COMPREHENSIVE ORGANIC SYNTHESIS

IN 9 VOLUMES

EDITORIAL BOARD

COMPREHENSIVE ORGANIC SYNTHESIS

Selectivity, Strategy & Efficiency in Modern Organic Chemistry

Editor-in-Chief
BARRY M. TROST
Stanford University, CA, USA

Deputy Editor-in-Chief
IAN FLEMING
University of Cambridge, UK

Volume 4

ADDITIONS TO AND SUBSTITUTIONS AT C—C π-BONDS

Volume Editor
MARTIN F. SEMMELHACK
Princeton University, NJ, USA

PERGAMON PRESS
OXFORD • NEW YORK • SEOUL • TOKYO

ELSEVIER SCIENCE Ltd
The Boulevard, Langford Lane
Kidlington, Oxford OX5 1GB, UK

First edition 1991
Second impression 1993
Third impression 1999

Library of Congress Cataloging in Publication Data

Comprehensive organic synthesis: selectivity, strategy and efficiency in modern organic chemistry/editor[s] Barry M. Trost, Ian Fleming.
p. cm.
Includes indexes.
Contents: Vol. 1.–2. Additions to C-X[pi]-Bonds — v. 3. Carbon–carbon sigma-Bond formation — v. 4. Additions to and substitutions at C-C[pi]-Bonds — v. 5. Combining C-C[pi]-Bonds — v. 6. Heteroatom manipulation — v. 7. Oxidation — v. 8. Reduction — v. 9. Cumulative indexes.
4. Organic Compounds — Synthesis I. Trost, Barry M. 1941–
II. Fleming, Ian, 1935–
QD262.C535 1991
547.2—dc20 90-26621

British Library Cataloguing in Publication Data

Comprehensive organic synthesis
4. Organic compounds. Synthesis
I. Trost, Barry M. (Barry Martin) 1941–
547.2

ISBN 0-08-040595-9 (Vol. 4)
ISBN 0-08-035929-9 (set)

™ The paper used in this publication meets the minimum requirements of American National Standard for Information Sciences — Permanence of Paper for Printed Library Materials, ANSI Z39.48-1984.

Contents

Preface

The emergence of organic chemistry as a scientific discipline heralded a new era in human development. Applications of organic chemistry contributed significantly to satisfying the basic needs for food, clothing and shelter. While expanding our ability to cope with our basic needs remained an important goal, we could, for the first time, worry about the quality of life. Indeed, there appears to be an excellent correlation between investment in research and applications of organic chemistry and the standard of living. Such advances arise from the creation of compounds and materials. Continuation of these contributions requires a vigorous effort in research and development, for which information such as that provided by the *Comprehensive* series of Pergamon Press is a valuable resource.

Since the publication in 1979 of *Comprehensive Organic Chemistry*, it has become an important first source of information. However, considering the pace of advancements and the ever-shrinking timeframe in which initial discoveries are rapidly assimilated into the basic fabric of the science, it is clear that a new treatment is needed. It was tempting simply to update a series that had been so successful. However, this new series took a totally different approach. In deciding to embark upon *Comprehensive Organic Synthesis*, the Editors and Publisher recognized that synthesis stands at the heart of organic chemistry.

The construction of molecules and molecular systems transcends many fields of science. Needs in electronics, agriculture, medicine and textiles, to name but a few, provide a powerful driving force for more effective ways to make known materials and for routes to new materials. Physical and theoretical studies, extrapolations from current knowledge, and serendipity all help to identify the direction in which research should be moving. All of these forces help the synthetic chemist in translating vague notions to specific structures, in executing complex multistep sequences, and in seeking new knowledge to develop new reactions and reagents. The increasing degree of sophistication of the types of problems that need to be addressed require increasingly complex molecular architecture to target better the function of the resulting substances. The ability to make such substances available depends upon the sharpening of our sculptors' tools: the reactions and reagents of synthesis.

The Volume Editors have spent great time and effort in considering the format of the work. The intention is to focus on transformations in the way that synthetic chemists think about their problems. In terms of organic molecules, the work divides into the formation of carbon–carbon bonds, the introduction of heteroatoms, and heteroatom interconversions. Thus, Volumes 1–5 focus mainly on carbon–carbon bond formation, but also include many aspects of the introduction of heteroatoms. Volumes 6–8 focus on interconversion of heteroatoms, but also deal with exchange of carbon–carbon bonds for carbon–heteroatom bonds.

The Editors recognize that the assignment of subjects to any particular volume may be arbitrary in part. For example, reactions of enolates can be considered to be additions to C—C π-bonds. However, the vastness of the field leads it to be subdivided into components based upon the nature of the bond-forming process. Some subjects will undoubtedly appear in more than one place.

In attacking a synthetic target, the critical question about the suitability of any method involves selectivity: chemo-, regio-, diastereo- and enantio-selectivity. Both from an educational point-of-view for the reader who wants to learn about a new field, and an experimental viewpoint for the practitioner who seeks a reference source for practical information, an organization of the chapters along the theme of selectivity becomes most informative.

The Editors believe this organization will help emphasize the common threads that underlie many seemingly disparate areas of organic chemistry. The relationships among various transformations becomes clearer and the applicability of transformations across a large number of compound classes becomes apparent. Thus, it is intended that an integration of many specialized areas such as terpenoid, heterocyclic, carbohydrate, nucleic acid chemistry, *etc.* within the more general transformation class will provide an impetus to the consideration of methods to solve problems outside the traditional ones for any specialist.

In general, presentation of topics concentrates on work of the last decade. Reference to earlier work, as necessary and relevant, is made by citing key reviews. All topics in organic synthesis cannot be treated with equal depth within the constraints of any single series. Decisions as to which aspects of a topic require greater depth are guided by the topics covered in other recent *Comprehensive* series. This new treatise focuses on being comprehensive in the context of synthetically useful concepts.

The Editors and Publisher believe that *Comprehensive Organic Synthesis* will serve all those who must face the problem of preparing organic compounds. We intend it to be an essential reference work for the experienced practitioner who seeks information to solve a particular problem. At the same time, we must also serve the chemist whose major interest lies outside organic synthesis and therefore is only an occasional practitioner. In addition, the series has an educational role. We hope to instruct experienced investigators who want to learn the essential facts and concepts of an area new to them. We also hope to teach the novice student by providing an authoritative account of an area and by conveying the excitement of the field.

The need for this series was evident from the enthusiastic response from the scientific community in the most meaningful way — their willingness to devote their time to the task. I am deeply indebted to an exceptional board of editors, beginning with my deputy editor-in-chief Ian Fleming, and extending to the entire board — Clayton H. Heathcock, Ryoji Noyori, Steven V. Ley, Leo A. Paquette, Gerald Pattenden, Martin F. Semmelhack, Stuart L. Schreiber and Ekkehard Winterfeldt.

The substance of the work was created by over 250 authors from 15 countries, illustrating the truly international nature of the effort. I thank each and every one for the magnificent effort put forth. Finally, such a work is impossible without a publisher. The continuing commitment of Pergamon Press to serve the scientific community by providing this *Comprehensive* series is commendable. Specific credit goes to Colin Drayton for the critical role he played in allowing us to realize this work and also to Helen McPherson for guiding it through the publishing maze.

A work of this kind, which obviously summarizes accomplishments, may engender in some the feeling that there is little more to achieve. Quite the opposite is the case. In looking back and seeing how far we have come, it becomes only more obvious how very much more we have yet to achieve. The vastness of the problems and opportunities ensures that research in organic synthesis will be vibrant for a very long time to come.

BARRY M. TROST
Palo Alto, California

Contributors to Volume 4

Professor E. Block
Department of Chemistry, SUNY-Albany, 1400 Washington Avenue, Albany, NY 12222, USA

Professor M. Brookhart
Department of Chemistry, University of North Carolina at Chapel Hill, Chapel Hill, NC 27599, USA

Dr M. Chapdelaine
Stuart Pharmaceuticals Division, ICI Americas Inc, Wilmington, DE 19897, USA

Professor D. P. Curran
Department of Chemistry, University of Pittsburgh, Pittsburgh, PA 15260, USA

Professor H. M. L. Davies
Department of Chemistry, Wake Forest University, Box 7486, Reynolds Station, Winston-Salem, NC 27109, USA

Dr S. A. Godleski
Senior Research Chemist, Kodak Research Laboratories, Building 82, 6th Floor, Rochester, NY 14650, USA

Professor K. E. Harding
Department of Chemistry, Texas A&M University, College Station, TX 77843, USA

Professor R. F. Heck
Department of Chemistry, University of Delaware, Newark, DE 19716, USA

Professor L. S. Hegedus
Department of Chemistry, Colorado State University, Fort Collins, CO 80523, USA

Professor P. Helquist
Department of Chemistry & Biochemistry, University of Notre Dame, Notre Dame, IN 46556, USA

Professor M. Hulce
Department of Chemistry, University of Maryland, Baltimore County Campus, Catonsville, MD 21228, USA

Professor M. E. Jung
Department of Chemistry & Biochemistry, University of California at Los Angeles, 405 Hilgard Avenue, Los Angeles, CA 90024, USA

Professor S. V. Kessar
Department of Chemistry, Panjab University, Chandigarh 160014, India

Professor P. Knochel
Department of Chemistry, University of Michigan, Ann Arbor, MI 48109, USA

Dr J. A. Kozlowski
Schering-Plough Research, 60 Orange Street, Bloomfield, NJ 07003, USA

Professor R. C. Larock
Department of Chemistry, Iowa State University, Ames, IA 50011, USA

Dr V. J. Lee
Building 65-355, Medical Research Division, American Cyanamid Company, Lederle Laboratories, Pearl River, NY 10965, USA

Dr W. W. Leong
Chemical Development, Schering-Plough Research, 60 Orange Street, Bloomfield, NJ 07003, USA

Dr V. Nair
Deputy Director, Regional Research Laboratory (CSIR), Industrial Estate P. O. Trivandrum 695019, Kerala State, India

Professor R. K. Norris
Department of Organic Chemistry, University of Sydney, Sydney, NSW 2006, Australia

Professor A. Padwa
Department of Chemistry, Emory University, Atlanta, GA 30332, USA

Dr C. Paradisi
Dipartimento Chimica Organica, Centro Meccanismi Reazioni Organiche del CNR, Università Degli Studi di Padova, via Marzolo 1, I-35131 Padova, Italy

Professor A. J. Pearson
Department of Chemistry, Case Western University, Cleveland, OH 44106, USA

Dr H.-G. Schmalz
Institut für Organische Chemie, Universität Frankfurt, Niederurseler Hang, D-6000 Frankfurt-Niederursel, Germany

Dr A. L. Schwan
Department of Chemistry, SUNY-Albany, 1400 Washington Avenue, Albany, NY 12222, USA

Professor M. F. Semmelhack
Department of Chemistry, Princeton University, Princeton, NJ 08544, USA

Professor J. K. Stille
Department of Chemistry, Colorado State University, Fort Collins, CO 80523, USA

Dr T. H. Tiner
Department of Chemistry, Texas A&M University, College Station, TX 77843, USA

Dr A. F. Volpe, Jr
Department of Chemistry, University of North Carolina at Chapel Hill, Chapel Hill, NC 27599, USA

Professor P. A. Wade
Department of Chemistry, Drexel University, Philadelphia, PA 19104, USA

Dr J. Yoon
Department of Chemistry, University of North Carolina at Chapel Hill, Chapel Hill, NC 27599, USA

Abbreviations

The following abbreviations have been used where relevant. All other abbreviations have been defined the first time they occur in a chapter.

Techniques

CD	circular dichroism
CIDNP	chemically induced dynamic nuclear polarization
CNDO	complete neglect of differential overlap
CT	charge transfer
GLC	gas–liquid chromatography
HOMO	highest occupied molecular orbital
HPLC	high-performance liquid chromatography
ICR	ion cyclotron resonance
INDO	incomplete neglect of differential overlap
IR	infrared
LCAO	linear combination of atomic orbitals
LUMO	lowest unoccupied molecular orbital
MS	mass spectrometry
NMR	nuclear magnetic resonance
ORD	optical rotatory dispersion
PE	photoelectron
SCF	self-consistent field
TLC	thin layer chromatography
UV	ultraviolet

***Reagents, solvents,* etc.**

Ac	acetyl
acac	acetylacetonate
AIBN	2,2′-azobisisobutyronitrile
Ar	aryl
ATP	adenosine triphosphate
9-BBN	9-borabicyclo[3.3.1]nonyl
9-BBN-H	9-borabicyclo[3.3.1]nonane
BHT	2,6-di-*t*-butyl-4-methylphenol (butylated hydroxytoluene)
bipy	2,2′-bipyridyl
Bn	benzyl
t-BOC	*t*-butoxycarbonyl
BSA	*N*,*O*-bis(trimethylsilyl)acetamide
BSTFA	*N*,*O*-bis(trimethylsilyl)trifluoroacetamide
BTAF	benzyltrimethylammonium fluoride
Bz	benzoyl
CAN	ceric ammonium nitrate
COD	1,5-cyclooctadiene
COT	cyclooctatetraene
Cp	cyclopentadienyl
Cp*	pentamethylcyclopentadienyl
18-crown-6	1,4,7,10,13,16-hexaoxacyclooctadecane
CSA	camphorsulfonic acid
CSI	chlorosulfonyl isocyanate
DABCO	1,4-diazabicyclo[2.2.2]octane
DBA	dibenzylideneacetone
DBN	1,5-diazabicyclo[4.3.0]non-5-ene
DBU	1,8-diazabicyclo[5.4.0]undec-7-ene

DCC	dicyclohexylcarbodiimide
DDQ	2,3-dichloro-5,6-dicyano-1,4-benzoquinone
DEAC	diethylaluminum chloride
DEAD	diethyl azodicarboxylate
DET	diethyl tartrate (+ or –)
DHP	dihydropyran
DIBAL-H	diisobutylaluminum hydride
diglyme	diethylene glycol dimethyl ether
dimsyl Na	sodium methylsulfinylmethide
DIOP	2,3-*O*-isopropylidene-2,3-dihydroxy-1,4-bis (diphenylphosphino)butane
DIPT	diisopropyl tartrate (+ or –)
DMA	dimethylacetamide
DMAC	dimethylaluminum chloride
DMAD	dimethyl acetylenedicarboxylate
DMAP	4-dimethylaminopyridine
DME	dimethoxyethane
DMF	dimethylformamide
DMI	*N,N'*-dimethylimidazolone
DMSO	dimethyl sulfoxide
DMTSF	dimethyl(methylthio)sulfonium fluoroborate
DPPB	1,4-bis(diphenylphosphino)butane
DPPE	1,2-bis(diphenylphosphino)ethane
DPPF	1,1'-bis(diphenylphosphino)ferrocene
DPPP	1,3-bis(diphenylphosphino)propane
E^+	electrophile
EADC	ethylaluminum dichloride
EDG	electron-donating group
EDTA	ethylenediaminetetraacetic acid
EEDQ	*N*-ethoxycarbonyl-2-ethoxy-1,2-dihydroquinoline
EWG	electron-withdrawing group
HMPA	hexamethylphosphoric triamide
HOBT	hydroxybenzotriazole
$IpcBH_2$	isopinocampheylborane
Ipc_2BH	diisopinocampheylborane
KAPA	potassium 3-aminopropylamide
K-selectride	potassium tri-*s*-butylborohydride
LAH	lithium aluminum hydride
LDA	lithium diisopropylamide
LICA	lithium isopropylcyclohexylamide
LITMP	lithium tetramethylpiperidide
L-selectride	lithium tri-*s*-butylborohydride
LTA	lead tetraacetate
MCPBA	*m*-chloroperbenzoic acid
MEM	methoxyethoxymethyl
MEM-Cl	β-methoxyethoxymethyl chloride
MMA	methyl methacrylate
MMC	methylmagnesium carbonate
MOM	methoxymethyl
Ms	methanesulfonyl
MSA	methanesulfonic acid
MsCl	methanesulfonyl chloride
MVK	methyl vinyl ketone
NBS	*N*-bromosuccinimide
NCS	*N*-chlorosuccinimide

NMO	*N*-methylmorpholine *N*-oxide
NMP	*N*-methyl-2-pyrrolidone
Nu^-	nucleophile
PPA	polyphosphoric acid
PCC	pyridinium chlorochromate
PDC	pyridinium dichromate
phen	1,10-phenanthroline
Phth	phthaloyl
PPE	polyphosphate ester
PPTS	pyridinium *p*-toluenesulfonate
Red-Al	sodium bis(methoxyethoxy)aluminum dihydride
SEM	β-trimethylsilylethoxymethyl
Sia_2BH	disiamylborane
TAS	tris(diethylamino)sulfonium
TBAF	tetra-*n*-butylammonium fluoride
TBDMS	*t*-butyldimethylsilyl
TBDMS-Cl	*t*-butyldimethylsilyl chloride
TBHP	*t*-butyl hydroperoxide
TCE	2,2,2-trichloroethanol
TCNE	tetracyanoethylene
TES	triethylsilyl
Tf	triflyl (trifluoromethanesulfonyl)
TFA	trifluoroacetic acid
TFAA	trifluoroacetic anhydride
THF	tetrahydrofuran
THP	tetrahydropyranyl
TIPBS-Cl	2,4,6-triisopropylbenzenesulfonyl chloride
TIPS-Cl	1,3-dichloro-1,1,3,3-tetraisopropyldisiloxane
TMEDA	tetramethylethylenediamine [1,2-bis(dimethylamino)ethane]
TMS	trimethylsilyl
TMS-Cl	trimethylsilyl chloride
TMS-CN	trimethylsilyl cyanide
Tol	tolyl
TosMIC	tosylmethyl isocyanide
TPP	*meso*-tetraphenylporphyrin
Tr	trityl (triphenylmethyl)
Ts	tosyl (*p*-toluenesulfonyl)
TTFA	thallium trifluoroacetate
TTN	thallium(III) nitrate

Contents of All Volumes

Volume 3 Carbon–Carbon σ-Bond Formation

Volume 5 Combining C—C π-Bonds

Volume 6 Heteroatom Manipulation

Displacement by Substitution Processes

Acylation-type Reactions

Protecting Groups

Functional Group Interconversion

Elimination Reactions

Volume 7 Oxidation

Oxidation of Unactivated C—H Bonds

Oxidation of Activated C—H Bonds

Oxidation of C=C Bonds

Oxidation of C—X Bonds

Oxidation of C—C Bonds

Oxidation of Heteroatoms

Special Topics

Volume 8 Reduction

Reduction of C=X Bonds

Volume 9

1.1
Stabilized Nucleophiles with Electron Deficient Alkenes and Alkynes

MICHAEL E. JUNG
University of California at Los Angeles, CA, USA

1.1.1 OVERVIEW

1.1.1.1 History

The addition of stabilized nucleophiles to activated π-systems is one of the oldest and most useful constructive methods in organic synthesis, dating back more than a hundred years. In 1887 Michael, after whom the most well-known example of this reaction was named, published the first of a series of papers on this reaction describing the addition of the sodium salt of both diethyl malonate (**1**) and ethyl acetoacetate (**2**) to ethyl cinnamate (**3**) to give the products (**4**) and (**5**), respectively (equation 1).[1] This Michael reaction, in which a carbanion stabilized by one or usually two electron-withdrawing groups adds to the β-carbon of an α,β-unsaturated carbonyl derivative, represents one of the earliest and most highly utilized methods for carbon–carbon bond formation. In that same 1887 paper,[1a] Michael also reported the synthesis of triethyl cyclopropane-1,1,2-tricarboxylate (**9**) from (**1**) and ethyl α-bromoacrylate (**6**) *via* the intermediates (**7**) and (**8**), a synthetic route to cyclopropanes used often today (Scheme 1). In 1894 he showed that an alkynic ester could also function as the electrophilic component in these reactions, *e.g.* (**1**) and (**10a**) or (**10b**) giving (**11a**) or (**11b**), respectively (equation 2).[2] This reaction and its variants were in wide use around the turn of the century to prepare straight chain polycarboxylates, *e.g.* the reaction of (**12**) to give (**13**; equation 3) and cyclohexane-1,3-diones, *e.g.* (**15**; equation 4) (*via* Knoevenagel and Claisen condensations).[3] The combination of Michael addition followed by intramolecular aldol condensation and decarboalkoxylation furnishes cyclohexenones, *e.g.* isophorone (**17**; equation 5), from enones such as mesityl oxide (**16**) and the salt of ethyl acetoacetate (**2**).[4] This process of annulation was made much more general in the 1930s by Robinson and his coworkers,[5] who showed that vinyl ketones and their derivatives, *e.g.* Mannich bases, could be used with simple ketone enolates to produce cyclohexenones of a wide variety, *e.g.* (**20**; equation 6), a process which has come to be known as the Robinson annulation.[6]

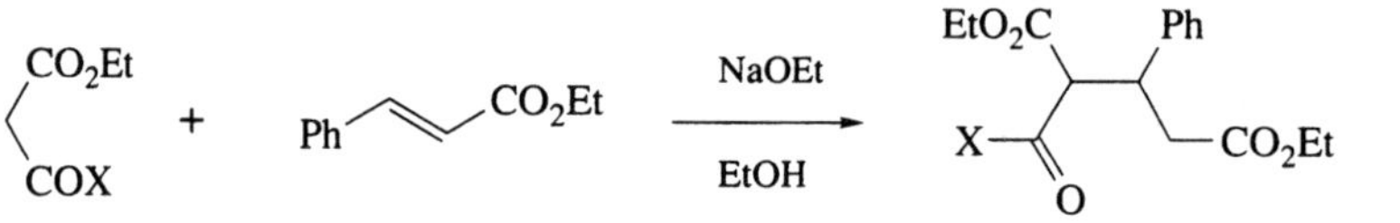

CO_2Et
COX
+
Ph
CO_2Et
NaOEt
EtOH
EtO_2C
Ph
X
O
CO_2Et

(1)

(**1**) X = OEt (**3**) (**4**) X = OEt
(**2**) X = Me (**5**) X = Me

CO_2Et
CO_2Et
+
Br
CO_2Et
NaOEt
EtOH
EtO_2C
Br
EtO_2C
CO_2Et
(**1**)
(**6**)
(**7**)
EtO_2C
Br
EtO_2C
CO_2Et
EtO_2C
EtO_2C
CO_2Et

(**8**) (**9**)

Scheme 1

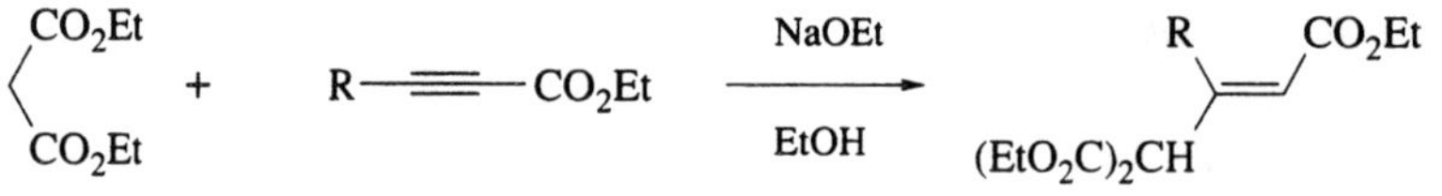

CO_2Et
CO_2Et
+
R
CO_2Et
NaOEt
EtOH
R
CO_2Et
$(EtO_2C)_2CH$

(2)

(**1**) (**10a**) R = CO_2Et (**11a**) R = CO_2Et
(**10b**) R = Ph (**11b**) R = Ph

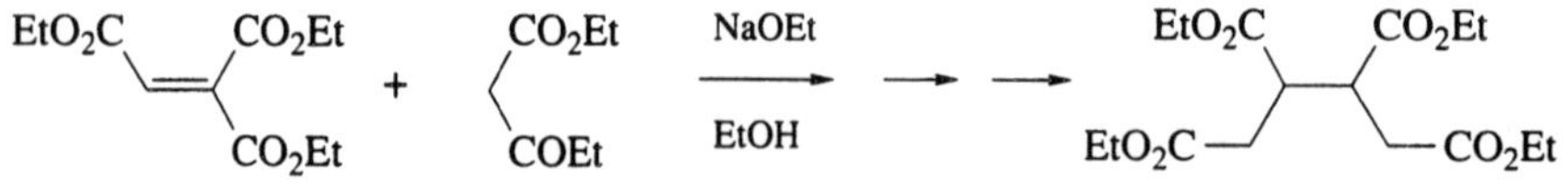

EtO_2C
CO_2Et
CO_2Et
+
CO_2Et
COEt
NaOEt
EtOH
EtO_2C
CO_2Et
EtO_2C
CO_2Et

(3)

(**12**) (**1**) (**13**)

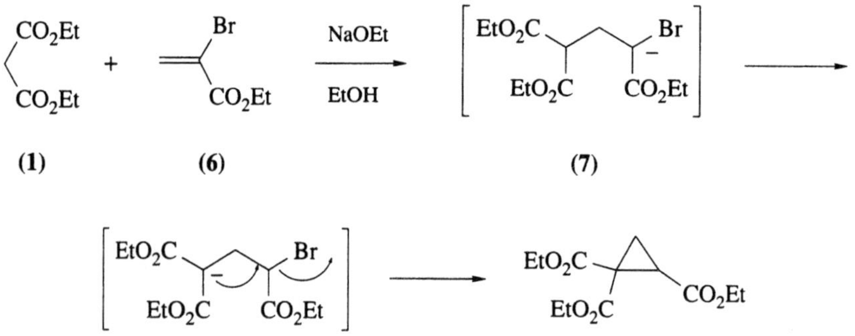

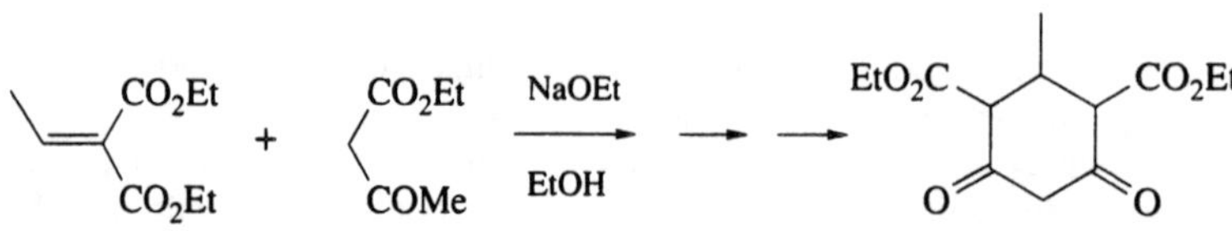

CO_2Et
CO_2Et
+
CO_2Et
COMe
NaOEt
EtOH
EtO_2C
CO_2Et
O
O

(4)

(**14**) (**2**) (**15**)

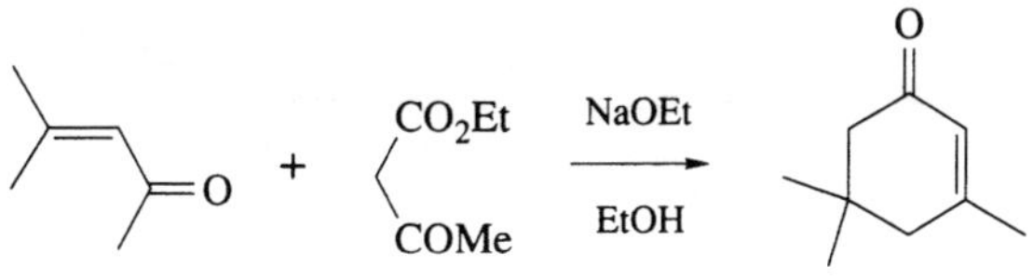

CO_2Et / COMe — NaOEt, EtOH — (5)

(16) **(2)** **(17)**

Ph — base — Ph — (6)

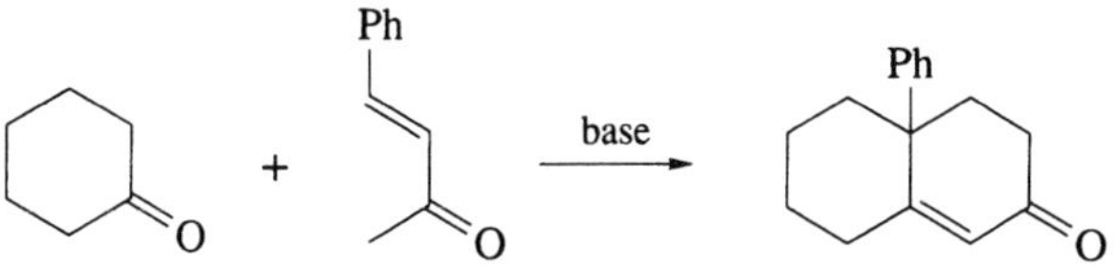

(18) **(19)** **(20)**

These early results paved the way for a myriad of applications of this process in organic synthesis, as described in detail in this chapter. It should be pointed out that only additions of stabilized nucleophiles are included here, with the addition of reactive nucleophiles (organolithiums, Grignards, cuprates, *etc.*), Lewis acid promoted additions and asymmetric nucleophilic additions, among others, being covered in Chapters 1.3 and 1.5, respectively, in this volume.

1.1.1.2 Other Reviews

The literally thousands of examples involving the addition of stabilized nucleophiles to activated alkenes and alkynes have spawned many reviews over the years. The Michael reaction has been discussed in detail several times[7] and reviews on other related topics, such as annulation[8] and Mannich base methiodides,[9] have also appeared, as have several reviews on nucleophilic additions to activated alkynes.[10]

It is not the goal of this chapter to present a complete or comprehensive view of these nucleophilic additions, but rather a selective and illustrative one. In particular, emphasis will be given to those methods with real potential for organic synthesis.

1.1.2 ALKENIC π-SYSTEMS

1.1.2.1 Carbon Nucleophiles

This section describes the additions of stabilized carbon nucleophiles, such as cyanide, malonate, ketone enolates, enamines, *etc.*, to alkenic π-systems. These reactions are highly useful in organic synthesis since they are all carbon–carbon bond-forming reactions, and therefore have been used extensively in organic chemistry.

1.1.2.1.1 Intermolecular additions

The original work of Michael[1] showed that highly stabilized carbanions from diactivated methylenes, such as malonate or acetoacetate, could add in a conjugate or 1,4-fashion to α,β-unsaturated esters in good yield. Over the years, innumerable variations on this theme have been exploited in organic synthesis. A typical example involves the condensation of ethyl acetoacetate (**2**) with methyl acrylate (**21**) in the presence of sodium ethoxide to give the product (**22**) in 73% yield (Scheme 2).[11] The mechanism of the reaction involves the formation of the stabilized anion (**23**) (from **2** and ethoxide), which can then add in a reversible fashion to the Michael acceptor (**21**) to give the stabilized anion (**24**). In hydroxylic solvents this anion is converted to the more stable anion (**25**) which is eventually protonated to give (**22**).

Several points arise from consideration of this mechanism: (i) unlike alkylation of stabilized enolates, the Michael reaction regenerates a basic species and thus usually catalytic quantities of base can be utilized;[7a,7c,12] (ii) the high acidity of the doubly activated methylene permits the use of weaker bases, *e.g.* simple amines (piperidine, triethylamine *etc.*) or hydroxides (Triton B, *i.e.* benzyltrimethylammonium hydroxide); (iii) it was recognized early that, when a full equivalent of base was used, the final more

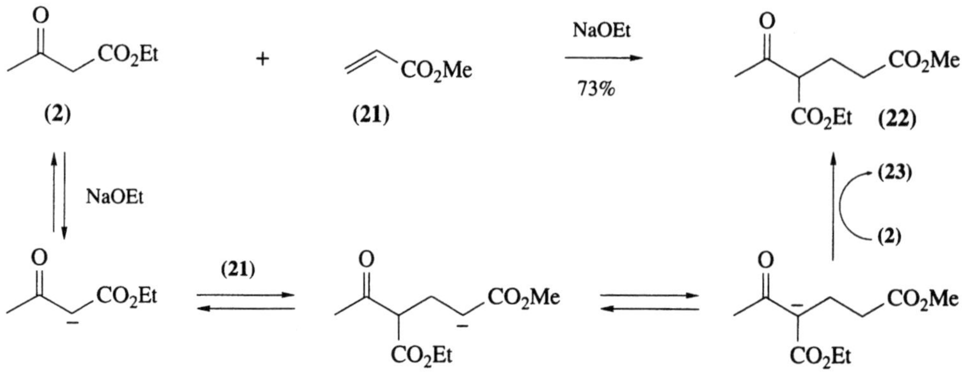

(2) (21) NaOEt 73% (22)

NaOEt (23) (2)

(23) (21) (24) (25)

Scheme 2

stable anion, *e.g.* (**25**), could be utilized synthetically and alkylated to give the product of overall addition and alkylation, *e.g.* the formation of (**29**; Scheme 3) by alkylation of (**28**), which is formed from (**26**) and (**27**);[13] (iv) multiple Michael addition is a potential side reaction, *e.g.* formation of (**32**) from (**1**) and (**30**), as is a Michael reaction followed by a subsequent Claisen or aldol condensation of the resultant 1,5-dicarbonyl compound,[14] both of which are shown (formation of **34** and **35**) in Scheme 4;[14a] (v) a rearrangement is often observed, the abnormal Michael reaction, resulting in transfer of an activating group from the nucleophilic species to the α-carbon of the Michael acceptor, *e.g.* formation of (**37**; equation 7);[15] and (vi) the key carbon–carbon bond formation step is reversible, so that starting materials and unusual products can sometimes result,[16] *e.g.* the products of arylmethylene transfer (**40**; equation 8)[16a] or malononitrile transfer (**43**; equation 9).[16b]

A large part of the usefulness of the Michael reaction in organic synthesis derives from the fact that almost any activated alkene can serve as an acceptor[7]—α,β-unsaturated ketones, esters, aldehydes, amides, acids, lactones, nitriles, sulfoxides, sulfones, nitro compounds, phosphonates, phosphoranes, quinones,

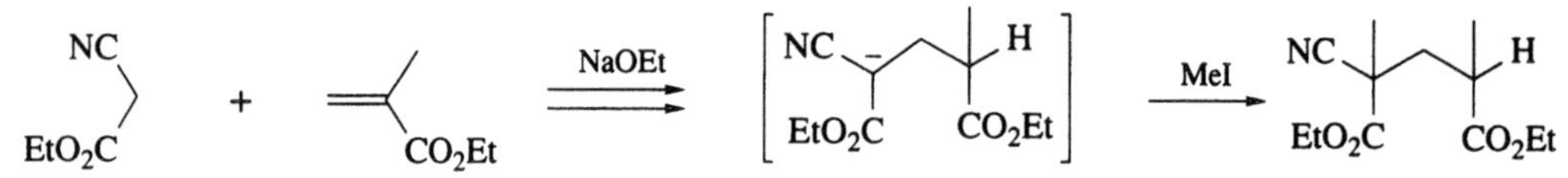

(26) (27) NaOEt (28) MeI (29)

Scheme 3

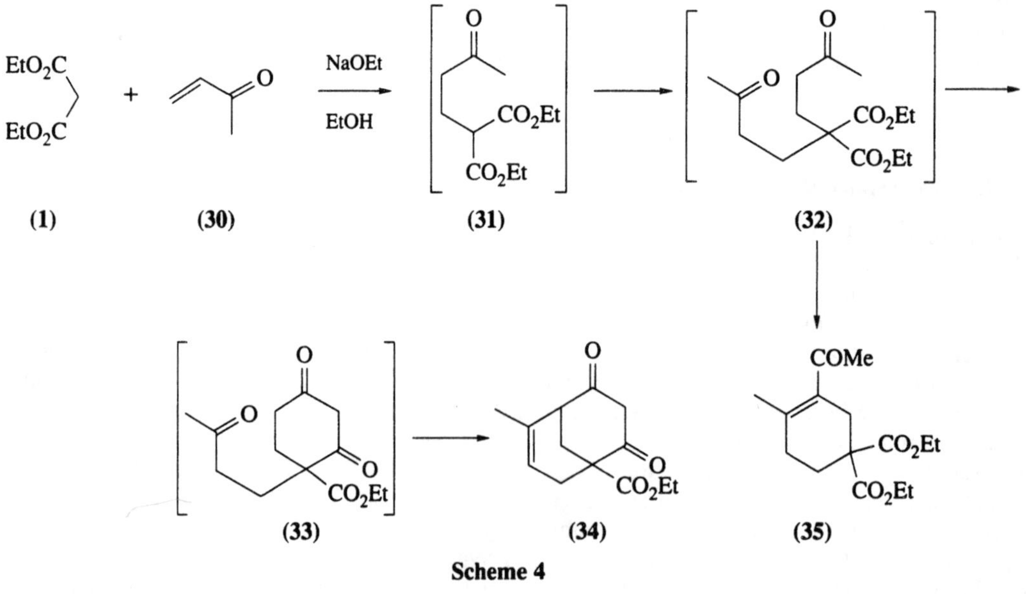

(1) (30) NaOEt EtOH (31) (32)

(33) (34) (35)

Scheme 4

(36) + (26) —NaOEt→ (37) (7)

(38) + (39) —EtOH, 71–91%→ (40) + (26) (8)

(41) + (42) —EtOH, 84%→ (43) + (39) (9)

coumarins, vinylpyridines, *etc.* Since good compilations of these reactions have appeared in various reviews, only a few typical examples are given in equations (10) to (14).[17] Not only is 1,4-addition possible but also 1,6- and 1,8-addition to dienyl and trienyl systems, such as (**58a,b**; equation 15) giving (**59a,b**).[18] Some 1,4-addition is also observed as a minor reaction pathway. The use of enamines leads to a final cyclization of the intermediate zwitterion, resulting in a six-membered ring (equation 16).[19] The use of a simple enamine of a cyclic ketone (**64**) with (**61**) followed by distillation causes loss of the amine and production of the cross-conjugated dienoate system (**65**; equation 17).[19e] A similar process is seen with 2-formylcycloalkanones.[19g]

(44) + (1) —NaOEt, EtOH, 90%→ (45) (10)

(46) + (47) —i, Δ; ii, H_3O^+; 68%→ (48) (11)

(49) + (50) —Et_2NH, 97%→ (51) (12)

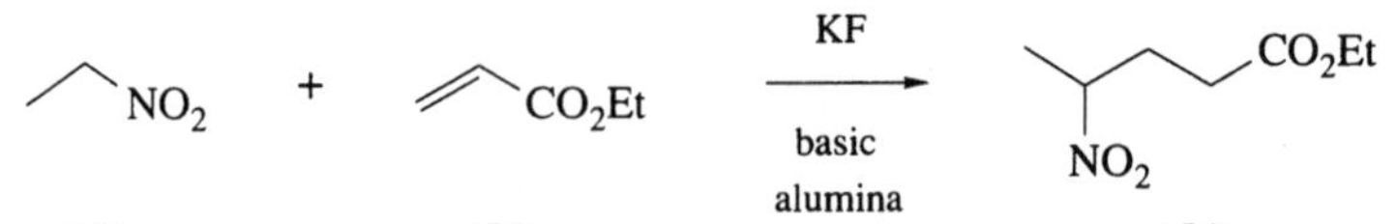

(52) (53) (54) (13)

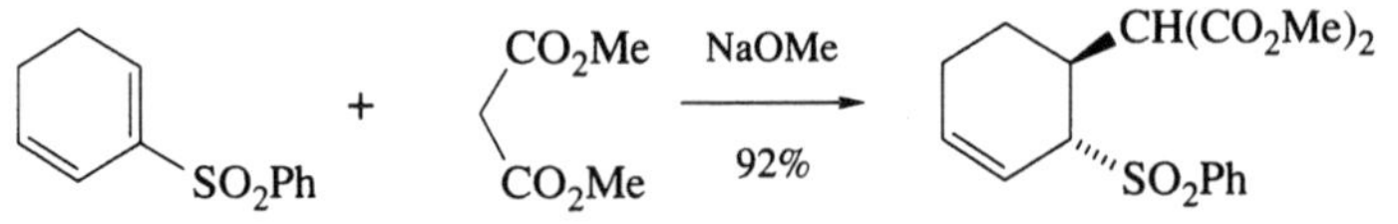

(55) (56) (57) (14)

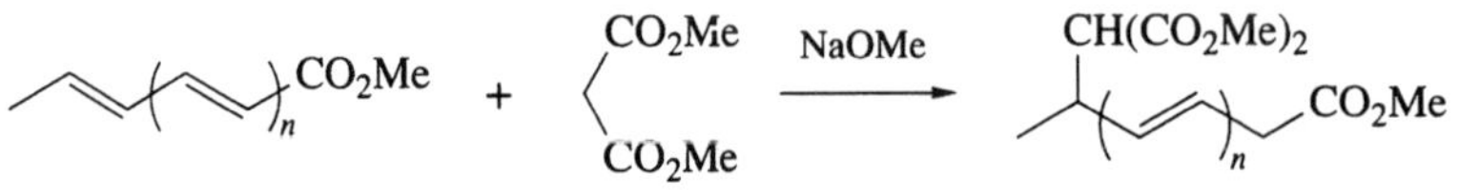

(58a) $n = 1$
(58b) $n = 2$
(56)
(59a) $n = 1$
(59b) $n = 2$ (15)

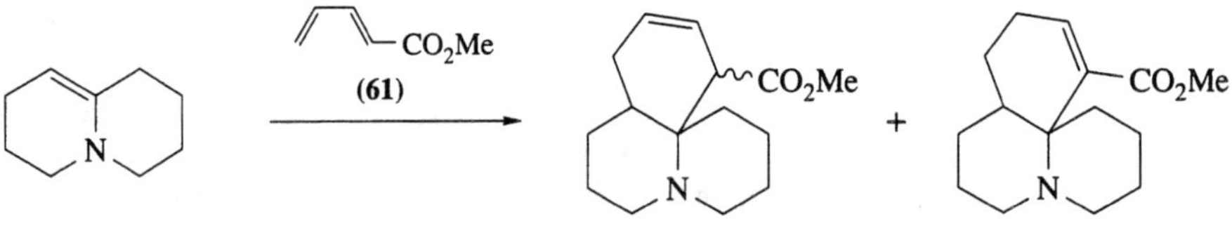

(60) (62) (63) (16)

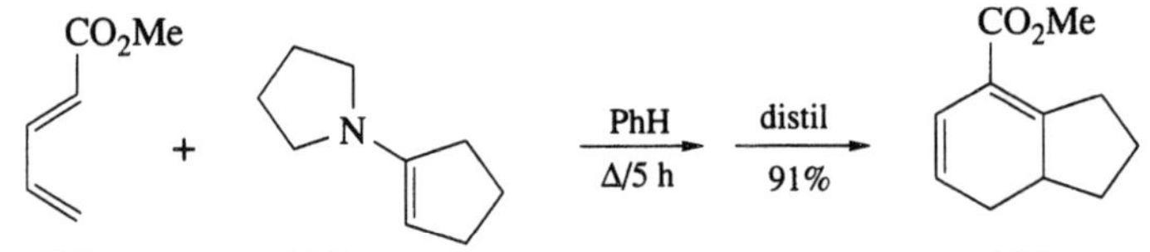

(61) (64) (65) (17)

Michael additions of unsymmetrical ketones usually give the products of addition at the more-substituted α-carbon,[20] *e.g.* the reaction of 2-methylcyclohexanone (**66**; equation 18) to give (**67**) in preference to (**68**) (the ratio is solvent dependent)[20a] and reaction of 2-butanone (**69**; equation 19) at the substituted carbon to give (**71**).[14f] This preference is reversed by using enamines,[21] namely the enamine (**72**) of 2-methylcyclohexanone (**66**) reacts with acrylonitrile (**70**) to give, after hydrolysis, the 2,6-disubstituted product (**73**) in 55% yield (Scheme 5).[21a] The combination of a Michael addition followed by a Claisen condensation permits the easy construction of cyclic ketone derivatives,[22] *e.g.* dimedone (**75**; Scheme 6) from mesityl oxide (**16**) and malonate (**1**) *via* the diketo ester (**74**), and 4-pentylcyclohexane-1,3-dione (**78**; Scheme 7) from 2-heptanone (**76**) and ethyl acrylate (**53**) *via* the keto ester (**77**).[20c] Other further cyclizations are also possible,[23] *e.g.* the production of cyclooctane[23a] or cycloheptane[23b] systems (Scheme 8) *via* an initial Michael reaction, aldol-type ring closure, quaternization of the amine and deacylative elimination.

A much more generally useful process was developed by Robinson to prepare cyclohexenones from ketones and methyl vinyl ketone or its derivatives. Again, because good compilations of the Robinson annulation exist,[8] only a few examples are given here. The first step of this process, the Michael addition, is carried out by normal base catalysis, while the second step, the aldol condensation, is best accomplished by the use of a secondary amine to form the enamine of the acyclic ketone, which then cyclizes

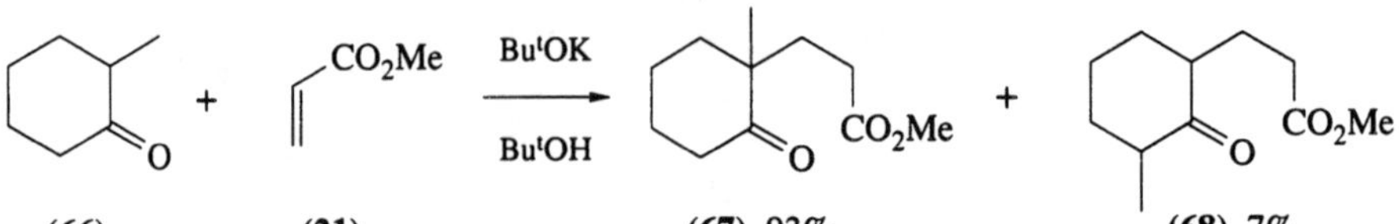

(66) (21) (67) 93% (68) 7% (18)

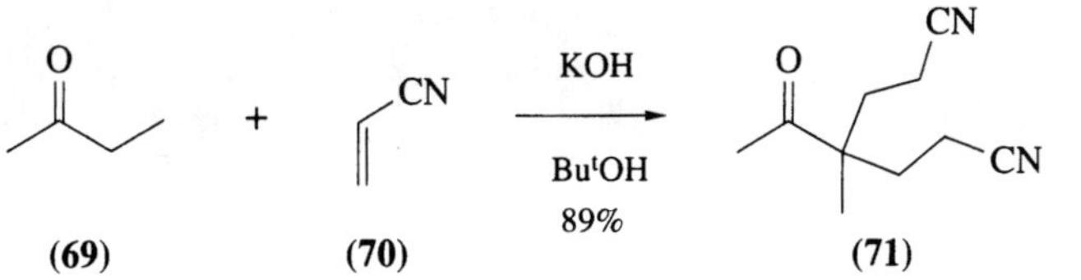

(69) + (70) $\xrightarrow[\text{Bu}^t\text{OH, 89\%}]{\text{KOH}}$ (71) (19)

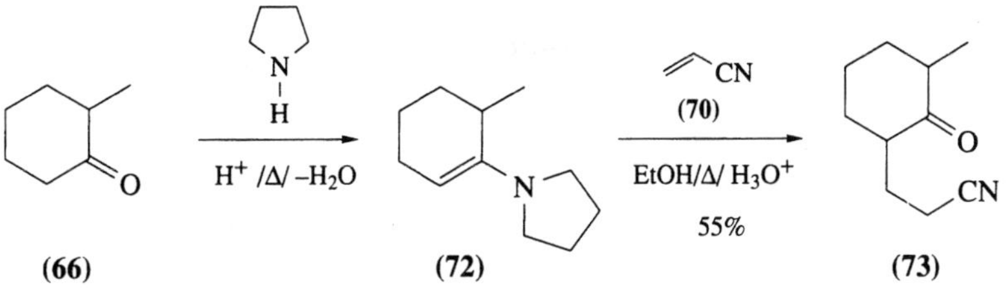

(66) $\xrightarrow[\text{H}^+/\Delta/-\text{H}_2\text{O}]{\text{pyrrolidine}}$ (72) $\xrightarrow[\text{EtOH}/\Delta/\text{H}_3\text{O}^+,\ 55\%]{(70)}$ (73)

Scheme 5

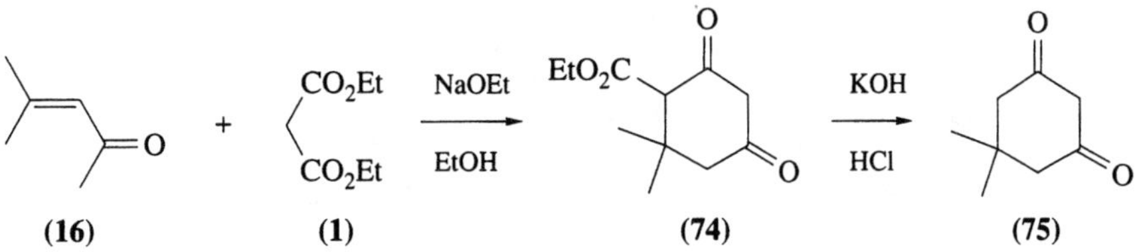

(16) + (1) $\xrightarrow[\text{EtOH}]{\text{NaOEt}}$ (74) $\xrightarrow[\text{HCl}]{\text{KOH}}$ (75)

Scheme 6

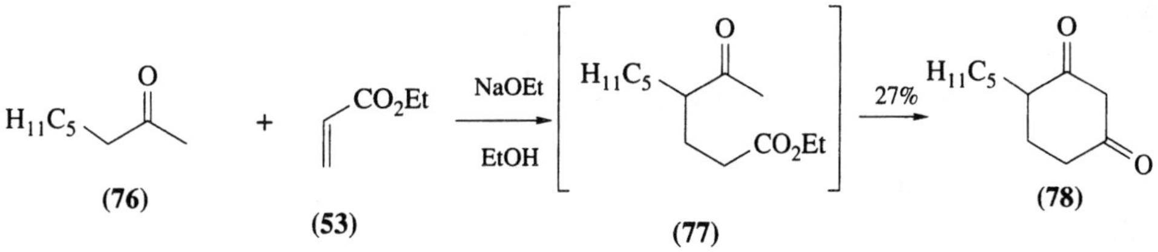

(76) + (53) $\xrightarrow[\text{EtOH}]{\text{NaOEt}}$ [(77)] $\xrightarrow{27\%}$ (78)

Scheme 7

(79) + (64) $\xrightarrow[25\ ^\circ\text{C}/2\ \text{h}]{\text{dioxane}}$ [(80)] $\xrightarrow[\text{ii, OH}^-,\ 25\%]{\text{i, MeI}}$ (81)

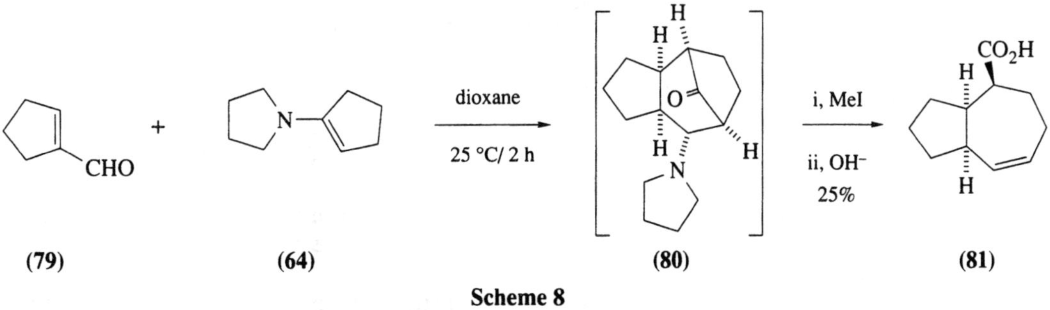

Scheme 8

and dehydrates (Scheme 9).[24] When β-keto esters are used, all of the key steps (Michael, aldol and decarboalkoxylation) can be done in one pot.[24c] The use of optically active amines in either of these two steps affords products of very high enantiomeric excess (equations 20 and 21).[25] Simple Robinson annulations can be carried out under acidic as well as basic conditions (equation 22).[26] The use of enamines of 2-substituted ketones such as (**72**) permits the production of the isomeric octalone (**91**; equation 23).[21a]

For years most Michael reactions were carried out under protic conditions so that rapid proton transfer was possible. However, in the early 1970s several groups performed Michael reactions under aprotic conditions and these processes are now quite common. Early attempts at trapping a kinetic enolate in aprotic solvents with simple enones such as methyl vinyl ketone or an α,β-unsaturated ester such as acrylate led to a scrambling of the enolate (using the Michael product as proton source).[8a] However, the introduction of an α-trialkylsilyl group in the enone (**93**; Scheme 10) permitted the trapping of kinetic

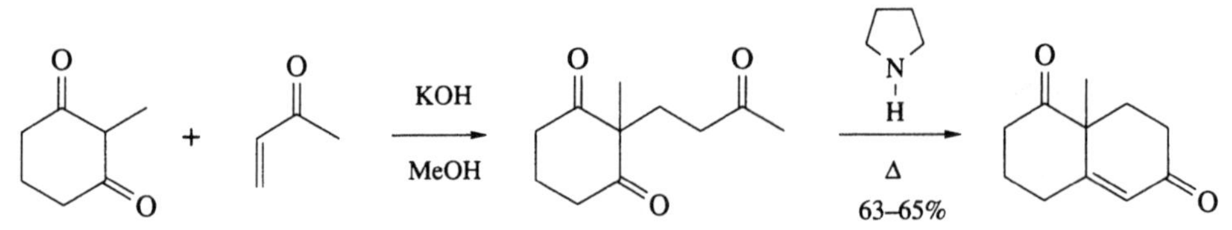

(82) (30) KOH MeOH (83) Δ 63–65% (84)

Scheme 9

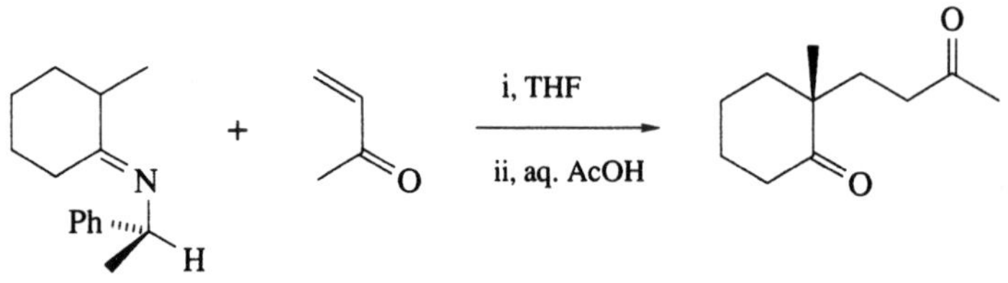

(85) (30) i, THF ii, aq. AcOH (86) 91% *ee* (20)

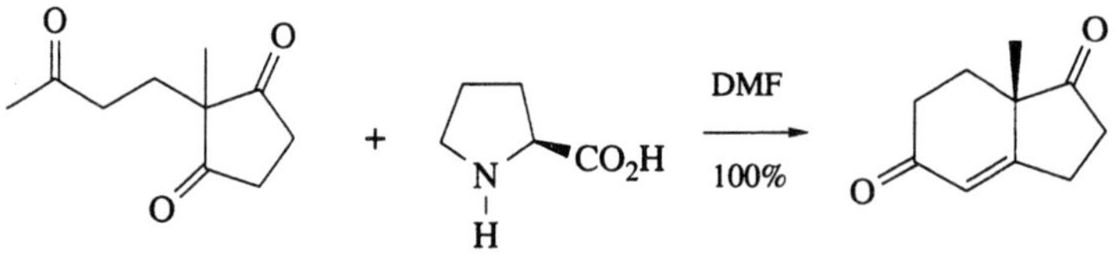

(87) (88) DMF 100% (89) 93% *ee* (21)

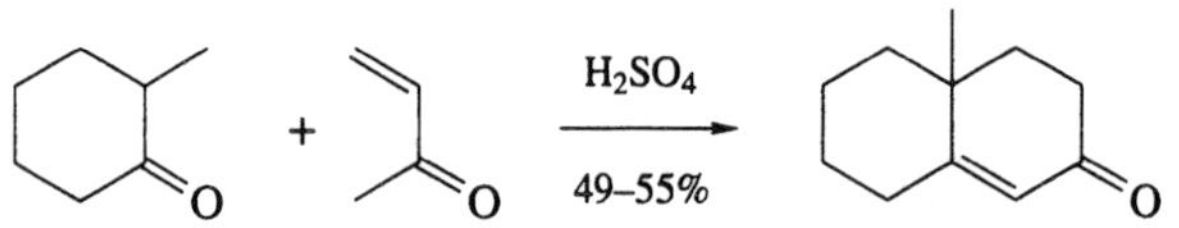

(66) (30) H_2SO_4 49–55% (90) (22)

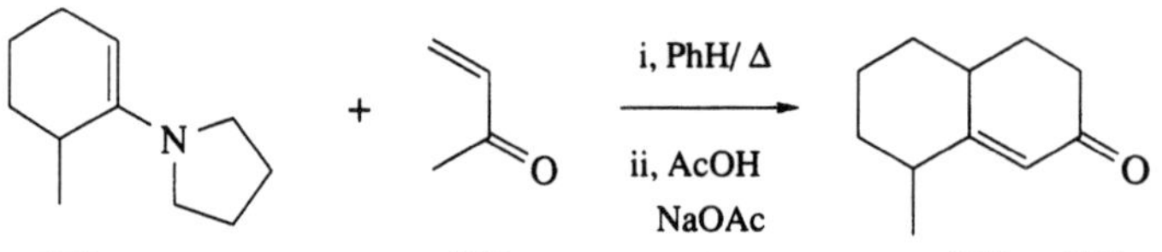

(72) (30) i, PhH/ Δ ii, AcOH NaOAc H_2O (91) 45% (23)

enolates such as (**92**).[27] Further treatment of (**94**) with base caused both aldol condensation and removal of the α-silyl group to give the desired enone (**95**).[27a]

As in simple Michael additions, dienones and trienones can be used in the Robinson annulation with an initial 1,6- or 1,8-addition.[8a] Dienoates can also be used,[28] *e.g.* (**82**; equation 24) and (**96**) give an 11:1 mixture of isomers (**97a,b**) in 47% yield.[28a]

Bicyclic keto esters can easily be prepared by a process called α,α′-annulation.[29] Thus, treatment of the enamine of cyclopentanone (**64**) with ethyl α-(bromomethyl)acrylate (**98**) affords, after work-up, the bicyclic keto ester (**99**) in 80% yield (equation 25).[29b] The mechanism probably involves an initial Michael addition and elimination (or a simple S_N2 or S_N2' alkylation) followed by an intramolecular Michael addition of the less-substituted enamine on the acrylate unit. The use of the enamine of 4,4-bis(ethoxycarbonyl)cyclohexanone (**100**; equation 26) with (**98**) gives a 45% yield of the adamantanedione diester (**101**) (yield based on **100**; 70% when based on **98**) *via* α,α′-annulation followed by Dieckmann condensation.[29e] Enamines of heterocyclic ketones can also serve as the initial nucleophiles, *e.g.* (**102**) and (**103**) give (**105**) *via* (**104**), formed *in situ*, in 70% yield (Scheme 11).[29g]

The use of α-thiophenyl enones (**106**; Scheme 12) allows the preparation of phenols such as (**107**) from cyclic ketones (**18**).[30a] The same product can also be obtained by normal Robinson annulation of methyl vinyl ketone (**30**) and the β-keto sulfoxide (**108**).[30b] Acceptors other than α,β-unsaturated carbonyls have been used in both the Michael reaction and the Robinson annulation process. For example, nu-

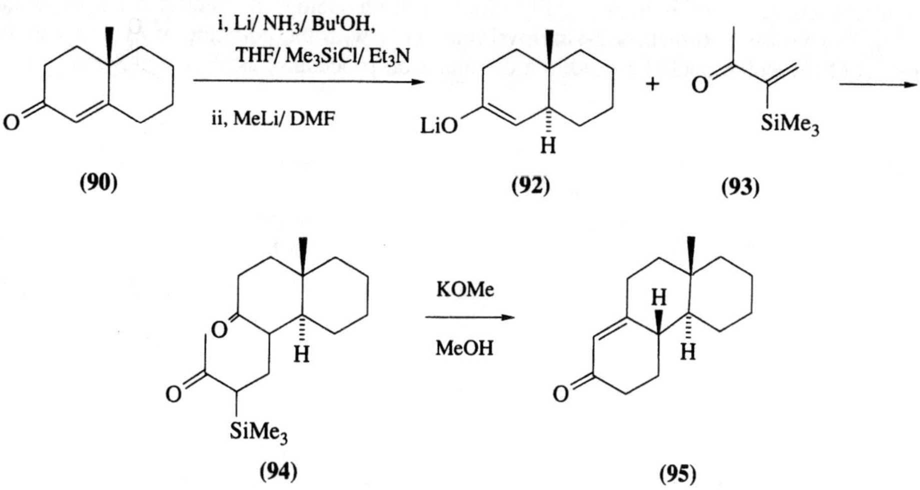
i, Li/ NH3/ ButOH,
THF/ Me3SiCl/ Et3N
ii, MeLi/ DMF
LiO
H
+
O
SiMe3
(90)
(92)
(93)
O
H
O
SiMe3
KOMe
MeOH
H
H
O
(94)
(95)

Scheme 10

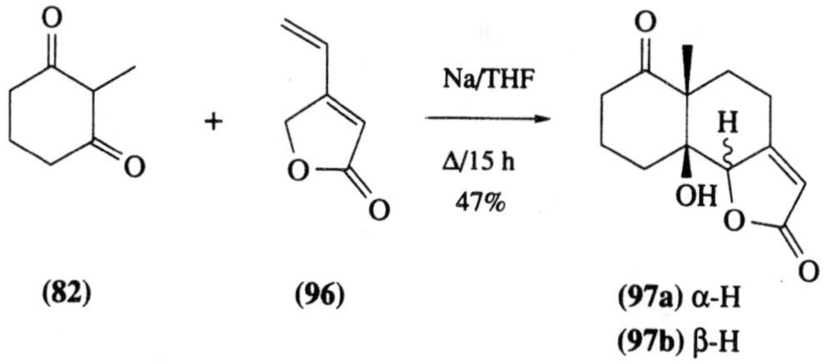
O
O
+
O
O
Na/THF
Δ/15 h
47%
O
H
OH
O
O
(82)
(96)
(97a) α-H
(97b) β-H

(24)

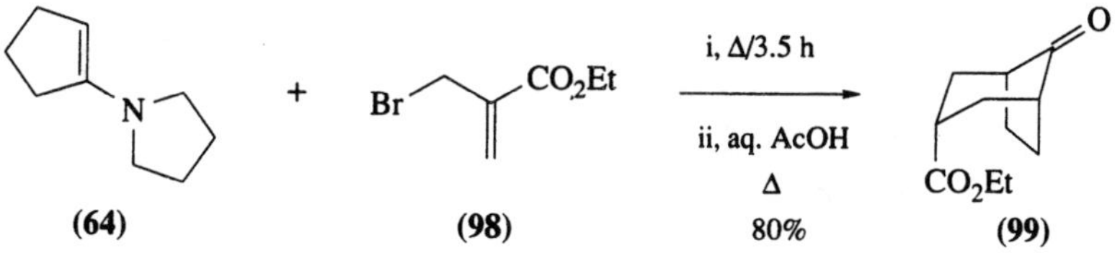
N
+
Br
CO2Et
i, Δ/3.5 h
ii, aq. AcOH
Δ
80%
O
CO2Et
(64)
(98)
(99)

(25)

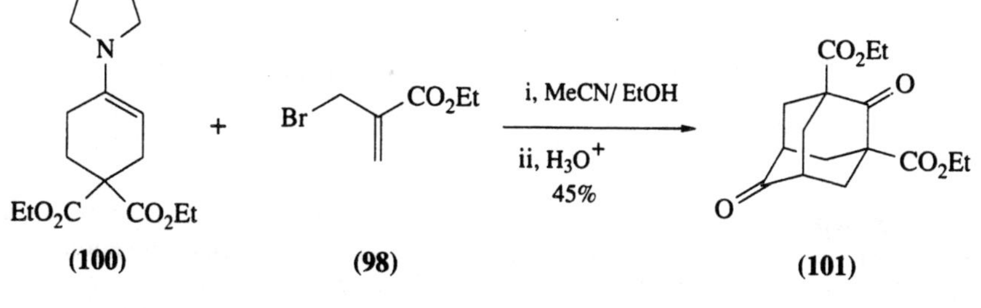
N
EtO2C
CO2Et
+
Br
CO2Et
i, MeCN/ EtOH
ii, H3O+
45%
CO2Et
O
CO2Et
O
(100)
(98)
(101)

(26)

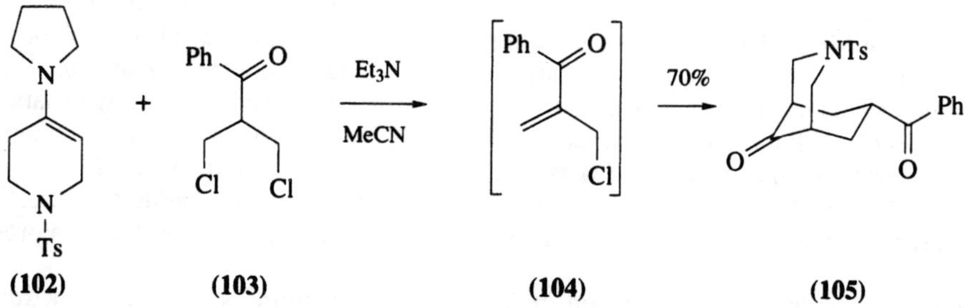
N
N
Ts
+
Ph
O
Cl
Cl
Et3N
MeCN
Ph
O
Cl
70%
NTs
Ph
O
O
(102)
(103)
(104)
(105)

Scheme 11

cleophiles have been added to 2- or 4-vinylpyridines in high yield;[31] a specific example is shown in equation (27).[31b] The use of 6-methyl-2-vinylpyridine (**112**) with the enamine of cyclohexanone (**113**) gives a product (**114**) which could be used in a diannulation procedure, as shown in Scheme 13, to give the dienone (**116**).[31c]

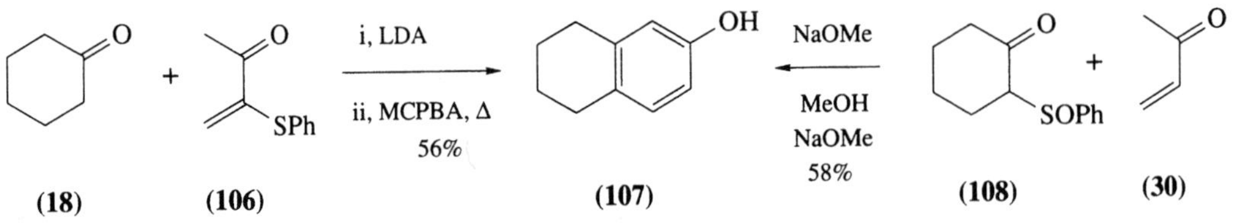

i, LDA; ii, MCPBA, Δ 56% — NaOMe, MeOH, NaOMe 58%

(**18**) (**106**) (**107**) (**108**) (**30**)

Scheme 12

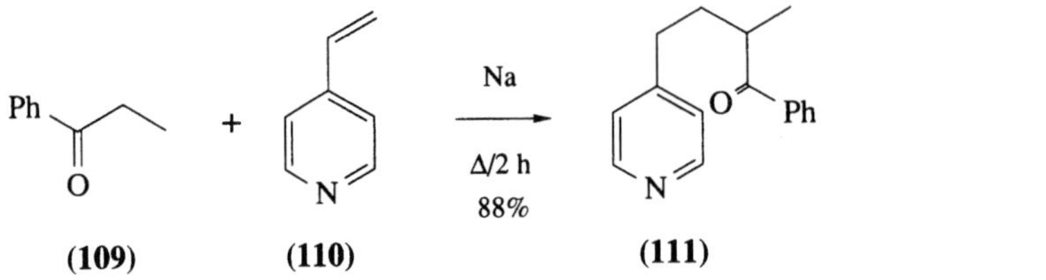

Na, Δ/2 h, 88% (27)

(**109**) (**110**) (**111**)

Δ, H_3O^+ — i, $HOCH_2CH_2OH/H^+$; ii, $Na/NH_3/EtOH$; iii, NaOH/EtOH; iv, H_3O^+

(**112**) (**113**) (**114**)

TsOH, AcOH, Δ — 52% from (**113**)

(**115**) (**116**)

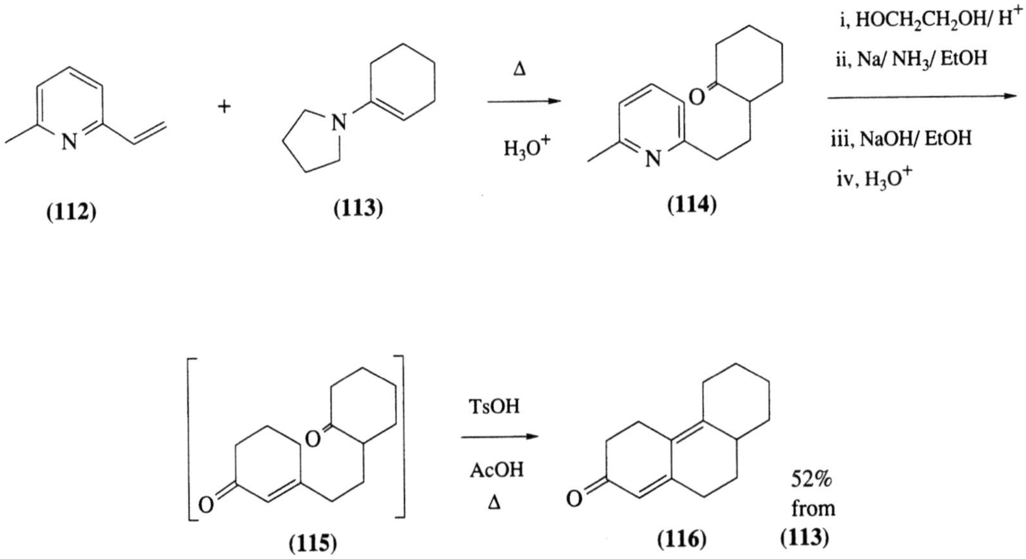

Scheme 13

In 1973, a series of papers from Schlessinger's group showed that Michael additions could be carried out in high yield under aprotic conditions.[32] In particular, alkyl (dithioalkyl)acetate enolates (Scheme 14) gave high yields of 1,4-addition products and, after hydrolysis, α-keto esters[32a–32d] (and is thus the 1,4-addition of an acyl anion equivalent), while α-(thiomethyl)acrylate as an acceptor resulted in high yields of the corresponding Michael adducts.[32c] A second acyl anion equivalent was also developed for use in Michael additions, the thioacetal monoxide; thus the anion of (**120**; Scheme 15) adds to MVK (**30**) to give a 96% yield of (**121**) which is easily hydrolyzed to (**122**).[32d,32e] The corresponding unsaturated compounds, namely ketene thioacetal monoxides (**124**), were developed as α-formyl cation synthons (Scheme 16). Addition of an enolate to (**124**) gave in high yield the expected adduct (**125**) which, on hydrolysis, produced the 1,4-dicarbonyl system (**126**).[32f,32g] This process has found further use,[33] *e.g.* in an efficient synthesis of allethrolone[33a] and also for trapping a kinetically formed enolate in a prostaglandin synthesis.[33b]

A large number of workers have examined the effects of temperature, solvent and reaction times on the ratio of the products of 1,4-addition *versus* 1,2-addition of dithiane and other α-thioalkyl carbanions.[34] Their results show that in general 1,2-addition is kinetically favored and therefore predominates at low temperatures and in less polar solvents,[34a,34b] whereas 1,4-addition is thermodynamically favored and predominates at 25 °C and in THF containing HMPA,[34c–34g] as the examples in equations (28) and (29) indicate. If a chiral ligand is used in these additions, fair enantiomeric excess (32–67%) can be obtained.[34f] It is possible to trap the enolates so generated with electrophiles,[32e,35] as shown in Scheme 17. The alkoxycarbonyl anion equivalent (**136**; Scheme 18) has also been added 1,4- and hydrolyzed to the ester (**138**).[36]

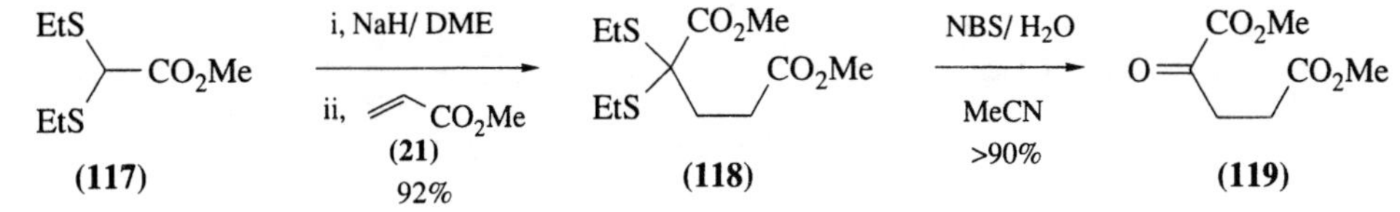

EtS, EtS, CO_2Me (**117**); i, NaH/ DME; ii, CO_2Me (**21**) 92%; (**118**); NBS/ H_2O, MeCN, >90%; (**119**)

Scheme 14

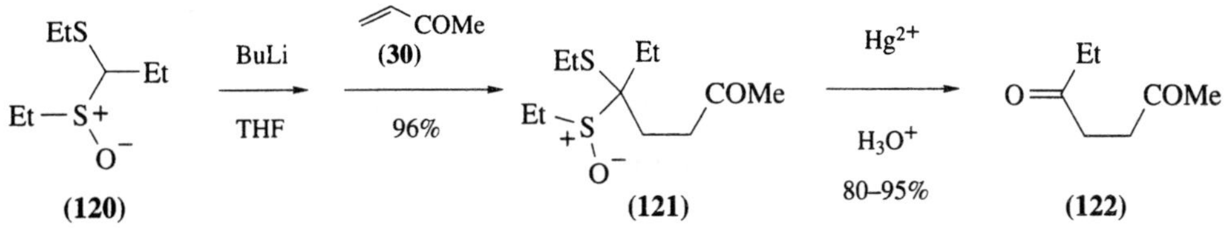

(**120**); BuLi, THF; COMe (**30**) 96%; (**121**); Hg^{2+}, H_3O^+, 80–95%; (**122**)

Scheme 15

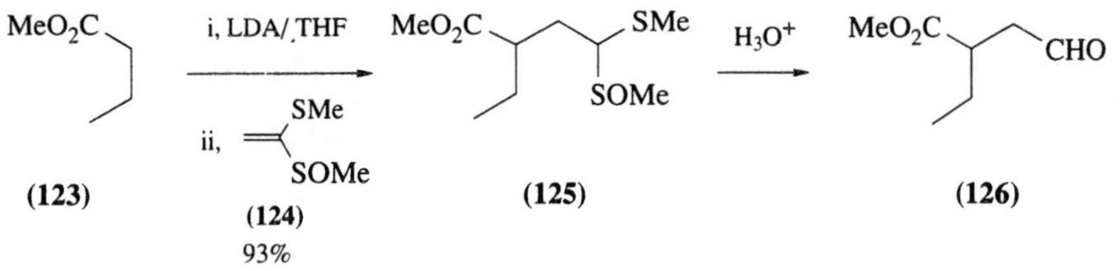

MeO_2C (**123**); i, LDA/ THF; ii, SMe, SOMe (**124**) 93%; (**125**); H_3O^+; CHO (**126**)

Scheme 16

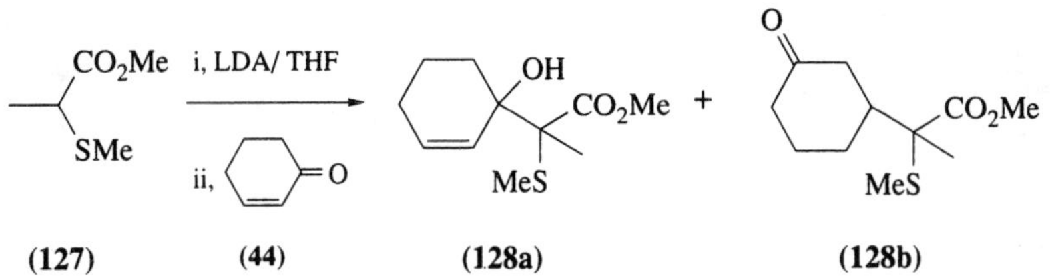

CO_2Me, SMe (**127**); i, LDA/ THF; ii, (**44**); OH, CO_2Me, MeS (**128a**) + O, CO_2Me, MeS (**128b**) (28)

	(**128a**)	(**128b**)
–78 °C	63%	7%
25 °C	0%	85%

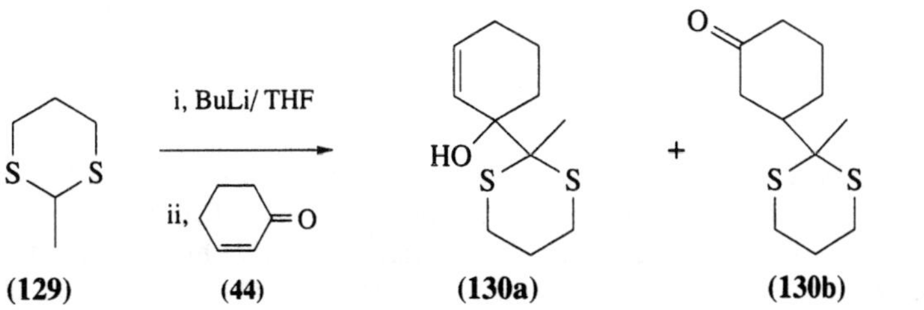

(**129**); i, BuLi/ THF; ii, (**44**); HO, S, S (**130a**) + O, S, S (**130b**) (29)

	(**130a**)	(**130b**)
no HMPA	99%	0%
1 equiv. HMPA	7%	93%
2 equiv. HMPA	3%	97%

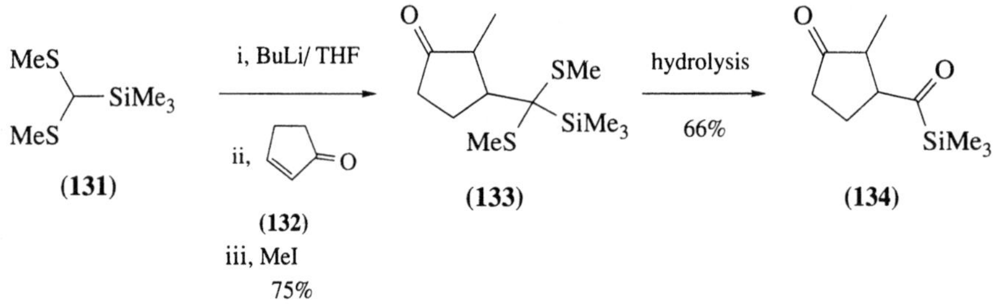

Scheme 17

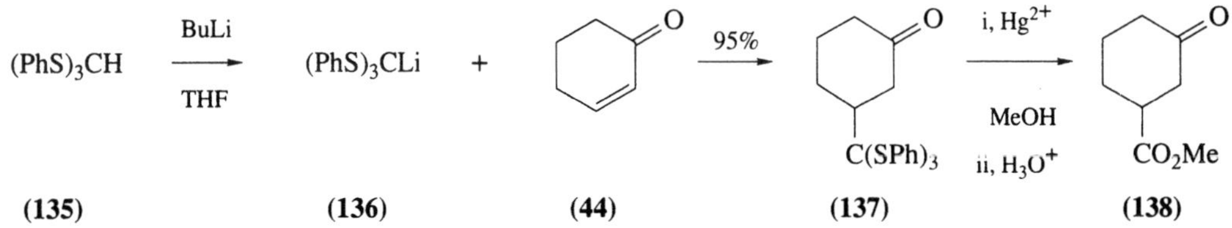

Scheme 18

An interesting difference in regiochemistry of addition has been observed with sulfur-substituted allyl anions.[37] The anion of the sulfide (**139**; equation 30) adds to cyclopentenone (**132**) in THF/HMPA (without HMPA 1,2-addition predominates, as expected from the above arguments) to give mainly the product of α-addition (**140a**) rather than that of γ-addition (**140b**).[37a,37b] However, addition of the anion of the corresponding sulfoxide (**141**) affords only the product of γ-addition, namely (**142b**).[37c] The use of the corresponding optically active allyl sulfoxides[37d–37f] causes the new carbon–carbon bond to be formed with virtually complete diastereoselectivity, *e.g.* (**143**; equation 31) gives (**144**) with 96% *de*.[37f]

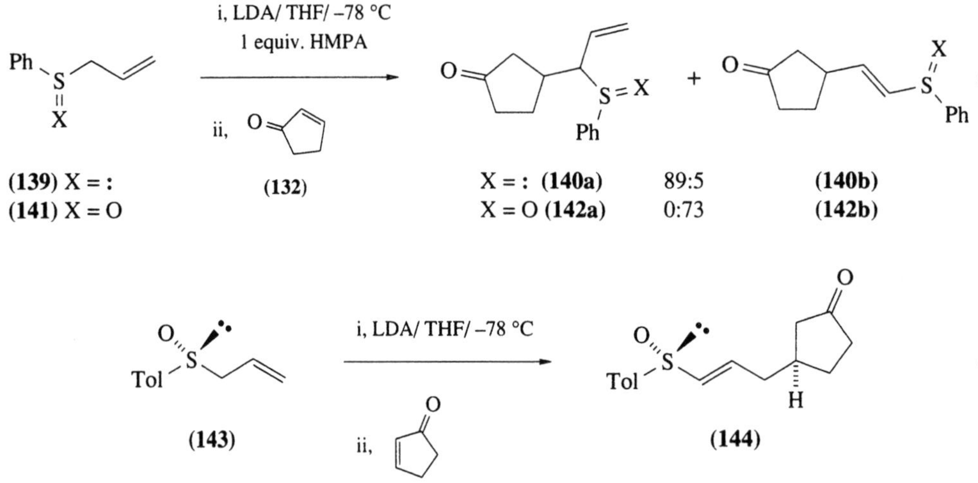

(30)

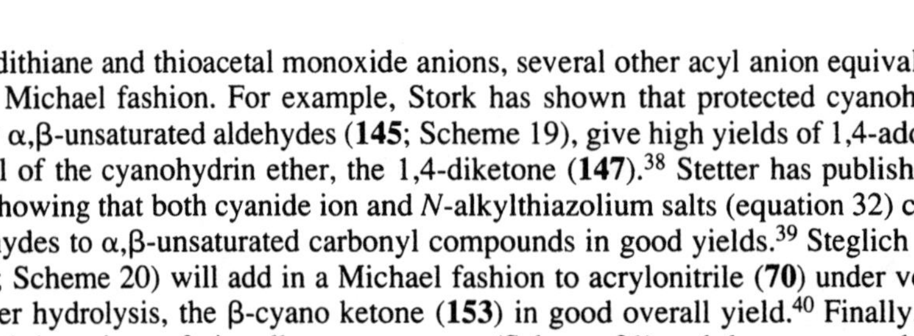

(31)

In addition to dithiane and thioacetal monoxide anions, several other acyl anion equivalents[35b,35c] have been added in a Michael fashion. For example, Stork has shown that protected cyanohydrins of aldehydes, especially α,β-unsaturated aldehydes (**145**; Scheme 19), give high yields of 1,4-addition and, after two-step removal of the cyanohydrin ether, the 1,4-diketone (**147**).[38] Stetter has published more than a score of papers showing that both cyanide ion and *N*-alkylthiazolium salts (equation 32) catalyze the 1,4-addition of aldehydes to α,β-unsaturated carbonyl compounds in good yields.[39] Steglich has shown that oxazolones (**152**; Scheme 20) will add in a Michael fashion to acrylonitrile (**70**) under very mild conditions to give, after hydrolysis, the β-cyano ketone (**153**) in good overall yield.[40] Finally, McMurry and others have added the anions of nitroalkanes to enones (Scheme 21) and then converted the nitro group to a carbonyl by reduction and hydrolysis.[41]

Since nitro groups are readily converted to carbonyls by various procedures,[41,42] nitroalkenes can serve as α-carbonyl cation equivalents in Michael reactions.[43] For example, addition of the amide enolate (**158**; Scheme 22) to the nitroalkene (**157**) affords, in 60% yield, the Michael adduct (**159**), which can be con-

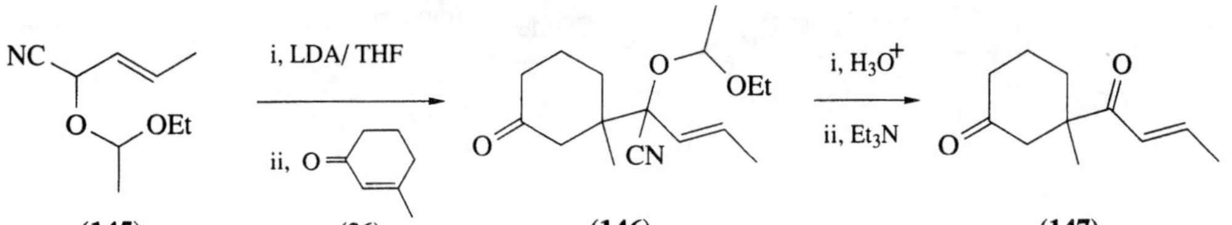

(145) **(36)** 70–85% **(146)** **(147)**

Scheme 19

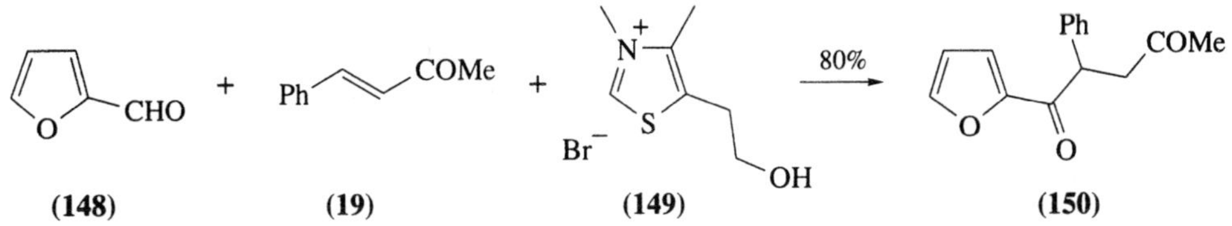

(148) **(19)** **(149)** **(150)** (32)

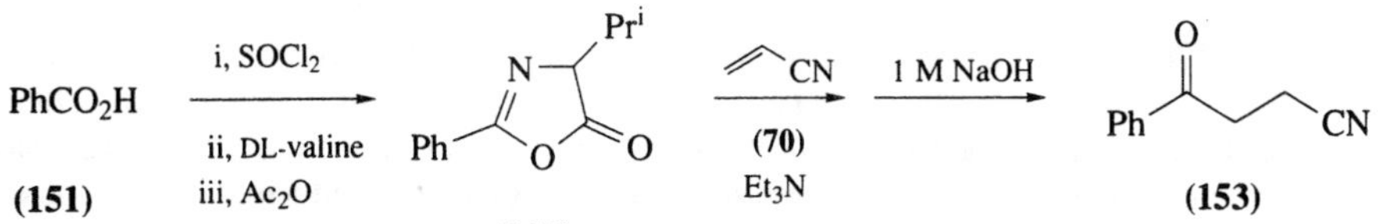

(151) **(152)** **(153)**

Scheme 20

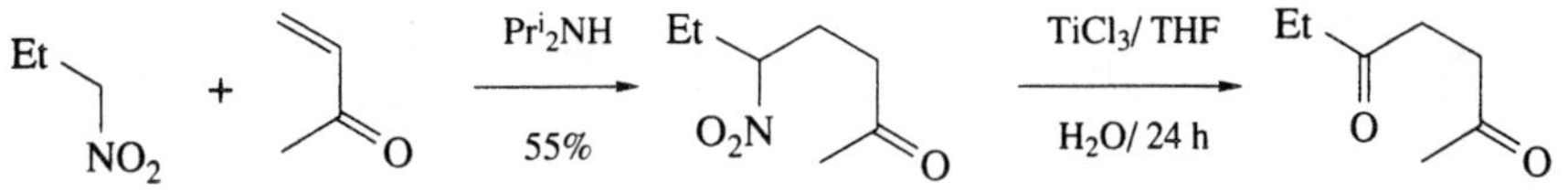

(154) **(30)** **(155)** **(156)**

Scheme 21

verted to the ketone (**160**) in 90% yield.[43a,43b] The highly stabilized anions of nitroalkanes are useful in Michael additions not only as acyl anion or α-carbonyl cation equivalents (*via* conversion to carbonyls) but also, *via* reductive removal of the nitro group, as alkyl anion equivalents or as alternatives to simple alkyl halides to give overall alkylation *via* a Michael addition process.[44] For example, treatment of a simple nitroalkane (**161**; Scheme 23) with DBU or tetramethylguanidine (TMG) in the presence of a Michael acceptor, such as MVK (**30**), affords the Michael adduct (**162**) which can be readily reduced with tributylstannane and AIBN to the ketone (**163**).[44b] Likewise, a vinylnitro compound can serve as a surrogate alkylating agent. Thus the nitroalkene (**164**; equation 33) can be used to trap the enolate (**165**, generated by cuprate addition to the corresponding cyclopentenone) to give the Michael adduct (**166**), the nitro group of which is removed reductively to give (**167**), a protected form of prostaglandin E_1.[44c]

In a like manner, vinyl sulfoxides and sulfones can serve as the equivalents of vinyl and alkenyl cations *via* a process involving a Michael addition followed by an elimination.[45] For example, addition of the enolate of the β-keto ester (**168**; Scheme 24) to phenyl vinyl sulfoxide (**169**) furnished, in 50% yield,

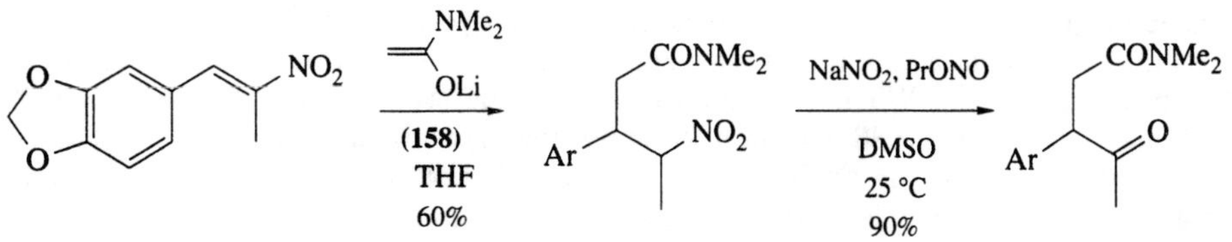

(157) **(159)** **(160)**

Scheme 22

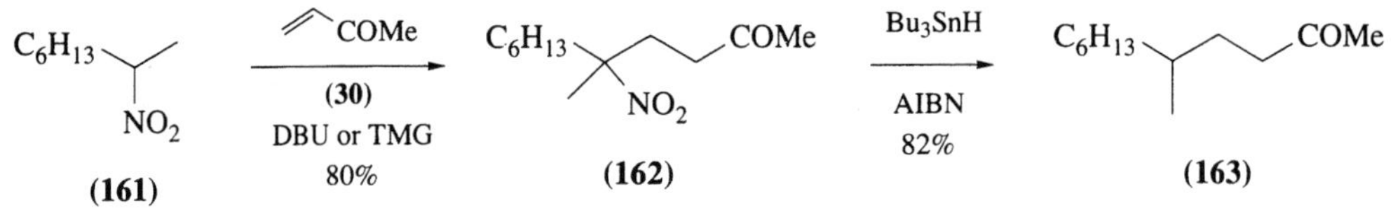

Scheme 23

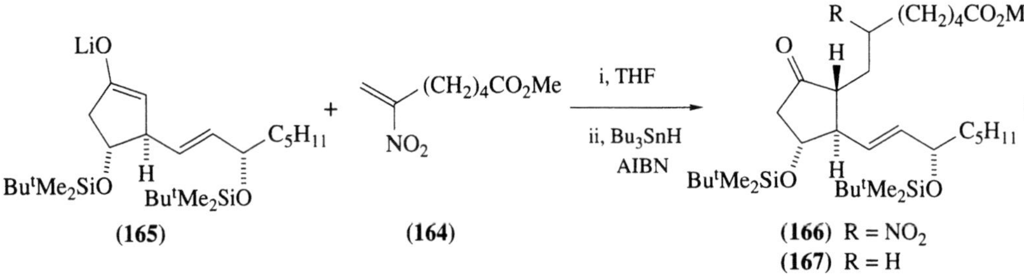

(33)

the Michael adduct (**170**) which lost phenylsulfenic acid on pyrolysis to give the vinyl-substituted β-keto ester (**171**) in 60% yield, the vinyl sulfoxide (**169**) being a vinyl cation equivalent in this process.[45a] In a similar addition–elimination process, treatment of alkenyl sulfones (**172**) with cyanide and a crown ether in refluxing *t*-butyl alcohol afforded the α-methylene nitrile (**173**) in good yield *via* Michael addition, proton transfer and elimination of phenylsulfinic acid (equation 34).[45b]

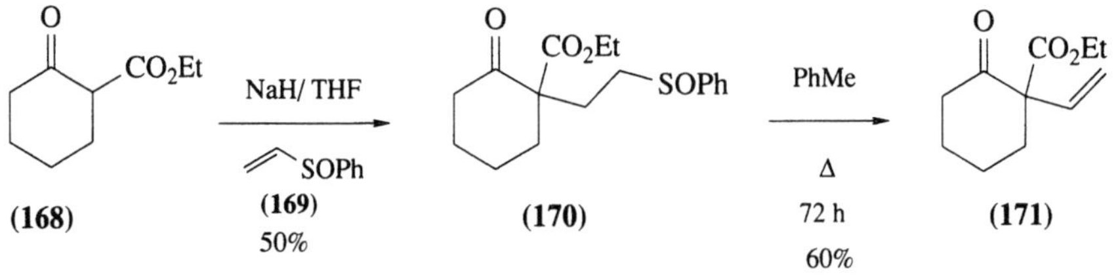

Scheme 24

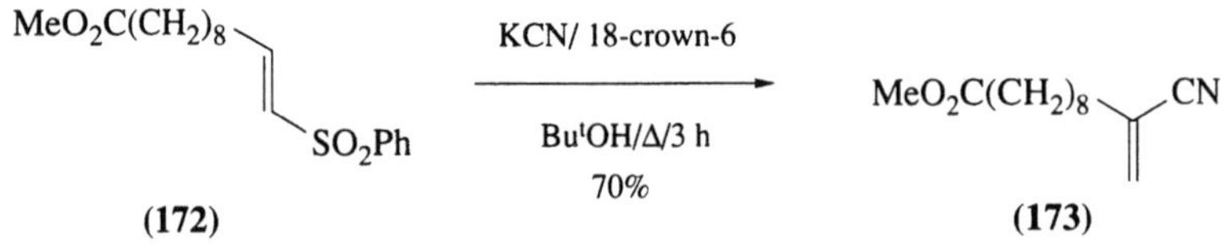

(34)

Michael additions followed by intramolecular acylation of the enolate so generated have been used often in organic synthesis to generate cycloalkanones.[46] For example, treatment of the dianion of the diester (**174**; equation 35) with 2,6-dichlorophenyl acrylate (**175**) affords the product of Michael addition–Dieckmann condensation (**176**) in 67% yield.[46a] Similarly, the anion (**178**; equation 36) of the carbonate ester produces a high yield of the lactone (**179**) when added to the unsaturated lactam (**180**).[46b] This process has been widely used in anthracycline and other polyketide aromatic syntheses.[47] The addition of a phthalide anion bearing a good leaving group that is also acidifying (arylsulfonyl or cyano) to a cyclohexenone or quinone monoketal affords the polycyclic hydroquinone (**182**; equation 37) or (**185**; equation 38) in one step *via* addition, intramolecular acylation, loss of the cyano or arylsulfonyl anion and final aromatization.[47a–47m] When the anion of the *o*-toluate (**186**) is added to the β-methoxy enone system (**187**), the *o*-acylphenol (**188**; equation 39) is obtained *via* a Michael addition, elimination and intramolecular Claisen condensation.[47o] The simple Michael addition of a naphthylacetonitrile (**189**; equation 40) to a cyclohexenecarboxylate (**190**) has also been used to produce intermediates, like (**191**), for anthracycline synthesis.[48]

Several other procedures for trapping the initially generated enolate of the Michael addition have also been used synthetically. Foremost among them is elimination of a good leaving group,[49] an overall addition–elimination process, illustrated by a specific example in equation (41).[49a] Using the 2-nitropropenyl pivalate (**195**; Scheme 25) Seebach[49c] was able to carry out a double Michael process, an initial addi-

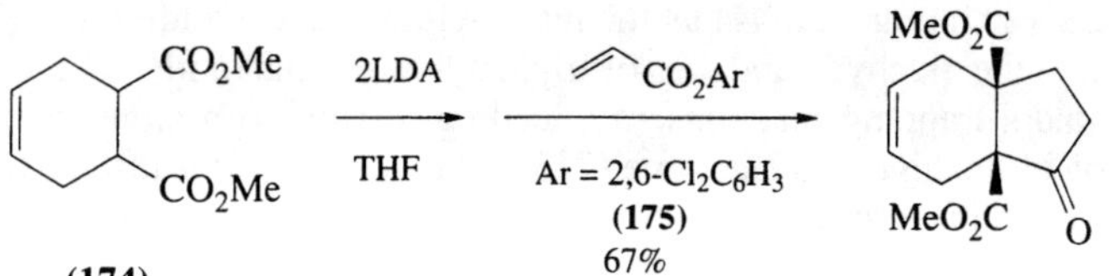

(174) (175) (176) (35)

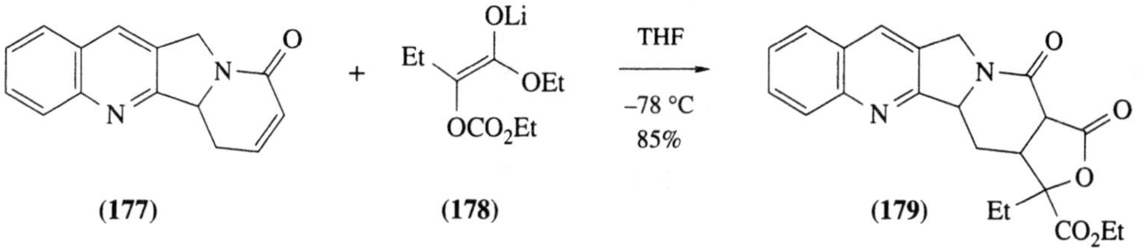

(177) (178) (179) (36)

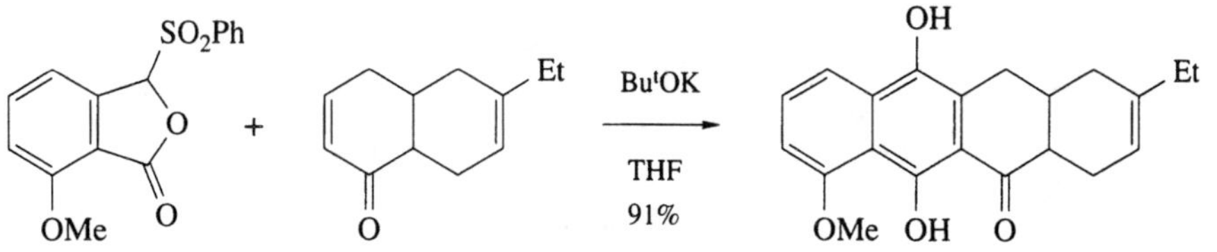

(180) (181) (182) (37)

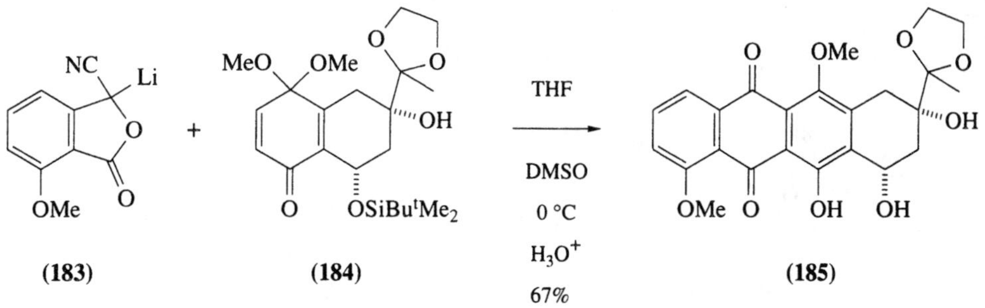

(183) (184) (185) (38)

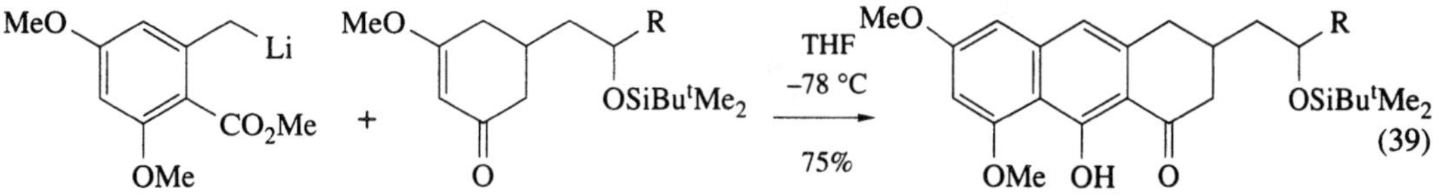

(186) (187) (188) (39)

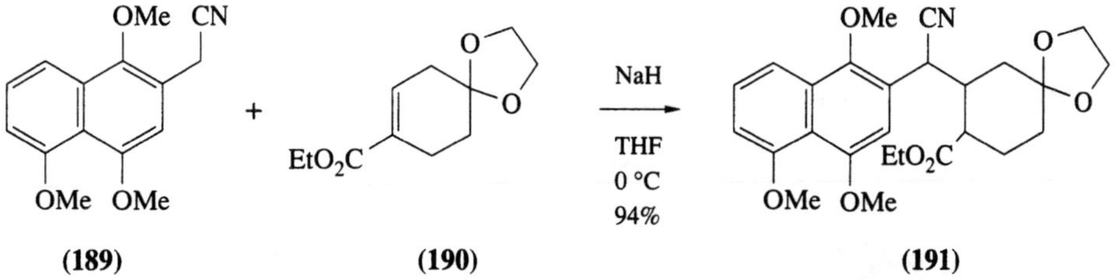

(189) (190) (191) (40)

tion–elimination followed by a simple Michael addition, to give (**197**). If a vinyl phosphonate or phosphonium salt is used as the Michael acceptor, the anionic intermediate can be trapped by an intramolecular Wittig reaction with a nearby ketone to give (**199**; equation 42) or (**202**; equation 43).[50] In certain cases, this process can lead to cyclohexenone products which are structurally transposed from those of the normal Robinson annulation (Scheme 26).[50c] An interesting variant on this approach uses the anion of a formylphosphorane (**207**; equation 44) as the nucleophile in a 1,6-addition to (**208**), followed by an aldol of the enolate on to the aldehyde and loss of triphenylphosphine oxide to give (**209**) in fair yield.[50e] Other intramolecular aldol-trapping reactions (not leading to Robinson annulation) have also been reported,[51] *e.g.* formation of the hindered enedione (**213**; Scheme 27).[51a] It is also possible to trap the initial enolate with an allylic leaving group or an isonitrile to produce cyclopentenes (**215**; Scheme 28),[52a,b] pyrrolines (**217**; Scheme 29)[52c] or pyrroles (**220**; Scheme 30).[52d]

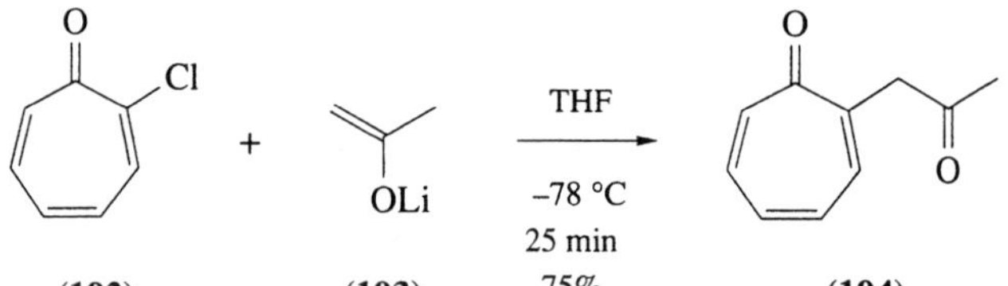

(**192**) + (**193**) → (**194**); THF, −78 °C, 25 min, 75% (41)

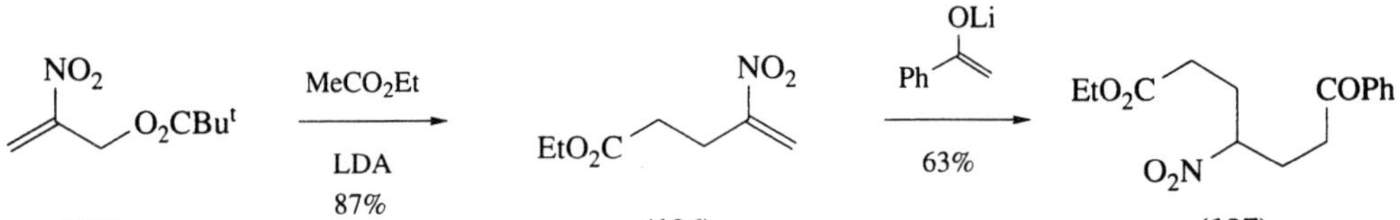

(**195**) → (**196**) → (**197**); $MeCO_2Et$, LDA, 87%; 63%

Scheme 25

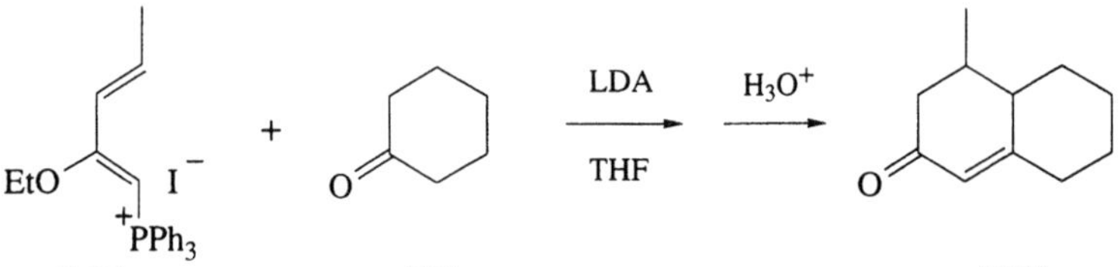

(**198**) + (**18**) → (**199**); LDA, THF; H_3O^+ (42)

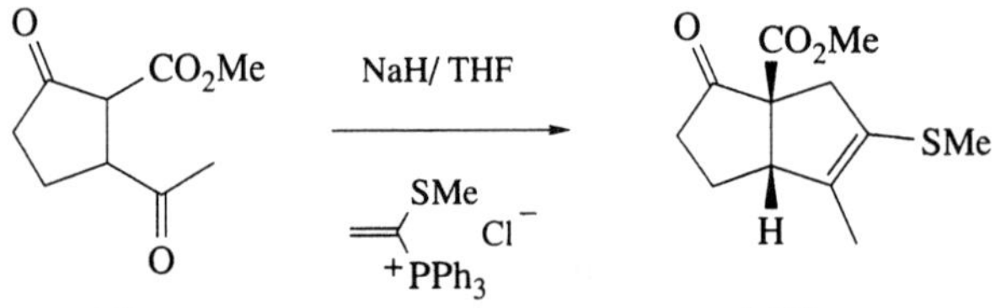

(**200**) + (**201**) → (**202**); NaH/ THF, 97% (43)

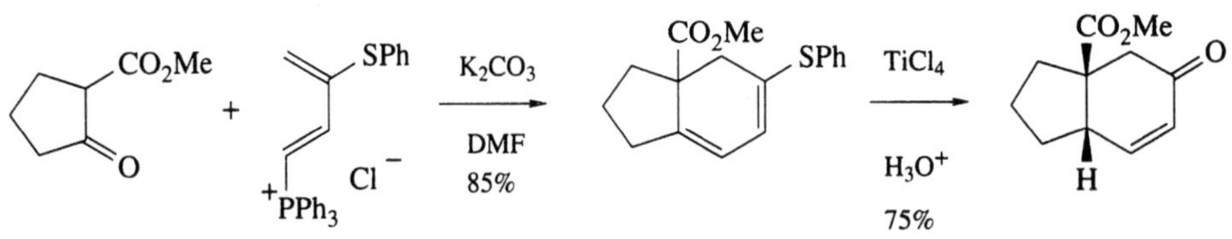

(**203**) + (**204**) → (**205**) → (**206**); K_2CO_3, DMF, 85%; $TiCl_4$, H_3O^+, 75%

Scheme 26

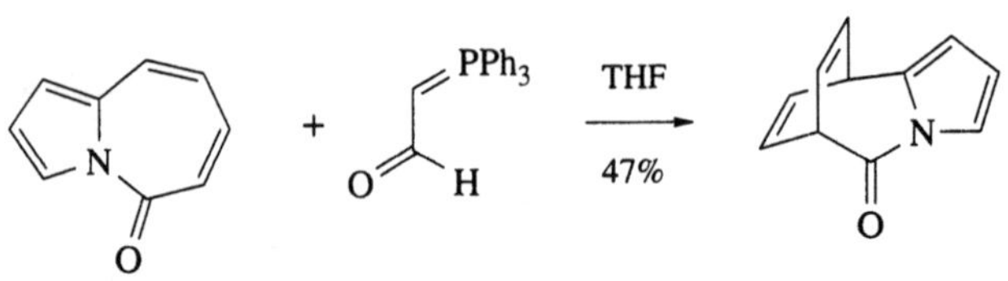

(**208**) + (**207**) → (**209**); THF, 47% (44)

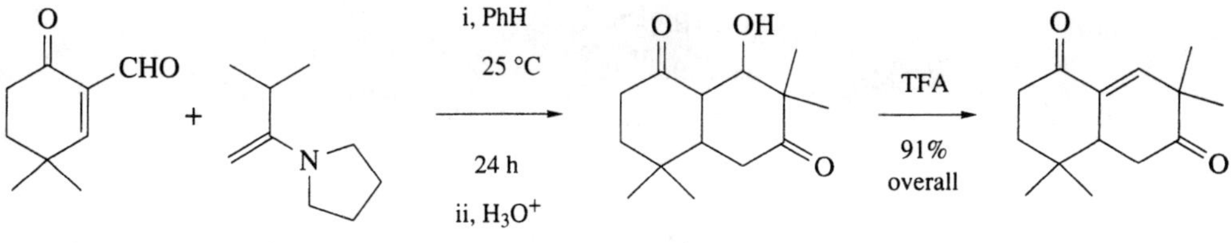

i, PhH 25 °C 24 h ii, H_3O^+ 92%; TFA 91% overall

(210) **(211)** **(212)** **(213)**

Scheme 27

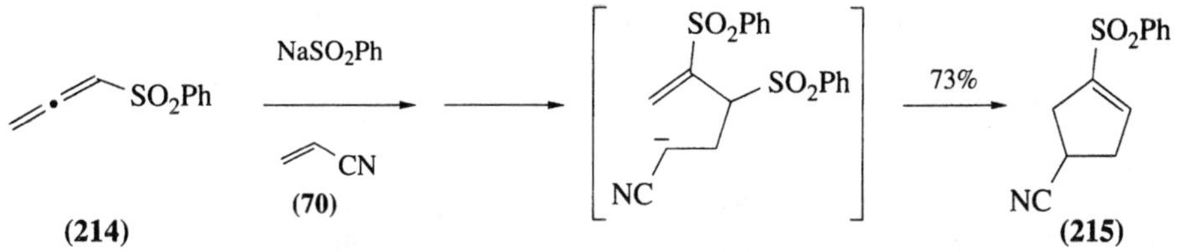

$NaSO_2Ph$; **(70)** CN; 73%

(214) **(215)**

Scheme 28

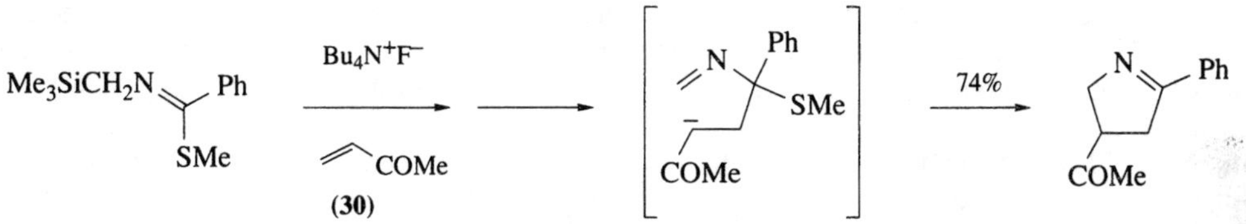

Me_3SiCH_2N Ph SMe; $Bu_4N^+F^-$; **(30)** COMe; 74%

(216) **(217)**

Scheme 29

Ph COPh **(219)** $KOBu^t$; 95%

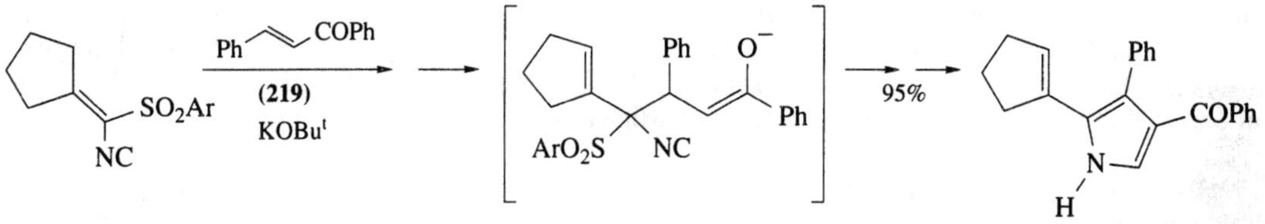

(218) **(220)**

Scheme 30

With certain exceptions, simple intermolecular Michael additions often proceed in poor yield when the acceptor is a β,β-disubstituted enone or ester, presumably due to steric hindrance of nucleophilic attack. However, this lack of reactivity can be overcome by two techniques. The first one involves the use of high pressures, which can greatly increase the yields of these hindered additions.[53] Thus, addition of diethyl malonate (**1**) to 3-methylcyclohexenone (**36**) in the presence of DBN at 15 kbar (1 bar = 100 kPa) affords the desired addition product (**221**; equation 45) in 71% yield.[53a] Alternatively, if the acceptor is sufficiently activated (usually with two strongly electron-withdrawing groups), then normal Michael addition is possible, even when contiguous quaternary centers are being formed. Holton has used this process to prepare very hindered compounds (**224**; equation 46) in high yield.[54]

The anion generated by a Michael addition can also be trapped by another Michael addition. Obviously, continued repetition of this process produces polymeric esters, nitriles, *etc.*, in a typical anionic poly-

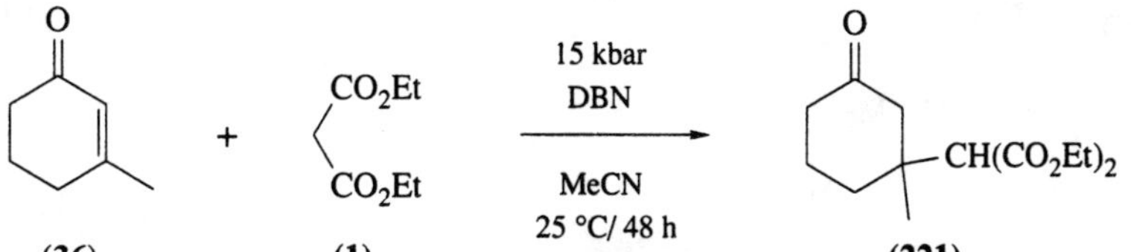

15 kbar DBN MeCN 25 °C/ 48 h 71% $CH(CO_2Et)_2$ (45)

(36) **(1)** **(221)**

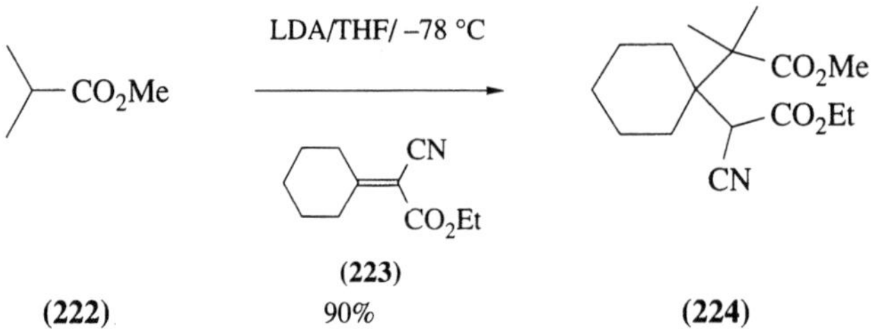

LDA/THF/ –78 °C; (223) 90%; (222) → (224) (46)

merization process. However, in certain cases, other chain-terminating steps can intervene to produce quite useful nonpolymeric material from these multi-Michael processes. Since an excellent review of these multicomponent one-pot annulations has recently appeared,[55] only a few key examples are noted here. Addition of excess lithium dimethylcuprate (**225**; Scheme 31) in THF to methyl crotonate (**226**) at 10 °C afforded the single stereoisomer of the keto diester (**227**) in 49% yield *via* cuprate addition, two Michael additions and final Dieckmann condensation.[56] Cyclopentanones and their bicyclic derivatives (**228**; equation 47) and (**230**; equation 48) are available from acrylates using cyanide as the initial nucleophile, again *via* two Michael additions followed by Dieckmann-like condensation.[57] A similar process using the thiomethyl sulfone (**231**; equation 49) and acrylate gives the β-keto ester (**232**) in 81% yield.[58] When the activating group of the Michael acceptor is also a good leaving group, as in nitroalkenes or vinylphosphonium salts, an S_N2 reaction is the final step. For example, treatment of the anion of isophorone (**17**) with 1-nitropropene (**233**) gave the tricyclooctanones (**236a,b**; Scheme 32) in 63% yield, by way of the intermediates (**234**) and (**235**).[59] A synthetically useful example of these one-pot annulations involves the condensation of an enolate ion with an enone bearing an α-bromoacrylate unit.[60] Treatment of (**237a,b**; Scheme 33) with (**238**) gives rise to the tricyclic compounds (**239a,b**), which are converted into the enals (**240a,b**). Curiously, the major products obtained in the parent and methyl-substituted cases have the opposite stereochemistry at one of the new ring junctions.[60a,60b] Posner has explored extensively the synthetic utility of these multi-Michael reactions.[61] The reaction of the enolate of cyclohexanone (**18**; Scheme 34) with α-substituted acrylates (**241**) or (**242**) gave rise to the cyclohexanol products (**243**) and (**244**), respectively, in good yield.[61a] The dibromo compound (**244**) could be converted easily into the aromatic diester (**245**). When vinyltriphenylphosphonium bromide (**247**; equation 50) is used as the acceptor with the enol borate from benzoylcyclopentane (**246**), the phosphine oxide (**248**) is produced *via* two Michael additions and a final Wittig reaction.[61b] Other multiple Michael reactions, especially those involving intramolecular Michaels, are discussed in the section on intramolecular additions.

Finally, a discussion of the relative stereochemistry of such reactions completes this section on intermolecular Michael additions. It should be pointed out that Chapter 1.5, this volume, is devoted wholly to asymmetric induction in Michael reactions and thus this topic is not covered here.

The diastereoselectivity of intermolecular Michael additions was first examined in two areas: the Robinson annulation using (*E*)-3-penten-2-one (**249**; equation 51)[62] and the reaction of enamines of cyclic

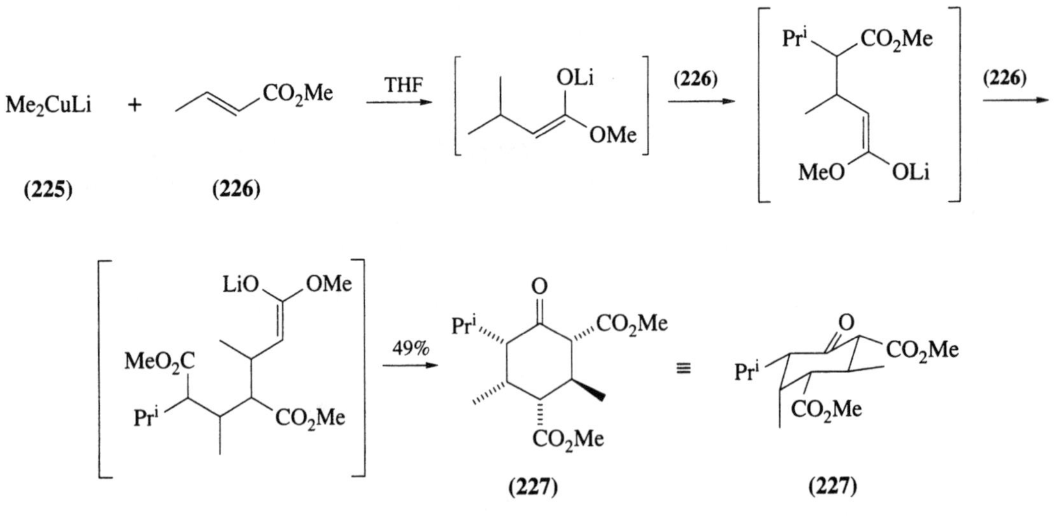

Me_2CuLi (225) + (226) —THF→ [OLi, OMe] —(226)→ [Pr^i, CO_2Me, MeO, OLi] —(226)→ [LiO, OMe, MeO_2C, Pr^i, CO_2Me] —49%→ (227) ≡ (227)

Scheme 31

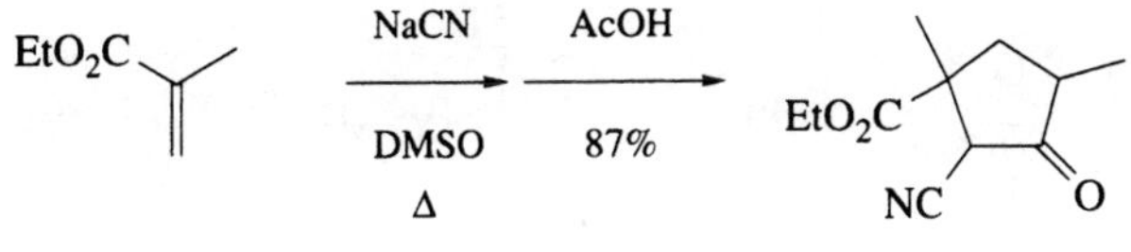

EtO_2C NaCN DMSO Δ AcOH 87% EtO_2C NC O (47)

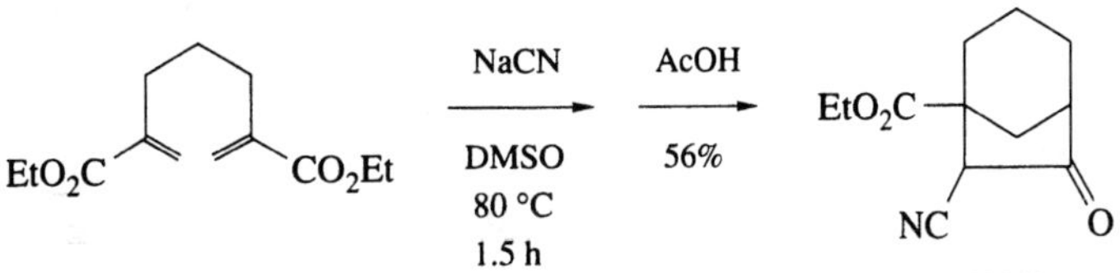

(27) **(228)**

EtO_2C CO_2Et NaCN DMSO 80 °C 1.5 h AcOH 56% EtO_2C NC O (48)

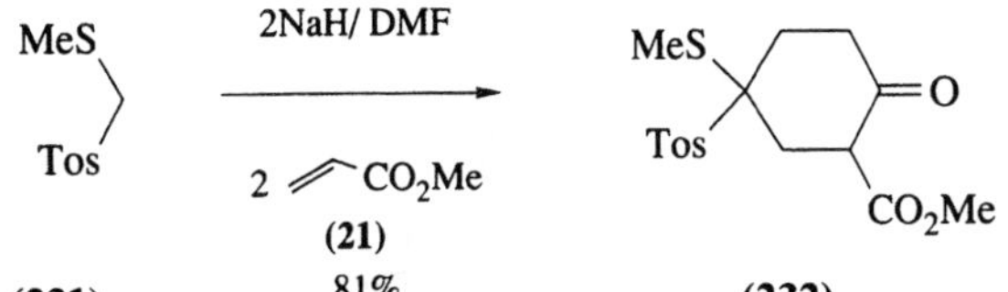

(229) **(230)**

MeS Tos 2NaH/ DMF 2 CO_2Me **(21)** 81% MeS Tos O CO_2Me (49)

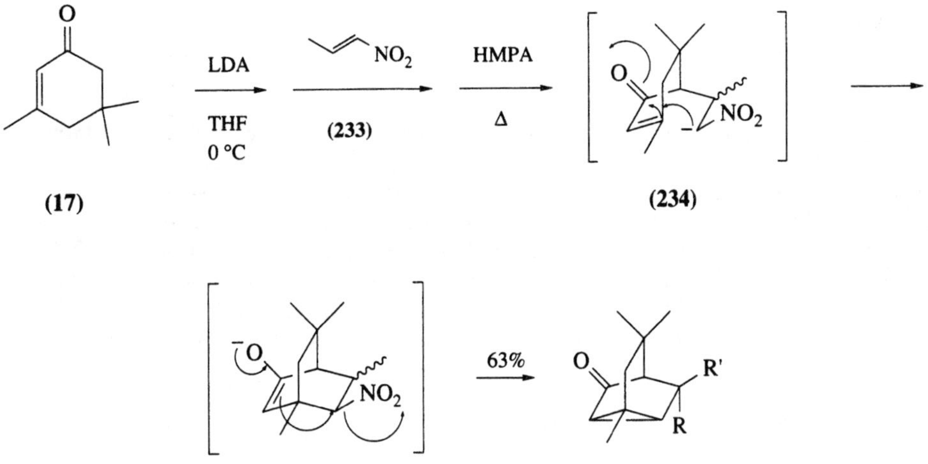

(231) **(232)**

O LDA THF 0 °C NO_2 **(233)** HMPA Δ O NO_2

(17) **(234)**

O NO_2 63% O R' R

(235) **(236a)** R = Me, R' = H

(236b) R = H, R' = Me

Scheme 32

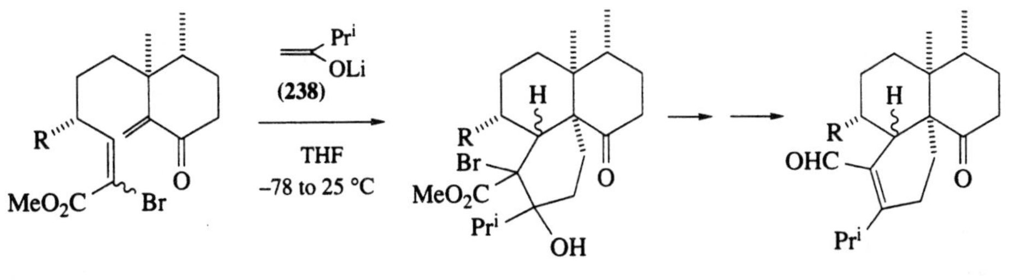

R MeO$_2$C Br O Pr^i OLi **(238)** THF –78 to 25 °C H R Br MeO$_2$C Pr^i OH O H R OHC Pr^i O

(237a) R = H 25% **(239a)** **(240a)** R = H, β-H

(237b) R = Me 30% **(239b)** **(240b)** R = Me, α-H

Scheme 33

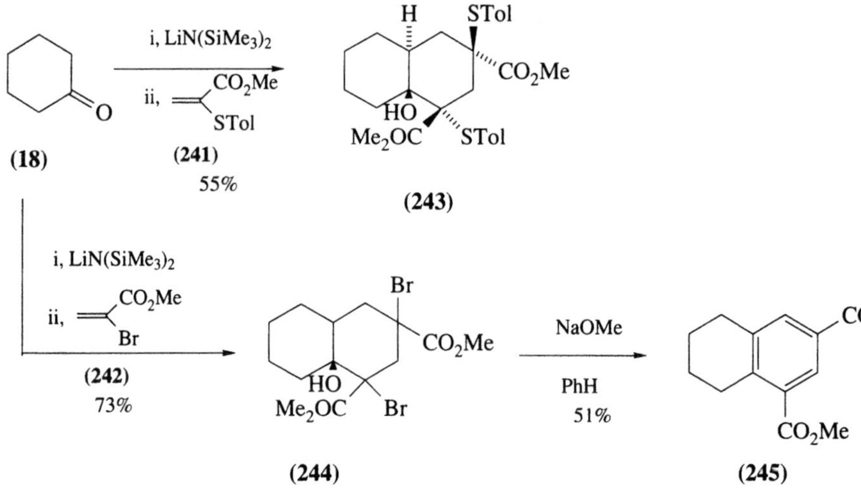

(18) i, $LiN(SiMe_3)_2$; ii, $CH_2{=}C(CO_2Me)STol$ **(241)** 55% → **(243)**

(18) i, $LiN(SiMe_3)_2$; ii, $CH_2{=}C(CO_2Me)Br$ **(242)** 73% → **(244)** NaOMe, PhH, 51% → **(245)**

Scheme 34

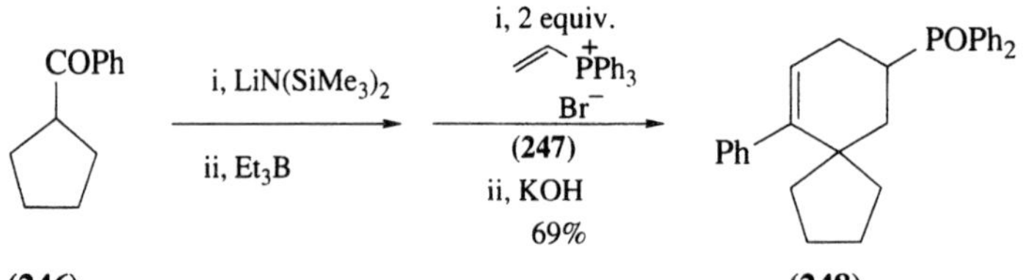

(246) i, $LiN(SiMe_3)_2$; ii, Et_3B → i, 2 equiv. $CH_2{=}CHP^+Ph_3\ Br^-$ **(247)**; ii, KOH, 69% → **(248)** (50)

ketones with β-substituted-α,β-unsaturated nitro compounds and sulfones.[63,64] Several groups have reported the stereoselective Michael additions of 2-substituted cyclohexanone enolates (**250**) and (**66**) with (**249**) to give, after aldol and dehydration, the corresponding *cis*- or *trans*-substituted octalones, (**251a, 251b**) and (**252a, 252b**), respectively (equation 51). In general, the use of apolar or alcoholic solvents affords a mixture in which the *cis* isomer (**251a**) predominates while the *trans* isomer (**252a**) is major in DMSO.[62a–62c,62f,62h,62i] The best example of this solvent dichotomy involves the production of only the *cis*-(**252a**) or only the *trans*-(**252b**) octalones from 2-methylcyclohexanone (**66**) and (**249**) in dioxane or DMSO, respectively.[62g] With 2-methylcyclohexane-1,3-dione (**82**) or its enamine (**253**) the *trans* isomer

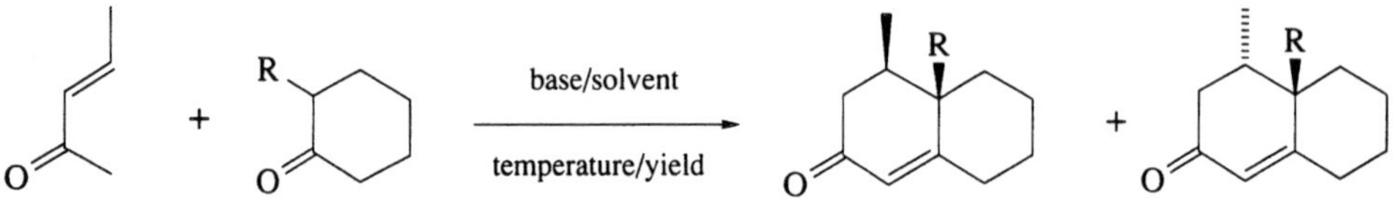

(**249**) + (**250**)/(**66**) —base/solvent, temperature/yield→ (**251**) + (**252**) (51)

		base/solvent/temperature/yield		ratio	
(**249**)	(**250**) R = CO_2Me	$KOt\text{-}C_5H_{11}/t\text{-}C_5H_{11}OH/–15$ °C/65%	(**251a**)	3.08:1	(**252a**)
		NaOMe/DMSO/29 °C/72%		0.61:1	
	(**66**) R = Me	NaH/dioxane/25 °C/65%	(**251b**)	>95:<5	(**252b**)
		NaH/DMSO/25 °C/72%		<5:>95	

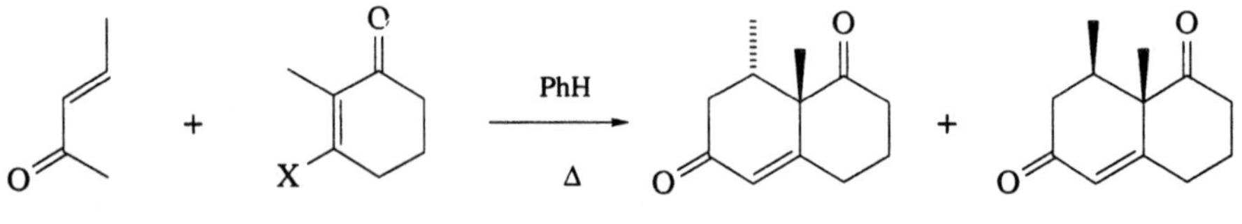

(**249**) + (**82**)/(**253**) —PhH, Δ→ (**254a**) + (**254b**) (52)

(**249**)	(**82**) X = OH	KOH/EtOH/pyrrolidine/71%	(**254a**)	6:1	(**254b**)
	(**253**) X = N(pyrrolidinyl)	AcOH/NaOAc/59%		10:1	

(**254a**) is the major product in benzene with ethanol or acetic acid, while in DMF a 1:1 mixture results (equation 52).[62d,62e]

The reaction of the morpholine enamine of cyclohexanone (**255**; equation 53) with β-nitrostyrene (**49**) or 1-nitropropene (**233**) followed by hydrolysis affords the *erythro* products (**256a,b**) almost exclusively; similar selectivity was also seen for the corresponding sulfones.[63] Seebach has extended these results to enamines of acyclic ketones and various enolates with little or no loss in selectivity and has proposed a 'topological rule' for such carbon–carbon bond-forming processes, as well as others.[64] As applied to the Michael addition, the best approach of the two reagents is represented by transition structure (**258**; equation 54) which has three important features: (i) all of the existing bonds are staggered; (ii) the donor (enolate, enamine) and acceptor (α,β-unsaturated system) units are in a *gauche* arrangement; and (iii) the actual donor and acceptor atoms, *e.g.* X and Z in (**258**), are situated close to each other (to minimize charge separation or to allow for chelation of metal ions). This approach, using a so-called closed transition state, predicts that the *erythro* product (**259**) would be formed preferentially, and thus correctly rationalizes most of the experimental results. For instance, Mulzer[65] reported that reaction of the lithium enolate of various β-lactones (**260a–260d**) with dimethyl maleate (**261**; equation 55) gives predominantly the corresponding diastereomer (**262a–262d**) rather than (**263a–263d**), which conforms to Seebach's topological rule *via* (**264**). However, the stereochemical results with ester and amide enolates are less easily rationalized by this model.[66,67] For example, Yamaguchi reported that the lithium enolates of simple esters (**266**; equation 56) react with β-substituted acrylates (**3**) in THF–HMPA at –78 °C to afford high yields of the *erythro*-glutarates (**267a**).[66a] The *threo* product was favored even when the HMPA was added after the lithium enolate was formed or when no HMPA was used at all! The lithium enolate of *t*-butyl propionate (**268**; equation 57) in THF without HMPA adds to ethyl crotonate (**269**) to give mainly the *threo* product (**270b**).[66a] It is hard to correlate these results with those of Heathcock,[67b] who showed that reaction of the (*E*)- and (*Z*)-enolates (**272**; Scheme 35) of *t*-butyl propionate (**268**) with the enone (**273**) gave stereospecifically the *threo*-glutarate (**274a**) from the (*E*)-enolate and the *erythro*-glutarate (**274b**) from the (*Z*)-enolate (results in accord with the Seebach model). The results with amide enolates are even harder to explain. Yamaguchi reports high *threo* selectivity in the addition of the enolate of the amide (**275**; equation 58) to ethyl crotonate (**269**),[66c] while Heathcock reports very low selectivity for reaction of the same enolate with a series of enones like (**273**).[67a] With very hindered amides (**277**; equation 59), Yamaguchi reports high *erythro* selectivity and good diastereomeric excess.[66c] Heathcock could achieve high *erythro* selectivity by using the cyclic amide (**279**; equation 60).[67a] These results and others[68] are not consistent with the Seebach model. Heathcock suggests an 'open transition state' model (Scheme 36) in which the interaction between the alkenic methyl and the pyrrolidine ring causes (**281b**) to be destabilized relative to (**281a**).[67a] Furthermore, when a halide is present on the alkyl group of the β-alkylacrylate moiety, as in (**282**; equation 61) a stereospecific alkylation of the kinetically generated enolate occurs to give high yields of the *trans*-substituted products, with only (**283a**; *threo* Michael adduct) being formed in the absence of HMPA and only (**283b**; *erythro* Michael adduct) being formed in the presence of HMPA.[66b]

Whatever the explanation for the stereocontrol, these processes are quite useful synthetically, enabling one to prepare nearly pure *erythro* or *threo* diastereomers in high yield. A number of groups have shown that by using a chiral auxiliary in the enamine, ester or amide unit, products of very high enantiomeric excess can be obtained.[69]

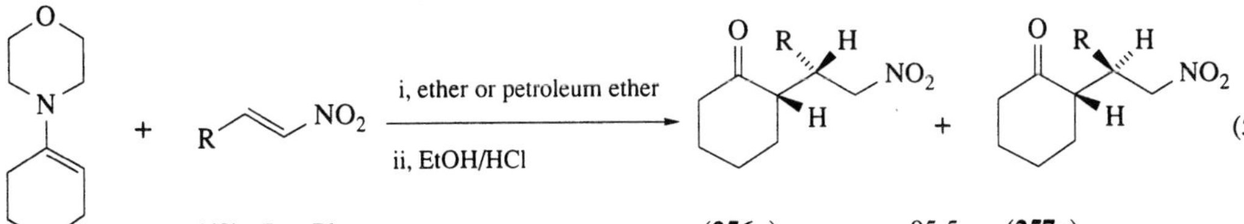

(**255**) + (**49**) R = Ph, (**233**) R = Me → i, ether or petroleum ether; ii, EtOH/HCl → (**256a**) + (**257a**) 95:5; (**256b**) + (**257b**) 99:1 (53)

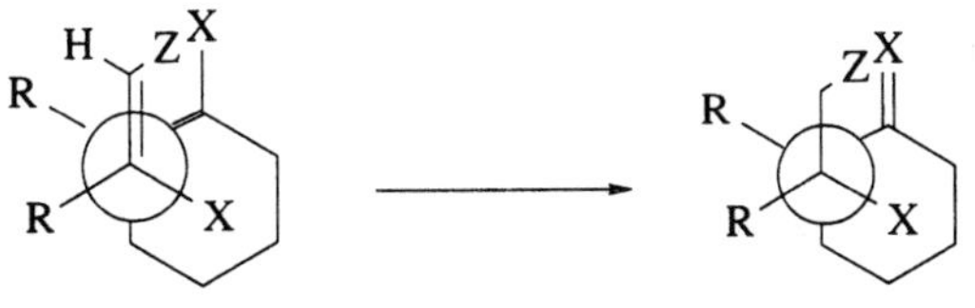

(**258**) → (**259**) (54)

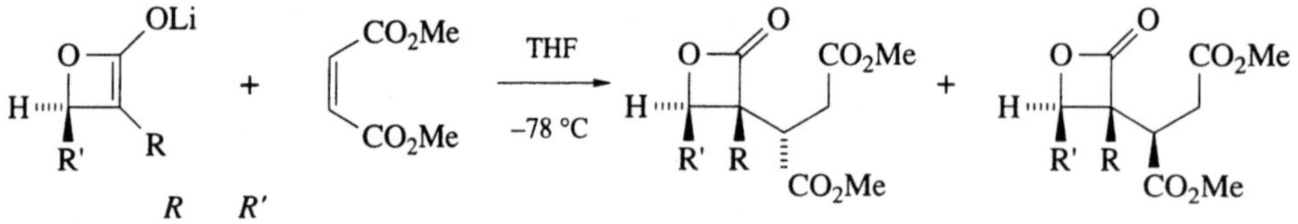

THF, −78 °C (55)

	R	R′						
(260a)	Ph	Bu^t	(261)	73%	(262a)	89:11	(263a)
(260b)	Ph	Pr^i		68%	(262b)	82:18	(263b)
(260c)	Ph	Me		56%	(262e)	85:15	(263e)
(260d)	Me	Bu^t		45%	(262d)	100:0	(263d)

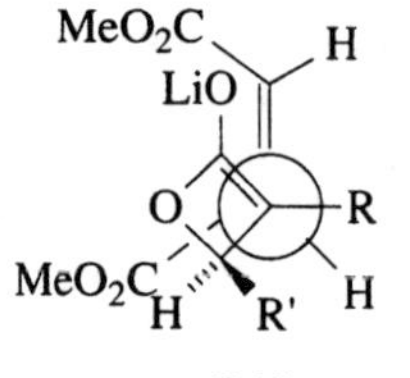

(264)

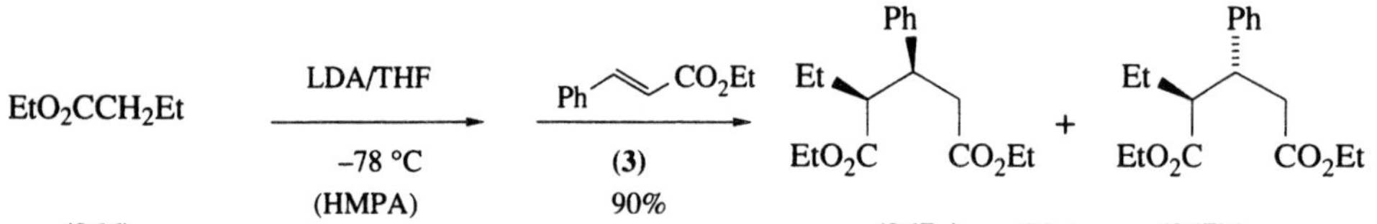

EtO_2CCH_2Et (266) —LDA/THF, −78 °C (HMPA)→ (3), 90% → (267a) >20:1 (267b) (56)

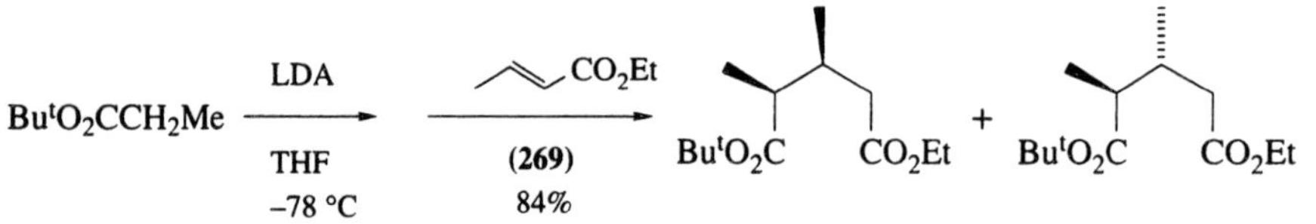

$Bu^tO_2CCH_2Me$ (268) —LDA, THF, −78 °C→ (269), 84% → (270a) <1:20 (270b) (57)

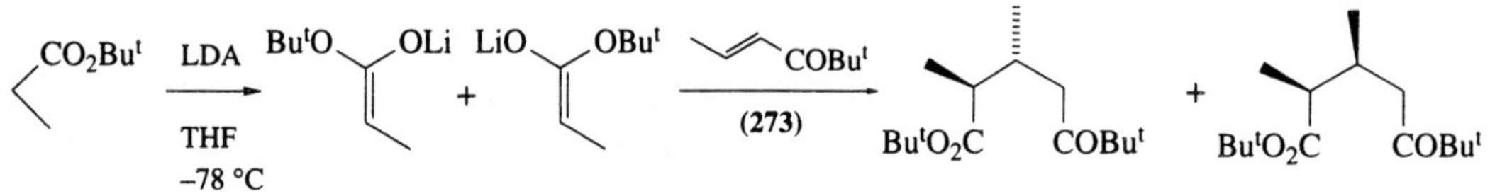

(268) —LDA, THF, −78 °C→ (Z)-(272) + (E)-(272) —(273)→ (274a) + (274b)

	(Z)-(272):(E)-(272)		(274a):(274b)
with HMPA	88:12	73%	13:87
no HMPA	6:94	85%	95:5

Scheme 35

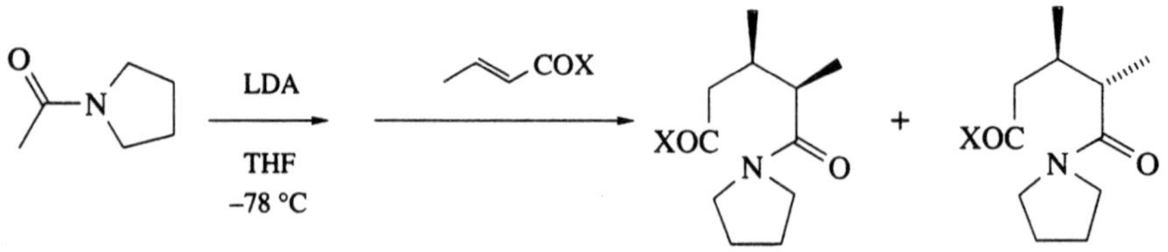

(275) —LDA, THF, −78 °C→ COX → (58)

(269) X = OEt, 92%	(275a)	1:20	(276a)
(273) X = Bu^t, 90%	(275b)	55:45	(276b)

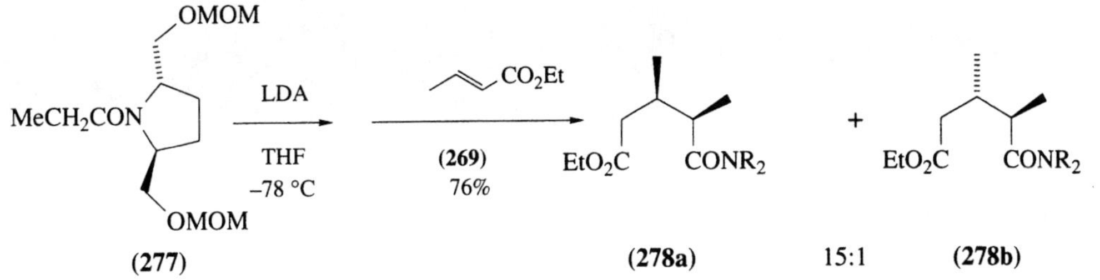

OMOM; $MeCH_2CON$; OMOM **(277)** — LDA, THF, –78 °C → CO_2Et **(269)** 76% → EtO_2C ... $CONR_2$ **(278a)** + EtO_2C ... $CONR_2$ **(278b)**, 15:1 (59)

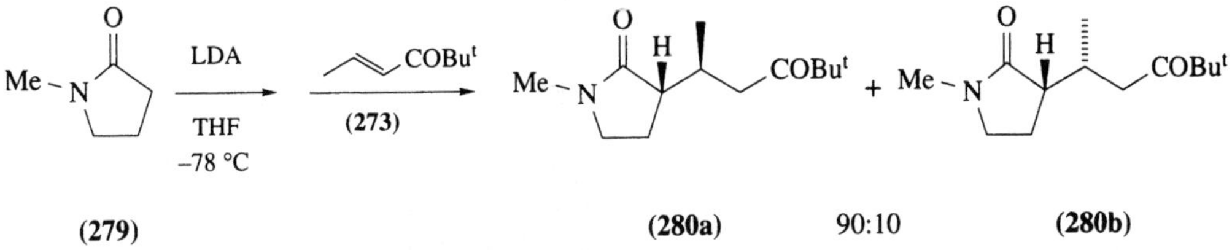

Me–N (O) **(279)** — LDA, THF, –78 °C → COBu^t **(273)** → **(280a)** (COBu^t) + **(280b)** (COBu^t), 90:10 (60)

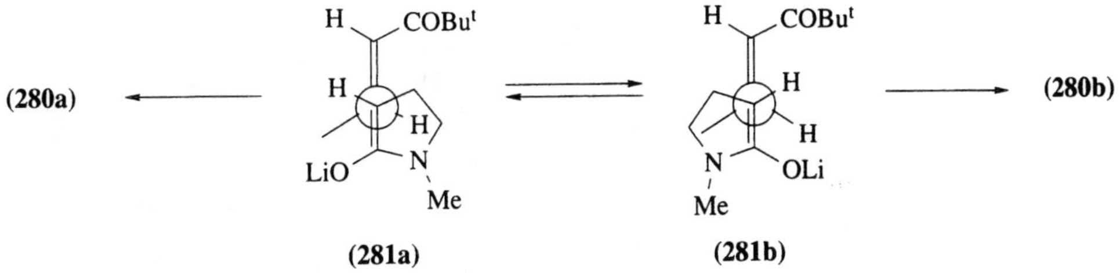

(280a) ← **(281a)** (H, COBu^t, H, H, LiO, N, Me) ⇌ **(281b)** (H, COBu^t, H, H, N, OLi, Me) → **(280b)**

Scheme 36

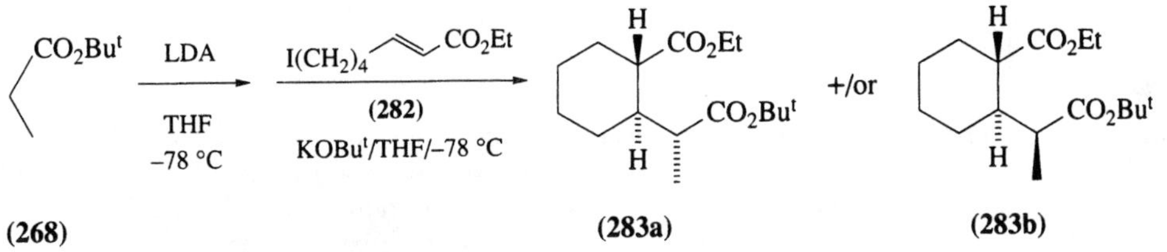

CO_2Bu^t **(268)** — LDA, THF, –78 °C → $I(CH_2)_4$ ... CO_2Et **(282)**, KOBu^t/THF/–78 °C → **(283a)** (CO_2Et, CO_2Bu^t) +/or **(283b)** (CO_2Et, CO_2Bu^t) (61)

	(283a)	(283b)
no HMPA	100%	0%
with HMPA	0%	100%

With cyclic Michael acceptors, *e.g.* cycloalkenones and unsaturated lactones, the relative stereochemistry is determined by other substituents on the ring. Generally an allylic substituent causes the addition to occur from the opposite face for steric reasons, giving the *trans* product (equation 62).[33d] The addition to cyclohexenone systems (equation 63) generally occurs in an antiparallel sense to give the axial product.[70] In the reaction of cyanide with octalones such as (**90**; equation 64) under basic conditions, a mixture of *cis* and *trans* products (**289a**,**289b**) is obtained which can be made to favor the *cis* isomer (**289a**).[71] However, the use of Nagata's reagent (HCN/Et_3Al) affords the *trans* isomer (**289b**) as the major product.[71a] This reagent has been used in several steroid syntheses to give predominately the desired *trans* C/D-ring juncture (**291b**; equation 65).[71a]

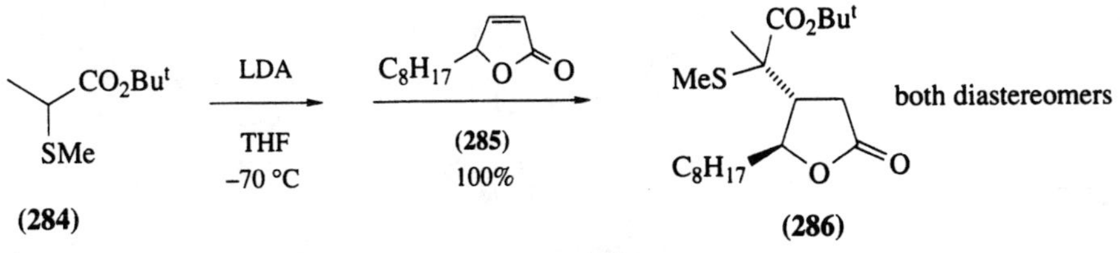

CO_2Bu^t, SMe **(284)** — LDA, THF, –70 °C → C_8H_{17} (lactone) **(285)** 100% → CO_2Bu^t, MeS, C_8H_{17} **(286)** both diastereomers (62)

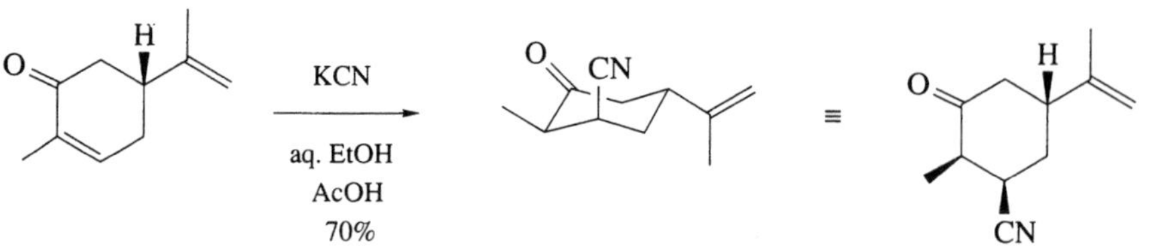

(90)		(289a)		(289b)
	KCN/NH_4Cl		43:57	
	$Me_2C(OH)CN/KCN/18$-crown-6		68:17	
	HCN/Et_3Al		29:71	

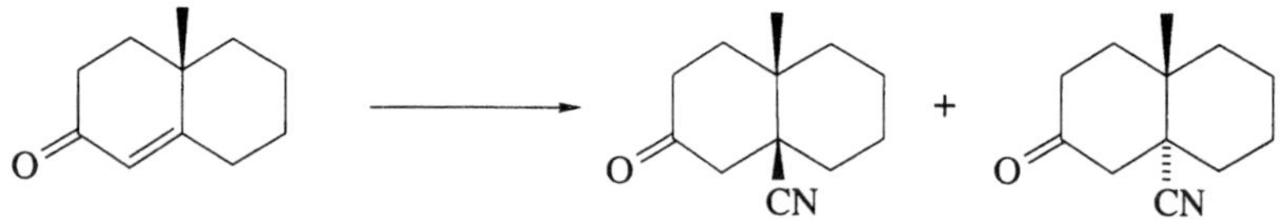

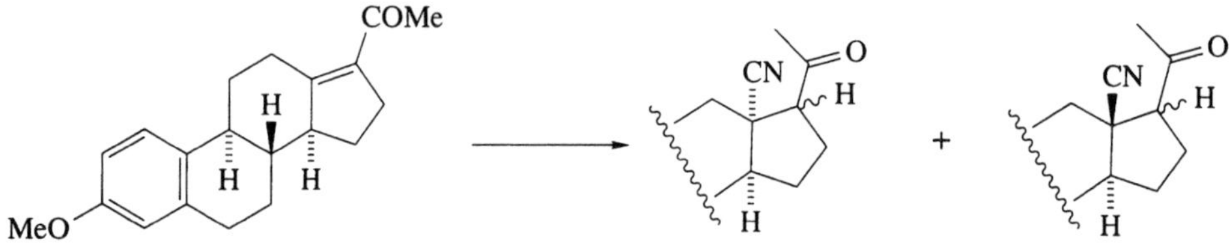

(290)		(291a)		(291b)
	KCN/NH_4Cl		57:22	
	HCN/Et_3Al		9:67	

1.1.2.1.2 Intramolecular additions

(i) Five-membered rings

Several groups have prepared cyclopentane systems by intramolecular Michael addition.[72] Reaction of the triester (**292**; Scheme 37) with phenyl vinyl ketone (**293**) and base produces the cyclopentane (**295**) in good yield *via* an intermolecular (giving **294**) and subsequent intramolecular Michael addition.[72a] When the Michael acceptor is a cyclohexenone (Scheme 38), the *cis*-fused hydrindanone is produced (**298** or **297**).[72b–72f] Spiro systems can also be formed by these reactions (**300a,b**; equation 66) in which the Michael addition gives a spiro ring fusion.

Stork reported excellent stereochemical control in the intramolecular Michael addition to produce methyl-substituted cyclopentane systems.[73] For example, base-catalyzed conjugate addition of (**301a,b**; equation 67) followed by intramolecular aldol condensation affords the hydrindenones (**302a,b**) in high yields with excellent stereoselectivity.[73a] The corresponding β-keto esters (**304a,b**; equation 68) are cyclized with sodium hydride in benzene to give again predominantly the *trans*-cyclopentanes (**305a,b**).[73b] The use of chiral esters permits high absolute stereocontrol.[73c] This work has been extended to systems with allylic substituents, *e.g.* (**307**; equation 69) gives mainly (**308**).[73d]

The introduction of heteroatoms in the chain connecting the nucleophile and the acceptor does not affect the reaction. Bicyclic butyrolactones, which are of interest in the forskolin area, have been prepared by this approach, *e.g.* (**314**; Scheme 39) and (**317**; Scheme 40).[74] Pyrrolidines such as (**319**; equation 70) are also easily formed by an internal Michael reaction.[75] This chemistry has been used by Büchi in the

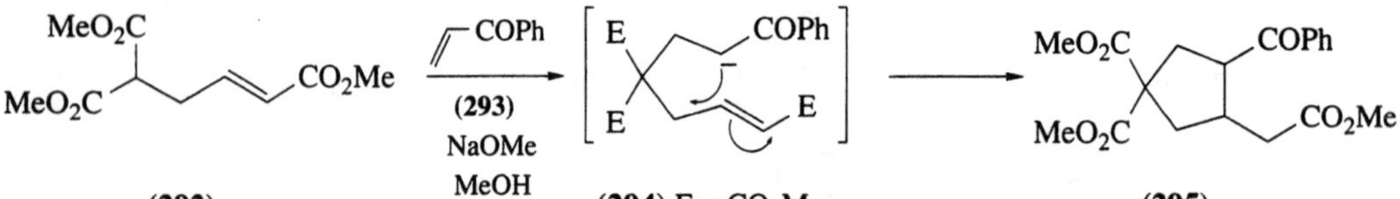

Scheme 37

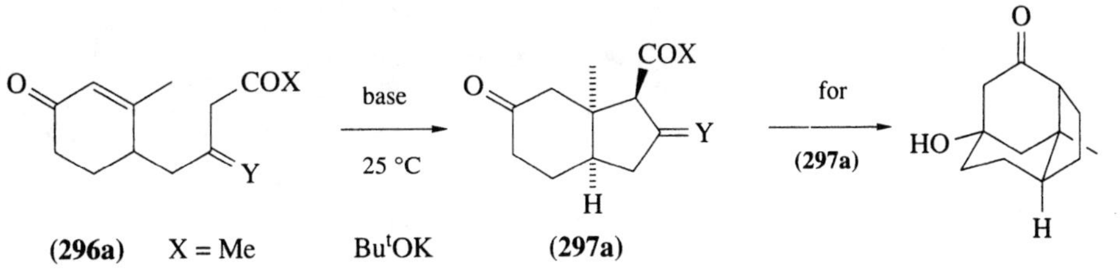

(296a)	X = Me	Bu^{t}OK	**(297a)**	**(298)** 75%
	Y = H_2	Bu^{t}OH/0.5 h		
(296b)	X = OEt	K_2CO_3	**(297b)** 45%	
	Y = O	EtOH/40 h		

Scheme 38

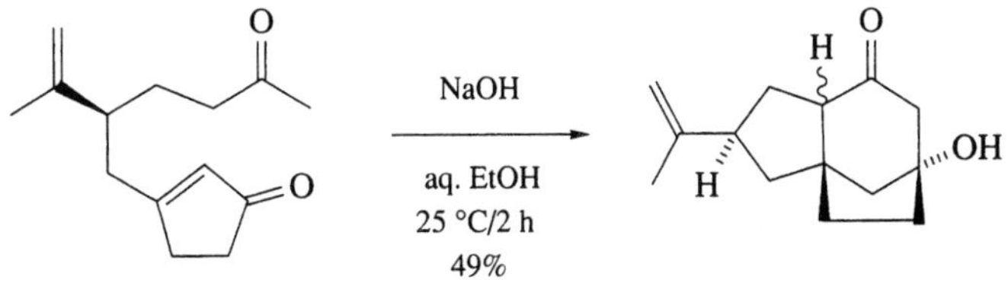

(299) → NaOH, aq. EtOH, 25 °C/2 h, 49% → **(300)** α-H 38%, β-H 11% (66)

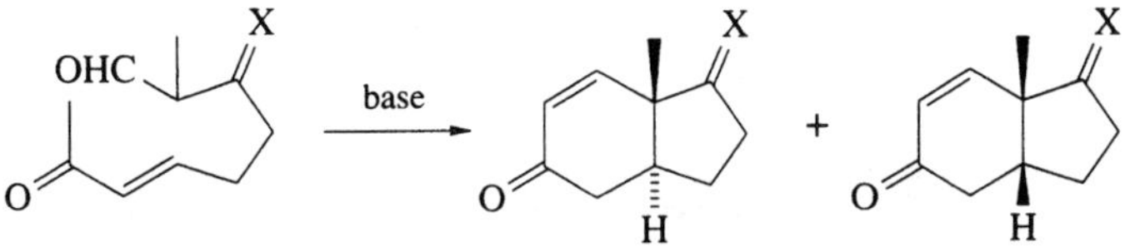

(301a) X = H_2	NaOMe	**(302a)**	3:1	**(303a)**
	Zr(OPr)$_4$		40:1	
(301b) X = O	LiOH	**(302b)**	4:1	**(303b)**
	Zr(OPr)$_4$		25:1	

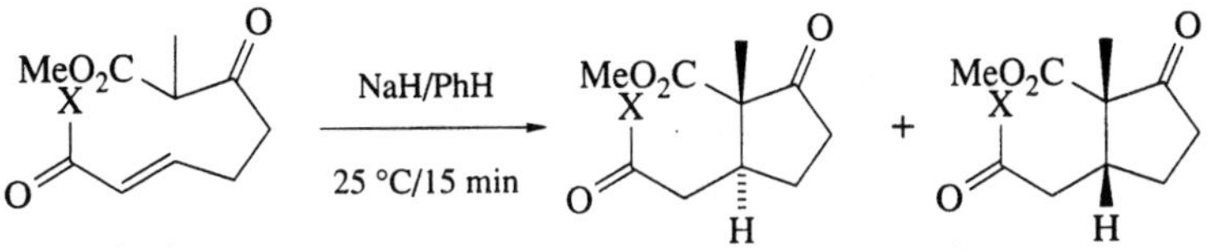

(67)

NaH/PhH, 25 °C/15 min (68)

(304a) X = Me	90%	**(305a)**	100:0	**(306a)**
(304b) X = OMe	90%	**(305b)**	22:1	**(306b)**

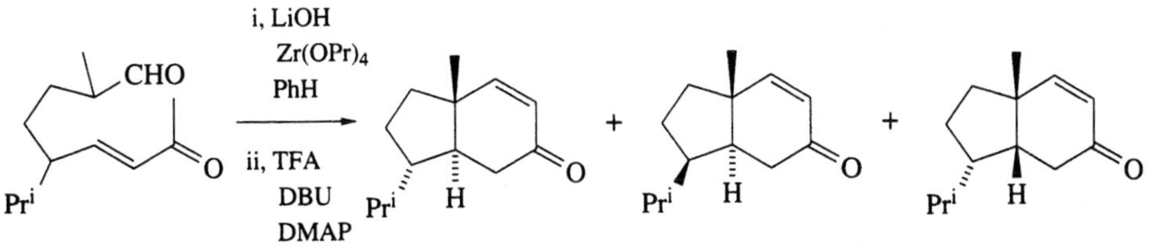

(307) → i, LiOH, Zr(OPr)$_4$, PhH; ii, TFA, DBU, DMAP, 80% → **(308)** + **(309)** + **(310)** (69)

10 : 1 : 1

synthesis of pentacyclic indole alkaloids such as vindoline and vindorosine, in which an acid-catalyzed Michael addition is followed by trapping of the spiroindolenium ion to give the products of an overall annulation process (**321a,b**; equation 71).[75a,75b] Finally, an intramolecular version of the thiazolium salt catalyzed Michael addition of an aldehyde has been used to generate a cyclopentanone system (**323**; equation 72) in a synthesis of hirsutic acid C.[76]

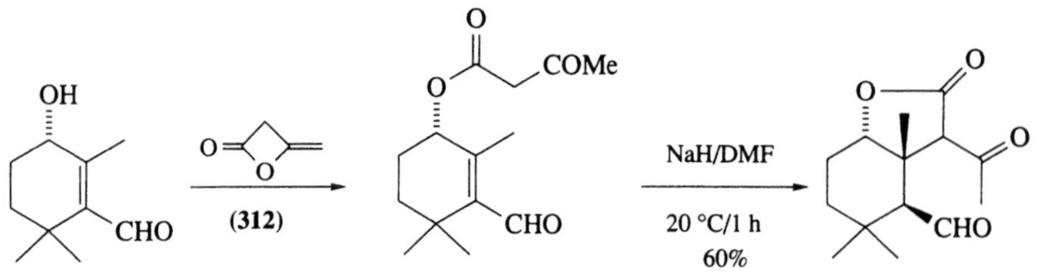

(311) (313) (314)

Scheme 39

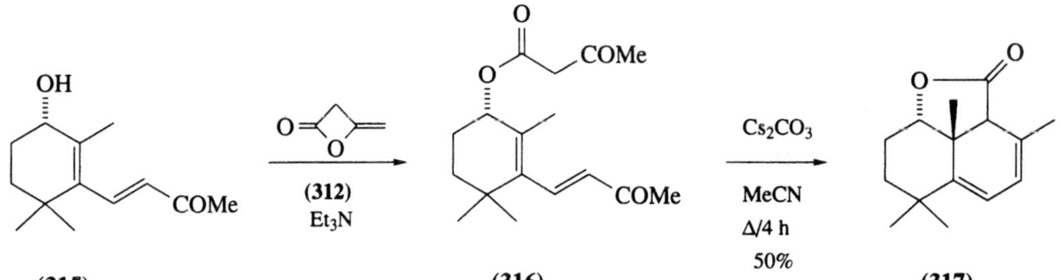

(315) (316) (317)

Scheme 40

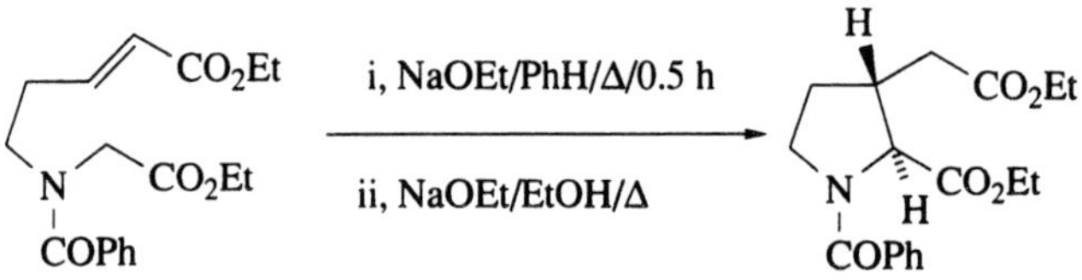

(70)

(318) (319)

(71)

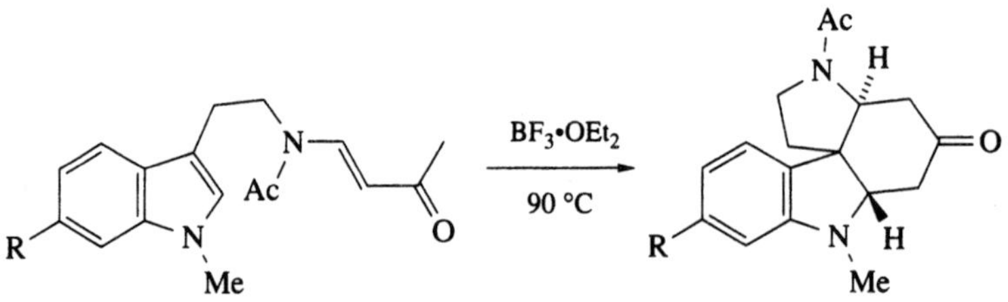

(320a) R = H 27 min, 38% (321a)
(320b) R = OTs 16 min, 89% (321b)

(72)

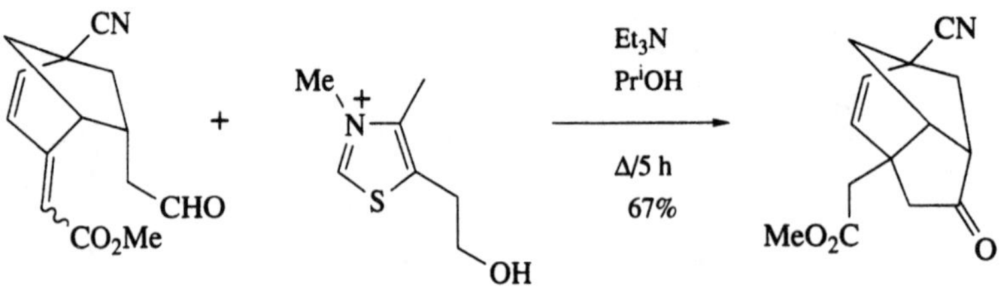

(322) (323)

(ii) Six-membered rings

4,4-Disubstituted cyclohexanones (**325**; equation 73) are easily prepared by a double Michael process, in which the intramolecular addition forms the six-membered ring.[77] Either the *cis* or *trans* compound can be obtained selectively. The enamines of cyclic ketones produce similar products.[77b] By using enynones 4,4-disubstituted cyclohexenones can be prepared by this route, *e.g.* griseofulvin (**328**; equation 74) (and its analogs) from the benzofuranone (**326**) and the enynone (**327**).[78] In fused systems, the *cis* ring juncture is again usually favored,[79] (**329**; equation 75) giving (**330**).[79a] The trione (**332**; Scheme 41) prepared by acidic hydrolysis of (**331**), also gives the *cis*-decalindione (**333**).[79c] Ivie showed that an intermediate similar to (**332**), namely (**336**; Scheme 42), could be prepared by intermolecular Michael addition to the dienone (**334**) and could then undergo intramolecular addition to give the *cis*-fused product (**337**).[79d,79e] The use of β-keto amides or esters as the nucleophilic components does not change the stereospecificity, the *cis* ring juncture again being preferred (equation 76).[80] However, there is a recent report of the opposite stereoselectivity, namely the formation of the *trans*-decalone (**341**; equation 77) from the keto triester (**340**).[81] Cyclization of the acyclic keto ester (**342**; equation 78), *via* its chiral enamine, gives exclusively the *trans*-substituted piperidine (**343**) in 80–83% yield and up to 90% *ee*.[82] This process of forming cyclohexane systems *via* an intramolecular Michael addition has been used often to form the A-ring in anthracycline syntheses (equations 79 and 80).[83] In these cases, the Michael acceptor is formed *in situ* by prototropic rearrangement; the products revert to anthraquinones by either loss of nitrous acid or air oxidation.

Bridged systems can also be prepared in high yield by this process.[84] For example, treatment of the keto diester (**348**; Scheme 43) with (**61**) in basic DMSO gave a 41% yield of the diketo diester (**352**) by way of the intermediates (**349**) to (**351**), involving two Michael additions and final Dieckmann condensation.[84a] Similar processes have also been used to make other bicyclo[3.3.1]nonanes and related com-

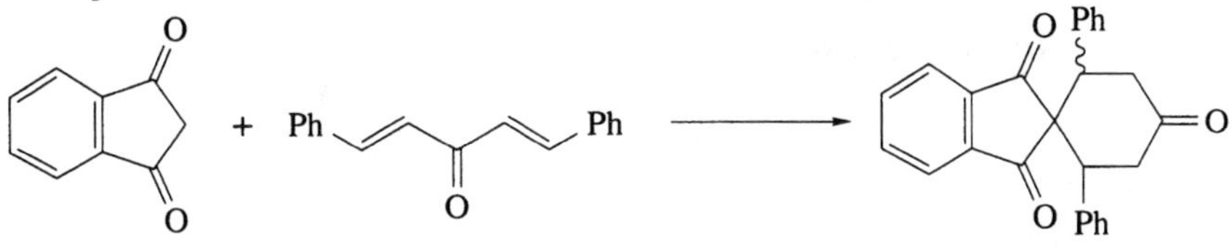

(73)

(**41**) (**324**) AcOH/Δ/1 h 63% NaOEt/EtOH 86% (**325a**) *trans* (**325b**) *cis*

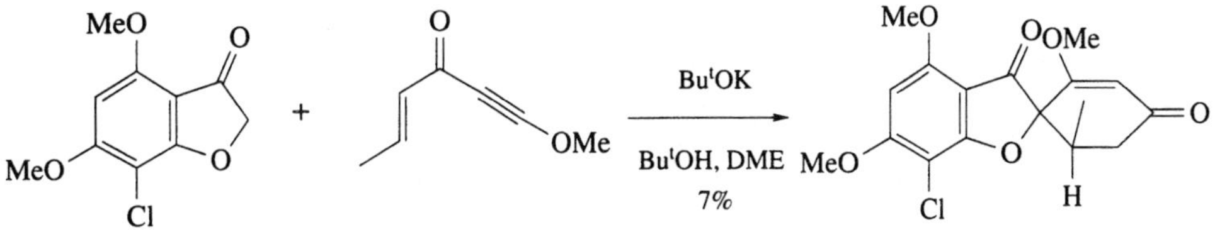

(74)

(**326**) (**327**) Bu^tOK, Bu^tOH, DME, 7% (**328**)

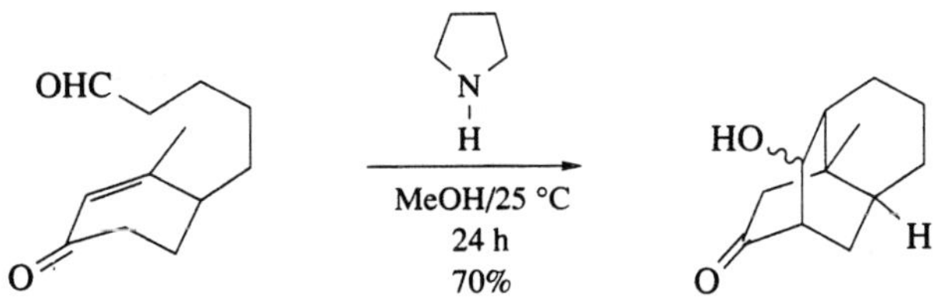

(75)

(**329**) MeOH/25 °C, 24 h, 70% (**330**)

(**331**) $HClO_4$, H_2O (**332**) (**333**)

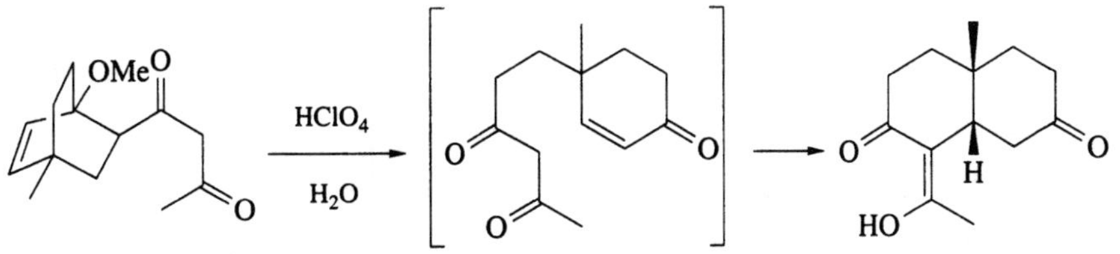

Scheme 41

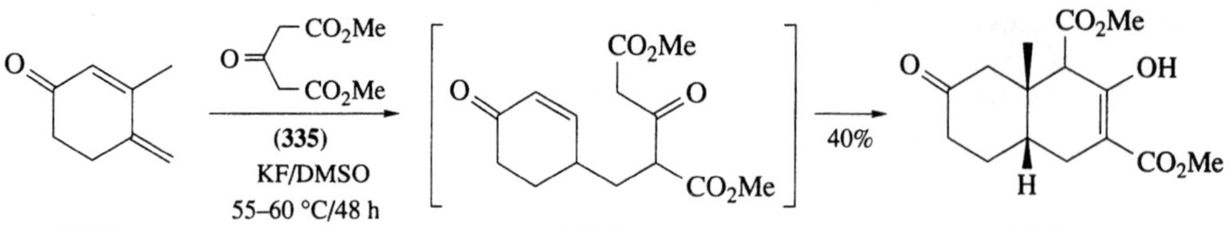

(334) (335) KF/DMSO 55–60 °C/48 h (336) 40% (337)

Scheme 42

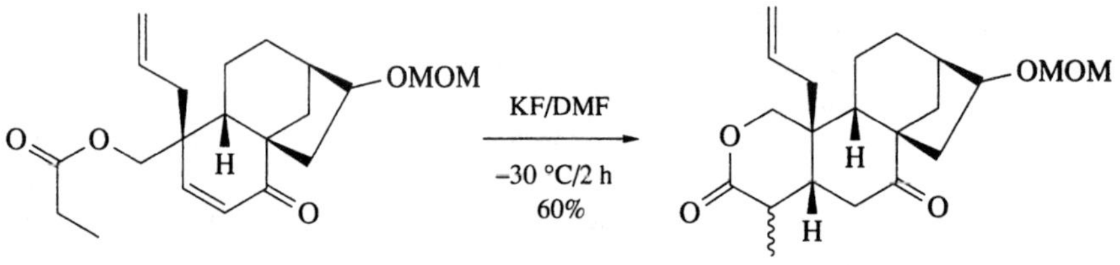

(338) KF/DMF −30 °C/2 h 60% (339) (76)

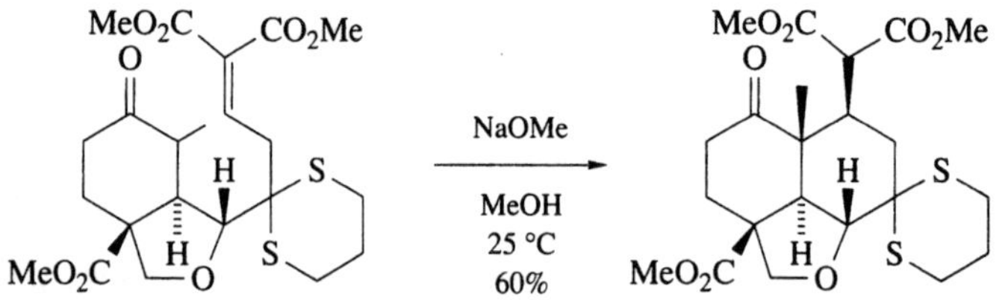

(340) NaOMe MeOH 25 °C 60% (341) (77)

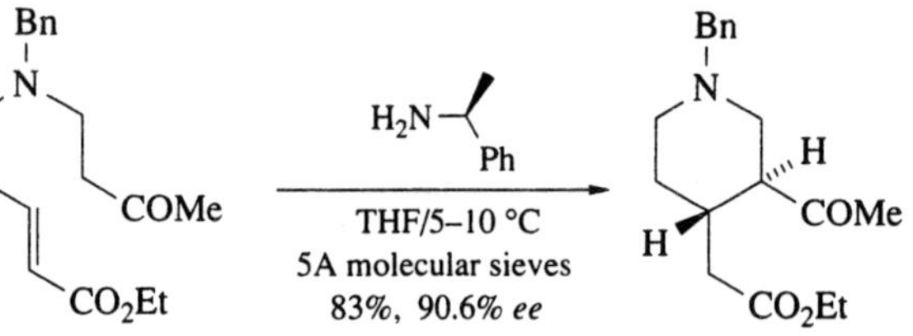

(342) THF/5–10 °C 5A molecular sieves 83%, 90.6% *ee* (343) (78)

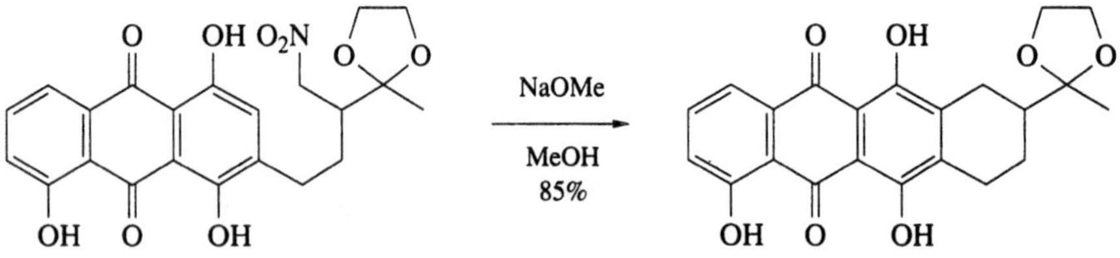

(344) NaOMe MeOH 85% (345) (79)

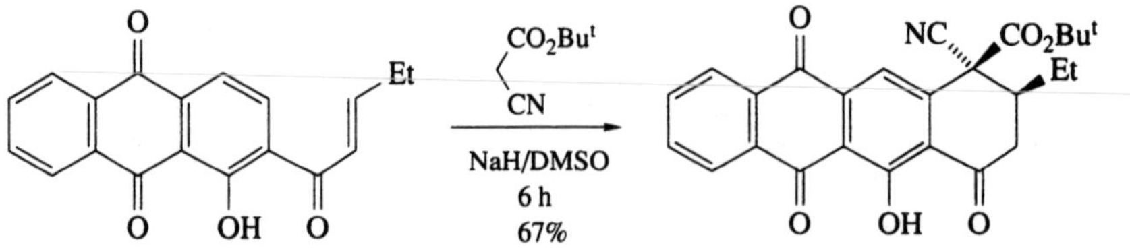

(346) NaH/DMSO 6 h 67% (347) (80)

pounds.[84b,84c] A synthesis of adrene used an intermolecular Michael to produce, after conversion of nitro to carbonyl, the key intermediate (**356**; Scheme 44) for an intramolecular Michael addition to afford the tricyclic dione (**357**).[84d] Deslongchamps showed that only *cis*-fused products were formed in cyclizations to give six-, seven- and eight-membered rings (equation 81), with poor acyclic stereocontrol except for the seven-membered case.[85] Finally, it is perhaps instructive to point out the failure of a simple internal Michael addition, namely the reluctance of the amide (**360**; equation 82) to cyclize to (**361**).[86]

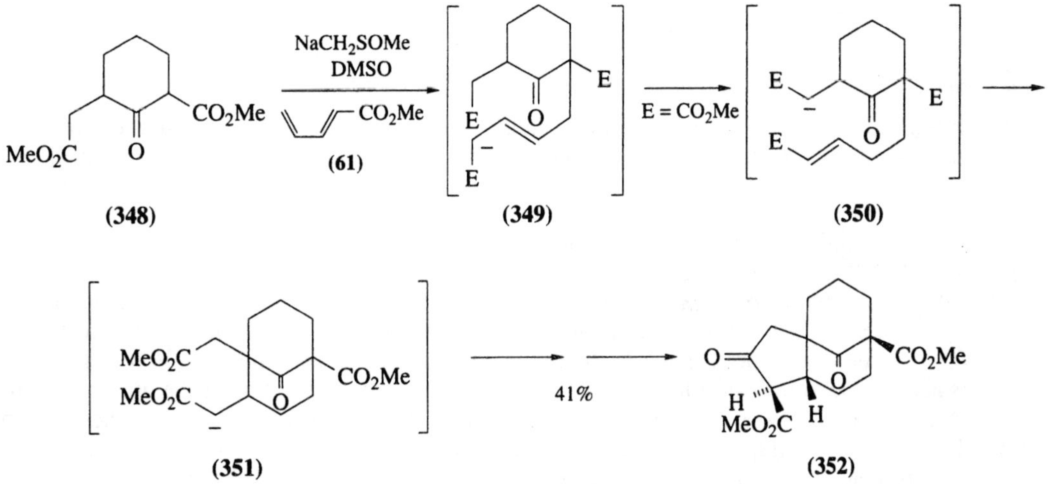

(**348**) NaCH$_2$SOMe, DMSO, (**61**) → (**349**) → E = CO_2Me (**350**) → (**351**) → → 41% (**352**)

Scheme 43

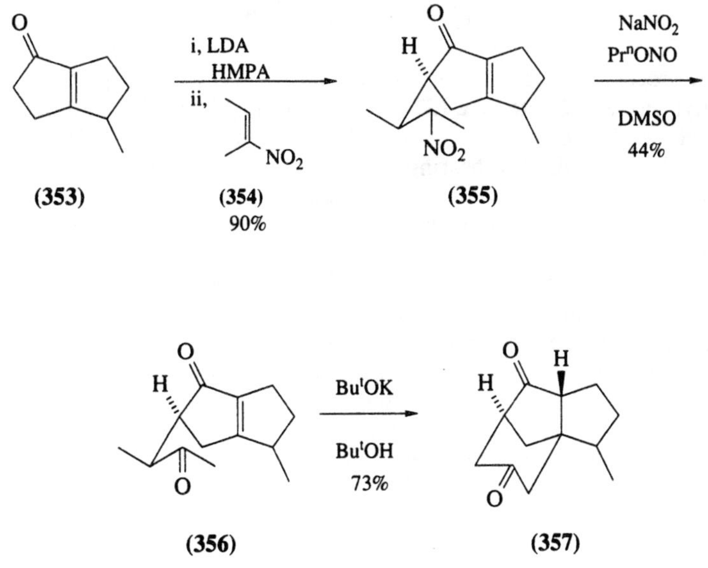

(**353**) i, LDA, HMPA; ii, (**354**) 90% → (**355**) → $NaNO_2$, Pr^{n}ONO, DMSO, 44%

(**356**) → Bu^{t}OK, Bu^{t}OH, 73% → (**357**)

Scheme 44

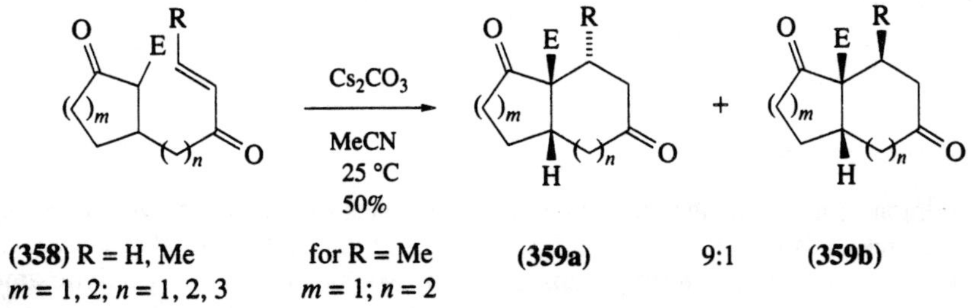

(**358**) R = H, Me; *m* = 1, 2; *n* = 1, 2, 3 → Cs_2CO_3, MeCN, 25 °C, 50% for R = Me, *m* = 1; *n* = 2 → (**359a**) + (**359b**) 9:1 (81)

(82)

(iii) Dienolate double Michael additions

In the 1960s and early 1970s, a highly useful synthetic technique was developed, for the formation of cyclohexanones with an electron-withdrawing substituent in the 4-position, and their bicyclic analogs, by a simple double Michael process using a dienolate ion as the initial nucleophile and a normal unsaturated system as Michael acceptor.[87] For instance, treatment of isophorone (**17**; Scheme 45) with lithium isopropyl(cyclohexyl)amide (LICA) in THF at –23 °C followed by addition of methyl acrylate (**21**) afforded the *endo*-bicyclo[2.2.2]octanone ester (**365**) as a single diastereomer in 98% yield.[87g] The reaction proceeds by addition of the dienolate (**362**) to the acrylate to give the ester enolate (**363**) which adds back *via* an internal Michael to the enone to generate the enolate (**364**); protonation gives the product (**365**). The *endo* stereochemistry is determined in the intramolecular Michael addition of (**363**) to give (**364**), and is presumably due to minimization of charge separation or partial chelation of the lithium cation by the oxygen atoms, as proposed earlier in the intermolecular additions. This reaction sequence has since been used many times to prepare compounds of use in natural product synthesis,[88] a few examples of which are given here. A substituent on the dienolate unit causes the initial Michael addition to occur *trans* to that substituent, thus setting an additional stereochemical center, *e.g.* (**366a,b**; equation 83) gave only (**367a,b**) when reacted with the acrylates (**269**) or (**223**).[88g,88h] Likewise, a substituent on the acceptor can also cause the initial Michael addition to be highly stereoselective (equation 84).[88e] The use of a pentenoate bearing an allylic chiral center as the acceptor affords the expected product with high diastereoselectivity.[88j] The initial dienolate can be generated by other reactions, *e.g.* a Michael reaction, so that a triple Michael process is observed, *e.g.* (**371**) and (**372**; Scheme 46) generate the dienolate (**373**) which adds to methyl acrylate (**21**) to give (**374**) in good yield.[88p] Finally, an intramolecular version of this process has been used in the synthesis of steroids and steroidal alkaloids.[88k–88o] For example, treatment of the enone (**375**; equation 85), bearing an appropriately positioned acrylate unit, with base in ether–hexane at low temperature produced, in 43% yield, the desired polycyclic keto ester (**376**), which was then taken on to atisine.[88o] It should be pointed out that these reactions could be formally viewed as Diels–Alder reactions of the dienolate ion and the Michael acceptor, since cycloadditions of similar cross-conjugated dienolate derivatives are known.[89] However, they are almost certainly correctly considered as double Michael additions since: (i) in certain cases the reaction can be carried out stepwise, by isolating the first Michael product and cyclizing it under similar conditions;[88q] (ii) the TLC of the reaction mixture under silylating conditions shows that the cyclization precedes silylation and thus the reaction does not involve the silyloxydiene;[88m,88n] and (iii) the reluctance of similar Diels–Alder reactions to proceed well.[88g,88q,90] Finally, it should be pointed out that by using a different set of conditions for dienolate formation, one can produce a through-conjugated dienolate, *e.g.* (**377**; Scheme 47) from (**36**), and effect a different double Michael process,[91] as shown, to give a product isomeric to that expected from the above references, namely (**380**).[91a]

1.1.2.2 Heteronucleophiles

The addition of heteroatomic nucleophiles, *e.g.* oxygen, nitrogen, sulfur, halogen, phosphorus, *etc.*, to activated alkenes is covered in this section.

1.1.2.2.1 Intermolecular additions

The conjugate addition of heteronucleophiles to activated alkenes has been used very often in organic synthesis to prepare compounds with heteroatoms β to various activating functional groups, *e.g.* ketones, esters, nitriles, sulfones, sulfoxides and nitro groups. As in the Michael reaction, a catalytic amount of a weak base is usually used in these reactions (with amines as nucleophiles, no additional base is added).

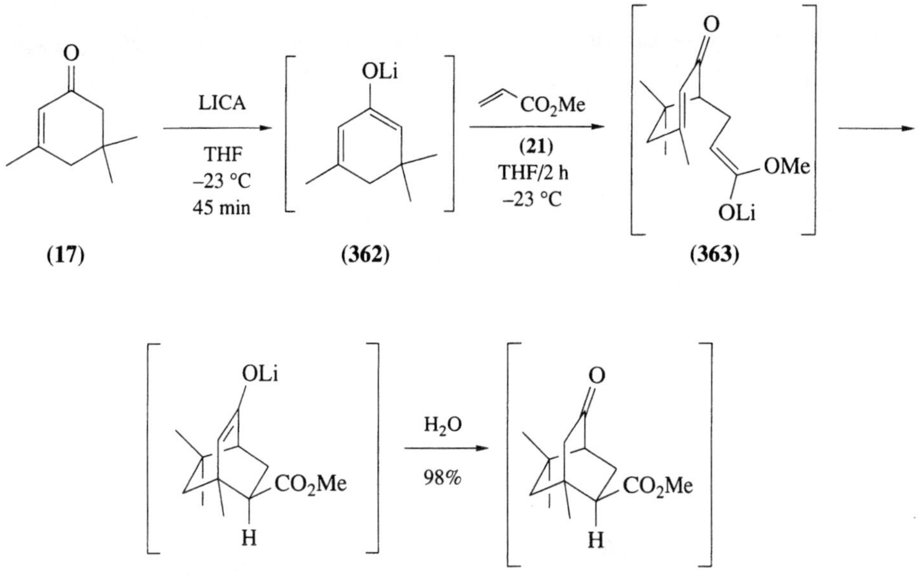

LICA, THF, –23 °C, 45 min; CO_2Me (21), THF/2 h, –23 °C

(17) (362) (363)

H_2O, 98%

(364) (365)

Scheme 45

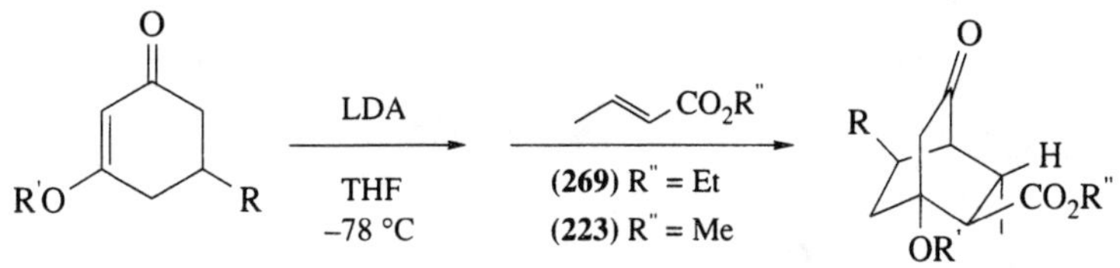

LDA, THF, –78 °C; CO_2R'' (269) R'' = Et, (223) R'' = Me (83)

(366a) R = Me, R' = MOM 94% **(367a)** R = Me, R' = MOM, R'' = Et

(366b) R = CH_2OMOM, R' = $SiMe_3$ 74% **(367b)** R = CH_2OMOM, R' = $SiMe_3$, R'' = Me

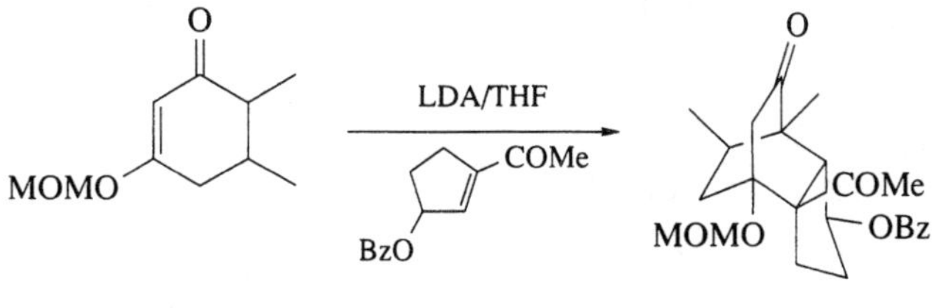

LDA/THF (84)

(368) **(369)** 62% **(370)**

Several representative examples of this large class of reactions are given in equations (86) to (89) and Scheme 48.[92] The reaction works quite well with good acceptors such as acrylonitrile (**70**; equation 86)[7f,92c,92f,92g] or nitroethylene (**387**; Scheme 48)[92i–92k] and with good nucleophiles such as thiols or amines.[92a,92b,92d,92e] As with Michael additions, the initially formed adduct can add a second time to give dialkylated products (**384**; equation 88), sometimes in excellent yield.[92b] Even with α-amidoacrylates as acceptors (equation 90), good yields of addition are obtained.[93]

This addition of a heteronucleophile to an enone has often been used as evidence for the formation of strained alkenes such as anti-Bredt's bridgehead enones.[94] Treatment of the β-chloro ketone (**393**; Scheme 49) with basic methanol produced the β-methoxy ketone (**395a**) in 91% yield by way of the bridgehead enone (**394**).[94a] Trapping with a thiol or amine also proceeded in excellent yield to give (**395b**) or (**395c**), respectively. The bicyclic enones can also be prepared by an intramolecular Wittig reaction and trapped *in situ* with alcohols.[94d] An intramolecular version of this process has been used in a clever synthesis of lycopodine (**400**; Scheme 50).[94e] Treatment of the β-bromo ketone (**396**) with the amine (**397**) in the presence of DBU afforded the desired tricyclic amino ketone (**399**) in quantitative

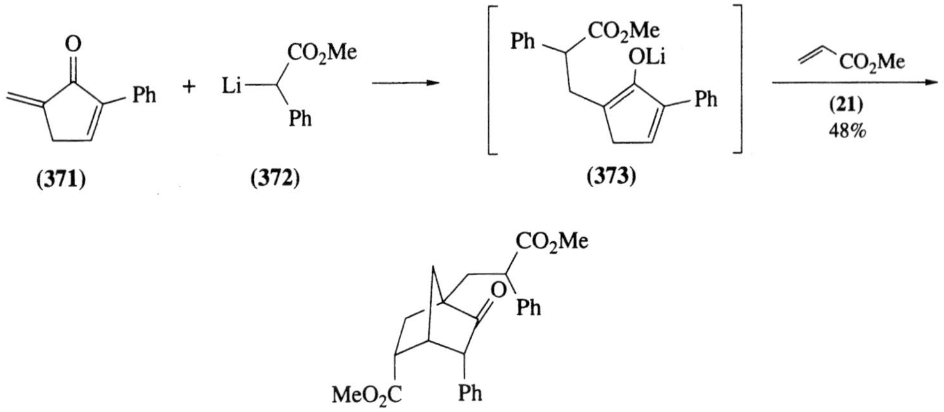

(371) + (372) → [(373)] —(21), 48%→ (374)

Scheme 46

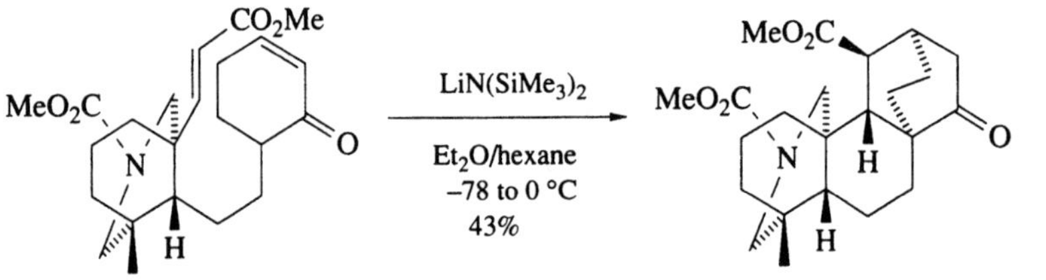

(375) —$LiN(SiMe_3)_2$, Et_2O/hexane, –78 to 0 °C, 43%→ (376) (85)

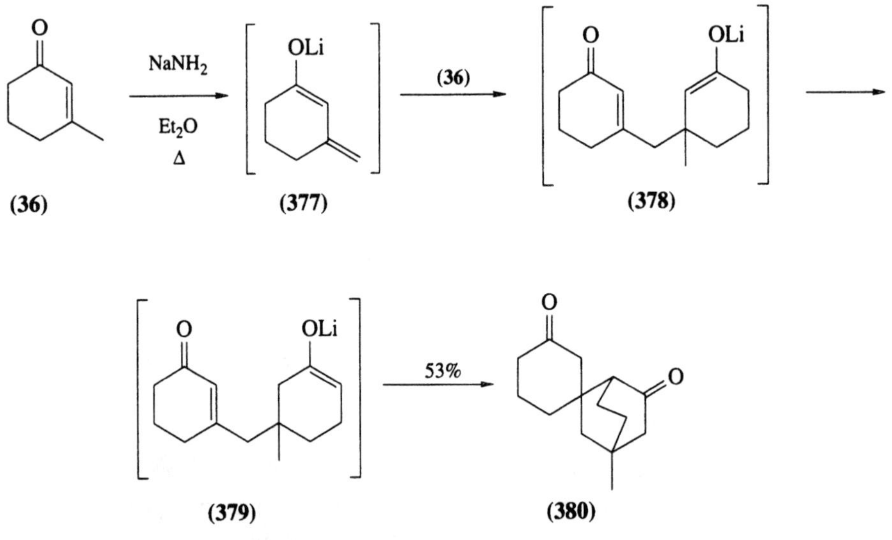

(36) —$NaNH_2$, Et_2O, Δ→ [(377)] —(36)→ [(378)] → [(379)] —53%→ (380)

Scheme 47

yield by way of the amino enone (**398**). The final conversion of (**399**) to lycopodine (**400**) was carried out in only two steps by Heathcock's route.[94f]

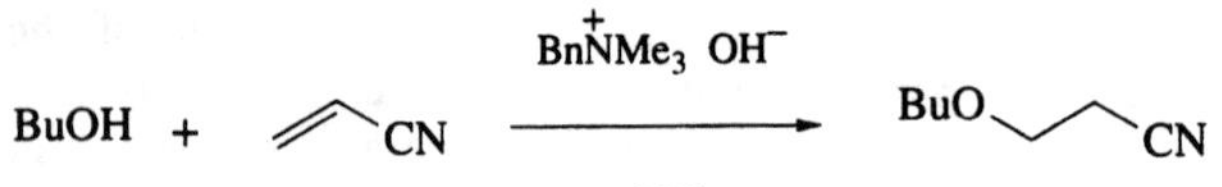

$$\text{BuOH} + \text{CH}_2{=}\text{CHCN}\ \textbf{(70)} \xrightarrow[86\%]{\text{Bn}\overset{+}{\text{N}}\text{Me}_3\ \text{OH}^-} \text{BuOCH}_2\text{CH}_2\text{CN}\ \textbf{(381)} \tag{86}$$

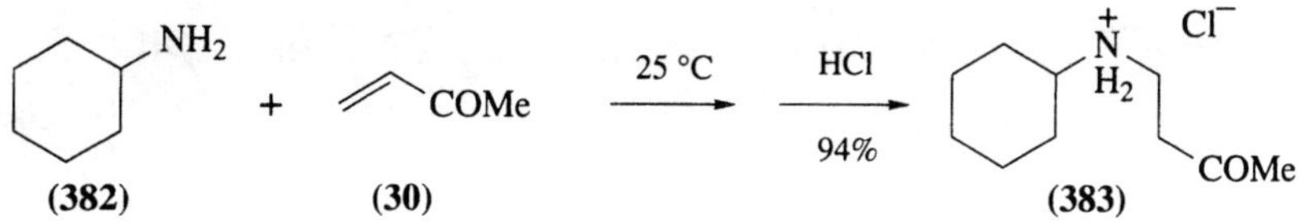

(382) + (30) $\xrightarrow{25\ °C}$ $\xrightarrow[94\%]{HCl}$ (383) (87)

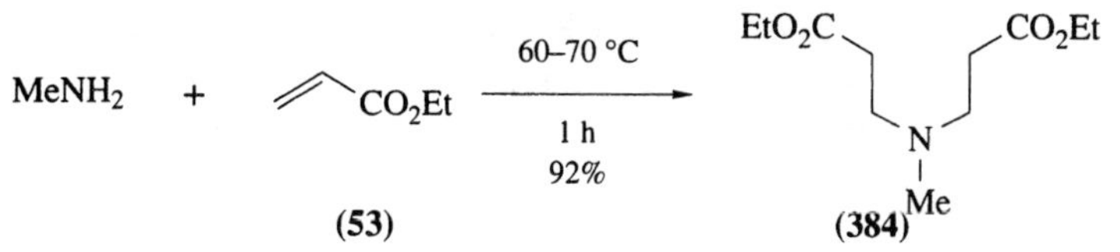

$MeNH_2$ + (53) $\xrightarrow[1\ h,\ 92\%]{60–70\ °C}$ (384) (88)

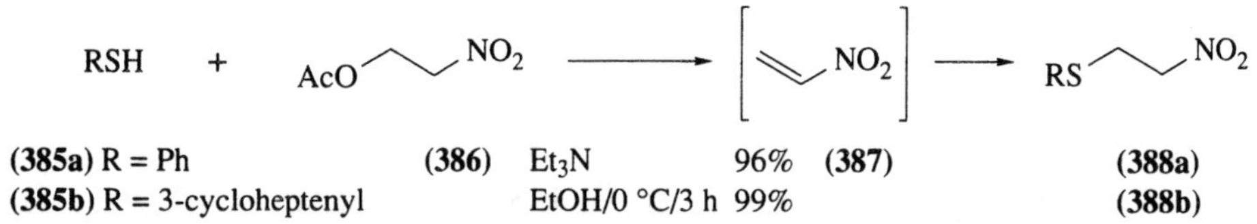

RSH + AcO–CH2CH2–NO_2 ⟶ [(387)] ⟶ RS–CH2CH2–NO_2

(385a) R = Ph **(386)** Et_3N 96% **(387)** **(388a)**
(385b) R = 3-cycloheptenyl EtOH/0 °C/3 h 99% **(388b)**

Scheme 48

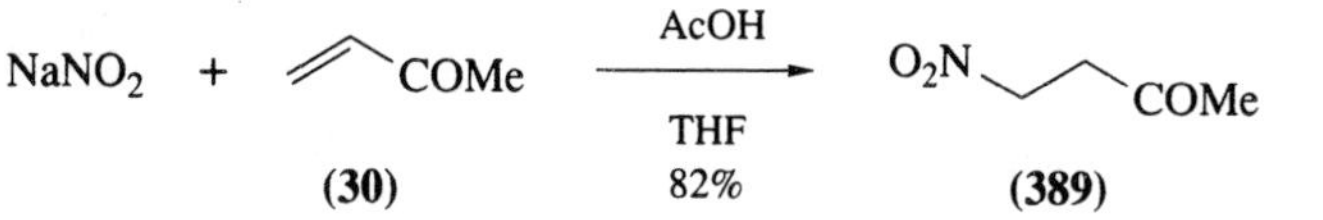

$NaNO_2$ + (30) $\xrightarrow[THF,\ 82\%]{AcOH}$ (389) (89)

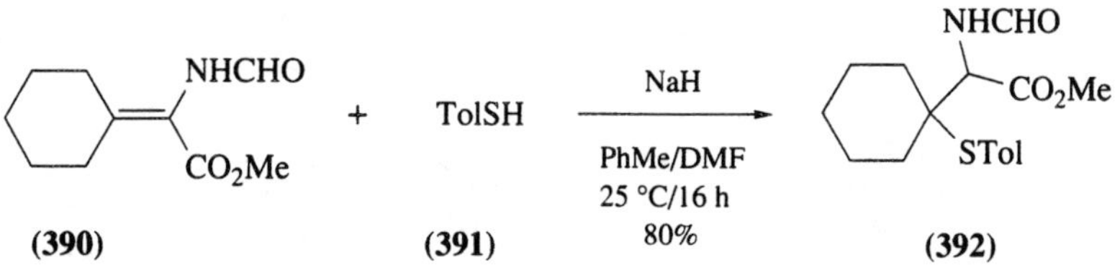

(390) + TolSH (391) $\xrightarrow[PhMe/DMF,\ 25\ °C/16\ h,\ 80\%]{NaH}$ (392) (90)

As in the Michael reaction, various methods for trapping the stabilized anion of the initial adduct have been developed. If, for example, there is a leaving group (thioalkyl, halogen), then an addition–elimination process occurs readily to regenerate the alkene.[95] The sequential addition of two different amines to the β,β-disubstituted nitro compound (**401**; Scheme 51) gives the disubstituted nitroalkene (**403**) and finally the pyrrole (**404**).[95a] The addition of a tertiary amine such as (**405**; Scheme 52) to a β-chloroacrylate (**406**) produces the alkenylammonium salt (**407**) which can undergo cycloaddition to give (**408**).[95b]

Another common trapping method is an intramolecular aldol reaction of the initially formed anion, as shown in equation (91) and Schemes 53 and 54.[96] In the first case, an aldol-like trapping of the iminium salt produced (**411**; equation 91).[96b] The initial heteronucleophile in the other two cases is ultimately lost from the product by oxidation and elimination, so that the overall process is C—C bond formation at the α-center of an enone. Thus, treatment of the formyl enone (**412**; Scheme 53) with an aluminum thiolate afforded in 60% yield the trapped product (**413**) which could be oxidized and eliminated to give (**414**).[96c] Addition of the corresponding aluminate species to the ketoacrylate (**415**; Scheme 54) produced only one diastereomer of the aldol product (**416**) which was converted into the alkene (**417**) in excellent yield.[96d]

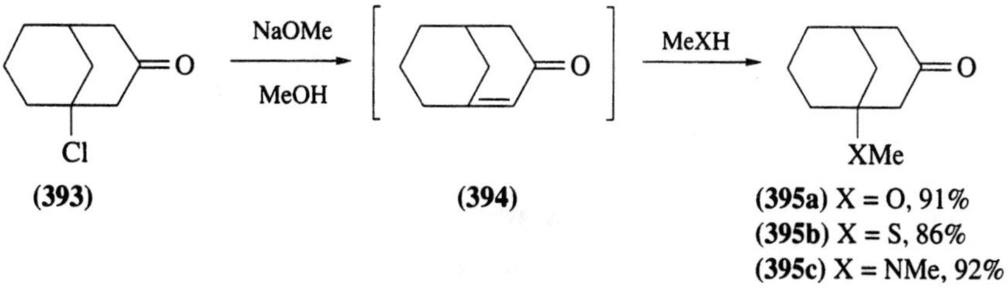

(393) $\xrightarrow[MeOH]{NaOMe}$ [(394)] $\xrightarrow{MeXH}$ (395)

(395a) X = O, 91%
(395b) X = S, 86%
(395c) X = NMe, 92%

Scheme 49

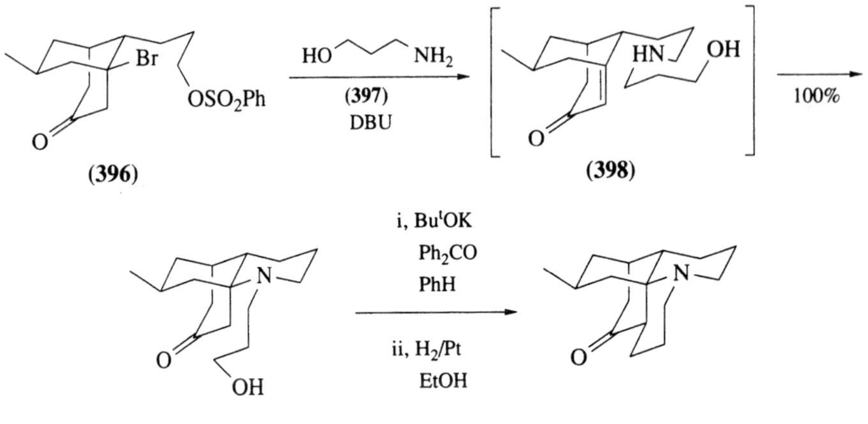

(396) (398) (399) (400)

Scheme 50

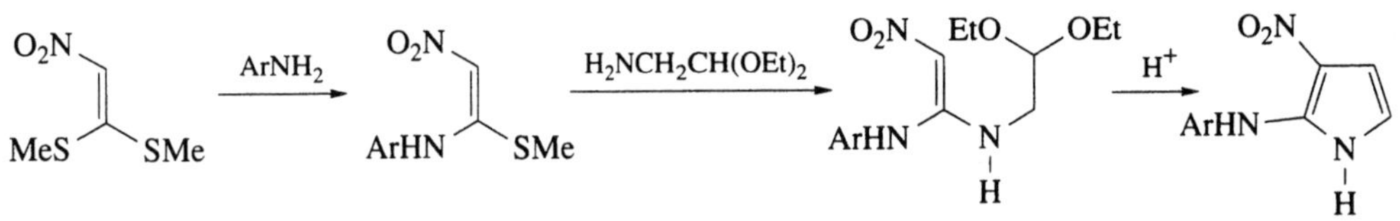

(401) (402) (403) (404) 44% overall

Scheme 51

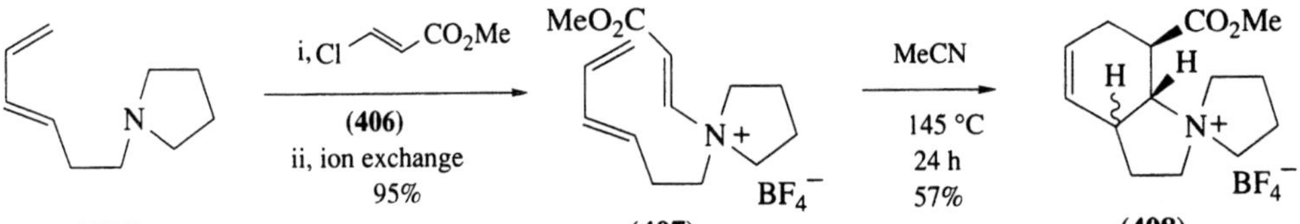

(405) (407) (408)

Scheme 52

This method of forming the oxahydrindene ring system was later applied to an elegant total synthesis of avermectin A_{1a}.[96e]

This C—C bond-forming process has been used very often in the intermolecular case[97] to produce α-hydroxyalkyl-α,β-unsaturated esters, ketones, nitriles, sulfones, *etc.* by treatment of the activated alkene

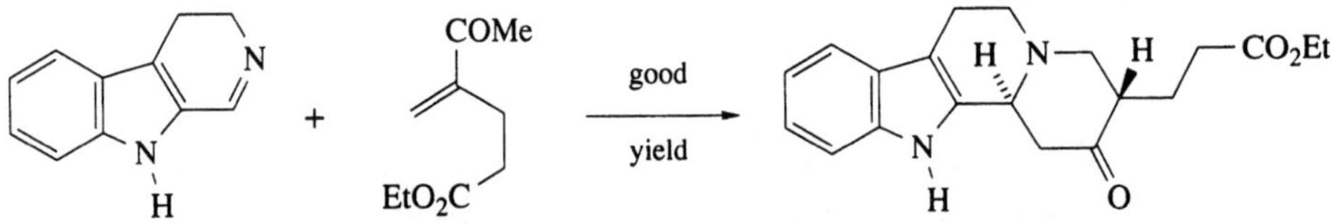

(409) + (410) → (411) (91)

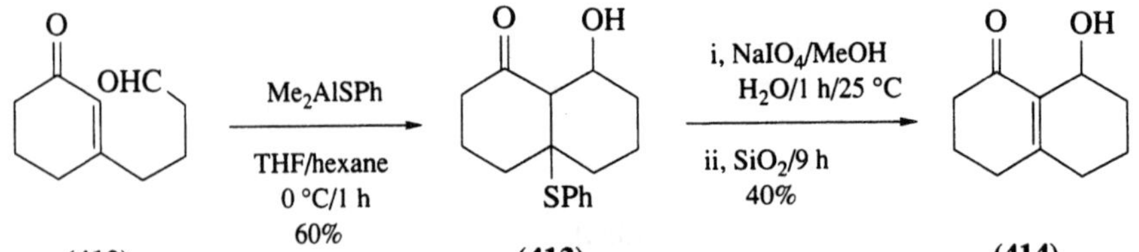

(412) (413) (414)

Scheme 53

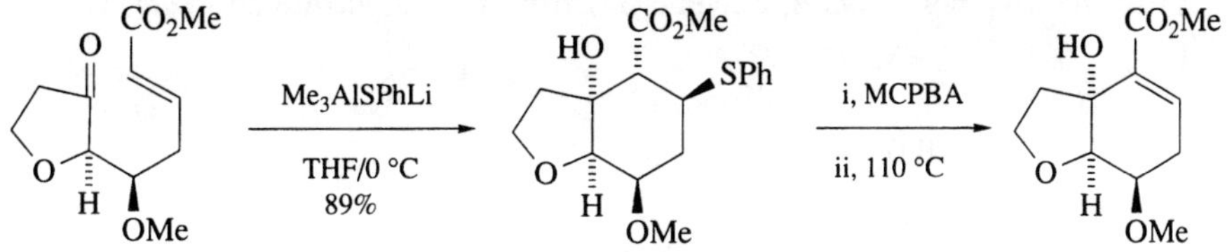

$Me_3AlSPhLi$, THF/0 °C, 89%; i, MCPBA; ii, 110 °C

(415) **(416)** **(417)**

Scheme 54

with a catalytic amount of 1,4-diazabicyclo[2.2.2]octane (DABCO) and any aldehyde (or even imine).[97d] The mechanism involves addition of DABCO to the activated alkene to form the zwitterion (**420**) which reacts with the aldehyde to give the alkoxide (**421**; Scheme 55). Proton transfer produces the zwitterion (**422**) which then eliminates DABCO to give the observed products (**419a–419e**) in high yields under mild conditions.[97] This process has also been catalyzed by tricyclohexylphosphine.[97i] Another interesting process for trapping the initial anion is an intramolecular proton transfer (Scheme 56), as in the formation of the ylides (**424a**) and (**424b**) from the addition of triphenylphosphine to maleic anhydride (**423a**) or maleimide (**423b**).[98a–98e] These stabilized ylides, and the esters derived from them, have been used in synthesis,[98f–98j] both in Wittig reactions and alkylations.

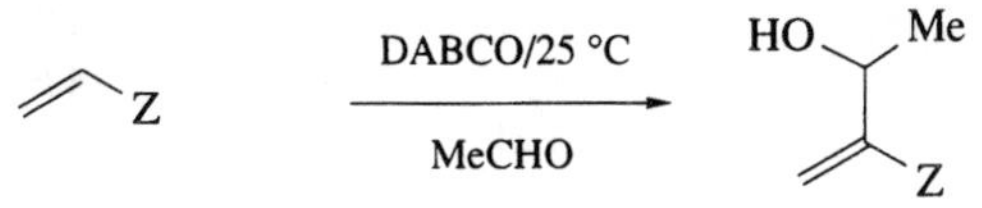

DABCO/25 °C, MeCHO

	Z				Z
(418a)	SO_2Ph	2 weeks	84%	**(419a)**	SO_2Ph
(418b)	CO_2Bu^t	7 d	89%	**(419b)**	CO_2Bu^t
(53)	CO_2Et	7 d	94%	**(419c)**	CO_2Et
(30)	COMe	24 h	81%	**(419d)**	COMe
(70)	CN	40 h	76%	**(419e)**	CN

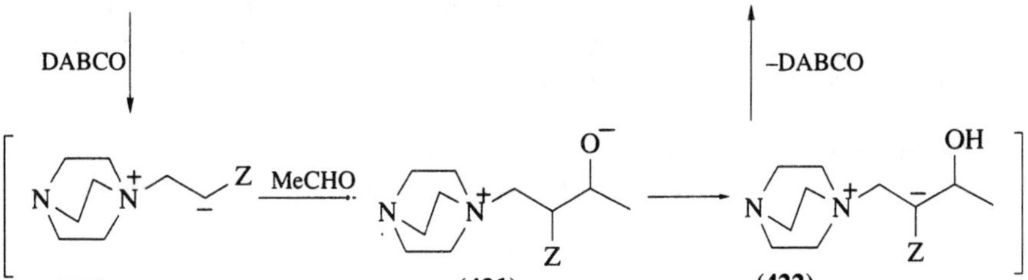

DABCO; MeCHO; –DABCO

(420) **(421)** **(422)**

Scheme 55

Ph_3P

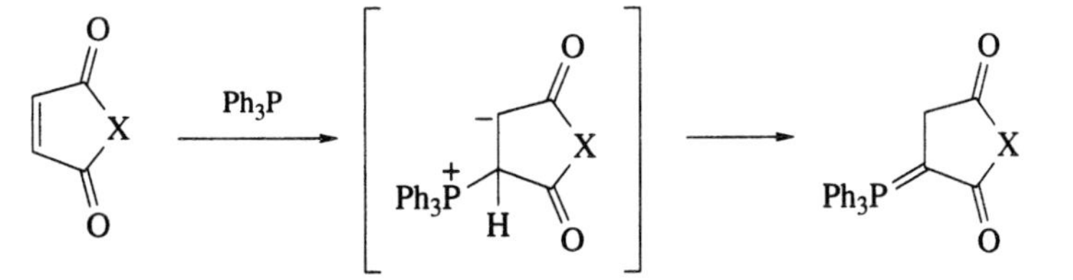

(423a) X = O
(423b) X = NH

(424a) X = O
(424b) X = NH

Scheme 56

Epoxidation of enones on treatment with basic hydrogen peroxide or *t*-butyl hydroperoxide, or with bleach, might be viewed as another form of trapping the initially produced anion.[99] In this case the enolate, *e.g.* (**425**; Scheme 57), attacks the oxygen of the hydroperoxide to eject hydroxide and yield the epoxy ketone (**426**).[99a] Finally, the initial anion can also be trapped by a sigmatropic rearrangement, as in the production of the azepines (**429a,b**; Scheme 58) from the vinylaziridine (**427**), *via* the zwitterion (**428**).[100]

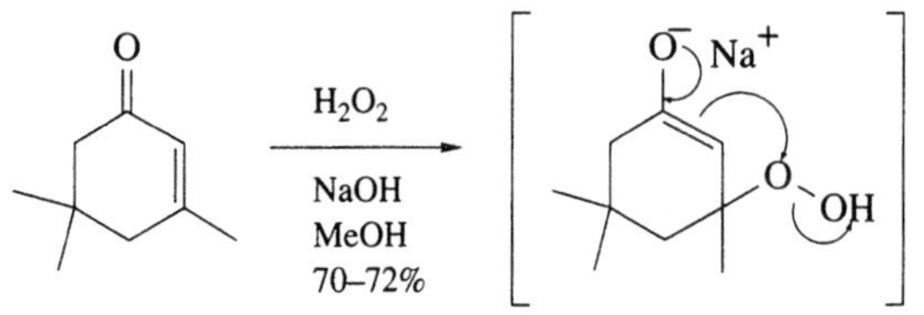

H$_2$O$_2$, NaOH, MeOH, 70–72%

(17) **(425)** **(426)**

Scheme 57

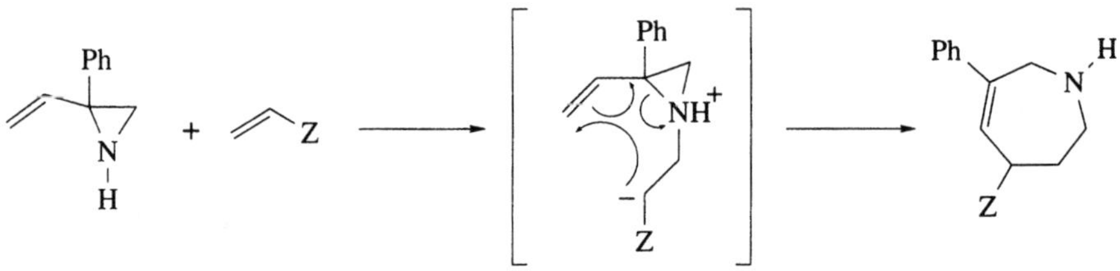

(427) **(70)** Z = CN; **(21)** Z = CO_2Me; **(418a)** Z = SO_2Ph **(428)**

(429a) Z = CN 70%
(429b) Z = CO_2Me 55%
(429c) Z = SO_2Ph 50%

Scheme 58

A discussion of the relative stereochemistry of these additions completes this section on intermolecular heteronucleophile additions.

In cyclic cases, the normal rules for relative stereocontrol described earlier for carbon nucleophiles apply to heteronucleophiles as well. An allylic substituent usually causes the addition to occur from the opposite face for steric reasons, resulting in the *trans* product (equation 92).[101] Addition of heteronucleophiles to cyclohexenone or dihydropyrone systems generally occurs in an antiparallel sense[102] to give the axial product (equation 93).[102a] This kinetic axial product is often easily rearranged to the thermodynamically more stable equatorial adduct when subjected to the reaction conditions for an extended period.[102b,102c,102e] The use of alkoxide as the nucleophile causes the lactone to be opened and, in certain cases, allows further base-catalyzed processes to occur (equation 94).[102g]

In acyclic systems, high diastereoselectivity can often be achieved.[103] Additions of amines or alkoxides to the (*Z*)-enoate (**436**; equation 95) at low temperature afforded mainly the single diastereomer (**437a,b**) in good yield.[103a,103b] Similar results were obtained with the enone corresponding to (**436**).[103c]

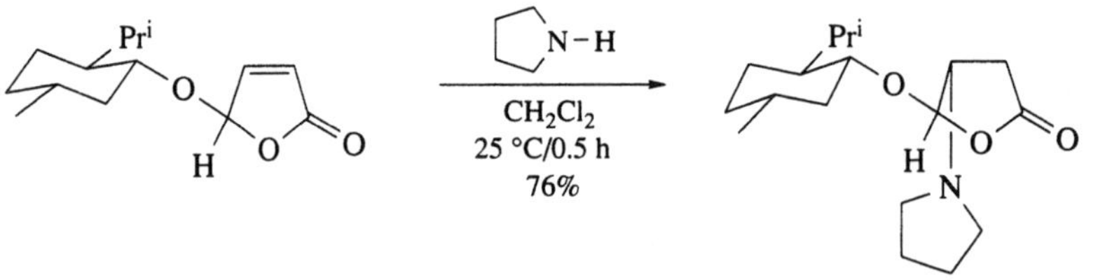

CH$_2$Cl$_2$, 25 °C/0.5 h, 76% (92)

(430) **(431)**

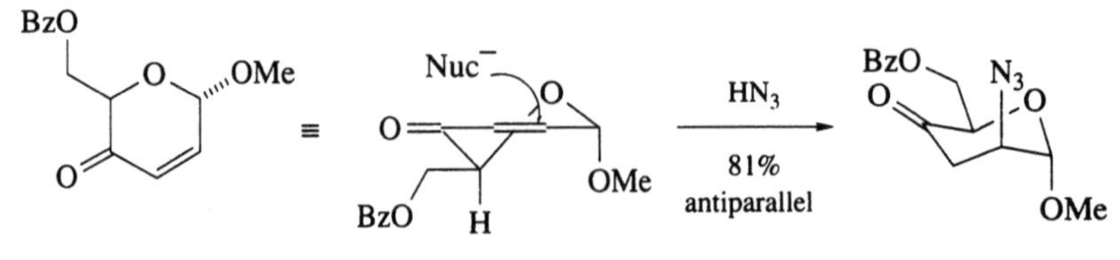

HN$_3$, 81%, antiparallel (93)

(432) **(432)** **(433)**

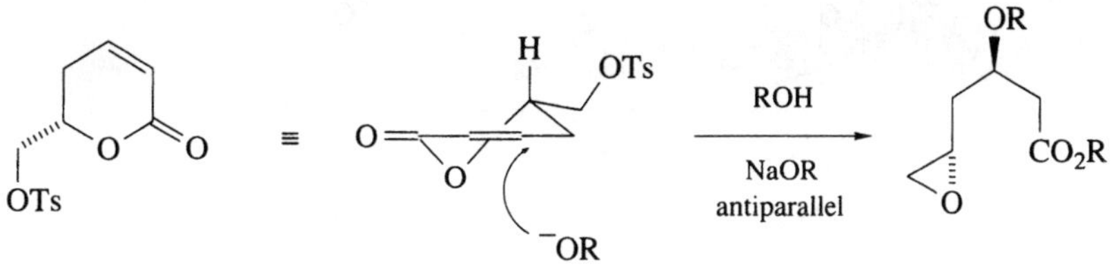

(94)

(434) **(434)** **(435)** R = allyl, 87%

Chiral α,β-unsaturated sulfoxides also produce predominately one diastereomer on addition of heteronucleophiles (equation 96).[104] Kinetic protonation of the anion (**440**; Scheme 59), generated by addition of thiophenoxide to the nitroalkene (**233**), followed by trapping with formaldehyde, produced a 87:13 mixture of diastereomers favoring (**441**).[105] Finally, the use of high pressure can increase the diastereoselectivity of the addition of amines to chiral crotonates.[106a]

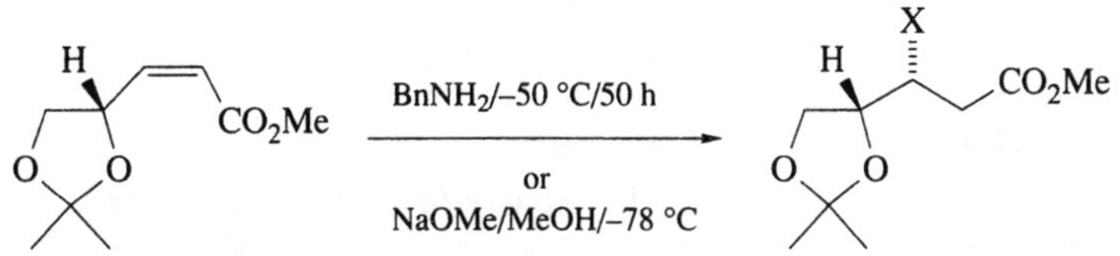

(95)

(436) **(437a)** X = NHBn
(437b) X = OMe

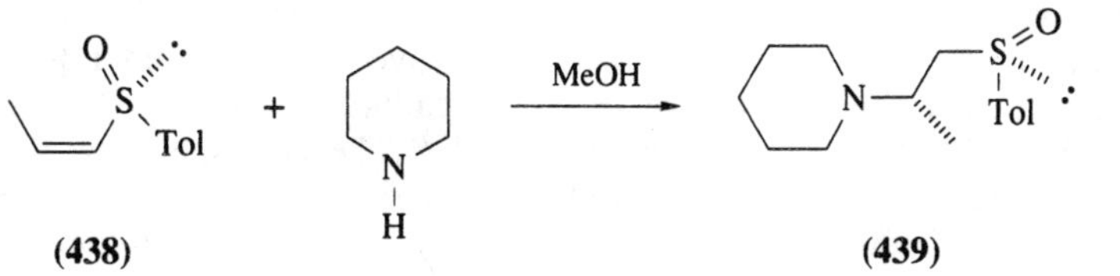

(96)

(438) **(439)**

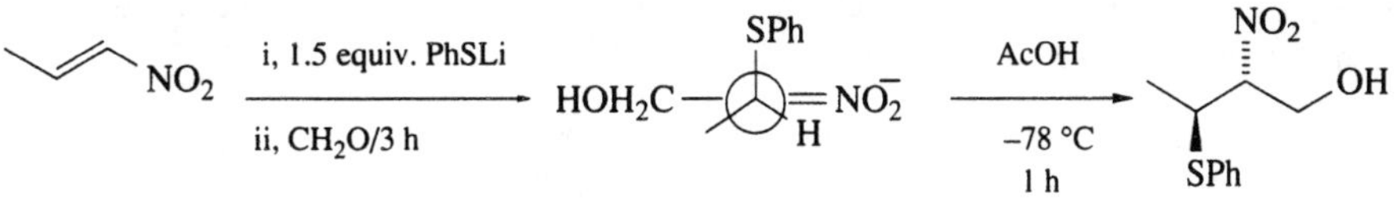

(233) **(440)** **(441)**

Scheme 59

1.1.2.2.2 Intramolecular additions

(i) Baldwin's rules for ring closure

Many intramolecular versions of heteronucleophilic additions have been published. However, a major contribution in this area is due to Baldwin who, in 1976, published a paper describing the rules for ring closure.[107a] In that paper and the succeeding ones,[107] he showed that 5-endo-trigonal cyclizations of hydroxy enones and unsaturated esters, such as (**442a**; equation 97) and (**442b**; equation 98), do not occur under basic conditions to give the expected products, (**443a**) and (**443b**), respectively (although the former reaction proceeds well under acidic conditions).[107b,107e] In order to demonstrate conclusively that the reluctance of the anion of (**442a**) to cyclize was not due to an unfavorable equilibrium, Baldwin treated (**443a**) (prepared *via* acid-catalyzed cyclization of **442a**) with NaOMe in MeOD. The α-protons of (**443a**) were exchanged for deuterium but none of the retro-Michael product (**442a**) was produced. The rule that 5-endo-trigonal cyclizations are disfavored has generally been followed, with few exceptions, *e.g.* the thiol (**442c**; equation 98) cyclizes in warm basic methanol to give (**443c**).[107b] The other favored cyclizations—5-exo, 6-endo, 6-exo, *etc.*—have been reported often, as described below. There are obviously many examples of each class, but due to space limitations only a few recent representative examples of each are presented.

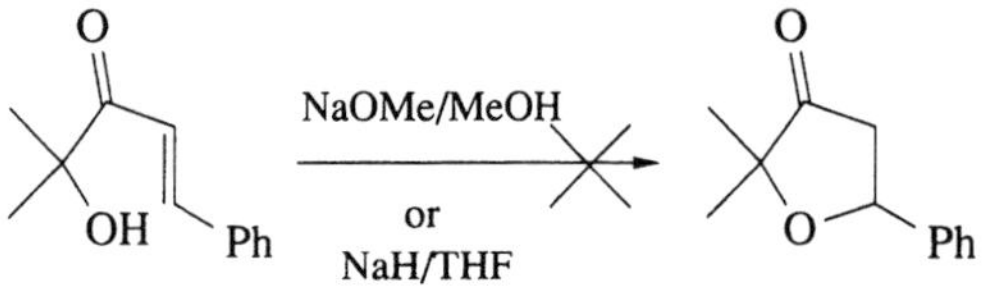

(442a) **(443a)** (97)

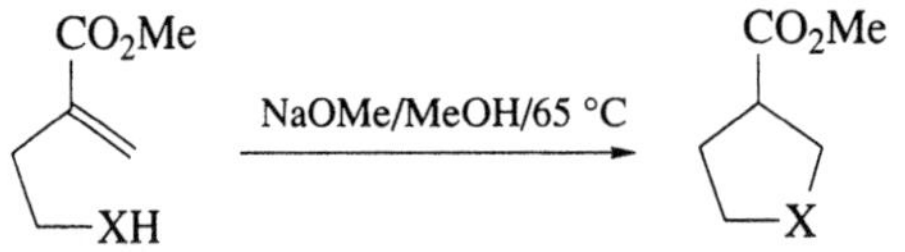

(442b) X = O no reaction **(443b)** X = O
(442c) X = S good yield **(443c)** X = S

(98)

(ii) 5-Exo-trigonal cyclizations

Baldwin showed that oxidation of the allylic alcohol (**444**; Scheme 60) gave the five-membered ring (**446**) in quantitative yield by way of the unisolable hydroxy enone (**445**).[107e] A useful 5-exo-trig cyclization has been reported often in the synthesis of carbohydrates from protected aldoses by a Wittig condensation followed by an intramolecular conjugate addition of the alkoxide.[108] As an example, reaction of the aldose (**447**; Scheme 61) with the stabilized Wittig reagents (**448a,b**) afforded a 3:1 mixture of β:α anomers of (**450a,b**), in which the key step is an intramolecular addition of the alkoxide of (**449**) to the enoate.[108a] The corresponding fully protected arabinose analog gave predominately the α-anomer.[108c] Hirama has used a similar cyclization of carbamates to give good 1,2- and 1,3-diastereoselectivity.[109a] For example, cyclization of either the (*E*)- or (*Z*)-allylic carbamate esters (**451a,b**; equation 99) gave the *trans*-oxazolidinones (**452a,b**) as the major products.[109a] The corresponding homoallylic carbamate, (**453**; equation 100) gave mainly the *cis* six-membered ring (**454**).[109a] Finally, an interesting example of a 5-exo-trig cyclization[109b] followed by a retro-Mannich reaction has been reported in the synthesis of a series of isoindole systems.[110] Treatment of the bromide (**455**; Scheme 62) with methylamine gave the pyrrolothiophene (**459**), the mechanism of which involves a 5-exo-trig cyclization of the amine (**456**) to (**457**), which undergoes a retro-Mannich reaction to give (**458**) and finally (**459**).[110]

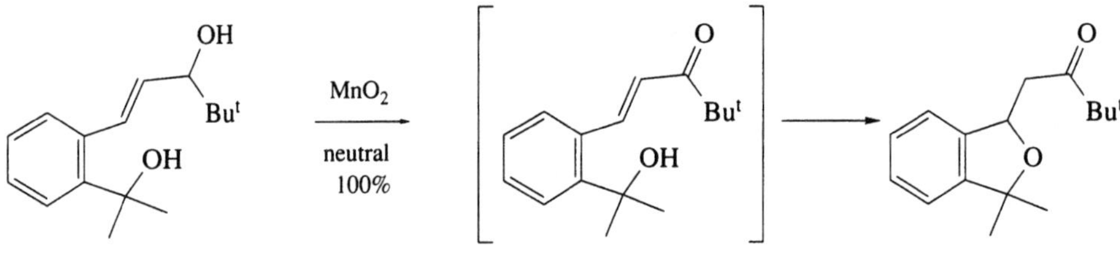

(444) **(445)** **(446)**

Scheme 60

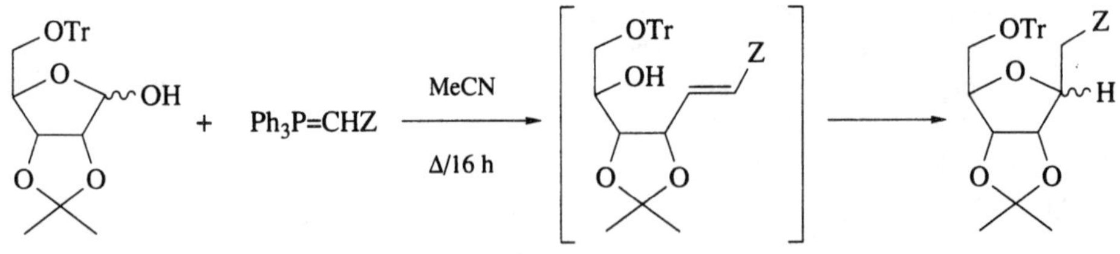

(447) **(448a)** Z = CO_2Me 98% **(449a, b)** **(450a, b)**
(448b) Z = CN 92%

Scheme 61

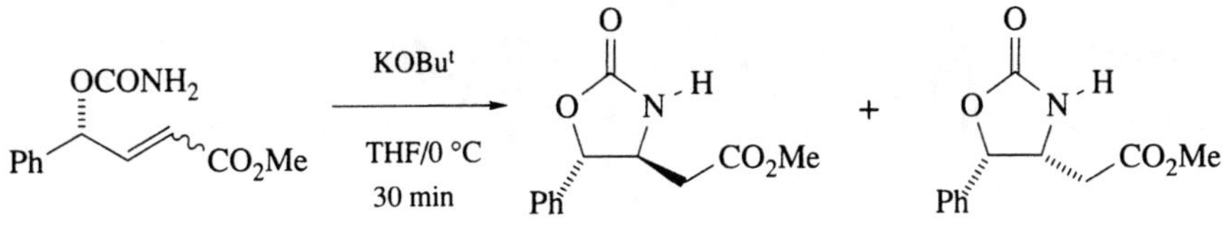

KOBu^t, THF/0 °C, 30 min

(*E*)-(**451a**) → (**452a**) 12:1 (**453a**)

(*Z*)-(**451b**) → (**452b**) >100:1 (**453b**) (99)

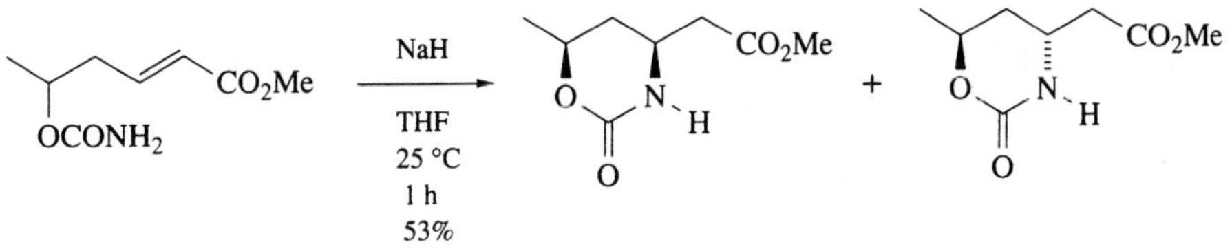

NaH, THF, 25 °C, 1 h, 53%

(**453**) → (**454a**) 10:1 (**454b**) (100)

$MeNH_2$, EtOH, 25 °C/1 h, 88%

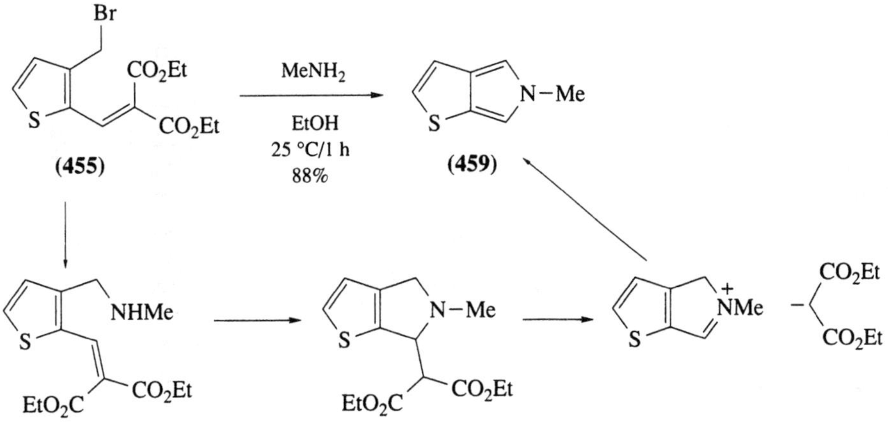

(**455**) → (**459**)

(**456**) → (**457**) → (**458**)

Scheme 62

(iii) 6-Exo-trigonal cyclizations

Baldwin has shown that the 6-exo-trig process is a favorable one,[107c,107e] *e.g.* cyclization of the phenolic enone (**459**; equation 101) in base furnished (**460**) in good yield.[107e] An intramolecular addition of an amine to a chiral vinyl sulfoxide, a 6-exo-trig process, was used in the synthesis of canadine.[111] Reaction of (**461**; equation 102) produced a 3–4:1 mixture of (**462a**) and (**462b**), in which the former predominated.[111c] An interesting 6-exo-trig cyclization was thought to be responsible for the racemization of the chiral hydroxy acid (*R*)-(**463**) under base-catalyzed alkylation conditions to give (±)-(**464**).[112] It was suggested that the dianion (**465**; Scheme 63) racemized *via* a 6-exo-trig cyclization to give the bridged acid dianion (**466**) which opened up to a mixture of (**465**) and (**467**) and was then alkylated to give racemic (**464**).

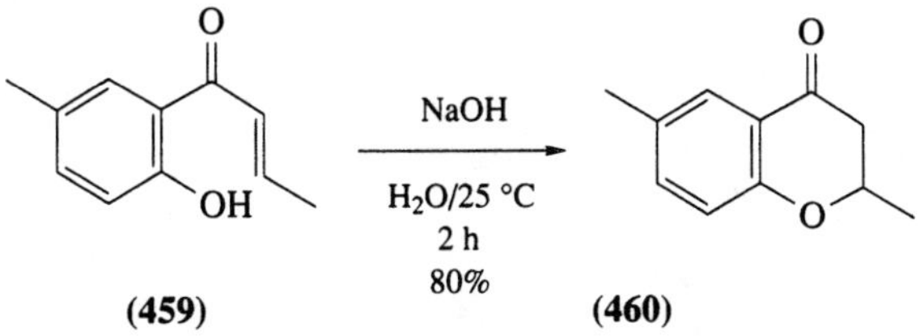

NaOH, H_2O/25 °C, 2 h, 80%

(**459**) → (**460**) (101)

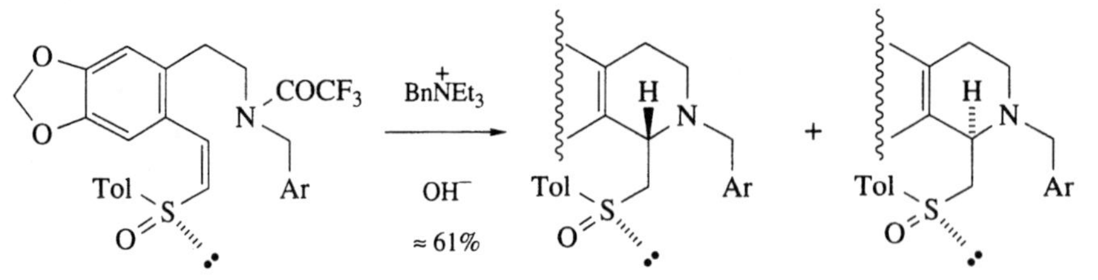

(461) (462a) 3–4:1 (462b) (102)

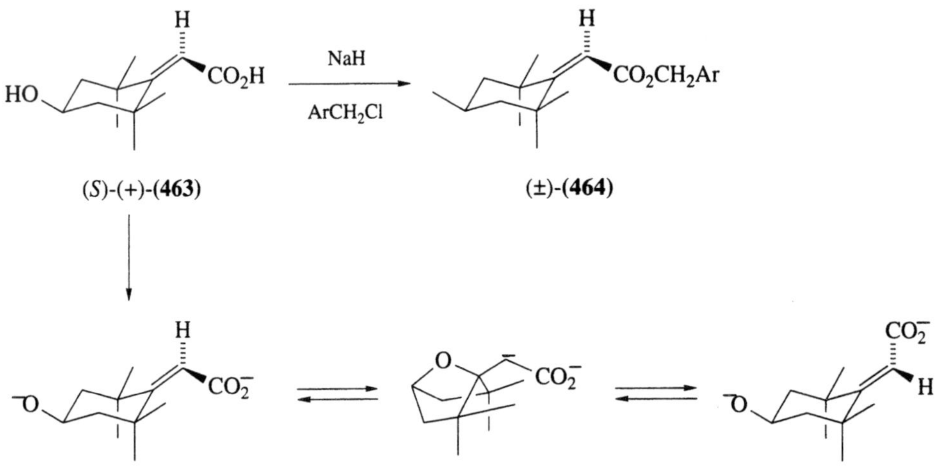

(*S*)-(+)-**(463)** (±)-**(464)**

(465) **(466)** **(467)**

Scheme 63

(iv) 6-Endo-trigonal cyclizations

These cyclizations are somewhat less common than those presented above but they are occasionally reported. In a recent forskolin synthesis, Ziegler reported the cyclization of (**468**; Scheme 64) to give (**470**) *via* a 6-endo-trig addition–elimination process involving the β-methoxy enone (**469**) as an intermediate.[113] A second 6-endo-trig addition–elimination process involves cyclization of (**471**; equation 103) with loss of benzenesulfinic acid to give (**472**).[114] Finally, a retro-6-endo-trig process was used to prepare a silyloxy diene for an intramolecular Diels–Alder reaction.[115] Treatment of *N*-methylpiperidone (**473**; Scheme 65) with acryloyl chloride and then Hünig's base produced the enone acrylamide (**474**), *via* a retro-6-endo-trig reaction. Silylation and thermal cyclization then afforded (**475**) in 72% yield.

(v) 5-Endo-trigonal cyclizations

Finally, it might be of interest to present one example of a formal violation of Baldwin's rules (since Baldwin uses the terms 'favored' and 'disfavored', there can be no violations of these rules, but only normally 'disfavored' reactions which proceed). We have reported[116] the formation of the cyclic keto alk-

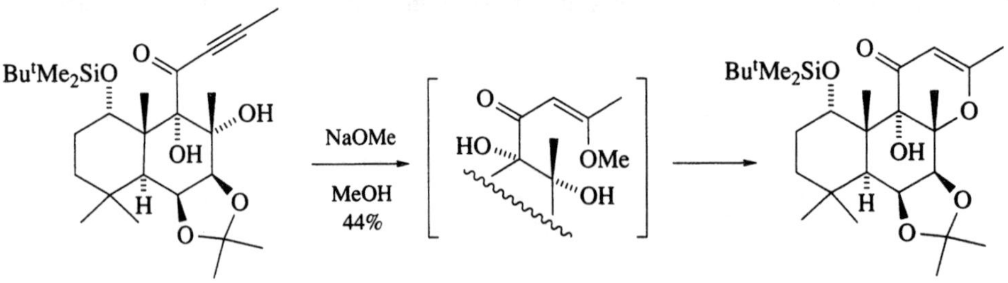

(468) **(469)** **(470)**

Scheme 64

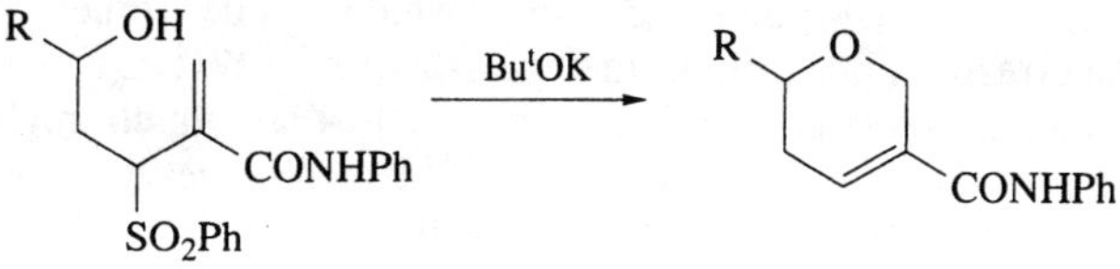

(103)

(471) (472)

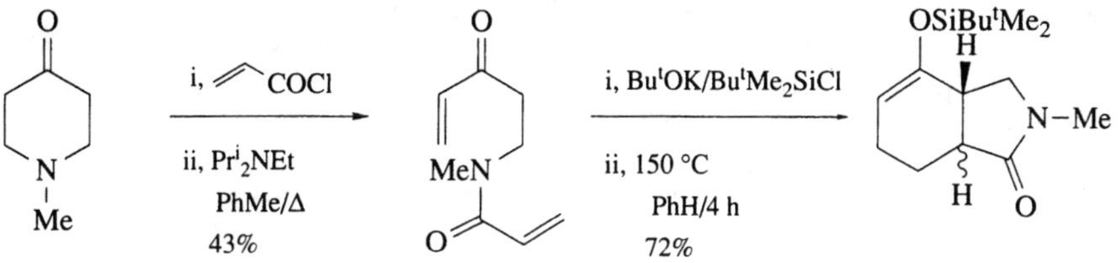

(473) (474) (475)

Scheme 65

enylammonium salt (**478**; Scheme 66) by treatment of the β-keto aldehyde anion (**476**) with tosyl (or mesyl) chloride in acetonitrile, a reaction that we suggest proceeds *via* the tosylate (**477**), which undergoes a 5-endo-trig cyclization and then elimination of the tosylate. The exact mechanism has not yet been proven.

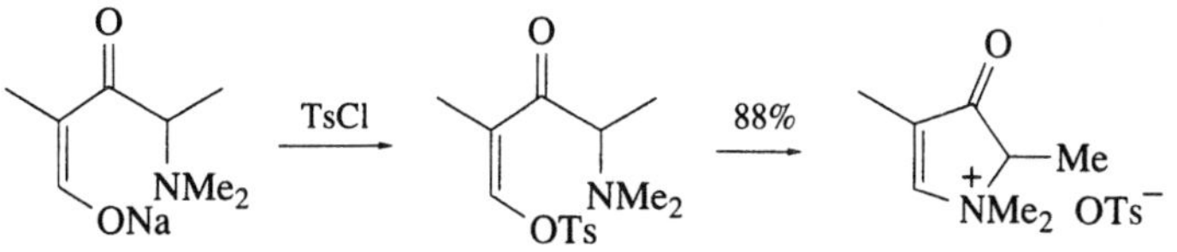

(476) (477) (478)

Scheme 66

1.1.3 ALKYNIC π-SYSTEMS

The addition of stabilized nucleophiles—both carbon and heteronucleophiles—to activated alkynes has been used far less often in organic synthesis than the corresponding addition to activated alkenes. Since several excellent reviews on nucleophilic additions to alkynes appeared in the 1960s and 1970s,[10] this section includes only a few representative examples from the early literature and some more recent applications.

1.1.3.1 Carbon Nucleophiles

1.1.3.1.1 Intermolecular additions

As mentioned early in this chapter, the first example of an addition of a stabilized carbon nucleophile to an activated alkyne was published by Michael in 1894 (see equation 2).[2] The addition of a β-keto ester (**2**) or β-diketone (**479**) to the alkynic ester (**10b**; equation 104) afforded the pyrone (**480a**) or (**480b**), respectively, by cyclization of the initial adduct.[117] Similarly, the addition of diethyl malonate (**1**) to an alkynic ketone (**481**; equation 105) afforded the pyrone-3-carboxylate (**482**) in good yield.[118] The yields of additions to alkynes are generally lower than with the corresponding alkenes, although in certain cases quite high yields of the initial Michael adduct can be obtained,[119] *e.g.* the cyanoacrylate (**485**; equation 106) is obtained from the addition of diphenylacetonitrile (**483**) to ethyl propiolate (**484**) in 92% yield[119a] and the ketoacrylate (**487**; Scheme 67) is produced from the tetrahydrophenanthrenone (**486**) and ethyl propiolate (**484**) in 83% yield.[119b] On heating in acid, the ketoacrylate (**487**) cyclizes to the pyrone (**488**). Depending on the reaction conditions, either the *cis* (kinetic) or *trans* (thermodynamic) enones can be obtained. For example, addition of the anion of the oxazolone (**152**; Scheme 68) to phenyl ethynyl ketone (**489**) produces initially the (*Z*)-product (*Z*)-(**490**), which can be isomerized thermally to the (*E*)-isomer (*E*)-(**490**).[120] The corresponding trifluoromethyl oxazolone (**491**; equation 107) adds to (**489**) at the aminal carbon rather than at the α-keto carbon to give (*Z*)-(**492**), which might find use as a method for the 1,4-addition of a trifluoroacetyl anion equivalent.[120] Dichloroacetylene (**493a**) (either preformed or formed *in situ* by reaction of trichloroethylene **493b** with base) is also sufficiently activated to react with stabilized nucleophiles, *e.g.* diethyl ethylmalonate (**494**), to give the dichlorovinyl product (**495**; Scheme 69).[121] Treatment with *t*-butyllithium affords the alkynic product (**496**), thus allowing for α-ethynylation of stabilized anions. The ethynyl iodonium salt (**498a**) reacts similarly with the enolate of β-diketone (**497**) to give the ethynylated product (**499**) in good yield (equation 108).[122a] However, a quite different product, namely (**500**; Scheme 70) has been obtained in 84% yield by the reaction of the corresponding 1-decynylphenyliodonium salt (**498b**) with (**497**) under similar conditions.[122b] In this latter reaction the initial addition occurs at the alkyl-substituted carbon to give the anion (**501**), which α-eliminates to the alkylidene carbene (**502**); insertion into the C—H bond then gives the cyclopentene (**500**). The reaction of the phenylsulfonylacetophenone (**503**; equation 109) with (**498b**) under the same conditions afforded the furan (**504**) in 67% yield *via* an analogous mechanism.[122b] The formation of (**499**) has now been shown to occur *via* the corresponding alkylidene carbene which undergoes a 1,2-phenyl migration.[122b]

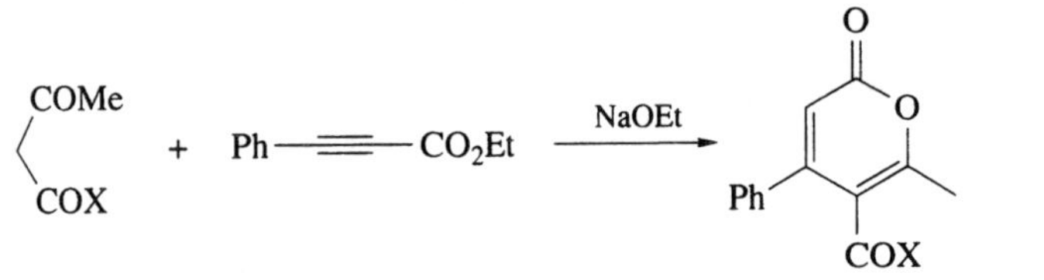

COMe / COX + Ph—≡—CO_2Et —NaOEt→ (pyranone with Ph, COX) (104)

(**2**) X = OEt (**10b**) (**480a**) X = OEt
(**479**) X = Me (**480b**) X = Me

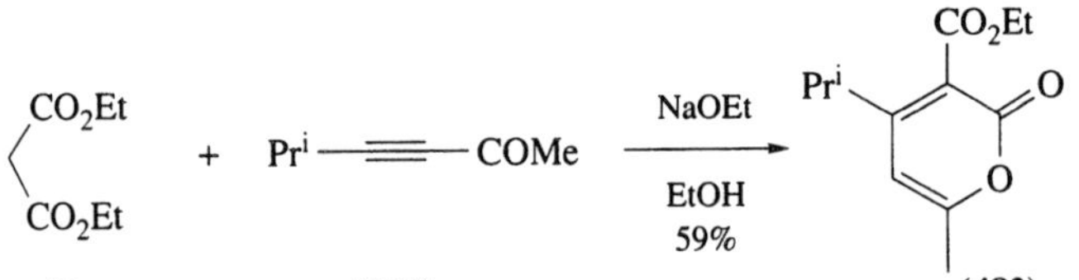

CO_2Et / CO_2Et + Pr^i—≡—COMe —NaOEt, EtOH, 59%→ (pyranone with Pr^i, CO_2Et) (105)

(**1**) (**481**) (**482**)

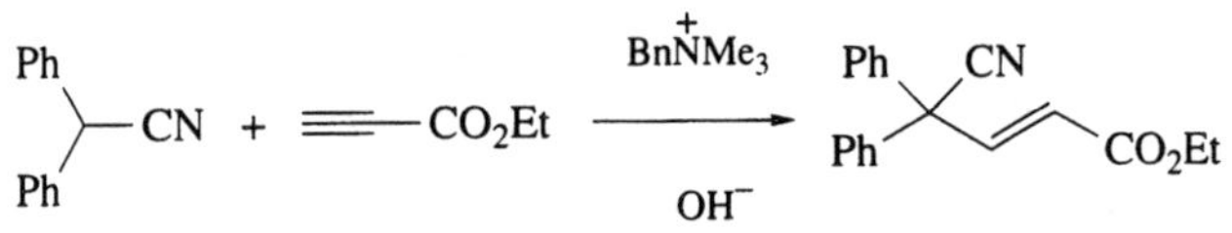

Ph_2CH—CN + ≡—CO_2Et —$Bn\overset{+}{N}Me_3$, OH^-, 92%→ $Ph_2C(CN)CH{=}CHCO_2Et$ (106)

(**483**) (**484**) (**485**)

(**486**) —≡—CO_2Et (**484**), $NaNH_2$, NH_3, 83%→ (**487**) —HCl, AcOH, 25 °C/12 h, Δ/3 h→ (**488**)

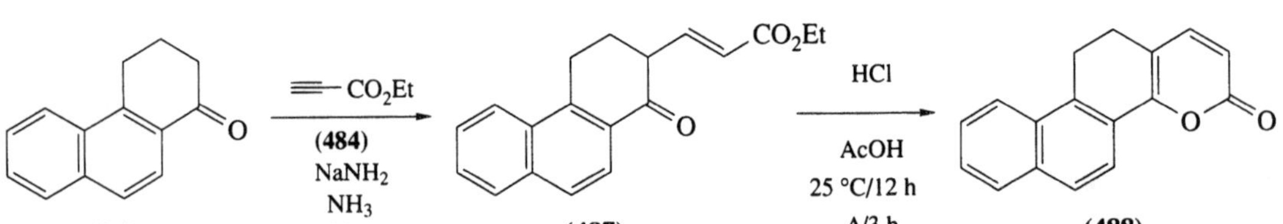

Scheme 67

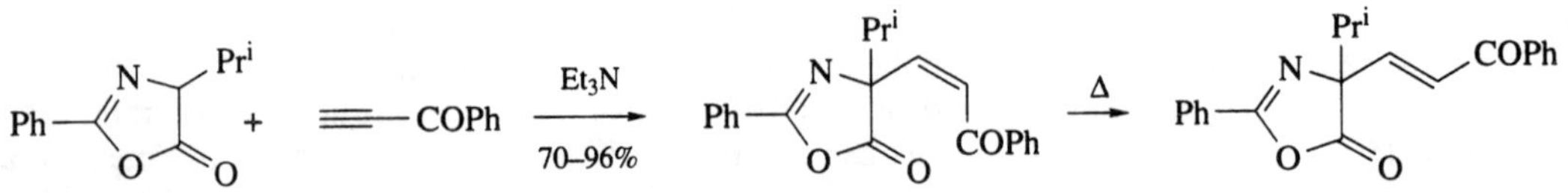

(**152**) + ≡—COPh (**489**) —Et_3N, 70–96%→ (*Z*)-(**490**) —Δ→ (*E*)-(**490**)

Scheme 68

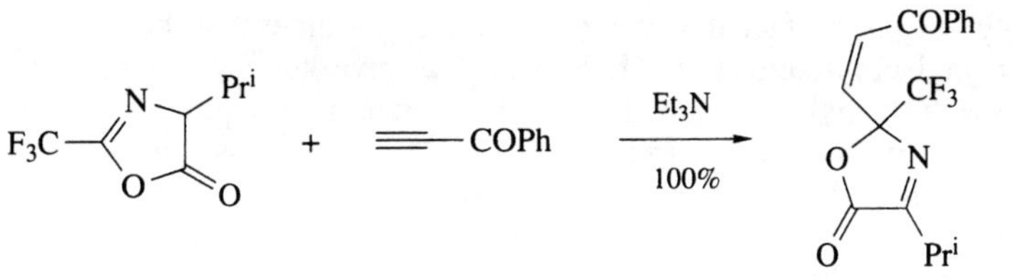

(491) + (489) $\xrightarrow[100\%]{Et_3N}$ (Z)-(492) (107)

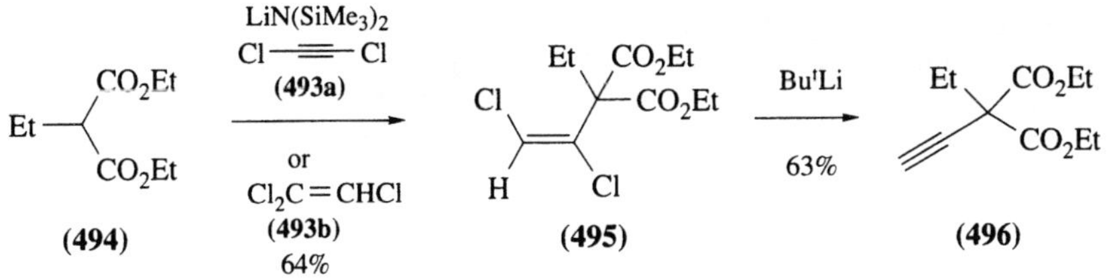

(494) $\xrightarrow[\text{or } Cl_2C{=}CHCl \text{ (493b), } 64\%]{LiN(SiMe_3)_2,\ Cl{-}{\equiv}{-}Cl \text{ (493a)}}$ (495) $\xrightarrow[63\%]{Bu^tLi}$ (496)

Scheme 69

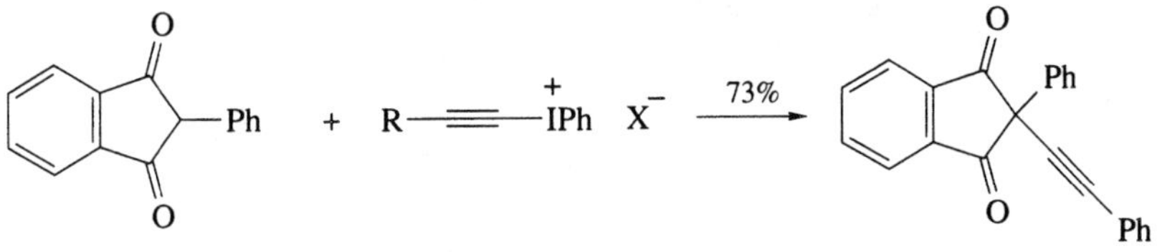

(497) + (498) $\xrightarrow{73\%}$ (499) (108)

(498a) R=Ph, X = Cl
(498b) R=C_8H_{17}, X = BF_4

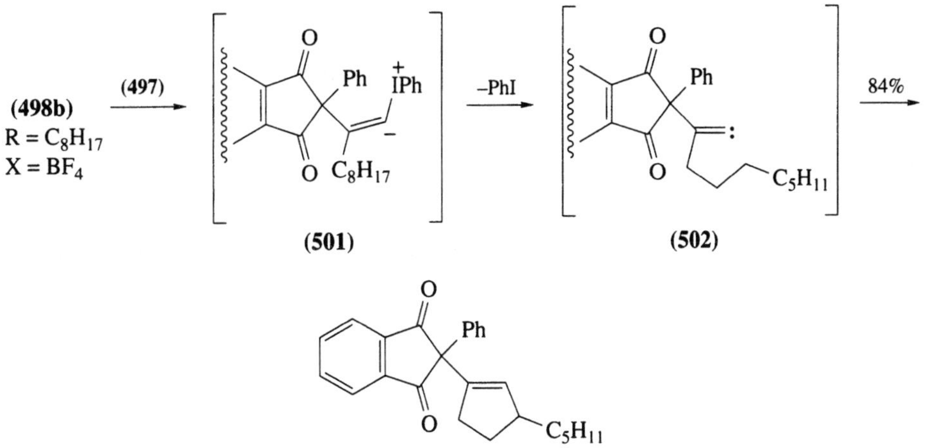

(498b) R = C_8H_{17}, X = BF_4 $\xrightarrow{(497)}$ [(501)] $\xrightarrow{-PhI}$ [(502)] $\xrightarrow{84\%}$ (500)

Scheme 70

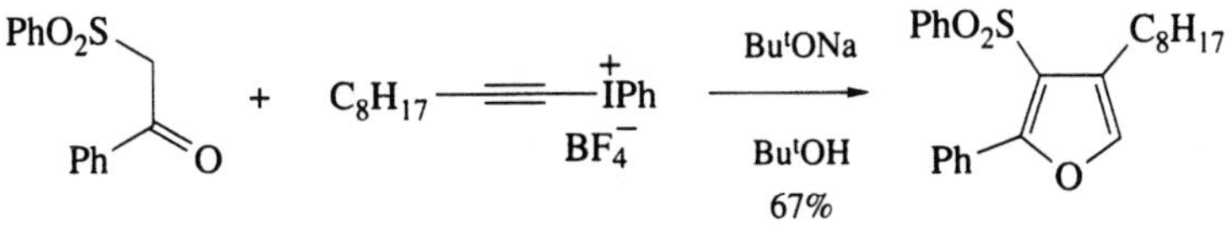

(503) + (498b) $\xrightarrow[Bu^tOH,\ 67\%]{Bu^tONa}$ (504) (109)

As with Michael additions to activated alkenes, the initial adducts with activated alkynes can be trapped by various processes. An aldol reaction can occur if a carbonyl is properly situated in the starting material (Scheme 71).[123] However, the use of methyl ethynyl ketone (**509**) and its homologs in the Robinson annulation process to give cyclohexadienones (**510**; equation 110) usually proceeds in poor yield,[124] presumably due to the fact that the enedione intermediate may be largely the (*E*)-isomer.[124b] However, alkoxyethynyl vinyl ketones (**511**; Scheme 72) give good yields of cyclic β-alkoxy enones (**513**) when reacted with diethyl malonate (**1**) in base, *via* internal trapping by proton transfer and an intramolecular Michael addition of the intermediate (**512**).[78a] This process was used in a synthesis of griseofulvin, as described earlier. A Dieckmann condensation can also serve as an internal trapping step. For example, reaction of dimethyl acetylenedicarboxylate (DMAD) (**514**; equation 111) with malononitrile (**39**) or ethyl cyanoacetate (**26**) in the presence of pyridine and acetic acid affords the salts (**515a**) or (**515b**), respectively, in good yield *via* an alkynic multiple Michael process followed by Dieckmann condensation.[125] However, the analogous condensation of (**514**) with dimethyl malonate (**56**; equation 112) under the same conditions afforded the seven-membered rings (**516a**) and (**516b**) in 63% and 4% yield, respectively.[125] These compounds are presumably formed by three consecutive intermolecular Michael additions followed by an intramolecular Michael addition to give the seven-membered rings (the last Michael addition being a 7-endo-trig cyclization).

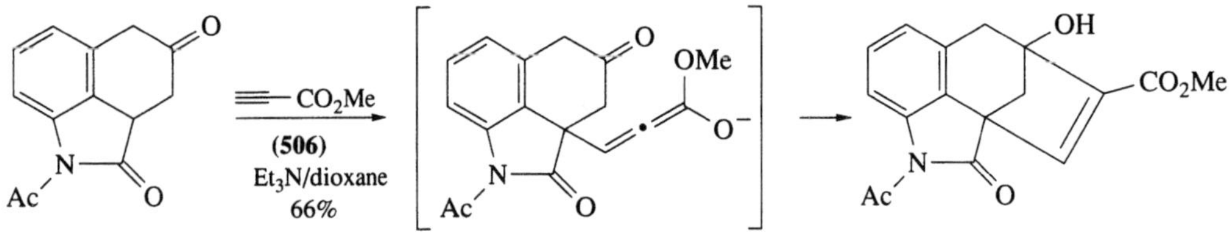

(**505**) —[$\equiv$—CO_2Me (**506**), Et_3N/dioxane, 66%]→ [(**507**)] → (**508**)

Scheme 71

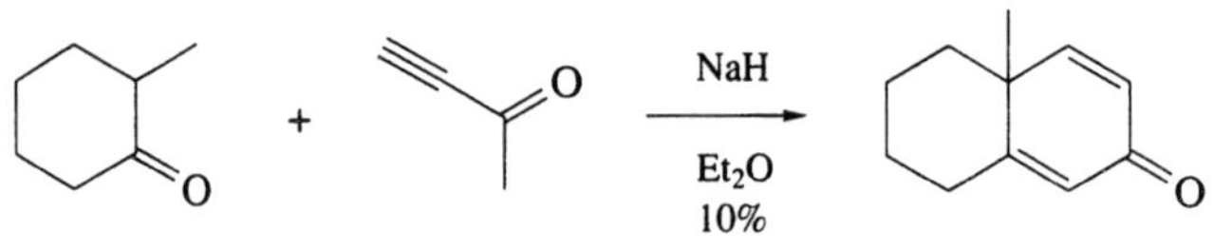

(**66**) + (**509**) —[NaH, Et_2O, 10%]→ (**510**) (110)

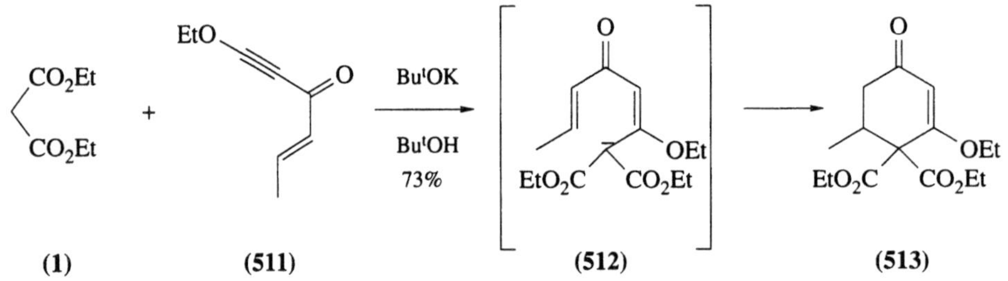

(**1**) + (**511**) —[Bu^tOK, Bu^tOH, 73%]→ [(**512**)] → (**513**)

Scheme 72

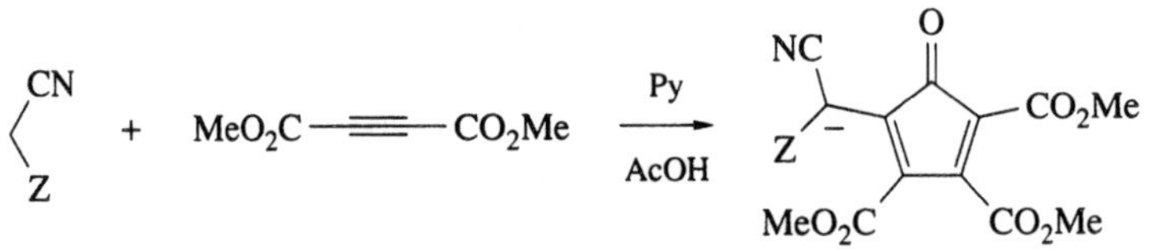

(**39**) Z = CN; (**26**) Z = CO_2Et + MeO_2C—≡—CO_2Me (**514**) —[Py, AcOH]→ (**515a**) Z = CN 60%; (**515b**) Z = CO_2Et 68% (111)

(**56**) —[MeO_2C—≡—CO_2Me (**514**), Py/AcOH]→ (**516a**) 63% + (**516b**) 4%; E = CO_2Me (112)

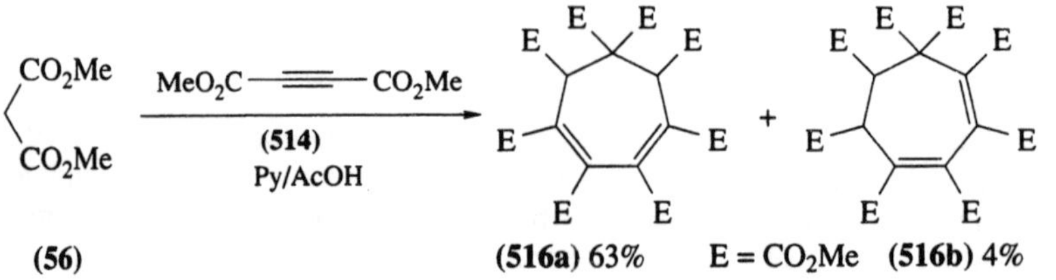

The reaction of enamines with alkynic esters has been used often in organic synthesis.[126] Enamines of aldehydes (**517**; Scheme 73) react with methyl or ethyl propiolate to give the 1-aminobutadiene-2-carboxylates (**519a,b**) by way of the cyclobutenes (**518a,b**).[126a] A similar reaction pathway is followed in the reaction of (**517**) with DMAD (**514**).[126b] Even enamines of β-diketones (β-amino enones) undergo this reaction to give analogous products which can then undergo further reactions.[126c] The reaction of enamines of cyclic ketones (**64**; Scheme 74) with methyl propiolate (**506**) or DMAD (**514**) at 90–110 °C affords the cyclic dienoates (**521a,b**), in which the ring has undergone a two-carbon expansion *via* the bicycloheptene intermediates (**520a,b**).[126a,126c,126e] By carrying out the addition at lower temperatures, the intermediate (**520a**) can be isolated.[126c,126e] Enamines of heterocyclic ketones, *e.g.* 3-piperidinoindole (**522a**; equation 113)[127a] or the 3-pyrrolidinodihydrothiophene (**522b**; equation 114),[127b] afford the corresponding seven-membered rings, (**523a**) and (**523b**), respectively, in excellent yields when heated with DMAD.[127] This type of two-carbon ring expansion of cyclic ketones has been used quite often in natural product synthesis.[128] It has also been applied to a variety of heterocyclic amino compounds to give products derived from similar reaction pathways.[129] Moreover, other nucleophilic alkenes can be used in place of enamines.[130] For example, the ynamine (**524**; equation 115) reacts with two equivalents of DMAD to give the aromatic product (**525**).[130a] The alkenic iminophosphoranes (**526**; equation 116), also, react with methyl propiolate or DMAD to give the ring-enlarged product (**527**).[130c,130d] Finally, the iminophosphoranes of certain ring systems, *e.g.* uracils, can give either simple Michael adducts or structurally rearranged products.[130e]

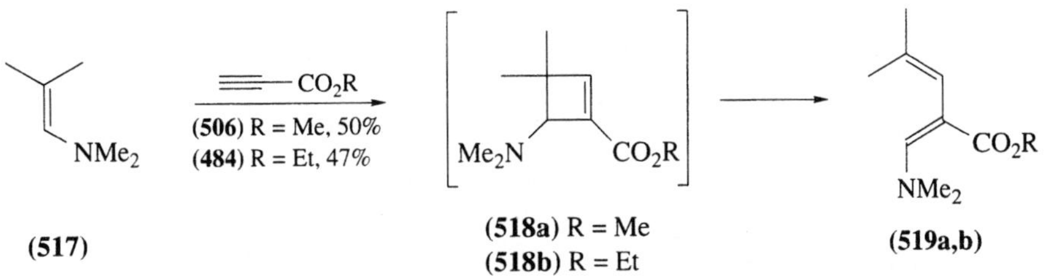

NMe_2 (**517**) —— ≡—CO_2R; (**506**) R = Me, 50%; (**484**) R = Et, 47% ——→ [Me_2N, CO_2R (**518a**) R = Me; (**518b**) R = Et] ——→ CO_2R, NMe_2 (**519a,b**)

Scheme 73

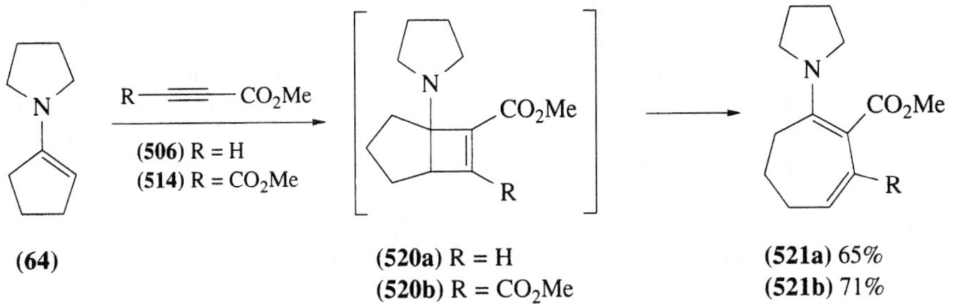

(**64**) —— R—≡—CO_2Me; (**506**) R = H; (**514**) R = CO_2Me ——→ [N, CO_2Me, R (**520a**) R = H; (**520b**) R = CO_2Me] ——→ N, CO_2Me, R (**521a**) 65%; (**521b**) 71%

Scheme 74

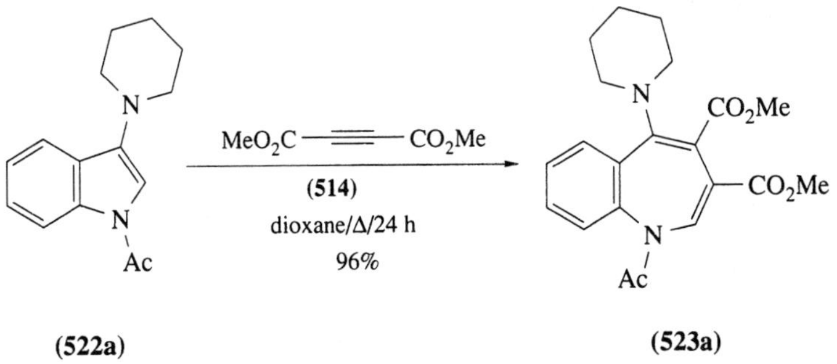

N, N, Ac (**522a**) —— MeO_2C—≡—CO_2Me (**514**) dioxane/Δ/24 h 96% ——→ N, CO_2Me, CO_2Me, N, Ac (**523a**) (113)

1.1.3.1.2 *Intramolecular additions*

There are only a few examples of intramolecular addition of a stabilized carbon nucleophile to an activated alkyne, one of these being in Trost's synthesis of hirsutic acid C, where basic treatment of the ketopropiolate (**528**; equation 117) at 110 °C resulted in a 65–70% yield of the bicyclic ketoacrylate (**529**), *via* a 5-exo-digonal cyclization.[76] Even though 5- and 6-endo-digonal cyclizations would be expected to be disfavored because of the inhibition of resonance stabilization of the enolate, due to angle strain (an allenic system in a five- or six-membered ring), under the right conditions they can be made to occur in high yields (*via* the vinyl anion not stabilized by resonance). For example, Deslongchamps has reported the formation of the hydrindenediones and the octalinediones, (**531a**) and (**531b**), respectively, from the corresponding β-keto ester ynones (**530a,b**) in very high yields (equation 118).[131] The reaction works poorly for the formation of the corresponding seven- and eight-membered systems.[131] However, this type of *endo* cyclization has not generally been seen in other systems. For instance, treatment of the ester ynamide (**532**; Scheme 75) with base did not afford the desired Michael adduct (**533a**) but rather the so-called 'anti-Michael' product (**533b**) and its endocyclic double-bonded isomer (**533c**).[86] Likewise, treatment of bis(phenylethynyl) ketone (**534**; equation 119) with ethyl acetoacetate (**2**) under basic conditions affords the 'anti-Michael' product (**535**) in 68% yield *via* a normal intermolecular Michael addition followed by an intramolecular 'anti-Michael' addition.[132]

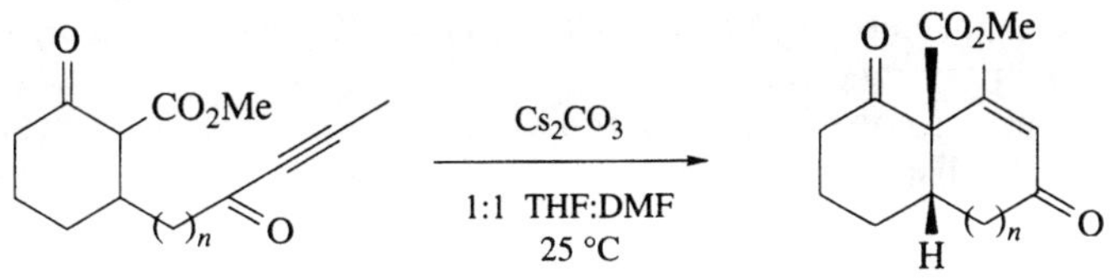

Scheme 75

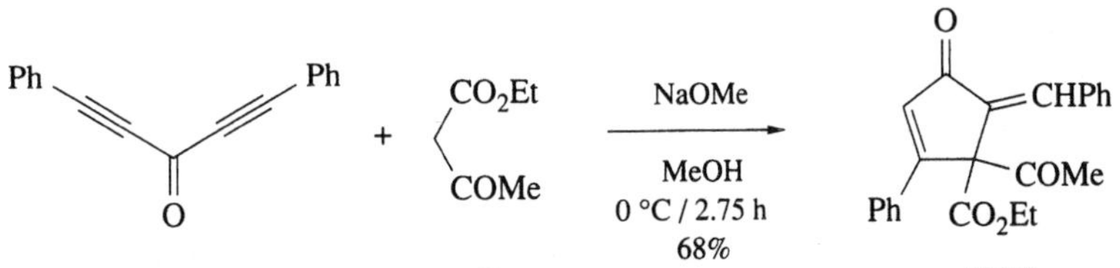

1.1.3.2 Heteronucleophiles

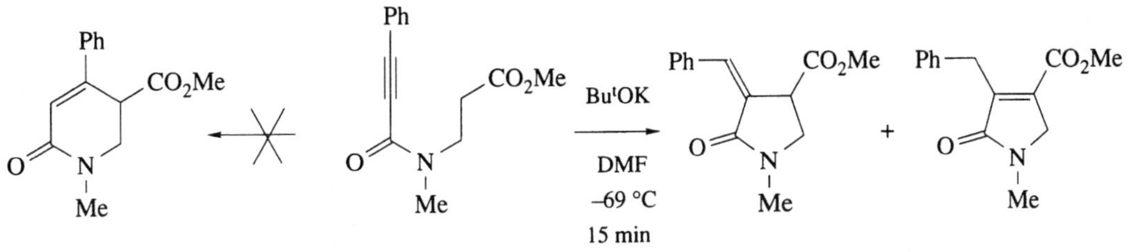

Since several excellent reviews summarize the literature in this area up to about 1970,[10] only a few representative examples from the older literature and some more recent applications are presented here.

1.1.3.2.1 Intermolecular additions

Michael was to first report, in 1896, the 1,4-addition of a heteronucleophile, namely the addition of ethoxide to diethyl acetylenedicarboxylate (**10a**) to give (**536**; equation 120).[133] As with the Michael addition, many electron-withdrawing groups have been used to activate the alkyne for addition—esters, aldehydes, ketones, nitriles, sulfones and so on. The number of heteronucleophiles used is quite extensive, including alcohols, amines (primary, secondary and tertiary), thiols, selenols, tellurols, halides, phosphines, phosphites, hydrazines and so on. The addition can be highly stereoselective or nearly stereorandom depending on the reaction conditions. The best mechanistic explanation for the large body of stereochemical results is as follows. Initial attack of the nucleophile on an activated alkyne (**A**; Scheme 76) occurs in an *anti* fashion to give the (*Z*)-vinyl anion (**B**) in preference to the (*E*)-vinyl anion (**C**).[134] If protonation is faster under the reaction conditions than equilibration, this compound leads to the (*Z*)-isomer (**D**) *via* overall *anti* addition.[135] This initially formed (*Z*)-vinyl anion (**B**) can isomerize to the (*E*)-vinyl anion (**C**) by way of the linear *sp*-hybridized anion (**E**) and its various resonance contributors (*e.g.* for Z = CO_2R, COR, *etc.*). If protonation is slow compared to equilibration, the more stable of the two products will be obtained—usually (**F**) when the nucleophile is alkoxide or a secondary or tertiary amine, but (**D**) when the nucleophile is a primary amine (due to internal hydrogen bonding). The relative rates of isomerization versus protonation will be directly dependent on the ability of the activating group Z to stabilize by resonance (*i.e.* to stabilize the linear anion **E**)[134d] and on the ability of the solvent to donate a proton to the anion. Thus, additions carried out in protic solvents tend to give a higher proportion of *anti* addition than those done in aprotic solvents, and alkynes activated by inductively stabilizing groups (sulfonyl, cyano) tend to give more *anti* addition. Finally, there is a second equilibration mechanism possible in certain cases, namely the direct isomerization of (**C**) and (**F**) *via* the intermediacy of the zwitterionic form (**G**) which can rotate about the central C—C single bond to lead to either product. A good discussion of this process for the addition of trialkylamines to various activated alkynes has recently been published.[136] This mechanistic picture can readily explain otherwise somewhat contradictory results.

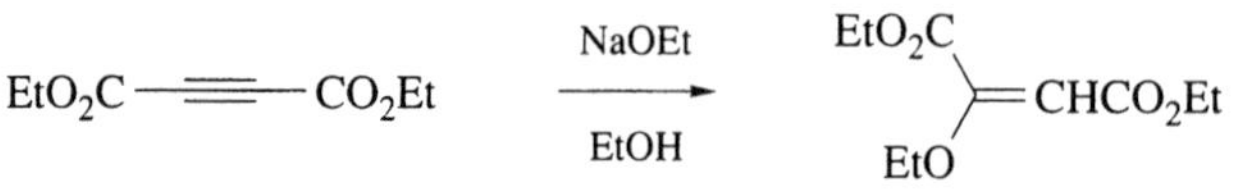

EtO_2C—≡—CO_2Et —(NaOEt, EtOH)→ EtO_2C(EtO)C=$CHCO_2Et$ (120)

(**10a**) (**536**)

H—≡—Z (**A**); (**B**), (**E**), (**C**), (**D**), (**G**), (**F**); Nu⁻; H-S

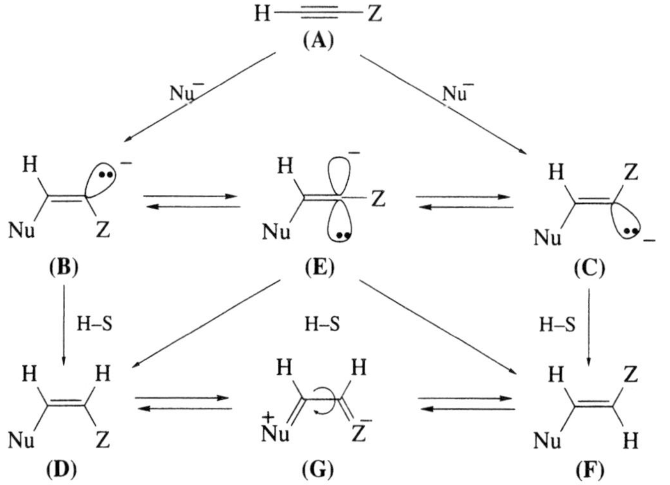

Scheme 76

Base-catalyzed addition of an alcohol[137] to methyl propiolate (**506**; equation 121) usually gives nearly exclusively the (*E*)-isomer (*E*)-(**537**),[137a] although under very weakly basic conditions a 2:1 mixture favoring the (*Z*)-isomer has been obtained.[137c] However, with DMAD (**514**) under certain conditions the (*Z*)-isomer (*Z*)-(**538**) predominates.[137a] An interesting rearrangement was initiated by addition of the alkoxide of the 3-hydroxy-*N*-phenylpyridine betaine (**539**; Scheme 77) with DMAD to give the furan (**543**), *via* the initial 1,4-adduct (**540**).[137g]

The addition of amines to activated alkynes leads to either the (*E*)- or (*Z*)-isomer with fairly high stereoselectivity based on the substitution pattern of the amine.[138] Primary amines add to alkynic ketones and propiolates (equation 122) to give an equilibrium mixture favoring the (*Z*)-isomer,[138a,138c] although there are reports of selective formation of the (*E*)-isomers.[138f] Secondary amines also afford an equilibrium mixture of (*E*)- and (*Z*)-isomers, but greatly favoring the (*E*)-isomer often to the exclusion of the (*Z*)-isomer (equation 123).[139] Addition of the chiral C(2)-symmetric pyrrolidine (**550**; Scheme 78) to the ester (**551**) gave in essentially quantitative yield the (*E*)-isomer which was cyclized with fluoride to furnish the lactone (**552**) in 94% overall yield. Alkylation of the anion of (**552**) proceeded with very high diastereoselectivity to give (**553**) as the major isomer.[139f] The use of aziridine[140] in protic solvents furnishes a higher proportion of the (*Z*)-isomer (*Z*)-(**554**), due to inhibition of the isomerization of the kinetic (*Z*)-isomer into the (*E*)-isomer *via* a zwitterionic intermediate (equation 124). With alkynic sulfones the proportion of the (*Z*)-isomer can be very high (equation 125).[138d] The addition of the 2-vinylaziridine (**427**; Scheme 79) to DMAD (**514**) at –20 °C produced the dihydroazepine (**558**) in 95% yield,

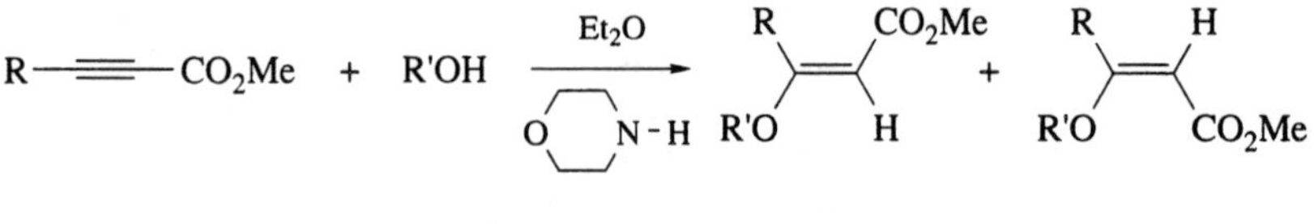

R—≡—CO_2Me + R'OH —(Et_2O, morpholine O⌒N-H)→ R(R'O)C=CH(CO_2Me) + R(R'O)C=C(H)CO_2Me (121)

(**506**) R = H	R' = Pr^i	87%	(*E*)-(**537**)	100:0	(*Z*)-(**537**)
(**514**) R = CO_2Me	R' = Me	92%	(*E*)-(**538**)	10:90	(*Z*)-(**538**)

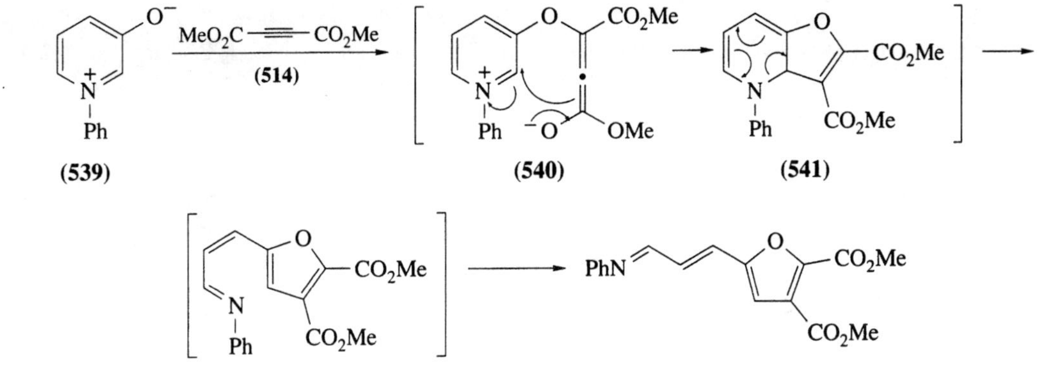

(539) (540) (541)

(542) (543)

Scheme 77

presumably *via* a [3,3] sigmatropic rearrangement of the initially formed zwitterion (**557**) from 1,4-addition.[100a] The addition of trialkylammonium salts[141] to activated alkynes in methanol can give either exclusive (*E*)- or (*Z*)-isomers or a mixture, depending on the activating group. With the ketone (**509**; equation 126) only the (*E*)-isomer (*E*)-(**560**) is formed, while with the nitrile (**559**), only the (*Z*)-isomer (*Z*)-(**561**) is produced. With propiolate (**506**), a nearly 1:1 mixture of (*E*)- and (*Z*)-(**562**) is obtained.[136] These results are easily explained by analysis of the ability of the activating group to stabilize the linear form of the initially formed Zq(*Z*)-vinyl anion.[134d] Thus the inductively stabilizing nitrile group isomerizes slowly and is directly protonated before equilibration to give the (*Z*)-isomer, while the resonance-stabilizing ketone allows rapid and complete equilibration and only the (*E*)-isomer is formed. The ester lies between these two extremes. This work[136] furnishes the best experimental evidence for the theoretical calculations of vinyl anion stability.[134d]

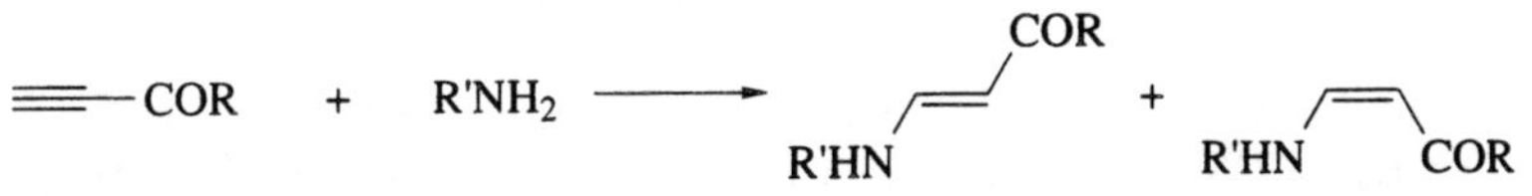

≡—COR + R'NH₂ ⟶ (*E*) + (*Z*) (122)

(**506**) R = OMe	R' = Bu^t	64%	(*E*)-(**544**)	0:100	(*Z*)-(**544**)
(**506**) R = OMe	R' = *c*-hexyl		(*E*)-(**545**)	15:85	(*Z*)-(**545**)
(**489**) R = Ph	R' = Bn	98%	(*E*)-(**546**)	0:100	(*Z*)-(**546**)

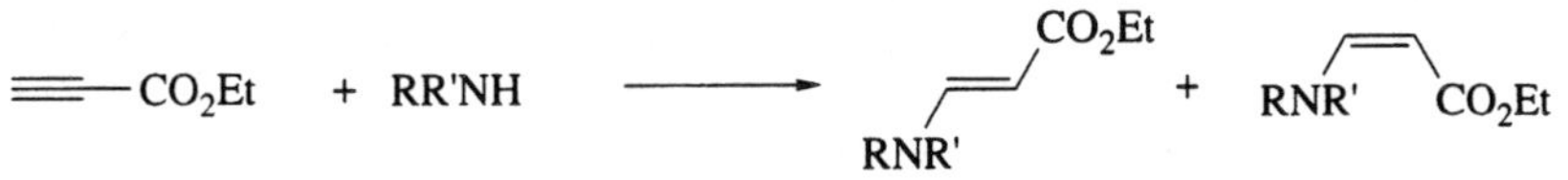

≡—CO_2Et + RR'NH ⟶ (*E*) + (*Z*) (123)

(**484**)	R = R' = $(CH_2)_4$	>95%	(*E*)-(**547**)	100:0	(*Z*)-(**547**)
	R = R' = Pr^i	90%	(*E*)-(**548**)	100:0	(*Z*)-(**548**)
	R = Me, R' = CH_2CO_2Et	90%	(*E*)-(**549**)	100:0	(*Z*)-(**549**)

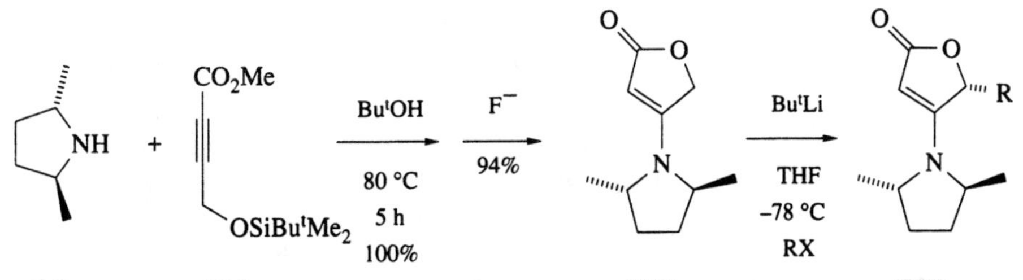

(550) (551) (552) (553) R = Me, 97:3

Scheme 78

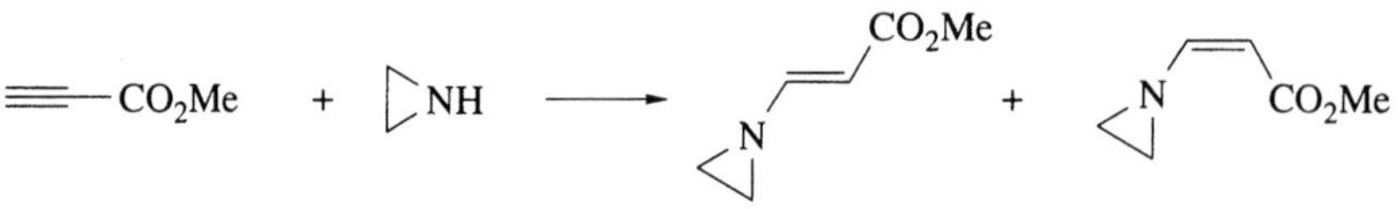

(124)

(506) DMF (E)-(554) 97:3 (Z)-(554)
MeOH 47:53

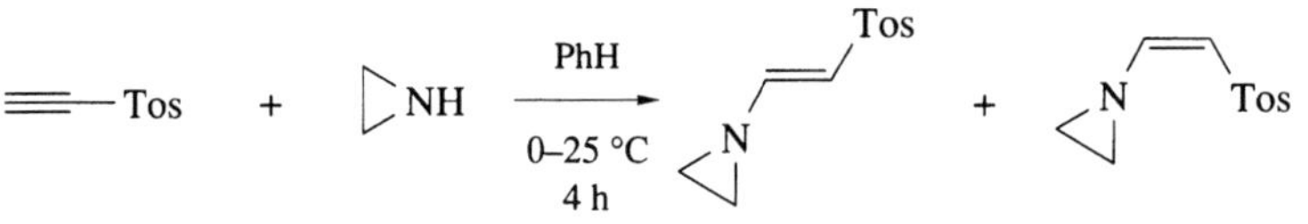

(125)

PhH, 0–25 °C, 4 h

(555) (E)-(556) 5:95 (Z)-(556)

−20 °C; 95%

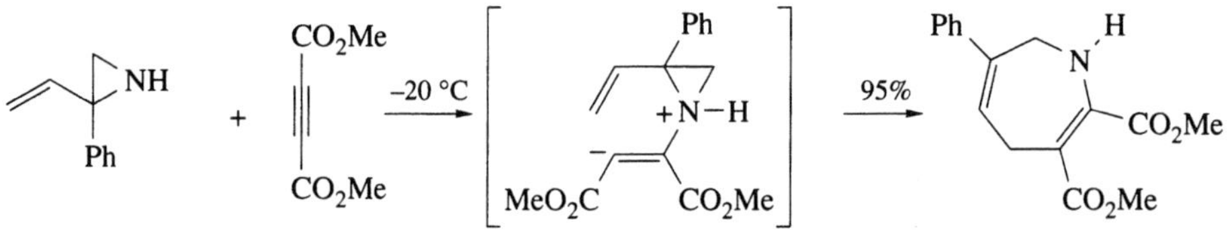

(427) (514) (557) (558)

Scheme 79

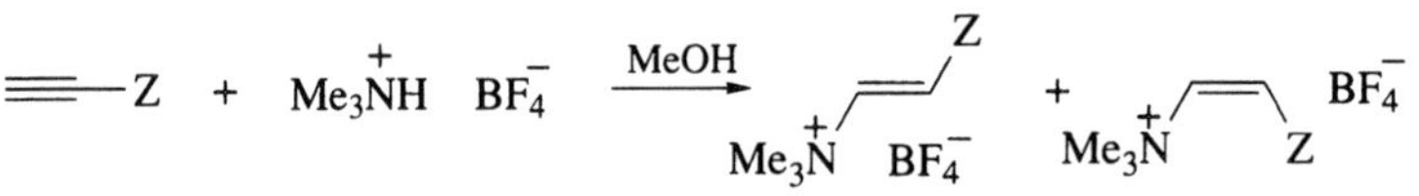

(126)

MeOH

(509) Z = COMe	25 °C/8 h	(E)-(560)	100:0	(Z)-(560)
(559) Z = CN	25 °C/18 h	(E)-(561)	0:100	(Z)-(561)
(506) Z = CO_2Me	65 °C/24 h	(E)-(562)	57:43	(Z)-(562)

The additions of thiol anions, selenides and tellurides are all highly selective for the (*Z*)-isomers, exclusively so for the selenides and tellurides.[142] For example, addition to (**10b**) gives primarily the (*Z*)-isomer (equation 127).[142a,142e,142l] The addition of dithiolate, dithiocarbamate and dithiocarbonate salts also occurs with high (*Z*)-selectivity in acetic acid/acetonitrile[142p] to give (*Z*)-(**566**; equation 128), but dithioic acids give mixtures enriched in the (*E*)-isomer (*E*)-(**567**) in carbon tetrachloride.[142g] The addition of thiols to activated alkynes has been used in a clever stereospecific trisubstituted alkene synthesis.[142b] Addition of thiophenoxide to the alkynic ester (**568**; Scheme 80) gives exclusively the (*Z*)-isomer (**569**), which is converted into the (E)-trisubstituted alkene (**570**) *via* copper-promoted Grignard addition. The kinetic (*Z*)-vinyl anion initially produced can be trapped with benzaldehyde to afford mainly the product of *anti* addition (*Z*)-(**571**; equation 129).[142k]

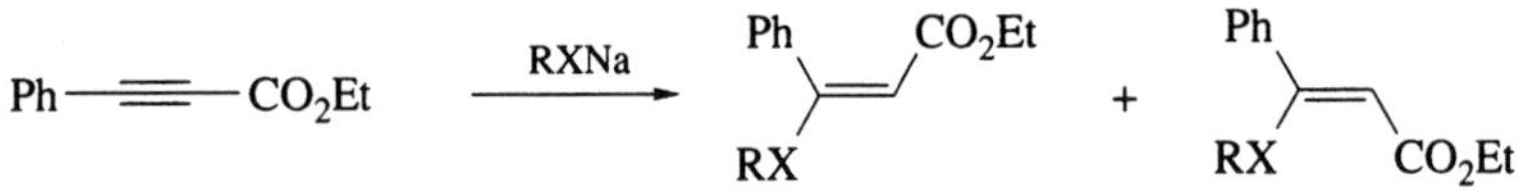

(127)

RXNa

(10b)	RX = PhS	85%	(E)-(563)	1:5	(Z)-(563)
	TolS	33%	(E)-(564)	0:100	(Z)-(564)
	PhSe	85%	(E)-(565)	0:100	(Z)-(565)

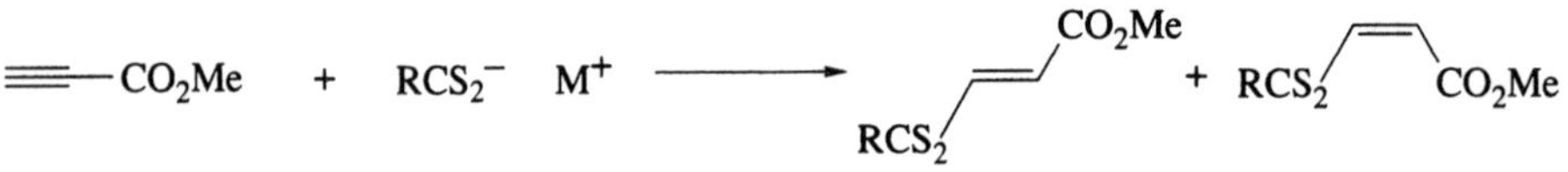

(128)

(506)	R = Ph	M = Et_3NH	AcOH/MeCN	88%	(E)-(566)	0:100	(Z)-(566)
	R = Me	M = H	CCl_4/Δ		(E)-(567)	95:5	(Z)-(567)

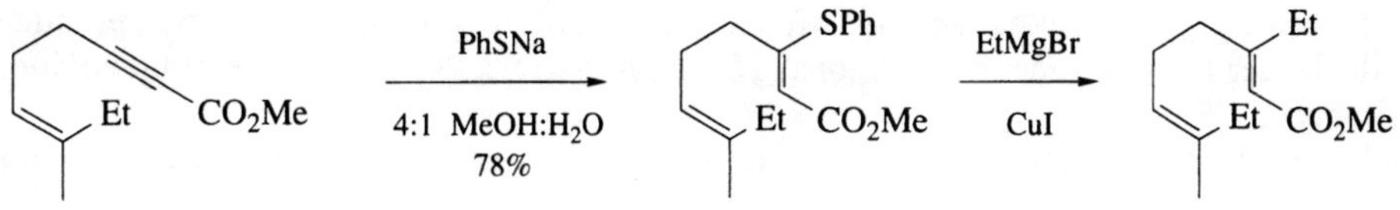

PhSNa, 4:1 MeOH:H_2O, 78%; EtMgBr, CuI

(568) (569) (570)

Scheme 80

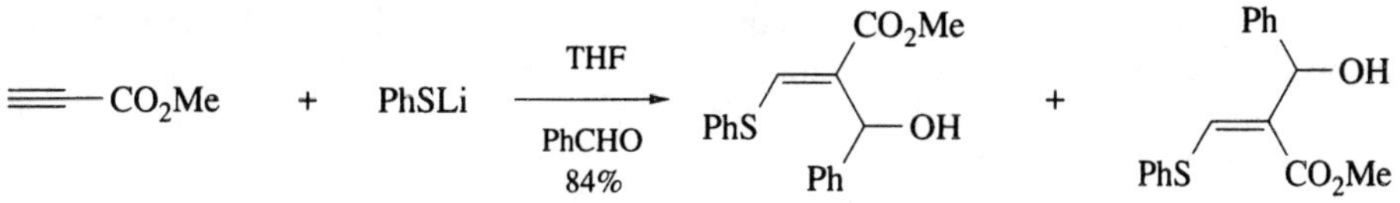

THF, PhCHO, 84%

(506) (E)-(571) 17:83 (Z)-(571) (129)

The addition of hydrogen halides to propiolic acid (**572**; equation 130) occurs by overall *anti* addition to furnish cleanly the (*Z*)-isomer of (**573**).[143] This product can also be formed in 80% yield by refluxing the acid (**572**) with methylmagnesium iodide in THF and quenching with glacial acetic acid.[143c] *N*,*N*-Dialkylhydrazines add to propiolic acid (**572**) to give the betaine (**574**),[144a] but when added to methyl propiolate (**506**) or DMAD (**514**) afford the cyclic betaines (**575**; equation 131).[144b] Hydrazine and *N*-alkylhydrazines produce the corresponding pyrazolin-5-one.[144c]

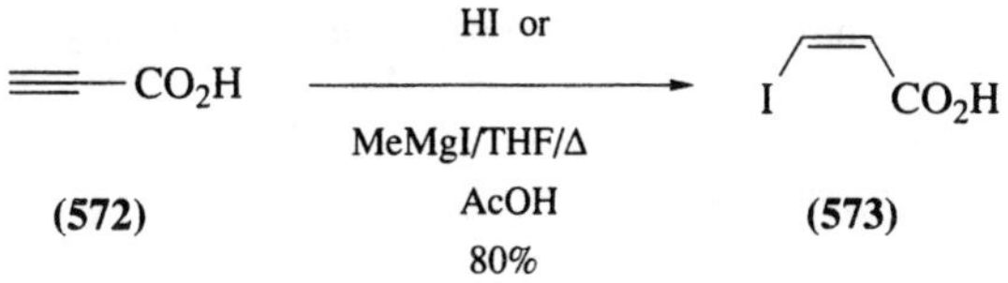

HI or MeMgI/THF/Δ, AcOH, 80%

(572) (573) (130)

R—≡—CO_2R' + Me_2NNH_2 → (574) or (575) (131)

(**572**) R = R' = H — MeOH — 57% (**574**)
(**506**) R = H, R' = Me — 1:1 MeOH:H_2O — 39% (**575a**)
(**514**) R = CO_2Me, R' = Me — 40% (**575b**)

Trialkylphosphines add *via anti* addition to give the expected salts (**577**; Scheme 81) which decompose upon work-up to products (**578**) not containing phosphorus.[145a] Triethyl phosphite (equation 132) reacts with propiolic acid (**572**) or methyl propiolate (**506**) to give predominately (>90%) the β-phosphono ester (**579**) with the (*E*)-configuration.[145b]

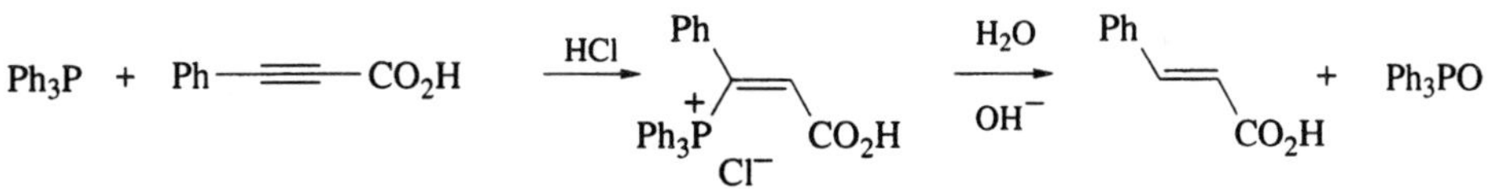

Ph_3P + Ph—≡—CO_2H —HCl→ (577) —H_2O, OH^-→ (578) + Ph_3PO

(576) (577) (578)

Scheme 81

≡—CO_2R + $P(OEt)_3$ → $(EtO)_2OP$—CH=CH—CO_2R (132)

(**572**) R = H
(**506**) R = Me

(**579**)
R = H
R = Me

Finally, a recent paper reports that under kinetic conditions (25 °C/6 h) phenoxide and thiophenoxide add to benzoyl(trifluoromethyl)acetylene (**580**; equation 133), 1,4 to the trifluoromethyl group rather than 1,4 to the benzoyl group (so-called 'anti-Michael' addition), to give the (*Z*)-addition products (**581**) in high yields. Under thermodynamic conditions (150 °C/24 h) a nearly 1:1 mixture of the regioisomeric adducts is produced.[146]

The initial vinyl anion produced can be trapped in several ways. One particularly useful intramolecular trapping involves the condensation of an α-amino, α-hydroxy or α-mercapto ketone with an activated alkynes to produce the heterocyclic aromatic compounds in one or two steps in good yield.[147] For example, condensation of the anilino ketone (**582**; equation 134) with methyl propiolate or DMAD produces the pyrroles (**583a,b**) in good yield.[147b] Similar processes in which the vinyl anion condenses on a ketone to form the C(3)–C(4) bond have been described often in the literature.[147a,147c,147d] However, the condensation of methyl *N*-ethylglycinate (**584**; equation 135) and DMAD affords the 3-hydroxypyrrole (**585**) *via* a Dieckmann condensation to form the C(2)–C(3) bond rather than the C(3)–C(4) bond.[147e] In this case and the following, one of the DMAD ester groups serves as the electrophile in the subsequent cyclization process. Quinolones (**588**; Scheme 82) can also be formed in two steps by condensing the anthranilate (**586**) with DMAD to give the anilinofumarate (**587**), which on pyrolysis affords the quinolone (**588**).[147c] Condensation of the enamino ketone (**589**; Scheme 83) with DMAD gives the amino fumarate (**590**) which on heating in methanol produces the pyrrolone (**591**).[147a] Furans can also be prepared by this general route. Condensation of benzoin (**592**; Scheme 84) with DMAD in the presence of potassium carbonate yields the hydroxydihydrofuran (**593**) which can be dehydrated to the furan (**594**) in 90% yield on heating with acidic methanol.[147a] The formation of thiophenes by this process shows this dichotomy of cyclization modes. Treatment of mercaptoacetone (**595**; Scheme 85) with methyl propiolate and molar quantities of potassium *t*-butoxide in DMSO furnishes only 2-acetyl-3-hydroxythiophene (**597**) by way of a Dieckmann-like condensation of the (*Z*)-isomer (*Z*)-(**596**). The use of catalytic quantities of base, however, gives a very different mixture of products in which the (*E*)-isomer of the initial adduct (*E*)-(**596**) and its condensation product (anion α to ester condensing on to the ketone) (**598**) predominate.[147f] Several other heterocyclic systems have been produced by similar chemistry.[147g]

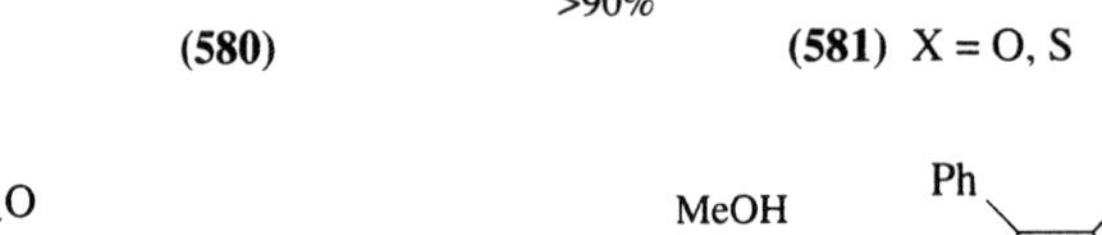

PhCO—≡—CF_3 (**580**) → [PhX⁻, 25 °C/6 h, >90%] (**581**) X = O, S (PhCO, PhX, CF_3) (133)

(**582**) + R—≡—CO_2Me → [MeOH, Δ] (**583**) (134)

(**506**) R = H 2 h/79% (**583a**) R = H

(**514**) R = CO_2Me 16 h/70% (**583b**) R = CO_2Me

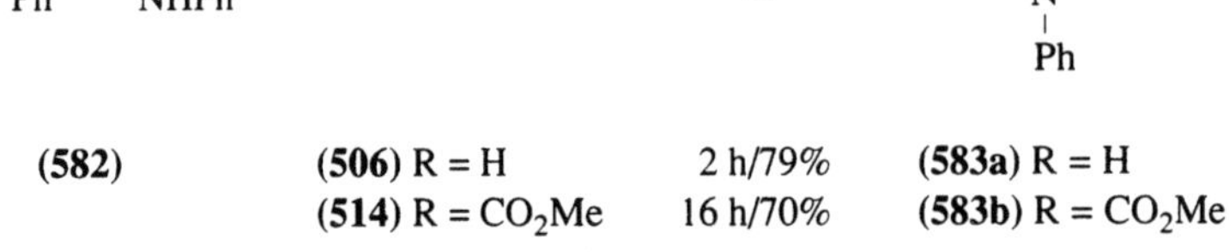

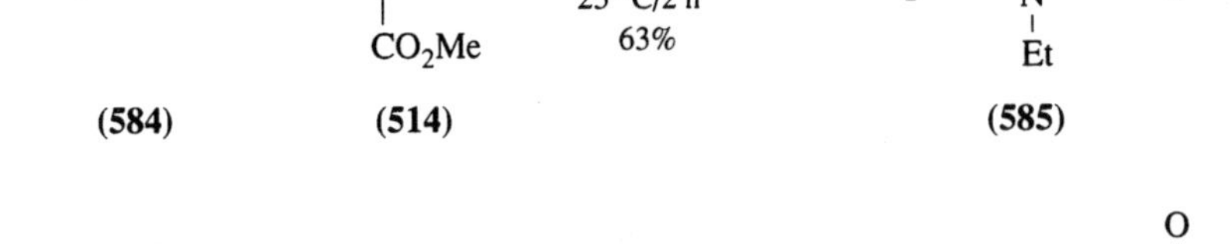

MeO_2C⌒NHEt (**584**) + (**514**) → [Et_2O, 25 °C/2 h, 63%] (**585**) (135)

(**586**) + (**514**) → [MeOH, 52%] (**587**) → [250 °C, 70%] (**588**)

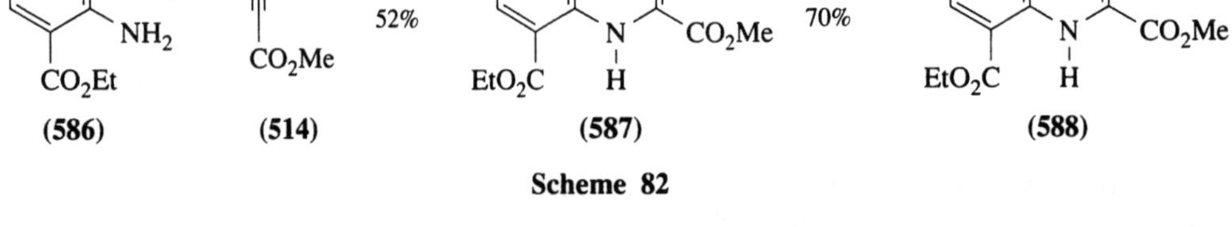

Scheme 82

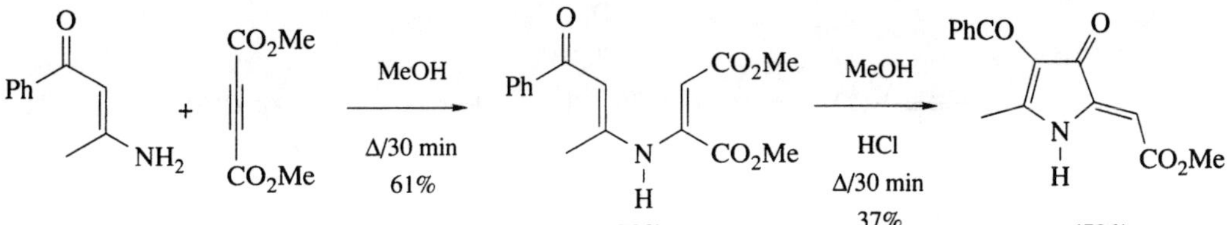

Scheme 83

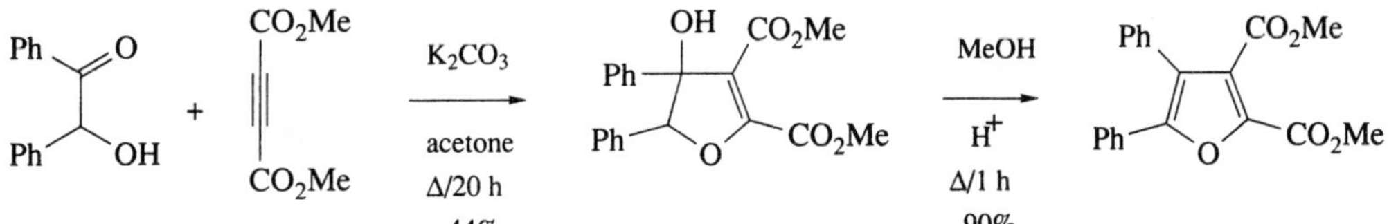

Scheme 84

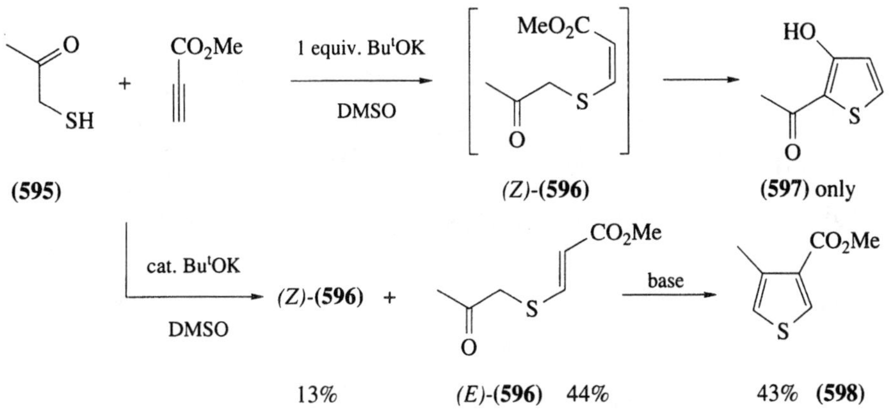

Scheme 85

1.1.3.2.2 Intramolecular additions

A number of intramolecular additions of heteronucleophiles to activated alkynes have been reported to give heterocyclic systems. For example, the addition of water, hydrogen sulfide or primary amines to the disubstituted diethynyl ketones (**598**; equation 136) furnish the pyrones (**599**) in good yields, *via* a 6-endo-digonal cyclization.[132,148] In certain cases (especially when R = Ph), the 'anti-Michael' products (**600**) can predominate.[132a] Reaction of the ethynyl amide or ketone, (**601a**) or (**601b**), respectively, with isothiocyanate or carbon disulfide (quenching with methyl iodide) gives exclusively the 'anti-Michael' products (**602a**) and (**602b**) from the amide (**601a**) but mainly the Michael products (**603**), rather than (**604**), from the ketone (**601b**; Scheme 86).[149] Treatment of the dihydroxy alkynic ketone (**605**; equation 137) with cesium carbonate in acetonitrile gives exclusively the furanone (**606**) *via* a 5-endo-digonal cyclization, rather than the product of the 6-endo-digonal cyclization (**607**).[113] An interesting 8-endo-digonal cyclization of an alkoxide is a key step in the formation of the oxocenone (**610**; Scheme 87) from the lactone (**608**), which presumably proceeds *via* the anion (**609**).[150]

1.1.4 ALLENIC π-SYSTEMS

Nucleophilic additions to activated allenes have been reported fairly often. Since there are several good recent reviews of this area,[151] only a brief overview is given here.

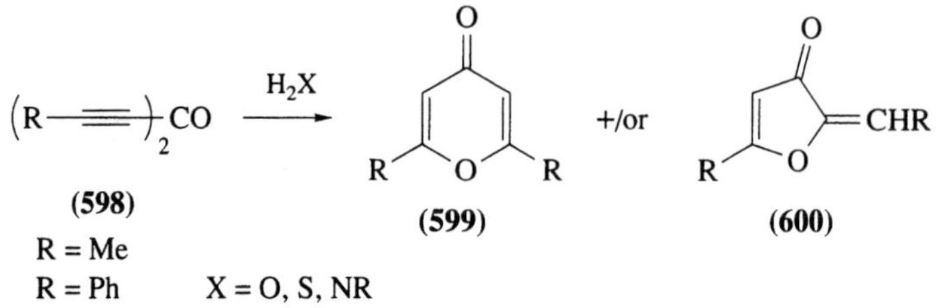

(136)

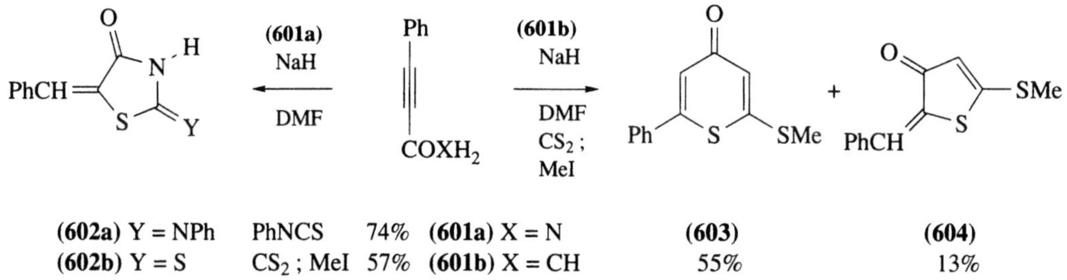

(598) R = Me R = Ph X = O, S, NR (599) (600)

(602a) Y = NPh PhNCS 74% (601a) X = N (603) 55% (604) 13%
(602b) Y = S CS_2 ; MeI 57% (601b) X = CH

Scheme 86

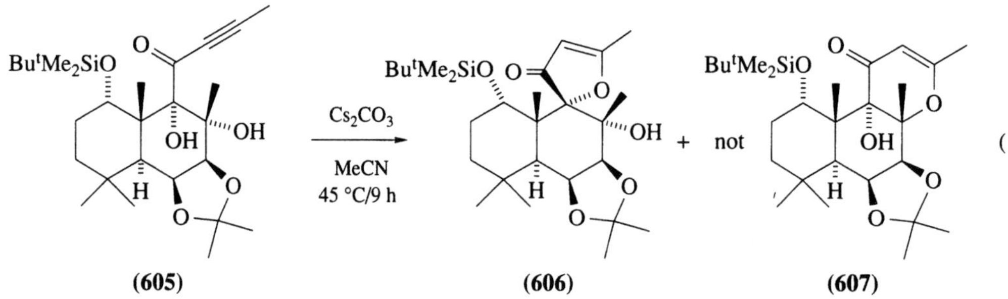

(137)

(605) (606) (607)

(608) (609) (610)

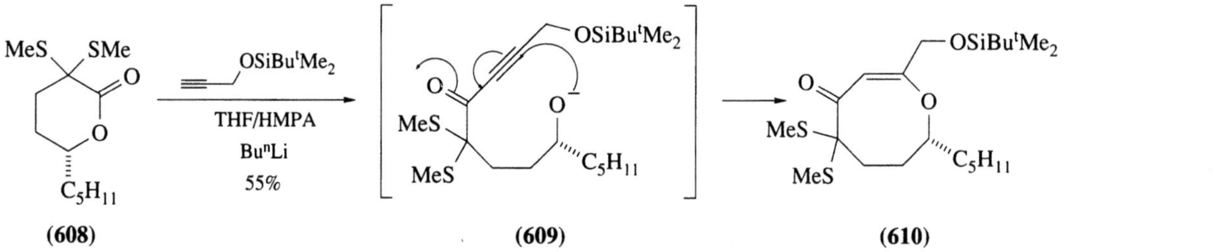

Scheme 87

1.1.4.1 Carbon Nucleophiles

Stabilized carbon nucleophiles add to the central carbon of activated allenes.[152] For example, addition of the enamine (**611**; Scheme 88) to diethyl allenedicarboxylate (**612**), followed by heating with acetic acid, affords the pyridone (**614**) in 70% yield *via* the intermediate (**613**).[152a] A more substituted fumarate was used in this process in the key step of a synthesis of camptothecin.[152b] A second approach to camptothecin used a pyridone synthesis based on the reaction of an imine of pyruvate (**615**; equation 138) with dimethyl allenedicarboxylate (**616**) to give (**617**).[152c] The addition of simple nucleophiles also works well, *e.g.* the formation of (**619**; equation 139) in 71% yield from (**1**) and (**618**).[153a] The additions of acylaminomalonates (**620**; equation 140) to the nitrile (**618**) or the ester (**621**) afford the unconjugated products (**622**) in very high yields.[153b] Addition of β-keto esters or β-diketones to the allenic dimethylsulfonium salt (**623**; equation 141) (prepared *in situ* from the corresponding propargylic salt) affords 4-methylfurans with electron-withdrawing substituents in the 3-position (**624a,b**).[154] However, malonates and substituted malonates (Scheme 89) react with (**623**) to give the rearranged α-alkenylmalonates (**628a**) and (**628b**), *via* a [2,3] sigmatropic shift of the sulfur ylides (**627a**) and (**627b**), formed from the initial Michael adducts (**626a**) and (**626b**). Finally, it has been reported that phosphonium ylides add to activated allenes to give new vinyl ylides and further products.[155]

1.1.4.2 Heteronucleophiles

Many heteronucleophiles have been added to allenic ketones, esters, nitriles, sulfones, *etc.*[156] In general, the product is the α,β-unsaturated system, as the following examples show (equation 142, Schemes 90 to 92). The β,γ-unsaturated system is formed first (and can sometimes be isolated)[156c,156g] and then rearranges to the α,β-unsaturated system.[156c] In general, the thermodynamically more stable alkene isomer is produced in this process, although mixtures are common. With dimethyl allenedicarboxylate (**616**), addition of alcohols and secondary amines is reported to give only the (*E*)-isomers (**635**), while addition of cyclohexylamine gives mainly the (*Z*)-isomer (*Z*)-(**636**) due to internal H-bonding (Scheme 93).[156n]

Cyclization of the initial acyclic adducts is also possible and several heterocyclic ring systems can be formed.[157] Addition of hydrazine or hydroxylamine to the nitrile (**618**; equation 143) produces the ami-

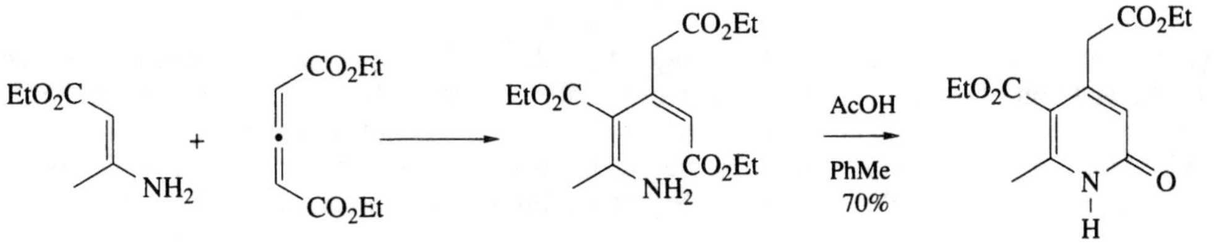

(**611**) + (**612**) → (**613**) —AcOH, PhMe, 70%→ (**614**)

Scheme 88

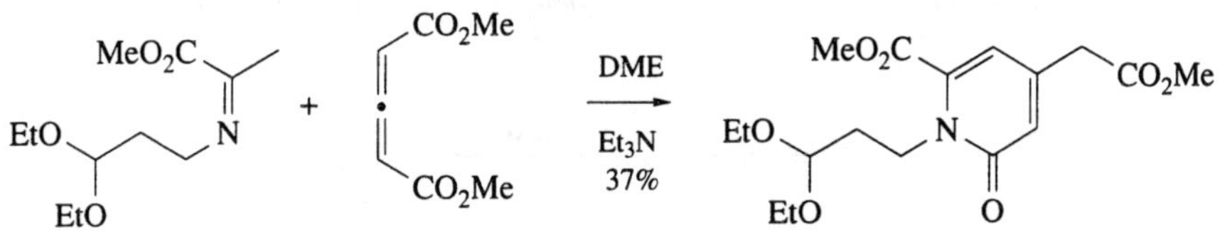

(**615**) + (**616**) —DME, Et_3N, 37%→ (**617**) (138)

(**618**) + (**1**) —NaOEt, 71%→ (**619**) (139)

(**620**) + Z-allene —NaOEt, EtOH, 1 h→ (**622**) (140)

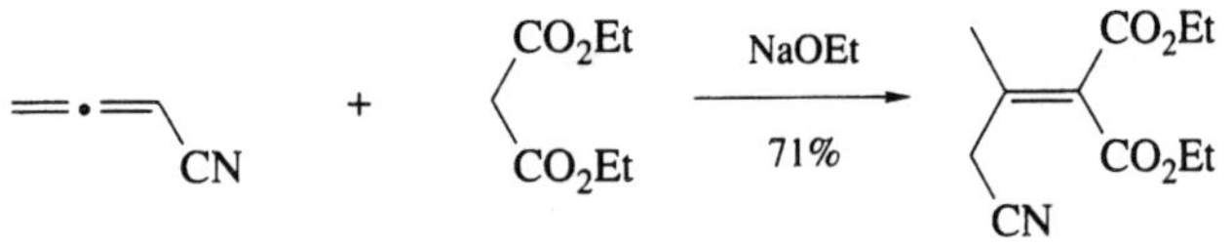

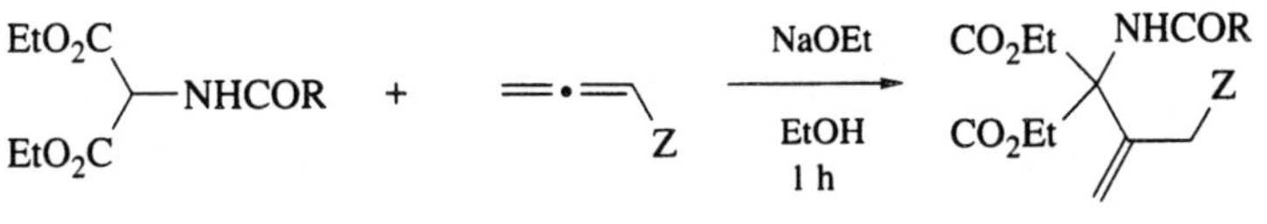

(**620**)			(**622**)
R = Bu^{t}O	(**618**) Z = CN	89%	R = Bu^{t}O, Z = CN
R = Me	(**621**) Z = CO_2Et	94%	R = Me, Z = CO_2Et

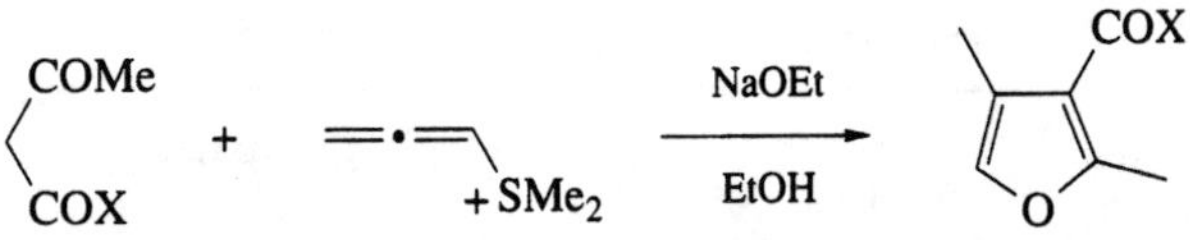

(**2**)/(**479**) + (**623**) —NaOEt, EtOH→ (**624**) (141)

(**2**) X = OEt	(**623**)	86%	(**624a**) X = OEt
(**479**) X = Me		89%	(**624b**) X = Me

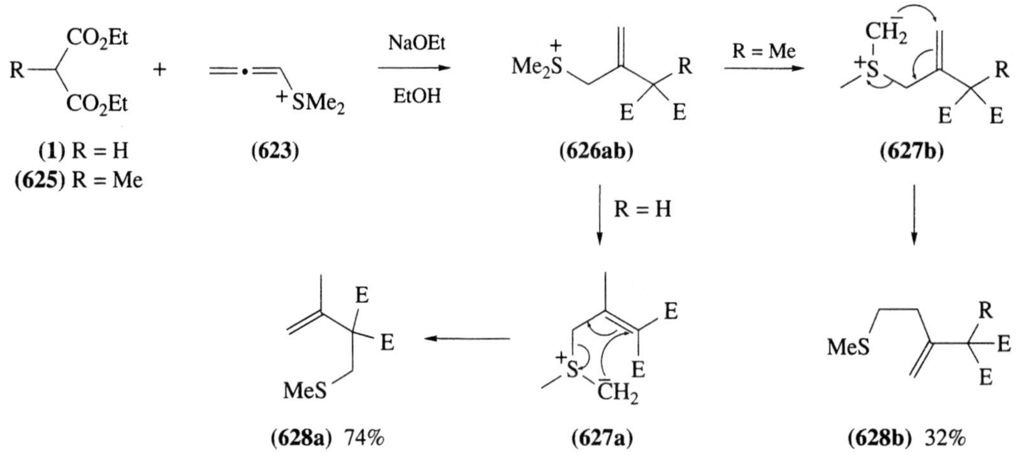

(1) R = H
(625) R = Me
(623)
(626ab)
(627b)
(628a) 74%
(627a)
(628b) 32%

Scheme 89

nopyrazole or isoxazole, (**637a**) or (**637b**) in good yield.[153a,157a] A synthetically quite useful process involves the addition of an *N*-acylarylhydroxylamine (**638**; Scheme 94) to an activated allene (**639**) under basic conditions to give the anion (**640**), which undergoes a [3,3] sigmatropic rearrangement to afford in high yield the 2-alkylated aniline (**642**) by way of the intermediate (**641**).[158] Heating (**642**) with formic acid produces the indole in this oxygen analog of the Fischer indole synthesis.[158] Addition of ethyl *N*-

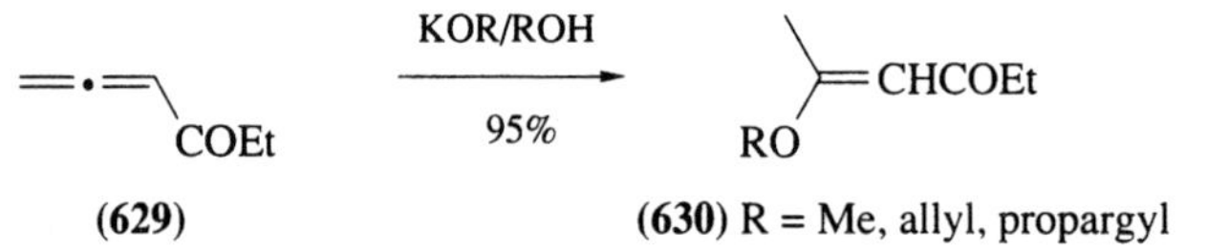

KOR/ROH, 95% (142)

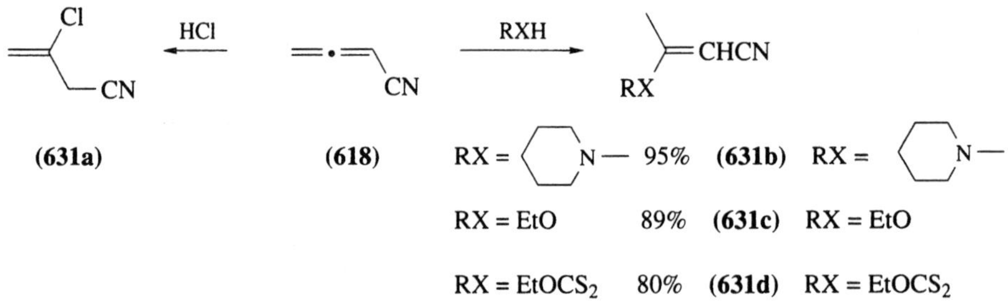

(629)
(630) R = Me, allyl, propargyl

(631a) ← HCl — (618) — RXH →

RX = piperidino 95% (631b) RX = piperidino
RX = EtO 89% (631c) RX = EtO
RX = $EtOCS_2$ 80% (631d) RX = $EtOCS_2$

Scheme 90

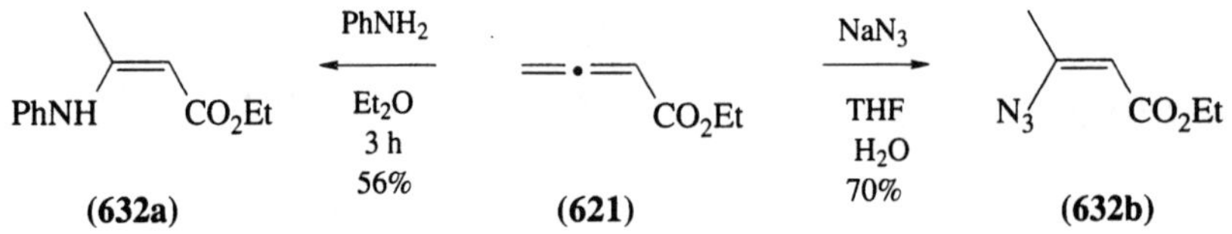

(632a) ← $PhNH_2$, Et_2O, 3 h, 56% — (621) — NaN_3, THF, H_2O, 70% → (632b)

Scheme 91

(634a) ← PhSH/PhH, Et_3N, MeOH/2 h, 93% — (633) — $PhCO_2H$, Et_3N/DMSO, 60% → (634b)

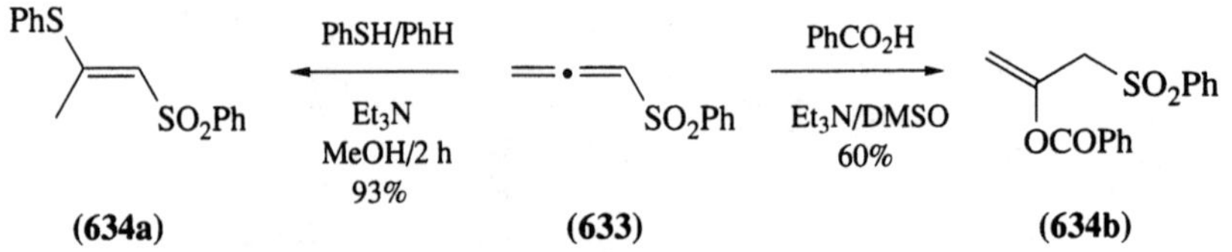

Scheme 92

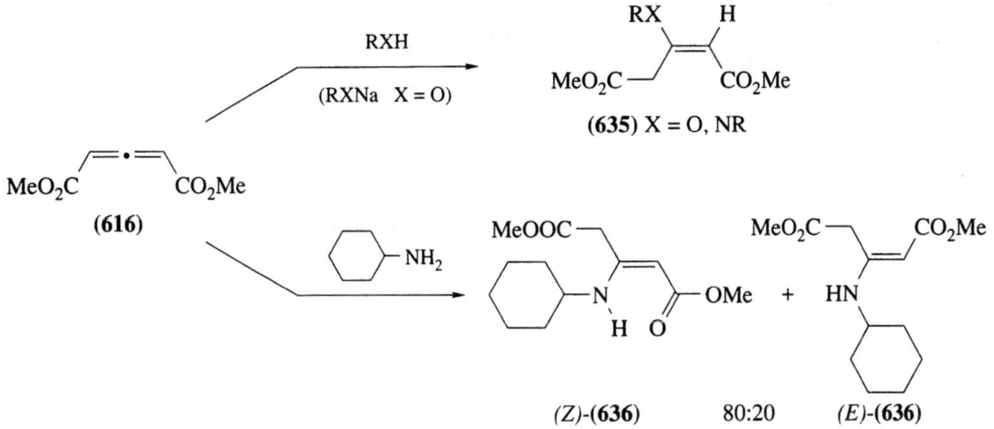

RXH (RXNa X = O) → (635) X = O, NR

(616) → cyclohexylamine → (Z)-(636) 80:20 (E)-(636)

Scheme 93

methylglycinate (**643a**) or methyl α-mercaptoacetate (**643b**) to the allenic nitrile (**644**; equation 144) affords in good yield the corresponding heterocycle (**645a**) or (**645b**).[157f] Similarly, reaction of the salicylate (**646a**), thiosalicylate (**646b**) and anthranilate (**646c**) with diethyl allenedicarboxylate (**612**) produces the corresponding heterocycle (equation 145) in moderate yield.[157h] The addition of ethylenediamine to the nitrile (**644**; equation 146) produces the imidazole (**648**) *via* two Michael additions and loss of acetonitrile.[157c] Similar reaction of ethylene glycol with cyanoallene gives the ethylene ketal of cyanoacetone.[157o] Reaction of *o*-phenylenediamine (**649a**) or 2-aminothiazole (**649b**) with dimethyl allenedicarboxylate (**616**) produces the corresponding heterocycle (Scheme 95) in moderate yield.[157b,157d,157j] Several other heterocyclic systems have been prepared by analogous applications of this general scheme.[157j] Finally, additions of trialkyl phosphites to allenic ketones (**629**; equation 147) give the methyleneoxaphospholenes in high yield (**651**).[157k,157l]

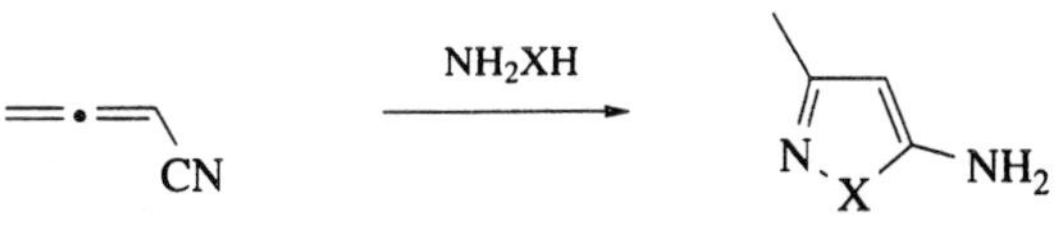

(618) → NH_2XH → (637a) X = NH; (637b) X = O (143)

X = NH	68%	(**637a**) X = NH
X = O	91%	(**637b**) X = O

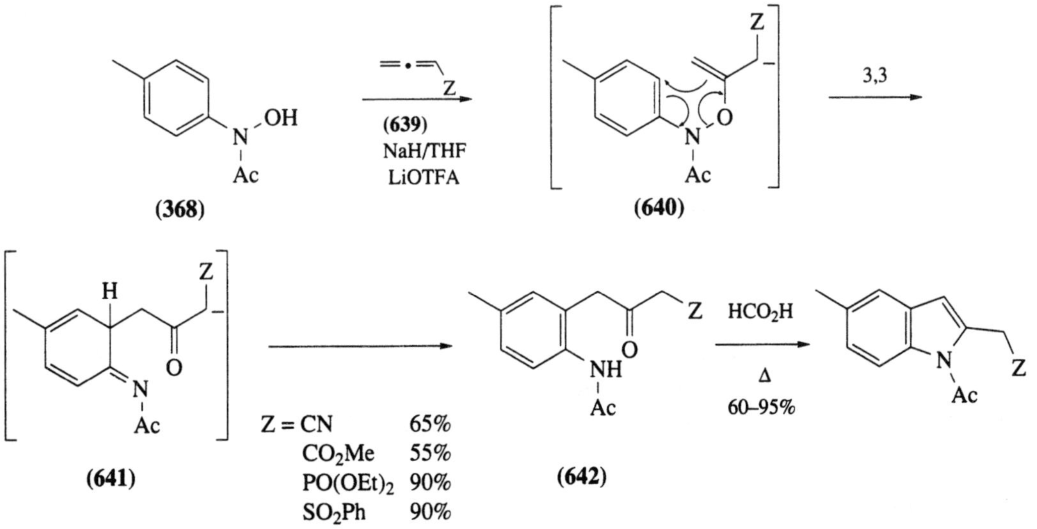

(368) → (639), NaH/THF, LiOTFA → [(640)] → 3,3 → [(641)] → (642) → HCO_2H, Δ, 60–95%

Z	Yield
CN	65%
CO_2Me	55%
$PO(OEt)_2$	90%
SO_2Ph	90%

Scheme 94

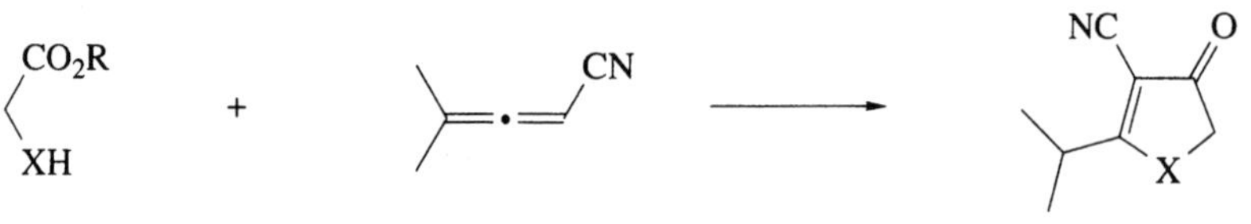

(144)

(643a) R = Et, X = NMe
(643b) R = Me, X = S
(644)
100 °C/2 h/84%
NaOEt/EtOH/89%
(645a) X = NMe
(645b) X = S

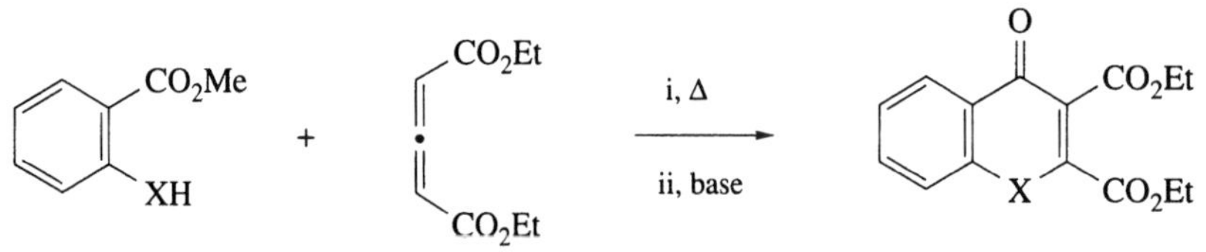

(145)

(646a) X = O
(646b) X = S
(646c) X = NMe
(612)
i, Δ; ii, base
29%
66%
65%
(647a) X = O
(647b) X = S
(647c) X = NMe

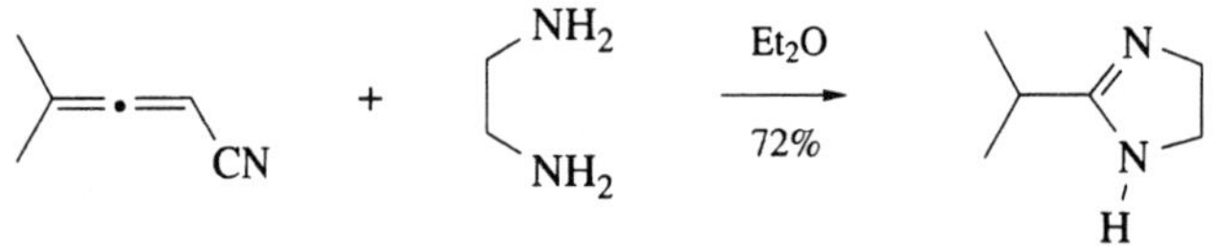

(146)

(644) + ethylenediamine, Et_2O, 72% → **(648)**

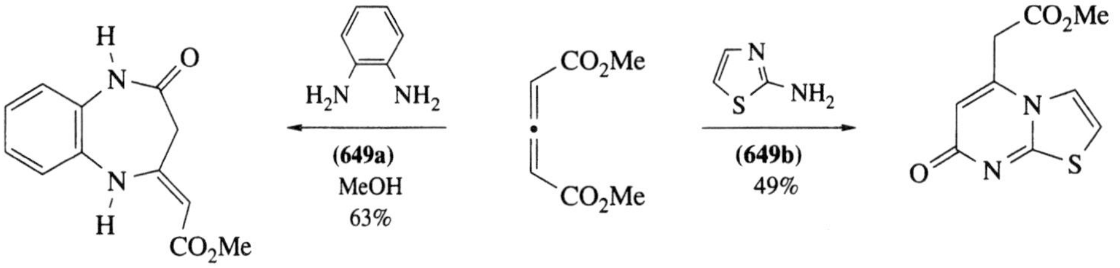

(650a) ← **(649a)** MeOH 63% — **(616)** — **(649b)** 49% → **(650b)**

Scheme 95

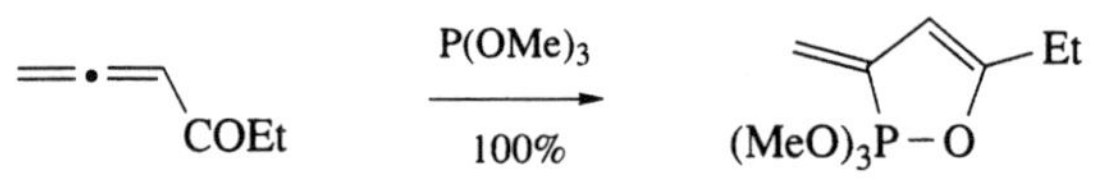

(147)

(629) $P(OMe)_3$, 100% → **(651)**

1.1.5 CONCLUSION

The addition of stabilized nucleophiles to activated π-systems is one of the most widely used constructive methods in organic synthesis. This chapter has provided a selective overview of this still burgeoning area where further progress continues, especially in the area of stereocontrol.

1.1.6 REFERENCES

1. (a) A. Michael, *J. Prakt. Chem.*, 1887, **35**, 349; (b) A. Michael, *Am. Chem. J.*, 1887, **9**, 115; (c) it is interesting to note that the addition of malonate anion to ethylidene malonate had been observed several years earlier but not exploited: T. Komnenos, *Justus Liebigs Ann. Chem.*, 1883, **218**, 145.
2. A. Michael, *J. Prakt. Chem.*, 1894, **49**, 20.

3. (a) K. Auwers, *Chem. Ber.*, 1891, **24**, 307; 1893, **26**, 364; 1895, **28**, 1130; (b) E. Knoevenagel, *Chem. Ber.*, 1894, **27**, 2345; *Justus Liebigs Ann. Chem.*, 1894, **281**, 25; 1896, **289**, 131.
4. (a) E. Knoevenagel, *Justus Liebigs Ann. Chem.*, 1897, **297**, 185; (b) G. Merling, *Chem. Ber.*, 1905, **38**, 979.
5. (a) W. S. Rapson and R. Robinson, *J. Chem. Soc.*, 1935, 1285; (b) J. R. Hawthorne and R. Robinson, *J. Chem. Soc.*, 1936, 763; (c) E. C. du Feu, F. J. McQuillin and R. Robinson, *J. Chem. Soc.*, 1937, 53; (d) P. S. Adamson, F. J. McQuillin, R. Robinson and J. L. Simonsen, *J. Chem. Soc.*, 1937, 1576; (e) F. J. McQuillin and R. Robinson, *J. Chem. Soc.*, 1938, 1097; (f) for earlier work using phenyl alkenyl ketones and an opposite aldol condensation see, H. Stobbe, *J. Prakt. Chem.*, 1912, **86**, 209.
6. For a good discussion of the various spellings of this term see, A. Nickon and E. F. Silversmith, 'Organic Chemistry: The Name Game', Pergamon Press, New York, 1987, p. 8.
7. For good reviews or discussions of the Michael reaction, see (a) R. Connor and W. R. McClellan, *J. Org. Chem.*, 1938, **3**, 570; (b) E. D. Bergmann, D. Ginsburg and R. Pappo, *Org. React. (N.Y.)*, 1959, **10**, 179; (c) H. O. House, 'Modern Synthetic Reactions', 2nd edn., Benjamin, Menlo Park, 1972, p. 595; (d) J. C. Stowell, 'Carbanions in Organic Synthesis', Wiley-Interscience, New York, 1979; (e) S. Patai and Z. Rappoport, in 'The Chemistry of Alkenes', ed. S. Patai, Interscience, London, 1964, vol. 1, chap. 8, p. 469; (f) H. A. Bruson, *Org. React.*, 1949, **5**, 79.
8. (a) M. E. Jung, *Tetrahedron*, 1976, **32**, 3; (b) R. E. Gawley, *Synthesis*, 1976, 777; (c) A. J. Waring, in 'Comprehensive Organic Chemistry', ed. D. H. R. Barton and W. D. Ollis, Pergamon Press, Oxford, 1979, vol. 1, p. 1043, 1049.
9. J. H. Brewster and E. L. Eliel, *Org. React. (N.Y.)*, 1953, **7**, 99.
10. (a) S. I. Miller and R. Tanaka, in 'Selective Organic Transformations', ed. B. S. Thyagarajan, Wiley-Interscience, New York, 1970, vol. 1, p. 143; (b) J. I. Dickstein and S. I. Miller, in 'The Chemistry of the Carbon–Carbon Triple Bond', ed. S. Patai, Wiley, New York, 1978, part II, p. 813; (c) E. Winterfeldt, *Angew. Chem., Int. Ed. Engl.*, 1967, **6**, 423; (d) I. A. Chekulaeva and L. V. Kondrat'eva, *Russ. Chem. Rev. (Engl. Transl.)*, 1965, **34**, 669; (e) E. Winterfeldt, in 'Chemistry of Acetylenes', ed. H. G. Viehe, Dekker, New York, 1969, chap. 4, p. 267; (f) T. F. Rutledge, 'Acetylenes and Allenes', Reinhold, New York, 1969, p. 232.
11. N. F. Albertson, *J. Am. Chem. Soc.*, 1948, **70**, 669.
12. For good early discussions of the best expermental conditions (solvent, base, concentration, *etc.*) for the Michael reaction, see (a) ref. 7a; (b) W. J. Hickinbottom, 'Reactions of Organic Compounds', 2nd edn., Longmans, London, 1948, p. 35; see also ref. 7c.
13. F. H. Howles, J. F. Thorpe and W. Udall, *J. Chem. Soc.*, 1900, **77**, 942.
14. (a) T. A. Spencer, M. D. Newton and S. W. Baldwin, *J. Org. Chem.*, 1964, **29**, 787; for other examples of double Michael addition, see (b) ref. 7f and refs. therein; (c) ref. 7c; (d) K. Balasubramanian, J. P. John and S. Swaminathan, *Synthesis*, 1974, 51; (e) L. Herzog, M. H. Gold and R. D. Geckler, *J. Am. Chem. Soc.*, 1951, **73**, 749; (f) H. A. Bruson and T. W. Reiner, *J. Am. Chem. Soc.*, 1953, **75**, 3585; (g) R. Bertocchio and J. Dreux, *Bull. Soc. Chim. Fr.*, 1962, 823, 1809.
15. (a) R. K. Hill and N. D. Ledford, *J. Am. Chem. Soc.*, 1975, **97**, 666, and refs. therein; (b) P. R. Shafer, W. E. Loeb, and W. S. Johnson, *J. Am. Chem. Soc.*, 1953, **75**, 5963.
16. (a) S. Patai and Z. Rappoport, *J. Chem. Soc.*, 1962, 377; (b) H. Junek and H. Sterk, *Tetrahedron Lett.*, 1968, 4309; (c) for a recent example of a retro-Michael process see, A. A. Chiu, R. R. Gorby, J. E. H. Hancock and E. J. Hustedt, *J. Org. Chem.*, 1984, **49**, 4313; 1985, **50**, 3245.
17. (a) P. D. Bartlett and G. F. Woods, *J. Am. Chem. Soc.*, 1940, **62**, 2933; (b) W. H. Bunnelle and L. A. Meyer, *J. Org. Chem.*, 1988, **53**, 4038; (c) A. Dornow and A. Frese, *Justus Liebigs Ann. Chem.*, 1952, **578**, 122; (d) D. E. Bergbreiter and J. J. Lalonde, *J. Org. Chem.*, 1987, **52**, 1601; (e) J. E. Backväll and S. K. Juntunen, *J. Am. Chem. Soc.*, 1987, **109**, 6396; (f) D. Mackay, E. G. Neeland and N. J. Taylor, *J. Org. Chem.*, 1986, **51**, 2351; (g) E. J. Corey and H. Estreicher, *J. Am. Chem. Soc.*, 1978, **100**, 6294; (h) A. R. Daniewski, *J. Org. Chem.*, 1975, **40**, 3135; (i) N. T. Boggs, III, R. E. Gawley, K. A. Koehler and R. G. Hiskey, *J. Org. Chem.*, 1975, **40**, 2850; (j) P. Kocovsky and D. Dvorák, *Tetrahedron Lett.*, 1986, **27**, 5015; (k) L. Fuentes, J. J. Vaquero, M. I. Ardid, A. Lorente and J. L. Soto, *Heterocycles*, 1988, **27**, 2125.
18. (a) J. W. Ralls, *Chem. Rev.*, 1959, **59**, 329; (b) E. P. Kohler and F. R. Butler, *J. Am. Chem. Soc.*, 1926, **48**, 1036; (c) E. H. Farmer and T. N. Mehta, *J. Chem. Soc.*, 1931, 1904; 1930, 1610; (d) E. H. Farmer and A. T. Healey, *J. Chem. Soc.*, 1927, 1060; (e) E. H. Farmer and S. R. W. Martin, *J. Chem. Soc.*, 1933, 960; (f) S. J. Danishefsky, W. E. Hatch, M. Sax, E. Abola and J. Pletcher, *J. Am. Chem. Soc.*, 1973, **95**, 2410; (g) J. H. Rigby, C. H. Senanayake and S. Rege, *J. Org. Chem.*, 1988, **53**, 4596.
19. (a) This reaction can also be thought of as a concerted Diels–Alder cycloaddition, although it is probably more correctly considered as a Michael reaction followed by ring closure; (b) F. Bohlmann, D. Schumann and O. Schmidt, *Chem. Ber.*, 1966, **99**, 1652; (c) H. O. House and T. H. Cronin, *J. Org. Chem.*, 1965, **30**, 1061; (d) S. J. Danishefsky and R. Cunningham, *J. Org. Chem.*, 1965, **30**, 3676; (e) G. A. Berchtold, J. Ciabattoni and A. A. Tunick, *J. Org. Chem.*, 1965, **30**, 3679; (f) S. J. Danishefsky and R. Cavanaugh, *J. Org. Chem.*, 1968, **33**, 2959; (g) S. J. Danishefsky, J. Eggler and G. A. Koppel, *Tetrahedron Lett.*, 1969, 4333.
20. (a) H. O. House, W. L. Roelofs and B. M. Trost, *J. Org. Chem.*, 1966, **31**, 646; (b) R. L. Frank and R. C. Pierle, *J. Am. Chem. Soc.*, 1951, **73**, 724; (c) J. J. Miller and P. L. de Benneville, *J. Org. Chem.*, 1957, **22**, 1268; (d) M. Yanagita and K. Yamakawa, *J. Org. Chem.*, 1957, **22**, 291.
21. (a) G. Stork, A. Brizzolara, H. K. Landesman, J. Szmuszkovicz and R. Terrell, *J. Am. Chem. Soc.*, 1963, **85**, 207, and refs. therein; (b) H. O. House and M. Schellenbaum, *J. Org. Chem.*, 1963, **28**, 34; for other uses of enamines in Robinson annulations, see (c) A. J. Birch, E. G. Hutchinson and G. Subba Rao, *J. Chem. Soc., Chem. Commun.*, 1970, 657; (d) J. E. Saxton, A. J. Smith and G. Lawton, *Tetrahedron Lett.*, 1975, 4161; for a good review of enamine chemistry, see (e) A. G. Cook, 'Enamines: Synthesis, Structure, and Reactions', Dekker, New York, 1969.
22. R. L. Shriner and H. R. Todd, *Org. Synth.*, 1943, **2**, 200.
23. (a) G. Stork and H. K. Landesman, *J. Am. Chem. Soc.*, 1956, **78**, 5129; (b) J. B. Hendrickson and R. K. Boeckman, Jr., *J. Am. Chem. Soc.*, 1971, **93**, 1307; (c) V. Dressler and K. Bodendorf, *Tetrahedron Lett.*, 1967,

4243; (d) R. D. Allan, B. G. Cordiner and R. J. Wells, *Tetrahedron Lett.*, 1968, 6055; (e) R. A. Appleton, K. H. Baggaley, S. C. Egan, J. M. Davies, S. H. Graham and D. O. Lewis, *J. Chem. Soc. C*, 1968, 2032.

24. (a) S. Ramachandran and M. S. Newman, *Org. Synth.*, 1973, **5**, 486; (b) R. L. Augustine and J. A. Caputo, *Org. Synth.*, 1973, **5**, 869; (c) P. M. McCurry, Jr. and R. K. Singh, *Synth. Commun.*, 1976, **6**, 75; (d) U. O'Connor and W. Rosen, *Tetrahedron Lett.*, 1979, 601; (e) F. T. Sher and G. A. Berchtold, *J. Org. Chem.*, 1977, **42**, 2569; (f) C. D. DeBoer, *J. Org. Chem.*, 1974, **39**, 2426.
25. (a) M. Pfau, G. Revial, A. Guingant and J. d'Angelo, *J. Am. Chem. Soc.*, 1985, **107**, 273; (b) T. Volpe, G. Revial, M. Pfau and J. d'Angelo, *Tetrahedron Lett.*, 1987, **28**, 2367; (c) J. d'Angelo, G. Revial, T. Volpe and M. Pfau, *Tetrahedron Lett.*, 1988, **29**, 4427; (d) Z. G. Hajos and D. R. Parrish, *J. Org. Chem.*, 1974, **39**, 1612, 1615; (e) J. Gutzwiller, P. Buchschacher and A. Fürst, *Synthesis*, 1977, 167.
26. C. H. Heathcock, J. E. Ellis, J. E. McMurry and A. Coppolino, *Tetrahedron Lett.*, 1971, 4995.
27. (a) G. Stork and J. Singh, *J. Am. Chem. Soc.*, 1974, **96**, 6181; (b) R. K. Boeckman, Jr., *J. Am. Chem. Soc.*, 1974, **96**, 6179; 1973, **95**, 6867; (c) G. Stork and B. Ganem, *J. Am. Chem. Soc.*, 1973, **95**, 6152; (d) T. Takahashi, Y. Naito and J. Tsuji, *J. Am. Chem. Soc.*, 1981, **103**, 5261.
28. (a) F. Kido, T. Fujishita, K. Tsutsumi and A. Yoshikoshi, *J. Chem. Soc., Chem. Commun.*, 1975, 337; (b) F. Kido, K. Tsutsumi, R. Maruta and A. Yoshikoshi, *J. Am. Chem. Soc.*, 1979, **101**, 6420; (c) A. G. Schultz and J. D. Godfrey, *J. Am. Chem. Soc.*, 1980, **102**, 2414; (d) F. Kido, Y. Noda, T. Maruyama, C. Kabuto and A. Yoshikoshi, *J. Org. Chem.*, 1981, **46**, 4264; (e) F. Kido, Y. Noda and A. Yoshikoshi, *J. Am. Chem. Soc.*, 1982, **104**, 5509.
29. (a) R. P. Nelson and R. G. Lawton, *J. Am. Chem. Soc.*, 1966, **88**, 3884; (b) R. P. Nelson, J. M. McEuen and R. G. Lawton, *J. Org. Chem.*, 1969, **34**, 1225; (c) J. M. McEuen, R. P. Nelson and R. G. Lawton, *J. Org. Chem.*, 1970, **35**, 690; (d) D. J. Dunham and R. G. Lawton, *J. Am. Chem. Soc.*, 1971, **93**, 2074; (e) H. Stetter and H. G. Thomas, *Chem. Ber.*, 1968, **101**, 1115; *Angew. Chem., Int. Ed. Engl.*, 1967, **6**, 554; (f) H. Stetter and K. Elfert, *Synthesis*, 1974, 36; (g) Th. R. Bok and W. N. Speckamp, *Tetrahedron*, 1979, **35**, 267.
30. (a) K. Takaki, M. Okada, M. Yamada and K. Negoro, *J. Chem. Soc., Chem. Commun.*, 1980, 1183; *J. Org. Chem.*, 1982, **47**, 1200; (b) D. L. Boger and M. D. Mullican, *J. Org. Chem.*, 1980, **45**, 5002.
31. (a) W. E. Doering and R. A. N. Weil, *J. Am. Chem. Soc.*, 1947, **69**, 2461; (b) G. Magnus and R. Levine, *J. Org. Chem.*, 1957, **22**, 270; (c) S. J. Danishefsky, P. Cain and A. Nagel, *J. Am. Chem. Soc.*, 1975, **97**, 380, and refs. therein.
32. (a) R. J. Cregge, J. L. Herrmann, J. E. Richman, R. F. Romanet and R. H. Schlessinger, *Tetrahedron Lett.*, 1973, 2595; (b) J. L. Herrmann, J. E. Richman and R. H. Schlessinger, *Tetrahedron Lett.*, 1973, 2599; (c) R. J. Cregge, J. L. Herrmann and R. H. Schlessinger, *Tetrahedron Lett.*, 1973, 2603; (d) J. L. Herrmann, J. E. Richman and R. H. Schlessinger, *Tetrahedron Lett.*, 1973, 3271; (e) J. L. Herrmann, J. E. Richman and R. H. Schlessinger, *Tetrahedron Lett.*, 1973, 3275; (f) J. L. Herrmann, G. R. Kieczykowski, R. F. Romanet, P. J. Wepplo and R. H. Schlessinger, *Tetrahedron Lett.*, 1973, 4711; (g) J. L. Herrmann, G. R. Kieczykowski, R. F. Romanet and R. H. Schlessinger, *Tetrahedron Lett.*, 1973, 4715; (h) for the use of α-phenylselenylacrylonitrile and other similar compounds as Michael acceptors see, D. L. J. Clive, T. L. B. Boivin and A. G. Angoh, *J. Org. Chem.*, 1987, **52**, 4943; (i) for an example of the addition of an ester enolate to β-substituted but not β,β-disubstituted acrylates see, M. Yamaguchi, M. Tsukamoto and I. Hirao, *Chem. Lett.*, 1984, 375.
33. (a) R. F. Romanet and R. H. Schlessinger, *J. Am. Chem. Soc.*, 1974, **96**, 3701; (b) R. Davis and K. B. Untch, *J. Org. Chem.*, 1979, **44**, 3755; (c) K. Ogura, M. Yamashita and G. Tsuchihashi, *Tetrahedron Lett.*, 1978, 1303; (d) J. L. Herrmann, M. H. Berger and R. H. Schlessinger, *J. Am. Chem. Soc.*, 1979, **101**, 1544.
34. For a good discussion of this process, see ref. 7d, p. 32. (a) A. G. Schultz and Y. K. Yee, *J. Org. Chem.*, 1976, **41**, 4044; (b) P. C. Ostrowski and V. V. Kane, *Tetrahedron Lett.*, 1977, 3549; (c) C. A. Brown and A. Yamaichi, *J. Chem. Soc., Chem. Commun.*, 1979, 100; (d) J. Lucchetti, W. Dumont and A. Krief, *Tetrahedron Lett.*, 1979, 2695; (e) L. Wartski and M. El-Bouz, *Tetrahedron*, 1982, **38**, 3285; (f) T. Rosen, M. J. Taschner, J. A. Thomas and C. H. Heathcock, *J. Org. Chem.*, 1985, **50**, 1190; (g) D. R. St. Laurent and L. A. Paquette, *J. Org. Chem.*, 1986, **51**, 3861; (h) F. Z. Basha, J. F. DeBernardis and S. Spanton, *J. Org. Chem.*, 1985, **50**, 4160; (i) K. Tomioka, M. Sudani, Y. Shinmi and K. Koga, *Chem. Lett.*, 1985, 329; (j) F. E. Ziegler and J. A. Schwartz, *Tetrahedron Lett.*, 1975, 4643; (k) for other α-hetero enolates see, R. E. Damon and R. H. Schlessinger, *Tetrahedron Lett.*, 1975, 4551; G. Neef and U. Eder, *Tetrahedron Lett.*, 1977, 2825.
35. (a) D. Seebach and R. Bürstinghaus, *Angew. Chem., Int. Ed. Engl.*, 1975, **14**, 57; for reviews of acyl anion equivalents, see (b) D. Seebach, *Angew. Chem., Int. Ed. Engl.*, 1969, **8**, 639; (c) O. W. Lever, Jr., *Tetrahedron*, 1976, **32**, 1943.
36. (a) A.-R. B. Manas and R. A. J. Smith, *J. Chem. Soc., Chem. Commun.*, 1975, 216; (b) W. D. Woessner, *Chem. Lett.*, 1976, 43.
37. (a) M. R. Binns and R. K. Haynes, *J. Org. Chem.*, 1981, **46**, 3790; (b) M. R. Binns, R. K. Haynes, T. L. Houston and W. R. Jackson, *Tetrahedron Lett.*, 1980, **21**, 573; (c) M. R. Binns, R. K. Haynes, T. L. Houston and W. R. Jackson, *Aust. J. Chem.*, 1981, **34**, 2465; (d) M. R. Binns, R. J. Goodridge, R. K. Haynes and D. D. Ridley, *Tetrahedron Lett.*, 1985, **26**, 6381; (e) D. H. Hua, G. Sinai-Zingde and S. Venkataraman, *J. Am. Chem. Soc.*, 1985, **107**, 4088; (f) D. H. Hua, S. Venkataraman, M. J. Coulter and G. Sinai-Zingde, *J. Org. Chem.*, 1987, **52**, 719; (g) D. H. Hua, M. J. Coulter and I. Badejo, *Tetrahedron Lett.*, 1987, **78**, 5465; (h) for a similar process using an allylic phosphine oxide anion see, R. K. Haynes and S. C. Vonwiller, *J. Chem. Soc., Chem. Commun.*, 1987, 92.
38. (a) G. Stork and L. A. Maldonado, *J. Am. Chem. Soc.*, 1974, **96**, 5272; (b) A. M. Casares and L. A. Maldonado, *Synth. Commun.*, 1976, **6**, 11; (c) J. A. Noguez and L. A. Maldonado, *Synth. Commun.*, 1976, **6**, 39.
39. (a) H. Stetter, *Angew. Chem., Int. Ed. Engl.*, 1976, **15**, 639, and refs. therein; (b) H. Stetter and H. Kuhlmann, *Chem. Ber.*, 1976, **109**, 2890; *Synthesis*, 1975, 379; (c) H. Stetter, H. Kuhlmann and G. Lorenz, *Org. Synth.*, 1980, **59**, 53; (d) E. Leete, M. R. Chedekel and G. B. Bodem, *J. Org. Chem.*, 1972, **37**, 4465; (e) for a recent example of the use of this process for the preparation of poly-1,4-diketones see, K. L. Pouwer, T. R. Vries, E. E. Havinga, E. W. Meijer and H. Wynberg, *J. Chem. Soc., Chem. Commun.*, 1988, 1432.

40. W. Steglich and P. Gruber, *Angew. Chem., Int. Ed. Engl.*, 1971, **10**, 655.
41. (a) J. E. McMurry and J. Melton, *J. Am. Chem. Soc.*, 1971, **93**, 5309; *J. Org. Chem.*, 1973, **38**, 4367; (b) J. E. McMurry, J. Melton and H. Padgett, *J. Org. Chem.*, 1974, **39**, 259; (c) W. T. Monte, M. M. Baizer and R. D. Little, *J. Org. Chem.*, 1983, **48**, 803; for the addition of less stabilized acyl anion equivalents, see (d) E. J. Corey and L. S. Hegedus, *J. Am. Chem. Soc.*, 1969, **91**, 4926; (e) R. Scheffold and R. Orlinski, *J. Am. Chem. Soc.*, 1983, **105**, 7200.
42. (a) D. St. C. Black, *Tetrahedron Lett.*, 1972, 1331; (b) R. M. Jacobson, *Tetrahedron Lett.*, 1974, 3215; (c) N. Kornblum and P. A. Wade, *J. Org. Chem.*, 1973, **38**, 1418; (d) J. R. Hanson, *Synthesis*, 1974, 1; (e) T.-L. Ho and C. M. Wong, *Synthesis*, 1974, 196; (f) S. Ranganathan, D. Ranganathan and A. K. Mehrotra, *J. Am. Chem. Soc.*, 1974, **96**, 5261; (g) A. H. Pagano and H. Schechter, *J. Org. Chem.*, 1970, **35**, 295; (h) W. E. Noland, J. H. Cooley and P. A. McVeigh, *J. Am. Chem. Soc.*, 1959, **81**, 1209.
43. (a) D. Seebach, H. F. Leitz and V. Ehrig, *Chem. Ber.*, 1975, **108**, 1924; (b) D. Seebach, V. Ehrig, H. F. Leitz and R. Henning, *Chem. Ber.*, 1975, **108**, 1946; (c) V. Ehrig and D. Seebach, *Chem. Ber.*, 1975, **108**, 1961; (d) D. Seebach and H. F. Leitz, *Angew. Chem., Int. Ed. Engl.*, 1971, **10**, 501; 1969, **8**, 983; (e) for approaches using Lewis acid promoted additions of silyl enol ethers see, M. Miyashita, T. Yanami and A. Yoshikoshi, *J. Am. Chem. Soc.*, 1976, **98**, 4679; M. Miyashita, T. Yanami, T. Kumazawa and A. Yoshikoshi, *J. Am. Chem. Soc.*, 1984, **106**, 2149.
44. (a) For a comprehensive review, see G. Rosini and R. Ballini, *Synthesis*, 1988, 833; (b) N. Ono, A. Kamimura, H. Miyake, I. Hamamoto and A. Kaji, *J. Org. Chem.*, 1985, **50**, 3692; (c) T. Tanaka, T. Toru, N. Okamura, A. Hazuto, S. Sugiura, K. Manabe, S. Kurozumi, M. Suzuki, T. Kawagishi and R. Noyori, *Tetrahedron Lett.*, 1983, **24**, 4103; (d) for other Michael additions of nitroalkyl anions followed by further transformations see, N. Ono, H. Miyake, R. Tanikaga and A. Kaji, *J. Org. Chem.*, 1982, **47**, 5017; N. Ono, H. Miyake, A. Kamimura, N. Tsukui and A. Kaji, *Tetrahedron Lett.*, 1982, **23**, 2957.
45. (a) G. A. Koppel and M. D. Kinnick, *J. Chem. Soc., Chem. Commun.*, 1975, 473; (b) D. F. Taber and S. A. Saleh, *J. Org. Chem.*, 1981, **46**, 4817; (c) for a similar use of phenyl vinyl sulfone as a vinyl cation equivalent for acyl anion addition see, H. Stetter and H.-J. Bender, *Chem. Ber.*, 1981, **114**, 1226.
46. (a) E. J. Corey, W.-G. Su and I. N. Houpis, *Tetrahedron Lett.*, 1986, **27**, 5951; (b) G. Stork and A. G. Schultz, *J. Am. Chem. Soc.*, 1971, **93**, 4074; (c) K. Schank and N. Moell, *Chem. Ber.*, 1969, **102**, 71; (d) W. Eisenhuth, H. B. Renfroe and H. Schmid, *Helv. Chim. Acta*, 1965, **48**, 375; (e) B. L. Chenard, D. K. Anderson and J. S. Swenton, *J. Chem. Soc., Chem. Commun.*, 1980, 932; (f) it is surprising that a similar base-catalyzed cyclization of the anion of homophthalic anhydride proceeds *via* a Diels–Alder cycloaddition rather than a Michael-trapping sequence, Y. Tamura, M. Sasho, K. Nakagawa, T. Tsugoshi and Y. Kita, *J. Org. Chem.*, 1984, **49**, 473; (g) H. Stetter and K. Marten, *Liebigs Ann. Chem.*, 1982, 250.
47. (a) F. M. Hauser and V. M. Baghdanov, *Tetrahedron*, 1984, **40**, 4719; (b) F. M. Hauser and S. Prasanna, *Tetrahedron*, 1984, **40**, 4711; *J. Org. Chem.*, 1979, **44**, 2596; (c) F. M. Hauser and D. Mal, *J. Am. Chem. Soc.*, 1984, **106**, 1862; (d) F. M. Hauser and R. P. Rhee, *J. Org. Chem.*, 1980, **45**, 3061; 1978, **43**, 178; (e) J. S. Swenton, J. N. Freskos, G. W. Morrow and A. D. Sercel, *Tetrahedron*, 1984, **40**, 4625 and refs. therein; (f) B. L. Chenard, M. G. Dolson, A. D. Sercel and J. S. Swenton, *J. Org. Chem.*, 1984, **49**, 318; (g) M. G. Dolson, B. L. Chenard and J. S. Swenton, *J. Am. Chem. Soc.*, 1981, **103**, 5263; (h) T.-T. Li, Y.-L. Wu and T. C. Walsgrove, *Tetrahedron*, 1984, **40**, 4701; (i) T.-T. Li and Y.-L. Wu, *J. Am. Chem. Soc.*, 1981, **103**, 7007; (j) G. A. Kraus and H. Sugimoto, *Tetrahedron Lett.*, 1978, 2263; (k) J. Wildeman, P. C. Borgen, H. Pluim, P. H. F. M. Rouwette and A. M. van Leusen, *Tetrahedron Lett.*, 1978, 2213; (l) R. A. Russell and R. N. Warrener, *J. Chem. Soc., Chem. Commun.*, 1981, 108; (m) A. I. Meyers and W. B. Avila, *J. Org. Chem.*, 1981, **46**, 3881; (n) J. H. Dodd and S. M. Weinreb, *Tetrahedron Lett.*, 1979, 3593; (o) J. H. Dodd, R. S. Garigipati and S. M. Weinreb, *J. Org. Chem.*, 1982, **47**, 4045; (p) G. E. Evans, F. J. Leeper, J. A. Murphy and J. Staunton, *J. Chem. Soc., Chem. Commun.*, 1979, 205; (q) F. J. Leeper and J. Staunton, *J. Chem. Soc., Chem. Commun.*, 1979, 206; (r) N. J. P. Broom and P. G. Sammes, *J. Chem. Soc., Chem. Commun.*, 1978, 162; *J. Chem. Soc., Perkin Trans. 1*, 1981, 465.
48. (a) K. A. Parker and E. A. Tallman, *Tetrahedron*, 1984, **40**, 4781; (b) A. S. Kende, J. Rizzi and J. Riemer, *Tetrahedron Lett.*, 1979, 1201; (c) K. A. Parker and J. L. Kallmerten, *Tetrahedron Lett.*, 1979, 1197; 1977, 4557; *J. Am. Chem. Soc.*, 1980, **102**, 5881; *J. Org. Chem.*, 1980, **45**, 2614.
49. (a) see ref. 18g; (b) R. J. Chorvat and R. Pappo, *Tetrahedron Lett.*, 1975, 623; (c) P. Knochel and D. Seebach, *Nouv. J. Chim.*, 1981, **5**, 75.
50. (a) S. F. Martin and S. R. Desai, *J. Org. Chem.*, 1978, **43**, 4673; 1977, **42**, 1664; (b) A. T. Hewson and D. T. MacPherson, *Tetrahedron Lett.*, 1983, **24**, 5807; (c) R. J. Pariza and P. L. Fuchs, *J. Org. Chem.*, 1983, **48**, 2304; 1983, **48**, 2306; for a similar process using an enamine, see (d) S. D. Darling and N. Subramanian, *J. Org. Chem.*, 1975, **40**, 2851; (e) W. Flitsch and E. R. Gesing, *Tetrahedron Lett.*, 1979, 4529.
51. (a) W. L. Meyer, M. J. Brannon, A. Merritt and D. Seebach, *Tetrahedron Lett.*, 1986, **27**, 1449; (b) W. L. Meyer, M. J. Brannon, C. da G. Burgos, T. E. Goodwin and R. W. Howard, *J. Org. Chem.*, 1985, **50**, 438; (c) A. S. Kende, D. Constantinides, S. J. Lee and L. S. Liebeskind, *Tetrahedron Lett.*, 1975, 405.
52. (a) A. Padwa and P. E. Yeske, *J. Am. Chem. Soc.*, 1988, **110**, 1617; (b) O. Tsuge, S. Kanemasa and M. Yoshioka, *J. Org. Chem.*, 1988, **53**, 1384; (c) O. Tsuge, S. Kanemasa, T. Yamada and K. Matsuda, *J. Org. Chem.*, 1987, **52**, 2523; (d) J. Moskal and A. M. van Leusen, *J. Org. Chem.*, 1986, **51**, 4131; (e) for the cyclization of (2-carbamoylallyl)lithium systems see, P. Beak and K. D. Wilson, *J. Org. Chem.*, 1987, **52**, 218; 1986, **51**, 4627; D. J. Kempf, K. D. Wilson and P. Beak, *J. Org. Chem.*, 1982, **47**, 1610.
53. (a) W. G. Dauben and J. M. Gerdes, *Tetrahedron Lett.*, 1983, **24**, 3841; (b) W. G. Dauben and R. A. Bunce, *J. Org. Chem.*, 1983, **48**, 4642; (c) K. Matsumoto, *Angew. Chem., Int. Ed. Engl.*, 1980, **19**, 1013; 1981, **20**, 770; (d) K. Matsumoto and T. Uchida, *Chem. Lett.*, 1981, 1673.
54. R. A. Holton, A. D. Williams and R. M. Kennedy, *J. Org. Chem.*, 1986, **51**, 5480.
55. G. H. Posner, *Chem. Rev.*, 1986, **86**, 831.
56. T. Olsson, M. T. Rahman and C. Ullenius, *Tetrahedron Lett.*, 1977, 75.
57. (a) H. Stetter and H. Kuhlmann, *Justus Liebigs Ann. Chem.*, 1979, 303, 944, 1122; (b) H. Stetter and K. Marten, *Liebigs Ann. Chem.*, 1982, 240.

58. K. Ogura, N. Yahata, M. Minoguchi, K. Ohtsuki, K. Takahashi and H. Iida, *J. Org. Chem.*, 1986, **51**, 508.
59. (a) R. M. Cory, P. C. Anderson, F. R. McLaren and B. R. Yamamoto, *J. Chem. Soc., Chem. Commun.*, 1981, 73; (b) R. M. Cory and R. M. Renneboog, *J. Org. Chem.*, 1984, **49**, 3898 and refs. therein; (c) R. M. Cory, D. M. T. Chan, Y. M. A. Naguib, M. H. Rastall and R. M. Renneboog, *J. Org. Chem.*, 1980, **45**, 1852; (d) G. H. Posner, J. P. Mallamo and A. Y. Black, *Tetrahedron*, 1981, **37**, 3921; for the similar loss of a nitro group in a furan synthesis, see (e) M. Miyashita, T. Kumazawa and A. Yoshikoshi, *J. Org. Chem.*, 1984, **49**, 3728; *Chem. Lett.*, 1981, 593; for the reaction of α-halo acrylates with dienolates to give cyclopropyl ketones, see (f) H. Hagiwara, H. Uda and T. Kodama, *J. Chem. Soc., Perkin Trans. 1*, 1980, 963; (g) H. Hagiwara, T. Kodama, H. Kosugi and H. Uda, *J. Chem. Soc., Chem. Commun.*, 1976, 413; (h) H. Hagiwara, K. Nakayama and H. Uda, *Bull. Chem. Soc. Jpn.*, 1975, **48**, 3769; (i) D. Spitzner, A. Engler, T. Liese, G. Splettstösser and A. de Meijere, *Angew. Chem., Int. Ed. Engl.*, 1982, **21**, 791; for a different trapping process in the reaction of the anion of methyl α-bromocrotonate and enones see, (j) T. Hudlicky, L. Radesca, H. Luna and F. E. Anderson, III, *J. Org. Chem.*, 1986, **51**, 4746.
60. (a) S. J. Danishefsky, S. Chackalamannil, M. Silvestri and J. P. Springer, *J. Org. Chem.*, 1983, **48**, 3615; (b) S. J. Danishefsky, P. Harrison, M. Silvestri and B. E. Segmuller, *J. Org. Chem.*, 1984, **49**, 1321.
61. (a) G. H. Posner, S.-B. Lu, E. Asirvatham, E. F. Silversmith and E. M. Shulman, *J. Am. Chem. Soc.*, 1986, **108**, 511; (b) G. H. Posner and S.-B. Lu, *J. Am. Chem. Soc.*, 1985, **107**, 1424; (c) G. H. Posner, K. S. Webb, E. Asirvatham, S. Jew and A. Degl'Innocenti, *J. Am. Chem. Soc.*, 1988, **110**, 4754; (d) G. H. Posner, E. Asirvatham, K. S. Webb and S. Jew, *Tetrahedron Lett.*, 1987, **28**, 5071; (e) G. H. Posner and E. Asirvatham, *Tetrahedron Lett.*, 1986, **27**, 663; (f) G. H. Posner, S.-B. Lu and E. Asirvatham, *Tetrahedron Lett.*, 1986, **27**, 659.
62. (a) J. A. Marshall, H. Faubl and T. M. Warne, Jr., *Chem. Commun.*, 1967, 753; (b) J. A. Marshall and R. A. Ruden, *Tetrahedron Lett.*, 1970, 1239; (c) J. A. Marshall and T. M. Warne, Jr., *J. Org. Chem.*, 1971, **36**, 178; (d) R. M. Coates and J. E. Shaw, *Chem. Commun.*, 1968, 47; (e) R. L. Hale and L. H. Falkow, *Chem. Commun.*, 1968, 1249; (f) H. C. Odom, Jr. and A. R. Pinder, *Chem. Commun.*, 1969, 26; *J. Chem. Soc., Perkin Trans. 1*, 1972, 2193; (g) C. J. V. Scanio and R. M. Starrett, *J. Am. Chem. Soc.*, 1971, **93**, 1539; (h) A. van der Gen, L. M. van der Linde, J. G. Wittereen and H. Boelens, *Recl. Trav. Chim. Pays-Bas*, 1971, **90**, 1034, 1945; (i) Y. Takagi, Y. Nakahara and M. Matsui, *Tetrahedron*, 1978, **34**, 517.
63. (a) A. Risaliti, M. Forchiassin and E. Valentin, *Tetrahedron Lett.*, 1966, 6331; (b) A. Risaliti, M. Forchiassin and E. Valentin, *Tetrahedron*, 1968, **24**, 1889; (c) F. P. Colonna, E. Valentin, G. Pitacco and A. Risaliti, *Tetrahedron*, 1973, **29**, 3011; (d) E. Valentin, G. Pitacco, F. P. Colonna and A. Risaliti, *Tetrahedron*, 1974, **30**, 2741; (e) M. Calligaris, G. Manzini, G. Pitacco and E. Valentin, *Tetrahedron*, 1975, **31**, 1501; (f) F. Benedetti, S. Fabrissin and A. Risaliti, *Tetrahedron*, 1984, **40**, 977; (g) S. Fabrissin, S. Fatutta, N. Malusà and A. Risaliti, *J. Chem. Soc., Perkin Trans. 1*, 1980, 686; (h) S. Fabrissin, S. Fatutta and A. Risaliti, *J. Chem. Soc., Perkin Trans. 1*, 1981, 109; (i) S. Fatutta and A. Risaliti, *J. Chem. Soc., Perkin Trans. 1*, 1974, 2387; (j) F. Benedetti, F. Fabrissin, S. Pricl and A. Risaliti, *Gazz. Chim. Ital.*, 1987, **117**, 391; (k) ref. 23e; (l) H. M. Hellman, R. A. Jerussi and J. Lancaster, *J. Org. Chem.*, 1967, **32**, 2148; (m) for the use of chiral vinylic sulfoxide in asymmetric Michael additions, see G. Tsuchihashi, S. Mitamura, S. Inoue and K. Ogura, *Tetrahedron Lett.*, 1973, 323.
64. (a) D. Seebach and J. Golinski, *Helv. Chim. Acta*, 1981, **64**, 1413; (b) R. Häner, T. Laube and D. Seebach, *Chimia*, 1984, **38**, 255.
65. (a) J. Mulzer, A. Chucholowski, O. Lammer, I. Jibril and G. Huttner, *J. Chem. Soc., Chem. Commun.*, 1983, 869; (b) J. Bertrand, L. Gorrichon and P. Maroni, *Tetrahedron*, 1984, **40**, 4127.
66. (a) M. Yamaguchi, M. Tsukamoto, S. Tanaka and I. Hirao, *Tetrahedron Lett.*, 1984, **25**, 5661; (b) M. Yamaguchi, M. Tsukamoto and I. Hirao, *Tetrahedron Lett.*, 1985, **26**, 1723; (c) M. Yamaguchi, K. Hasebe, S. Tanaka and T. Minami, *Tetrahedron Lett.*, 1986, **27**, 959; (d) M. Yamaguchi, M. Hamada, H. Nakashima and T. Minami, *Tetrahedron Lett.*, 1987, **28**, 1785; (e) M. Yamaguchi, M. Hamada, S. Kawasaki and T. Minami, *Chem. Lett.*, 1986, 1085.
67. (a) C. H. Heathcock, M. A. Henderson, D. A. Oare and M. A. Sanner, *J. Org. Chem.*, 1985, **50**, 3019; (b) C. H. Heathcock and D. A. Oare, *J. Org. Chem.*, 1985, **50**, 3022.
68. L. Z. Viteva and Y. N. Stefanovsky, *Monatsh. Chem.*, 1982, **113**, 181; 1981, **112**, 125; 1980, **111**, 1287.
69. (a) S. J. Blarer, W. B. Schweizer and D. Seebach, *Helv. Chim. Acta*, 1982, **65**, 1637; (b) S. J. Blarer and D. Seebach, *Chem Ber.*, 1983, **116**, 2250, 3086; (c) G. Calderai and D. Seebach, *Helv. Chim. Acta*, 1985, **68**, 1592; (d) D. Enders and K. Papadopoulos, *Tetrahedron Lett.*, 1983, **24**, 4967; (e) D. Enders, K. Papadopoulos, B. E. M. Rendenbach, R. Appel and F. Knoch, *Tetrahedron Lett.*, 1986, **27**, 3491; (f) K. Tomioka, W. Seo, K. Ando and K. Koga, *Tetrahedron Lett.*, 1987, **28**, 6637; (g) K. Tomioka, K. Yasuda and K. Koga, *Tetrahedron Lett.*, 1986, **27**, 4611; (h) K. Tomioka, K. Ando, K. Yasuda and K. Koga, *Tetrahedron Lett.*, 1986, **27**, 715; (i) B. de Jéso and J.-C. Pommier, *Tetrahedron Lett.*, 1980, **21**, 4511; (j) B. Nebout, B. de Jéso and J.-C. Pommier, *J. Chem. Soc., Chem. Commun.*, 1985, 504; (k) E. J. Corey and R. T. Peterson, *Tetrahedron Lett.*, 1985, **26**, 5025; (l) W. Oppolzer, R. Pitteloud, G. Bernardinelli and K. Baettig, *Tetrahedron Lett.*, 1983, **24**, 4975; (m) ref. 66c.
70. (a) P. Deslongchamps, 'Stereoelectronic Effects in Organic Chemistry', Pergamon Press, Oxford, 1983, p. 221; (b) C. Djerassi, R. A. Schneider, H. Vorbrueggen and N. L. Allinger, *J. Org. Chem.*, 1963, **28**, 1632; (c) C. W. Alexander and W. R. Jackson, *J. Chem. Soc., Perkin Trans. 2*, 1972, 1601; (d) C. Agami, M. Fadlallah and J. Levisalles, *Tetrahedron*, 1981, **37**, 903; *Tetrahedron Lett.*, 1980, 59; (e) R. A. Abramovitch and D. L. Struble, *Tetrahedron*, 1968, **24**, 357; (f) M. Yanagita, S. Inayama, M. Hirakura and F. Seki, *J. Org. Chem.*, 1958, **23**, 690; (g) T. Sakakibara and R. Sudoh, *J. Org. Chem.*, 1975, **40**, 2823.
71. (a) For an excellent discussion of cyanide additions, see W. Nagata and M. Yoshioka, *Org. React. (N.Y.)*, 1977, **25**, 255, and refs. therein; (b) for the use of trimethylsilyl cyanide in these additions, see K. Utimoto, Y. Wakabayahsi, T. Horiie, M. Inoue, Y. Shishiyama, M. Obayashi and H. Nozaki, *Tetrahedron*, 1983, **39**, 967.
72. (a) R. A. Bunce, E. J. Wamsley, J. D. Pierce, A. J. Shellhammer, Jr. and R. E. Drumright, *J. Org. Chem.*, 1987, **52**, 464; (b) W. S. Johnson, S. Schulman, K. L. Williamson and R. Pappo, *J. Org. Chem.*, 196, **27**, 2015; (c) D. H. R. Barton, A. da S. Campos-Neves and A. I. Scott, *J. Chem. Soc.*, 1957, 2698; (d) G. Stork, D. F.

Taber and M. Marx, *Tetrahedron Lett.*, 1978, 2445; (e) G. Stork, R. K. Boeckman, Jr., D. F. Taber, W. C. Still and J. Singh, *J. Am. Chem. Soc.*, 1979, **101**, 7107; (f) S. D. Burke, C. W. Murtiashaw and M. S. Dike, *J. Org. Chem.*, 1982, **47**, 1349; (g) A. R. Pinder, S. J. Price and R. M. Rice, *J. Org. Chem.*, 1972, **37**, 2202; (h) A. Alexakis, M. J. Chapdelaine and G. H. Posner, *Tetrahedron Lett.*, 1978, 4209.

73. (a) G. Stork, C. S. Shiner and J. D. Winkler, *J. Am. Chem. Soc.*, 1982, **104**, 310; (b) G. Stork, J. D. Winkler and N. A. Saccomano, *Tetrahedron Lett.*, 1983, **24**, 465; (c) G. Stork and N. A. Saccomano, *Nouv. J. Chim.*, 1986, **10**, 677; (d) S. K. Attah-Poku, F. Chau, V. K. Yadav and A. G. Fallis, *J. Org. Chem.*, 1985, **50**, 3418.
74. (a) T.-T. Li and Y.-L. Wu, *Tetrahedron Lett.*, 1988, **29**, 4039; (b) E. R. Koft, A. S. Kotnis and T. A. Broadbent, *Tetrahedron Lett.*, 1987, **28**, 2799; (c) for earlier examples of intramolecular cyclizations leading to heterocyclic five-membered rings see, ref. 7b, p. 502, Table XVIA.
75. (a) M. Ando, G. Büchi and T. Ohnuma, *J. Am. Chem. Soc.*, 1975, **97**, 6880; (b) G. Büchi, K. Matsumoto, K. Matsumoto and H. Nishimura, *J. Am. Chem. Soc.*, 1971, **93**, 3299; (c) S.-E. Yoo, S.-H. Lee and N.-J. Kim, *Tetrahedron Lett.*, 1988, **29**, 2195.
76. B. M. Trost, C. D. Shuey, F. DiNinno, Jr. and S. M. McElvain, *J. Am. Chem. Soc.*, 1979, **101**, 1284.
77. (a) W. ten Hoeve and H. Wynberg, *J. Org. Chem.*, 1979, **44**, 1508; (b) H. A. P. de Jongh, J. F. J. Gerhartl and H. Wynberg, *J. Org. Chem.*, 1965, **30**, 1409; (c) H. A. P. de Jongh and H. Wynberg, *Tetrahedron*, 1965, **21**, 515; (d) I. Ya. Shternberg and Ya. F. Freimanis, *J. Org. Chem. USSR (Engl. Transl.)*, 1968, **4**, 1081.
78. (a) G. Stork and M. Tomasz, *J. Am. Chem. Soc.*, 1964, **86**, 471; (b) T. P. C. Mulholland, R. I. W. Honeywood, H. D. Preston and D. T. Rosevear, *J. Chem. Soc.*, 1965, 4939; (c) simple enynones lacking the activating methoxy group on the alkyne give normal Michael addition to the enone system with enamines, G. Stork and J. Hill, private communication.
79. (a) K. Yamada, M. Aratani, Y. Hayakawa, H. Nakamura, H. Nagase and Y. Hirata, *J. Org. Chem.*, 1971, **36**, 3653; (b) K. Yamada, Y. Kyotani, S. Manabe and M. Suzuki, *Tetrahedron*, 1979, **35**, 293; (c) A. J. Birch and J. S. Hill, *J. Chem. Soc. C*, 1966, 2324; (d) H. Irie, J. Katakawa, Y. Mizuno, S. Udaka, T. Taga and K. Osaki, *J. Chem. Soc., Chem. Commun.*, 1978, 717; (e) H. Irie, Y. Mizuno, T. Taga and K. Osaki, *J. Chem. Soc., Perkin Trans. 1*, 1982, 25; (f) see also ref. 8 in ref. 72d; (g) for an interesting cyclization of a stabilized nucleophile to a cyclohexadienone (prepared by the trapping of an α-diazo ketone by an anisole unit) see, L. N. Mander and S. G. Pyne, *J. Am. Chem. Soc.*, 1979, **101**, 3373; *Aust. J. Chem.*, 1981, **34**, 1899; I. A. Blair, L. N. Mander, P. H. C. Mundill and S. G. Pyne, *Aust. J. Chem.*, 1981, **34**, 1887; (h) for the cyclization of ester enolates generated by the opening of silyloxycyclopropyl esters see, J. P. Marino and J. K. Long, *J. Am. Chem. Soc.*, 1988, **110**, 7916; E. L. Grimm, R. Zschiesche and H.-U. Reissig, *J. Org. Chem.*, 1985, **50**, 5543.
80. (a) M. J. Kenny, L. N. Mander and S. P. Sethi, *Tetrahedron Lett.*, 1986, **27**, 3927; (b) G. Stork and D. A. Livingston, *Chem. Lett.*, 1987, 105; (c) for the use of a β-diester as the internal nucleophile in forming a seven-membered ring, see S. Hanessian and A. G. Pernet, *Can. J. Chem.*, 1974, **52**, 1280.
81. (a) M. G. Brasca, H. B. Broughton, D. Craig, S. V. Ley, A. A. Somovilla and P. L. Toogood, *Tetrahedron Lett.*, 1988, **29**, 1853; (b) examination of molecular models indicates that the rigidity of the *trans*-fused oxahydrindanone system coupled with the stereochemistry of the chain bearing the Michael acceptor (*cis* to the α-H at the ring juncture) permits the new C–C bond to be formed much more easily on the α-face rather than the β-face.
82. Y. Hirai, T. Terada and T. Yamazaki, *J. Am. Chem. Soc.*, 1988, **110**, 958.
83. (a) A. E. Ashcroft, D. T. Davies and J. K. Sutherland, *Tetrahedron*, 1984, **40**, 4579; (b) M. P. Cava, Z. Ahmed, N. Benfaremo, R. A. Murphy, Jr. and G. J. O'Malley, *Tetrahedron*, 1984, **40**, 4767; (c) H. Uno, Y. Naruta and K. Maruyama, *Tetrahedron*, 1984, **40**, 4725; (d) for the intramolecular addition of a protected cyanohydrin anion to prepare intermediates for tetracycline synthesis, see E. Aufderhaar, J. E. Baldwin, D. H. R. Barton, D. J. Faulkner and M. Slaytor, *J. Chem. Soc. C*, 1971, 2175.
84. (a) S. J. Danishefsky, W. E. Hatch, M. Sax, E. Abola and J. Pletcher, *J. Am. Chem. Soc.*, 1973, **95**, 2410; (b) S. J. Danishefsky, G. A. Koppel and R. Levine, *Tetrahedron Lett.*, 1968, 2257; (c) S. J. Danishefsky and B. H. Migdalof, *Tetrahedron Lett.*, 1969, 4331; (d) M. Horton and G. Pattenden, *Tetrahedron Lett.*, 1983, **24**, 2125.
85. G. Berthiaume, J.-F. Lavallée and P. Deslongchamps, *Tetrahedron Lett.*, 1986, **27**, 5451.
86. S. H. Rosenberg and H. Rapoport, *J. Org. Chem.*, 1985, **50**, 3979.
87. (a) I. N. Nazarov and A. N. Elizarova, *J. Gen. Chem. USSR (Engl. Transl.)*, 1960, **30**, 474; (b) A. N. Elizarova, *J. Gen. Chem. USSR (Engl. Transl.)*, 1964, **34**, 3251; (c) H. L. Brown, G. L. Buchanan, A. F. Cameron and G. Ferguson, *J. Chem. Soc., Chem. Commun.*, 1967, 399; (d) A. J. Bellamy, *J. Chem. Soc. B*, 1969, 449; (e) J. Wiemann, L. Bobic-Korejzl and Y. Allamagny, *C. R. Hebd. Seances Acad. Sci., Ser. C*, 1969, **268**, 2037; (f) T. Ohnuma, T. Oishi and Y. Ban, *J. Chem. Soc., Chem. Commun.*, 1973, 301; (g) R. A. Lee, *Tetrahedron Lett.*, 1973, 3333.
88. (a) H. Hagiwara, K. Nakayama and H. Uda, *Bull. Chem. Soc. Jpn.*, 1975, **48**, 3769; (b) H. Hagiwara and H. Uda, *J. Chem. Soc., Perkin Trans. 1*, 1986, 629; (c) D. Spitzner, *Tetrahedron Lett.*, 1978, 3349; (d) K. B. White and W. Reusch, *Tetrahedron*, 1978, **34**, 2439; (e) E. G. Gibbons, *J. Org. Chem.*, 1980, **45**, 1540; *J. Am. Chem. Soc.*, 1982, **104**, 1767; (f) M. L. Quesada, R. H. Schlessinger and W. H. Parsons, *J. Org. Chem.*, 1978, **43**, 3968; (g) M. R. Roberts and R. H. Schlessinger, *J. Am. Chem. Soc.*, 1981, **103**, 724; (h) H. Nagaoka, K. Ohsawa, T. Takata and Y. Yamada, *Tetrahedron Lett.*, 1984, **25**, 5389; (i) H. Nagaoka, K. Kobayashi, T. Matsui and Y. Yamada, *Tetrahedron Lett.*, 1987, **28**, 2021; (j) H. Nagaoka, K. Kobayashi, T. Okamura and Y. Yamada, *Tetrahedron Lett.*, 1987, **28**, 6641; (k) M. Ihara, M. Toyota, M. Abe, Y. Ishida, K. Fukumoto and T. Kametani, *J. Chem. Soc., Perkin Trans. 1*, 1986, 1543; (l) M. Ihara, M. Toyota, K. Fukumoto and T. Kametani, *J. Chem. Soc., Perkin Trans. 1*, 1986, 2151; (m) M. Ihara, M. Katogi, K. Fukumoto and T. Kametani, *J. Chem. Soc., Chem. Commun.*, 1987, 721; (n) M. Ihara, T. Takahashi, N. Shimizu, Y. Ishida, I. Sudow, K. Fukumoto and T. Kametani, *J. Chem. Soc., Chem. Commun.*, 1987, 1467; (o) M. Ihara, M. Suzuki, K. Fukumoto, T. Kametani and C. Kabuto, *J. Am. Chem. Soc.*, 1988, **110**, 1963; (p) C. Thanupran, C. Thebtaranonth and Y. Thebtaranonth, *Tetrahedron Lett.*, 1986, **27**, 2295; (q) F. Richter and H.-H. Otto, *Tetrahedron Lett.*, 1987, **28**, 2945; 1985, **26**, 4351; (r) see also ref. 54.
89. (a) H. Nozaki, T. Yamuguti, S. Ueda and K. Kondo, *Tetrahedron*, 1968, **24**, 1445; (b) C. M. Cimarusti and J. Wolinsky, *J. Am. Chem. Soc.*, 1968, **90**, 113; (c) T. W. Scott, W. Vetter, W. E. Oberhänsli and A. Fürst,

Tetrahedron Lett., 1972, 1719; (d) S. J. Danishefsky and T. Kitahara, *J. Am. Chem. Soc.*, 1974, **96**, 7807; (e) M. E. Jung and C. A. McCombs, *Tetrahedron Lett.*, 1976, 2935; (f) G. M. Rubottom and D. S. Krueger, *Tetrahedron Lett.*, 1977, 611.

90. The mechanistically quite similar double Michael reaction of 2-silyloxy dienes and enones has been shown recently to go *via* a similar two-step mechanism rather than a Diels–Alder cycloaddition: S. Kobayashi, Y. Sagawa, H. Akamatsu and T. Mukaiyama, *Chem. Lett.*, 1988, 1777.
91. (a) G. Büchi, J. H. Hansen, D. Knutson and E. Koller, *J. Am. Chem. Soc.*, 1958, **80**, 5517; (b) for the Michael–aldol trapping of a similar unsubstituted dienolate, see G. A. Kraus and H. Sugimoto, *Tetrahedron Lett.*, 1977, 3929.
92. (a) N. C. Ross and R. Levine, *J. Org. Chem.*, 1964, **29**, 2346; (b) S. M. McElvain and K. Rorig, *J. Am. Chem. Soc.*, 1948, **70**, 1820; (c) M. Freifelder, *J. Am. Chem. Soc.*, 1960, **82**, 2386; (d) A. Pohland and H. R. Sullivan, *J. Am. Chem. Soc.*, 1955, **77**, 2817; (e) W. S. Johnson, E. L. Woroch and B. G. Buell, *J. Am. Chem. Soc.*, 1949, **71**, 1901; (f) L. L. Gershbein and C. D. Hurd, *Org. Synth., Coll. Vol.*, 1955, **3**, 458; (g) S. R. Buc, *Org. Synth., Coll. Vol.*, 1955, **3**, 93; (h) P. R. Haeseler, *Org. Synth., Coll. Vol.*, 1932, **1**, 196; (i) N. Ono, A. Kamimura and A. Kaji, *J. Org. Chem.*, 1986, **51**, 2139; (j) P. N. Confalone, G. Pizzolato, D. L. Confalone and M. R. Uskokovic, *J. Am. Chem. Soc.*, 1980, **102**, 1954; (k) M. Marx, F. Marti, J. Reisdorff, R. Sandmeier and S. Clark, *J. Am. Chem. Soc.*, 1977, **99**, 6754; (l) T. Miyakoshi, S. Saito and J. Kumanotani, *Chem. Lett.*, 1981, 1677.
93. (a) N. C. F. Yim, H. Bryan, W. F. Huffman and M. L. Moore, *J. Org. Chem.*, 1988, **53**, 4605; (b) R. Labia and C. Morin, *J. Org. Chem.*, 1986, **51**, 249.
94. (a) H. O. House, W. A. Kleschick and E. J. Zaiko, *J. Org. Chem.*, 1978, **43**, 3653; (b) H. O. House and T. V. Lee, *J. Org. Chem.*, 1979, **44**, 2819; (c) H. O. House, R. J. Outcalt, J. L. Haack and D. Van Derveer, *J. Org. Chem.*, 1983, **48**, 1654; (d) H. J. Bestmann and G. Schade, *Tetrahedron Lett.*, 1982, **23**, 3543; (e) G. A. Kraus and Y.-S. Hon, *J. Am. Chem. Soc.*, 1985, **107**, 4341; (f) C. H. Heathcock, E. F. Kleinman and E. S. Binkley, *J. Am. Chem. Soc.*, 1982, **104**, 1054.
95. (a) R. C. Young, R. C. Mitchell, T. H. Brown, C. R. Ganellin, R. Griffiths, M. Jones, K. K. Rana, D. Saunders, I. R. Smith, N. E. Sore and T. J. Wilks, *J. Med. Chem.*, 1988, **31**, 656; (b) M. E. Jung and K. R. Buszek, *J. Org. Chem.*, 1985, **50**, 5440; K. R. Buszek, Ph.D. Thesis, UCLA, 1978; (c) see also ref. 92i.
96. (a) Y. Yamada, T. Hirata and M. Matsui, *Tetrahedron Lett.*, 1969, 101; (b) C. S. Szántay, L. Töke and K. Honti, *Tetrahedron Lett.*, 1965, 1665; (c) A. Itoh, S. Ozawa, K. Oshima and N. Nozaki, *Tetrahedron Lett.*, 1980, **21**, 361; *Bull. Chem. Soc. Jpn.*, 1981, **54**, 274; (d) D. M. Armistead and S. J. Danishefsky, *Tetrahedron Lett.*, 1987, **28**, 4959; (e) S. J. Danishefsky, D. M. Armistead, F. E. Wincott, H. G. Selnick and R. Hungate, *J. Am. Chem. Soc.*, 1987, **109**, 8117.
97. (a) A. B. Baylis and M. E. Hillman, *Ger. Pat.* 2 155 113 (1972) (*Chem. Abstr.*, 1972, **77**, 34 174q); (b) S. E. Drewes and N. D. Emslie, *J. Chem. Soc., Perkin Trans. 1*, 1982, 2079; (c) H. M. R. Hoffmann and J. Rabe, *Angew. Chem., Int. Ed. Engl.*, 1983, **22**, 795, 796; *J. Org. Chem.*, 1985, **50**, 3849; (d) P. Perlmutter and C. C. Teo, *Tetrahedron Lett.*, 1984, **25**, 5951; (e) D. Basavaiah and V. V. L. Gowriswari, *Tetrahedron Lett.*, 1986, **27**, 2031; (f) H. Amri and J. Villieras, *Tetrahedron Lett.*, 1986, **27**, 4307; (g) P. Auvray, P. Knochel and J. F. Normant, *Tetrahedron*, 1988, **44**, 6095; (h) V. Calo, L. Lopez and G. Pesce, *J. Organomet. Chem.*, 1988, **353**, 405; (i) K. Morita, Z. Suzuki and H. Hirose, *Bull. Chem. Soc. Jpn.*, 1968, **41**, 2815.
98. (a) G. Aksnes, *Acta Chem. Scand.*, 1961, **15**, 692; (b) R. F. Hudson and P. A. Chopard, *Helv. Chim. Acta*, 1963, **46**, 2178; *Z. Naturforsch., Teil B*, 1963, **18**, 509; (c) C. Osuch, J. E. Franz and F. B. Zienty, *J. Org. Chem.*, 1964, **29**, 3721; (d) J. M. J. Tronchet and B. Gentile, *Helv. Chim. Acta*, 1979, **62**, 977; (e) J. M. J. Tronchet and M. J. Valero, *Helv. Chim. Acta*, 1979, **62**, 2788; for representative uses of these compounds in synthesis, see (f) A. G. M. Barrett and H. B. Broughton, *J. Org. Chem.*, 1984, **49**, 3673; (g) J. E. McMurry and S. F. Donovan, *Tetrahedron Lett.*, 1977, 2869; (h) M. P. Cooke, Jr., *Tetrahedron Lett.*, 1981, **22**, 381; (i) T. H. Kim and S. Isoe, *J. Chem. Soc., Chem. Commun.*, 1983, 730; (j) R. C. Cookson and N. J. Liverton, *J. Chem. Soc., Perkin Trans. 1*, 1985, 1589.
99. (a) R. L. Wasson and H. O. House, *Org. Synth., Coll. Vol.*, 1963, **4**, 552; (b) S. Marmor, *J. Org. Chem.*, 1963, **28**, 250; (c) N. C. Yang and R. A. Finnegan, *J. Am. Chem. Soc.*, 1958, **80**, 5845; (d) E. Weitz and A. Scheffer, *Chem. Ber.*, 1921, **54**, 2327.
100. (a) A. Hassner, R. D'Costa, A. T. McPhail and W. Butler, *Tetrahedron Lett.*, 1981, **22**, 3691; (b) A. Hassner and W. Chau, *Tetrahedron Lett.*, 1982, **23**, 1989.
101. B. L. Feringa and B. de Lange, *Tetrahedron Lett.*, 1988, **29**, 1303.
102. (a) T. Sakakibara, T. Kawahara and R. Sudoh, *Carbohydr. Res.*, 1977, **58**, 39; (b) T. Sakakibara and R. Sudoh, *Carbohydr. Res.*, 1976, **50**, 191; (c) E. Jegou, J. Cleophax, J. Leboul and S. D. Gero, *Carbohydr. Res.*, 1975, **45**, 323; (d) K. Torssell and M. P. Tyagi, *Acta Chem. Scand., Ser. B*, 1977, **31**, 297; (e) N. Gregersen and C. Pedersen, *Acta Chem. Scand.*, 1972, **26**, 2695; (f) B. Fraser-Reid, *Acc. Chem. Res.*, 1975, **8**, 192; (g) B. D. Roth and W. H. Roark, *Tetrahedron Lett.*, 1988, **29**, 1255; (h) W. Streicher, H. Reinshagen and F. Turnowsky, *J. Antibiot.*, 1978, **31**, 725; (i) M. Chmielewski and S. Maciejewski, *Carbohydr. Res.*, 1986, **157**, C1.
103. (a) H. Matsunaga, T. Sakamaki, H. Nagaoka and Y. Yamada, *Tetrahedron Lett.*, 1983, **24**, 3009; *Heterocycles*, 1984, **21**, 428; (b) J. Mulzer, M. Kappert, G. Huttner and I. Jibril, *Angew. Chem., Int. Ed. Engl.*, 1984, **23**, 704; (c) G. Fronza, C. Fuganti, P. Grasselli, L. Majori, G. Pedrocchi-Fantoni and F. Spreafico, *J. Org. Chem.*, 1982, **47**, 3289.
104. (a) D. J. Abbott, S. Colonna and C. J. M. Stirling, *J. Chem. Soc., Chem. Commun.*, 1971, 471; *J. Chem. Soc., Perkin Trans. 1*, 1976, 492; (b) G. Tsuchihashi, S. Mitamura and K. Ogura, *Tetrahedron Lett.*, 1973, 2469; (c) S. G. Pyne, R. Griffith and M. Edwards, *Tetrahedron Lett.*, 1988, **29**, 2089.
105. A. Kamimura and N. Ono, *J. Chem. Soc., Chem. Commun.*, 1988, 1278.
106. (a) J. d'Angelo and J. Maddaluno, *J. Am. Chem. Soc.*, 1986, **108**, 8112; (b) for an example where high pressure *decreases* the enantioselectivity of a Michael addition see, A. Sera, K. Takagi, H. Katayama, H. Yamada and K. Matsumoto, *J. Org. Chem.*, 1988, **53**, 1157.
107. (a) J. E. Baldwin, *J. Chem. Soc., Chem. Commun.*, 1976, 734, 738; (b) J. E. Baldwin, J. Cutting, W. Dupont, L. I. Kruse, L. Silberman and R. C. Thomas, *J. Chem. Soc., Chem. Commun.*, 1976, 736; (c) J. E. Baldwin and

J. A. Reiss, *J. Chem. Soc., Chem. Commun.*, 1977, 77; (d) J. E. Baldwin and L. I. Kruse, *J. Chem. Soc., Chem. Commun.*, 1977, 233; (e) J. E. Baldwin, R. C. Thomas, L. I. Kruse and L. Silberman, *J. Org. Chem.*, 1977, **42**, 3846; (f) for two different readings of these rules see, M. J. Perkins, P. C. Wong, J. Barrett and G. Dhaliwal, *J. Org. Chem.*, 1981, **46**, 2196; G. W. L. Ellis, C. D. Johnson and D. N. Rogers, *J. Am. Chem. Soc.*, 1983, **105**, 5090.

108. (a) H. Ohrui, G. H. Jones, J. G. Moffatt, M. L. Maddox, A. T. Christensen and S. K. Byram, *J. Am. Chem. Soc.*, 1975, **97**, 4602; (b) H. Ohrui and J. J. Fox, *Tetrahedron Lett.*, 1973, 1951; (c) B. E. Maryanoff, S. O. Nortey, R. R. Inners, S. A. Campbell, A. B. Reitz and D. Liotta, *Carbohydr. Res.*, 1987, **171**, 259; (d) in certain cases, elimination of the 3-alkoxy group occurs in preference to cyclization to give dienoates: F. Nicotra, F. Ronchetti, G. Russo and L. Toma, *Tetrahedron Lett.*, 1984, **25**, 5697; A. B. Reitz, A. D. Jordan, Jr. and B. E. Maryanoff, *J. Org. Chem.*, 1987, **52**, 4800; (e) for a similar process in a noncarbohydrate system, see R. Bloch and M. Seck, *Tetrahedron Lett.*, 1987, **28**, 5819; (f) for an analogous stereoselctive process in the formation of a tetrahydropyran *via* a 6-exo-trig cyclization see, A. P. Kozikowski, R. J. Schmiesing and K. L. Sorgi, *J. Am. Chem. Soc.*, 1980, **102**, 6577.

109. (a) M. Hirama, T. Shigemoto, Y. Yamazaki and S. Ito, *J. Am. Chem. Soc.*, 1985, **107**, 1797; (b) for the 5-*exo*-trig cyclization of an amino dienolate, see S. Boulaajaj, T. Le Gall, M. Vaultier, R. Grée, L. Toupet and R. Carrié, *Tetrahedron Lett.*, 1987, **28**, 1761.

110. C.-K. Sha, C.-P. Tsou, Y.-C. Li, R.-S. Lee, F.-Y. Tsai and R.-H. Yeh, *J. Chem. Soc., Chem. Commun.*, 1988, 1081.

111. (a) S. G. Pyne, *J. Chem. Soc., Chem. Commun.*, 1986, 1686; (b) S. G. Pyne and S. L. Chapman, *J. Chem. Soc., Chem. Commun.*, 1986, 1688; (c) S. G. Pyne, *Tetrahedron Lett.*, 1987, **28**, 4737.

112. S. M. Reddy and H. M. Walborsky, *J. Org. Chem.*, 1986, **51**, 2605.

113. F. E. Ziegler, B. H. Jaynes and M. T. Saindane, *J. Am. Chem. Soc.*, 1987, **109**, 8115.

114. K. Tanaka, H. Yoda and A. Kaji, *Tetrahedron Lett.*, 1985, **26**, 4747.

115. J. Lévy, J.-Y. Laronze and J. Sapi, *Tetrahedron Lett.*, 1988, **29**, 3303.

116. M. E. Jung and B. E. Love, *J. Chem. Soc., Chem. Commun.*, 1987, 1288.

117. (a) F. Feist and G. Pomme, *Justus Liebigs Ann. Chem.*, 1909, **370**, 72; (b) S. Ruhemann, *J. Chem. Soc.*, 1899, **75**, 245, 411.

118. (a) L. I. Smith and R. E. Kelly, *J. Am. Chem. Soc.*, 1952, **74**, 3305; (b) C. L. Bickel, *J. Am. Chem. Soc.*, 1950, **72**, 1022; (c) J. Metler, A. Uchida and S. I. Miller, *Tetrahedron*, 1968, **24**, 4285.

119. (a) E. Urech, E. Tagmann, E. Sury and K. Hoffmann, *Helv. Chim. Acta*, 1953, **36**, 1809; (b) W. E. Bachmann, G. I. Fujimoto and E. K. Raunio, *J. Am. Chem. Soc.*, 1950, **72**, 2533; (c) A. N. Pudovik, N. G. Khusainova and R. G. Galeeva, *J. Gen. Chem. USSR (Engl. Transl.)*, 1966, **36**, 73.

120. W. Steglich, P. Gruber, G. Höfle and W. König, *Angew. Chem., Int. Ed. Engl.*, 1971, **10**, 653.

121. A. S. Kende, P. Fludzinski, J. H. Hill, W. Swenson and J. Clardy, *J. Am. Chem. Soc.*, 1984, **106**, 3551, and refs. therein.

122. (a) F. M. Beringer and S. A. Galton, *J. Org. Chem.*, 1965, **30**, 1930; (b) M. Ochiai, M. Kunishima, Y. Nagao, K. Fuji, M. Shiro and E. Fujita, *J. Am. Chem. Soc.*, 1986, **108**, 8281.

123. H. Kappeler and E. Renk, *Helv. Chim. Acta*, 1961, **44**, 1541.

124. (a) R. B. Woodward and T. Singh, *J. Am. Chem. Soc.*, 1950, **72**, 494; (b) for a discussion, see ref. 7b, p. 215 and ref. 8a, p. 5.

125. (a) O. Diels, *Chem. Ber.*, 1942, **75**, 1452; (b) O. Diels and U. Kock, *Justus Liebigs Ann. Chem.*, 1944, **556**, 38; (c) E. Le Goff and R. B. LaCount, *J. Org. Chem.*, 1964, **29**, 423; (d) R. C. Cookson, J. Hudec and B. Whitear, *Proc. Chem. Soc., London*, 1961, 117.

126. (a) G. A. Berchtold and G. F. Uhlig, *J. Org. Chem.*, 1963, **28**, 1459; (b) K. C. Brannock, R. D. Burpitt, V. W. Goodlett and J. G. Thweatt, *J. Org. Chem.*, 1963, **28**, 1464; (c) C. F. Huebner, L. Dorfman, M. M. Robison, E. Donoghue, W. G. Pierson and P. Strachan, *J. Org. Chem.*, 1963, **28**, 3134; (d) A. K. Bose, G. Mina, M. S. Manhas and E. Rzucidlo, *Tetrahedron Lett.*, 1963, 1467; (e) K. C. Brannock, R. D. Burpitt, V. W. Goodlett and J. G. Thweatt, *J. Org. Chem.*, 1964, **29**, 818; (f) see ref. 21e, p. 370; (g) P. W. Hickmott, *Tetrahedron*, 1982, **38**, 3363; (h) D. N. Reinhoudt, *Adv. Heterocycl. Chem.*, 1977, **21**, 253; (i) G. J. M. Vos, P. H. Benders, D. N. Reinhoudt, R. J. M. Egberink, S. Harkema and G. J. van Hummel, *J. Org. Chem.*, 1986, **51**, 2004; (j) for a case where a simple Michael addition occurs rather than ring expansion see, M. D. Menachery, J. M. Saá and M. P. Cava, *J. Org. Chem.*, 1981, **46**, 2584; (k) for the use of silyl enol ethers instead of enamines in the reaction with alkynic esters (catalyzed by Lewis acids) see, R. D. Clark and K. G. Untch, *J. Org. Chem.*, 1979, **44**, 248, 253; (l) reaction of the enolate of 5-cyclohexenone-2-carboxylate with ethyl propiolate affords the corresponding cyclooctadienonecarboxylate in 60–65% yield: M. M. Abou-Elzahab, S. N. Ayyad and M. T. Zimaity, *Z. Naturforsch., Teil B*, 1986, **41**, 363.

127. (a) M.-S. Lin and V. Snieckus, *J. Org. Chem.*, 1971, **36**, 645; (b) D. N. Reinhoudt and C. G. Kouwenhoven, *Recl. Trav. Chim. Pays-Bas*, 1973, **92**, 865; (c) D. N. Reinhoudt and C. G. Leliveld, *Tetrahedron Lett.*, 1972, 3119.

128. See, among others (a) G. Stork and T. L. Macdonald, *J. Am. Chem. Soc.*, 1975, **97**, 1264; (b) W. G. Dauben and D. J. Hart, *J. Org. Chem.*, 1977, **42**, 922; (c) D. Becker, L. R. Hughes and R. A. Raphael, *J. Chem. Soc., Perkin Trans. 1*, 1977, 1674.

129. For a discussion, see (a) M. V. George, S. K. Khetan and R. K. Gupta, *Adv. Heterocycl. Chem.*, 1976, **19**, 279; (b) ref. 10a, p. 189; (c) ref. 10e, p. 299.

130. (a) J. Ficini and C. Barbara, *Bull. Soc. Chim. Fr.*, 1964, 871; (b) H. G. Viehe, R. Fuks and M. Reinstein, *Angew. Chem., Int. Ed. Engl.*, 1964, **3**, 581; (c) H. Wamhoff, G. Haffmanns and H. Schmidt, *Chem. Ber.*, 1983, **116**, 1691; (d) H. Wamhoff and G. Hendrikx, *Chem. Ber.*, 1985, **118**, 863; (e) H. Wamhoff and W. Schupp, *J. Org. Chem.*, 1986, **51**, 2787, and refs. therein; (f) for the reaction of silylated ynamines with propiolates to give ynenoates see, Y. Sato, Y. Kobayashi, M. Sugiura and H. Shirai, *J. Org. Chem.*, 1978, **43**, 199.

131. J.-F. Lavallée, G. Berthiaume, P. Deslongchamps and F. Grein, *Tetrahedron Lett.*, 1986, **27**, 5455.

132. (a) T. Metler, A. Uchida and S. I. Miller, *Tetrahedron*, 1968, **24**, 4285; (b) J. Chauvelier, *Bull. Soc. Chim. Fr.*, 1954, **21**, 734.
133. (a) A. Michael and J. E. Bucher, *Chem. Ber.*, 1896, **29**, 1792; for other early references to the addition of heteronucleophiles to activated alkynes, see (b) S. Ruhemann and K. C. Browning, *J. Chem. Soc.*, 1898, **74**, 723; (c) S. Ruhemann and A. V. Cunnington, *J. Chem. Soc.*, 1899, **75**, 954; (d) E. R. Watson, *J. Chem. Soc.*, 1904, **85**, 1319; (e) E. André, *C. R. Hebd. Seances Acad. Sci., Ser. C*, 1911, **152**, 525; (f) C. Moureu and I. Lazennec, *Bull. Soc. Chim. Fr.*, 1906, **35**, 1906; (g) F. Straus and W. Voss, *Chem. Ber.*, 1926, **59**, 1681.
134. (a) R. W. Strozier, P. Caramella and K. N. Houk, *J. Am. Chem. Soc.*, 1979, **101**, 1340; (b) C. E. Dykstra, J. E. Arduengo and T. Fukunaga, *J. Am. Chem. Soc.*, 1978, **100**, 6007; (c) K. N. Houk, R. W. Strozier, M. D. Rozeboom and S. Nagase, *J. Am. Chem. Soc.*, 1982, **104**, 323; (d) P. Caramella and K. N. Houk, *Tetrahedron Lett.*, 1981, **22**, 819.
135. Years earlier Truce proposed a general 'rule of *trans* nucleophilic addition' for alkynes based on his studies of thiol additions: W. E. Truce and J. A. Simms, *J. Am. Chem. Soc.*, 1956, **78**, 2756; see also S. I. Miller, *J. Am. Chem. Soc.*, 1956, **78**, 6091.
136. M. E. Jung and K. R. Buszek, *J. Am. Chem. Soc.*, 1988, **110**, 3965; for an earlier, less comprehensive study, see R. G. Kostyanovskii and Yu. I. El'natanov, *Bull. Acad. Sci. USSR, Div. Chem. Sci.*, 1983, 2322.
137. (a) E. Winterfeldt and H. Preuss, *Chem. Ber.*, 1966, **99**, 450; (b) E. Winterfeldt, W. Krohn and H. Preuss, *Chem. Ber.*, 1966, **99**, 2572; (c) J. Biougne, F. Theron and R. Vessiere, *Bull. Soc. Chim. Fr.*, 1972, 2882; (d) E. Ciganek, *J. Org. Chem.*, 1980, **45**, 1497; (e) D. J. Bates, M. Rosenblum and S. B. Samuels, *J. Organomet. Chem.*, 1981, **209**, C55; (f) Y. Vo-Quang, D. Marais, L. Vo-Quang and F. Le Goffic, *Tetrahedron Lett.*, 1983, **24**, 5209; (g) G. Ferguson, K. J. Fisher, B. E. Ibrahim, C. Y. Ishag, G. M. Iskander, A. R. Katritzky and M. Parvez, *J. Chem. Soc., Chem. Commun.*, 1983, 1216.
138. (a) R. Huisgen, K. Herbig, A. Siegl and H. Huber, *Chem. Ber.*, 1966, **99**, 2526; (b) K. Herbig, R. Huisgen and H. Huber, *Chem. Ber.*, 1966, **99**, 2546; (c) C. H. McMullen and C. J. M. Stirling, *J. Chem. Soc. B*, 1966, 1217, 1221; (d) W. E. Truce and D. G. Brady, *J. Org. Chem.*, 1966, **31**, 3543; (e) M. S. Sinsky and R. G. Bass, *J. Heterocycl. Chem.*, 1984, **21**, 759; (f) Z. F. Solomko, T. S. Chmilenko, V. I. Avramenko and T. I. Petrun'kova, *Ukr. Khim. Zh. (Russ. Ed.)*, 1976, **42**, 206.
139. (a) P. Walter and T. M. Harris, *J. Org. Chem.*, 1978, **43**, 4250; (b) J. Cossy and J. P. Pète, *Bull. Soc. Chim. Fr., Part 2*, 1979, 559; (c) J. J. Talley, *Tetrahedron Lett.*, 1981, **22**, 823; (d) C. B. Kanner and U. K. Pandit, *Tetrahedron*, 1982, **38**, 3597; (e) F. Johnson, K. M. R. Pillai, A. P. Grollman, L. Tseng and M. Takeshita, *J. Med. Chem.*, 1984, **27**, 954; (f) R. H. Schlessinger, E. J. Iwanowicz and J. P. Springer, *Tetrahedron Lett.*, 1988, **29**, 1489; (g) for earlier references, see refs. 138a–138d and ref. 10.
140. (a) W. E. Truce and D. W. Onken, *J. Org. Chem.*, 1975, **40**, 3200; (b) J. E. Dolfini, *J. Org. Chem.*, 1965, **30**, 1298; (c) R. Huisgen, B. Giese and H. Huber, *Tetrahedron Lett.*, 1967, 1883; (d) B. Giese and R. Huisgen, *Tetrahedron Lett.*, 1967, 1889; (e) the addition of 2-substituted aziridines gives predominantly the (*E*)-isomer, see: Y. Gelas-Mialhe, E. Tourund and R. Vessiere, *Can. J. Chem.*, 1982, **60**, 2830; P. T. Trapentsier, I. Ya. Kalvin'sh, E. E. Liepin'sh and E. Ya. Lukevits, *Chem. Heterocycl. Compd. (Engl. Transl.)*, 1983, **19**, 391.
141. (a) M. E. Jung and K. R. Buszek, *Tetrahedron Lett.*, 1986, **27**, 6165; (b) M. E. Jung and K. R. Buszek, *J. Org. Chem.*, 1985, **50**, 5440; (c) F. E. Herkes and H. E. Simmons, Jr., *J. Org. Chem.*, 1973, **38**, 2845; (d) A. W. McCulloch and A. G. McInnes, *Can. J. Chem.*, 1974, **52**, 3569.
142. (a) W. E. Truce and D. L. Goldhamer, *J. Am. Chem. Soc.*, 1959, **81**, 5795, and refs. therein; (b) S. Kobayashi and T. Mukaiyama, *Chem. Lett.*, 1974, 1425; (c) M. T. Omar and M. N. Basyouni, *Bull. Chem. Soc. Jpn.*, 1974, **47**, 2325; (d) H. R. Pfaendler, J. Gosteli and R. B. Woodward, *J. Am. Chem. Soc.*, 1979, **101**, 6306; (e) D. H. Wadsworth and M. R. Detty, *J. Org. Chem.*, 1980, **45**, 4611; (f) D. F. Corbett, *J. Chem. Soc., Chem. Commun.*, 1981, 803; (g) A. N. Volkov, K. A. Volkova, E. P. Levanova and B. A. Trofimov, *Bull. Acad. Sci. USSR, Div. Chem. Soc.*, 1983, 186; (h) S. M. Proust and D. D. Ridley, *Aust. J. Chem.*, 1984, **37**, 1677; (i) M. Renard and L. Hevesi, *Tetrahedron*, 1985, **41**, 5939; (j) O. De Lucchi, V. Lucchini, C. Marchioro and G. Modena, *Tetrahedron Lett.*, 1985, **26**, 4539; (k) A. Bury, S. D. Joag and C. J. M. Stirling, *J. Chem. Soc., Chem. Commun.*, 1986, 124; (l) M. R. Detty and B. J. Murray, *J. Am. Chem. Soc.*, 1983, **105**, 883; (m) M. R. Detty, B. J. Murray, D. L. Smith and N. Zumbulyadis, *J. Am. Chem. Soc.*, 1983, **105**, 875; (n) L. A. Tsoi, A. K. Patsaev, V. Zh. Ushanov and L. V. Vyaznikovtsev, *J. Org. Chem. USSR (Engl. Transl.)*, 1984, **20**, 1897; (o) V. N. Drozd, M. L. Petrov, O. A. Popova and A. S. Vyazgin, *J. Org. Chem. USSR (Engl. Transl.)*, 1984, **20**, 1082; (p) G. Levesque and A. Mahjoub, *Tetrahedron Lett.*, 1980, **21**, 2247; (q) J. L. Garraway, *J. Chem. Soc.*, 1962, 4077; (r) A. J. Luxen, L. E. E. Christiaens and M. J. Renson, *J. Organomet. Chem.*, 1985, **287**, 81.
143. (a) K. Bowden and M. J. Price, *J. Chem. Soc. B*, 1970, 1466, 1470, and refs. therein; (b) J. Biougne and F. Theron, *C. R. Hebd. Seances Acad. Sci., Ser. C*, 1971, **272**, 858; (c) M. E. Jung, J. A. Hagenah and L.-M. Zeng, *Tetrahedron Lett.*, 1983, **24**, 3973.
144. (a) L. K. Dalton, S. Demerac and B. C. Elmes, *Aust. J. Chem.*, 1980, **33**, 1365; (b) W. Sucrow and M. Slopianka, *Chem. Ber.*, 1978, **111**, 780; (c) R. H. Wiley and P. Wiley, in 'The Chemistry of Heterocyclic Compounds', vol. 20, ed. E. C. Taylor and A. Weissberger, Interscience, New York, 1964.
145. (a) H. Hoffman and H. J. Diehr, *Chem. Ber.*, 1965, **98**, 363; (b) J. P. J. van der Holst, C. van Hooidonk and H. Kienhuis, *Recl. Trav. Chim. Pays-Bas*, 1974, **93**, 40; (c) for earlier work in this area see refs. 10a, 10e.
146. C. L. Bumgardner, J. E. Bunch and M.-H. Whangbo, *J. Org. Chem.*, 1986, **51**, 4083; for the similar 'anti-Michael' additions of alkyllithiums to 3-phenyl and 3-trimethylsilyl propiolamides, see G. W. Klumpp, A. J. C. Mierop, J. J. Vrielink, A. Brugman and M. Schakel, *J. Am. Chem. Soc.*, 1985, **107**, 6740.
147. (a) J. B. Hendrickson, R. Rees and J. F. Templeton, *J. Am. Chem. Soc.*, 1964, **86**, 107; (b) D. S. James and P. E. Fanta, *J. Org. Chem.*, 1962, **27**, 3346; (c) S. K. Khetan, J. G. Hiriyakkanavar and M. V. George, *Tetrahedron*, 1968, **24**, 1567; (d) U. K. Pandit and H. O. Huisman, *Recl. Trav. Chim. Pays-Bas*, 1966, **85**, 311; 1964, **83**, 50; (e) E. Winterfeldt and H. J. Dillinger, *Chem. Ber.*, 1966, **99**, 1558; (f) F. Bohlmann and E. Bresinsky, *Chem. Ber.*, 1964, **97**, 2109; (g) see also refs. 10a, 10e, 10f.
148. (a) J. Chauvelier, *Ann. Chim. (Paris)*, 1948, **3**, 393; (b) F. Gaudemar-Bardone, *Ann. Chim. (Paris)*, 1958, **3**, 52; (c) F. Bardone, *C. R. Hebd. Seances Acad. Sci., Ser. C*, 1954, **238**, 1716.

149. W.-D. Rudorf and R. Schwarz, *Heterocycles*, 1986, **24**, 3459; *Tetrahedron Lett.*, 1987, **28**, 4267; for similar cyclizations on activated alkenes, see M. Augustin, G. Jahreis and W.-D. Rudorf, *Synthesis*, 1977, 472.
150. S. L. Schreiber, S. E. Kelly, J. A. Porco, Jr., T. Sammakia and E. M. Suh, *J. Am. Chem. Soc.*, 1988, **110**, 6210.
151. (a) S. R. Landor, in 'The Chemistry of the Allenes', ed. S. R. Landor, Academic Press, London, 1982, vol. 2, p. 361; (b) M. C. Caserio, in 'Selective Organic Transformations', ed. B. S. Thyagarajan, Wiley-Interscience, New York, 1970, vol. 1, p. 239; (c) W. T. Brady, in 'The Chemistry of Ketenes, Allenes, and Related Compounds', ed. S. Patai, Wiley, New York, 1980, vol. 1, p. 279; (d) H. F. Schuster and G. M. Coppola, 'Allenes in Organic Synthesis', Wiley-Interscience, New York, 1984; (e) ref. 10f, p. 73.
152. (a) S. J. Danishefsky, S. J. Etheredge, R. Volkmann, J. Eggler and J. Quick, *J. Am. Chem. Soc.*, 1971, **93**, 5575; (b) R. Volkmann, S. J. Danishefsky, J. Eggler and D. M. Solomon, *J. Am. Chem. Soc.*, 1971, **93**, 5576; (c) J. Quick, *Tetrahedron Lett.*, 1977, 327; (d) for the addition of tetrakis(dimethylamino)allene and other nucleophiles to allenetetracarboxylate to give zwitterionic adducts see, R. Gompper, J. Schelble and C. S. Schneider, *Tetrahedron Lett.*, 1978, 3897; R. Gompper and U. Wolf, *Justus Liebigs Ann. Chem.*, 1979, 1406.
153. (a) P. Kurtz, H. Gold and H. Disselnkötter, *Justus Liebigs Ann. Chem.*, 1959, **624**, 1; (b) Y. H. Paik and P. Dowd, *J. Org. Chem.*, 1986, **51**, 2910.
154. (a) J. W. Batty, P. D. Howes and C. J. M. Stirling, *J. Chem. Soc., Perkin Trans. 1*, 1973, 65; (b) B. S. Ellis, G. Griffiths, P. D. Howes, C. J. M. Stirling and B. R. Fishwick, *J. Chem. Soc., Perkin Trans. 1*, 1977, 286; (c) G. Griffiths, P. D. Howes and C. J. M. Stirling, *J. Chem. Soc., Perkin Trans. 1*, 1977, 912; *J. Chem. Soc., Chem. Commun.*, 1976, 296.
155. (a) G. Buono, G. Peiffer and A. Guillemonat, *C. R. Hebd. Seances Acad. Sci., Ser. C*, 1970, **271**, 937; (b) H. Strzelecka, M. Dupre and M. Simalty, *Tetrahedron Lett.*, 1971, 617.
156. (a) M. Bertrand and J. Le Gras. *C. R. Hebd. Seances Acad. Sci.*, 1965, **260**, 6926; (b) G. Eglinton, E. R. H. Jones, G. H. Mansfield and M. C. Whiting, *J. Chem. Soc.*, 1954, 3197; (c) C. J. M. Stirling, *J. Chem. Soc.*, 1964, 5856, 5863; (d) S. T. McDowell and C. J. M. Stirling, *J. Chem. Soc. B*, 1967, 351; (e) G. D. Appleyard and C. J. M. Stirling, *J. Chem. Soc. C*, 1967, 2686; (f) J. C. Chalchat, F. Théron and R. Vessière, *Bull. Soc. Chim. Fr.*, 1970, 711; (g) F. Théron and R. Vessière, *Bull. Soc. Chim. Fr.*, 1968, 2994; (h) M. Verny and R. Vessière, *Bull. Soc. Chim. Fr.*, 1967, 2508; *Tetrahedron*, 1969, **25**, 263; (i) A. N. Pudovik, N. G. Khusainova and I. M. Aladzheva, *J. Gen. Chem. USSR (Engl. Transl.)*, 1964, **34**, 2484; (j) R. L. Bol'shedvorskaya, G. A. Pavlova, L. D. Gavrilov, N. V. Alekseeva and L. I. Vereshchagin, *J. Org. Chem. USSR (Engl. Transl.)*, 1972, **8**, 1927; (k) F. Gaudemar, *C. R. Hebd. Seances Acad. Sci.*, 1956, **242**, 2471; (l) G. R. Harvey and K. W. Ratts, *J. Org. Chem.*, 1966, **31**, 3907; (m) H. J. Cristau, J. Viala and H. Christol, *Tetrahedron Lett.*, 1982, **23**, 1569; (n) E. Winterfeldt and J. M. Nelke, *Chem. Ber.*, 1968, **101**, 2381; (o) F. W. Nader, A. Brecht and S. Kreisz, *Chem. Ber.*, 1986, **119**, 1196.
157. (a) I. Tamnefors, A. Claesson and M. Karlsson, *Acta Pharm. Suec.*, 1975, **12**, 435; (b) R. M. Acheson, M. G. Bite and M. W. Cooper, *J. Chem. Soc., Perkin Trans. 1*, 1976, 1908; (c) S. R. Landor, P. D. Landor, Z. T. Fomum and G. B. Mpango, *J. Chem. Soc., Perkin Trans. 1*, 1979, 2289; (d) J. Ackroyd and F. Scheinmann, *J. Chem. Soc., Chem. Commun.*, 1981, 339; *J. Chem. Res. (S)*, 1982, 89; (e) P. M. Greaves and S. R. Landor, *J. Chem. Soc., Chem. Commun.*, 1966, 322; (f) I. T. Kay and N. Punja, *J. Chem. Soc. C*, 1970, 2409; (g) M. Bertrand, J. Elguero, R. Jacquier and J. Le Gras, *C. R. Hebd. Seances Acad. Sci., Ser. C*, 1966, **262**, 782; (h) Y. Tamura, T. Tsugoshi, S. Mohri and Y. Kita, *J. Org. Chem.*, 1985, **50**, 1542; (i) N. S. Nixon, F. Scheinmann and J. L. Suschitzky, *Tetrahedron Lett.*, 1983, **24**, 597; *J. Chem. Res. (S)*, 1984, 380; (j) R. M. Acheson and J. D. Wallis, *J. Chem. Soc., Perkin Trans. 1*, 1982, 1905; (k) G. Buono and G. Peiffer, *Tetrahedron Lett.*, 1972, 149; (l) G. Buono and J. R. Llinas, *J. Am. Chem. Soc.*, 1981, **103**, 4532 and refs. therein; (m) H. J. Bestmann, G. Graf, H. Hartung, S. Kolewa and E. Vilsmaier, *Chem. Ber.*, 1970, **103**, 2794; (n) C. Santelli, *Tetrahedron Lett.*, 1980, 2893; (o) Yu. M. Skvortsov, E. B. Oleinikova, A. N. Volkov, B. A. Trofimov, A. G. Mal'kina and M. V. Sigalov, *Izv. Akad. Nauk SSSR, Ser. Khim.*, 1979, 398.
158. S. Blechert, *Helv. Chim. Acta*, 1985, **68**, 1835; for a similar pyrrole-forming process by addition of a ketoxime to acetylenedicarboxylate, see T. Sheradsky, *Tetrahedron Lett.*, 1970, 25.

1.2
Conjugate Additions of Reactive Carbanions to Activated Alkenes and Alkynes

VING J. LEE

American Cyanamid, Pearl River, NY, USA

1.2.1 INTRODUCTION

The 1,4-conjugate addition of stabilized carbanions (1,3-dicarbonyl class) to α,β-unsaturated carbonyl compounds was reported by Michael in 1887 and quickly became established as an efficient method for carbon–carbon bond formation.[1] Numerous stabilized nucleophiles (donors) have been found to participate efficiently in 1,4-conjugate additions to other activated alkenes (acceptors).[2] In general, the reactivity of acceptors (a^3-synthons) towards various stabilized carbanions follows: α,β-unsaturated aldehydes >> α,β-unsaturated ketones > α,β-unsaturated nitriles > α,β-unsaturated esters > α,β-unsaturated amides.[3] Although the basic synthetic principles and reactivity were investigated in early years, limitations to this versatile form of carbon–carbon bond formation were observed. For example, the addition of organolithiums or organomagnesium halides to α,β-unsaturated aldehydes affords exclusive 1,2-addition, while additions to α,β-unsaturated ketones afford predominant 1,2-addition. During the last four decades intensive research in the use of unstabilized carbon nucleophiles for 1,4-conjugate additions was stimulated by three events. First, Kharasch and Tawney reported that preferential 1,4-conjugate addition of alkylmagnesium halides to isophorone was effected by catalytic amounts of copper(I) chloride,[4] and second, House, Respess and Whitesides showed that the actual reactive species in the earlier work was an organocopper species.[5] Third, Gilman and Kirby reported a comparative study of the addition of various aryl metallics to benzalacetophenone (1,3-diphenylpropene-1-one) in which several Group II (R_2Cd and R_2Zn) and Group III (R_3Al) organometallics afforded exclusive 1,4-conjugate addition.[6]

The ever-expanding application of 1,4-conjugate additions and, more recently, the tandem 1,4-conjugate addition–electrophile trapping protocol (triple convergent syntheses), especially with prostaglandin syntheses, and the Michael initiated ring closure protocol (MIRC)[7,8] has prompted the development and characterization of additional organometallic reagents for accomplishing these transformations (equation 1). In contrast to the robust organolithiums and Grignard reagents, more selective organometallic reagents have been introduced that are permissive of additional functionalities on either the acceptor or donor. Alternatively, modified a^3-synthons, *e.g.* α-silylvinyl carbonyl species and α,β-unsaturated thioamides, have been introduced in order to minimize undesirable carbanion-catalyzed processes, namely polymerization. Methods have been developed for the conjugate addition of organometallics to alkenes (a^2-synthons) with functionality capable of carbanion stabilization, *e.g.* NO_2, R_3P^+, R_2S^+. For example, 2-(*N*-methylanilino)acrylonitrile, ketene thioacetal monoxide, α-nitroalkenes, vinyl sulfoxides, vinyl sulfones and vinyl phosphonates serve as α-carbonyl cation or (and) vinyl cation equivalents. In this survey, the 1,4-conjugate addition of various organometallic reagents to activated alkenes and alkynes will be discussed and, where appropriate, the synthesis of these reagents will be discussed. In addition, sections will be devoted to (a) enolate and homoenolate additions, (b) acyl anion (and equivalents) additions, (c) multiple conjugate additions and (d) heteroconjugate additions, *i.e.* carbometallation. Where appropriate, asymmetric and diastereoselectivity aspects will also be discussed. This survey is not intended to be comprehensive, but the transformations cited are illustrative of their synthetic potential. Lewis acid promoted additions and stoichiometric organocopper reagents will be reviewed separately (Volume 4, Chapters 1.3 and 1.4, respectively).

A survey of the 1,4-conjugate addition of organometallic reagents should include a discussion of the interdependence of (a) electronegativity of various metals, (b) hard and soft acid and base (HSAB) theory, (c) the structure of the acceptor and (d) reaction conditions and solvent effects. The difference in electronegativity between a metal and carbon is reflected in the degree of ionic character imparted to the carbon–metal σ-bond and the direction of polarization.[9] Thus the highly ionic Group IA and IIA organometallics (RK, RNa, RLi and RMgX) are extremely reactive in contrast to the less ionic Group IIB organometallics (R_2Cd and R_2Zn) with the Group IB (RCu) and Group IIIB organometallics (R_3Al and R_3B) being intermediate. However with the Group IB and IIIB organometallics, carbon–carbon bond formation is mediated by either carbanion, free radical or electron-transfer processes. Typically, the nucleophilicity of the Group IIB and Group IIIB organometallics is enhanced by conversion to the corresponding stable 'ate' complexes which comprise of a Lewis acid (electron-deficient organometallic) and a Lewis base (organolithium) (Volume 4, Chapter 1.3).[10]

Chemoselectivity can also be partially explained by a hard and soft acid and base (HSAB) theory.[11] From HSAB theory, the lithium cation is a harder acid than the magnesium cation and, with α,β-unsaturated ketones, C-1 (carbonyl carbon) is a harder base than C-3; thus reactions of organolithiums are preferred at the the hard site, C-1, affording 1,2-addition products. The structure of the carbanion is also

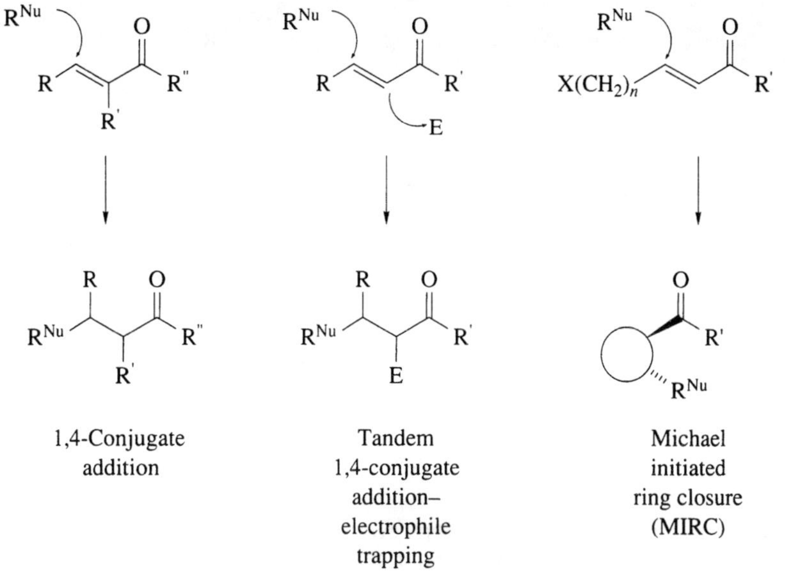

(1)

1,4-Conjugate addition

Tandem 1,4-conjugate addition–electrophile trapping

Michael initiated ring closure (MIRC)

a determinant in additions to α,β-unsaturated carbonyl acceptors; for example, benzylic lithiums[12] and lithio-2-phenyl-1,3-dithiane[13] undergo preferential, but not exclusive, 1,4-addition to α,β-unsaturated ketones, while alkyllithiums and lithio-1,3-dithiane[13b] show preferential 1,2-addition mode. The additional phenyl stabilization affords softer carbanions and thus higher preference for C-3; in addition, the higher the degree of substitution or bulkiness on the carbanion, the greater the tendency for 1,4-addition. It is instructive to note that the additions of *t*-butyllithium (–78 °C, ether)[14] and *t*-butylmagnesium chloride (25 °C, ether)[15] to 2-cyclohexen-1-one afford exclusively the 1,2- and 1,4-addition products (**1**) and (**2**), respectively (Scheme 1).

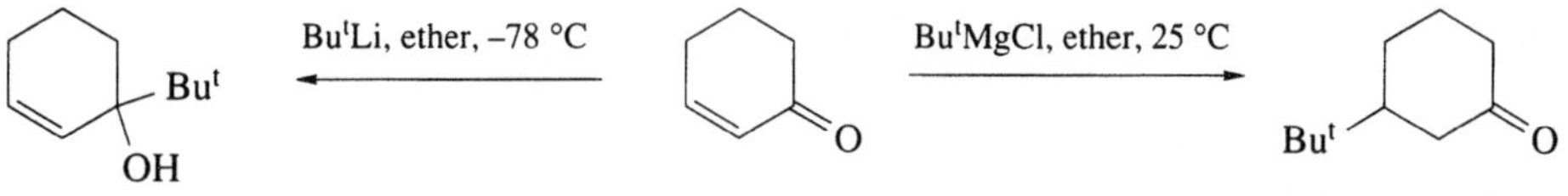

Bu^{t}Li, ether, –78 °C; Bu^{t}MgCl, ether, 25 °C

(**1**) (**2**)

Scheme 1

The electronics and structure of the acceptors, especially α,β-unsaturated ketones, is also a determinant in 1,2- *vs.* 1,4-addition processes. In general, substitution of aryl or large groups at the carbonyl unit increases the preference for 1,4-addition,[16a,b] while α,β-unsaturated aldehydes afford exclusive 1,2-addition and β,β-disubstitution suppresses 1,4-addition,[16c] presumably due to steric hinderance. House and Seyden-Penne have established good correlations between chemoselectivity and either the half-wave electrolytic reduction potentials,[17a] or the energy levels of the LUMO of various α,β-unsaturated ketones.[17b]

Finally, the effect of reaction conditions and solvents on chemoselectivity should also be considered.[18] Empirically, polar, more basic solvents, *e.g.* HMPA, DMF, serve to minimize counterion effects by formation of solvent-separated ion pairs and promote electron-transfer processes which are conducive for 1,4-additions,[14,19] while low polarity solvents afford contact ion pairs.[20] Numerous researchers have studied the effects of reaction conditions and solvents on the addition of α-thiosubstituted carbanions to α,β-enones;[21] in general, low polarity solvents afford kinetic 1,2-addition products while at higher temperatures the thermodynamic 1,4-addition products predominate. However, by using highly polar solvents, *e.g.* HMPA, and low temperatures, 1,4-adducts are formed predominantly and further warming to 25 °C results in exclusive formation of 1,4-adducts; thus the intermediate allylic alkoxide undergoes rearrangement to the more stable enolate (Scheme 2). In contrast, the addition of *t*-butyllithium to 2-cyclohexen-1-one affords a thermally noninterconvertible mixture of 1,2- and 1,4-addition products (**1** and **2**), but increasing quantities of 1,4-adduct are obtained with increasing amounts of HMPA in the reaction mixture.[14] In the absence of HMPA, low reaction temperatures (<–50 °C) also favor kinetic 1,4-addition of sulfur-stabilized organolithiums to α,β-enones.[20]

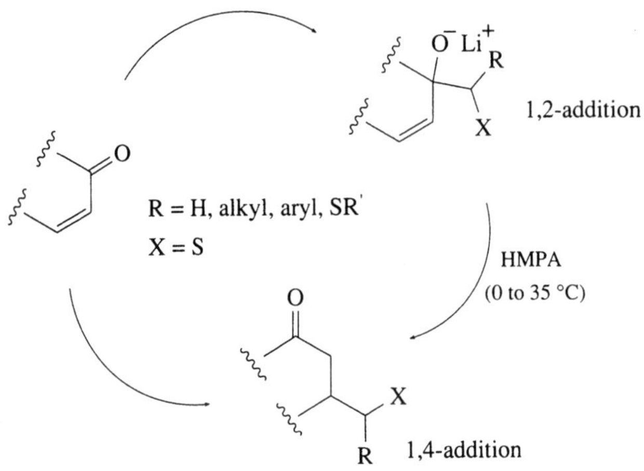

Scheme 2

Where applicable in this survey, emphasis will be given to the mode of synthesis of various organometallic reagents and optimal reaction conditions. Four general methods are employed for the synthesis of organometallic reagents: (a) oxidative metallation of organic halides, (b) metal–halogenation exchange, (c) metal–hydrogen exchange and (d) transmetallation.[22] Of these synthetic methods, the transmetallation protocol involving exchange of an organolithium (or Grignard reagent) with anhydrous metal halides, especially Group II metals, generates byproduct lithium halides (or magnesium halides) which modify the chemoselectivity of the newly formed organometallic reagent. The reactivity of the Group IIB organometallics (R_2Cd and R_2Zn) is highly influenced by the presence of magnesium halides ($MgI_2 > MgBr_2 > MgCl_2$) and lesser so with lithium halides (LiI > LiBr).[23] Where appropriate data exist, the differential reactivity of 'salt-free' organometallics *vs.* 'salt-containing' organometallics will be presented.

1.2.2 ADDITIONS TO ALKENIC π-SYSTEMS

1.2.2.1 Additions of Alkyl, Aryl, Alkenyl and Alkynyl Groups

1.2.2.1.1 Organolithiums

The reaction of organolithiums with α,β-unsaturated aldehydes, α,β-unsaturated ketones and α,β-unsaturated esters generally affords 1,2-addition products, which is attributed to the high ionicity of the organolithiums.[24] However, 1,4-chemoselectivity is attained when the acceptor carbonyl group is sterically hindered or deactivated towards 1,2-addition, as exemplified by the trityl enones (**3**; equation 2),[25] the N^1,N^2,N^2-(trimethyl)acrylhydrazides (**4**; equation 3),[26] hindered tertiary acrylamides (**5**; equation 4),[27] BHA (2,6-di-*t*-butyl-4-methoxyphenol) acrylates (**6**)[28] or BHT (2,6-di-*t*-butyl-4-methylphenol) acrylates (**7**; Scheme 3).[29] The tandem 1,4-conjugate addition–electrophile trapping protocol (aldehydes, alkyl halides, allylic halides, benzylic halides and dialkyl disulfides) has been extended to (**5**)[27,30] and (**6**) and (**7**).[31] In contrast to the other acceptors, the enones (**3**) also undergo 1,4-addition with lithium acetylides (–45 to 25 °C, THF) to afford the tritylalkynones.[25b]

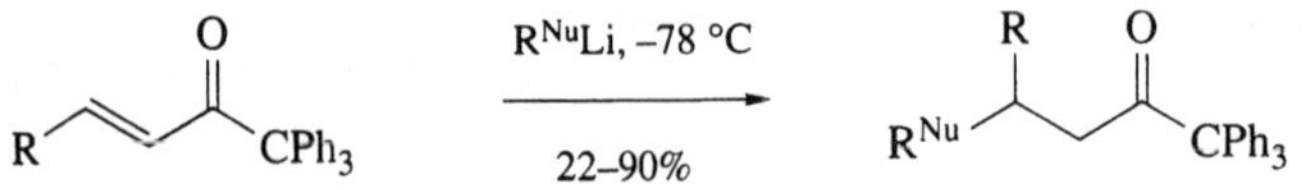

(2)

R = H, Me, Ph

R^{Nu} = Me, Bu^n, Bu^t, Ph, RC≡C

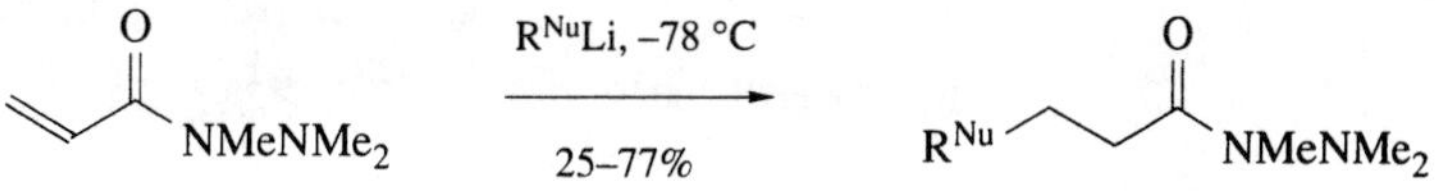

(4) $\xrightarrow[25\text{–}77\%]{R^{Nu}Li,\ -78\ °C}$ $R^{Nu}CH_2CH_2C(O)NMeNMe_2$ (3)

R^{Nu} = Me, Bu^n, Bu^t

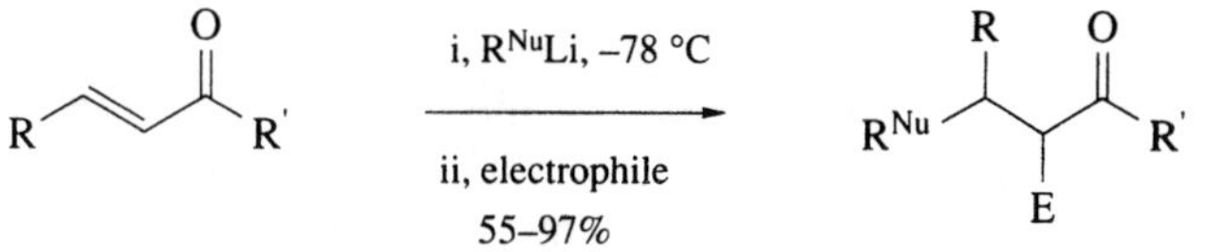

(5) R' = Me_2N, piperidino — i, $R^{Nu}Li$, –78 °C; ii, electrophile; 55–97% (4)

R^{Nu} = Bu^n, Bu^t, Ph, 4-MeC_6H_4, dithiane, enolates

E = H, alkyl, allyl, R"CH(OH), EtO_2C, PhS

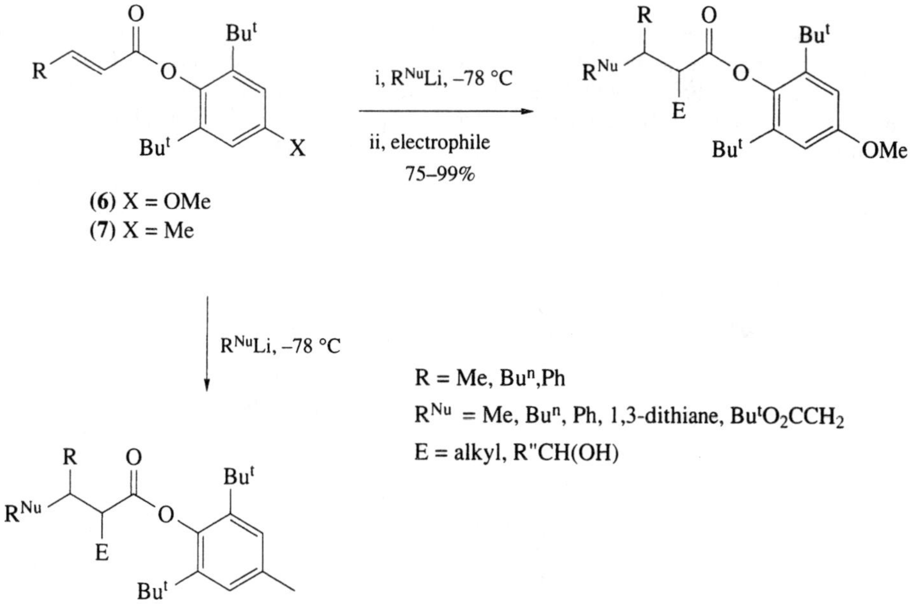

(6) X = OMe

(7) X = Me

i, $R^{Nu}Li$, –78 °C; ii, electrophile; 75–99%

$R^{Nu}Li$, –78 °C

R = Me, Bu^n,Ph

R^{Nu} = Me, Bu^n, Ph, 1,3-dithiane, $Bu^tO_2CCH_2$

E = alkyl, R"CH(OH)

Scheme 3

In general, β,β-disubstituted α,β-unsaturated carbonyl systems are poor acceptors for carbanions, due to steric hinderance, and the above acceptors are no exception;[16c,32] in contrast, for intramolecular cyclizations, the substituent limitations are not as stringent.[33] For example, Cooke reports that the intramolecular cyclization of *t*-butyl ω-iodo-α,β-unsaturated esters (**8**) to *t*-butyl cycloalkylacetates (**10**) *via* the ω-lithio-α,β-unsaturated esters (**9**) is highly efficient (equation 5).[34] However, the corresponding lower alkyl esters, ω-iodo-α,β-unsaturated tertiary amides or ω-iodo-α,β-unsaturated ketones afforded low yields of cyclized product. The reaction conditions necessary for optimal yields include *n*-butyllithium as the lithiation reagent and reaction temperatures of –100 °C. Similarly, Rodrigo reports that the intramolecular cyclization of ethyl (*o*-iodophenoxy)crotonates (**11**) affords dihydrobenzofurans (**12**; equation 6).[35]

In contrast, chemoselectivity can be attained by a 'charge directed conjugate addition' strategy in which 1,2-additions are suppressed, not by steric constraints, but by the presence of an adjacent negative

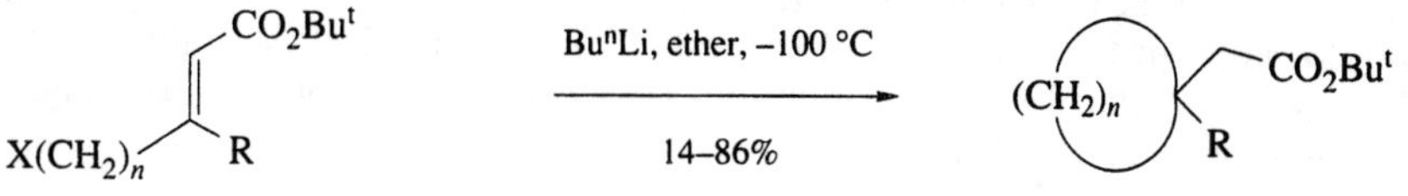

Bu^nLi, ether, –100 °C; 14–86% (5)

(**8**) X = I; *n* = 3–5

(**9**) X = Li; *n* = 3–5

(**10**) R = H, Me; *n* = 3–5

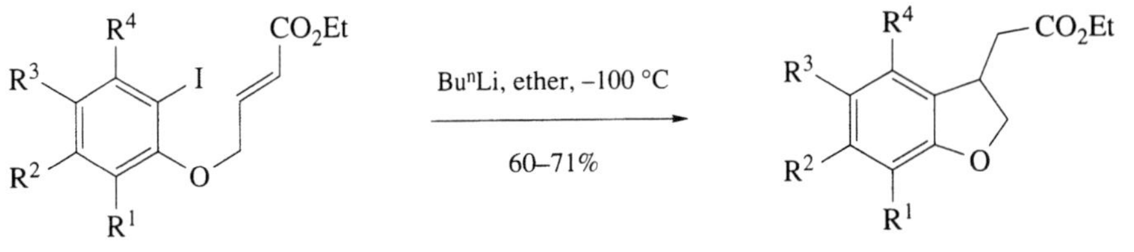

Bu^nLi, ether, −100 °C; 60–71% (6)

(11) → **(12)**

$R^1 = OMe, R^2 = R^3 = H, R^4 = Me$
$R^1 = OMe, R^2 = R^3 = R^4 = H$
$R^1 = R^2 = H, R^3 + R^4 = OCH_2O$
$R^1 = R^4 = H, R^2 = R^3 = Me$
$R^1 = OMe, R^3 = Me, R^2 = R^4 = H$
$R^1 = R^2 = H, R^3 = R^4 = (CH{=}CH)_2$

charge (on either carbon or nitrogen), which deactivates the carbonyl group to either initial or secondary reactions with organometallic reagents.[36] Typical acceptors are the secondary amides (**13**)[27] or anilides (**14**)[37] and the acryl(alkoxycarbonyl)methylenetriphenylphosphoranes (**15**; Scheme 4);[38a] the latter are obtained by acylation of (alkoxycarbonyl)methylenetriphenylphosphoranes.[38b] The adducts (**16**) are conveniently transformed in high yields to the corresponding carboxylic acids,[39a] esters[39b] or methyl ketones[39c] under mild conditions.

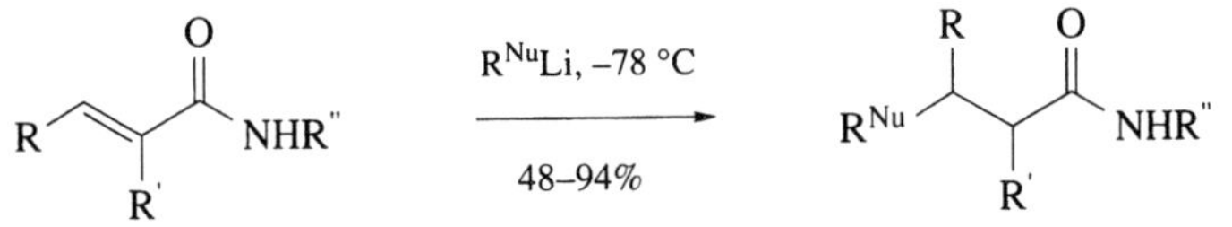

$R^{Nu}Li$, −78 °C; 48–94%

(**13**) R" = Me
(**14**) R" = Ph

R = H, Me
$R^{Nu} = Bu^t$, Ph

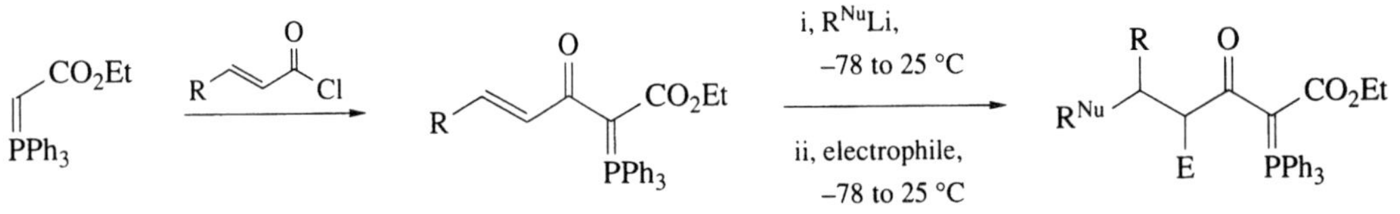

i, $R^{Nu}Li$, −78 to 25 °C; ii, electrophile, −78 to 25 °C; 72–98%

(15) **(16)**

$R = H$, alkyl, $CH_2{=}CH$; R^{Nu} = alkyl, Ph, $CH_2{=}CH$, $Me_3SiC{\equiv}CCH_2$, 1,3-dithiane, $Bu^tO_2CCH_2$
E = H, alkyl, R"CH(OH)

Scheme 4

It should be noted with (**3**), (**4**), (**13**) and (**14**) only hard lithium nucleophiles are tolerated, while for (**5**), (**6**), (**7**) and (**15**) the range of useful lithium nucleophiles, *e.g.* including lithio-1,3-dithianes and lithium enolates, is broader. In addition, none of these Michael acceptors react with organocuprates, Grignard reagents or stabilized carbanions.

Two intramolecular variants of the 'charge directed conjugate addition' strategy are reported to afford high yields of precursors of cycloalkylacetic esters (**20**),[40a] *trans*-2-substituted cycloalkylcarboxylate esters (**21**) and 2-cycloalkenylacetic esters (**22**).[40b,c] Both processes involve identical carbon skeletons, but different leaving groups at the ω-position. Low temperature lithiation (−100 °C, *n*-butyllithium) or oxidative metallation (Mg) of ω-iodides (**17**) and (**18**) affords the intramolecular 1,4-addition esters (**20** or **22**). Alternatively, the MIRC protocol for cyclization of a ω-chloride (or ω-bromide) (**19**) requires an initial intermolecular 1,4-addition of a carbanionic species, *e.g.* organolithium or lithium enolate, followed by ring closure to ester (**21**; Scheme 5).

In a variant of the 'charge directed conjugate addition' strategy, the incorporation of an α-trimethylsilyl unit in an acceptor serves to stabilize the intermediate carbanion and minimize anionic polymerizations.[41] This is exemplified by the highly efficient 1,4-conjugate addition of organolithiums (or Grignard reagents) to α-(trimethylsilyl)-α,β-unsaturated carboxylic acids (**23**); in contrast, α,β-unsaturated carb-

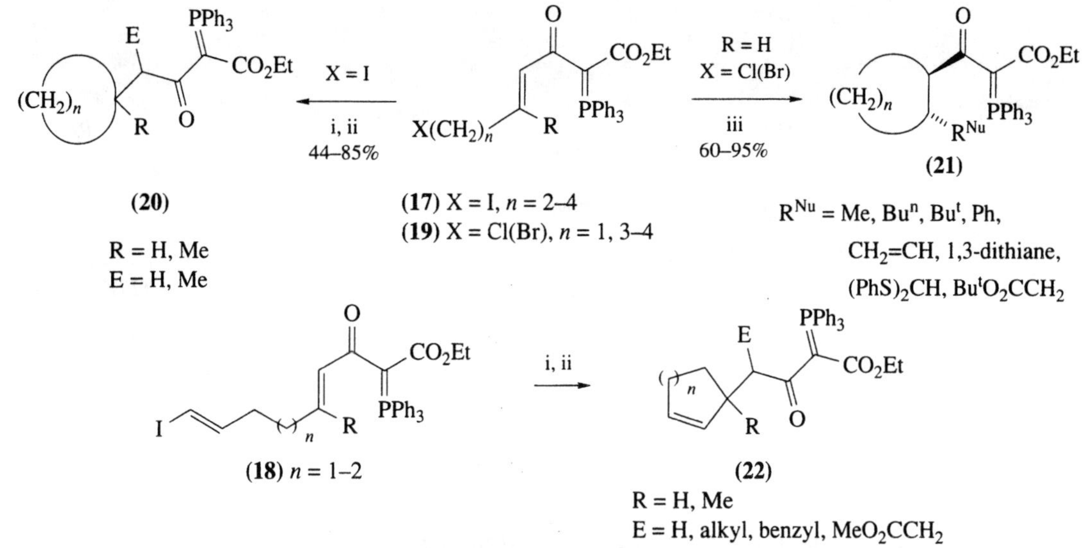

X = I; i, ii; 44–85%

R = H; X = Cl(Br); iii; 60–95%

(20)

R = H, Me
E = H, Me

(17) X = I, n = 2–4
(19) X = Cl(Br), n = 1, 3–4

(21)

R^{Nu} = Me, Bu^n, Bu^t, Ph, CH_2=CH, 1,3-dithiane, $(PhS)_2CH$, $Bu^tO_2CCH_2$

(18) n = 1–2

i, ii

(22)

R = H, Me
E = H, alkyl, benzyl, MeO_2CCH_2

i, Bu^nLi (Bu^tLi), –100 to –78 °C; ii, electrophile; iii, $R^{Nu}Li$, –78 °C

Scheme 5

oxylic acids typically afford exclusive 1,2-addition products in low yields. The resultant α-(trimethylsilyl)carboxylic acids (**24**) are desilylated with sodium hydroxide; thus, this is the equivalent of the 1,4-conjugate addition of an organometallic reagent to an α,β-unsaturated carboxylic acid.[42a] Two aspects of the reaction conditions should be highlighted: (a) excellent yields of (**24**) are obtained when acid (**23**) is added to the organometallic donor (–78 °C) and (b) in contrast to the intermediate lithium dianions, the intermediate magnesium dianions react with aldehydes to yield α,β-unsaturated acids (**25**; Scheme 6).[42b]

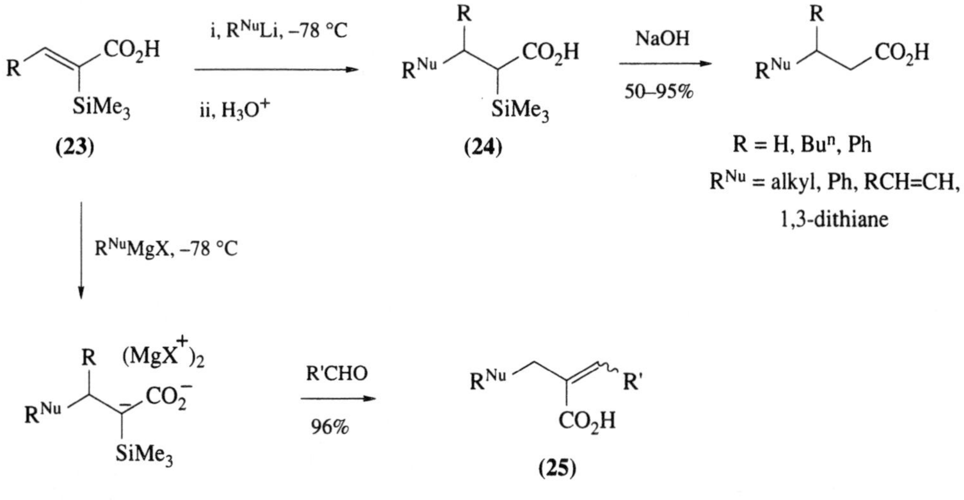

i, $R^{Nu}Li$, –78 °C; ii, H_3O^+

NaOH; 50–95%

(23) **(24)**

R = H, Bu^n, Ph
R^{Nu} = alkyl, Ph, RCH=CH, 1,3-dithiane

$R^{Nu}MgX$, –78 °C

R'CHO; 96%

(25)

R^{Nu} = Bu^t; R' = n-C_8H_{17}

Scheme 6

Similarly, organolithiums (or Grignard reagents) add to ester (**26**) to afford silyl enolates which are intercepted with aldehydes and ketones to give α,β-unsaturated esters (**27**),[43a] while addition of the bifunctional aryllithium (**28**) (MIRC protocol) affords the 2-(trimethylsilyl)tetrahydronaphthoate ester (**29**; Scheme 7).[43b]

The addition of organolithiums to α,β-unsaturated *s*-thioamides (**30**) occurs exclusively in a 1,4-mode which is a consequence of the high stability of the intermediate thioimidate anion.[44a,b] The thioamide acceptors are obtained by the reaction of an alkyl (or aryl) isothiocyanate with a vinylic lithium (or vinylic Grignard reagent) or an allylic Grignard reagent followed by a base-catalyzed conjugation. Typically, hard organolithiums add to afford the corresponding thioamide dianion which can be α-alkylated effi-

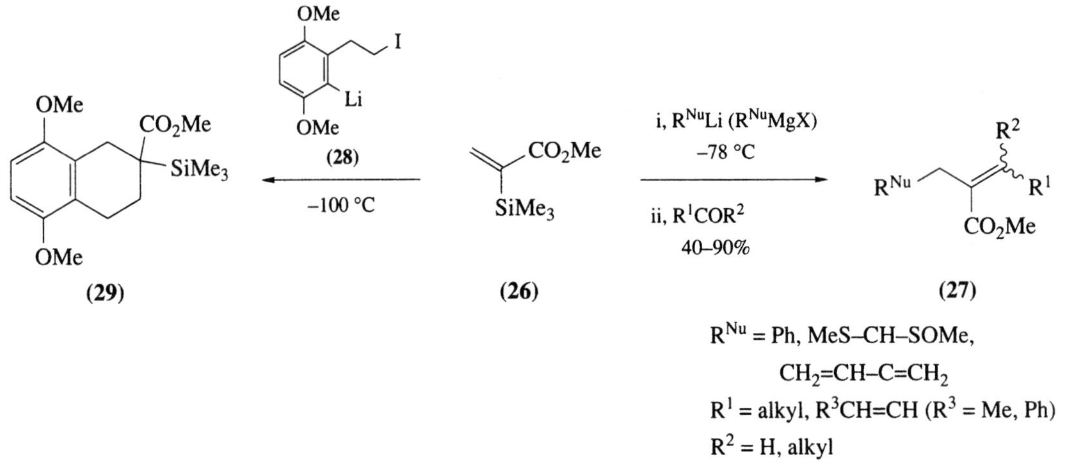

OMe
CO2Me
SiMe3
OMe
(29)
OMe
I
Li
OMe
(28)
−100 °C
CO2Me
SiMe3
(26)
i, RNuLi (RNuMgX)
−78 °C
ii, R1COR2
40–90%
R2
RNu
R1
CO2Me
(27)
RNu = Ph, MeS–CH–SOMe,
CH2=CH–C=CH2
R1 = alkyl, R3CH=CH (R3 = Me, Ph)
R2 = H, alkyl

Scheme 7

ciently; while with the α,β-unsaturated *N*-(trimethylsilyl)thioamides (**31**), both soft lithium anions, *e.g.* enolates, and hard lithium anions and Grignard reagents add efficiently (Scheme 8). Similarly, the corresponding α,β-unsaturated *t*-thioamides (**32**), including *N,N*-dimethylthiosorbamide, undergo exclusive 1,4-conjugate additions with organolithiums to afford an intermediate (*Z*)-enethiolate, which is a softer anion and thus not subject to further organometallic attack, but can undergo further α-alkylation or α-phenylsulfenylation.[44c]

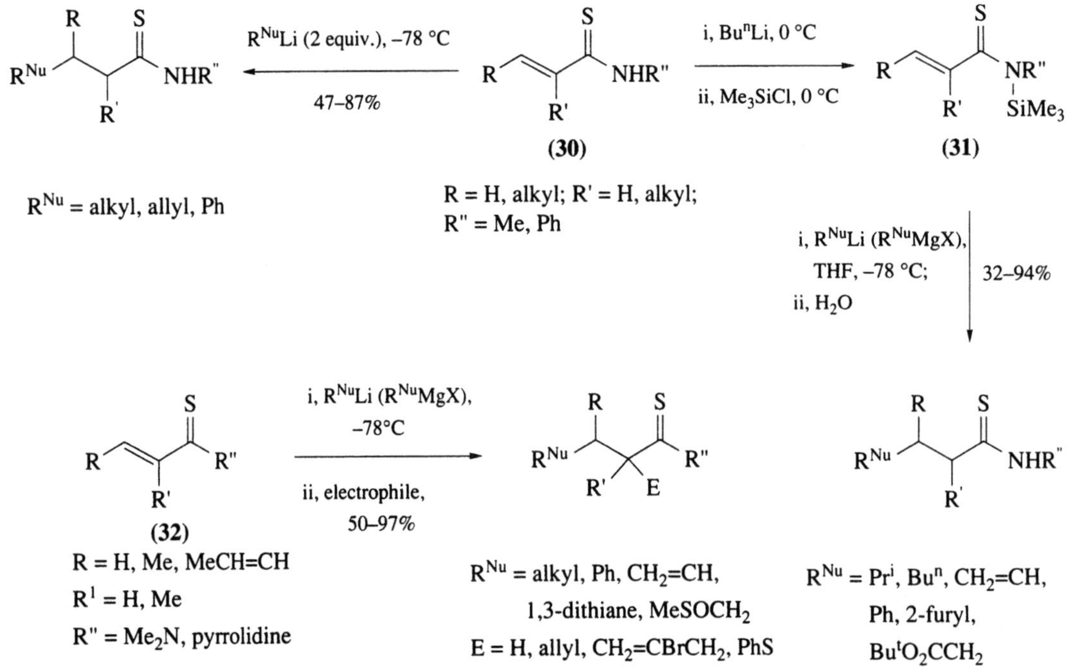

R
S
RNu
NHR"
R'
RNuLi (2 equiv.), −78 °C
47–87%
S
R
NHR"
R'
(30)
i, BunLi, 0 °C
ii, Me3SiCl, 0 °C
S
R
NR"
R'
SiMe3
(31)
RNu = alkyl, allyl, Ph
R = H, alkyl; R' = H, alkyl;
R" = Me, Ph
i, RNuLi (RNuMgX),
THF, −78 °C;
ii, H2O
32–94%
S
R
R"
R'
(32)
R = H, Me, MeCH=CH
R1 = H, Me
R" = Me2N, pyrrolidine
i, RNuLi (RNuMgX),
−78°C
ii, electrophile,
50–97%
R
S
RNu
R"
R'
E
RNu = alkyl, Ph, CH2=CH,
1,3-dithiane, MeSOCH2
E = H, allyl, CH2=CBrCH2, PhS
R
S
RNu
NHR"
R'
RNu = Pri, Bun, CH2=CH,
Ph, 2-furyl,
ButO2CCH2

Scheme 8

While α,β-unsaturated aldehydes react exclusively with organolithiums in the 1,2-mode, the corresponding imines or imidates, *e.g.* α,β-unsaturated azomethines, are excellent surrogates for '—CH═CH—CHO' [$^+CH_2CH_2CHO$ synthon] that undergo exclusive 1,4-additions. For example, the 2-styryldihydro-1,3-oxazines (**33**; equation 7),[45] 2-vinylbenzothiazoles (**34**; equation 8),[46] 2-alkenyl-2-imidazolines (**35**; equation 9)[47] and α-(phenylthio)vinyloxazolines (**36**; equation 10)[48] afford addition products that can be converted to the corresponding aldehydes, ketones or carboxylic acids (or esters). Similarly, Meyers has reported that *N*-cyclohexylnaphthylimines (**37**) are transformed to the corresponding 1,2-dihydronaphthalene aldehydes, as shown in Scheme 9.[49]

Notable is the general enantioselective addition of organolithiums to chiral vinyloxazolines (**38**) in which the intermediate azaenolate (**39**) can be further alkylated with high diastereoselectivity (Scheme

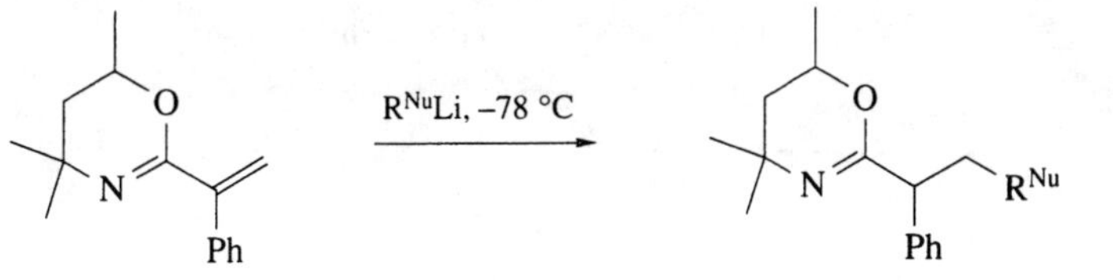

$R^{Nu}Li$, –78 °C (7)

(33) R^{Nu} = c-C_6H_{11}, Bu^s, Bu^i, Bu^t

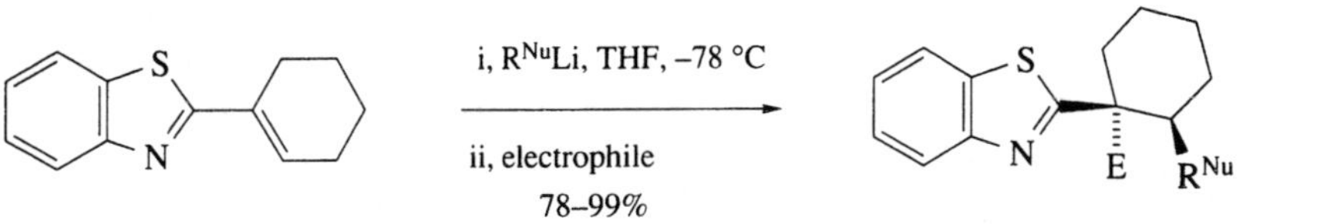

i, $R^{Nu}Li$, THF, –78 °C; ii, electrophile; 78–99% (8)

(34) E = H, allyl

R^{Nu} = Me, Bu^n, Bu^t, Ph, CH_2=$CHCH_2$, CH_2=CH, HC($NNMe_2$)CH_2

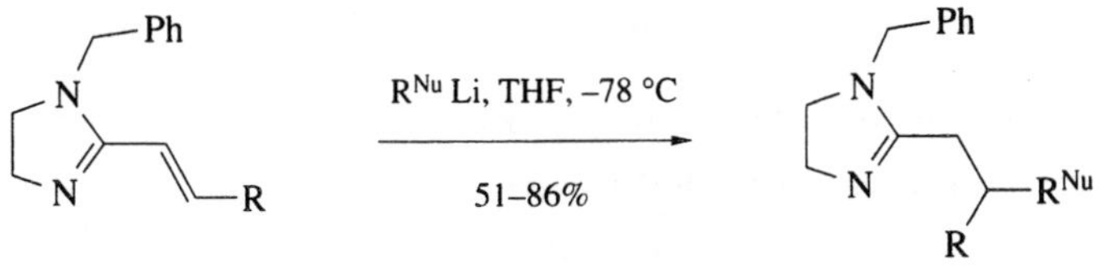

R^{Nu} Li, THF, –78 °C; 51–86% (9)

(35) R = Et, Pr^n, Ph; R^{Nu} = Bu^n, Ph

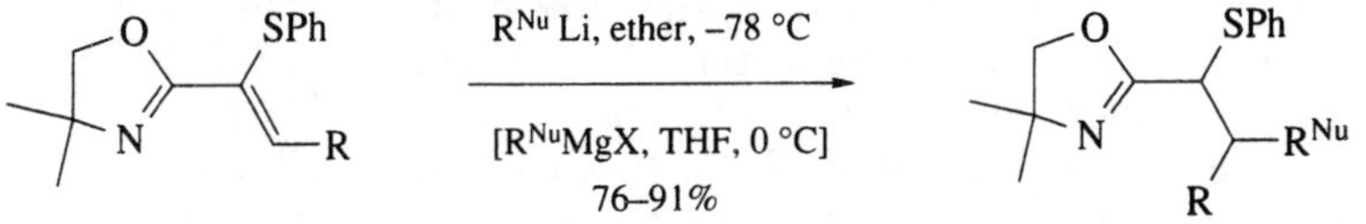

R^{Nu} Li, ether, –78 °C; [$R^{Nu}MgX$, THF, 0 °C]; 76–91% (10)

(36) R = Me, Ph; R^{Nu} = alkyl, Ph, RC≡C

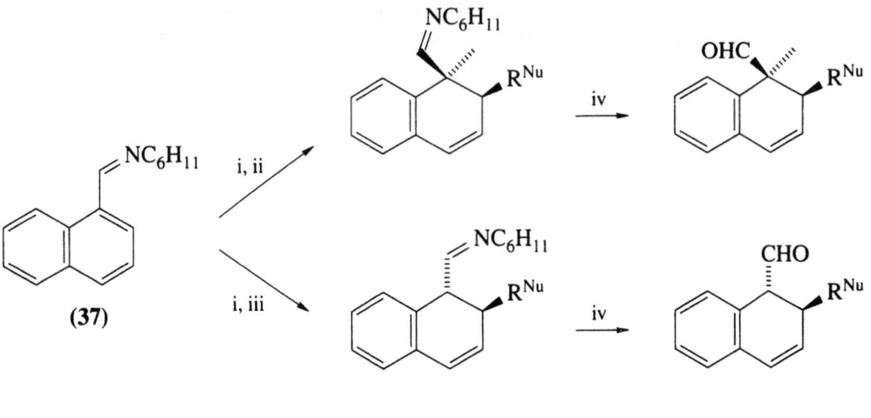

(37)

R^{Nu} = alkyl, benzyl, MeC=CH_2

i, R^{Nu} Li, –78 °C; ii, MeI; iii, Pr^nOH; iv, H_3O^+

Scheme 9

10).[50] The initial selectivity is attributed to coordination of the organolithium to the methyl ether substituent followed by delivery of the ligand to the *syn* face of the alkene.

The low temperature (–110 to –78 °C)[51] addition of organolithiums to β-substituted-α-nitroalkenes (**40**) affords intermediate nitronate anions which can be quenched with either a proton source or tetra-

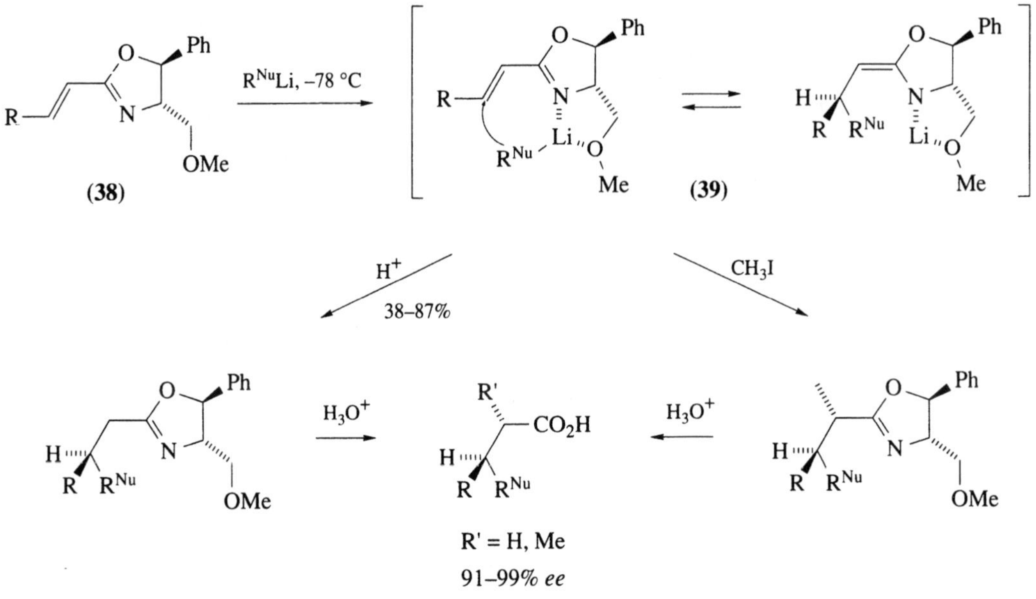

(38) $R^{Nu}Li$, –78 °C **(39)**

H^+ 38–87% CH_3I

H_3O^+ H_3O^+

R' = H, Me

91–99% *ee*

R = alkyl (C_1–C_3), Pr^i, Bu^t, c-C_6H_{11}, Ph, 2-$MeOC_6H_4$, $MeOCH_2CH_2$

R^{Nu} = alkyl (C_2–C_6), Ph, 4-MeC_6H_4

Scheme 10

nitromethane to afford nitroalkanes[52a] or geminal dinitroalkanes,[52b] respectively (Scheme 11). This process is the cornerstone to the synthesis of (+)-lycoricidine in which the addition of aryllithium (**41**) (–100 °C) to α-nitroalkene (**42**) affords both the *ido-* (**43**) and *gluco*-configured (**44**) addends (equation 11).[52c]

Seebach has shown that the nitroallylating reagent, 2-nitroallyl pivalate (**45**, NPP) can be used for triply convergent Michael coupling of dissimilar organometallic reagents.[53] As shown in Scheme 12, various organometallics add efficiently to reagent (**45**) to afford new α-nitroalkenes.

Conjugate additions of organolithiums to cycloalkenyl sulfones have become a pivotal point in various synthetic strategies.[54] Due to the sulfone activation, competing β- (primarily) and γ-deprotonation occurs; however, γ-deprotonation is suppressed by a γ-heteroatom (N or O) substituent, while β-deprotonation is retarded by the use of either *t*-butyl (preferably) or phenyl sulfones. The γ-heteroatom substituent also influences the addition of the ligand to the β-position so that either of two orientations are obtained by the expedient use of protecting groups.[54c] Typically the sulfonyl activating group is re-

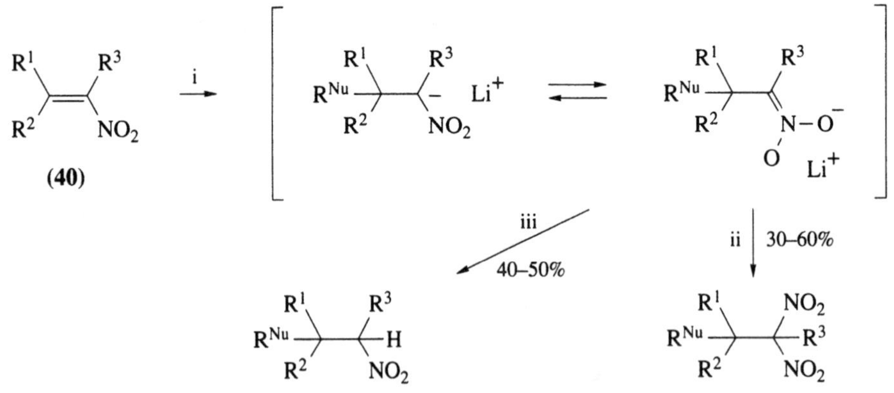

(40) i Li^+ Li^+

iii 40–50% ii 30–60%

R^1 = Ph; R^2 = H; R^3 = Me

R^{Nu} = alkyl

R^1 = Ph; R^1 = R^3 = H

R^{Nu} = alkyl, allyl, 1,3-dithiane, enolates

i, $R^{Nu}Li$, THF, –100 to –78 °C; ii, $C(NO_2)_4$; iii, H_3O^+

Scheme 11

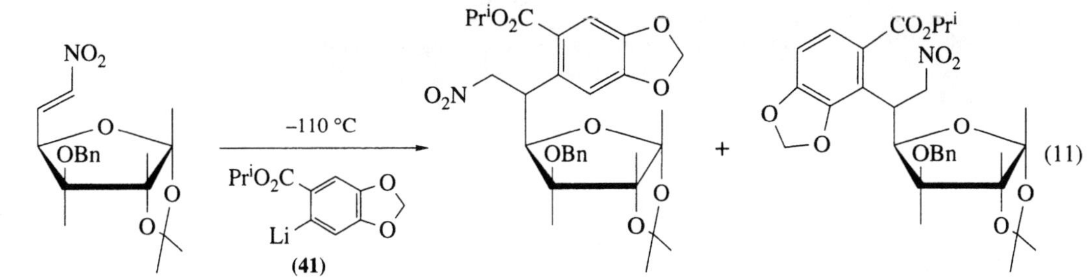

(42) → (43) + (44) (11)

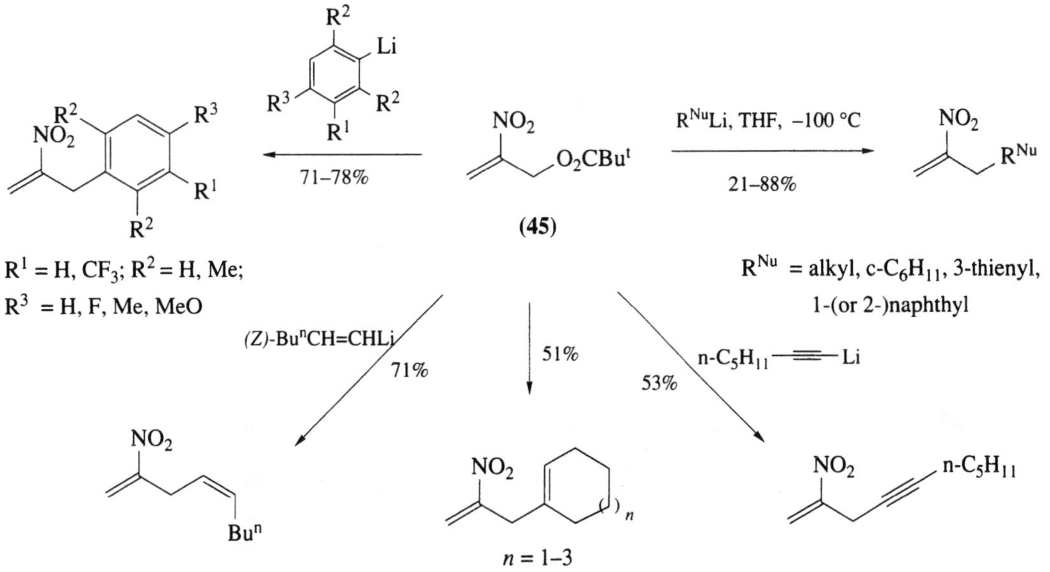

Scheme 12

ductively removed in a subsequent step; thus, the additions are the equivalent of adding to an unactivated alkene. The elaboration of cycloalkenyl sulfones, by an intermolecular tandem 1,4-conjugate addition–intramolecular (or intermolecular) electrophile trapping protocol, results in the formation of three (or four) contiguous asymmetric centers. Examples are shown in Scheme 13 for the conversion of sulfones (**46** and **47**) to (–)-PGE_2[55a] and (+)-carbacyclin (MIRC protocol).[55b] (±)-Morphine was synthesized by a tandem intramolecular 1,4-conjugate addition–intramolecular electrophile trapping protocol on the cyclohexenyl sulfone (**48**; Scheme 14).[56]

Eisch and Isobe have reported that open-chain vinyl sulfones (*e.g.* **49**) undergo exclusive α-deprotonation with organolithiums;[57] however, with α-(trimethylsilyl)vinyl phenyl sulfones (**50**), 1,4-conjugate addition occurs exclusively (Scheme 15). Desilylation is readily accomplished with anhydrous potassium fluoride. In contrast, the diastereoselective (>99% *threo*) addition of alkyllithiums to γ-hydroxy-(*Z*)-α-(trimethylsilyl)vinyl phenyl sulfones (**51**) occurs with concomitant desilylation, while the corresponding (*E*)-isomer (**52**) affords exclusive 1,4-silyl migration with no conjugate addition.[58a] Both (*E* and *Z*) γ-MEM ether isomers (**53** and **54**) afford *threo* (>99%) addition products (Scheme 16).[58b] The high diastereoselectivity is attributed to initial coordination of the organolithium to the γ-oxygen substituent (**55** and **56**) followed by delivery of the ligand to the *syn* face of the alkene; however, with (**52**), the intermediate alkoxide (**57**) undergoes 1,4-silyl transfer. As shown in Scheme 17, numerous organolithium reagents add to α-(trimethylsilyl)vinyl phenyl sulfones and the sequence of addition and trapping may be reversed to afford opposite diastereomers.[59a,b] Similarly, asymmetric induction is attained with organolithium additions to the chiral hemithioacetal α-(trimethylsilyl)vinyl phenyl sulfone (**58**) in the presence of lithium bromide (Scheme 18).[59c]

In spite of the high reactivity of the organolithiums, other compatible acceptors have also shown synthetic utility. For example, select organolithium reagents add exclusively to 2-(*N*-methylanilino)acrylonitrile (**59**), an α-carbonyl cation equivalent, to afford an intermediate α-aminonitrile anion (Section

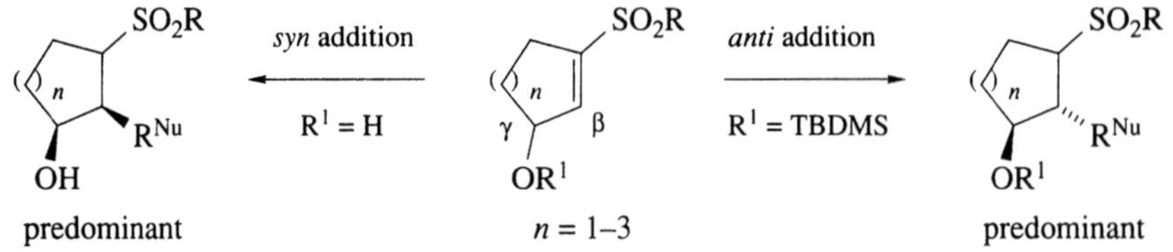

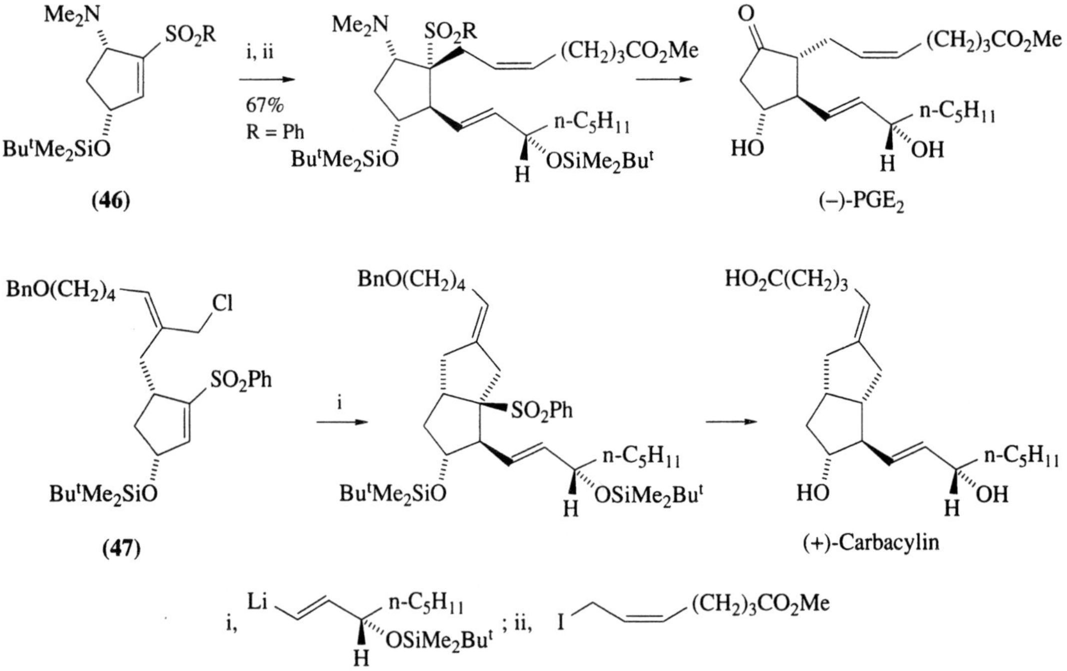

(46) (−)-PGE$_2$

(47) (+)-Carbacylin

Scheme 13

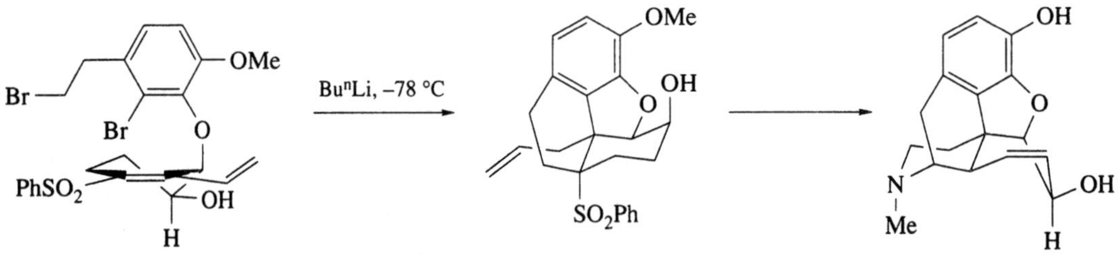

(48) (±)-Morphine

Scheme 14

1.2.2.3.1), which is quenched with either a proton source or further alkylated. Hydrolysis of the resultant α-aminonitrile (**60**) affords aldehydes or ketones (Scheme 19).[60]

Recently, η^6-arenechromium complexes and α,β-unsaturated iron acyls have shown utility as acceptors for organolithiums; in both examples, the role of the organometallic ligand is to stabilize the inci-

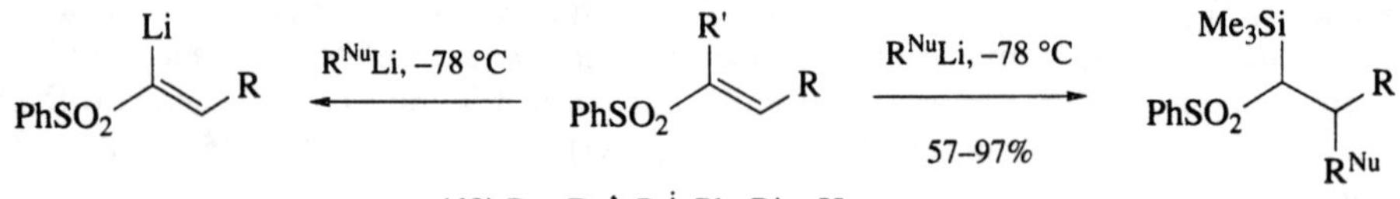

(**49**) R = Bu^t, Pr^i, Ph; R' = H
(**50**) R = Bu^t, Pr^i, Ph; R' = Me_3Si

Scheme 15

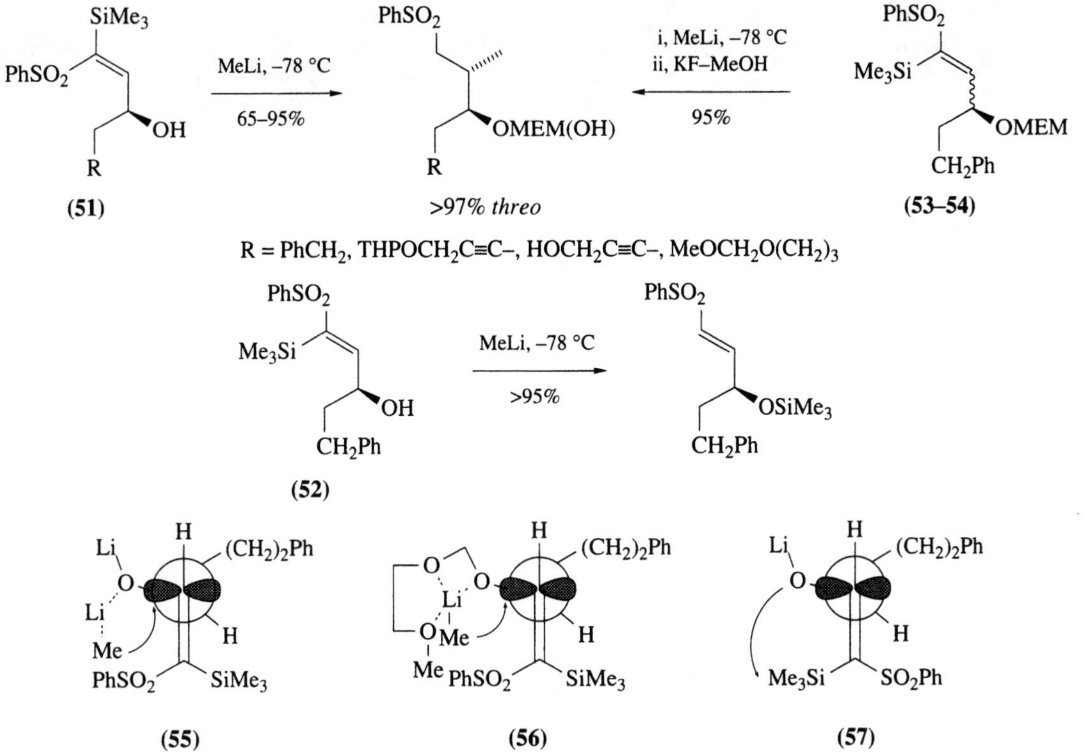

SiMe3
PhSO2
OH
R
MeLi, –78 °C
65–95%
PhSO2
OMEM(OH)
R
>97% threo
i, MeLi, –78 °C
ii, KF–MeOH
95%
PhSO2
Me3Si
OMEM
CH2Ph
(51)
(53–54)
R = PhCH2, THPOCH2C≡C–, HOCH2C≡C–, MeOCH2O(CH2)3
PhSO2
Me3Si
OH
CH2Ph
(52)
MeLi, –78 °C
>95%
PhSO2
OSiMe3
CH2Ph
(55)
(56)
(57)

Scheme 16

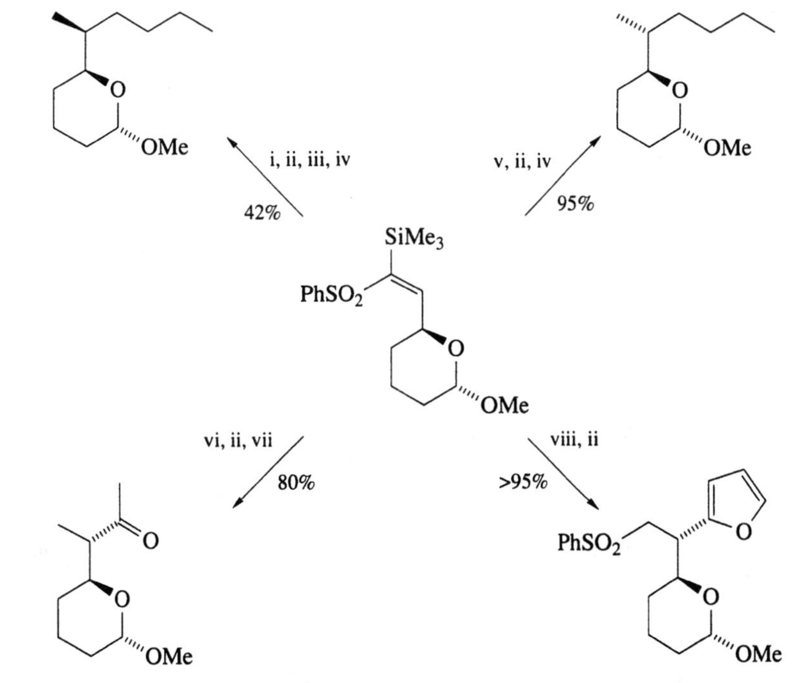

OMe
i, ii, iii, iv
42%
v, ii, iv
95%
SiMe3
PhSO2
O
OMe
vi, ii, vii
80%
viii, ii
>95%
PhSO2

i, MeLi, –78 °C; ii, KF–MeOH, 58 °C; iii, Bu^nLi, Pr^nBr, –78 °C; iv, Na–Hg;
v, Bu^nLi, –78 °C; vi, CH_2=C(Li)-OEt, –78 °C; vii, H_3O^+; viii, 2-furyllithium, –78 °C

Scheme 17

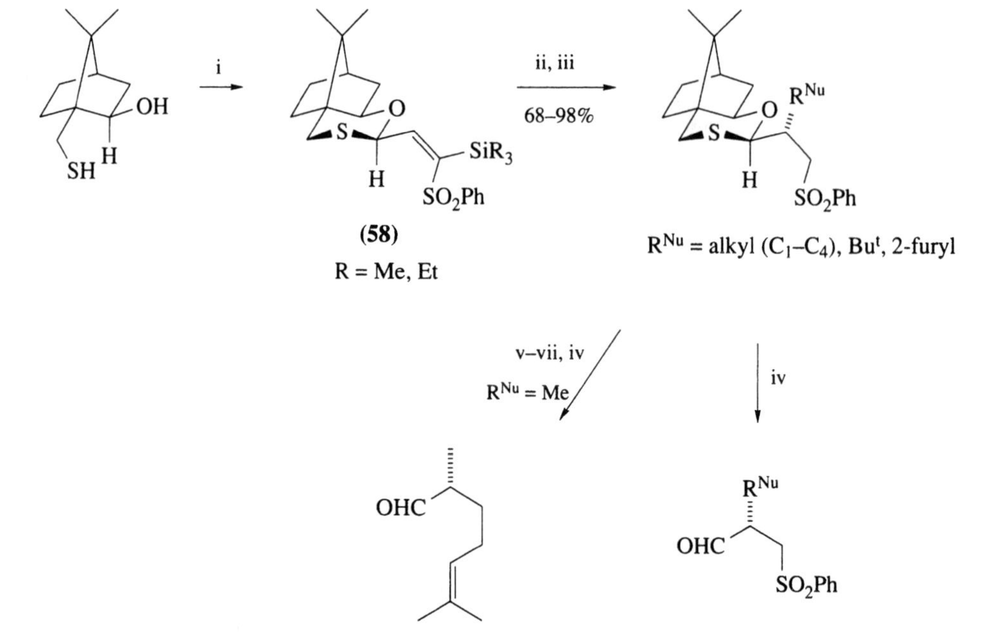

(58)

R = Me, Et

R^{Nu} = alkyl (C_1–C_4), Bu^t, 2-furyl

v–vii, iv R^{Nu} = Me

iv

i, [enal–SiR_3, SO_2Ph] ; ii, $R^{Nu}Li$, LiBr–Et_2O, n-C_6H_{14}, –78 °C;

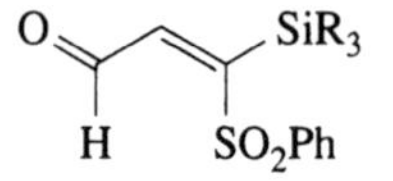

iii, $Bu^n{}_4NF$; iv, $HgCl_2$; v, Bu^nLi, –78 °C; vi, prenyl bromide; vii, Na–Hg

Scheme 18

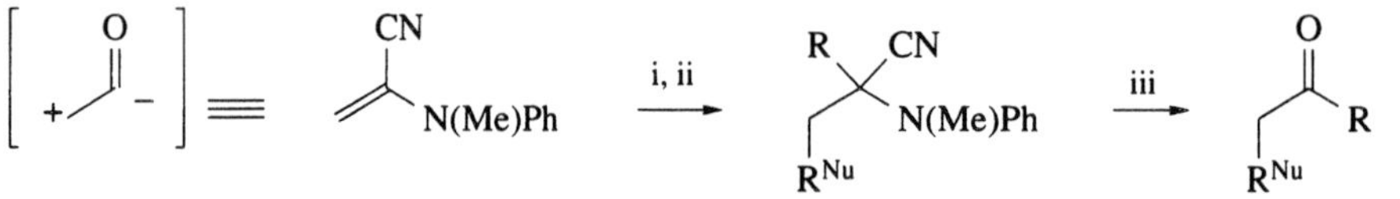

(59)

(60)

R^{Nu} = Bu^t, Ph

R = alkyl (C_1–C_3), benzyl, H_2O

i, $R^{Nu}Li$, THF, –78 °C; ii, RX (or H_2O); iii, 3N HCl

Scheme 19

pient benzylic carbanion or enolates. Knox and Semmelhack have shown that the regiospecific β-alkylation (arylation) of isolated styrene units is readily accomplished with complexes (**61**; Scheme 20).[61] Davies and Liebeskind have independently reported that the addition of organolithiums (–78 °C, THF) to chiral (*E*)-α,β-unsaturated iron acyls (**62** and **63**) and alkylation of the resultant enolate occurs with high diastereoselectivity (Scheme 21).[62a–f] The selectivity is attributed to a preferred *anti* conformation of the carbon monoxide ligand and acyl carbonyl and steric shielding of one face of the alkene by the triphenylphosphine ligand.[62g]

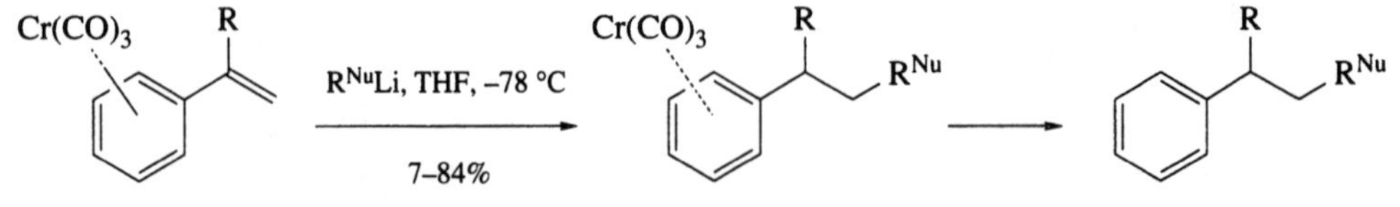

$R^{Nu}Li$, THF, –78 °C

7–84%

(61)

R = H, Me, MeCO, SEt

R^{Nu} = Me, Ph, Bu^n, 1,3-dithiane, Me_2CCN, $Me_2CCO_2Bu^t$

Scheme 20

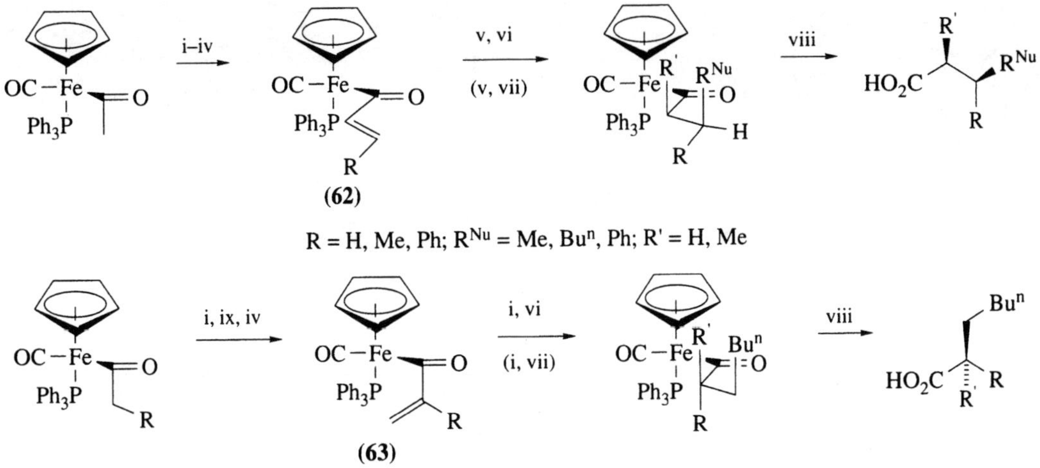

(62)

R = H, Me, Ph; R^{Nu} = Me, Bu^n, Ph; R' = H, Me

(63)

i, Bu^nLi, THF, –78 °C; ii, RCHO, –40 °C; iii, NaH, MeI, THF, 50 °C; iv, NaH, THF; v, $R^{Nu}Li$, THF, –78 °C; vi, R'X; vii, MeOH; viii, Br_2, H_2O; ix, $ClCH_2OR''$ (R" = Me, menthyl), –40 °C

Scheme 21

1.2.2.1.2 Organomagnesiums

The Grignard reagents have shown prevalent usage in carbon–carbon bond formation due to their ease of formation,[63] but some Grignard reagents are reported to be inaccessible because of synthetic complications, namely coupling. However, with newer methods for activation of magnesium, numerous sensitive Grignard reagents can be prepared quantitatively.[64] In addition, the greater tendency of Grignard reagents *vs.* organolithiums to form cyclic complexes with suitably juxtaposed unshared electron pairs on either nitrogen or oxygen, on a chiral auxiliary, is advantageous for diastereoselective additions.

The addition of Grignard reagents to α,β-unsaturated ketones generally affords 1,2-addition products, but examples of exclusive 1,4-additions to α,β-enones, with unique structural characteristics, are precedented. For example, the addition of *m*-methoxybenzylmagnesium chloride to α-methylene ketones (**64** and **65**) and subsequent acetylation affords the enol acetates (**66**) and (**67**) exclusively (Scheme 22);[65a–c] the absence of 1,2-addition products is due to severe 1,3-diaxial interactions of the incoming ligand with the axial juncture methyl.[65d] Similarly, as shown in Scheme 23, the anomalous 1,4-addition of Grignard reagents to sterically hindered aromatic ketones (**68**) affords *ortho*-alkylated ketones.[66]

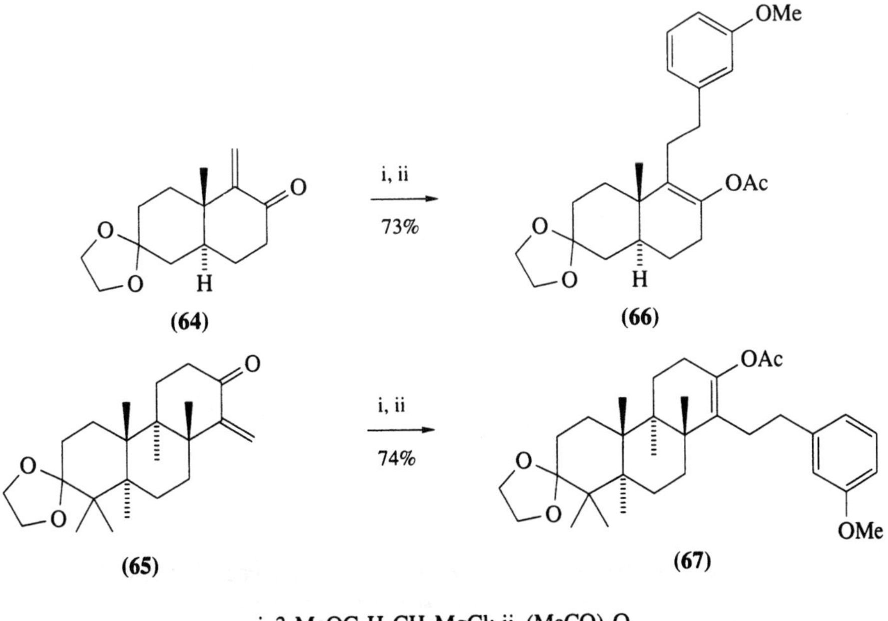

i, ii 73%

(64) **(66)**

i, ii 74%

(65) **(67)**

i, $3\text{-}MeOC_6H_4CH_2MgCl$; ii, $(MeCO)_2O$

Scheme 22

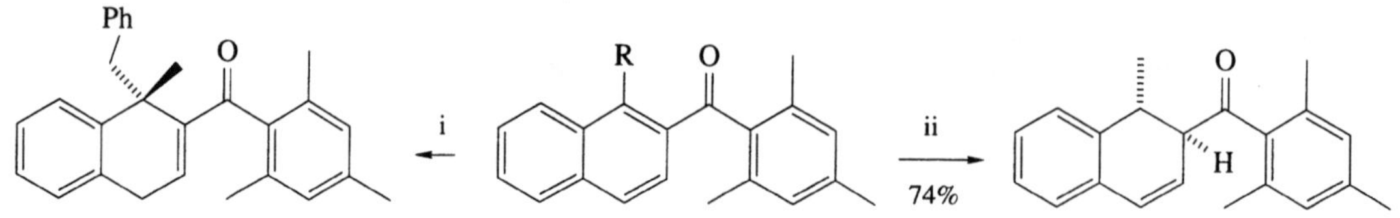

Ph

O R O O

i ii

74%

H

(68) R = H or Me

i, $PhCH_2MgCl$, Et_2O, 38 °C; ii, MeMgBr, Et_2O, 38 °C

Scheme 23

The 1,4-conjugate addition of Grignard reagents to *N,N*-disubstituted cinnamamides (**69**) was reported by Kohler in 1905;[67] subsequently numerous examples of Grignard conjugate additions to *N,N*-dialkylcrotonamides (**70**; Scheme 24),[68a] *N,N*-dialkylsorbamides (**71**)[68b] and higher analogs were reported. Notable is the exclusive 1,4-addition of Grignard reagents to amide (**71**), while organocuprates add exclusively in a 1,6-addition mode.[68b] The addition of allylic Grignard reagents affords allylic transposition adducts (**72**) exclusively, while both allenic and isomeric propargylic Grignard reagents afford predominantly allenic adducts (**73**; Scheme 25).[68c,d] In contrast, Grignard reagents add to *trans*-β-aroyl-*N*-alkylpropionamides (**75**) to afford exclusively β-aroyl-α-alkyl-*N*-alkylpropionamides (**76**; equation 12).[69]

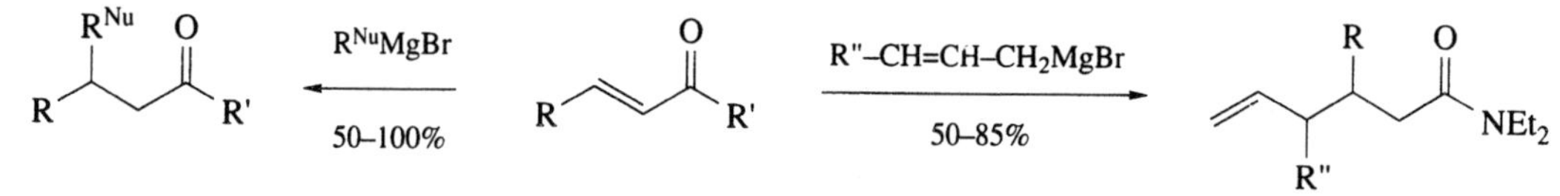

R^{Nu} O $R^{Nu}MgBr$ O R''–CH=CH–CH_2MgBr R O

R R' 50–100% R R' 50–85% NEt_2

R''

R' = Me_2N, Et_2N, piperidino, $(c\text{-}C_6H_{11})_2N$

R^{Nu} = Et, Pr^n, Ph

(69) R = Ph

(70) R = Me (Et)

(71) R = MeCH=CH

(72) R'' = H, Me, Et

Scheme 24

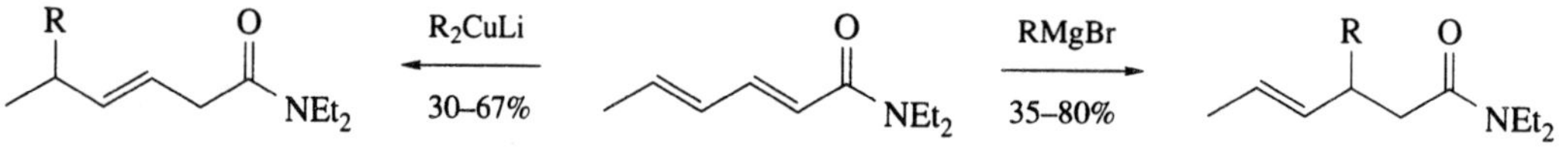

R O R_2CuLi O RMgBr R O

NEt_2 30–67% NEt_2 35–80% NEt_2

R = alkyl (C_3-C_7), CH_2=C(Me)–CH_2, MeCH–CH=CH_2, EtCH–CH=CH_2

(71)

R = alkyl (C_3-C_7), MeCH–CH=CH_2, EtCH–CH=CH_2

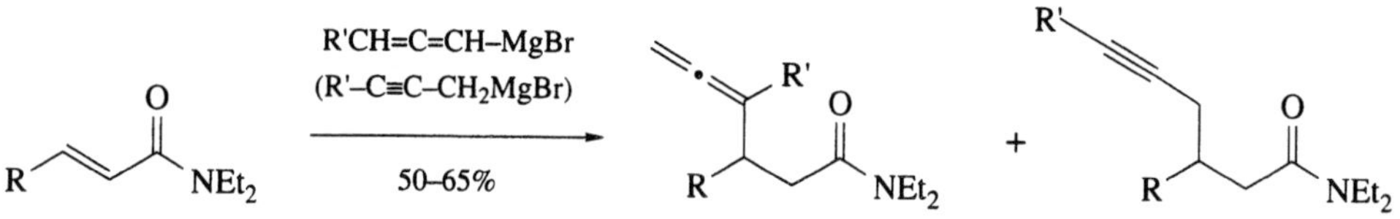

R'CH=C=CH–MgBr

(R'–C≡C–CH_2MgBr)

O R' O R' O

R NEt_2 50–65% R NEt_2 + R NEt_2

(70) R = Me, Et

(73) 81–90%

(74) 10–19%

Scheme 25

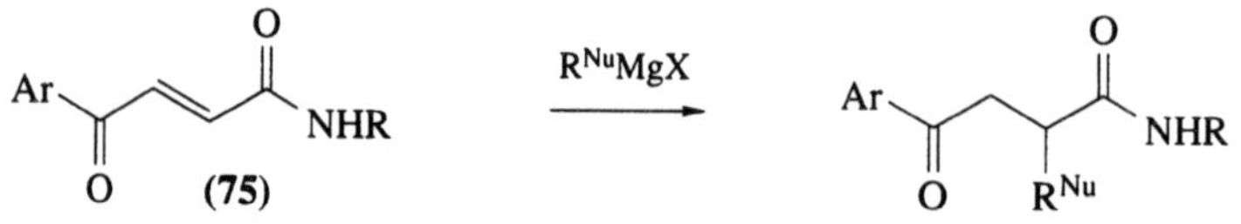

O $R^{Nu}MgX$ O

Ar NHR Ar NHR (12)

O **(75)** O R^{Nu}

Ar = Ph, 4-MeC_6H_4, 4-ClC_6H_4

R = Bu^n, Bu^s, Bu^t, benzyl

(76) R^{Nu} = Me, Et, Ph

Variable levels of asymmetric induction or diastereoselectivity have been found with additions of organometallics to α,β-unsaturated chiral amides (chiral auxiliary). For example, as shown in Scheme 26, Mukaiyama reports that the diastereoselective addition of Grignard reagents to β-substituted α,β-unsaturated *N*-methyl-(–)-ephedrine amides (**77**) occurs in moderate yields with >85% *ee*, while the use of either organolithiums or strong coordinating solvents negates the intramolecular ligand effects.[70] Similarly, Soai reports that Grignard reagents add to α,β-unsaturated (*S*)-prolinol amides (**78**; R′ = H) or α,β-unsaturated (*S*)-2-(1-hydroxy-1-methylethyl)pyrrolidine amides (**78**; R′ = Me) with 50–89% *ee* in the presence of a tertiary amine, *e.g.* DBN, DBU, (–)-sparteine or TMEDA.[71] The selectivity is attributed to a combination of the intrinsic aggregative properties and chelation of the Grignard reagent with the carbinol and carbonyl functionality. Similarly, Oppolzer reports that alkylmagnesium chlorides add to *N*-enoylsultams (**79** and **80**) with high diastereoselectivity (Scheme 27) which is attributed to initial chelation of the alkylmagnesium chloride to both the sultam and carbonyl groups followed by facial selective delivery of a second chelated alkylmagnesium chloride to the β-alkenic carbon (**81**).[72]

The addition of Grignard reagents to α,β-unsaturated *N,N*-dialkylthioamides (**32**) affords an intermediate (*Z*)-enethiolate which is a soft anion and resistant to further organometallic attack. Thus, in a tandem conjugate addition–electrophile trapping protocol with aldehydes, the enethiolate affords primarily *threo* aldols (**82**). The 1,4-conjugate addition of Grignard reagents to *N,N*-dimethyl-α-methacrylothioamide (**83**) affords the intermediate (*Z*)-enethiolate which undergoes aldol condensation to afford either the corresponding *erythro* aldols (**84**) or *threo* aldols (**85**) depending on the steric bulk of the aldehyde (Scheme 28).[73]

The α,β-unsaturated *N*-phenylthioimidates (**86**) have also shown utility as acceptors for nonallylic Grignard reagents (–5 to –10 °C) and alkyllithiums (–78 °C) while allylic Grignard reagents add exclusively in a 1,2-mode (equation 13). These adducts (**87**) can be converted quantitatively to the corresponding dithioesters with hydrogen sulfide. In contrast, α,β-unsaturated dithioesters (**88**) react with allylic Grignards in a 1,4-mode exclusively with allylic transposition (equation 14).[74]

Koga reports that α,β-unsaturated aldimines (**89**), derived from L-*t*-leucine *t*-butyl esters, undergo asymmetric additions with Grignard reagents which is due to formation of a highly ordered magnesium chelate (Scheme 29).[75]

The addition of Grignard reagents to α-nitroalkenes occurs in moderate to excellent yields with byproducts, from overreaction, being hydroxylamines and oximes.[76] In contrast to the optimal reaction temperatures of <–78 °C for organolithium additions, the optimal reaction temperature for Grignard reagent additions is <15 °C.[77a–c] The addition of phenylethynylmagnesium bromide to α-nitroalkenes (**40**) affords the intermediate magnesium nitronates (**90**) which can be hydrolyzed or halogenated to afford γ-alkynic nitroalkanes (**91**) or γ-alkynic α-bromonitroalkanes (**92**), respectively (Scheme 30).[77d] Vinylmagnesium bromide adds to the α-nitroalkenes (**93**) to afford exclusively the L-*ido*-furanosides (**94**; equation 15) (*cf.* **42** → **43** and **44**).[77e]

In contrast to the α-nitroalkenes, Grignard reagents add to mononitroarenes in a bimodal manner,[78a] with alkyl Grignard reagents adding efficiently in 1,4- (predominant) and 1,6-addition modes to afford

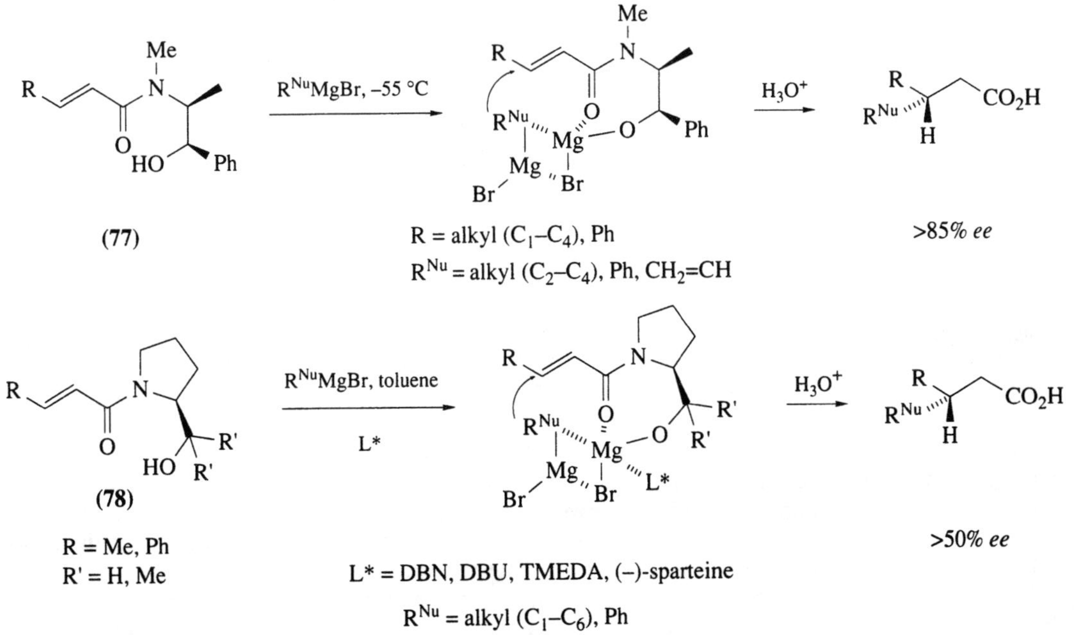

(**77**) $R^{Nu}MgBr$, –55 °C → H_3O^+ → >85% *ee*

R = alkyl (C_1–C_4), Ph

R^{Nu} = alkyl (C_2–C_4), Ph, CH_2=CH

(**78**) $R^{Nu}MgBr$, toluene, L* → H_3O^+ → >50% *ee*

R = Me, Ph

R' = H, Me

L* = DBN, DBU, TMEDA, (–)-sparteine

R^{Nu} = alkyl (C_1–C_6), Ph

Scheme 26

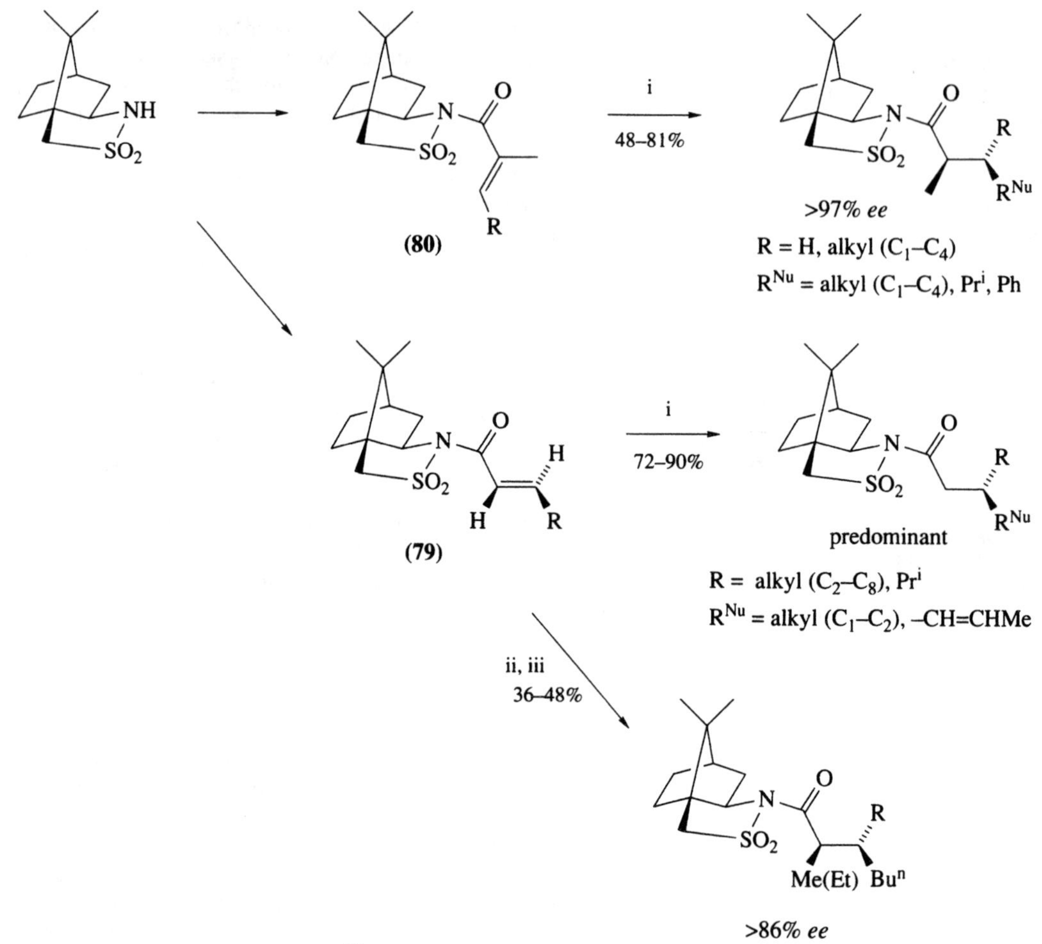

(80)

48–81%

>97% ee

R = H, alkyl (C_1–C_4)

R^{Nu} = alkyl (C_1–C_4), Pr^i, Ph

(79)

72–90%

predominant

R = alkyl (C_2–C_8), Pr^i

R^{Nu} = alkyl (C_1–C_2), –CH=CHMe

ii, iii 36–48%

Me(Et) Bu^n

>86% ee

i, R^{Nu}MgCl; ii, Bu^nMgCl; iii, MeI (EtI)

Scheme 27

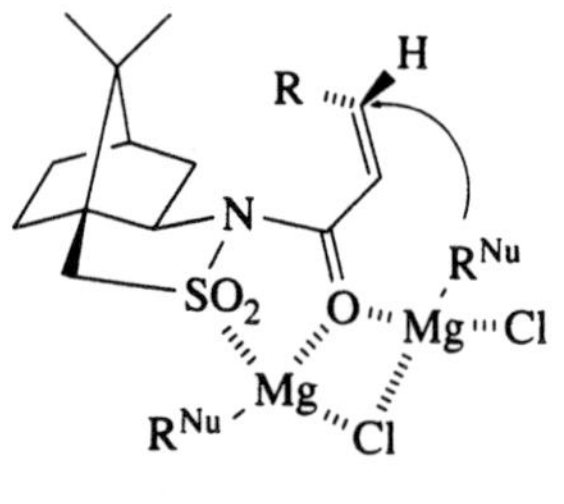

(81)

the intermediate nitronate salts,[78b,c] while aryl Grignard reagents add in a 1,2-addition mode to afford *N,N*-diarylamines and *N,N*-diarylhydroxylamines.[78d,e] The intermediate nitronate salts are subsequently oxidized with dichlorodicyano-1,4-benzoquinone (DDQ) or potassium permanganate,[79a] or treated with boron trifluoride to afford the *ortho*-alkylated nitroarenes or *ortho*-alkylated nitrosoarenes, respectively (Scheme 31).[79b]

Examples of Grignard reagent or other organometallic additions to α,β-unsaturated sulfoxides are rare, typically kinetic α-deprotonation occurs;[54b] however, the addition of methylmagnesium bromide to methylenecephem sulfoxide (**95**) affords the epimeric C-2 ethylcephems (**96**) in moderate yields (equation 16).[80] In contrast, Posner reports that Grignard reagents add to enantiomerically pure 2-(arylsulfinyl)-2-cycloalkenones with chirality transfer from the sulfoxide sulfur atom to the β-carbon; coupled with reductive removal of the sulfinyl group, this is equivalent to the enantioselective conjugate addition of Grignard reagents to α,β-enones.[81] The simple expedient of reacting (*S*)-(+)-(**97**) with an alkylmagnesium chloride or dialkylmagnesium, followed by reductive cleavage of the sulfinyl group affords the 3-

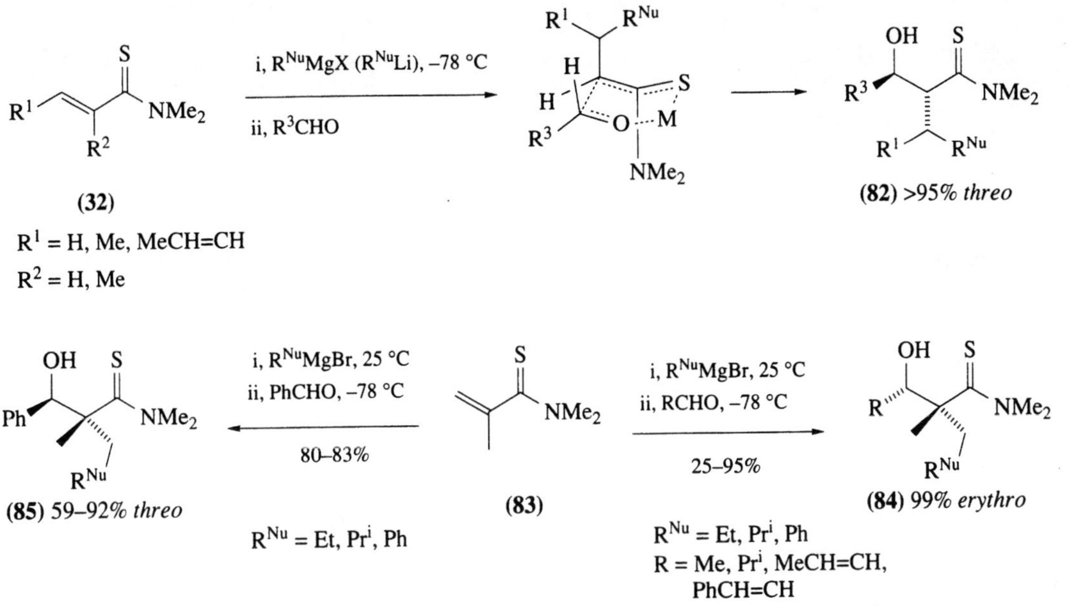

i, $R^{Nu}MgX$ ($R^{Nu}Li$), –78 °C
ii, R^3CHO

(32)

R^1 = H, Me, MeCH=CH
R^2 = H, Me

(82) >95% *threo*

i, $R^{Nu}MgBr$, 25 °C
ii, PhCHO, –78 °C
80–83%

(85) 59–92% *threo*

(83)

R^{Nu} = Et, Pr^i, Ph

i, $R^{Nu}MgBr$, 25 °C
ii, RCHO, –78 °C
25–95%

(84) 99% *erythro*

R^{Nu} = Et, Pr^i, Ph
R = Me, Pr^i, MeCH=CH, PhCH=CH

Scheme 28

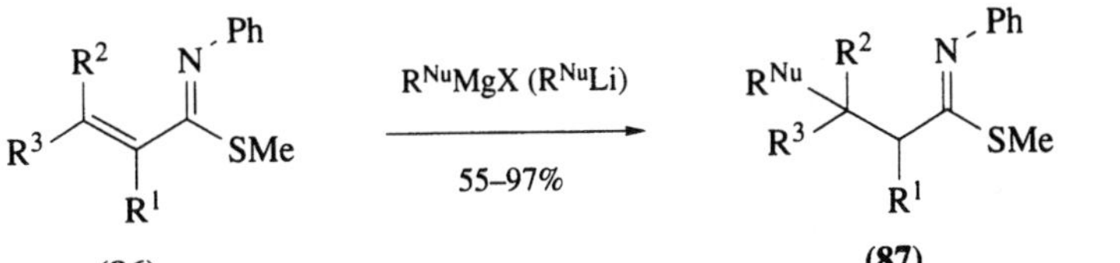

$R^{Nu}MgX$ ($R^{Nu}Li$)
55–97%

(86) → (87) (13)

R^1 = H, Me; R^2 = H, Me; R^3 = H, Me

R^{Nu} = alkyl (C_1–C_4), Bu^t, Bu^s, Pr^i, 4-MeC_6H_4, CH_2=CH

$R^1(R^2)C$=$C(R^3)CH_2MgX$
53–63%

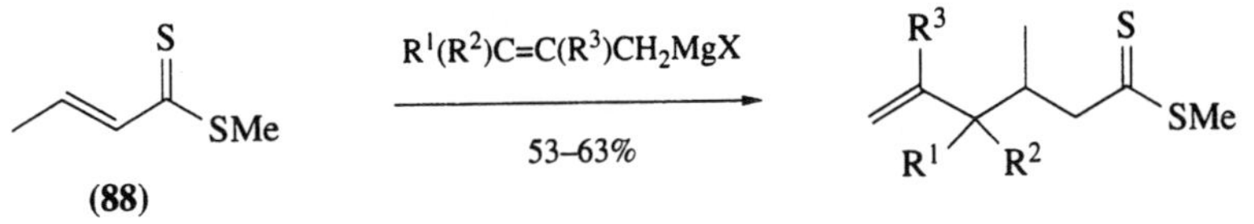

(88) (14)

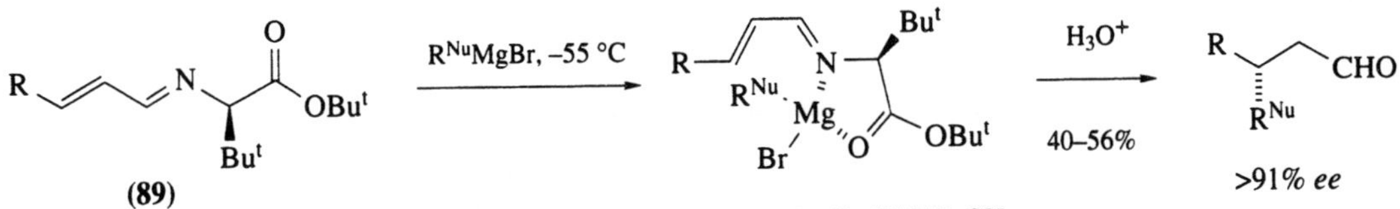

$R^{Nu}MgBr$, –55 °C

(89)

H_3O^+
40–56%

>91% *ee*

R^{Nu} = alkyl (C_2–C_4), c-C_6H_{11}, Ph, Me_2C=$CHCH_2CH_2$

Scheme 29

substituted cycloalkenones, (*R*)-(+)-(**99**) and (*S*)-(–)-(**100**), respectively. Similarly, 6-methoxy-2-naphthylmagnesium bromide reacts with (*S*)-(+)-(**97**) to afford the steroidal intermediate (**101**, >98% *ee*; Scheme 32).

The stereoselectivity of addition is a consequence of the intrinsic differences in chelation tendencies between various alkylmagnesium halides and dialkylmagnesium reagents and the preferred *anti*-conformation of the sulfoxide and carbonyl groups. Organometallic reagents with low Lewis acidity or low chelation tendency, *i.e.* less electrophilic (RMgI or R_2Mg), add to the more accessible face of the enone double bond (**97**). In contrast, higher Lewis acidity organometallics (RMgCl or RMgX–$ZnBr_2$) form an initial chelate (**98**) with both the sulfoxide and carbonyl groups, followed by addition to the double bond *anti* to the aryl group.[82]

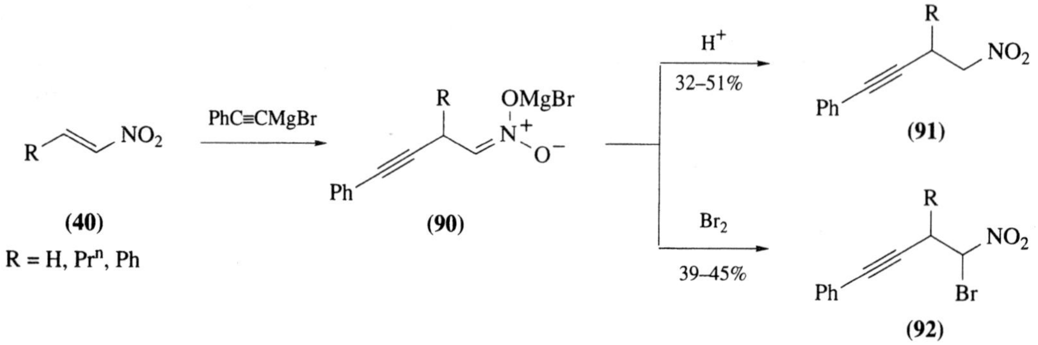

PhC≡CMgBr; H^+ 32–51%; Br_2 39–45%

(40) R = H, Pr^n, Ph (90) (91) (92)

Scheme 30

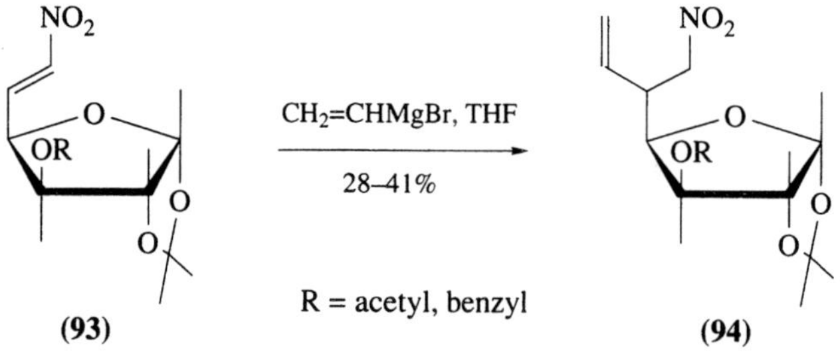

CH_2=CHMgBr, THF 28–41% (15)

R = acetyl, benzyl

(93) (94)

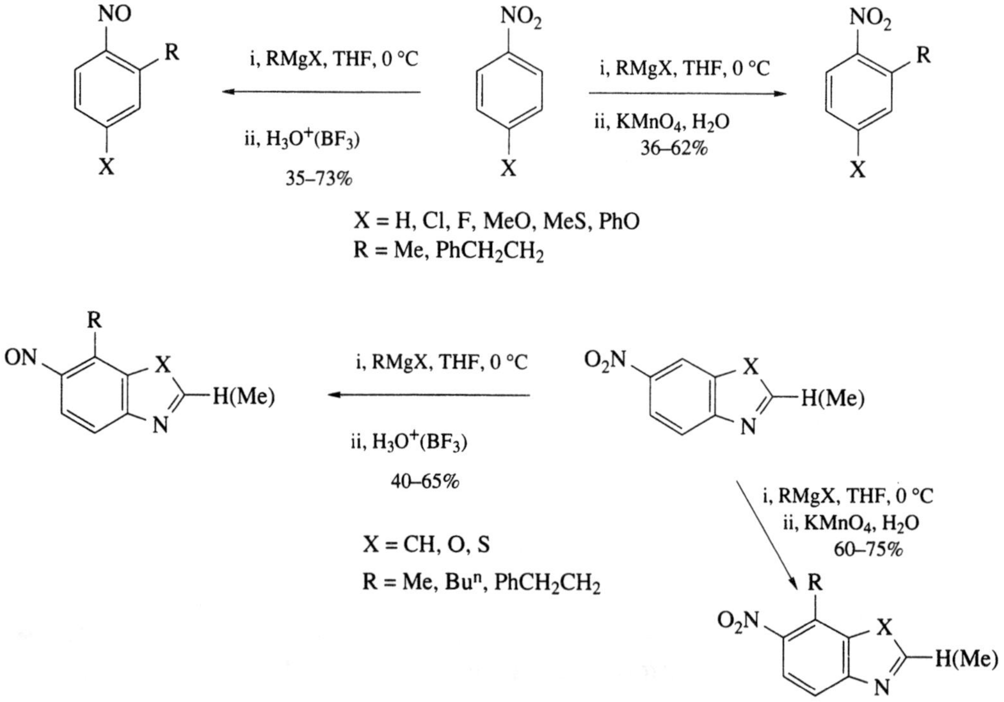

i, RMgX, THF, 0 °C; ii, $H_3O^+(BF_3)$ 35–73%

i, RMgX, THF, 0 °C; ii, $KMnO_4$, H_2O 36–62%

X = H, Cl, F, MeO, MeS, PhO
R = Me, $PhCH_2CH_2$

i, RMgX, THF, 0 °C; ii, $H_3O^+(BF_3)$ 40–65%

i, RMgX, THF, 0 °C; ii, $KMnO_4$, H_2O 60–75%

X = CH, O, S
R = Me, Bu^n, $PhCH_2CH_2$

Scheme 31

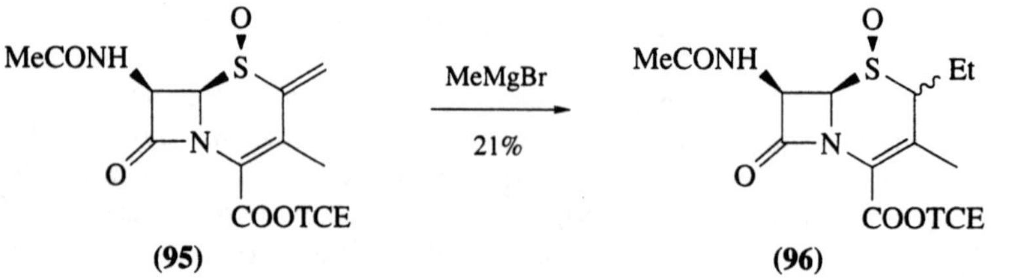

MeMgBr 21% (16)

(95) (96)

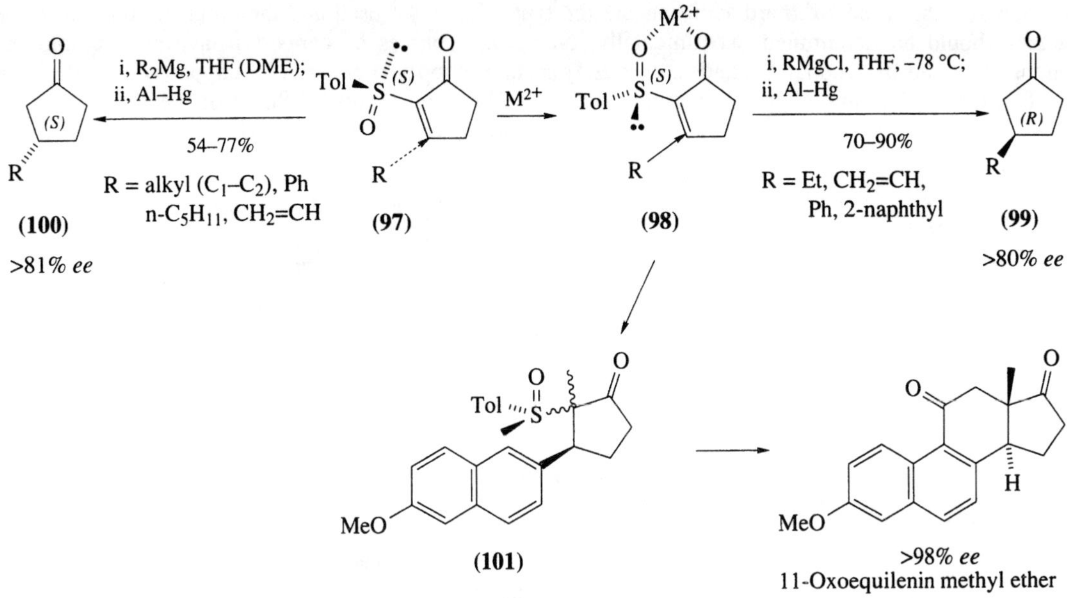

Scheme 32

Eisch has reported that alkyl Grignard reagents add to γ-hydroxy-α,β-unsaturated phenyl sulfones (**102a**) to afford γ-hydroxysulfones (**103**) (*cf.* **49**) while nonalkyl Grignard reagents add to γ-bromo-α,β-unsaturated phenyl sulfones (**102b**) to afford *trans*-(phenylsulfonyl)cyclopropanes (**104**; Scheme 33).[83]

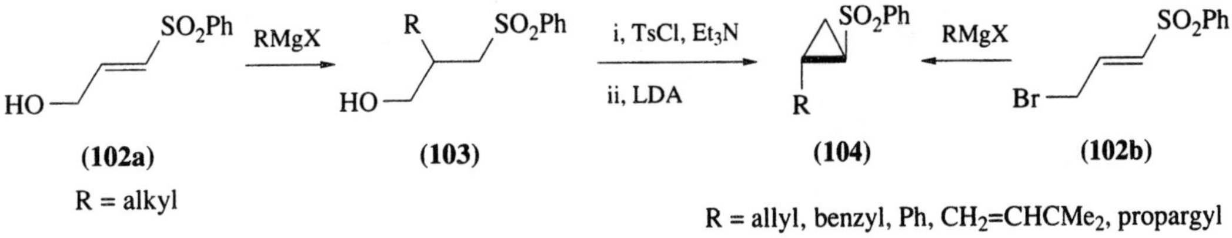

Scheme 33

The addition of Grignard reagents to 1,1-doubly activated alkenes, *e.g.* alkylidenemalonates (**105**),[68a,84a–d] alkylidenecyanoacetates (**106**),[84e,f] alkylideneisocyanatoacetates (**107**),[84g] alkylidenephosphonoacetates (**108**),[84h] isopropylidenemethylenemalonates (**109**)[84i,j] and their arylidene analogs, occurs in a 1,4-addition mode exclusively (Schemes 34 and 35). Similarly, as shown in Scheme 36, Mukaiyama reports that the addition of Grignard reagents to chiral oxazepines (**110**) and (**111**) affords an enantioselective synthesis of β-substituted carboxylic acids.[85]

Few comparative studies of the 1,4-addition capabilities of organomagnesium halides *vs.* diorganomagnesiums are available, although substrate and ligand variability is observed. Ethylmagnesium bromide shows a tendency to afford 1,4-addition products with 2-cyclohexen-1-one and *trans*-3-pentenone while diethylmagnesium affords exclusive 1,2-addition products.[86a] The addition of the homoenolate equivalent Grignard reagent (**112**) to alicyclic α,β-enones shows a temperature dependency in the ratios of 1,4-addition:1,2-addition, which varies from 1:1, at 0 °C, to 20:1, at –78 °C. The reversal of selectivity is attributed to low temperature alterations in the Schlenk equilibrium in which magnesium halides precipitate (or crystallize) from the reaction mixture, thus enriching the concentration of diorganomagnesium (**113**) in solution (Scheme 37).[86b]

1.2.2.1.3 Copper-catalyzed organomagnesium additions

The first example of a copper(I)-catalyzed Grignard conjugate addition was that described by Kharasch and Tawney;[4] subsequently this protocol was the cornerstone for various natural products syntheses, but it lacked predictability and dependability.[87] Experimental aspects that affect the efficiency of copper(I)-catalyzed Grignard additions are the type of catalyst used and the order of addition; these factors should be determined experimentally. Numerous sources of copper(I) have been employed, among them are the copper(I) halides, copper(I) acetate, copper(I) *t*-butoxide, copper(I) cyanide, copper(I) thiophenolate and copper(I) bromide–dialkyl sulfide complexes.[87d] Some of the conflicting experimental results from earlier studies were due to contamination with copper(II) species which oxidize

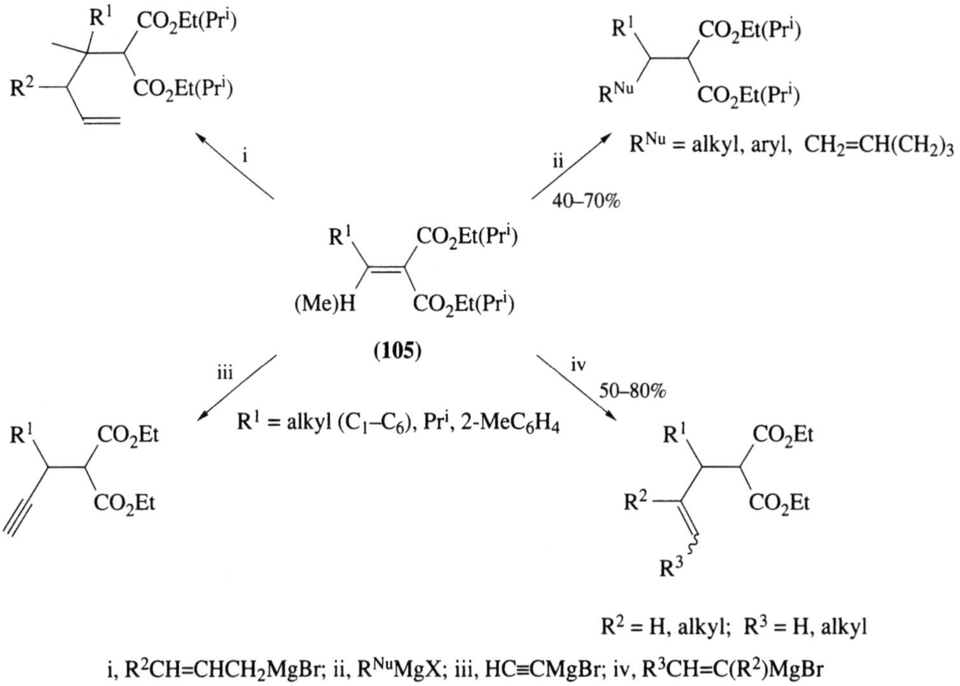

(105)

R^1 = alkyl (C_1–C_6), Pr^i, 2-MeC_6H_4

R^{Nu} = alkyl, aryl, CH_2=$CH(CH_2)_3$

ii, 40–70%; iv, 50–80%

R^2 = H, alkyl; R^3 = H, alkyl

i, R^2CH=$CHCH_2MgBr$; ii, $R^{Nu}MgX$; iii, HC≡CMgBr; iv, R^3CH=$C(R^2)MgBr$

Scheme 34

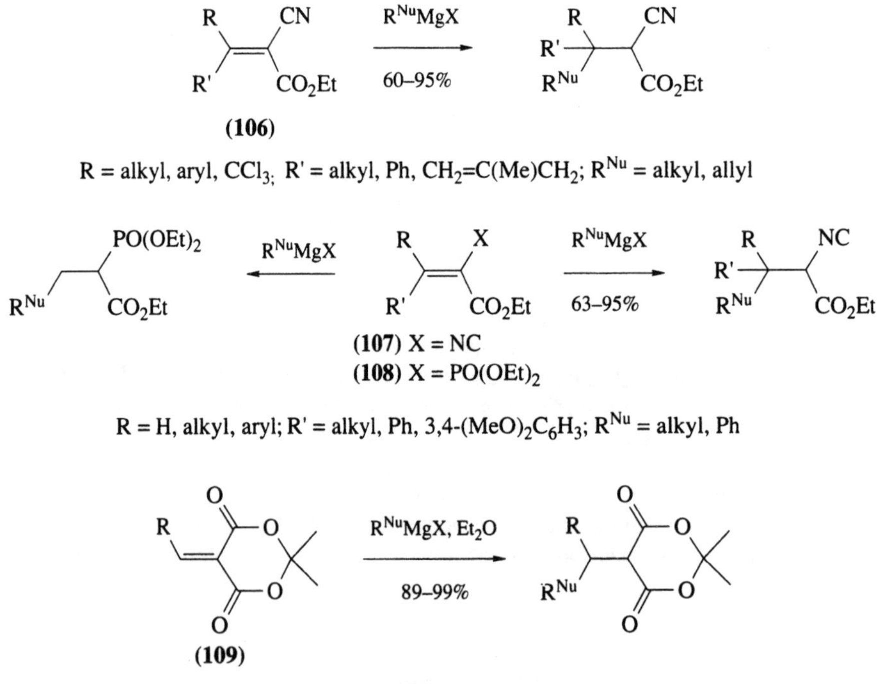

(106)

$R^{Nu}MgX$, 60–95%

R = alkyl, aryl, CCl_3; R' = alkyl, Ph, CH_2=$C(Me)CH_2$; R^{Nu} = alkyl, allyl

(107) X = NC
(108) X = $PO(OEt)_2$

$R^{Nu}MgX$, 63–95%

R = H, alkyl, aryl; R' = alkyl, Ph, 3,4-$(MeO)_2C_6H_3$; R^{Nu} = alkyl, Ph

(109)

$R^{Nu}MgX$, Et_2O, 89–99%

R = Ph, 2-thienyl, 1-naphthyl, (*E*)-PhCH=CH; R^{Nu} = alkyl, Ph

Scheme 35

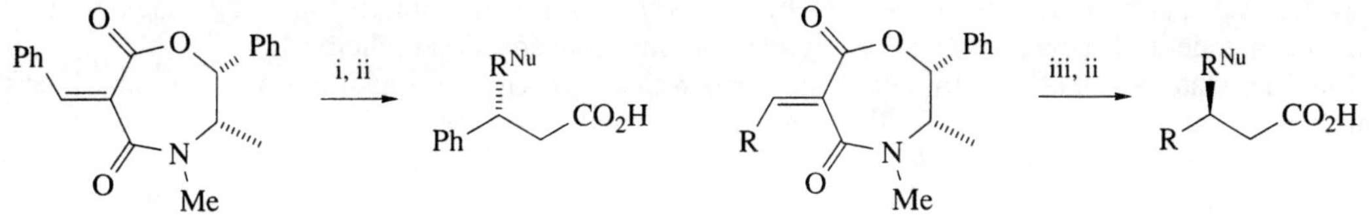

(Z)-(**110**) → (i, ii) → >99% *ee*

R^{Nu} = alkyl (C_1–C_4)

(E)-(**111**) → (iii, ii) → >58% *ee*

R = alkyl (C_1–C_4), Pr^i, Ph

R^{Nu} = alkyl (C_1–C_4), Pr^i, Ph

i, $R^{Nu}MgBr–NiCl_2$; ii, H_3O^+; iii, $R^{Nu}MgBr$

Scheme 36

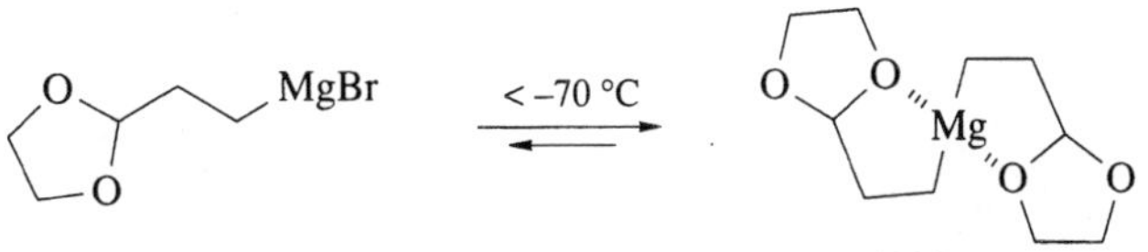

(**112**) ⇌ (< –70 °C) (**113**)

(**112**)–(**113**) (2 equiv.)

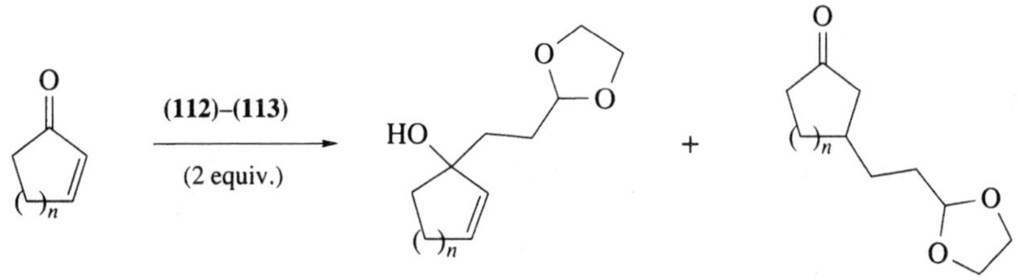

n	*Temp.* (°C)	*Yield* (%)	*1,2 vs. 1,4*
1	–78	73	5:95
1	25	69	78:22
2	–78	80	13:87
2	25	98	88:12

Scheme 37

the reactive intermediates, and the low solubility of the copper(I) salts which lead to heterogeneous reactions. Because of the advent of the stoichiometric organocopper reagents (Volume 4, Chapter 1.4), this method of C—C bond formation is rarely employed; however, under certain conditions the copper(I)-catalyzed Grignard additions offer advantages over the stoichiometric organocuprates. The stereoselectivity of copper-catalyzed Grignard additions is similar to that found with stoichiometric organocopper reagents. The examples highlighted have been arbitrarily chosen where the catalyst comprises <25 mol %.

Copper-catalyzed Grignard additions to α,β-enones, α,β-unsaturated esters and 1,1-doubly activated alkenes have been extensively studied. For example, catalyzed additions of alkyl, aryl and vinylic Grignard reagents to alkylidenecycloalkanones and cycloalkenones proceed in excellent yields (Scheme 38).[88a–e] As exemplified in Scheme 39, the conjugate addition of the homoenolate equivalent Grignard reagent (**112**) and bishomoenolate equivalent Grignard reagent (**114**) to cycloalkenones is the basis of general cyclopentenocycloalkanone and cyclohexenocycloalkanone syntheses.[88f] Similarly, the synthesis of a prototypical prostaglandin (**115**) was accomplished by a copper(I)-catalyzed conjugate addition of a Grignard reagent to an appropriate cyclopentenone.[88g] The presence of a proximal nitrogenous functionality on an α,β-enone acceptor is not detrimental to catalyzed Grignard additions, as shown in Scheme 40, by the use of basic alkaloid intermediates and β-enaminones.[89]

The catalyzed conjugate addition of Grignard reagents to α,β-enones affords, in principle, regiospecific enolates which can be intercepted with electrophiles under non-equilibrating, kinetically controlled conditions; however, few examples have been reported.[90] The conjugate addition of Grignard reagents to

2-cyclohexen-1-one and 3-methyl-2-cyclohexen-1-one affords the intermediate cyclohex-1-enolates which can undergo *C*-acylation,[90b] alkylation[90c,d] or aldol condensation (Scheme 41).

Ethyl acrylate reacts efficiently under the copper-catalyzed Grignard addition protocol, in which the combination of low temperatures (–20 → –50 °C) and catalytic copper(I) chloride (2.6%) was necessary to afford moderate to excellent yields of 1,4-addition products, while elaboration of complex α,β-unsaturated esters is not predictable.[91a] In contrast, addition to the α,β-unsaturated ester (**116**) affords the

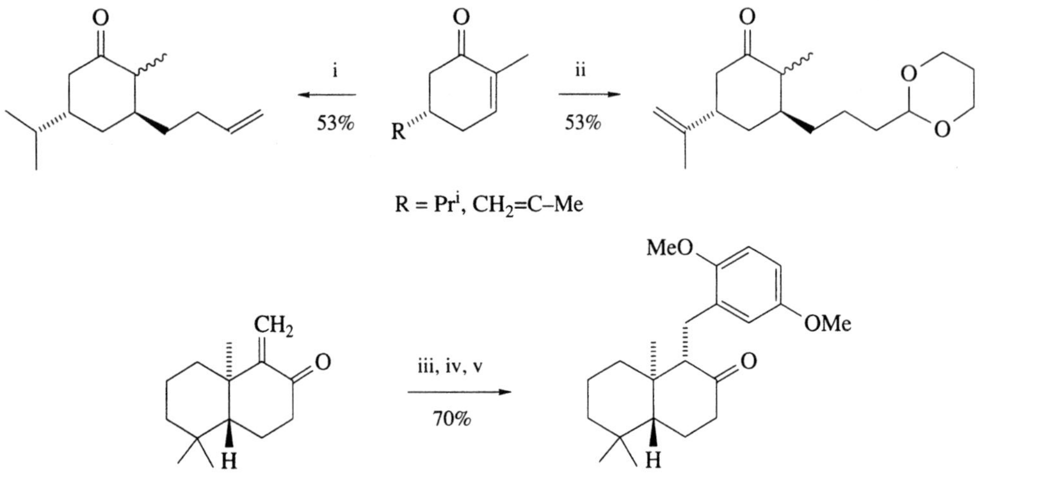

R = Pr^i, CH_2=C–Me

i, CH_2=$CHCH_2CH_2MgBr$, CuI (5%), THF, 0 °C; ii, (1,3-dioxan-2-yl)–$(CH_2)_3MgCl$, CuI (5%), THF, 0 °C;
iii, 2,5-$(MeO)_2C_6H_3MgBr$, CuI (20%), DME, 0 °C; $(MeCO)_2O$; v, HO^-

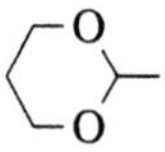

Scheme 38

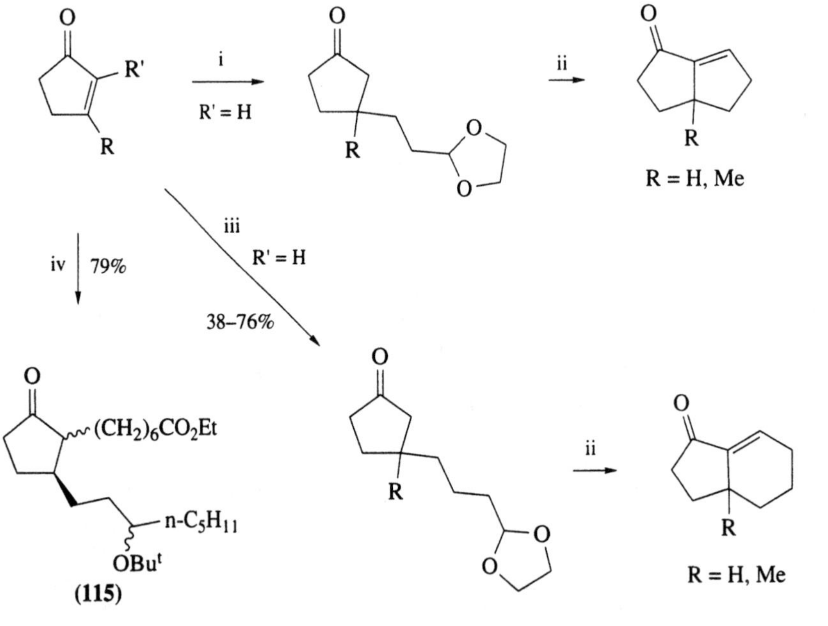

R = H, Me

R = H, Me

(115)

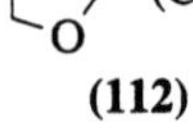

(1,3-dioxolan-2-yl)–$(CH_2)_nMgCl$
(112) $n = 2$
(114) $n = 3$

i, **(112)**, CuBr–Me_2S (25%), THF, –78 °C; ii, H_3O^+;
iii, **(114)**, CuBr–Me_2S (25%), THF, –78 °C;
iv, n-$C_5H_{11}CH(OBu^t)CH_2CH_2MgBr$, CuI–$(Bu^n)_3P$ (5%), THF, 0 °C

Scheme 39

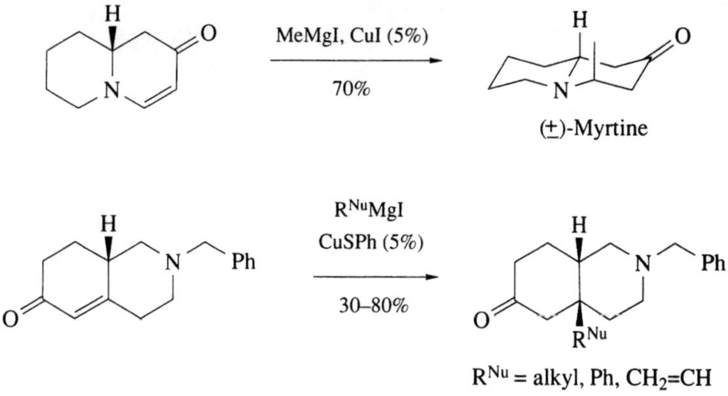

MeMgI, CuI (5%)

70%

(±)-Myrtine

$R^{Nu}MgI$
CuSPh (5%)

30–80%

R^{Nu} = alkyl, Ph, CH_2=CH

Scheme 40

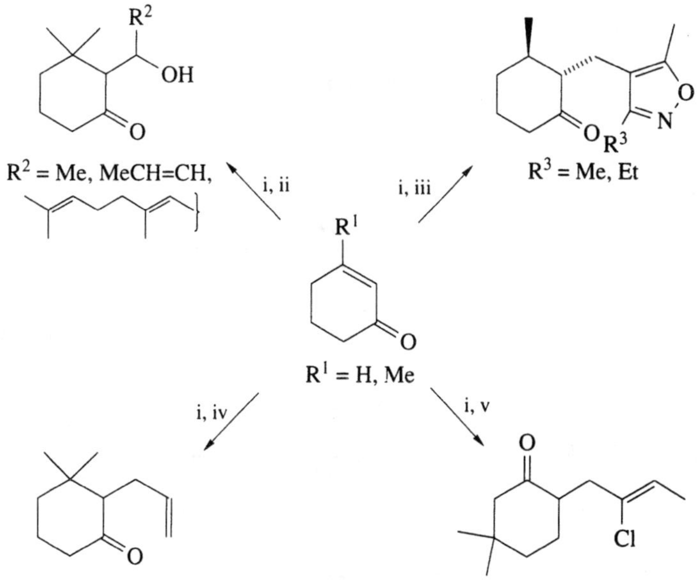

R^2 = Me, MeCH=CH,

R^3 = Me, Et

R^1 = H, Me

i, MeMgX, CuCl (5%); ii, R^2CHO; iii, chloromethylisoxazole;
iv, CH_2=$CHCH_2Br$; v, CH_2=C(Cl)CH_2Cl

Scheme 41

manno adduct (**117**) exclusively (Scheme 42).[91b] Asami and Mukaiyama have reported that enantioselective catalyzed Grignard additions to the chiral β-aminal-α,β-unsaturated ester (**118**) affords 3-alkylsuccinaldehydic methyl esters in >85% *ee* and moderate yields (Scheme 43).[92]

The catalyzed addition of Grignard reagents to 1,1-doubly activated alkenes (**105–108**) is a highly efficient process;[93a–c] notable are additions to *N*-acyldehydroamino acid ester (**119**), which cannot be accomplished with lithium dialkylcuprates (Scheme 44).[93d]

Itoh reports that copper-catalyzed Grignard additions to *N*-tosyl-*N*-alkyl-α,β-unsaturated amides and *N*-tosyl-α,β-unsaturated lactams proceed efficiently while deprotonation occurs exclusively with the corresponding *N*-alkyl-α,β-unsaturated lactams (Scheme 45).[94]

1.2.2.1.4 Organozincs

The organozincs were one of the first organometallic reagents synthesized, but interest in these reagents waned with the discovery of the more reactive and easier to handle Grignard reagents.[95] Within this class of organometallics there are four groups: (a) diorganozincs (**120**), (b) organozinc halides (**121**), (c) lithium triorganozincates ('ate' complex) (**122**) and (d) α-(alkoxycarbonyl)alkylzinc halides (**123**),

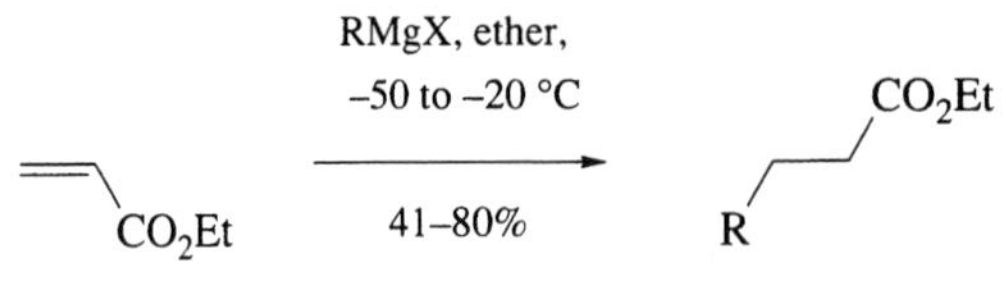

RMgX, ether, –50 to –20 °C, 41–80%

$R = n\text{-}C_5H_{11}$, $c\text{-}C_6H_{11}$, Ph, $CH_2{=}CH(CH_2)_4$, $CH_2{=}CH$, $PhCH_2$

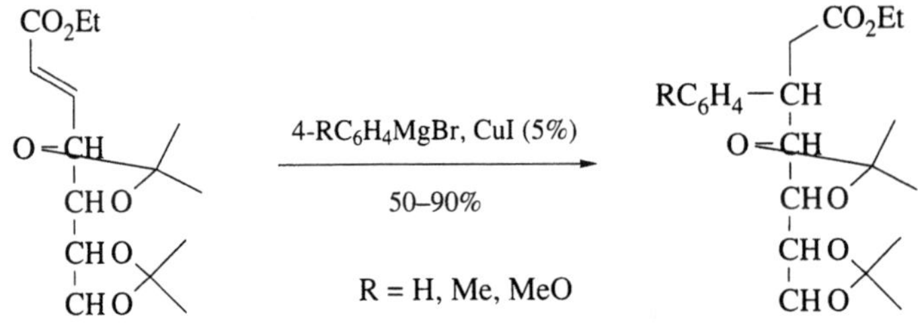

$4\text{-}RC_6H_4MgBr$, CuI (5%), 50–90%

R = H, Me, MeO

(116) **(117)**

Scheme 42

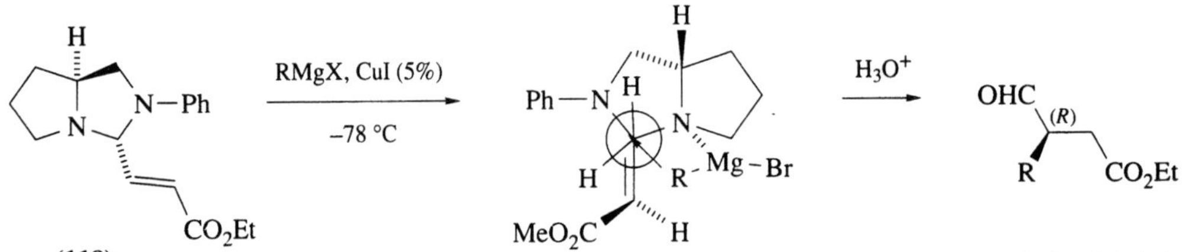

(118)

RMgX, CuI (5%), –78 °C; H_3O^+

R = alkyl (C_2–C_5), Pr^i

>85% *ee*

Scheme 43

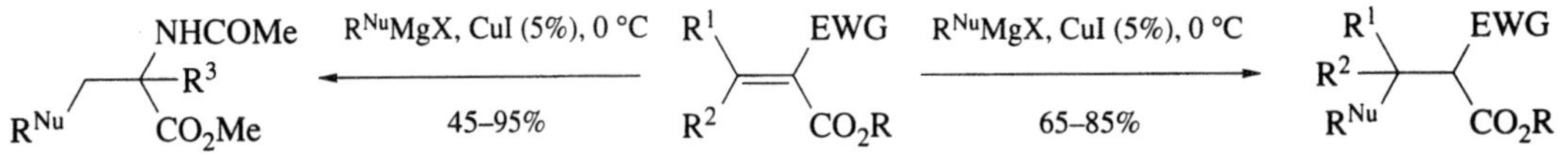

$R^{Nu}MgX$, CuI (5%), 0 °C, 45–95%

$R^{Nu}MgX$, CuI (5%), 0 °C, 65–85%

R^{Nu} = alkyl (C_2–C_4), Pr^i, Bu^t, phenyl, naphthyl

R^3 = H, Me

(105) EWG = CO_2R
(106) EWG = CN
(107) EWG = NC
(108) EWG = $PO(OEt)_2$
(119) EWG = NHCOMe

R^{Nu} = alkyl, phenyl

Scheme 44

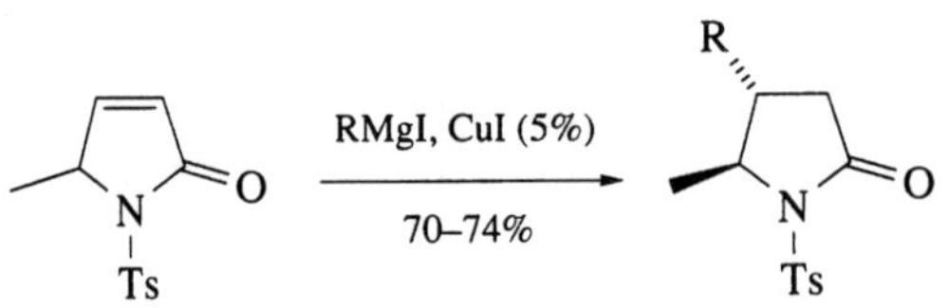

RMgI, CuI (5%), 70–74%

R = Me, Ph, $4\text{-}MeOC_6H_4$

$R^{Nu}MgI$, CuI (5%), 76–81%

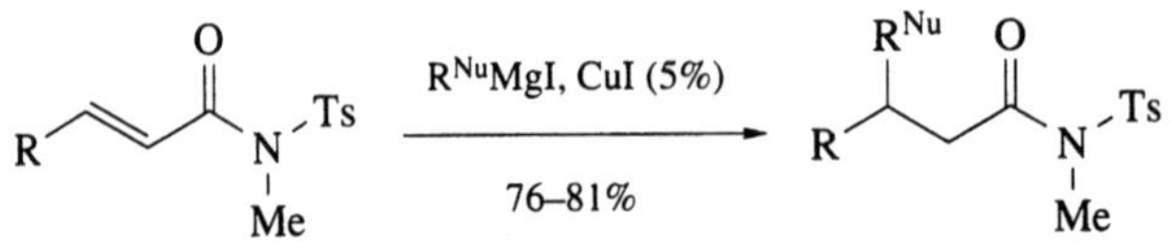

R = Me, Ph; R^{Nu} = Me, Ph

Scheme 45

i.e. Reformatsky reagents. In general, the organozincs and organocoppers have similar reactivity profiles, but the organozincs are tolerant of higher reaction temperatures.

R_2Zn **(120)** RZnX **(121)** LiR_3Zn **(122)** BrZnCH(R)CO_2R'' **(123)**

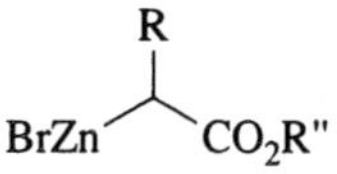

The diorganozincs are classified into two subgroups based on contrasting reactivity profiles: (a) the dialkylzincs and the diarylzincs and (b) the allylic, benzylic and propargylic zincs. The dialkylzincs (diarylzincs) are characterized by their low reactivity towards carbonyl groups[96] and other unsaturated species, while the allylic (benzylic or propargylic) zincs are highly reactive. Traditionally, they are prepared by several routes: (a) thermal recombination of organozinc halides or (b) transmetallation of organolithiums (or Grignard reagents) with an anhydrous zinc halide. The transmetallation process affords byproduct magnesium halides (or lithium halides) which modify the reactivity of the diorganozincs (MgX_2 >> LiX).[23,97] Recent developments in ultrasonic reaction methodology[98] have permitted rapid, quantitative one-pot syntheses of diverse diorganozinc reagents from alkyl bromides (iodides), allylic or benzylic bromides (iodides), or vinyl bromides (iodides) with minimal competing Wurtz coupling. The dialkylzincs (diarylzincs) react easily with α,β-enones (0–25 °C) and α,β-unsaturated aldehydes (–40 °C) in the presence of nickel acetylacetonate [$Ni(acac)_2$] (1%), as shown in Scheme 46;[99a,b] further alkylation of the intermediate zinc enolates is readily accomplished. In contrast to the stoichiometric organocuprates or the copper(I)-catalyzed Grignard reagents, these reagents add efficiently to β,β-disubstituted enones, as shown for two syntheses of (±)-β-cuparenone (**124**).[99c]

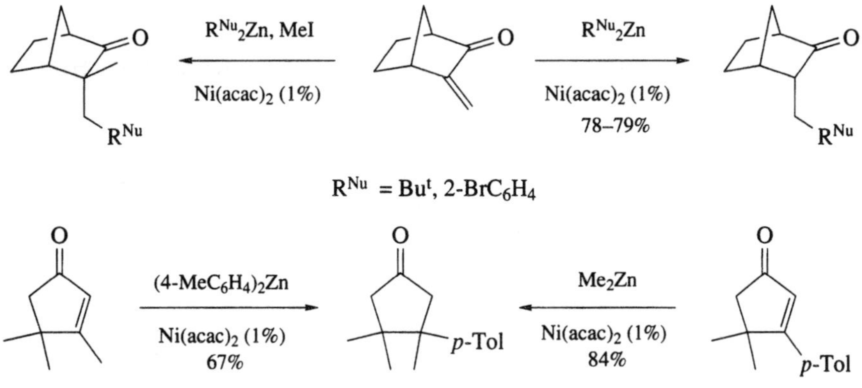

$R^{Nu}{}_2Zn$, MeI; $Ni(acac)_2$ (1%)

$R^{Nu}{}_2Zn$; $Ni(acac)_2$ (1%); 78–79%

R^{Nu} = Bu^t, 2-BrC_6H_4

$(4\text{-}MeC_6H_4)_2Zn$; $Ni(acac)_2$ (1%); 67%

Me_2Zn; $Ni(acac)_2$ (1%); 84%

p-Tol

(124)

Scheme 46

The organozinc halides are prepared by either of two processes: (a) selective transmetallation between an organolithium (or Grignard reagent) (1 equiv.) and an anhydrous zinc halide or (b) oxidative metallation of an organic halide. Typically these reagents are generated *in situ* because of thermal instability, which affords the less reactive diorganozincs and zinc halides. Few examples of syntheses with alkylzinc halides are known, but they retain their inertness towards various functionalities and are characterized by their exclusive 1,4-addition chemistry. A prime example, as shown in Scheme 47, is an alkylzinc halide promoted tandem 1,4-conjugate addition to α-halo-α,β-unsaturated esters to afford exclusively 1-alkyl-2-halo-*cis*-cyclopropanedicarboxylate esters (**125**).[100] Another variation is exemplified by the zinc-promoted three component coupling of alkyl halides, activated alkenes and ketones to afford β-hydroxyesters (nitriles) (**126**).[101] This process, in essence, is the equivalent of coupling a nitrile (or ester) with a carbonyl species (Scheme 47).

The addition of allylic (or benzylic) zinc halides to alkylidenemalonates (**105**),[102a–c] alkylidenecyanoacetates (**106**),[102d,e] alkylidenephosphonoacetates (**108**),[84h] alkylidene barbiturates[102f] and their arylidene analogs occurs exclusively in a 1,4-addition mode. However, the addition of allylic zinc halides to alkylidenemalonates (**105**) is temperature dependent; at low temperatures (–15 °C), the homoallylic malonate (**128**) is obtained, while at higher temperatures (68 °C) the isomeric 2-cyclopentenedicarboxylate esters (**129**) are obtained by a zinc halide promoted electrocyclic closure of the intermediate ester (**127**; Scheme 48).[102g]

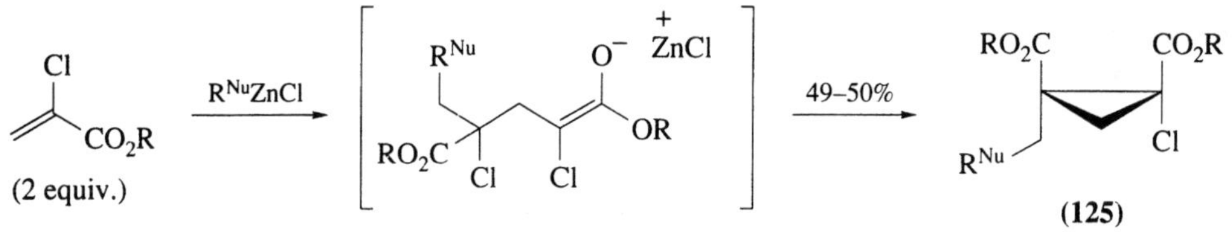

R^{Nu} = Et, Bu^n

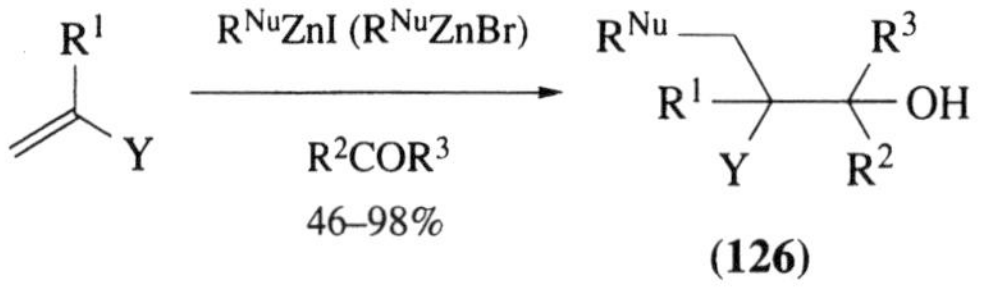

R^1 = H, Me; R^2 = alkyl, Ph; R^3 = H, alkyl; R^{Nu} = alkyl, c-C_6H_{11}, benzyl; Y = CN, CO_2Me

Scheme 47

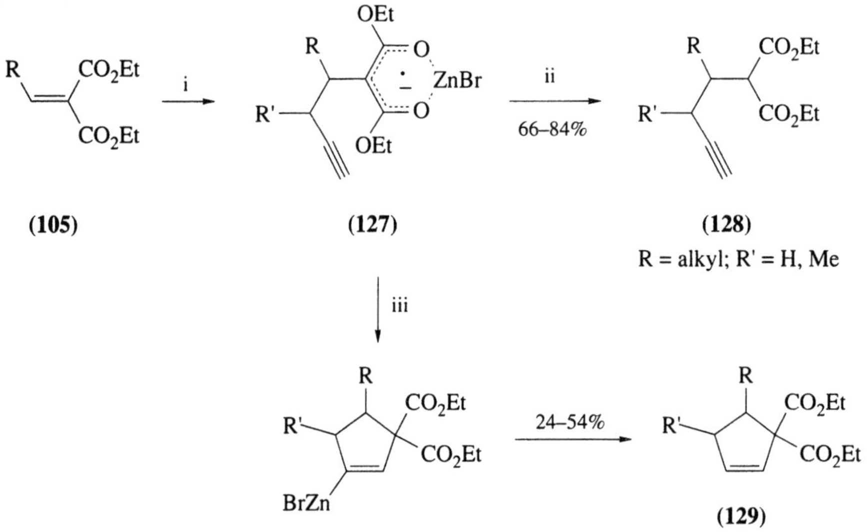

R = alkyl; R' = H, Me

R = alkyl; R' = H, Me

i, R'CH(Br)C≡CH, Zn, THF; ii, –15 °C, 48 h; iii, 65 °C, 48 h

Scheme 48

In contrast to the parent diorganozincs, the lithium triorganozincates (**122**) are more reactive, but few examples of their use in synthesis are available. The first lithium triorganozincates, lithium triphenylzincate and lithium diethyl(1,1-diphenyl-*n*-hexyl)zincate, were prepared by the reaction (1:1) of an organolithium with a symmetrical diorganozinc (Method A) and characterized by either reaction with benzalacetophenone or spectroscopically.[103a,b] Subsequently, two expedient syntheses of lithium and magnesium triorganozincates were reported in which either anhydrous zinc chloride–*N,N,N',N'*-tetramethylethylenediamine complex ($ZnCl_2$·TMEDA) (Method B)[104] or anhydrous zinc chloride (Method C) is treated with an alkyllithium or Grignard reagent (3 equiv.). The symmetrical triorganozincates effect 1,4-additions to α,β-enones with indiscriminate transfer of ligands, but with unsymmetrical triorganozincates, alkynyl and aryl ligands transfer slowly due to stabilized covalent carbon–zinc bonds. Method A is the method of choice for the synthesis of 'salt-free' unsymmetrical organozincates[103c] while method B affords zincates complexed with TMEDA. In general, organozincates made by method B afford higher yields of conjugate addition products.

By analogy to the mixed cuprates, unsymmetrical organozincates (**130**) have been introduced to overcome the ligand wastage in which methyl groups do not transfer as readily as other alkyl ligands.[105] The unsymmetrical organozincates exhibit selectivity for ligand transfer with the lithium (or magnesium) alkyldimethylzincates, RMe_2ZnLi, being optimal with the order of ligand transfer preferences being:

$PhMe_2Si$ >> CH_2=CH, Et, Bu^n, Pr^i > Bu^i, Ph > Me, Bu^t >> Bu^tCH_2. This is exemplified in Scheme 49 by an expedient prostaglandin synthesis in which the preferential transfer of vinyl ligands *vs.* methyl groups occurs with zincate (**132**).[105b] In contrast to the diorganozincs, the triorganozincates are sensitive to β-substitution on the α,β-enones and transition metal catalysis has negligible effect on either the chemoselectivity or the rate of reaction of these reagents.

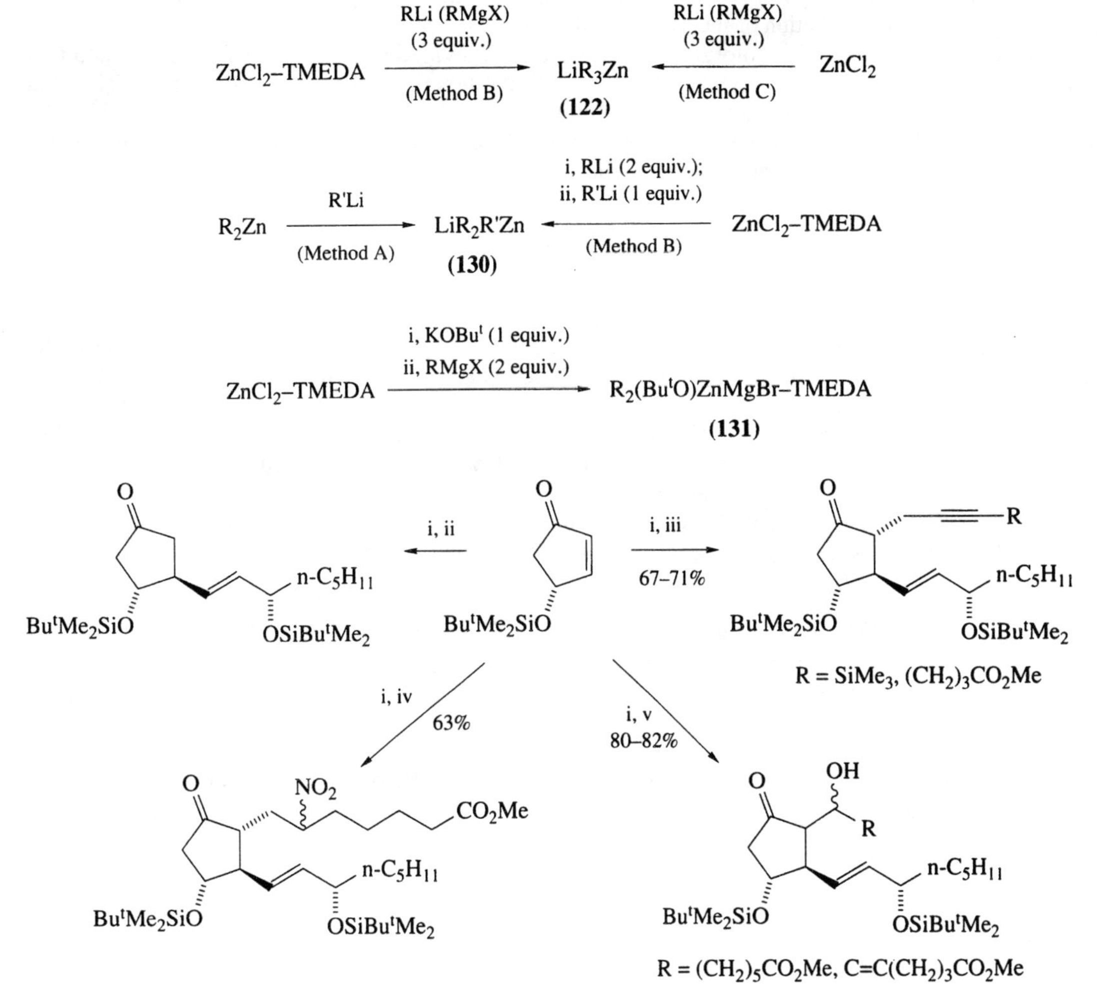

$ZnCl_2$–TMEDA —[RLi (RMgX) (3 equiv.), (Method B)]→ LiR_3Zn (**122**) ←[RLi (RMgX) (3 equiv.), (Method C)]— $ZnCl_2$

R_2Zn —[R'Li, (Method A)]→ $LiR_2R'Zn$ (**130**) ←[i, RLi (2 equiv.); ii, R'Li (1 equiv.), (Method B)]— $ZnCl_2$–TMEDA

$ZnCl_2$–TMEDA —[i, $KOBu^t$ (1 equiv.) ii, RMgX (2 equiv.)]→ $R_2(Bu^tO)ZnMgBr$–TMEDA (**131**)

i, ii; i, iii 67–71%; i, iv 63%; i, v 80–82%

Bu^tMe_2SiO, n-C_5H_{11}, $OSiBu^tMe_2$, NO_2, CO_2Me, OH, R

R = $SiMe_3$, $(CH_2)_3CO_2Me$

R = $(CH_2)_5CO_2Me$, C=C$(CH_2)_3CO_2Me$

i, Li *(E)*-Me_2ZnCH=CHCH($OSiBu^tMe_2$)-n-C_5H_{11} (**132**); ii, H_3O^+; iii, $BrCH_2C≡CR$, HMPA, –78 to –40 °C; iv, CH_2=C(NO_2)$(CH_2)_3CO_2Me$; v, RCHO, $Et_2O•BF_3$

Scheme 49

Feringa reports that alkyl heterozincates (**131**), made from zinc chloride·TMEDA complex, potassium *t*-butoxide and alkyl Grignard reagents in ether (or tetrahydrofuran), are highly effective in 1,4-additions to α,β-enones (>85%) while the aryl heterozincates are less efficient (<50%).[106]

Reformatsky first introduced electron-withdrawing substituents on the α-carbon of an organozinc halide, leading to the more reactive and thermally stable α-(alkoxycarbonyl)alkylzinc halides (**123**).[107a] Typically these reagents react with aldehydes or ketones to afford β-hydroxy esters while nitriles afford β-keto esters (Blaise reaction),[107b,c] but 1,4-conjugate additions to select α,β-unsaturated ketones are precedented (Section 1.2.2.2.2).

1.2.2.1.5 Organocadmiums

The reactivity of the organocadmiums has been the subject of numerous conflicting reports.[108a] Many of these organocadmium reagents are typically prepared by transmetallation of a Grignard reagent with anhydrous cadmium chloride (or bromide) in ether which generates byproduct magnesium halides which enhance ('activate') the nucleophilicity of the organocadmiums *via* a four-centered complex.[108b–d] Removal of the incipient magnesium halides is accomplished in two ways: (a) distillation or sublimation of the reagent or (b) precipitation of the salts with dioxane.

The reactivity of the 'salt-free' diorganocadmiums can be divided into two groups, (a) alkyl- and aryl-cadmiums and (b) allylic and benzylic cadmiums, with the latter group being more reactive. The 'salt-free' alkyl- or aryl-cadmiums participate in 1,4-conjugate additions to α,β-unsaturated ketones,[103a,109a–e] α-nitroalkenes (**40**),[109f] alkylidenemalonates (**105**),[109g] alkylideneacetoacetates[109b] and alkylidene-cyanoacetates (**106**).[109g,h] In contrast, the allylic and benzylic cadmiums react exclusively in a 1,2-mode.[110]

1.2.2.1.6 Organomanganese

Organomanganese reagents were first reported to undergo exclusive 1,4-addition to benzalacetophenone, but interest in these reagents was minimal.[6] Depending on the stoichiometry, the transmetallation of organolithiums (or Grignard reagents) with anhydrous manganese(II) iodide affords organomanganese halides (RMnX), diorganomanganese (R_2Mn) or manganate complexes (R_3MnLi or R_3MnMgX) which are poor reactants for the conjugate addition process. However, in the presence of copper(I) iodide (<5.0%), conjugate additions of organomanganese halides to α,β-enones occur efficiently. Even β,β-disubstituted-α,β-enones, *e.g.* 4-methyl-3-penten-2-one, which react with difficulty with organocopper reagents and copper-catalyzed Grignard reagents, undergo conjugate additions efficiently (Scheme 50).[111]

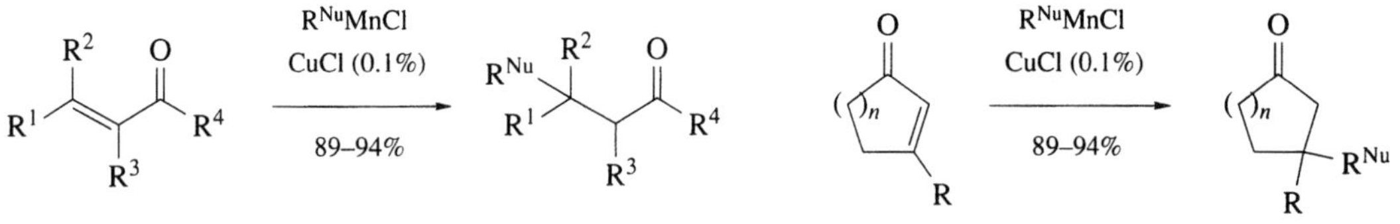

$R^{Nu}MnCl$, CuCl (0.1%), 89–94%

R^1 = H, Me, Pr^n
R^2 = H, Bu^n, n-C_5H_{11}
R^3 = H, Me
R^4 = Bu^n, n-C_5H_{11}

R^{Nu} = Bu^n, n-C_5H_{11}

$R^{Nu}MnCl$, CuCl (0.1%), 89–94%

R = H, Me
n = 1, 2

R^{Nu} = alkyl (C_1-C_4), Bu^t, Ph, Me_2C=CH, PhCH=CH

Scheme 50

1.2.2.1.7 Organosilanes

The cleavage of organosilanes to form anionic species was first reported by Brook;[112] subsequently, Sakurai and coworkers reported that the fluoride ion mediated desilylation of allyltrimethylsilane in the presence of α,β-enones afforded mixtures of 1,2-addition (predominant) and 1,4-addition products.[113] Less electrophilic Michael acceptors, *e.g.* α,β-unsaturated nitriles, α,β-unsaturated esters or α,β-unsaturated *N,N*-dialkylamides, were found to afford exclusive 1,4-addition by the fluoride ion mediated desilylation of allyltrimethylsilanes, benzyltrimethylsilanes or heteroaryltrimethylsilanes.[114] The use of either cesium fluoride or tetra-*n*-butylammonium fluoride in HMPA affords optimal yields of 1,4-adducts (Scheme 51). This allylation procedure complements the original Lewis acid catalyzed procedure (Volume 4, Chapter 1.3) of α,β-enones which fails with the above substrates.[113b] Typically, fluoride-mediated conjugate allylation yields are comparable to those obtained by the addition of lithium diallylcuprate.[114b] Data point towards the reactive species being an intermediate hypervalent fluorosilicon species (**133**). Similarly, Panek found that the fluoride ion catalyzed addition of 1-acyloxy-2-propenyltrimethylsilanes (**134**) to α,β-enones affords exclusive $1,4_{(\alpha)}$-addition of the 1-acyloxy-2-propenyl moiety with moderate diastereoselectivity (Scheme 52).[115]

Majetich reports that the fluoride ion catalyzed carbocyclization of allyl anion equivalents to α,β-enones is sensitive to the structure of the α,β-enone in which competing 1,2- and 1,4-addition processes occur.[116] For example, α,β-enones (**135**) afford predominantly the *syn*-bicyclo[4.3.0]nonan-3-ones (**136**)

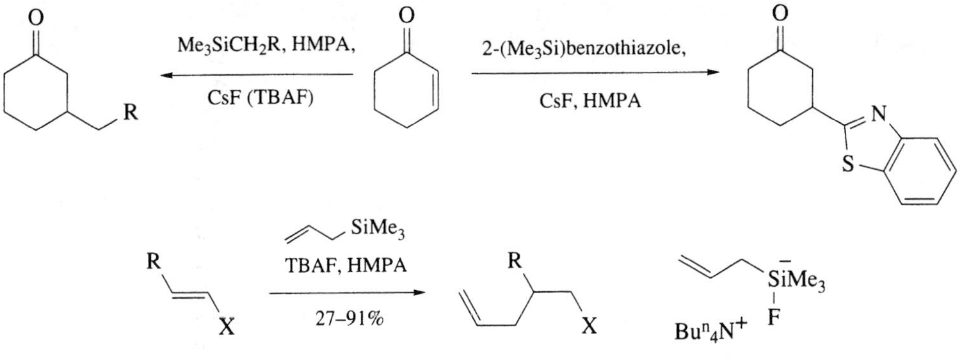

Me_3SiCH_2R, HMPA, CsF (TBAF)

2-(Me_3Si)benzothiazole, CsF, HMPA

SiMe$_3$ TBAF, HMPA 27–91%

$Bu^n{}_4N^+$ (133)

R = H, Bu^t, CO_2Me, Ph, 3-furyl, PhCH=CH; X = CN, CO_2Me, CO_2CH_2Ph, $CONEt_2$

Scheme 51

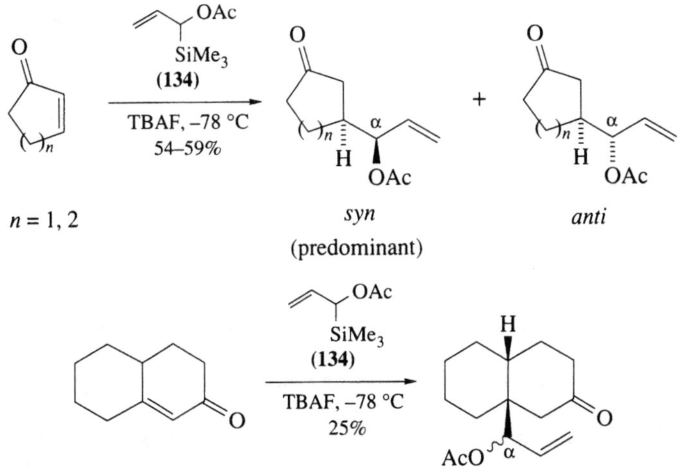

(134) TBAF, –78 °C 54–59%

n = 1, 2 *syn* (predominant) *anti*

(134) TBAF, –78 °C 25%

Scheme 52

while vinyl enones (**138**) afford both bicyclo[6.4.0]dodecen-3-ones (**139**) and bicyclo[2.2.2]octenes (**140**). In contrast, the vinyl enones (**141**) afford both bicyclo[6.3.0]undecen-3-ones (**142**) and vinylbicyclo[4.3.0]nonene-3-ones (**143**; Scheme 53).

1.2.2.2 Additions of Enolates, Azaenolates and α-Nitrile Anions

The conjugate addition of unstabilized enolates to various acceptors was conceptually recognized by early researchers; however, complications were encountered depending on the enolates and acceptors employed. Reexamination of this strategy was made possible by the development of techniques for kinetic enolate formation. This discussion is divided into three enolate classes (a) aldehyde and ketone enolates, azaenolates or equivalents, (b) ester and amide enolates, dithioenolates and dienolates and (c) α,*O*-carboxylic dianions and α-nitrile anions, in order to emphasize the differential reactivity of various enolates with various acceptors.[117] The α-nitrile anions are included because of their equivalence to the hypothetical α-carboxylic acid anion.

1.2.2.2.1 Aldehyde and ketone enolates, azaenolates and equivalents

The 1,4-conjugate addition of ketones to α,β-unsaturated ketones, under protic conditions, was reported by Stobbe in 1912;[118a] further developments afforded protocols for the synthesis of fused cyclohexenones, *i.e.* Robinson annulation.[118b–d] Robinson also reported that aprotic 1,4-conjugate additions of select reagents, *e.g.* sodiocyclohexanone with styryl methyl ketone,[119] was preferable to the protic addi-

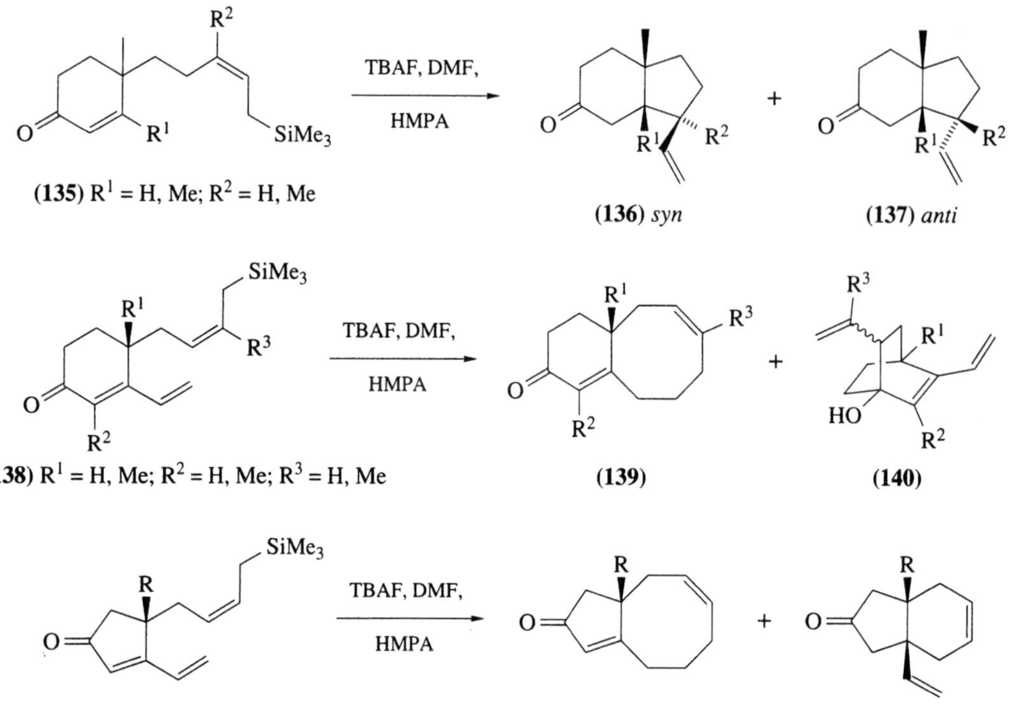

(135) R^1 = H, Me; R^2 = H, Me — TBAF, DMF, HMPA → (136) *syn* + (137) *anti*

(138) R^1 = H, Me; R^2 = H, Me; R^3 = H, Me — TBAF, DMF, HMPA → (139) + (140)

(141) R = H, Me — TBAF, DMF, HMPA → (142) + (143)

Scheme 53

tion protocol. However, complications with kinetic enolate additions include (a) enolate scrambling during addition of the acceptor, (b) self-aldolization and (c) polymerization of certain acceptors, *e.g.* acrylonitrile or methyl vinyl ketone. Enolate equivalents and modified acceptors, primarily a^3-synthons, have been introduced that obviate these complications. For example, azaenolates, *e.g.* metallated alkylidene amines, *N,N*-dimethylhydrazones and oximes (oxime ethers), do not suffer from anion scrambling or self-aldolization,[120] while the modified a^3-synthons, *e.g.* α-silyl vinyl ketones,[41,121] α-silylacrylate esters[43] and α,β-unsaturated thioamides,[122] are characterized by the presence of functionalities that stabilize an intermediate anion without proton transfer.

Gorrichon and Heathcock have reported that three factors affect the chemoselectivity of kinetic ketone enolate additions to α,β-enones. These factors are (a) enolate geometry and substitution pattern, (b) the cation and (c) reaction conditions. In general, under kinetic reaction conditions (–78 °C), lithium enolates and magnesium dienolates afford preferential 1,4-addition products while halomagnesium enolates afford substantial amounts of 1,2-addition products, but the intermediate allylic halomagnesium alkoxides undergo 1,2 to 1,4 equilibration at higher temperatures. In addition, increasing substitution on the enolates enhances the formation of 1,4-addition products.[123]

Heathcock[123b] and Seebach[124] have shown that the diastereoselectivity of the conjugate additions is highly dependent on both the lithium enolate geometry and geometry of the acceptor. For example, with (*E*)-acceptors, the (*Z*)-enolates (**144**) afford high *anti*-selectivity while the (*E*)-enolates (**145** and **146**) afford high *syn*-selectivity (Scheme 54). In contrast, Mukaiyama reports that tin(II) enolates of alicyclic ketones, *e.g.* (**147**), with (*E*)-β-nitrostyrene, afford predominant *anti*-selectivity (Scheme 55).[125]

Corriu and coworkers have reported an alternative procedure for the conjugate addition of ketones to α,β-unsaturated acceptors which employs CsF–$(RO)_4Si$ (Scheme 56);[126] this procedure affords adducts with α,β-enones, α,β-unsaturated esters and α,β-unsaturated amides. Mechanistically, silyl enol ether formation occurs initially, followed by fluoride ion catalyzed enolate formation.

Stork and Boeckmann have reported that enolates react with α-(trialkylsilyl)vinyl ketones (**148**) without proton scrambling to afford adducts in excellent yields; subsequent alcoholysis affords the desilylated ketone (Scheme 57).[41,121] Yoshida reports that various enolates undergo conjugate additions efficiently with α,β-unsaturated *N,N*-dialkylthioamides (**32**; Scheme 58),[122] while preferential 1,4-addition to *N,N*-dimethylthiosorbamide is observed (*cf.* Schemes 8 and 28).

The adducts of ketone enolates to various a^2-synthons, *e.g.* α-nitroalkenes (**40**), vinyl sulfoxides, 2-(*N*-methylanilino)acrylonitrile (**59**) and ketene dithioacetal monoxide (**151**), are versatile synthetic inter-

anti

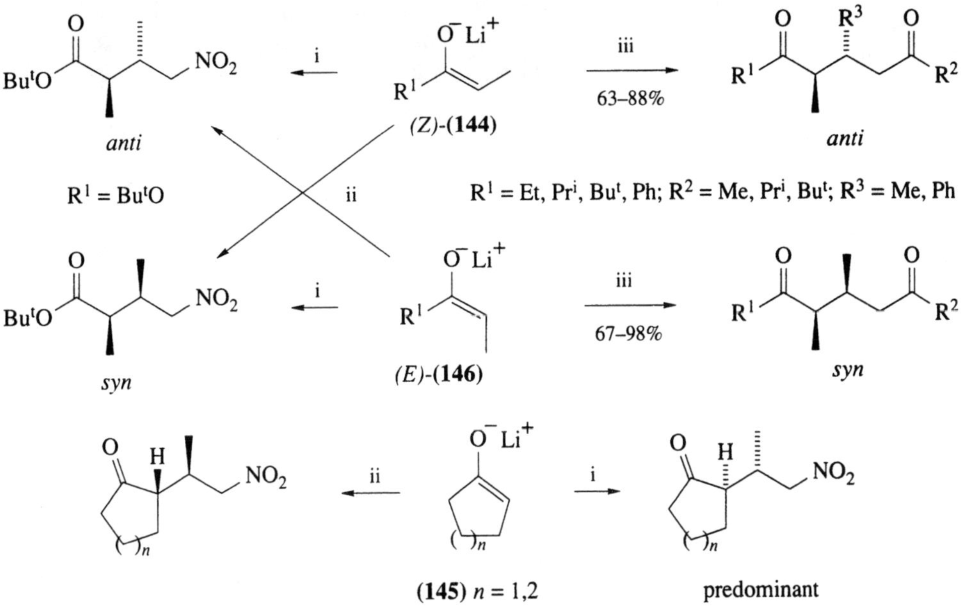

R^1 = Bu^{t}O

(Z)-(**144**)

63–88%

anti

R^1 = Et, Pr^i, Bu^t, Ph; R^2 = Me, Pr^i, Bu^t; R^3 = Me, Ph

syn

(E)-(**146**)

67–98%

syn

(**145**) n = 1,2

predominant

i, (E)-MeCH=CHNO$_2$; ii, (Z)-MeCH=CHNO$_2$; iii, (E)-R^{2}COCH=CHR^3

Scheme 54

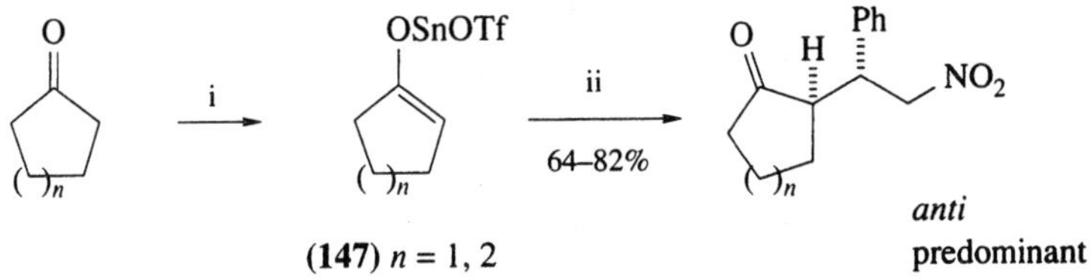

(**147**) n = 1, 2

64–82%

anti
predominant

i, Sn(OTf)$_2$, N-ethylpiperidine, –78 °C; ii, (E)-PhCH=CHNO$_2$

Scheme 55

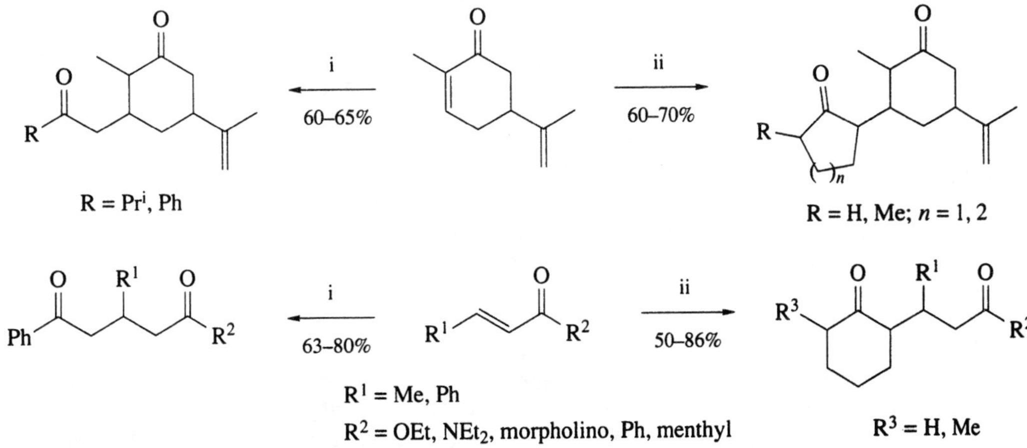

60–65%

60–70%

R = Pr^i, Ph

R = H, Me; n = 1, 2

63–80%

50–86%

R^1 = Me, Ph
R^2 = OEt, NEt$_2$, morpholino, Ph, menthyl

R^3 = H, Me

i, R^1COMe, CsF–Si(OMe)$_4$, 25–80 °C; ii, cycloalkanone, CsF–Si(OMe)$_4$, 25–100 °C

Scheme 56

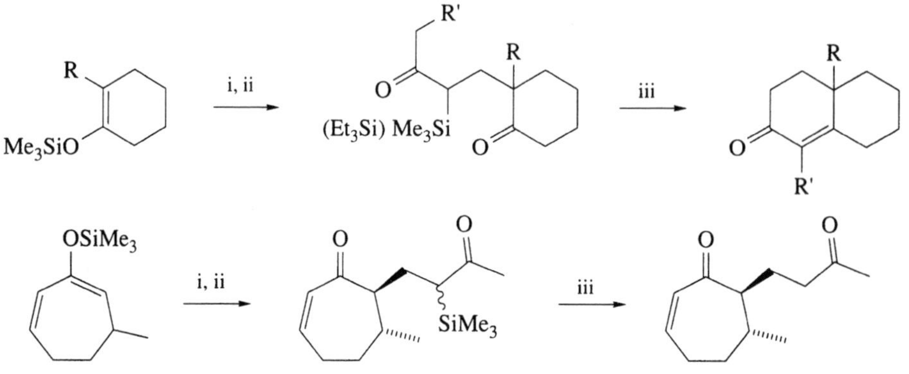

i, MeLi, –78 °C; ii, $R'CH_2COC(Me_3Si)=CH_2$ [(**148**), R' = H, Me]; iii, NaOEt, EtOH

Scheme 57

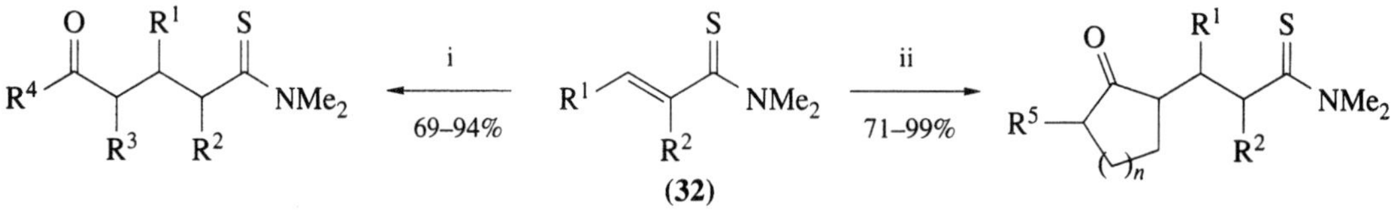

R^3 = H, Me
R^4 = Pr^i, Bu^t, Ph, NMe_2
Bu^{t}O, EtO_2CCH_2

R^1 = H, Me, Ph, MeCH=CH
R^2 = H, Me

R^5 = H, Me
n = 1, 2

i, $R^3CH_2COR^4$, LDA, –78 °C; ii, cycloalkanone, LDA, –78 °C

Scheme 58

mediates. Notable are the α-nitroalkenes which serve as α-carbonyl cations (*via* Nef reaction), β-amino-cations (*via* reduction) or vinyl cation equivalents (*via* elimination).[127] For example, Pattenden reports that the addition of the bicyclo[3.3.0]octenone (**149**) to 2-nitrobut-2-ene and subsequent oxidation affords the 1,4-diketone (**150**).[127b] Similarly, adducts with either 2-(*N*-methylanilino)acrylonitrile (**59**)[128a–c] or ketene dithioacetal monoxide (**151**),[128d] α-formyl cation equivalents, afford 1,4-diketones (Scheme 59).

In contrast to the paucity of citations concerning additions of organolithiums or Grignard reagents to vinyl sulfoxides, ketone enolate additions are highly precedented. Vinyl sulfoxides and vinyl sulfones are considered as equivalent a^2-synthons; however, further transformation of the sulfone unit is restricted to reductive cleavage. With the sulfoxide moiety, additional transformations are permissive, *i.e.* Pummerer rearrangement and eliminations. Jones has reported several prototypical prostaglandin syntheses in which ketone enolates and ethyl acetoacetate dianion are added to vinyl sulfoxides (**152** and **153**; Scheme 60).[129] Further, additions to chiral β-substituted vinyl sulfoxides result in chirality transfer from sulfur. For example, Yamazaki and Ishikawa report that the addition of ketone and ester enolates to (*R*,*E*)-3,3,3-trifluoroprop-1-enyl *p*-tolyl sulfoxide (**154**) occurs with high diastereoselectivity (equation 17).[130]

Takaki reports that ketone enolates add to dimethylstyryl sulfonium perchlorate (**155**) or methyl styryl sulfone (**156**) in a Robinson-type annulation sequence to afford the corresponding β-hydroxythiadecalin (**157**) or *S*-dioxide (**158**), respectively; subsequent reductive desulfonation of (**158**) affords diene (**159**).[131] However, additions to acceptor (**155**) suffer from competing cyclopropanation which is dependent on the electrophilicity of the carbonyl group and the ring size of the ketone (Scheme 61). As an aside, DeLucchi reports that 1,1-bis(benzenesulfonyl)ethylene (**160**) adds to ketones at the more substituted α-carbon *under neutral conditions* in refluxing acetonitrile (equation 18).[132]

In contrast to α,β-unsaturated phosphonium salts, α,β-unsaturated phosphonates (**161**)[133a] alkylate alicyclic ketone enolates efficiently, while the homologous diethyl butadiene phosphonate (**162**)[133b,c] undergoes 1,6-addition exclusively at the exclusion of intramolecular cyclization. Heathcock reports that

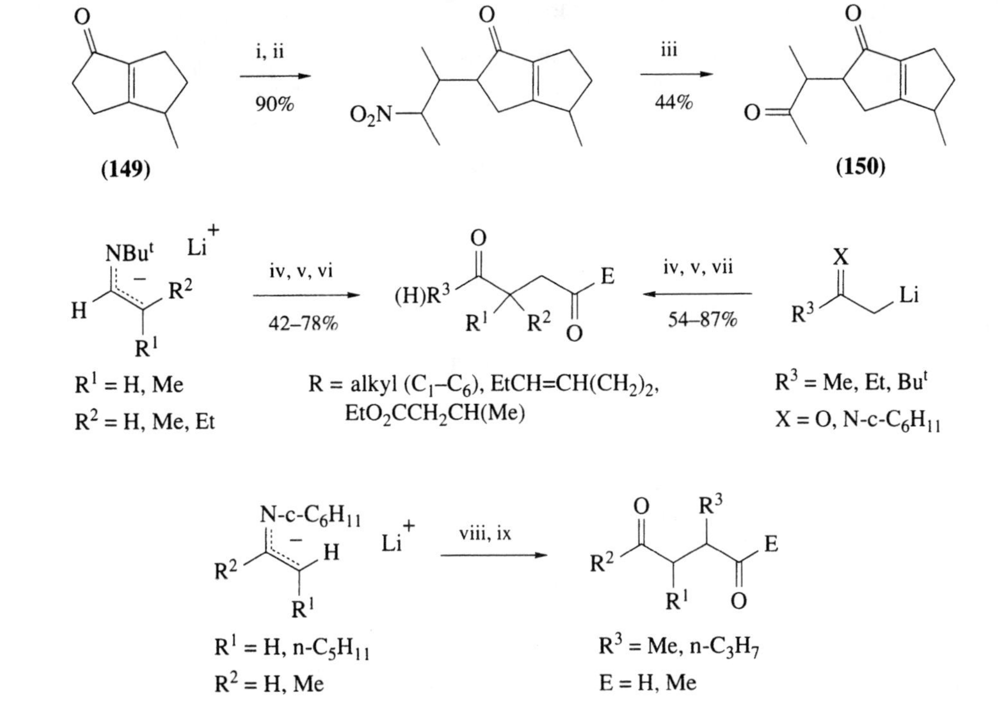

(149) (150)

i, ii 90%; iii 44%

R^1 = H, Me
R^2 = H, Me, Et

iv, v, vi 42–78%; iv, v, vii 54–87%

R = alkyl (C_1–C_6), EtCH=CH(CH_2)$_2$, $EtO_2CCH_2CH(Me)$

R^3 = Me, Et, Bu^t
X = O, N-c-C_6H_{11}

viii, ix

R^1 = H, n-C_5H_{11}
R^2 = H, Me

R^3 = Me, n-C_3H_7
E = H, Me

i, LDA, HMPA, –78 °C; ii, $NO_2(Me)C{=}CHMe$; iii, $NaNO_2$, Pr^{n}ONO, DMSO; iv, $CH_2{=}C(CN)[N(Me)Ph]$ (**59**); v, electrophile; vi, $CuSO_4$, H_2O; vii, H_2SO_4 (3N); viii, $R^3CH{=}C(SMe)SOMe$ (**151**); ix, Amberlite IR-120, 18–50 °C

Scheme 59

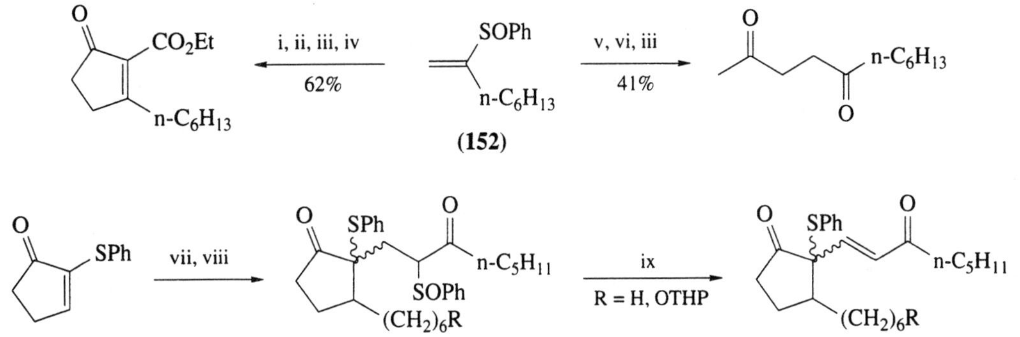

i, ii, iii, iv 62%; v, vi, iii 41%

(152)

vii, viii; ix R = H, OTHP

i, $LiCH_2COCH(Li)CO_2Et$; ii, MeSSMe; iii, CF_3CO_2H; iv, NaOH; v, acetone, LiTMP, –78 °C; vi, TFAA, pyridine; vii, $R(CH_2)_6MgBr$, CuCl (1%); viii, $CH_2{=}C(SOPh)CO$-n-C_5H_{11} (**153**); ix, $(MeO)_3P$

Scheme 60

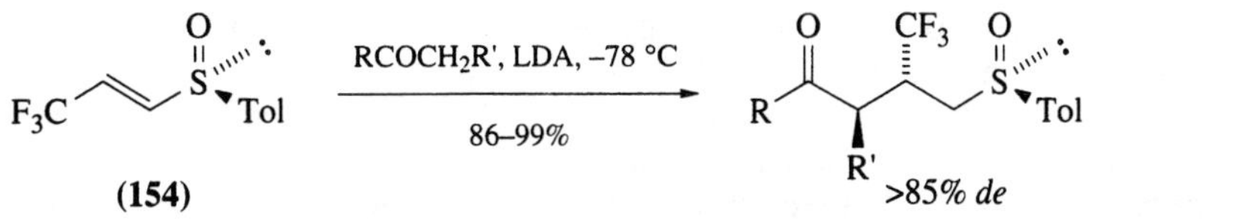

(154)

$RCOCH_2R'$, LDA, –78 °C; 86–99%

>85% *de* (17)

R = Et, Bu^t, Ph, OEt; R' = H, Me, CO_2Et

ester and ketone enolates add to ethyl 2-(diethylphosphono)acrylate (**163**) to afford the intermediate keto ester phosphonates which can react with aldehydes or ketones to afford α,β-unsaturated esters (Scheme 62).[134]

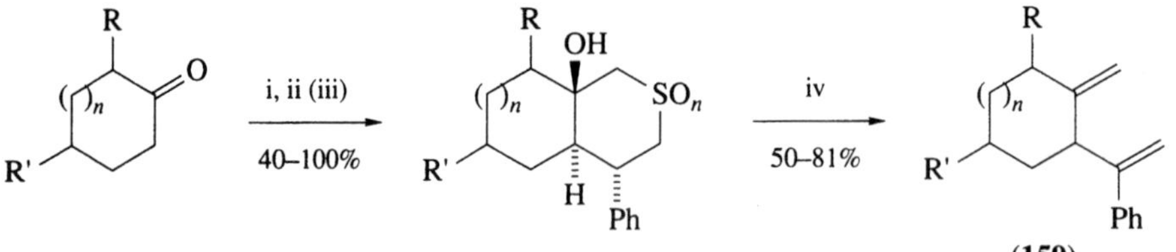

(157) $n = 0$
(158) $n = 2$
(159)

R = H, Me; R' = H, Bu^t; $n = 0 - 2$

i, LDA, –78 °C; ii, $PhCH{=}CH\overset{+}{S}Me_2\ ClO_4^-$ (155), DMF, 80 °C; iii, $PhCH{=}CHSO_2Me$ (156), DMF, 80 °C; iv, Na, EtOH, THF

Scheme 61

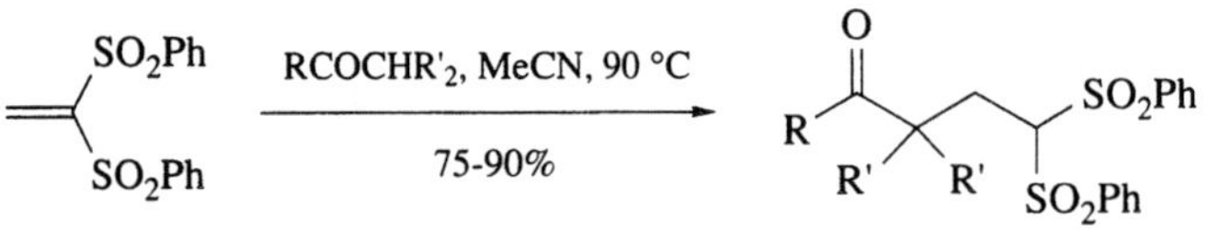

(160) — $RCOCHR'_2$, MeCN, 90 °C, 75-90% — (18)

R = Me, Ph; R' = H, Me

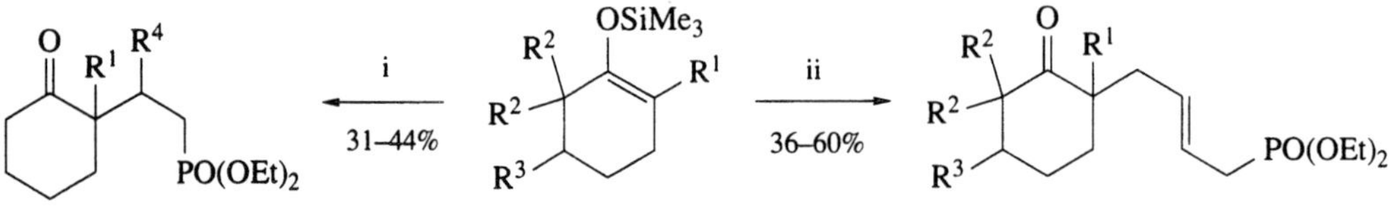

R^4 = H, Me R^1 = H, Me; R^2 = H, Me; R^3 = H, Pr^i

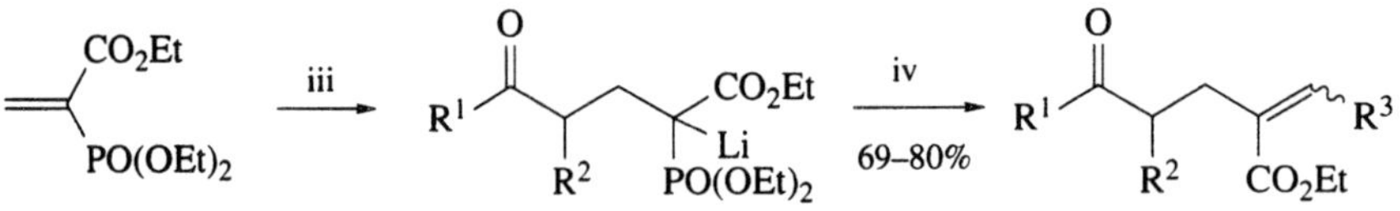

(163) R^1 = Et, Bu^t, Bu^tO; R^2 = H, Me R^3 = Et, Ph, MeCH(Ph)

i, $R^4CH{=}CHPO(OEt)_2$ (161), $PhCH_2\overset{+}{N}Me_3\ F^-$; ii, $CH_2{=}CH{-}CH{=}CHPO(OEt)_2$ (162), $PhCH_2\overset{+}{N}Me_3\ F^-$; iii, $R^1COCH_2R^2$, LDA, –78 °C; iv, R^3CHO

Scheme 62

Casey and coworkers have shown that ketone enolates add efficiently to α,β-unsaturated vinyl carbene complexes (**164**), irrespective of β,β-disubstitution on the complex or high substitution on the enolate;[135a] thus, contiguous β and γ quaternary centers are easily assembled. When coupled with the ease of release of the carbene ligand from the complexes by either oxidation to the ester functionality[135b–d] or elimination to the corresponding enol ether,[135a] the vinyl carbene complexes are synthetic equivalents for α,β-unsaturated esters or α,β-unsaturated aldehydes, respectively (Scheme 63).

The use of aldehyde enolates for conjugate additions is precluded by competing polymerization and aldolization processes; however, introduction of the 'CH_2CHO' moiety is accomplished with aldehyde enolate equivalents. For example, dianions of nitroethanes, *e.g.* β-phenylnitroethane (**165**) or methyl β-nitropropionate (**166**), add exclusively in the 1,4-mode to α,β-enones.[136a,b] Similarly, the dianion of 4-nitro-1-butene (**167**) adds in a 1,4-mode exclusively; unlike typical dienolates (Section 1.2.2.2.2) which react at the α-position, this dianion is formally equivalent to the crotonaldehyde γ-enolate (Scheme 64).[136c]

The introduction of a chiral auxiliary on the nitrogen of an azaenolate offers the capability to perform asymmetric conjugate additions of chiral nucleophiles to achiral acceptors. This is in contrast to the typical asymmetric conjugate addition of achiral nucleophiles to chiral acceptors.[137] For example, the con-

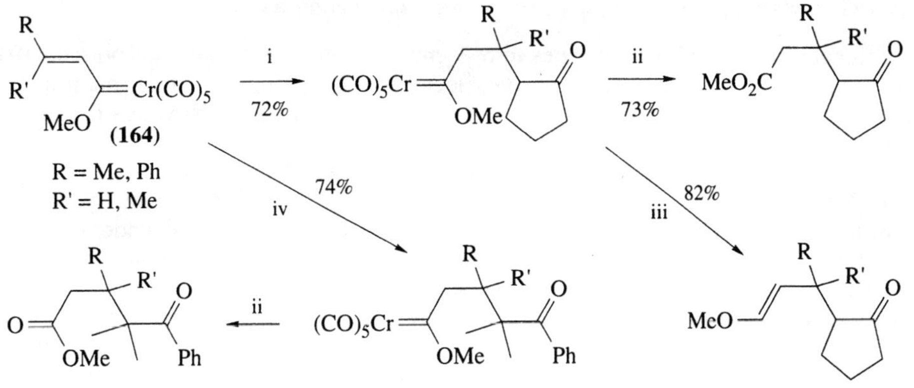

R = Me, Ph
R' = H, Me

i, lithium cyclopent-1-enolate; ii, $(NH_4)_2Ce(NO_2)_6$; iii, pyridine; iv, PhC(OLi)=CMe_2

Scheme 63

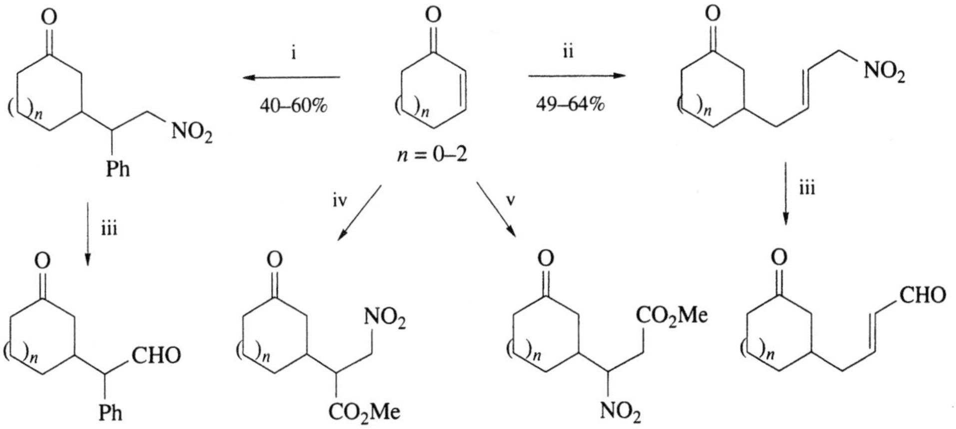

n = 0–2

i, $PhCH_2CH_2NO_2$ (**165**), Bu^{n}Li, HMPA, –90 °C; ii, CH_2=$CH(CH_2)_2NO_2$ (**167**), Bu^{n}Li, HMPA, –90 °C; iii, $TiCl_3$; iv, $MeO_2CCH_2CH_2NO_2$ (**166**); LDA, –78 °C; v, $MeO_2CH_2CH_2NO_2$ (**166**), DBU, 25 °C

Scheme 64

jugate addition of lithiated SAMP/RAMP hydrazones (**168**) of ketones to β-substituted-α,β-unsaturated esters occurs with excellent asymmetric induction (Scheme 65).

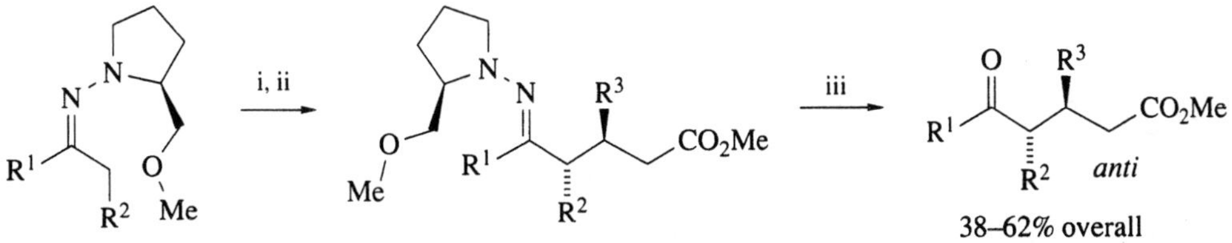

38–62% overall
$R^2 \neq$ H: >90% *de*, >92% *ee*
R^2 = H: >96% *ee*

R^1 = H, alkyl (C_1–C_6), Ph; R^2 = H, alkyl (C_1–C_3), benzyl; R^3 = alkyl (C_1–C_3), Ph

i, LDA, THF, –78 °C; ii, *(E)*-R^3CH=$CHCO_2Me$; iii, O_3, CH_2Cl_2, –78 °C

Scheme 65

1.2.2.2.2 Amide enolates, ester enolates, dithioenolates and dienolates

The 1,4-conjugate addition of ester enolates to α,β-enones was first reported by Kohler in 1910,[138a–c] as an anomalous Reformatsky reaction, but chemoselectivity was dependent on the structure of the α,β-enone and restricted to bromozinc enolates obtained from either α-bromoisobutyrate or bromomalonate esters (Scheme 66).[138d,e] Further evaluation, with lithio ester enolates and lithio amide enolate additions, has resulted in identification of four factors that affect the chemoselectivity and diastereoselectivity of additions to α,β-enones.[139] These factors are (a) enolate geometry, (b) acceptor geometry, (c) steric bulk of the β-substituent on the acceptor enone and (d) reaction conditions. In general, under kinetic reaction conditions (–78 °C), (*E*)-ester enolates afford preferential 1,2-addition products while (*Z*)-ester enolates afford substantial amounts of 1,4-addition products; however, 1,2 to 1,4 equilibration occurs at 25 °C in the presence of HMPA. The stereostructure of the 1,4-adducts is dependent on the initial enolate structure; for example, with (*E*)-enones, (*Z*)-ester enolates afford *anti* adducts, while (*E*)-ester enolates afford *syn* adducts (Scheme 54). In contrast, amide enolates show a modest preference for *anti* diastereomer formation.

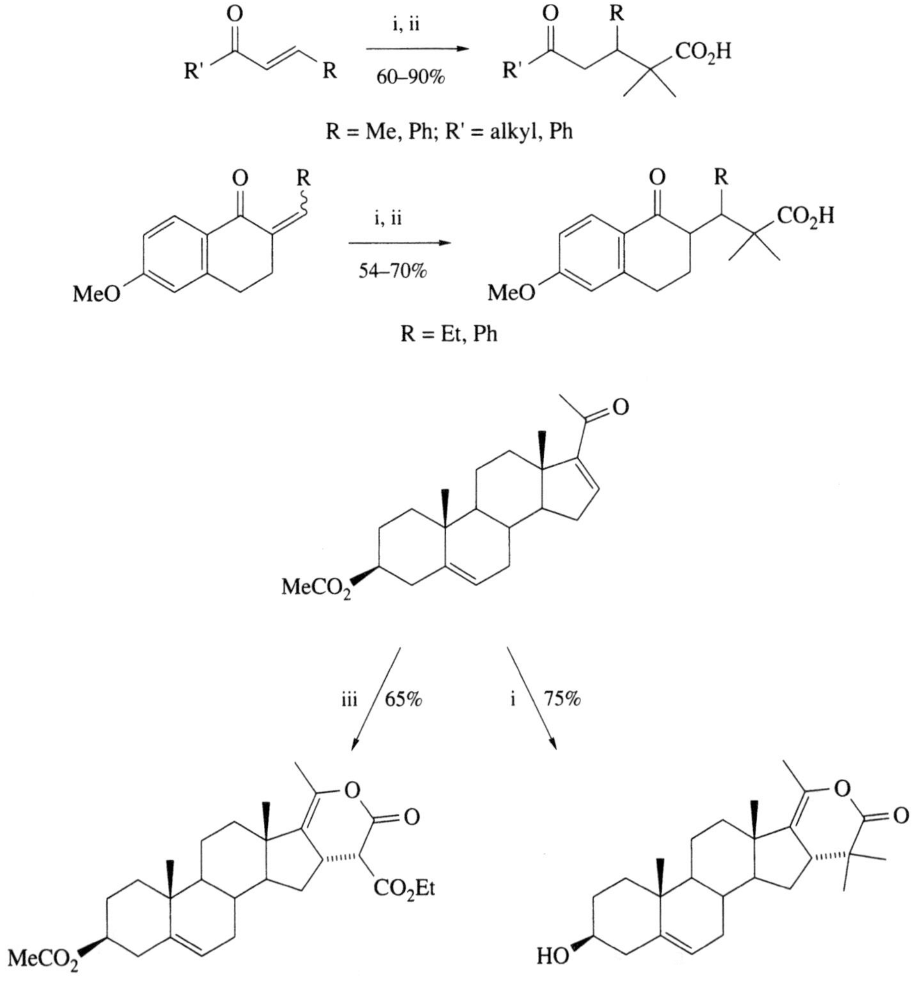

R = Me, Ph; R' = alkyl, Ph

R = Et, Ph

i, $BrZnC(Me)_2CO_2Et$, THF, 68 °C; ii, HO^-, H_3O^+; iii, $BrZnHC(CO_2Et)_2$, THF, 68 °C

Scheme 66

Various groups have reported that additions of α-hetero substituted ester enolates to α,β-enones are temperature dependent,[140a–c] but, in general, the 1,2 to 1,4-equilibration of α-seleno and α-thio substituted ester adducts occurs at lower temperature than the α-oxo substituted ester adducts. In contrast to the simple ester enolates, the α-hetero substituted ester enolates are extremely useful for functionalization of alicyclic α,β-enones with the tandem conjugate addition–electrophile trapping protocol, as shown in Scheme 67.[141] However, α-silyl ester enolates undergo conjugate additions to cyclopentenones only. Yamamoto reports that conjugate additions of α-thio substituted esters (or nitriles) can be achieved with catalytic 18-crown-6- potassium *t*-butoxide complex (5%) in toluene, while modest enantioselectivity is obtained with chiral crown ethers (>41% *ee*).[140d] McMurry and coworkers have reported that the conjugate addition of magnesium monoethyl malonate (**169**) to β-substituted-α,β-enones and β-substituted nitroalkenes introduces an acetic acid ester moiety which complements the above lithium enolate procedures which are not reliable for simple ester enolates.[142]

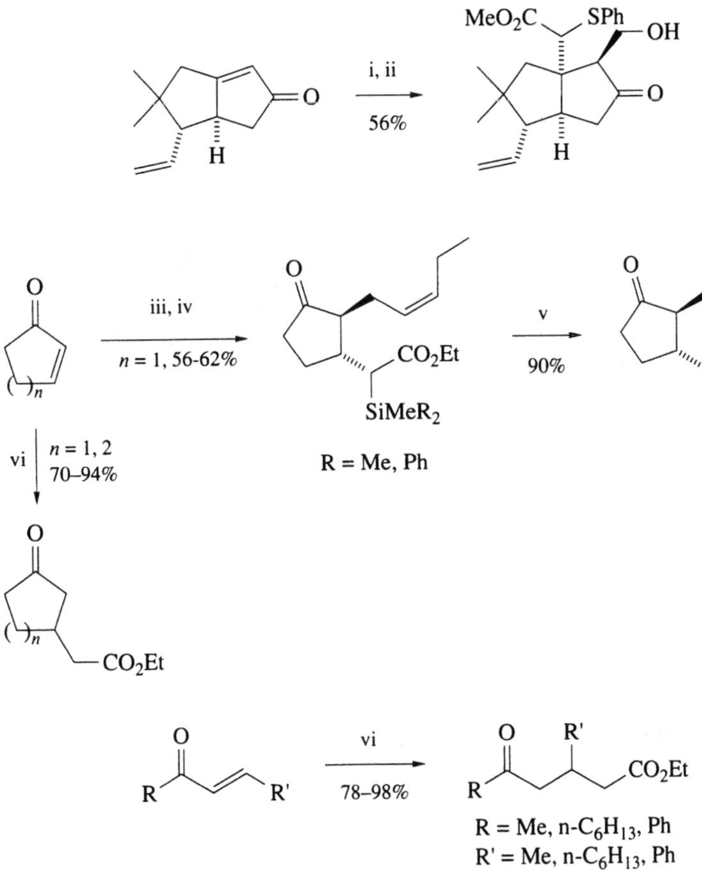

i, ii 56%

iii, iv n = 1, 56-62%

v 90%

vi n = 1, 2 70–94%

$SiMeR_2$ R = Me, Ph

vi 78–98%

R = Me, n-C_6H_{13}, Ph
R' = Me, n-C_6H_{13}, Ph

i, $PhSCH_2CO_2Me$, LDA, THF, –60 to –35 °C; ii, CH_2O; iii, $MeR_2SiCH_2CO_2Et$, LDA, THF, –20 °C; iv, $EtCH{=}CHCH_2Br$, –40 °C; v, KF, MeOH, 25 °C; vi, $Mg^{2+}(EtO_2CCHCO_2)^{2-}$ **(169)**

Scheme 67

Metzner reports that lithium dithioenolates, which are softer nucleophiles than the corresponding carbonyl enolates, add kinetically in a 1,4-mode exclusively to α,β-enones and the diastereoselectivity preferences are similar to that for ester enolate additions.[143a–d] Typically, kinetic deprotonation of dithioesters affords predominantly the (*Z*)-enethiolate (**170**) which is opposite to that for esters;[139] thus, with (*E*)-enones, *anti* adducts (**172**) are obtained. In contrast, the addition of methyl dithioacetate to α,β-disubstituted enones affords predominantly *syn* adducts (**174** and **175**) which is a consequence of intramolecular protonation of the resultant enolate (Scheme 68).[143e]

Few examples exist for the conjugate addition of ester enolates to α,β-unsaturated esters; typically the incipient enolate undergoes decomposition and secondary reactions. The first examples, described by Schlessinger,[144] are the addition of *t*-butyl lithioacetate and *t*-butyl α-lithio-α-(methylthio)propionate to butenolide (**176**; Scheme 69). Similarly, Normant reported that cyclopropanes are obtained from α-haloesters (**177**) and ethyl acrylate or acrylonitrile.[145]

The addition of ester enolates to α,β-unsaturated esters occurs with identical diastereoselectivity as shown for α,β-enones; this is exemplified by the diastereoselective synthesis of *erythro*- and *threo*-2,3-disubstituted glutarate esters (**178** and **179**).[146a–c] Similarly, Yamaguchi reports a general synthesis of *trans*-2-alkoxycarbonyl-1-cycloalkanepropionates (**181–184**) from ω-halo-α,β-unsaturated esters (**180**),

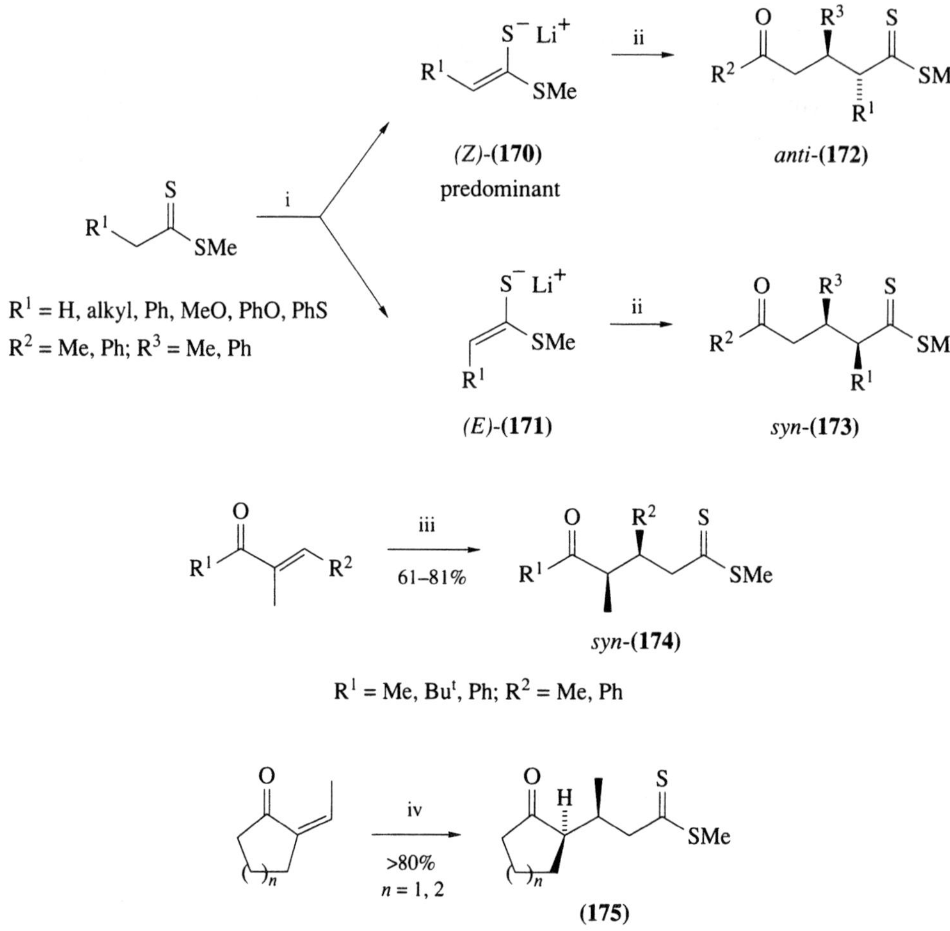
(Z)-(170)
predominant
anti-(172)
R¹ = H, alkyl, Ph, MeO, PhO, PhS
R² = Me, Ph; R³ = Me, Ph
(E)-(171)
syn-(173)
61–81%
syn-(174)
R¹ = Me, Buᵗ, Ph; R² = Me, Ph
>80%
n = 1, 2
(175)

i, LDA, THF, –78 °C; ii, (E)-$R^2COCH{=}CHR^3$; iii, $MeCS_2Me$, LDA, THF, –5 to 0 °C; iv, $MeCS_2Me$, LDA, THF, –45 to –25 °C

Scheme 68

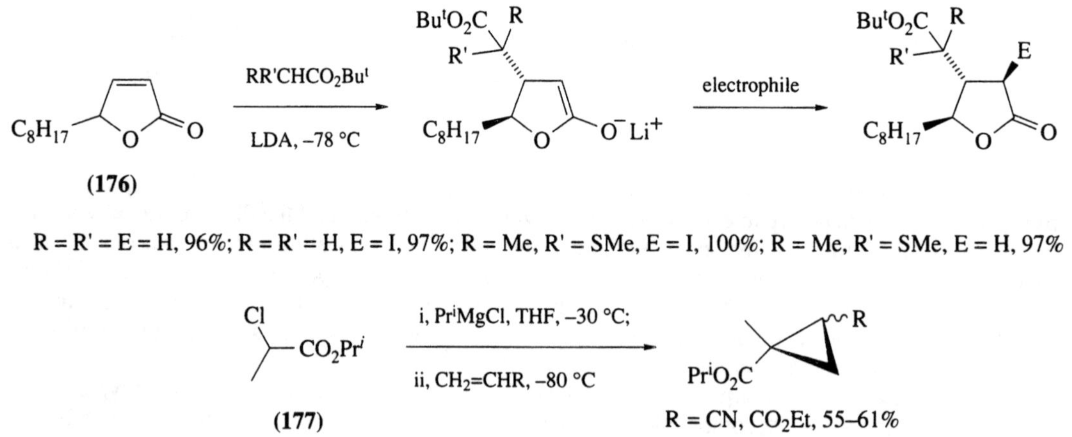
RR'CHCO₂Buᵗ
LDA, –78 °C
electrophile
(176)
R = R' = E = H, 96%; R = R' = H, E = I, 97%; R = Me, R' = SMe, E = I, 100%; R = Me, R' = SMe, E = H, 97%
i, PrⁱMgCl, THF, –30 °C;
ii, CH₂=CHR, –80 °C
(177)
R = CN, CO₂Et, 55–61%

Scheme 69

in a MIRC protocol, in which three contiguous asymmetric centers are formed (Scheme 70) (*cf.* **17**→**21**).[146d–h]

Posner has shown that ester enolates add to enantiomerically pure 2-(arylsulfinyl)-α,β-unsaturated lactones (**185**) and 2-(arylsulfinyl)cycloalkenones (**97**) to afford β-(carboxymethyl)lactones or β-(carboxy-

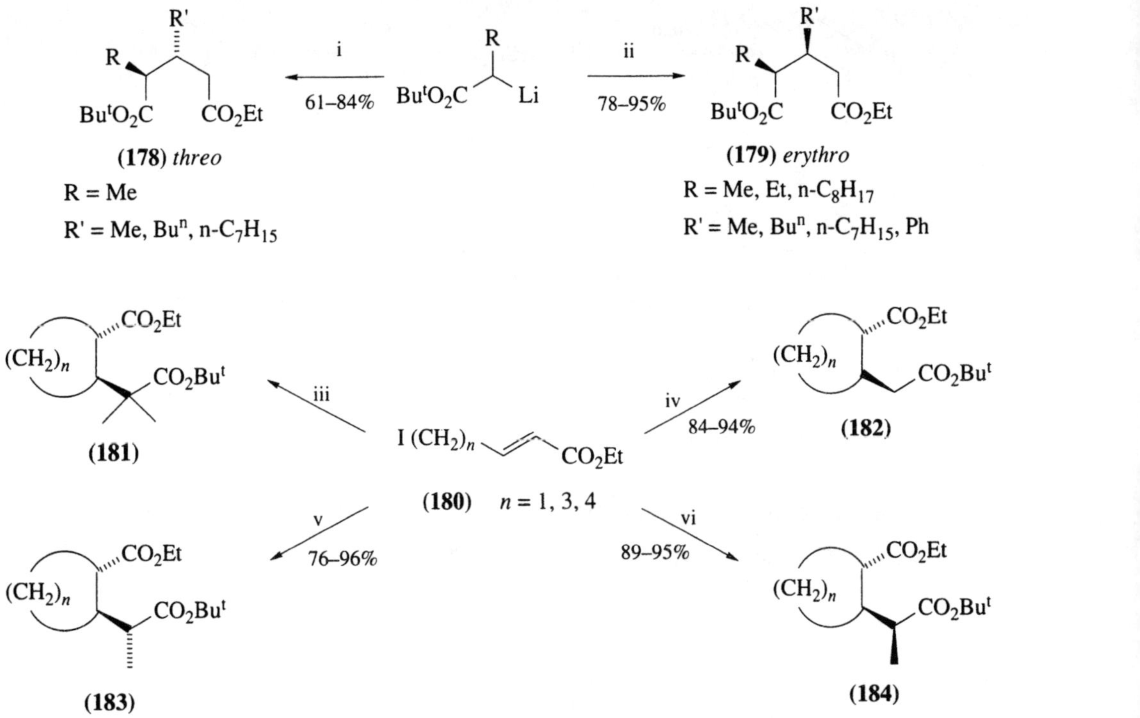

(178) *threo* R = Me R' = Me, Bu^n, n-C_7H_{15}

(179) *erythro* R = Me, Et, n-C_8H_{17} R' = Me, Bu^n, n-C_7H_{15}, Ph

(180) n = 1, 3, 4

i, *(Z)*-R'CH=CHCO$_2$Et, THF, –78 °C; ii, *(E)*-R'CH=CHCO$_2$Et, THF, HMPA, –78 °C; iii, $Pr^iCO_2Bu^t$, LDA, Bu^tOK, THF, –78 °C; iv, $MeCO_2Bu^t$, LDA, Bu^tOK, THF, –78 °C; v, $EtCO_2Bu^t$, LDA, Bu^tOK, THF, –78 °C; vi, $EtCO_2Bu^t$, LDA, Bu^tOK, HMPA, THF, –78 °C

Scheme 70

methyl)cycloalkanones, respectively, after reductive cleavage of the sulfinyl unit. The best enantiomeric selectivity was obtained with α-thio substituted ester enolates (Scheme 71).[147]

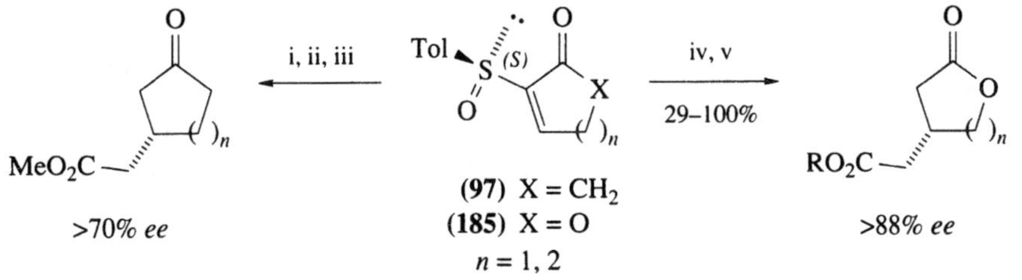

>70% *ee* — **(97)** X = CH_2 **(185)** X = O n = 1, 2 — >88% *ee*

i, $Me_3SiCH_2CO_2Me$, LDA, –78 °C; ii, Al–Hg; iii, KF; iv, YCH_2CO_2R (Y = H, PhS, 4-MeC_6H_4S); v, Raney Ni

Scheme 71

Schlessinger has shown that the addition of ester enolates to sulfur stabilized acceptors, *e.g.* ketene dithioacetal monoxide (**151**) and methyl α-(methylthio)acrylate (**187**), is highly efficient for the synthesis of γ-ketoesters.[148] Similarly, Ahlbrecht and Seebach have reported that amide and ester enolate additions to nitrogen stabilized acceptors, *e.g.* nitroalkenes (**40**) and 2-(*N*-methylanilino)acrylonitrile (**59**; Scheme 72), are highly efficient.[149]

Masked chiral α-hetero substituted carboxylic acid enolates have also shown utility in diastereoselective additions to nitroalkenes. For example, derivatives of α-hydroxycarboxylic acids, *e.g.* 1,3-dioxolan-4-ones (**187**); α-amino acids, *e.g.* 1,3-imidazolidin-4-ones (**188**); and α-amino-β-hydroxycarboxylic acids, *e.g.* methyl 1,3-oxazolidin-4-carboxylates (**189**) and methyl 1,3-oxazolin-4-carboxylates (**190**), have been employed.[150a] Further, diastereoselective additions of chiral β-hydroxyesters (**191**), *via* the enediolates, to nitroalkenes (**40**) afford predominant *anti*-β-hydroxyesters (**192**; Scheme

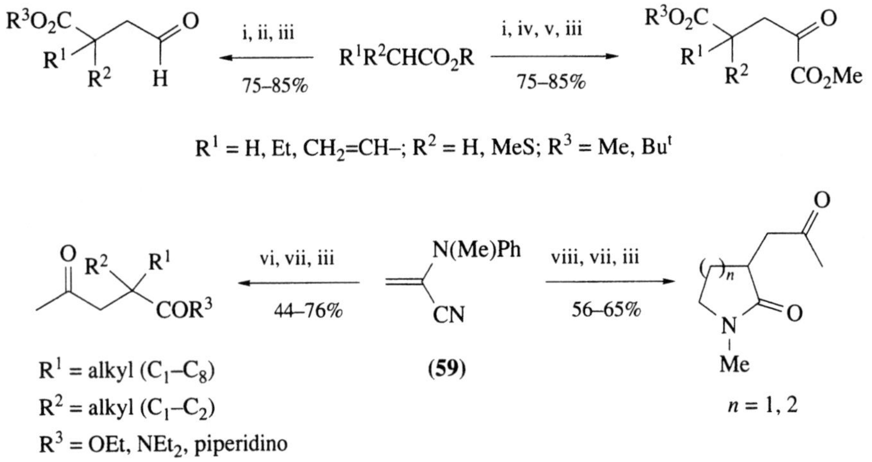

R^1 = H, Et, CH_2=CH–; R^2 = H, MeS; R^3 = Me, Bu^t

R^1 = alkyl (C_1–C_8)
R^2 = alkyl (C_1–C_2)
R^3 = OEt, NEt_2, piperidino

(59)

n = 1, 2

i, LDA, THF, –78 °C; ii, CH_2=C(SOMe)SMe (**151**); iii, H_3O^+; iv, CH_2=C(SMe)CO_2Me (**187**); v, LDA, MeSSMe, –78 °C; vi, amide or ester enolate, THF, –78 °C; vii, MeI; viii, lactam enolate, THF, –78 °C

Scheme 72

73).[150b] Hanessian reports a carbapenem synthesis in which intramolecular cyclization of the β-lactam nitroalkene (**193**) affords β-lactams (**194** and **195**; Scheme 74).[151]

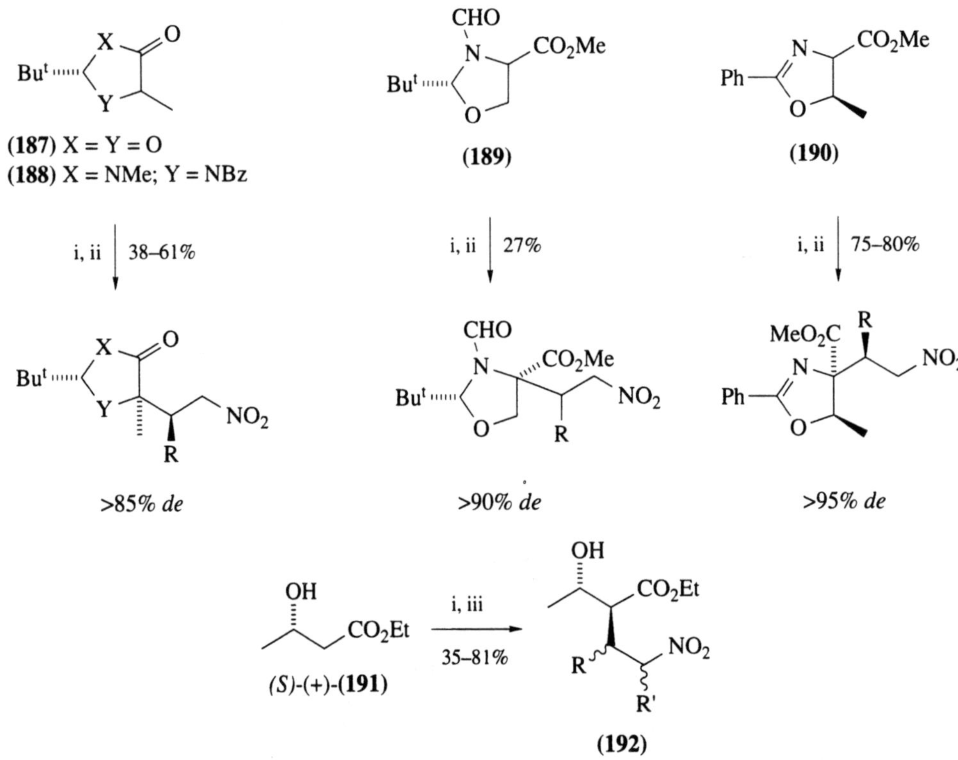

(**187**) X = Y = O
(**188**) X = NMe; Y = NBz

(**189**)

(**190**)

i, ii 38–61% — >85% *de*

i, ii 27% — >90% *de*

i, ii 75–80% — >95% *de*

(*S*)-(+)-(**191**) → i, iii, 35–81% → (**192**)

R = H, Me, Ph; R' = H, Me, $BrCH_2$, Ph, 3,4-(OCH_2)C_6H_3

i, LDA, –78 °C; ii, (*E*)-RCH=$CHNO_2$; iii, (*E*)-RCH=C(R')NO_2

Scheme 73

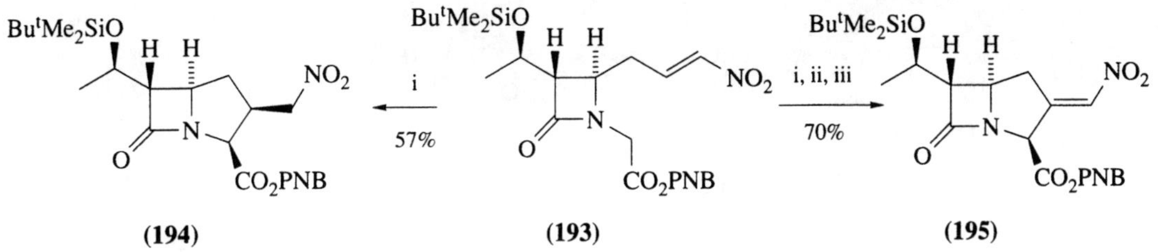

(194) **(193)** **(195)**

i, LiHMDS, THF, –100 °C; ii, PhSeCl; iii, H_2O_2

Scheme 74

Alternatively, enantioselective additions of chiral Schiff base glycine anion equivalents can also be used, as illustrated in Scheme 75. (+)-Phosphinothricin (**199**) and (+)-2-amino-4-phosphonobutyric acid (**200**) are obtained by the conjugate addition of the chiral *N*-alkylideneglycinate (**196**) to the phosphorus acceptors (**197**) and (**198**),[152a–c] while addition of the dihydropyrazine glycine equivalent (**201**)[152d] to α,β-unsaturated esters affords an asymmetric synthesis of glutamic acids and pyroglutamate esters.[152e]

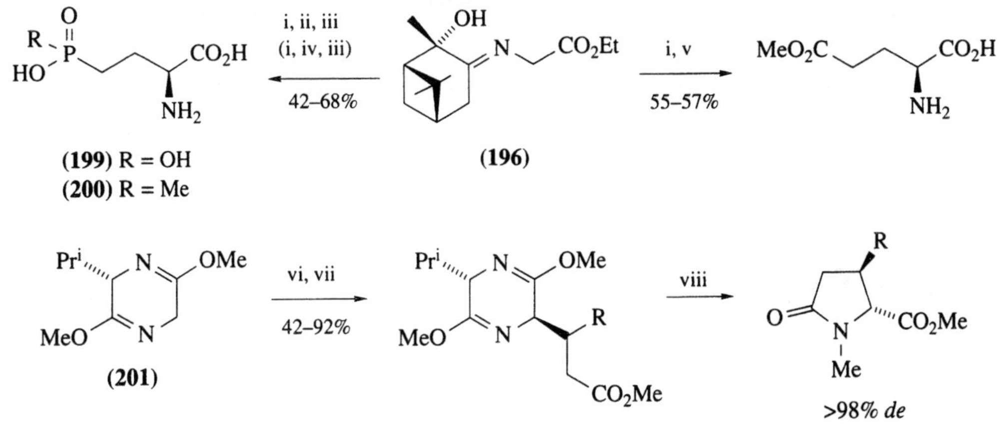

(199) R = OH
(200) R = Me

(196)

(201)

>98% *de*

i, Bu^tOK, THF, –78 °C; ii, $(EtO)_2P(O)CH{=}CH_2$ (**197**); iii, 6N HCl, Δ; iv, $Me(MeO)P(O)CH{=}CH_2$ (**198**); v, $CH_2{=}CHCO_2Me$; vi, Bu^nLi, THF, –78 °C; vii, $RCH{=}CHCO_2Me$, –78 to 0 °C; viii, 0.25 N HCl, 25 °C

Scheme 75

The addition of lithium alkoxydienolates to α,β-enones occurs exclusively in the 1,4(α)-mode. For example, alkoxydienolate (**202**), obtained from ethyl senecioate, adds efficiently, in a tandem conjugate addition–allylation protocol, to cyclopentenone to afford the α,β-functionalized cyclopentanone (**203**).[153] In contrast, the lithium dienolate (**204**), from 5-methylbutenolide, affords exclusive γ-alkylation,[154a,b] while the analogous phthalide enolates (**206**) can be exploited to accomplish regiospecific polynuclear aromatic syntheses (Scheme 76).[154c–g]

1.2.2.2.3 α,O-Carboxylic dianions and α-nitrile anions

In contrast to the ester enolates, the α,*O*-carboxylic dianions are intrinsically more reactive and their use in conjugate reactions is thus limited. Typically, α-substituted-α,*O*-carboxylic dianions add exclusively to α,β-unsaturated esters[155a] and nitroalkenes,[155b] while additions to α,β-enones are sensitive to the substitution pattern of the enones.[155c,d] Notable is the conjugate addition of dihydrobenzoic acid dianions (**207**), from Birch reduction of benzoic acids, to α,β-unsaturated esters (Scheme 77).[155e]

The addition of unstabilized α-nitrile carbanions to α,β-unsaturated carbonyl acceptors affords predominantly 1,2-addition products,[156] while lithiated acetonitrile derivatives having α-alkoxy, α-aromatic, α-dialkylamino, α-phenylselenyl, α-phenylthio or α-trimethylsilyl substituents afford 1,4-adducts. However, some of these are acyl anion equivalents (Section 1.2.2.3.2) so this discussion is limited to α-stabilized nitriles in which the nitrile function is retained after removal of the activating group. Notable examples are trimethylsilylacetonitrile (**208**),[157] phenylthioacetonitriles (**209**),[158a,b] phenylselenylacetoni-

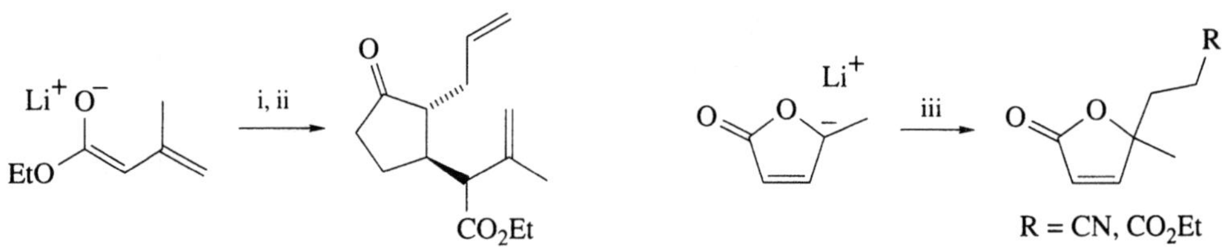

Li$^+$ O$^-$ EtO (202) —i, ii→ (203) CO_2Et; (204) Li$^+$ —iii→ (205) R = CN, CO_2Et

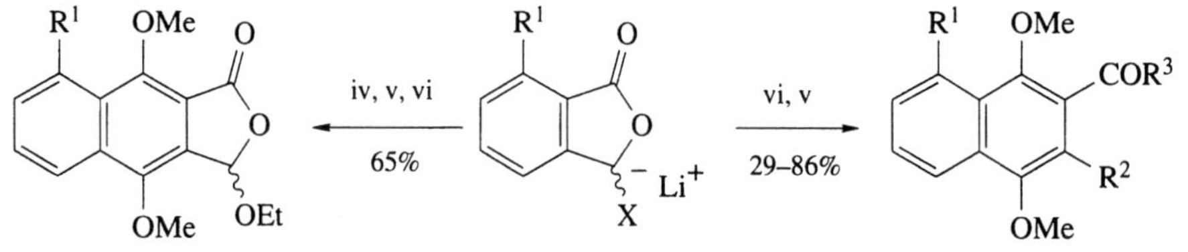

(206) R^1 = H, OMe; X = H, CN, $PhSO_2$, PhS; iv, v, vi 65%; vi, v 29–86%; R^2 = H, Me, CH_2SMe; R^3 = Me, EtO

i, 2-cyclopenten-1-one, THF, –78 °C; ii, CH_2=$CHCH_2Br$; iii, CH_2=CHR, THF, –78 °C; iv, 5-ethoxy-2(5*H*)-furanone; v, $(MeO)_2SO_2$, K_2CO_3, acetone; vi, (*E*)-R^2CH=$CHCOR^3$

Scheme 76

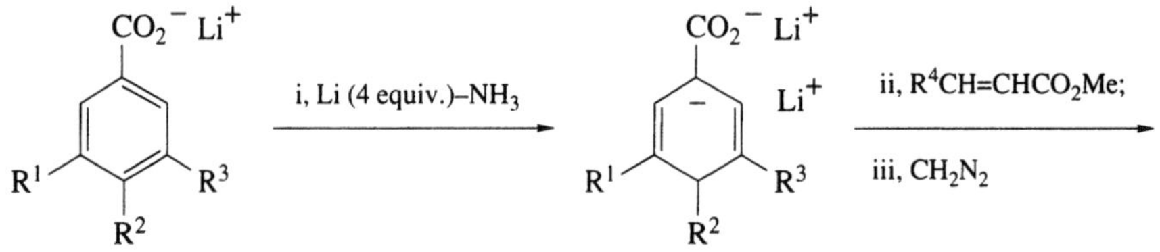

i, Li (4 equiv.)–NH_3 → (207) → ii, R^4CH=$CHCO_2Me$; iii, CH_2N_2

$R^1 = R^2 = R^3$ = H; $R^1 = R^3$ = H, R^2 = Me; $R^1 = R^3$ = OMe, R^2 = H

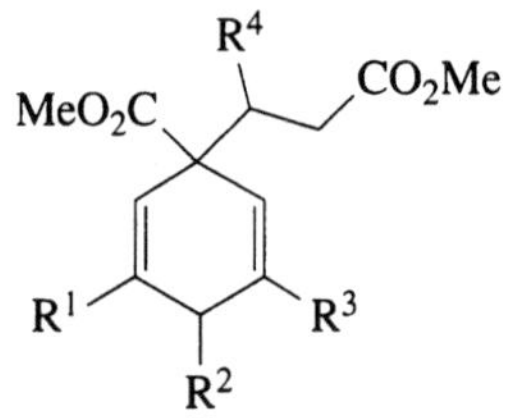

R^4 = H, Me

Scheme 77

trile (**210**)[158c] and phenylsulfonylacetonitrile (**211**).[158d] Seyden-Penne has reported that α-arylacetonitriles (**212**) add to α,β-cycloalkenones in a tandem conjugate addition–alkylation protocol to afford *trans* 2,3-disubstituted cycloalkanones (Scheme 78).[159] In contrast, Ahlbrecht has reported a versatile γ-ketonitrile synthesis in which unstabilized α-nitrile carbanions add to acceptor (**59**; equation 19).[160]

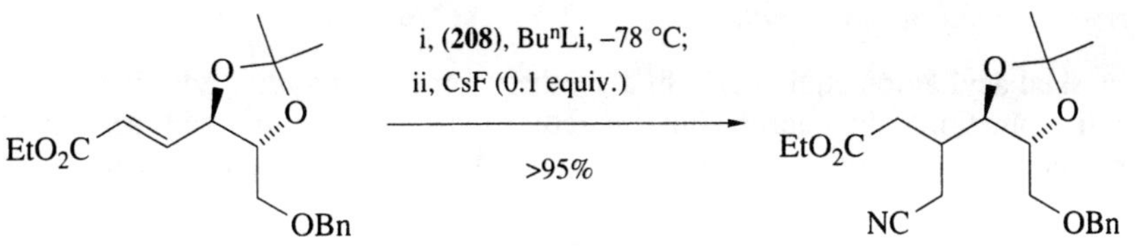

i, (**208**), Bu^nLi, –78 °C;
ii, CsF (0.1 equiv.)
>95%

R(X)CHCN

(**208**) X = $SiMe_3$; R = H
(**209**) X = PhS; R = H, alkyl
(**210**) X = PhSe; R = n-C_4H_9
(**211**) X = $PhSO_2$; R = H
(**212**) X = aryl; R = H

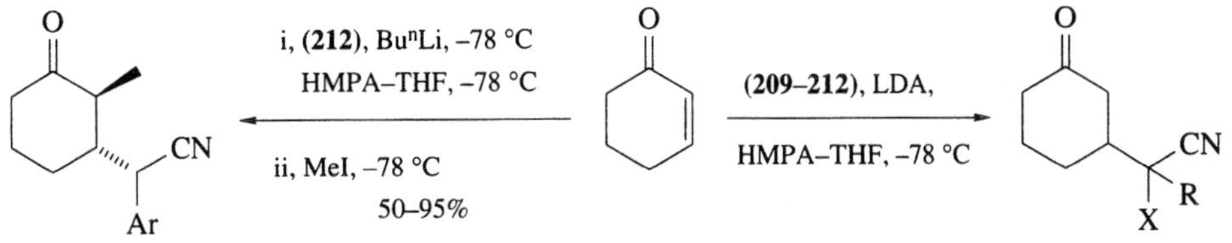

i, (**212**), Bu^nLi, –78 °C HMPA–THF, –78 °C
ii, MeI, –78 °C
50–95%

(**209–212**), LDA, HMPA–THF, –78 °C

Scheme 78

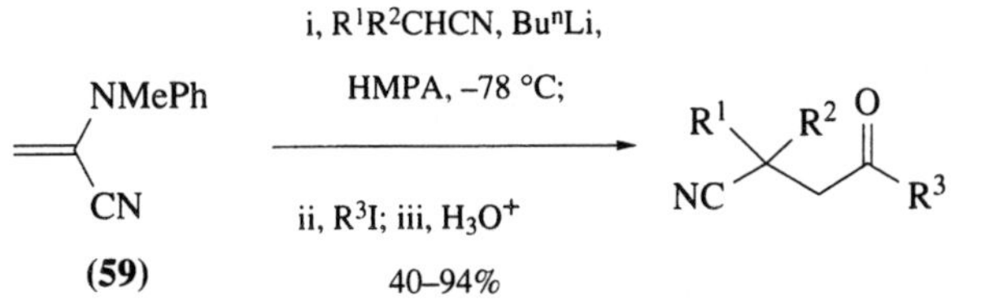

i, R^1R^2CHCN, Bu^nLi, HMPA, –78 °C;
ii, R^3I; iii, H_3O^+
40–94%

(**59**) (19)

R^1 = alkyl (C_1–C_4); R^2 = H, Me; R^3 = alkyl (C_1–C_4), allyl, benzyl

1.2.2.3 Additions of Acyl Anion Equivalents

1.2.2.3.1 Masked acyl anion equivalents

The realization of the principle of 'umpolung', in which reactivity at a normally electrophilic site is inverted to that of a nucleophilic species,[161] has resulted in the development of numerous masked acyl anion equivalents for carbon–carbon bond formation.[162] The use of masked acyl anion equivalents for conjugate additions was made possible by several observations. First, Stork reported that protected cyanohydrins afforded preferential kinetic 1,4-addition with α,β-enones[16b] and second, Brown reported that preferential kinetic 1,4-addition of 1,3-dithianes was effected by the use of HMPA.[163]

The masked acyl anion equivalents discussed in this section are divided into two classes, based on their unmasked functional groups, *i.e.* acyl and carboxylate. Representative acyl group synthons include the 1,3-dithianes (**214**),[164] bis(arylthio)acetals (**215**), [165] bis(arylseleno)acetals (**216**),[165] di(ethylthio)acetal monoxides (**217**),[166] α-silyl sulfides (**218**),[167] α-(dialkylamino)nitriles (**219**),[168] protected cyanohydrins (**220** and **221**)[16b,159b,168d–e,169] and *t*-butylhydrazone aldimines (**222**).[170] The carboxylate group synthons include the trithioorthoformates (**223**),[171] α-(trialkylmetallo)dithioacetals (**224** and **225**),[172] α-cyanodithioacetals (**226**),[173] methoxy(phenylthio)(trimethylsilyl)methane (**227**)[174] and the benzothiazoles (**228**; Scheme 79).[46,114,175] Numerous groups have reported additions of (**214–228**) to α,β-enones, α,β-unsaturated esters and α,β-unsaturated nitriles. The addition of the protected cyanohydrin (**229**),[169c] carboxylate synthon (**223**)[171e–i] or 1,3-dithiane (**230**)[176] to butenolides in a tandem conjugate addition–electrophile trapping protocol is representative of their synthetic versatility (Scheme 80).

1.2.2.3.2 Formal acyl anion equivalents

The use of masked acyl anion equivalents in a synthetic protocol requires additional steps to unmask the carbonyl unit. Sometimes the deprotection procedures are incompatible with sensitive compounds; thus, a direct nucleophilic acylation protocol is desirable. While C-nucleophilic carbonyl groups do not

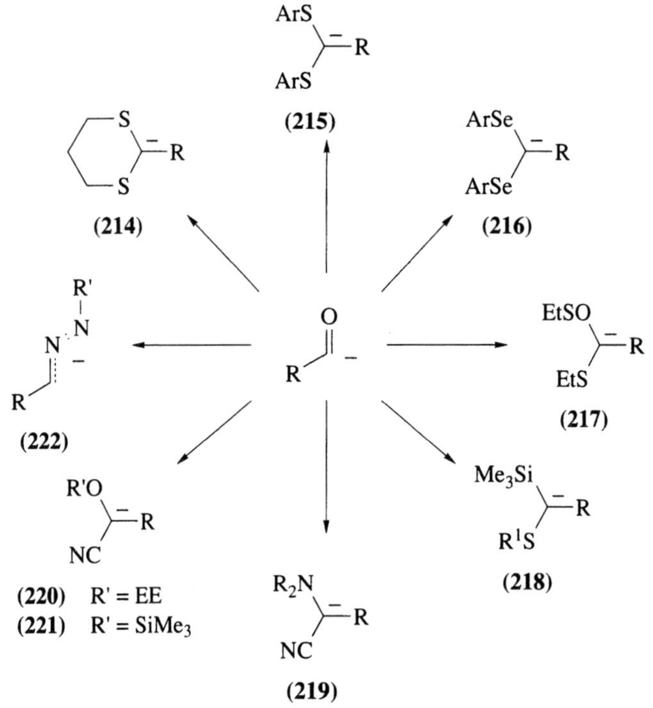

(214) (215) (216) (217) (218) (219) (220) R' = EE (221) R' = $SiMe_3$ (222)

Acyl Anion Equivalents

(223) (224) (225) (226) (227) (228)

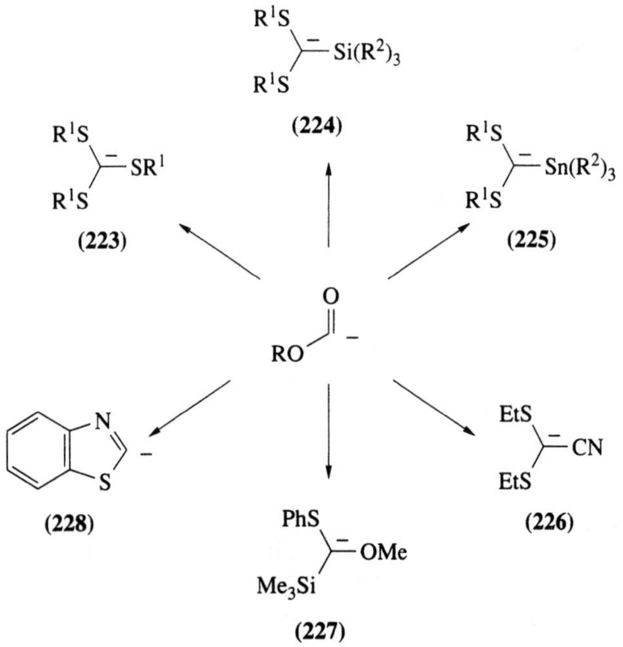

Carboxylate Equivalents

Scheme 79

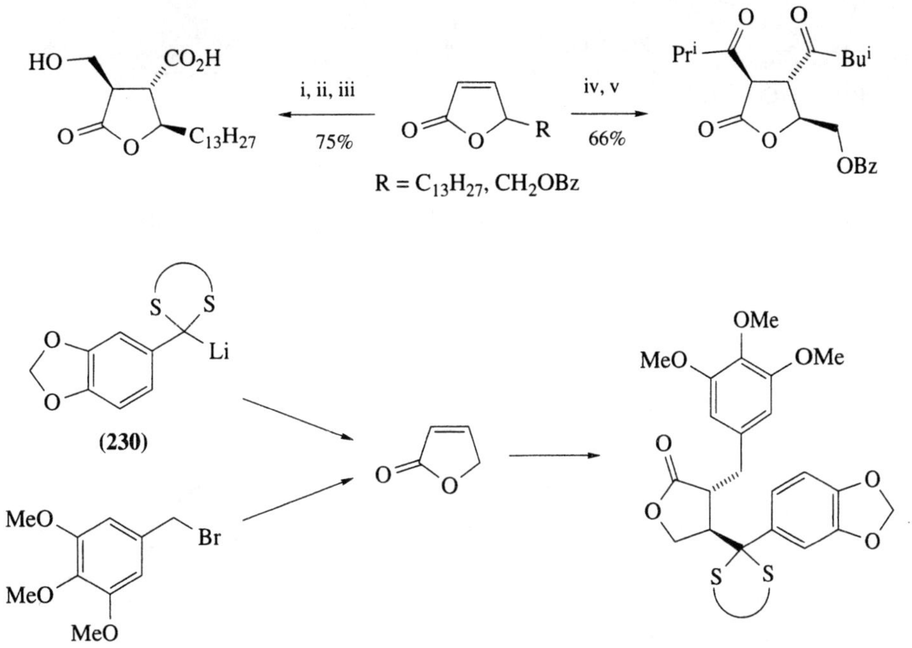

i, $(MeS)_3CH$ (**223**), LDA, THF, HMPA, –78 °C; ii, CH_2O, –40 °C; iii, HgO, BF_3–Et_2O;
iv, $Pr^iCH_2CH(OEE)CN$ (**229**), LDA, THF, HMPA, –78 °C; v, Pr^iCOCl, –78 to 25 °C

Scheme 80

exist *per se*,[177] the organotetracarbonylferrates (**231**) or acylnickel complexes (**232**) afford formal 1,4-addition of β-acyl units to α,β-unsaturated esters, α,β-enones or α,β-unsaturated nitriles. The two reagents differ mechanistically and in synthetic versatility; for example, the organoferrate system is sensitive to steric effects on the acceptor and tandem trapping at the α-position is not effective.[178] In contrast, the acyl nickel complexes are insensitive to steric effects on the acceptor and amenable to tandem electrophilic trapping protocols,[179a] as reported by Semmelhack, in an efficient synthesis of pyranonaphthoquinone antibiotic intermediates from quinone monoketals (**233**; Scheme 81).[179b]

Variations to the above processes have been introduced to obviate handling of the volatile and toxic metal carbonyls. For example, the tricarbonyl(α,β-enone)iron complexes (**234**),[180a–c] on treatment with organolithiums (or Grignard reagents), afford the corresponding 1,4-diketones[180d,e] or γ-ketoamides.[180f] The highly stable cobalt complex, $Co(NO)(CO)_2(Ph_3P)$,[181a] on reaction with alkyllithiums, affords an intermediate carbonylate complex which adds exclusively to α,β-enones; however, data is lacking as to the applicability of this complex in a tandem conjugate addition–electrophile trapping protocol.[181b]

Alternatively, the expedient cyanide (or fluoride) ion desilylation of acylsilanes (**235**), in the presence of α,β-enones or α,β-unsaturated esters, affords γ-dicarbonyl adducts;[182] even bis(trimethylsilyl)ketone (**236**), a formyl anion equivalent, adds to 2-cyclohexen-1-one to afford 3-formylcyclohexanone in moderate yields.[182e]

1.2.2.4 Additions of α-Heteroatom-stabilized Carbanions (Non-acyl Anion Equivalents)

1.2.2.4.1 Phosphorus- and sulfur-stabilized carbanions

The synthetic utility of α-phosphorus- and α-thio-stabilized carbanions is the subject of numerous reviews.[21] Notable are additions of phosphonium ylides (**237**),[183] sulfonium ylides (**238**),[184a–c] oxosulfonium ylides (**239**)[184a–c] and sulfoximine ylides (**240**)[184d] to electron-deficient alkenes which afford nucleophilic cyclopropanation products. In contrast, with α-(phenylthio)-stabilized carbanions, which are not acyl anion equivalents, either nucleophilic cyclopropanation or retention of the hetero substituent occurs, depending on the acceptor and reaction conditions used. For example, carbanion (**241**) adds to 1,1-

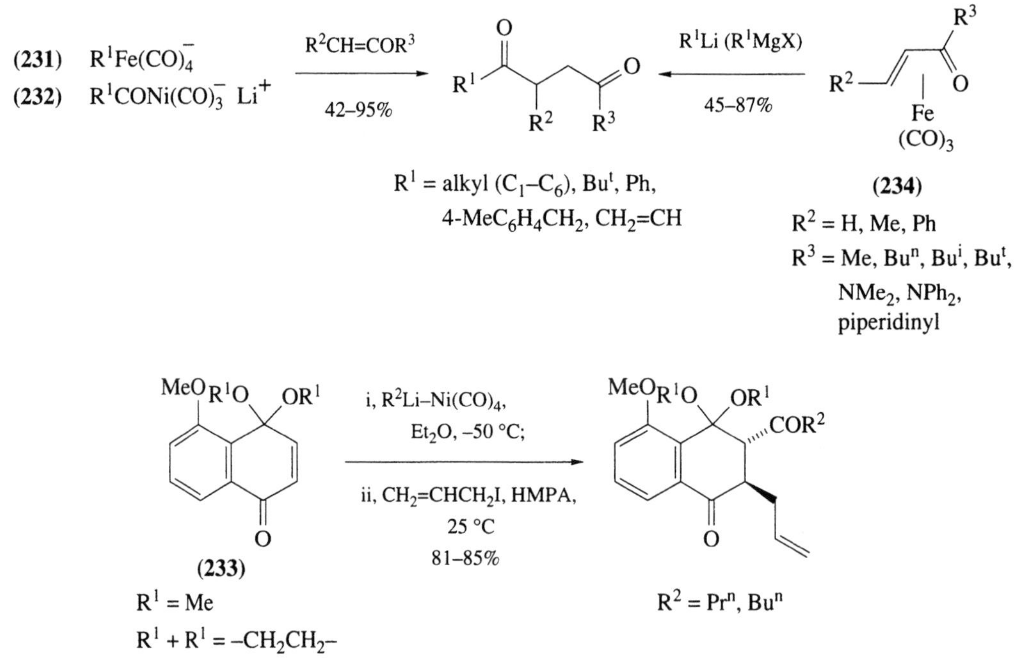

(231) $R^1Fe(CO)_4^-$
(232) $R^1CONi(CO)_3^-\ Li^+$

$R^2CH{=}COR^3$, 42–95%

R^1Li (R^1MgX), 45–87%

R^1 = alkyl (C_1–C_6), Bu^t, Ph, 4-$MeC_6H_4CH_2$, $CH_2{=}CH$

(234)
R^2 = H, Me, Ph
R^3 = Me, Bu^n, Bu^i, Bu^t, NMe_2, NPh_2, piperidinyl

(233)
R^1 = Me
$R^1 + R^1 = -CH_2CH_2-$

i, R^2Li–$Ni(CO)_4$, Et_2O, –50 °C;
ii, $CH_2{=}CHCH_2I$, HMPA, 25 °C
81–85%

R^2 = Pr^n, Bu^n

(235) $RCOSiMe_3$
(236) $Me_3SiCOSiMe_3$

Scheme 81

doubly stabilized a^2-synthons, *e.g.* *N*-(methylanilino)acrylonitrile (**59**),[185a] α-TMS vinyl phenyl sulfide (**242**),[185b] 1,1-bis(phenylthio)ethylene (**243**),[185c] with concomitant S_N2 displacement, while normal addition occurs with α,β-enones (Scheme 82).[185d] The initial products can be further desulfurized as shown for adduct (**244**).[186]

In contrast, few examples of conjugate additions of nonallylic α-sulfinyl (or α-sulfonyl) carbanions have been reported (for allylic α-sulfinyl carbanion additions, see Section 1.2.2.5.1). Notable is the diastereoselective addition of alkyl *t*-butyl sulfoxides (**245**) to α,β-unsaturated esters (equation 20)[187] which is complementary to the diastereoselective addition of enolates to β-substituted-α,β-unsaturated sulfoxides (equation 20).

1.2.2.4.2 Oxygen- and silicon-stabilized carbanions

While there are few examples of conjugate additions of either α-oxygen- or α-silyl-stabilized carbanions, Tamao and Posner have reported two hydroxymethyl synthons [$^-CH_2OH$] (**246** and **247**) which show synthetic promise. Additions with the silicon-based synthon (**246**) is restricted to 2-cyclohexen-1-ones and work-up requires a successive acid and base procedure that is incompatible with sensitive molecules,[188a,b] while the tin-based synthon (**247**) is more versatile and the hydroxyl group is obtained under neutral conditions (Scheme 83).[188c]

1.2.2.4.3 Nitrogen-stabilized carbanions

Aside from the nitronates, cyanide anion and acyl anion equivalents, *e.g.* (**219**), examples of conjugate additions of α-aza-stabilized carbanions are rare. The aminomethyl synthon [$^-CHRNH_2$] is typically introduced with either nitronates or cyanide; however, α-metallomethyl isocyanides (**248**) also show synthetic promise in conjugate additions. In addition, depending on hydrolytic conditions employed, they also serve as equivalents for the *N*-formamidomethyl anion [$^-CHRNHCHO$] or the isocyanatomethyl anion [$^-CHRN{=}C{:}$] (Scheme 84).[189]

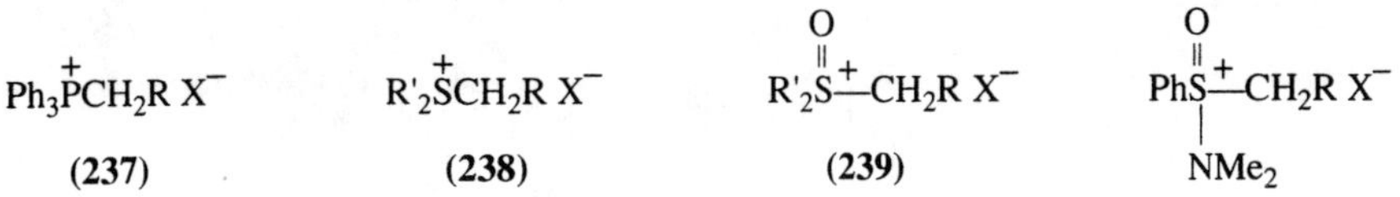

$Ph_3\overset{+}{P}CH_2R\ X^-$ (237) $R'_2\overset{+}{S}CH_2R\ X^-$ (238) $R'_2S(O)^+CH_2R\ X^-$ (239) $PhS(O)(NMe_2)^+CH_2R\ X^-$ (240)

RCH(Li)SPh (241)

i (ii) (iii), 76–100%; iv, 56–59%, n = 0, 1; v, vi, vii, viii, ix, 56–91%

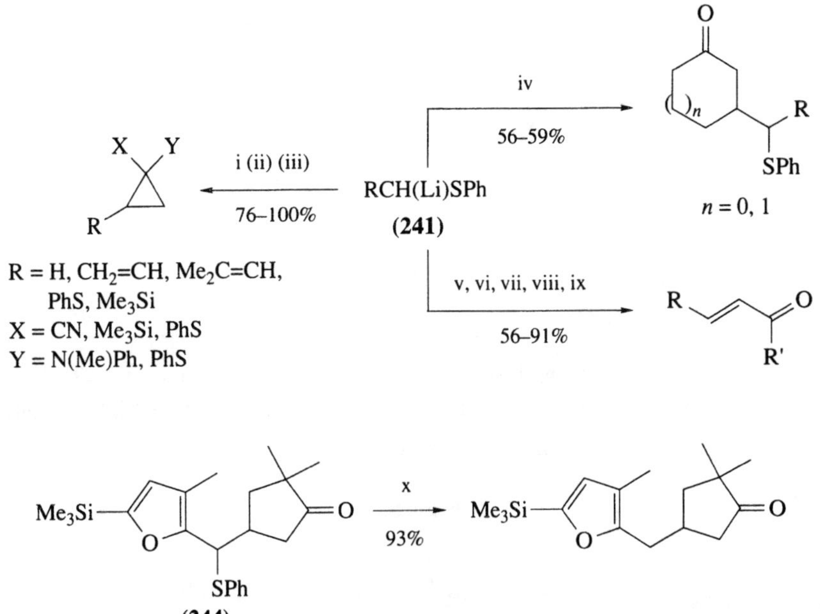

R = H, CH_2=CH, Me_2C=CH, PhS, Me_3Si
X = CN, Me_3Si, PhS
Y = N(Me)Ph, PhS

(244) x, 93%

i, CH_2=C(CN)N(Me)Ph (**59**), 25 °C, 30 h; ii, CH_2=C($SiMe_3$)SPh (**242**), TMEDA, –25 to 25 °C; iii, CH_2=C$(SPh)_2$ (**243**), TMEDA; iv, cycloalkenones, –78 °C; v, CH_2=C(CN)N(Me)Ph (**59**), –78 °C; vi, R'X; vii, 3N HCl; viii, MCPBA; ix, Δ, CCl_4; x, Raney Ni

Scheme 82

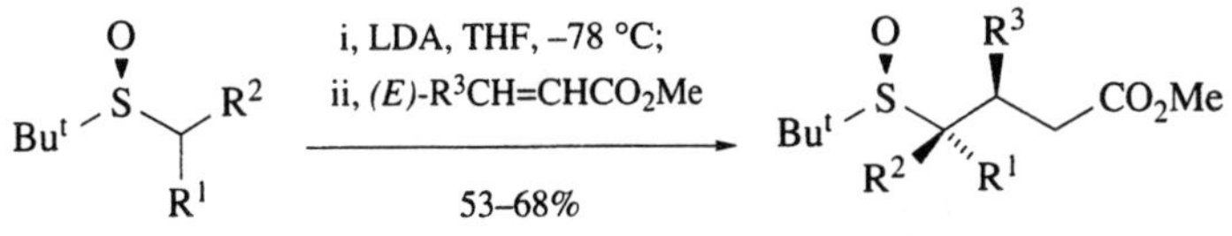

i, LDA, THF, –78 °C; ii, (E)-R^3CH=$CHCO_2Me$, 53–68% (20)

R^1 = alkyl (C_1–C_4), $PhCH_2$
R^2 = H, Me
(245)
R^3 = H, Me, Ph

1.2.2.5 Additions of Homoenolates or Homoenolate Equivalents

1.2.2.5.1 Homoenolate equivalents

The concept of homoenolization was recognized by Nickon in the 1960s but attempts at direct formation of homoenolates were frustrated by cyclopropanolate formation. This lack of success has prompted the development of homoenolate equivalents[190] of which the first example, the β-propionaldehyde anion equivalent (**112**), was previously discussed (Sections 1.2.2.1.2 and 1.2.2.1.3). Ghosez has shown that α-cyanoenamines (**249** and **250**) add preferentially in the 1,4(γ)-mode to cycloalkenones. The versatility of (**250**) which serves as either a β-carboxyvinyl anion equivalent [$^-CH{=}CHCO_2R$] or β-propionate anion equivalent [$^-CH_2CH_2CO_2R$] (Scheme 85) is notable.[191]

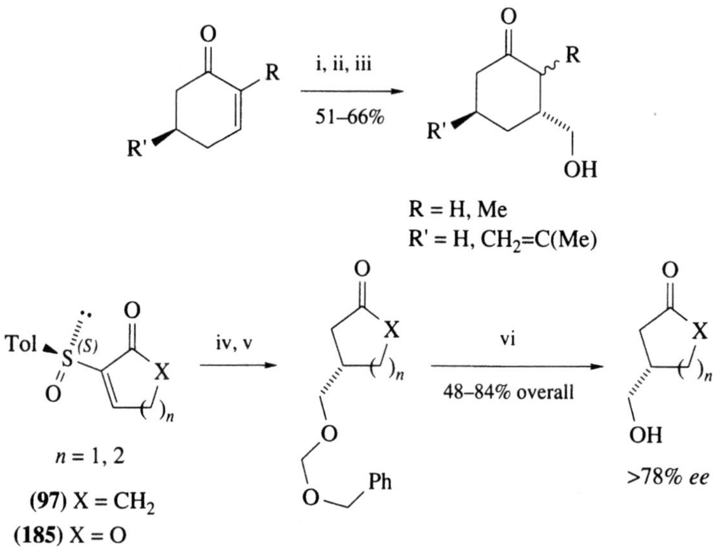

i, ii, iii
51–66%

R = H, Me
R' = H, CH_2=C(Me)

iv, v
vi
48–84% overall
n = 1, 2
(**97**) X = CH_2
(**185**) X = O
>78% *ee*

i, (CH_2=$CHCH_2$)Me_2SiCH_2MgCl (**246**), Et_2O, CuI (10%), 25 °C; ii, KHF_2 (2 equiv.), CF_3CO_2H, 50 °C; iii, 30% H_2O_2, $NaHCO_3$, MeOH, THF, 68 °C; iv, $PhCH_2OCH_2OCH_2SnBu^n_3$ (**247**), Bu^nLi (2 equiv.), –78 °C; v, Raney Ni, EtOH; vi, H_2, 10% Pd–C

Scheme 83

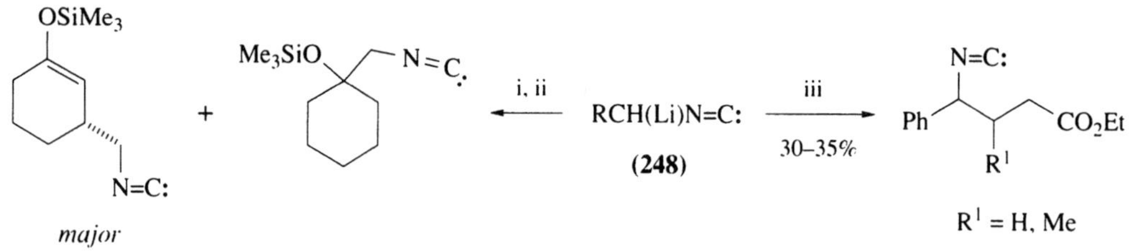

major
i, ii
RCH(Li)N=C:
(**248**)
iii
30–35%
R^1 = H, Me

i, cyclohexenone, –78 °C; ii, Me_3SiCl; iii, (*E*)-R^1CH=$CHCO_2Et$

Scheme 84

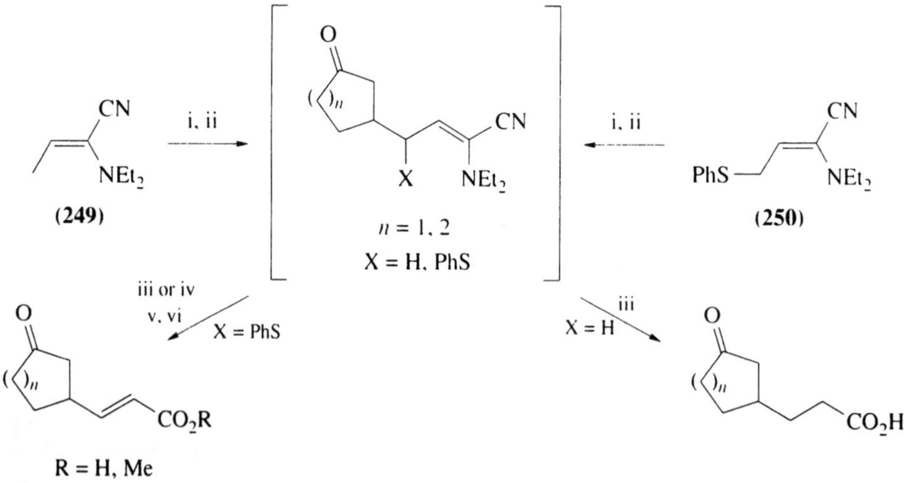

(**249**)
(**250**)
n = 1, 2
X = H, PhS
R = H, Me

i, LDA, THF, –60 to –30 °C; ii, cycloalkenones; iii, 3N HCl, Δ; iv, HCl, MeOH; v, MCPBA; vi, Δ, CCl_4

Scheme 85

Alternatively, the ambident α-hetero substituted allyl anions have been utilized as homoenolate equivalents. For example, in the presence of HMPA, allyl phenyl sulfides (**251**),[192a] allyl phenyl sulfones (**252**)[192b,c] and allyl phenyl selenides (**253**)[192d,e] add to α,β-enones in a $1,4_{(\alpha)}$-mode, while allyl phenyl sulfoxides (**254**) and allyl phosphine oxides (**255**) afford $1,4_{(\gamma)}$-addition exclusively, irrespective of solvent used.[193] Hua has shown that additions of either chiral sulfoxide (**254**; $R^1 = R^2 = R^3 = R^4 = H$, $R^5 = p$-tolyl) or allyl oxazaphospholidine oxide (**256**) occur with excellent enantioselectivity (>95% *ee*).[194] Similarly, Ahlbrecht reports that the α-azaallyl (**257**) adds exclusively in a $1,4_{(\gamma)}$-mode to acceptor (**59**) to afford 1,5-diketones (Scheme 86).[195]

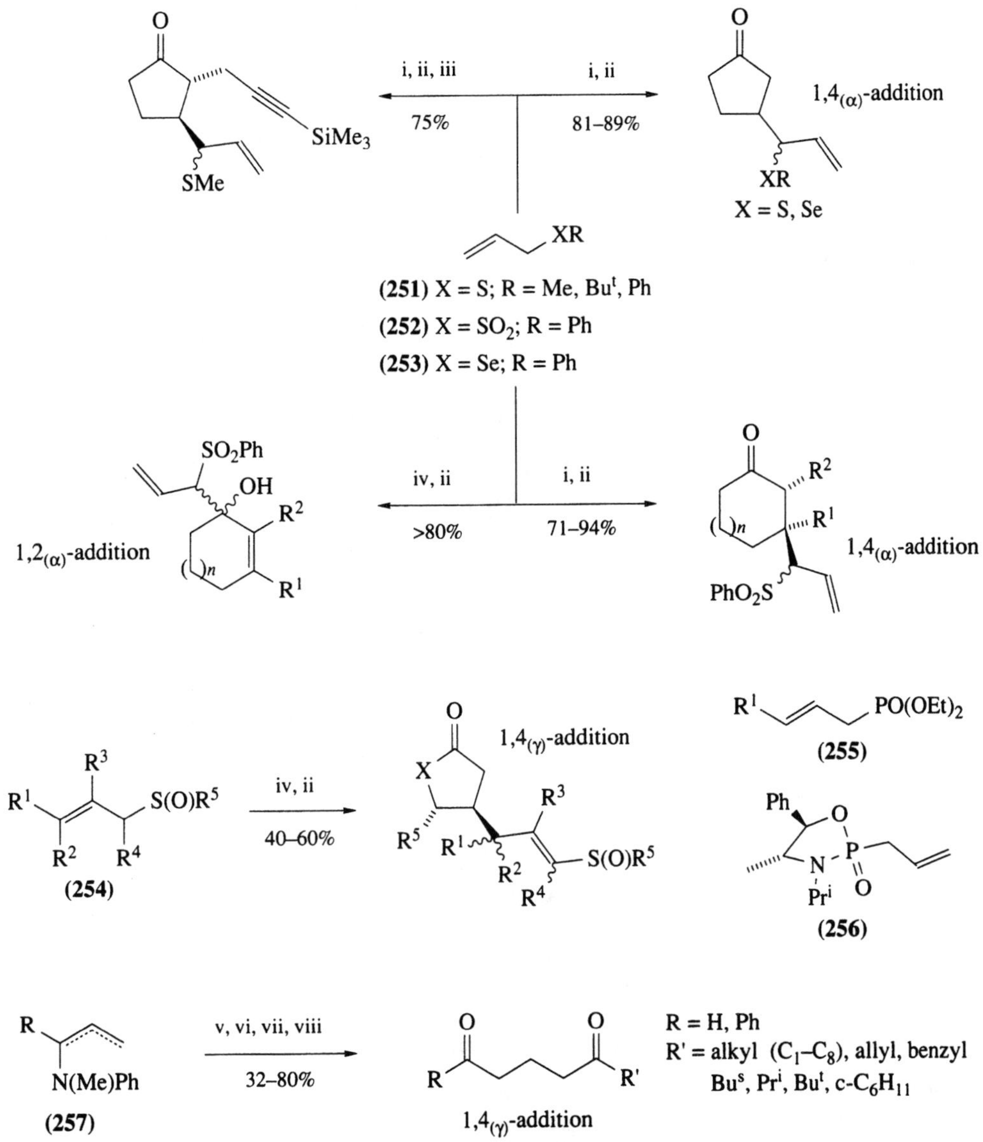

(**251**) X = S; R = Me, Bu^t, Ph
(**252**) X = SO_2; R = Ph
(**253**) X = Se; R = Ph

i, Bu^sLi (Bu^nLi), HMPA, –78 °C to –50 °C; ii, cycloalkenone or α,β-unsaturated lactone; iii, $BrCH_2C{\equiv}CSiMe_3$; iv, Bu^nLi, THF, –78 °C; v, Bu^nLi, TMEDA, THF, 0 °C; vi, $CH_2{=}C(CN)N(Me)Ph$ (**59**), –78 °C; vii, R'Br (R'I); viii, 3N HCl

Scheme 86

1.2.2.5.2 Formal homoenolates

Marino has reported that the fluoride-induced ring cleavage of 1-silyloxy-2-carboalkoxycyclopropanes affords γ-oxo-α-ester enolates, modified homoenolates, which add to Michael acceptors to afford cyclopentanes and cyclopentenes, *e.g.* (**258**→**259**);[196] alternatively, a carboannulation procedure affords octalin-1-ones (**260**→**261**; Scheme 87) (note the intramolecular enolate addition to the α,β-unsaturated sulfone).

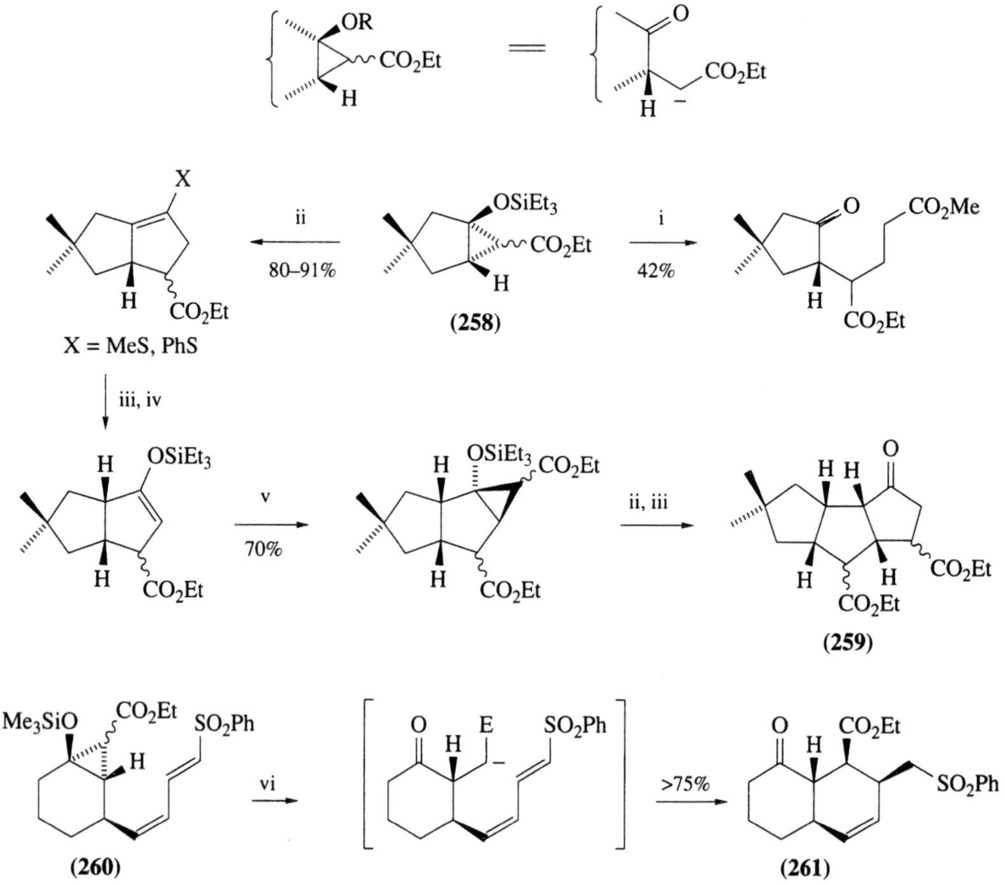

i, CH_2=$CHCO_2Me$, CsF; ii, CH_2=CX–$\overset{+}{P}Ph_3$ BF_4^-; iii, TFA; iv, LDA–THF, –78 °C, Et_3SiCl;

v, N_2=$CHCO_2Et$, $CuSO_4$; vi, CsF, DMF, 60 °C

Scheme 87

1.2.2.6 Heteroconjugate Additions (Carbometallation)

During the last three decades, the addition of organolithiums or Grignard reagents to vinyl derivatives of either second row (P, S or Si) or third row (Se) elements in their normal oxidation state has been recognized as an important procedure for the *in situ* generation of α-heteroatom stabilized carbanions. Carbometallation was initially recognized with vinylsilanes (**262**)[197] and subsequently for vinyl diarylphosphines (**263**),[198] aryl vinyl sulfides (**264**),[199] aryl vinyl selenides (**265**)[200] and 2-methylene-1,3-dithianes (**266**).[201] The driving force for these carbometallation reactions arises from the adjacent heteroatom stabilization of the incipient carbanion which is attributed to either charge dispersion or *d*-orbital interactions,[202] but it is not the role of this survey to discuss the issue.

Intermolecular carbometallations are typically accomplished with organolithiums; however, diastereocontrolled intramolecular carbometallation can be attained with appropriate Grignard reagents, as exemplified in Scheme 88, the suprafacial and 5-*exo*-trig closure of (*E*)-6-bromo-3-methyl-1-trimethylsilyl-1-hexene (**267**) to *trans*-1-methyl-2-(trimethylsilyl)methylcyclopentane (**268**).[203]

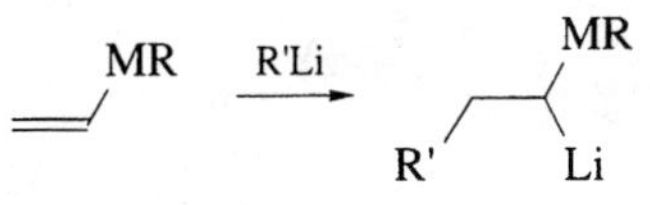

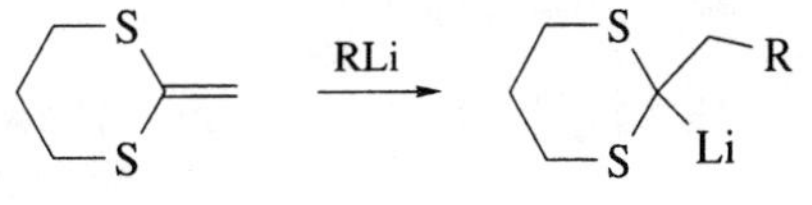

(**262**) MR = $SiMe_3$
(**263**) MR = PPh_2
(**264**) MR = SPh
(**265**) MR = SePh

(**266**)

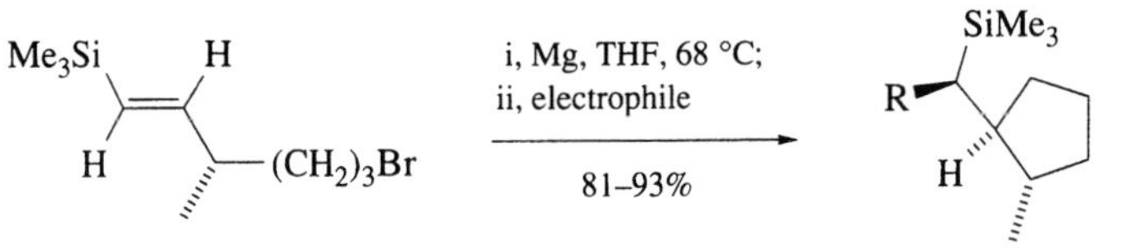

i, Mg, THF, 68 °C; ii, electrophile; 81–93%

(**267**) (**268**) R = H, D, allyl

Scheme 88

1.2.3 ANNULATIONS AND MULTIPLE CONJUGATE ADDITIONS (TANDEM MICHAEL REACTIONS)

Examples of the intermolecular tandem conjugate addition–electrophile trapping protocol with simple electrophiles have been cited earlier in this review, but the initial nucleophile can also possess a suitably juxtaposed electrophilic site, *e.g.* either a carbonyl group or Michael acceptor.[204] For example, the reaction of β-oxo-stabilized vinyllithiums, derived from β-heteroatom functionalized α,β-unsaturated esters or β-heteroatom vinyl ketones, *e.g.* (**269**), respectively, with α,β-unsaturated esters affords cyclopentenones (Scheme 89).[205a–c] Alternatively, Beak reports that (2-carbamoylallyl)lithiums [β′-lithiomethacrylate synthon] (**270**) react with Michael acceptors to afford cyclopentenecarboxamides (**271**);[205d–f] formally this would be a [3 + 2] anionic cycloaddition, but data show that this occurs in discrete stages: (a) initial conjugate addition, (b) 5-*endo*-trig cyclization and (c) sulfinate elimination. The α′-dienolates, derived from α,β-enones, are also excellent tandem Michael donor-acceptors and have been used extensively for the synthesis of complex bicyclo[2.2.1]heptanones (**272**),[206] bicyclo[2.2.2]octan-2-ones (**273**)[207] and tricyclo[3.2.1.0^{2,7}]octan-6-ones (**274**; Scheme 90).[208]

Examples of intramolecular transformations[209] include dienone–α,β-unsaturated ester cyclizations, *e.g.* (**275**→**276**),[209a,b] and ester–α,β-unsaturated ester annulations, *e.g.* (**277**→**278**; Scheme 91).[209c] Pertinent to the success of these multibond formation protocols is the selective use of the initiating carbanion (or enolates), the counter cation and the initial acceptor.

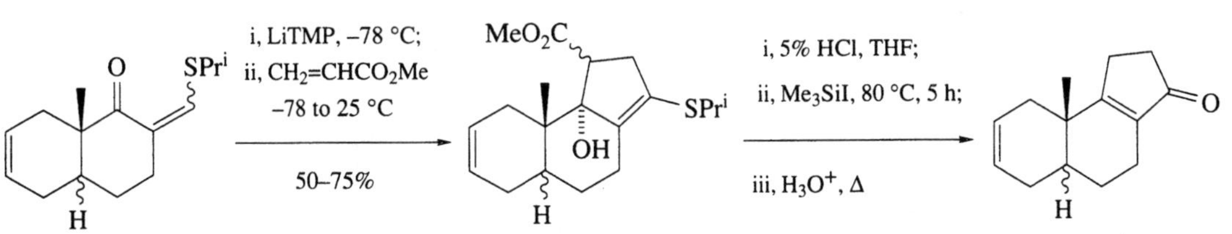

i, LiTMP, –78 °C; ii, CH_2=$CHCO_2Me$, –78 to 25 °C; 50–75%

i, 5% HCl, THF; ii, Me_3SiI, 80 °C, 5 h; iii, H_3O^+, Δ

(**269**)

LiTMP, THF, –78 °C

CH_2=CR^3(EWG), THF, –78 to 25 °C, 96 h; 22–89%

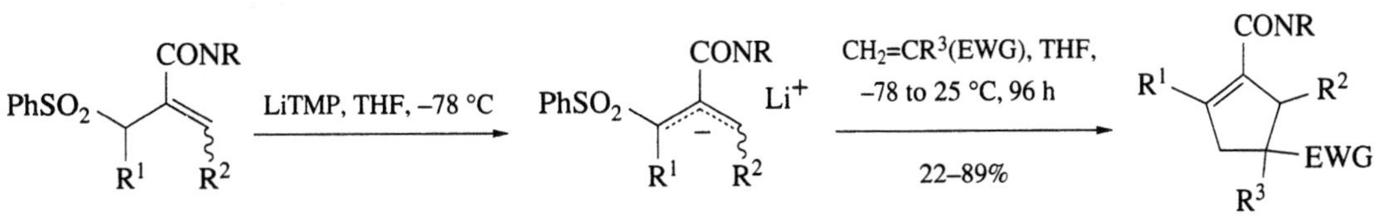

NR = NPr^i_2, NMe_2, N(Li)Ph
R^1 = H, Me; R^2 = H, Me

(**270**) EWG = CO_2alkyl (C_1–C_4), CO_2Ph, CON(Me)Ph, SO_2Ph
R^3 = H, Me, CN, $SiMe_3$

(**271**)

Scheme 89

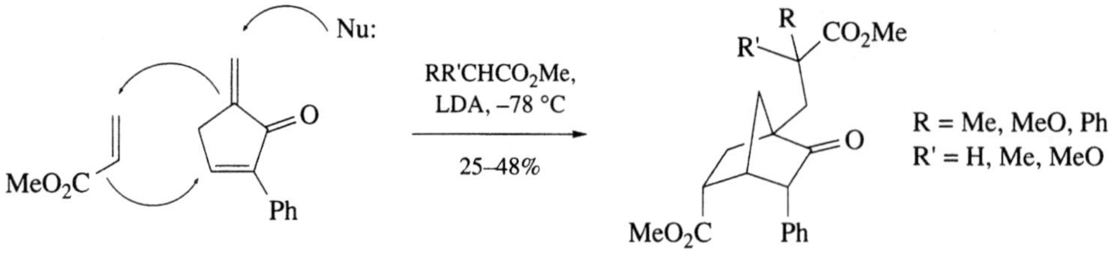

Nu:

MeO_2C O Ph

$RR'CHCO_2Me$, LDA, –78 °C

25–48%

R CO_2Me R' O MeO_2C Ph

(272)

R = Me, MeO, Ph
R' = H, Me, MeO

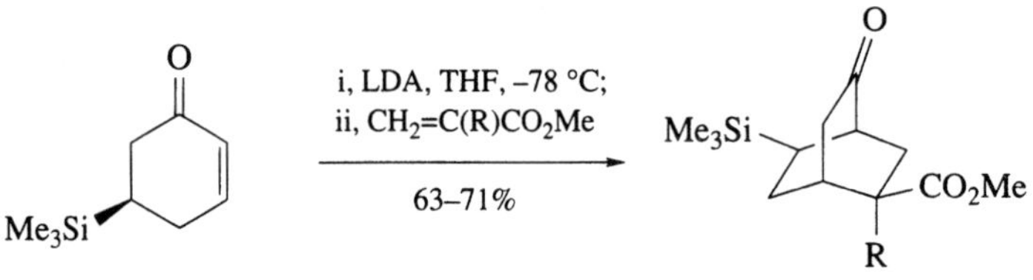

O Me_3Si

i, LDA, THF, –78 °C; ii, CH_2=C(R)CO_2Me

63–71%

O Me_3Si CO_2Me R

(273)

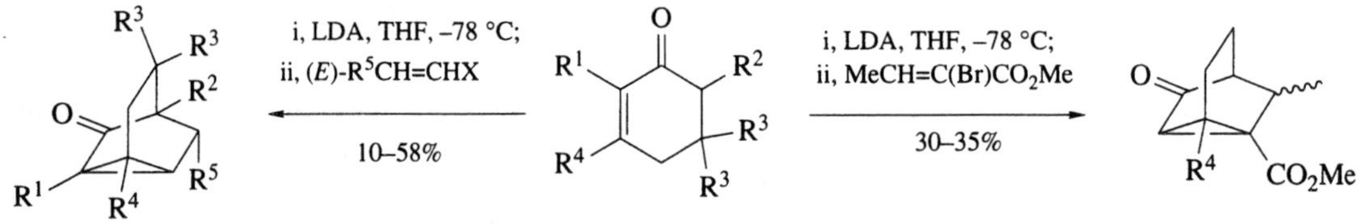

R^3 R^3 R^2 O R^5 R^1 R^4

i, LDA, THF, –78 °C; ii, (*E*)-R^5CH=CHX

10–58%

O R^1 R^2 R^3 R^4 R^3

i, LDA, THF, –78 °C; ii, MeCH=C(Br)CO_2Me

30–35%

O R^4 CO_2Me

R^1 = H, Me; R^2 = H, Me; R^3 = H, Me; R^4 = H, Me; R^5 = alkyl

(274b)

(274a) X = NO_2, $PhSO_2$, $Ph_3P^+Br^-$

Scheme 90

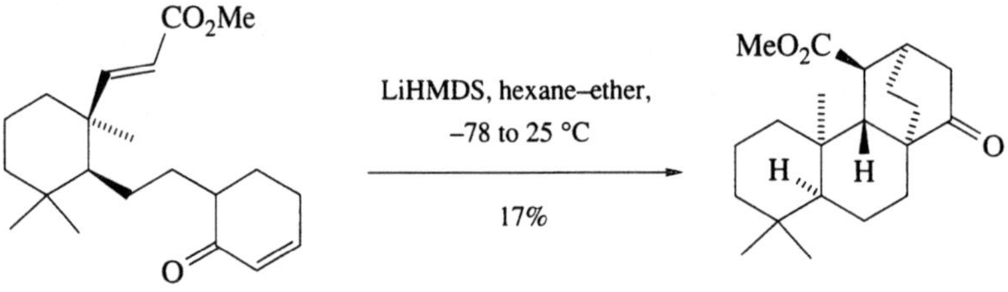

CO_2Me O

(275)

LiHMDS, hexane–ether, –78 to 25 °C

17%

MeO_2C H H O

(276)

O O OH H MeO_2C O O

(277)

KH, DMF, –20 °C, 2 h

>95%

O O H OH MeO_2C O O

(278)

Scheme 91

1.2.4 ADDITIONS TO 'PUSH–PULL' ALKENES

The addition of organometallic reagents to 'push–pull' alkenes is a synthetically important process in which retention of an alkenic unit in the product occurs; this process complements the direct conjugate addition of organometallic reagents to activated alkynes (Section 1.2.5) which is complicated primarily by multiple additions. 'Push–pull' alkenes, which are formally vinyl cation equivalents, are characterized by the presence of an electron-withdrawing functionality on one terminus of an alkene and an electron-donating functionality at the other terminus. Examples of 'push–pull' alkenes include enaminones (**279**), β-amino-α,β-unsaturated esters (**280**), β-nitroenamines (**281**), β-halovinyl ketones (**282**), β-halo-α,β-unsaturated esters and amides (**283**), β-alkoxy-α,β-unsaturated ketones (**284**), β-alkoxy-α,β-unsaturated esters (**285**), β-(alkylthio)enones (**286**), β-(alkylthio)-α,β-unsaturated esters (**287**) and β-halovinyl sulfones and sulfoxides (**288**). In contrast to additions to activated alkynes additions of organometallics to enaminones and nitroenamines are free of overaddition because the initial adducts do not eliminate the secondary amine functionality until work-up. Even select α,α-difluoroalkenes (**289**) react with organometallic reagents to afford fluoroalkenes.

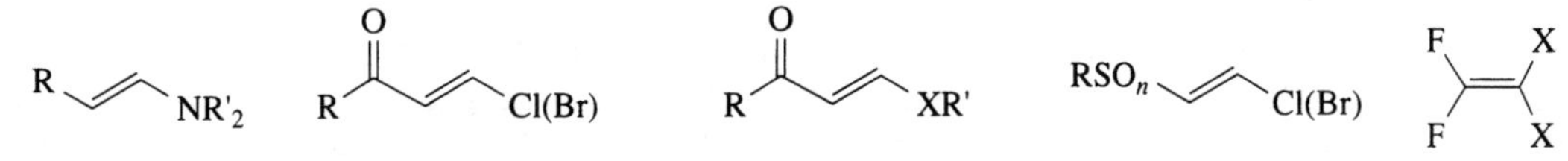

(**279**) R = COR"
(**280**) R = CO_2R"
(**281**) R = NO_2

(**282**) R = alkyl, aryl
(**283**) R = alkoxy

(**284**) R = alkyl, aryl; X = O
(**285**) R = alkoxy; X = O
(**286**) R = alkyl, aryl; X = S
(**287**) R = alkoxy; X = S

(**288a**) n = 2
(**288b**) n = 1

(**289**)

1.2.4.1 Enaminones

The addition of Grignard reagents[210a] and organolithium[210b] to enaminones (**279**) is an efficient process for the synthesis of α,β-unsaturated esters and α,β-enones. Typically, the enaminones are obtained by the treatment of activated methylene or methyl groups with dialkylformamide dialkyl acetals.[211] A variety of organolithiums have been added to enaminones, while Grignard reagents have been utilized

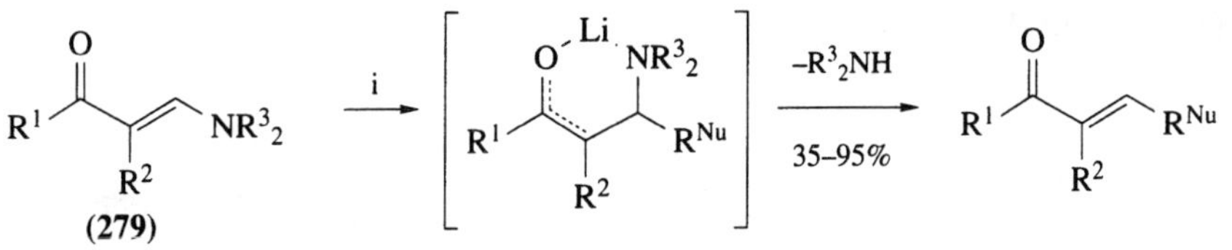

(**279**) —i→ [ ] —$-R^3_2NH$, 35–95%→

R' = H, Me, Ph, 4-ClC_6H_4, 2-thienyl, 2,6-$(MeO)_2C_6H_3$, 3,4,5-$(MeO)_3C_6H_2$
R^2 = H, Me, F, Ph, 3-$CF_3C_6H_4$; $R^1 + R^2 = (CH_2)_{3-4}$
R^3_2N = Me_2N, morpholino, piperidino, pyrrolidino

R^{Nu} = Me, Et, Bu^n, Bu^t, benzyl

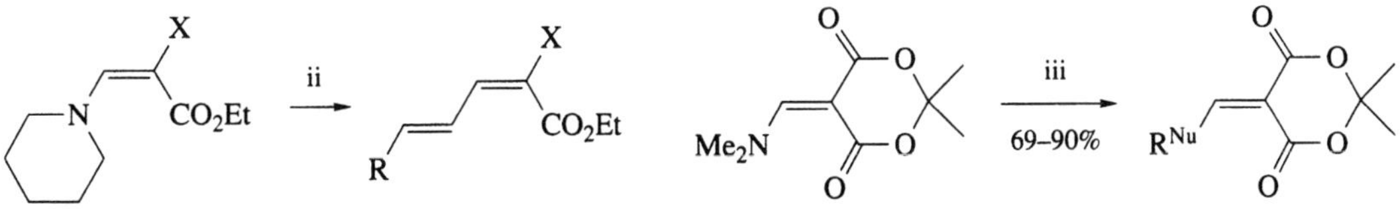

(**290**) —ii→ R = Me, Et; X = CN, CO_2Et

(**291**) —iii, 69–90%→ R^{Nu} = Me, Et, Pr^i, Ph

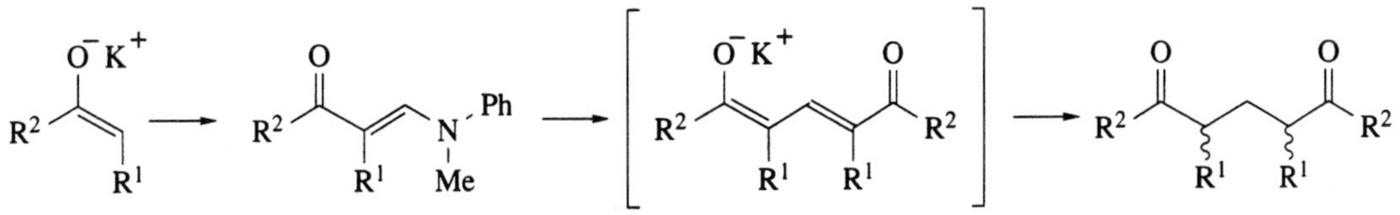

(**279a**) → [(**292**)] → (**293**)

i, R^{Nu}Li, ether (THF), –30 to 0 °C; ii, (Z)-RCH=CHMgX, THF, 25 °C; iii, R^{Nu}MgX, THF, 25 °C

Scheme 92

for additions to doubly activated enaminones (**290** and **291**; Scheme 92). While numerous examples of direct additions of organometallics to enaminones have been discussed, few examples of enolate additions to enaminones are reported. Recently, Kiyooka reported a synthesis of symmetrical 1,5-diketones permits a ketone (2 equiv.), *N*-methyl-*N*-phenylformamide (1 equiv.) and potassium metal (3 equiv.) react at 25 °C (THF). The enolate adds to enaminone (**279a**) to afford intermediate (**292**) and subsequent reduction affords the symmetrical 1,5-diketone (**293**); however, with preformed *N*-methyl-*N*-phenyl-enaminones (**279a**), cross-coupling with ketones permits the synthesis of unsymmetrical 1,5-diketones (Scheme 92).[212]

1.2.4.2 Nitroenamines

The addition of Grignard reagents or organolithiums (alkenyl, alkyl, alkynyl, allyl or aryl) to nitroenamines (**281**)[213] was reported by Severin to afford β-substituted-α-nitroalkenes.[214a,b] Similarly, ketone enolates (sodium or potassium), ester enolates (lithium) and lactone enolates (lithium) react to afford *aci*-nitroethylidene salts (**294**) which, on hydrolysis with either silica gel or dilute acid, afford γ-keto-α,β-unsaturated esters or ketones (**295**)[214c,d] or acylidene lactones (**296**).[214e] Alternatively, the salts (**294**, X = CH_2) can be converted to γ-ketoketones (**297**) with ascorbic acid and copper catalyst.

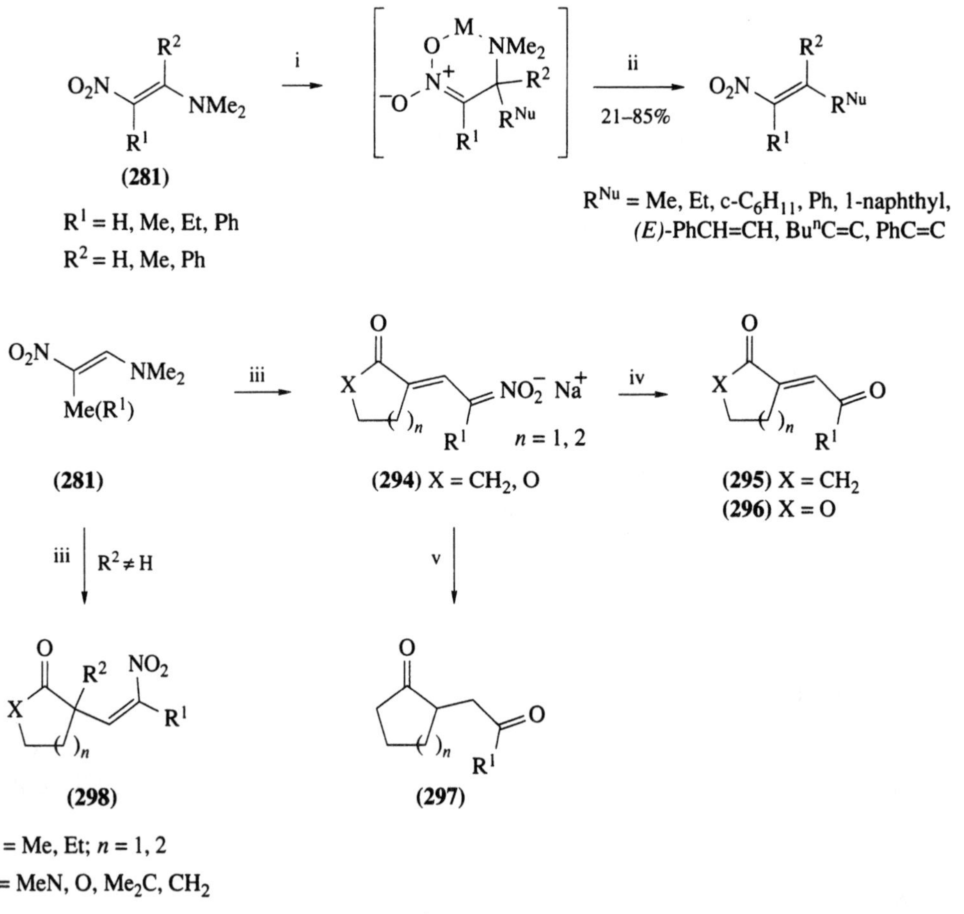

(**281**) R^1 = H, Me, Et, Ph; R^2 = H, Me, Ph

ii: 21–85%

R^{Nu} = Me, Et, c-C_6H_{11}, Ph, 1-naphthyl, (*E*)-PhCH=CH, $Bu^nC{\equiv}C$, PhC≡C

(**281**) iii; iii, $R^2 \neq H$

(**294**) X = CH_2, O; n = 1, 2

(**295**) X = CH_2
(**296**) X = O

(**298**) (**297**)

R^2 = Me, Et; n = 1, 2
X = MeN, O, Me_2C, CH_2

i, $R^{Nu}MgX$ (Li), ether (THF), 0 °C; ii, H_3O^+; iii, lactone enolate or ketone enolate; iv, SiO_2; v, ascorbic acid, Cu^{II}

Scheme 93

In contrast, α-nitroalkenation adducts (**298**) are obtained with thermodynamic enolates of α-substituted ketones or esters; in addition, Fuji reports that enantioselective addition of α-substituted lactone enolates to chiral nitroenamines is cation dependent with zinc affording maximum enantioselectivity.[215]

1.2.4.3 β-Haloenones and β-Halo-α,β-unsaturated Esters (Amides)[216]

Rupe and Iselin reported in 1916 the first example of an addition–elimination reaction in which Grignard reagents added to chloromethylenecamphor (**299**) to afford ketones (**300**).[216a] Subsequently, additions of dialkylcadmiums,[210b,216b] sodium dimsyl,[216c,d] allylidenetriphenylphosphorane (**303**),[216e] acyl anion equivalents, *e.g.* (**304**),[216f] and enolates, *e.g.* (**305**),[216g,h] to β-chloroenones and β-halo-α,β-unsaturated esters, *e.g.* (**301** and **302**), have been reported. The allylidenetriphenylphosphorane adducts afford a new Wittig reagent for polyene syntheses (Scheme 94).

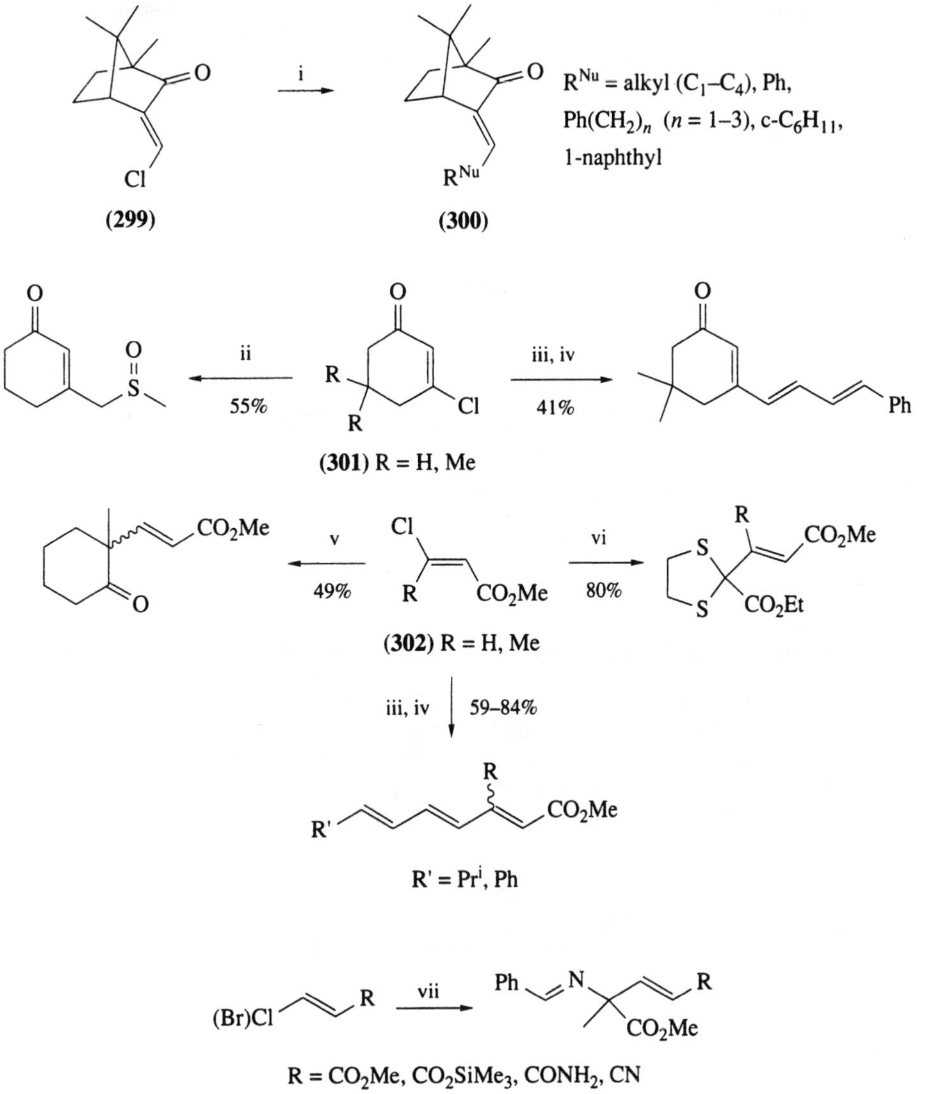

R^{Nu} = alkyl (C_1–C_4), Ph, Ph$(CH_2)_n$ (n = 1–3), c-C_6H_{11}, 1-naphthyl

(**299**) (**300**)

(**301**) R = H, Me

(**302**) R = H, Me

R' = Pr^i, Ph

R = CO_2Me, CO_2SiMe_3, $CONH_2$, CN

i, $R^{Nu}MgBr$, ether, 25 °C; ii, $NaCH_2SOMe$, THF; iii, CH_2=CH–CH=PPh_3 (**303**); iv, PhCHO (R'CHO); v, lithium 2-methylcyclohexen-1-olate, THF, –78 °C; vi, glyoxylate anion equivalent (**304**), THF, –78 °C; vii, PhCH=NCH(Li)CO_2Me (**305**), THF, –78 °C

Scheme 94

1.2.4.4 β-Alkoxy-α,β-unsaturated Esters

Kraus has shown that organolithiums add to diethyl alkoxymethylenemalonate in excellent yields to afford diethyl alkylidenemalonates; similarly, *ortho*-lithiated phenol ethers and ketone enolates afford alkoxy-carbonyl coumarins (**306** and **307**).[217a,b] In contrast, the treatment of diethyl alkoxymethylenemalonates with alkynyllithiums affords the adduct diesters (**308**) in which loss of the alkoxy group does not occur, but alkaline hydrolysis affords ylidenemalonic acids (**308**; Scheme 95).[217c,d]

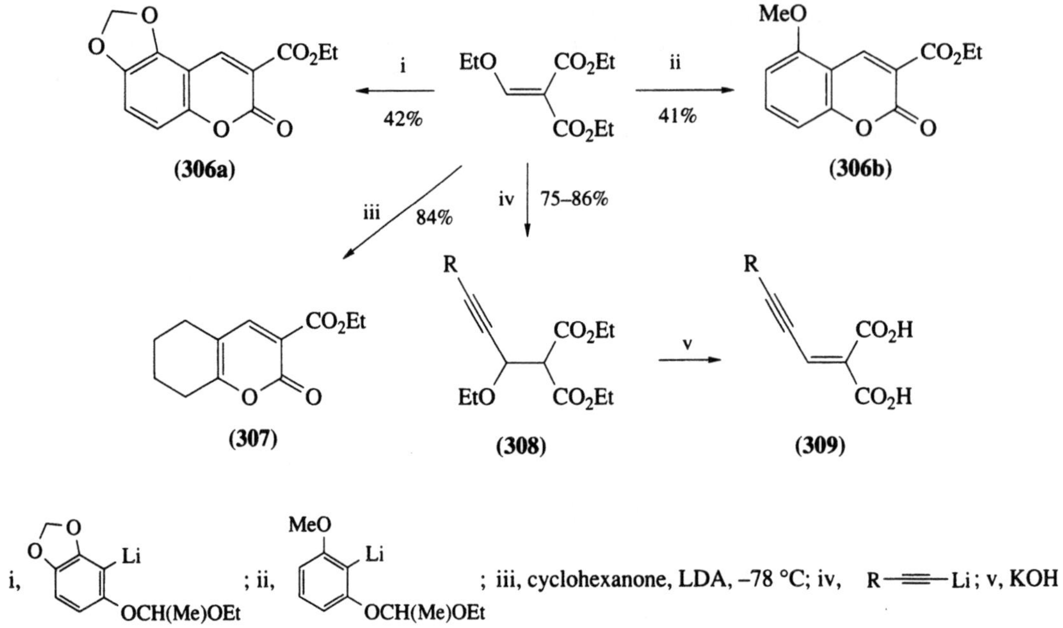

i, ; ii, ; iii, cyclohexanone, LDA, –78 °C; iv, R—≡—Li ; v, KOH

Scheme 95

1.2.4.5 β-(Alkylthio)enones, β-(Alkylthio)-α,β-unsaturated Esters and β-(Alkylthio)nitroalkenes

The copper(I)-catalyzed addition of Grignard reagents to β-(alkylthio)-α,β-unsaturated esters is the preferred protocol for substitution of the alkylthio substituent, *e.g.* (**310** → **311**),[218a] while potassium enolates add to α-ketoketene dithioacetals, *e.g.* (**312** → **313**), without dialkylation (Scheme 96).[218b,c] Further, Hanessian reports an expedient synthesis of penam (**315**) *via* intramolecular cyclization of the β-lactam nitroalkene (**314**), but loss of the methylthio moiety does not occur spontaneously (Scheme 97).[219]

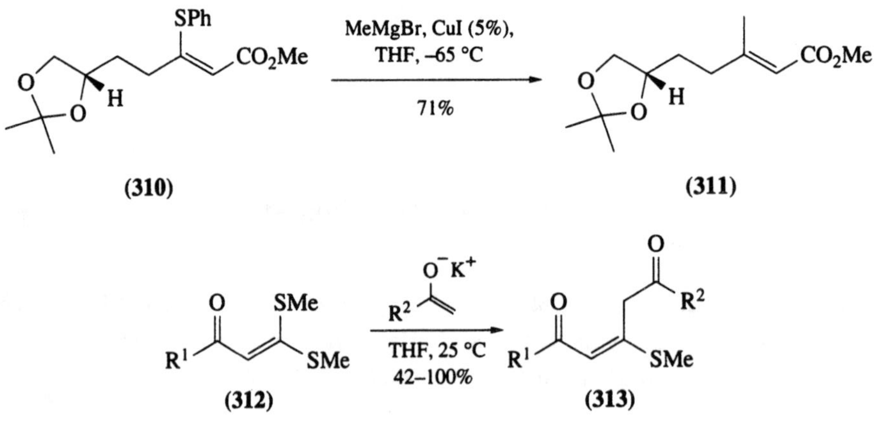

MeMgBr, CuI (5%), THF, –65 °C, 71%

(**310**) (**311**)

THF, 25 °C, 42–100%

(**312**) (**313**)

Scheme 96

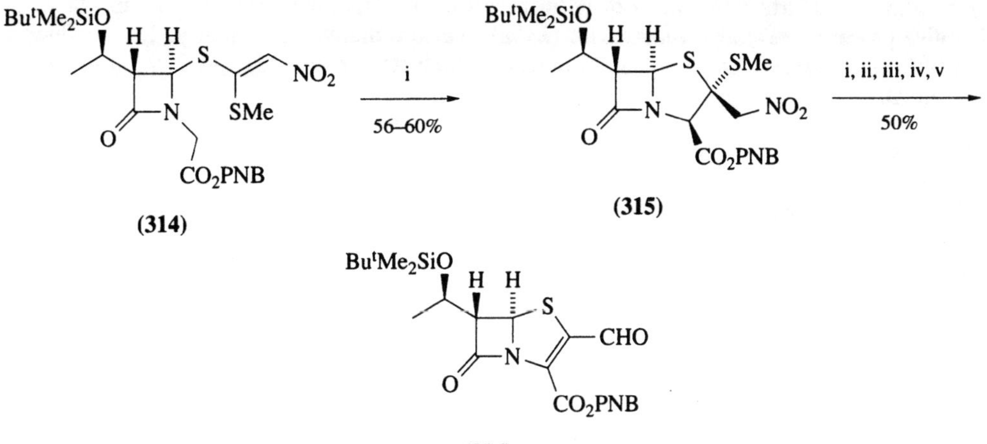

(314) (315) (316)

i, LiHMDS, THF, –78 °C; ii, MeI; iii, O_3; iv, MCPBA; v, $NaHCO_3$ (aq.)

Scheme 97

1.2.4.6 β-Halovinyl Sulfones

Enolate additions to β-halovinyl sulfones (**289a**) show synthetic promise; notable is the procedure, by Metcalf, for vinylation of α-amino acid derivatives, *e.g.* (**289a** → **317** → **318**).[220a] Recently, Ban reported that alicyclic ketone enolates add to the acetyl cation equivalent (**319**) to afford 1,3-diketones (Scheme 98).[220b] In general, β-halovinyl sulfones are synthetically superior to the β-halovinyl sulfoxides.

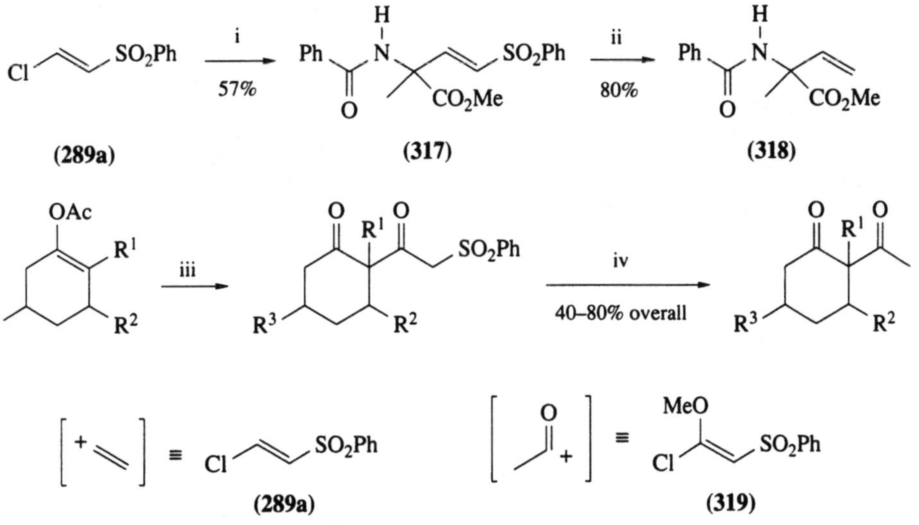

(289a) (317) (318)

(289a) (319)

i, $PhCONHCH(Me)CO_2Me$, LDA, THF, –78 °C; ii, Al–Hg; iii, (**319**), HMPA, THF, –78 °C; iv, Raney Ni

Scheme 98

1.2.4.7 α,α-Difluoroalkenes

The reaction of organolithiums with vinyl halides generally results in metal–halogen exchange; however, with fluoroalkenes, S_NV (nucleophilic vinylic substitution) processes occur. For example, organolithiums (or Grignard reagents) add to 1,1-dichloro-2,2-difluoroethylene (**290a**) and difluoroketene dithioacetals (**290b**) to afford α,α-dichloro-β-fluoroalkenes (**320a**) and fluoroketene thioacetals (**320b**), respectively; however, further treatment of (**320a**) with *n*-butyllithium affords lithium acetylides (**321**). In an iterative process, (**321**) also reacts with (**290a**) to afford dienynes. Similarly, 1,1,1-trifluoroethyl ethers (or sulfides) (**324**) react with organolithiums in a three-step process (**324** → **290c–d** → **320c–d** → **325**) to afford alkynic ethers (or sulfides) (Scheme 99).[221]

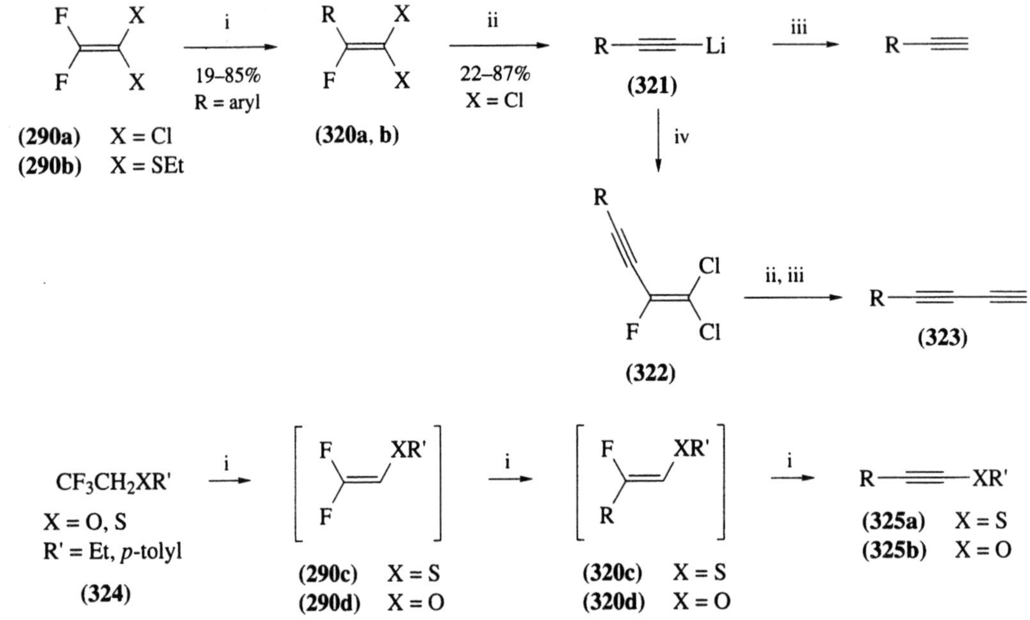

F X
i
19–85%
R = aryl
R X
ii
22–87%
X = Cl
R—≡—Li
(321)
iii
R—≡
(290a) X = Cl
(290b) X = SEt
(320a, b)
iv
R
Cl
F Cl
(322)
ii, iii
R—≡—≡
(323)
CF3CH2XR'
X = O, S
R' = Et, p-tolyl
(324)
F XR'
(290c) X = S
(290d) X = O
F XR'
R
(320c) X = S
(320d) X = O
R—≡—XR'
(325a) X = S
(325b) X = O

i, RLi, ether, –78 to 38 °C; ; ii, Bu^{n}Li (2 equiv.), –78 to 40 °C; iii, H_3O^+; iv, (**290a**), 20 to 38 °C

Scheme 99

1.2.5 ADDITIONS TO ALKYNIC π-SYSTEMS

In contrast to the well documented conjugate addition of carbon nucleophiles to activated alkenes, similar intermolecular attempts with activated alkynes with non-cuprate reactants are typically non-productive due to competing multiple addition processes.[87a,b] However, protic intramolecular conjugate additions of ketones as shown for the syntheses of griseofulvin and hirsutic acid,[222] are successful. Recently, several aprotic intramolecular conjugate additions to activated alkynes have been reported, as

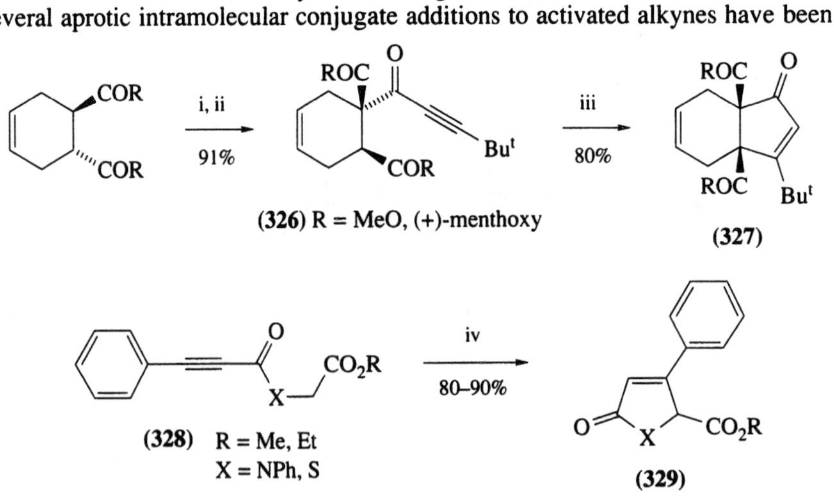

COR
COR
i, ii
91%
ROC
O
Bu^t
COR
(326) R = MeO, (+)-menthoxy
iii
80%
ROC
O
ROC
Bu^t
(327)
O
CO2R
X
(328) R = Me, Et
X = NPh, S
iv
80–90%
O
X
CO2R
(329)

i, LDA, THF, –78 °C; ii, Bu^{t}C≡CCO$_2$Ph; iii, KHMDS, THF, –45 °C; iv, LiH (NaH)

Scheme 100

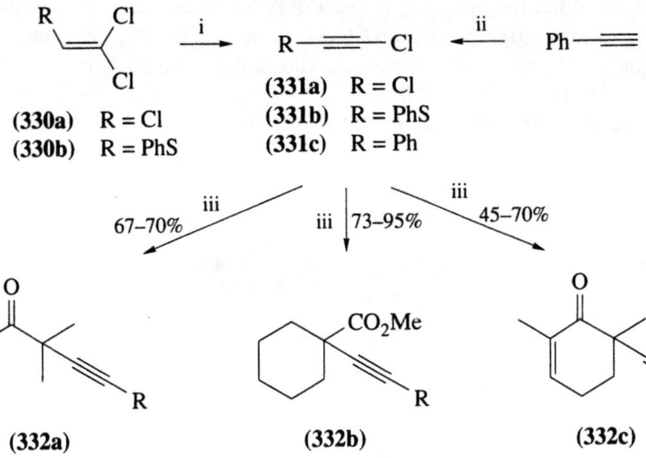

(330a) R = Cl
(330b) R = PhS

(331a) R = Cl
(331b) R = PhS
(331c) R = Ph

(332a) (332b) (332c)

i, LiHMDS, ether, –78 to 25 °C; ii, NaOH, Cl_2; iii, ester or ketone enolate, –78 to 25 °C

Scheme 101

shown for (**326** → **327**)[223a] and (**328** → **329**; Scheme 100).[223b] In contrast, Kende reports that α-disubstituted enolates add efficiently to chloroacetylenes (**331**) to afford α-ethynylated ketones (and esters) (**332**; Scheme 101).[224]

1.2.6 ELECTROCHEMICAL MEDIATED ADDITIONS

Electrochemical techniques in organic synthesis have become a burgeoning research area.[225] Several of the synthetically promising electrochemical intramolecular and intermolecular conjugate addition processes, catalyzed by either vitamin B_{12a} or CoHDP, are mediated by either [Co–R] complexes or [acyl–

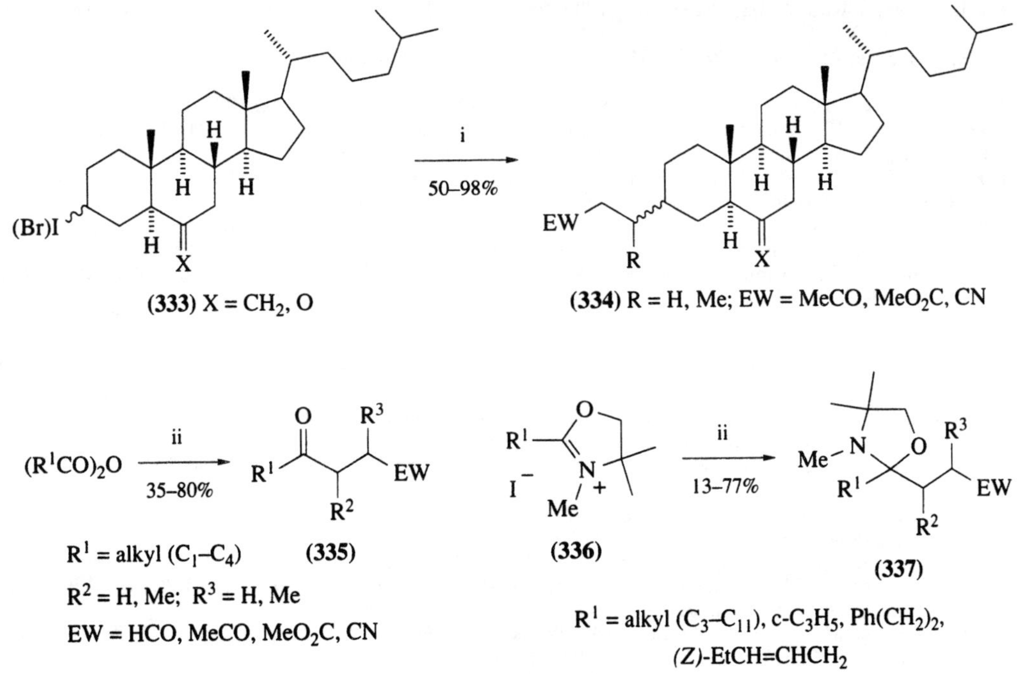

(**333**) X = CH_2, O

(**334**) R = H, Me; EW = MeCO, MeO_2C, CN

$(R^1CO)_2O$ → (**335**)

R^1 = alkyl (C_1–C_4)
R^2 = H, Me; R^3 = H, Me
EW = HCO, MeCO, MeO_2C, CN

(**336**) → (**337**)

R^1 = alkyl (C_3–C_{11}), c-C_3H_5, $Ph(CH_2)_2$, (*Z*)-$EtCH{=}CHCH_2$

i, (*E*)-$RCH{=}CH(EW)$, e^-; ii, $R^2CH{=}CR^3(EW)$, e^-

Scheme 102

Co(III)] complexes. For example, the coupling of 3-halocholestanes (**333**) and Michael acceptors affords epimeric mixtures of the 3-homologated steroids (**334**). The electrochemical nucleophilic acylation of α,β-unsaturated aldehydes, α,β-unsaturated ketones and α,β-unsaturated nitriles with acyl anhydrides affords adducts (**335**) in moderate yields.[226a,b] Similarly, the reduction of *N*-methyloxazolinium salts (**336**) affords the *N,O*-acetal intermediates (**337**) which are readily hydrolyzed (Scheme 102).[226c]

1.2.7 REFERENCES

1. (a) A. Michael, *J. Prakt. Chem.*, 1887, **35**, 349; (b) A. Michael, *Am. Chem. J.*, 1887, **9**, 115; (c) A. Michael, *J. Prakt. Chem.*, 1894, **49**, 20; (d) A. Michael, *Ber.*, 1894, **27**, 2126; (e) A. Michael, *Ber.*, 1900, **33**, 3731; (f) A. Michael and W. Schulthess, *J. Prakt. Chem.*, 1892, **45**, 55.
2. (a) E. D. Bergmann, D. Ginsburg and R. Pappo, *Org. React.*, 1959, **10**, 179; (b) H. O. House, 'Modern Synthetic Reactions', 2nd edn., Benjamin, Menlo Park, 1972, p. 595.
3. J. Fuhrhop and G. Penzlin, 'Organic Synthesis: Concepts, Methods, Starting Materials', Verlag Chemie, Weinheim, 1983, p. 1.
4. M. S. Kharasch and P. O. Tawney, *J. Am. Chem. Soc.*, 1941, **63**, 2308.
5. H. O. House, W. L. Respess and G. M. Whitesides, *J. Org. Chem.*, 1966, **31**, 3128.
6. H. Gilman and R. Kirby, *J. Am. Chem. Soc.*, 1941, **63**, 2046.
7. (a) R. D. Little and J. R. Dawson, *J. Am. Chem. Soc.*, 1978, **100**, 4607; (b) R. D. Little and J. R. Dawson, *Tetrahedron Lett.*, 1980, **21**, 2609; (c) R. D. Little, R. Verhé, W. T. Monte, S. Nugent and J. R. Dawson, *J. Org. Chem.*, 1982, **47**, 362.
8. G. H. Posner, *Chem. Rev.*, 1986, **86**, 831.
9. (a) L. Pauling, 'The Nature of the Chemical Bond', 3rd edn., Cornell University Press, Ithaca, NY, 1960; (b) M. Schlosser, 'Struktur und Reactivitat Polarer Organometalle', Springer, Berlin, 1973; (c) J. P. Oliver, in 'The Chemistry of the Metal–Carbon Bond. Volume 2: The Nature and Cleavage of Metal–Carbon Bonds', ed. F. R. Hartley and S. Patai, Wiley, New York, 1985, p. 789.
10. (a) G. Wittig, *Q. Rev. Chem. Soc.*, 1966, **20**, 191; (b) T. Tochtermann, *Angew. Chem., Int. Ed. Engl.*, 1966, **5**, 351; *Angew. Chem.*, 1966, **78**, 355.
11. (a) T.-L. Ho, *Chem. Rev.*, 1975, **75**, 1; (b) N. T. Anh, *Top. Curr. Chem.*, 1980, **88**, 145; (c) T.-L. Ho, *Tetrahedron*, 1985, **41**, 1; (d) J. M. Lefour and A. Loupy, *Tetrahedron*, 1978, **34**, 2597.
12. M. Clarembeau and A. Krief, *Tetrahedron Lett.*, 1985, **26**, 1093.
13. P. C. Ostrowski and V. V. Kane, *Tetrahedron Lett.*, 1977, 3549.
14. W. C. Still and A. Mitra, *Tetrahedron Lett.*, 1978, 2659.
15. F. C. Whitmore and G. W. Pedlow, Jr., *J. Am. Chem. Soc.*, 1941, **63**, 758.
16. (a) M. Cossentini, B. Deschamps, N. T. Ahn and J. Seyden-Penne, *Tetrahedron*, 1977, **33**, 409; (b) G. Stork and L. A. Maldonado, *J. Am. Chem. Soc.*, 1974, **96**, 5272; (c) R. Sauvetre and J. Seyden-Penne, *Tetrahedron Lett.*, 1976, 3949.
17. (a) H. O. House, *Acc. Chem. Res.*, 1976, **9**, 59; (b) B. Deschamps and J. Seyden-Penne, *Tetrahedron*, 1977, **33**, 413.
18. J. C. Stowell, 'Carbanions in Organic Synthesis', Wiley Interscience, New York, 1979, p. 32.
19. (a) C. Reichardt, 'Solvents and Solvent Effects in Organic Chemistry', 2nd edn., VCH, Weinheim, 1988; (b) for an in-depth discussion of conjugate additions to α,β-enones, see: D. Duval and S. Geribaldi, in 'The Chemistry of Enones—Part 1', ed. S. Patai and Z. Rappoport, Wiley, New York, 1989, p. 355.
20. T. Cohen, W. D. Abraham and M. Myers, *J. Am. Chem. Soc.*, 1987, **109**, 7923.
21. For a review of α-heterosubstituted carbanions, see: A. Krief, *Tetrahedron*, 1980, **36**, 2531.
22. E. Negishi, 'Organometallics in Organic Synthesis', Wiley-Interscience, New York, 1980.
23. (a) R. G. Jones and H. Gilman, *Chem. Rev.*, 1954, **54**, 835; (b) P. R. Jones, E. J. Goller and W. J. Kauffman, *J. Org. Chem.*, 1969, **34**, 3566.
24. J. L. Wardell, in 'The Chemistry of the Carbon–Metal Bond. Volume 4: The Use of Organometallic Compounds in Organic Synthesis', ed. F. R. Hartley, Wiley, New York, 1987, p. 1.
25. (a) D. Seebach and R. Locher, *Angew. Chem., Int. Ed. Engl.*, 1979, **18**, 957; *Angew. Chem.*, 1979, **91**, 1024; (b) R. Locher and D. Seebach, *Angew. Chem., Int. Ed. Engl.*, 1981, **20**, 569; *Angew. Chem.*, 1981, **93**, 614; (c) D. Seebach, M. Ertas, R. Locher and W. B. Schweizer, *Helv. Chim. Acta*, 1985, **68**, 264.
26. S. Knapp and J. Calienni, *Synth. Commun.*, 1980, **10**, 837.
27. G. B. Mpango, K. K. Mahalanabis, Z. Mahdavi-Damghani and V. Snieckus, *Tetrahedron Lett.*, 1980, **21**, 4823.
28. M. P. Cooke, Jr., *J. Org. Chem.*, 1986, **51**, 1637.
29. N. E. Schore and E. G. Rowley, *J. Am. Chem. Soc.*, 1988, **110**, 5224.
30. G. B. Mpango and V. Snieckus, *Tetrahedron Lett.*, 1980, **21**, 4827.
31. (a) C. H. Heathcock, M. C. Pirrung, S. H. Montgomery and J. Lampe, *Tetrahedron*, 1981, **37**, 4087; (b) R. Häner, T. Laube and D. Seebach, *J. Am. Chem. Soc.*, 1985, **107**, 5396.
32. (a) M. El-Bouz, M.-C. Roux-Schmitt and L. Wartski, *J. Chem. Soc., Chem. Commun.*, 1979, 779; (b) N. Wang, S. Su and L. Tsai, *Tetrahedron Lett.*, 1979, 1121.
33. (a) C. R. Hauser, R. S. Yost and B. I. Ringler, *J. Org. Chem.*, 1949, **14**, 261; (b) J. Munch-Petersen, *J. Org. Chem.*, 1957, **22**, 170.
34. M. P. Cooke, Jr., *J. Org. Chem.*, 1984, **49**, 1144.
35. G. Weeratunga, A. Jaworska-Sobiesiak, S. Horne and R. Rodrigo, *Can. J. Chem.*, 1987, **65**, 2019.
36. T. P. Murray and T. M. Harris, *J. Am. Chem. Soc.*, 1972, **94**, 8253.
37. J. E. Baldwin and W. A. Dupont, *Tetrahedron Lett.*, 1980, **21**, 1881.

38. (a) M. P. Cooke, Jr. and R. Goswami, *J. Am. Chem. Soc.*, 1977, **99**, 642; (b) P. A. Chopard, R. J. G. Searle and F. H. Devitt, *J. Org. Chem.*, 1965, **30**, 1015.
39. (a) M. P. Cooke, Jr., *J. Org. Chem.*, 1983, **48**, 744; (b) M. P. Cooke, Jr. and D. L. Burman, *J. Org. Chem.*, 1982, **47**, 4955; (c) M. P. Cooke, Jr., *J. Org. Chem.*, 1982, **47**, 4963.
40. (a) M. P. Cooke, Jr. and R. K. Widener, *J. Org. Chem.*, 1987, **52**, 1381; (b) M. P. Cooke, Jr., *Tetrahedron Lett.*, 1979, 2199; (c) M. P. Cooke, Jr. and J. Y. Jaw, *J. Org. Chem.*, 1986, **51**, 758.
41. (a) R. K. Boeckman, Jr., *J. Am. Chem. Soc.*, 1974, **96**, 6179; (b) G. Stork and J. Singh, *J. Am. Chem. Soc.*, 1974, **96**, 6181.
42. (a) M. P. Cooke, Jr., *J. Org. Chem.*, 1987, **52**, 5729; (b) aldehydes and ketones react efficiently with the lithio dianion of trimethylsilylacetic acid, see: P. A. Grieco, C.-L. J. Wang and S. D. Burke, *J. Chem. Soc., Chem. Commun.*, 1975, 537.
43. (a) O. Tsuge, S. Kanemasa and Y. Ninomiya, *Chem. Lett.*, 1984, 1993; (b) A. S. Narula and D. I. Schuster, *Tetrahedron Lett.*, 1981, **22**, 3707.
44. (a) Y. Tamaru, M. Kagotani and Z. Yoshida, *Tetrahedron Lett.*, 1981, **22**, 3409; (b) Y. Tamura, M. Kagotani, Y. Furukawa, Y. Amino and Z. Yoshida, *Tetrahedron Lett.*, 1981, **22**, 3413; (c) Y. Tamaru, T. Harada, H. Iwamoto and Z. Yoshida, *J. Am. Chem. Soc.*, 1978, **100**, 5221.
45. A. I. Meyers and A. C. Kovelesky, *Tetrahedron Lett.*, 1969, 4809.
46. (a) E. J. Corey and D. L. Boger, *Tetrahedron Lett.*, 1978, 9; (b) E. J. Corey and D. L. Boger, *Tetrahedron Lett.*, 1978, 13.
47. (a) R. C. F. Jones, M. W. Anderson and M. J. Smallridge, *Tetrahedron Lett.*, 1988, **29**, 5001; (b) M. W. Anderson, R. C. F. Jones and J. Saunders, *J. Chem. Soc., Perkin Trans. 1*, 1986, 205; (c) M. W. Anderson, R. C. F. Jones and J. Saunders, *J. Chem. Soc., Perkin Trans. 1*, 1986, 1995; (d) M. W. Anderson, M. J. Begley, R. C. F. Jones and J. Saunders, *J. Chem. Soc., Perkin Trans. 1*, 1984, 2599.
48. J. C. Clinet, *Tetrahedron Lett.*, 1988, **29**, 5901.
49. (a) A. I. Meyers, J. D. Brown and D. Laucher, *Tetrahedron Lett.*, 1987, **28**, 5279; (b) for enantioselective conjugate addition to α,β-unsaturated aldimines in the presence of a C-2 symmetric chiral diether, see: K. Tomioka, M. Shindo and K. Koga, *J. Am. Chem. Soc.*, 1989, **111**, 8266.
50. (a) A. I. Meyers and C. E. Whitten, *J. Am. Chem. Soc.*, 1975, **97**, 6266; (b) A. I. Meyers and C. E. Whitten, *Heterocycles*, 1976, **4**, 1687; (c) A. I. Meyers, R. K. Smith and C. E. Whitten, *J. Org. Chem.*, 1979, **44**, 2250; (d) A. I. Meyers, G. P. Roth, D. Hoyer, B. A. Barner and D. Laucher, *J. Am. Chem. Soc.*, 1988, **110**, 4611.
51. For a discussion of low temperature (<–80 °C) techniques, see: D. Seebach and A. Hidber, *Chimia*, 1983, **37**, 449.
52. (a) D. Seebach, E. W. Colvin, F. Lehr and T. Weller, *Chimia*, 1979, **33**, 1; (b) C. D. Bedford and A. T. Nielsen, *J. Org. Chem.*, 1978, **43**, 2460; (c) H. Paulsen and M. Stubbe, *Tetrahedron Lett.*, 1982, **23**, 3171.
53. D. Seebach and P. Knochel, *Helv. Chim. Acta*, 1984, **67**, 261.
54. (a) For a review of select aspects of sulfone chemistry, see: P. D. Magnus, *Tetrahedron*, 1977, **33**, 2019; (b) K. Tanaka and A. Kaji in 'The Chemistry of Sulphones and Sulphoxides', ed. S. Patai, Z. Rappoport and C. J. M. Stirling, Wiley, New York, 1988, p. 759; (c) for a review of cycloalkenyl sulfones, see: P. L. Fuchs and T. F. Braish, *Chem. Rev.*, 1986, **86**, 903.
55. (a) R. E. Donaldson, J. C. Saddler, S. Byrn, A. T. McKenzie and P. L. Fuchs, *J. Org. Chem.*, 1983, **48**, 2167; (b) D. K. Hutchinson and P. L. Fuchs, *J. Am. Chem. Soc.*, 1987, **109**, 4755.
56. (a) P. R. Hamann, J. E. Toth and P. L. Fuchs, *J. Org. Chem.*, 1984, **49**, 3865; (b) J. E. Toth and P. L. Fuchs, *J. Org. Chem.*, 1987, **52**, 473; (c) J. E. Toth, P. R. Hamann and P. L. Fuchs, *J. Org. Chem.*, 1988, **53**, 4694.
57. (a) J. J. Eisch and J. E. Galle, *J. Org. Chem.*, 1979, **44**, 3279; (b) M. Isobe, M. Kitamura and T. Goto, *Chem. Lett.*, 1980, 331.
58. (a) M. Isobe, M. Kitamura and T. Goto, *Tetrahedron Lett.*, 1980, **21**, 4727; (b) M. Isobe, M. Kitamura and T. Goto, *Tetrahedron Lett.*, 1979, 3465.
59. (a) M. Isobe, Y. Funabashi, Y. Ichikawa, S. Mio and T. Goto, *Tetrahedron Lett.*, 1984, **25**, 2021; (b) M. Isobe, M. Kitamura and T. Goto, *J. Am. Chem. Soc.*, 1982, **104**, 4997; (c) M. Isobe, J. Obeyama, Y. Funabashi and T. Goto, *Tetrahedron Lett.*, 1988, **29**, 4773.
60. (a) H. Ahlbrecht and K. Pfaff, *Synthesis*, 1978, 897; (b) J.-M. Fang and H.-T. Chang, *J. Chem. Soc., Perkin Trans. 1*, 1988, 1945.
61. (a) G. R. Knox, D. G. Leppard, P. L. Pauson and W. E. Watts, *J. Organomet. Chem.*, 1972, **34**, 347; (b) M. F. Semmelhack, W. Seufert and L. Keller, *J. Am. Chem. Soc.*, 1980, **102**, 6584.
62. (a) N. Aktogu, H. Felkin, G. J. Baird, S. G. Davies and O. Watts, *J. Organomet. Chem.*, 1984, **262**, 215; (b) S. G. Davies and J. C. Walker, *J. Chem. Soc., Chem. Commun.*, 1985, 209; (c) L. S. Liebeskind, R. W. Fengl and M. E. Welker, *Tetrahedron Lett.*, 1985, **26**, 3075; (d) L. S. Liebeskind and M. E. Welker, *Tetrahedron Lett.*, 1985, **26**, 3079; (e) S. G. Davies, I. M. Dordor-Hedgecock, K. H. Sutton and J. C. Walker, *Tetrahedron*, 1986, **42**, 5123; (f) S. G. Davies, *Pure Appl. Chem.*, 1988, **60**, 13; (g) J. W. Herndon, C. Wu and H. L. Ammon, *J. Org. Chem.*, 1988, **53**, 2873.
63. C. L. Raston and G. Salem, in 'The Chemistry of the Carbon–Metal Bond. Volume 4: The Use of Organometallic Compounds in Organic Synthesis', ed. F. R. Hartley, Wiley, New York, 1987, p. 159.
64. (a) R. D. Rieke, *Acc. Chem. Res.*, 1977, **10**, 301; (b) Y.-H. Lai, *Synthesis*, 1981, 585.
65. (a) R. E. Ireland, S. W. Baldwin, D. J. Dawson, M. I. Dawson, J. E. Dolfini, J. Newbould, W. S. Johnson, M. Brown, R. J. Crawford, P. F. Hudrlik, G. H. Rasmussen and K. K. Schmiegel, *J. Am. Chem. Soc.*, 1970, **92**, 5743; (b) R. E. Ireland, S. W. Baldwin and S. C. Welch, *J. Am. Chem. Soc.*, 1972, **94**, 2056; (c) S. P. Modi, J. O. Gardner, A. Milowsky, M. Wierzba, L. Forgione, P. Mazur, A. J. Solo, W. L. Duax, Z. Galdecki, P. Grochulski and Z. Wawrzak, *J. Org. Chem.*, 1989, **54**, 2317; (d) for discussion of vector approach analysis of α,β-enone reactivity, see: J. E. Baldwin, *J. Chem. Soc., Chem. Commun.*, 1976, 738.
66. For a review of conjugate additions of Grignard reagents to aromatic systems, see: R. C. Fuson, *Adv. Organomet. Chem.*, 1964, **1**, 221.
67. (a) E. P. Kohler and G. L. Heritage, *Am. Chem. J.*, 1905, **33**, 21; (b) N. Maxim and N. Ioanid, *Bul. Soc. Chim. Rom.*, 1928, **10**, 29; (c) G. Gilbert, *J. Am. Chem. Soc.*, 1955, **77**, 4413.

68. (a) T. Cuvigny and H. Normant, *Bull. Soc. Chim. Fr.*, 1961, 2423; (b) F. Barbot, A. Kadib-Elban and P. Miginiac, *Tetrahedron Lett.*, 1983, **24**, 5089; (c) G. Daviaud and P. Miginiac, *Tetrahedron Lett.*, 1971, 3251; (d) R. Epsztein, in 'Comprehensive Carbanion Chemistry—Part B: Selectivity in Carbon–Carbon Bond Forming Reactions', ed. E. Buncel and T. Durst, Elsevier, New York, 1984, p. 107.
69. W. I. Awad, M. F. Ismail and V. R. Selim, *Indian J. Chem.*, 1975, **13**, 658.
70. (a) T. Mukaiyama and N. Iwasawa, *Chem. Lett.*, 1981, 913; (b) M. Whittaker, *Chem. Ind. (London)*, 1986, 463.
71. (a) K. Soai, H. Machida and A. Ookawa, *J. Chem. Soc., Chem. Commun.*, 1985, 469; (b) K. Soai, H. Machida and N. Yokota, *J. Chem. Soc., Perkin Trans. 1*, 1987, 1909; (c) for an analogous bimodal addition of organolithiums to α,β-unsaturated (*S*)-proline amides, see: K. Soai and A. Ookawa, *J. Chem. Soc., Perkin Trans. 1*, 1986, 759.
72. W. Oppolzer, G. Poli, A. J. Kingma, C. Starkemann and G. Bernardinelli, *Helv. Chim. Acta*, 1987, **70**, 2201.
73. Y. Tamaru, T. Hioki, S. Kawamura, H. Satomi and Z. Yoshida, *J. Am. Chem. Soc.*, 1984, **106**, 3876.
74. (a) M. El-Jazouli, S. Masson and A. Thuillier, *Bull. Soc. Chim. Fr.*, 1988, 875; (b) M. El-Jazouli, N. Lage, S. Masson and A. Thuillier, *Bull. Soc. Chim. Fr.*, 1988, 883.
75. (a) S. Hashimoto, S. Yamada and K. Koga, *J. Am. Chem. Soc.*, 1976, **98**, 7450; (b) H. Kogen, K. Tomioka, S. Hashimoto and K. Koga, *Tetrahedron Lett.*, 1980, **21**, 4005; (c) for kinetic resolution *via* conjugate additions to a chiral α,β-unsaturated aldimine, see: L. A. Paquette, D. Macdonald, L. G. Anderson and J. Wright, *J. Am. Chem. Soc.*, 1989, **111**, 8037.
76. (a) E. P. Kohler and J. F. Stone, Jr., *J. Am. Chem. Soc.*, 1930, **52**, 761; (b) G. D. Buckley, *J. Chem. Soc.*, 1947, 1494.
77. (a) G. D. Buckley and E. Ellery, *J. Chem. Soc.*, 1947, 1497; (b) E. J. Corey and H. Estreicher, *J. Am. Chem. Soc.*, 1978, **100**, 6294; (c) M. S. Ashwood, L. A. Bell, P. G. Houghton and S. H. B. Wright, *Synthesis*, 1988, 379; (d) T. D. Mechkov, I. G. Sulimov, N. V. Usik, V. V. Perekalin and I. Mladenov, *Zh. Org. Khim.*, 1978, **14**, 733; (e) T. Iida, M. Funabashi and J. Yoshimura, *Bull. Chem. Soc. Jpn.*, 1973, **46**, 3202.
78. (a) G. Bartoli, *Acc. Chem. Res.*, 1984, **17**, 109; (b) G. Bartoli, R. Leardini, M. Lelli and G. Rosini, *J. Chem. Soc., Perkin Trans. 1*, 1977, 884; (c) for synthesis of seven-substituted indoles *via* vinyl Grignard additions to *ortho*-substituted nitroarenes, see: G. Bartoli, G. Palmieri, M. Bosco and R. Dalpozzo, *Tetrahedron Lett.*, 1989, **30**, 2129; (d) P. Buck and G. Köbrich, *Tetrahedron Lett.*, 1967, 1563; (e) G. Bartoli, E. Marcantoni, M. Bosco and R. Dalpozzo, *Tetrahedron Lett.*, 1988, **29**, 2251.
79. (a) G. Bartoli, M. Bosco and G. Baccolini, *J. Org. Chem.*, 1980, **45**, 522; (b) G. Bartoli, R. Leardini, A. Medici and G. Rosini, *J. Chem. Soc., Perkin Trans. 1*, 1978, 692.
80. D. O. Spry, *Tetrahedron Lett.*, 1980, **21**, 1293.
81. (a) G. H. Posner, J. P. Mallamo and K. Miura, *J. Am. Chem. Soc.*, 1981, **103**, 2886; (b) G. H. Posner, J. P. Mallamo, M. Hulce and L. L. Frye, *J. Am. Chem. Soc.*, 1982, **104**, 4180; (c) G. H. Posner and M. Hulce, *Tetrahedron Lett.*, 1984, **25**, 379.
82. For an alternative explanation of the stereoselectivity, see: S. D. Kahn, K. D. Dobbs and W. J. Hehre, *J. Am. Chem. Soc.*, 1988, **110**, 4602.
83. J. J. Eisch and J. E. Galle, *J. Org. Chem.*, 1979, **44**, 3278.
84. (a) B. Riegel, S. Siegel and W. M. Lilienfeld, *J. Am. Chem. Soc.*, 1946, **68**, 984; (b) M. S. Newman and M. Wolf, *J. Am. Chem. Soc.*, 1952, **74**, 3225; (c) G. A. Holmberg and R. Sjoholm, *Acta Chem. Scand.*, 1970, **24**, 3490; (d) Y. Yamamoto and S. Nishii, *J. Org. Chem.*, 1988, **53**, 3597; (e) E. Van Heyningen, *J. Am. Chem. Soc.*, 1954, **76**, 2241; (f) F. Gaudemar-Bardone, M. Mladenova and M. Gaudemar, *Synthesis*, 1988, 611; (g) U. Schöllkopf and R. Meyer, *Justus Liebigs Ann. Chem.*, 1977, 1174; (h) F. Barbot, E. Paraiso and P. Miginiac, *Tetrahedron Lett.*, 1984, **25**, 4369; (i) M. L. Haslego and F. X. Smith, *Synth. Commun.*, 1980, **10**, 421; (j) for diastereoselective addition of organolithiums or Grignard reagents to isopropylidene α-alkoxyalkylidenemalonates, see: M. Larchevêque, G. Tamagnan and Y. Petit, *J. Chem. Soc., Chem. Commun.*, 1989, 31.
85. (a) T. Mukaiyama, T. Takeda and M. Osaka, *Chem. Lett.*, 1977, 1165; (b) T. Mukaiyama, T. Takeda and K. Fujimoto, *Bull. Chem. Soc. Jpn.*, 1978, **51**, 3368.
86. (a) M. Gocmen and S. Soussan, *J. Organomet. Chem.*, 1974, **80**, 303; (b) M. Sworin and W. L. Neumann, *Tetrahedron Lett.*, 1987, **28**, 3217.
87. (a) G. H. Posner, *Org. React.*, 1972, **19**, 1; (b) G. H. Posner, 'An Introduction to Syntheses Using Organocopper Reagents', Wiley-Interscience, New York, 1980; (c) E. Erdik, *Tetrahedron*, 1984, **40**, 641; (d) S. H. Bertz, C. P. Gibson and G. Dabbagh, *Tetrahedron Lett.*, 1987, **28**, 4251.
88. (a) D. N. Brattesani and C. H. Heathcock, *J. Org. Chem.*, 1975, **40**, 2165; (b) H. O. House, C.-Y. Chu, J. M. Wilkins and M. J. Umen, *J. Org. Chem.*, 1975, **40**, 1460; (c) H. E. Ensley, C. A. Parnell and E. J. Corey, *J. Org. Chem.*, 1978, **43**, 1610; (d) K. J. Shea and P. Q. Pham, *Tetrahedron Lett.*, 1983, **24**, 1003; (e) M. Asaoka, K. Shima and H. Takei, *J. Chem. Soc., Chem. Commun.*, 1988, 430; (f) S. A. Bal, A. Marfat and P. Helquist, *J. Org. Chem.*, 1982, **47**, 5045; (g) R. E. Schaub and M. J. Weiss, *Tetrahedron Lett.*, 1973, 129.
89. (a) Z. Horii, K. Morikawa and I. Ninomiya, *Chem. Pharm. Bull.*, 1969, **17**, 846; (b) P. Slosse and C. Hootelé, *Tetrahedron Lett.*, 1979, 4587; (c) W. Bartmann, E. Guntrum, H. Urbach and J. Wunner, *Synth. Commun.*, 1988, **18**, 711.
90. (a) G. Stork, G. L. Nelson, F. Rouessac and O. Gringore, *J. Am. Chem. Soc.*, 1971, **93**, 3091; (b) F. Näf and R. Decorzant, *Helv. Chim. Acta*, 1974, **57**, 1317; (c) R. A. Kretchmer, E. D. Mihelich and J. J. Waldron, *J. Org. Chem.*, 1972, **37**, 4483; (d) R. A. Kretchmer and W. M. Schafer, *J. Org. Chem.*, 1973, **38**, 95.
91. (a) S.-H. Liu, *J. Org. Chem.*, 1977, **42**, 3209; (b) I. W. Lawston and T. D. Inch, *J. Chem. Soc., Perkin Trans. 1*, 1983, 2629.
92. M. Asami and T. Mukaiyama, *Chem. Lett.*, 1979, 569.
93. (a) N. Rabjohn, L. V. Phillips and R. J. DeFeo, *J. Org. Chem.*, 1959, **24**, 1964; (b) N. Rabjohn and R. J. DeFeo, *J. Org. Chem.*, 1960, **25**, 1307; (c) R. M. Schisla and W. C. Hammann, *J. Org. Chem.*, 1970, **35**, 3224; (d) C. Cardellicchio, V. Fiandanese, G. Marchese, F. Naso and L. Ronzini, *Tetrahedron Lett.*, 1985, **26**, 4387.
94. H. Nagashima, N. Ozaki, M. Washiyama and K. Itoh, *Tetrahedron Lett.*, 1985, **26**, 657.

95. (a) E. Frankland, *Justus Liebigs Ann. Chem.*, 1849, **71**, 171; (b) E. Frankland, *Justus Liebigs Ann. Chem.*, 1853, **85**, 329; (c) J. Furukawa and N. Kawabata, *Adv. Organomet. Chem.*, 1974, **12**, 83; (d) L. Miginiac, in 'The Chemistry of The Metal–Carbon Bond, Volume 3: Carbon–Carbon Bond Formation Using Organometallic Compounds', ed. F. R. Hartley and S. Patai, Wiley, New York, 1985, p. 99.
96. Y. Kawakami, Y. Yasuda and T. Tsuruta, *J. Macromol. Sci. Chem.*, 1969, **A3**, 205.
97. B. Marx, E. Henry-Basch and P. Freon, *C. R. Hebd. Seances Acad. Sci., Ser. C*, 1967, **264**, 527.
98. (a) K. S. Suslick, J. J. Gawienowski, P. R. Schubert and H. H. Wang, *Ultrasonics*, 1984, **22**, 33; (b) K. S. Suslick, *Adv. Organomet. Chem.*, 1986, **25**, 73.
99. (a) C. Petrier, J.-L. Luche and C. Dupuy, *Tetrahedron Lett.*, 1984, **25**, 3463; (b) C. Petrier, J. C. de Souza Barbosa, C. Dupuy and J.-L. Luche, *J. Org. Chem.*, 1985, **50**, 5761; (c) A. E. Greene, J.-P. Lansard, J.-L. Luche and C. Petrier, *J. Org. Chem.*, 1984, **49**, 931.
100. (a) Y. Kawakami and T. Tsuruta, *Tetrahedron Lett.*, 1971, 1173; (b) Y. Kawakami and T. Tsuruta, *Tetrahedron Lett.*, 1971, 1959; (c) T. Tsuruta and Y. Kawakami, *Tetrahedron*, 1973, **29**, 1173.
101. T. Shono, I. Nishiguchi and M. Sasaki, *J. Am. Chem. Soc.*, 1978, **100**, 4314.
102. (a) R. Malzieu, *Bull. Soc. Chim. Fr.*, 1966, 2682; (b) G. Daviaud and P. Miginiac, *Bull. Soc. Chim. Fr.*, 1970, 1617; J.-L. Moreau, Y. Frangin and M. Gaudemar, *Bull. Soc. Chim. Fr.*, 1970, 4512; (c) G. Daviaud, M. Massy and P. Miginiac, *Tetrahedron Lett.*, 1970, 5169; (d) G. Daviaud, M. Massy-Bardot and P. Miginiac, *C. R. Hebd. Seances Acad. Sci., Ser. C*, 1971, **272**, 969; (e) F. Gaudemar-Bardone, M. Mladenova and M. Gaudemar, *Synthesis*, 1988, 611; (f) Y. Frangin, G. Guimbal, F. Wissocq and H. Zamarlik, *Synthesis*, 1986, 1046; (g) M. Bellasoued, Y. Frangin and M. Gaudemar, *Synthesis*, 1978, 150.
103. (a) G. Wittig, F. J. Meyer and G. Lange, *Justus Liebigs Ann. Chem.*, 1951, **571**, 167; (b) R. Waack and M. A. Doran, *J. Am. Chem. Soc.*, 1963, **85**, 2861; (c) W. Tückmantel, K. Oshima and H. Nozaki, *Chem. Ber.*, 1986, **119**, 1581.
104. M. Isobe, S. Kondo, N. Nagasawa and T. Goto, *Chem. Lett.*, 1977, 679.
105. (a) R. A. Watson and R. A. Kjonaas, *Tetrahedron Lett.*, 1986, **27**, 1437; (b) Y. Morita, M. Suzuki and R. Noyori, *J. Org. Chem.*, 1989, **54**, 1785; (c) M. Suzuki, H. Koyano, Y. Morita and R. Noyori, *Synlett.*, 1989, 22.
106. J. F. G. A. Jansen and B. L. Feringa, *Tetrahedron Lett.*, 1988, **29**, 3593.
107. (a) M. W. Rathke, *Org. React.*, 1975, **22**, 423; (b) E. E. Blaise, *C. R. Hebd. Seances Acad. Sci.*, 1901, **132**, 478; (c) J. Cason, K. L. Rinehart, Jr. and S. D. Thornton, Jr., *J. Org. Chem.*, 1953, **18**, 1594.
108. (a) P. R. Jones and P. J. Desio, *Chem. Rev.*, 1978, **78**, 491; (b) J. Kollonitsch, *Nature (London)*, 1960, **188**, 140; (c) J. Kollonitsch, *J. Chem. Soc. A*, 1966, 453, 456; (d) H. Coudane, E. Henry-Basch, J. Michel, B. Marx, F. Huet and P. Freon, *C. R. Hebd. Seances Acad. Sci., Ser. C*, 1966, **262**, 861.
109. (a) M. Langlais and P. Freon, *C. R. Hebd. Seances Acad. Sci., Ser. C*, 1965, **261**, 2920; (b) M. Gocmen, G. Soussan and P. Fréon, *Bull. Soc. Chim. Fr.*, 1973, 562; (c) M. Gocmen and G. Soussan, *J. Organomet. Chem.*, 1973, **61**, 19; (d) E. Henry-Basch, J. Deniau, G. Emptoz, F. Huet, B. Marx and J. Michel, *C. R. Hebd. Seances Acad. Sci., Ser. C*, 1966, **262**, 598; (e) E. Henry-Basch, J. Michel, F. Huet, B. Marx and P. Fréon, *Bull. Soc. Chim. Fr.*, 1965, 927; (f) J. Michel and E. Henry-Basch, *C. R. Hebd. Seances Acad. Sci., Ser. C*, 1966, **262**, 1274; (g) B. Riegel, S. Siegel and W. M. Lilienfeld, *J. Am. Chem. Soc.*, 1946, **68**, 984; (h) F. S. Prout, *J. Am. Chem. Soc.*, 1952, **74**, 5915.
110. (a) D. A. Evans, D. J. Baillargeon and J. V. Nelson, *J. Am. Chem. Soc.*, 1978, **100**, 2242; (b) P. R. Jones, P. D. Sherman and K. Schwarzenberg, *J. Organomet. Chem.*, 1967, **10**, 521; (c) C. Bernardon, *Tetrahedron Lett.*, 1979, 1581.
111. (a) G. Cahiez and M. Alami, *Tetrahedron Lett.*, 1989, **30**, 3541; (b) G. Cahiez and B. Laboue, *Tetrahedron Lett.*, 1989, **30**, 3545.
112. A. G. Brook, *J. Am. Chem. Soc.*, 1957, **79**, 4373.
113. (a) For a comprehensive review of the reactions of allylsilanes and their use in organic synthesis, see: H. Sakurai, *Pure Appl. Chem.*, 1982, **54**, 1; (b) A. Hosomi, A. Shirahata and H. Sakurai, *Tetrahedron Lett.*, 1978, 3043.
114. (a) A. Ricci, M. Fiorenza, M. Antonietta Grifagni, G. Bartolini and G. Seconi, *Tetrahedron Lett.*, 1982, **23**, 5079; (b) G. Majetich, A. M. Casares, D. Chapman and M. Behnke, *Tetrahedron Lett.*, 1983, **24**, 1909.
115. J. S. Panek and M. A. Sparks, *Tetrahedron Lett.*, 1987, **28**, 4649.
116. (a) G. Majetich, R. Desmond and A. M. Casares, *Tetrahedron Lett.*, 1983, **24**, 1913; (b) G. Majetich, K. Hull, J. Defauw and R. Desmond, *Tetrahedron Lett.*, 1985, **26**, 2747; (c) D. Schinzer, *Synthesis*, 1988, 263.
117. For a review of enolate chemistry, see: D. Seebach, *Angew. Chem., Int. Ed. Engl.*, 1988, **27**, 1624; *Angew. Chem.*, 1988, **100**, 1685.
118. (a) H. Stobbe, *J. Prakt. Chem.*, 1912, **86**, 209; (b) M. E. Jung, *Tetrahedron*, 1976, **32**, 3; (c) R. E. Gawley, *Synthesis*, 1976, 777; (d) A. J. Waring, in 'Comprehensive Organic Chemistry', ed. D. H. R. Barton and W. D. Ollis, Pergamon Press, Oxford, 1979, vol. 1, p. 1043.
119. W. S. Rapson and R. Robinson, *J. Chem. Soc.*, 1935, 1285.
120. For a review on azaenolates in synthesis, see: (a) J. K. Whitesell and M. A. Whitesell, *Synthesis*, 1983, 517; (b) R. R. Fraser, in 'Comprehensive Carbanion Chemistry. Part B: Selectivity in Carbon–Carbon Bond Forming Reactions', ed. E. Buncel and T. Durst, Elsevier, New York, 1984, p. 65.
121. R. Urech, *J. Chem. Soc., Chem. Commun.*, 1984, 989.
122. Y. Tamaru, T. Harada and Z. Yoshida, *J. Am. Chem. Soc.*, 1979, **101**, 1316.
123. (a) J. Bertrand, L. Gorrichon and P. Maroni, *Tetrahedron*, 1984, **40**, 4127; (b) D. A. Oare and C. H. Heathcock, *Tetrahedron Lett.*, 1986, **27**, 6169.
124. R. Häner, T. Laube and D. Seebach, *Chimia*, 1984, **38**, 255.
125. R. W. Stevens and T. Mukaiyama, *Chem. Lett.*, 1985, 855.
126. (a) J. Boyer, R. J. P. Corriu, R. Perz and C. Reye, *Tetrahedron*, 1983, **39**, 117; (b) C. Chuit, R. J. P. Corriu, R. Perz and C. Reye, *Tetrahedron*, 1986, **42**, 2293.

127. (a) D. Seebach and V. Ehrig, *Angew. Chem., Int. Ed. Engl.*, 1974, **13**, 400; *Angew. Chem.*, 1974, **86**, 446; (b) M. Horton and G. Pattenden, *Tetrahedron Lett.*, 1983, **24**, 2125; (c) D. Seebach and P. Knochel, *Helv. Chim. Acta*, 1984, **67**, 261.
128. (a) H. Ahlbrecht and K. Pfaff, *Synthesis*, 1980, 413; (b) H. Ahlbrecht and A. von Daacke, *Synthesis*, 1984, 610; (c) H. Ahlbrecht and A. von Daacke, *Synthesis*, 1987, 24; (d) B. Cazes, C. Huynh, S. Julia, V. Ratovelomanana and O. Ruel, *J. Chem. Res. (S)*, 1978, 68.
129. (a) P. J. Brown, D. N. Jones, M. A. Khan, N. A. Meanwell and P. J. Richards, *J. Chem. Soc., Perkin Trans. 1*, 1984, 2049; (b) D. N. Jones, N. A. Meanwell and S. M. Mirza, *J. Chem. Soc., Perkin Trans. 1*, 1985, 145; (c) K. Seki, T. Ohnuma, T. Oishi and Y. Ban, *Tetrahedron Lett.*, 1975, 723.
130. T. Yamazaki, N. Ishikawa, H. Iwatsubo and T. Kitazume, *J. Chem. Soc., Chem. Commun.*, 1987, 1340.
131. (a) K. Takaki, H. Takahashi, Y. Oshshiro and T. Agawa, *J. Chem. Soc., Chem. Commun.*, 1977, 675; (b) K. Takaki, K. Nakagawa and K. Negoro, *J. Org. Chem.*, 1980, **45**, 4789.
132. O. De Lucchi, L. Pasquato and G. Modena, *Tetrahedron Lett.*, 1984, **25**, 3647.
133. (a) S. D. Darling, F. N. Muralidharan and V. B. Muralidharan, *Synth. Commun.*, 1979, **9**, 915; (b) S. D. Darling, F. N. Muralidharan and V. B. Muralidharan, *Tetrahedron Lett.*, 1979, 2757; (c) S. D. Darling, F. N. Muralidharan and V. B. Muralidharan, *Tetrahedron Lett.*, 1979, 2761.
134. W. A. Kleschick and C. H. Heathcock, *J. Org. Chem.*, 1978, **43**, 1256.
135. (a) C. P. Casey and W. R. Brunsvold, *Inorg. Chem.*, 1977, **16**, 391; (b) F. A. Cotton and C. M. Lukehart, *J. Am. Chem. Soc.*, 1971, **93**, 2672; (c) C. P. Casey, R. A. Boggs and R. L. Anderson, *J. Am. Chem. Soc.*, 1972, **94**, 8947; (d) E. O. Fischer and S. Riedmüller, *Chem. Ber.*, 1974, **107**, 915.
136. (a) D. Seebach, R. Henning and J. Gonnermann, *Chem. Ber.*, 1979, **112**, 234; (b) P. Bakuzis, M. L. F. Bakuzis and T. F. Weingartner, *Tetrahedron Lett.*, 1978, 2371; (c) D. Seebach, R. Henning and F. Lehr, *Angew. Chem., Int. Ed. Engl.*, 1978, **17**, 45; *Angew. Chem.*, 1978, **90**, 479.
137. (a) D. Enders and K. Papadopoulos, *Tetrahedron Lett.*, 1983, **24**, 4967; (b) D. Enders and B. E. M. Rendenbach, *Tetrahedron*, 1986, **42**, 2235; (c) D. Enders, K. Papadopoulos, B. E. M. Rendenbach, R. Appel and F. Knoch, *Tetrahedron Lett.*, 1986, **27**, 3491; (d) D. Enders, G. Bachstädter, K. A. M. Kremer, M. Marsch, K. Harms and G. Boche, *Angew. Chem., Int. Ed. Engl.*, 1988, **27**, 1522; *Angew. Chem.*, 1988, **100**, 1580; (e) for conjugate additions of zincates or cuprates of chiral azaenolates, see: K. Yamamoto, M. Kanoh, N. Yamamoto and J. Tsuji, *Tetrahedron Lett.*, 1987, **28**, 6347.
138. (a) E. P. Kohler and G. L. Heritage, *Am. Chem. J.*, 1910, **43**, 475; (b) E. P. Kohler, G. L. Heritage and A. L. McLeod, *Am. Chem. J.*, 1911, **46**, 221; (c) E. P. Kohler and H. Gilman, *J. Am. Chem. Soc.*, 1919, **41**, 683; (d) J. C. Dubois, J. P. Guette and H. B. Kagan, *Bull. Soc. Chim. Fr.*, 1966, 3008; (e) C. Gandolfi, G. Doria, M. Amendola and E. Dradi, *Tetrahedron Lett.*, 1970, 3923.
139. (a) C. H. Heathcock, M. A. Henderson, D. A. Oare and M. A. Sanner, *J. Org. Chem.*, 1985, **50**, 3019; (b) C. H. Heathcock and D. A. Oare, *J. Org. Chem.*, 1985, **50**, 3022.
140. (a) A. G. Schultz and Y. K. Yee, *J. Org. Chem.*, 1976, **41**, 4045; (b) S. Yamagiwa, N. Hoshi, H. Sato, H. Kosugi and H. Uda, *J. Chem. Soc., Perkin Trans. 1*, 1978, 214; (c) J. Luchetti and A. Krief, *Tetrahedron Lett.*, 1978, 2697; (d) M. Takasu, H. Wakabayashi, K. Furuta and H. Yamamoto, *Tetrahedron Lett.*, 1988, **29**, 6943; (e) for 1,8-addition of *t*-butyl lithioacetate to tropone, see: J. H. Rigby and C. Senanayake, *J. Am. Chem. Soc.*, 1987, **109**, 3147.
141. (a) W. K. Bornack, S. S. Bhagwat, J. P. Ponton and P. Helquist, *J. Am. Chem. Soc.*, 1981, **103**, 4647; (b) W. Oppolzer, M. Guo and K. Baettig, *Helv. Chim. Acta*, 1983, **66**, 2140; (c) H. Nishiyama, K. Sakuta and K. Itoh, *Tetrahedron Lett.*, 1984, **25**, 2487.
142. J. E. McMurry, W. A. Andrus and J. H. Musser, *Synth. Commun.*, 1978, **8**, 53.
143. (a) K. Kpegba, P. Metzner and R. Rakotonirina, *Tetrahedron*, 1989, **45**, 2041; (b) P. Metzner and R. Rakotonirina, *Tetrahedron Lett.*, 1983, **24**, 4203; (c) S. H. Bertz, L. W. Jelinski and G. Dabbagh, *J. Chem. Soc., Chem. Commun.*, 1983, 388; (d) P. Metzner and R. Rakotonirina, *Tetrahedron*, 1985, **41**, 1289; (e) S. Berrada and P. Metzner, *Tetrahedron Lett.*, 1987, **28**, 409.
144. J. L. Herrmann, M. H. Berger and R. H. Schlessinger, *J. Am. Chem. Soc.*, 1979, **101**, 1544.
145. H. Normant, *J. Organomet. Chem.*, 1975, **100**, 189.
146. (a) M. Yamaguchi, M. Tsukamoto and I. Hirao, *Chem. Lett.*, 1984, 375; (b) M. Yamaguchi, M. Tsukamoto, S. Tanaka and I. Hirao, *Tetrahedron Lett.*, 1984, **25**, 5661; (c) M. Yamaguchi, K. Hasebe, S. Tanaka and T. Minami, *Tetrahedron Lett.*, 1986, **27**, 959; (d) M. Yamaguchi, M. Tsukamoto and I. Hirao, *Tetrahedron Lett.*, 1985, **26**, 1723; (e) for a diastereoselective addition of β-lactone enolates to dimethyl maleate, see: J. Mulzer, A. Chucholowski, O. Lammer, I. Jibril and G. Huttner, *J. Chem. Soc., Chem. Commun.*, 1983, 8; (f) P. Prempree, S. Radviroongit and Y. Thebtaranonth, *J. Org. Chem.*, 1983, **48**, 3553; (g) for a diastereoselective and enantioselective addition of (–)-phenmenthyl propionate to methyl (*E*)-crotonate, see: E. J. Corey and R. T. Peterson, *Tetrahedron Lett.*, 1985, **26**, 5025; (h) for enantioselective conjugate addition of ester enolates to α,β-unsaturated esters with chiral catalyst, see: M. Alonso-Lôpez, J. Jimenez-Barbero, M. Martín-Lomas and S. Penadés, *Tetrahedron*, 1988, **44**, 1535.
147. G. H. Posner, *Acc. Chem. Res.*, 1987, **20**, 72.
148. (a) R. J. Cregge, J. L. Herrmann and R. H. Schlessinger, *Tetrahedron Lett.*, 1973, 2603; (b) J. L. Herrmann, G. R. Kieczykowski, R. F. Romanet, P. J. Wepplo and R. H. Schlessinger, *Tetrahedron Lett.*, 1973, 4711.
149. (a) H. Ahlbrecht and M. Dietz, *Synthesis*, 1985, 417; (b) D. Seebach and H. F. Lietz, *Angew. Chem., Int. Ed. Engl.*, 1971, **10**, 501; *Angew. Chem.*, 1971, **83**, 542.
150. (a) G. Calderari and D. Seebach, *Helv. Chim. Acta*, 1985, **68**, 1592; (b) M. Züger, T. Weller and D. Seebach, *Helv. Chim. Acta*, 1980, **63**, 2005.
151. (a) S. Hanessian and D. Desilets, in 'Trends in Medicinal Chemistry', ed. H. van der Goot, G. Domány, L. Pallos and H. Timmerman, Elsevier, Amsterdam, 1989, p. 165; (b) S. Hanessian, D. Desilets and Y. L. Bennani, *J. Org. Chem.*, 1990, **55**, 3098.
152. (a) G. Stork, A. Y. W. Leong and A. M. Touzin, *J. Org. Chem.*, 1976, **41**, 3491; (b) N. Minowa, M. Hirayama and S. Fukatsu, *Tetrahedron Lett.*, 1984, **25**, 1147; (c) S. Kanemasa, A. Tatsukawa, E. Wada and O. Tsuge,

Chem. Lett., 1989, 1301; (d) U. Schöllkopf, D. Pettig, U. Busse, E. Egert and M. Dyrbusch, *Synthesis*, 1986, 737; (e) J. E. Baldwin, R. M. Adlington and N. G. Robinson, *Tetrahedron Lett.*, 1988, **29**, 375.
153. (a) W. Oppolzer and R. Pitteloud, *J. Am. Chem. Soc.*, 1982, **104**, 6478; (b) for enantioselective addition, see: W. Oppolzer, R. Pitteloud, G. Bernardinelli and K. Baettig, *Tetrahedron Lett.*, 1983, **24**, 4975.
154. (a) G. A. Kraus and B. Roth, *Tetrahedron Lett.*, 1977, 3129; (b) G. A. Kraus and H. Sugimoto, *Tetrahedron Lett.*, 1977, 3929; (c) F. M. Hauser and R. P. Rhee, *J. Org. Chem.*, 1978, **43**, 179; (d) F. M. Hauser and S. Prasanya, *J. Org. Chem.*, 1979, **44**, 2597; (e) F. M. Hauser and D. W. Combs, *J. Org. Chem.*, 1980, **45**, 4071; (f) N. J. P. Broom and P. G. Sammes, *J. Chem. Soc., Perkin Trans. 1*, 1981, 465; (g) G. A. Kraus, H. Cho, S. Crowley, B. Roth, H. Sugimoto and S. Prugh, *J. Org. Chem.*, 1983, **48**, 3439.
155. (a) Y. N. Kuo, J. A. Yahner and C. Ainsworth, *J. Am. Chem. Soc.*, 1971, **93**, 6321; (b) M. Miyashita, R. Yamaguchi and A. Yoshikoshi, *Chem. Lett.*, 1982, 1505; (c) J. Mulzer, G. Hartz, U. Kühl and G. Brüntrup, *Tetrahedron Lett.*, 1978, 2949; (d) for discussion of conjugate additions of dienediolates, see: P. Ballester, A. Costa, A. G. Raso, A. Gomez-Solivellas and R. Mestres, *J. Chem. Soc., Perkin Trans. 1*, 1988, 1711; (e) G. S. Subba Rao, H. Ramanathan and K. Raj, *J. Chem. Soc., Chem. Commun.*, 1980, 315.
156. R. Sauvetre, M.-C. Roux-Schmitt, and J. Seyden-Penne, *Tetrahedron*, 1978, **34**, 2135.
157. K. Tomioka and K. Koga, *Tetrahedron Lett.*, 1984, **25**, 1599.
158. (a) N. Wang, S. Su and L. Tsai, *Tetrahedron Lett.*, 1979, 1121; (b) D. Morgans, Jr. and G. B. Feigelson, *J. Org. Chem.*, 1982, **47**, 1131; (c) P. A. Grieco and Y. Yokoyama, *J. Am. Chem. Soc.*, 1977, **99**, 5210; (d) E. Hatzigrigoriou and L. Wartski, *Synth. Commun.*, 1983, **13**, 319.
159. (a) E. Hatzigrigoriou, M.-C. Roux-Schmitt, L. Wartski, J. Seyden-Penne and C. Merienne, *Tetrahedron*, 1983, **39**, 3415; (b) M.-C. Roux-Schmitt, N. Seuron and J. Seyden-Penne, *Synthesis*, 1983, 494; (c) For conjugate additions to α,β-unsaturated aldehydes (imines), see: M. El-Bouz, M.-C. Roux-Schmitt and L. Wartski, *J. Chem. Soc., Chem. Commun.*, 1979, 779.
160. H. Ahlbrecht and M. Ibe, *Synthesis*, 1985, 421.
161. (a) D. Seebach and M. Kolb, *Chem. Ind. (London)*, 1974, 687; (b) D. Seebach, *Angew. Chem., Int. Ed. Engl.*, 1979, **18**, 239; *Angew. Chem.*, 1979, **91**, 259.
162. (a) For a review on nucleophilic acylation, see: O. W. Lever, Jr., *Tetrahedron*, 1976, **32**, 1943; (b) T. A. Hase, 'Umpoled Synthons', Wiley Interscience, New York, 1987.
163. (a) C. A. Brown and A. Yamaichi, *J. Chem. Soc., Chem. Commun.*, 1979, 100; (b) C. A. Brown, O. Chapa and A. Yamaichi, *Heterocycles*, 1982, **18**, 187.
164. (a) B.-T. Gröbel and D. Seebach, *Synthesis*, 1978, 357; (b) for conjugate additions to α,β-unsaturated aldehydes, see: L. Wartski and M. El-Bouz, *Tetrahedron*, 1982, **38**, 3285; (c) for enantioselective conjugate additions, see: K. Tomioka, M. Sudani, Y. Shinmi and K. Koga, *Chem. Lett.*, 1985, 329; (d) for additions to α-nitroalkenes, see: D. Seebach, H. F. Leitz and V. Ehrig, *Chem. Ber.*, 1975, **108**, 1924, 1946; (e) for additions to vinyltriphenylphosphonium salts, see: I. Kawamoto, S. Muramatsu and Y. Yura, *Tetrahedron Lett.*, 1974, 4223; (f) for additions to α,β-unsaturated nitriles, see: F. Z. Basha, J. F. DeBernardis and S. Spanton, *J. Org. Chem.*, 1985, **50**, 4160; (g) for discussion of additions of 2-vinyl-1,3-dithianes and 2-(1-(*E*)-propenyl)-1,3-dithianes, see: (i) F. E. Ziegler, J.-M. Fang and C. C. Tam, *J. Am. Chem. Soc.*, 1982, **104**, 7174; (ii) F. E. Ziegler and J. J. Mencel, *Tetrahedron Lett.*, 1983, **24**, 1859.
165. J. Lucchetti and A. Krief, *Synth. Commun.*, 1983, **13**, 1153.
166. (a) J. E. Richman, J. H. Herrmann and R. H. Schlessinger, *Tetrahedron Lett.*, 1973, 3267; (b) J. E. Richman, J. H. Herrmann and R. H. Schlessinger, *Tetrahedron Lett.*, 1973, 3271; (c) M. Mikolajczyk and P. Balczewski, *Synthesis*, 1987, 659; (d) for diastereoselective conjugate additions, see: L. Colombo, C. Gennari, G. Resnati and C. Scolastico, *J. Chem. Soc., Perkin Trans. 1*, 1981, 1284.
167. D. J. Ager and M. B. East, *J. Org. Chem.*, 1986, **51**, 3983.
168. (a) G. Stork, A. A. Ozorio and A. Y. W. Leong, *Tetrahedron Lett.*, 1978, 5175; (b) V. Reutrakul, S. Nimgirawath, S. Panichanun and P. Ratananukul, *Chem. Lett.*, 1979, 399; (c) H. Ahlbrecht and H.-M. Kompter, *Synthesis*, 1983, 645; (d) J. D. Albright, *Tetrahedron*, 1983, **39**, 3207; (e) M. Zervos, L. Wartski and J. Seyden-Penne, *Tetrahedron*, 1986, **42**, 4963.
169. (a) α-Silyloxynitriles: (i) S. Hünig and G. Wehner, *Chem. Ber.*, 1980, **113**, 302; (ii) S. Hünig and M. Öller, *Chem. Ber.*, 1980, **113**, 3803; (b) α-cyanocarbonates: A. T. Au, *Synth. Commun.*, 1984, **14**, 749; (c) α-(1-ethoxyethoxy)nitriles: (i) R. K. Boeckman, Jr., D. K. Heckendorn and R. L. Chinn, *Tetrahedron Lett.*, 1987, **28**, 3551; (ii) A. M. Casares and L. A. Maldonado, *Synth. Commun.*, 1976, **6**, 11; (iii) for additions to dienic sulfoxides, see: E. Guittet and S. Julia, *Synth. Commun.*, 1981, **11**, 709.
170. J. E. Baldwin, R. M. Adlington, A. U. Jain, J. N. Kolhe and M. W. D. Perry, *Tetrahedron*, 1986, **42**, 4247.
171. (a) A.-R. B. Manas and R. A. J. Smith, *J. Chem. Soc., Chem. Commun.*, 1975, 216; (b) R. A. J. Smith and A. R. Lal, *Aust. J. Chem.*, 1979, **32**, 353; (c) T. Cohen and S. M. Nolan, *Tetrahedron Lett.*, 1978, 3533; (d) T. Cohen and L.-C. Yu, *J. Org. Chem.*, 1985, **50**, 3266; (e) R. E. Damon and R. H. Schlessinger, *Tetrahedron Lett.*, 1976, 1561; (f) S. Hanessian, P. J. Murray and S. P. Sahoo, *Tetrahedron Lett.*, 1985, **26**, 5627; (g) S. Hanessian, S. P. Sahoo and M. Botta, *Tetrahedron Lett.*, 1987, **28**, 1147; (h) S. Hanessian and P. J. Murray, *Tetrahedron*, 1987, **43**, 5055; (i) S. Hanessian and P. J. Murray, *J. Org. Chem.*, 1987, **52**, 1170.
172. R. Bürstinghaus and D. Seebach, *Chem. Ber.*, 1977, **110**, 841.
173. G. S. Bates and S. Ramaswamy, *Can. J. Chem.*, 1983, **61**, 2006.
174. (a) J. Otera, Y. Niibo and H. Aikawa, *Tetrahedron Lett.*, 1987, **28**, 2147; (b) J. Otera, Y. Niibo and H. Nozaki, *J. Org. Chem.*, 1989, **54**, 5003.
175. J. V. Metzger, in 'Comprehensive Heterocyclic Chemistry', ed. A. R. Katritzky and C. W. Rees, Pergamon Press, Oxford, 1984, vol. 6, p. 235.
176. (a) F. E. Ziegler and J. A. Schwartz, *J. Org. Chem.*, 1978, **43**, 985; (b) A. Pelter, R. S. Ward, P. Satyanarayana and P. Collins, *J. Chem. Soc., Perkin Trans. 1*, 1983, 643.
177. (a) D. Seebach, *Angew. Chem., Int. Ed. Engl.*, 1969, **8**, 639; *Angew. Chem.*, 1969, **81**, 690; (b) J. P. Collman, *Acc. Chem. Res.*, 1975, **8**, 342; (c) D. Seyferth and R. M. Weinstein, *J. Am. Chem. Soc.*, 1982, **104**, 5534; (d) S. Murai, I. Ryu, J. Iriguchi and N. Sonoda, *J. Am. Chem. Soc.*, 1984, **106**, 2440; (e) D. Seyferth and R. C. Hui, *J. Am. Chem. Soc.*, 1985, **107**, 4551.

178. (a) M. P. Cooke, Jr. and R. M. Parlman, *J. Am. Chem. Soc.*, 1977, **99**, 5222; (b) M. Yamashita, H. Tashika, T. Nakanobo, M. Itokawa and R. Suemitsu, *Yakugaku Zasshi*, 1988, **37**, 250.
179. (a) E. J. Corey and L. S. Hegedus, *J. Am. Chem. Soc.*, 1969, **91**, 4926; (b) M. F. Semmelhack, L. Keller, T. Sato, E. J. Spiess and W. Wulff, *J. Org. Chem.*, 1985, **50**, 5566.
180. (a) J. A. S. Howell, B. F. G. Johnson, P. L. Josty and J. Lewis, *J. Organomet. Chem.*, 1972, **39**, 329; (b) A. N. Nesmeyanov, L. V. Rybin, N. T. Gubenko, M. I. Rybinskaya and P. V. Petrovskii, *J. Organomet. Chem.*, 1974, **71**, 271; (c) M. Brookhart, G. W. Koszalka, G. O. Nelson, G. Scholes and R. A. Watson, *J. Am. Chem. Soc.*, 1976, **98**, 8155; (d) T. N. Danks, D. Rakshit and S. E. Thomas, *J. Chem. Soc., Perkin Trans. 1*, 1988, 2091; (e) H. Kitahara, Y. Tozawa, S. Fujita, A. Tajiri, N. Morita and T. Asao, *Bull. Chem. Soc. Jpn.*, 1988, **61**, 3362; (f) A. Pouilhès and S. E. Thomas, *Tetrahedron Lett.*, 1989, **30**, 2285.
181. (a) R. F. Heck, *J. Am. Chem. Soc.*, 1963, **85**, 657; (b) L. S. Hegedus and R. J. Perry, *J. Org. Chem.*, 1985, **50**, 4955.
182. (a) A. G. Brook, *J. Am. Chem. Soc.*, 1957, **79**, 4373; (b) A. Degl'Innocenti, S. Pike, D. R. M. Walton and G. Seconi, *J. Chem. Soc., Chem. Commun.*, 1980, 1201; (c) D. Schinzer and C. H. Heathcock, *Tetrahedron Lett.*, 1981, **22**, 1881; (d) A. Ricci, A. Degl'Innocenti, S. Chimichi, M. Fiorenza, G. Rossini and H. J. Bestmann, *J. Org. Chem.*, 1985, **50**, 130; (e) A. Ricci, M. Fiorenza, A. Degl'Innocenti, G. Seconi, P. Dembech, K. Witzgall and H. J. Bestmann, *Angew. Chem., Int. Ed. Engl.*, 1985, **24**, 1068; *Angew. Chem.*, 1985, **97**, 1068.
183. (a) J. P. Freeman, *J. Org. Chem.*, 1966, **31**, 538; (b) P. A. Grieco and R. S. Finkelhor, *Tetrahedron Lett.*, 1972, 3781; (c) for additions of allylidenephosphoranes to (*S*)-*cis* configuration α,β-enones, see: (i) G. Büchi and H. Wuest, *Helv. Chim. Acta*, 1971, **54**, 1767; (ii) W. G. Dauben, D. J. Hart, J. Ipaktschi and A. P. Kozikowski, *Tetrahedron Lett.*, 1973, 4425; (ii) W. G. Dauben and J. Ipaktschi, *J. Am. Chem. Soc.*, 1973, **95**, 5088.
184. (a) B. M. Trost and L. S. Melvin, Jr., 'Sulfur Ylides—Emerging Synthetic Intermediates', Academic Press, New York, 1975; (b) F. Cook, P. Magnus and G. L. Bundy, *J. Chem. Soc., Chem. Commun.*, 1978, 714; (c) 'The Chemistry of the Sulphonium Group', ed. C. J. M. Stirling, Wiley, New York, 1981, vol. 1, 2; (d) H. G. Corkins, L. Veenstra and C. R. Johnson, *J. Org. Chem.*, 1978, **43**, 4233, and refs. therein.
185. (a) H. Ahlbrecht and M. Ibe, *Synthesis*, 1988, 210; (b) T. Cohen, J. P. Sherbine, S. A. Mendelson and M. Myers, *Tetrahedron Lett.*, 1985, **26**, 2965; (c) T. Cohen, R. B. Weisenfeld and R. E. Gapinski, *J. Org. Chem.*, 1979, **44**, 4744; (d) T. M. Dolak and T. A. Bryson, *Tetrahedron Lett.*, 1977, 1961.
186. Y. Takano, A. Yasuda, H. Urabe and I. Kuwajima, *Tetrahedron Lett.*, 1985, **26**, 6225.
187. M. Casey, A. C. Manage and L. Nezhat, *Tetrahedron Lett.*, 1988, **29**, 5821.
188. (a) K. Tamao and N. Ishida, *Tetrahedron Lett.*, 1984, **25**, 4249; (b) R. T. Taylor and J. G. Galloway, *J. Organomet. Chem.*, 1981, **220**, 295; (c) G. H. Posner, M. Weitzberg and S. Jew, *Synth. Commun.*, 1987, **17**, 611.
189. (a) U. Schöllkopf and K. Hantke, *Justus Liebigs Ann. Chem.*, 1973, 1571; (b) M. Westling and T. Livinghouse, *Synthesis*, 1987, 391.
190. (a) For a review of three-carbon homologating reagents, see: J. C. Stowell, *Chem. Rev.*, 1984, **84**, 409; (b) for a review of homoenolate anions and homoenolate anion equivalents, see: N. H. Werstiuk, *Tetrahedron*, 1983, **39**, 205.
191. (a) B. Lesur, J. Toye, M. Chantrenne and L. Ghosez, *Tetrahedron Lett.*, 1979, 2835; (b) S. De Lombaert, B. Lesur and L. Ghosez, *Tetrahedron Lett.*, 1982, **23**, 4251.
192. (a) M. R. Binns and R. K. Haynes, *J. Org. Chem.*, 1981, **46**, 3790; (b) M. R. Binns, R. K. Haynes, A. G. Katsifis, P. A. Schober and S. C. Vonwiller, *J. Org. Chem.*, 1989, **54**, 1960; (c) M. Hirama, *Tetrahedron Lett.*, 1981, **22**, 1905; (d) G. A. Kraus and K. Frazier, *Synth. Commun.*, 1978, **8**, 483; (e) I. H. Sanchez and A. M. Aguiar, *Synthesis*, 1981, 55.
193. (a) M. R. Binns, R. K. Haynes, A. G. Katsifis, P. A. Schober and S. C. Vonwiller, *J. Am. Chem. Soc.*, 1988, **110**, 5411; (b) R. K. Haynes, A. G. Katsifis, S. C. Vonwiller and T. W. Hambley, *J. Am. Chem. Soc.*, 1988, **110**, 5423.
194. (a) D. H. Hua, G. Sinai-Zingde and S. Venkataraman, *J. Am. Chem. Soc.*, 1985, **107**, 4088; (b) D. H. Hua, R. Chan-Yu-King, J. A. McKie and L. Myer, *J. Am. Chem. Soc.*, 1987, **109**, 5026; (c) D. H. Hua, S. Venkataraman, R. A. Ostrander, S. Z. Gurudas, P. J. McCann, M. J. Coulter and M. R. Xu, *J. Org. Chem.*, 1988, **53**, 507.
195. H. Ahlbrecht, M. Dietz and L. Weber, *Synthesis*, 1987, 251.
196. (a) J. P. Marino and E. Laborde, *J. Org. Chem.*, 1987, **52**, 1; (b) J. P. Marino, C. Silveira, J. V. Comasseto and N. Petragnani, *J. Org. Chem.*, 1987, **52**, 4140; (c) J. P. Marino and J. K. Long, *J. Am. Chem. Soc.*, 1988, **110**, 7916.
197. (a) L. F. Cason and H. G. Brooks, *J. Am. Chem. Soc.*, 1952, **74**, 4582; (b) L. F. Cason and H. G. Brooks, *J. Org. Chem.*, 1954, **19**, 1278; (c) J. E. Mulvaney and Z. G. Gardlund, *J. Org. Chem.*, 1965, **30**, 917; (d) P. F. Hudrlik and D. Peterson, *Tetrahedron Lett.*, 1974, 1133; (e) T. H. Chan and E. Chang, *J. Org. Chem.*, 1974, **39**, 3264; (f) G. R. Buell, R. J. P. Corriu, C. Guerin and L. Spialter, *J. Am. Chem. Soc.*, 1970, **92**, 7424; (g) A. G. Brook, J. M. Duff and D. G. Anderson, *Can. J. Chem.*, 1970, **48**, 561; (h) for additions to α-silylvinyl sulfides and α-silylvinyl sulfoxides, see: (i) D. J. Ager, *J. Chem. Soc., Perkin Trans. 1*, 1983, 1131; (ii) D. J. Ager, *J. Chem. Soc., Perkin Trans. 1*, 1986, 195; (iii) S. Kanemasa, H. Kobayashi, J. Tanaka and O. Tsuge, *Bull. Chem. Soc. Jpn.*, 1988, **61**, 3957.
198. D. J. Peterson, *J. Org. Chem.*, 1966, **31**, 950.
199. (a) W. E. Parham and R. F. Motter, *J. Am. Chem. Soc.*, 1959, **81**, 2146; (b) W. E. Parham, M. A. Kalnins and D. R. Theissen, *J. Org. Chem.*, 1962, **27**, 2698.
200. (a) T. Kauffmann, H. Ahlers, H.-J. Tilhard and A. Woltermann, *Angew. Chem., Int. Ed. Engl.*, 1977, **16**, 710; *Angew. Chem.*, 1977, **89**, 760; (b) M. Sevrin, J. N. Denis and A. Krief, *Angew. Chem., Int. Ed. Engl.*, 1978, **17**, 526; *Angew. Chem.*, 1978, **90**, 550; (c) S. Raucher and G. A. Koolpe, *J. Org. Chem.*, 1978, **43**, 4252.
201. (a) R. M. Carlson and P. Helquist, *Tetrahedron Lett.*, 1969, 173; (b) D. Seebach, M. Kolb and B. I. Gröbel, *Angew. Chem., Int. Ed. Engl.*, 1973, **12**, 69; *Angew. Chem.*, 1973, **85**, 42; (c) N. H. Anderson, P. F. Duffy, A. D. Denniston and D. B. Grotjahn, *Tetrahedron Lett.*, 1978, 4315; (d) for additions of organolithiums to

butadienyl-1,3-dithianes, see: D. Seebach, M. Kolb and B.-T. Gröbel, *Angew. Chem., Int. Ed. Engl.*, 1973, **12**, 69; *Angew. Chem.*, 1973, **85**, 42; (e) for diastereoselective addition of organolithiums to 2-(α-hydroxyalkylidene)-1,3-dithianes, see: T. Sato, M. Nakakita, S. Kimura and T. Fujisawa, *Tetrahedron Lett.*, 1989, **30**, 977.

202. (a) R. B. Woodward and R. H. Eastman, *J. Am. Chem. Soc.*, 1946, **68**, 2229; (b) D. S. Tarbell and M. A. McCall, *J. Am. Chem. Soc.*, 1952, **74**, 48; (c) W. J. Brehm and T. Levenson, *J. Am. Chem. Soc.*, 1954, **76**, 5389; (d) K. Tamao, R. Kanatani and M. Kumada, *Tetrahedron Lett.*, 1984, **25**, 1905.
203. K. Utimoto, K. Imi, H. Shiragami, S. Fujikura and H. Nozaki, *Tetrahedron Lett.*, 1985, **26**, 2101.
204. For a review of multicomponent annulations, see: G. H. Posner, *Chem. Rev.*, 1986, **86**, 831.
205. (a) R. R. Schmidt and J. Talbiersky, *Angew. Chem., Int. Ed. Engl.*, 1978, **17**, 204; *Angew. Chem.*, 1978, **90**, 220; (b) J. P. Marino and L. C. Katterman, *J. Chem. Soc., Chem. Commun.*, 1979, 946; (c) K. Isobe, M. Fuse, H. Kosugi, H. Hagiwara and H. Uda, *Chem. Lett.*, 1979, 785; (d) P. Beak and K. D. Wilson, *J. Org. Chem.*, 1986, **51**, 4627; (e) P. Beak and K. D. Wilson, *J. Org. Chem.*, 1987, **52**, 218; (f) P. Beak and D. A. Burg, *J. Org. Chem.*, 1989, **54**, 1647.
206. C. Thanupran, C. Thebtaranonth and Y. Thebtaranonth, *Tetrahedron Lett.*, 1986, **27**, 2295.
207. (a) E. G. Gibbons, *J. Org. Chem.*, 1980, **45**, 1540; (b) M. Asaoka and H. Takei, *Tetrahedron Lett.*, 1987, **28**, 6343; (c) H. Nagaoka, K. Kobayashi, T. Matsui and Y. Yamada, *Tetrahedron Lett.*, 1987, **28**, 2021; (d) for enantioselective synthesis, see: D. Spitzner, P. Wagner, A. Simon and K. Peters, *Tetrahedron Lett.*, 1989, **30**, 547; (e) for synthesis of tricyclo[5.2.2.0^{1,5}]undecanes, see: D. Spitzner and G. Sawitzki, *J. Chem. Soc., Perkin Trans. 1*, 1988, 373; (f) for synthesis of bicyclohydrocarbazolones, see: D. Schinzer and M. Kalesse, *Synlett.*, 1989, 34.
208. (a) R. M. Cory, D. M. T. Chan, Y. M. A. Naguib, M. H. Rastall and R. M. Renneboog, *J. Org. Chem.*, 1980, **45**, 1852; (b) R. M. Cory, P. C. Anderson, M. D. Bailey, F. R. McLaren, R. M. Renneboog and B. R. Yamamoto, *Can. J. Chem.*, 1985, **63**, 2618.
209. (a) M. Ihara, Y. Ishida, M. Abe, M. Toyota, K. Fukumoto and T. Kametani, *J. Chem. Soc., Perkin Trans. 1*, 1988, 1155; (b) M. Ihara, M. Toyota, K. Fukumoto and T. Kametani, *Tetrahedron Lett.*, 1984, **25**, 3235; (c) L. Lombardo and L. N. Mander, *J. Org. Chem.*, 1983, **48**, 2298; (d) S. Danishefsky, S. Chackalamannil, P. Harrison, M. Silvestri and P. Cole, *J. Am. Chem. Soc.*, 1985, **107**, 2474.
210. (a) For Grignard additions, see: E. Benary, *Ber.*, 1931, **64**, 2543; (b) for organolithium additions, see: (i) J.-P. Pradere and H. Quiniou, *C. R. Hebd. Seances Acad. Sci.*, 1971, **273**, 1013; (ii) P. Schiess, *Helv. Chim. Acta*, 1972, **55**, 2365; (iii) E. Elkik and M. Imbeaux-Oudotte, *C. R. Hebd. Seances Acad. Sci.*, 1973, **276**, 1203; (iv) E. Elkik and M. Imbeaux-Oudotte, *Bull. Soc. Chim. Fr.*, 1976, 439; (v) R. F. Abdulla and K. H. Fuhr, *J. Org. Chem.*, 1978, **43**, 4248.
211. R. F. Abdulla, *Tetrahedron*, 1979, **35**, 1675.
212. (a) S. Kiyooka, T. Yamashita, J. Tashiro, K. Takano and Y. Uchio, *Tetrahedron Lett.*, 1986, **27**, 5629; (b) S. Kiyooka and T. Yamashita, *Chem. Lett.*, 1987, 1775.
213. For a review of the chemistry of nitroenamines, see: S. Rajappa, *Tetrahedron*, 1981, **37**, 1453.
214. (a) T. Severin, D. Scheel and P. Adhikary, *Chem. Ber.*, 1969, **102**, 2966; (b) T. Severin and T. Wieland, *Synthesis*, 1973, 613; (c) T. Severin and B. Brück, *Chem. Ber.*, 1965, **98**, 3847; (d) T. Severin and H. Kullmer, *Chem. Ber.*, 1971, **104**, 440; (e) H. Lerche, D. König and T. Severin, *Chem. Ber.*, 1974, **107**, 1509.
215. K. Fuji, M. Node, H. Nagasawa, Y. Naniwa and S. Terada, *J. Am. Chem. Soc.*, 1986, **108**, 3855.
216. (a) H. Rupe and M. Iselin, *Ber.*, 1916, 25; (b) G. Martin, *Ann. Chim. (Paris)*, 1959, **4**, 541; (c) Y. Tamura, T. Nishimura, J. Eiho and T. Miyamoto, *Chem. Ind. (London)*, 1971, 1199; (d) J. P. Marino and T. Kaneko, *J. Org. Chem.*, 1974, **39**, 3176; (e) E. Vedejs and J. Bershas, *Tetrahedron Lett.*, 1975, 1359; (f) J. L. Herrmann, J. E. Richman and R. H. Schlessinger, *Tetrahedron Lett.*, 1973, 2599; (g) G. Dionne and C. R. Engel, *Can. J. Chem.*, 1978, **56**, 419; (h) P. Bey and J. P. Vevert, *J. Org. Chem.*, 1980, **45**, 3249; (i) for additions of organometallics to α-(difluoromethylene)-γ-lactones, see: M. Suda, *Tetrahedron Lett.*, 1981, **22**, 1421.
217. (a) G. A. Kraus and J. O. Pezzanite, *J. Org. Chem.*, 1979, **44**, 2480; (b) D. L. Boger and M. D. Mullican, *J. Org. Chem.*, 1984, **49**, 4033; (c) N. G. Clemo and G. Pattenden, *J. Chem. Soc., Perkin Trans. 1*, 1986, 2133; (d) C. A. Weber-Schilling and H.-W. Wanzlick, *Chem. Ber.*, 1971, **104**, 1518.
218. (a) K. Mori and H. Mori, *Tetrahedron*, 1987, **43**, 4097; (b) K. T. Potts, M. J. Cipullo, P. Ralli and G. Theodoridis, *J. Am. Chem. Soc.*, 1981, **103**, 3584; (c) X. Huang and B. Chen, *Synthesis*, 1987, 480.
219. S. Hanessian, A. Bedeschi, C. Battistini and N. Mogelli, *J. Am. Chem. Soc.*, 1985, **107**, 1438.
220. (a) B. W. Metcalf and E. Bonilavri, *J. Chem. Soc., Chem. Commun.*, 1978, 914; (b) H. Kinoshita, I. Hori, T. Oishi and Y. Ban, *Chem. Lett.*, 1984, 1517.
221. (a) K. Okuhara, *J. Org. Chem.*, 1976, **41**, 1487; (b) K. Tanaka, T. Nakai and N. Ishikawa, *Chem. Lett.*, 1979, 175; (c) A. E. Feiring, *J. Org. Chem.*, 1980, **45**, 1962; (d) N. Ishikawa, S. Butler and M. Maruta, *Bull. Chem. Soc. Jpn.*, 1981, **54**, 3084; (e) Z. Rappoport, *Recl. Trav. Chim. Pays-Bas*, 1985, **104**, 309; (f) R. P. Gajewski, J. L. Jackson, N. D. Jones, J. K. Swartzendruber and J. B. Deeter, *J. Org. Chem.*, 1989, **54**, 3311.
222. (a) G. Stork and M. Tomasz, *J. Am. Chem. Soc.*, 1962, **84**, 310; (b) B. M. Trost, C. D. Shuey and F. DiNinno, Jr., *J. Am. Chem. Soc.*, 1979, **101**, 1284.
223. (a) E. J. Corey and W. Su, *Tetrahedron Lett.*, 1988, **29**, 3423; (b) W.-D. Rudorf and R. Schwarz, *Z. Chem.*, 1988, **28**, 101.
224. A. S. Kende and P. Fludzinski, *Tetrahedron Lett.*, 1982, **23**, 2369, 2373.
225. (a) M. R. Rifi, in 'Techniques of Electroorganic Synthesis', ed. N. L. Weinberg, Wiley, New York, 1975; (b) M. R. Rifi and F. H. Covitz, 'Introduction to Organic Electrochemistry', Marcel Decker, New York, 1974.
226. (a) R. Scheffold, G. Rytz and L. Walder, in 'Modern Synthetic Methods', ed. R. Scheffold, Wiley, New York, 1983; (b) R. Scheffold, M. Dike, S. Dike, T. Herold and L. Walder, *J. Am. Chem. Soc.*, 1980, **102**, 3642; (c) T. Shono, S. Kashimura, Y. Yamaguchi and F. Kuwata, *Tetrahedron Lett.*, 1987, **28**, 4411.

1.3
Conjugate Additions of Carbon Ligands to Activated Alkenes and Alkynes Mediated by Lewis Acids

VING J. LEE
American Cyanamid, Pearl River, NY, USA

1.3.1 INTRODUCTION

The conjugate addition of stabilized carbanions, nonstabilized carbanions and heteroanions to activated alkenes and alkynes has been discussed extensively in previous chapters in this volume (Chapters 1.1 and 1.2). In summary, the regioselectivity of carbanion additions is sensitive to: (i) the electronegativity of the metal center;[1] (ii) the Hard and Soft Acid–Base (HSAB) characteristics of the metal center and the transferring ligand;[2] (iii) the structure of the acceptor;[3] and (iv) reaction conditions and solvent effects.[4] The difference in electronegativity between a metal and carbon is reflected in the degree of ionic character imparted to the carbon–metal σ-bond and the direction of polarization.[5] Thus the highly ionic Groups IA and IIA organometallics (RK, RNa, RLi and RMgX; Volume 4, Chapter 1.2) are extremely reactive, in contrast to the less ionic Group IIB organometallics (R_2Cd and R_2Zn; Volume 4, Chapter 1.2).

In contrast to the robust organolithiums and Grignard reagents, more selective and less basic organometallic reagents effective in transferring carbon ligands in a 1,4-mode have been introduced that are compatible with additional functionalities on either the acceptors or donors. These reagents are characterized by metal centers either with electron deficient *p*-orbitals which enhance their Lewis acidity, *e.g.* R_3Al or R_3B (Group IIIB), or that are electropositive *vis-à-vis* carbon, *e.g.* R_4Si or R_4Sn (Group IVB). Typically, carbon ligand transferrals with the latter class of organometallics are facilitated with a Lewis acid catalyst. However, with the Group IB organometallics, *e.g.* RCu (Volume 4, Chapter 1.4), and Group IIIB organometallics, carbon–carbon bond formation is mediated by either carbanion, free radical or electron transfer processes. For example, trialkylborane additions to α,β-enones are free radical mediated, while similar alkenylborane additions are not. The nucleophilicity of the Group IIB and Group IIIB organometallics is also enhanced with the corresponding stable 'ate' complexes which consist of a Lewis acid (electron deficient organometallic) and a Lewis base (organolithium).[6] In this chapter, the conjugate addition of carbon ligands to activated alkenes and alkynes mediated by Lewis acidic reagents will be discussed.

1.3.2 ADDITIONS TO ALKENIC π-SYSTEMS

1.3.2.1 Additions of Alkyl, Aryl, Alkenyl and Alkynyl Groups

1.3.2.1.1 Organoaluminums

The reactivity of the organoaluminums was described by Gilman[7a] and Wittig[7b] during the 1940s and subsequently they have become a widely used class of reagents in organic synthesis. In spite of the Lewis acid characteristics of the organoaluminums, their carbon–carbon bond formation reactions with electrophiles is similar to that of the organolithiums or organomagnesium halides. Fundamental to the chemistry of the organoaluminums is their heterophilic character in which the formation of coordination complexes with Lewis bases, *e.g.* carbonyl or ether groups, enhances the nucleophilicity of the attached ligands or moderates the reaction rates.[8] For example, trimethylaluminum, in either aromatic or hydrocarbon solvents, adds in a 1,2-mode to α,β-enones. In diethyl ether, both 1,2-additions and 1,4-additions are suppressed, but with catalytic nickel(II) acetylacetonate [$Ni(acac)_2$ (3%)] 1,4-addition products predominate. The trimethylaluminum reduces the nickel(II) to nickel(0), which undergoes oxidative addition to trimethylaluminum to afford the aluminum–nickel reagent $MeNiAlMe_2$ (**1**). The aluminum–nickel reagent undergoes initial addition to the carbonyl to afford an allylic methylnickel complex (**2**), which disproportionates to the corresponding 1,2- and 1,4-addition products (Scheme 1).[9] Alternatively, Kabalka reports that tri-*n*-propylaluminum reacts with α,β-enones (**3**; Scheme 1) at –78 °C in 1,4-mode by radical initiation, *e.g.* photochemically or with oxygen; in the absence of radical initiation, addition products are not observed.[10]

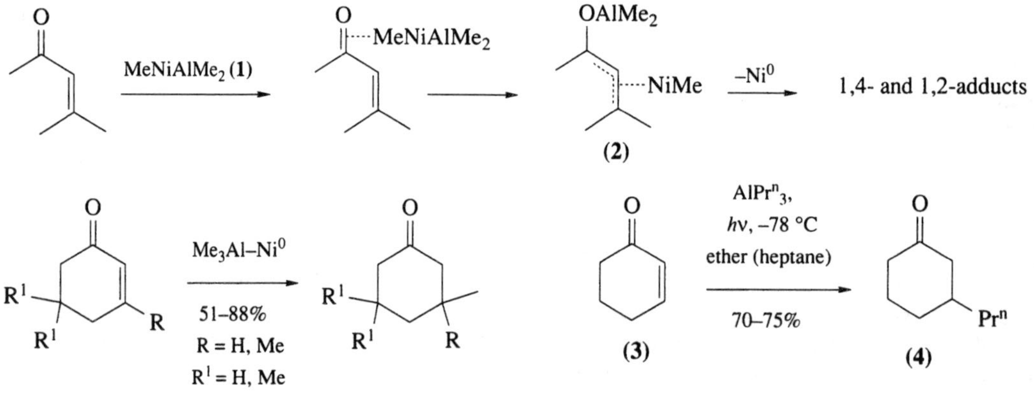

MeNiAlMe$_2$ (**1**) · MeNiAlMe$_2$ · OAlMe$_2$ · NiMe · (**2**) · –Ni^0 · 1,4- and 1,2-adducts

Me$_3$Al–Ni^0, 51–88%, R = H, Me, R^1 = H, Me · (**3**) · AlPr$^n_3$, *h*ν, –78 °C, ether (heptane), 70–75% · (**4**)

Scheme 1

While α,β-unsaturated aldehydes react exclusively with triorganoaluminums in the 1,2-mode, the identical reaction of the corresponding chiral α,β-unsaturated acetals (**5**), obtained from (*R,R*)-(+)-*N,N,N',N'*-tetramethyltartaric acid amides, is solvent dependent.[11] For example, in 1,2-dichloroethane, preferential γ-addition (S_N2') occurs (**5** giving **7**; Scheme 2), while in dichloromethane, preferential α-addition occurs (**5** giving **8**; Scheme 2). Subsequent hydrolysis of (**7**) affords the β-substituted aldehydes in high enantiomeric purity (Scheme 2); thus the acetals (**5**) serve as effective surrogates for '—CH═CH—CHO' ($^+CH_2CH_2CHO$).

Recently, Kunz reported that dialkylaluminum chlorides add in a 1,4-mode to *N*-(α,β-unsaturated acyl)oxazolidones (**9**; Scheme 3),[12] while high enantioselectivity is obtained with the chiral *N*-(α,β-unsaturated acyl)oxazolinone (**11**). Typically, higher dialkylaluminum halides (R* = ethyl, isobutyl) do not

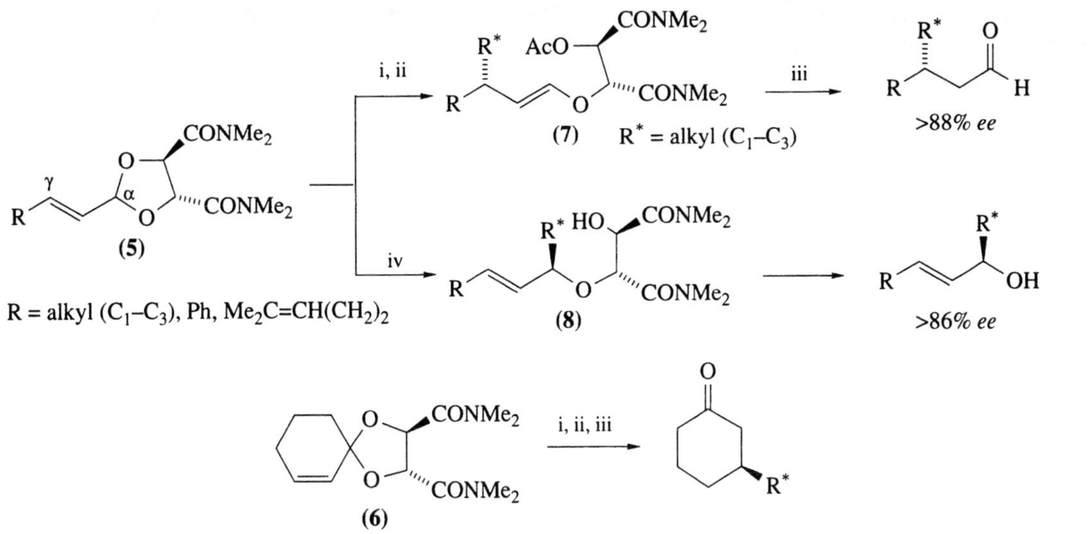

i, $R^*{}_3Al$, $ClCH_2CH_2Cl$; ii, $(MeCO)_2O$, pyridine; iii, H_3O^+; iv, $R^*{}_3Al$, $CHCl_3$

Scheme 2

add, but in combination with dimethylaluminum chloride, ligand transfer occurs exclusively with the higher alkyl ligand. These additions are specific for the described substrates, while *N,N*-disubstituted acrylamides do not react (Scheme 3).

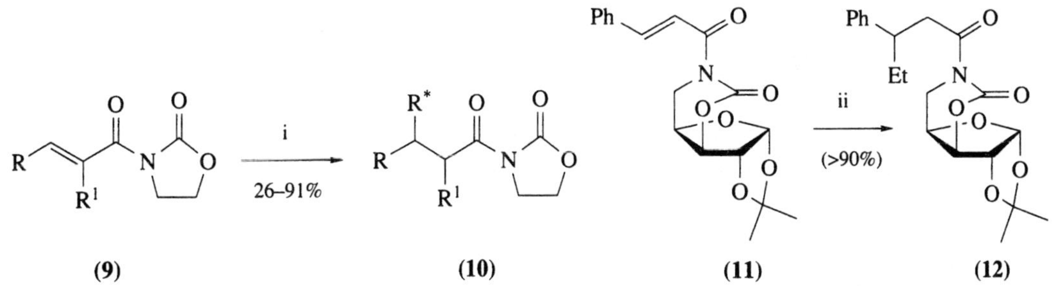

(3*S*):(3*R*) = 87:13

R^1 = H; R^2 = Me, Bu^i, Ph, 2-$MeOC_6H_4$

$R^1 + R^2 = (CH_2)_4$

i, Me_2AlCl–$R^*{}_2AlCl$ (1:1); ii, Et_2AlCl (2 equiv.)

Scheme 3

The conjugate addition of vinyl groups to α,β-enones is the cornerstone of various synthetic challenges, especially with prostaglandin syntheses. These transformations are typically effected with organocuprates which necessitate initial formation of stereochemically defined precursor vinyllithiums;[13] alternatively, the alkenylalanes effect alkenyl transfer to *s-cis* α,β-enones exclusively with retention of alkene stereochemistry.[14] The alkenylalanes are prepared by the regioselective and stereoselective hydroalumination of alkynes with dialkylaluminum hydrides, generally diisobutylaluminum hydride. Remote substituents on the alkynes may impact the hydroalumination process; for example, the hydroalumination of 1-octyn-3-ol ethers (**13**) to (**15**) affords the *cis*-octenylalanes (**17**) to (**19**), while the sterically encumbered trityl ether (**16**) affords *trans*-octenylalane (**20**) in moderate yields (Scheme 4). The unfavorable yield of alane (**20**) is attributed to competing Lewis acid catalyzed reactions, *i.e.* reductive propargylic ether cleavage (*cf.* alane **24**).[15] The specificity for 1,4-additions of alkenylalanes to *s-cis* α,β-enones (**22**) infers a highly ordered cyclic transition state.[14b] In contrast, the addition of either (*E*)-1-hexenyldiisobutylalane (**21**; Scheme 5) or (*E*)-6-triphenylmethoxy-1-heptenyldiisobutylalane (**24**) to hydroxyenone

(**25**) occurs exclusively *anti* to the hydroxy group, but isomeric at the acetyl position.[16] Thus, the initially formed alane alkoxide shields the *syn* face of the alkene.

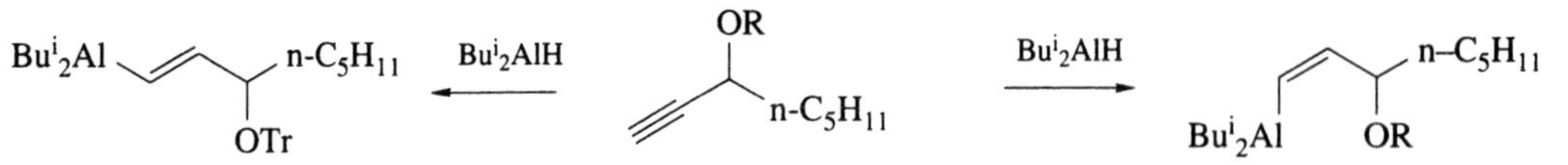

Bu^i_2Al n-C_5H_{11} OTr (**20**) ← Bu^i_2AlH — OR n-C_5H_{11} — Bu^i_2AlH → n–C_5H_{11} Bu^i_2Al OR

(**13**) R = Bu^t (**14**) R = Me
(**15**) R = 1-methylcyclohexyl
(**16**) R = Ph_3C

(**17**) R = Bu^t
(**18**) R = Me
(**19**) R = 1-methylcyclohexyl

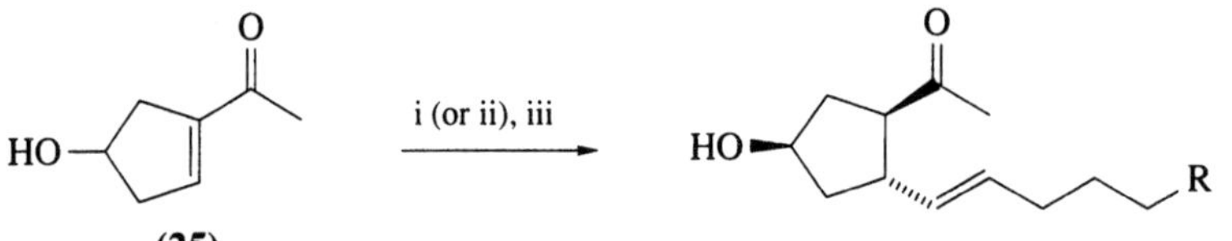

Scheme 4

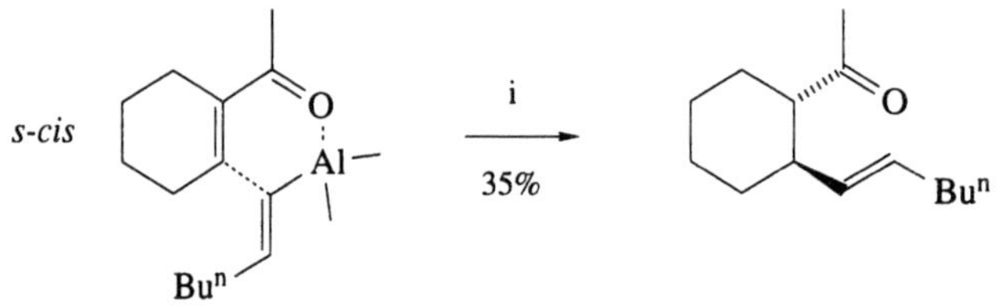

s-cis O Al Bu^n (**22**) — i, 35% → O Bu^n (**23**)

HO O (**25**) — i (or ii), iii → HO O R

(**26**) R = Me
(**27**) R = CH(OTr)Me

i, *(E)*-$Bu^i_2AlCH{=}CH(CH_2)_3Me$ (**21**); ii, *(E)*-$Bu^i_2AlCH{=}CH(CH_2)_3CH(OTr)Me$ (**24**); iii, NaOMe

Scheme 5

Irrespective of the stereochemistry of the α,β-enones, lithium alkenyltrialkylalanates, *e.g.* (**28**) and (**29**), obtainable by treatment of alkenylalanes with methyllithium (or *n*-butyllithium), are also excellent 1,4-conjugate alkenyl transfer reagents. The cornerstone of a general prostaglandin synthesis, as exemplified in Scheme 6, employs the alanate conjugate addition process exclusively.[15]

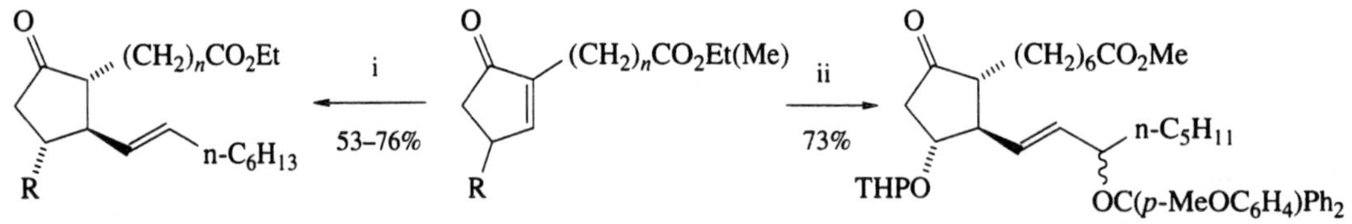

O $(CH_2)_nCO_2Et$ n-C_6H_{13} R (**31**) ← i, 53–76% — O $(CH_2)_nCO_2Et(Me)$ R (**30**) — ii, 73% → O $(CH_2)_6CO_2Me$ n-C_5H_{11} THPO OC(*p*-$MeOC_6H_4)Ph_2$ (**32**)

R = H, $MeCO_2$, THPO; n = 1 or 6

i, Li^+ *(E)*-$Me(CH_2)_5CH{=}CHAlMeBu^i_2{}^-$ (**28**); ii, Li^+ *(E)*-$Me(CH_2)_4(MTrO)CHCH{=}CHAlMeBu^i_2{}^-$ (**29**)

Scheme 6

Pecunioso reports that trialkylaluminums, alkenyldialkylalanes and alkynyldialkylalanes add efficiently to α-nitroalkenes (**33**) to (**35**) to afford nitroalkanes,[17a] γ-nitroalkenes[17b] and γ-nitroalkynes,[17c] respectively (Scheme 7). This method complements the organolithium and Grignard reagent mediated processes in which terminal α-nitroalkenes do not undergo efficient addition (*cf.* Volume 4, Chapter 1.2, Sections 1.2.2.1.1 and 1.2.2.1.2). Depending on the work-up conditions employed, either the nitroalkanes or the corresponding ketones (Nef reaction) are obtained. The latter work-up procedure affords an efficient synthesis of cyclopenten-3-ones (**36**), as shown in Scheme 7.[17b]

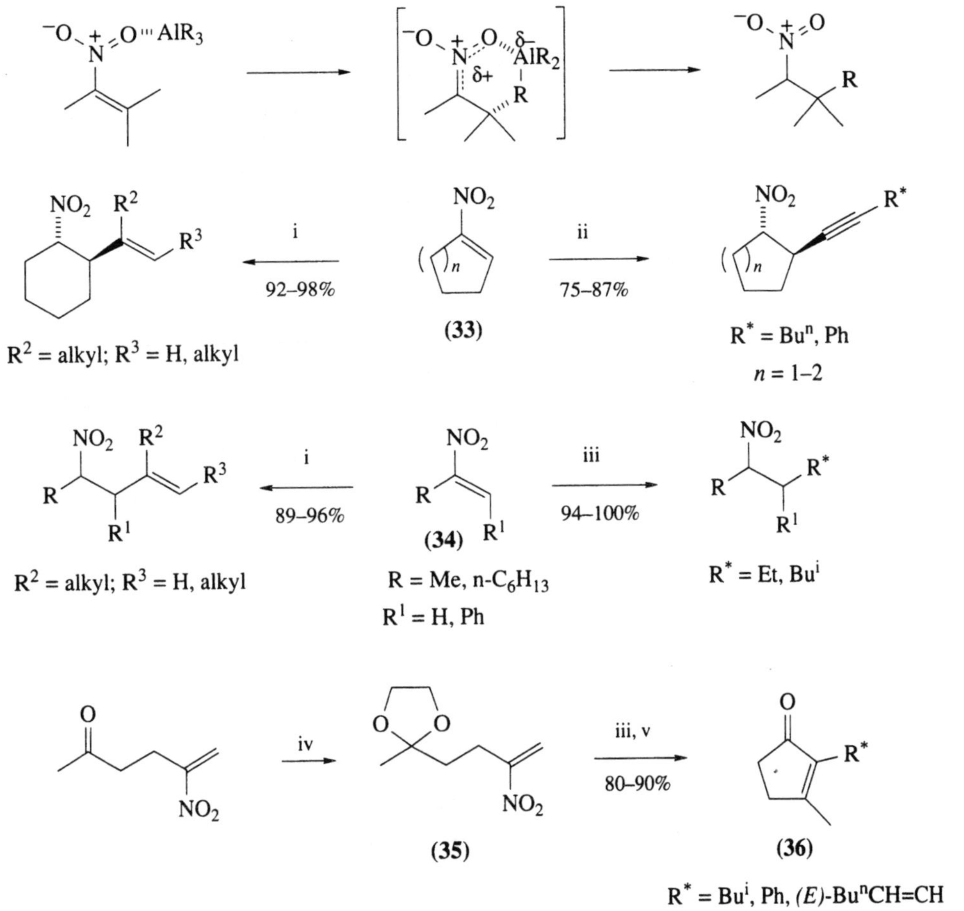

i, $Bu^i_2AlC(R^2)$=CHR^3; ii, $R^*C{\equiv}CAlEt_2$; iii, R^*_3Al; iv, $HOCH_2CH_2OH$, *p*-TsOH; v, 3 M HCl

Scheme 7

The direct conjugate addition of alkynyl groups to α,β-unsaturated systems is a synthetic challenge since virtually all alkynylmetallic reagents are ineffective at introducing alkynic moieties. The more popular stoichiometric organocuprates or copper-catalyzed Grignards do not transfer alkynyl ligands because of the strong alkynyl–copper bond. Typically, introduction of an alkynyl moiety is accomplished with a precursor functionality, *e.g.* (tri-*n*-butylstannyl)vinylcuprate.[18] The alkynyldialkylalanes transfer alkynyl ligands effectively to all α,β-enones in the presence of a nickel(0) catalyst, like $Ni(acac)_2$·DIBAL-H (1:1) (Scheme 8);[19] in the absence of the catalyst, only *s-cis* α,β-enones (**22**) are alkynylated, while fixed *s-trans* α,β-enones undergo 1,2-addition. Thus, a cyclic transition state is inferred for the noncatalyzed conjugate addition reactions. In contrast, the noncatalyzed conjugate addition of the trialkynylaluminum (**39**) to the 4-hydroxycyclopentenone (fixed *s-trans* configuration) (**40**) necessitates prior coordination to the 4-hydroxy group in order to direct the alkynyl ligand *syn* to the β-position.[20] Protection of the 4-hydroxy group, *e.g.* with THP, retards this conjugate addition process (equation 1).

While conjugate additions with organoaluminums typically occur *via* intramolecular ligand transferral from the aluminum coordination complexes, intermolecular ligand transferral from the more basic, nucleophilic reagents, *e.g.* RLi or RMgX, can be accomplished with unusual chemoselectivity and stereoselectivity in the presence of oxophilic organoaluminum reagents possessing bulky nontransferable ligands. For example, the bis(2,6-di-*t*-butyl-4-alkylphenoxy)methylaluminums (MAD **42** and MAT **43**; Scheme 9) facilitate conjugate additions of organolithiums to select cycloalkenones, quinone monoketals (**44**) and quinol ethers (**45**).[21]

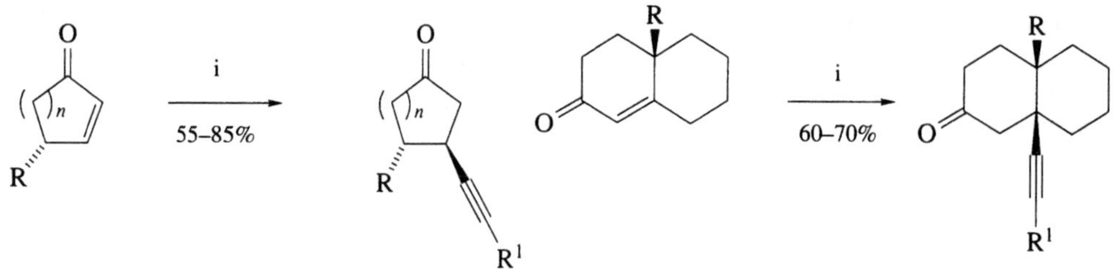

(37) R = H, $PhCMe_2O$; *n* = 1–2 **(38)**

R = H, Me; R^1 = Bu^n, Bu^t, $SiMe_3$

i, $Me_2AlC{\equiv}CR^1$, Ni^0 catalyst

Scheme 8

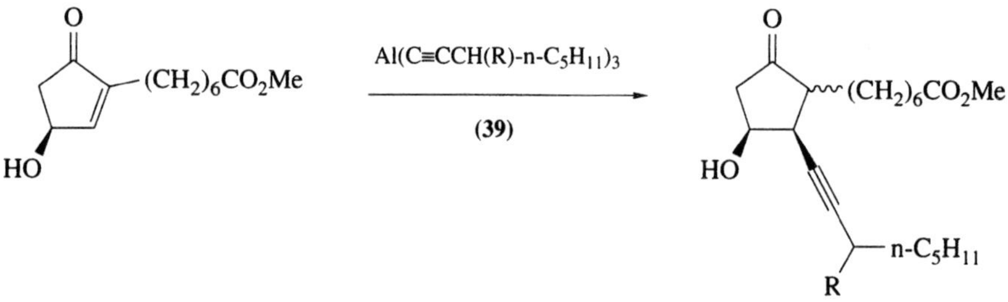

$Al(C{\equiv}CCH(R)\text{-}n\text{-}C_5H_{11})_3$ **(39)** (1)

(40)

(41) R = H, OTHP

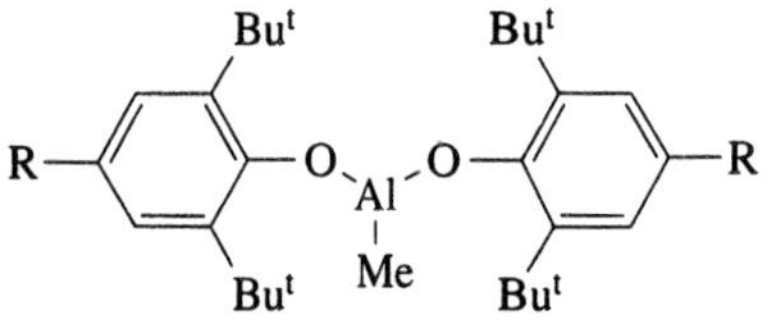

(42) MAD: R = Me
(43) MAT: R = Bu^t

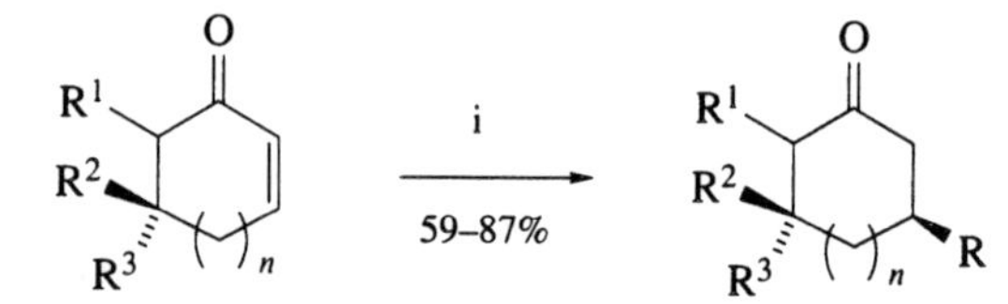

R^1 = H, Me; R^2 = H, Me
R^3 = H, Me, $OSiPh_3$

n = 0–1

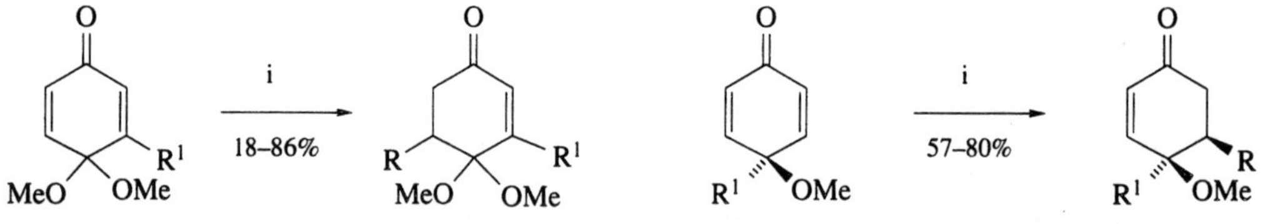

(44) R^1 = H, Me, Cl

(45) R^1 = Ph, $CH_2{=}CH$, 4-$Bu^tMe_2SiOC_6H_4$

i, MAD **(42)** 2 equiv., RLi 2 equiv., PhMe, –78 °C
[R = alkyl (C_1–C_4), Ph, 1,3-dithiane, $Me_3SiC{\equiv}C$, ester enolate]

Scheme 9

1.3.2.1.2 Organoboranes

The Lewis acidic organoboranes, on the other hand, are electrophilic in their chemical behavior.[22] This is attributed to the high degree of covalent character in the boron–carbon bond, which results in the low reactivity of organoboranes *versus* organolithiums towards electrophiles. In spite of the low reactivity of the triorganoboranes with simple aldehydes and ketones, formal conjugate addition of alkyl (or aryl) groups occurs with α,β-unsaturated aldehydes and α,β-unsaturated ketones to afford the alkenyloxydialkylboranes (vinyloxyboranes, **46**; Scheme 10) regiospecifically.[23] The ease of addition is dependent on the substituents on the α,β-unsaturated aldehyde or ketone; in the absence of β-substituents spontaneous addition occurs, while with β-substituted systems a radical initiator is required. Initiators include diacyl peroxides, light or oxygen.[24] Similar additions to α-nitroalkenes,[25] α,β-unsaturated imines,[26] α,β-unsaturated nitriles[27] (secondary alkyls only) and 1,4-benzoquinones (naphthoquinones)[28] also occur efficiently (equation 2). The free radical nature of these additions is shown by the inhibition of these reactions with galvinoxyl, a highly efficient free radical scavenger,[29a] and secondary product formation arising from solvents capable of radical formation, *e.g.* tetrahydrofuran and 2-propanol.[29b]

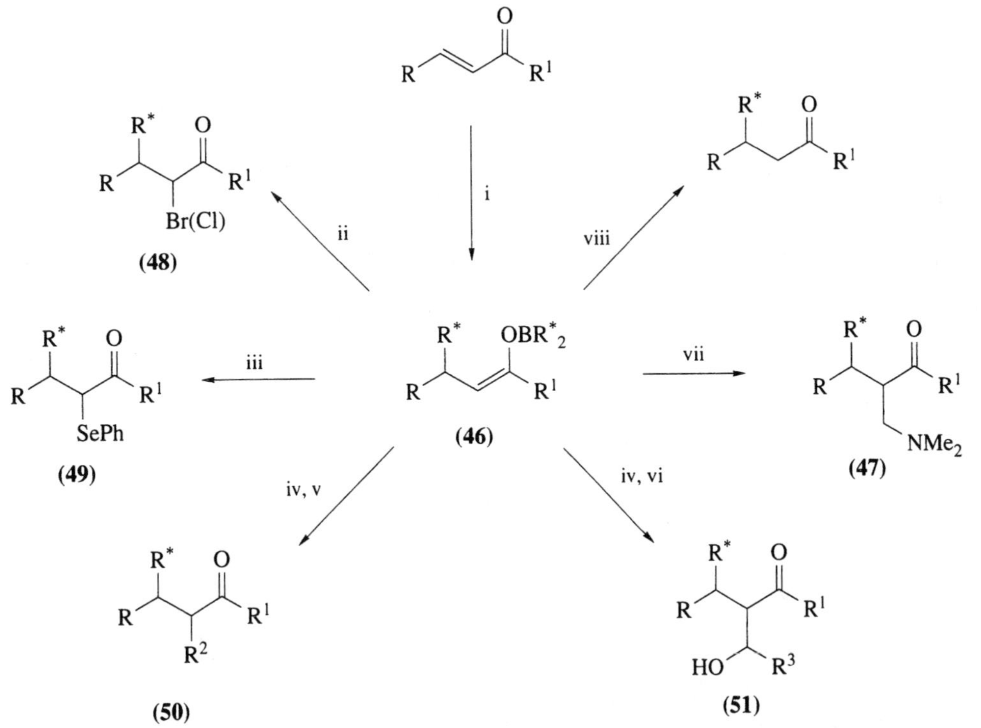

i, $R^*{}_3B$; ii, NBS (NCS); iii, PhSeCl; iv, $Me_2NCH_2CH_2OLi$; v, R^2X; vi, R^3CHO; vii, $CH_2{=}NMe_2{}^+\ I^-$; viii, H_3O^+

Scheme 10

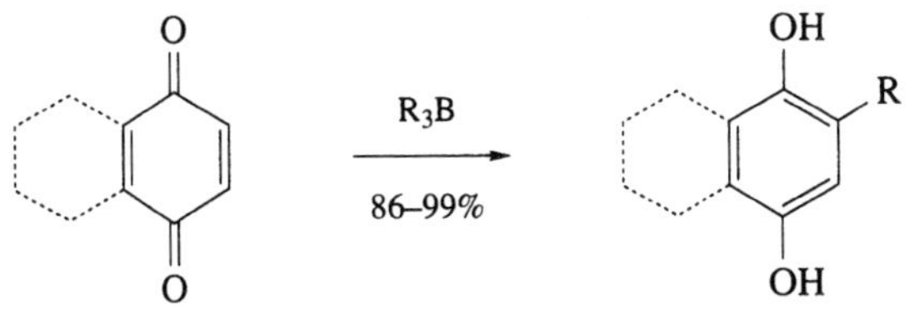

R_3B, 86–99% (2)

R = alkyl (C_1–C_6), c-C_6H_{11}, c-C_8H_{15}, $PhCH_2$

Further transformations of the intermediate alkenyloxydialkylboranes (**46**), as shown in Scheme 10, include: aqueous hydrolysis to the homologated aldehydes or ketones;[30] electrophilic additions to afford the corresponding α-dialkylaminomethyl aldehydes and ketones (**47**),[31] and α-halogenated (**48**)[32] or α-arylselenylated aldehydes or ketones (**49**);[33] transmetallation–alkylation to afford α-substituted aldehydes or ketones (**50**);[34] and aldol condensations with aldehydes (**51**).[35] These reagents offer the equivalent of the tandem 1,4-conjugate addition–electrophile-trapping protocol. A prototypical prostaglandin synthesis by an organoborane-mediated double Michael addition protocol is shown in Scheme 11.[36]

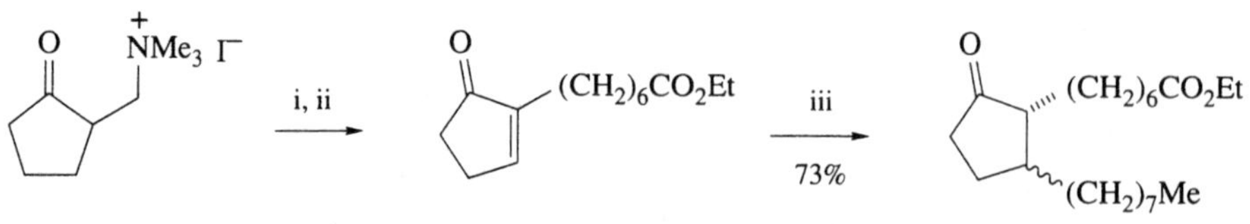

(52) (53) (54)

i, B{$(CH_2)_5CO_2Et$}$_3$, K_2CO_3; ii, Br_2, base; iii, B{$(CH_2)_7Me$}$_3$

Scheme 11

However, conjugate additions with trialkylboranes are inefficient in that only one of the three organic groups is transferred. The inefficiencies of group transfer are surmounted by the use of either the *B*-alkylboracyclanes (**55**)[37a,37b] or *B*-alkyldiphenylboranes (**56**),[37c] in which the *B*-alkyl groups are preferentially transferred under radical conditions. The *B*-alkylboracyclanes are obtained by sequential

(**55**) n = 1–2 (**56**)

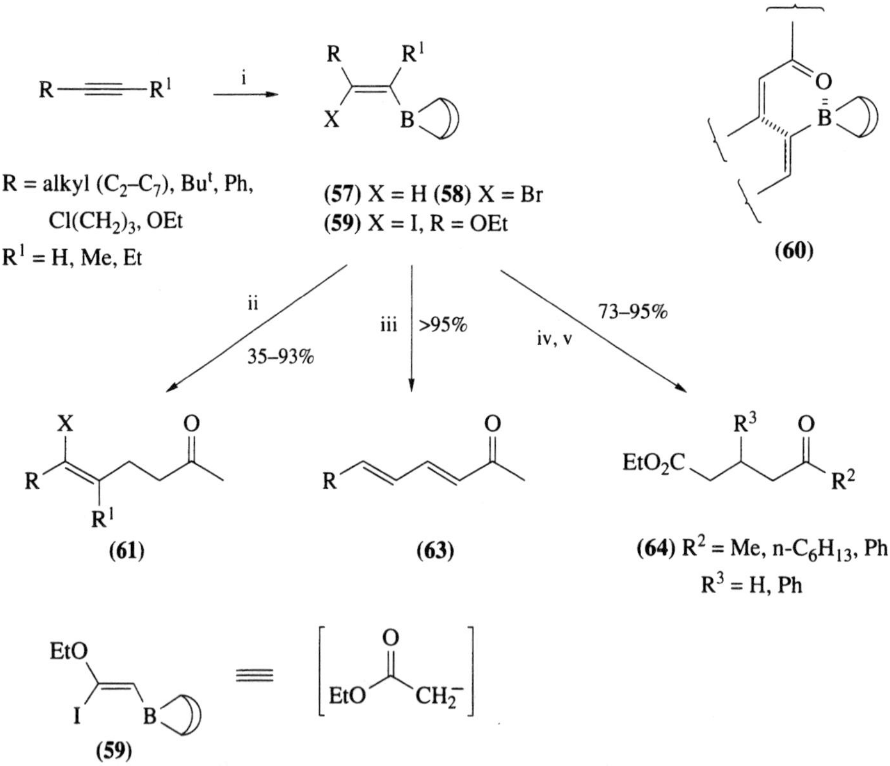

R = alkyl (C_2–C_7), Bu^t, Ph, $Cl(CH_2)_3$, OEt

R^1 = H, Me, Et

(**57**) X = H (**58**) X = Br

(**59**) X = I, R = OEt

(**60**)

(**61**) (**63**) (**64**) R^2 = Me, n-C_6H_{13}, Ph

R^3 = H, Ph

(**59**)

i, 9-BBN (*B*-Br-9-BBN) (*B*-I-9-BBN); ii, CH_2=CHCOMe; iii, MeOCH=CHCOMe (**62**); iv, R^{3}CH=CHCOR^2; v, EtOH

Scheme 12

hydroboration of either 2,4-dimethyl-1,4-pentadiene or 2,5-dimethyl-1,5-hexadiene followed by addition to appropriate alkenes. Hydroborations with the boracyclanes occur with high regioselectivity which is crucial for the subsequent conjugate additions in which the minor hydroboration isomers (Markownikov) will transfer preferentially with respect to the major hydroboration isomers (*anti*-Markownikov isomer).

In contrast, the *B*-1-alkenyl-9-borabicyclo[3.3.1]nonanes, (**57**) to (**59**) and (**65**), and *B*-1-alkynyl-9-borabicyclo[3.3.1]nonanes (**72**) effect alkenyl and alkynyl transfer, respectively, to *s-cis* α,β-enones exclusively *via* a cyclic transition state (**60**), with the 9-BBN moiety serving as a nontransferable ligand. Thus, the alkenylation of α,β-enones and β-methoxy-α,β-unsaturated ketones (**62**) affords γ,δ-enones (**61**) and α,β,δ,γ-dienones (**63**), respectively (Scheme 12). The *B*-(*E*)-1-alkenyl-9-borabicyclo[3.3.1]nonanes (**57**) to (**59**) are prepared by the regioselective and stereoselective hydroboration (9-BBN)[38a] or haloboration (*B*-X-9-BBN; X = Br, I)[38b] of alkynes, while the *B*-(*Z*)-1-alkenyl-9-borabicyclo[3.3.1]nonanes (**65**) and *B*-1-alkynyl-9-borabicyclo[3.3.1]nonanes (**72**) are prepared from *B*-methoxy-9-BBN and (*Z*)-lithio-1-alkenes (Scheme 13)[39] and lithium alkynides (Scheme 15),[40] respectively. Suzuki reports that *B*-1-(2-ethoxy-2-iodovinyl)-9-borabicyclo[3.3.1]nonane (**59**), an acetate ester enolate equivalent, adds to α,β-enones to afford δ-keto esters (**64**).[41] In contrast, α,β-enone alkenylations with the more stable *B*-(1-alkenyl)dialkoxyboranes (**68**) and (**71**) are facilitated by the presence of boron trifluoride etherate which generates *in situ* the reactive *B*-(1-alkenyl)alkoxyfluoroboranes (**69**). This procedure allows for the introduction of alkenyl groups which are not accessible by direct hydroboration, *e.g.* 1,2- or 2,2-disubstituted-1-alkenyl substituents, irrespective of alkene stereochemistry (Scheme 14).[42]

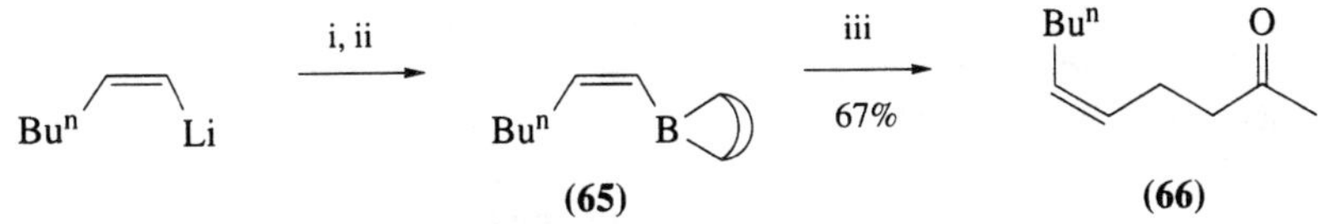

Bu^n Li → (i, ii) → Bu^n B (**65**) → (iii, 67%) → Bu^n O (**66**)

i, *B*-MeO-9-BBN; ii, BF_3•Et_2O; iii, CH_2=CHCOMe

Scheme 13

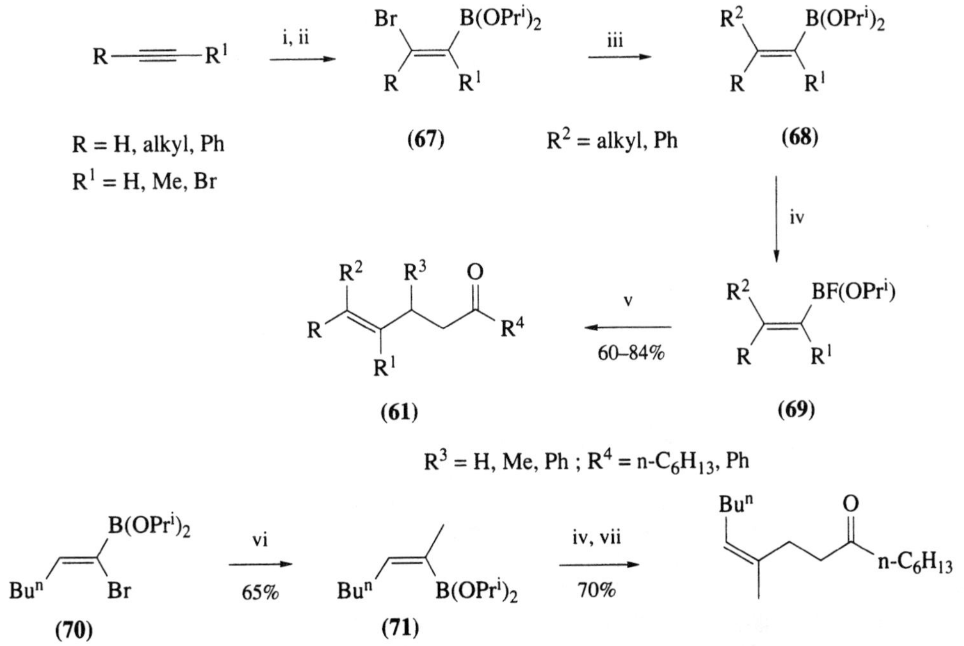

R—≡—R^1 → (i, ii) → Br, $B(OPr^i)_2$, R, R^1 (**67**) → (iii) → R^2, $B(OPr^i)_2$, R, R^1 (**68**)

R = H, alkyl, Ph
R^1 = H, Me, Br

R^2 = alkyl, Ph

(**68**) → (iv) → R^2, $BF(OPr^i)$, R, R^1 (**69**) → (v, 60–84%) → R^2, R^3, O, R, R^4, R^1 (**61**)

R^3 = H, Me, Ph ; R^4 = n-C_6H_{13}, Ph

$B(OPr^i)_2$, Bu^n, Br (**70**) → (vi, 65%) → Bu^n, $B(OPr^i)_2$ (**71**) → (iv, vii, 70%) → Bu^n, O, n-C_6H_{13}

i, BBr_3; ii, $Pr^i{}_2O$; iii, R^2ZnCl, Pd^0 catalyst; iv, BF_3•Et_2O; v, R^3CH=CHCOR^4; vi, MeLi;
vii, CH_2=CHCO-n-C_6H_{13}

Scheme 14

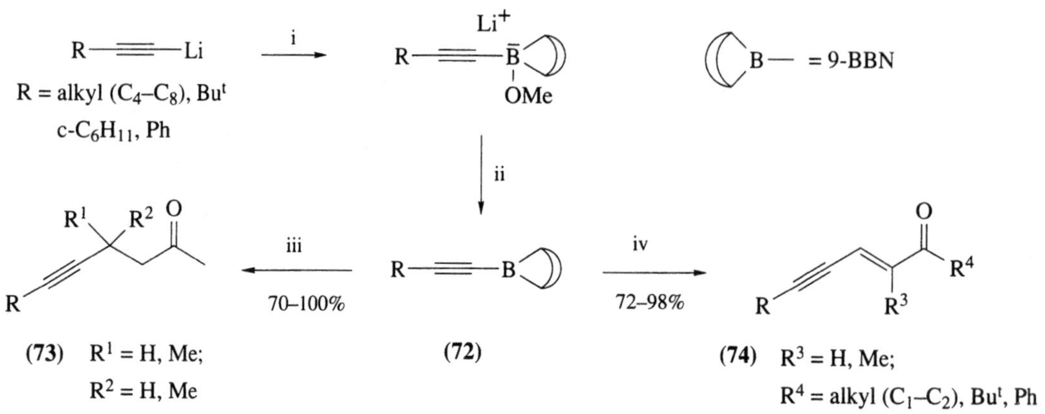

R—≡—Li
R = alkyl (C_4–C_8), Bu^t
c-C_6H_{11}, Ph

i, *B*-OMe-9-BBN; ii, $BF_3 \cdot Et_2O$, –78 °C; iii, R^1R^2C=CHCOMe; iv, MeOCH=C(R^3)COR^4

(73) R^1 = H, Me; R^2 = H, Me

(72)

(74) R^3 = H, Me; R^4 = alkyl (C_1–C_2), Bu^t, Ph

Scheme 15

The reactivity profiles of the boronate complexes are also diverse.[43] For example, the lithium methyltrialkylboronates (**75**) are inert, but the more reactive copper(I) methyltrialkylboronates (**76**) afford conjugate adducts with acrylonitrile and ethyl acrylate (Scheme 16).[44] In contrast, the lithium alkynylboronates (**77**) are alkylated by powerful acceptors, such as alkylideneacetoacetates, alkylidenemalonates and α-nitroethylene, to afford the intermediate vinylboranes (**78**) to (**80**), which on oxidation (peracids) or protonolysis yield the corresponding ketones or alkenes, respectively (Scheme 17).[45a] Similarly, titanium tetrachloride-catalyzed alkynylboronate (**77**) additions to methyl vinyl ketone afford 1,5-diketones (**81**).[45b] Mechanistically, the alkynylboronate additions proceed by initial β-attack of the electrophile and simultaneous alkyl migration from boron to the α-carbon.

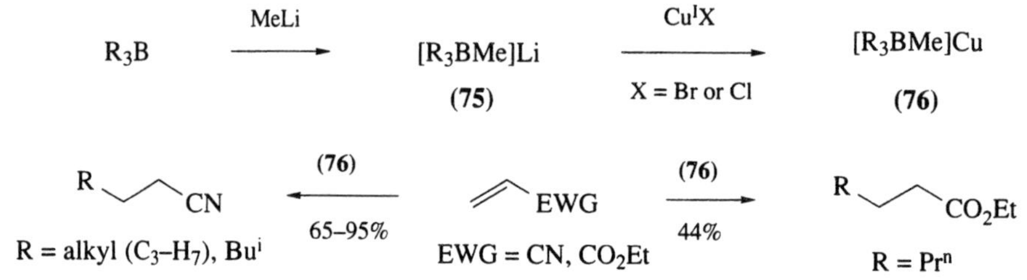

R_3B —MeLi→ [R_3BMe]Li (**75**) —Cu^IX, X = Br or Cl→ [R_3BMe]Cu (**76**)

R = alkyl (C_3–H_7), Bu^i ← (**76**), 65–95% — EWG = CN, CO_2Et — (**76**), 44% → R = Pr^n

Scheme 16

1.3.2.1.3 Lewis acid modified organocoppers and organocuprates

Enhancement of the chemoselectivity in the reaction of Grignard reagents with α,β-enones, in the presence of copper(I) salts, was first described by Kharasch and Tawney.[46] Subsequently, this conjugate addition protocol became the cornerstone for various natural products syntheses, but it lacked predictability and dependability. House and coworkers demonstrated that an organocopper (RCu, **82**) was the reactive species;[47a] independently, Gilman had reported on the synthesis and reactivity of lithium dimethylcuprate.[47b] With the development of general procedures for the synthesis of stoichiometric organocoppers and homocuprates (R_2CuM, **83**), their synthetic utility was soon established, but limitations on their thermal stability, ligand transfer efficiencies and acceptor reactivity profiles were soon recognized.[48] Additional copper-based reagents, *e.g.* mixed homocuprates (RR^1CuM, **84**), heterocuprates (RCuMX, **85**) and higher order cuprates (R_3CuLi, **86**),[49] were developed in order to address the above deficiencies. Some interesting variants are the Lewis acid modified organocoppers ($RCu \cdot MX_n$, **87**), Lewis acid modified organocuprates ($R_2CuLi \cdot MX_n$, **88**) and Lewis acid modified cyanocuprates ($R_2Cu(CN)Li_2 \cdot MX_n$, **89**).

Yamamoto[50] and Ibuka[51] have reported that ligand transferrals from organocoppers to a broad range of acceptors are facilitated in the presence of aluminum trichloride or boron trifluoride, at –78 °C, with high

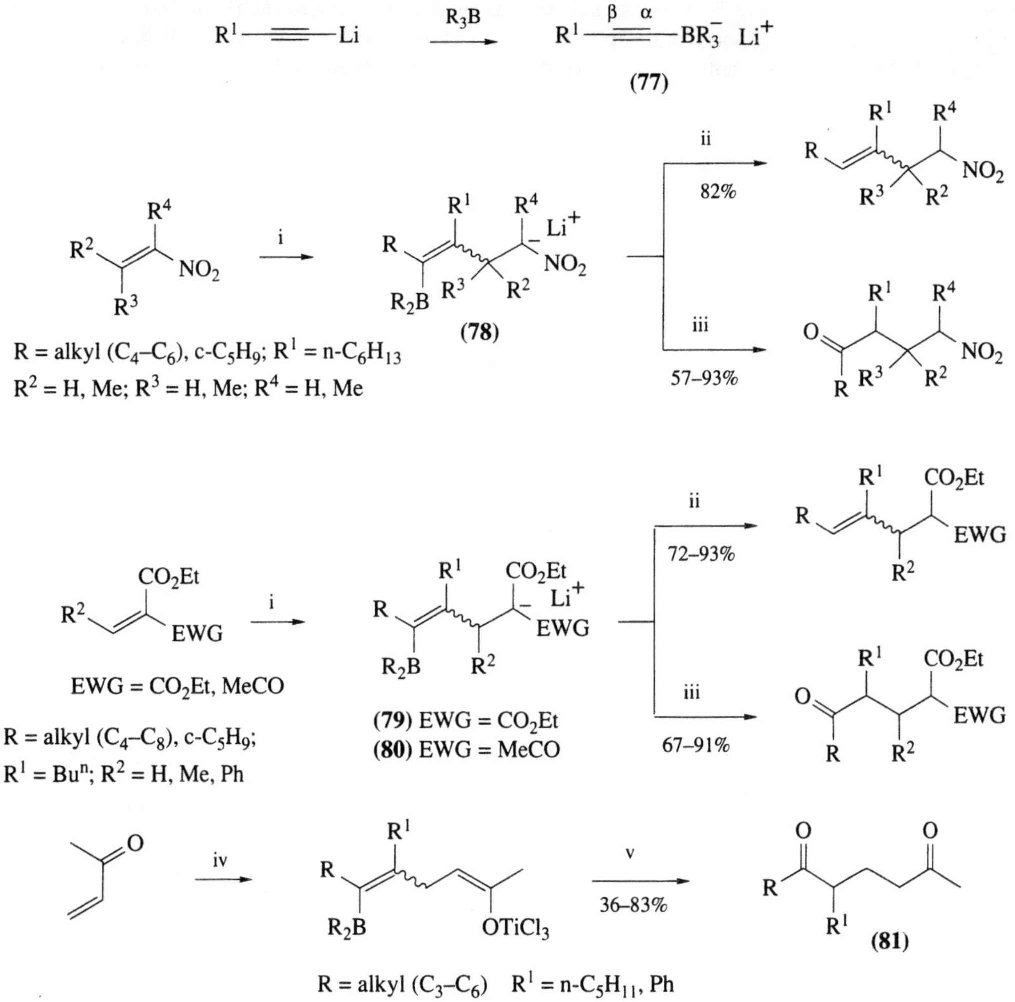

R^1—≡—Li $\xrightarrow{R_3B}$ R^1—≡—BR_3^- Li^+ (β, α)

(77)

R = alkyl (C_4–C_6), c-C_5H_9; R^1 = n-C_6H_{13}

R^2 = H, Me; R^3 = H, Me; R^4 = H, Me

(78)

ii, 82%; iii, 57–93%

EWG = CO_2Et, MeCO

R = alkyl (C_4–C_8), c-C_5H_9;

R^1 = Bu^n; R^2 = H, Me, Ph

(79) EWG = CO_2Et

(80) EWG = MeCO

ii, 72–93%; iii, 67–91%

iv; v, 36–83%

(81)

R = alkyl (C_3–C_6) R^1 = n-C_5H_{11}, Ph

i, $RC{\equiv}CBR^1{}_3{}^-$ Li^+ **(77)**; ii, $EtCO_2H$; iii, MCPBA; iv, **(77)**, $TiCl_4$, CH_2Cl_2, –78 °C; v, H_2O_2

Scheme 17

RCu	R_2CuM	RR^1CuM	RCuMX	R_3CuLi
(82)	**(83)**	**(84)**	**(85)**	**(86)**

RCu•MX_n	R_2CuLi•MX_n	R_2Cu(CN)Li_2•MX_n
(87) a–c	**(88)**	**(89)**

chemoselectivity and stereoselectivity. The acceptors include α,β-disubstituted-α,β-unsaturated esters, β,β-disubstituted-α,β-unsaturated esters, α,β-disubstituted-α,β-enones, β,β-disubstituted-α,β-enones, α,β,β-trisubstituted-α,β-enones and α,β-unsaturated carboxylic acids, which typically do not react with organocoppers or lower order organocuprates. However, characterization of the organocopper–Lewis acid complexes is lacking. It is instructive to compare the reactivity of the organocopper–Lewis acid complexes **(87)** *vis-à-vis* the homocuprates **(83)**; for example, organocopper–aluminum trichloride (**87b**; Scheme 18) is highly effective in conjugate additions to γ-*t*-butyldimethylsilyloxy-α,β-enones, **(94)** and **(96)** (exclusive *anti* addition to the silyl ether), and γ-cyclopropyl-α,β-enones, **(98)** and **(100)**.[51] In contrast, with the homocuprates **(83)** competing S_N2' reaction processes, reduction processes or cyclopropane cleavage typically occurs. The lack of cyclopropane cleavage with **(87b)** implies that carbon–carbon bond formation does not involve an electron transfer process. Chemoselectivity differences are also observed with methyl sorbate **(102)**, where (**87b**, R = Bu^n) affords predominantly 1,4-adduct **(103)**,[50b] while (**83**, R = Bu^n) affords 1,6-adduct (**104**; Scheme 18).[52] Other subtle differences include the diastereoselective addition of (**87a**) to γ-benzyloxy-α,β-unsaturated esters (**105**) and (**106**). For example, the (*E*)-ester (**105**) affords predominantly the *anti* adduct (**109**), while the (*Z*)-ester (**106**) affords the *syn* adduct (**110**; Scheme 19). In contrast, the corresponding homocuprate additions proceed exclusively in an S_N2' mode.[53]

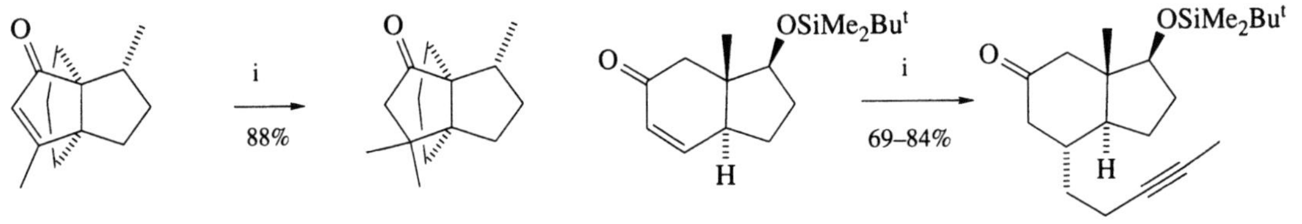

(90) —i, 88%→ (91)

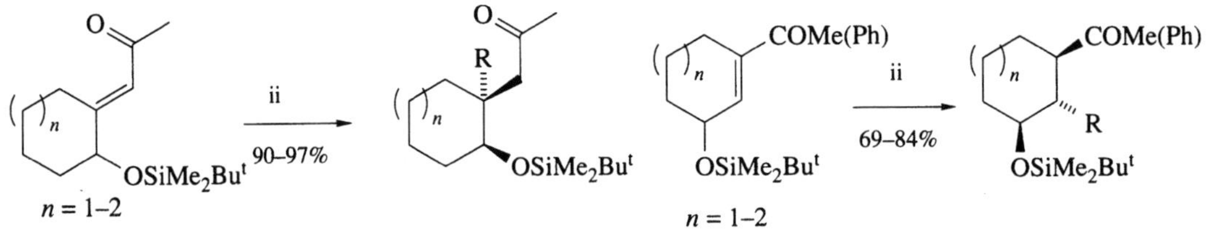

(92) —i, 69–84%→ (93)

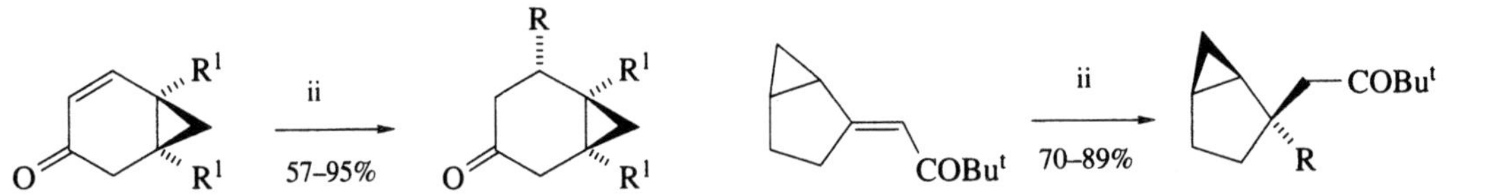

(94) n = 1–2 —ii, 90–97%→ (95)

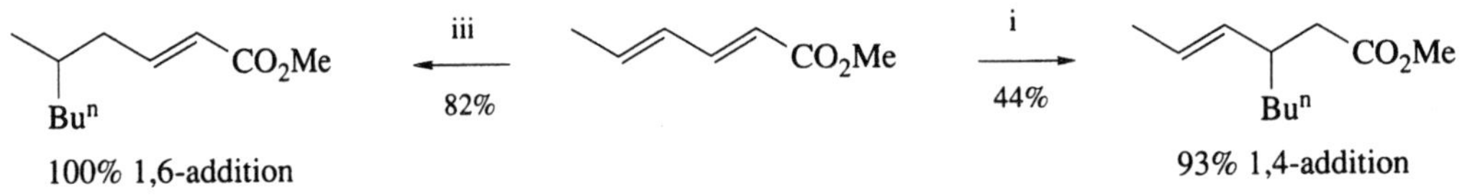

(96) n = 1–2 —ii, 69–84%→ (97)

(98) —ii, 57–95%→ (99)

(100) —ii, 70–89%→ (101)

(104) 100% 1,6-addition ←iii, 82%— (102) —i, 44%→ (103) 93% 1,4-addition

i, $RCu{\bullet}BF_3$ (**87a**), Et_2O, –78 to –25 °C; ii, $RCu{\bullet}AlCl_3$ (**87b**), Et_2O, –78 to –25 °C; iii, $Bu^n{}_2CuLi$ (**83**)

Scheme 18

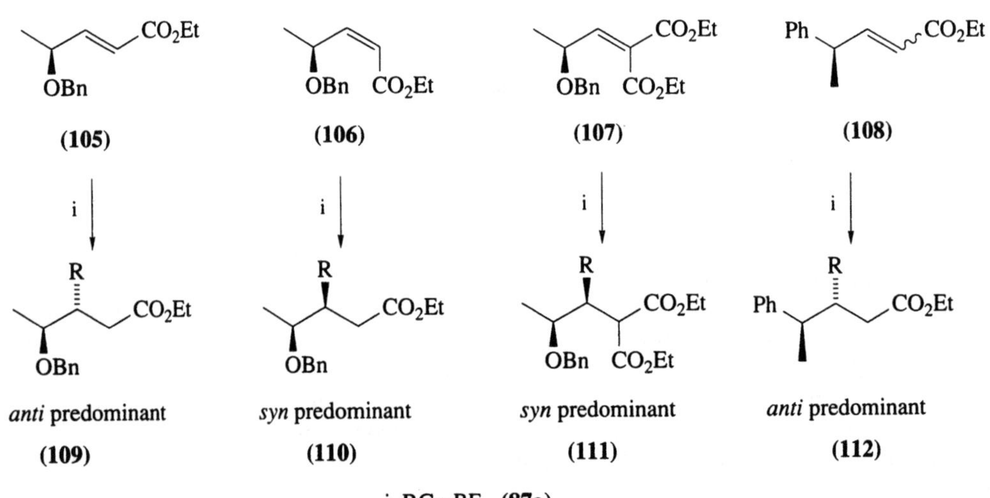

i, $RCu{\bullet}BF_3$ (**87a**)

Scheme 19

Oppolzer reports that similar additions of the ether soluble organocopper–boron trifluoride–tri-*n*-butylphosphine complexes (**87c**) to chiral α,β-unsaturated esters (Scheme 20) proceed with high diastereoselectivity. Subsequent hydrolysis or reduction of the chiral adducts, *e.g.* (**115**) and (**116**), affords

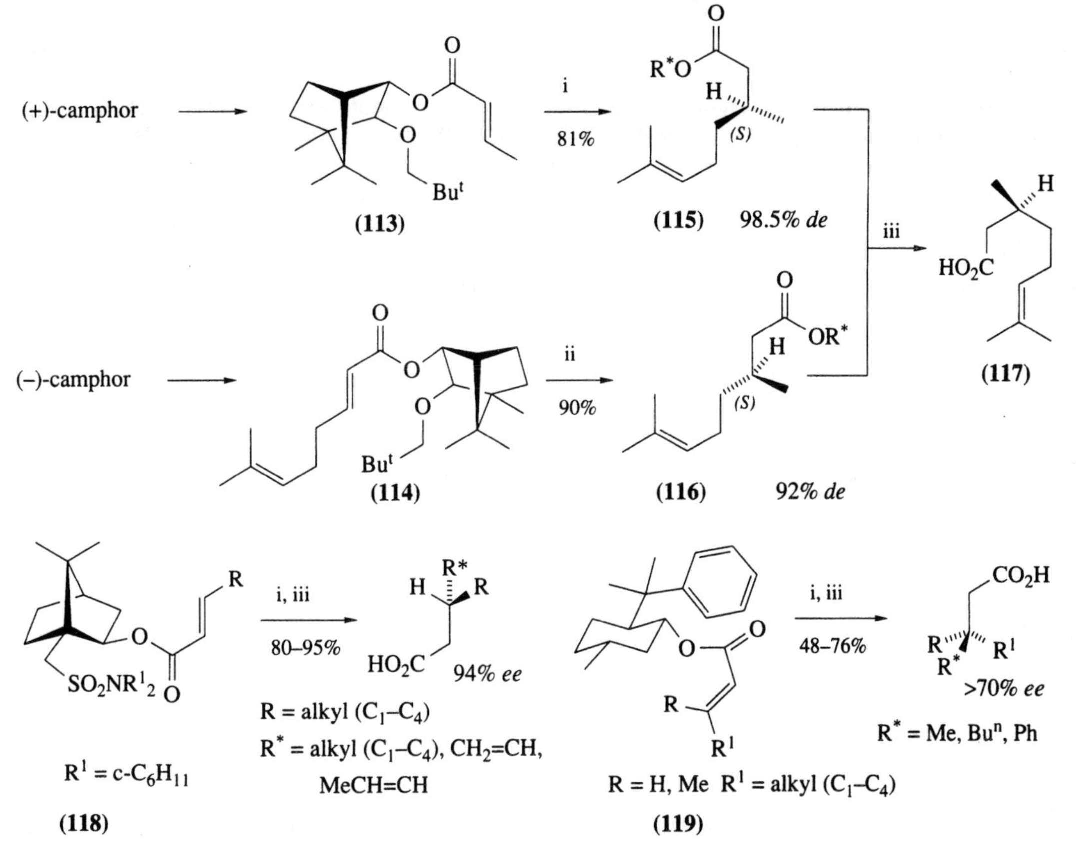

(+)-camphor → (**113**) —i, 81%→ (**115**) 98.5% *de*

(–)-camphor → (**114**) —ii, 90%→ (**116**) 92% *de*

(**115**), (**116**) —iii→ (**117**)

(**118**) R^1 = c-C_6H_{11} —i, iii, 80–95%→ 94% *ee*; R = alkyl (C_1–C_4); R^* = alkyl (C_1–C_4), CH_2=CH, MeCH=CH

(**119**) R = H, Me R^1 = alkyl (C_1–C_4) —i, iii, 48–76%→ >70% *ee*; R^* = Me, Bu^n, Ph

i, MeLi (R^*Li), CuI, Bu^n_3P, BF_3, –78 °C; ii, Me_2C=$CHCH_2CH_2$Li, CuI, Bu^n_3P, BF_3, –78 °C; iii, NaOH (MeONa), MeOH

Scheme 20

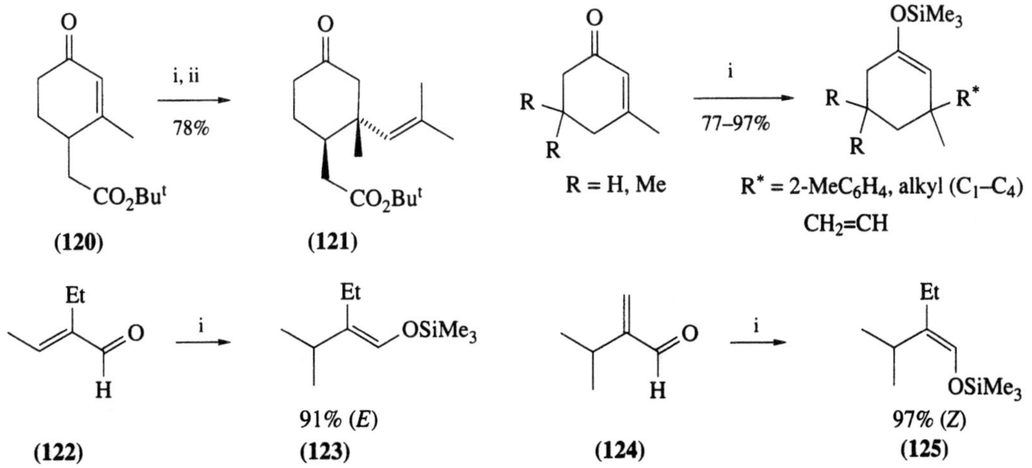

(**120**) —i, ii, 78%→ (**121**)

R = H, Me —i, 77–97%→ R^* = 2-MeC_6H_4, alkyl (C_1–C_4) CH_2=CH

(**122**) —i→ (**123**) 91% (*E*)

(**124**) —i→ (**125**) 97% (*Z*)

i, RMgBr, $CuBr{\cdot}SMe_2$ (3%), HMPA, –78 °C, TMSCl; ii, H_3O^+

Scheme 21

chiral β-substituted carboxylic acids or β-substituted alcohols, respectively.[54] As exemplified in Scheme 20, for the synthesis of (*S*)-citronellic acid (**117**), several combinations of enoate acceptor unit and chiral auxiliary will result in identical chiral products.

The combinations of chlorotrimethylsilane–hexamethylphosphoramide (HMPA) or chlorotrimethylsilane–4-(dimethylamino)pyridine (DMAP) are also powerful accelerants for copper(I)-catalyzed Grignard conjugate additions,[55] and stoichiometric organocopper and homocuprate additions (Scheme 21).[56] However, these reactions must be performed in tetrahydrofuran instead of ether.[57] These procedures are noted for their high yields with stoichiometric quantities of Grignard reagents, excellent chemoselectivity and efficiency with α,β-unsaturated amides and esters and enals.[58] Typically, additions to enals proceed *via* the *S-trans* conformers to afford stereo-defined silyl enol ethers; for example, enals (**122**) and (**124**) give the (*E*)-silyl enol ether (**123**) and (*Z*)-silyl enol ether (**125**), respectively.

Alternatively, as exemplified in Scheme 22, Normant reports that the chiral α,β-unsaturated acetals (**126**) and (**127**), obtained from (*R,R*)-2,3-butanediol, undergo exclusive γ-addition (S_N2') by complex (**87c**) with high diastereoselectivity. For example, (*E*)-acetals (**126**) and (*Z*)-acetals (**127**) afford (*S*)-β-substituted aldehydes and (*R*)-β-substituted aldehydes, respectively (*cf.* Scheme 2).[59]

Dramatic reactivity enhancements have also been reported for the Lewis acid modified cyanocuprates (**89**);[60] for example, additions of the α-hydroxyalkyl synthon ($^-$CH(R)OH) reagents, (**89a**) and (**89b**), to

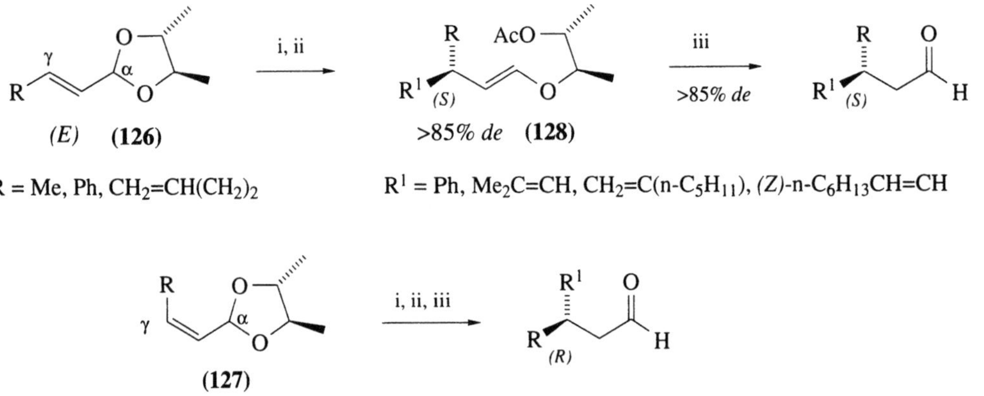

(*E*) (**126**) —i, ii→ >85% *de* (**128**) —iii, >85% *de*→ (*S*)

R = Me, Ph, CH_2=CH$(CH_2)_2$

R^1 = Ph, Me_2C=CH, CH_2=C(n-C_5H_{11}), (*Z*)-n-C_6H_{13}CH=CH

(**127**) —i, ii, iii→ (*R*)

i, $R^1Cu{\bullet}R^*{}_3P{\bullet}BF_3$ (**87c**); ii, $(MeCO)_2O$, C_5H_5N, DMAP; iii, H_3O^+

Scheme 22

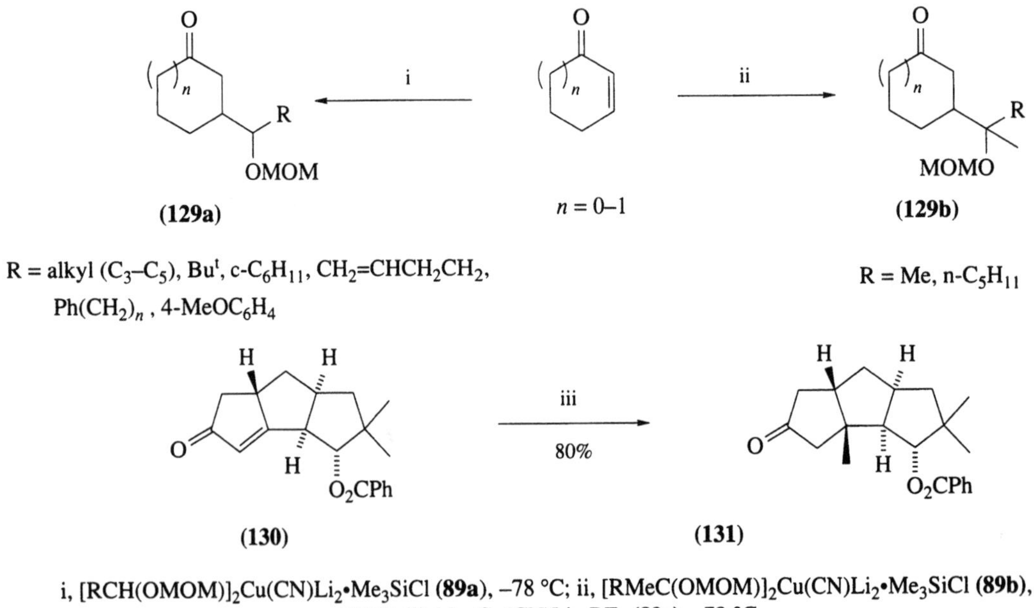

(**129a**) ←i— n = 0–1 —ii→ (**129b**)

R = alkyl (C_3–C_5), Bu^t, c-C_6H_{11}, CH_2=$CHCH_2CH_2$, Ph$(CH_2)_n$, 4-MeOC_6H_4

R = Me, n-C_5H_{11}

(**130**) —iii, 80%→ (**131**)

i, [RCH(OMOM)]$_2$Cu(CN)$Li_2{\bullet}Me_3SiCl$ (**89a**), –78 °C; ii, [RMeC(OMOM)]$_2$Cu(CN)$Li_2{\bullet}Me_3SiCl$ (**89b**), –78 °C; iii, $Me_2Cu(CN)Li_2{\bullet}BF_3$ (**89c**), –78 °C

Scheme 23

α,β-cycloenones proceed efficiently to afford homoaldol products (**129**).[61a–61d] Similarly, additions to sterically congested β,β-disubstituted-α,β-enones (**130**) proceed in the 1,4-mode exclusively, at low temperatures (Scheme 23).[61e]

1.3.2.1.4 Organozirconiums

The use of organozirconium reagents in synthesis was prompted by Wailes' discovery that alkynes and alkenes undergo hydrozirconation efficiently with bis(η^5-cyclopentadienyl)chlorohydridozirconium, [(η^5-C_5H_5)$_2$Zr(H)Cl].[62a,62b] Terminal alkynes undergo regiospecific *cis* addition, while unsymmetrically substituted alkynes show good regioselectivity.[62c] The alkenylzirconiums (**132**) are notable for their compatibility with various functionalities without masking; for example, the alkenylzirconium (**134**) is prepared in the presence of an ester functionality. In the presence of catalytic amounts of nickel acetylacetonate [Ni(acac)$_2$], excellent yields of 1,4-addition products with α,β-enones are obtained (Scheme 24).[63a] By-product dienes arising from alkene dimerization are minimized when a prereduced catalyst

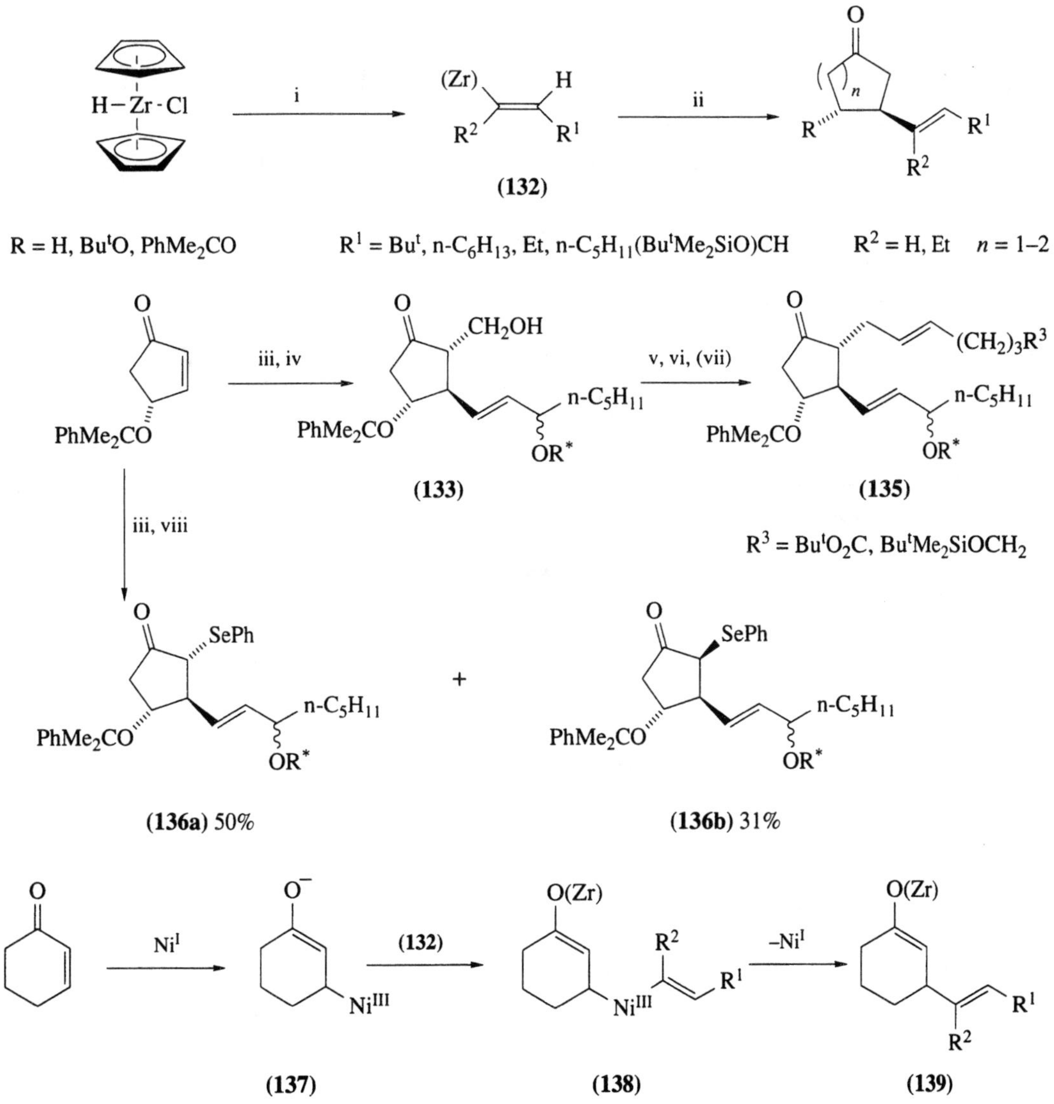

H−Zr−Cl
i
(Zr)
H
R²
R¹
(132)
ii
O
n
R
R¹
R²
R = H, ButO, PhMe₂CO
R¹ = But, n-C₆H₁₃, Et, n-C₅H₁₁(ButMe₂SiO)CH
R² = H, Et n = 1–2
O
PhMe₂CO
iii, iv
CH₂OH
n-C₅H₁₁
OR*
(133)
v, vi, (vii)
(CH₂)₃R³
(135)
R³ = ButO₂C, ButMe₂SiOCH₂
iii, viii
SePh
(136a) 50%
+
(136b) 31%
Niᴵ
O⁻
Niᴵᴵᴵ
(137)
(132)
O(Zr)
(138)
−Niᴵ
(139)

i, $R^1C{\equiv}CR^2$; ii, cycloalkenone, Ni^{II} catalyst; iii, *(E)*-(Zr)CH=CHCH(OR*)-n-C_5H_{11}, Ni^{II} catalyst; iv, CH_2O (g); v, $MeSO_2Cl$, Et_3N; vi, *(E)*-(Zr)CH=CH(CH_2)$_3$$CO_2Bu^t$ (**134**), Ni^{II} catalyst; vii, *(E)*-(Zr)CH=CH(CH_2)$_4$$OSiMe_2Bu^t$, Ni^{II} catalyst; viii, PhSeCl

Scheme 24

complex of nickel acetylacetonate and diisobutylaluminum hydride (DIBAL-H) is utilized. In contrast to the nickel acetylacetonate catalyzed dialkylalkenylalane conjugate additions (*cf.* Section 1.3.2.1.1), the alkenylzirconium transfers are highly efficient with terminal alkenylzirconiums and moderate yielding with internal alkenylzirconiums. The latter may be due to steric hindrance of the intermediate complex to conjugate addition and subsequent destruction of the nickel catalyst. Quenching of the intermediate zirconium enolates with either anhydrous formaldehyde or phenylselenyl chloride affords α-hydroxymethyl ketones (**133**)[64a] or α-phenylselenyl ketones (**136**),[64b] respectively; dehydration of ketol (**133**) and further alkenylzirconium conjugate addition affords the corresponding prostaglandin analogs (**135**). Similar conjugate additions have been successfully accomplished with α,β-unsaturated esters and α,β-enynones.

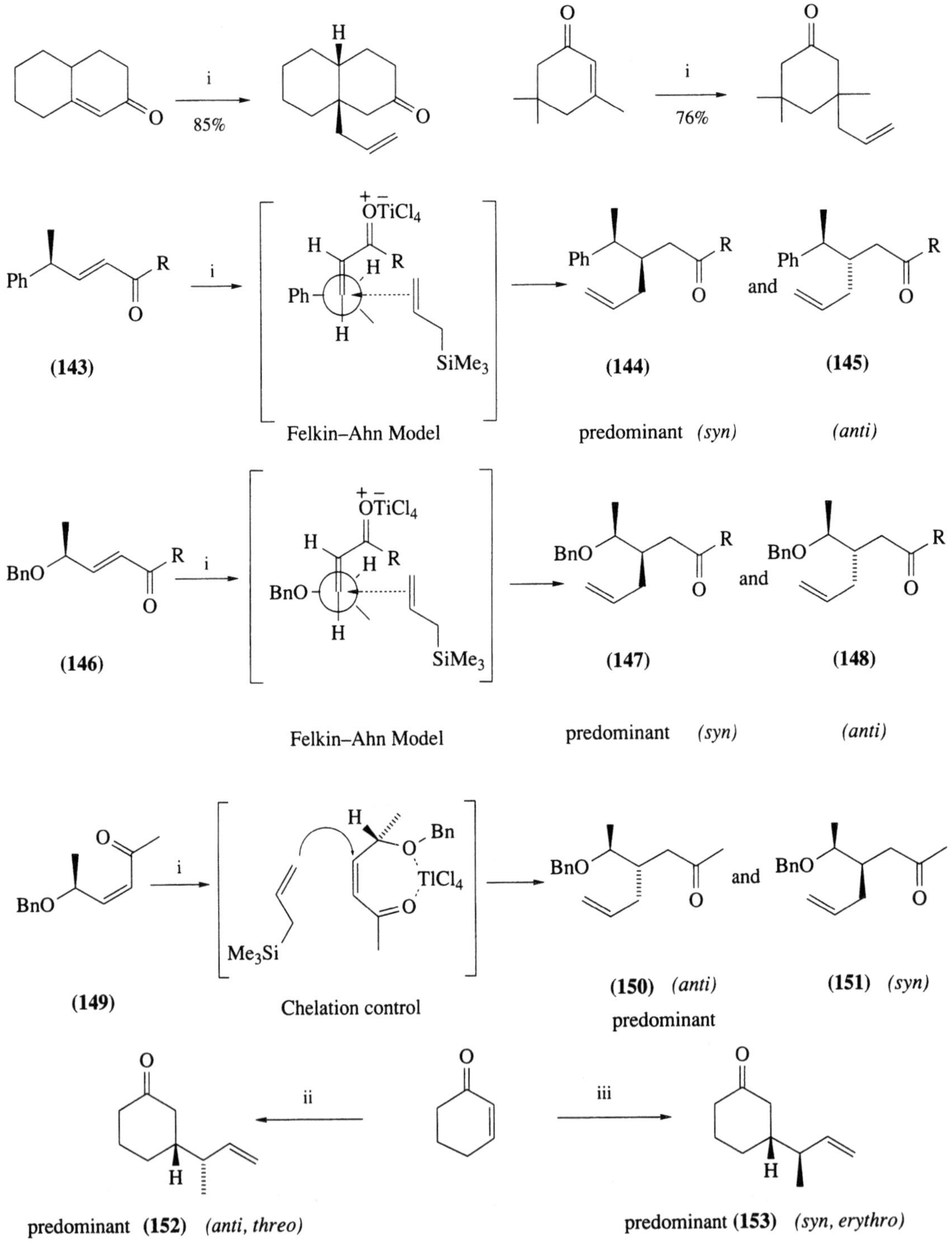

i, 85%; i, 76%

(**143**) → [Felkin–Ahn Model: $\overset{+\,-}{O}TiCl_4$, $SiMe_3$] → (**144**) predominant *(syn)* and (**145**) *(anti)*

(**146**) → [Felkin–Ahn Model: $\overset{+\,-}{O}TiCl_4$, $SiMe_3$] → (**147**) predominant *(syn)* and (**148**) *(anti)*

(**149**) → [Chelation control: $TiCl_4$, Me_3Si] → (**150**) *(anti)* predominant and (**151**) *(syn)*

predominant (**152**) *(anti, threo)* ←ii— cyclohexenone —iii→ predominant (**153**) *(syn, erythro)*

Scheme 25

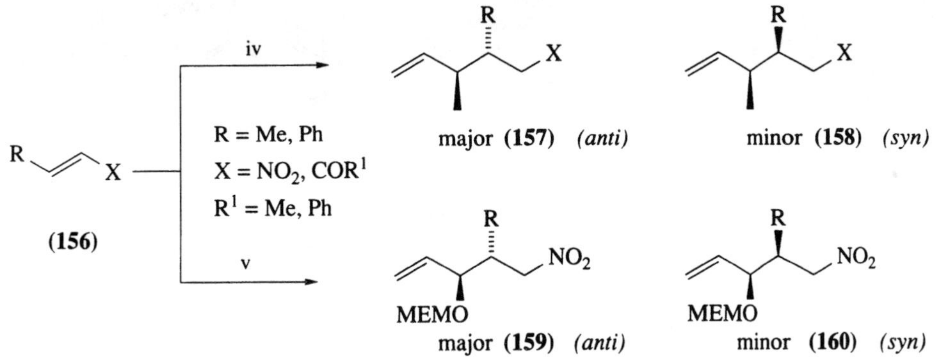

R = Me, Ph
X = NO_2, COR^1
R^1 = Me, Ph

(156)

major (157) *(anti)* minor (158) *(syn)*

MEMO major (159) *(anti)* MEMO minor (160) *(syn)*

i, $CH_2=CHCH_2SiMe_3$ (**140**), $TiCl_4$, CH_2Cl_2, –78 °C; ii, *(Z)*-$MeCH=CHCH_2SiMeR^1R^2$ (**141**), $TiCl_4$, CH_2Cl_2, –78 to –40 °C; iii, *(E)*-$MeCH=CHCH_2SiMeR^1R^2$ (**142**), $TiCl_4$, CH_2Cl_2, –78 °C; iv, *(E)*-$MeCH=CHCH_2SnBu^n_3$ (**153**), $(Et_2Al)_2SO_4$, C_6H_6, 80 °C; v, *(Z)*-$MOMOCH=CHCH_2SnBu^n_3$ (**155**), $(Et_2Al)_2SO_4$, C_6H_6, 80 °C

Scheme 25 *(continued)*

Mechanistically, the addition of a nickel(I) species to an α,β-enone generates a nickel(III) enolate (**137**) which undergoes transmetallation with an alkenylzirconium (**132**) and reductive elimination of nickel(I) to afford a zirconium enolate (**139**).[63b]

1.3.2.1.5 Organosilanes and organostannanes

The reactivity of the organosilanes[65] and organostannanes[66] towards electrophiles is dependent on the characteristics of the organic ligands. Typically, the alkylsilanes and alkylstannanes are unreactive, which is a consequence of the weakly polarized carbon–silicon and carbon–tin σ-bonds ($C^{\delta-}$—$M^{\delta+}$). However, allylsilanes[67] and allylstannanes are highly reactive to electrophiles because of extensive σ–π (C—Si or C—Sn) conjugation in the allyl metals and the β-carbonium ion stabilization effect of the metal center. Consequently, electrophiles add exclusively with allylic transposition.

The Lewis acid catalyzed conjugate addition of allylsilanes (**140**) to (**142**) and allylstannanes (**154**) and (**155**) to α,β-enones, described by Sakurai,[68a,68b] is highly efficient and experimentally simple in contrast to the allylcuprate additions. Various substituents can be incorporated into the allylsilanes (allylstannanes), *e.g.* alkoxy, alkoxycarbonyl and halogen, some of which are incompatible with cuprate reagents.[69] In addition, Heathcock and Yamamoto report that diastereoselectivity is correlated to the alkene geometry of both the allylmetals and the acceptor units; for example, allylation of (*E*)-enones (**143**) and (**146**) affords predominantly the *syn* adducts (**144**) and (**147**), while (*Z*)-enone (**149**) gives predominantly the *anti* adduct (**150**; Scheme 25).[68c] On the other hand, with cyclohexen-2-one the (*Z*)-silane (**141**) affords predominantly the *threo* adduct (**152**), while (**142**) affords *erythro* adduct (**153**).[68d] The more reactive allylstannanes (**154**) and (**155**) also afford similar diastereoselectivity.[68e,f]

Typically, α,β-unsaturated esters, α,β-unsaturated aldehydes and α,β-unsaturated nitriles are poor acceptors for the Lewis acid catalyzed silylallylation procedure, but they are excellent acceptors for the complementary fluoride ion mediated allylation procedure (*cf.* Volume 4, Chapter 1.2, Section 1.2.2.1.7). Other suitable acceptors include 1,4-quinones,[70] α,β-unsaturated acyl cyanides (**162**),[71a] silyl α,β-enoates (**163**)[71b] and nitroalkenes (Scheme 26);[72] reduction (titanium(III) trichloride) of the intermediate nitronates arising from nitroalkene allylation affords γ,δ-enones (**166**).

Majetich reports a general intramolecular Lewis acid allylation protocol for the synthesis of bicyclo[5.4.0]undecen-3-ones (**168**) and bicyclo[4.4.0]decen-3-ones (**170**), which are 1,6-addition products (Scheme 27). The same precursors, (**167**) and (**169**), when submitted to the fluoride ion cyclization protocol, also afford 1,2- and 1,4-addition products.[73] Typically, ethylaluminum dichloride, a 'proton sponge' Lewis acid, is used in order to minimize adventitious protonic desilylation. Other β,γ-unsaturated silanes also undergo similar intramolecular Lewis acid catalyzed additions; for example, the silylpropargylic enones (**171**) undergo intramolecular cyclization to the allenylspiro system (**172**).[74]

Subtle reactivity differences are reported for the reaction of allenylsilanes *vis-à-vis* the allenylstannanes with α,β-cycloalkenones. The stannane (**175**) affords the alkynic ketone (**176**)[75] exclusively, while

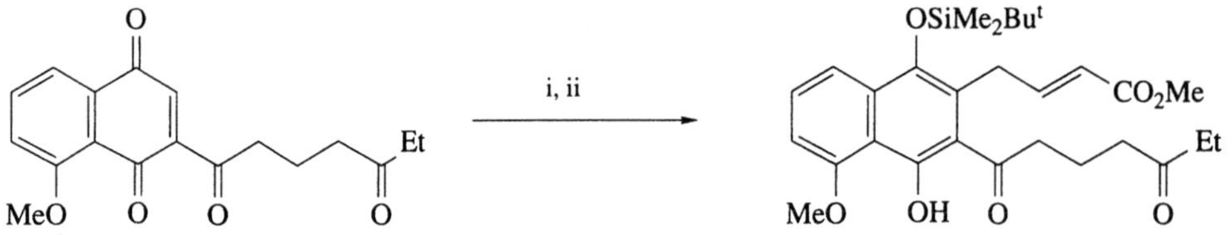

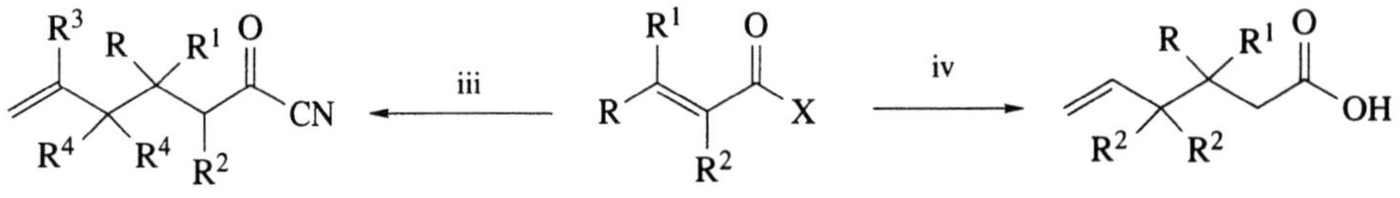

i, ii

MeO O O O Et → $OSiMe_2Bu^t$ CO_2Me Et MeO OH O O

(164) iii ← (162) X = CN
(163) X = $OSiMe_3$ → iv (165)

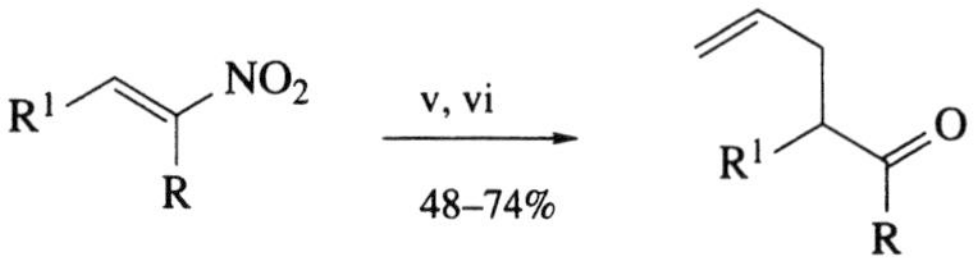

(34) v, vi 48–74% (166)

R = Me, Et; R^1 = n-$C_{10}H_{21}$, Ph, 4-ClC_6H_4, 4-$MeOC_6H_4$, $PhCH_2CH_2$

i, CH_2=CHCH[$SiMe_2Ph$]CO_2Me **(161)**, $SnCl_4$, CH_2Cl_2, –78 to –30 °C; ii, Bu^tMe_2SiCl, imidazole, DMF; iii, $R^4{}_2C$=C(R^3)CH_2SiMe_3, $TiCl_4$, CH_2Cl_2, –78 °C; iv, $R^2{}_2C$=$CHCH_2SiMe_3$, $TiCl_4$, CH_2Cl_2, –78 °C; v, CH_2=$CHCH_2SiMe_3$ **(140)**, $AlCl_3$, –20 °C; vi, $TiCl_3$

Scheme 26

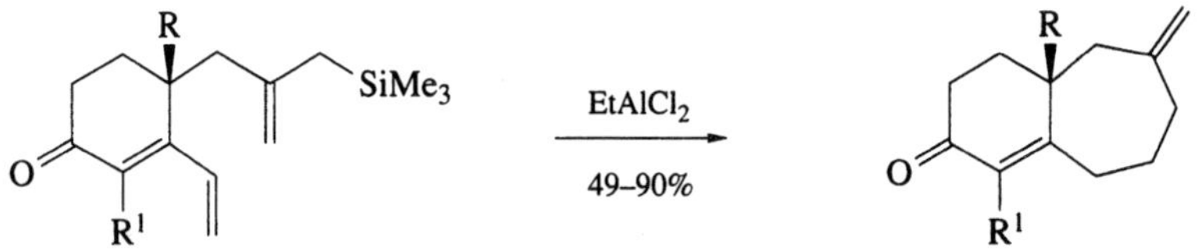

(167) $EtAlCl_2$ 49–90% **(168)**

R = H, Me; R^1 = H, Me; R^2 = H, Me

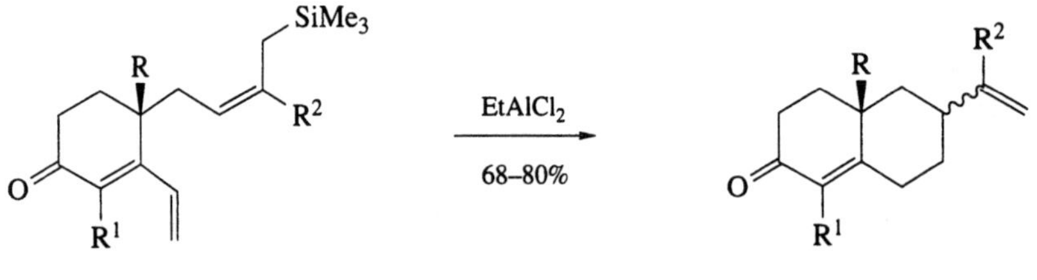

(169) $EtAlCl_2$ 68–80% **(170)**

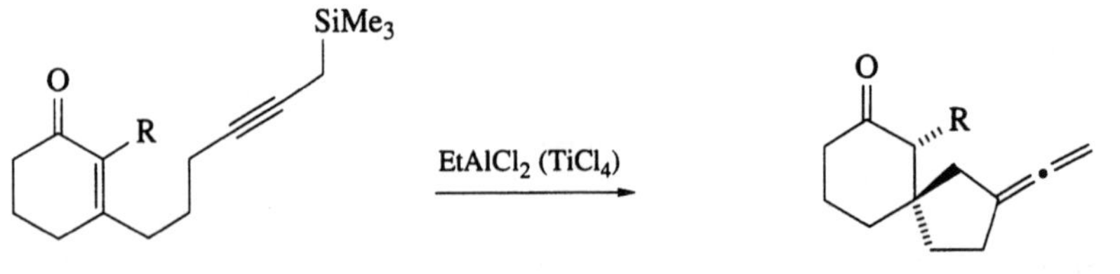

(171) $EtAlCl_2$ ($TiCl_4$) **(172)** R = H, Me

Scheme 27

the silane (**173**) affords the bicyclo[4.3.0]nonenylsilane (**174**; Scheme 28). The initial step, *i.e.* Lewis acid activated enone addition of the allenylmetallic, is identical for both reactions, but the subsequent

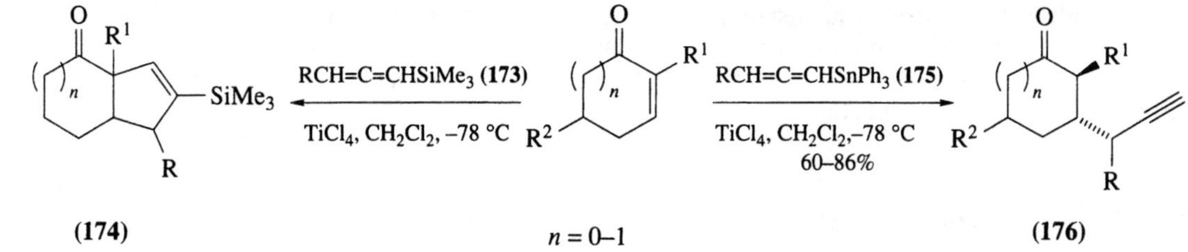

SiMe3
RCH=C=CHSiMe3 (173)
TiCl4, CH2Cl2, –78 °C
RCH=C=CHSnPh3 (175)
TiCl4, CH2Cl2,–78 °C
60–86%
(174)
n = 0–1
(176)

R = H, Me; R^2 = H, Me; R^1 = H, Me

Scheme 28

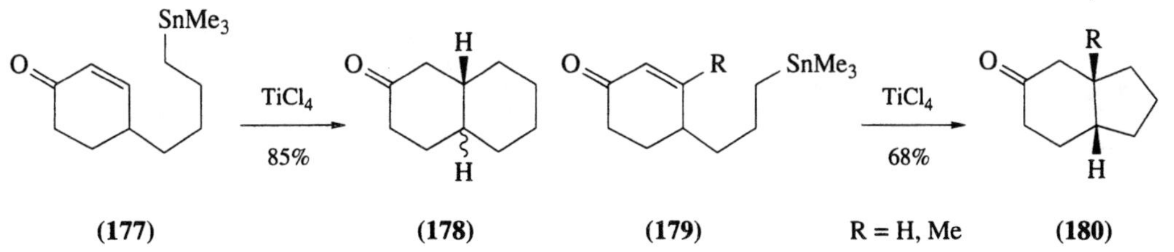

SnMe3
TiCl4
85%
SnMe3
TiCl4
68%
(177)
(178)
(179)
R = H, Me
(180)

Scheme 29

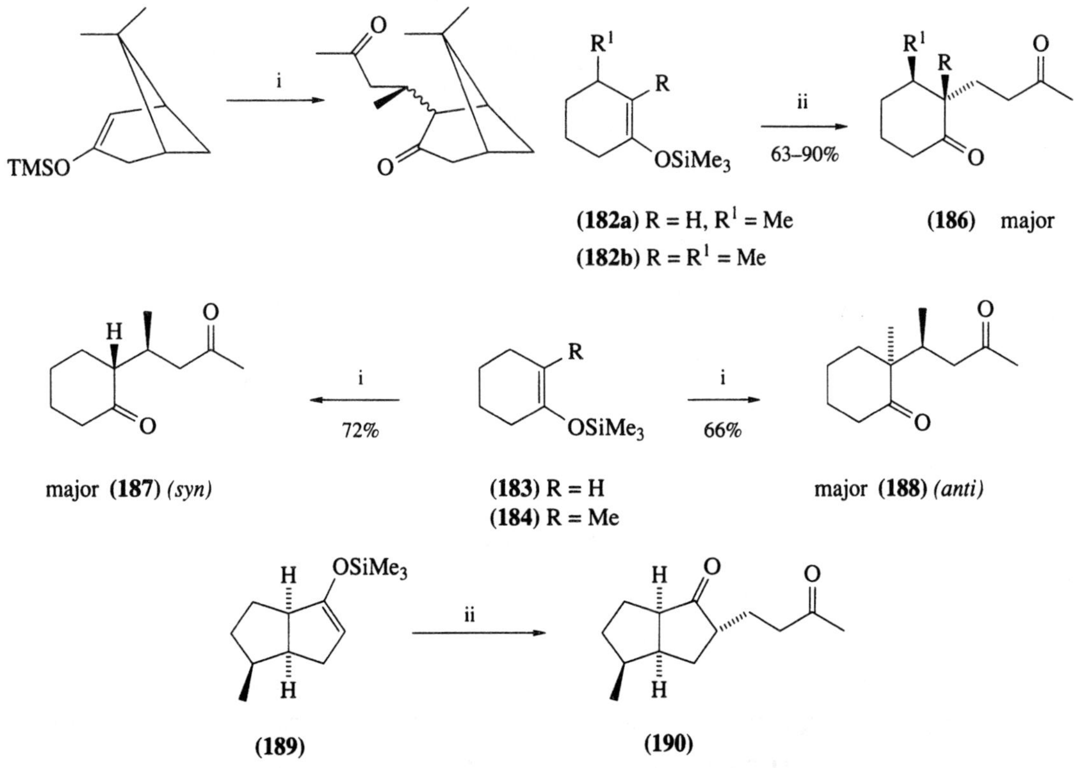

TMSO
i
OSiMe3
ii
63–90%
(182a) R = H, R1 = Me
(182b) R = R1 = Me
(186) major
i
72%
OSiMe3
i
66%
major (187) (syn)
(183) R = H
(184) R = Me
major (188) (anti)
OSiMe3
ii
(189)
(190)

i, *(E)*-MeCH=CHCOMe (**181**), $TiCl_4$, CH_2Cl_2, –78 °C;

ii, H_2C=CH-C(OCH_2CH_2O)Me (**185**), $TiCl_4$–Ti(OPr^i)$_4$, CH_2Cl_2, –80 to –95 °C

Scheme 30

step differs. The intermediate vinylstannane cation collapses to the alkyne, while the trimethylsilylvinyl cation undergoes a 1,2-shift to form an isomeric vinyl cation followed by titanium enolate capture.[76]

In contrast to the numerous examples of conjugate allyl transfers, the analogous alkyl transfers mediated by either alkylsilanes or alkylstannanes are scarce. However, in the presence of a β-electrophilic center cleavage of stereoproximate carbon–tin σ-bonds is facilitated; for example, the Lewis acid catalyzed carbocyclization of select 2-cyclohexenones with alkylstannane appendages, *e.g.* (**177**) or (**179**), affords bicyclic systems (Scheme 29).[77]

1.3.2.2 Additions of Enol Silyl Ethers and Silyl Ketene Acetals

The titanium(IV) chloride-catalyzed conjugate addition of enol silyl ethers (**182**) to (**184**) and (**189**; Scheme 30), and silyl ketene acetals, (**191**) to (**194**), to α,β-enones is the key feature in various synthetic strategies (Mukaiyama–Michael) (Scheme 31).[78,79] In contrast to the earlier described enolate addition

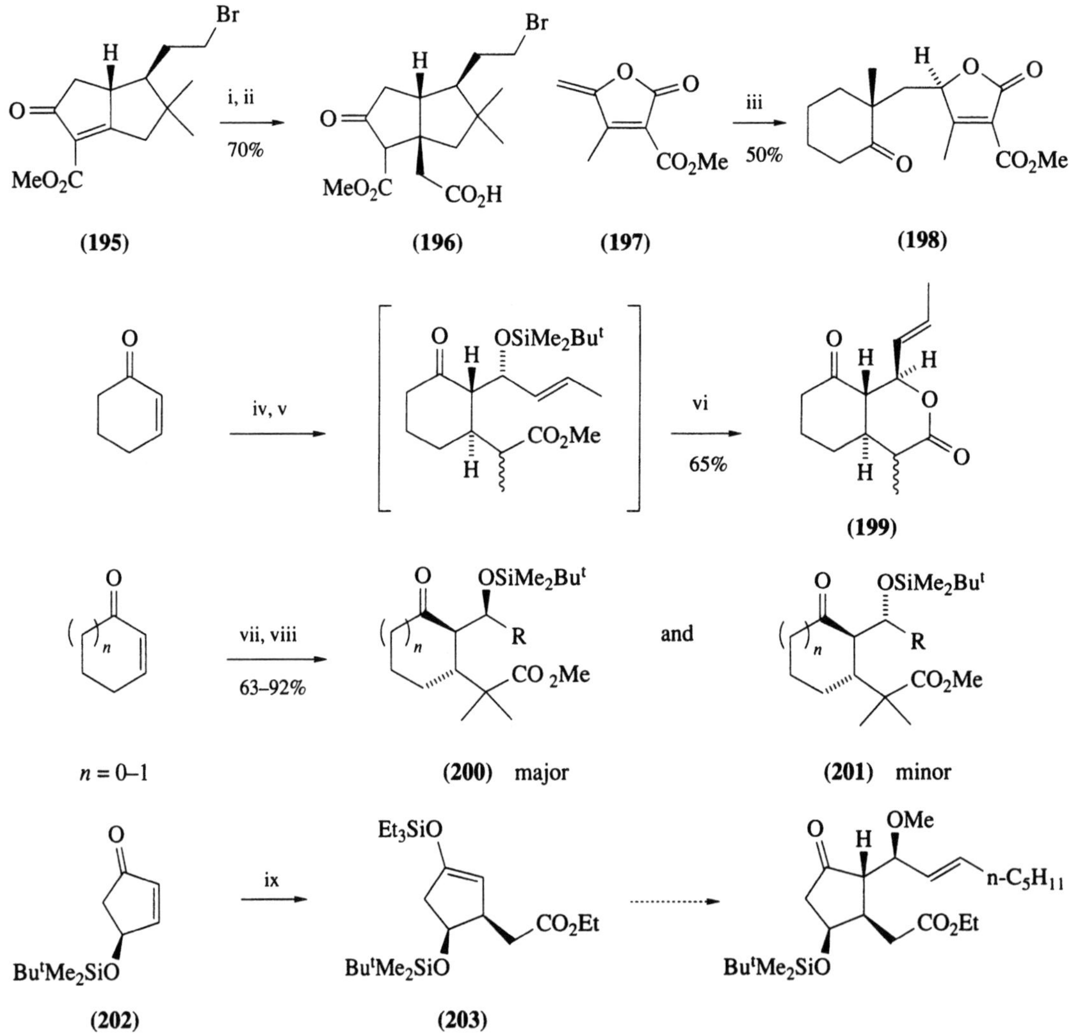
(195)
(196)
(197)
(198)
(199)
(200) major
(201) minor
n = 0–1
(202)
(203)
i, ii
70%
iii
50%
iv, v
vi
65%
vii, viii
63–92%
and
ix

i, CH_2=C(OSiMe$_2$Bu^t)OBu^t (**191a**), $TiCl_4$, CH_2Cl_2; ii, Bu$^n_4$N$^+$F$^-$; iii, 2-methylcyclohexanone trimethylsilyl enol ether (**186**), $TiCl_4$, CH_2Cl_2; iv, MeCH=C(OSiMe$_2$Bu^t)OMe (**192**), $Ph_3C^+ClO_4^-$ (5 mol%), CH_2Cl_2, −78 °C; v, *(E)*-MeCH=CHCHO; vi, $MeCO_2H$, THF; vii, Me_2C=C(OSiMe$_2$Bu^t)OMe (**193**), $Ph_3C^+ClO_4^-$ (5 mol%), CH_2Cl_2, −78 °C; viii, RCHO; ix, CH_2=C(OSiEt$_3$)OEt (**194**), HgI_2

Scheme 31

procedures (*cf.* Volume 4, Chapter 1.2, Section 1.2.2.2), this protocol is synthetically advantageous; for example, base sensitive acceptors may be used, β-quaternary centers are assembled in high yields (**196**; Scheme 31)[80] and by-products due to 1,2-addition or proton transferrals, *i.e.* enolate scrambling, are non-existent. Thus enol silyl ethers arising from either kinetic or thermodynamic enolates do not undergo scrambling. However, the highly Lewis acidic nature of these reactions precludes the use of methyl vinyl ketone (MVK) as an acceptor, while the corresponding ethylene ketal (**185**) is an excellent acceptor with bis(isopropoxy)titanium dichloride catalysis, *e.g.* the conversion of (**189**) to (**190**) in Scheme 30.[81]

The choice of catalyst determines the range of suitable acceptors and work-up procedures employed. For example, trityl cation promoted additions of acetals (**192**) and (**193**) to α,β-enones typically afford regiochemically pure enol silyl ethers that can be utilized *in situ* for aldol condensations (Tandem Michael–aldol) (Scheme 31).[82] With similar reaction conditions, additions of silyl thioketene acetals (**204**) to α,β-cycloenones and α-substituted-α,β-cycloenones afford *anti* (**205**) and *syn* adducts (**206**), respectively. The diastereoselectivity differences are attributed to a highly ordered transition state (Scheme 32).[83] Danishefsky reports that mercury(II) iodide catalyzes the exclusive *syn* addition of silyl ketene acetals (**194**) to enone (**202**) (*cf.* the formation of **203**; Scheme 31),[84] while the corresponding carbanion additions give exclusive *anti* addition.[85]

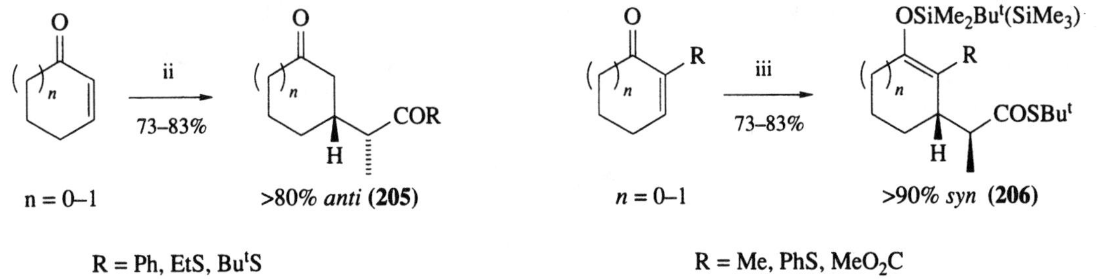

i, $MeCH{=}C(OSiMe_3)R$ [$MeCH{=}C(OSiMe_2Bu^t)R$], $Ph_3CCl{-}SnCl_2$ (5–10 mol %), CH_2Cl_2, –78 °C;
ii, $MeCH{=}C(OSiMe_2Bu^t)SBu^t$ [$MeCH{=}C(OSiMe_3)SBu^t$] (**204**), 2-$(C_5H_4N)CH_2OH$, $Ph_3C^+SbCl_6^-$, –78 °C

Scheme 32

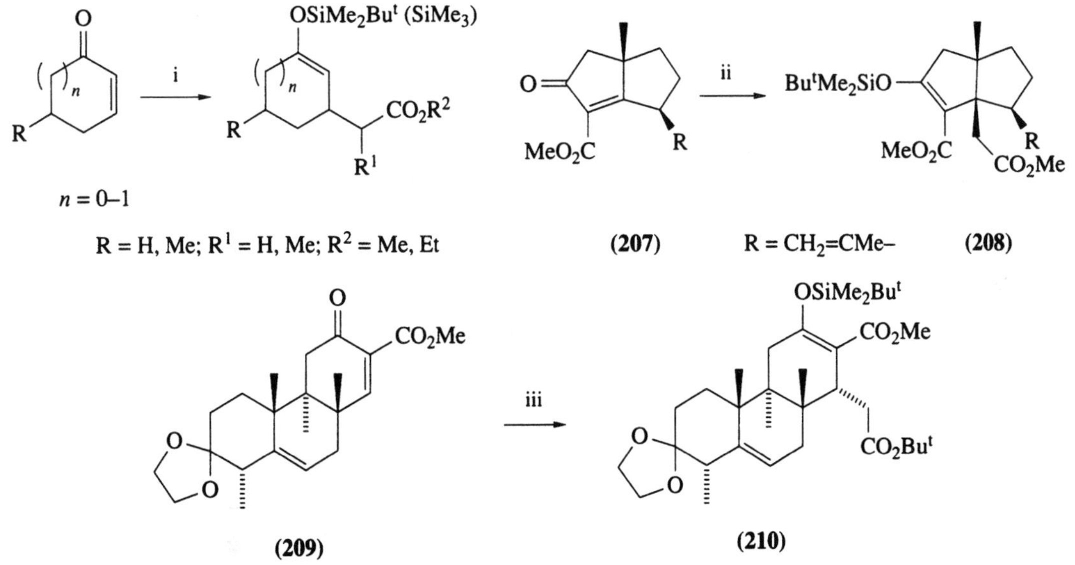

i, $R^1CH{=}C(OSiMe_2Bu^t)OR^2$ [$R^1CH{=}C(OSiMe_3)OR^2$] (**191**) and (**192**), MeCN, 55 °C;
ii, $CH_2{=}C(OSiMe_2Bu^t)OMe$, (**191b**), 15kbar; iii, $CH_2{=}C(OSiMe_2Bu^t)OBu^t$ (**191a**), 15 kbar

Scheme 33

Remarkable solvent effects have been reported for select silyl ketene acetal–enone additions; for example, acetonitrile suffices to promote additions in the absence of a Lewis acid.[86a] Alternatively, addi-

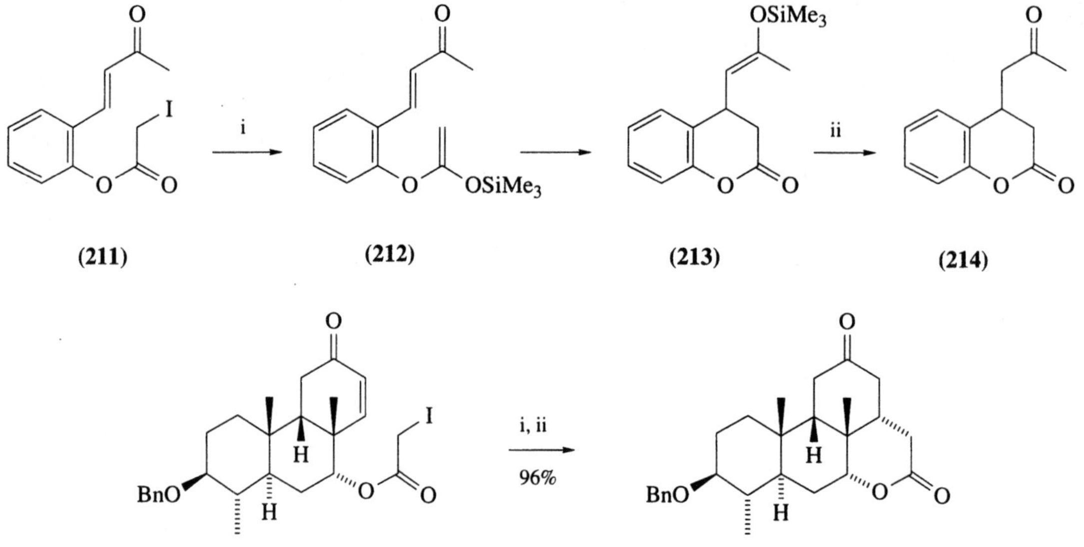

(211)
(212)
(213)
(214)
i
ii
OSiMe3
i, ii
96%
BnO

i, Me_3SiI; ii, H_3O^+

Scheme 34

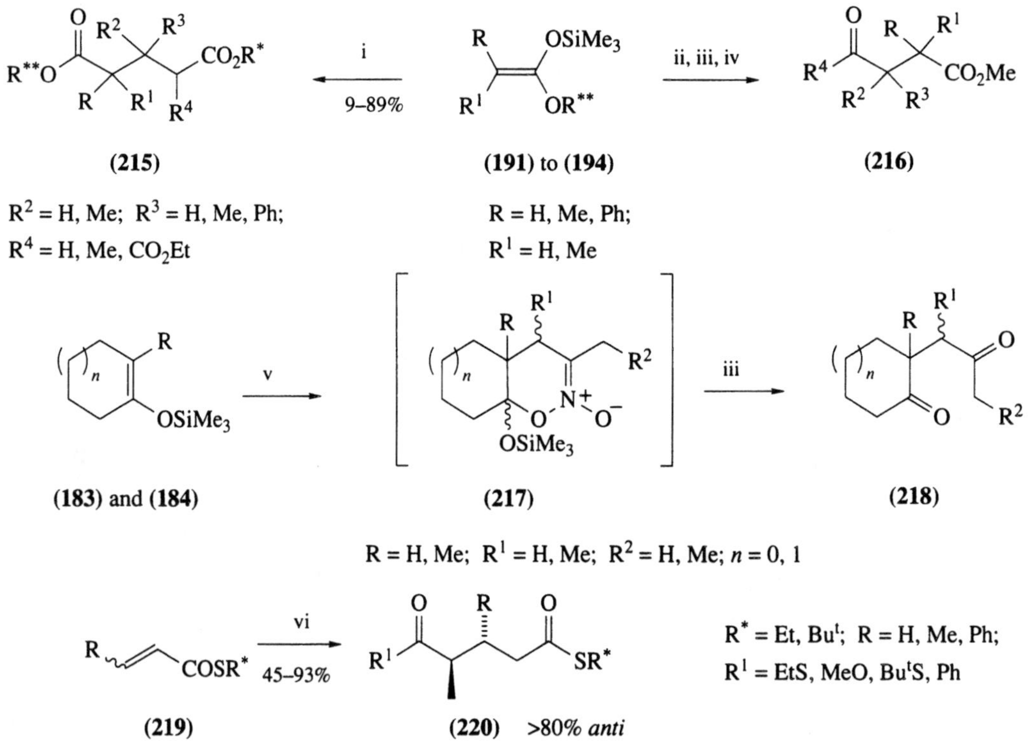

(215)
(191) to (194)
(216)
i
9–89%
ii, iii, iv
R2 = H, Me; R3 = H, Me, Ph;
R4 = H, Me, CO2Et
R = H, Me, Ph;
R1 = H, Me
(183) and (184)
(217)
(218)
v
iii
R = H, Me; R1 = H, Me; R2 = H, Me; n = 0, 1
(219)
(220) >80% anti
vi
45–93%
R* = Et, But; R = H, Me, Ph;
R1 = EtS, MeO, ButS, Ph

i, $R^2R^3C{=}C(R^4)CO_2R^*$, $Al(SO_3CF_3)_3$ or Al–montmorillonite; ii, $R^2R^3C{=}C(NO_2)R^4$, $TiCl_4$–$Ti(OPr^i)_4$; iii, H_2O; iv, CH_2N_2; v, $R^1CH{=}C(NO_2)CH_2R^2$, $TiCl_4$ ($SnCl_4$), CH_2Cl_2; vi, $MeCH{=}C(OSiMe_3)R^1$ [$MeCH{=}C(OSiMe_2Bu^t)R^1$], $SbCl_5$–$Sn(OTf)_2$ (5 mol %), –78 °C

Scheme 35

tions of silyl ketene acetals to sterically encumbered doubly activated α,β-enones, (**207**) or (**209**), which are not amenable to Lewis acid catalysis, proceed under high pressure (15 kbar, 1 bar = 100 kPa) (Scheme 33).[86b,c]

Few examples of intramolecular enol silyl ether or silyl ketene acetal cyclizations to α,β-enones have been reported. Notable, as exemplified in Scheme 34, is the iodotrimethylsilane-mediated intramolecular cyclization of δ-(iodoacetoxy)-α,β-enones (**211**) to δ-lactones (**214**). These cyclizations proceed with *in situ* generated silyl ketene acetals (**212**) arising from iodotrimethylsilane reduction of the iodoacetoxy moiety.[87]

In contrast to titanium(IV) tetrachloride, which causes polymerization of α,β-unsaturated esters, aluminum triflate[88] or aluminum-impregnated montmorillonite[87b] are excellent promoters of silyl ketene acetal additions to α,β-unsaturated esters (Scheme 35). Similarly, the addition of silyl ketene acetals and enol silyl ethers to nitroalkenes, followed by Nef-type work-up, affords γ-keto esters (**216**) and γ-diketones (**218**), respectively (Scheme 35).[89a,89b] Mechanistically, the γ-diketones (**218**) arise from Nef-type hydrolysis of an initial nitronate ester (**217**).[89c,89d] Mukaiyama reports that $SbCl_5$–$Sn(OTf)_2$ catalyzes diastereoselective *anti* additions of silyl ketene acetals, silyl thioketene acetals and enol silyl ethers to α,β-unsaturated thioesters (**219**).[90]

Vinyl sulfoxides (**221**), which are aldehyde α-cation equivalents, and vinylthiolium ions (**230**), which are α,β-unsaturated carbonyl β-cation equivalents, are also suitable acceptors for silyl ketene acetals and enol silyl ethers (Scheme 36). Kita reports that the bulky *t*-butyldimethylsilyl ketene acetals and trimethylsilyl ketene acetals form 1:1 adducts (**224**) and 1:2 adducts (**225**) with (**221**), respectively;[91] mechanistically, these additions proceed *via* an initial Pummerer rearrangement. The vinylthiolium ion additions are notable for their synthetic flexibility; for example, additions to the ketene dithioacetal (**229**) proceed with higher diastereoselectivity than the corresponding enolate additions to α,β-unsaturated esters.[91c,91d]

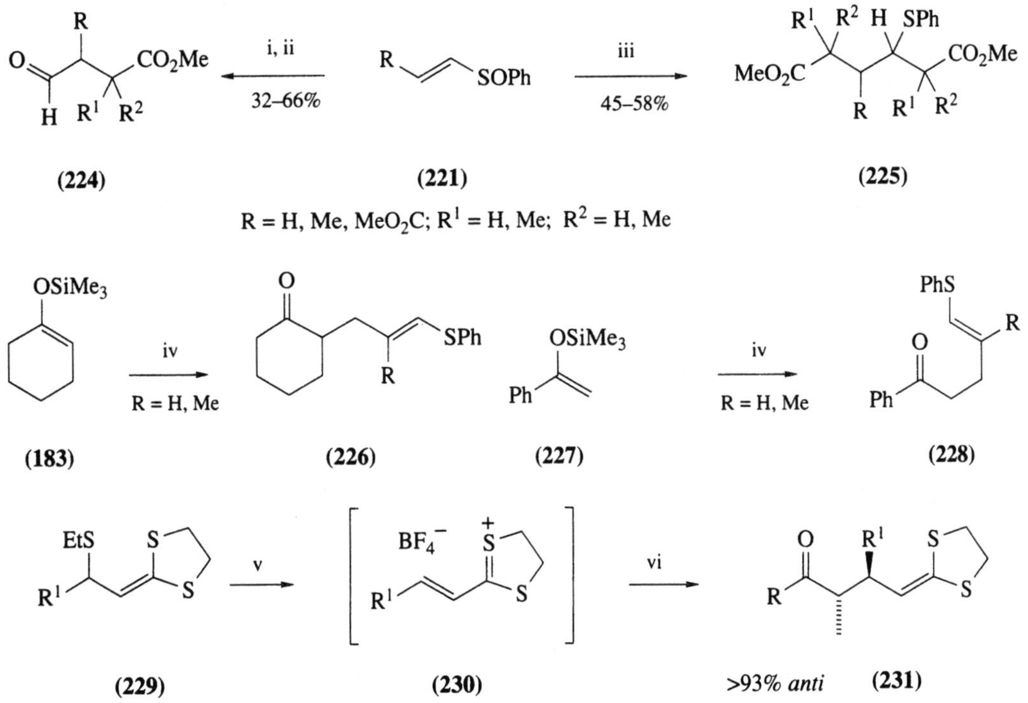

(**224**) ← i, ii; 32–66% — (**221**) — iii; 45–58% → (**225**)

R = H, Me, MeO_2C; R^1 = H, Me; R^2 = H, Me

(**183**) — iv; R = H, Me → (**226**)   (**227**) — iv; R = H, Me → (**228**)

(**229**) — v → [(**230**)] — vi → >93% *anti* (**231**)

R = Ph, Bu^t, EtO, MeO; R^1 = alkyl (C_1–C_3), Ph

i, $R^1R^2CH{=}C(OSiMe_2Bu^t)OMe$, ZnI_2, MeCN, 0 °C; ii, TBAF; iii, $R^1R^2CH{=}C(OSiMe_3)OMe$, ZnI_2, MeCN; iv, *(E)*-Me(R)C=CHSOPh (**222**), Me_3SiOTf, Pr^i_2NEt; v, $Ph_3C^+BF_4^-$; vi, $MeCH{=}C(OSiMe_2Bu^t)R$

Scheme 36

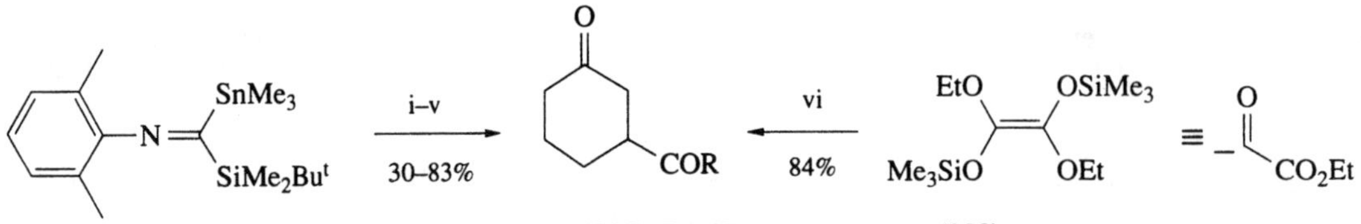

(232) **(235)** R = H **(232)**
(236) R = CO_2Et

i, Bu^{n}Li, THF, –78 °C; ii, CuBr–DMS; iii, Me_3SiCl (Bu^{t}Me$_2$SiCl); iv, cycloalkenone; v, H_3O^+;
vi, cycloalkenone, $ZnCl_2$

Scheme 37

1.3.2.3 Additions of Acyl Anion Equivalents

Except for the well-documented conjugate additions of diethylaluminum cyanide,[92] triethylaluminum–hydrogen cyanide and Lewis acid–tertiary alkyl isonitriles,[93] examples of Lewis acid catalyzed conjugate additions of acyl anion equivalents are scant. Notable examples are additions of copper aldimines (**233**),[94a,94b] prepared from (**232**), and silyl ketene acetals (**234**)[94c] to α,β-enones which afford 1,4-ketoaldehydes (**235**) and 2,5-diketo esters (**236**), respectively (Scheme 37). The acetal (**234**) is considered a glyoxylate ester anion equivalent.

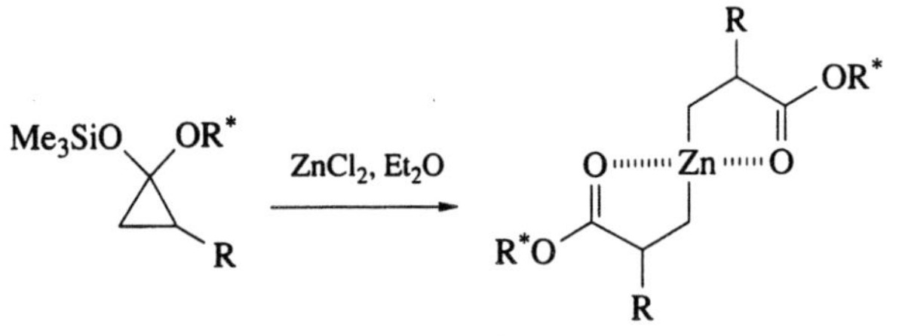

(237) **(238)**

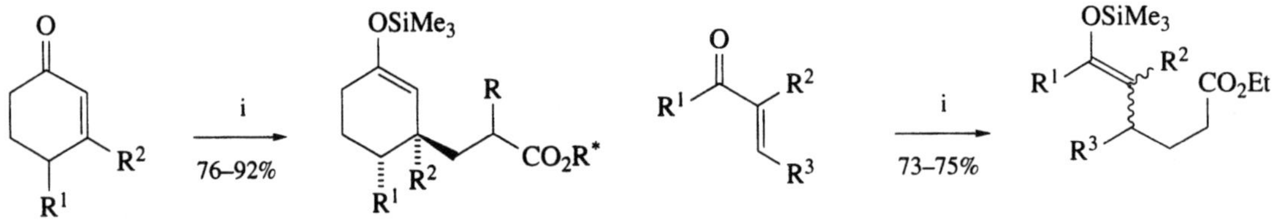

R^1 = H, Me; R^2 = H, Me R = H, Me R^1 = H, Me; R^2 = H, Et; R^3 = H, Me

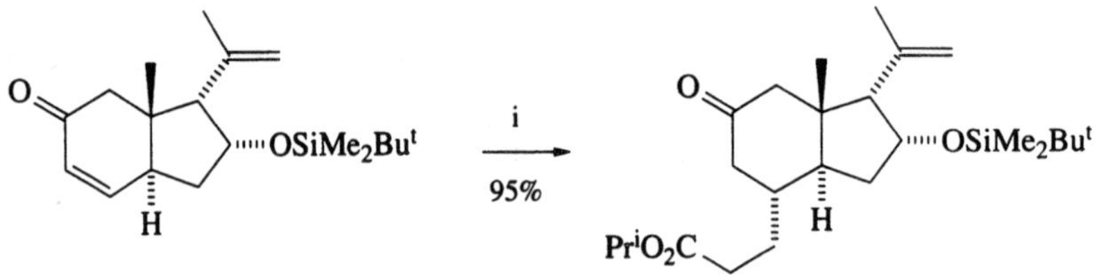

(240) **(241)**

i, (**238**), $CuBr{\bullet}SMe_2$ (0.5 mol %), HMPA, 0 °C

Scheme 38

1.3.2.4 Additions of Homoenolates

The concept of homoenolization was recognized during the 1960s. However, attempts at direct formation of homoenolates were complicated by their spontaneous cyclopropanolate formation. *In lieu* of direct methods, various groups have focussed on the development of homoenolate equivalents.[95] Kuwajima reports that the treatment of cyclopropanone silyl hemiketals (**237**) with zinc chloride, in the presence of HMPA, affords a convenient source for homoenolates (**238**). *In situ* transmetallation with a copper bromide–dimethyl sulfide complex affords the low reactivity copper homoenolate (**239**) which on activation with chlorotrimethylsilane–HMPA undergoes conjugate additions with α,β-enones and enals (Scheme 38).[96] Recently, Yoshida reported a versatile procedure for *in situ* generation of zinc homoenolates and homologous carbanions *via* a zinc–copper couple in the presence of HMPA (Scheme 39).[97] As an aside, the trimethylsilyl triflate catalyzed addition of γ-triphenylphosphorylidene *t*-butyldimethylsilyl enol ether (**248**), an α,β-enone β-anion equivalent, to activated alkenes affords β-substituted cycloalkenones (**249**).[98] Conceptually, this protocol is equivalent to the addition–elimination of a homoenolate or equivalent to a β-halo-α,β-enone (*cf.* Volume 4, Chapter 1.2, Section 1.2.4.3).

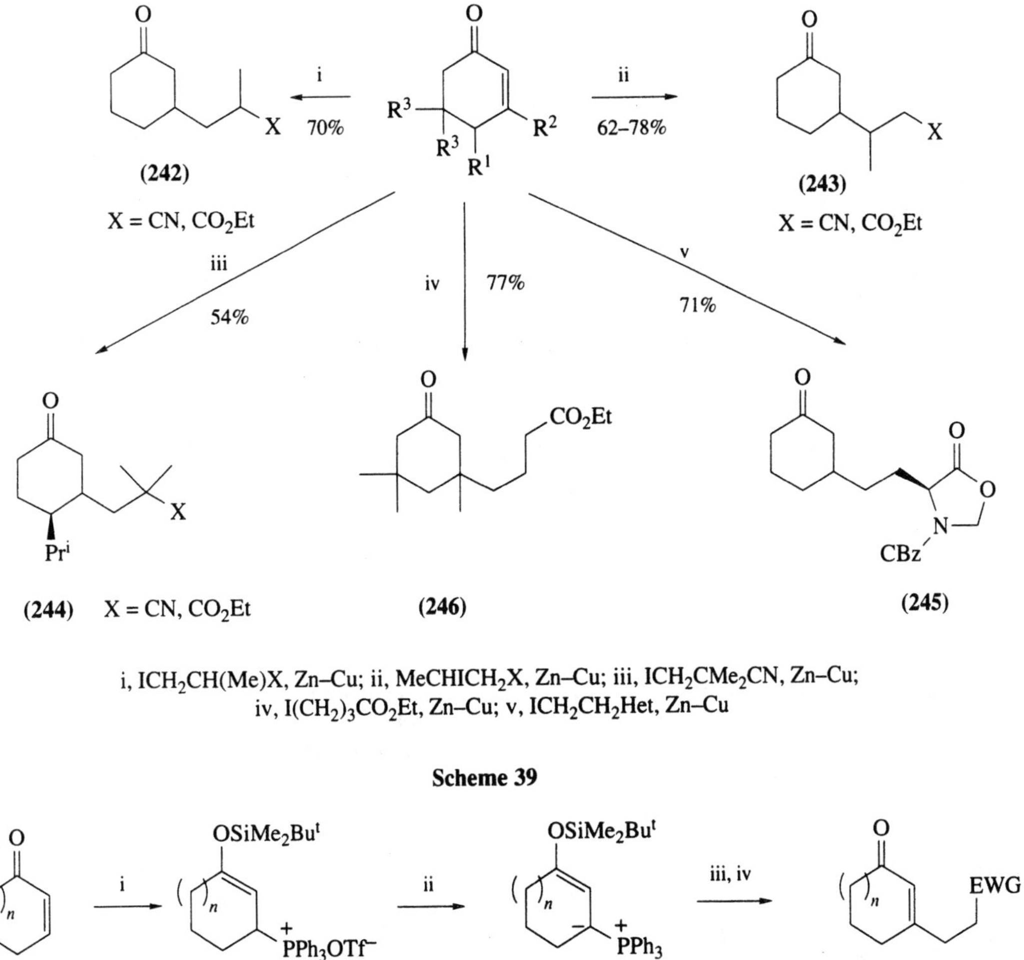

(**242**) X = CN, CO_2Et (**243**) X = CN, CO_2Et

(**244**) X = CN, CO_2Et (**246**) (**245**)

i, $ICH_2CH(Me)X$, Zn–Cu; ii, $MeCHICH_2X$, Zn–Cu; iii, ICH_2CMe_2CN, Zn–Cu; iv, $I(CH_2)_3CO_2Et$, Zn–Cu; v, ICH_2CH_2Het, Zn–Cu

Scheme 39

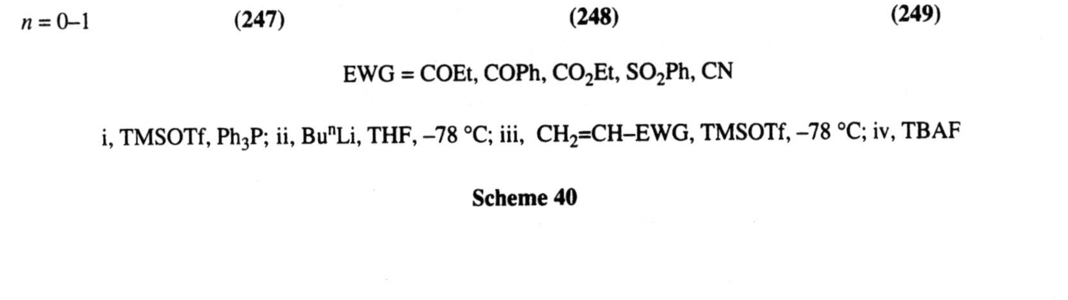

n = 0–1 (**247**) (**248**) (**249**)

EWG = COEt, COPh, CO_2Et, SO_2Ph, CN

i, TMSOTf, Ph_3P; ii, Bu^nLi, THF, –78 °C; iii, CH_2=CH-EWG, TMSOTf, –78 °C; iv, TBAF

Scheme 40

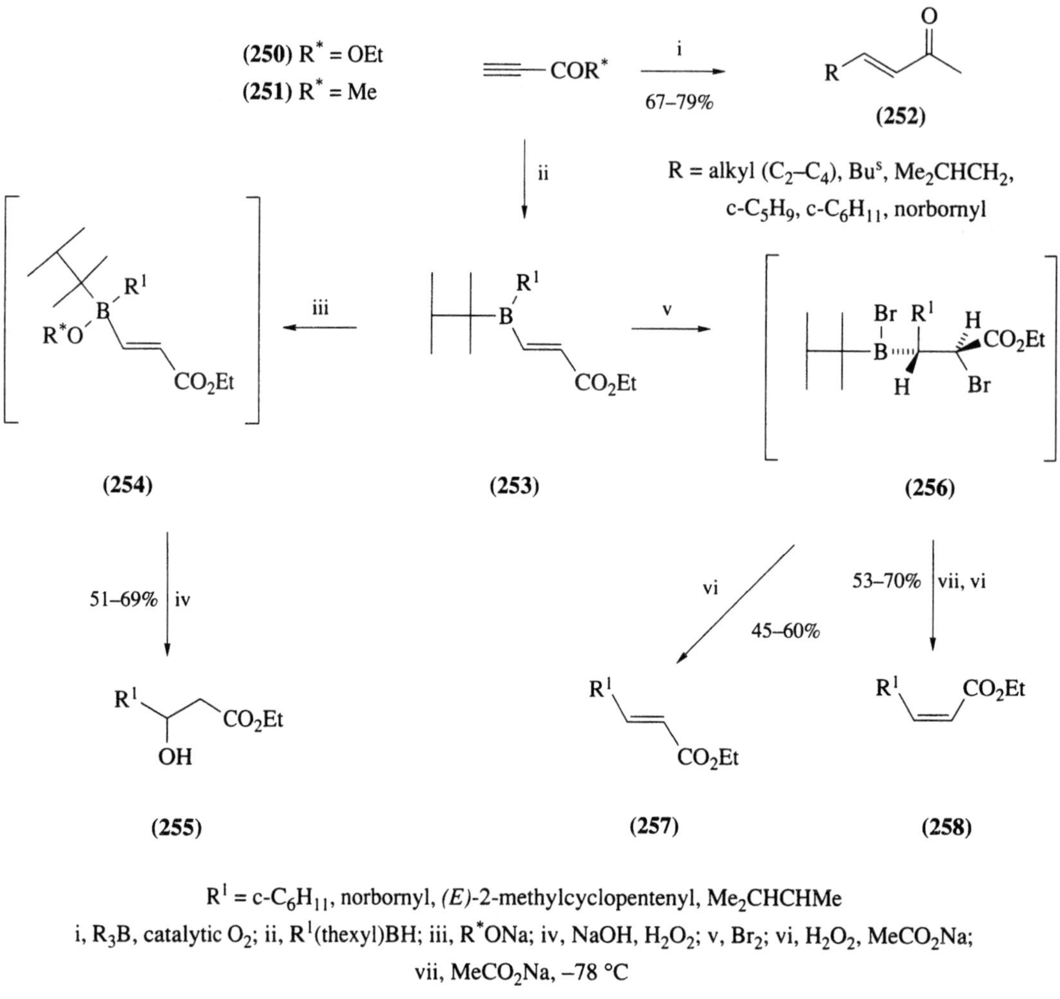

(**250**) R* = OEt
(**251**) R* = Me

≡—COR* —i (67–79%)→ (**252**)

R = alkyl (C_2–C_4), Bu^s, Me_2CHCH_2, c-C_5H_9, c-C_6H_{11}, norbornyl

(**254**) (**253**) (**256**)

51–69% iv; vi 45–60%; 53–70% vii, vi

CO_2Et CO_2Et CO_2Et OH

(**255**) (**257**) (**258**)

R^1 = c-C_6H_{11}, norbornyl, (*E*)-2-methylcyclopentenyl, $Me_2CHCHMe$

i, R_3B, catalytic O_2; ii, R^1(thexyl)BH; iii, R*ONa; iv, NaOH, H_2O_2; v, Br_2; vi, H_2O_2, $MeCO_2Na$; vii, $MeCO_2Na$, –78 °C

Scheme 41

1.3.3 ADDITIONS TO ALKYNIC π-SYSTEMS

The intermolecular conjugate addition of carbon ligands to activated alkynes is typically accomplished with organocopper reagents,[48,96] while similar additions with Lewis acids are rare. Notable examples are the organoborane[99] and silyl ketene acetal additions[100] to propynoate esters. The addition of trialkylboranes, catalyzed by oxygen, to butyn-3-one proceeds directly to afford (*E*)-enones (**252**). On the other hand, alkynes are functionalized *via* alkenylboronates or alkenylborates (**254**) which have a propensity to undergo alkyl migration from boron to α-carbon. For example, the hydroboration of ethyl propynoate with alkyl(thexyl)boranes (Scheme 41) affords the intermediate (*E*)-boraenoates (**253**), which can be transformed to either β-hydroxy esters (**255**),[99a] (*E*)-enoates (**257**) or (*Z*)-enoates (**258**).[99b,99c]

Further versatility of this approach has been realized with contrasting Lewis acid promoted additions of silyl ketene acetals, (**191**) to (**194**), to ethyl propynoate (Scheme 42). In fact, the tandem 1,4-conjugate addition–electrophile trapping protocol is feasible when titanium(IV) tetrachloride is employed. *In situ* functionalization of the intermediate titanate enoate (**259**), with select electrophiles, affords α-substituted enoates (**260**) to (**262**). On the other hand, the zinc iodide and zirconium(IV) tetrachloride protocols afford directly γ-alkoxycarbonyl-α-trimethylsilylenoates (**263**) and [2 + 2] adducts (**264**), respectively.[100]

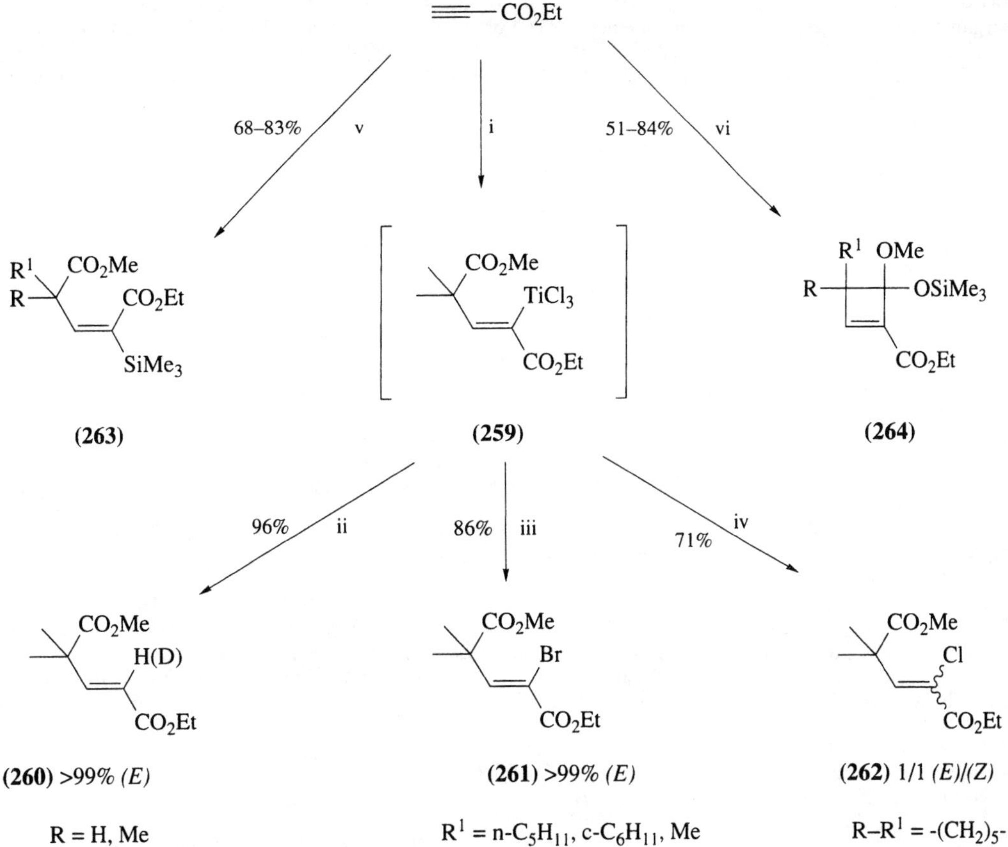

i, $Me_2C{=}C(OSiMe_3)OMe$ (**193**), $TiCl_4$, CH_2Cl_2, –78 °C; ii, $H_2O(D_2O)$; iii, NBS; iv, NCS;
v, (**191**)–(**194**), ZnI_2, CH_2Cl_2, 20 °C; vi, (**191**)–(**194**), $ZrCl_4$, CH_2Cl_2, 20 °C

Scheme 42

1.3.4 REFERENCES

1. (a) L. Pauling, 'The Nature of the Chemical Bond', 3rd edn., Cornell University Press, Ithaca, NY, New York, 1960; (b) M. Schlosser, 'Struktur and Reactivitat Polarer Organometalle', Springer, Berlin, 1973; (c) J. P. Oliver, in 'The Chemistry of the Metal–Carbon Bond', ed. F. R. Hartley and S. Patai, Wiley, New York, 1985, vol. 2, p. 789.
2. (a) T.-L. Ho, *Chem. Rev.*, 1975, **75**, 1; (b) N. T. Ahn, *Top. Curr. Chem.*, 1980, **88**, 145; (c) T.-L. Ho, *Tetrahedron*, 1985, **41**, 1; (d) J. M. Lefour and A. Loupy, *Tetrahedron*, 1978, **34**, 2597.
3. (a) M. Cossentini, B. Deschamps, N. T. Ahn and J. Seyden-Penne, *Tetrahedron*, 1977, **33**, 409; (b) G. Stork and L. A. Maldonado, *J. Am. Chem. Soc.*, 1974, **96**, 5272; (c) R. Sauvetre and J. Seyden-Penne, *Tetrahedron Lett.*, 1976, 3949; (d) H. O. House, *Acc. Chem. Res*, 1976, **9**, 59; (e) B. Deschamps and J. Seyden-Penne, *Tetrahedron*, 1977, **33**, 413.
4. J. C. Stowell, 'Carbanions in Organic Synthesis', Wiley-Interscience, New York, 1979, p. 32.
5. E. Negishi, 'Organometallics in Organic Synthesis', Wiley-Interscience, New York, 1980.
6. (a) G. Wittig, *Q. Rev., Chem. Soc.*, 1966, **20**, 191; (b) T. Tochtermann, *Angew. Chem., Int. Ed. Engl.*, 1966, **5**, 531; *Angew. Chem.*, 1966, **78**, 355.
7. (a) H. Gilman and R. Kirby, *J. Am. Chem. Soc.*, 1941, **63**, 2046; (b) G. Wittig and O. Bub, *Justus Liebigs Ann. Chem.*, 1949, **566**, 113.
8. (a) K. Maruoka and H. Yamamoto, *Tetrahedron*, 1988, **44**, 5001; (b) P. A. Chaloner, in 'The Chemistry of the Metal–Carbon Bond', ed. F. R. Hartley, Wiley, New York, 1987, vol. 4, p. 411; (c) K. Maruoka and H. Yamamoto, *Angew. Chem., Int. Ed. Engl.*, 1985, **24**, 668; *Angew. Chem.*, 1985, **97**, 670.
9. (a) L. Bagnell, E. A. Jeffery, A. Meisters and T. Mole, *Aust. J. Chem.*, 1975, **28**, 801; (b) E. C. Ashby and G. Heinsohn, *J. Org. Chem.*, 1974, **39**, 3297.
10. G. W. Kabalka and R. F. Daley, *J. Am. Chem. Soc.*, 1973, **95**, 4428.
11. (a) J. Fujiwara, Y. Fukutani, M. Hasegawa, K. Maruoka and H. Yamamoto, *J. Am. Chem. Soc.*, 1984, **106**, 5004; (b) Y. Fukutani, K. Maruoka and H. Yamamoto, *Tetrahedron Lett.*, 1984, **25**, 5911; (c) K. Maruoka, S. Nakai, M. Sakurai and H. Yamamoto, *Synthesis*, 1986, 130.
12. H. Kunz and K. J. Pees, *J. Chem. Soc., Perkin Trans. 1*, 1989, 1168.

13. (a) G. H. Posner, *Org. React. (N.Y.)*, 1972, **19**, 1; (b) G. H. Posner, 'An Introduction to Syntheses Using Organocopper Reagents', Wiley-Interscience, New York, 1980; (c) E. Erdik, *Tetrahedron*, 1984, **40**, 641; (d) S. H. Bertz, C. P. Gibson and G. Dabbagh, *Tetrahedron Lett.*, 1987, **28**, 4251.
14. (a) G. Zweifel and J. A. Miller, *Org. React. (N.Y.)*, 1984, **32**, 375; (b) J. Hooz and R. B. Layton, *Can. J. Chem.*, 1973, **51**, 2098.
15. K. F. Bernady, M. B. Floyd, J. F. Poletto and M. J. Weiss, *J. Org. Chem.*, 1979, **44**, 1438.
16. P. A. Bartlett and F. R. Green, III, *J. Am. Chem. Soc.*, 1978, **100**, 4858.
17. (a) A. Pecunioso and R. Menicagli, *J. Org. Chem.*, 1988, **53**, 45; (b) A. Pecunioso and R. Menicagli, *J. Org. Chem.*, 1988, **53**, 2614; (c) A. Pecunioso and R. Menicagli, *J. Org. Chem.*, 1989, **54**, 2391.
18. E. J. Corey and R. H. Wollenberg, *J. Am. Chem. Soc.*, 1974, **96**, 5581.
19. J. Schwartz, D. B. Carr, R. T. Hansen and F. M. Dayrit, *J. Org. Chem.*, 1980, **45**, 3053.
20. R. Pappo and P. W. Collins, *Tetrahedron Lett.*, 1972, 2627.
21. (a) K. Maruoka, K. Nonoshita and H. Yamamoto, *Tetrahedron Lett.*, 1987, **28**, 5723; (b) A. J. Stern, J. J. Rohde and J. S. Swenton, *J. Org. Chem.*, 1989, **54**, 4413.
22. (a) J. Weill-Raynal, *Synthesis*, 1976, 633; (b) D. S. Matteson, in 'The Chemistry of the Metal–Carbon Bond', ed. F. R. Hartley, Wiley, New York, 1987, vol. 4, p. 307.
23. (a) A. Suzuki, A. Arase, M. Matsumoto, M. Itoh, H. C. Brown, M. M. Rogic and M. W. Rathke, *J. Am. Chem. Soc.*, 1967, **89**, 5708; (b) H. C. Brown, M. M. Rogic, M. W. Rathke and G. W. Kabalka, *J. Am. Chem. Soc.*, 1967, **89**, 5709; (c) H. C. Brown, M. M. Rogic, M. W. Rathke and G. W. Kabalka, *J. Am. Chem. Soc.*, 1968, **90**, 4165, 4166; (d) W. Fenzl, R. Köster and H.-J. Zimmermann, *Justus Liebigs Ann. Chem.*, 1975, 2201; R. Köster, H.-J. Zimmermann and W. Fenzl, *Justus Liebigs Ann. Chem.*, 1976, 1116.
24. H. C. Brown and G. W. Kabalka, *J. Am. Chem. Soc.*, 1970, **92**, 712, 714.
25. O. P. Shitov, S. L. Ioffe, L. M. Leont'eva and V. A. Tarttakovskii, *Zh. Obshch. Khim.*, 1973, **43**, 1127.
26. N. Miyaura, M. Kashiwaga, M. Itoh and A. Suzuki, *Chem. Lett.*, 1974, 395.
27. G. W. Kabalka, *Intra-Sci. Chem. Rep.*, 1973, **7**, 57.
28. (a) M. F. Hawthorne and M. Reintjes, *J. Am. Chem. Soc.*, 1964, **86**, 951; (b) M. F. Hawthorne and M. Reintjes, *J. Am. Chem. Soc.*, 1965, **87**, 4585; (c) G. W. Kabalka, *J. Organomet. Chem.*, 1971, **33**, C25; (d) B. M. Mikhailov, G. S. Ter-Sarkisyan and N. A. Nikolaeva, *Zh. Obshch. Khim.*, 1971, **41**, 1721.
29. (a) G. W. Kabalka, H. C. Brown, A. Suzuki, S. Honma, A. Arase and M. Itoh, *J. Am. Chem. Soc.*, 1970, **92**, 710; (b) A. A. Akhrem, I. S. Levina, Y. A. Titov, V. A. Khripach, Y. N. Bubnov and B. M. Mikhailov, *Zh. Obshch. Khim.*, 1973, **43**, 2565.
30. J. Hooz and D. M. Gunn, *J. Am. Chem. Soc.*, 1969, **91**, 6195.
31. J. Hooz and J. N. Bridson, *J. Am. Chem. Soc.*, 1973, **95**, 602.
32. J. Hooz and J. N. Bridson, *Can. J. Chem.*, 1972, **50**, 2387.
33. J. Hooz and J. Oudenes, *Synth. Commun.*, 1980, **10**, 667.
34. (a) D. J. Pasto and P. Wojtkowski, *J. Org. Chem.*, 1971, **36**, 1790; (b) J. Hooz and J. Oudenes, *Synth. Commun.*, 1980, **10**, 139.
35. (a) D. E. Van Horn and S. Masamune, *Tetrahedron Lett.*, 1979, 2229; (b) Y. Yamamoto, H. Yatagai and K. Maruyama, *J. Chem. Soc., Chem. Commun.*, 1980, 1073; (c) D. A. Evans, J. V. Nelson, E. Vogel and T. R. Taber, *J. Am. Chem. Soc.*, 1981, **103**, 3099; (d) D. A. Evans, in 'Asymmetric Synthesis', ed. J. D. Morrison, Academic Press, New York, 1984, vol. 3, p. 1; (e) A. Pelter, *Chem. Soc. Rev.*, 1982, **11**, 191.
36. O. Attanasi, G. Baccolini, L. Caglioti and G. Rosini, *Gazz. Chim. Ital.*, 1973, **103**, 31.
37. (a) H. C. Brown and E. Negishi, *J. Am. Chem. Soc.*, 1971, **93**, 3777; (b) E. Negishi and H. C. Brown, *J. Am. Chem. Soc.*, 1973, **95**, 6757; (c) P. Jacobs, III, *J. Organomet. Chem.*, 1978, **101**, 156.
38. (a) P. Jacobs, III and H. C. Brown, *J. Am. Chem. Soc.*, 1976, **98**, 7832; (b) Y. Satoh, H. Serizawa, S. Hara and A. Suzuki, *J. Am. Chem. Soc.*, 1985, **107**, 5225.
39. H. C. Brown, N. G. Bhat and S. Rajagopalan, *Organometallics*, 1986, **5**, 816.
40. (a) J. A. Sinclair, G. A. Molander and H. C. Brown, *J. Am. Chem. Soc.*, 1977, **99**, 954; (b) G. A. Molander and H. C. Brown, *J. Org. Chem.*, 1977, **42**, 3106.
41. F. Kawamura, T. Tayano, Y. Satoh, S. Hara and A. Suzuki, *Chem. Lett.*, 1989, 1723.
42. S. Hara, S. Hyuga, M. Aoyama, M. Sato and A. Suzuki, *Tetrahedron Lett.*, 1990, **31**, 247.
43. A. Suzuki, *Acc. Chem. Res.*, 1982, **15**, 178.
44. N. Miyaura, M. Itoh and A. Suzuki, *Tetrahedron Lett.*, 1976, 255.
45. (a) A. Pelter, L. Hughes and J. M. Rao, *J. Chem. Soc., Perkin Trans. 1*, 1982, 719; (b) S. Hara, K. Kishimura and A. Suzuki, *Chem. Lett.*, 1980, 221.
46. M. S. Kharasch and P. O. Tawney, *J. Am. Chem. Soc.*, 1941, **63**, 2308.
47. (a) H. O. House, W. L. Respess and G. M. Whitesides, *J. Org. Chem.*, 1966, **31**, 3128; (b) H. Gilman, R. G. Jones and L. A. Woods, *J. Org. Chem.*, 1952, **17**, 1630.
48. (a) G. H. Posner, *Org. React. (N.Y.)*, 1972, **19**, 1; (b) G. H. Posner, 'An Introduction to Syntheses Using Organocopper Reagents', Wiley-Interscience, New York, 1980; (c) J. F. Normant, *Synthesis*, 1972, 63; (d) G. H. Posner, *Org. React. (N.Y.)*, 1975, **22**, 253; (e) J. P. Marino, *Ann. Rep. Med. Chem.*, 1975, **10**, 327; (f) H. O. House, *Acc. Chem. Res.*, 1976, **9**, 59.
49. (a) B. H. Lipshutz, *Synthesis*, 1987, 325; (b) B. H. Lipshutz, R. S. Wilhelm and J. A. Kozlowski, *Tetrahedron*, 1984, **40**, 5005.
50. (a) Y. Yamamoto and K. Maruyama, *J. Am. Chem. Soc.*, 1978, **100**, 3240; (b) Y. Yamamoto, S. Yamamoto, H. Yatagai, Y. Ishihara and K. Maruyama, *J. Org. Chem.*, 1982, **47**, 119; (c) Y. Yamamoto, *Angew. Chem., Int. Ed. Engl.*, 1986, **25**, 941; *Angew. Chem.*, 1986, **98**, 945; (d) for addition of a bifunctional organocopper reagent from (*E*)-5-chloro-3-lithio-2-pentene, see: E. Piers and A. V. Gavai, *J. Org. Chem.*, 1990, **55**, 2380; (e) M. Karpf and A. S. Dreiding, *Helv. Chim. Acta*, 1981, **64**, 1123.
51. (a) T. Ibuka, H. Minakata, Y. Mitsui and K. Kinoshita, *Tetrahedron Lett.*, 1980, **21**, 4073; (b) T. Ibuka and E. Tabushi, *J. Chem. Soc., Chem. Commun.*, 1982, 703.
52. (a) F. Barbot, A. Kadib-Elban and P. Miginiac, *Tetrahedron Lett.*, 1983, **24**, 5089; (b) C. Bretting, J. Munch-Petersen, P. M. Jorgensen and S. Refn, *Acta Chem. Scand.*, 1960, **14**, 151.

53. (a) Y. Yamamoto, S. Nishii and T. Ibuka, *J. Chem. Soc., Chem. Commun.*, 1987, 464; (b) Y. Yamamoto, S. Nishii and T. Ibuka, *J. Chem. Soc., Chem. Commun.*, 1987, 1572.
54. (a) W. Oppolzer, P. Dudfield, T. Stevenson and T. Godel, *Helv. Chim. Acta*, 1985, **68**, 212; (b) G. Helmchen and G. Wegner, *Tetrahedron Lett.*, 1985, **26**, 6051; (c) W. Oppolzer, R. Moretti, T. Godel, A. Meunier and H. J. Löher, *Tetrahedron Lett.*, 1983, **24**, 4971; (d) W. Oppolzer and H. J. Löher, *Helv. Chim. Acta*, 1981, **64**, 1981.
55. (a) M. Matsuzawa, Y. Horiguchi, E. Nakamura and I. Kuwajima, *Tetrahedron*, 1989, **45**, 349; (b) Y. Horiguchi, E. Nakamura and I. Kuwajima, *J. Org. Chem.*, 1986, **51**, 4323; (c) M. Asaoka, K. Shima and H. Takei, *Tetrahedron Lett.*, 1987, **28**, 5669; (d) M. Asaoka, K. Takenouchi and H. Takei, *Chem. Lett.*, 1988, 921; (e) M. Asaoka, K. Takenouchi and H. Takei, *Chem. Lett.*, 1988, 1225; (f) M. Asaoka, K. Shima, N. Fujii and H. Takei, *Tetrahedron*, 1988, **44**, 4757; (g) P. T. W. Chang and S. McLean, *Tetrahedron Lett.*, 1988, **29**, 3511; (h) J. Brendel and P. Weyerstahl, *Tetrahedron Lett.*, 1989, **30**, 2371.
56. (a) For chlorotrimethylsilane–*N,N,N′,N′*-tetramethylethylenediamine-accelerated organocopper additions, see: C. R. Johnson and T. J. Marren, *Tetrahedron Lett.*, 1987, **28**, 27; (b) for iodotrimethylsilane-accelerated organocopper additions, see: M. Bergdahl, E.-L. Lindstedt, M. Nilsson and T. Olsson, *Tetrahedron*, 1989, **45**, 535; M. Bergdahl, E.-L. Lindstedt, M. Nilsson and T. Olsson, *Tetrahedron*, 1988, **44**, 2055.
57. (a) Y. Horiguchi, M. Komatsu and I. Kuwajima, *Tetrahedron Lett.*, 1989, **30**, 7087; (b) E. J. Corey and N. W. Boaz, *Tetrahedron Lett.*, 1985, **26**, 6015, 6019.
58. (a) For additions to chiral β-oxazolidinyl-α,β-unsaturated esters, see: A. Bernardi, S. Cardani, T. Pilati, G. Poli, C. Scolastico and R. Villa, *J. Org. Chem.*, 1988, **53**, 1600; (b) A. Alexakis, J. Berlan and Y. Besace, *Tetrahedron Lett.*, 1986, **27**, 1047.
59. (a) P. Mangeney, A. Alexakis and J. F. Normant, *Tetrahedron Lett.*, 1986, **27**, 3143; (b) P. Mangeney, A. Alexakis and J. F. Normant, *Tetrahedron Lett.*, 1987, **28**, 2363.
60. B. H. Lipshutz, *Synthesis*, 1987, 325.
61. (a) R. J. Linderman, A. Godfrey and K. Horne, *Tetrahedron*, 1989, **45**, 495; (b) R. J. Linderman and J. R. McKenzie, *Tetrahedron Lett.*, 1988, **29**, 3911; (c) R. J. Linderman and J. R. McKenzie, *J. Organomet. Chem.*, 1989, **361**, 31; (d) R. J. Linderman and A. Godfrey, *J. Am. Chem. Soc.*, 1988, **110**, 6249; (e) L. Van Hijfte and R. D. Little, *J. Org. Chem.*, 1985, **50**, 3940.
62. (a) B. Kautzner, P. C. Wailes and H. Weigold, *J. Chem. Soc., Chem. Commun.*, 1969, 1105; (b) T. Gibson, *Organometallics*, 1987, **6**, 918; (c) D. W. Hart, T. F. Blackburn and J. Schwartz, *J. Am. Chem. Soc.*, 1975, **97**, 679.
63. (a) M. J. Loots and J. Schwartz, *J. Am. Chem. Soc.*, 1977, **99**, 8045; (b) F. M. Dayrit, D. E. Gladkowski and J. Schwartz, *J. Am. Chem. Soc.*, 1980, **102**, 3976.
64. (a) J. Schwartz, M. J. Loots and H. Kosugi, *J. Am. Chem. Soc.*, 1980, **102**, 1333; (b) J. Schwartz and Y. Hayashi, *Tetrahedron Lett.*, 1980, **21**, 1497.
65. (a) E. W. Colvin, in 'The Chemistry of the Metal–Carbon Bond', ed. F. R. Hartley, Wiley, New York, 1987, vol. 4, p. 539; (b) E. W. Colvin, 'Silicon in Organic Synthesis', Butterworths, London, 1981; (c) W. P. Weber, 'Silicon Reagents for Organic Synthesis', Springer-Verlag, Berlin, 1983; (d) Z. N. Parnes and G. I. Bolestova, *Synthesis*, 1984, 993; (e) T. H. Chan and I. Fleming, *Synthesis*, 1979, 761.
66. V. G. Kumar Das and C.-K. Chu, in 'The Chemistry of the Metal–Carbon Bond', ed. F. R. Hartley and S. Patai, Wiley, New York, 1985, vol. 3, p. 1.
67. A. Hosomi, *Acc. Chem. Res.*, 1988, **21**, 200.
68. (a) A. Hosomi and H. Sakurai, *J. Am. Chem. Soc.*, 1977, **99**, 1673; (b) A. Hosomi, H. Iguchi, M. Endo and H. Sakurai, *Chem. Lett.*, 1979, 977; (c) C. H. Heathcock, S. Kiyooka and T. A. Blumenkopf, *J. Org. Chem.*, 1984, **49**, 4214; (d) T. Tokoroyama and L. Pan, *Tetrahedron Lett.*, 1989, **30**, 197; (e) Y. Yamamoto and S. Nishii, *J. Org. Chem.*, 1988, **53**, 3597; (f) for anomalous additions of allylsilanes to 1-acetylcycloalkenes, see: R. Pardo, J.-P. Zahra and M. Santelli, *Tetrahedron Lett.*, 1979, 4557; (g) for trityl-catalyzed allylation protocol which affords the enol silyl ether, see: M. Hayashi and T. Mukaiyama, *Chem. Lett.*, 1987, 289.
69. S. Knapp, U. O'Connor and D. Mobilio, *Tetrahedron Lett.*, 1980, **21**, 4557.
70. K. Maruyama, H. Uno and Y. Naruta, *Chem. Lett.*, 1983, 1767.
71. (a) M. Santelli, D. El Abed and A. Jellal, *J. Org. Chem.*, 1986, **51**, 1199; (b) R. L. Danheiser and D. M. Fink, *Tetrahedron Lett.*, 1985, **26**, 2509.
72. M. Ochiai, M. Arimoto and E. Fujita, *Tetrahedron Lett.*, 1981, **22**, 1115.
73. (a) D. Schinzer, *Synthesis*, 1988, 263; (b) G. Majetich, J. Defauw and C. Ringold, *J. Org. Chem.*, 1988, **53**, 50; (c) G. Majetich and J. Defauw, *Tetrahedron*, 1988, **44**, 3833; (d) G. Majetich and K. Hull, *Tetrahedron Lett.*, 1988, **29**, 2773; (e) for discussion of 1,6-additions to androstanedienones, see: K. Nickisch and H. Laurent, *Tetrahedron Lett.*, 1988, **29**, 1533.
74. (a) D. Schinzer, J. Steffen and S. Solyom, *J. Chem. Soc., Chem. Commun.*, 1986, 829; (b) D. Schinzer, S. Solyom and S. Becker, *Tetrahedron Lett.*, 1985, **26**, 1831.
75. J. Haruta, K. Nishi, S. Matsuda, Y. Tamura and Y. Kita, *J. Chem. Soc., Chem. Commun.*, 1989, 1065.
76. R. L. Danheiser, D. J. Carini and A. Basak, *J. Am. Chem. Soc.*, 1981, **103**, 1604.
77. T. L. Macdonald and S. Mahalingam, *J. Am. Chem. Soc.*, 1980, **102**, 2113.
78. (a) N. Narasaka, K. Soai, Y. Aikawa and T. Mukaiyama, *Bull. Chem. Soc. Jpn.*, 1976, **49**, 779; (b) T. Yanami, M. Miyashita and A. Yoshikoshi, *J. Org. Chem.*, 1980, **45**, 607; (c) M. E. Jung and Y.-G. Pan, *Tetrahedron Lett.*, 1980, **21**, 3127; (d) for review of enol silyl ethers in synthesis, see: P. Brownbridge, *Synthesis*, 1983, **1**, 85.
79. For discussion of diastereofacial preferences in Lewis acid mediated additions of enol silyl ethers, see: (a) C. H. Heathcock and D. E. Uehling, *J. Org. Chem.*, 1986, **51**, 279; (b) C. H. Heathcock, M. H. Norman and D. E. Uehling, *J. Am. Chem. Soc.*, 1985, **107**, 2797; (c) T. Mukaiyama, M. Tamura and S. Kobayashi, *Chem. Lett.*, 1986, 1017.
80. (a) S. J. Danishefsky, K. Vaughan, R. C. Gadwood and K. Tsuzuki, *J. Am. Chem. Soc.*, 1980, **102**, 4262; (b) A. G. Schultz and J. D. Godfrey, *J. Org. Chem.*, 1976, **41**, 3494.

81. (a) O. Takazawa, H. Tamura, K. Kogami and K. Hayashi, *Chem. Lett.*, 1980, 1257; (b) J. W. Huffman, S. M. Potnis and A. V. Satish, *J. Org. Chem.*, 1985, **50**, 4266; (c) M. Tanaka, H. Suemune and K. Sakai, *Tetrahedron Lett.*, 1988, **29**, 1733; (d) for boron trifluoride etherate catalyzed additions of silyl enol ethers to methyl vinyl ketone in alcohols, see: P. Duhamel, L. Hennequin, N. Poirier and J. M. Poirier, *Tetrahedron Lett.*, 1985, **26**, 6201.
82. (a) S. J. Danishefsky and J. E. Audia, *Tetrahedron Lett.*, 1988, **29**, 1371; (b) S. Kobayashi and T. Mukaiyama, *Chem. Lett.*, 1986, 1805; (c) for [1,2-benzenediolato(2-)-*O,O'*]oxotitanium-catalyzed additions of silyl ketene acetals to α,β-unsaturated ketones, see: T. Mukaiyama and R. Hara, *Chem. Lett.*, 1989, 1171; (d) for tris(dimethylamino)sulfonium difluorotrimethylsiliconate (TASF) catalyzed additions of silyl ketene acetals, see: T. V. RajanBabu, *J. Org. Chem.*, 1984, **49**, 2083; (e) for electrochemically mediated additions of silyl ketene acetals and silyl enol ethers, see: T. Inokuchi, Y. Kurokawa, M. Kusumoto, S. Tanigawa, S. Takagishi and S. Torii, *Bull. Chem. Soc. Jpn.*, 1989, **62**, 3739; (f) for additions of silyl ketene acetals of *N*-methylephedrine, see: C. Gennari, L. Colombo, G. Bertolini and G. Schimperna, *J. Org. Chem.*, 1987, **52**, 2754; (g) for additions of aminoketene dithioacetals to α,β-enones, see: P. C. B. Page, S. A. Harkin and A. P. Marchington, *Synth. Commun.*, 1989, **19**, 1655; (h) for additions of silyl ketene acetals to quinones, see: M. Aso, K. Hayakawa and K. Kanematsu, *J. Org. Chem.*, 1989, **54**, 5597.
83. (a) T. Mukaiyama, S. Kobayashi, M. Tamura and Y. Sagawa, *Chem. Lett.*, 1987, 491; (b) T. Mukaiyama, M. Tamura and S. Kobayashi, *Chem. Lett.*, 1987, 743; (c) T. Mukaiyama, M. Tamura and S. Kobayashi, *Chem. Lett.*, 1986, 1817.
84. (a) S. J. Danishefsky, M. P. Cabal and K. Chow, *J. Am. Chem. Soc.*, 1989, **111**, 3456; (b) S. J. Danishefsky and B. Simoneau, *J. Am. Chem. Soc.*, 1989, **111**, 2599.
85. (a) Y. Morita, M. Suzuki and R. Noyori, *J. Org. Chem.*, 1989, **54**, 1785; (b) M. Suzuki, H. Koyano, Y. Morita and R. Noyori, *Synlett*, 1989, 22; (c) M. R. Binns, R. K. Haynes, A. G. Katsifis, P. A. Schober and S. C. Vonwiller, *J. Am. Chem. Soc.*, 1988, **110**, 5411; (d) R. K. Haynes, A. G. Katsifis, S. C. Vonwiller and T. W. Hambley, *J. Am. Chem. Soc.*, 1988, **110**, 5423; (e) R. Davis and K. G. Untch, *J. Org. Chem.*, 1979, **44**, 3755; (f) B. M. Trost, J. M. Timko and J. L. Stanton, *J. Chem. Soc., Chem. Commun.*, 1978, 436; (g) T. Toru, S. Kurozumi, T. Tanaka, S. Miura, M. Kobayashi and S. Ishimoto, *Tetrahedron Lett.*, 1976, 4087; (h) T. Tanaka, S. Kurozumi, T. Toru, M. Kobayashi, S. Miura and S. Ishimoto, *Tetrahedron*, 1977, **33**, 1105.
86. (a) Y. Kita, J. Segawa, J. Haruta, H. Yasuda and Y. Tamura, *J. Chem. Soc., Perkin Trans. 1*, 1982, 1099; (b) R. A. Bunce, M. F. Schlecht, W. G. Dauben and C. H. Heathcock, *Tetrahedron Lett.*, 1983, **24**, 4943; (c) for a review of high pressure organic synthesis, see: K. Matsumoto, A. Sera and T. Uchida, *Synthesis*, 1985, 1.
87. (a) A. S. Demir, R. S. Gross, N. K. Dunlap, A. Bashir-Hashemi and D. S. Watt, *Tetrahedron Lett.*, 1986, **27**, 5567; (b) M. Voyle, N. K. Dunlap, D. S. Watt and O. P. Anderson, *J. Org. Chem.*, 1983, **48**, 3242.
88. (a) N. Minowa and T. Mukaiyama, *Chem. Lett.*, 1987, 1719; (b) M. Kawai, M. Onaka and Y. Izumi, *Bull. Chem. Soc. Jpn.*, 1988, **61**, 2157.
89. (a) M. Miyashita, T. Kumazawa and A. Yoshikoshi, *Chem. Lett.*, 1980, 1043; (b) M. Miyashita, T. Yanami and A. Yoshikoshi, *J. Am. Chem. Soc.*, 1976, **98**, 4679; (c) M. A. Brook and D. Seebach, *Can. J. Chem.*, 1987, **65**, 836; (d) A. F. Mateos and J. A. de la Fuente Blanco, *J. Org. Chem.*, 1990, **55**, 1349.
90. (a) S. Kobayashi, M. Tamura and T. Mukaiyama, *Chem. Lett.*, 1988, 91; (b) T. Mukaiyama, S. Kobayashi, M. Tamura and Y. Sagawa, *Chem. Lett.*, 1987, 491; (c) T. Mukaiyama, M. Tamura and S. Kobayashi, *Chem. Lett.*, 1987, 743.
91. (a) Y. Kita, O. Tamura, F. Itoh, H. Yasuda, T. Miki and Y. Tamura, *Chem. Pharm. Bull.*, 1987, **35**, 562; (b) R. Hunter, L. Carlton, P. F. Cirillo, J. P. Michael, C. D. Simon and D. S. Walter, *J. Chem. Soc., Perkin Trans. 1*, 1989, 1631; (c) Y. Hashimoto, H. Sugumi, T. Okauchi and T. Mukaiyama, *Chem. Lett.*, 1987, 1691; (d) Y. Hashimoto and T. Mukaiyama, *Chem. Lett.*, 1986, 755; (e) Y. Hashimoto, H. Sugumi, T. Okauchi and T. Mukaiyama, *Chem. Lett.*, 1987, 1695.
92. (a) W. Nagata and M. Yoshioka, *Org. React. (N.Y.)*, 1977, **25**, 255; (b) K. Utimoto, Y. Wakabayashi, T. Horiie, M. Inoue, Y. Shishiyama, M. Obayashi and H. Nozaki, *Tetrahedron*, 1983, **39**, 967; (c) S. N. Suryawanshi and P. L. Fuchs, *J. Org. Chem.*, 1986, **51**, 902; (d) G. Stork and D. H. Sherman, *J. Am. Chem. Soc.*, 1982, **104**, 3758; (e) J. N. Marx and G. Minaskanian, *Tetrahedron Lett.*, 1979, 4175.
93. Y. Ito, H. Kato, H. Imai and T. Saegusa, *J. Am. Chem. Soc.*, 1982, **104**, 6449.
94. (a) Y. Ito, H. Imai, T. Matsuura and T. Saegusa, *Tetrahedron Lett.*, 1984, **25**, 3091; (b) M. Murakami, T. Matsuura and Y. Ito, *Tetrahedron Lett.*, 1988, **29**, 355; (c) M. T. Reetz, H. Heimbach and K. Schwellnus, *Tetrahedron Lett.*, 1984, **25**, 511.
95. (a) Review of three-carbon homologating reagents, see: J. C. Stowell, *Chem. Rev.*, 1984, **84**, 409; (b) review of homoenolate anions and homoenolate equivalents, see: N. H. Werstiuk, *Tetrahedron*, 1983, **39**, 205.
96. E. Nakamura, S. Aoki, K. Sekiya, H. Oshino and I. Kuwajima, *J. Am. Chem. Soc.*, 1987, **109**, 8056.
97. Y. Tamura, H. Tanigawa, T. Yamamoto and Z. Yoshida, *Angew. Chem., Int. Ed. Engl.*, 1989, **28**, 351; *Angew. Chem.*, 1989, **101**, 358.
98. S. Kim and P. H. Lee, *Tetrahedron Lett.*, 1988, **29**, 5413.
99. (a) A. Suzuki, S. Nozawa, M. Itoh, H. C. Brown, G. W. Kabalka and G. W. Holland, *J. Am. Chem. Soc.*, 1970, **92**, 3503; (b) E. Negishi and T. Yoshida, *J. Am. Chem. Soc.*, 1973, **95**, 6837; (c) E. Negishi, G. Lew and T. Yoshida, *J. Org. Chem.*, 1974, **39**, 2321.
100. (a) T. R. Kelly and M. Ghoshal, *J. Am. Chem. Soc.*, 1985, **107**, 3879; (b) A. Quendo and G. Rousseau, *Tetrahedron Lett.*, 1988, **29**, 6443.

1.4
Organocuprates in the Conjugate Addition Reaction

JOSEPH A. KOZLOWSKI
Schering-Plough Research, Bloomfield, NJ, USA

1.4.1 INTRODUCTION

One of the distinguishing features of the reaction of organocuprates with α,β-unsaturated carbonyl compounds is the overwhelming preference for the 1,4 *versus* the 1,2 mode of addition.[1] Dramatic changes in chemoselectivity are often observed when a copper(I) salt is added to a solution of an organomagnesium or organolithium reagent and an enone. For example, when *trans*-3-penten-2-one is treated with MeLi, a selective 1,2-addition takes place (99% 1,2). Use of methylmagnesium bromide leads to a mixture of 1,4- and 1,2-addition products (30:70). However, if the same enone is treated with Me_2CuLi, prepared from 2MeLi and CuI, only the conjugate addition product is formed (99% 1,4).[2]

$$\text{RLi} + \text{CuX} \longrightarrow \text{RCu} + \text{LiX} \qquad (1)$$

Typically organocuprates are formed by the addition of either an organolithium or organomagnesium reagent to a copper(I) salt under an inert atmosphere, where the relative stoichiometry is important in determining the type of reagent formed.[1,3,4] Monoorganocopper species, RCu (equation 1), are ether-insoluble compounds which tend to be thermally unstable. They can be solubilized with additives such as tributylphosphine[1] and are used successfully with Lewis acids such as boron trifluoride etherate ($BF_3{\cdot}Et_2O$).[5]

$$2\,RLi \quad + \quad CuX \quad \longrightarrow \quad R_2CuLi \quad + \quad LiX \qquad (2)$$

Diorganocuprates are ether-soluble reagents which react with a wide range of substrates.[1,6–10] These reagents can be prepared directly from a copper(I) salt (equation 2), or from a monoorganocopper species by addition of 1 mol equiv. of an RLi species.

Triorganocuprates are ether-soluble reagents which are often referred to as 'higher order cuprates'. Generally prepared from copper(I) cyanide, these reagents undergo the same reactions as diorganocuprates but offer a different reactivity profile (equation 3).[3,4]

$$2RLi \quad + \quad CuCN \quad \longrightarrow \quad R_2Cu(CN)Li_2 \qquad (3)$$

The many uses of organocuprates have been the subject of numerous reviews,[1,3,4,6–10] especially the conjugate addition reaction which has received the most attention[1] and is the subject of this discussion.

1.4.2 MECHANISTIC CONSIDERATIONS

Diorganocuprates such as dimethylcopperlithium are routinely written as R_2CuLi (R = Me), but this is simply a stoichiometric representation of the reagent and is not indicative of its actual structure or aggregation state. Common structural probes[11] such as X-ray crystallography have been hampered by difficulties in growing and maintaining a crystal of the cuprate. Although several X-ray structures of cuprate clusters have been reported which indicate a dimeric form,[12,13] characterization of the monomeric complexes $[Cu(DPPE)_2]$ $[Cu(C_6H_2Me_3)_2]$[14] and $[Li(THF)_4]$ $[Cu(C(SiMe_3)_3)_2]$[15] have also been noted. Discrepancies concerning the nature of bonding between copper, lithium and the alkyl group further complicate the issue.[12,16,17] An unequivocal picture of the structure and aggregation state of organocuprates does not exist.

Two general mechanisms are considered for the conjugate addition of organocuprates: (i) an electron-transfer process[9] or (ii) direct nucleophilic addition.[18] Both mechanisms are seen as proceeding through an oxidative addition to give a Cu^{III} species,[19] followed by reductive elimination. There is, however, no direct evidence for the Cu^{III} intermediate, and the direct transfer of the alkyl group cannot be ruled out. Attempts to discover free radicals in the conjugate addition have failed,[20–22] and the two mechanistic schemes differ only in the sequence or mode by which the oxidative addition occurs (Scheme 1).

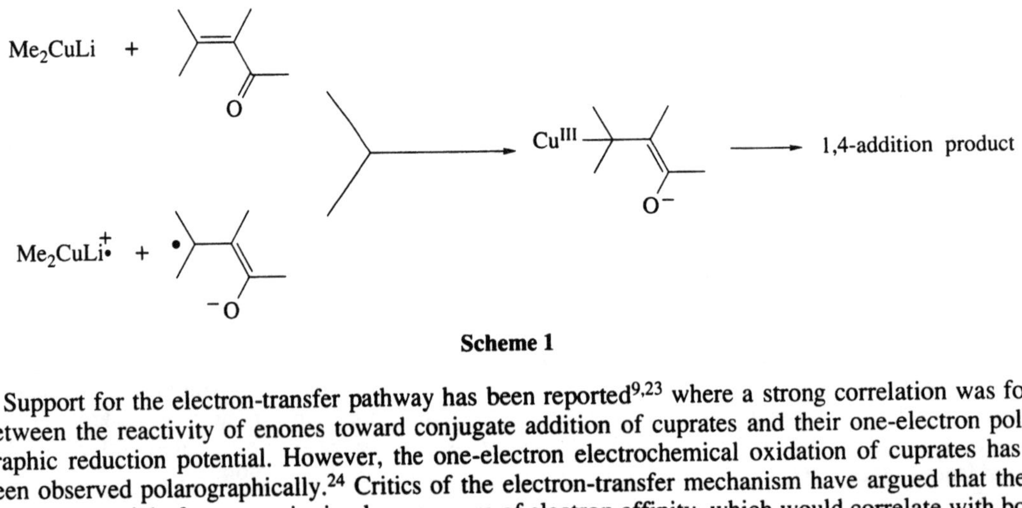

Me₂CuLi + [enone] / $Me_2CuLi^{+\bullet}$ + [radical enolate] ⟶ Cu^{III}–[enolate] ⟶ 1,4-addition product

Scheme 1

Support for the electron-transfer pathway has been reported[9,23] where a strong correlation was found between the reactivity of enones toward conjugate addition of cuprates and their one-electron polarographic reduction potential. However, the one-electron electrochemical oxidation of cuprates has not been observed polarographically.[24] Critics of the electron-transfer mechanism have argued that the reduction potential of an enone is simply a measure of electron affinity, which would correlate with both a one-electron (electron-transfer) and/or two-electron (nucleophilic addition) process.[18] In countering this argument House has pointed out that there is no obvious correlation between the reduction potentials of enones and reactivity in the Michael addition, and that β-substituents play a major role in carbanion conjugate addition.[9]

Further support for the electron-transfer argument was seen in the reaction of the γ-acetoxy enone (**1**) and Me_2CuLi.[25] This compound underwent elimination of acetoxy while reaction of the tetrahydropyranyl analog (**2**), with Me_2CuLi, resulted in the conjugate addition product (Scheme 2).[25] These results have been interpreted as arising from an electron-transfer step, which produces a radical anion intermediate that can undergo either a 1,2-elimination if the leaving group is good (*i.e.* acetoxy), or continue down the conjugate addition pathway if the leaving group is poor.

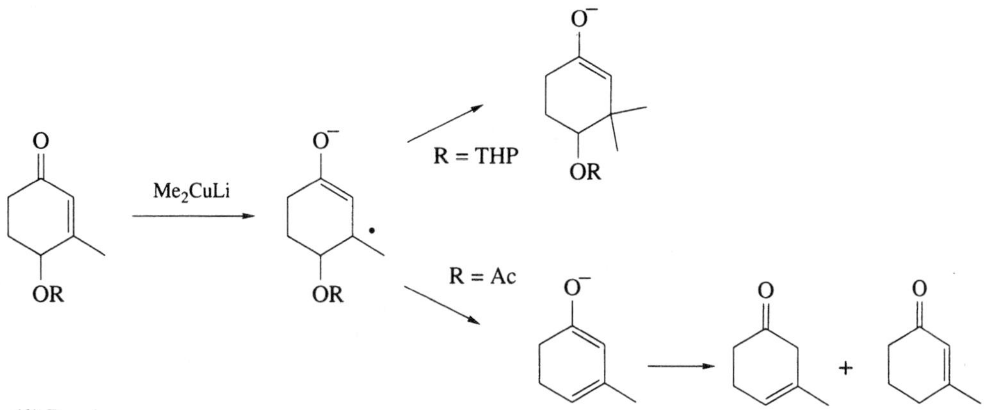

(**1**) R = Ac
(**2**) R = THP

Scheme 2

Two spectroscopic investigations support an enone–cuprate complex as an early intermediate in the conjugate addition mechanism. When cinnamate esters were treated with Me_2CuLi at temperatures where conjugate addition is slow (–70 °C), shifts in both the ^{13}C and ^{1}H signals of the π-system were observed.[26,27] An alkene–copper π-complex was proposed as an explanation. In a separate account, stopped flow spectroscopy was utilized to identify an intermediate complex in reactions of enones with Me_2CuLi.[28] The data were consistent with equilibrium formation of an intermediate alkene–copper complex which unimolecularly rearranges to a Cu^{III} intermediate (Scheme 3). Reductive elimination gives the β-alkylated ketone. Further evidence of an intermediate alkene–copper complex arose from the study of the reaction of the bicyclic enone (**3**) with Me_2CuLi. (Scheme 4).[29] It was reported that an ether solution of the two reagents gave a yellow precipitate, which was collected by centrifugation. The precipitate was then shown to be transformed into the β-substituted product by the addition of THF and TMS-Cl. This is the first example of an isolated complex, presumed to be a cuprate–enone complex, which was later converted to the conjugate addition product. An equilibrium between the *cis* and *trans* complexes, (**4**) and (**5**), was proposed to explain the observed isomeric mixture. In a THF solution, the *trans* complex (copper and oxygen of furan) was assumed to be favored, and was offered as an explanation for the formation of a single isomer.

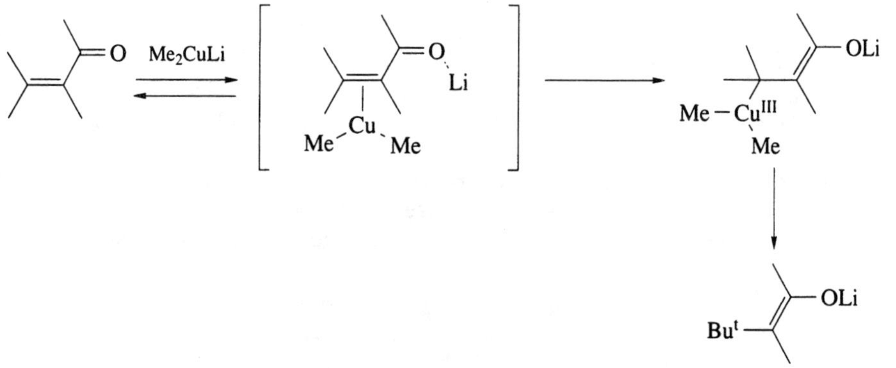

Scheme 3

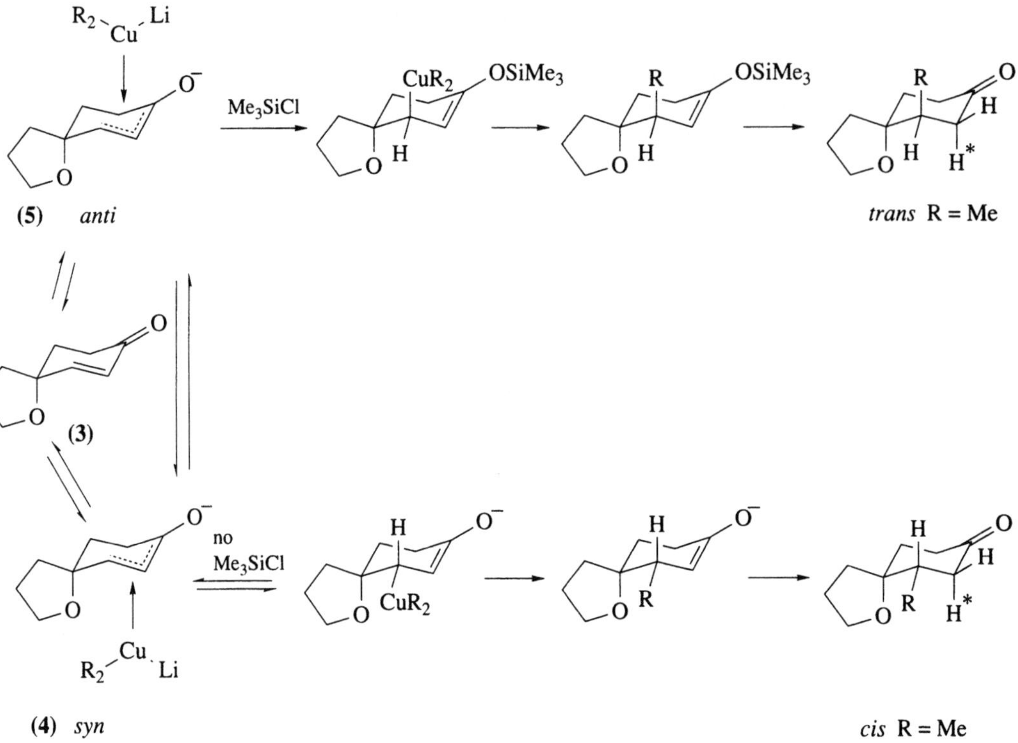

Scheme 4

The idea of a reversible alkene–copper complex had been used to explain the enantioselective alkylation of 2-cyclohexenone using a chiral auxilliary ligand.[30] Interaction of the cuprate with the *re* face of C-3 in 2-cyclohexenone was presumed to be favored over complexation with the *si* face of C-3 for steric reasons (equation 4).

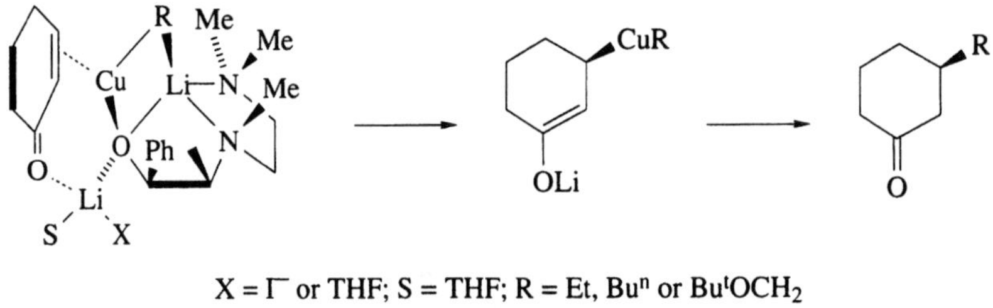

(4)

$X = I^-$ or THF; S = THF; R = Et, Bu^n or Bu^{t}OCH$_2$

1.4.3 STOICHIOMETRIC *VERSUS* CATALYTIC REAGENTS

An impressive feature of the copper-catalyzed reaction of organomagnesium and organolithium compounds is the wide range of conjugate additions achieved. Several reviews have documented these results,[1,31] and the reaction is considered to be catalytic when less than 10 mol % of a copper(I) salt is used. The solubility of the copper(I) salt influences the yield of conjugate addition, with more soluble copper(I) salts giving higher yields. For example, the yield in conjugate addition of isopropylmagnesium bromide to enoate (**6**; equation 5) was found to vary with the copper catalyst: CuCl (29–48%, least soluble); CuBr (46%); CuI (56%). Attempts to solubilize the copper(I) salt led to the use of copper(I) halides complexed with phosphines or sulfides (Me_2S), and good success has been achieved with this method.[1,31]

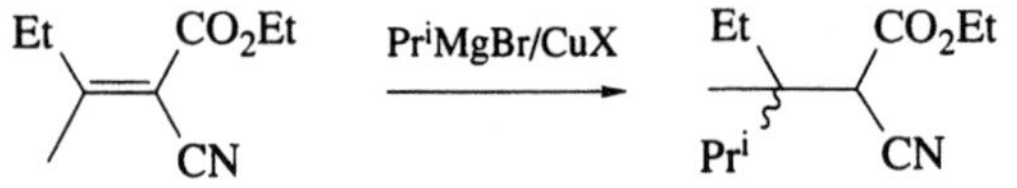

(5)

The order of addition of reagents has a significant impact on the yield of conjugate addition.[1] Commonly the substrate is added to a mixture of the copper(I) salt and the organo-magnesium or -lithium reagent, but inverse addition can often be beneficial (inverse refers to adding the organo-lithium or -magnesium reagent to a mixture of the substrate and copper catalyst). Enoate (**7**), for example,[1] reacts with Bu^nMgBr and CuCl to give (**8**) in 42% yield by the normal route and 89% yield by the inverse mode (equation 6).

CN; CO_2Et — $Bu^nMgBr/CuCl$ → CN; Bu^n; CO_2Et (6)

(**7**) (**8**)

In spite of the good success achieved in catalytic organocopper reactions, recent developments have emphasized the stoichiometric reagents. Presumably this is due to a greater reproducibility in results, with higher yields and greater stereoselectivity reported.[1,31]

1.4.4 VARIATIONS IN THE REAGENT

There is a wide tolerance for variations in the transferable R-group of organocuprates. Many of the early examples involved commercially available or easily synthesized (*i.e.*, methyl, phenyl, butyl) organolithium reagents. Organocuprates are usually formed by the addition of 2 mol equiv. of an RLi to a copper(I) salt, and a commercially available RLi simplifies the process. In spite of the limited variation of R for much of the early work, the methodology became accepted and established in the synthetic community. The deliveries of phenyl, methyl, or benzyl to enone (**9**; equation 7) were all shown to proceed in good yield.[1] Further scope of the methodology was demonstrated with cyclopropyl[32] and alkenyl[33] ligands (the alkenyl stereochemistry is retained).[34] These were also observed to be transferred in high yield as demonstrated by the reaction of 2-cyclohexenone with R_2CuLi (equation 8).

(**9**) + R_2CuLi → (7)

R = Ph, 84%; Me, 68%; CH_2Ph, 45%

+ R_2CuLi → (8)

R = cyclopropyl, 83%; R = 3,4-dihydro-2H-pyran-5-yl, 90%

Interesting and useful variations in R involve more complex alkyllithium reagents. For example, the treatment of β-iodoenone (**10**) with lithium phenylthio(2-vinylcyclopropyl)cuprate gave the conjugate addition product (**11**) as an intermediate *en route* to the bicyclic dienone (**12**; 82%; equation 9).[35] The ability of a cuprate to deliver a functionalized group to the enone was a definite advantage in this synthetic scheme.

Heteroatoms were also tolerated in the cuprate, as demonstrated by the reaction of methyl vinyl ketone with (**13**; equation 10).[36] A good yield of the conjugate addition product was reported (85%), without evidence of competing side reactions.

(9)

(10) **(11)** **(12)**

THF, $Pr_2{}^iS$
–78 to 20 °C, 14 h
85%

(10)

(13)

In a special example of the tolerance of R, Seyferth and Hui described a procedure for the direct nucleophilic β-acylation of α,β-unsaturated carbonyl compounds by the combination of CO with higher order organocuprates (formed from 2RLi and CuCN).[37] This methodology was successful with primary, secondary, and tertiary alkyllithium reagents, producing high yields of the corresponding β-acylated ketones (equation 11). Surprisingly, the methodology was not successful when primary alkyls and the traditional lower order reagents were examined.[38] Conjugate addition without acylation was found to occur. These cases necessitated use of the higher order cuprates to obtain the desired acylated product (equation 12).

$R_2Cu(CN)Li_2/CO$ (11)

R = Bu^n, 86%; R = Bu^s, 75%; R = Bu^t, 78%

$Bu^nCu(CN)Li/CO$ (12)

80%

Although a similar methodology was reported earlier,[39] in which an excess of the 1:1 $RLi/Ni(CO)_4$ reagent was used, the organocuprate method is advantageous in that it avoids use of the highly toxic $Ni(CO)_4$.

Triorganotin groups can also be transferred from the cuprate. β-Iodoenones are transformed into β-trimethylstannylenones by reaction of a cuprate prepared from Me_3SnLi and PhSCu (equation 13).[40]

Me_3SnLi + PhSCu (13)

The direct formation of functionalized organocuprates by oxidative addition of activated zerovalent copper, prepared from the lithium naphthalide reduction of $CuIPR_3$ complexes, into carbon–halogen bonds was found to be a mild method for the generation of cuprates.[41] Since formation of the cuprate by this method does not involve prior formation of an organolithium species, the range of R can be expanded to include functional groups which are sensitive to these highly basic reagents (equation 14). To date, organocuprates containing ester, nitrile, and chloride functionalities have been prepared directly from the corresponding alkyl bromide.[41]

+ activated Cu^0 + Br⌒⌒CO_2Et ⟶ (14)

Similarly compatible are organozinc compounds, RCu(CN)ZnI, which are formed under mild conditions.[42] The organozinc compounds were prepared from the corresponding primary and secondary iodides by treatment with zinc (activated by treatment with 4 mol % of 1,2-dibromoethane and 3 mol % TMS-Cl). Formation of the cuprate follows by addition of the soluble copper salt CuCN–2LiX (X = Cl, Br) to the zinc organometallic (equation 15). Functional groups such as ester and nitrile are stable to this reagent and the presence of TMS-Cl is necessary for the conjugate addition reaction to occur (equation 16).

$$\text{RI} \xrightarrow[\text{THF, 25–40 °C, 4–12 h}]{\text{activated Zn}} \text{RZnI} \xrightarrow[\text{0 °C, 10 min}]{\text{CuCN–2LiX}} \text{RCu(CN)ZnI} \quad (15)$$

$$+ \ \text{EtO}_2\text{C}\!\frown\!\frown\text{Cu(CN)ZnI} \xrightarrow[\text{r.t., overnight}]{-78\ °\text{C, 3 h, Me}_3\text{SiCl}} \quad (16)$$

A novel mode of formation of higher order cuprates has been reported *via* transmetallation between the cuprate and a vinylstannane.[43] Since vinyltin reagents are prepared directly from alkynes, a vinyllithium intermediate is bypassed. The mixed higher order cuprate is formed from the admixture of $Me_2Cu(CN)Li_2$ and a vinylstannane (equation 17). The exchange is quantitative, and is suggested to be driven by the preferential release of a vinyl group from tin.[43,44] The transfer of vinyl over methyl from the cuprate was highly selective, and the generality of the method was demonstrated with several substrates (equation 18).

$$\text{Me}_2\text{Cu(CN)Li}_2 + \text{R}'\text{CH=CHSnR}_3 \xrightarrow{\text{THF, r.t.}} \text{R}'\text{CH=CHCu(Me)(CN)Li}_2 + \text{MeSnR}_3 \quad (17)$$

$$+ \ \text{Bu}^n{}_3\text{Sn}\text{-cyclopentenyl} + \text{Me}_2\text{Cu(CN)Li}_2 \xrightarrow{92\%} \quad (18)$$

1.4.5 NONTRANSFERABLE LIGANDS

The preparation of traditional organocuprates such as R_2CuLi necessitates use of 2 mol equiv. of an RLi and 1 mol equiv. of a copper(I) salt. In the normal course of reaction only 1 equiv. of R is consumed in transfer, thereby leaving 1 equiv. unused in the form of RCu. Generally the second equivalent is lost in work-up as the product of reduction, RH. With respect to the R-group, the maximum efficiency for this process can only be 50%. If the RLi is commercially available (*e.g.* butyl, methyl, phenyl, vinyl), then this issue is usually not of concern. However, when one considers an organolithium which is not readily available, such as the chiral vinyllithium intermediate used by Corey *en route* to aplasmomycin,[45] the concept of ligand conservation becomes significant (equation 19). Rarely has an effort been made to recover the reduced ligand RH, but considerable attention has been given to the pursuit of a convenient nontransferable ligand R_R for the cuprate.

Bu_3Sn … $OSiPr^i_3$ ⟶ Li … $OSiPr^i_3$ —0.5 equiv. CuCN⟶ $Li_2(CN)Cu$ (… $OSiPr^i_3$)$_2$ (19)

Various R-groups differ in their ease of transfer from copper. For example, reaction of the mixed cuprate $Bu^n(Ph)CuLi$ with methyl vinyl ketone gave a 89:11 ratio of Bu^n to Ph transfer.[46] An improvement in the ratio of the transferable ligand, R_T, to the residual ligand, R_R, was realized by use of 1-pentynyllithium as R_R which led to a >97:3 ratio of *n*-butyl to 1-pentynyl transfer (equation 20).[46] An improved version of the alkyne scheme, which avoids the solubility problems associated with pentynylcopper, was reported[47] using 3-methyl-3-methoxy-1-butynyllithium as R_R. In spite of the many practical applications of alkynes as R_R ligands in organocuprates, they have been hampered by the lower reactivity which they impart to the reagent. As Whitesides has pointed out, 'The reactivity of organometallic groups in mixed cuprates is qualitatively intermediate between the reactivity of the constituent components'.[46]

$$\text{MVK} + Bu^n(Pr^n-C\equiv C)CuLi \longrightarrow Bu^n\text{-ketone} + Pr^n C\equiv C\text{-ketone} \quad (>97:3) \qquad (20)$$

The commercial availability of copper(I) cyanide has also provided an attractive alternative to the *in situ* preparation of copper alkynides, as cyanide functions very well as a nontransferable ligand (equation 21).[48] Unfortunately, the cyanocuprates (**14**) suffer from a diminished reactivity, similar to alkynylcuprates, and have not been widely used.[3] Other useful residual ligands are SPh,[49] Bu^tS,[49] and Bu^tO.[49]

$$RLi + CuCN \longrightarrow \underset{(\mathbf{14})}{RCu(CN)Li} \qquad (21)$$

The low reactivity associated with cyanocuprates was solved by introduction of the higher order reagents $R_2Cu(CN)Li_2$.[3,4] A dramatic increase in reactivity over the lower order organocuprates, RCu(CN)Li, was observed in applications of these reagents, but the higher order reagents are also subject to the same problem of selective ligand transfer. A satisfactory solution is the use of 2-lithiothiophene as the nontransferable ligand (equation 22).[50,51] A stoichiometric amount of reagent is generally required and good reactivity is observed with most substrates. For example, the transfer of phenyl, which is normally slow, is completely selective from the mixed thienylcuprate (**15**; equation 23). The high selectivity of ligand transfer[52] was attributed to stabilizing *d*–π* and *d*–*d* backbonding between copper and the thiophene ligand. That stabilizing interaction is also possible in $Me(Ph)Cu(CN)Li_2$, but phenyl transfer was observed.[53] The 2-furyl ligand couples smoothly with various substrates[54] which suggests that sulfur plays a key role in the selective retention of thiophene. This reagent combination has met with such success that (2-thienyl)Cu(CN)Li (**16**) is now commercially available.[55] To form the active reagent, all that is required is the addition of 1 mol equiv. of R_T to (**16**).

(22)

(**16**)

(23)

(**15**)

An unusual result was noted with this reagent. If steric demands are pushed to a severe level, then 1,2-addition of the thienyl group is observed.[50] The reasons are not clear, but a solution to the problem is the addition of $BF_3{\cdot}Et_2O$ to the reagent.[56]

Another class of sulfur-containing residual ligands utilized the dimethyl sulfoxide anion as the R_RLi (Table 1).[57] While the procedure is simple and general, 2 mol equiv. of the reagent and/or the addition of $BF_3{\cdot}Et_2O$ is necessary to obtain a good yield of β-alkylated product. Increased yields were obtained when the dimethyl sulfoxide anion was used in conjunction with the higher order cyanocuprate.[57] For example, the β-alkylation of enone (**17**) occurred in 95% yield with the higher order reagent and in 78% yield with the lower order cuprate. If only a single mole equivalent of reagent was used, then the disparity in yield was even greater; 22% for the lower order, and 67% for the higher order reagent (Table 1).

Table 1 Reaction of Isophorone (**17**) with Various Organometallic Reagents

Reagent	*Product yield (%)*
(**17**)	
$Li(MeSOCH_2)Cu(Bu^n)$	22 (1 equiv.) 78 (2.3 equiv.)
$Li(MeSOCH_2)Cu(CN)(Bu^n)$	67 (1 equiv.) 95 (2 equiv.)
$Bu^n(Ph_2P)CuLi$	75–88
$Bu^n(Cy_2N)CuLi$	29
$Bu^nCu/LiI/5$ equiv. $Bu^n{}_3P$	80–100

Cuprates based upon diphenylphosphidocopper(I) and dicyclohexylamidocopper(I) as R_RCu have also been introduced in the literature.[58] Addition of a mole equivalent of R_TLi is necessary to form the active reagents $R_T(Ph_2P)CuLi$ and $R_T(Cy_2N)CuLi$ respectively. These cuprates are reported to show enhanced thermal stability, and display a moderate reactivity with enones, but have not yet seen widespread use (Table 1). The handling and preparation of $^-PPh_2$ may detract from their appeal, and there is no clear advantage in reactivity over the reagents discussed above.

Phosphine-complexed cuprates, formed from copper(I) iodide, 1 mol equiv. of R_TLi, and 2–5 mol equiv. of tri-*n*-butylphosphine, show good reactivity in conjugate addition reactions with enones (Table 1).[59] The hazards surrounding the use of excess $Bu^n{}_3P$, and the availability of other methodologies lower the popularity of this method, but efficiency can be realized with this methodology.

1.4.6 SOLVENT EFFECTS

Good general success is observed with diethyl ether as solvent[1] for the reaction of Me_2CuLi, and other cuprates, with α,β-unsaturated carbonyl compounds. Many other solvents have been examined with variable success,[23,60] including tetrahydrofuran (THF), pyridine, acetonitrile, *N,N*-dimethylformamide (DMF), and dimethyl sulfoxide (DMSO). Additives such as hexamethylphosphoramide (HMPA) and dimethyl sulfide are also common.[23] Generally these solvents are analyzed with regard to their ability as donor ligands (*e.g.* HMPA is a good donor) and it is recognized that while good donor solvents enhance reaction rates of cuprates with alkyl halides, they generally retard or inhibit the reaction of cuprates with enones.[23,49,61] The effect can be dramatic as demonstrated by the reaction of isophorone with the higher order vinylcuprate (equation 24), and of the higher order phenyl reagent with enone (**18**; equation 25).[62]

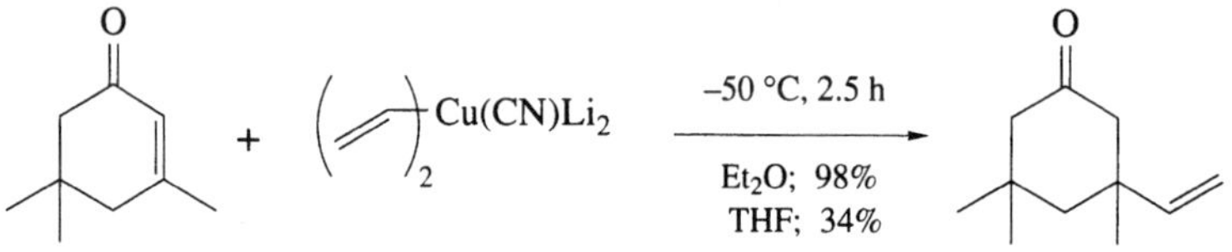

$(CH_2{=}CH)_2Cu(CN)Li_2$; −50 °C, 2.5 h; Et_2O; 98%; THF; 34% (24)

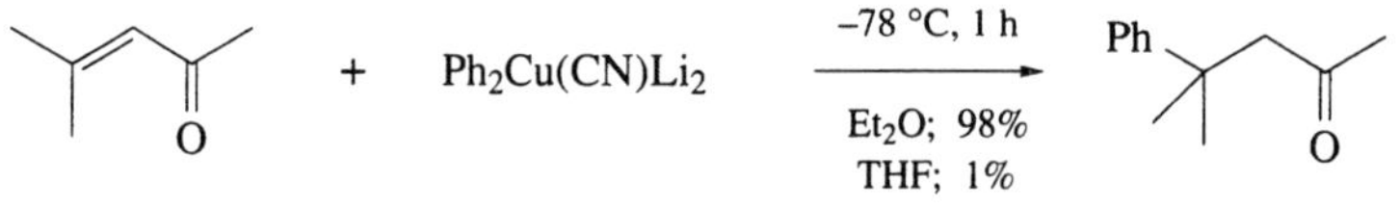

(18) + $Ph_2Cu(CN)Li_2$; −78 °C, 1 h; Et_2O; 98%; THF; 1% (25)

The inhibiting effects of THF (or other donor solvents) can often be used to advantage.[3] When a highly reactive enone such as 2-cyclopentenone is treated with $(vinyl)_2Cu(CN)Li_2$ in Et_2O, the yield is *ca.* 80%, while in THF the reaction is slower and more selective, giving the product in 90% yield. The chemoselectivity of organocuprates can be controlled by the choice of reaction medium (Scheme 5).[63] When bromoenone (**19**) was treated with Me_2CuLi in a Et_2O–Me_2S solution, the conjugate addition product (**20**) formed. Indeed this was not surprising as conjugate addition reactions of cuprates often proceed at a rate much faster than the displacement reaction of alkyl halides. However, when HMPA was included in the solvent mixture, the selectivity reversed and the methylated enone (**21**) was formed without a trace of the conjugate addition product.

Br ... Bu^t **(19)** — Me_2CuLi:
- Et_2O, Me_2S, 0–3 °C, 1.5 h; 83–92% → **(20)**
- Et_2O, Me_2S, HMPA, 25 °C, 7.5 h; 72–97% → **(21)**

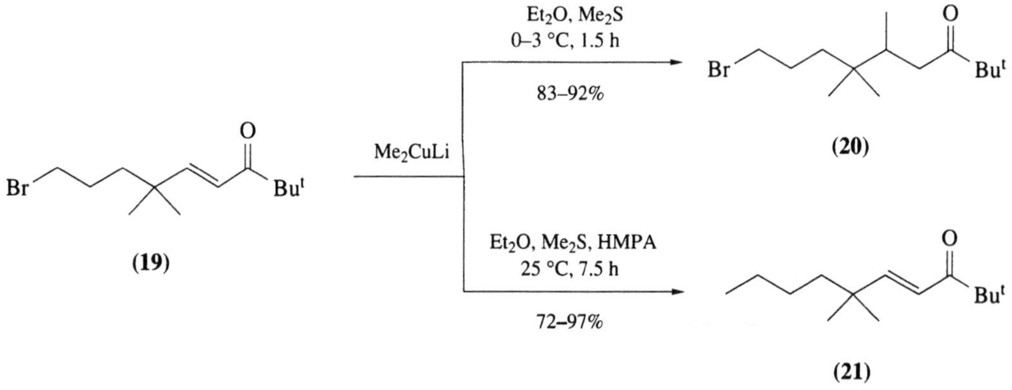

Scheme 5

Donor solvents might participate in at least two ways: complexation with Li^+, from LiI or LiBr; or complexation to the cuprate itself. Equal amounts of conjugate addition products were formed from the reaction of an enone with an Et_2O solution of Me_2CuLi + LiBr, or from an Et_2O solution of Me_2CuLi from which 95% of the LiBr had been removed. It was concluded that a good donor ligand would compete with the enone for coordination sites on the cuprate, and thus reduce the concentration of an enone–cuprate complex (a prerequisite for reaction in House's mechanism). A similar conclusion was reached in a separate study where fast conjugate addition reactions in hydrocarbon solvents (dichloromethane and Et_2O) were observed, while the rates were retarded in better coordinating solvents (THF, pyridine, acetonitrile).[60]

1.4.7 LEWIS ACID EFFECTS

Although organocuprates have seen great success as reagents for conjugate addition reactions, there are substrates encountered for which reagent reactivity is not sufficient. Poor results are often obtained in the reaction of cuprates with β,β-disubstituted acrylic esters, α,β-unsaturated nitriles, highly hindered enones, and α,β-unsaturated carboxylic acids. The reactivity can be increased by redesigning the reagent (*e.g.* higher order *versus* lower order cuprates), or by an additive such as a Lewis acid.[5] The additive must be chosen carefully to avoid direct reaction with the cuprate.

Good success has been reported with $BF_3{\cdot}Et_2O$. The combination of an organocopper species and $BF_3{\cdot}Et_2O$ gave a new complex,[64] $RCu{\cdot}BF_3$, which was far superior to RCu in its reactivity profile, and in many instances was also superior to R_2CuLi (equation 26).[5] There are many other examples of Lewis acid activation in the literature, with the details discussed in Volume 1, Chapter 1.12 and Volume 4, Chapter 1.3.

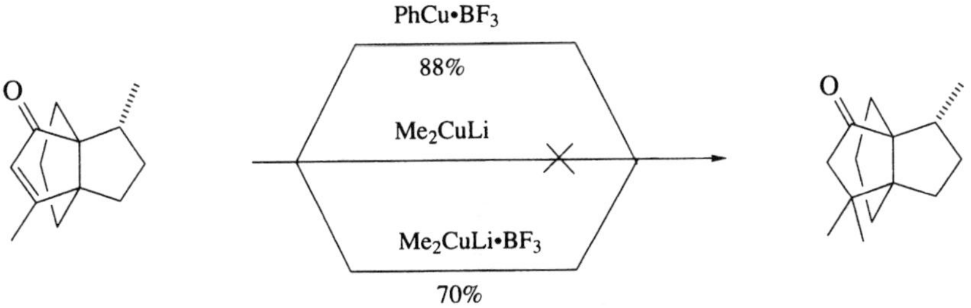

PhCu•BF₃ 88%; Me₂CuLi ✕; Me₂CuLi•BF₃ 70% (26)

1.4.8 THE α,β-UNSATURATED CARBONYL COMPOUND AS A SUBSTRATE

1.4.8.1 γ-Alkoxy-α,β-unsaturated Ketones

The reaction of organocuprates with enones is often described as being generally applicable and easily carried out. With reactive enones such as cyclopentenone, cyclohexenone, or similar acyclic enones, this is the case; it is difficult to find an organocuprate which will not add effectively to these substrates. However, there are many enone substrates for which little or no organocuprate conjugate addition occurs. For example, the reaction of γ-acetoxy-α,β-unsaturated ketones with cuprates results in reduction of acetoxy (*vide supra*) and not conjugate addition. Although this example provided some interesting mechanistic data, the class of substrates has potential synthetic application and a methodology to allow β-alkylation without reduction would be of value. Such a methodology was realized with the alkylcopper·$AlCl_3$ system (Scheme 6).[65] Use of 5–7 mol equiv. of an equimolar mixture of methylcopper and aluminum trichloride, in ether, was found to produce good yields of the conjugate addition product with no indication of reduction. The methodology was also found to be general, and was demonstrated with several substrates.

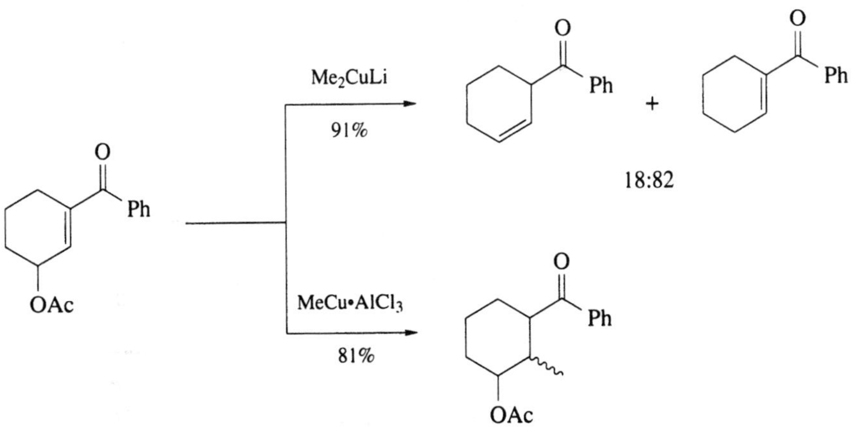

Me₂CuLi 91%; 18:82; MeCu•AlCl₃ 81%; OAc; Ph

Scheme 6

1.4.8.2 β-Cyclopropyl-α,β-unsaturated Ketones

β-Cyclopropyl-α,β-unsaturated ketones are another class of compounds for which simple conjugate addition is not the major pathway in reaction with conventional cuprates such as R_2CuLi. Generally a mixture of conjugate addition and cyclopropane ring-opened products are formed. When either geometrical or steric constraints cause the cyclopropyl group to be approximately perpendicular to the plane of the enone, the ring-opening reaction may dominate.[66] Again, use of an $RCu{\cdot}AlCl_3$ combination disfavors the process characteristic of electron transfer (ring opening) and the conjugate addition product appears in high yield (equation 27).[67] A marked increase in reactivity of $MeCu{\cdot}AlCl_3$ over Me_2CuLi was also noted. Me_2CuLi was found to be unreactive with the bicyclic enone (**22**) at –70 to –25 °C while the $MeCu{\cdot}AlCl_3$ complex readily underwent conjugate addition to give β-alkylated product in high yield (Scheme 7).

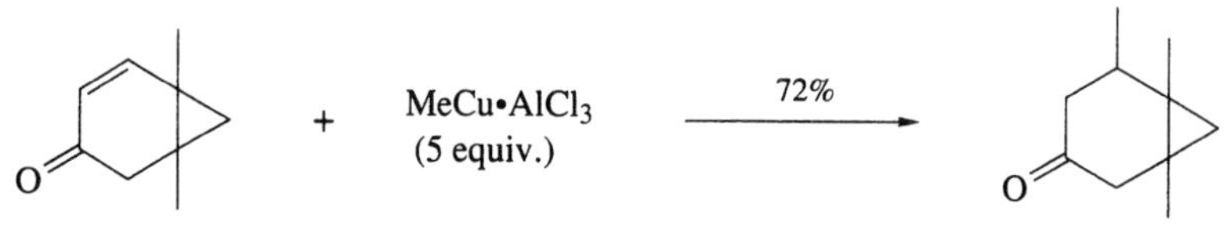

+ $MeCu{\bullet}AlCl_3$ (5 equiv.) → 72% (27)

Me_2CuLi, –70 to –25 °C (no reaction)

5 equiv. $MeCu{\bullet}AlCl_3$, –70 to –25 °C, 89%

Bu^t Bu^t O O

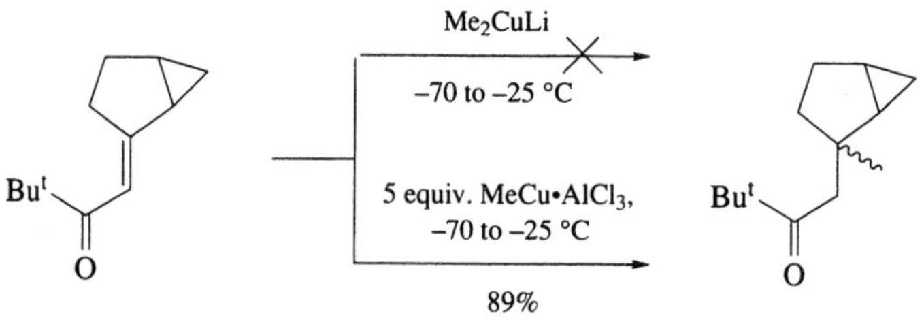

Scheme 7

1.4.8.3 β-Substituted-α,β-unsaturated Ketones

The simple substrates which are often chosen to demonstrate new methodologies often give an incomplete picture of the reactivity profile of the reagents. It is the more difficult cases which highlight the similarities and differences among cuprates. Enones which are either β,β-disubstituted or sterically congested may be troublesome in conjugate addition reactions. Treatment of isophorone (Scheme 8) with $(vinyl)_2CuLi{\cdot}Bu^n{}_3P$ (0 °C, 1 h) led to the conjugate addition product in 60% yield.[68] The higher order reagent $(vinyl)_2Cu(CN)Li_2$ gave the same product (–50 °C, 2.5 h) in 98% yield without the need for $Bu^n{}_3P$, and at a reaction temperature which was 50 °C lower.[62] Vinyldicyclohexylamido- and vinyldiphenylphosphido-cuprates were found to be similar to or less reactive than R_2CuLi, and both gave lower yields of the 1,4-addition product with isophorone compared to the higher order reagent (Scheme 8).[58a]

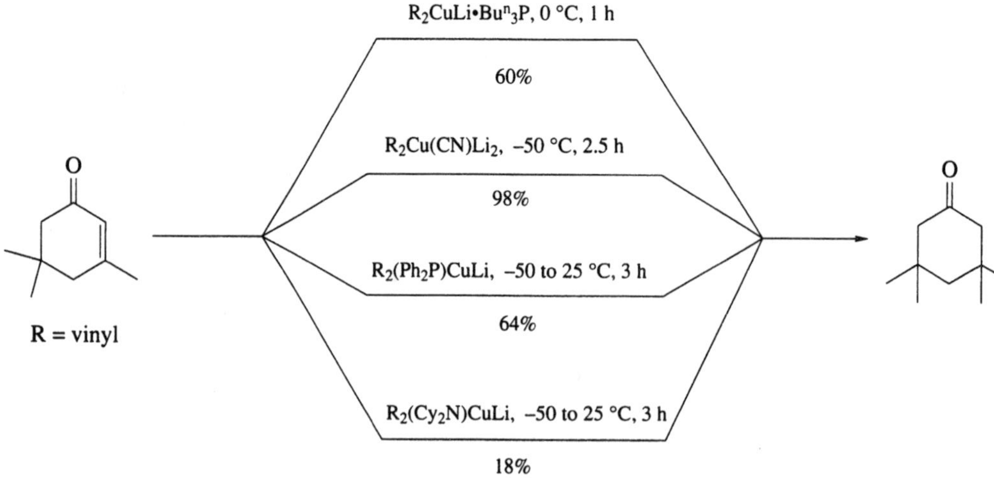

$R_2CuLi{\bullet}Bu^n{}_3P$, 0 °C, 1 h — 60%

$R_2Cu(CN)Li_2$, –50 °C, 2.5 h — 98%

$R_2(Ph_2P)CuLi$, –50 to 25 °C, 3 h — 64%

$R_2(Cy_2N)CuLi$, –50 to 25 °C, 3 h — 18%

O O

R = vinyl

Scheme 8

1.4.8.4 Polycyclic α,β-Unsaturated Carbonyl Compounds

Bicyclic or polycyclic enones may give slow or inefficient reaction with organocuprates. In particular, it is those enones which contain substituents arranged so as to create a steric hindrance on the carbon β to the carbonyl that see the greatest effect. The yields are often low to moderate, unless additional activation is provided. For example, when the bicyclic enone (**23**) was treated with R_2CuLi (R = 3-pentynyl), the β-substituted product was formed in only 46% yield,[69] while exposure of the same enone to $RCu{\cdot}BF_3$ (R = 3-pentynyl) gave the same β-alkylated product in 76% yield (Scheme 9).[70]

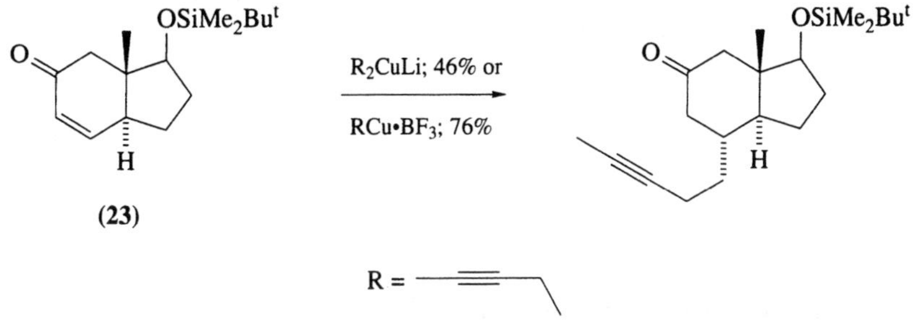

OSiMe₂Buᵗ → R_2CuLi; 46% or $RCu{\cdot}BF_3$; 76%

(**23**)

R =

Scheme 9

In many cases a Lewis acid has been added to a cuprate to enhance its reactivity with an enone, but there are also examples for which the Lewis acid–organocopper reagents do not work well, (Scheme 10). Reaction of the bicyclic enonate (**24**) with Me_2CuLi led to smooth conjugate addition,[71] but the use of either $Me_2CuLi{\cdot}BF_3$ or $MeCu{\cdot}BF_3$ resulted in formation of a dark resinous material. It is often difficult to predict when the reaction will go astray, but it should be recognized that a Lewis acid–cuprate complex is not always an effective solution to a reactivity problem.

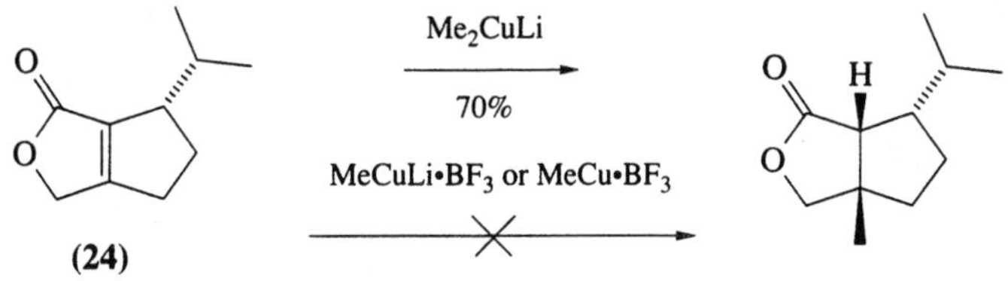

Me_2CuLi, 70%; $MeCuLi{\cdot}BF_3$ or $MeCu{\cdot}BF_3$ (no reaction)

(**24**)

Scheme 10

1.4.8.5 Polyalkenic α,β-Unsaturated Carbonyl Compounds

The conjugate addition of organocopper reagents to polyalkenic carbonyl compounds presents an opportunity in which either 1,4-, 1,6- and sometimes 1,8-addition can occur. Much of the early work in this area was conducted in the study of the reaction of a dienoate or dienone with an alkylcuprate or a catalytic copper/organomagnesium combination. The predominant mode of addition for the reaction of a dienoate and a copper-based organomagnesium reagent was found to be 1,6 not 1,4.[72] If the opportunity of 1,8-addition was present, then this mode prevailed.[73] However, a truly systematic study has not been carried out with the various reagents.

The reaction of a conjugated dienone and a catalytic copper/organomagnesium reagent produced a mixture of 1,2-, 1,4-, and 1,6-addition products.[1,74] In contrast, the stoichiometric cuprates showed a marked preference for 1,6- over 1,4-addition (equation 28).[1,75] However, both reagents are sensitive to steric demands, and the selectivity is often affected (Scheme 11).[1]

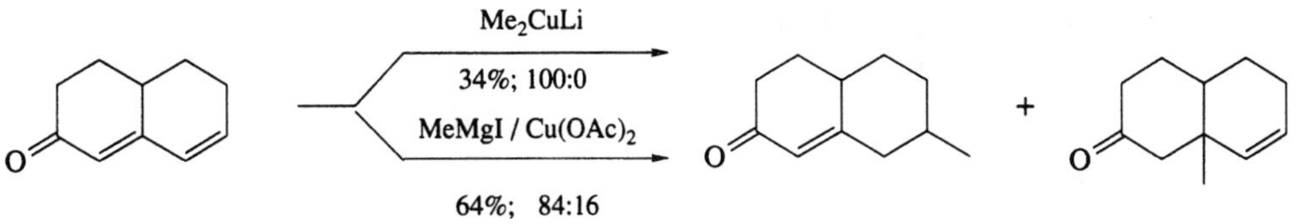

Me_2CuLi, 34%; 100:0 / $MeMgI$ / $Cu(OAc)_2$, 64%; 84:16 (28)

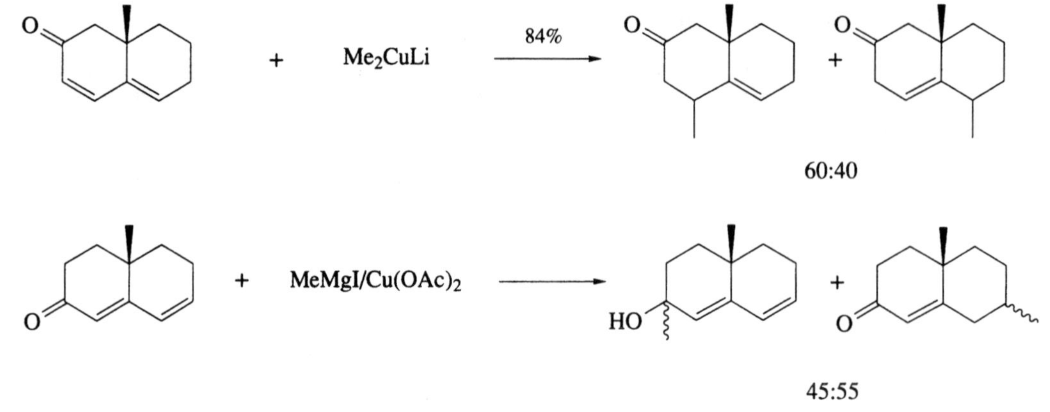

+ Me_2CuLi —84%→ +

60:40

+ MeMgI/Cu(OAc)$_2$ → HO +

45:55

Scheme 11

An interesting competition experiment was conducted on the steroid trienone (**25**), where the possibility for conjugate addition existed in three separate sites in the same molecule, C-1, C-5, and C-7 (equation 29). The catalytic reagent, MeMgI/CuI, gave only addition at C-1.[76]

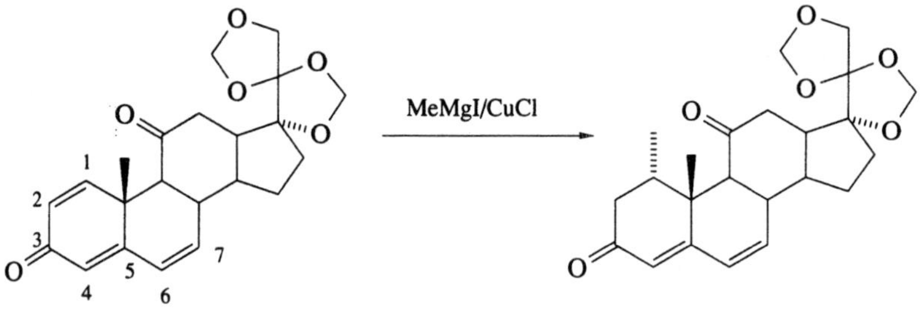

MeMgI/CuCl (29)

(**25**)

With Lewis acid–organocopper complexes, the question of chemoselectivity in polyalkenic carbonyl systems was further investigated. The reaction of BuCu·BF_3 with dienoate (**26**) led to selective 1,4-addition.[64a] This result was in contrast with the diorganocuprate, where 1,6-addition was reported (equation 30).[64c]

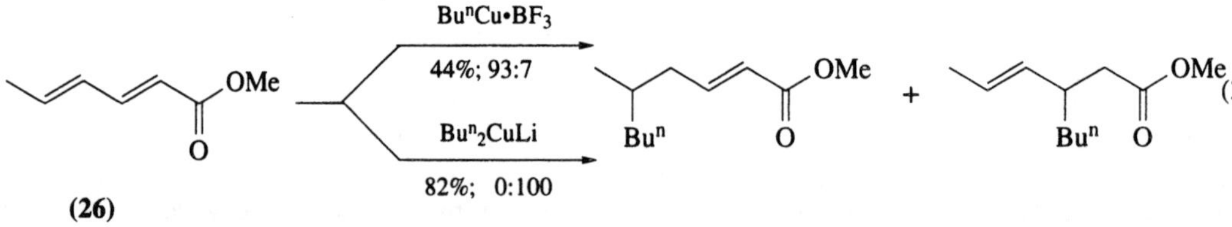

$Bu^nCu{\bullet}BF_3$ 44%; 93:7; $Bu^n{}_2CuLi$ 82%; 0:100 (30)

(**26**)

The corresponding methyl ketone showed a similar trend[77] but the selectivity was not as high. Use of $Bu^nCu{\cdot}BF_3$ gave 1,4- and 1,6-addition products in a ratio of 60:40, while the diorganocuprate selectivity remained high at 100% 1,6-addition. It was intriguing to see a change in selectivity with $Bu^nCu{\cdot}BF_3$ when the free acid was chosen as substrate (equation 31).[64a,c] The 14:86 ratio of 1,4 to 1,6 was a significant reversal from the 93:7 ratio seen with the ester.

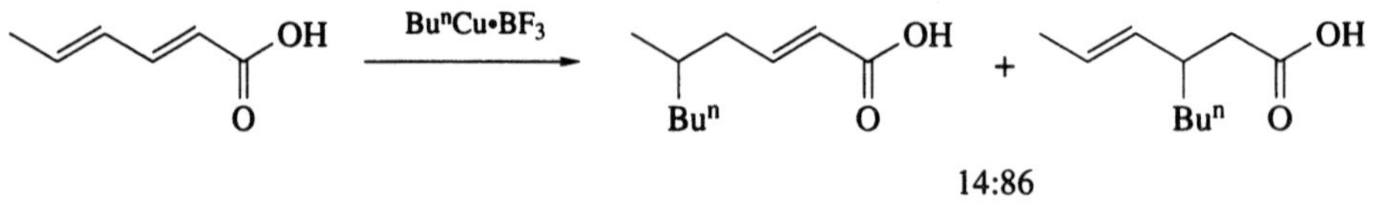

$Bu^nCu{\bullet}BF_3$ (31)

14:86

The conjugate addition of organomagnesium reagents to *N*,*N*-diethylsorbamide led to only 1,4-addition (equation 32).[78] Yields for the addition were good (35–60%) with alkylmagnesium compounds, and higher with allylic magnesium reagents (60–80%). With organocuprates, only the 1,6-addition was observed, in yields similar to those of the organomagnesium reagents.

Pr^iMgX; 55% — Bu^n_2CuLi; 40% — NEt_2, R, O (32)

1.4.8.6 α,β-Unsaturated Aldehydes

The first observations of reaction of lithiocuprates with α,β-unsaturated aldehydes indicated 1,2-addition was competing with 1,4-addition to a significant degree.[79] For example, lithium dimethylcuprate adds to enals to give a mixture of 1,2- and 1,4-addition products.[80] More recently, the addition of TMS-Cl to the reaction mixture was found to give the 1,4-addition product in reproducibly high yield (equation 33).[81] Good results were reported with acrolein and β-monosubstituted acroleins; however, lower ratios of 1,4- to 1,2-selectivity were observed with α,β,β-trisubstituted enals. For the more highly substituted enal, $Me_5Cu_3Li_2$ was used in place of Me_2CuLi to improve the conjugate addition selectivity to an acceptable 78:22 ratio (equation 34). Good results with other less-substituted substrates are also reported for this reagent.[80]

Bu^n_2CuLi — i, CHO, −100 °C; ii, Me_3SiCl; 91% → Bu^n … $OSiMe_3$ (33)

1,4:1,2 = 98:2

CHO — $Me_5Cu_3Li_2$, 86%: 1,4:1,2 78:22; Me_2CuLi, 86%: 36:64 (34)

Significant solvent effects have also been reported for the reaction of organocuprates with enals.[81] Less polar solvents are preferable, and a dramatic difference in the 1,4- to 1,2-selectivity was reported (equation 35).

Et, CHO — Me_2CuLi — ether/pentane, 75%: 1,4:1,2 95:5; THF, 55%: 40:60 (35)

Cuprates formed from organomagnesium reagents are acceptable reagents for the conjugate addition reaction with α,β-unsaturated aldehydes.[81] However, these cuprates are highly dependent upon the choice of copper(I) salt. With $CuBr{\cdot}Me_2S$, a dramatic increase in 1,4-:1,2-selectivity was reported over

CuBr (equation 36). This result must be taken with caution as the effects of using $CuBr{\cdot}Me_2S$ are not easily generalized.[82]

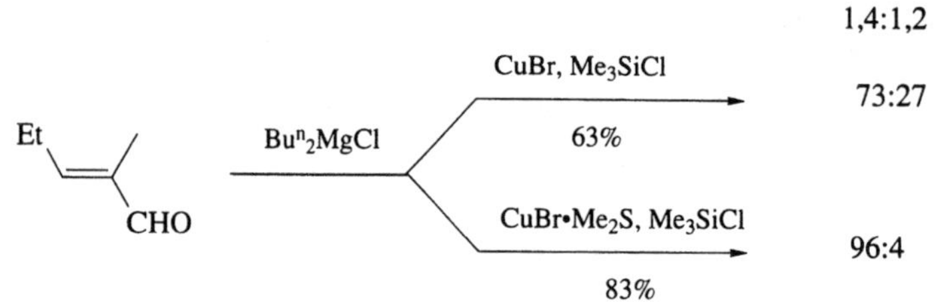

(36)

The direct nucleophilic acylation of enals by organocuprates is also successful.[37,38] With either the lower order RCuCNLi/CO or higher order $R_2CuCNLi_2$/CO reagents, high yields of the β-acylated product were observed (equation 37). A wide variation in R was also reported to be tolerated. Unfortunately, only β-monosubstituted enals were examined, so the extent to which substitution in the enal can be tolerated has yet to be determined.

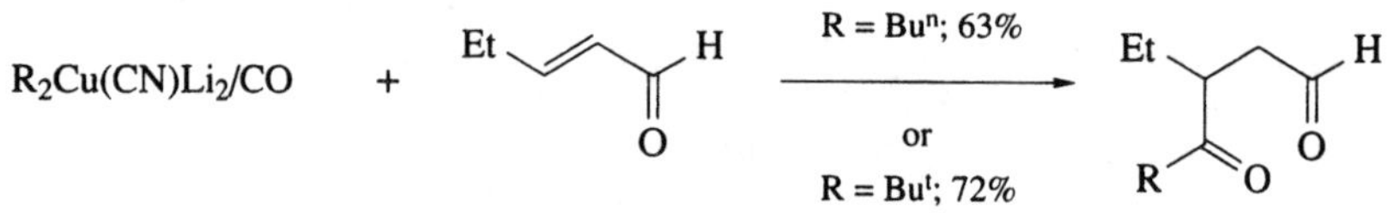

(37)

1.4.8.7 α,β-Unsaturated Esters

Relative to the reaction of cuprates with enones, the reaction of organocuprates with enoates has not been extensively studied.[64a,c,83–85] The lack of attention may be related to the unavailability of a reagent which could consistently produce high yields of the 1,4-addition product due to slow reactions and competition from 1,2-addition. Esters of hindered alcohols were prepared and bulky magnesiocuprates were employed, but without general success. An effective solution to this problem was realized with the introduction of $Bu^nCu{\cdot}BF_3$ as a reagent.[64a,b] Enoates with β-mono-, α,β-di-, and β,β-di-substituted structures underwent conjugate addition with this reagent while the corresponding diorganocuprates produced the dialkylation product (1,2- then 1,4-addition to the enone; equation 38). The success of $RCu{\cdot}BF_3$ in the conjugate addition reaction of enoates is well documented,[5] and for enoates such as (**27**) these appear to be the reagents of choice (equation 39).

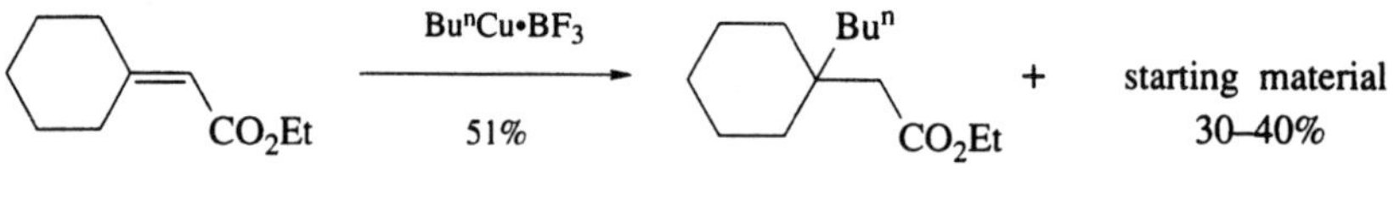

(38)

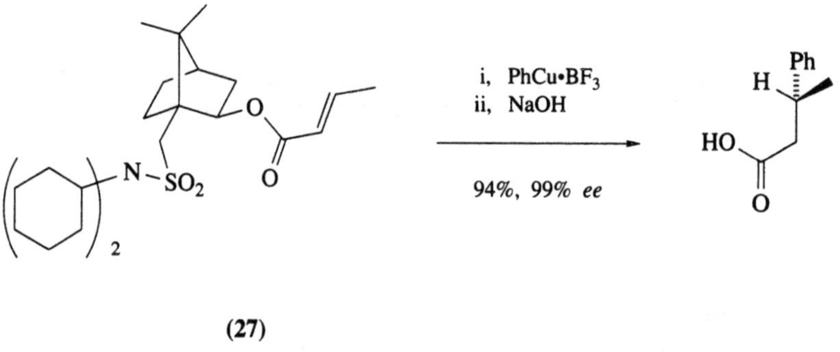

(39)

(**27**)

Higher order cuprates were also found to be effective in the conjugate addition reaction of α,β-unsaturated esters.[84] Primary alkyl, allyl, phenyl, vinyl, and trialkylsilyl reagents have all been used successfully with β-monosubstituted enoates. An increase in yield of the higher order reagent over the diorganocuprates order was demonstrated by the reaction of ethyl crotonate (equation 40). Under identical conditions, the 1,4-adduct was obtained in 75% yield with the higher order cuprate while the other diorganocuprate could only muster 38%.

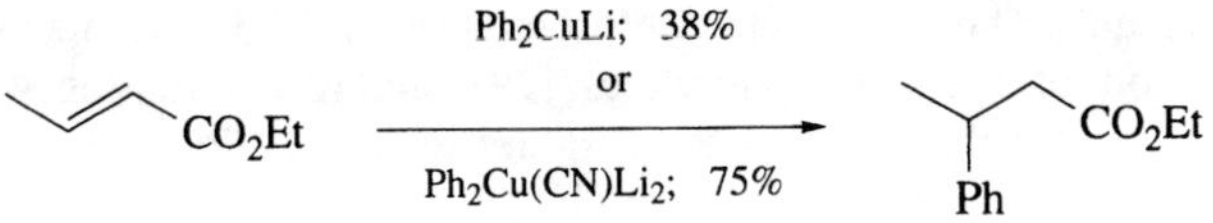

Ph_2CuLi; 38% or $Ph_2Cu(CN)Li_2$; 75% — CO_2Et → CO_2Et (Ph) (40)

It was found that with increasing substitution about the enoate, the efficiency of higher order reagents is lower due to competing side reactions. For example, treatment of ethyl tiglate with $Bu^n{}_2CuCNLi_2$ led to formation of a dialkylated ketone from 1,2- then 1,4-addition of the cuprate. Fortunately, we have already seen that good results can be obtained with enoates of this type if $RCu{\cdot}BF_3$ is employed (Scheme 12).

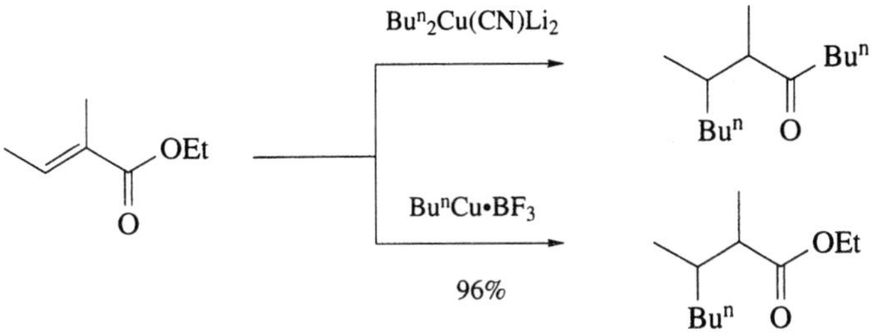

OEt, O — $Bu^n{}_2Cu(CN)Li_2$ → Bu^n, Bu^n, O; $Bu^nCu{\bullet}BF_3$, 96% → OEt, Bu^n, O

Scheme 12

Good chemoselectivity was demonstrated with the higher order reagent in the reaction of epoxy enoate (**28**) with $Bu^n{}_2CuCNLi_2$. Conjugate addition was reported[84] to occur in high yield without disturbing the epoxide (equation 41).

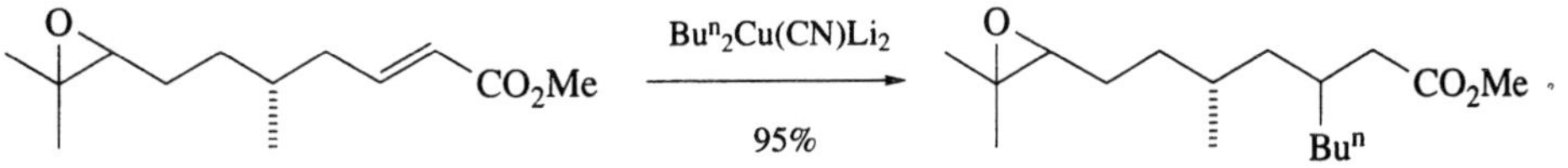

(**28**) CO_2Me —$Bu^n{}_2Cu(CN)Li_2$, 95%→ CO_2Me, Bu^n (41)

Cuprates derived from copper thiophenoxide and 3 mol equiv. of organomagnesium reagent were also shown to undergo conjugate addition with β-monosubstituted enoates.[86] The large molar excess of RMgX is necessary and ether is the preferred solvent for the reaction (equation 42). The results with disubstituted enoates have not been reported, and should provide a further test of the utility of this reagent.

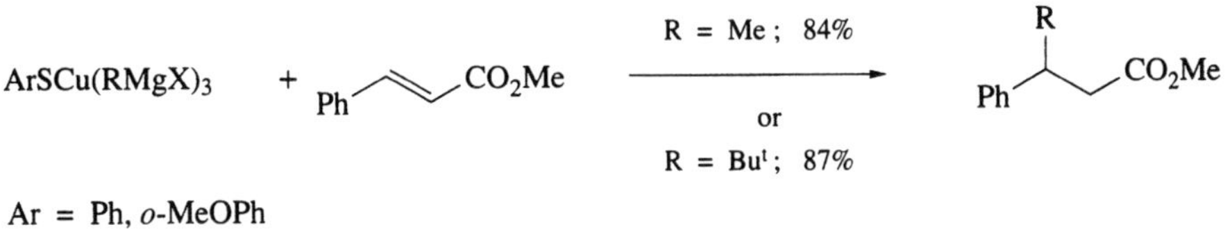

$ArSCu(RMgX)_3$ + Ph⁄⁄CO_2Me —R = Me ; 84% or R = Bu^t; 87%→ Ph(R)CO_2Me (42)

Ar = Ph, *o*-MeOPh

1.4.8.8 α,β-Alkynic Carbonyl Compounds

The conjugate addition of organocopper reagents to α,β-alkynic esters, ketones, aldehydes and acids is a useful method for the preparation of various tri- and tetra-substituted alkenes,[87] although addition to aldehydes is less common.[88] Use of a vinylcuprate results in the formation of a conjugated dienone, which further highlights the importance of this methodology (equation 43).[87]

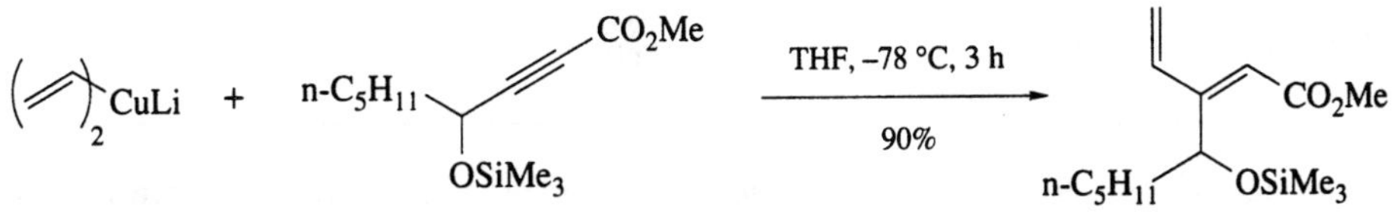

$(CH_2{=}CH)_2CuLi$ + n-C_5H_{11}(OSiMe$_3$)C≡CCO_2Me —THF, –78 °C, 3 h, 90%→ CO_2Me, n-C_5H_{11}, $OSiMe_3$ (43)

Early work in this area[89,90] established that the conjugate addition of cuprates to alkynic esters, in THF, occurred stereoselectively at –78 °C *via cis* addition. If the reaction is carried out at 0 °C, a mixture of (*E*)- and (*Z*)-isomers results. The change in stereoselectivity at higher temperatures was attributed to configurational equilibration of the copper enolates (**29**) and (**30**; equation 44). Quenching experiments[90,91] confirmed that the equilibration was temperature sensitive (Scheme 13) and also demonstrated that prolonged stirring at room temperature, followed by protonolysis, led to a predominance of the *trans* isomer (**31**). The *trans* configuration is defined according to the relative orientation of the carbonyl and the phenyl group in (**31**). In diethyl ether as solvent,[91] the addition was again *cis*, but the enolates were reported to equilibrate more rapidly (even at –78 °C).

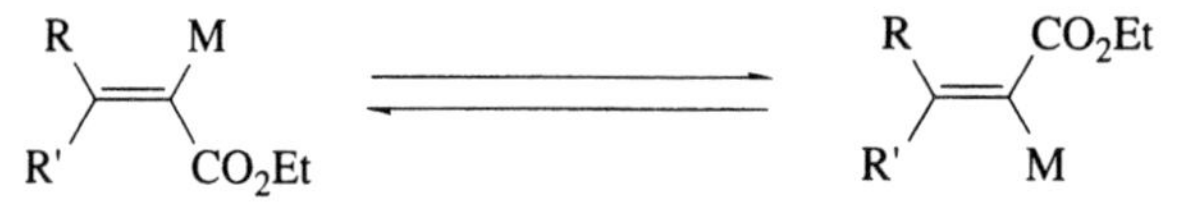

R, M / R', CO_2Et ⇌ R, CO_2Et / R', M (44)

(**29**) (**30**)

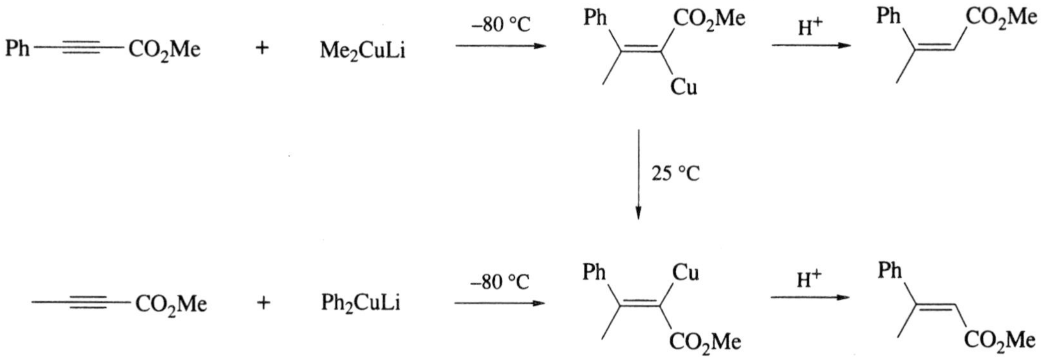

Ph—≡—CO_2Me + Me_2CuLi —(–80 °C)→ Ph, CO_2Me / Cu —(H^+)→ Ph, CO_2Me

↓ 25 °C

—≡—CO_2Me + Ph_2CuLi —(–80 °C)→ Ph, Cu / CO_2Me —(H^+)→ Ph / CO_2Me (**31**)

Scheme 13

Higher order cuprates have also been used in the addition to alkynic esters with good success. For example, ethyl tetrolate was found to react with $(Me_3Si)_2CuCNLi_2$ selectively to give the (*E*)-alcohol in 90% yield after reduction (equation 45).[92,93] Other organocopper reagents such as $Me_3SiCu·Me_2S·LiI$ and the cyanocuprate, $Me_3SiCuCNLi$, gave mixtures of (*E*)- and (*Z*)-isomers.[92]

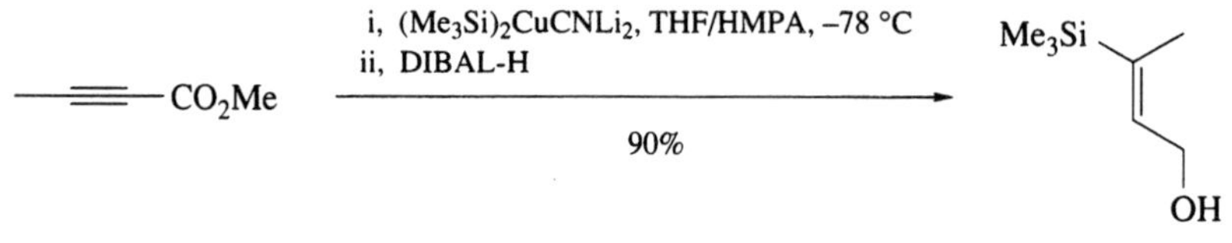

—≡—CO_2Me —(i, $(Me_3Si)_2CuCNLi_2$, THF/HMPA, –78 °C; ii, DIBAL-H; 90%)→ Me_3Si … OH (45)

Table 2 Reaction of (**32**) with Organocopper Reagents

Substrate	Reagent	Products			Yield (%)
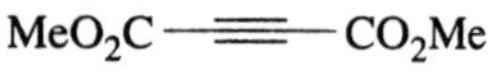 MeO_2C—≡—CO_2Me (**32**)		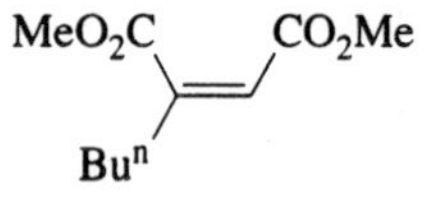 MeO_2C, CO_2Me / Bu^n		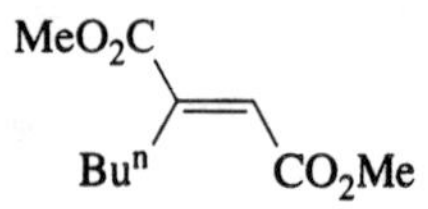 MeO_2C / Bu^n, CO_2Me	
	$Bu^nCu·BEt_3$	>99	:	<1	93
	$Bu^nCu·BF_3$	87	:	13	85
	Bu^n_2CuLi	60	:	40	98
	$Bu^nCu·9\text{-}BBN$	~100	:	0	92

Highly stereoselective *cis* additions of organocuprates to alkynic esters have also been reported with alkylcopper–Lewis acid complexes.[5] The choice of the proper Lewis acid was found to be important in determining the stereoselection as was demonstrated with enyne (**32**; Table 2).[5,88] While both the diorganocuprate and the alkylcopper·BF_3 complex gave mixtures of stereoisomers, the $Bu^nCu·BEt_3$ provided much higher stereoselectivity and a good chemical yield as well. The use of $Bu^nCu·9\text{-}BBN$ gave selectivity results similar to $Bu^nCu·BEt_3$ (100% *cis*).[5]

1.4.9 STEREOSELECTIVITY

The 1,4-addition of a cuprate or organomagnesium reagent to an α,β-unsaturated ketone differs from the traditional Michael addition in that it is an irreversible process.[94] As a consequence, the products should be governed by kinetic control,[94] which has stereochemical implications. For example, the reaction of methylmagnesium iodide/catalytic copper(I) chloride with 5-methyl-2-cyclohexenone produces a mixture of *trans:cis* 3,5-dimethylcyclohexenone in the ratio 95:5 (equation 46),[95] consistent with attack of the nucleophile from the less-hindered face of the enone.[96] Alternatively, reactions of this type can be analyzed in terms of the relative energies of the transition states.[95] As Allinger has pointed out, each of the two conformations of the 5-methyl-2-cyclohexenone has two possible reaction paths available to it, which leads to four transition states and two products.[95] Since the methyl group offers an obvious hindrance in only one of four pathways, the least-hindered face argument is an incomplete explanation. Of the two modes of attack available to each cyclohexenone, one leads to a chair and the other to a boat conformation, with the chair expected to be favored by 2–3 kcal mol^{-1} (1 cal = 4.18 J). Since pathway D is disfavored due to a 1,3-interaction of the methyl groups, the preferred route is pathway A, which leads to the *trans* product in agreement with the experimental result (Scheme 14).

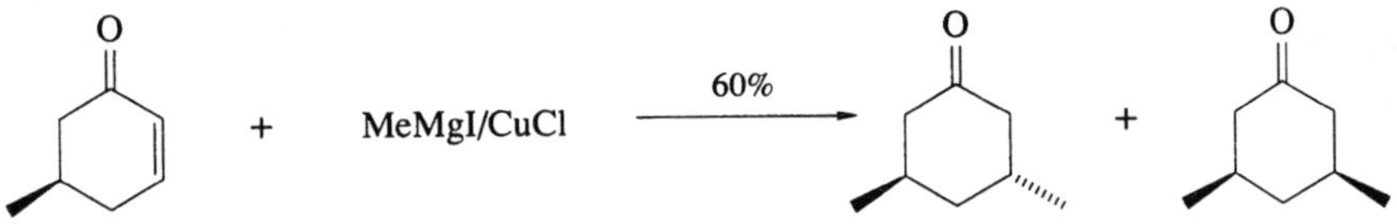

(46)

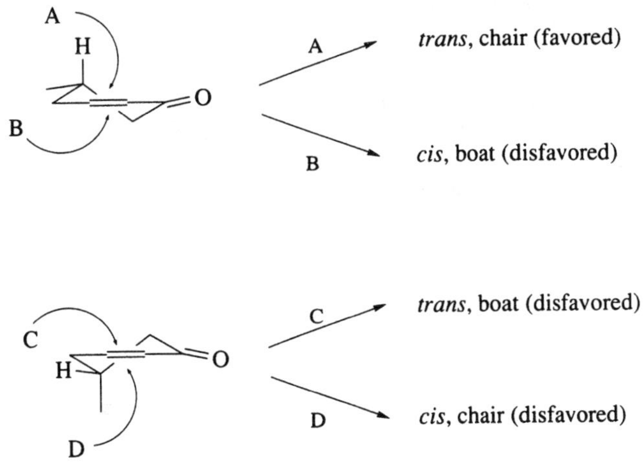

Scheme 14

In a similar experiment, enone (**33**) reacted with the diorganocuprate formed from copper(I) chloride and vinylmagnesium bromide to produce (**34**) and (**35**) as a 80:20 mixture.[97] The preference is consistent with a 1,2-interaction (*gauche*) of vinyl with ethoxycarbonyl, being of lower energy than a 1,3-interaction (diaxial) of vinyl with methyl (Scheme 15).

When a cuprate undergoes conjugate addition to an enone containing an α-substituent, a preference for the 2,3-*trans* product is generally observed.[1,98] If the α-substituent is 'large', as in (**36**), the preference is strong (>95:5; equation 47). If another electrophile is substituted for the proton,[99] the same stereoselectivity is observed. This important phenomenon has often been used to advantage in the synthesis of prostaglandins.[100]

In substrates where the carbonyl is exocyclic to the ring, a *cis* preference is reported for the 1,2-stereoselectivity.[5] The protonolysis has been reasoned to be the determining feature of the stereoselectivity. For example, when enone (**37**) was treated with dimethylcopperlithium, a *ca.* 70:30 *cis:trans* preference

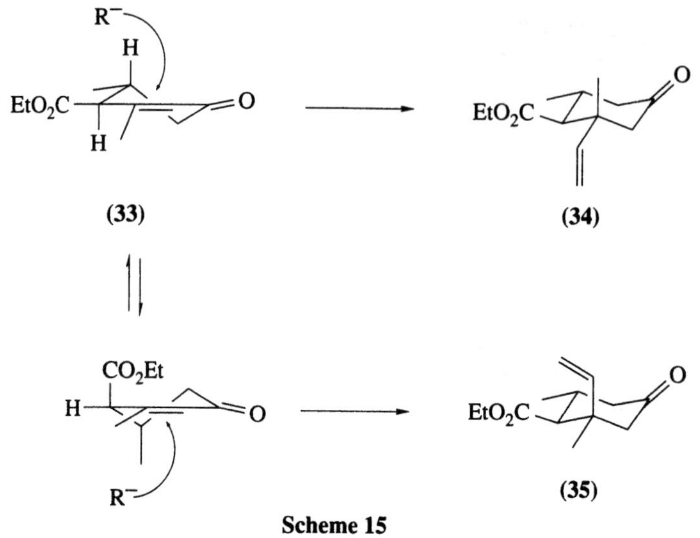

(33) (34)

(35)

Scheme 15

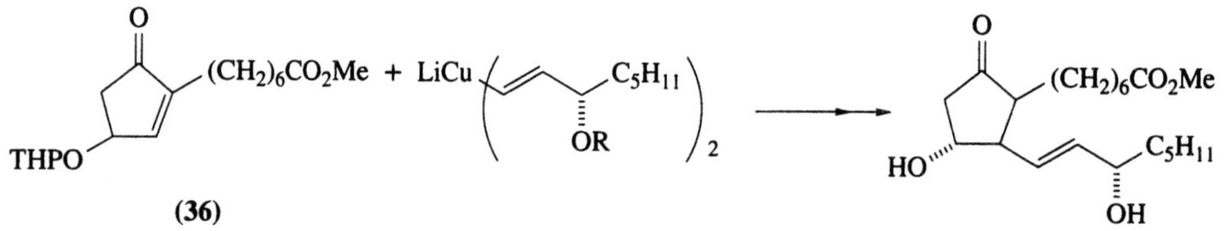

(36) (47)

was observed (equation 48). The proton was believed to arrive from axial direction on the preferred conformer (**40**; Scheme 16). Addition of $MeCu{\cdot}BF_3$ led to an 86:14 *cis:trans* ratio, not greatly different from the above example.

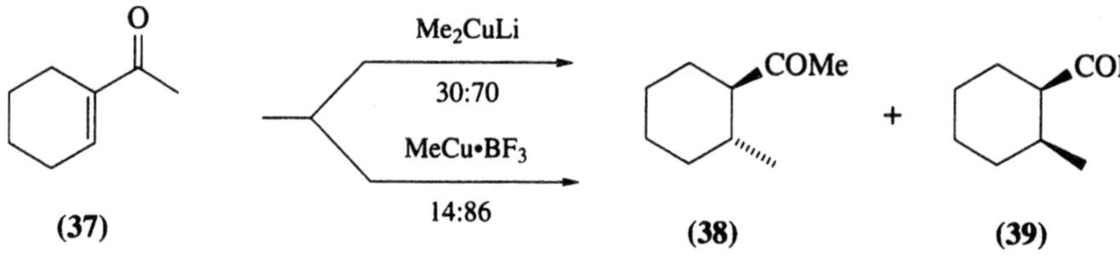

(37) (38) (39) (48)

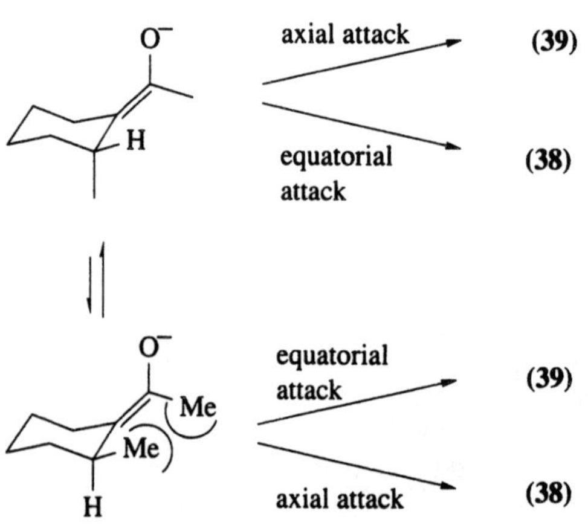

Scheme 16

A stereoselectivity dependence on the cuprate was also reported in the β-alkylation of γ-substituted acyclic enoates (Scheme 17).[101] The *trans* isomer (**41**) gave preferential formation of the *anti* isomer (**42**), regardless of the organocopper reagent employed. When the *cis* isomer (**43**) was treated with

$Bu^nCu{\cdot}BF_3$, compound (**42**) was formed with similar selectivity. However, if the *cis* compound (**43**) was treated with a BF_3-complexed diorganocuprate, then the *syn* isomer (**44**) was formed. The diastereoselectivities were explained in terms of a modified Felkin–Anh model, and electron transfer from the cuprate was used to account for the different stereochemical preferences of RCu and R_2CuLi with (**43**).[101]

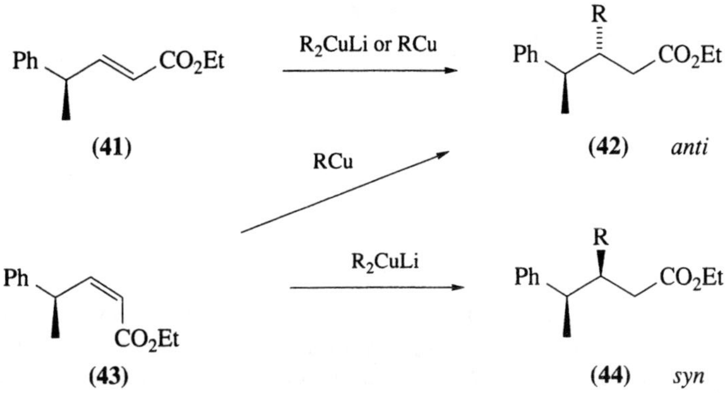

Reagent	*Substrate*	*Yield (%)*	*Anti:syn*
$Bu^nCuLi{\bullet}BF_3$	(**41**)	90	70:30
$Bu^nCuLi{\bullet}BF_3$	(**43**)	89	30:70
$Bu^nCu{\bullet}BF_3$	(**41**)	82	88:12
$Bu^nCu{\bullet}BF_3$	(**43**)	84	74:26

Scheme 17

With acyclic substrates such as (**45**), simple steric arguments are a suitable rationale for the observed stereoselectivity.[85a] For example, in the reaction of $PhCu{\cdot}BF_3$ with enoate (**45**), acid (**46**) was obtained, after hydrolysis, in 99.5% *ee*. This preference is consistent with an attack of the reagent from the least sterically encumbered side of the carbon–carbon double bond (shown by the arrow; equation 49).

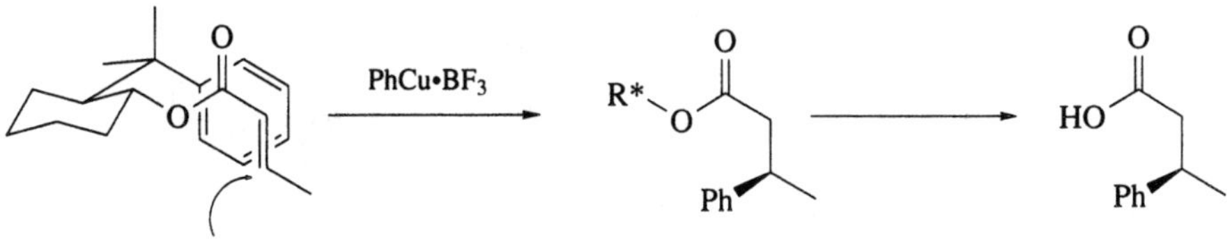

(49)

(**45**) (**46**) 99.5% *ee*

A stereoselectivity question also arose during addition/elimination of enones and enoates containing a good leaving group at the β-carbon.[102] A wide range of cuprates and leaving groups (L = halide,[103] acetate,[104] alkoxy,[105] alkylthio[106] *etc.*) have been used successfully, and this transformation constitutes a versatile synthesis of β-substituted enones and enoates with some obvious stereochemical implications (equation 50).

(50)

In many of the early reports, ester derivatives were extensively studied[105a,106b,107] and a preference for retention *versus* inversion was observed, where retention refers to introduction of the nucleophile to the same side of the carbon–carbon double bond as the leaving group it replaced (Scheme 18).

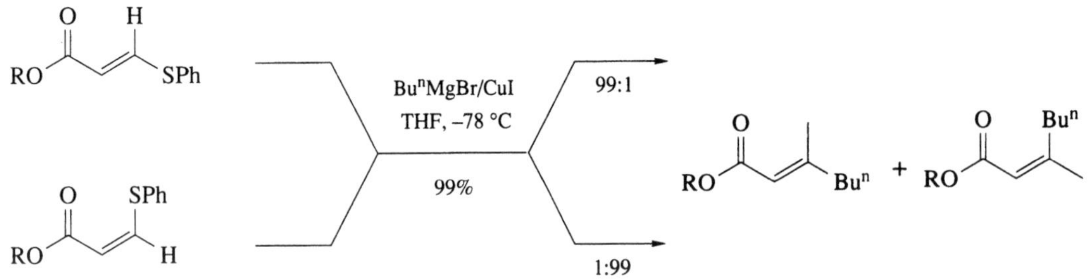
O
H
RO
SPh
O
SPh
RO
H
Bu^{n}MgBr/CuI
THF, –78 °C
99%
99:1
1:99
O
RO
Bu^n
+
O
Bu^n
RO

Scheme 18

The stereoselectivity in reactions of β-alkylthio-α,β-enones with organocuprates is dependent upon both the solvent and bulk of the ligand in the cuprate. Ether solutions favor inversion, while THF solutions favor retention (Scheme 19). It was proposed that in coordinating solvents such as THF, a more reactive enolate is formed, which eliminates before rotation occurs, thereby giving retention of configuration.[108]

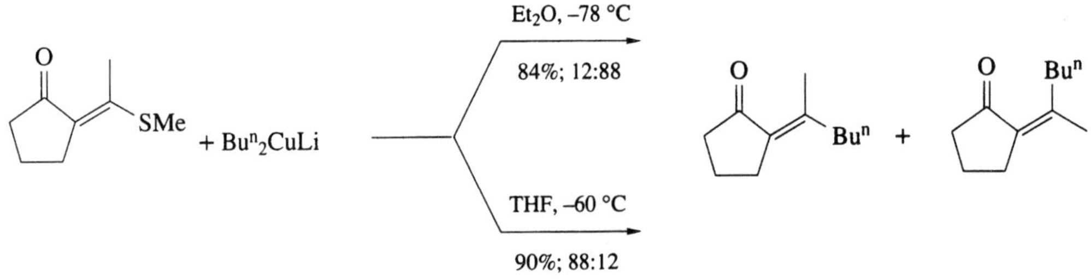
O
SMe
+ Bu$^n_2$CuLi
Et$_2$O, –78 °C
84%; 12:88
THF, –60 °C
90%; 88:12
O
Bu^n
+
O
Bu^n

Scheme 19

Bulky ligands (*e.g. t*-butyl) end up preferentially *anti* to the carbonyl, which is independent of the configuration of the β-alkylthio enone reactant (Scheme 20).

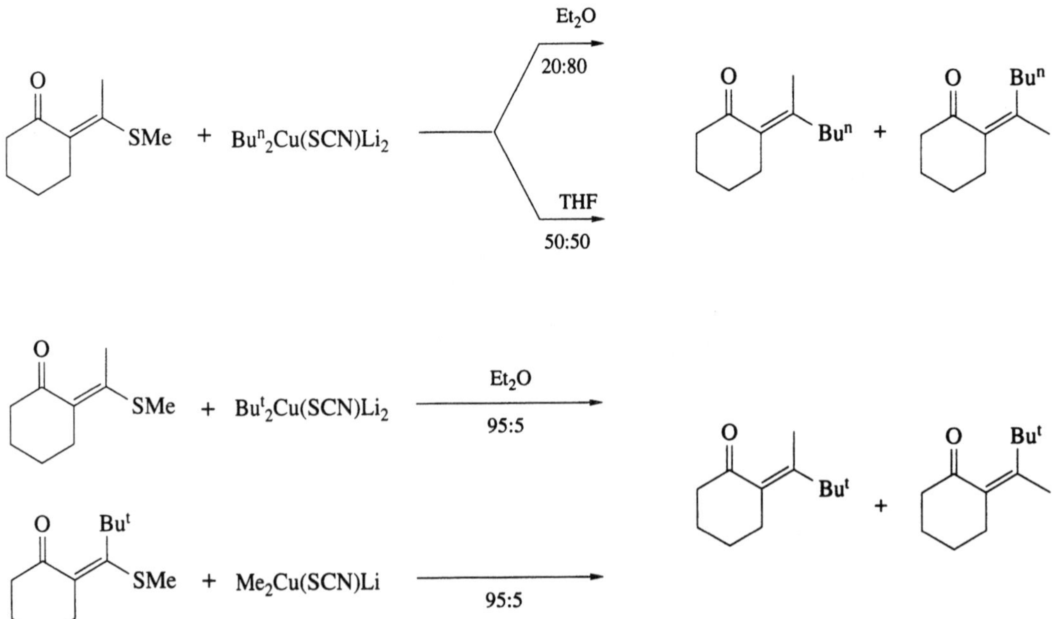
O
SMe
+ Bu$^n_2$Cu(SCN)Li$_2$
Et$_2$O
20:80
THF
50:50
O
Bu^n
+
O
Bu^n
O
SMe
+ Bu$^t_2$Cu(SCN)Li$_2$
Et$_2$O
95:5
O
Bu^t
SMe
+ Me$_2$Cu(SCN)Li
95:5
O
Bu^t
+
O
Bu^t

Scheme 20

Use of α-oxoketene dithioacetals and organocuprates provided a stereoselective synthesis of tri- and tetra-substituted alkenes.[109] In fact, it was reported that either the (*E*)- or (*Z*)-alkene could be prepared by the proper choice of cuprate and of the sequence of addition (Scheme 21).

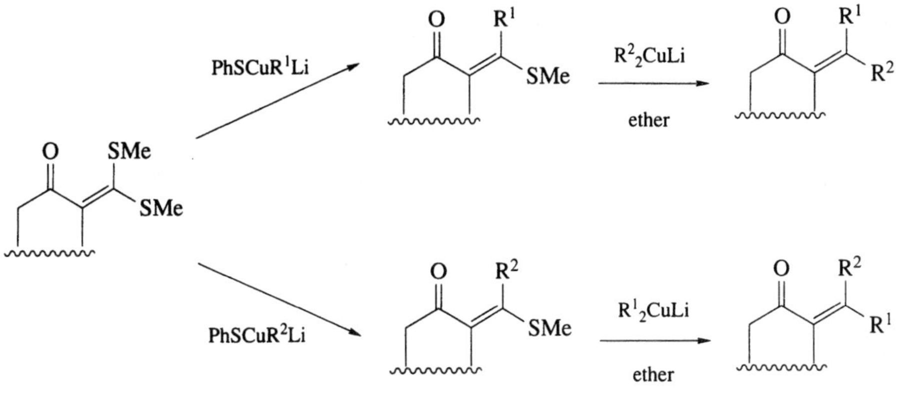

Scheme 21

1.4.10 APPLICATIONS

Conjugate addition of organocuprates has seen many applications in complex synthesis. For example, in the preparation of an opiate alkaloid it was necessary to introduce a methyl group at a late stage of the synthesis.[110] The transformation of (**47**) to (**48**) was successfully carried out with Me_2CuLi in a stereoselective manner (equation 51).

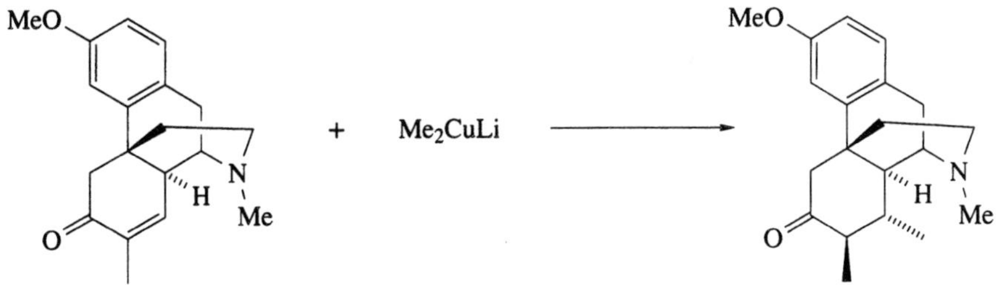

(51)

(**47**) (**48**)

The synthesis of pipitzol[111] also called for conjugate addition of a methyl group to a highly substituted enone. Good stereoselectivity (>95:5) was observed in the conversion of (**49**) to (**50**) with Me_2CuLi where attack was reasoned to occur from the less-hindered α-face (equation 52).

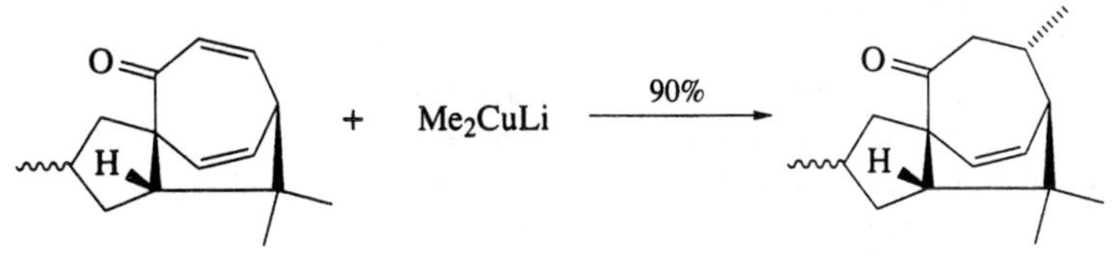

(52)

(**49**) (**50**)

Ligands other than methyl are also routinely delivered as organocuprates. For example, the transfer of vinylsilanes provided a methodology for conjugate addition of an enolate anion equivalent (equation 53).[113] Vinylsilanes are transformed to carbonyl groups by epoxidation followed by acid treatment.[112] There are no examples of simple enolates which undergo conjugate addition *via* cuprate formation, and simple enolates are generally poor nucleophiles for Michael reactions.[114]

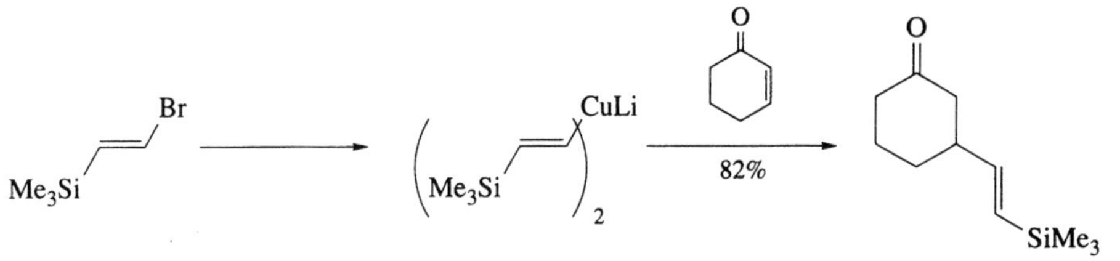

(53)

Ligands functioning as acyl anion equivalents also proceed efficiently in the conjugate addition to enones, producing the β-substituted product in high yield (equation 54).[113] Removal of the protecting group results in the free keto unit, and provides an alternative to the direct acylation (*vide supra*).[37,38]

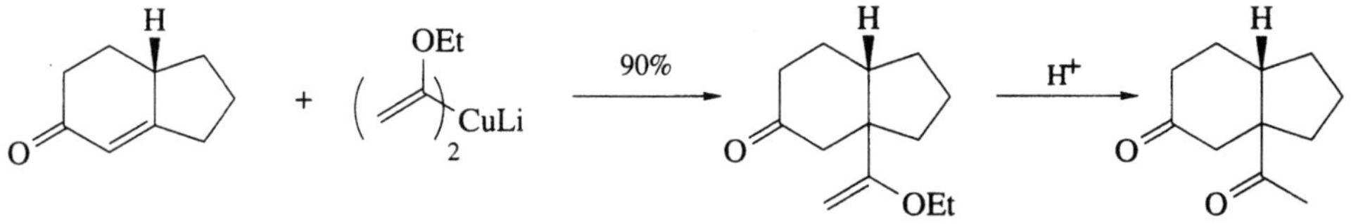

(54)

A synthesis of α-pyrones has been reported where a key step involved the chemoselective addition of an organocuprate to an α-oxoketene dithioacetal.[115] The methodology requires four steps and is quite general since the cuprate can deliver a range of ligands in a selective manner (Scheme 22).

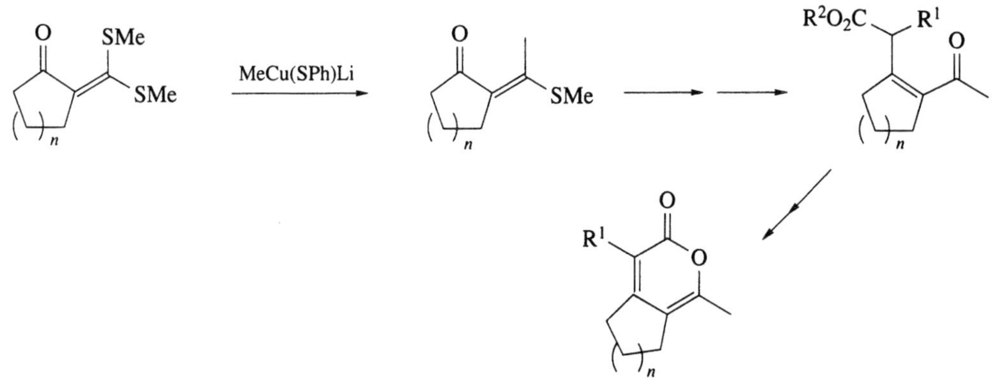

Scheme 22

The presence of a leaving group at the β-position of an enone provides the opportunity for double addition and creation of a quarternary carbon. If 2 mol equiv. of cuprate are employed, then the β,β-dialkylated product is expected. However, if a bisorganocuprate is used as the nucleophile, then a spiroannulation results.[116] Scheme 23 shows an example of a spiroannulation procedure with a biscuprate which formed spiro[4,4], [4,5] and [5,5] ring systems.[116] The bisorganocuprate was prepared from an organodilithium and copper thiophenoxide, with the obvious restriction being the limited range of dilithium species which can be prepared. Several such species were reported to undergo the spiroannulation, which provided for the rapid assembly of some interesting products in high yield (equation 55).

A nontransferable ligand is used to advantage in the above spiroannulation, and is generally important when a highly valued ligand must be transferred. In the synthesis of a steroid nucleus, the nucleophilic addition of a cuprate to enone (**51**) was one of two key steps (equation 56).[117] A good yield of the β-alkylated product was obtained, and a high stereochemical preference was reported as well.

From a study of the addition of organometallic reagents to β-substituted cyclopentyl sulfones, the reactivity of cuprates and organolithium species was compared (Scheme 24).[118] Organolithium reagents add to amine (**52**) to produce the *trans* adduct (**53**) in high stereoselectivity, while the cuprate adds to (**52**) (amine directed) to produce only the *cis* adduct (**55**). The amine was subsequently removed by quaternization and elimination, leaving the *cis*- or the *trans*-vinyl sulfones. Substrate (**52**) is apparently a special case since in earlier work vinyl sulfones were reported to be poor substrates for the reaction with organocuprates.[119,120] The ammonium ion is critical to the success of the addition; a related substrate with no amine present failed to give smooth addition.[118b]

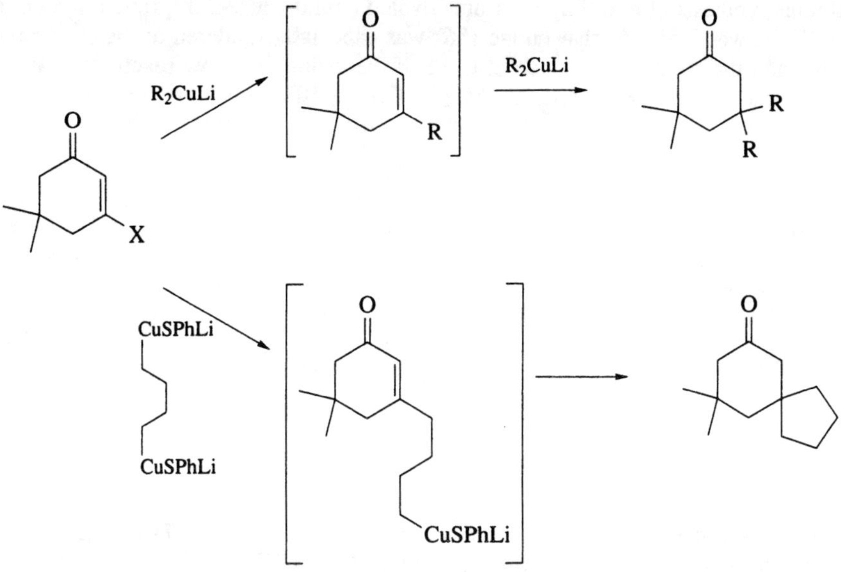

R_2CuLi R_2CuLi R R R X O O O

CuSPhLi CuSPhLi CuSPhLi O O

Scheme 23

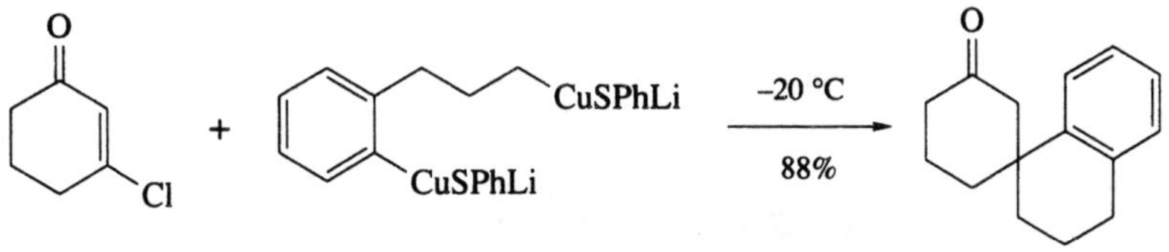

O Cl + CuSPhLi CuSPhLi −20 °C 88% O (55)

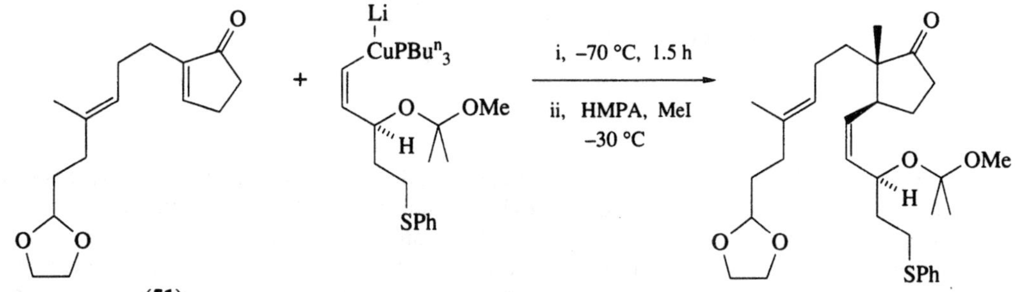

(51) O O O + Li CuPBu$^n_3$ O OMe H SPh i, −70 °C, 1.5 h ii, HMPA, MeI −30 °C O O O O OMe H SPh (56)

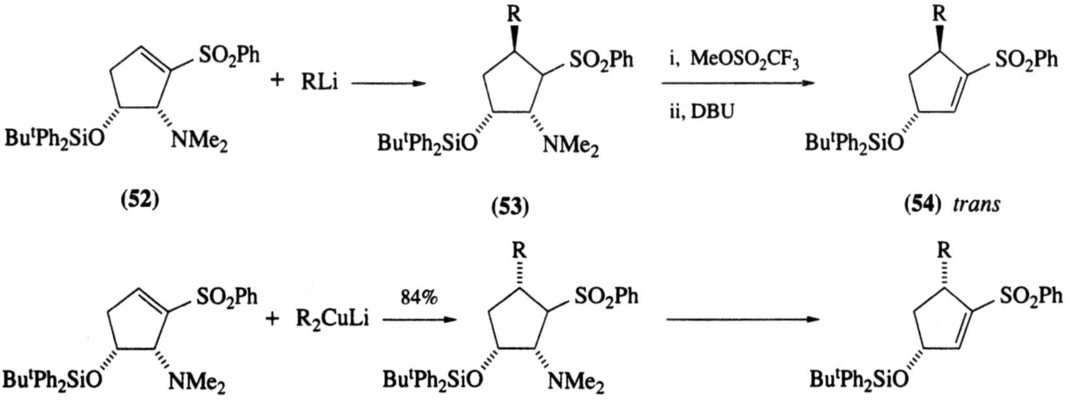

(52) SO$_2$Ph Bu^{t}Ph$_2$SiO NMe$_2$ + RLi (53) R SO$_2$Ph Bu^{t}Ph$_2$SiO NMe$_2$ i, MeOSO$_2$CF$_3$ ii, DBU (54) *trans* R SO$_2$Ph Bu^{t}Ph$_2$SiO

(52) SO$_2$Ph Bu^{t}Ph$_2$SiO NMe$_2$ + R_2CuLi 84% (55) R SO$_2$Ph Bu^{t}Ph$_2$SiO NMe$_2$ R SO$_2$Ph Bu^{t}Ph$_2$SiO

cis R = Me
cis:trans > 99:1

Scheme 24

In a formal total synthesis of coriolin,[121] an unactivated cuprate, Me_2CuLi, failed in a conjugate addition (Scheme 25). It was reasoned that enone (**56**) was especially hindered at the β-carbon due to the β,β-substitution, and the methine hydrogen at C-11 of the c-ring. The low reactivity was overcome by use of the higher order reagent–BF_3 complex, $Me_2Cu(CN)Li_2{\cdot}BF_3$, which produced the β-alkylated product in an isolated yield of 80%. The advantage in reactivity realized by addition of a Lewis acid can be substantial.

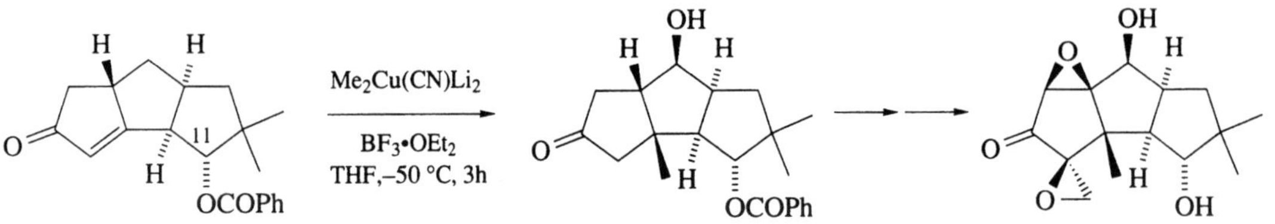

$Me_2Cu(CN)Li_2$, $BF_3{\cdot}OEt_2$, THF, –50 °C, 3h

(**56**) Coriolin

Scheme 25

In the total synthesis of mevinolin,[122] a stereoselective conversion of (**57**) into (**58**) was called for. Several organocopper reagents were examined, but it was $MeCu{\cdot}BF_3$ which exhibited the highest selectivity (Scheme 26). The reason for the selectivity is not obvious, but the results do indicate that there can be significant differences between the reactivity profile of the various organocopper reagents available.

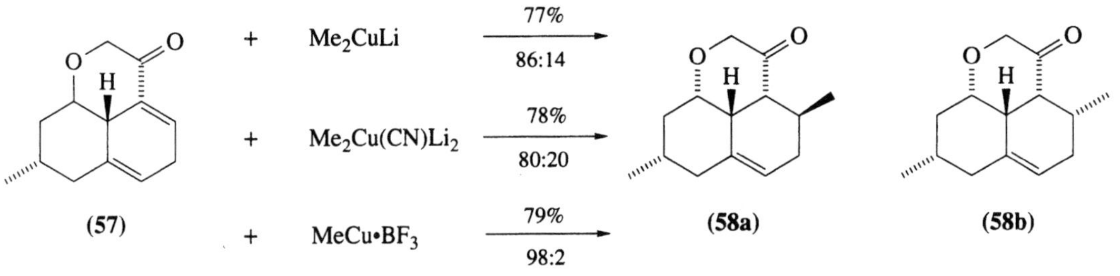

(**57**) + Me_2CuLi → 77%, 86:14

(**57**) + $Me_2Cu(CN)Li_2$ → 78%, 80:20

(**57**) + $MeCu{\cdot}BF_3$ → 79%, 98:2

(**58a**) (**58b**)

Scheme 26

In a good test of the regioselectivity of organocuprates, addition to trifluoromethyl alkynyl ketones was examined.[123] The methodology was used to provide a simple route to α,β-unsaturated trifluoromethyl ketones, and due to the electrophilic character of the fluoro ketone, 1,2-addition was expected to be a competing reaction. Surprisingly good 1,4-/1,2-selectivity was reported with the organocuprates, with cyanocuprates (higher and lower order) demonstrating the highest ratios (Scheme 27). In all cases the (*E*)-isomer prevailed, which indicated that isomerization of the products had occurred under the reaction conditions.

(**59**) + cuprates →

	1,4:1,2	*(E):(Z)*	
Me_2CuLi	88:12	76:24	
$MeCu(CN)Li_2$	100:0	53:47	(75%)
$Me_2Cu(CN)Li_2$	100:0	71:29	(93%)

Scheme 27

Two highly unusual reactions were noted in this study which merit attention. First, the treatment of (**59**) with cuprate (**60**) gave rise to 1,2-addition of the alkyne as the only product (67%; equation 57). Second, when trifluoromethyl ketone (**59**) was treated with either the higher or lower order methylcuprates, the cyanohydrin (**61**) was isolated in addition to the normal 1,4-addition product (equation 58).

This may be a rare example of CN transfer from the cuprate to a substrate, but more likely is a result of HCN addition formed from the acidic work-up.[124]

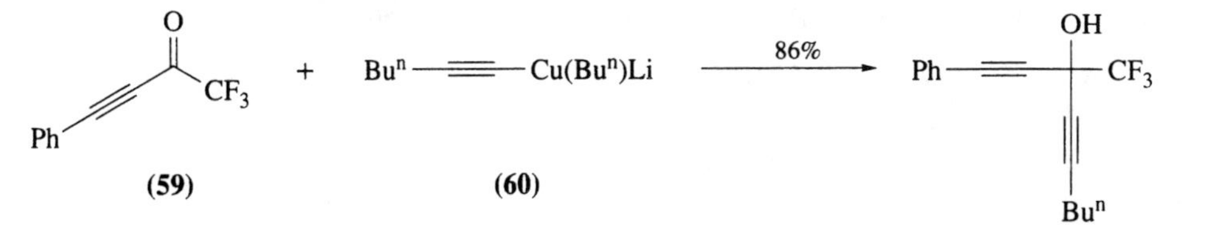

(57)

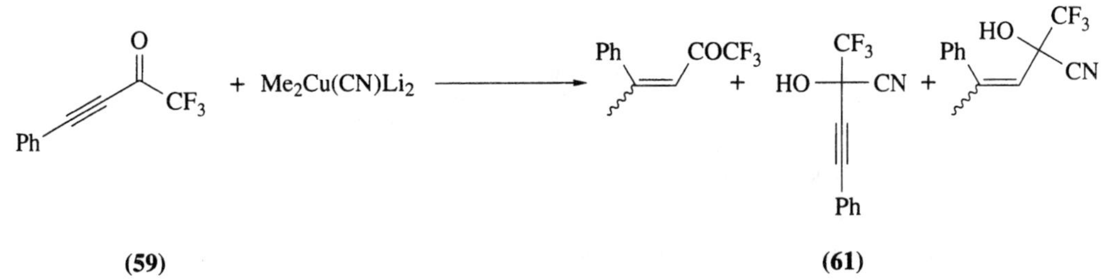

(58)

A Michael-initiated ring closure, begun by the conjugate addition of a cuprate to an alkyne, has also been reported.[125] Use of the higher order reagent was crucial to the success of this transformation as the lower order reagent led only to conjugate addition. It is likely that addition to the alkyne was followed by isomerization (facile at 0 °C) to attain a conformation suitable for ring opening of the epoxide (Scheme 28). The intermediacy of a vinylcuprate (**62**) is consistent with the reactivity difference between the lower and higher order reagents, where different species would undoubtedly be present.

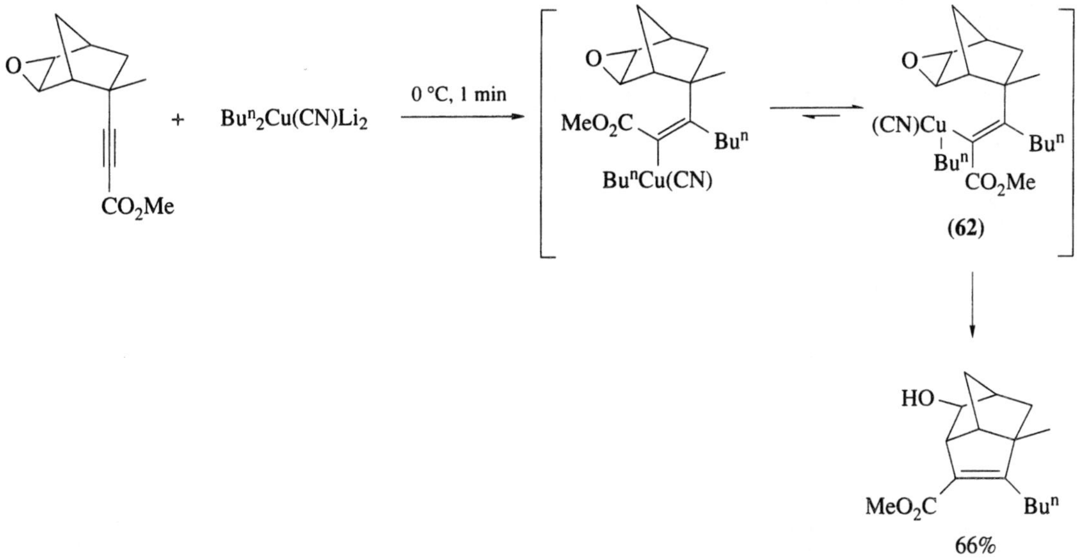

Scheme 28

1.4.11 CONCLUDING REMARKS

Organocuprates are easily prepared and undergo conjugate addition with a wide spectrum of substrates. A large variation in the R-group of R_2CuLi is tolerated and alternative modes of formation are available for ligands which contain sensitive functional groups. The emergence of technology which permits the selective transfer of a single equivalent of an R-group from the cuprate has increased the efficiency of the conjugate addition. Lewis acid additions have dramatically increased the reactivity of many cuprates, and transformations are now possible which were previously unattainable. Some mechanistic aspects are still unresolved but a more unified picture has emerged. The structure and aggregation state of the reagent is a topic which is also unsettled, and future work will undoubtedly be carried out here.

1.4.12 REFERENCES

1. G. H. Posner, *Org. React. (N.Y.)*, 1972, **19**, 1.
2. H. O. House, W. L. Respess and G. M. Whitesides, *J. Org. Chem.*, 1966, **31**, 3128.
3. B. H. Lipshutz, R. S. Wilhelm and J. A. Kozlowski, *Tetrahedron*, 1984, **40**, 5005.
4. B. H. Lipshutz, *Synthesis*, 1987, 325.
5. For recent review see Y. Yamamoto, *Angew. Chem., Int. Ed. Engl.*, 1986, **25**, 947.
6. G. H. Posner, *Org. React. (N.Y.)*, 1975, **22**, 253.
7. J. F. Normant, *Synthesis*, 1972, 63.
8. J. F. Normant, *J. Organomet. Chem. Lib.*, 1976, **1**, 219.
9. H. O. House, *Acc. Chem. Res.*, 1976, **9**, 59.
10. G. H. Posner, 'An Introduction to Synthesis Using Organocopper Reagents', Wiley, New York, 1980.
11. A. Camus, N. Marsich, G. Nardin and L. Randaccio, *Inorg. Chim. Acta*, 1977, **23**, 131.
12. G. van Koten, J. T. B. H. Jastrzebski, F. Muller and C. H. Stam, *J. Am. Chem. Soc.*, 1985, **107**, 607.
13. H. Hope, D. Oram and P. P. Power, *J. Am. Chem. Soc.*, 1984, **106**, 1149.
14. P. Leoni, M. Pasquali and C. A. Ghilardi, *J. Chem. Soc., Chem. Commun.*, 1983, 240.
15. C. Eaborn, P. B. Hitcock, J. D. Smith and A. C. Sullivan, *J. Organomet. Chem.*, 1984, **263**, C23.
16. R. G. Pearson and C. D. Gregory, *J. Am. Chem. Soc.*, 1976, **98**, 4098.
17. K. R. Steward, J. R. Lever and M.-H. Whangbo, *J. Org. Chem.*, 1982, **47**, 1472.
18. C. R. Johnson and G. A. Dutra, *J. Am. Chem. Soc.*, 1973, **95**, 7777.
19. C. R. Johnson, R. W. Herr and D. M. Wieland, *J. Org. Chem.*, 1973, **38**, 4263.
20. (a) G. M. Whitesides, W. F. Fischer, J. S. San Filipo, R. W. Basche and H. O. House, *J. Am. Chem. Soc.*, 1969, **91**, 4871; (b) G. M. Whitesides and P. E. Kendall, *J. Org. Chem.*, 1972, **37**, 3718.
21. C. P. Casey and R. A. Boggs, *Tetrahedron Lett.*, 1971, 2455.
22. F. Näf and P. Degen, *Helv. Chim. Acta*, 1971, **71**, 1939, and refs. therein.
23. H. O. House and J. M. Wilkins, *J. Org. Chem.*, 1978, **43**, 2443.
24. For a more detailed discussion see C. P. Casey and M. C. Cesa, *J. Am. Chem. Soc.*, 1979, **101**, 4236.
25. R. A. Ruden and W. E. Litterer, *Tetrahedron Lett.*, 1975, 2043.
26. G. Hallnemo, T. Olsson and C. Ullenius, *J. Organomet. Chem.*, 1985, **282**, 133.
27. G. Hallnemo, T. Olsson and C. Ullenius, *J. Organomet. Chem.*, 1984, **265**, C22.
28. S. R. Krauss and S. G. Smith, *J. Am. Chem. Soc.*, 1981, **103**, 141.
29. E. J. Corey and N. W. Boaz, *Tetrahedron Lett.*, 1985, **26**, 6015.
30. E. J. Corey, R. Naef and F. J. Hannon, *J. Am. Chem. Soc.*, 1986, **108**, 7114.
31. E. Erdik, *Tetrahedron*, 1984, **40**, 641.
32. J. P. Marino and L. J. Browne, *J. Org. Chem.*, 1976, **41**, 3629.
33. J. Ficini, P. Kanh, S. Falou and A. M. Touzin, *Tetrahedron Lett.*, 1979, 67.
34. A. Alexakis, G. Cahiez and J. F. Normant, *Tetrahedron*, 1980, **36**, 1961.
35. E. Piers, I. Nagakura and H. E. Morton, *J. Org. Chem.*, 1978, **43**, 3630.
36. E. J. Corey and D. Enders, *Chem. Ber.*, 1978, **111**, 1362.
37. D. Seyferth and R. C. Hui, *J. Am. Chem. Soc.*, 1985, **107**, 4551.
38. D. Seyferth and R. C. Hui, *Tetrahedron Lett.*, 1986, **27**, 1473.
39. E. J. Corey and L. S. Hegedus, *J. Am. Chem. Soc.*, 1969, **91**, 4926.
40. E. Piers and H. E. Morton, *J. Chem. Soc., Chem. Commun.*, 1978, 1033
41. R. D. Rieke and R. M. Wehmeyer, *J. Org. Chem.*, 1987, **52**, 5057.
42. P. Knochel, M. C. P. Yeh, S. C. Berk and J. Talbert, *J. Org. Chem.*, 1988, **53**, 2390.
43. J. R. Behling, K. A. Babiak, J. S. Ng, A. L. Campbell, R. Moretti, M. Koerner and B. H. Lipshutz, *J. Am. Chem. Soc.*, 1988, **110**, 2641, and refs. therein.
44. J. W. Labadie and J. K. Stille, *J. Am. Chem. Soc.*, 1983, **105**, 6129.
45. E. J. Corey, B. C. Pan, D. H. Hug and D. R. Deardorff, *J. Am. Chem. Soc.*, 1982, **104**, 6816.
46. W. H. Mandeville and G. M. Whitesides, *J. Org. Chem.*, 1974, **39**, 400.
47. E. J. Corey, D. Floyd and B. H. Lipshutz, *J. Org. Chem.*, 1978, **43**, 3418.
48. J. P. Gorlier, L. Hamon, J. Levisalles and J. Wagnon, *J. Chem. Soc., Chem. Commun.*, 1973, 88.
49. G. H. Posner, C. E. Whitten and J. J. Sterling, *J. Am. Chem. Soc.*, 1973, **95**, 7788.
50. B. H. Lipshutz, J. A. Kozlowski, D. A. Parker, K. E. McCarthy and S. L. Nguyen, *J. Organomet. Chem.*, 1985, **285**, 437.
51. The use of a 2-thienyl group as a nontransferable ligand in lower order cuprates has been described, H. Malmberg, M. Nilsson and C. Ullenius, *Tetrahedron Lett.*, 1982, **23**, 3823.
52. Ligands containing sulfur attached directly to copper in mixed lower order cuprates have been preferentially transferred (a) T. G. Back, S. Collins and K. W. Law, *Tetrahedron Lett.*, 1984, **25**, 1689; (b) E. Piers, K. F. Cheng and I. Nagakura, *Can. J. Chem.*, 1982, **60**, 1256.
53. For an example see ref. 3, p. 5030.
54. Y. Kojima and N. Kato, *Tetrahedron Lett.*, 1980, **21**, 4365, and refs. therein.
55. B. H. Lipshutz, M. Koerner and D. A. Parker, *Tetrahedron Lett.*, 1987, **28**, 945.
56. B. H. Lipshutz, D. A. Parker, J. A. Kozlowski and S. L. Nyuyen, *Tetrahedron Lett.*, 1984, **25**, 5959.
57. (a) C. R. Johnson and D. S. Dhanoa, *J. Org. Chem.*, 1987, **52**, 1885; (b) C. R. Johnson and D. S. Dhanoa, *J. Chem. Soc., Chem. Commun.*, 1982, 358.
58. (a) S. H. Bertz, G. Dabbagh and G. M. Villacorta, *J. Am. Chem. Soc.*, 1982, **104**, 5824; (b) S. H. Bertz and G. Dabbagh, *J. Chem. Soc., Chem. Commun.*, 1982, 1030; (c) S. H. Bertz and G. Dabbagh, *J. Org. Chem.*, 1984, **49**, 1119.
59. M. Suzuki, T. Suzuki, T. Kawagishi and R. Noyori, *Tetrahedron Lett.*, 1980, **21**, 1247.
60. G. Hallnemo and C. Ullenius, *Tetrahedron*, 1983, **39**, 1621, and refs. therein.
61. G. H. Posner, J. J. Sterling, C. E. Whitten, C. M. Lentz and D. J. Brunelle, *J. Am. Chem. Soc.*, 1975, **97**, 107.
62. B. H. Lipshutz, J. A. Kozlowski and R. S. Wilhelm, *J. Org. Chem.*, 1984, **49**, 3938.

63. H. O. House and T. V. Lee, *J. Org. Chem.*, 1978, **43**, 4369.
64. (a) Y. Yamamoto and K. Maruyama, *J. Am. Chem. Soc.*, 1977, **99**, 8068; (b) Y. Yamamoto and K. Maruyama, *J. Am. Chem. Soc.*, 1978, **100**, 3240; (c) Y. Yamamoto, S. Yamamoto, H. Yatagai, Y. Ishihara and K. Maruyama, *J. Org. Chem.*, 1982, **47**, 119.
65. T. Ibuka, H. Minakata, Y. Mitsui, K. Kinoshita and Y. Kawami, *J. Chem. Soc., Chem. Commun.*, 1980, 1193.
66. (a) C. Frejaville and R. Jullien, *Tetrahedron Lett.*, 1971, 2039; (b) R. Jullien, H. S. Lariviere and D. Zann, *Tetrahedron*, 1981, **37**, 3159.
67. (a) T. Ibuka and E. Tabushi, *J. Chem. Soc., Chem. Commun.*, 1982, 703; (b) T. Ibuka, E. Tabushi and M. Yasuda, *Chem. Pharm. Bull.*, 1983, **31**, 128.
68. J. Hooz and R. B. Layton, *Can. J. Chem.*, 1970, **48**, 1626.
69. B. B. Snider and T. C. Kirk, *J. Am. Chem. Soc.*, 1983, **105**, 2364.
70. H. Nemoto, H. Kurobe, K. Fukumoto and T. Kametani, *Tetrahedron Lett.*, 1984, **25**, 4669.
71. R. A. Roberts, V. Schüll and L. A. Paquette, *J. Org. Chem.*, 1983, **48**, 2076.
72. For an example see C. Bretting, J. Munch-Petersen, P. Moller Jorgensen and S. Refn, *Acta Chem. Scand.*, 1960, **14**, 151.
73. J. Munch-Petersen, C. Bretting, P. Moller Jorgensen, S. Refn and V. K. Andersen, *Acta Chem. Scand.*, 1961, **15**, 277.
74. A. J. Birch and M. Smith, *Proc. Chem. Soc., London*, 1962, 356.
75. J. A. Marshall, R. A. Ruden, L. D. Hirsch and M. Phillippe, *Tetrahedron Lett.*, 1971, 3795.
76. R. Wiechert, U. Kerb and K. Kieslich, *Chem. Ber.*, 1963, **96**, 2765.
77. F. Barbot, A. Kadib-Elban and P. Miginiac, *J. Organomet. Chem.*, 1983, **255**, 1.
78. F. Barbot, A. Kadib-Elban and P. Miginiac, *Tetrahedron Lett.*, 1983, **24**, 5089.
79. (a) E. J. Corey and D. L. Boger, *Tetrahedron Lett.*, 1978, 9; (b) J. P. Marino and D. M. Floyd, *Tetrahedron Lett.*, 1975, 3897.
80. D. L. J. Clive, V. Farina and P. L. Beaulieu, *J. Chem. Soc., Chem. Commun.*, 1981, 643.
81. A. Alexakis, C. Chuit, M. Commercon-Bourgain, J. P. Foulon, M. Jabri, P. Mangeney and J. F. Normant, *Pure Appl. Chem.*, 1984, **56**, 91.
82. B. H. Lipshutz, S. Whitney, J. A. Kozlowski and C. M. Breneman, *Tetrahedron Lett.*, 1986, **27**, 4273.
83. J. Munch-Petersen, *Bull. Soc. Chim. Fr.*, 1966, 471, and refs. therein.
84. B. H. Lipshutz, *Tetrahedron Lett.*, 1983, **24**, 127.
85. (a) W. Oppolzer and H. J. Loher, *Helv. Chim. Acta*, 1981, **64**, 2808; (b) W. Oppolzer, R. Moretti, T. Godel, A. Meunier and H. J. Loher, *Tetrahedron Lett.*, 1983, **24**, 4971.
86. M. Behforouz, T. T. Curran and J. L. Bolan, *Tetrahedron Lett.*, 1986, **27**, 3107.
87. For a review of this topic see, (a) J. F. Normant, *Mod. Synth Methods*, 1983, **3**, 139; (b) J. F. Normant and A. Alexakis, *Synthesis*, 1981, 841.
88. Y. Yamamoto, H. Yatagai and K. Maruyama, *J. Org. Chem.*, 1979, **44**, 1744.
89. E. J. Corey and J. A. Katzenellenbogen, *J. Am. Chem. Soc.*, 1969, **91**, 1851.
90. J. B. Siddall, M. Biskup and J. H. Fried, *J. Am. Chem. Soc.*, 1969, **91**, 1853.
91. J. Klein and R. Levene, *J. Chem. Soc., Perkin Trans. 2*, 1973, 1971.
92. J. E. Audia and J. A. Marshall, *Synth. Commun.*, 1983, **13**, 531.
93. In consideration of the literature precedent (refs. 3 and 4) the correct representation of the reagent formed by the addition of 2 mol equiv. TMSLi to CuCN should be $(TMS)_2Cu(CN)Li_2$, not $(TMS)_2CuLi$–LiCN; see I. Fleming and F. Roessler, *J. Chem. Soc., Chem. Commun.*, 1980, 276.
94. H. O. House and H. W. Thompson, *J. Org. Chem.*, 1963, **28**, 360.
95. For a more detailed discussion see N. L. Allinger and C. K. Riew, *Tetrahedron Lett.*, 1966, 1269.
96. (a) E. D. Gergmann, D. Ginsburg and R. Rappo, *Org. React. (N.Y.)*, 1959, **10**, 179; (b) E. Toromanoff, *Bull. Soc. Chim. Fr.*, 1962, 708; (c) E. L. Eliel, N. L. Allinger, S. J. Angyal and G. A. Morrison, 'Conformational Analysis', Wiley, New York, 1965.
97. T. R. Hoye, A. S. Magee and R. E. Rosen, *J. Org. Chem.*, 1984, **49**, 3224.
98. C. J. Sih, R. G. Salomon, P. Price and R. Sood, *J. Am. Chem. Soc.*, 1975, **97**, 857.
99. For a recent review see R. J. K. Taylor, *Synthesis*, 1985, 364; see also 'Comprehensive Organic Synthesis', ed. B. M. Trost, Pergamon Press, 1991, vol. 4, chap. 1.6.
100. R. Davis and K. G. Untch, *J. Org. Chem.*, 1979, **44**, 3755; see also ref. 99.
101. Y. Yamamoto, S. Nishii and T. Ibuka, *J. Chem. Soc., Chem. Commun.*, 1987, 1572.
102. R. K. Dieter, J. R. Fishpaugh and L. A. Silks, *Tetrahedron Lett.*, 1982, **23**, 3751.
103. (a) E. Piers and H. E. Morton, *J. Org. Chem.*, 1979, **44**, 3437; (b) J. L. Coke, H. J. Williams and S. Natarajan, *J. Org. Chem.*, 1977, **42**, 2380; (c) F. Leyendecker, J. Drouin and J. M. Coria, *Tetrahedron Lett.*, 1974, 2931.
104. (a) C. Ouannes and Y. Langlois, *Tetrahedron Lett.*, 1975, 3461. (b) C. P. Casey and D. F. Marten, *Synth. Commun.*, 1973, **3**, 321.
105. (a) S. Cacchi, A. Caputo and D. Misitri, *Indian J. Chem.*, 1974, **12**, 325; (b) G. H. Posner and D. J. Brunelle, *J. Chem. Soc., Chem. Commun.*, 1973, 907.
106. (a) R. M. Coates and L. O. Sandefur, *J. Org. Chem.*, 1974, **39**, 275; (b) S. Kobayashi, O. Meguro-Ku and T. Mukaiyama, *Chem. Lett.*, 1974, 705; (c) see ref. 119b.
107. (a) C. P. Casey, C. R. Jones and H. Tukada, *J. Org. Chem.*, 1981, **46**, 2089; (b) K. D. Richards, J. K. Aldean, A. Srinivasan, R. W. Stephenson and R. K. Olsen, *J. Org. Chem.*, 1976, **41**, 3674; (c) C. P. Casey, D. F. Marten and R. A. Boggs, *Tetrahedron Lett.*, 1973, 2071.
108. R. K. Dieter and L. A. Silks, *J. Org. Chem.*, 1986, **51**, 4687.
109. (a) R. K. Dieter, L. A. Silks, J. R. Fishpaugh and M. R. Kastner, *J. Am. Chem. Soc.*, 1985, **107**, 4679; (b) R. K. Dieter and L. A. Silks, *J. Org. Chem.*, 1983, **48**, 2786.
110. D. L. Leland and M. P. Kotick, *J. Med. Chem.*, 1980, **23**, 1427.
111. R. L. Funk and G. L. Bolton, *J. Org. Chem.*, 1987, **52**, 3174.
112. G. Stork and E. W. Colvin, *J. Am. Chem. Soc.*, 1971, **93**, 2080.
113. R. K. Boeckman, Jr. and K. J. Bruza, *J. Org. Chem.*, 1979, **44**, 4781.

114. (a) C. P. Casey and W. R. Brunsvold, *Inorg. Chem.*, 1977, **16**, 391; (b) J. Colonge and J. Dreux, *Bull. Soc. Chim. Fr.*, 1972, 4187.
115. R. K. Dieter and J. R. Fishpaugh, *J. Org. Chem.*, 1983, **48**, 4439.
116. (a) P. A. Wender and S. L. Eck, *Tetrahedron Lett.*, 1977, 1245; (b) P. A. Wender and A. W. White, *J. Am. Chem. Soc.*, 1988, **110**, 2218.
117. T. Takahashi, K. Shimizu, T. Doi and T. Tsuji, *J. Am. Chem. Soc.*, 1988, **110**, 2674.
118. (a) D. K. Hutchinson and P. L. Fuchs, *Tetrahedron Lett.*, 1986, **27**, 1429; (b) D. K. Hutchinson, S. A. Hardinger and P. L. Fuchs, *Tetrahedron Lett.*, 1986, **27**, 1425.
119. (a) G. H. Posner and D. J. Brunelle, *Tetrahedron Lett.*, 1973, 938; (b) G. H. Posner and D. J. Brunelle, *Tetrahedron Lett.*, 1973, 2747.
120. For acyclic sulfones see (a) J. J. Eisch and J. E. Galle, *J. Org. Chem.*, 1979, **44**, 3278; (b) P. Knochel and J. F. Normant, *Tetrahedron Lett.*, 1985, **26**, 425.
121. L. Van Hijfte, R. D. Little, J. L. Petersen and K. D. Moeller, *J. Org. Chem.*, 1987, **52**, 4647.
122. S. J. Hecker and C. H. Heathcock, *J. Org. Chem.*, 1985, **50**, 5159.
123. R. J. Linderman and M. S. Lonikar, *Tetrahedron Lett.*, 1987, **28**, 5271.
124. R. J. Linderman and M. S. Lonikar, *J. Org. Chem.*, 1988, **63**, 6013.
125. D. E. Lewis and H. L. Rigby, *Tetrahedron Lett.*, 1985, **26**, 3437.

1.5
Asymmetric Nucleophilic Addition to Electron Deficient Alkenes

HANS-GÜNTHER SCHMALZ
Universität Frankfurt, Germany

1.5.1 INTRODUCTION

Additions of nucleophiles to electron-deficient double bonds, *e.g.* the classical Michael addition or the Cu^I-mediated 1,4-addition of Grignard reagents to α,β-unsaturated carbonyl compounds (see previous chapters of this volume), constitute some of the very valuable items in the toolbox of the synthetic organic chemist. No wonder that they have found widespread application in organic synthesis. While early interest in these reactions was based particularly on their reliable chemo- and regio-selectivity, in recent years much progress has been made in the development of highly diastereo- and even enantio-selective methods, allowing the control of absolute stereochemistry.[1–4]

It is the aim of this chapter to survey these latter methods, independent of the special type of reagent (nucleophile) or substrate (electron-deficient alkene). The treatise will mainly focus on C—C bond-forming reactions, and will be restricted to such methods which: (i) allow the preparation of optically active, and, if possible, of enantiomerically pure (homochiral) compounds (EPC-synthesis);[5] (ii) involve the formation of at least one new stereogenic center during the addition step with a significant degree of asymmetric induction; and (iii) in general, give products containing only those stereogenic centers which are generated during the nucleophilic addition and, eventually, during subsequent alkylation steps (occasionally after removal of a chiral auxiliary group). Synthetic applications will be emphasized.

Most of the reactions which will be discussed lead to carbonyl compounds with a stereogenic center in the β-position. This is illustrated in Scheme 1: a substrate molecule (**1**; X = heteroatom or heteroatom-based functional group), having an electron-deficient double bond, is attacked by a nucleophilic reagent (possibly in the presence of a coordinating ligand or a catalyst) to form an anionic intermediate (**2**), which is then converted to the product (**3**) on hydrolytic work-up.

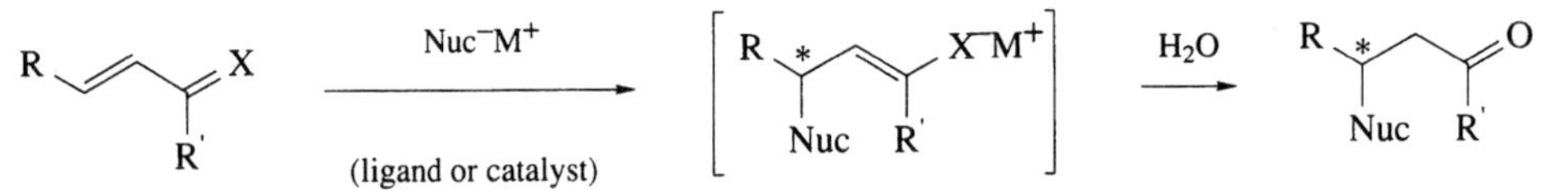

Nuc^-M^+ (ligand or catalyst); H_2O

(**1**) (**2**) (**3**)

Scheme 1

To achieve the desired asymmetric induction, chirality must be introduced. There are essentially three ways to do this. One can employ: (i) chirally modified substrates; (ii) chirally modified nucleophiles; or (iii) a chiral reaction medium (chiral coordinating cosolvents, ligands or catalysts). This chapter is organized according to these three approaches.

1.5.2 DIASTEREOSELECTIVE ADDITIONS OF ACHIRAL CARBON NUCLEOPHILES TO CHIRAL SUBSTRATES

1.5.2.1 Additions to Chiral Derivatives of α,β-Unsaturated Carboxylic Acids

1.5.2.1.1 Additions to chiral esters

The conjugate addition of (achiral) nucleophiles to α,β-unsaturated esters of chiral alcohols (**4**) generally produces optically active β-branched acids, (**7**):(*ent*-**7**) ≠ 1, after saponification of the diastereomeric ester intermediates (**5**) and (**6**; Scheme 2).[6]

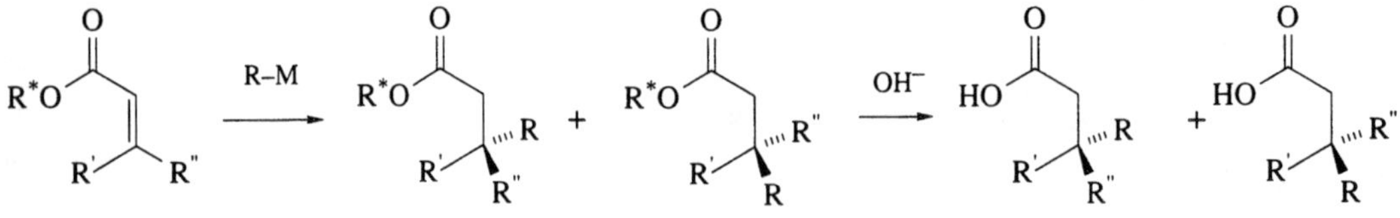

R–M; OH^-

(**4**) (**5**) (**6**) (**7**) (*ent*-**7**)

Scheme 2

Following this concept, the 1,4-addition of phenylmagnesium bromide to (–)-menthyl crotonate was investigated, however with quite disappointing results.[7] By using sugar-derived chiral auxiliaries, somewhat better selectivities were obtained,[8] but the results could not be reproduced.[9] Attempts to alkylate enamines with menthyl crotonate led also to unsatisfying yields and selectivities.[10]

The introduction of RCu/BF_3-type reagents[11] prompted investigation of the conjugate addition of these reagents to chiral enoates prepared from (–)-8-phenylmenthol (**8**).[12] While additions to the (*E*)-crotonate (**9**) were highly selective (Scheme 3) and proceeded in reasonable yields, (*Z*)-configured or tri- and tetra-substituted enoates gave only moderate results. These difficulties could be overcome by employing the Bu_3P-stabilized RCu/BF_3 reagents, which lead to increased reaction rates, improved yields and, most important, generally high selectivities.[13–16] Addition of these reagents to the enoates (**11**), accessible from the camphor-derived alcohol (**10**), gave carboxylic acids (**13**) of high enantiomeric purity, after hydrolysis of the initially formed esters (**12**; Scheme 4). The even more practical auxiliary (**14**), available in only two steps from camphor-10-sulfonyl chloride, also gave rise to high diastereoselectivities during 1,4-addition to its enoates (**15**). The results, summarized in Table 1, demonstrate the broad scope of this highly useful methodology. The chiral auxiliary alcohols are available in both enantiomeric forms and can efficiently be recycled.

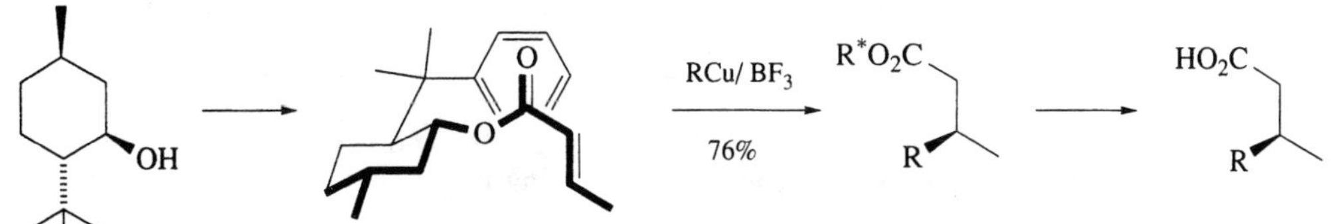
OH
Ph
O
RCu/ BF3
76%
R*O2C
R
HO2C
R

(**8**) (**9**) R = Ph, Bu^n >99% *ee*

Scheme 3

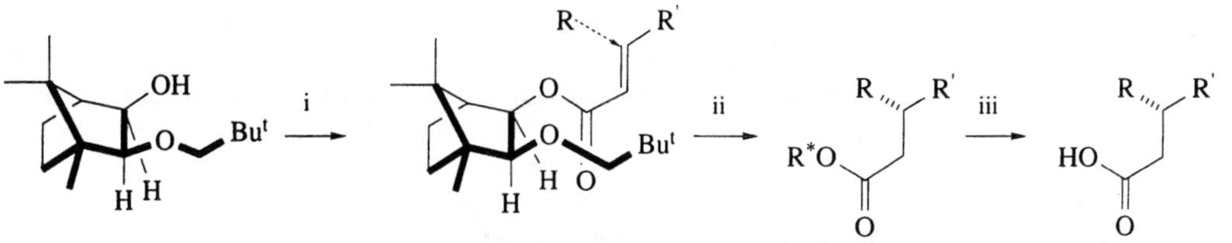
OH
O
Bu^t
H H
i
R
R'
O
O
Bu^t
H O
H
ii
R
R'
R*O
O
iii
R
R'
HO
O

(**10**) (**11**) (**12**) (**13**)

i, acid chloride, AgCN, benzene, reflux, 4 h; ii, add excess of an equimolar mixture of RLi, CuI, Bu_3P and BF_3•OEt_2, –78 to –35 °C; iii, NaOH, aq. EtOH, reflux

Scheme 4

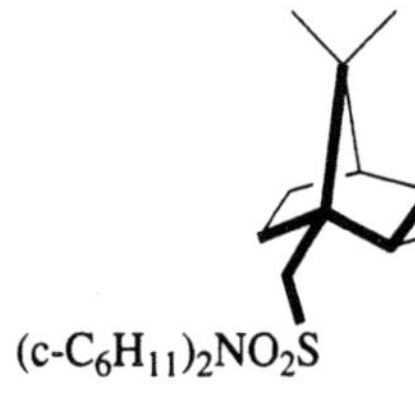
OH
(c-C6H11)2NO2S

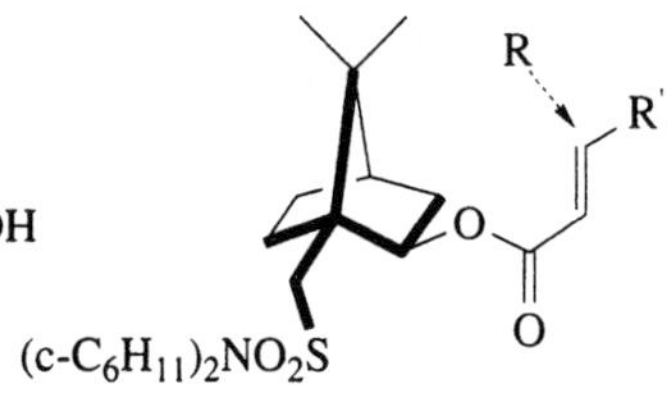
R
R'
O
O
(c-C6H11)2NO2S

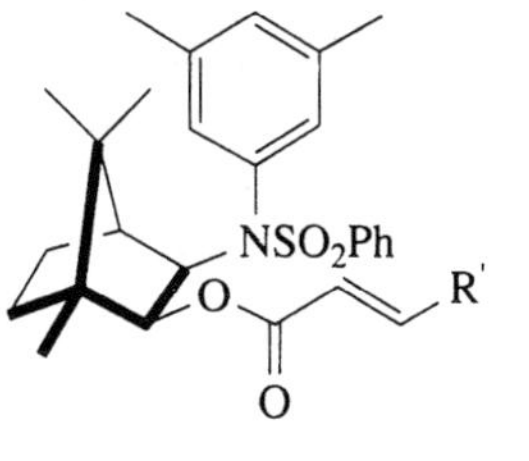
NSO2Ph
O
R'
O

(**14**) (**15**) (**16**)

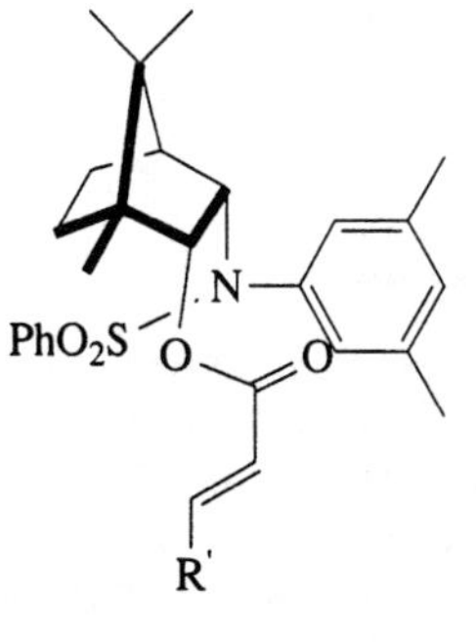
N
PhO2S
O
O
R'

(**17**)

Table 1 Preparation of Chiral Carboxylic Acids (**13**) or (*ent*-**13**) *via* Ester Intermediates by Addition of Bu_3P-Stabilized RCu/BF_3 Reagents to Chiral Enoates Followed by Hydrolysis (Scheme 4)

Enoate	*R′*	*R*	*Ester yield (%)*	*Carboxylic acid*	*ee (%)*	*Configuration*	*Ref.*
(**11**)	Bu^n	Me	82	(**13**)	94	(*R*)	13
(**11**)	Et	Me	85	(**13**)	92	(*R*)	13
(**11**)	Me	4-Methyl-3-pentenyl	81	(**13**)	98	(*S*)	13
(**11**)	Me	Vinyl	85	(**13**)	94	(*R*)	14
(**11**)	Me	2-Propenyl	86	(**13**)	99	(*R*)	14
(**11**)	1,5-Hexadien-2-yl	2-Propenyl	89	(**13**)	98	(*R*)	14
(*ent*-**11**)	n-C_8H_{17}	Me	90	(*ent*-**13**)	98	(*S*)	13
(*ent*-**11**)	4-Methyl-3-pentenyl	Me	90	(*ent*-**13**)	92	(*S*)	13
(**15**)	Me	Pr	98	(**13**)	97	(*S*)	15
(**15**)	Me	Bu^n	89	(**13**)	97	(*S*)	15
(**15**)	Me	Vinyl	80	(**13**)	98	(*R*)	15
(**15**)	Me	2-Propenyl	84	(**13**)	94	(*R*)	15
(**15**)	Pr	Me	89	(**13**)	94	(*R*)	15
(**15**)	Bu^n	Me	93	(**13**)	97	(*R*)	15

A related approach involves the diastereomeric chiral enoates (**16**) and (**17**), which are both prepared from (+)-camphor, but behave like pseudo-enantiomers, leading to products of opposite absolute configuration at the newly formed stereogenic center.[17] Conjugate addition of RCu/BF_3 reagents to these substrates at low temperatures generally proceeds with almost complete diastereoselectivity (>99:1). The right choice of the solvent, depending on the metal species involved, is crucial: while THF works well for the copper compounds prepared from Grignard reagents, the alkyllithium-derived species show best results in ether.

In all cases where secondary alcohols are used as chiral auxiliaries, one can understand (and predict) the stereochemical outcome of these reactions, if one takes into account: (i) that esters of secondary alcohols always try to adopt a conformation with a coplanar (horseshoe-type) arrangement of the carbonyl group, the ether oxygen and the alkoxy C—H bond;[18,19] and (ii) that in the presence of Lewis acids, (*E*)-enoates appear to prefer the *s-trans* conformation (antiplanar position of the carbonyl group and the C=C double bond).[20] Depending on the structure of the chiral auxiliary, one of the two (diastereotopic) faces of the double bond will now be more or less shielded, forcing the nucleophile to attack on the opposite face.

The homochiral β-branched carboxylic acids (**13** or *ent*-**13**; Table 1) have been used for the synthesis of a variety of natural products[21] such as California red scale pheromone,[14] α-skytanthine,[22] the southern corn rootworm pheromone,[15] and norpectinatone.[16] Pure (*S*)-3-trichloromethylbutyric acid (a structural subunit of some sponge metabolites) was prepared *via* conjugate addition of Cl_3CMgCl to the chiral crotonate (**16**) followed by hydrolysis (Scheme 5).[23]

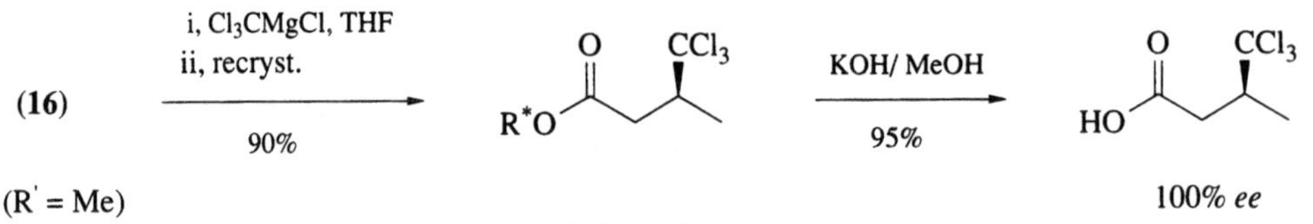

(**16**) (R′ = Me) —[i, Cl_3CMgCl, THF; ii, recryst.; 90%]→ R*O–C(=O)–CH₂–CH(CCl_3)–CH₃ —[KOH/ MeOH; 95%]→ HO–C(=O)–CH₂–CH(CCl_3)–CH₃ (100% *ee*)

Scheme 5

In an isolated case, a tertiary alcohol was used as chiral auxiliary: the crotonate (**19**), prepared from the (+)-camphor-derived alcohol (**18**), undergoes highly diastereoselective 1,4-addition of a cuprate reagent (Scheme 6).[24]

1.5.2.1.2 Additions to chiral amides and imides

Analogous to the use of chiral enoates (see previous section), α,β-unsaturated carboxylic amides, prepared from chiral amines, may be utilized in asymmetric 1,4-additions. When Grignard reagents are added to unsaturated amides (**21**), derived from (–)-ephedrine (**20**),[25] highly optically active β-substituted alkanoic acids (**22**; R and R′ = alkyl or phenyl) are obtained in a variety of cases, after hydrolysis of the initially formed adducts (Scheme 7). This method was used for the synthesis of the antibiotic (–)-malyngolide and its stereoisomers.[26] Recrystallization of the intermediate (saturated) amide was necess-

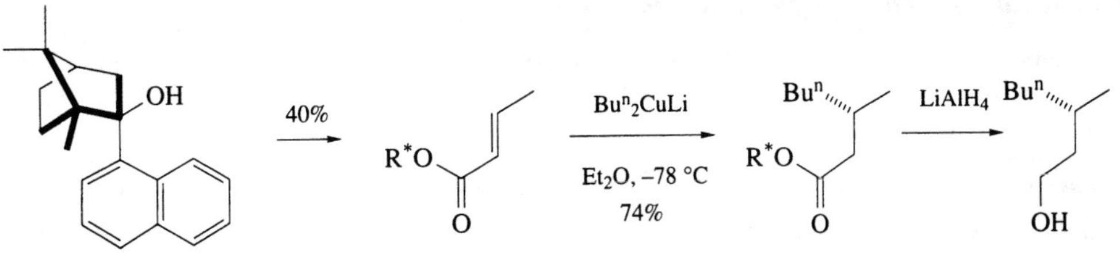

(18) (19) 95% *ee*

40% Bu^n_2CuLi Et_2O, –78 °C 74% $LiAlH_4$

Scheme 6

ary in this case to raise the enantiomeric excess (*ee*) of the hydrolysis product from 87 to 97%. A closely related approach involves addition of organometallic reagents to chiral (proline-derived) unsaturated amides in the presence of certain tertiary amines.[27–30] Best results are obtained with Grignard reagents in toluene in the presence of DBU, and when amino alcohol (**23**) is used as the chiral auxiliary (Scheme 8). The method was applied to the preparation of (*S*)-(–)-citronellol, however with only 63% *ee*.[30] With the lactam (**24**) as chiral auxiliary, conjugate addition of dialkylcuprates to the corresponding unsaturated imide (**25**) gives, after hydrolysis, carboxylic acids (**26**) with predictable absolute configuration and with reasonably high *ee* (Scheme 9).[31] In contrast to the reactions of chiral esters, these methods (based on chiral carboxylic amides or imides) seem to rely on rigid substrate conformations, which are due to chelation with a metal ion.

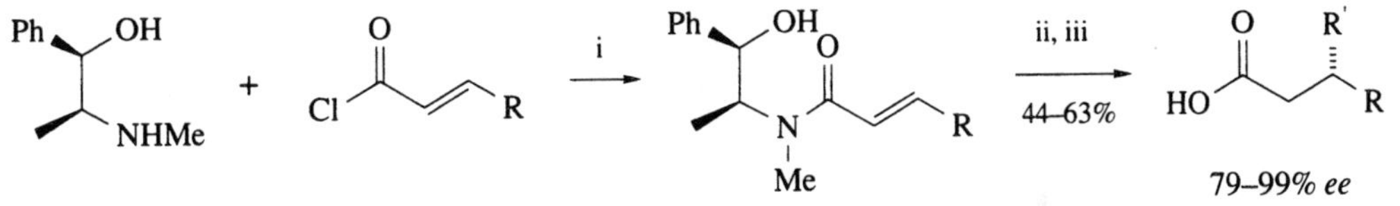

(20) (21) (22) 44–63% 79–99% *ee*

i, proton sponge; ii, 6 equiv. of R'MgBr, ether, –40 °C, 48 h, then add buffer pH 7; iii, AcOH, 3 M H_2SO_4, reflux, 3 h

Scheme 7

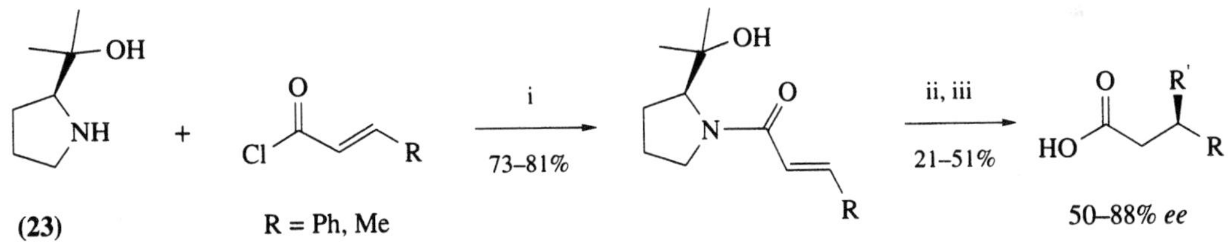

(23) R = Ph, Me 73–81% 21–51% 50–88% *ee*

i, 1 M aq. NaOH, ether, 1.5 h, r.t.; ii, R'MgBr, DBU, toluene, –40 °C to r.t.; iii, 3 M HCl, dioxane, reflux, 40 h

Scheme 8

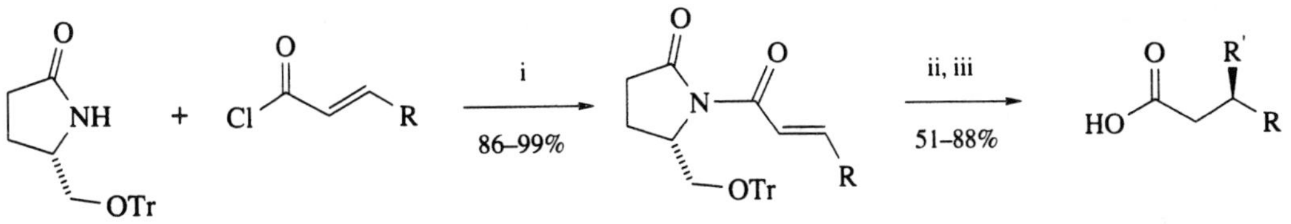

(24) R = Me, Bu (25) (26) 77–97% *ee* 86–99% 51–88%

i, Bu^nLi, THF, then add acid chloride; ii, R'_2CuLi, THF, –23 °C; iii, HCl, MeOH, then aq. KOH

Scheme 9

1.5.2.1.3 Additions to chiral **N-enoylsultams**

Another type of chiral auxiliary is the bornane-10,2-sultams (**27**) and (*ent*-**27**), which are quite easily accessible from the (+)-and (–)-camphor-10-sulfonyl chlorides.[21,32] They are readily converted into the corresponding *N*-enoylsultams (**28**) by successive treatment with sodium hydride and acyl chlorides. These substrates smoothly undergo 1,4-addition when treated with 2 mol equiv. of alkylmagnesium chlorides.[33] The β-substituted derivatives (**28**; R″ = H) lead predominantly to isomers (**29**) with reasonably high diastereoselectivity (Scheme 10). The diastereomeric purity of the products can be dramatically enhanced in most cases by simple recrystallization. Mild hydrolysis affords the free carboxylic acids (**30**) under recovery of the chiral auxiliary (**27**). The method also allows the diastereo- and enantioselective preparation of α,β-disubstituted alkanoic acids in two different ways: (A) conjugate addition to β-substituted derivatives (**28**; R″ = H) followed by *in situ* alkylation of the intermediate enolate leads preferentially into the same diastereomeric series (**31**) as does (B) the 1,4-addition to α,β-disubstituted derivatives (**28**; R″ ≠ H) with subsequent (kinetically controlled) protonation (Scheme 11). The steric course of the conjugate addition of Grignard reagents to *N*-enoylsultams (**28**) can be explained by the assumption of a reactive conformation as shown in (**32a**). A similar topological picture applies when β-silyl-substituted-*N*-enoylsultams are treated with PBu_3-stabilized RCu reagents in the presence of ethylaluminum dichloride (Scheme 12).[34] Methanolysis of the adducts (**33**) followed by successive fluorodephenylation and oxidation of the C—Si bond leads to homochiral β-hydroxy esters (**34**). Gilman reagents also smoothly add to (**28**),[35] leading preferentially to isomers (**35**; Scheme 13). The steric course of the addition step can be attributed in this case to a reactive conformation as shown in (**32b**). Addition of Me_2CuLi to the 1-cyclohexenoylsultam (**36**) and subsequent protonation gave the *cis* compound (**37**) as the sole product, which was then hydrolyzed to the homochiral acid (**38**; Scheme 14). An application of this methodology for the synthesis of β-necrodol (**39**)[36] is outlined in Scheme 15.

NH SO$_2$

(27)

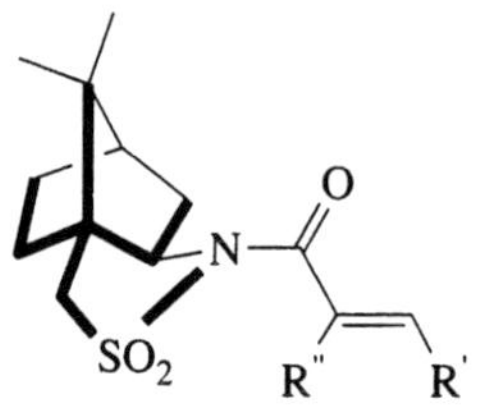

N O SO$_2$ R″ R′

(28)

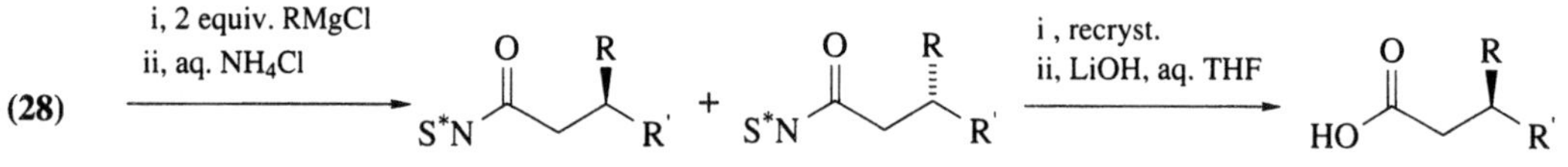

(**28**) → i, 2 equiv. RMgCl; ii, aq. NH_4Cl → S*N(C=O)...R R′ (**29**) 86–95% + S*N(C=O)...R R′ 5–14% → i, recryst.; ii, LiOH, aq. THF → HO(C=O)...R R′ (**30**)

R = alkyl
R′ = Me, Et
R″ = H

Scheme 10

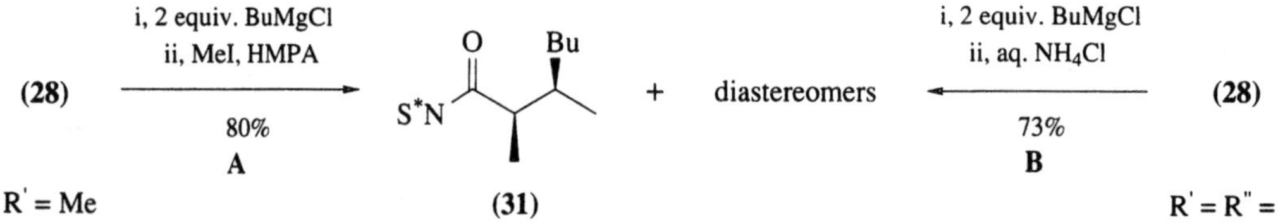

(**28**) → i, 2 equiv. BuMgCl; ii, MeI, HMPA; 80%; **A** → S*N(C=O)...Bu (**31**) + diastereomers ← i, 2 equiv. BuMgCl; ii, aq. NH_4Cl; 73%; **B** ← (**28**)

R′ = Me
R″ = H

R′ = R″ = Me

A:	86.7%	13.3%
B:	98.2%	1.8%

Scheme 11

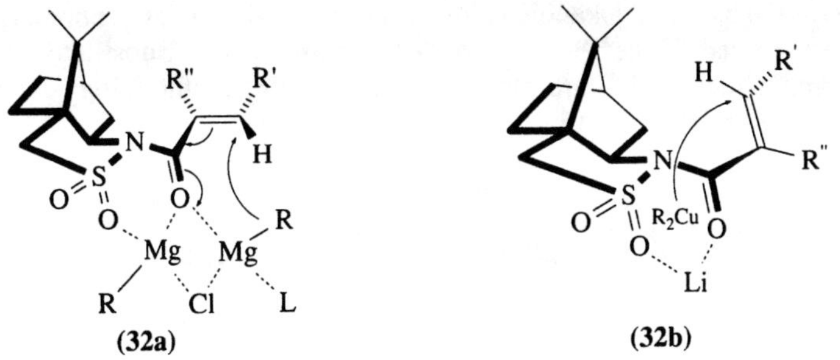
R''
R'
N
H
S
O
O
O
R
Mg
Mg
R
Cl
L
(32a)
H
R'
N
R''
S
O
R_2Cu
O
O
Li
(32b)

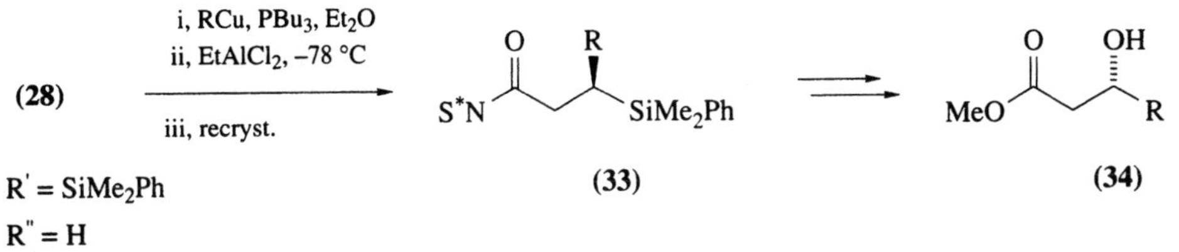
(28)
i, RCu, PBu_3, Et_2O
ii, $EtAlCl_2$, –78 °C
iii, recryst.
O
R
S*N
$SiMe_2Ph$
O
OH
MeO
R
R' = $SiMe_2Ph$
R'' = H
(33)
(34)

Scheme 12

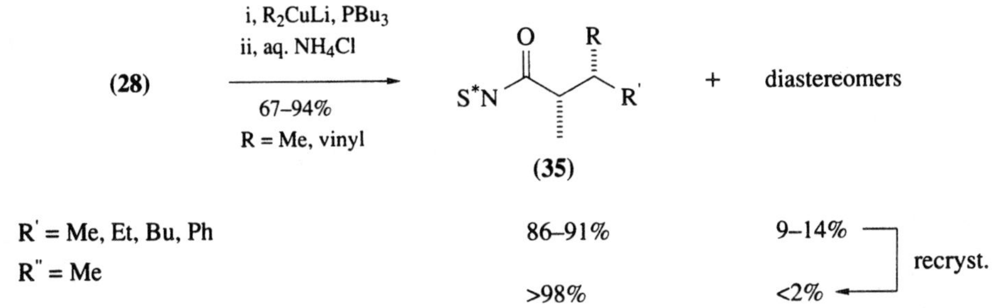
(28)
i, R_2CuLi, PBu_3
ii, aq. NH_4Cl
67–94%
R = Me, vinyl
O
R
S*N
R'
+
diastereomers
(35)
R' = Me, Et, Bu, Ph
R'' = Me
86–91%
9–14%
recryst.
>98%
<2%

Scheme 13

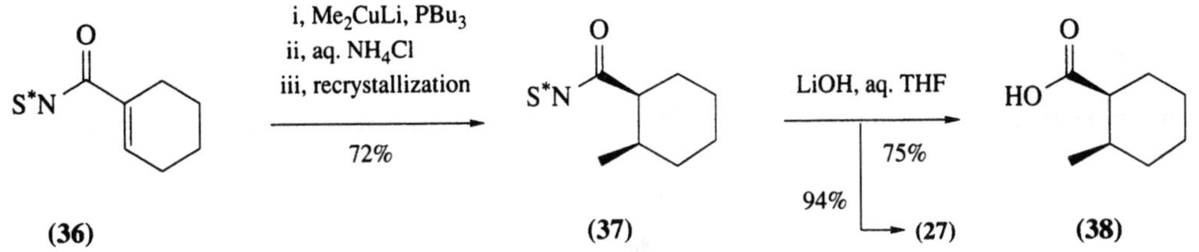
O
S*N
i, Me_2CuLi, PBu_3
ii, aq. NH_4Cl
iii, recrystallization
72%
O
S*N
LiOH, aq. THF
75%
94%
(27)
O
HO
(36)
(37)
(38)

Scheme 14

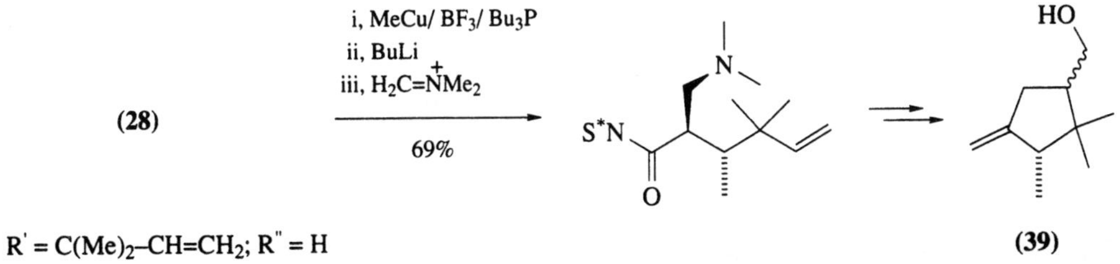
(28)
i, MeCu/ BF_3/ Bu_3P
ii, BuLi
iii, $H_2C{=}\overset{+}{N}Me_2$
69%
N
S*N
O
HO
R' = $C(Me)_2$–CH=CH_2; R'' = H
(39)

Scheme 15

1.5.2.1.4 Additions to chiral oxazolines

The first general method allowing the preparation of optically active β,β-dialkylpropionic acids *via* asymmetric synthesis is based on chiral oxazolines.[37–41] The α,β-unsaturated derivatives **(41)**, accessible as pure (*E*)-isomers from **(40)** and the respective aldehydes (RCHO), undergo highly selective 1,4-additions in a variety of cases (Scheme 16) when treated with alkyl- or aryl-lithium reagents. The products

(**42**) are hydrolyzed to β-substituted alkanoic acids (**43**) of generally high enantiomeric purity (>90% *ee*). The method has been applied to the synthesis of (+)-*ar*-turmerone.[42] Substituted dihydropyridines[43,44] and chiral 1,2-dihydronaphthalins[45–48] are also accessible by this methodology. The lignane, (+)-phyltetralin (**44**), was thus synthesized as outlined in Scheme 17.[49]

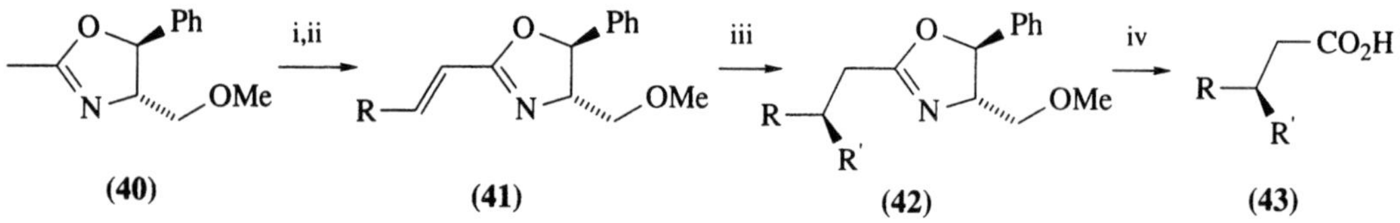

(**40**) (**41**) (**42**) (**43**) 91–99% *ee*

i, LDA, $(Pr^iO)_2POCl$, THF, –78 °C; ii, Bu^tOK, RCHO; iii, R′Li, THF, –78 °C, 3 h; iv, 1.5 M H_2SO_4, reflux, 24–36 h

Scheme 16

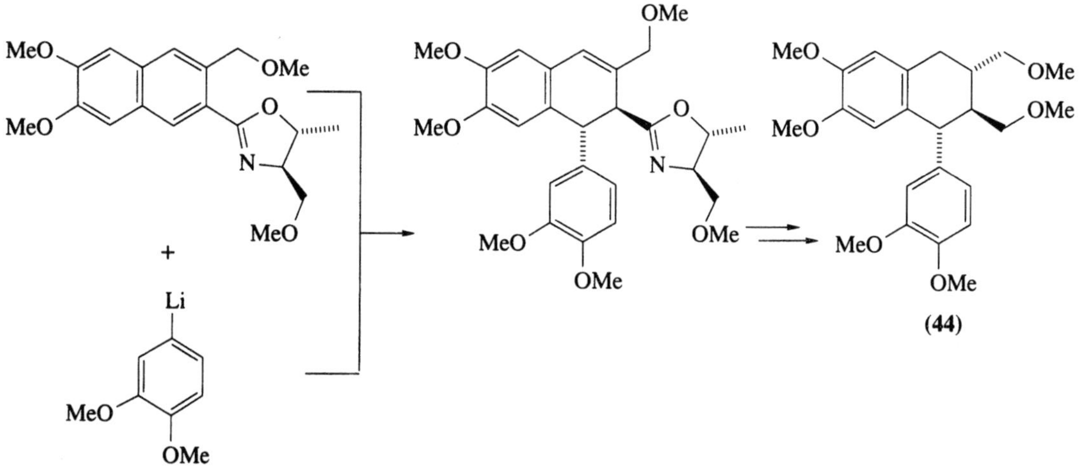

(**44**)

Scheme 17

The stereochemical outcome of all these reactions was rationalized in terms of a chelate complex formed between the reactants, from which the R′-group is then transferred to the alkene as drawn in (**45**). Nevertheless, in the case of substrates containing additional stereogenic centers and other coordinating functional groups, a reliable prediction of the product configuration is not always possible.[50]

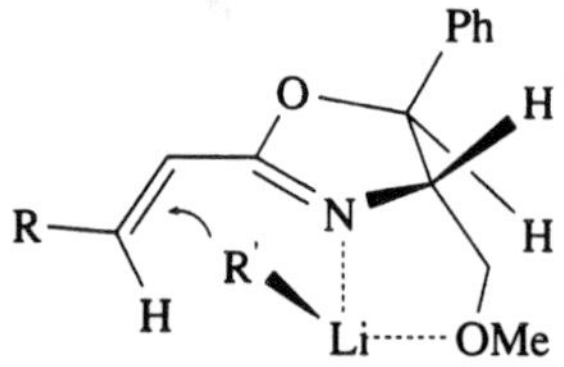

(**45**)

1.5.2.1.5 Additions to other chiral derivatives

Another type of chiral Michael acceptor, the oxazepine derivatives (**47**), is prepared by condensation of the (–)-ephedrine-derived malonic acid derivative (**46**) with aldehydes (Scheme 18).[51,52] Treatment of (**47**) with a variety of Grignard reagents in the presence of $NiCl_2$ affords, after hydrolysis and decarboxylation, the 3-substituted carboxylic acids (**48**), in most cases with more than 90% *ee*. Diastereoselective Michael additions to (**47**) were also used for the preparation of optically active cyclopropane derivatives (**49**)[53] and β-substituted-γ-butyrolactones (**50**; Scheme 18).[54] A total synthesis of indolmycin is based on this methodology.[55]

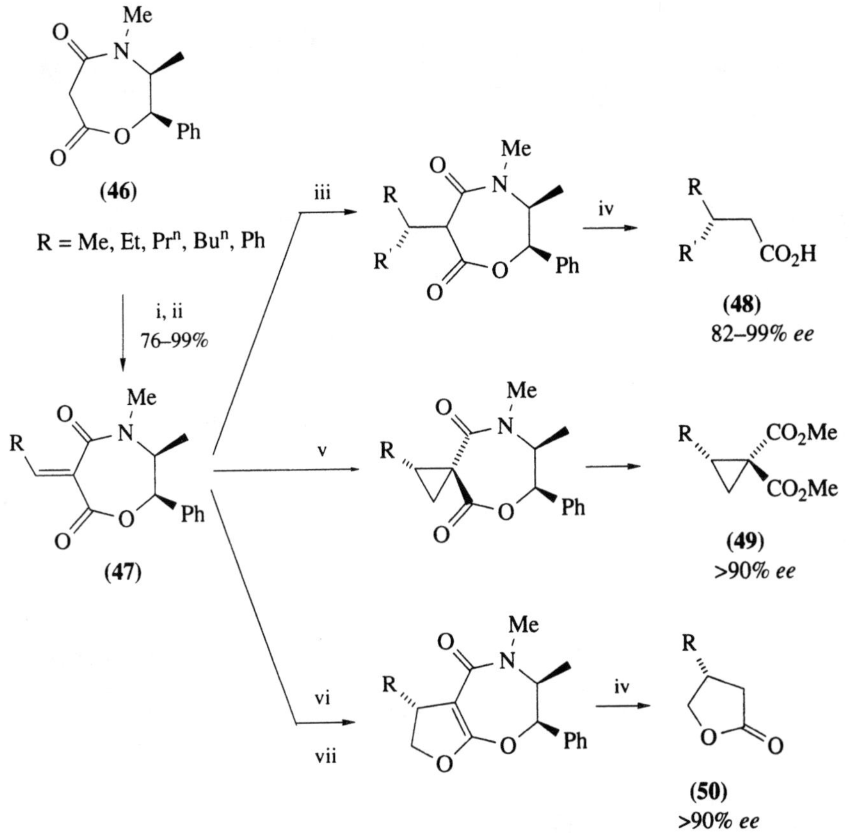

i, RCHO, $TiCl_4$, pyridine, THF, –78 °C to r.t., 12 h; ii, recryst.; iii, R'MgBr, $NiCl_2$, THF, –78 °C, 3 h; iv, 3 M H_2SO_4, AcOH, reflux, 6 h; v, $CH_2{=}SOMe_2$; vi, $PhSCH_2Li$, $NiCl_2$, THF/ toluene, –78 °C, 2 h; vii, $Me_3O^+BF_4^-$, CH_2Cl_2, 0 °C to r.t., 12 h

Scheme 18

A method for the preparation of 3-substituted succinaldehydic esters (**53**) involves conversion of the methyl ester of fumaraldehydic acid into the corresponding imidazoline (**52**) by reaction with the proline-derived chiral diamine (**51**; Scheme 19).[56] Cu^{I}-catalyzed conjugate addition of alkyl Grignard reagents and subsequent acidic hydrolysis yields the products (**53**), generally with high *ee*. A very closely related method uses a norephedrine-derived chiral auxiliary (Scheme 20).[57,58] Another related approach employs the chiral imidazoline (**54**), which on treatment with R_2CuLi reagents (R = Me, Bu, Ph, 1-butenyl) followed by hydrolysis gives chiral aldehydic esters (**55**; Scheme 21).[59]

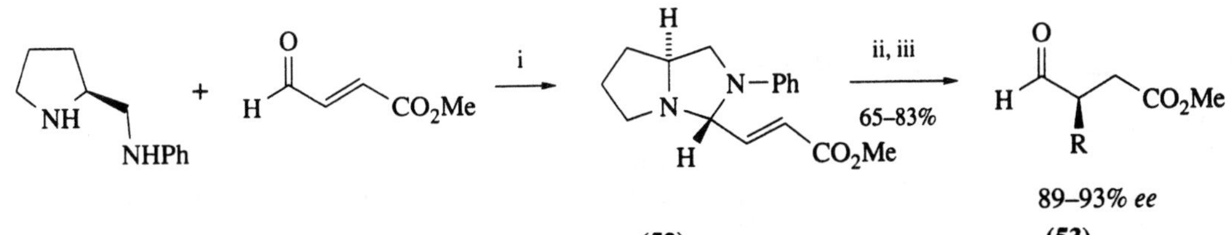

i, cat. H^+; ii, RMgBr, CuI, THF; iii, H_3O^+

Scheme 19

Conjugate addition of dialkylcuprates to the homochiral dioxinone (**57**), prepared from (*R*)-3-hydroxybutyric acid (**56**), produces single diastereomers (**58**), which can be hydrolyzed to the homochiral products (**59**).[60,61] The overall process (Scheme 22) constitutes an example for 'self-regeneration of a stereogenic center'. Interestingly, the high stereoselectivity can not be attributed to simple steric effects.

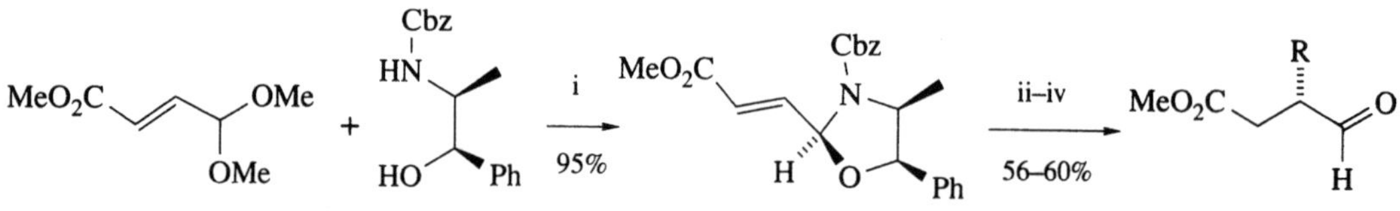

(R = Me, Et, Bu, Vinyl)

78–93% *ee*

i, BF_3•Et_2O, benzene; ii, R_2CuLi, Et_2O, –25 °C; iii, $HSCH_2CH_2SH$, BF_3•Et_2O, CH_2Cl_2; iv, MeI, $CaCO_3$, H_2O

Scheme 20

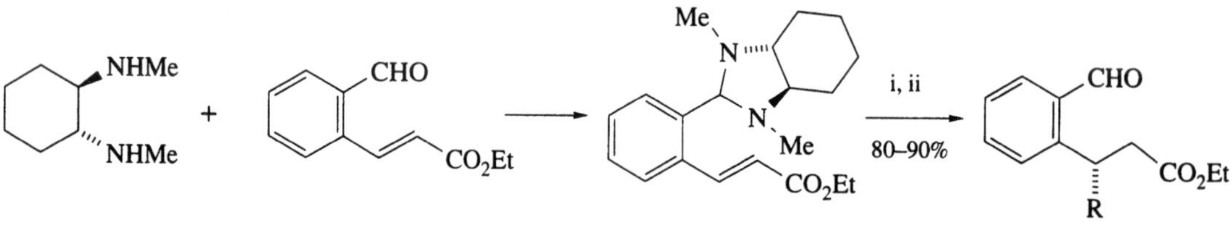

i, R_2CuLi, ether, 0 °C; ii, H_3O^+

(54) **(55)** 90–96% *ee*

Scheme 21

The pyramidalization of the trigonal centers (a common feature found in the X-ray structures of several related dioxinones) is probably responsible for the high selectivity.[61] The nucleophile preferentially attacks the electrophilic center from that (convex) face into which it is pyramidalized, in accordance with the principle of minimization of torsional strain. A related example for asymmetric 1,4-addition to chiral dioxinones was reported.[62]

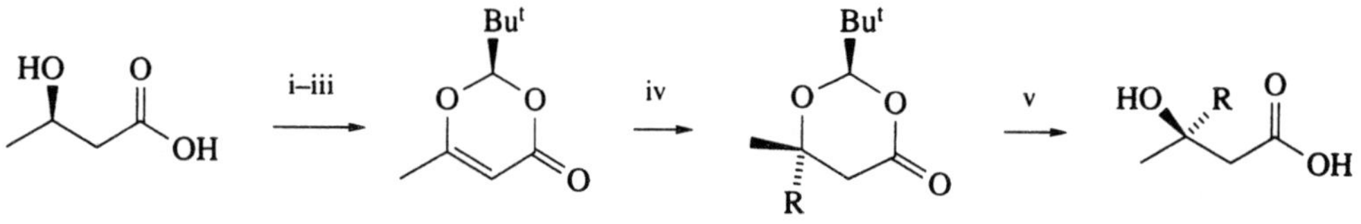

(56) **(57)** **(58)** **(59)** >99% *ee*

i, Bu^tCHO, CH_2Cl_2, Dowex 50 W, azeotropic removal of water; ii, LDA, THF, PhSeCl, –75 °C; iii, H_2O_2, pyridine, CH_2Cl_2, 0 °C; iv, R_2CuLi, ether, –75 °C, 2 h; v, 1 M HCl, THF, r.t., 45 min

Scheme 22

1.5.2.2 Additions to Chiral Derivatives of α,β-Unsaturated Aldehydes and Ketones

1.5.2.2.1 Additions to chiral acetals and ketals

The conversion of an α,β-unsaturated aldehyde or ketone into an allylic acetal or ketal, followed by S_N2'-type attack of a nucleophile, leads, after hydrolysis of an initially formed enol ether, to a β-substituted carbonyl compound. The overall sequence (Scheme 23) is equivalent to a direct conjugate addition, but has the advantage that it allows the temporary introduction of a chiral auxiliary group: if a chiral (C_2-symmetric) diol is used in the acetalization step, the subsequent nucleophilic addition leads to a mix-

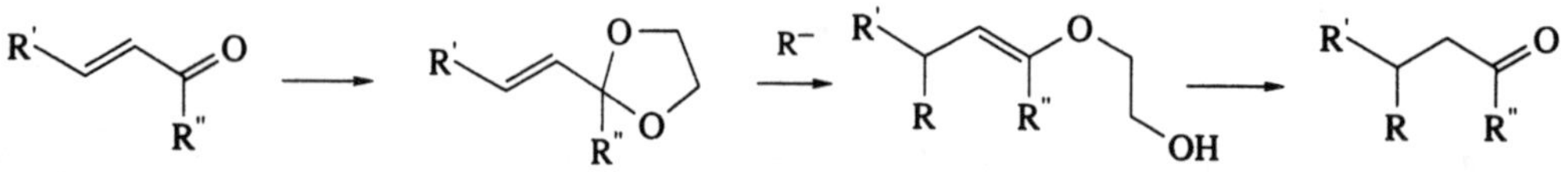

Scheme 23

ture of two diastereomeric enol ethers (generally in unequal amounts) which may be hydrolyzed to give the product in optically active form.

As a first (isolated) example of such a transformation, some asymmetric induction (26% *ee*) was observed during the addition of Me_2CuLi/BF_3 to the chiral ketal, derived from cyclohexenone and (*R*,*R*)-butane-2,3-diol.[63] The addition of trialkylaluminum compounds to α,β-unsaturated ketals and acetals derived from (*R*,*R*)-*N*,*N*,*N'*,*N'*-tetramethyltartaric acid diamide (**60**) or its enantiomer (*ent*-**60**), leads to much better results.[64–66] The initially formed enol ethers are first isolated as their corresponding acetates and then hydrolyzed to the actual β-substituted carbonyl compounds. While the ketals, *e.g.* (**61**), react cleanly and with reasonable diastereoselectivities to (**62**; Scheme 24), the acetals (**63**) generally lead to mixtures of regioisomers (**64**) and (**65**), however with an excellent degree of asymmetric induction (Scheme 25). The choice of the solvent is crucial: the desired isomer (**64**) is formed predominantly in less polar solvents like toluene or 1,2-dichloroethane, while more polar solvents (*e.g.* chloroform) strongly favor the direct S_N2 attack at the acetal center, leading to the (undesired) allylic ether derivative (**65**). The usefulness of this methodology is demonstrated by the synthesis of the alcohol (**66**), a structural subunit of vitamins K and E (Scheme 26).[65]

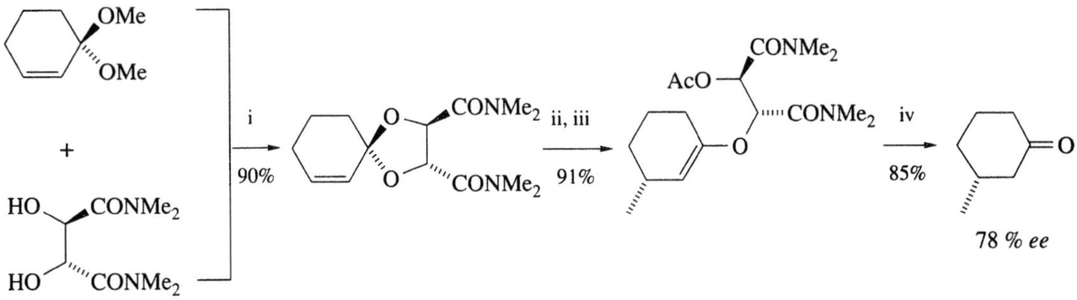

(**60**) (**61**) (**62**) 78 % *ee*

i, cat. TsOH; ii, 5 equiv. Me_3Al, toluene, 3.5 h, r.t.; iii, Ac_2O, pyridine, DMAP, 1 h, r.t.; iv, 6 M HCl, dioxane

Scheme 24

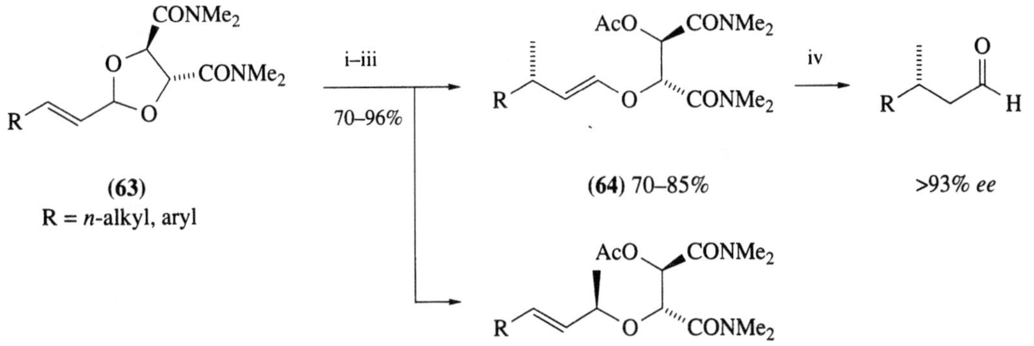

(**63**) R = *n*-alkyl, aryl

(**64**) 70–85% >93% *ee*

(**65**) 15–30%

i, 5 equiv. Me_3Al, 1,2-dichloroethane or toluene, 0 °C to r.t., 12 h; ii, Ac_2O, pyridine, DMAP, 1 h, r.t.; iii, separation of isomers by chromatography; iv, 6 M HCl, dioxane

Scheme 25

The stereochemical outcome of all these reactions may be explained in terms of a stereoelectronically assisted *syn* S_N2' process, assuming a substrate conformation as shown in (**67a**), and a preferred coordination of the Lewis acid to the less hindered acetal oxygen.[67]

In contrast, the corresponding reactions of RCu/BF_3 reagents (R = phenyl or alkenyl) with allylic acetals obviously proceed in an *anti* S_N2' fashion, as shown in (**67b**). The unsaturated acetals (**68**), prepared from (*R*,*R*)-butane-1,2-diol, are opened by the PBu_3-stabilized reagents to give, after hydrolysis, the β-substituted aldehydes (**69**) with high *ee* (Scheme 27).[68,69] The method has been applied to the preparation of a key intermediate (85% *ee*) for the synthesis of the California red scale pheromone.[69]

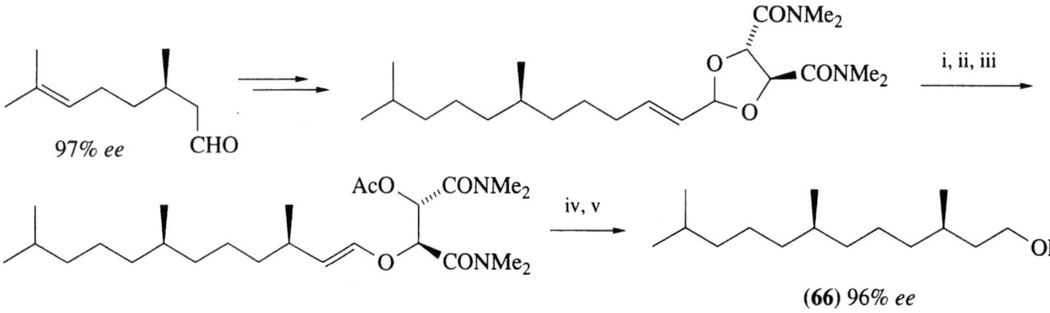

97% *ee* CHO

i, ii, iii

iv, v

(66) 96% *ee*

i, $AlMe_3$, toluene; ii, Ac_2O, pyridine; iii, chromatography; iv, 6M HCl, dioxane; v, $NaBH_4$

Scheme 26

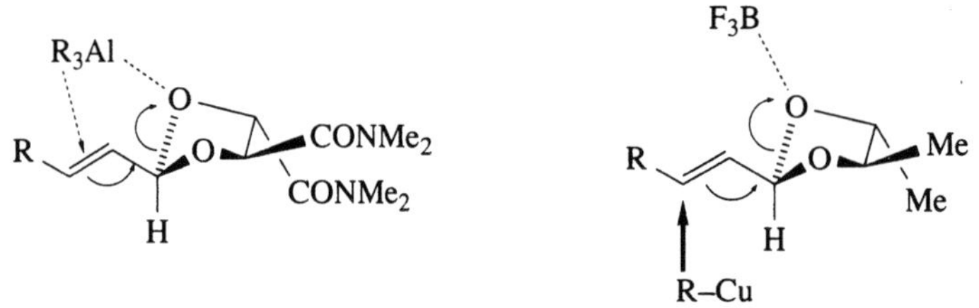

(67a) **(67b)**

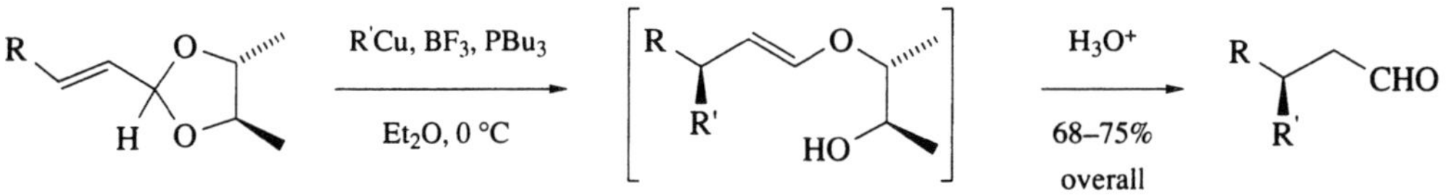

$R'Cu$, BF_3, PBu_3; Et_2O, 0 °C; H_3O^+, 68–75% overall

(68) R = Me, Pr R' = Ph, alkenyl **(69)** 85–95% *ee*

Scheme 27

1.5.2.2.2 Additions to chiral N,O*-acetals*

Analogous to the use of chiral acetals one can employ chiral *N,O*-acetals, accessible from α,β-unsaturated aldehydes and certain chiral amino alcohols, to prepare optically active β-substituted aldehydes *via* subsequent S_N2' addition and hydrolysis. However, the situation is more complicated in this case, since the *N,O*-acetal center constitutes a new stereogenic center which has to be selectively established. The addition of organocopper compounds to α,β-ethylenic oxazolidine derivatives prepared from unsaturated aldehydes and ephedrine was studied.[70–78] The (diastereo) selectivities were rather low (<50% *ee* after hydrolysis) in most cases, the highest value being 80% *ee* in a single case.[75] There is a strong solvent effect in these reactions, *e.g.* in the addition of lithium dimethylcuprate to the (*E*)-cinnamaldehyde-derived oxazolidine (**70**; Scheme 28):[73] the (*R*)-aldehyde (**71**) is formed preferentially in polar solvents, while the (*S*)-enantiomer (*ent*-**71**) is the major product in nonpolar solvents like hexane. This approach was utilized in the preparation of citronellal (80% *ee*) from crotonaldehyde (40% overall yield).[78]

1.5.2.2.3 Additions to chiral imines

The 1,4-addition of carbon nucleophiles to chiral α,β-unsaturated aldimines gives optically active β-substituted aldehydes.[79–86] Best results are obtained with imines prepared from *t*-leucine *t*-butyl ester (**72**). Addition of Grignard reagents to the acyclic imines (**73**) followed by acidic hydrolysis and reduction gives the alcohols (**74**) with high *ee* (Scheme 29).[80,83] The potassium salt of diethyl malonate also adds smoothly to (**73**), however providing a product of opposite configuration (48% yield, up to 86% *ee*).[81,84] Addition of Grignard reagents to cyclic compounds of type (**75**) leads initially to the intermedi-

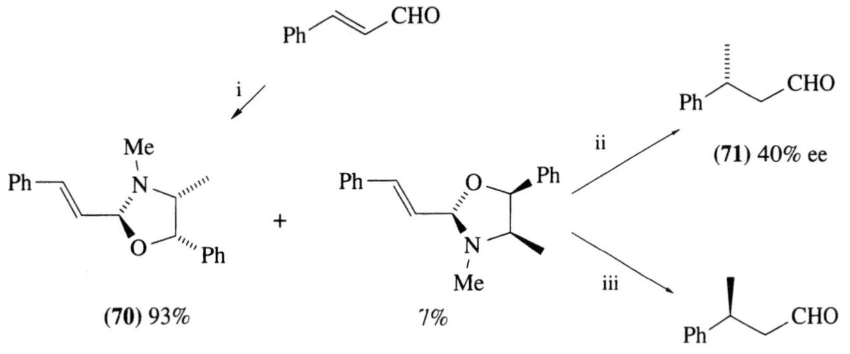

i, (+)-ephedrine, 4 Å mol. sieve, r.t.; ii, Me_2CuLi, Et_2O–HMPA (1:1), –42 °C, H_3O^+;
iii, Me_2CuLi, hexane, –42 °C, H_3O^+

Scheme 28

ate (*Z*)-enamides (**76**), which can either be (kinetically) protonated or alkylated (possibly after isomerization to the thermodynamically more stable (*E*)-isomer), allowing a selective entry into each class of compounds (**77**), (**78**) or (**79**; Scheme 30).[82,85] The products are obtained reliably with high *ee* (82–93%). The method was applied in a synthesis of (+)-ivalin (**80**; Scheme 31).[87]

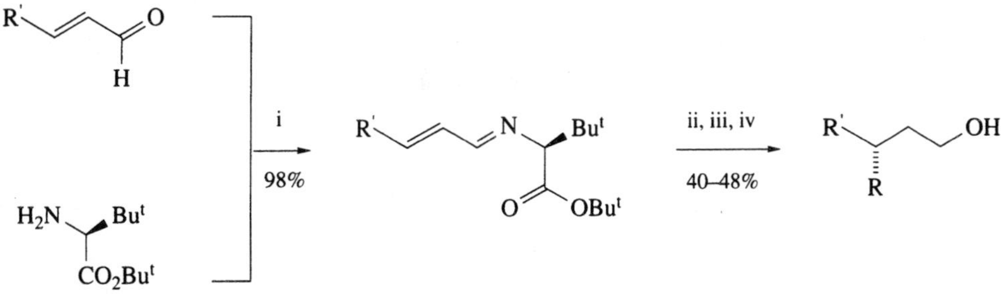

i, 4 Å mol. sieve, hexane, 0 °C to r.t.; ii, 2 equiv. RMgBr (R = Ph, c-C_6H_{11}, Bu^n, Et, $Me_2C{=}CH(CH_2)_2$), Et_2O–THF (5:1), –55 °C, 1.5 h; iii, 2 M HCl; iv, $NaBH_4$

Scheme 29

1.5.2.2.4 Additions to other chiral derivatives

Although desirable, the recovery of a chiral auxiliary is not always crucial. Some techniques employ α,β-unsaturated carbonyl compounds containing a 'disposable' stereogenic center, which is removed after the (asymmetric) conjugate addition step has been performed. This concept appears to be useful only if the chiral substrate is easily accessible in high enantiomeric purity.

Conjugate additions to 4-oxo-2-cyclopentenyl acetate (**81**), a compound readily available in both enantiomeric forms,[90] lead (after elimination of the acetoxy group) to substituted cyclopentenones (**82**) (Scheme 32).[88,89] For instance, this methodology was used, though employing (*rac*-**81**), in a synthesis of brefeldin A.[91] The chiral aldehyde (**83**), which is accessible from arabinose, was used in syntheses of botryodiplodin (**84**) and of lignans like (–)-burseran (**85**; Scheme 33).[92,93] In both syntheses, the conjugate addition constitutes a key step and proceeds with high diastereoselectivity. The preparation of homochiral 5-substituted cyclohexenones is possible *via* copper-catalyzed addition of Grignard reagents to 5-trimethylsilyl-2-cyclohexenone (**86**; or *ent*.-**86**) followed by oxidative elimination of the TMS group.[94–96] Multigram amounts of both enantiomers of (**86**) may be prepared in essentially pure form quite easily from anisole. According to Scheme 34, the method was used for the synthesis of (+)-α-curcumene (**87**) and of (–)-β-vetivone (**88**). Additions of Gilman reagents to 5-methoxy-2-cyclopentenone

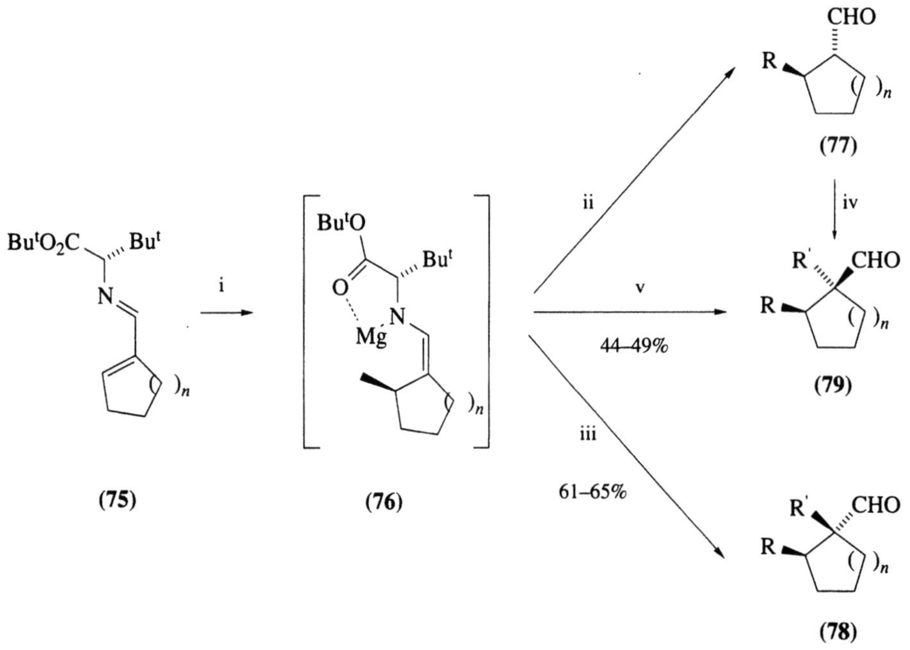

$n = 1$ or 2   R = Ph, vinyl   R' = Me, Et, benzyl, allyl, MOM

i, RMgBr, THF, –23 °C; ii, H_3O^+; iii, R'X, HMPA, THF, –23 °C to r.t., 15 h, then H_3O^+; iv, KH, R'X, THF, –23 °C; v, reflux for 3 to 6 h, then R'X, HMPA, THF, –23 °C to r.t., 15 h, then H_3O^+

Scheme 30

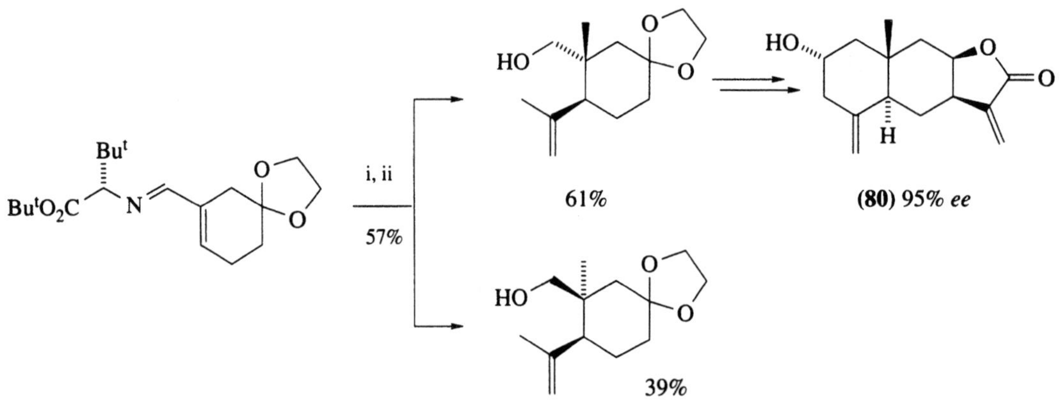

i, 2-propenylMgBr, THF, then MeI, HMPA, then H_3O^+; ii, $NaBH_4$, MeOH

Scheme 31

(**89**) also proceed with high diastereoselectivity to give 1,4-adducts, from which the methoxy group can be removed by treatment with samarium iodide to afford 3-substituted cyclopentanones of high enantiomeric purity (Scheme 35).[97,98] Another method, which does not seem to be of high preparative value though, involves diastereoselective complexation of the (enantiomerically pure) enone (**90**) by treatment with tricarbonyl(η^4-benzylideneacetone)iron to give a single complex (**91**). Reaction with α-lithioisobutyronitrile leads (after TFA quench) to compound (**92**), which is subsequently subjected to periodate cleavage to afford the pure carboxylic acid (**93**; Scheme 36).[99]

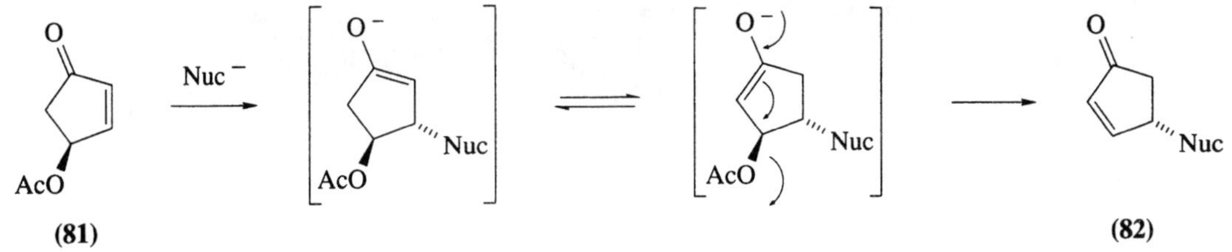

(81) (82)

Scheme 32

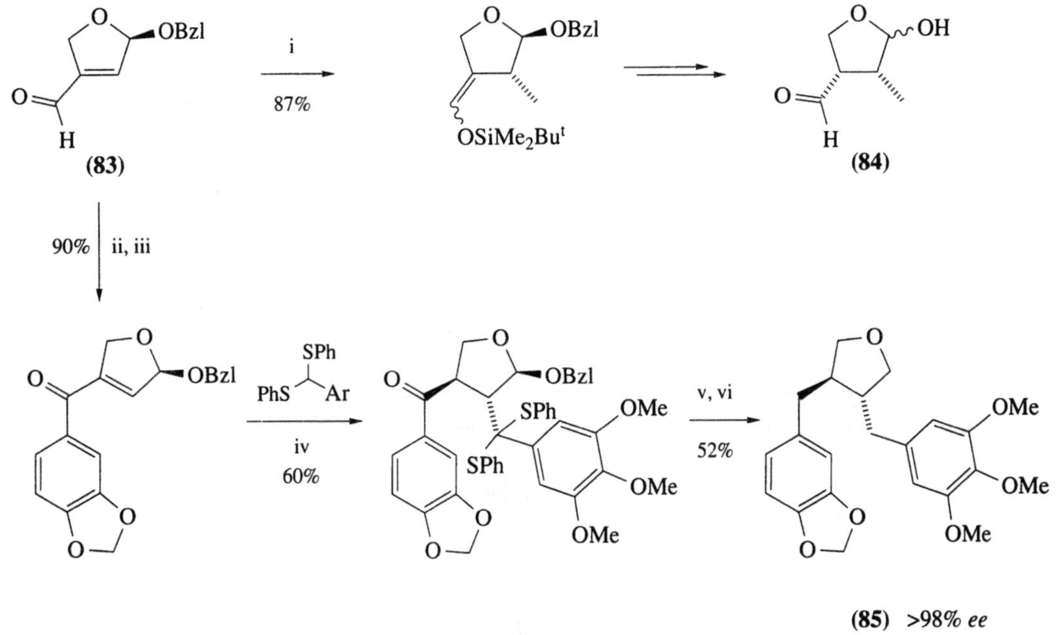

(83) (84)

(85) >98% *ee*

i, LiMeCNCu, Bu^tMe_2SiCl, THF, –78 °C to r.t.; ii, ArMgBr, THF, 0 °C; iii, CrO_3/pyridine, CH_2Cl_2, r.t., 15 min.; iv, Bu^nLi, THF, –78 °C to r.t.; v, Raney Ni, DME, 0 °C; vi, H_2, Pd/C, $AcOH/H_2O$, 6.5 h

Scheme 33

1.5.2.3 Additions to Chiral Vinylic Sulfur Compounds

Since early investigations about the asymmetric addition of diethyl sodiomalonate to optically active vinylic sulfoxides,[100,101] Posner and his coworkers[102–117] have developed a highly useful methodology based on the conjugate addition of carbon nucleophiles to homochiral α-arylsulfinyl-α,β-unsaturated carbonyl compounds. While acyclic derivatives still lead only to moderate results,[105] the strength of this method is for cyclic systems. For example, the 2-sulfinyl-2-cycloalkenones (**94**) and (**95**), the 2-sulfinyl-2-alkenolides (**96**) and (**97**), as well as their respective enantiomers are excellent substrates. All these compounds are quite readily accessible in enantiomeric purities of >98% and are configurationally stable, at least for several months at 0 °C.

Reaction of a variety of organometallic reagents (RM) with (**94**) or (**95**) leads diastereoselectively to addition products of type (**98**), which are then desulfurized yielding 3-substituted cycloalkanones (**99**) of reasonable to high enantiomeric purities (Scheme 37 and Table 2).[106,107,110] The steric course of these addition reactions can be understood in terms of an intermediate chelate, which is attacked by the nucleophile from the less hindered face, as shown in (**100a**). In contrast, dialkylmagnesium[108] and dialkylcuprate[109] reagents afford products with an absolute configuration opposite to that obtained with other reagents. This in turn can be rationalized as the result of an attack on the nonchelated substrate molecule, which should prefer the conformation shown in (**100b**). The role of the metal in controlling the stereochemistry of these reactions (chelate *versus* nonchelate mode) was the subject of a recent theoretical study.[119,120] Ester enolates may also be added as donors to these substrates (attack in the nonchelate addition mode),[115] especially with α-substituted derivatives leading to extraordinary high levels of asymmetric induction in some cases (*e.g.* Scheme 39).

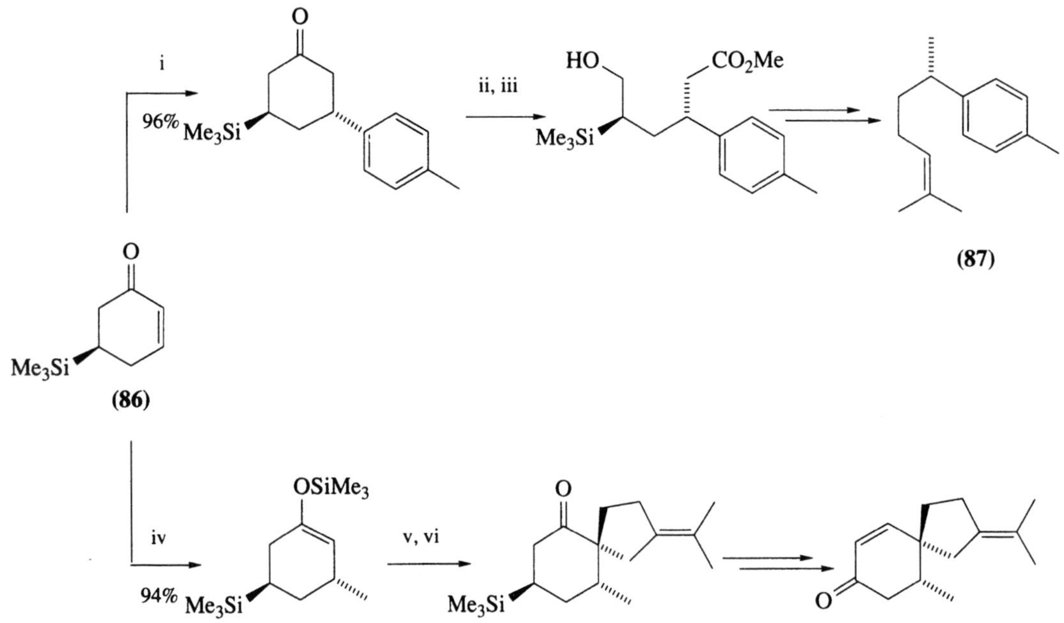

O
i
96% Me_3Si
ii, iii
HO
CO_2Me
Me_3Si
O
(87)
Me_3Si
(86)
$OSiMe_3$
iv
94% Me_3Si
v, vi
O
Me_3Si
O
(88)

i, *p*-TolMgBr, $CuBr{\bullet}Me_2S$, Me_3SiCl, HMPA; ii, MCPBA; iii, NaOMe, MeOH; iv, MeMgI, $CuBr{\bullet}Me_2S$, Me_3SiCl, HMPA; v, 1,4-dibromo-2-isopropylidenebutane, $ZnBr_2$; vi, NaOMe, THF, r.t. to 50 °C, 2.5 h

Scheme 34

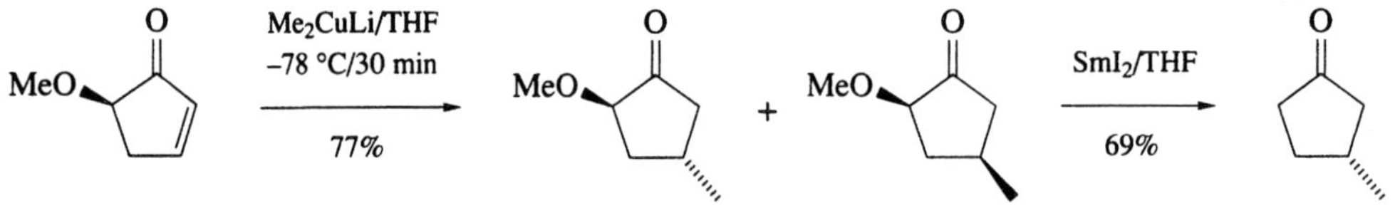

O
MeO
(89)
Me_2CuLi/THF
–78 °C/30 min
77%
O
MeO
98%
+
O
MeO
2%
SmI_2/THF
69%
O

Scheme 35

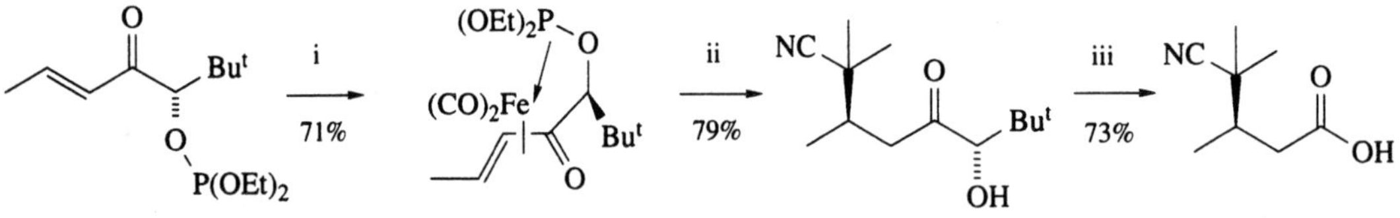

O
Bu^t
O
$P(OEt)_2$
(90)
i
71%
$(OEt)_2P$
O
$(CO)_2Fe$
Bu^t
O
(91)
ii
79%
NC
O
Bu^t
OH
(92)
iii
73%
NC
O
OH
(93) >99% *ee*

i, tricarbonyl(benzylideneacetone)iron, THF, 60–65 °C, 20 h; ii, α-lithioisobutyronitrile, THF, HMPA, –78 °C, then TFA; iii, H_5IO_6, MeOH, H_2O

Scheme 36

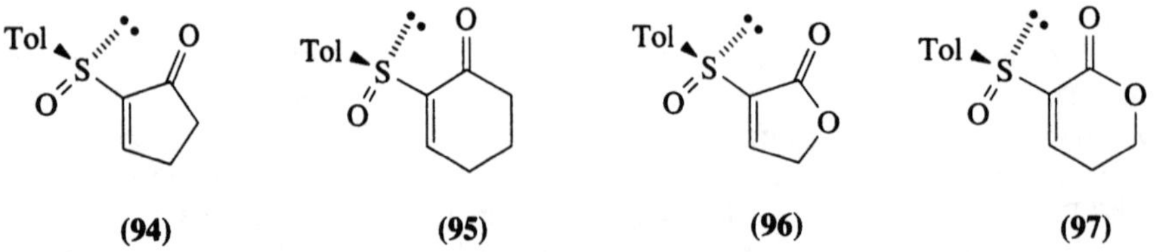

Tol S O O
(94)
Tol S O O
(95)
Tol S O O O
(96)
Tol S O O O
(97)

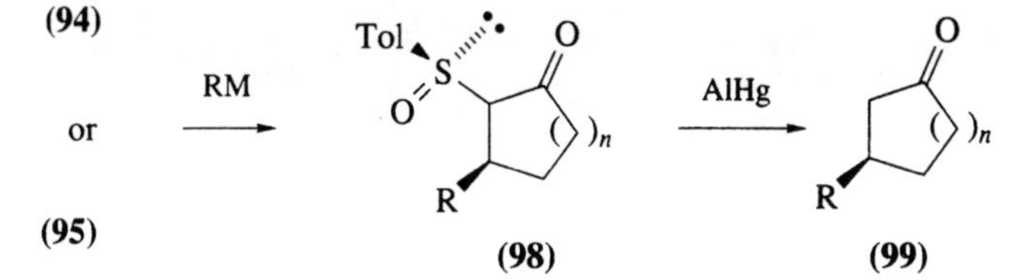

Scheme 37

Table 2 Preparation of Cyclic β-Substituted Carbonyl Compounds (**99**) *via* Addition of Organometallic Reagents to Unsaturated Sulfoxides (**94**) or (**95**) Followed by Desulfurization of the Intermediates (**98**; Scheme 37)

Starting material	*RM*	*R*	*Overall yield (%)*	*ee of* (**99**) *(%)*
(**94**)	$ZnBr_2$/MeMgBr	Me	89	87
(**94**)	$MeTi(Pr^iO)_3$	Me	90	90
(**94**)	MeMgCl	Me	91	95–100
(**95**)	$ZnBr_2$/MeMgBr	Me	95	62
(**95**)	$MeTi(Pr^iO)_3$	Me	85	86
(**94**)	$ZnBr_2$/EtMgCl	Et	90	80
(**94**)	$EtTi(Pr^iO)_3$	Et	67	>98
(**94**)	$ZnBr_2$/VinylMgBr	Vinyl	75	92
(**94**)	$ZnBr_2$/PhMgCl	Ph	70	>98
(**94**)	2-NaphthylMgBr	2-Naphthyl	90	>98

(**100a**) (**100b**)

Several applications in total syntheses exemplify the value of this methodology; 11-oxoequilenin methyl ether (**101**; Scheme 38),[105,107] (+)-α-cuparenone,[109] (–)-podorhizon,[112] (–)-methyl jasmonate (**102**; Scheme 39),[114] (+)-estrone methyl ether,[116] and the so called (+)-A-factor (**103**; Scheme 40)[117] were all prepared in high enantiomeric purity. Other applications constitute preparations of 2-alkylchroman-4-ones,[118] and of 3-vinylcyclopentanones, highly valuable intermediates for steroid total synthesis.[106,107]

In the course of a total synthesis of aphidicolin (**107**), the conjugate addition of the dienolate (**104**) to the chiral butenolide derivative (**105**) serves as a key step. A 7.4:1 mixture of diastereomeric products is obtained, from which the major isomer (**106**) can be isolated in pure form after recrystallization (Scheme 41).[121] The selectivity of this remarkable reaction, in which two quaternary stereogenic centers are simultaneously generated in a highly selective manner, can be explained by the assumption that the reactants approach each other in the chelate-mode indicated in (**108**).

Another application of additions to chiral vinylic sulfoxides constitutes the Hantzsch-type reaction of methyl 3-aminocrotonate with compound (**109**), yielding the dihydropyridine derivative (**110**) as a single

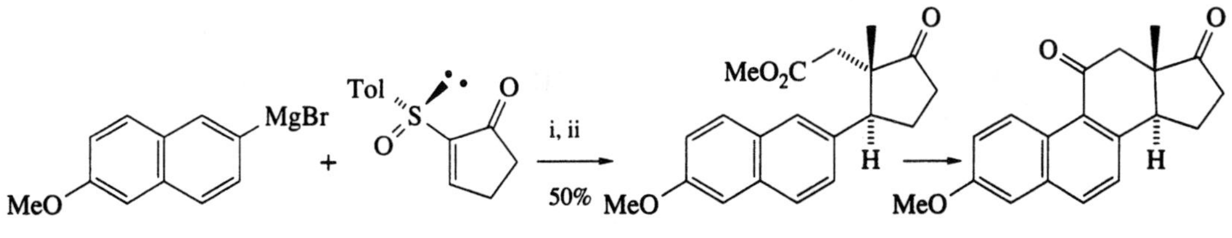

(**94**) >98% *ee* (**101**)

i, THF, –78 °C, then MeI, HMPA; ii, 3 equiv. Me_2CuLi, Et_2O, THF, 0 °C, 2 h, then $BrCH_2CO_2Me$, 25 °C

Scheme 38

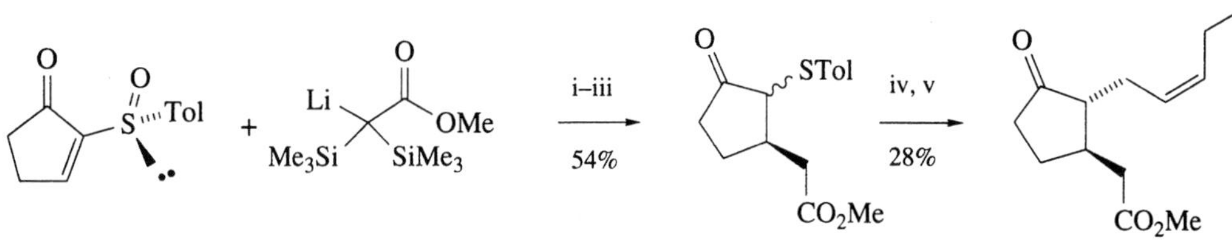

(*ent*-**94**) (**102**) >98% *ee*

i, THF, –78 °C, 5 min; ii, P_2I_4; iii, KF, iv, NaH, *(Z)*-2-pentenyl bromide; v, Raney Ni

Scheme 39

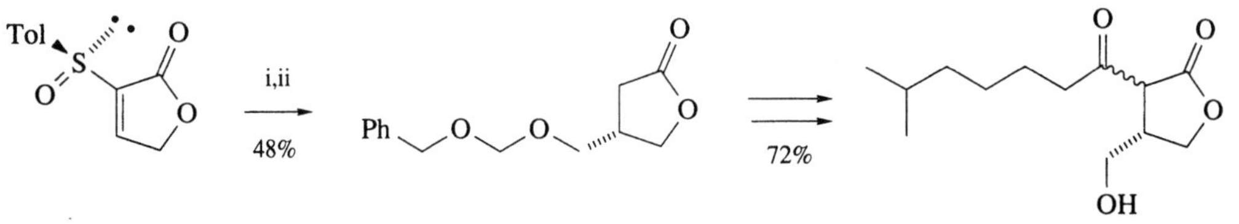

(**96**) 87% *ee* (**103**)

i, $PhCH_2OCH_2OCH_2Li$, 2,5-dimethyl-THF, –78 °C, 10 min; ii, Raney Ni, acetone

Scheme 40

(detectable) diastereomer, which was then oxidized to give the sulfone (**111**) with high *ee* (Scheme 42).[122]

Optically active organic compounds, bearing a trifluoromethyl substituent at a stereogenic center, are accessible *via* addition of enolates to the chiral sulfoxides (**112**) or (**113**).[123] Ephedrine-derived sulfoximines of type (**114**) also undergo diasteroselective addition of organometallic reagents, allowing the preparation of 3-alkylalkanoic acids with high *ee*.[124,125] The chirally modified vinylic sulfone (**115**) was utilized in the asymmetric preparation of α-substituted aldehyde derivatives (Scheme 43).[126]

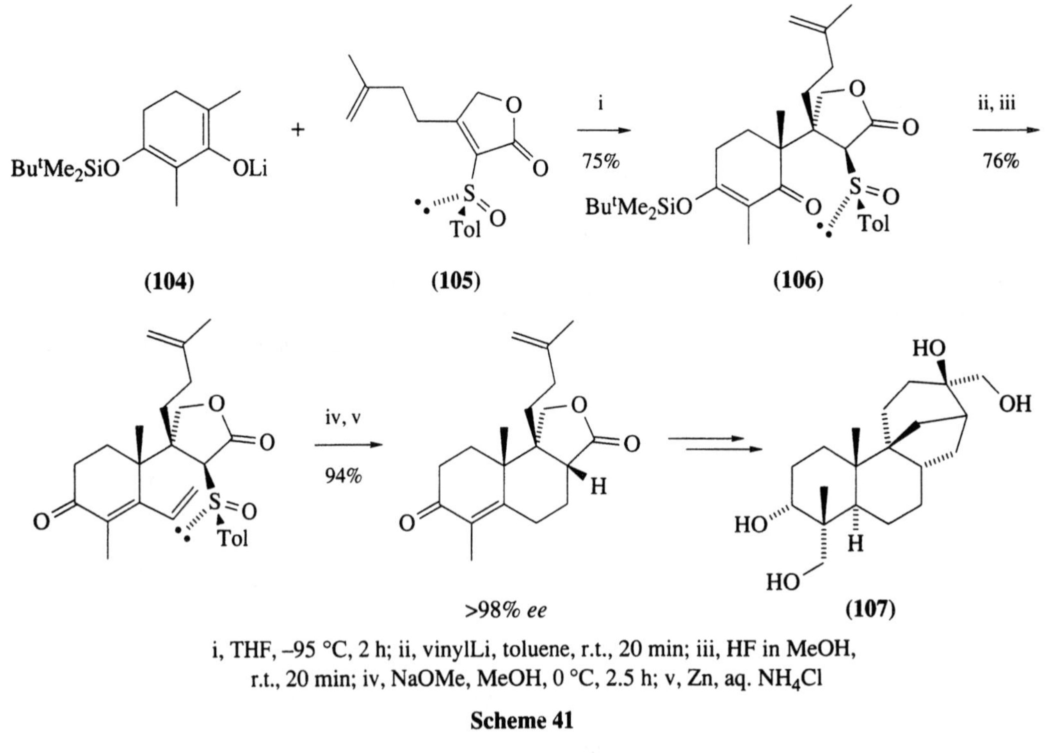

(**104**) (**105**) (**106**)

>98% *ee* (**107**)

i, THF, –95 °C, 2 h; ii, vinylLi, toluene, r.t., 20 min; iii, HF in MeOH, r.t., 20 min; iv, NaOMe, MeOH, 0 °C, 2.5 h; v, Zn, aq. NH_4Cl

Scheme 41

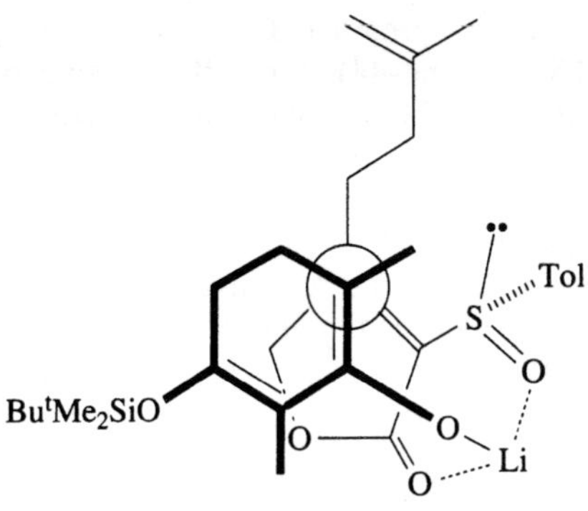

(108)

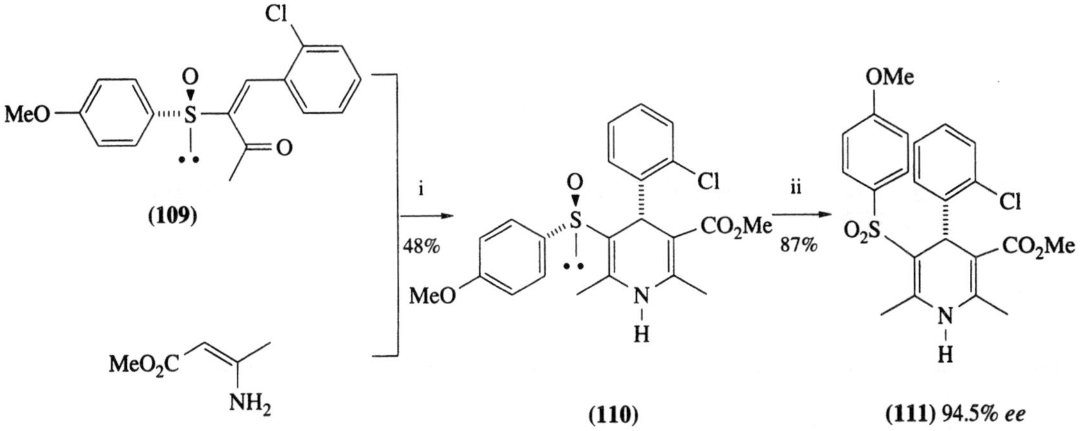

(110)

(111) 94.5% *ee*

i, MeOH, reflux; ii, Bu^{t}O$_2$H, 18-crown-6, KOH, EtOH

Scheme 42

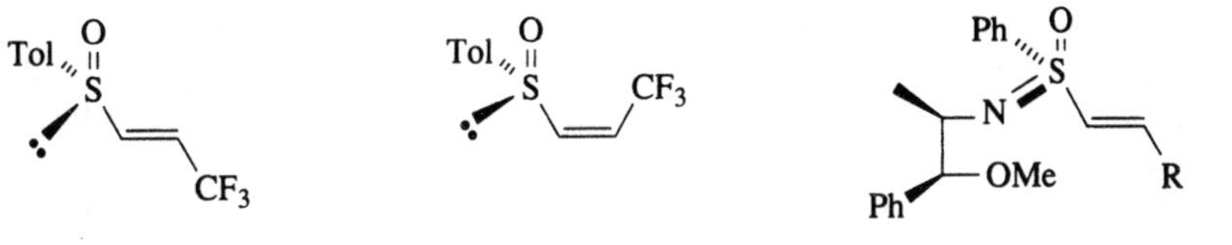

(112) **(113)** **(114)**

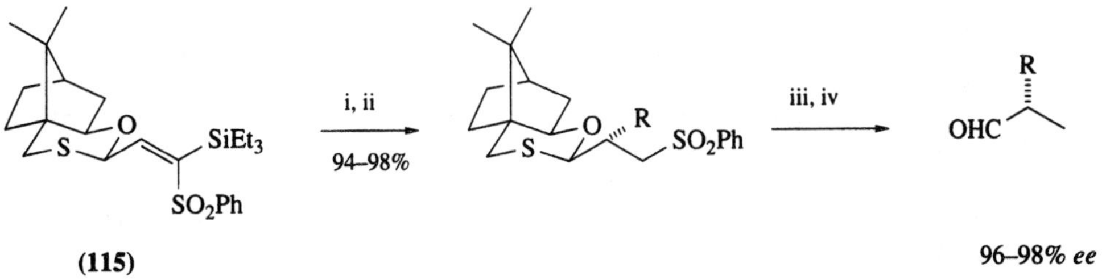

(115)

96–98% *ee*

i, RLi, LiBr–Et$_2$O, hexane, –78 °C; ii, Bu$^n{}_4$F; iii, Na–Hg; iv, HgCl$_2$

Scheme 43

1.5.2.4 Additions to α,β-Unsaturated Acyl Ligands Bound to a Chiral Metal Center

Addition of organolithium compounds to chiral (*E*)-configured α,β-unsaturated iron acyl complexes **(116)** proceeds with an exceptionally high degree of asymmetric induction.[127–132] The initially formed anionic intermediates (**117**) were either protonated or trapped with methyl iodide yielding diastereomerically pure products (**118**; Scheme 44). Since the starting complexes are configurationally stable under ambient conditions and accessible in enantiomerically pure form, this method allows (after cleavage of the metal–carbon bond) the preparation of a variety of homochiral organic compounds. The selectivity observed for these reactions can be explained in terms of a quite rigid substrate conformation as indicated in Scheme 44, where the acyl C═O group is *anti* with respect to the carbonyl ligand. It was proposed that more than simple steric effects are responsible for this conformational preference of the acyl group, since the diastereoselectivity of the conjugate addition reaction does not change much with the steric bulk of the phosphine ligand.[133]

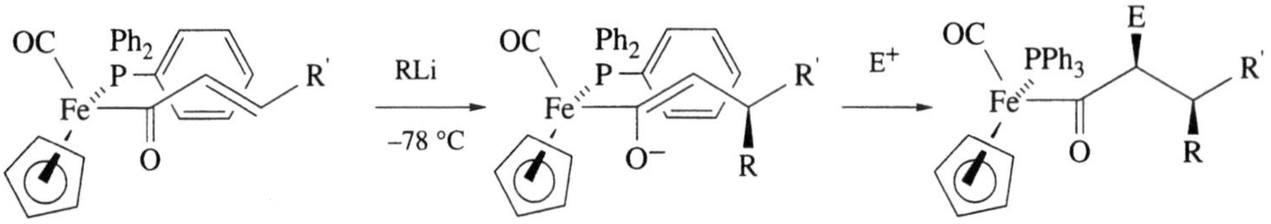

(**116**) R' = alkyl, Ph RLi, –78 °C (**117**) R = Me, Bu E^+ (**118**) E = H, Me

Scheme 44

1.5.3 DIASTEREOSELECTIVE ADDITIONS OF CHIRAL CARBON NUCLEOPHILES TO ACHIRAL SUBSTRATES

1.5.3.1 Reactions of Chiral Enolates

The chiral, nonracemic oxazepine derivative (**46**; Scheme 18) was studied as donor in the Michael addition to prochiral α,β-unsaturated carbonyl compounds.[134,135] The products were obtained with 44–55% *ee* after removal of the chiral auxiliary group. With 1-nitrocyclohexene as acceptor, somewhat better selectivities (62% *ee*) were observed.[136]

A double π-face selective aprotic Michael addition of the lithium dienolate, derived from the chiral senecioate (**119**), to cyclopentenone served for the total synthesis of (–)-kushimone (**120**; Scheme 45).[137,138] The selectivity of the key reaction was rationalized by the assumption of a transition state as shown in (**121a**). A related approach involves the Michael addition of enolates, derived from chiral propionates, to methyl (*E*)-crotonate (Scheme 46).[139] The formation of the '*threo*' isomer (**122**) as the major product indicates a transition state structure as shown in (**121b**). This method was utilized in the synthesis of the marine natural product 7,20-diisocyanoadociane (**123**; Scheme 47).[140]

The intramolecular Michael reaction of chiral acetoacetate derivatives (**124**) proceeds with excellent selectivity when the alcohol (**125**) is used as the chiral auxiliary (Scheme 48),[141] leading to valuable intermediates for the synthesis of homochiral 11-keto steroids.[142] The stereochemical outcome suggests a reactive conformation as shown in (**126**).

Diastereoselective Michael additions of enolates, prepared from the chiral amides (**127**) and (**128**), to prochiral α,β-unsaturated esters were utilized in the synthesis of (+)-dehydroiridodiol (**129**) and its isomer (**130**; Scheme 49).[143]

When Michael additions of chiral enolates to nitroalkenes were studied, it was found that lithium enolates (**132**) of 1,3-dioxolan-4-ones (**131**), derived from the corresponding α-hydroxy acids, afford the adducts (**133**) with high diastereoselectivity (Scheme 50).[144] Recrystallization leads, in general, to diastereomerically pure products, which in turn can efficiently be converted to homochiral compounds like (**134**), (**135**) or (**136**). A number of other chiral enolates (**137**)–(**140**) were also shown to undergo highly selective additions to nitroalkenes; however, product configurations were not determined in these cases.

Mukaiyama–Michael addition of a chiral ketene acetal to nonprochiral vinyl ketones gives products of 72–75% *ee*.[145] A chirally modified glycine derivative (Schiff-base) adds to vinylic phosphorus compounds to yield, after hydrolysis, products with 54–85% *ee*.[146] Another chiral glycine equivalent was used for the preparation of homochiral proline derivatives *via* diastereoselective addition to α,β-unsaturated aldehydes and ketones.[147,148]

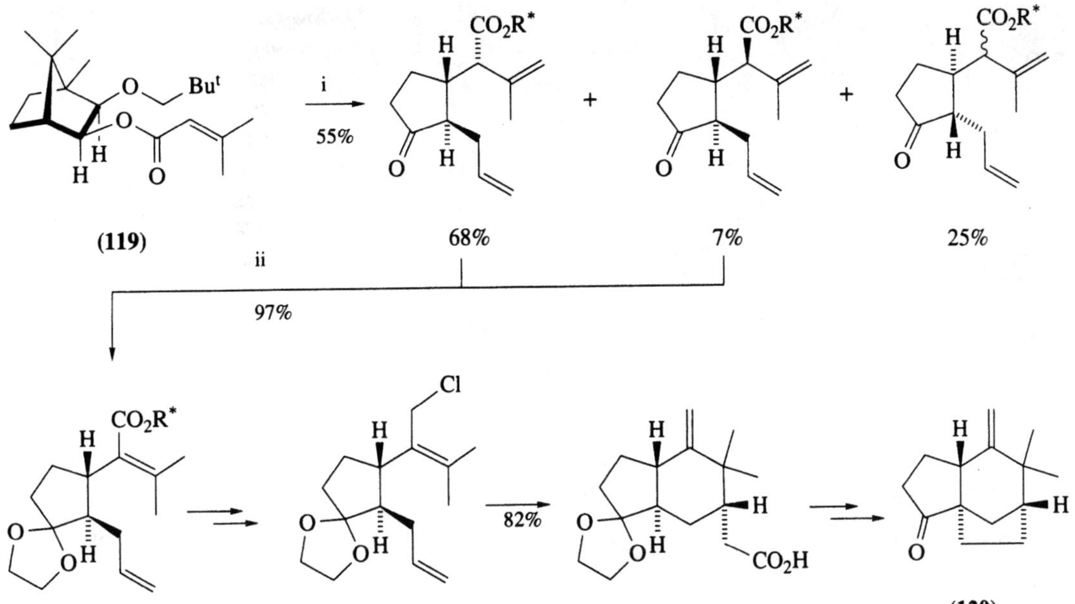

i, LDA, THF, –78 °C, then cyclopentenone, 5 min, then allyl bromide; ii, ethylene glycol, cat. TsOH, C_6H_6, reflux

Scheme 45

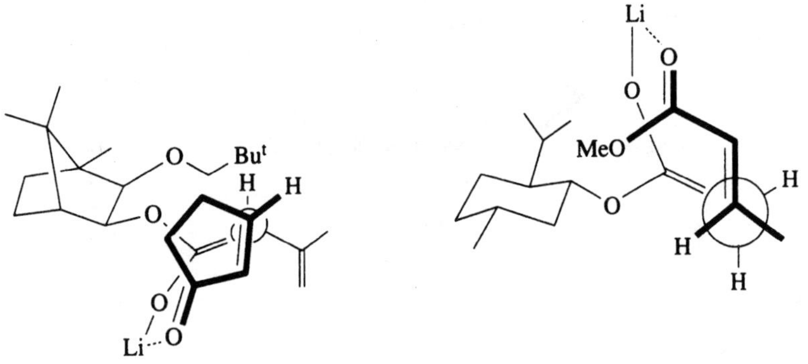

	(122)		'erythro' diastereomers
R^*OH = (–)-8-Ph-menthol	86%	4%	10%
R^*OH = (–)-menthol	69%	19%	12%

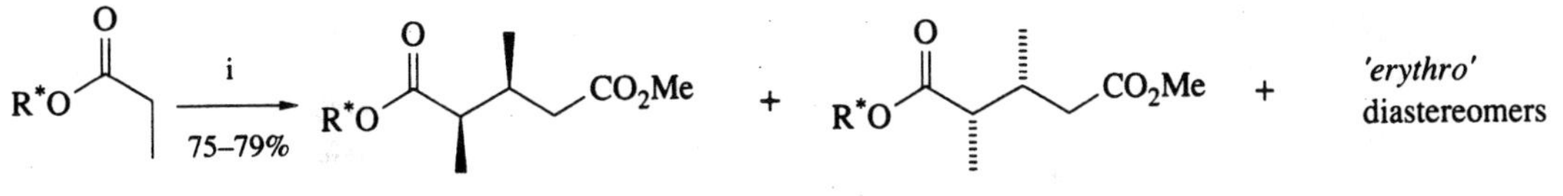

i, LDA, THF, –78 °C, 30 min, then *(E)*-methyl crotonate, –100 °C, 3 h, then AcOH quench

Scheme 46

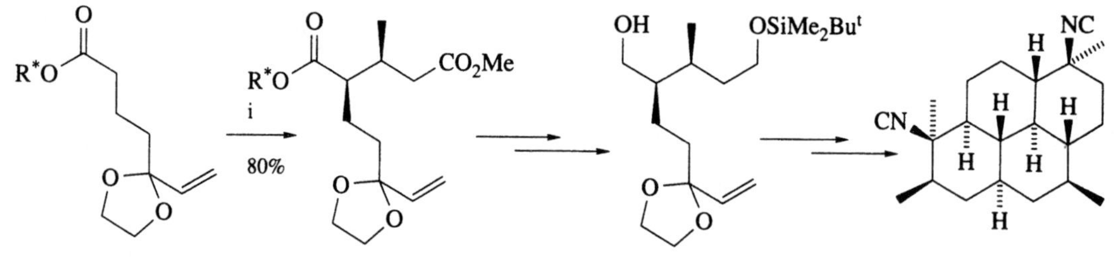

R^*OH = (–)-menthol + diastereomers 60% *ee* **(123)**

i, LDA, THF, –78 °C, 30 min, then *(E)*-methyl crotonate, –78 °C, 1 h, then AcOH quench

Scheme 47

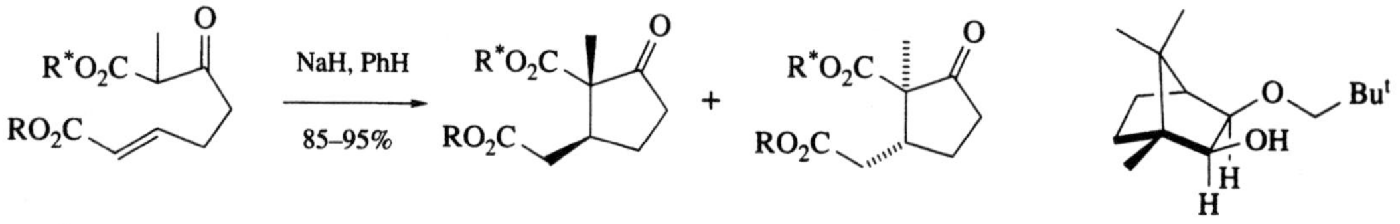

NaH, PhH 85–95%

(124)	R = Me	94%	6%
	R = Bu^t	>98%	<2%

(125)

Scheme 48

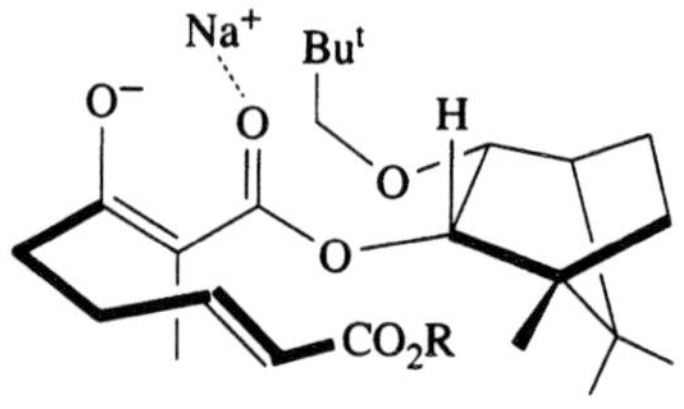

(126)

(127) **(128)**

(129) 85% *ee*

(130) 79% *ee*

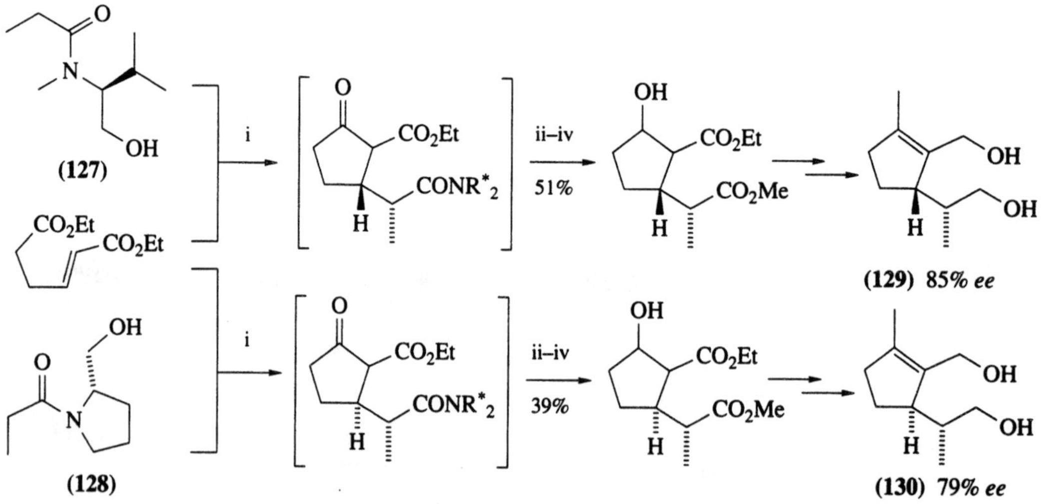

i, LDA, THF, Bu^tOK, –78 °C; ii, $NaBH_4$, EtOH, 0 °C, 30 min; iii, 2 M HCl, reflux, 2 h; iv, CH_2N_2, Et_2O, 0 °C, 30 min

Scheme 49

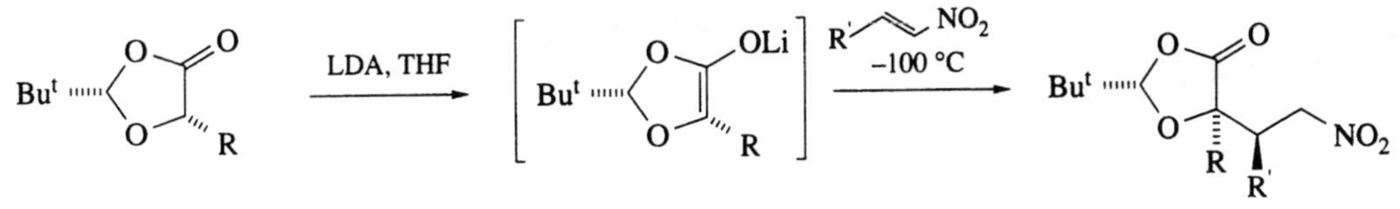

(131) (132) (133) 85 to >98% *ee*

R = Me, Ph, CH_2CO_2Li R' = Me, Et, Ph

Scheme 50

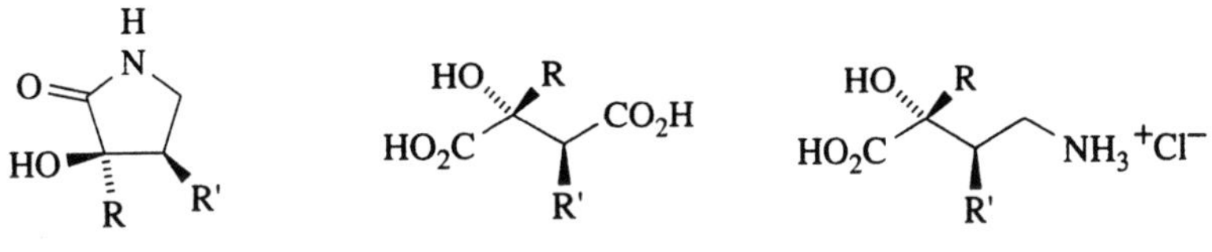

(134) (135) (136)

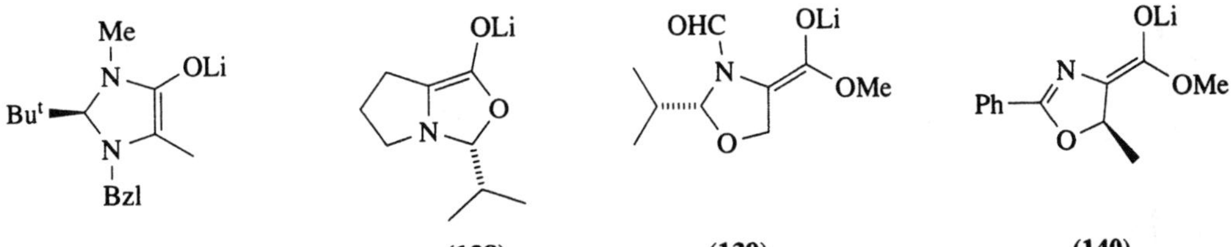

(137) (138) (139) (140)

1.5.3.2 Reactions of Chiral Azaenolates and Enamines

Early investigations have demonstrated that aldehydes and ketones can be enantioselectively α-alkylated *via* Michael reactions of the corresponding enamines, prepared from proline-derived secondary amines.[149–156] However, optical purities of the products were generally low and never exceeded 59% *ee*.[157] This kind of asymmetric α-alkylation could later be improved, allowing for example the preparation of compound (**141**) with high *ee* (Scheme 51).[158–160]

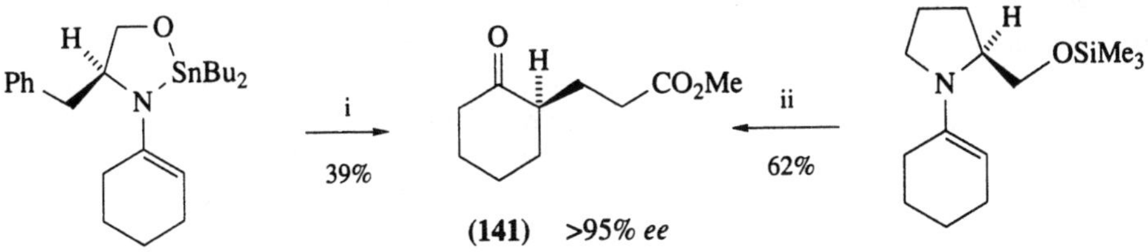

(141) >95% *ee*

i, $H_2C{=}CHCO_2Me$, pentane, 20 °C, 2.5 h; ii, $H_2C{=}CHCO_2Me$, $MgCl_2$, THF, 0 °C, 24 h, r.t., 24 h

Scheme 51

A method, based on a Michael-type (deracemizing) alkylation reaction, allows the enantioselective generation of quaternary carbon centers (Scheme 52).[161] The chiral imine (**143**), prepared in good yield from racemic (**142**) and (*S*)-(–)-1-phenylethylamine, smoothly reacts with (nonprochiral) methyl vinyl ketone to give, with quantitative recovery of the chiral auxiliary, the valuable diketone (**144**) in high yield and high enantiomeric purity. This technique was applied to the synthesis of steroid intermediates[162,163] and to the preparation of optically active oxa-spiro compounds.[164] Whereas 'normal' β-substituted Michael acceptors do not react with (**143**), (*E*)-crotyl cyanide (**145**) was shown to be a sufficiently strong acceptor.[165] Reaction of (*ent*-**143**) with (**145**) in cyclohexane leads diastereoselectively to a mixture of the bicyclic compounds (**147**) and (**148**), both secondary products of the initially formed Michael adduct (**146**; Scheme 53). The selectivity can be rationalized in terms of a transition state as drawn in (**149a**). An intramolecular application of this method involves the cyclization of

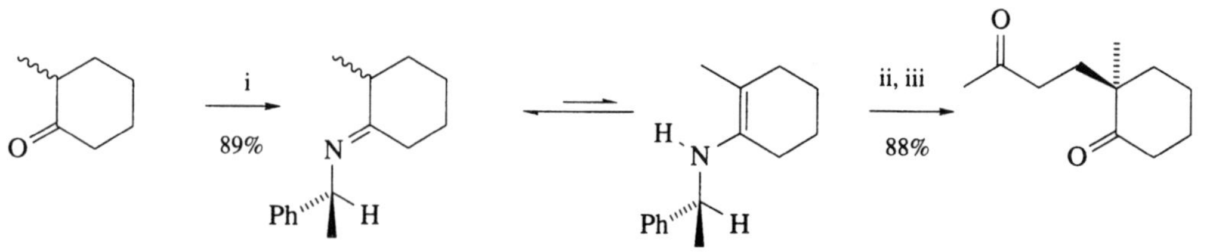

(142) **(143)** **(144)** 91% *ee*

i, *(S)*-(–)-1-phenylethylamine, TsOH, toluene, $-H_2O$; ii, MVK, THF, r.t., 3 d; iii, 10% aq. AcOH, r.t., 1 h

Scheme 52

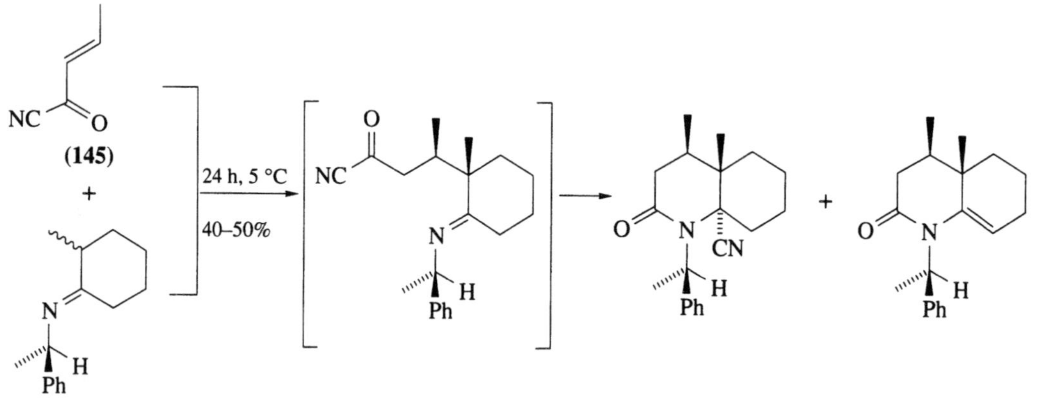

(145) + (*ent*-**143**) 24 h, 5 °C 40–50% **(146)** **(147)** + **(148)**

Scheme 53

compound (**150**) in a highly enantioselective manner (Scheme 54).[166] The product (**151**) can be transformed into the piperidine derivative (**152**), which in turn serves for the preparation of valuable intermediates for alkaloid total synthesis. The observed selectivity is in accordance with a transition state structure as drawn in (**149b**).

A highly selective method for the preparation of optically active β-substituted or β,γ-disubstituted-δ-keto esters and related compounds is based on asymmetric Michael additions of chiral hydrazones (**156**), derived from (*S*)-1-amino-2-methoxymethylpyrrolidine (SAMP) or its enantiomer (RAMP), to unsaturated esters (**154**).[167–172] Overall, a carbonyl compound (**153**) is converted to the Michael adduct (**155**) as outlined in Scheme 55. The actual asymmetric 1,4-addition of the lithiated hydrazone affords the adduct (**157**) with virtually complete diastereoselection in a variety of cases (Table 3). Some of the products were used for the synthesis of pheromones,[169] others were converted to δ-lactones.[170] The Michael acceptor (**158**) also reacts selectively with SAMP hydrazones.[171] Tetrahydroquinolindiones of type (**159**) are prepared from cyclic 1,3-diketones *via* SAMP derivatives like (**160**), as indicated in Scheme 56.[172]

Michael reactions of chiral lithioenamines of β-oxo esters with dimethyl alkylidenemalonates were studied.[173–175] Especially the α-alkyl-substituted compounds (**161**) and (**163**), derived from L-valine *t*-butyl ester, afford, after hydrolysis, the adducts (**162**) and (**164**), respectively, diastereoselectively and with high *ee* (Scheme 57).[175] In the presence of TMS-Cl, even weaker acceptors like acrylates or MVK were shown to react.[176] A somewhat related diastereoselective 1,4-addition, followed by a Pictet–Spengler-type cyclization, allows the preparation of compound (**165**; Scheme 58),[177] a central intermediate for the synthesis of several alkaloids.[177,178]

The anion of the bislactim ether (**166**) successfully acts as a chiral donor in Michael additions to prochiral substrates.[179–181] With methyl enoates, the reaction proceeds generally with high diastereoselectivity producing the isomer (**167**) as the major, and (**168**) as the minor product (Scheme 59).[179] This allows the preparation of virtually enantiomerically pure (*R*)-glutamic acid derivatives, since complete stereocontrol is usually obtained at the newly formed ring–stereogenic center. Conjugated dienoates are attacked under the same conditions in a 1,6-fashion with even higher selectivities to produce, after hydrolysis of the bislactim ether moiety, the adducts (**169**) as single isomers, however with unknown relative configuration (Scheme 60).[180] Enones, especially cyclic ones, are also suitable substrates

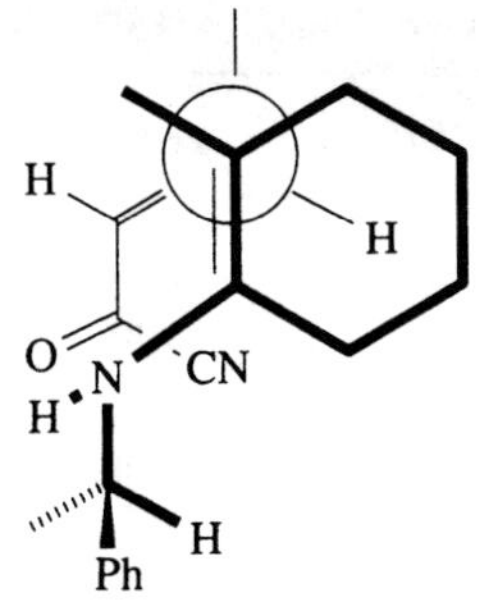

(149a) 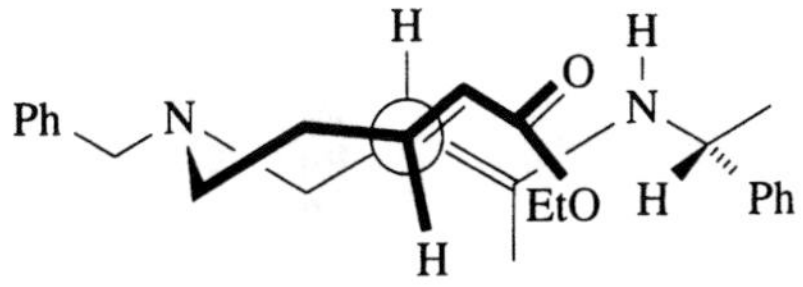 (149b)

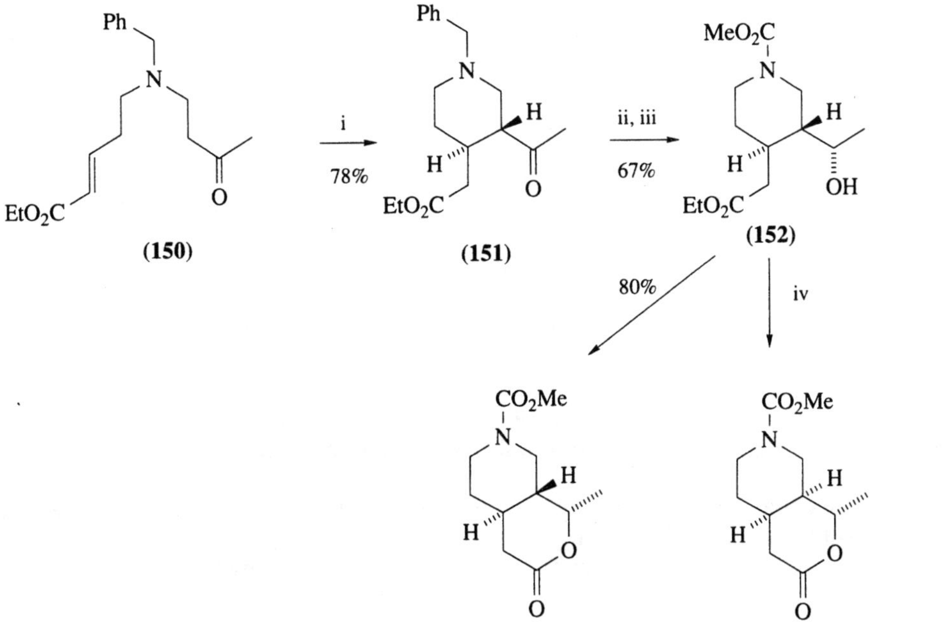

(150) (151) (152)

i, 1 equiv. (*R*)-(+)-1-phenylethylamine, 5 Å molecular sieve, THF, 5–10 °C; ii, $ClCO_2Me$, benzene, 60 °C; iii, $NaBH_4$, MeOH, –10 °C; iv, 5 steps

Scheme 54

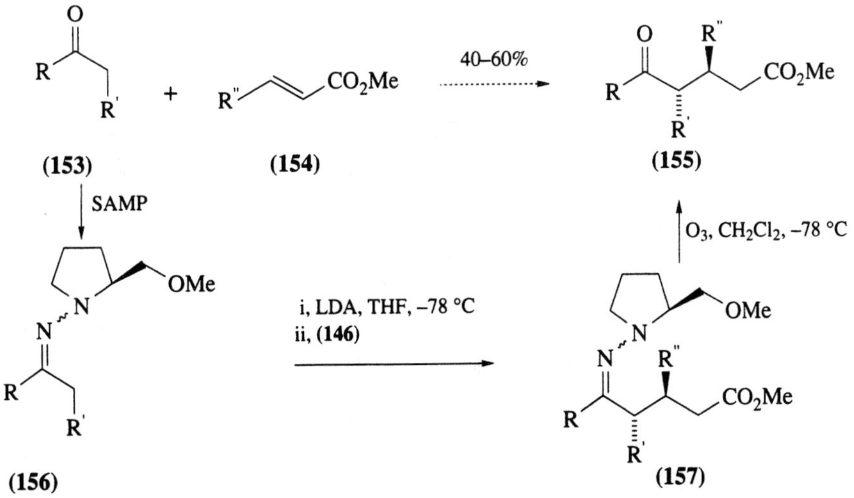

(153) (154) (155) (156) (157)

Scheme 55

Table 3 Preparation of Optically Active Substituted 5-Oxoalkanoates (**155**; Scheme 55)

R	*R'*	*R''*	*Overall yield (%)*	*ee (%)*	*Ref.*
Me	H	Me	50	>96	167
Pr^i	H	Me	46	>96	167
n-C_6H_{13}	H	Me	53	>99	167
Ph	H	Me	55	>96	167
Me	H	Ph	62	>99	167
n-C_5H_{11}	H	Ph	59	>99	167
H	Me	Me	58	>96	168
Et	Me	Me	45	92	168
Ph	Me	Ph	43	>99	168
Bu^n	Pr^n	Ph	40	>96	168
H	H	Me	43	>96	170
H	H	Et	37	90	170
H	H	Ph	51	>96	170
H	H	*p*-MeOPh	30	93	170

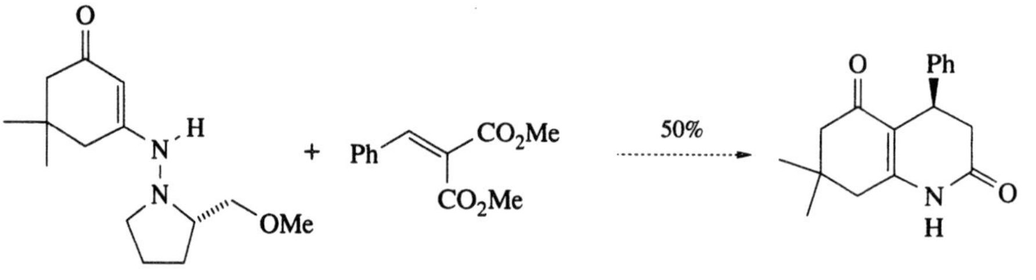

(160) + **(158)** —50%→ **(159)** >98% *ee*

Scheme 56

if the lithiated bislactim ether is first converted to the corresponding dialkylcuprate reagent (**170**).[181] The reaction with cyclopentenone, for instance, allows the preparation of compound (**171**) in pure form (Scheme 61).

The alkylation of the chiral enamine (**172**) by β-nitrostyrenes gives diastereomerically pure Michael adducts (**173**) with high *ee* (Scheme 62).[182] Comparable selectivities were observed with α-(methoxycarbonyl)cinnamates as substrates.[183] The same method was used for the enantioselective alkylation of β-tetralones in the 3-position by β-nitrostyrenes.[184]

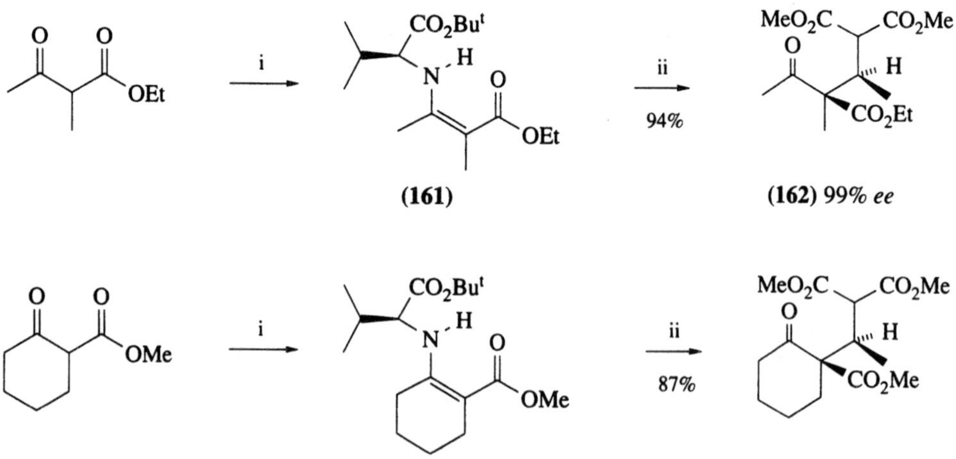

(161) —ii, 94%→ **(162)** 99% *ee*

(163) —ii, 87%→ **(164)** 99% *ee*

i, L-valine *t*-butyl ester; ii, LDA, then dimethyl ethylidenemalonate, toluene, 2 equiv. HMPA, –78 °C

Scheme 57

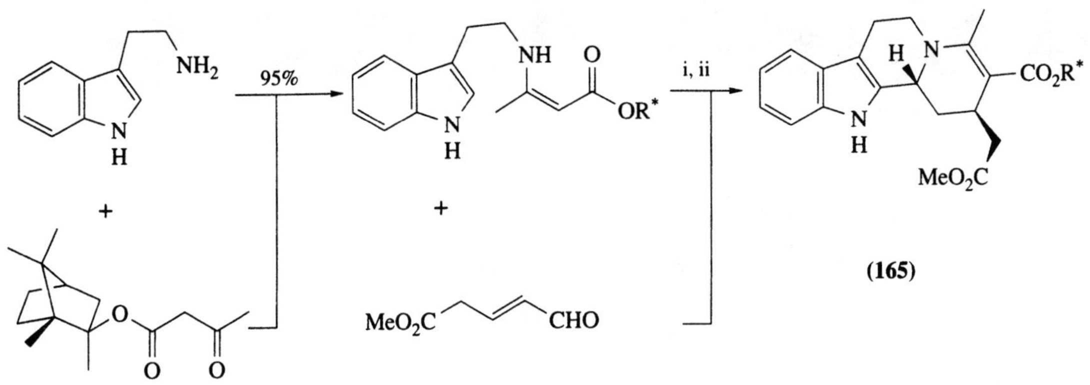

i, CH_2Cl_2, –78 °C; ii, TFA, CH_2Cl_2, –78 °C, 24 h, –20 °C, 14 d

Scheme 58

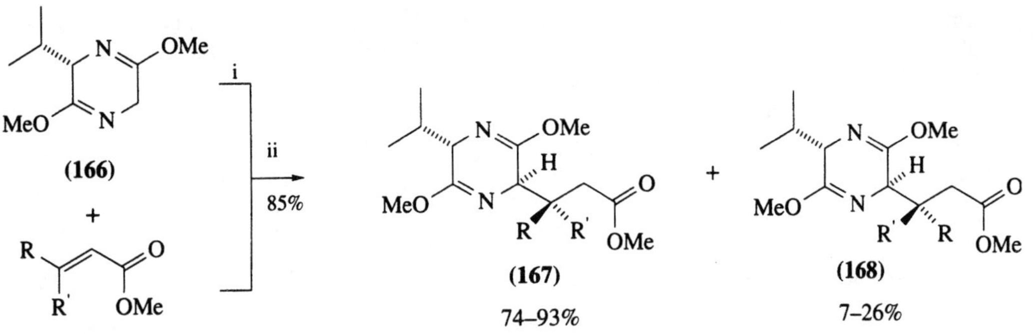

R = H, Me, Ph; R' = H, Ph

i, Bu^nLi, THF, –78 °C, 10 min; ii, –70 °C, 2–3 h, then 1 equiv. AcOH, –70 °C to r.t.

Scheme 59

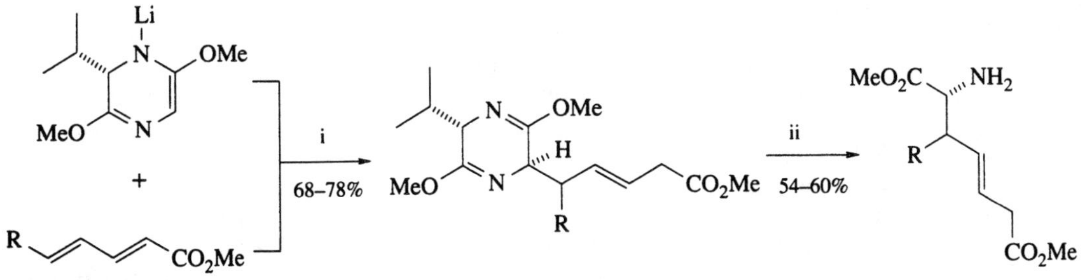

R = Me, Ph, 4-pyridyl

(169) >98% *ee*

i, THF, –70 °C, 2–3 h, then 1 equiv. AcOH, –70 °C to r.t.; ii, 0.25 M aq. HCl, r.t., 4 d

Scheme 60

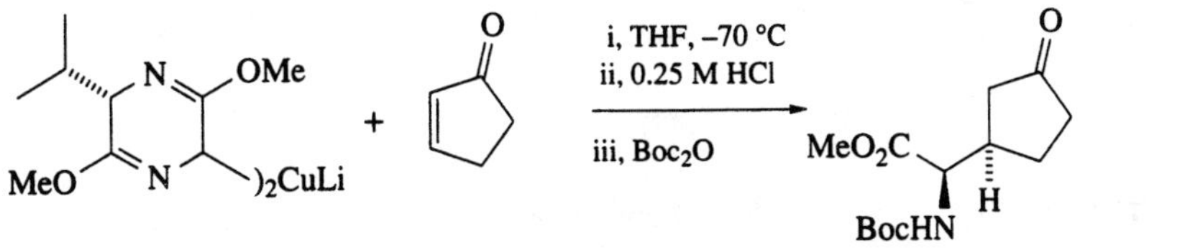

(170)

(171) >99% *ee*

Scheme 61

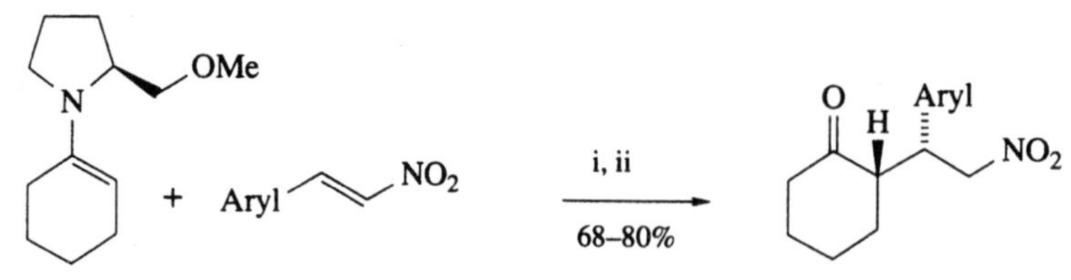

(172) (173) >90% *ee*

i, Et_2O, –80 °C, 30 min, r.t., 5–6 h; ii, aq. HCl, EtOH, 60 °C, 30 min

Scheme 62

1.5.3.3 Reactions of Chiral Sulfinyl and Phosphonyl Anions

In the first attempts to use a chiral α-sulfinyl ester enolate as donor in Michael additions to α,β-unsaturated esters, only low selectivities were observed.[185,186] Better results are obtained when the α-lithio sulfoxide (**174**), a chiral acyl anion equivalent, is employed. Conjugate addition of (**174**) to cyclopentenone derivatives occurs with reasonably high degrees of asymmetric induction, as exemplified by the preparation of the 11-deoxy prostanoid (**175**; Scheme 63).[187,188] Chiral oxosulfonium ylides and chiral lithiosulfoximines can be used for the preparation of optically active cyclopropane derivatives (up to 49% *ee*) from α,β-unsaturated carbonyl compounds.[189]

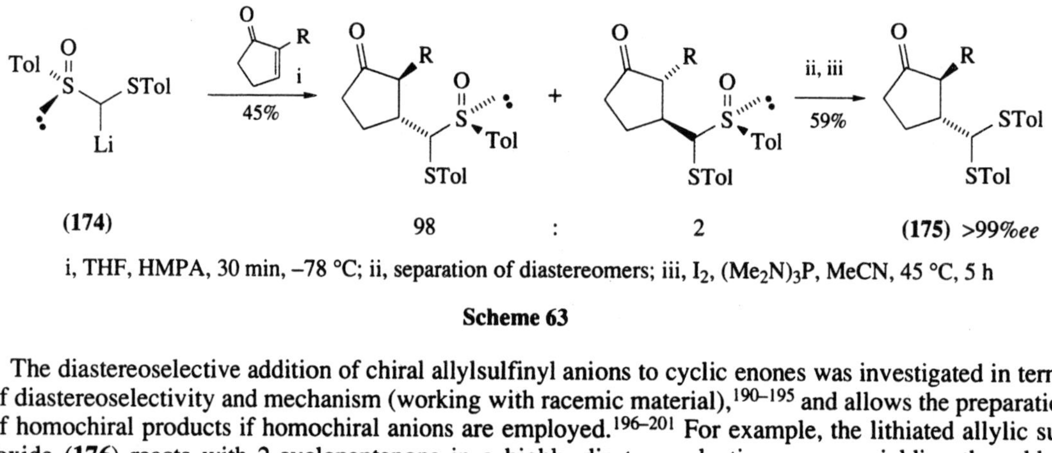

(174) 98 : 2 (175) >99%*ee*

i, THF, HMPA, 30 min, –78 °C; ii, separation of diastereomers; iii, I_2, $(Me_2N)_3P$, MeCN, 45 °C, 5 h

Scheme 63

The diastereoselective addition of chiral allylsulfinyl anions to cyclic enones was investigated in terms of diastereoselectivity and mechanism (working with racemic material),[190–195] and allows the preparation of homochiral products if homochiral anions are employed.[196–201] For example, the lithiated allylic sulfoxide (**176**) reacts with 2-cyclopentenone in a highly diastereoselective manner yielding the adduct (**177**), which can be converted to the optically active keto acid (**178**; Scheme 64). Application of this methodology allows the synthesis of cyclopentanoid natural products like (+)-hirsutene (**179**), which was prepared from (*ent*-**176**) according to Scheme 65.[196,199] The synthesis of (+)-pentalene (**182**; Scheme 66) utilizes two kinetic resolution steps:[197] at first, compound (*rac*-**180**) is (partially) resolved by treatment with 0.5 equiv. of lithiated (**176**); subsequent reaction of (**180**) with 2 equiv. of a racemic lithiated allyl sulfoxide affords the intermediate (**181**) diastereoselectively and in high yield. All these reactions are believed to proceed *via* a 10-membered '*trans*-decalyl'-like or '*trans*-fused chair–chair'-like transition state as shown in (**183**).

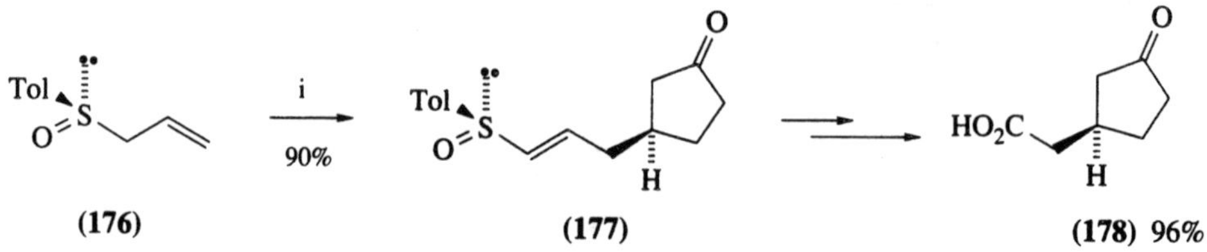

(176) (177) (178) 96% *ee*

i, LDA, THF, HMPA, –78 °C, 1 h, then 1 equiv. of 2-cyclopentenone, –78 °C, 5 min

Scheme 64

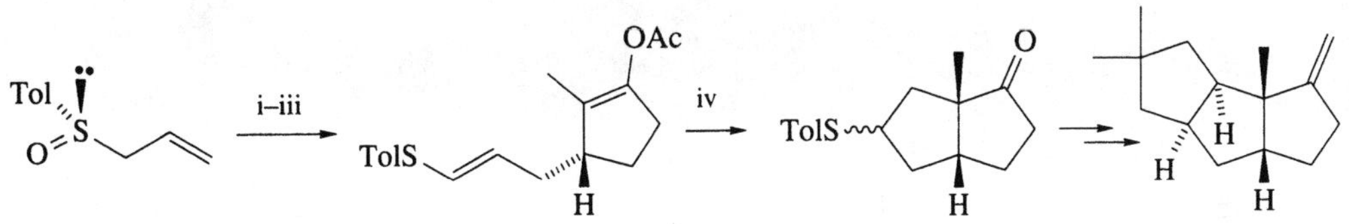

i, LDA, THF, HMPA, –78 °C, 1 h, then 2-methyl-2-cyclopentenone, –78 °C, 5 min; ii, Ac_2O; iii, Zn, AcOH, r.t., 24 h: iv, 1 equiv. $TiCl_4$, AcOH, H_2O, r.t.

Scheme 65

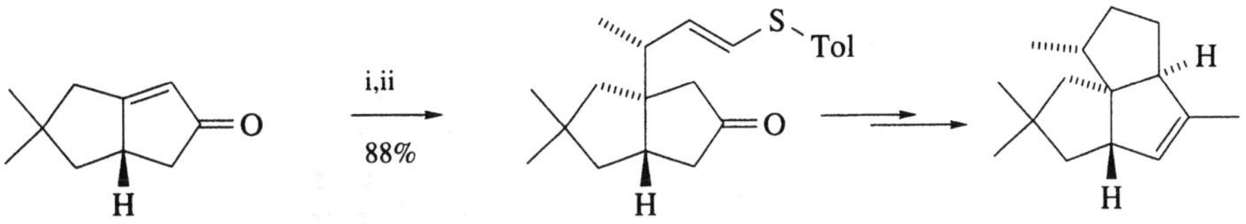

i, 2 equiv. rac-(Z)-2-butenyl phenyl sulfoxide anion, THF, –78 °C, 45 min; ii, Zn, AcOH, r.t., 24 h

Scheme 66

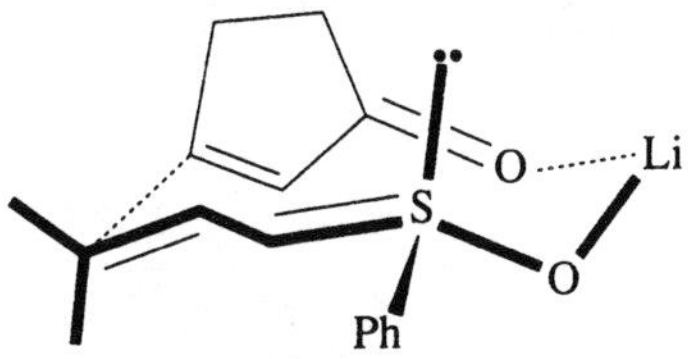

(183)

Since allyl sulfoxides may quite easily undergo racemization at the sulfur atom *via* a reversible [2,3] sigmatropic process, the configurationally more stable chiral allylic phosphine oxides were also investigated.[201] Compounds (**184**) and (**185**), prepared as a 1:1 mixture from allylphosphonyl dichloride and (–)-ephedrine, were shown to add to cycloalkenones with reasonably high diastereoselectivities. Ozonolysis of the initially formed 1,4-adducts affords the respective optically active ketoaldehydes (Scheme 67). With a *N*-isopropyl-substituted derivative even higher selectivities (88–98% *ee*) could be obtained.

1.5.3.4 Reactions of Organocopper Compounds Prepared from Chiral Carbanions

The chiral organocopper compound (**186**) adds diastereoselectively to 2-methyl-2-cyclopentenone, allowing the preparation of optically active steroid CD-ring building blocks (Scheme 68).[202–204] A related method was applied to a synthesis of the steroid skeleton *via* an intramolecular (transannular) Diels–Alder reaction of a macrocyclic precursor.[205] Chiral acetone anion equivalents based on copper azaenolates derived from acetone imines were shown to add to cyclic enones with good selectivity (60–80% *ee*, after hydrolysis).[206–208] Even better *ee* values are obtained with the mixed zincate prepared from (**187**) and dimethylzinc (Scheme 69). Other highly diastereoselective but synthetically less important 1,4-additions of chiral cuprates to prochiral enones were reported.[209,210]

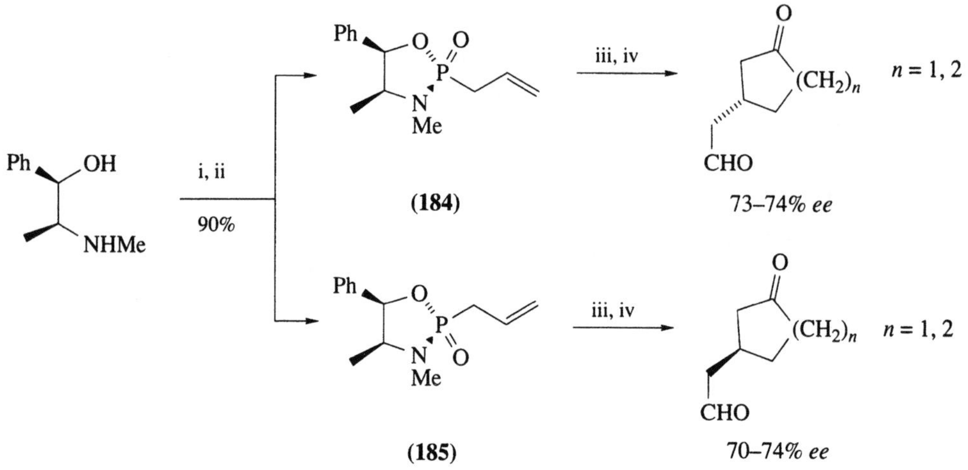

(184)

73–74% *ee*

$n = 1, 2$

(185)

70–74% *ee*

i, allyl-$POCl_2$, 2 equiv. Et_3N, toluene, –40 °C to r.t., 12 h; ii, column chromatography; iii, Bu^nLi, THF, –78 °C, then cycloalkenone, THF, –78 °C, 30 min; iv, O_3, CH_2Cl_2, –78 °C

Scheme 67

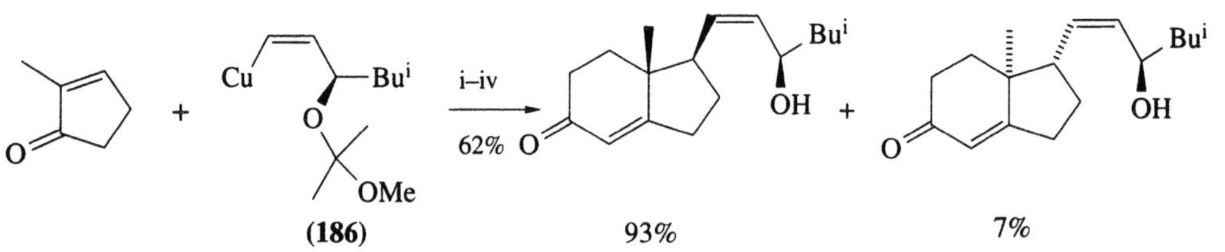

(186) 93% 7%

i, Bu_3P, Et_2O, –78 °C to –20 °C, 2 h, then add $H_2C{=}C(SiMe_3)COMe$, Et_2O, –20 °C, 2 h; ii, NaOMe, MeOH, reflux, 3 h; iii, 3 M HCl, 0 °C, 5 min; iv, chomatography

Scheme 68

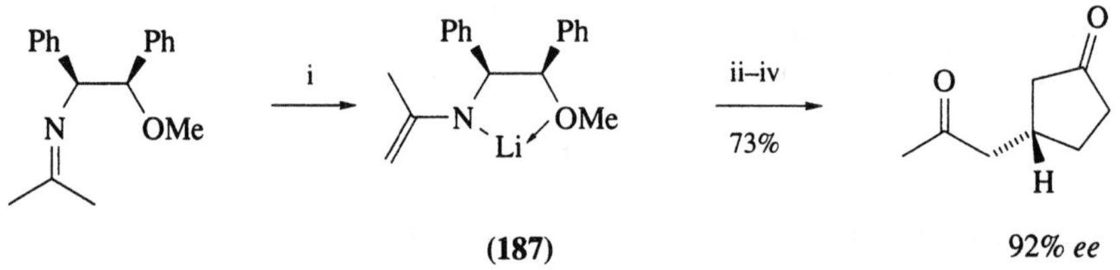

(187) 92% *ee*

i, Bu^nLi, THF, –40 °C, 30 min; ii, add $ZnMe_2$, –78 °C; iii, add 2-cyclopentenone, THF, –78 °C, 5 h, iv, H_3O^+

Scheme 69

1.5.4 ENANTIOSELECTIVE ADDITIONS OF ACHIRAL CARBON NUCLEOPHILES TO ACHIRAL SUBSTRATES

1.5.4.1 Reactions Employing Stoichiometric Amounts of Chiral Ligands or Coordinating Cosolvents

1.5.4.1.1 Reactions of organocopper compounds

Asymmetric induction in the conjugate addition of achiral residues to α,β-unsaturated carbonyl compounds can be achieved with chiral ligand-modified organocuprate reagents. Early experiments, employing a variety of chiral homo- and hetero-cuprates, gave very low enantioselectivities (<6% *ee*),[211–219] only in some cases were somewhat better *ee* values (15–34%) obtained.[220–223] A reasonably high asymmetric induction (68% *ee*) was observed with (*S*)-*N*-methylprolinol as the chiral (nontransferable) ligand in the copper-mediated addition of methyl Grignard reagents to 1,3-diphenyl-2-propene-1-one (chalcone).[224] Proline-derived ligands were also employed in other studies on the addition of magnesium- and lithium-cuprates to (mainly) the same substrate, leading to up to 94% *ee*.[225–229] It was found that the selectivity (up to 50% *ee*) during the addition of chiral organo(hetero)cuprates to 2-cyclohexenone depended upon virtually all experimental variables like the ligand, the solvent, the temperature, the counterion in the Cu^I precursor, and the cation (Li^+ or Mg^{2+}).[230] Stoichiometric amounts of the tridentate chelate (**188**) effect high asymmetric induction in conjugate additions of alkylcuprate reagents to cyclohexenone, allowing, for instance, the preparation of 3-butylcyclohexanone with 92% *ee* (Scheme 70).[231] Optical yields of 41–83% were obtained in the addition of alkyl(hetero)cuprates to cyclic and acyclic alkenones, the extent of asymmetric induction being a complex function of all experimental parameters.[232] Only a single example exists for the enantioselective transfer of an alkenyl residue from a chirally modified cuprate reagent: (**189**), a chiral building block for the synthesis of the pseudoguaianolide (+)-confertin (**190**),[235,236] was prepared from 2-methyl-2-cyclopentenone according to Scheme 71.[233,234]

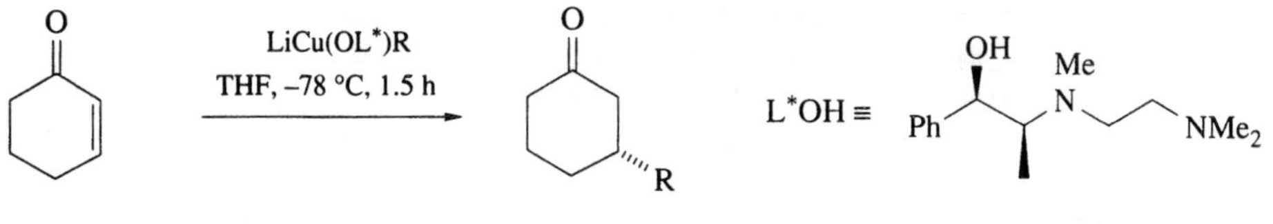

LiCu(OL*)R
THF, –78 °C, 1.5 h

85–92% *ee*

L*OH ≡ (**188**)

Scheme 70

76%

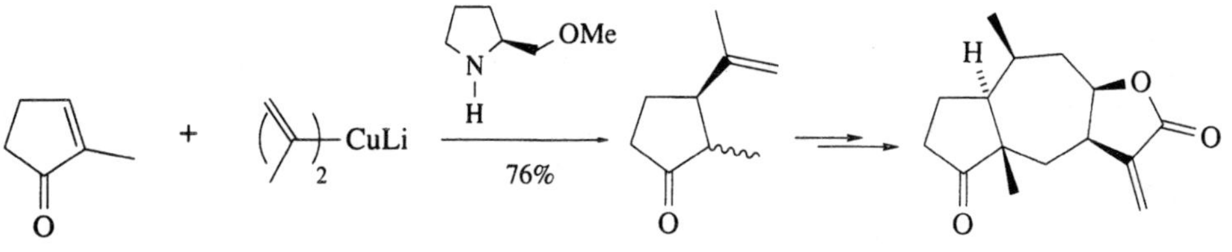

(**189**) 76% *ee* (**190**)

Scheme 71

1.5.4.1.2 Reactions of organolithium and organozinc compounds

Asymmetric induction in the 1,4-addition of organometallic compounds to prochiral electron-deficient alkenes can be achieved in the presence of chiral coordinating cosolvents.[237] With the tartaric acid-derived amino ether (**191**), addition of achiral Li-, Cu-, and Zn-organic compounds to α,β-unsaturated aldehydes, ketones, nitro compounds and ketene thioacetals led to adducts with 10 to 26% *ee*, and, in a single case, with 43% *ee*. The related chiral amino ether (**192**) effected the addition of BuLi to nitropropene with even higher enantioselectivity (58% *ee*).[238] Comparable selectivities (32 to 67% *ee*) were obtained in the conjugate addition of lithiated dithioacetals to prochiral unsaturated esters mediated by the chiral ligands (**193**) and (**194**; Scheme 72).[239] The conjugate addition of Grignard reagents to enones mediated by chiral diamine zinc(II) monoalkoxides gave the adducts with low *ee* (<15%).[240]

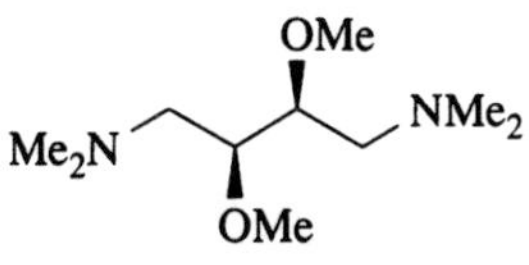

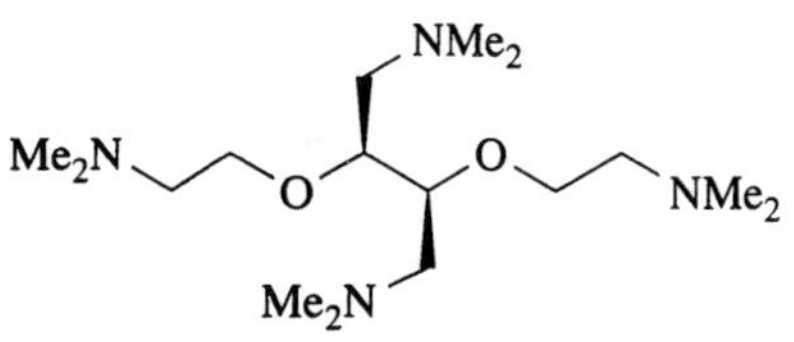

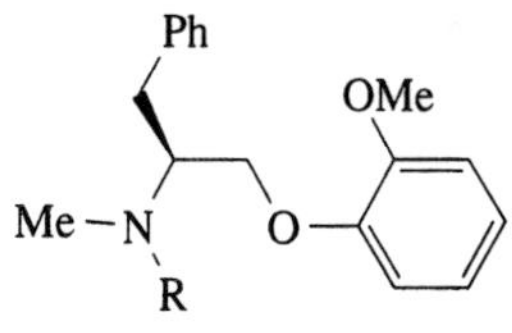

(**191**) (**192**)

(**193**) R = Me
(**194**) R = $(CH_2)_2NMe_2$

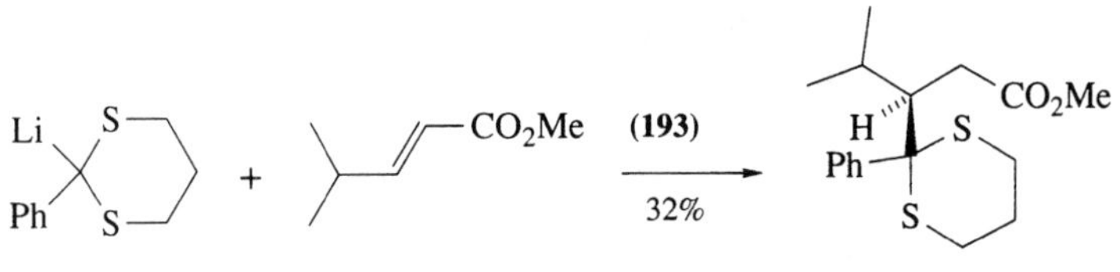

67% *ee*

Scheme 72

1.5.4.2 Reactions Mediated by Chiral Catalysts

1.5.4.2.1 Reactions of CH-acidic compounds

Chinchona alkaloids catalyze asymmetric alkylations of highly CH-acidic compounds with (nonprochiral) Michael acceptors.[241–243] For instance, the quinine-mediated reaction of methyl 1-oxo-2-indancarboxylate (**195**) with MVK affords the adduct (**196**) with high *ee* (Scheme 73). A polymer-bound catalyst may also be used without loss of selectivity.[244] The same reaction is catalyzed by a chiral Co^{II} diamine complex (66% *ee*).[245] Chiral crown ether complexed potassium bases catalyze the Michael addition of phenyl acetic esters and β-keto esters (*e.g.* **195**) to MVK and to methyl acrylate, yielding products with up to 99% *ee*.[246] Other chiral cation complexants were also tested.[247,248] Chiral phase-transfer catalysts in the Michael reaction were used to achieve high asymmetric inductions (up to 92% *ee*) in related alkylations of (less acidic) 2-alkyl-1-indanone derivatives.[249–251] The enantioselectivity of such reactions appears to be generally lower at higher pressures.[252]

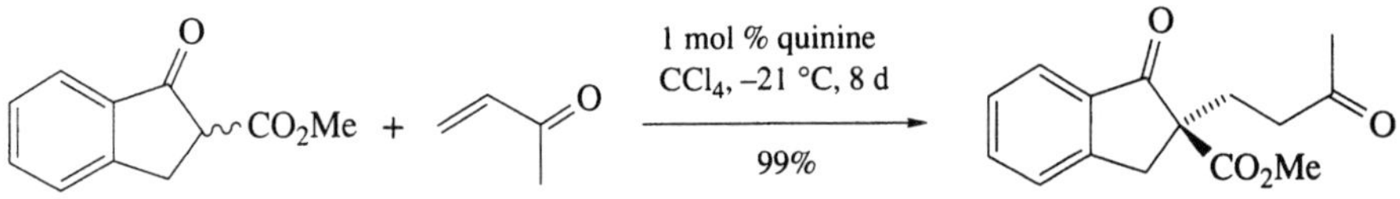

(**195**) (**196**) 76% *ee*

Scheme 73

1.5.4.2.2 Reactions of organometallic compounds

There are three techniques for true catalytic enantioselective 1,4-addition of organometallic reagents to prochiral α,β-unsaturated carbonyl compounds available so far: (i) the reaction of Grignard reagents with 2-cyclohexenone, mediated by catalytic amounts of the chiral binuclear Cu^{I} complex derived from the tropocoronand ligand H[CHIRAMT] (**197**) leads to rather low *ee* (4–14%); however, the authors mention in a footnote that substantially higher selectivities (up to 80% *ee*) can be obtained with use of an improved catalyst system;[253] (ii) the asymmetric conjugate addition of dialkylzinc compounds to certain enones is catalyzed by a chiral Ni complex prepared *in situ* from a nickel(II) salt and *N,N*-dibutylnorephedrine (**198**) and leads to moderate selectivities (<50% *ee*; Scheme 74);[254,255] (iii) the addition of tin(II) enethiolates, generated *in situ* from trimethylsilyl enethiolates (**199**) with catalytic amounts of $Sn(OTf)_2$ and the bidentate ligand (**200**), to enones yields chiral dithioesters (**201**) with up to 70% *ee* (Scheme 75).[256,257]

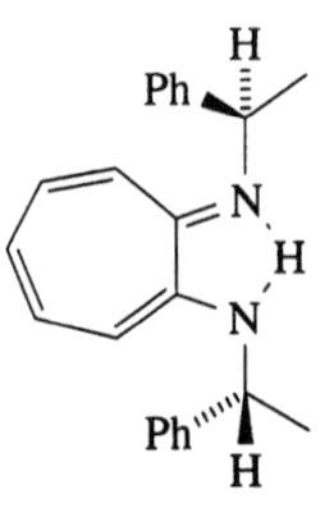

(**197**)

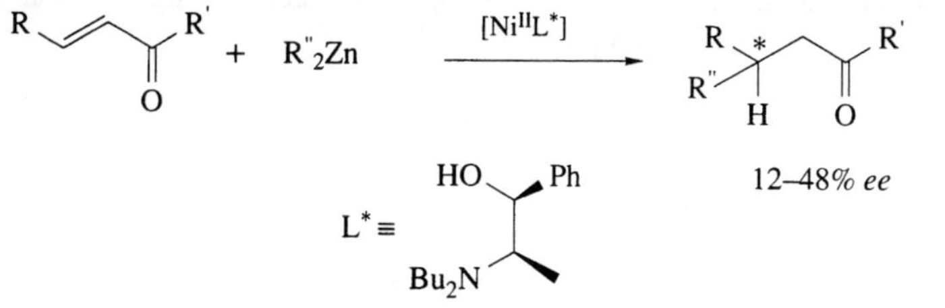

$[Ni^{II}L^*]$

12–48% *ee*

(198)

Scheme 74

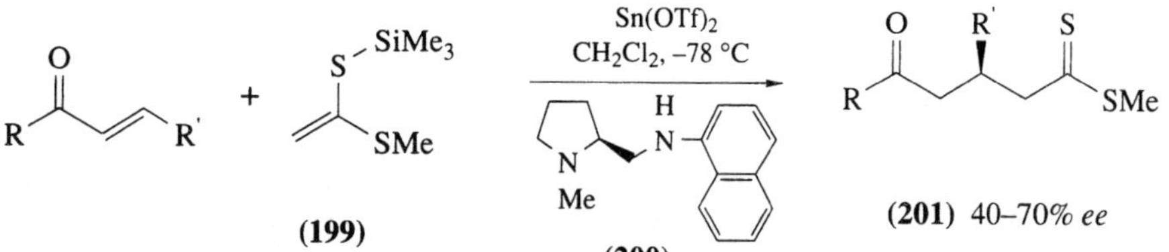

$Sn(OTf)_2$, CH_2Cl_2, –78 °C

(199) **(200)** **(201)** 40–70% *ee*

Scheme 75

1.5.5 ADDITIONS OF NONCARBON NUCLEOPHILES

The (diastereoselective) conjugate addition of silylcuprate reagents to a variety of chiral derivatives of α,β-unsaturated carboxylic acids can be used to prepare optically active β-silyl esters.[258,259] Best results are obtained with substrates of type (**25**). The (related) β-silyl ketones, which also constitute valuable building blocks for (acyclic) stereoselective synthesis, are now accessible in high *ee via* palladium-catalyzed enantioselective 1,4-disilylation of α,β-unsaturated ketones (Scheme 76).[260]

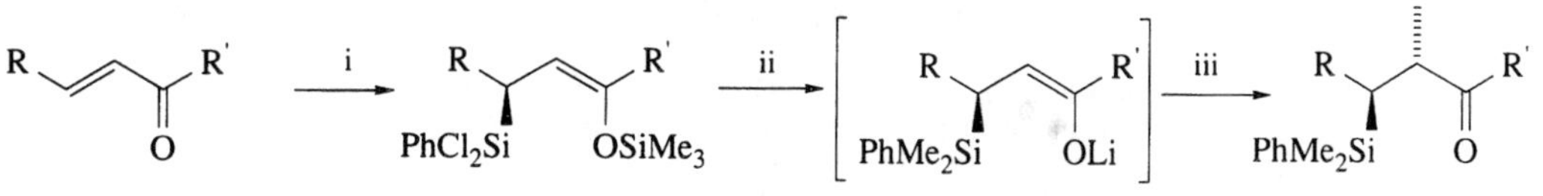

74–92% *ee*

i, $PhCl_2SiSiMe_3$, benzene, 0.005 mol % $PdCl_2$-(+)-BiNAP; ii, MeLi, Et_2O; iii, MeI, THF

Scheme 76

Thiols may be enantioselectively added in a conjugate fashion to α,β-unsaturated carbonyl compounds in the presence of chiral hydroxyamine catalysts (*e.g.* chinchona alkaloids).[242,244,249,252,261–269] In some cases *ee* of up to >80% were achieved (*e.g.* Scheme 77).[242,261,262] This methodology was utilized for the kinetic resolution of compound (*rac*-**86**; Scheme 34) in a multigram scale.[94] Related enantioselective 1,4-additions of thioacetates[270,271] and selenophenols[272] to enones are also known. Epoxidations, based on the asymmetric nucleophilic addition of peroxide anions to enones, are discussed separately.[273]

While sluggish under thermal conditions,[274,275] the asymmetric conjugate addition of amines to alkyl crotonates is achieved at room temperature under high pressure (15 kbar).[276] Thus, benzylamine can be added to the crotonate derived from 8-β-naphthyl menthol, with virtually complete diastereoselectivity. A related intramolecular 1,4-addition of an amine to a chiral enoate was used in a total synthesis of the alkaloid (–)-tylophorine.[277] Additions of amines to chiral iron complexes of type (**116**) proceed with excellent selectivity and allow the preparation of homochiral β-lactams.[127,128,130,132] In contrast, the addition of amine nucleophiles to chiral vinylic sulfoxides[278–280] and to chiral vinylsulfoximines[281] proceeds with comparably low selectivities.

Asymmetric conjugate addition of hydride to *N*-enoylbornane-10,2-sultams (**28**) was used to prepare homochiral β-branched alkanoic acids.[282,283] Another method involves the preparation of such compounds *via* a catalytic enantioselective 1,4-reduction:[284] in the presence of only 1 mol % of a chiral cobalt semicorrin ligand (readily available from pyroglutamic acid), β-methyl-2-alkenoic acids were reduced by $NaBH_4$ in high yields and with up to 96% *ee*.

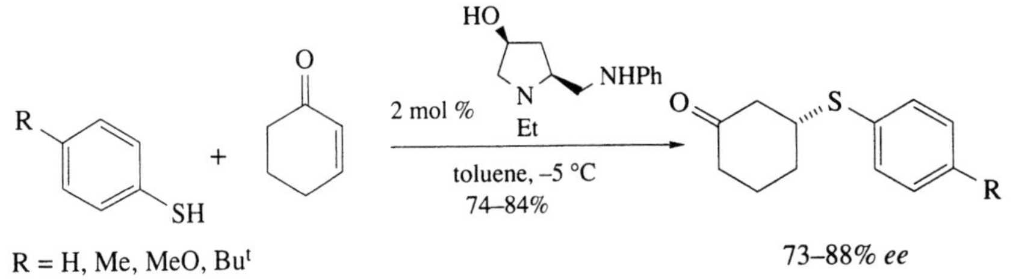

R = H, Me, MeO, Bu^t

2 mol % (HO-pyrrolidine, N-Et, NHPh); toluene, –5 °C; 74–84%

73–88% *ee*

Scheme 77

1.5.6 CONCLUSIONS

A number of techniques are now available allowing the preparation of enantiomerically pure (or at least enriched) compounds *via* asymmetric nucleophilic addition to electron-deficient alkenes. Some of these transformations have already been successfully applied in total synthesis. In most cases, the methods are based on diastereoselective reactions, employing chirally modified substrates or nucleophiles. There are only very few useful enantioselective procedures accessible so far. The search for efficient enantioselective methods, especially for those which are catalytic and do not require the use of stoichiometric amounts of chiral auxiliaries, remains a challenging task for the future.

1.5.7 REFERENCES

1. K. Tomioka and K. Koga, in 'Asymmetric Synthesis', ed. J. D. Morrison, Academic Press, New York, 1983, vol. 2, p. 201.
2. J. W. ApSimon and T. L. Collier, *Tetrahedron*, 1986, **42**, 5157.
3. M. Nógrádi, 'Stereoselective Syntheses', VCH, Weinheim, 1987, p. 221.
4. E. Winterfeldt, 'Prinzipien und Methoden der stereoselektiven Synthese', Vieweg, Braunschweig, 1988, chap. 6 and 7.
5. D. Seebach and E. Hungerbühler, in 'Modern Synthetic Methods 1980', ed. R. Scheffold, Salle-Sauerländer, Frankfurt, 1980, p. 93.
6. J. D. Morrison and H. S. Mosher, in 'Asymmetric Organic Reactions', Prentice-Hall, Englewood Cliffs, NJ, 1971; reprinted by the American Chemical Society, Washington, DC, 1976, p. 272.
7. Y. Inouye and H. M. Walborsky, *J. Org. Chem.*, 1962, **27**, 2706.
8. M. Kawana and S. Emoto, *Bull. Chem. Soc. Jpn.*, 1966, **39**, 910.
9. B. Gustafsson, A.-T. Hansson and C. Ullenius, *Acta Chem. Scand., Ser. B*, 1980, **34**, 113.
10. K. Igarashi, J. Oda, K. Inouye and M. Ohno, *Agric. Biol. Chem.*, 1970, **34**, 811
11. Y. Yamamoto and K. Maruyama, *J. Am. Chem. Soc.*, 1978, **100**, 3240.
12. W. Oppolzer and H. J. Löher, *Helv. Chim. Acta*, 1981, **64**, 2808.
13. W. Oppolzer, R. Moretti, T. Godel, A. Meunier and H. J. Löher, *Tetrahedron Lett.*, 1983, **24**, 4971.
14. W. Oppolzer and T. Stevenson, *Tetrahedron Lett.*, 1986, **27**, 1139.
15. W. Oppolzer, P. Dudfield, T. Stevenson and T. Godel, *Helv. Chim. Acta*, 1985, **68**, 212.
16. W. Oppolzer, R. Moretti and G. Bernardinelli, *Tetrahedron Lett.*, 1986, **27**, 4713.
17. G. Helmchen and G. Wegner, *Tetrahedron Lett.*, 1985, **26**, 6051.
18. A. McL. Mathieson, *Tetrahedron Lett.*, 1965, 4137.
19. W. B. Schweizer and J. Dunitz, *Helv. Chim. Acta*, 1982, **65**, 1547.
20. R. J. Loncharich, T. R. Schwartz and K. N. Houk, *J. Am. Chem. Soc.*, 1987, **109**, 14.
21. For a general review about 'Camphor Derivatives as Chiral Auxiliaries in Asymmetric Synthesis', see W. Oppolzer, *Tetrahedron*, 1987, **43**, 1969.
22. W. Oppolzer and E. J. Jacobsen, *Tetrahedron Lett.*, 1986, **27**, 1141.
23. G. Helmchen and G. Wegner, *Tetrahedron Lett.*, 1985, **26**, 6047.
24. P. Somfai, D. Tanner and T. Olsson, *Tetrahedron*, 1985, **41**, 5973.
25. T. Mukaiyama and N. Iwasawa, *Chem. Lett.*, 1981, 45.
26. T. Kogure and E. L. Eliel, *J. Org. Chem.*, 1984, **49**, 576.
27. K. Soai, A. Ookawa and Y. Nohara, *Synth. Commun.*, 1983, **13**, 27.
28. K. Soai, H. Machida and A. Ookawa, *J. Chem. Soc., Chem. Commun.*, 1985, 469.
29. K. Soai and A. Ookawa, *J. Chem. Soc., Perkin Trans. 1*, 1986, 759.
30. K. Soai, H. Machida and N. Yokota, *J. Chem. Soc., Perkin Trans. 1*, 1987, 1909; 1988, 415.
31. K. Tomioka, T. Suenaga and K. Koga, *Tetrahedron Lett.*, 1986, **27**, 369.
32. For a brief review, see W. Oppolzer, *Pure Appl. Chem.*, 1988, **60**, 39.
33. W. Oppolzer, G. Poli, A. J. Kingma, C. Starkemann and G. Bernardinelli, *Helv. Chim. Acta*, 1987, **70**, 2201.
34. W. Oppolzer, R. J. Mills, W. Pachinger and T. Stevenson, *Helv. Chim. Acta*, 1986, **69**, 1542.
35. W. Oppolzer, A. J. Kingma and G. Poli, *Tetrahedron*, 1989, **45**, 479.

36. W. Oppolzer and P. Schneider, *Helv. Chim. Acta*, 1986, **69**, 1817.
37. A. I. Meyers and C. E. Whitten, *J. Am. Chem. Soc.*, 1975, **97**, 6266.
38. A. I. Meyers and C. E. Whitten, *Heterocycles*, 1976, **4**, 1687.
39. A. I. Meyers and C. E. Whitten, *Tetrahedron Lett.*, 1976, 1947.
40. A. I. Meyers, R. K. Smith and C. E. Whitten, *J. Org. Chem.*, 1979, **44**, 2250.
41. For a review, see K. A. Lutomski and A. I. Meyers, in 'Asymmetric Synthesis', ed. J. D. Morrison, Academic Press, Orlando, 1983, vol. 3, p. 213.
42. A. I. Meyers and R. K. Smith, *Tetrahedron Lett.*, 1979, 2749.
43. A. I. Meyers, N. R. Natale, D. G. Wettlaufer, S. Rafii and J. Clardy, *Tetrahedron Lett.*, 1978, 227.
44. A. I. Meyers and N. R. Natale, *Heterocycles*, 1982, **18**, 13.
45. B. A. Barner and A. I. Meyers, *J. Am. Chem. Soc.*, 1984, **106**, 1865.
46. A. I. Meyers and D. Hoyer, *Tetrahedron Lett.*, 1984, **25**, 3667.
47. A. I. Meyers and B. A. Barner, *J. Org. Chem.*, 1986, **51**, 120.
48. A. I. Meyers, K. A. Lutomski and D. Laucher, *Tetrahedron*, 1988, **44**, 3107.
49. A. I. Meyers, G. P. Roth, D. Hoyer, B. A. Barner and D. Laucher, *J. Am. Chem. Soc.*, 1988, **110**, 4611.
50. F. E. Ziegler and P. J. Gilligan, *J. Org. Chem.*, 1981, **46**, 3874.
51. T. Mukaiyama, T. Takeda and M. Osaki, *Chem. Lett.*, 1977, 1165.
52. T. Mukaiyama, T. Takeda and K. Fujimoto, *Bull. Chem. Soc. Jpn.*, 1978, **51**, 3368.
53. T. Mukaiyama, K. Fujimoto and T. Takeda, *Chem. Lett.*, 1979, 1207.
54. T. Mukaiyama, K. Fujimoto, T. Hirose and T. Takeda, *Chem. Lett.*, 1980, 635.
55. T. Takeda and T. Mukaiyama, *Chem. Lett.*, 1980, 163.
56. M. Asami and T. Mukaiyama, *Chem. Lett.*, 1979, 569.
57. A. Bernardi, S. Cardani, G. Poli and C. Scolastico, *J. Org. Chem.*, 1986, **51**, 5041.
58. A. Bernardi, S. Cardani, T. Pilati, G. Poli, C. Scolastico and R. Villa, *J. Org. Chem.*, 1988, **53**, 1600.
59. A. Alexakis, R. Sedrani, P. Mangeney and J. F. Normant, *Tetrahedron Lett.*, 1988, **29**, 4411.
60. D. Seebach and J. Zimmermann, *Helv. Chim. Acta*, 1986, **69**, 1147.
61. D. Seebach, J. Zimmermann, U. Gysel, R. Ziegler and T.-K. Ha, *J. Am. Chem. Soc.*, 1988, **110**, 4763.
62. M. Sato, K. Takayama, T. Furuya, N. Inukai and C. Kaneko, *Chem. Pharm. Bull.*, 1987, **35**, 3971.
63. A. Ghribi, A. Alexakis and J. F. Normant, *Tetrahedron Lett.*, 1984, **25**, 3083.
64. Y. Fukutani, K. Maruoka and H. Yamamoto, *Tetrahedron Lett.*, 1984, **25**, 5911.
65. J. Fujiwara, Y. Fukutani, M. Hasegawa, K. Maruoka and H. Yamamoto, *J. Am. Chem. Soc.*, 1984, **106**, 5004.
66. K. Maruoka, S. Nakai, M. Sakurai and H. Yamamoto, *Synthesis*, 1986, 130.
67. D. Seebach, R. Imwinkelried and T. Weber, in "Modern Synthetic Methods 1986', ed. R. Scheffold, Springer-Verlag, Berlin, 1986, p. 213.
68. P. Mangeney, A. Alexakis and J. F. Normant, *Tetrahedron Lett.*, 1986, **27**, 3143.
69. A. Alexakis, P. Mangeney, A. Ghribi, I. Marek, R. Sedrani, C. Guir and J. F. Normant, *Pure Appl. Chem.*, 1988, **60**, 49.
70. M. Huche, J. Aubouet, G. Pourcelot and J. Berlan, *Tetrahedron Lett.*, 1983, **24**, 585.
71. J. Berlan, Y. Besace, G. Pourcelot and P. Cresson, *J. Organomet. Chem.*, 1983, **256**, 181.
72. Y. Besace, J. Berlan, G. Pourcelot and M. Huche, *J. Organomet. Chem.*, 1983, **247**, C11.
73. J. Berlan, Y. Besace, D. Prat and G. Pourcelot, *J. Organomet. Chem.*, 1984, **264**, 399.
74. J. Berlan, Y. Besace, E. Stephan and P. Cresson, *Tetrahedron Lett.*, 1985, **26**, 5765.
75. J. Berlan, Y. Besace, G. Pourcelot and P. Cresson, *Tetrahedron*, 1986, **42**, 4757.
76. J. Berlan and Y. Besace, *Tetrahedron*, 1986, **42**, 4767.
77. P. Mangeney, A. Alexakis and J. F. Normant, *Tetrahedron Lett.*, 1983, **24**, 373.
78. P. Mangeney, A. Alexakis and J. F. Normant, *Tetrahedron*, 1984, **40**, 1803.
79. For a review, see K. Koga, *ACS Symp. Ser.*, 1982, **185**, 73.
80. S. Hashimoto, S. Yamada and K. Koga, *J. Am. Chem. Soc.*, 1976, **98**, 7450.
81. S. Hashimoto, N. Komeshima, S. Yamada and K. Koga, *Tetrahedron Lett.*, 1977, 2907.
82. S. Hashimoto, H. Kogen, K. Tomioka and K. Koga, *Tetrahedron Lett.*, 1979, 3009.
83. S. Hashimoto, S. Yamada and K. Koga, *Chem. Pharm. Bull.*, 1979, **27**, 771.
84. S. Hashimoto, N. Komeshima, S. Yamada and K. Koga, *Chem. Pharm. Bull.*, 1979, **27**, 2437.
85. H. Kogen, K. Tomioka, S. Hashimoto and K. Koga, *Tetrahedron Lett.*, 1980, **21**, 4005.
86. H. Kogen, K. Tomioka, S. Hashimoto and K. Koga, *Tetrahedron*, 1981, **37**, 3951.
87. K. Tomioka, F. Masumi, T. Yamashita and K. Koga, *Tetrahedron Lett.*, 1984, **25**, 333.
88. M. Harre, P. Raddatz, R. Walenta and E. Winterfeldt, *Angew. Chem.*, 1982, **94**, 496; *Angew Chem., Int. Ed. Engl.*, 1982, **21**, 480.
89. E. Winterfeldt, 'Prinzipien und Methoden der stereoselektiven Synthese', Vieweg, Braunschweig, 1988, p. 101.
90. M. Suzuki, A. Yanagisawa and R. Noyori, *J. Am. Chem. Soc.*, 1988, **110**, 4718, and references cited therein.
91. Y. Köksal, P. Raddatz and E. Winterfeldt, *Angew. Chem.*, 1980, **92**, 486; *Angew. Chem., Int. Ed. Engl.*, 1980, **191**, 472.
92. N. Rehnberg and G. Magnusson, *Tetrahedron Lett.*, 1988, **29**, 3599.
93. N. Rehnberg, T. Frejd and G. Magnusson, *Tetrahedron Lett.*, 1987, **28**, 3589.
94. M. Asaoka, K. Shima and H. Takei, *Tetrahedron Lett.*, 1987, **28**, 5669.
95. M. Asaoka, K. Shima and H. Takei, *J. Chem. Soc., Chem. Commun.*, 1988, 430.
96. M. Asaoka, K. Takenouchi and H. Takei, *Chem. Lett.*, 1988, 1225.
97. A. B. Smith, III, N. K. Dunlap and G. A. Sulikowski, *Tetrahedron Lett.*, 1988, **29**, 439.
98. A. B. Smith, III and P. K. Trumper, *Tetrahedron Lett.*, 1988, **29**, 443.
99. W.-Y. Zhang, D. J. Jakiela, A. Maul, C. Knors, J. L. Lauher, P. Helquist and D. Enders, *J. Am. Chem. Soc.*, 1988, **110**, 4652.
100. G. Tsuchihashi, S. Mitamura, S. Inoue and K. Ogura, *Tetrahedron Lett.*, 1973, 323.
101. G. Tsuchihashi, S. Mitamura and K. Ogura, *Tetrahedron Lett.*, 1976, 855

102. For a review, see G. H. Posner, in 'Asymmetric Synthesis', ed. J. D. Morrison, Academic Press, New York, 1983, vol. 2, p. 225.
103. For a brief review, see G. H. Posner, *Chem. Scr.*, 1985, **25**, 157.
104. For a review, see G. H. Posner, *Acc. Chem. Res.*, 1987, **20**, 72.
105. G. H. Posner, J. P. Mallamo and K. Miura, *J. Am. Chem. Soc.*, 1981, **103**, 2886.
106. G. H. Posner, M. Hulce, J. P. Mallamo, S. A. Drexler and J. Clardy, *J. Org. Chem.*, 1981, **46**, 5244.
107. G. H. Posner, J. P. Mallamo, M. Hulce and L. L. Frye, *J. Am. Chem. Soc.*, 1982, **104**, 4180.
108. G. H. Posner and M. Hulce, *Tetrahedron Lett.*, 1984, **25**, 379.
109. G. H. Posner, T. P. Kogan and M. Hulce, *Tetrahedron Lett.*, 1984, **25**, 383.
110. G. H. Posner, L. L. Frye and M. Hulce, *Tetrahedron*, 1984, **40**, 1401.
111. G. H. Posner and L. L. Frye, *Isr. J. Chem.*, 1984, **24**, 88.
112. G. H. Posner, T. P. Kogan, S. R. Haines and L. L. Frye, *Tetrahedron Lett.*, 1984, **25**, 2627.
113. G. H. Posner and W. Harrison, *J. Chem. Soc., Chem. Commun.*, 1985, 1786.
114. G. H. Posner and E. Asirvatham, *J. Org. Chem.*, 1985, **50**, 2589.
115. G. H. Posner, M. Weitzberg, T. G. Hamill, E. Asirvatham, H. Cun-Heng and J. Clardy, *Tetrahedron*, 1986, **42**, 2919.
116. G. H. Posner and C. Switzer, *J. Am. Chem. Soc.*, 1986, **108**, 1239.
117. G. H. Posner and M. Weitzberg, *Synth. Commun.*, 1987, **17**, 611.
118. S. T. Saengchantara and T. W. Wallace, *J. Chem. Soc., Chem. Commun.*, 1986, 1592.
119. S. D. Kahn and W. J. Hehre, *J. Am. Chem. Soc.*, 1986, **108**, 7399.
120. S. D. Kahn, K. D. Dobbs and W. J. Hehre, *J. Am. Chem. Soc.*, 1988, **110**, 4602.
121. R. A. Holton, R. M. Kennedy, H.-B. Kim and M. Krafft, *J. Am. Chem. Soc.*, 1987, **109**, 1597.
122. R. Davis, J. R. Kern, L. J. Kurz and J. R. Pfister, *J. Am. Chem. Soc.*, 1988, **110**, 7873.
123. T. Yamazaki, N. Ishikawa, H. Iwatsubo and T. Kitazume, *J. Chem. Soc., Chem. Commun.*, 1987, 1340.
124. S. G. Pyne, *J. Org. Chem.*, 1986, **51**, 81.
125. S. G. Pyne, *Tetrahedron Lett.*, 1986, **27**, 1691.
126. M. Isobe, J. Obeyama, Y. Funabashi and T. Goto, *Tetrahedron Lett.*, 1988, **29**, 4773.
127. L. S. Liebeskind and M. E. Welker, *Tetrahedron Lett.*, 1985, **26**, 3079.
128. L. S. Liebeskind, M. E. Welker and R. W. Fengl, *J. Am. Chem. Soc.*, 1986, **108**, 6328.
129. S. G. Davies and J. C. Walker, *J. Chem. Soc., Chem. Commun.*, 1985, 209.
130. S. G. Davies, I. M. Dordor-Hedgecock, K. H. Sutton and J. C. Walker, *Tetrahedron Lett.*, 1986, **27**, 3787.
131. S. G. Davies, R. J. C. Easton, J. C. Walker and P. Warner, *Tetrahedron*, 1986, **42**, 175.
132. S. G. Davies, I. M. Dordor-Hedgecock, K. H. Sutton, J. C. Walker, R. H. Jones and K. Prout, *Tetrahedron*, 1986, **42**, 5123.
133. J. W. Herndon, C. Wu and H. L. Ammon, *J. Org. Chem.*, 1988, **53**, 2873.
134. T. Mukaiyama, Y. Hirako and T. Takeda, *Chem. Lett.*, 1978, 461.
135. T. Takeda and T. Mukaiyama, *Chem. Lett.*, 1980, 163.
136. T. Takeda, T. Hoshito and T. Mukaiyama, *Chem. Lett.*, 1981, 797.
137. W. Oppolzer, R. Pitteloud, G. Bernardinelli and K. Bättig, *Tetrahedron Lett.*, 1983, **24**, 4975.
138. W. Oppolzer and R. Pitteloud, *J. Am. Chem. Soc.*, 1982, **104**, 6478.
139. E. J. Corey and R. T. Peterson, *Tetrahedron Lett.*, 1985, **26**, 5025.
140. E. J. Corey and P. A. Magriotis, *J. Am. Chem. Soc.*, 1987, **109**, 287.
141. G. Stork and N. A. Saccomano, *Nouv. J. Chim.*, 1986, **10**, 677.
142. G. Stork and N. A. Saccomano, *Tetrahedron Lett.*, 1987, **28**, 2087.
143. M. Yamaguchi, K. Hasebe, S. Tanaka and T. Minami, *Tetrahedron Lett.*, 1986, **27**, 959.
144. G. Calderari and D. Seebach, *Helv. Chim. Acta.*, 1985, **68**, 1592.
145. C. Gennari, L. Colombo, G. Bertolini and G. Schimperna, *J. Org. Chem.*, 1987, **52**, 2754.
146. N. Minowa, M. Hirayama and S. Fukatsu, *Tetrahedron Lett.*, 1984, **25**, 1147.
147. Y. N. Belokon', A. G. Bulychev, M. G. Ryzhov, S. V. Vitt, A. S. Batsanov, Y. T. Struchkov, V. I. Bakhmutov and V. M. Belikov, *J. Chem. Soc., Perkin Trans. 1*, 1986, 1865.
148. Y. N. Belokon', A. G. Bulychev, V. A. Pavlov, E. B. Fedorova, V. A. Tsyryapkin, V. I. Bakhmutov and V. M. Belikov, *J. Chem. Soc., Perkin Trans. 1*, 1988, 2075.
149. S. Yamada, K. Hiroi and K. Achiwa, *Tetrahedron Lett.*, 1969, 4233.
150. S. Yamada and G. Otani, *Tetrahedron Lett.*, 1969, 4237.
151. S. Yamada and G. Otani, *Tetrahedron Lett.*, 1971, 1133.
152. G. Otani and S. Yamada, *Chem. Pharm. Bull.*, 1973, **21**, 2112.
153. G. Otani and S. Yamada, *Chem. Pharm. Bull.*, 1973, **21**, 2115.
154. G. Otani and S. Yamada, *Chem. Pharm. Bull.*, 1973, **21**, 2130.
155. T. Sone, K. Hiroi and S. Yamada, *Chem. Pharm. Bull.*, 1973, **21**, 2331.
156. T. Sone, S. Terashima and S. Yamada, *Chem. Pharm. Bull.*, 1976, **24**, 1288.
157. For a review, see D. E. Bergbreiter and M. Newcomb, in 'Asymmetric Synthesis', ed. J. D. Morrison, Academic Press, New York, 1983, vol. 2, p. 243.
158. B. de Jéso and J.-C. Pommier, *Tetrahedron Lett.*, 1980, **21**, 4511.
159. C. Stetin, B. de Jéso and J.-C. Pommier, *J. Org. Chem.*, 1985, **50**, 3863.
160. Y. Ito, M. Sawamura, K. Kominami and T. Saegusa, *Tetrahedron Lett.*, 1985, **26**, 5303.
161. M. Pfau, G. Revial, A. Guingant and J. d'Angelo, *J. Am. Chem. Soc.*, 1985, **107**, 273.
162. T. Volpe, G. Revial, M. Pfau and J. d'Angelo, *Tetrahedron Lett.*, 1987, **28**, 2367.
163. J. d'Angelo, G. Revial, T. Volpe and M. Pfau, *Tetrahedron Lett.*, 1988, **29**, 4427.
164. D. Desmaële and J. d'Angelo, *Tetrahedron Lett.*, 1989, **30**, 345.
165. J. d'Angelo, A. Guingant, C. Riche and A. Chiaroni, *Tetrahedron Lett.*, 1988, **29**, 2667.
166. Y. Hirai, T. Terada and T. Yamazaki, *J. Am. Chem. Soc.*, 1988, **110**, 958.
167. D. Enders and K. Papadopoulos, *Tetrahedron Lett.*, 1983, **24**, 4967.

168. D. Enders, K. Papadopoulos, B. E. M. Rendenbach, R. Appel and F. Knoch, *Tetrahedron Lett.*, 1986, **27**, 3491.
169. D. Enders and B. E. M. Rendenbach, *Tetrahedron*, 1986, **42**, 2235.
170. D. Enders and B. E. M. Rendenbach, *Chem. Ber.*, 1987, **120**, 1223.
171. D. Enders and B. E. M. Rendenbach, *Chem. Ber.*, 1987, **120**, 1731.
172. D. Enders, A. S. Demir, H. Puff and S. Franken, *Tetrahedron Lett.*, 1987, **28**, 3795.
173. K. Tomioka, K. Ando, K. Yasuda and K. Koga, *Tetrahedron Lett.*, 1986, **27**, 715.
174. K. Tomioka, K. Yasuda and K. Koga, *Tetrahedron Lett.*, 1986, **27**, 4611.
175. K. Tomioka, K. Yasuda and K. Koga, *J. Chem. Soc., Chem. Commun.*, 1987, 1345.
176. K. Tomioka, W. Seo, K. Ando and K. Koga, *Tetrahedron Lett.*, 1987, **28**, 6637.
177. C. Bohlmann, R. Bohlmann, E. G. Rivera, C. Vogel, M. D. Manadhar and E. Winterfeldt, *Liebigs Ann. Chem.*, 1985, 1752.
178. E. Winterfeldt and R. Freund, *Liebigs Ann. Chem.*, 1986, 1262.
179. U. Schöllkopf, D. Pettig, U. Busse, E. Egert and M. Dyrbusch, *Synthesis*, 1986, 737.
180. D. Pettig and U. Schöllkopf, *Synthesis*, 1988, 173.
181. U. Schöllkopf, D. Pettig, E. Schulze, M. Klinge, E. Egert, B. Benecke and M. Noltemeyer, *Angew. Chem.*, 1988, **100**, 181.
182. S. J. Blarer, W. B. Schweizer and D. Seebach, *Helv. Chim. Acta.* 1982, **65**, 1637.
183. S. J. Blarer and D. Seebach, *Chem. Ber.*, 1983, **116**, 2250.
184. S. J. Blarer and D. Seebach, *Chem. Ber.*, 1983, **116**, 3086.
185. F. Matloubi and G. Solladié, *Tetrahedron Lett.*, 1979, 2141.
186. G. Solladié, *Synthesis*, 1981, 185.
187. L. Colombo, C. Gennari, G. Resnati and C. Scolastico, *Synthesis*, 1981, 74.
188. L. Colombo, C. Gennari, G. Resnati and C. Scolastico, *J. Chem. Soc., Perkin Trans. 1*, 1981, 1284.
189. For a review, see M. R. Barbachyn and C. R. Johnson, in 'Asymmetric Synthesis', ed. J. D. Morrison, Academic Press, Orlando, 1984, vol. 4, p. 227.
190. M. R. Binns, R. K. Haynes, A. A. Katsifis, P. A. Schober and S. C. Vonwiller, *Tetrahedron Lett.*, 1985, **26**, 1565.
191. M. R. Binns, O. L. Chai, R. K. Haynes, A. A. Katsifis, P. A. Schober and S. C. Vonwiller, *Tetrahedron Lett.*, 1985, **26**, 1569.
192. M. R. Binns, R. J. Goodridge, R. K. Haynes and D. D. Ridley, *Tetrahedron Lett.*, 1985, **26**, 6381.
193. R. K. Haynes and S. C. Vonwiller, *J. Chem. Soc., Chem. Commun.*, 1981, 1284.
194. M. R. Binns, R. K. Haynes, A. G. Katsifis, P. A. Schober and S. C. Vonwiller, *J. Am. Chem. Soc.*, 1988, **110**, 5411.
195. R. K. Haynes, A. G. Katsifis, S. C. Vonwiller and T. W. Hambley, *J. Am. Chem. Soc.*, 1988, **110**, 5423.
196. D. H. Hua, G. Sinai-Zingde and S. Venkataraman, *J. Am. Chem. Soc.*, 1985, **107**, 4088.
197. D. H. Hua, *J. Am. Chem. Soc.*, 1986, **108**, 3835.
198. D. H. Hua, S. Venkataraman, M. J. Coulter and G. Sinai-Zingde, *J. Org. Chem.*, 1987, **52**, 719, 2962.
199. D. H. Hua, S. Venkataraman, R. A. Ostrander, G.-Z. Sinai, P. J. McCann, M. J. Coulter and M. R. Xu, *J. Org. Chem.*, 1988, **53**, 507.
200. D. H. Hua, S. Venkataraman, R. Chan-Yu-King and J. V. Paukstelis, *J. Am. Chem. Soc.*, 1988, **110**, 4741.
201. D. A. Hua, R. Chan-Yu-King, J. A. McKie and L. Myer, *J. Am. Chem. Soc.*, 1987, **109**, 5026.
202. T. Takahashi, Y. Naito and J. Tsuji, *J. Am. Chem. Soc.*, 1981, **103**, 5261.
203. T. Takahashi, H. Okumoto and J. Tsuji, *Tetrahedron Lett.*, 1984, **25**, 1925.
204. T. Takahashi, H. Okumoto, J. Tsuji and N. Harada, *J. Org. Chem.*, 1984, **49**, 948.
205. T. Takahashi, K. Shimizu, T. Doi and J. Tsuji, *J. Am. Chem. Soc.*, 1988, **110**, 2674.
206. K. Yamamoto, M. Iijima and Y. Ogimura, *Tetrahedron Lett.*, 1982, **23**, 3711.
207. K. Yamamoto, M. Iijima, Y. Ogimura and J. Tsuji, *Tetrahedron Lett.*, 1984, **25**, 2813.
208. K. Yamamoto, M. Kanoh, N. Yamamoto and J. Tsuji, *Tetrahedron Lett.*, 1987, **28**, 6347.
209. H. Malmberg, M. Nilsson and C. Ullenius, *Tetrahedron Lett.*, 1982, **23**, 3823.
210. S. Andersson, S. Jagner, M. Nilsson and F. Urso, *J. Organomet. Chem.*, 1986, **301**, 257.
211. R. A. Kretchmer, *J. Org. Chem.*, 1972, **37**, 2744.
212. J. S. Zweig, J.-L. Luche, E. Barreiro and P. Crabbé, *Tetrahedron Lett.*, 1975, 2355.
213. B. Gustafsson, M. Nilsson and C. Ullenius, *Acta Chem. Scand., Ser. B*, 1977, **31**, 667.
214. B. Gustafsson and C. Ullenius, *Tetrahedron Lett.*, 1977, 3171.
215. A. Takeda, T. Sakai, S. Shinohara and S. Tsuboi, *Bull. Chem. Soc. Jpn.*, 1977, **50**, 1133.
216. B. Gustafsson, *Tetrahedron*, 1978, **34**, 3023.
217. A.-T. Hansson, M. T. Rahman and C. Ullenius, *Acta Chem. Scand., Ser. B*, 1978, **32**, 483.
218. H. Malmberg, M. Nilsson and C. Ullenius, *Acta Chem. Scand., Ser. B*, 1981, **35**, 625.
219. A.-T. Hansson and M. Nilsson, *Tetrahedron*, 1982, **38**, 389.
220. F. Ghozland, J.-L. Luche and P. Crabbé, *Bull. Soc. Chim. Belg.*, 1978, **87**, 369.
221. B. Gustafsson, G. Hallnemo and C. Ullenius, *Acta Chem. Scand., Ser. B*, 1980, **34**, 443.
222. H. Malmberg and M. Nilsson, *J. Organomet. Chem.*, 1983, **243**, 241.
223. M. Huché, J. Berlan, G. Pourcelot and P. Cresson, *Tetrahedron Lett.*, 1981, **22**, 1329.
224. T. Imamoto and T. Mukaiyama, *Chem. Lett.*, 1980, 45.
225. F. Leyendecker, F. Jesser and B. Ruhland, *Tetrahedron Lett.*, 1981, **22**, 3601.
226. F. Leyendecker, F. Jesser and D. Laucher, *Tetrahedron Lett.*, 1983, **24**, 3513.
227. F. Leyendecker and D. Laucher, *Tetrahedron Lett.*, 1983, **24**, 3517.
228. F. Leyendecker, F. Jesser, D. Laucher and B. Ruhland, *Nouv. J. Chim.*, 1985, **9**, 7.
229. F. Leyendecker and D. Laucher, *Nouv. J. Chim.*, 1985, **9**, 13.
230. S. H. Bertz, G. Dabbagh and G. Sundararajan, *J. Org. Chem.*, 1986, **51**, 4953.
231. E. J. Corey, R. Naef and F. J. Hannon, *J. Am. Chem. Soc.*, 1986, **108**, 7114.
232. R. K. Dieter and M. Tokles, *J. Am. Chem. Soc.*, 1987, **109**, 2040.

233. G. Quinkert, *Chimia*, 1988, **42**, 207.
234. T. Müller, Dissertation, Universität Frankfurt/Main, 1988.
235. G. Quinkert, H.-G. Schmalz, E. Walzer, T. Kowalczyk-Przewloka, G. Dürner and J. W. Bats, *Angew. Chem.*, 1987, **99**, 82; *Angew. Chem., Int. Ed. Engl.*, 1987, **26**, 61.
236. G. Quinkert, H.-G. Schmalz, E. Walzer, S. Gross, T. Kowalczyk-Przewloka, C. Schierloh, G. Dürner, J. W. Bats and H. Kessler, *Liebigs Ann. Chem.*, 1988, 283.
237. W. Langer and D. Seebach, *Helv. Chim. Acta*, 1979, **62**, 1710.
238. D. Seebach, G. Crass, E.-M. Wilka, D. Hilvert and E. Brunner, *Helv. Chim. Acta*, 1979, **62**, 2695.
239. K. Tomioka, M. Sudani, Y. Shinmi and K. Koga, *Chem. Lett.*, 1985, 329.
240. J. F. G. A. Jansen and B. L. Feringa, *Tetrahedron Lett.*, 1988, **29**, 3593.
241. H. Wynberg and R. Helder, *Tetrahedron Lett.*, 1975, 4057.
242. H. Wynberg and B. Greijdanus, *J. Chem. Soc., Chem. Commun.*, 1978, 427.
243. K. Hermann and H. Wynberg, *J. Org. Chem.*, 1979, **44**, 2238.
244. M. Inagaki, J. Hiratake, Y. Yamamoto and J. Oda, *Bull. Chem. Soc. Jpn.*, 1987, **60**, 4121.
245. H. Brunner and B. Hammer, *Angew. Chem.*, 1984, **96**, 305.
246. D. J. Cram and G. D. Y. Sogah, *J. Chem. Soc., Chem. Commun.*, 1981, 625.
247. M. Alonso-López, M. Martín-Lomas and S. Penadés, *Tetrahedron Lett.*, 1986, **27**, 3551.
248. B. Raguse and D. D. Ridley, *Aust. J. Chem.*, 1984, **37**, 2059.
249. S. Colonna, A. Re and H. Wynberg, *J. Chem. Soc., Perkin Trans. 1*, 1981, 547.
250. R. S. E. Conn, A. V. Lovell, S. Karady and L. M. Weinstock, *J. Org. Chem.*, 1986, **51**, 4710.
251. A. Bhattacharya, U.-H. Dolling, E. J. J. Grabowski, S. Karady, K. M. Ryan and L. M. Weinstock, *Angew. Chem.*, 1986, **98**, 442.
252. A. Sera, K. Takagi, H. Katayama, H. Yamada and K. Matsumoto, *J. Org. Chem.*, 1988, **53**, 1157.
253. G. M. Villacorta, C. P. Rao and S. J. Lippard, *J. Am. Chem. Soc.*, 1988, **110**, 3175.
254. K. Soai, S. Yokoyama, T. Hayasaka and K. Ebihara, *J. Org. Chem.*, 1988, **53**, 4148.
255. K. Soai, T. Hayasaka, S. Shoji and S. Yokoyama, *Chem. Lett.*, 1988, 1571.
256. T. Yura, N. Iwasawa and T. Mukaiyama, *Chem. Lett.*, 1988, 1021.
257. T. Yura, N. Iwasawa, N. Narasaka and T. Mukaiyama, *Chem. Lett.*, 1988, 1025.
258. I. Fleming and N. D. Kindon, *J. Chem. Soc., Chem. Commun.*, 1987, 1177.
259. I. Fleming, *Pure Appl. Chem.*, 1988, **60**, 71.
260. T. Hayashi, Y. Matsumoto and Y. Ito, *J. Am. Chem. Soc.*, 1988, **110**, 5579.
261. H. Hiemstra and H. Wynberg, *J. Am. Chem. Soc.*, 1981, **103**, 417.
262. T. Mukaiyama, A. Ikegawa and K. Suzuki, *Chem. Lett.*, 1981, 165.
263. H. Yamashita and T. Mukaiyama, *Chem. Lett.*, 1985, 363.
264. R. R. Ahuja, A. A. Natu and V. N. Gogte, *Tetrahedron Lett.*, 1980, **21**, 4743.
265. S. I. Bkole and V. N. Gogte, *Indian J. Chem., Sect. B*, 1981, **20**, 218, 222.
266. N. Kobayashi and K. Iwai, *Tetrahedron Lett.*, 1980, **21**, 2167.
267. N. Kobayashi and K. Iwai, *J. Org. Chem.*, 1981, **46**, 1823.
268. A. Papagni, S. Colonna, S. Julia and J. Rocas, *Synth. Commun.*, 1985, **15**, 891.
269. P. Hodge, E. Khoshdel and J. Waterhouse, *J. Chem. Soc., Perkin Trans. 1*, 1985, 2327.
270. J. Gawronski, K. Gawronska and H. Wynberg, *J. Chem. Soc., Chem. Commun.*, 1981, 307.
271. J. Gawronski, K. Gawronska, H. Kolbon and H. Wynberg, *Recl. Trav. Chim. Pays-Bas*, 1983, **102**, 479.
272. H. Pluim and H. Wynberg, *Tetrahedron Lett.*, 1979, 1251.
273. K. B. Sharpless and R. A. Johnson, in 'Comprehensive Organic Synthesis', ed. B. M. Trost, Pergamon Press, Oxford, 1991, vol. 7, chap. 3.2.
274. R. Kinas, K. Pankiewicz, W. J. Stec, P. B. Farmer, A. B. Foster and M. Jarman, *J. Org. Chem.*, 1977, **42**, 1650.
275. M. Furukawa, T. Okawara and Y. Terawaki, *Chem. Pharm. Bull.*, 1977, **25**, 1319.
276. J. d'Angelo and J. Maddaluno, *J. Am. Chem. Soc.*, 1986, **108**, 8112.
277. M. Ihara, Y. Takino, K. Fukumoto and T. Kametani, *Tetrahedron Lett.*, 1988, **29**, 4135.
278. S. G. Pyne, R. C. Griffith and M. Edwards, *Tetrahedron Lett.*, 1988, **29**, 2089, 5042.
279. S. G. Pyne, *Tetrahedron Lett.*, 1988, **29**, 4737.
280. M. Hirama, H. Hioki, S. Ito and C. Kabuto, *Tetrahedron Lett.*, 1988, **29**, 3121.
281. S. G. Pyne, *J. Chem. Soc., Chem. Commun.*, 1986, 1686.
282. W. Oppolzer, G. Poli, C. Starkemann and G. Bernardinelli, *Tetrahedron Lett.*, 1988, **29**, 3559.
283. W. Oppolzer and G. Poli, *Tetrahedron Lett.*, 1986, **27**, 4717.
284. U. Leutenegger, A. Madin and A. Pfaltz, *Angew. Chem.*, 1989, **101**, 61.

1.6
Nucleophilic Addition–Electrophilic Coupling with a Carbanion Intermediate

MARTIN HULCE
University of Maryland, Baltimore, MD, USA

and

MARC J. CHAPDELAINE
ICI Americas, Wilmington, DE, USA

1.6.1 INTRODUCTION

Vicinal difunctionalization reactions have been extensively exploited in synthetic organic chemistry providing rapid, convergent access to complex structures in a stereocontrolled fashion. Examples of these reactions are numerous;[1] classical reactions such as Diels–Alder, [2 + 2] and 1,3-dipolar cycloadditions (Volume 5), epoxidation–functionalization of alkenes,[2] carbenoid additions to alkenes[3] and the additions of acyl[4] and alkyl[5] halides to alkenes, using Friedel–Crafts catalysts, have been joined by many new reactions of considerable synthetic potential, including, for instance, inverse electron demand Diels–Alder reactions[6] and radical cyclization–trapping (Volume 4, Part 4).

1.6.1.1 Tandem Vicinal Difunctionalization

Over the past 25 years, a specific vicinal difunctionalization sequence referred to as tandem vicinal difunctionalization has been developed.[7–10] The process consists of two reactions, one enabling the other to proceed. An initial Michael-type addition (also referred to as a conjugate or 1,4-addition) of a nucleophile, NuM, to a substrate (**1**), the 'Michael acceptor', which usually is an α,β-unsaturated carbonyl-containing compound, transforms both the α- and β-carbons. The β-carbon, changing from an sp^2- to an sp^3-hybridization state, becomes further substituted, and the α-carbon takes on nucleophilic character, usually as the carbanionic center of an ambident enolate ion (**2**), the 'conjugate enolate' (Scheme 1). The conjugate enolate ion subsequently can be trapped *in situ* using an appropriate electrophile, EX, thus further substituting the α-carbon. Conceptually, the process can be envisaged as a vinylogous substitution reaction; by means of a 'third party' two-carbon extension, nucleophile and electrophile react, forming two vicinal σ-bonds.

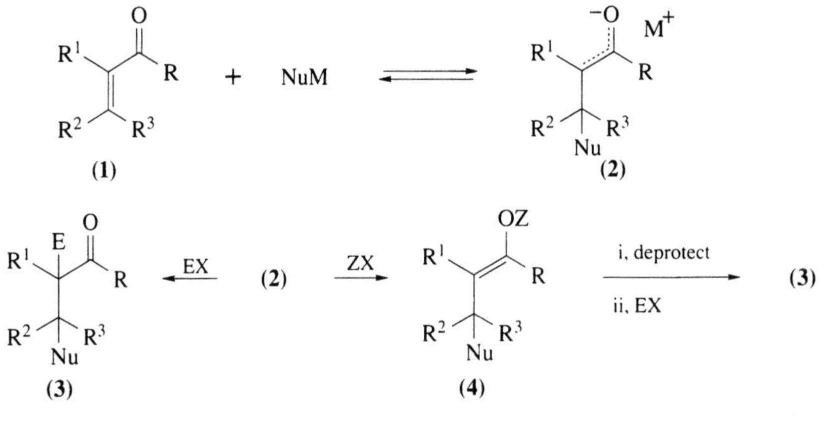

Scheme 1

The enolate ion generated by the conjugate addition process need not be α-functionalized *in situ*. It is an ambident anion and can be isolated as a neutral, uncharged species (**4**), an 'enolate equivalent', by straightforward *O*-functionalization using an appropriate protecting reagent, ZX, or by simple protonation. Isolation of enolate equivalent (**4**) allows two courses of action. The most frequently followed is regeneration of the enolate, with subsequent functionalization at the α-carbon to give the vicinally disubstituted product, (**3**). Alternatively, extensive chemical manipulation of the intervening species (**4**) is possible; a series of chemical transformations can be performed between initial β-functionalization and the ultimate α-functionalization.

The general reaction sequence has been named more specifically as tandem vicinal dialkylation, nucleophilic–electrophilic carbacondensation,[11] dicarbacondensation[12,13] or conjugate addition–enolate trapping,[14] usually in reference to the fact that most of the reaction examples available create two new vicinal carbon–carbon bonds. Many noncarbon nucleophiles and electrophiles also are known, resulting in vicinal carbon–heteroatom bonds in the products of the reaction sequence. For this reason, the broader term used here is more descriptive.

Tandem vicinal difunctionalizations by nucleophilic addition–enolate trapping sequences have developed into tools of considerable utility in the total synthesis of natural products, including various terpenoids,[15] steroids,[16,17] prostaglandins,[9] anthraquinones[18] and lactones.[19–21] The method can be used as a cyclization strategy to form cyclopropanes, cyclobutanes[22] and complex polycyclic products,[23] as well as to append new rings onto a substrate molecule.[24] Other applications of the reaction include the preparation of useful intermediates such as 1,3- and 1,5-diones, stereoisomerically pure α,β-unsaturated esters, γ,δ-unsaturated ketones and a variety of heterocyclic systems.[25–27] The methodology continues to widen in scope: an umpolung version, for instance, is available, allowing addition of the nucleophile to occur at the α-carbon and that of the electrophile to occur at the β-carbon.[28]

1.6.1.2 Historical Background

Early research optimizing reaction conditions for the 1,2-addition of Grignard reagents to α,β-unsaturated ketones noted the formation of undesired secondary products unless a large excess of the Grignard reagent was used.[29] Recognition that an alternate 1,4-addition pathway was possible, generating a conjugate enolate[30] capable of reacting with the starting ketone when present in significant concentrations, led to the identification of the observed secondary products as dimers formed by a tandem conjugate addition–aldol condensation process.[31]

The synthetic implications of this discovery were slow to be exploited. Base-initiated dimerizations of 2-cycloalkenones, known to give crystalline solids,[32,33] remained puzzling for some time before conjugate additions were suggested to account for some of the possible products;[34] indeed, the product of base-catalyzed dimerization of 4,4-dimethyl-2-cyclopentenone, which proceeds *via* a double Michael addition sequence, was not identified until 1969 (Scheme 2).[35] An unanticipated cyclopropanation reaction of acrylaldehyde[36,37] using ethyl bromomalonate and proceeding by means of a similar Michael addition–S_Ni enolate alkylation represents an early synthetic use of tandem vicinal difunctionalization.

H H + $Bu^tO^- K^+$ ⇌ O⁻ +

⇌ ⁻O O $\overset{Bu^tOH}{\rightleftharpoons}$ O O

40%

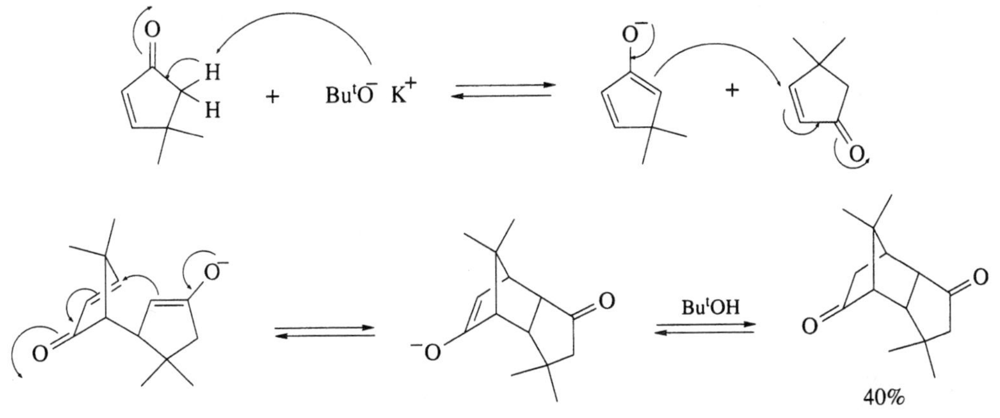

Scheme 2

As the synthetic value[38] of regiospecifically generated enolates became apparent in the 1960s, new methods were discovered and developed to prepare them. Dissolving metal conjugate reduction of α,β-

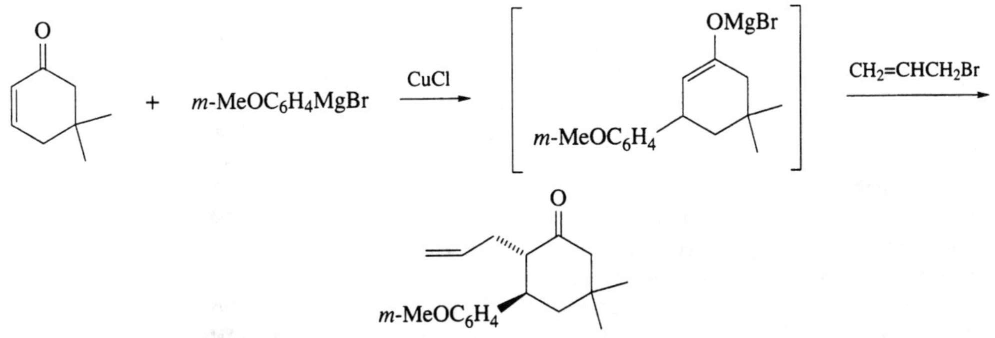

O + m-$MeOC_6H_4MgBr$ $\xrightarrow{CuCl}$ [OMgBr m-$MeOC_6H_4$] $\xrightarrow{CH_2=CHCH_2Br}$

O m-$MeOC_6H_4$

Scheme 3

unsaturated ketones to produce enolates, which could be *C*-alkylated using appropriate electrophiles,[39] comprises the first exploitation of a one-pot, three-component tandem vicinal difunctionalization reaction. The method was extended to conjugate additions of nucleophiles in aprotic media and applied to natural product synthetic strategy, the initial example being a key step in the total synthesis of lycopodine (Scheme 3).[40]

1.6.2 MECHANISM AND STEREOCHEMISTRY

The two distinct bond-forming steps in tandem vicinal difunctionalization have been studied extensively. The first step consists of a nucleophilic addition to a π-system; the nucleophile is almost invariably an organometal. 1,4-Addition to an α,β-unsaturated carbonyl substrate concomitantly generates a new σ-bond at the β-carbon and an enolate ion. The second step constitutes *C*-functionalization of the enolate intermediate, forming a new σ-bond between the nucleophilic α-carbon of the enolate and an electrophilic reagent.

The mechanism of the inital conjugate addition step undoubtedly varies according to the identity of the nucleophile.[41,42] The most common nucleophiles used are organocopper reagents. The fundamental aspect of organocopper conjugate additions to α,β-unsaturated carbonyl substrates is agreed[43] to be oxidative addition of a d^{10} cuprate, producing a transient copper(III) (d^8) intermediate. Reductive elimination follows to generate a new chemical bond at the β-carbon of the substrate, a conjugate enolate and a copper(I) species (Scheme 4). Copper(I)–π-bond coordination clearly precedes bond formation;[44] an alternative pathway involving carbocupration[45] instead of a copper(III) intermediate has been proposed and the specific nature of the reaction continues to be investigated.

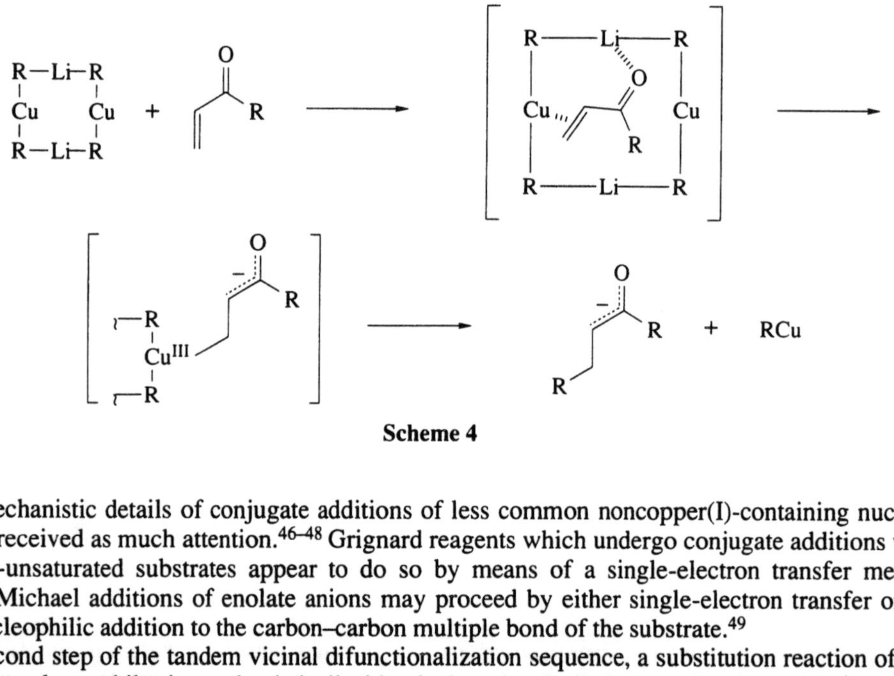

Scheme 4

The mechanistic details of conjugate additions of less common noncopper(I)-containing nucleophiles have not received as much attention.[46–48] Grignard reagents which undergo conjugate additions with specific α,β-unsaturated substrates appear to do so by means of a single-electron transfer mechanism, whereas Michael additions of enolate anions may proceed by either single-electron transfer or a traditional nucleophilic addition to the carbon–carbon multiple bond of the substrate.[49]

The second step of the tandem vicinal difunctionalization sequence, a substitution reaction of an enolate with an electrophile, is mechanistically identical to the *C*-alkylation of regiospecifically generated enolates.[38] For electrophiles of relatively low reduction potentials,[50] such as alkyl iodides, the substitution reaction appears to proceed *via* a single-electron transfer mechanism. Electrophiles with higher reduction potentials such as alkyl bromides, on the other hand, appear to undergo bond formation according to an S_N2 mechanism (Scheme 5).[51,52] Often it is found that the identity of the counterion profoundly influences the reactivity of the enolate towards the electrophile.[53,54] The counterion typically is predetermined by the first step of the tandem vicinal difunctionalization and can be a limiting factor in reactivity of the enolate towards reaction with an electrophilic reagent.

As with mechanism, the stereochemical outcome of tandem vicinal difunctionalization is dependent upon the individual bond-forming steps in the sequence. The conjugate addition reaction is quite sensitive to the steric environment of the α,β-unsaturated substrate, so that the bond-forming process at the β-carbon adheres to steric approach control factors: the 5-methoxycarbonyl substituent of (**5**) directs axial attack of a bis(dimethylphenylsilyl)copperlithium reagent so that only the 3,5-*trans* adduct is formed

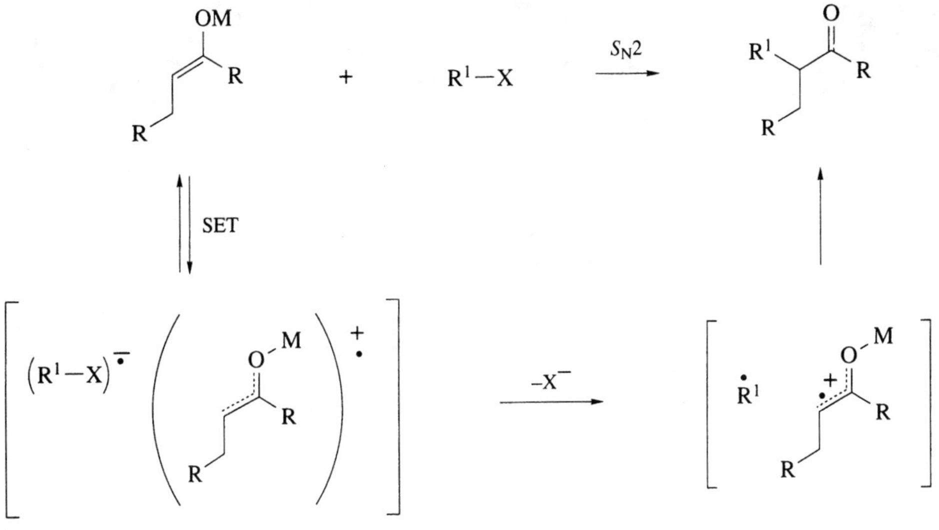

OM R R + R^1–X —S_N2→ R^1 O R R

SET

[$(R^1–X)^{\overline{\bullet}}$ (O M R R)$^{+\bullet}$] —$-X^-$→ [$\dot{R}^1$ O M + R R]

Scheme 5

(equation 1);[55] smaller substituents have essentially the same effect (equation 2).[56] Subsequent α-functionalization of the conjugate enolate does not always proceed with such high stereoselectivity, as these two examples indicate. Thermodynamically more stable *trans* products usually predominate, as predicted both by steric approach and product development control factors,[57,58] but a complex combination of other factors (the enolate counterion, the nature of the enolate itself, the reaction conditions, the nature of the electrophile) make predictions difficult and occasionally unreliable. When (**5**) undergoes conjugate addition with diphenylcopperlithium and the resulting adduct is α-methylated, the *cis* and not the *trans* product predominates (equation 3).[59] A relatively stereochemically remote substituent on an electrophile clearly influences the stereoselectivity of α-functionalization during tandem vicinal difunctionalization of cyclopentenone (**6**; equation 4).[60]

If the product of α-functionalization has a tertiary α-carbon, equilibration can occur under the conditions of tandem vicinal difunctionalization, as in equation (2). The equilibration process can proceed at

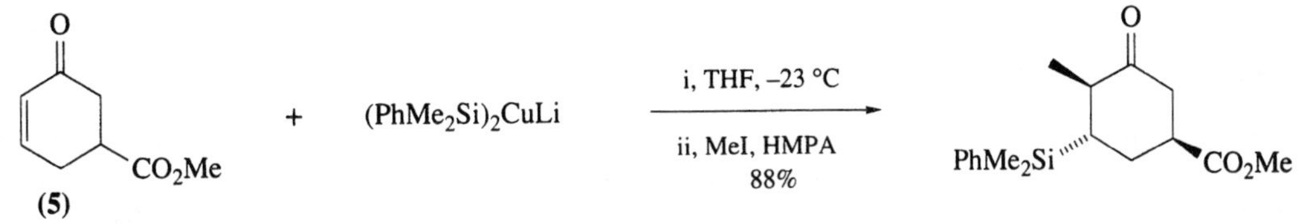

O CO_2Me (**5**) + $(PhMe_2Si)_2CuLi$ —i, THF, –23 °C; ii, MeI, HMPA; 88%→ O $PhMe_2Si$ CO_2Me (1)

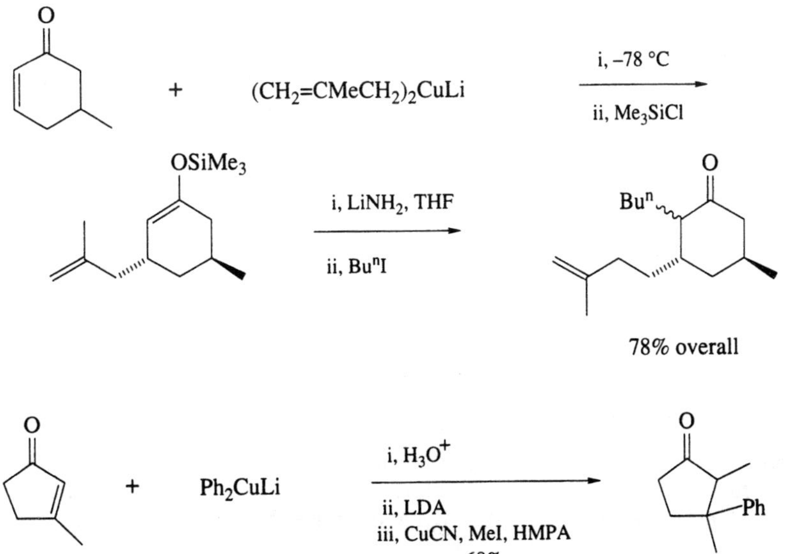

O + $(CH_2{=}CMeCH_2)_2CuLi$ —i, –78 °C; ii, Me_3SiCl→

$OSiMe_3$ —i, $LiNH_2$, THF; ii, Bu^nI→ O Bu^n (2)

78% overall

O + Ph_2CuLi —i, H_3O^+; ii, LDA; iii, CuCN, MeI, HMPA; 68%→ O Ph (3)

trans:*cis* = 35:65

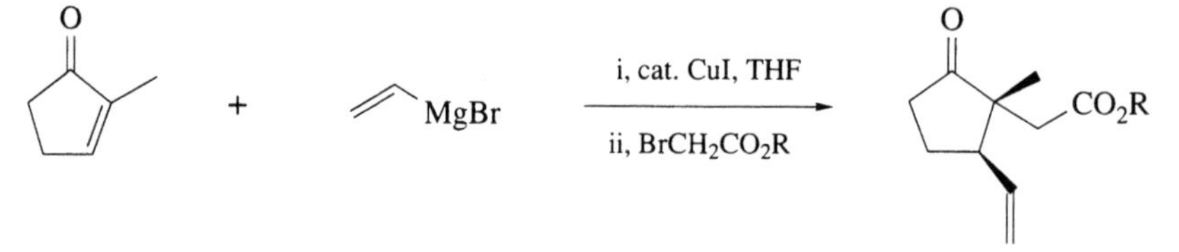

i, cat. CuI, THF
ii, $BrCH_2CO_2R$
(4)

(6)

R = Et, *trans*:*cis* = 88:12
R = Bu^t, >99% *trans*

rates sufficiently fast to obscure the original stereopreference of the α-bond-forming step: methylenecyclohexane annulation of cycloalkenones provides the expected kinetic *cis*-fused adducts after intramolecular α-alkylation of the conjugate addition products, but if there is no 2-substituent on the starting cycloalkenone, the kinetic product equilibrates to give substantial quantities of the thermodynamically more stable *trans* adducts (equation 5).[61]

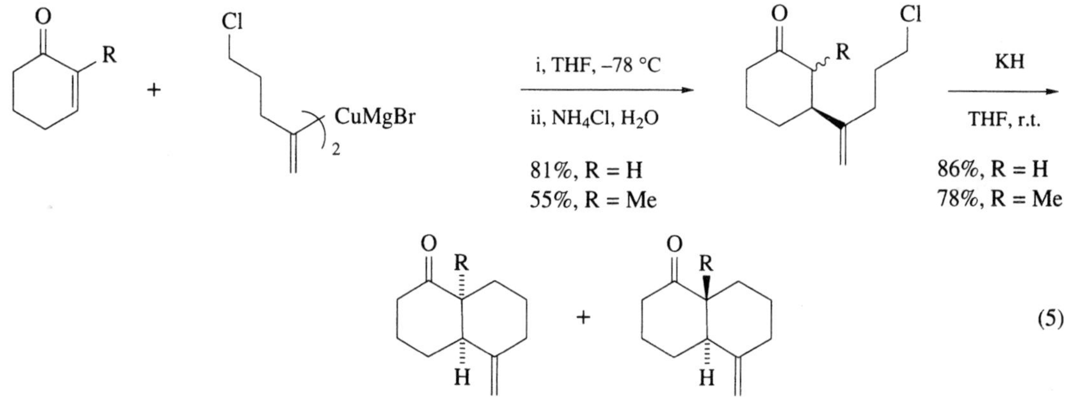

i, THF, –78 °C
ii, NH_4Cl, H_2O
81%, R = H
55%, R = Me

KH
THF, r.t.
86%, R = H
78%, R = Me

(5)

R = H	33%	66%
R = Me	100%	0%

1.6.3 THE α,β-UNSATURATED SUBSTRATE

The variety of conjugated unsaturated substrate types that can be used in tandem vicinal difunctionalization sequences allows considerable latitude in substrate choice. Generally, the only strict requirements that a potential substrate must meet are that the α-carbon of the α,β-unsaturated carbon–carbon bond moiety be substituted with an electron-withdrawing group, usually a carbonyl-containing group or some other unsaturated group capable of making the β-carbon electrophilic, and that there be no functional groups present in the substrate that would be incompatible with the nucleophile to be added. The latter requirement usually precludes the use of unprotected, relatively acidic functional groups in the substrate, such as carboxylic acid, hydroxy, thiol or β-keto ester groups. Exceptions to the former requirement are possible; especially noteworthy is the ability to use alkynes as substrates. A synthesis of spirovetevanes exploits tandem vicinal difunctionalization of a fulvene (equation 6).[62]

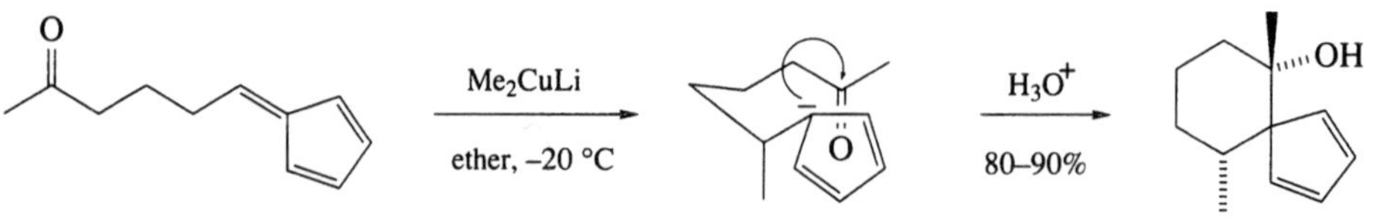

Me_2CuLi
ether, –20 °C

H_3O^+
80–90%

(6)

1.6.3.1 Aldehydes and Ketones

Ketones, specifically 2-alkenones and 2-cycloalkenones, have been used extensively as substrates for tandem vicinal difunctionalization, allowing delineation of various reactivity patterns based upon the structural elements present in the enone. Aldehyde substrates have been used less widely; comparison with analogous enone substrates indicates that they may be less effective Michael acceptors, resulting in lower isolated yields of reaction sequence products.[63]

1.6.3.1.1 Acyclic aldehydes and ketones

Aldehyde substrates behave in a manner similar to the analogous ketone substrates in organocopper-initiated 1,4-addition–conjugate enolate alkylation reactions (equation 7);[64] yields are comparable and the diastereoselectivity of the reaction sequence resulting from overall *trans* difunctionalization is good. In some cases, such as in the synthesis of a degradation product of chlorothricin,[65] complete regio- and stereo-control is possible. Cyclopropanations are possible using bromomalonates as Michael donors to α,β-unsaturated aldehydes,[37] as noted earlier; aldimine derivatives of α,β-unsaturated aldehydes serve as aldehyde functional group equivalents and permit naphthaldehydes to undergo organometal 1,4-addition–α-alkylation sequences.[66]

Ph–CH=CH–COR → (i, $(PhMe_2Si)_2CuLi$; ii, MeI) → Ph–CH($SiMe_2Ph$)–CH(Me)–C(O)R (7)

74%, R = H — *threo*:*erythro* = 12:1
57%, R = Me — *threo*:*erythro* = 49:1
70%, R = Ph — *threo*:*erythro* = 100:0

Acyclic enone substrates have been studied in detail: as with most reactions involving 1,4-addition, substrate substitution has a large influence on reaction outcome. Substituents on the α′-position of the enone act as steric directors, shielding the carbonyl carbon from the nucleophile and thereby promoting the 1,4-addition mode. Understandably, the degree of influence exhibited by an α′-substituent will vary, depending upon the identity of the nucleophile. For organocopper nucleophiles, phenyl and benzyl vinyl ketones are superior Michael acceptors compared to methyl vinyl ketones[67] (equation 7).[64] Methyl vinyl ketones are also the poorest substrates for enolate and acyl anion nucleophiles; interestingly, the homologous ethyl vinyl ketones are best, as illustrated by equation (8).[68,69]

CH_2=CH–COR + 2-lithio-2-(CO_2Et)-1,3-dithiolane → (i, THF, HMPA, –78 °C; ii, CH_2O) → 2-(CO_2Et)-1,3-dithiolan-2-yl–CH_2–CH(CH_2OH)–COR (8)

71%, R = Me
79%, R = Et
54%, R = Pr^n

Large steric requirements and charge at the α′-position of the enone substrate are combined in highly effective charge-directed conjugate addition–enolate functionalization reactions.[70] Michael acceptor acylphosphoranes such as (**7**; equation 9) react with a wide range of nucleophiles, including many alkyllithium reagents that normally add to enone substrates by 1,2- and not 1,4-addition;[71] the intermediate conjugate enolates are α-functionalized easily to give acylphosphorane adducts such as (**8**).[72] Conversion of compounds such as (**8**) to ketones can be accomplished by decarboxylation and hydrolysis; the method can be used to initiate intramolecular cyclizations leading to three-, five- and six-membered rings. In a related vein is the use of α,β-unsaturated acyliron complexes as Michael acceptors in tandem vicinal difunctionalization reactions (equation 10).[73,74] The [(η^5-C_5H_5)Fe(CO)PPh_3] moiety of (*E*)-crotonyl complex (**9**) activates the substrate towards 1,4-addition and provides excellent diastereoselectivity during the reaction sequence,[75] resulting in completely diastereospecific and usually rare *cis* dialkylation.[76] The substrate has been used in the synthesis of β-lactams and a single enantiomer of the pseudo-octahedral complex can be used in the difunctionalization sequence: (*R*)-(**9**) yields (2*R*,3*S*)-acids, -amides and -esters after decomplexation.

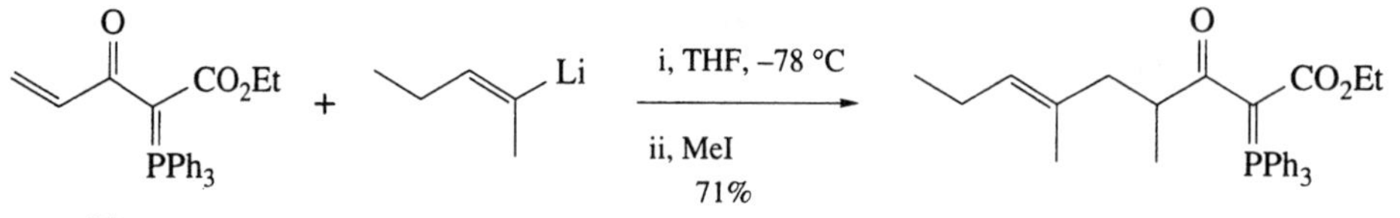

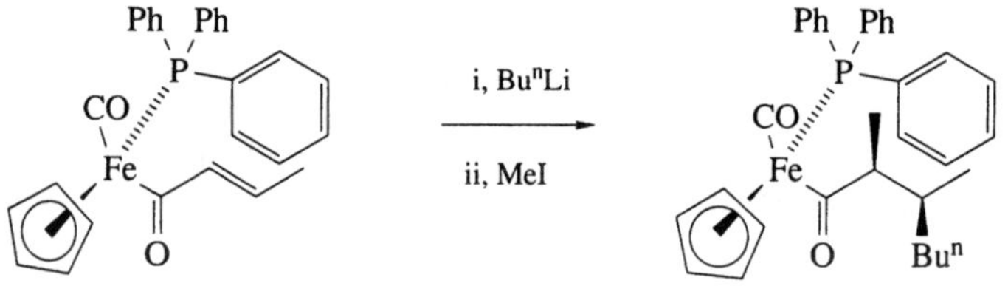

(9)

(10)

Enhancements of the chemical yields of tandem vicinal difunctionalizations are realized when the substrate enone bears an α-substituent. Replacement of the α-hydrogen of an α,β-unsaturated enone with a methyl group almost always provides such enhancement (equation 11)[77] by stabilizing the conjugate enolate, thereby reducing the amount of equilibration before α-functionalization occurs. Larger substituents, especially those that are able to stabilize the conjugate enolate by inductive or resonance effects, should provide similar results, although cycloalkenone difunctionalization examples indicate that this may not always be the case.

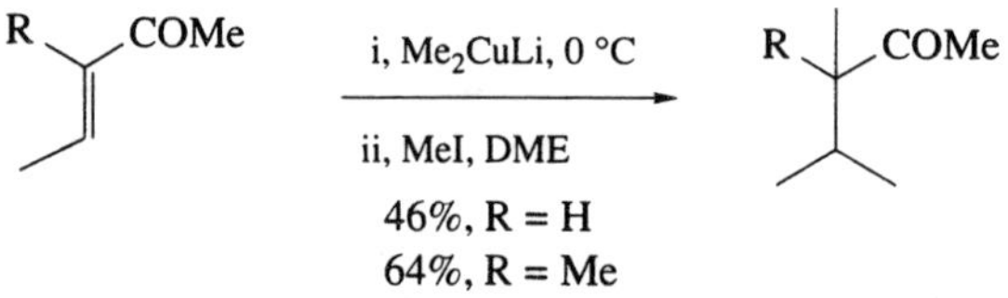

(11)

The sensitivity of the conjugate addition reaction to steric factors would indicate that β-substitution of the α,β-unsaturated substrate leads to lower overall reactivity of a prospective Michael acceptor towards tandem vicinal condensation. In the absence of Lewis acids[78,79] this prediction is borne out: β,β-disubstituted enones usually are poor substrates for difunctionalizations using bulky, stabilized Michael donors. Intermediate combinations of steric bulk in the initial nucleophile and the substrate are not as discriminating, especially when the nucleophile is an organocopper reagent or a simple enolate, as illustrated in equations (12)[67] and (13).[80] The combination of steric and electronic factors that deactivate the Michael acceptor towards conjugate addition can substantially reduce chemical yields of the difunctionalized product (equation 14).[69]

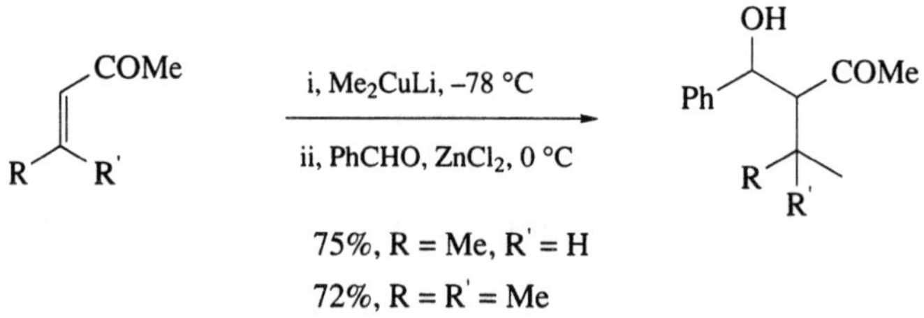

(12)

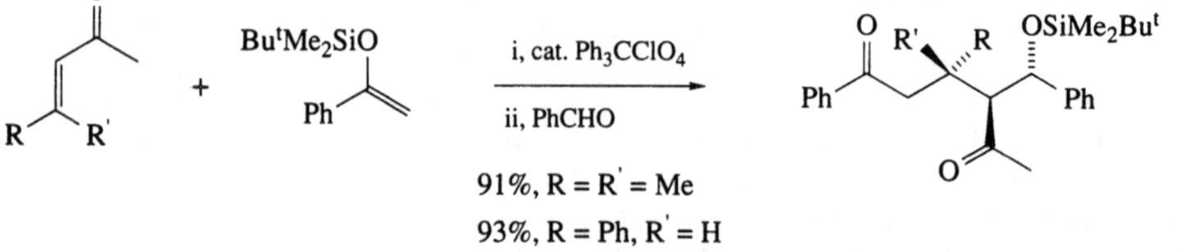

(13)

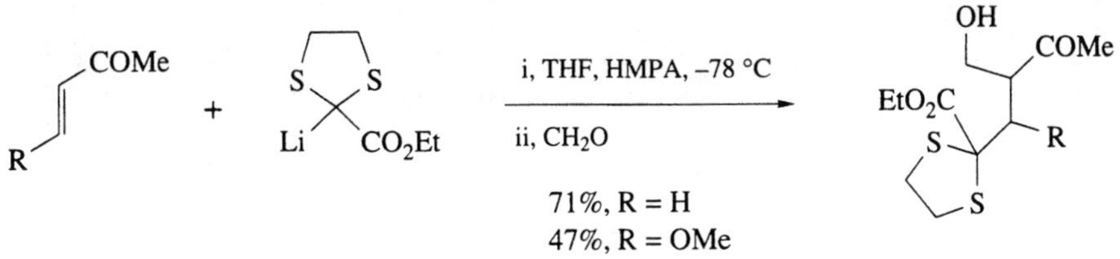

i, THF, HMPA, –78 °C; ii, CH_2O

71%, R = H
47%, R = OMe

(14)

For the most part, alkynic and allenic ketones have found limited use in conjugate addition–enolate trapping sequences;[69,81–83] their analogous esters have been used with far greater frequency (*vide infra*). Alkynic ketones, in particular, have found use in development of a new anionic polycyclization method consisting of intramolecular Michael addition followed by intramolecular alkylation (equation 15).[84]

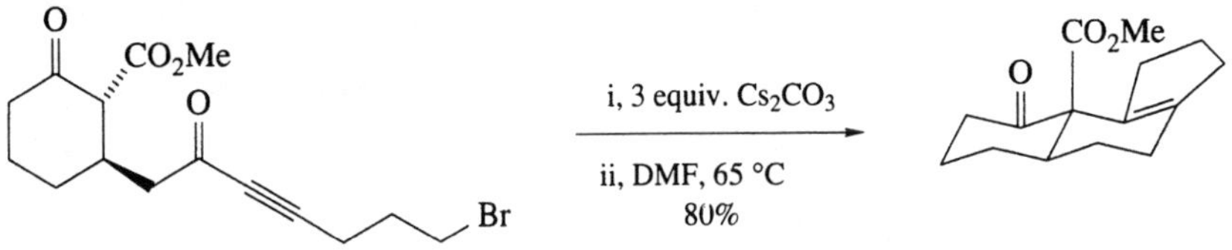

i, 3 equiv. Cs_2CO_3; ii, DMF, 65 °C
80%

(15)

1.6.3.1.2 Cyclic ketones

The prevalence of α,β-disubstituted cycloalkanones and substituted cycloalkanone substructures in nature has resulted in the use of a large number of moderate (5–8) ring size 2-cycloalkenones in total syntheses exploiting a tandem vicinal difunctionalization strategy. Equation (16) provides a representative example for a total synthesis of glycinoeclepin A,[11] in which (*R*)-carvone undergoes diastereospecific *trans* dialkylation to yield the 2,2,3,5-tetrasubstituted cyclohexanone (**10**); the method also has found extensive use in three-component coupling syntheses of prostaglandins.[9,85] General trends, especially for cyclopentenones and cyclohexenones, are discernable. On the whole, cyclopentenones are more reactive substrates than higher homologous cycloalkenones. Isolated yields usually are better, as illustrated by equation (17).[86] The rate of equilibration of the conjugate enolate intermediate varies with ring size, cyclopentanone enolates equilibrating more rapidly than cyclohexenone enolates.[87] Often, α-substituents must be employed to prevent equilibration: thus, 2-methyl-2-cyclopentenone is superior to 2-cyclopentenone as a substrate for difunctionalization. Arylhetero α-substituents which both stabilize the conjugate enolate toward equilibration and enhance subsequent α-functionalization can be used as substrate auxiliaries, resulting in 2-cyclopentenone synthetic equivalents which exhibit superior stereo- and regiochemical results in the difunctionalization sequence. Common examples include 2-phenylthio-[88] and 2-phenylselenyl-2-cyclopentenone.[89] Enantiomerically pure 2-arylsulfinyl-2-cycloalkenones offer additional, directable diastereofacial bias during difunctionalization to produce cycloalkanones of high enantiomeric purities (equation 18; NMP, *N*-methyl-2-pyrrolidone).[90] Substituents at the α-position of the substrate which activate it toward conjugate addition can promote regiospecific attack of the Michael donor when other competing modes of conjugate addition are possible (equation 19).[91]

Cycloalkenones are intrinsically less reactive than non-β-substituted acyclic enones. Cycloalkenones with β-substituents often undergo reaction with Michael donors quite slowly; a lower degree of stereo- and regio-control as well as lower chemical yield result.[92] These observations can be synthetically useful, allowing discrimination between two possible Michael acceptor sites in a molecule.[93,94] Lewis acid catalysis[78,79] of the conjugate addition process usually can greatly enhance the rate at which β-substituted cycloalkenones react.[95] Substituents at positions other than the α- and β-positions typically neither promote nor interfere with difunctionalization. In the case of particularly bulky α′-substitution, it is reasonable to assume that 1,4-addition might be enhanced when using nucleophiles that otherwise would show some

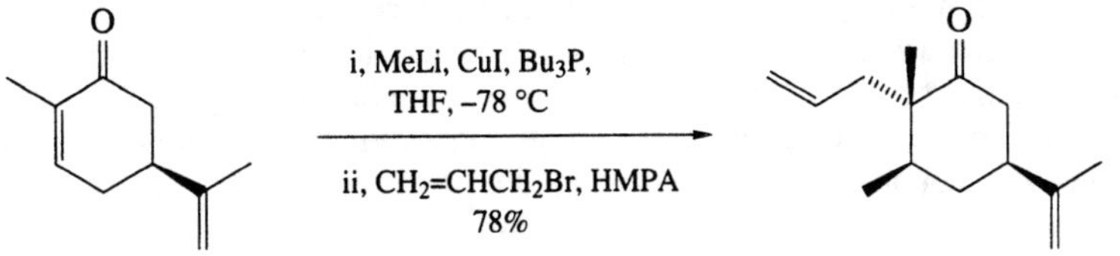

i, MeLi, CuI, Bu_3P, THF, –78 °C; ii, $CH_2{=}CHCH_2Br$, HMPA
78%

(16)

(10)

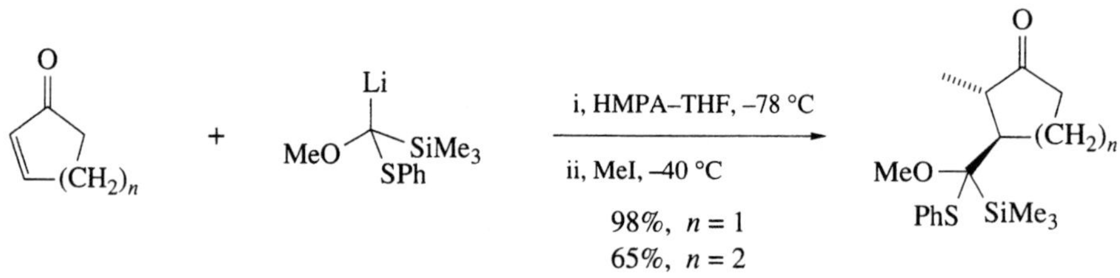

i, HMPA–THF, –78 °C; ii, MeI, –40 °C; 98%, $n = 1$; 65%, $n = 2$ (17)

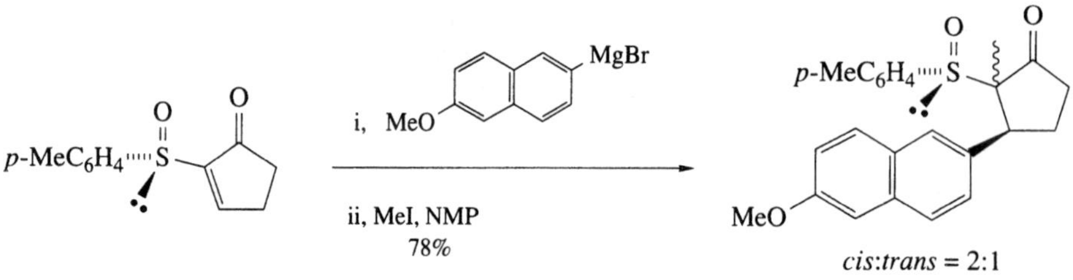

i, MeO-naphthyl-MgBr; ii, MeI, NMP; 78%; *cis*:*trans* = 2:1 (18)

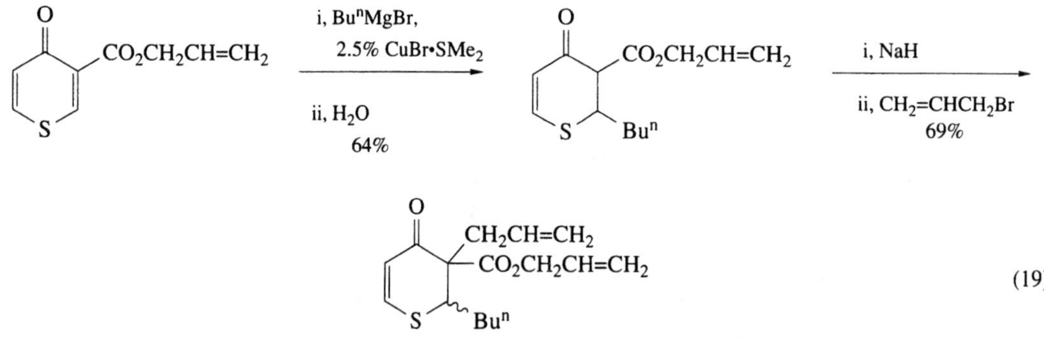

i, Bu^nMgBr, 2.5% $CuBr{\cdot}SMe_2$; ii, H_2O; 64%. i, NaH; ii, $CH_2{=}CHCH_2Br$; 69% (19)

THF, reflux, 70% (20)

preference for 1,2-addition; similarly, α′-disubstitution would eliminate any problems arising from equilibration of the conjugate enolate.

It is possible to use α,β-disubstituted cycloalkenone substrates for the construction of vicinal, quaternary carbon centers. Equation (20)[96] illustrates such a successful application; typically, however, enolate equilibration and steric congestion prevent straightforward application of the method.

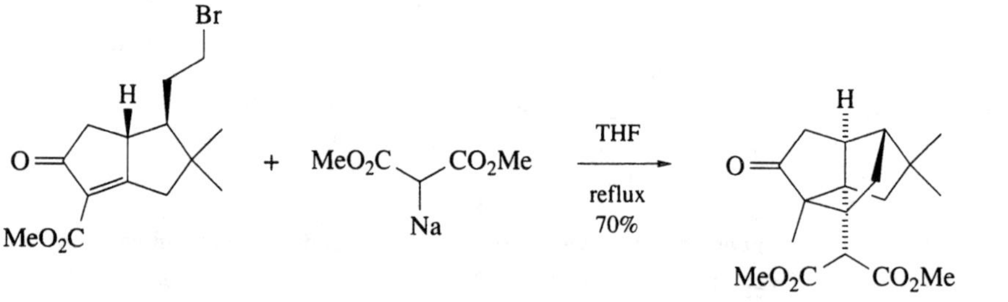

1.6.3.2 Esters and Amides

Conjugated, carbonyl-containing unsaturated compounds bearing a carbonyl carbon of oxidation state III are very good substrates for tandem vicinal difunctionalization and have been used widely. Anionic polymerization of acrylates, providing high molecular weight addition polymers, such as poly(methyl acrylate) and poly(methyl methacrylate), is an iterative tandem vicinal difunctionalization sequence consisting of sequential Michael additions and is indicative of the broad utility of the technique in applied chemistry.

1.6.3.2.1 Acyclic esters and amides

Alkenoates and alkenamides most commonly are prepared from the respective carboxylic acids or carboxylic acid halides, so it is interesting to inquire as to the suitability of these later compounds as substrates in difunctionalization reaction sequences. Under most reaction conditions, unsaturated acids and acid halides will react with the initial Michael donor to give either a salt or will result in acylation instead of the desired 1,4-adduct. A few rare examples demonstrate successful difunctionalization using such substrates, however. In equation (21), acrylic acid equivalent (**11**) undergoes 1,4-addition with *tert*-butylmagnesium chloride; the intermediate dianion undergoes regiospecific α-alkylation in excellent yield.[70] In equation (22), transient acylsilane (**12**) undergoes a 1,4-addition-initiated dimerization to give lactone (**13**).[97]

SiMe$_3$ / CO$_2$H (**11**) —[2.2 equiv. Bu^{t}MgCl, THF, –78 °C]→ [Bu^t … O$^-$, SiMe$_3$] —[i, C$_8$H$_{17}$CHO; ii, H$^+$; 96%]→ Bu^t … CO$_2$H, C$_8$H$_{17}$ (21)

O, Cl —[2 equiv. (Me$_3$Si)$_2$Cu(CN)Li$_2$, THF, –23 °C to r.t.]→ [O, SiMe$_3$ (**12**)] —[NH$_4$Cl, 69%]→ Me$_3$Si, O, O, SiMe$_3$ (**13**) (22)

In most cases[98] conjugate addition–enolate alkylation reaction sequences do not exhibit particular sensitivity with respect to the identity of the alkyl group present in the alkyl alkenoate substrate. When a Michael donor has been chosen that reacts in both the 1,4- and 1,2-addition modes, it may be possible to choose an alkyl group for the ester substrate that forces the Michael donor to undergo exclusive 1,4-addition by sterically shielding the carbonyl carbon from attack by the nucleophile (equation 23).[99]

OMe, Bu^t, Bu^t, O, O, Ph —[i, PhLi; ii, MeI; 88%]→ OMe, Bu^t, Bu^t, O, O, Ph, Ph (23)

Simple substitution at the α-carbon of the enoate substrate also does not appear to have much effect on the course of the reaction: methyl acrylate and methyl methacrylate both function well as substrates in the reaction sequence and isolated yields of the products are similar.[100] In certain sterically demanding reaction sequences, α-substitution does reduce the chemical yields of the products but the presence of an α-substituent appears to be of greater importance than the particular identity of the substituent.[101] As with ketones, enoate substrates are activated towards tandem vicinal difunctionalization when electron-withdrawing α-substituents are present.[102–104]

Enoate β-substitution and β-disubstitution cause a decrease in the rate of the initial conjugate addition step of the reaction sequence that is directly related to the steric bulk of the substituent.[103,105] Equation (24) provides a representative case in the α-alkylation of enoates by means of conjugate amination–enolate alkylation followed by dehydroamination.[106] When β-substitution results in stereoisomeric (*E*)- and (*Z*)-alkenoate substrates, tandem difunctionalization typically proceeds with greater facility for (*E*)-isomers.[64,103] Obviously, when the double bond of the ester is part of a medium-sized ring, an (*E*)-alkenoate geometry is mandated; in such cases, tandem vicinal difunctionalization proceeds with uniformly excellent results (equation 25).[25]

When alkynic esters are used as substrates in the reaction sequence, activated alkenes of good isomeric purity can be prepared. Initiation of the difunctionalization usually is by means of organocopper-mediated 1,4-addition, which proceeds *via* carbometallation of the alkynyl moiety of the ester to give a vinylcopper intermediate, (**14**). Trapping by an electrophile affords *cis*-difunctionalized alkenoates

O
OEt
R
i, LDA
ii, MeI
NPri2
SiO2
–Pri2NH

(24)

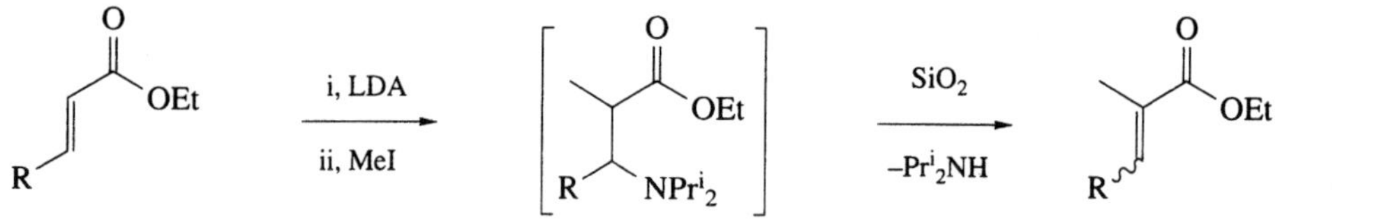

69%, R = n-C_7H_{15} *(E):(Z)* = 88:12
33%, R = Pr^i *(E):(Z)* = 95:5

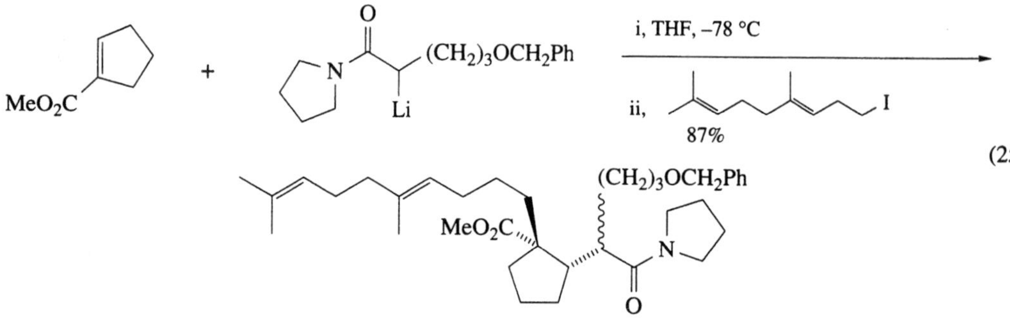

MeO2C
(CH2)3OCH2Ph
N
Li
i, THF, –78 °C
ii,
I
87%
MeO2C

(25)

(Scheme 6). Alternatively, (**14**) may equilibrate *via* an allenolate species, (**15**),[107] in which case the stereochemistry of the resulting alkenoate can be (*Z*) or (*E*), depending upon the steric interactions experienced between the electrophile and (**15**) during α-functionalization.[108] Loss of the stereointegrity of intermediate (**14**) is common[82,108–110] and clearly temperature- and reagent-dependent;[111,112] ester substrates appear to be somewhat less prone to equilibration than their corresponding ketones. Reactive electrophiles, such as acid chlorides, compete with equilibration effectively[111] to result in net *cis* difunctionalization, whereas less reactive or bulky electrophiles result in net *trans* difunctionalization. Equations (26)[113] and (27)[111] are indicative of the degree of control that can be obtained over the stereochemical outcome of the reaction.

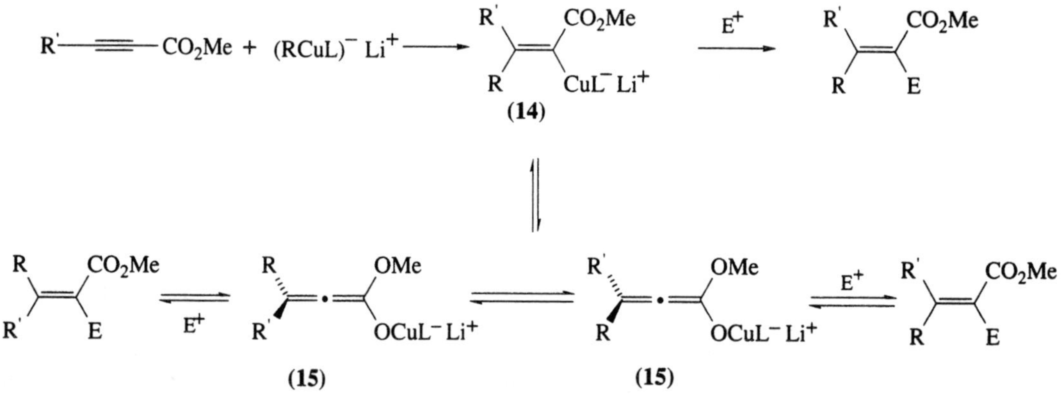

R'—≡—CO2Me + (RCuL)⁻ Li⁺
CO2Me
CuL⁻ Li⁺
(14)
E+
E
OMe
OCuL⁻ Li⁺
(15)
(15)

Scheme 6

HC≡CCO2Et
+
(MeCu—≡—Bun)Li
i, ether, –78 °C
ii,
COCl
, –78 °C
92%
O
CO2Et

(26)

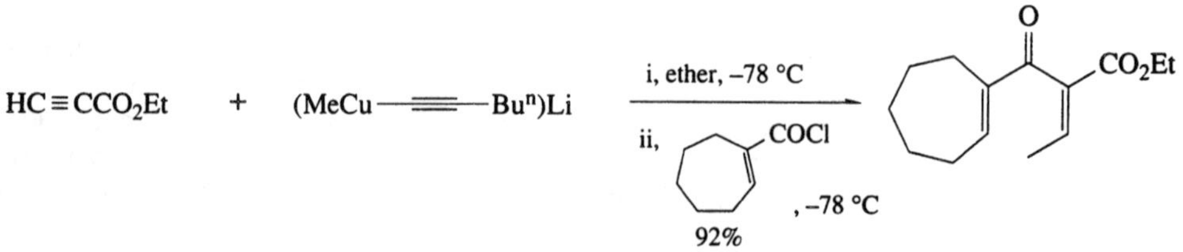

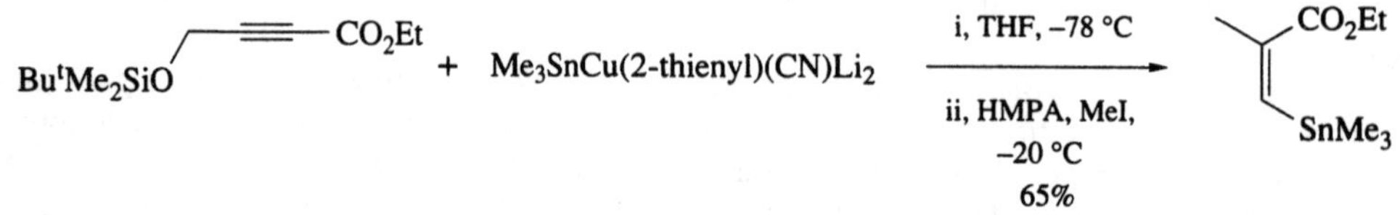

ButMe2SiO
CO2Et
+ Me3SnCu(2-thienyl)(CN)Li2
i, THF, –78 °C
ii, HMPA, MeI,
–20 °C
65%
SnMe3

(27)

Allenic esters react as enoates in tandem difunctionalization reactions using organocopper reagents as Michael donors; β,γ-unsaturated esters result (equation 28).[81] These substrates have found application in the synthesis of a variety of heterocyclic systems.[114]

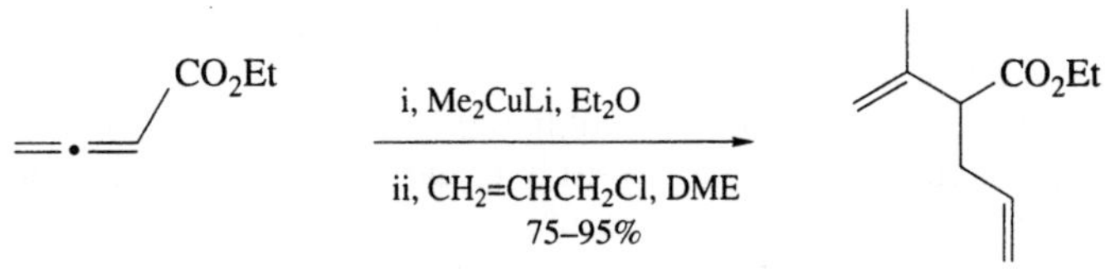

i, Me_2CuLi, Et_2O; ii, $CH_2{=}CHCH_2Cl$, DME; 75–95% (28)

Secondary and tertiary enamides and tertiary thioenamides undergo conjugate addition–α-functionalization reactions initiated by alkyllithium and Grignard reagent nucleophiles.[115–118] Secondary amides must be protected before the reaction by trimethylsilylation[118] or by amination;[119] alternatively, 2 equiv. of the organometal Michael donor can be employed, the first to deprotonate the amide before conjugate addition proceeds with the second equivalent of donor reagent.[115] Enantiomerically pure enoylsultams, such as (**16**) in equation (29), serve as very effective substrates for chelation-controlled, diastereodiscriminating 1,4-addition–enolate alkylation;[120,121] isolated yields and percent asymmetric induction usually are excellent. Conjugated, alkynic amides also have been noted to serve as substrates in a conjugate stannylation–α-alkylation sequence.[122,123]

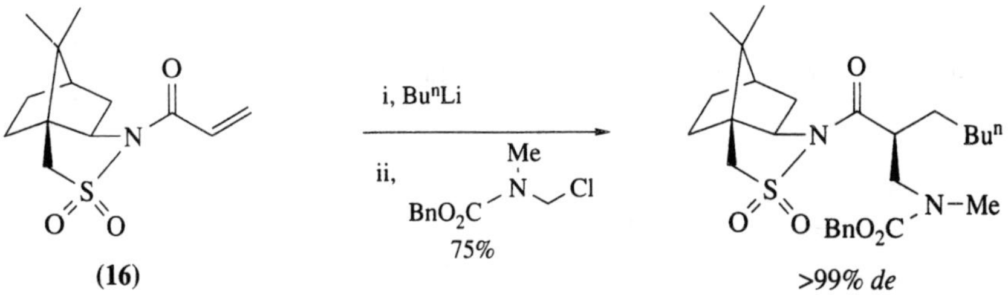

(**16**) → i, Bu^nLi; ii, BnO_2C–N(Me)–CH_2Cl; 75%; >99% *de* (29)

1.6.3.2.2 *Cyclic esters and amides*

While α,β-unsaturated lactams have not been exploited as substrates in tandem vicinal difunctionalization, the corresponding lactones serve as key substrates in the convergent preparation of lignans,[124–127] analogs of prostaglandins,[128,129] anthraquinone-derived antibiotics[18] and macrolides.[130] γ-Butenolides and 4-substituted γ-butenolides have found the most frequent use and react with a variety of combinations of Michael donor and electrophilic reagents to give uniformly high yields of *trans* α,β-disubstituted lactone adducts. Equation (30) is illustrative of the use of γ-butenolide in the synthesis of lignans related to the podophyllotoxins.[127]

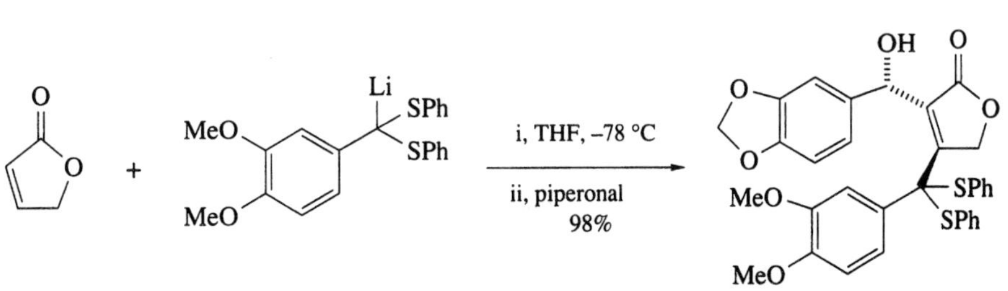

i, THF, –78 °C; ii, piperonal; 98% (30)

1.6.3.3 Alkynes

Alkynes are unusual in that they are able to undergo vicinal, nucleophilic addition–electrophilic coupling reactions without prerequisite activation of the triple bond by substitution with an electron-withdrawing group on one or both of the *sp*-hybridized carbons. Such tandem difunctionalization reactions proceed by carbometallation,[10,131–133] producing a vinylic organometal intermediate by *syn* Markovnikov addition of a nucleophilic organometal reagent across the triple bond. Highly stereoselective *syn* difunctionalization occurs after subsequent reaction with an electrophile (*cf.* Scheme 6 and Volume 4, Chapter

4.4). In particular, when the nucleophilic reagent is carbanionic in nature, the method serves as an eminently satisfactory way to prepare stereoisomerically pure di-, tri- and tetra-substituted alkenes. Most commonly, the nucleophilic organometal of choice is an organocopper reagent[10] or an organozinc reagent;[133] usually, terminal alkynes are used in the reaction, but some internal alkynes are reactive as well. There is considerable tolerance to a variety of functional groups in the substrate; again, the only cautions to be observed are that highly acidic or otherwise electrophilically reactive groups are not present. Equation (31) provides a typical example resulting in (*E*)-enyne (**17**);[134] iterative use of the method (Scheme 7) results in the rapid synthesis of a variety of insect pheromones, such as the codling moth constituent (**18**).[135] While heteroatom-containing substituents that are distant to the triple bond have no effect on the course of the reaction sequence, proximal heteroatoms can alter its regiochemistry, presumably through chelation-controlled attack of the nucleophile (equation 32).[136]

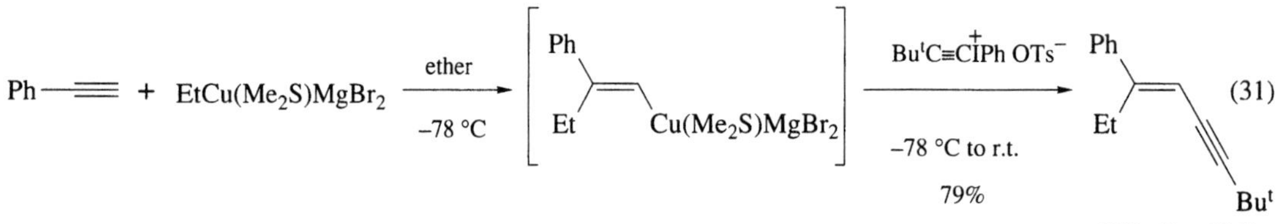

Ph—≡ + $EtCu(Me_2S)MgBr_2$ —(ether, –78 °C)→ [Ph, Et / $Cu(Me_2S)MgBr_2$] —($Bu^tC{\equiv}\overset{+}{C}IPh$ OTs^-, –78 °C to r.t., 79%)→ Ph, Et / ≡–Bu^t (31)

(**17**) (*E*) >98%

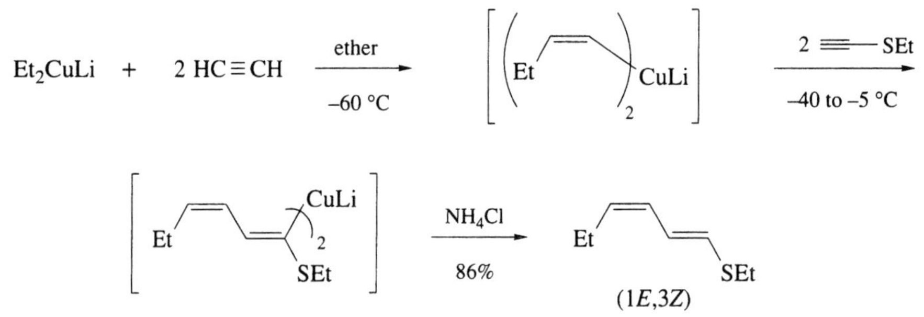

Et_2CuLi + 2 HC≡CH —(ether, –60 °C)→ [(Et⌒)$_2$CuLi] —(2 ≡—SEt, –40 to –5 °C)→ [(Et⌒⌒(SEt))$_2$CuLi] —(NH_4Cl, 86%)→ Et⌒⌒SEt (1*E*,3*Z*) (32)

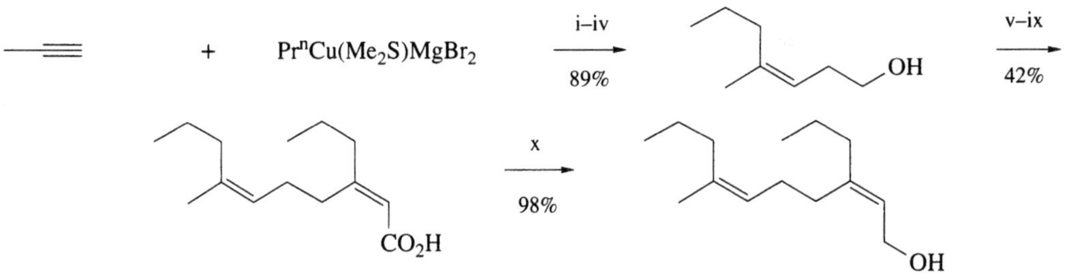

—≡ + $Pr^nCu(Me_2S)MgBr_2$ —(i–iv, 89%)→ ⌒OH —(v–ix, 42%)→ CO_2H —(x, 98%)→ OH (**18**)

i, –23 °C, 2 h; ii, $Pr^nC{\equiv}CLi$, HMPA, –78 °C, 1 h; iii, ethylene oxide, –78 to 20 °C; iv, aq. NH_4Cl; v, TsCl; vi, NaI; vii, Mg; viii, $CuBr{\bullet}Me_2S$; ix, $Pr^nC{\equiv}CH$; x, $LiAlH_4$

Scheme 7

A similar but conceptually distinct approach to difunctionalization of terminal alkynes consists of sequential carboboration–palladium-catalyzed cross-coupling;[137] equation (33) illustrates that this method also provides alkenes of high stereochemical purity by net *syn* Markovnikov addition. Benzyne-containing molecules can act as highly activated substrates for vicinal difunctionalizations initiated by nucleophiles;[138–140] thus, nucleophilic addition–electrophilic trapping can serve as an alternative to sequential directed metallation for the production of 1,2-disubstituted and 1,2,3-trisubstituted aromatic systems (equation 34).[141]

n-C_6H_{13}—≡ → i, BBr_3, CH_2Cl_2; ii, $PdCl_2(PPh_3)_2$; iii, Bu^nZnCl; iv, MeOLi, CH_2=$CHCH_2Br$; 56% → n-C_6H_{13}, Bu^n diene, (*E*) 98% (33)

i, 3 equiv. Bu^nLi, pentane, –78 °C; ii, MeI; 68%; N, O, Cl, Bu^n (34)

1.6.3.4 Other Substrates

Most substrates that do not contain a carbonyl moiety conjugated with an alkene or alkyne fall into two reactant types: those bearing a masked carbonyl equivalent, in conjugation with the alkene through a carbon atom of oxidation state II or III, and those bearing other activating substituents conjugated or coordinated with the alkene through a heteroatom.

1.6.3.4.1 Nitriles

As carboxylic acid synthetic equivalents, α,β-unsaturated nitriles provide an alternative to esters for use in carboxylic acid synthesis by tandem difunctionalization. Use of these substrates has been limited, however: when alkylzinc halides are used as Michael donors, difunctionalization proceeds well and products can be isolated in high yields;[142] sodium enolates also serve as Michael donors.[143] Alkylidenemalononitriles react with alkoxides[26] and thiolates[144] to initiate cyclizations, leading to tetrahydrothiofurans, pyrones and pyridines. In a related difunctionalization sequence, allenic and alkynic nitriles undergo conjugate addition reactions with *o*-aminobenzyl alcohol; subsequent dehydration of the conjugate adducts produce *o*-xylylene intermediates which undergo cyclization to give quinolines.[145]

1.6.3.4.2 Sulfoxides and sulfones

Vinylic sulfoxides are utilized in asymmetric synthesis; their role as Michael acceptors has been the topic of considerable study.[146] Only a limited number of Michael donor reagents can be used to initiate the conjugate addition–conjugate anion trapping sequence, and usually special reaction conditions are employed. The addition of dialkylcopperlithium reagents to *p*-chlorophenyl vinyl sulfoxide is successful at –60 °C in a mixed solvent system of ether and dimethyl sulfide; electrophilic trapping provides the vicinally dialkylated adducts in fair to good yield.[147] The reaction conditions appear to be specific for the particular substrate. A related vicinal difunctionalization of a vinylic sulfoxide has been used in the total synthesis of (*R*)-(+)-canadine,[148] consisting of an intramolecular 1,4-addition of an amine nucleophile; the conjugate adduct subsequently is subjected to cyclization *via* intramolecular Pummerer rearrangement, thereby alkylating the sulfoxide at the α-carbon.

Vinylic sulfones are much more reactive as Michael acceptors in tandem difunctionalization reactions and consequently have less stringent requirements as to initiating Michael donor and reaction conditions, as shown in equation (35).[149] They have been used to prepare polycyclic hydrocarbons[150] and heterocycles,[151] as well as a considerable number of prostaglandin analogs.[152,153] Uncommon stereochemically constrained *syn* conjugate addition–conjugate anion alkylation of a phenyl cyclohexenyl sulfone provides an important ring-forming step in a synthesis of morphiene (equation 36).[154]

i, MeLi, THF, –78 °C; ii, CH_2=$CHCH_2Br$; 55% (35)

i, Bu^nLi; ii, –78 to 4 °C; 63% (36)

1.6.3.4.3 Phosphorus-containing substrates

Vinylphosphonium salts and dialkyl vinylphosphonates generate Wittig-type conjugate anions upon 1,4-addition of a nucleophile; they, therefore, serve as substrates for tandem vicinal nucleophilic addition–Wittig alkenation when reacted with a Michael donor and a carbonyl-containing electrophile, resulting in net allylation of the donor reagent upon completion of the reaction sequence (equation 37).[102,155–157] The substrates can be used to prepare a variety of heterocycles[158] and carbocycles[159] when Wittig alkenation proceeds intramolecularly; 1,2-vinylenebisphosphonium salts function similarly.[160,161] The conjugate anions are reactive towards other electrophilic reagents as well, enabling, for instance, the production of relatively complex phosphonates,[162] cyclopropanes and cyclohexanes.[23]

$BnNH_2$ + $[CH_2{=}CHPPh_3]^+Br^-$ —(MeCN, reflux, 95%)→ $BnNH\diagdown\diagup\overset{+}{P}Ph_3\ Br^-$ —(i, 2 equiv. Bu^nLi, 0 °C, THF; ii, PhCHO; 73%)→ BnNH–CH=CH–Ph (37)

(E):(Z) = 71:29

1.6.3.4.4 Imine-containing substrates

α,β-Unsaturated aldehydes protected as imines by reaction with primary amines undergo tandem vicinal dialkylation;[163] when enantiomerically pure amino acid esters are used to generate these imines, the reaction sequence can proceed with good levels of asymmetric induction from the diastereodifferentiating effects of the imino moiety upon the constituent carbon–carbon bond-forming reactions.[164] Similarly, conjugated carboxylic acids protected as benzothiazoles act as substrates for conjugate addition–conjugate anion trapping reactions.[165] Perhaps the most synthetically useful imine-containing substrates are oxazolines prepared from 1- and 2-naphthoic acids.[166,167] These compounds allow direct introduction of vicinal substituents into the naphthalene π-system by addition of organolithium reagents followed by electrophilic trapping (equation 38). Imines of the corresponding 1-naphthaldehyde behave similarly.[66]

i, Bu^sLi, THF, –45 °C; ii, MeI; 94% (38)

1.6.3.4.5 Miscellaneous substrates

As previously illustrated in equation (6), fulvenes can undergo Michael donor-initiated tandem difunctionalization.[62] Nitroalkenes, superior heterodienes in [4 + 2] cycloaddition reactions,[168] would appear to be very good candidates for nucleophilic addition–conjugate anion trapping reactions. These substrates do undergo Michael additions[169] but so far have found scant use,[170] likely because the Nef-type[315] conjugate anion intermediates that form are not inherently nucleophilic at the α-carbon. (η-Arene) tricarbonylchromium complexes[171] offer a synthetic alternative to oxazoline-mediated vicinal dialkylation of arenes, undergoing nucleophilic addition to give an alkylchromium intermediate; carbonyl insertion into the chromium–carbon bond and subsequent alkylation affords ketones (equation 39).[172] Allenes undergo effective silylcupration–alkylation reactions.[173] Lastly, polyunsaturated ketones and esters, capable of undergoing 'extended' 1,(4 + 2*n*)-addition reactions, have received limited attention as substrates for conjugate addition–enolate trapping sequences.[65,105,174,175] 1,4-Addition-initiated vicinal difunctionalizations generally predominate, as exemplified in equation (40).[174] Related substrates (**19**) and (**20**), wherein one of the two double bonds in a conjugated alkadienoate is replaced by a cyclopropyl moiety, undergo net 2,6-dialkylation and 2,3-dialkylation, respectively (equations 41 and 42).[176]

$Cr(CO)_3$ — i, Bu^tLi, THF, −78 to 0 °C; ii, MeI, HMPA; iii, PPh_3; 67% → Bu^t, $COCH_3$ (39)

COMe + NO_2, Ph — LDA, THF, −78°C, 75% → O, Ph, NO_2 — NaOMe, MeOH, T → (40)

H, O, H, Ph, NO_2 + H, O, H, Ph, NO_2

T = r.t. 7% 93%

T = reflux 20% 80%

CO_2Et, CO_2Et (**19**) — i, $Bu^n{}_2CuLi$; ii, $CH_2{=}CHCH_2Br$; 88% → Bu^n, CO_2Et, CO_2Et (41)

CO_2Et, CO_2Et — i, $Bu^n{}_2CuLi$; ii, $CH_2{=}CHCH_2Br$; 98% → Bu^n, CO_2Et, CO_2Et (42)

1.6.4 THE NUCLEOPHILE

A wide range of nucleophilic reagents has been used to initiate tandem vicinal difunctionalizations of α,β-unsaturated carbonyl-containing substrates. Dominant among these reagents are organocopper (Gilman) reagents[8–10,43] and, to a lesser extent, enolates;[7,38] their use reflects the normally high regioselectivity of these reagents for 1,4- *versus* 1,2-addition as well as their reliably high relative nucleophilicities compared to their basicities.

1.6.4.1 Hydrides

As previously mentioned, the first exploitation of a one-pot, three-component tandem vicinal difunctionalization reaction employed a dissolving metal conjugate reduction of α,β-unsaturated ketones to produce enolates, which were *C*-alkylated using various electrophiles.[39] Such dissolving metal conjugate reduction–enolate trapping remains the most popular route to prepare α-alkylated ketones from enones. Other reagents have been developed to act as regioselective, 1,4-hydride donor reagents,[177] including potassium triphenylborane,[178] complex copper hydrides and stoichiometric and catalytic copper(I)–metal hydride mixtures,[179] DIBAL-H–HMPA with methylcopper[180] and 1,3-dimethyl-2-phenylbenzimidazoline;[181] applications to difunctionalizations are rare, however.[182] Reductive alkylation of enones and enoates is possible using lithium and potassium tri-*s*-butylbororanes (equation 43)[183] and the copper hydride cluster $(Ph_3PCuH)_6$;[184] copper hydride[185] and copper-containing sodium hydride[186] cause hydrodimerizations of enoates and alkynes, respectively. In general, such hydrodimerizations should be considered as possible products, especially when highly reactive intermediate enolates are generated. Related regioselective hydrosilylation of α,β-unsaturated carbonyl compounds results in silyl enol ethers, from which an enolate may be regenerated.[187]

i, $LiBu^s{}_3BH$, THF, –70°C; ii, MeI, –70 °C to r.t.; 95% (43)

1.6.4.2 Unstabilized Anions

Relatively simple carbon sp^3- and sp^2-hybridized organometals are frequently used to initiate conjugate alkylation–enolate trapping sequences by carbon–carbon σ-bond formation at the β-carbon of the substrate. While organocopper reagents have proven to be the most versatile of these reagents, many other organometals also can be used, depending upon the nature of the substrate.

1.6.4.2.1 Organocopper reagents

The first example of a three-component tandem vicinal dialkylation reaction employed copper(I) chloride-catalyzed 1,4-addition of a Grignard reagent to a cyclohexenone (Scheme 3).[40] Catalytic organocopper protocols are used occasionally (*e.g.* equations 4[60] and 19[91]), but usually an improvement in the yield of the conjugate adduct produced can be realized when stoichiometric organocopper reagents are employed. Copper(I)-catalyzed conjugate addition can be very convenient; typically, a Grignard reagent is used as nucleophile and copper(I) halides such as copper(I) iodide, copper(I) bromide or their dimethyl sulfide or trialkylphosphine complexes are present in amounts ranging from 2–10 mol %. In one instance, copper(I) catalysis was imperative for the success of a difunctionalization, the concentration of copper(I) present being able to effect both the reactivity and selectivity of the ambident conjugate enolate towards the electrophilic reagent.[188] In a complementary manner, activation of the substrate with an α-electron-withdrawing substituent can make copper(I) catalysis superfluous.[189]

In spite of the fact that one type of stoichiometric organocopper reagent may display chemical behavior very different from that of another, a variety of the reagents find common use in initiation of vicinal difunctionalizations. The more common of these reagents are: alkylcopper(I) reagents, with and without ligating agents that can be essential to their reactivity; dialkylcopper(I)metal reagents, with both alkyl groups identical; mixed dialkylcopper(I)metal reagents, wherein both alkyl groups are not identical; alkyl(alkylhetero)copper(I)metal reagents containing one copper–heteroatom bond; and the higher order organocopper reagents, exemplified by the cyanodialkylcopper(I)dilithium reagents. The choice of the appropriate reagent for the substrate is compounded by the fact that the reactivity and selectivity of the organocopper reagent is affected by the identity of the metal of the oganometal used to form it. The initial counterion of the copper(I) salt chosen to generate the organocopper species also can influence the outcome of the reaction.[190,191]

Simple alkylcopper(I) reagents are less reactive than their corresponding dialkylcopper(I)metals; this comparitive lack of reactivity in 1,4-addition reactions results in relatively rare use in vicinal difunctionalizations of α,β-unsaturated carbonyl substrates. Equation (44) illustrates that vinylcopper initiates dialkylation of a cyclopentenone,[192] but most organocoppers in fact are inert. In part, this inertness may be explained to the lack of solubility of the copper reagents in common solvents for conjugate addition–enolate trapping reactions, such as THF or diethyl ether. Indeed, when putatively solubilized *via* ligation with organophosphorus or organosulfur ligands, alkylcopper(I) reagents are activated towards 1,4-addition (*e.g.* equation 16).[11] Solubilizing ligands include trialkyl phosphites,[88] boron trifluoride,[193] dimethyl sulfide[194] and trialkylphosphines.[195,196] The latter reagent finds efficient use in three-component coupling syntheses of prostaglandins (equation 45).[197]

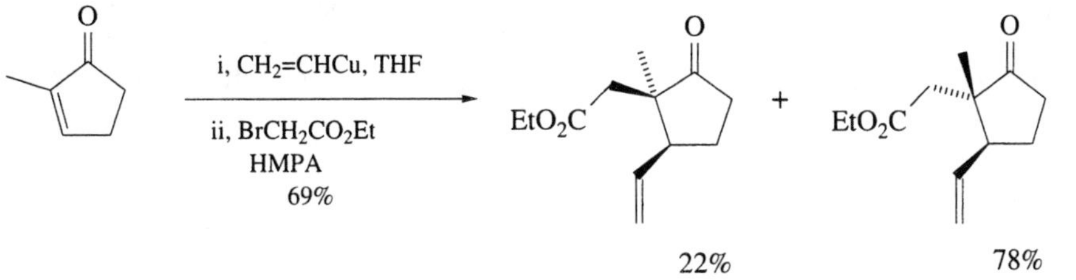

(44)

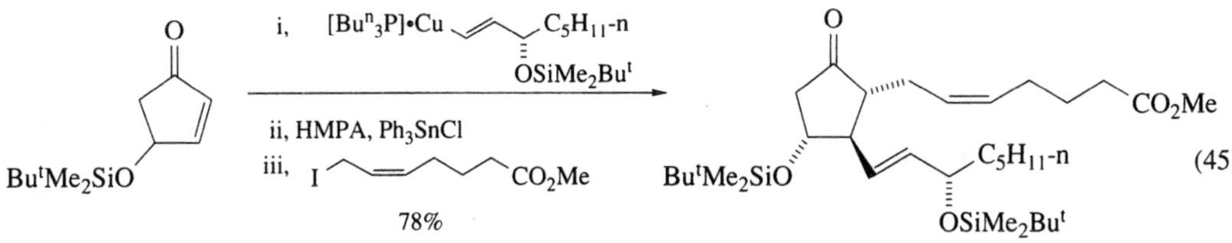

(45)

When terminal alkynes are used as substrates, alkylcoppermagnesium dihalides prepared from a copper(I) bromide–dimethyl sulfide complex typically are used (*e.g.* equation 31[134] and Scheme 7).[135] The choice of a Grignard reagent, from which the alkylcopper(I) species is prepared, is important; in this instance, alkyllithium-derived alkylcopper reagents usually are less satisfactory.[10]

Dialkylcopper(I)metal reagents, also referred to as homocuprates,[43] are the most popular Michael donors for tandem vicinal difunctionalization reactions of α,β-unsaturated carbonyl substrates. The alkyl groups are derived from 2 equiv. of an alkyllithium or Grignard reagent (*e.g.* equations 2,[56] 3,[59] 5,[61] 11[77] and 41[176]); the alkyl group can be methyl, primary or secondary alkyl, alkenyl, allyl, benzyl or aryl. Instances of tertiary dialkylcopper(I)metal conjugate addition–enolate trapping are rare. The reagents can possess additional, even complex functionality: in addition to the example of equation (5),[61] homocuprates have been generated with ether moieties present, including acetal groups;[198] trialkylsilyl groups[199] and imines[200] can be accomodated. Trialkylsilylmetal-derived homocuprate reagents are excellent Michael donors (*e.g.* equations 1[55] and 7[64]). As is the case with alkylcopper(I) reagents, solubilizing ligands can be used to facilitate difunctionalization. Often, these additional ligands do not appear essential for the success of the reaction sequence; occasionally, however, they are mandatory (equation 46).[201]

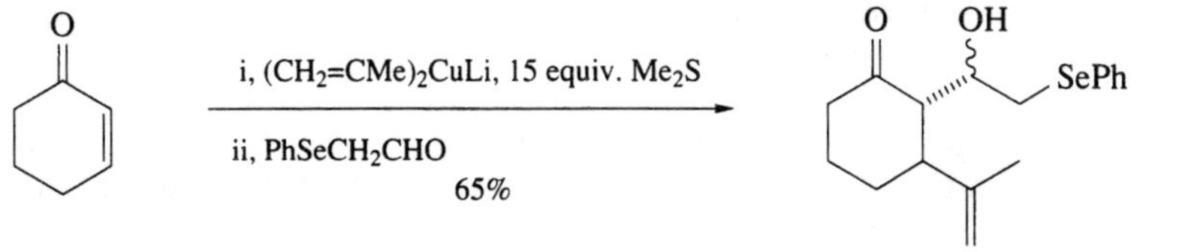

(46)

Enhanced yields result when a Lewis acid catalyst is used during addition of a dialkylcopper(I)metal reagent to a substrate.[78,79] In particular, boron trifluoride etherate has been used as an activating catalyst for conjugate addition in 1,4-addition–enolate trapping sequences.[78,202,203] Halotrimethylsilane-modified organocopper reagents can cause remarkable enhancements in the rates of conjugate additions to normally unreactive substrates:[204] trimethylsilyl enol ethers are the primary products, which can be employed in a subsequent α-functionalizing step.

Mixed or unsymmetrical dialkylcopper(I)metal reagents possess two chemically distinct alkyl groups, only one of which exhibits nucleophilicity and acts as a Michael donor. The two groups almost always differ in the formal hybridizations of the carbon atoms bonded to the copper nucleus and that group whose carbon–copper bond contains the lesser *s*-character is the group that acts as the nucleophile,[205] although a few exceptions to this rule have been noted.[206,207] Generally, the reagents are generated from an alkynylcopper(I) species and an alkylmetal (*e.g.* equation 26),[113] with pentynyl- and hexynyl-copper(I) being the reagents of choice; the net 1:1 nucleophile to electrophile stoichiometry of the Michael donor and substrate is preferable to that of a homocuprate reagent when the parent alkylmetal is difficult to obtain. For this reason, the reagents compete with the previously discussed trialkylphosphine–alkylcopper(I) complexes as preferred reagents with which to introduce the β-chain in conjugate addition–enolate trapping syntheses of prostanoids.[208] Normally alkylalkynylcopper(I)metal reagents are less reactive than their corresponding dialkylcopper(I)metal reagents. Rarely, the opposite can prove to be the case, as in equation (47):[209] the corresponding diarylcopper(I)magnesium halide of (**21**) fails to undergo 1,4-addition with the substrate. Mixed dialkylcopper(I)metal reagents also can serve as interesting and novel synthons: organocopper (**22**; equation 48) functions as a methyl acrylate synthetic equivalent which effectively undergoes vicinal dialkylation in a reverse order, α-bond formation preceding β-bond formation.[113]

$Cu(C{\equiv}CPr^n)M$, MeO; i, THF, r.t.; ii, $BrCH_2CO_2R$, HMPA; iii, saponify; HO_2C, MeO (47)

(**21a**) M = Li 57%
(**21b**) M = MgBr 0%

COCl + MeO_2C CuMeLi (**22**); –78 °C, Et_2O, 60%; CO_2Me (48)

Alkyl(heteroalkyl)copper(I)metal reagents are much like mixed dialkylcopper(I)metal reagents in that they contain only one alkyl group that is active as a nucleophile. The other group, bonded to the copper nucleus through a heteroatom, is retained by the copper complex almost without exception.[210] Most common among this class of reagents are those derived from addition of an alkylmetal reagent to phenylthiocopper at low temperature. The resulting reagents are thermally labile[211] but usually are as reactive as their corresponding dialkylcopper(I)metal reagents. A comparison between alkyl(phenylthio)- and alkylpentynyl-copper(I)metal reagents indicate that neither may offer particular advantage over the other.[212]

The increasingly popular higher order organocopper reagents[213] have been considered unsuitable for use in direct tandem vicinal difunctionalization reactions.[214,215] These reagents can be exploited by utilization of an indirect conjugate addition–enolate trapping sequence,[216] as illustrated by the synthesis of spiroketone (**23**) for a total synthesis of ginkolide B (equation 49);[21] examples of direct difunctionalization recently have been reported using trialkylsilylmetal- (*e.g.* equation 22)[97] and trialkylstannylmetal-derived (equation 27)[111] higher order organocoppers as well. Occasionally, these reagents offer unique

OMe, OMe; i, $Bu^t_2Cu(CN)Li_2$; ii, Me_3SiCl, Et_3N, 93%; $OSiMe_3$, OMe, OMe, Bu^t; i, trioxane, $TiCl_4$; ii, CH_2Cl_2, 65%; OMe, O, Bu^t (49)

(**23**)

reactivities unavailable using other Michael donor species, as evidenced by silylcupration–functionalization of alkynes[217] and allenes[173] using cyanobis(dimethylphenylsilyl)copperdilithium.

1.6.4.2.2 Other organometal reagents

Organometallic reagents can be used directly or with a transition metal catalyst in tandem vicinal difunctionalization reactions. Relatively basic reagents find greater use with substrate molecules that are not enolizable so that competitive α-deprotonation is avoided and with substrates that have been activated towards conjugate addition by an electron-withdrawing α-substituent.

A wide variety of Grignard reagents can be used as Michael donors without complication, including primary, secondary, tertiary, vinyl- and aryl-magnesium halides. These nucleophiles have been used most often with activated enoates and enones possessing α-alkoxylcarbonyl[189] or α-arylsulfinyl (*e.g.* equation 18)[90] substituents. Salts of unsaturated carboxylic acids with α-trimethylsilyl substituents also undergo Grignard reagent-initiated difunctionalization (equation 21).[70] *N,N*-dialkylthioenamides[118] and *N,N*-dialkylenamides[115] are quite reactive towards Grignard reagents, as illustrated in equation (50);[115] such amides have no acidic α-hydrogens and cannot be deprotonated by the organomagnesium reagent. Presumably, substituents on the amide nitrogen also hinder attack of the nucleophile at the carbonyl carbon.

i, MgBr; ii, MeI; 65% (50)

Alkyllithium reagents are used more frequently than their organomagnesium analogs; they exhibit the same reactivity trends. They also initiate tandem difunctionalizations of α,β-unsaturated amides[115,119] (*e.g.* equation 29)[121] and thioamides.[117] Alkyllithium reagents are preferred for use in charge-directed difunctionalizations[70,72–74] (*e.g.* equations 9[72] and 10[75]), additions to vinyl sulfones (equations 35[149] and 36[154]) and vicinal dialkylations of aryloxazolines (equation 38).[167] The latter application illustrates that the method by which the alkyllithium reagent is prepared can be important: 1-naphthyloxazolines undergo well-behaved dialkylation reactions initiated by commercially available butyllithium reagents, but allylic, benzylic and vinylic organolithiums are suitable only if generated by transmetallation[218] of tetraalkylstannanes with methyllithium. The explanation of this effect may lie in a change of the state of the aggregation of organolithium reagents, making them much more reactive.[219] Only rarely do simple organolithium reagents undergo 1,4-addition–enolate trapping reactions with substrates that otherwise would be anticipated to undergo nearly exclusive 1,2-addition, as illustrated in the synthesis of a tetralone by a benzyllithium-initiated conjugate addition–Dieckmann cyclization of methyl crotonate (equation 51).[18]

CO_2Me + MeO, Li, O, OMe, O —(−78 °C, THF, 52%)→ MeO, OH, CO_2Me, OMe, O (51)

Less common carbanionic Michael donors include organoaluminum and organozirconium reagents, which react with enones using nickel(II) catalysis[220–223] and acylate–nickel complexes.[224] Noncarbanionic organometal reagents that have been used in difunctionalization sequences include alkoxides,[26,98,114,225] sulfides,[100,114,144,225–227] trimethylsilyllithium,[47] trialkylstannyllithium reagents[130,228] and alkyl-[227] and aryl-selenides.[229,230] Perhaps the most commonly used heteroatom nucleophiles are metal amide reagents such as LDA (*e.g.* equation 24).[27,105,106,115]

1.6.4.3 Stabilized Anions

Many carbanionic nucleophiles that would be considered too 'hard' to react as Michael donors can be made into effective reagents for conjugate addition reactions by appending resonance or inductively stabilizing groups to 'soften' their intrinsic Lewis basicity. Such stabilized anionic Michael donors include enolates, alkylthio-substituted carbanions, ylides and nitro-substituted carbanions.

1.6.4.3.1 Enolates

Michael addition frequently is used as a generic descriptor of 1,4- or conjugate addition, but in fact refers to specific 1,4-addition of an enolate anion to an α,β-unsaturated carbonyl substrate, resulting in a 1,5-dione adduct.[231] Classical Michael additions are conducted in protic media; for tandem vicinal difunctionalization reaction sequences, this dictates that the α-functionalizing reagent be intramolecular in order to compete with proton capture by the conjugate enolate from the solvent,[232] or that the Michael adduct be isolated and α-functionalized under a different set of reaction conditions.[233,234] Generally, these reaction sequences suffer from side reactions. Amongst the undesired reactions are those with the base catalysts used, self-condensation of the substrate and the possiblility of 'retro-Michael' reaction; for these reasons, enolate-initiated difunctionalization sequences invariably are run in aprotic solvents.

Enolates are excellent Michael donor reagents, the initial step in difunctionalization proceeding with ease at low temperatures. The enolate can be generated directly from reaction with a hindered base such as LDA (*e.g.* equation 40)[175] or some other base (equation 15).[84] Alternatively, it can be generated from a neutral enolate synthetic equivalent, such as a silyl enol ether species, by ether cleavage using methyllithium,[235] fluoride,[236] trityl perchlorate (equation 13),[80,237,238] titanium(IV) tetrachloride,[239] alkylaluminum(III) dichlorides[240] and other Lewis acids.[241] Enol triflates may also serve this purpose.[242] The enolate can be derived from esters, especially β-keto esters and malonates (*e.g.* equation 20),[96] ketones or amides (*e.g.* equation 25).[25] An interesting extension of enolate addition–Dieckmann cyclization (equation 52) uses the *o*-xylylenolate (**24**) in a synthesis of tetralones, thereby mimicking a [2 + 4] cycloaddition reaction.[24]

MeO OLi OMe; CO_2Me + (**24**) — i, 0 °C, 5 h; ii, HCl, MeOH; 41% → MeO O (52)

1.6.4.3.2 Sulfur-containing anions

The synthesis of 4-alkyl thioketones is possible by exploiting the stabilizing effect of a sulfur atom upon an adjacent carbanionic center. Ambident allylic anions react so that conjugate addition proceeds exclusively with the α-carbon of the nucleophile,[129,243,244] as illustrated in equation (53);[245] arylsulfinyl and arylsulfonyl groups normally[246] behave similarly.[247–249] Sulfur-stabilized vinylic carbanions can be prepared and function as Michael donors in difunctionalization sequences.[250]

O + MeS, Li — i, THF, HMPA, –78 °C; ii, $ICH_2C{\equiv}CSiMe_3$; 75% → O, $SiMe_3$, SMe (53)

Organosulfur substituents can play dual roles when sulfur ylides are used as nucleophiles in conjugate addition–enolate trapping reactions; the conjugate enolate undergoes intramolecular displacement of the trialkylsulfonium moiety, resulting in net cyclopropanation.[251,252] Metallated dithianes, containing two geminal sulfur atoms to stabilize the adjacent negative charge, act as acyl anion equivalents to provide a route to 1,4-diketones.[253] As previously shown in equation (30),[127] these Michael donors have been used to prepare a number of lignan antibiotics.[20,124–126,254,255] The dithiane precursor can be further acidified by substitution with an electron-withdrawing group, such as the alkoxycarbonyl group present in equations (8) and (14).[69] Vinylogous dithianylidene anions can react by either an α-1,4- or a γ-1,4-addition mode, selectable by appropriate choice of counterion or by use of HMPA as a solvent adjuvant.[256] Metalated dithianes with one or both of the sulfur atoms oxidized to the sulfoxide or sulfone level also are efficient Michael donors.[102,257]

Orthothioformates[258,259] and their analogs[86,260] function as masked carboxylic acid Michael donors: methoxy(phenylthio)(trimethylsilyl)methyllithium, as illustrated in equation (17), can be used in a simple 1,4-addition–enolate trapping synthesis of sarkomycin.[86] A particularly interesting example[258] of difunctionalization using tri(phenylthio)methyllithium demonstrates umpolung of its carbanionic center, that carbon acting both as nucleophile and electrophile in the cyclopropanation of cyclohexenone (equation 54).

$$\text{cyclohexenone} + (PhS)_3CLi \xrightarrow[\text{iii, MeOH; iv, MeI, HMPA; 32\%}]{\text{i, THF; ii, Bu}^s\text{Li}} \text{SPh-substituted bicyclic ketone (H, H)} \quad (54)$$

1.6.4.3.3 Other anions

A limited number of other anionic species have been employed as Michael donors in tandem vicinal difunctionalizations. In a manner similar to sulfur ylides described above, phosphonium ylides can be used as cyclopropanating reagents by means of a conjugate addition–α-intramolecular alkylation sequence. Phosphonium ylides have been used with greater frequency[261–263] than sulfur ylides and display little steric sensitivity.[264] Phosphorus-stabilized allylic anions can display regiospecific γ-1,4-addition when used as Michael donors.[265]

The relative acidity of the α-hydrogens of nitroalkanes allows their conjugate bases to be formed easily and to be used as nucleophiles to initiate difunctionalizations of activated α,β-unsaturated esters[104,266] and amides[226] as well as 2-cyclopentenone.[267] Cyanide has been used as a Michael donor;[226,268,269] cyano-[270–272] and trimethylsilyl-stabilized[273] benzylic anions are very good Michael donors and allow the opportunity for further elaboration of the 1,5-difunctionalized adduct molecule.[274,275] *p*-Toluenesulfonylacetonitrile acts as a double Michael donor in equation (55)[276] to afford a cyclohexanone intermediate; subsequent γ-elimination of the sulfonyl group results in the product ketonitrile. Some very good Michael donors are incompatible with the difunctionalization strategy: enolates of thioesters yield conjugate enolates upon addition to α,β-unsaturated ketones, but the conjugate enolates readily equilibrate so that when an electrophile is added exclusive sulfur–electrophile bond formation is observed.[277]

$$\text{dienone} + p\text{-}MeC_6H_4SO_2CH_2CN \xrightarrow[\text{ii, reflux; 53\%}]{\text{i, cat. NaOEt, 20 °C, EtOH}} \text{CN-substituted bicyclic ketone} \quad (55)$$

1.6.5 THE ELECTROPHILE

Many different types of electrophilic α-functionalizing reagents are used in tandem vicinal difunctionalization reactions. The most common classes are alkyl iodides, allyl and propargyl bromides, aldehydes, ketones and to a lesser extent, nitroalkenes. Generally, in the case of difunctionalizations of α,β-unsaturated carbonyl substrates, softer, more polarizable electrophiles[278] show the best reactivity and highest *C*-regioselectivity when adding to conjugate enolate intermediates;[38] difunctionalizations of alkynes use the standard electrophiles of organocopper chemistry.[279] When organocopper reagents are used to initiate the reaction sequence for unsaturated carbonyl substrates, α-functionalization can be (and often is) sluggish[8,197] and an astute selection of electrophile, solvent or solvent adjuvant and counterion may be required for satisfactory adduct formation.

1.6.5.1 Reactive Alkyl Halides

The most common alkyl halide used is methyl iodide, illustrated by equations (7),[64] (9),[72] (10),[75] (23),[99] (24),[106] (38),[166] (43)[183] and (50).[115] Other alkyl halides that are reliable α-alkylating electrophiles include *n*-alkyl iodides (*e.g.* equation 25),[25] allyl bromide (equation 35),[149] propargylic bromides and iodides (equation 53),[245] benzylic halides and α-halo esters (equation 4);[60] other alkyl halides, especially secondary and tertiary alkyl halides, are far less reactive. When simple, more reactive alkyl halides are compared directly (equation 56), methyl and propargylic halides are better electrophiles than allylic and primary alkyl halides.[86] In many cases of organocopper-initiated dialkylation, the intermediate conjugate enolates do not α-alkylate at sufficient rates to compete with enolate equilibration or other side reactions;[280] often, the rate and amount of α-alkylation can be enhanced by addition of a polar aprotic adjuvant such as HMPA. The adjuvant can be present at the onset of the reaction sequence, as in equation (56), but more frequently it is added subsequent to the conjugate addition step (*e.g.* equations 16,[11] 27[111] and 44).[192] Organocopper 1,4-additions preferentially occur in relatively nonpolar aprotic solvents; a complete change of solvents[77,92,281] after the initial step of a difunctionalization sequence also can be effective (*e.g.* equations 11[77] and 18).[90] In difficult cases, the conjugate enolate can be trapped as a neutral intermediate, usually by simple protonation or by addition of chlorotrimethylsilane to generate a trimethylsilyl enol ether. Regeneration of the conjugate enolate to provide a copper(I)-free lithium, sodium or (rarely) potassium enolate that is much more reactive and is α-alkylated smoothly to give the dialkylated adduct in greater yield than direct tandem difunctionalization would provide (*e.g.* equations 2,[56] 3,[59] 5[61] and 53).[245] This effective transmetallation of the conjugate enolate also can be accomplished directly by addition of the appropriate metal salt to the reaction before adding the electrophile: excellent results have been achieved using triphenyltin(IV) chloride, for instance (*e.g.* equation 45).[197] Alkenyl and alkynyl halides normally do not function as electrophiles in tandem vicinal difunctionalizations; an exception is the alkynylphenyliodonium tosylate[134] reagent class (equation 31).

Cyclopentenone + MeO(SPh)C(Li)SiMe$_3$ —(i, HMPA/THF, –78 °C; ii, RX, –40 °C)→ 2-R-3-[C(OMe)(SPh)SiMe$_3$]cyclopentanone (56)

98%, RX = MeI
48%, RX = EtI
95%, RX = CH≡CCH$_2$Br
82%, RX = CH$_2$=CHCH$_2$Br

The alkyl halide electrophile can be intramolecular in nature, causing ring formation upon α-alkylation of the conjugate enolate. Such *Mi*chael *r*ing *c*losure (MIRC) reactions[282] have become a valuable addition to the repertoire of annulation reactions available in organic synthesis. The alkyl halide-containing moiety may originate either as part of the substrate (*e.g.* equations 15[84] and 20)[96] or, more commonly, may be present as part of the Michael donor (equation 5).[61] When both the Michael donor and the alkyl halide are present as substituents of the substrate molecule, double annulation occurs (equation 36).[154] The chemical yields of adducts from MIRC reactions usually are quite good, due to rate acceleration and decreased by-product formation inherent in the intramolecular nature of the process.

1.6.5.2 Carbonyl-containing Electrophiles

Aldehydes and ketones typically are superior electrophiles in conjugate addition–enolate trapping sequences when compared to alkyl halides. Aldehydes are the more reactive reagent class: formaldehyde (*e.g.* equation 14), *n*-alkyl aldehydes (equations 21 and 46) and aromatic aldehydes (equation 13 and 30) all are equally reactive. Branching α to the carbonyl moiety has a detrimental effect: as equation (57) illustrates, chemical yields of adducts can be significantly lowered.[67] This α-branching effect is clearly substrate and reaction condition dependent.[100] Ketones are less reactive electrophilic reagents and have been used less often. Nonetheless, they are quite efficient as α-alkylating reagents for conjugate enolate intermediates. Acetone and cyclohexanone are representative of the class.[100,108,142] As with the alkyl halide reagents, MIRC reactions are observed when a carbonyl moiety is present in the conjugate enolate (*e.g.* equation 6);[62] if the carbonyl moiety is masked as an acetal, the conjugate adduct can be isolated as a trimethylsilyl enol ether and ring closure accomplished[283] by a Mukaiyama reaction (equation 49).[21]

i, Me_2CuLi; ii, RCHO. $ZnCl_2$ (57)

77% R = Me
50% R = Et

Esters are far less reactive as electrophiles when compared to aldehydes and ketones. Successful tandem vicinal dialkylations are possible using alkyl formates,[67] but most esters lack the needed reactivity. More reactive thioesters can serve as electrophiles in these sequences.[208] Presence of a potentially electrophilic ester group as a substituent in the conjugate enolate permits very efficient Dieckmann cyclization to take place as the second step of a MIRC sequence (*e.g.* equations 51[18] and 52).[24] Ortho esters are far more reactive, giving β-keto esters as adducts when used in sequences that employ enones as substrates.[230]

Acyl halides have been investigated extensively as α-functionalizing reagents in conjugate addition–enolate trapping sequences.[8] Equation (26)[113] indicates that they can be highly effective trapping agents. It is not unusual for *O*-acylation to compete with *C*-acylation; the ratio of products depends upon the acyl halide itself,[113,284] the substrate chosen[285,286] and the reaction conditions.[188,287,288] Sometimes *O,C*-diacylated products are obtained,[289] an observation which has been used to synthetic advantage.[287,290] Acyl cyanides have been used as α-acylating reagents in a tandem difunctionalization sequence leading to heterocycles.[291]

In addition to the simple one-carbon electrophile formaldehyde, carbon dioxide[67,292] and carbonyl sulfide[273] also are excellent one-carbon electrophiles for use in conjugate enolate trapping.

1.6.5.3 Michael Acceptors

The addition of conjugate enolate anions derived from initial reaction between a nucleophile and an unsaturated substrate to a new Michael acceptor molecule should follow trends similar to those already discussed for the initial reaction in Sections 1.6.3 and 1.6.4.3.1. Indeed this is the case for these double Michael or sequential Michael reactions. An appropriate combination of Michael acceptor and conjugate anion is required for sequential Michael addition to take place; otherwise, Michael addition–aldol-type condensation is observed. Thus, with conjugate enolates derived from initial 1,4-addition of an organocopper nucleophile or a simple enolate to an α,β-unsaturated ketone or ester substrate, α-functionalization occurs by means of 1,4-addition to ambident electrophiles such as enoates[293–295] and enones.[296–298] Only 1,2-addition is observed when enals are used as electrophiles,[13,227,248,287,295] as can be seen when equations (58) and (59) are compared.[295] Sequential Michael addition reactions can be used in MIRC-type processes (equation 59). Typically, the initial nucleophile is a conjugated enolate which transfers its nucleophilicity upon 1,4-addition to the substrate, either intermolecularly[170,238,240] or intramolecularly.[17,27,299–301] The newly generated conjugate enolate then reaches back to the nascent enone, cyclization

i, ether; ii, acrylaldehyde (58)

i, ether; ii, methyl acrylate 47% (59)

$$\text{1-(trimethylsilyloxy)vinylcyclohexene} + \text{CH}_2\text{=CHCOMe} \xrightarrow[56\%]{\text{EtAlCl}_2} \text{decalone (H, H, COMe)} \tag{60}$$

ensues and the second, cyclized conjugate enolate can be quenched with a proton source. An example is given in equation (60).[240]

In a similar manner, alkynes can undergo sequential carbometallation reactions, previously illustrated in equation (32).[136] It is possible to combine carbocupration of alkynes with the use of a Michael acceptor as electrophile; when that electrophile is a vinylphosphonium salt, carbocupration can be coupled to Wittig alkenation to result in a stereospecific synthesis of dienes *via* a one-pot, four-component, four carbon–carbon bond-forming reaction sequence.[302]

1.6.5.4 Other Reagents

A limited number of other electrophile types have been used in difunctionalization reaction sequences. Disulfides,[105,212] sulfenyl chlorides[268,212] and sulfinyl chlorides[303] permit α-thiolation of the conjugate anion; selenyl halides behave similarly.[105,222,304] Molecular bromine[59,305] and iodine[82,110,306,307] trap conjugate enolates to give α-halo ketones, which may be used as electrophiles in other reactions or can be used to regenerate the conjugate enolate regiospecifically. Nitroalkenes appear to be quite reactive Michael-type acceptors;[308] also, unusual uses of lead tetraacetate[309] and an iminium triflate[310] have been reported. Lastly, when tandem vicinal difunctionalizations of acrylates or crotonates are mediated by their corresponding tetracarbonyliron complexes, their intermediate conjugate iron enolates undergo carbonyl insertion into the α-carbon–iron enolate bond; subsequent alkylation affords β-keto esters.[311]

1.6.6 SEQUENTIAL MICHAEL RING CLOSURE REACTIONS

A powerful modification of the Michael addition–enolate functionalization sequence involves use of a second Michael acceptor as the electrophilic α-functionalizing reagent. By way of example, ketone (**25**) in Scheme 8 undergoes 1,4-addition of the lithium enolate of methyl 2-methylpropanoate; the conjugate enolate that results then is *C*-methylated to generate adduct (**26**).[94] If methyl acrylate is substituted as the electrophilic reagent, α-functionalization generates a new conjugate ester enolate, (**27**), which undergoes yet a third conjugate addition reaction with the 2-phenyl-2-cyclopentenone moiety remaining from the original substrate (**25**). Norbornanone (**28**) is generated in 40% overall yield, the equivalent of 74% chemical yield per carbon–carbon bond-forming step in the sequence.[93] An analogous process determines the formation of the product diester of equation (59).[295] One-pot, multibond-forming, multiple tandem vicinal difunctionalization sequences with ultimate ring closure compose a reaction class referred to as *s*equential *Mi*chael *r*ing *c*losure (SMIRC) reactions.[23] Controlled anionic codi-, cotri- and cotetra-polymerization reactions like these can proceed in notably high yields and with excellent control of stereochemistry to generate complex polycyclic structures.

Both inter- and intra-molecular cyclopropanations are possible using SMIRC methodology,[98,312,313] but cyclohexane-forming sequences by sequential reaction of three- or four-components with ultimate 1,6-bond closure are more common.[23,101,207,314] Usually, the initial conjugate anion reacts in such a way that it is α-functionalized exclusively; in this way it acts as a two-carbon fragment in the annulation process. Equation (61) illustrates a four-component, tri-*n*-stannyllithium-initiated SMIRC sequence that yields a γ-hydroxystannane in 53% overall chemical yield, equivalent to a yield of 85% per chemical bond formed. Oxidative dehydrostannylation results in a macrolide.[130] By varying the identity of the various components in this sequential Michael–aldol ring closure reaction, various naphthalenes, benzothiophenes, benzofurans and quinolines can be generated.

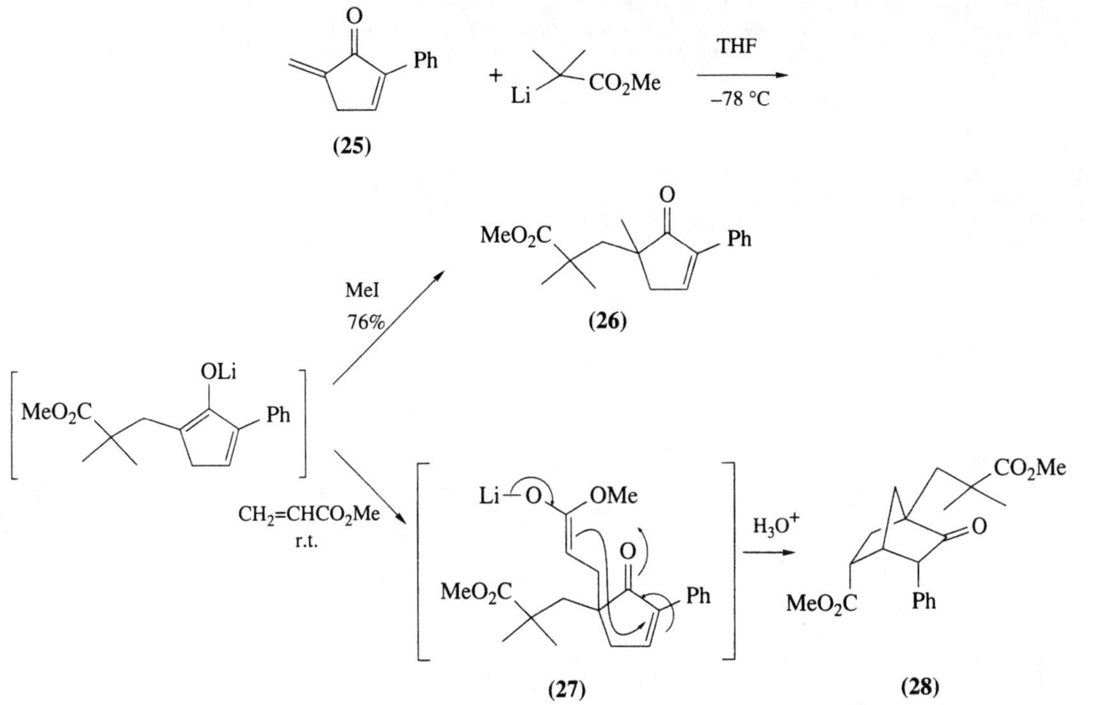

THF

−78 °C

(25)

MeI

76%

(26)

CH_2=$CHCO_2Me$

r.t.

H_3O^+

(27)

(28)

Scheme 8

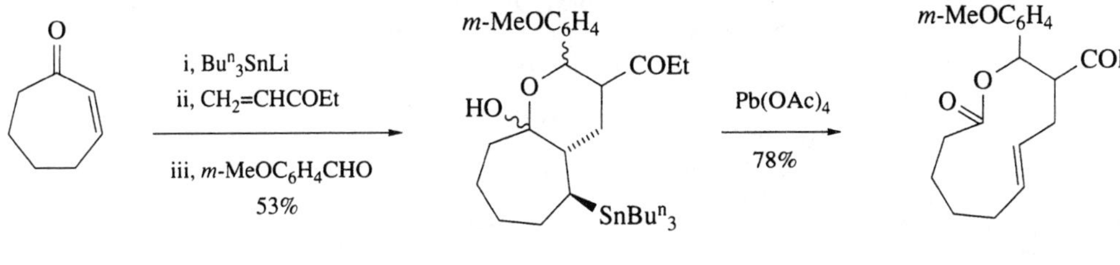

i, $Bu^n{}_3SnLi$

ii, CH_2=CHCOEt

iii, m-$MeOC_6H_4CHO$

53%

$Pb(OAc)_4$

78%

(61)

1.6.7 REFERENCES

1. M. Ono, *J. Synth. Org. Chem. Jpn.*, 1980, **10**, 923.
2. A. Schaap and J. F. Arens, *Recl. Trav. Chim. Pays-Bas*, 1968, **87**, 1249.
3. A. P. Marchand, in 'The Chemistry of the Functional Groups', ed. S. Patai, Wiley, New York, 1977, suppl. A, part I, pp. 534 and 625.
4. J. K. Groves, *Chem. Soc. Rev.*, 1972, **1**, 73.
5. H. Mayr and W. Striepe, *J. Org. Chem.*, 1983, **48**, 1159.
6. D. L. Boger, *Chem. Rev.*, 1986, **86**, 781.
7. M. J. Chapdelaine and M. Hulce, *Org. React. (N.Y.)*, 1990, **38**, 225.
8. R. J. K. Taylor, *Synthesis*, 1985, 364.
9. R. Noyori and M. Suzuki, *Chemtracts Org. Chem.*, 1990, **3**, 173.
10. J. F. Normant and A. Alexakis, *Synthesis*, 1981, 841.
11. A. Muria, N. Tanimoto, N. Sakamoto and T. Masamune, *J. Am. Chem. Soc.*, 1988, **110**, 1985.
12. H. Nishiyama, M. Sakuta and K. Itoh, *Tetrahedron Lett.*, 1984, **25**, 2487.
13. M. Suzuki, T. Kawagishi, T. Suzuki and R. Noyori, *Tetrahedron Lett.*, 1982, **23**, 4057.
14. N. Asao, T. Uyehara and Y. Yamamoto, *Tetrahedron*, 1988, **44**, 4173.
15. M. Vandewalle and P. DeClercq, *Tetrahedron*, 1985, **41**, 1767.
16. R. L. Funk and K. P. C. Vollhardt, *Chem. Soc. Rev.*, 1980, **9**, 41.
17. M. Ihara, T. Takahashi, N. Shimizu, Y. Ishida, I. Sudow, K. Fukumoto and T. Kametani, *J. Chem. Soc., Chem. Commun.*, 1987, 1467.
18. R. W. Franck, V. Bhat and C. S. Subramanian, *J. Am. Chem. Soc.*, 1986, **108**, 2455.
19. M. Asaoka, K. Ishibashi, N. Yanagida and H. Takei, *Tetrahedron Lett.*, 1983, **24**, 5127.
20. R. S. Ward, *Tetrahedron*, 1990, **46**, 5029.
21. E. J. Corey, M. Kang, M. C. Desai, A. Ghosh and I. N. Houpis, *J. Am. Chem. Soc.*, 1988, **110**, 649.
22. M. P. Cooke, Jr. and R. K. Widener, *J. Org. Chem.*, 1987, **52**, 1381.
23. G. H. Posner, *Chem. Rev.*, 1986, **86**, 831.
24. B. Tarnchompo, C. Thebtaranonth and Y. Thebtaranonth, *Synthesis*, 1986, 785.

25. R. B. Ruggeri, M. M. Hansen and C. H. Heathcock, *J. Am. Chem. Soc.*, 1988, **110**, 8734.
26. M. Igarashi, Y. Nakano, K. Takezawa, T. Watanabe and S. Sato, *Synthesis*, 1987, 68.
27. M. Ihara, T. Kirihara, A. Kawaguchi, M. Tsuruta, F. Keichiro and T. Kametani, *J. Chem. Soc., Perkin Trans. 1*, 1987, 1719.
28. A. Yanagisawa, S. Habaue and H. Yamamoto, *J. Am. Chem. Soc.*, 1989, **111**, 366.
29. E. P. Kohler, *J. Am. Chem. Soc.*, 1903, **29**, 352.
30. E. P. Kohler and M. Tishler, *J. Am. Chem. Soc.*, 1932, **54**, 1594.
31. E. P. Kohler and W. D. Peterson, *J. Am. Chem. Soc.*, 1933, **55**, 1073.
32. E. Knoevenagel and E. Reinecke, *Ber. Dtsch. Chem. Ges.*, 1899, **32**, 418.
33. O. Wallach, *Justus Liebigs Ann. Chem.*, 1908, **359**, 265.
34. L. Ruzicka, *Helv. Chim. Acta*, 1920, **3**, 781.
35. A. Bellamy, *J. Chem. Soc. B*, 1969, 449.
36. D. T. Warner and O. A. Moe, *J. Am. Chem. Soc.*, 1948, **70**, 3470.
37. D. T. Warner, *J. Org. Chem.*, 1959, **24**, 1536.
38. J. d'Angelo, *Tetrahedron*, 1976, **32**, 2979.
39. G. Stork, P. Rosen, N. Goldman, R. V. Coombs and J. Tsuji, *J. Am. Chem. Soc.*, 1956, **87**, 275.
40. G. Stork, *Pure Appl. Chem.*, 1968, **17**, 383.
41. E. C. Ashby, R. S. Smith and A. B. Goel, *J. Org. Chem.*, 1981, **46**, 5133.
42. F. Leyendecker, J. Drouin, J. J. DeBesse and J. M. Conia, *Tetrahedron Lett.*, 1977, 1591.
43. G. H. Posner, 'An Introduction to Synthesis Using Organocopper Reagents', Wiley, New York, 1980.
44. C. Ullenius and B. Christenson, *Pure Appl. Chem.*, 1988, **60**, 57.
45. J. Berlan, J.-P. Battioni and K. Koosha, *Bull. Soc. Chim. Fr., Part 2*, 1979, 183.
46. E. C. Ashby and T. L. Wiesemann, *J. Am. Chem. Soc.*, 1978, **100**, 3101.
47. W. C. Still, *J. Org. Chem.*, 1976, **41**, 3063.
48. L. Wartski, M. El-Bouz and J. Seyden-Penne, *J. Organomet. Chem.*, 1979, **177**, 17.
49. T. H. Lowry and K. S. Richardson, 'Mechanism and Theory in Organic Chemistry', 3rd edn., Harper and Row, New York, 1987, p. 620.
50. E. C. Ashby and J. N. Argyropoulos, *Tetrahedron Lett.*, 1984, **25**, 7.
51. E. C. Ashby and J. N. Argyropoulos, *J. Org. Chem.*, 1985, **50**, 3274.
52. T. Lund and H. Lund, *Tetrahedron Lett.*, 1986, **27**, 95.
53. D. G. Bhatt, N. Takamura and B. Ganem, *J. Am. Chem. Soc.*, 1984, **106**, 3353.
54. S. D. Kahn, K. D. Dobbs and W. J. Hehre, *J. Am. Chem. Soc.*, 1988, **110**, 4602.
55. D. J. Ager, I. Fleming and S. K. Patel, *J. Chem. Soc., Perkin Trans. 1*, 1981, 2520.
56. E. S. Binkley and C. H. Heathcock, *J. Org. Chem.*, 1975, **40**, 2156.
57. W. G. Dauben, G. J. Fonken and D. S. Noyce, *J. Am. Chem. Soc.*, 1956, **78**, 2579.
58. A. V. Kamernitzky and A. A. Akhrem, *Russ. Chem. Rev. (Engl. Transl.)*, 1961, 43.
59. G. H. Posner and C. M. Lentz, *J. Am. Chem. Soc.*, 1979, **101**, 934.
60. Y. Ito, M. Nakatsuka and T. Saegusa, *J. Am. Chem. Soc.*, 1982, **104**, 7609.
61. E. Piers, *Pure Appl. Chem.*, 1988, **60**, 107.
62. G. Büchi, D. Berthet, R. Decorzant, A. Grieder and A. Hauser, *J. Org. Chem.*, 1976, **41**, 3208.
63. E. G. Gibbons, *J. Org. Chem.*, 1980, **45**, 1540.
64. W. Bernhard, I. Fleming and D. Waterson, *J. Chem. Soc., Chem. Commun.*, 1984, 28.
65. J. A. Marshall, J. E. Audia and B. G. Shearer, *J. Org. Chem.*, 1986, **51**, 1730.
66. A. I. Meyers, J. D. Brown and D. Laucher, *Tetrahedron Lett.*, 1987, **28**, 5279, 5283.
67. K. K. Heng and R. A. J. Smith, *Tetrahedron*, 1979, **35**, 425.
68. G. H. Posner, S.-B. Lu and E. Asirvatham, *Tetrahedron Lett.*, 1986, **27**, 659.
69. M. Kato, H. Saito and A. Yoshikoshi, *Chem. Lett.*, 1984, 213.
70. M. P. Cooke, Jr., *J. Org. Chem.*, 1987, **52**, 5729.
71. T. Cohen, W. D. Abraham and M. Meyers, *J. Am. Chem. Soc.*, 1987, **109**, 7923.
72. M. P. Cooke, Jr., and D. L. Burman, *J. Org. Chem.*, 1982, **47**, 4955.
73. S. G. Davies, *Pure Appl. Chem.*, 1988, **60**, 13.
74. S. G. Davies, I. M. Dordor-Hedgecock, R. J. C. Easton, S. C. Preston, K. H. Sutton and J. C. Walker, *Bull. Soc. Chim. Fr.*, 1987, 608.
75. S. G. Davies, I. M. Dordor-Hedgecock, K. H. Sutton and J. C. Walker, *Tetrahedron Lett.*, 1986, **27**, 3787.
76. J. W. Herndon, D. Wu and H. L. Ammon, *J. Org. Chem.*, 1988, **53**, 2873.
77. R. M. Coates and L. O. Sandefur, *J. Org. Chem.*, 1974, **39**, 275.
78. B. H. Lipshutz, E. L. Ellsworth and T. J. Siahaan, *J. Am. Chem. Soc.*, 1988, **110**, 4834; 1989, **111**, 1351.
79. Y. Yamamoto, *Angew. Chem., Int. Ed. Engl.*, 1986, **25**, 947.
80. T. Mukaiyama and S. Kobayashi, *Heterocycles*, 1987, **25**, 205.
81. M. Bertrand, G. Gil and J. Viala, *Tetrahedron Lett.*, 1977, 1785.
82. J. Klein and R. Levene, *J. Chem. Soc., Perkin Trans. 2*, 1973, 1972.
83. F. A. Fouli, A. S. A. Youssef and J. M. Vernon, *Synthesis*, 1988, 291.
84. J.-F. Lavallée and P. Deslongchamps, *Tetrahedron Lett.*, 1987, **28**, 3457.
85. C. R. Johnson and T. D. Penning, *J. Am. Chem. Soc.*, 1988, **110**, 4726.
86. J. Otera, Y. Niibo and H. Aikawa, *Tetrahedron Lett.*, 1987, **28**, 2147.
87. G. H. Posner, J. J. Sterling, C. E. Whitten, C. M. Lentz and D. J. Brunelle, *J. Am. Chem. Soc.*, 1975, **97**, 107.
88. J.-M. Fang, *J. Org. Chem.*, 1982, **47**, 3464.
89. D. Liotta, M. T. Saindane, C. Barnum and G. Zima, *Tetrahedron*, 1985, **41**, 4881.
90. G. H. Posner, in 'Asymmetric Synthesis', ed. J. D. Morrison, Academic Press, New York, 1983, vol. 2, chap. 8.
91. G. Casy and R. J. K. Taylor, *J. Chem. Soc., Chem. Commun.*, 1988, 454.
92. R. K. Boeckman, Jr., *J. Org. Chem.*, 1973, **38**, 4450.
93. C. Thanupran, C. Thebtaranonth and Y. Thebtaranonth, *Tetrahedron Lett.*, 1986, **27**, 2295.

94. T. Siwapinyoyos and Y. Thebtaranonth, *Tetrahedron Lett.*, 1984, **25**, 353.
95. M. Zervos and L. Wartski, *Tetrahedron Lett.*, 1986, **27**, 2985.
96. S. J. Danishefsky, K. Vaughan, R. C. Gadwood, K. Tsuzuki and J. P. Springer, *Tetrahedron Lett.*, 1980, **21**, 2625.
97. A. Capperucci, A. Del'Innocenti, C. Faggi, A. Ricci, P. Dembech and G. Seconi, *J. Org. Chem.*, 1988, **53**, 3612.
98. M. Joucla, B. Fouchet, J. LeBrun and J. Hamelin, *Tetrahedron Lett.*, 1985, **26**, 1221.
99. M. P. Cooke, Jr., *J. Org. Chem.*, 1986, **51**, 1637.
100. T. Shono, Y. Matsumura, S. Kashimura and K. Hatanaka, *J. Am. Chem. Soc.*, 1979, **101**, 4752.
101. G. H. Posner, S.-B. Lu, E. Asirvatham, E. F. Silversmith and E. M. Shulman, *J. Am. Chem. Soc.*, 1986, **108**, 511.
102. T. Minami, K. Nishimura, J. Hirao, H. Suganuma and T. Agawa, *J. Org. Chem.*, 1982, **47**, 2360.
103. B. M. Trost and D. M. T. Chan, *J. Am. Chem. Soc.*, 1979, **101**, 6429.
104. J. H. Babler and K. P. Spina, *Tetrahedron Lett.*, 1985, **26**, 1923.
105. T. A. Hase and P. Kukkola, *Synth. Commun.*, 1980, **10**, 451.
106. T. Uyehara, N. Asao and Y. Yamamoto, *J. Chem. Soc., Chem. Commun.*, 1987, 1410.
107. I. Fleming and D. A. Perry, *Tetrahedron*, 1981, **37**, 4027.
108. J. P. Marino and R. J. Linderman, *J. Org. Chem.*, 1983, **48**, 4621.
109. J. B. Siddall, M. Biskup and J. H. Fried, *J. Am. Chem. Soc.*, 1969, **91**, 1853.
110. E. J. Corey and J. A. Katzenellenbogen, *J. Am. Chem. Soc.*, 1969, **91**, 1851.
111. E. Piers and R. D. Tillyer, *J. Org. Chem.*, 1988, **53**, 5366.
112. R. J. P. Corriu, J. J. E. Moreau and C. Vernhet, *Tetrahedron Lett.*, 1987, **28**, 2963.
113. J. P. Marino and R. J. Linderman, *J. Org. Chem.*, 1981, **46**, 3696.
114. Y. Tamura, T. Tsugoshi, S. Mohri and Y. Kita, *J. Org. Chem.*, 1985, **50**, 1542.
115. G. B. Mpango, K. K. Mahalanabis, Z. Mahdavi-Damghani and V. Snieckus, *Tetrahedron Lett.*, 1980, **21**, 4823.
116. J. E. Baldwin and W. A. DuPont, *Tetrahedron Lett.*, 1980, **21**, 1881.
117. Y. Tamaru, T. Harada, H. Iwamoto and Z. Yoshida, *J. Am. Chem. Soc.*, 1978, **100**, 5221.
118. Y. Tamaru, T. Hioki, S. Kawamura, H. Satomi and Z. Yoshida, *J. Am. Chem. Soc.*, 1984, **106**, 3876.
119. S. Knapp and J. Calienni, *Synth. Commun.*, 1980, **10**, 837.
120. W. Oppolzer, G. Poli, A. J. Kigma, C. Starkemann and G. Bernardinelli, *Helv. Chim. Acta*, 1987, **70**, 2201.
121. W. Oppolzer, *Pure Appl. Chem.*, 1988, **60**, 39.
122. E. Piers, J. M. Chong and B. A. Keay, *Tetrahedron Lett.*, 1985, **26**, 6265.
123. E. Piers and R. T. Skerlj, *J. Chem. Soc., Chem. Commun.*, 1986, 626.
124. F. E. Ziegler and J. A. Schwartz, *Tetrahedron Lett.*, 1979, 4643.
125. R. E. Damon, R. H. Schlessinger and J. F. Blount, *J. Org. Chem.*, 1976, **41**, 3772.
126. K. Tomioka, T. Ishiguro and K. Koga, *J. Chem. Soc., Chem. Commun.*, 1979, 652.
127. A. Pelter, R. S. Ward, M. C. Pritchard and I. T. Kay, *J. Chem. Soc., Perkin Trans. 1*, 1988, 1603; 1615.
128. A. G. Pernet, H. Nakamoto, N. Ishizuka, M. Aburatani, K. Nakahashi, K. Sakamoto and T. Takeuchi, *Tetrahedron Lett.*, 1979, 3933.
129. R. K. Haynes and P. A. Schober, *Aust. J. Chem.*, 1987, **40**, 1249.
130. G. H. Posner, K. S. Webb, E. Asirvatham, S. Jew, and A. Del'Innocenti, *J. Am. Chem. Soc.*, 1988, **110**, 4754.
131. J. V. N. Prasad and C. N. Pillai, *J. Organomet. Chem.*, 1983, **259**, 1.
132. E. Negishi, *Pure Appl. Chem.*, 1981, **53**, 2333.
133. E. Negishi, D. E. van Horn, T. Yoshida and C. L. Rand, *Organometallics*, 1983, **2**, 563.
134. P. J. Stang and T. Kitamura, *J. Am. Chem. Soc.*, 1987, **109**, 7561.
135. A. Marfat, P. R. McGuirk and P. Helquist, *J. Org. Chem.*, 1979, **44**, 3888.
136. A. Alexakis, G. Cahiez and J. F. Normant, *Tetrahedron*, 1980, **36**, 1961.
137. Y. Satoh, H. Serizawa, N. Miyaura, S. Hara and A. Suzuki, *Tetrahedron Lett.*, 1988, **29**, 1811.
138. M. C. Carre, B. Jamart-Gregoire, P. Geoffroy and P. Caubere, *Tetrahedron*, 1988, **44**, 127.
139. C. A. Townsend, S. B. Christensen and S. G. Davies, *J. Chem. Soc., Perkin Trans. 1*, 1988, 839.
140. M. Reuman and A. I. Meyers, *Tetrahedron*, 1985, **41**, 837.
141. P. D. Pansegrau, W. F. Rieker and A. I. Meyers, *J. Am. Chem. Soc.*, 1988, **110**, 7178.
142. T. Shono, I. Nishiguchi and M. Sasaki, *J. Am. Chem. Soc.*, 1979, **101**, 4752.
143. L. L. McCoy, *J. Org. Chem.*, 1960, **25**, 2078.
144. E. Anklam and P. Margaretha, *Helv. Chim. Acta*, 1984, **67**, 2206.
145. Z. T. Fomum, A. E. Nkengfack, S. R. Landor and P. D. Landor, *J. Chem. Soc., Perkin Trans. 1*, 1988, 277.
146. K. Takai, T. Maeda and M. Ishikawa, *J. Org. Chem.*, 1989, **54**, 58.
147. H. Sugihara, R. Tanikaga, K. Tanaka and A. Kaji, *Bull. Chem. Soc. Jpn.*, 1978, **51**, 655.
148. S. G. Pyne, *Tetrahedron Lett.*, 1987, **28**, 4737.
149. M. Isobe, M. Kitamura and T. Goto, *Chem. Lett.*, 1980, 331.
150. J. P. Ponton, P. Helquist, P. C. Conrad and P. L. Fuchs, *J. Org. Chem.*, 1981, **46**, 118.
151. P. R. Hamann, J. E. Toth and P. L. Fuchs, *J. Org. Chem.*, 1984, **49**, 3865.
152. R. E. Donaldson, J. C. Saddler, S. Byrn, A. T. McKenzie and P. L. Fuchs, *J. Org. Chem.*, 1983, **48**, 2167.
153. P. L. Fuchs and T. F. Braish, *Chem. Rev.*, 1986, **86**, 903.
154. J. E. Toth, P. R. Hamann and P. L. Fuchs, *J. Org. Chem.*, 1988, **53**, 4694.
155. A. I. Meyers, J. P. Lawson and D. R. Carver, *J. Org. Chem.*, 1981, **46**, 3119.
156. R. J. Linderman and A. I. Meyers, *Heterocycles*, 1983, **20**, 1737.
157. R. J. Linderman and A. I. Meyers, *Tetrahedron Lett.*, 1983, **24**, 3043.
158. E. Zbiral, *Synthesis*, 1974, 775.
159. K. B. Becker, *Tetrahedron*, 1980, **36**, 1717.
160. H. J. Cristau, H. Christol and D. Bottaro, *Synthesis*, 1978, 826.
161. V. W. Flitsch and E. R. Gesing, *Tetrahedron Lett.*, 1976, 1997.

162. R. Bodalski, T. Michalski, J. Monkiewicz and K. M. Pietrusiewicz, in 'Phosphorus Chemistry, Proceedings of the 1981 International Conference', ed. L. D. Quin and J. G. Verkade, American Chemical Society, Washington, DC, 1981, p. 243.
163. K. Tomioka, F. Masumi, T. Yamashita and K. Koga, *Tetrahedron Lett.*, 1984, **25**, 333.
164. K. Tomioka and K. Koga, in 'Asymmetric Synthesis', ed. J. D. Morisson, Academic Press, New York, 1983, vol. 2.
165. E. J. Corey and D. L. Boger, *Tetrahedron Lett.*, 1978, 9.
166. A. I. Meyers and G. Licini, *Tetrahedron Lett.*, 1989, **30**, 4049.
167. A. I. Meyers, K. A. Lutomski and D. Laucher, *Tetrahedron*, 1988, **44**, 3107.
168. S. E. Denmark, J. A. Sternberg and R. Lueoend, *J. Org. Chem.*, 1988, **53**, 1251.
169. T. Tanaka, T. Toru, N. Okamura, H. Hazato, S. Sugiura, K. Manabe, S. Kurozumi, M. Suzuki, T. Kawagishi and R. Noyori, *Tetrahedron Lett.*, 1983, **24**, 4103.
170. F. Richter and H.-H. Otto, *Tetrahedron Lett.*, 1987, **28**, 2945.
171. E. P. Kundig, *Pure Appl. Chem.*, 1985, **57**, 1855.
172. E. P. Kundig and D. P. Simmons, *J. Chem. Soc., Chem. Commun.*, 1983, 1320.
173. I. Fleming and M. Rowley, *Tetrahedron Lett.*, 1989, **45**, 413.
174. M. P. Cooke, Jr. and R. Goswami, *J. Am. Chem. Soc.*, 1977, **99**, 642.
175. H. Nagaoka, K. Kobayashi, T. Matsui and Y. Yamada, *Tetrahedron Lett.*, 1987, **28**, 2021.
176. P. A. Grieco and R. S. Finkelhor, *J. Org. Chem.*, 1973, **38**, 2100.
177. K. Nonoshita, K. Maruoka and H. Yamamoto, *Bull. Chem. Soc. Jpn.*, 1988, **61**, 2241.
178. K. E. Kim, S. B. Park and N. M. Yoon, *Synth. Commun.*, 1988, **18**, 89.
179. T. Tsuda, T. Fuji, K. Kawasaki and T. Seagusa, *J. Chem. Soc., Chem. Commun.*, 1980, 1013.
180. T. Tsuda, T. Kawamoto, Y. Kumamoto and T. Seagusa, *Synth. Commun.*, 1986, **16**, 639.
181. H. Chikashita and K. Itoh, *Bull. Chem. Soc. Jpn.*, 1986, **59**, 1747.
182. T. Tsuda, T. Yoshida, T. Kawamoto and T. Seagusa, *J. Org. Chem.*, 1987, **52**, 1624.
183. J. M. Fortunato and B. Ganem, *J. Org. Chem.*, 1976, **41**, 2194.
184. W. S. Mahoney, D. M. Brestensky and J. M. Stryker, *J. Am. Chem. Soc.*, 1988, **110**, 291.
185. M. F. Semmelhack and R. D. Stauffer, *J. Org. Chem.*, 1975, **40**, 3619.
186. S. A. Rao and M. Perisamy, *J. Chem. Soc., Chem. Commun.*, 1987, 495.
187. I. Ojima and T. Kogure, *Organometallics*, 1982, **1**, 1390.
188. W. P. Jackson and S. V. Ley, *J. Chem. Soc., Perkin Trans. I*, 1981, 1516.
189. S. Torii, H. Tanaka and Y. Nagai, *Bull. Chem. Soc. Jpn.*, 1977, **50**, 2825.
190. S. H. Bertz, C. P. Gibson and G. Dabbagh, *Tetrahedron Lett.*, 1987, **28**, 4251.
191. S. H. Bertz, C. P. Gibson and G. Dabbagh, *Organometallics*, 1988, **7**, 227.
192. K. C. Nicolaou, W. E. Barnette and P. Ma, *J. Org. Chem.*, 1980, **45**, 1463.
193. C. H. Heathcock, C. M. Tice and T. C. Germroth, *J. Am. Chem. Soc.*, 1982, **104**, 6081.
194. H. Nishiyama, M. Sasaki and K. Itoh, *Chem. Lett.*, 1981, 905.
195. M. Suzuki, A. Yanagisawa and R. Noyori, *J. Am. Chem. Soc.*, 1985, **107**, 3348.
196. T. Takahashi, K. Shimizu, T. Doi, J. Tsuji and Y. Fukazawa, *J. Am. Chem. Soc.*, 1988, **110**, 2674.
197. M. Suzuki, A. Yanagisawa and R. Noyori, *J. Am. Chem. Soc.*, 1988, **110**, 4718.
198. M. T. Crimmins, S. W. Mascarella and J. A. DeLoach, *J. Org. Chem.*, 1984, **49**, 3033.
199. S. E. Denmark and J. P. Germanas, *Tetrahedron Lett.*, 1984, **25**, 1231.
200. K. Yamamoto, M. Iijima, Y. Ogimura and J. Tsuji, *Tetrahedron Lett.*, 1984, **25**, 2813.
201. C. J. Kowalski and J.-S. Dung, *J. Am. Chem. Soc.*, 1980, **102**, 7950.
202. E. Piers and J. S. M. Wai, *J. Chem. Soc., Chem. Commun.*, 1987, 1342.
203. E. Piers and V. Karunaratne, *Can. J. Chem.*, 1984, **62**, 629.
204. M. Bergdahl, E.-L. Lindstedt, M. Nilsson and T. Olsson, *Tetrahedron*, 1988, **44**, 2055.
205. W. H. Mandeville and G. M. Whitesides, *J. Org. Chem.*, 1974, **39**, 400.
206. G. H. Posner, C. E. Whitten, J. J. Sterling and D. J. Brunelle, *Tetrahedron Lett.*, 1974, 2591.
207. T. Olsson, M. T. Rahman and C. Ullenius, *Tetrahedron Lett.*, 1977, 75.
208. T. Tanaka, S. Kurozumi, T. Toru, M. Kobayashi, S. Miura and S. Ishimoto, *Tetrahedron*, 1977, **33**, 1105.
209. G. H. Posner, M. J. Chapdelaine and C. M. Lentz, *J. Org. Chem.*, 1979, **44**, 3661.
210. E. Piers and J. M. Chong, *J. Org. Chem.*, 1982, **47**, 1602.
211. G. H. Posner, C. E. Whitten and J. J. Sterling, *J. Am. Chem. Soc.*, 1973, **95**, 7788.
212. S. Kurozumi, T. Toru, T. Tanaka, M. Kobayashi, S. Miura and S. Ishimoto, *Tetrahedron Lett.*, 1976, 4091.
213. B. H. Lipshutz, *Synthesis*, 1987, 325; *Synlett*, 1990, 119.
214. B. H. Lipshutz, R. S. Wilhelm and J. A. Kozlowski, *J. Org. Chem.*, 1984, **49**, 3938.
215. B. H. Lipshutz, R. S. Wilhelm and J. A. Kozlowski, *Tetrahedron*, 1984, **40**, 5005.
216. F.-T. Luo and E. Negishi, *J. Org. Chem.*, 1985, **50**, 4762.
217. I. Fleming, T. W. Newton and F. Roessler, *J. Chem. Soc., Perkin Trans. I*, 1981, 2527.
218. B. J. Wakefield, 'The Chemistry of Organolithium Compounds', Pergamon Press, New York, 1974, p. 14.
219. J. F. McGarrity, C. A. Ogle, Z. Brich and H. R. Loosli, *J. Am. Chem. Soc.*, 1985, **107**, 1810.
220. E. C. Ashby and G. Heinsohn, *J. Org. Chem.*, 1974, **39**, 3297.
221. E. A. Jeffery, A. Meisters and T. Mole, *J. Organomet. Chem.*, 1974, **174**, 365, 373.
222. J. Schwartz and Y. Hayashi, *Tetrahedron Lett.*, 1980, **21**, 1497.
223. J. Schwartz, M. J. Loots and H. Kosugi, *J. Am. Chem. Soc.*, 1980, **102**, 1333.
224. M. F. Semmelhack, L. Keller, T. Sato, E. J. Spiess and W. Wulff, *J. Org. Chem.*, 1985, **50**, 5566.
225. M. A. Gianturco, P. Friedel and A. S. Giammarino, *Tetrahedron*, 1964, **20**, 1763.
226. E. Campaigne and R. K. Mehra, *J. Heterocycl. Chem.*, 1977, **14**, 1337.
227. A. Itoh, S. Ozawa, K. Oshima and H. Nozaki, *Tetrahedron Lett.*, 1980, **21**, 361.
228. W. C. Still, *J. Am. Chem. Soc.*, 1977, **99**, 4836.
229. W. R. Leonard and T. Livinghouse, *J. Org. Chem.*, 1985, **50**, 730.
230. M. Suzuki, T. Kawagishi and R. Noyori, *Tetrahedron Lett.*, 1981, **22**, 1809.

231. E. D. Bergmann, D. Ginsburg and R. Pappo, *Org. React. (N.Y.)*, 1959, **10**, 179.
232. A. Alexakis, M. J. Chapdelaine and G. H. Posner, *Tetrahedron Lett.*, 1978, 4209.
233. S. J. Danishefsky and S. J. Etheredge, *J. Org. Chem.*, 1982, **47**, 4791.
234. H. J. Monteiro, *J. Org. Chem.*, 1977, **42**, 2324.
235. G. Stork and P. F. Hudrlik, *J. Am. Chem. Soc.*, 1968, **96**, 4462.
236. T. V. Rajanbabu, *J. Org. Chem.*, 1984, **49**, 2083.
237. S. Kobayashi and T. Mukaiyama, *Chem. Lett.*, 1986, 1805.
238. T. Mukaiyama, Y. Sagawa and S. Kobayashi, *Chem. Lett.*, 1986, 1821.
239. M. Asoka, K. Ishibashi, W. Takahashi and H. Takei, *Bull. Chem. Soc. Jpn.*, 1987, **60**, 2259.
240. H. Hagiwara, A. Okano, T. Akama and H. Uda, *J. Chem. Soc., Chem. Commun.*, 1987, 1333.
241. H. Hagiwara, A. Okano and H. Uda, *J. Chem. Soc., Chem. Commun.*, 1985, 1047.
242. W. J. Scott and J. E. McMurry, *Acc. Chem. Res.*, 1988, **21**, 47.
243. D. N. Jones and M. R. Peels, *J. Chem. Soc., Chem. Commun.*, 1986, 216.
244. R. K. Haynes, D. E. Lambert, P. A. Schober and S. G. Turner, *Aust. J. Chem.*, 1987, **40**, 1211.
245. M. R. Binns and R. K. Haynes, *J. Org. Chem.*, 1981, **46**, 3790.
246. R. K. Haynes and A. G. Katsifis, *J. Chem. Soc., Chem. Commun.*, 1987, 340.
247. F. M. Hauser and D. Mal, *J. Am. Chem. Soc.*, 1984, **106**, 1098.
248. J. Nokami, T. Ono, S. Wakabayashi, A. Hazato and S. Kurozumi, *Tetrahedron Lett.*, 1985, **26**, 1985.
249. F. M. Hauser, P. Hewawassam and D. Mal, *J. Am. Chem. Soc.*, 1988, **110**, 2919.
250. Y. Takahashi, K. Isobe, H. Hagiwara, H. Kosugi and H. Uda, *J. Chem. Soc., Chem. Commun.*, 1981, 714.
251. F. Cooke, P. Magnus and G. L. Bundy, *J. Chem. Soc., Chem. Commun.*, 1978, 714.
252. T. Cohen and M. Meyers, *J. Org. Chem.*, 1988, **53**, 457.
253. G. B. Mpango and V. Snieckus, *Tetrahedron Lett.*, 1980, **21**, 4827.
254. Y. Asano, T. Kamikawa and T. Tokoroyama, *Bull. Chem. Soc. Jpn.*, 1976, **49**, 3232.
255. R. Dahl, Y. Nabi and E. Brown, *Tetrahedron*, 1986, **42**, 2005.
256. F. E. Ziegler, J.-M. Fang and C. C. Tam, *J. Am. Chem. Soc.*, 1982, **104**, 7174.
257. K. Ogura, N. Yahata, M. Minoguchi, K. Ohtsuki, K. Takahashi and H. Iida, *J. Org. Chem.*, 1986, **51**, 508.
258. T. Cohen and L.-C. Yu, 1986, *J. Org. Chem.*, 1985, **50**, 3266.
259. G. H. Posner and E. Asirvatham, *Tetrahedron Lett.*, 1986, **27**, 663.
260. R. Bürstinghaus and D. Seebach, *Chem. Ber.*, 1977, **110**, 841.
261. H. J. Bestmann and F. Seng, *Angew. Chem., Int. Ed. Engl.*, 1962, **1**, 116.
262. J. P. Freeman, *J. Org. Chem.*, 1966, **31**, 538.
263. W. G. Dauben, D. J. Hart, J. Ipaktschi and A. P. Kozikowski, *Tetrahedron Lett.*, 1973, 4425.
264. P. A. Grieco and R. S. Finkelhor, *Tetrahedron Lett.*, 1972, 3781.
265. S. K. Haynes and S. C. Vonwiller, *J. Chem. Soc., Chem. Commun.*, 1987, 92.
266. A. Krief, M. J. Devos and M. Sevrin, *Tetrahedron Lett.*, 1986, **27**, 2283.
267. G. A. McAlpine, R. A. Raphael, A. Shaw, A. W. Taylor and H.-J. Wild, *J. Chem. Soc., Perkin Trans. 1*, 1976, 410.
268. M. Samson, H. DeWilde and M. Vandewalle, *Bull. Soc. Chim. Belg.*, 1977, **86**, 329.
269. M. J. Devos and A. Krief, *Tetrahedron Lett.*, 1979, 1891.
270. E. Hatzigrigoriou, M.-C. Roux-Schmitt, L. Wartski and J. Seyden-Penne, *Tetrahedron*, 1983, **39**, 3415.
271. E. Hatzigrigoriou and L. Wartski, *Synth. Commun.*, 1983, **13**, 319.
272. E. Hatzigrigoriou and L. Wartski, *Bull. Soc. Chim. Fr., Part 2*, 1983, 313.
273. E. Vedejs and B. Nader, *J. Org. Chem.*, 1982, **47**, 3193.
274. J. A. Noguez, and L. A. Maldonado, *Synth. Commun.*, 1976, **6**, 39.
275. N. Seuron and J. Seyden-Penne, *Tetrahedron*, 1984, **40**, 635.
276. M. Britten-Kelly, B. J. Willis and D. H. R. Barton, *Synthesis*, 1980, 27.
277. S. Berrada, S. Desert and P. Metzner, *Tetrahedron*, 1988, **44**, 3575.
278. H. O. House, 'Modern Synthetic Reactions', 2nd edn., Benjamin, Menlo Park, 1972, p. 526.
279. G. H. Posner, *Org. React. (N.Y.)*, 1972, **19**, 1; 1975, **22**, 253.
280. D. M. Lawler and N. S. Simpkins, *Tetrahedron Lett.*, 1988, **29**, 1207.
281. A. J. Dixon, R. J. K. Taylor and R. F. Newton, *J. Chem. Soc., Perkin Trans. 1*, 1981, 1407.
282. R. D. Little and J. R. Dawson, *Tetrahedron Lett.*, 1980, **21**, 2609.
283. G. D. Vite and T. A. Spencer, *J. Org. Chem.*, 1988, **53**, 2560.
284. T. Touru, S. Kurozumi, T. Tanaka, S. Miura, M. Kobayashi and S. Ishimoto, *Tetrahedron Lett.*, 1976, 4087.
285. S. Bernasconi, M. Ferrari, P. Gariboldi, G. Jommi, M. Sisti and R. Destro, *J. Chem. Soc., Perkin Trans. 1.*, 1981, 1994.
286. Y. Tamura, A. Wada, S. Okuyama, S. Fukumori, Y. Hayashi, M. Gohda and Y. Kita, *Chem. Pharm. Bull.*, 1981, **29**, 1312.
287. F. Naf and R. Decorzant, *Helv. Chim. Acta*, 1974, **57**, 1317.
288. S. Bernasconi, P. Gariboldi, G. Jommi and M. Sisti, *Tetrahedron Lett.*, 1980, **21**, 2337; T. H. Black, *Org. Prep. Proced. Int.*, 1989, **21**, 179.
289. R. G. Salomon and M. F. Salomon, *J. Org. Chem.*, 1975, **40**, 1488.
290. S. Bernasconi, P. Gariboldi, G. Jommi, M. Sisti and P. Tavecchia, *J. Org. Chem.*, 1981, **46**, 3719.
291. T. R. Kelly and H. Liu, *J. Am. Chem. Soc.*, 1985, **107**, 4998.
292. K. Narasaka, T. Sakakura, T. Uchimaru and C. Guedin-Vuong, *J. Am. Chem. Soc.*, 1984, **106**, 2954.
293. F. Naf and R. Decorzant, *Helv. Chim. Acta*, 1971, **54**, 1939.
294. R. A. Kretchmer, E. D. Mihelich and J. J. Waldron, *J. Org. Chem.*, 1972, **37**, 4483.
295. R. Davis and K. G. Untch, *J. Org. Chem.*, 1979, **44**, 3755.
296. R. K. Boeckman, Jr., *J. Am. Chem. Soc.*, 1973, **95**, 6867.
297. T. Takahashi, Y. Naito and J. Tsuji, *J. Am. Chem. Soc.*, 1981, **103**, 5261.
298. T. Takahashi, H. Okumoto and J. Tsuji, *Tetrahedron Lett.*, 1984, **25**, 1925.

299. M. Ihara, Y. Ishida, M. Abe, M. Toyota, K. Fukumoto and T. Kametani, *J. Chem. Soc., Perkin Trans. 1*, 1988, 1155.
300. M. Ihara, M. Katogi, K. Fukumoto and T. Kametani, *J. Chem. Soc., Chem. Commun.*, 1987, 721.
301. M. Ihara, M. Suzuki, K. Fukumoto, T. Kametani and C. Kabuto, *J. Am. Chem. Soc.*, 1988, **110**, 1963.
302. B. O'Conner and G. Just, *J. Org. Chem.*, 1987, **52**, 1801.
303. T. Fujisawa, A. Noda, T. Kawara and T. Sato, *Chem. Lett.*, 1981, 1159.
304. H. J. Reich, J. M. Renga and I. L. Reich, *J. Org. Chem.*, 1974, **39**, 2133.
305. T. Kitahara, Y. Takagi and M. Matsui, *Agric. Biol. Chem.*, 1979, **43**, 2359.
306. J. L. Herrmann, M. H. Berger and R. H. Schlessinger, *J. Am. Chem. Soc.*, 1973, **95**, 7923.
307. R. M. Carlson, A. R. Oyler and J. R. Peterson, *J. Org. Chem.*, 1975, **40**, 1610.
308. T. Tanaka, A. Hazato, K. Bannai, N. Okamura, S. Sugiura, K. Manabe, T. Toru, S. Kurozumi, M. Suzuki, T. Kawagishi and R. Noyori, *Tetrahedron*, 1987, **43**, 813.
309. J. W. Ellis, *J. Chem. Soc. (D)*, 1970, 406.
310. N. L. Holy and Y. F. Wang, *J. Am. Chem. Soc.*, 1974, **99**, 944.
311. B. W. Roberts, M. Ross and J. Wong, *J. Chem. Soc., Chem. Commun.*, 1980, 428.
312. R. M. Cory and R. M. Renneboog, *J. Org. Chem.*, 1984, **49**, 3898.
313. G. H. Posner, J. P. Mallamo and A. Y. Black, *Tetrahedron*, 1981, **37**, 3921.
314. S. J. Danishefsky, P. Harrison, M. Silvestri and B. E. Segmuller, *J. Org. Chem.*, 1984, **49**, 1319.
315. W. E. Noland, *Chem. Rev.*, 1955, **55**, 137.

1.7
Addition of H—X Reagents to Alkenes and Alkynes

RICHARD C. LAROCK and WILLIAM W. LEONG
Iowa State University, Ames, IA, USA

1.7.1 INTRODUCTION

The direct electrophilic addition of the elements H—X to alkenes, dienes and alkynes, where X = halogen, nitrogen, oxygen, sulfur and selenium, is among the oldest known and most important synthetic methods used to prepare the corresponding addition compounds. Unfortunately, much of the early literature from the late 1800s and early 1900s tends to be more confusing than helpful. Rather surprisingly though, relatively little in the way of major new synthetic advances have occurred since that time. However, a number of alternative, indirect procedures, such as solvomercuration–demercuration, have more recently been developed that offer important advantages in chemo-, regio- or stereo-selectivity. These processes are the focus of this chapter.

1.7.2 HYDROGEN HALIDE ADDITION

1.7.2.1 Introduction

The addition of H—X, where X = F, Cl, Br and I, to alkenes, dienes and alkynes has been extensively studied from both mechanistic[1–3] and synthetic standpoints. While it is one of the earliest methods employed for the synthesis of organic halides, it can often lead to mixtures of products, and the direct conversion of the corresponding alcohols is more commonly used nowadays. The early literature abounds with examples of this reaction in which either mixtures of products are formed, or the products are not well characterized. Few major advances have been reported in more recent times which overcome the synthetic disadvantages of this direct process. It can nevertheless be the method of choice for the synthesis of certain substrates.

In general, the ease of addition of H—X to simple alkenes follows their relative acidity, HI > HBr > HCl, but HF addition is often surprisingly easy. A diversity of mechanisms appear to be involved in these processes, ranging from relatively pure carbocation processes to those more reminiscent of four-center addition. Markovnikov addition is commonly observed when precautions are taken to exclude peroxide or other possible free radical initiators. Though strong acids are involved and rearrangements are not uncommon, the reaction conditions are otherwise quite mild and yields can be high.

1.7.2.2 H—F Addition

The addition of HF to alkenes to form alkyl fluorides is difficult to effect in the laboratory, because of the low reactivity of aqueous HF, the difficulties in handling anhydrous HF, the need for pressure equipment, the ease of polymerization of alkenes by HF, and the instability of the resulting alkyl fluorides, especially where secondary, or particularly tertiary, fluorides are concerned.[4–6]

While aqueous HF is unreactive towards alkene addition, anhydrous HF is surprisingly reactive. The addition of HF to simple alkenes, such as ethylene, propene and cyclohexene, has been effected by mixing the reagents in an appropriate metal container at temperatures of –78 to –45 °C, and gradually heating the mixture from room temperature to 90 °C.[7–9] Representative yields are 60–80%. Catalysts are unnecessary. Markovnikov addition is observed, but the stereochemistry of addition to norbornene is not clear.[10] With bornylene[11] and camphene,[12] HF addition gives excellent yields of a mixture of products.

With the great commercial interest in chlorofluorocarbons, the addition of HF to unsaturated organic chlorides, particularly vinylic chlorides has received considerable industrial attention.[4,5,13–19] While these reactions are more difficult to effect than those of simple alkenes and increasing halogen substitution about the C—C double bond decreases the alkene reactivity, the reactions of simple monochloroalkenes proceed at temperatures from –23 to +120 °C and can give good yields. Trichloroethylene or 1,1,2-trichloropropene, on the other hand, require temperatures in excess of 200 °C, and tetrahaloethylenes fail to react in the absence of a catalyst.[5] However, substitution of fluorine for chlorine tends to facilitate HF addition.[5]

While HF addition can on occasion efficiently produce addition products as in equation (1),[13] mixtures of halogen exchange products are often observed (equation 2).[18]

Cl HF Cl F
75% (1)

Cl F F
HF Cl Cl
Cl Cl + F (2)
65% 35%

Indeed, the exchange product can sometimes be obtained directly in high yield (equation 3),[15] or be favored by the addition of excess HF and $SbCl_5$, or SbF_3 plus Cl_2 (equation 4).[5,16] With the more highly halogenated starting materials, the higher reaction temperatures required lead to increasing halogen exchange.

Cl; HF; 64%; F F (3)

Cl; Cl; HF; SbF_3; Cl_2; F F F; 50%; + ; Cl F F; 50% (4)

Unsaturated fatty acids and steroids also react with HF, but the structure of the products has not always been clearly established.[4–6]

While procedures other than those described above, including the use of a variety of catalysts, have apparently been explored industrially, few experimental details are readily available.[5] It is known that BF_3 facilitates HF addition, but it also promotes alkene decomposition and halogen exchange and is, therefore, limited to the less reactive polyhaloalkenes.[17,19]

A significant, recent advance in this area involves the use of HF–pyridine or HF–trialkylamine reagents.[20] These stable poly(hydrogen fluoride) reagents react with simple alkenes at atmospheric pressure and room temperature in polyethylene containers using THF as the solvent to give secondary and tertiary alkyl fluorides in yields of 35–90% (equation 5).

$(HF)_x$•C_5H_5N; 80%; F (5)

There appear to be only two reports in the literature of the reaction of HF with dienes (equation 6).[21,22] The yields are quite low probably due to polymerization.

HF; 10%; F; 87%; + ; F; 13% (6)

There are, however, several publications which describe the addition of HF to alkynes to form either vinylic fluorides or difluoroalkanes.[4,23–26] Only the addition of HF to acetylene,[23,24] vinylacetylene[23] and 3,3,3-trifluoropropyne[25] have been reported to form vinylic fluorides (equations 7–9).

HC≡CH; HF; 15% conversion; F; 35%; + ; F F; 65% (7)

HF; 70–80%; 20–45% conversion; F (8)

F_3C—≡; HF; 92%; F_3C F (9)

While the direct reaction of HF and acetylene requires high pressure, proceeds in only low conversion and gives mixtures, a number of catalyst systems have been reported to give improved results.[23] Addition of boron trifluoride, however, favors the formation of 1,1-difluoroethane. Similar results are reported for vinylacetylene. On the other hand, 3,3,3-trifluoropropyne reacts readily to afford the anti-Markovnikov vinylic fluoride in 92% yield.[25]

The reaction of alkynes and HF more commonly gives geminal difluoroalkanes, usually in high yield. Terminal alkynes afford 2,2-difluoroalkanes (equation 10) and internal alkynes yield geminal difluorides (equation 11).[4,23,24,26]

$$\text{R}-\equiv \xrightarrow{\text{HF}} \text{RCF}_2\text{CH}_3 \quad (10)$$

$$\text{R}-\equiv-\text{R} \xrightarrow{\text{HF}} \text{RCF}_2\text{CH}_2\text{R} \quad (11)$$

These reactions are generally carried out at temperatures of –70 to +15 °C using either liquid HF,[24] or HF in ether or acetone.[26] The latter process affords regioisomeric mixtures of geminal difluorides when unsymmetrical internal alkynes, such as 2-pentyne, are employed.

Yields of geminal difluorides of 70–75% are also obtained from the reaction of terminal or internal alkynes with the $(HF)_x\cdot C_5H_5N$ reagent.[21]

On the other hand, the use of tetra-*n*-butylammonium or polymer-supported dihydrogentrifluoride at temperatures of 60–120 °C affords 57–90% yields of vinylic fluorides when alkynes activated by the presence of one or two strong electron-withdrawing groups (Ph, CHO, COR, CO_2R, CN) are employed (equation 12).[27,28] The (*Z*)-isomer tends to predominate. These reagents are substantially more reactive towards addition to these electron-deficient alkynes than $(Bu^n{}_4N)HF_2$, pyridine–HF, or $Et_3N\cdot 3HF$. It has been suggested that these reactions actually proceed by nucleophilic attack of fluoride anion.

$$n\text{-C}_7\text{H}_{15}-\equiv-\text{CN} \xrightarrow[\text{110 °C, 7 h} \\ 95\%]{(\text{Bu}^n{}_4\text{N})\text{H}_2\text{F}_3} n\text{-C}_7\text{H}_{15}\text{C(F)}=\text{CHCN} \quad (12)$$

(*E*):(*Z*) 70:30

1.7.2.3 H—Cl Addition

There are numerous examples of the direct addition of HCl to alkenes, even though HCl is less reactive than HBr or HI.[29,30] Several different experimental procedures have been employed. Most useful of these involves the passage of dry HCl gas into the alkene either neat or in an inert solvent such as pentane, dichloromethane or diethyl ether. Most successful of these reactions are those which generate tertiary and/or benzylic alkyl chlorides (equations 13–17).[31–35]

$$\text{(CH}_3)_2\text{C}=\text{CHCH}_3 \xrightarrow[100\%]{\text{HCl}} \text{(CH}_3)_2\text{CClCH}_2\text{CH}_3 \quad \text{(ref. 31)} \quad (13)$$

$$\text{methylenecyclobutane} \xrightarrow[89\%]{\text{HCl}} \text{1-chloro-1-methylcyclobutane} \quad \text{(ref. 32)} \quad (14)$$

$$\text{indene} \xrightarrow{\text{HCl}} \text{1-chloroindane} \quad \text{(ref. 33)} \quad (15)$$

Since these addition reactions appear to proceed through carbocation intermediates, carbon skeleton rearrangements can be expected (equations 18 and 19).[36–39]

The stereochemistry of HCl or DCl addition to a wide variety of alkenes has been examined. Addition of HCl to *cis*-2,3-dideutero-2-butene affords a mixture of *erythro* and *threo* 2-chlorobutanes.[40] 1-Methylcyclopentene[37] and 1,2-dimethylcyclopentene[41] give almost exclusively tertiary chlorides formed by *anti*

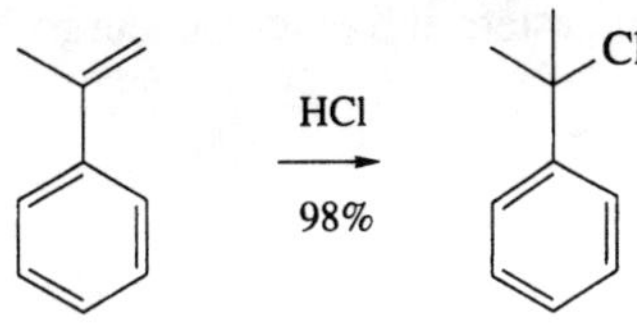

HCl, 98% (ref. 34) (16)

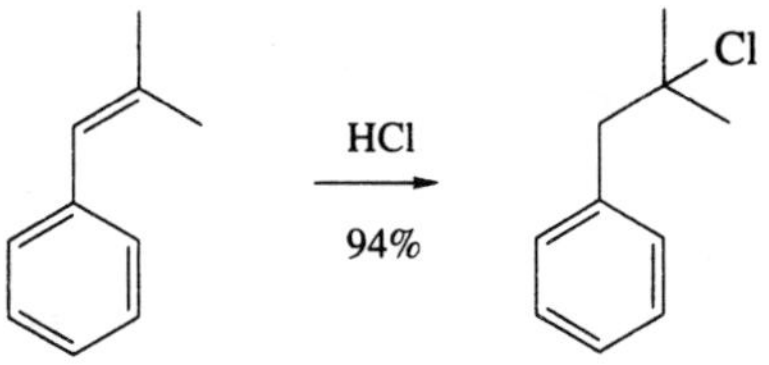

HCl, 94% (ref. 35) (17)

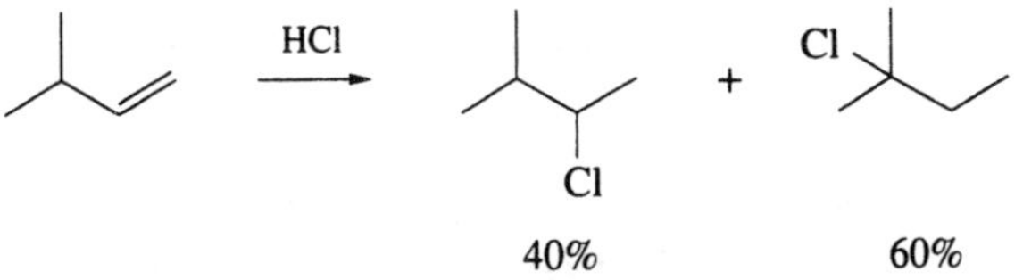

HCl → Cl 40% + Cl 60% (refs. 36, 37) (18)

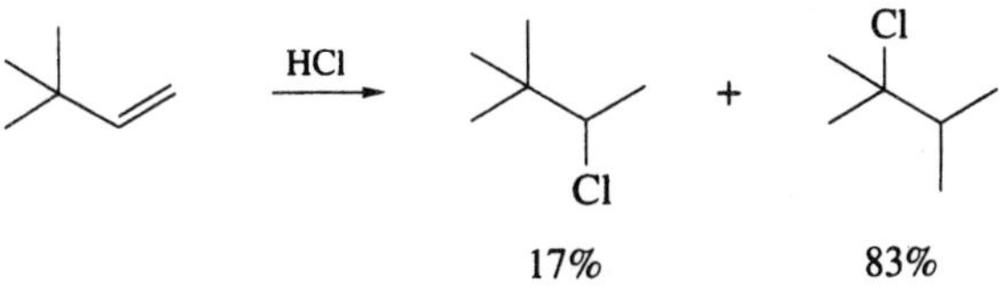

HCl → Cl 17% + Cl 83% (refs. 37–39) (19)

addition. With cyclohexene[42] and the tetrasubstituted alkenes (**1**)–(**3**),[43,44] products of both *syn* and *anti* addition are observed and the ratio can be dramatically reversed simply by changing the solvent and temperature of the reaction (equation 20).[44]

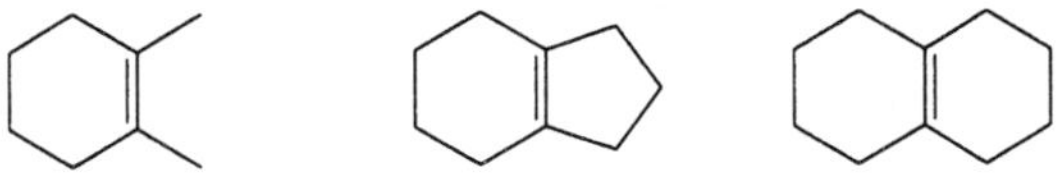

(**1**) (**2**) (**3**)

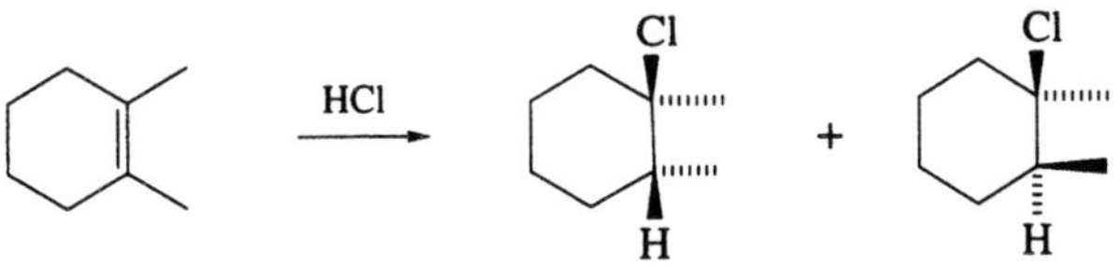

HCl (20)

CH_2Cl_2 (–98 °C)	88%	12%
Et_2O (0 °C)	5%	95%

With acenaphthene[45] and 4-*t*-butyl-1-phenylcyclohexene,[46] HCl or DCl addition in nonpolar solvents affords predominantly the products of *syn* addition.

Numerous bicyclic alkenes have been allowed to react with HCl to generate the corresponding *exo* chlorides.[47–53] Skeletal rearrangements, either during addition or soon thereafter, are observed in a number of these additions (equation 21).[47]

HCl, Et_2O → Cl + Cl (21)

Electron-rich alkenes, such as vinylic esters and ethers, undergo HCl addition in high yields (equations 22 and 23).[29]

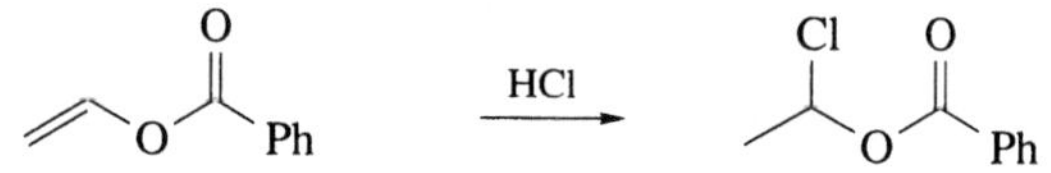

(22)

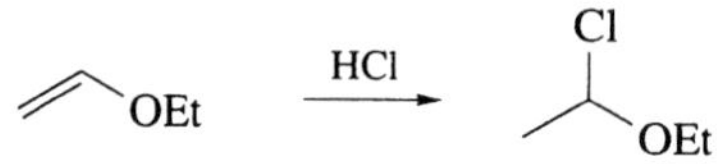

(23)

The addition of HCl to alkenes bearing electron-withdrawing groups also provides a useful approach to functionalized organic chlorides (equations 24–31).[29] Predominant *anti* addition has been observed in all of these examples where stereochemistry can be observed.[54–60]

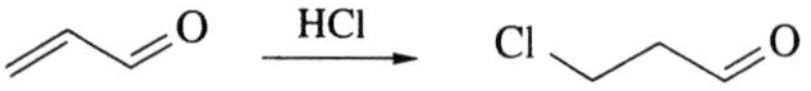

(ref. 54) (24)

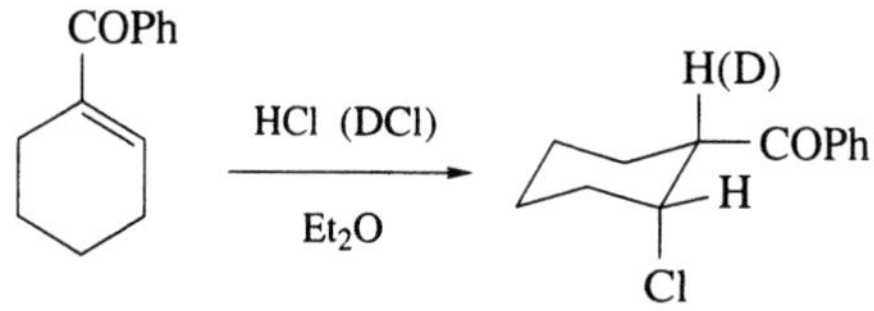

(ref. 55) (25)

(ref. 56) (26)

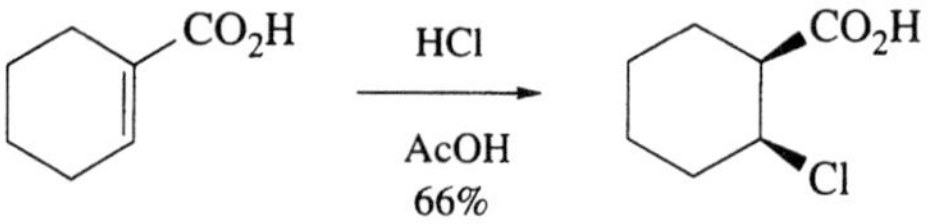

(ref. 57) (27)

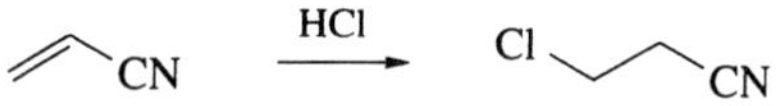

(ref. 58) (28)

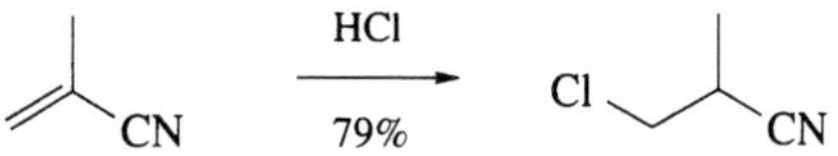

(ref. 59) (29)

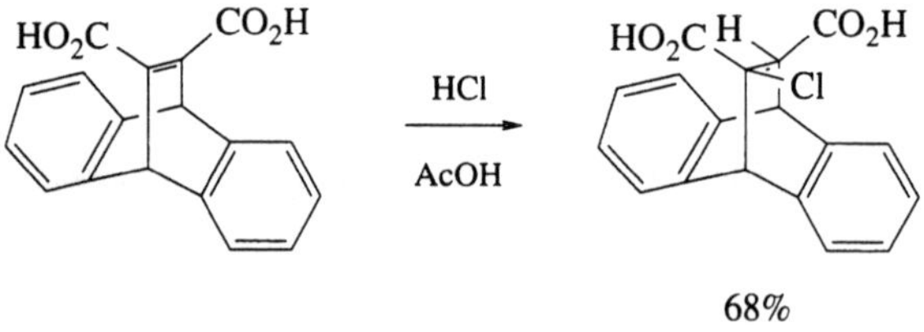

(ref. 57) (30)

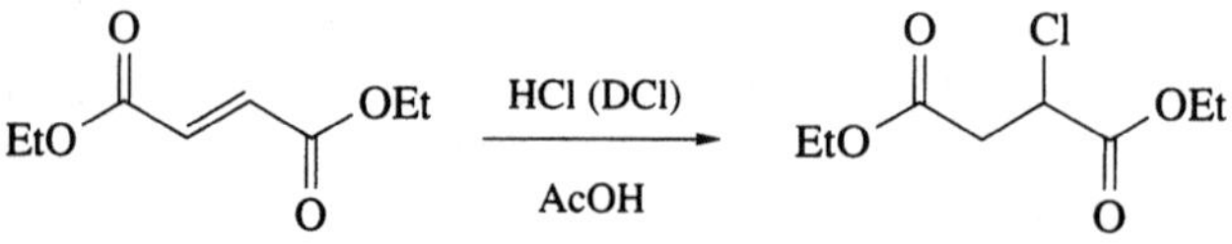

(ref. 60) (31)

With less reactive alkenes, particularly those bearing one or more halogens, it is often advisable to add a Lewis acid catalyst to the reaction (equations 32–41).[61–66] The metal halides $AlCl_3$, $FeCl_3$, $SnCl_4$ and $BiCl_3$ have been most commonly employed as catalysts.[29]

HCl / $FeCl_3$ / 79–89% (ref. 61) (32)

HCl / $AlCl_3$ / 23% (ref. 62) (33)

HCl / 67% (ref. 63) (34)

HCl / $AlCl_3$ / 68% (ref. 63) (35)

HCl / 56% (ref. 63) (36)

HCl / 91% (ref. 64) (37)

HCl / $FeCl_3$ / 77–94% (ref. 61) (38)

HCl / $FeCl_3$ / 90–100% (ref. 65) (39)

HCl / $AlCl_3$ / 72% (ref. 66) (40)

HCl / $AlCl_3$ 64% (ref. 66) / $FeCl_3$ 49% (ref. 61) (41)

Occasionally aqueous solutions of HCl have been employed in the hydrochlorination of alkenes (equations 42 and 43).[67,68] The latter procedure appears quite general and avoids the difficulties of working with gaseous HCl.

conc. HCl / H_2O → 80% (ref. 67) (42)

$$\text{n-C}_6\text{H}_{13}\text{CH=CH}_2 \xrightarrow[\text{115 °C}]{\text{HCl, H}_2\text{O} + \text{cat. n-C}_{16}\text{H}_{33}\text{PBu}^n_3\text{ Br}^-} \text{n-C}_6\text{H}_{13}\text{CHClCH}_3 \;\; 68\%$$ (ref. 68) (43)

Very little work on the addition of HCl to nonconjugated dienes has been reported. The reaction of divinyl sulfide with HCl can be controlled to give either the mono- or di-chloride (equation 44).[69,70]

(44)

Norbornadiene affords chiefly the alkenyl chloride, but significant amounts of nortricyclyl product are observed (equation 45).[71]

HCl, 72%; >80% + <20% (45)

A few examples of HCl addition to conjugated dienes have been reported.[29] 1,3-Butadiene[29,72] and isoprene[29,73] produce allylic chloride mixtures of 1,2- and 1,4-addition, while 1,3-cyclopentadiene[74] affords 3-chlorocyclopentene in excellent yield (equation 46).[74]

HCl, 70–90% (46)

Considerably more work has been reported on the addition of HCl to allenes.[75,76] Bismuth trichloride has frequently been added to catalyze these reactions. 1,2-Propadiene affords mixtures of 2-chloro-1-propene, 2,2-dichloropropane, propyne, and 1,3-dichloro-1,3-dimethylcyclobutane.[75–77]

Monosubstituted allenes afford predominantly 2-chloro-2-alkenes (equation 47),[73,75,76,78] although 1-phenyl-1,2-propadiene forms exclusively cinnamyl chloride (equation 48).[79]

HCl (47)

HCl, >98% (48)

1,3-Disubstituted allenes generally produce mixtures of vinylic and allylic chlorides (equation 49),[75,76,80,81] although 1,2-cyclononadiene forms the allylic chloride exclusively (equation 50).[80]

HCl (49)

HCl, $BiCl_3$ (50)

1,1-Disubstituted[73,76] and trisubstituted[76,82] allenes give rise to mixtures of regioisomeric allylic chlorides (equations 51 and 52).

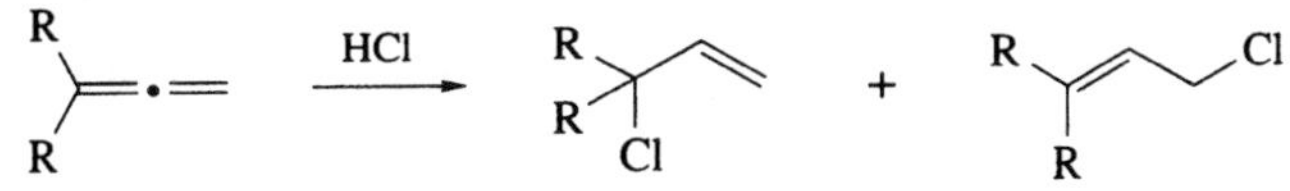

(51)

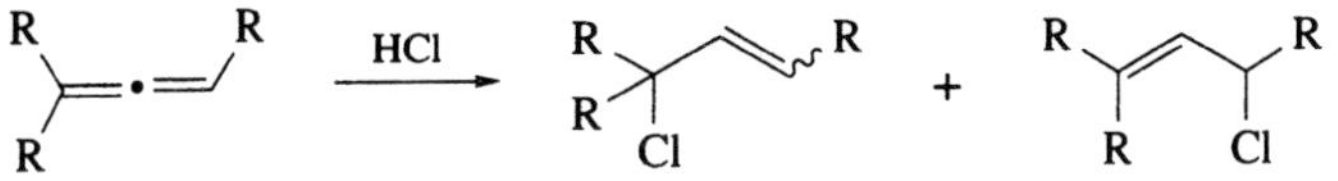

(52)

A few functionally substituted allenes have been reported to undergo HCl addition. Tetrafluoroallene and tetrachloroallene produce allylic chlorides (equation 53), while hydrochlorination of Bu^{t}C-Me=C=CCl$_2$ and (p-ClC$_6$H$_4$)$_2$C=C=CCl$_2$ yields Bu^{t}CMeClCH=CCl$_2$ and (p-ClC$_6$H$_4$)$_2$C=CHCCl$_3$ respectively.[83]

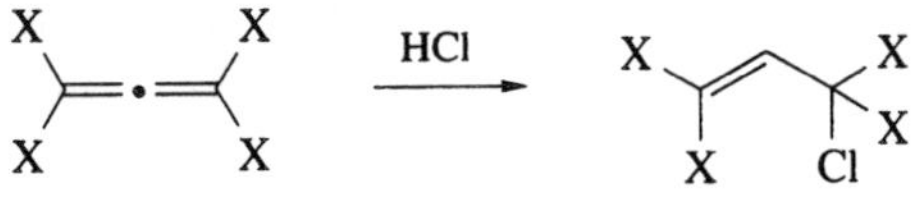

X = F, Cl

(53)

Allenic nitriles and esters provide nonconjugated products (equations 54 and 55).[76]

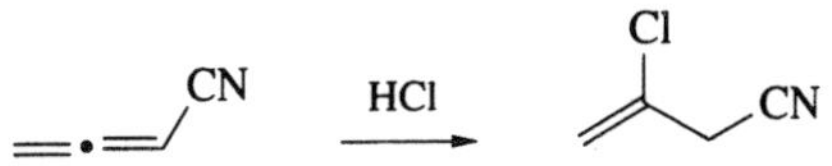

(54)

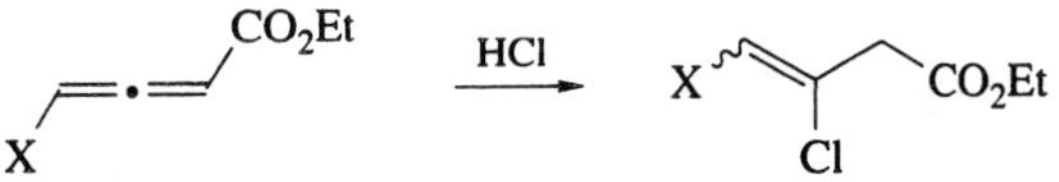

X = Br, I

(55)

The addition of HCl to alkynes has received considerable attention.[29] Numerous procedures, usually involving a metal chloride catalyst, have been developed for the industrially important conversion of acetylene to vinyl chloride.[29] The addition of HCl or DCl to terminal alkylalkynes generally produces 2-chloro-1-alkenes plus 2,2-dichloroalkanes (equation 56).

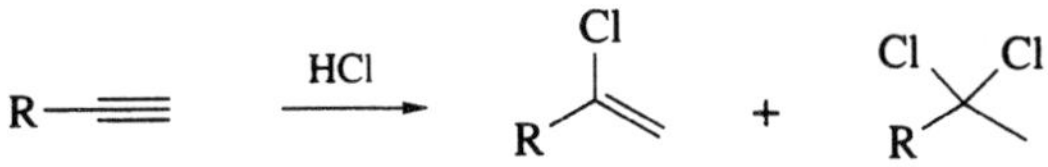

(56)

The reaction has been carried out using HCl either neat or in acetic acid,[84] or in the presence of HgCl$_2$,[85] BiCl$_3$[86] or Me$_4$NCl.[84] In acetic acid, mixtures of *syn* and *anti* adducts are produced,[84] but reactions catalyzed by HgCl$_2$ proceed by stereospecific *anti* addition.[85,87] The reaction of *t*-butylacetylene affords the expected vinylic chloride, plus a mixture of isomeric dichloride adducts.[84,88] The hydrochlorination of phenylacetylene produces mixtures of *syn* and *anti* adducts in which the ratio is significantly affected by the solvent and catalyst used.[84,89,90]

Symmetrical dialkylalkynes react with HCl in acetic acid to produce *anti* addition products plus ketone (equation 57).[30,91,92] The *anti* adducts are favored by higher HCl concentrations, or the addition of water or Me$_4$NCl.

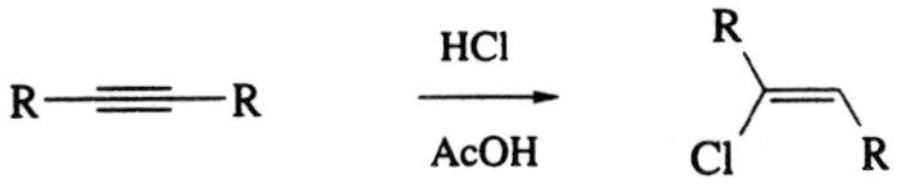

(57)

Unsymmetrical dialkylalkynes react with HCl in acetic acid plus Me$_4$NCl to give regioisomeric *anti* adducts,[84] while 1-phenyl-1-alkynes afford predominantly *syn* adducts (equation 58).[84,90–93]

$$\mathrm{Ph{-}C{\equiv}C{-}R} \xrightarrow{\mathrm{HCl}} \mathrm{Ph(Cl)C{=}CHR} \tag{58}$$

Vinylacetylene reacts with HCl, CuCl and NH_4Cl to give 2-chloro-1,3-butadiene.[29,94] Halogenated alkynes also undergo HCl addition (equations 59 and 60).[25]

$$\mathrm{F_3C{-}C{\equiv}CH} \xrightarrow[100\%]{\mathrm{HCl}} \mathrm{F_3C{-}CH{=}CH{-}Cl} \tag{59}$$

$$\mathrm{F_3C{-}C{\equiv}C{-}CF_3} \xrightarrow[\mathrm{AlCl_3},\ 78\%]{\mathrm{HCl}} \mathrm{F_3C{-}CH{=}C(Cl){-}CF_3} \tag{60}$$

A wide variety of functionally substituted alkynes react with HCl to produce vinylic chlorides (equations 61–65).[29,95,96] Both *syn*[29] and *anti*[97,98] adducts have been reported to predominate in the reaction of alkynoic acids.

$$\mathrm{HOCH_2C{\equiv}CH} \xrightarrow[\mathrm{HgCl_2}]{\mathrm{HCl}} \mathrm{HOCH_2C(Cl){=}CH_2} \quad \text{(ref. 29)} \tag{61}$$

$$\mathrm{RO{-}C{\equiv}CH} \xrightarrow{\mathrm{HCl}} \mathrm{RO{-}C(Cl){=}CH_2} \quad \text{(ref. 29)} \tag{62}$$

$$\mathrm{Bu^n{}_2P{-}C{\equiv}C{-}R} \xrightarrow{\mathrm{HCl}} \mathrm{Bu^n{}_2P{-}CH{=}C(Cl)R} \quad \text{(ref. 95)} \tag{63}$$

$$\mathrm{EtO_2C{-}C{\equiv}C{-}CO_2Et} \xrightarrow[\mathrm{AcOH},\ 76\%]{\mathrm{LiCl}} \mathrm{Et_2OC(Cl)C{=}CH{-}CO_2Et} \quad \text{(ref. 96)} \tag{64}$$

$$\mathrm{R{-}C{\equiv}C{-}CO_2H} \xrightarrow[\mathrm{H_2O}]{\mathrm{HCl}} \mathrm{R(Cl)C{=}CH{-}CO_2H} \tag{65}$$

A wide variety of alkynes have also been reported to react with $(Et_3NH)HCl_2$ either neat or in chloroform (equation 66).[99] *Anti* adducts predominate (R^1, R^2 = CO_2Me, CO_2Me; CH_2Cl, H; Ph, H; Ph, Me), except when R^1 = Ph and R^2 = Bu^t.

$$\mathrm{R^1{-}C{\equiv}C{-}R^2} \xrightarrow{\mathrm{(Et_3NH)HCl_2}} \mathrm{R^1(Cl)C{=}CHR^2}\ (anti) + \mathrm{R^1(Cl)C{=}CHR^2}\ (syn) \tag{66}$$

The addition of HCl across the triple bond of PhC≡CX (X = Cl, Br) can even be effected using Et_4NCl in aqueous DMSO (equation 67).[100]

$$\mathrm{Ph{-}C{\equiv}C{-}X} \xrightarrow[\mathrm{H_2O,\ DMSO}]{\mathrm{Et_4NCl}} \mathrm{Ph(Cl)C{=}CHX} \qquad X = \mathrm{Cl, Br} \tag{67}$$

1.7.2.4 H—Br Addition

The electrophilic addition of HBr to alkenes has been extensively studied.[30,101] The reaction is quite facile, but care must be taken to avoid competing anti-Markovnikov free radical chain processes by omitting light and free radical initiators such as peroxides.[102–104] The addition of HBr to alkenes is exothermic and considerably more rapid than HCl addition. Hydrobromination has usually been carried out neat, in inert solvents, or in water or acetic acid. Recently, good results have also been reported using excess aqueous HBr at 115 °C in the presence of a catalytic amount of n-$C_{16}H_{33}PBu^n{}_3Br$.[68] Catalysts are generally unnecessary except with the less reactive alkenes, where Lewis acids might be added.

The literature prior to the observation of competitive free radical processes initiated by peroxides and/or light is often contradictory. When the reaction is properly carried out, however, high yields of Markovnikov addition products can be achieved from a wide variety of simple alkenes (equations 68–71).[61,101,105–111] Unsymmetrical alkenes such as 2-pentene give mixtures of regioisomers.[112–114]

HBr (68)

R	*Yield (%)*	*Ref.*	*R*	*Yield (%)*	*Ref.*
Me	89–98	61	$Ph(CH_2)_2$	80–85	108
Et	100	105	$PhCH_2$	92	108
Pr^n	≥84	106	Ph	65–85	109
Me_3CCH_2	100	107			

$H_2C{=}C$ $(CH_2)_n$ —HBr→ Br–C $(CH_2)_n$ (69)

n	*Yield (%)*	*Ref.*
3	61, 72	67, 110
4	53	110
5	65	110
6	63	110

HBr, 83% (ref. 31) (70)

HBr (ref. 111) (71)

X	*Y*	*Yield (%)*
H	H	100
Me	H	89
Me	Me	79
Ph	H	91

On occasions rearrangements have been observed during hydrobromination (equations 72 and 73).[31,48,115]

The stereochemistry of HBr addition to alkenes has been thoroughly studied. The reaction of *cis*- or *trans*-2-butene with DBr in DOAc gives a mixture consisting of 60% *threo* and 40% *erythro* products.[116] While cyclohexene[117] and alkene **(1)**[118] undergo predominant *anti* addition, the alkenes **(4)**–**(6)** give

HBr, 65%; products: 52% (Br on secondary carbon) + 48% (Br on tertiary carbon) (ref. 31) (72)

DBr, D_2O, >80%; products: Br, D 46% + Br, D 3% + Br, D 51% (73) (refs. 48, 115)

stereoisomeric mixtures. Acenaphthalene,[45] indene,[119] and *cis*- or *trans*-1-phenyl-1-propene[120] undergo 74–90% *syn* addition of DBr, while 1-phenyl-4-*t*-butylcyclohexene gives a mixture of stereoisomers.[46]

(1) **(4)** (ref. 118) **(5)** (ref. 118) **(6)** (ref. 41)

The hydrobromination of a number of halogenated alkenes has been examined. Allyl bromide reacts with HBr in the absence of peroxides, or upon addition of antioxidants[121] or $FeCl_3$,[61] or in acetic acid,[122] to produce primarily 1,2-dibromopropane (equation 74). Allyl chloride behaves similarly.[123]

X —HBr→ Br, X (74)

X = Cl, Br

On the other hand, 3,3,3-trifluoro-1-propene,[62,101] 3,3,3-trifluoro-2-trifluoromethyl-1-propene[63] and 3,3,4,4,5,5,5-heptafluoro-1-pentene[124] react with HBr to afford exclusively the terminal bromides (equations 75–77).

CF_3 —HBr, $AlBr_3$, 35%→ Br CF_3 (75)

CF_3, CF_3 —HBr, $AlBr_3$, 81%→ Br CF_3, CF_3 (76)

CF_2 CF_2 CF_3 —HBr, act. C, 85%→ Br CF_2 CF_2 CF_3 (77)

Vinylic halides are less reactive than simple alkenes, but generally afford clean addition products. The addition of HBr to vinyl chloride[125] or vinyl bromide[61,122,126,127] in the presence of iron(III) chloride or bromide affords primarily the geminal dihalides (equation 78).

With other vinylic halides, the substitution pattern determines the direction of HBr addition (equations 79–83).[61,65,128–130]

Polyhalogenated alkenes also undergo HBr addition (equations 84–87).[61,63,131] Even tetrahaloethylenes undergo HBr addition in a continuous flow reactor in the presence of a catalyst.[132]

HBr, FeX_3; X = Cl, Br (78)

HBr, $FeCl_3$, 92–100%; X = Cl, Br (refs. 61, 65) (79)

HBr → (7) + (8) (refs. 61, 65) (80)

X	*Additive*	*Yield* (%)	*(7):(8)*
Cl	–	60–100	≥98:≤2
Cl	$FeCl_3$	90	≥90:≤10
Br	–	70	≥92:≤8
Br	$FeCl_3$	61–95	~66:~34

HBr, $FeCl_3$; 85–90% + 10–15% (ref. 128) (81)

HBr, $FeCl_3$ (ref. 129) (82)

X	*Yield* (%)
Cl	69–90
Br	75–80

HBr, AcOH (ref. 130) (83)

HBr (ref. 131) (84)

HBr, $AlBr_3$, 53% (ref. 63) (85)

HBr, $AlBr_3$, 45% (ref. 63) (86)

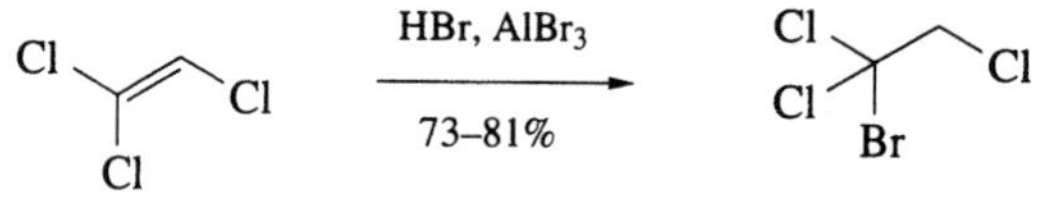

(ref. 61) (87)

A number of functional groups including sulfones, ethers and even alcohols (occasionally) can be accommodated during hydrobromination. Functional groups in close proximity to the C—C double bond can have an important directing effect (equations 88 and 89).[133,134]

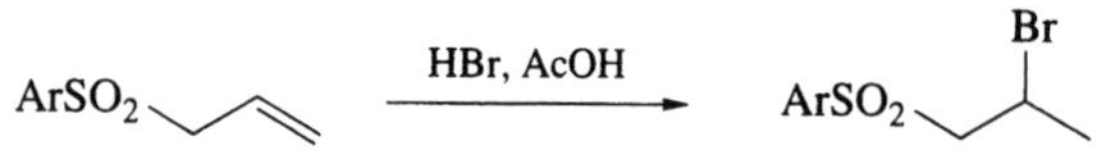

(ref. 133) (88)

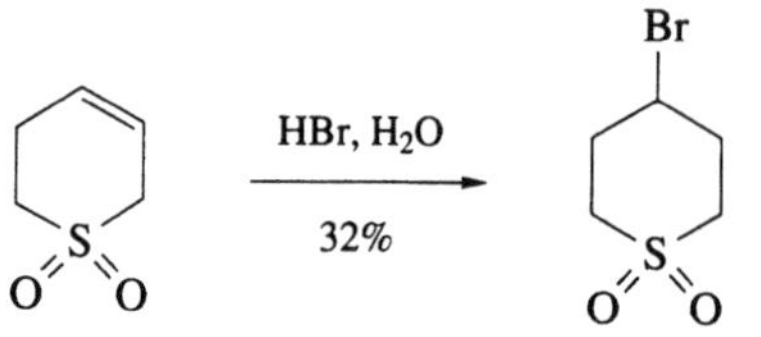

(ref. 134) (89)

α,β-Unsaturated aldehydes,[101] ketones[101] and esters[109,135,136] afford β-bromo carbonyl products, often in excellent yield (equation 90). Even tribenzoylethylene reacts in fair yield (equation 91).[137]

(90)

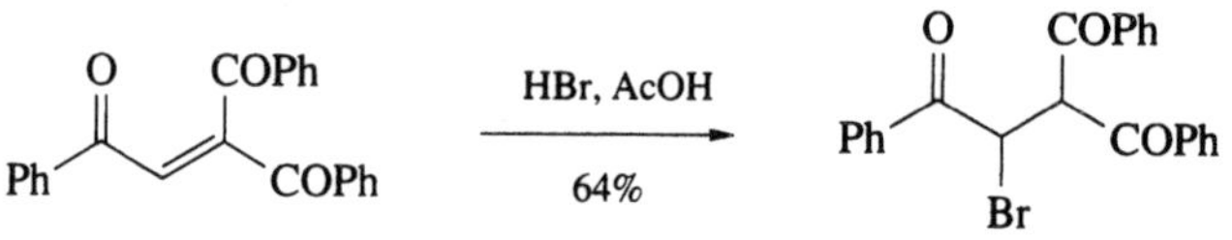

(91)

A wide variety of unsaturated carboxylic acids have been allowed to react with HBr.[101] Carboxylic acids with remote C—C double bonds react as simple alkenes.[138,139] 4-Pentenoic acid reacts with HBr neat or in a polar solvent to give exclusively 4-bromopentanoic acid, but the reaction in nonpolar solvents affords only 5-bromopentanoic acid.[136] On the other hand, 5-methyl-4-hexenoic acid produces only the 5-bromo acid. A similar pattern is followed by 3-butenoic, 3-pentenoic and 4-methyl-3-pentenoic acids. No matter what the substitution pattern, 2-alkenoic acids always favor the 3-bromo acid.[113,136,140,141] Addition of HBr to cyclic α,β-unsaturated acids initially forms predominantly the product of *trans* diaxial addition which upon longer reaction time or higher temperature isomerizes to the *trans* product (equation 92).[57,141,142] Similar observations have been made on bicyclic α,β-unsaturated acids.[141]

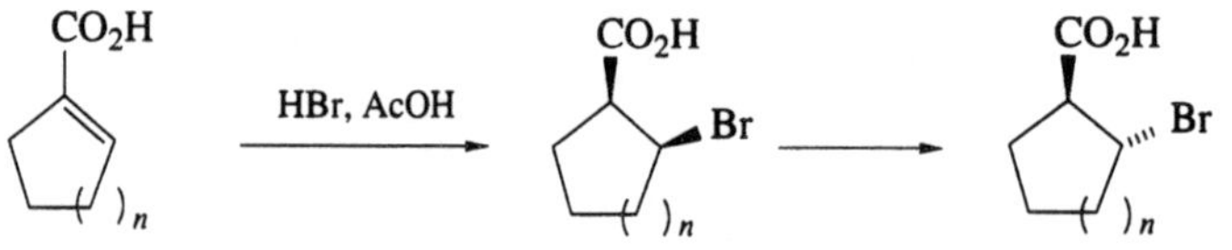

(92)

The stereochemistry of addition to unsaturated dicarboxylic acids has also been examined (equations 93 and 94).[130,143]

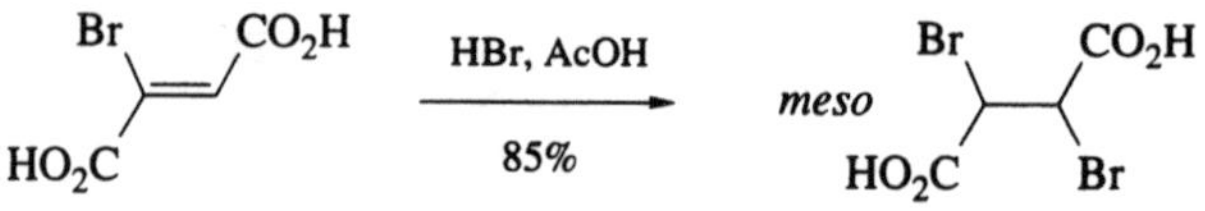

(ref. 130) (93)

Addition of HBr to unsaturated nitriles can also proceed in good yield (equation 95).[59]

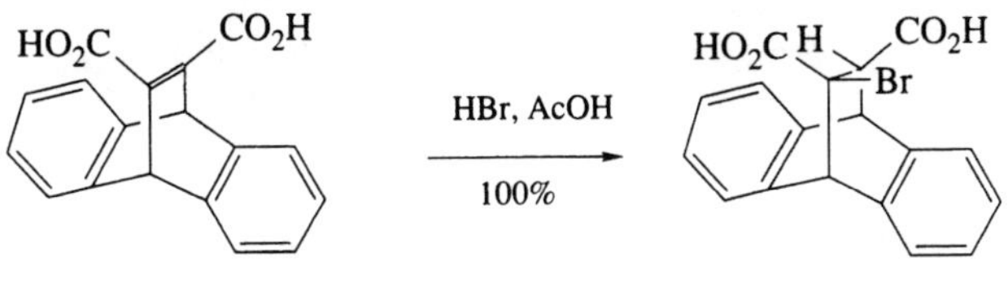

HBr, AcOH, 100% (ref.143) (94)

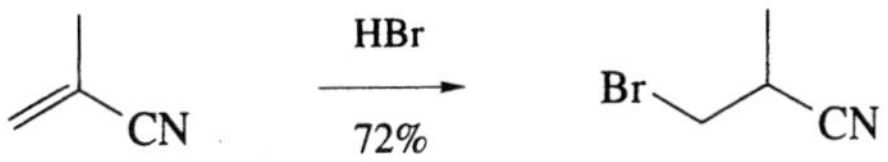

HBr, 72% (95)

Nonconjugated dienes react as expected to give mono- or di-addition products (equations 96–98).[70,144,145] Norbornadiene also produces nortricyclyl product (equation 99).[71]

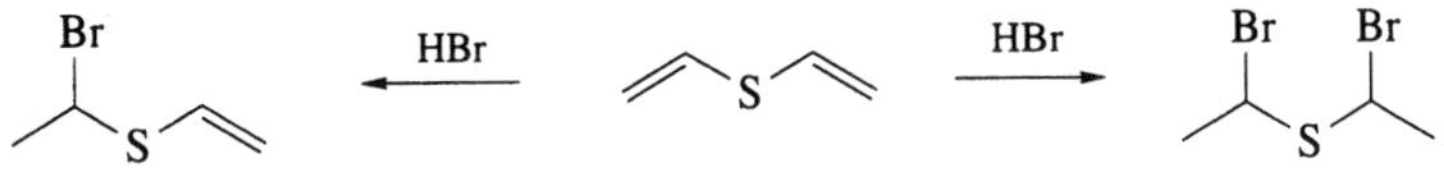

HBr (ref. 70) (96)

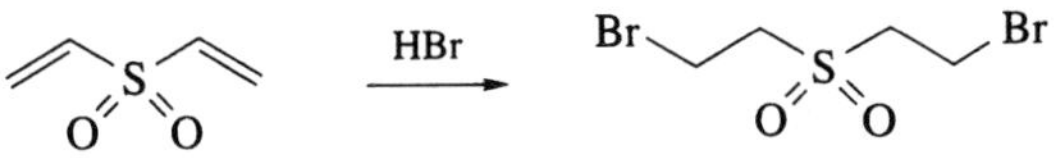

HBr (ref. 144) (97)

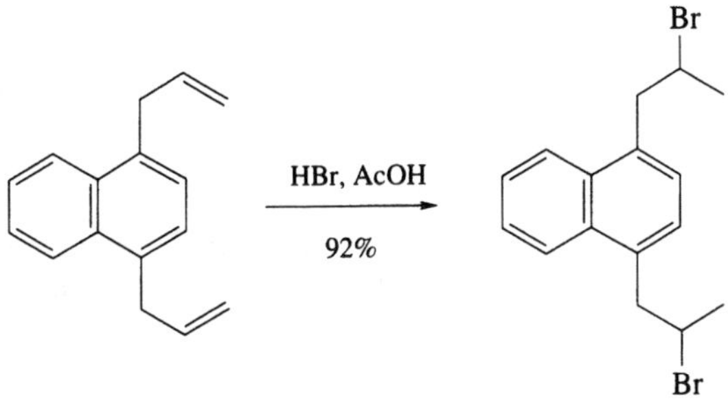

HBr, AcOH, 92% (ref. 145) (98)

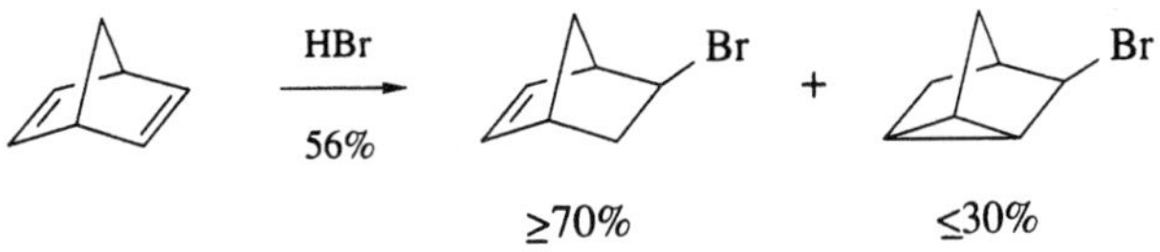

HBr, 56% ≥70% + ≤30% (99)

Conjugated dienes generally produce mixtures of 1,2- and 1,4-addition compounds.[101] 1,3-Butadiene reacts rapidly with HBr even at low temperatures and gives mixtures in which either 1-bromo-2-butene or 3-bromo-1-butene predominate, depending on the reaction conditions (equation 100.)[72,146,147] Further HBr addition gives mixtures of 1,3-dibromobutane and 2,3-dibromobutane.[147]

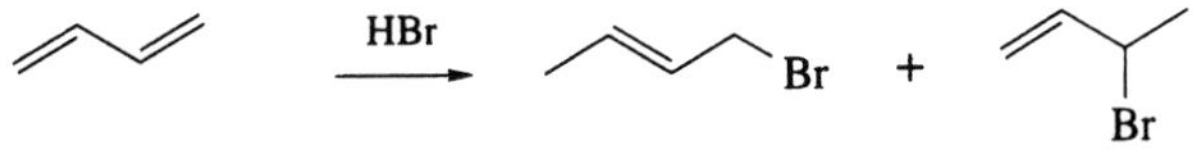

HBr (100)

Isoprene initially forms 3-bromo-3-methyl-1-butene, which rapidly rearranges to 1-bromo-3-methyl-2-butene (equation 101).[148,149]

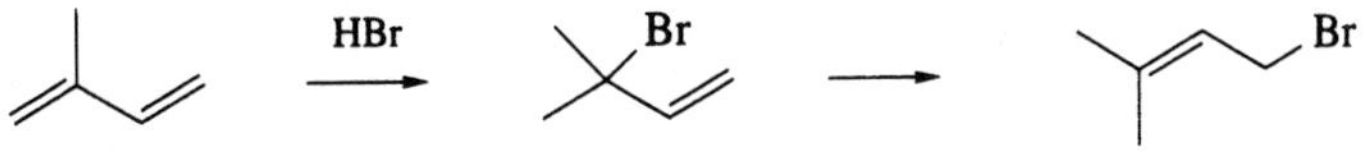

HBr (101)

2,3-Dimethyl-1,3-butadiene cleanly affords 1-bromo-2,3-dimethyl-2-butene,[150] but further HBr addition produces a mixture of dibromides (equation 102).[151]

The addition of HBr to 2-methyl-1,3-pentadiene, 1-bromo-1,3-butadiene, 1-phenyl-1,3-butadiene and 2,4-hexadiene produces 2,4-dibromo-2-methylpentane, 1,3-dibromo-1-butene, 3-bromo-1-phenyl-1-

(102)

butene and 2-bromo-3-hexene plus 4-bromo-2-hexene, respectively.[101] The addition of DBr to 1,3-cyclohexadiene provides allylic bromide in 85% yield; both 1,4-*syn* and 1,2-*anti* addition are suggested to occur, followed by stereospecific rearrangement.[152]

Numerous studies of the addition of HBr to allenes have been reported.[75,76] The low temperature addition of HBr to allene produces a mixture of products (equation 103).[77] However, gas phase photocatalyzed addition generates only the vinylic bromide, in near quantitative yield.[153]

(103)

Terminal monosubstituted allenes afford primarily (*E*)- and (*Z*)-2-bromo-2-alkenes using either approach (equation 104).[153–155] However, terminal disubstituted allenes give quite different products (equation 105).[82,153,155]

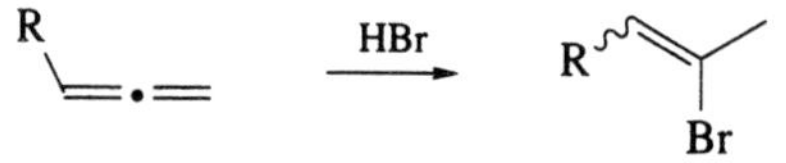

(104)

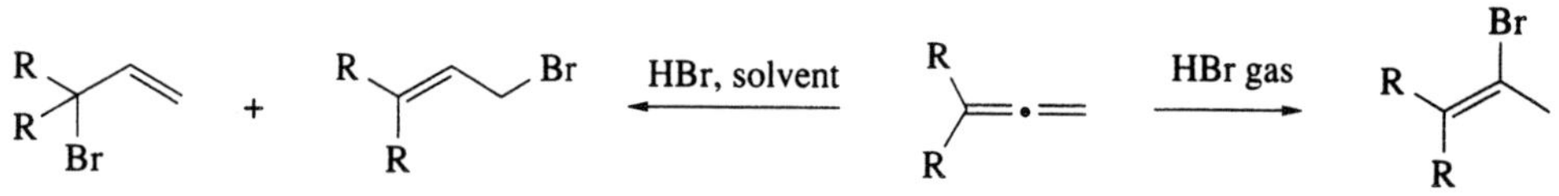

(105)

Symmetrical 1,3-disubstituted allenes give predominantly vinylic bromide or allylic bromide depending upon reaction conditions (equation 106).[82,153,154] Analogous unsymmetrical allenes give mixtures of all four possible bromides.[81,82]

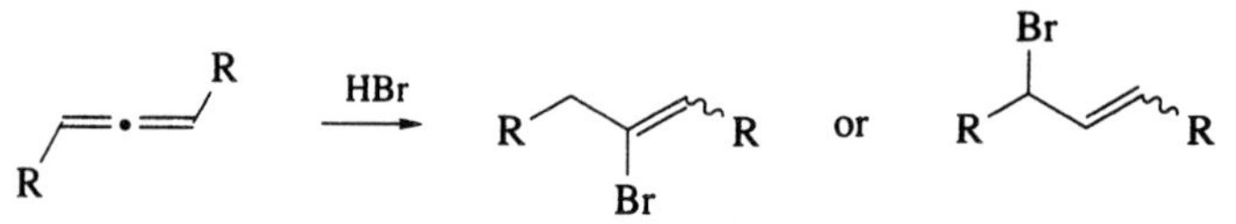

(106)

Totally different products are observed from aqueous[82] and gas phase[153] HBr addition to tetrasubstituted allenes (equation 107).

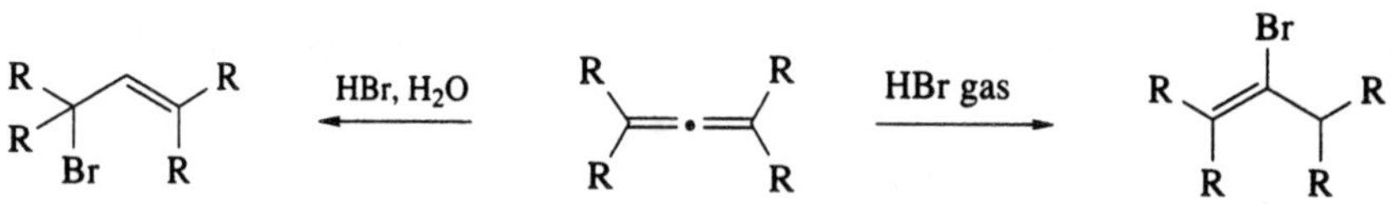

(107)

The products of addition of HBr to cyclic allenes is highly dependent on the size of the ring and the solvent used (equation 108).[154,156] 1,2-Cyclononadiene affords rather selectively either product.

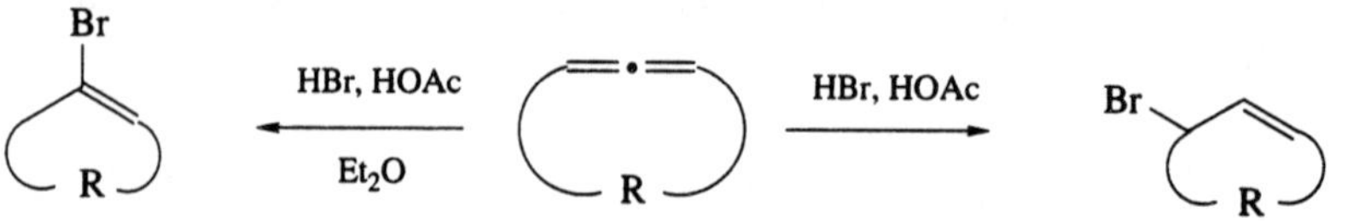

(108)

1,2-Cyclodecadiene and 1,2,6-cyclononatriene have been converted to the corresponding allylic bromide and vinylic bromide respectively, while 1,2-cyclotridecadiene reacts with HBr in HOAc to give a mixture of bromides.

The hydrobromination of functionally substituted allenes has received little attention. Vinylic allenes generate mixtures of bromides,[76] while tetrafluoroallene affords the allylic bromide,[83] and allenic acids produce 3-bromo-3-alkenoic acids (equation 109).[157,158]

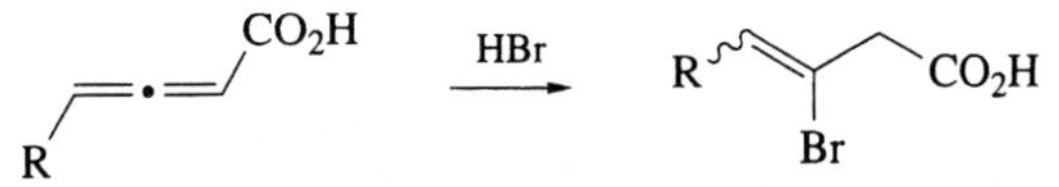

(109)

There are numerous examples of the addition of HBr to alkynes.[101] Addition to acetylene is difficult, tends to produce mixtures of bromide products, and requires a catalyst.[94,101,122] Simple terminal alkynes react with HBr in the absence of peroxides to produce mixtures of 2-bromo-1-alkenes and geminal dibromides (equation 110).[85,159] The latter can be produced in high yield when excess HBr is used.[160]

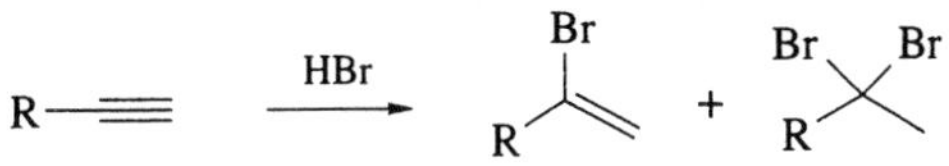

(110)

Mercury(II) bromide[85] and iron(III) bromide[159] have been employed as catalysts in these reactions. The former catalyst affords stereospecific *trans* adducts.[85]

In the presence of peroxides, 1-bromo-1-alkenes or 1,2-dibromoalkanes are produced depending on the amount of HBr employed (equation 111).[85,159,160]

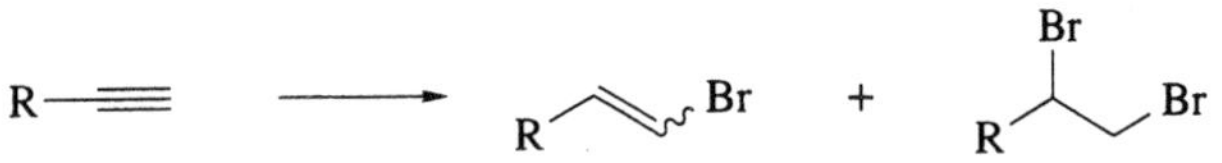

(111)

The hydrobromination of arylalkynes in acetic acid produces good yields of vinylic bromides in which *syn* addition is favored over *anti* addition by ~3 to 1 (equation 112).[90,161,162]

(112)

3,3,3-Trifluoro-1-propyne readily adds HBr in an anti-Markovnikov fashion (equation 113).[25,163]

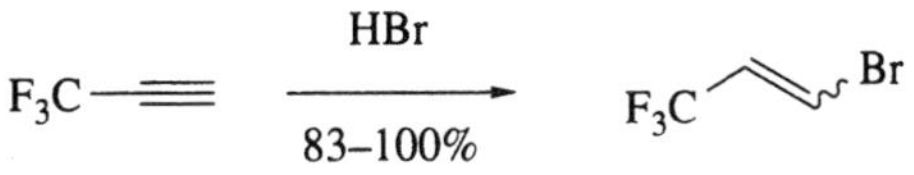

(113)

The stereochemistry of HBr addition to 1-alkyn-3-ones can be controlled by proper choice of reagents (equation 114).[164]

(114)

	% *cis*	% *trans*
LiBr, CF_3CO_2H (30 min)	1	89
LiBr, CH_3CO_2H (overnight)	68	12

Alkynoic acids in which the C—C triple bond is far removed from the acid moiety behave like simple alkynes.[165,166] Propiolic acid, on the other hand, adds HBr to form almost exclusively (*Z*)-3-bromopropenoic acid (equation 115).[98]

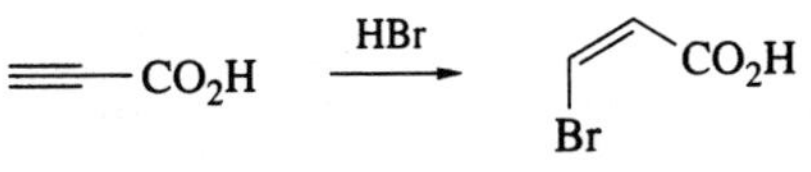

(115)

Surprisingly, no work has apparently been carried out on the addition of HBr to simple dialkylalkynes. Hexafluoro-2-butyne readily adds HBr in the presence of $AlBr_3$ (equation 116).[25]

$$F_3C{-}{\equiv}{-}CF_3 \xrightarrow{\text{HBr}} F_3C{-}CH{=}C(Br){-}CF_3 \quad (116)$$

Diarylalkynes add HBr in an *anti* fashion (equation 117).[130]

$$Ar{-}{\equiv}{-}Ar \xrightarrow[\text{90–95\%}]{\text{HBr, AcOH}} Ar(Br)C{=}CH(Ar) \quad (117)$$

1-Halo-1-alkynes react with HBr–$HOAc$[167] or HBr–Cu_2Br_2[168,169] to afford primarily *anti* adducts (equation 118).

$$R{-}{\equiv}{-}X \xrightarrow{\text{HBr}} R(Br)C{=}CH(X) \quad (118)$$

X = Cl, Br

Even aqueous Et_4NBr will hydrobrominate PhC≡CX, where X = Cl or Br, although the yield is low (equation 119).[100]

$$Ph{-}{\equiv}{-}X \xrightarrow[\text{DMSO}]{Et_4NBr,\ H_2O} Ph(Br)C{=}CH(X) \quad (119)$$

X = Cl, Br

Anti addition products have also been reported for a wide variety of functionally substituted alkynes (equations 120 and 121).[95,97,170,171]

$$Bu^n{}_2P{-}{\equiv}{-} \xrightarrow{\text{HBr}} Bu^n{}_2P{-}CH{=}C(Br){-}CH_3 \quad \text{(ref. 95)} \quad (120)$$

$$R{-}{\equiv}{-}CO_2H \xrightarrow{HBr,\ H_2O} R(Br)C{=}CH(CO_2H) \quad \text{(refs. 97, 170, 171)} \quad (121)$$

R = H, Me, Ar, CO_2H

However, HBr addition to tetrolic acid and phenylpropiolic acid in benzene is reported to afford the opposite regioisomers (equation 122),[170] and the stereochemistry of HBr addition to $PhC{\equiv}CCO_2Et$[171] and NCC≡CCN[172] is unclear.

$$R{-}{\equiv}{-}CO_2H \xrightarrow{\text{HBr, PhH}} R{-}CH{=}C(Br){-}CO_2H \quad (122)$$

R = Me, Ph

Recently, two very useful procedures for the conversion of terminal alkynes to 2-bromo-1-alkenes have been reported (equations 123 and 124).[173,174] Both procedures accommodate alcohol groups as well.

$$R{-}{\equiv} \xrightarrow{(Et_4N)HBr_2} R(Br)C{=}CH_2 \quad \text{(ref. 173)} \quad (123)$$

$$\text{R}{-}{\equiv} \xrightarrow[\text{ii, AcOH}]{\text{i, BrBR'}_2} \text{R(Br)C=CH}_2 \qquad \text{(ref. 174)} \quad (124)$$

The addition of HBr to alkynes *via* hydrometallation–halogenation is another important approach to vinylic halides which will be discussed elsewhere in this series (see Volume 8, Chapters 3.9–3.12).

1.7.2.5 H—I Addition

Although HI addition to alkenes and alkynes is faster than that of the other hydrohalides and free radical anti-Markovnikov additions are not a problem, this reaction has received less attention than the others.[175] The hydroiodination of alkenes is most commonly run using concentrated HI in water or acetic acid at or below room temperature. While the early literature suggests that simple terminal alkenes afford small amounts of anti-Markovnikov products, only Markovnikov products have been reported in the more recent literature (equations 125–129).[67,176–179]

$$\text{RCH=CH}_2 \xrightarrow{\text{HI}} \text{RCHICH}_3 \qquad (125)$$

R	*Yield (%)*	*Ref.*
Me	79, >90	176, 177
Et	73	176
Bu^n	94	178
Me_3CCH_2	92	176

$$\text{methylenecyclobutane} \xrightarrow[94\%]{\text{HI, AcOH}} \text{1-iodo-1-methylcyclobutane} \qquad \text{(ref. 67)} \quad (126)$$

$$\text{cyclohexene} \xrightarrow[88\text{–}90\%]{\text{KI, H}_3\text{PO}_4} \text{iodocyclohexane} \qquad \text{(ref. 178)} \quad (127)$$

$$\text{Me}_2\text{C=CMe}_2 \xrightarrow[91\%]{\text{KI, H}_3\text{PO}_4} \text{Me}_2\text{CI–CHMe}_2 \qquad \text{(ref. 178)} \quad (128)$$

$$\text{vinylcyclopropane} \xrightarrow[85\%]{\text{HI, AcOH}} \text{1-cyclopropyl-1-iodoethane} \qquad \text{(ref. 179)} \quad (129)$$

While cyclopropylethylene is reported not to rearrange,[179] this appears questionable since other alkenes prone to rearrangements have been reported to do so (equation 130).[31,38,39,180]

$$\text{Me}_3\text{CCH=CH}_2 \xrightarrow{\text{HI}} \text{Me}_3\text{CCHICH}_3 + \text{Me}_2\text{CICHMe}_2 \qquad (130)$$

$$\text{CH}_2\text{=CHCH}_2\text{X} \xrightarrow[90\text{–}100\%]{\text{HI}} \text{CH}_3\text{CHICH}_2\text{X} \qquad (131)$$

X = Cl, Br

Allyl chloride and bromide afford high yields of HI addition products (equation 131).[176,177]

Even vinylic halides add HI readily, though 1-bromo-1-propene gives a mixture of products (equations 132–135).[125,131,177] However, perfluoroalkenes have proven unreactive towards HI addition.

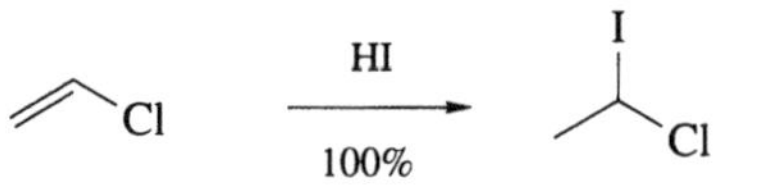

(ref. 125) (132)

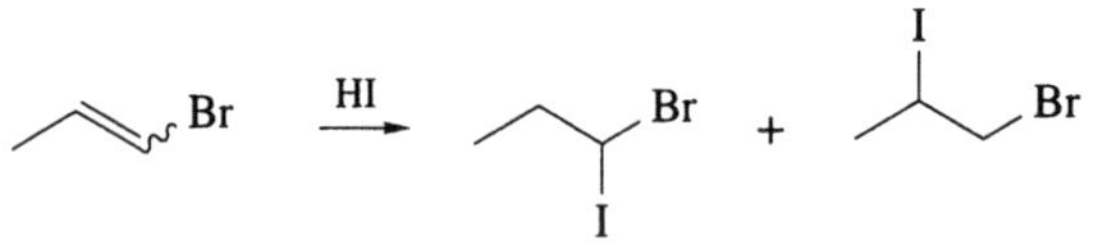

(ref. 177) (133)

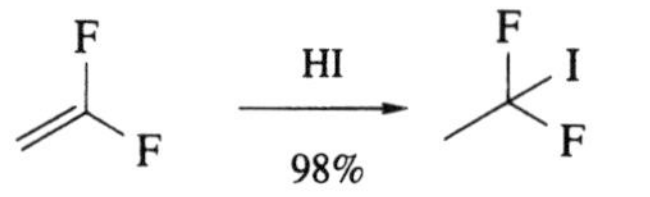

(ref. 131) (134)

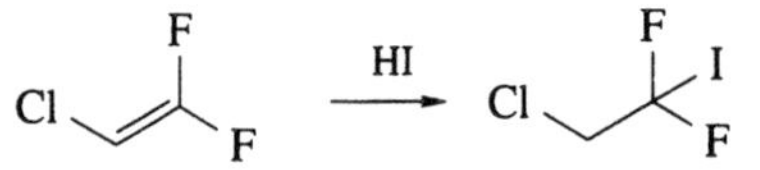

(ref. 131) (135)

Certain functional groups can be accommodated during HI addition. While alcohols are converted to alkyl iodides under the usual reaction conditions,[181] sulfones present no problems.[133,134] Remote carboxylic acid functionality has no effect on the addition of HI,[182] and α,β-unsaturated carboxylic acids produce good yields of exclusively the 3-iodocarboxylic acids, no matter what the substitution pattern about the C—C double bond (equation 136).[114,140,183–186]

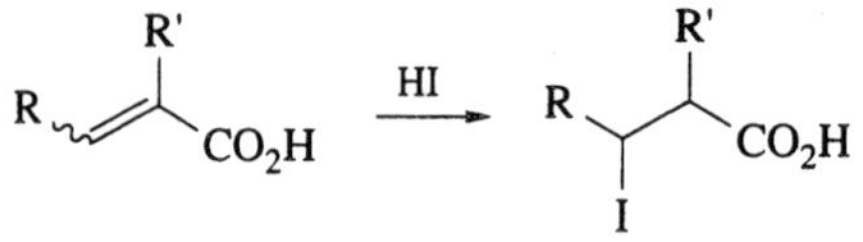

(136)

The addition of ammonium and phosphonium salts as catalysts for HI addition to alkenes has been recommended as a useful alternative procedure.[68]

While the reaction of alkenes with I_2 at temperatures of 125–130 °C reportedly produces regioisomeric mixtures of iodoalkanes,[187] in the presence of dehydrated alumina in refluxing petroleum ether, this reaction gives low to reasonable yields of Markovnikov adducts.[188,189] This approach does not appear very general, however, and rearrangement has been observed.

There are very few examples of the addition of HI to dienes (equations 137–139).[72,77,144]

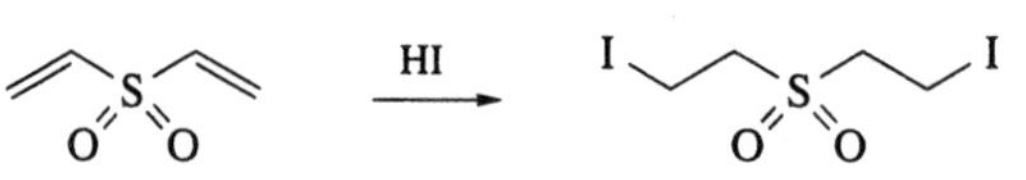

(ref. 144) (137)

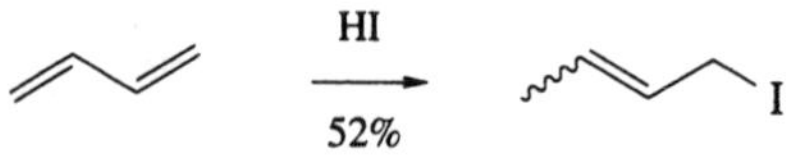

(ref. 72) (138)

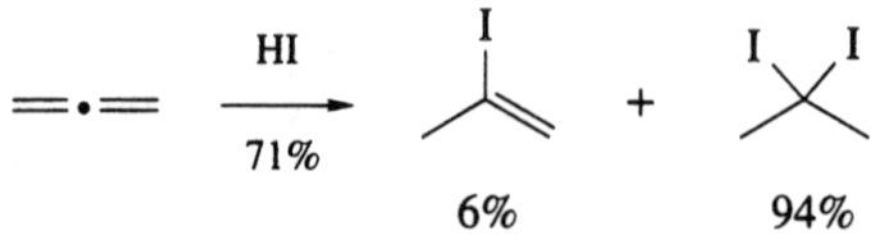

(ref. 77) (139)

Relatively little work has also been reported on the addition of HI to alkynes. 1-Propyne affords a mixture of iodides (equation 140).[77,85]

The reaction of 3,3,3-trifluoro-1-propyne requires a temperature of 100 °C or the addition of AlI_3 to produce the terminal vinylic iodide (equation 141).[25]

HI, 86%; 35%; 65% (ref. 77) (140)

F_3C—≡ HI, 65% (AlI_3, 80%) → F_3C–CH=CH–I (141)

A variety of haloenynes react with HI and Cu_2I_2 to afford products of *anti* addition (equation 142).[168,169]

$$C{=}C{-}C{\equiv}C{-}X \xrightarrow[Cu_2I_2]{HI} (C{=}C)(I)C{=}C(H)(X) \qquad X = Cl, Br, I \tag{142}$$

A variety of functionally substituted alkynes also undergo facile HI addition (equations 143–145).[95,164,190]

Bu^n_2P—≡—R → (HI) Bu^n_2P, I, R (ref. 95) (143)

RCO—≡ → RCO–CH=CH–I (*Z*) + RCO–CH=CH–I (*E*) (ref. 164) (144)

NaI, CF_3CO_2H	0%	95%
NaI, CH_3CO_2H	70%	17%

≡—CO_2Me → (HI) I–CH=CH–CO_2Me (ref. 190) (145)

While 4-pentynoic acid reportedly yields the geminal diiodide (equation 146),[165] 2-alkynoic acids generally produce almost exclusively the 3-iodo-2-alkenoic acids of *anti* addition (equation 147).[97,98,191]

≡–CH_2CH_2–CO_2H → (HI) $CH_3CI_2CH_2CH_2CO_2H$ (146)

R—≡—CO_2H → (HI) I, R, CO_2H (147)

R = H, Me

3-Bromo- and 3-iodo-propiolic acids are also reported to give the corresponding 3,3-dihalo-2-propenoic acids of unknown stereochemistry.[191]

Alkynes bearing two strong electron-withdrawing groups are still capable of undergoing HI addition at or below room temperature (equation 148).[172,192,193]

X—≡—X → (HI) XCH=C(X)I (148)

X = CO_2Me, CO_2H, CN

Terminal alkynes also undergo HI addition in good yields upon reaction with I_2 and dehydrated alumina (equation 149),[189] or by iodoboron addition followed by protonolysis (equation 150).[174] These methods appear quite useful.

$$\mathrm{Bu^n{-}C{\equiv}CH} \xrightarrow[62\%]{I_2,\ Al_2O_3} \mathrm{Bu^nC(I){=}CH_2} \qquad (149)$$

$$\mathrm{R{-}C{\equiv}CH} \xrightarrow[\text{ii, AcOH; } 80\text{–}100\%]{\text{i, IBR'}_2} \mathrm{RC(I){=}CH_2} \qquad (150)$$

1.7.3 H—N ADDITION

1.7.3.1 Introduction

There are several methods available for the electrophilic addition of hydrogen and nitrogen to alkenes, dienes and alkynes. While the direct electrophilic addition of amines to these substrates is not feasible, aminomercuration–demercuration affords a very useful indirect approach to such amines. The addition of amides to C—C multiple bonds can be effected directly through the Ritter reaction or by the less direct, but equally useful, amidomercuration–demercuration process using either nitriles or amides. Similarly, H—N_3 addition to alkenes can be carried out directly or *via* mercuration to produce organic azides.

1.7.3.2 Aminomercuration–Demercuration

There do not appear to be any direct approaches for electrophilic addition of amines across C—C double or triple bonds. However, aminomercuration–demercuration affords a very useful process to effect this transformation (equation 151). This reaction has been thoroughly reviewed recently.[194]

$$\mathrm{RCH{=}CH_2} \xrightarrow[HgX_2]{R_2NH} \mathrm{RCH(NR_2)CH_2HgX} \xrightarrow[NaOH]{NaBH_4} \mathrm{RCH(NR_2)CH_3} \qquad (151)$$

Although the relative reactivity of mercuric salts is $Hg(ClO_4)_2 \approx Hg(NO_3)_2 >> Hg(OAc)_2 > HgCl_2$,[195] the latter two salts have been used in almost all examples of this reaction reported to date. The reaction can be reversible or irreversible depending on the mercury salt employed.[196]

Most work on aminomercuration has employed secondary aliphatic amines, or primary or secondary anilines. Primary amines have been successfully employed using mercuric perchlorate or mercuric nitrate.[197] No reactions utilizing ammonia have been reported. While excess amine or various ether solvents can be used, it is sometimes desireable to add water.[198,199]

A wide variety of alkenes have been shown to undergo aminomercuration–demercuration, including alkenes bearing alcohol, ester, acid, amide, imide, urea, urethane, ether, sulfide and aryl groups.[194]

Although Markovnikov addition is generally observed, little work on the regio- or stereo-selectivity of this reaction has been reported. Both *cis*- and *trans*-2-pentene afford mixtures of the 2- and 3-substituted amines.[199] α,β-Unsaturated esters yield β-amino esters.[200] The aminomercuration of *cis*- and *trans*-2-butenes gives $\geq 97\%$ of the *anti* adducts.[201]

While reductive demercuration of the intermediate β-aminomercurials can usually be carried out *in situ* using alkaline sodium borohydride,[202,203] regeneration of the starting alkene or nitrogen migration can be a problem. Alkene formation can often be circumvented by adding 10% NaOH prior to reduction[197] and using a phase transfer catalyst.[204] This reducing system also cuts down on the problem of nitrogen migration.

The intramolecular solvomercuration of unsaturated amines provides a very useful route to a variety of nitrogen heterocycles.[194,205] This reaction is particularly useful for the preparation of pyrrolidines and piperidines and their heteroatom-substituted analogs (equation 152).

Bicyclic[206] and spirocyclic[207] amines are also readily prepared by intramolecular aminomercuration–demercuration (equations 153 and 154).

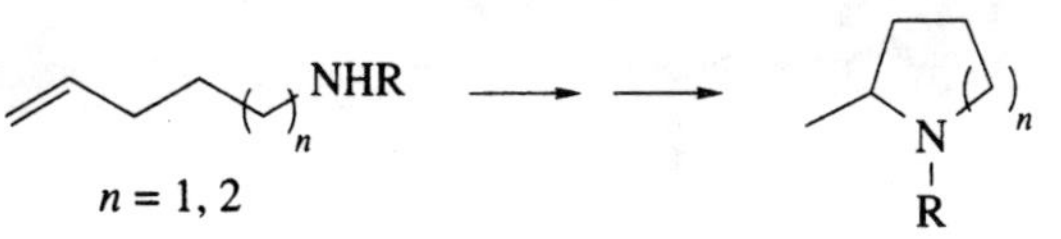

$n = 1, 2$ (152)

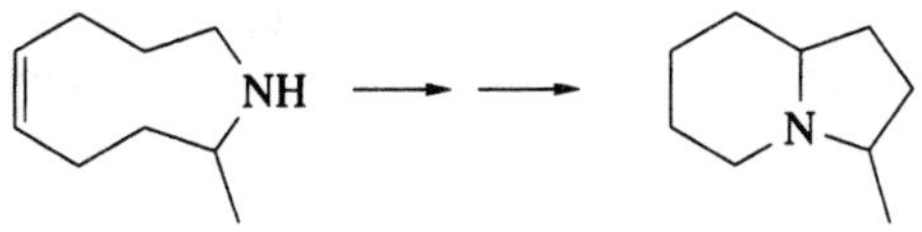

(153)

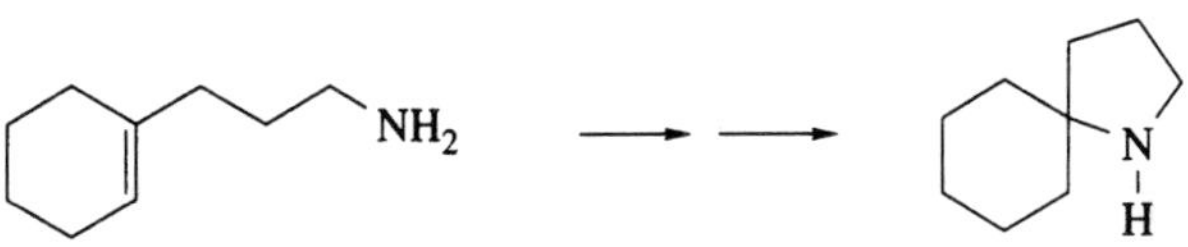

(154)

Analogous cyclic amines can also be prepared by aminomercuration–demercuration of unsaturated organic halides (equation 155).[208]

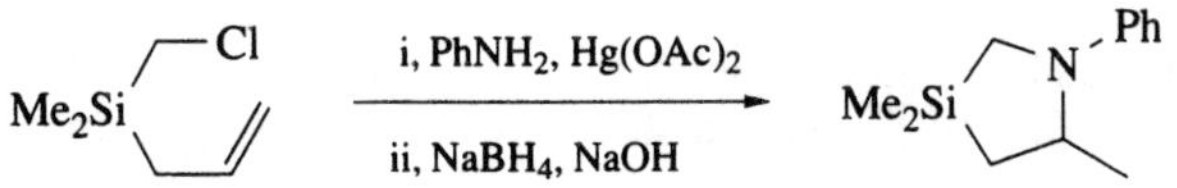

i, $PhNH_2$, $Hg(OAc)_2$; ii, $NaBH_4$, NaOH (155)

High stereoselectivity is frequently observed in these types of cyclization.

These cyclic aminomercurials are particularly prone to alkene formation and rearrangement during reduction. However, alkaline sodium borohydride under phase transfer conditions usually works well.[209,210]

Nonconjugated dienes also undergo aminomercuration–demercuration to generate pyrrolidines, piperidines and their heterocyclic analogs (equations 156–158).[211–218]

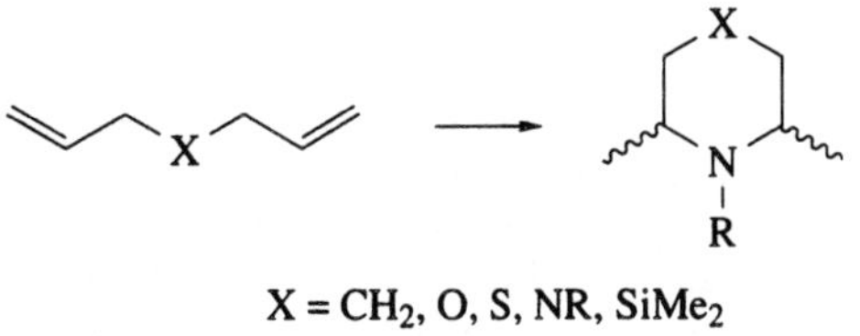

X = CH_2, O, S, NR, $SiMe_2$ (refs. 211–214) (156)

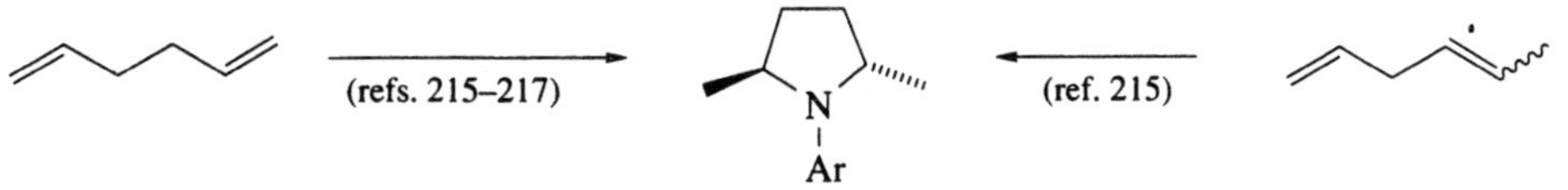

(refs. 215–217) (ref. 215) (157)

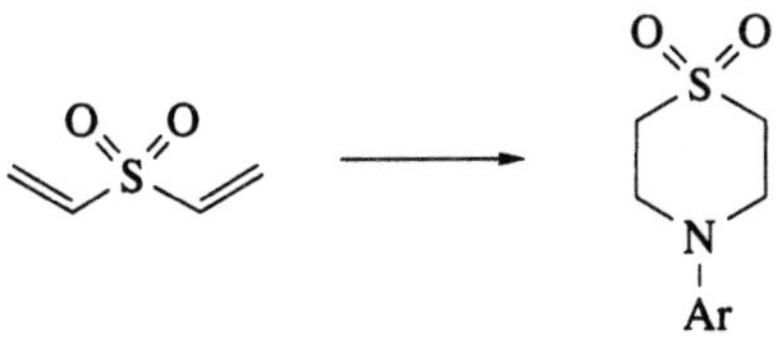

(ref. 218) (158)

Conjugated dienes afford either allylic amines or saturated diamines (equation 159).[204,219] When mercury(II) oxide plus tetrafluoroboric acid is employed, dihydropyrroles are formed instead (equation 160).[220]

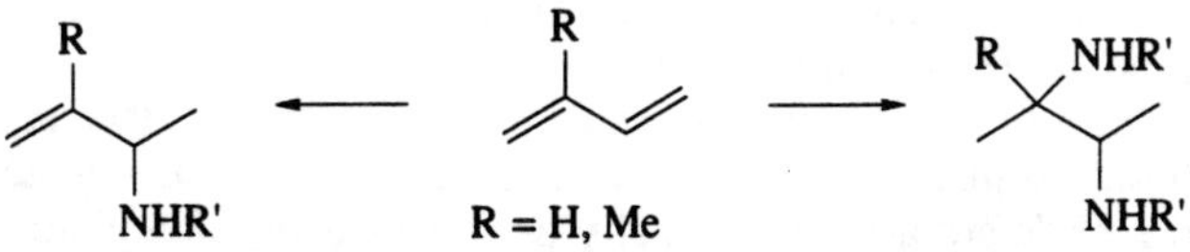

R = H, Me (159)

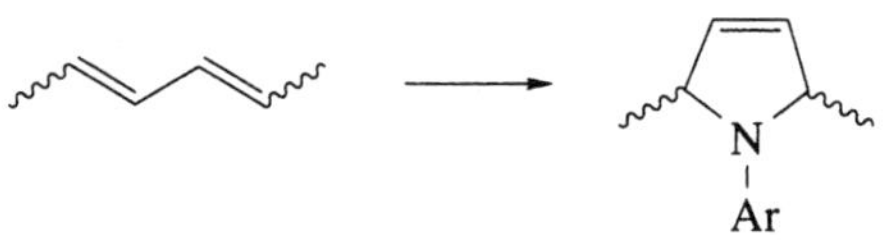

(160)

Allylic amines are usually obtained from the aminomercuration of allenes (equation 161),[204,221] while allenes with amino groups in a side chain produce the corresponding heterocycles (equation 162).[222]

(161)

$n = 3, 4$

(162)

The aminomercuration–demercuration of alkynes affords a variety of nitrogen-containing products. Most commonly imines are formed,[194] though there are a few examples of enamines being produced (equation 163).[223–228]

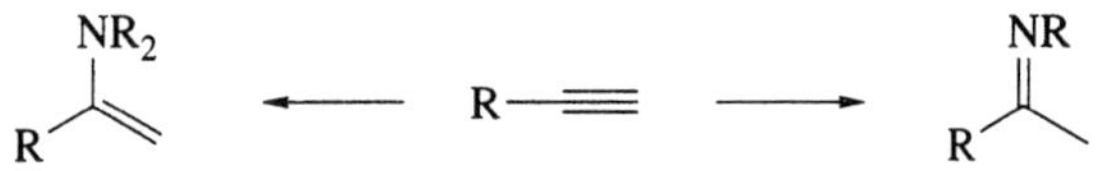

(163)

1.7.3.3 Amide Addition

1.7.3.3.1 Ritter reaction

The preparation of amides by the addition of hydrogen cyanide or alkyl nitriles to alkenes in the presence of acids, known as the Ritter reaction, has been reviewed.[229–232] The reaction may be considered simplistically as nucleophilic attack of a nitrile on a carbocation formed by the protonation of an alkene. Subsequent hydrolysis of the nitrilium intermediate gives the amide product (equation 164). The overall result is addition of a molecule of H—NHCOR to a C—C double bond.

(164)

Although terminal alkenes provide the best yields, the Ritter reaction is also successful when using trisubstituted alkenes and haloalkenes. Markovnikov addition is generally observed. Rearranged products arise occasionally, especially with alkenes that are prone to cationic rearrangements (equation 165).[233]

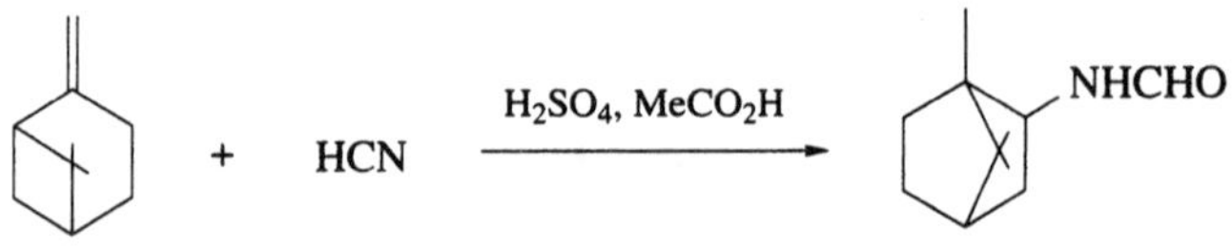

(165)

The Ritter reaction of internal alkenes may yield mixtures of regioisomers, and on occasion, small amounts of rearranged products (equation 166).[234]

The acid used in the Ritter reaction is usually sulfuric acid, although other acids such as perchloric, phosphoric, polyphosphoric, formic and sulfonic acids have been used. Lewis acids such as aluminum trichloride and boron trifluoride are also occasionally used. However, high yields are generally best obtained with sulfuric acid. The choice of solvents varies among sulfuric acid, glacial acetic acid, acetic an-

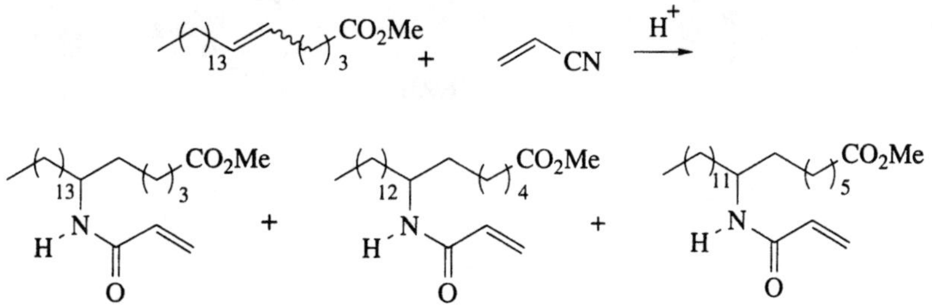

(166)

hydride, di-*n*-butyl ether, some chlorinated solvents, hexanes and nitrobenzene. Reaction times and temperatures are variable, but generally the reaction conditions are very mild.

The synthetic utility of the Ritter reaction is evident from surveying the literature. An early example of the Ritter reaction is found in the synthesis of *in vivo* metabolites of the drug mephentermine (equation 167).[235]

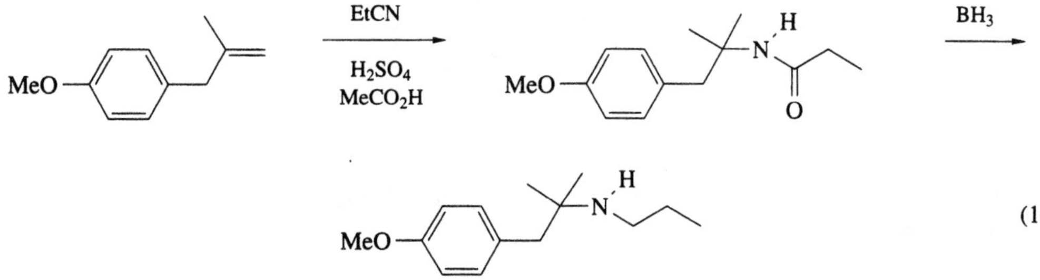

(167)

The synthesis of heterocycles is another common application of the Ritter reaction. Equations (168)–(170) illustrate the preparation of a lactam,[236] oxazoline[237] and dihydroisoquinoline,[235] respectively.

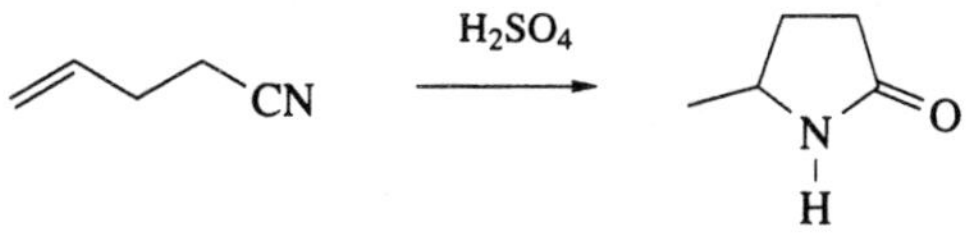

(168)

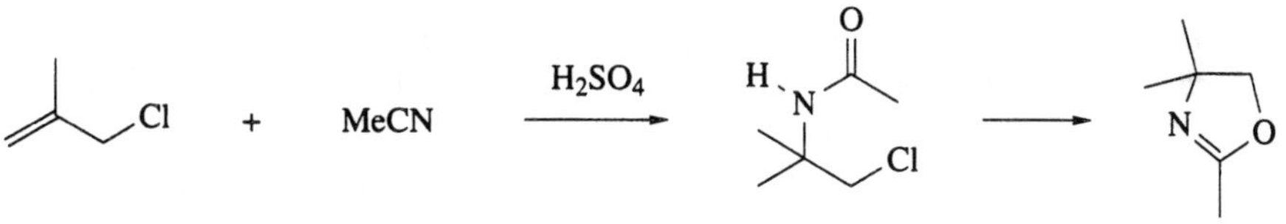

(169)

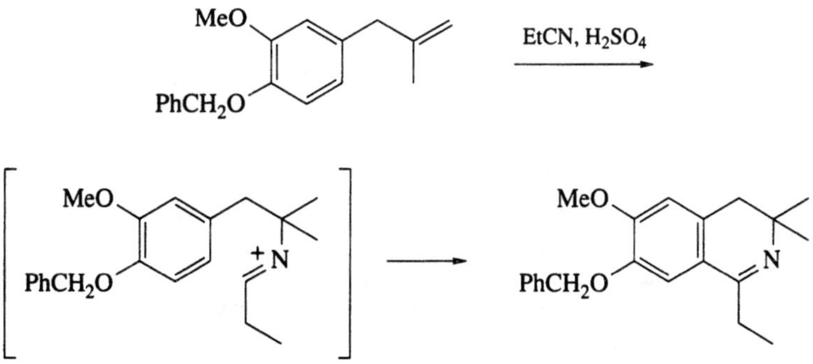

(170)

Recently, nonconjugated dienes have been successfully used in the Ritter reaction for the preparation of bridgehead amides[238] and azatricycloundecane amides (equations 171 and 172).[239]

Extension of the Ritter reaction to conjugated dienes, however, has been less successful. The reaction is often met with competing Diels–Alder reactions or extensive polymerization. The polymerization that occurs during the reaction of 1,3-dienes and nitriles has been used in the synthesis of linear

MeCN, H_2SO_4 / $MeCO_2H$ → NHCOMe (171)

MeCN / H_2SO_4 → [$^+N{\equiv}CMe$] →

[N] → N, NHCOMe (172)

polyamides.[240] However, satisfactory Ritter reactions with conjugated dienes can provide pyrroline derivatives (equation 173).[241]

+ R—≡N $\xrightarrow{H_2SO_4}$ [N, R, +] → N, R (173)

1.7.3.3.2 Amidomercuration–demercuration

The reaction of alkenes or dienes, mercuric salts, and either nitriles or amides generates β-amidomercurials, which can be easily reduced to the corresponding amides (equation 174). This reaction effects H—NHCOR addition to alkenes and dienes, and has been thoroughly reviewed recently.[242]

$$\gt C{=}C\lt \xrightarrow[\text{RCN or RCONH}_2]{\text{HgX}_2} \text{XHg}-\overset{|}{\underset{|}{C}}-\overset{|}{\underset{|}{C}}-\text{NHCOR} \xrightarrow[\text{NaOH}]{\text{NaBH}_4} \text{H}-\overset{|}{\underset{|}{C}}-\overset{|}{\underset{|}{C}}-\text{NHCOR} \quad (174)$$

Amidomercuration using nitriles has received the most attention. This reaction commonly employs mercury(II) nitrate with the nitrile as solvent. Acetonitrile is the most widely used nitrile, but RSCN and R_2NCN can also be employed.[243]

A variety of alkenes undergo successful amidomercuration, but alkenes of the type $R_2C{=}CH_2$ and $R_2C{=}CHR$ fail to react in the anticipated fashion.[243–245] While rearrangements have been observed with α- and β-pinene,[245] 3,3-dimethyl-1-butene reacts normally.[244]

Most useful of the demercuration procedures reported is the use of alkaline sodium borohydride[244] or sodium–mercury amalgam.[243,246] Reduction with $LiAlH_4$ affords the corresponding amines.[243]

There appears to be only one example of intramolecular amidomercuration using an unsaturated nitrile (equation 175),[247] and no examples of the amidomercuration of dienes or alkynes using nitriles.

CN → HgCl, O, NH (175)

Amides can also be utilized for amidomercuration.[242,248,249] The reaction is usually run using equivalent amounts of the alkene and $Hg(NO_3)_2$, plus 10 equiv. of amide in refluxing dichloromethane (equation 176).

$$\text{>C=C<} + H_2NCOR \xrightarrow[CH_2Cl_2]{Hg(NO_3)_2} \xrightarrow[Bu^nNH_2]{NaBH_4,\ NaOH} H-\overset{|}{\underset{|}{C}}-\overset{|}{\underset{|}{C}}-NHCOR \quad (176)$$

The reaction may be subject to limitations in the alkene structure similar to those of the nitrile process. Besides simple amides, one can also utilize urea and urethanes in this reaction. Demercuration is best effected by using alkaline sodium borohydride in the presence of a primary amine such as Bu^nNH_2.

Intramolecular amidomercuration of unsaturated amides and subsequent reduction is a very useful approach to cyclic amides (equations 177 and 178).[242] Mercuric acetate has been employed in all such reactions to date. The reaction is often highly regio- and stereo-selective.[250–253]

R–C(O)–NH ... → N-COR pyrrolidine ... HgOAc (refs. 250–252) (177)

NH → → (ref. 253) (178)

Sulfonamides can also be employed in amidomercuration when mercury(II) nitrate is utilized as the mercury salt (equation 179).[254]

$$\text{>C=C<} + H_2NSO_2R \xrightarrow[CH_2Cl_2]{Hg(NO_3)_2} \xrightarrow[Bu^nNH_2]{NaBH_4,\ NaOH} H-\overset{|}{\underset{|}{C}}-\overset{|}{\underset{|}{C}}-NHSO_2R \quad (179)$$

While no intramolecular examples of this reaction exist, dienes undergo reaction to generate *cis*-pyrrolidines (equation 180).

→ N–$SO_2C_6H_4Me$ ← (180)

1.7.3.4 Azide Addition

1.7.3.4.1 Direct H—N_3 addition

The chemistry of alkyl and alkenyl azides has been well summarized in several recent reviews.[255–257] The azides can be prepared *via* numerous methods, of which the addition of hydrazoic acid to C—C multiple bonds is one. With the exception of cyclopropenes,[258] most alkenes are unreactive towards hydrazoic acid itself. However, the addition can be catalyzed by acids (phosphoric acid,[259] sulfuric acid[260] or trifluoroacetic acid[261]) or Lewis acids (aluminum trichloride, boron trifluoride or titanium tetrachloride).[262]

The acid-catalyzed reactions proceed through a carbocation intermediate formed by protonation of the double bond, followed by nucleophilic attack of HN_3 and subsequent loss of a proton (equation 181).

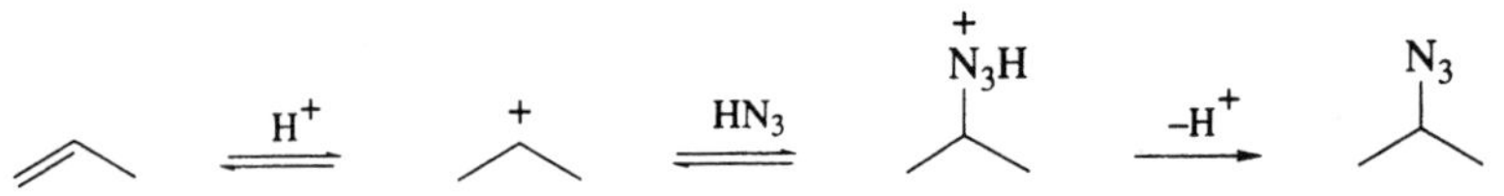

(181)

This reaction works well for simple alkenes and cycloalkenes. Attempts to add hydrazoic acid to diarylalkenes in the presence of sulfuric acid, however, leads to considerable decomposition due to Schmidt rearrangement.[260]

The addition of HN_3 to alkenes in the presence of Lewis acids also involves carbocationic intermediates. This is nicely illustrated by the boron trifluoride-catalyzed addition of HN_3 to α- or β-pinene in which carbocation rearrangements are observed (equation 182).[263]

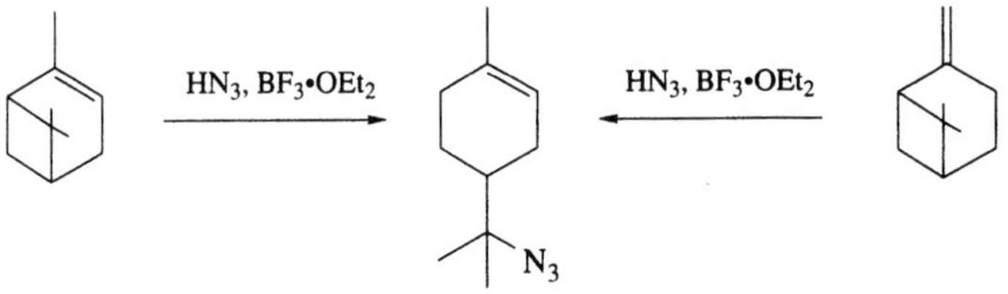

(182)

The combination of hydrazoic acid and boron trifluoride has also been used to convert 5β-hydroxypregnane into the corresponding azido compound (equation 183).[264] This reaction is presumed to involve elimination of water to form a Δ^5-pregnene which then adds HN_3.

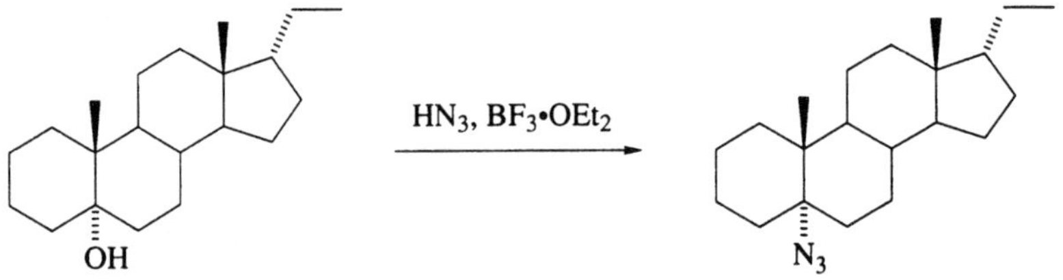

(183)

The addition of hydrazoic acid to numerous acyclic and cyclic alkenes also proceeds well in the presence of titanium tetrachloride or aluminum trichloride.[262] Dichloromethane or chloroform are ideal solvents for these catalysts. The addition is regiospecific and tolerates the presence of primary and secondary alcohols, as well as esters (equation 184).

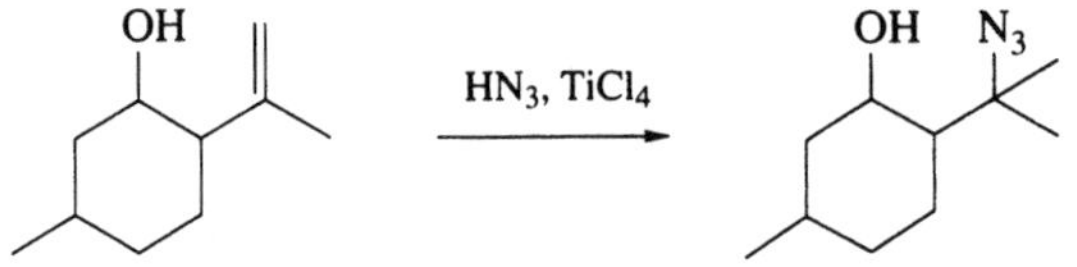

(184)

These Lewis acids may also be used to catalyze additions of hydrazoic acid to enol ethers and silyl enol ethers (equation 185). In an earlier report, the addition of HN_3 to enol ethers was accomplished with trifluoroacetic acid.[261] Curiously, the same additions are observed in comparable yields even in the absence of a catalyst.[262]

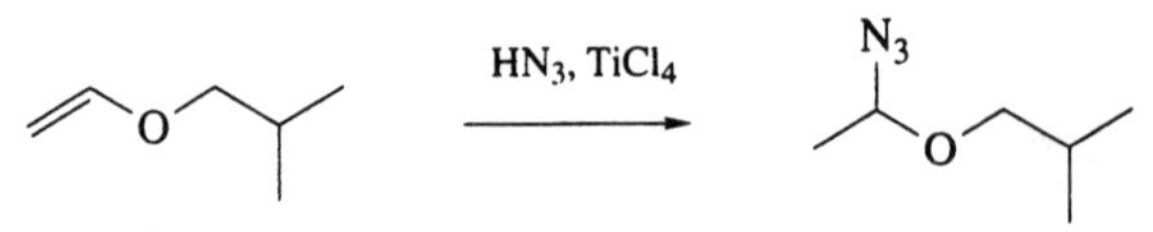

(185)

The reaction of hydrazoic acid with alkynes generally provides triazoles instead of azides (equation 186).[265]

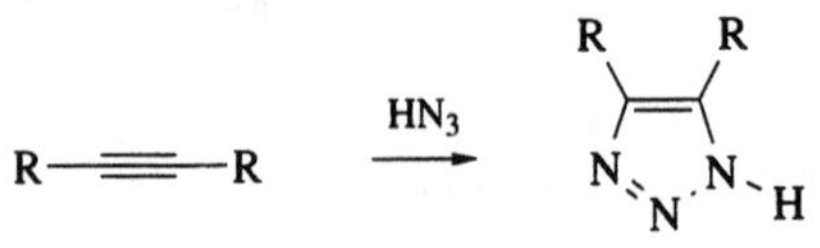

(186)

Alkoxyalkynes are an exception, but unfortunately they undergo further additions to yield geminal diazides as the only product (equation 187).[266]

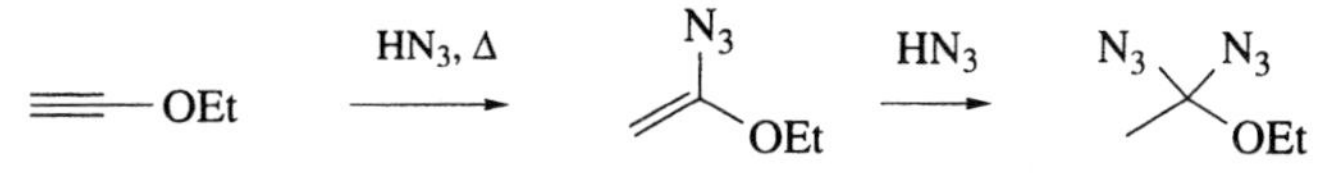

(187)

1.7.3.4.2 Azidomercuration–demercuration

The formal addition of H—N_3 across C—C double bonds can also be brought about *via* azidomercuration–demercuration, a process recently reviewed.[267] In the presence of NaN_3, $Hg(OAc)_2$, $Hg(O_2CCF_3)_2$ or $Hg(NO_3)_2$, in aqueous THF, methanol or DMF, alkenes undergo azidomercuration (equation 188).[268–274]

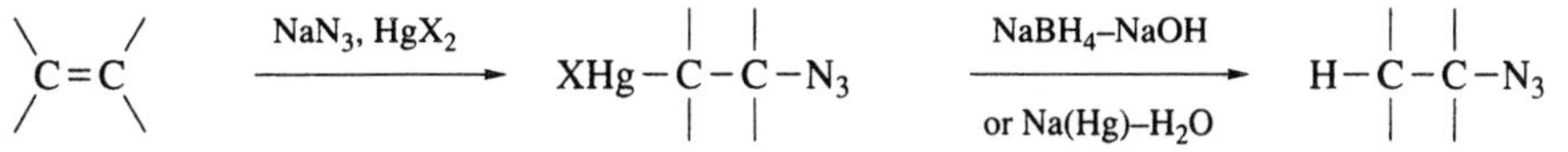

(188)

While highly substituted alkenes react poorly, strained alkenes like cyclopropene[268] and norbornene[270] undergo a facile reaction. 3,3-Dimethyl-1-butene does not rearrange during addition.[270] The reaction conditions are mild enough that unsaturated carbohydrates can be utilized.[271–274]

Alkaline sodium borohydride is the preferred demercuration reagent, but sodium–mercury amalgam in D_2O is best for stereospecific reduction with retention.[268]

Only two dienes, and no alkynes, have been subjected to azidomercuration (equations 189 and 190).[275–277]

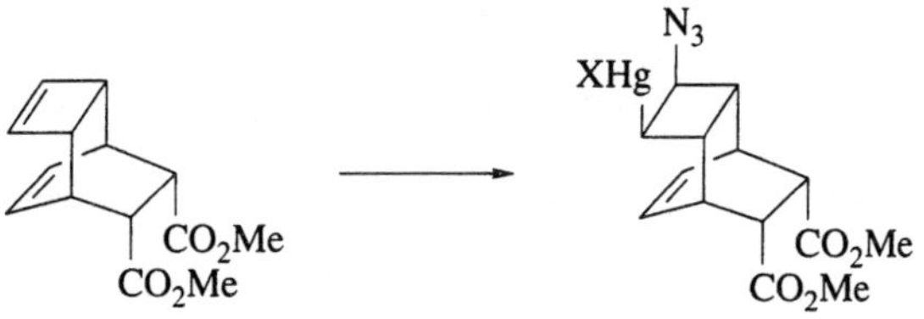

(refs. 275, 276) (189)

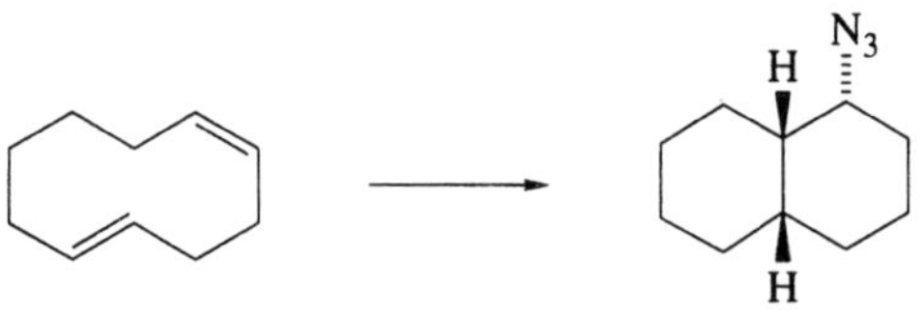

(ref. 277) (190)

1.7.4 H—O ADDITION

1.7.4.1 H—OH Addition

1.7.4.1.1 Direct hydration

The addition of H—OH to alkenes is a well studied reaction for the synthesis of alcohols with important industrial applications.[278,279] It is seldom used as a laboratory procedure. The hydration is commonly achieved under acid or metal catalysis, but may also be accomplished photochemically under acidic or neutral conditions. The acid-catalyzed addition of water to the C—C double bond occurs in a Markovnikov fashion, and is typically carried out using sulfuric acid, phosphoric acid or perchloric acid. Yields are highly variable and depend on exact reaction conditions. The large number of mechanistic studies reported concerning this reaction has convincingly provided evidence for an $A_{SE}2$ mechanism involving a cationic species (Scheme 1). As anticipated, rearrangements of the cationic species are observed.

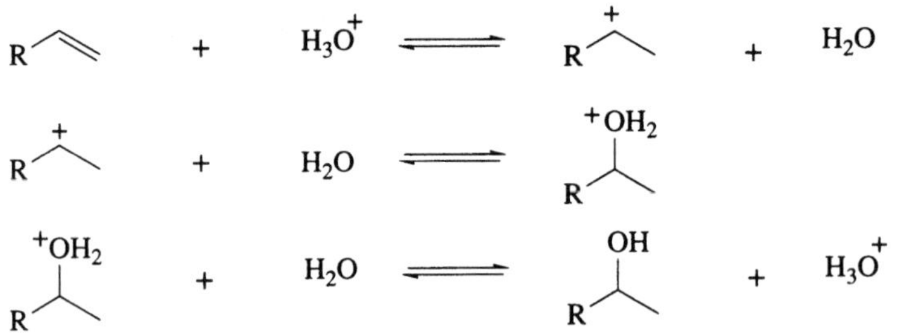

Scheme 1

The rate of hydration is dependent on the substitution and hence the nucleophilicity of the C—C double bond.[280] In general, trisubstituted alkenes are hydrated faster than terminal alkenes. Thus, the reaction is most commonly applied to the preparation of tertiary alcohols (equation 191).[281]

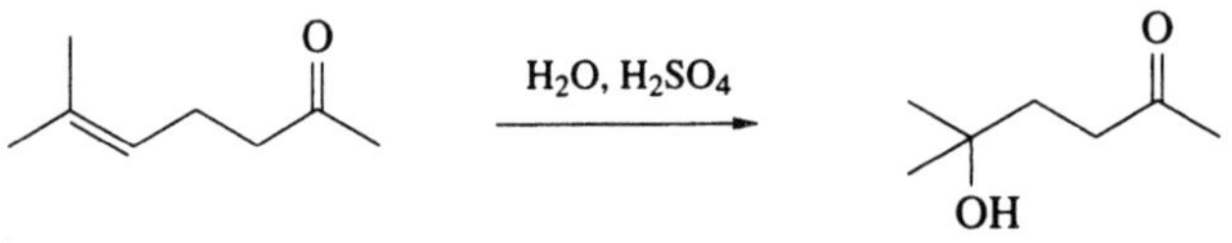

H_2O, H_2SO_4 (191)

Furthermore, *cis* isomers are more easily hydrated than their *trans* counterparts, although exceptions are known. Thus, the hydration of *cis*-1,2-dicyclopropylethylene is 2.5 times faster than its *trans* isomer.[282] Strain introduced into a ring by the incorporation of a *trans* double bond reverses the trend, thereby making the hydration of *trans*-cyclooctene more rapid than the *cis* compound.[283] Smaller ring alkenes are also rather sluggish towards hydration.[284]

Recently, there have been many new catalysts developed for the hydration of alkenes, although most are more suited for industrial purposes. These catalysts include zeolites such as pentasil,[285] mordenite[286] and ferrierite,[185] as well as those containing heavy metals,[287] heteropolyacids[288] and sulfonic acid exchange resins.[289] It is reported that some of these new catalysts perform as high as 99.6% conversions and 99.4% selectivity, as illustrated in the hydration of propene into 2-propanol (equation 192).[290]

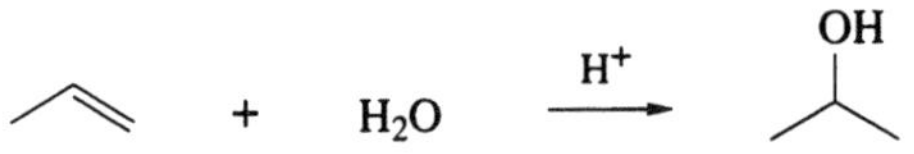

(192)

Hydration by synthetic zeolites is also applicable to structurally more complicated alkenes (equation 193).[291]

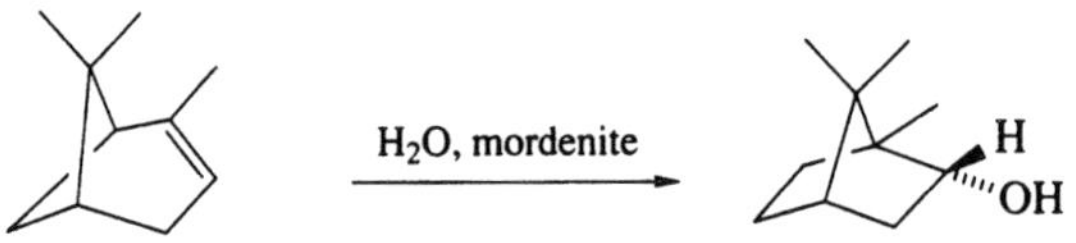

H_2O, mordenite (193)

The metal ion promoted hydration of alkenes that mimic biochemical reactions has also received attention.[292]

The rate of acid-catalyzed hydration of styrenes can be enhanced photochemically (Scheme 2).[293,294] Except for nitro-substituted styrenes, such photohydrations undergo exclusive Markovnikov addition.

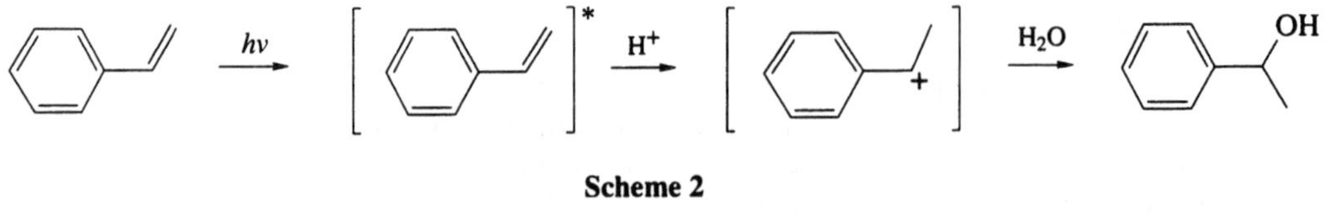

Scheme 2

The products obtained *via* the acid-catalyzed hydration of conjugated dienes are dependent on the structure of the diene. Thus, through the use of deuterated reagents, the sulfuric acid-catalyzed hydration of 1,3-cyclohexadiene was shown to proceed exclusively by a 1,2 addition (equation 194).[295]

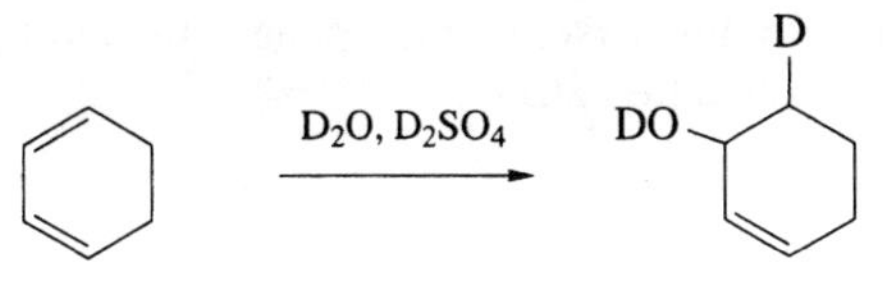

$$\text{cyclohexa-1,3-diene} \xrightarrow{D_2O,\ D_2SO_4} \text{DO–C}_6\text{H}_8\text{D} \tag{194}$$

On the other hand, the hydration of acyclic dienes such as 1- or 2-substituted 1,3-butadienes affords an equilibrium mixture of allylic alcohols resulting from both 1,2- and 1,4-addition (equation 195).[296]

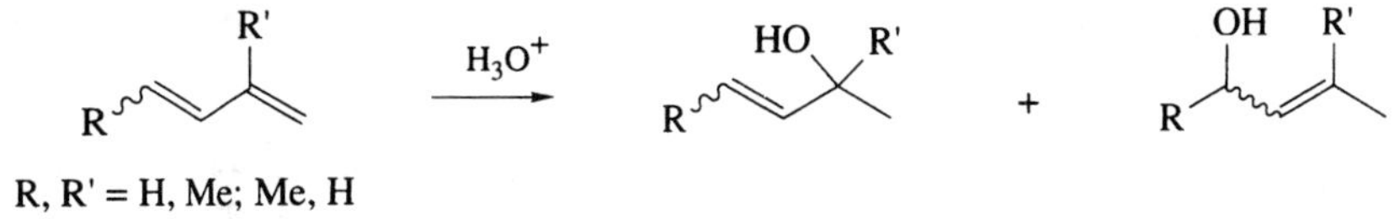

(195)

R, R' = H, Me; Me, H

Interestingly, stereochemistry also plays a role in the mode of addition.[297] Thus, hydration of *cis*-1-ethoxy-1,3-butadiene shows deuterium incorporation at both the 2 and 4 positions (equation 196), while the *trans* isomer incorporates deuterium only at the 4 position (equation 197).

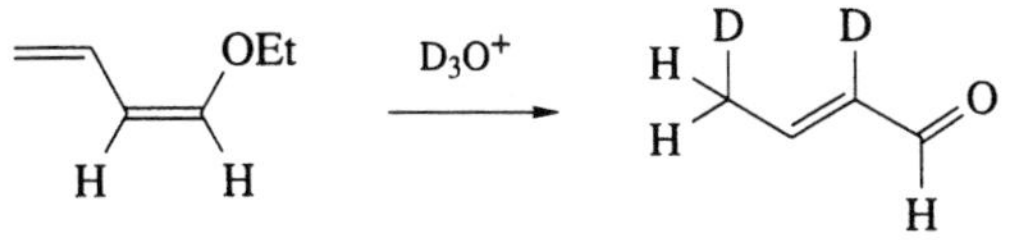

(196)

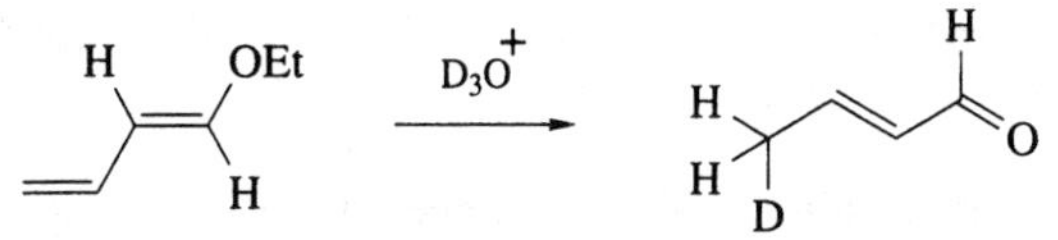

(197)

The addition of water to allenes has been reviewed recently.[298,299] The product obtained from the acid-catalyzed hydration of allene, alkylallenes and 1,3-dialkylallenes is usually the ketone. The intermediate is a vinyl cation formed by protonation on the terminal carbon of the allene moiety (equation 198).

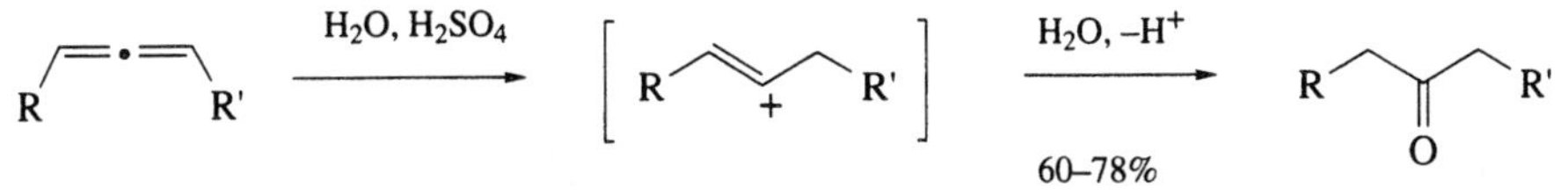

(198)

However, the site of protonation changes to the central carbon when electron-donating substituents such as acetates, alkoxides, arenes and fluoride are attached to the allene moiety. Thus, the hydration of allenyl ethers provides unsaturated aldehydes showing deuterium incorporation at the central carbon (equation 199).[300]

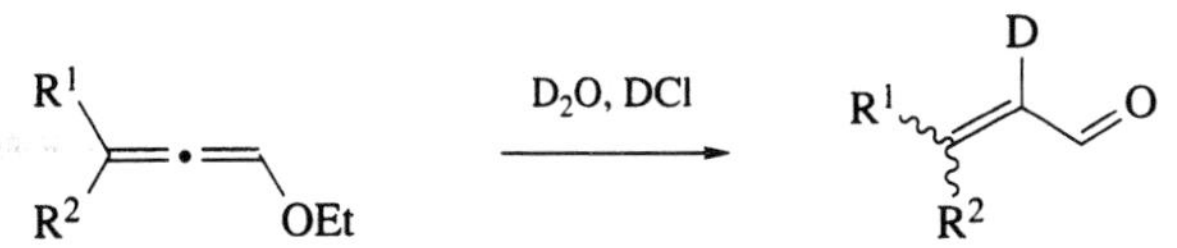

(199)

Carboxylic acids are obtained when water is added to the carbon–carbon double bond of ketenes (equation 200).[301] The addition is acid catalyzed, and involves protonation at the β-carbon.

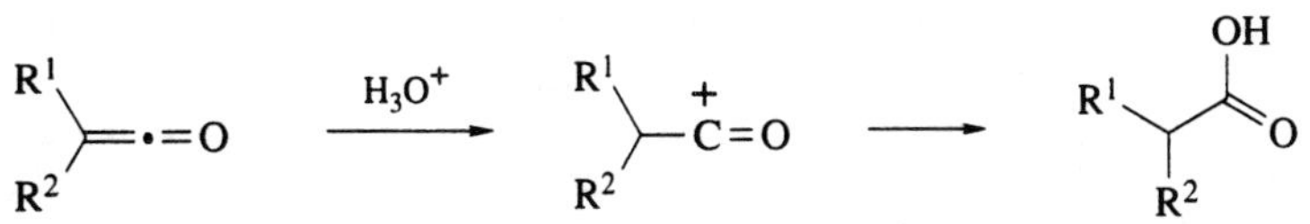

(200)

The direct hydration of a C—C triple bond has been recently reviewed.[302,303] Hydration is possible under acid or metal catalysis, as well as by photohydration. Hydration under acid catalysis is generally done with sulfuric acid, although other acid systems such as phosphoric acid/boron trifluoride have been reported.[304] It is established that acid catalysis occurs by a vinyl cationic intermediate which reacts with water to form an enol, followed by tautomerism into a ketone. The addition of water to terminal alkynes is governed by Markovnikov's rule (equation 201).

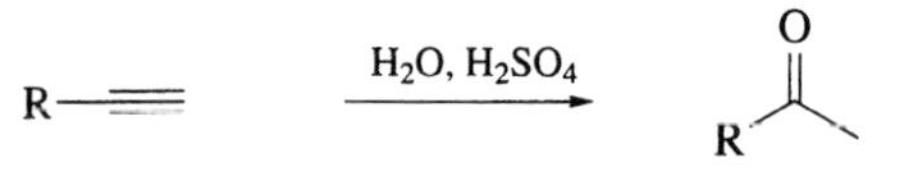

$$R—\equiv \xrightarrow{H_2O,\ H_2SO_4} RC(O)CH_3 \quad (201)$$

Except for very reactive alkynes, acid-catalyzed hydrations are usually sluggish. This slow hydration can be overcome by the addition of catalytic amounts of mercury(II) salts. Such hydrations are generally mild and will tolerate the presence of other functional groups. Specific examples of mercury-catalyzed hydrations are discussed in the next section.

The hydration of alkynes is also accomplished by use of catalytic amounts of palladium and gold salts.[305] The mildness of this reaction is demonstrated by the preparation of 5-oxo-prostaglandin derivatives (equation 202). In this connection, it should be noted that attempted use of other metal salts to catalyze C—C triple bond hydrations has met with little success.[306]

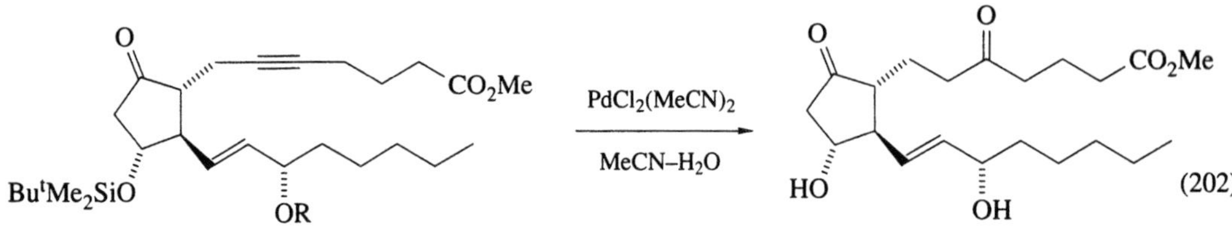

(202) Reagents: $PdCl_2(MeCN)_2$, $MeCN–H_2O$. Starting material labels: O, CO_2Me, Bu^tMe_2SiO, OR. Product labels: O, O, CO_2Me, HO, OH.

Analogous to the photohydration of styrenes, the acid-catalyzed hydration of phenylalkynes is also possible photochemically.[293,294,307] Except for nitro-substituted arylalkynes, the reaction follows Markovnikov's rule. This reaction has thus far received limited synthetic attention.

The hydration of heteroatom-substituted alkynes proceeds well under general acid catalysis. This reaction provides a convenient synthesis of amides,[308] esters[309] and thioesters.[308] Equation (203) illustrates the utility of this reaction for the regiospecific synthesis of 3-alkenamides.

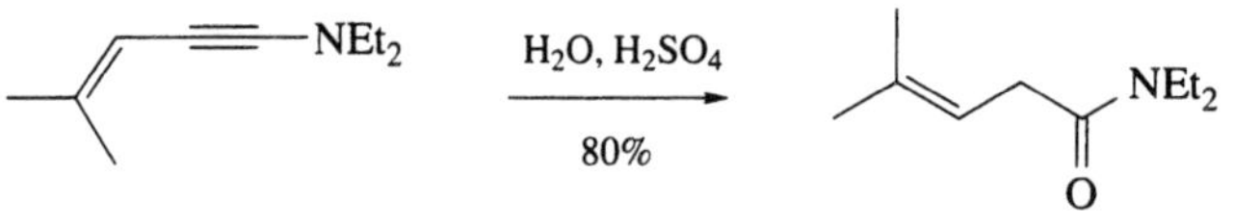

(203) $Me_2C{=}CH–C{\equiv}C–NEt_2 \xrightarrow[80\%]{H_2O,\ H_2SO_4} Me_2C{=}CH–CH_2–C(O)NEt_2$

The hydration of conjugated diynes readily affords 1,3-diketones. The hydration may be acid- or mercuric sulfate-catalyzed. The Rupe rearrangement is observed under general acid catalysis (equation 204).[310]

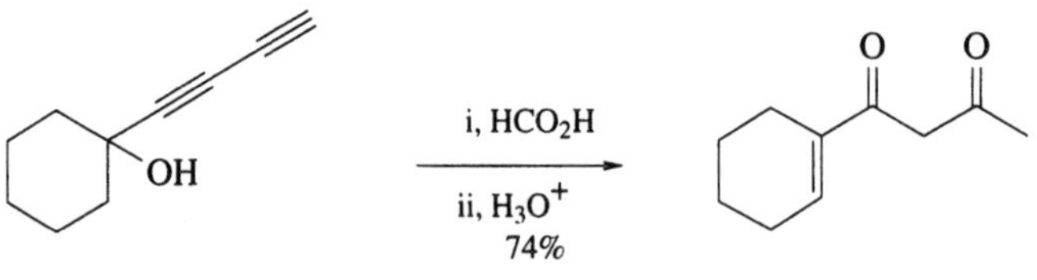

(204) 1-(buta-1,3-diynyl)cyclohexanol (OH) → i, HCO_2H; ii, H_3O^+; 74% → 1-(cyclohex-1-enyl)butane-1,3-dione (O, O)

1.7.4.1.2 Hydroxymercuration–demercuration

In view of the many limitations inherent in the direct acid-catalyzed hydration of alkenes, indirect hydration *via* hydroxymercuration–demercuration has become a very valuable method for the preparation of alcohols. This process has recently been thoroughly reviewed.[311]

While many different procedures have been reported for the hydroxymercuration–demercuration of alkenes, the most useful procedure uses mercury(II) acetate in aqueous THF, followed by *in situ* alkaline sodium borohydride reduction (equation 205).[312,313] Virtually all substitution patterns about the C—C double bond are accommodated.

The following relative rates have been observed: $R_2C{=}CH_2 > RCH{=}CH_2 > (Z)\text{-}RCH{=}CHR > (E)\text{-}RCH{=}CHR$. Only certain aryl-substituted alkenes, tetrasubstituted alkenes or sterically hindered alkenes

$$\mathrm{C{=}C} \xrightarrow[\mathrm{H_2O{-}THF}]{\mathrm{Hg(OAc)_2,}} \mathrm{AcOHg{-}C{-}C{-}OH} \xrightarrow[\mathrm{NaOH}]{\mathrm{NaBH_4}} \mathrm{H{-}C{-}C{-}OH} \qquad (205)$$

fail to react. The reaction rate is increased by electron donation and decreased by electron withdrawal, as expected for an electrophilic process.

The overall process is generally highly regiospecific. Mono- or di-substituted terminal alkenes and tri-substituted alkenes afford >99% of the Markovnikov adduct (equation 206).[312–314]

R ⟶ R, OH (206)

The direction of hydration is disubstituted internal alkenes is controlled by steric factors (equation 207).[312]

R ⟶ R, OH + R, OH (207)

R	%	%
Et	64	36
Pr^i	91	9
Bu^t	98	2

Functional groups in the vicinity of the C—C double bond can also have a substantial effect on the regiochemistry of hydration (equations 208 and 209).[315–317]

OH —93%⟶ OH, OH (94%) + OH, OH (6%) (ref. 315) (208)

R^1, CO_2R, R^2 ⟶ R^1, OH, CO_2R, R^2 (refs. 316, 317) (209)

R = H, Me

The regio- and stereo-selectivity of various substituted cycloalkenes have been examined (equations 210–212).[318–324]

Unfortunately, these reactions tend to afford mixtures. Similar mixtures are observed when using 2-cyclohexen-1-ol,[315] 3-cyclohexen-1-ol,[315,325] or the corresponding ethers or esters.[326,327]

The regio- and stereo-selectivity of the hydration of bicyclic alkenes has also been carefully examined (equation 213).[328–330]

⟶ OH (2%) + OH (18%) + OH (11%) + OH (69%) (210)

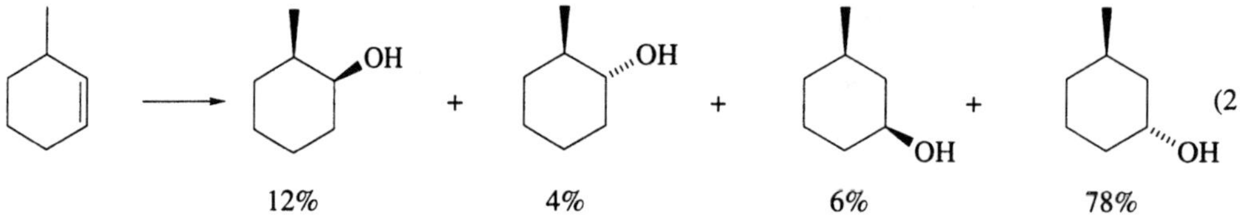

OH OH OH OH (211)

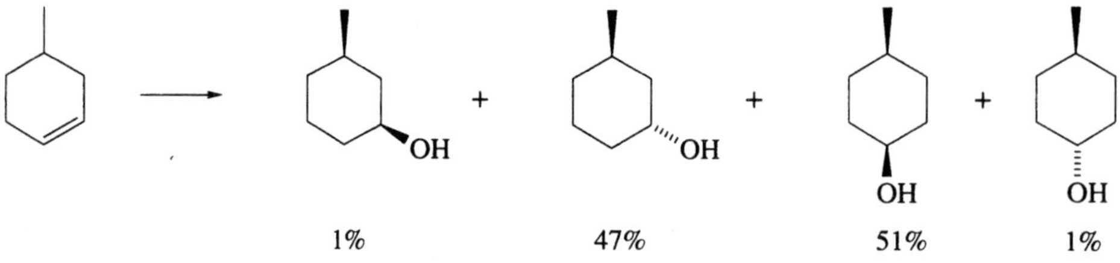

12% 4% 6% 78%

OH OH OH OH (212)

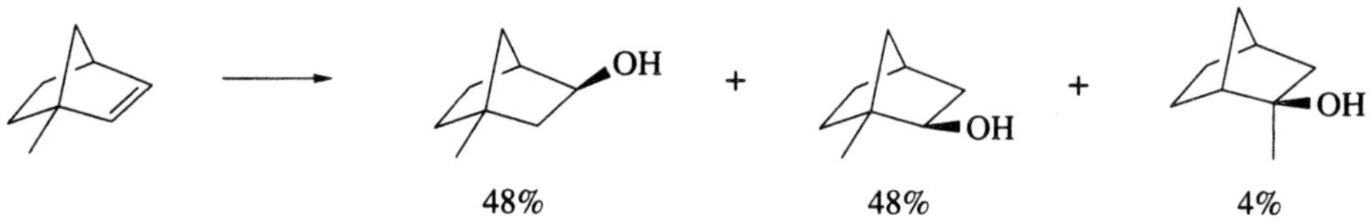

1% 47% 51% 1%

OH OH OH (213)

48% 48% 4%

The stereoselectivity of this hydration process has been studied on a number of alkenes. Acyclic alkenes and most cyclic alkenes give exclusively *trans* addition (equations 214 and 215).[331,332] However, some strained cyclic alkenes, such as *trans*-cyclooctene and *trans*-cyclononene, give *cis* adducts.[332]

D H H D → HO H D D H HgX (ref. 331) (214)

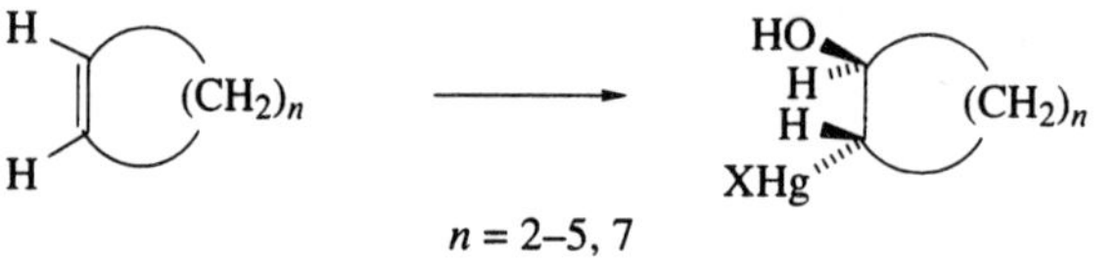

H H $(CH_2)_n$ → HO H H XHg $(CH_2)_n$ (ref. 332) (215)

$n = 2–5, 7$

Various substituted cyclic alkenes have been reported to give mixtures of stereoisomers (equations 216 and 217).[319,320,322,323,333–336]

R → HO R (refs. 323, 333–335) (216)

R → HO R (refs. 319, 320, 322, 336) (217)

As noted earlier, many bicyclic and polycyclic alkenes undergo hydroxymercuration–demercuration. Norbornene and derivatives[328] give *cis* adducts, but less strained alkenes afford the expected *trans* adducts.

Attempts at asymmetric induction using mercury salts of optically active carboxylic acids have met with limited success.[337,338]

Relatively few functional groups have been reported to interfere in the reaction, and rearrangements during hydroxymercuration or reduction are rare.

While reduction is usually best carried out using alkaline sodium borohydride, this reaction proceeds by free radicals and is not stereospecific. By employing Na(Hg) in D_2O, reduction with retention is observed. However, the stereospecific reduction of β-hydroxymercurials in which the mercury moiety is α to a carbonyl group has been reported using alkaline $NaBH_4$–EtOH,[316] alkaline H_2S[317] and $HS(CH_2)_3SH$ (equation 218).[316]

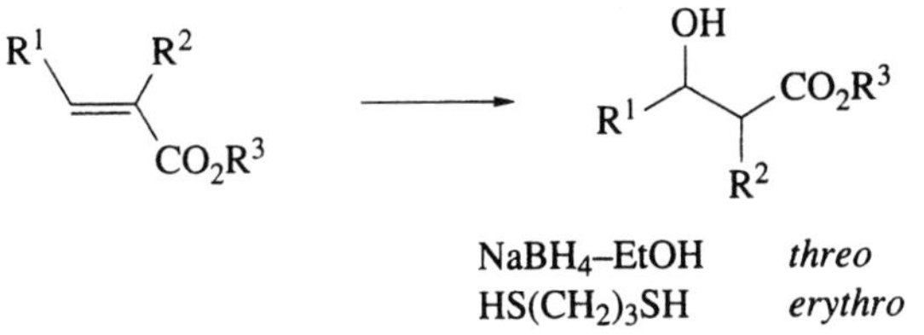

(218)

$NaBH_4$–EtOH *threo*
$HS(CH_2)_3SH$ *erythro*

Numerous dienes and polyenes have been subjected to hydroxymercuration–demercuration.[311] With nonconjugated dienes, the products can usually be predicted by applying what one has learned from the corresponding simple alkenes. Isolated double bonds are more reactive than conjugated double bonds and frequently one of the double bonds is sufficiently more reactive than the others that monohydroxylated products can be obtained. Improvements in selectivity have been reported by using mercury(II) trifluoroacetate[339] or by adding sodium lauryl sulfate.[340]

When five- or six-membered ring ethers can readily be formed by intramolecular alkoxymercuration of the initially formed alkenol, cyclic ethers are often the observed product (equation 219).[341] This process has recently proven useful in the synthesis of spirocyclic acetals (equation 220).[342]

(219)

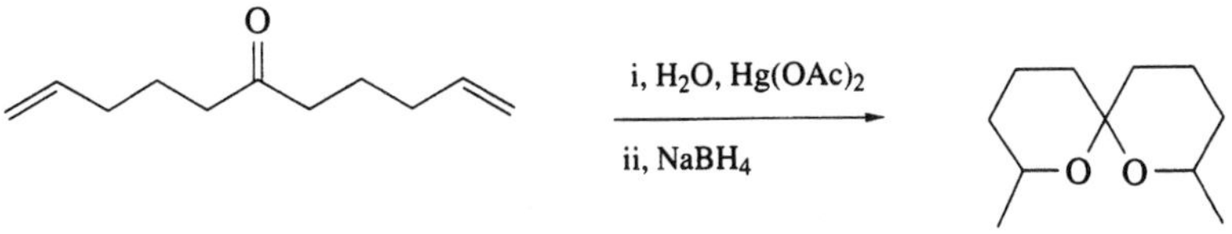

i, H_2O, $Hg(OAc)_2$
ii, $NaBH_4$
(220)

Carbon–carbon bond formation is another side reaction occasionally observed (equation 221).[343]

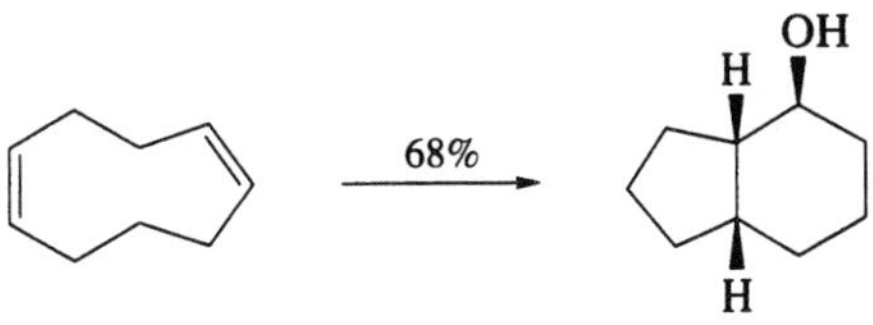

68%
(221)

Carbon skeleton rearrangements during either hydroxymercuration or demercuration are also more common with dienes.

The hydroxymercuration–demercuration of conjugated dienes generally does not afford monohydration products selectivity, but diols can sometimes be obtained in reasonable yield. The direction of addition of H—OH is that expected by extrapolation from simple alkenes and allylic alcohols.

Allenes undergo mercury-catalyzed hydration to give a variety of products depending upon the substituents and the substitution pattern (equations 222 and 223).[344–347]

Alkynes undergo mercury-catalyzed hydration to afford ketones (equation 224).[311,348–350]

A wide variety of mercury(II) salts have been employed as catalysts, though mercury(II) sulfate is the most widely used.[351] Phenylmercury(II) hydroxide is useful for the selective conversion of terminal al-

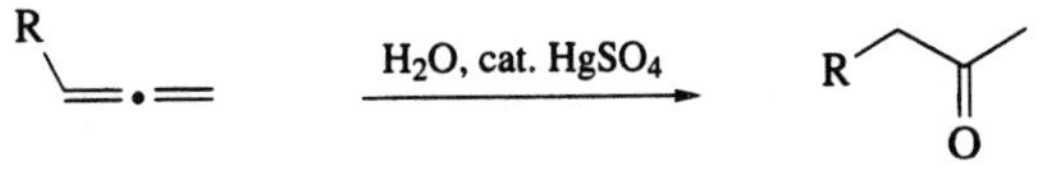

H_2O, cat. $HgSO_4$
(refs. 344, 345) (222)

(refs. 346, 347) (223)

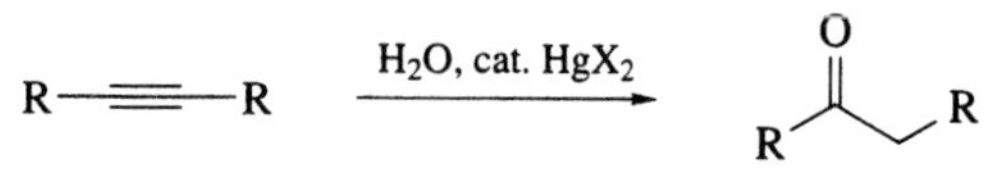

H_2O, cat. HgX_2 (224)

kynes to methyl ketones (equation 225).[352] Internal alkynes or conjugated alkynes fail to react in this process.

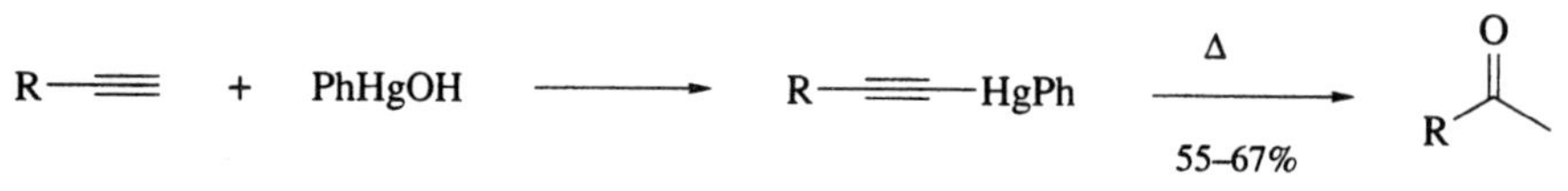

R—≡ + PhHgOH ⟶ R—≡—HgPh ⟶ (Δ, 55–67%) (225)

Mercury-impregnated polymeric resins have also proven valuable for the hydration of alkynes.

These mercury-catalyzed processes are commonly effected in water, aqueous sulfuric acid, acetic acid or alcohol solutions.

The hydration of terminal alkynes produces methyl ketones. Simple unsymmetrical internal alkynes generally afford mixtures of regioisomeric ketones, but alkynes bearing certain functional groups exhibit high regioselectivity (equations 226 and 227).[353–355]

(ref. 353) (226)

$n = 2, 3$ (refs. 354, 355) (227)

While many functional groups are accommodated by this process, acid-promoted rearrangements and eliminations are not uncommon (equations 228–231).[356–360] In fact, the hydration of certain unsaturated alkynes produces oxygen heterocycles (equations 232 and 233).[361,362]

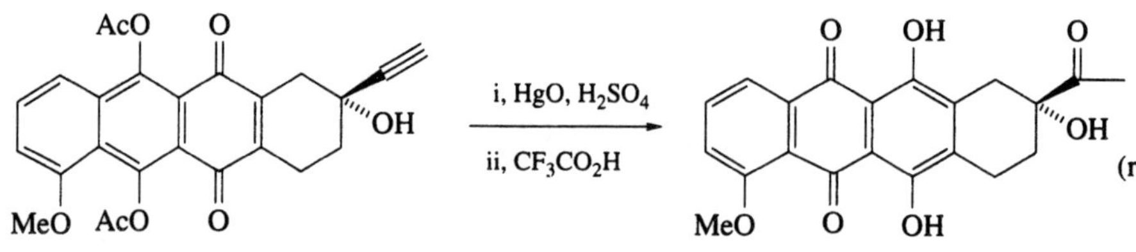

i, HgO, H_2SO_4; ii, CF_3CO_2H (228) (ref. 356)

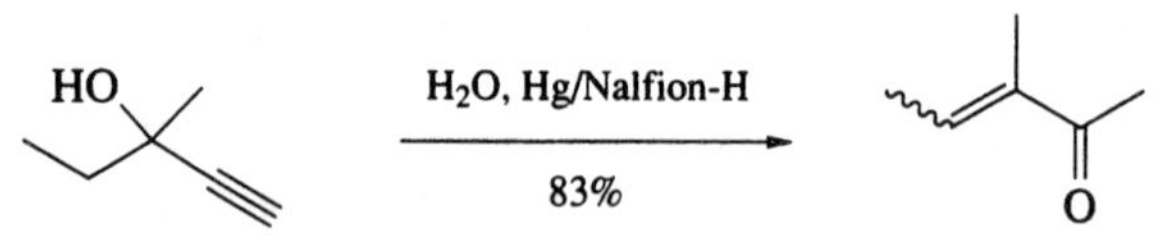

H_2O, Hg/Nalfion-H, 83% (ref. 357) (229)

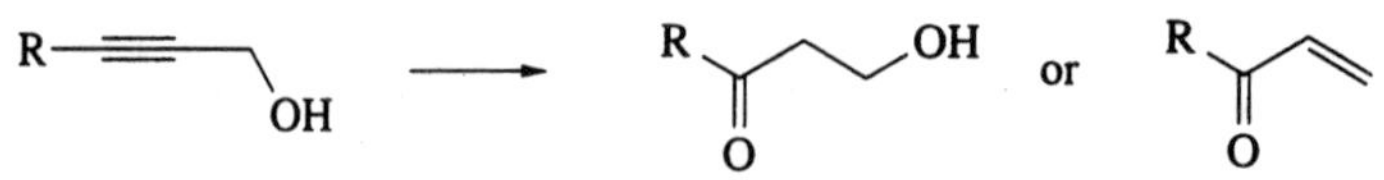

or (230) (refs. 358, 359)

(ref. 360) (231)

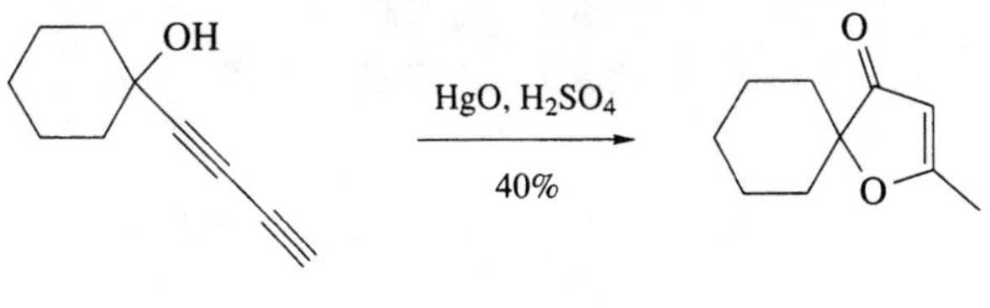
OH
HgO, H_2SO_4
40%
O
O

(ref. 361) (232)

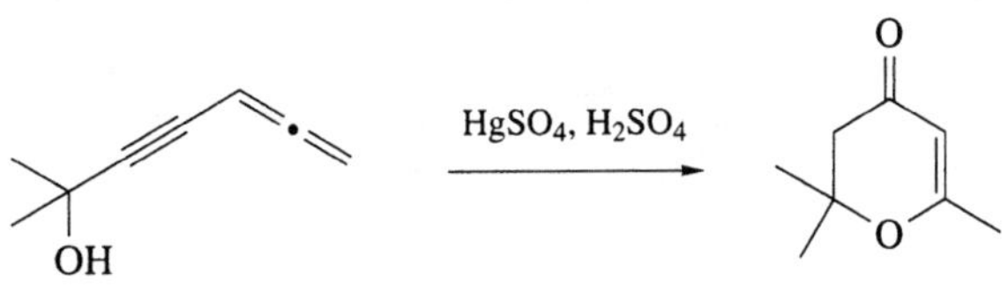
OH
$HgSO_4$, H_2SO_4
O
O

(ref. 362) (233)

1.7.4.2 H—O_2H and H—O_2R Additions

1.7.4.2.1 Acid-catalyzed addition

There are very few examples in the literature of electrophilic additions of hydrogen peroxide or alkyl hydroperoxides to C—C multiple bonds. The products of these additions, alkyl hydroperoxides or dialkyl peroxides respectively, can be explosively unstable[363] and are more commonly obtained by peroxymercuration–demercuration methods (*vide infra*).

Early work on the electrophilic addition of hydrogen peroxide to alkenes was performed in the presence of an acid catalyst, usually sulfuric acid[364] or *p*-toluenesulfonic acid.[365] The reaction proceeds *via* Markovnikov-directed protonation of the double bond (Scheme 3). Subsequent nucleophilic attack of hydrogen peroxide on the carbocation, followed by loss of a proton, furnishes the alkyl hydroperoxide.[366]

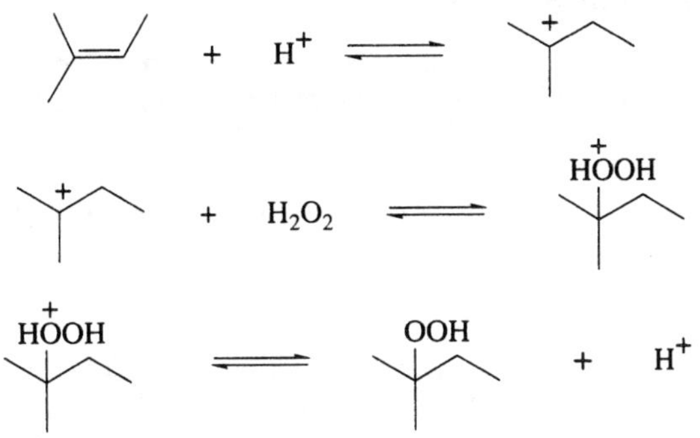
+ H^+
+ H_2O_2
$\overset{+}{H}OOH$
$\overset{+}{H}OOH$
OOH
+ H^+

Scheme 3

Since the reaction involves a carbocation, it is subject to the normal substituent effects. Hence, the addition of hydrogen peroxide to a mixture of 3-methyl-2-hexene and 3-methyl-3-hexene yields the tertiary alkyl hydroperoxide as the only product (equation 234).[367] On the other hand, cyclohexene does not react under similar conditions.

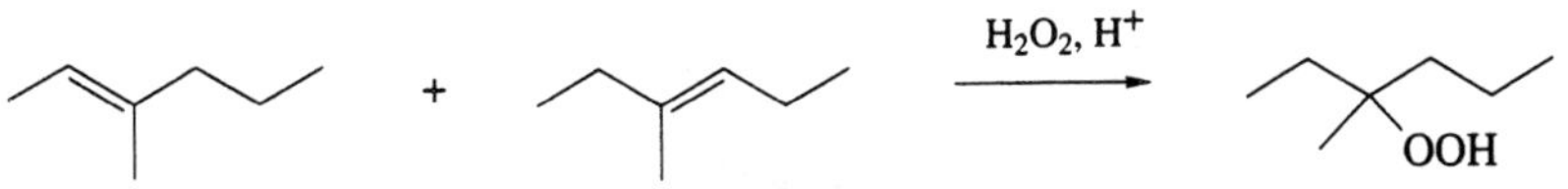
+
H_2O_2, H^+
OOH

(234)

Hydrogen peroxide also adds well to alkenes with heteroatom substituents on the double bond, as exemplified by the reactions of vinyl ethers (equation 235)[368,369] and enamines (equation 236).[370]

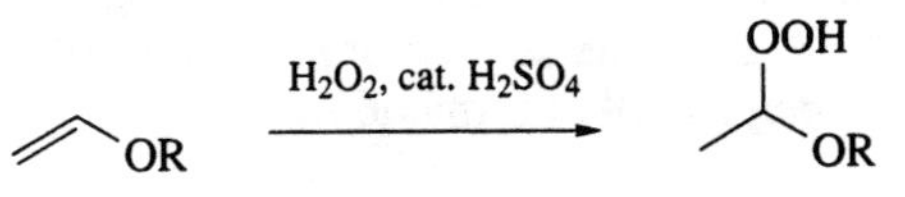
OR
H_2O_2, cat. H_2SO_4
OOH
OR

(235)

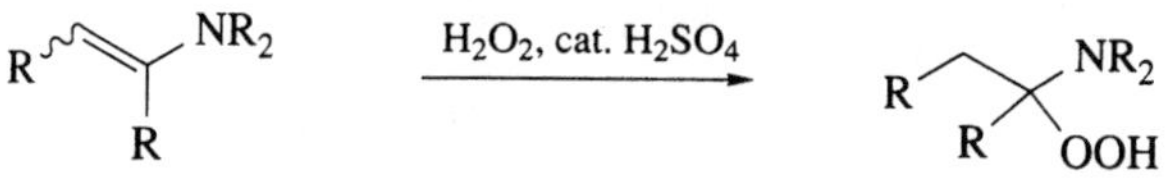

$$\text{R(R)C=CH-NR}_2 \xrightarrow{H_2O_2,\ \text{cat. } H_2SO_4} \text{R}_2\text{C(OOH)(NR}_2) \qquad (236)$$

Recently, it has been proposed that alkyl hydroperoxides are formed from the Co^{III}-catalyzed Markovnikov addition of hydrogen peroxide to alkenes.[371] The alkyl hydroperoxides thus formed are immediately decomposed into ketones and alcohols under the reaction conditions (equation 237).

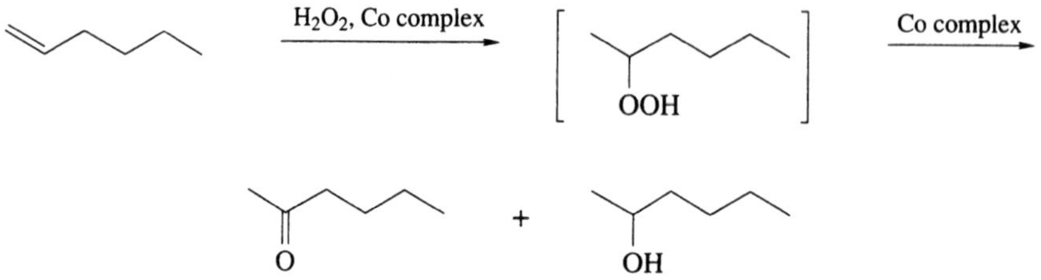

$$\xrightarrow{H_2O_2,\ \text{Co complex}} \left[\ \text{OOH}\ \right] \xrightarrow{\text{Co complex}} \text{O} + \text{OH} \qquad (237)$$

In a manner analogous to the acid-catalyzed addition of hydrogen peroxide, dialkyl peroxides are formed by the addition of alkyl hydroperoxides to alkenes (equation 238).[366]

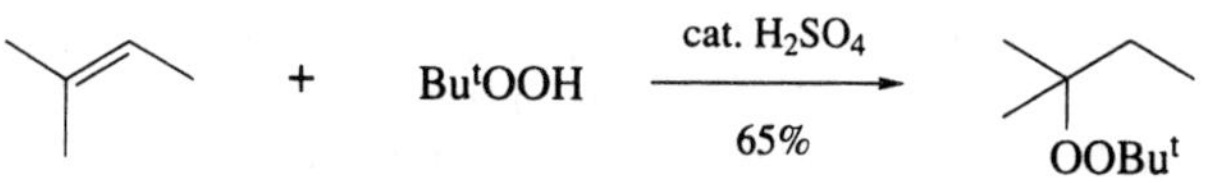

$$+ \ \text{Bu}^t\text{OOH} \xrightarrow[65\%]{\text{cat. } H_2SO_4} \text{OOBu}^t \qquad (238)$$

Surprisingly, there is no literature precedent for the addition of either hydrogen peroxide or alkyl hydroperoxides to dienes, polyenes or alkynes.

1.7.4.2.2 Peroxymercuration–demercuration

The peroxymercuration–demercuration of alkenes provides a more versatile approach to the corresponding hydroperoxides or dialkyl peroxides than direct acid-catalyzed addition (equation 239).[372–374]

$$\text{C=C} \xrightarrow[\text{HgX}_2]{\text{HOOH or ROOH}} \text{XHg}-\text{C}-\text{C}-\text{OOH(R)} \longrightarrow \text{H}-\text{C}-\text{C}-\text{OOH(R)} \qquad (239)$$

The most widely used mercury salts for this transformation are $Hg(OAc)_2$ in the presence of catalytic amounts of $HClO_4$[375] or $Hg(O_2CCF_3)_2$.[376] While hydrogen peroxide itself can be used, mono- and dimercuration products have been observed.[377] Alkyl hydroperoxides generally give cleaner reactions. Dichloromethane is commonly employed as the solvent.

A wide variety of alkenes undergo Markovnikov peroxymercuration. Even α,β-unsaturated carbonyl compounds afford peroxymercurials whose regiochemistry is dependent on the substitution pattern about the C—C double bond.[378–381] Selective *anti* addition to the double bond of simple alkenes is the general rule.[375,382]

While successful demercuration using alkaline sodium borohydride has been reported, this process can also give rise to epoxides (equation 240).[383] Tri-*n*-butyltin hydride often gives improved results.[384]

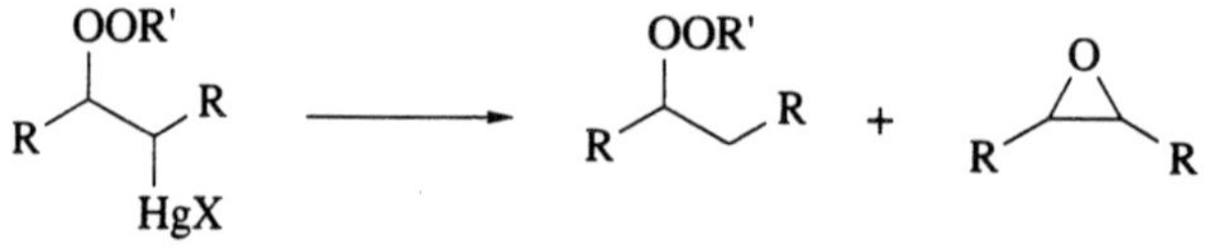

$$\text{RCH(OOR')CH(HgX)R} \longrightarrow \text{RCH(OOR')CH}_2\text{R} + \text{R-epoxide-R} \qquad (240)$$

Several alkenyl hydroperoxides have been successfully cyclized to five-, six- and seven-membered ring peroxides (equation 241).[385–388] Alkaline sodium borohydride reduction of these mercurials is frequently accompanied by epoxide or cyclic ether formation.

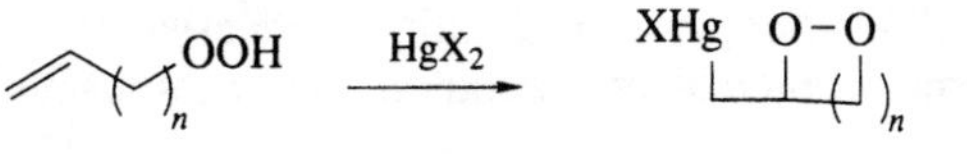

(241)

Dienes can afford either mono- or di-mercurated products (equations 242 and 243).[385,389–391] Mercury(II) trifluoroacetate or mercury(II) nitrate are the reagents of choice for this latter reaction. Alkaline sodium borohydride reduction is often accompanied here by formation of unsaturated alcohols.

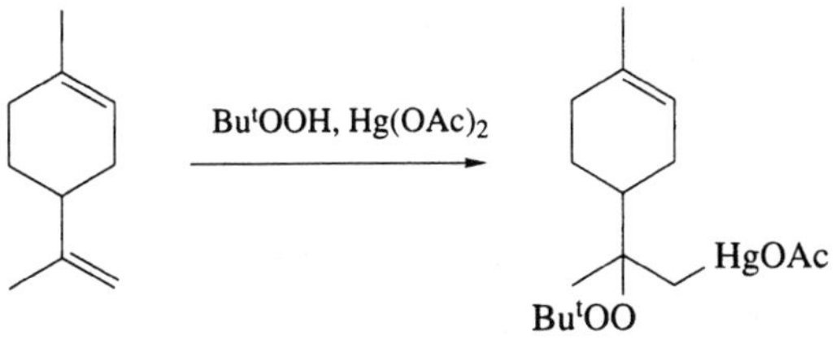

(ref. 389) (242)

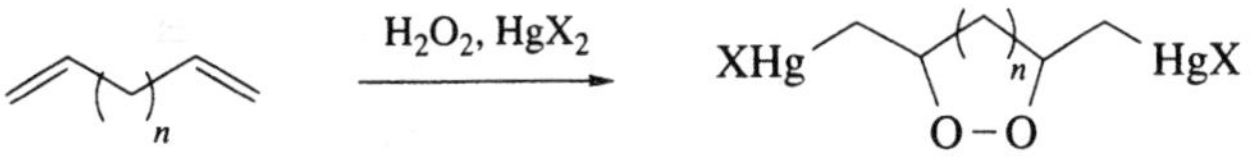

(243) (refs. 385, 390, 391)

1.7.4.3 H—OR Addition

1.7.4.3.1 Direct H—OR addition

Alcohols add directly to alkenes under acid or metal catalysis, or under photoinitiated conditions to give dialkyl ethers. Although these additions provide certain ethers under rather mild conditions, alkoxymercuration–demercuration is generally a superior method.

As in the hydration of alkenes, sulfuric acid is the preferred catalyst. The reaction occurs in accordance with Markovnikov's rule (equation 244).

(244)

As might be anticipated, isomeric products resulting from cationic rearrangements have been observed. The yield of ether is dependent on the amount of the catalyst used, the strength of the acid, and the reaction temperature.[392] Optimum conditions generally involve trace amounts of acid catalyst and room temperature reaction. Acids with weakly nucleophilic conjugate bases are recommended so as not to compete with the alcohols for the cationic intermediate.[393]

The addition of alcohols to alkenes is also catalyzed by boron trifluoride–hydrogen fluoride, but the reaction proceeds only under vigorous conditions and the yields are generally low.[394] Recently, the etherification of alkenes has also been achieved industrially by using acidic montmorillonite[395] and macroporous sulfuric acid resins (ion-exchange resins).[396]

Alcohol addition to alkenes can also be achieved photochemically under neutral conditions (equation 245).[397] This reaction appears to be a photochemically induced ionic addition to the double bond. An arene photosensitizer is required. Unfortunately, this reaction is limited to cyclic alkenes and double bond isomerization is significant.

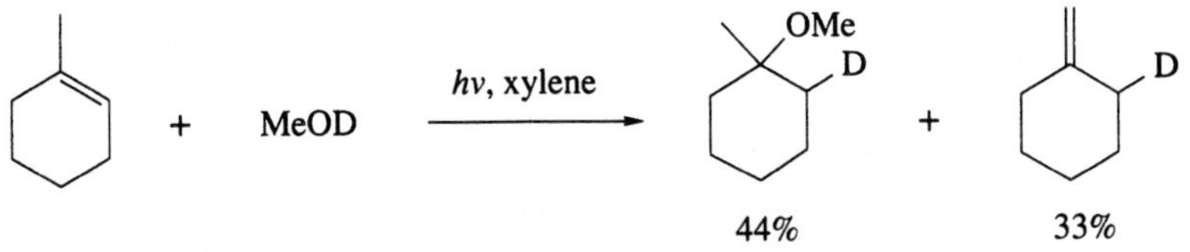

(245)

The presence of electron-donating groups on an alkene enhances the nucleophilicity of the double bond. Thus, the addition of alcohols to vinylic ethers proceeds rapidly by either acid[398] or mercury(II) acetate catalysis.[399] Equation (246) illustrates the addition of an allylic alcohol to 2-methoxypropene, followed by a Claisen rearrangement which affords good yields of γ,δ-unsaturated ketones.

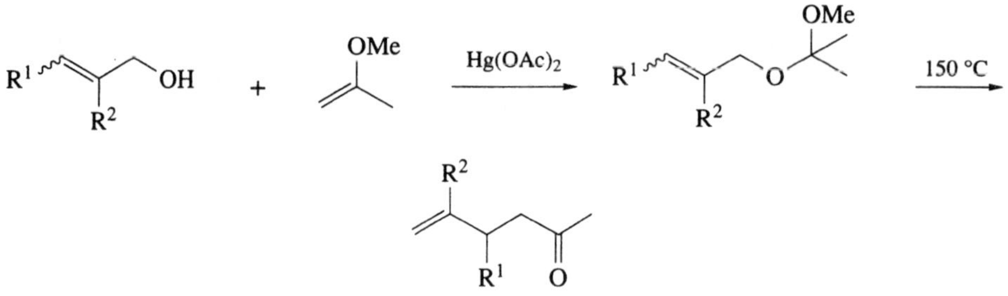

(246)

Under acidic conditions, dihydropyran will undergo additions with alcohols at room temperature to form 2-tetrahydropyranyl ethers (equation 247).[398] This reaction constitutes an important method for the protection of primary and secondary alcohols.[400]

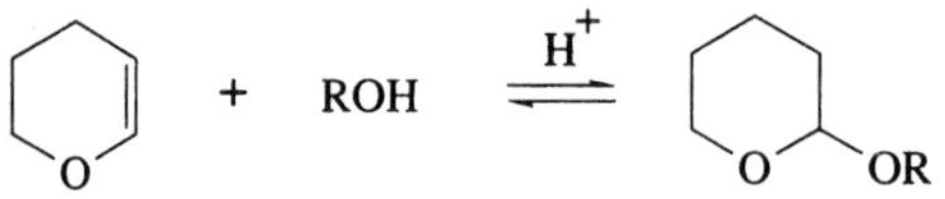

(247)

Alcohols will add to allenes in the presence of trace amounts of acids to give vinyl ethers (or acetals) or allylic ethers.[401] Analogous to the hydration of allenes, protonation occurs on the terminal carbon of the allenic functionality in 1,2-propadiene, 3-alkyl-1,2-propadienes and 1,3-dialkyl-1,2-propadienes (equation 248).[402] Addition of an alcohol to the resulting vinylic cation produces a vinylic ether, which may on further reaction form an acetal of the corresponding ketone.

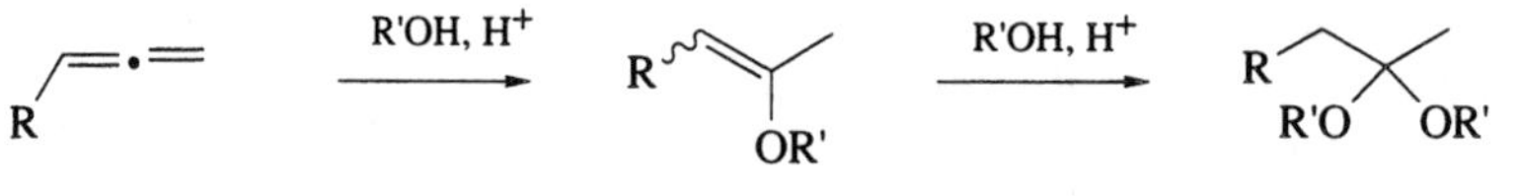

(248)

On the other hand, allylic ethers are obtained from 3,3-dialkyl-1,2-propadienes and tetraalkyl allenes (equation 249).[403]

When treated with catalytic amounts of acid,[404,405] silver salts[406–410] or mercury salts,[409,410] allenic alcohols will cyclize to ethers (equations 250–253). By proper choice of reaction conditions, the yields of products obtained in these cyclizations are generally high. Interestingly, the stereoselectivity observed in equation (253) is proposed to arise from a chair conformation of the intermediate.[408]

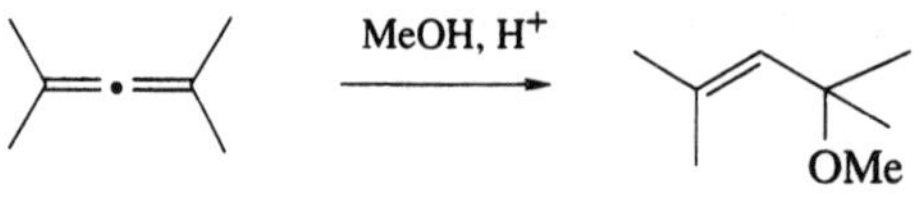

(249)

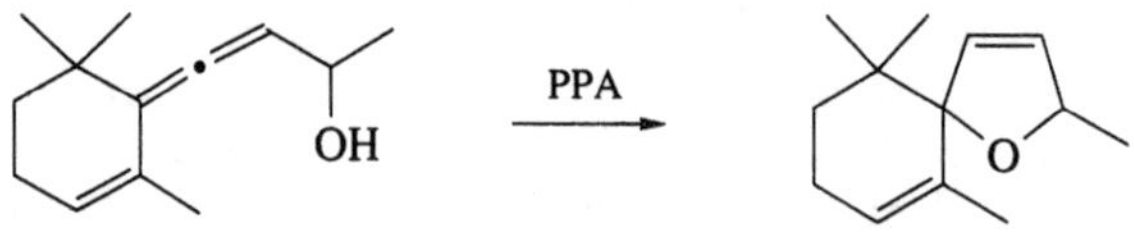

(250)

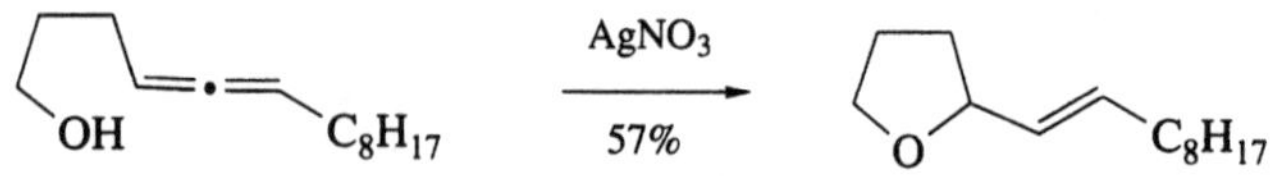

(251)

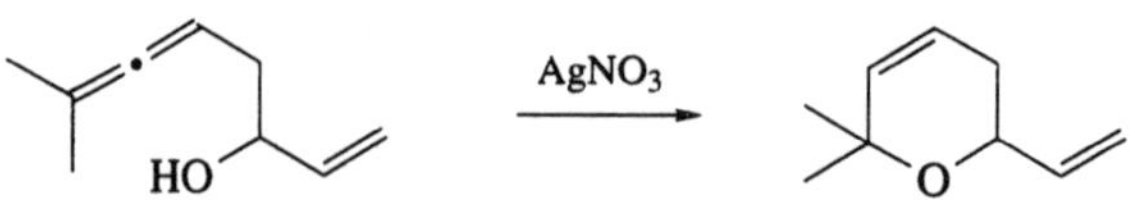

(252)

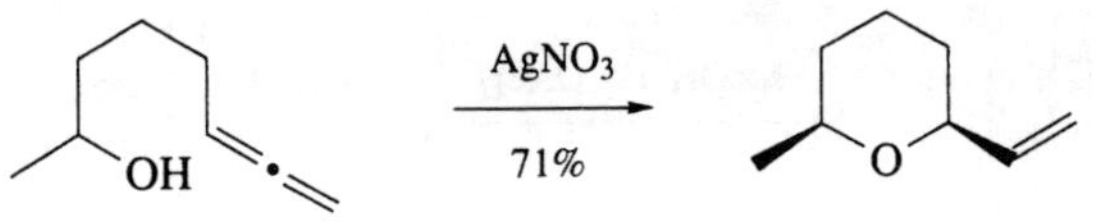

$$\xrightarrow[71\%]{AgNO_3} \quad (253)$$

The addition of alcohols to alkynes is catalyzed by bases,[411,412] or salts of palladium,[413] silver[414] or mercury[413] (equations 254–256). The mercury-catalyzed processes are summarized in the following section.

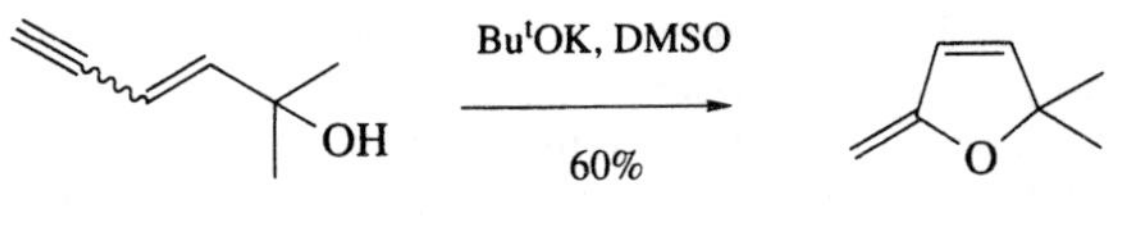

$$\xrightarrow[60\%]{Bu^tOK,\ DMSO} \quad \text{(ref. 412)} \quad (254)$$

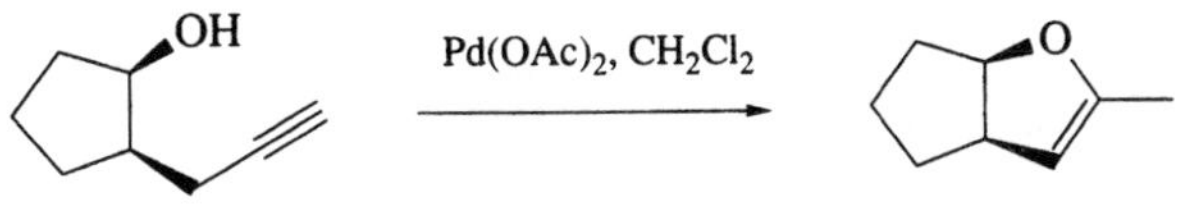

$$\xrightarrow{Pd(OAc)_2,\ CH_2Cl_2} \quad (255)$$

$$\xrightarrow[90\%]{Ag_2CO_3,\ \text{benzene}} \quad (256)$$

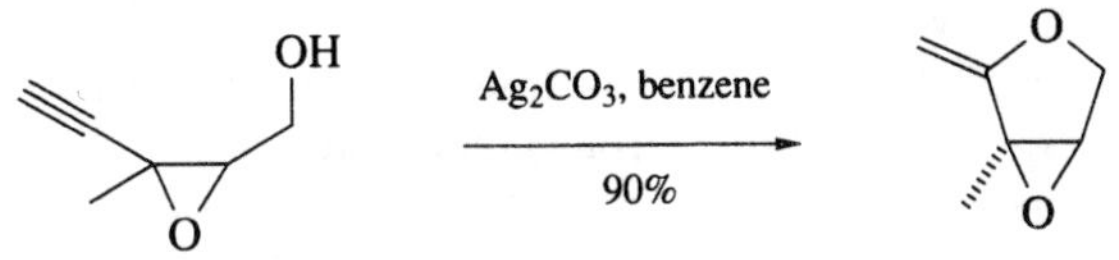

1.7.4.3.2 Alkoxymercuration–demercuration

The alkoxymercuration–demercuration of alkenes, dienes and alkynes in the presence of alcohols provides an even more versatile approach to the corresponding ethers than the acid-catalyzed process. This reaction has been extensively studied and thoroughly reviewed recently.[415] The reaction of alkenes is best carried out using mercury(II) acetate or, for more highly substituted alcohols or alkenes, mercury(II) trifluoroacetate (equation 257).[416,417]

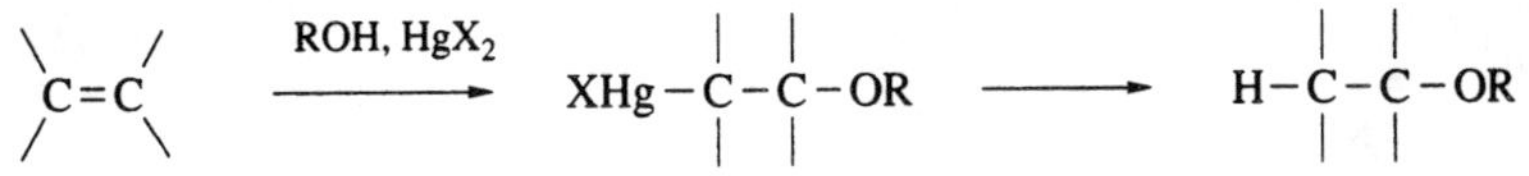

$$\text{C=C} \xrightarrow{ROH,\ HgX_2} XHg-\text{C}-\text{C}-OR \longrightarrow H-\text{C}-\text{C}-OR \quad (257)$$

The reaction is commonly run at room temperature using the alcohol as the solvent. For hindered alcohols, better results can sometimes be obtained at 0 °C.[417] There are few examples in which other solvents or only stoichiometric amounts of alcohols have been employed (equation 258).[418]

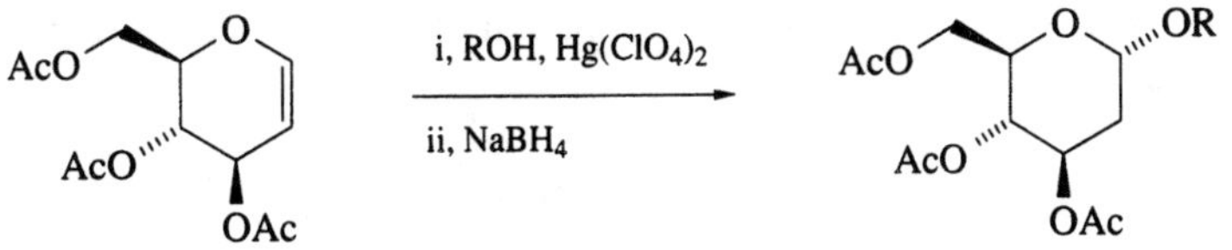

$$\xrightarrow[\text{ii, } NaBH_4]{\text{i, } ROH,\ Hg(ClO_4)_2} \quad (258)$$

A wide variety of alcohols and alkenes can be utilized in this process. Numerous functional groups in either the alcohol or the alkene can be accommodated during alkoxymercuration. For example, alkenes differing in their nucleophilicity as much as vinylic ethers[419] and α,β-unsaturated aldehydes or ketones[420,421] react smoothly (equations 259 and 260).

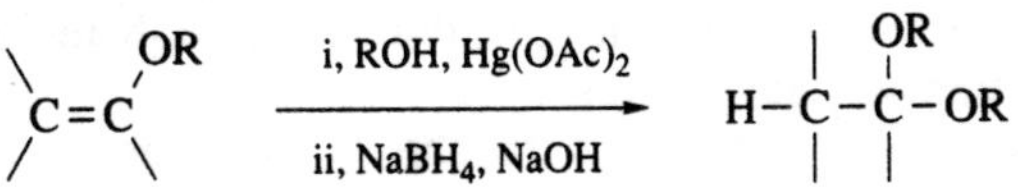

$$\text{C=C(OR)} \xrightarrow[\text{ii, } NaBH_4,\ NaOH]{\text{i, } ROH,\ Hg(OAc)_2} H-\text{C}-\text{C(OR)}-OR \quad (259)$$

$$\text{CH}_2\!=\!\text{C(CH}_3\text{)COR} \xrightarrow{\text{MeOH, Hg(OAc)}_2} \text{AcOHgCH}_2\text{C(OMe)(CH}_3\text{)COR} \qquad (260)$$

R = H, Me

The reaction is extremely versatile with regard to the substitution pattern of the alkene. All but the most hindered of alkenes tend to react and very few functional groups interfere. The regio- and stereo-selectivity is very similar to that reported for hydroxymercuration (see Section 1.7.4.1.2). The reaction is generally highly regioselective in the Markovnikov sense. Carbonyl groups (acids, esters, aldehydes and ketones) can, however, reverse the direction of addition depending on the substitution pattern about the C—C double bond (equation 261). The more remote the functional group, the less its influence.[422–424]

$$\text{C=C–C(=O)–} \xrightarrow{\text{ROH, HgX}_2} \text{RO–C–C(HgX)–C(=O)–} \quad \text{or} \quad \text{XHg–C–C(OR)–C(=O)–} \qquad (261)$$

The stereochemistry of addition to acyclic, cyclic and polycyclic alkenes is essentially identical to that of hydroxymercuration wherever the two processes have been compared. Fewer data have been accumulated for alkoxymercuration however.

One major advantage of the alkoxymercuration–demercuration approach to ethers over the acid-catalyzed process is the fact that carbon skeleton rearrangements are seldom observed. Only unsaturated cyclopropanes,[425,426] or aryl-substituted alkenes[427,428] in the presence of highly electrophilic mercury salts afford rearranged products.

The kinetics of alkoxymercuration have received considerable attention.[415] The relative reactivity of various alkenes parallels that reported for hydroxymercuration.

The demercuration of β-alkoxymercurials is usually best effected using alkaline sodium borohydride. Few rearrangements during this free radical reduction process have been observed.[429] When the mercury moiety is positioned α to a carbonyl group, alkaline H_2S[430,431] or 1,3-propanedithiol[316] provide alternatives that afford complementary stereochemical results (equation 262). The use of sodium–mercury amalgam is also useful for stereospecific reduction.[432,433]

$$\text{R}^1\text{CH=C(R}^2\text{)CO}_2\text{R}^3 \xrightarrow{\text{R}^4\text{OH, Hg(OAc)}_2} \xrightarrow{\text{i or ii}} \text{R}^1\text{CH(OR}^4\text{)CH(R}^2\text{)CO}_2\text{R}^3 \qquad (262)$$

i, $NaBH_4$ — *threo*
ii, $HS(CH_2)_3SH$ — *erythro*

The intramolecular alkoxy- or phenoxy-mercuration of unsaturated alcohols or phenols, respectively, provides an exceptionally useful process for the formation of cyclic ethers, particularly those bearing a five- or six-membered ring (equation 263).[415]

$$\text{C=C–C}_n\text{–OH} \xrightarrow{\text{HgX}_2} \xrightarrow{\text{NaBH}_4} \text{H–C–C(–O–)–C}_n \qquad (263)$$

The choice of mercury salt for this reaction appears to be less critical than intermolecular variants and many salts (X = Cl, OAc, NO_3, SO_4, ClO_4, CF_3CO_2) have been successfully utilized. The electrophilicity of the salt can effect the regio- and stereo-chemistry of addition.[434–438] The regioselectivity is also determined by the substitution pattern of the unsaturated alcohol. For cycloalkenes, steric factors, ring strain, and other reaction variables become important (equation 264). High regio- and stereo-selectivity are often encountered in these intramolecular processes.

The demercuration of these cyclic mercurials is fraught with more problems than analogous mercurials formed by intermolecular processes. Alkaline sodium borohydride is once again the most common reducing agent, but elimination to the starting unsaturated alcohol is not unusual. The extent of elimination varies with the mercury ligand, the pH and the solvent used.[434] Phase transfer approaches offer advant-

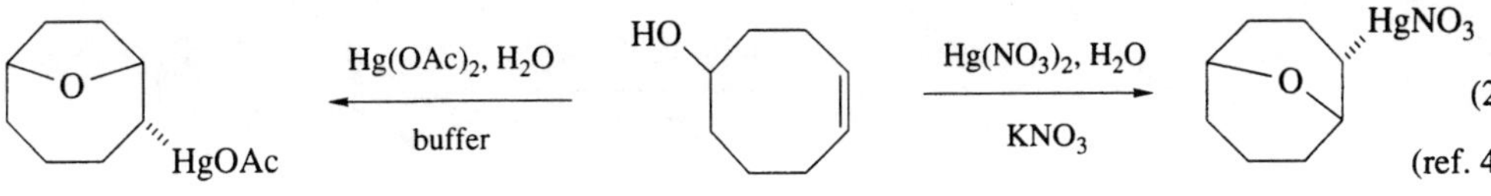

(264)
(ref. 434)

ages here.[439–441] Sodium trithiocarbonate reduction of cyclic alkoxymercurials bearing a neighboring carboxylic acid moiety provides stereoselectivity complementary to $NaBH_4$.[442]

The alkoxymercuration of dienes and polyenes has been extensively studied. Mono-, di- or poly-mercurated adducts are formed as expected based on results from the alkoxymercuration of analogous alkenes, but the product distribution varies depending on the nature of the diene or polyene.

Nonconjugated dienes or polyenes can be monomercurated if the diene or polyene is employed in excess, but more frequently these compounds are allowed to react with an excess of the mercury salt and di- or poly-mercurated products are obtained. The relative reactivity of isolated double bonds is basically that expected from studies on simple alkenes.

The alkoxymercuration of conjugated dienes is more complex since these compounds are less reactive than simple alkenes and the products often react further to afford dimercurated products whose regio- or stereo-chemistry is strongly affected by the initial product. For example, 1,3-butadiene, 1,3-pentadiene, 2,3-dimethyl-1,3-butadiene and 1,3-cyclohexadiene react with mercury(II) acetate in methanol to produce 1,2-adducts (equation 265).[443–446] These adducts sometimes rearrange with time.[446] 1,3-Butadiene reacts further to afford primarily the *meso* adduct of double 1,2-additions.[447] Surprisingly, little additional work has been reported on reactions of conjugated dienes.

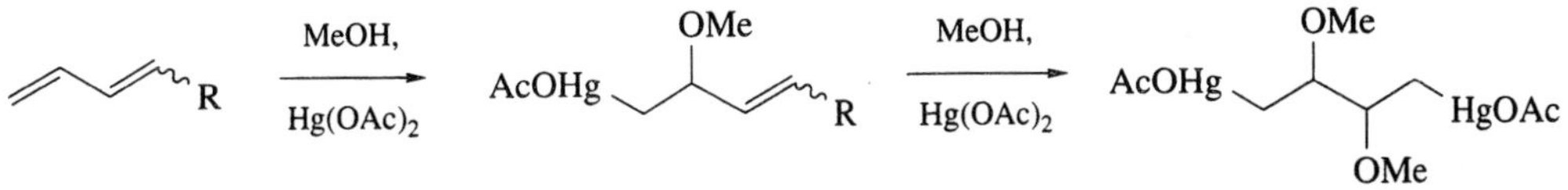

(265)

Considerable work on the alkoxymercuration of allenes has been reported.[415] Mono- or di-mercuration products are observed depending on the substitution pattern of the allene. While allene and 1,2-butadiene afford dimercurated acetals, more highly substituted allenes produce vinylmercurials (equations 266 and 267).[448] Cyclic allenes also afford vinylmercurials.[449,450] Alkaline sodium borohydride reduction of these vinylmercurials produces the expected allylic ethers.

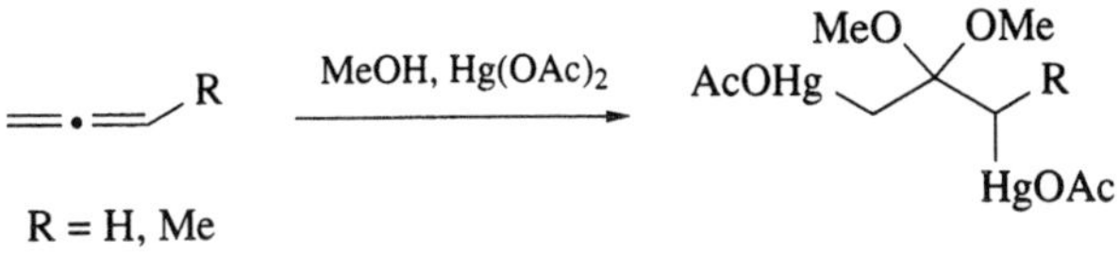

(266)

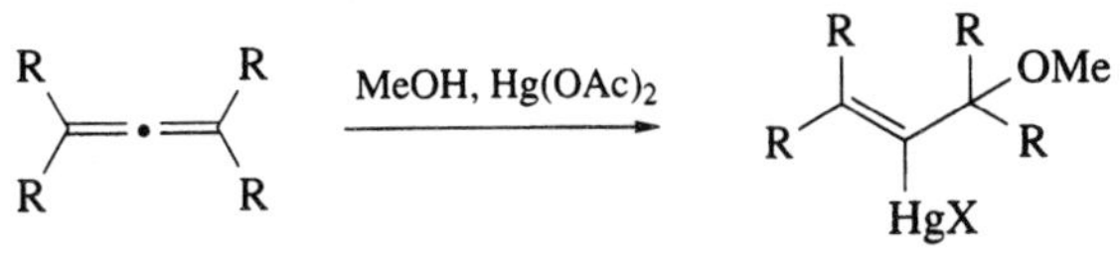

R = H, Me

(267)

Nonconjugated dienols undergo intramolecular cyclization in the same manner as simple dienols (equation 268).[451]

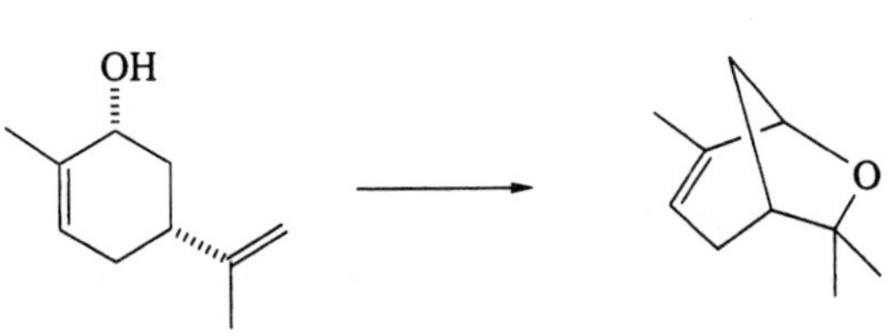

(268)

The alkoxymercuration of conjugated dienols does not appear to have been examined, while allenic alcohols provide unsaturated ethers (equation 269).[452]

The alkoxymercuration of alkynes generally provides vinylic ethers[453] or acetals[454] depending on the reaction conditions (equations 270 and 271). Only catalytic amounts of mercury salts are generally

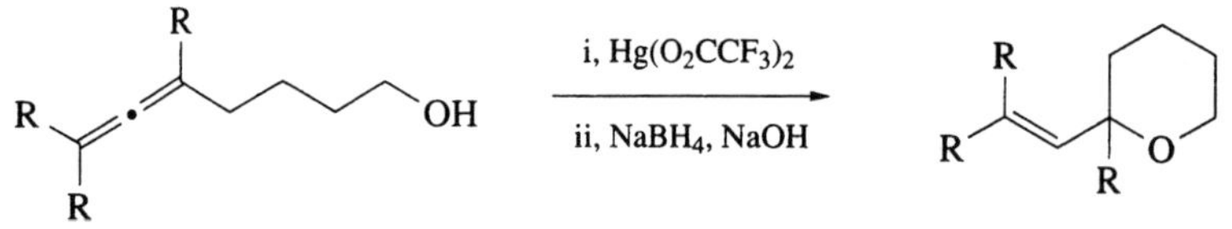

(269)

necessary in these reactions. Internal alkynes usually afford acetals exclusively. However, the intramolecular alkoxymercuration of alkynols can be stopped at the vinyl ether stage (equation 272).[455]

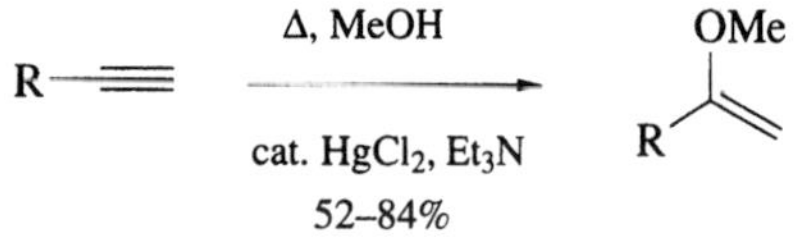

(270)

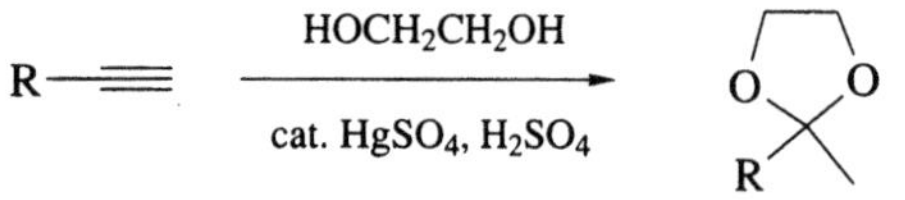

(271)

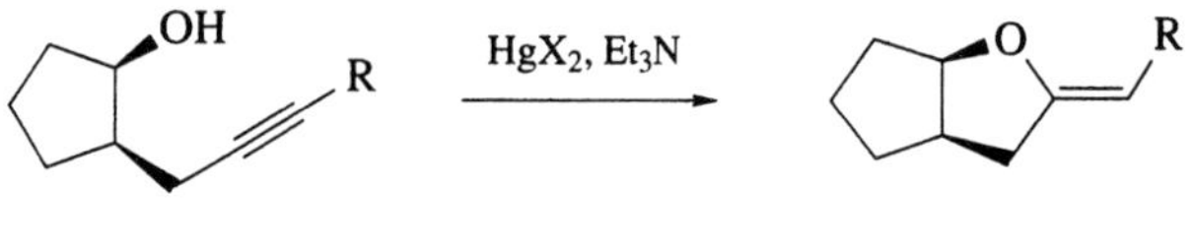

(272)

1.7.4.4 H—O_2CR Addition

1.7.4.4.1 Direct addition of H—O_2CR

Although a seldom used laboratory method, the addition of carboxylic acids to alkenes provides carboxylic esters. The addition is catalyzed by protic acids or Lewis acids and obeys Markovnikov's rule (equation 273).[456–458]

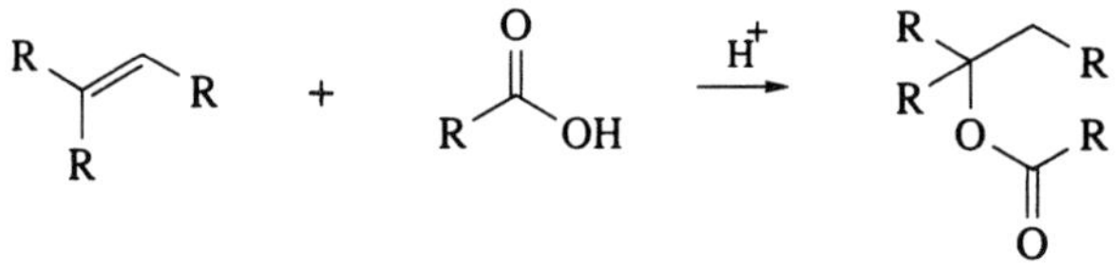

(273)

The best yields of ester are generally obtained with alkenes which provide highly stabilized carbocations. This reaction is particularly effective for the preparation of hindered esters (equation 274).[459]

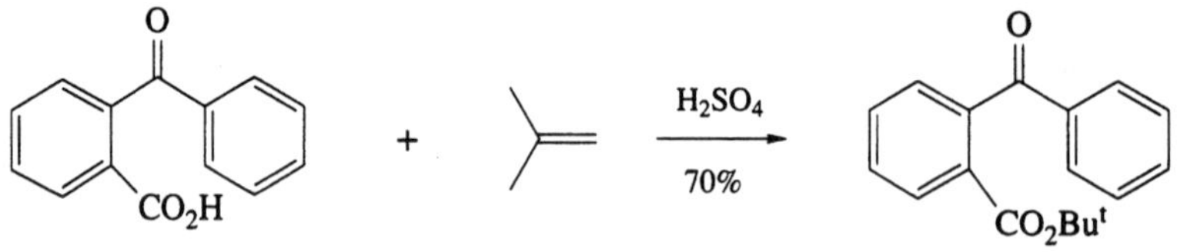

(274)

To obtain high yields from the esterification of terminal alkenes, boron trifluoride–hydrogen fluoride is reported to be more effective than sulfuric acid as a catalyst (equation 275).[456]

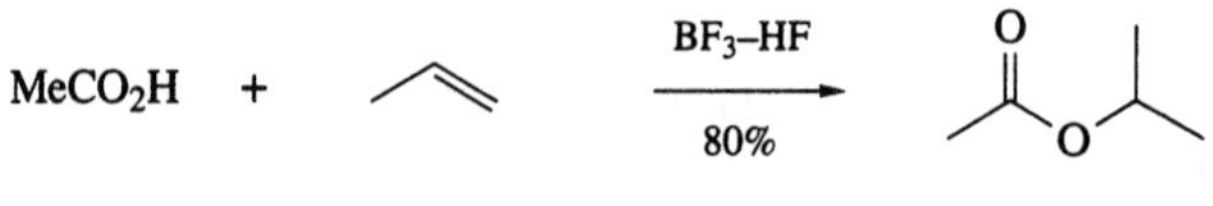

(275)

In the presence of a strong acid as catalyst, alkenoic acids will add internally to furnish a valuable route to lactones (Scheme 4).[460] Regardless of the position of the double bond in the chain, the product obtained is always the γ- and/or δ-lactone due to acid-catalyzed double bond migration to a position favorable for ring closure.[461] The choice of catalyst and the reaction temperature are important in minimizing the amount of cycloalkenone side product that usually accompanies lactone formation.[462]

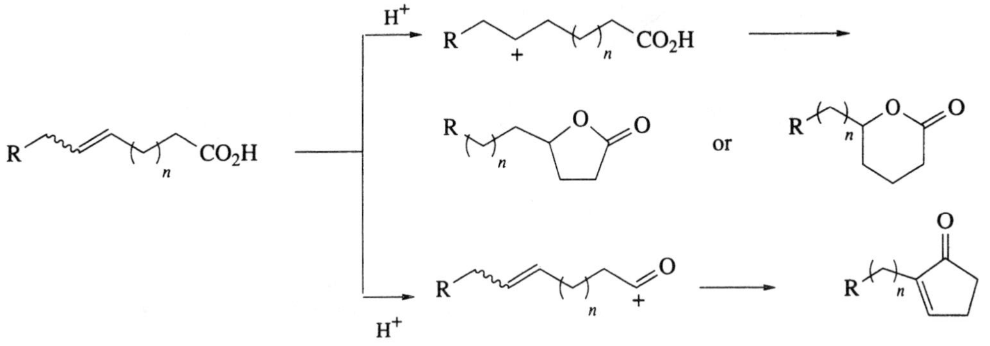

Scheme 4

In an earlier study, interesting stereochemical aspects of this lactonization reaction were demonstrated, as illustrated in the synthesis of (±)-ambreinolide (equation 276).[463]

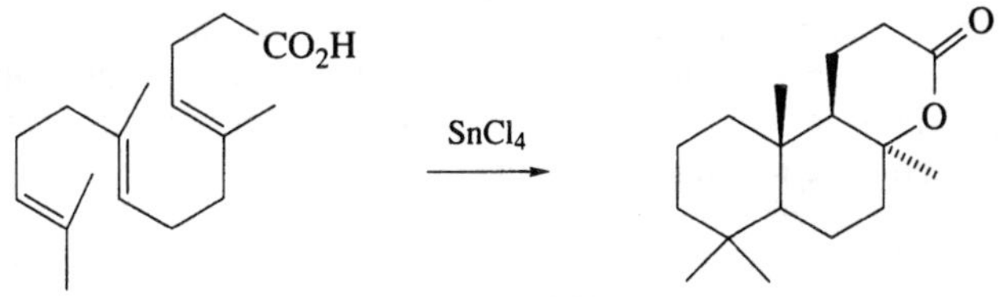

(276)

The intermolecular addition of carboxylic acids to dienes is not very efficient. Conjugated dienes generally form polymeric products,[456] whereas nonconjugated dienes tend to give polycyclic products (equation 277).[464] The addition to allenes, on the other hand, is marred by regio- or stereo-isomerism (equation 278).[465]

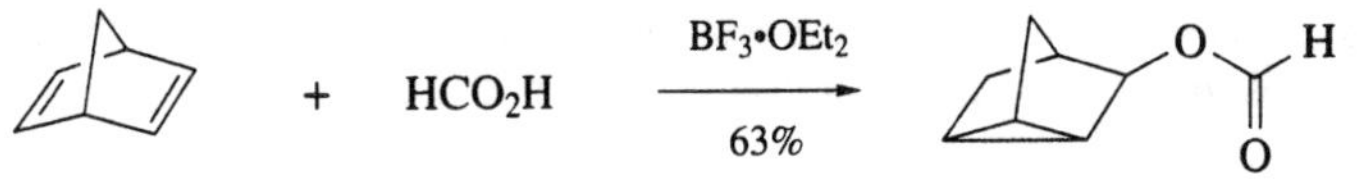

(277)

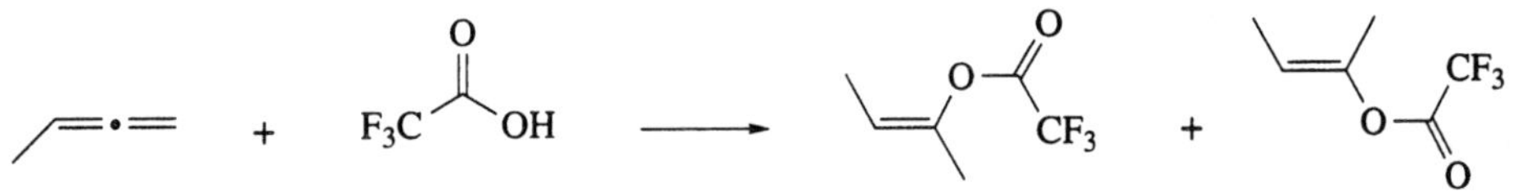

(278)

More recently, interest in the addition of carboxylic acids to alkenes has focused on the development of new catalytic systems such as clay,[466] sheet silicate (montmorillonites),[467] wofatit[468] and zeolites.[469]

The addition of carboxylic acids to alkynes affords enol esters which are useful as intermediates in organic synthesis.[470] As in the addition to alkenes, a catalyst is usually required for high conversions of alkynes to enol esters. Simple acid catalysis has been employed (equation 279),[471] but the more common catalysts are Lewis acids, such as boron trifluoride etherate,[472] silver nitrate,[473] zinc acetate[474] and zinc oxide (equations 280 and 281).[475,476]

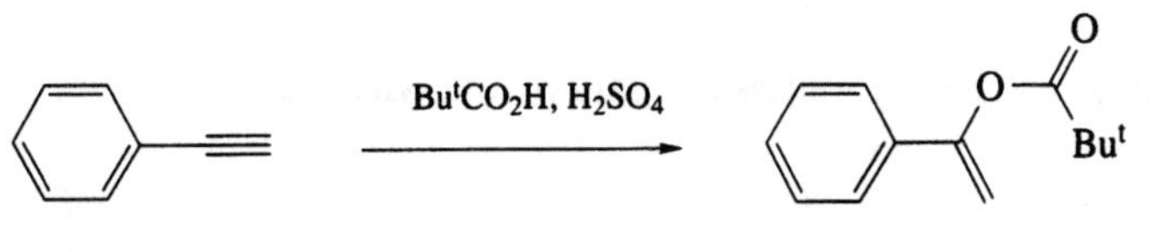

(279)

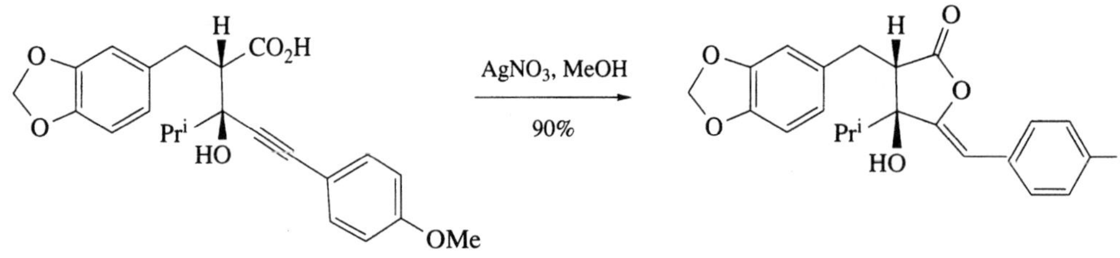

$AgNO_3$, MeOH; 90% (280)

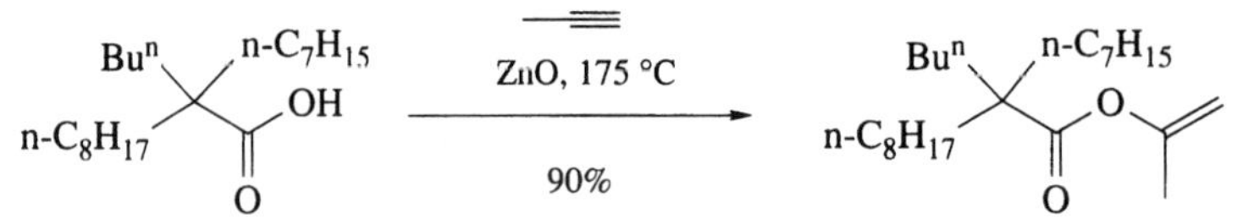

ZnO, 175 °C; 90% (281)

Mercury(II) salts and transition metals are also important catalysts for this reaction. Mercury catalysis will be discussed in the next section, while transition metal catalysis is discussed elsewhere in this series.

1.7.4.4.2 Acyloxymercuration–demercuration

The acyloxymercuration–demercuration of alkenes provides an alternative route to esters which is probably less prone to carbon skeleton rearrangements than the direct addition of carboxylic acids to alkenes (equation 282). This reaction has recently been reviewed.[477] The reaction is most commonly run using mercury(II) acetate in acetic acid, though other mercury salts may be used and aprotic solvents can also be employed. Equilibria have been measured for the reaction of mercury(II) trifluoroacetate and alkenes in tetrahydrofuran, and were found to be solvent dependent.[478]

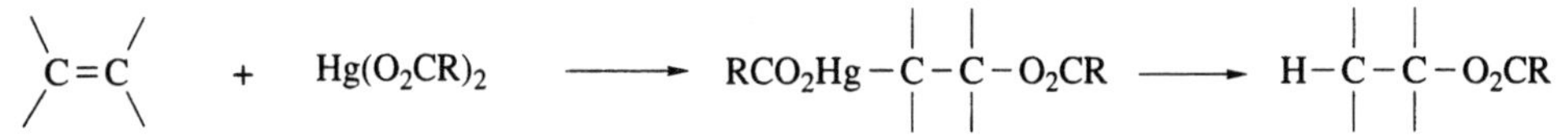

$$\mathrm{>C{=}C<} + \mathrm{Hg(O_2CR)_2} \longrightarrow \mathrm{RCO_2Hg{-}\overset{|}{\underset{|}{C}}{-}\overset{|}{\underset{|}{C}}{-}O_2CR} \longrightarrow \mathrm{H{-}\overset{|}{\underset{|}{C}}{-}\overset{|}{\underset{|}{C}}{-}O_2CR} \qquad (282)$$

Acyloxymercuration follows Markovnikov's rule and involves *anti* addition with most cyclic alkenes.[479–482] However, certain strained bicyclic alkenes, such as norbornene, afford *cis exo* adducts.[483] With substituted cyclic alkenes, the regio- and stereo-selectivity of acetoxymercuration is very similar to hydroxy- and methoxy-mercuration.[484,485] While simple alkenes, and vinylic ethers[486,487] and esters[488–491] undergo acetoxymercuration, there does not appear to be any example of alkenes bearing electron-withdrawing groups undergoing this reaction (equation 283).

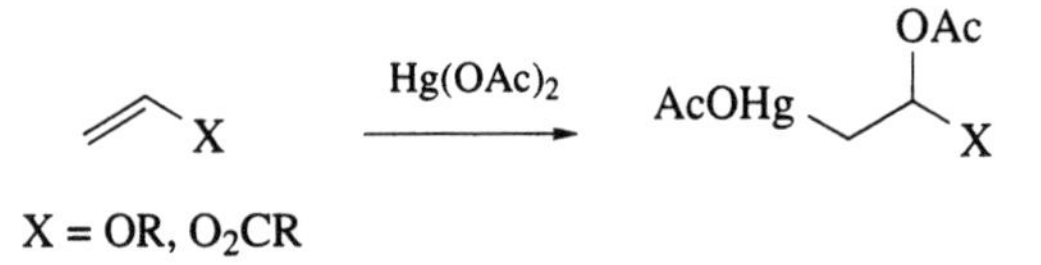

$Hg(OAc)_2$ (283)

X = OR, O_2CR

The reaction accommodates halides, esters, ethers, nitriles, cyclopropanes, epoxides, alcohols and nitro groups. Even carbohydrates can be used.[492–494] However, vinylic halides afford aldehydes and ketones (equation 284).[495–498]

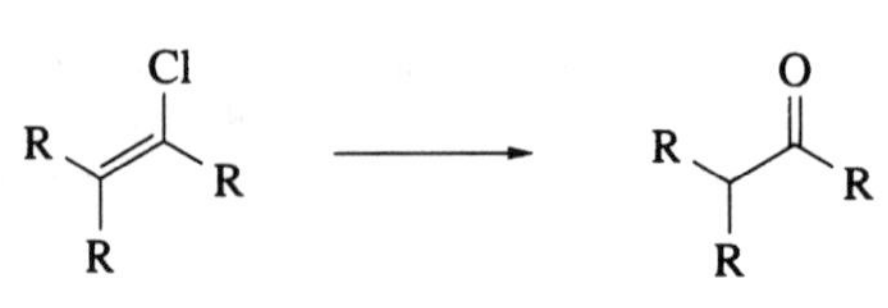

(284)

Mercury(II) salts also catalyze the transesterification of vinylic esters (equation 285).[477]

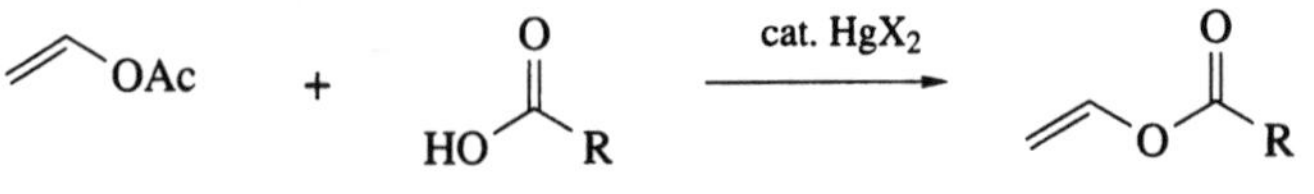

cat. HgX_2 (285)

While the demercuration of β-acyloxymercurials has not received much attention, sodium borohydride or alkaline sodium borohydride have been most widely employed for this task. Hydrolysis to the corresponding alcohol or reversion to alkene are significant side reactions, however, and they are sensitive to the solvent used.[499,500] Sodium amalgam and tri-*n*-butyltin hydride[501] have also been utilized as reducing agents.

A number of alkenoic acids undergo intramolecular acyloxymercuration when treated with mercury(II) acetate or mercury(II) chloride (equations 286 and 287).[477]

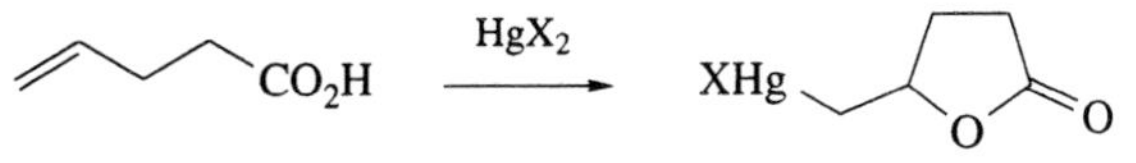

CO_2H —HgX_2→ XHg (286)

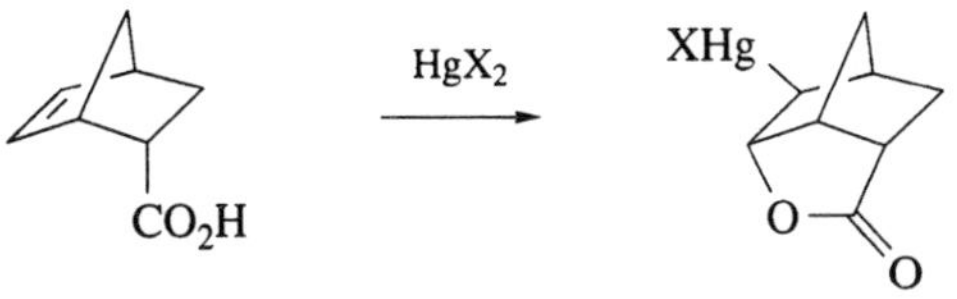

CO_2H —HgX_2→ XHg (287)

It appears that all but one[502] (eleven-membered ring) of the examples of this reaction in the literature involve the synthesis of monocyclic or tricyclic five-membered ring lactones.[503–507] Sodium borohydride,[508,509] sodium trimethoxyborohydride[510] and sodium amalgam[483] have been employed for demercuration of these substrates.

Dienes and polyenes undergo acyloxymercuration. Nonconjugated dienes afford either mono- or dimercurated species, depending on the stoichiometry of the reaction (equation 288).[511–513] Products involving carbon–carbon bond formation are occasionally observed during this process.[512] Allenes afford vinylmercurials (equation 289).[514,515]

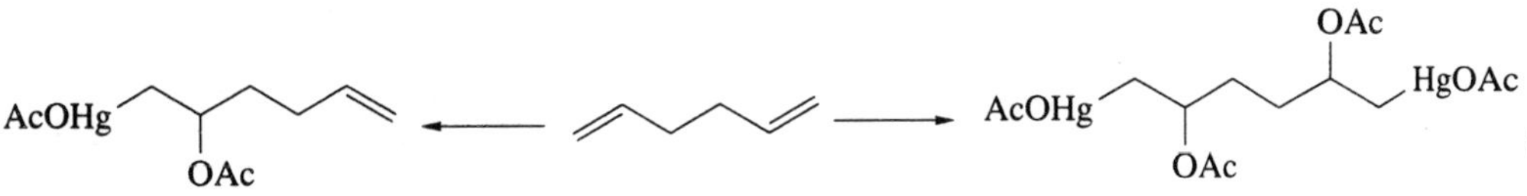

AcOHg, OAc ← → AcOHg, OAc, OAc, HgOAc (288)

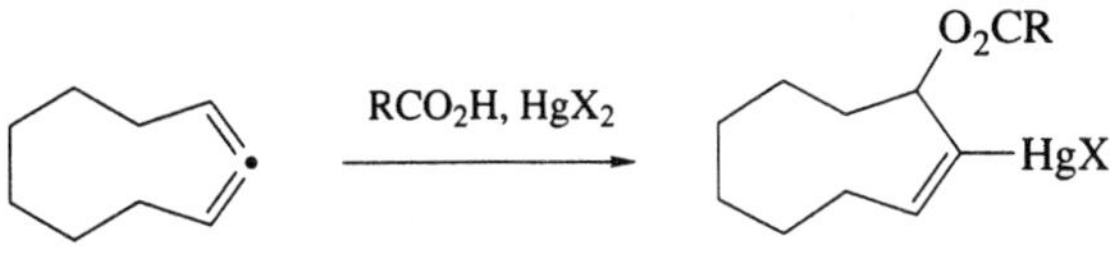

—RCO_2H, HgX_2→ O_2CR, HgX (289)

The acyloxymercuration of alkynes has been reported to produce a wide variety of products. Terminal alkynes afford either dialkynylmercurials[516,517] or polymercurated products whose structures have not been well established (equation 290). Internal alkynes usually afford vinylic mercurials in which the mercury(II) salt has added in an *anti* fashion (equation 291).[518–520] Only sodium borohydride has been used to demercurate a few of these mercurials.[520]

$$2\ R{-}{\equiv} \quad + \quad Hg(OAc)_2 \longrightarrow R{-}{\equiv}{-}Hg{-}{\equiv}{-}R \qquad (290)$$

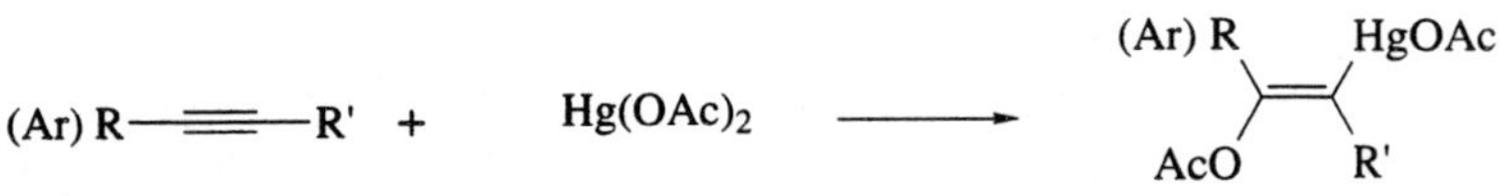

$$(Ar)\ R{-}{\equiv}{-}R' \quad + \quad Hg(OAc)_2 \longrightarrow (Ar)\ R(AcO)C{=}C(HgOAc)R' \qquad (291)$$

Mercury salts, such as mercury(II) acetate,[521–525] mercury(II) oxide,[524,526–528] mercury(II) trifluoroacetate,[529,530] mercury(II) sulfate[524,531] and mercury(II) phosphate[531] catalyze the addition of carboxylic acids to alkynes. Acetic anhydride in the presence of boron trifluoride etherate can also be effectively used in this reaction (equation 292).[521,522] Alkynoic acids undergo mercury-catalyzed cyclization to lactones (equation 293).[523,532,533]

Applications of the lactonization reaction are illustrated in the syntheses of obtusilactones,[530] benzamidopyranone,[529] oxaspirolactone,[527] diaxospirodione,[528] and morphohinones[526] (equations 294–298).

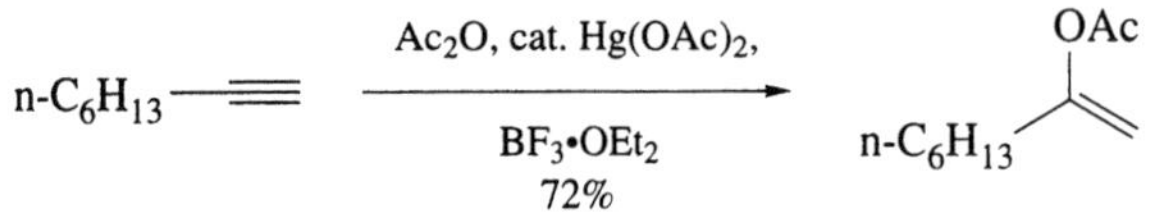
n-C6H13
Ac2O, cat. Hg(OAc)2,
BF3•OEt2
72%
OAc
n-C6H13
(292)

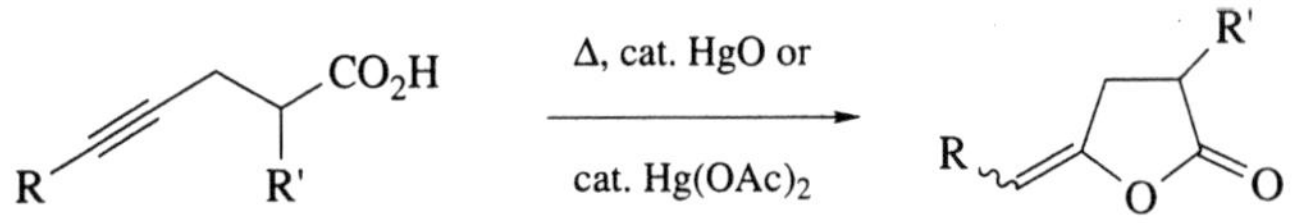
CO2H
R
R'
Δ, cat. HgO or
cat. Hg(OAc)2
R'
R
O
O
(293)

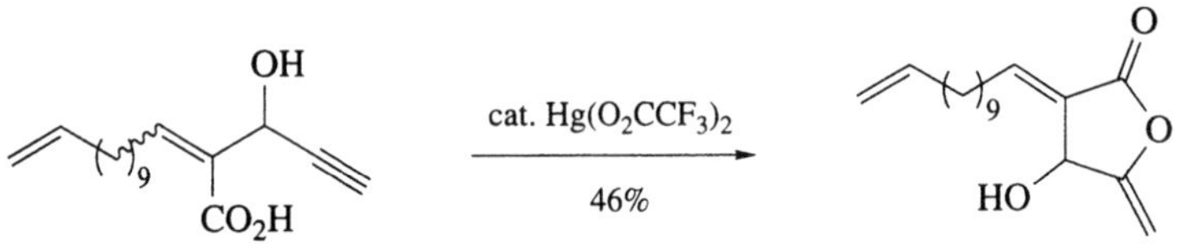
OH
9
CO2H
cat. Hg(O2CCF3)2
46%
O
9
O
HO
(294)

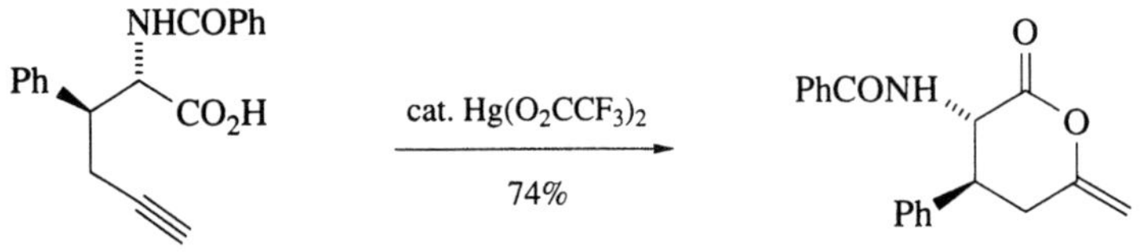
NHCOPh
Ph
CO2H
cat. Hg(O2CCF3)2
74%
O
PhCONH
O
Ph
(295)

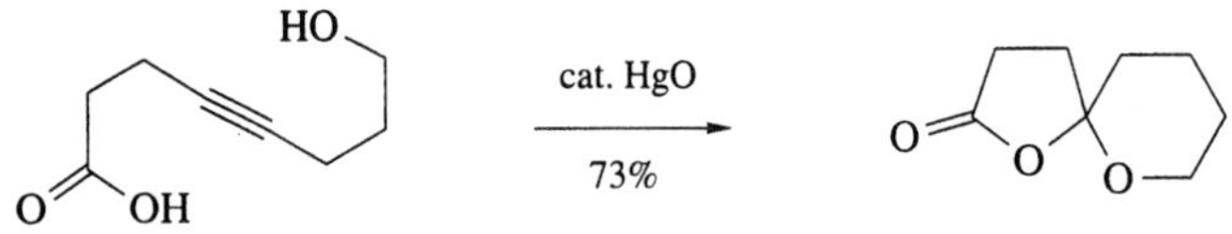
HO
O
OH
cat. HgO
73%
O
O
O
(296)

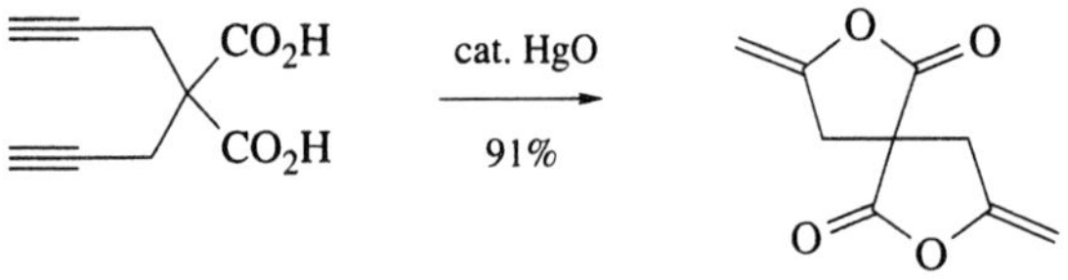
CO2H
CO2H
cat. HgO
91%
O
O
O
O
(297)

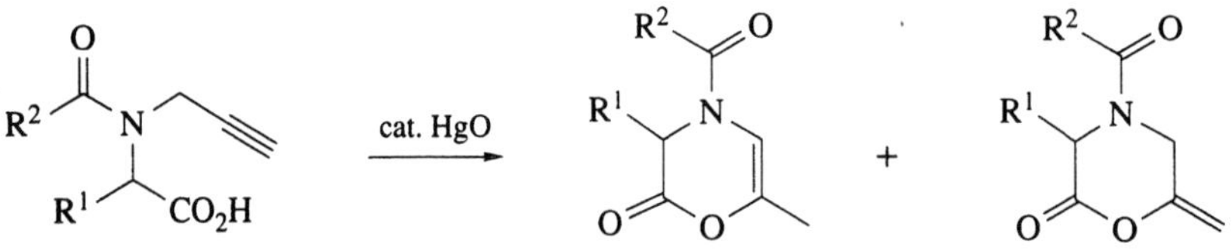
O
R2
N
R1
CO2H
cat. HgO
R2
O
R1
N
O
O
+
R2
O
R1
N
O
O
(298)

1.7.5 H—SR ADDITION

The addition of thiols to C—C multiple bonds may proceed *via* an electrophilic pathway involving ionic processes or a free radical chain pathway. The main emphasis in the literature has been on the free radical pathway, and little work exists on electrophilic processes.[534–537] The normal mode of addition of the relatively weakly acidic thiols is by the electrophilic pathway in accordance with Markovnikov's rule (equation 299). However, it is established that even the smallest traces of peroxide impurities, oxygen or the presence of light will initiate the free radical mode of addition leading to anti-Markovnikov products. Fortunately, the electrophilic addition of thiols is catalyzed by protic acids, such as sulfuric acid[538] and *p*-toluenesulfonic acid,[539] and Lewis acids, such as aluminum chloride,[540] boron trifluoride,[536] titanium tetrachloride,[540] tin(IV) chloride,[536,540] zinc chloride[536] and sulfur dioxide.[541]

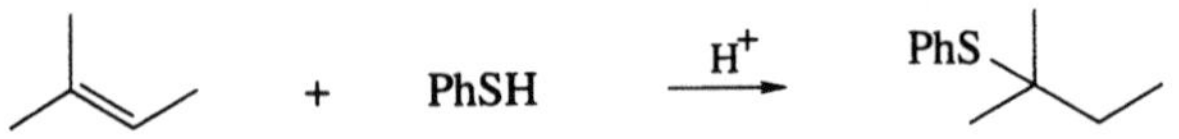
+
PhSH
H+
PhS
(299)

More acidic H—SR reagents, such as dialkyldithiophosphoric acids, may add to alkenes without the aid of a catalyst to yield exclusively the product of Markovnikov addition (Scheme 5).[542] The anti-Markovnikov product is obtained if the alkene used contains even slight amounts of peroxide impurities.[543]

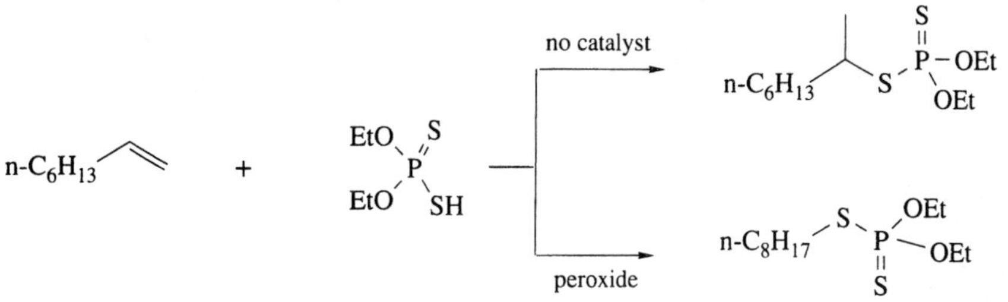

no catalyst

peroxide

Scheme 5

The reaction of H—SR compounds with vinylic ethers and vinylic sulfides occurs more readily than with unactivated alkenes (equations 300 and 301).[544–547] An application of this reaction is found in the synthesis of organophosphorus insecticides.[548]

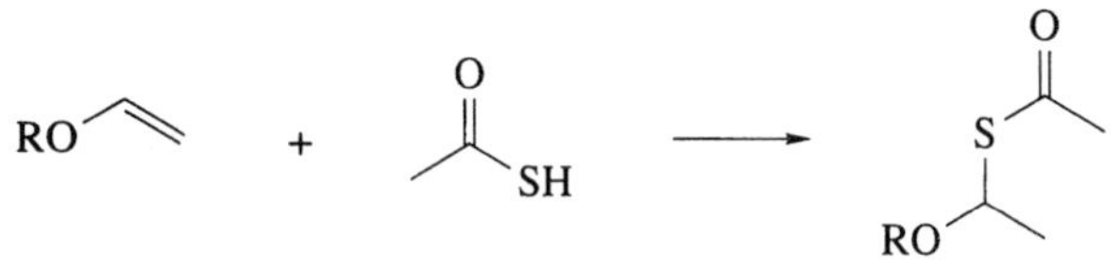

(300)

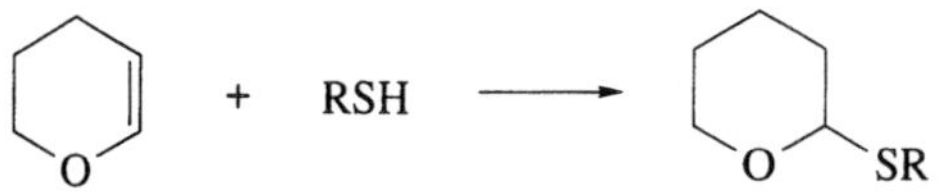

(301)

The intramolecular addition of thiols to alkenes provides a novel entry into heterocycles. One example of this is the reaction of hydrogen sulfide with various nonconjugated dienes to form six-membered rings (equation 302).[549,550]

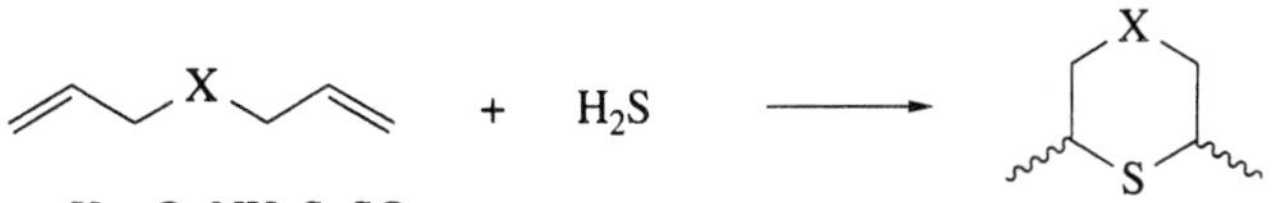

$X = O, NH, S, SO_2$ (302)

An unusual example of intramolecular thiol addition to double bonds is illustrated in the synthesis of spirothiazines.[551] The key step is described as an acid-catalyzed Markovnikov addition of a thiol to a double bond (Scheme 6).

The thiylation of conjugated dienes and allenes proceeds as expected. Thus, the acid-catalyzed reaction of thiols with 1,3-dienes affords the 1,4-addition product,[552] whereas the nucleophilic addition to allene provides the product arising from sulfur attack at the central carbon (equations 303 and 304).[553]

In contrast to the reaction of alkenes, the addition of H—SR reagents to alkynes generally requires higher temperatures and the presence of a base.[534] Terminal alkynes generally react to give anti-Markovnikov products. Substituted alkynes, however, will provide terminal alkenes if the reaction is carried out in sodium and liquid ammonia (equation 305).[554] Polyynes react with H_2S to produce thiophenes (equation 306).[555]

Recently, it has been reported that sulfenic acids add readily to 1-alkynes to give α,β-unsaturated sulfoxides,[556] themselves useful for the synthesis of various cyclopentenones (equation 307).[557]

1.7.6 H—SeR ADDITION

There are several reviews that enumerate the many methods that effect the addition of H—SeR to C—C multiple bonds.[558,559] Of these, only a few methods involve true electrophilic additions of selenols to C—C multiple bonds. High yields of the Markovnikov product are reported when the relatively acidic

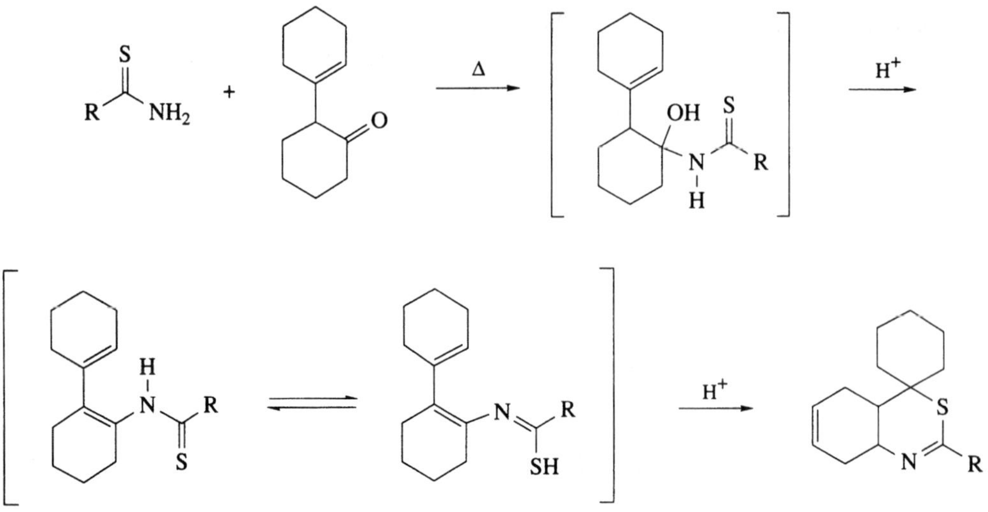

Scheme 6

(303)

(304)

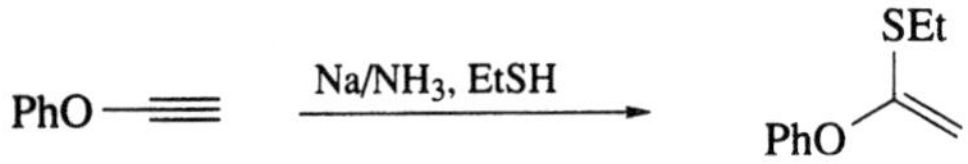

(305)

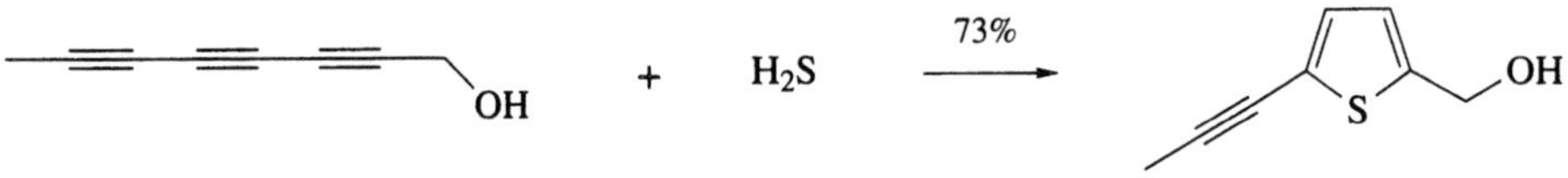

(306)

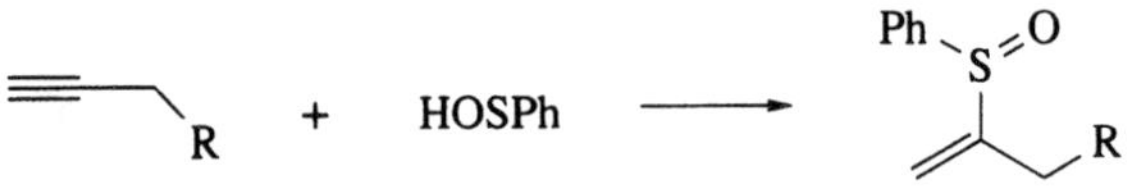

(307)

selenophenol reacts with vinylic sulfides, vinylic ethers or vinylic selenides (equation 308).[560] In the reaction with vinylic selenides, $BF_3{\cdot}OEt_2$ is required as a catalyst, because of the relatively nonnucleophilic double bond. Recently, it has been reported that the acid-catalyzed intramolecular selenol addition to double bonds affords spiroimino selenides (equation 309).[561]

An example of adding selenols to triple bonds in a Markovnikov fashion is the reaction of selenophenol and 1-hexyne (equation 310).[562]

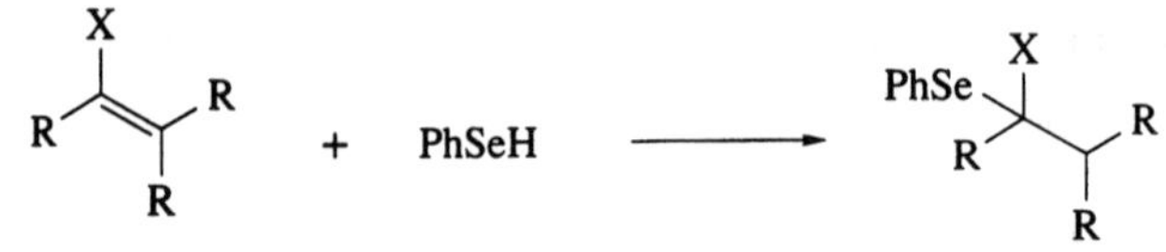

(308)

X = OR, SR, SeR

Se ‖ R–C–NH$_2$ + [2-(cyclohex-1-enyl)cyclohexanone] —(Δ, H$^+$, toluene)→ [N=C(R)SeH enamine] —(H$^+$)→ [spiro Se,N heterocycle, R] (309)

$$\text{Bu}^n\text{—}\equiv \ + \ \text{PhSeH} \xrightarrow[80\%]{\text{NaOEt}} \text{Bu}^n(\text{PhSe})\text{C}{=}\text{CH}_2 \quad (310)$$

1.7.7 REFERENCES

1. M. J. S. Dewar and R. C. Fahey, *Angew. Chem., Int. Ed. Engl.*, 1964, **3**, 245.
2. R. C. Fahey, *Top. Stereochem.*, 1968, **3**, 237.
3. P. B. D. de la Mare and R. Bolton, in 'Electrophilic Additions to Unsaturated Systems', Elsevier, New York, 1966, chap. 5.
4. A. L. Henne, *Org. React. (N.Y.)*, 1944, **2**, 50.
5. E. Forche, *Methoden Org. Chem. (Houben-Weyl)*, 1962, **V/3**, 99.
6. C. M. Sharts and W. A. Sheppard, *Org. React. (N.Y.)*, 1974, **21**, 125.
7. A. V. Grosse and C. B. Linn, *J. Org. Chem.*, 1938, **3**, 26.
8. A. V. Grosse, R. C. Wackher and C. B. Linn, *J. Phys. Chem.*, 1940, **44**, 275.
9. S. M. McElvain and J. W. Langston, *J. Am. Chem. Soc.*, 1944, **66**, 1759.
10. K. O. Alt and C. D. Weis, *Helv. Chim. Acta*, 1969, **52**, 812.
11. M. Hanack and R. Hähnle, *Chem. Ber.*, 1962, **95**, 191.
12. M. Hanack and W. Keberle, *Chem. Ber.*, 1961, **94**, 62.
13. J. L. Webb and J. E. Corn, *J. Org. Chem.*, 1973, **38**, 2091.
14. A. L. Henne and F. W. Haeckl, *J. Am. Chem. Soc.*, 1941, **63**, 2692.
15. M. W. Renoll, *J. Am. Chem. Soc.*, 1942, **64**, 1115.
16. A. L. Henne and A. M. Whaley, *J. Am. Chem. Soc.*, 1942, **64**, 1157.
17. A. L. Henne and R. C. Arnold, *J. Am. Chem. Soc.*, 1948, **70**, 758.
18. A. L. Henne and E. P. Plueddeman, *J. Am. Chem. Soc.*, 1943, **65**, 1271.
19. A. L. Henne and R. C. Arnold, *J. Am. Chem. Soc.*, 1948, **70**, 758.
20. G. A. Olah, M. Nojima and I. Kerekes, *Synthesis*, 1973, 779.
21. M. Hanack, H. Eggensperger and R. Hähnle, *Justus Liebigs Ann. Chem.*, 1962, **652**, 96.
22. M. Hanack and W. Kaiser, *Justus Liebigs Ann. Chem.*, 1962, **657**, 12.
23. E. Forche, *Methoden Org. Chem. (Houben-Weyl)*, 1962, **V/3**, 108.
24. A. V. Grosse and C. B. Linn, *J. Am. Chem. Soc.*, 1942, **64**, 2289.
25. R. N. Haszeldine, *J. Chem. Soc.*, 1952, 3490.
26. A. L. Henne and E. P. Plueddeman, *J. Am. Chem. Soc.*, 1943, **65**, 587.
27. P. Albert and J. Cousseau, *J. Chem. Soc., Chem. Commun.*, 1985, 961.
28. J. Cousseau and P. Albert, *Bull. Soc. Chim. Fr.*, 1986, 910.
29. R. Stroh, *Methoden Org. Chem. (Houben-Weyl)*, 1962, **V/3**, 812.
30. H. O. House, in 'Modern Synthetic Reactions,' 2nd edn., Benjamin, Menlo Park, CA, 1972, p. 446.
31. A. Michael and F. Zeidler, *Justus Liebigs Ann. Chem.*, 1911, **385**, 269.
32. H. C. Brown and M. Borkowski, *J. Am. Chem. Soc.*, 1952, **74**, 1894.
33. R. A. Pacaud and C. F. H. Allen, *Org. Synth., Coll. Vol.*, 1943, **2**, 336.
34. H. C. Brown and M.-H. Rei, *J. Org. Chem.*, 1966, **31**, 1090.
35. R. H. Hall, R. G. Pyke and G. F. Wright, *J. Am. Chem. Soc.*, 1952, **74**, 1597.
36. F. C. Whitmore and F. Johnston, *J. Am. Chem. Soc.*, 1933, **55**, 5020.
37. Y. Pocker and K. D. Stevens, *J. Am. Chem. Soc.*, 1969, **91**, 4205.
38. G. G. Ecke, N. C. Cook and F. C. Whitmore, *J. Am. Chem. Soc.*, 1950, **72**, 1511.
39. G. G. Ecke, N. C. Cook and F. C. Whitmore, *J. Am. Chem. Soc.*, 1950, **72**, 1511.
40. J. Tierney, F. Costello and D. R. Dalton, *J. Org. Chem.*, 1986, **51**, 5191.
41. G. S. Hammond and C. H. Collins, *J. Am. Chem. Soc.*, 1960, **82**, 4323.
42. R. C. Fahey and M. W. Monahan, *J. Am. Chem. Soc.*, 1970, **92**, 2816.
43. R. C. Fahey and C. A. McPherson, *J. Am. Chem. Soc.*, 1971, **93**, 2445.
44. K. B. Becker and C. A. Grob, *Synthesis*, 1973, 789.
45. M. J. S. Dewar and R. C. Fahey, *J. Am. Chem. Soc.*, 1963, **85**, 2245.
46. K. D. Berlin, R. O. Lyerla, D. E. Gibbs and J. P. Devlin, *J. Chem. Soc., Chem. Commun.*, 1970, 1246.
47. H. C. Brown and K.-T. Liu, *J. Am. Chem. Soc.*, 1967, **89**, 3898.
48. J. K. Stille, F. M. Sonnenberg and T. H. Kinstle, *J. Am. Chem. Soc.*, 1966, **88**, 4922.
49. H. C. Brown and K.-T. Liu, *J. Am. Chem. Soc.*, 1967, **89**, 466.
50. S. J. Cristol and R. Caple, *J. Org. Chem.*, 1966, **31**, 2741.
51. F. T. Bond, *J. Am. Chem. Soc.*, 1968, **90**, 5326.
52. H. C. Brown and K.-T. Liu, *J. Am. Chem. Soc.*, 1967, **89**, 3900.

53. P. K. Freeman, F. A. Raymond and M. F. Grostic, *J. Org. Chem.*, 1967, **32**, 24.
54. C. Moureu and R. Chaux, *Org. Synth., Coll. Vol.*, 1941, **1**, 166.
55. C. Armstrong, J. A. Blair and J. Homer, *J. Chem. Soc., Chem. Commun.*, 1969, 103.
56. E. Schjånberg, *Ber. Dtsch. Chem. Ges. B*, 1937, **70**, 2385.
57. W. R. Vaughan, R. L. Craven, R. Q. Little, Jr. and A. C. Schoenthaler, *J. Am. Chem. Soc.*, 1955, **77**, 1594.
58. R. Stewart and R. H. Clark, *J. Am. Chem. Soc.*, 1947, **69**, 713.
59. C. L. Stevens, *J. Am. Chem. Soc.*, 1948, **70**, 165.
60. R. C. Fahey and H.-J. Schneider, *J. Am. Chem. Soc.*, 1970, **92**, 6885.
61. M. S. Kharasch, S. C. Kleiger and F. R. Mayo, *J. Org. Chem.*, 1939, **4**, 428.
62. A. L. Henne and S. Kaye, *J. Am. Chem. Soc.*, 1950, **72**, 3369.
63. R. N. Haszeldine, *J. Chem. Soc.*, 1953, 3565.
64. R. N. Haszeldine and B. R. Steele, *J. Chem. Soc.*, 1957, 2193.
65. M. S. Kharasch, H. Engelmann and F. R. Mayo, *J. Org. Chem.*, 1937, **2**, 288.
66. H. J. Prins, *Recl. Trav. Chim. Pays-Bas*, 1926, **45**, 80.
67. W. Shand, Jr., V. Schomaker and J. R. Fischer, *J. Am. Chem. Soc.*, 1944, **66**, 636.
68. D. Landini and F. Rolla, *J. Org. Chem.*, 1980, **45**, 3527.
69. S. H. Bales and S. A. Nickelson, *J. Chem. Soc.*, 1922, **121**, 2137.
70. S. H. Bales and S. A. Nickelson, *J. Chem. Soc.*, 1923, **123**, 2486.
71. L. Schmerling, J. P. Luvisi and R. W. Welch, *J. Am. Chem. Soc.*, 1956, **78**, 2819.
72. R. Voigt, *J. Prakt. Chem.*, 1938, **151**, 307.
73. T. L. Jacobs and R. N. Johnson, *J. Am. Chem. Soc.*, 1960, **82**, 6397.
74. R. B. Moffett, *Org. Synth., Coll. Vol.*, 1963, **4**, 238.
75. W. Smadja, *Chem. Rev.*, 1983, **83**, 263.
76. T. L. Jacobs, in 'The Chemistry of the Allenes', ed. S. R. Landor, Academic Press, London, 1982, vol. 2, p. 419.
77. K. Griesbaum, W. Naegele and G. G. Wanless, *J. Am. Chem. Soc.*, 1965, **87**, 3151.
78. G. F. Hennion and J. J. Sheehan, *J. Am. Chem. Soc.*, 1949, **71**, 1964.
79. T. Okuyama, K. Izawa and T. Fueno, *Tetrahedron Lett.*, 1970, 3295.
80. R. K. Sharma, B. A. Shoulders and P. D. Gardner, *J. Org. Chem.*, 1967, **32**, 241.
81. A. V. Fedorova, *J. Gen. Chem. USSR (Engl. Transl.)*, 1963, **33**, 3508.
82. J.-P. Bianchini and A. Guillemonat, *Bull. Soc. Chim. Fr.*, 1968, 2120.
83. T. L. Jacobs, in 'The Chemistry of the Allenes', ed. S. R. Landor, Academic Press, London, 1982, vol. 2, p. 501.
84. R. C. Fahey, M. T. Payne and D.-J. Lee, *J. Org. Chem.*, 1974, **39**, 1124.
85. H. Hunziker, R. Meyer and H. H. Günthard, *Helv. Chim. Acta*, 1966, **49**, 497.
86. G. F. Hennion and C. E. Welsh, *J. Am. Chem. Soc.*, 1940, **62**, 1367.
87. H. Hunziker, *Chimia*, 1963, **17**, 391.
88. K. Griesbaum and Z. Rehman, *J. Am. Chem. Soc.*, 1970, **92**, 1416.
89. F. Marcuzzi, G. Melloni and G. Modena, *Tetrahedron Lett.*, 1974, 413.
90. F. Marcuzzi and G. Melloni, *J. Am. Chem. Soc.*, 1976, **98**, 3295.
91. R. C. Fahey and D.-J. Lee, *J. Am. Chem. Soc.*, 1968, **90**, 2124.
92. R. C. Fahey and D.-J. Lee, *J. Am. Chem. Soc.*, 1967, **89**, 2780.
93. R. C. Fahey and D.-J. Lee, *J. Am. Chem. Soc.*, 1966, **88**, 5555.
94. F. Karlsson and L. Granberg, *Chem. Scr.*, 1979, **13**, 147.
95. G. Borkent and W. Drenth, *Recl. Trav. Chim. Pays-Bas*, 1970, **89**, 1056.
96. K. Mai and G. Patil, *Chem. Ind. (London)*, 1986, 670.
97. K. Bowden and M. J. Price, *J. Chem. Soc. (B)*, 1970, 1472.
98. K. Bowden and M. J. Price, *J. Chem. Soc. (B)*, 1970, 1466.
99. J. Cousseau, *J. Chem. Soc., Perkin Trans. 1*, 1977, 1797.
100. R. Tanaka, S.-Q. Zheng, K. Kawaguchi and T. Tanaka, *J. Chem. Soc., Perkin Trans. 2*, 1980, 1714.
101. A. Roedig, *Methoden Org. Chem. (Houben-Weyl)*, 1960, **V/4**, 102.
102. F. R. Mayo and C. Walling, *Chem. Rev.*, 1940, **27**, 351.
103. F. W. Stacy and J. F. Harris, Jr., *Org. React. (N.Y.)*, 1963, **13**, 150.
104. B. A. Bohn and P. I. Abell, *Chem. Rev.*, 1962, **62**, 599.
105. M. S. Kharasch and J. A. Hinckley, Jr., *J. Am. Chem. Soc.*, 1934, **56**, 1212.
106. M. S. Kharasch, J. A. Hinckley, Jr. and M. Gladstone, *J. Am. Chem. Soc.*, 1934, **56**, 1642.
107. M. S. Kharasch, C. W. Hannum and M. Gladstone, *J. Am. Chem. Soc.*, 1934, **56**, 244.
108. H. E. Carter, *J. Biol. Chem.*, 1935, **108**, 619.
109. C. Walling, M. S. Kharasch and F. R. Mayo, *J. Am. Chem. Soc.*, 1939, **61**, 2693.
110. J. G. Traynham and O. S. Pascual, *J. Org. Chem.*, 1956, **21**, 1362.
111. P. J. Card, F. E. Friedli and H. Schechter, *J. Am. Chem. Soc.*, 1983, **105**, 6104.
112. H. J. Lucas and H. W. Moyse, *J. Am. Chem. Soc.*, 1925, **47**, 1459.
113. W. M. Lauer and F. H. Sodola, *J. Am. Chem. Soc.*, 1934, **56**, 1215.
114. H. J. Lucas and A. N. Prater, *J. Am. Chem. Soc.*, 1937, **59**, 1682.
115. H. W. Kwart and J. L. Nyce, *J. Am. Chem. Soc.*, 1964, **86**, 2601.
116. D. J. Pasto, G. R. Meyer and S.-Z. Kang, *J. Am. Chem. Soc.*, 1969, **91**, 2163.
117. R. C. Fahey and R. A. Smith, *J. Am. Chem. Soc.*, 1964, **86**, 5035.
118. G. S. Hammond and T. D. Nevitt, *J. Am. Chem. Soc.*, 1954, **76**, 4121.
119. M. J. S. Dewar and R. C. Fahey, *J. Am. Chem. Soc.*, 1963, **85**, 2248.
120. M. J. S. Dewar and R. C. Fahey, *J. Am. Chem. Soc.*, 1963, **85**, 3645.
121. M. S. Kharasch and F. R. Mayo, *J. Am. Chem. Soc.*, 1933, **55**, 2468.
122. J. P. Wibaut, *Recl. Trav. Chim. Pays-Bas*, 1931, **50**, 313.
123. J. G. Traynham and J. S. Conte, *J. Org. Chem.*, 1957, **22**, 702.

124. O. R. Pierce, E. T. McBee and R. E. Cline, *J. Am. Chem. Soc.*, 1953, **75**, 5618.
125. M. S. Kharasch and C. W. Hannum, *J. Am. Chem. Soc.*, 1934, **56**, 712.
126. M. S. Kharasch, M. C. McNab and F. R. Mayo, *J. Am. Chem. Soc.*, 1933, **55**, 2521.
127. G. N. Burkhardt and W. Cocker, *Recl. Trav. Chim. Pays-Bas*, 1931, **50**, 837.
128. P. D. Readio and P. S. Skell, *J. Org. Chem.*, 1966, **31**, 753.
129. H. L. Boering and L. L. Sims, *J. Am. Chem. Soc.*, 1979, **79**, 6270.
130. G. Drefahl and C. Zimmer, *Chem. Ber.*, 1960, **93**, 505.
131. R. N. Haszeldine and J. E. Osborne, *J. Chem. Soc.*, 1956, 61.
132. J. D. Park, M. L. Sharrah and J. R. Lacher, *J. Am. Chem. Soc.*, 1949, **71**, 2339.
133. J. Troeger and K. Artmann, *J. Prakt. Chem.*, 1896, **53**, 489.
134. E. A. Fehnel and P. A. Lackey, *J. Am. Chem. Soc.*, 1951, **73**, 2473.
135. R. Mozingo and L. A. Patterson, *Org. Synth., Coll. Vol.*, 1955, **3**, 576.
136. E. J. Boorman, R. P. Linstead and H. N. Rydon, *J. Chem. Soc.*, 1933, 568.
137. C.-K. Dien and R. E. Lutz, *J. Am. Chem. Soc.*, 1956, **78**, 1989.
138. R. Ashton and J. C. Smith, *J. Chem. Soc.*, 1934, 435.
139. E. P. Abraham, E. L. R. Mowat and J. C. Smith, *J. Chem. Soc.*, 1937, 948.
140. H. P. Talbot, *Justus Liebigs Ann. Chem.*, 1900, **313**, 231.
141. W. R. Vaughan and R. Caple, *J. Am. Chem. Soc.*, 1964, **86**, 4928.
142. W. R. Vaughan, R. Caple, J. Csapilla and P. Scheiner, *J. Am. Chem. Soc.*, 1965, **87**, 2204.
143. W. R. Vaughan and K. M. Milton, *J. Am. Chem. Soc.*, 1952, **74**, 5623.
144. J. R. Alexander and H. McCombie, *J. Chem. Soc.*, 1931, 1913.
145. K. J. Clark, *J. Chem. Soc.*, 1957, 463.
146. L. W. J. Newman and H. N. Rydon, *J. Chem. Soc.*, 1936, 261.
147. M. S. Kharasch, E. T. Margolis and F. R. Mayo, *J. Org. Chem.*, 1936, **1**, 393.
148. L. Claisen, *J. Prakt. Chem.*, 1922–23, **105**, 65.
149. H. Staudinger, W. Kreis and W. Schilt, *Helv. Chim. Acta*, 1922, **5**, 750.
150. Y.-R. Naves, A. V. Grampoloff and P. Bachmann, *Helv. Chim. Acta*, 1947, **30**, 1599.
151. R. G. Kelso, K. W. Greenlee, J. M. Derfer and C. E. Boord, *J. Am. Chem. Soc.*, 1952, **74**, 287.
152. G. S. Hammond and J. Warkentin, *J. Am. Chem. Soc.*, 1961, **83**, 2554.
153. R. Y. Tien and P. I. Abell, *J. Org. Chem.*, 1970, **35**, 956.
154. M. S. Baird, *Synthesis*, 1976, 385.
155. A. V. Fedorova and A. A. Petrov, *J. Gen. Chem. USSR (Engl. Transl.)*, 1961, **31**, 3273.
156. S. N. Moorthy, A. Singh and D. Devaprabhakara, *J. Org. Chem.*, 1975, **40**, 3452.
157. N. R. Rosenquist and O. L. Chapman, *J. Org. Chem.*, 1976, **41**, 3326.
158. J. M. Schwab and D. C. T. Lin, *J. Am. Chem. Soc.*, 1985, **107**, 6046.
159. C. A. Young, R. R. Vogt and J. A. Nieuwland, *J. Am. Chem. Soc.*, 1936, **58**, 1806.
160. M. S. Kharasch, J. G. McNab and M. C. McNab, *J. Am. Chem. Soc.*, 1935, **57**, 2463.
161. P. G. Gassman and C. K. Harrington, *J. Org. Chem.*, 1984, **49**, 2258.
162. C. A. Grob and H. R. Pfaendler, *Helv. Chim. Acta*, 1971, **54**, 2060.
163. A. L. Henne and M. Nager, *J. Am. Chem. Soc.*, 1952, **74**, 650.
164. M. Taniguchi, S. Kobayashi, M. Nakagawa and T. Hino, *Tetrahedron Lett.*, 1986, **27**, 4763.
165. W. H. Perkin and J. L. Simonsen, *J. Chem. Soc.*, 1907, **91**, 816.
166. P. L. Harris and J. C. Smith, *J. Chem. Soc.*, 1935, 1572.
167. J. A. Pincock and C. Somawardhana, *Can. J. Chem.*, 1978, **56**, 1164.
168. Y. N. Shmatov, Y. I. Porfir'eva and A. A. Petrov, *J. Org. Chem. USSR (Engl. Transl.)*, 1976, **12**, 297.
169. Y. N. Shmatov and Y. I. Porfir'eva, *J. Org. Chem. USSR (Engl. Transl.)*, 1980, **16**, 1715.
170. A. Michael, *J. Org. Chem.*, 1939, **4**, 128.
171. J. J. Sudborough and K. J. Thompson, *J. Chem. Soc.*, 1903, **83**, 1153.
172. C. Moureu and J. C. Bongrand, *Ann. Chim. (Paris)*, 1920, **14**, 33.
173. J. Cousseau, *Synthesis*, 1980, 805.
174. S. Hara, H. Dojo, S. Takinami and A. Suzuki, *Tetrahedron Lett.*, 1983, **24**, 731.
175. A. Roedig, *Methoden Org. Chem. (Houben-Weyl)*, 1960, **V/4**, 535.
176. M. S. Kharasch and C. W. Hannum, *J. Am. Chem. Soc.*, 1934, **56**, 1782.
177. M. S. Kharasch, J. A. Norton and F. R. Mayo, *J. Am. Chem. Soc.*, 1940, **62**, 81.
178. H. Stone and H. Shechter, *Org. Synth., Coll. Vol.*, 1963, **4**, 543.
179. G. Gustavson, *J. Prakt. Chem.*, 1896, **54**, 104.
180. H. Cohn, E. D. Hughes, M. H. Jones and M. G. Peeling, *Nature (London)*, 1952, **169**, 291.
181. N. Polgar and R. Robinson, *J. Chem. Soc.*, 1945, 389.
182. E. P. Abraham and J. C. Smith, *J. Chem. Soc.*, 1936, 1605.
183. A. Michael and P. Freer, *J. Prakt. Chem.*, 1889, **40**, 95.
184. W. C. Young, R. T. Dillon and H. J. Lucas, *J. Am. Chem. Soc.*, 1929, **51**, 2528.
185. M. L. Sherrill and E. S. Matlack, *J. Am. Chem. Soc.*, 1937, **59**, 2134.
186. J. Corse and E. F. Jansen, *J. Am. Chem. Soc.*, 1955, **77**, 6632.
187. R. Kh. Freidlina, O. P. Bondarenko, R. A. Amriev, F. K. Velichko and R. P. Rilo, *Bull. Acad. Sci. USSR, Div. Chem. Sci.*, 1985, 1781.
188. R. M. Pagni, G. W. Kabalka, R. Boothe, K. Gaetano, L. J. Steward, R. Conaway, C. Dial, D. Gray, S. Larson and T. Luidhardt, *J. Org. Chem.*, 1988, **53**, 4477.
189. L. J. Stewart, D. Gray, R. M. Pagni and G. W. Kabalka, *Tetrahedron Lett.*, 1987, **28**, 4497.
190. G. F. Dvorko, T. F. Karpenko and E. A. Shilov, *Kinet. Catal. (Engl. Transl.)*, 1965, **6**, 731.
191. F. Stolz, *Ber. Dtsch. Chem. Ges.*, 1886, **19**, 536.
192. E. Baudrowski, *Ber. Dtsch. Chem. Ges.*, 1882, **15**, 2697.
193. G. F. Dvorko and E. A. Shilov, *Kinet. Katal.*, 1965, **6**(1), 37 (*Chem. Abstr.*, 1965, **62**, 16 009h).

194. R. C. Larock, in 'Solvomercuration/Demercuration Reactions in Organic Synthesis', Springer Verlag, Berlin, 1986, chap. 6.
195. A. Lattes, *Afinidad*, 1972, **29**, 153 (*Chem. Abstr.*, 1972, **77**, 48 544x).
196. J. Barluenga, J. J. Pérez-Prieto, A. M. Bayón and G. Asensio, *Tetrahedron*, 1984, **40**, 1199.
197. R. C. Griffith, R. J. Gentile, T. A. Davidson and F. L. Scott, *J. Org. Chem.*, 1979, **44**, 3580.
198. H. Hodjat-Kachani, J. J. Périé and A. Lattes, *Chem. Lett.*, 1976, 409.
199. M. B. Gasc, J. J. Périé and A. Lattes, *Tetrahedron*, 1978, **34**, 1943.
200. J. Barluenga, J. Villamaña and M. Yus, *Synthesis*, 1981, 375.
201. J. E. Backväll and B. Åkermark, *J. Organomet. Chem.*, 1974, **78**, 177.
202. A. Lattes and J. J. Périé, *Tetrahedron Lett.*, 1967, 5165.
203. J. J. Périé and A. Lattes, *Bull. Soc. Chim. Fr.*, 1971, 1378.
204. G. Etemad-Moghadam, M. C. Benhamou, V. Spéziale, A. Lattes and A. Bielawska, *Nouv. J. Chim.*, 1980, **4**, 727.
205. M. Tokuda, Y. Yamada, T. Takagi and H. Suginome, *Tetrahedron*, 1987, **43**, 281.
206. S. R. Wilson and R. A. Sawicki, *J. Org. Chem.*, 1979, **44**, 330.
207. H. Hodjat, A. Lattes, J. P. Laval, J. Moulines and J. J. Périé, *J. Heterocycl. Chem.*, 1972, **9**, 1081.
208. M. G. Voronkov, S. V. Kirpichenko, A. T. Abrosimova, A. I. Albanov and V. V. Keiko, *J. Organomet. Chem.*, 1987, **326**, 159.
209. V. P. Lopatinskii, E. E. Sirotkina, Y. P. Shekhirev, N. G. Men'shikova and L. F. Kovaleva, *Tr. Tomsk. Gos. Univ., Ser. Khim.*, 1964, **170**, 35 (*Chem. Abstr.*, 1965, **63**, 1763f).
210. M. C. Benhamou, G. Etemad-Moghadam, V. Spéziale and A. Lattes, *Synthesis*, 1979, 891.
211. J. Barluenga, C. Nájera and M. Yus, *Synthesis*, 1979, 896.
212. J. Barluenga, C. Nájera and M. Yus, *Synthesis*, 1978, 911.
213. J. Barluenga, C. Nájera and M. Yus, *J. Heterocycl. Chem.*, 1980, **17**, 917.
214. J. Barluenga, C. Jiménez, C. Nájera and M. Yus, *Synthesis*, 1982, 414.
215. J. Barluenga, C. Nájera and M. Yus, *J. Heterocycl. Chem.*, 1981, **18**, 1297.
216. V. Gómez-Aranda, J. Barluenga and M. Yusá, *An. Quim.*, 1972, **68**, 221.
217. V. Gómez-Aranda, J. Barluenga, M. Yus and G. Asensio, *Synthesis*, 1974, 806.
218. J. Barluenga, C. Jiménez, C. Nájera and M. Yus, *Synthesis*, 1982, 417.
219. V. Gómez-Aranda, J. Barluenga, M. Yus-Astiz and F. Aznar, *Rev. Acad. Cienc. Exactas, Fis.-Quim. Nat. Zaragoza*, 1974, **29**, 321 (*Chem. Abstr.*, 1976, **85**, 20 716w).
220. J. Barluenga, J. Perez-Prieto and G. Asensio, *J. Chem. Soc., Chem. Commun.*, 1982, 1181.
221. H. Hodjat-Kachani, J. J. Périé and A. Lattes, *Chem. Lett.*, 1976, 405.
222. S. Arseniyadis and J. Gore, *Tetrahedron Lett.*, 1983, **24**, 3997.
223. J. A. Loritsch and R. R. Vogt, *J. Am. Chem. Soc.*, 1939, **61**, 1462.
224. J. Barluenga, F. Aznar, R. Liz and R. Rodes, *J. Chem. Soc., Perkin Trans. 1*, 1980, 2732.
225. J. Barluenga and F. Aznar, *Synthesis*, 1975, 704.
226. P. F. Hudrlik and A. M. Hudrlik, *J. Org. Chem.*, 1973, **38**, 4254.
227. J. J. P. Staudinger and K. H. W. Tuerck, *Br. Pat.* 573 752 (1945) (*Chem. Abstr.*, 1949, **43**, 3031i).
228. J. Barluenga, F. Aznar, R. Liz and C. Postigo, *J. Chem. Soc., Chem. Commun.*, 1986, 1465.
229. L. I. Krimen and D. J. Cota, *Org. React. (N.Y.)*, 1969, **17** 213.
230. Z. Rappoport (ed.), 'The Chemistry of the Cyano Group', Wiley-Interscience, New York, 1970.
231. F. Johnson and R. Madroñero, *Adv. Heterocycl. Chem.*, 1966, **6**, 95.
232. E. N. Zil'berman, *Russ. Chem. Rev. (Engl. Transl.)*, 1960, **26**, 331.
233. J. J. Ritter and P. P. Minieri, *J. Am. Chem. Soc.*, 1948, **70**, 4045.
234. S. Blum, S. Gertler, S. Sarel and D. Sinnreich, *J. Org. Chem.*, 1972, **37**, 3114.
235. R. T. Coults, A. Benderly, A. L. C. Mak and W. G. Taylor, *Can. J. Chem.*, 1978, **56**, 3054.
236. H. Schnell and J. Nentwig, *Methoden Org. Chem. (Houben-Weyl)*, 1958, **XI/2**, 561.
237. S. Julia and C. Papantoniou, *C. R. Hebd. Seances Acad. Sci.*, 1965, **260**, 1440.
238. R. Bishop and G. Burgess, *Tetrahedron Lett.*, 1987, **28**, 1585.
239. R. Bishop, S. C. Hawkins and I. C. Ibana, *J. Org. Chem.*, 1988, **53**, 427.
240. A. I. Meyers and J. C. Sirear, in 'The Chemistry of the Cyano Group', ed. Z. Rappoport, Wiley-Interscience, New York, 1970, p. 341.
241. A. I. Meyers and J. J. Ritter, *J. Org. Chem.*, 1958, **23**, 1918.
242. R. C. Larock, in 'Solvomercuration/Demercuration Reactions in Organic Synthesis', Springer Verlag, Berlin, 1986, chap. 7.
243. J. Beger and D. Vogel, *J. Prakt. Chem.*, 1969, **311**, 737.
244. H. C. Brown and J. T. Kurek, *J. Am. Chem. Soc.*, 1969, **91**, 5647.
245. A. J. Fry and J. A. Simon, *J. Org. Chem.*, 1982, **47**, 5032.
246. D. Chow, J. H. Robson and G. F. Wright, *Can. J. Chem.*, 1965, **43**, 312.
247. A. Factor and T. G. Traylor, *J. Org. Chem.*, 1968, **33**, 2607.
248. J. Barluenga, C. Jiménez, C. Nájera and M. Yus, *J. Chem. Soc., Chem. Commun.*, 1981, 670.
249. J. Barluenga, C. Jiménez, C. Nájera and M. Yus, *J. Chem. Soc., Perkin Trans. 1*, 1983, 591.
250. K. E. Harding and S. R. Burks, *J. Org. Chem.*, 1981, **46**, 3920.
251. S. J. Danishefsky, E. Taniyama and R. R. Webb, II, *Tetrahedron Lett.*, 1983, **24**, 11.
252. K. E. Harding and S. R. Burks, *J. Org. Chem.*, 1984, **49**, 40.
253. S. R. Wilson and R. A. Sawicki, *J. Org. Chem.*, 1979, **44**, 330.
254. J. Barluenga, C. Jiménez, C. Nájera and M. Yus, *J. Chem. Soc., Chem. Commun.*, 1981, 1178.
255. E. F. V. Scriven and K. Turnbull, *Chem. Rev.*, 1988, **88**, 351.
256. E. P. Kyba and A. Hassner, in 'Azides and Nitrenes: Reactivity and Utility', ed. E. F. V. Scriven, Academic Press, Orlando, 1984, pp. 2, 35.
257. M. E. C. Biffin, J. Miller and D. B. Paul, in 'The Chemistry of the Azido Group', ed. S. Patai, Wiley-Interscience, London, 1971.

258. A. Hassner and J. E. Galle, *J. Am. Chem. Soc.*, 1972, **94**, 3930.
259. R. E. Schaad, *US Pat.* 2 557 924 (1951) (*Chem. Abstr.*, 1952, **46**, 1028).
260. S. N. Ege and K. W. Sherk, *J. Am. Chem. Soc.*, 1953, **75**, 354.
261. E. P. Kyba and A. M. John, *Tetrahedron Lett.*, 1977, 2739.
262. A. Hassner, R. Fibiger and D. Andisik, *J. Org. Chem.*, 1984, **49**, 4237.
263. A. Pancrazi, I. Kabore, B. Delpech and Q. Khuong-Huu, *Tetrahedron Lett.*, 1979, 3729.
264. A. Pancrazi and Q. Khuong-Huu, *Tetrahedron*, 1974, **30**, 2337.
265. L. W. Hartzel and F. R. Benson, *J. Am. Chem. Soc.*, 1954, **76**, 667.
266. Y. A. Sinnema and J. F. Arens, *Recl. Trav. Chim. Pays-Bas*, 1955, **74**, 901.
267. R. C. Larock, in 'Solvomercuration/Demercuration Reactions in Organic Synthesis', Springer Verlag, Berlin, 1986, chap. 8.
268. J. E. Galle and A. Hassner, *J. Am. Chem. Soc.*, 1972, **94**, 3930.
269. V. I. Sokolov and O. A. Reutov, *Izv. Akad. Nauk SSSR, Ser. Khim.*, 1967, 1632.
270. C. H. Heathcock, *Angew. Chem.*, 1969, **81**, 148; *Angew. Chem., Int. Ed. Engl.*, 1969, **8**, 134.
271. S. Czernecki, C. Georgoulis and C. Provelenghiou, *Tetrahedron Lett.*, 1979, 4841.
272. J. S. Brimacombe, J. A. Miller and U. Zakir, *Carbohydr. Res.*, 1975, **41**, C3.
273. J. S. Brimacombe, J. A. Miller and U. Zakir, *Carbohydr. Res.*, 1975, **44**, C9.
274. J. S. Brimacombe, J. A. Miller and U. Zakir, *Carbohydr. Res.*, 1976, **49**, 233.
275. G. Mehta and P. N. Pandey, *J. Org. Chem.*, 1975, **40**, 3631.
276. T. Sasaki, K. Kanematsu and A. Kondo, *Tetrahedron*, 1975, **31**, 2215.
277. J. G. Traynham and H. H. Hsieh, *J. Org. Chem.*, 1973, **38**, 868.
278. F. Asinger, in 'Mono-Olefins: Chemistry and Technology', Pergamon Press, Oxford, 1968, p. 628.
279. V. J. Nowlan and T. T. Tidwell, *Acc. Chem. Res.*, 1977, **10**, 252.
280. V. Lucchini, F. Marcuzzi and G. Modena, *Stud. Org. Chem. (Amsterdam)*, 1987, **31**, 287 (*Chem. Abstr.*, 1988, **108**, 186 112q).
281. J. Meinwald, *J. Am. Chem. Soc.*, 1955, **77**, 1617.
282. P. Knittel and T. T. Tidwell, *J. Am. Chem. Soc.*, 1977, **99**, 3408.
283. Y. Chiang and A. J. Kresge, *J. Am. Chem. Soc.*, 1985, **107**, 6363.
284. G. Mihaila, C. Luca and D. Dranga, *Rev. Roum. Chim.*, 1985, **30**, 1019.
285. K. Eguchi, T. Tokiai and H. Arai, *Appl. Catal.*, 1987, **34**, 275.
286. C. Liu, Q. Chen and B. Shen, *Huaxue Xuebao*, 1986, **44**, 545 (*Chem. Abstr.*, 1987, **107**, 40 086s).
287. D. Kallo and G. Onyestyak, *Acta Chim. Hung.*, 1987, **34**, 605 (*Chem. Abstr.*, 1988, **109**, 92 034j).
288. A. Aoshima, S. Yamatomo and T. Yamaguchi, *Nippon Kagaku Kaishi*, 1987, 976 (*Chem. Abstr.*, 1988, **108**, 39 642w).
289. A. Delion, B. Torck and M. Hellin, *J. Catal.*, 1987, 103.
290. R. R. Carls, G. Osterbury, M. Prezeli and W. Webers (Deutsch Texaco A.-G.), *Ger. (East) Pat.* 3 628 007 (1987) (*Chem. Abstr.*, 1988, **108**, 133 847b).
291. M. Nomura and Y. Fujihara, *Nippon Kagaku Kaishi*, 1983, **12**, 1818 (*Chem. Abstr.*, 1984, **100**, 192 083g).
292. L. R. Gahan, J. M. Harrowfield, A. J. Herlt, L. F. Lindoy, P. O. Whimp and A. M. Sargeson, *J. Am. Chem. Soc.*, 1985, **107**, 6231.
293. P. Wan, S. Culshaw and K. Yates, *J. Am. Chem. Soc.*, 1982, **104**, 2509.
294. J. McEwen and K. Yates, *J. Am. Chem. Soc.*, 1987, **109**, 5800.
295. J. L. Jensen and V. Uaprasert, *J. Org. Chem.*, 1976, **41**, 649.
296. W. K. Chwang, P. Knittel, K. M. Koshy and T. T. Tidwell, *J. Am. Chem. Soc.*, 1977, **99**, 3395.
297. T. Okuyama, T. Sakagami and T. Fueno, *Tetrahedron*, 1973, **29**, 1503.
298. W. S. Madja, *Chem Rev.*, 1983, **83**, 263.
299. T. L. Jacobs, in 'The Chemistry of Allenes', ed. S. R. Landor, Academic Press, New York, 1982, vol. 2, p. 417.
300. D. J. Scheffel, A. R. Cole, D. M. Jung and M. D. Schiavelli, *J. Am. Chem. Soc.*, 1980, **102**, 267.
301. A. D. Allen and T. T. Tidwell, *J. Am. Chem. Soc.*, 1982, **104**, 2515.
302. V. Jager and H. G. Viehe, *Methoden Org. Chem. (Houben-Weyl)*, 1977, **V/2a**, 726.
303. P. F. Hudrlik and A. M. Hudrlik, in 'The Chemistry of the Carbon–Carbon Triple Bond', ed. S. Patai, Wiley, New York, 1978, vol. 1, p. 199.
304. A. A. Pourzal and P.-H. Bonnet, *Monatsh. Chem.*, 1983, **114**, 809 (*Chem. Abstr.*, 1983, **99**, 194 553v).
305. K. Imi, K. Imai and K. Utimoto, *Tetrahedron Lett.*, 1987, **28**, 3127.
306. M. Bassetti and B. Floris, *Gazz. Chim. Ital.*, 1986, **116**, 595.
307. Y. Chiang, A. J. Kresge, J. A. Santaballa and J. Wirz, *J. Am. Chem. Soc.*, 1988, **110**, 5506.
308. M. V. Kormer and S. E. Tolchinskii, *Zh. Org. Khim.*, 1983, **19**, 1166 (*Chem. Abstr.*, 1983, **99**, 121 783k).
309. N. Banait, M. Hojatti, P. Findlay and A. J. Kresge, *Can. J. Chem.*, 1987, **65**, 441.
310. M. G. Constantino, P. M. Donate and N. Petragnani, *Tetrahedron Lett.*, 1982, **23**, 1051.
311. R. C. Larock, in 'Solvomercuration/Demercuration Reactions in Organic Synthesis', Springer Verlag, Berlin, 1986, chap. 2.
312. H. C. Brown and P. J. Geoghegan, Jr., *J. Org. Chem.*, 1970, **35**, 1844.
313. H. C. Brown and P. J. Geoghegan, Jr., *J. Am. Chem. Soc.*, 1967, **89**, 1522.
314. H. C. Brown, P. J. Geoghegan, Jr. and J. T. Kurek, *J. Org. Chem.*, 1981, **46**, 3810.
315. H. C. Brown, P. J. Geoghegan, Jr., J. T. Kurek and G. J. Lynch, *Organomet. Chem. Synth.*, 1970, **1**, 7.
316. F. H. Gouzoules and R. A. Whitney, *J. Org. Chem.*, 1986, **51**, 2024.
317. K. Maskens and N. Polgar, *J. Chem. Soc., Perkin Trans. 1*, 1973, 109.
318. H. C. Brown, G. J. Lynch, W. J. Hammar and L. C. Liu, *J. Org. Chem.*, 1979, **44**, 1910.
319. D. J. Pasto and J. A. Gontarz, *J. Am. Chem. Soc.*, 1970, **92**, 7480.
320. D. J. Pasto and J. A. Gontarz, *J. Am. Chem. Soc.*, 1971, **93**, 6902.
321. P. Chamberlain and G. H. Witham, *J. Chem. Soc. (B)*, 1970, 1382.
322. S. Bentham, P. Chamberlain and G. H. Whitham, *J. Chem. Soc. (D)*, 1970, 1528.

323. D. Jasserand, R. Granger, J.-P. Girard and J.-P. Chapat, *C. R. Hebd. Seances Acad. Sci., Ser. C*, 1971, **272**, 1693.
324. D. J. Pasto and J. A. Gontarz, *J. Am. Chem. Soc.*, 1971, **93**, 6909.
325. H. B. Henbest and B. Nicholls, *J. Chem. Soc.*, 1959, 227.
326. M. R. Johnson and B. Rickborn, *Chem. Commun.*, 1968, 1073.
327. M. R. Johnson and B. Rickborn, *J. Org. Chem.*, 1969, **34**, 2781.
328. H. C. Brown and J. H. Kawakami, *J. Am. Chem. Soc.*, 1973, **95**, 8665.
329. H. C. Brown, W. J. Hammar, J. H. Kawakami, I. Rothberg and D. L. Vander Jagt, *J. Am. Chem. Soc.*, 1967, **89**, 6381.
330. H. C. Brown, J. H. Kawakami and S. Ikegami, *J. Am. Chem. Soc.*, 1967, **89**, 1525.
331. J. E. Backväll, B. Åkermark and S. O. Ljunggren, *J. Am. Chem. Soc.*, 1979, **101**, 2411.
332. W. L. Waters, T. G. Traylor and A. Factor, *J. Org. Chem.*, 1973, **38**, 2306.
333. Y. Senda, S. Kamiyama and S. Imaizumi, *J. Chem. Soc., Perkin Trans. 1*, 1978, 530.
334. D. Jasserand, J.-P. Girard, J. C. Rossi and R. Granger, *Tetrahedron*, 1976, **32**, 1535.
335. Y.-H. Suen and H. Kagan, *Bull. Soc. Chim. Fr.*, 1970, 2270.
336. H. B. Henbest and R. S. McElhinney, *J. Chem. Soc.*, 1959, 1834.
337. T. Sugita, Y. Yamasaki, O. Itoh and K. Ichikawa, *Bull. Chem. Soc. Jpn.*, 1974, **47**, 1945.
338. J. Barluenga, J. M. Martinez-Gallo, C. Najera and M. Yus, *J. Chem. Res. (S)*, 1985, 266.
339. H. C. Brown, P. J. Geoghegan, Jr., G. J. Lynch and J. T. Kurek, *J. Org. Chem.*, 1972, **37**, 1941.
340. J. K. Sutter and C. N. Sukenik, *J. Org. Chem.*, 1982, **47**, 4174.
341. V. Gómez-Aranda, J. Barluenga and M. Yus-Astiz, *Rev. Acad. Cienc. Exactas, Fis.-Quim. Nat. Zaragoza*, 1973, **28**, 225 (*Chem. Abstr.*, 1974, **80**, 83 164g).
342. W. Kitching, J. A. Lewis, M. T. Fletcher, J. J. DeVoss, R. Drew and C. J. Moore, *J. Chem. Soc., Chem. Commun.*, 1986, 855.
343. G. Nagendrappa and D. Devaprabhakara, *Tetrahedron Lett.*, 1970, 4687.
344. T. L. Jacobs and R. N. Johnson, *J. Am. Chem. Soc.*, 1960, **82**, 6397.
345. G. F. Hennion and J. J. Sheehan, *J. Am. Chem. Soc.*, 1949, **71**, 1964.
346. R. K. Sharma, B. A. Shoulders and P. D. Gardner, *J. Org. Chem.*, 1967, **32**, 241.
347. G. Mehta, *Org. Prep. Proced.*, 1970, **2**, 245.
348. M. Miocque, N. M. Hung and V. Q. Yen, *Ann. Chim. (Paris)*, 1963, **8**, 157.
349. V. Jager and H. G. Viehe, *Methoden Org. Chem. (Houben-Weyl)*, 1977, **V/2a**, 726.
350. P. F. Hudrlik and A. M. Hudrlik, in 'The Chemistry of the Carbon–Carbon Triple Bond', ed. S. Patai, Wiley, New York, 1978, vol. 1, p. 199.
351. R. R. Vogt and J. A. Nieuwland, *J. Am. Chem. Soc.*, 1921, **43**, 2071.
352. V. Janout and S. L. Regen, *J. Org. Chem.*, 1982, **47**, 3331.
353. R. N. Haszeldine and K. Leedham, *J. Chem. Soc.*, 1954, 1261.
354. G. Stork and R. Borch, *J. Am. Chem. Soc.*, 1964, **86**, 935.
355. S. Padmanabhan and K. M. Nicholas, *Synth. Commun.*, 1980, **10**, 503.
356. Y. Tamura, H. Annoura, H. Yamamoto, H. Kondo, Y. Kita and H. Fujioka, *Tetrahedron Lett.*, 1987, **28**, 5709.
357. G. A. Olah and A. P. Fung, *Synthesis*, 1981, 473.
358. E. D. Venus-Danilova and S. N. Danilov, *Zh. Obshch. Khim.*, 1932, **2**, 645 (*Chem. Abstr.*, 1933, **27**, 1624).
359. M. Koulkes, *Bull. Soc. Chim. Fr.*, 1957, 127.
360. B. S. Kupin and A. A. Petrov, *Zh. Obshch. Khim.*, 1959, **29**, 2281 (*Chem. Abstr.*, 1960, **54**, 9722i).
361. M. G. Constantino, P. M. Donate and N. Petragnani, *Tetrahedron Lett.*, 1982, **23**, 1051.
362. A. P. Khrimyan, A. V. Kavapetyan and S. O. Badanyan, *Khim. Geterotsikl. Soedin.*, 1984, 619 (*Chem. Abstr.*, 1984, **101**, 72 562m).
363. A. I. Meyers, S. Schwartzman, G. L. Olson and H.-C. Cheung, *Tetrahedron Lett.*, 1976, 2417.
364. A. G. Davis and A. M. White, *Nature (London)*, 1952, **170**, 668.
365. E. S. Huyer and D. Eordway, *J. Org. Chem.*, 1979, **44**, 777.
366. A. G. Davis, R. V. Foster and A. M. White, *J. Chem. Soc.*, 1954, 2200.
367. A. G. Davis, R. V. Foster and A. M. White, *J. Chem. Soc.*, 1953, 1541.
368. N. A. Milas, R. L. Peeler and O. L. Magelli, *J. Am. Chem. Soc.*, 1954, **76**, 2322.
369. F. McCapra, P. Leeson, V. Donovan and G. Perry, *Tetrahedron*, 1986, **42**, 3223.
370. A. Rieche, *Angew. Chem.*, 1961, **73**, 57.
371. D. E. Hamilton, R. S. Drago and A. Zombeck, *J. Am. Chem. Soc.*, 1987, **109**, 374.
372. R. C. Larock, in 'Solvomercuration/Demercuration Reactions in Organic Synthesis', Springer Verlag, Berlin, 1986, chap. 4.
373. V. I. Sokolov and O. A. Reutov, *Zh. Org. Khim.*, 1969, **5**, 174; *J. Org. Chem. USSR (Engl. Transl.)*, 1969, **5**, 168.
374. D. H. Ballard, A. J. Bloodworth and R. J. Bunce, *J. Chem. Soc. (D)*, 1969, 815.
375. A. J. Bloodworth and J. L. Courtneidge, *J. Chem. Soc., Perkin Trans. 1*, 1981, 3258.
376. A. J. Bloodworth and I. M. Griffin, *J. Organomet. Chem.*, 1974, **66**, C1.
377. A. J. Bloodworth and M. E. Loveitt, *J. Chem. Soc., Perkin Trans. 1*, 1977, 1031.
378. A. J. Bloodworth and R. J. Bunce, *J. Chem. Soc. (D)*, 1970, 753.
379. A. J. Bloodworth and R. J. Bunce, *J. Organomet. Chem.*, 1973, **60**, 11.
380. A. J. Bloodworth and R. J. Bunce, *J. Chem. Soc. (C)*, 1971, 1453.
381. A. J. Bloodworth and I. M. Griffin, *J. Chem. Soc., Perkin Trans. 1*, 1974, 688.
382. A. J. Bloodworth and I. M. Griffin, *J. Chem. Soc., Perkin Trans. 2*, 1975, 531.
383. A. J. Bloodworth and J. L. Courtneidge, *J. Chem. Soc., Perkin Trans. 1*, 1982, 1797.
384. A. J. Bloodworth and J. L. Courtneidge, *J. Chem. Soc., Chem. Commun.*, 1981, 1117.
385. A. J. Bloodworth and J. A. Khan, *J. Chem. Soc., Perkin Trans. 1*, 1980, 2450.
386. N. A. Porter and J. R. Nixon, *J. Am. Chem. Soc.*, 1978, **100**, 7116.
387. J. R. Nixon, M. A. Cudd and N. A. Porter, *J. Org. Chem.*, 1978, **43**, 4048.

388. N. A. Porter, M. A. Cudd, R. W. Miller and A. T. McPhail, *J. Am. Chem. Soc.*, 1980, **102**, 414.
389. E. Schmitz, A. Rieche and O. Brede, *J. Prakt. Chem.*, 1970, **312**, 30.
390. A. J. Bloodworth and M. E. Loveitt, *J. Chem. Soc., Perkin Trans. 1*, 1978, 522.
391. A. J. Bloodworth and M. E. Loveitt, *J. Chem. Soc., Chem. Commun.*, 1976, 94.
392. D. R. Steven, *J. Org. Chem.*, 1955, **20**, 1232.
393. S. Pavlov, M. Bogavac and V. Arsenijevic, *Bull. Soc. Chim. Fr.*, 1974, 2985.
394. R. D. Morin and A. E. Bearse, *Ind. Eng. Chem.*, 1951, **43**, 1596.
395. M. P. Atkins and I. M. Wren (British Petroleum), *Eur. Pat.* 249 352 (1987) (*Chem. Abstr.*, 1988, **108**, 188 886u)
396. F. Ancillotti, M. M. Mauri, E. Pescarollo and L. Romaguoni, *J. Mol. Catal.*, 1974, **4**, 37.
397. J. A. Marshall, *Acc. Chem. Res.*, 1969, **2**, 33.
398. M. Miyashita, A. Yoshikoshi and P. A. Grieco, *J. Org. Chem.*, 1977, **42**, 3772.
399. T. C. McKenzie, *Org. Prep. Proced. Int.*, 1987, **19**, 435.
400. T. W. Green, 'Protecting Groups in Organic Synthesis', Wiley-Interscience, New York, 1981.
401. W. Smadja, *Chem. Rev.*, 1983, **83**, 263.
402. T. L. Jacobs, in 'The Chemistry of Allenes', ed. S. R. Landor, Academic Press, New York, 1982, vol. 2, p. 417.
403. J. K. Crandall, W. H. Machleder and S. A. Sojka, *J. Org. Chem.*, 1973, **38**, 1149.
404. R. Pellicciari, E. Castagnino, R. Fringuelli and S. Corsano, *Tetrahedron Lett.*, 1979, 481.
405. R. M. Carlson, R. W. Jones and A. S. Hatcher, *Tetrahedron Lett.*, 1975, 1741.
406. P. Audin, A. Doutheau, L. Ruest and J. Gore, *Bull. Soc. Chim. Fr., Ser. II*, 1981, 313.
407. L.-I. Olsson and A. Claesson, *Synthesis*, 1979, 743.
408. G. Balme, A. Dontheau, J. Gore and M. Malacria, *Synthesis*, 1979, 508.
409. P. Audin, A. Dontheau and J. Gore, *Tetrahedron Lett.*, 1982, **23**, 4337.
410. T. Gallagher, *J. Chem. Soc., Chem. Commun.*, 1984, 1554.
411. B. A. Trofimov, *Zh. Org. Khim.*, 1987, **23**, 1788 (*Chem. Abstr.*, 1987, **107**, 77 247w).
412. P. H. M. Schreurs, W. G. Galesloot and L. Brandsma, *Recl. Trav. Chim. Pays-Bas*, 1975, **94**, 70.
413. M. Riediker and J. Schwartz, *J. Am. Chem. Soc.*, 1982, **104**, 5842.
414. P. Pale and J. Chuche, *Tetrahedron Lett.*, 1987, **28**, 6447.
415. R. C. Larock, in 'Solvomercuration/Demercuration Reactions in Organic Synthesis', Springer Verlag, Berlin, 1986, chap. 3.
416. H. C. Brown, J. T. Kurek, M.-H. Rei and K. L. Thompson, *J. Org. Chem.*, 1984, **49**, 2551.
417. H. C. Brown, J. T. Kurek, M.-H. Rei and K. L. Thompson, *J. Org. Chem.*, 1985, **50**, 1171.
418. S. Honda, K. Kakehi, H. Takai and K. Takiura, *Carbohydr. Res.*, 1973, **29**, 477.
419. R. K. Boeckman, Jr. and C. J. Flann, *Tetrahedron Lett.*, 1983, **24**, 4923.
420. A. J. Bloodworth and R. J. Bunce, *J. Organomet. Chem.*, 1973, **60**, 11.
421. A. J. Bloodworth and R. J. Bunce, *J. Chem. Soc. (D)*, 1970, 753.
422. C. H. Lam and M. S. F. Lie Ken Jie, *Chem. Phys. Lipids*, 1976, **16**, 181.
423. F. D. Gunstone and R. P. Inglis, *Chem. Phys. Lipids*, 1973, **10**, 73.
424. F. D. Gunstone and R. P. Inglis, *Chem. Phys. Lipids*, 1973, **10**, 89.
425. R. M. Babb and P. D. Gardner, *Chem. Commun.*, 1968, 1678.
426. V. S. Dombrovskii, N. I. Yakushkina and I. G. Bolesov, *Zh. Org. Khim.*, 1979, **15**, 1325; *J. Org. Chem. USSR (Engl. Transl.)*, 1979, **15**, 1184.
427. I. V. Bodrikov, V. R. Kartashov, L. I. Koval'ova and N. S. Zefirov, *J. Organomet. Chem.*, 1974, **82**, C23.
428. A. J. Bloodworth and I. M. Griffin, *J. Chem. Soc., Perkin Trans. 1*, 1975, 195.
429. R. Sterzycki and W. Sobotka, *Symp. Pap. — IUPAC Int. Symp. Chem. Nat. Prod., 11th*, 1978, **3**, 142 (*Chem. Abstr.*, 1980, **92**, 94 581q).
430. H. D. West and H. E. Carter, *J. Biol. Chem.*, 1937, **119**, 103.
431. K. Maskens and N. Polgar, *J. Chem. Soc., Perkin Trans. 1*, 1973, 109.
432. J. T. Slama, H. W. Smith, C. G. Willson and H. Rapoport, *J. Am. Chem. Soc.*, 1975, **97**, 6556.
433. M. J. Aberchrombie, A. Rodgman, K. R. Bharucha and G. F. Wright, *Can. J. Chem.*, 1959, **37**, 1328.
434. F. G. Bordwell and M. L. Douglass, *J. Am. Chem. Soc.*, 1966, **88**, 993.
435. T. Hosokawa, S. Miyagi, S. Murahashi, A. Sonoda, Y. Matsuura, S. Tanimoto and M. Kakudo, *J. Org. Chem.*, 1978, **43**, 719.
436. R. K. Summerbell, G. Lestina and H. Waite, *J. Am. Chem. Soc.*, 1957, **79**, 234.
437. R. K. Summerbell and J. R. Stephens, *J. Am. Chem. Soc.*, 1954, **76**, 6401.
438. R. K. Summerbell and J. R. Stephens, *J. Am. Chem. Soc.*, 1955, **77**, 6080.
439. M. C. Benhamou, G. Etemad-Moghadam, V. Spéziale and A. Lattes, *Synthesis*, 1979, 891.
440. G. Etemad-Moghadam, M. C. Benhamou, V. Spéziale, A. Lattes and A. Bielawska, *Nouv. J. Chim.*, 1980, **4**, 727.
441. J.-R. Pougny, M. A. M. Nassr and P. Sinay, *J. Chem. Soc., Chem. Commun.*, 1981, 375.
442. P. A. Bartlett and J. L. Adams, *J. Am. Chem. Soc.*, 1980, **102**, 337.
443. K. H. McNeely and G. F. Wright, *J. Am. Chem. Soc.*, 1955, **77**, 2553.
444. A. J. Bloodworth, M. G. Hutchings and A. J. Sotowicz, *J. Chem. Soc., Chem. Commun.*, 1976, 578.
445. B. Giese, K. Hueck and U. Lüning, *Tetrahedron Lett.*, 1981, **22**, 2155.
446. J. Barluenga, J. Pérez-Prieto and G. Asensio, *J. Chem. Soc., Perkin Trans. 1*, 1984, 629.
447. J. R. Johnson, W. H. Jobling and G. W. Bodamer, *J. Am. Chem. Soc.*, 1941, **63**, 131.
448. W. L. Waters and E. F. Kiefer, *J. Am. Chem. Soc.*, 1967, **89**, 6261.
449. W. H. Pirkle and C. W. Boeder, *J. Org. Chem.*, 1977, **42**, 3697.
450. R. Vaidyanathaswamy, D. Devaprabhakara and V. V. Rao, *Tetrahedron Lett.*, 1971, 915.
451. L. Garver, P. van Eikeren and J. E. Byrd, *J. Org. Chem.*, 1976, **41**, 2773.
452. P. Audin, A. Doutheau and J. Gore, *Tetrahedron Lett.*, 1982, **23**, 4337.
453. J. Barluenga, F. Aznar and M. Bayod, *Synthesis*, 1988, 144.

454. H. S. Hill and H. Hibbert, *J. Am. Chem. Soc.*, 1923, **45**, 3108.
455. M. Riediker and J. Schwartz, *J. Am. Chem. Soc.*, 1982, **104**, 5842.
456. R. D. Morin and A. E. Bearse, *Ind. Eng. Chem.*, 1951, **43**, 1596.
457. J. Guenzet and M. Camps, *Tetrahedron*, 1974, **30**, 849.
458. P. E. Peterson and E. V. P. Tao, *J. Org. Chem.*, 1964, 2322.
459. W. S. Johnson, A. L. McCloskey and D. A. Dunnigan, *J. Am. Chem. Soc.*, 1950, **72**, 514.
460. T. Fujita, S. Watanabe, K. Suga, T. Miura, K. Sugahara and H. Kikuchi, *J. Chem. Technol. Biotechnol.*, 1982, **32**, 476.
461. L. Jalander and M. Broms, *Acta Chem. Scand. Ser. B*, 1982, **36**, 371.
462. M. F. Ansell and M. H. Palmer, *Q. Rev., Chem. Soc.*, 1964, 211.
463. G. Stork and A. W. Burgstahler, *J. Am. Chem. Soc.*, 1955, **79**, 5068.
464. L. Schmerling, J. P. Luvisi and R. W. Welch, *J. Am. Chem. Soc.*, 1956, **78**, 2819.
465. P. Cramer and T. T. Tidwell, *J. Org. Chem.*, 1981, **46**, 2683.
466. J. A. Ballantine, P. A. Diddams, W. Jones, J. H. Purnell and J. T. Thomas (British Petroleum), *Br. Pat.* 2 179 563 (1987) (*Chem. Abstr.*, 1987, **107**, 98 587x).
467. J. A. Ballantine, M. Davis, J. H. Purnell, M. Rayanakorn, J. M. Thomas and K. J. Williams, *J. Chem. Soc., Chem. Commun.*, 1981, 8.
468. M. Malewski and S. Lewandowski (Instytut. Chemii. Przemyslowej), *Pol. Pat.* 123 461 (1982) (*Chem. Abstr.*, 1985, **102**, 184 838h).
469. L. B. Young (Mobil Oil Corp.), *US Pat.* 4 448 983 (1984) (*Chem. Abstr.*, 1984, **101**, 54 583q).
470. For leading references see T. Mitsudo, Y. Hori, Y. Yamakawa and Y. Watanabe, *J. Org. Chem.*, 1987, **52**, 2230.
471. J.-P. Montheard, M. Camps and A. Benzaid, *Synth. Commun.*, 1983, **13**, 663.
472. P. F. Hudrlik and A. M. Hudrlik, *J. Org. Chem.*, 1973, **38**, 4254.
473. P. G. Willard, T. T. Jong and J. P. Porwell, *J. Org. Chem.*, 1984, **47**, 736.
474. B. A. Morrow, *J. Catal.*, 1984, **86**, 328.
475. E. S. Rothman, S. S. Hecht, P. E. Pfeffer and L. S. Silbert, *J. Org. Chem.*, 1972, **37**, 3551.
476. W. Reppe, *Justus Liebigs. Ann. Chem.*, 1956, **601**, 81.
477. R. C. Larock, in 'Solvomercuration/Demercuration Reactions in Organic Synthesis', Springer Verlag, Berlin, 1986, chap. 5.
478. H. C. Brown, M.-H. Rei and K.-T. Liu, *J. Am. Chem. Soc.*, 1970, **92**, 1760.
479. R. F. Richter, C. Philips and R. D. Bach, *Tetrahedron Lett.*, 1972, 4327.
480. A. J. Bloodworth and J. L. Courtneidge, *J. Chem. Soc., Perkin Trans. 1*, 1981, 3258.
481. S. Wolfe, P. G. C. Campbell and G. E. Palmer, *Tetrahedron Lett.*, 1966, 4203.
482. S. Wolfe and P. G. C. Campbell, *Can. J. Chem.*, 1965, **43**, 1184.
483. A. Factor and T. G. Traylor, *J. Org. Chem.*, 1968, **33**, 2607.
484. D. J. Pasto and J. A. Gontarz, *J. Am. Chem. Soc.*, 1971, **93**, 6909.
485. M. R. Johnson and B. Rickborn, *J. Org. Chem.*, 1969, **34**, 2781.
486. I. F. Lutsenko, R. M. Khomutov and L. V. Eliseeva, *Izv. Akad. Nauk SSSR, Ser. Khim.*, 1956, 181; *Bull. Acad. Sci. USSR, Div. Chem. Sci.*, 1956, 173.
487. M. N. Popova, D. E. Stepanov, V. S. Yakovenko, E. S. Stepuk, I. A. Komarovskaya and V. Donskikh, *Dokl. Vses. Konf. Khim. Atsetilena, 4th*, 1972, **2**, 360 (*Chem. Abstr.*, 1973, **79**, 78 910x).
488. H. Lüssi, *Helv. Chim. Acta*, 1966, **49**, 1684.
489. G. Slinckx and G. Smets, *Tetrahedron*, 1966, **22**, 3163.
490. H. Hopff and M. A. Osman, *Tetrahedron*, 1968, **24**, 2205.
491. H. Hopff and M. A. Osman, *Tetrahedron*, 1968, **24**, 3887.
492. Y. A. Zhdanov, G. A. Korol'chenko and L. A. Kubasskaya, *Russ. Akad. Nauk SSSR*, 1959, **128**, 1185; *Russ. Chem. Rev. (Engl. Transl.)*, 1959, **128**, 887.
493. Y. A. Zhdanov, G. A. Korol'chenko, L. A. Kubasskaya and R. M. Krivoruchko, *Dokl. Akad. Nauk SSSR*, 1959, **129**, 1049; *Russ. Chem. Rev. (Engl. Transl.)*, 1959, **129**, 1101.
494. K. Takiura and S. Honda, *Carbohydr. Res.*, 1972, **23**, 369.
495. M. Julia and C. Blasioli, *Bull. Soc. Chim. Fr.*, 1976, 1941.
496. H. Yoshioka, K. Takasaki, M. Kobayashi and T. Matsumoto, *Tetrahedron Lett.*, 1979, 3489.
497. S. F. Martin and T. Chou, *Tetrahedron Lett.*, 1978, 1943.
498. G. Fráter, *Tetrahedron Lett.*, 1981, **22**, 425.
499. A. Lethbridge, R. O. C. Norman and C. B. Thomas, *J. Chem. Soc., Perkin Trans. 1*, 1973, 2763.
500. A. Lethbridge, R. O. C. Norman and C. B. Thomas, *J. Chem. Soc., Perkin Trans. 1*, 1975, 231.
501. E. J. Corey and H. L. Pearce, *Tetrahedron Lett.*, 1980, **21**, 1823.
502. Y. Hagihara and N. Iritani, *Jpn. Pat.* 4222 (1955) (*Chem. Abstr.*, 1957, **51**, 16 520i).
503. O. A. El Seoud, A. T. do Amaral, M. Moura Campos and L. do Amaral, *J. Org. Chem.*, 1974, **39**, 1915.
504. R. L. Rowland, W. L. Perry and H. L. Friedman, *J. Am. Chem. Soc.*, 1951, **73**, 1040.
505. S. V. Arakelyan, M. T. Dangyan and A. A. Avetisyan, *Izv. Akad. Nauk Arm. SSR, Khim. Nauki*, 1962, **15**, 435 (*Chem. Abstr.*, 1963, **59**, 5186a).
506. S. V. Arakelyan, L. G. Rashidyan and M. T. Dangyan, *Izv. Akad. Nauk Arm. SSR, Khim. Nauki*, 1964, **17**, 173 (*Chem. Abstr.*, 1964, **61**, 8331b).
507. S. V. Arakelyan, L. S. Gareyan, S. O. Titanyan and M. T. Dangyan, *Khim. Geterotsikl. Soedin*, 1970, 13 (*Chem. Abstr.*, 1972, **77**, 88 614b).
508. H. B. Henbest and B. Nicholls, *J. Chem. Soc.*, 1959, 227.
509. H. Christol, F. Plénat and J. Revel, *Bull. Soc. Chim. Fr.*, 1971, 4537.
510. F. R. Jensen and J. J. Miller, *Tetrahedron Lett.*, 1966, 4861.
511. V. Gómez-Aranda, J. Barluenga, M. Yus and G. Asensio, *Synthesis*, 1974, 806.
512. M. Julia and E. Colomer Gasquez, *Bull. Soc. Chim. Fr.*, 1973, 1796 (*Chem. Abstr.*, 1973, **79**, 104 790e).

513. V. Gómez-Aranda, J. Barluenga and M. Yus-Astiz, *Rev. Acad. Cienc. Exactas, Fis.-Quim. Nat. Zaragoza*, 1973, **28**, 225 (*Chem. Abstr.*, 1974, **80**, 83 164g).
514. R. D. Bach, U. Mazur, R. N. Brummel and L.-H. Lin, *J. Am. Chem. Soc.*, 1971, **93**, 7120.
515. W. S. Linn, W. L. Waters and M. C. Caserio, *J. Am. Chem. Soc.*, 1970, **92**, 4018.
516. M. Camps and J.-P. Montheard, *C. R. Hebd. Seances Acad. Sci., Ser. C*, 1976, **283**, 215.
517. W. H. Carothers, R. A. Jacobson and G. J. Berchet, *J. Am. Chem. Soc.*, 1933, **55**, 4665.
518. A. E. Borisov, V. D. Vil'chevskaya and A. N. Nesmeyanov, *Dokl. Akad. Nauk SSSR*, 1953, **90**, 383 (*Chem. Abstr.*, 1954, **48**, 4434f).
519. R. C. Larock, K. Oertle and K. M. Beatty, *J. Am. Chem. Soc.*, 1980, **102**, 1966.
520. S. Uemura, H. Miyoshi and M. Okano, *J. Chem. Soc., Perkin Trans. 1*, 1980, 1098.
521. P. F. Hudrlik and A. M. Hudrlik, *J. Org. Chem.*, 1973, **38**, 4254.
522. G. F. Hennion, D. B. Killian, T. H. Vaughn and J. A. Nieuwland, *J. Am. Chem. Soc.*, 1934, **56**, 1130.
523. R. A. Amos and J. A. Katzenellenbogen, *J. Am. Chem. Soc.*, 1978, **43**, 560.
524. S. R. Sandler, *J. Chem. Eng. Data*, 1973, **18**, 445 (*Chem. Abstr.*, 1973, **79**, 115 076c).
525. Z. A. Chobanyan, S. Z. Davtyan and S. O. Badanyan, *Arm. Khim. Zh.*, 1980, **33**, 589 (*Chem. Abstr.*, 1981, **94**, 46 728m).
526. M. Yamamoto, S. Tanaka, K. Naruchi and K. Yamada, *Synthesis*, 1982, 850.
527. M. Yamamoto, M. Yoshitake and K. Yamada, *J. Chem. Soc., Chem. Commun.*, 1983, 991.
528. M. Yamamoto, *J. Chem. Soc., Perkin Trans. 1*, 1981, 582.
529. M. A. Sofia and J. A. Katzenellenbogen, *J. Org. Chem.*, 1985, **50**, 2331.
530. S. W. Rollinson, R. A. Amos and J. A. Katzenellenbogen, *J. Am. Chem. Soc.*, 1981, **103**, 4114.
531. J. H. Werntz, *US Pat.* 1 963 108 (1934) (*Chem. Abstr.*, 1934, **28**, 4745).
532. D. M. T. Chan, T. B. Marder, D. Melstein and N. J. Taylor, *J. Am. Chem. Soc.*, 1987, **109**, 6385, and references therein.
533. G. A. Krafft and J. A. Katzenellenbogen, *J. Am. Chem. Soc.*, 1981, **103**, 5459.
534. S. Patai (ed.), 'The Chemistry of the Thiol Group', Wiley, New York, 1974.
535. G. C. Barrett, in 'Comprehensive Organic Chemistry', ed. D. H. R. Barton and W. D. Ollis, Pergamon Press, Oxford, 1978, vol. 3, p. 3.
536. E. N. Prilezhaeva and M. F. Stostakovskii, *Russ. Chem. Rev. (Engl. Transl.)*, 1963, **32**, 399.
537. F. R. Mayo and C. Walling, *Chem. Rev.*, 1940, **27**, 351.
538. V. N. Ipatieff, H. Pines and B. S. Friedman, *J. Am. Chem. Soc.*, 1938, **60**, 2731.
539. C. G. Screttas, and M. Micha-Screttas, *J. Org. Chem.*, 1979, **44**, 713.
540. M. J. Belley and R. Zamboni, *J. Org. Chem.*, 1989, **54**, 1230.
541. S. O. Jones and E. E. Reid, *J. Am. Chem. Soc.*, 1938, **60**, 2452.
542. W. E. Bacon and W. M. LeSuer, *J. Am. Chem. Soc.*, 1954, **76**, 670.
543. W. E. Bacon, N. A. Meinhardt and W. M. LeSuer, *J. Org. Chem.*, 1960, **25**, 1993.
544. M. F. Shostakovskii, E. N. Prilezhaeva and E. S. Shapiro, *Izv. Akad. Nauk SSSR Otdel. Khim. Nauk*, 1954, 303 (*Chem. Abstr.*, 1954, **48**, 10 535f).
545. E. N. Prilezhaeva, N. P. Petukhova and M. F. Shostakovskii, *Dokl. Akad. Nauk SSSR*, 1964, **154**, 160 (*Chem. Abstr.*, 1964, **60**, 9143g).
546. M. G. Missakian, R. Ketcham and A. R. Martin, *J. Org. Chem.*, 1974, **39**, 2010.
547. F. Kipnis and J. Ornfelt, *J. Am. Chem. Soc.*, 1951, **73**, 822.
548. M. I. Kabachnik, T. A. Mastryukova, M. F. Shostakovskii, E. N. Prilezhaeva, D. M. Paikin, M. P. Shabanova and N. M. Gamper, *Dokl. Akad. Nauk SSSR*, 1956, **109**, 777 (*Chem. Abstr.*, 1957, **51**, 4634g).
549. D. Harman and W. E. Vaughn, *J. Am. Chem. Soc.*, 1950, **72**, 631.
550. C. S. Marvel and E. D. Weil, *J. Am. Chem. Soc.*, 1954, **76**, 61.
551. C. Bochu, A. Couture and P. Grandclaudon, *J. Org. Chem.*, 1988, **53**, 4852.
552. B. Saville, *J. Chem. Soc.*, 1962, 5040.
553. W. H. Mueller and K. Griesbaum, *J. Am. Chem. Soc.*, 1967, **32**, 856.
554. A. I. Borisova, A. K. Filippova, V. K. Voronov and M. F. Shostakovskii, *Izv. Akad. Nauk SSSR, Ser. Khim.*, 1969, 2498 (*Chem. Abstr.*, 1970, **72**, 66 538z).
555. K. E. Schülte, J. Reich and L. Horner, *Chem. Ber.*, 1962, **95**, 1943.
556. D. N. Jones, in 'Perspectives in the Organic Chemistry of Sulfur', ed. B. Zwanenbury and A. J. H. Klunder, Elsevier, Amsterdam, 1987, p. 189.
557. P. J. Brown, D. N. Jones, M. A. Khan and N. A. Meanwell, *J. Chem. Soc., Perkin Trans. 1*, 1984, 2049.
558. D. Liotta (ed.), 'Organoselenium Chemistry', Wiley, New York, 1987.
559. K. C. Nicolaou and N. A. Petasis, 'Selenium in Natural Products Synthesis', CIS, Philadelphia, 1984.
560. A. J. Anciaux, A. Eman, W. Dumont, D. Van Ende and A. Krief, *Tetrahedron Lett.*, 1975, 1613.
561. C. Bouche, A. Couture and P. Grandclaudon, *J. Org. Chem.*, 1988, **53**, 4852.
562. E. G. Kataev and V. N. Petrov, *Zh. Obshch. Khim.*, 1962, **32**, 3699 (*Chem. Abstr.*, 1963, **59**, 481f).

1.8
Electrophilic Addition of X—Y Reagents to Alkenes and Alkynes

ERIC BLOCK and ADRIAN L. SCHWAN
State University of New York at Albany, NY, USA

1.8.1 INTRODUCTION

Electrophilic addition of the halogens and related X—Y reagents to alkenes and alkynes has been a standard procedure since the beginning of modern organic chemistry.[1] *Anti* electrophilic bromination of such simple compounds as cyclohexene and (*E*)- and (*Z*)-2-butene, and variants of this reaction when water or methanol are solvents (formation of halohydrin or their methyl ethers, respectively), are frequently employed as prototype examples of stereospecific reactions in elementary courses in organic chemistry. A simple test for unsaturation involves addition of a dilute solution of bromine in CCl_4 to the compound in question: alkenes and alkynes discharge the color. Prior to the development of modern analytical methods, quantitative analysis of alkenes was performed by titration with bromine, thiocyanogen or related interhalogens. Spurred by interest in biologically active organofluorine compounds, methods have recently been developed for controlled addition of highly reactive elemental fluorine and fluorine-containing interhalogens to alkenes and alkynes. Regio- and stereo-specific addition of sulfenyl halides and reagents with nitrogen–halogen bonds to alkenes and alkynes have been used for many years for controlled introduction of sulfur and nitrogen, respectively, as well as for other more general synthetic purposes. More recently, electrophiles containing selenium and tellurium have been added to the list of synthetically useful reagents that undergo stereocontrolled addition to π-bonds.

Mechanistic details for these electrophilic addition reactions have been determined by Ingold, Bartlett, Tarbell, Roberts, Kimball, Grob and Winstein among others; these studies represent an important chapter in the history of organic chemistry.[1a,c,d,2] All of the above electrophilic reagents contain a polar or polarizable bond which is cleaved during the addition so that the rate-determining transition state has a net positive charge. The electrophile can add in a one-step process or in a two-step process involving a discrete cationic intermediate. Bond formation in the adduct can be preceded by formation of complexes of various types. Thus, reaction of adamantylideneadamantane (Ad_2) with bromine has been shown to involve formation of an initial charge-transfer complex ($Ad_2 \rightarrow Br_2$), followed by formation of two bromonium–polybromide complexes ($Ad_2Br^+ \, Br_n^-$; $n = 3$ or 5).[3] Regioselection in attachment of X—Y to an alkene can be described as Markovnikov or anti-Markovnikov (with reference to the more electrophilic part of X—Y) while stereoselection can be described as *syn* (X and Y add to the same side of the plane of the double bond) or *anti*. When the π-system has two distinct faces, configurational selectivity is also possible. In the case of a bicyclic alkene such as bicyclo[2.2.1]hept-2-ene the attack of the electrophile from the less hindered *exo* face is favored; on the other hand bicyclo[2.2.1]hepta-2,5-diene shows a preference for *endo* attack (Scheme 1).[4]

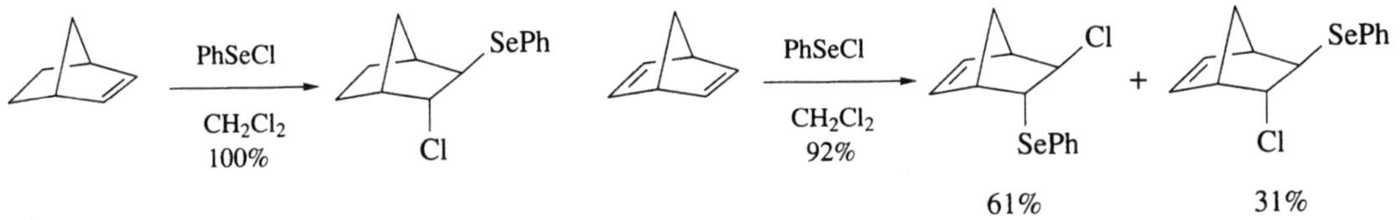

PhSeCl, CH_2Cl_2, 100% → SePh, Cl

PhSeCl, CH_2Cl_2, 92% → Cl, SePh (61%) + SePh, Cl (31%)

Scheme 1

The major focus in this chapter will be on synthesis, with emphasis placed on more recent applications, particularly those where regiochemistry and stereochemistry are precisely controlled. The reader is referred to the earlier reviews for full mechanistic information and details of historic interest. Electrophilic addition of X—Y to an alkene, where X is the electrophile, gives products with functionality Y β to the heteroatom X. Further transformations of X and/or Y provide the basis for diverse synthetic applications. These transformations include replacement of Y by hydrogen, elimination to form a π-bond (either including the carbon bonded to X or β to that carbon so that X is now in an allylic position), and nucleophilic or radical substitution. Representative examples of these synthetic methods will be given below. This chapter will include examples of heterocycles formed in one-pot reactions where the the initial alkene–electrophile adduct contains an electrophilic group that can react further. Examples of heterocycles formed in several steps from alkene–electrophile adducts will also be considered. Cases in which activation by an external electrophile directly results in addition of an internal heteroatom nucleophile are treated in Chapter 1.9 of this volume.

1.8.2 ELECTROPHILIC SULFUR, SELENIUM AND TELLURIUM

1.8.2.1 Sulfenyl Halides and Related Compounds

Early work on electrophilic addition of sulfur halides to alkenes stems from the development of bis(β-chloroethyl) sulfide ('mustard gas') as a chemical warfare agent. The original synthesis of mustard gas involved addition of sulfur dichloride (or S_2Cl_2) to ethylene according to the equation $2CH_2{=\!=}CH_2 + SCl_2 \rightarrow (ClCH_2CH_2)_2S$. Electrophilic addition of sulfenyl compounds to alkenes and alkynes, which has been widely studied from a mechanistic standpoint,[1c,d,5] finds considerable utility in organic synthesis due to the regio- and stereo-selectivity of the addition, as well as the numerous options for subsequent replacement of the sulfur or vicinal substituent with other functional groups. The most common type of agents for electrophilic addition of sulfur to carbon–carbon multiple bonds are sulfenyl halides (RS—X, X = halogen). Other possibilities include RS—OY and RS—NR_3^+, discussed following the halides, Y—S—X (Y is also a leaving group), and R_2S^+—X, discussed under the heading 'sulfonium salts' in Section 1.8.2.2.

Representative examples of addition of sulfenyl halides to alkenes include the *anti* addition of ethanesulfenyl chloride to cyclopentene and dihydropyran (Scheme 2)[6] and the diastereoselective addition of benzenesulfenyl chloride to optically active bornyl propenoates as in equation (1).[7] The chlorine can be replaced with retention of stereochemistry by other nucleophiles such as nitrogen (see below), hydrogen (as hydride), oxygen or carbon. Examples of the latter process, termed carbosulfenylation, are shown in Scheme 3,[8] where the regioselectivity is 85% and the diastereofacial selectivity is >99%, and Scheme 4.[9] In this latter sequence, acetonylation and allylation are accomplished using a silyl enol ether or allyltrimethylsilane as the carbon nucleophiles using catalytic zinc bromide. White and coworkers used the carbosulfenylation procedure in an enantioselective total synthesis of (–)-monic acid C (Scheme 5).[10] Here the triphenylmethanesulfenyl group undergoes spontaneous elimination. Replacement of the chlorine by oxygen is used in the synthesis of a 2α-glycoside of *N*-acetylneuraminic acid (Scheme 6).[11] In the final step the phenylthio group is removed by treatment with tri-*n*-butyltin hydride.

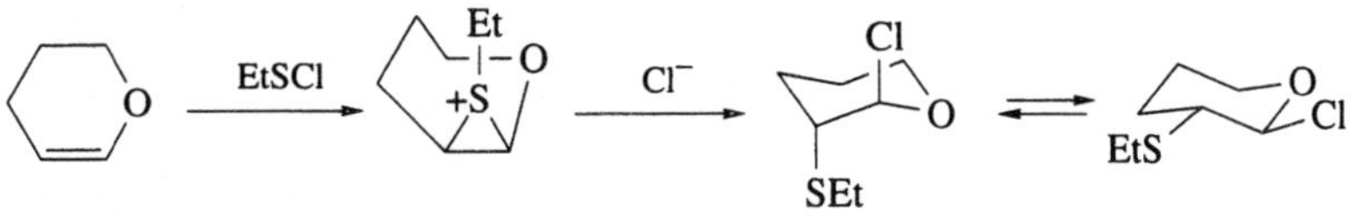

EtSCl; Cl^-

Scheme 2

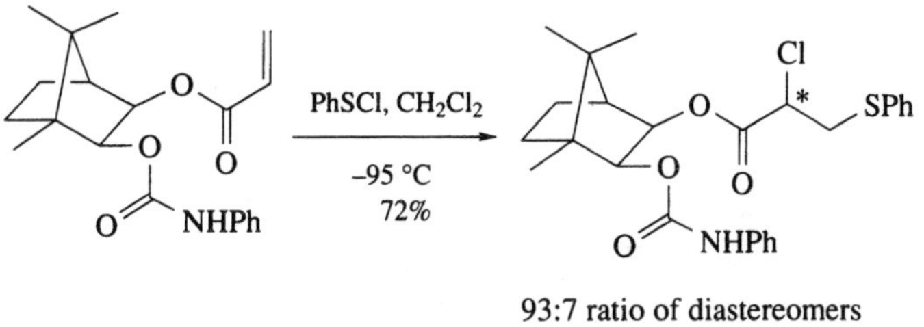

$PhSCl, CH_2Cl_2$; –95 °C; 72% (1)

93:7 ratio of diastereomers

i, ii; 61%

i, PhSCl; ii, Me_2Zn, 0.2 equiv. $TiCl_4$, CH_2Cl_2

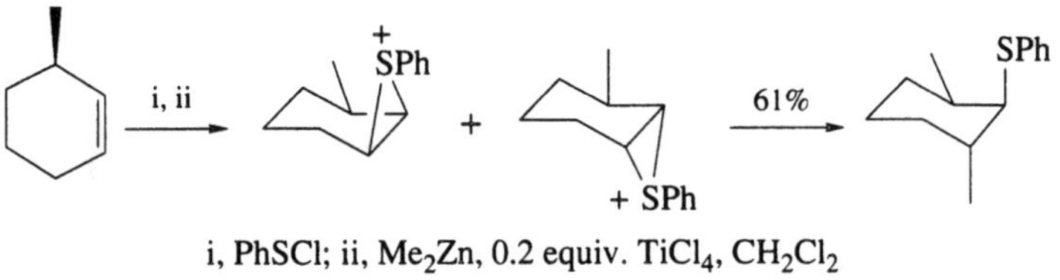

Scheme 3

Thiiranes can be formed directly and stereospecifically from 1,2-disubstituted alkenes by addition of trimethylsilylsulfenyl bromide, formed at –78 °C from reaction of bromine with bis(trimethylsilyl) sulfide (Scheme 7).[12] A two-step synthesis of thiiranes can be achieved by addition of succinimide-*N*-sulfenyl chloride or phthalimide-*N*-sulfenyl chloride to alkenes followed by lithium aluminum hydride cleavage of the adducts (Scheme 8).[13] Thiaheterocycles can also be formed by intramolecular electrophilic addition of sulfenyl chlorides to alkenes, *e.g.* as seen in Schemes 9[14] and 10.[15] Related examples involving sulfur dichloride are shown in Schemes 11[16] and 12.[17] In the former case addition of sulfur dichloride to 1,5-cyclooctadiene affords a bicyclic dichloro sulfide *via* regio- and stereo-specific intramolecular addition of an intermediate sulfenyl chloride. Removal of chlorine by lithium aluminum hydride reduction affords 9-thiabicyclo[3.3.1]nonane, which can be further transformed into bicyclo[3.3.0]oct-1,5-ene.[16]

Addition of a sulfenyl chloride to an alkene in the presence of silver fluoride affords β-fluoro sulfides, as shown in equation (2).[18] The same *trans* adducts can also be prepared from the sulfenyl chloride adducts and silver fluoride. An alternative procedure for the formal addition of methanesulfenyl fluoride to alkenes involves the use of dimethyl(methylthio)sulfonium tetrafluoroborate and triethylamine tris(hydrofluoride) as in equation (3).[19]

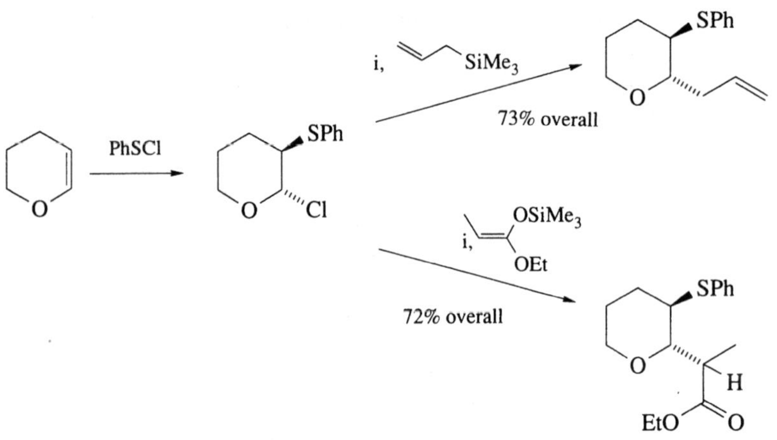

i, $ZnBr_2$ (0.05 equiv.), CH_2Cl_2, 20 °C

Scheme 4

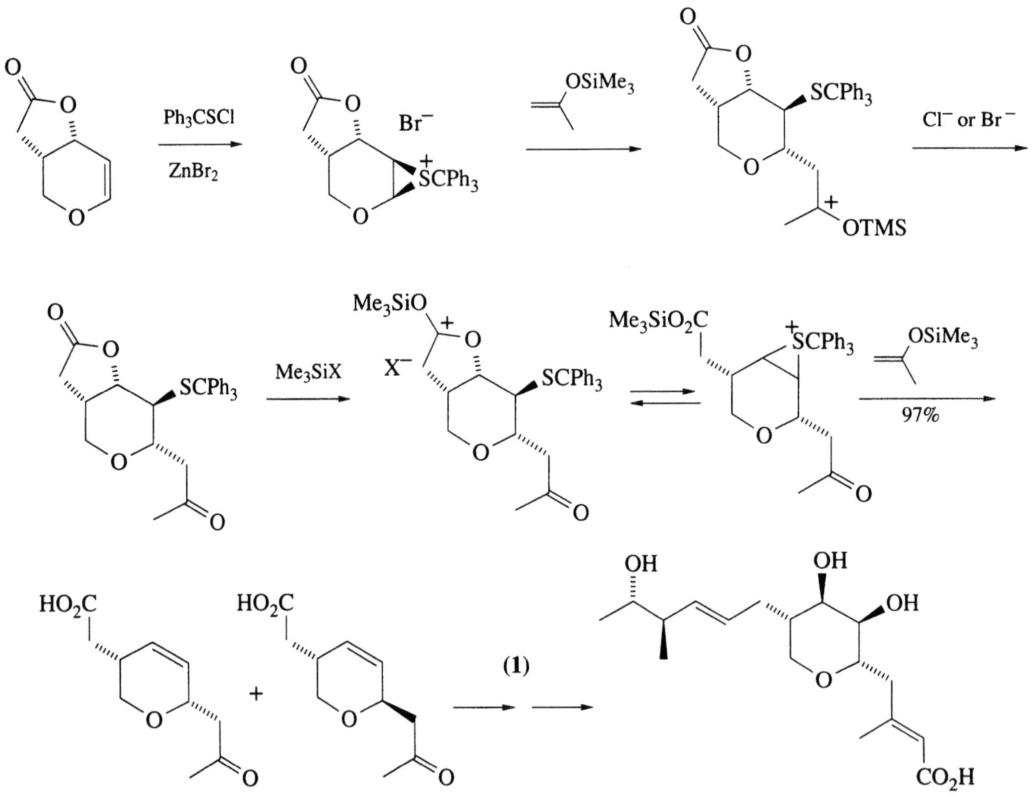

i, overall, Ph_3CSCl, $CH_2{=}C(Me)OSiMe_3$, cat. $ZnBr_2$, CH_2Cl_2, –78 °C, 15 h

Scheme 5

The groups of Trost and coworkers and Caserio and coworkers independently developed procedures for azasulfenylation. Caserio reported that when alkenes are treated with a sulfenamide such as $MeSNMe_2$ in the presence of the Meerwein reagent trimethyloxonium tetrafluoroborate, azasulfenylation of the alkene is observed, presumably *via* the thioammonium ion $MeSNMe_3^+$ (equation 4).[20] Boron trifluoride etherate can also serve as a catalyst for addition of sulfenamides to alkenes. The adduct stereochemistry is strictly *anti* as established by treatment of the dihydropyran methanesulfenyl chloride adduct (see Scheme 2) with $AgBF_4$ followed by trimethylamine.

Cycloazasulfenylation (or 'sulfenocycloamination') can be achieved by addition of benzenesulfenyl chloride to an alkene containing a remote nitrogen followed by treatment of the β-chloro sulfide adduct

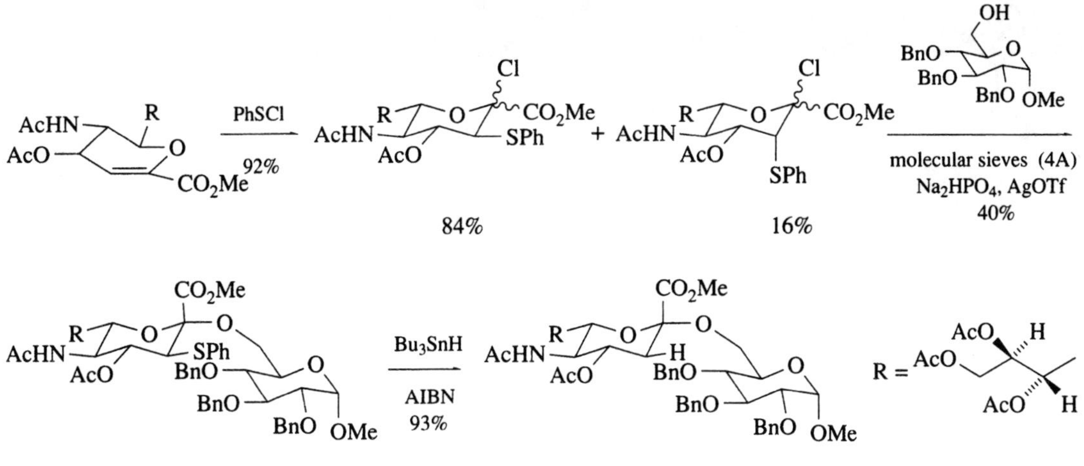

Scheme 6

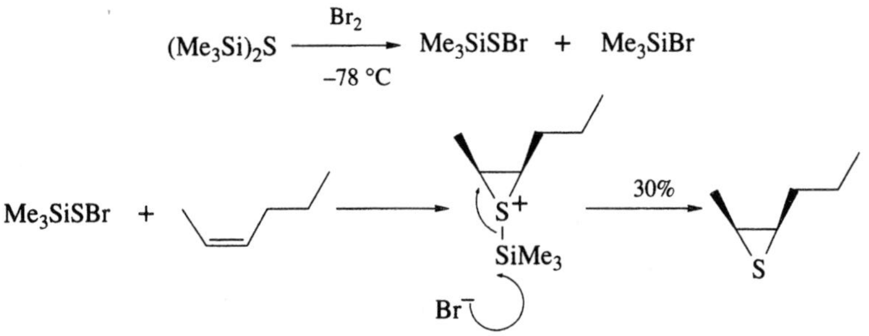

Scheme 7

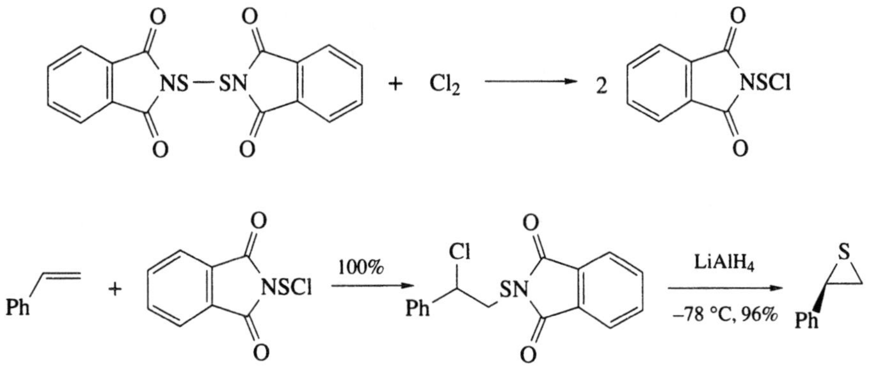

Scheme 8

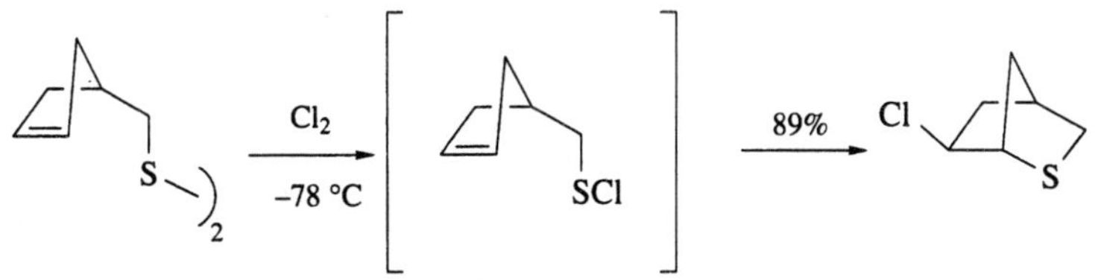

Scheme 9

with potassium carbonate/sodium iodide (Scheme 13).[21] The phenylthio group can be replaced with hydrogen using Raney nickel, or can be oxidized to a sulfoxide and benzenesulfenic acid eliminated by heating. These overall procedures have been used in the synthesis of the pyrrolizidine alkaloids retronecine and turneforcidine (Scheme 14)[22] as well as β-lactam antibiotics (Scheme 15).[23] If alkenes containing a remote nitrogen are treated with the methylsulfenylium equivalent bis(methylthio)methylsulfonium

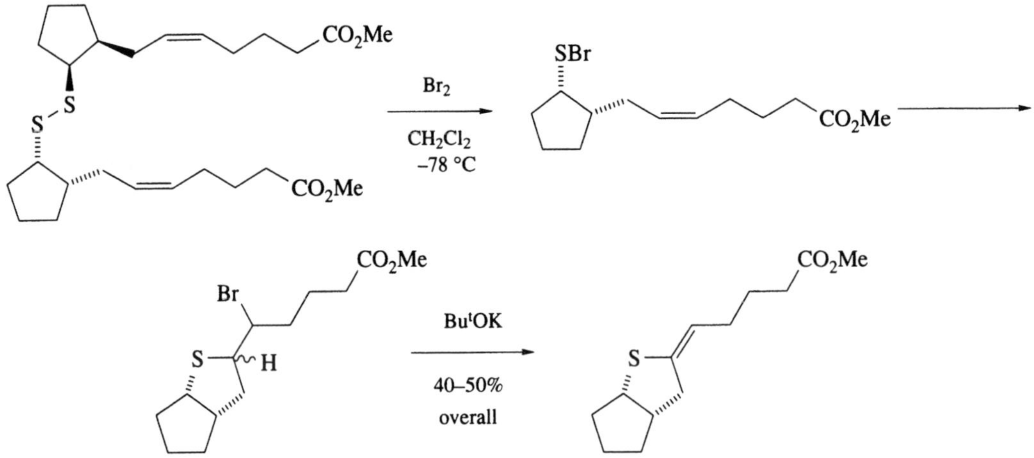

CO2Me
SBr
Br2
CH2Cl2
–78 °C
CO2Me
CO2Me
CO2Me
Br
S
H
ButOK
40–50%
overall
CO2Me
S

Scheme 10

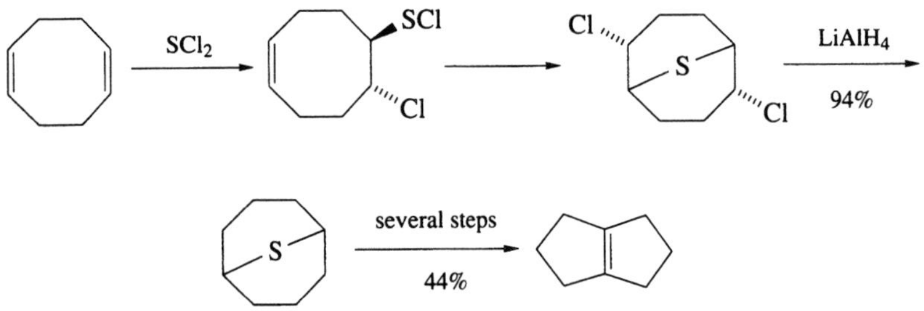

SCl2
SCl
Cl
Cl
S
Cl
LiAlH4
94%
S
several steps
44%

Scheme 11

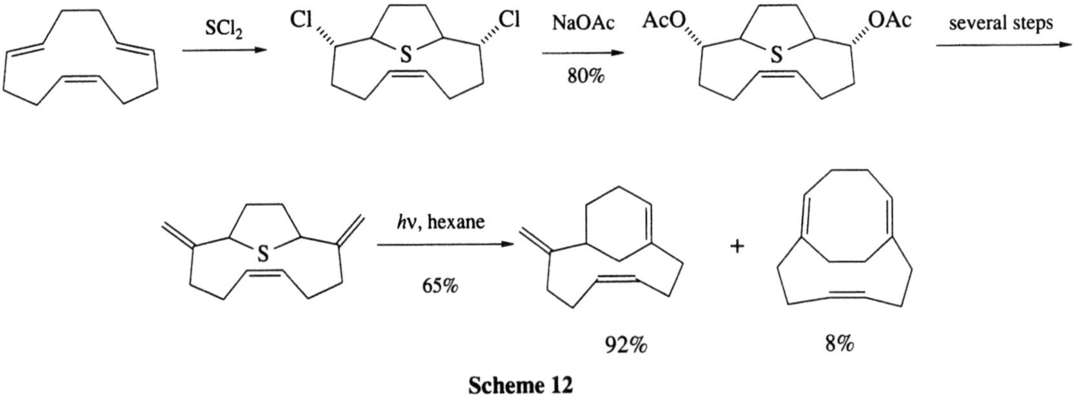

SCl2
Cl
Cl
S
NaOAc
80%
AcO
OAc
S
several steps
S
hν, hexane
65%
+
92%
8%

Scheme 12

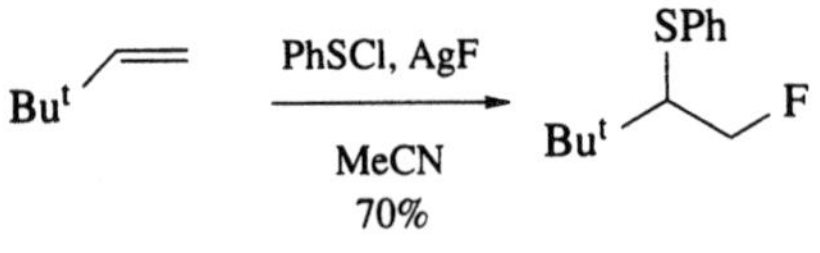

But
PhSCl, AgF
MeCN
70%
SPh
But
F

(2)

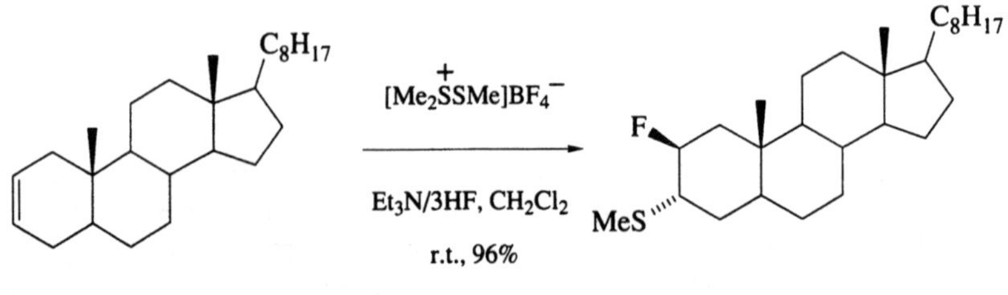

C8H17
[Me2SSMe]BF4
Et3N/3HF, CH2Cl2
r.t., 96%
C8H17
F
MeS

(3)

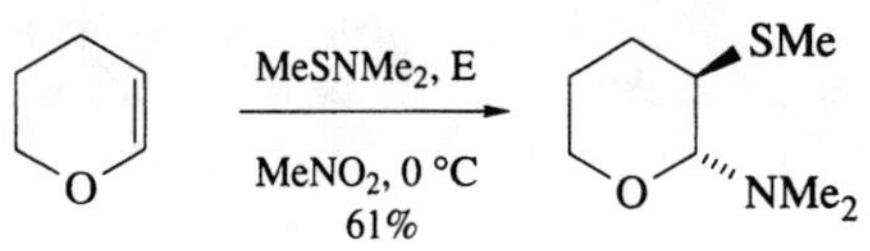

MeSNMe$_2$, E; MeNO$_2$, 0 °C; 61% (4)

E = Me$_2$O•BF$_3$, Me$_3$O$^+$BF$_4^-$

tetrafluoroborate [(MeS)$_2$SMe$^+$ BF$_4^-$], instead of benzenesulfenyl chloride, nitrogen heterocycles can be isolated in one step;[24] these types of reactions are treated in Chapter 1.9 of this volume.

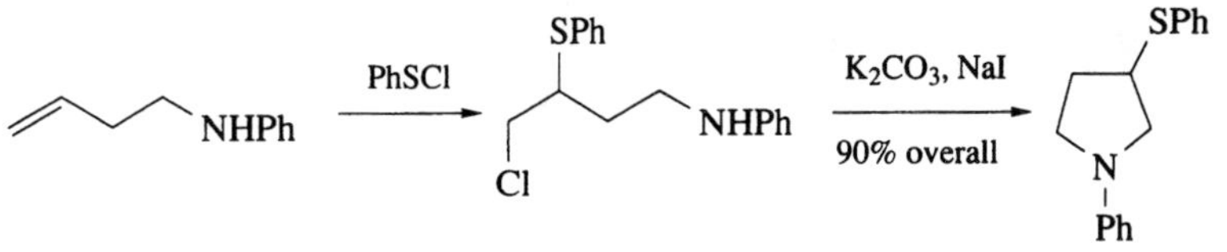

PhSCl; K$_2$CO$_3$, NaI; 90% overall

Scheme 13

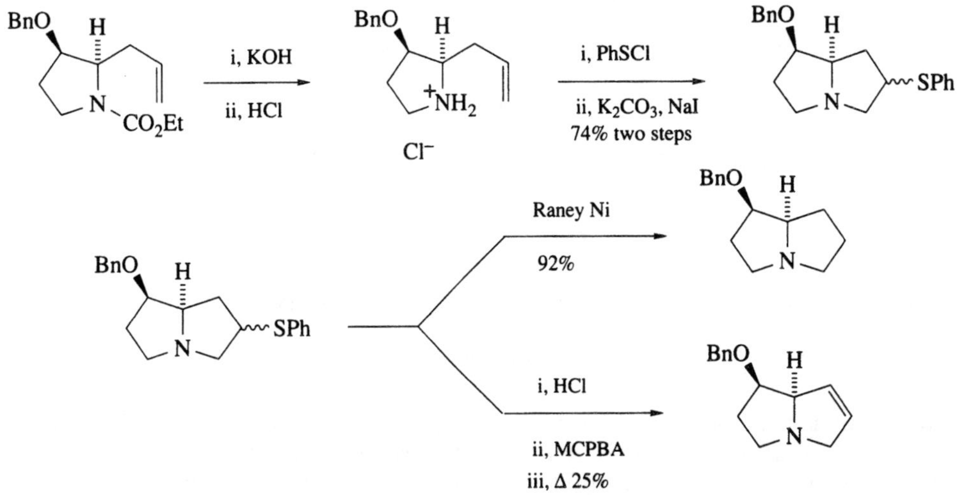

i, KOH; ii, HCl; i, PhSCl; ii, K$_2$CO$_3$, NaI; 74% two steps

Raney Ni; 92%

i, HCl; ii, MCPBA; iii, Δ 25%

Scheme 14

i; PhSCl; ii; X = Bu^{t}Me$_2$Si, 83%; X = SPh, 47%

Et$_3$N, K$_2$CO$_3$; NaI, MeCN; 82%

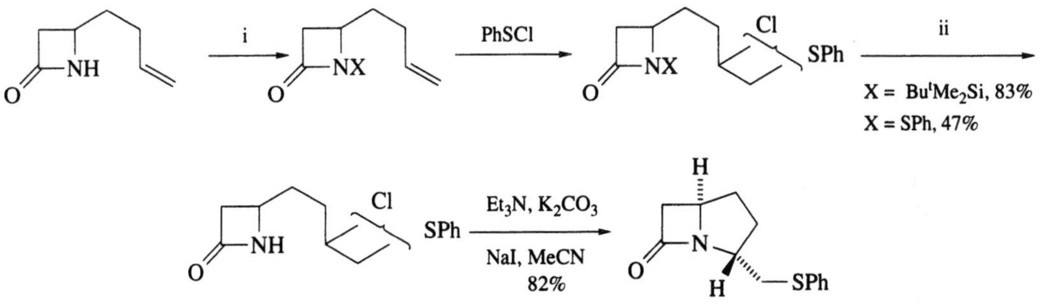

X = SPh: i, PhSCl; ii, NaBH$_4$, MeOH

X = Bu^{t}Me$_2$Si: i, Et$_3$N, Bu^{t}Me$_2$SiCl, DMF; ii, TBAF

Scheme 15

Intramolecular electrophilic addition of a sulfenate ester (RS—OR′) to a carbon–carbon double bond (intramolecular oxosulfenylation) is indicated in the work of Block and Wall (Scheme 16).[25] Zefirov and coworkers find that arenesulfenamides can be activated with sulfur trioxide to give arenesulfenyl sulfamates, which react with alkenes such as cyclohexene giving good yields of the oxosulfenylated *trans* 1,2-adducts, as shown in Scheme 17.[26] Sulfenyl trifluoro- or trichloro-acetates, prepared from reaction of thiosulfinates with the corresponding trihaloacetic anhydrides, give *trans* 1,2-diadducts with a variety of alkenes (Scheme 18).[27]

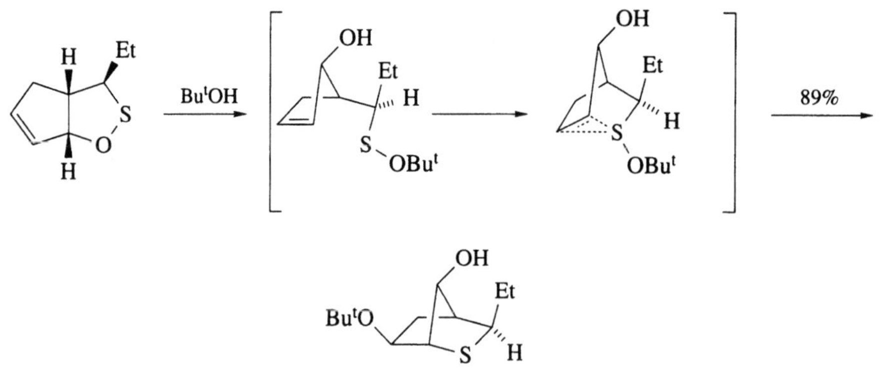

Scheme 16

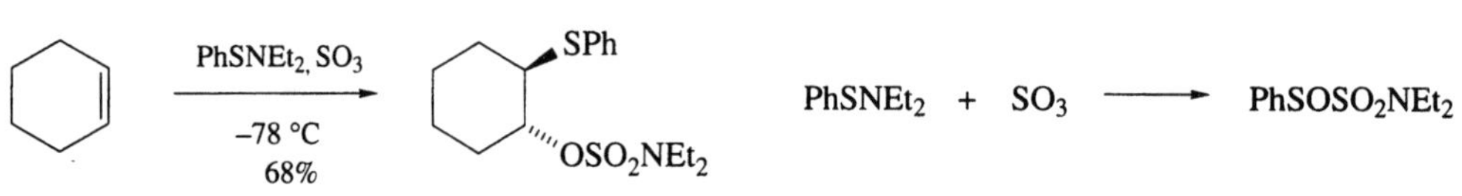

$PhSNEt_2 + SO_3 \longrightarrow PhSOSO_2NEt_2$

Scheme 17

$3\ MeS(O)SMe + (CX_3CO)_2O \longrightarrow 2\ MeSSMe + 2\ MeS(O)OC(O)CX_3$ *via* $MeSOC(O)CX_3$

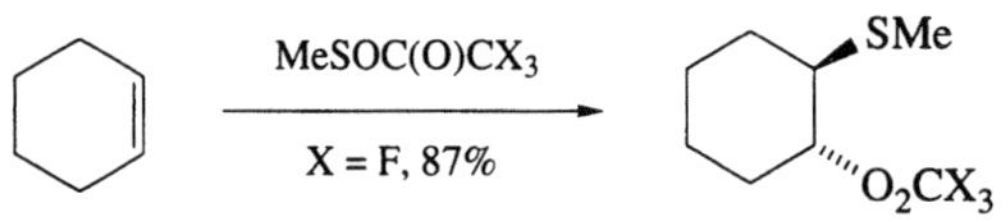

Scheme 18

An example of addition of sulfenyl halides to an alkyne is shown in Scheme 19[28] where the initial *trans* adduct of 1,4-dichlorobut-2-yne is further transformed by treatment with base or aluminum amalgam. Another example involves boron trifluoride-catalyzed addition of an arenesulfenamide to alkynes in the presence of acetonitrile, giving β-acetamidinovinyl sulfides (Scheme 20),[29] which can be hydrolyzed to β-keto phenyl sulfides.[30]

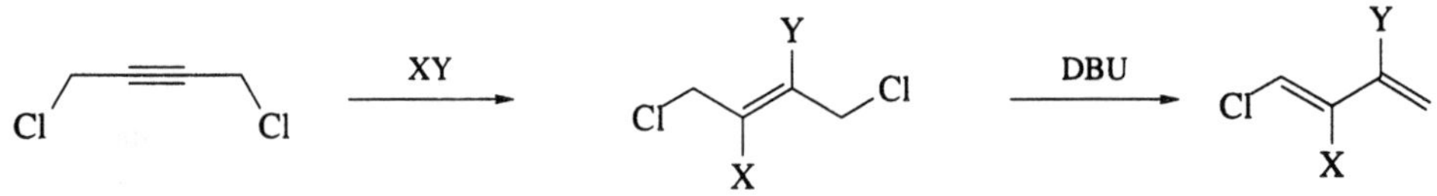

XY	(2)	DBU product
PhSCl	98 %	85%
PhSBr	80	63
PhSeCl	100	46
PhSeBr	93	68
Br_2	97	85
I_2	86	81

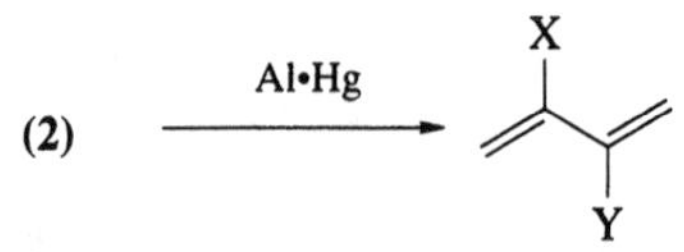

(2) —Al•Hg→

XY	*Yield* (%)
PhSCl	68
PhSBr	40
PhSeCl	75
PhSeBr	76

Scheme 19

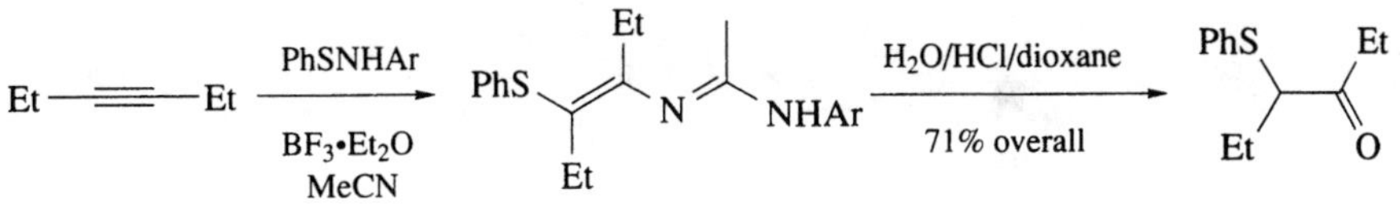

Et—≡—Et → (PhSNHAr; BF_3•Et_2O, MeCN) → PhS(Et)C=C(Et)N=C(Me)NHAr → (H_2O/HCl/dioxane; 71% overall) → PhS(Et)CH–C(O)Et

Ar = p-$NO_2C_6H_4$

Scheme 20

1.8.2.2 Sulfonium Salts

As seen above, β-chloroalkyl sulfides are generally prepared by addition of sulfenyl chlorides to alkenes. A new procedure involves reaction of alkenes with dimethyl sulfoxide activated with phenyl dichlorophosphate or phosphorus oxychloride (Scheme 21).[31] In this latter reaction the addition proceeds in a completely regiospecific manner with the methylthio group being added to the carbon bearing the greater number of hydrogen atoms. Exclusive *trans* addition is observed with simple cycloalkenes. It is suggested that the active electrophile is the oxysulfonium salt $[Me_2SOPO(R)Cl]^+ Cl^-$ (R = OPh or Cl). When the double bond is conjugated with a benzene ring, double addition of the methylthio group occurs (Scheme 21); dimethyl sulfide is suggested to be the source of the second methylthio group. In the above procedures chlorotrimethylsilane may be used in place of the phosphorus reagents to activate dimethyl sulfoxide.[32] Double addition of the methylthio (or other alkylthio groups) to unactivated alkenes can be achieved using the combination of dimethyl disulfide (or other dialkyl disulfides) and boron trifluoride etherate (equation 5).[33] Chlorosulfonium salts have been found to add to alkenes, giving the β-chloroalkylsulfonium salt adducts in good yield (Scheme 22).[34]

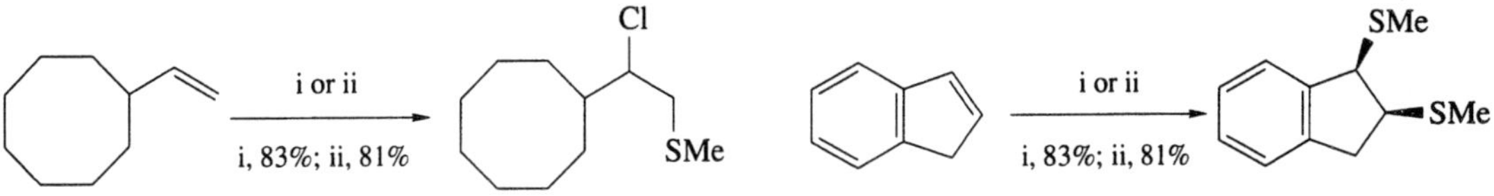

(vinylcyclooctane → (i or ii; i, 83%; ii, 81%) → cyclooctyl–CH(Cl)–CH₂SMe; indene → (i or ii; i, 83%; ii, 81%) → trans-1,2-bis(SMe)indane)

i, Me_2SO (10 equiv.), $PhOP(O)Cl_2$ (5 equiv.), CH_2Cl_2, –20 to 20 °C; ii, Me_2SO (6 equiv.), $POCl_3$ (3 equiv.), CH_2Cl_2, –20 to 20 °C

Scheme 21

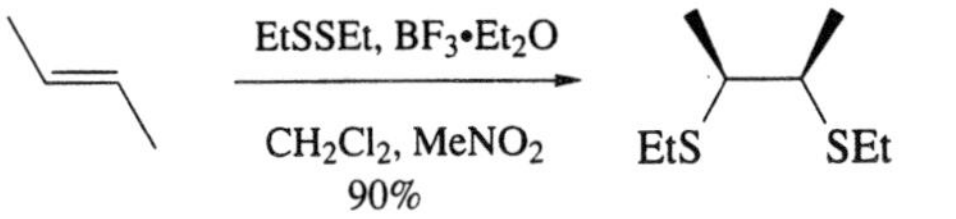

2-butene → (EtSSEt, BF_3•Et_2O; CH_2Cl_2, $MeNO_2$, 90%) → MeCH(SEt)CH(SEt)Me (5)

tetrahydrothiophene → (SO_2Cl_2; SO_2, –30 °C) → S-chlorotetrahydrothiophenium Cl^- → (styrene; 85%) → PhCH(Cl)CH₂–S⁺(tetrahydrothiophenium) Cl^-

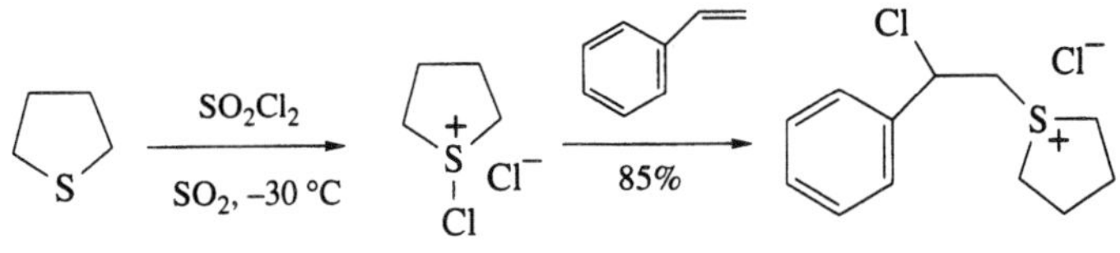

Scheme 22

An alternative route from alkenes to 2-azasulfides reported by the groups of Caserio and Trost involves addition of a thiosulfonium salt, *e.g.* dimethyl(methylthio)sulfonium tetrafluoroborate ($MeSSMe_2^+ BF_4^-$), followed by treatment of the resultant thiosulfenylated adduct with an amine or other nitrogen nucleophiles (Schemes 23[20] and 24).[35] Trost reports that the addition of the thiosulfonium salt can be followed by addition of an oxygen nucleophile, such as acetate, or a carbon nucleophile, such as cyanide, effecting oxosulfenylation and cyanosulfenylation, respectively (Scheme 25).[36]

The dimethyl sulfide–sulfur trioxide complex also functions as a thiosulfenylating agent, *e.g.* converting 1,5-cyclooctadiene into a single adduct, presumably of *trans* stereochemistry (Scheme 26),[37] giving the *trans* adduct of 2-butyne (equation 6),[38] and 1,4-adducts with acyclic 1,3-dienes (Scheme 26).[37] With cyclopentadiene both 1,2- and 1,4-addition occurs (Scheme 26).[37]

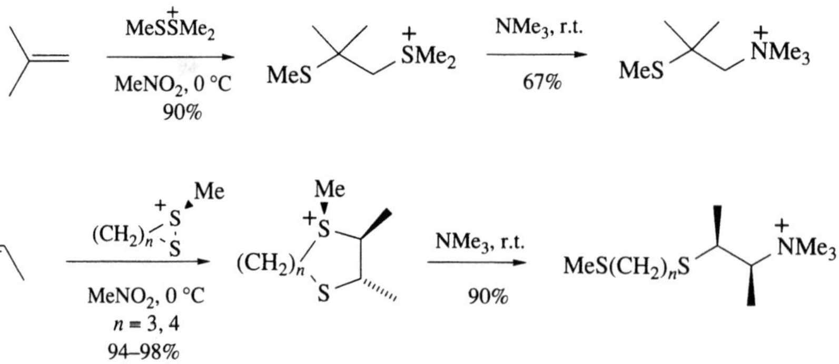

MeS$\overset{+}{S}$Me$_2$ MeNO$_2$, 0 °C 90% SMe$_2$ MeS NMe$_3$, r.t. 67% NMe$_3$ MeS

Me S S (CH$_2$)$_n$ MeNO$_2$, 0 °C n = 3, 4 94–98% Me S (CH$_2$)$_n$ S NMe$_3$, r.t. 90% MeS(CH$_2$)$_n$S NMe$_3$

Scheme 23

i 70% N$_3$ SMe Ph ii 98% NH$_2$ Ph SMe

i, Me$_2\overset{+}{S}$(SMe) BF$_4^-$, NaN$_3$, MeNO$_2$, Me$_2$S; ii, Me$_2\overset{+}{S}$(SMe) BF$_4^-$, NH$_3$, CH$_2$Cl$_2$, Me$_2$S

Scheme 24

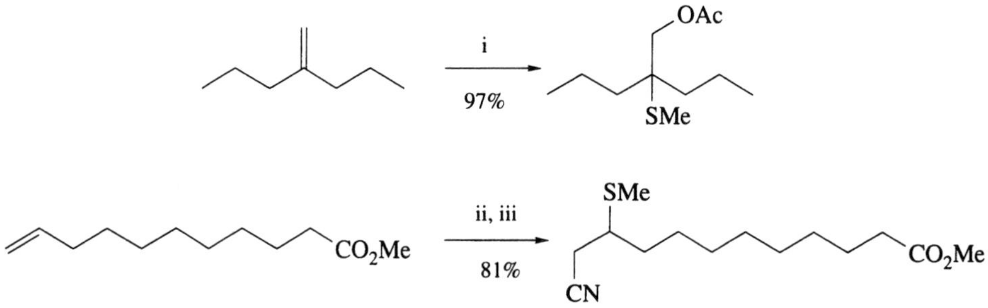

i 97% OAc SMe

CO$_2$Me ii, iii 81% SMe CN CO$_2$Me

i, Me$_2\overset{+}{S}$(SMe) BF$_4^-$, KOAc, MeCN; ii, Me$_2\overset{+}{S}$(SMe) BF$_4^-$, NaCN, HBF$_4$, MeNO$_2$; iii, MeOH

Scheme 25

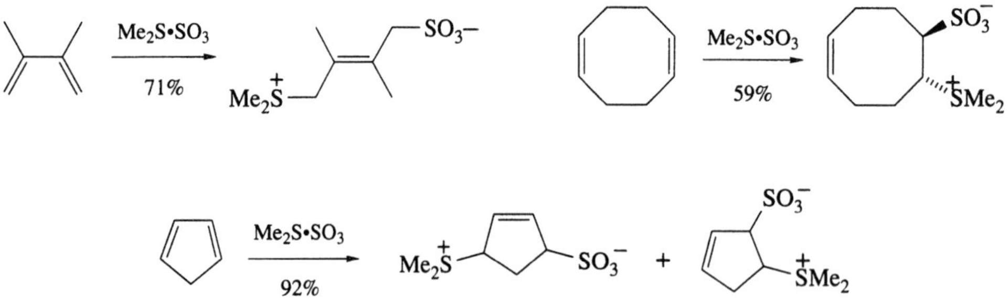

Me$_2$S•SO$_3$ 71% SO$_3^-$ Me$_2\overset{+}{S}$ Me$_2$S•SO$_3$ 59% SO$_3^-$ $\overset{+}{S}$Me$_2$

Me$_2$S•SO$_3$ 92% Me$_2\overset{+}{S}$ SO$_3^-$ + SO$_3^-$ $\overset{+}{S}$Me$_2$

Scheme 26

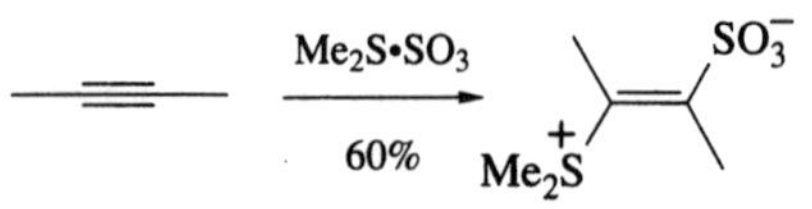

Me$_2$S•SO$_3$ 60% SO$_3^-$ Me$_2\overset{+}{S}$ (6)

1.8.2.3 Selenenyl Halides, Pseudohalides and Related Compounds.

The reaction of electrophilic selenium reagents with alkenes and alkynes is the subject of several recent reviews[39,40] and mechanistic studies.[41] The majority of electrophilic selenium reactions involve selenenic (Se^{II}) compounds. Reactions of Se^{IV} electrophiles, which have been less studied than their Se^{II} counterparts, are considered in Section 1.8.2.4. Aryl, rather than alkyl, selenium derivatives are favored for electrophilic reactions because of the less offensive odors, increased stability, lower volatility and, in general, greater ease of handling of the former, compared to the latter, compounds. Diphenyl diselenide (PhSeSePh), a commercially available crystalline solid, can be readily converted into benzeneselenenyl bromide (PhSeBr) or chloride (PhSeCl) by treatment with bromine or chlorine, respectively. These latter two reagents can in turn be converted to such useful Se^{II} electrophiles such as PhSeOAc, PhSeSCN, $PhSeN_3$, PhSeCN and $PhSeSO_2Ar$. Addition of selenenyl halides to alkenes forming β-halo selenides bears a superficial resemblance to addition of sulfenyl halides to alkenes, although there are significant differences in the reactivity, nature of the intermediates and the regioselectivity of product formation. The 1,2-additions of ArSeX to alkenes are generally highly *anti* stereospecific, consistent with the intermediacy of selenium-bridged intermediates (episelenuranes and seleniranium or episelenonium ions). The regiochemistry of addition varies with the temperature and solvent with anti-Markovnikov adducts favored at low temperatures (–78 °C) where kinetic control predominates and the Markovnikov adducts favored at room temperature under thermodynamic control. When ArSeX reacts with alkenes in the presence of other nucleophiles the latter are often incorporated in the β-position, instead of the halide ion. In cases where there exist two distinct faces of the π-system, attack of ArSeX is favored from the less sterically congested side of the double bond (configurational selectivity, *e.g. exo versus endo* attack in bicyclic systems; Scheme 1).

Regio- and stereo-specific addition of methaneselenenyl bromide in acetic acid to estra-1,3,5(10),16-tetraen-3-ol gave 17α-(methylseleno)estra-1,3,5(10)-triene-3,16β-diol 16-acetate in 80% yield (equation 7).[42] The latter compound and related seleno analogs of estradiol are of interest for biological studies using radioactive ^{75}Se. Regiospecific addition of benzeneselenenyl bromide to ethyl vinyl ether followed by treatment of the adduct with an allylic alcohol affords mixed acetals (Scheme 27).[43] These on oxidation and heating then give esters of δ,γ-unsaturated acids by sequential selenoxide elimination and Claisen rearrangement. A related alkoxyselenation process involving regiospecific addition of benzeneselenenyl bromide to 3,4-dihydro-2*H*-pyrans in the presence of alcohols followed by oxidative deselenation gives cyclic α,β-unsaturated acetals, which can be further transformed into hexopyranose derivatives or the C-glycoside pseudomonic acid C.[44] The phenylseleno group in 3,4-dihydro-2*H*-pyran alkoxyselenation adducts can also be removed by reductive deselenation with triphenyltin hydride. This latter procedure has been successfully employed in efficient syntheses of β-2-deoxyglycosides.[45] The useful homologating reagent phenylselenoacetaldehyde can be prepared by regiospecific addition of phenylselenenyl bromide to ethyl vinyl ether in the presence of ethanol, followed by hydrolysis (Scheme 28).[46]

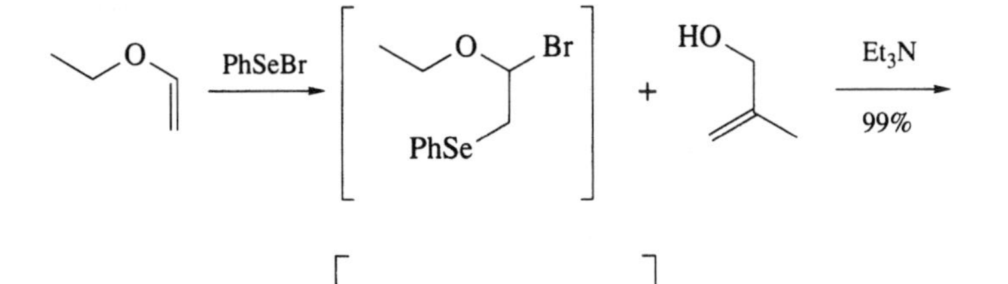

MeSeBr, AcOH, 80% (7)

PhSeBr; Et_3N, 99%; $NaIO_4$; 100%

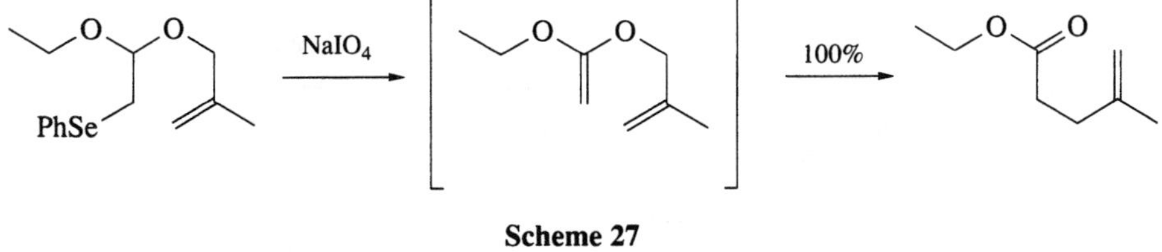

Scheme 27

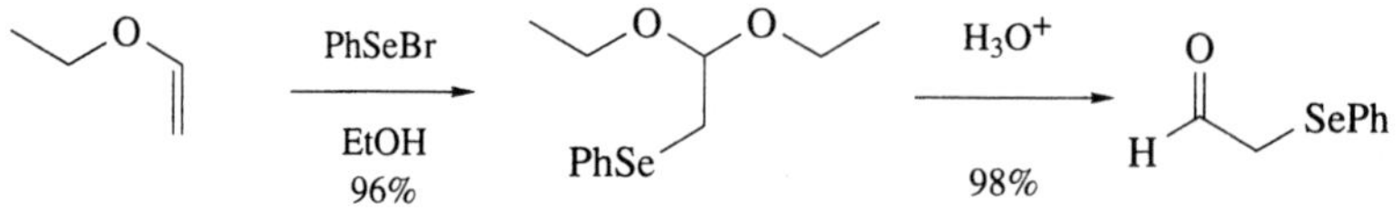

PhSeBr, EtOH, 96%; H_3O^+, 98%

Scheme 28

Addition of benzeneselenenyl chloride to conjugated dienes in methanol as solvent followed by treatment of the resultant methoxyselenation product with LDA affords 1-phenylseleno-1,3-dienes.[47] Regiospecific addition of benzeneselenenyl chloride to allylic alcohols is the key step in a 1,3-enone transposition sequence (Scheme 29).[49] *cis*-Vicinal dichlorides can be prepared from the corresponding alkenes by a process involving *trans* addition of phenylselenenyl chloride followed by treatment with chlorinating agents, ultimately leading to *trans* displacement of the seleno moiety (Scheme 30).[50] Chlorination with SO_2Cl_2 of the β-chloroalkyl phenyl selenides derived from terminal alkenes, followed by hydrolysis to selenoxide, followed by elimination leads instead to 2-chloro-1-alkenes (Scheme 31).[51] Arylselenenyl chloride, generated *in situ* from aryl diselenides and NCS, are effective as catalysts for chlorination of alkenes with NCS to give principally a rearranged allylic chloride (Scheme 32).[52] Addition of phenylselenenyl chloride and silver trifluoroacetate to nitroalkenes results in regiospecific formation of the phenylseleno–trifluoroacetoxy adducts which can be converted to nitroallylic alcohols (Scheme 33).[53] In a related example, successive addition of phenylselenenyl chloride and silver crotonate to alkenes or cycloalkenes affords β-phenylseleno crotonates which give γ-lactones upon reaction with triphenyltin hydride (Scheme 34).[54]

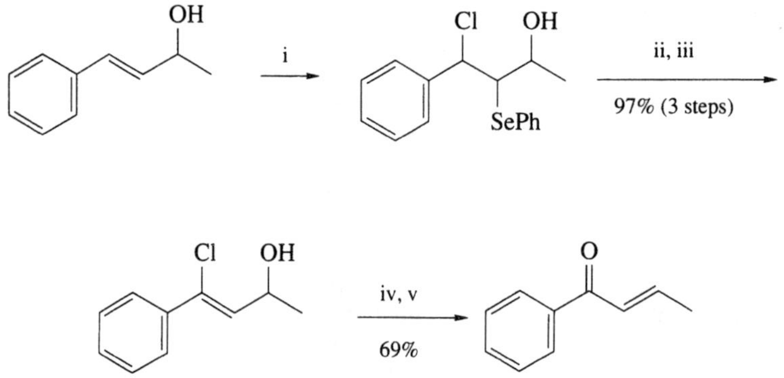

i; ii, iii, 97% (3 steps); iv, v, 69%

i, PhSeCl, CH_2Cl_2, –78 °C; ii, O_3, CH_2Cl_2; iii, Et_2NH, CH_2Cl_2, Δ; iv, AcCl, pyridine, Et_2O; v, $Hg(OAc)_2$, TFA

Scheme 29

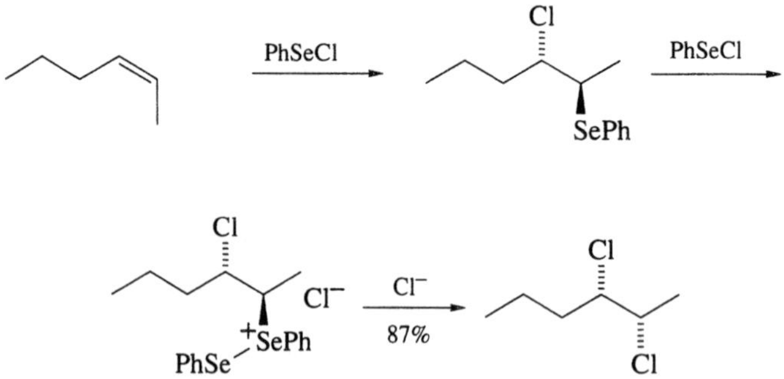

PhSeCl; PhSeCl; Cl^-, 87%

Scheme 30

i, PhSeCl, ii, SO_2Cl_2, 82%; i, aq. $NaHCO_3$, ii, PhH, Δ, 79%

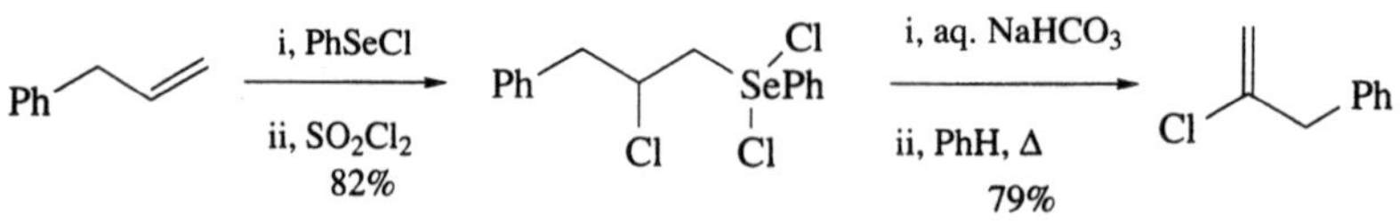

Scheme 31

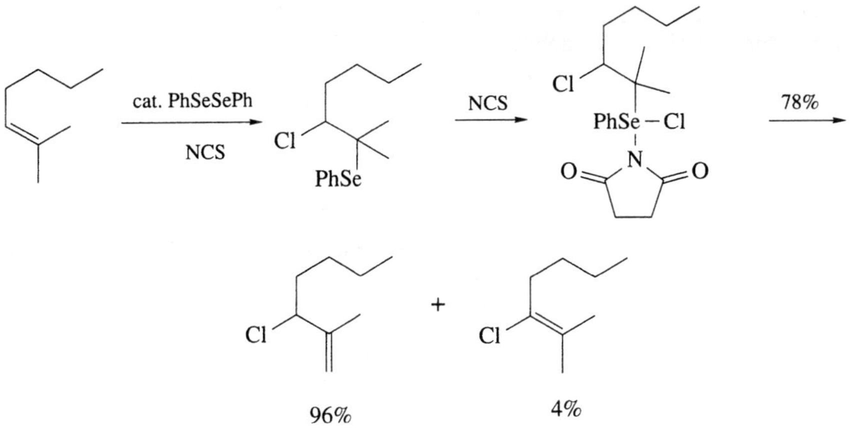

Scheme 32

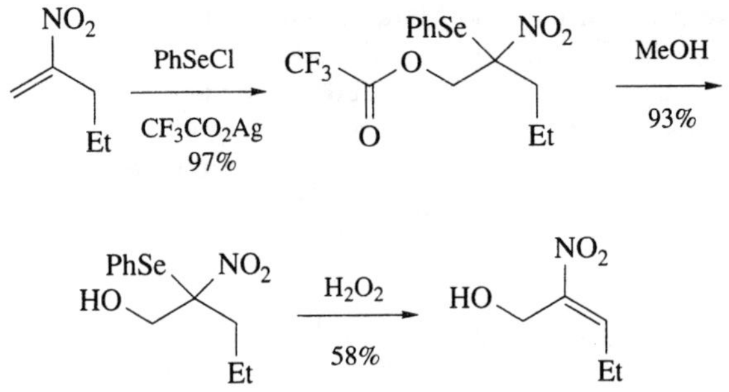

Scheme 33

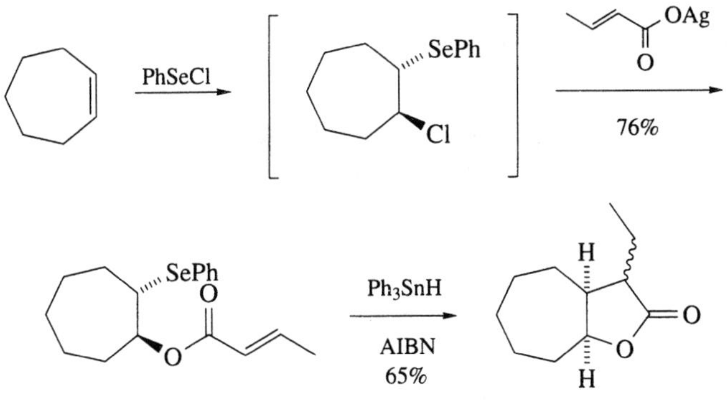

Scheme 34

Phenyl selenocyanate, PhSeCN, adds regio- and stereo-specifically to nucleophilic alkenes such as enamines and ketene acetals giving β-phenylselenocarbonitriles by a process termed vicinal cyanoselenenylation.[55] Phenyl selenocyanate also adds to simple alkenes in the presence of Lewis acids.[55] In alcohol in the presence of metal salts β-alkoxyalkyl phenyl selenides are formed.[56] Benzeneselenenyl thiocyanate, PhSeSCN, undergoes Markovnikov addition to alkenes affording either the β-phenylseleno thiocyanate or isothiocyanate, depending on the alkene.[57] Isothiocyanate formation dominates when mercury(II) salts are present.[58]

Se-Phenyl areneselenosulfonates, $PhSeSO_2Ar$, add to alkenes to give β-phenylseleno sulfones (selenosulfonation). The reaction may be conducted in the presence of a Lewis acid such as $BF_3{\cdot}Et_2O$ to give regioselectivity or regiospecifically the adduct of Markovnikov orientation arising from stereospecific ionic *anti* addition. Alternatively, the addition may be thermally induced to afford anti-Markovnikov products generated by a nonstereospecific free radical reaction.[59] Reaction of alkenes such as cyclooctene with phenylselenenyl chloride in the presence of sodium azide leads to the formation of phenylselenenyl azide which adds stereospecifically to the double bond (equation 8).[60] In a similar manner alkenes can be nitroselenated using a mixture of phenylselenenyl chloride and silver nitrite[53] or cyanamidoselenated using *N*-phenylselenophthalimide and cyanamide.[61]

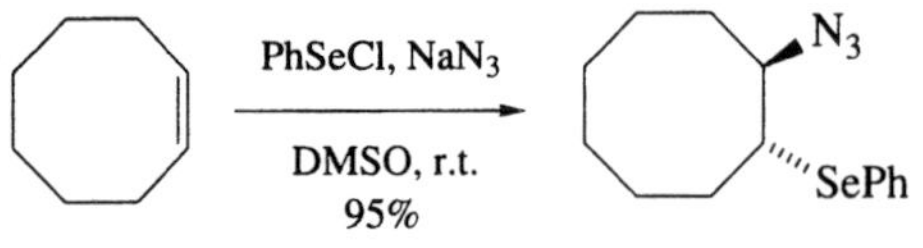

PhSeCl, NaN_3, DMSO, r.t., 95% (8)

Addition of a selenenyl halide to an alkyne takes a course analogous to that of the sulfenyl halide, as seen in the reaction of 1,4-dichlorobut-2-yne with PhSeCl or PhSeBr (Scheme 19).[28,62] Selenenyl halides also undergo facile addition to allenes with exclusive attack by selenium occurring at the central allenic carbon.[63]

1.8.2.4 Tetravalent Selenium

Brief reviews of electrophilic addition reactions of tetravalent selenium have appeared.[39,64] Alkenes react with selenium tetrabromide ($SeBr_4$) and tetrachloride ($SeCl_4$) to afford 2:1 adducts (Scheme 35 and equation 9)[65,66] with *anti* stereospecificity and Markovnikov regioselectivity. Dienes undergo intramolecular double addition with SeX_4 (Scheme 36).[67] Monoadducts are formed with alkyl- or aryl-selenium trichloride ($RSeCl_3$) and R_2SeCl_2 with *anti* stereospecificity (equation 10).[68,69] The $RSeCl_3$ adducts can be hydrolyzed with aqueous bicarbonate to selenoxides which spontaneously eliminate, affording vinylic or allylic chlorides depending on the regiochemistry of elimination (compare Scheme 31).[70] Methylselenium trichloride, $MeSeCl_3$, adds *anti* stereoselectively to disubstituted alkynes.[71]

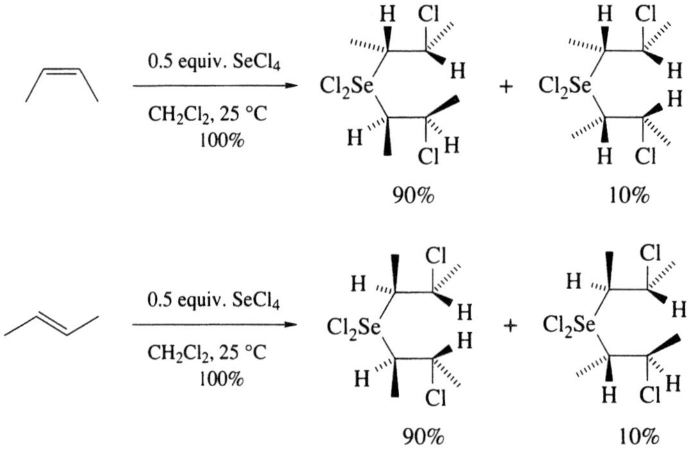

0.5 equiv. $SeCl_4$, CH_2Cl_2, 25 °C, 100% — 90% + 10%

0.5 equiv. $SeCl_4$, CH_2Cl_2, 25 °C, 100% — 90% + 10%

Scheme 35

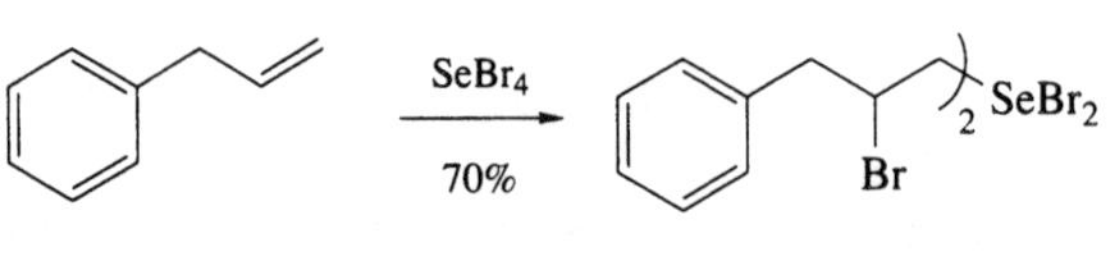

$SeBr_4$, 70% (9)

$SeBr_4$, 77%

$SeBr_4$, 99%

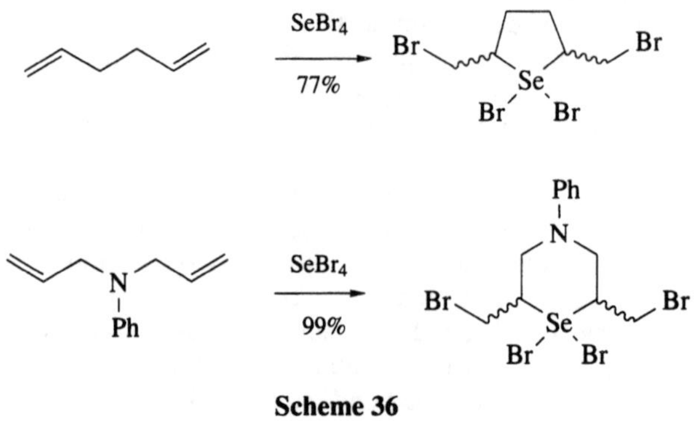

Scheme 36

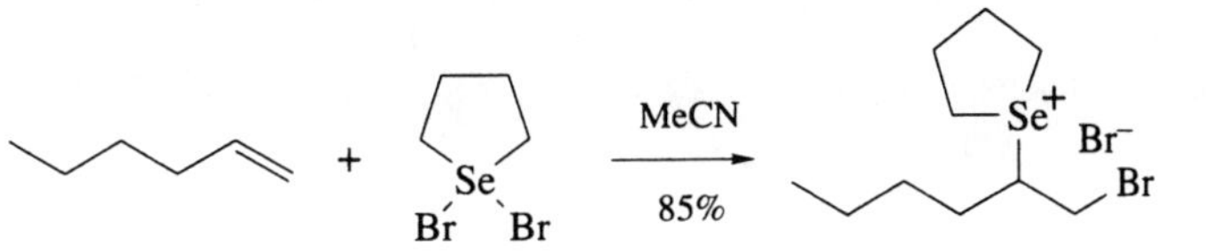

MeCN, 85% (10)

1.8.2.5 Tellurium Compounds

Electrophilic addition reactions of tetravalent tellurium compounds have been reviewed.[64] 2-Naphthyltellurium trichloride ($ArTeCl_3$) adds to alkenes in an *anti* stereospecific manner (equation 11), whereas tellurium tetrachloride gives mixtures of 2:1 adducts with both *syn* and *anti* addition.[72] A one-pot alkene inversion procedure has been developed, based upon $TeCl_4$ addition to alkenes followed by treatment of the β-chloroalkyltellurium trichloride adduct with aqueous Na_2S (Scheme 37).[73] Tellurium compounds such as tellurinyl acetates, ArTe(O)OAc, prepared *in situ* through reaction of tellurinic acid anhydrides with acetic acid, can be employed in oxytelluration and aminotelluration procedures (Schemes 38 and 39).[74] In the oxytelluration reaction intermediate triacetates of the type $RCH_2Te(OAc)_2Ph$ are reduced with hydrazine to the corresponding tellurides.

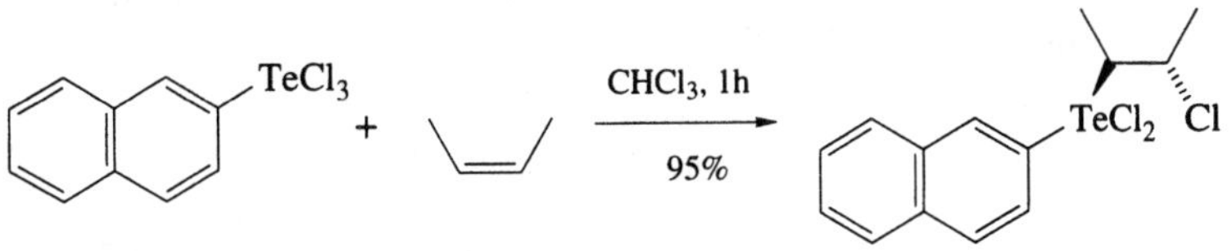

$TeCl_3$ + ; $CHCl_3$, 1h, 95%; $TeCl_2$ Cl (11)

$TeCl_4$, MeCN, 0 °C; Cl, $TeCl_3$; Na_2S, 100%; 97% + 3%

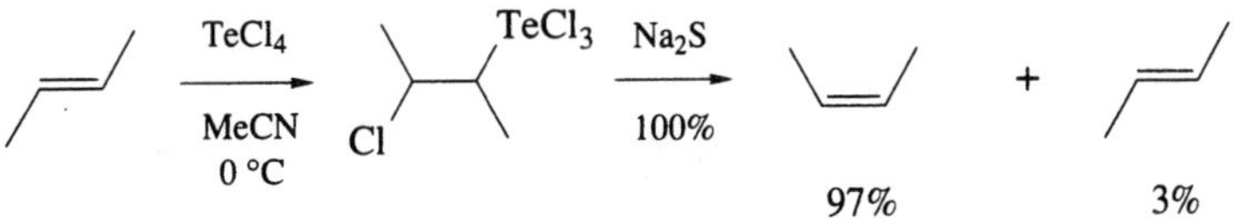

Scheme 37

$[ArTe(O)]_2O$ + AcOH ⟶ ArTe(O)OAc + H_2O

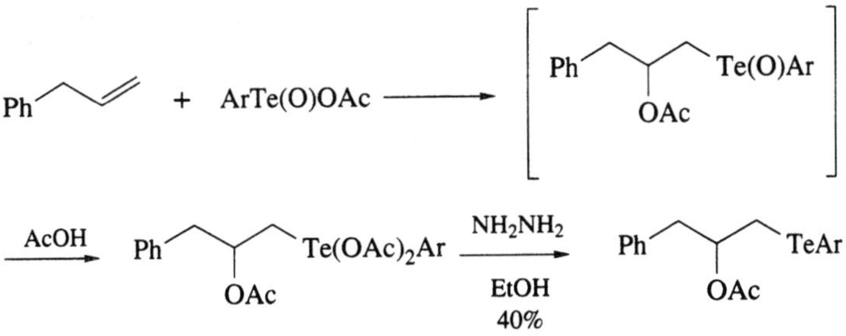

Ph + ArTe(O)OAc ⟶ [Ph, OAc, Te(O)Ar]

AcOH ⟶ Ph, OAc, $Te(OAc)_2Ar$; NH_2NH_2, EtOH, 40% ⟶ Ph, OAc, TeAr

Scheme 38

Ph; i, ii, 71% ⟶ Ph, OEt, TePh

Ph; iii, 97% ⟶ Ph, $NHCO_2Et$, TePh

i, PhTe(O)OAc, $SnCl_4$; ii, NH_2NH_2, EtOH; iii, PhTe(O)OAc, BF_3•OEt_2, NH_2CO_2Et

Scheme 39

1.8.3 HALOGENS

1.8.3.1 Fluorine

Until recently there have been few examples reported of the controlled addition of fluorine to alkenes and alkynes.[75] Many elementary texts dismiss the utility of F_2 in organic chemistry with statements such as 'fluorine is so reactive that it not only adds to the double bond but also rapidly replaces all the hydrogens with fluorines, often with considerable violence.'[76] Because of the weak F—F bond (37.7 kcal mol^{-1}, 1 kcal = 4.18 kJ) reactions of fluorine at ambient temperatures tend to be homolytic in nature, frequently leading to extensive decomposition. However by working at –78 °C with a 1% stream of fluorine in nitrogen in $CHCl_3$–$CFCl_3$ in the presence of a proton source such as ethanol to suppress radical processes, ionic addition of F_2 in a stereospecific *syn* manner is observed, as shown in Scheme 40.[77] The exclusive *syn* addition, which is opposite to the stereochemistry observed with other halogen additions, is characteristic of all electrophilic fluorination processes (*e.g.* with $FOCF_3$, $FOCF_2CF_3$, $FOCOCF_3$, FOCOMe) and is considered as a good criterion for their existence.[77] Rapid collapse of the tight ion pair involving the α-fluorocarbocation is believed to account for the stereospecificity. Fluorination of enones by the above procedure also proceeds without complications (Scheme 41).[77] Dehydrofluorination of the initial *syn* difluoro adduct affords α-fluoroenones.

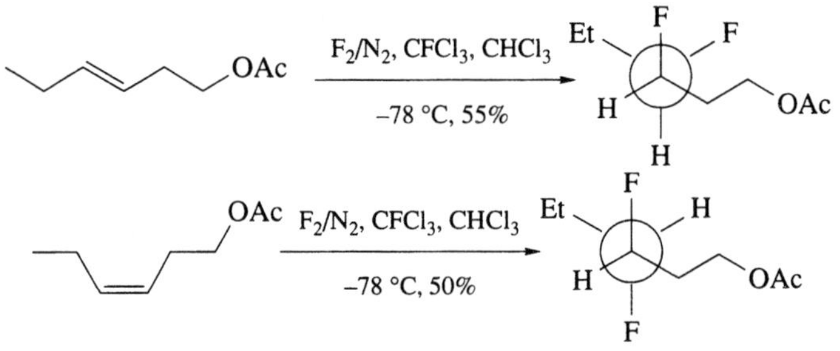

OAc F_2/N_2, $CFCl_3$, $CHCl_3$ –78 °C, 55% Et F F H OAc H

OAc F_2/N_2, $CFCl_3$, $CHCl_3$ –78 °C, 50% Et F H H OAc F

Scheme 40

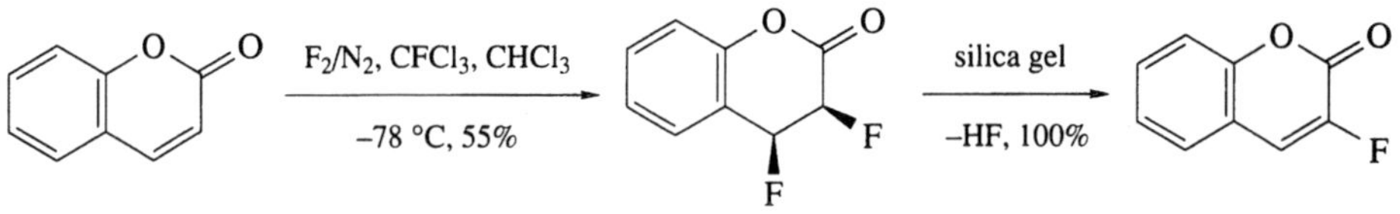

O O F_2/N_2, $CFCl_3$, $CHCl_3$ –78 °C, 55% O O F F silica gel –HF, 100% O O F

Scheme 41

1.8.3.2 Chlorine

Few recent studies have appeared on electrophilic addition reactions of chlorine which significantly update the earlier literature.[1,2] While simple alkenes react readily with elemental chlorine it is possible to carry out chlorinations of such groups as sulfur–sulfur bonds in the presence of C═C functions without competition from chlorination of the C═C bond (Scheme 9).[14] Alkenes have been chlorinated in good yield under mild conditions using chlorine incorporated into an anion-exchange resin.[78a] Isomerically pure vinyl silanes on low temperature chlorination followed by desilicohalogenation give vinyl chlorides with inverted geometry in good yield with high stereoselectivity.[78b] A comparative study of the chlorination, bromination and iodination of 1,5-cyclooctadiene is described in Section 1.8.3.4 below.

1.8.3.3 Bromine

What is the precise mechanistic process by which bromine (and by extension chlorine) adds to alkenes? Studies designed to probe this question are generally difficult to conduct due to the very fast product-forming steps. By applying stopped-flow spectrokinetics to sterically crowded alkenes such as adamantylideneadamantane (Ad_2), where steric hindrance retards the final addition step by hampering rearside nucleophilic attack of the counterion at the bromonium carbons of the intermediate ion pairs, it is possible to obtain useful mechanistic information on the initial steps in the interaction of bromine with a π-bond. Bellucci and coworkers[3] present a mechanism for the Ad$_2$–bromine reaction involving a sequence of multiple equilibria involving 1:1 and 1:2 Ad$_2$–Br$_2$ molecular charge-transfer complexes and bromonium–bromide, –tribromide and –pentabromide salts arising from solvent-assisted, Br$_2$-assisted or unassisted ionization of the molecular complexes and from the tribromide–pentabromide equilibrium. Charge-transfer complexes have already been indicated in the ionic bromination of cyclohexene.[79] It is concluded from these studies that a large part of the increase in bromination rate resulting from increasing substitution on the double bond (*e.g.* for mono-:di-:tri-:tetra-substituted double bonds the relative rate constants for bromine addition are 1:60:2000:20 000)[80] must be due to an increase in the alkene–Br$_2$ charge-transfer complex formation constant.

Another approach in the study of the mechanism and synthetic applications of bromination of alkenes and alkynes involves the use of crystalline bromine–amine complexes such as pyridine hydrobromide perbromide (PyHBr$_3$), pyridine dibromide (PyBr$_2$), and tetrabutylammonium tribromide (Bu$_4$NBr$_3$) which show stereochemical differences and improved selectivities for addition to alkenes and alkynes compared to Br$_2$ itself.[81] The improved selectivity of bromination by PyHBr$_3$ forms the basis for a synthetically useful procedure for selective monoprotection of the higher alkylated double bond in dienes by bromination (Scheme 42).[80] The less-alkylated double bonds in dienes can be selectively monoprotected by tetrabromination followed by monodeprotection at the higher alkylated double bond by controlled-potential electrolysis (the reduction potential of vicinal dibromides is shifted to more anodic values with increasing alkylation; Scheme 42).[80] The question of which diastereotopic face in chiral allylic alcohols reacts with bromine has been probed by Midland and Halterman as part of a stereoselective synthesis of bromo epoxides (Scheme 43).[82]

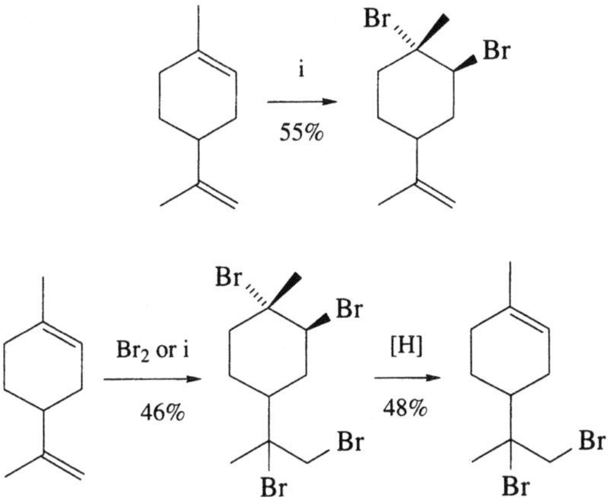

i, pyridinium hydrobromide perbromide

Scheme 42

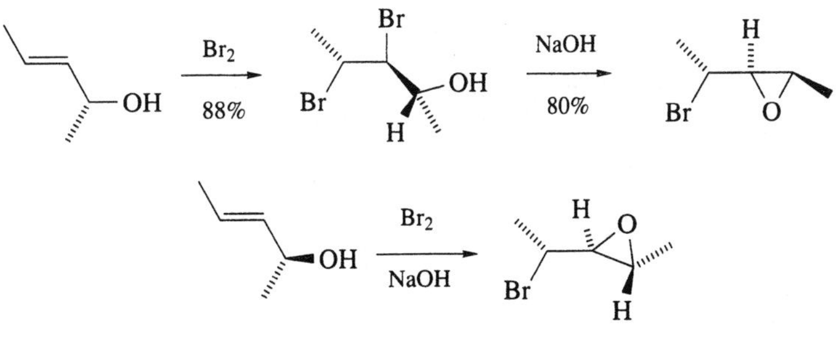

Scheme 43

White and coworkers used *trans* Addition of bromine and (–)-1-borneol to 3,4-dihydro-2*H*-pyran to afford the stereoisomeric bornyl bromotetrahydropyranyl ethers (**3**) and (**4**; Scheme 44) which are employed in an enantioselective total synthesis of (–)-monic acid C (Scheme 5).[10] A simple synthesis of 4-methylthio-1,2-dithiolane, the photosynthesis inhibitor of the green alga *Chara globularis* (Scheme 45),[83] exploits the observation that allylic sulfides undergo rearrangement on addition of bromine.[84] Block and Naganathan employed the *trans* addition of bromine to 4-thioacetoxycyclopentene as a key step in the synthesis of 2-bromo-5-thiabicyclo[2.1.1]hexane (Scheme 46).[85] *trans* Addition of bromine to 1,4-dichlorobut-2-yne followed by base- or aluminum amalgam-induced elimination is employed in the synthesis of halo-1,3-butadienes (Scheme 19).[28]

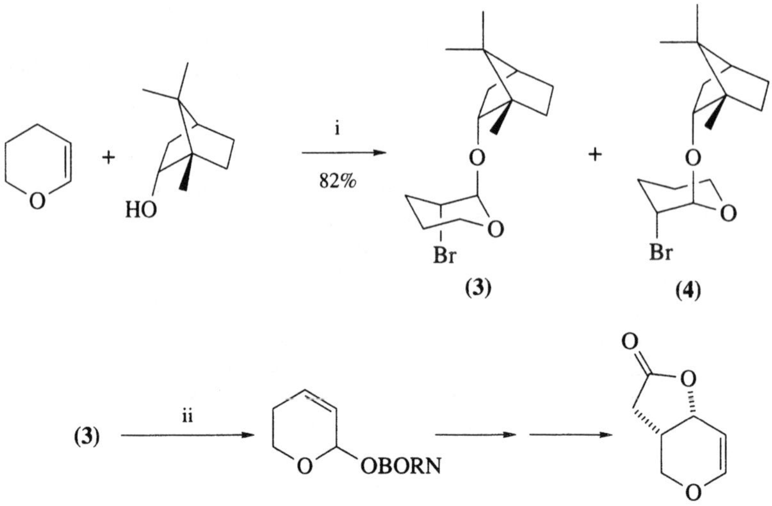

i, Br_2, $PhNMe_2$, CH_2Cl_2, –78 to 0 °C; ii, DBU, 95 °C, 20 h

Scheme 44

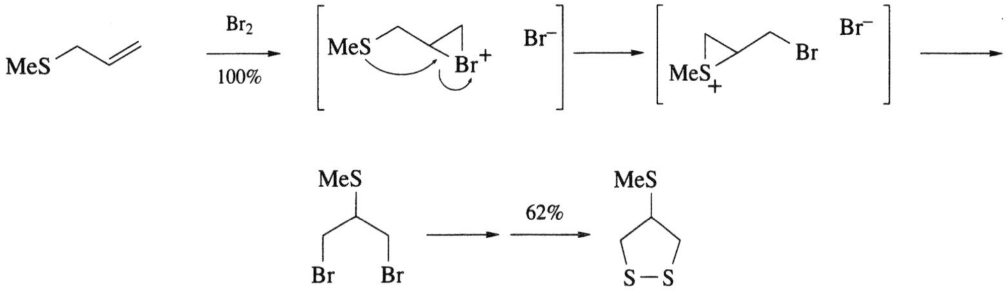

Scheme 45

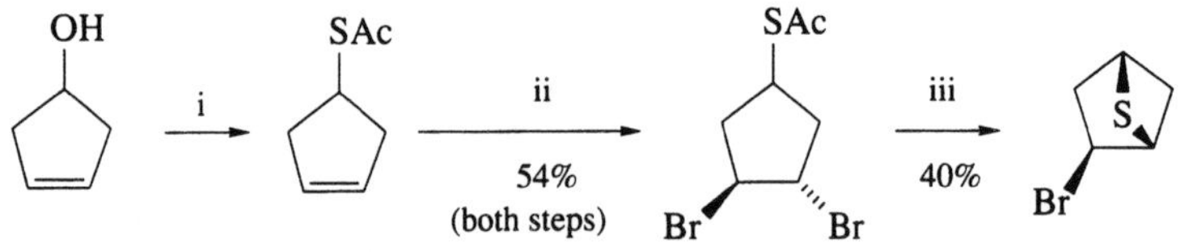

i, diisopropylazodicarboxylate, PPh_3, THF, MeC(O)SH, 0 °C to r.t., 3 h; ii, Br_2/CCl_4, 0 °C, 1 h; iii, KOH, $EtOH/H_2O$, reflux, 90 h

Scheme 46

1.8.3.4 Iodine

Iodine, less reactive than bromine, is best added to alkenes by use of ICl and IBr (see Section 1.8.3.5). Iodine itself adds rapidly but reversibly to alkenes forming diiodides by mechanisms that can be either ionic or radical. The position of the equilibrium depends upon the structure of the alkene, the solvent and the temperature. Simple vicinal diiodides survive distillation in the dark, but are unstable toward iodine or radicals. In the presence of functions containing free OH groups, such as alumina, HI generated from I_2 adds to alkenes irreversibly with the result that the HI adduct, rather than the I_2 adduct, is the exclusive product.[86a] A comparison of the reaction of 1,5-cyclooctadiene with chlorine, bromine and iodine in CH_2Cl_2 reveals that chlorine gas at –50 °C gives a 93:7 mixture of *trans*- and *cis*-5,6-dichlorocyclooctene, bromine gives *trans*-5,6-dibromocyclooctene in 95%, while iodine gives a 1:1 mixture of *endo,exo*- and *endo,endo*-2,6-diiodobicyclo[3.3.0]octanes in 70% yield.[87] Significant differences in product composition are seen in all cases with changes in solvent. Addition of iodine to terminal (*E*)-vinylsilanes in the presence of Lewis acids (iodesilylation) gives different ratios of (*E*)- and (*Z*)-vinyl iodides depending on the amount of Lewis acid.[88]

Unlike alkenes, which react reversibly with I_2, alkynes react in solution irreversibly with I_2, forming (*E*)-1,2-diiodoalkenes.[89] The reaction may involve radicals rather than ionic species.[90] Ordinary alumina appears to promote the addition, forming electrophilic species.[86b] The addition of iodine to 1,4-dichlorobut-2-yne is analogous to the addition found with bromine cited above (Scheme 19).[28]

1.8.3.5 Interhalogens

Halofluoridation has been reviewed.[91] Iodine fluoride, IF, and bromine fluoride, BrF, formed from the elements in $CFCl_3$ at –78 °C, are highly reactive toward π-systems, reacting at –75 °C in the dark in a matter of seconds. Additions to alkenes are regiospecific in the Markovnikov sense (X^+ F^-) and *anti* stereospecific, affording vicinal fluoro iodides or bromides.[92] Other sources of these two interhalogens, which have been added to alkenes and alkynes, include $I^+(collidine)_2BF_4^-$, $I^+(py)_2BF_4^-$, I_2/AgF, *N*-iodosuccinimide/HF–pyridine, and BrF_3.[93] Addition of IF or BrF to terminal alkynes gives adducts of type RCF_2CX_2 (equation 12).[94] A variety of mechanistic studies have appeared on the addition of BrCl, ICl, IBr and related reagents to alkenes and alkynes.[95]

XF, $CHCl_3$ → CF_2CX_2 (X = Br, 60%; X = I, 80%) (12)

1.8.3.6 Reagents with O—X Bonds

Electrophilic addition of hypohalites, HOX, formed through reaction of halogens with water, has been known for many years.[96] Even hypofluorous acid, HOF, will add to alkenes and alkynes.[97] A variety of other types of reagents with oxygen–halogen bonds are also known to add to alkenes and alkynes. Examples of such reagents, capable of undergoing low-temperature addition to alkenes, include acyl hypohalites, XOC(O)R (equations 13[48] and 14[98] and Scheme 47),[99] perchlorates $XOClO_3$,[100] sulfates, $XOSO_2Y$ (Scheme 48),[101] triflates, $XOSO_2CF_3$[102,103] and trifluoromethyl hypofluorite, CF_3OF (equation 15).[104,105] A related reagent, cesium fluoroxysulfate $FOSO_3Cs$, has also been found to add to alkenes under mild conditions, affording vicinal fluoroalkyl sulfates (Scheme 49).[103] The latter reactions exhibit low regio- and stereo-selectivities, with a preference for anti-Markovnikov and *syn* addition.

Zefirov and coworkers have developed procedures for chlorosulfamation of alkenes and alkynes using reagents of the type R_2NSO_2OCl formed by insertion of sulfur trioxide into the nitrogen–chlorine bond in *N*-chloroamines (R_2NCl).[106]

AcO, OAc, O, AcO — $MeCO_2F$; $CFCl_3$, AcOH; 5 min, –78 °C; 84% → AcO, O, OAc, AcO, OAc, F (13)

$MeCO_2F$; $CHCl_3$; 55% → F, OAc, O (14)

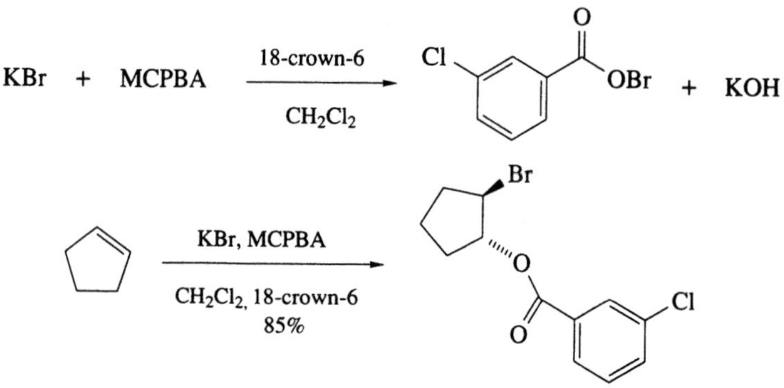

Scheme 47

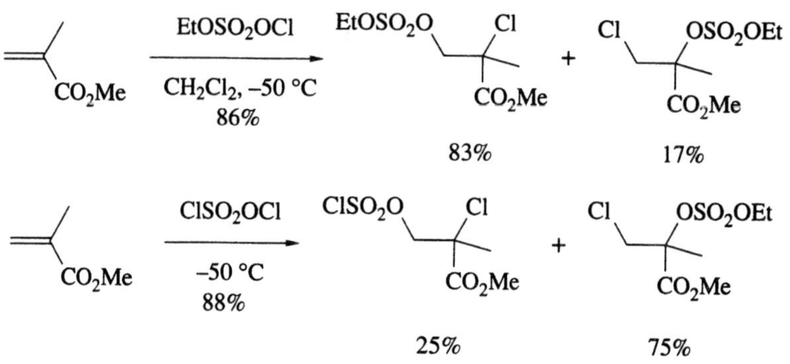

Scheme 48

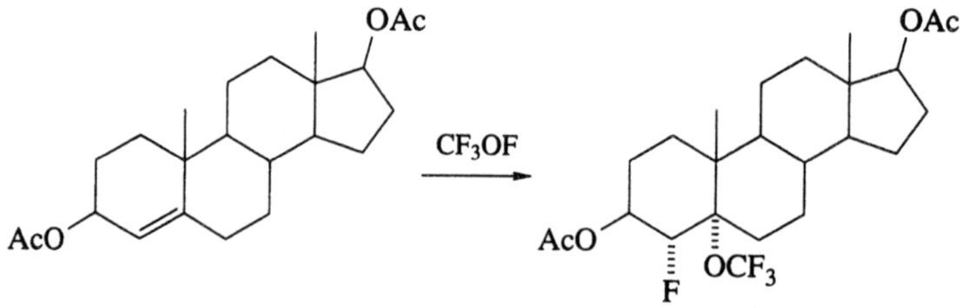

(15)

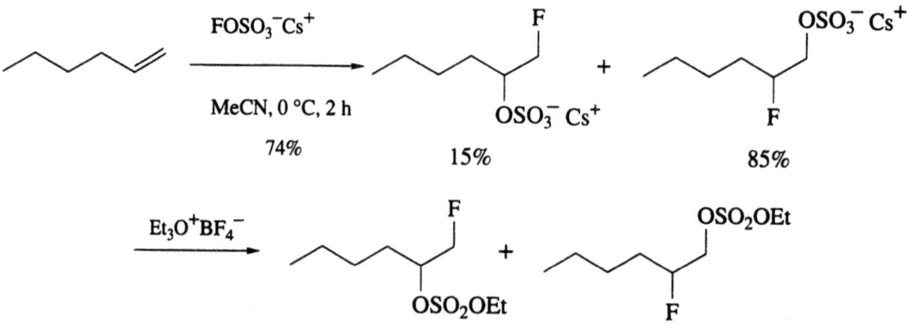

Scheme 49

1.8.4 PSEUDOHALOGENS

1.8.4.1 Thiocyanogen

The addition of thiocyanogen[107] to unsaturated compounds[108] forms the basis of a classical analytical method for determination of the unsaturation in fats and oils.[108a] The addition has been found to afford both α,β-dithiocyanates as well as isomeric α-isothiocyanato-β-thiocyanates in ratios dependent on solvent and catalysts.[108b] Metals such as iron[108b] or tin[108c] favor the α,β-dithiocyanate form by coordinating to the harder (HSAB concept) nitrogen terminus of NCS; in related studies use of mercury salts (a softer metal) favors isothiocyanate formation.[58] Thiocyanogen[108a] addition has been utilized in the inversion of alkene geometry (Scheme 50).[108b] More recently thiocyanogen addition to a silylated alkene has been used in the synthesis of various silylated thiols, unsaturated thiocyanates and *trans*-2,3-bis(trimethylsilyl)thiirane (Scheme 51).[109] (*E*)-1,2-Dithiocyanatoalkenes can be prepared by addition of thiocyanogen, formed *in situ* from $Hg(SCN)_2/I_2$, to alkynes (equation 16).[110]

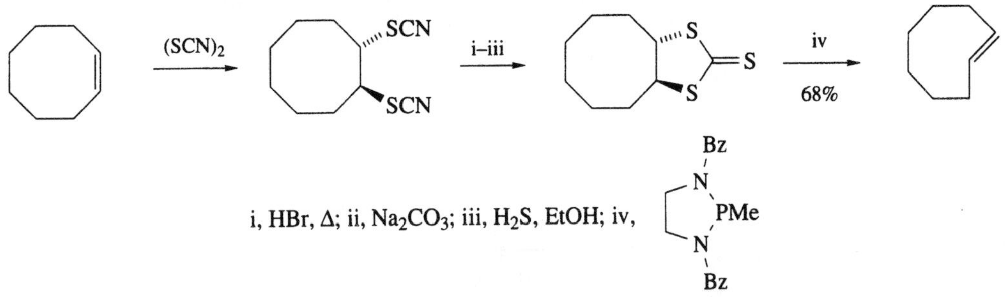

$(SCN)_2$ → SCN, SCN → i–iii → S, S, =S → iv, 68%

i, HBr, Δ; ii, Na_2CO_3; iii, H_2S, EtOH; iv, Bz–N, PMe, N–Bz

Scheme 50

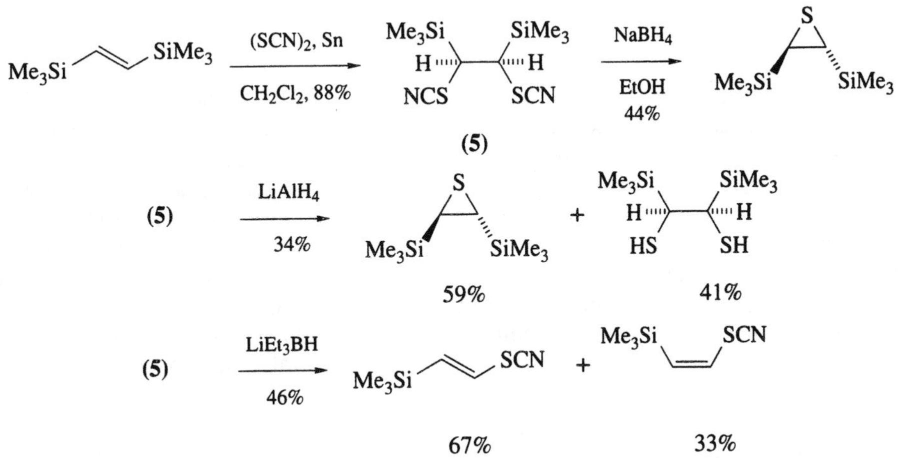

Me_3Si, $SiMe_3$ → $(SCN)_2$, Sn; CH_2Cl_2, 88% → Me_3Si, $SiMe_3$, H, H, NCS, SCN **(5)** → $NaBH_4$; EtOH 44% → S, Me_3Si, $SiMe_3$

(5) → $LiAlH_4$; 34% → S, Me_3Si, $SiMe_3$ (59%) + Me_3Si, $SiMe_3$, H, H, HS, SH (41%)

(5) → $LiEt_3BH$; 46% → Me_3Si, SCN (67%) + Me_3Si, SCN (33%)

Scheme 51

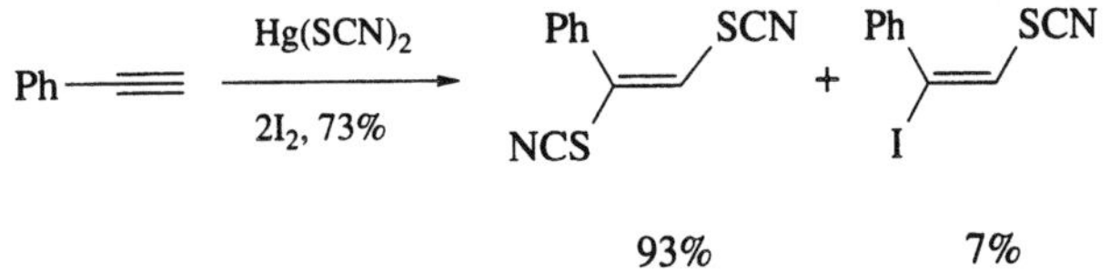

Ph—≡ → $Hg(SCN)_2$; $2I_2$, 73% → Ph, SCN, NCS (93%) + Ph, SCN, I (7%) (16)

1.8.4.2 Halogen Azides

Some of the most useful procedures for the stereospecific introduction of nitrogen functionalities into the carbon skeleton are based on reagents with a N—X bond such as halogen azides, nitrates and isocyanates together with haloamines and haloamides. Hassner has published an excellent overview of applications of the reagent, XN_3.[111] Special procedures must be used to prepare the explosive iodine azide, IN_3, *in situ* in organic solvents;[112] BrN_3 and ClN_3 can be prepared using a two-phase system (water–pentane) by the interaction of Br_2 or Cl_2 with NaN_3 in the presence of acid.[111] More recently BrN_3 has been prepared *in situ* from NBS and NaN_3 in the presence of the alkene in DME/H_2O. The BrN_3 adducts can be converted to aziridines by treatment with lithium aluminum hydride (Schemes 52[113] and 53).[114]

Iodine azide adds stereo- and regio-specifically to 2-cholestene giving a *trans*-diaxial adduct with azide at the 2-position, a consequence of attack by I^+ from the less hindered face of the steroid. With 1-hexene the addition proceeds regiospecifically to yield the secondary azide, most likely *via* a three-membered iodonium ion intermediate.[111] With (*Z*)-2-octene, BrN_3 addition gives a single adduct in 71% yield

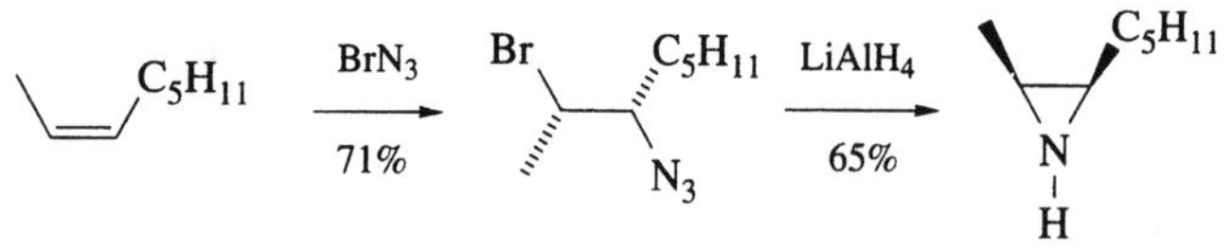

Scheme 52

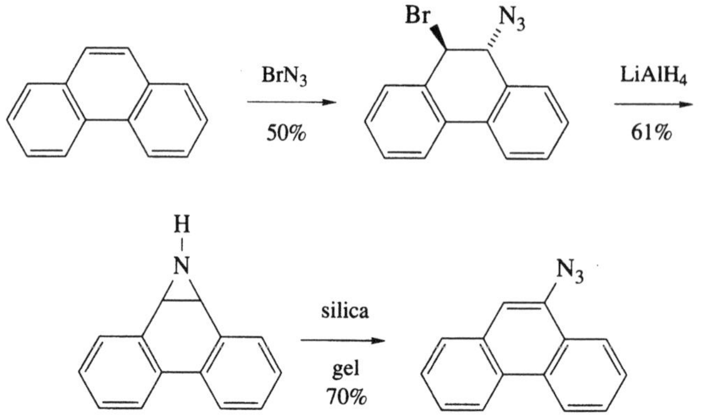

Scheme 53

in a process termed directiospecific.[113] A notable exception to the rule that azide is preferentially attached to the center best stabilizing a positive charge is seen with IN_3 addition to 3,3-dimethyl-1-butene, $Me_3CCH{=}CH_2$. The adduct has the structure $Me_3CCHICH_2N_3$. It is suggested that the sterically bulky *t*-butyl group forces the azide ion to attack at the primary carbon (compare equation 2). Another example in which steric effects dominate over polar effects is seen in the addition of IN_3 to 1-arylcyclohexenes; in this example addition of IN_3 is followed by HI elimination to afford an allylic azide (Scheme 54).[115] Addition of IN_3 to vinyl sulfones affords single regioisomers which eliminate either HI, giving β-azidovinyl sulfones, or HN_3, giving α-iodovinyl sulfones (Scheme 55).[116] The halogen in halogen azide adducts can also be displaced intermolecularly by nucleophiles (including azide ion), as in Schemes 56[117] and 57,[118] or removed using a tin hydride, as in the elegant synthesis of the alkaloid (±)-*O*-methylorantine by Wasserman and coworkers (Scheme 58).[119]

Addition of halogen azides to allenes affords a variety of useful products by regioselective processes (Scheme 59).[120] In the addition of BrN_3 and ClN_3 and even IN_3 to alkenes, both ionic and free radical addition are possible. Thus in polar solvents such as nitromethane, BrN_3 adds to 2-cholestene giving a single diaxial adduct. Under free radical conditions (pentane, light, peroxides) two stereoisomers are formed (Scheme 60).[111] Similar results have been reported for IN_3 (Scheme 61).[121]

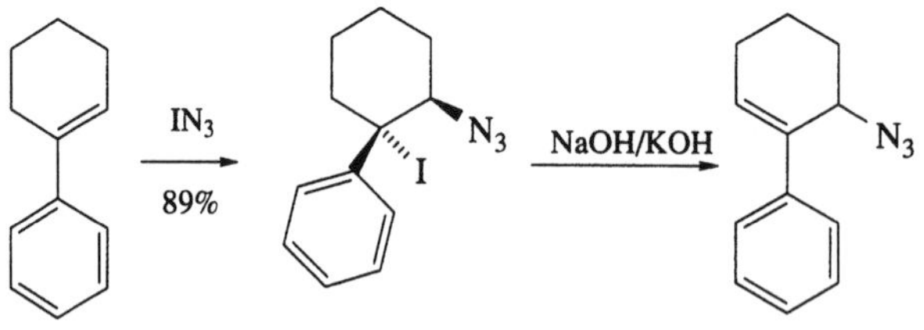

Scheme 54

1.8.4.3 Halogen Nitrates

The work of Lown and coworkers[122] on iodonium nitrates has been reviewed.[1] Addition of these reagents proceeds in a *trans* stereospecific manner, but with mediocre regiospecificity. The Lown group has also studied $BrNO_3$ and found it to be a more regiospecific reagent.[123]

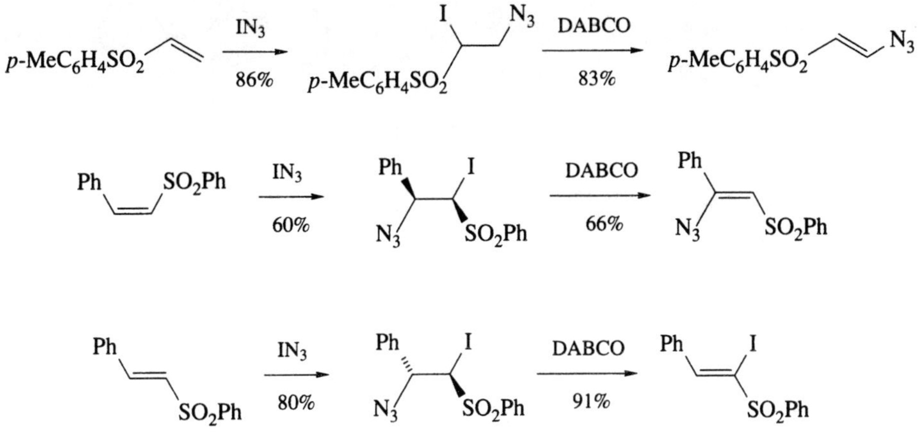

Scheme 55

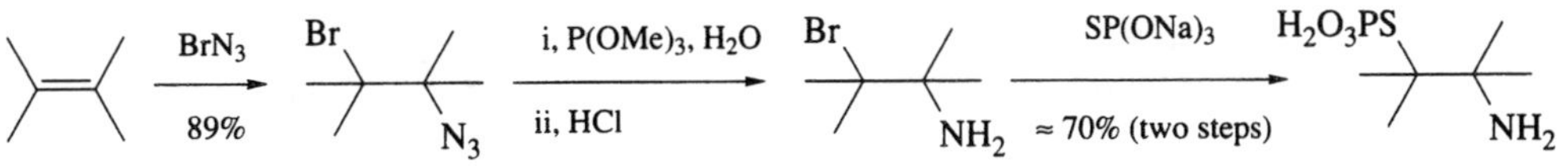

Scheme 56

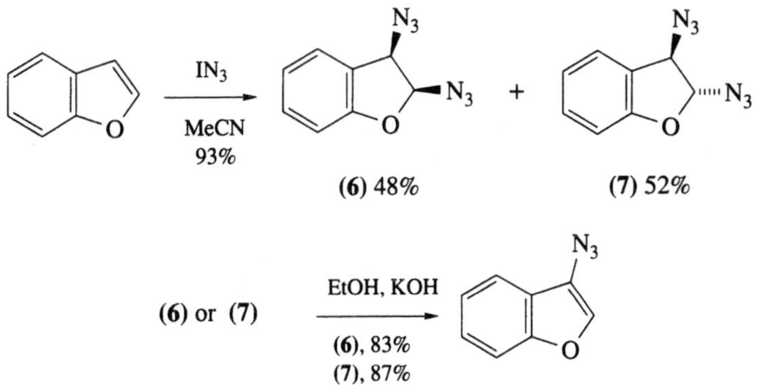

(6) 48% (7) 52%

Scheme 57

Barluenga has introduced a new reagent for the addition of XNO_3 across double bonds. Thus, $Hg(NO_3)_2{\cdot}X_2$, where X = Br or I, is a good reagent for the efficient generation of 1,2-halonitrates (Scheme 62).[124] Unfortunately this mercury reagent, like simple INO_3 or $BrNO_3$, is inert toward alkynes.

1.8.4.4 Halogen Isocyanates

The chemistry of halogen isocyanates and the mode by which they interact with double bonds has been reviewed.[1] The addition to alkenes has *anti* stereospecificity, but displays limited regiospecificity. Halogen isocyanates have received little attention in recent years, but there are two examples worth noting. An INCO addition was employed in an important step in the synthesis of methyl α-D-tetronitroside (Scheme 63).[125] The preparation allowed for a confident structural assignment relating to tetronitrose, one of the sugar components of the antibiotics tetrocarcins A and B, and kijanimicin. Barluenga's $I(py)_2^+$ BF_4^- reagent, introduced in Section 1.8.3.5,[93c] has been utilized with the OCN^- ion to afford 1-iodo-2-isocyanatoalkanes.[93c]

1.8.4.5 Halogen Thiocyanates

The chemistry of the XSCN reagents has been reviewed briefly.[1] When X = Br or Cl, the reagent is polarized $X^-(SCN)^+$, while when X = I the polarization is $I^+(SCN)^-$.[126] Examples using ISCN or its equivalents dominate the recent literature on XSCN reagents. Thiocyanate is often introduced as an ionic

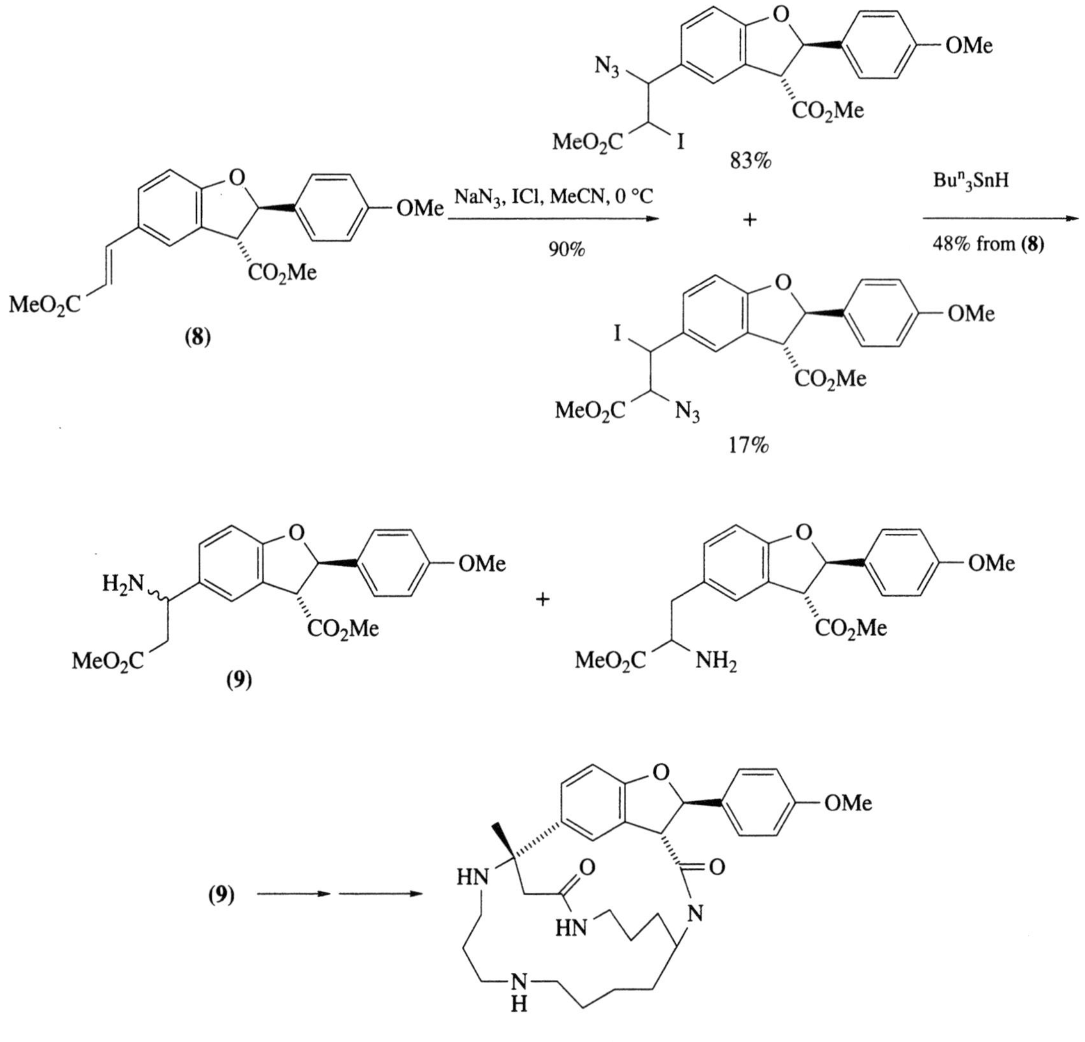

N_3 MeO$_2$C I CO$_2$Me OMe 83%

(8) MeO$_2$C CO$_2$Me OMe

NaN$_3$, ICl, MeCN, 0 °C 90%

Bu$^n{}_3$SnH 48% from (8)

I MeO$_2$C N_3 CO$_2$Me OMe 17%

H$_2$N MeO$_2$C (9) CO$_2$Me OMe + MeO$_2$C NH$_2$ CO$_2$Me OMe

(9) → → HN HN N N H O O OMe

(±)-*O*-Methylorantine

Scheme 58

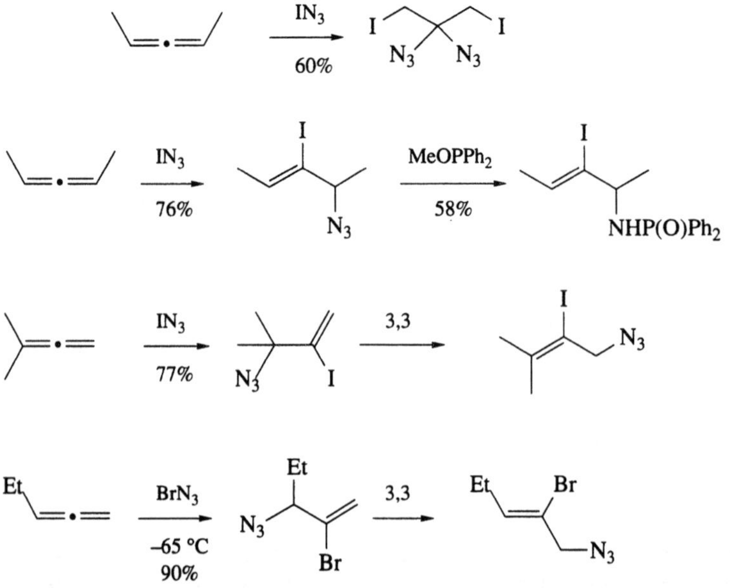

IN$_3$ 60% I I N$_3$ N$_3$

IN$_3$ 76% I N$_3$ MeOPPh$_2$ 58% I NHP(O)Ph$_2$

IN$_3$ 77% N$_3$ I 3,3 I N$_3$

Et BrN$_3$ –65 °C 90% Et N$_3$ Br 3,3 Et Br N$_3$

Scheme 59

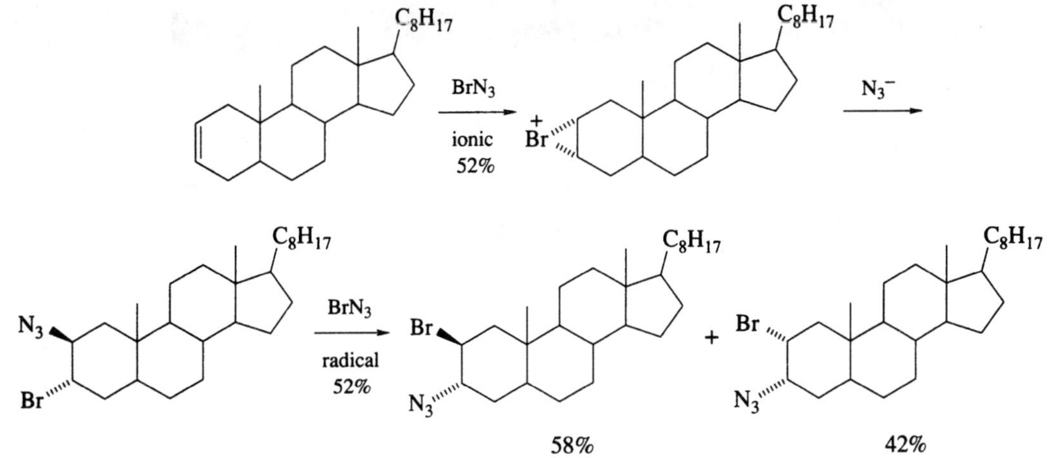
C8H17
BrN3
ionic
52%
Br
N3⁻
N3
Br
BrN3
radical
52%
58%
42%

Scheme 60

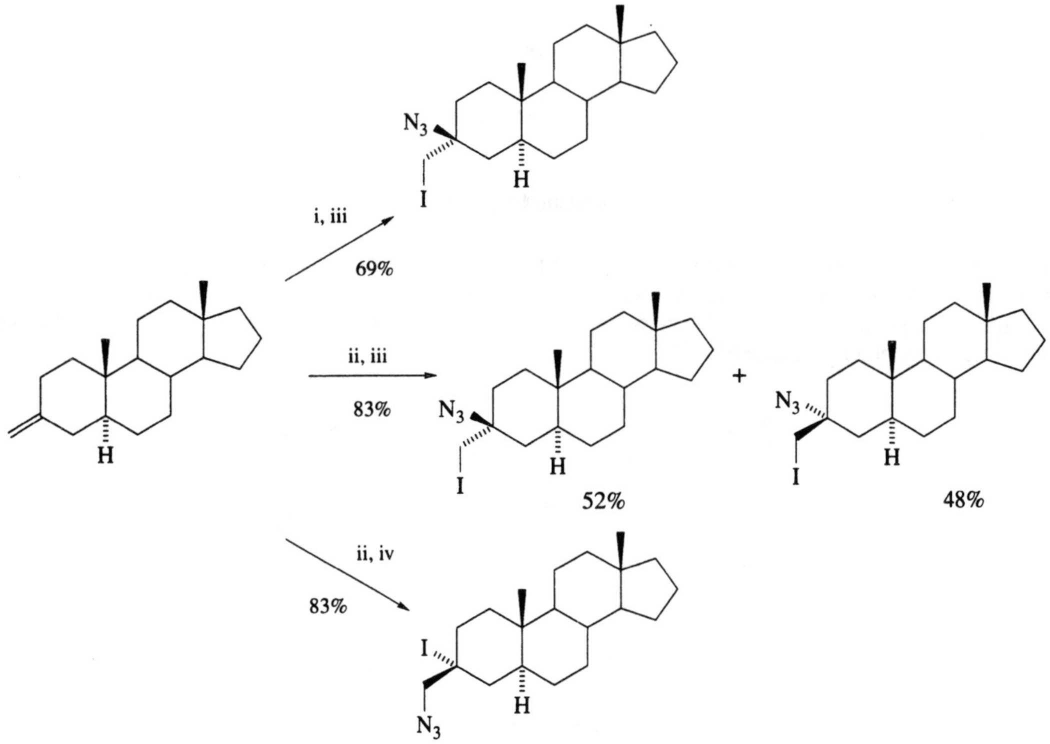
i, iii
69%
ii, iii
83%
ii, iv
83%
N3
I
H
52%
48%

i, *N*-iodosuccinimide•HN_3, CH_2Cl_2; ii, NaN_3•ICl, MeCN; iii, no removal of O_2 from reaction solution; iv, displacement of O_2 in reaction solution with N_2

Scheme 61

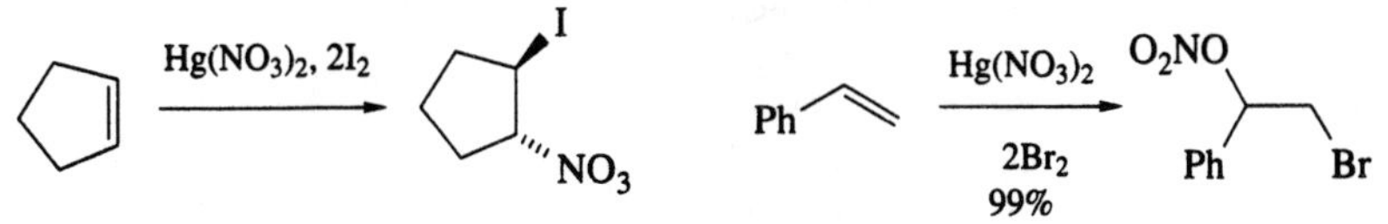
Hg(NO3)2, 2I2
I
NO3
Ph
Hg(NO3)2
2Br2
99%
O2NO
Ph
Br

Scheme 62

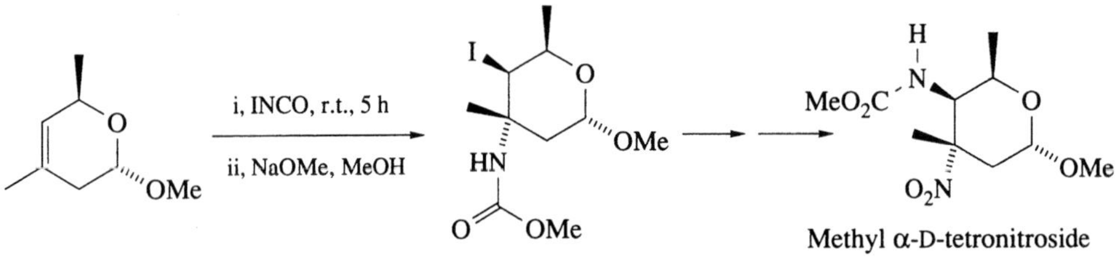

i, INCO, r.t., 5 h
ii, NaOMe, MeOH

Methyl α-D-tetronitroside

Scheme 63

salt. Addition product distribution is usually sensitive to the identity of the counterion of the thiocyanate salt. Another problem sometimes incurred is that the *vic*-halothiocyanates often isomerize to the thermodynamically more stable *vic*-haloisothiocyanates (Scheme 64).[127] Some of the factors governing this isomerism have been evaluated.[128] Adducts of BrSCN do not suffer from the problem of isomerism.[126]

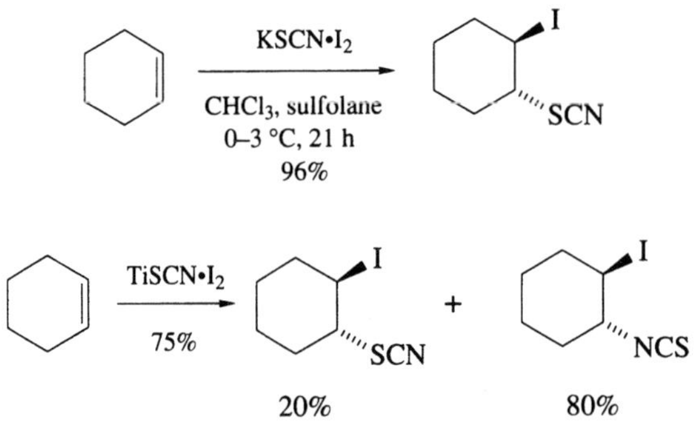

KSCN•I_2, $CHCl_3$, sulfolane, 0–3 °C, 21 h, 96%

TiSCN•I_2, 75%; 20% (SCN) + 80% (NCS)

Scheme 64

Grayson and Whitham have found the regiochemistry of addition of ISCN (from LiSCN and I_2) to vinyl silanes to be consistent with the intermediacy of an iodonium ion which is particularly stabilized at one of the carbons due to the β-silicon effect.[129] The products were a thiocyanate in one case and an isothiocyanate in the other (Scheme 65).[129]

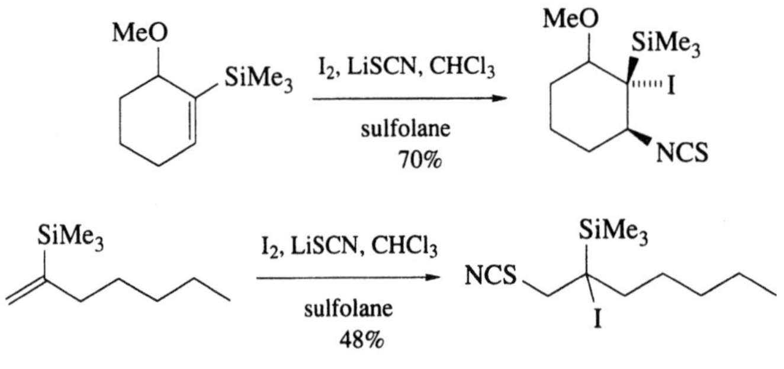

I_2, LiSCN, $CHCl_3$, sulfolane, 70%

I_2, LiSCN, $CHCl_3$, sulfolane, 48%

Scheme 65

The $I(py)_2^+$ BF_4^- reagent of Barluenga (see Sections 1.8.3.5 and 1.8.4.4) is a useful reagent for introduction of ISCN as well as other pseudohalides. The reagent was used for the iodothiocyanation of 3-hexyne.[93d] Other efficient iodothiocyanations include the use of $Hg(SCN)_2{\cdot}I_2$[110] (see Section 1.8.4.3) and the use of inorganic solid supported KSCN, as indicated by equation (17).[130]

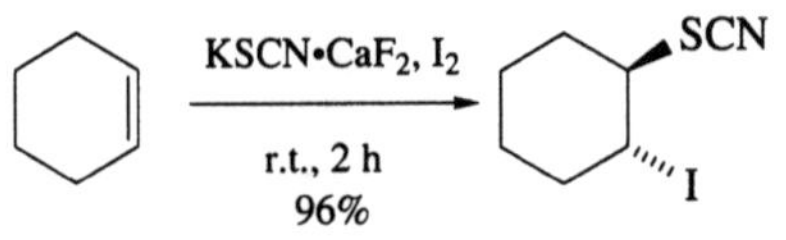

KSCN•CaF_2, I_2, r.t., 2 h, 96% (17)

1.8.4.6 Haloamides

Currently the most popular of the pseudohalides are *N*-haloamides. These reagents are used in conjunction with the solvent or an added nucleophilic reagent to give overall 1,2-addition of a halogen and a nucleophile to the alkenic substrate. The amides of choice seem to be *N*-halosuccinimides. The low nucleophilicity of the succinimdyl anion allows for a variety of nucleophiles to attack the initially formed halonium ion.

Alcohols have been shown to be suitable nucleophiles. Thus unsymmetrical ketene acetals can be prepared by treating vinyl ethers with *N*-iodosuccinimde (NIS) in the presence of functionalized alcohols (Scheme 66).[131] In an elegant example, McDougal *et al.* treated the 2-phenylthio-7-oxa[2.2.1]hept-2-ene (**10**) with NCS in MeOH to effect *syn, exo* chloromethoxylation of the double bond (Scheme 67).[132] Presumably the phenylthio group can bridge to form a thiiranium ion as the initial intermediate. This cation is then attacked on the *exo* face by MeOH. The preparation of a key intermediate along the pathway to the tricyclo[5.4.0.0^{2,8}]undecane ring system employs NBS in the presence of MeOH or NaOAc (equation 18).[133] The authors use molecular models and force-field calculations to help account for the high degree of regio- and stereo-selectivity of the addition.

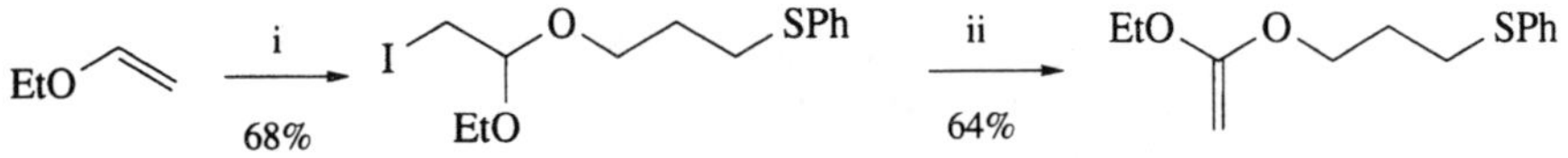

i, *N*-iodosuccinimide, $HO(CH_2)_3SPh$; ii, Bu^tOK, THF

Scheme 66

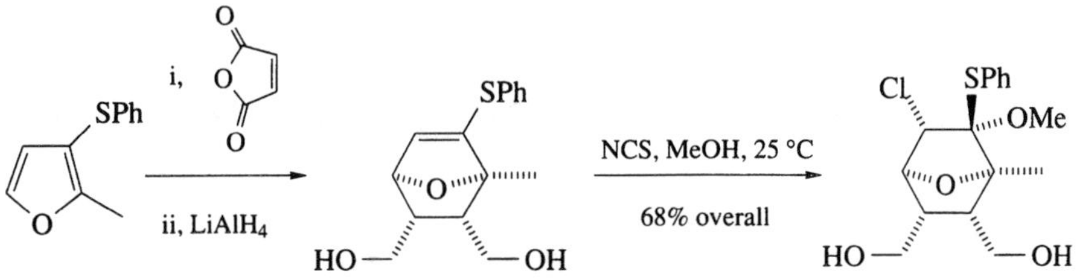

Scheme 67

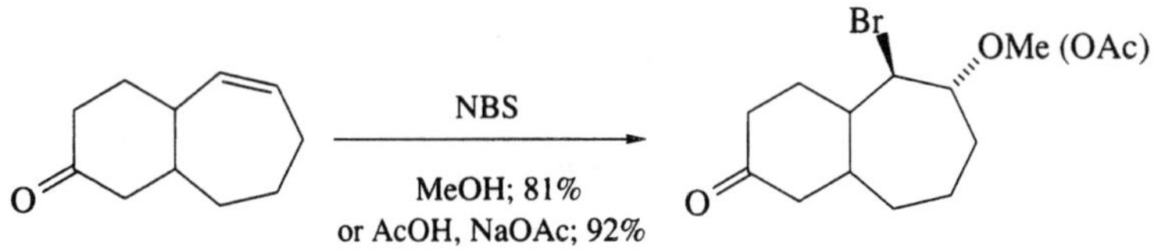

(18)

Jung and Kohn have demonstrated the use of an external nitrogen nucleophile. Haloamination of alkenes with NBS and cyanamide, followed by further manipulation, results in vicinal diamines in high yield (Scheme 68).[134]

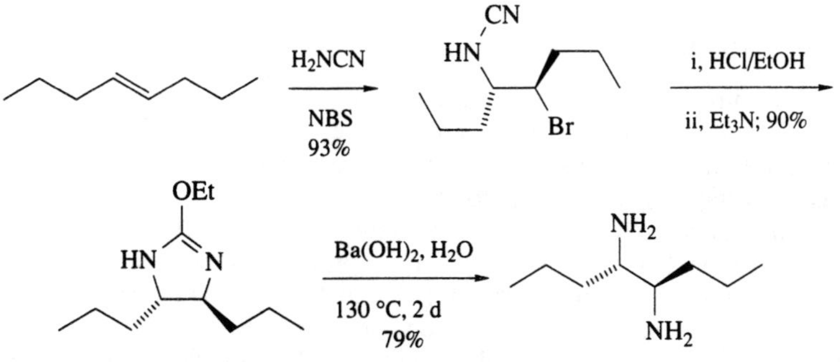

Scheme 68

N-Bromoacetamide (NBA) is also a useful positive bromine equivalent. Thus NBA in the presence of H_2O or LiOAc/HOAc has resulted in the functionalization of 11,12-dihydrobenzo[*e*]pyrene (Scheme 69).[135] NBA has also been employed in conjunction with H_2O for the regio- and stereo-selective hydrobromination of the 10,11 double bond of avermectin B1a, a complex antibiotic containing *five* double bonds.[136]

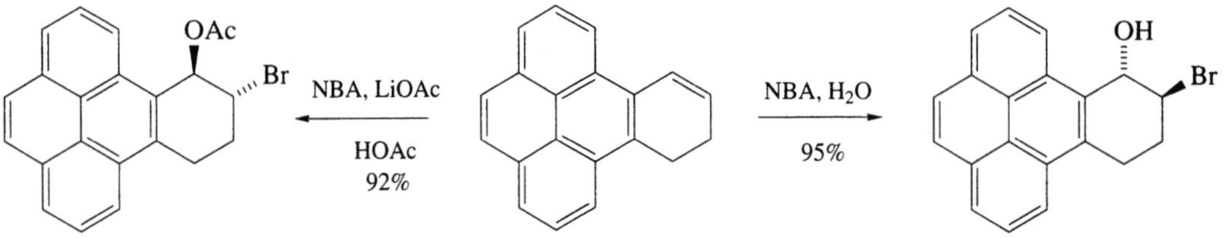

OAc, Br ← NBA, LiOAc / HOAc 92% — NBA, H_2O / 95% → OH, Br

NBA = *N*-bromoacetamide

Scheme 69

Several groups have employed *N*-halosuccinimdes for the halofluorination of alkenes. Laurent and co-workers have found that triethylamine tris(hydrofluoride) is a suitable fluoride source for the halofluorination of a number of alkenes (Scheme 70).[137] Other examples of halofluorinations include NBS and tetrabutylammonium fluoride as the halogen sources for the formal addition of BrF across funtionalized cyclohexenes[138] and the halofluorination of alkenes and alkynes by all three *N*-halosuccinimides and polymer-supported HF.[139]

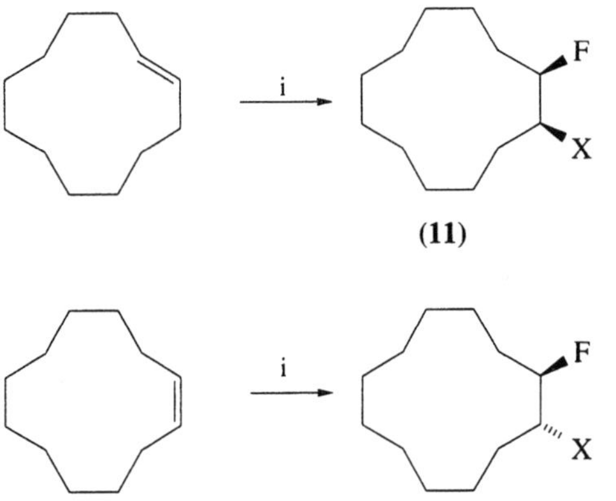

Product	*X*	*Yield (%)*
(11)	Cl	96
(11)	Br	95
(11)	I	82
(12)	Cl	92
(12)	Br	91
(12)	I	76

(11) **(12)**

i, $Et_3N/3HF$, CH_2Cl_2 or ether, *N*-halosuccinimide, 0 to 20 °C

Scheme 70

1.8.5 ELECTROPHILIC NITROGEN

1.8.5.1 Nitronium Tetrafluoroborate

Like *N*-haloamides, nitronium tetrafluoroborate (NO_2BF_4) is employed with a nucleophile to effect overall addition of a nitro group and the nucleophile to the double bond of an alkene. The nature of the cation resulting from the attack of an alkene on NO_2^+ is addressed in a recent review by Smit.[140] One solvent commonly used for trapping the initially formed cations is acetonitrile. The reaction of alkenes with NO_2BF_4 in dichloromethane/acetonitrile followed by aqueous work-up affords good yields of products of nitroacetamidation (equation 19).[141,142] Nitration in the presence of trifluoroacetic anhydride and subsequent elimination of trifluoroacetic acid is an effective method for the preparation of 1-nitro-1,3-dienes (equation 20).[143] It has also been found that nitration using NO_2BF_4 in the presence of lithium perchlorates and fluoride sources leads to 1,2-nitroperchlorates[144] and 1,2-nitrofluorides,[145] respectively.

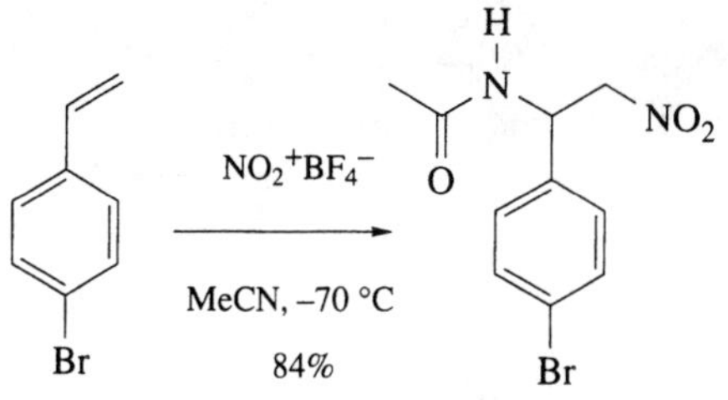

$NO_2^+BF_4^-$, MeCN, –70 °C, 84% (19)

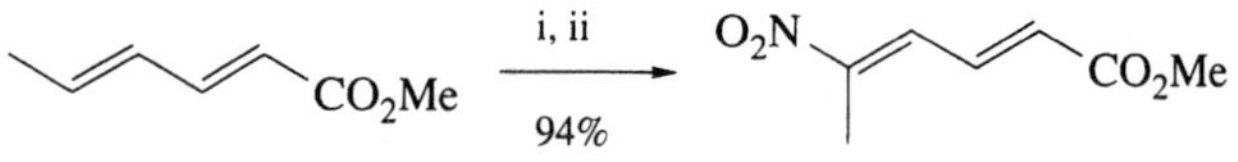

i, ii, 94% (20)

i, $NH_4^+NO_3^-$, TFAA, HBF_4, CH_2Cl_2; ii, KOAc, ether

1.8.5.2 Nitryl and Nitrosyl Halides

The nitrosochlorination of alkenes has been reviewed.[146] The products of nitrosochlorination are often dimers due to the propensity of nitroso groups to couple with one another. For this reason, and because of the advent of newer reagents, modern synthetic uses of NOCl are scarce. Instances where NOCl has proven to be a useful reagent include the preparation of α-oximinocarbonyl compounds (Scheme 71)[147] and the synthesis of precursors of insect growth regulators (Scheme 72).[148]

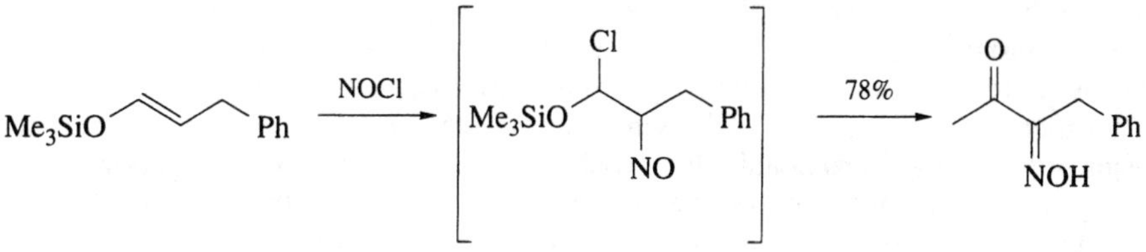

Scheme 71

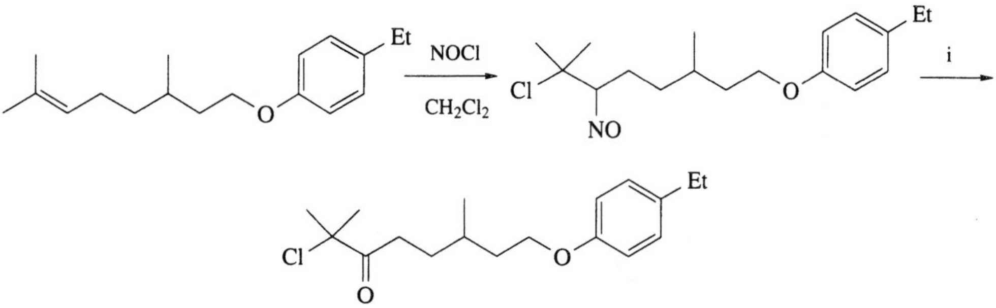

i, levulinic acid, acetone

Scheme 72

Nitryl halides have been more popular reagents. Treatment of substituted styrenes with iodine and sodium nitrite leads to Markovnikov addition of NO_2I (electropositive NO_2) to the double bond. Elimination of HI affords β-nitrostyrenes[149,150] as shown in Scheme 73.[149] (*E*)-2-nitro-1-(trimethylsilyl)ethylene has been prepared by addition of nitryl chloride to trimethylsilylethylene, followed by elimination of HI. The author advertises the use of NO_2Cl and the other reagents involved as more economical than some of the other approaches to the nitroalkene (Scheme 74).[151]

1.8.6 ELECTROPHILIC BORON

The electrophilic addition of hydrogen-bearing boranes to alkenes (hydroboration), now recognized as a vital component of organic synthesis, is covered in Chapter 1.7 of this volume. The boron compounds that do fall under the realm of this chapter are boron bromides. Suzuki and coworkers have demonstrated

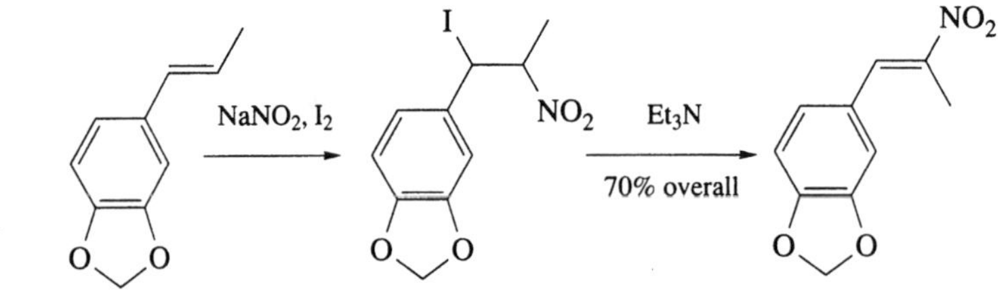

Scheme 73

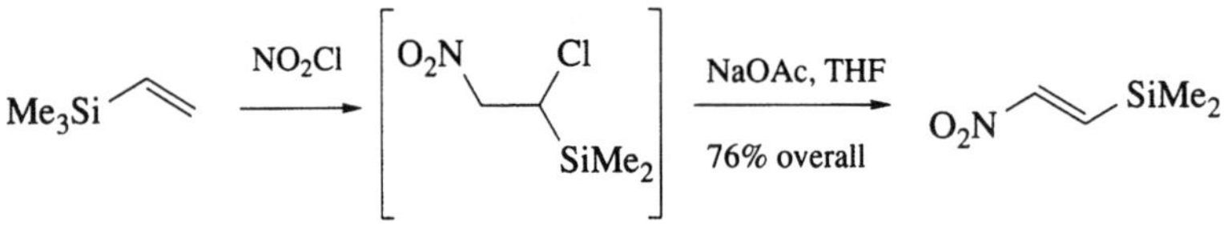

Scheme 74

that these species are synthetically useful X—Y addition reagents for the functionalization of terminal alkynes and allenes.[152] Addition of B-bromo- or B-iodo-9-borabicyclo[3.3.1]nonane (B-X-9-BBN) and BBr_3 was found to proceed in a Markovnikov manner to afford *cis* alkenes. The reagents are chemoselective for terminal alkynes and allenes, while internal alkynes and simple alkenes of either type remain unaffected.[152]

The synthetic utility of the products of haloborane addition, β-halovinylboranes, gives this chemistry its great value. For instance, the products of addition of BBr_3 can be treated with phenyl isocyanate or a positive halogen source to provide *N*-phenyl-β-bromo-α,β-unsaturated amides[153] or (*Z*)-1-halo-2-bromoalkenes,[154] respectively (Scheme 75). The vinyl boranes can also be oxidized under buffered conditions as a means of preparing α-bromoaldehydes (Scheme 75).[155] Alternatively, organozinc nucleophiles, when used along with a palladium catalyst, have been shown to displace the bromine of the bromovinyl boranes.[156] The boron substituent remains intact under these conditions.[156]

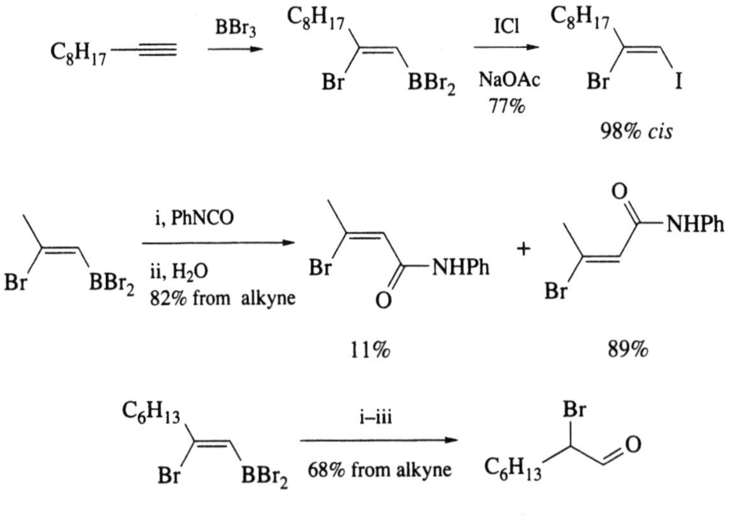

i, pH 5 buffer; ii, KOAc; iii, H_2O_2

Scheme 75

The synthetic utility of the alkyne/B-X-9-BBN adducts has also been explored.[152] (*Z*)-1-Alkynyl-2-halo-1-alkenes are the result of lithium alkynide action on these latter boranes (Scheme 76).[157] The adducts resulting from the treatment of alkynes with B-X-9-BBN have been shown to be labile under protic conditions. However the decomposition has been harnessed and is an efficient preparation of 2-halo-1-alkenes.[158]

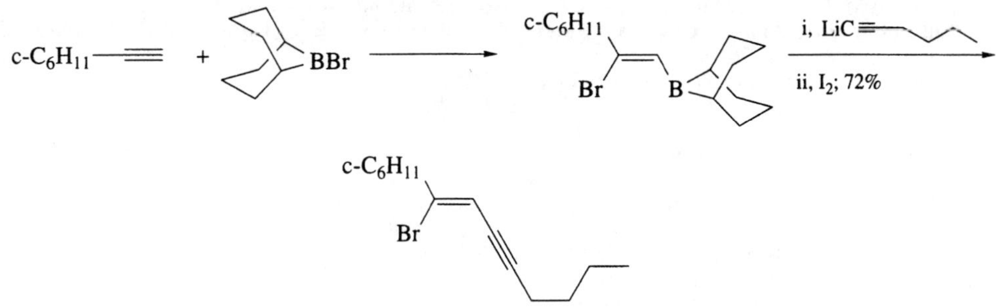

c-C_6H_{11} + BBr → c-C_6H_{11} Br B; i, LiC≡; ii, I_2; 72%

c-C_6H_{11} Br

Scheme 76

1.8.7 REAGENTS THAT ADD *VIA* ELECTROPHILIC AND RADICAL MECHANISMS

In the course of reaction with an alkene or alkyne many reagents of type X—Y can be dissociated into radical pairs X· Y· as well as, or instead of, ionic species X^+Y^-. In some cases the radical pair can add to the π-bond(s) (see Chapter 4.1 of this volume). Furthermore, in some cases both ionic and radical behavior can be observed in the same reaction. For example, while the sulfonyl bromide bromomethanesulfonyl bromide, $BrCH_2SO_2Br$, adds to alkenes and alkynes by processes that most likely involve free radicals (equation 21),[159] mechanisms are less clear cut in other related cases, (equation 22[160] and Scheme 77).[161]

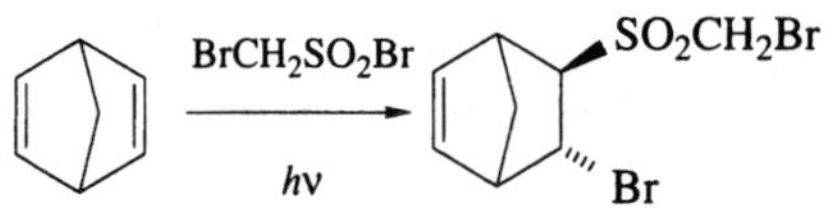

$BrCH_2SO_2Br$, hν → SO_2CH_2Br, Br (21)

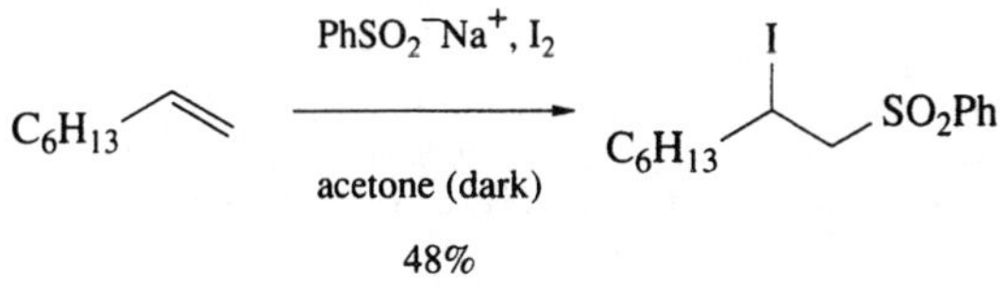

C_6H_{13} → $PhSO_2^-Na^+$, I_2; acetone (dark); 48% → C_6H_{13} I SO_2Ph (22)

i → [I Tol] → ii, 100% → Tol

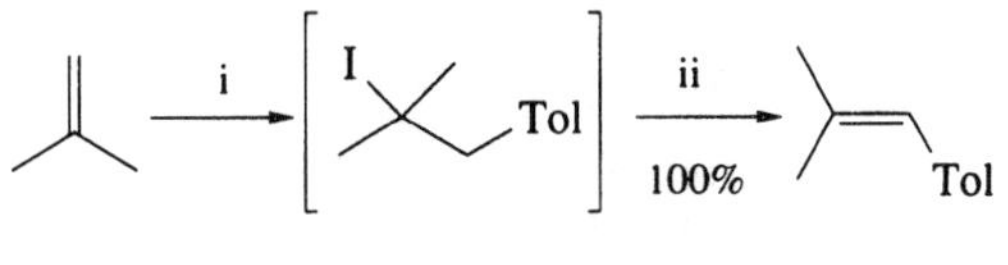

i, Ts^-Na^+•H_2O/I_2, MeOH; ii, Et_3N/MeCN

Scheme 77

ACKNOWLEDGEMENTS

A.L.S. wishes to thank NSERC of Canada for financial support in the form of a Postdoctoral Fellowship. We gratefully acknowledge support from the National Science Foundation, the National Institutes of Health, the donors of the Petroleum Research Fund (administered by the American Chemical Society) and the Herman Frasch Foundation.

1.8.8 REFERENCES

1. (a) P. B. D. De la Mare and R. Bolton, 'Electrophilic Additions to Unsaturated Systems', 2nd edn., Elsevier, New York, 1982; (b) R. C. Fahey, *Top. Stereochem.*, 1968, **3**, 237; (c) G. H. Schmid and D. G. Garratt, in 'The Chemistry of Doubly Bonded Functional Groups', ed. S. Patai, Wiley, Chichester, 1977, supplement A, p. 725; (d) G. H. Schmid in 'The Chemistry of Carbon-Carbon Triple Bond', ed. S. Patai, Wiley, Chichester, 1978, supplement A, p. 275; (e) K. A. V'yunov and A. I. Ginak, *Russ. Chem. Rev. (Engl. Transl.)*, 1981, **50**, 151; (f) L. S. Boguslavskaya. *Russ. Chem. Rev. (Engl. Transl.)*, 1972, **41**, 740.

2. L. F. Fieser and M. Fieser, 'Advanced Organic Chemistry', Reinhold, New York, 1961.
3. C. G. Bellucci, R. Bianchini, C. Chiappe, F. Marioni, R. Ambrosetti, R. S. Brown and H. Slebocka-Tilk, *J. Am. Chem. Soc.*, 1989, **111**, 2640.
4. D. G. Garratt and A. Kabo, *Can. J. Chem.*, 1980, **58**, 1030.
5. (a) G. H. Schmid, *Top. Sulfur Chem.*, 1977, **3**, 101; (b) W. A. Smit, N. S. Zefirov and I. V. Bodrikov, in 'Organic Sulfur Chemistry', ed., R. Kh. Freidlina and A. E. Skorova, Pergamon Press, Oxford, 1981, p. 159; (c) G. Capozzi and G. Modena, in 'Organic Sulfur Compounds: Theoretical and Experimental Advances', ed. F. Bernardi, I. G. Csizmadia and A. Mangini, Elsevier, New York, 1985, p. 246; (d) E. Kühle, 'The Chemistry of the Sulfenic Acids', Thieme, Stuttgart, 1973; (e) V. Lucchini, G. Modena, G. Valle and G. Capozzi, *J. Org. Chem.*, 1981, **46**, 4720; (f) G. H. Schmid, *Phosphorus Sulfur*, 1988, **36**, 197.
6. M. J. Baldwin and R. K. Brown, *Can. J. Chem.*, 1968, **46**, 1093.
7. F. Effenberger, T. Beisswenger and H. Isak, *Tetrahedron Lett.*, 1985, **26**, 4335.
8. M. T. Reetz and T. Seitz, *Angew. Chem., Int. Ed. Engl.*, 1987, **26**, 1028.
9. R. K. Alexander and I. Paterson, *Tetrahedron Lett.*, 1983, **24**, 5911.
10. J. D. White, P. Theramongkol, C. Kuroda, and J. R. Engebreecht, *J. Org. Chem.*, 1988, **53**, 5909.
11. T. Kondo, H. Abe and T. Goto, *Chem. Lett.*, 1988, 1657.
12. F. Capozzi, G. Capozzi and S. Menichetti, *Tetrahedron Lett.*, 1988, **29**, 4177.
13. M. U. Bombala and S. V. Ley, *J. Chem. Soc., Perkin Trans. 1*, 1979, 3013.
14. J. Ohishi, K. Tsuneoka, S. Ikegami and S. Akaboshi, *J. Org. Chem.*, 1978, **43**, 4013.
15. M. Shibasaki and S. Ikegami, *Tetrahedron Lett.*, 1977, 4037.
16. E. J. Corey and E. Block, *J. Org. Chem.*, 1966, **31**, 1663.
17. E. Block and S. Ahmad, *Phosphorus Sulfur*, 1985, **25**, 139.
18. S. T. Purrington and I. D. Correa, *J. Org. Chem.*, 1986, **51**, 1080.
19. G. Haufe, G. Alvernhe, D. Anker, A. Laurent and C. Saluzzo, *Tetrahedron Lett.*, 1988, **29**, 2311.
20. M. C. Caserio and J. K. Kim, *J. Am. Chem. Soc.*, 1982, **104**, 3231.
21. T. Ohsawa, M. Ihara, K. Fukumoto and T. Kametani, *Heterocycles*, 1982, **19**, 1605, 2075.
22. T. Ohsawa, M. Ihara, K. Fukumoto and T. Kametani, *J. Org. Chem.*, 1983, **48**, 3644.
23. M. Ihara, Y. Haga, M. Yonekura, T. Ohsawa, K. Fukumoto and T. Kametani, *J. Am. Chem. Soc.*, 1983, **105**, 7345.
24. G. Capozzi, *Pure Appl. Chem.*, 1987, **59**, 989.
25. E. Block and A. Wall, *J. Org. Chem.*, 1987, **52**, 809.
26. N. S. Zefirov, N. V. Zyk, A. G. Kutateladze and Y. A. Lapin, *J. Org. Chem. USSR (Engl. Transl.)*, 1987, **23**, 351.
27. T. Morishita, N. Furukawa and S. Oae, *Tetrahedron*, 1981, **37**, 2539.
28. A. J. Bridges and J. W. Fischer, *J. Org. Chem.*, 1984, **49**, 2954.
29. L. Benati, P. C. Montevecchi and P. Spagnolo, *J. Chem. Soc., Chem. Commun.*, 1987, 1050.
30. L. Benati, P. C. Montevecchi and P. Spagnolo, *Tetrahedron Lett.*, 1988, **29**, 2381.
31. H.-J. Liu and J. M. Nyangulu, *Tetrahedron Lett.*, 1988, **29**, 5467.
32. F. Bellesia, F. Ghelfi, U. M. Pagnoni and A. Pinetti, *J. Chem. Res. (S)*, 1987, 238.
33. M. C. Caserio, C. L. Fisher and J. K. Kim, *J. Org. Chem.*, 1985, **50**, 4390.
34. D. L. Schmidt, J. P. Heeschen, T. C. Klingler and L. P. McCarty, *J. Org. Chem.*, 1985, **50**, 2840.
35. B. M. Trost and T. Shibata, *J. Am. Chem. Soc.*, 1982, **104**, 3225.
36. B. M. Trost, T. Shibata and S. J. Martin, *J. Am. Chem. Soc.*, 1982, **104**, 3228.
37. T. V. Popkova, A. V. Shastin and E. S. Balenkova, *J. Org. Chem. USSR (Engl. Transl.)*, 1988, **24**, 1045.
38. T. V. Popkova, A. V. Shastin, E. I. Lazhko and E. S. Balenkova, *J. Org. Chem. USSR (Engl. Transl.)*, 1986, **22**, 2210.
39. T. G. Back, in 'Organoselenium Chemistry', ed. D. Liotta, Wiley-Interscience, New York, 1987, p. 1; T. G. Back, in 'The Chemistry of Organic Selenium and Tellurium Compounds', ed. S. Patai, Wiley-Interscience, New York, 1986, vol. 2, p. 136.
40. (a) K. C. Nicolaou and N. A. Petasis, in 'Selenium in Natural Product Synthesis', CIS, Philadelphia, 1984, p. 113; (b) C. Paulmier, 'Selenium Reagents and Intermediates in Organic Synthesis', Pergamon Press, Oxford, 1986, p. 182.
41. G. H. Schmid and D. G. Garratt, *J. Org. Chem.*, 1983, **48**, 4169; *Tetrahedron Lett.*, 1983, **24**, 5299, and refs. therein.
42. T. Arunachalam and E. Caspi, *J. Org. Chem.*, 1981, **46**, 3415.
43. M. Petrzilka, *Helv. Chim Acta*, 1978, **61**, 2286; R. Pitteloud and M. Petrzilka, *Helv. Chim. Acta*, 1979, **62**, 1319.
44. (a) G. Bérubé, E. Luce and K. Jankowski, *Bull. Chim. Soc. Fr., Part 2*, 1983, 109; (b) A. P. Kozikowski, K. L. Sorgi and R. J. Schmiesing, *J. Chem. Soc., Chem. Commun.*, 1980, 477; A. P. Kozikowski, R. J. Schmiesing and K. L. Sorgi, *J. Am. Chem. Soc.*, 1980, **102**, 6577.
45. M. Perez and J.-M. Beau, *Tetrahedron Lett.*, 1989, **30**, 75.
46. R. Baudat and M. Petrzilka, *Helv. Chim. Acta*, 1979, **62**, 1406.
47. A. Toshimitsu, S. Uemura and M. Okano, *J. Chem. Soc., Chem. Commun.*, 1982, 965.
48. M. J. Adam, B. D. Pate, J. R. Nesser and L. D. Hall, *Carbohydrate Res.*, 1983, **124**, 215.
49. D. Liotta and G. Zima, *J. Org. Chem.*, 1980, **45**, 2551.
50. A. M. Morella and A. D. Ward, *Tetrahedron Lett.*, 1984, **25**, 1197.
51. L. Engman, *Tetrahedron Lett.*, 1987, **28**, 1463.
52. T. Hori and K. B. Sharpless, *J. Org. Chem.*, 1979, **44**, 4204, 4208.
53. D. Seebach, G. Calderari and P. Knochel, *Tetrahedron*, 1985, **41**, 4861.
54. D. L. J. Clive and P. L. Beaulieu, *J. Chem. Soc., Chem. Commun.*, 1983, 307.
55. S. Tomoda, Y. Takeuchi and Y. Nomura, *J. Chem. Soc., Chem. Commun.*, 1982, 871; *Tetrahedron Lett.*, 1982, **23**, 1361; *Chem. Lett.*, 1982, 1733.
56. A. Toshimitsu, T. Aoai, S. Uemura and M. Okano, *J. Org. Chem.*, 1980, **45**, 1953.

57. D. G. Garratt, *Can. J. Chem.*, 1979, **57**, 2180.
58. A. Toshimitsu, S. Uemura, M. Okano and N. Watanabe, *J. Org. Chem.*, 1983, **48**, 5246.
59. T. G. Back and S. Collins, *J. Org. Chem.*, 1981, **46**, 3249.
60. A. Hassner and A. S. Amarasekara, *Tetrahedron Lett.*, 1987, **28**, 5185.
61. R. Hernández, E. I. León, J. A. Salazar and E. Suárez, *J. Chem. Soc., Chem. Commun.*, 1987, 312.
62. (a) G. H. Schmid and D. G. Garratt, *Chem. Scr.*, 1976, **10**, 76; (b) C. N. Filer, D. Ahern, R. Fazio and E. J. Shelton, *J. Org. Chem.*, 1980, **45**, 1313.
63. (a) D. G. Garratt, P. L. Beaulieu and M. D. Ryan, *Tetrahedron*, 1980, **36**, 1507; (b) S. Halazy and L. Hevesi, *Tetrahedron Lett.*, 1983, **24**, 2689.
64. J. Bergman, L. Engman and J. Siden, in 'The Chemistry of Organic Selenium and Tellurium Compounds', ed. S. Patai and Z. Rappoport, Wiley-Interscience, New York, 1986, vol. 1, p. 526.
65. D. G. Garratt, M. Ujjainwalla and G. H. Schmid, *J. Org. Chem.*, 1980, **45**, 1206.
66. Y. V. Migalina, S. V. Galla-Bobik, S. M. Khripak and V. I. Staninets, *Chem. Heterocycl. Compd. (Engl. Transl.)*, 1982, **18**, 690.
67. Y. V. Migalina, V. I. Staninets, V. G. Lendel, I. M. Balog, V. A. Palyulin, A. S. Koz'min and N. S. Zefirov, *Chem. Heterocycl. Compd. (Engl. Transl.)*, 1977, **13**, 49.
68. N. M. Magdesieva and M. F. Gordeev, *J. Org. Chem. USSR (Engl. Transl.)*, 1987, **23**, 857.
69. D. G. Garratt and G. H. Schmid, *J. Org. Chem.*, 1977, **42**, 1776.
70. L. Engman, *J. Org. Chem.*, 1987, **52**, 4086.
71. D. G. Garratt and G. H. Schmid, *Chem. Scr.*, 1977, **11**, 170.
72. J. E. Backväll, J. Bergman and L. Engman, *J. Org. Chem.*, 1983, **48**, 3918.
73. J. E. Backväll and L. Engman, *Tetrahedron Lett.*, 1981, **22**, 1919.
74. N. X. Hu, Y. Aso, T. Otsubo and F. Ogura, *Phosphorus Sulfur*, 1988, **38**, 177.
75. S. T. Purrington, B. S. Kagan and T. B. Patrick, *Chem. Rev.*, 1986, **86**, 997.
76. G. M. Loudon, 'Organic Chemistry', 2nd edn., Benjamin/Cummings, Menlo Park, CA, 1988.
77. S. Rozen and M. Brand, *J. Org. Chem.*, 1986, **51**, 3607; see also D. H. R. Barton, J. Lister-James, R. H. Hesse, M. M. Pechet and S. Rozen, *J. Chem. Soc., Perkin Trans. 1*, 1982, 1105.
78. (a) A. Bongini, G. Cainelli, M. Contento and F. Manescalchi, *J. Chem. Soc., Chem. Commun.*, 1980, 1278; (b) R. B. Miller and G. McGarvey, *J. Org. Chem.*, 1978, **43**, 4424.
79. C. G. Bellucci, R. Bianchini and R. Ambrosetti, *J. Am. Chem. Soc.*, 1985, **107**, 2464.
80. U. Husstedt and H. J. Schäfer, *Synthesis*, 1979, 964, 966.
81. (a) G. E. Heasley, J. M. Bundy, V. L. Heasley, S. Arnold, A. Gipe, D. McKee, R. Orr, S. L. Rodgers and D. F. Shellhamer, *J. Org. Chem.*, 1978, **43**, 2793; (b) J. Berthelot and M. Fournier, *Can. J. Chem.*, 1986, **64**, 603; (c) M. Fournier, F. Fournier and J. Berthelot, *Bull. Soc. Chim. Belg.*, 1984, **93**, 157.
82. M. M. Midland and R. L. Halterman, *J. Org. Chem.*, 1981, **46**, 1227.
83. E. Block and V. Eswarakrishnan, *Phosphorus Sulfur*, 1986, **26**, 101.
84. (a) M. Raasch, *J. Org. Chem.*, 1975, **40**, 161; (b) J. M. Bland and C. H. Stammer, *J. Org. Chem.*, 1983, **48**, 4393.
85. E. Block and S. Naganathan, in press.
86. (a) R. M. Pagni, G. W. Kabalka, R. Boothe, K. Gaetano, L. J. Stewart, R. Conway, C. Dial, D. Gray, S. Larson and T. Luidhardt, *J. Org. Chem.*, 1988, **53**, 4477; (b) S. Larson, T. Luidhardt, G. W. Kabalka and R. M. Pagni, *Tetrahedron Lett.*, 1988, **29**, 35; G. Hondrogiannis, L. C. Lee, G. W. Kabalka and R. M. Pagni, *Tetrahedron Lett.*, 1989, **30**, 2069.
87. S. Uemura, S. Fukuzawa, A. Toshimitsu, M. Okano, H. Tezuka and S. Sawada, *J. Org. Chem.*, 1983, **48**, 270.
88. T. H. Chan and K. Koumaglo, *Tetrahedron Lett.*, 1986, **27**, 883.
89. R. A. Hollins and M. P. A. Campos, *J. Org. Chem.*, 1979, **44**, 3931.
90. V. L. Heasley, D. F. Shellhamer, L. E. Heasley, D. B. Yaeger and G. E. Heasley, *J. Org. Chem.*, 1980, **45**, 4649.
91. L. S. Boguslavskaya, *Russ. Chem. Rev. (Engl. Transl.)*, 1984, **53**, 1178.
92. S. Rozen and M. Brand, *J. Org. Chem.*, 1985, **50**, 3342.
93. (a) R. D. Evans and J. H. Schauble, *Synthesis*, 1977, 551; (b) L. S. Boguslavskaya, N. N. Chuvatkin, A. V. Kartashov and L. A. Ternovskoi, *J. Org. Chem. USSR (Engl. Transl.)*, 1987, **23**, 230; (c) J. Barluenga, J. M. González, P. J. Campos and G. Asensio, *Angew. Chem., Int. Ed. Engl.*, 1985, **24**, 319; (d) J. Barluenga, M. A. Rodríguez, J. M. González, P. J. Campos and G. Asensio, *Tetrahedron Lett.*, 1986, **27**, 3303.
94. S. Rozen and M. Brand, *J. Org. Chem.*, 1986, **51**, 222.
95. (a) V. L. Heasley, D. W. Spaite, D. F. Shellhamer and G. E. Heasley, *J. Org. Chem.*, 1979, **44**, 2608; (b) V. L. Heasley, D. F. Shellhamer, J. A. Iskikian, D. L. Street and G. E. Heasley, *J. Org. Chem.*, 1978, **43**, 3139; (c) M. Adinolfi, G. Barone, R. Lanzetta, G. Laonigro, L. Mangoni and M. Parrilli, *Tetrahedron*, 1984, **40**, 2183; (d) T. Negoro and Y. Ikeda, *Bull. Chem. Soc. Jpn.*, 1984, **57**, 2111, 2116; T. Negoro and Y. Ikeda, *Bull. Chem. Soc. Jpn.*, 1985, **58**, 3655; T. Negoro and Y. Ikeda, *Bull. Chem. Soc. Jpn.*, 1986, **59**, 2547, 3515, 3519; (e) C. W. McCleland in 'Synthetic Reagents', ed. J. S. Pizey, Horwood, Chichester, 1983, vol. 5, p. 85.
96. R. C. Cambie, W. I. Noall, G. J. Potter, P. S. Rutledge and P. D. Woodgate, *J. Chem. Soc., Perkin Trans. 1*, 1977, 226.
97. K. G. Migliorese, E. H. Appelman and M. N. Tsangaris, *J. Org. Chem.*, 1979, **44**, 1711.
98. S. Rozen, O. Lerman, M. Kol and D. Hebel, *J. Org. Chem.*, 1985, **50**, 4753.
99. M. Srebnik, *Synth. Commun.*, 1989, **19**, 197.
100. C. J. Schack, D. Pilipovich and J. F. Hon, *Inorg. Chem.*, 1973, **12**, 897.
101. (a) N. S. Zefirov, A. S. Koz'min and V. D. Sorokin, *J. Org. Chem.*, 1984, **49**, 4086; (b) N. S. Zefirov, A. S. Koz'min, V. D. Sorokin and V. V. Zhdankin, *Sulfur Letters*, 1986, **4**, 45; (c) N. S. Zefirov, A. S. Koz'min, V. D. Sorokin and V. V. Zhdankin, *J. Org. Chem. USSR (Engl. Transl.)*, 1986, **22**, 802.
102. Y. Katsuhara and D. D. DesMarteau, *J. Org. Chem.*, 1980, **45**, 2441.
103. N. S. Zefirov, V. V. Zhdankin, A. S. Koz'min, A. A. Fainzilberg, A. A. Gakh, B. I. Ugrak and S. V. Romaniko, *Tetrahedron*, 1988, **44**, 6505.

104. D. H. R. Barton, L. J. Danks, A. K. Ganguly, R. H. Hesse, G. Tarzia and M. M. Pechet, *J. Chem. Soc., Perkin Trans. 1*, 1976, 101.
105. R. H. Hesse, *Isr. J. Chem.*, 1978, **17**, 60.
106. (a) N. S. Zefirov, N. V. Zyk, S. I. Kolbasenko and A. G. Kutateladze, *J. Org. Chem.*, 1985, **50**, 4539; (b) N. S. Zefirov, N. V. Zyk, S. I. Kolbasenko and E. M. Itkin, *J. Org. Chem. USSR (Engl. Transl.)*, 1986, **22**, 397.
107. R. P. Welcher and P. F. Cutrufello, *J. Org. Chem.*, 1972, **37**, 4478.
108. (a) J. L. Wood, *Org. React. (N.Y.)*, 1946, **3**, 240; (b) R. J. Maxwell, L. S. Silbert and J. R. Russell, *J. Org. Chem.*, 1977, **42**, 1510; (c) E. Block, A. J. Yencha, M. Aslam, V. Eswarakrishnan, J. Luo and A. Sano, *J. Am. Chem. Soc.*, 1988, **110**, 4748.
109. (a) E. J. Corey, F. A. Carey and R. A. E. Winter, *J. Am. Chem. Soc.*, 1965, **87**, 934; (b) E. J. Corey and J. I. Shulman, *Tetrahedron Lett.*, 1968, 3655.
110. J. Barluenga, J. M. Martínez-Gallo, C. Nájera and M. Yus, *J. Chem. Soc., Perkin Trans. 1*, 1987, 1017.
111. A. Hassner, *Acc. Chem. Res.*, 1971, **4**, 9.
112. K. Dehnicke, *Angew. Chem., Int. Ed. Engl.*, 1979, **18**, 507.
113. D. Van Ende and A. Krief, *Angew. Chem., Int. Ed. Engl.*, 1974, **13**, 279.
114. J. N. Denis and A. Krief, *Tetrahedron*, 1979, **35**, 2901.
115. S. Sivasubramanian, S. Aravind, L. T. Kumarasingh and A. Arumugam, *J. Org. Chem.*, 1986, **51**, 1985.
116. Y. Tamura, J. Haruta, S. M. Bayomi, M. W. Chun, S. Kwon and M. Ikeda, *Chem. Pharm. Bull.*, 1978, **26**, 784.
117. C. Lion, J.-P. Boukou-Poba and I. Saumtally, *Bull. Soc. Chim. Belg.*, 1987, **96**, 711.
118. Y. Tamura, M. W. Chun, S. Kwon, S. M. Bayomi, T. Okada and M. Ikeda, *Chem. Pharm. Bull.*, 1978, **26**, 3515.
119. H. H. Wasserman, R. K. Brunner, J. D. Buynak, C. G. Carter, T. Oku and R. P. Robinson, *J. Am. Chem. Soc.*, 1985, **107**, 519.
120. A. Hassner and J. Keogh, *J. Org. Chem.*, 1986, **51**, 2767.
121. R. C. Cambie, J. L. Jurlina, P. S. Rutledge and P. D. Woodgate, *J. Chem. Soc., Perkin Trans. 1*, 1982, 315; R. C. Cambie, J. D. Robertson, P. S. Rutledge and P. D. Woodgate, *Aust. J. Chem.*, 1982, **35**, 863.
122. J. W. Lown and A. V. Joshua, *Can. J. Chem.*, 1977, **55**, 122, and refs. therein.
123. J. W. Lown and A. V. Joshua, *Can. J. Chem.*, 1977, **55**, 508.
124. J. Barluenga, J. M. Martinez-Gallo, C. Nájera and M. Yus, *J. Chem. Soc., Chem. Commun.*, 1985, 1422.
125. K. Funaki, K. Takeda and E. Yoshii, *Tetrahedron Lett.*, 1982, **23**, 3069.
126. R. C. Cambie, D. S. Larsen, P. S. Rutledge and P. D. Woodgate, *J. Chem. Soc., Perkin Trans. 1*, 1981, 58.
127. R. C. Cambie, H. H. Lee, P. S. Rutledge and P. D. Woodgate, *J. Chem. Soc., Perkin Trans. 1*, 1979, 757.
128. P. D. Woodgate, H. H. Lee, P. S. Rutledge and R. C. Cambie, *Tetrahedron Lett.*, 1976, 1531.
129. E. J. Grayson and G. H. Whitham, *Tetrahedron*, 1988, **44**, 4087.
130. T. Ando, J. H. Clark, D. G. Cork, M. Fujita and T. Kimura, *J. Chem. Soc., Chem. Commun.*, 1987, 1301.
131. D. S. Middleton and N. S. Simpkins, *Synth. Commun.*, 1989, **19**, 21.
132. P. G. McDougal, Y.-I. Oh and D. Van Derveer, *J. Org. Chem.*, 1989, **54**, 91
133. R. Gleiter and G. Müller, *J. Org. Chem.*, 1988, **53**, 3912.
134. S.-H. Jung and H. Kohn, *J. Am. Chem. Soc.*, 1985, **107**, 2931.
135. S. K. Agarwal, D. R. Boyd, R. Dunlop and W. B. Jennings, *J. Chem. Soc., Perkin Trans. 1*, 1988, 3013.
136. T. L. Shih, H. Mrozik, J. Ruiz-Sanchez and M. H. Fisher, *J. Org. Chem.*, 1989, **54**, 1459.
137. G. Alvernhe, A. Laurent and G. Haufe, *Synthesis*, 1987, 562.
138. M. Maeda, M. Abe and M. Kojima, *J. Fluorine Chem.*, 1987, **34**, 337.
139. A. Gregorcic and M. Zupan, *Bull. Chem. Soc. Jpn.*, 1987, **60**, 3083; A. Gregorcic and M. Zupan, *J. Fluorine Chem.*, 1984, **24**, 291.
140. W. A. Smit, *Sov. Sci. Rev., Sect. B*, 1985, **7**, 156.
141. A. J. Bloom, M. Fleischman and J. M. Mellor, *J. Chem. Soc., Perkin Trans. 1*, 1984, 2357.
142. M. L. Scheinbaum and M. Dines, *J. Org. Chem.*, 1971, **36**, 3641.
143. A. J. Bloom and J. M. Mellor, *J. Chem. Soc., Perkin Trans. 1*, 1987, 2737.
144. N. S. Zefirov, A. S. Koz'min, V. V. Zhdankin, A. V. Nikulin and N. V. Zyk, *J. Org. Chem.*, 1982, **47**, 3679.
145. G. A. Olah and B. G. B. Gupta, *Synthesis*, 1980, 44; G. A. Olah and M. Nojima, *Synthesis*, 1973, 44.
146. P. P. Kadzyauskas and N. S. Zefirov, *Russ. Chem. Rev. (Engl. Transl.)*, 1968, **37**, 543.
147. J. K. Rasmussen and A. Hassner, *J. Org. Chem.*, 1974, **39**, 2558.
148. K. Derdzinski and A. Zabza, *Bull. Acad. Pol. Sci., Ser. Sci. Chem.*, 1983, **31**, 217.
149. W.-W. Sy and A. W. By, *Tetrahedron Lett.*, 1985, **26**, 1193.
150. S. Jew, H.-D. Kim, Y.-S. Cho and C.-H. Cook, *Chem. Lett.*, 1986, 1747.
151. R. F. Cunico, *Synth. Commun.*, 1988, **18**, 917.
152. A. Suzuki, in 'Reviews on Heteroatom Chemistry', ed. S. Oae, MYU, 1988, vol. 1, p. 291.
153. Y. Satoh, H. Serizawa, S. Hara and A. Suzuki, *Synth. Commun.*, 1984, **14**, 313.
154. S. Hara, T. Kato, H. Chimizu and A. Suzuki, *Tetrahedron Lett.*, 1985, **26**, 1065.
155. Y. Satoh, T. Tayano, H. Koshino, S. Hara and A. Suzuki, *Synthesis*, 1985, 406.
156. Y. Satoh, H. Serizawa, N. Miyaura, S. Hara and A. Suzuki, *Tetrahedron Lett.*, 1988, **29**, 1811.
157. S. Hara, Y. Satoh, H. Ishiguro and A. Suzuki, *Tetrahedron Lett.*, 1983, **24**, 735.
158. S. Hara, S. Takinami, and S. Hyuga and S. Suzuki, *Chem. Lett.*, 1984, 345; S. Hara, H. Dojo, S. Takinami and A. Suzuki, *Tetrahedron Lett.*, 1983, **24**, 731.
159. E. Block, M. Aslam, V. Eswarakrishnan, K. Gebreyes, J. Hutchinson, R. S. Iyer, J.-A. Laffitte and A. Wall, *J. Am. Chem. Soc.*, 1986, **108**, 4568.
160. L. M. Harwood, M. Julia and G. L. Thuillier, *Tetrahedron*, 1980, **36**, 2483.
161. K. Inomata, T. Kobayashi, S. Sasaoka, H. Kinoshita and H. Kotake, *Chem. Lett.*, 1986, 289; K. Inomata, S. Sasaoka, T. Kobayashi, Y. Tanaka, S. Igarashi, T. Ohtani, H. Kinoshita, Y. Tanaka, S. Igarashi, T. Ohtani, H. Kinoshita and H. Kotake, *Bull. Chem. Soc. Jpn.*, 1987, **60**, 1767.

1.9
Electrophilic Heteroatom Cyclizations

KENN E. HARDING and TAMMY H. TINER
Texas A&M University, College Station, TX, USA

1.9.1 INTRODUCTION

1.9.1.1 Coverage

This chapter on electrophilic heteroatom cyclizations covers reactions of carbon–carbon π-bonds in which activation by an *external* electrophilic reagent results in addition of an *internal* heteroatom nucleophile. The general reaction is illustrated in Scheme 1. These cyclization reactions generate heterocyclic products, but many synthetic applications involve subsequent cleavage of the newly formed heterocyclic ring.

The above definition excludes a number of types of heteroatom cyclofunctionalization reactions from discussion in this chapter. Examples include: neighboring group participation in which the cyclic structure is only a transient intermediate; cyclization reactions where the C—C π-bond is internally activated

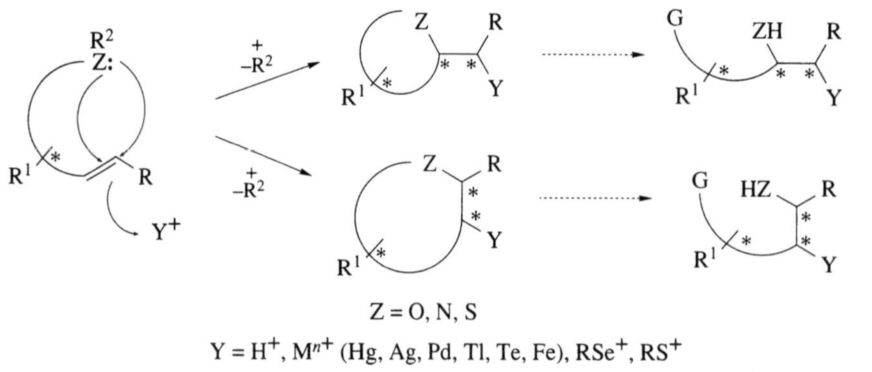

Z = O, N, S

Y = H^+, M^{n+} (Hg, Ag, Pd, Tl, Te, Fe), RSe^+, RS^+

Scheme 1

(conjugation with another functionality) *and* the reaction is best considered as an acid- or base-initiated Michael reaction; cationic π-cyclization where the cyclofunctionalization is the termination step; and reactions in which the heterocyclization involves the heteroatom acting as an electrophile or reactive radical.

Several recent reviews have included specific types of electrophilic cyclofunctionalization reactions.[1] Important areas covered in these reviews are: halolactonization;[1a] cyclofunctionalization of unsaturated hydroxy compounds to form tetrahydrofurans and tetrahydropyrans;[1b] cyclofunctionalization of unsaturated amino compounds;[1c] cyclofunctionalization of unsaturated sulfur and phosphorus compounds;[1d–1f] electrophilic heterocyclization of unconjugated dienes;[1g] synthesis of γ-butyrolactones;[1h] synthesis of functionalized dihydro- and tetrahydro-furans;[1j] cyclofunctionalization using selenium reagents;[1k,1m] stereocontrol in synthesis of acyclic systems;[1n] stereoselectivity in cyclofunctionalizations;[1p] and cyclofunctionalizations in the synthesis of α-methylenelactones.[1q] Previous reference works have also addressed this topic.[2]

1.9.1.2 Synthetic Considerations

The identity of the heteroatom present in a target molecule dictates the identity of the nucleophilic atom (Z) to be used in an electrophilic heteroatom cyclization reaction of the type shown in Scheme 1. However, successful application of this strategy to the synthesis of specific target molecules also requires selection of appropriate combinations of nucleophilic functionality (Z–R) and activating electrophile. Therefore, major subdivisions within this chapter are based on the identity of the heteroatom, although comparisons between results observed for the different heteroatoms will be made.

These heterocyclization reactions provide initial products with a functionality β to the heteroatom, except for cases where a proton is the electrophile. Synthetic applications often depend upon further transformation of this functionality. Useful transformations include: replacement by hydrogen, elimination to form a π-bond, nucleophilic substitution, and substitution *via* radical intermediates. These reactions will be discussed only when understanding the cyclization step requires inclusion of the functional group transformation.

Although these cyclofunctionalization reactions have been used extensively to generate specific heterocyclic structures in target molecules, they can also be used to control the introduction of functional groups which are not part of a heterocyclic system. Cyclofunctionalization followed by ring cleavage (see Scheme 1) has been used to introduce functionality into existing ring systems or acyclic systems with control of regiochemistry and stereochemistry, to control conformational mobility and to serve as a means for protecting the double bond and/or nucleophilic functional group against other reagents in a synthesis. The most recent extensions of this synthetic strategy involve the synthesis and cyclization of substrates in which the nucleophilic Z group is attached to the substrate by a transient 'tether', which is added to an initial substrate and removed after cyclization to generate the final product (Scheme 2).

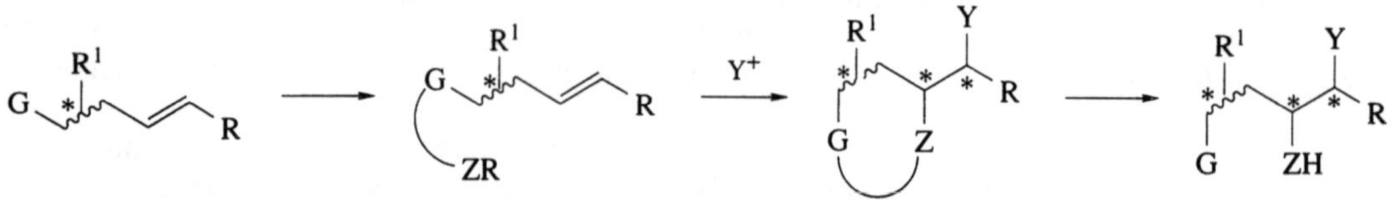

Scheme 2

1.9.1.3 Mechanistic Considerations

1.9.1.3.1 General mechanism

A generalized mechanism for electrophile-initiated cyclofunctionalization is shown in Scheme 3. (To simplify this initial discussion, factors affecting regioselectivity will not be considered until later.) The specific mechanistic pathway may vary from system to system, but it is generally considered that an activated intermediate must be involved. Reaction of the π-system with an electrophile can generate several potential intermediates for the cyclization (a π-complex **B**, a cyclic 'onium' ion **C**, an addition product **D**, a carbocation **E** or an intermediate with an electrophile–Z bond **F**), each of which could undergo attack by the internal nucleophile to give cyclic products. Each type of intermediate has been considered to be involved in some examples of cyclofunctionalization, but most cyclofunctionalizations are best rationalized with intermediates (**B**) and/or (**C**). Although Scheme 3 shows each intermediate being generated from the starting alkene, interconversions between intermediates (**B**) to (**F**) are expected in some cases.

All mechanistic pathways that generate product (**H**) represent examples of neighboring group participation, but anchimeric assistance is observed only if the rate-limiting step is that of ring closure.[3] Since each of the steps may be readily reversible under the conditions of the reaction, it is not always clear whether a given reaction was conducted under conditions of kinetic or thermodynamic control, which step was rate-limiting, or which step controlled the stereochemistry.

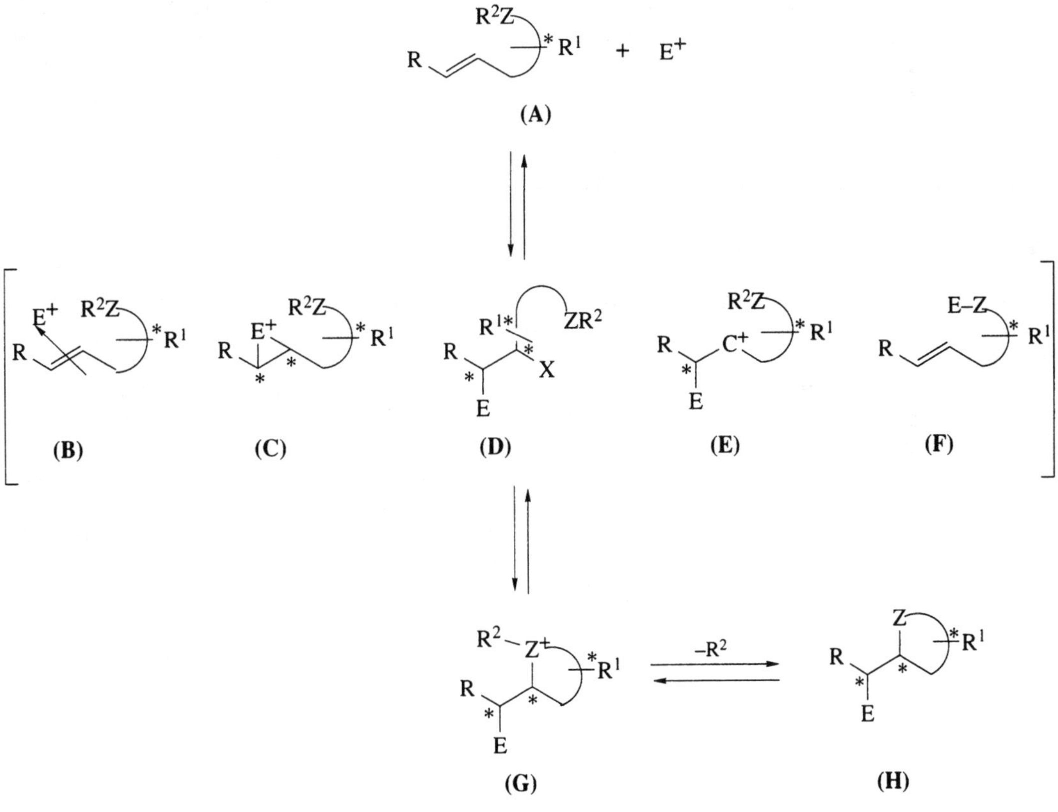

Scheme 3

Most of the types of cyclofunctionalization reaction discussed in this chapter have been shown to result in stereospecific *anti* addition across the π-bond. This result suggests that the important intermediates are π-complexes (**B**) or 'onium' ions (**C**) rather than carbocations (**E**). In the case of cyclofunctionalization with some electrophiles, such as phenylselenenyl chloride, it has been shown that the formation of addition products such as (**D**) occurs faster than the cyclization.[4] Stereospecific *trans* addition in these reactions then requires conversion to intermediates (**B**) or (**C**) before nucleophilic attack, since nucleophilic attack on intermediate (**D**) and substitution of X with inversion would result in *syn* addition. Thus, in the discussions below, intermediates (**B**) and/or (**C**) are considered to be the key cyclization precursors in most reactions. It should be noted that cyclization of intermediate (**F**) by alkene insertion would be expected to lead to stereospecific *syn* addition also.

In a few cases, detailed mechanistic studies have shown that the cyclization step is rate limiting.[5,6,7] One method has been to demonstrate that the overall rate of reaction is a function of the nucleophilicity of the ZR group or of the formed ring size.[5,6] However, the cyclization step need not be the rate-limiting step in all electrophilic heteroatom cyclizations.[8] The uncertainty about which step is rate limiting complicates attempts to derive general rationales for predicting the stereochemical results of these reactions.

1.9.1.3.2 Stereoselectivity

When the cyclization substrate (**A**; Scheme 3) contains stereogenic centers, and the formation of the C—Z bond generates a new stereogenic center, two diastereomers of the cyclization product (**H**) can be formed (stereospecific *anti* addition assumed). The factors which lead to high stereoselectivity in this process are of considerable importance and have been the subject of numerous studies in recent years. This reaction mechanism shows that all pathways leading to cyclic products are potentially reversible; thus, the ratios of products in these reactions may be the result of thermodynamic rather than kinetic control. Unfortunately, many studies have not determined which type of control was operating under the reaction conditions used.

The stereoselectivity of cyclization reactions conducted under conditions of thermodynamic control can often be reliably predicted by estimation or calculation of the energy differences between the diastereomers of the cyclization product (**H**) or its immediate precursor (**G**). It has been shown that, even in cases where the (**G**) to (**H**) step is not reversible, thermodynamic control of the diastereomer ratio can be influenced through the use of cyclization substrates in which the neutralization step (**G** to **H**) is the slow step, thus allowing for equilibration of the diastereomers of (**G**). Thermodynamic equilibration of diastereomeric products will occur only if the reaction reverses to starting materials (**A**), or interconversion of the diastereomers of the intermediates (**B**) or (**C**) occurs in some other way (*e.g.* the **C** to **D** interconversion can equilibrate diastereomers if E = X).

The factors controlling product stereochemistry under conditions of kinetic control depend upon which step of the reaction is rate limiting. The initial reaction of an electrophile with (**A**) leads to diastereomers of intermediates (**B**) and/or (**C**). If the formation of these intermediates is the slow step of the reaction, stereoselectivity will be determined by the diastereofacial discrimination in the attack of the electrophile on the π-system. The stereoselectivity of the reaction will be controlled in the cyclization step *only* if the ring closure is rate limiting and the diastereomeric precursor intermediates are interconverting rapidly. In any case, it is apparent that high stereoselectivity will be observed only if the chiral elements present in the starting substrate result in significant energy differences between the diastereomeric transition states in the product-determining step.

The most successful examples of stereochemical control in electrophilic heteroatom cyclizations are those in which the substitution pattern constrains the substrate so that the two diastereofaces of the π-system are significantly different. The most straightforward prediction of stereochemistry involves incorporating both the π-system and the directing chiral center into a ring such that rotation about the vinylic bond that attaches the nucleophile to the double bond is highly restricted. Comparison of equations (1) and (2) illustrates this difference. For this reason, in the sections on cyclizations to form five- and six-membered rings, examples with constrained C═C—C bonds will be discussed separately.

HO, O — PhSeCl, ether; 25 °C; *ca.* 65% → PhSe, O, O (1)

variety of conformations possible — 1:1 mixture of isomers

HO, O — PhSeCl, EtOAc; reflux; 82% → H, O, O, H, PhSe (2)

bond rotation severely restricted — single diastereoisomer

1.9.1.3.3 Regioselectivity

While the regiochemistry of simple electrophilic additions to double bonds is controlled by a combination of electronic (Markovnikov rule), stereoelectronic (*trans* diaxial addition to cyclohexenes) and steric factors,[9] the intramolecular nature of electrophilic heteroatom cyclizations introduces additional conformational, stereoelectronic and entropic factors. The combination of these factors in cyclofunctionalization reactions results in a general preference for *exo* cyclization over *endo* cyclization (Scheme 4).[3,10] However, *endo* closure may predominate in cases where electronic or ring strain factors strongly favor that mode of cyclization. The observed regiochemistry may differ under conditions of kinetic control from that observed under conditions of thermodynamic control.

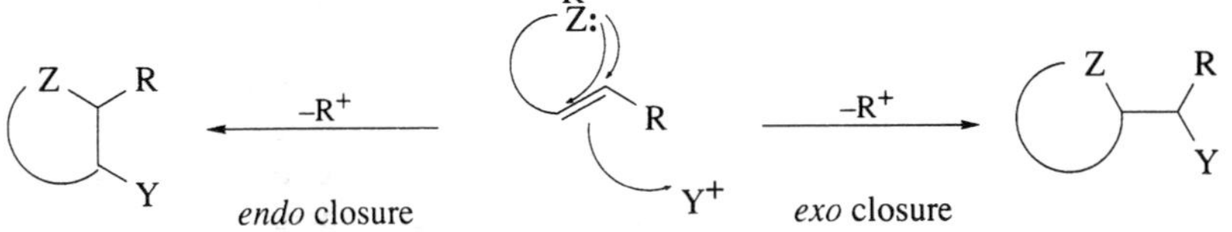

$-R^+$ *endo* closure; $-R^+$ *exo* closure

Scheme 4

The stereochemistry and regiochemistry of cyclofunctionalization reactions in the following sections are discussed in terms of the above mechanistic concepts. Emphasis is placed on the relationship between the products and the heteroatom-substituted substrates; thus, examples of electrophilic heterocyclization involving two sequential additions to nonconjugated dienes are not discussed. These reactions have been covered in a recent review.[1g]

1.9.2 OXYGEN NUCLEOPHILES

Cyclofunctionalization reactions involving oxygen nucleophiles have been studied more extensively than reactions involving any other heteroatom. Examples of functional groups used as oxygen nucleophiles are shown in Figure 1.

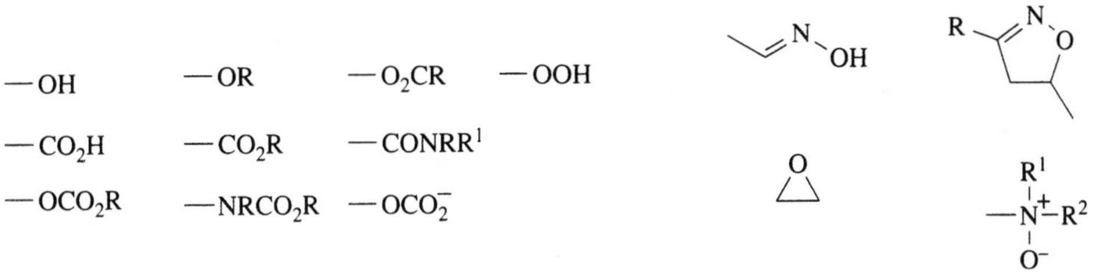

—OH	—OR	$—O_2CR$	—OOH
$—CO_2H$	$—CO_2R$	$—CONRR^1$	
$—OCO_2R$	$—NRCO_2R$	$—OCO_2^-$	

Figure 1

1.9.2.1 Small Rings

The formation of strained three- and four-membered rings by electrophile-initiated cyclization requires that the reaction be conducted under conditions which minimize the possibility of simple addition of the activating reagent across the double bond or reversal of the cyclization product to intermediates that can be trapped by external nucleophiles.

Cyclization of allylic alcohols to form epoxides has been particularly problematical, and the reactions have been more of mechanistic than of synthetic interest. For reactions conducted under basic conditions, it is possible that epoxide formation involves initial halogen addition followed by nucleophilic displacement to form the epoxide. Early examples of direct formation of epoxides from allylic alcohols with sodium hypobromite,[11] bromine and 1.5 M NaOH,[12] and *t*-butyl hypochlorite[13] have been reviewed previously.[1p] Recently it has been shown that allylic alcohols can be cyclized effectively with bis(*sym*-collidine)iodine(I) perchlorate (equation 3).[14] An unusual example of epoxide formation competing with other cyclization types is shown in equation (4).[15] In this case, an allylic benzyl ether competes effectively with a γ-hydroxyl group as the nucleophile.

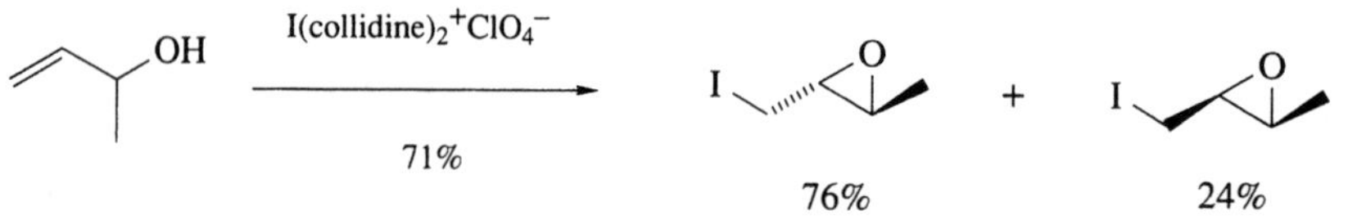

$I(collidine)_2^+ClO_4^-$, 71%: 76% + 24% (3)

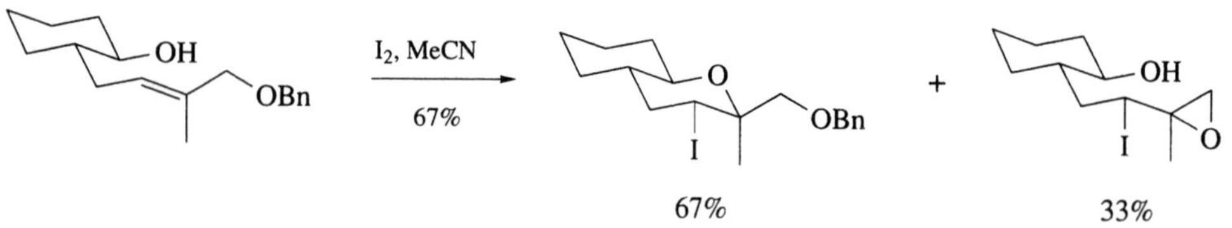

I_2, MeCN, 67%: 67% + 33% (4)

Cyclofunctionalization of homoallylic alcohols with bis(*sym*-collidine)iodine(I) perchlorate produces oxetanes in good yield if the 4-*exo* mode of cyclization is favored electronically by the alkene substitution pattern (Table 1).[14] Geminal substitution at the carbinol carbon also favors this mode of cyclization (compare entries 2 and 3 with entry 1). Substitution at C-3 leads to reaction only by the 4-*exo* mode (entry 4), while substitution at C-4 leads to cyclization only *via* the 5-*endo* mode (entry 5).

Table 1 Cyclization of Homoallylic Alcohols with $I(collidine)_2^+ ClO_4^-$

Entry	*Reactant*	*Yield (%)*	*Product(s) (% of total)*	
1	R = R' = H		50	50
2	R = R' = Me	62	100	
3	R, R' = $(CH_2)_5$	64	100	
4	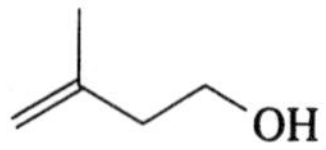	67	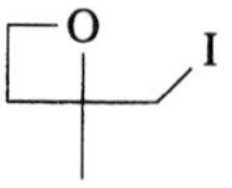	
5	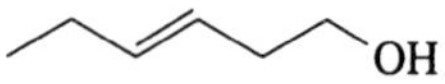	70	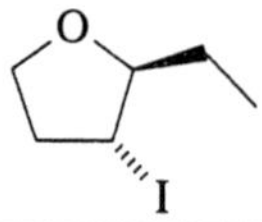	

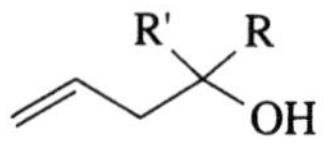

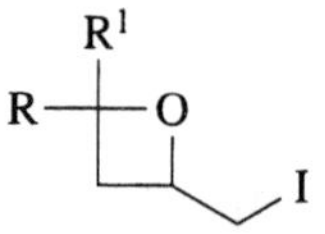

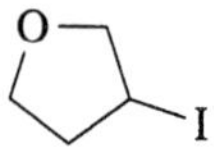

One example of oxetane formation proceeding by an intramolecular *syn* addition of a sulfenate ester intermediate has been reported (equation 5).[16] This mode of cyclofunctionalization appears to fail unless the substrate enforces the correct alignment of the sulfenate ester.[16b] However, related 7-oxanorbornen-2-ols give oxetane products with an *exo* phenylsulfenyl group.[17]

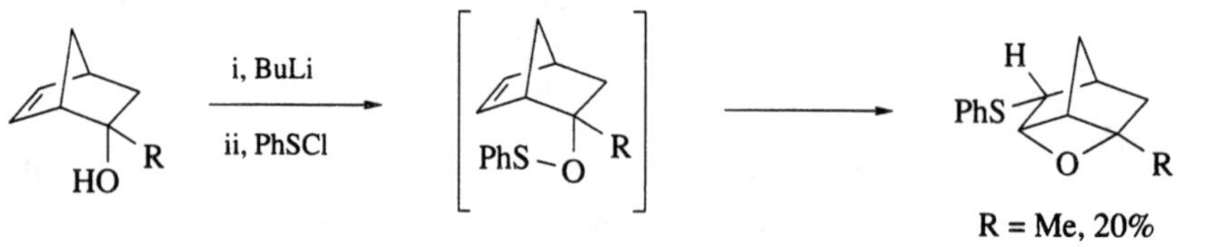

i, BuLi; ii, PhSCl (5)

R = Me, 20%
R = $CH=CH_2$, 60%

Cyclizations of β,γ-unsaturated acids form β-lactones (4-*exo* cyclization) when the reactions are conducted under conditions of kinetic control.[1a,1p] The most common procedure for β-lactone formation, developed by Barnett, involves halolactonization in a two-phase system using an aqueous solution of the carboxylate salt of the substrate with the halogen (Br_2 or I_2) added in an organic solvent.[18] Cyclization with bis(*sym*-collidine)iodine(I) perchlorate provides a higher yield than the Barnett procedure in cases where cyclization is not favored by geminal α-substitution (Table 2, entries 1 and 2).[14] Iodo- and bromolactonization of thallium salts of the carboxylic acids have also been shown to be useful procedures for formation of β-lactones.[21,22] It should be noted that the initially formed β-lactone products may be readily rearranged to thermodynamically more stable γ-lactones.[19,23,24]

Table 2 Cyclization of β,γ-Unsaturated Acids to β-Lactones

Entry	*Reactant*	*Conditions*	*Yield (%)*	*Product(s) (% of total)*		*Ref.*
	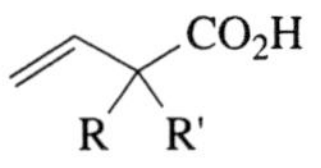			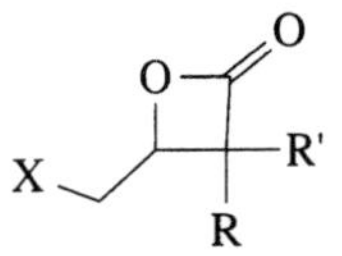		
1	R = R' = H	aq. $NaHCO_3$, I_2, Et_2O	25			18b
2	R = R' = H	$I(collidine)_2^+\ ClO_4^-$	75			14
3	R = R' = H	Tl_2CO_3, Br_2, CH_2Cl_2	51			21
4	R = R' = Me	aq. $NaHCO_3$, I_2, Et_2O	75			18b
	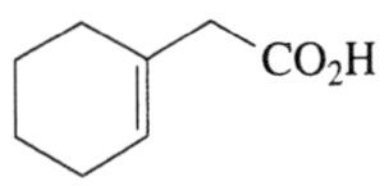			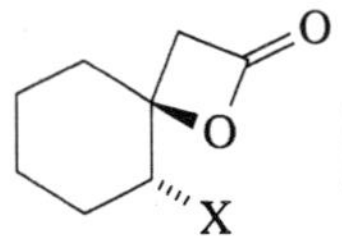	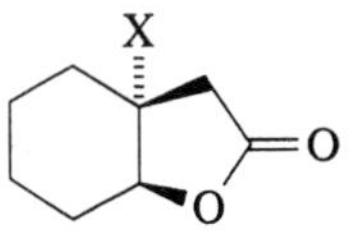	
5		NBS/DMF	53	100	0	21
6		aq. $NaHCO_3$, Br_2, CH_2Cl_2	66	67	33	21
	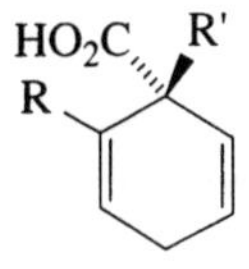			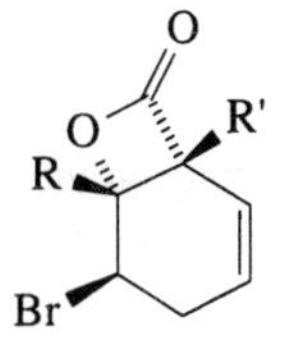		
7	R = R' = H	aq. $NaHCO_3$, Br_2, CH_2Cl_2	42–94			19,20

Formation of β-lactones by 4-*endo* cyclization has been observed in very few cases. Examples of such halolactonization reactions proceeding by stereospecific *anti* addition have been discussed in previous reviews.[1p,2b]

1.9.2.2 Five- and Six-membered Rings

The lack of significant ring strain and favorable entropic factors results in facile cyclofunctionalization to form five- or six-membered rings. Emphasis in this review is placed on examples which illustrate general principles of regiochemical and stereochemical control.

1.9.2.2.1 Cyclizations with C=C—C constrained by existing ring

Electrophile-initiated cyclizations of substrates where an existing ring constrains rotation of the vinylic bond which attaches the nucleophile to the double bond generally occur with quite predictable stereochemical and regiochemical control. These reaction types have been used extensively in synthesis and only representative examples and leading references will be presented here.

(i) Synthesis of fused ring lactones and ethers

The cyclization of 2-cycloalkeneacetic acids leads to *cis*-fused γ-lactones arising from *anti* addition across the double bond (equation 6 and Table 3). This type of reaction, effected with a wide variety of electrophiles and substrate substitution patterns, has been used both for generation of γ-lactone target molecules and selective introduction of a hydroxy group onto the original ring by cleavage of the lactone cyclization product. As shown in equation (7), the 5-*exo* mode of cyclization is preferred even when this mode of addition is anti-Markovnikov.[1q,37] However, one example of iodolactonization to a tricyclic product produced a mixture of a γ-lactone and a bridged ring δ-lactone (6-*endo* cyclization).[38]

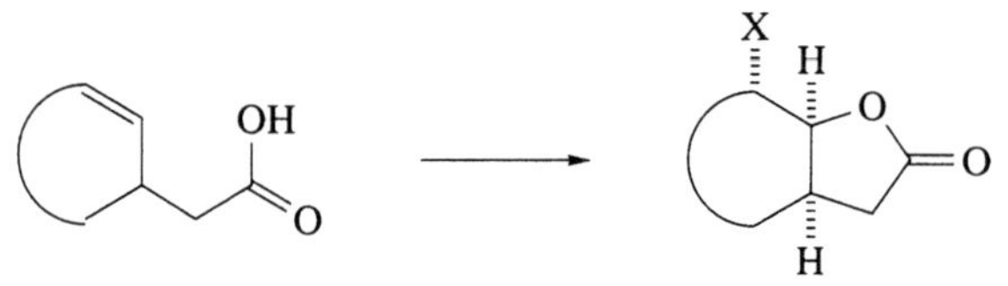

OH
O
X
H
O
O
H

(6)

Table 3 Fused Ring γ-Lactones from Cyclization of 2-Cycloalkenylacetic Acids (Equation 6)

X	*Refs.*	*X*	*Refs.*
I	22, 25–28	RS	16b, 32, 35
Br	21, 25a, 29, 30	H	1q, 33
PhSe	4, 31, 36	$ArCl_2Te$	34

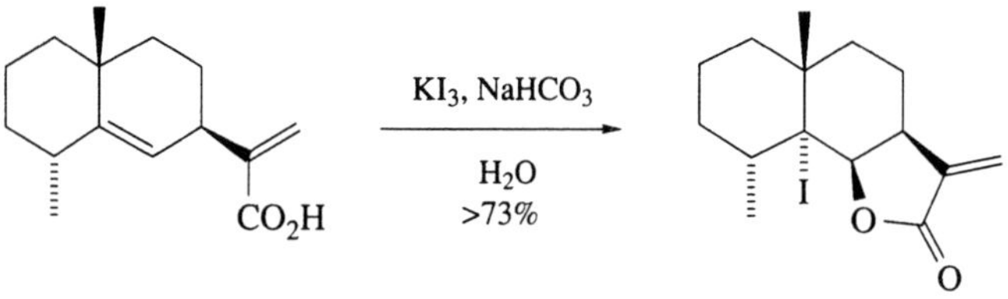

CO2H
KI3, NaHCO3
H2O
>73%
I
O
O

(7)

Cyclization to yield fused ring γ-lactones has also been effected using *N,N*-dialkylamide derivatives as the nucleophile, as shown in equation (8).[29,39,40] Isolation of the fused ring iminolactone from cyclization of an *N*-monoalkylamide derivative with phenylselenenyl chloride has been reported also.[41]

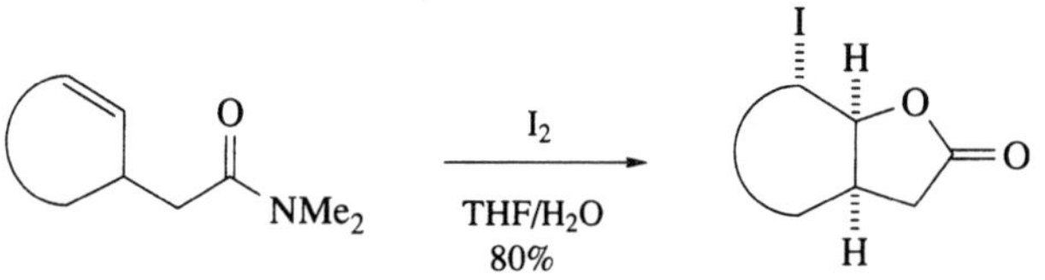

O
NMe2
I2
THF/H2O
80%
I
H
O
O
H

(8)

Cyclizations with some electrophiles generate products in which the C—E bond is solvolyzed under the reaction conditions. Cyclization of 2-cyclopenteneacetic acid with lead tetraacetate forms the acetoxy-substituted lactone from solvolysis of the plumbolactonization intermediate (equation 9).[42] When the substrate contains a second suitably located carboxylate group, bislactones are generated (equation 10).[43,44] Similar cyclizations to form ether–lactones have been reported.[45] Unsaturated fused ring lactones have been obtained from 2-cyclopenteneacetic acid upon treatment with hydroxy(tosyloxy)-iodobenzene or hydroxybis(phenyloxy)phosphoryloxyiodobenzene, as shown in equation (11).[46]

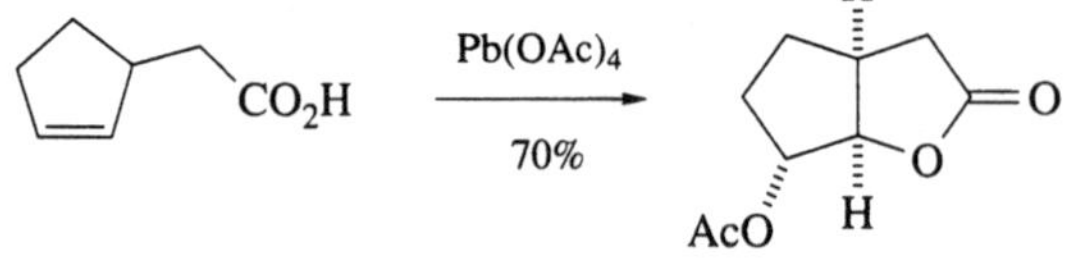

CO2H
Pb(OAc)4
70%
H
O
O
AcO
H

(9)

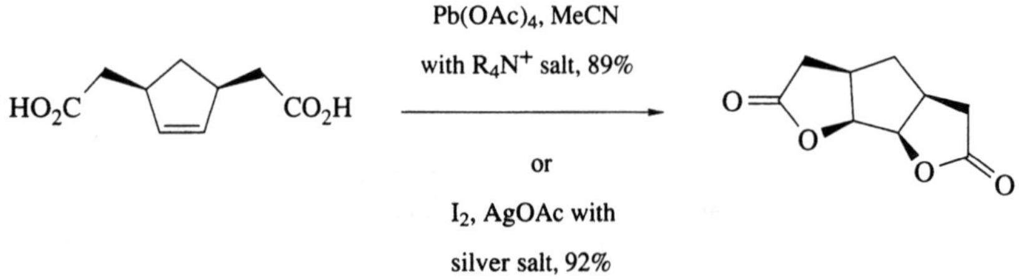

HO2C
CO2H
Pb(OAc)4, MeCN
with R4N+ salt, 89%
or
I2, AgOAc with
silver salt, 92%
O
O
O
O

(10)

TsO-I(OH)-Ph
CH_2Cl_2, 56%
or
$(RO)_2PO_2(HO)IPh$, 50%

(11)

Fused ring γ-lactones may also be formed by cyclizations of 1-cycloalkeneacetic acids under equilibrating conditions. Nicolaou obtained evidence for the presence of an unstable β-lactone in the phenylselenolactonization of 1-cyclohexeneacetic acid, but rearrangement occured readily at room temperature to the more stable γ-lactone (equation 12).[32] An example of a one-step conversion of a 2-methyl-1-cyclohexeneacetic acid to a fused butenolide by use of diphenyldiselenide and electrochemical oxidation has been reported.[47] Recent studies by Rutledge showed that simultaneous addition of bromine and thallium carbonate to 1-cyclohexeneacetic acid gave the γ-lactone as the exclusive product (compare to Table 2); however, the mechanism of this reaction may differ from other cyclofunctionalizations.[21]

(12)

PhSeCl or PhSCl	E = PhSe or PhS	(ref. 32)
Tl_2CO_3 and Br_2/CH_2Cl_2	E = Br	(ref. 21)

The cyclofunctionalization of cyclohexa-2,4-dieneacetic acids results in 1,4-addition to form *cis*-fused γ-lactones, as shown in equation (13) and Table 4. Most reaction conditions gave products with the electrophile *trans* to the lactone ring (entries 1–4), but the stereochemistry of the palladium-catalyzed reaction was reversed if an excess of a complexing ligand was added to the reaction (entries 5 and 6).[49,50] Results of lactonization in cyclohepta-2,4-dieneacetic acid systems were similar, but selenolactonization produced 1,2-addition products under some conditions.[51] It is possible that these products result from a 1,3-rearrangement of the initial allyl selenide.[52]

(13)

Table 4 Cyclofunctionalization of Cyclohexa-2,4-dieneacetic Acid (Equation 13)

Entry	*Conditions*	*Yield (%)*	*X*	*Ref.*
1	PhSeCl, Et_3N, CH_2Cl_2	70	β-SePh	48
2	I_2, MeCN	60	β-I	48
3	$Hg(OAc)_2$, BF_3/MeOH	60	β-HgOAc	48
4	5 mol % $Pd(OAc)_2$, HOAc, acetone, benzoquinone	77	β-OAc	49
5	Entry 4 plus 2 equiv. LiOAc and 0.5 equiv. LiCl	57	α-OAc (76%)	49
6	Entry 4 plus 2 equiv. LiOAc and 2 equiv. LiCl	75	α-Cl (>95%)	49

Cis-fused tetrahydrofurans are produced when 2-cycloalkenyl-substituted ethanol derivatives are cyclized *via* 5-*exo* ring closure (equation 14 and Table 5). The related 1-cycloalkenyl alcohol systems also yield *cis*-fused tetrahydrofurans upon reaction with phenylselenyl reagents, similar to their carboxylic acid analogs (see equation 12).[60] A sulfoetherification to a fused ring tetrahydrofuran from a system with an exocyclic methylene provided an 86:14 ratio of *cis*- and *trans*-fused isomers.[61]

E^+

(14)

Table 5 Cyclization to Fused Ring Tetrahydrofurans (Equation 14)

Entry	*Conditions*	*Yield (%)*	*Ref.*
1	I_2, Na_2CO_3, MeCN	>78	53
2	I^- or Br^- + MCPBA	85 and 78	30
3	NBS, acetone	85	54
4	PhSeCl	86–90	55, 57
5	PhSeNPht	83	31,60
6	$Hg(OAc)_2$	>66	59
7	$(ArTe{=}O)_2O$, HOAc	42–53	58
8	$ArTeCl_3$, $CHCl_3$, Δ	90	56

A recent study of group selectivity in iodocyclizations which could form either *cis*-fused tetrahydrofuran or γ-lactone products (equation 15) has shown that the observed selectivity correlates with the conformational bias of each isomeric substrate.[62] One isomer, with no significant conformational bias, produced a mixture of the products upon cyclization of the ester, but gave only the γ-lactone upon cyclization of the carboxylic acid.

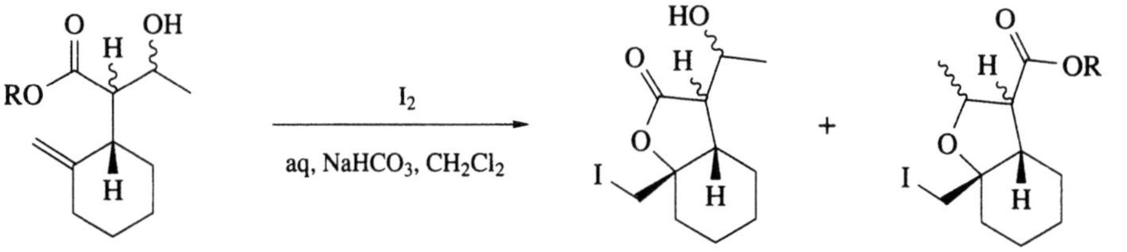

I_2, aq, $NaHCO_3$, CH_2Cl_2 (15)

In cyclofunctionalization of some substituted cyclohexadienylethyl alcohols, products of 1,4-addition were observed (equation 16).[63] However, the stereochemistry of the addition to the diene functionality depended upon the choice of electrophile.

(16)

Conditions	*Yield*	*X*
PhSCl, CH_2Cl_2		α-SPh
PhSeCl, $NaHCO_3$, CH_2Cl_2	>70%	α-SePh
NBS, CH_2Cl_2	94%	β-Br

Few applications of cyclizations to form fused ring δ-lactones or tetrahydropyrans are found. Two consecutive bromolactonizations were used to effect stereoselective dihydroxylation of a cyclohexadienone system in a total synthesis of erythronolide B (Scheme 5).[64a] Iodolactonization of an *N,N*-diethylbenzamide derivative to form a *cis*-fused benzolactone was a key step in a recent synthesis of pancratistatin.[64b] A *cis*-fused tetrahydropyran was produced in good yield by intramolecular oxymercuration as shown in equation (17),[59] although attempts to cyclize a more highly functionalized system have been reported to fail.[65] Formation of a fused ring tetrahydropyran *via* an anti-Markovnikov 6-*endo* selenoetherification has been reported in cases where steric and stereoelectronic factors disfavor a 5-*exo* cyclization to a spirocyclic structure.[38]

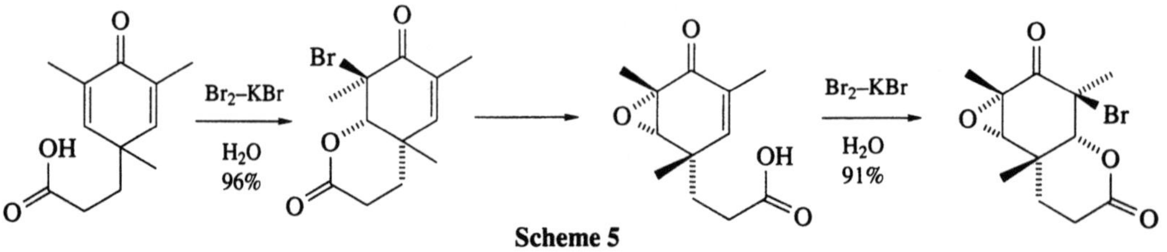

Br_2–KBr, H_2O, 96%; Br_2–KBr, H_2O, 91%

Scheme 5

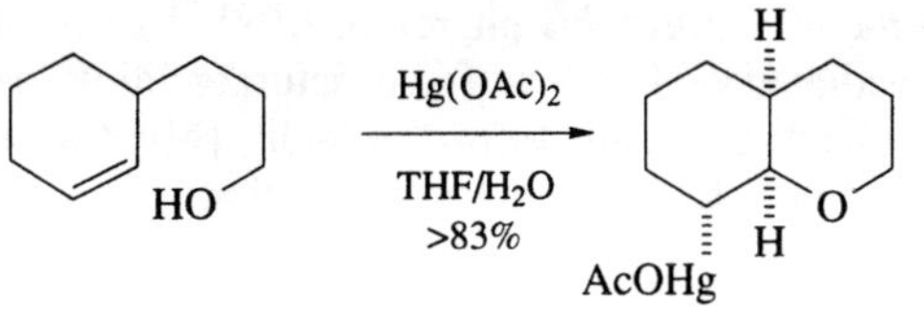

Hg(OAc)$_2$ THF/H_2O >83% HO AcOHg H H O (17)

(ii) Synthesis of bridged ring lactones and ethers

Cyclofunctionalization of cycloalkenyl systems to form bridged ring systems under kinetically controlled conditions generally results in products formed from *exo* cyclization. This type of cyclization has been applied to the synthesis of a variety of natural products and novel unnatural products including twistane,[66] dodecahedrane,[67] reserpine,[68] picrotoxin,[69] gibberellins[70] and antheridiogens,[71] kessanol,[72] incensole,[73] the quassinoid ring system,[74] isolineatin,[75] actinobolin,[76] aklavinone and ε-pyrromycinone,[77] shikimic acid derivatives,[78] *exo*-brevicomin and the *Mus musculus* pheromone,[79,80,81] compactin and mevinolin[82] and glycinoeclipin A.[83,84] Many examples have been covered in earlier reviews of halolactonization,[1a] intramolecular oxymercuration,[85] phenylselenoetherification and lactonization[1k,1m] and cyclizations to form tetrahydrofurans and tetrahydropyrans.[1b] Syntheses of diheterotricyclodecanes by a variety of cyclofunctionalization reaction types have been included in an earlier review (see Scheme 6).[86]

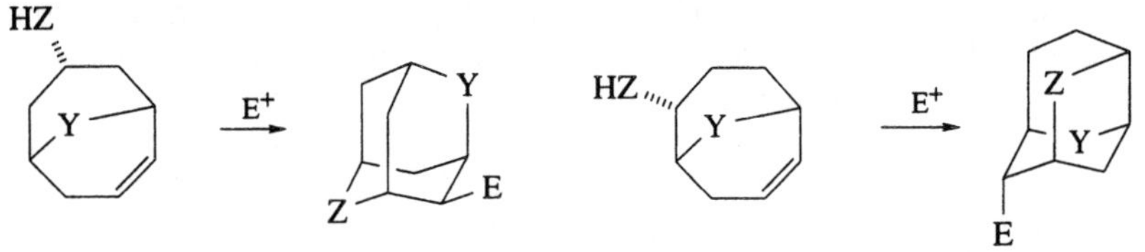

HZ Y E$^+$ Y Z E HZ Y E$^+$ Z Y E

2,6-diheteroadamantane 2,7-diheteroisotwistane

Scheme 6

Some unusual nucleophile functional groups that have been utilized in cyclofunctionalizations to form bridged ring systems include cyclic hemiacetals (equation 18)[75] and epoxides (equation 19).[87] The cyclizations in equation (20) involve a cyclic enol ether as the reactive π-system and have been used in syn-

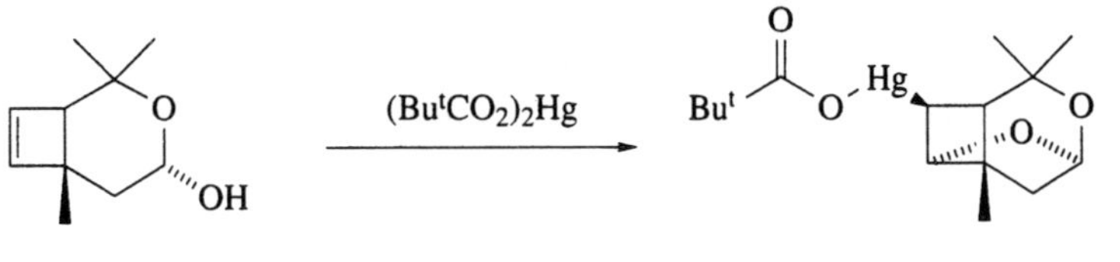

O OH $(Bu^tCO_2)_2Hg$ O Bu^t O Hg O O (18)

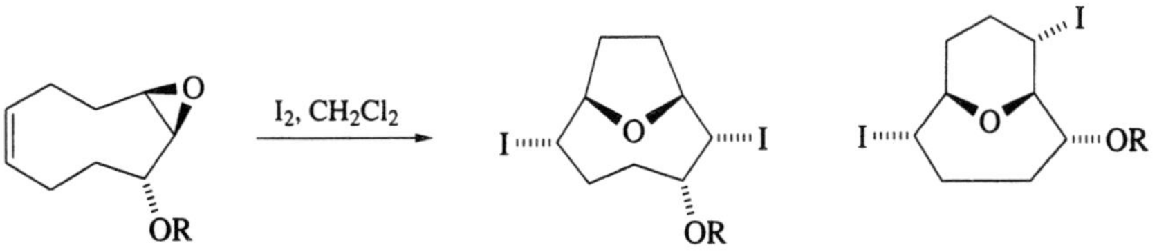

O OR I_2, CH_2Cl_2 I O I OR I O I OR (19)

R = OAc, >94% 40% 60%

R = $SiBu^tMe_2$, >89% 0% 100%

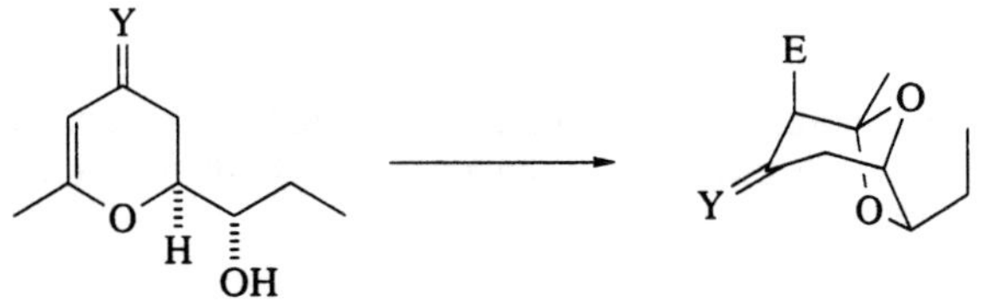

Y O H OH E O Y O (20)

Y = H, H	PhSeCl, Et_3N	91%	E = PhSe
Y = O	Hg(OAc)$_2$		E = HgOAc
Y = H, OH	Hg(OAc)$_2$		E = HgOAc

theses of *exo*-brevicomin and the *Mus musculus* pheromone.[79,80,81] Bridged spiroacetals have also been synthesized by palladium(II)-mediated cyclization of ω-unsaturated diols, as shown in equations (21) and (22).[88,89,90] These bis-heterocyclizations result from the facile β-hydrogen elimination of the initially formed σ-alkylpalladium intermediate to form an enol ether that serves as the substrate for the second cyclization.[50]

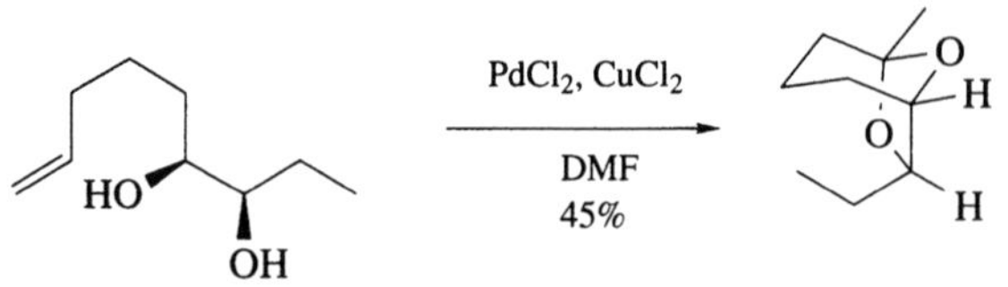
HO
OH
$PdCl_2$, $CuCl_2$
DMF
45%
O
H
O
H

(21)

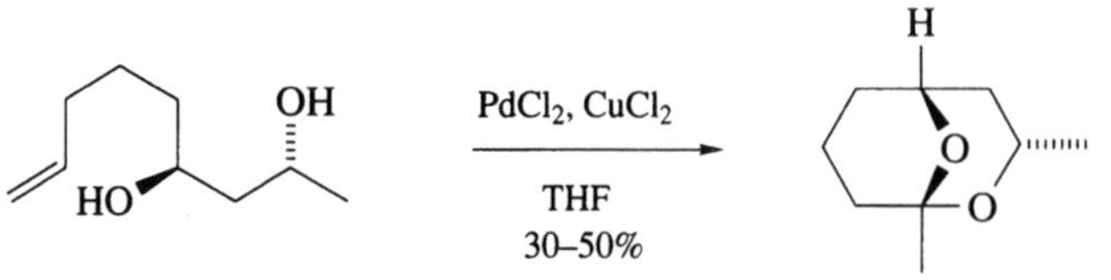
HO
OH
$PdCl_2$, $CuCl_2$
THF
30–50%
H
O
O

(22)

As mentioned earlier in the discussion of cyclizations leading to β-lactones, the β-lactones formed from halolactonization of 1,4-dihydrobenzoic acids readily rearrange to produce bridged ring γ-lactones.[19] In some cases, the substitution pattern favors formation of the γ-lactone even under conditions of kinetic control (equation 23).[20] Synthesis of a variety of γ-lactones by iodolactonization of dihydrobenzoic acid derivatives has been reported recently by Hart (equation 24).[91] Attempted iodolactonization of the acid in the case where R′ = H resulted primarily in an oxidative decarboxylation; however, iodolactonization was effected using the amide derivative.

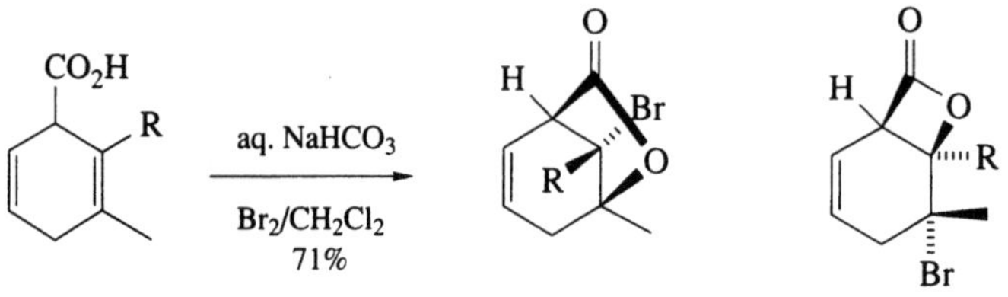
CO_2H
R
aq. $NaHCO_3$
Br_2/CH_2Cl_2
71%
O
H
Br
R
O
O
H
O
R
Br

(23)

R = H	93%	7%
R = Me	>80%	<20%

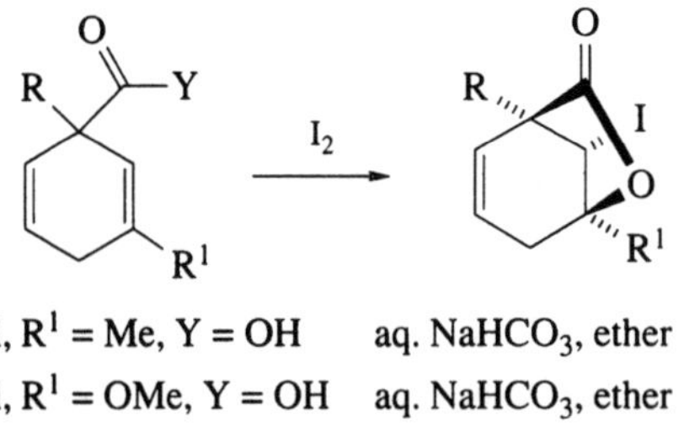
O
R
Y
R^1
I_2
O
R
I
O
R^1

(24)

R = alkenyl, R^1 = Me, Y = OH	aq. $NaHCO_3$, ether	61%
R = alkenyl, R^1 = OMe, Y = OH	aq. $NaHCO_3$, ether	>66%
R = alkenyl, R^1 = SPr^i, Y = OH	aq. $NaHCO_3$, ether	67%
R = alkyl, R^1 = H, Y = NR_2	THF, water	70–94%

The effect of conformational control on the regiochemistry of cyclizations in steroid systems has been studied extensively by Kočovsky. A cyclization which proceeds both in an anti-Markovnikov and 6-*endo* mode is shown in equation (25).[92]

(iii) Cyclization of heteroatom-tethered cyclic systems

The cyclofunctionalization strategy involving a transient 'tether' attached to a preexisting functional group, as outlined in Scheme 2, has been used to introduce a new oxygen functionality onto a ring with

NBS, 10% $HClO_4$

dioxane/water

90% 10% (25)

stereochemical and regiochemical control. This strategy has been particularly useful in stereoselective synthesis of amino sugars and aminocyclitols. Examples of cyclizations of derivatives of cyclic allyl alcohols and amines are given in Table 6. The attachment of the tether to an allylic functional group generally leads to *cis*-fused products upon cyclization. Exceptions are found in the cyclizations of the

Table 6 *O*-Cyclization of Derivatives of Cyclic Allylic Alcohols and Amines

Entry					*Ref.*
1		HgX_2, ~80%			93
2		I_2/THF or $I(collidine)_2^+\ ClO_4^-$, 67%			94
3	R = alkyl Y = OMe	I_2, aq. $NaSO_3$/THF	96% (X = I)		96b, 96c
4	R = alkyl Y = OMe	i, Br_2, $AgBF_4$ ii, H_2O, $NaHCO_3$	79% (X = Br)		96a
5	R = H Y = OEt	$I(collidine)_2^+ClO_4^-$ CH_2Cl_2		61% (X = I)	95a
6	R = Me Y = OEt	$I(collidine)_2^+ClO_4^-$ dioxane, Δ	71% (X = I)		95b
7	R = H Y = OEt	NBS/CH_2Cl_2, 0 °C		60% (X = Br)	97
8	R = H Y = NH_2	PhSeCl, CH2Cl_2 silica gel, 25 °C		76–92%	98a
9	Y = NH_2 Z = O	PhSeCl, CH_2Cl_2	91%		98a
10	Y = OMe Z = NH_2	PhSeCl, CF_3SO_3Ag CF_3SO_3H, silica gel		65%	98b

trans-cycloalkene derivatives shown in entries 9 and 10. Cyclization to a bridged ring product has been observed in the bromocyclization of the phthalimide derivative of cyclohexenylamine.[98c,98d] Cyclic amines which are allylic to an exocyclic methylene also generate *cis*-fused ring systems as shown by equation (26).[96b] The cyclization of carbamate derivatives of cyclic homoallylic amines generates bridged ring systems which are derivatives of 1,3-amino alcohols (equation 27).[99] Preparation of 1,3- and 1,4-diol derivatives *via* cyclofunctionalization of 19-carbamoyloxysteroids has been studied by Kočovsky.[9]

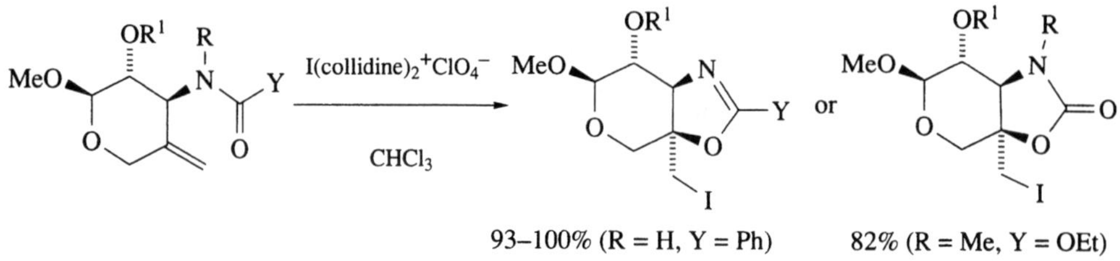
OR¹
R
MeO
N
Y
O
I(collidine)₂⁺ClO₄⁻
CHCl₃
MeO
OR¹
N
Y
O
I
or
MeO
OR¹
R
N
O
I
93–100% (R = H, Y = Ph)
82% (R = Me, Y = OEt)

(26)

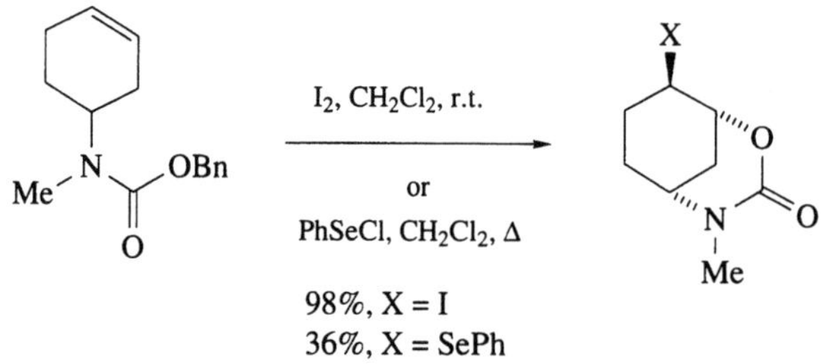
Me
N
OBn
O
I₂, CH₂Cl₂, r.t.
or
PhSeCl, CH₂Cl₂, Δ
98%, X = I
36%, X = SePh
X
O
N
Me
O

(27)

1.9.2.2.2 Other C=C cyclizations

The discussion of 'unconstrained' cyclization substrates in this section includes not only acyclic structures but also cyclic structures in which the vinylic bond which connects the nucleophile to the double bond is not constrained by the ring system. The factors which control the stereochemistry and regiochemistry in cyclizations of this type are much less obvious than those for the cases discussed in the previous section. The major synthetic utility of these reactions, however, comes from cases in which stereochemistry and regiochemistry can be predictably controlled. Thus, the initial discussion in this section will examine these two types of selectivity.

(i) Regioselectivity

A number of cyclizations of acyclic substrates illustrate that a balance of electronic and stereoelectronic factors control the regiochemical course of the reactions. Examples of such balance are found in cases where the preference for *exo* cyclization is opposed by ring strain or by electronic factors (Markovnikov rule). The ring strain generated in 4-*exo* ring closure often results in 5-*endo* cyclization predominating. The 4-*exo* mode of cyclization is normally found only when the substitution pattern also favors this mode and equilibration does not occur. Some examples of cyclizations of homoallylic alcohols are found in Table 1. Cyclizations of acyclic β,γ-unsaturated acids normally result in formation of γ-lactones *via* the 5-*endo* mode (equation 28 and Table 7). Attempts to effect selenolactonization of substrates with substitution heavily favoring 4-*endo* closure (R^3 = alkyl, $R^1 = R^2$ = H) resulted only in the formation of allyl selenides *via* a decarboxylative elimination from the acyclic addition product.[102] The cyclization of 2-vinylcyclohexanols with thallium triacetate produced tetrahydrofuran products of the 5-*endo* type (equation 29).[103]

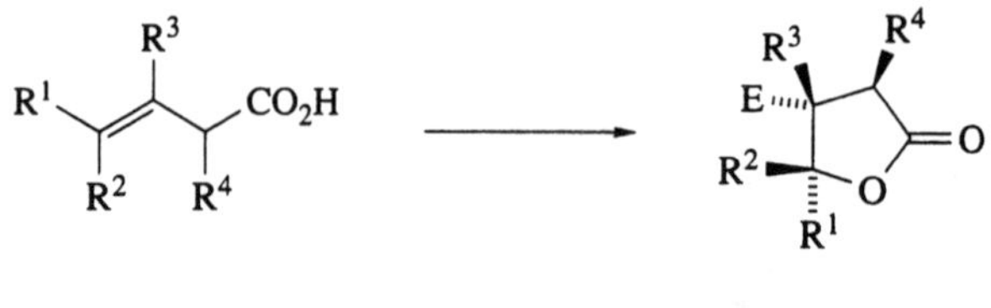
R³
R¹
CO₂H
R²
R⁴
R³
R⁴
E
O
R²
O
R¹

(28)

Table 7 Cyclofunctionalization of β,γ-Unsaturated Carboxylic Acids (Equation 28)

R^1	R^2	R^3	R^4	*E*	*Conditions*	*Yield (%)*	*Ref.*
Me	Me	H	H	I	TlOAc, I_2	65	22
Me	Me	H	H	Br	$Tl_2CO_3/Br_2/CH_2Cl_2$	74	21
Me	Me	H	H	I	$NaHCO_3/I_2/MeOH/H_2O$	95	100
Me	Me	H	H	Br	$TlX/Br_2/CH_2Cl_2$	85–90	100
Ph	H	H	H	I	$NaHCO_3/I_2$/ether	Only product	101
Me	H	Me	H	PhSe	PhSeCl, silica gel	86	102
Et	H	H	H	PhSe	PhSeCl, silica gel	95	102

Tl(OAc)$_3$, AcOH, 80 and 82% (29)

The 5-*exo* and 6-*endo* modes of cyclization are more closely balanced since ring strain is not a factor. Thus, in cyclizations of simple 4-alkenyl systems with a disubstituted internal double bond, the *exo* cyclization mode appears to be favored slightly when there is no Markovnikov preference (equation 30). In some systems, the ratio of 6-*endo* to 5-*exo* products has been found to vary with substrate structure (*e.g.* double-bond geometry), the electrophilic reagent used and reaction conditions (kinetic or thermodynamic control).[104,124b,145a] The 6-*endo* mode can be made to predominate by significantly increasing the electronic bias. The introduction of a strongly stabilizing group at the 6-position leads to preferential 6-*endo* cyclization, as shown in equations (31),[104a,105a] (32)[105b] and (33).[105c]

I(collidine)$_2$$^+ClO_4$$^-$ or NaI, ArCO$_3$H, 18-crown-6; 88% + 12% (30)

NIS, EtCN, –78 °C, 95% (31)

TsOH–Py, CH_2Cl_2 (32)

I_2, MeCN; 34%, Y = H, H; 36%, Y = O (33)

For systems in which 6-*endo* cyclization is favored only by alkyl substitution, the ratio of 5-*exo* to 6-*endo* cyclization appears to be sensitive to both reaction conditions and substrate structure. Cyclizations of a variety of 5,5-dimethyl-4-pentenols (equation 34) are shown in Table 8. Although the use of 2,4,4,6-

tetrabromocyclohexa-2,5-dienone (TBCD) promoted the 6-*endo* cyclization in some systems (Table 8, entries 4 and 5),[104a,104b,108] recent cyclization studies directed toward synthesis of oxacyclic squalenoids found 5-*exo* cyclization to predominate under all conditions examined (entries 6 and 7).[109,110]

(34)

Table 8 Cyclization of Some 4-Pentenols (Equation 34)

Entry	*R, R*[1]	*Conditions*	*Product yield (5-ring:6-ring) (%)*	*Ref.*
1	Me, CH═CH2	NBS/CCl_4	85	106
2	Me, steroid	NBS/CCl_4	Only product	107
3	Me, C≡C	NBS/CCl_4	Major:minor	108
4	Me, C≡C	TBCD	20:80	108
5	H, alkyl	TBCD	20:58	104a
6	Me, pyranopyran	$TBCD/MeNO_2$	61:31	109
7	Me, pyranopyran	NBS/CH_2Cl_2	62:22	110
8	H, phenyl	$Pd(OAc)_2$, $Cu(OAc)_2$	Trace:45	104f
9	H, Me	PhSCl, Pr^i_2NEt	42:42	16b

An interesting method for the generation of anti-Markovnikov products has been described by Bartlett.[111] Cyclization of γ-alkenyl alcohols of the type shown in equation (35) with thallium triacetate in a nucleophilic solvent generates *trans*-2,5-disubstituted tetrahydrofurans. The reaction is considered to proceed *via* an initial 6-*endo* cyclization to a tetrahydropyran, which undergoes heterolytic dethallation with concomitant 1,2-oxygen migration through a bridged oxonium ion.[104a,111] The products from this type of cyclization are equivalent to those expected from use of 'RO^+' as electrophile. Examples of *exo* mode cyclization followed by ring expansion have been described also.

(35)

72% (solvent = AcOH, R^1O = AcO)

73% (solvent = acetone/H_2O, R^1O = HO)

The lactonization of 5,7-dienoic acids with selenium and sulfur electrophiles generates δ-lactones by regioselective 1,4-addition to the diene (equation 36).[52a] With substitution at the 8-position, reaction with PhSeCl or PhSCl gave a 1:1 mixture of rapidly interconverting 1,2- and 1,4-addition products. This interconversion, presumably *via* a 1,3-rearrangement,[52b] was not observed with the alkylthio adduct. An example of conjugate cyclization of an enyne system to form an allenyl product has also been reported (equation 37).[112]

(36)

R	*Conditions*	*X*	*Yield (%)*
H	PhSeCl, –78 °C, dark	PhSe	95
H	PhSCl, 25 °C	PhS	82
Me	$MeO_2CCH_2CH_2SCl$	RS	69–72

NBS, 62% (37)

(ii) Stereoselectivity

The major use of cyclofunctionalization in stereochemical control has been in the area of relative asymmetric induction; that is, control of stereochemistry in the introduction of a new stereogenic center by stereogenic center(s) present in the substrate. Several types of structural features have been found to provide high levels of stereocontrol in cyclofunctionalizations to five- and six-membered rings.

Perhaps the most important structural feature that has been used extensively in recent times for controlling stereochemistry is the allylic oxygen functionality. Cyclization of derivatives of the type shown in equation (38) with a variety of electrophiles leads to stereoselective formation of the '*cis*' five- and six-membered ethers and lactones (Table 9).[113] Although different theoretical approaches have been taken to rationalize these examples of 1,2-relative asymmetric induction,[113,115e,117] each analysis leads to the conclusion that the cyclizations involve a transition state conformation with the allylic oxygen functionality oriented toward the alkene bond, either fully eclipsed (**2**; OR-in-plane conformation)[113,115e] or slightly staggered (**3**; inside alkoxy conformation),[117] and that electrophilic attack occurs from the side opposite the alkyl chain. The same *cis* stereoselectivity is found in the iodolactonizations of *N,N*-dimethylamide derivatives related to (**1**) ($n = 1$).[114b] Recent results have shown that an allylic fluorine substituent exerts even higher stereochemical control in this type of cyclization, a result consistent with calculations that the major control factor is orientation of the electronegative substituent in the plane of the alkene.[115e] Selectivity of the same type has been found in mercuricyclizations of substrates with an allylic *N*-benzylamine substituent, forming α-*C*-glucosides of *N*-benzyl-D-glucosamine.[118]

E^+; n = 1 or 2; $E^+ = I_2$, NBS, PdX_2, HgX_2, PhSeOTf (38)

R = H, alkyl, SiR_3, Ac; R^1 = H, alkyl, CO_2R; R^2 and R^3 = H or alkyl; R^4 = H, alkyl, benzyl; Y = H,H or O

Table 9 Control of Stereochemistry by Allylic Oxygen Substituents (Equation 38)

Product type	*Leading refs.*
γ-Lactones	104j, 114
Tetrahydrofurans and *C*-furanosides	104e, 115
Tetrahydropyrans and *C*-pyranosides	104e, 116

(**2**)
OR-in-plane conformation

(**3**)
'inside alkoxy' conformation

The stereochemical control exerted by the allylic oxygen functionality generally dominates control by other stereogenic centers in the ring, although substitution at other centers may affect reaction rate, the level of stereocontrol or regioselectivity.[115b,115e] One example of a dramatic lack of stereocontrol is found in the 2,4-dimethyl-3-hydroxy-4-pentenoic acid system with the 2-methyl group *syn* to the allylic hydroxy group (*cis*:*trans* = 42:58).[114a,115e] Although stereoselectivity in some systems is not affected by the change of the allylic substituent from OH to OR, some cyclizations with palladium salts show lower selectivity with the ethers. This change could be a result of direct complexation of the electrophilic metal with the hydroxy group.[116c]

Several studies indicate that the level of stereocontrol is greater when R^1 is an electron-withdrawing group,[104d,104e] and can vary with the electrophile used.[104e,115h] In particular, the variable selectivity observed with phenylselenyl reagents appears to depend upon reaction conditions and substrate structure.[104d,104e,115g,115h]

Stereocontrol opposite to that shown in Table 9 occurs when the double bond contains a nonhydrogen substituent *cis* to the carbinol carbon (equation 39). In this system the OR-in-plane conformation is disfavored by $A^{1,3}$ strain. In cases with a secondary allylic oxygen substituent, cyclization through an H-in-plane conformation gives the *trans* product with high selectivity.[113a,115e] Several examples demonstrating this type of stereocontrol have been described.[104e,114a,115b,115h,119]

E^+, n = 1 or 2 (39)

R^1 and R^2 = H or alkyl; R^3 = alkyl, OR, CO_2Et

The stereochemistry of products from cyclizations of systems with an allylic alkyl group and a *cis* substituent on the alkene is also controlled by steric interactions. Cyclization through an H-in-plane conformation generates the new stereogenic center *trans* to the allylic group (equation 40).[1n,115h,120,124b] An exception has been reported in a cyclization effected with phenylselenenyl chloride.[121] If the *cis* alkene substituent (R^2) is hydrogen, $A^{1,3}$ steric interactions are absent; the stereoselectivity in kinetically controlled cyclizations is generally low with electrophiles such as halogen, mercury(II) ion or phenylselenenyl chloride, and the *cis* isomer predominates in many cases (*cis:trans* ratios varying from 3.8:1 to 2:3).[4,104j,115e,115h,115j,120b,122] Exceptions include the iodolactonizations of some 2-alkyl-3-methylpentenoic acids, which have been found to proceed with very high *cis* 1,2-relative asymmetric induction (7.9–30:1) irrespective of the configuration at C-2.[123]

E^+, n = 1 or 2 (40)

R and R^2 = alkyl; R^1 = H or alkyl; Y = O or H, H

The effect of the nature of the electrophile on the stereoselectivity of reactions with substrates containing a terminal alkene and an allylic substituent is dramatically illustrated by some recent results with palladium electrophiles.[124] Cyclizations of 3-methyl- or 3-phenyl-5-hydroxyalkenes with palladium catalysts proceed with high selectivity (≥9:1) for the 2,3-*trans* isomer (equation 41).[50,124] It is suggested that the steric interactions of the palladium–alkene complex affects the stereochemistry of these cyclizations. In some related cyclizations to form tetrahydropyran products (equation 42 and Table 10), reaction with iodine in the presence of sodium bicarbonate gives a different major diastereomer from cyclization with mercury(II) trifluoroacetate or palladium chloride.[125]

cat. $PdCl_2$, $CuCl_2$; CO, MeOH (41)

90% (R^1 = Me, R^2 = H)	90%	10%
88% (R^1 = H, R^2 = Me)	87%	13%
87% (R^1 = R^2 = H)	only isomer observed	

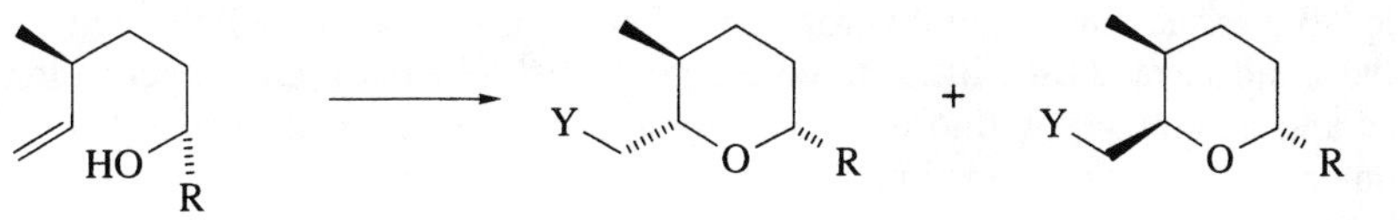

(42)

Table 10 Stereoselectivity in Cyclizations to 2,3,5-Trisubstituted Tetrahydropyrans (Equation 42)

Entry	*Reagent*	*Y*	*Product ratio*	*Ref.*
1	$PdCl_2$, $CuCl_2$, MeOH, CO	CO_2Me	7.9:1	125a
2	$Hg(OTFA)_2$, MeCN, $PdCl_2$, LiCl, $CuCl_2$, CO, MeOH	CO_2Me	2.5–3.0:1	125a
3	I_2, $NaHCO_3$	I	1:2.9	125b

One general solution to low levels of 1,2-asymmetric induction in iodolactonization reactions of unsaturated acids with an allylic methyl group was developed by Bartlett, who demonstrated that cyclization in a nonbasic solvent (CH_3CN) and in the absence of a proton scavenger allowed for equilibration to the thermodynamically favored *trans* product (equation 43).[1p,120b,122a] The *trans* isomers of both γ- and δ-lactones are generally preferred under these conditions by ratios of greater than 10:1. Similar equilibration has been effected with mercury(II) salts,[120b] phenylselenenyl triflate[126a] and *N*-(phenylseleno)phthalimide in the presence of $TiCl_4$.[126b] The low stereoselectivity in cyclizations with other phenylselenenyl reagents (*e.g.* equation 1) may be due to rapid nonselective addition of electrophile to the starting alkene combined with a lack of reversal of intermediates (**D**) or (**C**) back to starting alkene (Scheme 3).[4,120b] Cyclization of the ester derivatives (Scheme 7) with iodine also generates the thermodynamically more stable product even though the final lactone product would not be considered to be in equilibrium with starting ester.[8a] In this case, equilibration is achieved since the dealkylation of the cyclic intermediate (**4**) is the slow step, and equilibration of all prior intermediates can occur (*cf.* **G** to **H** in Scheme 3). Numerous examples of this strategy have been reported in previous reviews,[1p,127] and the method has found widespread use in synthesis.

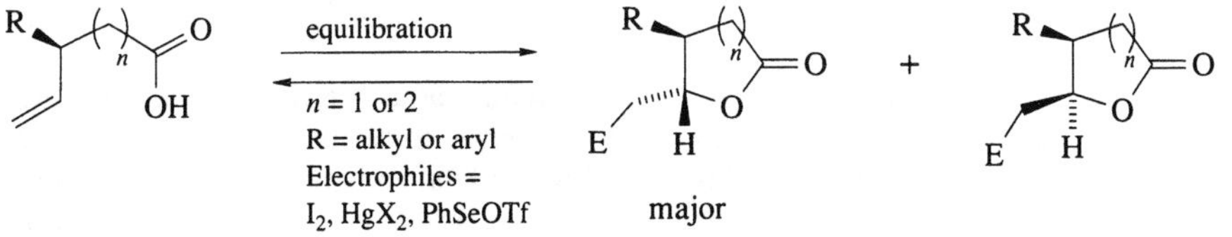

equilibration

n = 1 or 2
R = alkyl or aryl
Electrophiles =
I_2, HgX_2, PhSeOTf

major

(43)

slow step

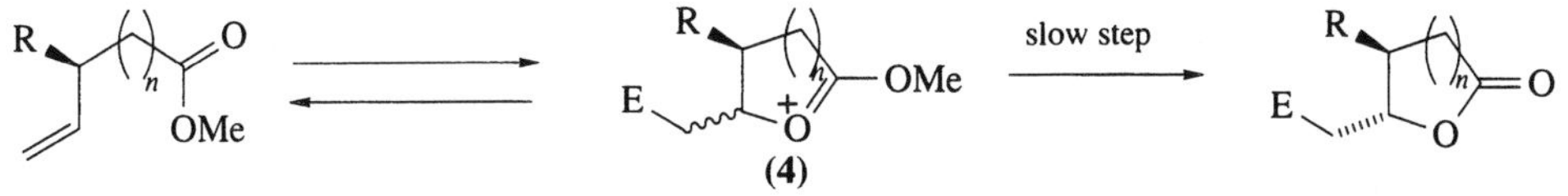

(**4**)

Scheme 7

Steric effects similar to those shown in equations (39) and (40) are found when the substitution pattern leads to tetrahydropyran systems through 6-*endo* cyclization. Cyclizations of systems with an allylic oxygen and a *syn* alkene substituent give products rationalized by cyclization through H-in-plane conformations as shown earlier in equations (31) and (32).[105,128] Examples with allylic methyl substitution have been reported also.[104a]

A few examples of high 1,3-asymmetric induction under kinetically controlled conditions have been reported.[1p] A recent study has shown that iodoetherification proceeds with high stereoselectivity on systems with an α,β-unsaturated ester as the π-participant and an electronegative homoallylic substituent (equation 44 and Table 11).[129] When the homoallylic substituent was a methyl group, the major diastereomer was the *cis* isomer. These results were rationalized using AM1 calculations of chair-like transition state models. The calculations predict an H-in-plane conformation for the iodonium ion or π-complex and either an *anti* or *gauche* conformation between the substituent, Y, and the hydroxy group (structures **5** and **6**). The normal preference for an equatorial methyl leads to cyclization through transition state (**5**) to produce the *cis* product (Table 11, entry 6). However, when Y is an electronegative substituent, the calculations predict a preference for the conformation where the substituent is *gauche* to the hydroxy group, which acquires positive charge as the C—O bond is formed (**6**; Table 11, entries 1–5). A selenoetherification of a related system with an internal (*Z*) double bond, but without an ester functionality, showed moderate stereoselectivity (3:1) for the product predicted by transition state (**6**).[130] Iodocyclizations to 2,3,5-trisubstituted tetrahydrofurans of systems with a homoallylic hydroxy group and a terminal double bond have shown little stereoselectivity,[115k,115m] but mercuricyclizations with a terminal double bond and a homoallylic fluorine have shown good stereoselectivities (4.5:1 to 20:1).[131a] The effect of a homoallyic sulfoxide substituent on the stereochemistry in mercuricyclizations of secondary alkenols has been reported recently.[131b]

I_2, $NaHCO_3$, solvent (44)

Table 11 Relative Asymmetric Induction by Homoallylic Substituent (Equation 44)[129]

Entry	*(E) or (Z)*	*Y*	*Solvent*	*Product ratio*
1	*E*	OH	THF	1:4.5
2	*E*	OH	Et_2O	1:8.6
3	*Z*	OH	Et_2O	1:11
4	*E*	OMe	THF	1:4.6
5	*E*	F	THF	1:6.3
6	*E*	Me	THF	3.6:1

(5) **(6)**

Iodolactonizations of *N*-acyl derivatives of 2-amino-4-pentenoic acids produce the *cis* γ-lactones selectively (equation 45).[132] These lactonizations have been proposed to result from thermodynamic control,[129] although chelation of the bromonium ion by the nitrogen has been proposed also.[132b] Formation of δ-lactones by 6-*exo* iodolactonization of 3-(*N*-acylamino)-5-hexenoic acids proceeds with little selectivity.[133]

NBS, THF (45)

R = $NHCO_2R$, NPht, NHTs 6–8.8:1

High levels of 1,3-asymmetric induction also have been obtained in some cyclizations to γ-lactones under conditions of thermodynamic control (equation 46 and Table 12).[114b,134] Cyclizations of carboxylic acid esters result in equilibration related to that shown in Scheme 7, and *cis* products predominate (Table 12, entry 1). Alternatively, cyclizations of *N,N*-dimethylamide derivatives produce the *trans* isomer with very high selectivity (entries 2–4).[114b,134a] A related cyclization of an *N*-acyloxazolidinone derivative has been reported to proceed with high stereoselectivity.[134b] The *trans* selectivity is attributed to equilibration of the initial cyclic product, with the steric interactions between an amide methyl group and the β-substituent ($A^{1,3}$ strain) disfavoring the *cis* isomer. Phenylselenenyl triflate and iodine cyclizations of the carboxylic acids gave low selectivity (entries 5 and 6).[114b,126]

Examples of 1,3-asymmetric induction in cyclizations to δ-lactones have been observed. Iodolactonization of 3-methyl-5-hexenoic acid to a δ-lactone under equilibrating conditions showed reasonable stereoselectivity (6:1 *cis:trans*).[120b] Recent studies have examined the formation of δ-lactones from cyclization of 5-hexenoic acids with a homoallylic oxygen substituent at C-3.[135] Selenolactonization of 3-hydroxy-5-hexenoic acid under conditions of kinetic control provided the *trans* lactone in modest yield (40%) and high stereoselectivity.[135b] Equilibrating conditions led to a slight preponderance of the *cis*

(46)

Table 12 Stereoselectivity in Cyclizations to α-Methyl-γ-lactones

Entry	*R*	*Y*	*Conditions*	*Yield (%)*	*trans:cis*	*Ref.*
1	Alkyl	OMe	3 equiv. I_2, MeCN		1:9	134a
2	Alkyl	NMe_2	3 equiv. I_2, MeCN		*Trans* only	134a
3	H	NMe_2	NBS	75	>99:1	114b
4	Me	NMe_2	NBS	74	94:6	114b
5	H	OH	3 equiv. I_2, MeCN	92	32:68	114b
6	H	OH	PhSeOTf	85	3:4 or 4:3	126

isomer in higher yields (65%). Iodolactonizations of the corresponding 3-silyloxy derivatives under kinetic control were also selective (5.5:1 *trans:cis* with a triisopropylsilyl derivative). Selenolactonization of a related system with an internal double bond (*Z* configuration) showed little selectivity.[135a]

Examples of high levels of 1,4-asymmetric induction in kinetically controlled cyclofunctionalizations are rare. Cyclizations of 5-hydroxyalkenes to form 2,5-disubstituted tetrahydrofurans (equation 47) proceed with low selectivity in most cases.[1p] Exceptions shown in Table 13 are the palladium-catalyzed cyclization to a *trans* 2-vinyl system (entry 1)[50,104f] and cyclization of a 2-phenyl system with mercury(II) chloride to give a preponderance of the *cis* isomer (entry 2), presumably through equilibration.[136] Equilibration with substituents other than phenyl (Me or *t*-butyl) resulted in much lower selectivity.[136,138]

(47)

Table 13 Stereoselectivity in Cyclofunctionalization to 2,5-Dialkyltetrahydrofurans (Equation 47)

Entry	R^1, R^2, R^3	*Conditions*	R^4	*Yield (%)* *cis*	*trans*	*Ref.*
1	Ph, Me, H	$Pd(OAc)_2$, $Cu(OAc)_2$	$CH{=}CH_2$	0	40	104f
2	Ph, H, H	$HgCl_2$	CH_2HgX	56	4	136
3	$CHMe_2$, H, H	I_2, MeCN	CH_2I	18	70	137
4	$CHMe_2$, H, DCB	I_2, MeCN	CH_2HgX	95	0	137

Bartlett developed a strategy for the highly stereoselective synthesis of *cis* 2,5-disubstituted tetrahydrofurans involving the use of bulky ether derivatives for the cyclization, such that dealkylation of the oxonium ion is the slow step.[137] This strategy results in a large *cis:trans* ratio (compare Table 13, entries 3 and 4) because steric interactions between the bulky substituent on oxygen and the substituents at C-2 and C-5 cannot be avoided in the *trans* intermediate.[137] The success of this strategy requires a substituent which is bulky, but does not prevent cyclization, and which is cleaved faster than either of the ring groups in the dealkylation step. The 2,6-dichlorobenzyl (DCB) group has been found to be useful in a number of such reactions.[115h,137,139] Application of the slow dealkylation strategy to stereoselective syntheses of *trans* 2,5-disubstituted tetrahydropyrans was not successful.[15] Also, cyclizations of 2-alkyl-5-hexenoic acids to δ-lactones proceed with very low levels of 1,4-asymmetric induction under conditions of either kinetic or thermodynamic control.[134a,140]

An interesting example of the apparent dependence of stereoselectivity on the rate of breakdown of the initial cyclization intermediate has been observed in cyclizations with isoxazolines as the nucleophilic functionality (equation 48 and Table 14).[141] The stereochemistry of products from cyclizations to form tetrahydrofuran products varies significantly upon changing the 3-substituent of the isoxazoline ring (Table 14, entries 1–3). The exact factors controlling these variations have not been determined, but the stereoselectivity may be synthetically useful. Particularly noteworthy is the high selectivity for the less stable *trans* isomer of 2,3-disubstituted tetrahydropyrans; this type of selectivity is not obtainable by other strategies.[15] Some cyclizations with Br_2 and PhSeBr also have been reported.[141a]

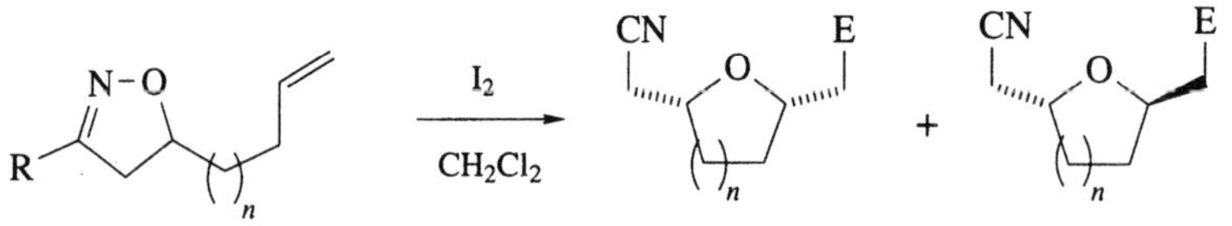

(48)

Table 14 Cyclic Ethers from Cyclofunctionalization of Unsaturated Isoxazolines (Equation 48)[141]

Entry	*n*	*R*	*Yield (%)*	*Ratio (cis:trans)*
1	1	Ph_3C	60	1:4
2	1	Me_3C	59	1:1
3	1	Me_3Si	40	6.6:1
4	2	Ph_3C	82	1:9
5	3	Ph_3C	80	1:3

The stereoselectivity in cyclizations of vinyl ether systems to form furanosyl disaccharides is controlled by both the side chain substituent (1,4-asymmetric induction) and the configuration of the existing pyranoside linkage (equation 49).[142] Substrates with no side chain substituent show no stereoselectivity.

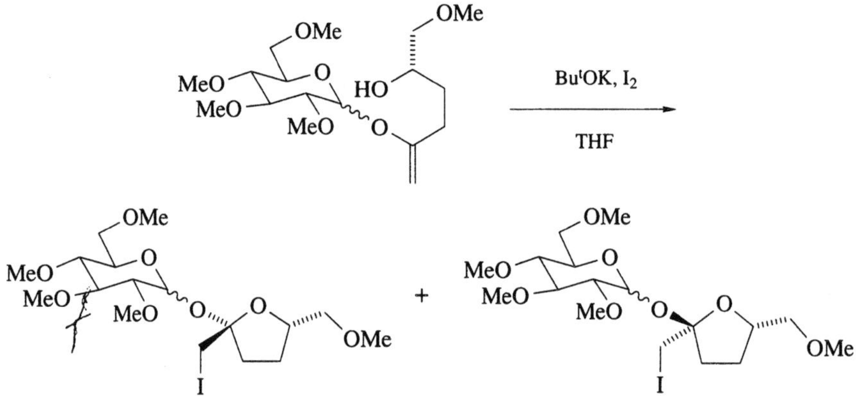

(49)

β-anomer 8:1

α-anomer ≥40:1

Formation of tetrahydropyran ring systems through 6-*endo* or 6-*exo* cyclization of substrates which lack 1,2-interactions of the type shown in equations (39) and (40), generally proceed with moderate selectivity for the diastereomer that would result from cyclization through chair conformations with equatorial substituents.[7,110,131b,142,143]

The levels of 1,5-asymmetric induction in the palladium-catalyzed alkoxy-carbonylations of alkenols to form 2,6-disubstituted tetrahydropyrans have been shown to be quite reasonable (Table 10 and equation 50).[144] Recent studies have shown that cyclization with palladium(II) acetate in DMSO in the absence of CO results in controlled β-hydride elimination to form vinyl-substituted tetrahydropyrans with high levels of 1,4- and 1,5-asymmetric induction (equation 51).[144b]

Although several of the stereoselective cyclizations discussed above have more than one chiral center which could affect the diastereoselectivity of the reaction, they have been discussed in terms of the factor which seems the most important. High stereoselectivity in cyclizations of some highly substituted systems results from cooperative effects of the substituents. Some examples of high stereoselectivity in cyclizations used in syntheses of macrolide antibiotics appear to result from a combination of substitution pattern *and* the (*Z*)-alkene configuration of the substrate. Examples include the formation of a 2,3,5-trisubstituted tetrahydrofuran in a synthesis of monensin[145a] and of a 2,2,5-trisubstituted tetrahydrofuran in a synthesis of ionomycin.[145b] Studies[7] on intramolecular oxymercuration to form substituted tetrahydropyrans (equation 52), conducted as part of a total synthesis of antibiotic X-206, demonstrated that formation of a mercuronium ion was reversible, and that extremely high levels of stereocontrol can result from additive contributions of several control elements, including extraannular allylic oxygen substituents.[113b]

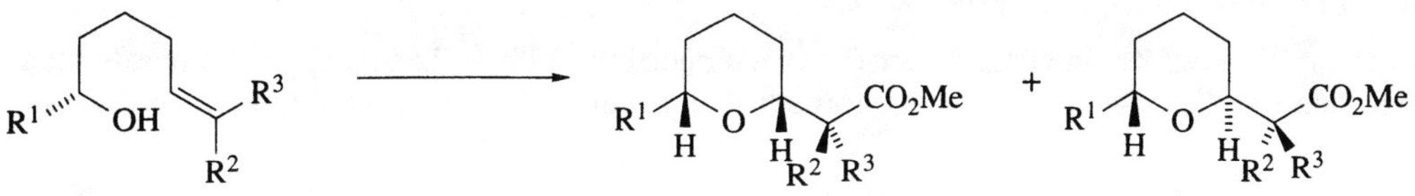

(50)

R^1	R^2	R^3	*cis:trans ratio*
Me	H	H	≈20:1
Bu^tCH_2	H	H	6:1
Bu^i	Me	H	83:10
Bu^i	H	Me	≥97:3

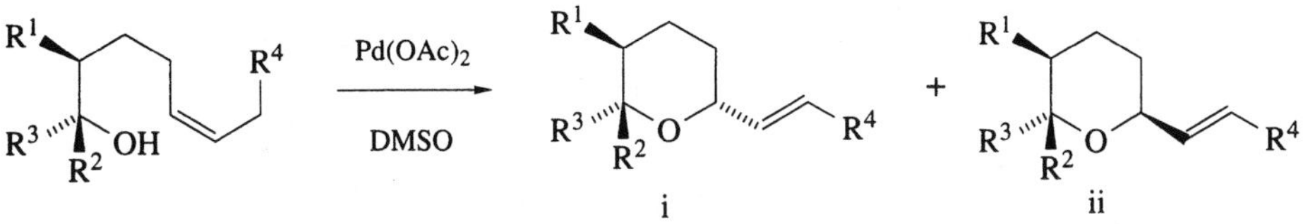

Pd(OAc)$_2$, DMSO (51)

R^1	R^2	R^3	R^4	*Ratio (i:ii)*
H	Me	H	Me	90:5
Me	H	H	Me	98:2
Me	H	Me	Me	94:3

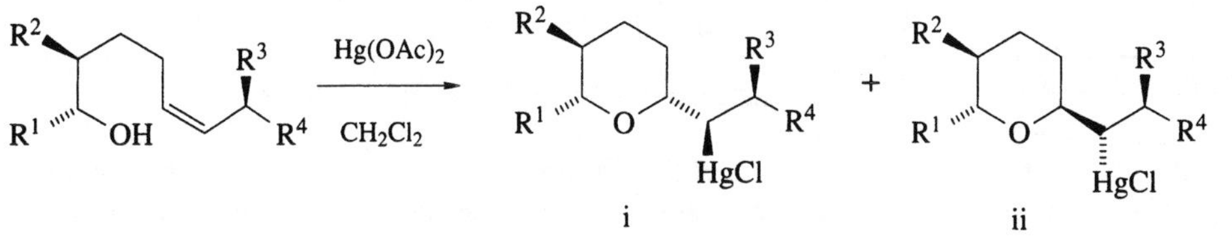

Hg(OAc)$_2$, CH_2Cl_2 (52)

R^1	R^2	R^3	R^4	*Ratio (i:ii)*
H	Me	Me	Me	85:15
secondary alkyl	Me	Me	Me	>97:3
H	H	OBn	secondary alkyl	95:5
H	Me	OBn	secondary alkyl	>97:3

Examples of 5-*exo* cyclizations in which the hydroxy group is constrained by a five-membered ring are shown by equation (53). The stereoselectivity in these cyclizations ranges from quite high in iodocyclization (9:1 in H_2O and >20:1 in CH_2Cl_2) to only moderate (2–3:1) with mercury and phenylselenenyl chloride.[1p] This type of cyclization has been widely applied to synthesis of prostacyclins.[146] Similar cyclizations with dimethyl(methylthio)sulfonium fluoroborate have been reported recently.[35]

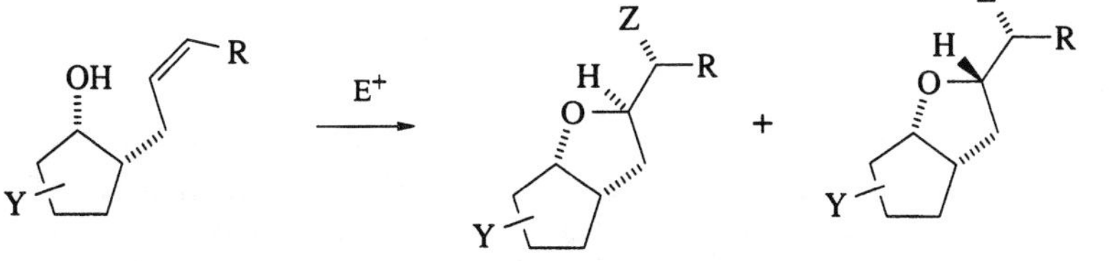

(53)

E^+ = I_2, Hg(OAc)$_2$, NBS, PhSeCl, DBDMH (1,3-dibromo-5,5-dimethylhydantoin)

(iii) Cyclization of heteroatom-tethered systems

Cyclizations of substrates in which internal oxygen nucleophiles have been attached to unsaturated alcohols provide a method for stereoselective formation of acyclic 1,2- and 1,3-diol functionalities, as outlined in Scheme 2.

Cyclizations of chloral hemiacetal derivatives of cyclic allyl alcohols were regio- and stereo-selective (Table 6, entry 1), but a mixture of regioisomers was obtained from analogous derivatives of acyclic allyl alcohols with a nonterminal double bond.[93] Hemiacetal derivatives of allyl alcohols with a terminal vinyl group have been cyclized with mercury(II) acetate to give acetal derivatives of *threo* 1,2-diols with moderate selectivities (equation 54 and Table 15, entries 1 and 2).[147] Moderate to excellent stereoselectivity has been observed in the iodocyclizations of carbonate derivatives of allyl alcohols (entries 3–5).[94a] The currently available results do not provide a rationale for the variation in observed stereoselectivity.

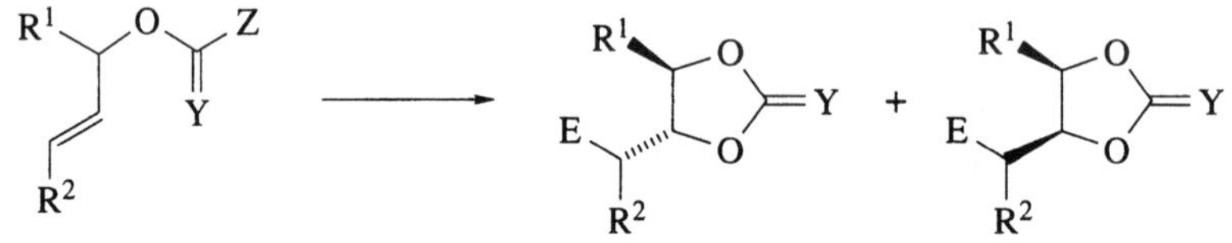

(54)

Table 15 *O*-Cyclizations of Derivatives of Acyclic Allylic Alcohols (Equation 54)

Entry	R^1, R^2	*Y*	*Z*	*Conditions*	*E*	*Ratio (trans:cis)*	*Ref.*
1	Et, H	H, CCl_3	OH	i, $Hg(OAc)_2$; ii, $NaBH_4$	H	81:19	147
2	Bu^t, H	H, CCl_3	OH	i, $Hg(OAc)_2$; ii, $NaBH_4$	H	91:9	147
3	Pr, H	O	OLi	I_2/THF, 12 h	I	60:40	94a
4	Pr, H	O	O	I_2/THF, 12 h	I	80:20	94a
5	Bu^n, Me	O	OLi	I_2/THF/ 12 h	I	93:7	94a

Cyclizations of derivatives of homoallylic alcohols have proven useful for 1,3-asymmetric induction (equation 55 and Table 16). This approach was initially examined with phosphate esters,[8b,151] but the resistance of the cyclic phosphate ester functionality to hydrolytic removal has limited the applications of this approach. Cyclic carbonates, which are readily cleaved, can be prepared by cyclization of *t*-butyl carbonate[148a,149] lithium carbonate,[94a,148b,148c] and carbamate derivatives of homoallylic alcohols,[125a,150] although the stereoselectivity is somewhat higher in the phosphate cyclizations.[148a] In most systems, the stereoselectivities obtained from the two types of carbonate derivatives did not differ significantly, although the *t*-butyl carbonate reactions were significantly slower, and the stereoselectivity of the lithium carbonate reaction was greatest at low temperature (Table 16, entries 1 and 3).[148c] Although the stereoisomer ratio did not change with time in cyclizations of *t*-butyl carbonate systems with a terminal vinyl group, cyclization of a *t*-butyl carbonate derivative of a 2-alkyl-4-hydroxyalkene was shown to give an extremely high stereoisomer ratio (>50:1) under mild conditions, but reaction for long periods at higher temperature (equilibration) resulted in a significantly lower selectivity (5:1; Table 16, entries 5 and 6).[149]

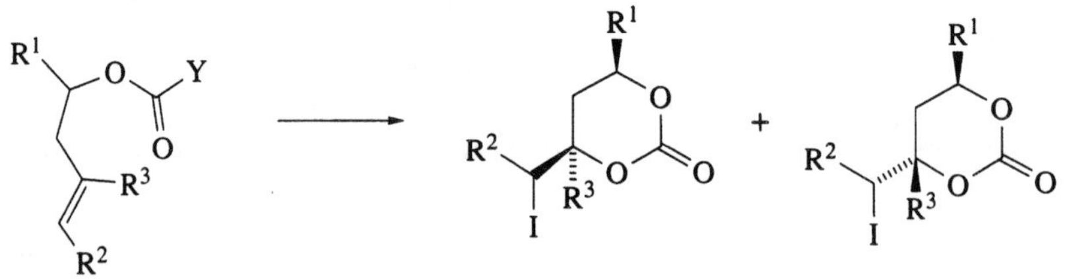

(55)

The synthesis of amino alcohol derivatives by cyclization of derivatives of allylic and homoallylic amines has received considerable attention recently. Early studies by McManus and Pittman[152a] demonstrated that allylic amides could be converted into 2-oxazolines by treating them with concentrated sulfuric acid and drowning the reaction mixture in cold dilute base. The results suggested that reaction occurred through carbocation intermediates, and cyclizations of allylic systems with substitution favoring 6-*endo* cyclization gave tetrahydrooxazine products.[152b] Allylic amides (equation 56) have also been cyclized with halogen,[153] selenium,[154] tellurium,[155] and sulfur[156] electrophiles. The halogen electrophile of choice appears to be *N*-iodosuccinimide.[153a,157] The regiochemistry is predicted by the Markovnikov rule, with 1,2-disubstituted alkene systems giving a mixture of oxazoline and tetrahydrooxazine products.[153b] The stereoselectivity of cyclization is variable, with 4,5-dialkyloxazolines obtained in *trans:cis*

Table 16 *O*-Cyclizations of Derivatives of Acyclic Homoallylic Alcohols (Equation 55)

Entry	R^1, R^2, R^3	*Y*	*Conditions*	*Ratio (cis:trans)*	*Ref.*
1	Alkyl, H, H	OLi	$I_2/THF/CO_2$, –78 °C to r.t.	19:1	94a, 148
2	Alkyl, H, H	OBu^t	I_2/MeCN, –20 °C	10:1	148a
3	Et, Me, H	OLi	$I_2/THF/CO_2$, 0 °C	4:1	148a
4	Et, Me, H	OBu^t	I_2/MeCN, –20 °C	4:1	148a
5	Alkyl, H, Me	OBu^t	I_2/EtCN, –40 °C, 30 min	50:1	149
6	Alkyl, H, Me	OBu^t	I_2/EtCN, –10 °C, 14 h	5:1	149
7	Alkyl, alkyl, H	NH_2	I_2/ether/saturated $NaHCO_3$, r.t.	10–17:1	150, 125a
8	Alkyl, alkyl, H	NMe_2	I_2/ether/aq. $NaHCO_3$, r.t.	>20:1	125a

ratios varying from 80:20 to 24:76 (equation 57).[153a,157] Examples of tetrahydrooxazine formation by 6-*exo* cyclization of homoallylic *N*-acylamines have also been reported.[156,158] The product from an iodocyclization in aqueous THF was shown to result from rearrangement of the initial product (equation 58).[158b] A related cyclization of the amide from a symmetrical homoallylic amine (R = allyl) and a chiral acid gave the pyrrolidine derivative in only 7–26% *ee*.[159] A synthesis of 5-methylene-2-oxazolines by cyclization of amide derivatives of propargylamine has been reported also (equation 59).[160]

The cyclization of carbamate derivatives of unsaturated amines has proven synthetically useful. Cyclizations of carbamates of allylamines containing a terminal vinyl group give oxazolidinone products (equation 60 and Table 17, entries 1 and 2).[99,161] Bromocyclizations of systems with a di- or tri-substituted alkene often give mixtures of oxazolidinones and tetrahydrooxazinones,[163a] while cyclization of an *N*-cinnamyl carbamate with phenylsulfenyl chloride gave only the oxazolidinone product.[163b,163c] The stereochemistry of the cyclization of primary carbamates of either allylic or homoallylic amines is low

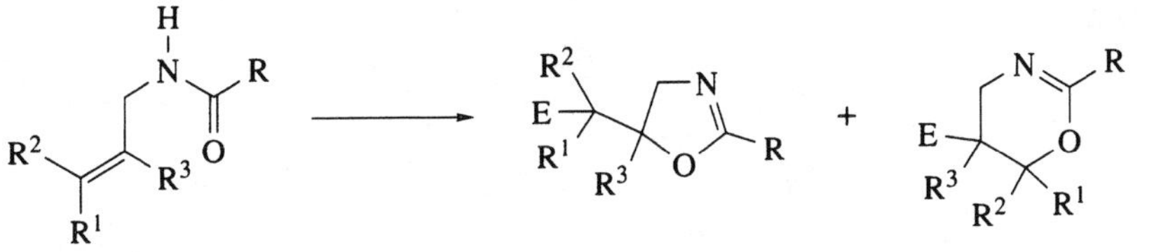

(56)

$E = H, Cl, Br, I, PhSe, PhS, TeX_4^-$

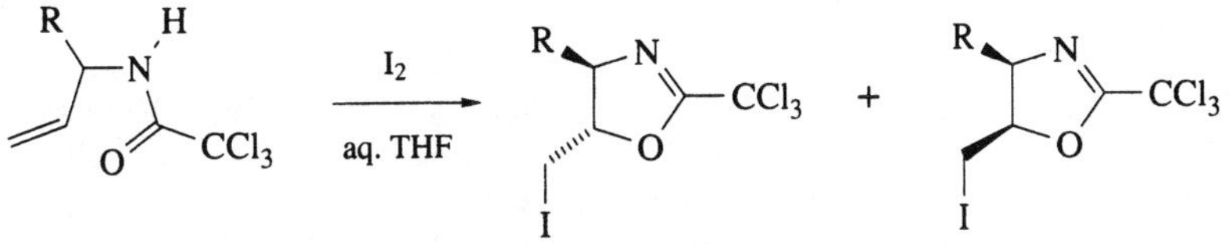

I_2, aq. THF

(57)

R	*Ratio* (trans:cis)	*Ref.*
CH_2OBn	80:20	157b
$C_{15}H_{31}$	55:45	153a
Pr^n	66:30	157a
Ph	24:76	157a

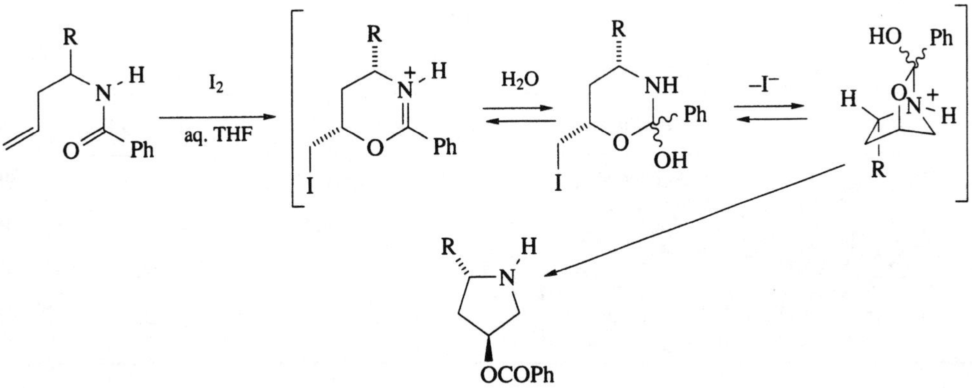

I_2, aq. THF; H_2O; $-I^-$

(58)

high yields

E = SMe, Br, I

(59)

(Table 17), but cyclization of *N*-substituted carbamates is highly selective (entries 2, 4 and 5), presumably by insuring reversible formation of the intermediate prior to benzyl cleavage and steric interactions between the *N*-substituent and the alkyl substituent favoring the axial orientation (equation 61). Cyclizations of this type, where R = H and Y = α-phenethyl, result in little asymmetric induction. However, the diastereomeric products can be separated and used for the synthesis of chiral nonracemic amino alcohol derivatives.[164]

(60)

Table 17 Stereoselectivity in *O*-Cyclizations of Carbamate Derivatives of Unsaturated Amines

Entry	*n*	*Y*	*Ratio (cis:trans)*	*Ref.*
1	0	H	1:1.5	161
2	0	CO_2Bn	1:6.6	161
3	1	H	2.3:1	133
4	1	CO_2Bn	1:2.3	133
5	1	TBDMS	1:14	133, 162

n = 1 or 2

(61)

Cyclizations of urea derivatives of allylamines with selenium reagents have been examined recently (equation 62 and Table 18).[98] Cyclization of the allylic ureas produces *trans*-2-oxazoline derivatives with high stereoselectivity when the double bond is internal (Table 18, entry 2). Although *O*-methylisourea derivatives can be cyclized to dihydroimidazoles [see Section 1.9.3.2.2(ii)], cyclization with phenylselenenyl trifluoromethanesulfonate and trifluoromethanesulfonic acid produces 5,6-dihydro-1,3-oxazines (6-*endo* products), even when the double bond is monosubstituted (entry 3).

(A) or (B)

(62)

Table 18 *O*-Cyclization of Urea Derivatives of Acyclic Allylic Amines.

Entry		*Conditions*	*Yield (%) (A)*	*Yield (%) (B)*	*Ref.*
1	Z = NH_2, Y = O R^1 = Ph, R^2 = H	PhSeCl, $CHCl_3$ Silica gel, 25 °C	96% (*trans:cis* = 2:1)		98a
2	Z = NH_2, Y = O R^1 = Bu^n, R^2 = Pr^n	PhSeCl, $CHCl_3$ Silica gel, 25 °C	81% (*trans* only)		98a
3	Z = OMe, Y = NH R^1 = Ph, R^2 = H	PhSeCl, CF_3SO_3Ag Silical gel, CF_3SO_3H		66%	98b
4	Z = OMe, Y = NH R^1 = Bu^n, R^2 = Pr^n	PhSeCl, CF_3SO_3Ag Silical gel, CF_3SO_3H		62%	98b

The cyclization of carbamate derivatives of 2-alkenylpyrrolidines has been examined (equation 63 and Table 19). Cyclization to oxazolidinones proceeds with high stereoselectivity through an H-in-plane transition state for systems with a terminal double bond. With phenyl substitution, the only product is the tetrahydrooxazinone (Table 19, entry 4). However, when the phenyl group contained an electron-withdrawing substituent *and* the alkene had (*Z*)-geometry, the major product was the oxazolidinone (entry 6). Thus, the regioselectivity is affected by small changes in substrate structure.

$$\xrightarrow[\text{CH}_2\text{Cl}_2]{\text{E(collidine)}_2^+\ \text{ClO}_4^-} \quad \textbf{(A)} + \textbf{(B)} \qquad (63)$$

Table 19 Cyclization of 2-Alkenylpyrrolidine Derivatives (Equation 63)

Entry	*R*	R^1	R^2	R^3	*E*	*Yield (%) (A)*	*Yield (%) (B*	*Ref.*
1	Me	H	H	H	Br	80		104g
2	Bn	H	H	Me	Br or I	83		165
3	Bu^t	H	Ph	H	I		73	104g
4	Bu^t	Ph	H	H	I		60	104g
5	Bu^t	*p*-NO_2Ph	H	H	Br		60	104g
6	Bu^t	H	*p*-NO_2Ph	H	Br	58	12	104g

Allylamines may be converted directly to oxazolidinones through condensation with CO_2, to form carbamate salts, and treatment with iodine (equation 64 and Table 20).[166] High stereoselectivity was obtained only in cases where the α-substituent was hydroxymethyl (entry 5) or the amine was secondary and the substituent was phenoxymethyl (entry 4). The results with primary amines are comparable to those for amide cyclizations shown in equation (57).

$$\longrightarrow \qquad (64)$$

Table 20 Cyclization to 2-Oxazolidinones from Allylic Amines and Carbon Dioxide (Equation 64)

Entry	*R*	R^1	*Conditions*	*Ratio (cis:trans)*	*Yield (%)*	*Ref.*
1	Pr	H	Amberlyst A26 (CO_3^{2-} form), I_2, $CHCl_3$, r.t.	45:55	95	166a
2	Pr	CH_2Ph	Same as entry 1	70:30	95	166a
3	CH_2OPh	H	Same as entry 1	70:30	94	166a
4	CH_2OPh	CH_2Ph	Same as entry 1	99:1	90	166a
5	CH_2OH	H	Same as entry 1	93:7	80	166a
6	H	H	CO_2, MeOH, I_2		60	166b

One example of the cyclization of a hemiaminal adduct of a homoallylic β-lactam has been reported to proceed with high 1,3-asymmetric induction (equation 65).[167]

$$\xrightarrow[25\text{–}50\%]{\text{i, Hg(OAc)}_2\text{; ii, NaCl}} \qquad (65)$$

(iv) Miscellaneous

The cyclofunctionalization of cycloalkenyl systems where the chain containing the nucleophilic functionality is attached at one end of the double bond leads to spirocyclic structures. Cyclizations of cyclic and acyclic enol ethers to generate spiroacetals are shown in equations (66)[168] and (67).[169] These reactions generate the thermodynamically more stable products based on anomeric and steric factors.[170] Spiroacetal products have also been obtained using isoxazolines as the nucleophilic functionality (*cf.* Table 14).[141b] Studies of steric and stereoelectronic control in selenoetherification reactions which form spirocyclic tetrahydrofurans have been reported.[38] An interesting example of stereoelectronic control in the formation of a spirocyclic lactone has been reported in a recent mevinolin synthesis (equation 68).[171]

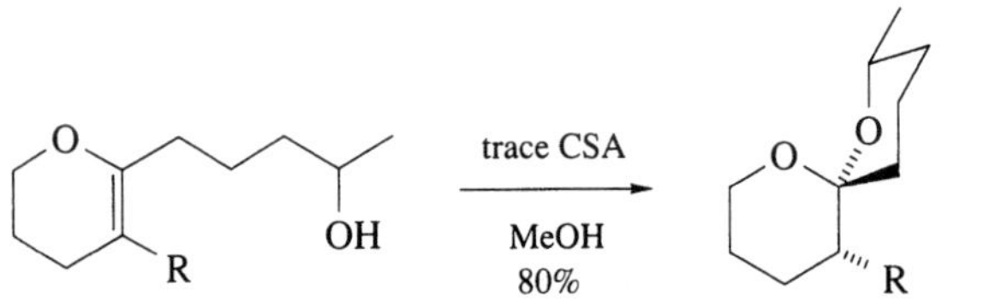

trace CSA; MeOH; 80% (66)

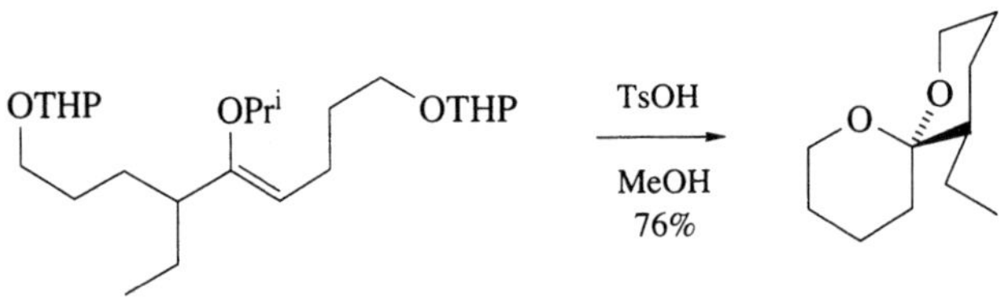

TsOH; MeOH; 76% (67)

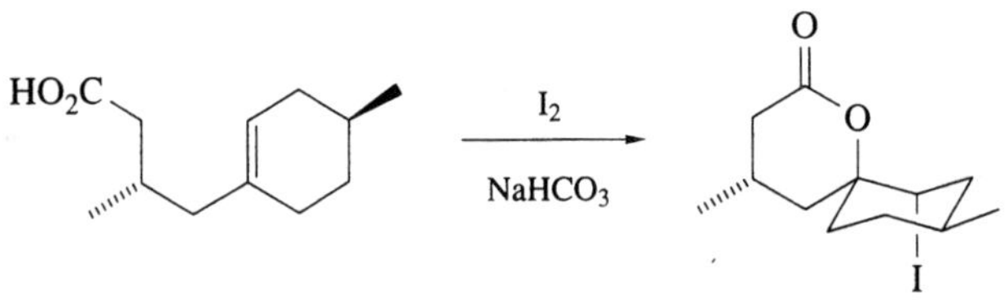

I_2; $NaHCO_3$ (68)

Early studies on mercuricyclization of hydroperoxides have been summarized by Bartlett.[1p] Recent applications include the generation of endoperoxide intermediates (**7**) for use in biomimetic syntheses of prostaglandins (equation 69).[172] These results, and others,[173a,173b] show that *cis*-3,5-dialkyl-1,2-dioxolanes are the predominant products from both 5-*exo* and 5-*endo* cyclizations. As shown in equation (70), where 5-*endo* cyclization involves electrophilic attack on a diene system, the ratio of the 5-*exo* and 5-*endo* modes of cyclization could be varied by changing the mercury(II) salt used. Studies on 6-*exo* cyclizations by Porter and Bartlett have shown that *trans*-3,6-disubstituted-1,2-dioxanes predominate (equation 71),[1p,173c] but 6-*exo* cyclizations of related tertiary hydroperoxides are not selective.[104b] Halocyclization of unsaturated hydroperoxides has also been reported.[173a,174] Recent studies have shown that cyclization of homoallylic hydroperoxides with NIS proceeds through a radical chain mechanism, whereas cyclization with NBS is mainly a polar process.[174a] A recent study of competition between 5-*exo* and 6-*exo* peroxycyclizations showed little regioselectivity.[174c]

Several examples of uncommon oxygen nucleophiles should be mentioned. Cyclizations with selenium electrophiles where the nucleophile is an OH group of a hemiacetal function have been used to generate monocyclic[104h] and spirocyclic[175a] acetals (equation 72). Related cyclizations to generate spirocyclic acetals with substituents on both rings using mercury(II) salts[175b] or NIS[175c] have been found to show significant stereoselectivity. Selenocyclization of dialkenyl ketones to fused and spirocyclic acetals

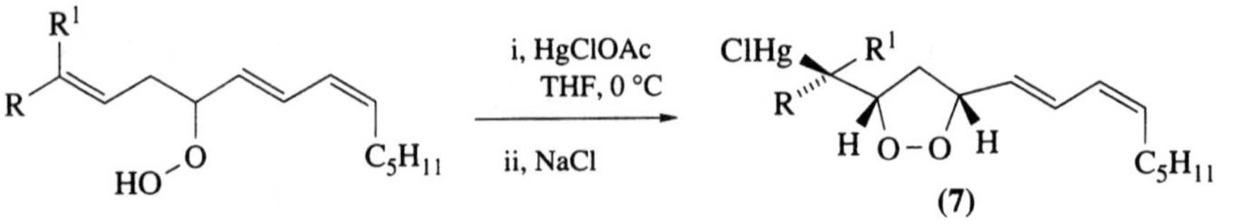

i, HgClOAc, THF, 0 °C; ii, NaCl (69)

(**7**)

80% [R = H, $R^1 = CH_2CH(OMe)_2$]; 93% [$R = CH_2CH(OMe)_2$, R^1 = H]

i, HgX_2

ii, aq. KBr

C_5H_{11} HO O R

R HgBr (8) + ClHg R + *trans*-(8) (70)

	(8)		*trans*-(8)
$Hg(O_2CBu^t)_2$	36%	55%	9%
$Hg(TFA)_2$	76%	21%	3%
$Hg(OMs)_2$	trace	85%	15%

i, $Hg(O_2CBu^t)_2$, CH_2Cl_2, –20 °C; ii, aq. KBr (71)

$R = R^1 = H$	90%	10%
$R = H, R^1 = Et$	88%	12%
$R = Et, R^1 = H$	79%	21%

using photooxidative (single electron transfer) cleavage of diphenyldiselenide has been reported recently.[175d]

PhSeNPht; $ZnBr_2$, CH_2Cl_2; *m* and *n* = 1 or 2 → + *cis* isomer when *n* = 1 (72)

Heterocyclizations involving the oxygen of enolizable β-dicarbonyl groups[115g,176] and of epoxides[177] as the nucleophile have also been reported. Iodocyclizations of unsaturated *N*-oxides of tertiary amines to form substituted tetrahydro-1,2-oxazines,[178a,178b] and iodocyclizations of *N,N'*-dimethyl-*N'*-allylbenzohydrazides to form substituted 4*H*-1,3,4-oxadiazinium salts[178c] have been reported.

Alkyl ethers can act as nucleophiles in cyclofunctionalization reactions,[179] and recently acetal oxygens have been shown to be excellent nucleophiles for participation in heteroatom cyclizations (Scheme 9). This reaction has been used as a nonhydrolytic procedure for unraveling complex acetals (conversion of **9** to **10**),[180] for specific liberation of an anomeric center (conversion of **11** to **12**),[181a] for preparation of glycosyl bromides[181b] and for saccharide coupling (reaction of **13** and **14** to give **15**).[182]

The selective activation of a mixture of (**13**) and (**14**) requires that iodonium ion formation be reversible. Then, the rate of cyclization to the activated intermediate is dependent upon the nucleophilicity of the glycosidic oxygen, which is found to be a function of the protecting group present on the C-2 oxygen. The method has also been used for the synthesis of oligosaccharides and for the synthesis of 2-deoxyglycosides with a 2-bromo group used to deactivate the *n*-pentenyl protecting group.[182a] Further studies have also demonstrated the utility of *n*-pentenyl acetals as a general protecting groups for alcohols and carbonyl groups (equations 73 and 74).[183]

1.9.2.2.3 Asymmetric induction with chiral auxiliaries

The asymmetric halolactonization reactions of unsaturated L-proline amides, developed by Terashima and coworkers,[184] has been extended to α-alkyl acrylic acid derivatives (equation 75 and Table 21).[185] This allows for the synthesis of either enantiomer of an α-methyl-α-hydroxy acid using L-proline as the auxiliary. Less successful approaches to asymmetric induction with a chiral auxiliary include iodolac-

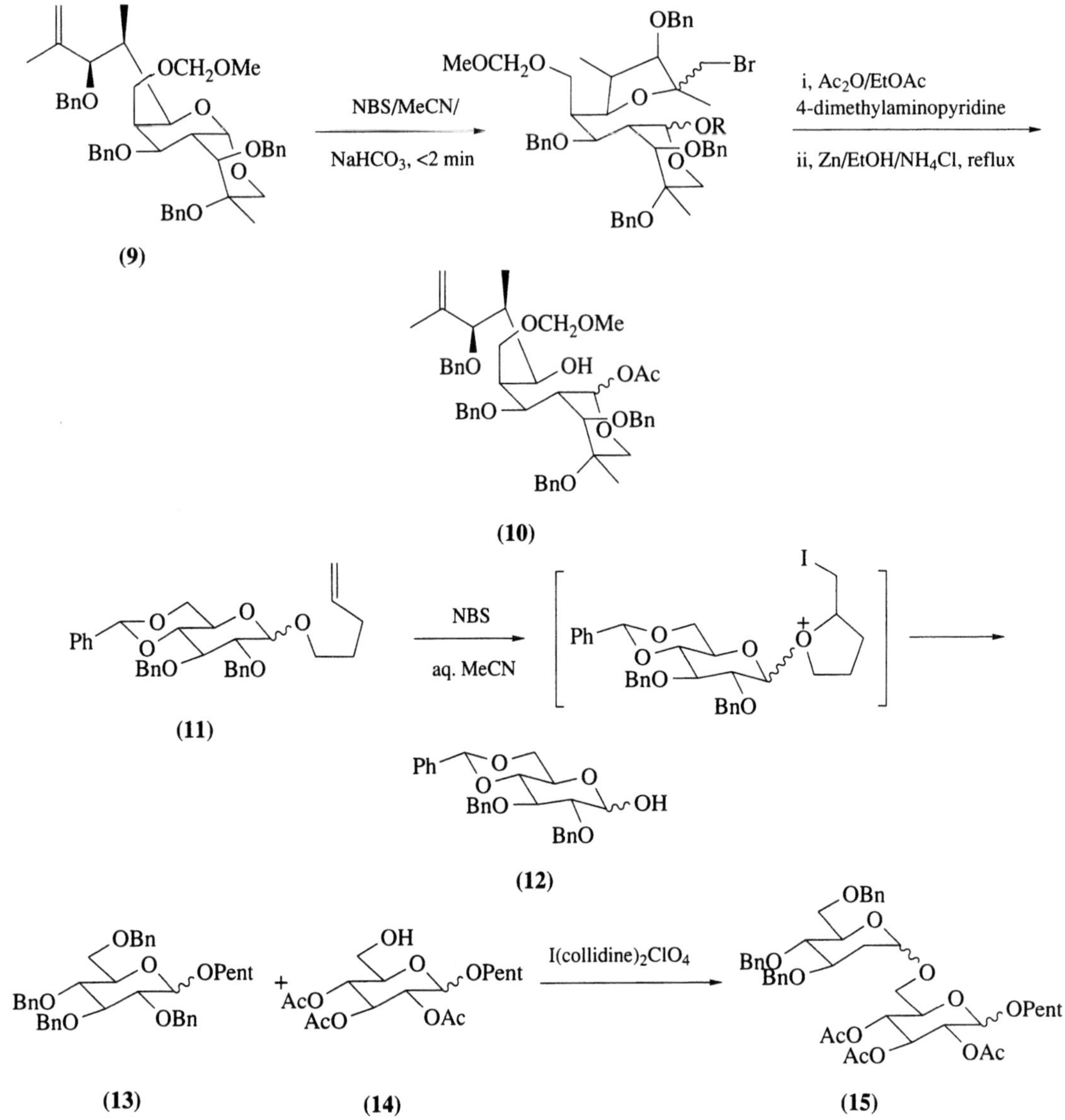

NBS/MeCN/ NaHCO$_3$, <2 min

i, Ac$_2$O/EtOAc 4-dimethylaminopyridine

ii, Zn/EtOH/NH$_4$Cl, reflux

(9) (10)

NBS aq. MeCN

(11) (12)

I(collidine)$_2$ClO$_4$

(13) (14) (15)

Scheme 9

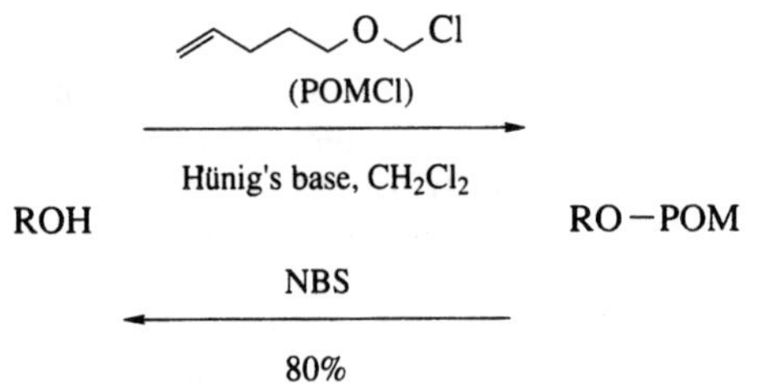

(POMCl) Hünig's base, CH$_2$Cl$_2$

ROH ⇌ RO–POM (73)

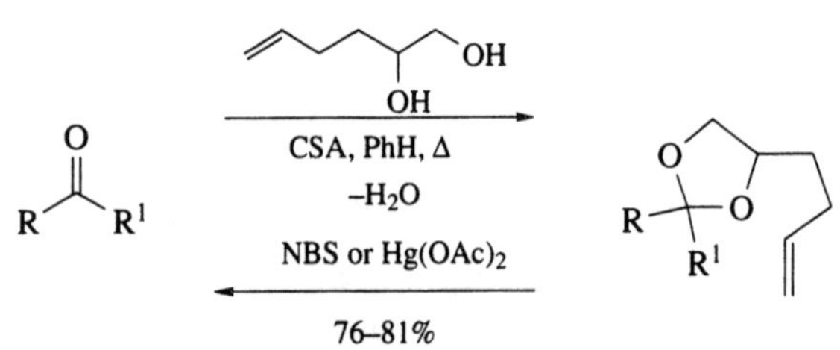

NBS 80%

CSA, PhH, Δ –H$_2$O

NBS or Hg(OAc)$_2$ 76–81% (74)

tonization of amides from symmetrical *bis*-γ,δ-unsaturated acids and chiral amines (<17% *ee*),[186] cyclization of amides of symmetrical homoallylic amines and chiral acids (<26% *ee*),[159] cyclization of carbamates of allylic amines with a chiral *N*-alkyl substituent,[164a,164c] use of mercury salts of chiral carboxylic acids (<5% *ee*)[187] and intramolecular oxypalladation with chiral palladium(II) complexes (<30% *ee*).[50,188] It is interesting to note that the one successful approach (Table 21) incorporates the chiral center of the auxiliary into the newly formed ring system.

Table 21 Asymmetric Bromolactonization of Proline Amides

Entry		*Conditions*	*Yield (%)*	*Diastereomer ratio*	*Crystalline yield (major isomer) (%)*	*Ref.*
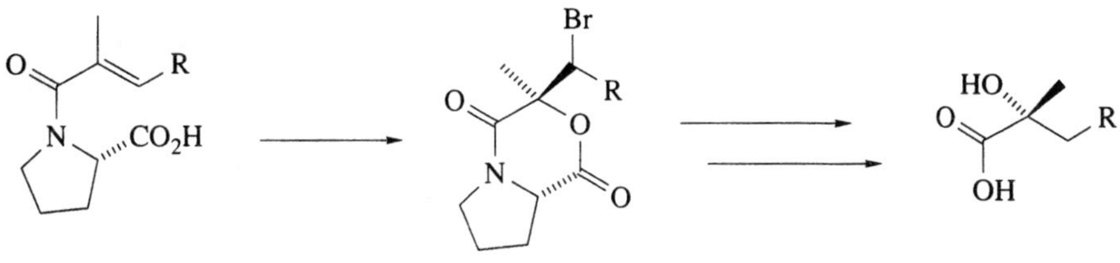		(75)				
1	R = Me	NBS/DMF	84	94:6	64	184a
2	R = Me (Z-isomer)	NBS/Bu^{t}OK/DMF	76	61:39		184b
3	R = Pr^n	NBS/CCl_4			76	185
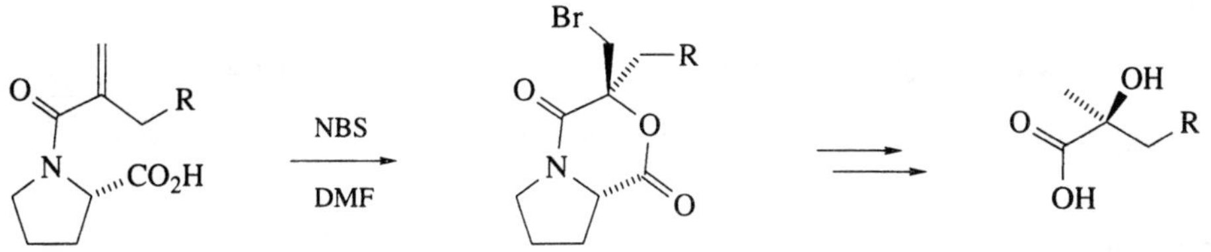		NBS, DMF				
4	R = Pr^n				76	185
5	R = H				64	185

1.9.2.2.4 Cyclizations with alkynes and allenes

Electrophilic heteroatom cyclizations of systems involving alkyne and allene π-systems have attracted significant attention. A major difference from alkene cyclizations is that the electrophilic group in the initial product may be a vinyl substituent, and, in the case of metal electrophiles, possess different reactivity patterns than when attached to a saturated carbon.

The cyclofunctionalization of 4-alkynoic and 5-alkynoic acids generates γ-alkylidene-γ-lactones and δ-alkylidene-δ-lactones, respectively (equation 76 and Table 22). The initial products from the reactions catalyzed by mercury or silver salts undergo protiodemetallation under the reaction conditions. The vinylpalladium intermediates undergo either protiodemetallation or coupling with an added allyl halide. The palladium(II)-catalyzed cyclization of 3-alkynoic acids proceeds by 5-*endo* closure to give 3-buten-4-olides (Table 22, entry 2).[50,190]

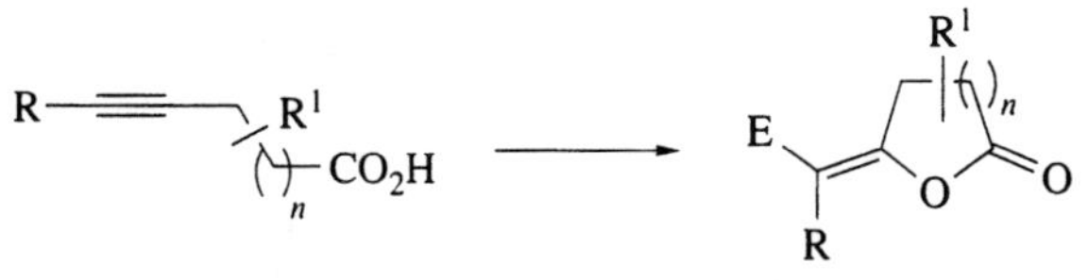

(76)

Several cyclofunctionalization reactions of alkynic alcohols are synthetically useful. Metal ion-promoted cyclofunctionalization of *cis*-2-propargylcyclopentanol systems proceeds by the 5-*exo* mode (equation 77 and Table 23).[197] Protiodemetallation or reductive demetallation provides the cyclic enol ether in high yields. This method has been used by Noyori in the synthesis of prostacyclin (PGI_2).[197b,197c] Reactions with catalytic amounts of mercury(II) or palladium(II) salts gave the endocyclic enol ether as the major product.[197a,198] A related cyclization with Ag_2CO_3 has been reported by Chuche.[191] Schwartz

Table 22 Lactonization of Acyclic Alkynic Acids (Equation 76)

Entry	*Reagent*	*n*	*E*	*Ref.*
1	Mercury(II) salts	1 or 2	H	189
2	$[PdCl_2(PhCN)_2]$, Et_3N	1 or 2	H	190
3	$AgCO_3$, PhH	1	H	191
4	Li carboxylate, $[PdCl_2(MeCN)_2]$, $CH_2{=}CHCH_2Cl$	1 or 2	Allyl	192
5	NCS, NIS or NBS, $KHCO_3$, CH_2Cl_2	1 or 2	Cl, I, Br	189b, 189c, 193
6	PhSeNPhth	1	PhSe	194
7	PhSCl, Et_3N	1	PhS	195
8	I_2, KI, H_2O	1 or 2	I	196

found that the organomercury intermediate could be intercepted by NBS or NIS to give the β-halo enol ether, and that the *trans*-cyclopentanol isomer cyclized *via* the 6-*endo* mode to give a fused dihydropyran.[197a]

(77)

Table 23 Cyclization of Propargyl-substituted Cyclopentanols (Equation 77)

Reagent	*Yield (%)*	*Ref.*
i, $Hg(TFA)_2$, Et_3N, –78 °C ii, LiI, ether	45	197a
i, $Hg(TFA)_2$, Et_3N, –78 °C ii, $NaBH_4$, NaOMe, MeOH	67	197b
i, $[PdCl_2(PhCN)_2]$ ii, $NH_4^+HCO_3^-$	71	197c

Successive intramolecular oxypalladation of dihydroxyalkynes has been used for the synthesis of spirocyclic and bridged acetals. Equations (78) to (80) show the syntheses of some insect pheromones by this method.[198] The results from several diols suggest that preferred modes for initial cyclization are 5-*endo* > 5-*exo* > 6-*endo*. A related conversion of 8-hydroxyoct-4-ynoic acid to an oxaspirolactone with mercury(II) oxide has been reported.[199]

cat. $PdCl_2$, aq. MeCN, 85% (78)

$PdCl_2(PhCN)_2$, Et_2O, r.t., 95% — 17% + 83% (79)

$PdCl_2(PhCN)_2$, Et_2O, r.t., 80% (80)

exo-Brevicomin

Several examples of intramolecular solvomercuration of arylalkynes have been reported (equations 81 to 83).[200] The mercury moiety can be removed by protiodemetallation or converted to a variety of other functional groups. The use of a methyl ether as the nucleophilic functionality is noteworthy, as is the change in the regioselectivity (equation 82 *versus* equation 83). A bromocyclization related to equation (83) using a phenylethynylbenzoate ester also gives an isocoumarin derivative *via* 6-*endo* cyclization.[201] Palladium-catalyzed cyclizations of β,γ-alkynic ketones or 2-methyl-3-alkyn-1-ols to form substituted furans are discussed in Chapter 1.4, Section 3.1.3 in this volume.

i, $Hg(OAc)_2$, AcOH; ii, NaCl — 5-*endo*; 60–70% (81)

i, $Hg(OAc)_2$, AcOH; ii, NaCl — 5-*exo*; 38% (82)

i, $Hg(OAc)_2$, AcOH; ii, NaCl — 6-*endo* (83)

Allenes can also act as the π-participant in electrophilic heteroatom cyclizations. Reviews of electrophilic additions to allenes discuss early examples of this type of cyclization.[1d,1e,202] Numerous examples of cyclizations of α-functionalized allenes, including carboxylic acids, phosphonates, sulfinates and alcohols, to form five-membered heterocycles (equation 84) are cited in these reviews. The silver nitrate-mediated conversion of α-allenic alcohols to 2,5-dihydrofurans[203] has recently been applied to trimethylsilyl-substituted systems.[204]

E^+ (84)

E^+ = HY, ArSeCl, ArSCl, Br_2, HgX_2, $AgNO_3$; R = H or alkyl

X = >CRR[1], >C=O, >S=O, O=PPh, R[1]OP=O,

Most cyclizations of β-allenic alcohols have produced six-membered ring products *via* a 6-*endo* process (equation 85).[168e,205] The exception is found with terminally unsubstituted allenes as shown in equation (86).[205d] The bicyclic acetal results from a 5-*exo* cyclization to an enol ether, which undergoes a second cyclization to form the acetal.

Cyclization of a variety of γ-allenic alcohols with silver nitrate proceeds by 5-*exo* cyclization to form 2-alkenyltetrahydrofurans (equation 87).[205c,206] Little stereoselectivity is seen in cyclizations of secondary alcohols. Cyclization by intramolecular oxypalladation/methoxycarbonylation or oxymercuration followed by transmetallation and methoxycarbonylation also showed no stereoselectivity (equation 88 and Table 24, entries 1 and 2).[50,207a] However, cyclization of the corresponding *t*-butyldimethylsilyl ether derivatives with mercury(II) trifluoroacetate followed by transmetallation/methoxycarbonylation pro-

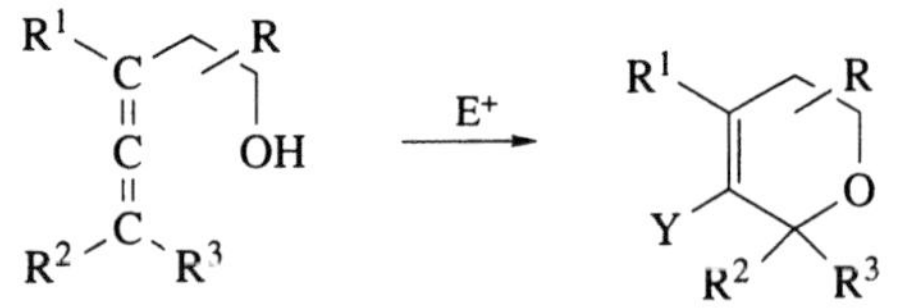

(85)

E^+ = HY, ArSeCl, BF_3, $HgCl_2$, $AgNO_3$; R^1 = H, alkyl or SPh; R^2 and R^3 = alkyl, H or alkyl, alkyl

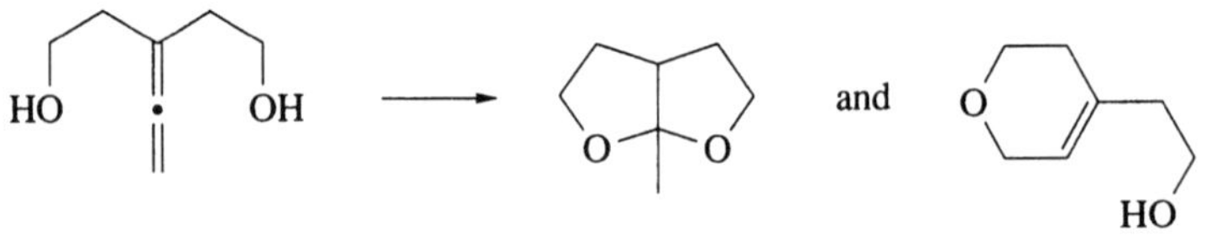

(86)

Reagent	*Yield*	*Product ratio*	
cat. $HgCl_2$, CH_2Cl_2, r.t.	90%	100%	0%
1.1 equiv. $AgNO_3$, aq. acetone	80%	60%	40%

ceeds with high stereoselectivity (entries 3 and 5).[207a,207b] This selectivity is presumed to derive from equilibrium control brought about by slow desilylation and the steric effect of the silyl group. Two exceptions to 5-*exo* cyclizations of systems with the allene γ to the nucleophilic atom are the mercury(II) oxide cyclization of 4,5-hexadienoic acid to a δ-lactone (equation 89)[189d] and the cyclization of β-allenic oximes to dihydro-1,2-oxazepines (equation 90).[208]

δ-Allenic alcohols with a variety of allene substitution patterns form 2-alkenyltetrahydropyrans by a 6-*exo* ring closure upon cyclization with silver or mercury salts (equation 91).[209] The organosilver intermediate undergoes protiodemetallation under the reaction conditions, while the vinylmercury

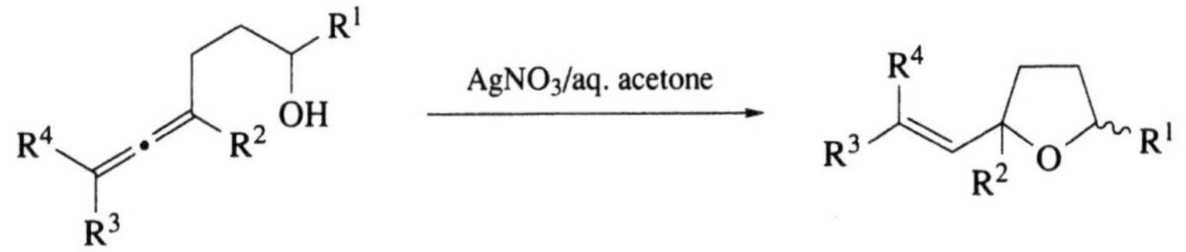

AgNO$_3$/aq. acetone (87)

(88)

Table 24 Effect of *O*-Silylation on Stereoselectivity of Cyclization of γ-Allenic Alcohols (Equation 88)

Entry	*R, R*^{1}	*Reagents*	*Yield (%)*	*cis:trans*
1	Me, H	i, $Hg(OAc)_2$, CH_2Cl_2; ii, cat. $PdCl_2$, $CuCl_2$, MeOH, CO (1 atm)	55	50:50
2	Me, H	Cat. $PdCl_2$, $CuCl_2$, MeOH, CO (1 atm)	72	50:50
3	Me, $SiMe_2Bu^t$	i, $Hg(TFA)_2$, CH_2Cl_2; ii, cat. $PdCl_2$, $CuCl_2$, MeOH, CO (1 atm)	53	94:6
4	Me, $SiMe_2Bu^t$	Cat. $PdCl_2$, $CuCl_2$, MeOH, CO (1 atm)	60	50:50
5	Pr, $SiMe_2Bu^t$	i, $Hg(TFA)_2$, CH_2Cl_2; ii, cat. $PdCl_2$, $CuCl_2$, MeOH, CO (1 atm)	89	>98:2

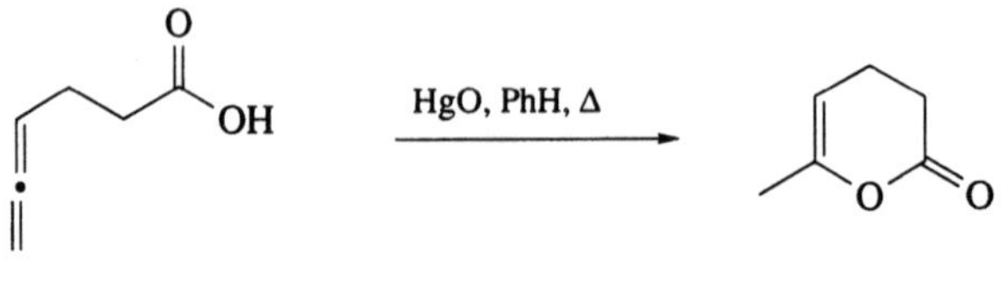

HgO, PhH, Δ (89)

AgBF$_4$, CHCl$_3$

70%

(90)

intermediate is stable and mercury is removed by subsequent reduction with sodium borohydride. In cyclizations of analogous secondary alcohols or of 3-alkyl-substituted systems, the *cis*-fused isomer predominates (equation 92).[209,210] The predominant formation of the *cis* isomer is rationalized in terms of a chair-like transition state. Enolizable γ- and δ-allenic ketones have been converted to 2-alkenyl-substituted dihydrofuran and dihydropyran structures in moderate yield using catalytic amounts of mercury(II) oxide and *p*-TsOH in refluxing cyclohexane.[211]

AgNO$_3$/aq. acetone or Hg(TFA)$_2$/CH$_2$Cl$_2$, NaBH$_4$ 40–90%

R^1, R^2, R^3 = H or alkyl

(91)

AgNO$_3$/aq. acetone or Hg(TFA)$_2$/CH$_2$Cl$_2$, NaBH$_4$ 75–87%

R^1, R^2, R^3 = H or alkyl

2,6-*cis*-isomer 81–95%

(92)

1.9.2.3 Large Rings

Electrophilic heteroatom cyclizations have been rarely applied to the synthesis of large rings. Although isolated examples of cyclization to seven-membered rings (*e.g.* equation 90) are known,[9,58,141b,184b,208] applications to synthesis have been limited. A selenoetherification of a vinyl ether to an eight-membered acetal has been reported.[212] Examples of phenylselenolactonizations to form 14- and 16-membered rings are shown in equation (93).[31] Reagents such as phenylselenenyl chloride lead to simple addition across the double bond with no cyclization.

PhSeNPht

CSA, CH$_2$Cl$_2$

(93)

64%, n = 1 — 78% — 12%

69%, n = 3

1.9.3 NITROGEN NUCLEOPHILES

The generation of nitrogen heterocycles by electrophilic heterocyclization has been reviewed recently.[1c] Examples of functional groups used as nitrogen nucleophiles are shown in Figure 2.

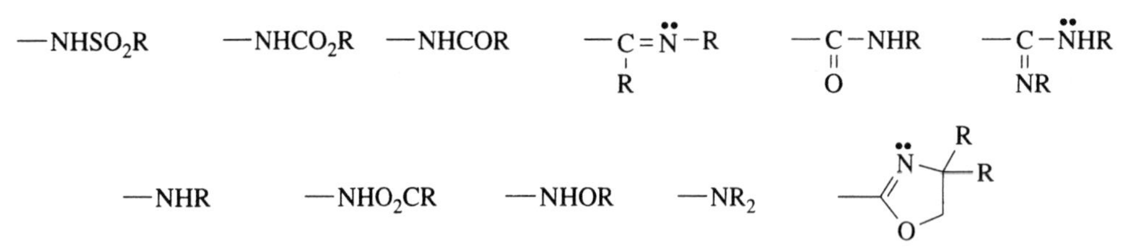

—$NHSO_2R$ —$NHCO_2R$ —NHCOR —C(R)=N–R —C(=O)–NHR —C(=NR)–NHR

—NHR —NHO_2CR —NHOR —NR_2

Figure 2

1.9.3.1 Small Rings

Few examples of cyclization to small ring nitrogen compounds are known. Cyclization of alkenic amides normally results in nucleophilic attack by the carbonyl oxygen and formation of lactone products (see Section 1.9.2, Oxygen Nucleophiles). However, cyclizations of some β,γ-unsaturated amide derivatives to β-lactams have proven successful (equation 94 and Table 25). Ganem demonstrated that the use of *N*-sulfonyl derivatives with a more acidic carboxamide nitrogen proton resulted in preferential cyclization to β-lactam derivatives.[213] Similar results were obtained from cyclization of *O*-acylhydroxamate derivatives.[214] The formation of β-lactams by cyclization of amides with phenylsulfenyl chloride is reported to result from rapid addition of the chalcogenide to the double bond, with cyclization resulting from subsequent *N*-deprotonation and attack of the amide anion on the episulfonium ion intermediate.[215]

(94)

Table 25 *N*-Cyclization of Amide Derivatives to Form β-Lactams (Equation 94)

Entry	*R*	*Y*	*Conditions*	*Yield (%)*	*Ref.*
1	H or alkyl	SO_2Ar or SO_2OR^1	I_2, CH_2Cl_2/aq. $NaHCO_3$	77–95	213
2	H	O_2CR^1	Br_2, K_2CO_3/10% aq. MeCN	75–90	214
3	H or alkyl	CH_2Ar	i, PhSCl; ii, KOH, Bu_4NBr	45–97	215

1.9.3.2 Five- and Six-membered Rings

1.9.3.2.1 Cyclizations with C=C—C constrained by existing ring

Cyclizations of substrates in which an existing ring constrains rotation of the vinylic bond that attaches the nucleophile to the double bond form fused ring nitrogen heterocycles with highly predictable regiochemistry and stereochemistry. Examples include cyclizations to fused ring pyrrolidines[216] and pyrrolidones[217] (equation 95 and Table 26).

(95)

Fused ring products also may be formed from the transannular cyclization of medium ring unsaturated amine derivatives, as shown in Table 27.[218] A transannular cyclization to the loline ring system was reported by Wilson (entry 4).[218c] Cyclization of a tertiary amine to a pyrrolizidine perchlorate has also been reported.[219]

One example of the formation of a bridged ring product *via* 6-*endo* closure has been reported using phenylselenenyl chloride (equation 96).[220] The regioselectivity is presumably controlled by conjugation of the alkene with the thiazole ring. Cyclizations to form bridged ring systems from monocyclic sub-

Table 26 Cyclization to Fused-ring Pyrrolidine Derivatives

Entry	R^1, R^2	*Y*	*Conditions*	*E*	*Yield (%)*	*Ref.*
1	H	CO_2R	PhSeBr, AgO_2CCF_3	PhSe	59	216a
2	H	CO_2R	NPhSeNPht, CH_2Cl_2	PhSe	>82	216b
3	$(CH_2)_4$	H	Br_2, CH_2Cl_2	Br	72	216d
4	alkyl, H	CO_2R	I_2, K_2CO_3	I	76	216e
5	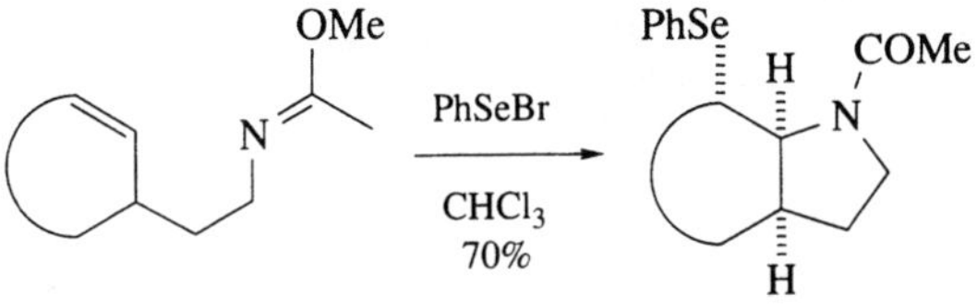		PhSeBr, $CHCl_3$, 70%			216e
6	 CH_2Ph	SMe	I_2		63	217b
7	$SiMe_3$	$OSiMe_3$	I_2		86–88	217a

For entries 1–4, see equation (95).

Table 27 Cyclization with Nitrogen Nucleophiles

Entry	*Y*	*X*	*Yield (%)*	*Product(s) (% of total)*		*Ref.*
1	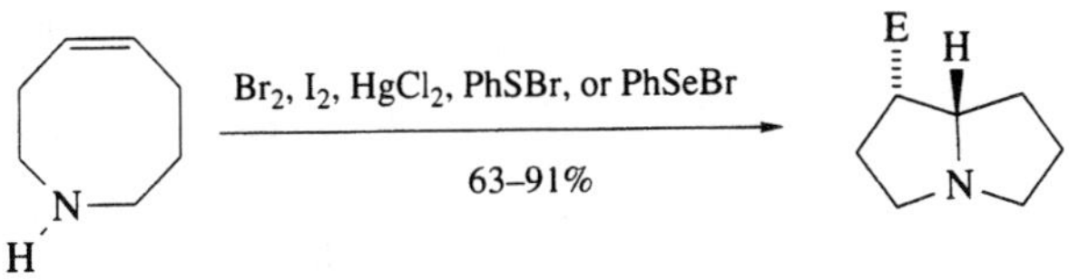		Br_2, I_2, $HgCl_2$, PhSBr, or PhSeBr; 63–91%			218a
	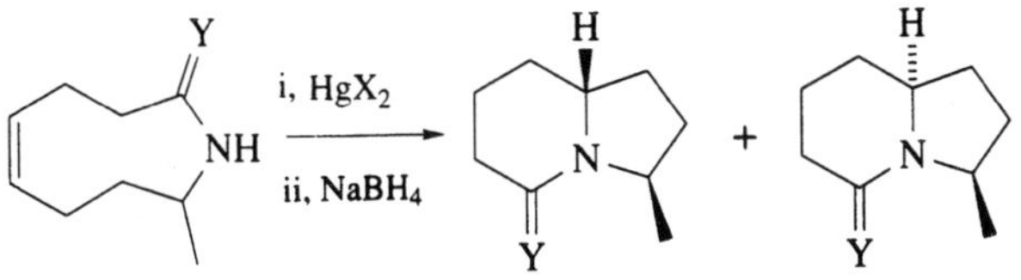		i, HgX_2; ii, $NaBH_4$			
2	O	OAc	83	100	0	218b
3	H, H	Cl	68	70	30	218c
4	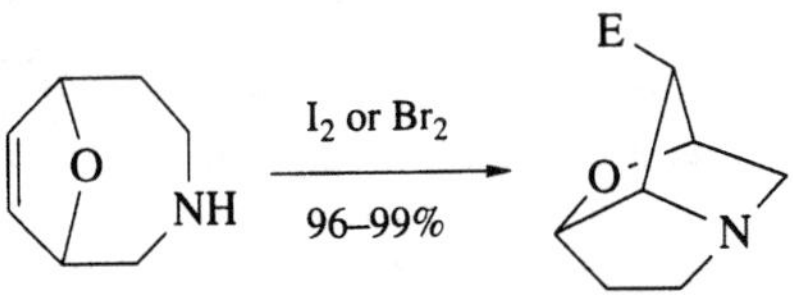		I_2 or Br_2; 96–99%			218c

strates (equation 97)[221] as well as from bridged ring substrates (equation 98)[222] have been covered in previous reviews.[1c,86] In equation (98) the organomercury intermediate undergoes substitution in the highly polar solvent with participation of the neighboring nitrogen. The bridged lactam (**16**), prepared by bromolactamization of a silyl imidate derivative,[223] was used in a recent synthesis of an aminouronic acid (Scheme 10).[217a] The synthesis of bridged ring systems from monocyclic dienes has been discussed in recent reviews.[1c,224]

Examples of highly stereoselective iodocyclizations of δ-aminocyclohexenes to generate azaspirocycles are shown in equations (99)[225a] and (100).[225b] The stereocontrol in these examples is attributed to the effect of the *t*-butyldimethylsilyloxy substituent.[225]

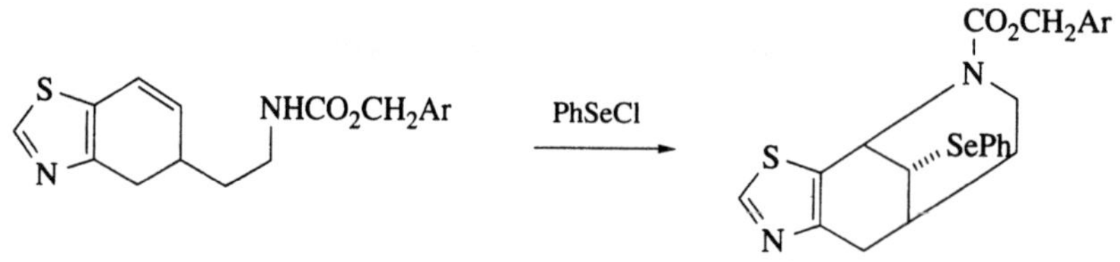

(96)

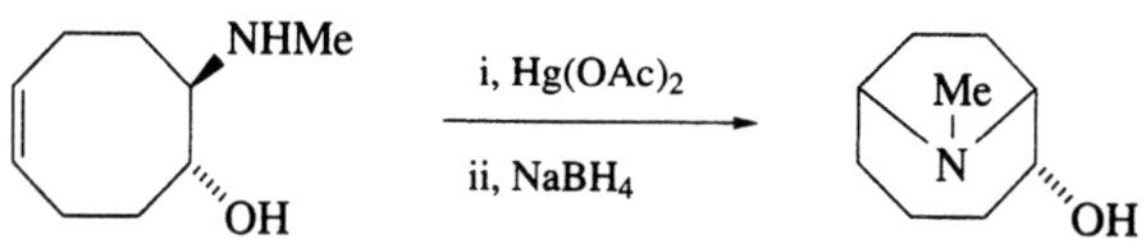

(97)

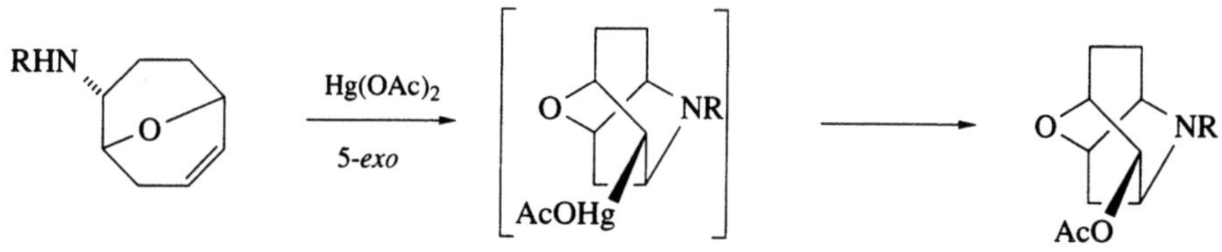

(98)

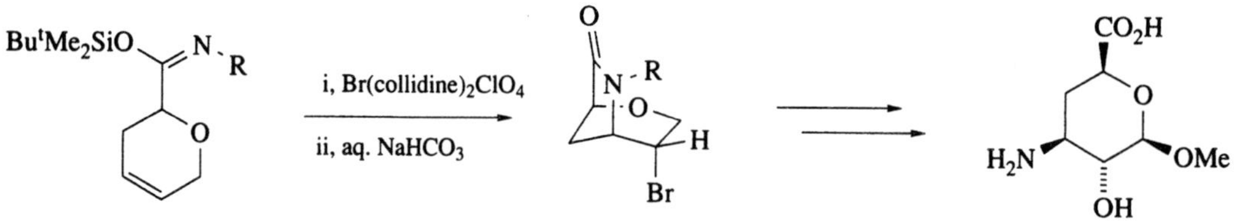

(16)

Scheme 10

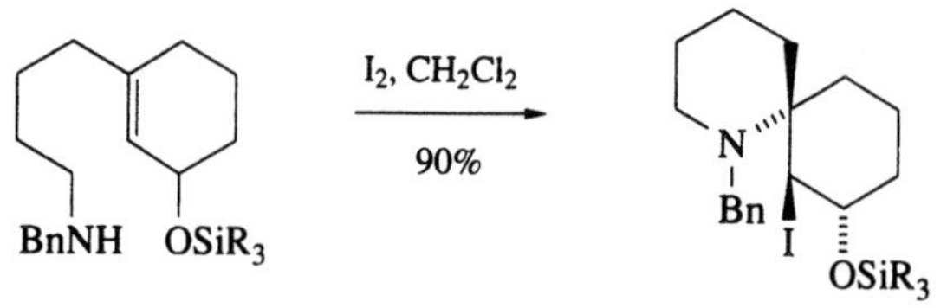

(99)

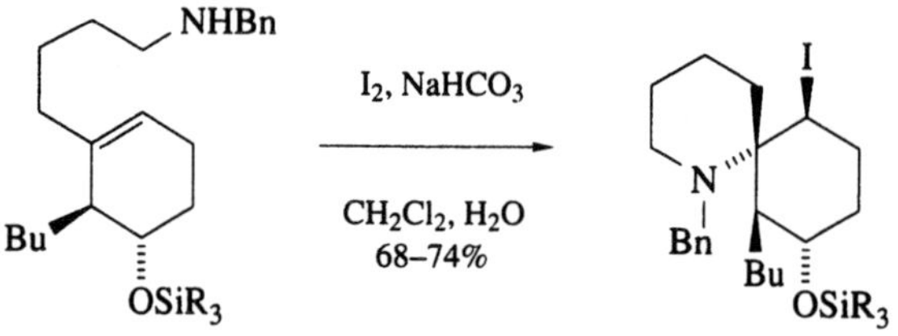

(100)

Fused ring heterocycles, prepared by cyclization of substrates with a 'tethered' nitrogen nucleophile, have been used in the synthesis of amino sugars and aminocyclitols. The examples shown in Table 28 make use of imidate-type functionality (equation 101) to insure nucleophilic attack by nitrogen. The bromocyclization of *N,N*-dialkylaminomethyl ethers of 2-cyclohexen-1-ol to form bicyclic oxazolidine derivatives has been reported also.[228b]

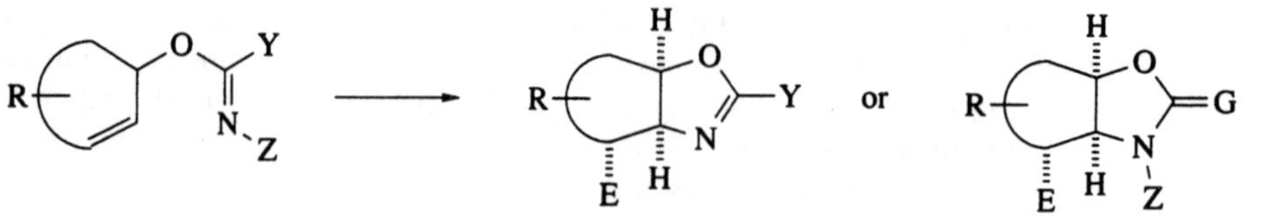

(101)

(A) **(B)**

Table 28 *N*-Cyclization of Derivatives of Cyclic Allylic Alcohols

Entry	*Y*	*Z*	*Conditions*	*Yield (%) (A)*	*Yield (%) (B)*	*Ref.*
1	CCl_3	H	NBS or NIS or I(*sym*-collidine)$_2BF_4$	70–95		94c, 226
2	SMe	Alkyl	Br(coll)$_2ClO_4$ or I_2		61–90 (O)	96b,227
3	Ph	COPh	NBS		>85 (OR, Ph)	226d,228a
4	NMe_2	H	$Hg(O_2CCF_3)_2$; $NaBH_4$	80		228c, 228d

1.9.3.2.2 Other C=C cyclizations

(i) Regioselectivity and stereoselectivity

Most cyclizations to form five-membered rings involve the 5-*exo* mode of cyclization. An exception has been reported in the intramolecular sulfenoamination of β-alkenylamines used in the synthesis of pyrrolidine alkaloids (equation 102).[215a,229] Other examples include the aminomercurations of *N*-substituted-1-amino-3-butenes[162a,230] and the methylsulfenilium ion-initiated cyclization of *O*-(2-propenyl)-*N*-tosylaniline.[231]

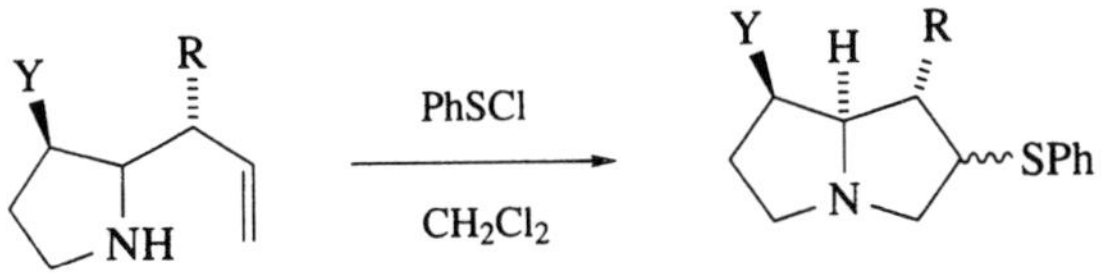

PhSCl, CH_2Cl_2 (102)

Y, R = H, H or OBn, CH_2OBn

Systems which can react by either 5-*exo* or 6-*endo* cyclization normally produce the five-membered ring system. Exceptions result when equilibration of the initially formed five-membered ring is facile and substitution electronically favors the 6-*endo* mode of cyclization. Several examples have been found in amidoselenation reactions.[41,158c,216a,216e,232] For example, halocyclization of thioimidate (**17**) produced only the pyrrolidone product,[217b,217c,233] while selenocyclization of amide (**18**) produced only the piperidone product.[232a] Note that the cyclization of (**18**) to a piperidone also involves regioselectivity in the cyclization of an amide functionality to a lactam rather than an imidate. Both the ring size and type of product can be explained by equilibration.

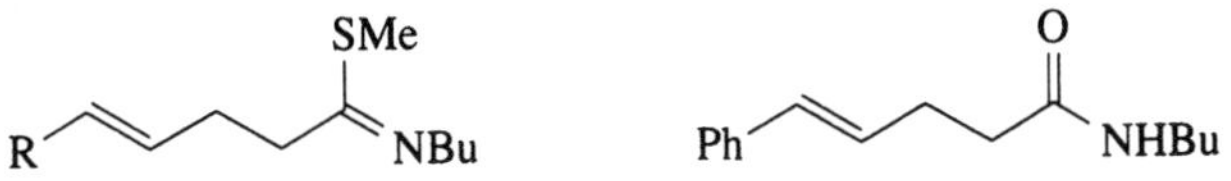

R= Ph or alkyl
(**17**) (**18**)

The regiochemistry of the palladium-catalyzed cyclizations of some alkenic sulfonamides was found to vary with the amount of palladium used (Scheme 11).[50,234a] The regioselectivity in cyclizations of ammonium salts of 1-amino-4-hexene with palladium chloride[234b] (5-ring:6-ring = 80:20) was significantly different from cyclizations of the same amine promoted with platinum salts[234b] (5-ring:6-ring = 9:91).

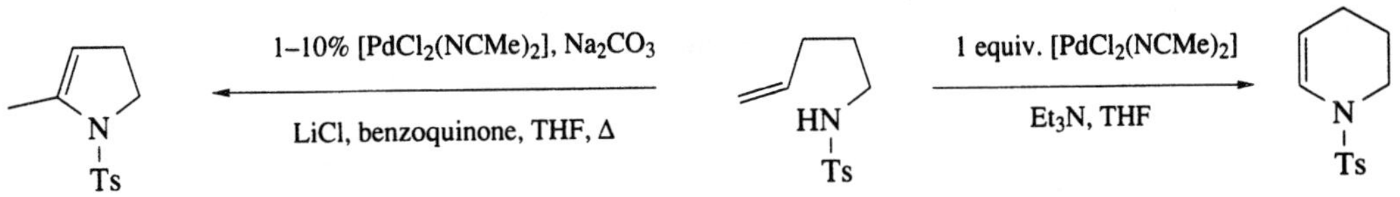

1–10% $[PdCl_2(NCMe)_2]$, Na_2CO_3; LiCl, benzoquinone, THF, Δ — 1 equiv. $[PdCl_2(NCMe)_2]$; Et_3N, THF

Scheme 11

Another aspect of regioselectivity with many nitrogen nucleophiles is the ambident nature of the nucleophile, which can lead to products of the same ring size through either *N*-cyclization or *O*-cyclization (see Scheme 12 for an example). As shown by entries 1 and 2 in Table 25, preferential *N*-cyclization to β-lactams has been effected by use of sulfonyl or acyloxy substitution on nitrogen. However, these substituents do not lead to *N*-cyclization in substrates which generate five- or six-membered products. Although isolated examples of the cyclization of amides to lactams have been observed (*cf.* cyclization of **18** described above), most examples of lactam formation have involved amide derivatives which direct the cyclization to nitrogen (see below). A recent study has shown that *N*-heterocyclic α-alkyl γ,δ-unsaturated amides form lactam products upon bromocyclization (equation 103 and Table 29).[235] No explanation for the anti-Markovnikov 6-*endo* cyclization in entry 3 is given.

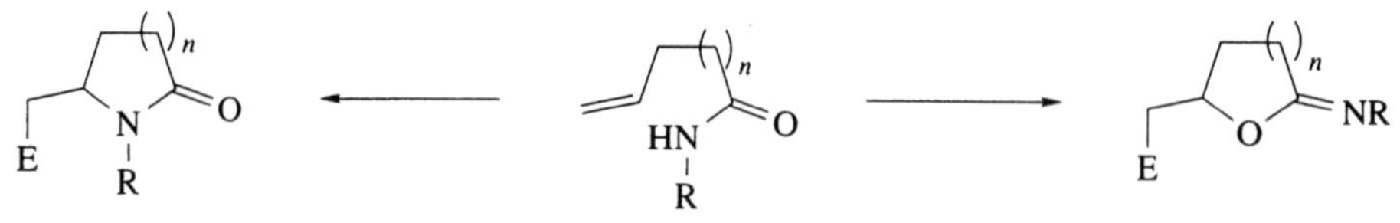

Scheme 12

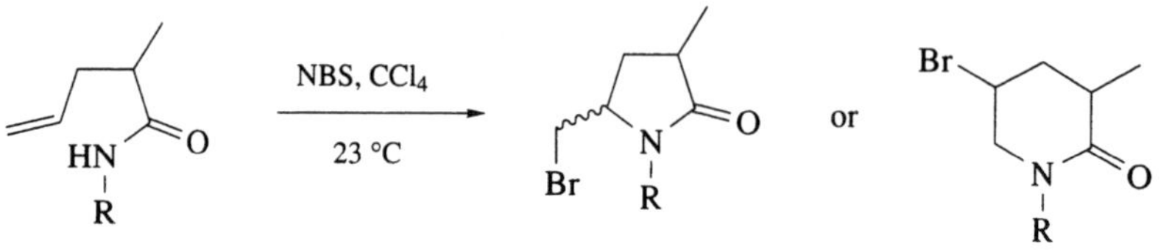

NBS, CCl_4, 23 °C (103)

Table 29 *N*-Cyclization of γ,δ-Unsaturated Amides (Equation 103)[235]

Entry	*R*	*Yield*
1	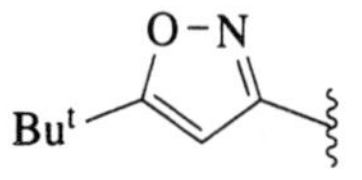 3-Bu^t-isoxazol-5-yl	30%
2	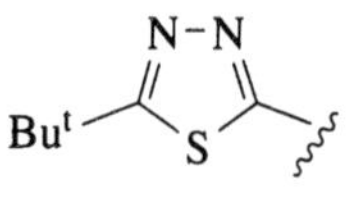 5-Bu^t-1,3,4-thiadiazol-2-yl	73%
3	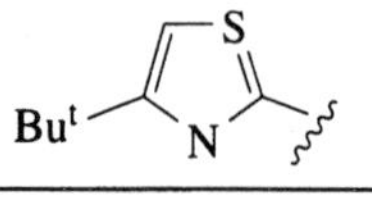 4-Bu^t-thiazol-2-yl	43%

Many of the factors controlling stereochemistry in cyclizations with a nitrogen nucleophile are analogous to those discussed earlier with oxygen as nucleophile. One example is found in the strong directing effect of an allylic hydroxy group, as shown in equation (104)[236] (compare with the results for oxygen cyclizations, shown in Table 9). The *cis* isomer (resulting from OH-in-plane cyclization) was formed with high selectivity in a variety of substituted systems. In particular, the configuration of a C-4 (homoallylic) alkyl group had little effect on stereoselectivity. The tosylamide system cyclized more readily than the carbamate system, which required higher temperatures and did not cyclize with NBS.

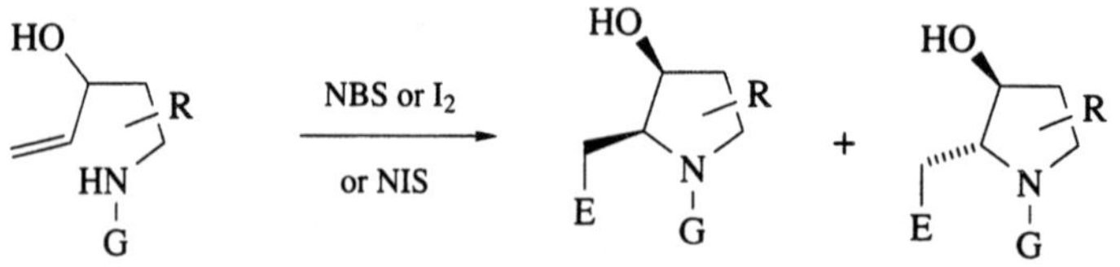

NBS or I_2 or NIS (104)

G	*Product ratio (cis:trans)*
O_2STos	93:7–97:3
CO_2Me	90:10–95:5

Similar stereochemical results have been reported in iodolactamizations of thioimidates (equation 105)[217b,217c] and the palladium(II)-catalyzed aminocarbonylation reaction (equation 106).[50,237] In some cases, the stereoselectivity in the aminocarbonylation reaction varied considerably with changes in the group on nitrogen or reaction conditions.

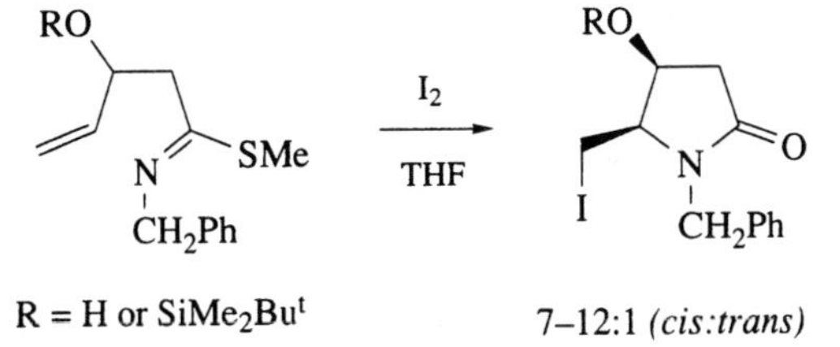
RO
SMe
N
CH_2Ph
I_2
THF
RO
O
N
I
CH_2Ph
R = H or $SiMe_2Bu^t$
7–12:1 *(cis:trans)*

(105)

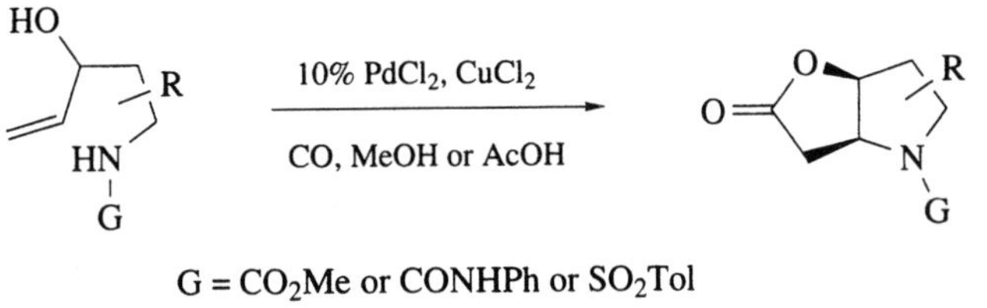
HO
R
HN
G
10% $PdCl_2$, $CuCl_2$
CO, MeOH or AcOH
O
O
R
N
G
G = CO_2Me or CONHPh or SO_2Tol

(106)

As found in oxygen cyclizations (equations 36 and 37), steric interactions result in a strong preference for the generation of *trans* isomers from the cyclization of systems with an allylic substituent and a double bond with the (*Z*)-configuration (equation 107).[238]

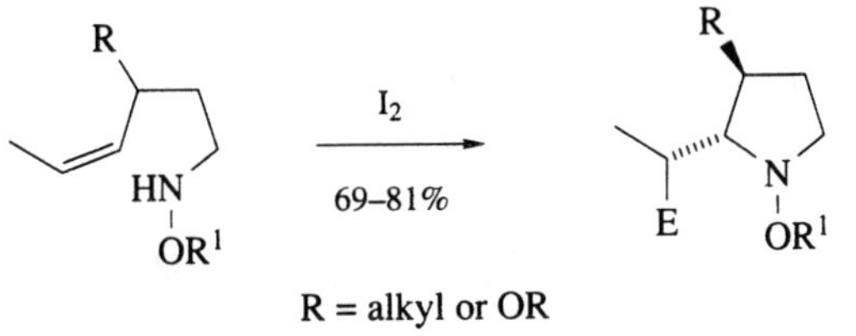
R
HN
OR^1
I_2
69–81%
R
N
E
OR^1
R = alkyl or OR

(107)

The stereochemistry of cyclizations involving a terminal double bond or an internal double bond with an (*E*)-configuration and an allylic alkyl group depends upon the substrate and reaction conditions. Iodocyclizations of *N,O*-bis(trimethylsilyl)imidates containing allylic substituents provide iodolactams with a small preference for the *cis* stereoisomers (equation 108).[223] Cyclization of 3-methyl-4-pentenylamine with $[PtCl_4]^{2-}$ in aqueous acid produced the *trans*- and *cis*-dimethylpyrrolidine in 85% yield and an 88:12 ratio, although the reaction required 21 d for completion.[239] Iodocyclization with an oxazoline as the nucleophile gave a slight preference for the *cis* isomer with R = Me, but a large preference for the *trans* isomer when R was the much more sterically demanding *t*-butyl group (equation 109).[240] Selenoamination of some arylamine systems with an (*E*) double bond and an allylic substituent to generate mitosane ring systems proceeded with a moderate (80:20) to high (>95:5) preference for the *trans* isomers.[241]

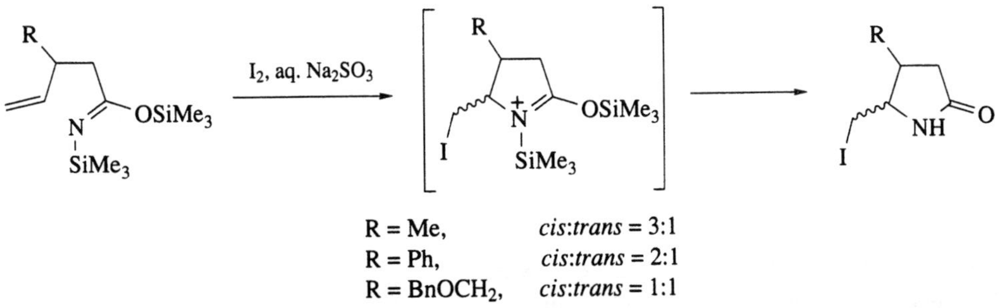
R
N
$SiMe_3$
$OSiMe_3$
I_2, aq. Na_2SO_3
R
I
N
$SiMe_3$
$OSiMe_3$
R
I
NH
O
R = Me, *cis:trans* = 3:1
R = Ph, *cis:trans* = 2:1
R = $BnOCH_2$, *cis:trans* = 1:1

(108)

High asymmetric induction has been observed in some cyclizations to 2,5-disubstituted pyrrolidines. The selectivity in these cases appears to be a function of the nitrogen derivative used and the reaction conditions (equation 110 and Table 30). Cyclization of carbamate derivatives with mercury salts or I(col-

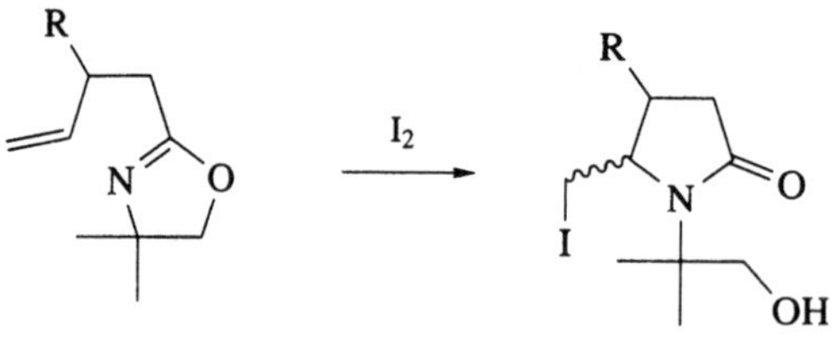

(109)

R= Me, *trans*:*cis*= 1:1–2; R= Bu^t, *trans*:*cis*= 9:1

lidine)$_2$ClO$_4$ provides *trans* products with high stereoselectivity.[242] The high selectivity in the mercuration reactions has been shown to be a result of kinetic control; equilibration of the products results in a predominance of the *cis* isomer.[246] Cyclization with *N*-phenylselenophthalamide (PhSeNPht) produced a 1:1 mixture of the two isomers (Table 30, entry 3).[216b] Cyclizations of substrates in which the C-5 substituent is part of a β-lactam ring have been reported to give only the *trans* isomer (entry 4).[243]

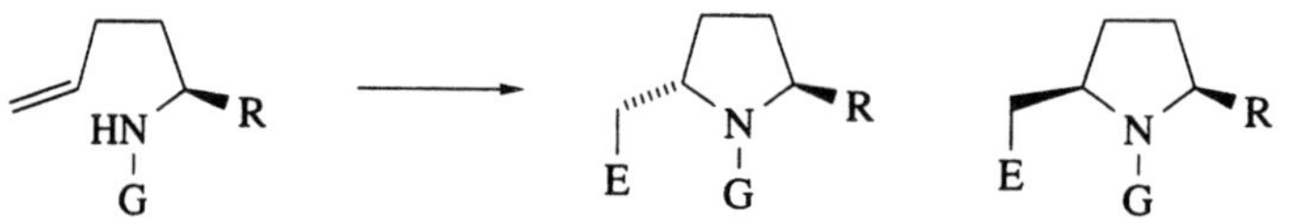

(110)

Table 30 Cyclization to 2,5-Disubstituted Pyrrolidine Derivatives (Equation 110)

Entry	*G*	*R*	*Electrophile*	*E*	*Ratio (trans:cis)*	*Ref.*
1	CO_2R or COMe	Me or Pr	$Hg(OAc)_2$	HgOAc	98:2	242a, 242b
2	CO_2R	Me	$I(coll)_2ClO_4$	I	>95:5	83a, 242b, 242c
3	CO_2R	Me	PhSeNPht	SePh	≈1:1	216b
4	G, R = —$COCH_2$—		$Hg(OAc)_2$ or I_2	HgOAc or I	>95:5	243
5	OEt	Me	I_2	I	≤75:25	244
6	Me	Me	$HgCl_2/H_2O$–THF	HgCl	93:7	245
7	Me	Me	$HgCl_2$/THF	HgCl	36:64	245
8	H	Me	$PtCl_4^{2-}$/aq. HCl	H	60:40	239

In contrast, cyclizations of *N*-alkoxy derivatives with iodine are much less selective (Table 30, entry 5), and similar results are observed with systems containing internal double bonds of either (*E*)- or (*Z*)-configuration.[238] The stereoselectivity of the aminomercuration reaction of *N*-methyl derivatives has been reported to be controlled by the choice of mercury(II) salt and solvent (entries 6 and 7).[244] Cyclization with platinum salts showed little selectivity (entry 8).[239] An iodocyclization of a complex *N*-alkyl system has been used in a recent synthesis of (+)-croomine.[238]

High stereoselectivity has been observed in the cyclizations to γ-lactams of thioimidate systems with a homoallylic substituent (equation 111).[217b,217c] The selectivity is considered to arise from $A^{1,2}$ strain favoring a quasiaxial orientation of the homoallylic substituent in (**17**; see above). In comparison, the cyclizations to γ-lactams shown in Table 29 were not stereoselective.[235]

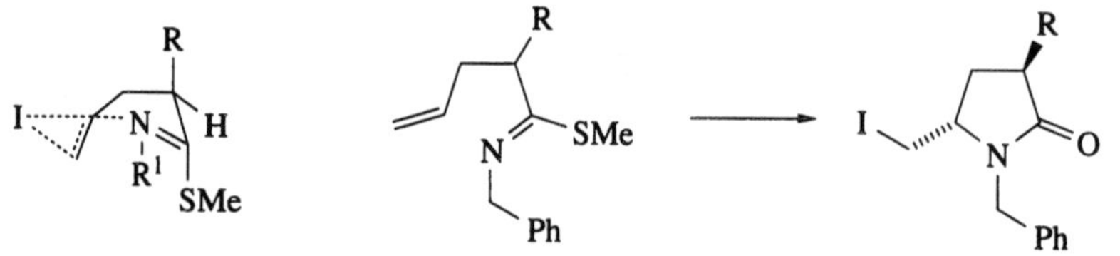

(111)

(17) R = Me, C_6H_{11}, Ph *trans*:*cis* = 12–14:1

Few examples of stereoselectivity in 6-*exo* cyclizations to piperidine derivatives have been reported. The aminocarbonylation reactions of systems with an allylic hydroxy substituent that are quite stereoselective in cyclization to pyrrolidine systems (equation 106) are nonselective in cyclizations to piperidine systems.[237] Cyclizations with mercury(II) acetate also proceed with low selectivity (59–69% *cis* isomer).[237] A series of aminoalditols have been synthesized by aminomercuration of oxygen-substituted 6-(*N*-benzylamino)hexenes. The stereochemistry of these cyclizations (equation 112)[247] does not appear to be controlled solely by the stereochemistry of the allylic oxygen substituent. Cyclization of the aminoalkene (**18**) derived from galactose (entry 1) with mercury(II) trifluoroacetate under conditions of kinetic control did give a predominance of the equatorial product after reductive oxygenation.[247b] Cyclization with mercury(II) bromide (thermodynamic control) gave a 1:1 mixture of the stereoisomeric organomercury compounds. Cyclization of the aminoalkene (**18**) derived from mannose (entry 2) with mercury(II) trifluoroacetate also gave predominantly the product with the hydroxymethyl *cis* to the allylic benzyloxy substituent.[247a] Iodocyclization gave the same ratio of stereoisomers. However, cyclization of the aminoalkene (**18**) derived from glucose (entry 3) produced the *trans* isomer as the predominant product.[247a]

(**18**) $\xrightarrow{Hg(OTFA)_2,\ THF}$ i + ii (112)

Entry	*R¹*	*R²*	*R³*	*R⁴*	*Ratio (i:ii)*
1	OBn	H	OBn	H	75:10
2	H	OBn	H	OBn	1:7
3	H	OBn	OBn	H	61:39

Intramolecular amidomercuration of carbamate derivatives to generate 2,6-disubstituted piperidine systems (equation 113 and Table 31) proceeds with low selectivity under conditions of kinetic control (*cis:trans* = 40:60),[246,248] but cyclization under conditions of thermodynamic control highly favors the *cis* isomer (*cis:trans* ≥ 98:2).[246] Interestingly, only the *cis* isomer was isolated from cyclization with phenylselenenyl chloride in the presence of silica gel.[216a]

(113)

Table 31 Cyclization of Carbamates to 2,6-Disubstituted Piperidine Derivatives (Equation 113)

Entry	*R*	*R¹*	*Conditions*	*Yield (%)*	*cis:trans*	*Ref.*
1	Me	Bn	$Hg(OAc)_2$, THF		40:60	246
2	Me	Bn	$Hg(O_2CCF_3)_2$, CH_3NO_2	93	>98:2	246
3	Me	Me	$Hg(OAc)_2$, THF	>62	~40:60	248
4	$C_{11}H_{23}$	Me	PhSeCl, silica	84	*cis* only	216a

Lower stereoselectivity has been observed with other nitrogen nucleophiles (equation 114 and Table 32). Iodocyclization of *N*-alkoxyamine derivatives gave a slight preponderance of the *trans* products with terminal vinyl groups (Table 32, entry 1), but the *cis* isomers predominated when the double bond was disubstituted with a (*Z*)-configuration (entry 2).[238] Aminomercuration of a related primary amine with the double bond an (*E*)/(*Z*) mixture provided a moderate excess of the *cis* isomer (entry 3).[249] The cyclization of a similar system with a homoallylic alkoxy substituent was also only moderately selective.[250a] However, the cyclization of a conformationally constrained system proceeded with quite high selectivity (equation 115).[250b] Aminomercuration reactions to form *N*-aryl-1-aza-4-silacyclohexanes showed variable stereoselectivities (*cis:trans* = 30:70 to 58:42).[251]

A stereoselective 6-*exo* selenoamination to form a tetrahydroisoquinoline ring system is probably a result of the specific substitution pattern.[252] Selenocyclizations which generate piperidine systems by 6-*endo* cyclization give products predicted by reaction through the more stable chair conformation (equation 116).[41,158c,216e,232a]

(114)

Table 32 Cyclization of Amines and *N*-Alkoxyamines to 2,6-Disubstituted Piperidine Derivatives (Equation 114)

Entry	R^1, R^2	*R*	*G*	*Conditions*	*Yield (%)*	*cis:trans*	*Ref.*
1	H, H	Alkyl	OCH_2OCH_3	I_2/CH_2Cl_2, Et_2O	~82	~36:64	238
2	H, alkyl	Alkyl	OCH_2OCH_3	I_2/CH_2Cl_2, Et_2O	80	67:33	238
3	alkyl, H	Alkyl	H	i, $Hg(OAc)_2$ ii, $NaBH_4$	52	79:21	249

i, $Hg(OAc)_2$, MeOH; ii, $NaBH_4$

3.3% 76% (115)

PhSeCl, MeCN (116)

Y	*Z*	*G*	*Yield (%)*	*Ref.*
O	H, Bu	Bu	73	232a
H, H	H, COMe	COMe	79	41, 158c
H, H	=C(OR)Me	COMe	30–40	216e

(ii) Cyclization of heteroatom-tethered systems

Cyclizations of systems in which nitrogen nucleophiles have been tethered to the oxygen atoms of unsaturated alcohols have been used for stereo- and regio-selective syntheses of 1,2- and 1,3-amino alcohols. A variety of nitrogen nucleophiles have been examined.

Bromocyclization of the thiocarbamidate generated from *t*-butylisothiocyanate and 3-buten-2-ol gave oxazolidinone products with slight selectivity for the *cis* isomer (*cis:trans* $\leq$ 1.9:1),[227a] while iodocyclizations of *N*-sulfonylated carbamates gave somewhat higher stereoselectivity (equation 117).[253]

(117)

Y	*Z*	*Conditions*	*Yield (%)*	*Ratio (c:t)*	*Ref.*
NBu^t	SMe	$Br(coll)_2ClO_4$, CH_2Cl_2, –78 °C	68	1.4:1	227a
O	$NHSO_2C_6H_4Me$	I_2, ether/K_2CO_3	78	2.7:1	253

Iodocyclization of trichloroacetimidates forms *trans*-1,3-oxazolines with moderate to good stereoselectivity (equation 118).[254] Cyclizations of systems with internal double bonds (equation 119) generate 5-*exo* products *only* if the double bond has a (*Z*)-configuration (entries 2 and 3) *or* an oxygen substituent is attached to the allylic carbon (entries 3 and 4).[104c,153a,157b,254c]

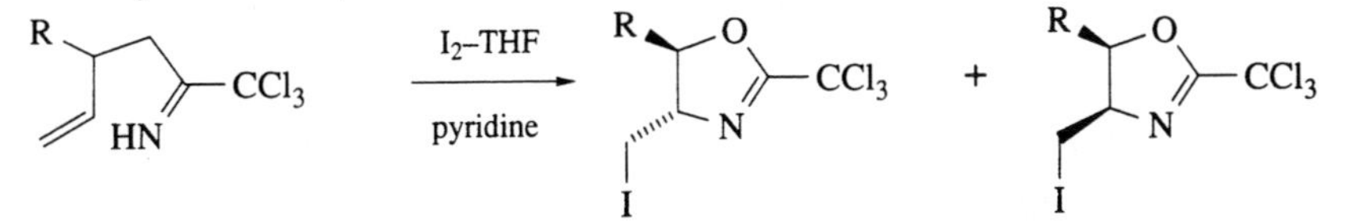

I_2–THF, pyridine (118)

Ratio (trans:cis)
75:25 (R = C_5H_{11})
95:5 (R = Pr^i)

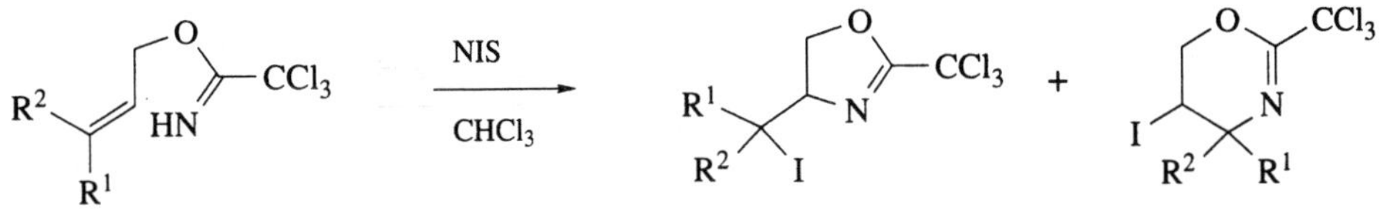

NIS, $CHCl_3$ (119)

Entry	R^1	R^2	*Yield*	
1	alkyl or aryl or alkenyl	H		
2	H	alkyl	88–90%	80–90%
3	H	CH_2OR	81–98%	
4	CH_2OR	H	55–75%	18–38%

The bromocyclization of *N,N*-dialkylaminomethyl ethers of allyl and propargyl alcohols to form oxazolidinium salts has been reported, but not used in synthesis.[255] The heterocyclization of *N*-acylaminomethyl ethers with mercury salts has been used for stereoselective synthesis of a variety of 1,2-amino alcohol systems. These cyclizations form *trans*-4,5-dialkyloxazolidine products with good to excellent stereoselectivities (equation 120 and Table 33). As shown by entry 5, 6-*endo* cyclization predominates (6:3) with an internal double bond of (*E*)-configuration, but this mode of cyclization is reduced with substrates containing a (*Z*) double bond and/or allylic oxygen substitution (Table 33, entries 6–9).

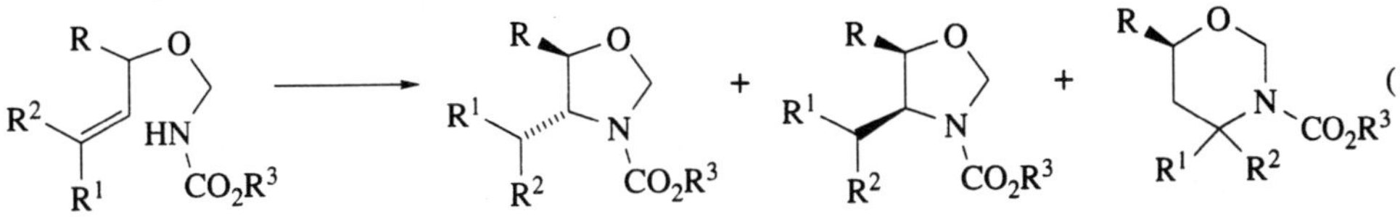

(120)

Table 33 *N*-Cyclization of *N*-Acylaminomethyl Ether Derivatives of Allylic Alcohols (Equation 120)

Entry	*R*	R^1	R^2	*Conditions*	*Ratio*	*Ref.*
1	Me	H	H	i, $Hg(OAc)_2$, MeCN; ii, $NaBH_4$, ^-OH	75:25:0	256a
2	$(CH_2)_3OBn$	H	H	i, $Hg(OAc)_2$, MeCN; ii, $NaBH_4$, ^-OH	81:19:0	256d
3	$CHMe_2$	H	H	i, $Hg(OAc)_2$, MeCN; ii, $NaBH_4$, ^-OH	>95:5:0	256a
4	Ph	H	H	i, $Hg(OAc)_2$, MeCN; ii, $NaBH_4$, ^-OH	>95:5:0	256a
5	Me	Me	H	i, $Hg(OTFA)_2$, MeCN; ii, $NaBH_4$, ^-OH	2:1:6	256b
6	Me	H	Me	i, $Hg(OTFA)_2$, MeCN; ii, $NaBH_4$, ^-OH	10:1:5.5	256b
7	$(CH_2)_2NPht$	H	CH_2OBn	i, $Hg(OTFA)_2$, EtOAc; ii, $NaBH_4$, ^-OH	65:0:17	256c
8	H	H	CH_2OBn	i, $Hg(NO_3)_2$, MeCN; ii, $NaBH_4$, ^-OH	Oxazolidine only	256b
9	H	CH_2OBn	H	i, $Hg(NO_3)_2$, MeCN; ii, $NaBH_4$, ^-OH	Oxazolidine only	256b

Quite recently, new methods for stereocontrol in cyclizations of *N*-acylaminomethyl ethers have been developed. *N,O*-acetals of type (**19**; Scheme 13) were prepared from the corresponding secondary allyl alcohol and the diastereomers were separated by chromatography and/or crystallization. Cyclization with mercury(II) salts and reduction of the organomercury intermediate proceeded with high stereocontrol exerted by the amidal stereogenic center, *not* the stereogenic center on the original alcohol.[257a]

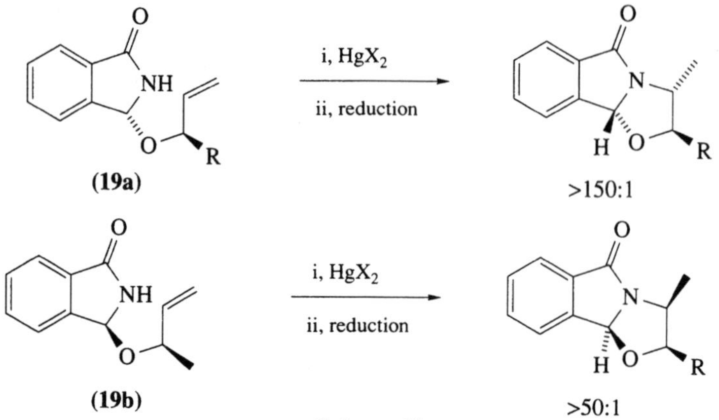

(**19a**) — i, HgX_2; ii, reduction → >150:1

(**19b**) — i, HgX_2; ii, reduction → >50:1

Scheme 13

A related strategy provides for control of absolute stereochemistry through the use of a chiral auxiliary (Scheme 14).[257b] Generation of the amidal (**20**) proceeds with reasonable stereoselectivity (87:13), and chromatography provides a pure diastereomer. Cyclization of either diastereomer gives an oxazolidine product in which the configuration of the new C—N bond is controlled solely by the configuration of the stereogenic center containing the —CCl_3 group. Interestingly, the product is a *trans*-2,5-disubstituted oxazolidine, while the cyclizations shown in Scheme 13 give products with a *cis* relationship between the groups at C-2 and C-5. Cyclization of a system similar to that shown in Scheme 14, but with stereocenters only on the chiral auxiliary (H replacing the —CCl_3 group) gave only a 60:40 ratio of diastereomers.

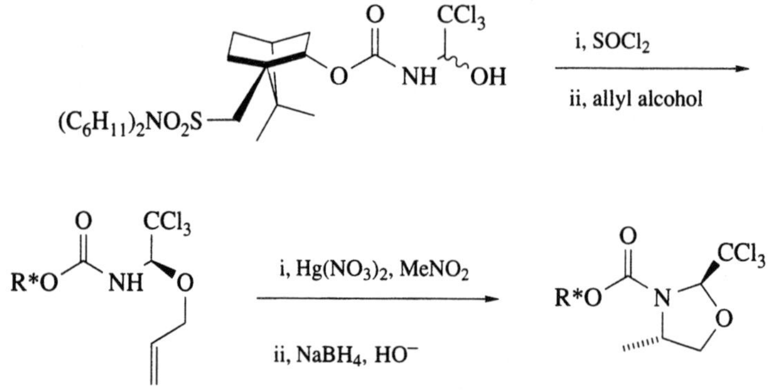

$(C_6H_{11})_2NO_2S$ — i, $SOCl_2$; ii, allyl alcohol →

(**20**) — i, $Hg(NO_3)_2$, $MeNO_2$; ii, $NaBH_4$, HO^- → only isomer observed

Scheme 14

Studies on tethered derivatives of homoallylic alcohols are fewer in number. Iodocyclization of trichloroacetimidate derivatives of 4-penten-2-ol (**21**) gave *cis*- and *trans*-dihydrooxazines in an 80:20 ratio,[254a,254b] while cyclization of the *N*-benzenesulfonylcarbamate derivatives (**22**) provided an 86:14 ratio of cyclic carbamates.[253] Cyclization of the *N*-acylaminomethyl derivatives (**23**) with mercury(II) trifluoroacetate under conditions of kinetic control gave a 75:25 ratio of *cis*- and *trans*-tetrahydrooxazine derivatives, but cyclization under conditions which allowed for equilibration of the organomercury intermediates produced a 6:94 ratio.[258]

A variety of nitrogen nucleophiles have been tethered to the nitrogen of allylic and homoallylic amines. Examples include halocyclizations of amidines (equation 121),[259a,259b] *O*-substituted isoureas (equations 122[259a,259c] and 123[98b]) and *S*-alkylisothioureas (equation 124),[259e,259f] aminopalladation/methoxycarbonylation of alkylurea derivatives (equation 125)[259d] and bromocyclization of substituted urea derivatives (equation 126).[235] Bromocyclization of an isothiourea has also been reported.[1c,259e] Although most studies have resulted in cyclization to five-membered rings, halocycliza-

HN CCl_3 O (21) HN O SO_2Ph (22) HN O CO_2Bn (23)

tion of amidine derivatives[259b] and selenocyclization of urea derivatives[98a] of cinnamylamine give products of 6-*endo* cyclization.

NIS, THF or I_2, py/THF; or Br_2, CCl_4; 53–100% (121)

R^1 = Ph or CCl_3, R^2 = H or Ph

NIS or I_2; 71–90% (122)

$R^1 = R^3 = SiMe_3$, R^2 = Me or Ph; R^1 = Et, R^3 = H, R^2 = Ph(Me)CH

PhSeCl, $CHCl_3$; silica gel (123)

I_2, CH_2Cl_2; 77–89% (124)

R = Ph or Ac

$PdCl_2$, $CuCl_2$; CO, MeOH; 47–92% (125)

NBS, CCl_4, 23 °C (126)

R^1 = aryl, aroyl, or EtO_2CCH_2; R^2 = Me or Ph

Related intramolecular aminomercurations of cyclic guanidine derivatives have been studied in some detail,[260a] but are primarily useful for the synthesis of bicyclic heterocycles (equations 127 and 128). Cyclizations of allylic isothiouronium salts have also been reported (equation 129).[260b,260c]

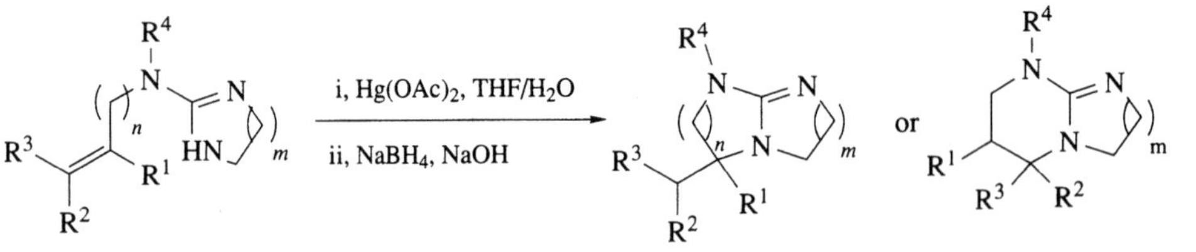

i, $Hg(OAc)_2$, THF/H_2O; ii, $NaBH_4$, NaOH — or (127)

n = 1 or 2; m = 1, 2 or 3

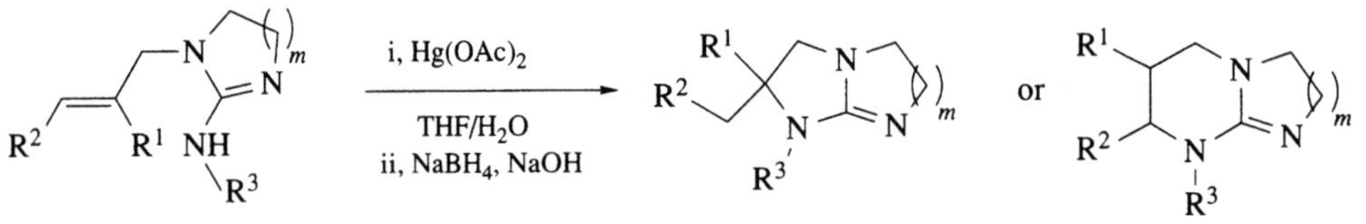

i, $Hg(OAc)_2$, THF/H_2O; ii, $NaBH_4$, NaOH — or (128)

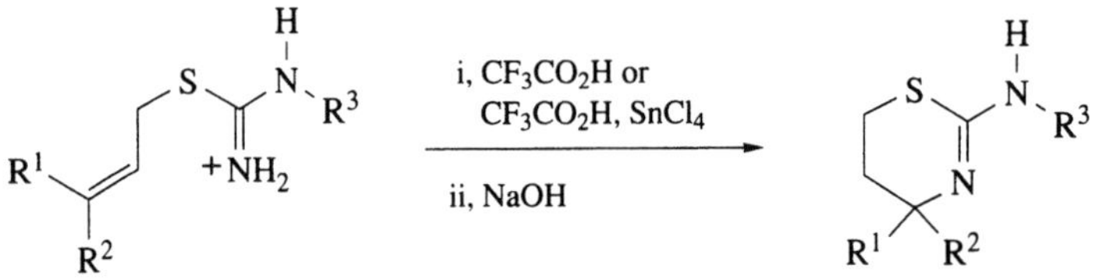

i, CF_3CO_2H or CF_3CO_2H, $SnCl_4$; ii, NaOH (129)

R^1, R^2 = alkyl or aryl, R^3 = H or aryl

(iii) Miscellaneous

Some synthetically useful reactions not covered in the above sections should be mentioned. An unusual heterocyclization to form indole derivatives is found in the aminomercuration of allylenaminones (equation 130).[261] Palladium(II)-catalyzed cyclizations of *o*-allyl- and *o*-vinyl-anilines and aminoquinones are discussed elsewhere.[50] Other electrophiles used to form dihydroindole derivatives include phenylselenenyl chloride[216a] and methyl(bismethylthio)sulfonium salts.[231]

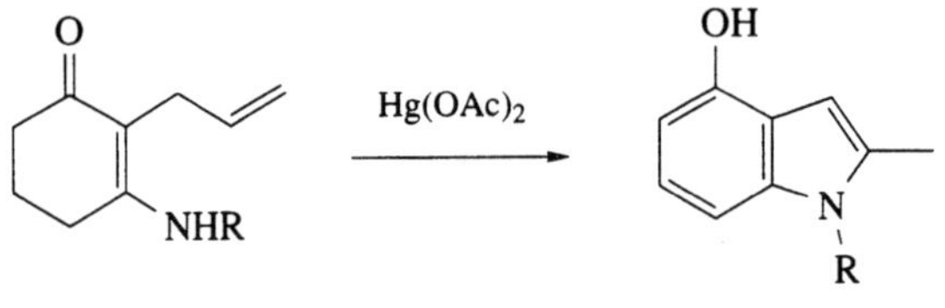

$Hg(OAc)_2$ (130)

Cyclizations of alkenic amines and imines using organoiron complexes to generate bicyclic β-lactams are discussed in Chapter 3.1 of this volume. Examples of heterocyclizations of alkenic *N,N*-dialkylamine and pyridine derivatives to form cyclic quaternary ammonium salts are cited in the Staninets review.[1c] A cyclization of an enol thioether has been used to generate a thiazolidine intermediate used in cephalosporin synthesis (equation 131).[262]

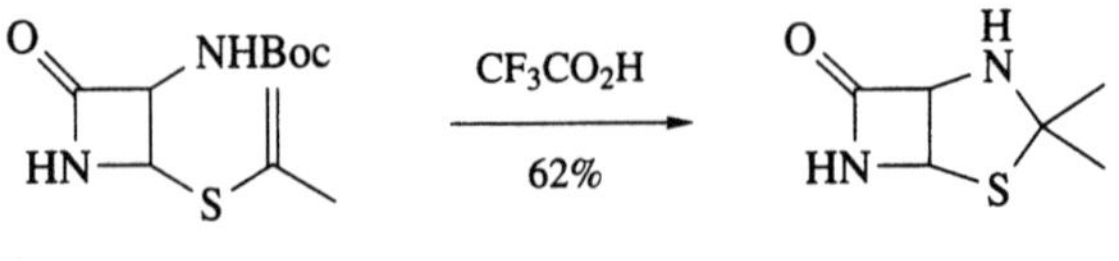

CF_3CO_2H, 62% (131)

Examples of cyclization through alkene insertion of a metal–nucleophile intermediate (Scheme 3, intermediate **F**) using organolanthanide catalysts, $(Cp'_2LaH)_2$, have been reported recently (equation 132), although turnover numbers rather than synthetic yields were reported.[263]

$$\xrightarrow{{}^1\!/_2(\mathrm{Cp'_2LaH})_2} \qquad (132)$$

$n = 1$, R = H
$n = 1$, R = Me (*trans*:*cis* = 5:1 at 25 °C or 8:1 at 0 °C)
$n = 2$, R = H

1.9.3.2.3 Cyclizations with alkynes and allenes

Cyclizations with nitrogen nucleophiles involving alkynes and allenes have received little attention until recently. The cyclizations of several α-aminoallenes to 3-pyrrolines with silver tetrafluoroborate was reported by Claesson and coworkers (equation 133).[264] A similar cyclization to form Δ^1-carbapenems has been reported (equation 134).[265a] Diastereomeric allenes ($R^1 \neq R^2$) were shown to cyclize with complete stereocontrol. Cyclization with palladium chloride in the presence of allyl bromide or electrophilic alkenes allowed for the intermediate vinylpalladium species to be trapped by the electrophile.[265b] A related product was obtained by cyclization of an alkynic substrate (equation 135).[265a] Other examples of 5-*endo* cyclization of β-aminoalkynes[50] include the formation of indoles by cyclization of 2-alkynylanilines with mercury salts[200] or palladium chloride,[266a,266b,266c] formation of 1-pyrrolines with catalytic palladium chloride (equation 136)[198] and formation of pyrroles by cyclization of hydroxy-substituted β-aminoalkynes.[198,266d]

$$\xrightarrow[\text{CHCl}_3,\ 85\%]{\text{AgBF}_4} \qquad (133)$$

$$\xrightarrow[\text{CH}_2\text{Cl}_2,\ 60\%]{\text{AgBF}_4} \qquad (134)$$

$$\xrightarrow[\text{H}_2\text{O/acetone},\ 45\%]{\text{AgNO}_3/\text{CaCO}_3} \qquad (135)$$

$$\xrightarrow[\text{MeCN},\ \Delta,\ 43\%]{\text{cat. PdCl}_2} \qquad (136)$$

Cyclization of β-allenylamine systems to six-membered ring products has been observed recently. Cyclization of terminally substituted β-allenylamides with silver tetrafluoroborate produced lactam products (equation 137).[267a] The cyclization shown in equation (138) was used to generate the indolizidine ring system of allopumiliotoxin alkaloids.[267b]

$$\xrightarrow[\text{CHCl}_3,\ 60\%]{\text{AgBF}_4} \qquad (137)$$

p-TsOH, MeCN, >40% (138)

A number of cyclizations of γ-allenylamines have been reported. Cyclization studies of N-alkyl and N-aryl systems with mercury and silver salts to form 2-alkenylpyrrolidines[268a] were followed by studies of stereoselectivity in the formation of 2,5-disubstituted pyrrolidines (equation 139 and Table 34).[268b] Although cyclization of the primary amine was nonselective, only the *cis* product could be observed with N-substituted amines. This reaction has been used in a stereoselective synthesis of anatoxin-a.[268d] Interestingly, the aminopalladation/methoxycarbonylation reaction provides a modest excess of the *trans* isomer (Table 34, entry 3).[268c] Recently, it has been demonstrated that reasonable stereochemical control can be exerted by a chiral group attached to nitrogen if a coordinating group is present on the N-substituent (equation 140).[268e]

(139)

Table 34 Cyclization of γ-Allenylamine Derivatives (Equation 139)

Entry	*R*	*Conditions*	*Cis:trans ratio*	*R*^{1}	*Ref.*
1	H	$AgBF_4$, CH_2Cl_2	1:1	H	268d
2	CO_2Bn or CH_2Ph or SO_2PhMe	$AgBF_4$, CH_2Cl_2	>50:1	H	268d
3	CH_2Ph or SO_2PhMe	$PdCl_2$, $CuCl_2$ MeOH, CO	1:3	CO_2Me	268c

0.5 equiv. $AgOSO_2CF_3$, CH_2Cl_2, 20 °C, 90–97% (140)

Y = H, H or O

80% *de*

Cyclization of δ-allenylamines is successful also. Synthesis of 2-alkenylpiperidines can be effected with mercury(II) or silver salts, with silver salts giving higher yields (equation 141 and Table 35).[268a] The significant asymmetric induction found in the cyclization of a chiral allene (78% *ee*, entry 2) suggests that the low stereoselectivity observed in the synthesis of the 2,6-disubstituted system (entry 3)[269] may be a result of starting with a diastereomeric mixture. Aminopalladation/methoxycarbonylation has been effected in moderate yield also (entry 4).

Silver ion catalysis has also been used with allenic oximes to generate cyclic nitrones, which are trapped by 1,2-dipolarophiles (equation 142).[270] This reaction was used in a synthesis of pyrrolizidine alkaloids.[270b]

(141)

Table 35 Cyclization of δ-Allenylamine Derivatives (Equation 141)

Entry	*R*[1]	*R*[2]	*R*	*Z*	*Conditions*	*Yield (%)*	*Ref.*
1	H	H or alkyl	Pr or Bn	H	$AgNO_3$, H_2O/acetone	76–95	268a
2	H	Me	Bn	H	$AgBF_4$, CH_2Cl_2, 20 °C	86 (78% *ee*)	269b
3	Me	Me	Bn	H	$AgNO_3$, H_2O/acetone	~100 (62:38)	269a
4	H	H or Me	Bn	CO_2Me	$PdCl_2$, MeOH/CO	42–52	268c

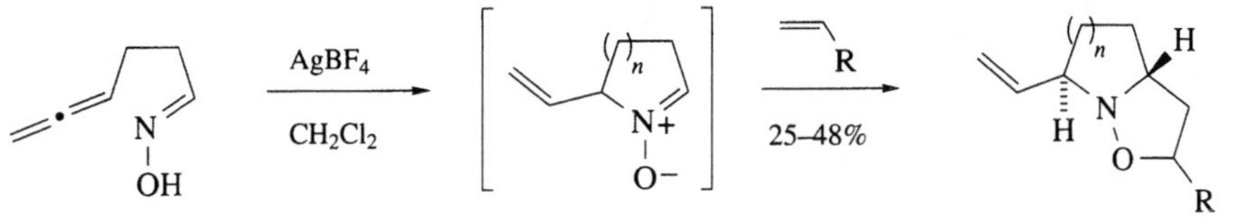

(142)

1.9.4 SULFUR NUCLEOPHILES

Reviews which cover heterocyclization of unsaturated sulfur compounds have been published recently.[1d,1f] Cyclizations of sulfur systems can also be effected by radical chain mechanisms or by reactions in which the sulfur acts as an electrophile. Examples of nucleophilic sulfur functional groups used in heterocyclization reactions are shown in Figure 3.

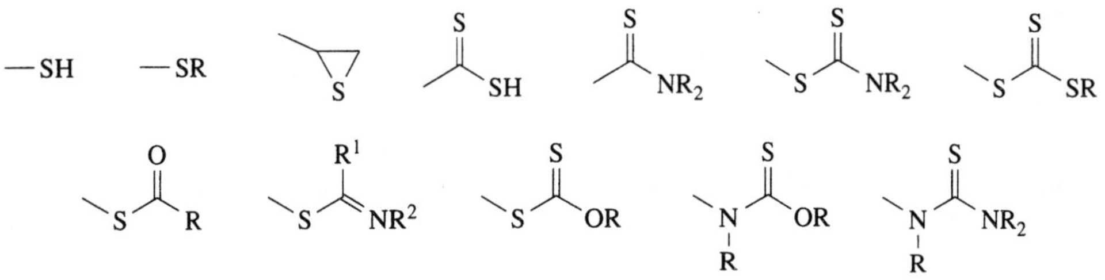

Figure 3

Sulfur heterocyclization reactions have been used to provide sulfur heterocycles of many types, but few examples have been used as general synthetic methods. One exception is found in the synthesis of thiaprostacyclins.[146] The key step in the synthesis of the thiaprostacyclins (**25**) was the cyclization of 9-thia-$PGF_{2\alpha}$ derivatives with phenylselenenyl chloride (Scheme 15).[271a] The cyclization reaction was shown to proceed with either the thiol (R = H) or thioacetate (R = COMe) as nucleophile and with high stereoselectivity (the (*Z*)-isomer gave only the *endo* product, while the (*E*)-isomer gave only the *exo* product).[271a,271b] The stereoselectivity is much higher than found in the corresponding selenoetherification reactions with oxygen as nucleophile.[272] Thiaprostacyclins have also been prepared by cyclization with halogens, but reactions of this type are considered to proceed through the sulfenyl halide with electrophilic addition of the sulfur to the alkene.[273] 9-Thiaprostacyclins have also been prepared by nucleophilic addition of a thiolate to an alkyne analog of (**24**).[274]

Cyclization of either of the thiols (**24**) with acid provided the saturated thia-PGI_1 analogs (**26**) as a 2.7:1 mixture of the *exo* and *endo* isomers.[271a] Although many addition reactions of thiols to alkenes are radical reactions catalyzed by light, oxygen, or other radical initiators, conversion to (**26**) was not effected under photolytic conditions or with AIBN. A radical chain mechanism for selenothiolactonization has also been reported recently.[275]

A recent paper[276] has shown that some halocyclizations involving sulfur functionalities other than thiols (*e.g.* thiazolidinones) may involve initial reaction at sulfur to generate a sulfenyl halide which cyclizes *via* attack of the electrophilic sulfur. This type of reaction would be an alternative to the nucleophilic heterocyclization mechanism previously proposed for a synthesis of a biotin intermediate from a vinylthiazolidine.[1p,277]

One recent study (Scheme 16) has shown that the iminothiolactones generated from iodocyclization of γ,δ-unsaturated thioamides can be converted to 2-acetamidothiophenes[278a,278b] in addition to thiolactones, as previously reported.[114b,278c] Iodocyclizations of allylic or homoallylic thioureas to generate dihydrothiazoles and dihydrothiazines (equation 143) have been reported recently.[259e]

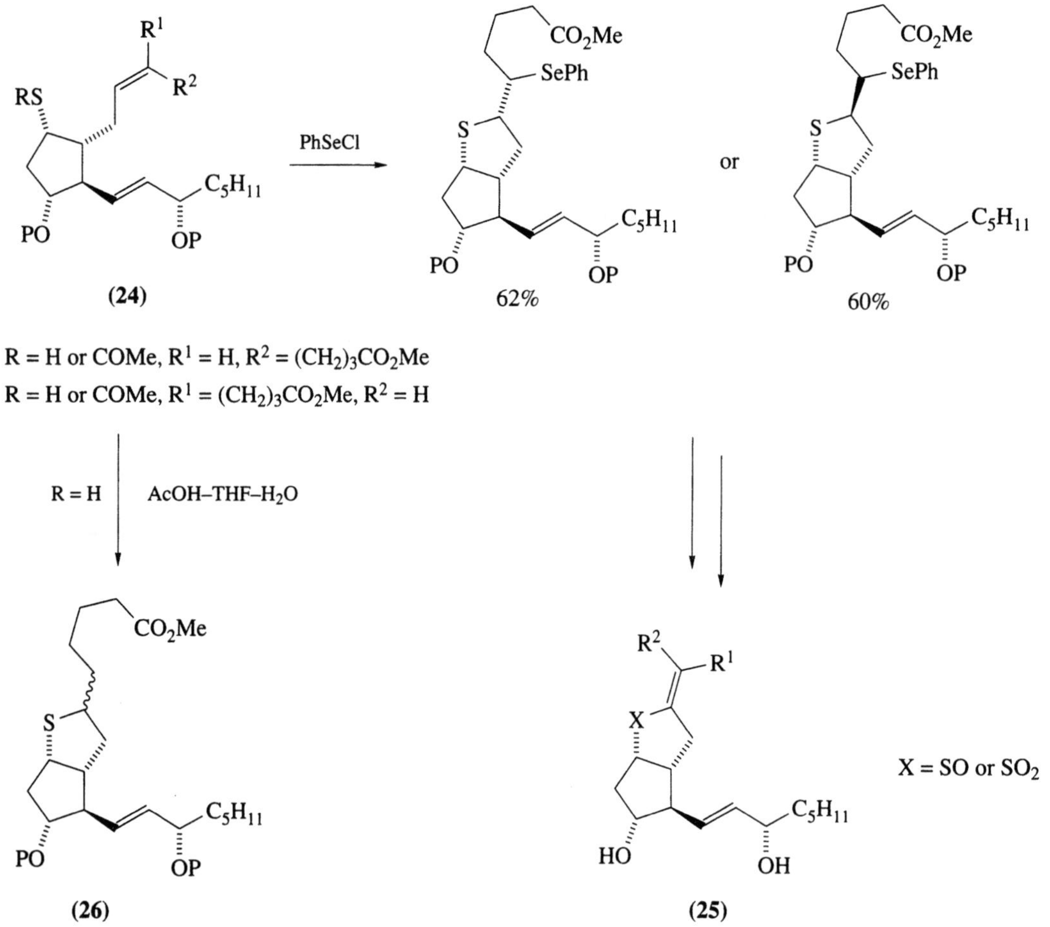

PhSeCl

CO_2Me SePh S PO OP C_5H_{11} 62%

or

CO_2Me SePh S PO OP C_5H_{11} 60%

(24)

R = H or COMe, R^1 = H, R^2 = $(CH_2)_3CO_2Me$
R = H or COMe, R^1 = $(CH_2)_3CO_2Me$, R^2 = H

R = H AcOH–THF–H_2O

(26)

X = SO or SO_2

(25)

Scheme 15

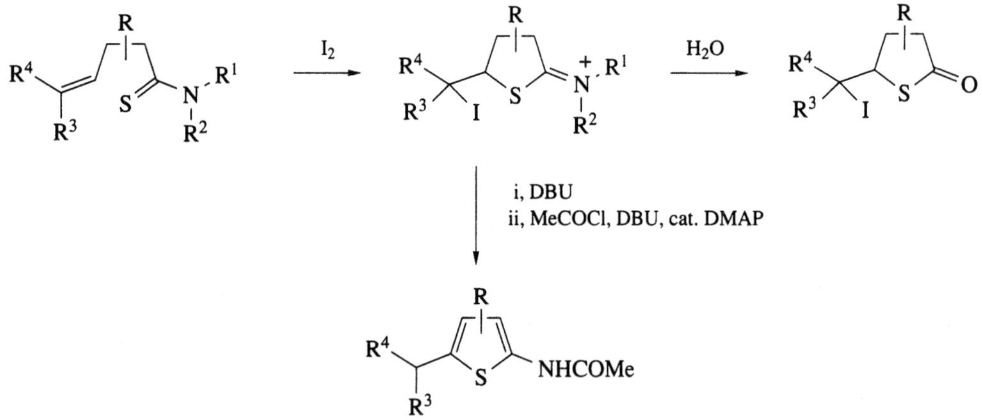

I_2 H_2O

i, DBU
ii, MeCOCl, DBU, cat. DMAP

NHCOMe

Scheme 16

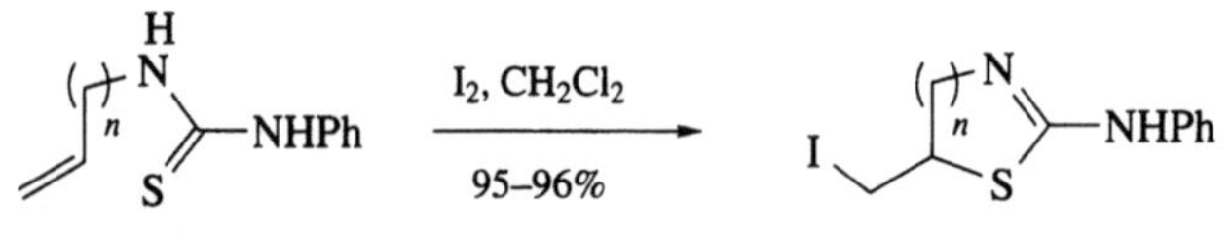

I_2, CH_2Cl_2
95–96%

NHPh (143)

n = 1 or 2

1.9.5 REFERENCES

1. (a) M. D. Dowle and D. I. Davies, *Chem. Soc. Rev.*, 1979, 171; Yu. I. Gevaza and V. I. Staninets, *Khim. Geterotsikl. Soedin.*, 1988, 1299; (b) Yu. I. Gevaza and V. I. Staninets, *Khim. Geterotsikl. Soedin.*, 1982, 1443; (c) V. I. Staninets and Yu. I. Gevaza, *Khim. Geterotsikl. Soedin.*, 1985, 435; (d) Yu. I. Gevaza and V. I. Staninets, *Khim. Geterotsikl. Soedin.*, 1986, 291; (e) C. M. Angelov, *Phosphorus Sulfur*, 1983, **15**, 177; (f) E. N. Karaulova, *Russ. Chem. Rev. (Engl. Transl.)*, 1987, **56**, 546; (g) Yu. I. Gevaza and V. I. Staninets, *Khim. Geterotsikl. Soedin.*, 1984, 867; (h) S. Kano, S. Shibuya and T. Ebata, *Heterocycles*, 1980, **14**, 661; (i) M. M. Joullie and J. E. Semple, *Heterocycles*, 1980, **14**, 1825; (j) D. L. J. Clive, *Aldrichimica Acta*, 1978, **11**, 43; (k) C. Paulmier, 'Selenium Reagents and Intermediates in Organic Synthesis', Pergamon Press, Oxford, 1986, p. 229; (l) P. A. Bartlett and J. L. Adams, *J. Am. Chem. Soc.*, 1980, **102**, 337; (m) P. A. Bartlett, in 'Asymmetric Synthesis', ed. J. D. Morrison, Academic Press, San Diego, 1984, vol. 3, chap. 6; (n) N. Petragnani and H. M. C. Ferraz, *Synthesis*, 1985, 27.
2. (a) P. Kočovsky, F. Tureček and J. Hájíček, 'Synthesis of Natural Products: Problems of Stereoselectivity', CRC Press, Boca Raton, FL, 1986, vol. 2; (b) P. B. D. de la Mare and R. Bolton, 'Electrophilic Additions to Unsaturated Systems', Elsevier, Amsterdam, 1982.
3. B. Capon and S. P. McManus, 'Neighboring Group Participation', Plenum Press, New York, 1976.
4. D. L. J. Clive, C. G. Russell, G. Chittattu and A. Singh, *Tetrahedron*, 1980, **36**, 1399.
5. D. L. H. Williams, *J. Chem. Soc. B*, 1969, 517.
6. R. G. Bernett, J. T. Doi and W. K. Musker, *J. Org. Chem.*, 1985, **50**, 2048.
7. S. L. Bender, Ph.D. Thesis, Harvard University, 1986 (*Chem. Abstr.*, 1987, **106**, 84 249f); D. A. Evans, S. L. Bender and J. Morris, *J. Am. Chem. Soc.*, 1988, **110**, 2506.
8. (a) P. A. Bartlett and J. Myerson, *J. Am. Chem. Soc.*, 1978, **100**, 3950; (b) P. A. Bartlett and K. K. Jernstedt, *Tetrahedron Lett.*, 1980, **21**, 1607.
9. P. Kočovsky and I. Stieborova, *J. Chem. Soc., Perkin Trans. 1*, 1987, 1969.
10. J. E. Baldwin, *J. Chem. Soc., Chem. Commun.*, 1976, 734.
11. S. Winstein and L. Goodman, *J. Am. Chem. Soc.*, 1954, **76**, 4368.
12. M. M. Midland and R. L. Halterman, *J. Org. Chem.*, 1981, **46**, 1227.
13. B. Ganem, *J. Am. Chem. Soc.*, 1976, **98**, 858.
14. R. D. Evans, J. W. Magee and J. H. Schauble, *Synthesis*, 1988, 862.
15. P. A. Bartlett and P. C. Ting, *J. Org. Chem.*, 1986, **51**, 2230.
16. (a) W. L. Brown and A. G. Fallis, *Can. J. Chem.*, 1987, **65**, 1828; (b) S. M. Tuladhar and A. G. Fallis, *Can. J. Chem.*, 1987, **65**, 1833.
17. O. Arjona, R. F. d. l. Pradilla, J. Plumet and A. Viso, *Tetrahedron Lett.*, 1989, **45**, 4565.
18. (a) W. E. Barnett and J. C. McKenna, *Tetrahedron Lett.*, 1971, 2595; (b) W. E. Barnett and W. H. Sohn, *Tetrahedron Lett.*, 1972, 1777.
19. B. Ganem, G. W. Holbert, L. B. Weiss and K. Ishizumi, *J. Am. Chem. Soc.*, 1978, **100**, 6483.
20. W. E. Barnett and L. L. Needham, *J. Org. Chem.*, 1975, **40**, 2843.
21. R. C. Cambie, P. S. Rutledge, R. F. Somerville and P. D. Woodgate, *Synthesis*, 1988, 1009.
22. R. C. Cambie, K. S. Ng, P. S. Rutledge and P. D. Woodgate, *Aust. J. Chem.*, 1979, **32**, 2793.
23. G. W. Holbert, L. B. Weiss and B. Ganem, *Tetrahedron Lett.*, 1976, 4435.
24. W. E. Barnett and W. H. Sohn, *J. Chem. Soc., Chem. Commun.*, 1972, 472.
25. (a) P. A. Bartlett and C. F. Pizzo, *J. Org. Chem.*, 1981, **46**, 3896; (b) P. A. Bartlett and J. F. Barstow, *J. Org. Chem.*, 1982, **47**, 3933.
26. E. J. Corey, N. M. Weinshenker, T. K. Schaaf and W. J. Huber, *J. Am. Chem. Soc.*, 1969, **91**, 5675.
27. S. Ohuchida, N. Hamanaka and M. Hayashi, *Tetrahedron*, 1983, **39**, 4263.
28. R. Bernini, E. Davini, C. Iavarone and C. Trogolo, *J. Org. Chem.* 1986, **51**, 4600.
29. J. C. Barrish, H. L. Lee, T. Mitt, G. Pizzolato, E. G. Baggiolini and M. R. Uskokovic, *J. Org. Chem.*, 1988, **53**, 4282.
30. M. Srebnik and R. Mechoulam, *J. Chem. Soc., Chem. Commun.*, 1984, 1070.
31. (a) K. C. Nicolaou, D. A. Claremon, W. E. Barnette and S. P. Seitz, *J. Am. Chem. Soc.*, 1979, **101**, 3704; (b) K. C. Nicolaou, *Tetrahedron*, 1981, **37**, 4097.
32. K. C. Nicolaou, S. P. Seitz, W. J. Sipio and J. F. Blount, *J. Am. Chem. Soc.*, 1979, **101**, 3884.
33. N. Petragnani and H. M. C. Ferraz, *Synthesis*, 1978, 476.
34. J. V. Comasseto and N. Petragnani, *Synth. Commun.*, 1983, **13**, 889.
35. G. J. O'Malley and M. P. Cava, *Tetrahedron Lett.*, 1985, **26**, 6159.
36. R. F. W. Jackson, R. A. Raphael, J. H. A. Stibbard and R. C. Tidbury, *J. Chem. Soc., Perkin Trans. 1*, 1984, 2159.
37. W. C. Still and M. J. Schneider, *J. Am. Chem. Soc.*, 1977, **99**, 948.
38. A. G. Schultz and P. Sundararaman, *J. Org. Chem.*, 1984, **49**, 2455.
39. E. J. Corey, M. Shibasaki and J. Knolle, *Tetrahedron Lett.*, 1977, 1625.
40. (a) G. W. J. Fleet and C. R. C. Spensley, *Tetrahedron Lett.*, 1982, **23**, 109; (b) G. W. J. Fleet and M. J. Gough, *Tetrahedron Lett.*, 1982, **23**, 4509.
41. A. Toshimitsu, K. Terao and S. Uemura, *J. Org. Chem.*, 1987, **52**, 2018.
42. Y. Guindon, Y. St. Denis, S. Daigneault and H. E. Morton, *Tetrahedron Lett.*, 1986, **27**, 1237.
43. E. J. Corey and A. W. Gross, *Tetrahedron Lett.*, 1980, **21**, 1819.
44. M. Kato, M. Kageyama, R. Tanaka, K. Kuwahara and A. Yoshikoshi, *J. Org. Chem.*, 1975, **40**, 1932.
45. M. Kato, M. Kageyama and A. Yoshikoshi, *J. Chem. Soc., Perkin Trans. 1*, 1977, 1305.
46. (a) M. Shah, M. J. Taschner, G. F. Koser and N. L. Rach, *Tetrahedron Lett.*, 1986, **27**, 4557; (b) G. F. Koser, J. S. Lodaya, D. G. Ray, III and P. B. Kokil, *J. Am. Chem. Soc.*, 1988, **110**, 2987.
47. S. Torii, K. Uneyama, M. Ono and T. Bannou, *J. Am. Chem. Soc.*, 1981, **103**, 4606.
48. A. J. Pearson and T. Ray, *Tetrahedron*, 1985, **41**, 5765.
49. J. E. Backväll, P. G. Andersson and J. O. Vagberg, *Tetrahedron Lett.*, 1989, **30**, 137.

50. For additional examples and a discussion of the mechanism of electrophilic addition to alkenes with palladium salts, see Volume 4, Chapter 3.4.
51. A. J. Pearson, S. L. Kole and T. Ray, *J. Am. Chem. Soc.*, 1984, **106**, 6060.
52. (a) M. R. Huckstep, R. J. K. Taylor and M. P. L. Caton, *Tetrahedron Lett.*, 1986, **27**, 5919; (b) P. Brownbridge and S. Warren, *J. Chem. Soc., Perkin Trans. 1*, 1976, 2125.
53. R. E. Ireland, D. Häbich and D. W. Norbeck, *J. Am. Chem. Soc.*, 1985, **107**, 3271.
54. R. E. Ireland and P. Maienfisch, *J. Org. Chem.*, 1988, **53**, 640.
55. J. L. J. Clive, G. Chittattu, N. J. Curtis, W. A. Kiel and C. K. Wong, *J. Chem. Soc., Chem. Commun.*, 1977, 725.
56. J. V. Comasseto, H. M. C. Ferraz and N. Petragnani, *Tetrahedron Lett.*, 1987, **28**, 5611.
57. K. C. Nicolaou and Z. Lysenko, *Tetrahedron Lett.*, 1977, 1257.
58. N. X. Hu, Y. Aso, T. Otsubo and F. Ogura, *Tetrahedron Lett.*, 1987, **28**, 1281.
59. F. Tureček, *Collect. Czech. Chem. Commun.*, 1982, **47**, 858.
60. K. C. Nicolaou, R. L. Magolda, W. J. Sipio, W. E. Barnette, Z. Lysenko and M. M. Joullie, *J. Am. Chem. Soc.*, 1980, **102**, 3784.
61. S. Toteberg-Kaulen and E. Steckhan, *Tetrahedron*, 1988, **44**, 4389.
62. M. J. Kurth, R. L. Beard, M. Olmstead and J. G. Macmillan, *J. Am. Chem. Soc.*, 1989, **111**, 3712.
63. (a) M. T. Crimmins and J. G. Lever, *Tetrahedron Lett.*, 1986, **27**, 291; (b) M. T. Crimmins, W. G. Hollis, Jr. and J. G. Lever, *Tetrahedron Lett.*, 1987, **28**, 3647.
64. (a) E. J. Corey, S. Kim, S. Yoo, K. C. Nicolaou, L. S. Melvin, Jr., D. J. Brunelle, J. R. Falck, E. J. Trybulski, R. Lett and P. W. Sheldrake, *J. Am. Chem. Soc.*, 1978, **100**, 4620; (b) S. J. Danishefsky and J. Y. Lee, *J. Am. Chem. Soc.*, 1989, **111**, 4829.
65. G. A. Kraus and K. Frazier, *J. Org. Chem.*, 1980, **45**, 4820.
66. H. W. Whitlock, *J. Am. Chem. Soc.*, 1962, **84**, 3412.
67. L. A. Paquette, M. J. Wyvratt, O. Schallner, J. L. Muthard, W. J. Begley, R. M. Blankenship and D. Balogh, *J. Org. Chem.*, 1979, **44**, 3616.
68. (a) R. B. Woodward, F. E. Bader, H. Bickel, A. J. Frey and R. W. Kierstead, *Tetrahedron*, 1958, **2**, 1; (b) G. Stork, *Pure Appl. Chem.*, 1989, **61**, 439.
69. E. J. Corey and H. L. Pearce, *J. Am. Chem. Soc.*, 1979, **101**, 5841.
70. (a) E. J. Corey, R. L. Danheiser, S. Chandrasekaran, G. E. Keck, B. Gopalan, S. D. Larsen, P. Siret and J. Gras, *J. Am. Chem. Soc.*, 1978, **100**, 8034; (b) H. Nagaoka, M. Shimano and Y. Yamada, *Tetrahedron Lett.*, 1989, **30**, 971.
71. L. N. Mander, *Pure Appl. Chem.*, 1989, **61**, 397.
72. N. H. Andersen and F. A. Golec, Jr., *Tetrahedron Lett.*, 1977, 3783.
73. T. Kato, C. C. Yen, T. Kobayashi and Y. Kitahara, *Chem. Lett.*, 1976, 1191.
74. D. G. Batt, N. Takamura and B. Ganem, *J. Am. Chem. Soc.*, 1984, **106**, 3353.
75. J. D. White, M. A. Avery and J. P. Carter, *J. Am. Chem. Soc.*, 1982, **104**, 5486.
76. M. Yoshioka, H. Nakai and M. Ohno, *J. Am. Chem. Soc.*, 1984, **106**, 1133.
77. F. M. Hauser and D. Mal, *J. Am. Chem. Soc.*, 1984, **106**, 1098.
78. P. A. Bartlett and L. A. McQuaid, *J. Am. Chem. Soc.*, 1984, **106**, 7854.
79. M. Bhupathy and T. Cohen, *Tetrahedron Lett.*, 1985, **26**, 2619.
80. S. J. Danishefsky, W. H. Pearson, D. F. Harvey, C. J. Maring and J. P. Springer, *J. Am. Chem. Soc.*, 1985, **107**, 1256.
81. B. P. Mundy and W. G. Bornmann, *J. Org. Chem.*, 1984, **49**, 5264.
82. D. L. J. Clive, K. S. K. Murthy, A. G. H. Wee, J. S. Prasad, G. V. J. da Silva, M. Majewski, P. C. Anderson, R. D. Haugen and L. D. Heerze, *J. Am. Chem. Soc.*, 1988, **110**, 6914.
83. K. Mori and H. Watanabe, *Pure Appl. Chem.*, 1989, **61**, 543.
84. A. Murai, *Pure Appl. Chem.*, 1989, **61**, 393.
85. H. Straub, K.-P. Zeller and H. Leditschke, *Methoden Org. Chem. (Houben-Weyl)*, 1974, 154.
86. C. Ganter, *Top. Curr. Chem.*, 1976, **67**, 15.
87. E. Alvarez, E. Manta, J. D. Martin, M. L. Rodriguez and C. Ruiz-Perez, *Tetrahedron Lett.*, 1988, **29**, 2093.
88. N. T. Byrom, R. Grigg and B. Kongkathip, *J. Chem. Soc., Chem. Commun.*, 1976, 216.
89. B. Kongkathip and N. Kongkathip, *Tetrahedron Lett.*, 1984, **25**, 2175.
90. G. Dauphin, A. Fauve and H. Veschambre, *J. Org. Chem.*, 1989, **54**, 2238.
91. (a) C.-P. Chuang, J. C. Gallucci and D. J. Hart, *J. Org. Chem.*, 1988, **53**, 3210; (b) C.-P. Chuang, J. C. Gallucci, D. J. Hart and C. Hoffman, *J. Org. Chem.*, 1988, **53**, 3218; (c) D. J. Hart, H.-C. Huang, R. Krishnamurthy and T. Schwartz, *J. Am. Chem. Soc.*, 1989, **111**, 7507.
92. P. Kočovsky and F. Tureček, *Tetrahedron Lett.*, 1981, **22**, 2699.
93. L. E. Overman and C. B. Campbell, *J. Org. Chem.*, 1974, **39**, 1474.
94. (a) A. Bongini, G. Cardillo, M. Orena, G. Porzi and S. Sandri, *J. Org. Chem.*, 1982, **47**, 4626; (b) G. Cardillo, M. Orena, G. Porzi and S. Sandri, *J. Chem. Soc., Chem. Commun.*, 1981, 465; (c) H. W. Pauls and B. Fraser-Reid, *J. Org. Chem.*, 1983, **48**, 1392.
95. (a) M. Georges, D. Mackay and B. Fraser-Reid, *Can. J. Chem.*, 1984, **62**, 1539; (b) M. Georges and B. Fraser-Reid, *Tetrahedron Lett.*, 1981, **22**, 4635.
96. (a) S. Knapp and D. V. Patel, *Tetrahedron Lett.*, 1982, **23**, 3539; (b) S. Knapp and D. V. Patel, *J. Org. Chem.*, 1984, **49**, 5072; (c) S. Knapp, M. J. Sebastian and H. Ramanathan, *Tetrahedron*, 1986, **42**, 3405.
97. J. E. Baldwin, D. R. Kelly and C. B. Ziegler, *J. Chem. Soc., Chem. Commun.*, 1984, 133.
98. (a) C. Betancor, E. I. León, T. Prange, J. A. Salazar and E. Suárez, *J. Chem. Soc., Chem. Commun.*, 1989, 450; (b) R. Freire, E. I. León, J. A. Salazar and E. Suårez, *J. Chem. Soc., Chem. Commun.*, 1989, 452; (c) P. G. Sammes and D. Thetford, *Tetrahedron Lett.*, 1986, **27**, 2275; (d) P. G. Sammes and D. Thetford, *J. Chem. Soc., Perkin Trans. 1*, 1989, 655.
99. S. Takano and S. Hatakeyama, *Heterocycles*, 1982, **19**, 1243.
100. E. J. Corey and T. A. Hase, *Tetrahedron Lett.*, 1979, 335.

101. R. C. Cambie, R. C. Hayward, J. L. Roberts and P. S. Rutledge, *J. Chem. Soc., Perkin Trans. 1*, 1974, 1864.
102. D. Goldsmith, D. Liotta, C. Lee and G. Zima, *Tetrahedron Lett.*, 1979, 4801.
103. H. M. C. Ferraz, *Tetrahedron Lett.*, 1986, **27**, 811.
104. (a) P. C. Ting and P. A. Barlett, *J. Am. Chem. Soc.*, 1984, **106**, 2668; (b) P. A. Bartlett and C. Chapuis, *J. Org. Chem.*, 1986, **51**, 2799; (c) A. Bongini, G. Cardillo, M. Orena, S. Sandri and C. Tomasini, *J. Org. Chem.*, 1986, **51**, 4905; (d) F. Freeman and K. D. Robarge, *Tetrahedron Lett.*, 1985, **26**, 1943; (e) F. Freeman and K. D. Robarge, *J. Org. Chem.*, 1989, **54**, 346; (f) T. Hosokawa, M. Hirata, S. Murahashi and A. Sonoda, *Tetrahedron Lett.*, 1976, 1821; (g) K. A. Parker and R. O'Fee, *J. Am. Chem. Soc.*, 1983, **105**, 654; (h) S. Current and K. B. Sharpless, *Tetrahedron Lett.*, 1978, 5075; (i) B. B. Snider and M. I. Johnston, *Tetrahedron Lett.*, 1985, **26**, 5497; (j) Y. Tamaru, S. Kawamura and Z. Yoshida, *Tetrahedron Lett.*, 1985, **26**, 2885.
105. (a) K. Suzuki and T. Mukaiyama, *Chem. Lett.*, 1982, 1525; (b) K. C. Nicolaou, R. A. Daines, T. K. Chakraborty and Y. Ogawa, *J. Am. Chem. Soc.*, 1988, **110**, 4685; (c) T. Morikawa, I. Kumadaki and M. Shiro, *Chem. Pharm. Bull.*, 1985, **33**, 5144.
106. E. Demole and P. Enggist, *Helv. Chim. Acta*, 1971, **54**, 456.
107. O. Tanaka, N. Tanaka, T. Ohsawa, Y. Iitaka and S. Shibata, *Tetrahedron Lett.*, 1968, 4235.
108. T. Kato, I. Ichinose, T. Hosogai and Y. Kitihara, *Chem. Lett.*, 1976, 1187.
109. E. J. Corey and D. Ha, *Tetrahedron Lett.*, 1988, **29**, 3171.
110. C. A. Broka and Y.-T. Lin, *J. Org. Chem.*, 1988, **53**, 5876.
111. J. P. Michael, P. C. Ting and P. A. Bartlett, *J. Org. Chem.*, 1985, **50**, 2416.
112. K. S. Feldman, *Tetrahedron Lett.*, 1982, **23**, 3031.
113. (a) A. R. Chamberlin, R. L. Mulholland, Jr., S. D. Kahn and W. J. Hehre, *J. Am. Chem. Soc.*, 1987, **109**, 672; (b) for an example of a stereoselective 5-*exo* iodolactonization in which the allylic oxygen substituent is external (extraannular) to the newly formed ring see: Y. G. Kim and J. K. Cha, *Tetrahedron Lett.*, 1989, **30**, 5721.
114. (a) A. R. Chamberlin, M. Dezube, P. Dussault and M. C. McMills, *J. Am. Chem. Soc.*, 1983, **105**, 5819; (b) Y. Tamaru, M. Mizutani, Y. Furukawa, S. Kawamura, Z. Yanagi, K. Yanagi and M. Minobe, *J. Am. Chem. Soc.*, 1984, **106**, 1079; (c) Y. Ohfune and N. Kurokawa, *Tetrahedron Lett.*, 1985, **26**, 5307; (d) G. Nakaminami, S. Shioi, Y. Sugiyama, S. Isemura, M. Shibuya and M. Nakagawa, *Bull. Chem. Soc. Jpn.*, 1972, **45**, 2624; (e) S. W. Rollinson, R. A. Amos and J. A. Katzenellenbogen, *J. Am. Chem. Soc.*, 1981, **103**, 4114.
115. (a) A. A. Akhrem, E. I. Kvasyuk, I. A. Mikhailopulo and T. I. Prikota, *Zh. Obshch. Khim.*, 1977, **47**, 1206; (b) Y. Tamaru, M. Hojo, S. Kawamura, S. Sawada and Z. Yoshida, *J. Org. Chem.*, 1987, **52**, 4062; (c) Y. Tamaru, T. Kobayashi, S. Kawamura, H. Ochiai, M. Hojo and Z. Yoshida, *Tetrahedron Lett.*, 1985, **26**, 3207; (d) Y. G. Kim and J. K. Cha, *Tetrahedron Lett.*, 1988, **29**, 2011; (e) M. Labelle and Y. Guindon, *J. Am. Chem. Soc.*, 1989, **111**, 2204; (f) F. J. L. Herrera, M. S. P. Gonzalez, M. N. Sampedro and R. M. D. Aciego, *Tetrahedron*, 1989, **45**, 269; (g) G. Brussani, S. V. Ley, J. L. Wright and D. J. Williams, *J. Chem. Soc., Perkin Trans. 1*, 1986, 303; (h) A. B. Reitz, S. O. Nortey, B. E. Maryanoff, D. Liotta and R. Monahan, III, *J. Org. Chem.*, 1987, **52**, 4191; (i) S. Murata and T. Suzuki, *Tetrahedron Lett.*, 1987, **28**, 4297; (j) F. Nicotra, R. Perego, F. Ronchetti, G. Russo and L. Toma, *Gazz. Chim. Ital.*, 1987, **114**, 193; (k) F. Nicotra, L. Panza, F. Ronchetti, G. Russo and L. Toma, *Carbohydr. Res.*, 1987, **171**, 49.
116. (a) A. B. Boschetti, F. Nicotra, L. Panza and G. Russo, *J. Org. Chem.*, 1988, **53**, 4181; (b) J.-M. Lancelin, J.-R. Pougny and P. Sinay, *Carbohydr. Res.*, 1985, **136**, 369; (c) M. F. Semmelhack, C. Bodurow and M. Baum, *Tetrahedron Lett.*, 1984, **25**, 3171.
117. K. N. Houk, S. R. Moses, Y.-D. Wu, N. G. Rondan, V. Jager, R. Schohe and F. R. Fronczek, *J. Am. Chem. Soc.*, 1984, **106**, 3880.
118. M. Carcano, F. Nicotra, L. Panza and G. Russo, *J. Chem. Soc., Chem. Commun.*, 1989, 297.
119. (a) P. D. Kane and J. Mann, *J. Chem. Soc., Perkin Trans. 1*, 1984, 657; (b) D. R. Williams and F. H. White, *J. Org. Chem.*, 1987, **52**, 5067.
120. (a) D. B. Collum, J. H. McDonald, III and W. C. Still, *J. Am. Chem. Soc.*, 1980, **102**, 2118; (b) P. A. Bartlett, D. P. Richardson and J. Myerson, *Tetrahedron Lett.*, 1984, **40**, 2317.
121. A. G. M. Barrett, H. B. Broughton, S. V. Attwood and A. A. L. Gunatilaka, *J. Org. Chem.*, 1986, **51**, 495.
122. (a) F. B. Gonzalez and P. A. Bartlett, *Org. Synth.*, 1985, **64**, 175; (b) H. J. Gunther, E. Guntrum and V. Jager, *Liebigs Ann. Chem.*, 1984, 15; (c) J. A. Marshall, M. J. Coghlan and M. Watanabe, *J. Org. Chem.*, 1984, **49**, 747.
123. (a) M. J. Kurth and E. G. Brown, *J. Am. Chem. Soc.*, 1987, **109**, 6844; (b) F. E. Ziegler and A. Kneisley, *Tetrahedron Lett.*, 1985, **26**, 263.
124. (a) M. F. Semmelhack and N. Zhang, *J. Org. Chem.*, 1989, **54**, 4483; (b) M. McCormick, R. Monahan, III, J. Soria, D. Goldsmith and D. Liotta, *J. Org. Chem.*, 1989, **54**, 4485.
125. (a) C. P. Holmes and P. A. Bartlett, *J. Am. Chem. Soc.*, 1989, **54**, 98; (b) formation of isomers with 'reversed' stereochemistry in iodolactonization and iodoetherification has been attributed in some cases to intramolecular delivery of the electrophile from a hypoiodite intermediate: P. A. Bartlett and J. Myerson, *J. Am. Chem. Soc.*, 1978, **100**, 3950.
126. (a) S. Murata and T. Suzuki, *Chem. Lett.*, 1987, 849; (b) S. V. Ley, N. J. Anthon, A. Armstrong, M. G. Brasca, T. Clarke, D. Culshaw, C. Greck, P. Grice, A. B. Jones, B. Lygo, A. Madin, R. N. Sheppard, A. M. Z. Slawin and D. J. Williams, *Tetrahedron*, 1989, **45**, 7161.
127. T. L. B. Boivin, *Tetrahedron*, 1987, **43**, 3309.
128. (a) F. Paquet and P. Sinay, *Tetrahedron Lett.*, 1984, **25**, 3071; (b) F. Paquet and P. Sinay, *J. Am. Chem. Soc.*, 1984, **106**, 8313.
129. M. Labelle, H. E. Morton, Y. Guindon and J. P. Springer, *J. Am. Chem. Soc.*, 1988, **110**, 4533.
130. S. Hatakeyama, K. Sakurai, K. Saigo and S. Takano, *Tetrahedron Lett.*, 1985, **26**, 1333.
131. (a) P. Bravo, G. Resnati, F. Viani and A. Arnone, *J. Chem. Soc., Perkin Trans 1*, 1989, 839; (b) A. Arnone, P. Bravo, M. Frigerio, G. Resnati and F. Viani, *J. Chem. Res. (S)*, 1989, 278.
132. (a) N. Kurokawa and Y. Ohfune, *J. Am. Chem. Soc.*, 1986, **108**, 6041; (b) Y. Ohfune, K. Hori and M. Sakaitani, *Tetrahedron Lett.*, 1986, **27**, 6079.

133. W. Yi-Fong. T. Izawa, S. Kobayashi and M. Ohno, *J. Am. Chem. Soc.*, 1982, **104**, 6465.
134. (a) P. A. Bartlett, K. H. Holm and A. Morimoto, *J. Org. Chem.*, 1985, **50**, 5179; (b) R. H. Bradbury, J. M. Revill, J. E. Rivett and D. Waterson, *Tetrahedron Lett.*, 1989, **30**, 3845; (c) P. Herold, R. Duthaler, G. Rihs and C. Angst, *J. Org. Chem.*, 1989, **54**, 1178.
135. (a) F. Bennett and D. W. Knight, *Tetrahedron Lett.*, 1988, **29**, 4625; (b) F. Bennett and D. W. Knight, *Tetrahedron Lett.*, 1988, **29**, 4865.
136. V. Speziale, J. Rossel and A. Lattes, *J. Heterocycl. Chem.*, 1974, **11**, 771.
137. S. D. Rychnovsky and P. A. Bartlett, *J. Am. Chem. Soc.*, 1981, **103**, 3963.
138. R. Amouroux, F. Chastrette and M. Chastrette, *J. Heterocycl. Chem.*, 1981, **18**, 565.
139. R. Amouroux, B. Gerin and M. Chastrette, *Tetrahedron Lett.*, 1982, **23**, 4341.
140. S. Torii, T. Inokuchi and K. Yoritaka, *J. Am. Chem. Soc.*, 1981, **46**, 5030.
141. (a) M. J. Kurth and M. J. Rodriguez, *J. Am. Chem. Soc.*, 1987, **109**, 7577; (b) M. J. Kurth, private communication; (c) The use of isoxazolines as nucleophiles in iodocyclizations to ring-fused ethers has been reported recently: M. J. Kurth and M. J. Rodriguez, *Tetrahedron*, 1989, **45**, 6693.
142. A. G. M. Barrett, B. C. B. Bezuidenhoudt, A. F. Gasiecki, A. R. Howell and M. A. Russell, *J. Am. Chem. Soc.*, 1989, **111**, 1392.
143. (a) E. J. Corey, J. W. Ponder and P. Ulrich, *Tetrahedron Lett.*, 1980, **21**, 137; (b) D. A. Evans, S. L. Bender and J. Morris, *J. Am. Chem. Soc.*, 1988, **110**, 2506; (c) S. Hanessian, J. Kloss and T. Sugawara, *J. Am. Chem. Soc.*, 1986, **108**, 2758; (d) R. A. Johnson and E. G. Nidy, *J. Org. Chem.*, 1980, **45**, 3802; (e) M. F. Semmelhack and A. Zask, *J. Am. Chem. Soc.*, 1983, **105**, 2034; (f) J. Gutzwiller, G. Pizzolato and M. R. Uskokovic, *J. Am. Chem. Soc.*, 1971, **93**, 5907.
144. (a) M. F. Semmelhack and C. Bodurow, *J. Am. Chem. Soc.*, 1984, **106**, 1496; (b) M. F. Semmelhack, C. R. Kim, W. Dobler and M. Meier, *Tetrahedron Lett.*, 1989, **30**, 4925.
145. (a) T. Fukuyama, C.-L. J. Wang and Y. Kishi, *J. Am. Chem. Soc.*, 1979, 260; (b) R. L. Dow, Ph.D. Thesis, Harvard University, 1985 (*Chem. Abstr.*, 1987, **106**, 49 833v). We thank Professor D. Evans for providing information on this dissertation.
146. R. F. Newton and S. M. Roberts, *Synthesis*, 1984, 449.
147. B. Giese and D. Bartmann, *Tetrahedron Lett.*, 1985, **26**, 1197.
148. (a) P. A. Bartlett, J. D. Meadows, E. G. Brown, A. Morimoto and K. K. Jernstedt, *J. Org. Chem.*, 1982, **47**, 4013; (b) M. Majewski, D. L. J. Clive and P. C. Anderson, *Tetrahedron Lett.*, 1984, **25**, 2101; (c) B. H. Lipshutz and J. A. Kozlowski, *J. Org. Chem.*, 1984, **49**, 1147.
149. W. S. Johnson and M. F. Chan, *J. Org. Chem.*, 1985, **50**, 2598.
150. M. Hirama and M. Uei, *Tetrahedron Lett.*, 1982, **23**, 5307.
151. P. A. Bartlett and K. K. Jernstedt, *J. Am. Chem. Soc.*, 1977, **99**, 4829.
152. (a) S. P. McManus and J. T. Carrol, *J. Org. Chem.*, 1970, **35**, 3768; (b) S. P. McManus, C. U. Pittman, Jr. and P. E. Fanta, *J. Org. Chem.*, 1972, **37**, 2353.
153. (a) A. Bongini, G. Cardillo, M. Orena, S. Sandri and C. Tomasini, *J. Chem. Soc., Perkin Trans. 1.*, 1986, 1345; (b) S. P. McManus, D. W. Ware and R. A. Hames, *J. Org. Chem.*, 1978, **43**, 4288; (c) T. A. Degurko and V. I. Staninets, *Dopov. Akad. Nauk Ukr. RSR, Ser. B*, 1973, **35**, 345 (*Chem. Abstr.*, 1973, **79**, 104 490g); (d) L. Goodman and S. Winstein, *J. Am. Chem. Soc.*, 1957, **79**, 4788.
154. F. Chretien and Y. Chapleur, *J. Org. Chem.*, 1988, **53**, 3615.
155. (a) L. Engman, *J. Am. Chem. Soc.*, 1984, **106**, 3977; (b) J. Bergman and J. Siden, *Tetrahedron*, 1984, **40**, 1607.
156. Z. K. M. Abd El Samii, M. I. Al Ashmawy and J. M. Mellor, *J. Chem. Soc., Perkin Trans. 1*, 1988, 2517.
157. (a) G. Cardillo, M. Orena and S. Sandri, *J. Chem. Soc., Perkin Trans. 1*, 1983, 1489; (b) A. Bongini, G. Cardillo, M. Orena, S. Sandri and C. Tomasini, *J. Chem. Soc., Perkin Trans. 1.*, 1985, 935.
158. (a) Yu. I. Gevaza, O. M. Dorokhova, V. I. Staninets and E. S. Levchenko, *Dopov. Akad. Nauk Ukr. RSR, Ser. B*, 1975, 423 (*Chem. Abstr.*, 1975, **83**, 58 731u); (b) S. Takano, Y. Iwabuchi and K. Ogasawara, *J. Chem. Soc., Chem. Commun.*, 1988, 1527; (c) A. Toshimitsu, K. Terao and S. Uemura, *J. Org. Chem.*, 1986, **51**, 1724.
159. S. Takano, C. Kasahara and K. Ogasawara, *Heterocycles*, 1982, **19**, 1443.
160. G. Capozzi, C. Caristi and M. Gattuso, *J. Chem. Soc., Perkin Trans. 1*, 1984, 255.
161. S. Kobayashi, T. Isobe and M. Ohno, *Tetrahedron Lett.*, 1984, **25**, 5079.
162. (a) Y. Yamamoto, T. Komatsu and K. Maruyama, *J. Org. Chem.*, 1985, **50**, 3115; (b) Y. Yamamoto, S. Nishii, K. Maruyama, T. Komatsu and W. Ito, *J. Am. Chem. Soc.*, 1986, **108**, 7778.
163. (a) M. Muhlstadt, B. Olk and R. Widera, *J. Prakt. Chem.*, 1986, **328**, 163; (b) M. Muhlstadt, B. Olk and R. Widera, *J. Prakt. Chem.*, 1986, **328**, 173; (c) M. Muhlstadt, R. Meusinger, B. Olk, L. Weber and R. Widera, *J. Prakt. Chem.*, 1986, **328**, 309.
164. (a) G. Cardillo, M. Orena, S. Sandri and C. Tomasini, *Tetrahedron*, 1987, **43**, 2505; (b) A. Bongini, G. Cardillo, M. Orena, G. Porzi and S. Sandri, *Tetrahedron*, 1987, **43**, 4377; (c) A. Bongini, G. Cardillo and M. Orena, *Chem. Lett.*, 1988, 87.
165. L. E. Overman and R. J. McCready, *Tetrahedron Lett.*, 1982, **23**, 4887.
166. (a) G. Cardillo, M. Orena and S. Sandri, *J. Org. Chem.*, 1986, **51**, 713; (b) T. Toda and Y. Kitagawa, *Angew. Chem., Int. Ed. Engl.*, 1987, **26**, 334.
167. M. L. Phillips, R. Bonjouklian, N. D. Jones, A. H. Hunt and T. K. Elzey, *Tetrahedron Lett.*, 1983, **24**, 335.
168. (a) R. Amouroux, *Heterocycles*, 1984, **22**, 1489; (b) C. G. Chavdarian, L. L. Chang and B. C. Onisko, *Heterocycles*, 1988, **27**, 651; (c) P. Kocienski and C. Yeates, *J. Chem. Soc., Perkin Trans. 1*, 1985, 1879; (d) S. V. Ley and B. Lygo, *Tetrahedron Lett.*, 1984, **25**, 113; (e) G. Pairaudeau, P. J. Parsons and J. M. Underwood, *J. Chem. Soc., Chem. Commun.*, 1987, 1718.
169. M. Mortimore and P. Kocienski, *Tetrahedron Lett.*, 1988, **29**, 3357.
170. P. Deslongchamps, 'Stereoelectronic Effects in Organic Chemistry,' Pergamon Press, Oxford, 1983, p. 7.
171. P. M. Wovkulich, P. C. Tang, N. K. Chadha, A. D. Batcho, J. C. Barrish and M. R. Uskokovic, *J. Am. Chem. Soc.*, 1989, **111**, 2596.
172. E. J. Corey, K. Shimoji and C. Shih, *J. Am. Chem. Soc.*, 1984, **106**, 6425.

173. (a) E. Bascetta and F. D. Gunstone, *J. Chem. Soc., Perkin Trans. 1*, 1984, 2207; (b) N. A. Porter, P. J. Zuraw and J. A. Sullivan, *Tetrahedron Lett.*, 1984, **25**, 807; (c) N. A. Porter and P. J. Zuraw, *J. Org. Chem.*, 1984, **49**, 1345.
174. (a) A. J. Bloodworth and R. J. Curtis, *J. Chem. Soc., Chem. Commun.*, 1989, 173; (b) L. A. Paquette, R. V. C. Carr and F. Bellamy, *J. Am. Chem. Soc.*, 1984, **100**, 6764; (c) A. J. Bloodworth, R. J. Curtis and N. Mistry, *J. Chem. Soc., Chem. Commun.*, 1989, 954.
175. (a) A. M. Doherty, S. V. Ley, B. Lygo and D. J. Williams, *J. Chem Soc., Perkin Trans. 1*, 1984, 1371; (b) W. Kitching, J. A. Lewis, M. V. Perkins, R. Drew, C. J. Moore, V. Schurig, W. A. König and W. Francke, *J. Org. Chem.*, 1989, **54**, 3893; (c) P. A. Bartlett, I. Mori and J. A. Bose, *J. Org. Chem.*, 1989, **54**, 3236; (d) G. Pandey, V. Jayathirtha Rao and U. T. Bhalerao, *J. Chem. Soc., Chem. Commun.*, 1989, 416.
176. (a) W. P. Jackson, S. V. Ley and J. A. Morton, *J. Chem. Soc., Chem. Commun.*, 1980, 1028; (b) W. P. Jackson, S. V. Ley and A. J. Whittle, *J. Chem. Soc., Chem. Commun.*, 1980, 1173; (c) S. V. Ley, B. Lygo and H. Molines, *J. Chem. Soc., Perkin Trans. 1*, 1984, 2403; (d) R. Antonioletti, F. Bonadies and A. Scettri, *Tetrahedron Lett.*, 1988, **29**, 4987.
177. I. P. Kupchik, E. B. Koryak, Yu. I. Gevaza and V. I. Staninets, *Ukr. Khim. Zh. (Russ. Ed.)*, 1988, **54**, 1180 (*Chem. Abstr.*, 1989, **111**, 77 279r).
178. (a) Yu. I. Gevaza, V. I. Staninets and E. V. Konovalov, *Zh. Org. Khim.*, 1975, **11**, 906; (b) Yu. I. Gevaza, V. I. Staninets and E. B. Koryak, *Khim. Geterotsikl. Soedin.*, 1976, 1340; (c) I. D. Brindle and M. S. Gibson, *Can. J. Chem.*, 1979, **57**, 3155.
179. (a) P. Kočovsky and V. Cerny, *Collect. Czech. Chem. Commun.*, 1980, **45**, 3023; (b) P. Kočovsky and V. Cerny, *Collect. Czech. Chem. Commun.*, 1980, **45**, 3030; (c) P. Kočovsky, *Collect. Czech. Chem. Commun.*, 1983, **48**, 3597; (d) P. Kočovsky, *Collect. Czech. Chem. Commun.*, 1983, **48**, 3606; (e) P. Kočovsky, *Collect. Czech. Chem. Commun.*, 1983, **48**, 3660; (f) R. W. Rickards and H. Ronneberg, *J. Org. Chem.*, 1984, **49**, 572; (g) H. Niwa, K. Wakamatsu, T. Hida, K. Niiyama, H. Kigoshi, M. Yamada, H. Nagase, M. Suzuki and K. Yamada, *J. Am. Chem. Soc.*, 1984, **106**, 4547.
180. (a) D. R. Mootoo and B. Fraser-Reid, *J. Chem. Soc., Chem. Commun.*, 1986, 1570; (b) D. R. Mootoo, V. Date and B. Fraser-Reid, *J. Chem. Soc., Chem. Commun.*, 1987, 1462; (c) D. R. Mootoo and B. Fraser-Reid, *J. Org. Chem.*, 1989, **54**, 5548.
181. (a) D. R. Mootoo, V. Date and B. Fraser-Reid, *J. Am. Chem. Soc.*, 1988, **110**, 2662; (b) P. Konradsson and B. Fraser-Reid, *J. Chem. Soc., Chem. Commun.*, 1989, 1124.
182. (a) D. R. Mootoo, P. Konradsson, U. Udodong and B. Fraser-Reid, *J. Am. Chem. Soc.*, 1988, **110**, 5583; (b) B. Fraser-Reid, P. Konradsson, D. R. Mootoo and U. Udodong, *J. Chem. Soc., Chem. Commun.*, 1988, 823; (c) D. R. Mootoo and B. Fraser-Reid, *Tetrahedron Lett.*, 1989, **30**, 2363.
183. A. Wu, D. R. Mootoo and B. Fraser-Reid, *Tetrahedron Lett.*, 1988, **29**, 6549.
184. (a) S. Terashima and S. Jew, *Tetrahedron Lett.*, 1977, 1005; (b) S. Jew, S. Terashima and K. Koga, *Tetrahedron*, 1979, **35**, 2345.
185. P. F. Corey, *Tetrahedron Lett.*, 1987, **28**, 2801.
186. S. Takano, C. Murakata and Y. Inamura, *Heterocycles*, 1981, **16**, 1291.
187. M. F. Grundon, D. Stewart and W. E. Watts, *J. Chem. Soc., Chem. Commun.*, 1973, 573.
188. T. Hosokawa, Y. Imada and S. Murahashi, *Bull. Chem. Soc. Jpn.*, 1985, **58**, 3282.
189. (a) R. A. Amos and J. A. Katzenellenbogen, *J. Org. Chem.*, 1978, **43**, 560; (b) G. A. Krafft and J. A. Katzenellenbogen, *J. Am. Chem. Soc.*, 1981, **103**, 5459; (c) R. W. Spencer, T. F. Tam, E. Thomas, V. J. Robinson and A. Krantz, *J. Am. Chem. Soc.*, 1986, **108**, 5589; (d) A. Jellal, J. Grimaldi and M. Santelli, *Tetrahedron Lett.*, 1984, **25**, 3179; (e) M. Yamamoto, *J. Chem. Soc., Perkin Trans. 1*, 1981, 582.
190. C. Lambert, K. Utimoto and H. Nozaki, *Tetrahedron Lett.*, 1984, **25**, 5323.
191. P. Pale and J. Chuche, *Tetrahedron Lett.*, 1987, **28**, 6447.
192. N. Yanagihara, C. Lambert, K. Iritani, K. Utimoto and H. Nozaki, *J. Am. Chem. Soc.*, 1986, **108**, 2753.
193. (a) M. J. Sofia, P. K. Chakravarty and J. A. Katzenellenbogen, *J. Org. Chem.*, 1983, **48**, 3318; (b) S. B. Daniels, E. Cooney, M. J. Sofia, P. K. Chakravarty and J. A. Katzenellenbogen, *J. Biol. Chem.*, 1983, **258**, 15046.
194. T. Toru, S. Fujita and E. Maekawa, *J. Chem. Soc., Chem. Commun.*, 1985, 1082.
195. T. Toru, S. Fujita, M. Saito and E. Maekawa, *J. Chem. Soc., Perkin Trans. 1*, 1986, 1999.
196. For results of kinetic studies of these cyclizations and of iodocyclizations of the related alkenoic acids, see: J. T. Doi, G. W. Luehr, D. del Carmen and B. C. Lippsmeyer, *J. Org. Chem.*, 1989, **54**, 2764.
197. (a) M. Riediker and J. Schwartz, *J. Am. Chem. Soc.*, 1982, **104**, 5842; (b) M. Suzuki, A. Yanagisawa and R. Noyori, *Tetrahedron Lett.*, 1983, **24**, 1187; (c) M. Suzuki, A. Yanagisawa and R. Noyori, *J. Am. Chem. Soc.*, 1988, **110**, 4718.
198. K. Utimoto, *Pure Appl. Chem.*, 1983, **55**, 1845.
199. M. Yamamoto, M. Yoshitake and K. Yamada, *J. Chem. Soc., Chem. Commun.*, 1983, 991.
200. R. C. Larock and L. W. Harrison, *J. Am. Chem. Soc.*, 1984, **106**, 4218.
201. M. A. Oliver and R. D. Gandour, *J. Org. Chem.*, 1984, **49**, 558.
202. (a) W. Smadja, *Chem. Rev.*, 1983, **83**, 263; (b) T. L. Jacobs, in 'The Chemistry of the Allenes', ed. S. R. Landor, Academic Press, New York, 1982, vol. 2, p. 417; (c) A. Claesson and L.-I. Olsson, in 'The Chemistry of the Allenes', ed. S. R. Landor, Academic Press, New York, 1982, vol. 3, p. 755.
203. L. Olsson and A. Claesson, *Synthesis*, 1979, 743.
204. S. S. Nikam, K.-H. Chu and K. K. Wang, *J. Org. Chem.*, 1986, **51**, 745.
205. (a) A. J. Bridges and R. D. Thomas, *J. Chem. Soc., Chem. Commun.*, 1984, 694; (b) R. C. Cookson and P. J. Parsons, *J. Chem. Soc., Chem. Commun.*, 1978, 822; (c) J.-J. Chilot, A. Doutheau and J. Gore, *Bull. Soc. Chim. Fr.*, 1984, 307; (d) J.-J. Chilot, A. Doutheau, J. Gore and A. Saroli, *Tetrahedron Lett.*, 1986, **27**, 849; (e) J. Grimaldi and A. Cormons, *C. R. Hebd. Seances Acad. Sci., Ser. C*, 1979, **289**, 373; (f) J. Grimaldi and A. Cormons, *Tetrahedron Lett.*, 1987, **28**, 3487; (g) T. L. Jacobs, R. Macomber and D. Zunker, *J. Am. Chem. Soc.*, 1967, **89**, 7001.
206. A. Audin, A. Doutheau, L. Ruest and J. Gore, *Bull. Soc. Chim. Fr.*, 1978, 313.

207. (a) R. D. Walkup and G. Park, *Tetrahedron Lett.*, 1987, **28**, 1023; (b) R. D. Walkup and G. Park, *Tetrahedron Lett.*, 1988, **29**, 5505.
208. J. Grimaldi and A. Cormons, *Tetrahedron Lett.*, 1985, **26**, 825.
209. P. Audin, A. Doutheau and J. Gore, *Bull. Soc. Chim. Fr.*, 1984, 297.
210. T. Gallagher, *J. Chem. Soc., Chem. Commun.*, 1984, 1554.
211. T. Delair, A. Doutheau and J. Gore, *Bull. Soc. Chim. Fr.*, 1988, 125.
212. M. Petrzilka, *Helv. Chim. Acta*, 1978, **61**, 3075.
213. A. J. Biloski, R. D. Wood and B. Ganem, *J. Am. Chem. Soc.*, 1982, **104**, 3233.
214. G. Rajendra and M. J. Miller, *Tetrahedron Lett.*, 1985, **26**, 5385.
215. (a) M. Ihara, K. Fukumoto and T. Kametani, *Heterocycles*, 1982, **19**, 1435; (b) M. Ihara, Y. Haga, M. Yonekura, T. Ohsawa, K. Fukumoto and T. Kametani, *J. Am. Chem. Soc.*, 1983, **105**, 7345.
216. (a) D. L. J. Clive, V. Farina, A. Singh, C. K. Wong, W. A. Kiel and S. M. Menchen, *J. Org. Chem.*, 1980, **45**, 2120; (b) R. R. Webb, II and S. J. Danishefsky, *Tetrahedron Lett.*, 1983, **24**, 1357; (c) D. J. Hart and J. A. McKinney, *Tetrahedron Lett.*, 1989, **30**, 2611; (d) I. Monkovic, T. T. Conway, H. Wong, Y. G. Perron, I. J. Pachter and B. Belleau, *J. Am. Chem. Soc.*, 1973, **95**, 7910; (e) K. Terao, A. Toshimitsu and S. Uemura, *J. Chem. Soc., Perkin Trans. 1*, 1986, 1837.
217. (a) S. Knapp and A. T. Levorse, *J. Org. Chem.*, 1988, **53**, 4006; (b) H. Takahata, T. Takamatsu, M. Mozumi, T.-S. Chen, T. Yamazaki and K. Aoe, *J. Chem. Soc., Chem. Commun.*, 1987, 1627. (c) H. Takahata, T. Takamatsu and T. Yamazaki, *J. Org. Chem.*, 1989, **54**, 4812.
218. (a) S. R. Wilson and R. A. Sawicki, *J. Org. Chem.*, 1979, **44**, 287; (b) S. R. Wilson and R. A. Sawicki, *J. Org. Chem.*, 1979, **44**, 330; (c) S. R. Wilson, R. A. Sawicki and J. C. Huffman, *J. Org. Chem.*, 1981, **46**, 3887.
219. R. A. Johnson, *J. Org. Chem.*, 1972, **37**, 312.
220. K. Katsuura and K. Mitsuhashi, *Chem. Pharm. Bull.*, 1983, **31**, 2094.
221. H. F. Campbell, O. E. Edwards, J. W. Elder and R. J. Kolt, *Pol. J. Chem.*, 1979, **53**, 27.
222. H. Szezepanski and C. Ganter, *Helv. Chim. Acta*, 1977, **60**, 1435.
223. S. Knapp and K. E. Rodriques, *Tetrahedron Lett.*, 1985, **26**, 1803.
224. J. Barluenga, J. Perez-Prieto, A. M. Bayon and G. Asensio, *Tetrahedron*, 1984, **40**, 1199.
225. (a) D. Tanner and P. Somfai, *Tetrahedron*, 1986, **42**, 5657; (b) D. Tanner, M. Sellén, and J. E. Backväll, *J. Org. Chem.*, 1989, **54**, 3374.
226. (a) A. Bongini, G. Cardillo, M. Orena, S. Sandri and C. Tomasini, *Tetrahedron*, 1983, **39**, 3801; (b) G. Cardillo, M. Orena, S. Sandri and C. Tomasini, *J. Org. Chem.*, 1984, **49**, 3951; (c) H. W. Pauls and B. Fraser-Reid, *J. Chem. Soc., Chem. Commun.*, 1983, 1031; (d) P. G. Sammes and D. Thetford, *J. Chem. Soc., Perkin Trans. 1*, 1988, 111.
227. (a) S. Knapp and D. V. Patel, *J. Am. Chem. Soc.*, 1983, **105**, 6985; (b) S. Knapp, G. S. Lal and D. Sahai, *J. Org. Chem.*, 1986, **51**, 380.
228. (a) P. G. Sammes and D. Thetford, *J. Chem. Soc., Chem. Commun.*, 1985, 352; (b) V. I. Staninets, T. A. Degurko and Yu. V. Melika, *Ukr. Khim. Zh. (Russ. Ed.)*, 1973, **39**, 1043 (*Chem. Abstr.*, 1974, **80**, 36459x); (c) R. M. Giuliano, T. W. Deisenroth and W. C. Frank, *J. Org. Chem.*, 1986, **51**, 2304; (d) a similar cyclization of an isourea with the double bond exocyclic to the ring has been reported: R. M. Giuliano and T. W. Deisenroth, *Carbohydr. Res.*, 1986, **158**, 249.
229. (a) T. Ohsawa, M. Ihara, K. Fukumoto and T. Kametani, *Heterocycles*, 1982, **19**, 2075; (b) T. Ohsawa, M. Ihara, K. Fukumoto and T. Kametani, *J. Org. Chem.*, 1983, **48**, 3644.
230. J. J. Perie, J. P. Laval, J. Roussel and A. Lattes, *Tetrahedron*, 1972, **28**, 675.
231. G. Capozzi, *Heterocycles*, 1986, **24**, 583.
232. (a) A. Toshimitsu, K. Terao and S. Uemura, *J. Chem. Soc., Chem. Commun.*, 1986, 530; (b) S. R. Berryhill and M. Rosenblum, *J. Org. Chem.*, 1980, **45**, 1984.
233. S. Kano, T. Yokomatsu, H. Iwasawa and S. Shibuya, *Heterocycles*, 1987, **26**, 359.
234. (a) L. S. Hegedus and J. M. McKearin, *J. Am. Chem. Soc.*, 1982, **104**, 2444; (b) B. Pugin and L. M. Venanzi, *J. Organomet. Chem.*, 1981, **214**, 125.
235. T. W. Balko, R. S. Brinkmeyer and N. H. Terando, *Tetrahedron Lett.*, 1989, **30**, 2045.
236. Y. Tamaru, S. Kawamura, T. Bando, K. Tanaka, M. Hojo and Z. Yoshida, *J. Org. Chem.*, 1988, **53**, 5491.
237. Y. Tamaru, M. Hojo and Z. Yoshida, *J. Org. Chem.*, 1988, **53**, 5731.
238. D. R. Williams, M. H. Osterhout and J. M. McGill, *Tetrahedron Lett.*, 1989, **30**, 1327.
239. (a) J. Ambuehl, P. S. Pregosin, L. M. Venanzi, G. Ughetto and L. Zambonelli, *J. Organomet. Chem.*, 1978, **160**, 329; (b) J. Ambuehl, P. S. Pregosin, L. M. Venanzi, G. Consiglio, F. Bachechi and L. Zambonelli, *J. Organomet. Chem.*, 1979, **181**, 255.
240. M. J. Kurth and S. H. Bloom, *J. Org. Chem.*, 1989, **54**, 411.
241. S. J. Danishefsky, E. M. Berman, M. Ciufolini, S. J. Etheredge and B. E. Segmuller, *J. Am. Chem. Soc.*, 1985, **107**, 3891.
242. (a) K. E. Harding and S. R. Burks, *J. Org. Chem.*, 1981, **46**, 3920; (b) K. E. Harding and S. R. Burks, *J. Org. Chem.*, 1984, **49**, 40; (c) S. R. Burks, Ph.D. Thesis, Texas A&M University, 1983 (*Chem. Abstr.*, 1984, **104**, 171408x).
243. T. Aida, R. Legault, D. Dugat and T. Durst, *Tetrahedron Lett.*, 1979, 4993.
244. D. R. Williams, D. L. Brown and J. W. Benbow, *J. Am. Chem. Soc.*, 1989, **111**, 1923.
245. M. Tokuda, Y. Yamada and H. Suginome, *Chem. Lett.*, 1988, 1289.
246. K. E. Harding and T. H. Marman, *J. Org. Chem.*, 1984, **49**, 2838.
247. (a) R. C. Bernotas and B. Ganem, *Tetrahedron Lett.*, 1985, **26**, 1123; (b) R. C. Bernotas, M. A. Pezzone and B. Ganem, *Carbohydr. Res.*, 1987, **167**, 305.
248. W. Carruthers, M. J. Williams and M. T. Cox, *J. Chem. Soc., Chem. Commun.*, 1984, 1235.
249. Y. Moriyama, D. Doan-Huynh, C. Monneret and Q. Khuong-Huu, *Tetrahedron Lett.*, 1977, 825.
250. (a) K. Kurihara, T. Sugimoto, Y. Saitoh, Y. Igarashi, H. Hirota, Y. Moriyama, T. Tsuyuki, T. Takahashi and Q. Khuong-Huu, *Bull. Chem. Soc. Jpn.*, 1985, **58**, 3337; (b) Y. Saitoh, Y. Moriyama, H. Hirota, T. Takahashi and Q. Khuong-Huu, *Bull. Chem. Soc. Jpn.*, 1981, **54**, 488.

251. J. Barluenga, C. Jimenez, C. Najera and M. Yus, *Synthesis*, 1982, 414.
252. S. J. Danishefsky, P. J. Harrison, R. R. Webb, II and B. O'Niel, *J. Am. Chem. Soc.*, 1985, **107**, 1421.
253. M. Hirama, M. Iwashita, Y. Yamazaki and S. Ito, *Tetrahedron Lett.*, 1984, **25**, 4963.
254. (a) G. Cardillo, M. Orena, G. Porzi and S. Sandri, *J. Chem. Soc., Chem. Commun.*, 1982, 1308; (b) G. Cardillo, M. Orena and S. Sandri, *Pure Appl. Chem.*, 1988, **60**, 1679; (c) A. Bongini, G. Cardillo, M. Orena, S. Sandri and C. Tomasini, *J. Chem. Soc., Perkin Trans. 1*, 1986, 1339.
255. (a) Yu. V. Melika, I. V. Smolanka and V. I. Staninets, *Ukr. Khim. Zh. (Russ. Ed.)*, 1973, **39**, 799 (*Chem. Abstr.*, 1973, **79**, 115481f); (b) Yu. V. Melika, V. I. Staninets and I. V. Smolanka, *Ukr. Khim. Zh. (Russ. Ed.)*, 1973, **39**, 280 (*Chem. Abstr.*, 1973, **78**, 159 500v).
256. (a) K. E. Harding, R. Stephens and D. R. Hollingsworth, *Tetrahedron Lett.*, 1984, **25**, 4631; (b) K. E. Harding and D. R. Hollingsworth, *Tetrahedron Lett.*, 1988, **29**, 3789; (c) K. E. Harding and D. Nam, *Tetrahedron Lett.*, 1988, **29**, 3793; (d) K. E. Harding and M. W. Jones, *Heterocycles*, 1989, **28**, 663.
257. (a) J. M. Takacs, M. A. Helle and L. Yang, *Tetrahedron Lett.*, 1989, **30**, 1777; (b) K. E. Harding, D. R. Hollingsworth and J. Reibenspies, *Tetrahedron Lett.*, 1989, **30**, 4775.
258. K. E. Harding, T. H. Marman and D. Nam, *Tetrahedron*, 1988, **44**, 5605.
259. (a) P. A. Hunt, C. May and C. J. Moody, *Tetrahedron Lett.*, 1988, **29**, 3001; (b) I. I. Ershova, Yu. I. Gevaza and V. I. Staninets, *Khim. Geterotsikl. Soedin.*, 1981, **43**, 32; (c) E. Bruni, G. Cardillo, M. Orena, S. Sandri and C. Tomasini, *Tetrahedron Lett.*, 1989, **30**, 1679; (d) Y. Tamaru, M. Hojo, H. Higashimura and Z. Yoshida, *J. Am. Chem. Soc.*, 1988, **110**, 3994; (e) P. I. Creeke and J. M. Mellor, *Tetrahedron Lett.*, 1989, **30**, 4435; (f) see also ref. 1c and T. A. Krasnitskaya, I. V. Smolanka and A. L. Vais, *Khim. Geterotsikl. Soedin.*, 1973, 424.
260. (a) F. Esser, *Synthesis*, 1987, 460; (b) N. Cohen, *Tetrahedron Lett.*, 1971, 3943; (c) N. Cohen and B. L. Banner, *J. Heterocycl. Chem.*, 1977, **14**, 717.
261. H. Ida, Y. Yuasa and C. Kibayashi, *Tetrahedron Lett.*, 1982, **23**, 3591.
262. K. Heusler, *Helv. Chim. Acta*, 1972, **55**, 388.
263. M. R. Gagné and T. J. Marks, *J. Am. Chem. Soc.*, 1989, **111**, 4108.
264. A. Claesson, C. Sahlberg and K. Luthman, *Acta Chem. Scand., Ser. B*, 1979, **33**, 309.
265. (a) J. S. Prasad and L. S. Liebeskind, *Tetrahedron Lett.*, 1988, **29**, 4253; (b) J. S. Prasad and L. S. Liebeskind, *Tetrahedron Lett.*, 1988, **29**, 4257.
266. (a) E. C. Taylor, A. H. Katz and H. Salgado-Zamora, *Tetrahedron Lett.*, 1985, **26**, 5963; (b) A. Arcadi, S. Cacchi and F. Marinelli, *Tetrahedron Lett.*, 1989, **30**, 2581; (c) D. E. Rudisill and J. K. Stille, *J. Org. Chem.*, 1989, **54**, 5856; (d) K. Utimoto, H. Miwa and H. Nozaki, *Tetrahedron Lett.*, 1981, **22**, 4277.
267. (a) J. Grimaldi and A. Cormons, *Tetrahedron Lett.*, 1986, **27**, 5089; (b) L. E. Overman and S. W. Goldstein, *J. Am. Chem. Soc.*, 1984, **106**, 5360.
268. (a) S. Arseniyadis and J. Gore, *Tetrahedron Lett.*, 1983, **24**, 3997; (b) T. Gallagher and P. Vernon, *J. Chem. Soc., Chem. Commun.*, 1987, 243; (c) T. Gallagher, *Tetrahedron Lett.*, 1986, **27**, 6009; (d) T. Gallagher and P. Vernon, *J. Chem. Soc., Chem. Commun.*, 1987, 245; (e) D. N. A. Fox, D. Lathbury, M. F. Mahon, K. C. Molloy and T. Gallagher, *J. Chem. Soc., Chem. Commun.*, 1989, 1073.
269. (a) S. Arseniyadis and J. Sartoretti, *Tetrahedron Lett.*, 1985, **26**, 729; (b) T. Gallagher and D. Lathbury, *J. Chem. Soc., Chem. Commun.*, 1986, 114.
270. (a) T. Gallagher and D. Lathbury, *Tetrahedron Lett.*, 1985, **26**, 6249; (b) T. Gallagher and D. Lathbury, *J. Chem. Soc., Chem. Commun.*, 1986, 1017.
271. (a) K. C. Nicolaou, W. E. Barnette and R. L. Magolda, *J. Am. Chem. Soc.*, 1981, **103**, 3486; (b) K. C. Nicolaou, W. E. Barnette and R. L. Magolda, *J. Am. Chem. Soc.*, 1978, **100**, 2567.
272. K. C. Nicolaou, W. E. Barnette and R. L. Magolda, *J. Am. Chem. Soc.*, 1981, **103**, 3480.
273. (a) K. C. Nicolaou, W. E. Barnette, G. P. Gasic and R. L. Magolda, *J. Am. Chem. Soc.*, 1977, **99**, 7736; (b) K. C. Nicolaou, W. E. Barnette and R. L. Magolda, *J. Am. Chem. Soc.*, 1981, **103**, 3472.
274. K. Shimoji, Y. Arai and M. Hayashi, *Chem. Lett.*, 1978, 1375.
275. T. Toru, T. Kanefusa and E. Maekawa, *Tetrahedron Lett.*, 1986, **27**, 1583.
276. E. H. M. Abd Ellal, M. I. Al Ashmawy and J. M. Mellor, *J. Chem. Soc., Chem. Commun.*, 1987, 1577.
277. P. N. Confalone, G. Pizzolato, E. G. Baggiolini, D. Lollar and M. R. Uskokovic, *J. Am. Chem. Soc.*, 1975, **97**, 5936.
278. (a) H. Takahata, K. Moriyama, M. Maruyama and T. Yamazaki, *J. Chem. Soc., Chem. Commun.*, 1986, 1671; (b) H. Takahata, T. Suzuki, M. Maruyama, K. Moriyama, M. Mozumi, T. Takamatsu and T. Yamazaki, *Tetrahedron*, 1988, **44**, 4777; (c) I. I. Ershova, V. I. Staninets and T. A. Degurko, *Dopov. Akad. Nauk Ukr. RSR, Ser. B*, 1975, 1097 (*Chem. Abstr.*, 1976, **84**, 58199x).

2.1
Arene Substitution *via* Nucleophilic Addition to Electron Deficient Arenes

CRISTINA PARADISI
Centro Meccanismi Reazioni Organiche del CNR, Padova, Italy

2.1.1 INTRODUCTION

This chapter reviews those reactions of nucleophiles with electrophilic aromatic systems which proceed *via* formation of a covalently bound intermediate (σ-adduct), and result in the substitution of either a nucleofugal ring substituent or of a ring hydrogen. A few different processes are classified in this category. In the S_NAr reaction the nucleophile attacks, in what is often the rate-limiting step, the carbon carrying the nucleofugal substituent, Y, and the substitution product forms *via* expulsion of Y^- from the σ-adduct.[1–4] When nucleophilic attack occurs at a ring carbon other than that bearing the leaving group, *cine* (*e.g.* the Von Richter reaction)[1] and/or *tele* substitution can take place.[5] The reaction of 2,3-dinitroaniline with piperidine, shown in equation (1), serves as a remarkable example since it gives the products of S_NAr (**1**), of *cine* (**2**) and of *tele* substitution (**3**) in comparable yields.[5]

Significant advances have recently been made in the development of synthetically useful procedures to achieve formal nucleophilic displacement of hydrogen.[6,7] The hydride ion being such a poor leaving group, processes other than H^- expulsion are effective in most cases. A common approach involves oxidation of the σ-adduct by inorganic (*e.g.* H_2O_2, halogens, $KMnO_4$, hypohalites) and organic (*e.g.* NBS, chloranil, DDQ, *p*-benzoquinone) reagents.[8] In the absence of an added oxidant 'spontaneous' oxidation[1] of the σ-adduct may occur *via* reduction of the substrate itself,[8–10] or *via* disproportionation of the σ-adduct.[11,12] In these cases the substitution product is normally obtained in poor yield (Scheme 1).[12] Special cases of 'spontaneous' oxidation have been observed in which rearomatization of the σ-adduct is accomplished *via* $NO_2 \rightarrow NO$ reduction in what can be formally viewed as an intramolecular redox process.[13,14] The reaction of 1,2,4-triazines with nitronate anions produces oximes in good yields, according to the mechanism proposed in Scheme 2.[14]

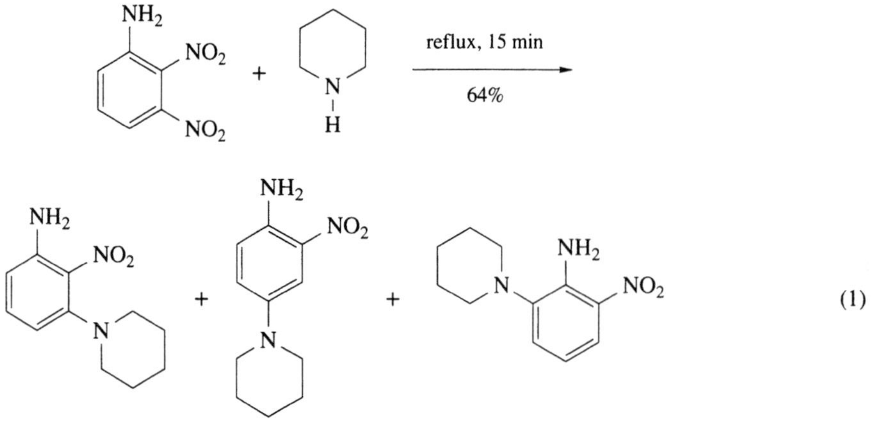

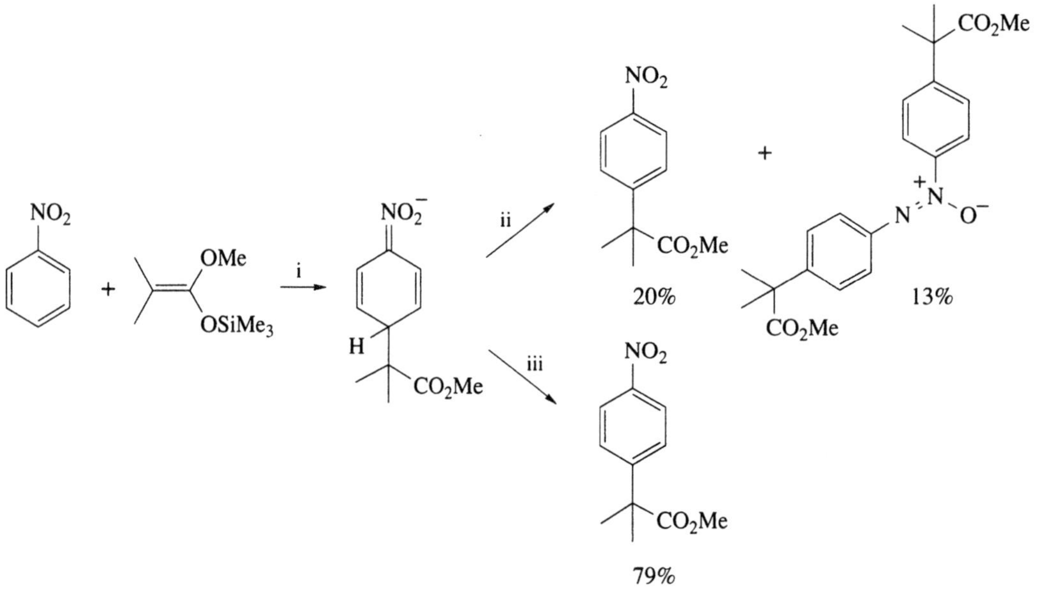

i, TASF, THF; ii, HCl; iii, Br_2 or DDQ

Scheme 1

When the attacking atom of the nucleophile carries a leaving group, replacement of hydrogen can take place *via* the VNS (vicarious nucleophilic substitution) reaction,[15] as shown in Scheme 3.[16] The key feature of this reaction is the rate-limiting base-induced β-elimination from the σ-adduct.

Other processes, limited to heteroaromatic systems, include the S_N(ANRORC) reaction (Scheme 4),[17,18] and ring transformation reactions.[18,19] Reactions which proceed *via* σ-adduct intermediates, but do not lead to substitution on the aromatic nuclei, such as the S_N(AEAE) reaction,[20] or give nonaromatic products, are not included. Also not covered are processes involving aryl–metal intermediates, such as most copper-catalyzed aromatic substitutions.

Activation of the aromatic ring towards nucleophilic attack is required for all these reactions to proceed at reasonable rates and is commonly provided by strongly electron-withdrawing substituents. Reactivity also strongly depends on the leaving group and on the nucleophile. Scales of activating power of

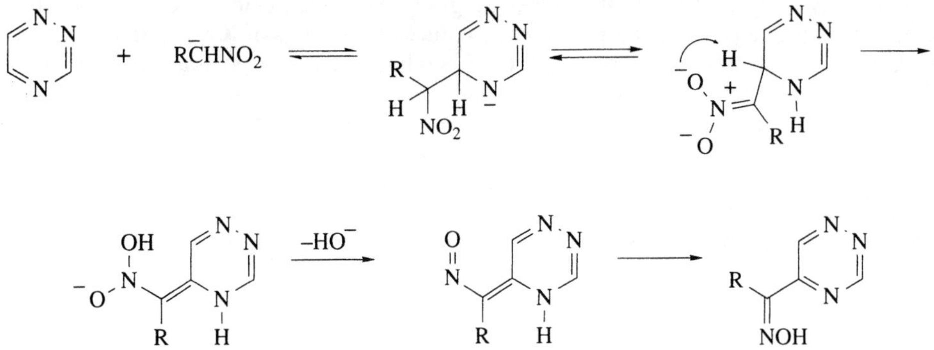

Scheme 2

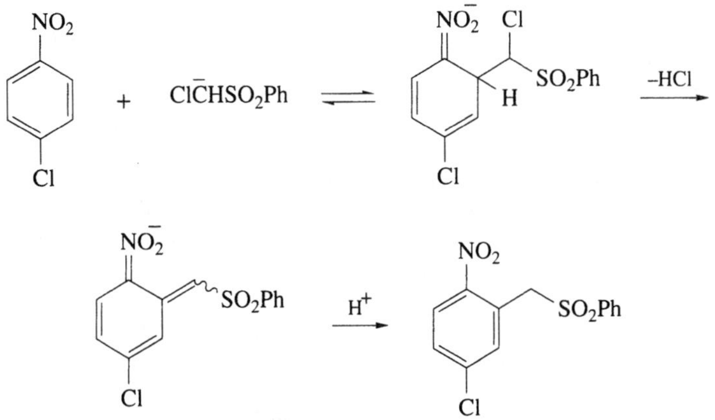

Scheme 3

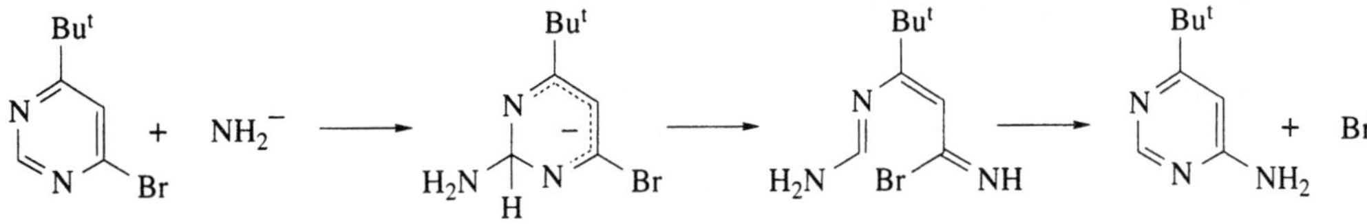

Scheme 4

substituents, of leaving group ability and of nucleophilicity, compiled in earlier reviews[1,2] on nucleophilic aromatic substitution, can serve as guidelines in predicting reactivity and selectivity. A partial listing is reported below for the reader's convenience. Ability as: (i) activating substituent, N_2^+ > ^+NR (heterocyclic) > NO > NO_2 > N (heterocyclic) > SO_2R > $^+NR_3$ > CF_3 > CN > CHO, COR > CO_2H > SO_3^- > Br, Cl, I; (ii) leaving group, F > NO_2 > OTs > SOPh > Cl, Br, I > N_3 > $^+NR_3$ > OAr, OR, SR, SO_2R, NH_2; (iii) nucleophile, RS^- > R_3C^- > RO^- > RNH_2 > ArO^- > HO^- > $ArNH_2$ > NH_3 > I^- > Br^- > Cl^-.

The use of oxazoline units to activate *ortho* positions has lately found many applications and has been reviewed.[21] Other activating groups recently employed include aryliodonium,[22] azopyridinium,[23] and, for nitrogen heterocycles, 2,6-dimethyl-4-oxopyridin-1-yl[24,25] and pyridinium[26] units. A rather extreme case of activation was reported in the reaction of polynitroarenes with HX (X = Cl, Br), which results in nucleophilic displacement of NO_2^- by X^-, and which proceeds *via* the highly electrophilic nitro-protonated form of the substrate.[27] Activation of the substrate towards nucleophilic attack is also achieved by π-complexing with transition metal ligands, a subject covered in Chapter 2.4 of this volume. Among the leaving groups, NO_2 has recently been exploited with success in many applications.[28–30]

Reference is also made to computer programs potentially useful in the prediction of reactivity/selectivity.[31] It is now well established that nucleophilic reactivity[32] is the result of a complex interplay of substrate–nucleophile–solvent interactions, with solvation[32,33] and steric effects[34,35] playing major roles. The relative abilities of leaving groups, for example, have been found to vary depending on the nucleophile.

Thus, nitrite displays enhanced reactivity as leaving group when the nucleophile is thiophenoxide.[36,37] Since the nucleophilic reactivity of anions is greatly reduced in solvents of low polarity, due to ionic association, or in protic solvents, due to solvation or H-bonding, the use of polar aprotic solvents (mostly DMF, DMSO and HMPA) and of phase transfer catalysis has become common practice in synthetic applications. PTC has proved of great value, especially in large-scale applications where the alternative use of polar aprotic solvents would be prohibitively expensive and product recovery and purification more difficult. Despite the considerable body of fundamental studies on micellar catalysis[38,39] and related molecular host–guest systems[40,41] in nucleophilic aromatic substitution, synthetic applications are lacking. The reasons probably include limited substrate solubilization and temperature range, as well as the difficulty of separating the products from the surfactants.

For any of the nucleophilic aromatic reactions covered in this chapter, regioselectivity, when more than one activated position is available, depends in most cases on the selectivity of attack by the nucleophile. However, when the conversion of the σ-adduct to product is the rate-limiting step, as in the VNS reaction, the final product distribution may differ from that expected, based on the relative electrophilic reactivity of the possible reaction sites.[15] Important roles are played by deactivating steric effects and by stabilizing specific interactions such as those, for example, between an ion-paired nucleophile and a nitro activating group, which favor attack at the *ortho* position.

Diastereo- and enantio-selectivity in nucleophilic aromatic substitution is limited to atropisomerism in binaphthyl- and biaryl-forming reactions.[21,42]

Nucleophilic attack is faster at unsubstituted ring positions than at similarly activated but substituted ring positions.[8,15,43–45] Since the addition is in most cases reversible, the opportunity exists for competing reactions. Indeed an extremely varied spectrum of reactivities is found in these systems, depending on reactants and reaction conditions. Examples are known of competing S_NAr and *cine* substitution,[46] S_NAr and *tele* substitution,[47] S_NAr and *cine* and *tele* substitution (equation 1),[5] S_NAr and VNS,[16] and S_NAr and S_N(ANRORC).[48]

Moreover, reactions at side chain centers, such as S_N2 displacement[49–51] and attack on an ester carbonyl,[50,52,53] may compete or prevail. Further complexity may result from the presence of ambident nucleophiles.[54,55]

A more intriguing type of competition is due to radical processes, which usually involve the substrate radical anion. These can fragment, if they carry a suitable substituent, *via* anion expulsion.[56] The resulting aryl radical, Ar·, can form the reduced product, ArH, by hydrogen abstraction[57] or the product of substitution *via* reaction with the nucleophile according to the $S_{RN}1$ mechanism, discussed in Chapter 2.2 of this volume. Examples of competition between S_NAr and radical processes of this type have been reported.[57–59]

In the case of nitro-substituted arenes, the formation of radical anions usually leads, in protic solvents, to products of nitro reduction, mostly azoxyarenes.[60] The challenge of avoiding reduction is commonly encountered, since the nitro group is a favorite activating substituent, easily introduced and readily transformed into a variety of functionalities. Early attempts to perform nucleophilic aromatic substitution on nitroaryl halides with alkoxides[61] and thiolate[62] ions, failed because of this complication. The presence of molecular oxygen[63] or other radical traps, such as azobenzene,[57] di- or tri-nitrobenzene[64] or excess thiol,[65] is often sufficient to inhibit reduction.

Ion-pairing effects may be crucial in determining the competition between radical and nonradical reactions. Recent investigations have shown that ion pairing accelerates the reduction of nitroarenes in alcoholic media.[60] Therefore, the higher yields of substitution products often obtained in the absence of ion pairing are due, at least in some cases, not only to the promotion of substitution, but also to the depression of radical pathways.

The material is organized in sections grouped according to the central atom of the attacking nucleophile. The coverage is not exhaustive, but hopefully representative of recent developments and trends.

2.1.2 CARBON NUCLEOPHILES

During the last decade the use of carbon nucleophiles in aromatic substitution reactions has increased greatly. Important advances concern particularly the alkylation of unsubstituted ring positions by formal nucleophilic displacement of hydrogen, according to the addition/oxidation sequence in its many variants and to the VNS reaction. In the following sections some of the most significant applications are described, grouped according to the nucleophile rather than to the reaction type.

2.1.2.1 Organometallics

A convenient route to biaryls involves nucleophilic displacement of *o*-methoxy, *o*-fluoro, or *o*-bromo in 2-aryloxazolines[21] and oxazoles[66] by ArLi or ArMgBr. Chiral binaphthyls have been obtained as a result of asymmetric induction by a chiral *o*-alkoxy leaving group[42] or by a chiral oxazoline *ortho*-activating substituent.[67] The latter method has also been applied to the synthesis of chiral biphenyls with moderate success.[68] Among the chiral leaving groups employed, derived from the naturally occurring alcohols 1-menthol, quinine, quinidine, α-fenchol and borneol, (–)-menthoxy gave the highest chiral transfer efficiencies (defined as % chemical yields × % optical yield), which ranged from 41% to 65% (Scheme 5).[42] The use of a chiral oxazoline *ortho*-activating group, exemplified in Scheme 6, led to a mixture of diastereomeric products, which upon acidic hydrolysis and LAH reduction yielded (hydroxymethyl)binaphthyls of enantiomeric excess in the range 87.4–96%.[67]

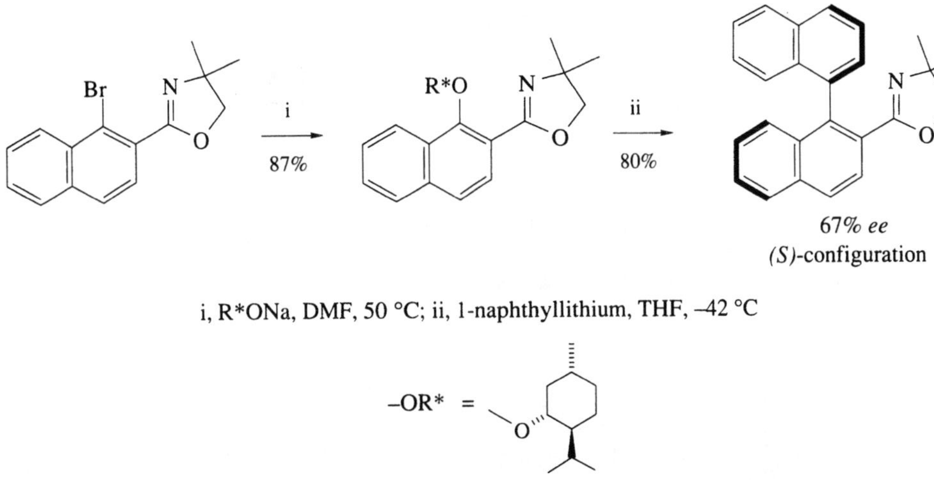

Br N O i 87% R*O N O ii 80% N O

67% *ee*
(*S*)-configuration

i, R*ONa, DMF, 50 °C; ii, 1-naphthyllithium, THF, –42 °C

–OR* =

Scheme 5

OMe CONH$_2$ i 75% OMe O Ph N OMe ii 71%

OMe Ph O N OMe iii, iv 65% OMe OH

diastereomeric ratio:
91.9:8.9

96% *ee*
(*R*)-configuration

i, $Et_3O{\bullet}BF_4$, DCE, 30 h, r.t, then (+)-1-MeO-2-NH_2-3-phenyl-3-hydroxypropane, reflux, 48 h;
ii, 2 equiv. 2-MeO-naphthylMgBr, THF; iii, HCl, EtOH, reflux; iv, $LiAlH_4$, THF, 25 °C, 6 h

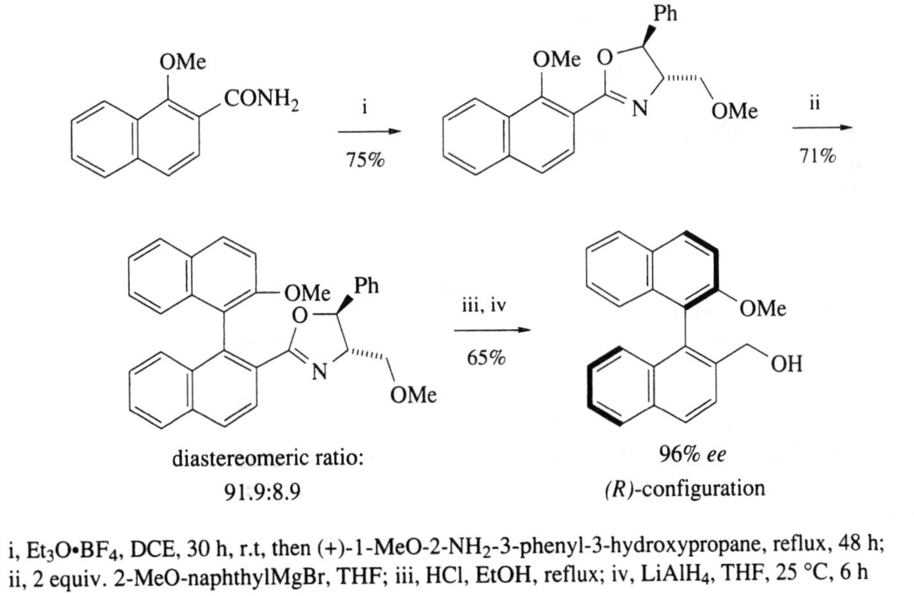

Scheme 6

Nucleophilic displacement of RSO_2^- (R = alkyl, aryl) by RMgX and ArMgX is a convenient route to alkylate the 2-position of pyridines.[69] The method is not applicable to the 4-isomers which give sluggish reactions and radical-derived products.[69]

Grignard and lithium reagents have also been reported to effect nucleophilic displacement of hydrogen. Treatment of 3-pyridyloxazoline with a variety of lithium and Grignard reagents gave the products of addition to the 4-position of the pyridine ring in good yields.[70,71] The dihydropyridines so obtained are readily oxidized to pyridines, as in the example of Scheme 7.[70] In contrast, treatment of the isomeric 4-pyridyloxazoline with lithium reagents gave *ortho* metallation.[21] The addition/oxidation sequence has been used to prepare optically active (*S*)-4-naphthylquinoline derivatives from a chiral 3-quinolinyloxazoline precursor.[72] The addition of Grignard reagents to 1-carboxy-[73,74] and 1-alkoxycarboxy-pyridinium salts,[75] the latter unstable species being generated *in situ* (*e.g.* **4**; Scheme 8), provides an alternative convenient route to alkylated pyridines. With 1-carboxypyridines, a catalytic amount of copper(I) iodide was required to attain regioselectivity and form the product of 1,4-addition almost exclusively.[73,74] 1-(2,6-Dimethyl-4-oxopyridin-1-yl) pyridinium cation reacted with aryl and heteroaryl Grignard reagents to give regiospecifically the corresponding 4-arylpyridines.[24]

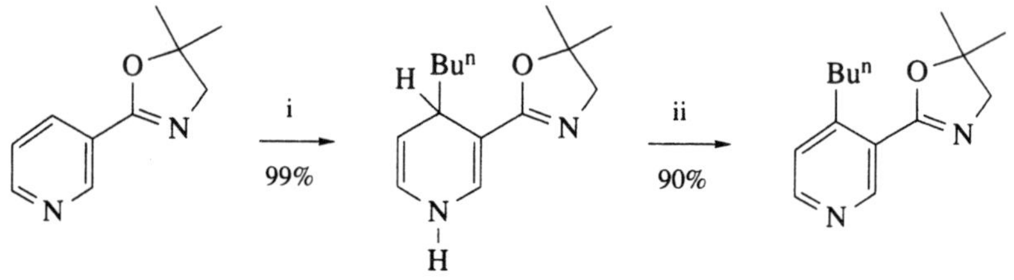

i, Bu^nLi, THF, –78 °C, 1 h; ii, chloranil, toluene, 2 h

Scheme 7

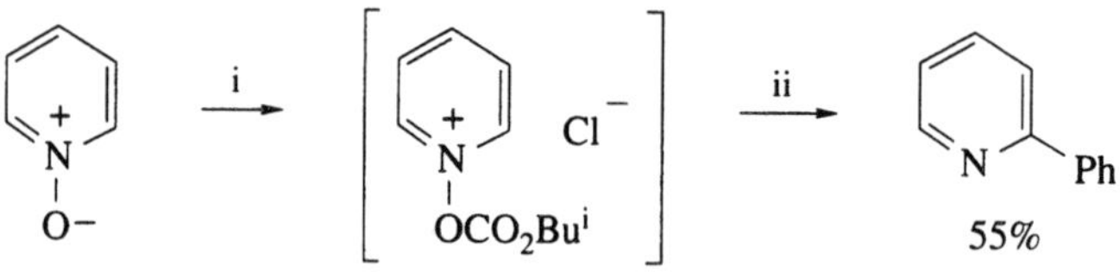

(4)

i, $ClCO_2Bu^i$, THF, 20 °C; ii, PhMgCl, –78 °C, 30 min

Scheme 8

Aromatic nitro compounds can be alkylated in *ortho* or *para* positions by treatment with alkyllithium or alkyl Grignard reagents followed by oxidation.[76] Thus, for example, the reaction of 2-methyl-1-nitronaphthalene with BuLi gave 3-butyl-2-methyl-1-nitronaphthalene in 70–75% yield when the adduct was oxidized with Br_2 or DDQ in ether.[76] Interesting redox processes have been identified in the reaction of nitroarenes with 2-lithium-1,3-dithiane.[11] A unified mechanism has been proposed for these reactions, comprising an initial single electron transfer step to form a radical anion–radical pair, followed by geminate radical combination to form the σ-adduct, or by diffusion from the solvent cage to give the radical anion and products derived therefrom.[11,77,78]

Alkyl (but not aryl) Grignard reagents add irreversibly[79] to activated unsubstituted positions of nitroarenes, including polycyclic and heterocondensed systems.[78] Reactive positions are pointed by the arrows in (**5**)–(**7**) (X = S, NR), and (**8**) and (**9**) (A = thiazole, oxazole, thiophene, pyrrole, pyridine, benzene).[78] The resulting nitronate intermediates can be: oxidized to alkylated nitroarenes with DDQ in THF, LTA in CH_2Cl_2, or $KMnO_4$ in alkaline acetone–water;[76,78] converted to alkylnitrosoarenes with HCl (37%) $BF_3{\cdot}Et_2O$,[78] or reduced to alkylanilines with $LiAlH_4$ or $NaBH_4$, Pd/C in THF.[79] At low temperatures high chemoselectivity was observed: keto, cyano and ester functions were not attacked by the

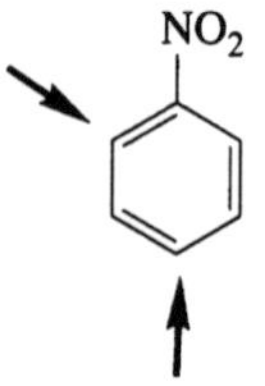

(5) 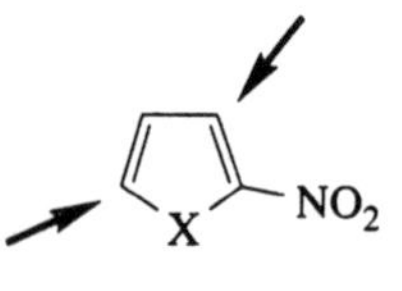 **(6)** 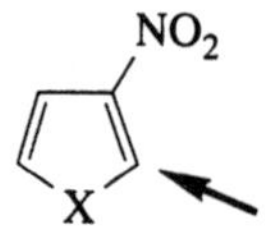 **(7)** 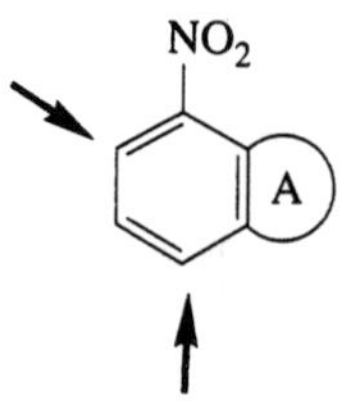 **(8)** 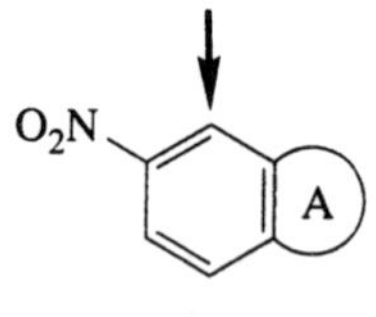 **(9)**

Grignard reagent.[80] Competition by nucleophilic aromatic substitution was not observed unless the only active position(s) was (were) substituted with a leaving group, as in the reaction of 1-methoxy-2-nitronaphthalene which gave 1-alkyl-2-nitronaphthalenes in 73–95% yields.[81] The use of Me_3SiCH_2MgCl (Peterson reagent) provides an entry to nitro-substituted benzyl anion intermediates, as shown in the example of Scheme 9.[82]

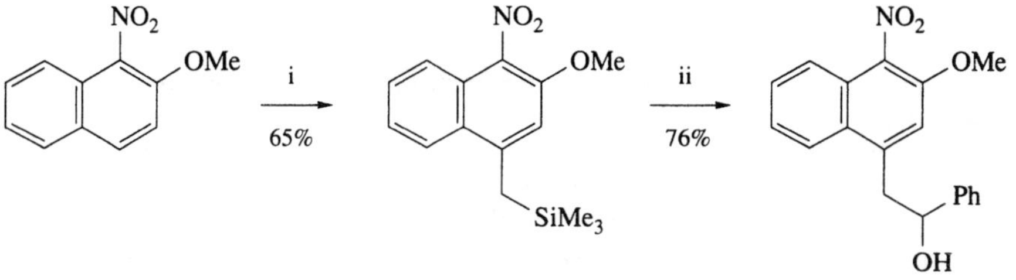

i, Me_3SiCH_2MgCl, THF, –30 °C, 1 h, then DDQ, –30 to 0 °C, 1 h;
ii, PhCHO, TBAF, THF, r.t., 30 min

Scheme 9

2.1.2.2 Active Hydrogen Compounds

Activated nitro and halo substituents have been efficiently replaced by a variety of alkyl groups *via* S_NAr reaction with carbanions. Examples include the displacement of the nitro group in compounds (**10**; X = 4-PhCO, 4-MeOCO, 4-CN, 4-NO_2, 4-$PhSO_2$, 3,5-$(CF_3)_2$) by the anion of 2-nitropropane in HMPA at room temperature (equation 2),[83] and the reaction of *p*-dinitrobenzene with several ketones, esters and nitriles (RH; equation 3) in Bu^{t}OK/liquid NH_3 at –70 °C.[84] Interestingly, under the latter reaction conditions, *p*-chloronitrobenzene gave the product of alkylation rather than of S_NAr displacement of chloride, as in equation (4).[85] Further examples include the dehalogenation of *p*-halonitrobenzenes by 9-fluorenyl anions in DMSO at room temperature,[34] and dehalogenation and denitration reactions by the carbanions of phenyl- and diphenyl-acetonitrile in DMSO or under PTC conditions.[86]

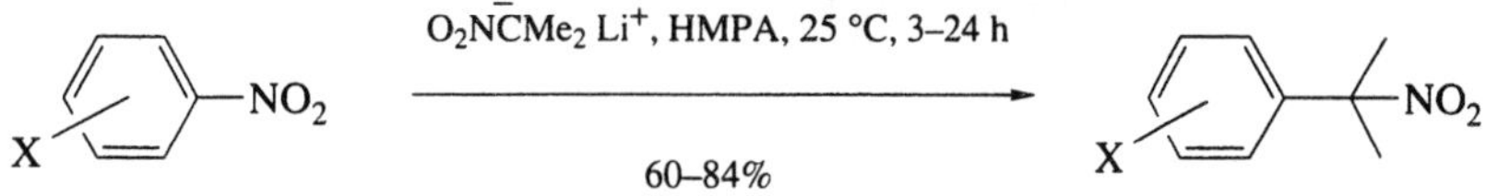

$O_2N\bar{C}Me_2$ Li^+, HMPA, 25 °C, 3–24 h; 60–84% (2)

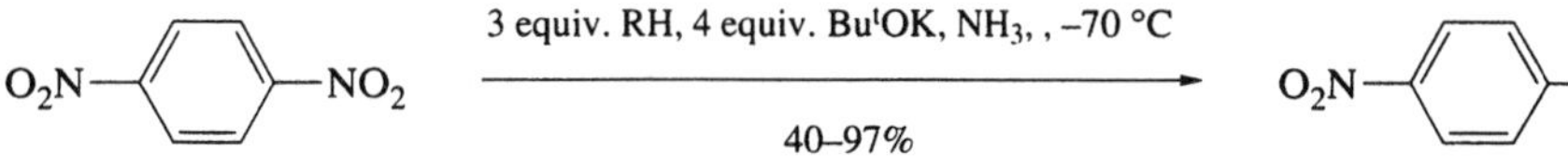

(**10**)

3 equiv. RH, 4 equiv. Bu^{t}OK, NH_3, , –70 °C; 40–97% (3)

Me_2CO, Bu^{t}OK, NH_3, –70 °C; 56.3% 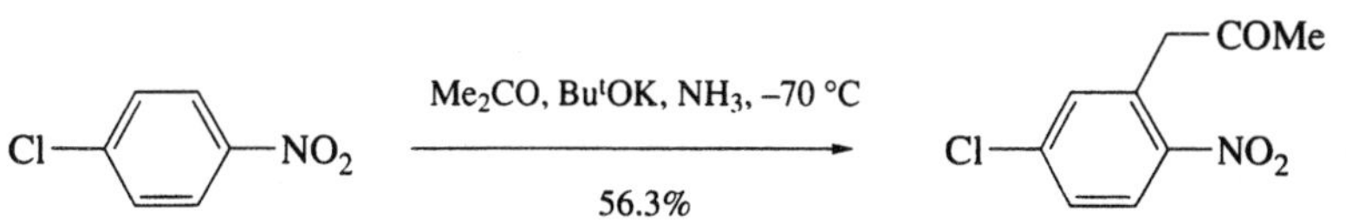 (4)

Unsymmetrical biphenyls have been prepared in good yields by the S_NAr reaction of nitroarenes, bearing a leaving group *ortho* or *para* to the nitro group, with the anion of 2,6-di-*t*-butylphenol, which behaves as a carbon nucleophile.[9] Reaction conditions and scope are summarized in equation (5; X = 2-F, 2-Cl, 2-Br, 2-I, 2-NO_2, 2-SO_2Ph, 4-NO_2, 4-SO_2Ph). Arylation took place at activated unsubstituted ring positions in substrates lacking a suitably located leaving group, such as *m*-dinitrobenzene, according to an addition/oxidation sequence.[9] Bis- and tris-(dialkylamino)benzenes also behave as carbon nucleophiles in displacing chloride from activated arenes. In this case, however, greater activation of the electrophile was required, such as that provided by at least two nitro groups.[87]

The reaction of nitroarenes with silyl enol ethers and ketene silyl acetals in MeCN/THF with 1 equiv. of TASF, followed by *in situ* oxidation with Br_2 or DDQ, provides an easy route to α-nitroaryl carbonyl compounds (Scheme 1).[12] The use of these compounds as reagents for the synthesis of arylacetic acids, propionic acids, indoles, 2-indolinones and other heterocyclic compounds has recently been described.[88]

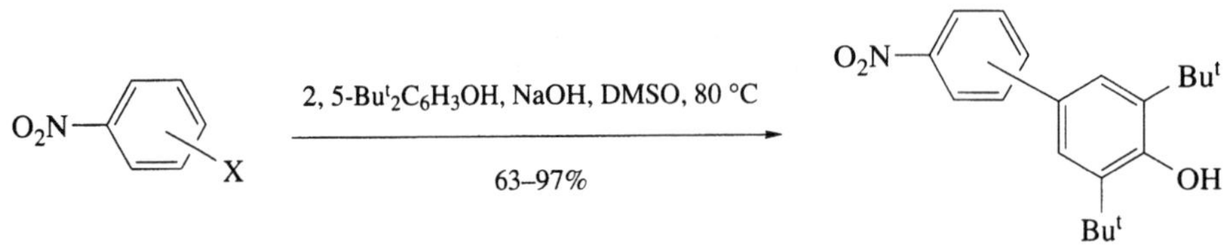

O_2N–C_6H_4–X $\xrightarrow[\text{63–97\%}]{\text{2, 5-Bu}^t_2\text{C}_6\text{H}_3\text{OH, NaOH, DMSO, 80 °C}}$ O_2N–C_6H_4–C_6H_2(Bu^t)$_2$OH (5)

The alkylation reaction is limited to nitro-substituted arenes and heteroarenes and is highly chemoselective; nucleophilic displacement of activated halogens, including fluorine, was not observed. The regioselectivity is determined by the bulkiness of the silicon reagent. With unhindered silyl derivatives a strong preference for *ortho* addition was observed, as in the example of equation (6). With bulkier reagents attack took place exclusively at the *para* position (Scheme 1). The success of this reaction, which could not be reproduced with alkali enolates, was attributed at least in part to the essentially nonbasic reaction conditions under which side processes due to base-induced reactions of nitroarenes can be effectively eliminated.[12]

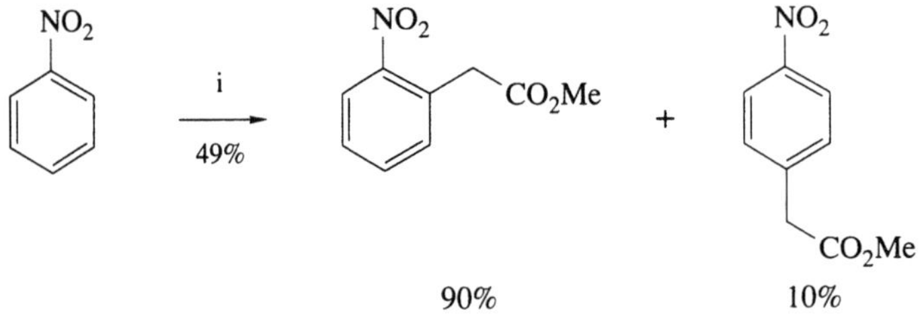

$PhNO_2$ $\xrightarrow[\text{49\%}]{\text{i}}$ o-$O_2NC_6H_4CH_2CO_2Me$ (90%) + p-$O_2NC_6H_4CH_2CO_2Me$ (10%) (6)

i, CH_2=C(OMe)$OSiMe_3$, TASF, THF, –78 °C, 18–21 h; Br_2

Base-catalyzed rearrangement of *N*-(aryloxy)pyridinium salts (**11**) leads to 2-arylpyridines (**12**) *via* intramolecular attack on the *ortho* position of the aryloxy ring by the 2-pyridyl carbanion. When X was 3-CO_2Me, pyrido[3,2-*d*]coumarins (**13**; R = NO_2, CN) were directly obtained in useful yields from this reaction. When in (**11**) X was 3-COMe and R was 4-NO_2, 10-hydroxy-10-methyl-6-nitropyrido[2,3-*d*]benzopyran (**14**) was obtained in 10.8% yield.[89]

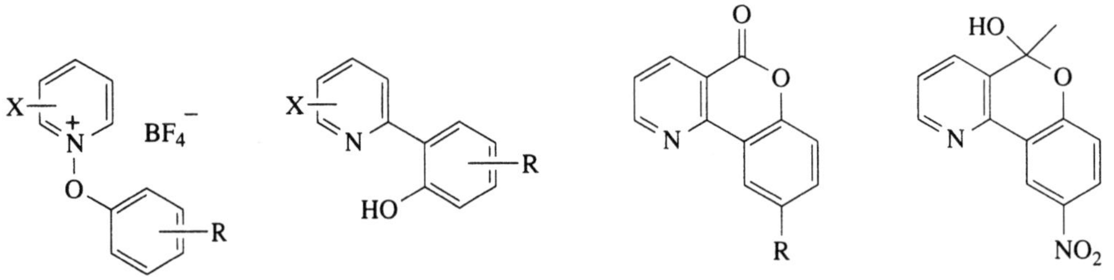

(**11**) (**12**) (**13**) (**14**)

The *N*-(2,6-dimethyl-4-oxopyridin-1-yl)pyridinium salts (**15**)[24] have proved to be versatile intermediates for the regiospecific synthesis of 4-substituted pyridines (**17**) *via* attack by the appropriate carbon nucleophiles, *e.g.* ionized ketones,[90] nitroalkanes,[91] esters and nitriles,[92] and α-diketones, α-keto esters, α-diesters, disulfones *etc.* (Scheme 10).[93] Aromatization of the intermediate 1,4-dihydro adduct (**16**) was generally achieved under free radical conditions.

Selective *ortho* alkylation of anilines and phenols can be carried out *via* the sequence of Scheme 11, which involves as the key step a Sommelet–Hauser-type [2,3] sigmatropic rearrangement of ylides (**19**).[94,95] The success of these reactions depends on the effective generation of the intermediate aza- and phenoxy-sulfonium salts (**18**). Conversion rates ranged from 37% to 82% for anilines (yields, based on unrecovered starting material, were 41–90%)[94] and were somewhat lower for the reactions of phenols.[95] High conversion rates have recently been reported for phenol alkylations using alkyl isopropyl sulfide, SO_2Cl_2 and triethylamine.[96] Useful elaborations of the alkyl side chain in the product have been reported, including desulfurization *via* Raney nickel reduction,[94,95] cyclization to heterocyclic compounds (*e.g.* indoles when sulfides with β-carbonyl groups are used,[97] chromans, chromenes and coumarins[96]) and hydrolysis of —CHR′SR″ to —CHO when R′ is an alkylthio group.[95,98]

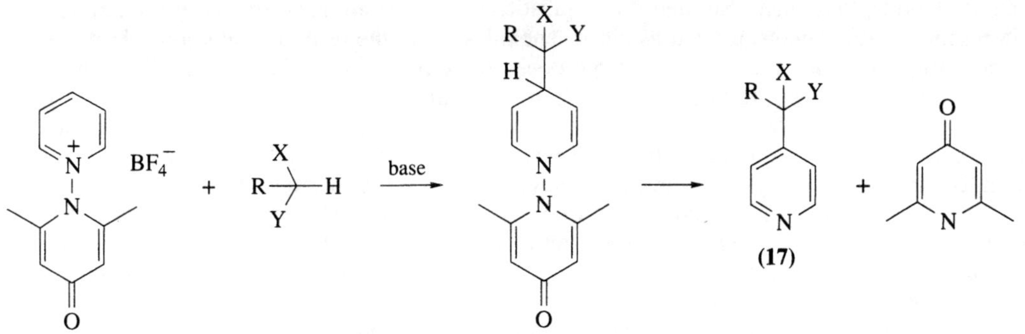

Scheme 10

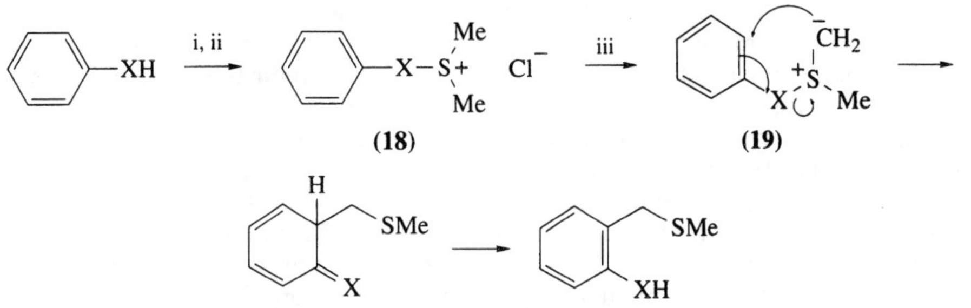

X = NH: i, Bu^{t}OCl, CH_2Cl_2, –65 °C; ii, 3 equiv. MeSMe; iii, MeONa, MeOH

X = O: i–ii, (succinimido-N-S^+Me_2 Cl^-) (prepared *in situ* from NCS and Me_2S), CH_2Cl_2, –25 °C,

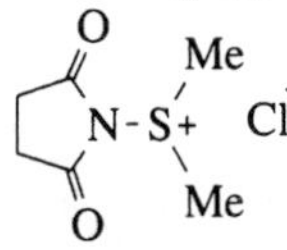

or $Me_2\overset{+}{S}Cl$ Cl^- (prepared *in situ* from Me_2S and Cl_2), CH_2Cl_2, –60 °C; iii, Et_3N

Scheme 11

A related process involves fluoride-induced desilylation in HMPA at room temperature of benzyldimethyl(trimethylsilylmethyl)ammonium halides (**20**) to (**21**; 62–84% yields; R = H, 2-Me, 4-Me, 2-Cl, 4-Cl, 4-OAc; X = Cl, Br).[99,100] Compunds (**20**) were prepared by reaction of the corresponding benzyl halide derivatives with (dimethylaminomethyl)trimethylsilane, $Me_3SiCH_2NMe_2$. The product of Stevens rearrangement (**22**) formed competitively and predominantly from precursors (**20**) having strong electron-withdrawing substituents (R = 2-COMe, 2-CN, 4-CN, 2-NO_2, and 4-NO_2).[100]

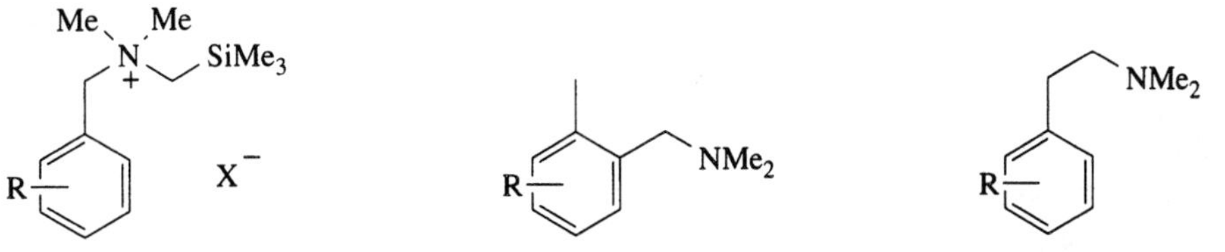

Several α-functionalized alkyl substituents can be introduced, often with good regioselectivity, into unsubstituted positions of nitro-substituted arenes by means of the VNS reaction (Scheme 3).[15,101] Useful nucleophiles are of the type RCHXY, where X is a leaving group (Cl, PhS, Me_2NCS_2, PhO, MeO), Y is an electron-withdrawing substituent (SO_2Ph, SO_2Bu^t, SO_2OPh, $SO_2CH_2Bu^t$, SOPh, $POPh_2$, $P(OEt)_2$, CN, CO_2R', PhS, Cl), and R is H, alkyl, aryl, PhS, or Cl.[15] The corresponding carbanions are relatively inert towards oxidation and self-condensation.[15] The nucleophile of greatest applicability appears to be the anion of chloromethyl phenyl sulfone. Most substituents on the aromatic ring do not interfere with the VNS reaction. Nitrophenols, present as nitrophenoxides under the reaction conditions, do not react, but nitrobenzoates, in which the negative charge does not conjugate with the ring, do.[102] Activation by a nitro substituent is a requisite to enter the VNS reaction also for electrophilic arenes, such as pyridine. The only reported exceptions are 1,2,4-triazines[103] and benzothiazoles.[104] The following nitro-substituted heteroarenes were found to undergo the VNS reaction: nitropyridines and 4-nitropyridine *N*-oxide (but not the 2- and 3-NO_2 isomers);[105] nitrothiophenes and nitro-*N*-alkylpyrroles (but not nitrofurans);[106] and 5-, 6- and 8-nitroquinolines.[107] Studies on nitrobenzene derivatives showed that in the absence of steric effects attack at *ortho* is favored. However, steric and ion-pairing effects are of major consequence in determining the reaction selectivity. In DMSO, DMF or NH_3, solvents in which the carbanions are not strongly associated with the cation, steric effects dominate and the product of *para* alkylation was obtained exclusively with tertiary carbanions and preferentially even with relatively unhindered secondary carbanions like the anion of bromo- and iodo-methyl phenyl sulfone.[16] With Bu^tOK in THF[108] and in DMF[109] at low temperatures, however, *ortho* alkylation appears to be favored, as a result of specific stabilizing interactions of the ion-paired K^+Nu^- with the nitro group, which favors approach and attack at the *ortho* position.

The VNS reaction generally proceeds only to the stage of monoalkylation since the products are anions (Scheme 3), inert towards attack by the nucleophile. In the presence of two or three nitro groups, however, products of di- and tri-substitution were obtained.[110] The products of VNS were obtained selectively also in the presence of activated leaving groups, as shown in the example of equation (7).[111] In other instances, like in the reaction with chloromethyl phenyl sulfone in DMSO/KOH, competing S_NAr displacement of F^- and NO_2^- was found in 4-F-, 2-NO_2- and 4-NO_2-substituted nitrobenzene.[16] In these cases, however, low combined yields of S_NAr and VNS products were obtained. In the reaction of chloromethyl phenyl sulfone with nitro-substituted benzophenones VNS ring alkylation is accompanied by Darzens' condensation (equation 8).[112] It has been pointed out that in these cases the VNS reaction can be promoted by increasing the concentration of the base, which has an effect on the rate-determining β-elimination step of the VNS reaction, but no effect on either the Darzens' or the S_NAr reactions.[15] The products of bisannulation rather than of VNS substitution were obtained from the reaction of chloromethyl phenyl sulfone with quinoxaline, several naphthyridines, and 1-cyano- and 1-(methylsulfonyl)-naphthalene.[15]

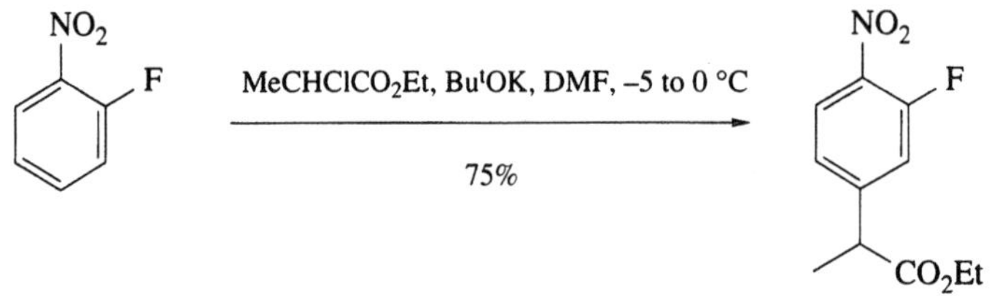

NO_2, F — $MeCHClCO_2Et$, Bu^tOK, DMF, –5 to 0 °C; 75% → NO_2, F, CO_2Et (7)

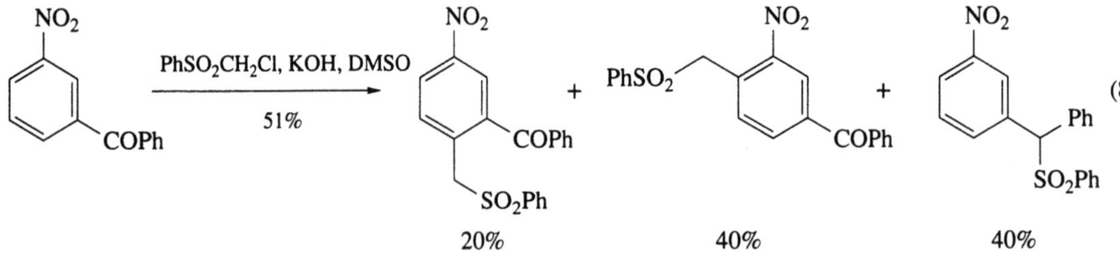

NO_2, COPh — $PhSO_2CH_2Cl$, KOH, DMSO; 51% → NO_2, COPh, SO_2Ph (20%) + $PhSO_2$, NO_2, COPh (40%) + NO_2, Ph, SO_2Ph (40%) (8)

Some useful elaboration of the initially introduced nucleophiles has also been reported.[113] Thus, formylation of nitro aromatic rings was achieved *via* VNS reaction of the nitroarene with the anions of trisphenylthiomethane[114] and chloroform,[115] followed by hydrolysis, as in the example of Scheme 12.

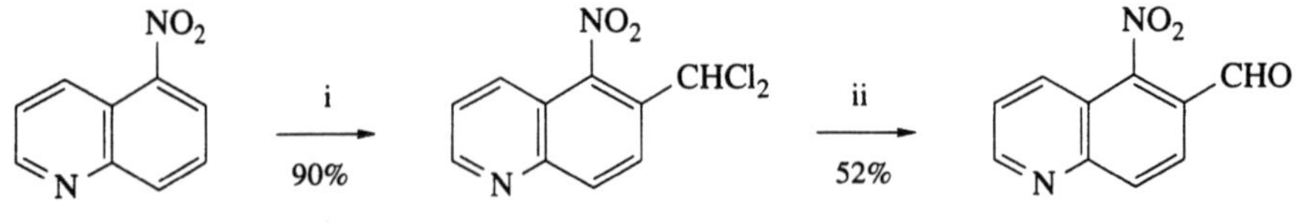

NO_2, N — i, 90% → NO_2, $CHCl_2$, N — ii, 52% → NO_2, CHO, N

i, $CHCl_3$, Bu^tOK, THF/DMF; ii, HCO_2H, H_2O

Scheme 12

Some synthetic applications involving an intramolecular VNS reaction have been developed.[116–118]

2.1.2.3 Cyanides

1-Cyanoisoquinolines and 2-cyanoquinolines were recently prepared *via* the reaction of isoquinolines and quinolines with *p*-TsCl/KCN.[119] The reaction proceeds in two steps, as shown in the example of Scheme 13, and involves Reissert intermediates, which upon treatment with base give the cyano aromatic products in good overall yields.[119]

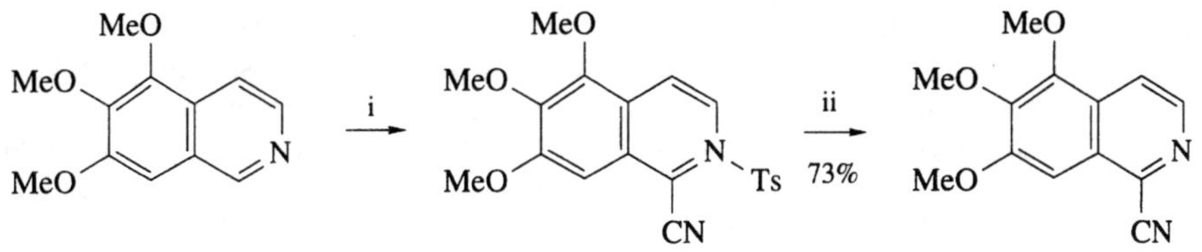

i, 1.5 equiv. *p*-TsCl, 6 equiv. KCN, THF, 25 °C, 12 h; ii, 1.2 equiv. DBU, THF, 25 °C, 1 h

Scheme 13

Regioselective cyanation in the α-position of pyridines and quinolines was achieved by treatment of the corresponding *N*-oxides with trimethylsilyl cyanide in the presence of triethylamine.[120] The silyl reagent could also be generated *in situ* by the reaction of chlorotrimethylsilane with KCN, as in the example reported in equation (9). In the presence of catalytic amounts of $Bu^n{}_4NF$ the reactions could be carried out at 5 °C in THF. The method has been extended to pyrimidine *N*-oxides, which underwent cyanation at the 2- and 6-position, attack at C-2 being generally favored.[121] Thus, 4-methoxypyrimidine 1-oxide gave the 2-carbonitrile as the sole product in 86% yield.[121]

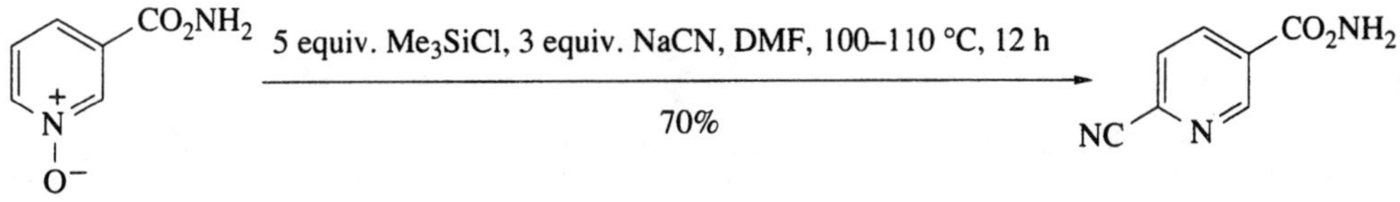

5 equiv. Me_3SiCl, 3 equiv. NaCN, DMF, 100–110 °C, 12 h; 70% (9)

2.1.3 NITROGEN NUCLEOPHILES

Mechanistic features of S_NAr displacement reactions involving amino nucleophiles have been the object of many investigations, a major point of interest being the occurrence of base-catalyzed paths. Strongly activated aryl halides react readily with ammonia and with primary and secondary amines to give the corresponding arylamines. Thus, for example, 2,4-dinitrofluorobenzene is used to tag the amino end of a peptide or protein chain.

Some of these reactions are characterized by high negative activation volumes.[122] Recently, high pressure conditions have been employed to activate some S_NAr reactions involving primary and secondary amines and activated aryl halides[123] and halopyridines.[124] The reaction of 4-chloronitrobenzene with Bu^nNH_2 in THF, for example, gave 4-*N*-butylnitroaniline in 76% yield after 20 h at 7.2 kbar and 50 °C (equation 10).[123] By contrast, at 1 atm and 80 °C no reaction took place during the same time.[123] The promotion of reactivity was much less pronounced with secondary amines, with the exception of cyclic ones, and absent with tertiary and aromatic amines.[123]

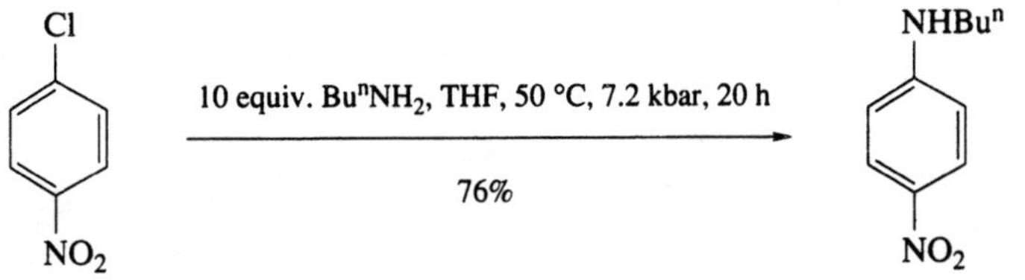

10 equiv. Bu^nNH_2, THF, 50 °C, 7.2 kbar, 20 h; 76% (10)

Ammonolysis of aryl halides has been performed under PTC conditions.[125] The reaction of 2,4-dinitrochlorobenzene with NH_3 (g) in toluene at ambient temperature and pressure gave, in the presence of 10% TBAB and after 3 h, 2,4-dinitroaniline in 16% yield (0.6% in the absence of TBAB). The reaction was complete after 24 h but the product final yield was not reported.[125]

The use of formamide as solvent in the ammonolysis of activated aryl chlorides has been recommended when the substrate carries cyano ring substituents which can undergo hydrolysis in water.[126]

Anilines (**23**; R = 2-Pr, $PhCH_2$; X = Cl, Br, I) have been obtained in high yields (40–95%) from the reaction of the primary amines 2-propylamine and benzylamine with the ethers (**24**; R′ = Me, $CH_2{=}CHCH_2$, CH_2-oxirane) in EtOH at room temperature.[127] With secondary amines dealkylation of the aromatic ether to the corresponding phenol took place rather than S_NAr dealkoxylation.

Diarylamines are readily obtained from the reaction of activated fluoroarenes with arylamines containing electron-releasing groups.[128] Anilines carrying electron-withdrawing substituents require deprotonation to enter S_NAr or nucleophilic hydrogen displacement reactions. Diarylamines (25; R′, R″ = *o*- or *p*-NO_2, CN, PhCO, $PhSO_2$) have been obtained in fair to good yields *via* defluorination or denitration of activated fluorobenzenes[129] and nitrobenzenes[130] respectively, using an excess of the aniline reagent and Bu^tOK in DMSO at 20–25 °C or K_2CO_3 in DMSO or DMF at 120–130 °C. Under the latter conditions (K_2CO_3 in DMF at 120–130 °C) the reaction of 4-cyanoaniline with 4-fluoronitrobenzene gave the triarylamine (**26**) in 64% yield whereas, with *p*-dinitrobenzene (**26**) only formed as a minor product (3% yield), the major product (50% yield) being the diamine (**25**; R′ = 4-CN; R″ = 4-NO_2). The difference in reactivity was attributed to the greater bulk of the nitro substituent with respect to fluorine.[129]

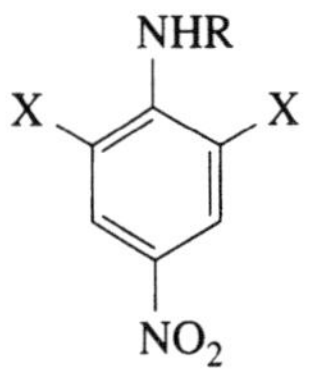

(23) 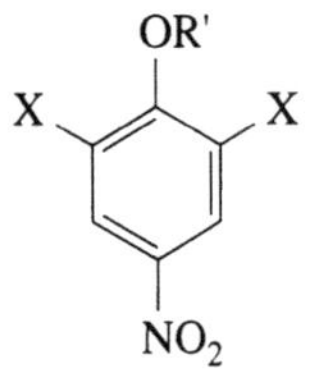 **(24)** 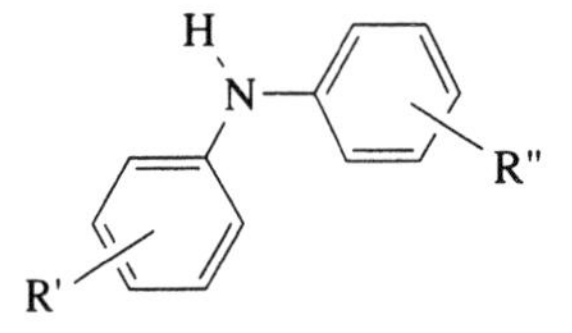 **(25)** 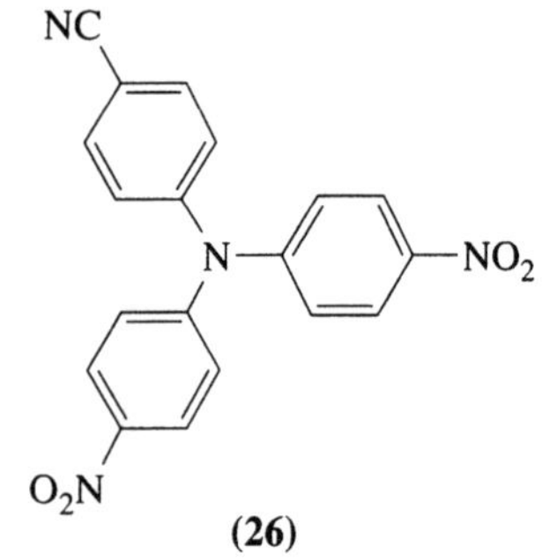 **(26)**

Treatment of the salicylic acid derivatives, dinitrolactone (**27**) and dinitro ester (**28**), with aqueous amines (NH_3, $MeNH_2$, Me_2NH) afforded salicylamides (**29**; R, R′ = H, Me) in quantitative yields.[131]

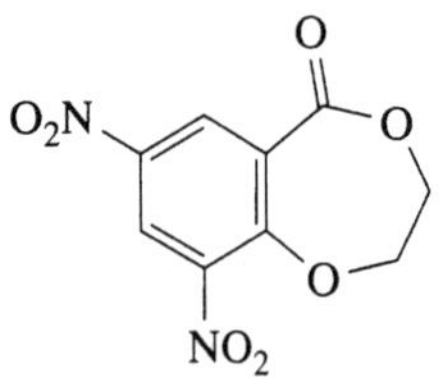

(27) 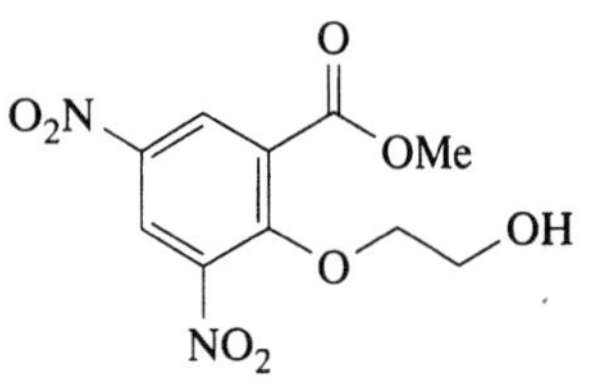 **(28)** 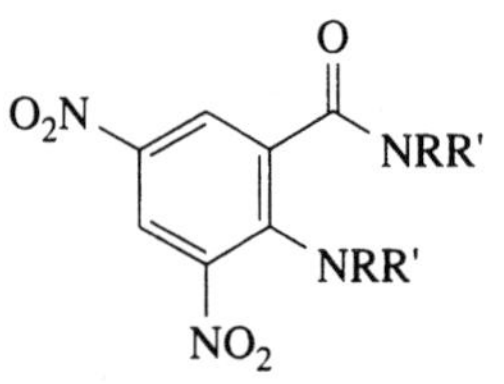 **(29)**

Nitroanions derived from phenothiazenes readily displace the halide in 4-halonitrobenzenes in DMSO at room temperature (equation 11).[34] On the other hand nucleophilic displacement of hydrogen as well as S_NAr displacement of a leaving group were observed in the reaction of nitrobenzene derivatives with *p*-nitroaniline and Bu^tOK in HMPA.[132] The oxidation with potassium permanganate of adducts formed from nitrogen heterocycles, such as pyridazines and 4-nitropyridazine 1-oxide, with liquid ammonia or with KNH_2 has been widely applied to the synthesis of aminoaza aromatics.[133]

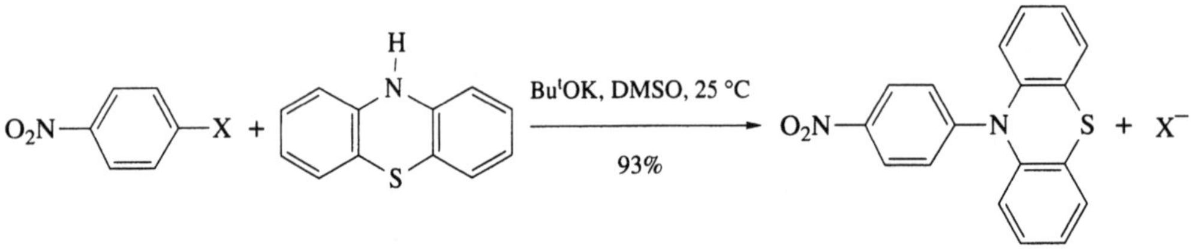

Bu^tOK, DMSO, 25 °C; 93% (11)

Amine–amine exchange reactions have recently been performed with high yields and under mild conditions (DMSO at 30 °C[134] or MeCN at room temperature[135]) on activated naphthylamines with primary amines or pyrrolidine. Equation (12), where R′, R″ = H, H; Me, H; Et, H; Pr^i, H; Bu^t, H; $PhCH_2$, H; 4-$MeOC_6H_4$, H; —$(CH_2)_4$—, summarizes the reactions of *N,N*-dimethyl-2,4-bistrifluoroacetyl-1-naphthylamine.[135]

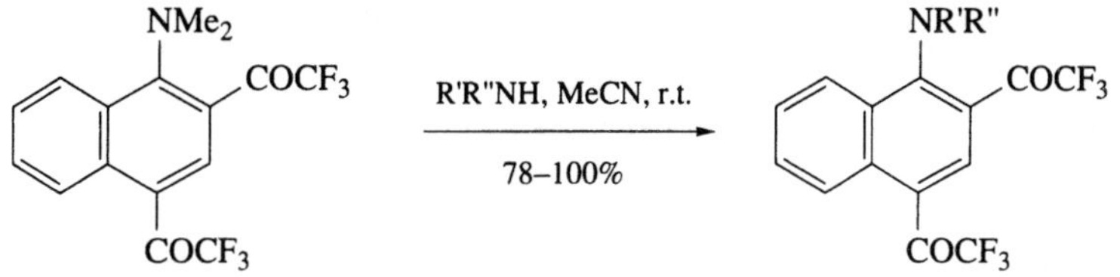

(12)

Several useful syntheses of polycondensed systems have been reported, based on an intramolecular displacement involving an amino nucleophile. Phenazine-1-carboxylic acids (**30**) have been prepared *via* reaction of *N*-phenyl-3-nitroanthranilic acids (**31a**; X = H; R = 6′-MeO, 3′-MeO) with $NaBH_4$ under alkaline conditions, a process involving reduction of NO_2 to NH_2, followed by intramolecular nucleophilic attack and aromatization.[136] An improved procedure utilizes *N*-(2-fluorophenyl)-3-nitroanthranilic acids (**31b**; X = F; R = H, 3′-F, 5′-F, 6′-F, 5′-Cl, 6′-Cl, 5′-Me) with the advantage that no ambiguity as to the carbon attacked exists in this case since ring closure occurs *via* S_NAr displacement of fluoride.[137] Reactions were conducted in H_2O, MeOH or EtOH and gave (**30**) in 47–92% yield.

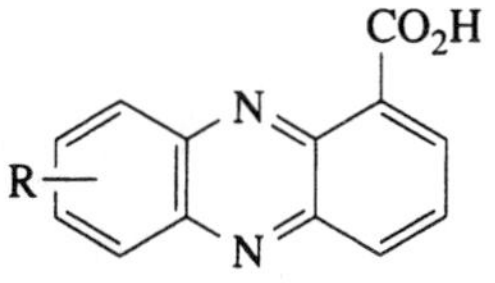

(30) 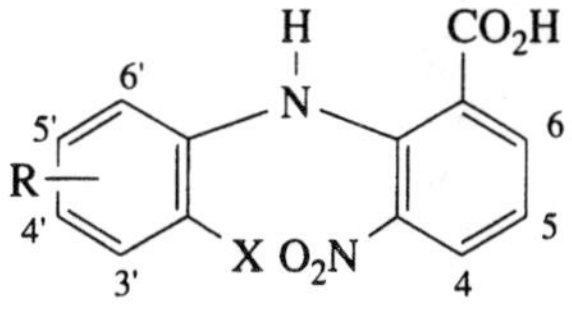 **(31a–b)**

A large number of 5-deazaflavins (**32**; R^1, R^2 = H, alkyl, aryl; R^3, R^4 = H, Cl, NO_2, OH; 48 examples in all), have been prepared in good yields *via* condensation of 6-substituted aminouracils with *o*-halobenzaldehydes in DMF under reflux. The mechanism shown in Scheme 14 was proposed for this reaction.[138] Several bis(5-deazaflavin-10-yl)alkanes (**33**; *n* = 6, 8, 10, 12) have also been prepared *via* the same route using bis(uracil-6-ylamino)alkanes.[138] By an analogous reaction the substituted quinolines (**34a**) and (**34b**) were obtained in 87% and 50% yield, respectively, from enaminones (**35a**; X = Y = NMe; Z = O) and (**35b**; X = Y = CH_2; Z = Me_2) and pentafluorobenzaldehyde in glacial acetic acid at reflux.[139]

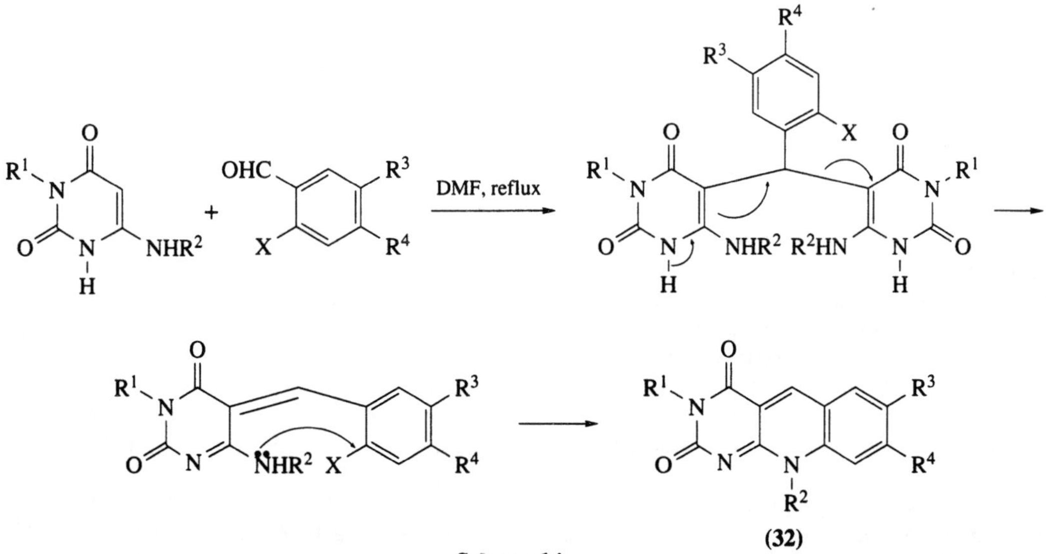

(32)

Scheme 14

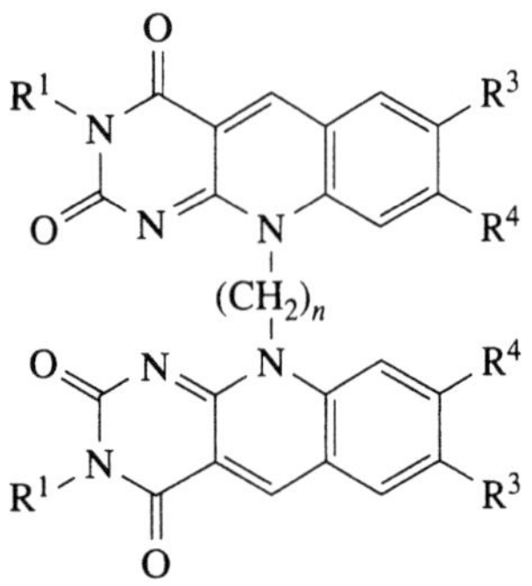

(33) 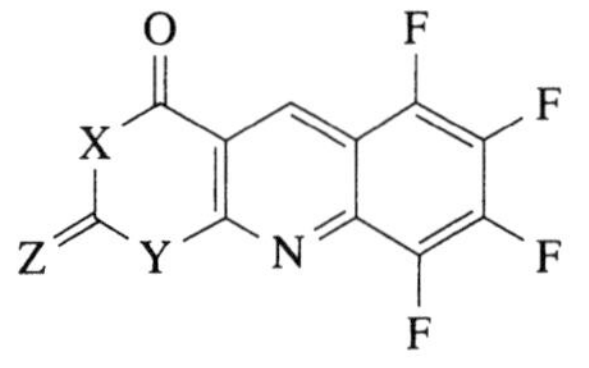 (34a–b) 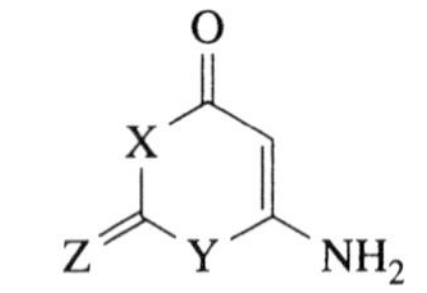 (35a–b)

Several 3-aryl-2-methylimidazo[4,5-*b*]pyridines (**36**; X = H, R = H, 4-F, 4-Et; X = Cl, R = H, 4-F, 3-CF_3, 2,4-Me_2, 2,4-Cl_2, 2-Me-6-Et, 3-CF_3-4-Cl) with pesticidal activity have been prepared in moderate yields by the reaction of *N*-(2-chloro-3-pyridyl)acetamides (**37**) with aromatic amines in the presence of P_2O_5 and $Et_3N{\cdot}HCl$ at 150 °C.[140] With X = H the cyclized products were isolated directly, whereas with X = Cl treatment of the intermediate amidines (**38**) with K_2CO_3 in DMF was required to arrive at the final cyclized product.

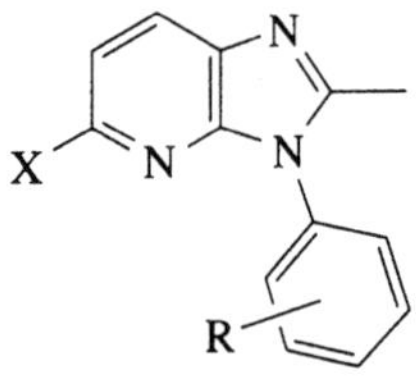

(36) 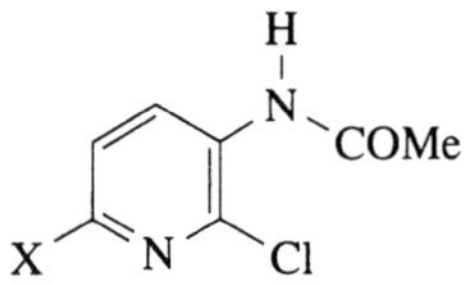 (37) 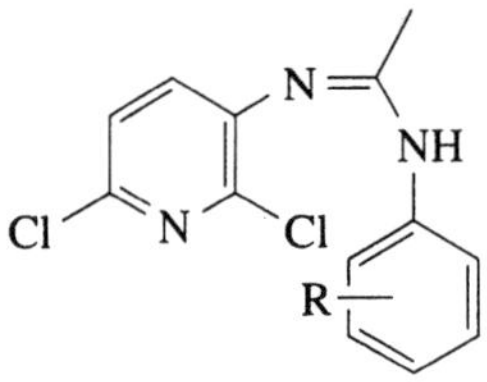 (38)

A one-pot synthesis of 3-amino-2,1-benzisoxazoles (**39**) has been reported, based on the initial S_NAr displacement of halogen by hydroxylamine in the precursor *o*-halobenzonitriles (**40**; X = F, Cl; R = 3-CN, 5-NO_2, 5-CN, 3,5-$(NO_2)_2$, 3-NO_2-5-CN, 5-NO_2-6-Cl), followed by cyclization and aromatization (Scheme 15).[141]

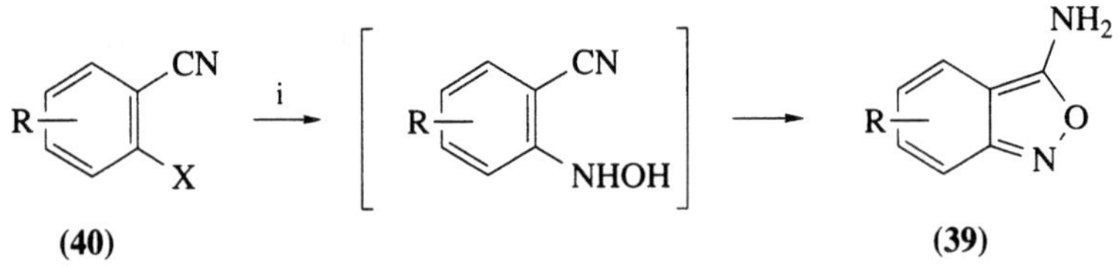

(40) (39)

i, $NH_2OH{\cdot}HCl$, THF, aq. NaOH, ≤40 °C or $NH_2OH{\cdot}HCl$, MeOH, NaOMe

Scheme 15

The one-pot conversion of pyrimidinediones (**41**; $R^1 = R^2$ = Me; R^1 = Me, R^2 = Ph; R^1 = H, R^2 = Me, Et, Ph, 2,6-$Me_2C_6H_3$) to isoalloxazines (**42**) has been reported to proceed in good yield *via* oxidative cyclization in DMF at 120 °C under oxygen.[142] The reaction was completely inhibited by degassing. The precursors (**41**) were in turn prepared in high yield *via* S_NAr debromination by anilines on the appropriate 5-Br-6(alkyl- or aryl-amino)pyrimidinediones (**43**).[143]

A clever application of the VNS reaction was recently reported to afford direct amination of nitrobenzenes, an otherwise prohibitive process, using 4-amino-1,2,4-triazole (Scheme 16).[144] Exclusive attack at position 4 was observed in all cases studied (R = H, Me, Cl, CO_2H, OMe, F, I, CN) with yields in the range 22–91%.

Several activated aryl and heteroaryl halides were converted into the corresponding *N,N*-dimethylamino derivatives by treatment with HMPA at 150 °C.[145] Chlorobenzenes activated by *o*- and/or *p*-NO_2, CN, and 4-$ClC_6H_4SO_2$ groups gave this reaction with 62–94% yields. Chlorine on pyrazine, tetrazole, thia-

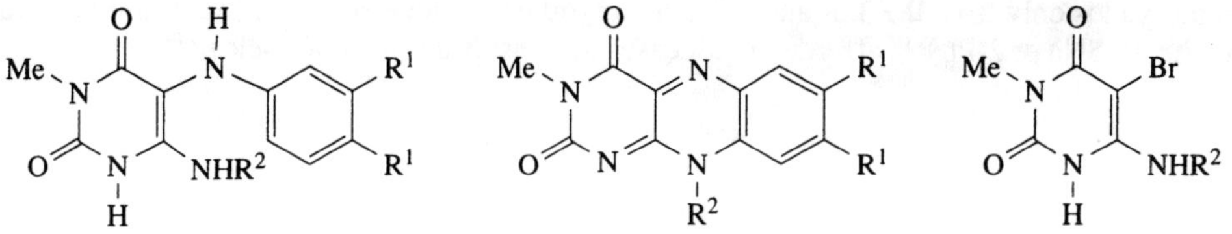

(41) **(42)** **(43)**

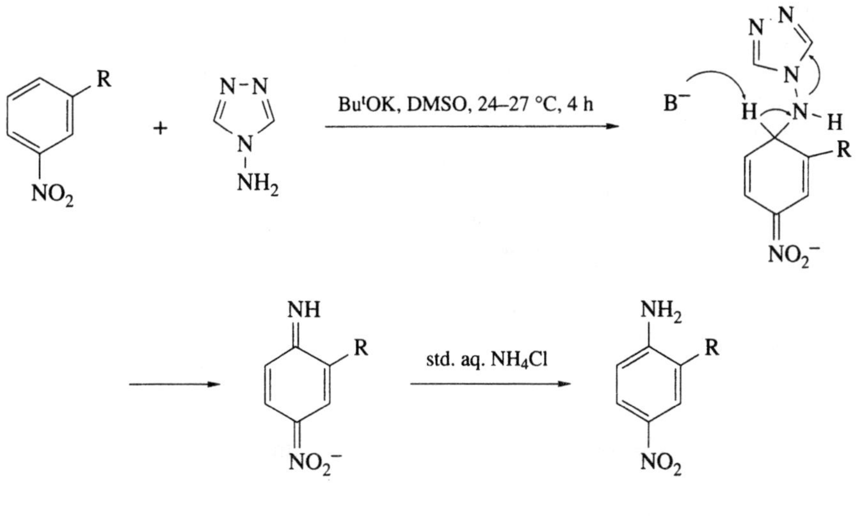

Scheme 16

zole, as in the example of equation (13), and quinoline rings, was sufficiently reactive to produce a clean substitution. It was proposed that the reaction proceeds *via* S_NAr displacement of the suitably activated leaving groups (halide and also nitrite) by *N,N*-dimethylamide formed by the solvent at elevated temperatures.[145]

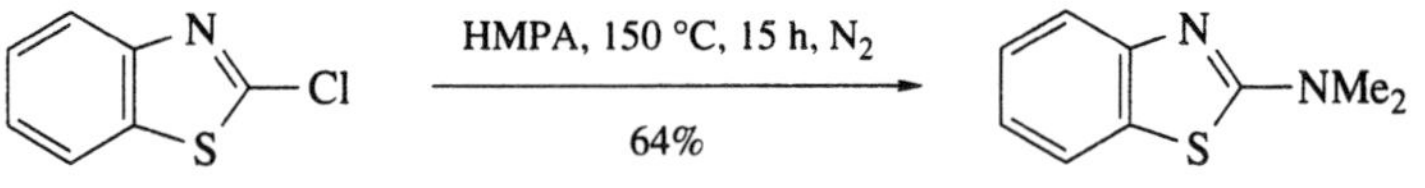

HMPA, 150 °C, 15 h, N_2; 64% (13)

2.1.4 OXYGEN NUCLEOPHILES

The great majority of synthetic applications with oxyanions involve S_NAr displacement of leaving groups other than hydrogen. Nucleophilic displacement of hydrogen by oxyanions has been observed, as, for example, in the reaction of nitrobenzene with Bu^{t}OK in THF to give 4-Bu^{t}O-nitrobenzene and potassium nitrobenzenide,[10] but has not found as yet general application in synthesis. In one example alkoxydehydrogenation *via* addition/oxidation has been proposed to be a step in the synthesis of 2-alkoxy-5,10,15,20-tetraphenylporphyrins from the corresponding 2-nitro derivative *via* a multistep *cine* substitution sequence.[146]

The use of solvents like HMPA has found many applications in S_NAr displacements by alkoxides since the report by Shaw and collaborators on the ability of this solvent to promote methoxylation *via* chloride displacement in unactivated aryl chlorides, including chlorobenzene.[147] Due to the deactivating effect of the OMe group, bissubstitution did not take place on the dichlorobenzenes even under prolonged reaction times and with a large excess of MeONa. S_N2 demethylation of chloroanisoles to the corresponding chlorophenols was instead observed.[51] The reaction of tri- and tetra-chlorobenzenes with 1 equiv. of MeONa is not regioselective, with the exception of 1,2,4-trichlorobenzene which gave 2,5-dichloroanisole (88%) as the sole product. When an excess of nucleophile was used, mixtures resulting from monosubstitution, bissubstitution and demethylation were obtained.[51] An analogous survey showed that in DMF methoxydehalogenation takes place regioselectively in dichloro- and dibromo-pyridines (equation 14).[148] Substitution of the second halide was more difficult with these substrates, and bisethers could be obtained in fair yields only from the 2,6- and 3,5-dihalopyridines using 8 equiv. of MeONa. In contrast, the reaction with MeSNa or 2-PrSNa afforded in all cases the bissulfides in good yields.[148]

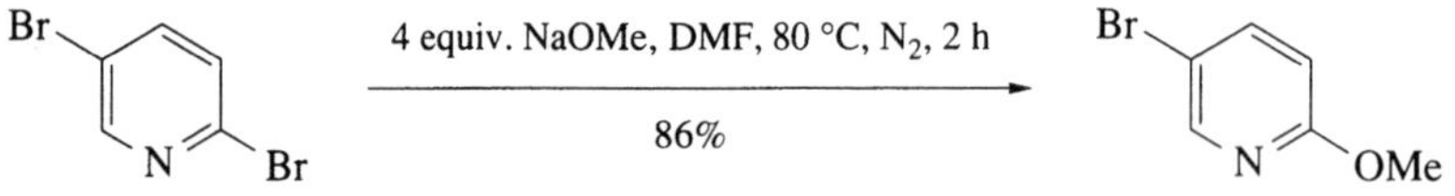

4 equiv. NaOMe, DMF, 80 °C, N_2, 2 h; 86% (14)

Aryl fluoroalkyl ethers have been prepared from the reaction, at room temperature in HMPA, of fluoro-substituted alkoxides with activated fluoro-,[149] nitro-,[149] and, at 150 °C, also chloro-arenes[150,151] and some chloro-substituted pyrazines (equation 15), pyrimidines, quinolines,[150,152] and pyridines.[152] Disubstitution was observed in the presence of comparably activated leaving groups such as in 2,4- and 2,6-dichloronitro- or cyano-benzenes, whereas regiospecific substitution took place at position 4 in 3,4-dichloronitro- or cyano-benzene and at position 2 in 2-fluoro-6-chlorocyanobenzene.[151] Steric hindrance and the number of fluorine substituents in the alkoxide pose limits to the reactivity. Thus, tertiary alkoxides, or alkoxides containing more than four fluorine substituents, displace activated nitro and fluoro, but not chloro substituents.[149,150] The secondary hexafluoro-2-propoxide anion does not react even with the more reactive nitro and fluoro derivatives.[149]

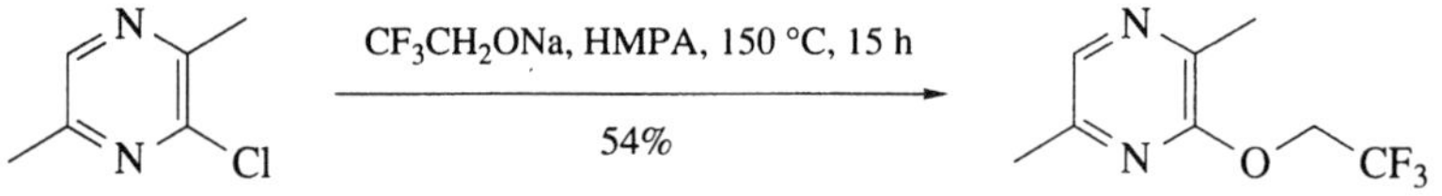

CF_3CH_2ONa, HMPA, 150 °C, 15 h; 54% (15)

A convenient one-step conversion of moderately activated nitroarenes to phenols was achieved in DMSO *via* nucleophilic nitrite displacement by the anion of an aldoxime.[153] The resulting *O*-arylaldoxime is rapidly cleaved to the phenol derivative under the reaction conditions. The reaction is also applicable to activated fluorides and even to 2-chloropyridine which, at 110 °C, is converted to 2-pyridone in 72% yield.[153] A somewhat related process concerns the synthesis, in 82–92% yield, of 4-alkoxybenzonitriles (**45**; R = Me, CH_2-oxirane, CH_2Ph, $CHMeCH_2Me$ from *O*-alkyl-4-nitrobenzaldoximes (**44**) *via* hydride-induced elimination of the alkoxide followed by alkoxy denitration (Scheme 17).[154]

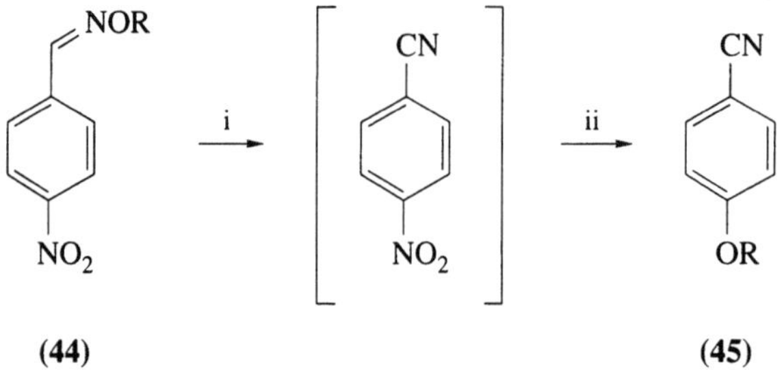

(**44**) (**45**)

i, NaH (excess), DMF, r.t., 1–4 h; ii, RONa (formed in i)

Scheme 17

The use of PTC conditions has proved beneficial in several cases and the method of choice in large scale operations. *m*-Dinitrobenzene was converted to *m*-nitroanisole in 81% yield on the mole scale by treatment for 2 h with MeONa in chlorobenzene at 80 °C in the presence of a catalytic amount of trioctylmethylammonium chloride (Aliquat 336).[155] No reaction occurred in the absence of the onium salt catalyst.[155] Analogous results were obtained by performing the reaction in HMPA.[83]

4-Nitrophenyl ethers (**46**; R = 2-Pr, 2-Bu, *n*-C_8H_{17}, Ph) were prepared in good yields from the reaction of 4-chloronitrobenzene in the appropriate alcohol (chlorobenzene for R = Ph) with KOH in the presence of TBAB.[156] Without the onium salt, the substitution products formed in poor yields due to reduced S_NAr rates and to competing nitro reduction. Reagents capable of complexing the K^+ cation (glymes, high molecular weight polyethers and, best of all, 18-crown-6) could be used in place of the onium salt.[156] Reflux conditions were found unsuitable for these reactions because of the removal of oxygen, a powerful inhibitor of the reduction process.

OR
NO_2
(46)

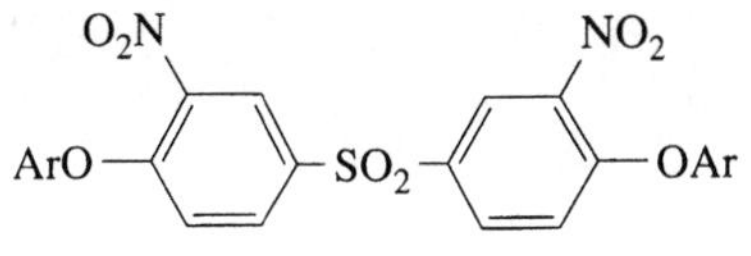

O_2N NO_2
ArO— —SO_2— —OAr
(47)

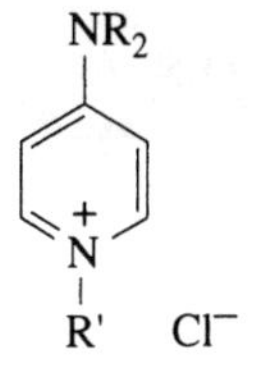

NR_2
N+
R' Cl^-
(48)

Potassium hydroxide in DMSO has proved a useful reagent to promote the reaction of activated aryl and heteroaryl chlorides with long chain primary alcohols, to form the corresponding ethers (equation 16).[157] The procedure failed with secondary alcohols.

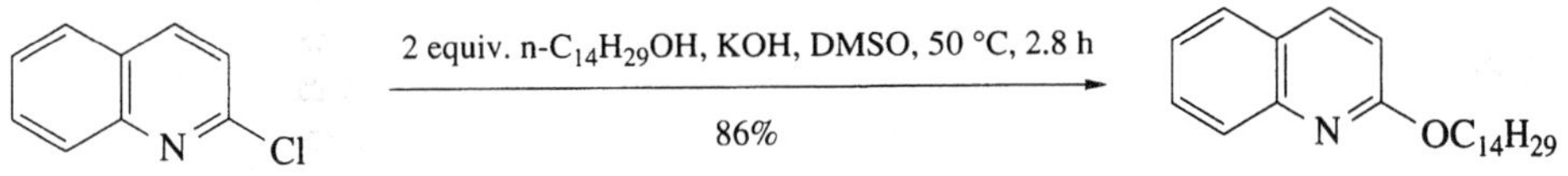

$$\text{2-chloroquinoline} \xrightarrow[\text{86\%}]{\text{2 equiv. n-C}_{14}\text{H}_{29}\text{OH, KOH, DMSO, 50 °C, 2.8 h}} \text{2-OC}_{14}\text{H}_{29}\text{-quinoline} \quad (16)$$

Perfluorotoluene and pentafluoropyridine have been proposed as reagents for protecting alcohols and phenols as perfluoroaryl ethers, a reaction which takes place under mild PTC conditions. Deprotection involves demethoxylation with NaOMe in DMF at 60 °C (Scheme 18; R = phenyl, steroidal).[158]

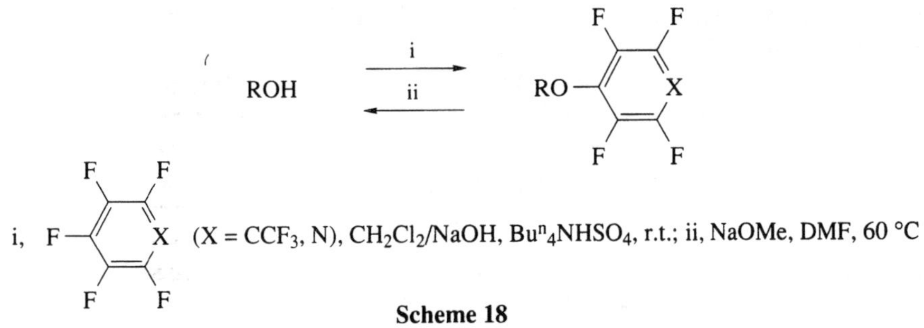

$$\text{ROH} \underset{\text{ii}}{\overset{\text{i}}{\rightleftarrows}} \text{RO}-\text{C}_6\text{F}_4\text{X}$$

i, C_6F_5X (X = CCF_3, N), CH_2Cl_2/NaOH, $Bu^n_4NHSO_4$, r.t.; ii, NaOMe, DMF, 60 °C

Scheme 18

PTC with onium salts (TBAB or TBPB) in a two-phase system (H_2O/NaOH/CH_2Cl_2) afforded bis(aryloxynitrophenyl) sulfones (**47**) in good yields at room temperature, from the reaction of the corresponding chlorides with several phenols.[159]

More elaborate catalysts have also been used, like tris(3,6-dioxaheptyl)amine (TDA-1) in dechlorinations of chloropyridines by benzyl alcohol[160] and of 4-chloronitrobenzene, 4-chlorobenzonitrile, and 2-chloro-5-(trifluoromethyl)nitrobenzene by phenols and methanethiol.[161]

In the reaction of activated aryl halides with phenoxide and thiophenoxide in chloro- or *o*-dichlorobenzene at reflux, pyridinium salts (**48**; R, R = —$CH_2CH_2CHMeCH_2CH_2$—; R = Me, Bu^n, *n*-C_6H_{13}; R′ = $CH_2CHEtBu$, CH_2Bu^t) proved to be superior catalysts to simpler onium salts, like TBAB, due mainly to their greater thermal stability.[162]

Electrochemical reduction of phenol and thiophenol in the presence of a suitable support electrolyte has been proposed as a method to produce solutions of reactive $PhX^{-}\,^{+}NR_4$ (X = O, S) to be used in S_NAr reactions.[163]

Investigations on the use of ionic fluorides as bases in organic synthesis have led to the interesting observation that fluoride hydrogen-bonds to a variety of H-bond electron-acceptor (protic) compounds and that such H-bound complexes may actually be more effective nucleophilic reagents than the salt of the protic.[164] Examples of fluoride-promoted nucleophilic aromatic substitutions range from the synthesis of aromatic polyethers *via* polycondensation of bis(4-chloro-3-nitrophenyl) sulfone with 2,2-bis(4-hydroxyphenyl)propane (equation 17),[165] to the synthesis of aryl-4-cyanophenyl ethers obtained from the reaction of activated nitro- or chloro-arenes with the hydrogen-bonded complex of 4-cyanophenol with KF (equation 18).[166] In the latter case, H-bonding results in charge localization on the oxygen, thus presumably increasing the selectivity of the nucleophile. However, no data were supplied to allow a comparison of the H-bonded complex and the phenoxide salt.

One method of synthesis of 1,2-benzisoxazoles involves cyclization of *o*-halo- or *o*-nitro-benzoyl oximes *via* intramolecular S_NAr.[167,168] A recently reported variant, which gives access to sterically constrained 3-phenyl-1,2-benzisoxazoles, employs a reversed sequence of steps, *e.g.* nucleophilic

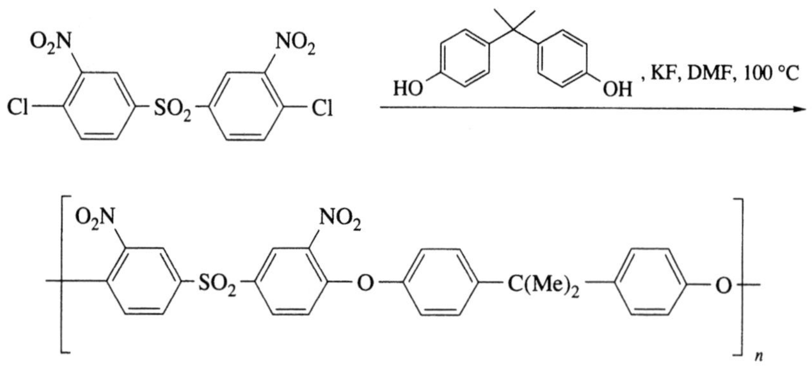

, KF, DMF, 100 °C

(17)

KF–HOC_6H_4CN, DMSO, 110 °C, 2 h

88%

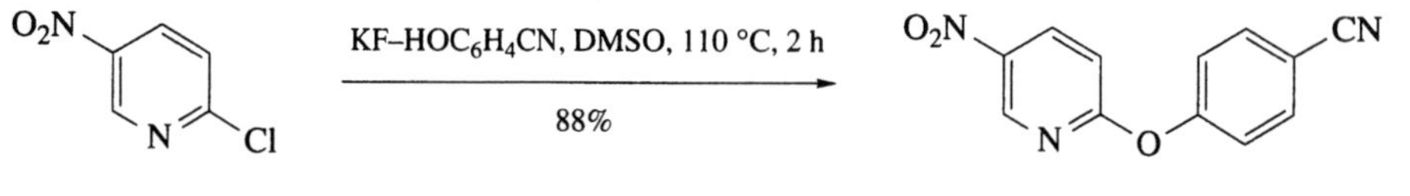

(18)

displacement by the anion of acetone oxime of an *o*-fluoro substituent in the precursor benzophenone, followed by cyclization *via* acid-induced transoxymation (Scheme 19; R = H; 2,3-Cl_2-4-MeO; 2,6-Me_2; 2,6-Cl_2).[169] In contrast to the intramolecular S_NAr route, which is not selective (Cl and F are displaced at comparable rates, so that mixtures of 3-(2-chlorophenyl)- and 3-(2-fluorophenyl)-1,2-benzisoxazoles were obtained from 2-chloro-2′-fluorobenzophenone oximes), the intermolecular S_NAr reaction is highly selective. Thus only 3-(2,3-dichloro-4-methoxyphenyl)-1,2-benzisoxazole (**49**; R = 2,3-Cl_2-4-OMe) was obtained in 73% overall yield from 2,3-dichloro-2′-fluoro-4-methoxybenzophenone.[169]

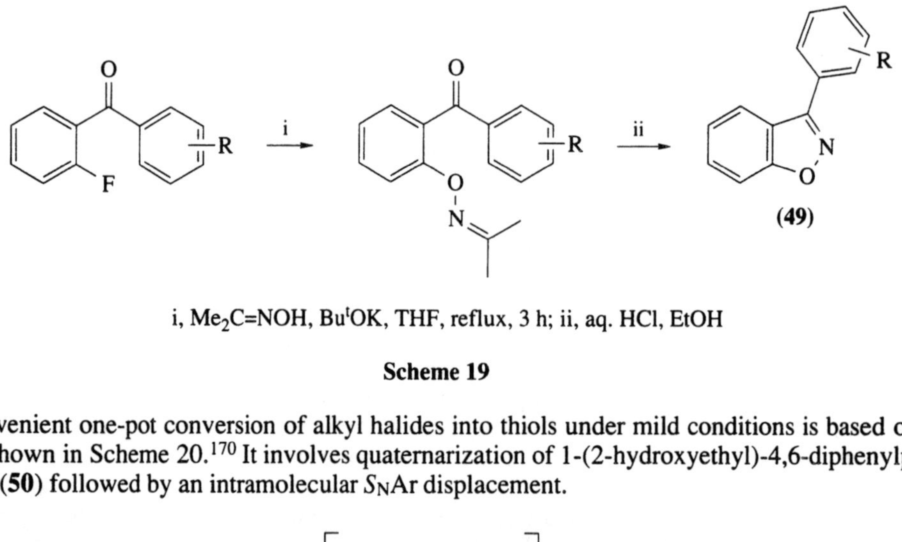

(49)

i, Me_2C=NOH, Bu^tOK, THF, reflux, 3 h; ii, aq. HCl, EtOH

Scheme 19

A convenient one-pot conversion of alkyl halides into thiols under mild conditions is based on the sequence shown in Scheme 20.[170] It involves quaternarization of 1-(2-hydroxyethyl)-4,6-diphenylpyridine-2-thione (**50**) followed by an intramolecular S_NAr displacement.

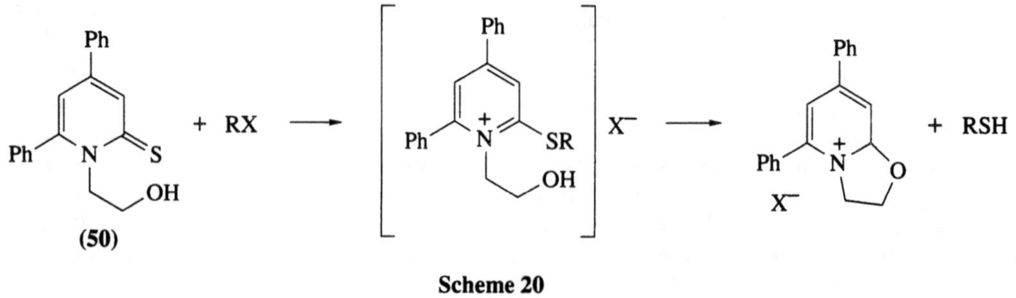

(50)

Scheme 20

2-Alkoxy- and 2-aryloxy-4-amino-5-pyrimidinecarbonitriles (**51**; X = MeO, EtO, PhO, 4-$MeOC_6H_4O$, 4-ClC_6H_4O, 4-BrC_6H_4O, 4-CNC_6H_4O) were prepared in good yields from the corresponding chloro or bromo derivatives (X = Cl, Br) by the reaction with RONa in refluxing acetone.[171] 2-Amino derivatives (**51**; X = NH_2, NHR, NR_2) were prepared in analogous fashion.[171]

(51)

2.1.5 SULFUR NUCLEOPHILES

Sulfur anions are generally good nucleophiles, more powerful than their oxygen analogs. While the nucleophilic displacement of hydrogen by thioanions has not been frequently observed,[6] the S_NAr displacement of nitrite, halides and other leaving groups constitutes a common and often easily accomplished step in synthesis. However, undesired redox processes occur frequently with thioanions due to their low oxidation potentials. To avoid oxidation of the nucleophile, reactions with thioanions are best carried out under an inert atmosphere. Under these conditions, however, simple nitrobenzene derivatives like 4-chloronitrobenzene readily undergo reduction. Thus, complex mixtures resulting from reduction of the nitro group were obtained from the reaction of 4-chloronitrobenzene with MeSNa in DMF[172] and with MeSNa and EtSNa in HMPA,[173] under conditions which with more substituted arenes or with secondary thioanions lead to the products of substitution in high yields.[173] 4-Chloromethylthiobenzene was instead obtained in 92% yield from the reaction in MeOH of 4-chloronitrobenzene with an excess of MeSH upon the dropwise addition of methanolic KOH.[65]

Displacement of the nitro group of moderately activated nitroarenes by thioanions occurs readily in polar aprotic solvents.[174–176] Kornblum and collaborators found that nitrobenzenes carrying one electron-withdrawing group react with thioanions, in HMPA at 25 °C, to give the products of nitrite displacement (**52**) in good to excellent yields (equation 19; RS = MeS, $PhCH_2S$, PhS, $PhSO_2$; W = 2-, 3-, and 4-NO_2, 4-CN, 4-PhCO, 4-EtOCO, 4-$PhSO_2$, 3,5-$(CF_3)_2$).[83]

RSNa, HMPA, 25 °C; 61–96% (19)

(**52**)

Thiol anions in polar aprotic solvents are also effective in displacing chloride in activated and even weakly activated arenes. In substrates with more than one replaceable chlorine regioselective displacement of one chlorine was seldom observed, the reaction usually yielding mixtures of mono- and polysubstituted products. It was thus concluded that alkylthio substituents exert an activating effect roughly comparable to that of a chlorine.[172,173] Exhaustive substitution was often observed when the thiolate reagent was used in excess.[172,173,177–179] Several bis-, tris-, tetrakis- and pentakis-(methylthio)benzoic acids and derivatives,[178] and poly(alkylthio)benzenes,[179] were prepared in good yields in DMF and HMPA, respectively. The extent of substitution could be controlled in some cases by proper choice of the reaction conditions, *e.g.* the amount of nucleophile and/or the temperature. Thus, the reaction of 4-chloro-3,5-dinitrotoluene (**53**) with excess methanethiolate in DMF at ice bath temperature for 15 min produced (**54**) in 78% yield, whereas the tris(sulfide) (**55**) was isolated in 60% yield when the reaction mixture was allowed to stand for 4.5 h at room temperature.[172] In the examples of Scheme 21, advantage was taken of the different rates of the alkylthio denitration and alkylthio dechlorination processes to prepare the isomeric tris(sulfides) (**56**) and (**57**) in 70% and 65.5% yield, respectively.[173] In contrast, the reaction of 1-chloro-2,4-dinitrobenzene with 1 and 2 equiv. of Me_2CHSNa gave mixtures of mono- and di-, and di- and tri-substitution products, respectively.[173] Reactions of polychloroarenes in HMPA with EtSNa, and, even more so, with MeSNa, proceed *via* S_NAr displacement of chloride followed by dealkylation *via* S_N2 displacement of thiophenoxides by the thiolate anion. The reaction has been developed into a one-pot synthetic route to thiophenols from aryl chlorides and bromides, as exemplified in equation (20).[180] Mercaptopyridines and mercaptoquinolines were prepared by the same procedure in DMF.[181] Alkoxyaryl alkyl sulfides were obtained from dichlorobenzenes and dichlorotoluenes by sequential treatment with RSNa and R′ ONa in HMPA.[182]

Treatment of activated diaryl sulfides with Na_2S in DMF at 130 °C results in the cleavage of the sulfide and in the formation of sodium aryl sulfides which can be used *in situ* to prepare mixed diaryl or alkyl aryl sulfides.[183] An example is reported in Scheme 22.

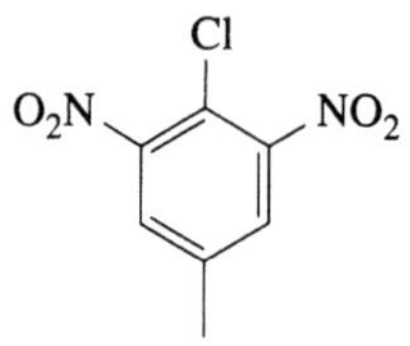

(53) 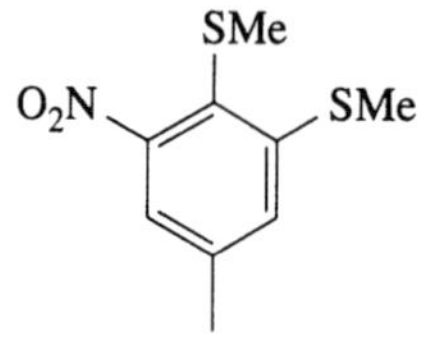 **(54)** 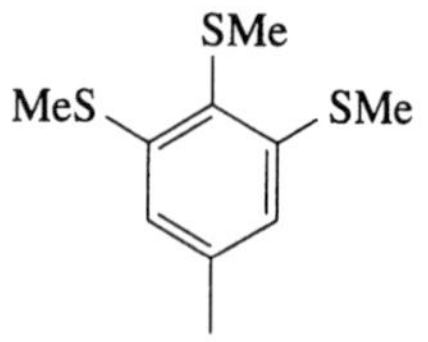 **(55)**

(a) **(56)**

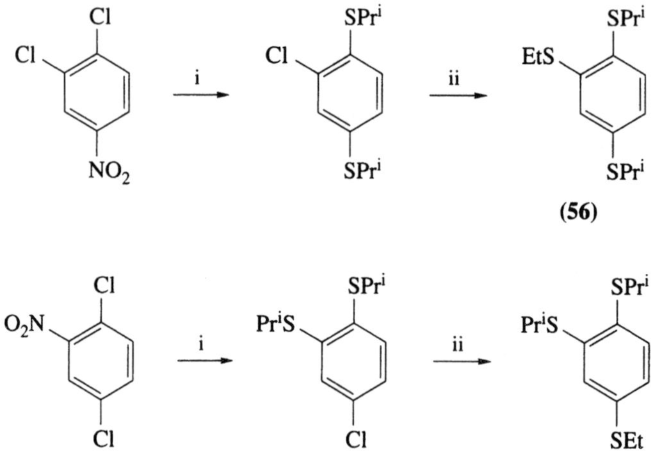

(b) **(57)**

i, 2 equiv. Pr^{i}SNa, HMPA, r.t.; ii, 1 equiv. EtSNa, 80 °C

Scheme 21

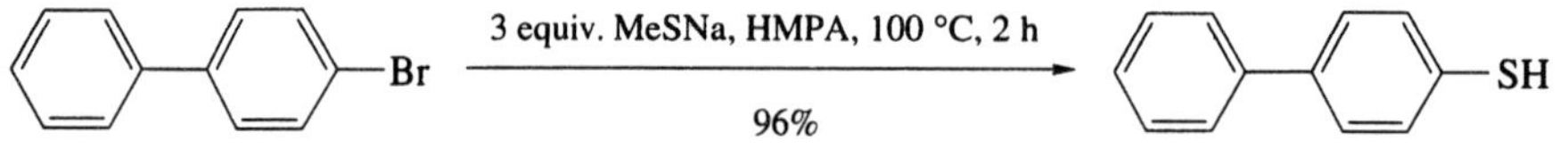

3 equiv. MeSNa, HMPA, 100 °C, 2 h; 96% (20)

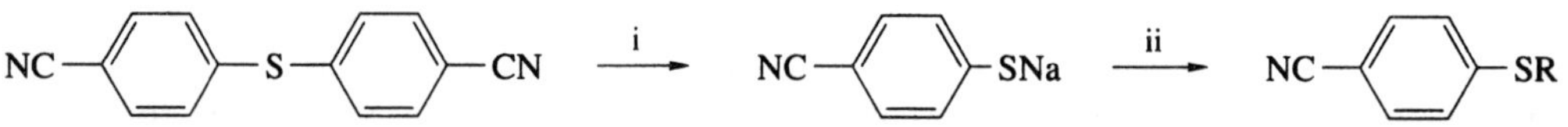

i, Na_2S, DMF, 130 °C, 12–18 h; ii, RX (alkyl or aryl halide)

Scheme 22

The reaction of EtSLi with the isomeric dinitrotoluenes (**58**) in HMPA at 20 °C gave regiospecific displacement of the nitro group adjacent to the alkyl group in excellent yields (equation 21).[184] With a bulky *o*-alkyl substituent, like in 2,4-dinitro-*t*-butylbenzene, regiospecificity was lost, the two nitro groups being displaced in equal proportions. Notably, when the same substrate was allowed to react with Bu^{t}Li slow displacement of the nonadjacent nitro group took place, accompanied by redox processes: azoxy and hydrazo compounds and di-*t*-butyl disulfide (80%) were isolated from the reaction mixture.

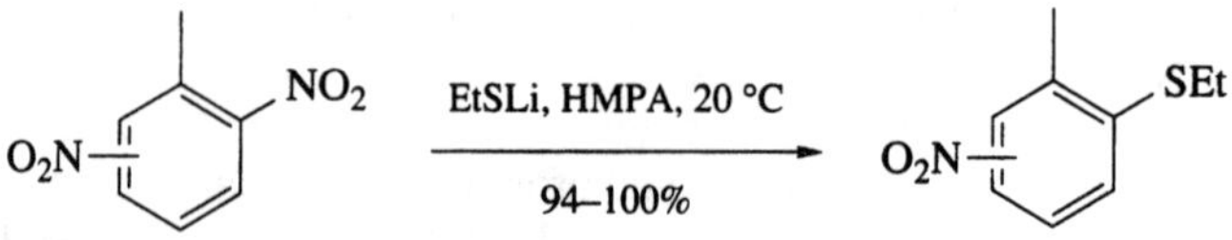

EtSLi, HMPA, 20 °C; 94–100% (21)

(58)

Dichloropyridines were converted in good yield to the corresponding bis(alkylthio)pyridines *via* reaction with sodium thiolates in DMF at 140 °C.[185] Room temperature oxidation of the sulfides with chlorax solution gave bis(alkylsulfonyl)pyridines in high yields.[185]

The convenient preparation of 2,7-dinitrothianthrene (**59**) *via* base-induced cyclization of 2-chloro-5-nitrobenzenethiol in acetone at room temperature (equation 22), provides an easy access to a number of 2,7-substituted thianthrenes *via* elaboration of the nitro groups in (**59**).[186]

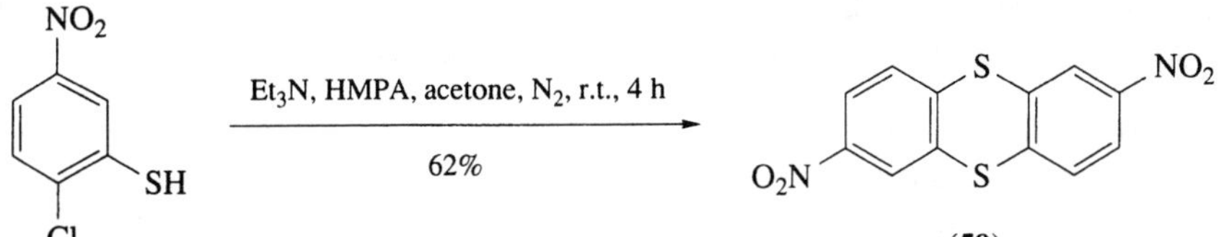

NO_2 SH Cl — Et_3N, HMPA, acetone, N_2, r.t., 4 h; 62% → S S NO_2 O_2N (22)

(**59**)

Polyglymes offer a viable alternative as solvents to the potentially carcinogenic HMPA.[187,188] Alkyl and phenyl sulfides were prepared from the reaction of suitable aryl halide precursors and an excess of sodium thiolates in tetraglyme.[188] By the same procedure 2-(alkylthio)-substituted (alkyl = $Me(CH_2)_n$; n = 3, 15, 17) quinolines, pyrimidines and pyrazines were prepared from the corresponding chlorides.[187]

Halide displacement by thioanions such as thioalkoxides, thiocyanate and sulfite, can be readily carried out under PTC conditions in activated arenes. The reaction of nitro- and dinitro-aryl halides with thiophenol and NaOH in the presence of an ammonium or phosphonium salt catalyst proceeds readily in toluene at room temperature to give the phenyl aryl sulfides in nearly quantitative yield.[189] Excellent yields of thiocyanoarenes were obtained from the reaction of the 2,4-dinitrohalobenzenes with KSCN in toluene at 90 °C with a tetraalkyl onium salt PTC catalyst.[190]

Protonated tertiary amines ($(Me(CH_2)_n)_3N$ (n = 2, 3, 4, 5 and 7) were found to be better catalysts than tetraalkyl ammonium salts for the sulfodechlorination of 1-chloro-2,4-dinitrobenzene.[191]

Unsymmetrical diaryl sulfides (**60**; R = 2-NO_2, 4-NO_2, 2,4-$(NO_2)_2$; R′ = H, 4-NMe_2) were obtained in 80–97% yields from the reaction under PTC conditions of nitro-activated aryl halides with arenethiolates generated *in situ* from the reduction of the corresponding diaryl disulfides with aminoiminomethanesulfinic acid (**61**).[192] Arylthiolates carrying electron-withdrawing substituents were not sufficiently reactive. The reaction could also be applied to the synthesis of diaryl selenides, but not of ditellurides.

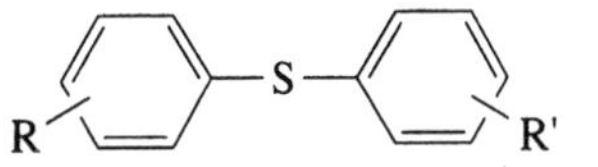

R S R' (**60**)   HN O S H_2N OH (**61**)

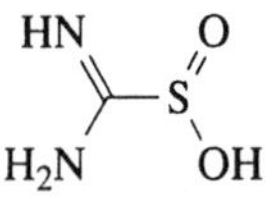

Symmetrical diaryl sulfides were produced in fair yield from the PTC reaction of sodium sulfide with molten aryl chlorides activated by a cyano, nitro, phthalimido or anhydrido group.[193] Typical conditions require use of a 3:1 mole ratio of aryl chloride to Na_2S and 10% of catalyst (crown ethers and onium salts) at 200 °C for 24 h.

Thioiminium salts (**62**; R′ = H, Me; R = H, Ph, CH=CH_2) can be used to generate *in situ* thiolates RCH_2S^- which under PTC conditions in $CHCl_3$, at room temperature, react with aryl halides to give the alkyl aryl sulfides (**63**; R = H, Ph, CH=CH_2; R″ = H, CO_2H, Cl, NO_2).[194] Yields are best with activated aryl halides. With an excess of (**62**), *p*-dichlorobenzene gave products of disubstitution (**63**; R″ = CH_2SR).

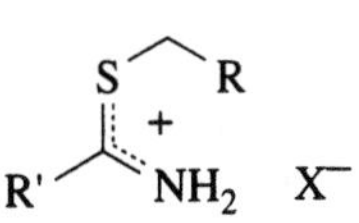

S R + R' NH_2 X^- (**62**)   R S R" (**63**)

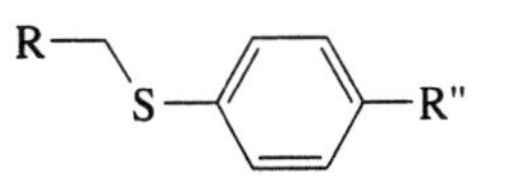

Catalysts of greater thermal stability have also been used, such as *N*-alkyl salts of 4-dialkylamino-pyridines (**48**)[162] and tris(polyoxaheptyl)amine (TDA-1).[161]

Displacement of chloride has also been accomplished on moderately activated substrates, such as $C_6H_nCl_{6-n}$ (n = 2, 3, 4), under PTC conditions at temperatures which depend on the number of chlorine substituents. The dichlorobenzenes react with aliphatic thiols at 110 °C in a heterogeneous mixture with concentrated aq. KOH containing dicyclohexano–18-crown-6 as PTC catalyst.[195] Reaction times were long, ranging from 14 h with primary thiols to >200 h with tertiary ones. The reactivity of the first chlorine atom followed the order 1,2- > 1,3- > 1,4-dichlorobenzenes. The reaction of 1,2-dichlorobenzene with secondary and tertiary thiolates was selective and gave exclusively the product of monosubstitution. With Bu^nSH some disulfide (5% by GC) was also formed. The reaction of 1,3- and 1,4-dichlorobenzene gave in all cases mixtures of mono- and di-sulfides of variable composition.

Several alkyl aryl sulfides in which the alkyl group is a linear primary C_nH_{2n} chain (n = 7, 8, 12), have been prepared from the corresponding thiols and di-, tri-, and tetra-chlorobenzenes under PTC conditions with KOH and a catalyst in toluene.[196] Ammonium and phosphonium salts, 18-crown-6, and polyethylene glycols were used, (tricyclohexyl-*n*-dodecyl)phosphonium bromide being the most efficient catalyst. The regioselectivity is determined by the prevalent influence of the –*I*-activating effect of the chloro substituent, *ortho* > *meta* > *para*. Thus the reaction of 1,2,4,5-tetrachlorobenzene (**64**; X = H, Y = Cl) gave disulfide (**65**; X = H, Y = SR) in 89% yield, and 1,2,4,6-tetrachlorobenzene (**64**; X = Cl, Y = H) gave monosulfide (**65**; X = Cl, Y = H) in 88% yield. The reaction of 3-chlorobromobenzene with *n*-heptanethiol gave exclusively the product of bromide displacement in 86% yield.

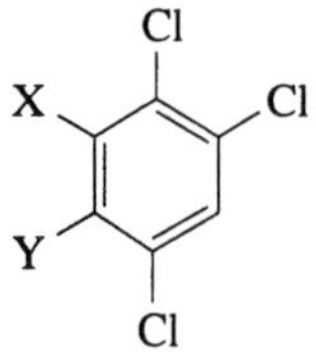

(64) 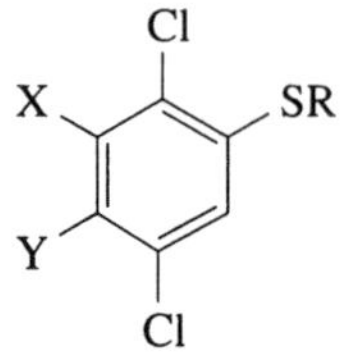 **(65)**

A one-pot synthesis of benzothiazolones from 2-halonitrobenzenes has been reported (equation 23).[197] The proposed mechanism involves the displacement of halide by some sulfide species (H_2S or HS_n^-, n = integer) generated *in situ*, followed by nitro reduction and condensation with SCO.

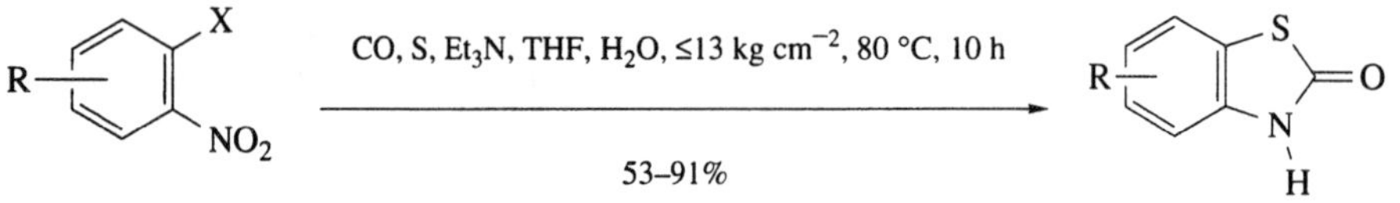

CO, S, Et_3N, THF, H_2O, ≤13 kg cm^{-2}, 80 °C, 10 h; 53–91% (23)

Mercaptobenzoic acids (*ortho, meta* and *para* isomers) were prepared in fair to good yields from the reaction of the chlorobenzoic acids with elemental sulfur and molten NaOH–KOH (1:1 molar ratio) at 270 °C for 3 min. Both 2- and 4-chloropyridine gave the mercaptopyridines by an analogous procedure.[198] Use of selenium instead of sulfur gave poor yields and mixtures of isomeric products.

2.1.6 OTHER NUCLEOPHILES

2.1.6.1 Hydride

Anionic σ-adducts with the hydride ion have been detected and some also isolated as stable salts.[8,43] A few examples of nucleophilic displacement by hydrogen have been reported. Displacement of NO_2^-, Cl^- and Br^- by hydride, *via* σ-adduct intermediates, was observed in DMSO solutions of $XC_6H_4NO_2$ (X = 2-Br, 2-Cl and 4-NO_2) and of 1-X-2,4-dinitrobenzene (X = Br, Cl) with $NaBH_4$.[199] Another example is the reaction of arene diazonium salts carrying an electron-withdrawing substituent with formamide and Et_3N to form arenes in moderate to good yields. The mechanism of this proto dediazoniation was shown to involve transfer of the formyl hydrogen as hydride within a 1-aryl-3-formyltriazene intermediate (Scheme 23).[200]

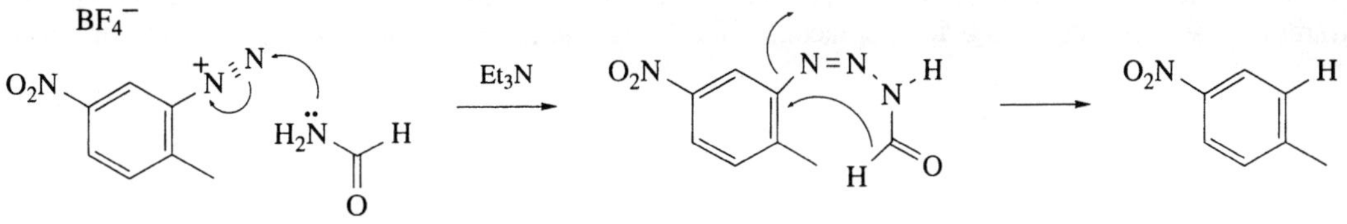

Scheme 23

2.1.6.2 Halides

Great effort has been given in the last decade to the synthesis of fluoro aromatics due to their growing importance as pesticides and pharmaceuticals.[201,202] One of the most useful reactions to introduce a fluorine or other halogen substituent into an arene is nucleophilic displacement by fluoride (halide) of an activated leaving group. Special areas of research deal with the selective introduction of fluorine into biologically active molecules and with the synthesis of radiopharmaceuticals labeled with short-lived radionuclides, particularly with the positron-emitting nuclides ^{18}F ($t_{1/2}$ = 110 min) and ^{122}I ($t_{1/2}$ = 3.6 min).

^{18}F-labeled aromatics have been prepared most efficiently from activated arenes in DMSO *via* displacement of nitrite with $Rb^{18}F$[203,204] and with $Cs^{18}F$,[204] and of fluoride ($^{19}F \rightarrow {}^{18}F$ exchange), chloride, and bromide with $Cs^{18}F$,[204] of iodide with $H^{18}F$–K_2CO_3,[205] and of Me_3N with $Cs^{18}F$.[206] The temperatures needed to obtain high conversions in the required short reaction times (<20 min) depend on the leaving group ability as well as on the activating group and range from 80–150 °C for the displacement of nitrite,[203] to 180 °C for the displacement of iodide.[205] The use of rapid heating in microwave ovens has been reported to significantly reduce the reaction times in these and related systems.[207] The use of polar aprotic solvents like DMSO can be avoided in these reactions by the choice of better nucleophiles like anhydrous tetrabutylammonium fluoride (TBAF) which can be used in THF at room temperature (equation 24).[35] The convenience of using the KF–Ph_4PBr system has also been pointed out, rate enhancements being especially significant in less polar solvents (1,2-dimethoxyethane, acetonitrile).[208]

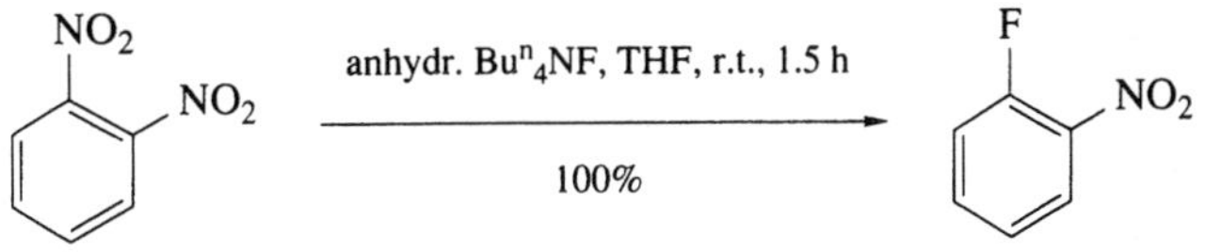

anhydr. $Bu^n{}_4NF$, THF, r.t., 1.5 h; 100% (24)

Fluorodesulfonylation reactions on arylfluorosulfonyl fluorides, carried out with KF in sulfolane or DMF, or without solvent in the presence of 18-crown-6, have shown that the FSO_2^- is a good activating group and a good leaving group (equation 25).[209]

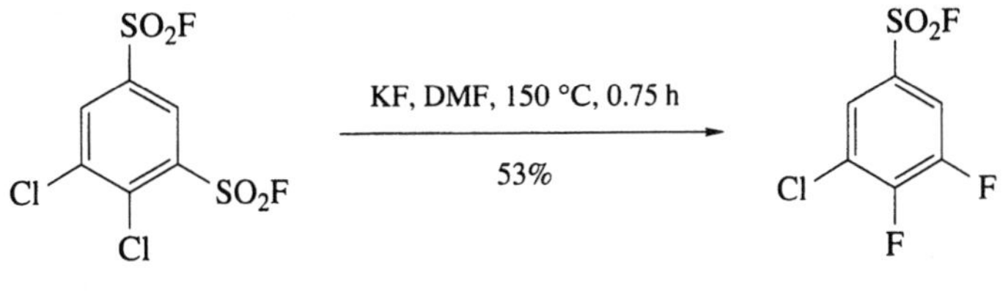

KF, DMF, 150 °C, 0.75 h; 53% (25)

The synthetic potential of 'solid-state' nucleophilic aromatic substitutions has been pointed out in a report on the displacement of *o*-nitro groups in arene diazonium ions by chloride from the chlorozincate counterion in the salt.[210] The reaction is selective for the *ortho* position. Chloroarenes and heteroarenes have been synthesized in good yields from the corresponding nitro compounds upon treatment with phenyltetrachlorophosphorane (PTCP) in phenylphosphonic dichloride at 170 °C for 5 h. Reactive substrates ranged from nitrobenzene to nitropyridines and nitropyrimidines, yields being in the range 66–94%. An

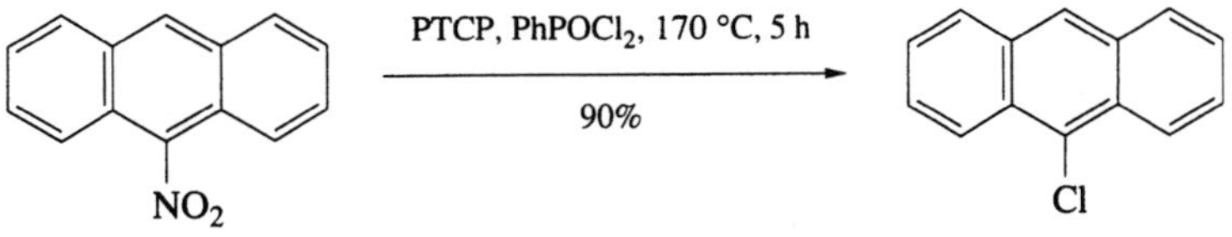

PTCP, $PhPOCl_2$, 170 °C, 5 h; 90% (26)

example is shown in equation (26). A stepwise mechanism (Scheme 24), involving in the final stage an intramolecular chlorine attack to displace NOCl from intermediate (**66**), was proposed for this reaction.[211]

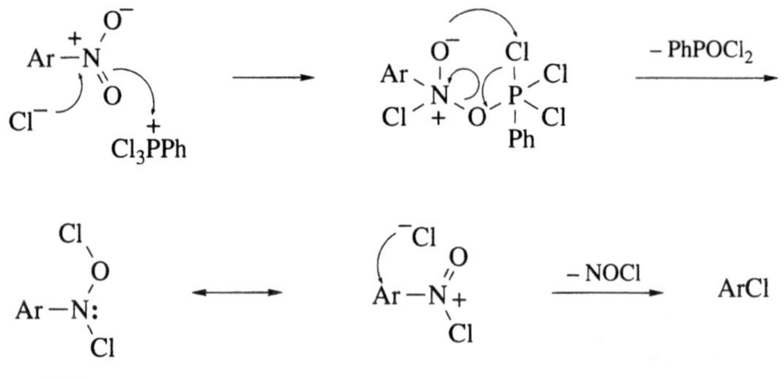

Scheme 24

2.1.6.3 Phosphorus Nucleophiles

Attack by phosphorus nucleophiles on pyridinium salts is well established, and is the basis for a method for the regioselective synthesis of 4-alkylpyridines,[212] and of 1-alkylisoquinolines.[213] Although the phosphonate adducts (**67**) were obtained in high yields in all cases examined (R = Me, Et, $CH_2{=}CHCH_2$, Bu, $PhCH_2$), the final step gave satisfactory yields only in the case R = Bu (Scheme 25).

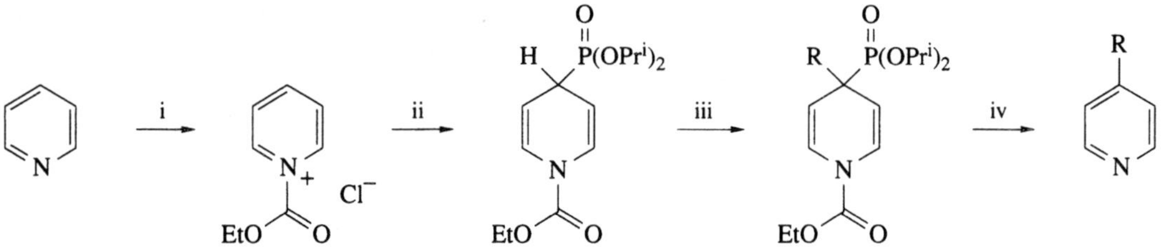

i, EtOCOCl; ii, $P(OPr^i)_3$; iii, BuLi, THF, –78 °C, RBr; iv, BuLi

Scheme 25

The reaction of nitronaphthalenes and nitroisoquinolines with dimethyl phosphite in MeONa/MeOH (equation 27), proceeded *via* nucleophilic substitution of hydrogen according to a redox stoichiometry and gave substituted dimethyl naphthalene- and isoquinoline-phosphonates and benzazepines.[214]

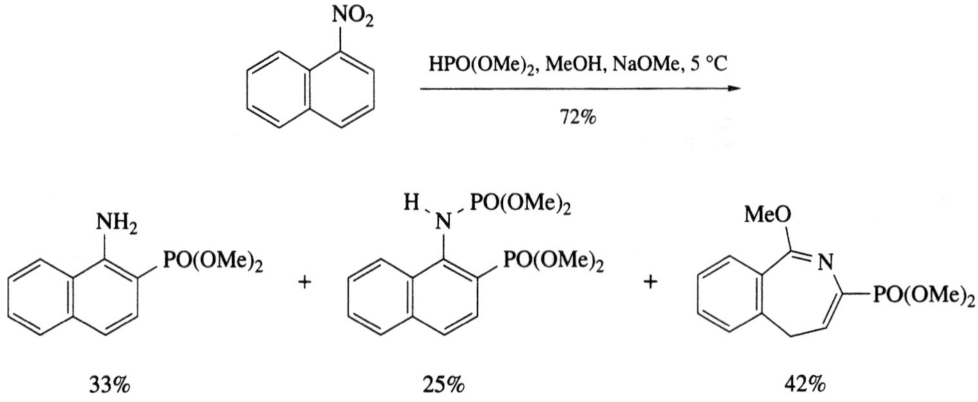

(27)

2.1.6.4 Selenium and Tellurium Nucleophiles

The methods of synthesis of diaryl and aryl alkyl selenides have been recently reviewed.[215] 4-Nitrophenyl methyl selenide was obtained in 93% yield from the reaction of 4-chloronitrobenzene in DMF with a suspension of MeSeLi in THF, prepared from powdered Se and MeLi.[216] Other mixed RSeAr selenides were synthesized *via* alkylation with RI of $ArSe^-$ resulting from the nucleophilic substitution of unactivated haloarenes ArCl with MeSeLi in DMF at 120 °C, followed by $MeSe^-$-induced demethylation.[217]

Diaryl selenides (**68**; X = Z = H, Y = NO_2; X = Y = H, Z = NO_2; X = H, Y = Z = NO_2; X = Cl, Y = NO_2, Z = H) were prepared in 75–95% yields from diaryl diselenides and chloro- or bromo-nitrobenzenes under PTC conditions, by the procedure already described in Section 2.1.5.[192]

The synthesis of aromatic tellurides has been performed *via* the reaction of iodoarenes (*o*-, *m*- and *p*-iodonitrobenzene, and di- and tri-methyliodonitrobenzenes) with phenyltelluride in HMPA at 80–90 °C, preferably in the presence of CuI. Yields range between 53 and 95%.[218] Symmetrical diaryl tellurides (**69**; R = 3-Me, 4-Me, 2,4-Me_2, 2,4,6-Me_3, 4-MeO) were prepared in good yields from reaction of nonactivated aryl iodides with Na_2Te in DMF at 60 °C.[219] The mechanisms of these reactions, which are limited to the iodoarenes, have not been discussed.

Y

X—⟨⟩—Se—⟨⟩—Z R—⟨⟩—Te—⟨⟩—R

(68) **(69)**

2.1.7 REFERENCES

1. J. F. Bunnett and R. E. Zahler, *Chem. Rev.*, 1951, **49**, 273.
2. J. Miller, 'Aromatic Nucleophilic Substitution', Elsevier, New York, 1968.
3. Th. J. de Boer and I. P. Dirkx, in 'The Chemistry of the Nitro and Nitroso Groups', ed. H. Feuer, Interscience, New York, 1969, part 1, chap. 8.
4. J. A. Zoltewicz, *Top. Curr. Chem.*, 1975, **59**, 33.
5. D. P. Self, D. E. West and M. R. Stillings, *J. Chem. Soc., Chem. Commun.*, 1980, 281.
6. O. N. Chupakhin and I. Ya. Postovskii, *Russ. Chem. Rev. (Engl. Transl.)*, 1976, **45**, 454.
7. O. N. Chupakhin, V. N. Charushin and H. C. van der Plas, *Tetrahedron*, 1988, **44**, 1.
8. G. A. Artamkina, M. P. Egorov and I. P. Beletskaya, *Chem. Rev.*, 1982, **82**, 427.
9. G. P. Stahly, *J. Org. Chem.*, 1985, **50**, 3091.
10. R. D. Guthrie and D. E. Nutter, *J. Am. Chem. Soc.*, 1982, **104**, 7478.
11. G. Bartoli, R. Dalpozzo, L. Grossi and P. E. Todesco, *Tetrahedron*, 1986, **42**, 2563.
12. T. V. RajanBabu, G. S. Reddy and T. Fukunaga, *J. Am. Chem. Soc.*, 1985, **107**, 5473.
13. R. B. Davies and L. C. Pizzini, *J. Org. Chem.*, 1960, **25**, 1884.
14. A. Rykowski and M. Makosza, *Tetrahedron Lett.*, 1984, **25**, 4795.
15. M. Makosza and J. Winiarski, *Acc. Chem. Res.*, 1987, **20**, 282.
16. M. Makosza, J. Golinski and J. Baran, *J. Org. Chem.*, 1984, **49**, 1488.
17. H. C. van der Plas, *Acc. Chem. Res.*, 1978, **11**, 462.
18. H. C. van der Plas, *Tetrahedron*, 1985, **41**, 237.
19. A. N. Kost, S. P. Gromov and R. S. Sagitullin, *Tetrahedron*, 1981, **37**, 3423.
20. P. J. Newcombe and R. K. Norris, *Aust. J. Chem.*, 1981, **34**, 1879.
21. M. Reuman and A. I. Meyers, *Tetrahedron*, 1985, **41**, 837.
22. S. Spyroudis and A. Varvoglis, *J. Chem. Soc., Perkin Trans. 1*, 1984, 135.
23. I. Onyido and C. I. Ubochi, *Heterocycles*, 1987, **26**, 313.
24. A. R. Katritzky and H. Beltrami, *J. Chem. Soc., Chem. Commun.*, 1979, 137.
25. A. R. Katritzky, H. Beltrami, J. G. Keay, D. N. Rogers, M. P. Sammes, C. W. F. Leung and C. Man Lee, *Angew. Chem., Int. Ed. Engl.*, 1979, **18**, 792.
26. M. Bruix, M. L. Castellanos, M. R. Martin and J. de Mendoza, *Tetrahedron Lett.*, 1985, **26**, 5485.
27. A. T. Nielsen, A. P. Chafin and S. L. Christian, *J. Org. Chem.*, 1984, **49**, 4575.
28. J. R. Beck, *Tetrahedron*, 1978, **34**, 2057.
29. W. Fischer and V. Kvita, *Helv. Chim. Acta*, 1985, **68**, 854.
30. W. Fischer and V. Kvita, *Helv. Chim. Acta*, 1985, **68**, 846.
31. C. E. Peishoff and W. L. Jorgensen, *J. Org. Chem.*, 1985, **50**, 1056.
32. F. G. Bordwell, T. A. Cripe and D. A. Hughes, *Adv. Chem. Ser.*, 1987, **215**, 137.
33. G. Modena, C. Paradisi and G. Scorrano, *Stud. Org. Chem. (Amsterdam)*, 1985, **19**, 568.
34. F. G. Bordwell and D. L. Hughes, *J. Am. Chem. Soc.*, 1986, **108**, 5991.
35. J. H. Clark and D. K. Smith, *Tetrahedron Lett.*, 1985, **26**, 2233.
36. J. F. Bunnett and W. D. Merritt, *J. Am. Chem. Soc.*, 1957, **79**, 5967.
37. G. Bartoli and P. E. Todesco, *Acc. Chem. Res.*, 1977, **10**, 125.

38. J. H. Fendler, 'Membrane Mimetic Chemistry', Wiley Interscience, New York, 1982, p. 327.
39. C. A. Bunton and G. Savelli, *Adv. Phys. Org. Chem.*, 1986, **22**, 213.
40. R. H. de Rossi, M. Barra and E. B. de Vargas, *J. Org. Chem.*, 1986, **51**, 2157.
41. M. Barra, R. H. de Rossi and E. de Vargas, *J. Org. Chem.*, 1987, **52**, 5004.
42. J. M. Wilson and D. J. Cram, *J. Org. Chem.*, 1984, **49**, 4930.
43. E. Buncel, M. R. Crampton, M. J. Strauss and F. Terrier, 'Electron Deficient Aromatic- and Heteroaromatic-Base Interactions', Elsevier, Amsterdam, 1984.
44. F. Terrier, *Chem. Rev.*, 1982, **82**, 78.
45. R. H. de Rossi and A. Nunez, *J. Org. Chem.*, 1982, **47**, 319.
46. P. Goldman and J. D. Wuest, *J. Am. Chem. Soc.*, 1981, **103**, 6224.
47. M. Novi, C. Dell'Erba and F. Sancassan, *J. Chem. Soc., Perkin Trans. 1*, 1983, 1145.
48. A. Rykowski and H. C. van der Plas, *J. Heterocycl. Chem.*, 1982, **19**, 653.
49. N. S. Nudelman and D. Palleros, *J. Chem. Soc., Perkin Trans. 2*, 1981, 995.
50. M. W. Logue and B. H. Han, *J. Org. Chem.*, 1981, **46**, 1683.
51. L. Testaferri, M. Tiecco, M. Tingoli, D. Chianelli and M. Montanucci, *Tetrahedron*, 1983, **39**, 193.
52. G. Guanti, C. Dell'Erba, F. Pero and G. Cevasco, *J. Chem. Soc., Perkin Trans. 2*, 1978, 422.
53. R. S. Dainter, H. Suschitzky, B. J. Wakefield, N. Hughes and A. J. Nelson, *Tetrahedron Lett.*, 1984, **25**, 5693.
54. W. M. Koppes, G. W. Lawrence, M. E. Sitzman and H. G. Adolph, *J. Chem. Soc., Perkin Trans. 1*, 1981, 1815.
55. L. Forlani, P. De Maria, E. Foresti and G. Pradella, *J. Org. Chem.*, 1981, **46**, 3178.
56. R. A. Rossi and R. H. de Rossi, *ACS Monogr.*, 1983, **178**.
57. J. A. Zoltewicz and T. M. Oestreich, *J. Am. Chem. Soc.*, 1973, **95**, 6863.
58. M. Novi, C. Dell'Erba, G. Garbarino and F. Sancassan, *J. Org. Chem.*, 1982, **47**, 2292.
59. D. R. Carver, J. S. Hubbard and J. F. Wolfe, *J. Org. Chem.*, 1982, **47**, 1036.
60. C. Paradisi and G. Scorrano, *Adv. Chem. Ser.*, 1987, **215**, 339.
61. E. Mitscherlich, *Justus Liebigs Ann. Chem.*, 1834, **12**, 311.
62. H. H. Hodgson and F. W. Handley, *J. Soc. Chem. Ind., London*, 1927, **46**, 435.
63. A. Bassani, M. Prato, P. Rampazzo, U. Quintily and G. Scorrano, *J. Org. Chem.*, 1980, **45**, 2263.
64. I. M. Sosonkin, D. A. Novokhatka, N. A. Lakomova, A. V. Golikov, T. K. Ponomareva, A. Sh. Glaz, V. V. Evstigneev and F. F. Lakomov, *Zh. Org. Khim.*, 1987, **23**, 234.
65. E. H. Gold, V. Piotrowski and B. Z. Weiner, *J. Org. Chem.*, 1977, **42**, 554.
66. D. J. Cram, J. A. Bryant and K. M. Doxsee, *Chem. Lett.*, 1987, 19.
67. A. I. Meyers and K. A. Lutomski, *J. Am. Chem. Soc.*, 1982, **104**, 879.
68. A. I. Meyers and R. J. Himmelsbach, *J. Am. Chem. Soc.*, 1985, **107**, 682.
69. N. Furukawa, M. Tsuruoka and H. Fujihara, *Heterocycles*, 1986, **24**, 3337.
70. A. I. Meyers and R. A. Gabel, *J. Org. Chem.*, 1982, **47**, 2633.
71. A. E. Hauck and C.-S. Giam, *J. Chem. Soc., Perkin Trans. 1*, 1980, 2070.
72. A. I. Meyers and D. G. Wettlaufer, *J. Am. Chem. Soc.*, 1984, **106**, 1135.
73. D. L. Comins and A. H. Abdullah, *J. Org. Chem.*, 1982, **47**, 4315.
74. D. L. Comins, R. K. Smith and E. D. Stroud, *Heterocycles*, 1984, **22**, 339.
75. T. R. Webb, *Tetrahedron Lett.*, 1985, **26**, 3191.
76. F. Kienzle, *Helv. Chim. Acta*, 1978, **61**, 449.
77. G. Bartoli, M. Bosco, R. Dalpozzo and L. Grossi, *NATO Adv. Study Inst. Ser., Ser. C*, 1989, **257**, 489.
78. G. Bartoli, *Acc. Chem. Res.*, 1984, **17**, 109.
79. G. Bartoli, M. Bosco, R. Dalpozzo and M. Petrini, *Tetrahedron*, 1987, **43**, 4221.
80. G. Bartoli, M. Bosco and R. Dalpozzo, *Tetrahedron Lett.*, 1985, **26**, 115.
81. G. Bartoli, M. Bosco, A. Melandri and A. C. Boicelli, *J. Org. Chem.*, 1979, **44**, 2087.
82. G. Bartoli, M. Bosco, R. Dalpozzo and P. E. Todesco, *J. Org. Chem.*, 1986, **51**, 3694.
83. N. Kornblum, L. Cheng, R. C. Kerber, M. M. Kestner, B. N. Newton, H. W. Pinnick, R. G. Smith and P. A. Wade, *J. Org. Chem.*, 1976, **41**, 1560.
84. G. Iwasaki, S. Sacki and M. Hamana, *Chem. Lett.*, 1986, 31.
85. G. Iwasaki, M. Hamana and S. Sacki, *Heterocycles*, 1982, **19**, 162.
86. M. Makosza, M. Jagusztyn-Grochowska, M. Ludwikow and M. Jawdosink, *Tetrahedron*, 1974, 3723.
87. F. Effenberger, W. Agster, P. Fischer, K. H. Jogun, J. J. Stezowski, E. Daltrozzo and G. Kollmannsberger-von Nell, *J. Org. Chem.*, 1983, **48**, 4649.
88. T. V. RajanBabu, B. L. Chenard and M. A. Petti, *J. Org. Chem.*, 1986, **51**, 1704.
89. R. A. Abramovitch, M. N. Inbasekaran, S. Kato, T. A. Radzkowska and P. Tomasik, *J. Org. Chem.*, 1983, **48**, 690.
90. C. M. Lee, M. P. Sammes and A. R. Katritzky, *J. Chem. Soc., Perkin Trans. 1*, 1980, 2458.
91. A. R. Katritzky, J. G. Keay, D. N. Rogers, M. P. Sammes and C. W. F. Leung, *J. Chem. Soc., Perkin Trans. 1*, 1981, 588.
92. M. P. Sammes, C. M. Lee and A. R. Katritzky, *J. Chem. Soc., Perkin Trans. 1*, 1981, 2476.
93. M. P. Sammes, C. W. F. Leung and A. R. Katritzky, *J. Chem. Soc., Perkin Trans. 1*, 1981, 2835.
94. P. G. Gassman and G. Gruetzmacher, *J. Am. Chem. Soc.*, 1973, **95**, 588.
95. P. G. Gassman and D. R. Amick, *J. Am. Chem. Soc.*, 1978, **100**, 7611.
96. S. Inoue, H. Ikeda, S. Sato, K. Horie, T. Ota, O. Miyamoto and K. Sato, *J. Org. Chem.*, 1987, **52**, 5497.
97. P. G. Gassman and T. J. van Bergen, *J. Am. Chem. Soc.*, 1973, **95**, 590, 591.
98. P. K. Claus, E. Jaeger and A. Setzer, *Monatsh. Chem.*, 1985, **116**, 1017.
99. M. Nakano and Y. Sato, *J. Chem. Soc., Chem. Commun.*, 1985, 1684.
100. M. Nakamo and Y. Sato, *J. Org. Chem.*, 1987, **52**, 1844.
101. M. Makosza, in 'Current Trends in Organic Synthesis', ed. H. Nozaki, Pergamon Press, New York, 1983, p. 401.
102. M. Makosza and S. Ludwiczak, *Synthesis*, 1986, 50.

103. A. Rykowski and M. Makosza, *Liebigs Ann. Chem.*, 1988, 627.
104. M. Makosza, J. Golinski and A. Rykowski, *Tetrahedron Lett.*, 1983, 3277.
105. M. Makosza, B. Chylinska and B. Mudryk, *Liebigs Ann. Chem.*, 1984, 8.
106. M. Makosza and E. Slonka, *Bull. Pol. Acad. Sci., Chem.*, 1984, **32**, 69.
107. M. Makosza, A. K. Kinowski, W. Danikiewicz and B. Mudryk, *Liebigs. Ann. Chem.*, 1986, 69.
108. M. Makosza, T. Glinka and A. K. Kinowski, *Tetrahedron*, 1984, **40**, 1863.
109. B. Mudryk and M. Makosza, *Tetrahedron*, 1988, **44**, 209.
110. M. Makosza and S. Ludwiczak, *J. Org. Chem.*, 1984, **49**, 4562.
111. J. P. Stahly, B. C. Stahly and K. C. Lilje, *J. Org. Chem.*, 1984, **49**, 578.
112. M. Makosza, J. Golinski, J. Baran and D. Dziewonska-Baran, *Chem. Lett.*, 1984, 1619.
113. M. Makosza and A. Tyrala, *Synthesis*, 1987, 1142.
114. M. Makosza and J. Winiarski, *Chem. Lett.*, 1984, 1623.
115. M. Makosza and Z. Owczarczyk, *Tetrahedron Lett.*, 1987, **28**, 3021.
116. R. A. Murphy, Jr. and M. P. Cava, *Tetrahedron Lett.*, 1984, **25**, 803.
117. M. Makosza and K. Wojciechowski, *Tetrahedron Lett.*, 1984, **25**, 4791.
118. K. Wojciechowski and M. Makosza, *Tetrahedron Lett.*, 1984, **25**, 4793.
119. D. L. Boger, C. E. Brotherton, J. S. Panek and D. Yohannes, *J. Org. Chem.*, 1984, **49**, 4056.
120. H. Vorbrüggen and K. Krolikiewicz, *Synthesis*, 1983, 316.
121. H. Yamanaka, S. Nishimura, S. Kaneda and T. Sakamoto, *Synthesis*, 1984, 681.
122. K. R. Brower, *J. Am. Chem. Soc.*, 1959, **81**, 3504.
123. T. Ibata, Y. Isogami and J. Toyoda, *Chem. Lett.*, 1987, 1187.
124. S. Hashimoto, S. Otani, T. Okamoto and K. Matsumoto, *Heterocycles*, 1988, **27**, 319.
125. G. Barak and Y. Sasson, *J. Chem. Soc., Chem. Commun.*, 1987, 1267.
126. H. J. Niclas, M. Bohle, J.-D. Rick, F. Zeuner and L. Zolch, *Z. Chem.*, 1985, **25**, 137.
127. J. F. Pilichowski and J. C. Gramain, *Synth. Commun.*, 1984, **14**, 1247.
128. J. J. Kulagowski and C. W. Rees, *Synthesis*, 1980, 215.
129. J. H. Gorvin, *J. Chem. Soc., Chem. Commun.*, 1985, 238.
130. J. H. Gorvin, *J. Chem. Soc., Perkin Trans. 1*, 1988, 1331.
131. P. R. Jones and S. D. Rothenberg, *J. Org. Chem.*, 1986, **51**, 3016.
132. H. Iida, M. Yamazaki, K. Takahashi and K. Yamada, *Nippon Kagaku Kaishi*, 1976, 138 (*Chem. Abstr.*, 1976, **84**, 134 795f).
133. H. Tondys and H. C. van der Plas, *J. Heterocycl. Chem.*, 1986, **23**, 621.
134. S. Sekiguchi, T. Horie and T. Suzuki, *J. Chem. Soc., Chem. Commun.*, 1988, 698.
135. M. Hojo, R. Masuda and E. Okada, *Tetrahedron Lett.*, 1987, **49**, 6199.
136. P. K. Brooke, S. R. Challand, M. E. Flood, R. B. Herbert, F. G. Holliman and P. N. Ibberson, *J. Chem. Soc., Perkin Trans. 1*, 1976, 2248.
137. G. W. Rewcastle and W. A. Denny, *Synth. Commun.*, 1987, **17**, 1171.
138. T. Nagamatsu, Y. Hashiguchi and F. Yoneda, *J. Chem. Soc., Perkin Trans. 1*, 1984, 561.
139. J. B. Jiang and J. Roberts, *J. Heterocycl. Chem.*, 1985, **22**, 159.
140. L. Andersen, F. E. Nielsen and E. B. Pedersen, *Chem. Scr.*, 1988, **28**, 191.
141. J. Wrubel and R. Mayer, *Z. Chem.*, 1984, **24**, 254.
142. M. Sako, Y. Kojima, K. Hirota and Y. Maki, *J. Chem. Soc., Chem. Commun.*, 1984, 1691.
143. M. Sako, Y. Kojima, K. Hirota and Y. Maki, *Heterocycles*, 1984, **22**, 1021.
144. A. R. Katritzky and K. S. Laurenzo, *J. Org. Chem.*, 1986, **51**, 5039.
145. J. T. Gupton, J. P. Idoux, G. Baker, C. Colon, A. D. Crews, C. D. Jurss and R. C. Rampi, *J. Org. Chem.*, 1983, **48**, 2933.
146. M. M. Catalano, M. J. Crossley and L. G. King, *J. Chem. Soc., Chem. Commun.*, 1984, 1537.
147. J. E. Shaw, D. C. Kunerth and S. B. Swanson, *J. Org. Chem.*, 1976, **41**, 732.
148. L. Testaferri, M. Tiecco, M. Tingoli, D. Bartoli and A. Massoli, *Tetrahedron*, 1985, **41**, 1373.
149. J. P. Idoux, M. L. Madenwald, B. S. Garcia, D.-L. Chu and J. T. Gupton, *J. Org. Chem.*, 1985, **50**, 1876.
150. J. P. Idoux, J. T. Gupton, C. McCurry, A. D. Crews, C. D. Jurss, C. Colon and R. C. Rampi, *J. Org. Chem.*, 1983, **48**, 3771.
151. J. T. Gupton, J. P. Idoux, G. De Crescenzo and C. Colon, *Synth. Commun.*, 1984, **14**, 621.
152. J. T. Gupton, G. Hertel, G. De Crescenzo, C. Colon, D. Baran, D. Dukesherer, S. Novick, D. Liotta and J. P. Idoux, *Can. J. Chem.*, 1985, **63**, 3037.
153. R. D. Knudsen and H. R. Snyder, *J. Org. Chem.*, 1974, **39**, 3343.
154. D. Mauleòn, R. Granados and C. Minguillòn, *J. Org. Chem.*, 1983, **48**, 3105.
155. F. Montanari, M. Pelosi and F. Rolla, *Chem. Ind. (London)*, 1982, 412.
156. C. Paradisi, U. Quintily and G. Scorrano, *J. Org. Chem.*, 1983, **48**, 3105.
157. G. Uray and I. Kriessmann, *Synthesis*, 1984, 679.
158. M. Jarman and R. McCague, *J. Chem. Soc., Chem. Commun.*, 1984, 125.
159. S. M. Andrews, C. Konstantinou and W. A. Feld, *Synth. Commun.*, 1987, **17**, 1041.
160. P. Ballesteros, R. M. Claramunt and J. Elguero, *Tetrahedron*, 1987, **43**, 2557.
161. G. Soula, *J. Org. Chem.*, 1985, **50**, 3717.
162. D. J. Brunelle and D. A. Singleton, *Tetrahedron Lett.*, 1984, **25**, 3383.
163. T. Fuchigami, T. Awata, T. Nonaka and M. M. Baizer, *Bull. Chem. Soc. Jpn.*, 1986, **59**, 2873.
164. J. H. Clark, *Chem. Rev.*, 1980, **80**, 429.
165. Y. Imai, M. Ueda and M. Ii, *Makromol. Chem.*, 1978, **179**, 2989.
166. J. H. Clark and N. D. S. Owen, *Tetrahedron Lett.*, 1987, **28**, 3627.
167. R. K. Smalley, *Adv. Heterocycl. Chem.*, 1981, **29**, 1.
168. G. M. Shutske, L. L. Setescak, R. C. Allen, L. Davis, R. C. Effland, K. Ranbom, J. M. Kitzen, J. C. Wilker and W. J. Novick, Jr., *J. Med. Chem.*, 1982, **25**, 36.
169. M. Shutske, *J. Org. Chem.*, 1984, **49**, 180.

170. P. Molina, M. Alajarin, M. J. Vilaplana and A. R. Katritzky, *Tetrahedron Lett.*, 1985, **26**, 467.
171. H.-W. Schmidt, G. Koitz and H. Junek, *J. Heterocycl. Chem.*, 1987, **24**, 1305.
172. J. R. Beck and J. A. Yahner, *J. Org. Chem.*, 1978, **43**, 2048.
173. P. Cogolli, L. Testaferri, M. Tingoli and M. Tiecco, *J. Org. Chem.*, 1979, **44**, 2636.
174. J. R. Beck, *J. Org. Chem.*, 1972, **37**, 3224.
175. J. R. Beck, *J. Org. Chem.*, 1973, **38**, 4086.
176. J. R. Beck and J. A. Yahner, *J. Org. Chem.*, 1974, **39**, 3440.
177. P. Cogolli, F. Maiolo, L. Testaferri, M. Tingoli and M. Tiecco, *J. Org. Chem.*, 1979, **44**, 2642.
178. J. R. Beck and J. A. Yahner, *J. Org. Chem.*, 1978, **43**, 2052.
179. L. Testaferri, M. Tingoli and M. Tiecco, *J. Org. Chem.*, 1980, **45**, 4376.
180. L. Testaferri, M. Tingoli and M. Tiecco, *Tetrahedron Lett.*, 1980, **21**, 3099.
181. L. Testaferri, M. Tiecco, M. Tingoli, D. Chianelli and M. Montanucci, *Synthesis*, 1983, 751.
182. D. Chianelli, L. Testaferri, M. Tiecco and M. Tingoli, *Synthesis*, 1982, 475.
183. T. L. Evans and R. D. Kinnard, *J. Org. Chem.*, 1983, **48**, 2496.
184. F. Benedetti, D. R. Marshal, C. J. M. Stirling and J. L. Leng, *J. Chem. Soc., Chem. Commun.*, 1982, 918.
185. S. G. Woods, B. T. Matyas, A. P. Vinogradoff and Y. C. Tong, *J. Heterocycl. Chem.*, 1984, **21**, 97.
186. I. W. J. Still and V. A. Sayeed, *Synth. Commun.*, 1983, **13**, 1181.
187. S. D. Pastor, H. K. Naraine and R. Sundar, *Phosphorus Sulfur*, 1988, **36**, 111.
188. S. D. Pastor and E. T. Hessell, *J. Org. Chem.*, 1985, **50**, 4812.
189. W. P. Reeves, T. C. Bothwell, J. A. Rudis and J. V. McClusky, *Synth. Commun.*, 1982, **12**, 1071.
190. W. P. Reeves, A. Simmons, Jr. and K. Keller, *Synth. Commun.*, 1980, **10**, 633.
191. M. Gisler and H. Zollinger, *Angew. Chem., Int. Ed. Engl.*, 1981, **20**, 203.
192. J. T. B. Ferreira, F. Simonelli and J. V. Comasseto, *Synth. Commun.*, 1986, **16**, 1335.
193. T. L. Evans, *Synth. Commun.*, 1984, **14**, 435.
194. P. Singh, M. S. Batra and H. Singh, *Indian J. Chem., Sect. B*, 1983, **22**, 729.
195. D. Landini, F. Montanari and F. Rolla, *J. Org. Chem.*, 1983, **48**, 604.
196. D. J. Brunelle, *J. Org. Chem.*, 1984, **49**, 1309.
197. K. Konishi, I. Nishiguchi, T. Hirashima, N. Sonoda and S. Murai, *Synthesis*, 1984, 254.
198. T. Kamiyama, S. Enomoto and M. Inoue, *Chem. Pharm. Bull.*, 1985, **33**, 5184.
199. V. Gold, A. Y. Miri and S. R. Robinson, *J. Chem. Soc., Perkin Trans. 2*, 1980, 243.
200. M. D. Threadgill and A. P. Gledhill, *J. Chem. Soc., Perkin Trans. 1*, 1986, 873.
201. M. R. C. Gerstenberger and A. Haas, *Angew. Chem., Int. Ed. Engl.*, 1981, **20**, 647.
202. C. D. Hewitt and M. J. Silvester, *Aldrichimica Acta*, 1988, **21**, 3.
203. M. Attinà, F. Cacace and A. P. Wolf, *J. Chem. Soc., Chem. Commun.*, 1983, 108.
204. C.-Y. Shiue, M. Watanabe, A. P. Wolf, J. S. Fowler and P. Salvadori, *J. Labelled Comp. Radiopharm.*, 1984, **21**, 533.
205. M. S. Berridge, C. Crouzel and D. Comar, *J. Labelled Comp. Radiopharm.*, 1985, **22**, 687.
206. G. Angelini, M. Speranza, A. P. Wolf and C.-Y. Shiue, *J. Fluorine Chem.*, 1985, **27**, 177.
207. D.-R. Hwang, S. M. Moerlein, L. Lang and M. J. Welch, *J. Chem. Soc., Chem. Commun.*, 1987, 1799.
208. J. H. Clark and D. J. Macquarrie, *Tetrahedron Lett.*, 1987, **28**, 111.
209. M. Van Der Puy, *J. Org. Chem.*, 1988, **53**, 4398.
210. R. W. Trimmer, L. R. Stover and A. C. Skjold, *J. Org. Chem.*, 1985, **50**, 3612.
211. E. Bay, P. E. Timony and A. Leone-Bay, *J. Org. Chem.*, 1988, **53**, 2858.
212. K. Akiba, H. Matsuoka and M. Wada, *Tetrahedron Lett.*, 1981, **22**, 4093.
213. K. Akiba, Y. Negishi, K. Kurumaya, N. Ueyama and N. Inamoto, *Tetrahedron Lett.*, 1981, **22**, 4977.
214. W. Danikiewicz and M. Makosza, *J. Chem. Soc., Chem. Commun.*, 1985, 1792.
215. C. Paulmier, 'Selenium Reagents and Intermediates in Organic Synthesis', Pergamon Press, Oxford, 1986, p. 84.
216. M. Tiecco, L. Testaferri, M. Tingoli, D. Chianelli and M. Montanucci, *Synth. Commun.*, 1983, **13**, 617.
217. M. Tiecco, L. Testaferri, M. Tingoli, D. Chianelli and M. Montanucci, *J. Org. Chem.*, 1983, **48**, 4289.
218. H. Suzuki, H. Abe, N. Ohmasa and A. Osuka, *Chem. Lett.*, 1981, 1115.
219. H. Suzuki and M. Inouye, *Chem. Lett.*, 1985, 389.

2.2
Nucleophilic Coupling with Aryl Radicals

ROBERT K. NORRIS
University of Sydney, Australia

2.2.1 INTRODUCTION

Electrophilic substitution allows replacement of a proton or another electrofuge with a different group and is the most-used process for functionalizing aromatic rings, whereas nucleophilic substitution is often considered to be more difficult or to require special substituents or reaction conditions. Prior to 1970, only two significant classes of nucleophilic substitution processes on aromatic rings had been identified. The first, formally termed the S_NAr reaction, requires the presence of one or more electron-withdrawing groups on the aromatic ring, and is a two step addition–elimination process (see Chapter 2.1 of this volume), whereas the second proceeds through the intermediacy of arynes in a two-step elimination–addition sequence and often is accompanied by *cine* substitution (see Chapter 2.3 of this volume). It was during a study of the latter reaction type involving treatment of 5- and 6-iodopseudocumenes with potassium amide in liquid ammonia that Kim and Bunnett[1] observed that more of the respective *ipso* substitution products were being formed than could be explained on the basis that the reaction was proceeding through a common aryne intermediate. It was shown that the *ipso* substitution reaction was radical in nature and involved single-electron transfer steps, being inhibited by radical scavengers and promoted by one-electron donors, such as potassium. The reaction was recognized as being an aromatic analog of substitution processes at the aliphatic carbon in *p*-nitrobenzylic and α-nitroalkyl halides, independently identified in 1966 by the research groups of Kornblum[2] and Russell.[3] Bunnett[1,4] suggested the term '$S_{RN}1$', standing for substitution, radical nucleophilic, unimolecular, for these reactions and this widely adopted description will be used throughout this chapter.

The utility of the $S_{RN}1$ process from a synthetic viewpoint was very quickly recognized. It allows substitution of nucleofugic substituents on *unactivated* aromatic rings (*cf.* the S_NAr reaction) and does not give rise to rearranged products (*cf. cine* substitution in the aryne mechanism). Detailed mechanistic studies have identified the nature of the individual steps occurring in $S_{RN}1$ reactions (see Section 2.2.2.1), but from a synthetic viewpoint, the key process is the trapping of an aryl radical by a nucleophile, establishing a new aryl carbon to carbon or heteroatom bond, depending on the nature of the nucleophile. The generality of trapping of radicals by nucleophiles was recognized in the late 1960s and an overview by Russell[5] in 1970 summarizes the early observations. The mechanistic and synthetic work performed on the $S_{RN}1$ reaction since that time, however, has served as the most useful source of information on the chemo- and regio-selectivity of this key step.

The $S_{RN}1$ reaction has been the subject of numerous reviews. The very comprehensive review by Kornblum[6] principally deals with the aliphatic version, in which the new bond is formed between an aliphatic carbon and a nucleophile,[6] whereas other more general reviews treat both the aliphatic and aromatic reactions[7–12] or discuss the reaction in the context of electron transfer processes and/or electrochemical processes.[13–15] Among the reviews principally dealing with the aromatic version, one by Bunnett[4] deals with the discovery and initial work on the aromatic $S_{RN}1$ reaction, the review by Wolfe and coworkers[16] summarizes the early synthetic uses of the reaction, and a later review by Beugelmans discusses the application of the reaction to heterocyclic synthesis.[17] The monograph by Rossi and Rossi[18] details many aspects of the aromatic version, with particular emphasis on the mechanistic intricacies of the reaction. The discussion of the $S_{RN}1$ reaction in this chapter will be restricted to the aromatic version of the reaction, except for a brief mention of reaction with vinylic substrates. (*i.e.* it is limited to trapping of nucleophiles by sp^2-hybridized radical sites). Reactions which are catalyzed by or involve electron transfer to transition metal ions, *e.g.* the Sandmeyer reaction or the recently discovered cobalt carbonyl catalyzed carbonylation of aryl and vinyl halides,[19] also will be omitted.

2.2.2 THE $S_{RN}1$ REACTION

2.2.2.1 Mechanistic Considerations

The steps in the general $S_{RN}1$ reaction as applied to aromatic systems are given in equations (1)–(4), where X^- is a nucleofuge. Equation (1) represents the generalized initiation step. The single electron necessary for initiating the whole reaction sequence has been shown to arise from the nucleophile (equation 5) in either photostimulated or thermal (spontaneous) processes, added nonparticipant nucleophiles or radical anions (entrainment), or from solvated electrons and/or dissolving alkali metals. Numerous possibilities exist for both the nature of the photostimulation of the $S_{RN}1$ reaction and the way in which electron transfer from nucleophile to substrate takes place. These are discussed at length in Chapter 7 of ref. 18, but some points are further elaborated below.

An example of transfer of electron from nucleophile to substrate is seen in the formation of the radical anions (observable by ESR) of 5-halo-2*H*,3*H*-benzo[*b*]thiophene-2,3-diones on treatment with nucleophiles.[20] It has been proposed in some cases, that this single-electron transfer step takes place through a charge transfer complex between the nucleophile and the aromatic substrate.[21,22] Some reactions occur spontaneously, *i.e.* without any catalysts or reagents other than the substrate and the nucleophile, but the initiation process is usually, although not invariably, photostimulated (near-ultraviolet radiation, 300–

$$ArX + e^- \longrightarrow ArX^{\bullet -} \quad (1)$$

$$ArX^{\bullet -} \longrightarrow Ar^{\bullet} + X^- \quad (2)$$

$$Ar^{\bullet} + Nu^- \longrightarrow ArNu^{\bullet -} \quad (3)$$

$$ArNu^{\bullet -} + ArX \longrightarrow ArNu + ArX^{\bullet -} \quad (4)$$

$$ArX + Nu^- \longrightarrow ArX^{\bullet -} + Nu^{\bullet} \quad (5)$$

400 nm). Indeed, the catalytic effect of irradiation has been used extensively as a mechanistic probe for the occurrence of aromatic substitution by the $S_{RN}1$ mechanism. It should be noted, however, that some powerful nucleophiles, such as the phenylated anions of the Group IVA elements[23–25] do not need photostimulation to bring about reaction. In similar fashion, reactions between the enolate ion from pinacolone ($^-CH_2COBu^t$) and iodobenzene or bromobenzene also proceed spontaneously in DMSO in the absence of irradiation, although irradiation certainly increases the reaction rate.[26]

A practical disadvantage of using dissolving alkali metals, first noted by Bunnett and Kim in their initial discovery of the aromatic $S_{RN}1$ reaction,[1] is formation of reduction products, *i.e.* replacement of the nucleofuge by a hydrogen atom (equation 6; Solv-H is solvent). The first example of an electrochemically induced aromatic substitution of the $S_{RN}1$-type was reported in 1970,[27] when it was demonstrated by cyclic voltammetry that the substitution of iodide by cyanide in *p*-iodonitrobenzene could be accomplished. In 1974, Pinson and Savéant[28] carried out the substitution of the bromine in *p*-bromobenzophenone with PhS^- ion under electrocatalytic conditions (0.2 F mol^{-1}) and since this beginning, electrochemically induced $S_{RN}1$ processes have been extensively studied, mainly by French chemists. In 1980 Savéant[13] reviewed the earlier literature, which principally consisted of mechanistic studies. In more recent work, especially since 1985, electrochemically initiated reactions have been used in preparative procedures and these would appear to be an important addition to the synthetic use of the $S_{RN}1$ reaction, particularly with some nucleophiles which fail to react or react very poorly under other conditions. These nucleophiles include phenoxides, CN^- and *aci*-nitronic ions (see Sections 2.2.3.1.4–6 respectively). In some cases these reactions only take place under electrochemical conditions with stoichiometric amounts of current (*i.e.* are noncatalytic processes) and are not really chain processes (*e.g.* reaction of iodobenzene with 2-nitropropan-2-ide ion).[29]

$$Ar^{\bullet} + e^- \longrightarrow Ar^- \xrightarrow{\text{Solv-H}} ArH \quad (6)$$

The propagation steps in the $S_{RN}1$ sequence are the dissociative step (equation 2), the associative step (equation 3), and the single-electron transfer step (equation 4). These processes are discussed extensively in Chapter 8 of ref. 18. The nature of the nucleofuge is critical in that poor nucleofuges retard the dissociation of the intermediate radical anion, $ArX^{-\cdot}$ (equation 2), and prevent the reaction from proceeding at a reasonable rate or stop it completely (see Section 2.2.2.3).

The associative step (equation 3) determines the nature of the product, since in this step the synthetically important bond formation between the aromatic moiety and the nucleophile takes place. The rates for the association of a number of nucleophiles with a variety of aryl and heteroaryl radicals has been measured in electrochemical studies[29–31] and competitive product studies.[32–35] The range of nucleophiles, classified according to the atom which becomes directly bonded to the aromatic ring, and, where the nucleophile is ambident, the regiochemistry of the association reaction shown in equation (3) are detailed in Section 2.2.3. The final propagating step (equation 4), that returns the chain-propagating electron from product radical anion to another molecule of substrate, is essential if a chain reaction is to continue.

Several alternative processes can impede or dramatically alter the overall substitution reaction. First, if the product radical anion, $ArNu^{-\cdot}$, is significantly more stable than the radical anion of the substrate, $ArX^{-\cdot}$, then the electron transfer step in equation (4) can become rate limiting and the reaction may become very sluggish. This may be one of the contributing factors towards the absence of examples of nitrite ion acting as a nucleophile in aromatic $S_{RN}1$ reactions. It is known from other studies that NO_2^- is readily trapped by aryl radicals,[36,37] so failure of the associative step, (equation 3; with $Nu^- = NO_2^-$) cannot account for lack of reaction. The nitroarene radical ion, $ArNO_2^{-\cdot}$, formed is relatively too stable and so propagation of the chain is stopped. Second, the radical anion, $ArNu^{-\cdot}$, can fragment (*i.e.* dissociate with breaking of a bond other than the newly formed Ar—Nu bond), rather than transferring an electron to ArX. The mode of fragmentation depends on the nature of the aryl group, including the attached substituents, and the nature of the group Nu. Radical ions bearing a second nucleofugic group can dissociate to give new aryl radicals whose subsequent reaction leads to overall disubstitution, whereas dissociation within the Nu substituent can give rise to alternative aryl radicals or to unproductive radicals that terminate the radical chain. Two examples will suffice to show how poor yields of the desired products result when these processes take place. The reaction of *p*-iodoanisole with $PhSe^-$ ion in liquid ammonia, in addition to the desired product (**1**) gives the selenides (**2**) and (**3**; equation 7). This mixture of products is explicable in terms of the reversible association/dissociation sequence given in equation (8), in which two different cleavages of the aryl–Se bond compete with the electron transfer propagation step.[35] This phenomenon, leading to 'aryl scrambling' also occurs with arsenic, antimony and tellurium derivatives (see Sections 2.2.3.2.5 and 2.2.3.3.2). As the result of a quite different cleavage, reaction of bromobenzene with $^-CH_2CN$ provoked by potassium metal, gives a poor yield of phenylacetonitrile, and gives rise to appreciable amounts of 1,2-diphenylethane (**4**; equation 9).[38] When the same reaction was carried out under photostimulation (less reducing reaction conditions), (**4**) is still the major product, phenylacetonitrile remains a minor product, and small amounts of 1,1,2-triphenylethane and 1,1,2,2-tetraphenylethane are detected.[39,40] Fragmentation of the initially formed radical anion of phenylacetonitrile to give cyanide ion and benzyl radicals (equation 10) explains the formation of (**4**) (by dimerization) and toluene (electron transfer/protonation). Similar behavior is shown by the intermediate radical anions formed in reaction of aryl radicals with *aci*-nitronate and alkanethiolate ions. The occurrence of alternative fragmentation processes in $S_{RN}1$ reactions has been reviewed by Rossi.[41]

$$p\text{-MeOC}_6\text{H}_4\text{I} \xrightarrow{\text{PhSe}^-/\text{NH}_3/h\nu} p\text{-MeOC}_6\text{H}_4\text{SePh} + (p\text{-MeOC}_6\text{H}_4)_2\text{Se} + \text{PhSePh} \quad (7)$$

(**1**) 25% (**2**) 19% (**3**) 20%

$$p\text{-MeOC}_6\text{H}_4^{\bullet} + \text{PhSe}^- \rightleftharpoons [p\text{-MeOC}_6\text{H}_4\text{SePh}]^{\bullet-} \rightleftharpoons p\text{-MeOC}_6\text{H}_4\text{Se}^- + \text{Ph}^{\bullet} \quad (8)$$

$$\text{PhBr} + {}^-\text{CH}_2\text{CN} \xrightarrow{\text{K/NH}_3} \underset{5\%}{\text{PhCH}_2\text{CN}} + \underset{(\mathbf{4})\ 14\%}{(\text{PhCH}_2)_2} + \underset{3\%}{\text{Ph}_2\text{CH}_2} + \underset{26\%}{\text{PhMe}} + \underset{43\%}{\text{PhH}} \quad (9)$$

$$\text{Ph}^{\bullet} + {}^-\text{CH}_2\text{CN} \longrightarrow [\text{PhCH}_2\text{CN}]^{\bullet-} \longrightarrow \text{PhCH}_2^{\bullet} + \text{CN}^- \quad (10)$$

Termination steps in $S_{RN}1$ reactions compete with the propagation steps and, although these processes have aroused considerable mechanistic and theoretical speculation (see Chapter 9 in ref. 18), their effects, with several important exceptions, are not significant. For example the self-coupling of aryl radicals (equation 11) does not appear to occur under the conditions used for the $S_{RN}1$ reaction. One potentially disruptive termination step is reduction of the intermediate aryl radical (equation 6). The source of the reducing electron can be a dissolving metal (or solvated electron), one of the radical anion intermediates in the reaction ($ArNu^{-\cdot}$ or $ArX^{-\cdot}$), an electrode, or the nucleophile itself. These termination processes can normally be minimized by suitable choice of reaction conditions. In the rare cases of reduction of the intermediate radical by the nucleophile (equation 12) competing successfully with the associative propagating step (equation 3) the $S_{RN}1$ reaction is doomed to failure. This phenomenon was used to explain the failure of *p*-nitrophenyl and *p*-cyanophenyl radicals to couple with *aci*-nitronic ions.[36] Another termination reaction is abstraction of a hydrogen from solvent (Solv-H; equation 13) to give an aromatic hydrocarbon and a solvent-derived radical, which invariably is nonchain propagating. Although rate constant data for abstraction of hydrogen atoms from DMSO, DMF and acetonitrile indicated that these solvents might lead to interference with the successful operation of $S_{RN}1$ reactions,[42] the abstraction of hydrogen from these dipolar aprotic solvents, or from the much poorer hydrogen atom donor liquid ammonia, most commonly used in these reactions, does not appear to interfere significantly. One process, which sometimes limits the scope of the $S_{RN}1$ reaction, involves the reaction of aryl radicals with enolate ions having abstractable β-hydrogens. The classic example of this side reaction occurs in the reaction of phenyl radical (from iodobenzene) with the enolate ion (**5**) from diisopropyl ketone, as outlined in Scheme 1. Hydrogen abstraction from the enolate ion by the phenyl radical competes with association of the phenyl radical with (**5**), and gives the relatively stable (and hence chain-terminating) radical anion (**6**). Disproportionation of (**6**) into the dianion (**7**) and the α,β-unsaturated ketone (**8**), followed by protonation of (**7**) by the solvent, regenerates the enolate ion (**5**). Michael addition of (**5**) to (**8**) gives the by-product (**9**) in yields as high as 20%.[43] Similar abstraction processes occur in intramolecular reactions, and the work of Semmelhack and coworkers with the deuterated ketone (**10**) showed that β-hydrogen (deuterium) abstraction was occurring by isolation of the deuterated α,β-unsaturated ketone (**11**).[44,45] Another significant termination phenomenon is fragmentation of the radical anion of the intended product, ($ArNu^{-}\cdot$), to give an unproductive (nonchain-propagating) radical, *e.g.* the benzyl radical (equation 10).

$$2Ar\bullet \longrightarrow Ar\text{—}Ar \tag{11}$$

$$Ar\bullet + Nu^{-} \longrightarrow Ar^{-} + Nu\bullet \tag{12}$$

$$Ar\bullet + Solv\text{-}H \longrightarrow ArH + Solv\bullet \tag{13}$$

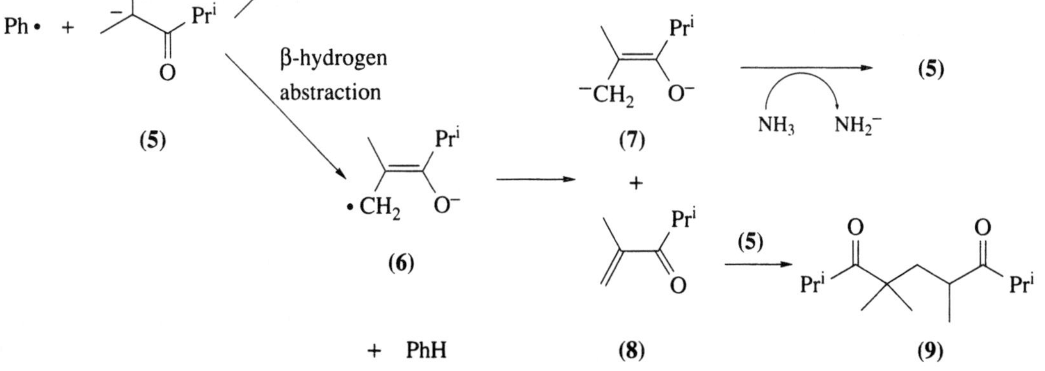

Ph• + (**5**) — association → [Ph-C(Me)$_2$-CO-Pr^i]$^{-\bullet}$

β-hydrogen abstraction → (**6**) + PhH → (**7**) + (**8**); (**7**) —NH_3/NH_2^-→ (**5**); (**8**) —(**5**)→ (**9**)

Scheme 1

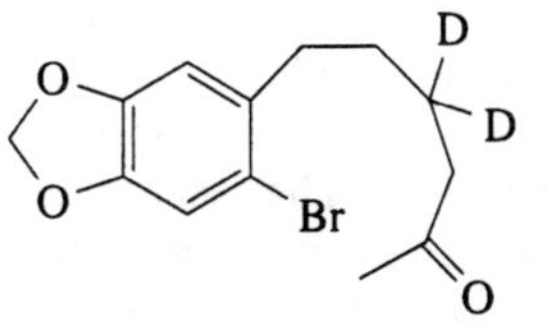

(**10**) 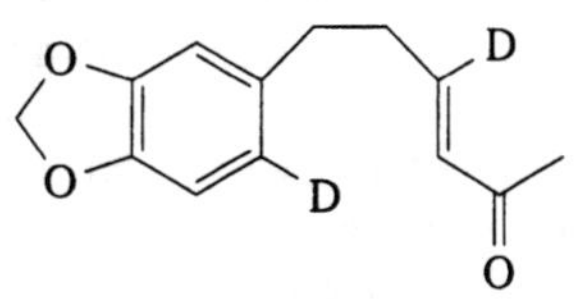 (**11**)

Inhibition of the $S_{RN}1$ reaction by a number of substances such as 2-methyl-2-nitrosopropane, tetraphenylhydrazine, oxygen, di-*t*-butyl nitroxide, and *m*- and *p*-dinitrobenzene has been demonstrated, but clearly for synthetic purposes all such substances must be excluded from the appropriate reactions. One important practical consequence is that all $S_{RN}1$ reactions should be carried out under an inert atmosphere to avoid inhibition by oxygen; nitrogen or argon are generally used. Anomalous cases are known, however, where oxygen did not affect the reaction rate or in fact, accelerated the reaction. Coupling of phenyl radicals with $(EtO)_2PO^-$ appears to be too rapid to be affected by oxygen.[21] The reaction of iodobenzene with $K^+\ ^-CH_2COBu^t$ in DMSO in the dark was actually accelerated by oxygen, but no useful yield of the expected $S_{RN}1$ product was obtained since it was destroyed by oxygen under the reaction conditions.[26]

2.2.2.2 Choice of Solvents

The number of solvents that have been used in $S_{RN}1$ reactions is somewhat limited in scope, but this causes no practical difficulties. Characteristics that are required of a solvent for use in $S_{RN}1$ reactions are that it should dissolve both the organic substrate and the ionic alkali metal salt (M^+Nu^-), not have hydrogen atoms that can be readily abstracted by aryl radicals (*cf.* equation 13), not have protons which can be ionized by the bases (*e.g.* NH_2^- or Bu^tO^- ions), or the basic nucleophiles (Nu^-) and radical ions ($RX^{-\cdot}$ or $RNu^{-\cdot}$) involved in the reaction, and not undergo electron transfer reactions with the various intermediates in the reaction. In addition to these characteristics, the solvent should not absorb significantly in the wavelength range normally used in photostimulated processes (300–400 nm), should not react with solvated electrons and/or alkali metals in reactions stimulated by these species, and should not undergo reduction at the potentials employed in electrochemically promoted reactions, but should be sufficiently polar to facilitate electron transfer processes.

A comparison of the suitability of solvents for use in $S_{RN}1$ reactions was made in benzenoid systems[46] and in heteroaromatic systems.[47] The marked dependence of solvent effect on the nature of the aromatic substrate, the nucleophile, its counterion and the temperature at which the reaction is carried out, however, often make comparisons difficult. Bunnett and coworkers[46] chose to study the reaction of iodobenzene with potassium diethyl phosphite, sodium benzenethiolate, the potassium enolate of acetone, and lithium *t*-butylamide. From extensive data based on the reactions with $K^+\ (EtO)_2PO^-$ (an extremely reactive nucleophile in $S_{RN}1$ reactions and a relatively weak base) the solvents of choice (based on yields of diethyl phenylphosphonate, given in parentheses) were found to be liquid ammonia (96%), acetonitrile (94%), *t*-butyl alcohol (74%), DMSO (68%), DMF (63%), DME (56%) and DMA (53%). The powerful dipolar aprotic solvents HMPA (4%), sulfolane (20%) and NMP (10%) were found not to be suitable. A similar but more discriminating trend was found in reactions of iodobenzene with the other nucleophilic salts listed above.[46] Nearly comparable suitability of liquid ammonia and DMSO have been found with other substrate/nucleophile combinations. For example, the reaction of *p*-iodotoluene with Ph_2P^- (equation (14) gives 89% and 78% isolated yields (of the corresponding phosphine oxide) in liquid ammonia and DMSO respectively.[48]

$$Ph_2P^- + p\text{-}MeC_6H_4I \xrightarrow[89\%]{NH_3/h\nu} p\text{-}MeC_6H_4PPh_2 \qquad (14)$$

As an example of the dangers associated with interpreting solvent effects in these reactions, it was found in reactions involving $K^+\ ^-CH_2COMe$ with 2-chloroquinoline, that liquid ammonia gave the most rapid reaction and the best yields (at –33 °C) and that the rate of consumption of 2-chloroquinoline at ambient temperature increased with solvent in the order benzene ≈ ether << DME < THF < DMF < DMSO. Moon and Wolfe wrote that, 'however, for preparative purposes, the more polar solvents DMF and especially DMSO, appear to offer no advantages over THF, in which substitution proceeds at very respectable rates with a minimum of side reactions'.[47] These same authors found in the same study, however, that iodobenzene failed to react in THF with $K^+\ ^-CH_2COMe$. The reaction of bromobenzene with $K^+\ ^-CH_2COBu^t$, as shown in equation (15), after 80 min gave the substitution product (**12**) in yields of 96% (liquid ammonia, –33 °C), 84% (DMSO, 35 °C) and 57% (DMF, 35 °C) and failed completely in

THF.[45] To confuse the situation further, THF is found to be a quite appropriate solvent for cyclization reactions in which the enolate ion from an amide attacks a benzene ring intramolecularly.[49,50]

$$\text{PhBr} + \overset{-}{\text{H}_2\text{C}}\text{C(=O)Bu}^t \xrightarrow{\text{NH}_3/h\nu} \text{PhCH}_2\text{C(=O)Bu}^t \quad \textbf{(12)} \qquad (15)$$

Aprotic solvents, such as DMSO and acetonitrile generally give good results, and the latter appears to be the solvent of choice (with appropriate supporting electrolytes) in electrochemically initiated reactions.

Protic solvents (other than ammonia) are generally unsuitable on account of their high acidity relative to that of most nucleophiles used in $S_{RN}1$ reactions. Water was found to be unsuitable, even with water-soluble substrates and weakly basic nucleophiles.[46] The reported reaction of halobenzenes with PhO^- in 50% aqueous Bu^tOH, catalyzed by sodium amlagam,[51] was shown to be unreproducible.[52] Methanol was used as solvent in an unusual reaction, believed to be occurring by the $S_{RN}1$ mechanism and catalyzed by MeO^- at 147 °C, in which PhS^- replaces the bromine in 3-bromoisoquinoline.[53] Other protic solvents have been reported to give acceptable yields in $S_{RN}1$ reactions, but on the whole these involve substrates which give good yields in other solvents as well.

In general, from among the protic solvents, only liquid ammonia (the first used)[1] is particularly useful, and is still used more than any other solvent despite the low temperature at which reactions have to be carried out (b.p. –33 °C) and the fact that solubilities of some aromatic substrates and salts (M^+Nu^-) are poor. Ammonia has the added advantage of being easily purified by distillation, being an ideal system for production of solvated electrons, and has very low reactivity with basic nucleophiles and radical anions, and aryl radicals. Also, poor solubilities can sometimes be ameliorated by use of cosolvents such as THF. In addition it can be used as a solvent for the *in situ* reductive generation of nucleophiles such as $ArSe^-$ and $ArTe^-$ ions, *e.g.* the formation of $PhTe^-$ from diphenyl ditelluride (equation 16).[54,55]

$$\text{PhTeTePh} + 2\text{Na} \xrightarrow[>80\%]{\text{NH}_3} 2\text{Na}^+\text{PhTe}^- \qquad (16)$$

2.2.2.3 Nature of the Nucleofuge

Practical considerations, namely the ready availability of halogenated aromatic compounds, normally lead to the use of a halogen as the nucleofuge (X) in aromatic $S_{RN}1$ reactions (see equation 2). Other leaving groups have been used, however, and the best comparative lists are given by Rossi and Bunnett, who studied the reductive cleavage of various aromatic substrates (ArX) with potassium metal in liquid ammonia, in the presence of the enolate from acetone.[56,57] The suitability of groups as nucleofuges was measured by the overall production of phenylacetone (**13**), together with its reduction product (**14**) and 1,1-diphenylacetone (**15**; equation 17). Unsuitable nucleofuges gave predominant formation of benzene, formed by the alternative cleavage/protonation sequence (equation 18), failed to react or formed other products. In addition to the halogens, other nucleofuges used in the above study[56,57] and other work, with varying degress of success, include: the $(EtO)_2(P{=}O)O$,[33,38,58,59] PhO,[1,58] Me_3N^+,[38,60] Ph_2S^+,[58] and PhI^+ groups. Although ArS and ArSe groups were used as nucleofuges in the earlier studies,[56–58,61] the use of these groups is not recommended since fragmentation of radical anions of the form $Ar{-}Y{-}Ar^{-\cdot}$ (Y = Se, Te) can introduce the problem of aryl scrambling (*e.g.* equation 8).[41] The reactions of arenediazonium ions with some nucleophiles almost certainly occur by a mechanism which has many of the characteristics of the $S_{RN}1$ reaction. These reactions will not be discussed in detail here, but two recent synthetic routes which present convenient preparations of arenecarbonitriles[62] and diaryl sulfides[63,64] and involve intermediacy of arenediazo sulfides, are summarized in Scheme 2. The radical anion of the diazo sulfide is believed to be formed and to dissociate as shown in equations (19) and (20)[62,64] to give aryl radicals which are trapped by either CN^- and/or arenethiolate ions.

The reactivity sequence in the the aryl halides, in decreasing order, has been found to be ArI > ArBr > ArCl >> ArF (with the same aryl group) and this is believed to result, among other factors, from the well-known greater nucleofugicity of the halogens in the sequence $I^- > Br^- > Cl^- >> F^-$, differences in reduction potentials of the corresponding substrates, and solvation effects (see Chapter 8 in ref. 18). Typical examples confirming the above reactivity order are found in the reaction of the halobenzenes with

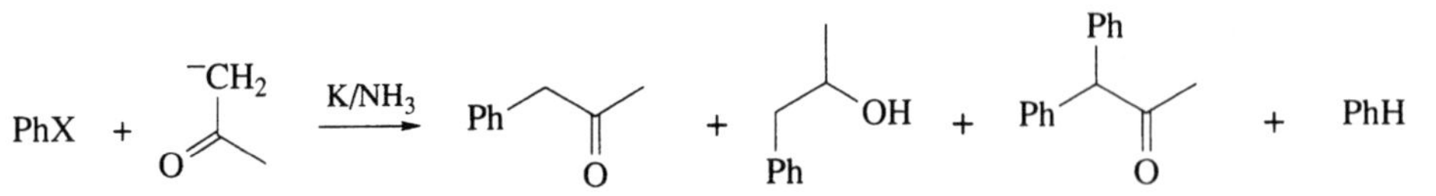

$$PhX + {}^{-}CH_2COCH_3 \xrightarrow{K/NH_3} PhCH_2COCH_3\ \textbf{(13)} + PhCH_2CH(OH)CH_3\ \textbf{(14)} + Ph_2CHCOCH_3\ \textbf{(15)} + PhH \quad (17)$$

$$PhX \xrightarrow{K/NH_3} PhX^{\bullet -} \longrightarrow Ph^{-} + X^{\bullet} \xrightarrow{NH_3} PhH \quad (18)$$

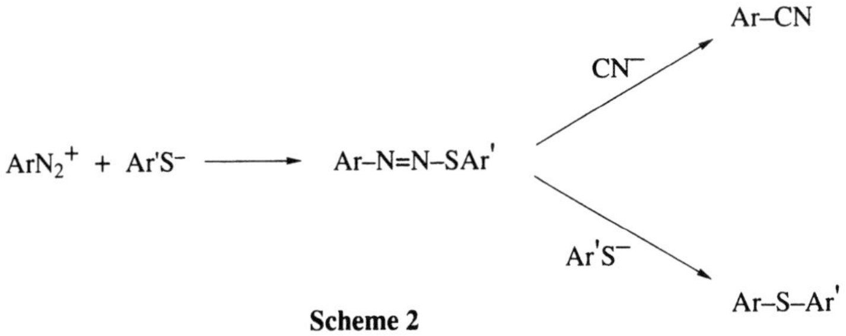

$$ArN_2^+ + Ar'S^- \longrightarrow Ar{-}N{=}N{-}SAr' \xrightarrow{CN^-} Ar{-}CN$$

$$Ar{-}N{=}N{-}SAr' \xrightarrow{Ar'S^-} Ar{-}S{-}Ar'$$

Scheme 2

$$Ar{-}N{=}N{-}S{-}Ar' + Nu^- \longrightarrow [Ar{-}N{=}N{-}S{-}Ar']^{\bullet -} + Nu^{\bullet} \quad (19)$$

$$[Ar{-}N{=}N{-}S{-}Ar']^{\bullet -} \longrightarrow Ar^{\bullet} + N_2 + Ar'{-}S^- \quad (20)$$

$^-CH_2COMe$,[58] $^-CH_2CN$[40] or, in competitive studies, with $^-CH_2COBu^t$ and $(EtO)_2PO^-$,[32] and in reactions of $^-CH_2COMe$ with 2-bromo-, 2-chloro- and 2-fluoro-pyridine.[65] Bunnett and coworkers have identified a peculiar leaving group effect in the reaction of halobenzenes with the enolate ion of acetone, in reactions stimulated by solvated electrons. They found that both the ratio of (**13**):(**14**) and the amount of benzene formed in the reaction given in equation (17) depended on the halogen, and both decreased as the halogen was changed from fluorine through chlorine and bromine to iodine. This phenomenon was interpreted as resulting from competition between rate of advance of solvated electrons through the solvent to the reactive intermediates and the rate of decomposition of the intermediate radical anions, $ArX^{-\cdot}$ (X = halogen).[66,67] This observation is of more than mechanistic interest, since it indicates that the aryl iodides (or bromides) are usually the substrates of choice in benzene derivatives because, not only are the reactions faster, but, in the case of reactions stimulated by solvated electrons, production of by-products is minimized. On account of the greater degree of reactivity in some heteroaromatic and polycyclic aromatic substrates, the use of aryl chlorides usually gives quite satisfactory results.

An exception to the above generalizations appears to be in reactions which are induced by electrochemical means. In cases where the intermediate radical anion, $ArX^{-\cdot}$, decomposes so rapidly that the aryl radical produced is still near the reducing electrode, a second electron transfer can take place with production of the aryl anion (see equation 6). Clearly a less nucleofugic halogen is preferable under these conditions. As confirmation of this principle, in the reaction of 2-halogenoquinolines with PhS^- and $PhCOCH_2^-$ ions the best electrochemically induced substitution reactions take place when the halogen is chlorine, less substitution occurs with bromine and significant, if not predominant reduction to give quinoline, occurs with the iodo compound.[68] The special selectivity problems which arise when a substrate has two nucleofuges attached is discussed in Section 2.2.2.4.2.

2.2.2.4 Nature of the Aromatic Substrate

The generalizations in the following discussion are based on reports of reactions carried out under ideal conditions, *i.e.* liquid ammonia or DMSO as solvent, photostimulation rather than 'dark' reactions, most appropriate nucleofuge, *etc.* The range of nucleophiles which can be used are itemized in Section 2.2.3, which also contains several extensive tables in which the range of substrates can be clearly seen. It must be noted, however, that reactions which proceed in high yield with one nucleophile may fail or proceed inefficiently with another if the nucleophile is incompatible with substituents on the aromatic substrate.

2.2.2.4.1 Benzene rings with one nucleofuge

The $S_{RN}1$ on benzene rings with only one nucleofuge, the best normally being either iodine or bromine, is an excellent method for replacing the nucleofuge without interference from *cine* substitution processes. The unsubstituted benzene ring reacts readily with most nucleophiles, but there are now sufficient examples of reactions which fail with the unsubstituted ring, but succeed when the ring has particular substituents, to modify the earlier generalization that there is no requirement for activation by substituents.[4] Bunnett and coworkers have summarized the effect of *ortho* substituents on $S_{RN}1$ reaction with $Bu^tCOCH_2^-$ and $(EtO)_2PO^-$ ions.[69] They showed that a wide range of *o*-substituents and, in an earlier study,[70] a large number of substituents in general, can be tolerated without deleterious effect. Beugelmans and coworkers, have more recently observed that *o*-, *m*- or *p*-cyano, *o*-carbonyl, *o*-methoxy or *o*-amino functions have a substantial facilitating effect on the reaction of halobenzenes with nucleophiles such as phenoxides, or the enolate ions of aldehydes and β-dicarbonyl compounds and that reactions with these nucleophiles generally fail to give substitution products without these specific substituents.[71–75]

In general, electron-donating groups such as alkyl, alkoxy or aryloxy groups do not interfere, with good results being obtained even with two *o*-methyl groups (equation 21).[70] Moderate steric hindrance does not impede substitution, but one Bu^t or two Pr^i groups *ortho* to the nucleofugic halogen reduce reactivity of the aromatic substrate and lead to substantial replacement of the halogen by hydrogen.[70] Ionized phenolic groups (—O^-) do prevent reaction,[70] but, contrary to the earlier report,[70] the dimethylamino group does not interfere.[76] *o*-Amino groups have the special facilitating effect discussed above, but *m*- and *p*-iodoanilines also react readily with $(EtO)_2PO^-$ ion.[77]

$$\xrightarrow{K^+\ ^-CH_2COMe/NH_3/h\nu} \qquad (21)$$

Electron-withdrawing groups, on the whole, do not interfere with the reaction as long as they do not react with the nucleophile, and, as mentioned above, facilitate the reactions with certain nucleophiles. This is also apparent in reactions carried out under electrochemical stimulation. Examples of substitution reactions on benzene rings bearing ionized carboxy (—CO_2^-), carbonylamino, acetyl, benzoyl, formyl and cyano groups are to be found in the references cited and the tables presented in Sections 2.2.3 and 2.2.4. There are also numerous examples of electrochemically stimulated reactions where benzoyl,[28,29,31,78–81] cyano[30,78,80,82–86] and *p*-acetyl groups[87] are present on the benzene ring.

The nitro group, which most successfully facilitates both the S_NAr reaction and the aliphatic version of the $S_{RN}1$ reaction, appears to prevent the aromatic $S_{RN}1$ reaction under nucleophile-initiated conditions,[39,70,77] with the possible exception of the reaction of *o*-iodonitrobenzene with $^-CH_2COBu^t$.[69]

Aromatic substrates with a CF_3 group either *ortho* or *para* to the halogen, on reaction with the the enolates from acetone or pinacolone, undergo reactions in which HF is eliminated from the initially formed $S_{RN}1$ product.[88] The normal substitution product is obtained, however, on treatment of *m*-CF_3-iodobenzene with ketone enolates[70,88] or on reaction of *o*-CF_3-iodobenzene with $(EtO)_2PO^-$ ion.[88]

2.2.2.4.2 Benzene rings with two nucleofuges

When the aromatic ring bears two nucleofugic groups, the decision as to which of the two is replaced or whether both are replaced depends on the nature of the two groups, their relative positions (*ortho*, *meta* or *para*) and the reaction conditions. The dependence on the above factors is best explained in terms of dissociation of the intermediate anion radical, $X—Ar—Y^{-\cdot}$, with two nucleofuges X and Y, and the subsequent reactions shown in equations (22)–(26). Initially, the better nucleofuge (Y^-), departs from the first-formed radical anion, as shown in equation (22), and consideration of all reactions with dihalobenzenes results in affirmation of the reactivity sequence $I^- > Br^- > Cl^- >> F^-$. In cases where one of the two halogens is fluorine only monosubstitution occurs (*e.g.* equation 27).[70,89–92] These results show that the C—F bond does not cleave in either equation (22) or (24). With the other halogens, the sequence of reactions in equations (22), (23) and (24) and (26) lead to disubstituted products, whereas the reactions in equations (22), (23) and (25) lead to monosubstituted products. The product distribution is determined by the relative rates of the reactions in equations (24) and (25).

For diiodo compounds the dissociation of the radical anion $IArNu^{-\cdot}$ appears always to be faster than the electron transfer step (equation 25) and disubstitution invariably takes place.[89,91,92] Bromoiodo and dibromo compounds behave similarly to the diiodo derivatives, with trace amounts of the monosubstituted monobromo compounds being found among the reaction products, but only when the halogens are in a *meta* relationship.[91,93] *o*-Dibromobenzene, for example, reacts readily with $^{-}CH_2COBu^t$ to give the disubstituted product (**16**) in 62% yield with no evidence of formation of the monosubstituted product.[94] Similarly, with $MeCOCH_2^-$, no monosubstituted product is isolated, but the initially formed disubstituted product (**17**) undergoes an internal aldol condensation to give a 64% yield of a mixture of the compounds (**18**) and (**19**) and also undergoes a further $S_{RN}1$ reaction with *o*-dibromobenzene to give (**20**) in 5% yield.[94]

$$XArY^{-\cdot} \longrightarrow XAr^{\cdot} + Y^- \qquad (22)$$

$$XAr^{\cdot} + Nu^- \longrightarrow XArNu^{-\cdot} \qquad (23)$$

$$XArNu^{-\cdot} \longrightarrow {}^{\cdot}ArNu + X^- \qquad (24)$$

OR

$$XArNu^{-\cdot} + XArY \longrightarrow XArNu + XArY^{-\cdot} \qquad (25)$$

$${}^{\cdot}ArNu + Nu^- \longrightarrow ArNu_2^{-\cdot} \qquad (26)$$

$$p\text{-}IC_6H_4F \xrightarrow[91\%]{(EtO)_2PO^-/NH_3/h\nu} p\text{-}(EtO)_2P(=O)C_6H_4F \qquad (27)$$

(**16**) R = Bu^t
(**17**) R = Me

(**18**) (**19**) (**20**)

Chloroiodo compounds give disubstituted products when the halogens are in an *ortho* or *para* relationship, with trace amounts of monosubstituted products,[89,90,92] whereas with *m*-chloroiodobenzene the ratio of monosubstituted to disubstituted product depends on the nucleophile. With $(EtO)_2PO^-$ ion, the monosubstituted product (**21**) can be isolated in 89% yield[91,92] and the disubstituted product (**22**) is formed in only 4% yield, whereas with PhS^- the disubstituted product (**23**; 91%) is by far the major product.[90] This difference in behavior has been attributed to stabilization, by the electron-withdrawing $P(O)(OEt)_2$ group, of the intermediate radical anion formed in the reaction with $(EtO)_2PO^-$, allowing the intermediate to undergo the electron transfer process in equation (25) in preference to the fragmentation reaction in equation (24).[92] By way of contrast, the *m*-chloroiodo relationship in some 5-chloro-7-iodoquinolines (both of the halogens are on the benzene rather than the pyridine ring), results in initial, almost exclusive substitution of the iodine by sulfur-, carbon- and phosphorus-based nucleophiles.[95–97] *o*-Dihalobenzenes (**24**) and (**25**) when treated with the dithiolate (**26**) give good yields of cyclic product (**27**; equation 28), and no monosubstituted product could be detected.[98]

Other substrates, in which one of the two nucleofuges is not a halogen, have been studied. Haloarenediazo sulfides on treatment with arenethiolate ions give substantial formation of bis(arylthio)arenes except that the monofluorophenyl sulfide is produced when the halogen is fluorine,[64] and *p*-iodophenyltrimethylammonium ion, on treatment with benzenethiolate ion in liquid ammonia, gives *p*-bis(phenylthio)benzene in 95% yield.[90]

Benzene derivatives with two nucleofuges have been used in the preparation of polymeric materials with varying degrees of success. Poly(1,4-phenylene sulfide) has been prepared by condensation of *p*-dichlorobenzene with sodium sulfide,[99,100] and in a related process, diazonium ions have been shown to initiate the polymerization of *p*-halobenzenethiolate ions.[101] In a preliminary study, poorly characterized polymers were obtained from reaction of equimolar amounts of *p*-dihalobenzenes and the enolate ions from ketones in the presence of excess base. When an excess of the ketone enolates was used, the normal *p*-disubstituted derivatives were formed.[102]

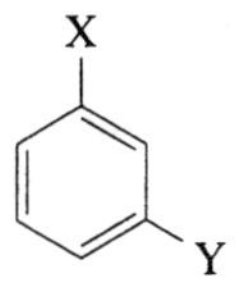

(21) X = Cl ; Y = $P(O)(OEt)_2$
(22) X = Y = $P(O)(OEt)_2$
(23) X = Y = SPh

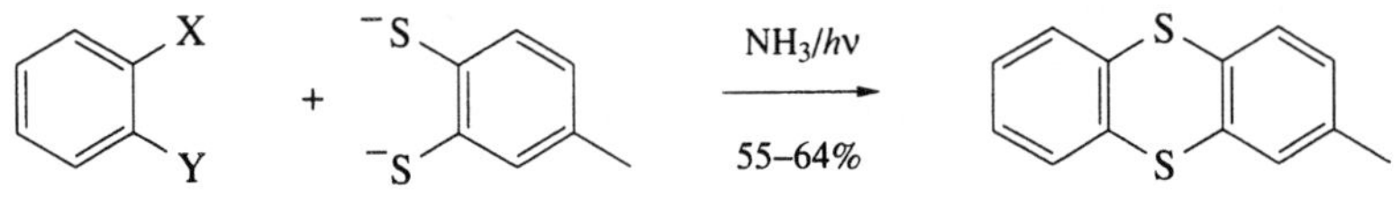

(24) X = Cl, Y = Br + **(26)** $\xrightarrow[55–64\%]{NH_3/h\nu}$ **(27)** (28)
(25) X = Y = I

2.2.2.4.3 *Polycyclic arenes*

The extension of the $S_{RN}1$ reaction to polycyclic aromatic substrates has been limited almost entirely to the simple halogenated derivatives (see tables and references in Section 2.2.3). The substrates used include 4-halobiphenyls, 1- and 2-halonaphthalenes, 9-bromoanthracene and 9-bromophenanthrene. There appears to be only one report on the effect of additional substituents on these ring systems; the ionized phenolic group in 1-bromo-2-naphthoxide causes substitution of the bromine by $^{-}CH_2CN$ to be completely supplanted by a reductive process.[39]

The increased stability of the radical anions of the products ($ArNu^{-}\cdot$), when the aryl moiety is a polycyclic system, has important practical consequences in that intermediates, which in the benzene series dissociate and give fragmented or rearranged products, are now sufficiently stable to allow chain-propagating electron transfer to take precedence.[41] The successful reaction of 4-chlorobiphenyl with $^{-}CH_2CN$ (equation 29),[39] for example, contrasts with that of bromobenzene (see equations 9 and 10),[38] and other halobenzenes.[38–40] Similar reduction in the occurrence of fragmentation processes and 'aryl scrambling' is seen when the benzene ring is replaced by polycyclic systems in reactions with alkanethiolate,[103] $ArSe^{-}$ and $ArTe^{-}$ ions,[54] and Ph_2As^{-} and Ph_2Sb^{-} ions.[25] Thus, butanethiolate gives a poor yield of substitution product (14%) with iodobenzene, but the reaction in the naphthalene system proceeds well (equation 30).[103]

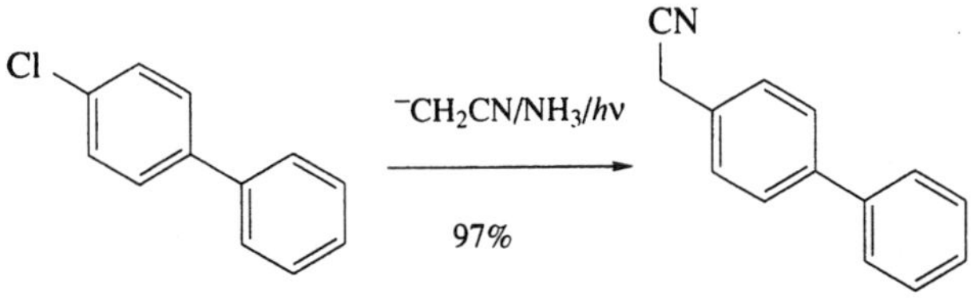

$$\xrightarrow[97\%]{^{-}CH_2CN/NH_3/h\nu} \quad (29)$$

$$\xrightarrow[81\%]{BuS^{-}/NH_3/h\nu} \quad (30)$$

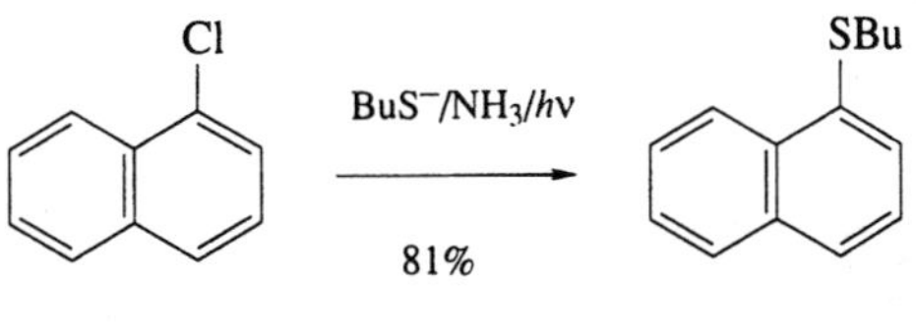

2.2.2.4.4 *Heteroaromatic compounds*

Numerous halogenated heteroaromatic substrates undergo $S_{RN}1$ reactions with a variety of nucleophiles (see tables and references in Section 2.2.3 for specific examples). The substrates used to date include all four 2-halopyridines, 3-bromo- and 3-iodo-pyridine, 4-bromopyridine, 2-chloro-, 2-bromo- and 2-iodo-quinoline, 3-bromoquinoline, 3-bromoisoquinoline, 2-chloropyrimidine, 2-chloropyrazine, 2-chloroquinoxaline, 4-chloroquinazoline, 2-chloro-, 2-bromo- and 2-iodo-thiophene, 3-bromothiophene, 2-chlorothiazole, 2-bromothiazole and 9-alkyl-6-iodopurines. Although the earlier studies were restricted to these monosubstituted derivatives (2-chloroquinoline often being used as a model system), more recent studies indicate that a similar range of substituents to that tolerated in the benzene system can be situated on the heteroaromatic ring. *o*-Substituents such as aryl, alkyl, alkoxy, fluoro, amino and functionalized amino, acyl, cyano and carbonylamino groups do not interfere. Alkyl and alkoxy groups in other positions similarly do not have a deleterious effect on reactivity. It would appear that the nitrogen-containing, π-deficient heterocycles give the best results, with yields from π-excessive substrates (*e.g.* halothiophenes) generally being poor. 2-Halothiazoles (π-excessive substrates) do react but S_NAr processes take place with some nucleophiles. Dihalopyridines behave in similar fashion to the benzene derivatives. Disubstituted derivatives are formed in good yields in the reaction of the 2,3-, 2,5-, 2,6- and 3,5-dihalo systems with $^-CH_2COBu^t$ (equation 31),[65,104] but with the anion of phenylacetonitrile mixtures of mono- and di-substituted products result.[104] It is noteworthy that under different reaction conditions, treatment of 2,6-dibromopyridine with the anion of 2-pyridylacetonitrile gives only the monosubstitution product in a reaction whose mechanism has not been delineated.[105]

$$X_2C_5H_3N + Bu^tCOCH_2^- \xrightarrow[43-89\%]{NH_3/h\nu} (Bu^tCOCH_2)_2C_5H_3N \qquad (31)$$

The relative reactivity of heteroaromatic compounds and their reactivity relative to the benzene system have not been extensively studied, but the reactivity sequences (decreasing) 2-chloroquinoline > 2-bromopyridine > bromobenzene,[65] 2-bromopyridine > 3-bromopyridine > 4-bromopyridine[65] and the reactivity order I > Br > Cl > F[65,106] allow a relative assessment of reactivity to be made. The relative ease of displacement of chlorine in differing environments is clearly shown in the dichloro compounds (**28**) and (**29**), in which the indicated chlorines are cleanly and preferentially replaced.[104]

The trend observed with the polycyclic hydrocarbons (see preceding section), namely that the product radical anions ($ArNu^{-\cdot}$) are more stable than those derived from the simple benzene analogs, is even more evident with the heteroaromatic substrates and, as a consequence, fragmentation processes are minimized.[41] For example, 2-chloroquinoline is the only substrate of many studied to undergo a substitution reaction with $PhCH_2S^-$ ion without fragmentation of the benzyl–S bond,[103] and to react with diphenylarsenide ion without scrambling of the aryl moieties.[25]

(**28**) (**29**)

2.2.2.4.5 *Vinylic compounds*

The report by Bunnett and coworkers[107] of substitution of halide by $MeCOCH_2^-$ or PhS^- ions in vinyl bromides or iodides appears to be the only synthetic application of the $S_{RN}1$ reaction at vinylic sites. The reactions, particularly with benzenethiolate ion, are slower than those of aryl halides and the yields are not as good. Tautomers or mixtures of products result in the reactions with $MeCOCH_2^-$ (*e.g.* equation 32).

$$PhCH{=}CHBr + {}^-CH_2COMe \xrightarrow{NH_3/h\nu} PhCH{=}CHCH_2Ac\ (48\%) + PhCH_2CH{=}CHAc\ (34\%) \qquad (32)$$

2.2.3 SUBSTITUTIONS BY THE $S_{RN}1$ REACTION

The $S_{RN}1$ reactions are grouped according to the attacking element in the nucleophilic species, *i.e.* according to the new aryl–element bond being formed. These reactions are convenient one-step substitution reactions leading from aryl halides (or arenes with appropriate nucleofuges) to a variety of classes of aromatic compounds.

With the majority of nucleophiles, regiochemical problems resulting from ambident behavior do not present any practical difficulties. Theoretical considerations indicate that in the key bond-forming reaction (equation 3) the aryl radical should attach itself to the site of greatest basicity.[108] This theory is consistent with the observation that enolates (including aryloxides) do not give rise to *O*-arylation products. With aryloxides, arylation *ortho* and *para* to the ionized hydroxy occurs (see Section 2.2.3.1.4). Ambident behavior of anions in aromatic $S_{RN}1$ reactions is very rare, and is observed in the arylation of the delocalized carbanions of conjugated alkenes, arylation of arylamides, where *o*- and *p*-attack usually predominates over *N*-attack,[109] and to a trivial extent in the heteroarylation (in the *o*- and *p*-positions on the benzene ring) of the α-carbanion from phenylacetonitrile. An example of this latter minor irritation is seen in the reaction of 4-bromopyridine with $^{-}$CH(Ph)CN, which gives 3% of (**30**) and 6% of (**31**).[110] The regiochemistry of the *C*-arylation of ketones, which form different enolate ions is discussed in Section 2.2.3.1.1.

(30) **(31)**

The preparation of cyclic compounds is discussed separately in Section 2.2.4.

2.2.3.1 Formation of C—C Bonds, Carbon-based Nucleophiles

2.2.3.1.1 Preparation of α-aryl ketones

The reactions between haloarenes and ketone enolates are by far the most extensively studied among $S_{RN}1$ reactions, and represent one of the few generally applicable methods for preparation of α-aryl ketones. The extensive list of the aromatic substrates and the ketones, from which the enolates are derived, together with the yields of α-arylated or α-heteroarylated ketones are collected in Tables 1 and 2 respectively. Most of these examples are taken from reactions which have been performed under photostimulated conditions, although several take place spontaneously ('in the dark'). Most of the reactions were carried out in liquid ammonia, and the enolate ion was generated from the ketone by treatment with an appropriate base, usually NH_2^- or Bu^tO^-. The alternative process, under alkali metal stimulation, is normally best avoided since it leads to formation of products in which either the carbonyl group has been reduced,[56,66,67] the aromatic ring has been reduced (*e.g.* with naphthalene derivatives),[111] or substantial replacement of halogen by hydrogen takes place.[66,67] This change in product distribution is evident when the yield of phenylacetone in the potassium-stimulated reaction (equation 33)[67] is compared with that obtained (60%) in the photostimulated reaction of PhF with $MeCOCH_2^-$.[58]

$$\text{PhF} + {}^{-}\text{CH}_2\text{COMe} \xrightarrow{\text{K/NH}_3} \underset{29\%}{\text{PhCH}_2\text{COMe}} + \underset{12\%}{\text{PhCH}_2\text{CH(OH)Me}} + \underset{42\%}{\text{PhH}} \quad (33)$$

Even when the reductive processes, so evident in metal-stimulated processes, are avoided, several side reactions can still cause reductions in the yield of the desired α-arylated ketones. The first, abstraction of β-hydrogen atoms from the enolate ion by the aryl radical, has already been mentioned (Section 2.2.2.1) and is sometimes a serious, chain-terminating process.[43–45] This abstraction reaction, however, appears to be quite unpredictable. β-Hydrogen abstraction from the enolate of 2,4-dimethyl-3-pentanone (Pr^iCOPr^i; Table 1) which severely disrupts the reaction with iodobenzene, does not prevent high-yielding reactions of the same enolate (and those from other ketones with α-branching) with many other substrates. In in-

Table 1 Yields of α-Arylated Ketones Formed in Photostimulated Reaction of Ketone Enolates with Aryl Substrates

ArX	*Ketone*[a]	*Yield (%)*[b]	*Ref.*
PhBr	MeCOMe	93	70
PhBr	3-Pentanone	80	112
PhBr	4-Heptanone	80	112
PhBr	Bu^tCOMe	90	112
PhBr	Cyclobutanone	90	112
PhBr	Cyclopentanone	64	112
PhBr	Cyclohexanone	72	112
PhBr	Cycloheptanone	58	112
PhBr	Cyclooctanone	95 (92)	112
PhBr	2-Indanone	90	112
PhI	MeCOMe	67	58
PhI	Pr^iCOPr^i	32	112
PhI	Bu^tCOMe	87[c]	113
PhI	PhCOMe	67	45
$4\text{-}AcC_6H_4Br$	2′-Acetonaphthone	56	114
$2\text{-}OMeC_6H_4I$	MeCOMe	67	71
$2\text{-}OMeC_6H_4I$	Pr^iCOMe	66	71
$2\text{-}OMeC_6H_4I$	Bu^tCOMe	100	71
$2,4\text{-}(MeO)_2C_6H_3Br$	MeCOMe	76	70
$3,5\text{-}(MeO)_2C_6H_3I$	MeCOMe	68	70
$2,4,6\text{-}(MeO)_3C_6H_2I$	MeCOMe	92	70
$2\text{-}(CO_2^-)C_6H_4Br$	MeCOMe	85	70
$3\text{-}(CO_2^-)C_6H_4Br$	MeCOMe	80	70
$4\text{-}(CO_2^-)C_6H_4Br$	MeCOMe	70	70
$2,5\text{-}Pr^i_2C_6H_3I$	MeCOMe	78	70
$2,5\text{-}Bu^t_2C_6H_3I$	MeCOMe	26	70
$2,4,6\text{-}Et_3C_6H_2Br$	MeCOMe	70	70
4-Bromobiphenyl	MeCOMe	69	70
2-Iodobiphenyl	Bu^tCOMe	83	69
$3\text{-}FC_6H_4I$	MeCOMe	56	70
$1,2\text{-}Br_2C_6H_4$	Bu^tCOMe	62[d]	94
$1,4\text{-}Br_2C_6H_4$	Bu^tCOMe	65[d]	102
$2\text{-}CF_3C_6H_4I$	MeCOMe	12	88
$3\text{-}CF_3C_6H_4I$	MeCOMe	35	70
$3\text{-}NH_2C_6H_4I$	MeCOMe	66	73
$4\text{-}NH_2C_6H_4I$	MeCOMe	33	73
$3\text{-}NMe_2C_6H_4I$	MeCOMe	82	76
$4\text{-}NMe_2C_6H_4I$	MeCOMe	90	76
$2\text{-}CF_3C_6H_4I$	Bu^tCOMe	12	88
$3\text{-}CF_3C_6H_4I$	Bu^tCOMe	40	88
$4\text{-}CF_3C_6H_4I$	Bu^tCOMe	21	88
1-Chloronaphthalene	MeCOMe	88	111
1-Iodonaphthalene	MeCOMe	76	70
2-Iodonaphthalene	MeCOMe	75	70
9-Bromoanthracene	MeCOMe	98 (84)	70
9-Bromophenanthrene	MeCOMe	62	70

[a]With unsymmetrical methyl ketones arylation occurred at the primary carbon. [b]Where two yields are stated the figure in parentheses is the isolated yield and the other is analytical (*e.g.* determined by GLC or ^1H NMR). [c]Reaction in the dark, catalyzed by iron(II) sulfate. [d]Both halogens replaced.

termolecular reactions between aryl radicals and enolate ions, which do not have α-branching, β-hydrogen abstraction is usually negligible, except in reactions with *o*-halobenzamides where β-hydrogen abstraction is significant and is even more evident in the *N*-alkylated benzamides.[123]

A second reaction which reduces the yield of α-arylated ketone is α,α-diarylation. This process also appears to be somewhat unpredictable. Although concentrations of the reacting species were a little different, the reaction of iodobenzene with acetone enolate appears to give higher proportions of 1,1-diphenylacetone than bromobenzene.[58] With bromobenzene, reactions with enolates of other ketones generally give diphenylated ketones in 10–20% yields.[112] 3-Bromothiophene with acetone enolate gives a 25:51 ratio of di- to mono-arylated acetones, whereas the 2-bromo isomer appears to give predominant monoarylation.[121] 2-Chlorothiazole, even with a fourfold excess of pinacolone enolate, gave mono- and di-arylation in a 53:25 ratio, but reaction of 2-bromothiazole gave no detectable diarylation.[118] It would appear from inspection of the references cited in Table 1, that yields of monoarylated compounds are good, that monoarylation normally greatly predominates over diarylation and is favored by use of excess of the enolate ion. In a deliberate attempt to improve diarylation, treatment of $MeCOCH_2^-$ with a three-

Table 2 Yields of α-Heteroarylated Ketones Formed in Photostimulated Reaction of Ketone Enolates with Heteroaryl Substrates

ArX	*Ketone*[a]	*Yield (%)*[b]	*Ref.*
2-Chloropyridine	MeCOMe	85	65
2-Bromopyridine	Pr^iCOMe	97	65
2-Bromopyridine	Bu^tCOMe	94	65
2-Bromopyridine	Cyclohexanone	47	65
3-Bromopyridine	MeCOMe	65	65
4-Bromopyridine	MeCOMe	28	65
2-Fluoro-3-iodopyridine	Bu^tCOMe	92	115
2-Methoxy-3-iodopyridine	Bu^tCOMe	84	115
2-Amino-3-iodopyridine	Bu^tCOMe	87	115
2,6-Dichloropyridine	Bu^tCOMe	86[c]	65
2,6-Dibromopyridine	Bu^tCOMe	89[c]	65
2,3-Dichloropyridine	Bu^tCOMe	63[c]	104
3,5-Dichloropyridine	Bu^tCOMe	43[c]	104
3,5-Dibromopyridine	Bu^tCOMe	85[c]	104
2-Chloropyrimidine	MeCOMe	61	116
2-Chloropyrimidine	Bu^tCOMe	32	116
2-Chloropyrimidine	Pr^iCOPr^i	88	116
2,6-$(MeO)_2$-4-Cl-Pyrimidine	Bu^tCOMe	98	116
4-Bu^t-5-Br-Pyrimidine	MeCOMe	70–75	117
4-Bu^t-5-Br-Pyrimidine	Bu^tCOMe	95	117
4-Bu^t-5-Br-Pyrimidine	PhCOMe	85–90	117
4-Ph-5-Br-Pyrimidine	MeCOMe	25–30	117
4-Ph-5-Br-Pyrimidine	Bu^tCOMe	60–65	117
4-Ph-5-Br-Pyrimidine	PhCOMe	63	117
3-Cl-6-OMe-Pyridazine	MeCOMe	60[d]	116
3-Cl-6-OMe-Pyridazine	Bu^tCOMe	72[d]	116
2-Chloropyrazine	MeCOMe	98[d]	116
2-Chloropyrazine	Bu^tCOMe	95[d]	116
2-Chloropyrazine	Pr^iCOPr^i	85[d]	116
2-Chloropyrazine	PhCOMe	82[d]	116
2-Chlorothiazole	Bu^tCOMe	53	118
4-Me-2-Cl-Thiazole	Bu^tCOMe	64	118
5-Me-2-Cl-Thiazole	Bu^tCOMe	67	118
2-Chloroquinoline	MeCOMe	90 (55)	119
2-Chloroquinoline	3-Pentanone	38 (68)	119
2-Chloroquinoline	Cyclopentanone	44 (63)	119
2-Chloroquinoline	Pr^iCOPr^i	68 (94)	119
2-Chloroquinoline	$MeCOCH(OMe)_2$	80	73
2-Chloroquinoline	PhCOMe	14	119
2-Chloroquinoline	$PhCOCH_2Me$	31 (50)	119
5-Cl-7-I-8-Pr^iO-Quinoline	MeCOMe	73[e]	96
5-Cl-7-I-8-Pr^iO-Quinoline	Bu^tCOMe	70[e]	96
5-Cl-7-I-8-Pr^iO-Quinoline	*p*-$MeOC_6H_4COMe$	70[e]	96
5-Cl-7-I-8-Pr^iO-Quinoline	2-Acetylfuran	80[e]	96
2-Chloroquinoxaline	Bu^tCOMe	70[d]	120
2-Bromothiophene	MeCOMe	31	121
3-Bromothiophene	MeCOMe	51	121
6-I-9-Et-Purine	MeCOMe	87	122
6-I-9-Et-Purine	Cyclopentanone	65	122
6-I-9-Et-Purine	PhCOMe	70	122
6-I-9-Et-Purine	2-Acetylfuran	67	122

[a]With unsymmetrical methyl ketones arylation occurred at the primary carbon. [b]Where two yields are stated the figure in parentheses is the isolated yield and the other is analytical (*e.g.* determined by GLC or ^1H NMR). [c]Both halogens replaced. [d]Photostimulation was not necessary. [e]Only the 7-iodine was replaced.

fold excess of bromobenzene gave a 58:32 ratio of di- to mono-arylation with only trace amounts of triarylation.[70] Other experiments also support the conclusion that triarylation is difficult.[102]

It is noticeable from the data in Table 1, that the number of examples of reactions involving the enolates of aryl alkyl ketones is limited. Bromo- and iodo-benzene were reported not to react with the enolate of acetophenone[112] but the reaction with iodobenzene under 'more intense and longer irradiation' was reported to give a 67% yield of phenacylbenzene.[45] Sufficient examples of the α-arylation of the enolates of aryl (and heteroaryl) alkyl ketones have been cited to indicate that these reactions are worth further investigation.

The regiochemistry of arylation of ketones has not been studied systematically, but there are sufficient data in the literature to allow reasonable predictions to be made. With enolates having one *t*-alkyl group attached to the carbonyl (*e.g.* $^-CH_2COBu^t$) or with those derived from aryl alkyl ketones, clearly no problem arises and attack at the sole available site takes place. Symmetrical ketones such as acetone, 3-pentanone and diisopropyl ketone also present no problems. With unsymmetrical dialkyl ketones, isomeric enolate ions can form and it would appear that the distribution of the two possible products formed by arylation of this mixture of enolate ions is determined principally by the equilibrium concentration of the various possible enolate ions.[124] Although the ratio of attack at the two possible α-positions could vary according to the relative reactivity of the aryl radical toward the various enolate ions, and indeed this is the case, the above generalization explains the observed distribution of products. Reactions with the enolates derived from 2-butanone give predominant formation of products from attack at the secondary α-position; the ratio of products (**32**; Ar = Ph):(**33**; Ar = Ph) in reactions with iodo- or bromo-benzene is in the range 1.3–3.2:1,[124] with *o*-iodoanisole the corresponding ratio (with Ar = *o*-anisyl) is 1.5:1,[71] and with 3-amino-2-chloropyridine the ratio is 2.3:1(Ar = 3-amino-2-pyridyl).[125] In the reaction of several aryl halides with the mixture of enolates from 3-methyl-2-butanone, the predominance of the less-substituted primary enolate is reflected in major formation of isomer (**34**) in all cases. The ratio (**34**):(**35**) varied with the aryl group as follows: 16:1[45] or 9:1[124] (Ar = Ph); 7:1 (Ar = 2-pyridyl);[65] and 5:1 (Ar = 2-quinolinyl).[119] When there is a substituent *ortho* to the halogen being displaced, the attack at the primary α-carbon is enhanced and vanishingly small proportions of products (or intermediates) corresponding to (**35**) were formed.[71,76,123,125,126] In the only reported case of a reaction with an α-substituted cyclohexanone, attack at the tertiary carbon of the thermodynamically more stable enolate occurred.[122,127]

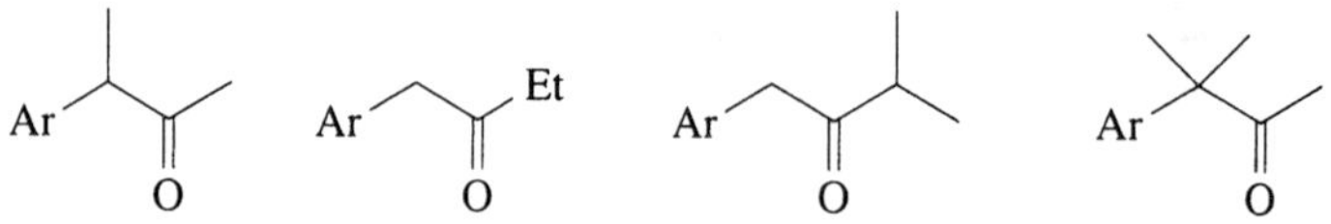

(**32**) (**33**) (**34**) (**35**)

A substantial number of important syntheses that depend on the attack of ketone enolates on *ortho*-functionalized aryl or heteroaryl halides, and that lead to cyclized products are discussed in Section 2.2.4.2.

2.2.3.1.2 Preparation of other α-arylated carbonyl compounds

The arylation of acetaldehyde enolate by iodobenzene[17,128] or *p*-iodoaniline[73] failed, and the reduced products, benzene or aniline were obtained. In the reaction of aldehyde enolates with *o*-substituted derivatives, however, arylation does occur and in these cases subsequent cyclization reactions usually take place (see Section 2.2.4.2). The special effect of the *o*-substituents in facilitating reactions with aldehyde enolates is shown by the rare example of simple substitution (*i.e.* without concomitant cyclization) in equation (34).[71]

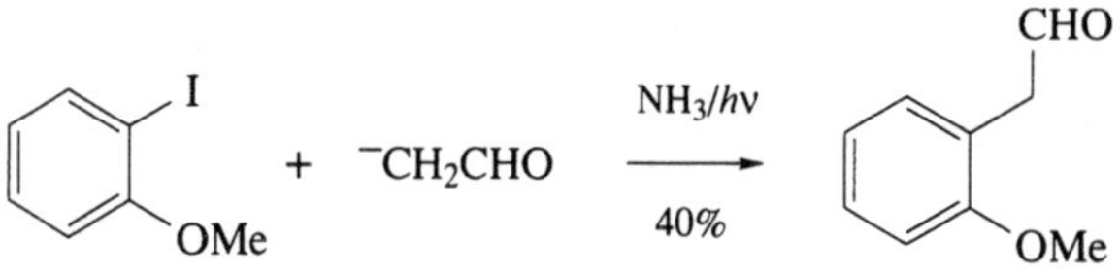

$$\text{2-IC}_6\text{H}_4\text{OMe} + {}^{-}\text{CH}_2\text{CHO} \xrightarrow[40\%]{\text{NH}_3/h\nu} \text{2-(OHCCH}_2\text{)C}_6\text{H}_4\text{OMe} \quad (34)$$

The α-arylation of the enolates of esters has only been studied superficially, and the results obtained with the *t*-butyl esters of acetic, propionic and isobutyric acids indicate that the method is not very promising in the benzene series. Arylation of the acetic ester with *p*-bromoanisole proceeds in 67%, but is accompanied by 29% α,α-diarylation,[44] and arylation of the isobutyric ester with either bromobenzene or *p*-bromoanisole gave low yields of arylated material (11% and 5%, respectively).[44,45] The principal process occurring with the isobutyric ester was β-hydrogen abstraction from the enolate ion, clearly demonstrated by deuterium-labelling experiments.[44] The reaction of the lithium enolate of the propionic ester with bromobenzene gave an encouraging result, with the α-phenylated ester being formed in 60% yield.[45] The potassium enolates of ethyl phenylacetate and ethyl α-phenylpropionate give 84% and 42% yields, respectively, of the products in which both of the bromines in 2,6-dibromopyridine have been replaced, with no trace of the monosubstituted monobromo compounds being detected.[104]

Most of the information on simple α-arylation reactions on the enolates of amides comes from a single study by Rossi and Alonso.[128] They showed that with acetamide itself no arylation reactions took place, presumably on account of ionization of the N—H rather than the C—H bond and the observation that the

$MeCONH^-$ anion (also $MeCONMe^-$), at least under electrochemically stimulated conditions,[68] does not undergo $S_{RN}1$ reactions. The use of acetamides with the nitrogen protected by *N,N*-disubstitution allowed formation of the enolate ion with sodium amide in liquid ammonia and the resulting enolates gave yields of 50–80% of the α-arylated products (*e.g.* equations 35 and 36).[128] α,α-Diarylation was a minor nuisance (the mono- and di-substituted compounds were separable) and the insolubility of some salts (*e.g.* the potassium enolate of *N*-acetylpiperidine) in liquid ammonia prevented satisfactory reaction. The enolate ion from *N*-(*p*-chlorobenzyl)-*N*-methylacetamide polymerizes on irradiation in liquid ammonia by the $S_{RN}1$ mechanism, but the enolate from *N*-(*p*-bromophenyl)-*N*-methylacetamide does not, a difference explained in terms of an intramolecular interaction in the latter enolate between the nucleophilic and the nucleofugic sites.[102] Examples of intramolecular attacks of amide enolates on *o*-substituted aryl halides are known, and these are discussed in Section 2.2.4.1.

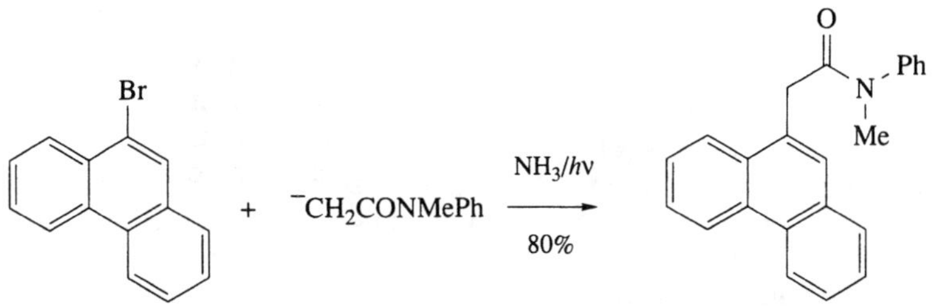

Br + $^-CH_2CONMePh$ —$NH_3/h\nu$, 80%→ (35)

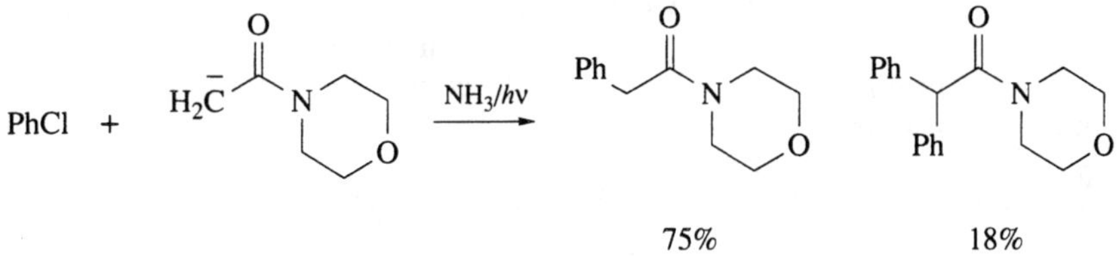

PhCl + $H_2\bar{C}$ —$NH_3/h\nu$→ 75% 18% (36)

Initial use of β-dicarbonyl compounds as sources of nucleophiles in aromatic $S_{RN}1$ reactions was disappointing, particularly in view of their successful use in the aliphatic case.[6,10] Monoanions from malonic esters,[45,112,119] acetoacetic esters[112,119] and β-diketones[112,119,129] all failed to react with halobenzenes or 2-chloroquinoline. The dianions could be arylated but these reactions took place at the site α to only one of the carbonyls. Thus the dipotassium salt of 2,4-pentanedione gave an 82% yield of 1-mesityl-2,4-pentanedione on treatment with 2-bromomesitylene[112] and the dialkali metal salts of benzoylacetone reacted readily with 2-chloroquinoline to give a 71% yield of 1-phenyl-4-(2-quinolinyl)-1,3-butanedione.[130] The latter salt was also able to entrain otherwise sluggish $S_{RN}1$ processes.[131] Beugelmans and coworkers found, however, that the presence of a cyano group modifies the reactivity of benzene (and pyridine) rings, facilitating reaction between the monoanions of a wide range of β-dicarbonyl compounds and leads to α-arylated-β-dicarbonyl derivatives in excellent yield, generally in excess of 70% and mostly near 90%.[72] Examples of these reactions are given in equations (37) and (38), and, as shown in equation (38), the reaction can be further modified, for example by basic work-up, to give the product resulting from a retro-Claisen reaction. The difficulty of predicting whether aromatic $S_{RN}1$ reactions will take place with β-dicarbonyl compounds is exemplified by the further observations by Beugelmans' group that although 2-chloro-, 4,7-dichloro-, 5,7-dichloro-8-methoxy-, and 7-iodo-8-methoxy-quinoline are unreactive, 5-chloro-7-iodo-8-methoxyquinoline reacts under photostimulation with the monoanions of diethyl malonate, ethyl acetoacetate and acetylacetone with displacement of only the 7-iodine in yields from 60–100%.[97]

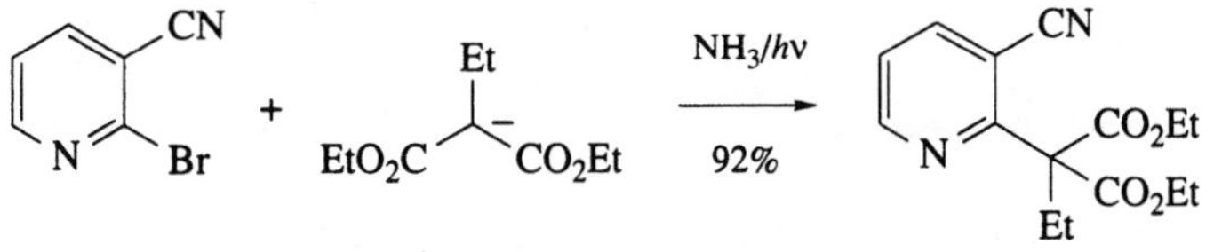

CN, N, Br + Et, EtO_2C, CO_2Et —$NH_3/h\nu$, 92%→ CN, N, CO_2Et, CO_2Et, Et (37)

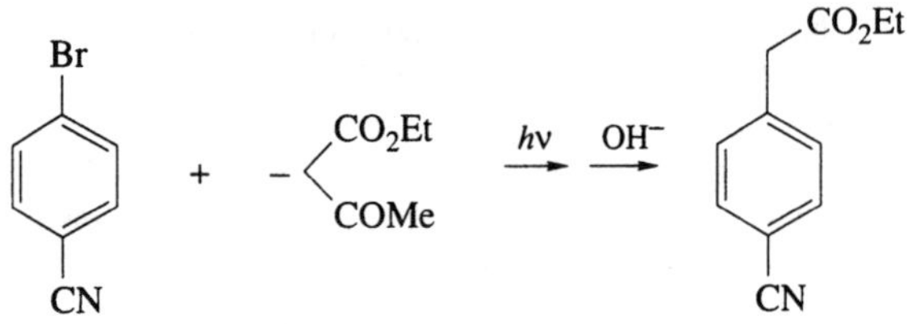

$$p\text{-}BrC_6H_4CN + {}^{-}CH(CO_2Et)(COMe) \xrightarrow{h\nu} \xrightarrow{OH^-} p\text{-}NCC_6H_4CH_2CO_2Et \quad (38)$$

2.2.3.1.3 Preparation of α-arylated nitriles

In general, a number of products can be obtained in the reaction of an aromatic substrate with an α-cyanoalkyl anion as shown in equation (39), and the detailed mechanism of formation of these products is discussed in the early paper by Bunnett and Gloor.[38] The product distribution from the reaction of bromobenzene with $^-CH_2CN$ is typical of that obtained with benzene derivatives, when carried out under metal-stimulated conditions (see equation 9). The reaction with halobenzenes, even under photostimulated (less reducing) conditions, gives less benzene, but still gives low yields of phenylacetonitrile on account of the fragmentation process, discussed earlier in Section 2.2.2.1 (equation 10). It has been suggested[18,38] that the fragmentation could be put to good use in the production of alkylbenzenes (*i.e.* $ArCH_2R$ in equation 39) since in the potassium-stimulated processes with a range of leaving groups and a number of α-cyanoalkyl ions, the principal product other than benzene (in yields up to 56%) is the alkane $PhCH_2R$. This proposal has the added potential that successful nucleofugal groups (X in equation 39) include Me_3N^+ and $(EtO)_2P(O)$, readily derived from amino and phenolic functionalities, respectively. This strategy, exemplified by the butylation sequence in Scheme 3,[38] which would allow halogen, amino or hydroxy groups on a benzene ring to be replaced by alkyl groups such as methyl, propyl, isopropyl or benzyl has not, as yet, been further developed.

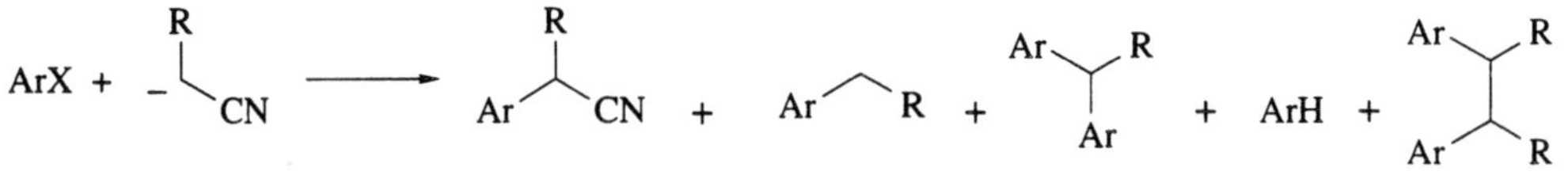

$$ArX + {}^{-}CH(R)CN \longrightarrow ArCH(R)CN + ArCH_2R + Ar_2CHR + ArH + ArCH(R)CH(R)Ar \quad (39)$$

$$PhOH \xrightarrow[87\%]{i} PhOP(O)(OEt)_2 \xrightarrow[38\%]{ii} PhBu$$

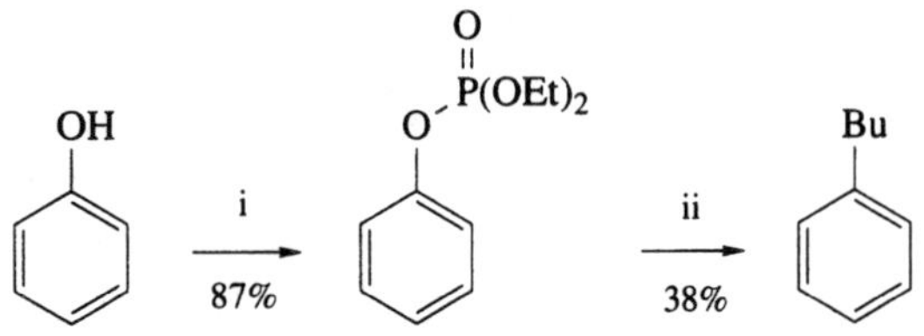

i, $(EtO)_2POCl/NaOH$; ii, $^-CH(CN)(CH_2)_2Me/K/NH_3$

Scheme 3

In contrast with the behavior of the benzene derivatives, the reaction of α-cyanoalkyl (principally cyanomethyl) ions with other aromatic substrates, leads mainly to the expected substitution products, ArCH(R)CN (equation 39). This difference in behavior (not as pronounced in the thiophene system)[132] has been rationalized by molecular orbital considerations.[18,41,133] Examples of these successful syntheses are collected in Table 3.

A useful synthesis of heteroaryl ketones uses the photostimulated α-cyanoalkylation reaction followed by oxidation (Scheme 4). The oxidation of products formed from phenylacetonitrile (R = Ph in Scheme 4) proceeded in over 90%.[110]

The reactions of 2,6-dibromopyridine with the carbanions formed from α-phenylbutyronitrile or α-ethyl-α-phenylbutyronitrile give mixtures of products in which one or both of the bromines have been replaced together with products in which one bromine has been replaced by hydrogen and the other by the carbanion.[104]

The reaction of *p*-iodo- or *p*-bromo-anisole or of 1-iodonaphthalene with the anion formed from the α,β-unsaturated nitrile (**36**) gives 60–70% combined yields of the isomeric nitriles (**37**) and (**38**) together with small amounts of the diarylated derivatives (**39**).[135]

Table 3 Yields of α-Arylated Nitriles Formed in the Photostimulated Reaction of α-Cyanoalkyl Anions with Aryl and Heteroaryl Substrates

ArX	*Nitrile*	*Yield (%)*	*Ref.*
Bromobenzene	MeCN	8[a]	38
Bromobenzene	MeCN	25	39
4-Chlorobiphenyl	MeCN	94	39
4-Chlorobenzophenone	MeCN	97	39
1-Chloronaphthalene	MeCN	89	39
2-Chloronaphthalene	MeCN	93	39
9-Bromophenanthrene	MeCN	70	39
2-Chloropyridine	MeCN	56	39
2-Bromopyridine	$PhCH_2CN$	76, 88	110, 134
3-Bromopyridine	$PhCH_2CN$	48	110
4-Bromopyridine	$PhCH_2CN$	15	110
2-Chloropyrimidine	$PhCH_2CN$	31	110
2-Chloropyrazine	$PhCH_2CN$	78[b]	116
2,4-Dichloropyrimidine	$PhCH_2CN$	58[c]	104
2-Chloroquinoline	$PhCH_2CN$	46, 88	110, 134
3-Bromoquinoline	$PhCH_2CN$	45	110
3-Bromothiophene	MeCN	37[d]	132

[a]Stimulated by potassium metal (see text). [b]In the dark. [c]Only the 4-chlorine is replaced. [d]Also obtained from 2-bromothiophene.

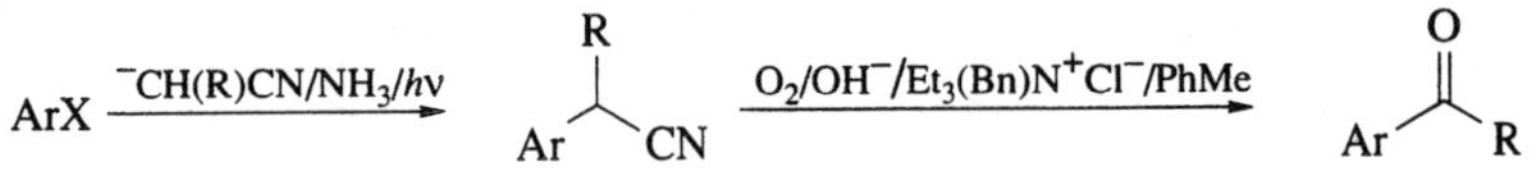

$$\text{ArX} \xrightarrow{^-\text{CH(R)CN/NH}_3/h\nu} \text{ArCH(R)CN} \xrightarrow{\text{O}_2/\text{OH}^-/\text{Et}_3(\text{Bn})\text{N}^+\text{Cl}^-/\text{PhMe}} \text{ArC(O)R}$$

Scheme 4

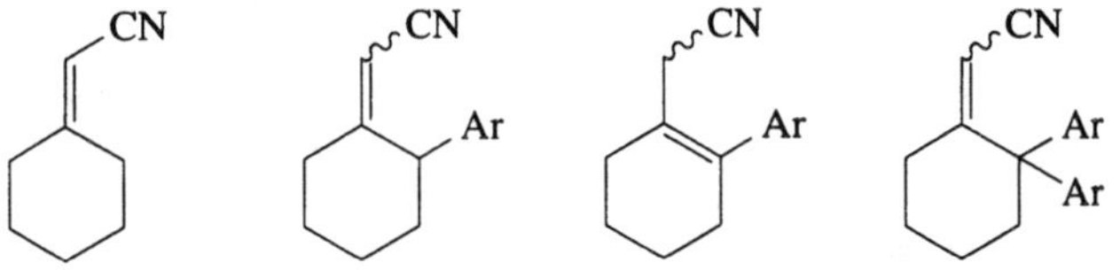

(36) **(37)** **(38)** **(39)**

2.2.3.1.4 Preparation of hydroxy- and amino-biaryls

The study of the reactivity of phenoxide ions with aryl halides under $S_{RN}1$ conditions has undergone dramatic changes since the end of 1987, after a negative and confused beginning. In 1973 products were not isolated in the reaction of phenoxide ion with bromobenzene under alkali metal stimulated conditions,[124] but four years later the claim was made that bromobenzene on reaction with phenoxide ion and sodium amalgam in 50% aqueous Bu^{t}OH was converted into diphenyl ether.[51] In 1979, the latter work was shown to be unreproducible,[52] and additionally, phenoxide ion was shown not to react with 2-chloroquinoline under electrochemical stimulus.[68] In 1980, *p*-cresol was recovered in over 90% yield on attempted photostimulated reaction under basic conditions with iodobenzene, although the iodobenzene was consumed.[45] At the end of 1987, it was clearly demonstrated that 2-, 3- and 4-chlorobenzonitrile reacted in electrochemically induced $S_{RN}1$ reaction with 2,6- and 2,4-di-*t*-butylphenoxide ions, as shown in the representative reaction given in equation (40).[86] Since this report the idea that phenoxides were not good substrates in $S_{RN}1$ reactions has been allayed. The electrochemically induced reaction of 4-bromobenzophenone with phenoxide ion was found to give a 40% yield of benzophenone and 60% of a 1:2 mixture of the *para*-coupled compound (**40**) and the *ortho* isomer (**41**), with no trace of any *O*-arylated compounds,[80,81] and the reaction of phenoxide ion with 4-chlorobenzonitrile gave a similar result.[80] The synthetic utility of these coupling reactions (clearly still in their infancy) leading to unsymmetrical hydroxylated biaryls, is seen in the further example in Scheme 5, wherein removable Bu^t groups are used to control the regiochemistry of coupling which takes place under electrochemical or photochemical stimulation in overall yields near 70%.[136] At the same time as the electrochemically induced reactions were being developed, Beugelmans and coworkers carefully reexamined the photostimulated reaction of a variety of phenoxides and 1- and 2-naphthoxides with aryl halides.[75] They confirmed that arylation of *p*-cresol by simple aryl halides was low yielding, with only 3% of *ortho*-phenylated *p*-cresols being detected, but noted that electron-withdrawing substituents *o*- or *p*- to the nucleofuge raised the yield of hydroxylated biaryls, as did electron-donating groups on the phenoxide ion. The presence of an electron-donating group, such as methoxy, *ortho* to the nucleofuge also had the same activating effect. They also recognized the usefulness of Bu^t groups as both blocking and activating groups on the phenoxide moiety (*cf.* Scheme 5). Examples of these processes are given in equations (41) and (42).[75] In the naphthalene series the 1- and 2-naphthoxides also readily undergo arylation in good yields, the former gives mixtures of 2- and 4-arylated products,[75] and the latter give arylation solely in the 2-position.[75,137] The increased activity of the naphthoxides is shown by the fact that substituent effects on the aryl halide are not as critical; 2-naphthoxide ion gives 48% and 85% 1-arylation with *p*-iodoanisole[137] and *p*-bromobenzonitrile[75] respectively, despite the large difference in electronic properties of the methoxy and cyano groups. By a simple extension of this reaction, reasonable yields of unsymmetrical 1,1′-,[137,138] 1,2′-[138] and 2,2′-binaphthyl[138] derivatives are also available in the one-step coupling of naphthoxides and iodonaphthalenes. The anions from *p*-cresol, *p*-methoxyphenol and 2,4-dimethoxyphenol were readily arylated by 5-chloro-7-iodo-8-isopropoxyquinoline in the vacant *o*- or *p*-positions with only the 7-iodine being replaced.[97] It can be seen, that in the elapse of little more than one year, the $S_{RN}1$ reaction of phenoxides with aryl halides (with judicious choice of substrates) has been developed as an excellent route to hydroxybiaryls.

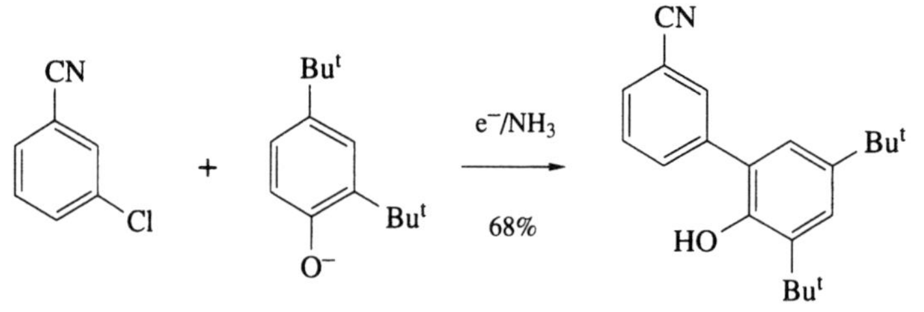

CN Bu^t Cl Bu^t O$^-$ + — e$^-$/NH$_3$, 68% → CN Bu^t HO Bu^t (40)

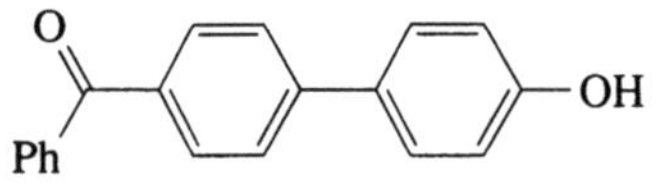

(40) 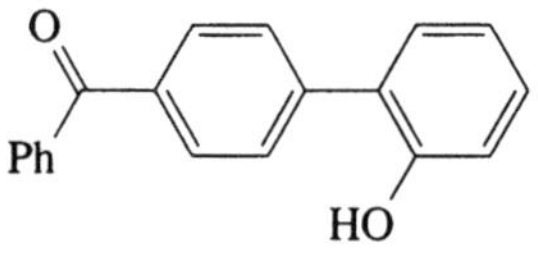 **(41)**

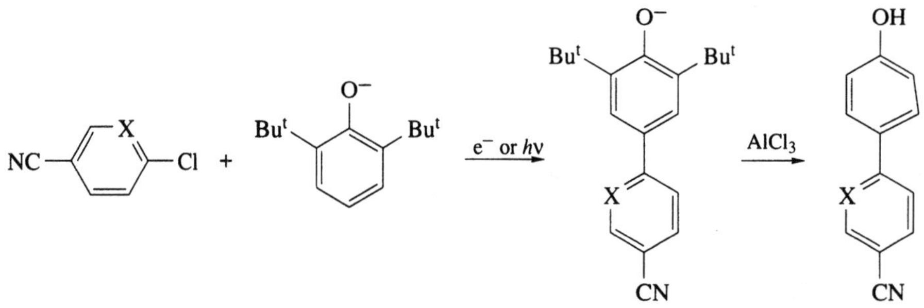

NC Cl X + O$^-$ Bu^t Bu^t — e$^-$ or $h\nu$ → O$^-$ Bu^t Bu^t X CN — $AlCl_3$ → OH X CN

Scheme 5

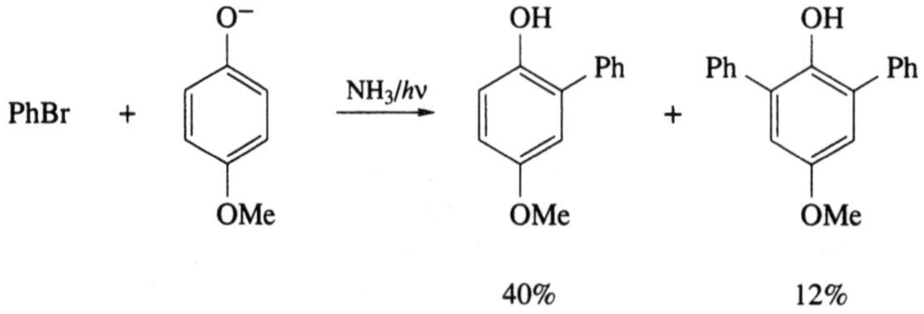

PhBr + O$^-$ OMe — $NH_3/h\nu$ → OH Ph OMe (40%) + OH Ph Ph OMe (12%) (41)

The study of the arylation of arylamide ions prior to 1987, in similar fashion to that of the analogous reaction with aryloxides discussed above, gave little promise of synthetically useful reactions. In Kim and Bunnett's original papers on the $S_{RN}1$ reaction,[1] they reported that the anilide ion ($PhNH^-$) behaves

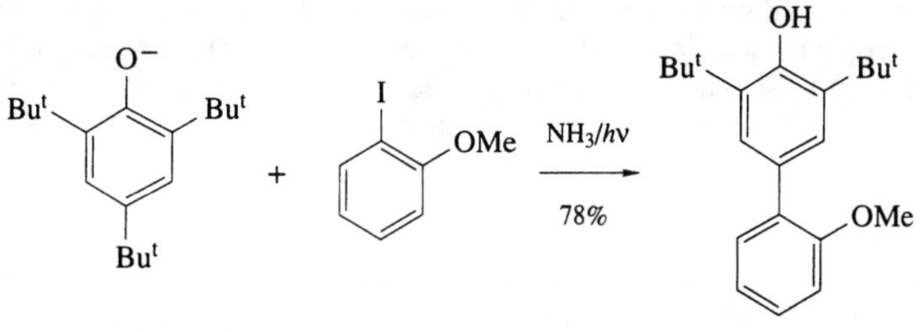

(42)

in ambident fashion and gives poor yields of diphenylamine (19%) and *o*- and *p*-aminobiphenyl (11% each). Rossi and coworkers have opened up another potentially useful biaryl synthesis by examining the reactions of the anion of 2-naphthylamine, which they believed should have a lower ionization potential to initiate the photostimulated reaction than the $PhNH^-$ anion, with aryl iodides and bromides. Indeed, arylation of the arylamide ion, generated from 2-naphthylamine by treatment with NH_2^- or Bu^tO^- in liquid ammonia, takes place at the 1-position of the naphthalene ring in 45–63% yields with iodoarenes and to a lower extent with bromoarenes. *N*-Arylation occurs to a maximum extent of only 8% with the substrates studied.[109] An example of this new aminobiaryl synthesis, clearly still in the developmental stage, is shown in equation (43).

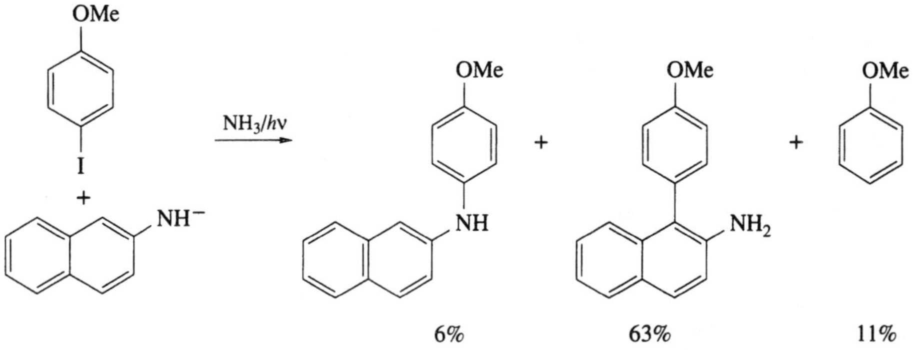

(43)

2.2.3.1.5 Preparation of arenecarbonitriles

Despite being the nucleophile in the electrochemically induced $S_{RN}1$-like process with *p*-iodonitrobenzene,[27] which predates the discovery of the aromatic $S_{RN}1$ reaction,[1] the use of cyanide ion has been almost totally neglected, except for electrochemical studies, which principally revealed that reactions of haloarenes with cyanide ion either do not take place with, for example, normally reactive substrates such as 2-chloroquinoline,[68] are 'not entirely catalytic'[78,139] or are 'noncatalytic' substitution processes.[140,141] More recent electrochemical studies have shown that coupling between various aryl radicals and cyanide is significantly slower than that with PhS^-, $^-CH_2COMe$ and $(EtO)_2PO^-$.[31] This factor, combined with the stability of the product of the association step (equation 3), namely $ArCN^{-\cdot}$, relative to that of the radical anion of the substrate, $ArX^{-\cdot}$, which makes the chain propagation step in equation (4) unfavorable, presumably explains the necessity for more than catalytic amounts of current, even in those reactions which do take place. In what appears to be the only reference to reactions of cyanide with aryl halides under nonelectrochemical conditions (see p. 42 in ref. 18), only failed reactions were reported. The photochemically or electrochemically stimulated reaction of aryldiazo phenyl sulfides, already discussed in Section 2.2.2.3 (see Scheme 2), gives moderate to good yields of arenecarbonitriles, presumably on account of the propagating step (equation 4) becoming thermodynamically favorable.[62]

2.2.3.1.6 Miscellaneous reactions with carbanions

Very little work has been done on reactions involving nucleophiles formed from hydrocarbons.[124,142] The limitation on basicity of the carbanion, so that it does not react with solvent, has led to use of conjugated hydrocarbons, such as dienes or alkenes conjugated with aromatic rings. When initiated by dissolving alkali metal in liquid ammonia, complex mixtures are often produced on account of reduction processes,[124] and regiochemistry and multiplicity of arylation in conjugated systems also create problems.[142] The photostimulated phenylation of indene anion in DMSO by bromobenzene leads to mixtures of mono-, di- and tri-phenylated indenes[124,142] and similar results are obtained with fluorene and anethole.[124] Phenylation of 1,3-pentadiene, accomplished by potassium-induced reaction with bromobenzene in liquid ammonia, followed by catalytic reduction of all aliphatic double bonds, does give a useful yield of 1-phenylpentane (74%), albeit contaminated with diphenylpentanes (7%) and triphenylpentanes (3%).

The 2- and 4-picolyl anions are phenylated or mesitylated on reaction with chlorobenzene, phenyltrimethylammonium ion and 2-bromomesitylene under stimulation by light or potassium metal. The mechanism of reaction with bromobenzene and iodobenzene is not certain, with the aryne mechanism almost certainly intruding, and with iodobenzene some diarylation of the picolinyl anion results. The reaction of the 2-picolyl anion with 2-bromomesitylene, where an aryne process is impossible, is shown in equation (44).[60] Similar reactions take place between the 4-picolyl anion and 2- or 4-bromopyridine or 2-chloroquinoline.[134]

Br + $H_2\bar{C}$ N $\xrightarrow[87\%]{NH_3/h\nu}$ N (44)

Although the anions from nitroalkanes have been shown by ESR[36] and electrochemical studies[29] to be very good traps for aryl radicals, the dissociation of the first-formed radical anions, *e.g.* as shown in equation (45),[29] produces nonchain-propagating radicals, accounting for the lack of photostimulated or metal-promoted examples of this reaction type.[46] The electrochemical reaction (two electrons per molecule) of iodobenzene with the anion from 2-nitropropane gives only 29% of isopropylbenzene and the corresponding reaction with 4-bromobenzophenone gives 4-isopropylbenzophenone in 50% yield.[29]

$$(ArCMe_2NO_2)^{\bullet -} \longrightarrow ArCMe_2^{\bullet} + NO_2^- \qquad (45)$$

Dimsyl anion fails to undergo $S_{RN}1$ reactions with aryl halides,[134,143] and the occurrence of fragmentation processes similar to those in equations (11) and (45) have been used to explain this failure,[143] correcting the earlier, falacious claim that dimsyl anion and halobenzenes react to give benzyl methyl sulfoxide in high yield.[144]

Acetylide ions have also been reported as unreactive in $S_{RN}1$ reactions with aryl[45] and heteroaryl[134] halides.

2.2.3.2 Formation of Aryl C—N, C—P, C—As and C—Sb Bonds

2.2.3.2.1 Preparation of arylamines

The advantage of the radical process, that substitution of nucleofuges by NH_2^- on an aromatic ring takes place without *cine* substitution (by the aryne mechanism), was clearly seen in the first aromatic $S_{RN}1$ reactions recognized.[1] The same regiochemical advantage is seen in the potassium-stimulated conversion of *o*-haloanisoles predominantly into *o*-anisidine rather than the *m*-isomer (the sole product of the aryne process). Also, unlike substitutions proceeding by the aryne mechanism, reactions readily take place when both *ortho* positions are blocked (equation 46),[1] and 2-bromomesitylene reacted with NH_2^- under photostimulation to give 2,4,6-trimethylaniline in 70% yield.[34]

I $\xrightarrow[64\%]{NH_2^-/K/NH_3}$ NH_2 (46)

Despite the abovementioned advantages, the only application of this useful synthetic procedure appears to be the otherwise difficult conversion of phenols into amines *via* the aryl diethyl phosphates in overall yields in excess of 55% (*cf.* Scheme 3).[56] Even this potentially useful reaction does not appear to have been used since the original report. The reaction of amide ion with bromothiophenes is complicated by rearrangement of the 2- to the 3-bromo isomer but 3-thiophenamine is the ultimate product in yields up to 79%.[121]

The reactions of the ambident $ArNH^-$ ions, which do not *N*-arylate to a great extent, are discussed in Section 2.2.3.1.4. Based on the negative results obtained with *N*-acyl anions such as $AcNH^-$ and $AcNMe^-$,[68] and the anion of phthalimide,[134] it would appear that these anions do not undergo *N*-arylation under $S_{RN}1$ conditions.

2.2.3.2.2 *Preparation of triarylphosphines*

Triarylphosphines were prepared by the reaction between lithium diphenylphosphide in THF and *m*- and *p*-iodotoluene (or the corresponding bromo compounds), 4-bromobiphenyl and *p*-dibromobenzene in yields of 70–80% (isolated after oxidation, as the phosphine oxides).[145] The absence of *cine* substitution products is a synthetic advantage and would have been taken as a *prima facie* indication that the displacements are examples of the $S_{RN}1$ reaction, had the mechanism been recognized at the time. Operation of the radical ion mechanism in DMSO, or liquid ammonia, in which marginally improved yields are obtained, was confirmed by Swartz and Bunnett,[48] but no extension to the scope of the reaction was made. Rossi and coworkers have developed a procedure for 'one-pot' preparation of triarylphosphines starting from elemental phosphorus (Scheme 6).[146] As an example of the synthesis of a symmetrical triarylphosphine, triphenylphosphine (isolated as its oxide) was obtained in 75% yield, with iodobenzene as the aryl halide (ArX in Scheme 6, steps i–iii only). Unsymmetrical phosphines result from the full sequence of reactions in Scheme 6, and *p*-anisyldiphenylphosphine (isolated as its oxide) was produced in 55% yield, based on the phosphorus used, when chlorobenzene (ArX) and *p*-methoxyanisole (Ar'X) were used.

$$P \xrightarrow{\text{i, ii}} P^{3-} \xrightarrow{\text{iii}} Ar_3P \xrightarrow{\text{i, ii}} Ar_2P^- \xrightarrow{\text{iv}} Ar_2PAr'$$

i, Na/NH_3; ii, Bu^tOH; iii, $ArX/h\nu$; iv, $Ar'X/h\nu$

Scheme 6

2.2.3.2.3 *Preparation of dialkyl arylphosphonates*

The greatest number of preparative (and mechanistic) studies with phosphorus-based nucleophiles have been carried out with the dialkyl phosphite ions. These reactions generally proceed in good to high yield in either liquid ammonia, THF or DMSO under photostimulation, and representative examples are collected in Table 4. A detailed experimental procedure for the preparation of diethyl phenylphosphonate by this method is available.[147] The reactions generally require photostimulation and the quantum yields have been evaluated.[21,22] In the absence of photostimulation, reactions are slower or do not proceed at all, and in the particular case of *o*-haloiodobenzenes, ionic processes occur to the virtual exclusion of the substitution reaction.[89] With the notable exception of aromatic rings bearing nitro groups, where the $S_{RN}1$ reaction failed, superior results were obtained using the photostimulated reaction in THF, than by using a variety of metal-catalyzed processes.[77] Electrochemically stimulated reactions with $(EtO)_2PO^-$ proceed with remarkably low current input and the reaction with *p*-chlorobenzonitrile proceeds with 100% conversion in 10 min with a current input of only 0.01 electron per molecule of substrate.[139]

2.2.3.2.4 *Miscellaneous reactions with phosphanions*

In addition to the dialkyl phosphite ions, discussed in the previous section, a number of other phosphanions readily undergo photostimulated *P*-arylation in liquid ammonia. The yields for these reactions, which have to date only been studied for *P*-phenylations are collected in Table 5.[23] Iodobenzene is more reactive than bromobenzene, which fails to react with the $Ph_2P—O^-$ ion.[23]

Table 4 Yields of Dialkyl Arylphosphonates Formed in the Photostimulated Reaction of Dialkyl Phosphite Ions with Aryl Halides: $ArX + (RO)_2PO^- \rightarrow (RO)_2P(=O)Ar$[a]

ArX	*Yield (%)*	*Ref.*	*ArX*	*Yield (%)*	*Ref.*
Iodobenzene	93[b]	90, 147	*m*-Chloroiodobenzene	91	92
Iodobenzene	88[c]	90	*p*-Chloroiodobenzene	59[d]	92
Iodobenzene	96	90	*m*-Bromoiodobenzene	87[d]	90
p-Iodotoluene	95	90	*p*-Bromoiodobenzene	56[d]	92
p-Iodoanisole	95	90	*m*-Diiodobenzene	94[d]	90
o-Iodoaniline	87	77	*p*-Diiodobenzene	87[d]	90
m-Iodoaniline	90	77	1-Iodonaphthalene	93	90
p-Iodoaniline	90	77	2-Iodopyridine	78	77
2-Iodo-*m*-xylene	87	90	3-Iodopyridine	70	77
m-Iodobenzotrifluoride	95	90	2-Methoxy-3-iodopyridine	78	115
m-Fluoroiodobenzene	96	90	3-Bromoquinoline	76	77
p-Fluoroiodobenzene	91	92	5-Cl-7-I-8-Pr^{i}O-Quinoline	70	95

[a]Unless otherwise stated, the alkyl group is ethyl; *i.e.* R = Et. [b]R = Me. [c]R = Bu. [d]Both halogens are replaced.

Table 5 Yields of *P*-Phenylated Phosphorus Derivatives Formed in the Photostimulated Reaction of Various Phosphanions with Phenyl Halides in Liquid Ammonia[a]

Halobenzene	*Phosphanion*[b]	*Product*	*Yield (%)*
Iodobenzene	$PhP(OBu)O^-$	$Ph_2P(=O)OBu$	95
Bromobenzene	$PhP(OBu)O^-$	$Ph_2P(=O)OBu$	72
Iodobenzene	$Ph_2P—O^-$	$Ph_3P=O$	95
Iodobenzene	$(EtO)_2P—S^-$	$PhP(=S)(OEt)_2$	95
Iodobenzene	$(Me_2N)_2P—O^-$	$PhP(=O)(NMe_2)_2$	64
Bromobenzene	$(Me_2N)_2P—O^-$	$PhP(=O)(NMe_2)_2$	65

[a]Taken from ref. 23. [b]The counterion in all reactions was K^+.

2.2.3.2.5 *Reactions with arsenide and stibide ions*

The earlier literature concerning the reaction of arsenide ions with aryl halides has been summarized by Rossi and coworkers in their paper investigating the $S_{RN}1$ reaction between Ph_2As^- and several haloarenes.[24] In this and subsequent work,[25] it was found that scrambling of aryl groups took place in the reactions of Ph_2As^- with all the substrates examined, except for 4-chlorobenzophenone (quantitative yield of substitution product)[24] and 2-chloroquinoline (60% yield).[25] Consequently, the use of the Ph_2As^- ion and also of the Ph_2Sb^- ion which gives gives scrambling of the aryl groups with all the substrates so far examined,[25] are not generally satisfactory for selective preparation of unsymmetrical triarylarsines or triarylstibines.

The 'one pot' procedure used for conversion of elemental phosphorus into triarylphosphines (see Scheme 6), with appropriate modification (*i.e.* replacement of P by As or Sb), has been used to prepare triphenylarsine (75%), 2-quinolyldiphenylarsine (90%) and triphenylstibine (45%).[146]

2.2.3.3 Formation of Aryl C—O, C—S, C—Se and C—Te Bonds

Alkoxide and aryloxide ions are not arylated on oxygen. Tertiary alkoxides do not react, thus making them appropriate bases for generation of nucleophiles, and primary and secondary alkoxides undergo α-hydrogen abstraction with aryl radicals normally making them unsuitable for use as bases; *e.g.* isopropoxide ion reacts with phenyl radicals to give benzene and acetone ketyl.[58] The *C*-arylation of aryloxides is discussed in Section 2.2.3.1.4.

2.2.3.3.1 *Preparation of diaryl and aryl alkyl sulfides*

The photostimulated reaction between aryl halides and arenethiolates gives diaryl sulfides (equation 47) and examples are given in Table 6. The yields obtained from substituted iodobenzenes are generally high but reactions with heteroaryl halides are less prominent and, for example, normally reactive 2-chloroquinoline fails to react with PhS^- under photostimulated conditions.[134] The recently reported reactions of 2- and 3-halothiophenes with benzenethiolate indicate that scrambling of the products, through fragmentation processes, led to complex mixtures.[148] The use of *p*-bifunctionalized thiols/halides in the preparation of poly(1,4-phenylene sulfides) has been mentioned in Section 2.2.2.4.2.[99–102]

$$ArX + Ar'S^- \xrightarrow{NH_3/h\nu} ArSAr' + X^- \qquad (47)$$

Table 6 Yields of Diaryl Sulfides Formed in Photostimulated Reaction of Arenethiolates with Aryl and Heteroaryl Halides

ArX	*Thiol*	*Yield (%)*	*Ref.*
Iodobenzene	PhSH	94	149
Iodobenzene	*p*-$MeOC_6H_4SH$	71	61
Bromobenzene	PhSH	23	149
o-Iodoanisole	PhSH	91	149
m-Iodoanisole	PhSH	88	149
p-Iodoanisole	PhSH	76	149
p-Iodoanisole	*p*-$MeOC_6H_4SH$	73	61
o-Iodotoluene	PhSH	68	149
m-Iodotoluene	PhSH	81	149
p-Iodotoluene	PhSH	72	149
m-Fluoroiodobenzene	PhSH	96	149
o-Chloroiodobenzene	PhSH	77[a]	91
m-Chloroiodobenzene	PhSH	91[a]	91
m-Chloroiodobenzene	2-Pyrimidinethiol	100[b]	95
p-Chloroiodobenzene	PhSH	89[a]	91
m-Dibromobenzene	PhSH	92[a]	91
m-Iodobenzotrifluoride	PhSH	71	149
4-I-$C_6H_4N^+Me_3$ I^-	PhSH	95[a]	91
2-Iodo-*m*-xylene	PhSH	19	149
4-Iododiphenyl ether	PhSH	92	149
1-Iodonaphthalene	PhSH	85	149
2-Iodopyridine	PhSH	58[c]	106
2-Bromopyridine	PhSH	21, 65[d]	106, 134
3-Bromopyridine	PhSH	24[c]	106
5-Cl-7-I-8-Pr^{i}O-Quinoline	PhSH	70[e]	95
5-Cl-7-I-8-Pr^{i}O-Quinoline	f	65–100[e]	95

[a]Both nucleofuges are replaced [b]Only the iodine was replaced. [c]In DMF at 80 °C. [d]In HMPA at 80 °C. [e]Only the 7-iodine was replaced. [f]Various heteroaromatic thiols.

Electrochemical studies indicate that the PhS^- ion is much less reactive (with regard to trapping Ar·) than enolate and phosphorus-centered nucleophiles[150] and 2-pyridyl and 2-quinolyl radicals have particularly low reactivity toward thiolate ions.[30] Hydrogen atom abstraction from DMSO (and also from the methyl groups in the Me_4N^+ ion of the supporting electrolyte), becomes a significant side reaction in the reaction of bromobenzene with PhS^-.[151]

The reaction of aryl halides with alkanethiolate ions is complicated by the problem of fragmentation of the first-formed radical ions (equation 48). The reaction between iodobenzene and ethanethiolate gives only 30% of ethyl phenyl sulfide and 47% of products derived from the benzenethiolate ion.[61] Fragmentation decreases in importance as the stability of the species $ArSR^{-}\cdot$ increases, but is more significant for the more stable aliphatic radicals (*e.g.* when R = $PhCH_2$ or $EtOCOCH_2$ in equation 48).[74,103] As a result, halobenzenes with electron-withdrawing substituents such as cyano or acyl groups,[74,78] or naphthyl halides,[103,111] react to give alkyl aryl sulfides in good yield with simple alkanethiolates or alkanedithiolates[152] with substantially reduced incursion of fragmentation processes (*e.g.* see equation 30). 2-Chloroquinoline and 2-bromopyridine (and its 3-substituted derivatives) behave in similar fashion, thus widening the scope of the reaction significantly.[152] 2-Chloroquinoline, alone among all the aryl halides examined to date, reacts cleanly with $PhCH_2S^-$ to give benzyl 2-quinolyl sulfide,[103] whilst all other substrates give significant, usually major proportions of products resulting from cleavage of the benzyl–S bond. The thiolate, $EtO_2CCH_2S^-$, on reaction with 2-chloroquinoline, however, gives 2-quinolinethiol in 75% yield, with no detectable amount of the expected sulfide.[74] The highly specific fragmentation occurring with this anion could perhaps be used as a method for replacement of halogens on aromatic rings by the thiol group.

$$\text{Ar}\cdot + {}^{-}\text{SR} \longrightarrow \text{ArSR}^{\cdot -} \longrightarrow \text{ArS}^{-} + \text{R}\cdot \quad (48)$$

2.2.3.3.2 Preparation of diaryl selenides and tellurides

The reactions of aryl halides with $PhSe^-$ and $PhTe^-$ ions are complicated by fragmentation of the intermediate radical anions which leads to scrambling of the aryl groups and production of mixtures (see equations 7 and 8).[35] Diphenyl selenide[54] and telluride[54,55] have been prepared in 73% and 90% respectively from iodobenzene, and *p*-iodobromobenzene gives 70% and 40% isolated yields of *p*-bis(phenylselenyl)- and *p*-bis(phenyltelluryl)-benzene respectively.[54] Unsymmetrical phenyl aryl chalcogenides can be prepared when the aryl moiety sufficiently stabilizes the intermediate radical anions, lowering the occurrence of fragmentation processes. 4-Chlorobiphenyl, 1-halonaphthalenes, 9-bromoanthracene and 2-chloroquinoline give 37–98% yields of the phenyl aryl selenides[54,153,154] and 1-bromonaphthalene and 2-chloroquinoline give 53% and 43% yields of the respective aryl phenyl tellurides.[54,55] In all but the reaction of $PhSe^-$ with 2-chloroquinoline, minor amounts of scrambled products are still formed, and prolonged irradiation or use of an excess of the phenyl chalcogenide further reduce the yield.[35] Several useful methods for *in situ* generation of aryl chalcogenide ions from the respective elements, aryl halides and alkali metals in dipolar aprotic solvents[155] or in liquid ammonia[156] (analogous to that used with phosphorus, see Scheme 6) also have been developed and these lead to diaryl chalcogenides and related derivatives in fair to good yields. Very recently electrochemically stimulated reactions with sonication in acetonitrile, between either $PhSe^-$ or $PhTe^-$ and 4-chlorobenzonitrile have been reported to give yields of 57 and 42% of the respective substitution products, and subsequently it was shown that all three bromobenzonitriles, give 36–70% yields with $PhSe^-$.[83] Similar yields have been obtained from electrochemical reactions of the same chlorobenzonitriles with $PhSe^-$ and $PhTe^-$ ions in the presence of redox catalysts,[82] and the same technique has been successfully applied (in 44–86% yields) in the replacement of either bromine or chlorine by PhTe or PhSe groups in *o*-, *m*- and *p*-halobenzophenones.[79] In reactions involving alkyl chalcogenide ions, fragmentation is expected, and accordingly, photostimulated reactions of adamantyl selenide or telluride ions with iodobenzene give most unpromising mixtures.[157]

2.2.4 CYCLIZATIONS UTILIZING THE $S_{RN}1$ REACTION

Cyclization processes taking advantage of the $S_{RN}1$ reaction as one of its steps have been increasingly used in the last decade, and fall into two general categories. The first type have an intramolecular $S_{RN}1$ reaction on an aryl halide by a nucleophile already attached to the aromatic ring, and the second class have an $S_{RN}1$ reaction on *ortho*-functionalized aryl halides followed by cyclization of the initial substitution product. Beugelmans recently summarized the use of both these cyclization sequences for the preparation of heterocyclic systems.[17] The majority of the reactions involve attack on aromatic rings by carbanions, almost exclusively enolate ions. The following sections are divided according to mode of ring closure and the nature of the products.

2.2.4.1 Ring Closure by Intramolecular $S_{RN}1$ Reactions

Formation of cyclized products, resulting from intramolecular attack of the intermediate aryl radical in the $S_{RN}1$ reaction on a second aromatic ring (*i.e.* intramolecular homolytic arylation) are rare and not synthetically useful.[64,158–161]

The first reported intramolecular $S_{RN}1$ reaction by Semmelhack and coworkers,[162,163] was the photostimulated ring closure of (**42**) with Bu^tO^- to give a 93% yield of cephalotaxinone (**43**), the immediate precursor of the antileukemia agent, cephalotoxin. As a consequence of this synthesis, Semmelhack and Bargar, studied the viability and the regiochemistry of the photostimulated ring closure of the series of methyl ketones (**44**; X = Br or I; n = 2, 3, 4 or 6) and their α- or α′-permethylated derivatives in the presence of Bu^tO^- in liquid ammonia.[44,45] It was possible to prepare six-, seven-, eight- and ten-membered rings, but competing β-hydrogen abstraction process interfered in some cases. The presence of α-methyl groups enforced regiochemical control on the reaction and prevented internal hydrogen transfer reactions. The regiochemistry of the ring closure for the ketones (**44**; X = Br; n = 3 or 4) appeared to be controlled by the proportion of the respective enolate ions present, and greater proportions of the cyclic ketones resulting from attack through the methyl group occurred to give the seven- and eight-membered cyclic ketones in preference to the products with five- and six-membered rings, respectively. Cyclization of the ketone (**44**; X = I; n = 2) failed completely on account of β-hydrogen abstraction processes.

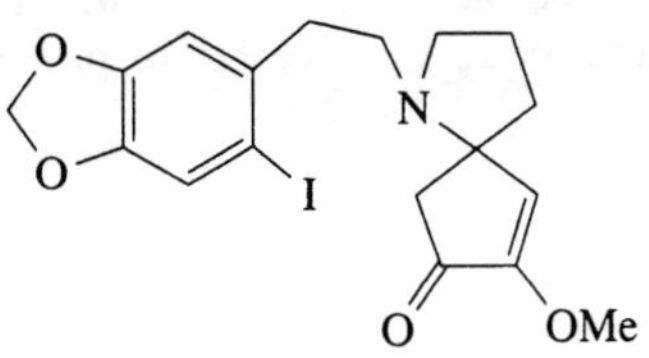

(42) 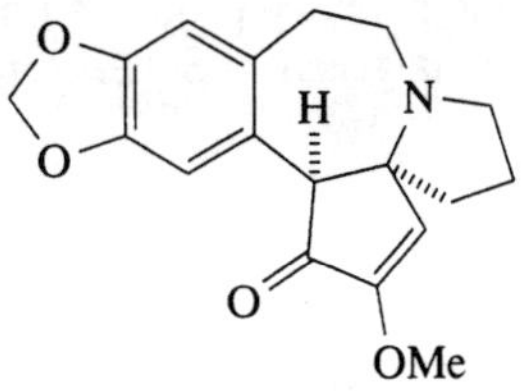 (43) 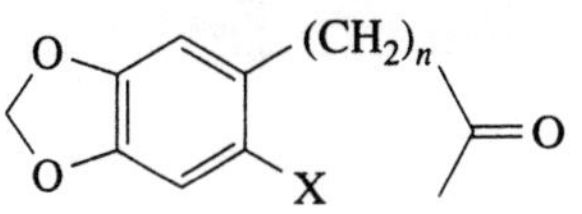 (44)

The photoinduced cyclization of mono- and di-anions of *N*-acyl-*o*-chloroanilines and *N*-acyl-*o*-chlorobenzylamines was developed by Wolfe and coworkers as a general method for the preparation of oxindoles[49,50] and 1,4-dihydro-3(2*H*)-isoquinolinones, respectively.[50] The generalized reaction for the oxindole synthesis is shown in equation (49; Z = CH), and the yields range from 32–83%, with most in excess of 60%. The reaction proceeded readily when substituents were *ortho* to the chlorine (*i.e.* not an aryne mechanism) or elsewhere on the ring, except that a CF_3 or additional Cl substituents caused the reactions to fail. α,β-Unsaturated-*N*-alkylanilides reacted, but only in liquid ammonia with KNH_2 as base, to give 3-alkylideneoxindoles in yields usually over 65%. The method was also extended to the preparation of azaindoles (equation 49; Z = N), but in this case there had to be an alkyl group on the amide nitrogen before successful reactions occurred. The analogous reactions of acylbenzylamines, to give isoquinolone derivatives proceeded in 40–65% yield.

LDA/THF/*hν* (49)

Z = CH, N

o-Iodothiobenzanilide and *o*-iodothioacetanilide undergo ring closure on treatment with Bu^tO^- in DMSO, but only when the reaction is performed with entrainment by $^-CH_2COMe$, to give 2-phenyl- and 2-methyl-1,3-benzothiazoles respectively.[164]

Two further quite different cyclization processes each have an initial intermolecular $S_{RN}1$ reaction followed by a second in which the ring closure is made. The reactions of *o*-dihalobenzenes (**24**) and (**25**) with the dithiolate (**26**) to give (**27**) have already been mentioned (see equation 28), but the yield of analogous product formed between (**26**) and 1-bromo-2-iodonaphthalene is only 24% and use of 1,2-ethanedithiolate with *o*-diiodobenzene gives only 13% of benzo-1,4-dithiane.[98] A rather convenient one-step synthesis of [*m.m*]-*meta*-cyclophanediones (*m* = 3–8) in most acceptable yields is obtained on tandem $S_{RN}1$ reactions of the ketones (**45**; equation 50).[165]

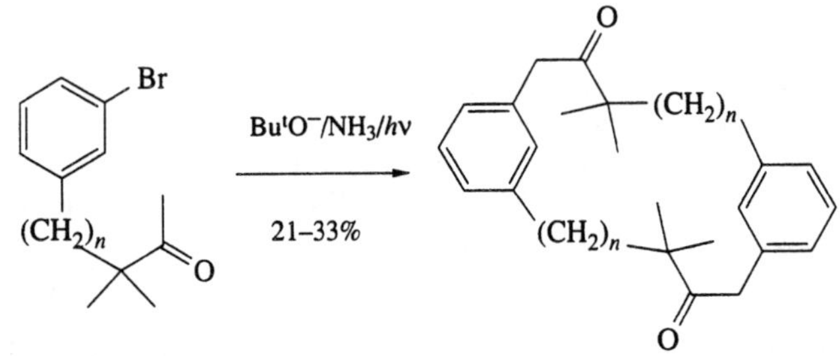

$Bu^tO^-/NH_3/h\nu$, 21–33% (50)

(**45**) *n*= 0–5

2.2.4.2 $S_{RN}1$ Reactions Followed by Cyclization

The only examples of carbocyclic products arising from cyclizations subsequent to $S_{RN}1$ reactions, are the isolated example reported by Bunnett and Singh,[94] which gives products (**18**)–(**20**; Section 2.2.2.4.2) and the recently reported synthesis of unsymmetrically hydroxylated 2,2′-binaphthyls.[114] The final products result from aldol condensation of the primary products under the basic reaction conditions. In the binaphthyl synthesis, an example of which is given in Scheme 7, the sequence of reactions proceeded in 59–83% overall yield.[114]

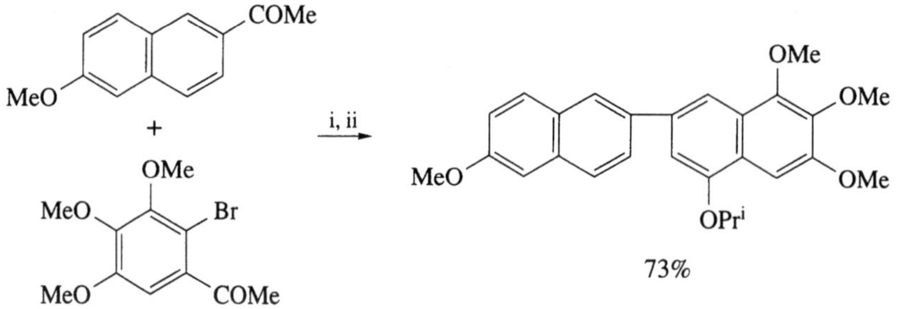

MeO-naphthyl-COMe + 2-bromo-3,4,5-trimethoxyacetophenone →(i, ii) binaphthyl product (OPr^i), 73%

i, Bu^{t}OK/DMSO/$h\nu$; ii, 2-bromopropane

Scheme 7

The first of many heterocyclization reactions, depending on an initial *S*RN1 reaction between the enolate of a carbonyl compound and an *o*-substituted haloarene, followed by cyclization, is represented by the formation of indoles from *o*-haloanilines and enolate ions, first reported by Beugelmans and Roussi,[166] and shortly afterwards by Bard and Bunnett.[76] This reaction (Scheme 8; Z = CH; X = Br or I) was used not only with the enolates of a wide range of ketones,[73,76,166,167] but also with enolates derived from aldehydes.[73,166,167] Symmetrical ketone enolates gave single products, and methyl alkyl ketones with branching at the α-carbon gave good yields of the 2-substituted indole. 2-Butanone gave mixtures of 2,3-dimethyl- and 2-ethyl-indole, with the former predominating. Aldehyde enolates (R^1 = H) led regiospecifically to 3-alkylindoles but, under these photostimulated reaction conditions, appreciable amounts of dehalogenated products were formed.[73] This competing process was virtually eliminated under electrochemical conditions.[167]

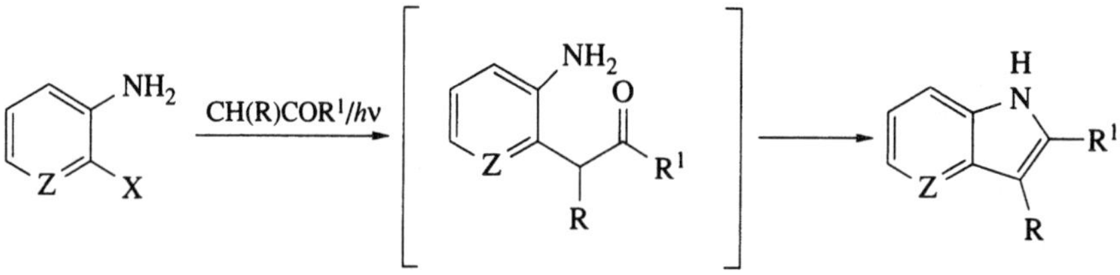

CH(R)COR^1/$h\nu$

Z = CH, N; R, R^1 = H, alkyl; X = I, Br, Cl

Scheme 8

4-Azaindoles (1*H*-pyrrolo[3,2-*b*]pyridines) were prepared by an entirely analogous method with that developed for indoles, starting with 3-amino-2-chloropyridine, and the same diversity of reacting enolates and similar regioselectivity effects were observed (Scheme 8; Z = N; X = Cl).[76,125,168] 1,2-Disubstituted pyrrolo[2,3-*b*]pyridines were also prepared by a related process, in which 2-fluoropyridine was first selectively *o*-lithiated and then iodinated to give 2-fluoro-3-iodopyridine. Preparation of 2-amino derivatives by *S*NAr replacement of the 2-fluorine gave the required substrates for the *S*RN1 ring closure sequence. Using similar strategies, 2-alkylpyrrolo[2,3-*b*]-, -[2,3-*c*]- and -[3,2-*c*]-pyridines were also prepared.[115]

The fusion of a furan ring onto isolated benzene rings, the benzene ring of quinolines, or pyridine rings, is readily brought about in good yield by reaction of *o*-halo ethers with ketone enolates followed by ether cleavage and furan ring formation, leading to benzo[*b*]furans,[71] furo[3,2-*h*]quinolines or furo[3,2-*b*]pyridines.[96] Examples of these syntheses are given in Scheme 9.

The isoquinoline system is conveniently prepared from treatment of *o*-iodobenzylamines with the enolate ions derived from symmetrical ketones (or ketones with one α-position blocked), aldehydes, or the dimethyl acetal of pyruvaldehyde, to give aminocarbonyl compounds which condensed *in situ* to give 2- and/or 3-substituted 1,2-dihydroisoquinolines. Catalytic dehydrogenation or borohydride reduction of these products then led to the corresponding isoquinolines or tetrahydroisoquinolines in moderate to high

i, MeCOR/KOBu^t/NH$_3$/$h\nu$; ii, Me$_3$SiCl/NaI; iii, 48% HBr/AcOH/100 °C

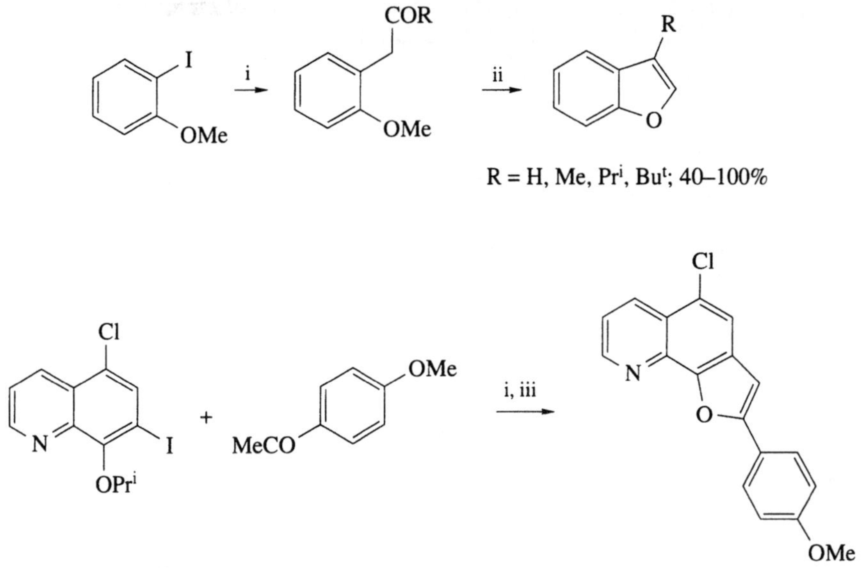

R = H, Me, Pr^i, Bu^t; 40–100%

Scheme 9

yields.[169,170] Natural products (and analogous compounds) with the benzo[*c*]phenanthridine skeleton, were prepared through this route by using substituted *o*-iodobenzylamines and substituted α-tetralones in the initial $S_{RN}1$ reaction, followed by oxidation and catalytic dehydrogenation of the subsequent condensation product.[171] The isoquinoline system, in the form of 1,2-dihydroisoquinolones is also readily accessible through the $S_{RN}1$ reaction of *o*-halobenzamides (and their *N*-alkyl derivatives) with the enolate ions of ketones, including those from acetophenones, followed by cyclization and dehydration (usually spontaneous) of the primary product, as outlined in Scheme 10.[123,172] An alternate route through isocoumarins is also possible. Treatment of *o*-halobenzoic acids with ketone enolates first gives $S_{RN}1$ products which can be isolated as esters,[70] or can also be cyclized on treatment with acid to give isocoumarins. Subsequent treatment of the isocoumarin with a primary amine leads to dihydroisoquinolones,[172] or when α-tetralones are used as the carbonyl component and catalytic dehydrogenation of the final cyclization product is carried out, the final products are benzo[*c*]phenanthridones.[171]

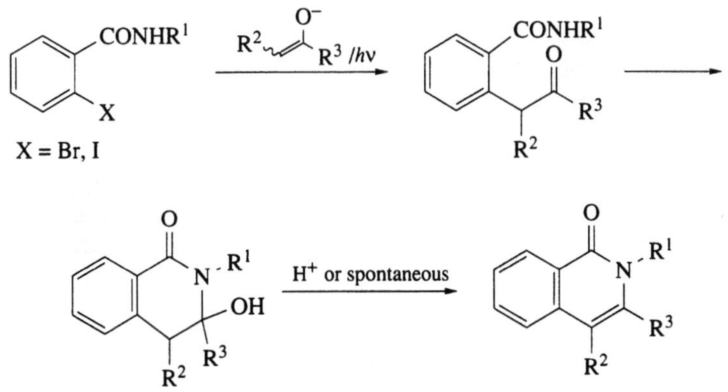

X = Br, I

H$^+$ or spontaneous

Scheme 10

The halogens in *o*-bromo- or *o*-iodo-veratric acids are replaced by ketone enolates to give primary products which on lactonization or lactamization lead to benzazepines or benzoxepines respectively.[126] There would appear to be no reason why this reaction could not be extended to reactions of *o*-halophenylacetic acids in general.

In one of the few examples not involving enolate ions, the substitution of Br in *o*-bromo-benzaldehyde or -acetophenone by $^-SCH_2CO_2Et$ ion followed by Claisen condensation of the resulting sulfides, **(46)** and **(47)**, under the reaction conditions ultimately gave substituted benzo[*b*]thiophenes **(48)** and **(49)** in

55 and 40% yields respectively. The relatively low yields are a result of competing radical anion fragmentation processes which led to the thiols (**50**; 10%) and (**51**; 48%).[74]

(**46**) R = H
(**47**) R = Me

(**48**) R = H
(**49**) R = Me

(**50**) R = H
(**51**) R = Me

2.2.5 REFERENCES

1. J. F. Bunnett and J. K. Kim, *J. Am. Chem. Soc.*, 1970, **92**, 7463, 7464.
2. N. Kornblum, R. E. Michel and R. C. Kerber, *J. Am. Chem. Soc.*, 1966, **88**, 5662.
3. G. A. Russell and W. C. Danen, *J. Am. Chem. Soc.*, 1966, **88**, 5663.
4. J. F. Bunnett, *Acc. Chem. Res.*, 1978, **11**, 413.
5. G. A. Russell, in 'Chemical Society Special Publication No. 24', The Chemical Society, London, 1970, p. 271.
6. N. Kornblum, in 'The Chemistry of the Functional Groups, Supplement F', ed. S. Patai, Wiley, Chichester, 1982, chap. 10.
7. I. P. Beletskaya and V. N. Drozd, *Russ. Chem. Rev. (Engl. Transl.)*, 1979, **48**, 431.
8. G. Simig, *Magy. Kem. Lapja*, 1980, 35, 347 (*Chem. Abstr.*, 1981, **94**, 64 649).
9. N. M. Alvarez, *Rev. Soc. Quim. Mex.*, 1983, **27**, 318 (*Chem. Abstr.*, 1984, **101**, 190 663).
10. R. K. Norris, in 'The Chemistry of the Functional Groups, Supplement D', ed. S. Patai and Z. Rappoport, Wiley, Chichester, 1983, chap. 16, p. 681.
11. Q. Chen, *Huaxue Tongbao*, 1984, **197**, 165 (*Chem. Abstr.*, 1984, **101**, 89 914).
12. C. Xia and Z. Chen, *Huaxue Tongbao*, 1986, 20 (*Chem. Abstr.*, 1987, **106**, 137 665).
13. J.-M. Savéant, *Acc. Chem. Res.*, 1980, **13**, 323.
14. L. E. Eberson, *J. Mol. Cat.*, 1983, **20**, 27.
15. J. Prousek, *Pokroky Chemie*, 1988, **19**, 1.
16. J. F. Wolfe and D. R. Carver, *Org. Prep. Proced. Int.*, 1978, **10**, 227.
17. R. Beugelmans, *Bull. Soc. Chim. Belg.*, 1984, **93**, 547.
18. R. A. Rossi and R. H. de Rossi, *ACS Monogr.*, 1983, **178**, 1.
19. J. J. Brunet, C. Sidot and P. Caubere, *J. Org. Chem.*, 1983, **48**, 1166; *Tetrahedron Lett.*, 1981, **22**, 1013; *J. Organomet. Chem.*, 1981, **204**, 229.
20. F. Ciminale, G. Bruno, L. Testaferri, M. Tiecco and G. Martelli, *J. Org. Chem.*, 1978, **43**, 4509.
21. S. Hoz and J. F. Bunnett, *J. Am. Chem. Soc.*, 1977, **99**, 4690.
22. M. A. Fox, J. Younathan and G. E. Fryxell, *J. Org. Chem.*, 1983, **48**, 3109.
23. J. E. Swartz and J. F. Bunnett, *J. Org. Chem.*, 1979, **44**, 4673.
24. R. A. Rossi, R. A. Ruben and S. M. Palacios, *J. Org. Chem.*, 1981, **46**, 2498.
25. R. A. Alonso and R. A. Rossi, *J. Org. Chem.*, 1982, **47**, 77.
26. R. G. Scamehorn and J. F. Bunnett, *J. Org. Chem.*, 1977, **42**, 1449.
27. D. E. Bartak, W. C. Danen and M. D. Hawley, *J. Org. Chem.*, 1970, **35**, 1206.
28. J. Pinson and J.-M. Savéant, *J. Chem. Soc., Chem. Commun.*, 1974, 933.
29. C. Amatore, M. Gareil, M. A. Oturan, J. Pinson, J.-M. Savéant and A. Thiébault, *J. Org. Chem.*, 1986, **51**, 3757.
30. C. Amatore, M. A. Oturan, J. Pinson, J.-M. Savéant and A. Thiébault, *J. Am. Chem. Soc.*, 1985, **107**, 3451.
31. C. Amatore, C. Combellas, S. Robveille, J.-M. Savéant and A. Thiébault, *J. Am. Chem. Soc.*, 1986, **108**, 4754.
32. C. Galli and J. F. Bunnett, *J. Am. Chem. Soc.*, 1979, **101**, 6137.
33. C. Galli and J. F. Bunnett, *J. Am. Chem. Soc.*, 1981, **103**, 7140.
34. R. A. Alonso, A. Bardon and R. A. Rossi, *J. Org. Chem.*, 1984, **49**, 3584.
35. A. B. Pierini, A. B. Penenory and R. A. Rossi, *J. Org. Chem.*, 1984, **49**, 486.
36. G. A. Russell and A. R. Metcalfe, *J. Am. Chem. Soc.*, 1979, **101**, 2359.
37. A. L. J. Beckwith and R. O. C. Norman, *J. Chem. Soc. B*, 1969, 403.
38. J. F. Bunnett and B. F. Gloor, *J. Org. Chem.*, 1973, **38**, 4156.
39. R. A. Rossi, R. H. De Rossi and A. F. Lopez, *J. Org. Chem.*, 1976, **41**, 3371.
40. R. A. Rossi, R. H. De Rossi and A. B. Pierini, *J. Org. Chem.*, 1979, **44**, 2662.
41. R. A. Rossi, *Acc. Chem. Res.*, 1982, **15**, 164.
42. B. Helgee and V. D. Parker, *Acta Chem. Scand., Ser. B*, 1980, **32**, 129.
43. J. F. Wolfe, M. P. Moon, M. C. Sleevi, J. F. Bunnett and R. R. Bard, *J. Org. Chem.*, 1978, **43**, 1019.
44. M. F. Semmelhack and T. M. Bargar, *J. Org. Chem.*, 1977, **42**, 1481.
45. M. F. Semmelhack and T. M. Bargar, *J. Am. Chem. Soc.*, 1980, **102**, 7765.
46. J. F. Bunnett, R. G. Scamehorn and R. P. Traber, *J. Org. Chem.*, 1976, **41**, 3677.
47. M. P. Moon and J. F. Wolfe, *J. Org. Chem.*, 1979, **44**, 4081.
48. J. E. Swartz and J. F. Bunnett, *J. Org. Chem.*, 1979, **44**, 340.
49. J. F. Wolfe, M. C. Sleevi and R. R. Goehring, *J. Am. Chem. Soc.*, 1980, **102**, 3646.

50. R. R. Goehring, Y. P. Sachdeva, J. S. Pisipati, M. C. Sleevi and J. W. Wolfe, *J. Am. Chem. Soc.*, 1985, **107**, 435.
51. S. Rajan and P. Sridaran, *Tetrahedron Lett.*, 1977, 2177.
52. R. A. Rossi and A. B. Pierini, *J. Org. Chem.*, 1980, **45**, 2914.
53. J. T. Zoltewicz and T. M. Oestreich, *J. Am. Chem. Soc.*, 1973, **95**, 6863.
54. A. B. Pierini and R. A. Rossi, *J. Org. Chem.*, 1979, **44**, 4667.
55. A. B. Pierini and R. A. Rossi, *J. Organomet. Chem.*, 1979, **168**, 163.
56. R. A. Rossi and J. F. Bunnett, *J. Am. Chem. Soc.*, 1972, **94**, 683.
57. R. A. Rossi and J. F. Bunnett, *J. Am. Chem. Soc.*, 1974, **96**, 112.
58. R. A. Rossi and J. F. Bunnett, *J. Org. Chem.*, 1973, **38**, 1407.
59. R. A. Rossi and J. F. Bunnett, *J. Org. Chem.*, 1972, **37**, 3570.
60. J. F. Bunnett and B. F. Gloor, *J. Org. Chem.*, 1974, **39**, 382.
61. J. F. Bunnett and X. Creary, *J. Org. Chem.*, 1975, **40**, 3740.
62. M. Novi, G. Petrillo and C. Dell'Erba, *Tetrahedron Lett.*, 1987, **28**, 1345.
63. G. Petrillo, M. Novi, G. Garbarino and C. Dell'Erba, *Tetrahedron Lett.*, 1985, **26**, 6365.
64. G. Petrillo, M. Novi, G. Garbarino and C. Dell'Erba, *Tetrahedron*, 1986, **42**, 4007.
65. A. P. Komin and J. F. Wolfe, *J. Org. Chem.*, 1977, **42**, 2481.
66. M. J. Tremelling and J. F. Bunnett, *J. Am. Chem. Soc.*, 1980, **102**, 7375.
67. R. R. Bard, J. F. Bunnett, X. Creary and M. J. Tremelling, *J. Am. Chem. Soc.*, 1980, **102**, 2852.
68. C. Amatore, J. Chaussard, J. Pinson, J.-M. Savéant and A. Thiébault, *J. Am. Chem. Soc.*, 1979, **101**, 6012.
69. J. F. Bunnett, E. Mitchel and C. Galli, *Tetrahedron*, 1985, **41**, 4119.
70. J. F. Bunnett and J. E. Sundberg, *Chem. Pharm. Bull.*, 1975, **23**, 2620.
71. R. Beugelmans and H. Ginsburg, *J. Chem. Soc., Chem. Commun.*, 1980, 508.
72. R. Beugelmans, M. Bois-Choussy and B. Boudet, *Tetrahedron*, 1982, **38**, 3479.
73. R. Beugelmans and G. Roussi, *Tetrahedron, Suppl. 1*, 1981, 393.
74. R. Beugelmans, M. Bois-Choussy and B. Boudet, *Tetrahedron*, 1983, **39**, 4153.
75. R. Beugelmans and M. Bois-Choussy, *Tetrahedron Lett.*, 1988, **29**, 1289.
76. R. R. Bard and J. F. Bunnett, *J. Org. Chem.*, 1980, **45**, 1546.
77. J. J. Bulot, E. E. Aboujaoude, N. Collignon and P. Savignac, *Phosphorus Sulfur*, 1984, **21**, 197.
78. J. Pinson and J.-M. Savéant, *J. Am. Chem. Soc.*, 1978, **100**, 1506.
79. C. Degrand, R. Prest and P. L. Compagnon, *J. Org. Chem.*, 1987, **52**, 5229.
80. N. Alam, C. Amatore, C. Combellas, J. Pinson, J.-M. Savéant, A. Thiébault and J.-N. Verpeaux, *J. Org. Chem.*, 1988, **53**, 1496.
81. C. Amatore, C. Combellas, J. Pinson, J.-M. Savéant and A. Thiébault, *J. Chem. Soc., Chem. Commun.*, 1988, 7.
82. C. Degrand, *J. Electroanal. Chem. Interfacial Electrochem.*, 1987, **238**, 239.
83. C. Degrand, *J. Org. Chem.*, 1987, **52**, 1421.
84. C. Degrand, *J. Chem. Soc., Chem. Commun.*, 1986, 1113.
85. C. Amatore, J. Pinson, J.-M. Savéant and A. Thiébault, *J. Electroanal. Chem. Interfacial Electrochem.*, 1980, **107**, 59.
86. N. Alam, C. Amatore, C. Combellas, A. Thiébault and J.-N. Verpeaux, *Tetrahedron Lett.*, 1987, **28**, 6171.
87. W. J. M. van Tilborg, C. J. Smit and J. J. Scheele, *Tetrahedron Lett.*, 1977, 2113.
88. J. F. Bunnett and C. Galli, *J. Chem. Soc., Perkin Trans. 1*, 1985, 2515.
89. R. R. Bard, J. F. Bunnett and R. P. Traber, *J. Org. Chem.*, 1979, **44**, 4918.
90. J. F. Bunnett and X. Creary, *J. Org. Chem.*, 1974, **39**, 3611.
91. J. F. Bunnett and X. Creary, *J. Org. Chem.*, 1974, **39**, 3612.
92. J. F. Bunnett and R. P. Traber, *J. Org. Chem.*, 1978, **43**, 1867.
93. J. F. Bunnett and S. J. Shafer, *J. Org. Chem.*, 1978, **43**, 1873.
94. J. F. Bunnett and P. Singh, *J. Org. Chem.*, 1981, **46**, 5022.
95. R. Beugelmans and M. Bois-Choussy, *Tetrahedron*, 1986, **42**, 1381.
96. R. Beugelmans and M. Bois-Choussy, *Heterocycles*, 1987, **26**, 1863.
97. R. Beugelmans and M. Bois-Choussy, *Eur. J. Med. Chem. — Chim. Ther.*, 1988, **23**, 539.
98. A. B. Pierini, M. T. Baumgartner and R. A. Rossi, *J. Org. Chem.*, 1987, **52**, 1089.
99. W. Koch, W. Risse and W. Heitz, *Makromol. Chem., Suppl. 12 (Polym. Specific Prop.)*, 1985, 105.
100. V. Z. Annenkova, L. M. Antonik, I. V. Shafeeva, T. I. Vakul'skaya, V. Y. Vitkovskii and M. G. Voronkov, *Vysokomol. Soedin., Ser. B*, 1986, **28**, 137 (*Chem. Abstr.*, 1986, **105**, 6848).
101. M. Novi, G. Petrillo and M. L. Sartirana, *Tetrahedron Lett.*, 1986, **27**, 6129.
102. R. A. Alonso and R. A. Rossi, *J. Org. Chem.*, 1980, **45**, 4760.
103. R. A. Rossi and S. M. Palacios, *J. Org. Chem.*, 1981, **46**, 5300.
104. D. R. Carver, T. D. Greenwood, J. S. Hubbard, A. P. Komin, Y. P. Sachdeva and J. F. Wolfe, *J. Org. Chem.*, 1983, **48**, 1180.
105. G. R. Newkome, Y. J. Joo, D. W. Evans, S. Pappaldo and F. R. Fronczek, *J. Org. Chem.*, 1988, **53**, 786.
106. S. Kondo, M. Nakanishi and K. Tsuda, *J. Heterocycl. Chem.*, 1984, **21**, 1243.
107. J. F. Bunnett, X. Creary and J. E. Sundberg, *J. Org. Chem.*, 1976, **41**, 1707.
108. L. M. Tolbert and S. Siddiqui, *Tetrahedron*, 1982, **38**, 1079.
109. A. B. Pierini, M. T. Baumgartner and R. A. Rossi, *Tetrahedron Lett.*, 1987, **28**, 4653.
110. C. K. F. Hermann, Y. P. Sachdeva and J. F. Wolfe, *J. Heterocycl. Chem.*, 1987, **24**, 1061.
111. R. A. Rossi, R. H. De Rossi and A. F. Lopez, *J. Am. Chem. Soc.*, 1976, **98**, 1252.
112. J. F. Bunnett and J. E. Sundberg, *J. Org. Chem.*, 1976, **41**, 1702.
113. C. Galli and J. F. Bunnett, *J. Org. Chem.*, 1984, **49**, 3041.
114. R. Beugelmans and M. Bois-Choussy and Q. Tang, *J. Org. Chem.*, 1987, **52**, 3880.
115. L. Estel, F. Marsais and G. Queguiner, *J. Org. Chem.*, 1988, **53**, 2740.
116. D. R. Carver, A. P. Komin, J. S. Hubbard and J. F. Wolfe, *J. Org. Chem.*, 1981, **46**, 294.

117. E. A. Oostveen and H. C. Van der Plas, *Recl. Trav. Chim. Pays-Bas*, 1979, **98**, 441.
118. S. C. Dillender, Jr., T. D. Greenwood, M. S. Hendi and J. F. Wolfe, *J. Org. Chem.*, 1986, **51**, 1184.
119. J. V. Hay and J. F. Wolfe, *J. Am. Chem. Soc.*, 1975, **97**, 3702.
120. D. R. Carver, J. S. Hubbard and J. F. Wolfe, *J. Org. Chem.*, 1982, **47**, 1036.
121. J. F. Bunnett and B. F. Gloor, *Heterocycles*, 1976, **5**, 377.
122. V. Nair and S. D. Chamberlain, *J. Org. Chem.*, 1985, **50**, 5069.
123. R. Beugelmans and M. Bois-Choussy, *Synthesis*, 1981, 729.
124. R. A. Rossi and J. F. Bunnett, *J. Org. Chem.*, 1973, **38**, 3020.
125. R. Beugelmans, B. Boudet and L. Quintero, *Tetrahedron Lett.*, 1980, **21**, 1943.
126. R. Beugelmans and H. Ginsburg, *Heterocycles*, 1985, **23**, 1197.
127. V. Nair and S. D. Chamberlain, *J. Am. Chem. Soc.*, 1985, **107**, 2183.
128. R. A. Rossi and R. A. Alonso, *J. Org. Chem.*, 1980, **45**, 1239.
129. R. G. Scamehorn, J. M. Hardacre, J. M. Lukanich and L. R. Sharpe, *J. Org. Chem.*, 1984, **49**, 4881.
130. J. F. Wolfe, J. C. Greene and T. Hudlicky, *J. Org. Chem.*, 1972, **37**, 3199.
131. J. V. Hay, T. Hudlicky and J. F. Wolfe, *J. Am. Chem. Soc.*, 1975, **97**, 374.
132. Y. L. Gol'dfarb, A. P. Yakubov and L. I. Belen'kii, *Chem. Heterocycl. Compd. (Engl. Transl.)*, 1979, 853.
133. R. A. Rossi, R. H. de Rossi and A. F. Lopez, *J. Org. Chem.*, 1976, **41**, 3367.
134. M. P. Moon, A. P. Komin, J. F. Wolfe and G. F. Morris, *J. Org. Chem.*, 1983, **48**, 2392.
135. R. A. Alonso, E. Austin and R. A. Rossi, *J. Org. Chem.*, 1988, **53**, 6065.
136. C. Combellas, H. Gautier, J. Simon, A. Thiébault, F. Tournilhac, M. Barzoukas, D. Josse, I. Ledoux, C. Amatore and J.-N. Verpeaux, *J. Chem. Soc., Chem. Commun.*, 1988, 203.
137. A. B. Pierini, M. T. Baumgartner and R. A. Rossi, *Tetrahedron Lett.*, 1988, **29**, 3451.
138. R. Beugelmans, M. Bois-Choussy and Q. Tang, *Tetrahedron Lett.*, 1988, **29**, 1705.
139. C. Amatore, J. Pinson, J.-M. Savéant and A. Thiébault, *J. Am. Chem. Soc.*, 1981, **103**, 6930.
140. C. Amatore, J. Pinson, J.-M. Savéant and A. Thiébault, *J. Am. Chem. Soc.*, 1982, **104**, 817.
141. C. Amatore, J.-M. Savéant, C. Combellas, S. Robveille and A. Thiébault, *J. Electroanal. Chem. Interfacial Electrochem.*, 1985, **184**, 25.
142. L. M. Tolbert and S. Siddiqui, *J. Org. Chem.*, 1984, **49**, 1744.
143. R. A. Alonso and R. A. Rossi, *Tetrahedron Lett.*, 1985, **26**, 5763.
144. S. Rajan and K. Muralimohan, *Tetrahedron Lett.*, 1978, 483.
145. A. M. Aguiar, H. J. Greenberg and K. E. Rubenstein, *J. Org. Chem.*, 1963, **28**, 2091.
146. E. R. Bornancini, R. A. Alonso and R. A. Rossi, *J. Organomet. Chem.*, 1984, **270**, 177.
147. J. F. Bunnett and R. H. Weiss, *Org. Synth.*, 1978, **58**, 134.
148. M. Novi, G. Garbarino, G. Petrillo and C. Dell'Erba, *J. Org. Chem.*, 1987, **52**, 5382.
149. J. F. Bunnett and X. Creary, *J. Org. Chem.*, 1974, **39**, 3173.
150. C. Amatore, C. Combellas, J. Pinson, M. A. Oturan, S. Robveille, J.-M. Savéant and A. Thiébault, *J. Am. Chem. Soc.*, 1985, **107**, 4846.
151. J. E. Swartz and T. T. Stenzel, *J. Am. Chem. Soc.*, 1984, **106**, 2520.
152. R. Beugelmans and H. Ginsburg, *Tetrahedron Lett.*, 1987, **28**, 413.
153. A. B. Pierini and R. A. Rossi, *J. Organomet. Chem.*, 1978, **144**, C12.
154. A. B. Penenory, A. B. Pierini and R. A. Rossi, *J. Org. Chem.*, 1984, **49**, 3834.
155. D. J. Sandman, J. C. Stark, L. A. Acampora and P. Gagne, *Organometallics*, 1983, **2**, 549.
156. R. A. Rossi and A. B. Penenory, *J. Org. Chem.*, 1981, **46**, 4580.
157. S. M. Palacios, R. A. Alonso and R. A. Rossi, *Tetrahedron*, 1985, **41**, 4147.
158. M. Novi, C. Dell'Erba, G. Garbarino and S. Fernando, *J. Org. Chem.*, 1982, **47**, 2292.
159. M. Novi, G. Garbarino and C. Dell'Erba, *J. Org. Chem.*, 1984, **49**, 2799.
160. M. Novi, G. Garbarino, C. Dell'Erba and G. Petrillo, *J. Chem. Soc., Chem. Commun.*, 1984, 1205.
161. M. Novi, G. Garbarino, G. Petrillo and C. Dell'Erba, *J. Chem. Soc., Perkin Trans. 2*, 1987, 623.
162. M. F. Semmelhack, R. D. Stauffer and T. D. Rogerson, *Tetrahedron Lett.*, 1973, 4519.
163. M. F. Semmelhack, B. P. Chong, R. D. Stauffer, T. D. Rogerson, A. Chong and L. D. Jones, *J. Am. Chem. Soc.*, 1975, **97**, 2507.
164. W. R. Bowman, H. Heaney and P. H. G. Smith, *Tetrahedron Lett.*, 1982, **23**, 5093.
165. S. Usui and Y. Fukazawa, *Tetrahedron Lett.*, 1987, **28**, 91.
166. R. Beugelmans and G. Roussi, *J. Chem. Soc., Chem. Commun.*, 1979, 950.
167. K. Boujlel, J. Simonet, G. Roussi and R. Beugelmans, *Tetrahedron Lett.*, 1982, **23**, 173.
168. R. Fontan, C. Galvez and P. Viladoms, *Heterocycles*, 1981, **16**, 1473.
169. R. Beugelmans, J. Chastanet and G. Roussi, *Tetrahedron Lett.*, 1982, **23**, 2313.
170. R. Beugelmans, J. Chastanet and G. Roussi, *Tetrahedron*, 1984, **40**, 311.
171. R. Beugelmans, J. Chastanet, H. Ginsburg, L. Quintero-Cortes and G. Roussi, *J. Org. Chem.*, 1985, **50**, 4933.
172. R. Beugelmans, H. Ginsburg and M. Bois-Choussy, *J. Chem. Soc., Perkin Trans. 1*, 1982, 1149.

2.3
Nucleophilic Coupling with Arynes

SATINDER V. KESSAR
Panjab University, Chandigarh, India

2.3.1 INTRODUCTION

Functional group substitution plays a pivotal role in the synthetic chemistry of aromatic compounds, and one important reaction serving this purpose proceeds through arynes. Formally derived from aromatic rings by removal of two *ortho* hydrogen atoms, and kinetically unstable, these intermediates find sporadic reference in the early literature — primarily to explain the strange phenomenon of *cine* substitution. The first decisive evidence came with the observation that equal amounts of anilines (**2**) and (**3**) are formed in the reaction of ^{14}C-labeled chlorobenzene with KNH_2 in liquid ammonia (equation 1).[1] This demonstrated the involvement of a symmetrical intermediate like (**1**). Additional support for aryne intermediacy was found in its reactions with phenyllithium and in Diels–Alder-type trapping.[2] A great deal of work on the generation and reactions of arynes was undertaken in the 1960s, and is covered in a comprehensive monograph by Hoffmann.[3] Much of the subsequent work is dealt with in some more recent reviews.[4,5]

Cl, KNH_2/NH_3 → (**1**) → (**2**) NH_2 + (**3**) NH_2 (1)

* = ^{14}C

The structure of benzyne, the parent member of the aryne group of intermediates, has been the focus of many experimental and theoretical studies. For instance, its IR spectrum has been recorded by low temperature solid matrix photolysis of precursors like benzocyclobutanedione (**4**) and diazalactone (**5**).[6–9] Force field calculations based on these observations indicate an extended triple bond in benzyne (**6a**).[10] Semiempirical and *ab initio* calculations also predict the 'arynic bond' to be longer than a normal triple bond and the other C—C bonds to be similar in length to those in benzene.[11,12] The observed microwave spectrum of benzyne is in accord with these conclusions.[13] So are the vertical ionization potentials obtained from its photoelectron spectrum,[14] although some doubts about the assignments have been raised.[15,16] A recent *ab initio* study, in which electron correlation is taken into account, indicates more cumulene (**6b**) like features in benzyne than predicted by earlier calculations at the Hartree–Fock level.[17] Irrespective of the finer details of the geometric dimensions and orbital energy levels of benzyne, it may be considered to have a weak third bond produced by lateral overlap of in-plane orbitals. The symmetric combination (**7**) of these two orbitals is lower in energy than the antisymmetric combination (**8**) and hence the ground state of benzyne should be a singlet. This inference is supported by all recent calculations and there is little evidence for triplet character in any reaction of arynes.

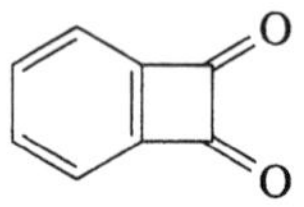

(**4**) 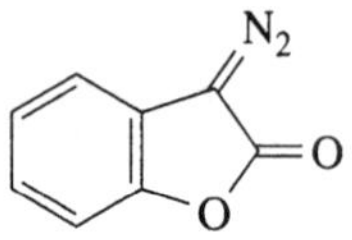 (**5**)  (**6a**)  (**6b**)

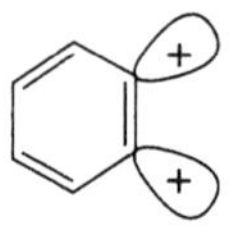

(**7**) 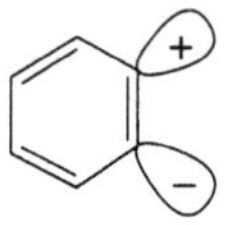 (**8**)

The reactivity of benzyne can be explained in terms of its high enthalpy of formation which has been experimentally determined to be around 120 kcal mol^{-1} (1 cal = 4.18 J).[18–20] The difference between enthalpies of formation of an alkyne and an alkene is usually about 37 kcal mol^{-1}; with an enthalpy value of 20 kcal mol^{-1} for benzene, the strain energy in benzyne comes out to be nearly 63 kcal mol^{-1}.[4e] Besides general reactivity, benzyne also exhibits marked electrophilicity and from this point of view, its LUMO energy level is of special import. The HOMO of benzyne is calculated to be at –9.58 eV, which is little different in energy from the corresponding bonding orbital of acetylene. However, the LUMO is estimated to be at 1.33 eV, which is much lower in energy than the acetylene LUMO. This has been attributed to 'bending' of the triple bond which results in an efficient mixing of the π^* orbital with a σ^* orbital lying only slightly higher in energy.[21] The lowering of the benzyne LUMO decreases its energy gap with the HOMO of an attacking nucleophile and makes for an easy reaction between the two. Thus, many nucleophilic reagents which are inert towards alkynes add readily to arynes. If electron-withdrawing substituents such as halogens are present on an aryne, its electrophilicity is further enhanced. Another point of interest in nucleophilic addition to arynes is the high polarizability of the formal triple bond. As a consequence, arynes behave as 'soft' electrophiles and their reactions with polarizable nucleophiles are specially facile. However, sometimes reactions of aryl halides with nucleophiles under seemingly benzyne-generating conditions may not involve these intermediates.[5a,22,23]

Other prominent reactions of arynes include cycloadditions and the ene reaction as shown in equations (2)–(5). The former have been used extensively for the synthesis of a variety of benzo-annulated systems and for the diagnosis of aryne intermediates.[1–5] Normally, these reactions do not fall in the category of nucleophilic coupling. However, in some cases the electrophilic character of the arynic partner is quite manifest and as such a brief discussion is included in Section 2.3.3.

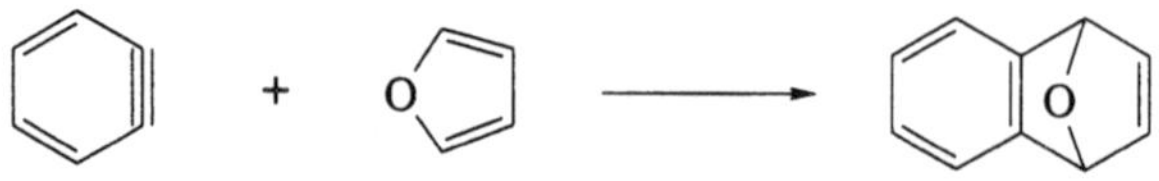

(**9**) + furan ⟶ (2)

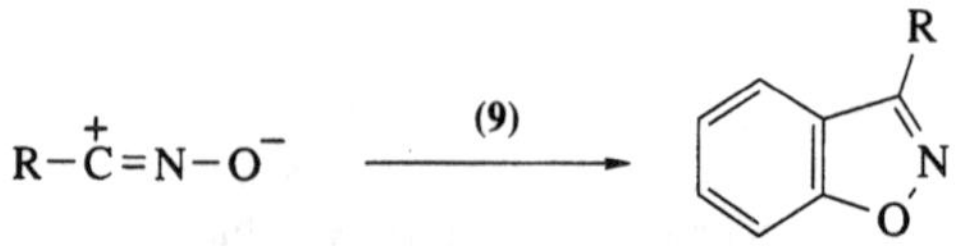

$R-\overset{+}{C}=N-O^-$ $\xrightarrow{(9)}$ (3)

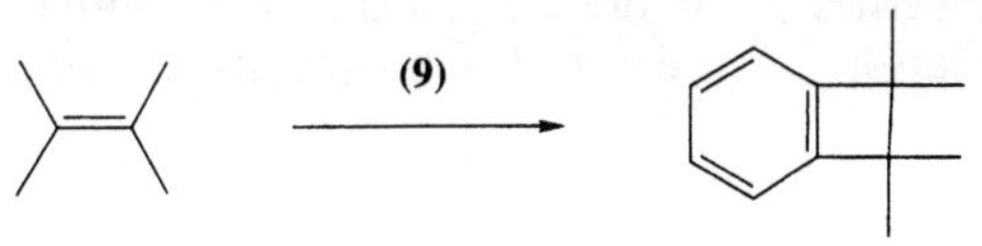

(4)

(5)

Dehydroheteroarenes like (**10**) and (**11**) have also been proposed as intermediates in nucleophilic substitution.[23–25] Some of these reactions were evaluated uncritically and operation of other mechanisms like addition–elimination (AE) and ring opening–ring closure (ANRORC) can now be demonstrated in many such cases. Nevertheless, there is conclusive evidence for heteroaryne intermediacy in some reactions of heterocyclic halides. From the preparative point of view, nucleophilic coupling of such intermediates has found only limited applications.[26–28] Reactive intermediates with an additional formal bond between nonadjacent atoms, like (**12**) and (**13**), have also been postulated but again hold little synthetic interest.

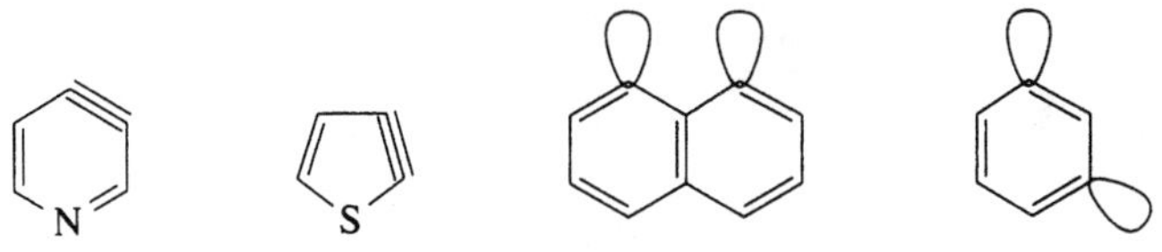

(**10**) (**11**) (**12**) (**13**)

Many stable metal complexes of arynes are known but in most of their reactions of synthetic interest, the yields are poor. For example, thermolysis of titanocene (Cp_2TiPh_2) at 80–100 °C gives rise to a titanium–benzyne complex which reacts with molecular nitrogen to afford aniline with low efficiency.[29] However, procedures are available for *in situ* generation of zirconium complexes (**14**) and for their coupling reactions to synthesize functionalized aromatic compounds in preparatively useful yields (Scheme 1).[30] Whether such complexes should be regarded as π-bonded benzynes or σ-bonded *o*-phenylenes, remains a debatable point.[31]

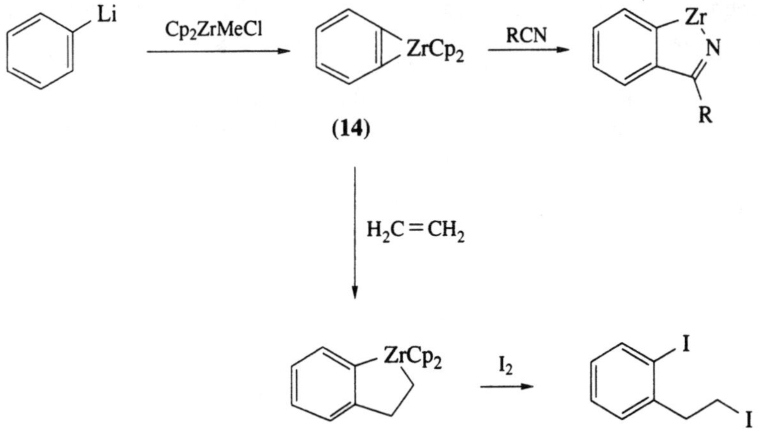

Scheme 1

2.3.2 GENERATION OF ARYNES

Much of the early work on arynes was undertaken using the aryl halide–strong base route (equation 6), but a variety of other procedures, discussed below, are also available. In practice, the choice of the method of generation is often governed by the nature of the reaction to be carried out subsequently. Since arynes are transitory intermediates used *in situ*, the conditions employed for their formation have to be compatible with the properties of the other reacting partner. Further, in preparative work, the accessibility of the benzyne precursor and its cost can be important considerations.

2.3.2.1 From Aryl Anions

Aryl anions having a good leaving group (X) in the *ortho* position readily afford arynes (equation 6). For this purpose, halides have been used widely but other electrofugic groups, like OPh, OTs, OTf, N^+_2, NR^+_3, SR^+_2, IR^+, have also been employed. Since both the deprotonation and the benzyne formation steps can be reversible,[32] many factors such as the nature of the leaving group, the base and the solvent play an important role in the rate and efficiency of aryne generation. The ease of halide expulsion from (**15**) is I > Br > Cl > F but the rates of initial proton removal are in the reverse order. In the case of aryl fluorides, reprotonation can compete effectively with benzyne formation, while with iodides, there is greater ease of reversal in the second step. For these reasons, and because of easier accessibility, chloro- and bromo-benzenes are used most commonly. In general leaving groups other than halogens offer no particular advantage. For example, phenyl benzenesulfonate reacts with lithium 2,2,6,6-tetramethylpiperidide (LTMP) to give benzyne but the yields of adducts are less than those obtained with bromobenzene.[33]

X base X −X⁻ (6)

(15) **(9)**

A wide variety of base–solvent combinations can be employed for the deprotonation of aryl halides. However, alkali metal aryls/alkyls in ether solvents or amides in liquid ammonia have been used most extensively. In these systems, reactions of the generated arynes with the solvent or the base can pose a serious problem, especially when coupling with relatively weak nucleophiles is desired. Metal aryls and alkyls can be used in ether solvents which have low reactivity towards arynes but these bases are very powerful nucleophiles. On the other hand, metal amides are somewhat less reactive but ammonia, the commonly employed solvent, has a greater proclivity for aryne capture. Thus, the use of amidic bases in aprotic solvents may be preferable, provided the solubility is acceptable. Sodamide in THF, alone or with HMPA, can function effectively when used in conjunction with a second anionic base such as sodium *t*-butoxide.[34] In reactions carried out with commercial sodamide in THF special precautions to exclude air and moisture are necessary.[35] More recently, LDA in THF is finding extensive use in coupling reactions of arynes.[36–38] For minimizing aryne reaction with the amide base used for its generation, the highly hindered LTMP has also been recommended.[39,40] With this base, another common side reaction, hydride transfer from the α-carbon leading to aryne reduction, is also precluded. It should be noted that whereas

+ Nu⁻ Nu H⁺ Nu H (7)

(9) **(16)**

(9)

Nu

(17)

the use of ethers minimizes reaction of the benzyne with the solvent, it also prolongs the life time of the adduct anion (**16**), which in ammonia is protonated promptly. Therefore, undesirable reactions of (**16**) occur to a greater extent under aprotic conditions.

At higher temperatures, weaker bases can be used for the generation of arynes from aryl halides. Sodium methanesulfinyl carbanions in DMSO are effective above room temperature,[41,42] while sodium hydroxide reacts with chlorobenzene at about 250 °C to give phenol, presumably through a benzyne intermediate.[1] With substrates having strongly activating groups, even mild bases lead to deprotonation, *e.g.* betain (**19**) can be generated from (**18**) with potassium acetate (equation 8).[43] Arynes can also be formed from diazonium carboxylates derived by the rearrangement of *N*-nitrosoacylarylamines such as (**20**).[44] However, these methods are more useful for [2 + 4] trapping reactions of arynes than for their coupling with strong nucleophiles.

$PhN_2^+BF_4^-$ (**18**) —KOAc→ (**19**) → (**9**) (8)

(**20**) → PhN_2^+ ^-O_2CR → (**19**)

A number of procedures for the generation of arynes using disubstituted precursors have also been developed. Metal–halogen exchange in dihalobenzenes with lithium amalgam or magnesium is one such useful method.[1] Electroreduction of dihalobenzenes,[45] reaction of Grignard reagents with aryl sulfoxides[46] or defluorosilylation of arylsilanes[47,48] have also been used for the generation of arynes (equation 9). Decarboxylation of 2-halobenzoates also leads to aryne formation but requires high temperature.[1] However, 1-(*o*-halophenyl)-2-benzenesulfonylhydrazides are decomposed under milder conditions.[49]

(9)

X = Cl, Br or Tf

A very popular route to arynes involves decomposition of benzenediazonium 2-carboxylates.[50] The explosive nature of diazonium compounds is a major disadvantage of this method, although procedures for safe preparation and handling have been described.[51] An alternative is to carry out the diazotization of anthranilic acid *in situ*, but the yields of benzyne adducts obtained in this manner are usually inferior due to side reactions.[52] In any case, the benzenediazonium carboxylate route is usually not suitable for coupling of arynes with anionic nucleophiles, since in a strongly basic medium dediazotization to benzoic acid eclipses the formation of the reactive intermediate.[53] This method can, however, be used in reactions with uncharged nucleophiles. Diphenyliodonium-2-carboxylate is a safe benzyne precursor but requires decomposition above 150 °C.[54] A new convenient procedure comprising thermolysis of silyl sulfoxides has been developed recently (Scheme 2).[5d]

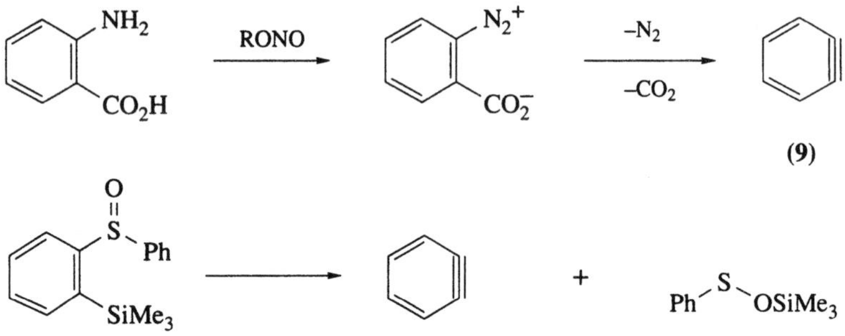

NH_2 / CO_2H —RONO→ N_2^+ / CO_2^- —($-N_2$, $-CO_2$)→ **(9)**

S(O)Ph / $SiMe_3$ → benzyne + Ph–S–$OSiMe_3$

Scheme 2

2.3.2.2 From Fragmentation of Cyclic Systems

Arynes can be readily generated by thermal or photolytic fragmentation of several benzo-fused systems.[1] The conditions required for the cleavage of such precursors vary widely. For instance, azosulfone (**21**), obtained by diazotization of 2-aminobenzenesulfinic acid, decomposes in organic solvents at room temperature to give nitrogen, sulfur dioxide and benzyne.[55] In the case of phthalic anhydride (**22**), gas phase pyrolysis at 700 °C is needed, while phthaloyl peroxide (**23**) affords benzyne on irradiation. Oxidative decomposition of 1-aminobenzotriazole can be effected at temperatures as low as –80 °C (equation 10).[56] Several variations of this method have been explored with a view to avoid oxidizing agents. The salt (**24**) furnishes benzyne on dissolving in polar solvents at room temperature.[57] Of the cyclic benzyne precursors, benzotriazoles have found more preparative use in general but not many examples involving coupling with nucleophiles have been reported.

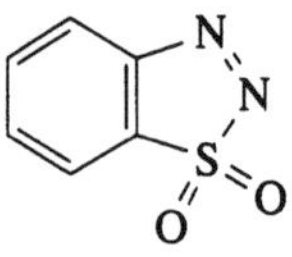

(21) 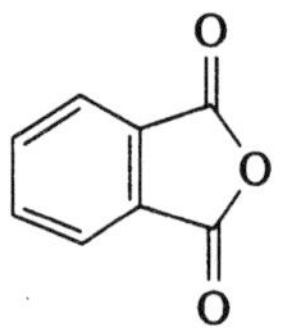 **(22)** 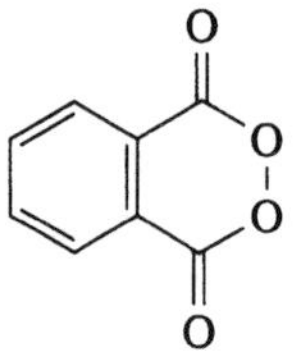 **(23)** 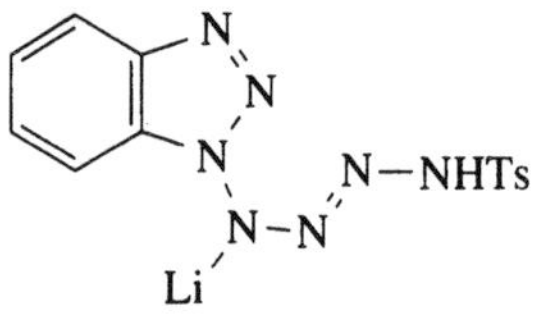 **(24)**

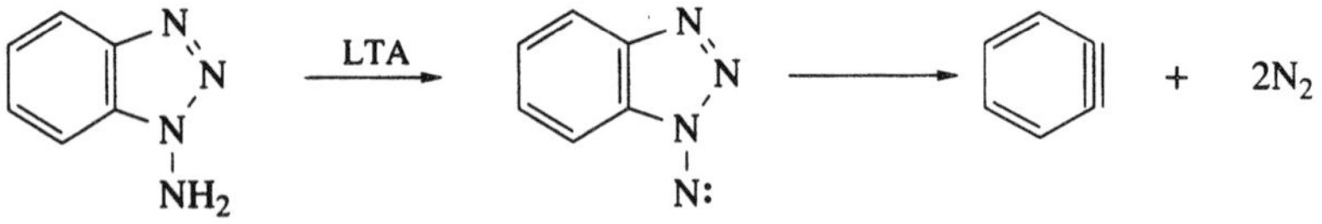

N–NHTs, N–N, Li

NH_2 —LTA→ N: → benzyne + $2N_2$ (10)

An important drawback of the aryne routes starting with bidentate or cyclic precursors can be the effort needed to prepare the precursor itself, especially for substituted arynes. However, these have the advantage that the arynic bond can be generated without positional ambiguity as illustrated in the following section.

2.3.2.3 Generation of Substituted Arynes

Generation of arynes from *ortho*- and *para*-substituted halides is straightforward. In the *meta* case, aryne formation can take place in two directions (Scheme 3). Regioselectivity in this situation is dependent on the relative rates and reversibility of the deprotonation and the dehalogenation steps. With substituents which exert an electron-withdrawing (*–I*) effect or favor *ortho* metalation by chelation, aryne formation in the adjacent position predominates, if the halide ion loss is fast. In the case of aryl halides with *+I* substituents the regioselectivity can be poor. For example, treatment of 3-bromoanisole (**25a**) with LDA/THF leads to 3-methoxybenzyne (**27a**) primarily.[36] In contrast, the reaction of 3-bromotoluene (**25b**) with KNH_2/NH_3 gives (**27b**) and (**29b**) in 40:60 ratio. Interestingly, with the corresponding chloride this ratio is reversed to 79:21. It has been suggested that in both halides, metalation *para* to the methyl group is faster ($k_3 > k_1$).[1] However, in the aryl bromide this step is virtually irreversible due to rapid bromide ion loss. With the chloride, equilibration between the metalated species (**28**) and (**26**) takes place and the chloride ion loss from the latter is faster ($k_2 > k_4$). When there is ambiguity in aryne formation from aryl halides, bidentate precursors can be used to get the arynic bond in a predetermined position. For instance, aryne (**27b**) can be exclusively generated from diazonium carboxylate (**30**). This route has been used for obtaining some other substituted arynes also.[52,58]

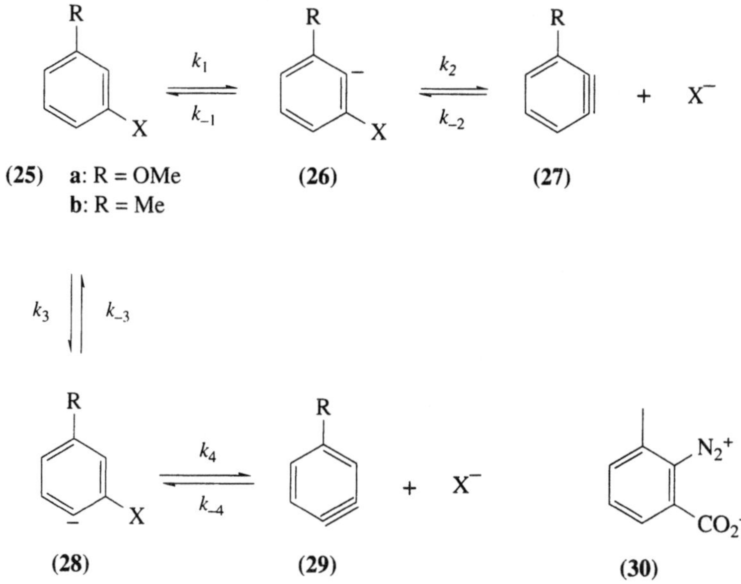

(**25**) **a**: R = OMe
b: R = Me

(**26**) (**27**) (**28**) (**29**) (**30**)

Scheme 3

In polycyclic systems with bond fixation, dehydrohalogenation tends to occur in one direction only, *e.g.* 2-bromonaphthalene gives 1,2-naphthalyne (**31**). The 2,3-isomer (**33**) can be obtained by the oxidation of aminotriazole (**32**).[59,60] Similarly, in some heterocyclic systems marked regioselectivity on dehydrohalogenation is observed. Thus, treatment of 3-bromopyridine with strong base affords (**34**). For the generation of 2,3-dehydropyridine, the dihalogenated substrate (**35**) is required.[24] The question of chemoselectivity comes up with substrates having two potential leaving groups, such as the mixed dihalobenzenes which can afford two different arynes on dehydrohalogenation. In such cases the heavier halogen is often lost preferentially,[61] *e.g.* (**36**) gives 3-chlorobenzyne. The generation of arynes from di- and tri-halogenated arenes can get complicated by prior isomerization, termed 'halogen dance'.[62] However, perchloro- and perfluoro-arynes can be obtained efficiently.[63,64]

Finally, the conditions used for generation of substituted arynes have to be compatible with the nature of the substituent group. In the aryl halide–strong base route, a wide range of substituents can be tolerated; alkyl, aryl, trifluoromethyl, halogen, alkoxy, dialkylamino, methylthio, carboxylic and cyano groups exert no deleterious effect. However, strongly electron-withdrawing substituents can so stabilize the aryl anion that the halide ion loss does not occur under the usual conditions.[1] Thus (**37**) has considerable lifetime at room temperature, while the corresponding bromo compound decomposes somewhat faster. Strongly electron-donating substituents, on the other hand, can retard the deprotonation step and it may be stopped altogether if a negatively charged atom is appended to the aromatic ring. For instance, halophenoxides are not metalated by phenyllithium in ether although *n*-butyllithium is effective.[1] The anions (**38**) and (**39**) are stable towards KNH_2/NH_3, whereas in (**40**) and (**41**) aryne formation takes place

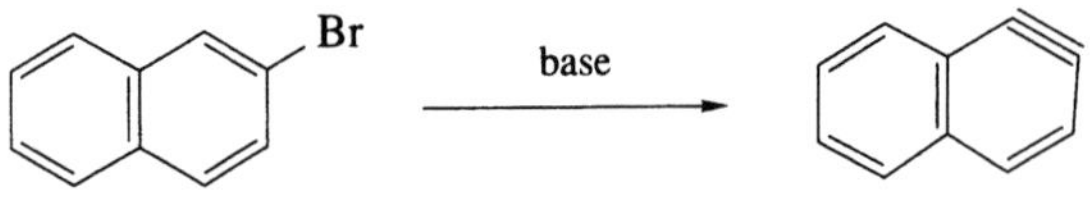

(31)

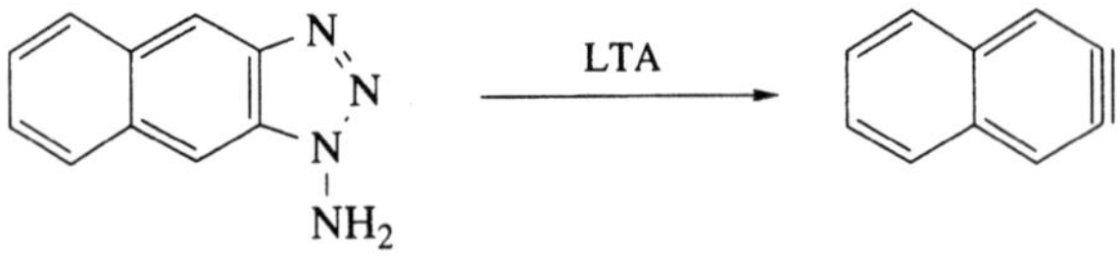

(32) (33)

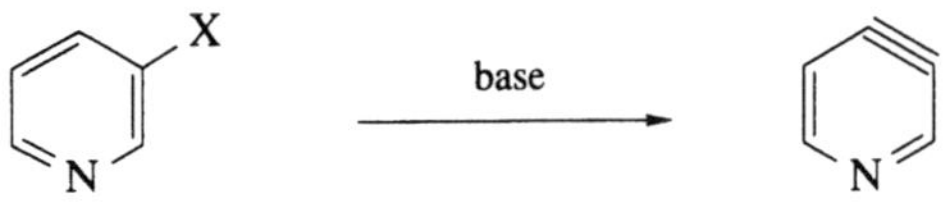

(34)

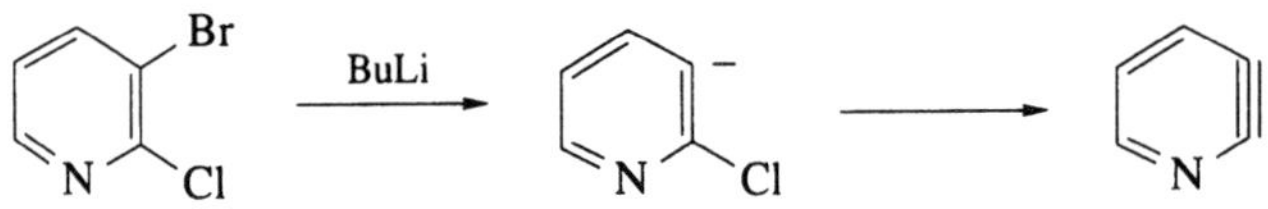

(35)

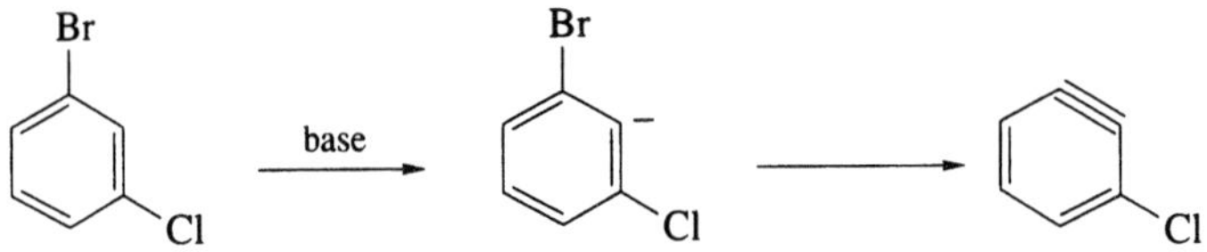

(36)

Scheme 4

under similar conditions.[65,66] It is thus clear that the extent of substituent charge delocalization into the aromatic ring imposes a delicate balance on aryne formation. Of course, if the negative charge bearing atom in the side chain is not attached directly to the aromatic ring, benzyne formation proceeds normally.

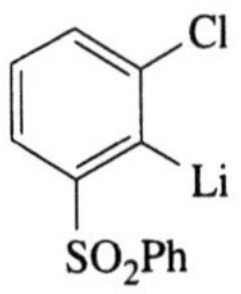

(37) 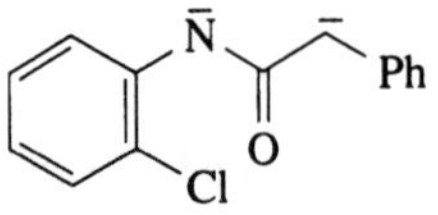 (38) 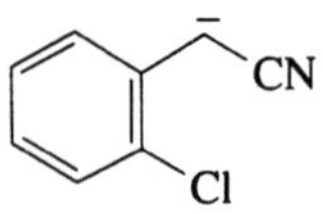 (39)

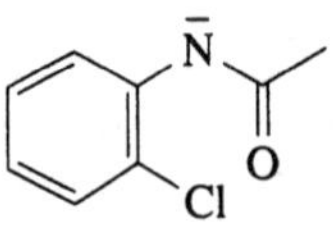

(40) 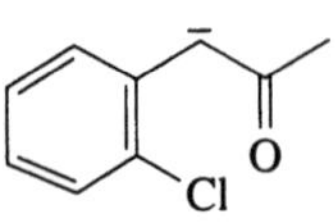 (41)

2.3.3 ADDITION OF NUCLEOPHILES TO ARYNES

A wide variety of anionic and uncharged nucleophiles add readily to arynes (Scheme 5). The adduct, anionic or zwitterionic, can abstract a proton from the surroundings or undergo an internal rearrangement. In the absence of such quenching, it can be trapped by addition of electrophiles. However, further aryne addition can also take place to give (**17**) and polymeric products derived therefrom (equation 7). Another complication in coupling of nucleophiles with arynes arises if the phenylation product is sufficiently acidic to undergo rapid deprotonation; further aryne reaction can then occur. Thus, considerable amounts of di- and tri-phenylation products are formed when α-picoline is reacted with 1 equiv. of an aryl halide in the presence of a strong base.[67]

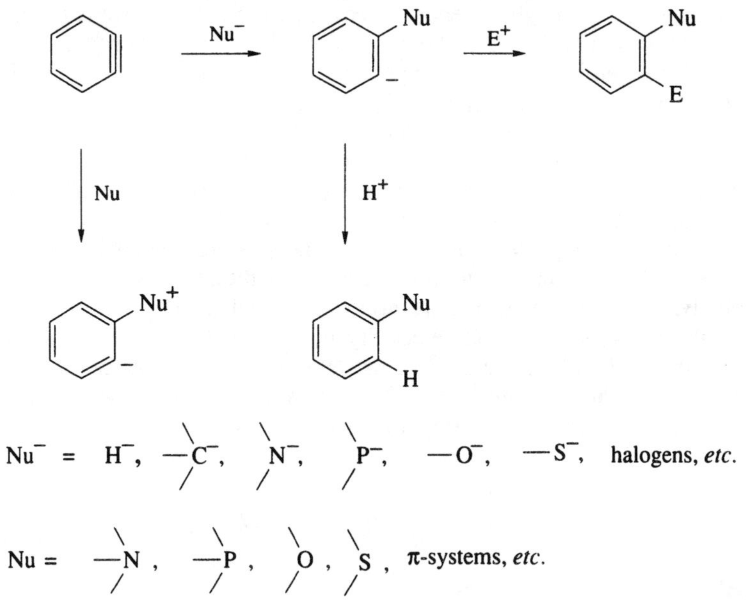

Nu⁻ = H⁻, C⁻, N⁻, P⁻, —O⁻, —S⁻, halogens, *etc.*

Nu = N, P, O, S, π-systems, *etc.*

Scheme 5

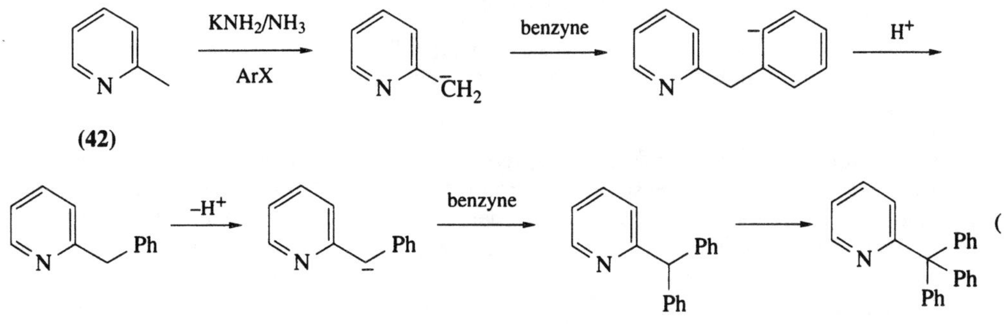

KNH₂/NH₃, ArX; benzyne; H⁺

(**42**)

−H⁺; benzyne (11)

24% 18% 18%

2.3.3.1 Relative Reactivity of Nucleophiles

As mentioned in the introductory section, arynes behave as 'soft acids'. Therefore, the relative reactivity of a nucleophile should be governed by basicity as well as polarizability. The following gradations, established through competitive reactions of arynes with different nucleophiles, are more or less in line with this expectation.[1] (i) BuLi > PhSLi > PhNMeLi > PhC≡CLi > ROLi > ArOLi, for 9,10-phenanthryne in ether; (ii) PhS^- > Ph_3C^- > $PhC{\equiv}C^-$ > enolates > PhO^- > RO^- > I^-, CN^-, for benzyne in liquid ammonia; (iii) I^- > Br^- > Cl^- > EtOH for benzyne in alcohol.

The above series perhaps do not exactly reflect the order of inherent nucleophilicities towards arynes, because factors like the size of the attacking reagent, its state of association and the nature of the counterion can play an important role. For instance, diethylamine is 18 times more reactive towards benzyne than the bulkier diisopropylamine.[68] Nevertheless, aryne preference for soft nucleophiles is clear in a general way. It is also manifested in reactions with ambident anions derived from phenol, pyrrole and indole as substantial *C*-phenylation is observed in base-promoted reactions with aryl halides.[69]

The fact that the rates for addition of different nucleophiles to arynes vary by a wide margin shows that these intermediates are sufficiently long lived to allow high chemoselectivity. The circumstance that the nucleophilic reactivity is not entirely governed by basicity is useful in reactions where the nucleophile is not strong enough a base to generate the aryne from its precursor. In such an eventuality, another strong base has to be used, and 2 mol equiv. may be needed for the deprotonation of the aryl halide and the conjugate acid of the nucleophile. In some cases, the reaction goes to completion only if a slight further excess of the base is used. This base can compete for the aryne, but good yields of the desired product can still be obtained if the nucleophile has higher reactivity toward the aryne. For example, benzenethiolate ion is not basic enough to generate benzyne from chlorobenzene in liquid ammonia and a metal amide is used for this purpose. Nevertheless, diaryl sulfides are obtained in acceptable yields because the polarizable benzenethiolate ion is a superior aryne nucleophile.[1,70] Under similar conditions, the yield of diaryl ethers is poor and amination predominates over the reaction of benzyne with the weakly nucleophilic phenoxide ions. In the case of addenda which are poor bases and are also weakly nucleophilic, it is customary to use a nonbasic route for aryne generation.

The structure of an aryne can have a marked effect on its selectivity and the efficiency of its coupling reactions.[1,71] In general, any factor which increases the stability and the life time of a reactive intermediate should enhance its discerning power. An example is the superior selectivity of 1,2-naphthalyne, as compared to benzyne, in phenyllithium *vs.* lithium piperidide addition. This was attributed to greater stability of 1,2-naphthalyne arising out of a more effective orbital overlap across the shorter 'dehydro bond'.[1] Detailed analysis revealed that the selectivity of 1,2-naphthalyne is higher at the α-position, which is shielded by the *peri* hydrogen atom.[72] This trend is more pronounced in the case of diethylamine/diisopropylamine pair due to appreciable difference in the size of the two competing nucleophiles (Scheme 6). Thus, steric factors also play an important role in aryne selectivity.

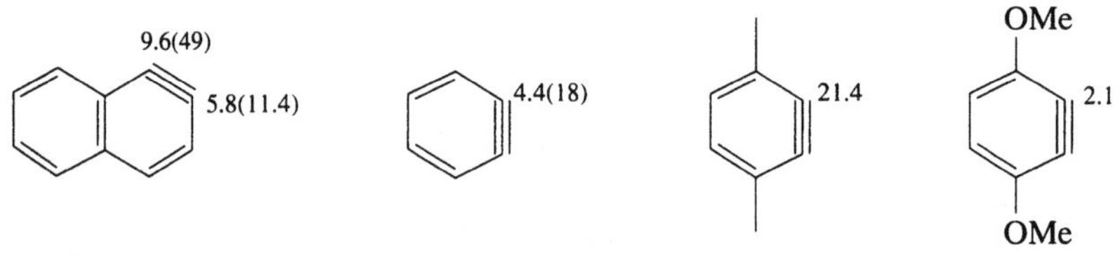

Scheme 6 Relative aryne selectivity values for phenyllithium/lithium piperidide pair and diethylamine/diisopropylamine pair (in parenthesis)

In connection with the substituent effects, the kinetic stability of benzyne is suggested to be increased by electron withdrawal (–*I*) and decreased by electron release (+*I*).[73] However, the inference cannot be extrapolated to selectivity of substituted arynes in general. For example, in additions involving competition between phenyllithium and lithium piperidide, the methyl substituents (+*I*) on benzyne increase its selectivity, whereas methoxy groups (–*I*) decrease it (Scheme 6). On the other hand, in reactions of carbanions derived from acetonitrile in alkylamine solvents both +*I* and –*I* benzyne substituents lower selectivity and cause predominant amination. Thus, the method was found unsuitable for preparation of many substituted benzyl nitriles.[74] In symmetrically disubstituted arynes there is partial cancellation of polarization, and in fact acceptable yields of acetonitrile adducts could be obtained from 3,6-dimethoxybenzyne.[75] The selectivity of substituted arynes varies with the set of nucleophiles in the competition and no comprehensive theory or simple generalization is available on this point.

2.3.3.2 Regioselectivity in Nucleophilic Addition to Unsymmetrical Arynes

A nucleophile can add to either end of the 'triple bond' in an unsymmetrical aryne. Since two products can be formed, the extent and the direction of regioselectivity observed in such a reaction may determine its preparative value. Substituents present on an aryne can orient the incoming nucleophile through electronic and steric effects. As the reacting 'triple bond' is orthogonal to the ring π-system, the inductive effect of the substituents is more important than their mesomeric effect. The inductive effect may be considered to influence the site of the nucleophilic attack by polarization of the 'triple bond' or by stabilization/destabilization of the negative charge in the transition state. Arguments based on either approach lead to similar predictions. The degree of selectivity is also expected to be sensitive to nucleophile reactivity, the stronger nucleophiles being less selective.[76] Significant regioselectivity is usually obtained in additions to arynes having *–I* substituents in position 3 (**43a**), as the developing negative charge can be stabilized to a greater extent in (**44a**) than in (**45a**). The repulsive interactions are also minimal for the *meta* addition. On the other hand, with the *+I* substituents (**43b**) the electronic and the steric effects may operate in opposition to each other. The transition state corresponding to the *ortho* adduct (**45b**) is destabilized to a lesser extent electronically but it is more congested. As a consequence only with powerful electron-releasing groups, like those having a negatively charged atom attached to the aromatic ring,[77] can high *ortho:meta* product ratios be obtained. With weakly electron-releasing alkyl groups, the *ortho:meta* ratio is rarely greater than unity and can be much less if large substituents and nucleophiles are involved. In addition of lithium piperidide to 3-isopropylbenzyne this ratio is 1:24, and it is 1:2 in addition to 3-methylbenzyne. This trend in favor of *meta* addition has been utilized for the synthesis of the *meta*-substituted toluene (**48**).[78] In coupling iodide (**46**) with (**47**), the *ortho* isomer (**49**) was only a minor product. Steric effects arising out of different degrees of association of the nucleophilic anion with its countercation have been invoked to explain the decreasing *ortho:meta* ratio in the reaction of 2-chlorotoluene with $LiNH_2$, $NaNH_2$ and KNH_2.[79] Similarly, the preponderance of β-adducts in 1,2-naphthalyne-mediated reactions can be ascribed to *peri* hydrogen shielding of the α-position.[1]

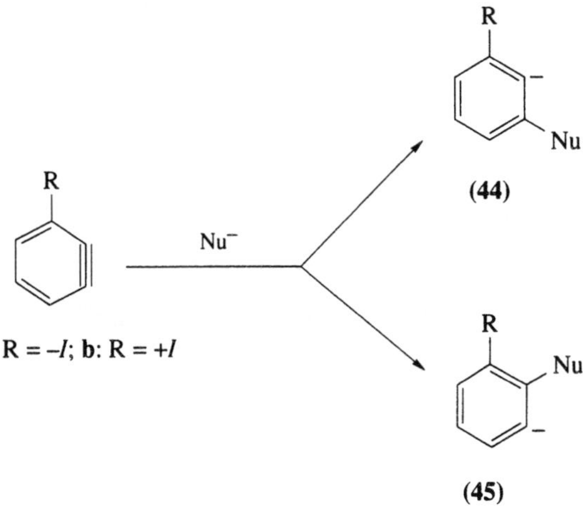

R
Nu⁻
R
Nu
(**44**)
(**43**) **a**: R = *–I*; **b**: R = *+I*
R
Nu
(**45**)

Scheme 7

MgX
+
SiMe₃
I
Cl
MgX
Cl
(**46**) (**47**)

(**47**)
62%
+
SiMe₃
SiMe₃
(**48**) 91% (**49**) 9%

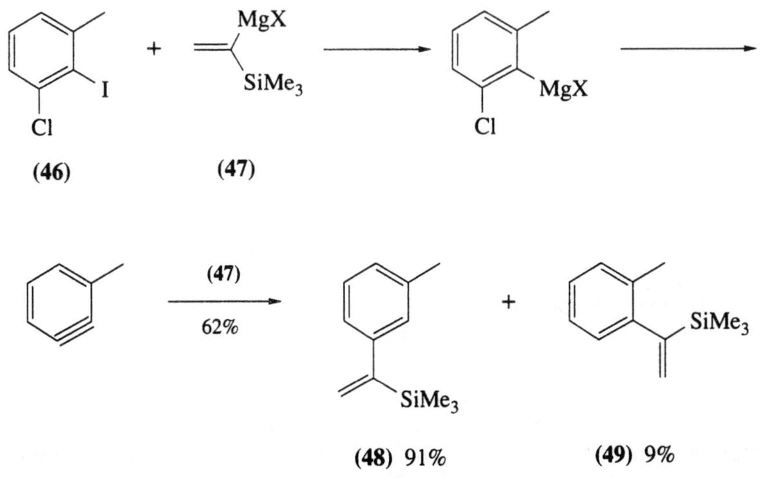

(12)

It seems that the repulsive steric interactions play a more dominant role in regioselectivity of aryne reactions than is sometimes realized. In fact, it has been argued that in nucleophilic addition to arynes, the transition state is reached early, while the incipient bond is still very much extended. Consequently, steric effects were considered not to be of great importance.[80,81] It should, however, be noted that the 'dehydro bond' orbitals are so oriented that the optimal approach trajectory for the nucleophile lies in the plane of the *ortho* substituent (**50**). Thus, if the nucleophile is large or is heavily solvated/associated, then considerable steric hindrance to attack at the aryne carbon adjacent to a substituent can be expected.

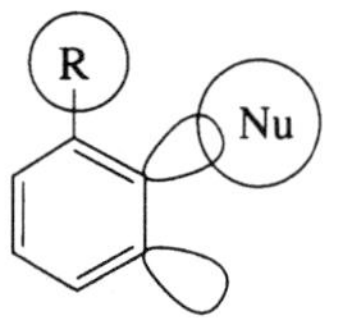

(50)

In addition of organometallic reagents to some arynes, prior counterion complexation with the substituent can direct the incoming group to the *ortho* position (kinetic control). Addition of alkyllithiums to oxazolinyl (OXZ) aryne (**51**) to give the *ortho* product (**52**) is explained in this manner. In contrast, lithium dialkylcuprates add to the aryne (**51**) exclusively at the *meta* position. This is ascribed to thermodynamic control of the reaction, which results in the formation of the more ligated and stable adduct (**53**).[82] Control of nucleophilic addition to arynes by complex-induced proximity effects has not been explored with substituents other than OXZ,[83] but has considerable synthetic potential if it can be achieved, say through solvent manipulation.

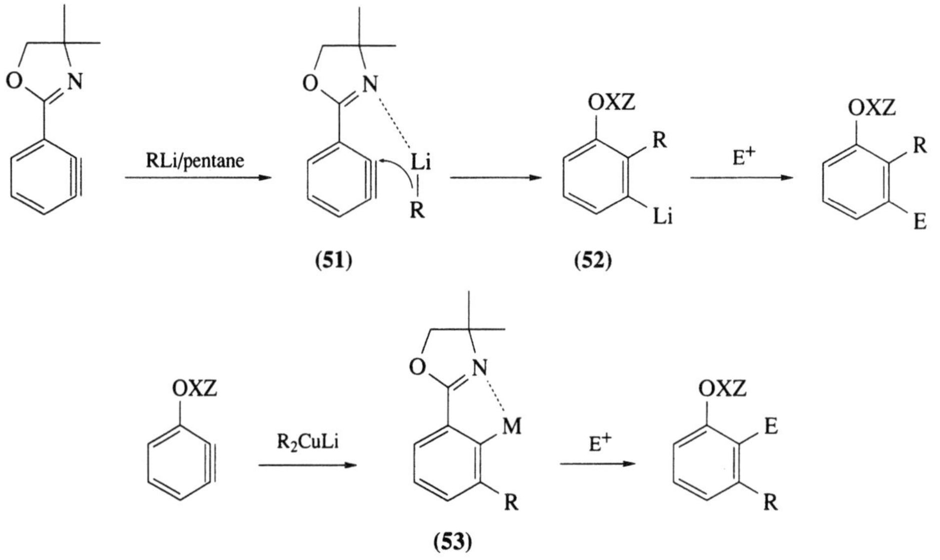

RLi/pentane

(51) **(52)**

R_2CuLi

(53)

Scheme 8

In reactions of 4-substituted benzynes, both the inductive and steric effects are less pronounced and nearly equal addition at the two ends of the 'dehydro bond' occurs with strong nucleophiles. Weak nucleophiles, however, do show some positional selectivity. For example, the *para:meta* ratio in addition of amide ions to 4-chlorobenzyne is near unity but is 4.9 for the addition of ammonia. It was suggested that in the transition state for the addition of weaker nucleophiles, the incipient bond is more developed and there is greater transfer of negative charge. Consequently, sensitivity to the electronic effects of the substituents is heightened.[76] The recent work of Hart provides the only example of high *meta/para* regioselectivity in the addition of a strong nucleophile. Reaction of tetrahalide (**54**) with arylmagnesium bromide gave (**57**) along with a negligible amount of *m*-terphenyl. The pronounced selectivity in the addition of the Grignard reagent to the 'dehydro bond' was traced to the necessity of keeping the two negative charges away from each other in (**55**) and (**56**).[84]

In the case of 3,4-pyridynes there is some preference for nucleophilic attack at the *meta* position, because the 4-pyridyl anion is more stable than its 3-isomer. The selectivity, however, is marginal and disappears altogether when strong nucleophiles are involved. On the other hand, 2,3-pyridynes afford only C-2 adducts. It has been argued that nucleophilic attack at C-3 is highly disfavored because the negative charge developing at C-2 experiences a strong coulombic repulsion from the nitrogen lone pair. How-

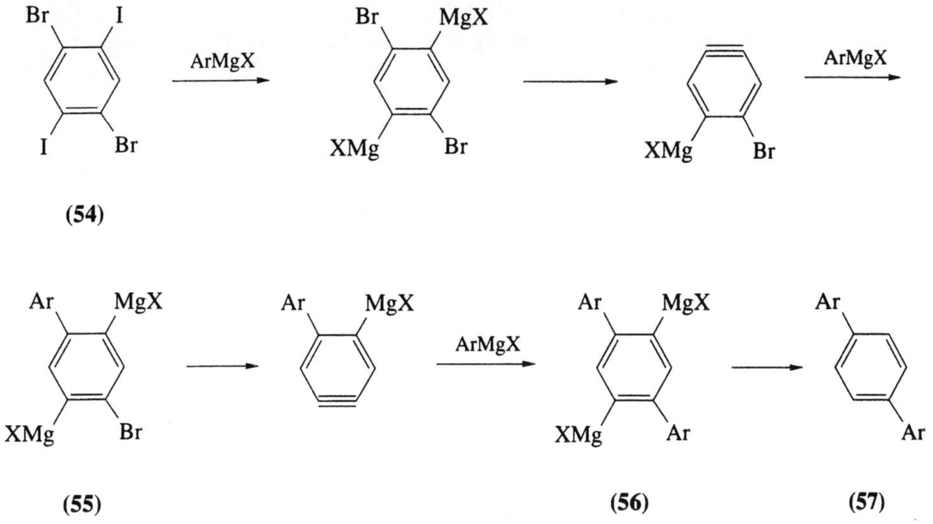

(54)

(55) **(56)** **(57)**

Scheme 9

ever, pyridyne intermediacy in substitution reactions of 2-halopyridines is not fully established. An addition–elimination mechanism can also explain the exclusive formation of 2-substituted products.[24]

2.3.3.3 Use of Arynic Substitution in Synthesis

Arynic substitution is a versatile technique of functional group transformation in aromatic systems and has found varied applications in preparation of simple compounds and in multi-step synthesis.[3,4,5] The present section comprises examples illustrative of its synthetic scope. Attention is also drawn to some allied strategies which, when used in conjunction with the nucleophilic coupling of arynes, have opened convenient routes to complex natural products.

2.3.3.3.1 Intermolecular addition of anionic nucleophiles

A variety of substituted aromatic compounds have been prepared through addition of anionic nucleophiles to arynes generated from readily accessible precursors.[1] Most of the laboratory preparations start with aryl halides. The coupling yields are usually good to modest (equations 13–15) but can be poor (equation 16).[85] Sometimes, a dramatic improvement in reaction efficiency can be achieved by the change of the base/solvent pair or other reaction conditions. For instance, in arylation of phenoxides and benzenethiolates, a switch over to DMSO as the solvent boosted the yield considerably (equation 17).[86] Another example, illustrative of this point, is the reaction of *N*-methylpyrrolidone with aryl halides where an acceptable yield could not be obtained under a variety of conditions except with LICA in THF (equation 18).[71]

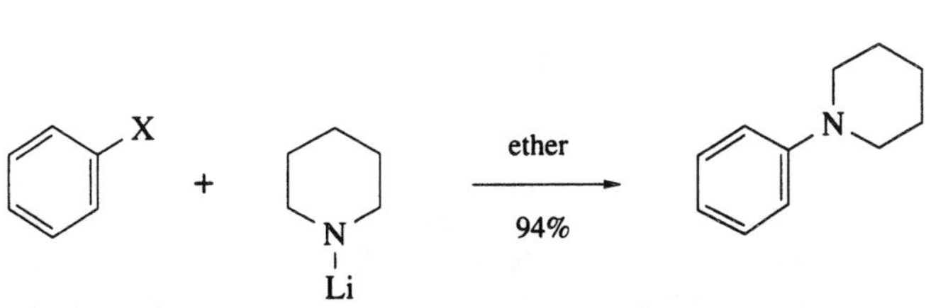

(13)

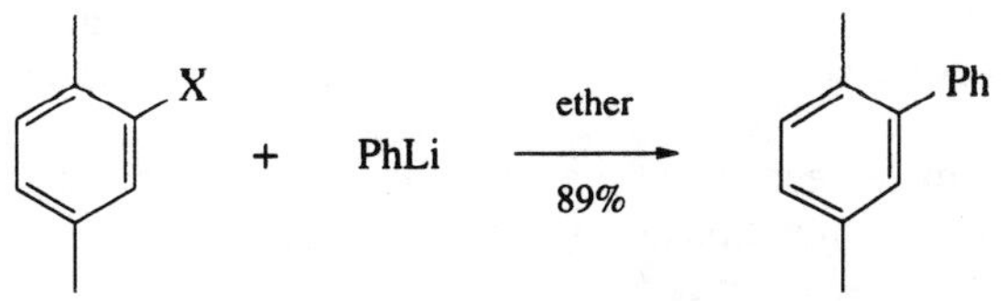

(14)

X + Ph_3CK —ether, 46%→ Ph_4C (15)

MeO, MeO, Br + O, O, O —$NaNH_2$/THF, 13%→ Ar, O, O, O (16)

X + PhSH —KNH_2/NH_3, 42%, or $KOBu^t$/DMSO, >90%→ PhSPh (17)

RO, RO, Br + N, Me, O —LICA/THF, 35%→ Ar, N, Me, O (18)

The synthesis of aromatic compounds through unsymmetrical arynic intermediates entails regioselectivity in the addition step. Possibility of *cine* substitution, however, can be turned into an advantage for getting isomers which are not easily accessible otherwise. Use of oxozolinyl aryl halides for such introduction of substituents was mentioned in Section 2.3.3.2. Some other examples are given in equations (19)–(21).[87–89]

RO, RO, N, Me, Br, MeO —KNH_2/NH_3→ RO, RO, N, Me, MeO, NH_2 ——→ Aporphine (19)

OMe, X —KNH_2/NH_3→ OMe, NH_2 (20)

F, F —$NaNH_2$/EtC(Me_2)OH, Et_2NH/THF→ F, NEt_2 (21)

Polyhalogenated arenes can function as diaryne equivalents *via* the tandem sequence illustrated in Scheme 10.[90,91] The predominant *meta* addition to aryne (**59**) can be ascribed to steric hindrance, while in addition to (**58**) both steric and electronic effects favor the observed regioselectivity.

If the moeity adding to the aryne carries a suitably placed electrophilic center, cyclization with the initially formed anion can occur as observed in the synthesis of dibenzothiophene (**60**).[92] Of wider synthetic import in this context is enolate addition to arynes. Here, after the initial ring closure many types of products can be formed (Scheme 11).[35] However, with some substrates, a particular reaction course can predominate;[93,94] for example, reaction of cyclodecanone with bromobenzene provides an efficient method for benzo-annulation and ring expansion. Similarly, high yields of benzocyclobutenols can be obtained from the monoketals of α-diketones.[95]

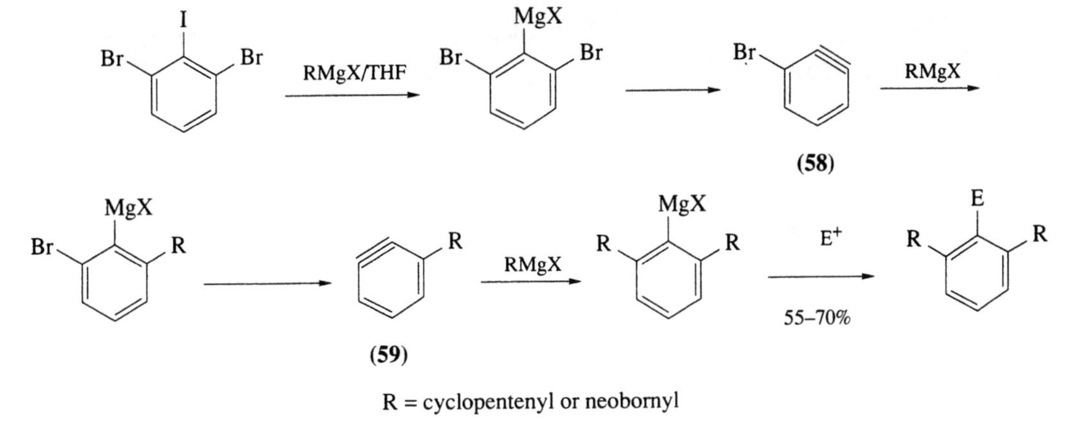

R = cyclopentenyl or neobornyl

Scheme 10

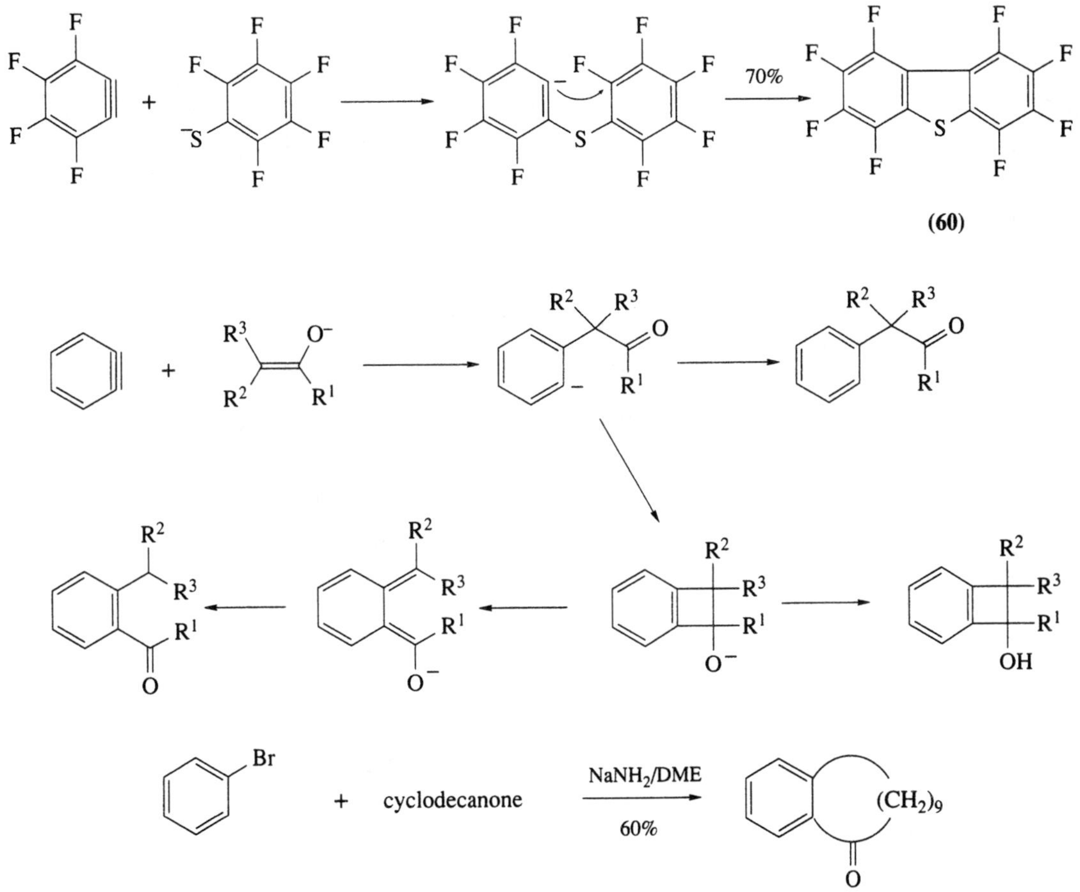

Scheme 11

The initial adducts formed from ester dienolates and arynes undergo ring closure to give naphthols in moderate yield (equation 22). Phthalides enter into a similar reaction to furnish 10-hydroxyanthrones which can undergo ready air oxidation to anthraquinones (equation 23).[96,97] The latter can be obtained directly from the phthalides bearing a leaving group at position 3. Thus using cyanophthalides, a number of naturally occurring anthraquinones and a precursor (**61**) of the antineoplastic drug 4-dimethoxyduanomycin have been synthesized (Scheme 12).[5b,98] Aryne reactions of anthranilic acid esters have been used to get acridones like acronycine (**62**).[99]

The nitrile group present on a nucleophile adding to an aryne can also undergo subsequent reactions. Simplest of these is the formation of an imine, which on hydrolytic work-up gives a cyclic ketone.[100] However, further imine rearrangement (**63** → **64**) can occur to give ring-opened products in which two

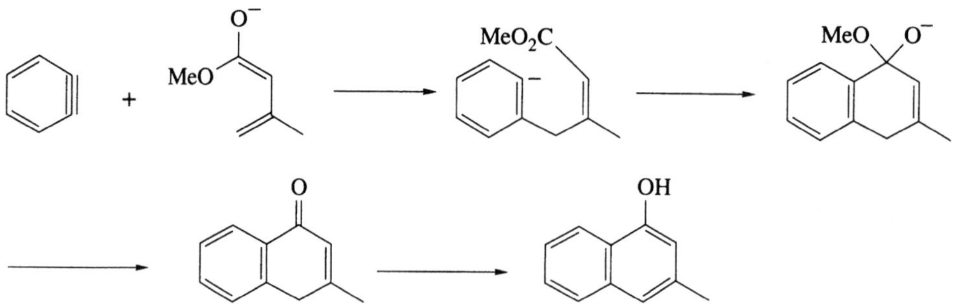

(22)

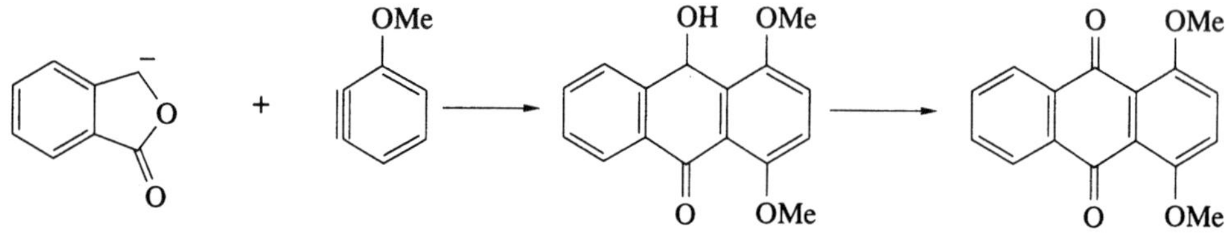

(23)

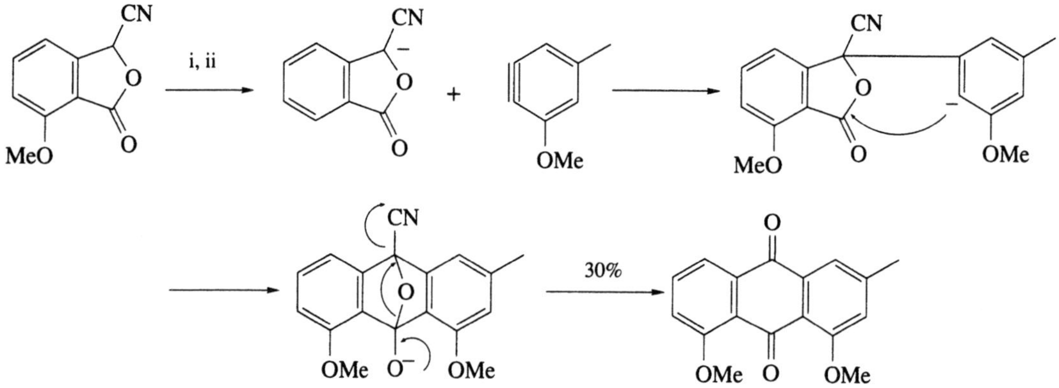

(61)

i, LDA (3 moles), THF, –78 to –40 °C; ii, 2-chloro-5-methylanisole

Scheme 12

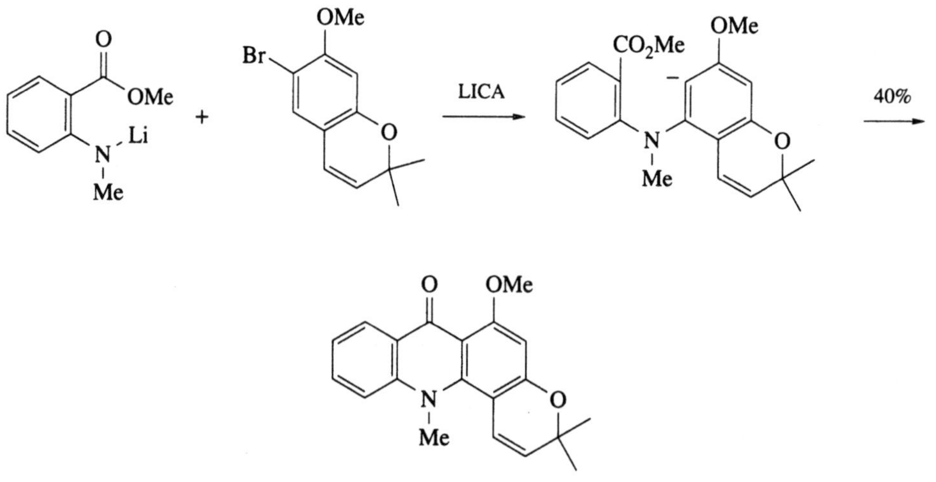

(24)

(62)

new groups are attached to the aromatic substrate (equation 25).[101] A number of substituted carboxylic acids and aldehydes have been synthesized in this manner and a new aryne annulation protocol has been developed.[5b,102] This rearrangement needs some driving force which may be provided by the ligation of lithium with the OXZ group. In the case of benzyl nitriles, which also give rearranged products, it may come from the additional stability of (**65**) due to charge delocalization. In the absence of such features, only simple adducts are obtained from the reactions of aryl halides with alkyl nitriles. Also, if ammonia is used as a solvent in these reactions, rapid protonation of the first-formed adduct anion may preclude all subsequent events.

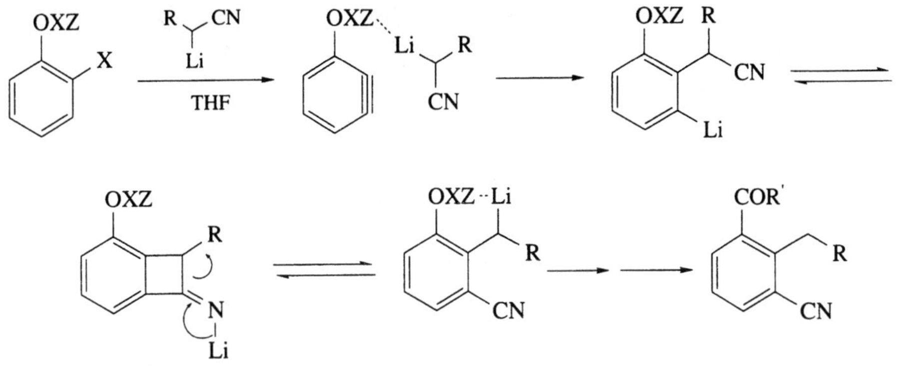

OXZ X R CN Li THF OXZ Li R CN OXZ R CN Li

OXZ R N Li OXZ Li R CN COR' R CN (25)

(**63**) (**64**) R' = H, OH

Ph CN Ph CN (26)

(**65**)

2.3.3.3.2 *Intramolecular addition of anionic nucleophiles*

One of the most productive applications of nucleophilic aryne coupling is for the construction of benzo-fused polycyclic systems (**66** → **68**).[66] This procedure, often called the benzyne cyclization, has been extensively used for making four-, five- and six-membered rings but larger systems have also been obtained. Its scope is extended because of the fact that some nucleophiles which are sluggish in intermolecular aryne reactions participate readily in ring formation due to favorable entropic factors. A *meta* substituted precursor like (**69**) can also be used, if it is more readily accessible than the *ortho* isomer, although the yield may be inferior due to some benzyne formation on the wrong side. Cyclizatioin of *m*-isomers can sometimes be more efficient, perhaps due to prior complexation of the proton-abstracting base with the side chain nucleophile. In such a case, not only abstraction of the correct proton (H^1 in **67**) occurs but also the nucleophile is kept poised for reaction at the time of aryne generation. Usually aryl

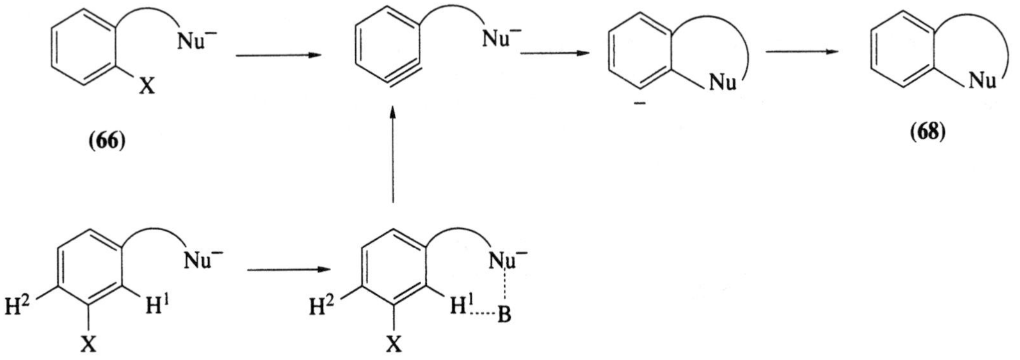

Nu^- X Nu^- Nu Nu

(**66**) (**68**)

H^2 H^1 Nu^- X H^2 H^1....B Nu^- X

(**69**) (**67**)

Scheme 13

bromides afford better yields of cyclic products than the chlorides. These general aspects of the benzyne cyclization are illustrated in equations (27)–(31).[103–107]

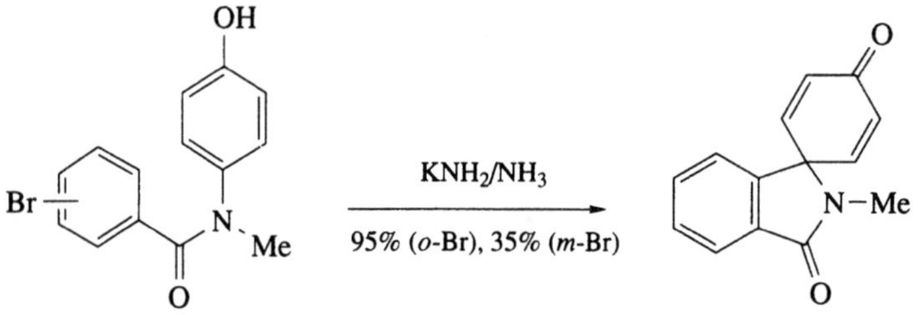

KNH$_2$/NH$_3$ 95% (*o*-Br), 35% (*m*-Br) (27)

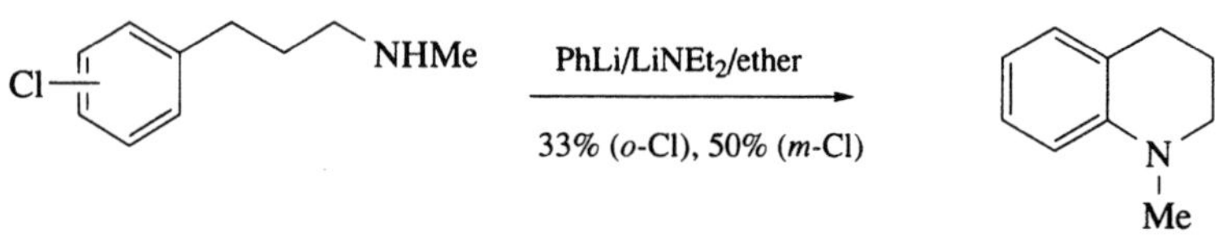

PhLi/LiNEt$_2$/ether 33% (*o*-Cl), 50% (*m*-Cl) (28)

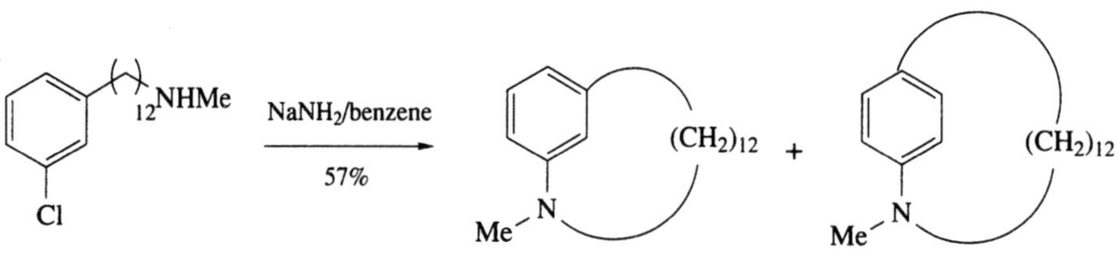

NaNH$_2$/benzene 57% (29)

90% 10%

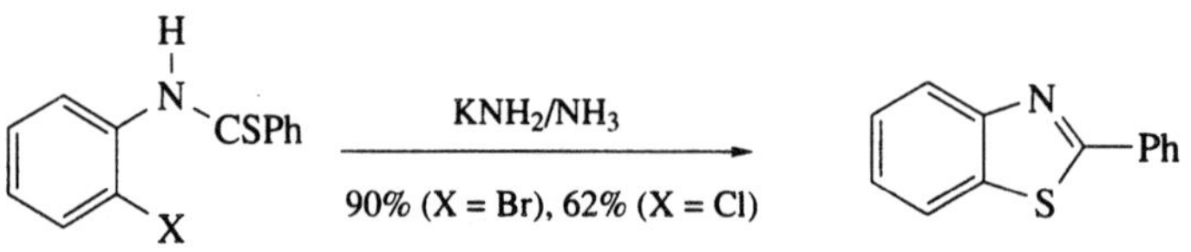

KNH$_2$/NH$_3$ 90% (X = Br), 62% (X = Cl) (30)

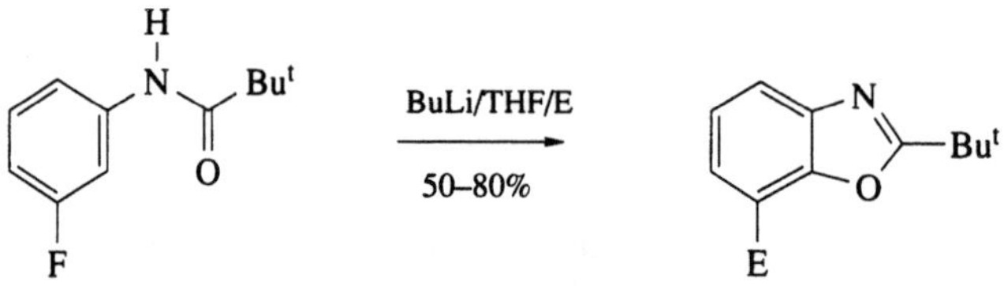

BuLi/THF/E 50–80% (31)

Cyclization of side chain nitriles has found extensive use in the synthesis of benzocyclobutenes (**70**; *n* = 2),[104] the versatile synthons which open on mild thermolysis to give *o*-quinodimethanes for inter- and intra-molecular [4 + 2] trapping.[108] The nitrile group in (**70**) can be manipulated into a variety of functionalities for appending the dienophile portion. For example, in the synthesis of chelidonine, the nitrile (**71**) was converted, by hydrolysis followed by Curtius degradation and reaction of the formed isocyanate with benzyl alcohol, to a urethane (**72**). The latter was then condensed with a benzyl bromide to get the compound (**73**), which was elaborated further as shown in Scheme 14.[109]

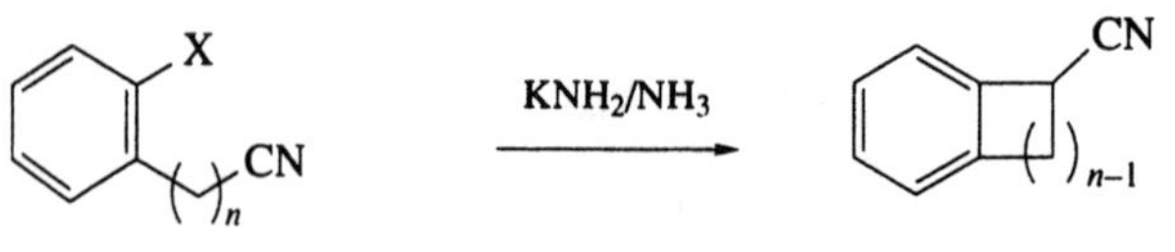

KNH$_2$/NH$_3$ (32)

(70)

The benzyne route to 1-cyanobenzocyclobutenes has been employed for the synthesis of a myriad of complex natural products. The cyano group in (**74**) was used to attach the dienophile as well as for building the imino bridge of (**75**), an important precursor for some diterpene alkaloids (Scheme 15).[110]

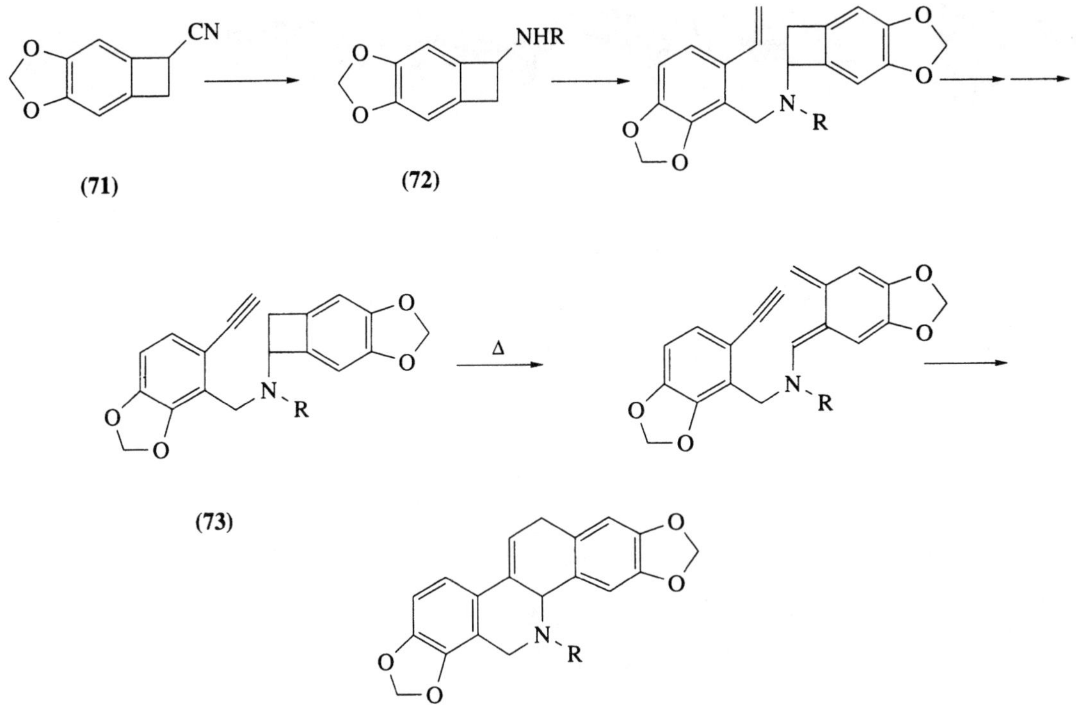

(71) (72)

(73)

Scheme 14

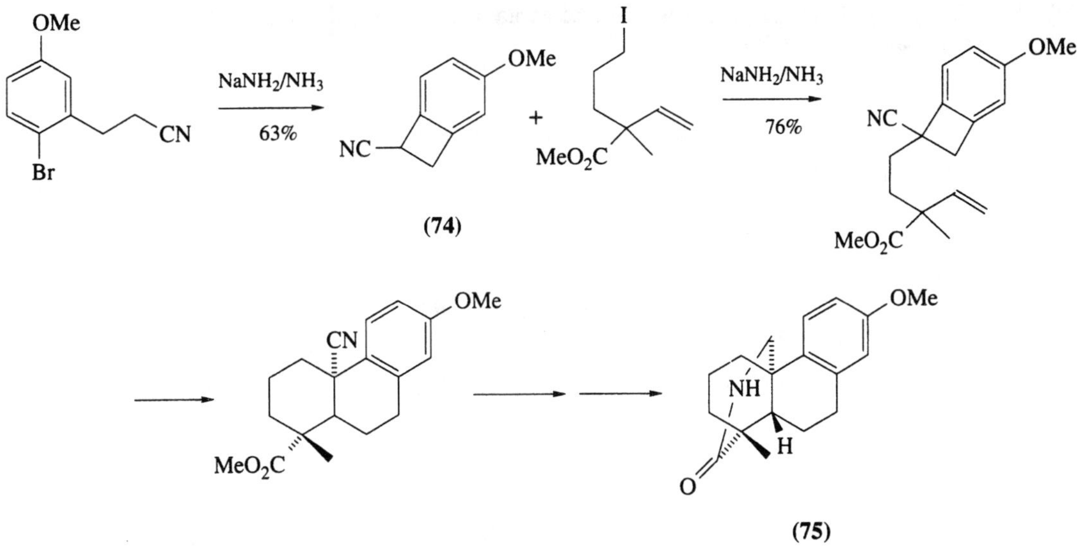

$NaNH_2/NH_3$ 63%

(74)

$NaNH_2/NH_3$ 76%

(75)

Scheme 15

Attachment of a crotonate chain to the alcohol (**78**), through a mixed carbonate ester, sets up the intramolecular Diels–Alder reaction in the synthesis of podophylotoxin.[111] The starting benzocyclobutenol (**78**) was procured by the sequence of steps shown in Scheme 16. A similar route using a carboxylic ester, in place of the nitrile (**76**), for the benzyne cyclization gave a mixture of *cis*- and *trans*-benzocyclobutenes but ester hydrolysis led to the pure *trans* acid (**77**).[112]

Among other elaborations of the cyano group of the benzocyclobutenes are Na–NH_3 decyanation, DIBAL reduction, and side chain elongation as used in the synthesis of estrogenic steroids.[113–115] Based on this methodology, synthesis of some tetracyclic triterpenes[116] and alkaloids like yohimbine[117] and xylopinine[118] have also been reported.

Benzyne cyclization of the side chain nitriles or carboxylic esters has also been employed for constructing other rings as shown in the synthesis of *N*-methylisoindole (**79**) and the lysergic acid precursor

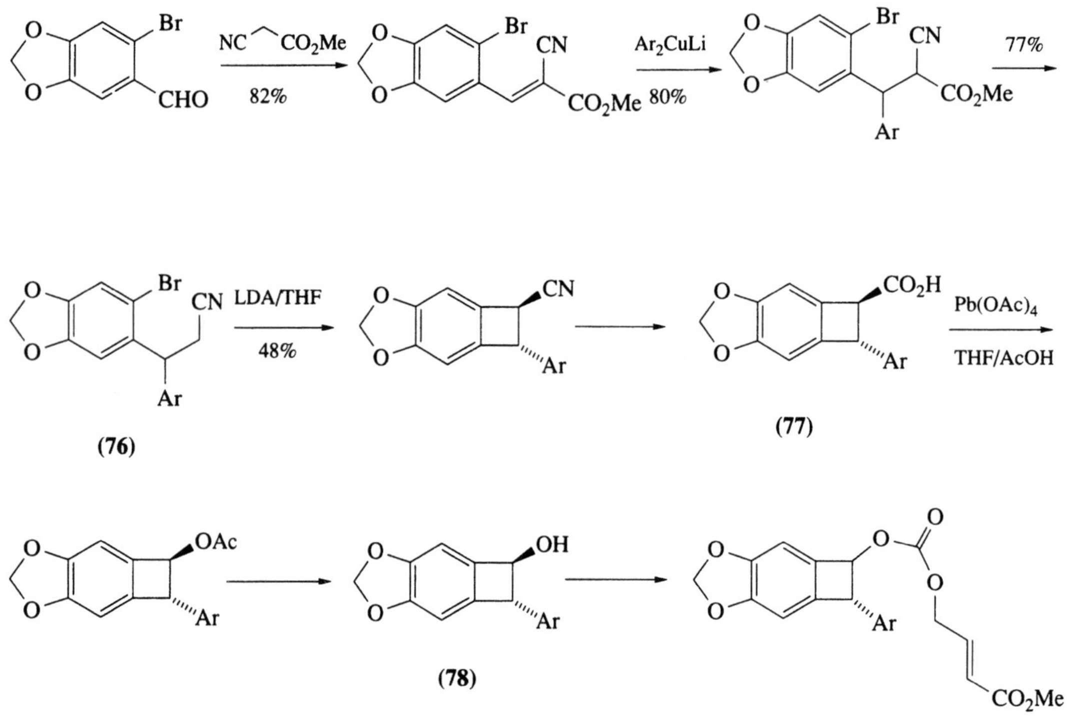

(76)

(77)

(78)

Scheme 16

(**80**).[119,120] The low yield in the latter case may be explained in terms of competitive aryne ring closure with the nitrogen atom to give (**81**).

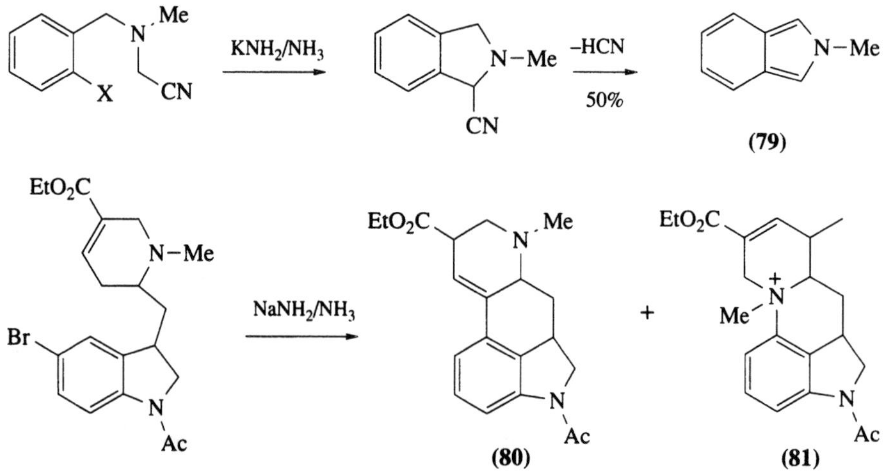

(79)

(80)

(81)

Scheme 17

Competition between two nucleophilic sites participating in an arynic cyclization depends upon their relative nucleophilicities as well as on the size of the rings being formed. For example, the bromo compound (**82**) could be cyclized to a lycorane smoothly because coupling with the nitrogen atom would lead to a four-membered ring.[121] In contrast, a similar reaction of enaminones (**83a**) and (**83b**) gave mostly five- or six-membered ring *N*-phenylation products (**84**).[122] Similarly, in cephalotaxine synthesis, closure to a seven-membered ring failed under the usual benzyne-generating conditions. The product (**86**) could be obtained with potassium triphenylmethide as the base, but still the yield was low. A nickel complex mediated cyclization was somewhat more efficient, while irradiation of (**85**) under $S_{RN}1$ reaction conditions[123] afforded (**86**) in 94% yield.[124] This is an example where a nonaryne procedure for ring closure was clearly advantageous.

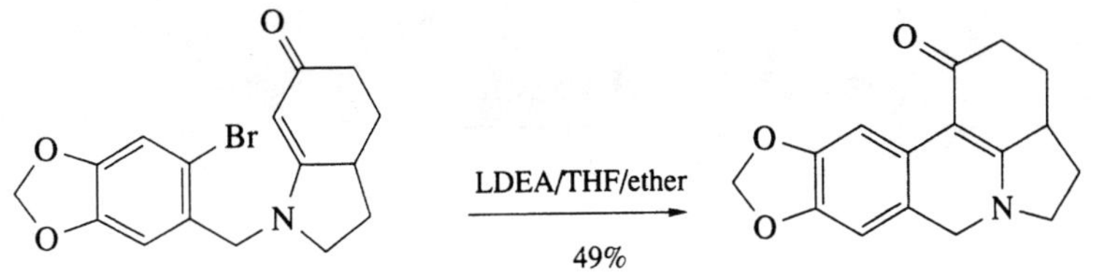

LDEA/THF/ether, 49% (33)

(82)

(34)

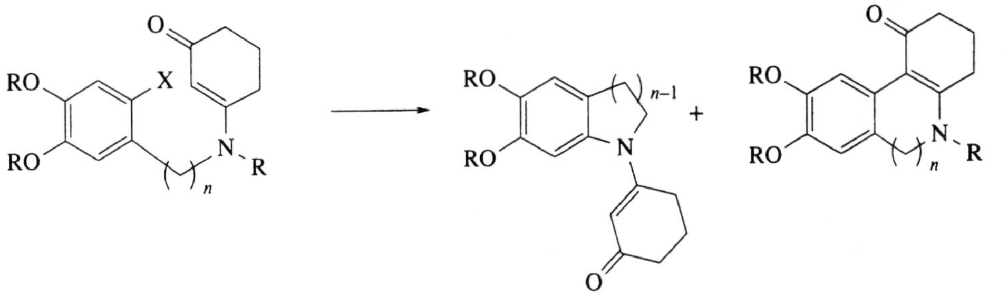

(83) **a**: $n = 2$; **b**: $n = 3$; R = H or alkyl **(84)**

(35)

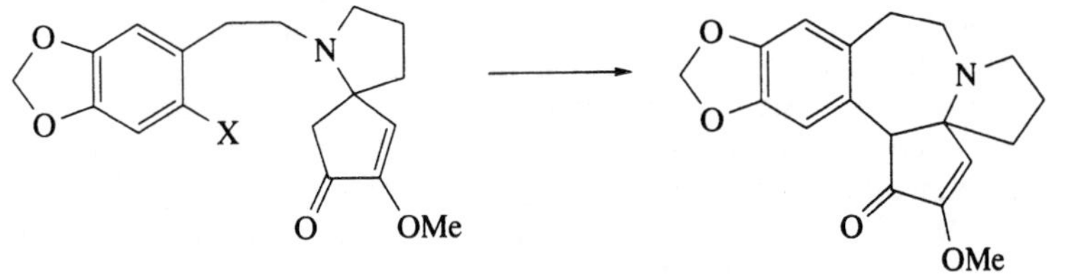

(85) **(86)**

Carbanions derived from side chain tertiary amides have also been cyclized to provide isoquinolones and isoindoles (equation 36).[125,126] While benzyne intermediacy in the formation of the former is likely, the latter seems to arise through a $S_{RN}1$ reaction pathway. Synthesis of indole from the *meta* bromo compound (**87**), on the other hand, clearly involves an aryne cyclization.[127] A more versatile route to indoles is based on intramolecular addition of aminyl anions to arynes (equation 38).[128] A somewhat similar dihydroindole preparation constitutes the first step in a synthesis of lycoranes (equation 39).[129] The synthesis of (**88**) also falls in the same category of reactions, but it is noteworthy because only a few examples of ring closure of heteroarynes are mentioned in literature.[27,28]

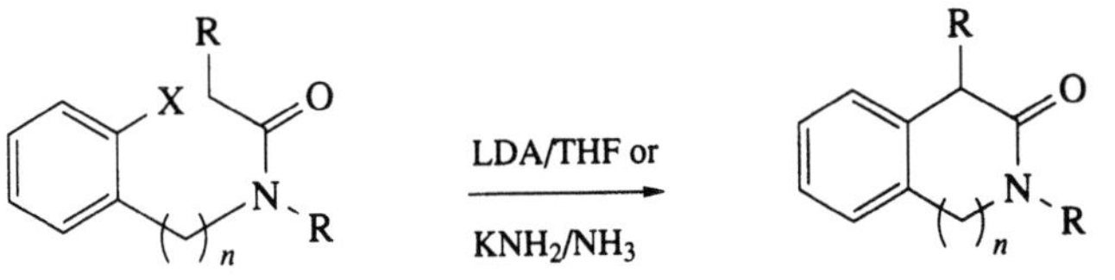

LDA/THF or KNH_2/NH_3 (36)

$n = 0$ or 1

$NaNH_2/Bu^tONa$, 56% (37)

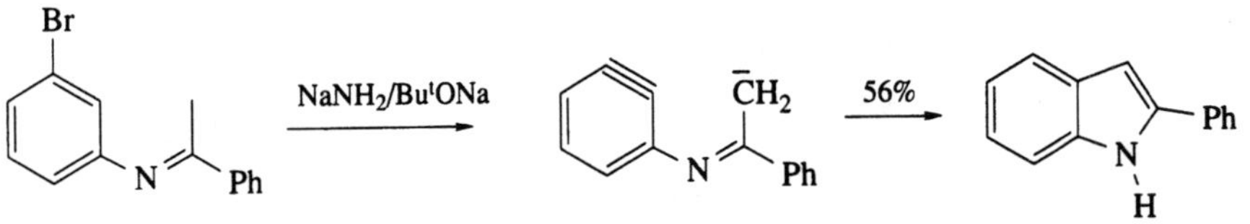

(87)

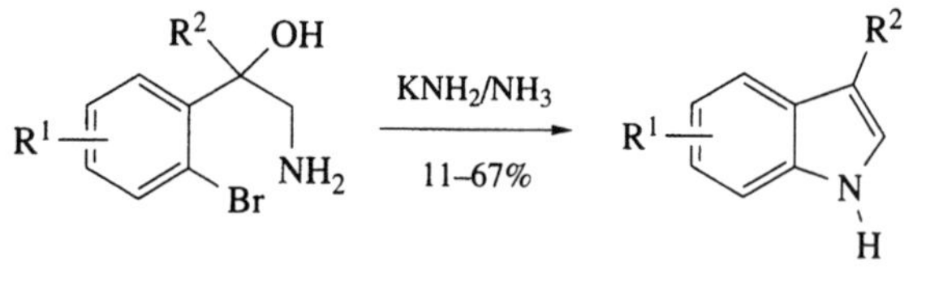

KNH$_2$/NH$_3$ 11–67% (38)

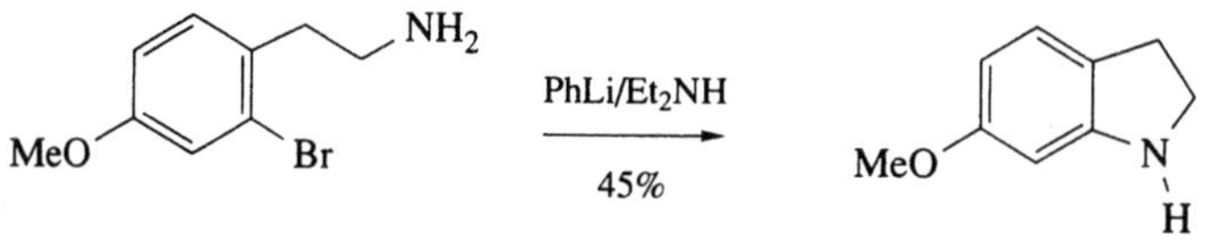

PhLi/Et$_2$NH 45% (39)

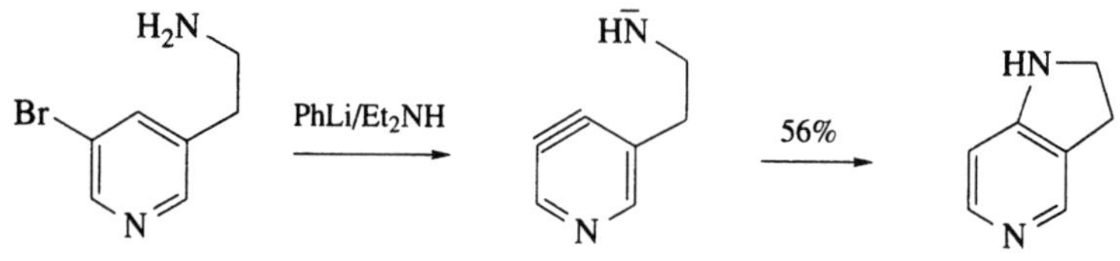

PhLi/Et$_2$NH 56% (40)

(88)

Normally aromatic systems do not function as effective nucleophiles towards arynes even in intramolecular reactions. For example, the aryne (**89a**), generated from the corresponding halide and MNH$_2$ in liquid ammonia, leads only to amine (**92**) and quaternary compound (**93**). Since (**93**) tends to undergo Hoffmann degradation under the basic reaction conditions, it is preferable to cyclize the *N*-demethyl substrates and carry out the quaternization subsequently to prepare dibenzoindolizidines. The alkaloids cryptausoline (**96a**) and cryptowoline (**96b**) have been synthesized in this manner.[130,131] To forge a biphenyl link by the benzyne route, advantage can be taken of the fact that appendage of a negatively charged atom to an aromatic ring confers strong nucleophilicity at the *ortho* and the *para* positions. Thus, phenolic isoquinolines afford, through the phenoxide (**89b**), aporphines (**90**) in 20–30% yield. In these reactions, dienones (**91**) arising from aryne attack at the *para* position are also formed, albeit in poor

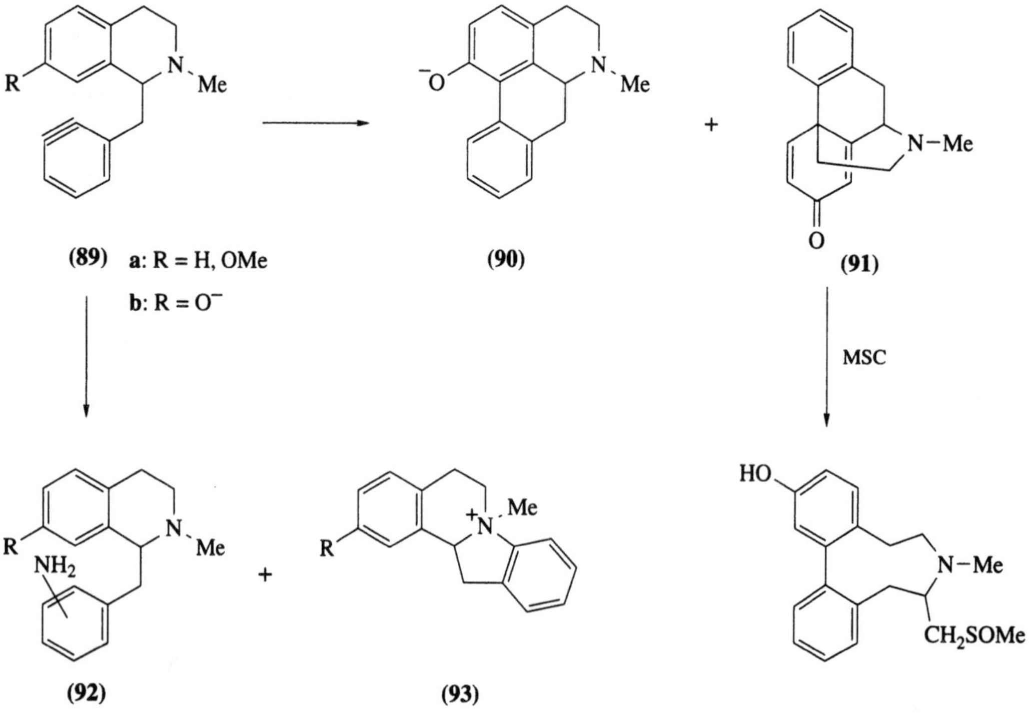

(89) **a**: R = H, OMe
b: R = O$^-$

(90)

(91)

MSC

(92)

(93)

Scheme 18

yield.[132–137] If methanesulfinyl carbanion (MSC) in DMSO is employed as the base, ring opening of (**91**) can take place.[138] This base–solvent system affords acceptable yields of aporphinoids from isoquinolinium methiodides even in the absence of anionic activation.[139] Similar conditions have also been used to get cularine alkaloids through a cyclization which involves addition of an oxyanion to the arynic bond (**97** → **98**).[140]

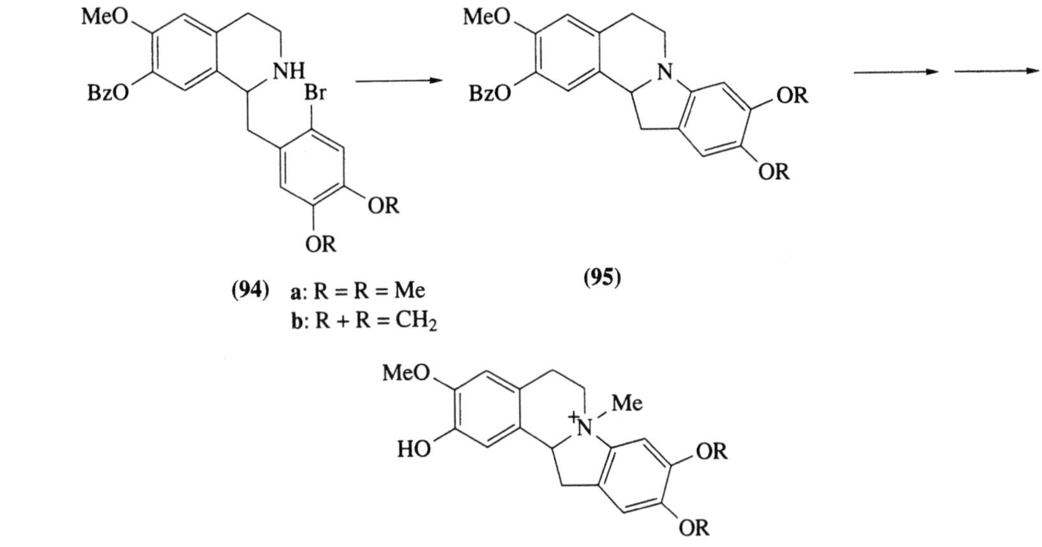

(**94**) **a**: R = R = Me
b: R + R = CH_2

(**95**)

(**96**) (41)

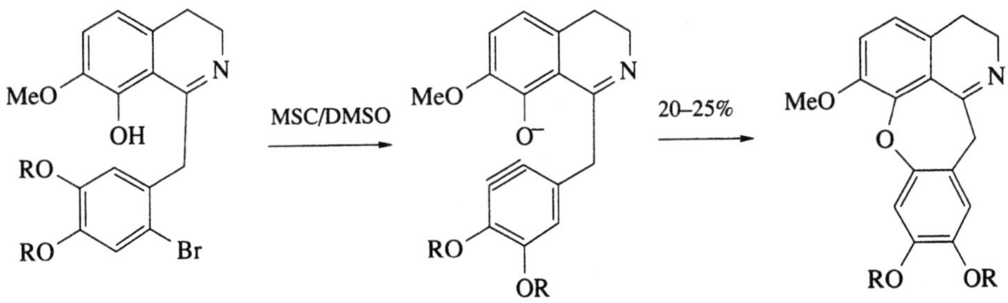

(**97**) (**98**) (42)

The above mentioned coupling of arynes with oxyanion-activated aromatic rings proceeds in modest yields. Since nitrogen has a greater proclivity to share negative charge, aminyl anions should serve as more efficient aryl activators. This concept forms the basis of a benzyne route to phenanthridines.[5a,141] Thus, treatment of the chloro compound (**99**) with KNH_2/NH_3 affords dihydrophenanthridine (**100**), almost quantitatively. Under similar conditions, phenanthridine (**103**) can be obtained directly from anil (**101**) even though the rings to be joined are *trans* disposed. Addition–elimination of amide ions across the azomethine bond has been postulated to explain this result.[141] Many substituents and heterocyclic rings are compatible with this cyclization and a variety of systems have been synthesized.[142,143] Pentacyclic compounds, like (**105**), can be obtained in one step through a double benzyne cyclization.[144] Pyridynes also enter into this reaction as shown in the synthesis of the grass alkaloid perlolidine (**104**).[145] Further, this route provides a convenient access to the benzo[*c*]phenanthridine nucleus present in some alkaloids and compounds of interest in cancer chemotherapy.[146–149] Low yields in synthesis of the 8,9-oxygenated alkaloids (**106b**) can be much improved by the use of LDA/THF in the benzyne cyclization step.[37]

2.3.3.3.3 Addition of uncharged heteroatom nucleophiles

Molecules having a nucleophilic heteroatom add readily to arynes. In fact, some of these exhibit reactivity comparable to anionic nucleophiles, *e.g.* the reactivity ratio between phenyllithium and triethyl-

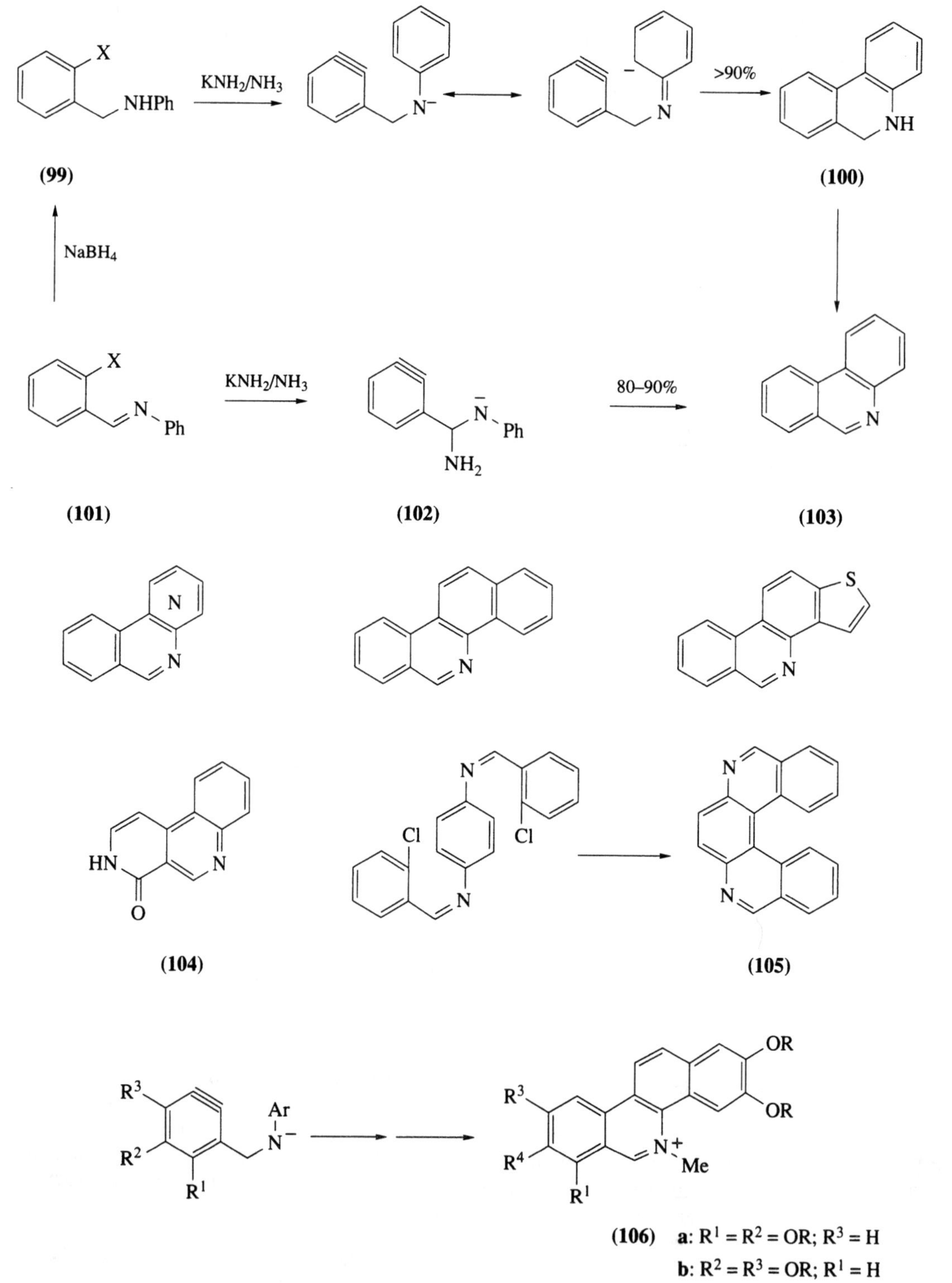

X
NHPh
KNH$_2$/NH$_3$
N$^-$
N
>90%
NH

(99)

(100)

NaBH$_4$

X
N
Ph
KNH$_2$/NH$_3$
N
Ph
NH$_2$
80–90%
N

(101)

(102)

(103)

N
N

N

S
N

HN
N
O

N
Cl
Cl
N

N
N

(104)

(105)

R^3
R^2
R^1
Ar
N$^-$

R^3
R^4
R^1
OR
OR
N$^+$
Me

(106) **a**: R^1 = R^2 = OR; R^3 = H
b: R^2 = R^3 = OR; R^1 = H

Scheme 19

amine[1] for 2,3-dehydroanisole is only 1.8. Initially a zwitterion (**107**) is formed which may abstract a proton from the surroundings. In an aprotic solvent, it can often be intercepted by addition of electrophiles or may evolve into stable products through a variety of rearrangement and elimination steps. If a hydrogen is available on the heteroatom itself (**108**), migration occurs to give neutral products. Otherwise, a proton can shift from the β- or the α-position, the former process being somewhat more facile (**109** → **110**). The new zwitterion thus formed can undergo further reorganization or fragmentation. Although such reactions normally lead to a mixture of products, structural features of the substrate can

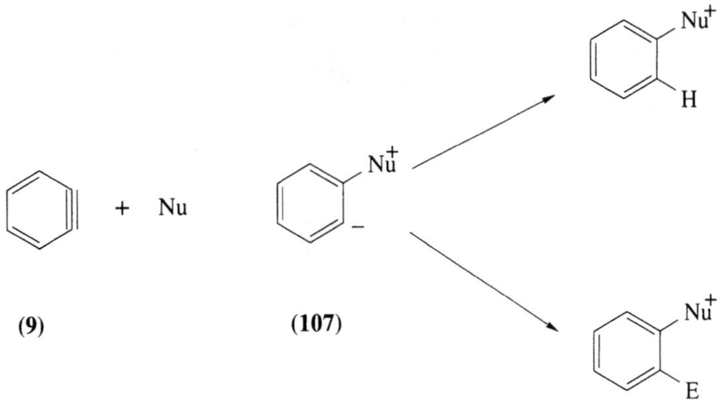

(43)

(44)

(45)

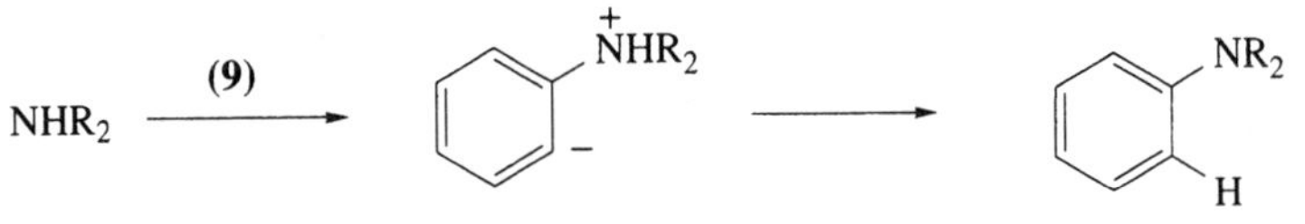

(46)

(9) (107) (108) (109) (110) (111) a: R = H b: R = SiMe 3 (112) (113)

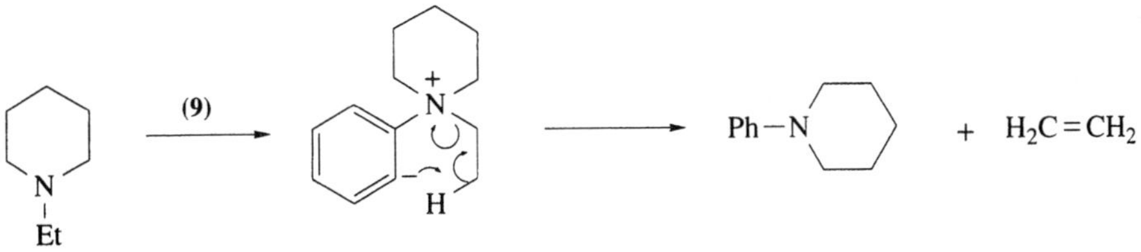

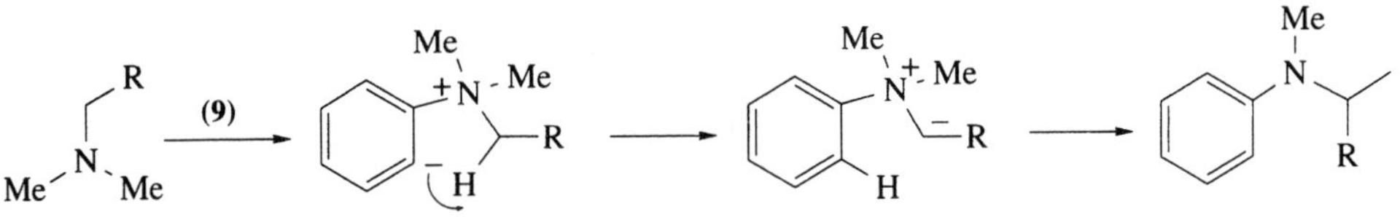

favor a particular pathway. For instance, in the reaction of trimethylamine (**111a**) with benzyne, very little product (**113a**) corresponding to Stevens rearrangement is obtained. But with silylamine (**111b**), the compound (**113b**) is formed in 60% yield;[150,151] the proton transfer leading to carbanion (**112b**) is obviously promoted by the neighboring silicon atom.

Like tertiary amines, dialkyl sulfides also react with arynes to give betaines prone to rearrangements and some of these have been put to ingenious preparative use.[152] The reaction of thiirane (**114**) with benzenediazonium carboxylate was employed to produce, in a stereospecific manner, unsaturated sulfides (**115**), which serve as valuable synthetic intermediates.[153] Certain betaines, derived from aryne reaction with sulfides, reorganize to give products in which the original sulfur ligands are linked directly. Since dialkyl sulfides are readily accessible and the sulfur pendant on the product can be removed or utilized for further functionalization, this sequence constitutes a powerful method for carbon–carbon bond formation. For example, a 3,2-sigmatropic shift in the ylide (**117**), formed through reaction of farnesyl nerolide sulfide with benzyne, gave (**118a**) which, on treatment with lithium in liquid ammonia, furnished squalene (**118b**) in good yield.[154] An attempt to bring about a similar rearrangement by forming the corresponding methylsulfonium cation, in place of the phenylsulfonium cation (**116**), led to a complex mixture of products. Another useful strategy based on aryne reaction with dialkyl sulfides (**119** → **120**) has been devised to synthesize a number of strained cyclophanedienes.[155,156]

Ethers do not exhibit marked nucleophilicity towards arynes but readily enter into reaction with more electrophilic halogenated analogs like (**121**). In the betaine derived from (**121**) and ethyl ether, a proton

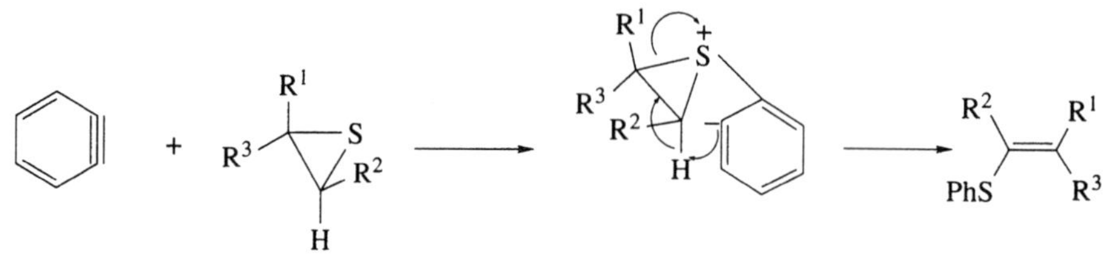

(9) (114) (115) (47)

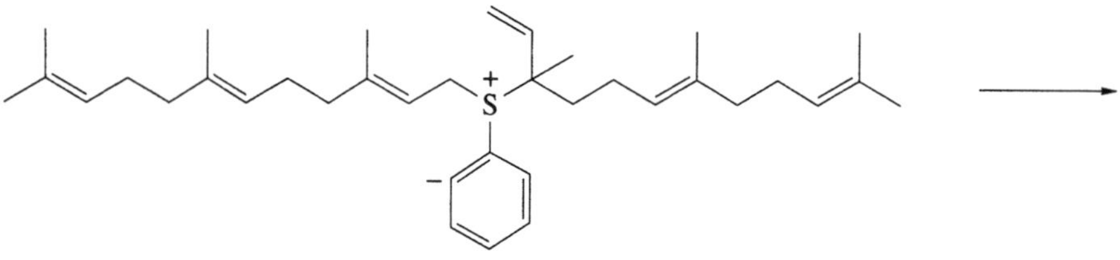

(116)

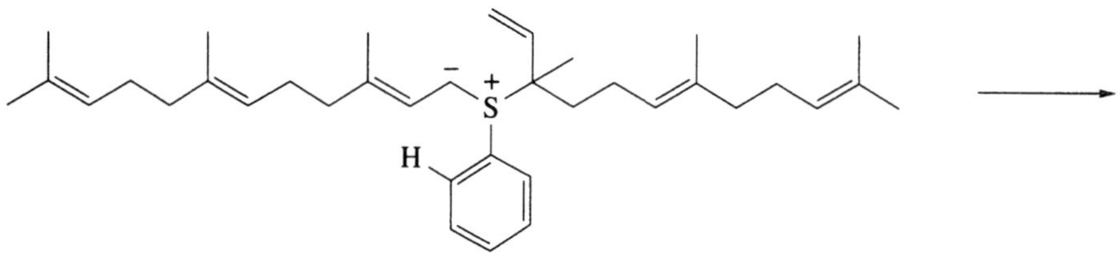

(117)

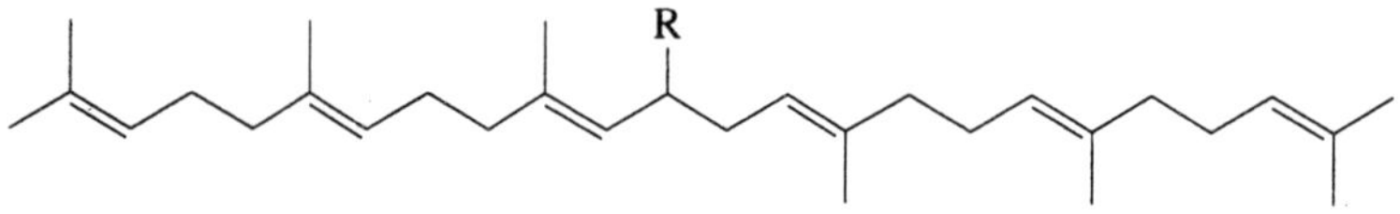

(118) **a**: R = SPh
b: R = H

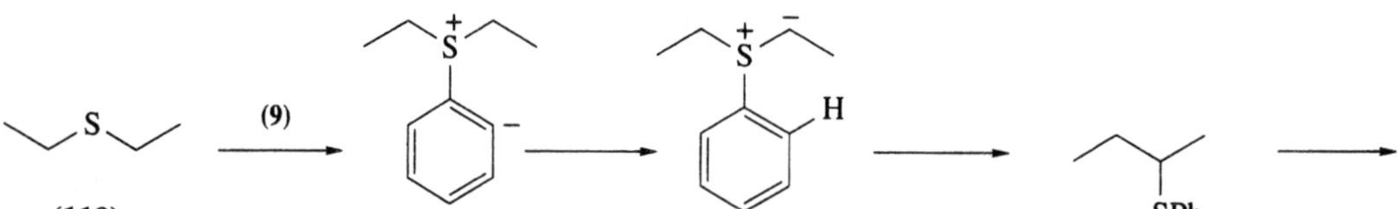

(48)

(119) (9) (120) (49)

shift is followed by alkene elimination.[157] With cyclic ethers such a shift is disfavored and good yields of products arising through ring opening by added nucleophiles can be obtained (**122** → **123**).[158] The oxygen of *N*-oxides is more reactive and even simple arynes are attacked (equation 50).[1]

Reactions of phosphines and phosphites have received some attention but their preparative value is limited. The zwitterion formed from diphenylmethylphosphine and benzyne rearranges to ylide (**124**) which can be captured by Wittig alkenation, with cyclohexanone, in about 20% yield.[159] Some synthetically useful reactions of tellurium and selenium compounds with arynes have been reported. For example, heating diphenyl iodonium carboxylate and bis(*p*-ethoxyphenyl) ditelluride in dichlorobenzene affords the compound (**125**).[160] The corresponding reactions with diphenyl selenide and diphenyl sulfide

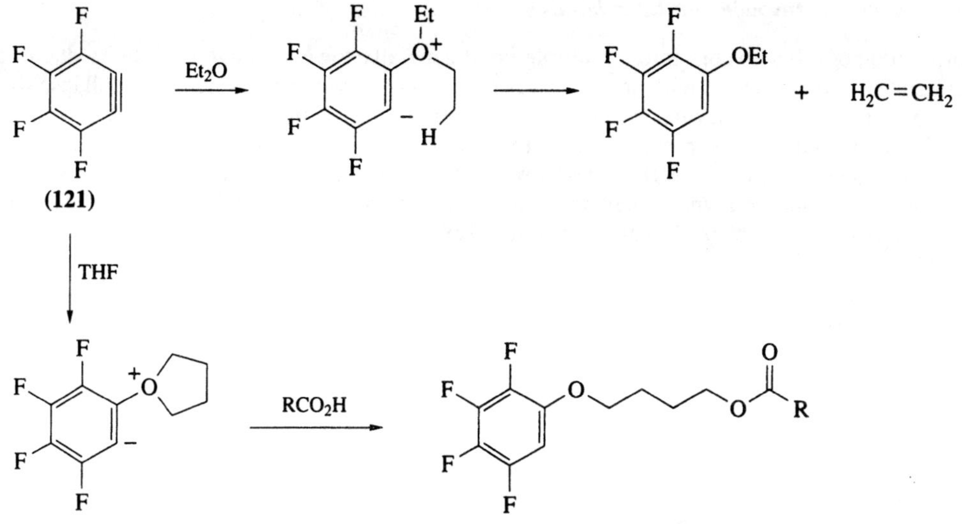

Et_2O

THF

RCO_2H

(121)

(122)

(123)

Scheme 20

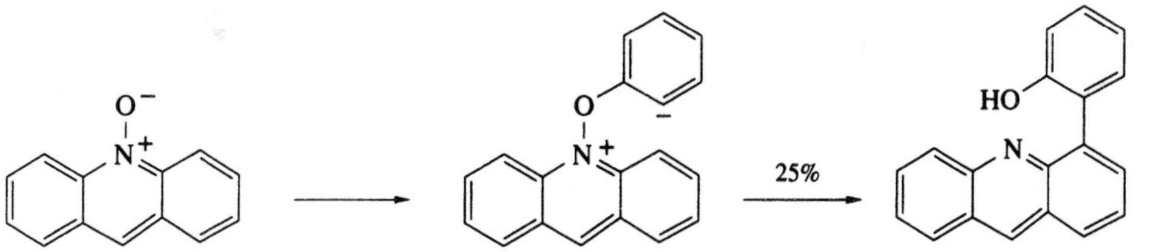

25%

(50)

proceed in slightly lower yield.[161] The selenadiazole (**126**) participates in an interesting aryne reaction to give nitrile (**127**) in excellent yield.[162]

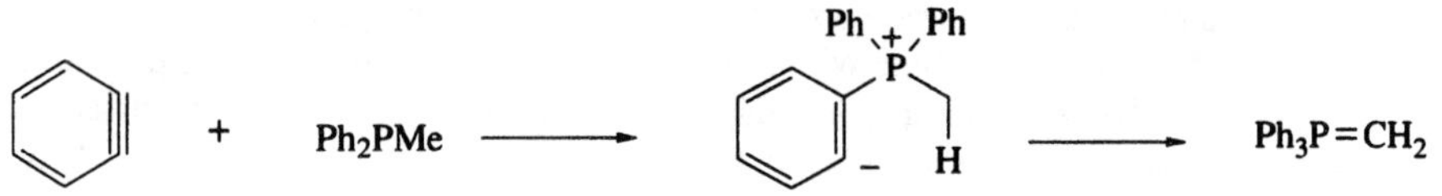

Ph_2PMe → $Ph_3P{=}CH_2$ (51)

(9) (124)

(9) 97% (52)

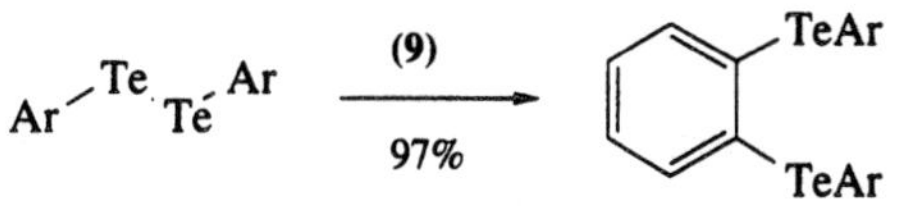

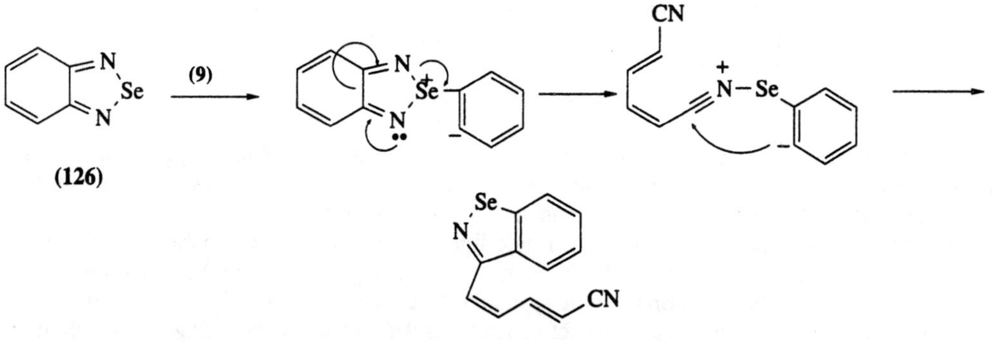

(125)

(9)

(126)

(53)

(127)

2.3.3.3.4 Additions involving multiple bonds

Incorporation of a heteroatom into a multiple bond generally lowers its nucleophilicity due to change in hybridization. Still, reaction with arynes can occur to give a zwitterion which may collapse to a four-membered ring or undergo further reactions. For example, interaction of azirine (**128**) with benzyne leads to 2,3-diphenylindole,[163] presumably by the pathway shown in the equation (54). Similarly, anils on reaction with benzyne afford acridines in low yield.[164] Even carbonyl compounds can react, but only with strongly electrophilic arynes. Generation of tetrachlorobenzyne in the presence of aldehydes and ketones was found to give benzodioxan derivatives (**129**).[165]

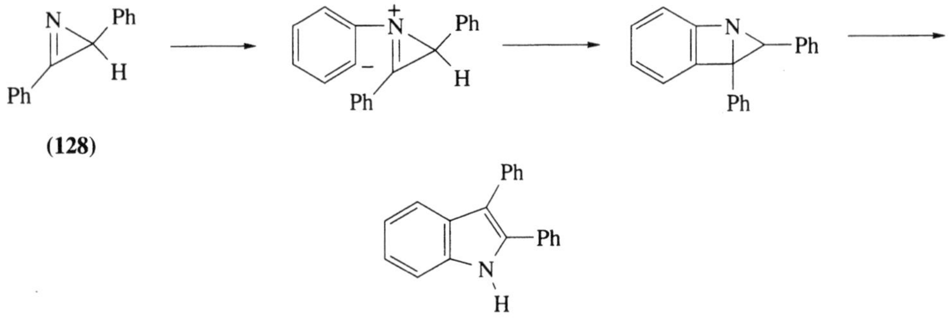

(**128**)

(54)

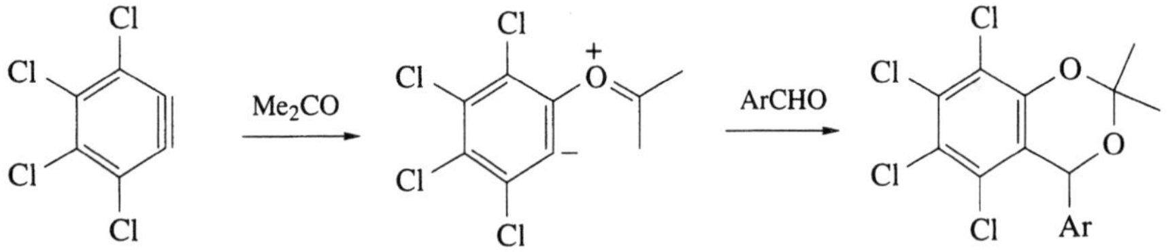

Me_2CO ArCHO

(55)

(**129**)

Alkenes on reaction with arynes afford benzocyclobutenes in low yield. Particularly, if an allylic hydrogen is present the ene reaction tends to predominate. The main interest in this area is perhaps mechanistic, because concerted [2 + 2] addition is forbidden by Woodward–Hoffmann rules. The fact that with 1,2-disubstituted alkenes mixtures of stereoisomers are obtained, although stereoretention predominates, indicates the involvement of an intermediate with a lifetime long enough to allow some bond rotation before ring closure. Extended Hückel calculations also show that the potential energy surface for this reaction has a 'valley' corresponding to an intermediate on the reaction pathway.[166,167] There has been much conjecture on the dipolar/diradical nature of this intermediate. Although calculations indicate a slight charge imbalance in it, the lack of rearrangements and the absence of solvent effects suggest diradical character. In the case of alkenes bearing strong electron donating groups, it is reasonable to consider the first step to be a nucleophilic addition. Good regioselectivity, in accordance with a dipolar intermediate, is observed in reactions of such alkenes with polarized benzynes and, importantly, the yields are high.[168,169] For example, treatment of *o*-bromoanisole with sodamide in the presence of 1,1-dimethoxyethylene gave (**130**) in 70% yield. A similar addition constitutes the key step in an elegant synthesis of taxodione (**131**).[168] In 3-methylbenzyne, inductive polarization of the dehydro bond is weak and the preferred orientation must rely on the steric effect to align the approaching nucleophile. Accordingly, a much lower selectivity is observed in the reaction given in equation (56). Reaction of *O*-silylated enolates of carboxylic esters with benzyne gives *ortho*-alkylbenzoic acids due to ring opening of the initially formed bicyclic product.[170] Monovinyl ethers and acetates also afford [2 + 2] adducts with arynes in acceptable yields (equation 57).[171] Benzocyclobutenes prepared in this manner were used for the synthesis of some protoberberine alkaloids.[172]

The carbon end of enamines should be even more nucleophilic than that of the oxygen analogs. The expected regioselectivity is observed in reactions with arynes (equation 58).[173] However, the yields of benzocyclobutenes are often low because of competitive initial reaction at the nitrogen end.[174] Silylynamines on interaction with benzyne give zwitterionic intermediates (**132**), which can undergo a 1,3-shift leading to stable products (equation 59).[175] In enamides the charge density on the β-carbon must be relatively low, yet intramolecular addition to arynes can take place (equation 60).[176]

Finally, electron rich aromatic rings can also act as nucleophiles towards arynes, *e.g.* 2-phenylanisole is the major product of Ag^+-catalyzed decomposition of benzenediazonium carboxylate in anisole (equa-

δ− δ+ MeO OMe OMe OMe OMe OMe OMe OMe OMe OMe

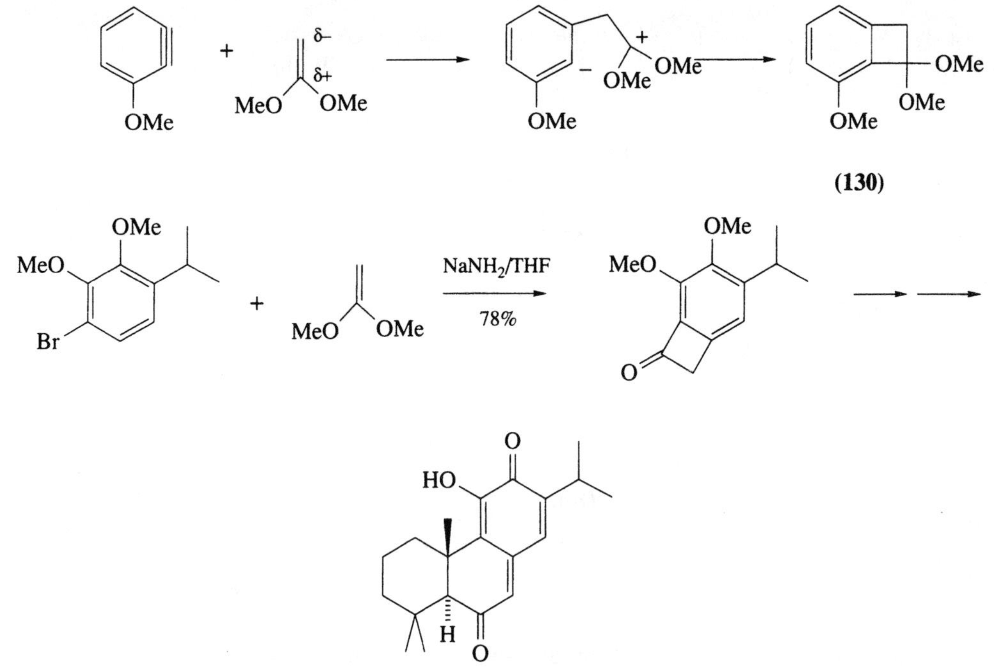

(130)

OMe MeO Br MeO OMe $NaNH_2$/THF 78% OMe MeO O

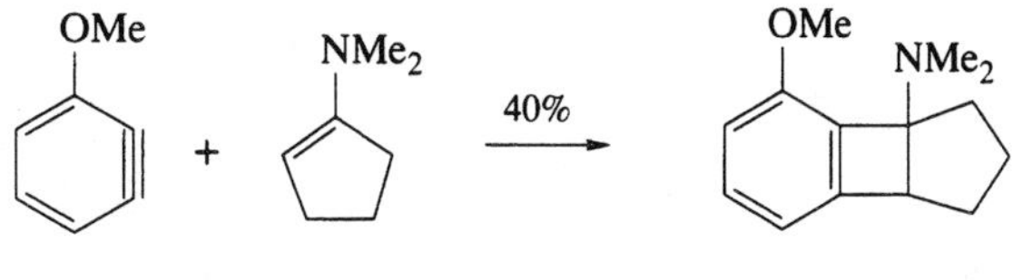

HO O H O

(131)

Scheme 21

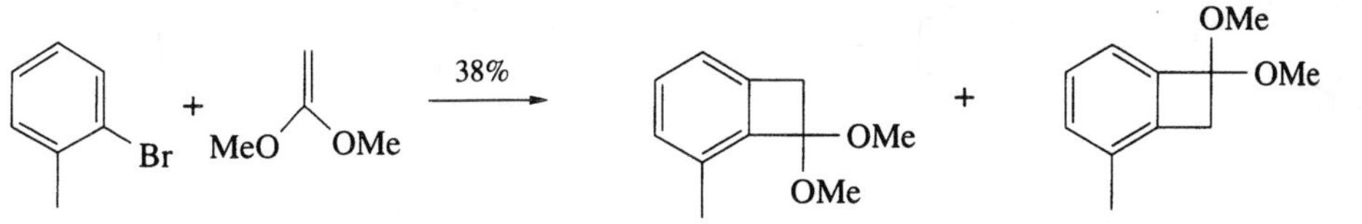

Br MeO OMe 38% OMe OMe OMe OMe 75% 25% (56)

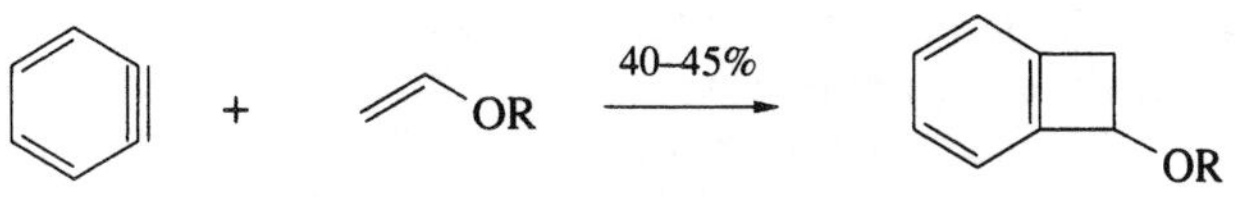

OR 40–45% OR (57)

R = Me or Ac

OMe NMe_2 40% OMe NMe_2 (58)

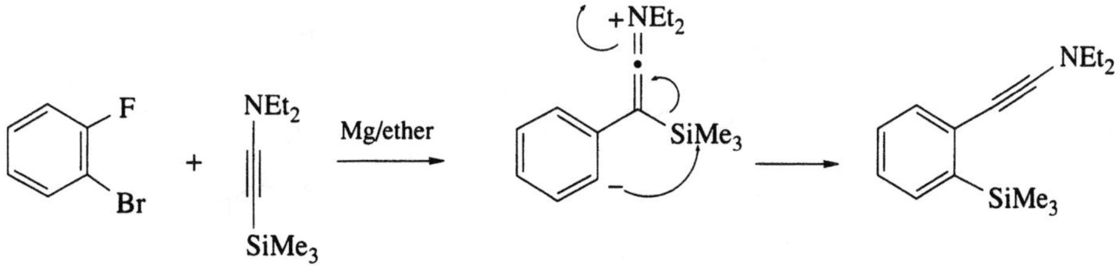

F Br NEt_2 $SiMe_3$ Mg/ether $+NEt_2$ $SiMe_3$ NEt_2 $SiMe_3$ (59)

(132)

tion 61).[177] In this reaction, benzyne complexation with the silver cation increases its nucleophilicity at the cost of [2 + 4] reactivity. The high regioselectivity obtained under these conditions is indeed remarkable (91.6% *ortho*, 8.4% *para*).[178] However, the yields in nucleophilic coupling of aromatic rings with

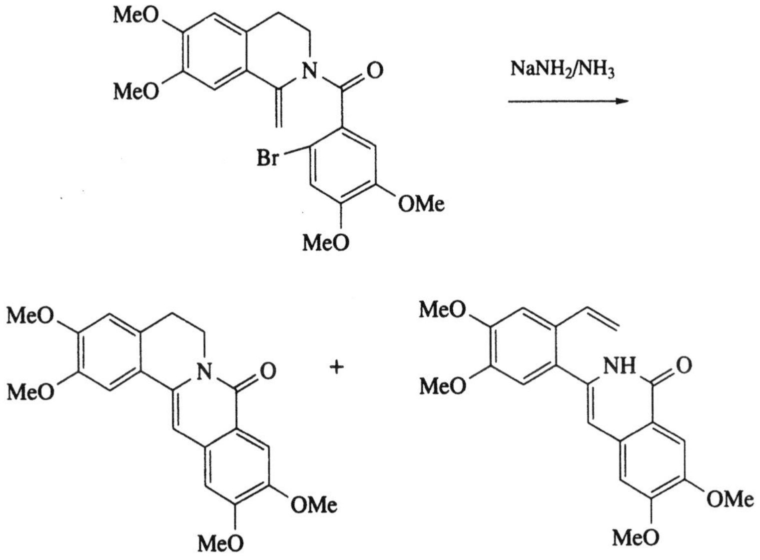

NaNH$_2$/NH$_3$

30% 20%

(60)

arynes are poor even under metal ion catalysis, and such reactions have found hardly any synthetic applications.

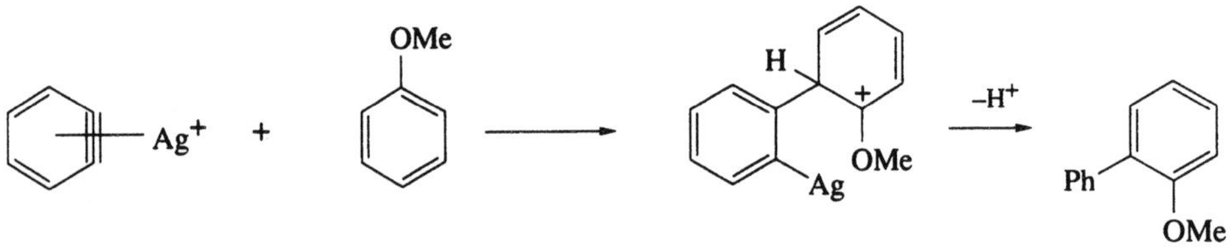

(61)

As mentioned in the introductory section, Diels–Alder and 1,3-dipolar additions of arynes have been used extensively in organic synthesis. In general, electron rich dienes and dipoles are more reactive in these reactions. This fact can be rationalized within the rubric of concerted cycloadditions in terms of a decrease in the HOMO–LUMO energy gap. However, asynchronous reaction, with bond formation at the nucleophilic end of the diene being more advanced and appreciable charge separation in the transition state (**133**), has been invoked to explain the regioselectivity observed in aryne addition to some enamides. For instance, the reaction of amide (**134**) with 3-methoxyarynes was found to give only one

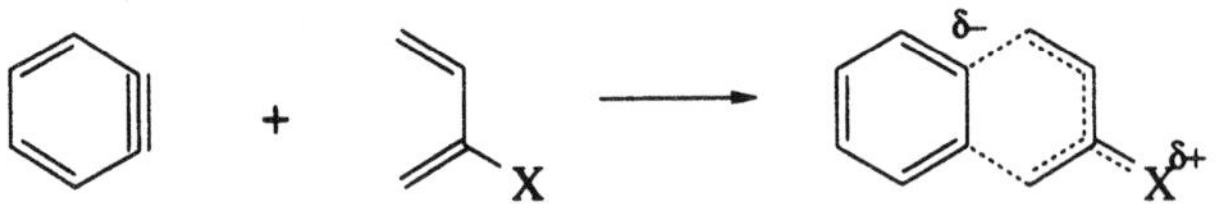

(**9**) (**133**)

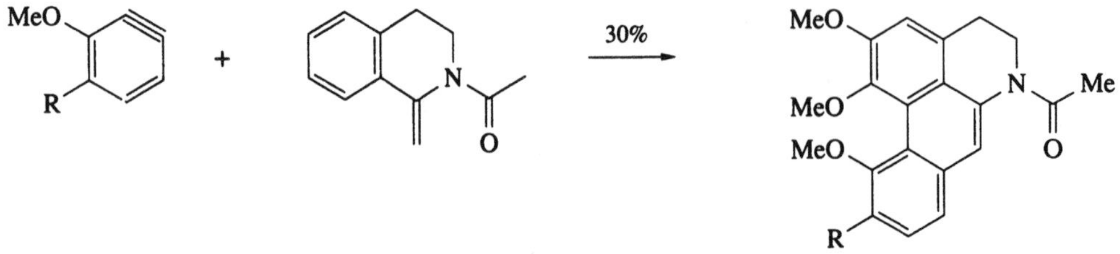

(**134**) (**135**)

Scheme 22

isomer, *i.e.* (**135**). The selectivity was attributed to a stabilization of the developing negative charge, in a transition state similar to (**133**), by the adjoining methoxy group although alternate explanations also seem tenable.[179] In any case, this reaction affords a useful access to 11-substituted aporphines which are difficult to synthesize otherwise.[180]

2.3.4 REFERENCES

1. J. D. Roberts, H. E. Simmons, Jr., L. A. Carlsmith and C. W. Vaughan, *J. Am. Chem. Soc.*, 1953, **75**, 3290.
2. G. Wittig and L. Pohmer, *Angew. Chem.*, 1955, **67**, 348.
3. R. W. Hoffmann, 'Dehydrobenzene and Cycloalkynes', Academic Press, New York, 1967.
4. (a) J. T. Sharp, in 'Comprehensive Organic Chemistry', ed. D. H. R. Barton and W. D. Ollis, Pergamon Press, Oxford, 1979, vol. 1, p. 477; (b) T. L. Gilchrist, in 'The Chemistry of Functional Groups', ed. S. Patai and Z. Rappoport, Wiley, New York, 1983, Suppl. C, part 1, p. 383; (c) L. Grundman, *Methoden Org. Chem. (Houben-Weyl)*, 1981, **2B**; (d) R. H. Levin, in 'Reactive Organic Intermediates', ed. M. Jones, Jr. and R. A. Moss, Wiley, New York, 1984; (e) C. Wentrup, 'Reactive Molecules', Wiley, New York, 1984, p. 288.
5. (a) S. V. Kessar, *Acc. Chem. Res.*, 1978, **11**, 283; (b) E. R. Biehl and S. P. Khanapure, *Acc. Chem. Res.*, 1989, **22**, 275; (c) L. Castedo and E. Guitian, in 'Studies in Natural Products Chemistry', ed. Atta-ur-Rahman, Elsevier, Amsterdam, 1989, vol. 3, part B; (d) S. V. Kessar, in 'Studies in Natural Products Chemistry', ed. Atta-ur-Rahman, Elsevier, Amsterdam, 1989, vol. 4, part C, p. 541.
6. O. L. Chapman, K. Mattes, C. L. McIntosh, J. Pacansky, G. V. Calder and G. Orr, *J. Am. Chem. Soc.*, 1973, **95**, 6134.
7. O. L. Chapman, C.-C. Chang, J. Kolc, N. R. Rosenquist and H. Tomioka, *J. Am. Chem. Soc.*, 1975, **97**, 6586.
8. I. R. Dukin and J. E. MacDonald, *J. Chem. Soc., Chem. Commun.*, 1979, 772.
9. H. H. Nam and G. E. Leroi, *J. Mol. Struct.*, 1987, **157**, 301.
10. J. W. Laing and R. S. Berry, *J. Am. Chem. Soc.*, 1976, **98**, 660.
11. M. J. S. Dewar, G. P. Ford and C. H. Reynolds, *J. Am. Chem. Soc.*, 1983, **105**, 3162.
12. J. O. Noell and M. D. Newton, *J. Am. Chem. Soc.*, 1979, **101**, 51.
13. R. D. Brown, P. D. Godfrey and M. Rodler, *J. Am. Chem. Soc.*, 1986, **108**, 1296.
14. M. J. S. Dewar and T. P. Tien, *J. Chem. Soc., Chem. Commun.*, 1985, 1243.
15. C. Wentrup, R. Blanch, H. Biehl and G. Gross, *J. Am. Chem. Soc.*, 1988, **110**, 1874.
16. I. H. Hillier, M. A. Vincent, M. F. Guest and W. von Niessen, *Chem. Phys. Lett.*, 1987, **134**, 403.
17. A. C. Scheiner, H. F. Schaefer and B. Lui, *J. Am. Chem. Soc.*, 1989, **111**, 3118.
18. H. F. Gruetzmacher and J. Lohmann, *Justus Liebigs Ann. Chem.*, 1967, **705**, 81.
19. S. K. Pollack and W. J. Hehre, *Tetrahedron Lett.*, 1980, **21**, 2483.
20. M. Moini and G. E. Leroi, *J. Phys. Chem.*, 1986, **90**, 4002.
21. N. G. Rondan, L. N. Domelsmith, K. N. Houk, A. T. Browne and R. H. Levin, *Tetrahedron Lett.*, 1979, **20**, 3237.
22. J. F. Bunnett, *Acc. Chem. Res.*, 1978, **11**, 413.
23. H. E. Van der Plas and F. Rocterdink, in 'The Chemistry of Functional Groups', ed. S. Patai and Z. Rappoport, Wiley, New York, 1983, Suppl. C, part 1, p. 421.
24. M. G. Reinecke, *Tetrahedron*, 1982, **38**, 427.
25. H. H. Nam and G. E. Leroi, *J. Am. Chem. Soc.*, 1988, **110**, 4096.
26. S. P. Khanapure and E. R. Biehl, *Heterocycles*, 1988, **27**, 2643.
27. I. Ahmed, G. W. H. Cheeseman and B. Jaques, *Tetrahedron*, 1979, **35**, 1145.
28. T. Kauffmann and H. Fischer, *Chem. Ber.*, 1973, **106**, 220.
29. E. G. Berkvich, V. B. Shur, M. E. Vol'pin, B. Lorenz, S. Rummel and M. Wahren, *Chem. Ber.*, 1980, **113**, 70.
30. S. L. Buchwald, B. T. Watson and R. T. Lum, *J. Am. Chem. Soc.*, 1987, **109**, 7137; S. L. Buchwald and R. B. Nielsen, *Chem. Rev.*, 1988, **88**, 1047.
31. S. J. McLain, R. R. Schrock, P. R. Sharp, M. R. Churchill and W. J. Young, *J. Am. Chem. Soc.*, 1979, **101**, 263.
32. J. D. Roberts, D. A. Semenow, H. E. Simmons, Jr. and L. A. Carlsmith, *J. Am. Chem. Soc.*, 1956, **78**, 601.
33. I. Fleming and T. Mah, *J. Chem. Soc., Perkin Trans. 1*, 1976, 1577.
34. P. Caubere, *Top. Curr. Chem.*, 1978, **73**, 49; *Acc. Chem. Res.*, 1974, **7**, 301.
35. L. S. Liebeskind, L. J. Lescosky and C. M. McSwain, Jr., *J. Org. Chem.*, 1989, **54**, 1435.
36. M. E. Jung and G. T. Lowen, *Tetrahedron Lett.*, 1986, **27**, 5319.
37. S. V. Kessar, Y. P. Gupta, P. Balakrishnan, K. K. Sawal, T. Mohammad and M. Dutt, *J. Org. Chem.*, 1988, **53**, 1708.
38. S. P. Khanpure, L. C. Crenshaw, R. T. Reddy and E. R. Biehl, *J. Org. Chem.*, 1988, **53**, 4915.
39. R. A. Olofson and C. M. Dougherty, *J. Am. Chem. Soc.*, 1973, **95**, 582.
40. K. L. Shepard, *Tetrahedron Lett.*, 1975, 3371.
41. E. J. Corey and M. Chaykovsky, *J. Am. Chem. Soc.*, 1965, **87**, 1345.
42. T. Kametani, S. Shibuya and S. Kano, *J. Chem. Soc., Perkin Trans. 1*, 1973, 1212.
43. C. Rüchardt and C. C. Tan, *Angew. Chem., Int. Ed. Engl.*, 1970, **9**, 522.
44. B. Baigrie, J. I. G. Cadogan, J. R. Mitchell, A. K. Robertson and J. T. Sharp, *J. Chem. Soc., Perkin Trans. 1*, 1972, 2563.
45. N. Egashira, J. Takenaga and F. Hori, *Bull. Chem. Soc. Jpn.*, 1987, **60**, 2671.
46. N. Furukawa, T. Shibutani and H. Fujihara, *Tetrahedron Lett.*, 1987, **28**, 2727.
47. R. F. Cunico and E. M. Dexheimer, *J. Organomet. Chem.*, 1973, **59**, 153.
48. Y. Himeshima, T. Sonoda and H. Kobayashi, *Chem. Lett.*, 1983, 1211.
49. J. F. Bunnett and H. Takayama, *J. Am. Chem. Soc.*, 1968, **90**, 5173.

50. M. Stiles and R. G. Miller, *J. Am. Chem. Soc.*, 1960, **82**, 3802.
51. F. M. Logullo, A. H. Seitz and L. Friedman, *Org. Synth., Coll. Vol.*, 1973, **5**, 54.
52. L. Friedman and F. M. Logullo, *J. Org. Chem.*, 1969, **34**, 3089.
53. J. F. Bunnett and C. Pyun, *J. Org. Chem.*, 1969, **34**, 2035.
54. D. D. Mazza and M. G. Reinecke, *J. Org. Chem.*, 1988, **53**, 5799.
55. G. Wittig and R. W. Hoffmann, *Org. Synth., Coll. Vol.*, 1973, **5**, 60.
56. C. D. Campbell and C. W. Rees, *J. Chem. Soc. (C)*, 1969, 742.
57. M. Keating, M. E. Peek, C. W. Rees and R. C. Storr, *J. Chem. Soc., Perkin Trans. 1*, 1972, 1315.
58. W. H. Best and D. Wege, *Aust. J. Chem.*, 1986, **39**, 635.
59. R. H. Hales, J. S. Bradshaw and D. R. Pratt, *J. Org. Chem.*, 1971, **36**, 314.
60. J. S. Bradshaw and R. H. Hales, *J. Org. Chem.*, 1971, **36**, 318.
61. J. F. Bunnett and F. J. Kearley, Jr., *J. Org. Chem.*, 1971, **36**, 184.
62. J. F. Bunnett, *Acc. Chem. Res.*, 1972, **5**, 139.
63. O. M. Nefedov, A. I. D'Yachenko and A. K. Prokof'ev *Russ. Chem. Rev. (Engl. Transl.)*, 1977, **46**, 941.
64. H. Heaney, *Fortschr. Chem. Forsch.*, 1970, **16**, 35.
65. J. F. Bunnett and B. F. Hrutford, *J. Am. Chem. Soc.*, 1961, **83**, 1691.
66. (a) J. F. Bunnett, T. Kato, R. R. Flynn and J. A. Skorcz, *J. Org. Chem.*, 1963, **28**, 1; (b) R. Huisgen and J. Sauer, *Angew. Chem.*, 1960, **72**, 91.
67. P. H. Dirstine and F. W. Bergstorm, *J. Org. Chem.*, 1946, **11**, 55.
68. T. Kauffmann, H. Fischer, R. Nurenberg, M. Vestweber and R. Wirthwein, *Tetrahedron Lett.*, 1967, 2911.
69. G. Wittig and B. Reichel, *Chem. Ber.*, 1963, **96**, 2851.
70. F. Scardiglia and J. D. Roberts, *Tetrahedron*, 1958, **3**, 197.
71. J. D. Stewart, S. C. Fields, K. S. Kochhar and H. W. Pinnick, *J. Org. Chem.*, 1987, **52**, 2110.
72. T. Kauffmann, H. Fischer, R. Nurenberg and R. Wirthwein, *Justus Liebigs Ann. Chem.*, 1970, **731**, 23.
73. F. Gavina, S. V. Luis and A. M. Costero, *Tetrahedron*, 1986, **42**, 155.
74. E. R. Biehl, E. Nieh and K. C. Hsu, *J. Org. Chem.*, 1969, **34**, 3595.
75. Y. X. Han, M. V. Jovanovic and E. R. Biehl, *J. Org. Chem.*, 1985, **50**, 1334.
76. J. F. Bunnett and J. K. Kim, *J. Am. Chem. Soc.*, 1973, **95**, 2254.
77. G. B. R. de Graaff, H. J. den Hertog and W. C. Melger, *Tetrahedron Lett.*, 1965, 963.
78. T. K. Vinod and H. Hart, *Tetrahedron Lett.*, 1988, **29**, 885.
79. R. Levine and E. R. Biehl, *J. Org. Chem.*, 1975, **40**, 1835.
80. R. Huisgen and L. Zirngibl, *Chem. Ber.*, 1958, **91**, 1438.
81. J. A. Zoltewicz and J. F. Bunnett, *J. Am. Chem. Soc.*, 1965, **87**, 2640.
82. A. I. Meyers and P. D. Pansegrau, *J. Chem. Soc., Chem. Commun.*, 1985, 690.
83. P. Beak and A. I. Meyers, *Acc. Chem. Res.*, 1986, **19**, 356.
84. H. Hart, K. Harada and C.-J. F. Du, *J. Org. Chem.*, 1985, **50**, 3104.
85. T. Kametani, K. Kigasawa, M. Hiiragi and O. Kusama, *J. Heterocycl. Chem.*, 1973, **10**, 31.
86. R. B. Bates and K. D. Janda, *J. Org. Chem.*, 1982, **47**, 4374.
87. M. S. Gibson, G. W. Prenton and J. M. Walthew, *J. Chem. Soc. (C)*, 1970, 2234.
88. E. L. Smithwick, Jr. and R. T. Shuman, *Synthesis*, 1974, 582.
89. M. F. Moreau-Hochu and P. Caubere, *Tetrahedron*, 1977, **33**, 955.
90. H. Hart and T. Ghosh, *Tetrahedron Lett.*, 1988, **29**, 881.
91. T. Ghosh and H. Hart, *J. Org. Chem.*, 1988, **53**, 3555.
92. R. D. Chambers and D. J. Spring, *Tetrahedron Lett.*, 1969, 2481.
93. P. Caubere and B. Loubinoux, *Bull. Soc. Chim. Fr.*, 1968, 3008.
94. P. Caubere and G. Guillaumet, *Bull. Soc. Chim. Fr.*, 1972, 4643.
95. B. Gregoire, M. C. Carre and P. Caubere, *J. Org. Chem.*, 1986, **51**, 1419.
96. (a) P. G. Sammes and T. W. Wallace, *J. Chem. Soc., Perkin Trans. 1*, 1975, 1377; (b) D. J. Dodsworth, M. P. Calcagno, E. U. Ehrmann, B. Devadas and P. G. Sammes, *J. Chem. Soc., Perkin Trans. 1*, 1981, 2120.
97. (a) C. A. Townsend, S. G. Davis, S. B. Christensen, J. C. Link and C. P. Lewis, *J. Am. Chem. Soc.*, 1981, **103**, 6885; (b) C. A. Townsend, P. R. O. Whittamore and S. W. Brobst, *J. Chem. Soc., Chem. Commun.*, 1988, 726.
98. S. P. Khanapure, R. T. Reddy and E. R. Biehl, *J. Org. Chem.*, 1987, **52**, 5685.
99. M. Watanabe, A. Kurosaki and S. Furukawa, *Chem. Pharm. Bull.*, 1984, **32**, 1264.
100. S. V. Kessar, D. Pal and M. Singh, *Tetrahedron*, 1973, **29**, 177.
101. A. I. Meyers and P. D. Panesgrau, *Tetrahedron Lett.*, 1984, **25**, 2941.
102. L. C. Crenshaw, S. P. Khanapure, U. Siriwardane and E. R. Biehl, *Tetrahedron Lett.*, 1988, **29**, 3777.
103. G. Wittig and L. Pohmer, *Chem. Ber.*, 1956, **89**, 1334.
104. J. F. Bunnett and J. A. Skorcz, *J. Org. Chem.*, 1962, **27**, 3836.
105. T. M. Harris and C. R. Hauser, *J. Org. Chem.*, 1964, **29**, 1391.
106. R. Huisgen, H. Konig and A. R. Lepley, *Chem. Ber.*, 1960, **93**, 1496.
107. R. D. Clark and J. M. Caroon, *J. Org. Chem.*, 1982, **47**, 2804.
108. W. Oppolzer, *Synthesis*, 1978, 793.
109. W. Oppolzer and K. Keller, *J. Am. Chem. Soc.*, 1971, **93**, 3836.
110. T. Kametani, Y. Sato, T. Honda and K. Fukumoto, *J. Am. Chem. Soc.*, 1976, **98**, 8185.
111. M. E. Jung and G. T. Lowen, *Tetrahedron Lett.*, 1986, **27**, 5319.
112. D. I. Macdonald and T. Durst, *Tetrahedron Lett.*, 1986, **27**, 2235; *J. Org. Chem.*, 1988, **53**, 3663.
113. T. Kametani, H. Nemoto, H. Ishikawa, K. Shiroyama, H. Matsumoto and K. Fukumoto, *J. Am. Chem. Soc.*, 1977, **99**, 3461.
114. D. F. Taber, K. Raman and M. D. Gaul, *J. Org. Chem.*, 1987, **52**, 28.
115. W. Oppolzer, K. Bättig and K. Petrzilka, *Helv. Chim. Acta*, 1978, **61**, 1945.
116. T. Kametani, Y. Hirai, Y. Shiratori, K. Fukumoto and F. Satoh, *J. Am. Chem. Soc.*, 1978, **100**, 554.
117. T. Kametani, M. Kajiwara and K. Fukumoto, *Tetrahedron*, 1974, **30**, 1053.
118. T. Kametani, K. Ogasawara and T. Takahashi, *Tetrahedron*, 1973, **29**, 73.

119. B. Jaques and R. G. Wallace, *Tetrahedron*, 1977, **33**, 581.
120. M. Julia, F. Le Goffic, J. Igolen and M. Baillarge, *Tetrahedron Lett.*, 1969, 1569.
121. H. Iida, Y. Yuasa and C. Kibayashi, *J. Org. Chem.*, 1979, **44**, 1074.
122. H. Iida, Y. Yuasa and C. Kibayashi, *J. Org. Chem.*, 1979, **44**, 3985.
123. R. A. Rossi and R. H. de Rossi, 'Aromatic Substitution by the $S_{RN}1$ Mechanism', American Chemical Society, Washington, DC, 1983.
124. M. F. Semmelhack, B. P. Chong, R. D. Stauffer, T. D. Rogerson, A. Chong and L. D. Jones, *J. Am. Chem. Soc.*, 1975, **97**, 2507.
125. R. R. Goehring, Y. P. Sachdeva, J. S. Pisipati, M. C. Sleevi and J. F. Wolfe, *J. Am. Chem. Soc.*, 1985, **107**, 435; J. F. Wolfe, M. C. Sleevi and R. R. Goehring, *J. Am. Chem. Soc.*, 1980, **102**, 3646.
126. S. V. Kesar, P. Singh, R. Chawla and P. Kumar, *J. Chem. Soc., Chem. Commun.*, 1981, 1074.
127. L. Lalloz and P. Caubere, *J. Chem. Soc., Chem. Commun.*, 1975, 745.
128. I. Fleming and M. Woolias, *J. Chem. Soc., Perkin Trans. 1*, 1979, 827.
129. H. Lida, S. Aoyagi and C. Kibayashi, *J. Chem. Soc., Perkin Trans. 1*, 1975, 2502.
130. T. Kametani and K. Ogasawara, *J. Chem. Soc. (C)*, 1967, 2208.
131. F. Benington and R. D. Morin, *J. Org. Chem.*, 1967, **32**, 1050.
132. S. V. Kessar, S. Batra, U. K. Nadir and S. S. Gandhi, *Indian J. Chem.*, 1975, **13**, 1109.
133. S. V. Kessar, R. Randhawa and S. S. Gandhi, *Tetrahedron Lett.*, 1973, 2923.
134. T. Kametani, A. Ujiie, K. Takahashi, T. Nakano, T. Susuki and K. Fukomoto, *Chem. Pharm. Bull.*, 1973, **21**, 766.
135. T. Kametani, K. Fukumoto and T. Nakano, *J. Heterocycl. Chem.*, 1972, **9**, 1363.
136. T. Kametani, S. Shibuya, K. Kigasawa, M. Hiiragi and O. Kusama, *J. Chem. Soc. (C)*, 1971, 2712.
137. R. J. Spangler, D. C. Boop and J. H. Kim, *J. Org. Chem.*, 1974, **39**, 1368.
138. S. Kano, T. Ogawa, T. Yokomatsu, W. E. Komiyama and S. Shibuya, *Tetrahedron Lett.*, 1974, 1063.
139. J. M. Boente, L. Castedo, A. R. de Lera, J. M. Saa, R. Suau and M. C. Vidal, *Tetrahedron Lett.*, 1983, **24**, 2295.
140. A. R. Lera, S. Aubourg, R. Suau and L. Castedo, *Heterocycles*, 1987, **26**, 675.
141. S. V. Kessar, R. Gopal and M. Singh, *Tetrahedron*, 1973, **29**, 167.
142. S. V. Kessar, M. Singh, P. Jit, G. Singh and A. K. Lumb, *Tetrahedron Lett.*, 1971, 471.
143. S. V. Kessar, N. Parkash and G. S. Joshi, *J. Chem. Soc., Perkin Trans. 1*, 1973, 1158.
144. S. V. Kessar, Y. P. Gupta, P. Singh, V. Jain and P. S. Pahwa, *J. Chem. Soc. Pak.*, 1979, **1**, 129.
145. S. V. Kessar, Y. P. Gupta, P. S. Pahwa and P. Singh, *Tetrahedron Lett.*, 1976, 3207.
146. S. V. Kessar, M. Singh and P. Balakrishnan, *Indian J. Chem.*, 1974, **12**, 323.
147. J. P. Gillespie, L. G. Amoros and F. R. Stermitz, *J. Org. Chem.*, 1974, **39**, 3239.
148. F. R. Stermitz, J. P. Gillespie, L. G. Amoros, R. R. Romero, T. A. Stermitz, K. A. Larson, S. Earl and J. E. Ogg, *J. Med. Chem.*, 1975, **18**, 708.
149. Y. Simanek, in 'The Alkaloids', ed. A. Brossi, Academic Press, New York, 1985, p. 26.
150. Y. Sato, T. Toshimasa, A. Toyohiko and S. Hideaki, *J. Org. Chem.*, 1976, **41**, 3559.
151. Y. Sato, T. Toyo'oka, T. Aoyama and H. Shirai, *J. Chem. Soc., Chem. Commun.*, 1975, 640.
152. B. M. Trost and L. S. Melvin, Jr., 'Sulfur Ylides', Academic Press, New York, 1975.
153. J. Nakayama, S. Takeue and M. Hoshino, *Tetrahedron Lett.*, 1984, **25**, 2679.
154. G. M. Blackburn, W. D. Ollis, C. Smith and I. O. Sutherland, *J. Chem. Soc., Chem. Commun.*, 1969, 99.
155. T. Otsubo and V. Boekelheide, *J. Org. Chem.*, 1977, **42**, 1085.
156. T. Otsubo and V. Boekelheide, *Tetrahedron Lett.*, 1975, 3881.
157. S. Hayashi and N. Ishikawa, *Bull. Chem. Soc. Jpn.*, 1972, **45**, 642.
158. R. S. Pal and M. M. Bokadia, *Pol. J. Chem.*, 1978, **52**, 1473.
159. D. Seyferth and J. M. Burlitch, *J. Org. Chem.*, 1963, **28**, 2463.
160. N. Petragnani and V. G. Toscano, *Chem. Ber.*, 1970, **103**, 1652.
161. J. Nakayama, T. Tajiri and M. Hoshino, *Bull. Chem. Soc. Jpn.*, 1986, **59**, 2907.
162. C. D. Campbell, C. W. Rees, M. R. Bryce, M. D. Cooke, P. Hanson and J. M. Vernon, *J. Chem. Soc., Perkin Trans. 1*, 1978, 1006.
163. V. Nair and K. H. Kim, *J. Org. Chem.*, 1975, **40**, 3784.
164. C. W. Fishwick, R. C. Gupta and R. C. Storr, *J. Chem. Soc., Perkin Trans. 1*, 1984, 2827.
165. H. Heaney and C. T. McCarty, *J. Chem. Soc. (D)*, 1970, 123.
166. D. M. Hayes and R. Hoffmann, *J. Phys. Chem.*, 1972, **76**, 656.
167. S. Inagaki and K. Fukui, *Bull. Chem. Soc. Jpn.*, 1973, **46**, 2240.
168. R. V. Stevens and G. S. Biacchi, *J. Org. Chem.*, 1982, **47**, 2393.
169. R. V. Stevens and G. S. Biacchi, *J. Org. Chem.*, 1982, **47**, 2396.
170. S. M. Ali and S. Tanimoto, *J. Chem. Soc., Chem. Commun.*, 1988, 1465.
171. H. H. Wasserman and J. Solodar, *J. Am. Chem. Soc.*, 1965, **87**, 4002.
172. T. Kametani, T. Kato and K. Fukumoto, *Tetrahedron*, 1974, **30**, 1043.
173. T. Kametani, K. Kigasawa, M. Hiiragi, T. Hayasaka and O. Kusama, *J. Chem. Soc. (C)*, 1971, 1051.
174. H. Heaney and S. V. Ley, *J. Chem. Soc., Perkin Trans. 1*, 1974, 2693.
175. Y. Sato, Y. Kobayashi, M. Sugiura and H. Shirai, *J. Org. Chem.*, 1978, **43**, 199.
176. T. Kametani, T. Sugai, Y. Shoji, T. Honda, F. Satoh and K. Fukumoto, *J. Chem. Soc., Perkin Trans. 1*, 1977, 1151.
177. L. Friedman, *J. Am. Chem. Soc.*, 1967, **89**, 3071.
178. I. Tabushi, H. Yamada, Z. Yoshida and R. Oda, *Bull. Chem. Soc. Jpn.*, 1977, **50**, 285.
179. L. Castedo, E. Guitian, C. Saa, R. Suau and J. M. Saa, *Tetrahedron Lett.*, 1983, **24**, 2107.
180. C. Saa, E. Guitian, L. Castedo and J. M. Saa, *Tetrahedron Lett.*, 1985, **26**, 4559.

2.4
Nucleophilic Addition to Arene–Metal Complexes

MARTIN F. SEMMELHACK
Princeton University, NJ, USA

2.4.1 INTRODUCTION

2.4.1.1 Coverage

This chapter covers reactions in which coordination of a transition metal to the π-system of an arene ring activates the ring toward addition of nucleophiles, to give η^5-cyclohexadienyl–metal complexes (**1**; Scheme 1). If an electronegative atom is present in the *ipso* position, elimination of that atom (X in **1**) leads to nucleophilic aromatic substitution (path **a**). Reaction of the intermediate with an electrophile (E^+) can give disubstituted 1,3-cyclohexadiene derivatives (path **b**). If a hydrogen occupies the *ipso* position, oxidation of the intermediate gives formal nucleophilic substitution for hydrogen (path **c**). These processes have been uncovered and developed relatively recently and have significant potential in synthesis methodology. General reviews have appeared[1–3] as well as others with an emphasis on [(η^6-arene)$Cr(CO)_3$] complexes,[4–6] on [(η^6-arene)FeCp]$^+$ (Cp = η^5-cyclopentadienyl) complexes[7,8] and on [(η^6-arene)$Mn(CO)_3$]$^+$ complexes,[9] but none with emphasis on synthesis scope and limitations. The purpose of this chapter is to illustrate the established methodology and to point to future possibilities; mechanistic questions are included when helpful in understanding the scope and limitations. The coverage is representative and not comprehensive.

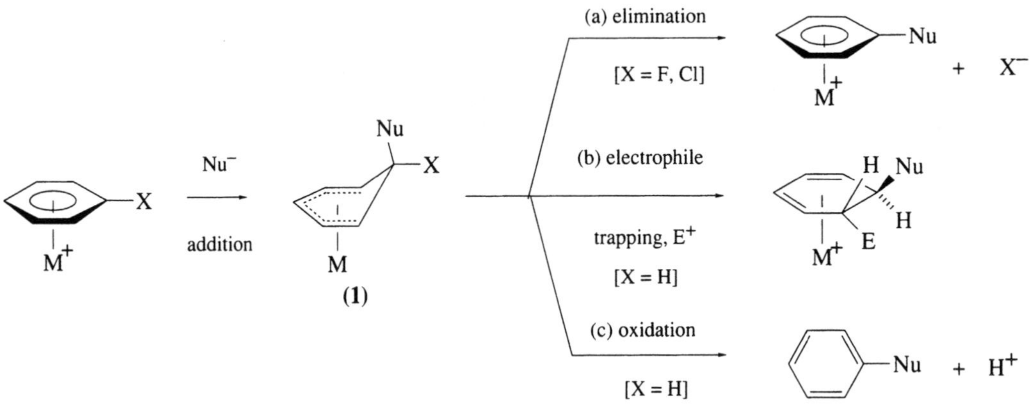

(a) elimination [X = F, Cl]; (b) electrophile trapping, E$^+$ [X = H]; (c) oxidation [X = H]; Nu$^-$ addition; (1)

Scheme 1

2.4.1.2 Relation to Nucleophilic Aromatic Substitution by the *S*NAr Mechanism

The reactions are related mechanistically to traditional nucleophilic aromatic substitution (*S*NAr), but offer several unique features. It may be well to summarize the comparison of general features for traditional arene ring substitution. Arene rings are polarizable and tend to react as electron donors, as in electrophilic aromatic substitution.[10] A cationic intermediate, the phenonium ion, is produced, which spontaneously loses a proton to give overall substitution of an electrophile for a proton. With suitable electron-withdrawing substituents attached to the arene *via* σ-bonds, nucleophiles may add to produce the corresponding anionic intermediate, a cyclohexadienyl anion, stabilized by interaction with the substituents as represented by (**2**) in Scheme 2.[11] Attack is kinetically favored at a position bearing a hydrogen (path **a**), but the intermediate thus formed (**2**) cannot proceed directly to substitution (by loss of a hydride ion). In the general version of the *S*NAr reaction, an electronegative atom (usually halide) is at-

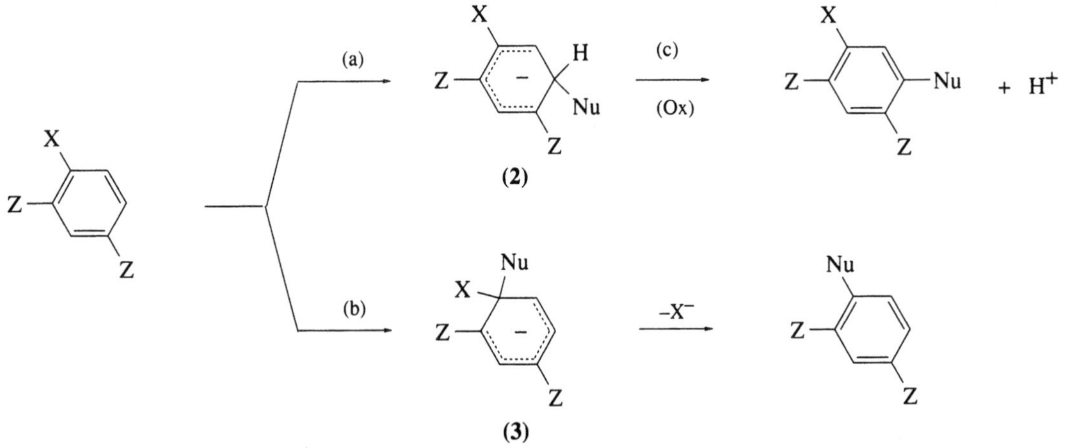

(a); (c) (Ox); (2); (b); $-X^-$; (3)

Z = electron-withdrawing group
X = electronegative atom (group)

Scheme 2

tached to the arene *ortho* or *para* to an activating group, and the addition of the nucleophile is reversible. A slower addition occurs at the carbon bearing the electronegative atom to produce an anionic intermediate (**3**), which can lose the electronegative atom in an irreversible step and give the product from nucleophilic substitution for halide (path **b**). In a few cases, loss of hydrogen from intermediates related to (**2**) has been induced by oxidation, which presumably allows departure of a proton, as shown in step **c**, Scheme 2.[11,12] The overall addition–oxidation pathway of nucleophilic aromatic substitution for hydrogen is too limited to be considered synthesis methodology.

2.4.1.3 Arene–Metal Complexes, Background

The first arene–transition metal complexes were prepared in the 1950s[13,14] and it was immediately recognized that the added polarizibility or electron deficiency would promote addition of nucleophiles to the arene ligand. A number of cyclohexadienyl complexes were characterized following nucleophilic addition, but the question of inducing the *ipso* hydrogen to depart was not answered (equation 1).

(**4**) + Nu^- ⟶ (**5**) $\xrightarrow{?}$ PhNu (1)

The isolation of the first halobenzene complex, (η^6-chlorobenzene)tricarbonylchromium(0), allowed a test for a direct analog of classical S_NAr reactivity.[15] The activating effect of the $Cr(CO)_3$ unit was found to be comparable to a single *p*-nitro substituent in reaction with methoxide in methanol and the substituted arene ligand was detached with mild oxidation (equation 2).

$$(PhCl)Cr(CO)_3 + NaOH \xrightarrow[50\ ^\circ C]{MeOH} (PhOMe)Cr(CO)_3 \xrightarrow{I_2} PhOMe \quad (2)$$

Three types of arene π-complexes have seen significant development in synthesis methodology: neutral [(η^6-arene)$Cr(CO)_3$] (**6**), the isoelectronic cationic [(η^6-arene)$Mn(CO)_3$] (**7**) and the cationic (η^6-arene)(η^5-cyclopentadienyl)iron(II) complexes (**8**). The [(arene)$Cr(CO)_3$] species are formed by simple displacement of neutral ligands (L) from [$Cr(CO)_3L_3$] by the arene and a large number of variously substituted [(arene)$Cr(CO)_3$] complexes are available.[14] Gentle and general syntheses of the cationic complexes (**7**) and (**8**) have only recently become available and the development of these complexes as synthesis intermediates is less developed. The focus of this chapter will be on the chromium system, but the iron and manganese analogs will be included for comparison and future potential.

(**6**) (**7**) (**8**)

2.4.2 PREPARATION OF ARENE–METAL COMPLEXES

2.4.2.1 [(Arene)$Cr(CO)_3$]

The general process is shown in equation (3). The L unit in the [$Cr(CO)_3L_3$] can be CO (most common),[15–20] MeCN,[21,22] 2-alkylpyridine,[23] ammonia[24–27] and other donor ligands. The rate (reaction temperature) is related to the nature of L; the most reactive readily available source of $Cr(CO)_3$ is [(η^6-naphthalene)$Cr(CO)_3$], which undergoes favorable arene exchange under mild conditions with many substituted arenes.[28] A technical problem in the use of $[Cr(CO)_6]$ is the crystallization of sublimed $[Cr(CO)_6]$ in the condenser. The original solution is still used, a fairly complicated heated condenser apparatus known as the Strohmeier apparatus,[16] but a simple air condenser is also effective.[17] The most general and convenient procedure also relies on continuous washing down of the $[Cr(CO)_6]$ with a mixture of THF and di-*n*-butyl ether at reflux.[18,19] A variety of polar and nonpolar aprotic solvents have been used, and, for some purposes, such as complexation of α-amino acids with aromatic side chains, water–THF mixtures are effective.[20]

$$\text{R–C}_6\text{H}_5 + [Cr(CO)_3L_3] \longrightarrow [(\text{R–C}_6\text{H}_5)Cr(CO)_3] + 3L \qquad (3)$$

L = CO, MeCN, NH_3, Py, *etc.*

The primary competing reaction is irreversible oligomerization of the coordinatively unsaturated $[Cr(CO)_n]$ species. Addition of small amounts of weak donor ligands, such as α-picoline[29] or THF,[18–20] appear to lower the temperature necessary for the complexation step from 100 °C or more to 60 °C, presumably by assisting in the dissociation of one or more CO ligands and retarding the rate of oligomerization of the reactive source of $Cr(CO)_3$.[18–20] After complexation, the yellow to red $[(arene)Cr(CO)_3]$ complex can be purified by crystallization from nonpolar organic solvents or chromatographed on silica gel. The complexes are somewhat sensitive to air while in solution, but the solutions can be handled in air briefly and crystalline solids can be stored without special precautions. Examples are known with a wide variety of arenes, including sensitive species such as $[(\eta^6\text{-indole})Cr(CO)_3]$.[13,14,30] Most common synthesis operations, such as acid and base hydrolysis, hydride reduction and carbanion addition to ketones, can be carried out on side chain functional groups without disturbing the arene–Cr bond.[31] However, even very mild oxidation conditions will detach the arene by oxidizing the metal. An important aspect of the complexation procedure is diastereoselectivity. Complexes of disubstituted (and more highly substituted) arenes can have molecular asymmetry; a stereogenic center in a side chain leads to diastereoisomers. Under favorable conditions, especially with low temperature conditions for complexation, significant diastereoselectivity is observed in the complexation step (equation 4).[32–35]

SiMe3, OH, R, MeO + $[Cr(CO)_3(naphth)]$ —(25 °C, 69%)→ [SiMe3, OH, R, MeO, $Cr(CO)_3$] + [SiMe3, OH, R, MeO, $Cr(CO)_3$] (4)

98:2

2.4.2.2 $[(Arene)Mn(CO)_3]^+$

A gentle procedure has made these complexes available with a variety of substituted arenes.[36] Direct displacement of CO from the perchlorate salts of $[Mn(CO)_5]^+$ or $[Mn(CO)_3(acetone)_3]^+$ with the arene in dichloromethane at reflux leads to precipitation of the $[(arene)Mn(CO)_3]$ salt. The conditions are milder than the $AlCl_3$-promoted procedure employed earlier.[37] Only a handful of substituted arenes have been attached to $[Mn(CO)_3]^+$, but the general synthesis method suggests few limitations (equation 5).

$$\text{R–C}_6\text{H}_5 + [Mn(CO)_3(acetone)_3]^+\ {}^-ClO_4 \xrightarrow{25\ ^\circ C} [(\text{R–C}_6\text{H}_5){}^+Mn(CO)_3]\ {}^-ClO_4 \qquad (5)$$

The complexes are air stable; indeed, a limitation is the need for powerful oxidizing agents, such as Jones reagent Cr^{VI}, to detach the arene ligand.[38] They are highly reactive toward nucleophiles. This limits the number of compatible synthesis manipulations that can be carried on in the presence of the $[(arene)Mn(CO)_3]$ unit but broadens the scope of effective nucleophiles.

2.4.2.3 $[(Arene)FeCp]^+$

One of the few useful reactions of ferrocene is the $AlCl_3$-catalyzed exchange of one Cp group for an arene ligand (equation 6).[39–44]

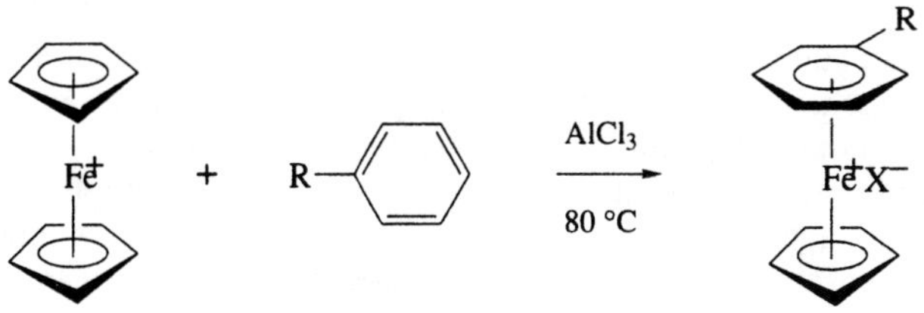

$$\mathrm{Cp_2Fe^+ + R\text{-}C_6H_5 \xrightarrow[80\ ^\circ C]{AlCl_3} [(R\text{-}C_6H_5)FeCp]^+X^-} \quad (6)$$

The requirement for $AlCl_3$ adds some obvious limitations, but a modest number of substituted arenes have been successfully coordinated to the $[FeCp]^+$ unit. Photoinduced arene exchange has been described[45] and is particularly effective with the η^6-chlorobenzene-η^5-cyclopentadienyliron cation without the need for Lewis acids.[42] More basic (electron rich) arenes readily replace chlorobenzene, leading to FeCp cation complexes of *O*-phenylethyl-*p*-toluenesulfonate and variously substituted thiophenes.[42] The charged complexes are not purified by conventional organic techniques, such as chromatography, but recrystallization is possible. The complexes are very air and heat stable; again, methods of removal of the arene from the Fe are few. The simplest is pyrolysis at >200 °C.[46]

2.4.3 NUCLEOPHILIC SUBSTITUTION FOR HETEROATOMS ON ARENE LIGANDS, S_NAr REACTION

The smooth replacement of a heteroatom (usually halide) from arene ligands requires reversible addition of the nucleophile, since the kinetic site of addition is usually at a position bearing a hydrogen substituent (Scheme 3, path k_1).

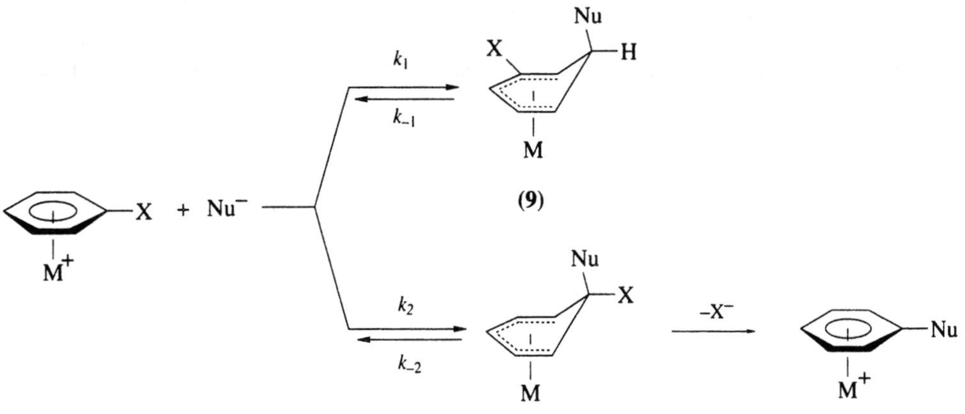

(Arene-X)M$^+$ + Nu$^-$ ⇌ (k_1, k_{-1}) Nu,H-cyclohexadienyl–M **(9)**

(Arene-X)M$^+$ + Nu$^-$ ⇌ (k_2, k_{-2}) Nu,X-cyclohexadienyl–M **(10)** → ($-X^-$) (Arene-Nu)M$^+$

Scheme 3

The relative rates of each step depend critically on the nature of M and of the nucleophile. More reactive nucleophiles and more reactive complexes disfavor equilibration, ($k_1 >> k_{-1}$) and the process can stop with formation of the first cyclohexadienyl intermediate (**9**). Equilibration leads through (**10**) to the substitution product. The overall order of reactivity for electron deficient arenes is:[2,47] $[(arene)Mn(CO)_3]^+$ > 2,4-$(NO_2)C_6H_3Cl$ > $[(arene)FeCp]^+$ >> 4-$(NO_2)C_6H_4Cl$ > $[(arene)Cr(CO)_3]$.

2.4.3.1 [(Halobenzene)Cr(CO)$_3$]

2.4.3.1.1 Intermolecular nucleophilic substitution with heteroatom nucleophiles

A patent issued in 1965 claims substitution for fluoride on [(fluorobenzene)Cr(CO)$_3$] in DMSO by a long list of nucleophiles, including alkoxides (from simple alcohols, cholesterol, ethylene glycol, pinacol, dihydroxyacetone), carboxylates, amines and carbanions (from triphenylmethane, indene, cyclohexanone, acetone, cyclopentadiene, phenylacetylene, acetic acid and propiolic acid).[48] Many of those results have not appeared in the primary literature. In the reaction of methoxide with [(halobenzene)Cr(CO)$_3$], the fluorobenzene complex is *ca.* 2000 times more reactive than the chlorobenzene complex.[47] The difference is taken as evidence for a rate-limiting attack on the arene ligand followed by fast loss of halide; the concentration of the cyclohexadienyl anion complex (*e.g.* **10**) does not build up. In the reaction of [(fluorobenzene)Cr(CO)$_3$] with amine nucleophiles, the coordinated aniline product appears rapidly at 25 °C and a careful mechanistic study suggests that the loss of halide is now rate limiting.[49]

Hydroxide, alkoxide and phenoxide nucleophiles react with [(chlorobenzene)Cr(CO)$_3$] at 25–50 °C in polar aprotic media to give high yields of the phenol or aryl ether chromium complexes; oxidation with excess iodine at 25 °C for a few hours releases the free arene, CO, chromium(III) and iodide anion, as summarized in equation (7).[15,50–52] Only the arene remains in the organic layer after a conventional aqueous extraction. An alternate general procedure for detaching the arene ligand from the Cr(CO)$_3$ unit is to expose a solution of the complex to air in the presence of normal room light or sunlight. The solution soon becomes cloudy with oxidized chromium derivatives and the organic solution can be processed to isolate the free arene.[19]

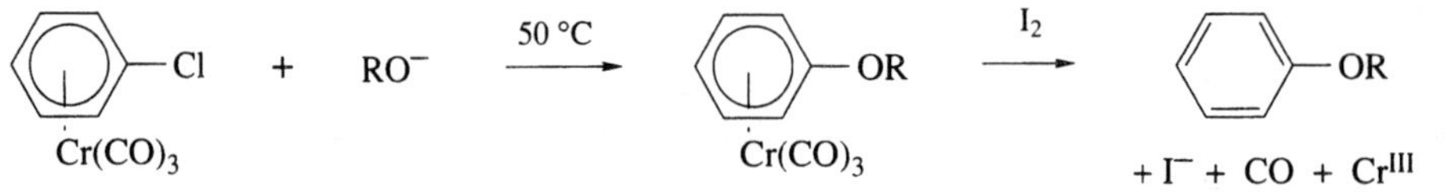

$$[(\eta^6\text{-C}_6\text{H}_5\text{Cl})\text{Cr(CO)}_3] + \text{RO}^- \xrightarrow{50\ ^\circ\text{C}} [(\eta^6\text{-C}_6\text{H}_5\text{OR})\text{Cr(CO)}_3] \xrightarrow{\text{I}_2} \text{C}_6\text{H}_5\text{OR} + \text{I}^- + \text{CO} + \text{Cr}^{\text{III}} \quad (7)$$

The reaction proceeds almost exclusively by direct substitution (*ipso*), as shown by reactions of isomeric chlorotoluene complexes (Scheme 4).[50] The relative rates of substitution of the isomeric ligands are similar, in the order 1.0:1.4:2.4 for *o:m:p*.[50] A version of *cine* substitution is possible under special conditions, as discussed below.

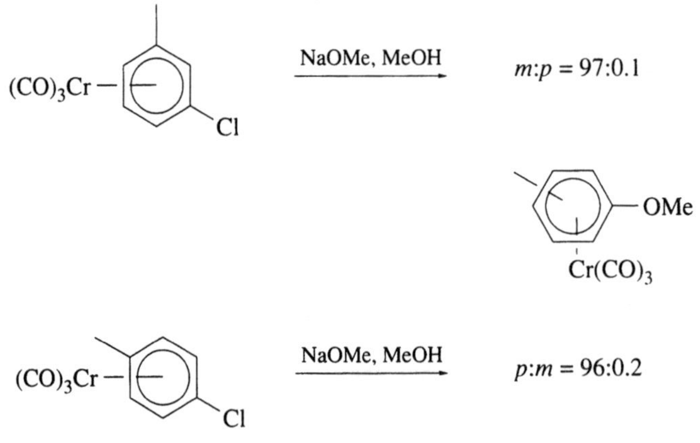

(CO)$_3$Cr–(*m*-chlorotoluene) $\xrightarrow{\text{NaOMe, MeOH}}$ *m*:*p* = 97:0.1

[(*m*-methylanisole)Cr(CO)$_3$]

(CO)$_3$Cr–(*p*-chlorotoluene) $\xrightarrow{\text{NaOMe, MeOH}}$ *p*:*m* = 96:0.2

Scheme 4

While polar protic solvents, such as MeOH, strongly retard reaction,[53] phase transfer catalysis using benzene[54] or addition of crown ethers to potassium alkoxides in benzene[53] allows reaction at 25 °C. Even with strong electron donors, such as alkyl, methoxy or dialkylamino in the *ortho, meta* or *para* positions, substitution for chloride by potassium methoxide proceeds smoothly using the crown ether activation in benzene (equation 8).[53]

In general, replacement of fluoride occurs under milder conditions and in higher yield than chloride; however, compared to chloride, a fluoride substituent slows the complexation of arenes and often leads to lower yields in the formation of the complexes. An excellent procedure for formation of the parent

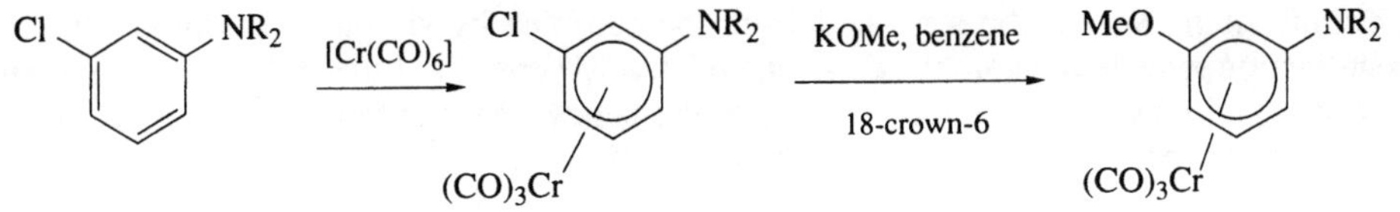

Cl NR2 [Cr(CO)6] Cl NR2 (CO)3Cr KOMe, benzene 18-crown-6 MeO NR2 (CO)3Cr (8)

[(fluorobenzene)$Cr(CO)_3$] complex has been reported.[55] The bromo- and iodo-arene complexes are known, but generally are not effective in the S_NAr reaction.[51]

Direct replacement of oxygen leaving groups, such as *p*-toluenesulfonate, is not effective, perhaps due to steric retardation of the departure of the *endo* leaving group.[49] One intriguing exception which bears further development is the reaction of the diphenyl ether mono-$Cr(CO)_3$ complex (**11**) with reactive anions (Scheme 5).[56] If the intermediate cyclohexadienyl anion complex (**12**) is quenched with an oxidizing agent at low temperature, *meta* substitution for hydrogen and loss of the $Cr(CO)_3$ unit are observed (see below for discussion of this pathway). However, if the same intermediate is allowed to warm to room temperature, anion equilibration can occur even with the very reactive 2-methyl-1,3-dithianyl carbanions and direct (*ipso*) substitution for phenoxide is observed.

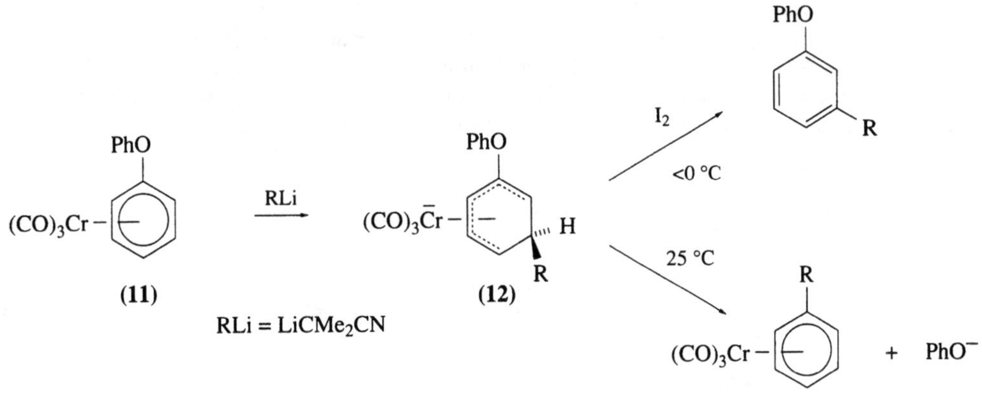

(11) RLi (12) RLi = $LiCMe_2CN$ I_2 <0 °C 25 °C + PhO^-

Scheme 5

Thiolate anions[57,58] and oxime alkoxides[58] react under phase transfer conditions to give aryl sulfides and *O*-aryl oximes, respectively; the *o*-dichlorobenzene complex can be converted selectively to the monosubstitution product (equation 9). The arylation of oximes leads to a simple process for benzofuran formation (equation 10). Simple primary and secondary amine nucleophiles react smoothly in the absence of added base, in a very general and efficient process for aniline derivatives.[49]

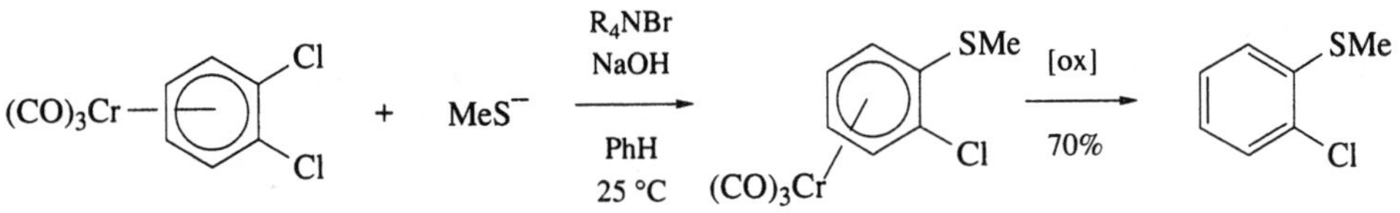

(CO)3Cr Cl Cl + MeS^- R_4NBr NaOH PhH 25 °C SMe Cl (CO)3Cr [ox] 70% SMe Cl (9)

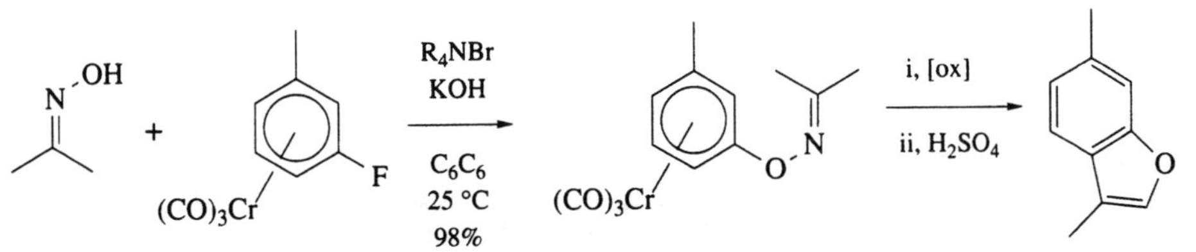

N OH + (CO)3Cr F R_4NBr KOH C_6C_6 25 °C 98% (CO)3Cr O N i, [ox] ii, H_2SO_4 O (10)

Chromium activation allows a strategy for aryl ether synthesis in four stages: (i) electrophilic chlorination; (ii) chromium coordination; (iii) alkoxide substitution for chloride; and (iv) oxidative decomplexation.[53] The process is effective for the synthesis of 6-methoxytetrahydroquinolines and 5-methoxydihydroindole derivatives, for example. Chlorination of *N*-acetyltetrahydroquinoline with SO_2Cl_2 followed by deacetylation provides 6-chlorotetrahydroisoquinoline in 77% yield (equation 11).[53] Complexation with [$Cr(CO)_6$] in diglyme–cyclohexane at 125 °C for 53 h using the Strohmeier apparatus gave the chromium complex (**13**) in 85% yield, based on 40% recovery of starting material. The unreacted tetrahydroisoquinoline can be separated from the complex by simple acid extraction, since the $Cr(CO)_3$ coordination renders the complexed amine relatively nonbasic. Indeed, the complexed amine is moderately acidic and must be protected (*in situ*) as the benzyl ether before methoxide treatment. The overall yield for protection, substitution and oxidative decomplexation is 80%.

i, SO_2Cl_2; ii, deacetylation; iii, $[Cr(CO)_6]$; 85% → **(13)** → i, KH, BnBr; ii, KOMe, 18-crown-6; iii, I_2 (11)

2.4.3.1.2 *Cyclizations* via *heteroatom substitution for halide*

Intramolecular substitution for chloride or fluoride is particularly effective. Oxygen heterocycles with fused benzo rings are obtained from $Cr(CO)_3$ complexes of fluorobenzene with an *o*-(hydroxyalkyl) side chain.[59,60] For example, complexation of 3-(2-fluorophenyl)-1-propanol with $[Cr(CO)_3(pyridine)_3]$ at 25 °C in ether (promoted by $BF_3 \cdot Et_2O$) followed by reaction in DMSO with excess potassium *t*-butoxide for 3 h at 25 °C gave the chroman complex (**14**).[60] The yield in the cyclization step was 75% and iodine decomplexation was quantitative (equation 12). Efforts to produce the dihydrobenzofuran under the same

$[Cr(CO)_3(Py)_3]$, $BF_3 \cdot Et_2O$ → KOBu^t, 3 h, 25 °C, 75% → **(14)** → I_2, 100% (12)

conditions failed; only intermolecular substitution products were obtained.[60]

From the *o*-dichlorobenzene complex, reaction with a dialkoxide can produce a cyclic bisether. This idea has been applied in the preparation of benzo-crown ethers, as outlined in Scheme 6.[61] The first substitution product (**15**) does not undergo cyclization, but a second intermolecular substitution to give (**16**) sets up a final cyclization to the bischromium complex (**17**); the overall yield is 27%. A bisthia analog was also prepared.[61]

Coordination of arenes with $Cr(CO)_3$ also activates the ring hydrogens toward abstraction with strong base (metallation).[59,62] Simple arene ligands can be metallated with alkyllithium reagents; alkoxy, amino and halo substituents on the arene direct the metallation to the *ortho* position with rates and regioselectivity higher than with the corresponding free arene.[55,63–65] This allows a strategy for annulating aromatic rings *via ortho* lithiation, trapping with a bifunctional electrophile, and finally nucleophilic substitution for the electronegative substituent (usually F) (equation 13).

(13)

E = electrophilic unit; N = nucleophile

Following an initial report[59] including carbon nucleophiles for the cyclization (see below), a series of papers have defined useful possibilities with heteroatom nucleophiles.[55,63–65] Although [(chlorobenzene)$Cr(CO)_3$] undergoes lithiation to give a moderately stable species which can be trapped with elec-

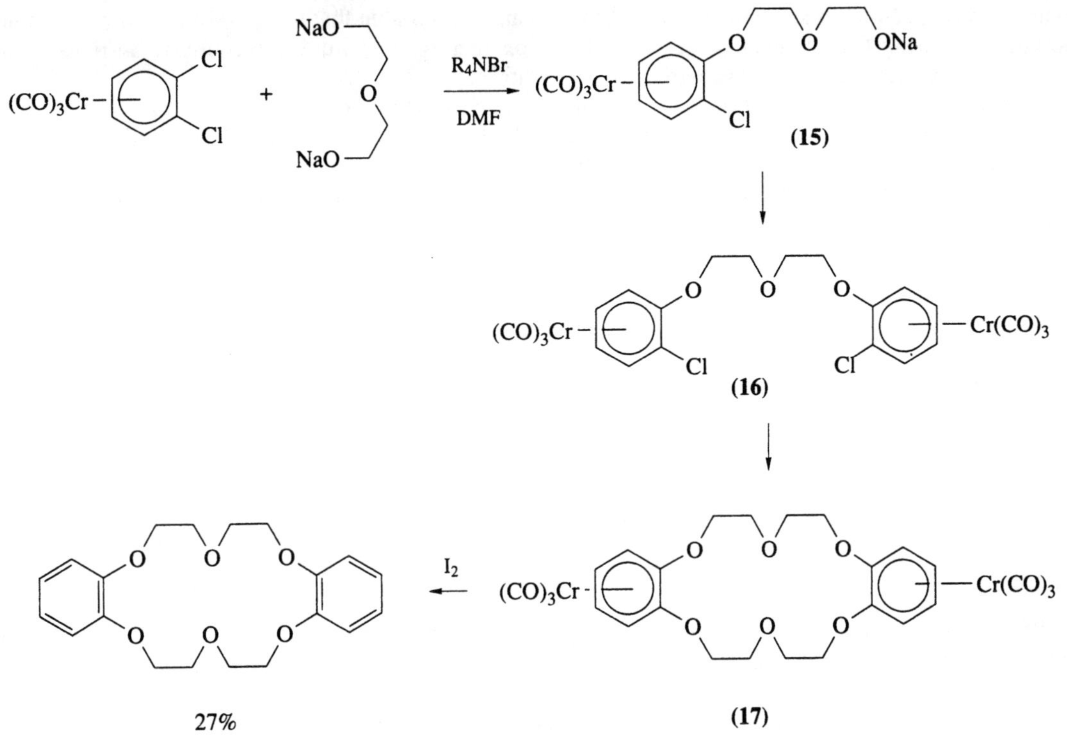

Scheme 6

trophiles to produce the *ortho*-substituted [(chloroarene)$Cr(CO)_3$] complexes,[66] the fluoro analog appears to be more efficient, in the same way. The *ortho* lithio haloarene complexes are highly basic and the electrophilic trapping species is restricted to those with low kinetic acidity and high electrophilicity. Good results are obtained with isocyanates, ketenes and acyl derivatives with α-protons of low acidity.[55,63–65] Examples (Scheme 7) include phenyl isocyanate, which reacts twice to give the the six-membered heterocycle (**18**), while the dimethyl-*N*-carboxy anhydride (**19**) proceeds directly to the five-

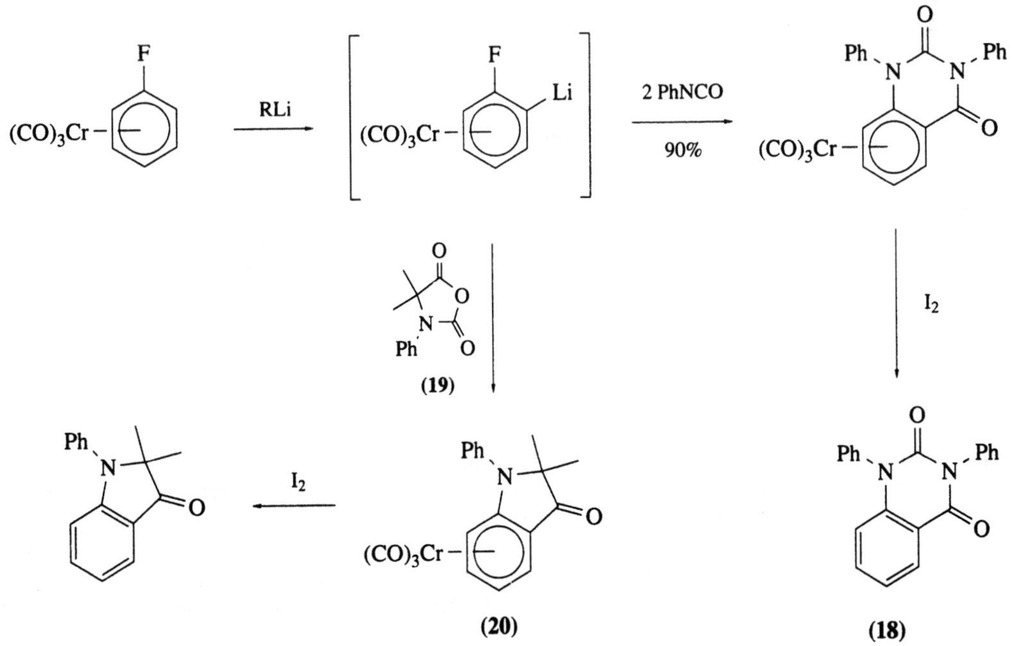

Scheme 7

membered ring (**20**) after spontaneous decarboxylation. It is possible that the *gem*-dimethyl group is important in favoring five-membered ring formation, considering the failure in forming furan rings by direct cyclization with a simple side chain hydroxy group.[55]

A variation is shown in equation (14), in which an arylcuprate is prepared and used to couple with a vinyl halide, leading eventually to the β-methylenedihydrobenzofuran complex (**21**). Cyclization occurs spontaneously upon fluoride-induced removal of the silyl protecting group.[64] The complex (**21**) is much more stable than the free ligand and can be used to advantage as a synthesis equivalent of 3-methylenedihydrobenzofuran.

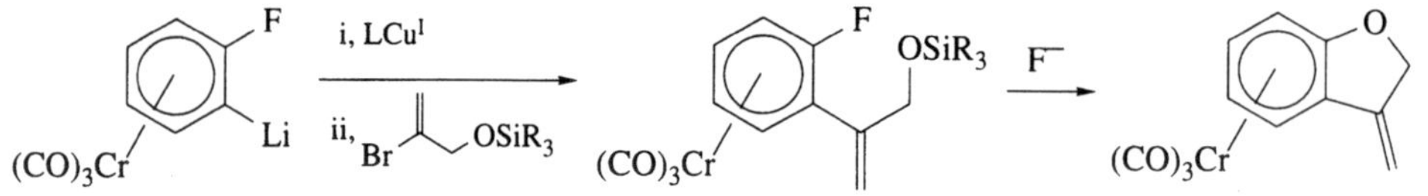

(CO)$_3$Cr F Li — i, LCu^I; ii, Br⌒OSiR$_3$ → (CO)$_3$Cr F OSiR$_3$ — F$^-$ → (CO)$_3$Cr O (**21**) (14)

2.4.3.1.3 Nucleophilic substitution with carbon nucleophiles

(i) Addition–elimination, S$_N$Ar

While carbon nucleophiles were suggested to be efficient in substitution for fluoride in the early patent,[48] the first examples in the primary literature appeared in 1974.[51,52] It is now clear that there are three reactivity classes of carbon nucleophiles: (i) stabilized carbanions (from carbon acids with pK_a < *ca.* 18); (ii) more reactive carbanions (pK_a > 20), which give complete conversion to cyclohexadienyl addition products prior to slow equilibration *via* reversible anion addition; and (iii) more reactive carbanions (pK_a > 20), which give irreversible addition to the arene ligand (Scheme 8).

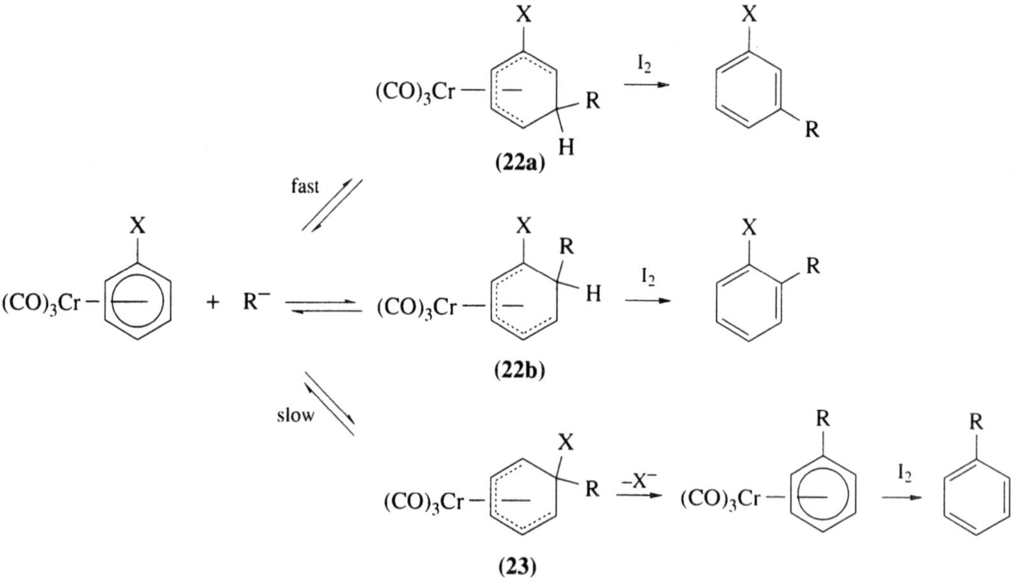

(CO)$_3$Cr–(C$_6$H$_5$X) + R$^-$ ⇌ fast: (CO)$_3$Cr–(cyclohexadienyl, X, R, H) (**22a**) —I$_2$→ X, R (meta); (CO)$_3$Cr–(cyclohexadienyl, X, R, H) (**22b**) —I$_2$→ X, R (ortho); slow: (CO)$_3$Cr–(cyclohexadienyl, X, R) (**23**) —$-X^-$→ (CO)$_3$Cr–(C$_6$H$_5$R) —I$_2$→ C$_6$H$_5$R

Scheme 8

Diethyl sodiomalonate is an example of type (i). Reaction with [(fluorobenzene)Cr(CO)$_3$] proceeds to completion after 20 h at 50 °C in HMPA to give the diethyl phenylmalonate complex in over 95% yield. Monitoring the reaction by NMR gave no evidence for an intermediate (*e.g.* the cyclohexadienyl anion complex); interruption of the reaction by addition of iodine at less than 20 h gave significant amounts of unreacted fluorobenzene. A satisfactory picture is the simple one, that the anion adds reversibly and unfavorably ($k_1 < k_{-1}$, as in Scheme 3), slowly finding itself at the *ipso* position; then irreversible loss of fluoride gives the substitution product.[51,52]

Lithioisobutyronitrile (from LDA and isobutyronitrile) is an example of anion type (ii). The initial addition to [(chlorobenzene)Cr(CO)$_3$] is over within minutes at –78 °C, but the substitution product does not appear until the mixture is warmed to 25 °C (Scheme 8). Quenching with iodine after short reaction time leads to a mixture of phenylisobutyronitrile and *o*- and *m*-chlorophenylisobutyronitrile. This appears to be a case of fast addition *ortho* and *meta* to the chloride to give cyclohexadienyl anionic complexes (*e.g.* **22**) followed by slow rearrangement to the *ipso* intermediate (**23**). Quenching with iodine before equilibration is complete leads to oxidation of the intermediates (**22**) and formal substitution for hydrogen (see below).

Carbon nucleophiles of type (iii) add to the arene ligand and do not rearrange; examples include the very reactive anions, such as 2-lithio-2-methyl-1,3-dithiane, and the less sterically encumbered anions, such as lithio acetonitrile and *t*-butyl lithioacetate. In these cases, the anion adds to an unsubstituted position (mainly *ortho* or *meta* to Cl, as in **22**) and does not rearrange. Then iodine quenching, even after a long period at 25 °C, gives almost exclusively the products from formal substitution for hydrogen, as from (**22**) in Scheme 8.

Hydrocarbon anions, such as cyclopentadienyl and analogs (fluorenyl, indenyl, pentadienyl), substitute for fluoride, leading, for example, to phenylcyclopentadiene as a $Cr(CO)_3$ complex on the arene (equation 15).[67]

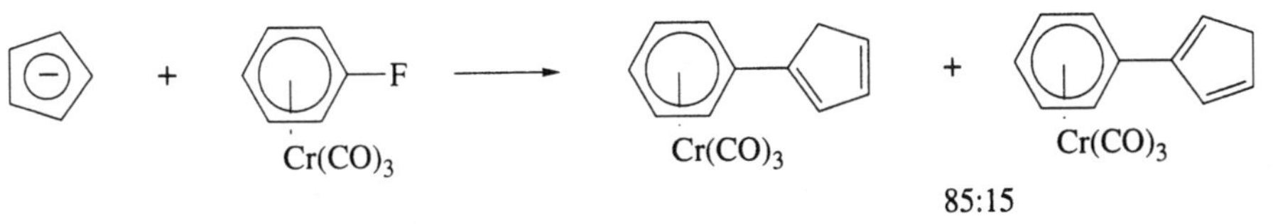

(15)

Scheme 8 summarizes the possibilities. Successful substitution for halogen by carbanions requires reversible addition and that sets upper limits on the reactivity of the anion, although steric effects can also favor equilibration (more-substituted equilibrate faster). As before, fluoride is a better leaving group than chloride. An example is shown in equation (16), producing methyl 2-phenyl-2-(*t*-butylthio)propionate (**24**) in 94% yield after 15 h at 25 °C. There is an important solvent effect on the rate of equilibration of the carbanions[68] and the conditions chosen for these early experiments are not necessarily the best; there may be room to expand the scope of useful anions by careful choice of media (less polar solvents should favor equilibration).

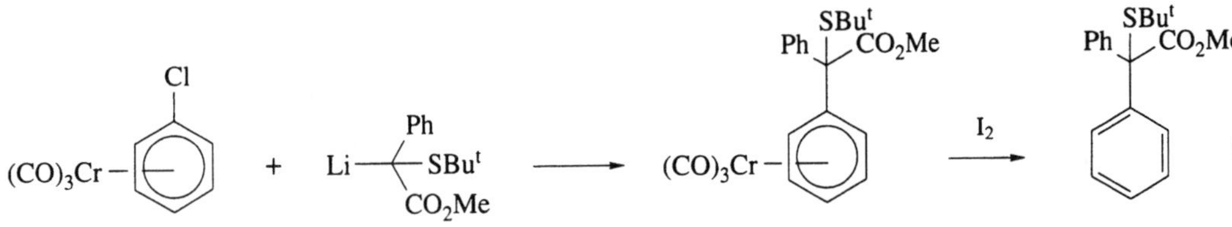

(16)

(**24**) 94% overall

Except in the formation of minor side products, the replacement of halogen by hydride as nucleophile is not observed. Halide exchange is also not a general process with [(haloarene)$Cr(CO)_3$] species.

(ii) Cine–tele *substitution*

There is an alternate mechanism for halide replacement, following the sequence of nucleophile addition, protonation and elimination of HX (Scheme 9). In this pathway, the addition of the nucleophile need not be at the *ipso* position; it can be *ortho* to halide leading to '*cine*' substitution or it can be at the *meta* or *para* positions, leading to '*tele*' substitution.[69,70] The mechanism is the same for both *cine* and *tele* substitution and the different names reflect a differentiation in the IUPAC naming schemes.

The processes depend on the formation of the cyclohexadienyl anion intermediates in a favorable equilibrium (carbon nucleophiles from carbon acids with $pK_a > 22$ or so), protonation (which can occur at low temperature with even weak acids, such as acetic acid) and hydrogen shifts in the proposed diene–chromium intermediates (**25**) and (**26**). Hydrogen shifts lead to an isomer (**26**), which allows elimination of HX and regeneration of an arene–chromium complex (**27**), now with the carbanion unit indirectly substituted for X (Scheme 9).

An important example of indirect substitution utilizes alkoxy leaving groups; the intermediate (**28**) from the reaction of the [(diphenyl ether)$Cr(CO)_3$] complex and the 2-methyl-1,3-dithiane anion can be induced to eliminate phenol by protonation at low temperature; the result is *tele* substitution (equation 17).

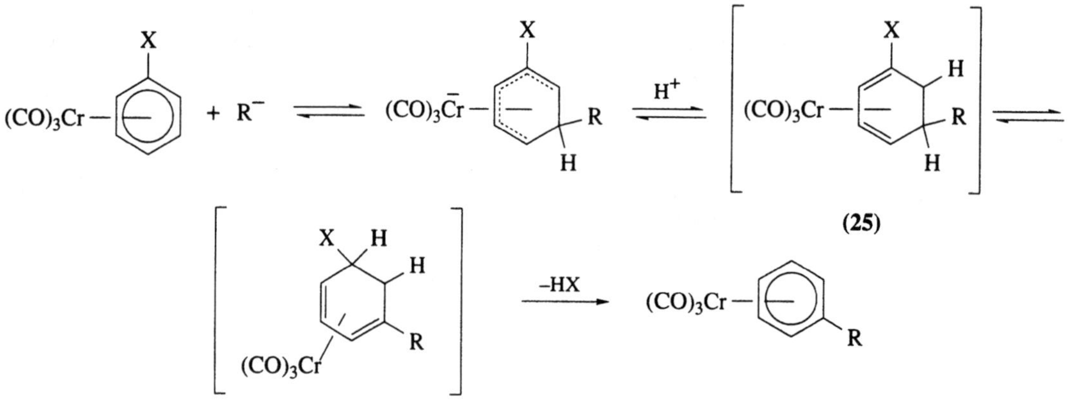

(25)

(26)

(27)

Scheme 9

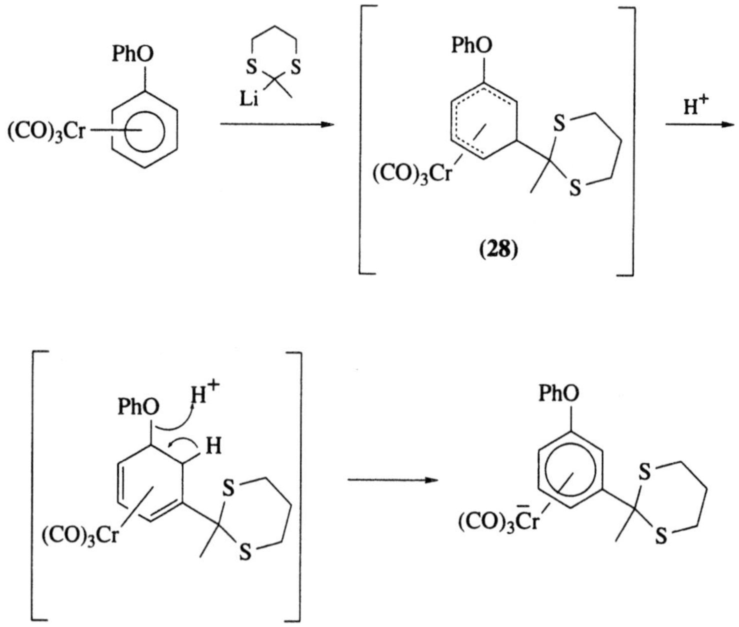

(28)

(17)

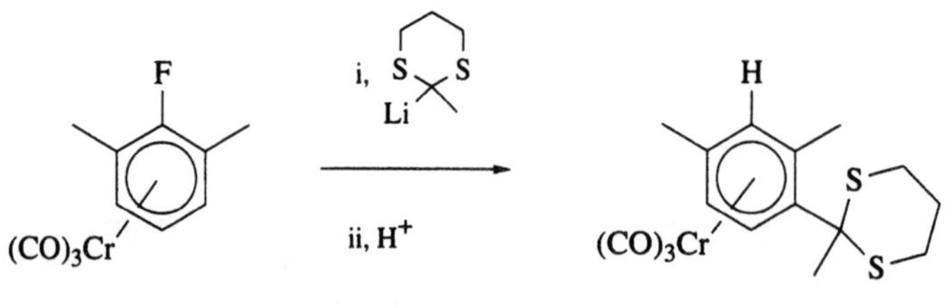

(18)

(29)

An impressive example is the reaction of [(2,6-dimethyl-1-fluorobenzene $Cr(CO)_3$] with 2-lithio-2-methyl-1,3-dithiane at –78 °C followed by treatment with trifluoroacetic acid at –78 °C.[69] Loss of HF leads to the 1,2,4-substitution (*tele*) product (**29**) in 62% yield (equation 18).[70]

The directing effects of F (strong *meta*) *versus* Cl (weak, *ortho–meta*; discussed below) allow control over the site of *tele* substitution; the chloro analog leads to (**30**), with the 1,3,5-substitution pattern (equation 19).[71]

Under certain conditions, apparently when the $Cr(CO)_3$ unit is spontaneously displaced from the product by a weak donor ligand (solvent), the simple addition product (1,3-cyclohexadiene derivative **31**) can be the major product (equation 20).[71]

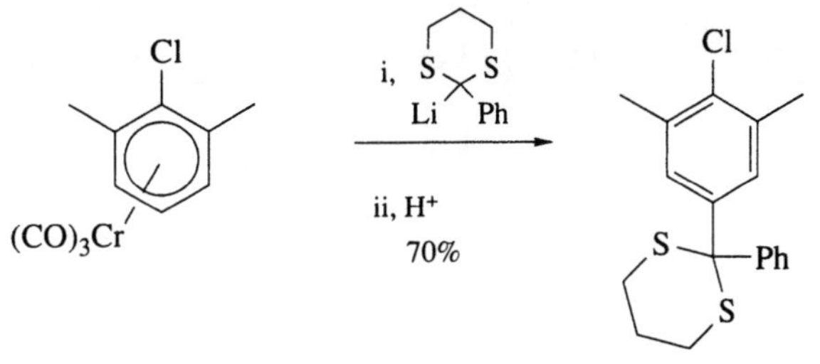

(19)

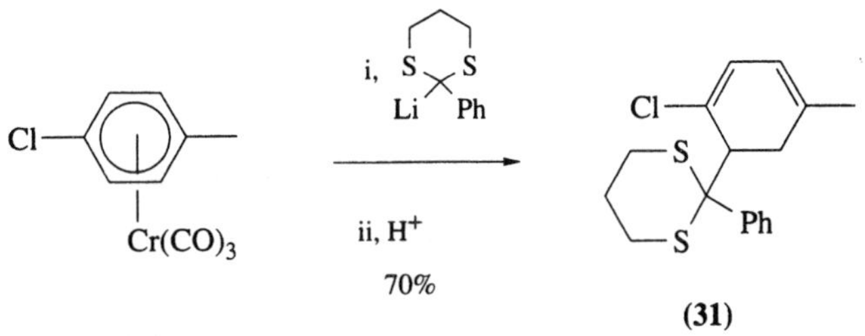

(30)

(20)

(31)

2.4.3.2 [(Halobenzene)CpFe]$^+$

2.4.3.2.1 Nucleophilic substitution with heteroatom nucleophiles

While the potential for these species in nucleophile substitution for chloride has been demonstrated, the processes have not been fully developed nor applied. The two-stage process of addition–substitution for chloride and arene detachment is exemplified for the iron system in equation (21).

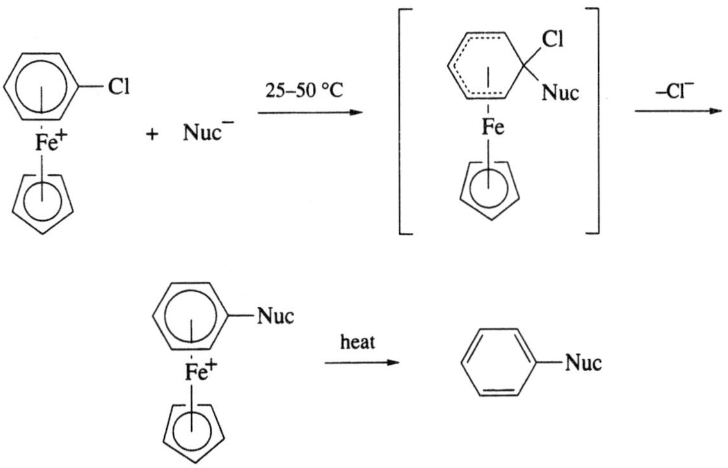

(21)

The first stage (addition–elimination) is well known with CN^- and a variety of N-, O- and S-nucleophiles.[7,72–75] The detachment of the product arene from the Fe is more difficult. The conversions in equation (22) demonstrate the stability of the arene–Fe bond toward oxidation, as a side chain methyl is converted to a carboxylic acid, and suggest the generality of heteroatom substitution under mild conditions.[76]

With [η^6-(*o*-dichlorobenzene)FeCp]$^+$, selective mono- and di-substitution can be achieved with a variety of N- and O- and stabilized C-nucleophiles.[77] With a bisheteroatom nucleophile, benzo-fused heterocycles are produced (equation 23).[74]

The analogous *p*-dichlorobenzene complex will react rapidly with phenoxide and alkoxide nucleophiles in the first substitution and more slowly in the second; monosubstitution can be readily achieved.[78] Simple amines give only monosubstitution, even in the presence of excess amine (equation 24).

$$\text{Cl-C}_6\text{H}_4\text{-Me(FeCp}^+) \xrightarrow[94\%]{KMnO_4} \text{Cl-C}_6\text{H}_4\text{-CO}_2\text{H(FeCp}^+) \xrightarrow[25\ °C]{Nuc^-} \text{Nuc-C}_6\text{H}_4\text{-CO}_2\text{H(FeCp}^+) \qquad (22)$$

Nuc = PhS (51%) and EtO (68%)

(23)

X, Y = O, S, NH; R = H, Me

(24)

Substitution with amine nucleophiles in the series of chlorotoluene complexes showed that the substitution is direct; no *cine* or *tele* substitution was observed.[75] The analogous [(fluoroarene)FeCp]$^+$ complexes are known, but less well developed. Kinetic studies show that the fluoro derivatives are more reactive compared to the chloro analogs.[47]

A special example is the replacement of a nitro group on [(η^6-nitrobenzene)FeCp]$^+$.[78,79] Reaction with O-, S-, N- and stabilized C-nucleophiles gives overall addition–elimination (equation 25). The nitroarene complexes are prepared by oxidation of the corresponding aniline complexes and are readily available.[78,79]

(25)

(32)

85–90 °C, DMSO

(33)

DMSO, 160 °C

+ [CpRu(DMSO)$_3$]$^+$ (26)

In the isoelectronic ruthenium series, nucleophilic substitution by the disodium salt of 4,4′-dihydroxybenzophenone on the complex (**32**) of *p*-dichlorobenzene with a $[RuCp]^+$ unit produced the aromatic polyether complex represented by (**33**). Displacement by DMSO at 160 °C led to the free polymer and the recoverable complex $[CpRu(DMSO)_3]^+$, which can be recycled directly by complexation with *p*-dichlorobenzene (equation 26). While ruthenium is not attractive to use in stoichiometric processes, this example appears to allow easy recycling.[80]

2.4.3.2.2 Nucleophile addition–substitution with carbon nucleophiles

Stabilized carbon nucleophiles (*e.g.* from β-diketones, β-keto esters, malonate esters, *etc.*) can be arylated by substitution for chloride on the arene in [Fe(arene)Cp] cation complexes.[72,78,81] A base is necessary and two heterogeneous systems are favored: potassium carbonate in DMF or potassium fluoride prepared on Celite-545. As usual in the $[FeCp]^+$ system, detachment of the substituted arene requires somewhat extreme conditions, usually pyrolytic sublimation at 200 °C.[46] An example is given in equation (27).

Cl; Fe^+ + EtO_2C (O) $\xrightarrow[25\ °C,\ 4–6\ h;\ 80\%]{K_2CO_3,\ DMF}$ O; CO_2Et; Fe^+ (27)

2.4.3.3 [(Halobenzene)$Mn(CO)_3$]$^+$

The most reactive of the common arene–metal complexes has been the least developed with regard to nucleophilic substitution for halide. It is only an improved, gentler method of complexation that allowed isolation of the fluorobenzene complex.[36] The chlorobenzene analog is easily prepared and undergoes substitution with alkoxy, phenoxy, thiolate, amine and azide nucleophiles (equation 28).[36,82] Reaction is complete with aniline (a weak nucleophile) within 3 min at room temperature. In examples with a methyl substituent on the arene as a label, the substitution is shown to be direct. The detachment of the substituted arene ligand from the $Mn(CO)_3$ unit is still limited by the requirement for strong oxidizing conditions, a serious limitation if this series is to be put forward in synthesis methodology.

Cl; $—\overset{+}{Mn}(CO)_3$ + Nu^- ⟶ Nu; $—\overset{+}{Mn}(CO)_3$ (28)

Nu^- = alkoxy, phenoxy, thiolate, amine and azide

2.4.4 ADDITION TO ARENE–METAL COMPLEXES TO GIVE η^5-CYCLOHEXADIENYL INTERMEDIATES

2.4.4.1 Addition–Oxidation: Formal Nucleophilic Substitution for Hydrogen

2.4.4.1.1 Addition–oxidation with [(arene)$Cr(CO)_3$] complexes

(i) General features

Once it is recognized that cyclohexadienyl anionic complexes of chromium (**34**) can be generated by addition of sufficiently reactive nucleophiles and that simple oxidizing techniques convert the anionic intermediates to free substituted arenes, a general substitution process becomes available, which does not depend on a specific leaving group on the arene (equation 29).[52,83] While no heteroatom nucleophile has been successfully utilized, the process is general for carbanions derived from carbon acids with $pK_a > 22$ or so.[83] Exceptions among the very reactive anions include organolithium reagents (deprotonation of the arene ligand and addition to the CO ligand are favored) and Grignard reagents (addition to the CO ligand is favored). With anions such as ester enolate anions, reaction occurs within minutes at –78 °C and the intermediate cyclohexadienyl complex can be observed spectroscopically.[83] The intermediates are exceedingly air sensitive and are generally quenched directly, without purification. In one case, from the addition of 2-lithio-1,3-dithiane, the adduct has been crystallized and fully characterized by X-ray diffraction analysis.[83] Oxidation with excess iodine (at least 2.5 mol equiv. of I_2) also proceeds at –78 °C and is complete within hours below room temperature. The products are the substituted arene, HI, Cr^{III} and CO. For acid sensitive products, such as trialkylsilyl-substituted arenes, an excess of amine (conveniently, diisopropylamine) can buffer the mixture. Table 1 provides a representative sample of the carbanions, which have been tested with $[(\eta^6\text{-benzene})Cr(CO)_3]$.

$$(CO)_3Cr\text{–}(\eta^6\text{-}C_6H_6) + R^- \xrightarrow{-70\ °C} [\text{cyclohexadienyl}(R)(H)Cr(CO)_3]^- \ (\mathbf{34}) \xrightarrow{I_2} C_6H_5R \quad (29)$$

Table 1 Reactivity of Carbanions (R—Li) Toward $[(Benzene)Cr(CO)_3]$

Unreactive	*Successful*	*Metallation*
1. $LiCH(CO_2Me)_2$	7. $LiCH_2CO_2Bu^t$	18. Bu^nLi
2. $LiCH_2COBu^t$	8. $LiCH_2CN$	19. MeLi
3. MeMgBr	9. KCH_2COBu^t	20. Bu^sLi
4. Bu^tMgBr	10. LiCH(CN)(OR)	
5. Me_2CuLi	11. $LiCH_2SPh$	
6. LiC(CN)(Ph)(OR)	12. 2-Li-1,3-dithiane	
	13. $LiCH{=}CH_2$	
	14. LiPh	
	15. $LiC{\equiv}CR$	
	16. $LiCH_2CH{=}CH_2$	
	17. $LiC(Me)_3$	

(ii) Regioselectivity

Regioselectivity has been the subject of numerous studies with a variety of mono- and poly-substituted arene ligands. The addition of synthetically interesting carbanions to the typical arene ligand, especially those bearing electron-donating substitutents, can be reversible at rates comparable to the rate of addition: it is not always obvious whether to assume kinetic or thermodynamic control.[68,84–88] However, correlations have been suggested which allow prediction or rationalization of regioselectivity with a modest degree of confidence. With some significant exceptions as discussed below, the difference between the kinetic and thermodynamic selectivity has not been determined or is small.

With arenes bearing a single resonance donor substituent (NR_2, OMe, F), the addition is strongly preferred at the *meta* position, with small amounts of *ortho* substitution (0–10%).[89,90] Representative examples are shown in Table 2 and equation (30).

The *meta* acylation of anisole, using a carbonyl anion equivalent as the nucleophile, illustrates the unique regioselectivity available with the $Cr(CO)_3$ activation (equation 31).

However, the selectivity is more complicated with a methyl or chloro substituent. Again, *meta* substitution is always significant, but *ortho* substitution can account for 50–70% of the mixture in some cases.[89–91] More reactive anions (1,3-dithianyl) and less substituted carbanions (*e.g.* *t*-butyl lithioacetate) tend to favor *ortho* substitution. Representative examples are shown in Table 2 and equation (30). Entries 2–4 show that variation of reaction temperatures from –100 to 0 °C has no significant effect in that highly selective system. The added activating effect of the Cl substitutent allows addition of the pinacolone enolate anion (entry 11), whereas no addition to the anisole nor toluene ligand is observed with the same anion. Table 3 and equation (32) also includes results with *para*-directing substituents, CF_3 and Me_3Si (entries 14 and 15).[90,91] In general, electron-withdrawing substituents such as acyl and cyano are also ac-

Table 2 Addition–Oxidation with Anisole, *N,N*-Dimethylaniline and [(Fluorobenzene)$Cr(CO)_3$] Complexes

$$(CO)_3Cr\text{-}(C_6H_5X) \xrightarrow[\text{ii, excess } I_2]{\text{i, LiY, }-78\ ^\circ C} o\text{-}XC_6H_4Y + m\text{-}XC_6H_4Y + p\text{-}XC_6H_4Y \quad (30)$$

Entry	*X*	*LiY*	*Ratio (o:m:p)*	*Combined yield (%)*
1	OMe	$LiCH_2CN$	3:97:0	38
2	OMe	$LiC(Me)_2CN$	3:97:0	93
3	OMe	$LiCH_2CO_2Bu^t$	6:94:0	86
4	OMe	$LiCH(Me)CO_2Bu^t$	4:96:0	93
5	OMe	$LiC(Me)_2CO_2Bu^t$	0:100:0	76
6	OMe	$LiC(CN)(OR)CH_2Ph$[a]	0:100:0	75
7	OMe	2-lithio-1,3-dithiane	10:90:0	35[b]
8	$N(Me)_2$	$LiC(Me)_2CN$	1:99:0	92
9	F	$LiC(Me)_2CN$	2:98:0	84

[a]The R group is CH(Me)(OEt). [b] A major side reaction is *ortho* deprotonation of the anisole ligand.

$$[(\text{anisole})Cr(CO)_3] \xrightarrow[\text{ii, } I_2\text{; iii, } H_3O^+\text{; iv, } HO^-]{\text{i, } LiC(CN)(CHMe_2)OCH(Me)OEt} 88\% \quad m\text{-}MeOC_6H_4COCH(CH_3)_2 \quad (31)$$

tivated toward nucleophile addition to the substituent (C═O or CN) and this process competes with addition to the ring itself.

Table 3 Addition–Oxidation with [(Chlorobenzene)$Cr(CO)_3$] and [(Toluene)$Cr(CO)_3$] Complexes

$$(CO)_3Cr\text{-}(C_6H_5X) \xrightarrow[\text{ii, excess } I_2]{\text{i, LiY, }-78\ ^\circ C} o\text{-}XC_6H_4Y + m\text{-}XC_6H_4Y + p\text{-}XC_6H_4Y \quad (32)$$

Entry	*X*	*LiY*	*Ratio (o:m:p)*	*Combined yield (%)*
1	Me	$LiCH_2CN$	35:63:2	88
2	Me	$LiC(Me)_2CN$ (–100 °C/1.5 min)	2:96:2	52
3	Me	$LiC(Me)_2CN$ (–78 °C/1.5 min)	1:97:2	95
4	Me	$LiC(Me)_2CN$ (0 °C/20 min)	1:97:2	86
5	Me	$LiCH_2CO_2Bu^t$	28:72:0	89
6	Me	$LiC(Me)_2CO_2Bu^t$	3:97:0	96
7	Me	2-Lithio-1,3-dithiane	52:46:2	94
8	Cl	$LiCH_2CO_2Bu^t$	54:45:1	98
9	Cl	$LiCH(Me)CO_2Bu^t$	53:46:1	88
10	Cl	$LiC(Me)_2CO_2Bu^t$	5:95:0	84
11	Cl	$LiCH_2COBu^t$	70:24:0	87
12	Cl	$LiC(Me)_2CN$	10:89:1	84
13	Cl	2-Lithio-1,3-dithiane	46:53:1	56
14	CF_3	LiC(CN)(Me)(OR)[a]	0:30:70	33
15	$SiMe_3$	$LiC(Me)_2CN$	0:2:98	65

[a] The R group is —CH(Me)(OEt).

Regioselectivity in *ortho*-disubstituted arenes is often high and useful. A series of examples is summarized in Figure 1. The resonance donor substituent (OMe) appears to dominate the directing influences, favoring addition at the less hindered position *meta* to itself.[91] But with two identical substituents, the addition is preferred in the adjacent position.[88,91,92] For example, with the complex of benzocyclobutene (**35**), six carbon nucleophiles were tested and each gave addition exclusively at C-3.[88,92] With the complex of indane (**36**), selective addition at C-3 was observed with dithianyl anions, but cyano-stabilized anions gave up to 20% of the isomeric product from addition at C-4.[88,92] Substituents at the benzylic carbons in the indane ligand have a strong effect on selectivity and can lead to the 1,2,4-substitution product.[88]

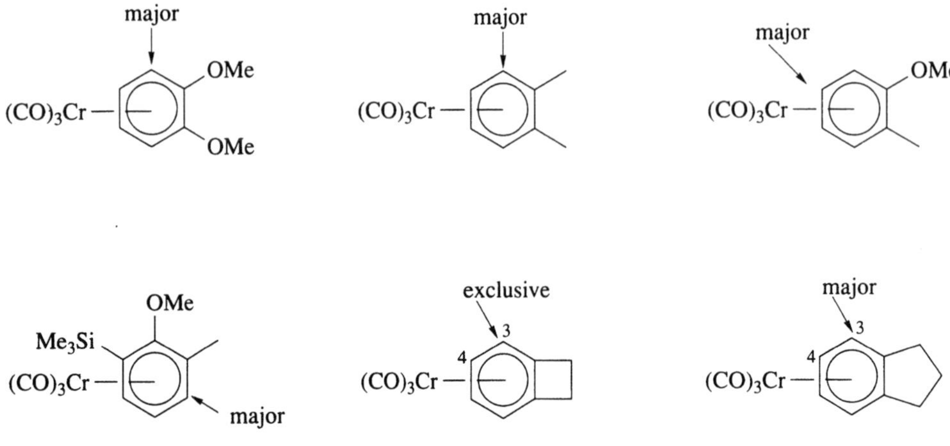

major OMe $(CO)_3Cr$ OMe major $(CO)_3Cr$ major OMe $(CO)_3Cr$

OMe Me_3Si $(CO)_3Cr$ major exclusive 3 4 $(CO)_3Cr$ **(35)** major 3 4 $(CO)_3Cr$ **(36)**

Figure 1 Summary of regioselectivity of substituted arene ligands

Addition to *o*-alkylaniline complexes has been the subject of a detailed study and presents a clear case of different products from kinetic *versus* thermodynamic control.[87,88] Addition of 2-lithio-2-cyanopropane to [(*N*-methyl(tetrahydroquinoline)$Cr(CO)_3$] (**37**; equation 33) with variation in time, temperature and solvent (THF or THF with 4 mol equiv. of HMPA added) gives product distributions ranging from 2:1 for C-5:C-3 at –78 °C for 1 min to >96% addition at C-5, when the adduct is allowed to equilibrate (8.6 h at –78 °C). The data show that equilibration is occurring within minutes at –78 °C and the product distribution does not change upon warming to 20 °C. The effect of added HMPA is an initial product ratio of 54:44 for C-5:C-3 and strongly inhibited equilibration; the effect of HMPA was also observed in halide substitution (addition–elimination)[90] and clearly established during addition to simple arenes.[68,84] With the less sterically demanding anion, $LiCH_2CN$, addition is incomplete after 22 h at –78 °C and favored at C-3 (82:18 for C-3:C-5).

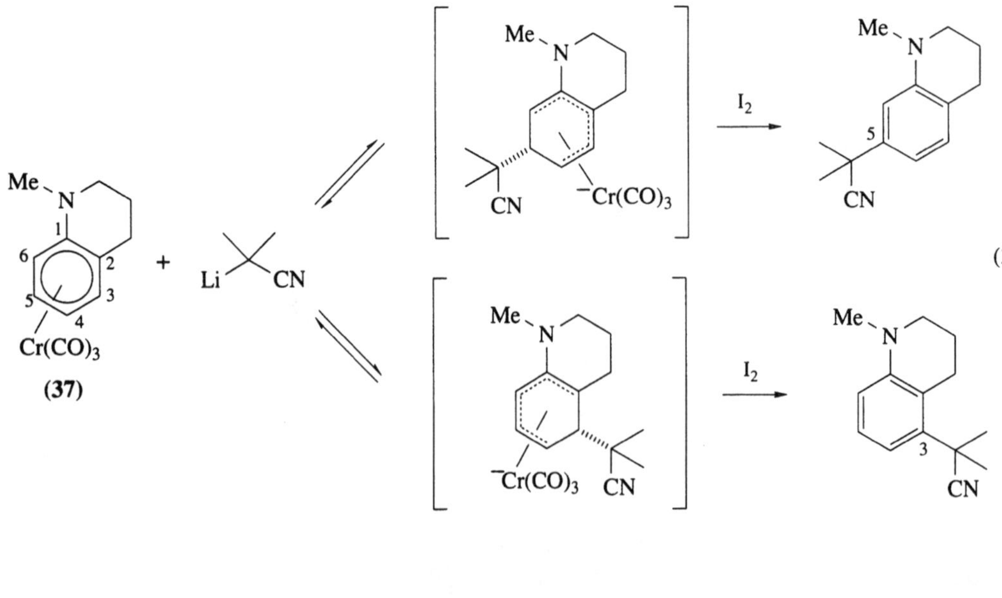

Me N 1 2 3 4 5 6 $Cr(CO)_3$ **(37)** + Li CN [Me N CN $^-Cr(CO)_3$] I_2 → Me N 5 CN [Me N $^-Cr(CO)_3$ CN] I_2 → Me N 3 CN

(33)

The same study shows that with the complex of *N,N*-dimethyl-*o*-toluidine (**38**), the selectivity depends on time, temperature and the nature of the anion (Table 4 and equation 34).[87] Again, equilibration occurs with the tertiary nitrile-stabilized anion, favoring the C-4 substitution product, while lithioacetonitrile favors addition at C-3. The 2-methyl-1,3-dithianyl anion gives precisely the same product mixture at –78 °C and at –30 °C; there is no evidence for equilibration with this anion.

Table 4 Addition–Oxidation with [(*N,N*-Dimethyl-*o*-Toluidine)$Cr(CO)_3$]

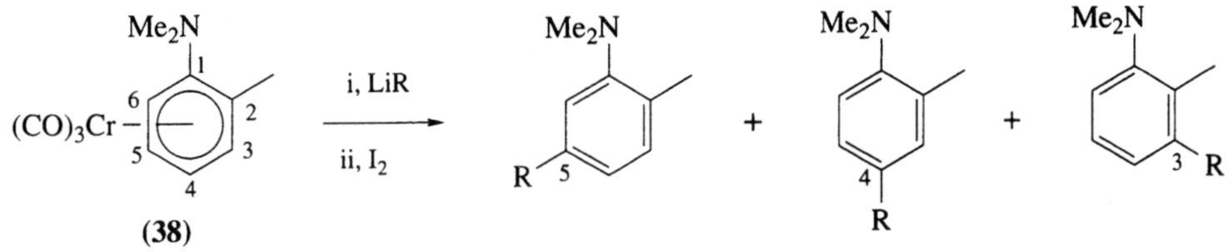

Me_2N 1 6 2 $(CO)_3Cr$ 5 3 4 (**38**) —i, LiR; ii, I_2→ Me_2N R 5 + Me_2N 4 R + Me_2N 3 R (34)

Carbanion	*Conditions*	*1,2,5 (%)*	*1,2,4 (%)*	*1,2,3 (%)*
$LiC(Me)_2CN$	–78 °C/2.5 h	32	32	36
$LiC(Me)_2CN$	–50 °C/24 h	8	81	11
$LiCH_2CN$	–40 °C/24 h	3	4	93
2-Li-1,3-Thianyl	–78 °C/2.5 h	25	10	65

The complex of *N*-methylindoline (**39**) gives similar behavior, favoring addition at C-4 (indole numbering), but with significant addition at C-7, depending on the anion and opportunity for equilibration.[87,93] With the 2-methyl-1,3-dithianyl nucleophile, a high selectivity for C-4 is observed (equation 35).

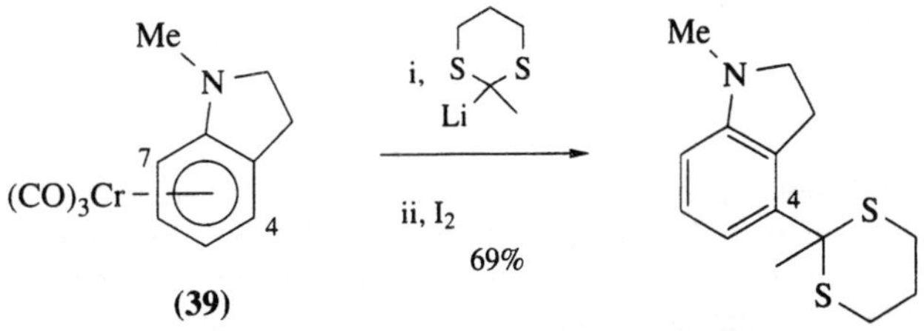

Me N 7 $(CO)_3Cr$ 4 (**39**) —i, S S Li; ii, I_2; 69%→ Me N 4 S S (35)

96% this isomer

Indole is a particularly interesting case, because the $Cr(CO)_3$ unit selectively activates the six-membered ring (**40**),[93,94] while in free indole the five-membered ring dominates the (electrophilic addition) reactivity. The selectivity in addition to the $Cr(CO)_3$ complexes of indole derivatives shows a preference for addition at C-4 (indole numbering) and C-7, with steric effects due to substituents at C-3 and N-1 as well as anion type influencing the selectivity.[93] In Scheme 10, example A, a hydrogen substituent at C-3 and a tertiary carbanion leads to selective C-4 substitution. In example B, the same substrate adds 2-lithio-1,3-dithiane predominantly at C-7. With a trimethylsilylmethyl substituent at C-3, the addition is preferred at C-7 (example C). Finally, even with the silylmethyl substituent at C-3, a sufficiently large *N*-protecting group can disfavor addition at C-7 (example D).

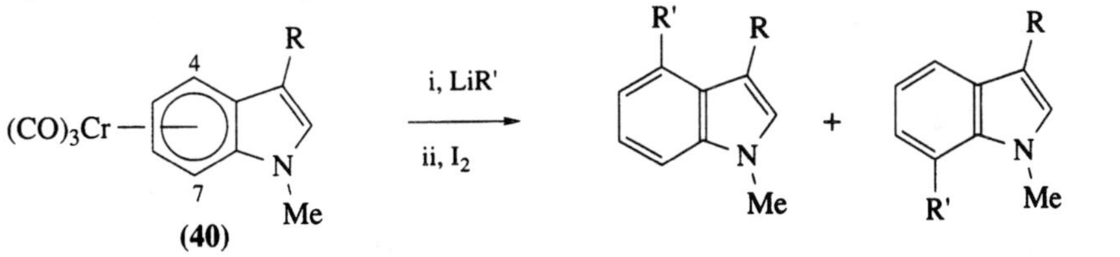

R 4 $(CO)_3Cr$ N 7 Me (**40**) —i, LiR'; ii, I_2→ R' R N Me + R N R' Me

A.	R' = CMe_2CN; R = H; Y = Me	99	:	1	(92% yield)
B.	R' = (1,3-dithianyl); R = H; Y = Me	14	:	86	(68%)
C.	R' = CMe_2CN; R = CH_2SiMe_3	17	:	83	(82%)
D.	R' = CMe_2CN; R = CH_2SiMe_3; Y = $SiBu^tMe_2$	95	:	5	(78%)

Scheme 10

Another example in which the regioselectivity of addition is different under kinetic *versus* thermodynamic control is the naphthalene series. In the addition of $LiCMe_2CN$ to [(naphthalene)$Cr(CO)_3$] (**41**), a mixture of products is observed from addition at C-α and C-β in the ratio 42:58 under conditions where equilibration is minimized (0.3 h, –65 °C, THF–HMPA). With the same reactants, but in THF and at 0 °C, the product is almost exclusively the α-substituted naphthalene (equation 36).[84,85,92]

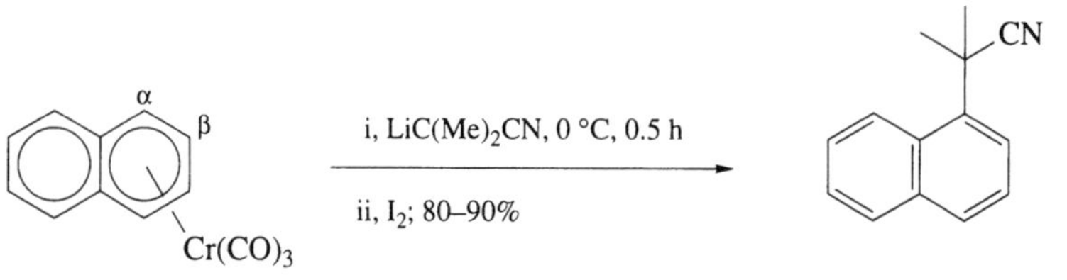

i, $LiC(Me)_2CN$, 0 °C, 0.5 h; ii, I_2; 80–90% (36)

(**41**) >98% selectivity

Using the standard procedures, 1,4-dimethoxynaphthalene is complexed at the less-substituted ring with high selectivity to give (**42**).[68] Again, under conditions of minimum equilibration of anion addition, $LiCMe_2CN$ gave a mixture (after iodine oxidation) of the 1,4-dimethoxy-β-substituted and 1,4-dimethoxy-α-substituted products in the ratio 78:22. After equilibration, the α-substitution product was essentially the only product found (equation 37).

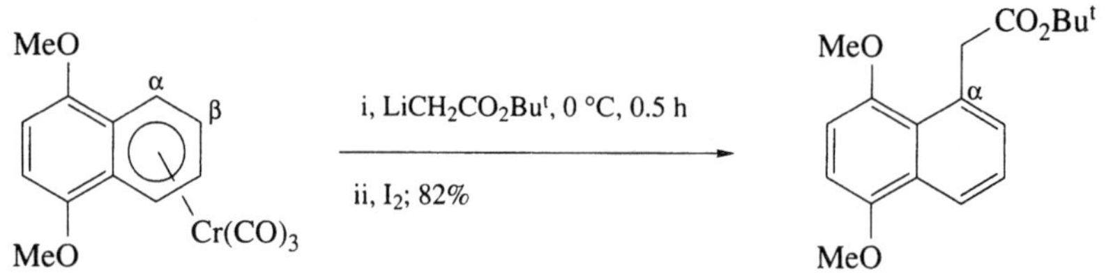

i, $LiCH_2CO_2Bu^t$, 0 °C, 0.5 h; ii, I_2; 82% (37)

(**42**)

The 1,4-dimethoxynaphthalene ligand was used to probe for the parameters which influence the rate of equilibration. The change from pure THF to a mixture of 3:1 THF:HMPA slows the rate of equilibration by a factor of 50 000.[68,85] Replacement of Li with K in the carbanion nucleophile also slows equilibration, by a factor of 500. The equilibration is shown to be intermolecular (dissociation of the nucleophile) by elegant 'crossover' experiments with doubly labeled substrates.[68,85] With the 1,4-dimethoxynaphthalene ligand, cyano-stabilized anions (including cyanohydrin acetal anions) and ester enolates equilibrate even at low temperature and strongly favor addition at the α-position (C-5), as in equation (37). The kinetic site of addition is also generally C-α. However, the 2-lithio-1,3-dithiane anion and phenyllithium do not equilibrate over the temperature range –78 to 0 °C, equations (38) and (39). The sulfur-stabilized anions favor addition at C-β, while phenyllithium gives a mixture favoring C-α.

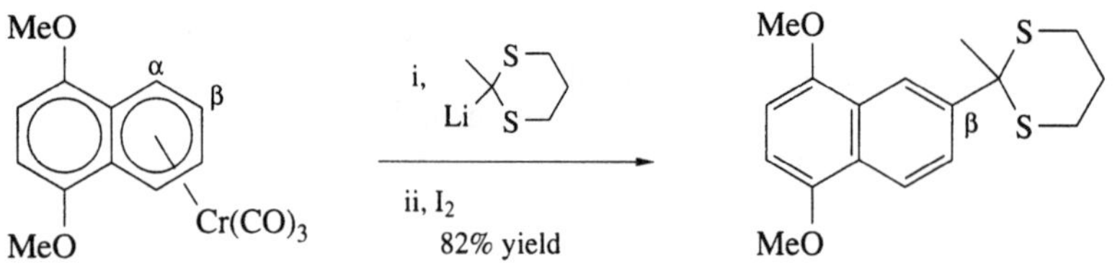

i, 2-lithio-2-methyl-1,3-dithiane; ii, I_2; 82% yield (38)

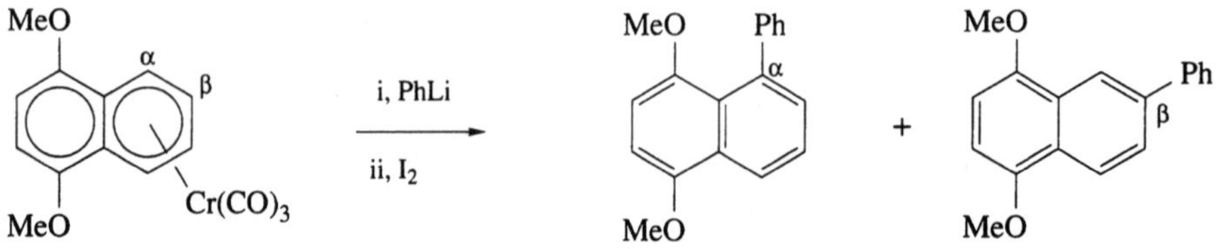

(**42**)

i, PhLi; ii, I_2 (39)

(**42**) 84:16

The steric effect on regioselectivity shows clearly in the series in Table 5 and equation (40).[95] Comparing similar anion type, except for size, entries 1, 3 and 7 show that *ortho* substitution is very significant with a primary carbanion but essentially absent with a tertiary cyano-stabilized anion. It is striking that as the size of the alkyl substitutent on the arene increases, not only is *ortho* substitution disfavored, but *meta* is as well: compare entries 3–6 and compare 7–9. With the very large $CH(Bu^t)_2$ group (entry 9), only *para* substitution is observed. Regioselectivity is also dependent on the electronic nature of the nucleophile. Most remarkably, addition of the primary sulfur-stabilized anion shows nearly equal amounts of *ortho* and *meta* substitution with the toluene ligand, but, as the size of the arene substituent increases, the *para* substitution product increases at the expense of the *meta* product (entries 10–12).

Table 5 Steric Effects on Regioselectivity

$$(CO)_3Cr-(C_6H_5X) \xrightarrow[\text{ii, } I_2]{\text{i, LiR}} o\text{-}XC_6H_4R + m\text{-}XC_6H_4R + p\text{-}XC_6H_4R \quad (40)$$

Entry	*Complex*	*Anion*	*Product Ratio (o:m:p)*	*Combined yield (%)*
1	X = Me	$LiCH_2CN$	35:63:2	88
2	X = Bu^t	$LiCH_2CN$	28:48:24	51
3	X = Me	LiC(OR)MeCN[a]	0:96:4	75
4	X = Et	LiC(OR)MeCN[a]	0:94:6	89
5	X = Pr^i	LiC(OR)MeCN[a]	0:80:20	88
6	X = Bu^t	LiC(OR)MeCN[a]	0:35:65	86
7	X = Me	$LiC(Me)_2CN$	1:97:2	95
8	X = Bu^t	$LiC(Me)_2CN$	0:55:45	78
9	X = $CH(Bu^t)_2$	$LiC(Me)_2CN$	0:0:100	63
10	X = Me	$LiCH_2SPh$	52:46:2	96
11	X = Pr^i	$LiCH_2SPh$	47:46:7	86
12	X = Bu^t	$LiCH_2SPh$	45:32:23	88

[a] R = C(Me)OEt.

As presented in the next section, the dependence of selectivity on anion type is understood as a change in the balance of charge *versus* orbital control. However, in the anions listed so far there is the possibility that differences in carbanion structure, extent of delocalization *etc.*, might contribute to the changes in regioselectivity. In an effort to compare addition of carbanions, which differ little in structure, a series of 2-(*p*-substituted)aryl-2-lithio-1,3-dithiane nucleophiles was allowed to react with the *t*-butylbenzene ligand.[95] The *para* substitutent in the carbanion was varied in the expectation of varying the reactivity of the anion and thereby varying the addition selectivity. Table 6 and equation (41) presents the results. More reactive anions tend to add to the *meta* position (almost *m*:*p* = 3:1 in entry 1; to *m*:*p* = 1:3 in entry 8; *ortho* addition is never significant with this ligand). Obviously, this change in regioselectivity must be due to an electronic factor in the anion; steric effects are constant.

(iii) Interpretation of regioselectivity

Selectivity in polar reactions can be discussed in terms of two features: charge control and orbital control.[96] Analysis of charge control requires a knowledge of charge distribution in the arene ligand. Analysis of orbital control can be approximated by emphasizing interaction of the frontier molecular orbitals, HOMO for the nucleophile and LUMO for the arene complex. It was recognized early that the *ortho–para* selectivity observed with weakly polar substituents, such as Me and Cl, correlated with the LUMO for the free arene and the assumption was made that the LUMO for the arene ligand has a distribution of coefficients similar to the free arene.[91] This has been supported by computation at the level of extended Hückel theory.[88,95,97] LUMOs for (uncoordinated) toluene and chlorobenzene are distributed nearly equally at the *ortho* and *meta* positions and show small density at the *para* position (Figure 2, A).[98–100] Similarly for the case of two donor substituents arranged *ortho*, the LUMO is predicted to be localized at adjacent positions (Figure 2, C), consistent with the observed selectivity. With electron acceptor substituents (R_3Si, CF_3), LUMO has large coefficients *ipso* and *para* (Figure 2, B), consistent with *para* addition selectivity.

Table 6 Electronic Effects in Regioselectivity

$(CO)_3Cr$–(Bu^t-benzene) — i, Y-C$_6$H$_4$-C(Li)(dithiane); ii, I_2 → meta-Bu^t product + para-Bu^t product (41)

Entry	*Para Y*	*Ratio (m:p)*	*Combined yield, (%)*	σ_p *for Y*
1	O^-	73:27	84	—
2	NMe_2	63:37	80	−0.83
3	NMe_2	57:43[a]	78	−0.83
4	OMe	54:46	71	−0.27
5	Me	43:57	75	−0.17
6	H	28:72	75	0
7	H	23:77[a]	85	0
8	Cl	21:79	80	+0.23

[a] HMPA was added to the medium; in all other cases, THF was the solvent.

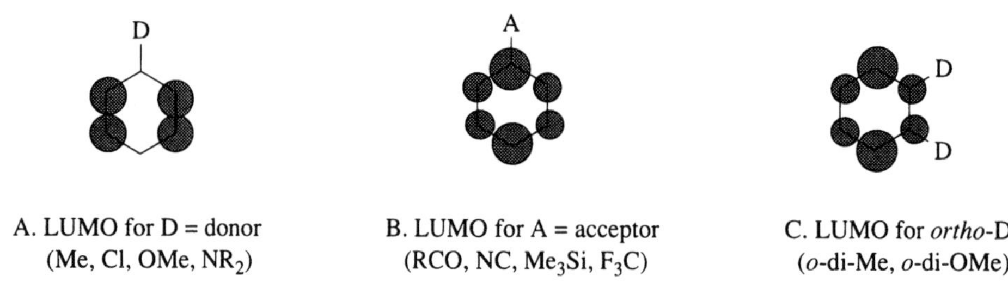

A. LUMO for D = donor (Me, Cl, OMe, NR_2)   B. LUMO for A = acceptor (RCO, NC, Me_3Si, F_3C)   C. LUMO for *ortho*-D (*o*-di-Me, *o*-di-OMe)

Figure 2 Orbital distributions for substituted benzenes

However, the frontier orbital picture based on the free arene does not account for nearly exclusive *meta* selectivity in addition to [(anisole)$Cr(CO)_3$]; LUMO for anisole shows essentially the same pattern as for toluene.[98–100] With a strong resonance electron donor the traditional electronic picture (deactivation of the *ortho* and *para* positions) is sufficient to account for the observed *meta* selectivity. In this case the balance of charge control and orbital control is pushed toward charge control by strong polarization. The same argument applies to the aniline and fluorobenzene complexes.

A delicate balance of charge and orbital effects can also account for the dependence of selectivity on anion type. The central assumption is that nucleophiles with a higher lying HOMO (softer) should give a better orbital energy match in the HOMO–LUMO interaction and increase the orbital control term. For the toluene ligand, this predicts strong *ortho–meta* selectivity for more reactive anions.

The surprising change in regioselectivity from *meta* to *para* substitution as the size of the alkyl group is increased in [(alkylbenzene $Cr(CO)_3$] complexes (Table 5 and equation 40) is not easily rationalized based on the HOMO–LUMO picture and the direct polarization by arene substituents. A parameter unique to the arene–metal complexes is the conformation of the $Cr(CO)_3$ group, and it has been emphasized that this unit tends to be eclipsed with electron donor substituents and *anti* with electron acceptors, in the absence of overwhelming steric effects (Figure 3).[101] In addition, the arene ligand carbons which are eclipsed with a CO ligand are predicted to be more electrophilic (Figure 3).[102] This factor should favor addition to arene positions eclipsed with a CO ligand, under charge control conditions. In this picture, *meta* selectivity with anisole is consistent with an eclipsed conformation and *meta* activation. In most cases with polar substituents on the arene, the direct polar effect of the substituent and the indirect effect through conformational preferences lead to the same predicted selectivity: donor substitutents favor *meta* substitution and acceptors favor *ortho* and *para*. The emphasis on conformational effects in regioselectivity has been applied in several other systems[87,88,93,103,104] and supported by direct NMR measurements.[104–107]

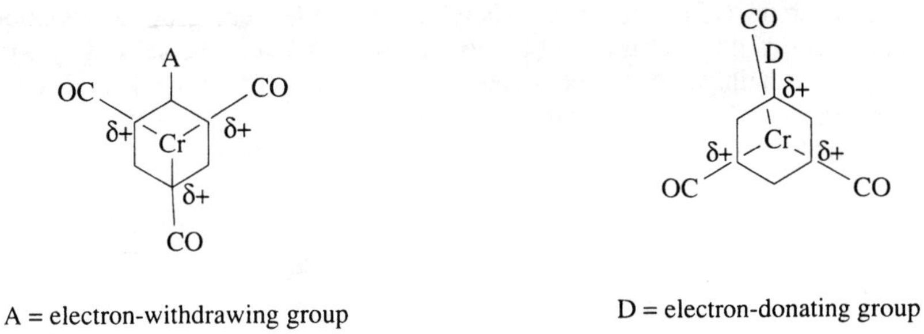

A = electron-withdrawing group

D = electron-donating group

Figure 3 Conformational preferences in [(arene)$Cr(CO)_3$] complexes

The picture emphasizing conformational effects[102] does not account for significant *ortho–meta* selectivity in some cases, nor preferential addition adjacent to *ortho* donor substituents. However, all of the existing regioselectivity results can be accommodated by analyzing orbital control *versus* charge control, assuming an arene-centered LUMO similar to that for the free arene, and assuming that conformational preferences for the arene ligand have a major influence on the charge distribution of the arene.

The examples in Table 5 and equation (40) were chosen to test exactly these parameters and to separate the direct effect of substituents on charge distribution from the indirect effect of the substituent through its influence on the conformational preferences. With a very large substituent on the arene (*e.g.* Bu^t and $CH(Bu^t)_2$), the favored conformation (Figure 4) now positions the CO ligand eclipsed with the *ortho* and *para* positions as steric interactions override electronic preferences.[104] Electron density is predicted to be reduced at these positions and a charge-controlled coupling with a nucleophile would be favored *para* (*ortho* is disfavored by direct steric interactions). The arene-centered LUMO, expected to be relatively independent of the conformation of the $Cr(CO)_3$ unit, has large density at the *ortho* and *meta* positions (Figure 4). Orbital-controlled coupling would be favored *meta*.

a. Favored conformation and resulting charge distribution

b. LUMO for an alkylbenzene

Figure 4 Charge distribution and LUMO for [(Bu^t_2CHPh)$Cr(CO)_3$]

The series of *para*-substituted 2-aryl-1,3-dithianyl carbanions shown in Table 6 and equation (41) are expected to differ in LUMO energy. The phenoxide-substituted anion (entry 1, Table 6 and equation 41) should have higher LUMO relative to, for example, the simple phenyl-substituted anion (entry 6) and gives a greater preponderance of *meta* substitution due to stronger orbital control. The smooth correlation of selectivity with reactivity for this series of closely related anions is consistent with a simple two-parameter analysis. Other regioselectivity results, such as those with *o*-alkylanilines (Table 2 and equation 30) and [(indole)$Cr(CO)_3$] also have been analyzed in terms of orbital control and conformational effects.[87,88,93]

(iv) Application in the synthesis of deoxyfrenolicin.[108]

The combination of *ortho* metallation and *meta* nucleophilic acylation was used to prepare a key intermediate in a synthesis of deoxyfrenolicin (**42**), as outlined in Scheme 11. The complex of anisole is *ortho*-metallated with *n*-butyllithium and quenched with chlorotrimethylsilane; the resulting [(*o*-(trimethylsilyl)anisole)$Cr(CO)_3$] (**43**) is then metallated again, converted to the arylcuprate, and coupled with (*E*)-2-hexenyl bromide to give the complex of 1-trimethylsilyl-2-methoxy-3-(2-hexenyl)benzene (**44**). Addition of the carbanion from the cyanohydrin acetal of 4-pentenal, followed by the standard iodine oxidation and subsequent hydrolysis of the cyanohydrin acetal to regenerate the carbonyl group gave the trisubstituted arene (**45**). The trimethylsilyl group underwent proto-desilylation during the iodine oxidation step. The yield overall from (**44**) to (**45**) was 61% and the selectivity for addition adjacent to the hexenyl side chain was >95%. The role of the trimethylsilyl group is to disfavor addition at the alternative activated *meta* position (*ortho* to Me_3Si). The ketone (**45**) serves as an effective intermediate for the construction of deoxyfrenolicin, using conventional methodology.

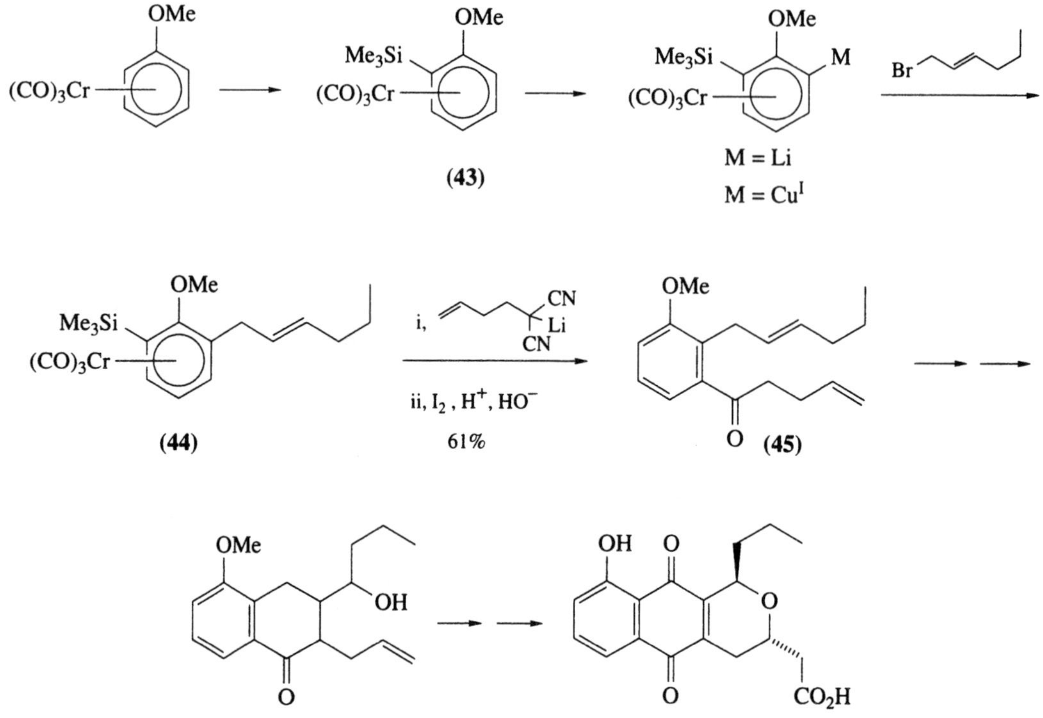

(**43**)

M = Li
M = Cu^I

(**44**)

i, [CN/Li/CN carbanion reagent]
ii, I_2, H^+, HO^-
61%

(**45**)

(**42**) Racemic deoxyfrenolicin

Scheme 11

(v) Intramolecular addition–oxidation

Intramolecular addition–oxidation with reactive carbanions is generally successful; most of the examples involve cyano-stabilized carbanions. Formation of a six-membered fused ring (**46**) is efficient (equation 42), but five-membered fused ring formation is not.[17] The only addition–oxidation product is a cyclic dimer, the [3.3]metacyclophane (**47**).[17]

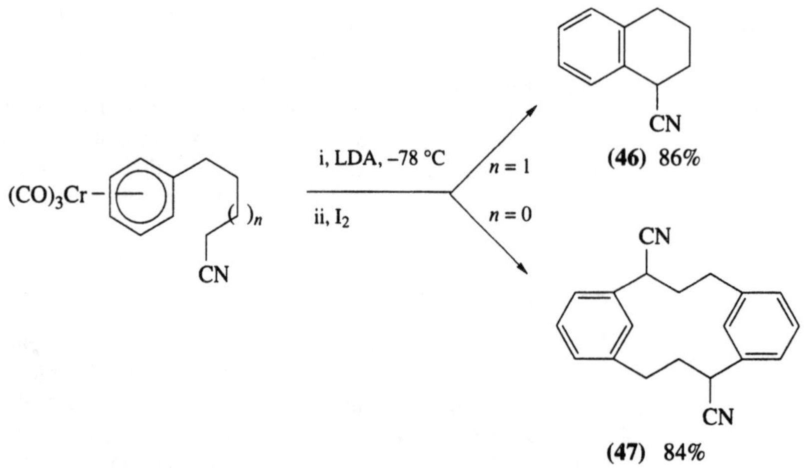

i, LDA, –78 °C
ii, I_2
$n = 1$ → (**46**) 86%
$n = 0$ → (**47**) 84%

(42)

Reversibility again is apparent with the higher homolog (**48**; Scheme 12). At low temperature (0.5 h, –78 °C), a mixture of cyclohexadienyl anionic intermediates is formed with the spiro ring isomer (**49**) preferred by a factor of almost 3:1. An alternative quenching procedure, using trifluoroacetic acid (see below), retains the spiroring and produces the spiro[5.5]cyclohexadiene product (**51**). If the initial adduct is allowed to warm to 0 °C or above, essentially complete rearrangement to the fused ring adduct (**50**) occurs and oxidative quenching gives the seven-membered ring (**52**) in good yield, Scheme 12.[17]

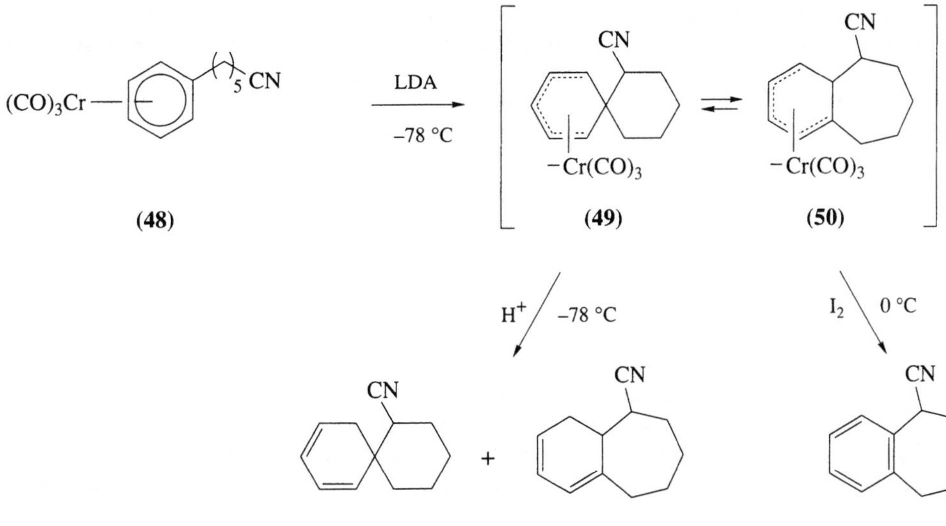

Scheme 12

2.4.4.1.2 Addition–oxidation with [(arene)FeCp]⁺ complexes

It has been recognized for many years that nucleophiles will add to [(η^6-benzene)FeCp]$^+$ complexes.[109–112] The nucleophiles range from cyanide anion[113] to metal hydrides and methyllithium; the intermediate neutral [(η^5-(6-substituted)cyclohexadienyl)FeCp] complexes can be isolated and characterized. The regioselectivity has been determined for a number of monosubstituted arene ligands. Chloro[110] and methoxycarbonyl[112] substituents are *ortho* directing, while methoxy[111] is a *meta* director, and a methyl group leads to similar amounts of addition at the *ortho*, *meta* and *para* positions.[109] In general, electron-withdrawing groups (nitro, halo, benzoyl, cyano and sulfonyl) favor *ortho* addition, as exemplified for the addition of the acetone enolate in equation (43).[113,114]

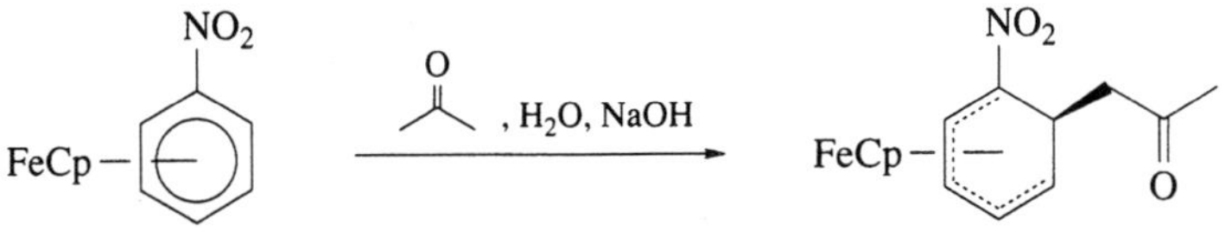

(43)

The primary limitations in applying these complexes are the two-stage detachment of the *endo* C-6 hydrogen (from **53**) and the detachment of the FeCp unit. Direct abstraction with trityl cation has been employed for the first stage,[110] and oxidation with *N*-bromosuccinimide completes both stages, leading to the substituted arene in the few cases tested.[115,116] In the cases with electron-withdrawing substituents on the starting arene, Ce^{IV} has been employed to aromatize and detach the arene unit after addition (equation 44).[113]

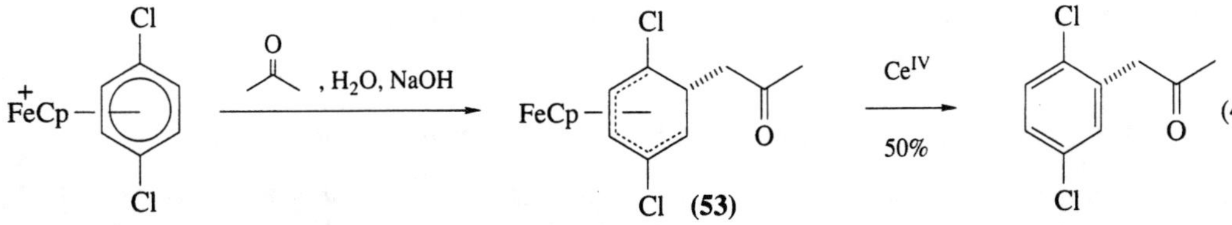

(44)

Because of the high reactivity of the [(arene)FeCp]$^+$ complexes, it can be expected that further developments and definition of selectivity features will be forthcoming.

2.4.4.1.3 Addition–oxidation with [(arene)Mn(CO)$_3$]$^+$ complexes

The reactivity of the [(η^6-arene)Mn(CO)$_3$]$^+$ complexes toward addition of nucleophiles is high and general, leading to neutral, isolable [(η^6-(6-substituted)cyclohexadienyl)Mn(CO)$_3$] derivatives, *e.g.* (**54**).[9,117] The addition of Grignard reagents to anisole and aniline ligands shows good *meta* selectivity (equation 45).[38,118,119] Addition to the chlorobenzene ligand gives similar amounts of *ortho* and *meta* (orbital control?), while the toluene ligand favors addition at all three positions (*o, m, p*).[38,119]

MeO (CO)$_3$Mn$^+$ — [arene] —MeLi, 0 °C→ (CO)$_3$Mn — [cyclohexadienyl, MeO, H] (45)

(**54**)

The completion of the substitution process to produce a substituted arene free of the metal has not been worked out in general. Direct abstraction of the *endo*-hydride of the η^5-cyclohexadienyl intermediate (**55**) with trityl cation is not effective, but an interesting thermal rearrangement has been observed at 130–150 °C to create an isomeric cyclohexadienyl ligand (in **56**), which bears an *exo* (easily abstracted) hydrogen substituent (equation 46).[119,120]

$^+$Mn(CO)$_3$ —MeLi→ Mn(CO)$_3$ (H_{endo}) (**55**) —150 °C→ (CO)$_3$Mn (H_{exo}, H_{endo}) (**56**) —Ph$_3$C$^+$→ (CO)$_3$Mn$^+$ (46)

A general process for addition of Grignard reagents and ketone enolates to [(arene)Mn(CO)$_3$]$^+$ complexes includes oxidative decomplexation of the substituted arene with Jones reagent.[38] An alternative is *endo*-hydride abstraction with strong acid, producing hydrogen and, using a coordinating solvent, displacing the arene ligand (equation 47).[38] This latter process allows easy recycling of the MnI reagent.

$^+$Mn(CO)$_3$ —RM→ Mn(CO)$_3$ (R, H) —(RM = PhMgBr), CF$_3$CO$_2$H–MeCN, 80%→ Ph-benzene + H$_2$; —(RM = LiCH$_2$COCMe$_3$), CrO$_3$ (Jones' reagent), 75%→ PhCH$_2$C(O)CMe$_3$ (47)

The [(arene)Mn(CO)$_3$]$^+$ system is very promising, but more development work is necessary in order to assess the full possibilities for overall addition–oxidation, substitution for hydrogen.

2.4.4.2 Addition–Protonation: Synthesis of Substituted 1,3-Cyclohexadienes

The intermediate η^5-cyclohexadienyl anionic species from nucleophilic addition to the [(arene)Cr(CO)$_3$] complexes are obviously highly electron rich and should be susceptible to reactions with electrophiles. Protonation is efficient at low temperature and is suggested to produce a labile η^4-1,3-cyclohexadiene complex (**57**), which can undergo H-migration to give the more stable 1,3-cyclohexadiene isomer (**58**); in simple systems, the major product is the 1-substituted 1,3-cyclohexadiene, Scheme 13.[83] The products can be aromatized by reaction with 2,3-dichloro-5,6-dicyanobenzoquinone.

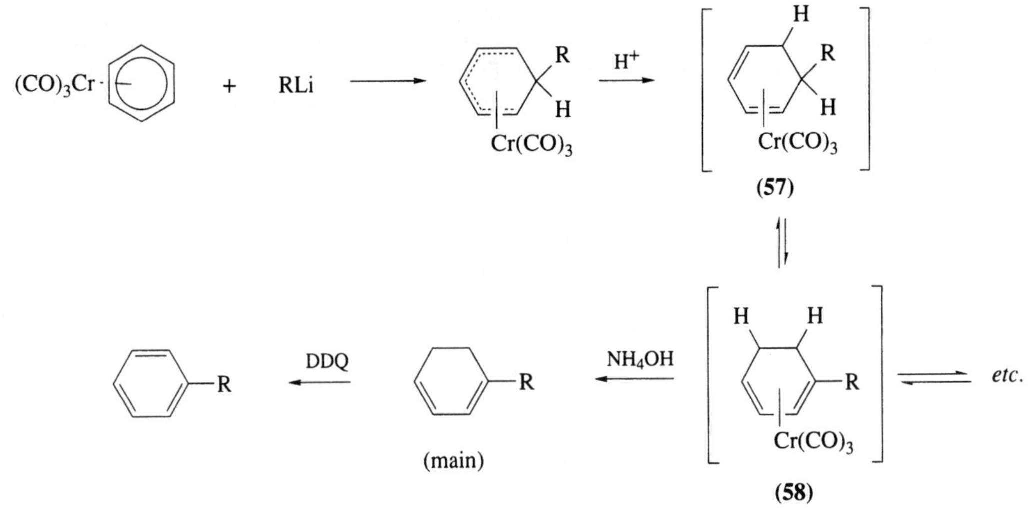

$(CO)_3Cr$ + RLi → ... $Cr(CO)_3$ $\xrightarrow{H^+}$ [... $Cr(CO)_3$] (**57**) ⇌ [... $Cr(CO)_3$] (**58**) ⇌ *etc.*; (**58**) $\xrightarrow{NH_4OH}$ (main) $\xrightarrow{DDQ}$ Ph–R

Scheme 13

The reaction appears to be general, although a limited number of examples have been reported. A particularly useful process begins with *meta* addition to $[(anisole)Cr(CO)_3]$, followed by protonation and hydrolysis of the enol ether unit (equation 48). The result is a 5-substituted cyclohex-2-en-1-one (**59**).[121] The intermediate dienol ether (**60**) can be isolated in high yield, before the aqueous hydrolysis. Other alkene positional isomers of the product enone can be obtained selectively depending on the conditions of the acid treatment and the acid hydrolysis.[121] The addition–protonation process does not change the oxidation state of the chromium and a procedure has been defined for recovery of the Cr^0 in order to allow direct recycling.[122] The addition of CO or other donor ligands during the protonation process does not influence the product distribution; CO insertion to produce formyl derivatives is not observed.

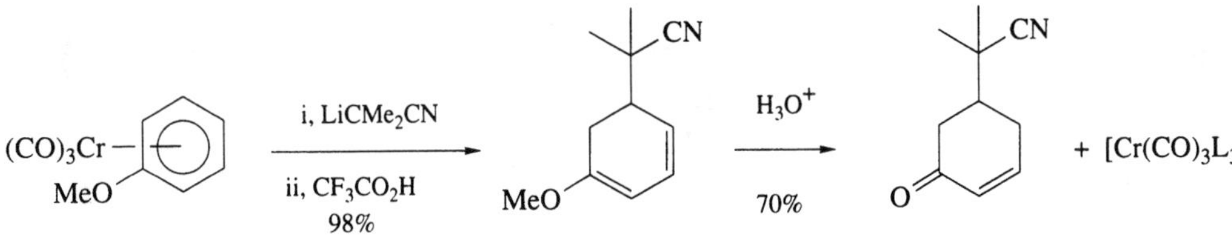

$(CO)_3Cr$–(anisole) $\xrightarrow[\text{ii, } CF_3CO_2H \text{ } 98\%]{\text{i, } LiCMe_2CN}$ (**60**) $\xrightarrow[70\%]{H_3O^+}$ (**61**) + $[Cr(CO)_3L_3]$ (48)

The intramolecular version of addition–protonation with (*m*-cyanoalkyl)anisole ligands produces spirocyclic enones, such as (**62**; equation 49).[121]

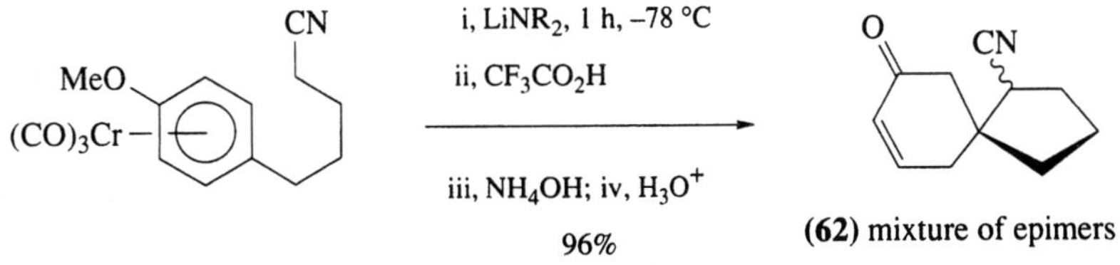

$\xrightarrow[\text{iii, } NH_4OH\text{; iv, } H_3O^+ \text{ } 96\%]{\text{i, } LiNR_2\text{, 1 h, } -78\ ^\circ C\text{; ii, } CF_3CO_2H}$ (**62**) mixture of epimers (49)

This process has been coupled with *meta* addition of a carbonyl anion equivalent and the controlled *exo* addition of the incoming nucleophile to generate acorenone and acorenone B stereospecifically from $[(o\text{-methylanisole})Cr(CO)_3]$ (**63**; Scheme 14).[123] The first step is addition of a cyanohydrin acetal anion (**64**) to the less-hindered *meta* position in $[(o\text{-methylanisole})Cr(CO)_3]$. Addition of allylMgBr to the resulting ketone, anti-Markovnikov addition of HBr to the alkene, substitution for Br by CN, and coordination of the arene with $Cr(CO)_3$ produces the two diastereoisomeric complexes (**65**) and (**66**); the arene–metal unit is a center of asymmetry. The diastereoisomers were separated by HPLC, and the separate isomers were carried through the sequence of side chain anion generation and acid quenching to lead, separately, to the spirocyclohexenones *e.g.* (**67**). The yields were lower than for the model (equation 49) and the best case (64% for the cyclization in the acorenone B series) is based on recovery of starting material (30%). The isopropyl substituent apparently disfavors the ring closure. Conventional procedures added the appropriate functionality for the natural products. The cyclization process is stereospecific, based on the *exo* addition of the nucleophile. The *meta*-directing effect of the methoxy group favors formation of two new bonds at the quaternary center; spirocyclizations are the only cases of anion addition to a substituted arene carbon in the series with $Cr(CO)_3$ activation.

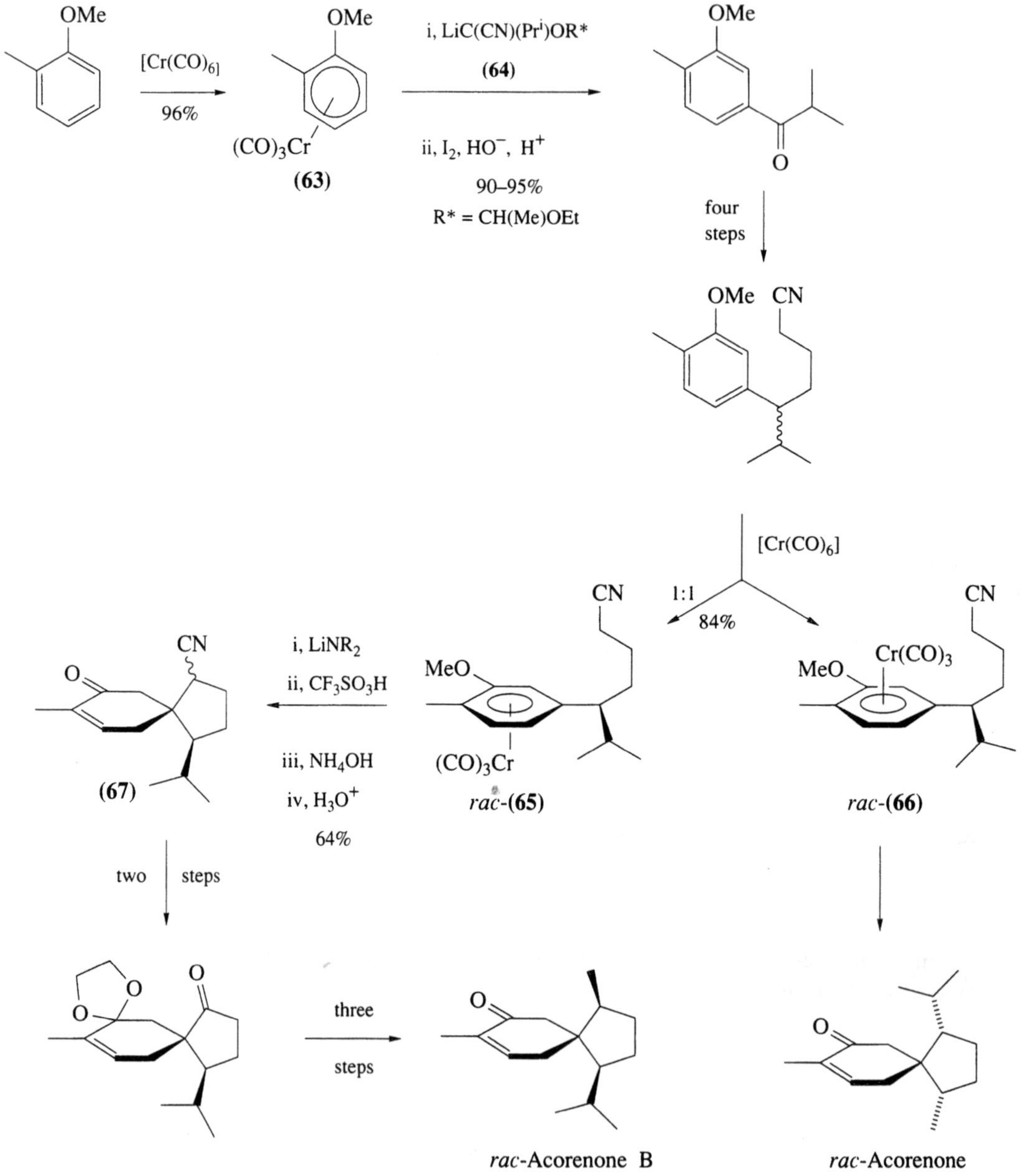

OMe; $[Cr(CO)_6]$, 96%; (63); i, $LiC(CN)(Pr^i)OR^*$ (64); ii, I_2, HO^-, H^+; 90–95%; $R^* = CH(Me)OEt$; four steps; OMe CN; $[Cr(CO)_6]$; 1:1; 84%; MeO; $(CO)_3Cr$; rac-(65); $Cr(CO)_3$; rac-(66); i, $LiNR_2$; ii, CF_3SO_3H; iii, NH_4OH; iv, H_3O^+; 64%; (67); two steps; three steps; *rac*-Acorenone B; *rac*-Acorenone

Scheme 14

2.4.4.3 Addition–Acylation

The efficient trapping of the cyclohexadienyl anionic intermediates with protons raises the possibility of quenching with carbon electrophiles. The process is not as general as the proton quench; early experiments suggested that reversal of the nucleophile addition and coupling of the electrophile with nucleophile was the preferred pathway.[83] However, when the nucleophile adds essentially irreversibly, quenching with a limited set of carbon electrophiles is successful.[84,124–126] For example, addition of 2-lithio-1,3-dithiane to [(benzene)Cr(CO)$_3$], followed by addition of methyl iodide and then oxidation or addition of a donor ligand (CO, PPh$_3$) produces a cyclohexa-1,3-diene (**68**) substituted by both acetyl (Me + CO) and the nucleophile (Scheme 15). The insertion of CO occurs without exception, with a variety of electrophiles. The insertion is efficient without added CO, but gives somewhat higher yields under a modest pressure (<4 bar) of CO during electrophile coupling. The attachment of the electrophile is from the *endo* direction, consistent with initial addition to the metal following by CO insertion (**70**) and migration up to the arene unit; the products then have the *trans* arrangement of the new substituents.

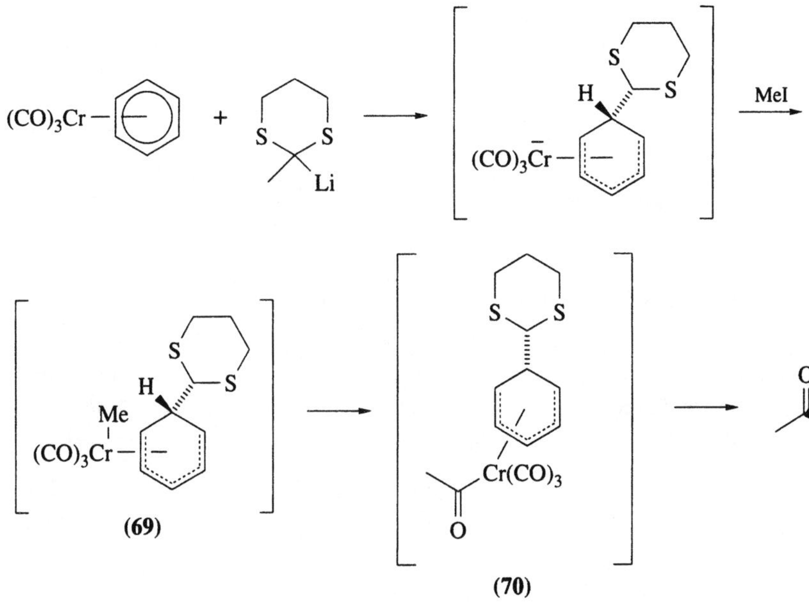

(CO)$_3$Cr–(benzene) + 2-lithio-1,3-dithiane → [(CO)$_3$Cr$^-$ complex] —MeI→

(**69**) → (**70**) → (**68**)

Scheme 15

While the addition–oxidation and the addition–protonation procedures are successful with ester enolates as well as more reactive carbon nucleophiles, the addition–acylation procedure requires more reactive anions and the addition of a polar aprotic solvent (HMPA has been used) to disfavor reversal of anion addition. Under these conditions, cyano-stabilized anions and ester enolates fail (simple alkylation of the carbanion) but cyanohydrin acetal anions are successful. The addition of the cyanohydrin acetal anion (**71**) to [(1,4-dimethoxynaphthalene)Cr(CO)$_3$] occurs by kinetic control at C-β in THF–HMPA and leads to the α,β-diacetyl derivative (**72**) after methyl iodide addition, and hydrolysis of the cyanohydrin acetal (equation 50).[84,124–126]

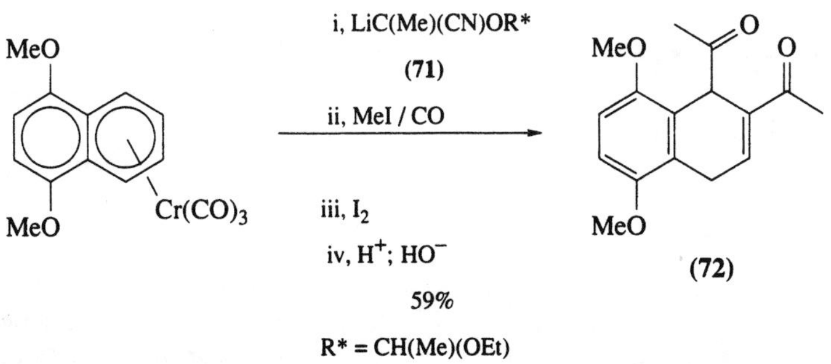

i, LiC(Me)(CN)OR* (**71**); ii, MeI / CO; iii, I_2; iv, H^+; HO^-; 59%; R* = CH(Me)(OEt) — (**72**) (50)

More reactive anions such as the 2-lithio-1,3-dithiane derivatives, phenyllithium and *t*-butyllithium do not require a special solvent and proceed in high yield in THF. While HMPA is known to suppress the migratory insertion to CO in anionic complexes,[127] it does not deter the CO insertion in these cases; no example of direct alkylation is reported. The only electrophile which adds without CO insertion is the proton, as discussed above. Good alkylating agents (primary iodides and triflates, allyl bromide, benzyl bromide) react below 0 °C, but ethyl bromide requires heating at 50 °C. The reaction is selective for a primary alkyl iodide in the presence of an ester or a ketone unit. Surprisingly, acyl chlorides are unreactive. A simple S_N2 process is proposed, based on the reactivity pattern of the electrophiles and the lack of rearrangement during alkylation with cyclopropylcarbinyl iodide (no long-lived radical intermediates).[84,125]

Addition of a cyanohydrin acetal anion to [(benzene)$Cr(CO)_3$] followed by reaction with allyl bromide produces the cyclohexadiene derivative (**73**) in 94% yield, which undergoes a Diels–Alder reaction rapidly to give a tricyclic framework (**74**). After quenching with methyl iodide and disassembling of the cyanohydrin group, the diketone (**75**) is obtained in 50% yield overall (equation 51).[125] These products are obviously interesting as potential intermediates for synthesis.

$(CO)_3Cr$–(benzene) → i, LiC(Me)(CN)OR*; ii, allyl bromide → (**73**) → 50% → (**74**) (51)

R* = CH(Me)(OEt)

2.4.4.4 Addition to Styrene-type Ligands Activated by $Cr(CO)_3$

Nucleophile addition to styrene derivatives (*e.g.* **75**) coordinated with $Cr(CO)_3$ is another example of addition–electrophile trapping.[23,128] Addition of reactive anions is selective at the β-position of the styrene ligand, leading to the stabilized benzylic anion (**76**). The intermediate reacts with protons and a variety of carbon electrophiles to give substituted alkylbenzene ligands (in **77**) (equation 52).

(**75**) → $LiMe_2CN$, –30 °C → [(**76**)] → E → (**77**) (52)

E = H^+, MeCOCl, MeI

In the dihydronaphthalene series, the selective *exo* addition of the nucleophile and *exo* addition of the electrophile (steric approach) results in exclusive formation of the *cis*-α,β-disubstituted tetralin (**78**; equation 53).[128]

$Cr(CO)_3$ complex → $LiCMe_2CN$, –30 °C → [$Cr(CO)_3$ anion] → i, MeI; ii, I_2; 65% → (**78**) (53)

The addition–protonation procedure maintains the arene–chromium bond and allows further application of the activating effect of the metal. In an approach to the synthesis of anthraquinone antibiotics, the dihydronaphthalene complex (**79**) was allowed to react with a cyanohydrin acetal anion and then quenched with acid.[129] The resulting tetralin complex (**80**) could be metallated effectively and carried on to a key intermediate (**81**) in anthraquinone construction (equation 54)

While it can be expected that the $[Mn(CO)_3]^+$ and $[FeCp]^+$ complexes of styrene derivatives would allow addition of a wider variety of nucleophiles, no useful methodology based on this possibility has been put forward.

$Cr(CO)_3$ OMe (79) —[i, LiC(Me)(CN)OR*; ii, H^+, HO^-]→ $Cr(CO)_3$ O OMe (80) → O O O OMe OMe (81) (54)

2.4.5 REFERENCES

1. J. P. Collman, L. S. Hegedus, J. R. Norton and R. G. Finke, 'Principles and Applications of Organotransition Metal Chemistry', University Science Books, Mill Valley, CA, 1987, pp. 424, 921.
2. W. E. Watts, in 'Comprehensive Organometallic Chemistry', ed. G. Wilkinson, F. G. A. Stone and E. W. Abel, Pergamon Press, Oxford, 1982, vol. 8, p. 1013.
3. L. A. P. Kane-Maguire, E. D. Honig and D. A. Sweigart, *Chem. Rev.*, 1984, **84**, 525.
4. M. F. Semmelhack, *J. Organomet. Chem. Libr.*, 1976, **1**, 361.
5. (a) M. F. Semmelhack, *Ann. N. Y. Acad. Sci.*, 1977, **295**, 36; (b) M. F. Semmelhack, G. R. Clark, J. L. Garcia, J. J. Harrison, Y. Thebtaranonth, W. Wulff and A. Yamashita, *Tetrahedron*, 1981, **37**, 3957.
6. M. F. Semmelhack, *Pure Appl. Chem.*, 1981, **53**, 2379.
7. R. G. Sutherland, M. Iqbal and A. Piorko, *J. Organomet. Chem.*, 1986, **302**, 307.
8. D. Astruc, *Tetrahedron*, 1983, **39**, 4027.
9. P. J. C. Walker and R. J. Mawby, *Inorg. Chim. Acta*, 1973, **7**, 621.
10. G. Pattenden, in 'Comprehensive Organic Synthesis', ed. B. M. Trost, Pergamon Press, Oxford, 1991, vol. 3, chap. 1.10.
11. C. Paradisi, in 'Comprehensive Organic Synthesis', ed. B. M. Trost, Pergamon Press, Oxford, 1991, vol. 4, chap. 2.1.
12. G. A. Artamkina, M. P. Egorov and I. P. Beletskaya, *Chem. Rev.*, 1982, **82**, 427.
13. H. H. Zeiss, P. J. Wheatley and H. J. S. Winkler, 'Benzenoid-Metal Complexes', Ronald Press, New York, 1966.
14. W. E. Silverthorn, *Adv. Organomet. Chem.*, 1975, **13**, 47.
15. B. Nicholls and M. C. Whiting, *J. Chem. Soc.*, 1959, 551.
16. W. Strohmeier, *Chem. Ber.*, 1961, **94**, 2490.
17. M. F. Semmelhack, L. Keller and Y. Thebtaranonth, *J. Am. Chem. Soc.*, 1977, **99**, 959.
18. C. A. L. Mahaffy and P. L. Pauson, *Inorg. Synth.*, 1979, **19**, 154.
19. S. Top and G. Jaouen, *J. Organomet. Chem.*, 1979, **182**, 381.
20. C. Sergheraert, J.-C. Brunet and A. Tartar, *J. Chem. Soc., Chem. Comm.*, 1982, 1417.
21. K. Ofele and E. Dotzauer, *J. Organomet. Chem.*, 1971, **30**, 211.
22. G. R. Knox, D. G. Leppard, P. L. Pauson and W. E. Watts, *J. Organomet. Chem.*, 1972, **34**, 347.
23. K. Ofele, *Chem. Ber.*, 1966, **99**, 1732.
24. M. D. Rausch, *Pure Appl. Chem.*, 1972, **30**, 523.
25. G. A. Moser and M. D. Rausch, *Synth. React. Inorg. Metal-Org. Chem.*, 1974, **4**, 37.
26. M. D. Rausch, G. A. Moser, E. J. Zaiko and A. L. Lipman, Jr., *J. Organomet. Chem.*, 1970, **23**, 185.
27. J. Verbrel, R. Mercier and J. Belleney, *J. Organomet. Chem.*, 1982, **235**, 197.
28. E. P. Kundig, C. Perret, S. Spichiger and G. Bernardinelli, *J. Organomet. Chem.*, 1985, **286**, 183.
29. (a) R. L. Pruett, *Prep. Inorg. React.*, 1965, **2**, 187; (b) M. D. Rausch, *J. Org. Chem.*, 1974, **39**, 1787.
30. E. O. Fischer, H. A. Goodwin, C. G. Kreiter, H. D. Simmons, Jr., K. Sonogashira and S. B. Wild, *J. Organomet. Chem.*, 1968, **14**, 359.
31. A. Meyer and G. Jaouen, *J. Chem. Soc., Chem. Commun.*, 1974, 787; for a summary, see ref. 1, p. 933.
32. S. G. Davies, *Chem. Ind. (London)*, 1986, 506.
33. M. Uemura, T. Kobayashi, K. Isobe, T. Minami and Y. Hayashi, *J. Org. Chem.*, 1986, **51**, 2859.
34. M. Uemura, T. Kobayashi, K. Isobe, T. Minami and Y. Hayashi, *J. Am. Chem. Soc.*, 1987, **109**, 5277.
35. M. Uemura, T. Minami, K. Hirotsu and Y. Hayashi, *J. Org. Chem.*, 1989, **54**, 469.
36. K. K. Bhasin, W. G. Balkeen and P. L. Pauson, *J. Organomet. Chem.*, 1981, **204**, C25.
37. T. H. Coffield, V. Sandel and R. D. Closson, *J. Am. Chem. Soc.*, 1957, **79**, 5826.
38. Y. K. Chung, P. G. Williard and D. A. Sweigart, *Organometallics*, 1982, **1**, 1053.
39. I. U. Khand, P. L. Pauson and W. E. Watts, *J. Chem. Soc., C*, 1968, 2257.
40. A. N. Nesmeyanov, N. A. Vol'kenau and I. N. Bolesova, *Dokl. Akad. Nauk SSSR*, 1963, **149**, 615.
41. A. N. Nesmeyanov, N. A. Vol'kenau and I. N. Bolesova, *Tetrahedron Lett.*, 1963, 1725.
42. C. C. Lee, M. Iqbal, U. S. Gill and R. G. Sutherland, *J. Organomet. Chem.*, 1985, **288**, 89.
43. D. Astruc and R. Dabard, *J. Organomet. Chem.*, 1976, **111**, 339.
44. J. F. Helling and W. A. Hendrickson, *J. Organomet. Chem.*, 1977, **141**, 99.
45. T. P. Gill and K. R. Mann, *Inorg. Chem.*, 1983, **22**, 1986.
46. R. G. Sutherland, W. J. Pannekock and C. C. Lee, *Can. J. Chem.*, 1978, **56**, 1782.
47. A. C. Knipe, J. McGuinness and W. E. Watts, *J. Chem. Soc., Chem. Commun.*, 1979, 842.
48. M. C. Whiting (Ethyl Corp.), *US Pat.* 3 225 071 (1966) (*Chem. Abstr.*, 1966, **64**, 6694h).
49. J. F. Bunnett and H. Hermann, *J. Org. Chem.*, 1971, **36**, 4081.

50. S. I. Rosca and S. Rosca, *Rev. Chem. (Bucharest)*, 1964, **25**, 461.
51. M. F. Semmelhack and H. T. Hall, Jr., *J. Am. Chem. Soc.*, 1974, **96**, 7091.
52. M. F. Semmelhack and H. T. Hall, Jr., *J. Am. Chem. Soc.*, 1974, **96**, 7092.
53. T. Oishi, M. Fukui and Y. Endo, *Heterocycles*, 1977, **7**, 947.
54. A. Alemagna, C. Baldoli, P. Del Buttero, E. Licandro and S. Maiorana, *J. Chem. Soc., Chem. Commun.*, 1985, 417.
55. M. Ghavshou and D. A. Widdowson, *J. Chem. Soc., Perkin Trans. 1*, 1983, 3065.
56. J. C. Boutonnet, F. Rose-Munch and E. Rose, *Tetrahedron Lett.*, 1985, 3899.
57. A. Alemagna, P. Del Buttero, C. Gorini, D. Landini, E. Licandro and S. Maiorana, *J. Org. Chem.*, 1983, **48**, 605.
58. A. Alemagna, P. Cremonesi, P. Del Buttero, E. Licandro and S. Maiorana, *J. Org. Chem.*, 1983, **48**, 3114.
59. M. F. Semmelhack, J. Bisaha and M. Czarny, *J. Am. Chem. Soc.*, 1979, **101**, 768.
60. R. P. Houghton, M. Voyle and R. Price, *J. Organomet. Chem.*, 1983, **259**, 183.
61. C. Baldoli, P. Del Buttero, S. Maiorana and A. Papagni, *J. Chem. Soc., Chem. Commun.*, 1985, 1181.
62. R. J. Card and W. S. Trayhanovsky, *J. Org. Chem.*, 1980, **45**, 2560.
63. J. P. Gilday and D. A. Widdowson, *Tetrahedron Lett.*, 1986, 5525.
64. P. J. Beswick and D. A. Widdowson, *Synthesis*, 1985, 492.
65. J. P. Gilday and D. A. Widdowson, *J. Chem. Soc., Chem. Commun.*, 1986, 1235.
66. M. F. Semmelhack and C. Ullenius, *J. Organomet. Chem.*, 1982, **235**, C10.
67. A. Ceccon, A. Gambaro, F. Gotthardi, F. Manoli and A. Venzo, *J. Organomet. Chem.*, 1989, **363**, 91.
68. E. P. Kundig, V. Desobry, D. P. Simmons and E. Wenger, *J. Am. Chem. Soc.*, 1989, **111**, 1804.
69. F. Rose-Munch, E. Rose, A. Semra, L. Mignon, J. Garcia-Oricain and C. Knobler, *J. Organomet. Chem.*, 1989, **363**, 297.
70. F. Rose-Munch, E. Rose and A. Semra, *J. Chem. Soc., Chem. Commun.*, 1987, 942.
71. F. Rose-Munch, E. Rose, A. Semra, Y. Jeannin, and F. Robert, *J. Organomet. Chem.*, 1988, **353**, 53.
72. R. M. Moriarty and U. S. Gill, *Organometallics*, 1986, **5**, 253.
73. A. N. Nesmeyanov, N. A. Vol'kenau and I. N. Bolesova, *Dokl. Akad. Nauk SSSR*, 1967, **175**, 606.
74. R. G. Sutherland, A. Piorko, U. S. Gill and C. C. Lee, *J. Heterocycl. Chem.*, 1982, **19**, 801.
75. C. C. Lee, U. S. Gill, M. Iqbal, C. I. Azogu and R. G. Sutherland, *J. Organomet. Chem.*, 1982, **231**, 151.
76. A. N. Nesmeyanov, N. A. Vol'kenau, E. I. Sirorkina and V. V. Deryabin, *Dokl. Akad. Nauk SSSR*, 1967, **177**, 1110.
77. C. C. Lee, A. S. Abd-el-aziz, R. L. Chowdhury, U. S. Gill, A. Piorko and R. G. Sutherland, *J. Organomet. Chem.*, 1986, **315**, 79.
78. C. C. Lee, A. S. Abd-El-Aziz, R. L. Chowdhury, A. Piorko and R. G. Sutherland, *Synth. React. Inorg. Metal-Org. Chem.*, 1986, **16**, 541.
79. R. L. Chowdhury, C. C. Lee, A. Piorko, and R. G. Sutherland, *Synth. React. Inorg. Metal-Org. Chem.*, 1985, **15**, 1237.
80. J. A. Segal, *J. Chem. Soc., Chem. Comm.*, 1985, 1338.
81. R. G. Sutherland, A. S. Abd-el-aziz, A. Piorko and C. C. Lee, *Synth. Commun.*, 1987, **17**, 393.
82. P. L. Pauson and J. A. Segal, *J. Chem. Soc., Dalton Trans.*, 1975, 1677.
83. M. F. Semmelhack, H. T. Hall, Jr., R. Farina, M. Yoshifuji, G. Clark, T. M. Bargar, K. Hirotsu and J. Clardy, *J. Am. Chem. Soc.*, 1979, **101**, 3535.
84. E. P. Kundig, V. Desobry and D. P. Simmons, *J. Am. Chem. Soc.*, 1983, **105**, 6962.
85. E. P. Kundig, A. F. Cunningham, Jr., P. Paglia and D. P. Simmons, *Helv. Chim. Acta*, 1990, **73**, 386.
86. B. Ohlsson and C. Ullenius, *J. Organomet. Chem.*, 1984, **267**, C34.
87. B. Ohlsson and C. Ullenius, *J. Organomet. Chem.*, 1988, **350**, 35.
88. B. Ohlsson, C. Ullenius, S. Jagner, C. Grivet, E. Wegner and E. P. Kundig, *J. Organomet. Chem.*, 1989, **365**, 243.
89. M. F. Semmelhack and G. Clark, *J. Am. Chem. Soc.*, 1977, **99**, 1675.
90. G. Clark, Ph.D. Thesis, Cornell University, 1977.
91. M. F. Semmelhack, G. R. Clark, R. Farina and M. Saeman, *J. Am. Chem. Soc.*, 1979, **101**, 217.
92. E. P. Kundig, *Pure Appl. Chem.*, 1985, **57**, 1855.
93. M. F. Semmelhack, W. Wulff and J. L. Garcia, *J. Organomet. Chem.*, 1982, **240**, C5.
94. A. P. Kozikowski and K. Isobe, *J. Chem. Soc., Chem. Comm.*, 1978, 1076.
95. M. F. Semmelhack, J. L. Garcia, D. Cortes, R. Farina, R. Hong and B. K. Carpenter, *Organometallics*, 1983, **2**, 467.
96. G. Klopmann, 'Chemical Reactivity and Reaction Paths', Wiley, New York, 1974.
97. See footnote 19 in ref. 92.
98. E. Heilbronner and P. A. Straub, 'HMO', Springer, New York, 1966.
99. K. W. Bowers, in 'Radical Ions', ed. E. T. Kaiser and L. Kevan, Interscience, New York, 1968, p. 211.
100. K. W. Bowers, *Adv. Magn. Reson.*, 1965, **1**, 317.
101. T. A. Albright, P. Hofmann and R. Hoffmann, *J. Am. Chem. Soc.*, 1977, **99**, 7546.
102. T. A. Albright and B. K. Carpenter, *Inorg. Chem.*, 1980, **19**, 3092.
103. (a) W. R. Jackson, I. D. Rae, M. G. Wong, M. F. Semmelhack and J. N. Garcia, *J. Chem. Soc., Chem. Commun.*, 1982, 1359; (b) W. R. Jackson, I. D. Rae, M. G. Wong, *Aust. J. Chem.*, 1986, **39**, 303.
104. A. Solladié-Cavallo and G. Wipff, *Tetrahedron Lett.*, 1980, 3047.
105. A. Solladié-Cavallo and J. Suffert, *J. Org. Magn. Reson.*, 1980, **14**, 426.
106. F. van Meurs, J. M. van der Toorn and H. van Bekkum, *J. Organomet. Chem.*, 1976, **113**, 341 and 353.
107. A. Solladié-Cavallo, *Polyhedron*, 1985, **11**, 901.
108. M. F. Semmelhack and A. Zask, *J. Am. Chem. Soc.*, 1983, **105**, 2034.
109. I. U. Khand, P. L. Pauson and W. E. Watts, *J. Chem. Soc. (C)*, 1968, 2257 and 2261.
110. I. U. Khand, P. L. Pauson and W. E. Watts, *J. Chem. Soc. (C)*, 1969, 116 and 2024.
111. F. Haque, J. Miller, P. L. Pauson, and J. B. P. Tripathi, *J. Chem. Soc. (C)*, 1971, 743.

112. J. F. McGreer and W. E. Watts, *J. Organomet. Chem.*, 1976, **110**, 103.
113. R. G. Sutherland, R. L. Chowdhury, A. Piorko and C. C. Lee, *J. Organomet. Chem.*, 1987, **319**, 379.
114. R. G. Sutherland, R. L. Chowdhury, A. Piorko and C. C. Lee, *J. Chem. Soc., Chem. Commun.*, 1985, 1296.
115. A. Zimniak, *Bull Acad. Pol. Sci., Ser. Chim.*, 1979, **27**, 743.
116. J. C. Boutonnet and E. Rose, *J. Organomet. Chem.*, 1981, **221**, 157.
117. L. A. P. Kane-Maguire and D. A. Sweigart, *Inorg. Chem.*, 1979, **18**, 700.
118. P. L. Pauson and J. A. Segal, *J. Chem. Soc., Dalton Trans.*, 1975, 1683.
119. G. A. M. Munro and P. L. Pauson, *J. Chem. Soc., Chem. Commun.*, 1976, 134.
120. G. A. M. Munro and P. L. Pauson, *Z. Anorg. Allg. Chem.*, 1979, **458**, 211.
121. M. F. Semmelhack, J. J. Harrison and Y. Thebtaranonth, *J. Org. Chem.*, 1979, **44**, 3275.
122. J. C. Boutonnet, J. Levisalles, J. M. Normant and E. Rose, *J. Organomet. Chem.*, 1983, **255**, C21.
123. M. F. Semmelhack and A. Yamashita, *J. Am. Chem. Soc.*, 1980, **102**, 5924.
124. E. P. Kundig and D. P. Simmons, *J. Chem. Soc., Chem. Commun.*, 1983, 1320.
125. E. P. Kundig, N. P. Do Thi, P. Paglia, D. P. Simmons, S. Spichiger and E. Wenger, in 'Organometallics in Organic Synthesis', ed. A. de Meijere and H. tom Dieck, Springer, Berlin, 1987, p. 265.
126. R. C. Cambie, G. R. Clark, S. R. Gallagher, P. S. Rutledge, M. J. Stone and P. D. Woodgate, *J. Organomet. Chem.*, 1988, **342**, 315.
127. J. P. Collman, *Acc. Chem. Res.*, 1975, **8**, 342.
128. M. F. Semmelhack, W. Seufert and L. Keller, *J. Am. Chem. Soc.*, 1980, **102**, 6584.
129. M. Uemura, T. Minami and Y. Hayashi, *J. Chem. Soc., Chem. Commun.*, 1984, 1193.

3.1
Heteroatom Nucleophiles with Metal-activated Alkenes and Alkynes

LOUIS S. HEGEDUS
Colorado State University, Fort Collins, CO, USA

3.1.1 INTRODUCTION

Electron-rich, unsaturated hydrocarbons, which are normally resistant to nucleophilic attack, become generally reactive towards nucleophiles upon complexation to an electrophilic transition metal such as palladium(II), platinum(II) or iron(II). Complexation also directs the regio- and stereo-chemistry of the nucleophilic attack, the result of which is a new organometallic complex, which can often be used to promote additional functionalization of the original substrate. Synthetically useful examples of such processes are presented in the following sections.

3.1.2 ADDITIONS TO ALKENES

Palladium(II) salts, in the form of organic solvent soluble complexes such as $PdCl_2(RCN)_2$, $Pd(OAc)_2$ or Li_2PdCl_4, are by far the most extensively utilized transition metal complexes to activate simple (unactivated) alkenes towards nucleophilic attack (Scheme 1). Alkenes rapidly and reversibly complex to palladium(II) species in solution, readily generating alkenepalladium(II) species (**1**) *in situ*. Terminal monoalkenes are most strongly complexed, followed by internal *cis* and *trans* (respectively) alkenes. Geminally disubstituted, trisubstituted and tetrasubstituted alkenes are only weakly bound, if at all, and intermolecular nucleophilic additions to these alkenes are rare.

Once complexed to palladium(II), the alkene is generally activated towards nucleophilic attack, with nucleophiles ranging from chloride to phenyllithium undergoing reaction. The reaction is, however, quite sensitive to conditions and displacement of the alkene by the nucleophile (path a) or oxidative destruction of the nucleophile can become an important competing reaction. Nucleophilic attack occurs predominately to exclusively at the more-substituted position of the alkene (the position best able to stabilize positive charge) and from the face opposite the metal (*trans* attack, path b) to produce a new carbon–nucleophile bond and a new carbon–metal bond. This newly formed σ-alkylmetal complex (**2**) is

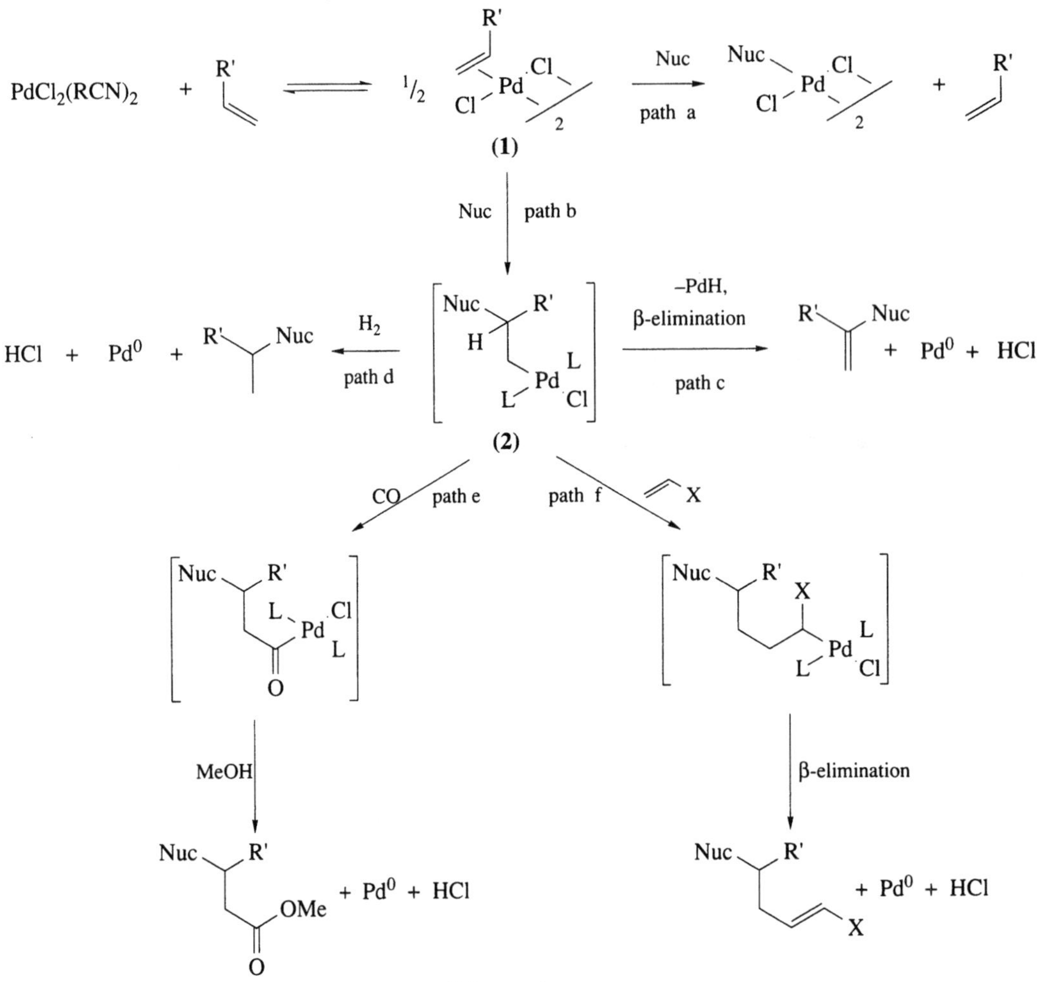

Scheme 1

usually quite reactive and unstable and it can undergo a number of synthetically useful transformations. β-Hydrogen elimination (path c) is facile above *ca.* –20 °C and results in an overall nucleophilic substitution on the alkene. The metal–carbon σ-bond is readily hydrogenolyzed and exposure to 1 atm of hydrogen gas at 25 °C results in an overall nucleophilic addition to the alkene (path d). Small molecules, such as CO or alkenes, readily insert into palladium–carbon σ-bonds permitting further functionalization of the original alkene (paths e and f). All of these transformations ultimately produce palladium(0), while palladium(II) is required to activate alkenes. Thus, for any of these processes to be run in a catalytic fashion, a way to rapidly reoxidize palladium(0) to palladium(II) in the presence of both substrate and product is required. Although many systems have been developed ($CuCl_2/O_2$; $K_2S_2O_8$; benzoquinone), it is often this redox step which is most problematical in palladium(II)-catalyzed nucleophilic additions to alkenes. Notwithstanding these problems, a number of very useful catalytic processes have been developed.

3.1.2.1 Addition of Oxygen Nucleophiles to Alkenes

One of the earliest uses of palladium(II) salts to activate alkenes towards additions with oxygen nucleophiles is the industrially important Wacker process, wherein ethylene is oxidized to acetaldehyde using a palladium(II) chloride catalyst system in aqueous solution under an oxygen atmosphere with copper(II) chloride as a co-oxidant.[1,2] The key step in this process is nucleophilic addition of water to the palladium(II)-complexed ethylene. As expected from the regioselectivity of palladium(II)-assisted addition of nucleophiles to alkenes, simple terminal alkenes are efficiently converted to methyl ketones rather than aldehydes under Wacker conditions.

This unique one-step transformation of alkenes — stable to acids, bases and nucleophiles — to methyl ketones has considerable synthetic potential, particularly when applied in a specific manner to more highly functionalized alkenes. Although traditional Wacker conditions are somewhat harsh, modified conditions utilizing DMF–water (7:1) as solvent and benzoquinone or copper(I) chloride–oxygen as reoxidant permit the efficient palladium(II)-catalyzed oxidation of terminal alkenes to ketones in the presence of a wide array of functional groups (Table 1).[3] Under these milder conditions, the reaction is specific for terminal alkenes and will tolerate virtually all common organic functional groups. From a synthetic point of view, this makes terminal alkenes functionally equivalent to methyl ketones. More recently developed variations of this oxidation procedure involving the use of phase transfer catalysis[4] or the use of electrochemical oxidation of palladium(0) back to palladium(II) for catalysis[5] may offer some advantages in specific systems.

Although tolerant of a wide variety of functionality, this process (and most other transition metal catalyzed processes) is sensitive to the presence of adjacent functional groups which can coordinate to the catalyst. This sensitivity often is manifest in a change in the expected regioselectivity or reactivity of a process. For example, catalytic oxidation of the terminal alkene in the β-lactams, shown in equation (1), led exclusively to the aldehyde rather than to the expected methyl ketone.[6] This change in regioselectivity is likely due to coordination of the amide carbonyl group to palladium to generate the more stable (*versus* six-membered) five-membered palladiocycle intermediate (Scheme 2).

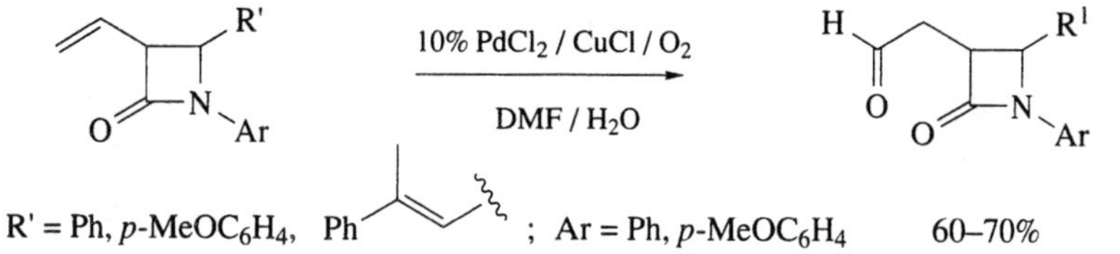

10% $PdCl_2$ / CuCl / O_2, DMF / H_2O (1)

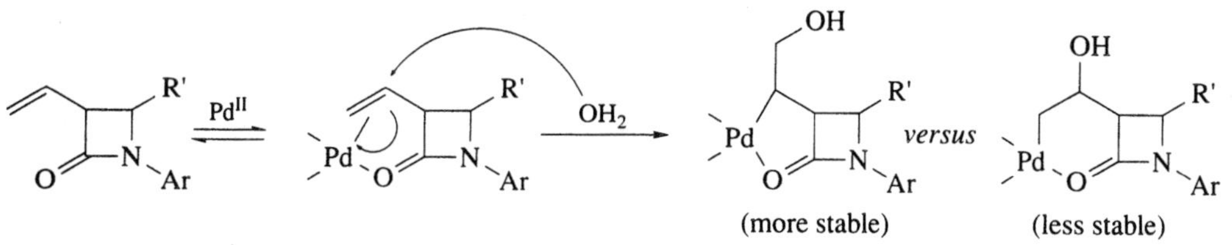

R' = Ph, *p*-$MeOC_6H_4$, Ph; Ar = Ph, *p*-$MeOC_6H_4$ 60–70%

Pd^{II} OH_2 *versus* (more stable) (less stable)

Scheme 2

Internal alkenes do not undergo this oxidation under normal circumstances, hence the excellent chemoselectivity observed. However, allyl ethers and acetates with internal double bonds undergo facile oxidation, giving exclusively the β-alkoxy ketone in 60–80% yield (equation 2).[7] In this case additional coordination of the palladium(II) catalyst to the ether oxygen may enhance its activation of the alkene, although the regioselectivity is different from that expected from the arguments advanced for the reaction in equation (1). Simple internal alkenes (*e.g.* 2-decene) were oxidized to mixtures of regioisomeric ketones by a 2% $PdCl_2$/$CuCl_2$/polyethylene glycol 400/H_2O catalyst system.[8]

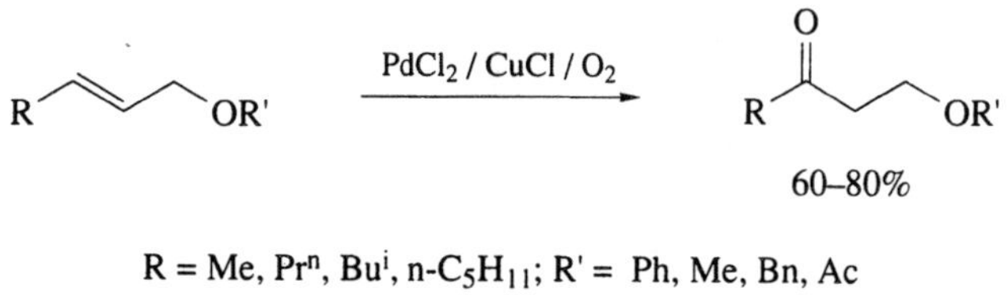

$PdCl_2$ / CuCl / O_2 (2)

60–80%

R = Me, Pr^n, Bu^i, n-C_5H_{11}; R' = Ph, Me, Bn, Ac

Finally, α,β-unsaturated carbonyl compounds are converted to β-keto systems when treated with 20% Na_2PdCl_4 catalyst in 50% acetic acid as solvent and *t*-butyl hydroperoxide or hydrogen peroxide as reoxidant (equation 3).[9] It is not clear if the mechanism of this process is related to the other palladium(II)-catalyzed addition of oxygen nucleophiles to alkenes.

Alcohols and carboxylic acids also readily add to metal-activated alkenes[2] and industrial processes for the conversion of ethylene to vinyl acetate, vinyl ethers and acetals are well established. However, very little use of intermolecular versions of this chemistry with more complex alkenes has been developed. In

Table 1 Palladium(II)-catalyzed Oxidation of Functionalized Terminal Alkenes to Methyl Ketones

Entry	*Alkene*	*Product*	*Yield (%)*
1	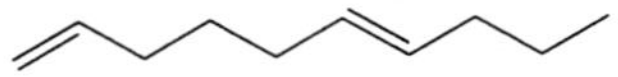	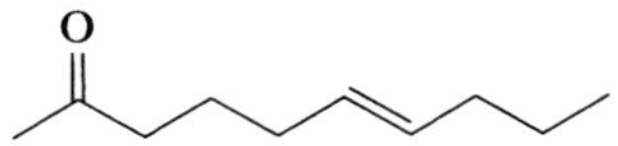 O	78
2	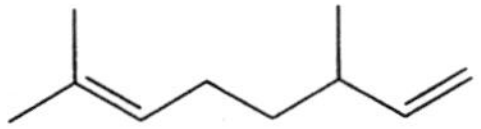	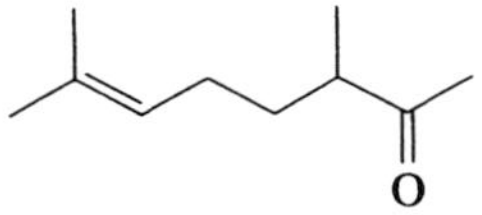 O	—
3	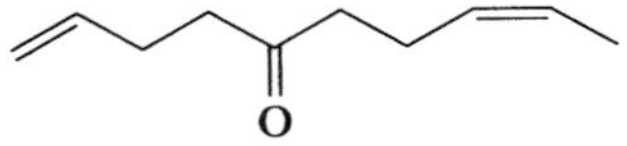 O	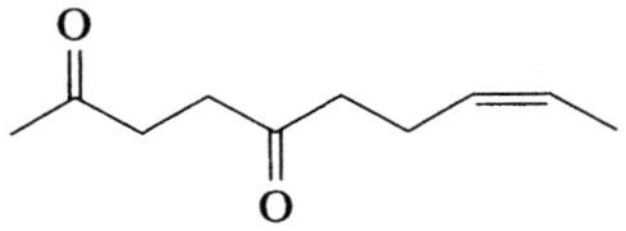 O O	70
4	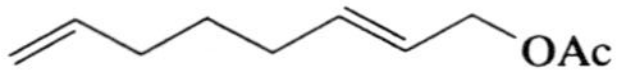 OAc	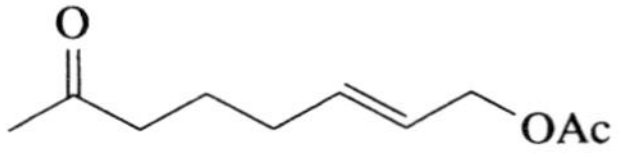 O OAc	83
5	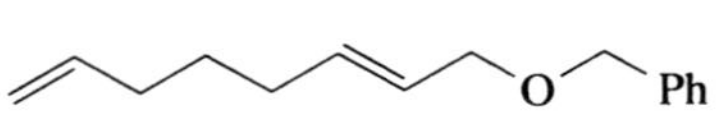 O Ph	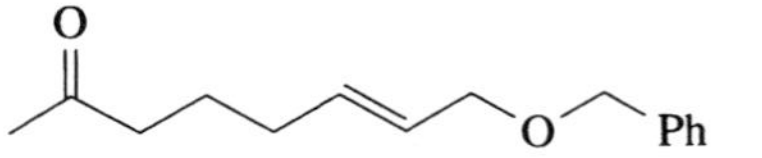 O O Ph	76
6	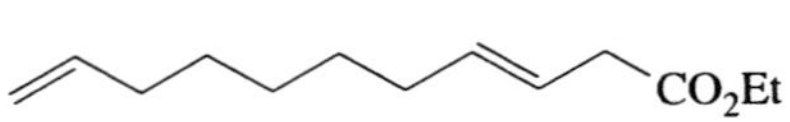 CO_2Et	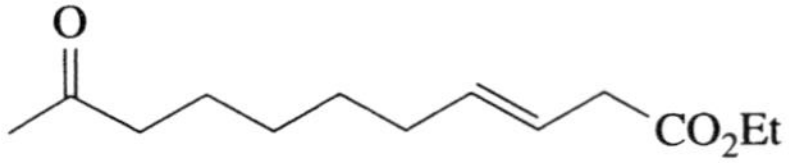 O CO_2Et	86
7	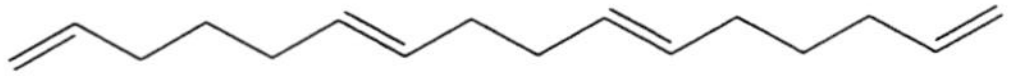	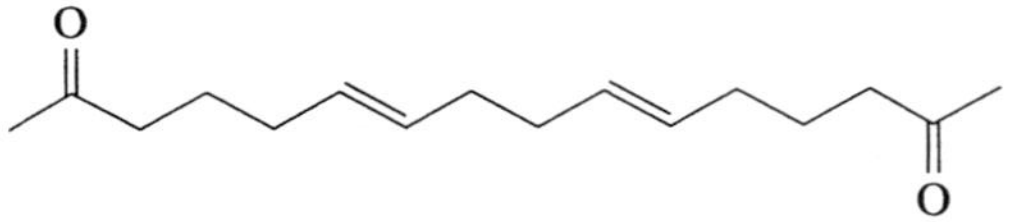 O O	62
8	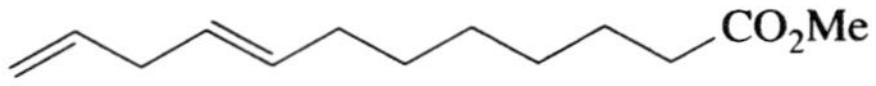 CO_2Me	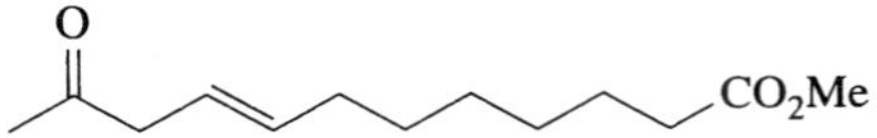 O CO_2Me	67

Table 1 *(continued)*

Entry	*Alkene*	*Product*	*Yield (%)*
9	Cl	O, Cl	62
10	OHC, O	OHC, O, O	59
11	$SiMe_3$	$SiMe_3$, O	74
12	Br	O, Br	74
13	OH	O, OH	79
14	OTs	O, OTs	82
15	OTHP	O, OTHP	81

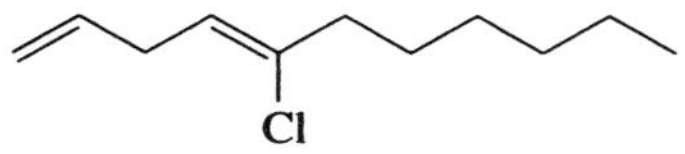

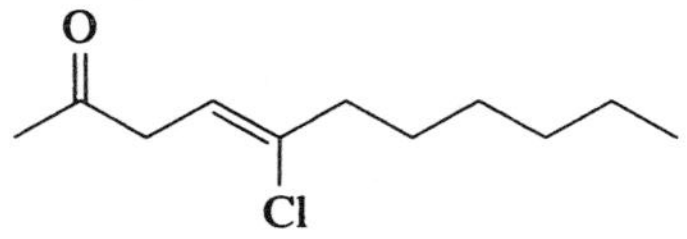

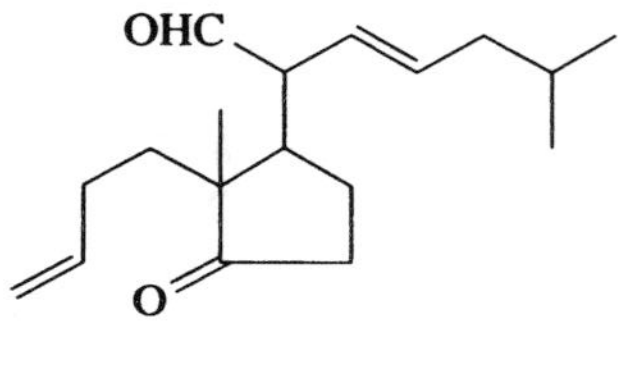

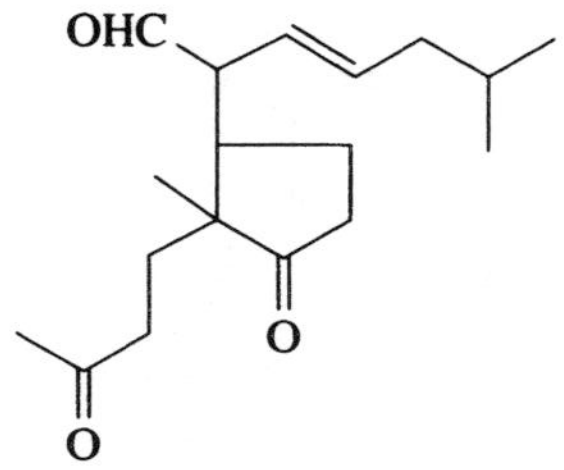

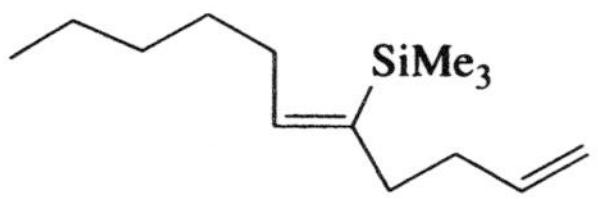

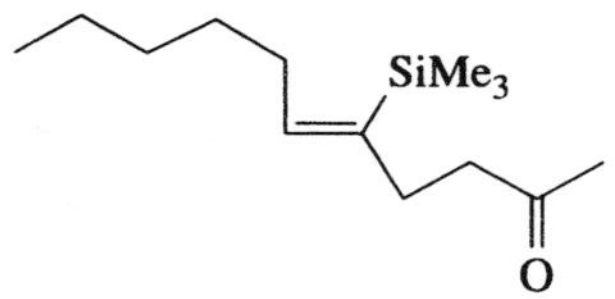

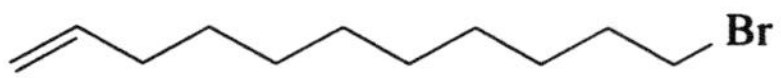

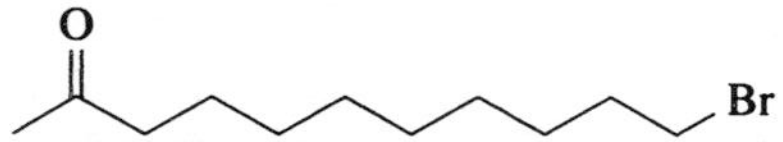

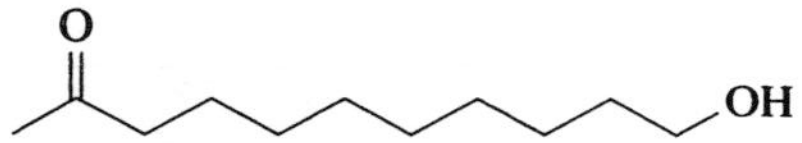

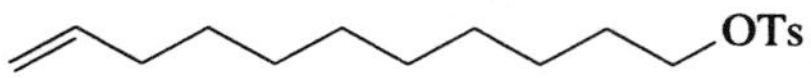

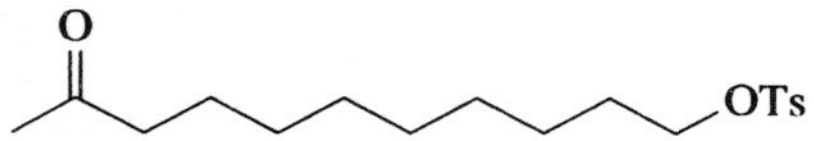

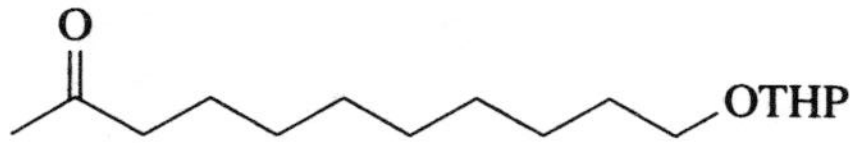

Table 1 *(continued)*

Entry	*Alkene*	*Product*	*Yield (%)*
16	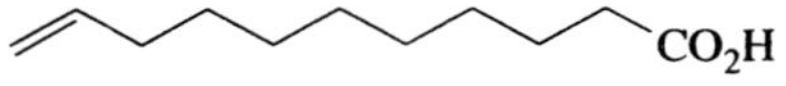 CO_2H	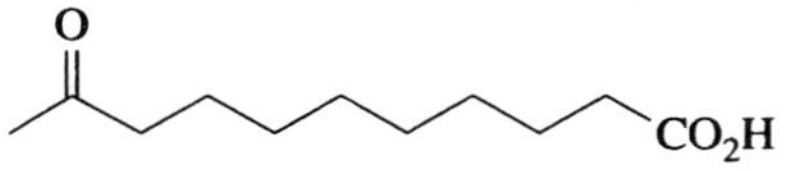 O, CO_2H	83
17	 CO_2Me	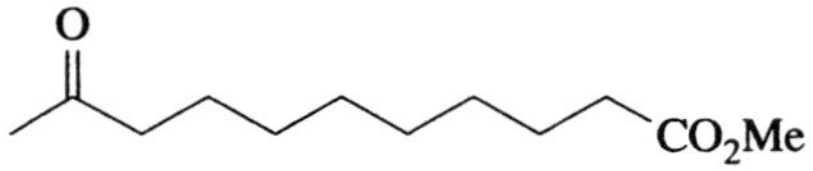 O, CO_2Me	86
18	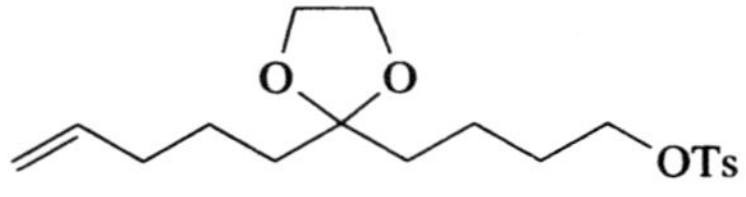 O, O, OTs	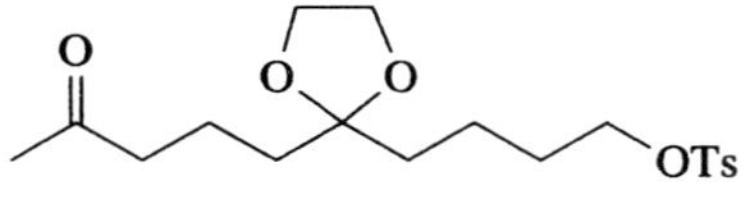 O, O, O, OTs	70
19	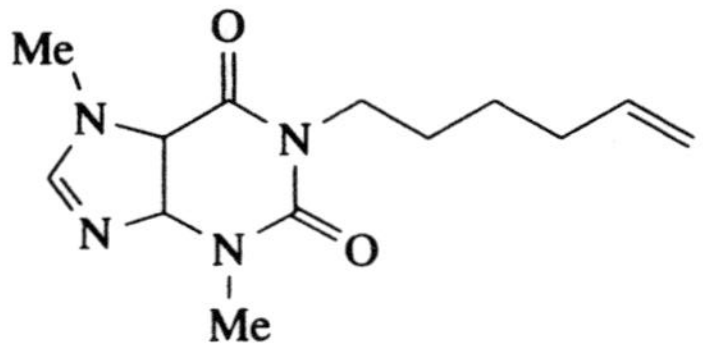 Me, N, N, O, N, N, O, Me	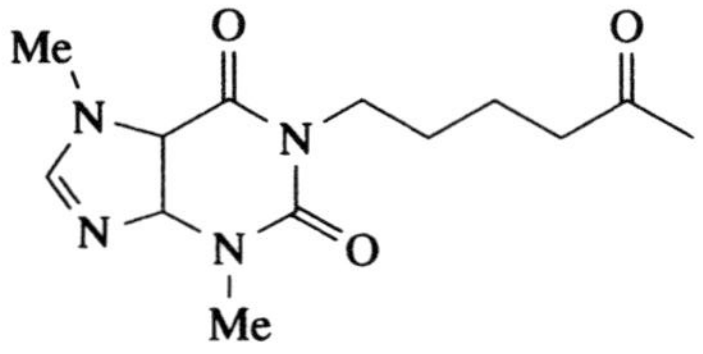 Me, N, N, O, N, N, O, Me, O	85

$$\text{R-CH=CH-C(O)X} \xrightarrow[50\ ^\circ\text{C}]{\text{Na}_2\text{PdCl}_4\ /\ 50\%\ \text{AcOH}/\ \text{Bu}^t\text{OOH}} \text{RC(O)CH}_2\text{C(O)X} \quad 60\text{–}80\% \tag{3}$$

R = Me, Et, Pr^i, n-C_7H_{15}, Ph; X = OMe, Me

contrast, intramolecular versions of this process, to form oxygen heterocycles, has been extensively developed.

The earliest studies centered on the cyclization of ω-hydroxyalkenes to furan and pyran ring systems (equations 4 and 5).[10] The regiochemistry (attack at the more substituted position) was that expected and the yields were only modest. However, modern advances in catalysis should permit much more efficient processes and many older systems probably warrant re-examination.

$$\xrightarrow[\text{Cu(OAc)}_2\ /\ \text{O}_2,\ \text{MeOH, H}_2\text{O}]{10\%\ \text{Pd(OAc)}_2} \quad 15\text{–}52\% \tag{4}$$

R^1 = Me, Ph; R^2 = H, Me, Et, Ph

$$\longrightarrow \quad 45\% \tag{5}$$

Palladium-catalyzed cyclization of *o*-allylphenols to benzofurans has been extensively studied.[11] As is usual, early systems were catalytically inefficient but continued studies led to substantial improvement. A wide range of catalyst systems work for this process. One of the most efficient, from a standpoint of catalyst turnover and chemical yield, was based on chiral π-allylpalladium catalysts (equation 6), although the optical yields were low. γ-Pyrones can also be efficiently synthesized by palladium(II)-catalyzed addition of phenolic OH groups to conjugated enones (equation 7).[12]

$$\xrightarrow[\text{Cu(OAc)}_2\ /\ \text{O}_2\ \text{or Bu}^t\text{OOH}]{[\text{PdCl}]_2} \quad 60\text{–}80\% \tag{6}$$

Y = H, Me, MeCO, Bu^t, SiMe$_3$,

80:20, 15–29% *ee*

$$\xrightarrow{10\%\ \text{Li}_2\text{PdCl}_4\ /\text{CuCl}_2\ /\text{H}_2\text{O}} \quad 65\text{–}80\% \tag{7}$$

As detailed in Scheme 1, all of these reactions proceed through unstable σ-alkylpalladium(II) complexes (*e.g.* **2**), which have a rich chemistry in their own right. Particularly useful synthetic transformations, which involve trapping of these intermediates with carbon monoxide (path e), have been developed (equations 8 and 9; Scheme 3).[13–15] The yields and catalytic efficiencies in these more recently developed processes are remarkably good and the procedures are simple enough for synthetic chemists with no experience in organometallic chemistry to use. Allenes underwent a similar hydroxylation–carbonylation process to produce functionalized tetrahydrofurans in good yield (equation 10).[16]

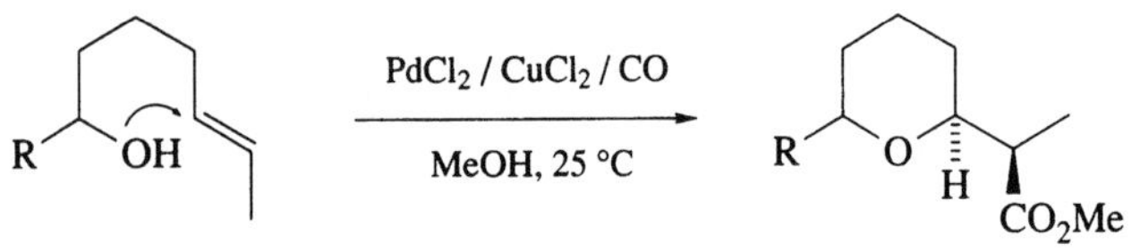

$PdCl_2$ / $CuCl_2$ / CO; MeOH, 25 °C (8)

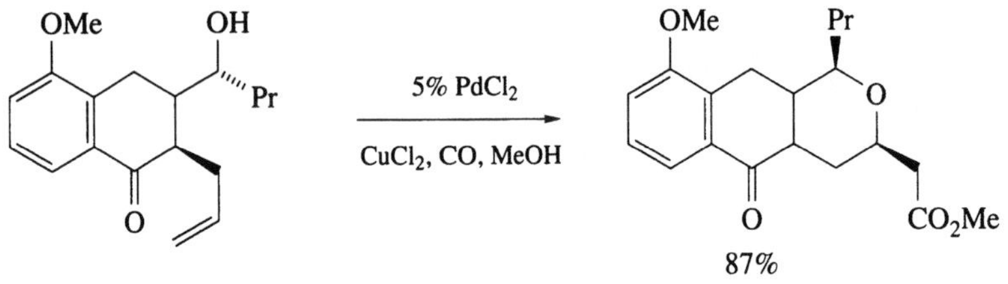

5% $PdCl_2$; $CuCl_2$, CO, MeOH; 87% (9)

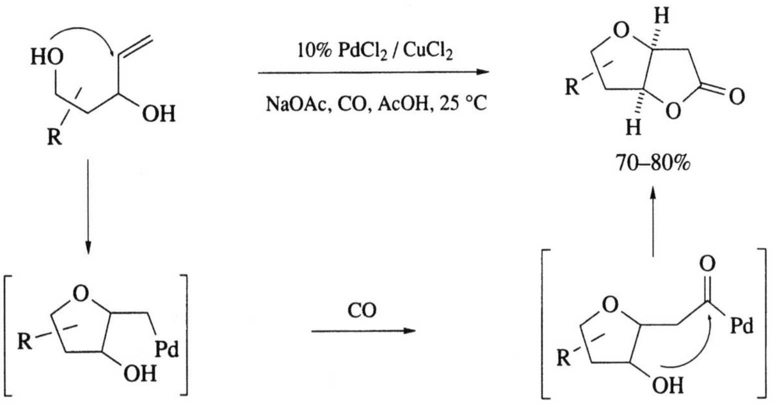

10% $PdCl_2$ / $CuCl_2$; NaOAc, CO, AcOH, 25 °C; 70–80%; CO

Scheme 3

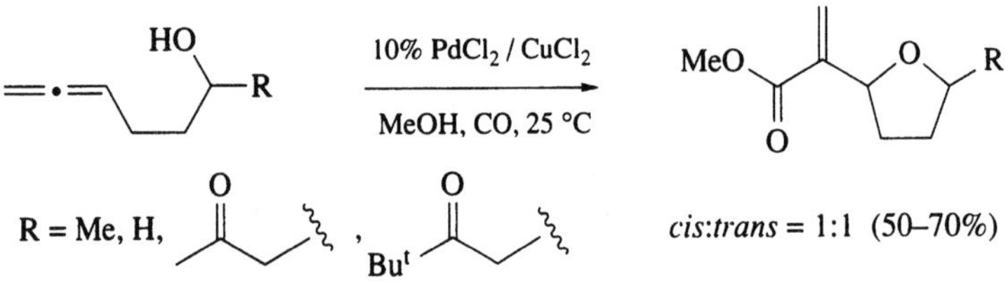

10% $PdCl_2$ / $CuCl_2$; MeOH, CO, 25 °C (10)

R = Me, H, ...; *cis*:*trans* = 1:1 (50–70%)

Carboxylate ions are also effective nucleophiles in palladium-catalyzed reactions of alkenes and several classes of lactones including γ-pyrones (equation 11)[17] and isocoumarins (equation 12)[18] have been made in this manner. These early studies used stoichiometric amounts of palladium salts, since efficient redox systems had not yet been developed. However, with more modern techniques catalysis in these systems should be relatively straightforward. The more recent catalytic cyclization–carbonylation process in equation (13) is indicative of this.[19]

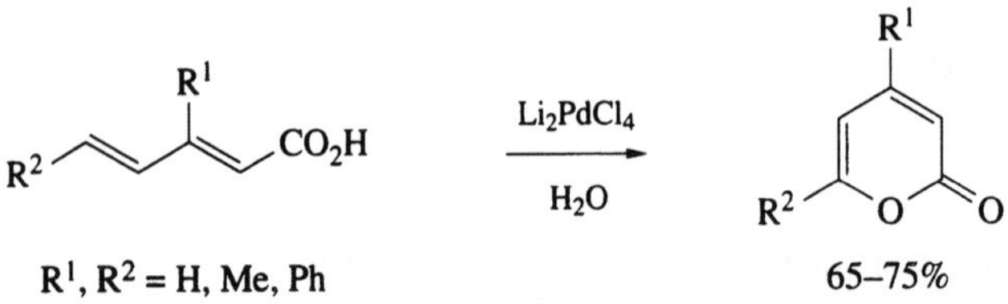

Li_2PdCl_4; H_2O (11)

R^1, R^2 = H, Me, Ph; 65–75%

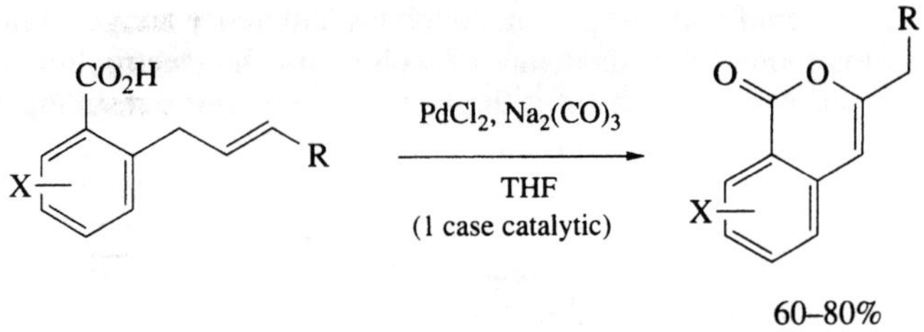

(12)

X = 4-Cl, 5-MeO, 6-Cl, 7-MeO

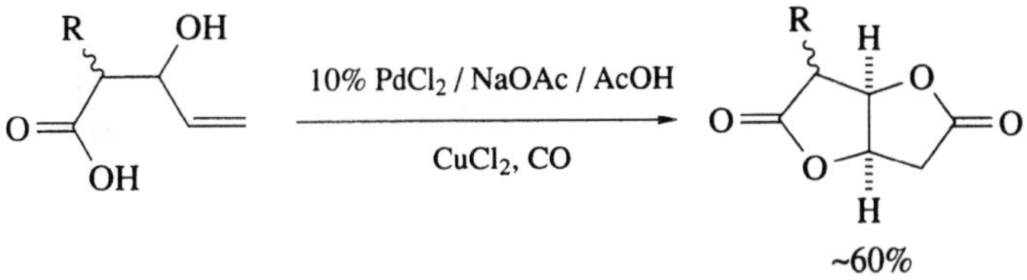

(13)

Chelating dialkenes such as 1,5-cyclooctadiene, norbornadiene and dicyclopentadiene, as well as allylamines, were among the first alkenes whose reactions with oxygen (and other) nucleophiles in the presence of palladium(II) salts were studied. These reactions were quite efficient, but led to stable σ-alkylpalladium(II) complexes, making the development of catalytic systems difficult. As a consequence, little synthetic use of this chemistry has developed. However, potentially useful synthetic transformations are possible, as evidenced by the reaction described in Scheme 4.[20] This reaction illustrates the importance of reaction conditions to the selectivity of the process, a phenomenon typical of organotransition metal chemistry. Nucleophilic attack of the alcohol on the coordinated dialkene produced stable σ-alkylpalladium(II) complex (**3**), in which the remote alkene remained coordinated. Carbonylation of this complex proceeded by an initial insertion of the alkene into the palladium–carbon σ-bond to give (**4**), followed by CO insertion to give (**5**). In contrast, decomplexation of the remote alkene by prior treatment with diisopropylamine, followed by carbonylation, led to the simple alkoxycarbonylation product (**6**). This ability to control selectivity by choice of ligand and conditions is a major feature of organotransition metal chemistry.

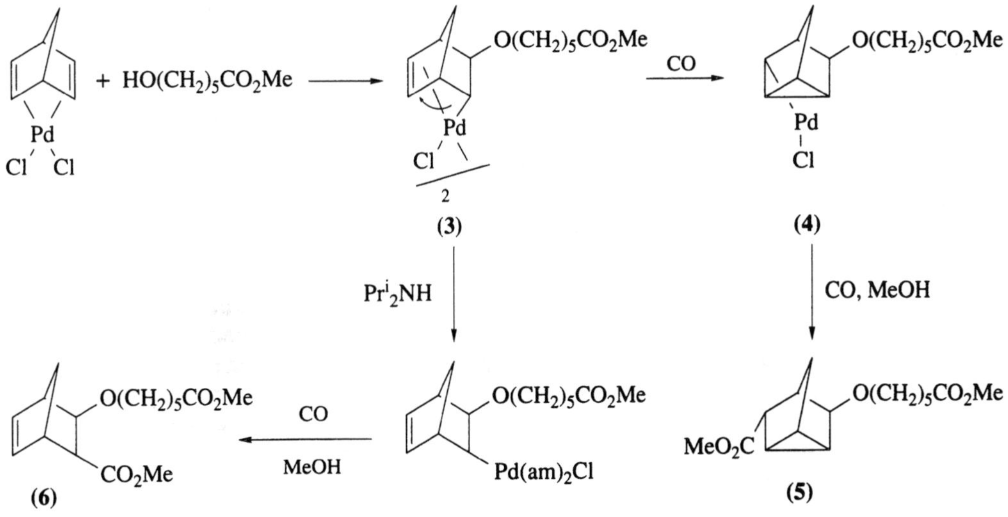

Scheme 4

3.1.2.2 Addition of Nitrogen Nucleophiles to Alkenes

Nitrogen nucleophiles such as amines (and in intramolecular cases, amides and tosamides) readily add to alkenes complexed to palladium(II) and iron(II) with reactivity and regiochemical features parallel to those observed for oxygen nucleophiles. However, these metal-assisted amination reactions are subject to several problems not encountered with oxygen nucleophiles and hence are generally less synthetically useful. Amines are much more potent ligands than are alcohols for the electrophilic metals used to activate alkenes and displacement of the alkene from the metal by the amine resulting in poisoning of the catalyst is a serious problem. Additionally, the product of successful amination is an enamine, which is not only a potent catalyst poison but also prone to Lewis acid catalyzed polymerization. Finally, in contrast to alcohols and ethers, amines and enamines are readily oxidized under conditions normally used to carry the requisite $Pd^0 \rightarrow Pd^{II}$ redox chemistry, making catalysis difficult. For these reasons, efficient catalytic intermolecular amination of alkenes has not yet been achieved. However, the stoichiometric amination of alkenes, followed by further reaction of the resulting σ-alkylpalladium(II) complex has been used to effect unconventional transformations of alkenes (Scheme 5).

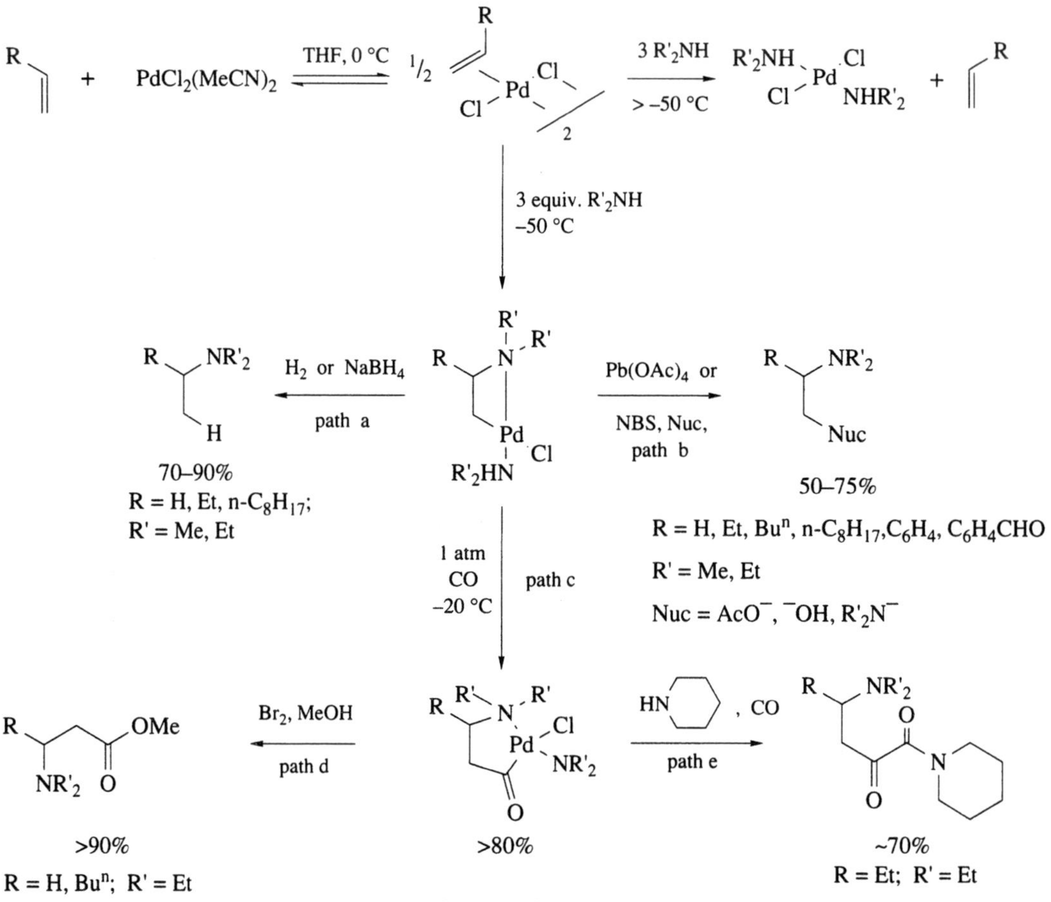

Scheme 5

Terminal alkenes could be efficiently aminated by nonhindered secondary amines in a process requiring 1 equiv. of palladium(II) chloride, 3 equiv. of amine and a reduction at temperatures below –20 °C (path a, Scheme 5);[21,22] however, primary amines and/or internal alkenes were less efficient, producing only 40–50% yields of amination product. Oxidative cleavage of the unstable σ-alkylpalladium(II) in the presence of a nucleophile resulted in vicinal oxamination or diamination of the alkene (path b).[23,24] Carbonylation resulted in the isolation of stable σ-acylpalladium(II) species (path c),[25] which were oxidatively cleaved to give β-amino esters (path d)[26] or further carbonylated to give γ-amino-α-ketoamides (path e).[27]

All of these processes are of limited synthetic utility because of the requirement of the use of stoichiometric amounts of palladium complexes. However, by judicious choice of reactants and conditions the above-mentioned impediments to catalysis can be overcome. For example, an efficient palladium(II)-catalyzed cyclization of *o*-allyl- and *o*-vinyl-anilines to indoles has been developed (equation 14).[28] Because arylamines are ~10^6 less basic than aliphatic amines, and because the cyclized product in this system gave an enamine (indole) stabilized by aromatization, the problems of catalyst poisoning by substrate or product were circumvented, and catalysis was successfully achieved. The system was quite tolerant of a variety of functional groups and was used to prepare indoloquinones in excellent yield (equation 15).[29] Since this process also proceeded through an unstable σ-alkylpalladium(II) complex both inter- (Scheme 6)[30] and intra-molecular (equation 16)[29,31] insertion processes were possible, resulting in extensive elaboration of the simple starting materials, utilizing two sequential palladium-assisted bond-forming processes.

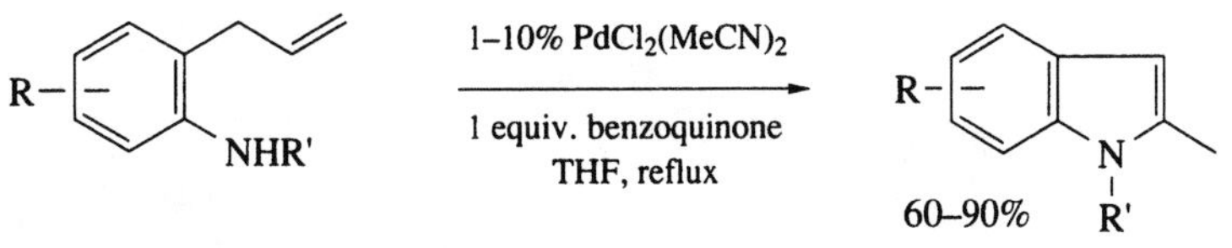

1–10% $PdCl_2(MeCN)_2$; 1 equiv. benzoquinone; THF, reflux; 60–90% (14)

R = H, 3-Me, 3-CO_2Et, 4-MeO, 5-MeO, 4,5-MeO_2, 4-Br; R' = H, Me, Ac, Ts

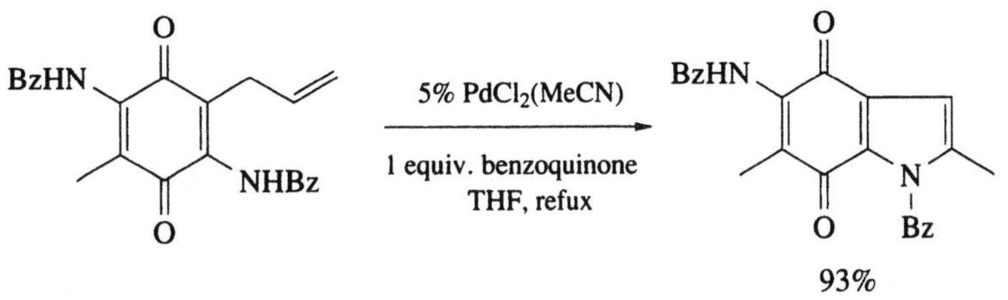

5% $PdCl_2(MeCN)$; 1 equiv. benzoquinone; THF, reflux; 93% (15)

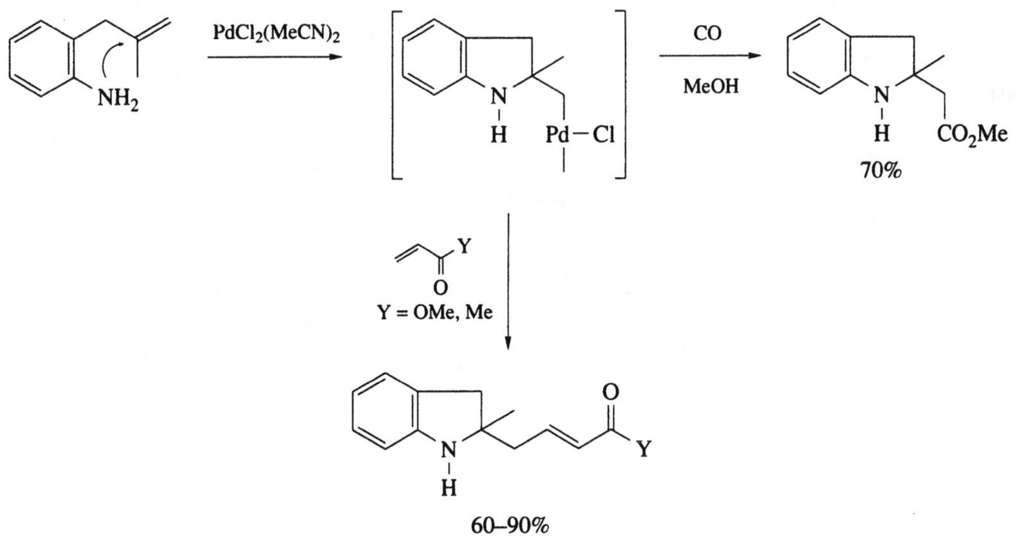

$PdCl_2(MeCN)_2$; CO, MeOH; 70%; Y = OMe, Me; 60–90%

Scheme 6

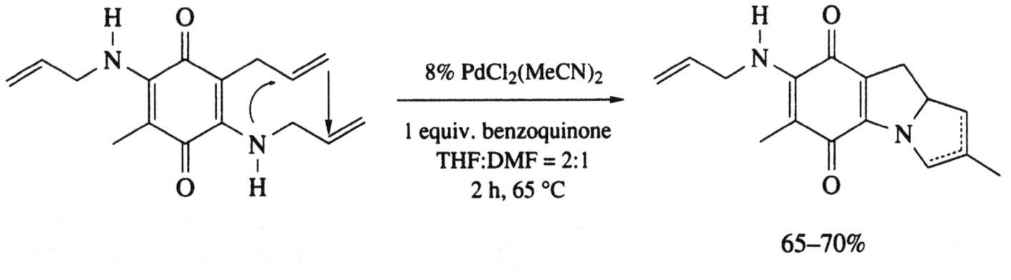

8% $PdCl_2(MeCN)_2$; 1 equiv. benzoquinone; THF:DMF = 2:1; 2 h, 65 °C; 65–70% (16)

As stated above, aliphatic amines are potent ligands for electrophilic transition metals and are efficient catalyst poisons in attempted alkene amination reactions. However, tosylation of the basic amino group greatly reduces its complexing ability, yet does not compromise its ability to nucleophilically attack complexed alkenes. Thus, a variety of alkenic tosamides efficiently cyclized under palladium(II) catalysis producing *N*-tosylenamines in excellent yield (equations 17 and 18).[32] Again, this alkene amination proceeded through an unstable σ-alkylpalladium(II) species, which could be intercepted by carbon monoxide, to result in an overall aminocarbonylation of alkenes. With ureas of 3-hydroxy-4-pentenylamines (Scheme 7), this palladium-catalyzed process was quite efficient but it was somewhat less so with

4-hydroxy-5-hexenylamine (equation 19).[33] ω-Aminoallenes underwent a related palladium-catalyzed aminocarbonylation (equation 20).[34]

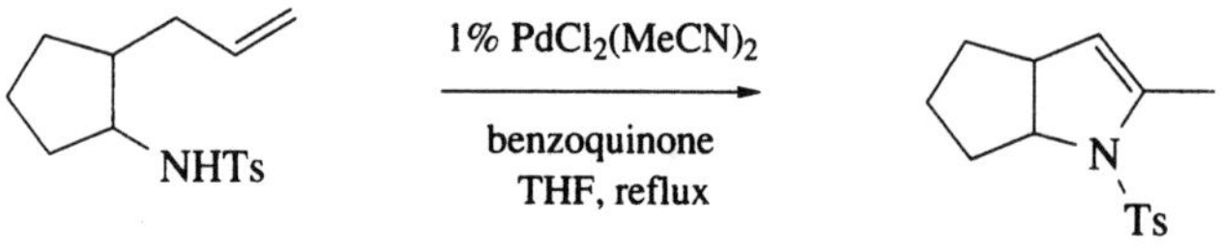

1% $PdCl_2(MeCN)_2$, benzoquinone, THF, reflux — 90% (17)

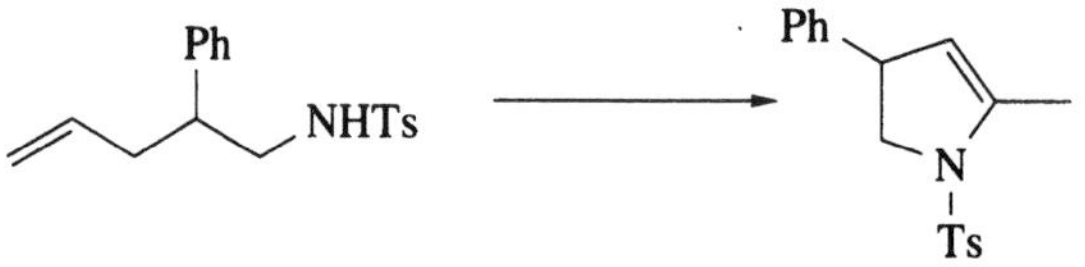

— 85% (18)

10% $PdCl_2$, $CuCl_2$, CO, MeOH

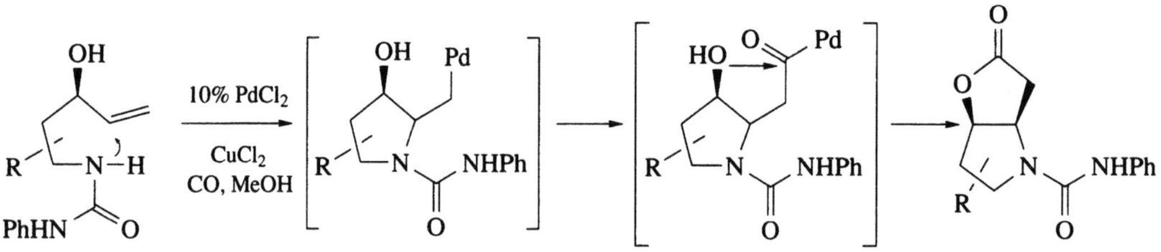

R = H, 2,2-Me_2, 1-Ph, 2-Me, 3-Me — 70–97%

Scheme 7

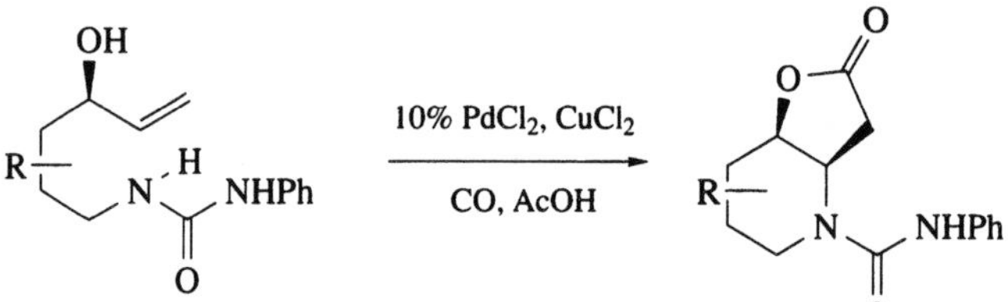

10% $PdCl_2$, $CuCl_2$, CO, AcOH (19)

R = H, 2-Me, 2,2-Me_2 — 30%

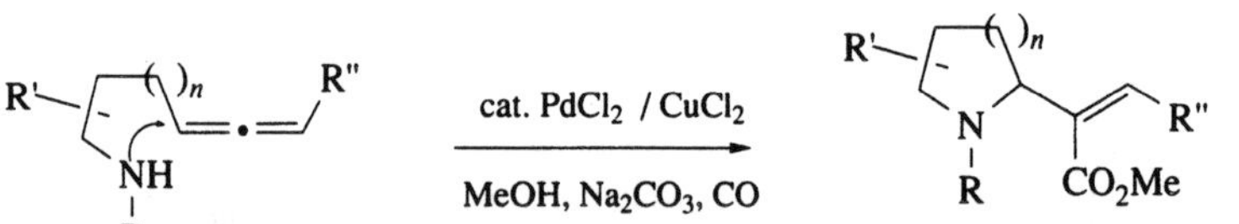

cat. $PdCl_2$ / $CuCl_2$, MeOH, Na_2CO_3, CO — 40–80% (20)

R = C_6H_4Me, Ts, CO_2Me; R' = H, 1-CO_2Et; R" = H, Me, n-C_5H_{11}; n = 1, 2

Chelating alkenes, such as 1,5-dienes and allylamines, complexed to palladium(II) undergo facile attack by nitrogen nucleophiles. However, the resulting σ-alkylpalladium(II) complexes are very stable, preventing efficient catalytic processes. These stable complexes undergo all the typical reactions of σ-alkylmetal complexes — hydrogenolysis, carbon monoxide insertion and alkene insertion — so that the organic compound can be freed from the metal and converted into a useful product. An example of this is the stoichiometric addition of phthalimide to tertiary allylic amines (Scheme 8).[35]

Cationic cyclopentadienyliron dicarbonyl (F_p) alkene complexes are generally reactive towards a wide variety of nucleophiles, including nitrogen nucleophiles, but they too generate stable σ-alkylmetal complexes from which the metal must be removed in a second chemical step (usually oxidation). This makes catalysis impossible and severely limits application of this methodology to organic synthesis (equation 21).[36] However, in contrast to palladium, iron is relatively inexpensive and stoichiometric procedures

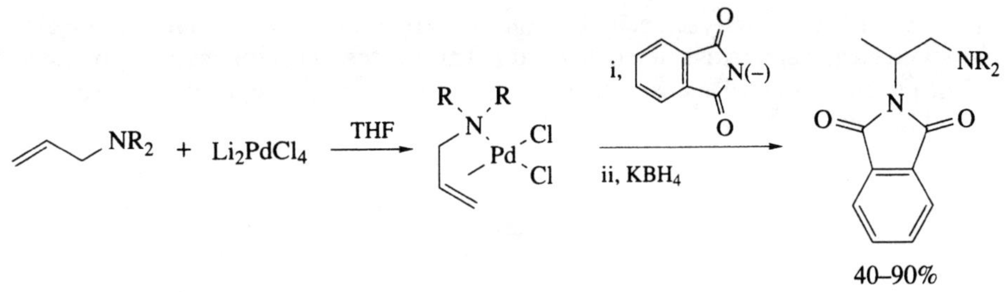

$$\text{CH}_2{=}\text{CHCH}_2\text{NR}_2 + \text{Li}_2\text{PdCl}_4 \xrightarrow{\text{THF}} [\text{R}_2\text{N}\cdots\text{PdCl}_2] \xrightarrow[\text{ii, KBH}_4]{\text{i, phthalimide N(–)}} \text{product}$$

40–90%

Scheme 8

may be justified when the use of this process greatly simplifies the desired synthesis. The bicyclic β-lactam synthesis in Scheme 9 is such an example.[37]

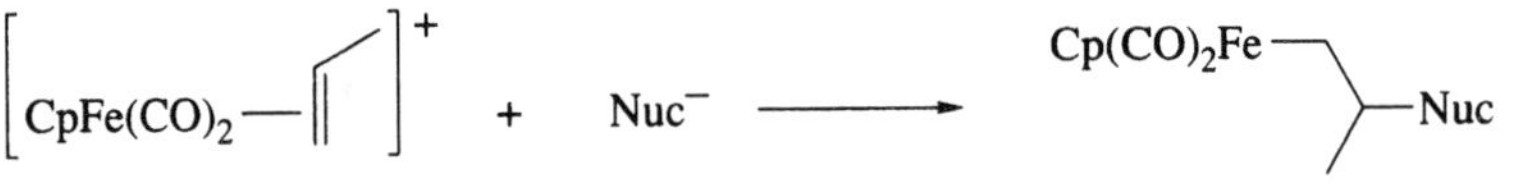

$$\left[\text{CpFe(CO)}_2\text{—}\|\right]^+ + \text{Nuc}^- \longrightarrow \text{Cp(CO)}_2\text{Fe—CH}_2\text{CH(Nuc)CH}_3 \qquad (21)$$

$[\text{Cp-Fe(CO)}_2\text{—alkene}]^+$ (≡ Fp) + CO_2Me keto ester ⟶ F_p^+–alkene ketoester 74% $\xrightarrow{\text{NH}_3}$

$[\text{F}_\text{p}^+ \ldots \text{CO}_2\text{Me}, \text{N-H}]$ ⟶ F_p pyrroline CO_2Me 95% $\xrightarrow{\text{NaBH}_4}$ F_p HN pyrrolidine CO_2Me $\xrightarrow{\Delta,\ \text{CO}}$

Cp-Fe(CO) acyl N CO_2Me 58% $\xrightarrow{\text{Ag}_2\text{O}}$ bicyclic β-lactam CO_2Me 32%

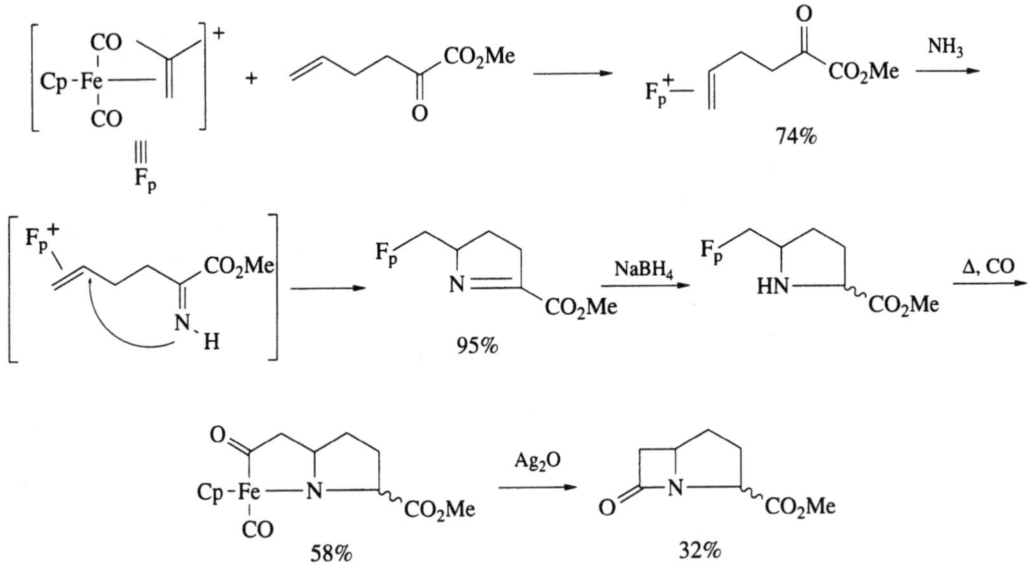

Scheme 9

3.1.2.3 Metal-catalyzed Allylic Transpositions, Oxy-Cope and Aza-Cope Reactions

Electrophilic transition metals, particularly palladium(II) salts, catalyze a number of heteroatom allylic transposition processes[38] by a mechanism which almost certainly involves nucleophilic attack of a heteroatom on a metal-bound alkene (Scheme 10), often termed 'cyclization-induced rearrangement'.

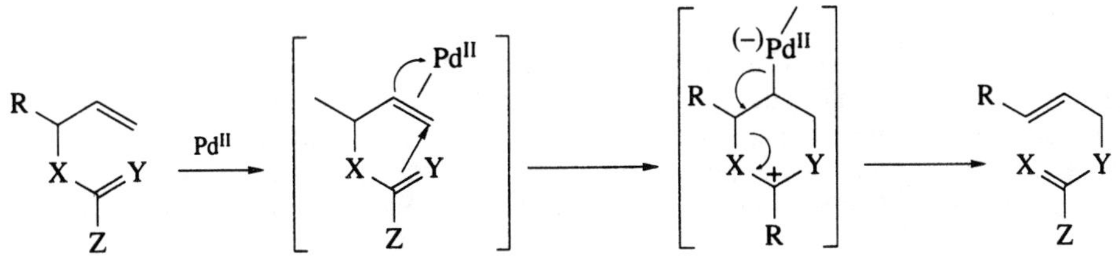

R, X, Y, Z $\xrightarrow{\text{Pd}^{\text{II}}}$ [Pd^{II}, X, Y, Z] ⟶ [$(-)\text{Pd}^{\text{II}}$, R, X, +, Y, R] ⟶ R, X, Y, Z

Scheme 10

Allylic esters equilibrated under very mild conditions (2 h, 25 °C) in the presence of 2–4% $PdCl_2$ catalyst. These rearrangements were not complicated by skeletal rearrangements, cyclizations or elimination, as is often observed with acid-catalyzed equilibrations. (*Z*)-Alkenes rearranged much more slowly than (*E*)-alkenes and with clean suprafacial stereochemistry. This process was limited to allylic esters unsubstituted at C-2 of the allyl group. Representative synthetically interesting examples are shown in equations (22)[38,39] and (23) to (27).[38]

OAc → OAc 88% (22)

OAc ⇌ OAc 91% (23)

C_7H_{15} OAc → C_7H_{15} OAc 96% (24)

steroid OAc → OAc 90% (25)

C_5H_{11} OAc R Br O O → 91% → OAc C_5H_{11} R Br O O (100% chirality transfer) (26)

^{18}O O ⇌ (AcOH solvent) (No loss of label) ^{18}O O 60% 40% (27)

Allylimidates underwent a clean Claisen-type rearrangement in the presence of 5 mol % palladium(II) chloride (equation 28),[40] as did allyl carbamates[39] and *S*-allylthioimidates (equation 29).[41] This S to N rearrangement has found application particularly in the synthesis of pyrimidines, systems for which thermal S to N allylic rearrangements were generally ineffective (equation 30).[42] Finally, *O*-allyl *S*-methyl dithiocarbonates cleanly underwent palladium(II)-catalyzed O to S allylic transposition (equation 31).[43]

R' O N R → $PdCl_2(PhCN)_2$ → R' O N R H (28)

1% $PdCl_2(PhCN)_2$, THF, reflux; 98% (12 cases) (29)

2% $PdCl_2(PhCN)_2$, THF; ~90% (30)

X = C–R^6, N; R^1 = Me; R^2–R^6 = H, *etc.*

5% $PdCl_2(MeCN)_2$, THF, 30 min; 100% (31)

3.1.3 ADDITION OF HETEROATOM NUCLEOPHILES TO METAL-ACTIVATED ALKADIENES

Nonconjugated dienes undergo metal-catalyzed additions of heteroatom nucleophiles as isolated double bonds and most are subject to the general reaction chemistry presented in the previous sections. Conjugated 1,3-dienes are fundamentally different in their reactivity, since addition of a nucleophile to the terminal carbon of a coordinated 1,3-diene produces a π-allylmetal species (Scheme 11), which itself has a very rich chemistry (Volume 4, Chapter 3.3). Most stable, neutral transition metal η^4-diene complexes are relatively inert to attack by heteroatom nucleophiles, although a few are attacked by relatively reactive carbanions (Volume 4, Chapter 3.2). However, electrophilic transition metals, particularly palladium(II) salts, which do not form stable complexes with 1,3-dienes, do activate these substrates to undergo a variety of synthetically useful reactions with heteroatom nucleophiles. Most useful of these is the palladium(II)-catalyzed 1,4-additions of nucleophiles to 1,3-dienes. The main features of this process are summarized in Scheme 12, exemplified in the palladium(II) acetate catalyzed 1,4-diacetoxylation, 1,4-acetoxychlorination and 1,4-acetoxy-trifluoroacetoxylation of 1,3-cyclohexadiene.[44] The process involves the regio- and stereo-selective *trans*-acetoxypalladation of a terminus of the diene, producing a *trans*-acetoxy-η^3-allylpalladium complex. When the reaction is run in the presence of excess chloride ion, external nucleophilic attack of chloride on the η^3-allylpalladium complex occurs, producing *cis*-1-acetoxy-4-chlorocyclohex-2-ene. In the absence of chloride ion, acetate transferred internally from the same face of the η^3-allyl complex as the metal, producing *trans*-1,4-diacetoxycyclohex-2-ene. With a catalytic amount of lithium chloride, *cis*-1,4-diacetoxycyclohex-2-ene was the exclusive product. Presumably, chloride displaced acetate from the metal center of the η^3-allylpalladium intermediate, permitting external nucleophilic attack of the excess acetate present. Finally, in the presence of trifluoroacetic acid, *trans*-1-acetoxy-4-trifluoroacetoxycyclohex-2-ene was formed, presumably by prior coordination of trifluoroacetate to the metal and internal transfer to the η^3-allyl ligand. The wide range of dienes that underwent this reaction, in good yield and with high regio- and stereo-selectivity are listed in Table 2. The real utility of this chemistry lies in the facile further functionalization of these systems using both conventional and palladium(0)-catalyzed substitution reactions (Volume 4, Chapter 3.3). With a slight change in reaction conditions, this same range of dienes was efficiently dialkoxylated by alcohols (equation 32).[45]

1,2-Dienes underwent facile chloropalladation producing an unstable σ-vinylpalladium(II) intermediate, which inserted another mole of allene to give an isolable 2-substituted η^3-allylpalladium complex.

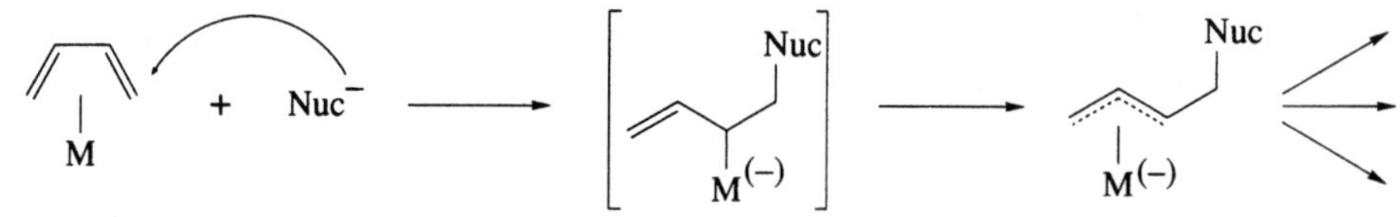

Scheme 11

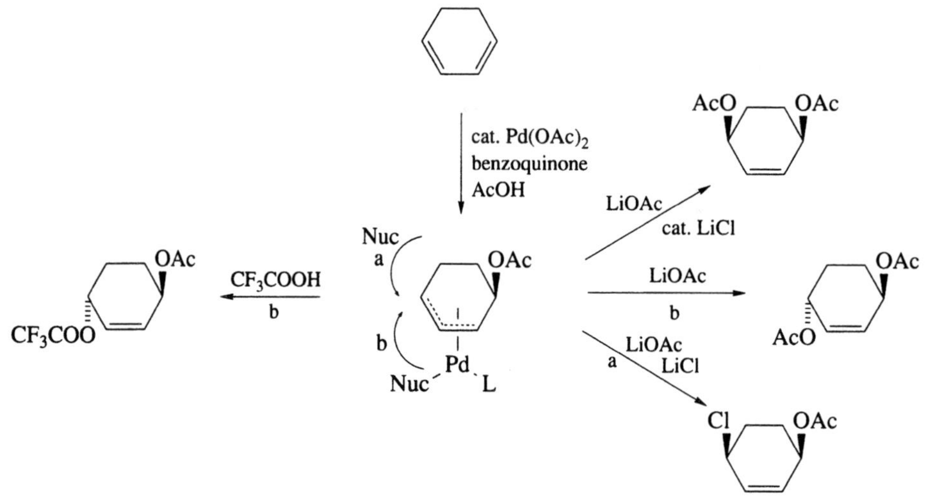

Scheme 12

Table 2 1,3-Dienes Catalytically Diacetoxylated

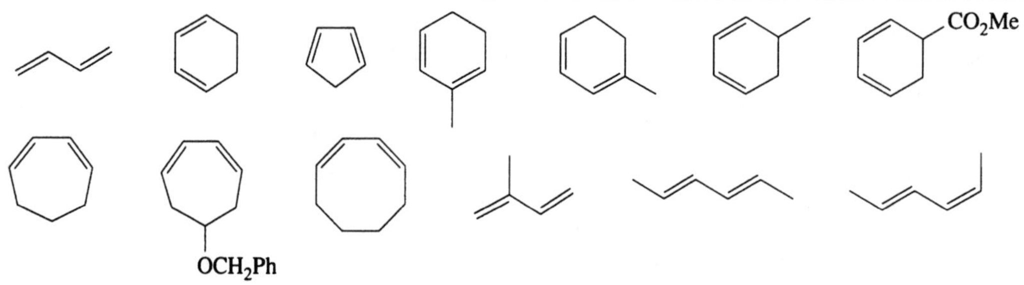

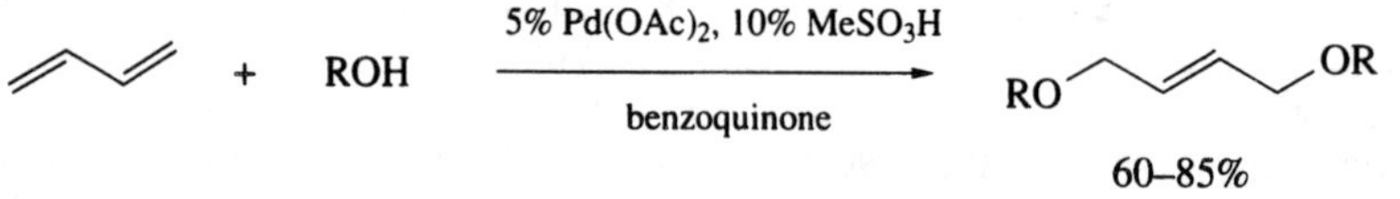

$$\text{butadiene} + \text{ROH} \xrightarrow[\text{benzoquinone}]{5\%\ Pd(OAc)_2,\ 10\%\ MeSO_3H} \text{RO}\diagup\!\!\diagdown\!\!=\!\!\diagup\text{OR} \quad 60\text{–}85\% \tag{32}$$

When this reaction was carried out in the presence of copper(II) chloride as oxidant, efficient catalytic production of bis-2,3-chloromethyl-1,3-butadiene resulted (Scheme 13).[46]

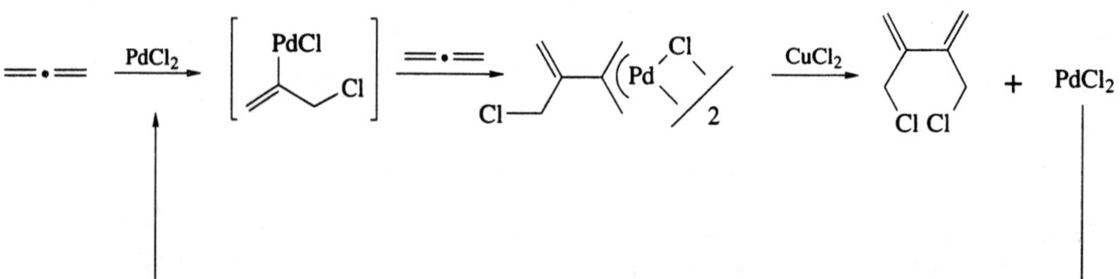

Scheme 13

3.1.4 ADDITION OF HETEROATOM NUCLEOPHILES TO METAL-ACTIVATED ALKYNES

Although alkynes are highly reactive toward a wide range of transition metals, very few instances of metal-catalyzed reactions of nucleophiles with alkynes are known. This is, in part, because most stable alkyne–metal complexes are inert to nucleophilic attack, while most unstable alkyne–metal complexes tend to oligomerize alkynes faster than anything else. Hence synthetic methodology involving this process is quite limited.

There are two notable exceptions to this. Palladium(II) salts catalyzed a number of cyclization reactions of alkynes bearing remote heteroatoms in a process thought to involve nucleophilic attack of the heteroatom on the palladium-complexed alkynes. Thus pyrroles[47] and indoles[48] (Scheme 14), furans and substituted furans (equation 33)[49] and unsaturated lactones (equations 34–36)[40,50] were efficiently produced by treatment of the appropriately disposed alkynic amines, alcohols or carboxylates with 1–5% palladium chloride catalyst. These reactions required a somewhat unusual protolytic cleavage of the σ-vinylpalladium(II) species to regenerate the palladium(II) chloride catalyst. Evidence for the intermediacy of this species was obtained by intercepting it with allyl chlorides, producing substituted unsaturated lactones (equations 37 and 38).[51]

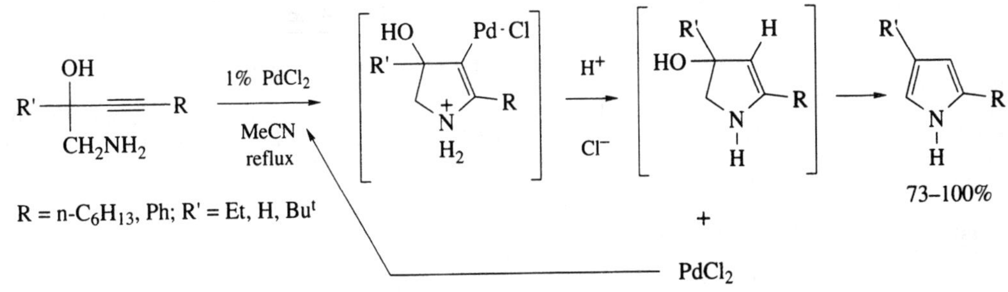

1% $PdCl_2$, MeCN, reflux; H^+, Cl^-; + $PdCl_2$; 73–100%

$R = n\text{-}C_6H_{13}$, Ph; R' = Et, H, Bu^t

Scheme 14

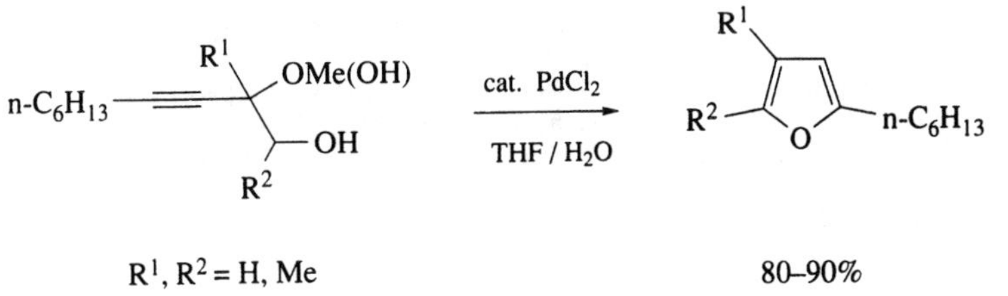

cat. $PdCl_2$, THF / H_2O; 80–90% (33)

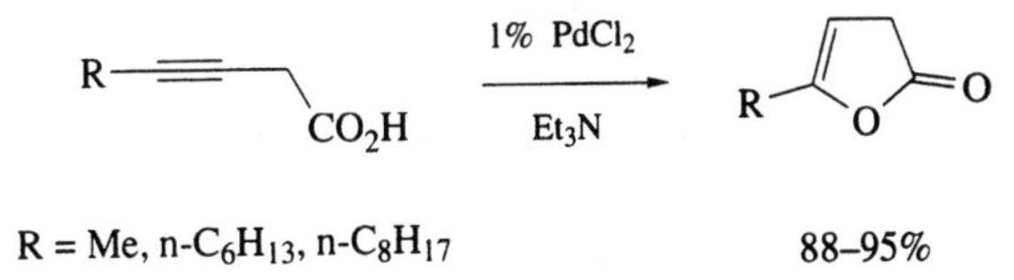

$R^1, R^2 = H, Me$

1% $PdCl_2$, Et_3N; 88–95% (34)

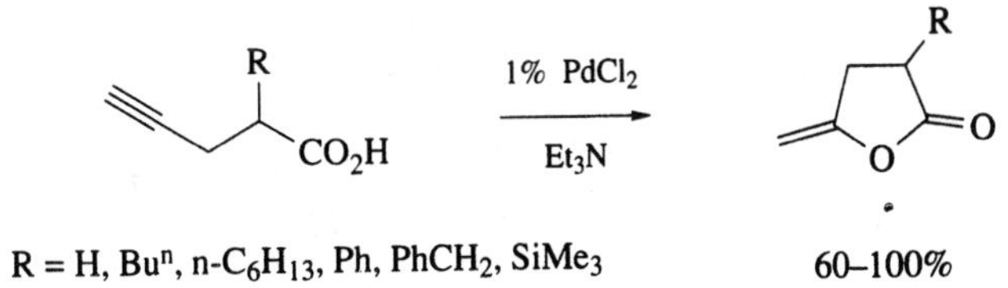

$R = Me, n\text{-}C_6H_{13}, n\text{-}C_8H_{17}$

1% $PdCl_2$, Et_3N; 60–100% (35)

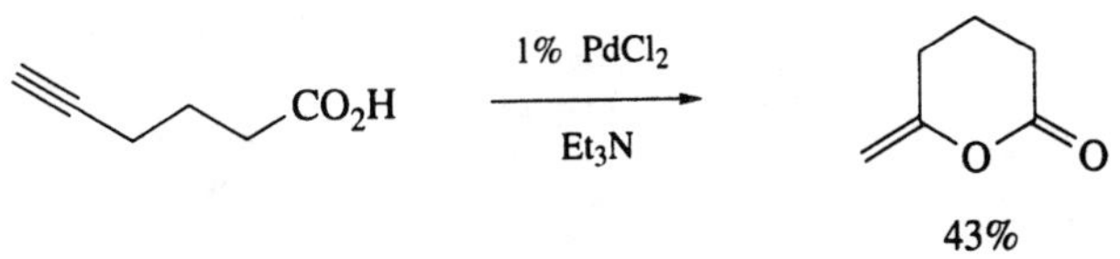

$R = H, Bu^n, n\text{-}C_6H_{13}, Ph, PhCH_2, SiMe_3$

1% $PdCl_2$, Et_3N; 43% (36)

In contrast to many stable transition metal–alkyne complexes, cationic cyclopentadienyliron–alkyne complexes are reactive toward a range of nucleophiles. However, since most of the nucleophiles studied were carbanions, discussion of this chemistry is deferred to Volume 4, Chapter 3.2, Section 3.2.3.

(37)

R^1 = H, n-C_6H_{13}, Ph, $SiMe_3$; R^2 = 1-Me, 2-Me, 3-Me

(38)

3.1.5 REFERENCES

1. (a) P. M. Henry, 'Palladium Catalyzed Oxidation of Hydrocarbons', Reidel, Dordrecht, 1980; (b) R. F. Heck, 'Palladium Reagents in Organic Synthesis', Academic Press, New York, 1985.
2. (a) J. Tsuji, 'Organic Synthesis with Palladium Compounds', Springer-Verlag, Berlin, 1980, p. 6; (b) For a general review on nucleophilic attack on transition metal complexes see: 'Chemistry of the Metal Carbon Bond', ed. F. R. Hartley and S. Patai, Wiley, New York, 1985, vol. 2, p. 401.
3. J. Tsuji, *Synthesis*, 1984, 369.
4. H. A. Zahalka, K. Januszkiewicz and H. Alper, *J. Mol. Catal.*, 1986, **35**, 249.
5. J. Tsuji and M. Minatodd, *Tetrahedron Lett.*, 1987, **28**, 3683.
6. A. J. K. Bose, L. Krishnan, D. R. Wagle and M. S. Manhas, *Tetrahedron Lett.*, 1986, **27**, 5955.
7. J. Tsuji, H. Nagashima and K. Hori, *Tetrahedron Lett.*, 1982, **23**, 2679.
8. H. Alper, K. Januszkiewicz and D. J. H. Smith, *Tetrahedron Lett.*, 1985, **26**, 2263.
9. J. Tsuji, H. Nagashima and K. Hori, *Chem. Lett.*, 1980, 257.
10. T. Hosokawa, M. Hirata, S. Murahashi and A. Sonoda, *Tetrahedron Lett.*, 1976, 1821.
11. (a) T. Hosokawa, Y. Inada and S. Murahashi, *Bull. Chem. Soc. Jpn.*, 1985, **58**, 3282; (b) T. Hosokawa, C. Okuda and S. Murahashi, *J. Org. Chem.*, 1985, **50**, 1282.
12. A. Kasahara, T. Izumi and M. Ooshima, *Bull. Chem. Soc. Jpn.*, 1974, **47**, 2526.
13. M. F. Semmelhack and C. Boudrow, *J. Am. Chem. Soc.*, 1984, **106**, 1496.
14. Y. Tamaru, T. Kobayashi, S. Kawamura, H. Ochiai, M. Hojo and Z. Yoshida, *Tetrahedron Lett.*, 1985, **26**, 3207.
15. M. F. Semmelhack and A. Zask, *J. Am. Chem. Soc.*, 1983, **105**, 2034.
16. R. D. Walkup and G. Park, *Tetrahedron Lett.*, 1987, **28**, 1023.
17. (a) A. Kasahara, T. Izumi, K. Sato, M. Masamura and T. Hayasaka, *Bull. Chem. Soc. Jpn.*, 1977, **50**, 1899; (b) T. Izumi and A. Kasahara, *Bull. Chem. Soc. Jpn.*, 1975, **48**, 1673.
18. D. E. Korte, L. S. Hegedus and R. K. Wirth, *J. Org. Chem.*, 1977, **42**, 1329.
19. Y. Tamaru, H. Higashimura, K. Naka, M. Hojo and Z. Yoshida, *Angew. Chem., Int. Ed. Engl.*, 1985, **24**, 1045.
20. R. C. Larock and D. R. Leach, *J. Org. Chem.*, 1984, **49**, 2144.
21. B. Åkermark, J. E. Backväll, K. Siirala-Hansen, K. Sjoberg, L. S. Hegedus and K. Zetterberg, *J. Organomet. Chem.*, 1974, **72**, 127.
22. L. S. Hegedus, B. Åkermark, K. Zetterberg and L. F. Olsson, *J. Am. Chem. Soc.*, 1984, **106**, 7122.
23. J. E. Backväll and E. E. Bjorkman, *J. Org. Chem.*, 1980, **45**, 2893.
24. J. E. Backväll, *Tetrahedron Lett.*, 1978, 163.
25. L. S. Hegedus and K. Siirala-Hansen, *J. Am. Chem. Soc.*, 1975, **97**, 1184.
26. L. S. Hegedus, O. P. Anderson, K. Zetterberg, A. B. Packard, G. F. Allen and K. Siirala-Hansen, *Inorg. Chem.*, 1977, **16**, 1887.
27. F. Ozawa, N. Nakano, I. Eoyama, T. Yamamoto and A. Yamamoto, *J. Chem. Soc., Chem. Commun.*, 1980, 382.
28. (a) L. S. Hegedus, G. F. Allen, J. J. Bozell and E. L. Waterman, *J. Am. Chem. Soc.*, 1978, **100**, 5800; (b) For a review on transition metals in the synthesis of indoles see: L. S. Hegedus, *Angew. Chem., Int. Ed. Engl.*, 1988, **27**, 1113.
29. (a) P. R. Weider, L. S. Hegedus, H. Asada and S. D'Andrea, *J. Org. Chem.*, 1985, **50**, 4276; (b) For a review on palladium catalysis in the synthesis of indoloquinones see: L. S. Hegedus, P. R. Weider, T. A. Mulhern, H. Asada and S. D'Andrea, *Gazz. Chim. Ital.*, 1986, **116**, 213.
30. L. S. Hegedus, G. F. Allen and D. J. Olsen, *J. Am. Chem. Soc.*, 1980, **102**, 3583.
31. S. J. Danishefsky and E. Taniyama, *Tetrahedron Lett.*, 1983, **24**, 15.
32. L. S. Hegedus and J. M. McKearin, *J. Am. Chem. Soc.*, 1982, **104**, 2444.
33. Y. Tamura, M. Hojo and Z. Yoshida, *J. Org. Chem.*, 1988, **53**, 5731.
34. D. Lathbury, P. Vernon and T. Gallagher, *Tetrahedron Lett.*, 1986, **27**, 6009.
35. K. A. Parker and R. P. Ofee, *J. Org. Chem.*, 1983, **48**, 1547.
36. P. Lennon, M. Madhavaro, A. Rosan and M. Rosenblum, *J. Organomet. Chem.*, 1976, **108**, 93; for a review see: M. Rosenblum, T. C. Chang, B. M. Foxman, S. B. Samuels and C. Stockmann, 'Organic Synthesis Today

and Tomorrow, Proceedings of the 3rd IUPAC Symposium on Organic Synthesis, 1980', Pergamon Press, Oxford, 1981, p. 47.
37. S. R. Berryhill, T. Price and M. Rosan, *J. Org. Chem.*, 1983, **48**, 158.
38. For a review of mercury(II)- and palladium(II)-catalyzed sigmatropic rearrangements see: L. E. Overman, *Angew. Chem., Int. Ed. Engl.*, 1984, **23**, 579.
39. A. C. Oehlschlager, P. Misra and S. Dhani, *Can. J. Chem.*, 1984, **62**, 791.
40. T. G. Schenck and B. Bosnich, *J. Am. Chem. Soc.*, 1985, **107**, 2058.
41. Y. Tamaru, M. Kogatani and Z. Yoshida, *J. Org. Chem.*, 1980, **45**, 5221; see R. F. Heck, 'Palladium Reagents in Organic Synthesis', Academic Press, New York, 1985, p. 36.
42. M. Mizutani, Y. Sanemitsu, Y. Tamaru and Z. Yoshida, *Tetrahedron*, 1985, **41**, 5289; *J. Org. Chem.*, 1985, **50**, 764.
43. P. R. Auburn, J. Whelan and B. Bosnich, *Isr. J. Chem.*, 1986, **27**, 250.
44. (a) J. E. Backväll, 'Organic Synthesis, An Interdisciplinary Challenge, 5th IUPAC Symposium', ed. J. Streith, H. Prinzbach and G. Schill, Pergamon Press, Oxford, 1985, p. 69; (b) J. E. Backväll, S. E. Bystrom and R. E. Nordberg, *J. Org. Chem.*, 1984, **49**, 4619.
45. J. E. Backväll and J. O. Vagberg, *J. Org. Chem.*, 1988, **53**, 5695.
46. L. S. Hegedus, N. Kambe, Y. Ishii and A. Mori, *J. Org. Chem.*, 1985, **50**, 2240.
47. K. Utimoto, H. Miwa and H. Nozaki, *Tetrahedron Lett.*, 1981, **22**, 4277.
48. E. C. Taylor, A. H. Katz and H. Salgado-Zamora, *Tetrahedron Lett.*, 1985, **26**, 5963.
49. Y. Wakabayashi, Y. Fukuda, H. Shiragami, K. Utimoto and H. Nozaki, *Tetrahedron*, 1985, **41**, 3655.
50. C. Lambert, K. Utimoto and H. Nozaki, *Tetrahedron Lett.*, 1984, **25**, 5323.
51. N. Yanagihara, C. Lambert, K. Iritani, K. Utimoto and H. Nozaki, *J. Am. Chem. Soc.*, 1986, **108**, 2753.

3.2
Carbon Nucleophiles with Alkenes and Alkynes

LOUIS S. HEGEDUS
Colorado State University, Fort Collins, CO, USA

3.2.1 INTRODUCTION

The same transition metal systems which activate alkenes, alkadienes and alkynes to undergo nucleophilic attack by heteroatom nucleophiles also promote the reaction of carbon nucleophiles with these unsaturated compounds, and most of the chemistry in Scheme 1 in Section 3.1.2 of this volume is also applicable in these systems. However two additional problems which seriously limit the synthetic utility of these reactions are encountered with carbon nucleophiles. Most carbanions are strong reducing agents, while many electrophilic metals such as palladium(II) are readily reduced. Thus, oxidative coupling of the carbanion, with concomitant reduction of the metal, is often encountered when carbon nucleophiles are studied. In addition, catalytic cycles invariably require reoxidation of the metal used to activate the alkene [usually palladium(II)]. Since carbanions are more readily oxidized than are the metals used, catalysis of alkene, diene and alkyne alkylation has rarely been achieved. Thus, virtually all of the reactions discussed below require stoichiometric quantities of the transition metal, and are practical only when the ease of the transformation or the value of the product overcomes the inherent cost of using large amounts of often expensive transition metals.

3.2.2 ALKYLATION OF MONOALKENES

Terminal monoalkenes were alkylated by stabilized carbanions ($pK_a \approx 10–18$) in the presence of 1 equiv. of palladium chloride and 2 equiv. of triethylamine, at low temperatures (Scheme 1).[1] The resulting unstable σ-alkylpalladium(II) complex was reduced to give the alkane (path a), allowed to β-hydride eliminate to give the alkene (path b), or treated with carbon monoxide and methanol to produce the ester (path c).[2] As was the case with heteroatom nucleophiles, attack at the more substituted alkene position predominated, and internal alkenes underwent alkylation in much lower (≈30%) yield. In the absence of triethylamine, the yields were very low (1–2%) and reduction of the metal by the carbanion became the major process. Presumably, the tertiary amine ligand prevented attack of the carbanion at the metal, directing it instead to the coordinated alkene. The regiochemistry (predominant attack at the more sub-

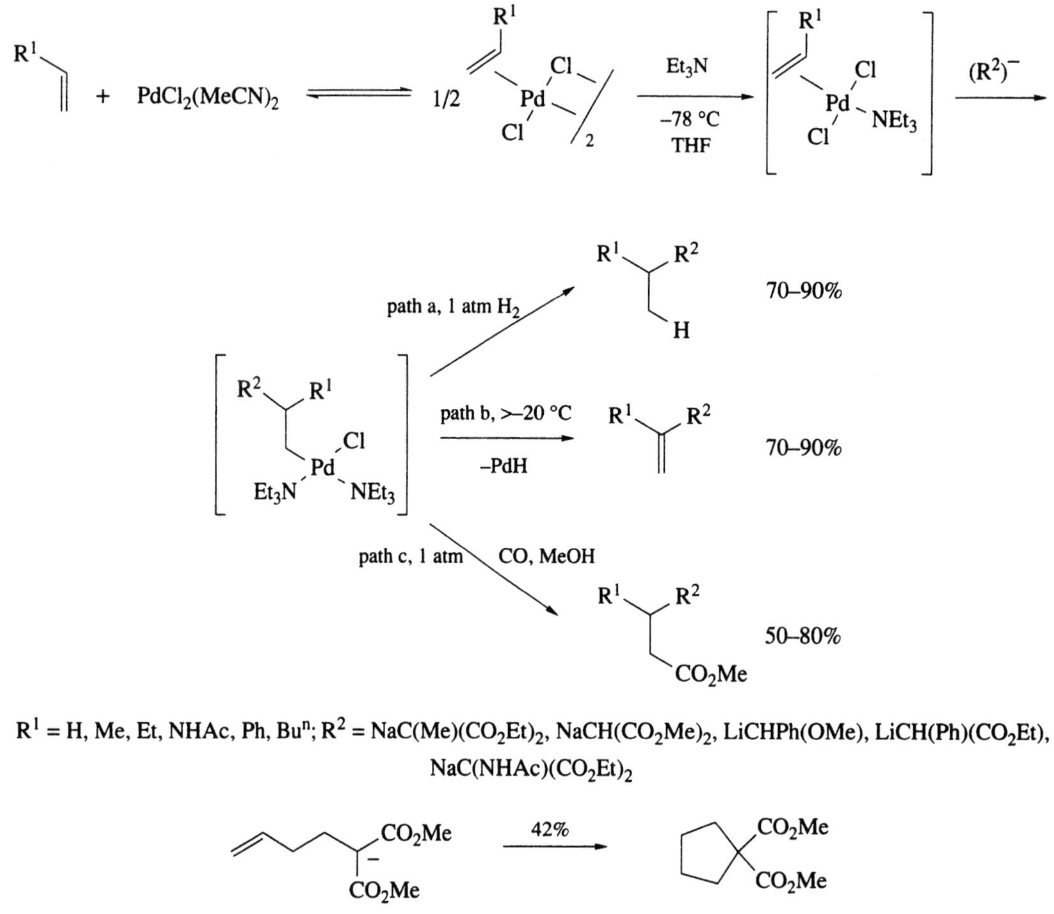
Et3N
−78 °C
THF
path a, 1 atm H2
70–90%
path b, >−20 °C
−PdH
70–90%
path c, 1 atm CO, MeOH
50–80%
R1 = H, Me, Et, NHAc, Ph, Bun; R2 = NaC(Me)(CO2Et)2, NaCH(CO2Me)2, LiCHPh(OMe), LiCH(Ph)(CO2Et), NaC(NHAc)(CO2Et)2
42%

Scheme 1

stituted alkene terminus) is consistent with attack by the nucleophile without prior coordination (see below).

Because it requires stoichiometric amounts of expensive palladium(II) salts, this process has found little use in the synthesis of more complex organic compounds. Two exceptions are seen in equation (1)[3] and Scheme 2.[4]

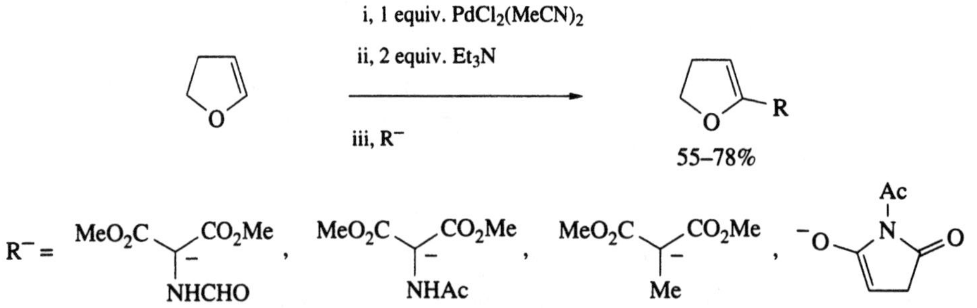
i, 1 equiv. PdCl2(MeCN)2
ii, 2 equiv. Et3N
iii, R−
55–78%
R− =

(1)

The above system failed entirely when nonstabilized carbanions such as ketone or ester enolates or Grignard reagents were used as carbon nucleophiles, leading to reductive coupling of the anions rather than alkylation of the alkene. However, the fortuitous observation that the addition of HMPA to the reaction mixture prior to addition of the carbanion prevented this side reaction[1] extended the range of useful carbanions substantially to include ketone and ester enolates, oxazoline anions, protected cyanohydrin anions, nitrile-stabilized anions[5] and even phenyllithium (Scheme 3).[5]

With these nonstabilized carbanions, attack occurred almost exclusively at the less-substituted terminus of the alkene, regioselectivity opposite that observed with stabilized carbanions. This regioselectivity

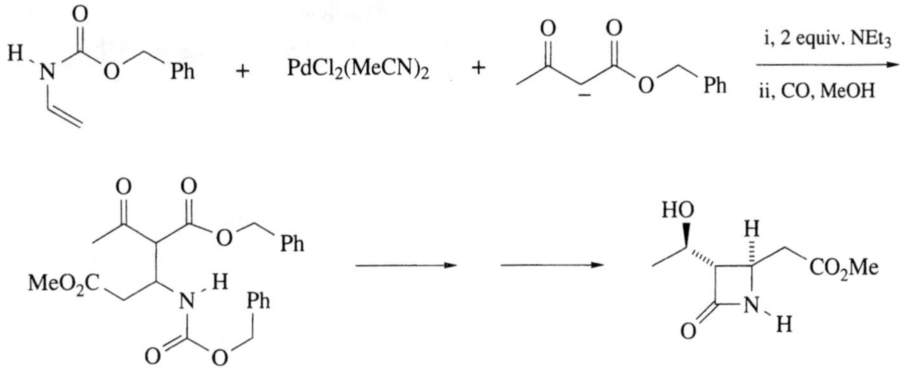

i, 2 equiv. NEt_3
ii, CO, MeOH

Scheme 2

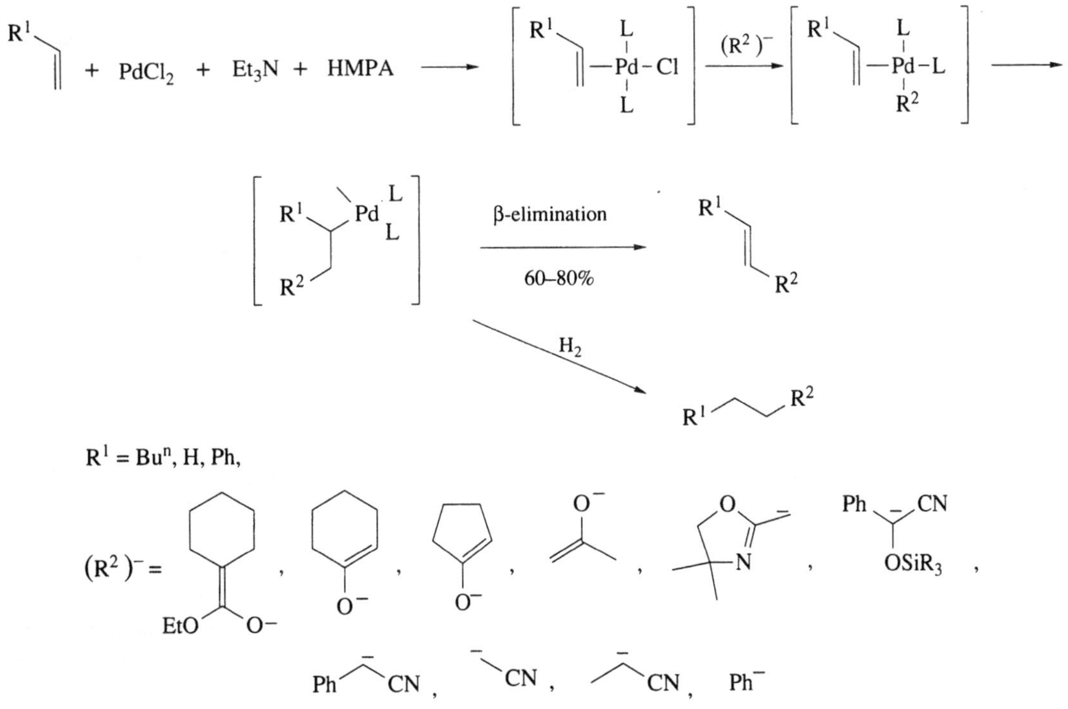

$R^1 = Bu^n$, H, Ph,

β-elimination
60–80%

Scheme 3

is that commonly observed in the insertion of alkenes into metal–carbon σ-bonds (Chapter 4.3, Volume 4), and implies a change in mechanism wherein direct alkylation of the metal followed by insertion of the alkene occurred rather than external nucleophilic attack of the carbanion on the metal-bound alkene. Again, the requirement of stoichiometric quantities of palladium salts has limited the synthetic applications of this chemistry.

In contrast, the closely related palladium acetate-promoted intramolecular alkylation of alkenes by trimethylsilyl enol ethers (Scheme 4)[6,7] has been used to synthesize a large number of bridged carbocyclic systems (Table 1). In principle, this process should be capable of being made catalytic in palladium(II), since silyl enol ethers are stable to a range of oxidants used to carry the $Pd^0 \rightarrow Pd^{II}$ redox chemistry required for catalysis. In practice, catalytically efficient conditions have not yet been developed, and the reaction is usually carried out using a full equivalent of palladium(II) acetate. This chemistry has been used in the synthesis of quadrone (equation 2).[8] With the more electrophilic palladium(II) trifluoroacetate, methyl enol ethers underwent this cyclization process (equation 3).[9]

Chelating alkenes such as allylic[10] and homoallylic[11] amines and sulfides underwent alkylation by a range of stabilized carbanions to produce stable σ-alkylpalladium(II) complexes. In these cases the regioselectivity was strictly governed by the inherent stability of a five (*versus* four or six) membered chelate σ-alkylpalladium complex, with allylic systems (Scheme 5) being alkylated at the more sub-

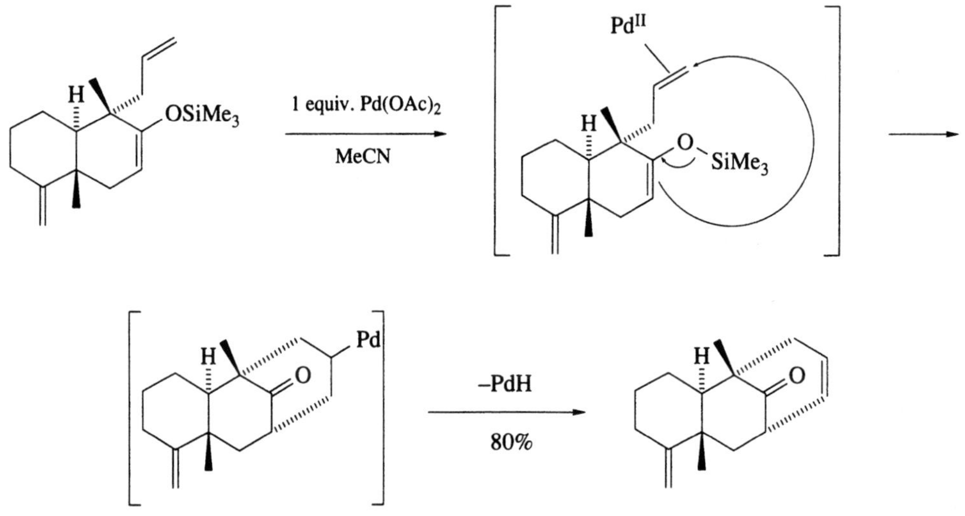

1 equiv. $Pd(OAc)_2$

MeCN

Pd^{II}

$OSiMe_3$

–PdH

80%

Scheme 4

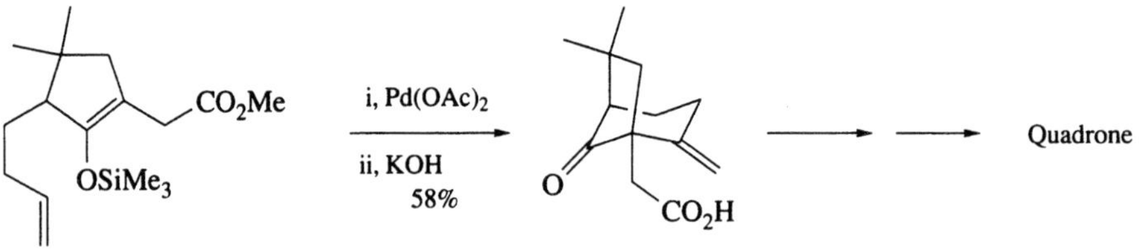

i, $Pd(OAc)_2$

ii, KOH

58%

Quadrone (2)

1 equiv. $Pd(TFA)_2$

8 cases, ≈50% yield

60%

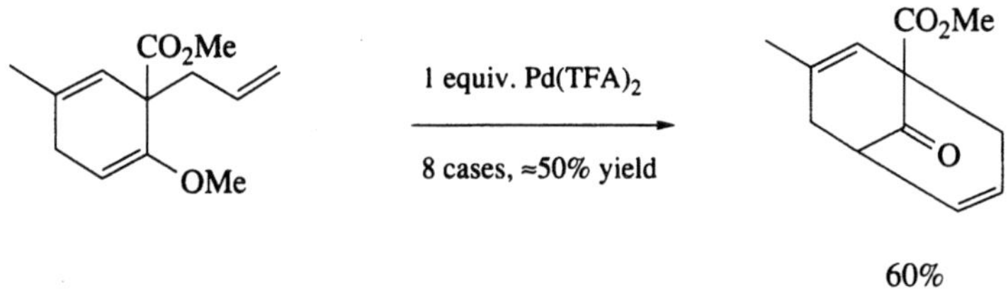

(3)

stituted position and homoallylic systems at the less substituted terminus (Scheme 6). Reduction of these complexes produced the saturated hydrocarbons, while treatment with conjugated enones resulted in insertion of the alkene.

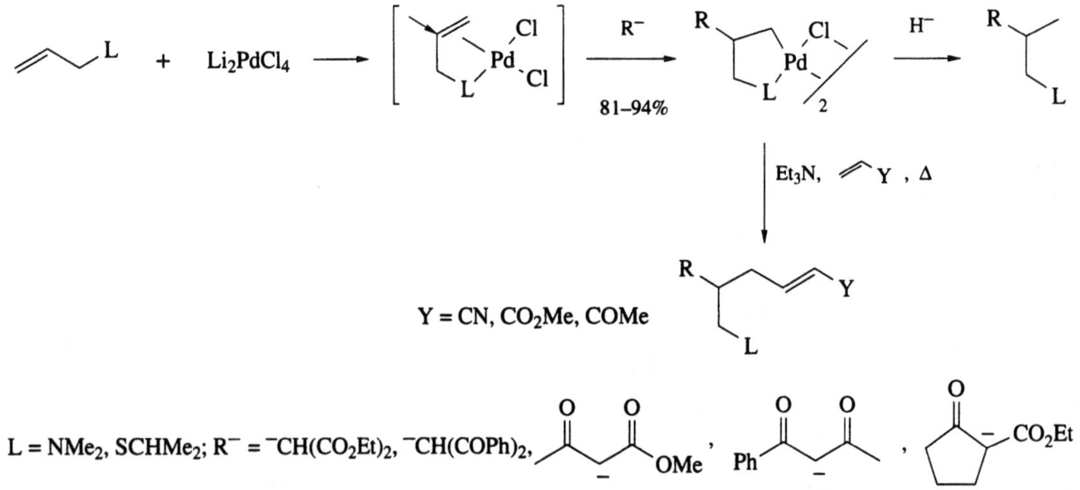

Li_2PdCl_4

R^-

81–94%

H^-

Et_3N, Y, Δ

Y = CN, CO_2Me, COMe

L = NMe_2, $SCHMe_2$; R^- = $^-CH(CO_2Et)_2$, $^-CH(COPh)_2$,

Scheme 5

Table 1 Palladium(II) Acetate Assisted Cyclization of Trimethylsilylenol Ethers

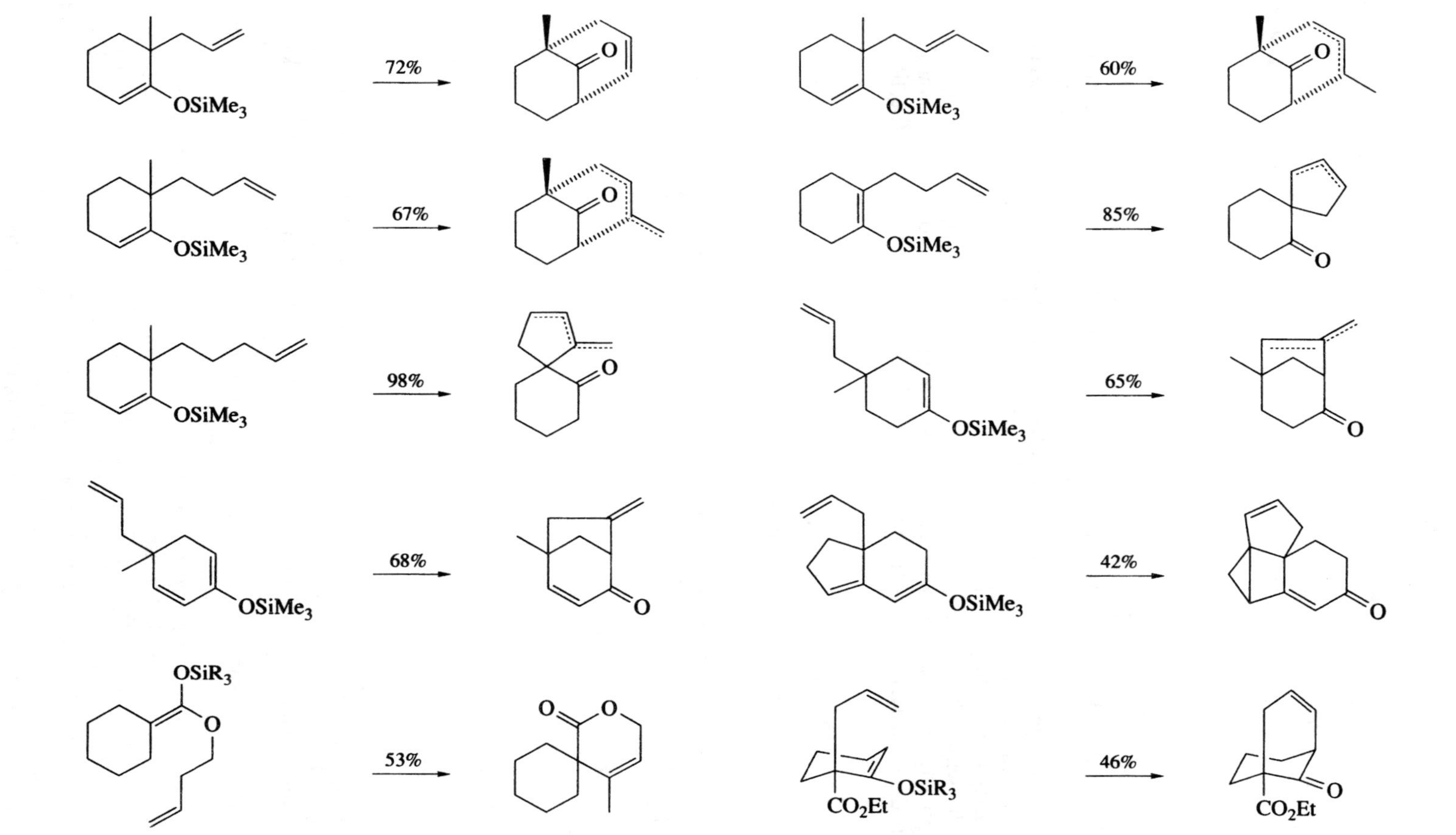

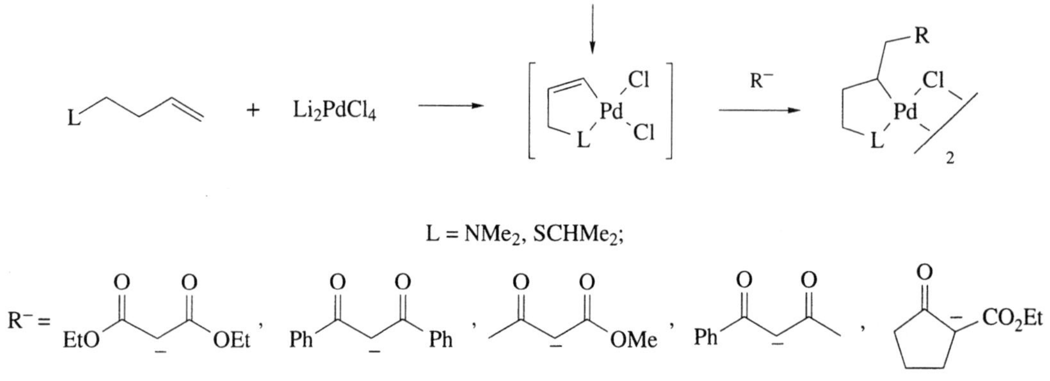

$L = NMe_2$, $SCHMe_2$;

$R^- =$

Scheme 6

In contrast to the above stoichiometric palladium-assisted reactions, palladium(II) chloride efficiently catalyzed a variety of Cope rearrangements (Scheme 7 and Table 2).[12] The mechanism of this process is not known, but is likely to involve 'cyclization-induced catalysis' as noted in Section 3.1.2.3 of this volume. This rearrangement has several important features. Successful catalysis required that C-2 or C-5 of the diene had an alkyl substituent present, perhaps to stabilize the developing positive charge at these positions during rearrangement. Alkyl substituents at both C-2 and C-5 suppressed the rearrangement, reflecting the steric inhibition of placing the bulky metal at a sterically hindered tertiary position. With chiral, optically active dienes, complete 1,4-transfer of chirality was observed, as was the case in the related palladium-catalyzed allylic transposition of allyl acetates discussed above (Section 3.1.2.3, this volume).

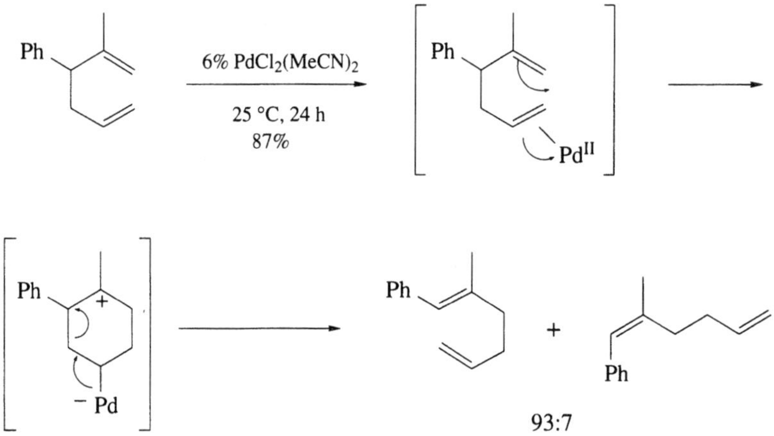

6% $PdCl_2(MeCN)_2$

25 °C, 24 h
87%

93:7

Scheme 7

Cationic cyclopentadienyliron dicarbonyl (F_p) complexes of alkenes are generally reactive toward a wide range of nucleophiles, including carbanions, producing very stable σ-alkyliron complexes (Scheme 8).[13,14] Again, attack at the more substituted alkene terminus predominated. These σ-alkyliron complexes required chemical removal of the iron to free the organic fragment, making catalytic processes impossible. As a consequence little use of these simpler systems in synthesis has resulted. However, a number of synthetically useful transformations involving alkoxyalkenes have been developed.

α-Methylene lactones were produced by the reaction of cyclohexanone enolate with the F_p–α-ethoxyacrylate complex (Scheme 9)[15] while ethoxyethylene was converted to the corresponding γ-lactones (Scheme 10).[16] Dimethoxyethylene–F_p complexes provided routes to efficiently dialkylate ethylene (Scheme 11).[17] Using this methodology, a variety of furans were synthesized (Scheme 12).[18] This chemistry was noteworthy in two respects. Firstly, all transformations were highly stereoselective, permitting the facile synthesis of complex molecules having multiple stereogenic centers. Secondly, the oxidative cleavage of the σ-alkyliron complexes led to nucleophilic displacement of the oxidized iron fragment, rather than the normal insertion of CO to form acyl complexes. Thus furans, rather than the expected lactones, were formed.

Table 2 Palladium(II)-catalyzed Cope Rearrangements

34% ⇌ 66%

R — 64–94% → R

R = CO_2Et, COMe, CN, CO_2H

O, OMe — 81% → O, OMe

MeO_2C — 90% → MeO_2C

HO — 100% → O

Ph — 86% → Ph + Ph

(3*R*)-(5*E*) — (2*Z*)-(5*R*) 7:3 (2*E*)-(5*S*)

H H

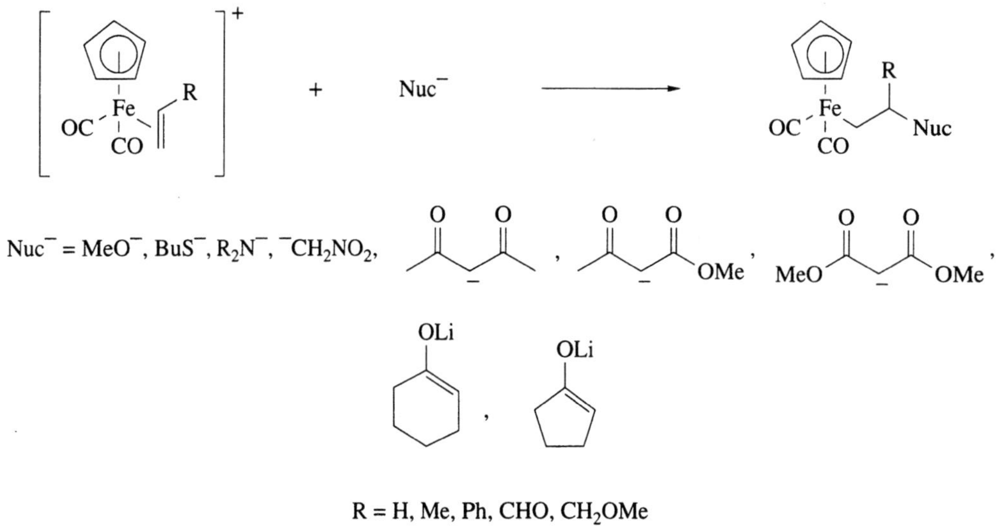

$Nuc^- = MeO^-, BuS^-, R_2N^-, ^-CH_2NO_2$,

R = H, Me, Ph, CHO, CH_2OMe

Scheme 8

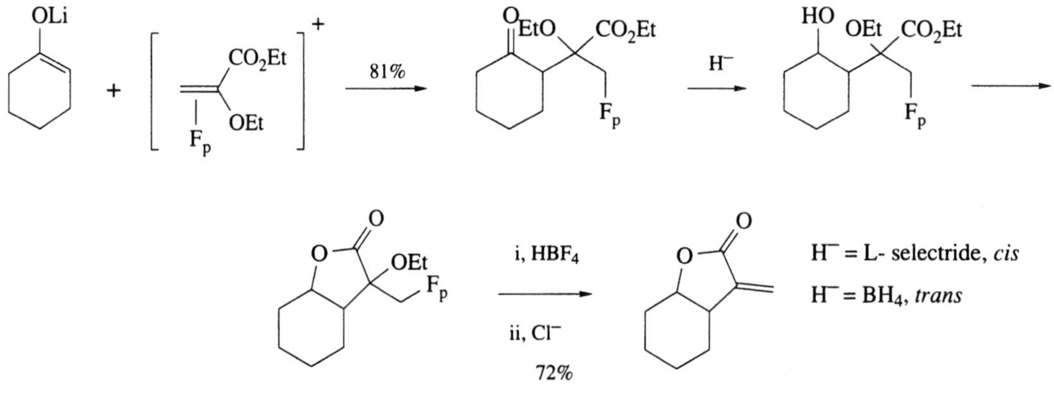

H^- = L- selectride, *cis*

$H^- = BH_4$, *trans*

Scheme 9

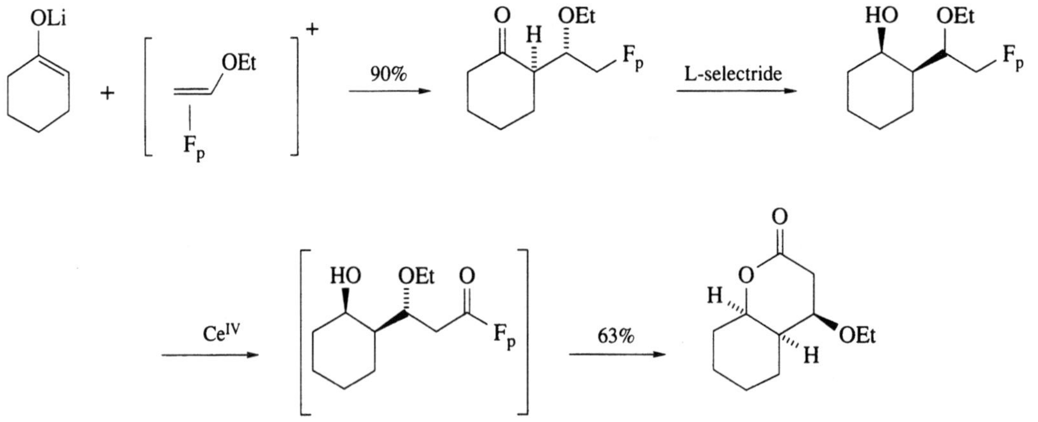

Scheme 10

Cationic iron–alkene complexes also participate in an unusual 'cycloaddition' process, wherein electron-deficient alkenes are attacked by nucleophilic σ-allylic F_p complexes, generating stabilized carbanions and cationic alkene–iron complexes. Attack of the carbanion on the alkene forming a five-membered ring completes this process (Scheme 13). Oxidative removal of the iron produces useful organic compounds.[19–21]

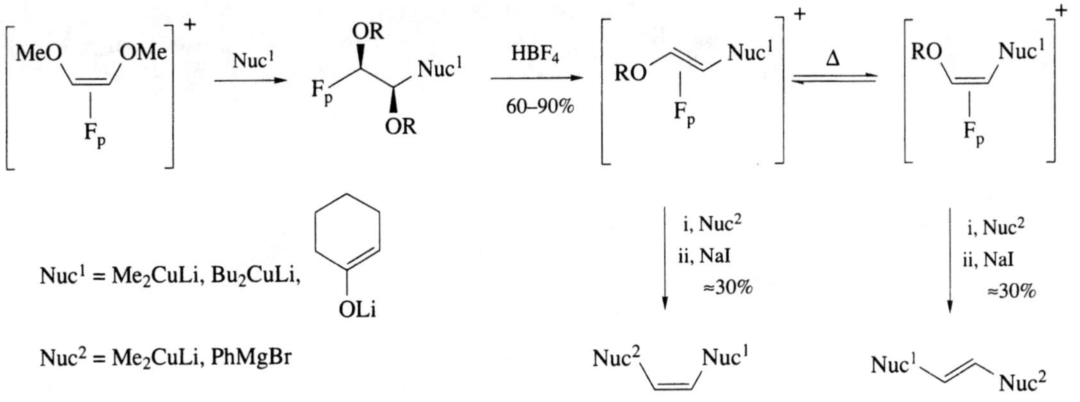

$Nuc^1 = Me_2CuLi, Bu_2CuLi,$ OLi

$Nuc^2 = Me_2CuLi, PhMgBr$

Scheme 11

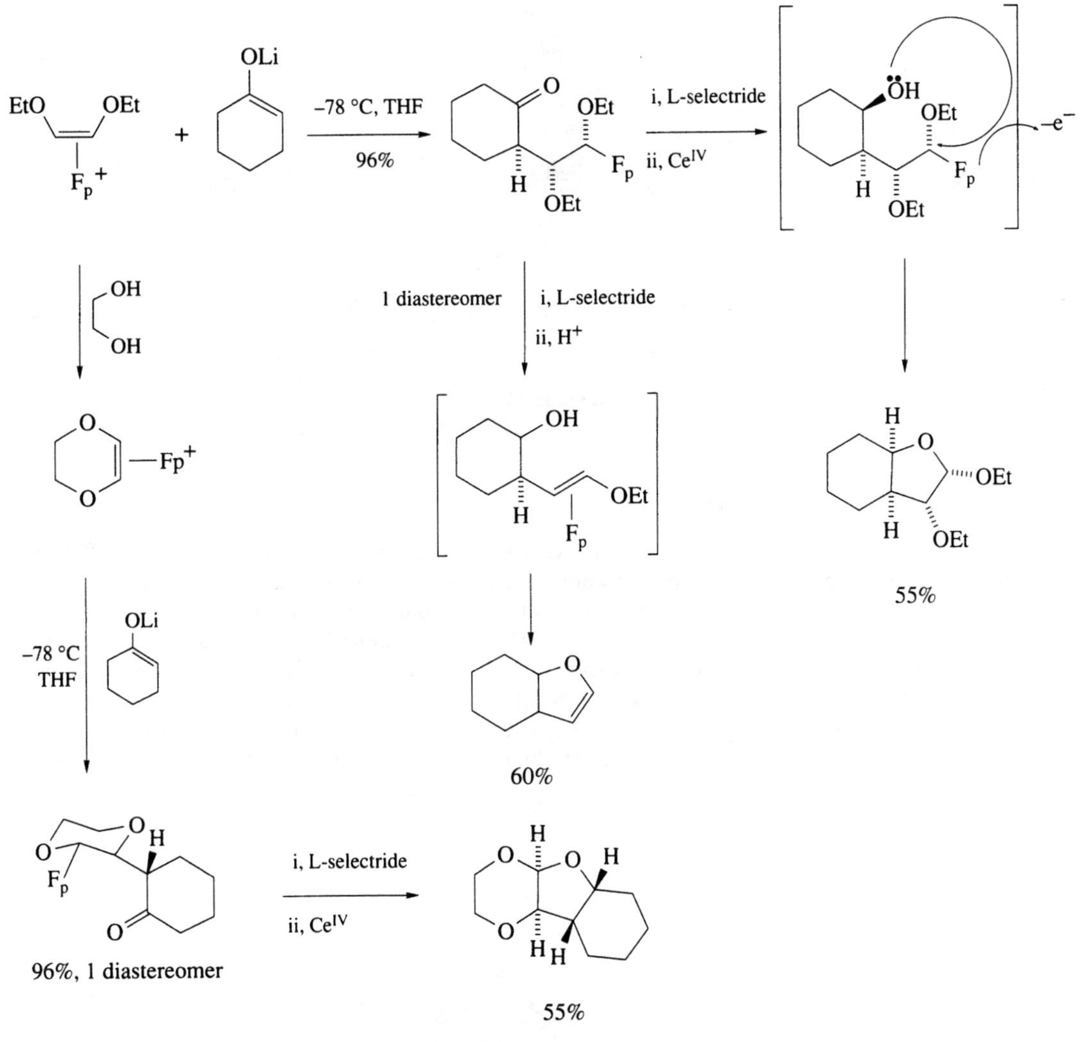

Scheme 12

This chemistry has been used to synthesize cyclopentanoid derivatives used in the synthesis of sarkomycin and brefeldin A.[22] Azulene derivatives were also synthesized using this chemistry (Scheme 14).[23]

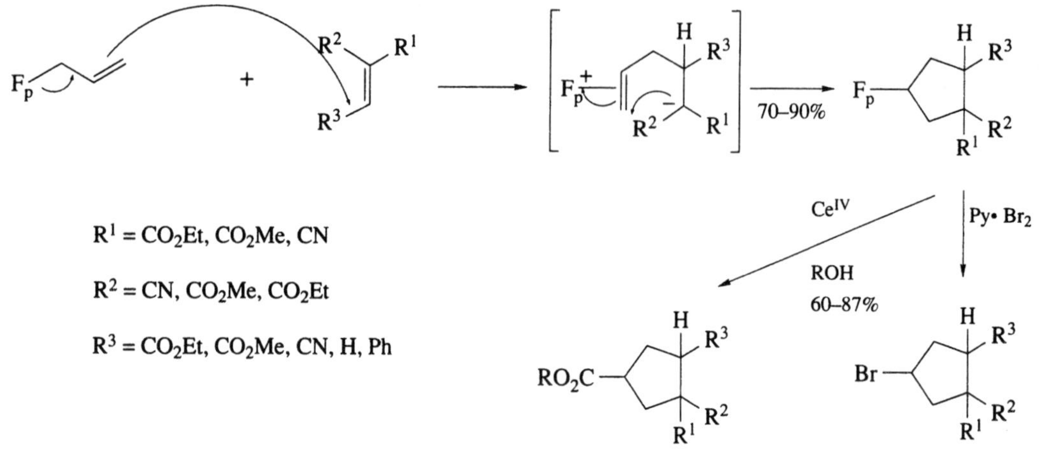

$R^1 = CO_2Et, CO_2Me, CN$

$R^2 = CN, CO_2Me, CO_2Et$

$R^3 = CO_2Et, CO_2Me, CN, H, Ph$

70–90%

Ce^{IV} ROH 60–87%

Py• Br_2

Scheme 13

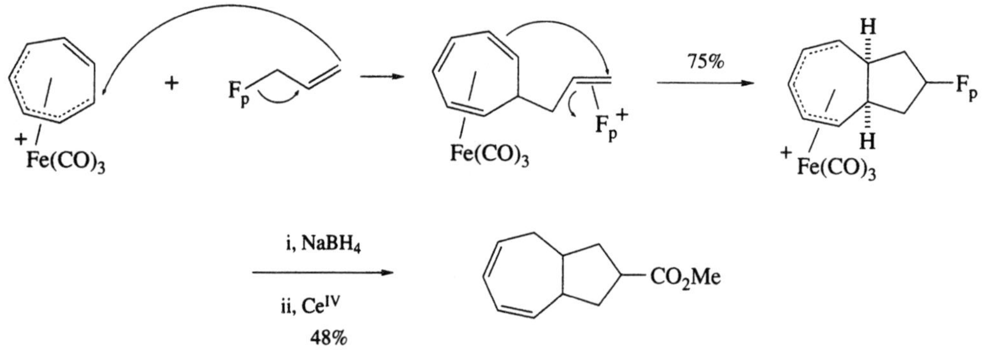

75%

i, $NaBH_4$

ii, Ce^{IV}

48%

Scheme 14

3.2.3 ALKYLATION OF ALKADIENES

1,3-Dienes form very stable complexes with a variety of metal carbonyls, particularly $Fe(CO)_5$, and the neutral η^4-diene metal carbonyl complexes are quite resistant to normal reactions of dienes (*e.g.* hydrogenation, Diels–Alder). However, they are subject to nucleophilic attack by a variety of nonstabilized carbanions. Treatment of η^4-cyclohexadiene iron tricarbonyl with nonstabilized carbanions, followed by protonolysis of the resulting complex, produced isomeric mixtures of alkylated cyclohexenes (Scheme 15).[24] With acyclic dienes, this alkylation was shown to be reversible, with kinetic alkylation occurring at an internal position of the complexed dienes but rearranging to the terminal position under thermodynamic conditions (Scheme 16).[25] By trapping the kinetic product with an electrophile, overall 'carbo-

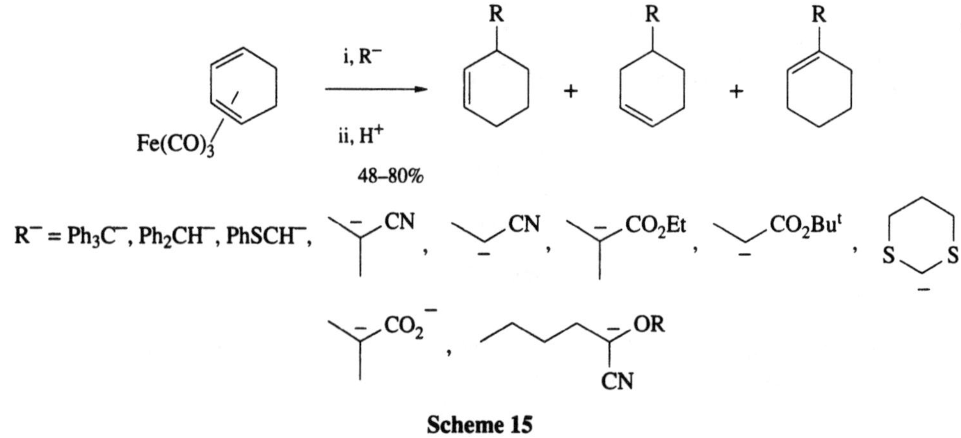

i, R^-

ii, H^+

48–80%

$R^- = Ph_3C^-, Ph_2CH^-, PhSCH^-$,

Scheme 15

acylation' was achieved (Schemes 17 and 18).[26,27] With cyclic dienes, alkylation occurred from the face opposite the metal, and acylation from the same face as the metal to produce exclusively *trans* product.

As expected, cationic diene complexes (as well as cationic dienyl complexes, see Chapter 3.4, of this

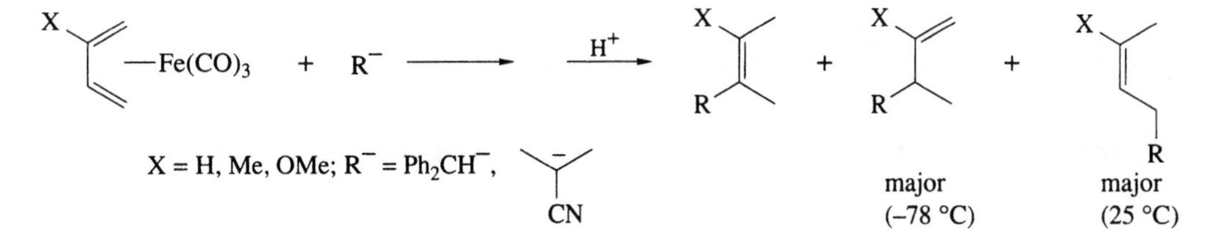

X = H, Me, OMe; R^- = Ph_2CH^-,

major (–78 °C) major (25 °C)

Scheme 16

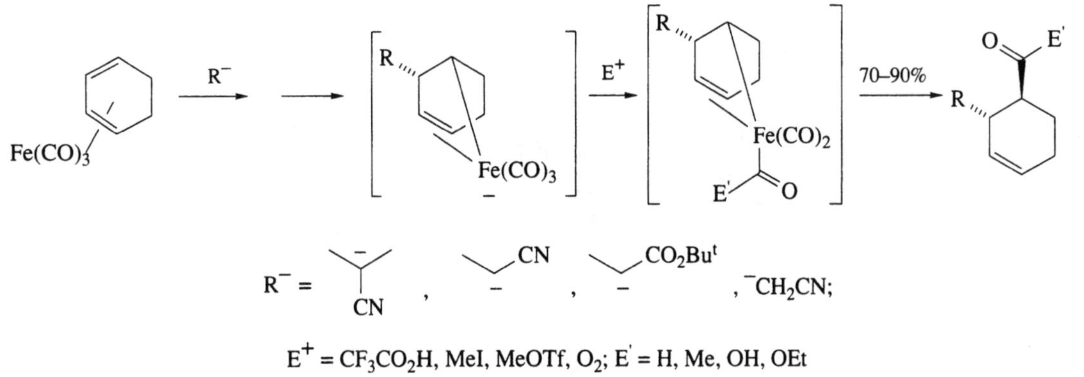

70–90%

R^- = , $^-CH_2CN$;

E^+ = CF_3CO_2H, MeI, MeOTf, O_2; E' = H, Me, OH, OEt

Scheme 17

57–85%

R^- = , $^-CH_2CN$

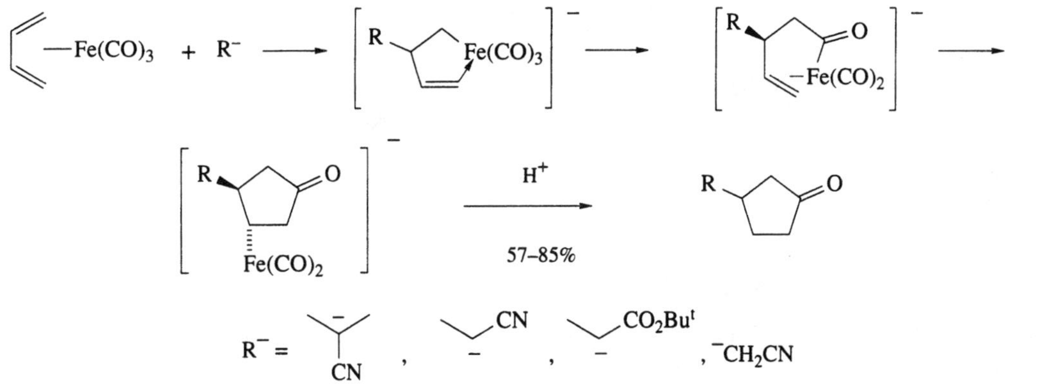

Scheme 18

volume) are considerably more reactive toward nucleophiles than are the corresponding neutral complexes. Using this feature, very efficient processes for stereocontrolled alkylation of cyclohexadiene (Scheme 19)[28] and cycloheptadiene (Scheme 20)[29] have been developed. They share many common features. A range of carbanions attacked, always from the face opposite the metal, producing a neutral η^3-allylmolybdenum complex. Further functionalization of the cyclohexadiene system was not studied. However in the cycloheptadiene series the initially formed η^3-allyl complex was reconverted into a cationic diene complex, which could be alkylated again. Both alkyl groups entered from the face opposite the metal, giving *cis*-1,2- and -1,4-dialkyl systems in high yield and with high diastereoselectivity. The potential for further functionalization of the resulting η^3-allyl systems is high, but has yet to be extensively explored.

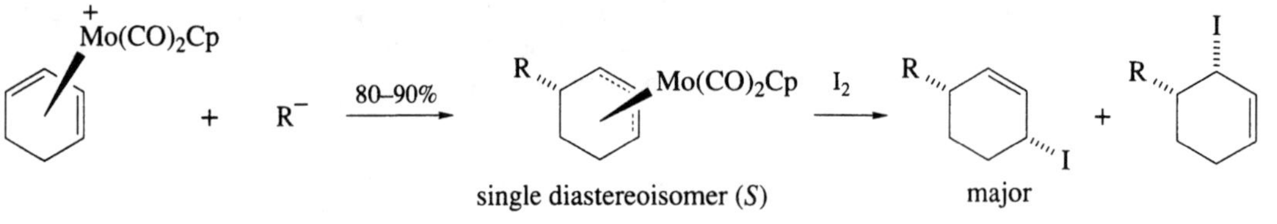

80–90%

single diastereoisomer (*S*)

major

$R^- = {}^-CH(CO_2Me)_2$, ${}^-CMe(CO_2Me)_2$, ${}^-CH(SO_2Ph)(CO_2Me)$, ${}^-CH(COMe)(CO_2Me)$, p-$MeOC_6H_4{}^-$, ${}^-CH_2CO_2Me$

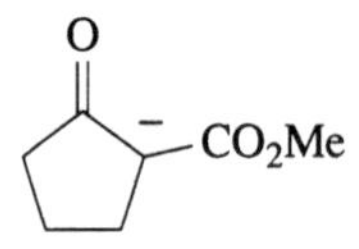

Scheme 19

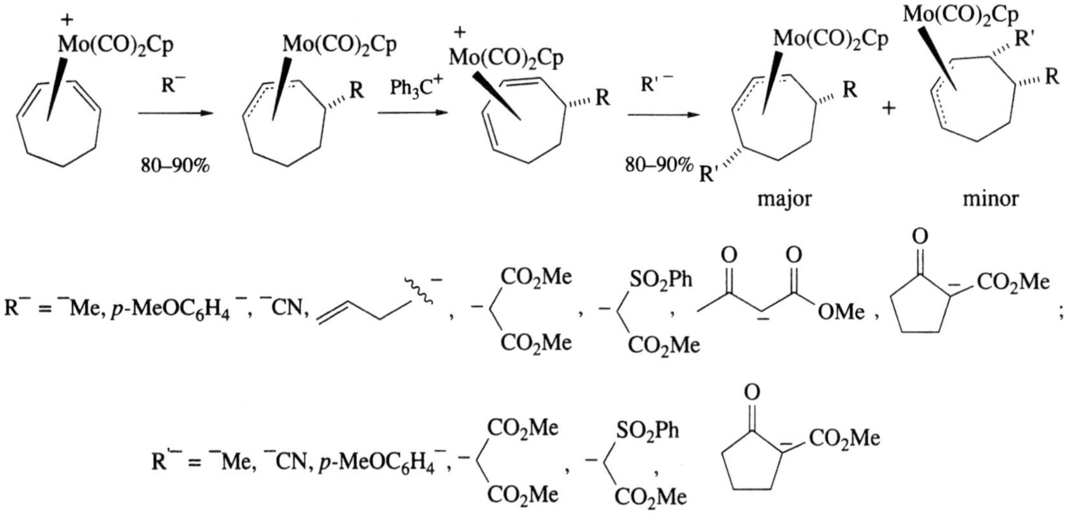

80–90%

80–90%

major minor

$R^- = {}^-Me$, p-$MeOC_6H_4{}^-$, ${}^-CN$, ...

$R'^- = {}^-Me$, ${}^-CN$, p-$MeOC_6H_4{}^-$, ...

Scheme 20

3.2.4 ALKYLATION OF ALKYNES

Although alkynes are highly reactive toward a wide range of transition metals, few instances of metal-catalyzed reactions of carbanions with alkynes are known. The most extensively developed system involves cationic iron complexes of internal alkynes. These complexes underwent alkylation by a range of carbanions to produce stable σ-vinyliron complexes (Scheme 21).[30] The addition was stereoselectively

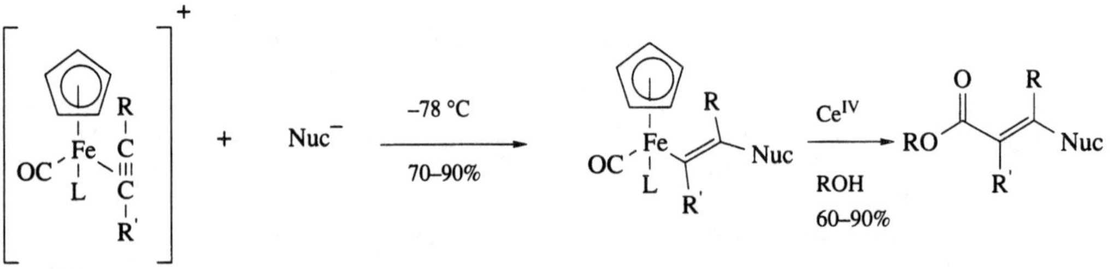

–78 °C

70–90%

Ce^{IV}

ROH

60–90%

L = PPh_3, $P(OPh)_3$; R, R' = Me; R = Me, R' = CO_2Me; R = Ph, R' = Me;

$Nuc^- = {}^-Me$, ${}^-Ph$, ${}^-CH(CO_2Et)_2$, ${}^-CN$, ${}^-SPh$, –C≡C–

Scheme 21

trans, and also regioselective, with the nucleophile adding away from ester groups on the alkyne, and on the alkyne carbon bearing a phenyl group. Oxidation promoted CO insertion with retention of alkene geometry (in most cases) producing acyliron complexes which upon further oxidation cleaved to give α,β-unsaturated esters. At present, the process is stepwise, and the metal complex must be destroyed to free the organic ligand. More efficient systems await development.

3.2.5 REFERENCES

1. L. S. Hegedus, R. E. Williams, M. A. McGuire and T. Hayashi, *J. Am. Chem. Soc.*, 1980, **102**, 4973.
2. L. S. Hegedus and W. H. Darlington, *J. Am. Chem. Soc.*, 1980, **102**, 4980.
3. L. V. Dunkerton and A. J. Sermo, *J. Org. Chem.*, 1982, **47**, 2814.
4. G. M. Wieber, L. S. Hegedus, B. Åkermark and A. Kramer, *J. Org. Chem.*, 1989, **54**, 4649; J. Montgomery, G. M. Wieber and L. S. Hegedus, *J. Am. Chem. Soc.*, 1990, **112**, 6255.
5. L. S. Hegedus and M. A. McGuire, *Organometallics*, 1982, **1**, 1175.
6. A. S. Kende, B. Roth and P. J. Sanfillippo, *J. Am. Chem. Soc.*, 1982, **104**, 1784.
7. A. S. Kende and D. J. Wustrow, *Tetrahedron Lett.*, 1985, **26**, 5411.
8. A. S. Kende, B. Roth, P. J. Sanfillippo and T. J. Blacklock, *J. Am. Chem. Soc.*, 1982, **104**, 5808.
9. A. S. Kende, R. A. Battista and S. B. Sandoval, *Tetrahedron Lett.*, 1984, **25**, 1341.
10. R. A. Holton and R. A. Kjonaas, *J. Am. Chem. Soc.*, 1977, **99**, 4177.
11. R. A. Holton and R. A. Kjonaas, *J. Organomet. Chem.*, 1977, **142**, C15.
12. L. E. Overman, *Angew. Chem., Int. Ed. Engl.*, 1984, **23**, 579.
13. P. Lennon, M. Madhavarso, A. Rosan and M. Rosenblum, *J. Organomet. Chem.*, 1976, **108**, 93.
14. P. Lennon, A. Rosan and M. Rosenblum, *J. Am. Chem. Soc.*, 1977, **99**, 8426.
15. T. C. Chang and M. Rosenblum, *Tetrahedron Lett.*, 1983, **24**, 695; *Isr. J. Chem.*, 1984, **24**, 99.
16. T. C. Chang, T. S. Coolbaugh, B. M. Foxman, M. Rosenblum, N. Simms and C. Stockmann, *Organometallics*, 1987, **6**, 2394.
17. M. Marsi and M. Rosenblum, *J. Am. Chem. Soc.*, 1984, **106**, 7264.
18. M. Rosenblum, B. M. Foxman and M. M. Turnbull, *Heterocycles*, 1987, **25**, 419.
19. T. S. Abraham, R. Baker, C. M. Exon and V. B. Rao, *J. Chem. Soc., Perkin Trans. 1*, 1982, 285.
20. R. Baker, C. M. Exon, V. B. Rao and R. W. Turner, *J. Chem. Soc., Perkin Trans. 1*, 1982, 295.
21. T. S. Abraham, R. Baker, C. M. Exon, V. B. Rao and R. W. Turner, *J. Chem. Soc., Perkin Trans. 1*, 1982, 301.
22. R. Baker, R. B. Keen, M. D. Morris and R. W. Turner, *J. Chem. Soc., Chem. Commun.*, 1984, 987.
23. N. Genco, D. F. Marten, S. Raghu and M. Rosenblum, *J. Am. Chem. Soc.*, 1976, **98**, 848.
24. M. F. Semmelhack and J. W. Herndon, *Organometallics*, 1983, **2**, 363.
25. M. F. Semmelhack and H. T. M. Lee, *J. Am. Chem. Soc.*, 1984, **106**, 2715.
26. M. F. Semmelhack, J. W. Herndon and J. P. Springer, *J. Am. Chem. Soc.*, 1983, **105**, 2497.
27. M. F. Semmelhack, J. W. Herndon and J. K. Lin, *Organometallics*, 1983, **2**, 1885.
28. A. J. Pearson, M. N. I. Khand, J. C. Clardy, and He. Chung-heng, *J. Am. Chem. Soc.*, 1985, **107**, 2748.
29. A. J. Pearson and M. N. I. Khand, *J. Org. Chem.*, 1985, **50**, 5276.
30. D. L. Reger, S. A. Klaeren and L. Lebioda, *Organometallics*, 1986, **5**, 1072; *J. Am. Chem. Soc.*, 1986, **108**, 1940; for a review see D. L. Reger, *Acc. Chem. Res.*, 1988, **21**, 279.

3.3
Nucleophiles with Allyl–Metal Complexes

STEPHEN A. GODLESKI

Kodak Research Laboratories, Rochester, NY, USA

3.3.1 INTRODUCTION

3.3.1.1 Scope

Interest in π-allylpalladium chemistry has undergone incredible growth, particularly over the last decade. This growth is exemplified by the significant number of reviews on this subject that have now appeared, including several within the last few years.[1–16] In light of this substantial coverage, this chapter is not intended to serve as a comprehensive review of this area of chemistry. Rather, the focus of this chapter will be on the selectivities, namely chemo-, diastereo-, regio- and enantio-selectivity, that are exhibited by π-allylpalladium chemistry. As possible, the mechanistic basis of these selectivities will be elucidated. In addition, the selection of examples provided to illustrate the nature of reactions involving π-allylpalladium complexes will be largely made from contributions that have appeared in the last decade.

3.3.1.2 Nature of π-Allylpalladium-mediated Functionalization

The essence of π-allylpalladium-mediated functionalization is the activation of an allylic system to attack by a nucleophilic compound. The regio- and stereo-selectivity associated with this reaction have been found to be functions of the nucleophile, the substitution on the allyl moiety, the nature of the ligands on palladium and the reaction medium.

The means of entry into π-allyl complexes, the elucidation of the range of nucleophiles that can be employed, a summary of the reaction types available for allylpalladium species and the diastereo-, regio- and enantio-selectivities of this reaction will be discussed in turn.

3.3.2 CHEMOSELECTIVITY

3.3.2.1 Precursors to π-Allylpalladium Complexes

3.3.2.1.1 Stoichiometric formation from alkenes

One of the first significant advances in the chemistry of π-allylpalladium complexes was the discovery that alkenes could be directly converted into the corresponding allyl complex by substitution into the allylic C—H bond. A variety of recipes have now been reported that can accomplish this transformation. Initially, palladium chloride[17–23] or its more soluble forms, sodium or lithium tetrachloropalladate[24–27] and bisacetonitrile palladium dichloride,[28–30] in alcohol or aqueous acetic acid solvent were employed. The use of palladium trifluoroacetate, followed by counterion exchange with chloride, represents the mildest and most effective means available to accomplish this reaction.[31]

This C—H substitution process results in a Markovnikov orientation, with the H that is allylic to the more substituted end of the alkene preferentially abstracted. The stereochemistry of the resulting π-allyl complex does not represent the stereochemistry of the starting alkene, as the complexes are capable of isomerization under the conditions in which they are formed. Typically, a thermodynamic mixture is obtained, with the *syn* form of the complex predominating over the *anti* form (equation 1). The *syn* form is more stable due to unfavorable steric interactions that the *anti* form encounters with the coordination sphere of the palladium.

R Pd H ⇌ R H Pd ⇌ H Pd R (1)

syn *anti*

In addition, the palladium will preferentially position itself on the sterically less demanding face of an allyl system. Both stereoselectivities are neatly demonstrated by reacting either geometric isomer of 2-ethylidenenorpinane (**1**) with $PdCl_2$, where only a single π-allylpalladium complex (**2**) that has a *syn* configuration is observed and has the Pd *trans* to the *gem*-dimethyl bridge (equation 2).[32]

or → Pd (2)

(**1**) (**2**)

3.3.2.1.2 Stoichiometric formation from other unsaturated precursors

π-Allylpalladium complexes can also be generated by the addition of $PdCl_2$ to 1,3-dienes,[33–39] allenes,[40,41] vinylcyclopropenes,[42,43] methylenecyclopropanes,[44,45] spirocyclopentanes,[46,47] and cyclopropenes (equations 3–8).[48,49]

$PdCl_2$ → Cl Pd (3)

PdCl2 Cl Pd (4)

R PdCl2 R Pd Cl (5)

PdCl2 Cl Pd (6)

Cl PdCl2 Pd (7)

PdCl2 Cl Pd (8)

3.3.2.1.3 Stoichiometric formation via *addition of alkenyl- and alkynyl-palladium to alkenes*

Alkynyl- or vinyl-palladiums can be generated *in situ* from the corresponding organomercury compounds and added to alkenes to form π-allylpalladium complexes after H-rearrangement (equation 9).[50]

Bu^t HgCl + PdCl2 Bu^t Pd (9)

3.3.2.1.4 Stoichiometric formation via *addition of palladium complexes to allyl–X substrates*

Allyl halides,[51–63] alcohols,[64] and ethers[65,66] have been converted to their respective π-allyl derivatives by reaction with a Pd^{II} salt in the presence of a reducing agent. Allyl-silanes[67] and -stannanes[68] have also been found to undergo reaction with lithium palladates in methanol to generate π-allyl chloride dimers.

3.3.2.1.5 Catalytic formation from allyl–X precursors

The initial investigation of the reactivity of π-allylpalladium complexes generated as their chloride dimers with various nucleophiles indicated that these species were quite limited as electrophiles.[2] In what may be the most significant advance in π-allylpalladium chemistry, it was subsequently noted that the addition of ligands, typically phosphines, dissociated the chloride dimers and greatly enhanced the electrophilicity of the allyl ligand.[2] Importantly, it was further discovered that a ratio of two phosphines per palladium gave optimal performance of the complex as an electrophile.[69–74] This ratio suggested the intermediacy of a bisphosphine cationic complex, with a chloride serving as a counterion (**3**). The final crucial realization that the same complexes could be generated by reaction of Pd^0 complexes[72] with alkenes allylically substituted with a leaving group, X,[73,74] was the true impetus for the explosive growth in this chemistry as these processes could now be conducted with catalytic, rather than stoichiometric, use of the metal. In this catalytic sequence, the palladium in its zero oxidation state initially acts as a nucleophile, displacing the X leaving group (oxidative addition) with the metal undergoing oxidation to its 2+ state in the process. The cationic metal center activates the allyl ligand to attack by nucleophiles. In the nucleophilic addition, the metal serves as a leaving group (reductive elimination) and returns to its zero oxidation state, thus allowing the process to be catalytic in metal. Catalytic use of Pd in the addition of nucleophiles to allyl intermediates in the dimerization of 1,3-butadiene had previously been observed,[75] but this use did not enjoy the potential generality of that defined by the allyl–X species.

Pd^+ Cl^- L L

(3)

A wide variety of groups have been found to be capable of serving the role of X in this process, including halogens,[9] esters,[9] alcohols,[9] phosphates,[76] amines,[77] ammonium salts,[78–83] selenides[84] and arsenites.[85] Among this group allyl alcohols and ethers are generally poor actors, and the vast majority of work has been done utilizing allyl esters, principally allyl acetates.

Several new leaving groups have been discovered recently which merit special discussion. Allyl sulfones, surprisingly, function as substrates for palladium catalysis.[86] As the sulfone group had previously been proven to be able to stabilize an adjacent carbanion, this result allowed allyl sulfones now to be considered as synthons for 1,1- and 1,3-dipoles (equation 10). That is, the allyl sulfone can be used alternately as a nucleophile and electrophile, greatly extending its synthetic utility.

– + ⟵⟶ SO_2Ph ⟵⟶ – + (10)

α-Nitroalkenes have been found to be equivalent to allyl nitro compounds upon treatment with Et_3N and a Pd^0 catalyst in DMF.[87,88] The significantly greater availability of the α-nitroalkenes over allyl nitro compounds makes this method of forming π-allyl complexes attractive.

A particularly important advance in the nature of the leaving group, X, used in π-allylpalladium precursors, has been the development of substrates that allow functionalization of the incipient allyl complex under neutral conditions. With these precursors, the leaving group, X, is, or can react further to become, sufficiently basic so as to generate *in situ* Nu^- from Nu—H. In another variation, the X itself is transformed into the nucleophilic partner for the allyl ligand. Examples include vinyl epoxides,[7,9–11] allyl esters of acetoacetate,[7–9,11] allyl carbonates[8,9,11] and carbamates (equation 11).[8,9,11]

O + Pd^0 ⟶ O^- Pd^+ L L $\xrightarrow{\text{Nu-H}}$

OH Pd^+ L L + Nu^- ⟶ Pd^0 + Nu OH (11)

Although allyl ethers are generally not good actors in the π-allylpalladium reaction, the added relief of ring strain associated with epoxide opening allows vinyl epoxides to become excellent substrates for palladium-mediated processes.[7,9–11] The initially formed alkoxymethylallylbisphosphine cation is capable of deprotonating dimethyl malonate, for example, and then is sufficiently electrophilic to undergo addition by malonate anion on the allyl ligand. The interesting functionality associated with this transformation and the ability to carry it out under neutral conditions has been widely exploited, as reported in the following sections.

Allyl carbonates[8,9,11] serve as excellent substrates with Pd^0 catalysts. In this process, the initially formed π-allyl-PdL_2^+ RO—CO_2^- undergoes decarboxylation to provide π-allyl-PdL_2^+ RO^-. The alkoxy group is sufficiently basic to deprotonate *in situ* a number of commonly used Nu—H species. Allyl carbamates react similarly to generate R_2N^- bases *in situ* (equation 12).

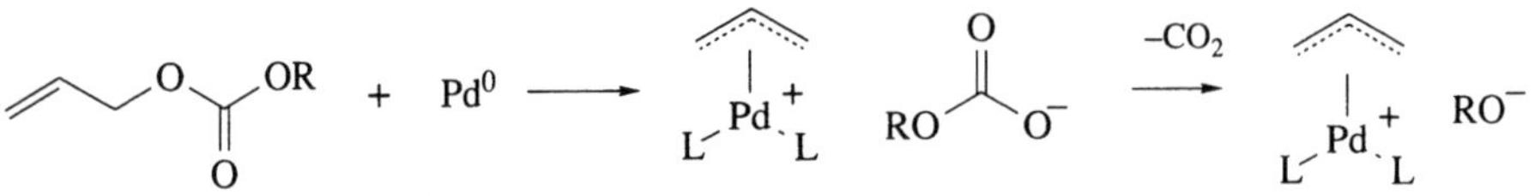

(12)

Allyl esters of acetoacetates[7,8,9,11] react with Pd^0 catalysts to generate initially a bisphosphine allylpalladium cation, with the β-ketocarboxylate serving as counterion. Under the reaction conditions the β-ketocarboxylate decarboxylates, yielding a π-allylpalladium ketone enolate complex. The required nucleophile is thus formed *in situ* and is capable of Pd-mediated alkylation. A wide spectrum of reactions have been based on this chemistry which will be discussed in later sections.

Allyl borates,[89] imidates,[90,91] alcohols,[77,92] arsenites[93] and *O*-allylisourea[94,95] have also been found to function as leaving groups that allow neutral functionalization by similar mechanisms, although the chemistry of these compounds has not been widely explored.

A final example of this class of allyl–X compounds is a vinylcyclopropane that is geminally substituted on the three-membered ring with electron withdrawing groups. Upon addition of a Pd^0 complex, the allylic C—C bond ruptures, with the stabilized carbanion serving as a leaving group (equation 13).[96] The complex can be independently reacted with electrophiles at the carbanion as well as with nucleophiles on the allyl moiety.

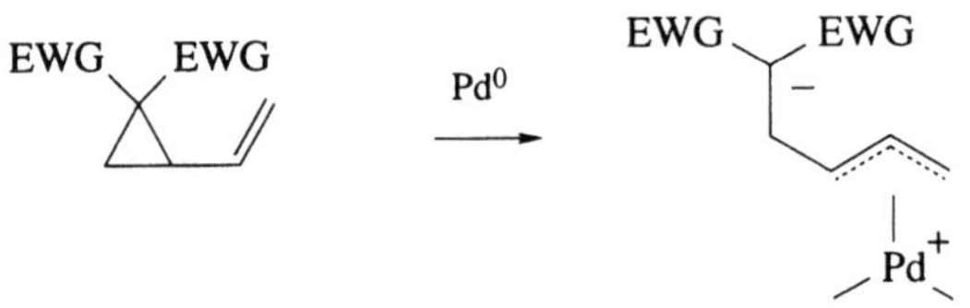

(13)

3.3.2.2 Range of Nucleophiles Employed in Palladium-catalyzed Allylic Alkylation

The reactivity of π-allylpalladium–phosphine complexes generated stoichiometrically or from alkenes allylically substituted with a leaving group, is essentially identical and, as a result, allyl species will be generally considered in this section without distinction as to the origin of the palladium complex.

3.3.2.2.1 Carbon nucleophiles

(i) Soft nucleophiles

The first examples demonstrating the electrophilicity of π-allyl ligands complexed to Pd involved stabilized 'soft' carbanions, namely the anion of diethyl malonate (equation 14).[97,98] Much of the early work that described the use of carbon nucleophiles in π-allylpalladium chemistry focused on such 'soft' carbanions, which can be generically defined as RCXY, where R = H, alkyl or aryl, and X and Y could be selected from the following list of functionalities known to stabilize adjacent carbanions so as to achieve an acidity of the conjugate acid in the pK_a range of 10–20:[1–16] —CO_2R^1, —C(O)R^1, —SR^1, —S(O)R^1, —SO_2R^1, —NO_2, —CN. Also included in this category are enamines,[1,99,100] the carbanion derived from nitroalkanes[101–104] and the cyclopentadiene anion.[105] Equations (15–18) show examples of their use.

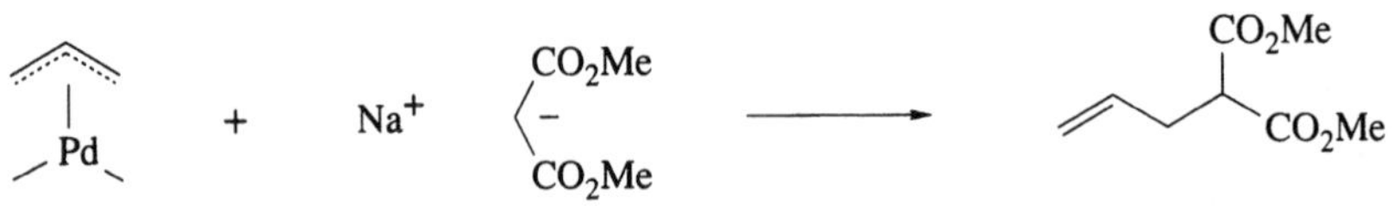

(14)

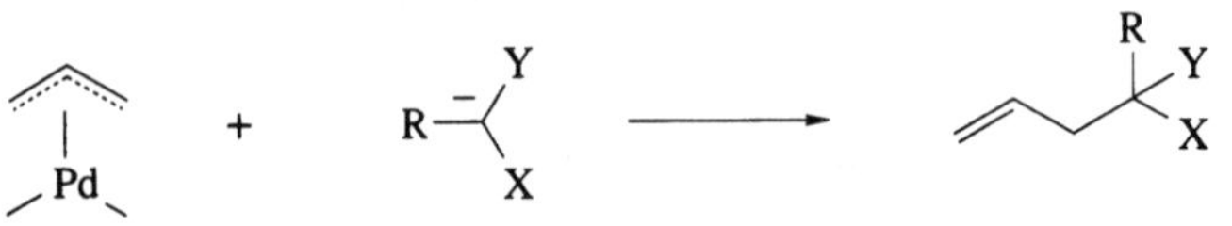

(15)

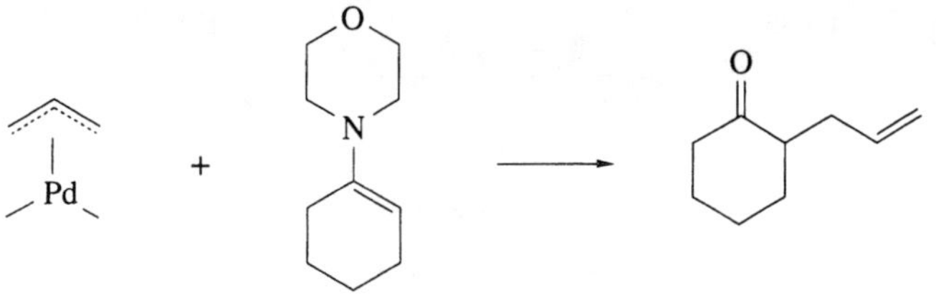

(16)

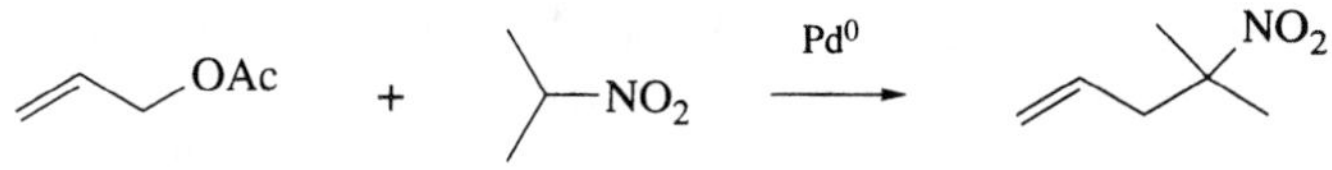

$\xrightarrow{Pd^0}$ (17)

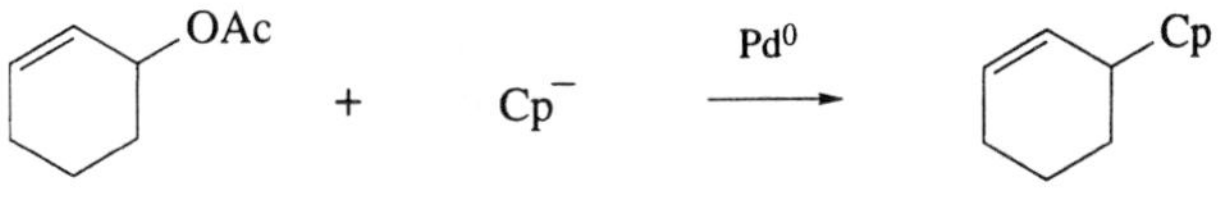

$\xrightarrow{Pd^0}$ (18)

(ii) Enolates

Initial reports on the use of simple enolates as nucleophiles in π-allylpalladium chemistry met with only limited success.[77,106] The enolate of acetophenone reacted with allyl acetate in the presence of $Pd(PPh_3)_4$, but gave predominantly dialkylated product.[106] The use of the enol silyl ether of acetophenone gave only monoalkylated product with allyl acetate and Pd^0 catalysis, but substituted allyl acetates did not function in this reaction.[106] Enol stannanes, however, have been found to give monoalkylated products with a wide variety of allyl acetates (equation 19).[106] *In situ* generation of enol stannanes from lithium enolates and trialkylstannyl trifluoroacetates followed by Pd^0-catalyzed allylation has been demonstrated.[107]

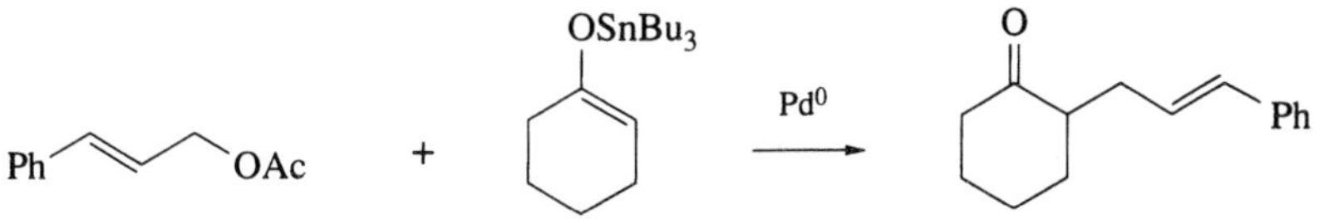

(19)

Similarly, the addition of triethylborane to lithium enolates allowed ready reaction with allyl nitro compounds catalyzed by palladium(0) complexes.[108,109]

The use of bis(dibenzylideneacetonato)palladium and 1,2-bis(diphenylphosphino)ethane as catalyst was reported to allow efficient monoalkylation of lithium enolates of ketones, with allyl acetates.[110] $Pd(PPh_3)_4$ and allylammonium salts have also been reacted successfully with lithium enolates.[80]

The addition of potassium enolates to preformed π-allylpalladium complexes has also been reported.[111]

α-Substituted ketones have also been prepared by the reaction of metallated ketimines with allyl–X compounds catalyzed by $Pd(acac)_2$–PPh_3.[79]

The *in situ* generation and alkylation of palladium enolates has gained considerable recent attention. A variety of precursors have been utilized to access the Pd enolate. The first reported use of this intermediate employed an allyl β-ketocarboxylate ester (**4**).[112,113] The suggested mechanism involves oxidative addition into the allylic C—O bond by Pd to yield an allylpalladium β-ketocarboxylate, which subsequently decarboxylates to give an allylpalladium enolate. The enolate then reacts to given an allylated ketone and Pd^0 (equation 20). A number of reports on this reaction have appeared,[114–116] including two recent reviews.[8,9,117] Allyl esters substituted in the α-position with esters, nitriles and nitro groups also function in this reaction (equation 21).[118]

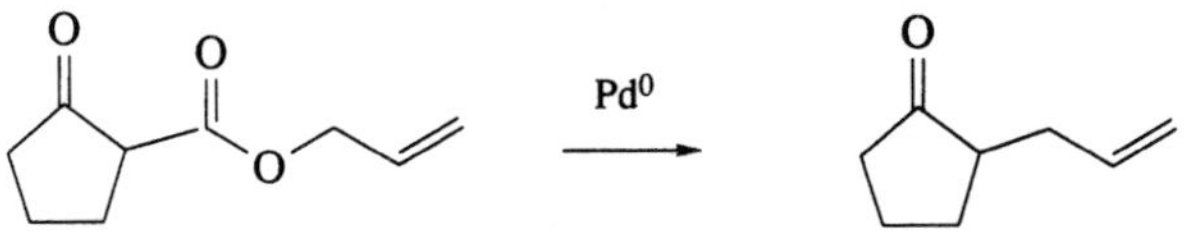

(20)

(4)

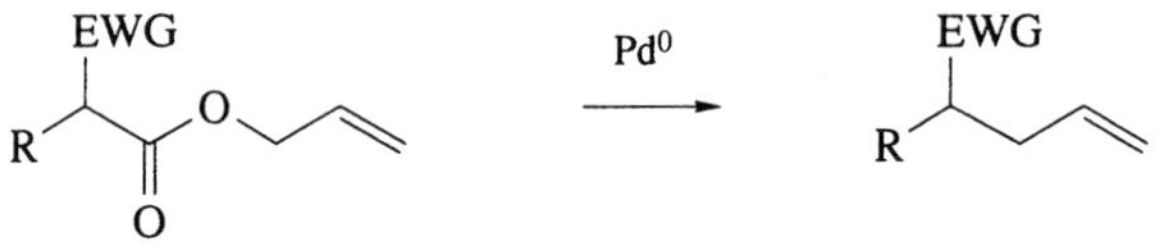

(21)

EWG = CO_2R', CN, NO_2

Considerable use has also been made of allyl carbonates as substrates for the allylation of Pd enolates.[9] The reaction of Pd^0 complexes with allyl enol carbonates[119,120] proceeds by initial oxidative addition into the allylic C—O bond of the carbonate followed by decarboxylation, yielding an allylpalladium enolate, which subsequently produces Pd^0 and the allylated ketone (equation 22). In like fashion, except now in an intermolecular sense, allyl carbonates have been found to allylate enol silyl ethers (equation 23),[121] enol acetates (with $MeOSnBu_3$ as cocatalyst) (equation 24),[122] ketene silyl acetals (equation 25)[123] and anions α to nitro, cyano, sulfonyl and keto groups.[115,124] In these cases, the alkoxy moiety liberated from the carbonate on decarboxylation serves as the key reagent in generating the Pd enolate.

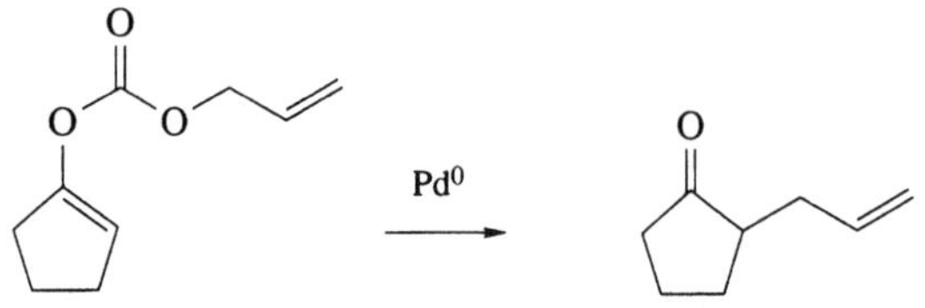

(22)

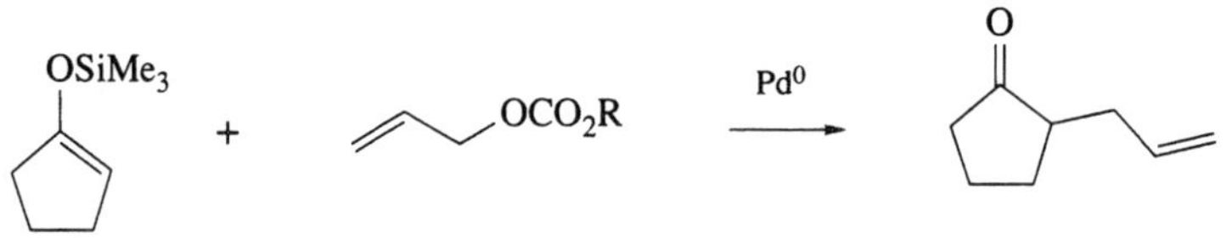

(23)

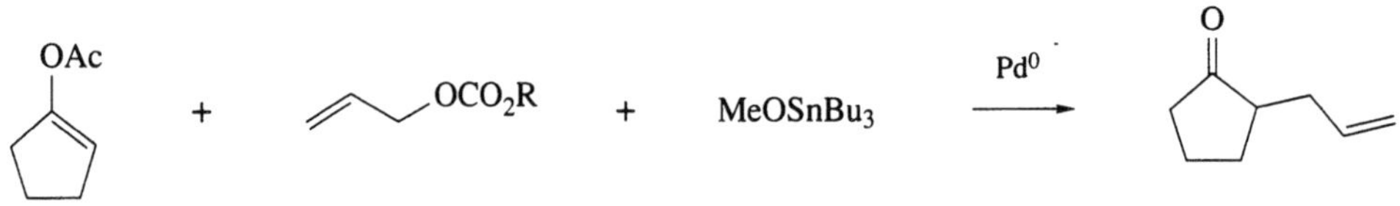

(24)

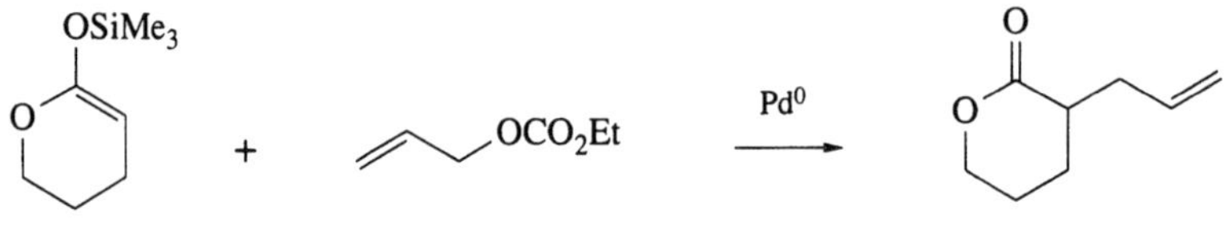

(25)

Allyl carbamates (equation 26)[125] and isoureas (equation 27)[95] generate allylpalladium enolates by virtually identical mechanisms.

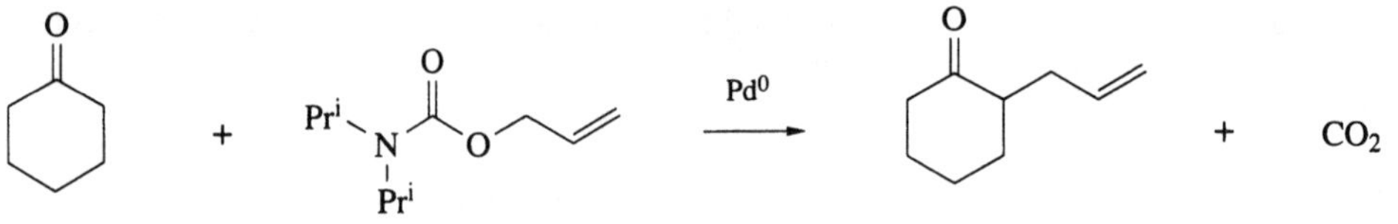

(26)

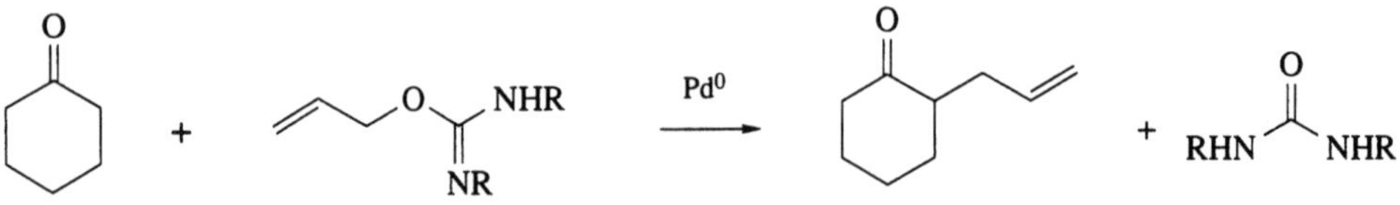

(27)

β-Ketoacids[126,127] form the same intermediates as the allyl β-ketoesters by nucleophilic addition of the carboxylate to a π-allylpalladium complex. Decarboxylation generates the allylpalladium enolate, which again yields Pd^0 and allylated ketone. Enol silyl ethers have also been employed with allyl arsenites[93] to provide allylated ketones.

The addition of simple ester or ketoenolates to π-allylpalladium complexes may constitute the second step of an ingenious [3 + 2] cycloaddition reaction. One substrate that undergoes this process is 2-(trimethylsilylmethyl)allyl acetate (**5**). The mechanism proposed involves initial formation of a 2-(trimethylsilylmethyl)allylpalladium cation followed by desilylation by the acetate liberated in the oxidative addition (Scheme 1). The dipolar intermediate can be envisioned as an η^3-trimethylenemethane–PdL_2 species (**6**) or, less likely, an η^4-complex (**7**).

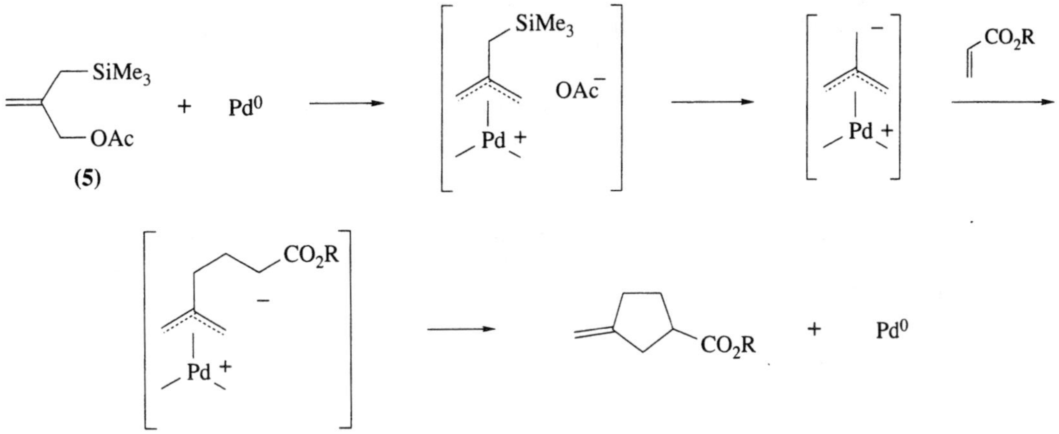

Scheme 1

(6) **(7)**

The addition of the trimethylenemethane–palladium complex to alkenes may proceed by a concerted process or *via* a stepwise mechanism in which the anion of the 1,3-dipole attacks Michael-fashion to generate an intermediate anion which collapses to form a five-membered ring by attack on the allylpalladium complex. This [3 + 2] cycloaddition reaction has been reviewed.[128] A number of additional reports of its use have appeared recently.[129–134]

Ethyl-2-(sulfonylmethyl)- and 2-(cyanomethyl)-allyl carbonates[135] as well as (methoxycarbonyl)methylallyl carbonates[136] serve as substrates for the [3 + 2] cycloaddition. Oxidative addition into the allylic C—O bond of the carbonate, followed by decarboxylation, gives a 2-substituted allylpalladium alkoxide. The alkoxide then deprotonates the C—H α to the electron-withdrawing substituent at the 2-position of the allyl. This anion then undergoes a Michael addition to an α,β-unsaturated ketone or ester, followed by intramolecular allylation of the anion of the Michael product (Scheme 2).

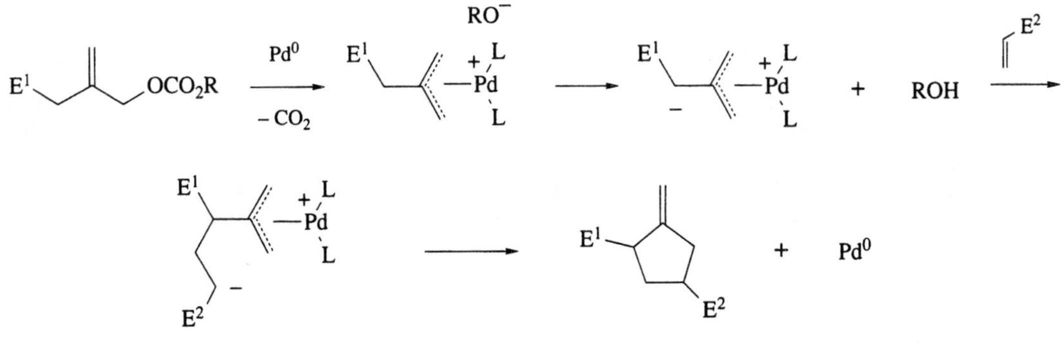

Scheme 2

Vinylcyclopropanes having two electron-withdrawing groups geminally substituted on the three-membered ring also have been found to serve as 1,3-dipolar equivalents capable of [3 + 2] cycloadditions.

The enolates of esters or ketones again are envisioned to add to π-allylpalladium intermediates in the second step of the proposed mechanism (Scheme 3).[96]

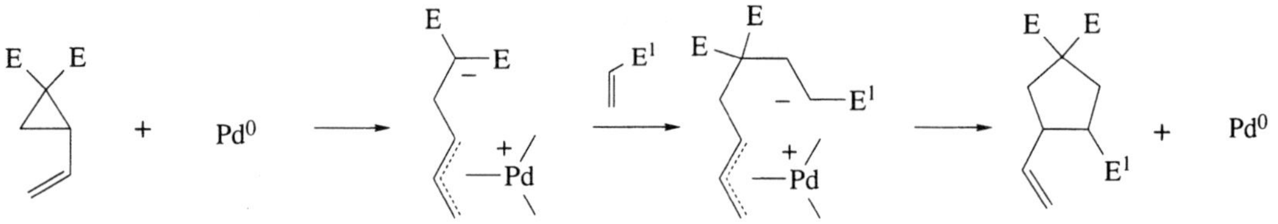

Scheme 3

(iii) Organotin nucleophiles

The range of nucleophiles that function in π-allylpalladium alkylations encompasses a variety of organotin reagents, including allyl,[137–140] vinyl,[141–143] aryl,[141–144] acetylenyl[144] and heteroaromatic[144–146] stannanes (equations 28–31).

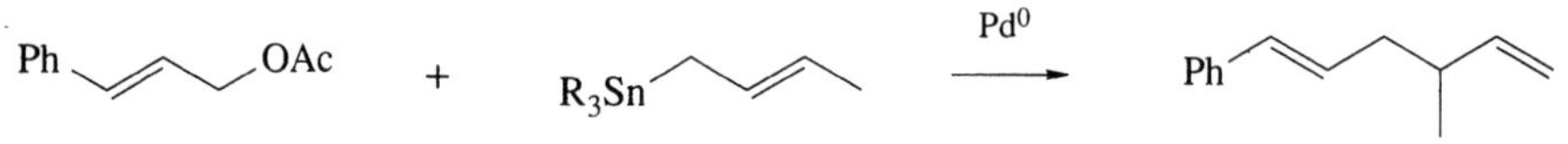

(28)

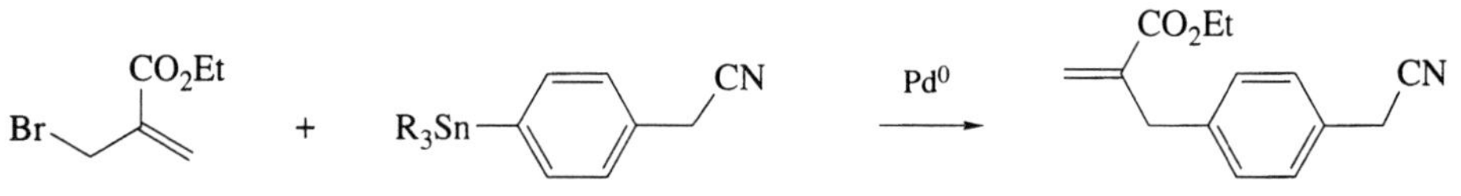

(29)

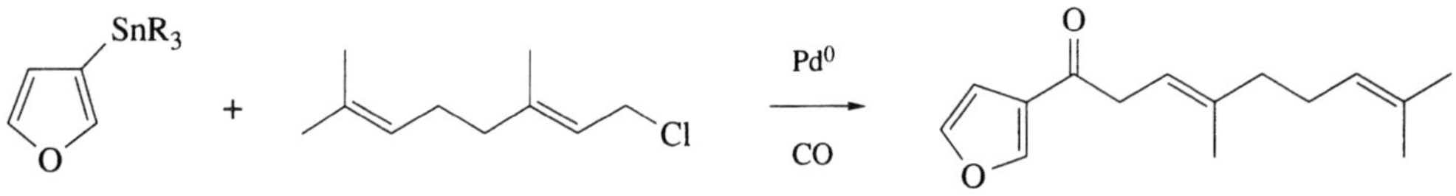

(30)

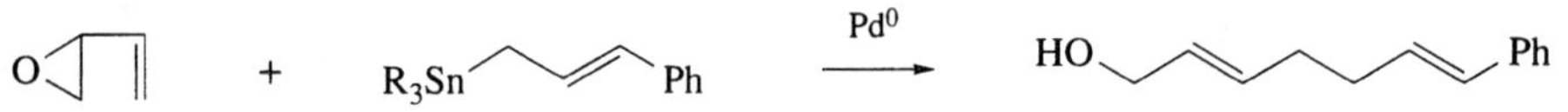

(31)

In the palladium-catalyzed addition of allylstannanes to allyl acetates, complete allyl inversion of the allylstannane is observed.[137] Symmetrical coupling is also possible, for example, by treatment of cinnamyl acetate with hexa-*n*-butyldistannane.[137] In this reaction, cinnamyl tri-*n*-butylstannane is generated *in situ* by Pd catalysis and it then couples with the allyl acetate with allyl inversion (equation 32).

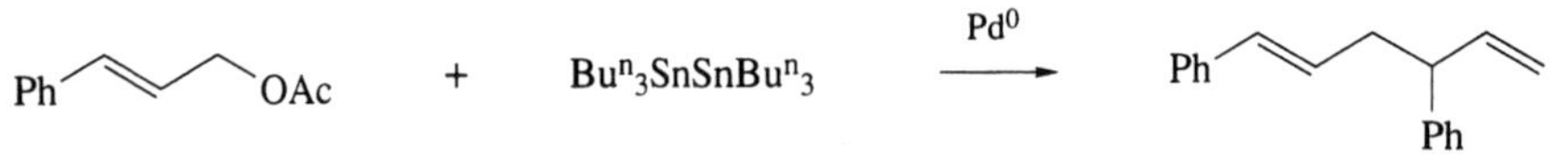

(32)

The mechanism of coupling of allyltin and allylpalladium can be altered by the addition of maleic anhydride to the reaction mixture (*vide infra*).[139] In this modified procedure, reaction of the allylpalladium chloride dimer with allyl-$Bu^n{}_3Sn$ now gives coupling without allyl inversion of the allyltin reagent.

If the coupling of the allyl halide and tin reagent is run under a CO atmosphere, an allyl R ketone is generated by insertion of CO into the allylpalladium complex, followed by coupling (equation 33).[145–147]

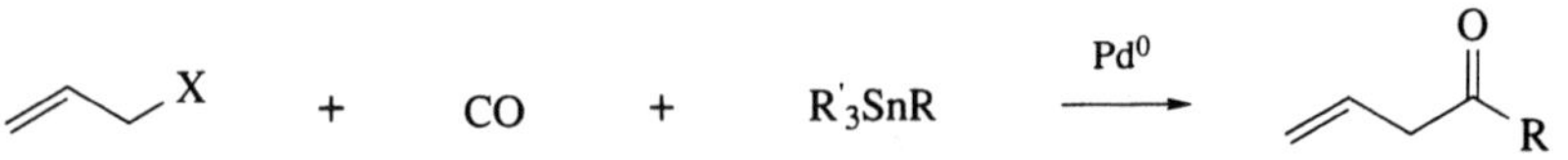

(33)

(iv) Organo-thallium and -zinc reagents

The π-allyl-σ-arylpalladium–phosphine complexes have been of particular interest in the study of the mechanism of their reductive elimination to yield allylated aromatics.[148,149] Their preparation can be achieved by the addition of a triarylthallium or arylzinc halide to the corresponding allylpalladium phosphino halides (equation 34).

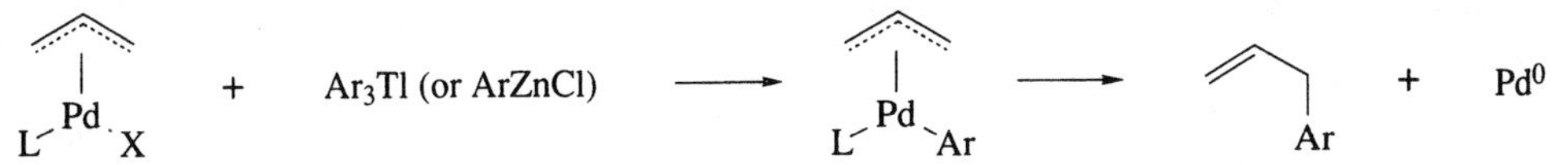

(34)

A variety of additional organozinc-mediated couplings to π-allylpalladium complexes have been reported, including those of perfluoroalkylzinc iodides,[150] vinylzinc bromides,[151] allylzinc acetates[152] and arylzinc bromides (equations 35–37).[153,154]

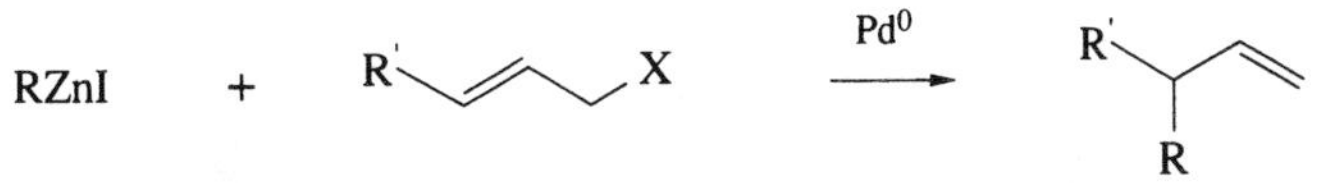

(35)

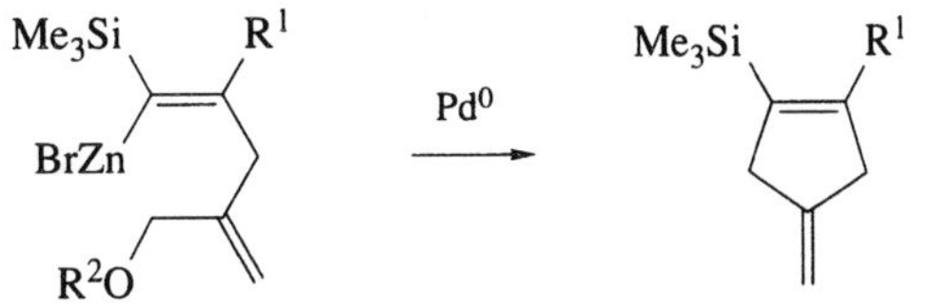

(36)

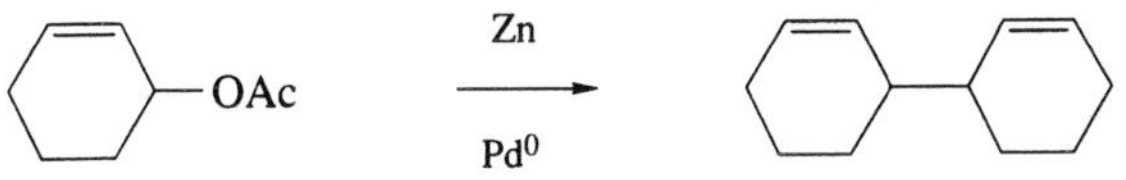

(37)

(v) Organo-aluminum and -zirconium reagents

Alkenyl-aluminum and -zirconium derivatives have been found to couple with allyl halides in the presence of Pd^0 catalysts (equation 38), although simple alkyl-aluminum and -zirconium reagents fail in the reaction.[154] The 1,4-dienes thus generated are important intermediates in organic synthesis.

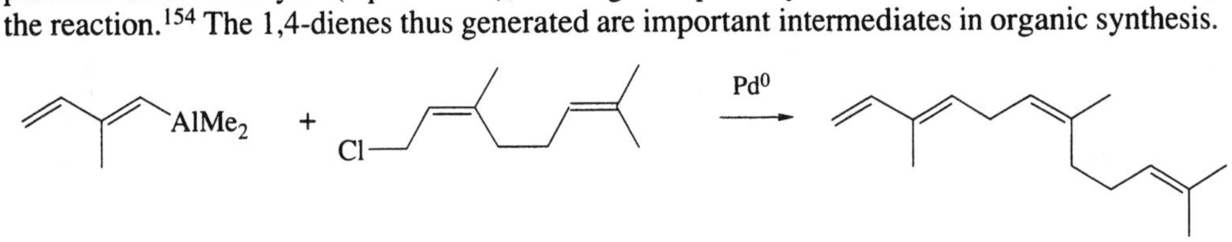

(38)

The coupling of alkenylzirconiums with π-allyl complexes derived from the D-ring of steroids has been demonstrated to be regiospecific and to yield products which possess the natural configuration at C-20 (equation 39).[155]

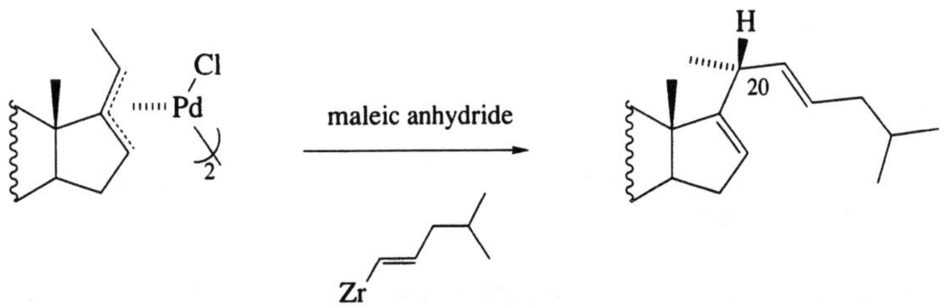

(39)

The effects on coupling efficiency and regiochemical control in nonsymmetrical allyl complexes as a function of added ligand in these reactions has been determined[155,156] (*vide infra*) and applied in the synthesis of flexibilene and humulene.[157]

(vi) Grignard and organolithium reagents

Reaction with even harder nucleophiles such as organolithiums and Grignard reagents is substantially limited by virtue of the fact that these carbon nucleophiles add by direct attack at the metal center, as opposed to the softer carbon nucleophiles which add by attack on the allyl ligand. Direct metal addition can lead to the opening of alternative reaction pathways, *e.g.* β-H elimination, in competition with reductive elimination which accomplishes nucleophile allylation (equation 40).

(40)

Successful use of such nucleophiles has been limited to those that are incapable of undergoing these competing reactions, namely, methyl and phenyl organometallics (equations 41–44).[83,158–162]

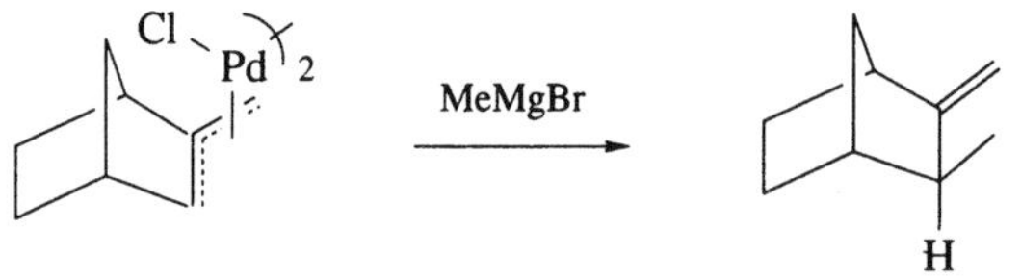

(41)

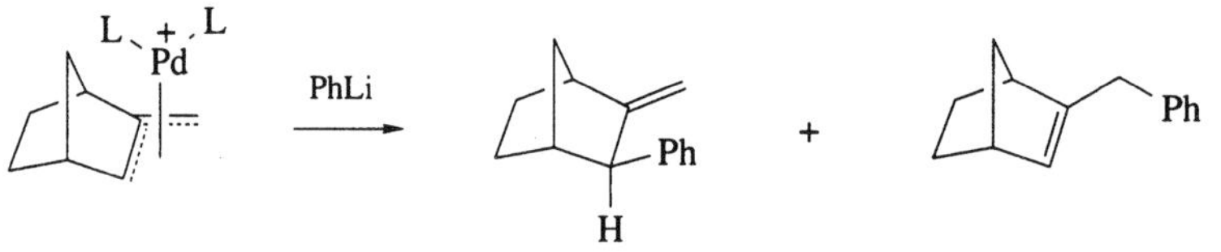

(42)

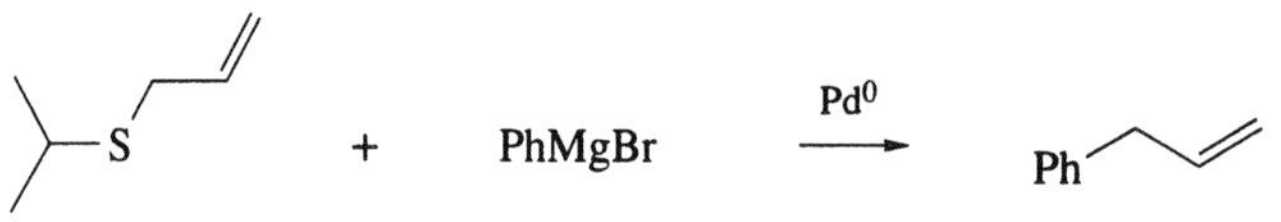

(43)

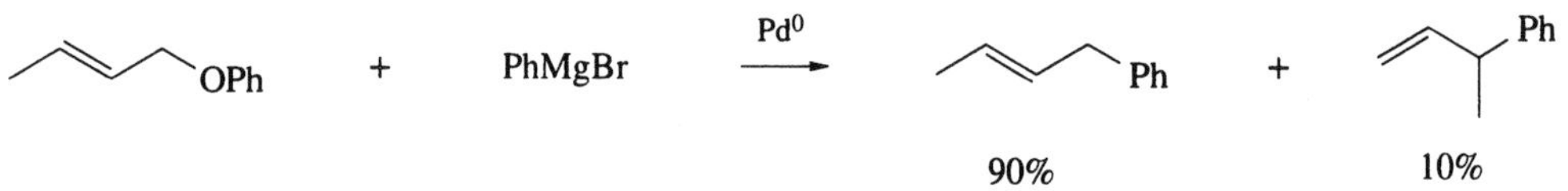

(44)

An interesting reaction in which the initial addition product can be isolated occurs with diallylpalladium and organolithiums (equation 45).[161]

(45)

3.3.2.2.2 Heteroatom nucleophiles

(i) Oxygen nucleophiles

Oxygen nucleophiles can be added successfully to π-allylpalladium complexes by various methodologies. Those additions that are fostered by the use of classical oxidizing agents will be treated separately in Section 3.3.2.3.2.

The most intense interest in the addition of oxygen nucleophiles to π-allylpalladium complexes has centered on the delivery of OAc^-. For example, in allylpalladium–OAc complexes, acetate can be induced to migrate to the allyl ligand by the addition of CO (equation 46).[162–164] Rearrangements and isomerizations of allyl acetates can also be readily accomplished *via* Pd catalysis (equations 47 and 48).[165–167]

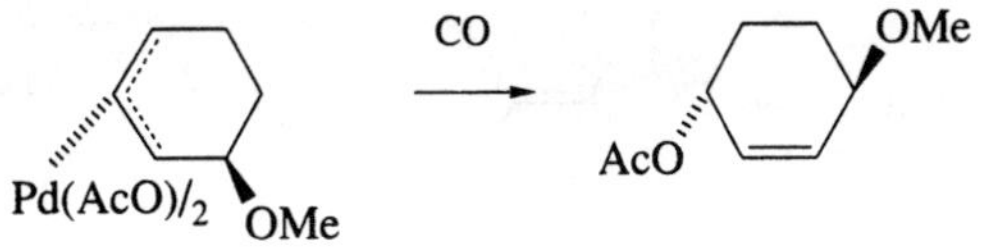

(46)

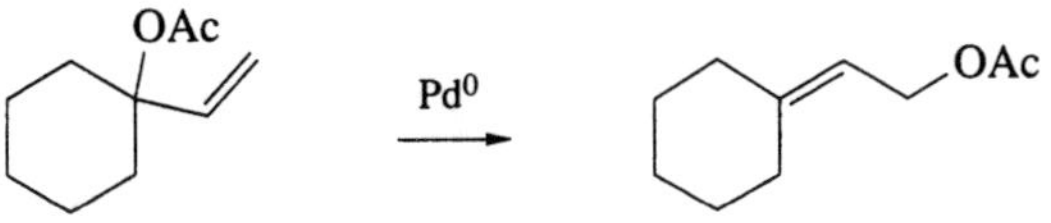

(47)

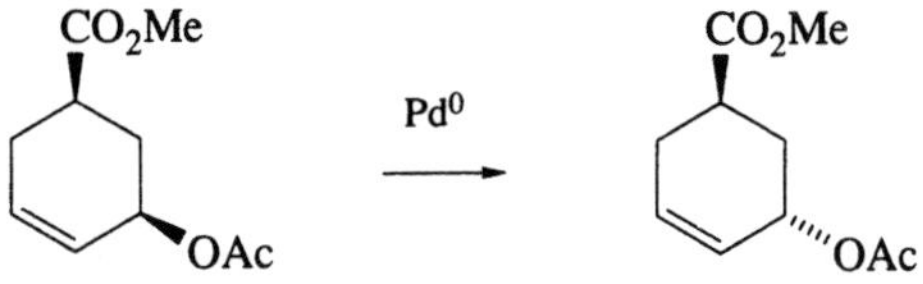

(48)

Stereocontrol (*vide infra*) can be efficiently engineered in the addition of acetate to π-allylpalladium complexes (equation 49).[168–171] The regioselective addition of acetate of allylpalladium complexes to prepare γ-acetoxy-(*E*)-α,β-unsaturated esters[172] and γ-acetoxy-α,β-unsaturated sulfones has also been achieved (equations 50 and 51).[173,174]

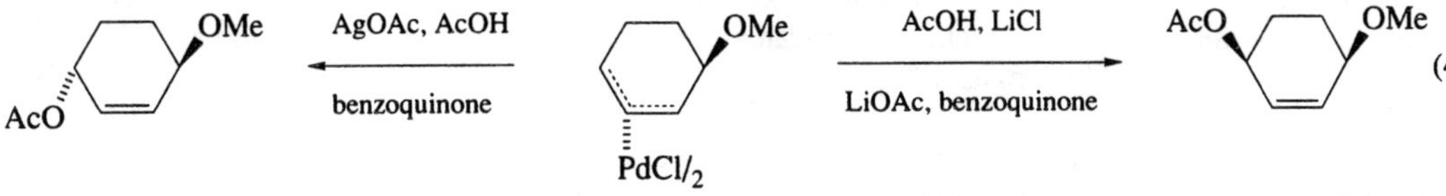

(49)

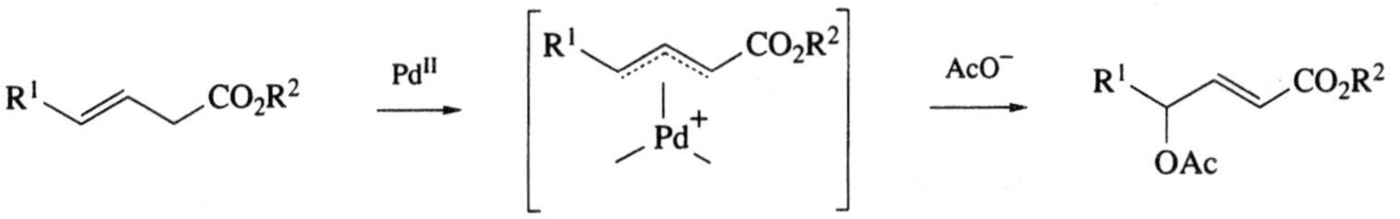

(50)

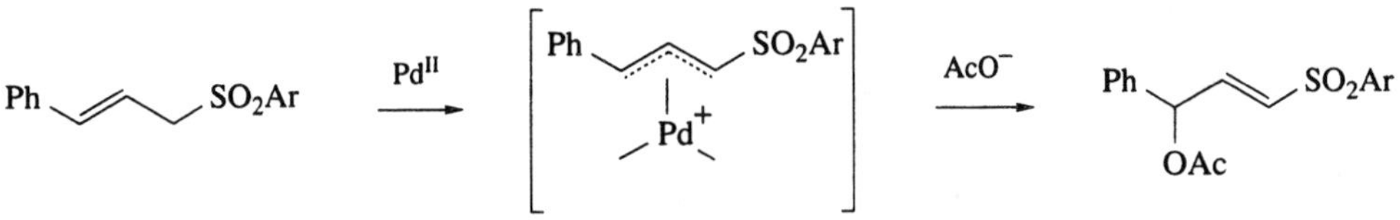

(51)

Other synthetically useful additions of carboxylate anions to π-allylpalladium complexes have been demonstrated, as illustrated in equations (52–54).[175–179] Harder oxygen nucleophiles, such as phenoxides[175,176] and alkoxides,[175–180] can also be added to π-allylpalladium complexes (equation 55). Finally, the addition of NO_2 to allylpalladium chloride dimer results in the production of allyl nitrite (equation 56).[181]

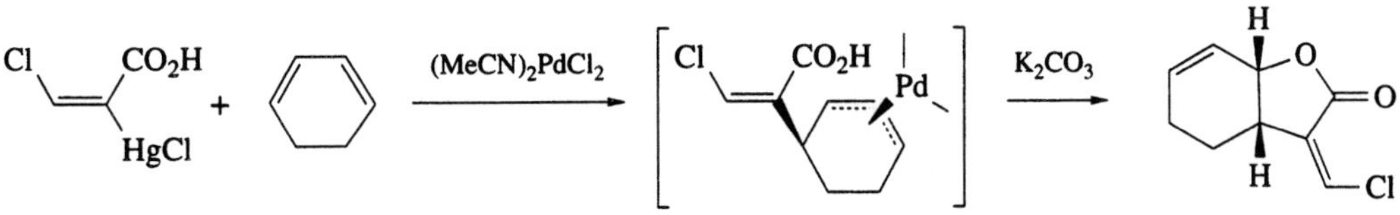

(52)

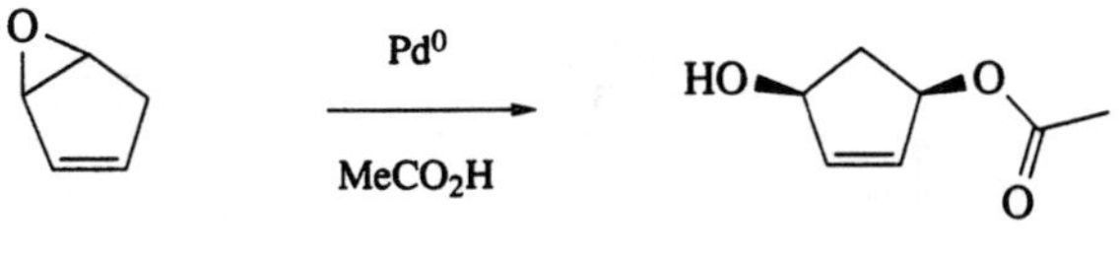

(53)

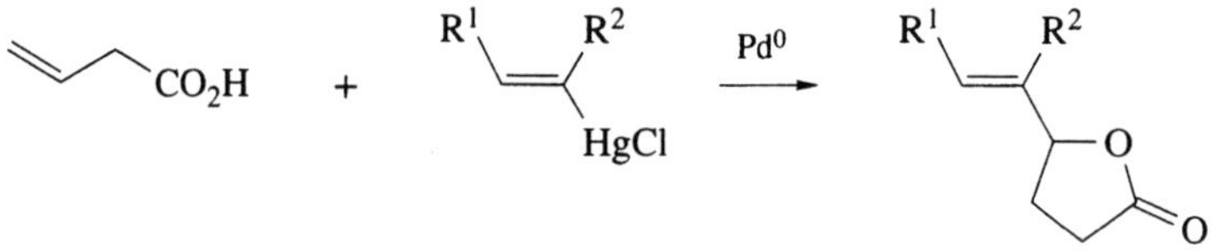

(54)

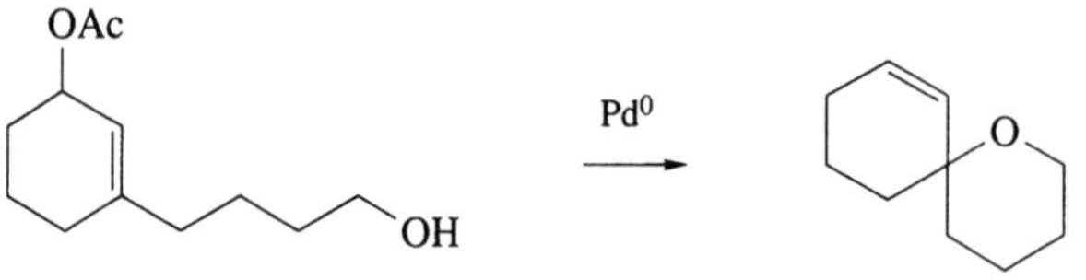

(55)

(56)

(ii) Nitrogen nucleophiles

Ammonia fails to act as an effective nucleophile for π-allylpalladium complexes. Some examples of primary amines[81,181–183] have been reported, although diallylation is often a problem in these reactions. Secondary amines are, however, excellent substrates for palladium-catalyzed allylation.[81,84,181,183–195] Amides,[176,196] sulfonamides,[196,197] azides[198] and magnesium amides[199] have also been shown to be effective nucleophiles.

Due to the failure of ammonia in this reaction, a variety of alternative strategies to prepare primary alkyl amines utilizing Pd catalysis have been investigated.[196–198,200] 4,4′-Dimethoxybenzhydrylamine (DMB; **8**) reacts with allyl acetates under Pd^0 catalysis and can be debenzylated with formic acid to yield a primary amine (equation 57).[200] Similarly, *p*-toluenesulfonamides[196,197] and azides can be conveniently converted to amines following their participation in the Pd-catalyzed alkylation.

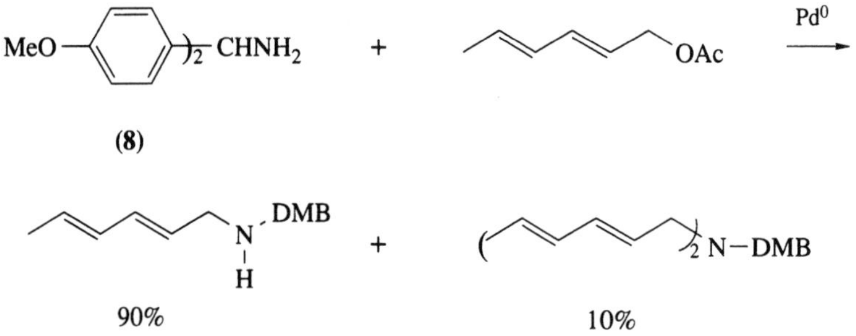

(57)

Some examples of the use of nitrogen nucleophiles as the key step in organic syntheses include the preparation of pyrroles (equation 58),[201,202] a number of alkaloids[203-207] including ibogamine,[204] catharanthine[205,206] and perhydrohistrionicotoxin,[207] amino sugars (equation 59),[208] 5-amino-1,3-pentadienes[209] and 1-azaspirocycles.[210]

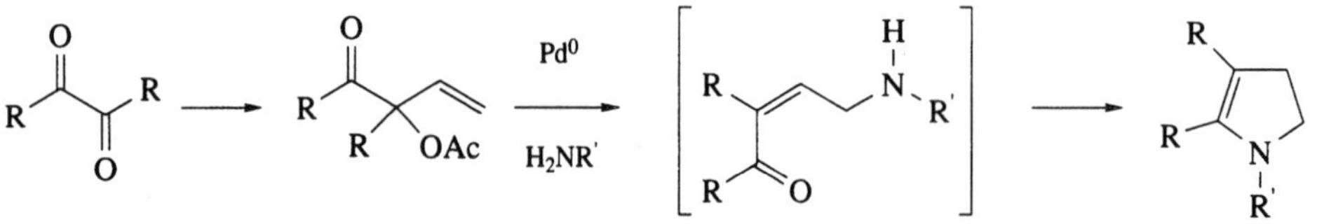

(58)

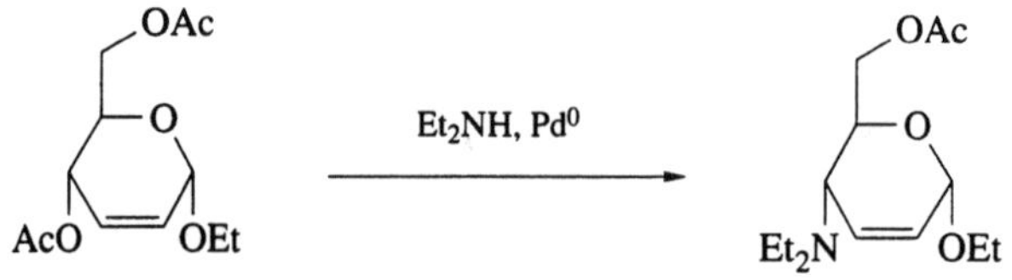

(59)

(iii) Other heteroatom nucleophiles

Sulfur nucleophiles have not been widely used in π-allylpalladium chemistry, principally due to their ability to interfere with the catalyst by coordination. This is a less serious concern for sulfinic acid and sulfinate salts than with sulfides, and as a result, several reports have appeared using RSO_2H and RSO_2^-.[211–219] Alkenes allylically functionalized with a leaving group, dienes[214] and vinyl nitro compounds[218,219] have all served as precursors for the required π-allyl species in these reactions (equations 60 and 61).

AcO–CH₂CH=CHCH₂–OAc → ($PhSO_2H$, Pd^0) $PhSO_2$–CH₂CH=CHCH₂–SO_2Ph (60)

1-nitrocyclohexene (NO_2) → ($PhSO_2Na$, Pd^0) cyclohexenyl–SO_2Ph (61)

Pd^0 has also been found to catalyze thiono–thiolo allylic rearrangement of *O*-allyl phosphoro- and phosphono-thionates *via* a π-allyl intermediate (equation 62).[220]

$(EtO)_2P(=S)$–O–allyl → (Pd^0) $(EtO)_2P(=O)$–S–allyl + $(EtO)_2P(=O)$–S–allyl (62)

The use of alkyl sulfides as nucleophiles has been realized by the employment of *O*-allyl *S*-alkylthiocarbonates.[221,222] In this process, oxidative addition by Pd^0 is followed by the liberation of COS and generation, *in situ*, of RS^-, which subsequently adds to the allylpalladium complex (equation 63).

allyl–O–C(=S)–SR → (Pd^0, –COS) π-allyl–Pd^+ RS^- → allyl–SR (63)

Trimethylsilyl alkyl and aryl sulfides were found to function as latent sources of a sulfide nucleophile when used in conjunction with allyl carbonates or vinyl epoxides with Pd catalysts (equation 64).[223]

geranyl–O–C(=O)–OMe → ($PhSSiMe_3$, Pd^0) geranyl–SPh (64)

The addition of phosphites to allyl halides catalyzed by Pd^0 has been found to yield alkyl phosphonates (equation 65).[215]

AcO–CH₂CH=CHCH₂–Cl + $P(OEt)_3$ + Pd^0 → AcO–CH₂CH=CHCH₂–P(=O)$(OEt)_2$ (65)

An additional example utilizing phosphorus nucleophiles employs lithium diphenylthiophosphides and allyl carboxylates and yields allylic diphenylphosphine sulfides (equation 66).[224]

cyclohexenyl–OAc + Li–P(=S)Ph₂ → (Pd^0) cyclohexenyl–P(=S)Ph₂ (66)

Palladium(0) catalysts can also function to transfer trimethylsilyl groups from aluminum to carbon *via* π-allyl intermediates (equation 67).[225]

Br, OAc + Me_3SiAl $\xrightarrow{Pd^0}$ Br, $SiMe_3$ (67)

(iv) Transition metal nucleophiles

Acyliron anionic complexes, such as, $MeCOFe(CO)_4^- Li^+$, and anionic acylnickel carbonyl complexes, $RCONi(CO)_x^- M^+$, react with allylpalladium complexes to give unsaturated ketones (equations 68 and 69).[226]

(68)

(69)

$NaCo(CO)_4$ was found to carboalkoxylate allylic acetates with Pd^0 catalysts under a CO atmosphere (equation 70).[226]

(70)

3.3.2.3 Other Reactions of π-Allylpalladium Complexes

3.3.2.3.1 Insertion into π-allylpalladium complexes

The insertion of CO into π-allylpalladium complexes has been used to prepare various β,γ-unsaturated carboxylic acid derivatives. The process can be initiated from preformed allyl complexes,[227–235] catalytically from appropriate π-allyl precursors[228,229,236,237] or from diene–Pd complexes.[238,239] Particularly mild conditions, involving low pressures and temperatures, to accomplish this insertion have been discovered.[235] The key reagent in allowing these mild conditions to be employed was found to be carboxylate anions.[235] For unsymmetrical allyl complexes, CO insertion occurs preferentially at the less-substituted allyl terminus (equations 71–73).

Isocyanides react similarly with π-allylpalladium complexes to generate β,γ-unsaturated iminoethers.[240–243] Comparable to the CO reaction, isocyanide insertion occurs preferentially in the less

(71)

(72)

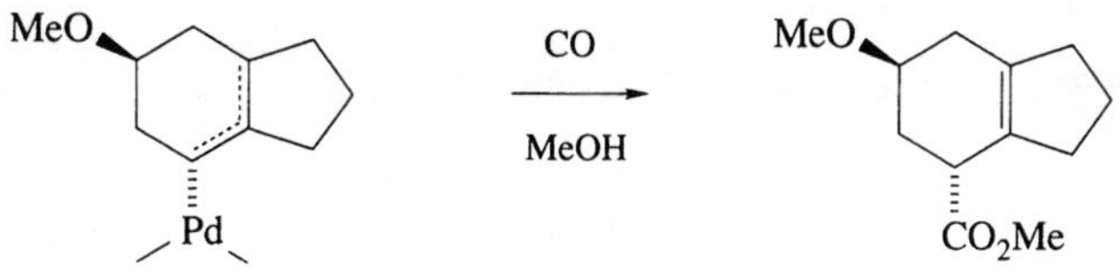

(73)

substituted allyl terminus (equation 74). CO_2[243–245] and SO_2[245] also insert efficiently into π-allylpalladium complexes (equations 75 and 76).

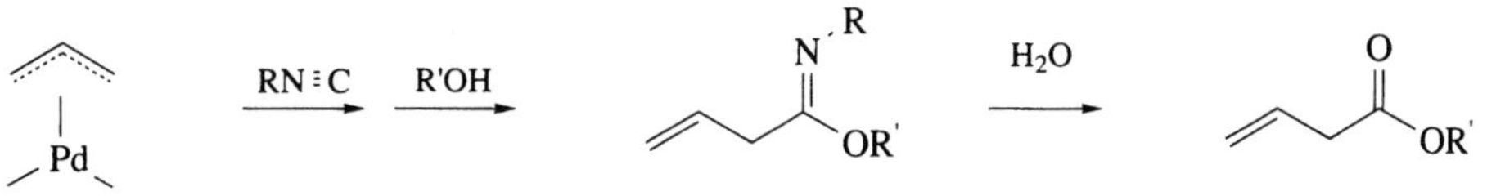

(74)

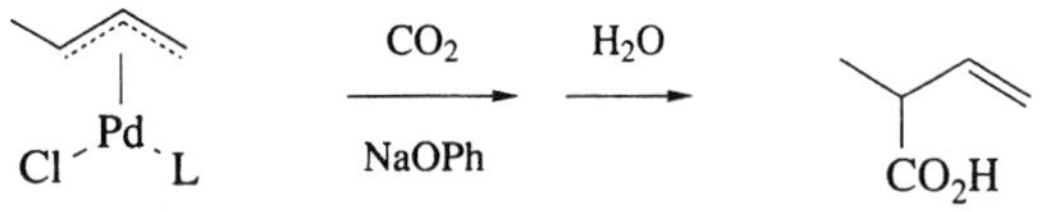

(75)

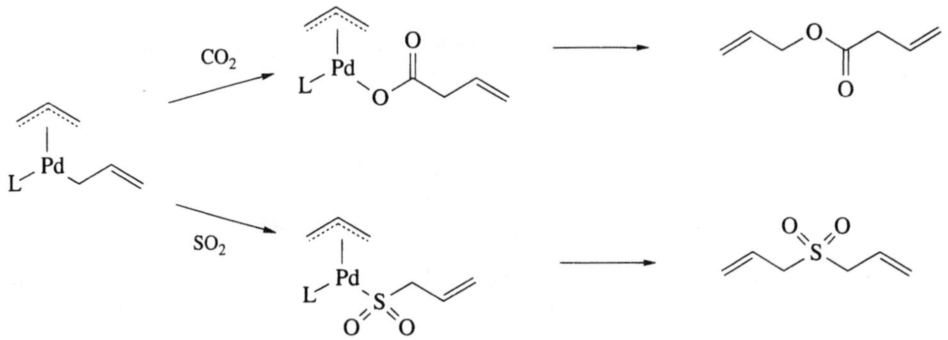

(76)

The insertion of 1,3-dienes into a π-allylpalladium complex is believed to proceed *via* an intermediate in which the metal is complexed to the less hindered double bond of an unsymmetrical diene, followed by an electrocyclic rearrangement which links the more substituted allyl terminus with the more substituted alkene (equation 77).[246–251] Electron-withdrawing substituents on the π-allyl fragment generally increase the rate of insertion,[248] whereas substituents on the diene generally slow the rate.[268]

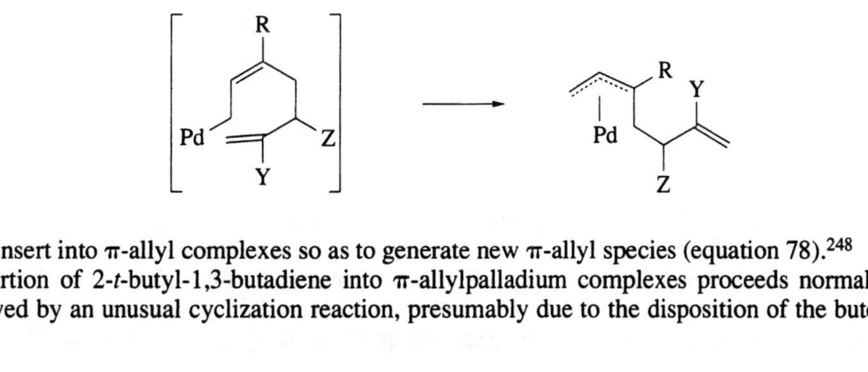

(77)

Allenes insert into π-allyl complexes so as to generate new π-allyl species (equation 78).[248]

The insertion of 2-*t*-butyl-1,3-butadiene into π-allylpalladium complexes proceeds normally, but is then followed by an unusual cyclization reaction, presumably due to the disposition of the butenyl frag-

(78)

ment in the *anti* configuration, caused by the preferential positioning of the *t*-butyl in the *syn* site (equation 79).[251]

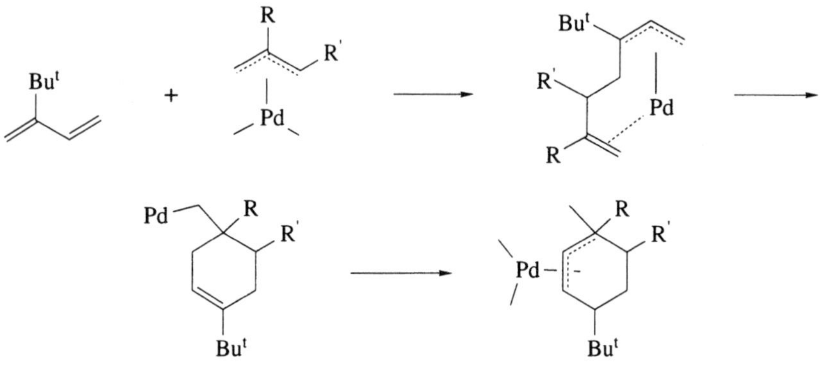

(79)

α-Dienyl-ω-allyl acetates cyclize by isomerization. The key step in this process entails insertion of a 1,3-diene into a π-allyl fragment (equation 80).[252]

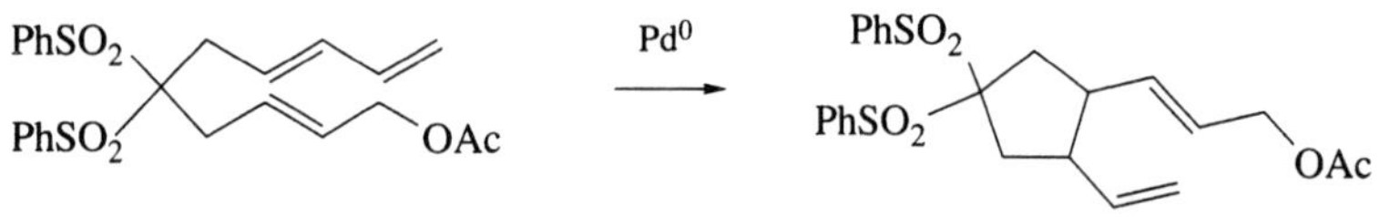

(80)

Strained alkenes such as norbornene, norbornadiene and bicyclo[2.2.2]octenes readily insert into π-allyl complexes with the less substituted allyl terminus linked to the alkene (equation 81).[253–261] This reaction has been utilized to prepare interesting prostaglandin analogs.[253,254]

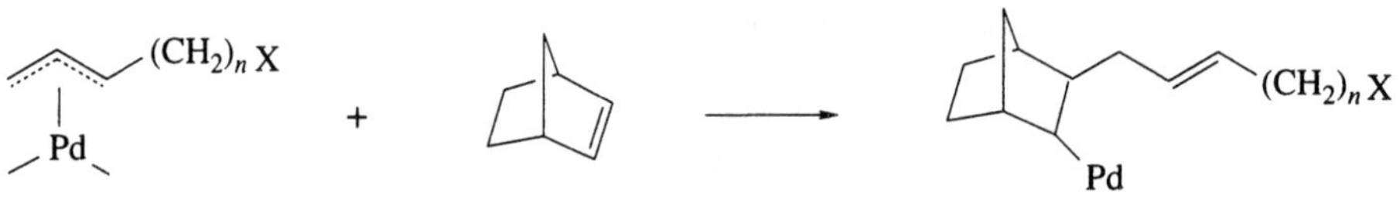

(81)

Intramolecular insertion of a simple alkene into a π-allyl complex has also been demonstrated, resulting in synthetically useful cyclization methodologies (equations 82 and 83).[262,263]

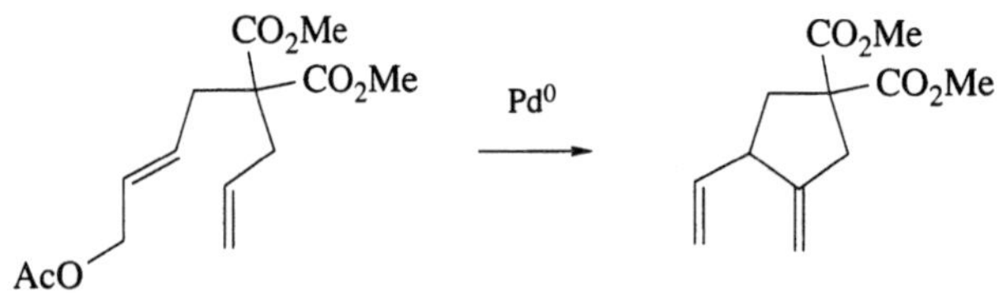

(82)

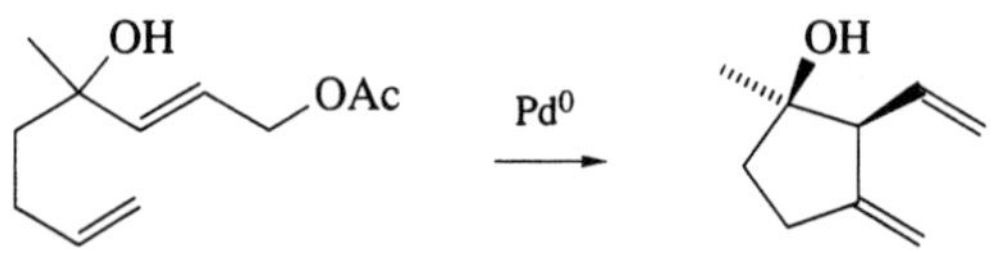

(83)

The course of the insertion of diphenylketene into π-allylpalladium intermediates has been found to depend on the nature of the π-allyl precursor. Allyl acetates give dienes, while allyl carbonates give carboxymethylated products (equations 84 and 85).[264] An intermediate allyl Pd—OR species (**9**) is believed to exist in both cases. When R = Ac, decarbonylation is followed by β-H elimination, whereas if R = Me, alkoxide attacks the acylpalladium intermediate and yields the methoxycarbonyl compound.

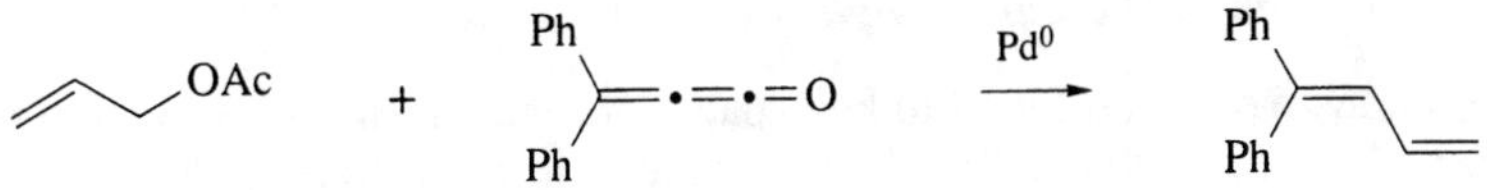

(84)

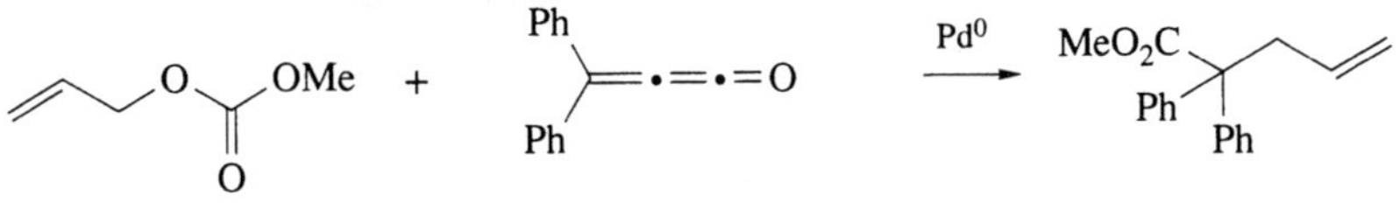

(85)

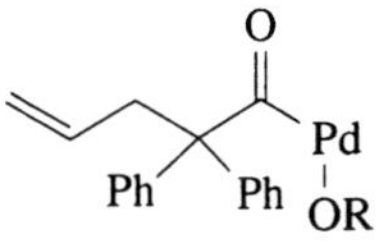

(9)

Finally, the reaction of hydrazones with π-allylpalladium complexes may involve insertion of C═N into the allyl ligand.[265,266]

3.3.2.3.2 *Oxidation of π-allylpalladium complexes*

Treatment of π-allylpalladium complexes with excess $PdCl_2$,[267–270] or MnO_2[267] results in the formation of α,β-unsaturated ketones or aldehydes (equation 86).

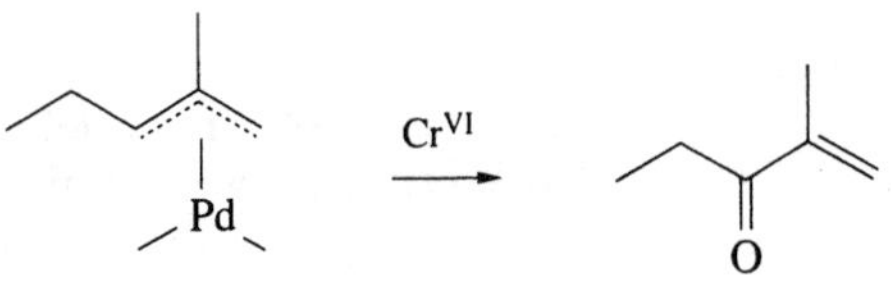

(86)

π-Allyl complexes derived from steroidal substrates have been found to undergo oxidation to allylic alcohols with MCPBA (equation 87).[271,272] Similarly, treatment of preformed allyl complexes with lead tetraacetate[273] or mercury(II) chloride in acetic acid[274] provides the corresponding allyl acetate.

(87)

A $MoO_2(acac)_2$–Bu^tO_2H reagent has been shown to yield mixtures of allyl alcohols and α,β-unsaturated ketones on reaction with π-allyl species (equation 88).[275] Photo-oxidation of preformed allyl complexes has also been reported (equation 89).[276,277]

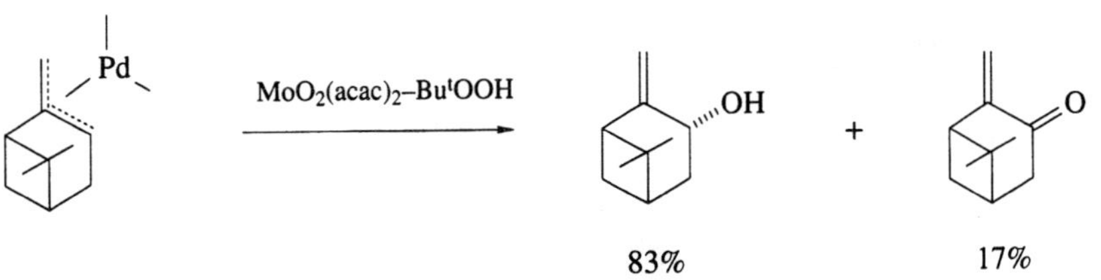

(88)

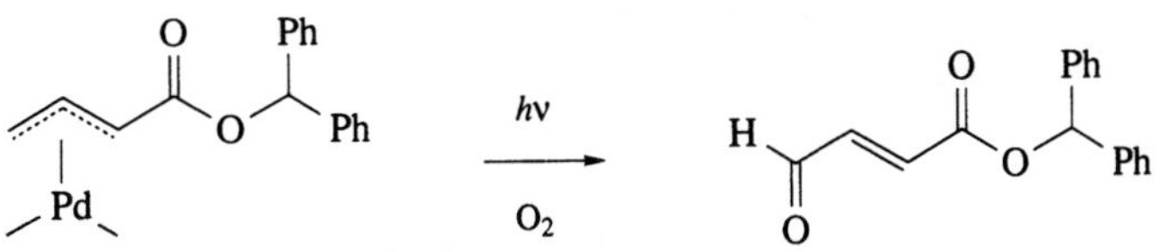

(89)

3.3.2.3.3 *Reduction of π-allylpalladium complexes*

Basic alcohol solutions have been found to be capable of reducing π-allylpalladium complexes to the corresponding alkene (equations 90 and 91).[278–283] In similar fashion, aqueous NaOH media has been reported to effect a disproportionation of the allylpalladium chloride dimer (equation 92).[284]

KCN / MeOH (90)

KCN / MeOH (91)

30% 70%

NaOH / H_2O (92)

A variety of metal hydride reducing agents can also function in this reaction. Sodium borohydride, for example, has demonstrated good regioselectivity, with preferential delivery of the hydride to the less substituted allyl terminus (equation 93).[285–287] Allyl sulfones are substituted by hydride under Pd catalysis and treatment with $NaBH_4$ (equation 94).[288,289] Sodium cyanoborohydride functions in like manner.[290]

$NaBH_4$ / Pd^0 (93)

5% 95%

$NaBH_4$ / Pd^0 (94)

$LiBHEt_3$ has been found to reduce allylic ethers, carboxylates, sulfides, sulfones, selenides and silyl ethers (equations 95 and 96).[291]

$LiBHEt_3$ / Pd^0 (95)

97% 2% 1%

$n\text{-}C_7H_{15}$ SePh → $LiBHEt_3$ / Pd^0 → $n\text{-}C_7H_{15}$ + $n\text{-}C_7H_{15}$ + $n\text{-}C_7H_{15}$ (96)

81% 18% 1%

Steroidal π-allylpalladium complexes are reduced by $LiAlH_4$ with high regio- and stereo-selectivity (equation 97).[292] Similar results were found with $LiAlH(OBu^t)_3$.[293]

LiAlH$_4$ 74%   26%   (97)

Considerable work has been done with trialkyltin hydride reducing agents.[287,294–299] These reagents are relatively mild in their reducing properties with respect to typical organic functionalities, yet function well with π-allylpalladium complexes, thereby allowing considerable chemoselectivity to be exercised. For example, allylic acetates and amines are reduced regioselectively with tri-*n*-butyltin hydride under Pd^0 catalysis (equation 98).[287] Vinyl chlorides and acetates have been prepared from the corresponding allyl *gem*-dichlorides and diacetates using this methodology (equations 99 and 100).[297,298]

OAc, CN; R_3SnH, Pd^0; CN (98)

Cl, Cl; R_3SnH, Pd^0; Cl 38% + Cl 62% (99)

OAc, OAc; R_3SnH, Pd^0; OAc 35% + OAc 65% (100)

Allyl carbonates can function as protecting groups for alcohols, with removal of the protecting group being effected by R_3SnH and Pd^0 (equation 101).[299]

ROH ⟶ RO(C=O)O-allyl $\xrightarrow[Pd^0]{R_3SnH}$ ROH + CO_2 + propene (H) (101)

Tetra-substituted tin enolates of ketones, which are otherwise difficult to prepare, can be formed *via* Pd-catalyzed tin hydride reduction of allyl β-ketocarboxylates (equation 102).[294] Allyl β-ketocarboxylates can also be transformed into α-bromo ketones using this method (equation 103).[296]

$R^1C(O)C(R^2)(R^3)C(O)O$-allyl $\xrightarrow[Pd^0]{R_3SnH}$ $R^1C(O)C(R^2)(R^3)C(O)OSnR_3$ + propene (H) ⟶

$R_3SnO(R^1)C{=}C(R^3)(R^2)$ + CO_2 (102)

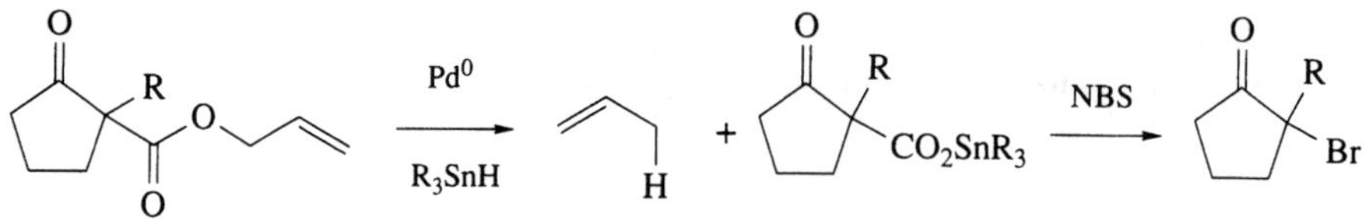

(103)

Polymethylhydrosiloxane (PMHS) has been reported to be a more selective reducing agent when coupled with Pd^0 catalysts than R_3SnH, permitting, for example, the reduction of allylic acetates in the presence of enones and acyl halides (equation 104).[300]

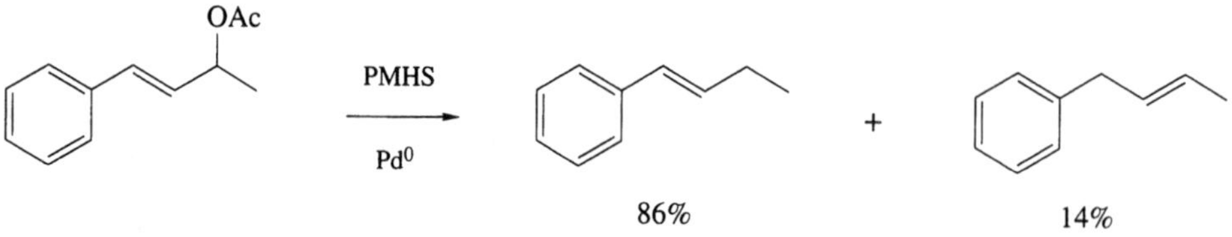

(104)

Formic acid and formate salts are also effective reducing agents for π-allylpalladium complexes (equation 105).[301–303]

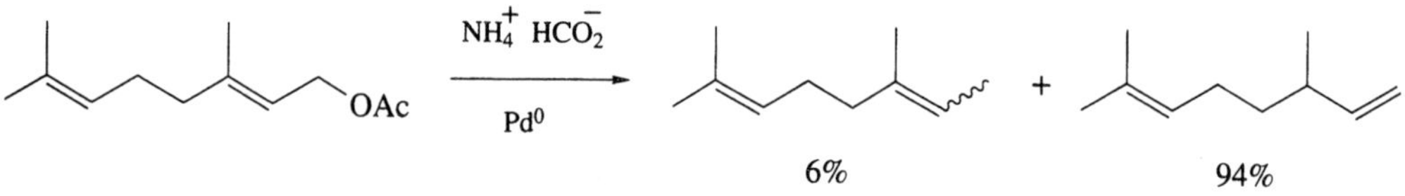

(105)

Two-phase systems utilizing water-soluble palladium phosphine catalysts, formate and allyl chlorides or acetates have also been developed.[303]

Allyl carboxylates can serve as protecting groups for acid functionalities removable by formate and Pd^0.[302]

Alkylzinc compounds bearing β-hydrogens reduce allyl acetates on treatment with Pd^0 (equation 106).[304]

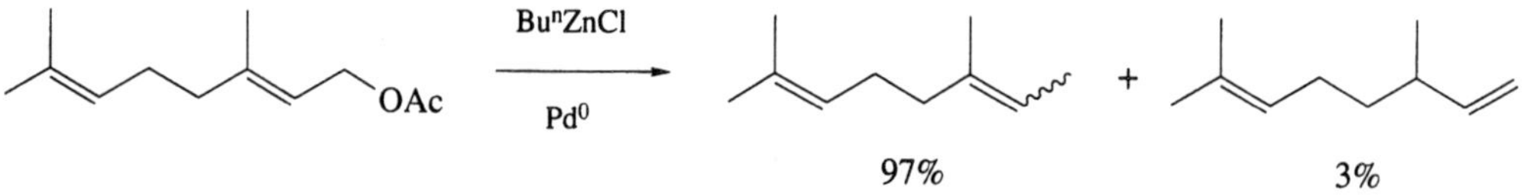

(106)

Allyl acetates can also be reduced by Pd^0 catalysts in conjunction with samarium iodide,[305] model NAD(P)H compounds[306] and under electrochemical conditions (equations 107 and 108).[307]

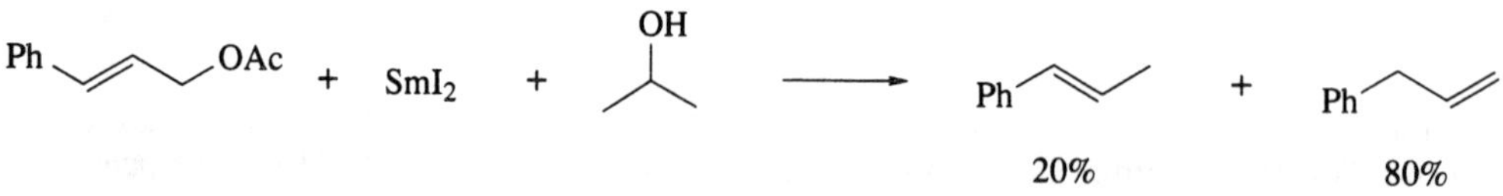

(107)

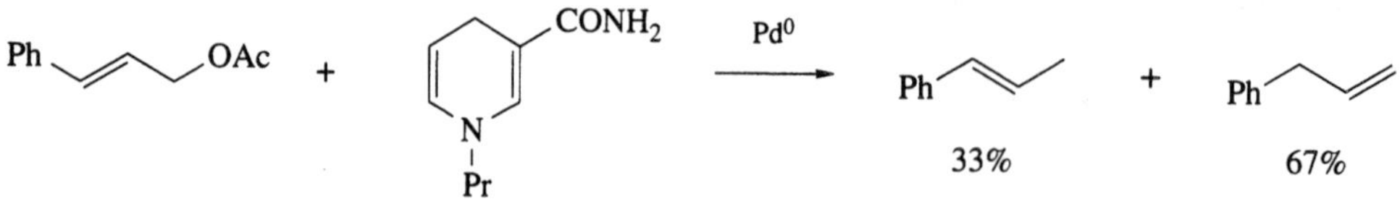

(108)

Finally, dimethylglyoxime has been reported to reduce allylpalladium complexes.[308]

3.3.2.3.4 *Umpolung of π-allylpalladium complexes*

The reversal in mode of reactivity or umpolung of π-allylpalladium complexes can be accomplished by two strategies. The first involves conversion of the π-allyl species to an allyl–M species, where M is a metal which will impart nucleophilic character to the allyl fragment, in contrast to the normal electrophilicity of the Pd species. Transfer of the allyl moiety from palladium to tin accomplishes such an umpolung, and can be achieved by the reaction of allyl acetates with $Et_2AlSnBu_3$,[309] allyl phosphonates with Bu_3SnLi and Et_2AlCl,[310] allyl acetates with Bu_3SnCl and SmI_2[311] or allyl acetates with Sn_2^+ salts,[312] all catalyzed by Pd^0 (equations 109–112). The tin is normally bonded at the less substituted allyl terminus as a result of these reactions.

Ph~~OAc + Et_2AlSnR_3 —Pd^0→ Ph~~SnR_3 (109)

~~O–P(O)(OPh)$_2$ + R_3SnLi + Et_2AlCl —Pd^0→ ~~SnR_3 (110)

Ph~~OAc + R_3SnCl + SmI_2 —Pd^0→ Ph~~SnR_3 (111)

~~OAc + Sn^{II} + RCHO —Pd^0→ ~~CH(R)OH (112)

An analogous process, which provides a nucleophilic allylsilane, has been reported, employing $(Me_3Si)_3Al$ etherate in conjunction with allyl acetates and Pd^0 (equation 113).[225]

Br-allyl OAc + $(Me_3Si)_3Al{\bullet}Et_2O$ —Pd^0→ Br-allyl $SiMe_3$ (113)

The conversion of allyl acetates into nucleophiles has also been accomplished by treatment with Pd^0 catalysts and SmI_2,[305–313] Zn[314] or by electrochemical means (equations 114–116).[307–315]

Ph~~OAc + RCHO + SmI_2 —Pd^0→ Ph~~CH(R)OH (114)

~~OAc + Zn + RCHO —Pd^0→ ~~CH(OH)R (115)

Ph~~OAc —Pd^0, +2e$^-$→ —Me_3SiCl→ Ph~~$SiMe_3$ (116)

An alternative strategy for the conversion of allylpalladium species from electrophiles to nucleophiles entails transformation from an η^3-allyl to an η^1-allyl. Although often implicated as an intermediate in allylpalladium reactions, there are only a very limited number of characterized, structurally rigid η^1-allyl complexes available. One such complex is an η^1-allylarylphosphinepalladium.[316,317] Interestingly, the η^1-species behaves as a nucleophile and reacts readily with a variety of electrophiles, including electron deficient alkenes (equations 117 and 118).[317]

R~~Pd(L)(L)Ar ——→ ~~CH(R)E (117)

HCl, E = Cl; Br_2, E = Br; NBS, E = Br; CCl_4, E = CCl_3; $HCCl_3$, E = $HCCl_2$

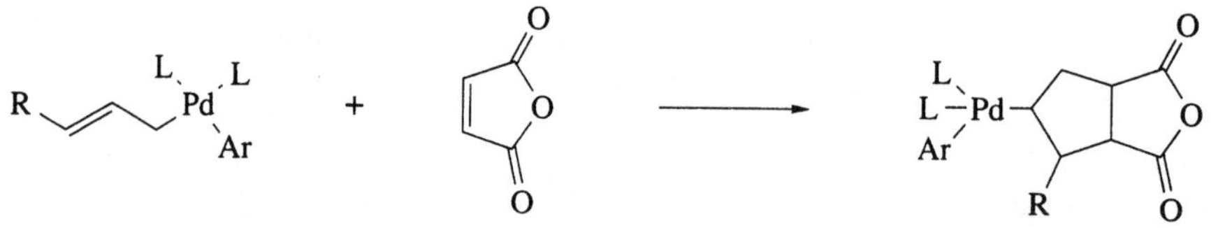

(118)

Diallylpalladium complexes derived from the dimerization of butadienes have been isolated and found to contain both an η^3- and an η^1-allyl unit. Although it is difficult to discern which allyl moiety is reacting, treatment of this complex with acetylenedicarboxylate or DOAc yields products consistent with nucleophilic participation by the η^1-fragment (equations 119 and 120).[318,319]

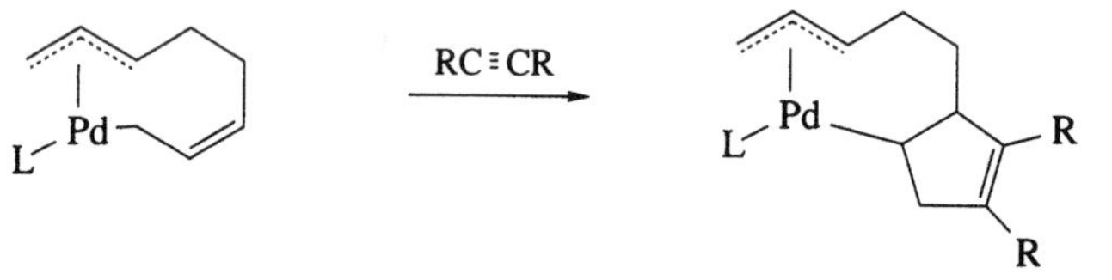

(119)

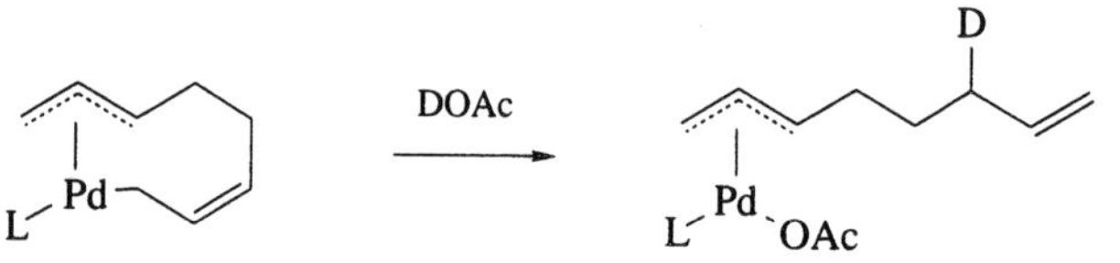

(120)

A likely explanation for the nucleophilicity induced by CN^- in a cyclooctenylallyl complex is its *in situ* conversion to an η^1-allylpalladium (equation 121).[320]

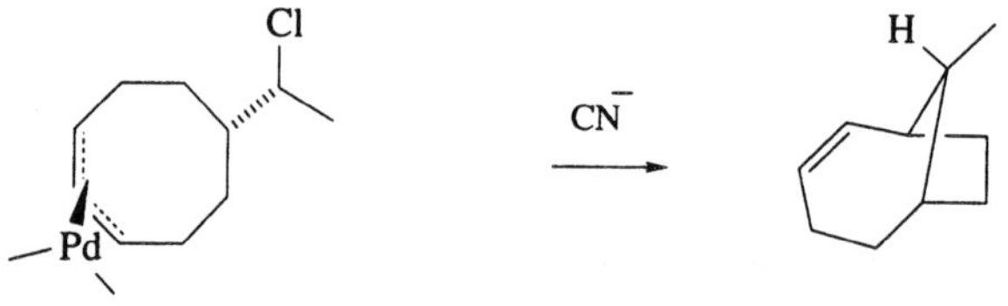

(121)

Dicarbonates of enediols have been converted to conjugated dienes on treatment with Pd^0 catalysts. Nucleophilic displacement of the allyl carbonate by an η^1-allyl complex may be responsible (equation 122).[321]

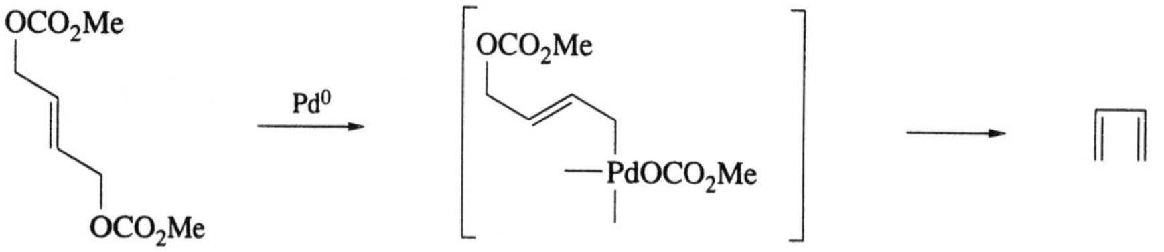

(122)

3.3.2.3.5 Transformation of π-allylpalladium complexes to dienes

Two methodologies for the conversion of π-allylpalladium complexes to dienes have been elucidated. The first, and most extensively studied, is the induced elimination of a β-hydrogen. Although π-allyl complexes are significantly less susceptible to the loss of a β-H than σ-alkylpalladium complexes, the loss can be accomplished from preformed π-allylpalladium complexes by thermolysis[322,323] or catalytically from allyl–X species by treatment with a weak base. The mild conditions that can be employed in the latter case allow preclusion of isomerization of the alkene into conjugation with a carbonyl[324] or aromatization of a generated cyclohexadiene (equations 123 and 124).[324]

Examples of this process have also been provided with steroidal substrates[325] (equation 125).

An interesting series of transformations have been reported which demonstrate a variety of selectivities of organopalladium chemistry. 1,3-Dienes can be regioselectively functionalized in a 1,4-fashion

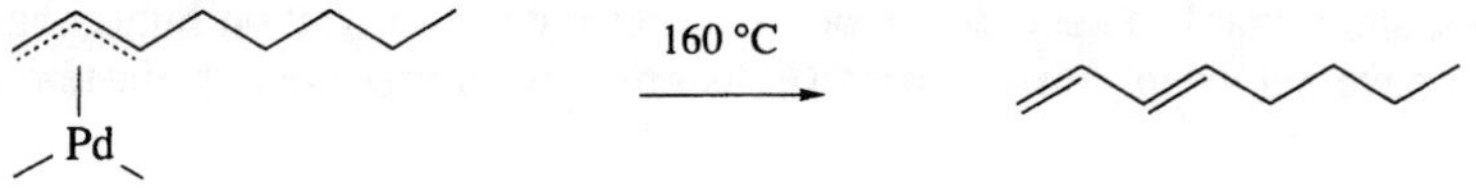

(123)

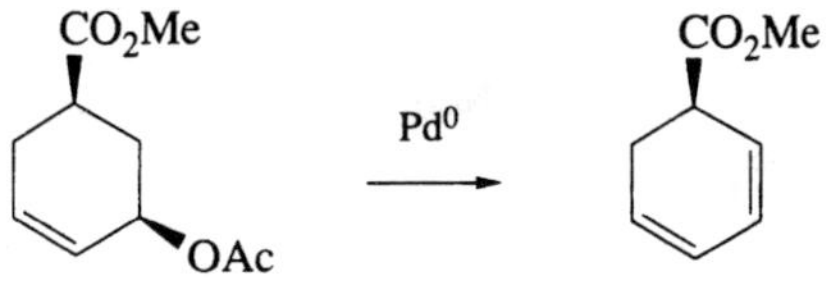

(124)

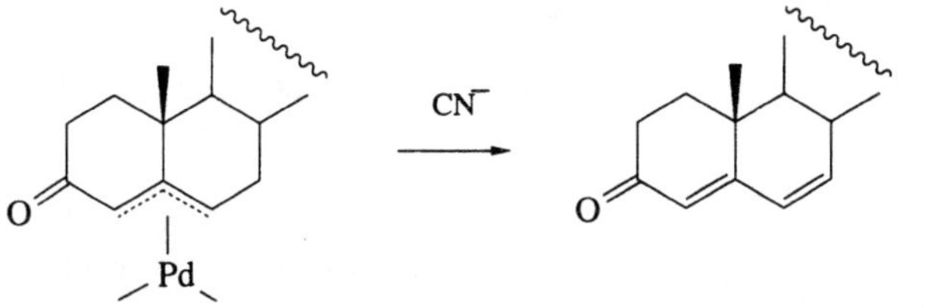

(125)

with Pd^{II} catalysts and Cl^- and OAc^-. The resulting alkene can be treated with Pd^0, and oxidative addition occurs selectively into the allyl C—Cl bond. The addition of a nucleophile gives exclusive attack remote from the OAc group. Further reaction with Pd^0 then ionizes the allyl acetate, and Et_3N induces β-H elimination to provide a nucleophilically substituted 1,4-diene (Scheme 4).[215]

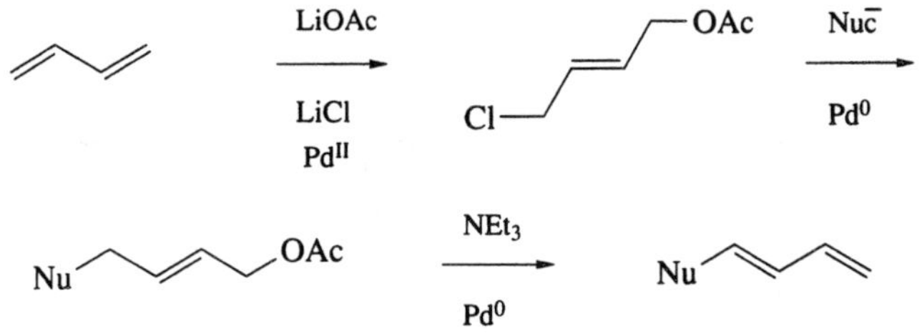

Scheme 4

The final step in the Pd-catalyzed reaction of diphenyl ketenes and allyl acetates (Section 3.3.2.3.1) involves β-H elimination to generate a diene (equation 126).[264]

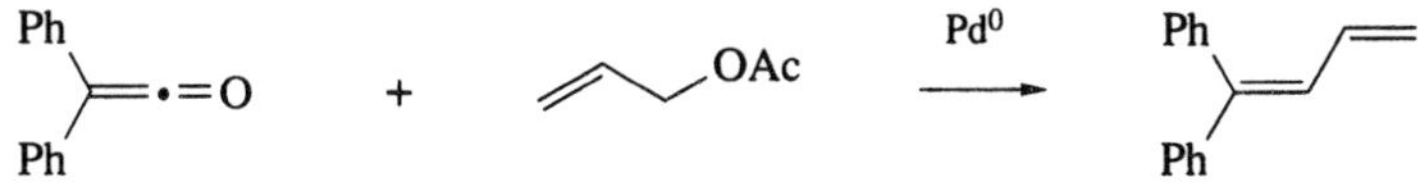

(126)

Vinyl epoxides have been found to eliminate a β-H on treatment with Pd^0 in the absence of nucleophiles. Acyclic vinyl epoxides give dienols, whereas cyclic vinyl epoxides yield β,γ-unsaturated ketones (equations 127 and 128).[326] Additional examples of β-H elimination of allylpalladium complexes to generate dienes have been reported.[327–332]

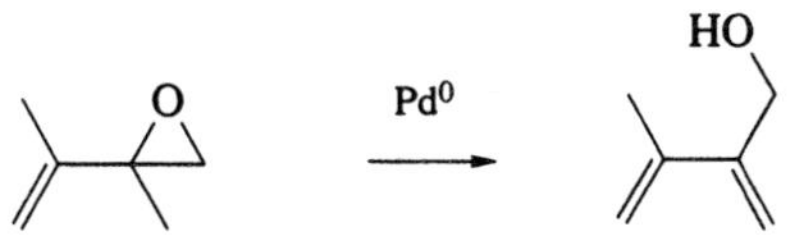

(127)

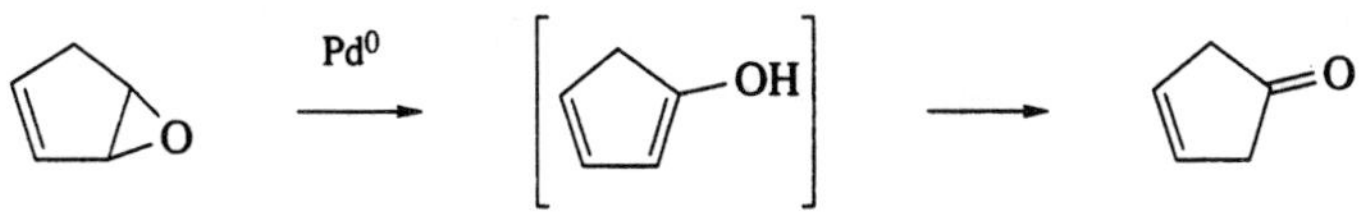

(128)

The second methodology that has been developed features a decarboxylation rather than a β-H elimination. This process gave exclusive formation of (*E*)-isomers from mixtures of diastereomeric starting materials (equation 129).[333]

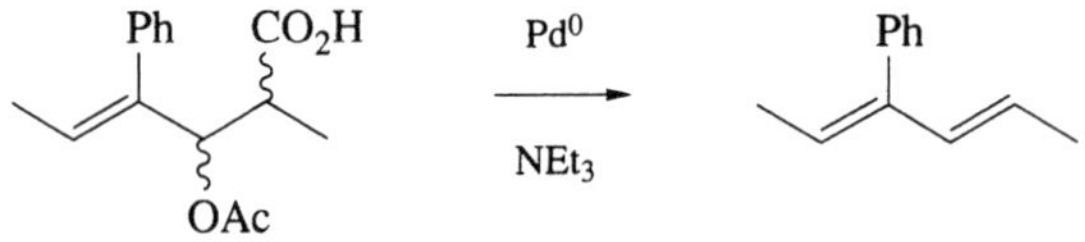

(129)

3.3.2.3.6 Photochemistry of π-allylpalladium complexes

The photochemistry of π-allylpalladium complexes has been studied to a limited extent. Two basic reactions have been observed. Irradiation at 366 nm of π-allylpalladium complexes produced 1,5-diene dimers, reportedly *via* a radical coupling mechanism.[334,335] Similar irradiations in the presence of species capable of trapping the presumed allyl radical intermediate, such as $BrCCl_3$, $BrCH_2Ph$ or allyl bromide, now yield alkylated and halogenated allyls, in addition to 1,5-diene dimer. This reaction fails for simple alkyl or aryl halides due to the instability of the associated radical (equations 130 and 131).[336]

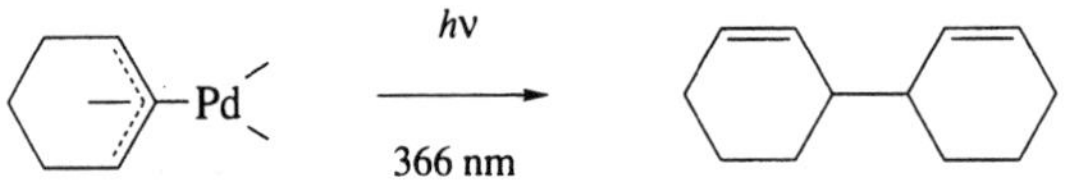

(130)

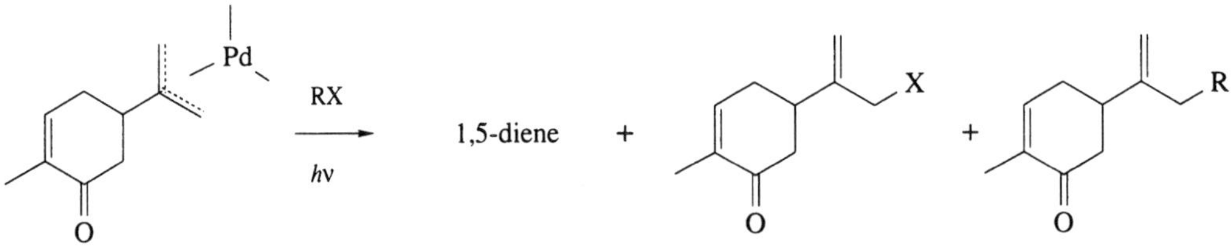

(131)

3.3.2.3.7 Addition reactions to π-allylpalladium complexes

π-Allylpalladium complexes can undergo reactions with weak σ-bonds which formally appear to result from addition to the metal, followed by coupling to the allyl fragment. The reaction of methyl- and chloromethyl-disilanes, $Me_nSi_2Cl_{6-n}$, $n = 2–6$, exemplifies the process. Treatment of, for example, hexamethyldisilane, allyl chloride and a Pd^0 catalyst produces allyltrimethylsilane and chlorotrimethylsilane as products. When unsymmetrical chloroalkyldisilanes were employed, the predominant product originated from the more highly chlorinated silicon moiety transferring to the allyl group. Unsymmetrical allyls position the silyl group at the less substituted terminus (equation 132).[337–339]

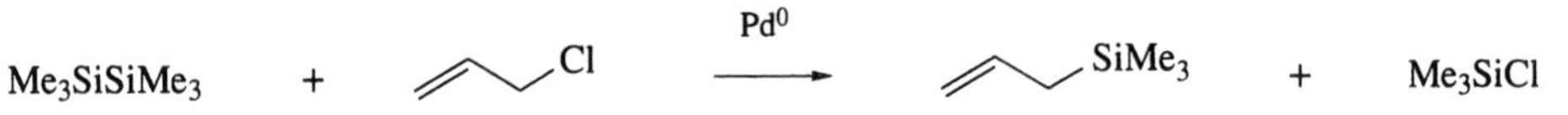

$Me_3SiSiMe_3$ + allyl chloride $\xrightarrow{Pd^0}$ allyl-$SiMe_3$ + Me_3SiCl (132)

Hexamethyldistannane reacts with allyl acetates or halides in a similar fashion.[340,341] Again, the trialkylstannane will preferentially couple to the less substituted allyl terminus (equation 133).

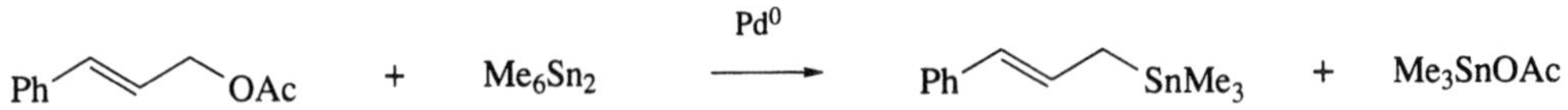

PhCH=CHCH$_2$OAc + Me_6Sn_2 $\xrightarrow{Pd^0}$ PhCH=CHCH$_2$SnMe$_3$ + Me_3SnOAc (133)

Both of these methods could be included as examples of umpolung of π-allyl species, as the electrophilic allylpalladium has now been transformed into a nucleophilic allyl moiety. Finally, Br_2 can also be added to π-allylpalladium complexes (equation 134).[342]

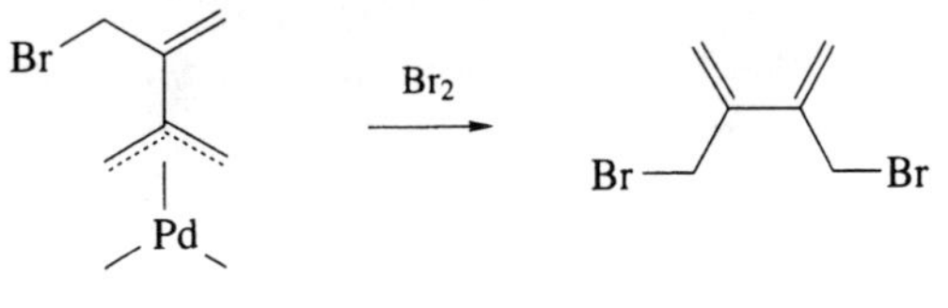

(134)

3.3.2.3.8 *Oxa-π-allylpalladium complexes*

Oxa-π-allylpalladium complexes (**10**), which can also be envisioned as palladium enolates (**11**), are susceptible to β-hydride elimination, and as such have been principally used in methodologies for the preparation of β,β-unsaturated carbonyl compounds.

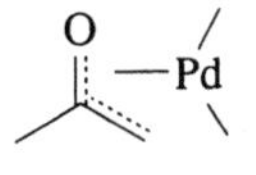

(**10**) 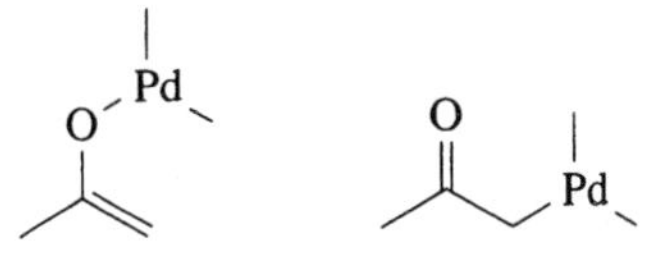 (**11**)

The direct reaction of ketones with Pd^{II} salts does effect the transformation, but the conversions are typically low and the process is largely nonregiospecific (equation 135).[343–345]

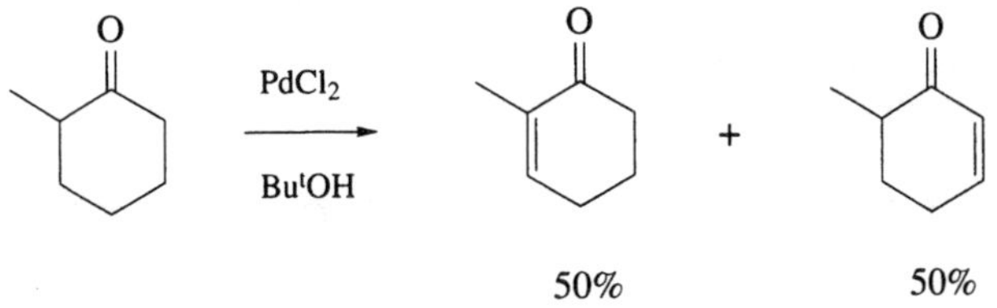

(135)

Enediones can successfully be generated from diones,[346,347] except for cyclohexandiones where aromatization can be a competitive process.[348,349] This reaction has been specifically used as an aromatization procedure with various heterocycles.[350–353]

A significant advance in this methodology was based on the realization that enol silyl ethers could be employed with Pd^{II} complexes to allow selective generation of oxa-π-allylpalladium complexes and, thereby, regiospecific dehydrogenation.[354] Considerable use of this process has been made in the synthesis of natural products (equation 136).[355–358] Decarboxylation of β-ketocarboxylates in the presence of Pd^{II} salts also specifically generates oxa-π-allyls, which subsequently dehydrogenate (equation 137).[359]

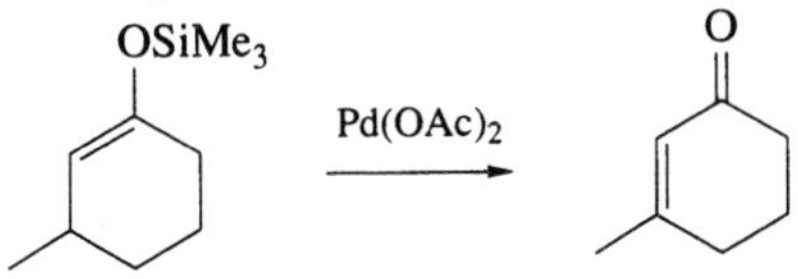

(136)

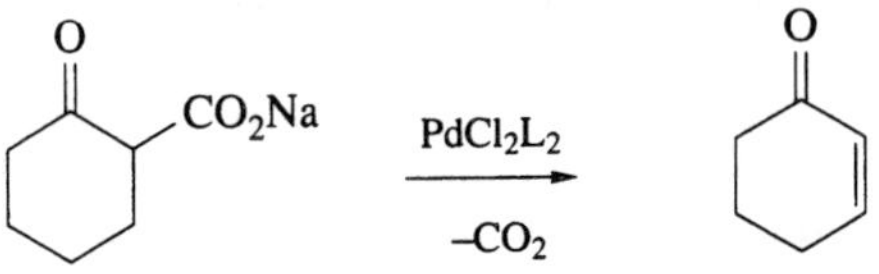

(137)

An extremely useful alternative method, employing catalytic amounts of Pd^0 complexes, rather than stoichiometric (or near stoichiometric) quantities of Pd^{II} salts has been developed. The oxa-π-allyl complex is accessed *via* allyl β-ketocarboxylates[360–363] or allyl alkenylcarbonates[360,361,363,364] in the 'intramolecular' cases and *via* enol acetates,[361,363,365] enol silyl ethers[361,366] or ketene silyl acetals[367] with allyl carbonates in the 'intermolecular' cases. The products of these reactions are the corresponding α,β-unsaturated ketones, aldehydes and esters (equations 138–142).

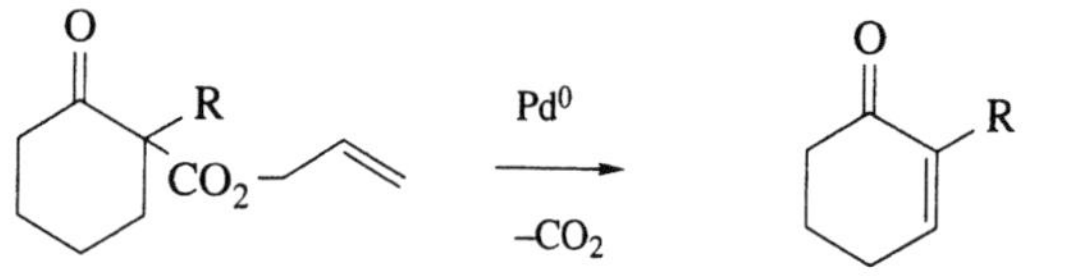

(138)

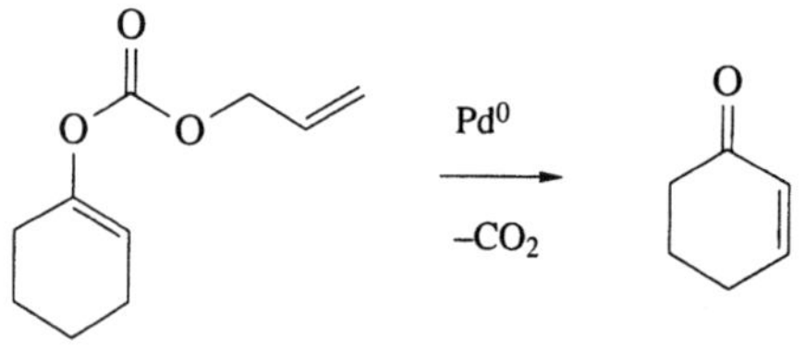

(139)

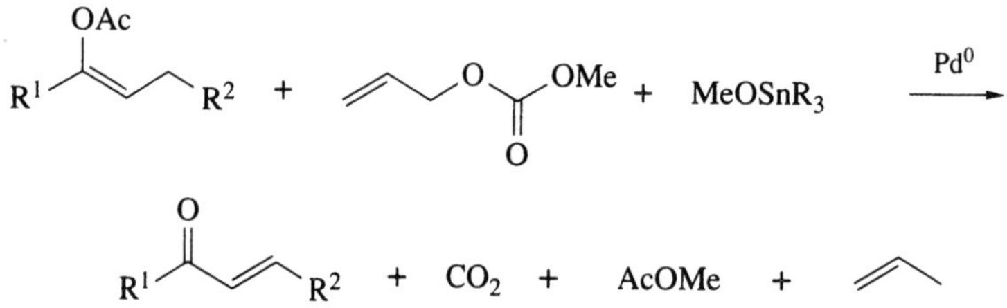

(140)

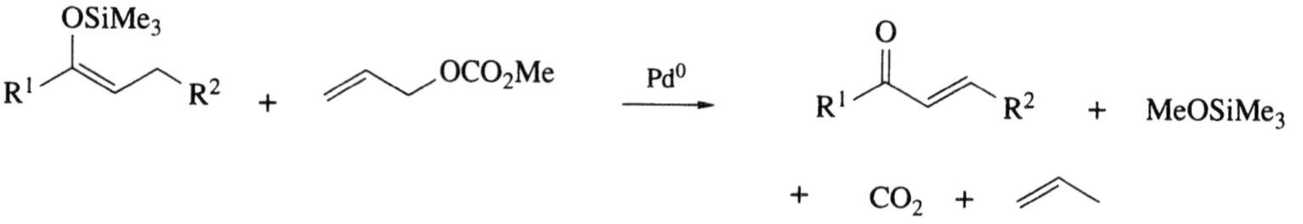

(141)

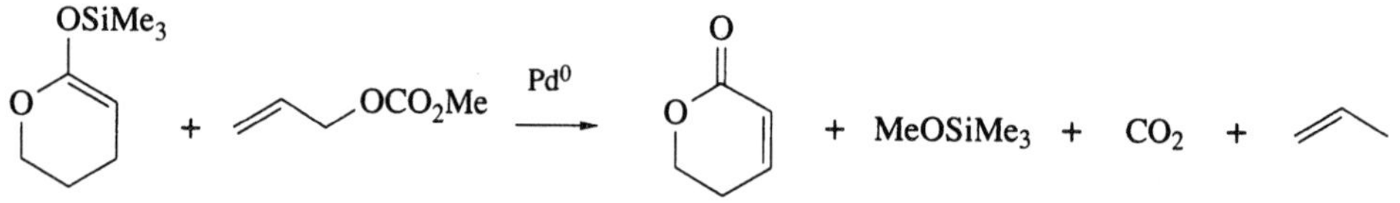

(142)

The mechanisms proposed for these reactions are all quite analogous, and only the intramolecular cases will be considered in detail (Scheme 5). Oxidative addition by Pd^0 into the allylic C—O bond of the allyl β-ketocarboxylate produces an allylpalladium carboxylate. This species then undergoes decarboxylation to yield an allylpalladium enolate (oxa-π-allyl), which subsequently eliminates a β-H to form the enone and provide an allyl-Pd-H. Reductive elimination from the allyl-Pd-H yields propene and returns Pd to its zero oxidation state. A similar mechanism can be imagined for the alkenyl allylcarbonate. Oxidative addition by the Pd^0 forms an allylpalladium carbonate, which decarboxylates again to give an allylpalladium enolate (oxa-π-allyl). β-Hydride elimination and reductive elimination complete the process. The intermolecular cases derive the same allylpalladium enolate intermediates, only now as the result of bimolecular processes.

Dehydrogenation of an allyl β-ketocarboxylate is a key step in a synthesis of methyl jasmonate that has been reported.[362]

The biologically significant α-methylene ketones have been prepared by a similar strategy, entailing a Pd^0-catalyzed decarboxylation–deacetoxylation of allyl-α-acetoxymethyl β-ketocarboxylates. The crucial allylpalladium enolate now β-eliminates OAc rather than H (equation 143).[368]

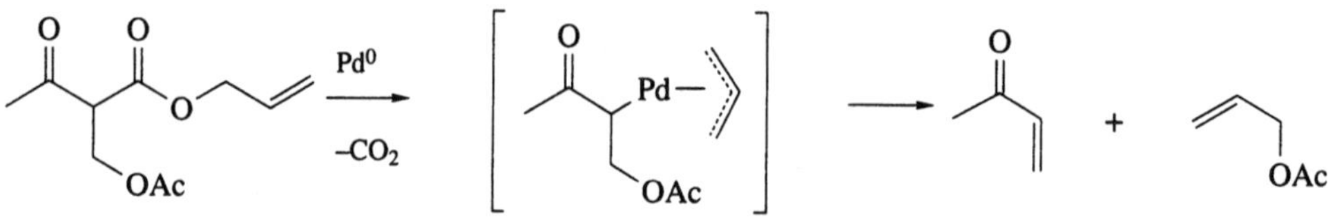

(143)

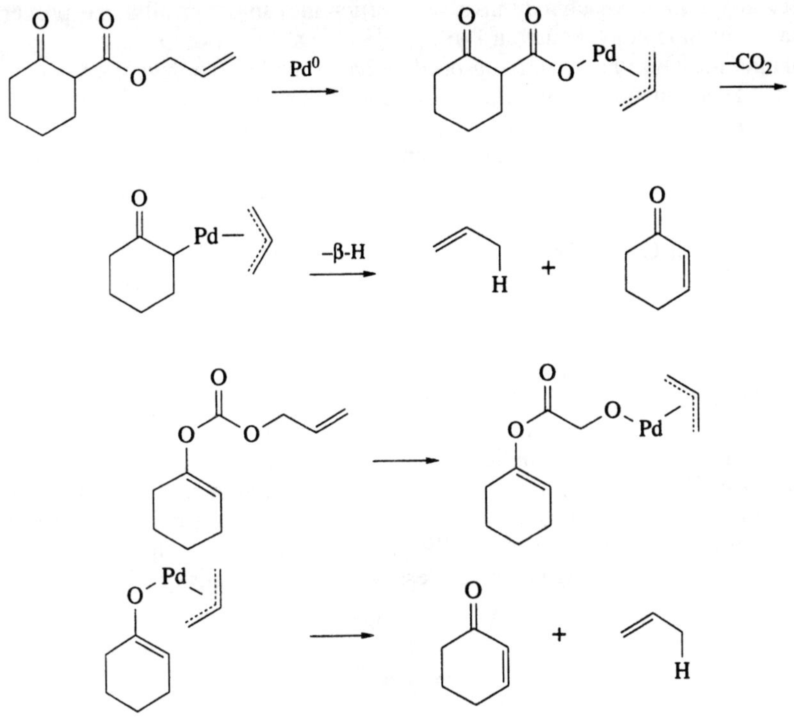

Pd^0 −CO_2 −β-H

Scheme 5

What may be characterized as an aza-π-allylpalladium complex has been proposed as an intermediate in the preparation of α,β-unsaturated nitriles by decarboxylation–dehydration of allyl α-cyanocarboxylates (equation 144).[369] These reactions have been summarized in the context of several reviews.[89,117] If the oxa-π-allyl species is blocked from the elimination of a β-H, then it can undergo typical π-allylpalladium chemistry, such as insertion by alkenes, dienes and CO (equation 145).[370,371]

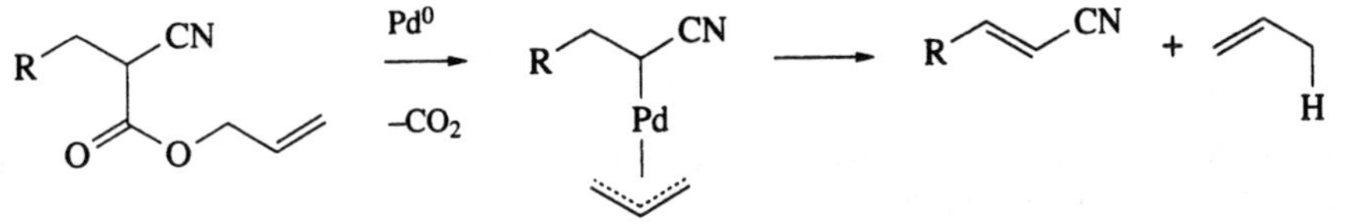

Pd^0, −CO_2 (144)

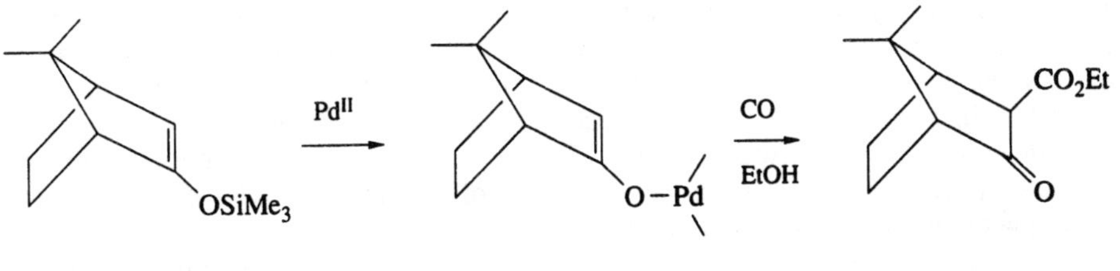

Pd^{II}; CO, EtOH; $OSiMe_3$; CO_2Et (145)

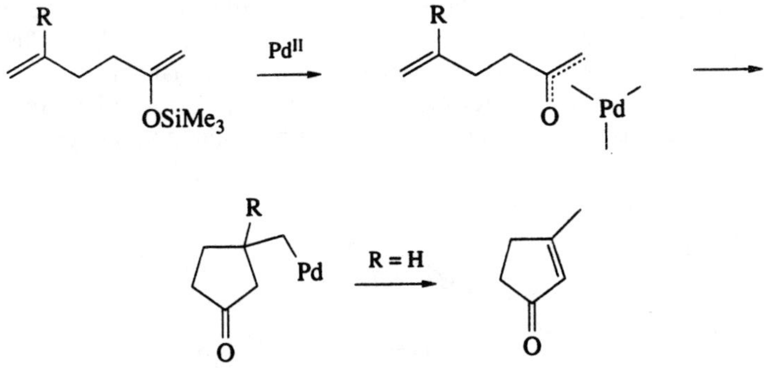

Pd^{II}; $OSiMe_3$ (146)

R = H

An intramolecular variation, which may involve insertion into an oxa-π-allyl, has proven to be a valuable new cyclization methodology (equation 146).[372–374]

A similar palladium-catalyzed cyclization procedure has recently been developed which involves enol ethers capable of β-H elimination.[375] Significant evidence has been accumulated suggesting that an oxa-π-allyl complex is not an intermediate in these reactions, but that it is better characterized as an enolate addition to a Pd^{II}–alkene complex.[376,377] Synthetic applications of this reaction have also appeared.[376–379]

3.3.3 DIASTEREOSELECTIVITY

3.3.3.1 Diastereoselectivity and Mechanism

In considering the diastereoselectivity associated with reactions of π-allylpalladium complexes, the stereochemical nature of two processes must be evaluated. The first is the formation of the π-allylpalladium complex, and the second is the allylic functionalization of a nucleophile reacting with the allyl complex. As previously described (Section 3.3.2.1.1) the generation of π-allyl complexes from alkenes by their reaction with stoichiometric amounts of palladium proceeds so as to leave the substituents on the allyl termini preferentially in the *syn* rather than the *anti* position and the palladium situated on the sterically less congested face of the allyl ligand, regardless of the stereochemistry of the starting alkene (**12**).

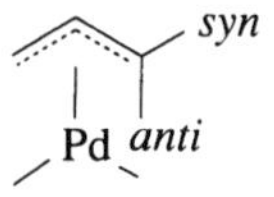

syn

Pd anti

(12)

The *syn–anti* interconversions and facial rearrangements necessary to achieve this preferred configuration are generally believed to be accomplished *via* η^1-intermediates (equation 147).

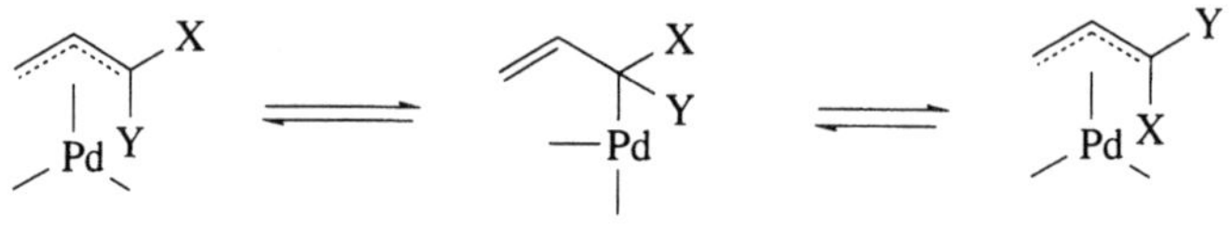

X Pd Y ⇌ X Y —Pd ⇌ Y Pd X (147)

In the catalytic process, Pd^0 complexes react with alkenes possessing an allylic C—X bond, displacing X and forming a π-allyl species. The stereochemistry of this oxidative addition reaction was initially determined indirectly by ascertaining the overall stereochemistry of the two-step process (formation of π-allyl; reaction with nucleophile) and by having previously established the stereochemical course of the attack of a given nucleophile on a stoichiometrically generated allyl complex. By this difference method, the oxidative addition was found to proceed with inversion of configuration of the initial C—X bond.[32,327,380] This determination was strongly supported by the analogous oxidative addition of Pd^0 into the C—X bond of benzyl halides, which was previously known to proceed with inversion of configuration.[381] Recently, the stereochemistry of the oxidative addition of Pd^0 into an allyl acetate was directly verified as proceeding with inversion of configuration.[382]

It is reasonable to assume that the identical complex will be generated whether it be done stoichiometrically from an alkene, to give a chloride or carboxylate dimer followed by the addition of 2 equiv. of a phosphine per Pd, or by the addition of an allyl–X compound to give a phosphine–Pd^0 complex. This assumption is supported by the fact that complexes generated in either manner have been found to exhibit identical reaction profiles.[380] Furthermore, for the vast majority of allylpalladium reactions studied, it is most likely that the reactive species is a cationic bisphosphine–palladium complex (**13**).[13] Calculations

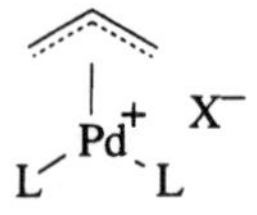

Pd^+ X^- L L

(13)

also support that the cationic bisphosphine complex is the most reactive species of the various allylpalladium species that can be reasonably considered.[383] As a result, all mechanistic rationales will be made assuming this to be the active catalyst.

Pd+ X− L L —Nuc−→ Nuc PdL$_2$ —L′→ Nuc + PdL$_2$L′ (148)

Attack by nucleophiles on the electrophilic π-allylpalladium complex can take place by two distinct mechanisms, which have opposite stereochemical consequences. Stereochemical inversion is achieved by attack of the nucleophile directly on the allyl ligand on the face opposite the palladium (equation 148). Retention is achieved by attack of the nucleophile at the metal center, followed by reductive elimination (equation 149). The reductive elimination step could proceed through an η^1-allylpalladium–Nuc species or directly from an η^3-allylpalladium–Nuc complex. Recent evidence strongly suggests the latter pathway.[384]

Pd+ L L X− —Nuc−→ Pd L Nuc ——→ L–Pd-Nuc L

Nuc + PdL$_2$ (149)

Nucleophiles partition between the two mechanisms based on their 'hard–soft' characteristics, with soft nucleophiles undergoing ligand attack and hard nucleophiles attacking at the metal. A limited class of nucleophiles appear capable of adding by either mechanism, with secondary factors controlling their choice of mode of addition.

The various nucleophiles enumerated in the chemoselectivity section (Section 3.3.2.2) will be considered in turn and classified as to the stereochemistry of addition and, thereby, the mechanism they follow.

In addition to direct determination of the stereochemistry of the allylation process, alternative means of classification of nucleophiles as to those that proceed by ligand or metal addition have been advanced. In one case the classification is made based on the regiochemistry of addition to a specific unsymmetrical allyl complex,[385] and in a second by the ability of a particular allyl complex to react only by ligand addition,[386] and therefore to be inert to nucleophiles that add *via* attack at the metal.

It must also be noted that despite the ability of stereochemical interconversions to take place in many π-allyl complexes *via* a variety of mechanisms, the addition of nucleophiles has been found to be sufficiently fast such that the allylation process is generally stereospecific.

3.3.3.2 Stereochemistry of Addition of Carbon Nucleophiles

3.3.3.2.1 Soft carbanions

Soft carbanions, $R\bar{C}XY$, as defined in Section 3.3.2.2.1, have been shown to add to π-allylpalladium complexes directly on the allyl ligand, on the face opposite the Pd (inversion) (equations 150 and 151).[32,72,153,327,380,387–392]

In combination with the inversion step in the oxidative addition, allyl–X π-allyl precursors show overall retention of configuration relative to the C—X bond *via* a double inversion process (equation 152).

The anion of cyclopentadiene has also been demonstrated to add *via* ligand addition (inversion).[105,385,386] Indenyl nucleophiles derived from the corresponding allylsilane have been classified as adding *via* ligand addition,[385] but the sodium salt of the indenyl anion has curiously been suggested to add *via* addition directly to the metal.[386]

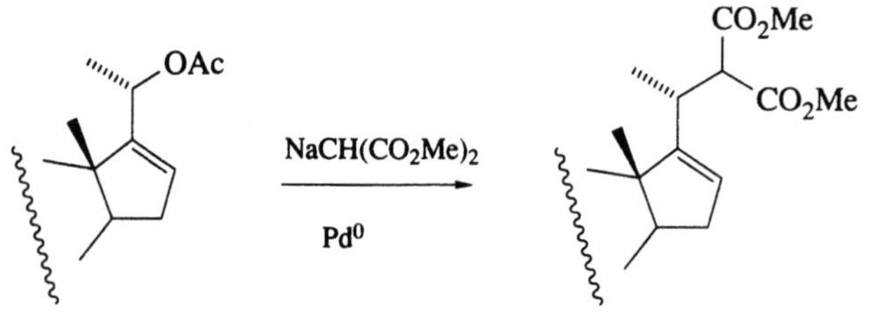

(150)

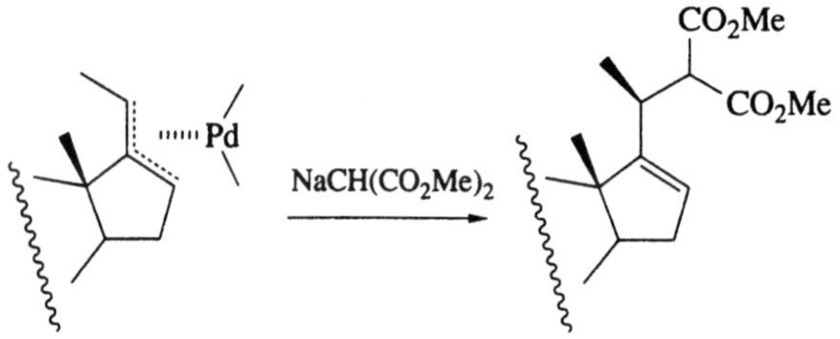

(151)

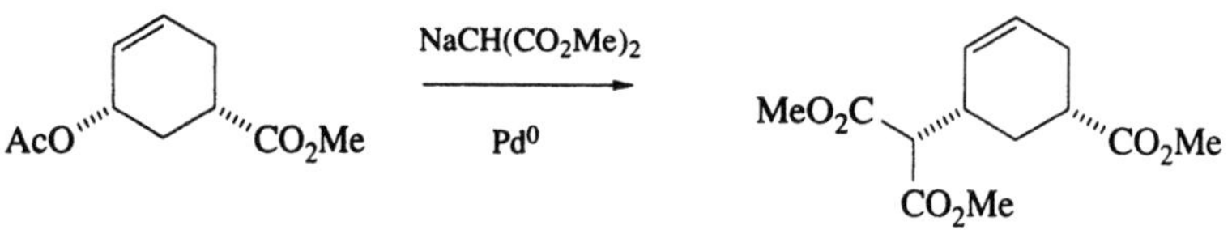

(152)

Highly acidic carbon nucleophiles, pK_a < 10, have also been investigated and found to add *via* ligand addition (equation 153).[393]

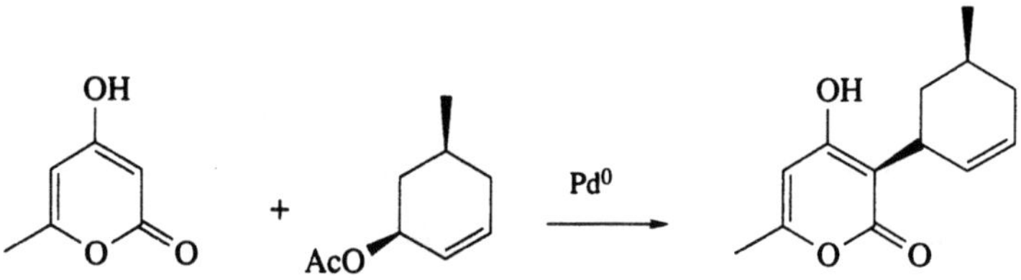

(153)

Loss of stereospecificity in the addition of soft carbon nucleophiles can occur if the rate of nucleophilic attack is slow, due, for example, to extreme steric bulk, *e.g.* $NaCH(SO_2Ph)_2$,[167] of the nucleophile (equation 154). In this case, the initially displaced OAc^- has sufficient time to return and attack the π-allyl complex. Acetate anions (*vide infra*) are capable of either ligand or metal addition, thus scrambling the stereochemistry of the starting allyl acetate.

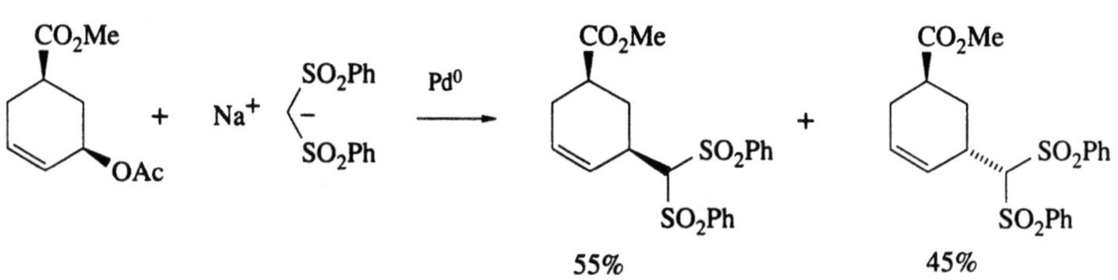

(154)

3.3.3.2.2 Enolates

Tin enolates add to π-allylpalladium complexes directly on the allyl ligand (inversion).[106,385] Therefore, in tandem with the inversion of configuration incurred in the oxidative addition of a Pd^0 catalyst into an allyl acetate, a net overall retention is observed (equation 155).

Similarly, lithium enolates of ketones add to allyl acetates *via* Pd^0 catalysis by a double inversion process,[110] and potassium ketone enolates have been shown to add to preformed π-allyl complexes with inversion (equations 156 and 157).[111]

OSnR$_3$ + CO$_2$Me OAc $\xrightarrow{Pd^0}$ CO$_2$Me O (155)

OAc Li O $\xrightarrow{Pd^0}$ O Ph Ph (156)

OMe K Ph O Pd → O OMe Ph (157)

The predominant stereochemical course of the allylation of Pd enolates follows that previously shown by the other enolates, but also consistently exhibits some loss of stereointegrity (equations 158–160).[114,115,126] This lessening of stereospecificity can be attributed either to scrambling of the stereochemistry of the starting material by attack of the initially ionized carboxylate at the metal (equation 161)

O O O Ph $\xrightarrow[-CO_2]{Pd^0}$ O Ph (158)

95:5 *cis:trans* 80:20 *cis:trans*

O O O Ph $\xrightarrow[-CO_2]{Pd^0}$ O Ph (159)

24:76 *cis:trans* 34:66 *cis:trans*

O O O OMe $\xrightarrow[-CO_2]{Pd^0}$ O OMe + O OMe

1,4 1,2

cis → 61% *cis*-1,4; 14% *trans*-1,4; 23% 1,2 (160)

trans → 51% *cis*-1,4; 26% *trans*-1,4; 23% 1,2

or by direct reductive elimination of an allylpalladium enolate intermediate (equation 162). Epimerization of the starting material has been noted in this reaction.[115]

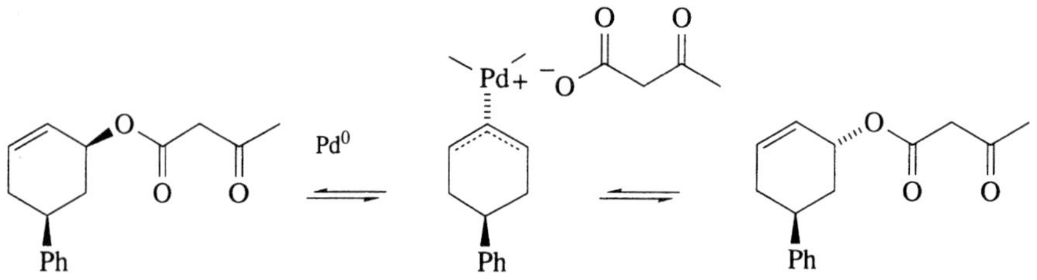

(161)

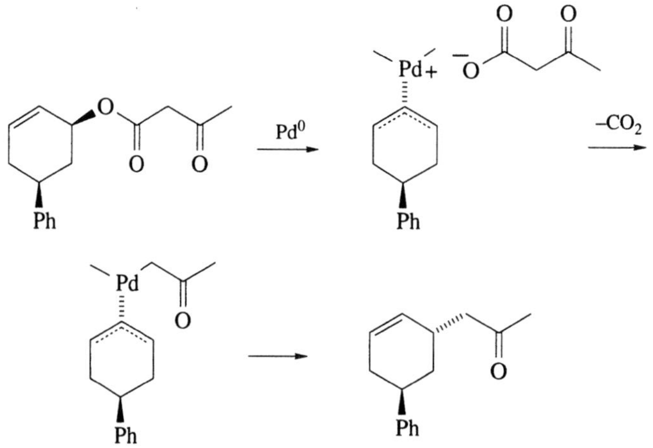

(162)

The reaction of β-ketoacids with allyl carboxylates is also believed to proceed *via* a palladium enolate intermediate.[126] Less than complete stereospecificity is also observed in these reactions (equation 163). Interestingly, the bicyclic lactone substrate employed to ascertain the stereointegrity of this reaction, in addition to being incapable of any *syn–anti* isomerization, cannot epimerize the starting material by carboxylate attack at the metal. The observed stereochemical leakage could be due to epimerization of the intermediate allyl complex (equation 164) or reductive elimination of an allylpalladium enolate (retention) (equation 165).

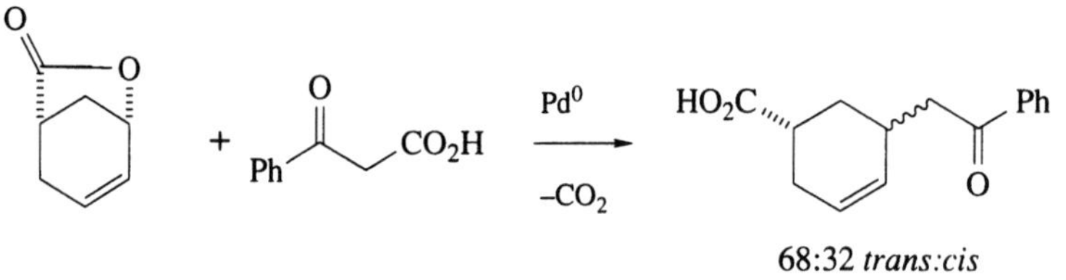

(163)

68:32 *trans:cis*

(164)

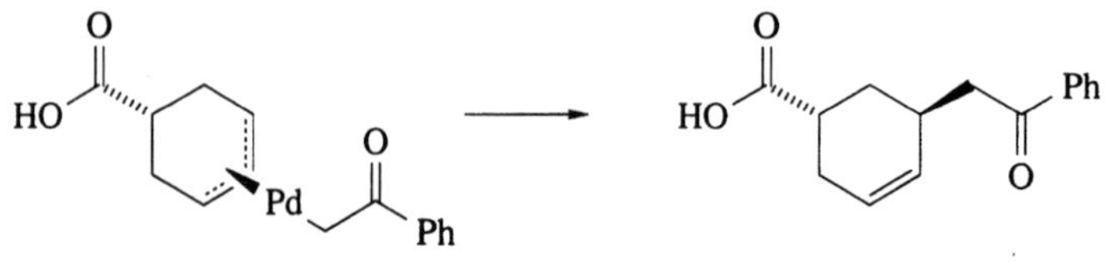

(165)

3.3.3.2.3 Organotins

The palladium-catalyzed reaction of aryl- and vinyl-tin reagents with stereochemically defined allyl chlorides proceeds with overall retention of configuration, indicating that the second step, entailing interaction of the π-allylpalladium complex and the organotin, proceeds by transmetallation and reductive elimination (attack at Pd, retention) (equations 166 and 167).[142,145] Comparable results were obtained with cyclic vinyl epoxides and aryltins.[143]

CO_2Me … Cl —(R_3SnPh, Pd^0)→ CO_2Me … Ph (166)

—(R_3SnPh, Pd^0)→ OH, Ph + OH, Ph (167)

Allyltin reagents were suggested to add to π-allylpalladium complexes directly on the allyl ligand,[137,138] although a definitive stereochemical experiment could not be successfully carried out. The fact that these reactions proceed with allyl inversion of the stannane (equation 168) would require an unlikely, sterically unfavorable allylpalladium–R intermediate if metal attack was followed.

Ph⁄⁀OAc + ⁀SnR_3 —(Pd^0)→ Ph (168)

However, the reaction of preformed allylpalladium complexes with allyltin reagents with coadded maleic anhydride does proceed *via* metal addition (equation 169).[139]

OMe, Pd —(R_3Sn, maleic anhydride)→ OMe + OMe (169)

75% 25%

It is clear that allyltin compounds do add to allylpalladium complexes at the metal center and that these intermediates can be induced to undergo reductive elimination with maleic anhydride. In view of the phenyl- and vinyl-tin results (metal attack), it is tempting to assume all organotin reagents proceed by metal addition, but this remains to be verified.

3.3.3.2.4 Organozincs

Direct assays of the stereochemistry of the addition of phenylzinc halides to allylic esters catalyzed by palladium uniformly agree that this process takes place with overall inversion of configuration, indicating direct addition to the metal by the zinc reagent (equations 170–172).[153,154,386] Methods of classification of the mode of nucleophilic addition based on regiochemical considerations also support this result.[385] A large number of studies have been conducted on the allylpalladium–Ar intermediate in this reaction. Two interesting aspects emerge from these studies. First, these species appear to be capable of reductive elimination directly from an η^3-allyl form without collapse to an η^1-allyl, and second, that the reductive elimination process is greatly accelerated by the addition of coordinating electron deficient alkenes, *e.g.* acrylates.[394]

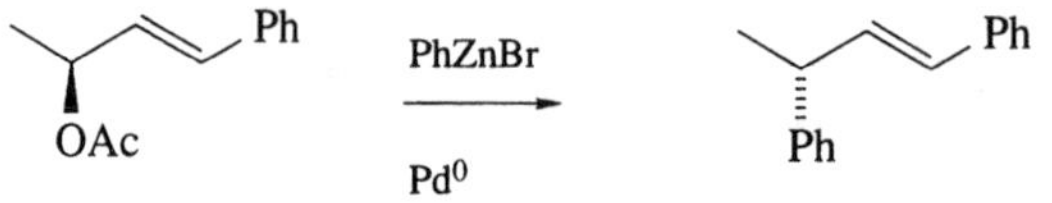

(170)

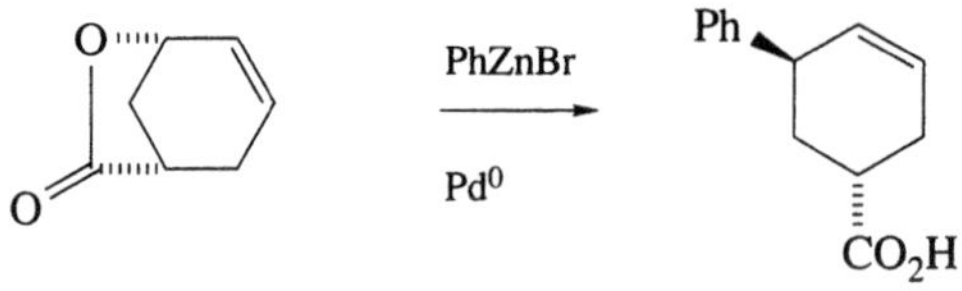

(171)

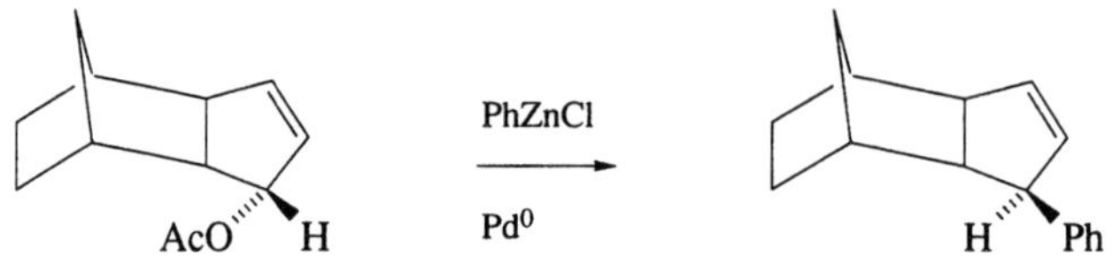

(172)

3.3.3.2.5 Organo-aluminum, -zirconium and -mercury compounds

Dialkylvinylaluminums transfer their vinyl substituent to stereodefined allylic acetates with Pd^0 catalysis with overall inversion of configuration. Metal attack of the organoaluminum is thereby indicated (equation 173).[154]

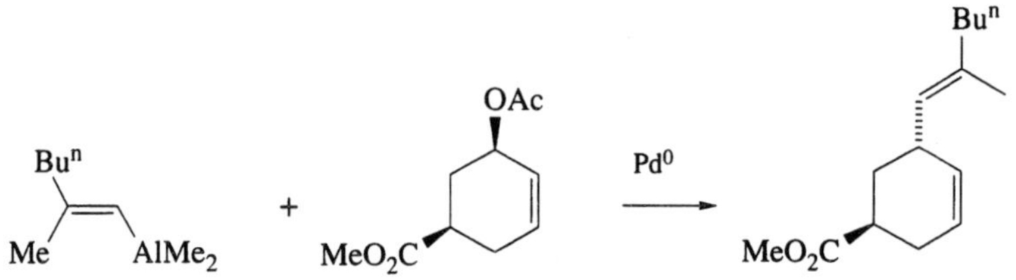

(173)

Alkenylzirconiums react with either preformed allylpalladium complexes[155] or allyl acetates and Pd^0 catalysts[156,395] *via* addition directly to the palladium, followed by reductive elimination. In the reaction with steroidal π-allylpalladium complexes (equation 174) the natural configuration at C-20 is now obtained, in contrast with the addition of malonates (ligand attack) which yielded the unnatural configuration.[74]

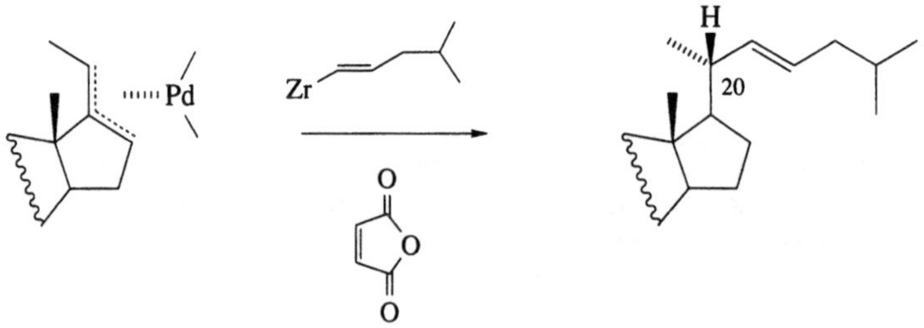

(174)

Similar to the tin- and zinc-based methodologies, electron deficient alkenes were found to strongly catalyze the reductive elimination step.[155,156,395]

Organomercury reagents are also believed to add *via* metal addition.[385]

3.3.3.2.6 Grignards and organolithiums

Grignard reagents have been shown to add to preformed π-allylpalladium complexes directly at the metal center.[158,387,396,397] Reductive elimination completes the allylation process with retention of configuration with respect to the metal. Again, electron deficient alkenes (maleic anhydride) were found to accelerate the reductive elimination (equations 175 and 176).

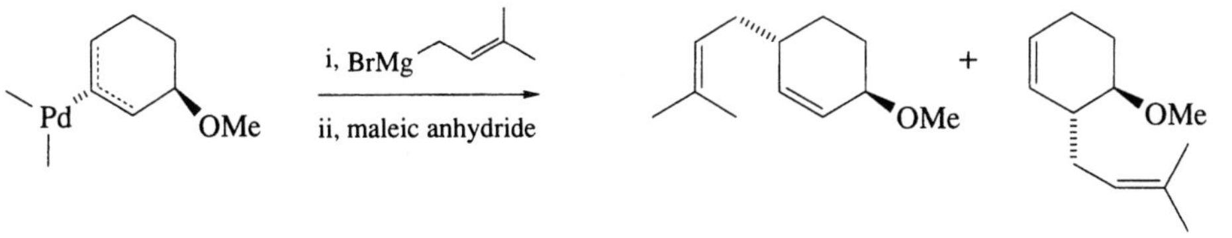

i, BrMg ; ii, maleic anhydride (175)

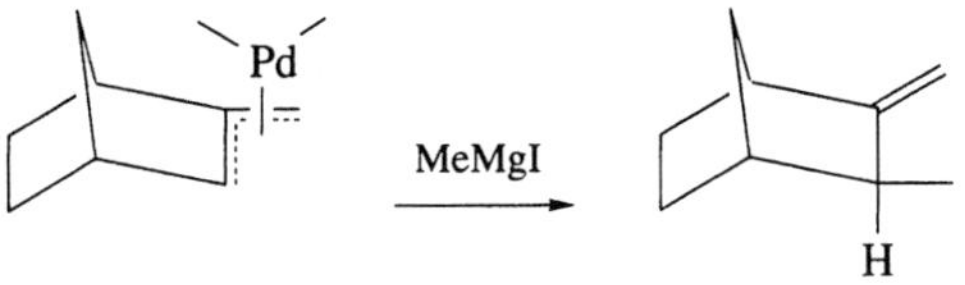

MeMgI (176)

Organolithiums[158,160,385] behave in identical fashion to the Grignard reagents, adding with retention of configuration with respect to the Pd.

3.3.3.3 Stereochemistry of Addition of Oxygen Nucleophiles

In choosing between the two possible modes of addition to a π-allylpalladium complex, namely ligand or metal attack, the overwhelming majority of nucleophiles opt for one mode to the total exclusion of the other. Oxygen nucleophiles, specifically carboxylates, however, appear to enjoy a relatively unique position in being able to add by both mechanisms.

The initial suggestion that acetates might be capable of migration directly from the metal to the allyl ligand came from the report that π-allylpalladium acetate dimers could be induced by treatment with CO to undergo acetoxylation of the allyl ligand.[163] Although the required stereochemical evidence was not determined in this early report, the use of CO to effect this reaction was highly reminiscent of other processes which occur by a *cis* reductive elimination mechanism.

The loss of stereospecificity in the addition of bis(sulfone) anions to cyclohexenyl allylic acetates was attributed to a scrambling of the stereochemistry of the starting acetate. The ability of Pd^0 catalyst to effect this epimerization was confirmed in the absence of added nucleophile. This epimerization was attributed to the ability of the acetate to return to add to the π-allyl complex *via* attack at the metal center (equation 177).[167] This suggestion was confirmed by treatment of a preformed allylpalladium acetate dimer with CO, which resulted in *cis* migration of the acetate from Pd to the allyl ligand (equation 178).[164]

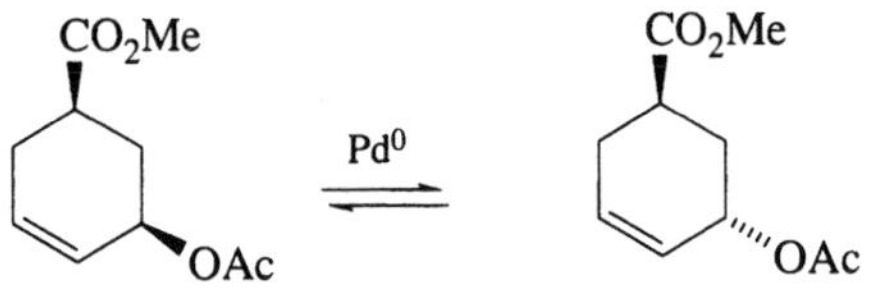

Pd^0 (177)

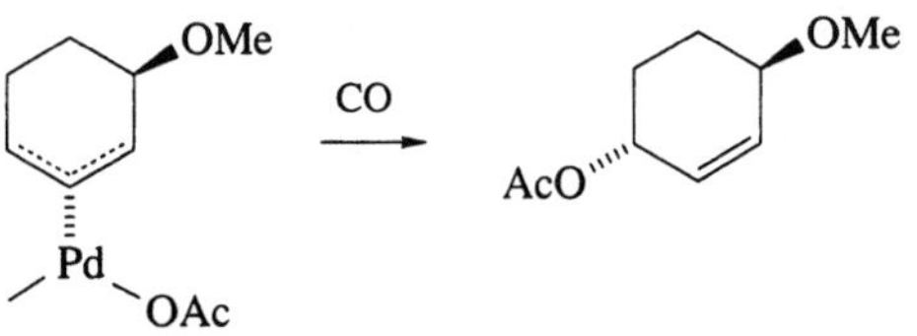

CO (178)

That acetate could add *via* either ligand or metal addition was subsequently verified.[168–170,175] Of crucial importance to the synthetic utility of this addition, conditions have been determined which allow virtually complete control of the mode of attack of acetate by manipulation of the ligands present in the reaction mixture (equation 179). Addition of acetate to a preformed allyl complex was shown to result in metal addition, followed by *cis* migration (reductive elimination) to the allyl ligand. When chloride ion was added, addition now proceeded *via* attack on the ligand on the face opposite the palladium. The chloride ion appears to block addition to the Pd by occupying its available coordination sites and forcing ligand addition as an alternative.[168]

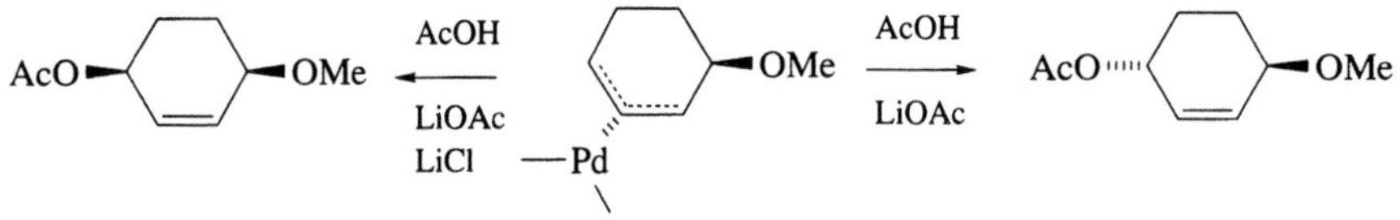

(179)

Carboxylates may enjoy their ability to undergo metal addition and *cis* migration because of their bidentate nature and could effect this migration *via* an oxygen isomerization process (equation 180). Carboxylate and phenolate nucleophiles can also add *via* ligand addition without Cl^- blocking of the metal (equation 181).[176] The reported loss of stereochemical integrity in the intramolecular addition of alkoxides to a π-allylpalladium complex could also be rationalized by competing metal–ligand addition processes (equation 182).[180]

(180)

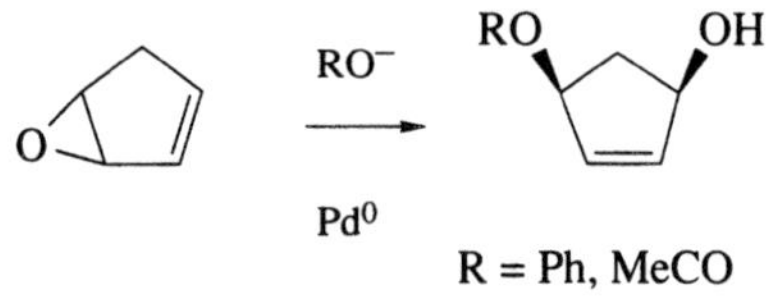

R = Ph, MeCO (181)

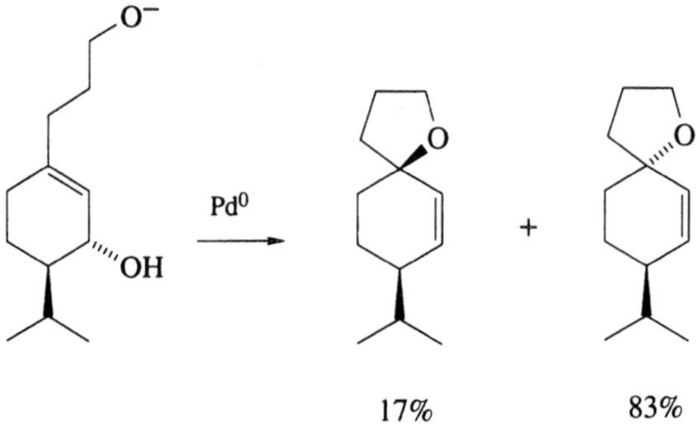

17% 83% (182)

3.3.3.4 Stereochemistry of Addition of Amine Nucleophiles

Amine nucleophiles appear, in general, to prefer addition to the allyl ligand on the face opposite the palladium.[185,195,208,386]

Addition of dimethylamine to 1,3-cyclohexadiene catalyzed by Pd^{II} complexes proceeds *via* an amino-π-allyl complex of known configuration, to which a second equivalent of $HNMe_2$ adds on the face opposite Pd to yield a *cis*-1,4-diaminocyclohexene (equation 183).[185] Similar results were obtained from an acetoxychlorocyclohexene (equation 184)[195] and in the preparation of amino sugars (equation 185).[208]

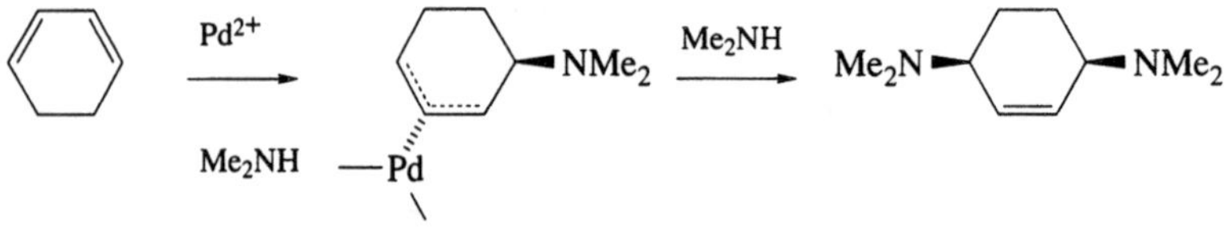

(183)

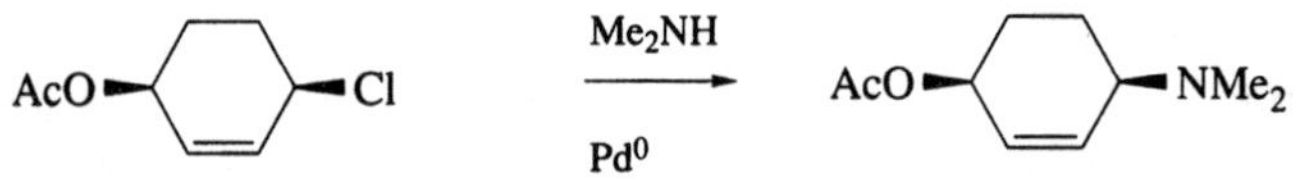

(184)

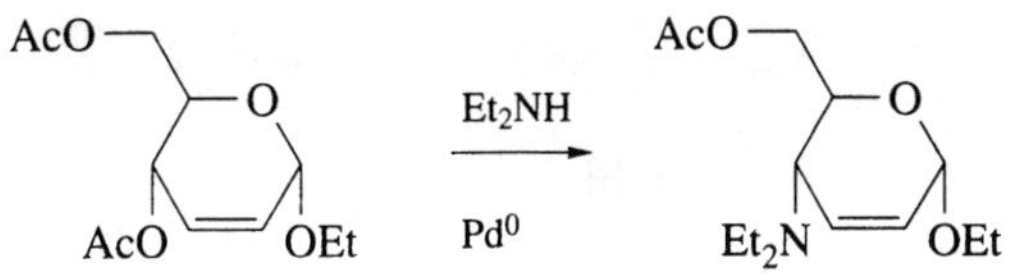

(185)

Loss of stereospecificity, however, has also been reported in the addition of amines. The use of homogeneous Pd^0 catalysts in the addition of dimethylamine to a cyclohexenyl acetate led to substantial stereochemical scrambling (equation 186). Employment of polymer-bound Pd^0 catalysts, however, gave complete stereospecificity *via* ligand addition.[398] The epimerization noted in this reaction is apparently due to acetate attack at the metal center, which is prohibited by steric congestion of the metal in the polymer matrix (equation 187).[398]

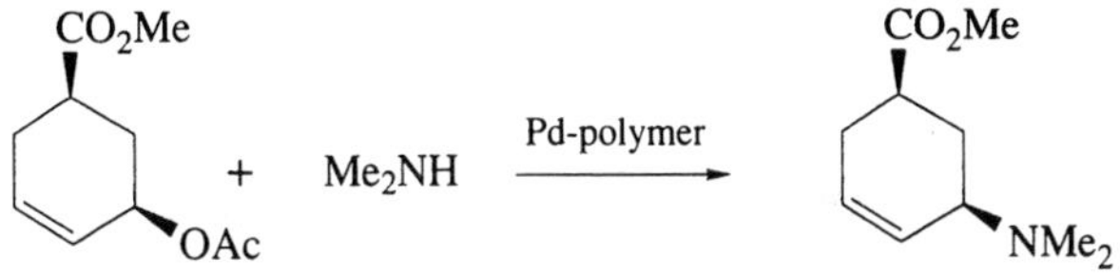

(186)

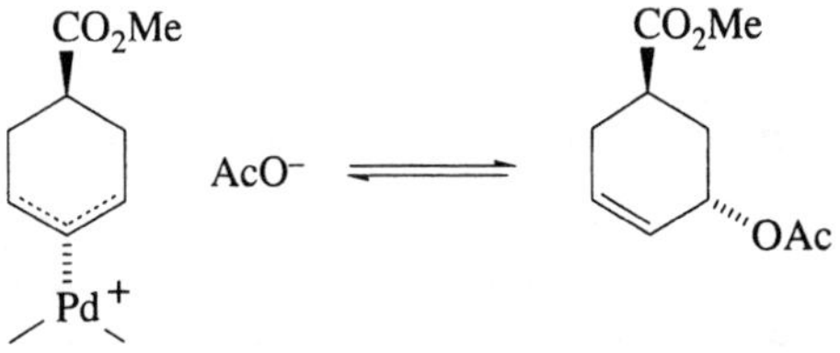

(187)

The reaction of dimethylamine with a preformed allylpalladium complex was found to yield products resulting from, on one extreme, complete ligand addition, on the other extreme, to up to 14% loss of stereospecificity, depending on the reaction conditions (equation 188).[190]

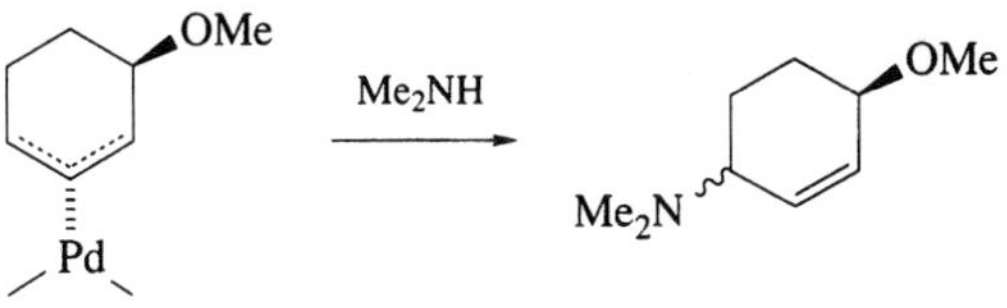

(188)

Dimethoxybenzhydrylamine (DMB) exhibited loss of stereospecificity in its Pd-catalyzed addition to acyclic allylic acetates. Pure (*Z*)-allylic acetates yielded only (*E*)-products in contrast to the complete stereospecificity in the addition of malonate anions to the same substrates, presumably due to the enhanced rate of *syn–anti* interconversion caused by the ability of the amine to coordinate to Pd (equation 189).[200]

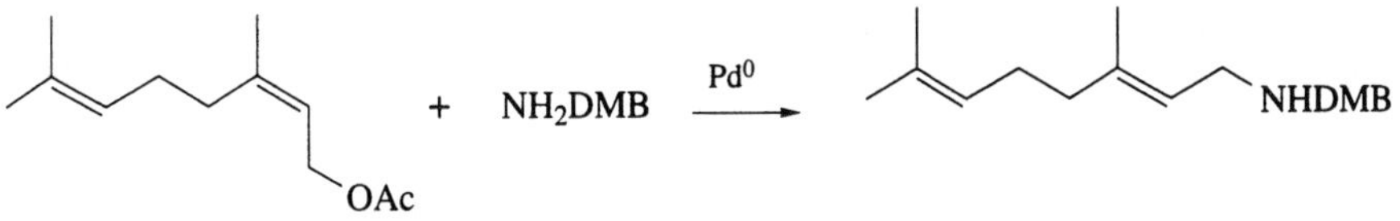

(189)

DMB-NH_2 also showed significant stereoscrambling in its addition to cyclohexenyl allylic acetates[200] (equation 190). This loss of stereointegrity could be due to either: (i) epimerization of the methoxycarbonyl group, although this seems unlikely under the reported reaction conditions; (ii) epimerization of the allylic acetate *via* metal addition, although no epimerized unreacted starting material could be detected; (iii) epimerization of the intermediate π-allyl complex by attack of uncomplexed PdL_2; or (iv) attack of the amine at both the ligand and metal center.[200]

Amides have been shown to add by both ligand (equation 191)[197] and metal attack (equation 192).[175]

CO_2Me + NH_2DMB $\xrightarrow{Pd^0}$ CO_2Me NHDMB (190)

60% *cis*
40% *trans*

AcO Cl + NH_2Ts $\xrightarrow{Pd^0}$ AcO NHTs (191)

Pd NHAc → Ac N H H (192)

3.3.3.5 Stereochemistry of Addition of Sulfur Nucleophiles

The addition of sodium phenylsulfinate nucleophiles to stereodefined acyclic allylic chlorides was reported to proceed with complete overall retention of configuration, indicating that this nucleophile adds with inversion of configuration, *i.e. via* attack at the allyl ligand (equation 193).[215] A cyclohexenyl acetate substrate also showed predominant ligand addition, but some isomeric product was also produced (equation 194).[216] This loss could be due to acetate epimerization of starting material, π-allyl epimerization by PdL_2, or by attack of the sulfur at the metal, followed by reductive elimination.

AcO Cl + $PhSO_2Na$ $\xrightarrow{Pd^0}$ AcO SO_2Ph (193)

OAc + $ArSO_2Na$ $\xrightarrow{Pd^0}$ SO_2Ar (194)

75:25 *cis*:*trans*

O-Allyl-*S*-alkyl dithiocarbonates react with Pd^0 catalysts to provide a π-allyl complex with a thiocarbonate serving as counterion. Under the reaction conditions the carbonate expels free methyl mercaptide, generating COS. Attack of the mercaptide was shown to occur exclusively *via* ligand addition (equation 195).[221]

O S MeS $\xrightarrow[-CO_2]{Pd^0}$ MeS (195)

A similar *in situ* generation of a sulfide nucleophile can be accomplished by the reaction of an allylic carbonate with a Pd^0 catalyst and $PhSSiMe_3$. Following oxidative addition by the Pd^0 into the allylic C—O bond to form the π-allyl complex, CO_2 is lost from the carbonate counterion, generating methoxide ion. The methoxide attacks the thiosilane, liberating thiophenoxide which subsequently adds *via* ligand addition to the Pd complex (equation 196).[223]

CO_2Me … OCO_2Me + $PhSSiMe_3$ $\xrightarrow{Pd^0}$ CO_2Me … SPh (196)

3.3.3.6 Stereochemistry of Addition of Other Heteroatom Nucleophiles

The stereochemistry of the Pd-catalyzed reaction of lithium diphenylthiophosphides with allyl acetates was determined using a 5-phenylcyclohex-2-enyl system. This phosphorus nucleophile was found to add exclusively *via* ligand addition (equation 197).[224,386]

Ph … OCOPh + $LiP(S)Ph_2$ $\xrightarrow{Pd^0}$ Ph … $P(S)Ph_2$ (197)

The silylation of allylic acetates has been accomplished utilizing tris(trimethylsilyl)aluminum and Pd^0 catalysts. The stereochemistry of this reaction has been determined and is consistent with exclusive ligand addition (equation 198).[225]

CO_2Me … OAc + $Al(SiMe_3)_3$ $\xrightarrow{Pd^0}$ CO_2Me … $SiMe_3$ (198)

3.3.3.7 Diastereoselectivity of other Reactions of π-Allylpalladium Complexes

3.3.3.7.1 Insertion into π-allylpalladium complexes

The insertion of CO[239] or alkenes[253] into π-allylpalladium complexes proceeds with complete retention of configuration with respect to the metal (equations 199 and 200).

MeO … Pd– $\xrightarrow[MeOH]{CO}$ MeO … CO_2Me (199)

+ Pd … CO_2Me ⟶ … CO_2Me … Pd … H (200)

3.3.3.7.2 Oxidation of π-allylpalladium complexes

Treatment of preformed, stereodefined π-allylpalladium complexes derived from steroids with MCPBA gave an allylic alcohol product with retention of configuration with respect to the metal (metal attack) (equation 201).[272] In like fashion, $MoO_2(acac)_2$–Bu^tO_2H oxidation of preformed π-allylpalladium complexes provided allylic hydroxy compounds with retention of configuration with respect to the Pd (equation 202).[275]

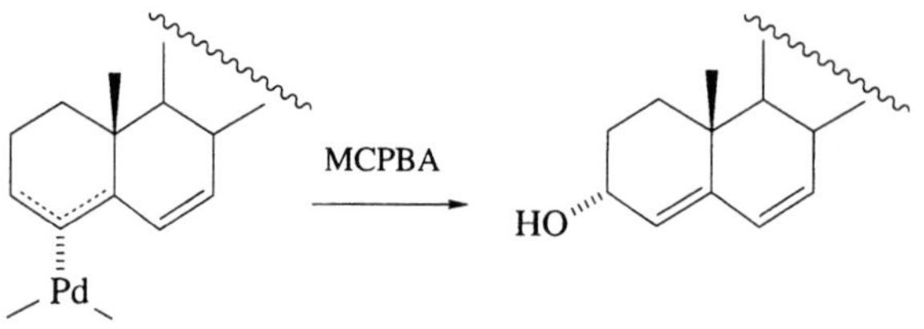

MCPBA (201)

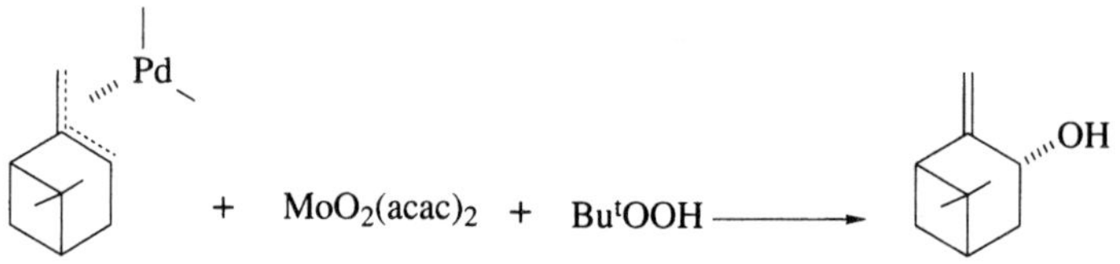

$+ \ MoO_2(acac)_2 \ + \ Bu^tOOH \longrightarrow$ (202)

3.3.3.7.3 *Reduction of π-allylpalladium complexes*

Addition of hydride to π-allylpalladium complexes uniformly proceeds by attack at the metal, followed by reductive elimination, irrespective of the source of the hydride, *e.g.* $NaBH_4$,[286,287] $LiAlH_4$,[292] R_3SnH[385] or $NH_4^+ \ HCO_2^-$ (equation 203).[385]

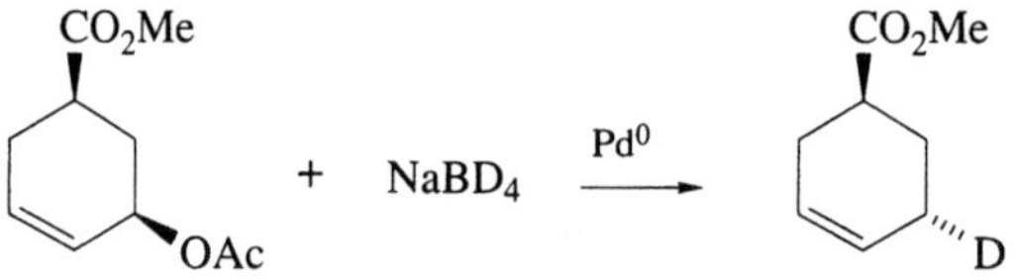

$+ \ NaBD_4 \xrightarrow{Pd^0}$ (203)

3.3.3.7.4 *Umpolung*

Conversion of allylic acetates of allylstannanes by treatment with $Et_2AlSnBu_3$ and Pd^0 catalysts proceeds by addition of the tin to the metal, followed by reductive elimination (equation 204).[309] The corresponding allyl phosphonate, however, showed some loss of stereochemical integrity (equation 205).[310] The Pd^0-catalyzed reaction of allylic acetates, trialkyltin chlorides and SmI_2 produced allylstannanes with no stereospecificity.[311]

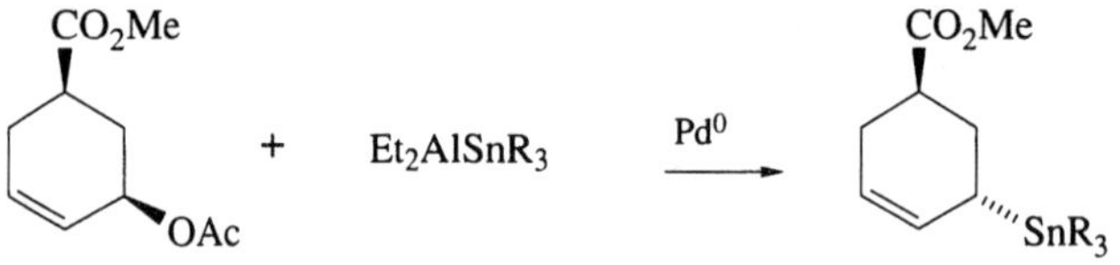

$+ \ Et_2AlSnR_3 \xrightarrow{Pd^0}$ (204)

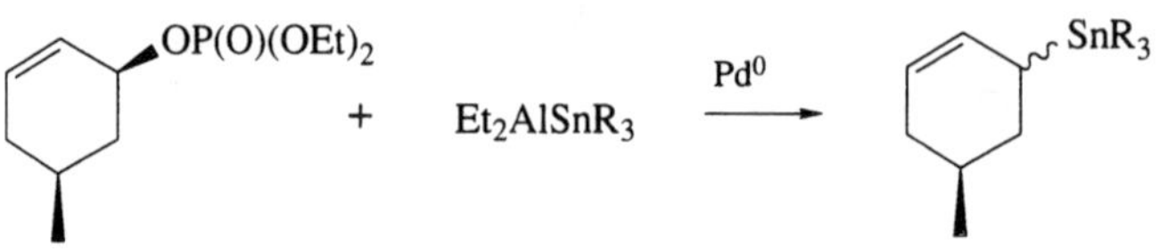

$+ \ Et_2AlSnR_3 \xrightarrow{Pd^0}$ (205)

78:22 *cis:trans*

3.3.3.7.5 *Addition reactions*

The addition of 1,1,1-trichloro-2,2,2-trimethyldisilane to preformed allyl complexes proceeded *via* retention of configuration with respect to the metal (equation 206).[339]

Ph Pd → Ph $SiMe_3$ (206)

3.3.4 REGIOSELECTIVITY

3.3.4.1 Regioselectivity in the Addition of Nucleophiles to π-Allylpalladium Complexes

3.3.4.1.1 Nucleophiles that add via *attack at the π-allyl ligand*

The factors that determine the regiochemistry of addition of nucleophiles that attack directly on the allyl ligand can be partitioned between those based on steric and those based on electronic influences. The most important factor overall in determining regioselectivity is the steric interaction between the incoming nucleophile and the allyl terminus to which it is bonding. The relative steric environments of the allyl termini, as well as the size of the nucleophile, are key in determining regiochemistry. A secondary steric factor that can be significant is based on consideration of the alkene–palladium complex that is formed as the reaction proceeds. In general, Pd complexes to less substituted alkenes are favored. The size of the ligands on the palladium will also affect the ability of the metal to interact with an alkene, with less sterically demanding ligands being favored.

The electronic factors in determining regiochemistry are based on the ability of substituents on the allyl moiety or ligands on the palladium to induce an asymmetric distribution of the electron deficiency present in the allyl ligand, thereby making one terminus more reactive with an electron-rich nucleophile.

Asymmetric bonding of the π-allyl ligand to the palladium, which is not a distinct consideration from these steric and electronic factors but rather a manifestation of them, must also be assimilated into an understanding of the regiochemistry of π-allyl alkylations.

3.3.4.1.2 Nucleophiles that add via *attack at the metal*

Since the coupling of the nucleophile and the allyl ligand constitutes a reductive elimination from the metal, it is believed to proceed with a *cis* geometry of these groups about the metal. The geometric distribution of ligands in the allylpalladium complex that would foster the formation of a *cis* orientation between the palladium–Nuc bond and a specific allyl terminus must then be considered in understanding the regiospecificity of this reaction.

3.3.4.2 Regioselectivity in the Addition of Soft Carbon Nucleophiles to π-Allylpalladium Complexes

Soft carbon nucleophiles (pK_a 10–20) have been shown to add to π-allylpalladium complexes by direct attack at the allyl ligand and to add preferentially to the sterically less congested allyl terminus. Although all the effects responsible for the regioselectivity demonstrated in a particular case cannot easily be accounted for, the discussion will be organized by the apparent dominant controlling factor being exhibited.

3.3.4.2.1 Effect of substituents on the allyl ligand

(i) Alkyl- and aryl-substituted π-allyl complexes

Asymmetry on the allyl ligand caused by the substitution of alkyl or aryl groups on the terminus results in preferential addition by the nucleophile at the less substituted end of the allyl group. As is often observed in these reactions, the relative steric interactions between the allyl termini and the nucleophile must be the dominating factor in determining regioselectivity. Arguments based on electronic factors would predict greater stabilization of cationic character in the allyl ligand at the more substituted terminus and preferential addition there. Similarly, if the relative stability of the incipient palladium–alkene complex dominated, then attack at the more substituted end should be observed, as this leads to a less substituted and more stable alkene complex.

Relatively subtle changes in the steric environment of the allyl termini result in substantial variations in regioselectivity, as shown in a series of π-allyl ligands having a methyl at one terminus and either Pr^n, Bu^i or Pr^i at the other terminus. This series varies from a 77:23 ratio favoring the less substituted terminus to >99:1 as the difference in group size increases, with dimethyl malonate as nucleophile (equation 207).[399]

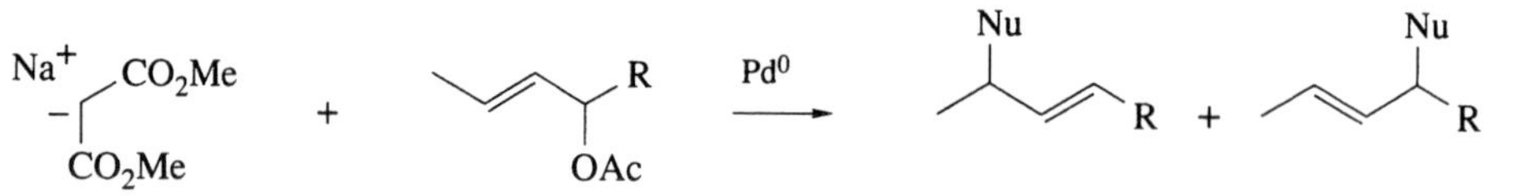

(207)

R = Pr^n	77	23
R = Bu^i	93	7
R = Pr^i	>99	1

The size of the nucleophile can also exert effects on the regiochemical outcome of the reaction. On switching from a dimethyl malonate to a more bulky methyl(methylsulfonyl) acetate, the preference for attack at the less substituted terminus increases, as seen by comparing equations (208) and (209).[380]

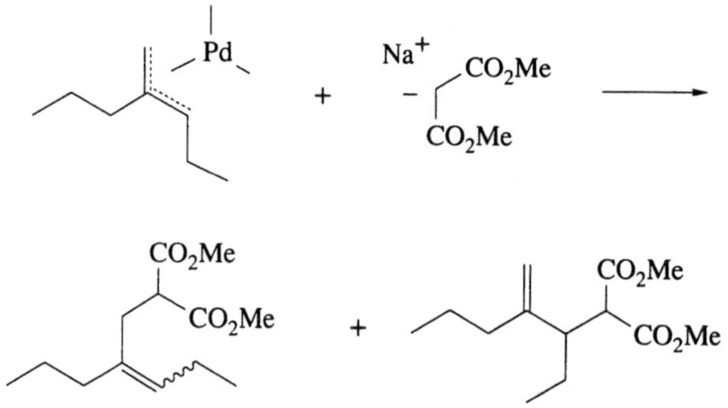

(208)

89% 11%

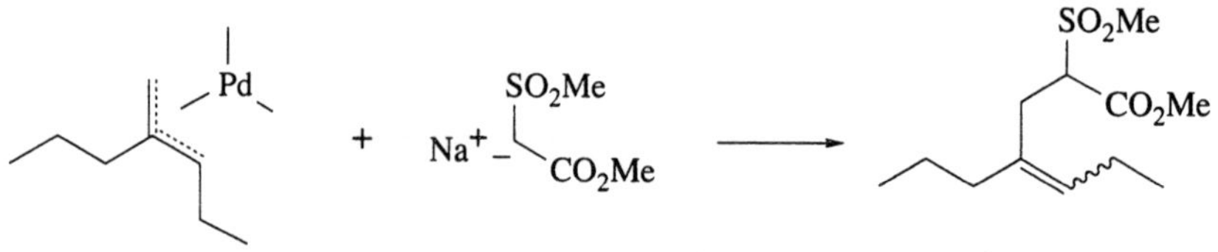

(209)

A fascinating control of regiochemistry has been discovered in the preparation of medium size rings. Intramolecular cyclization of phenylsulfonyl acetates, when confronted by a choice of attack at the allyl termini to create an *n* or *n* + 2 ring, *n* = 6–14, invariably opts for the larger ring.[400–402] Although this reflects the usual propensity for attack at the less substituted allyl terminus, other factors including the nature of the nucleophile, the ligands on the Pd and the chain configuration have also been cited as contributing to the observed selectivity (equation 210).

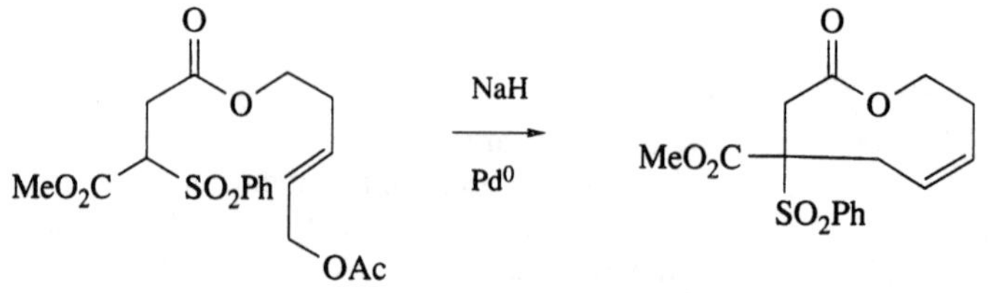

(210)

mixture of alkene isomers

Soft carbon nucleophiles other than malonate derivatives follow similar patterns of regioselectivity, as shown by the addition of cyclopentadiene anion (equation 211)[105] and the highly acidic 4-hydroxy-6-methyl-2-pyrone (equation 212).[393]

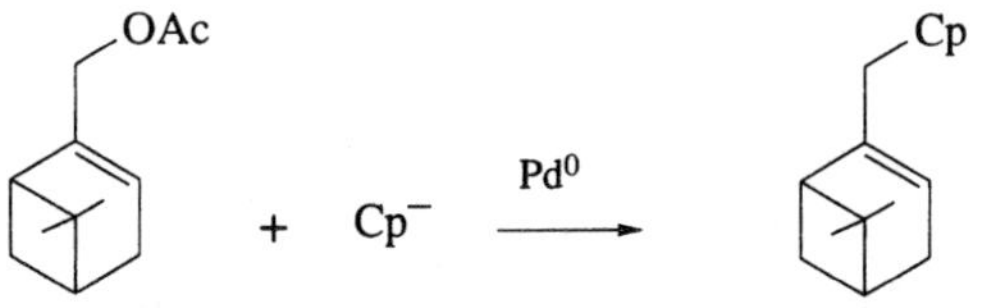

(211)

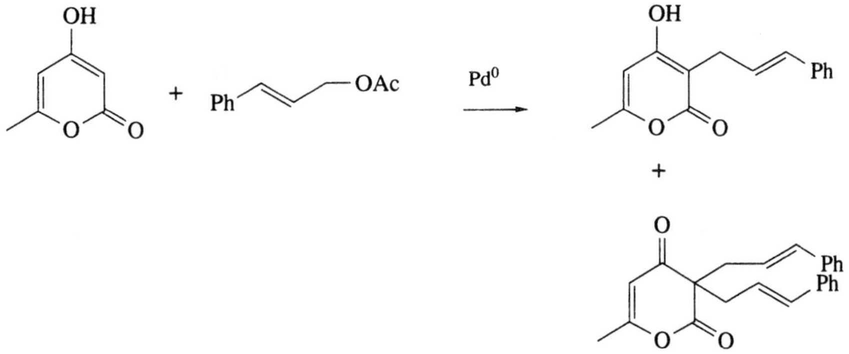

(212)

(ii) Functional group effects

The substitution of various functional groups on the terminus of a π-allyl ligand, including carbonyls (equation 213)[327,391,392,403,404] and nitriles (equation 214),[405] as well as heteroatoms such as —SR (equation 215)[406,407] and —P(O)R_2 (equation 216),[408] induces a strong tendency for addition of soft carbon nucleophiles to the remote terminus. This tendency is difficult to rationalize on steric grounds as isosteric alkyl groups do not exert nearly as effective regiocontrol. Analysis of electronic factors also does not offer an obvious explanation. Extrapolating the electrophilicity imparted to the allyl ligand by the Pd^{II} to the extreme limit of an allyl cation simplifies the analysis. The charge distribution created by a carbonyl group, *versus* an oxygen or sulfur group, must be vastly different, as judged by the relative contributions of the resonance forms pictured in equations (217)–(219). The only common electronic property enjoyed by these functionalities is their ability to withdraw electrons in the σ-sense. The relevance of this feature, and the irrelevance of their widely differing π-electron effects, remains unclear. The ability of these groups to promote asymmetry in the bonding of the allyl ligand,[409] either through electronic interactions or by internal coordination, remains largely unstudied by X-ray structure determination.

The effects created by these functionalities are, surprisingly, capable of being transmitted from positions even more remote to the allyl ligand. 1,3-Diene monoepoxides, for example, selectively undergo the addition of nucleophiles remote from the incipient hydroxymethyl group (equation 220).[410–413] Nitro,[414] malonate[415] and acetoxy[416,417] functionalities show similar tendencies (equations 221–223). Tri-

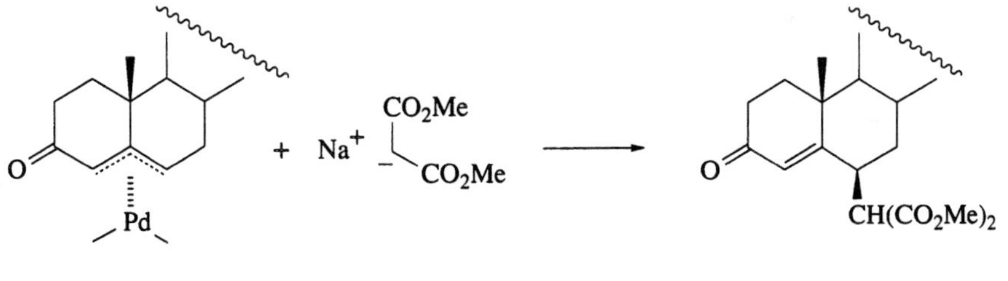

(213)

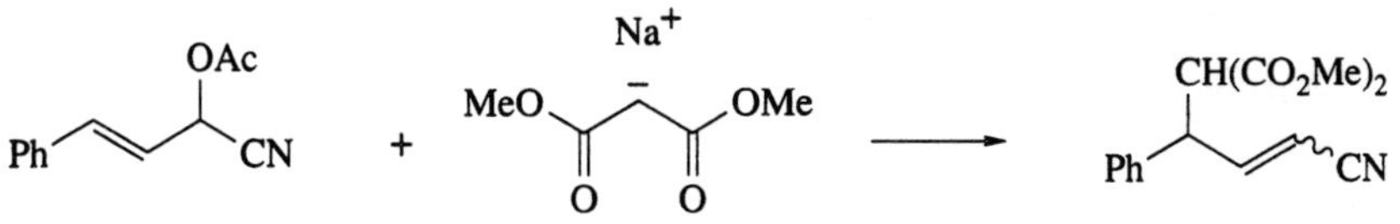

(214)

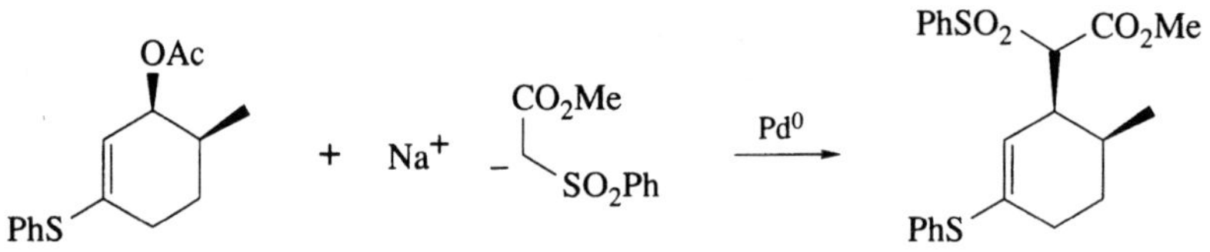

(215)

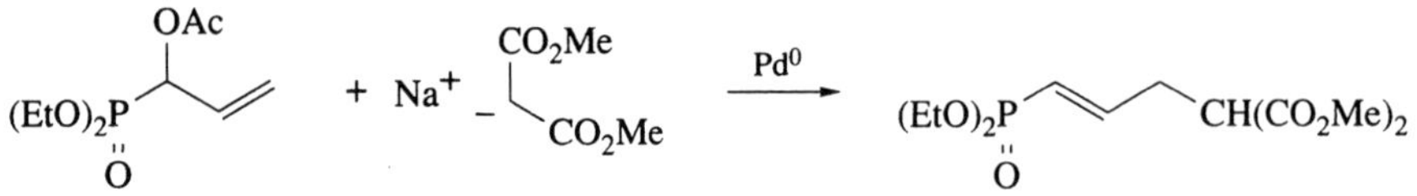

(216)

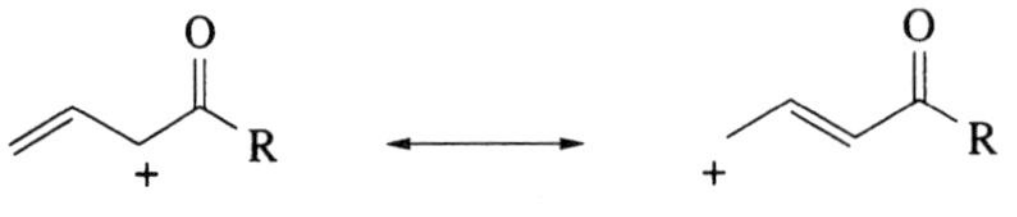

(217)

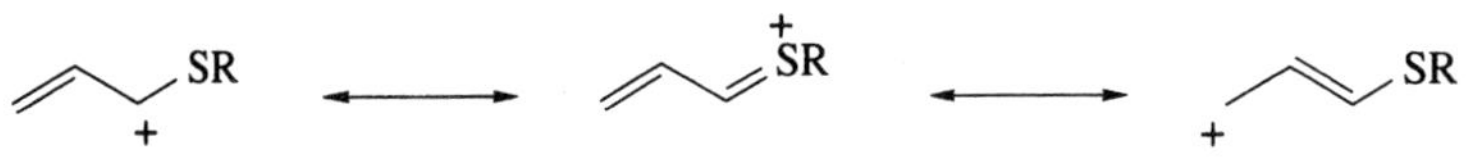

(218)

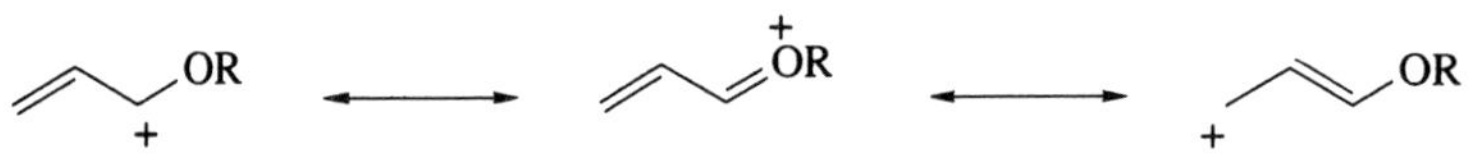

(219)

methylsilyl-substituted allyl ligands also direct carbon nucleophiles to the remote terminus, presumably based on simple steric considerations (equation 224).[418,419]

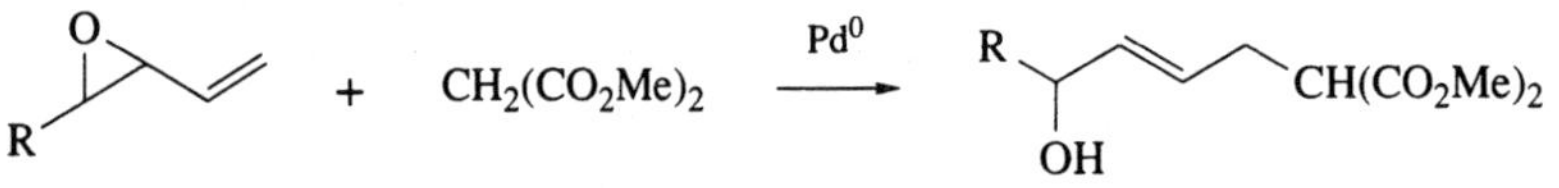

(220)

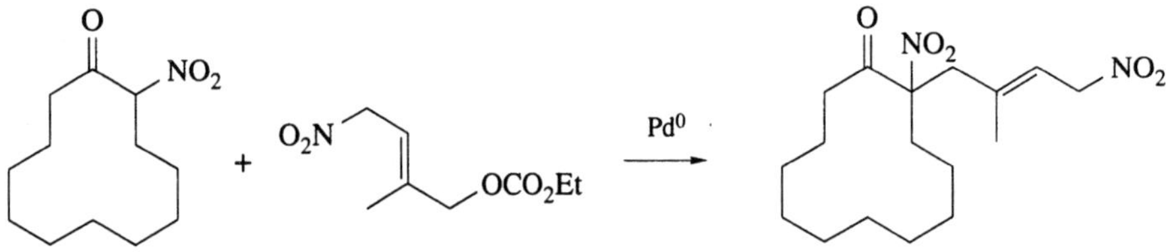

(221)

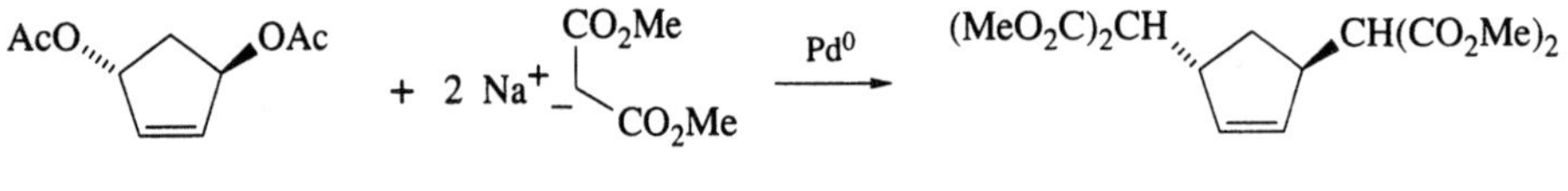

(222)

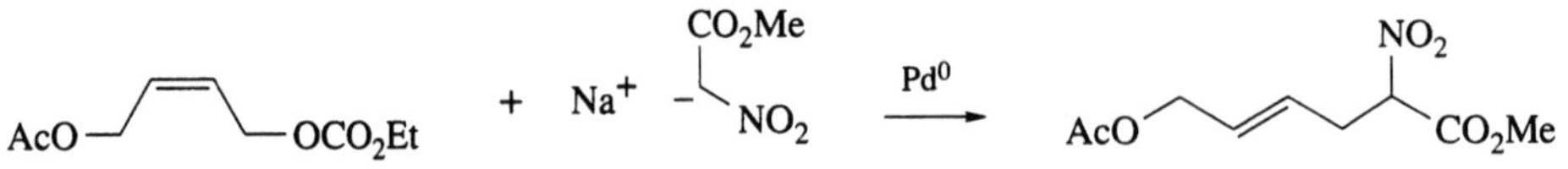

(223)

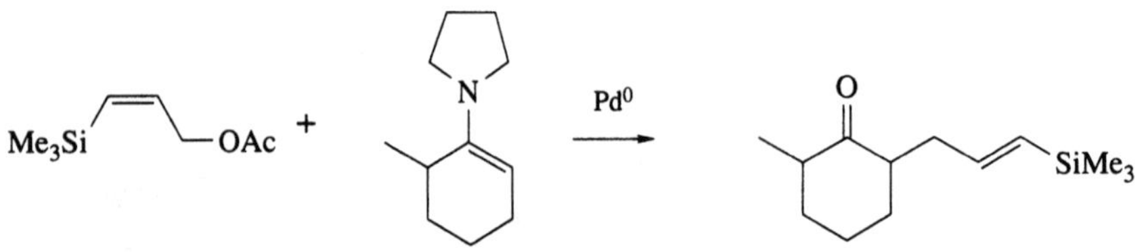

(224)

(iii) Ligand effects

The nature of the ligands on the palladium in π-allyl complexes can influence the regioselectivity exhibited by soft carbon nucleophiles. π-Allylpalladium complexes generated from methylenecycloalkanes provide an example of the effect of ligands on regiochemistry. The complexes derived from methylenecyclopentane and methylenecycloheptane both exhibit exclusive exocyclic addition by the anion of methyl(methylsulfonyl) acetate with triphenylphosphine ligands on the Pd (equation 225). In contrast, the complex derived from methylenecyclohexane yields a 62:38 ratio of exocyclic:endocyclic addition (equation 226).

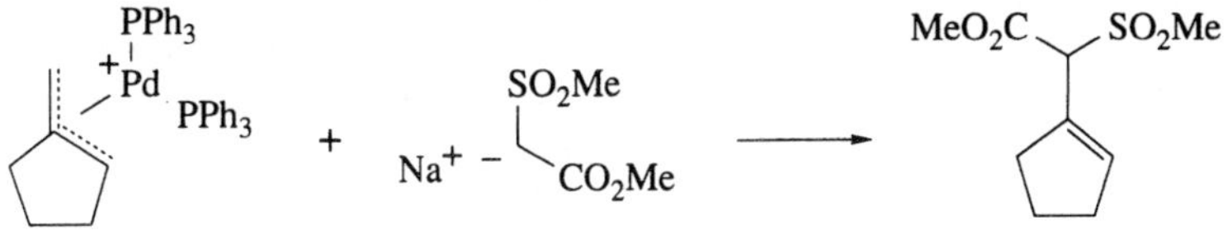

(225)

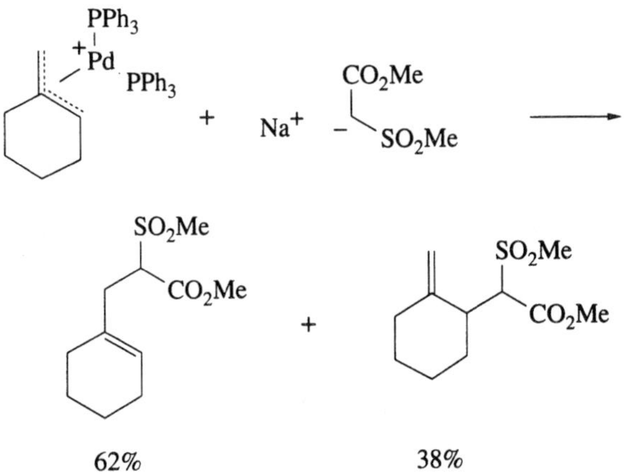

(226)

62% 38%

This result has been explained[72,390] by the particular instability of the alkene complex resulting from exocyclic addition (**14**) relative to endocyclic addition (**15**) in the cyclohexyl system. This rationale is supported by the results obtained by the use of the considerably more sterically bulky tri-*o*-tolylphosphine instead of triphenylphosphine. A 15:85 exocyclic:endocyclic ratio is obtained with the bulky phosphine. This result is nicely explained by the ability of the bulky phosphine to magnify the instability of the endocyclic alkene–palladium complex (**14**) relative to the exocyclic alkene complex (**15**) and favor endocyclic addition.

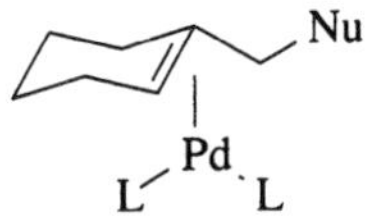

(**14**) 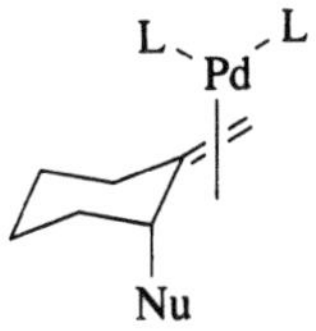 (**15**)

An elegant study on the influence of ligands on the regiochemistry of π-allylpalladium alkylations has recently appeared.[420] This study is based on the premise that acceptor ligands should induce greater positive charge at the more substituted terminus of the η^3-allyl ligand, leading to a greater preference for reaction at this terminus, and thereby modifying the usual steric preference for attack at the less substituted terminus.

The model reaction employed the allyl complex derived from 3-methylbutene and sodiodiethyl malonate as nucleophile (equation 227). The ratio of products observed was consistent with the relative induced charge as indicated by the ^{13}C NMR shift of C-3 and by the difference between the shifts at C-1 and C-3.[421] Thus, the ratio of product from attack at the less substituted terminus to attack at the more substituted terminus was small for the best acceptor ligands, 1.3:1 for L = $P(OPh)_3$ and 2.7:1 for L = PPh_3, and increases for less potent acceptors, *e.g.* 10:1 for L = $As(PPh_3)_3$. For donor ligands such as L = pyridine, only attack at the less substituted terminus was observed.

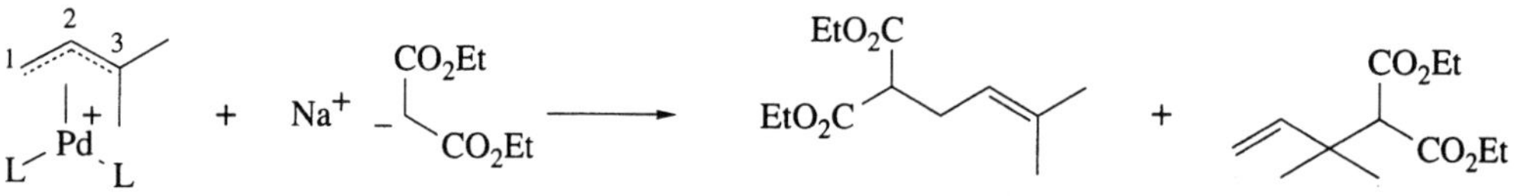

(227)

3.3.4.3 Regioselectivity in the Addition of Enolates to π-Allylpalladium Complexes

Since enolates also add *via* a ligand attack process, the regioselectivity that they exhibit is quite comparable to soft carbon nucleophiles. Alkyl or aryl substituents at the allyl termini direct attack to the less substituted terminus (equations 228–232); functional groups such as CO_2Me and halogen at one allyl terminus direct attack to the remote terminus (equations 233 and 234). Remote functionalities such as —OR also direct addition to the allyl terminus more removed from the substituent (equations 235 and 236).

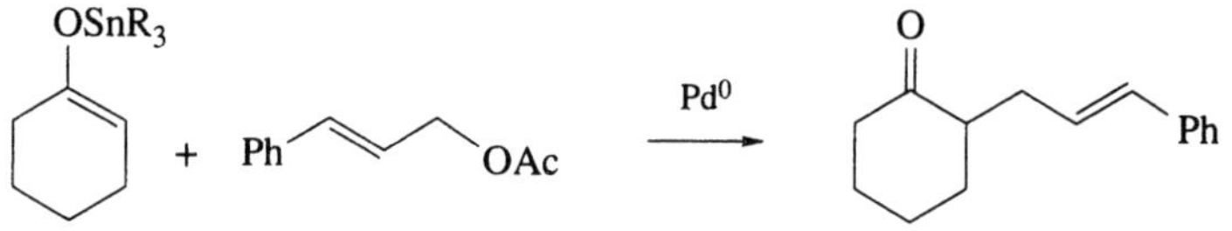

(228)

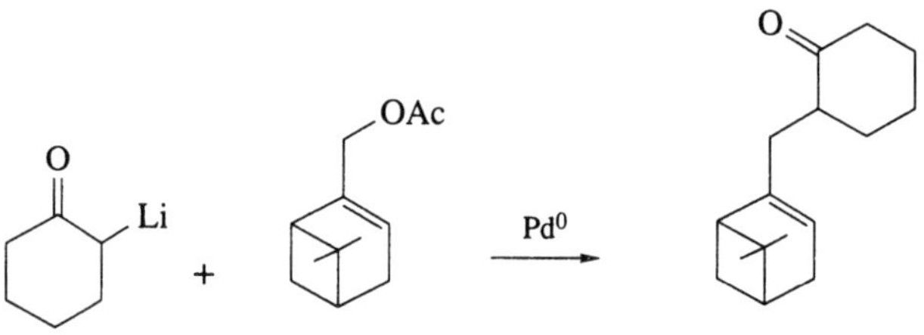

(229)

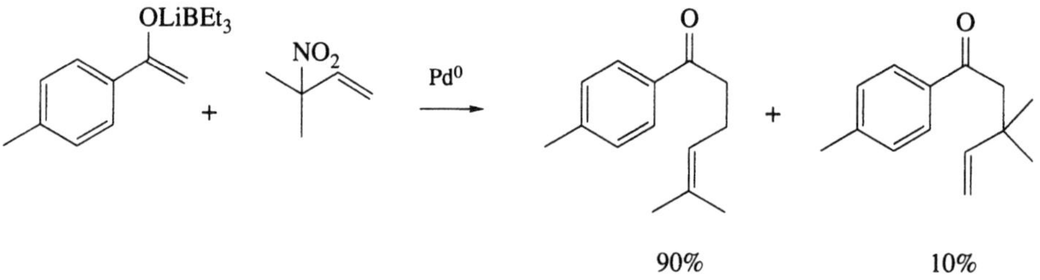

90% 10%

(230)

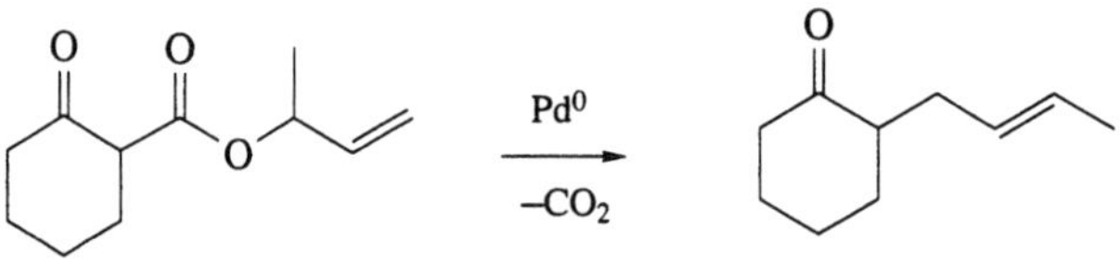

(231)

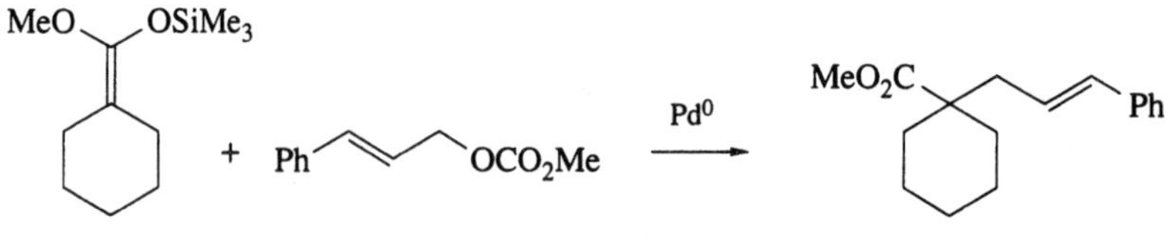

(232)

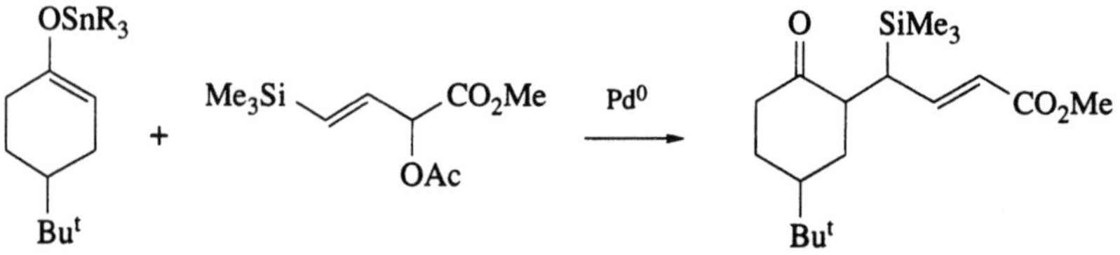

(233)

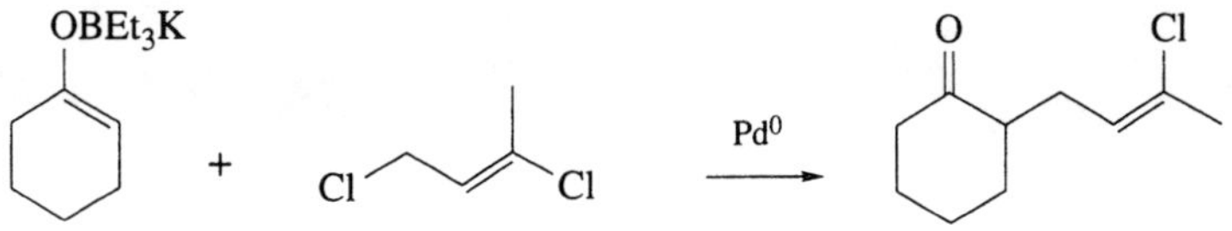

(234)

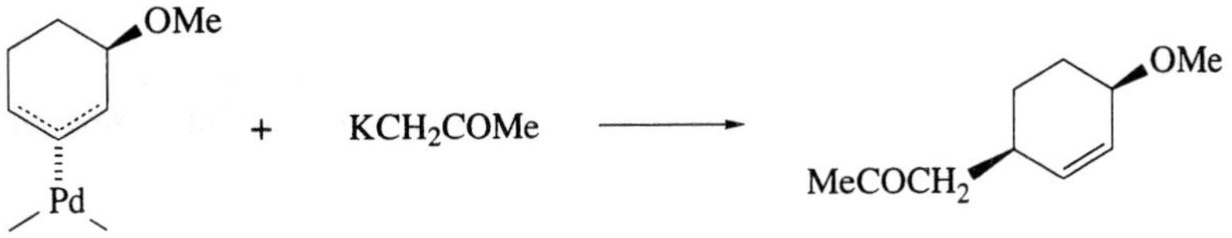

(235)

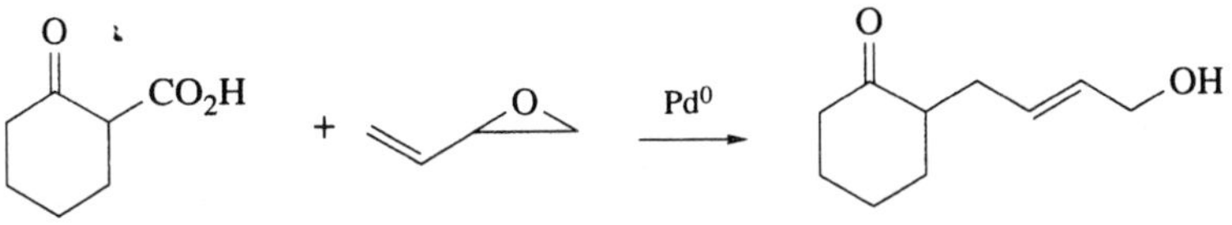

(236)

These regioselectivity patterns are observed for tin enol ethers (equations 228 and 233),[106,107,399] alkali metal enolates (equations 229 and 235),[111] lithium trialkylborate enolates (equations 230 and 234)[108,109] and palladium enolates (equations 231, 232 and 236).[95,111,112,115,126,127]

Two cases merit brief further discussion. In a competition between the regiochemical directing effects of a TMS and a CO_2Me group positioned on opposite termini of an allyl ligand, the CO_2Me group dominates and directs the enolate nucleophile to the TMS end (equation 233).[107] The palladium ester enolate derived from oxidative addition into a vinylcyclopropane, followed by Michael addition, closes to give only a five-membered ring product in preference to the alternative seven-membered ring (equation 237).[96]

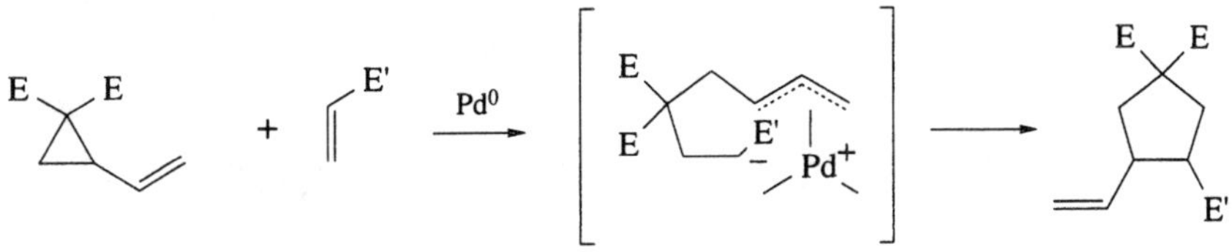

(237)

3.3.4.4 Regioselectivity of Organotins

Aryl- and vinyl-tins have been shown to add *via* attack at the metal, in contrast to the soft carbon nucleophiles and enolates. Although this change in mechanism is accompanied by a change in the factors controlling regiochemistry, these tin reagents behave comparably to the carbon nucleophiles that act *via* ligand addition. Alkyl- (equation 238),[145] alkoxycarbonyl- (equation 239),[142] and epoxy-substituted (equation 240)[143] allyls all efficiently direct the organotin to the remote terminus. Allyltins have been reported to add with inversion of the allyl moiety and suggested to attack directly the allyl ligand, except if maleic anhydride is added to the reaction mixture, when metal addition is observed.

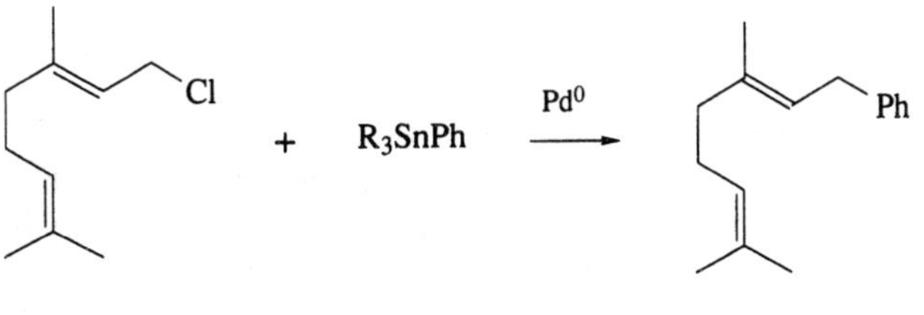

(238)

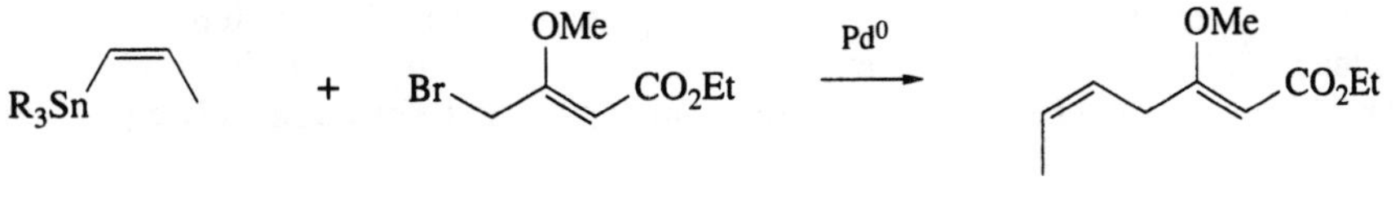

(239)

$$\text{Ph-vinyl epoxide} + PhSnR_3 \xrightarrow{Pd^0} \text{Ph-CH(OH)-CH=CH-CH}_2\text{Ph} \quad (240)$$

The opposite regioselectivity observed for the addition of methylbutenyltin to similar allyl substrates (equations 241 and 242),[137,139] caused by the presence of maleic anhydride in the latter case (equation 242), may be viewed as support for a change in mechanism when the electron deficient alkene ligand is present. Conversely, both reactions could proceed by metal addition, with the regiocontrol exercised by the maleic anhydride acting in the addition–reductive elimination sequence.

(241) SnR_3 + Ph~~OAc $\xrightarrow{Pd^0}$ Ph

(242) —Pd + R_3Sn $\xrightarrow{\text{maleic anhydride}}$

In situ generation of an allylstannane from an allylic acetate and hexamethyldistannane results in five-membered ring formation in preference to the allyl alternative seven-membered ring (equation 243).[140]

(243) OAc OAc + Me_6Sn_2 $\xrightarrow{Pd^0}$ $SnMe_3$ OAc ⟶

A remote methoxy substituent directs maleic anhydride-mediated allylation predominantly in the 1,4-sense, comparable to the phenyltin–vinyl epoxide example (equation 244).[139]

(244) OMe Pd + R^3Sn $\xrightarrow{\text{maleic anhydride}}$ OMe (75%) + OMe (25%)

3.3.4.5 Regioselectivity of Organozincs

In contrast to the organotin derivatives which proceed by metal addition but apparently respond to asymmetry in the allyl ligand with regiochemical outcomes similar to those displayed by soft carbon nucleophiles (ligand attack), organozinc derivatives add by metal addition and show the complementary regiochemistry.[155,399] For example, perfluoroalkylzinc halides and phenylzinc chloride preferentially add to the more substituted allyl terminus (equations 245 and 246). In addition to exhibiting complementary regiochemical selectivity to the soft carbon nucleophiles, PhZnCl is also pronouncedly more discriminating, showing a 99:1 selectivity in an allyl system possessing a methyl at one terminus and a Pr^n at the other (equation 246) (*cf.* diethyl malonate, 22:73). Explanations based on differential donor–acceptor properties of the allyl termini that have been advanced to explain regioselectivity in metal addition processes appear difficult to apply here, as the termini would appear to be electronically equivalent.

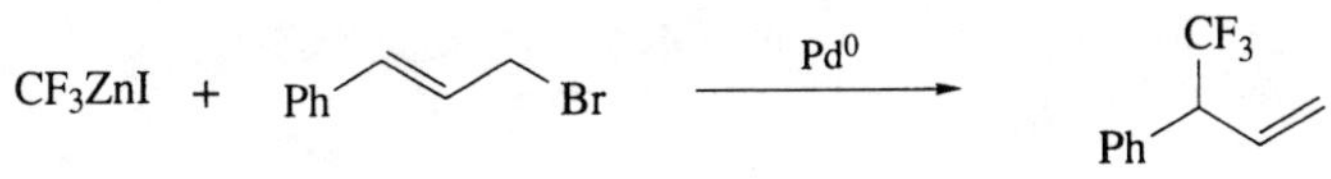

CF_3ZnI + Ph⌒⌒Br —Pd^0→ Ph–CH(CF_3)–CH=CH_2 (245)

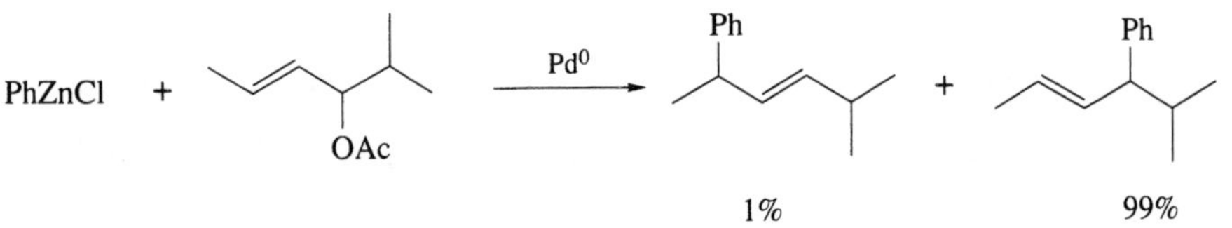

PhZnCl + (OAc) —Pd^0→ (Ph) 1% + (Ph) 99% (246)

3.3.4.6 Regioselectivity of Organoaluminums and Organozirconiums

Alkenylaluminums add *via* metal addition and prefer coupling to the less substituted allyl terminus (equation 247).[154]

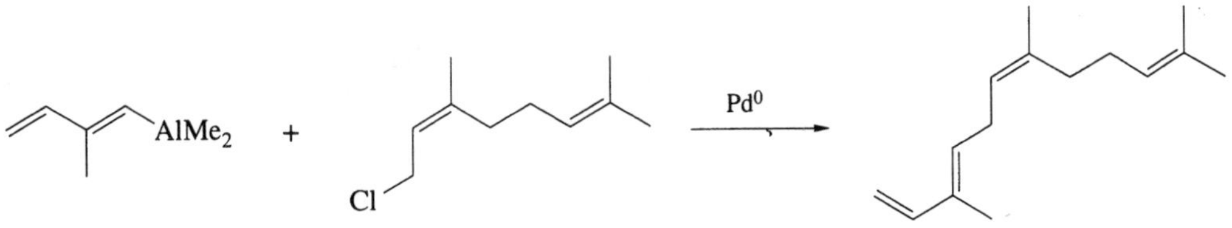

$AlMe_2$ + (Cl) —Pd^0→ (247)

The most elegant and comprehensive mechanistic work associated with controlling the regiochemistry of additions to π-allylpalladium complexes that occur *via* metal attack–reductive elimination has been done in the area of organozirconiums.[155,156,395] The coupling of the ligands of the palladium allyl–alkenyl complex, formed on addition of an alkenylzirconium to a π-allylpalladium, was assumed to be an intramolecular process which requires a *cis* orientation of the groups that bond. As a result, the geometric distribution of ligands in the allylpalladium–alkenyl complexes immediately preceding formation of the carbon–carbon bond should determine the regioselectivity.

The model systems employed in these studies were derived from steroids. In these systems, it was reasoned that an asymmetrical allyl ligand should exert unequal *trans* influences upon the ligands on the palladium. The favored situation would have the stronger acceptor ligand *trans* to the more strongly donating (closer) terminus of the allyl ligand. In the allyl ligands studied, it was suggested that the distance Pd–C(20) would be less than the distance Pd–C(16), based on relative steric environments. As a result, the exocyclic C-20 allyl terminus would exert a stronger electron donating interaction towards the Pd and prefer to have an acceptor ligand *trans*. That is, complex (**16**) would be favored over (**17**). A donor ligand L should alternatively favor (**17**). In agreement with this rationale, coupling in the absence of maleic anhydride (L = Cl) gave a predominance of C-16 addition and in the presence of the acceptor ligand (L = maleic anhydride) gave a >95:5 ratio favoring alkenylation at C-20.

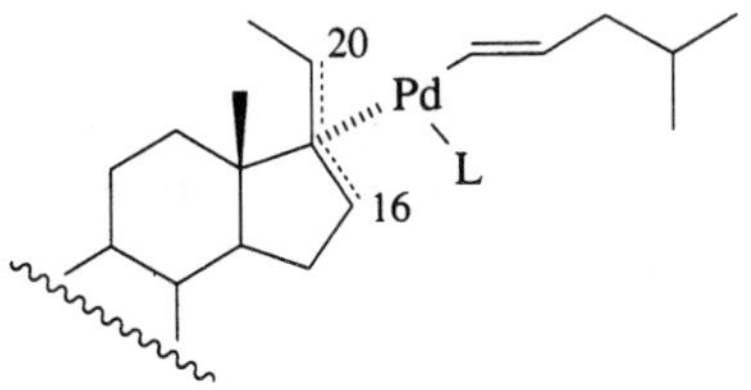

20 Pd L 16   Pd L

(**16**) 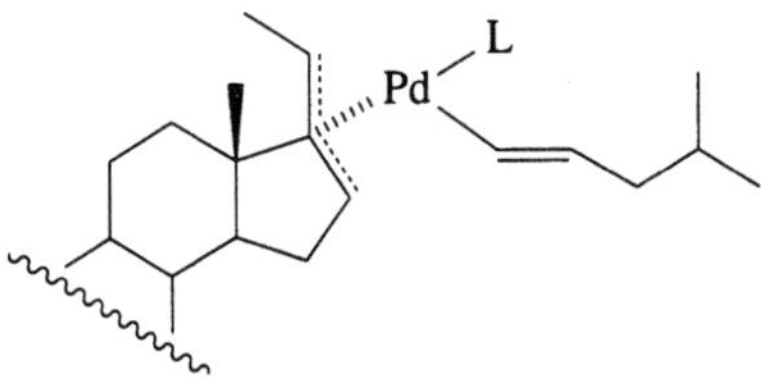   (**17**)

3.3.4.7 Regioselectivity of Grignards and Organolithiums

The addition of Grignards and organolithium reagents proceeds by attack at the metal center in π-allylpalladium complexes. The regiochemical selectivity exhibited by these hard carbon nucleophiles with π-allyl complexes substituted at the termini with alkyl or aryl groups is comparable to the soft carbon nucleophiles (ligand attack) in most cases, with addition occurring predominantly at the less substituted terminus (equations 248 and 249).[159,387]

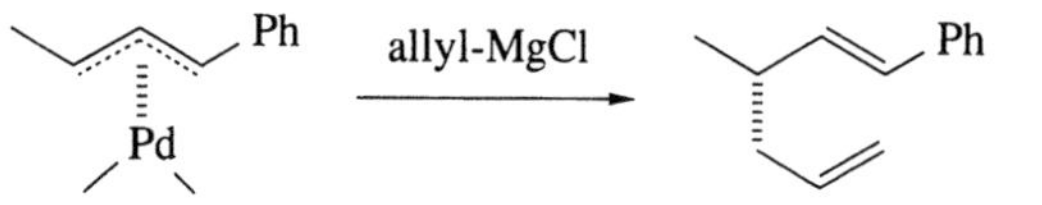

allyl-MgCl (248)

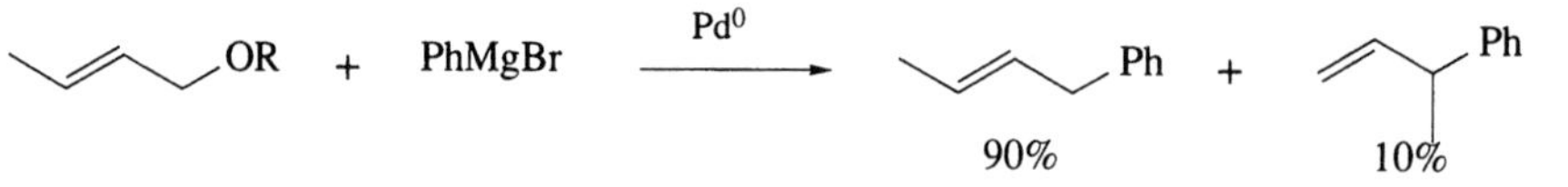

Pd⁰ (249)

90% 10%

An elegant procedure has been elucidated for allyl–allyl′ coupling which demonstrates the crucial need for a π-acceptor ligand, such as maleic anhydride, to promote the reductive elimination (C—C bond-forming step).[366,367] These allyl–allyl′ couplings proceed with predominant 'head to head' regiochemistry, that is, with C—C bond formation between the less substituted allyl termini (equation 250). This is in contrast to the addition of allyltin reagents which are suggested to proceed by ligand addition in the absence of maleic anhydride and give 'head to tail' allyl–allyl′ couplings.

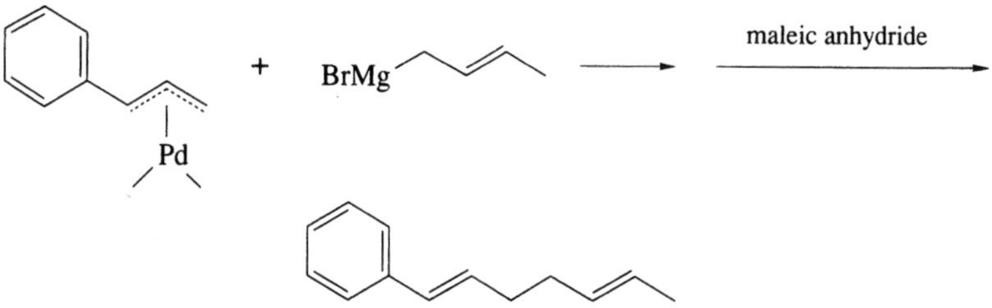

maleic anhydride (250)

Some additional examples of regioselectivity of Grignard reagents have been reported. In contrast to the addition of alkenylzirconiums to the identical steroidal π-allyl complexes which selectively coupled to C-20, the Grignard reagent in the presence of maleic anhydride shows complete selectivity for addition to C-16 (equation 251).[396,397]

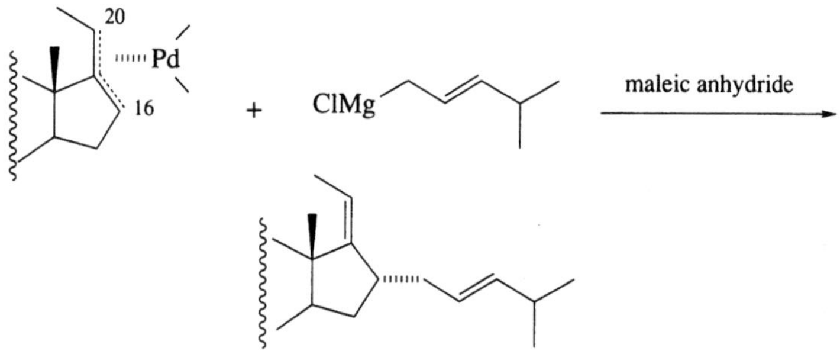

maleic anhydride (251)

Addition of MeMgI[158] or PhLi[160] to π-allylpalladium complexes derived from methylenenorbornane exhibits selective C—C bond formation at the more hindered allyl terminus. Soft carbon nucleophiles with the same allyl species show the opposite regioselectivity (equation 252).[158,160]

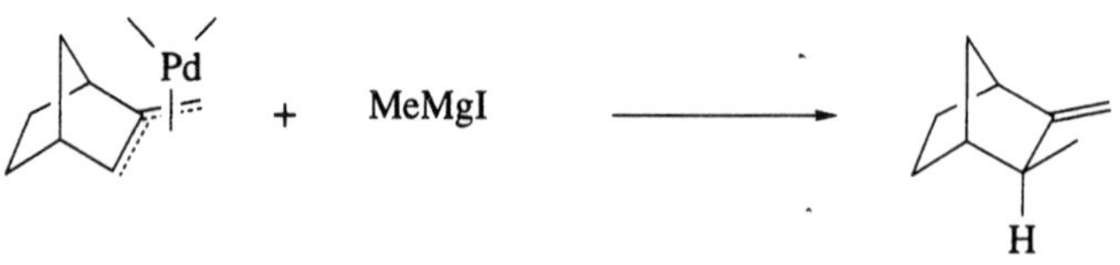

(252)

Regiodirection by remote oxygen functionality in the addition of Grignards parallels that observed by soft carbon nucleophiles, showing predominant attack at the more remote allyl terminus (equation 253).[397]

It is clear that although substantial progress has been made in the understanding and control of regioselectivity in the addition of nucleophiles to π-allylpalladium complexes *via* attack at the metal, considerable variability still exists in these reactions that is not readily mechanistically rationalized.

OMe OMe OMe
i, BrMg
ii, maleic anhydride
Pd
71% 29%
(253)

3.3.4.8 Regioselectivity of Oxygen Nucleophiles

Understanding the factors that control the addition of oxygen nucleophiles to π-allylpalladium complexes is complicated by the fact that: (i) oxygen nucleophiles, especially carboxylates, can add *via* ligand or metal addition; and (ii) the oxygen nucleophile can also serve as a leaving group, allowing equilibration of the products, thus superimposing thermodynamic over kinetic control of the addition.

Under conditions where metal attack is known to occur, an acetate nucleophile will preferentially add to the more substituted allyl terminus when directed by alkyl substitution (equation 254).[163] Remote oxygen-containing functionality directs acetate addition to the more distant terminus when metal attack is fostered (equations 255 and 256). Identical directing effects are observed from remote oxygen groups under conditions known to favor ligand addition (equations 257 and 258).[68,169,176] Reports on the addition of acetate where stereochemical information is not available and therefore the mechanism of attack is unknown indicate that CO_2R,[172] SO_2R[173,174] and CN[385] groups substituted directly on the allyl ligand all direct attack at the remote allyl terminus (equations 259–261).

H Pd OAc
CO
H OAc
(254)

MeOCH2 Pd OAc
CO
MeOCH2 CH2OAc
(255)

OMe Pd OAc
CO
OMe AcO
(256)

OAc Pd OAc
HOAc
LiOAc
LiCl
OAc AcO
(257)

O + PhOH
Pd0
OH PhO
(258)

CO2Et Pd
AcO−
CO2Et OAc
(259)

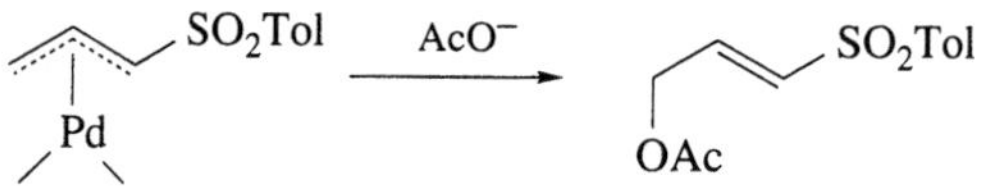

(260)

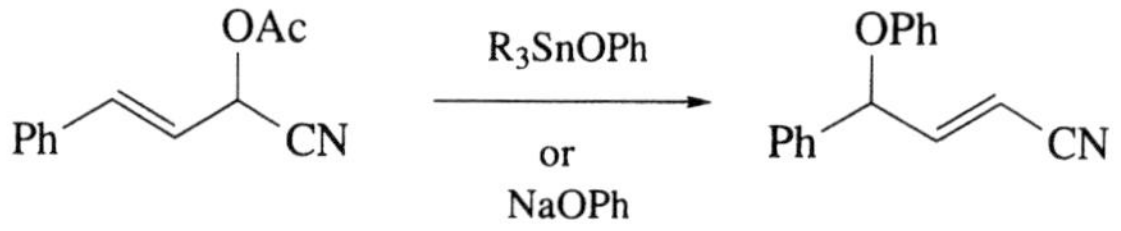

(261)

3.3.4.9 Regioselectivity of Amine Nucleophiles

Amine nucleophiles appear generally to add by attack at the allyl ligand. Similar to oxygen nucleophiles, however, a caveat must be noted in evaluating the observed regioselectivities in that the allyl amine products are capable of rearrangement by reentry into the π-allyl manifold.

Substitution on the terminus of the allyl ligand by alkyl[81,183,186,191,199,200] or aryl groups generally directs the addition of the amine to the less substituted allyl terminus (equations 262 and 263). The preference for reaction at the less substituted terminus of the allyl group increases as the amine nucleophile becomes more bulky (equations 264 and 265).[422]

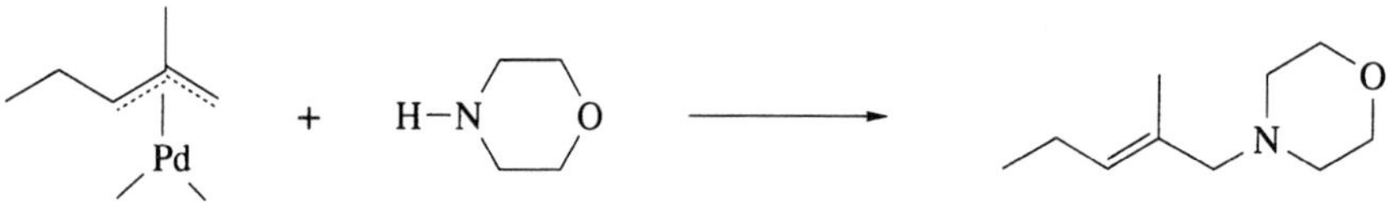

(262)

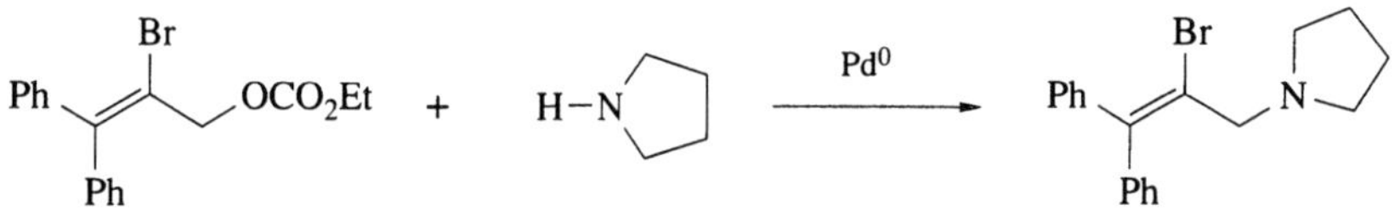

(263)

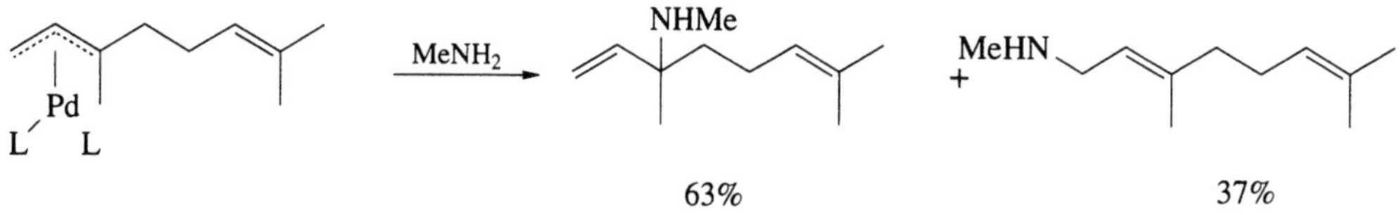

(264)

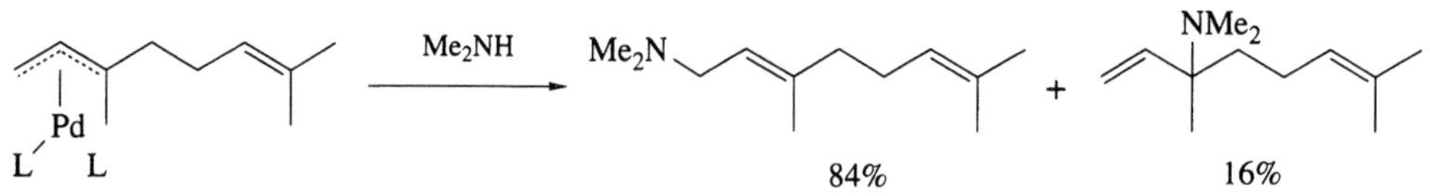

(265)

Carbonyl substitution on the allyl ligand leads to amine addition to the allyl terminus remote from the carbonyl moiety (equation 266).[202]

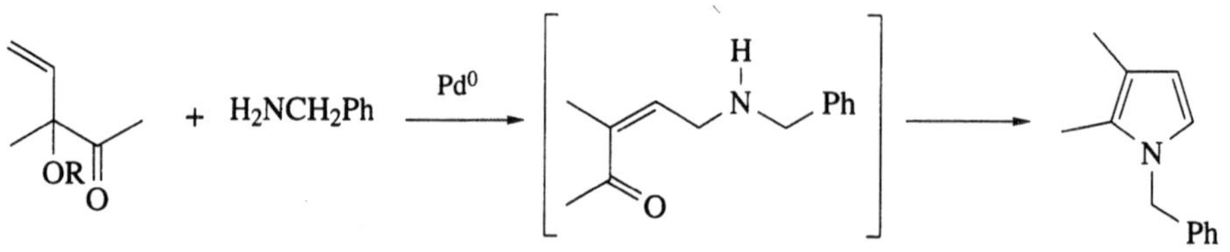

(266)

When presented with the choice between cyclization to give a five- or seven-membered ring, an amine nucleophile yields the five-membered ring.[203] However, reaction of an amine with the option to give either a bicyclo[4.2.0] or bicyclo[2.2.2] system produced the latter exclusively (equations 267 and 268).[204,205]

Functionality remote to the allyl ligand, whether it be oxygen,[181,183,190,193,195,201,410] nitrogen,[185] or phosphorus,[209] directs selective attack of an amine remote to the electronegative substituent (equations

OAc H N Ph $\xrightarrow{Pd^0}$ H N H Ph (267)

OAc H N Ph $\xrightarrow{Pd^0}$ Ph N (268)

269 and 270). Ligand effects can also be significant in directing the regioselectivity of the addition of amines to π-allylpalladium complexes. In a study of the reactivity of geranyl- **18**) and neryl-palladium (**19**) complexes, it was noted that donor ligands on the palladium, *e.g.* pyridine or simple amines, direct the addition to the less substituted terminus. Triphenylphosphine complexed to the palladium, however, gave preferential addition to the more substituted terminus.

NMe$_2$ Pd $\xrightarrow{Me_2NH}$ Me$_2$N NMe$_2$ (269)

Ph N H + O $\xrightarrow{Pd^0}$ Ph N OH (270)

The regioselectivity of addition in these complexes also seems to depend on whether the side chain occupies a *syn* (geranyl; **18**) or *anti* (neryl; **19**) position. The bisphosphine neryl complex gives exclusive attack at the more substituted terminus with $MeNH_2$, while the geranyl complex gives 70% attack at the more substituted end but also 30% addition to the less substituted allyl terminus. The difference is greater in the reactions with dimethylamine where the geranyl complex mainly gives attack at the less substituted end, while the neryl complex preferentially gives addition to the more substituted allyl terminus.

Pd L L **(18)** Geranyl Pd L L **(19)** Neryl Pd **(20)**

The possibility for internal coordination by the remote double bond exists for the geranyl complex (**20**). The alkene can act as an acceptor ligand and intensify the positive charge at the allyl terminus *trans* to it. This can account for the greater propensity of the geranyl system to add an amine to the less substituted terminus.

Amide nucleophiles can add to π-allylpalladium complexes by either ligand or metal addition. Only a limited number of examples have been examined, but these nucleophiles show a surprising tendency to add to the more substituted allyl terminus (equation 271).[197] However, in common with amine nucleophiles, they do add to the more distant terminus from remotely substituted polar functionalities (equation 272).[197]

OAc + NH_2Ts $\xrightarrow{Pd^0}$ NHTs (67%) + NHTs (33%) (271)

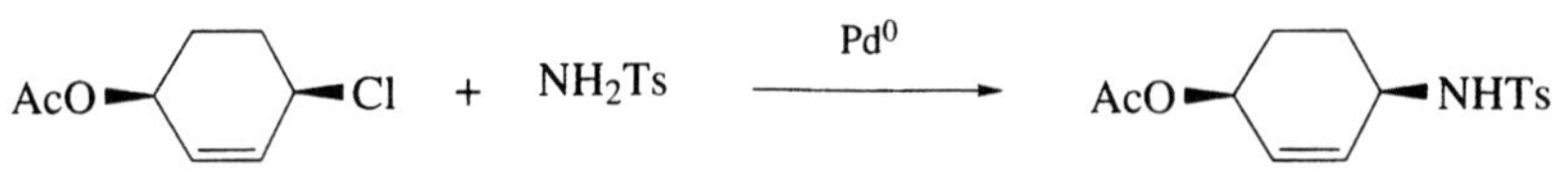

(272)

Azides have been found to add to π-allylpalladium complexes on the less alkyl- or aryl-substituted terminus (equations 273 and 274).[423]

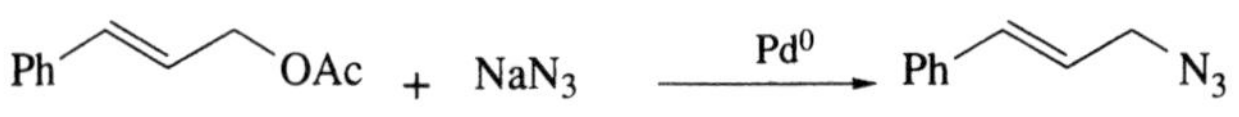

(273)

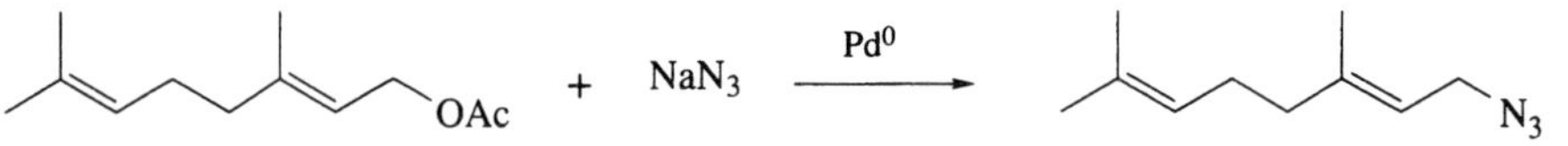

(274)

Magnesium amides also add to unsymmetrical π-allylpalladium complexes on the less substituted terminus (equation 275).[199]

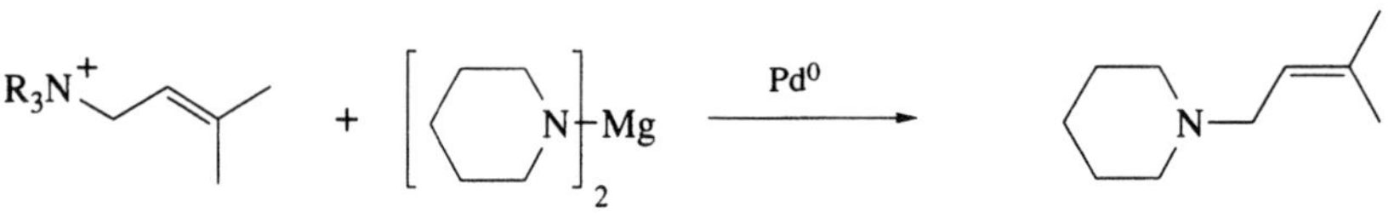

(275)

3.3.4.10 Regioselectivity of Sulfur Nucleophiles

All of the various sulfur nucleophiles that have been studied, including $PhSO_2$—, RS— and $(EtO)_2P(O)S$—, have been demonstrated to add to π-allylpalladium complexes by ligand addition. Interpretation of the regioselectivities of these nucleophiles must be tempered by the realization that the allyl–SO_2Ph, –SR and –$SP(O)(OEt)_2$ products are capable of isomerization by means of reentering the π-allylpalladium manifold.

The regioselectivity of *p*-toluenesulfonate nucleophiles as directed by alkyl groups asymmetrically substituted on the allyl ligand is very difficult to characterize, with preferential attack at both the more[211–213] and less[213,218] substituted allyl terminus alternately observed (equations 276–278).

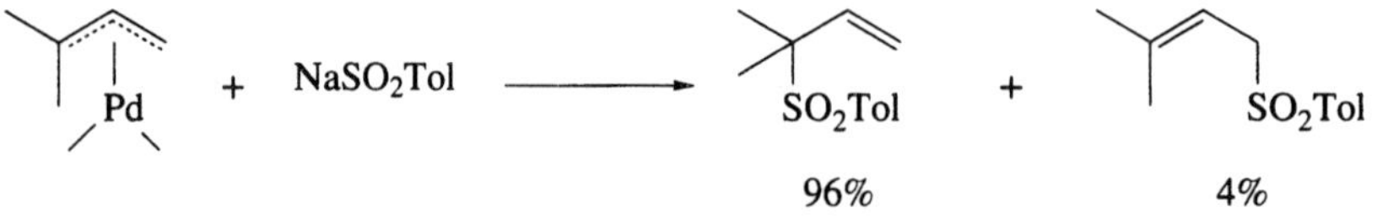

(276)

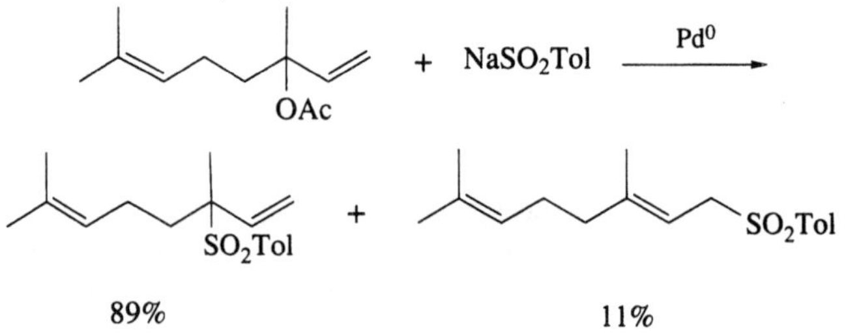

(277)

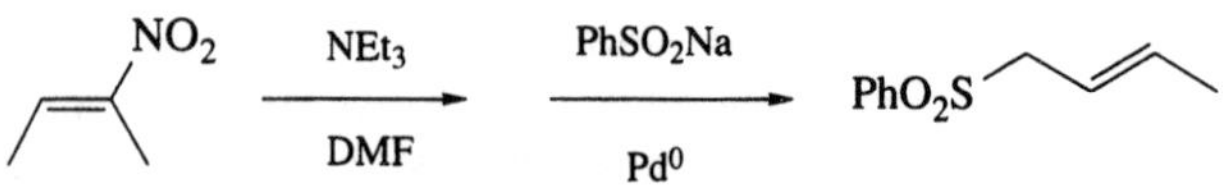

(278)

Particularly perplexing is a result obtained with the two possible crotyl precursors which should give the same π-allylpalladium complex, yet gave opposite regiochemical results (equations 279 and 280).[213]

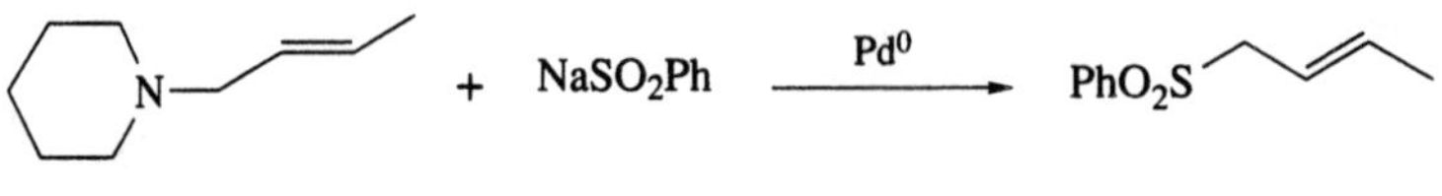

(279)

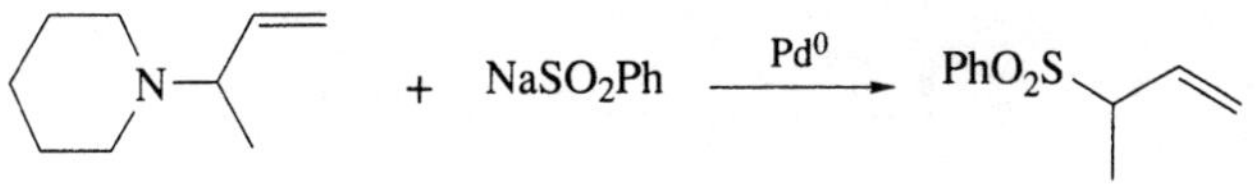

(280)

Aryl-substituted allyls give the addition of sulfonate at the remote allyl terminus (equation 281).[212]

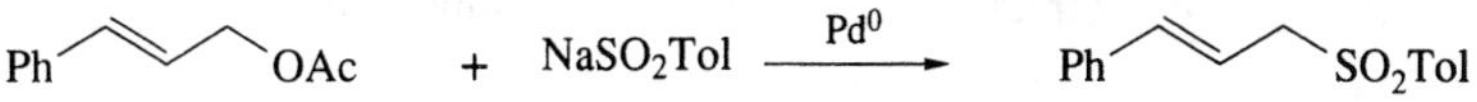

Ph⁀OAc + $NaSO_2Tol$ $\xrightarrow{Pd^0}$ Ph⁀SO_2Tol (281)

In contrast to the malonate additions to methylenecycloalkane-derived π-allyl complexes, where substantial endocyclic addition has been observed with the cyclohexyl analog, essentially exclusive exocyclic addition occurs for both the five- and six-membered ring cases with $PhSO_2Na$ (equations 282 and 283).[219]

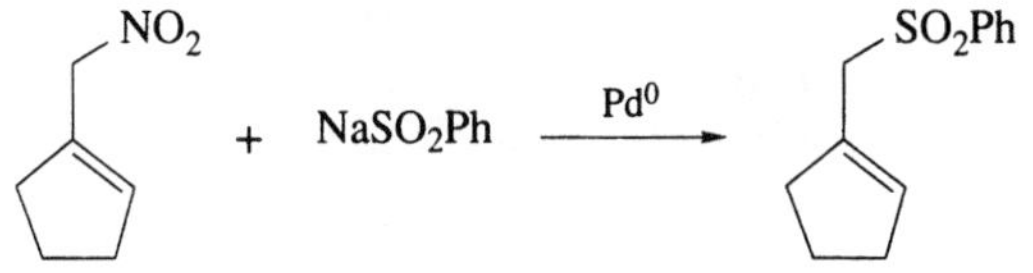

(282)

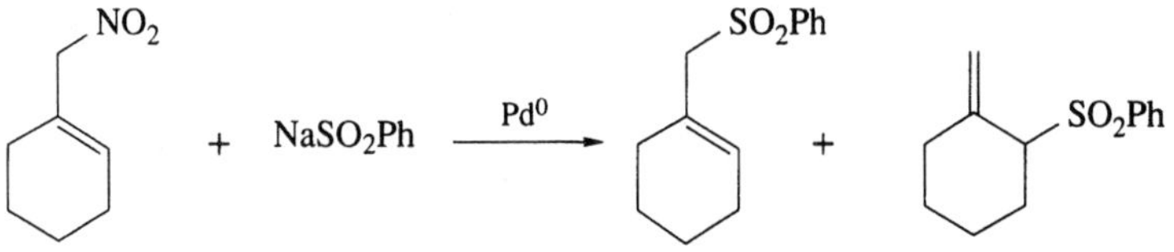

(283)

98% 2%

Ethoxycarbonyl groups show efficient regiodirecting powers in sulfonate additions, giving exclusive C—S bond formation at the allyl terminus remote to the carbonyl functionality (equation 284).[219] Remote oxygen functionality also directs incoming sulfonate nucleophiles to the distal allyl terminus (equation 285).[215]

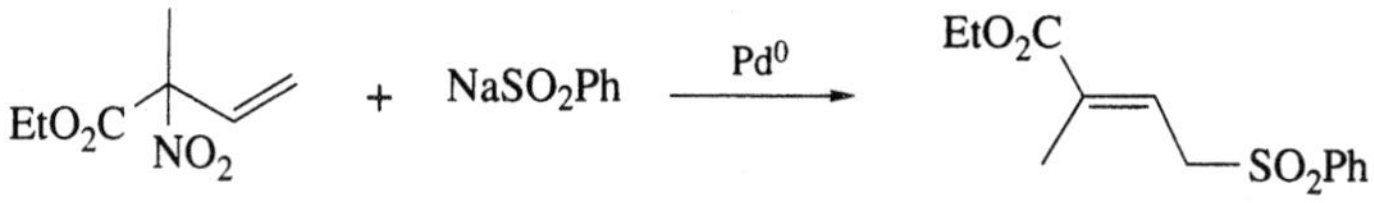

(284)

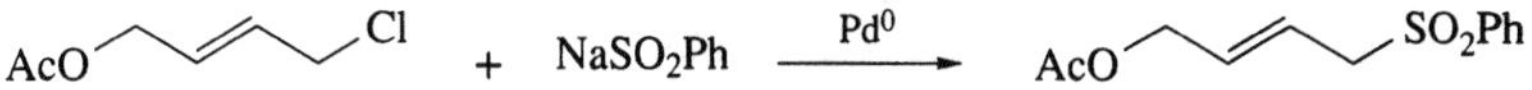

(285)

S-Allyl dithiocarbonates on treatment with Pd^0 catalyst expel COS and serve as precursors to RS^- nucleophiles. Alkyl-substituted precursors show preferential attack at the more substituted allyl terminus, while aryl-substituted precursors give exclusive attack at the less substituted allyl terminus (equation 286).[222]

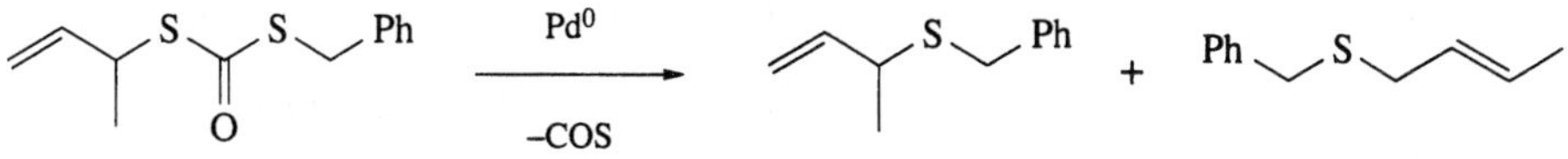

(286)

70% 30%

O-Allyl-*S*-alkyl dithiocarbonates show attack at the less substituted allyl end when directed by alkyl or aryl groups appended to the allyl ligand (equations 287 and 288).[221]

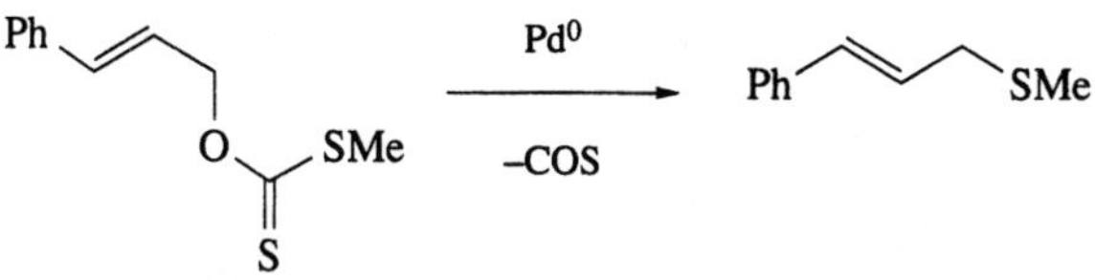

(287)

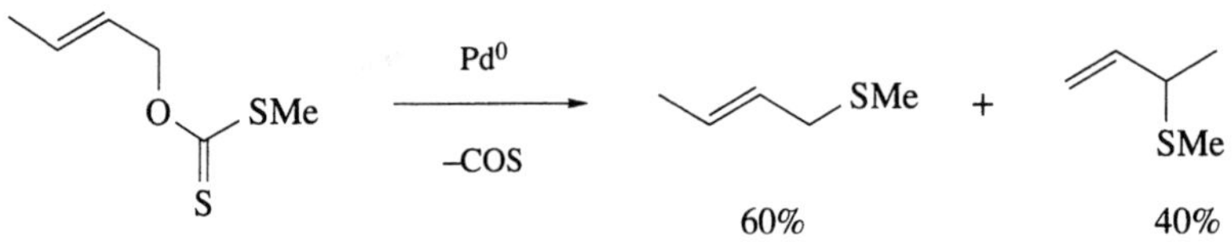

$$\text{(288)}$$

O-Allyl phosphonothionates rearrange on treatment with Pd^0 catalysts to the *S*-allyl derivatives.[220] Alkyl substituents on the allyl group direct attack to the less substituted allyl terminus (equation 289).

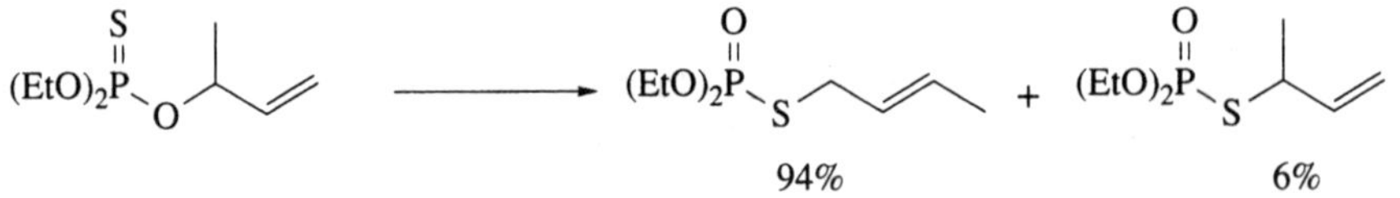

$$\text{(289)}$$

Trimethylsilyl alkyl and aryl sulfides react with allyl carbonates and vinyl epoxides to deliver the alkyl or aryl sulfide to the allyl unit. These species show typical regioselectivities by: (i) adding to the less substituted end of alkyl substituted allyls (equation 290); (ii) adding to the allyl terminus more distant from remote oxygen functionality (equation 291); and (iii) showing substantial endocyclic addition to methylenecyclohexane-derived allyls (equation 292).[223]

$$\text{(290)}$$

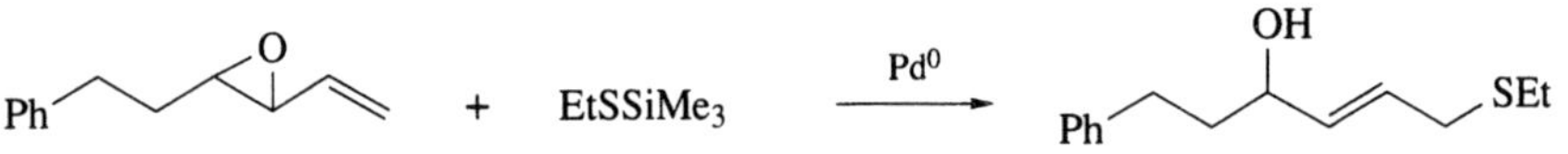

$$\text{(291)}$$

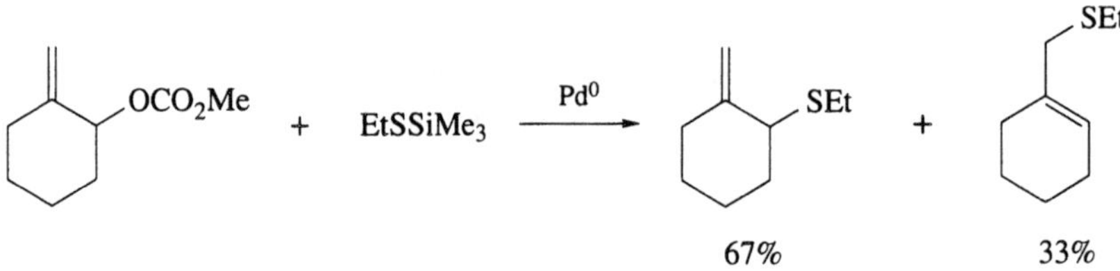

$$\text{(292)}$$

3.3.4.11 Regioselectivity in Other Heteroatom Nucleophiles

Lithium diphenylthiophosphides add *via* ligand addition and, in the single case studied, couple preferentially to the more substituted allyl terminus (equation 293).[224]

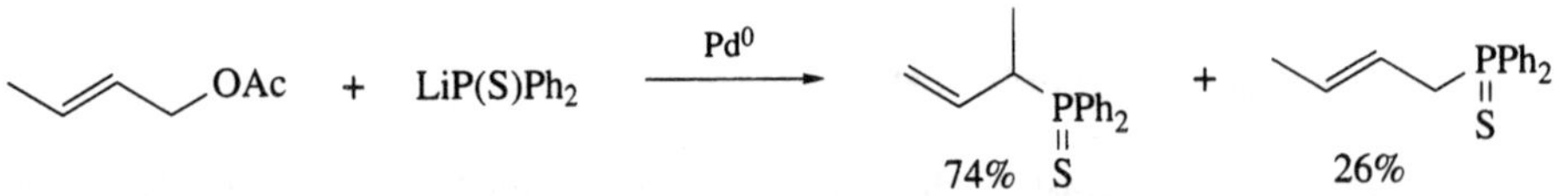

$$\text{(293)}$$

Tris(trimethylsilyl)aluminum reacts with π-allylpalladium complexes to yield allylsilanes. The regiochemistry of this addition is highly dependent on reaction conditions, and no trends can be readily discerned.[225]

Acyliron and acylnickel anions, as well as $NaCo(CO)_4$, acylate π-allylpalladium complexes selectively at the less substituted allyl terminus (equation 294).[225]

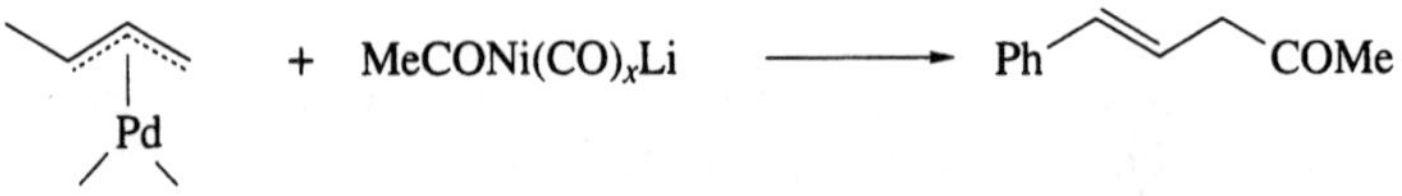

$$\text{(294)}$$

3.3.4.12 Regioselectivity in Other π-Allylpalladium Reactions

3.3.4.12.1 Insertion of CO, CO_2, C=N

The reaction of CO with π-allylpalladium complexes shows highly regioselective insertion into the less substituted allyl terminus[235,237–239] as illustrated in equations (295)–(297). One interesting exception is shown in an allyl complex possessing a Cl group on its terminus, where insertion takes place into the more substituted allyl end (equation 298).[234]

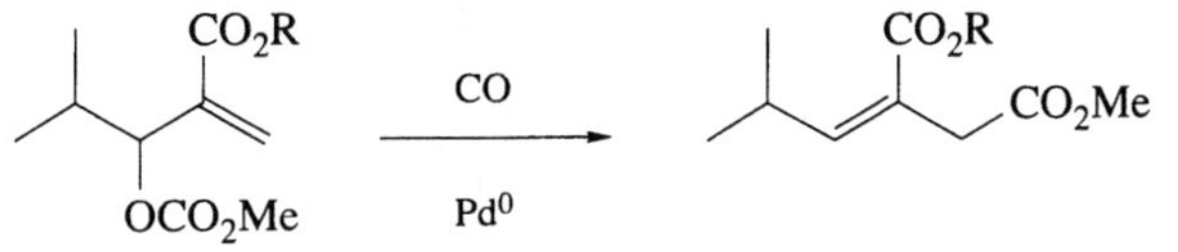

(295)

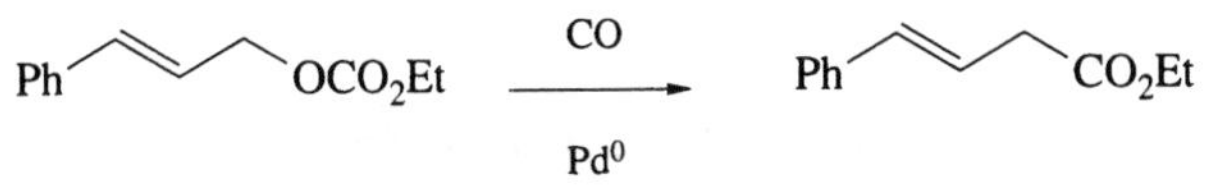

(296)

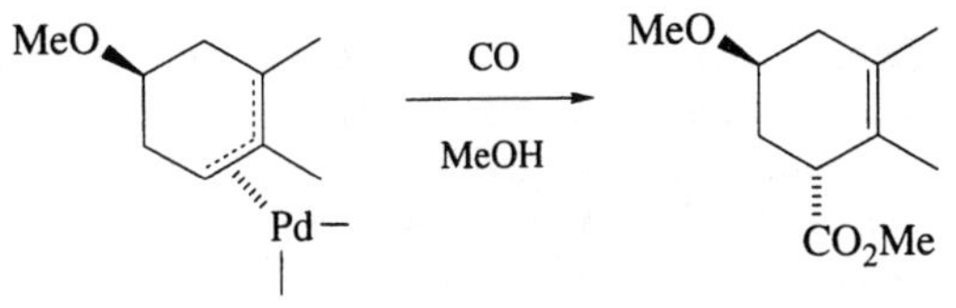

(297)

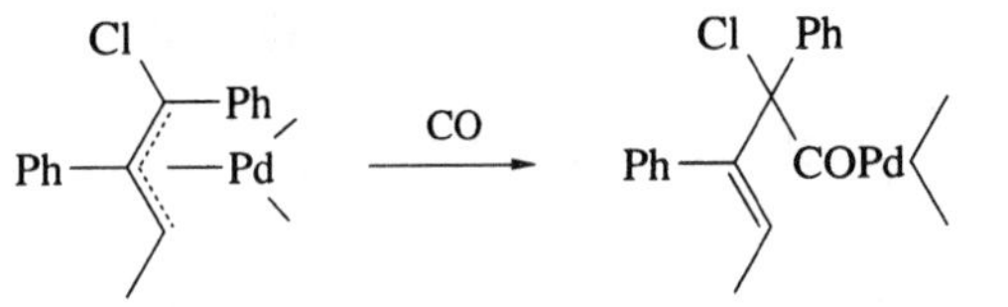

(298)

CO_2 insertion into π-allylpalladium complexes exhibits the opposite regioselectivity, with, for example, exclusive reaction at the more substituted terminus of a butenylpalladium complex (equation 299).[244]

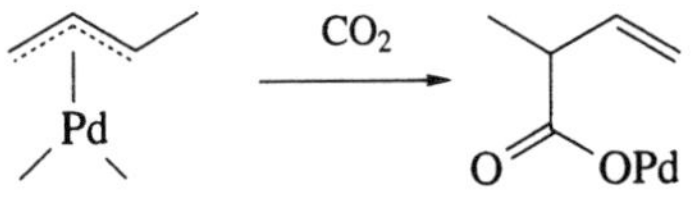

(299)

Isocyanide insertion into π-allylpalladium complexes generally occurs regioselectivity into the less substituted allyl terminus (equation 300).[240]

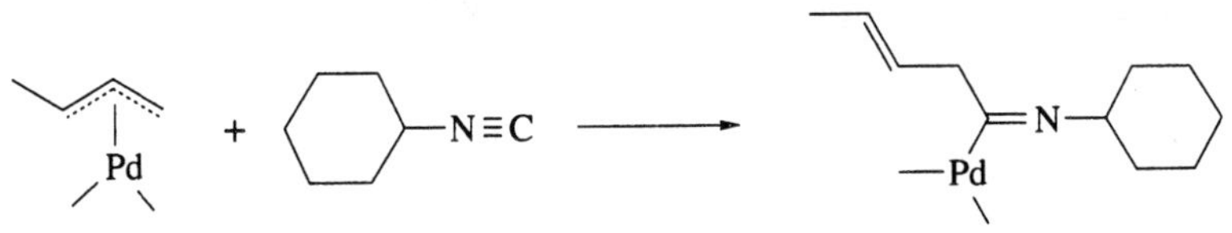

(300)

3.3.4.12.2 Insertion of dienes

The regioselectivity of the insertion of 1,3-dienes into π-allylpalladium complexes possesses two components. First, with respect to the allyl unit, the insertion takes place at the more substituted terminus.[248,249,251,252] Second, with regard to the diene it has been shown that 2-substituted-1,3-dienes insert at the more substituted alkene.[248,249] These interesting features have been explained by the intermediacy of an η^1-allyl species which is bound to the palladium at the less substituted terminus and with the Pd complexed to the less substituted alkene of the diene. The insertion process proceeds from this intermediate by an electrocyclic reaction which occurs outside the coordination sphere of the Pd (equation 301). This diene insertion has been recently employed as the key step in a potentially highly useful cyclization process (equation 302).[252]

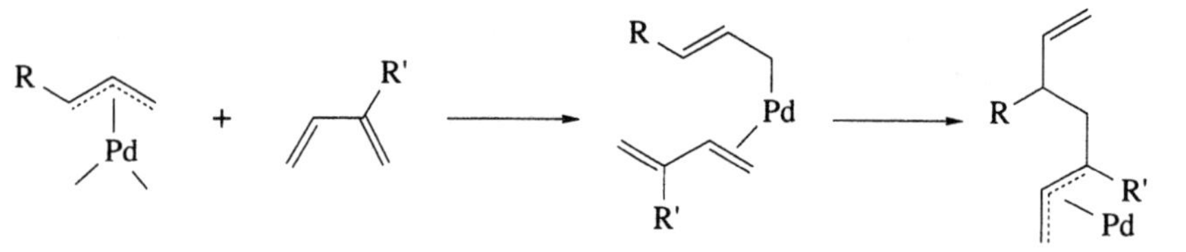

(301)

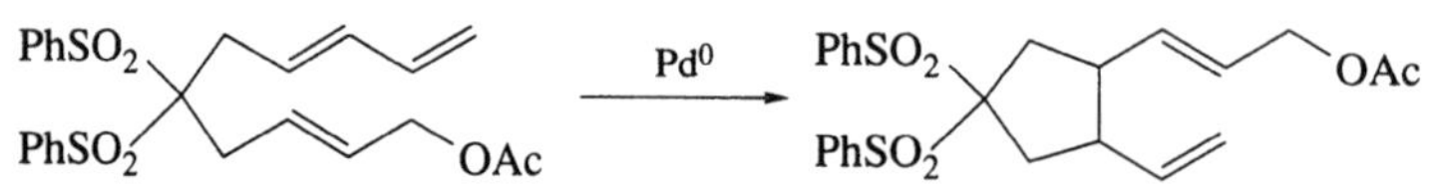

(302)

3.3.4.12.3 *Insertion of alkenes into π-allylpalladium complexes*

Intramolecular insertion of alkenes into π-allylpalladium complexes has been shown to proceed regioselectively to yield five-membered rings.[262,263] This transformation is equivalent to a palladium ene reaction and is completed by a β-hydride elimination. The C=C bond of ketenes is also capable of undergoing insertion into π-allylpalladium intermediates.[264] The final course of this reaction is dependent on the leaving group in the initial allyl–X precursor. For X = OAc, dienes are observed; for X = $MeOCO_2$, a methoxycarbonyl adduct is obtained (equation 303).[264]

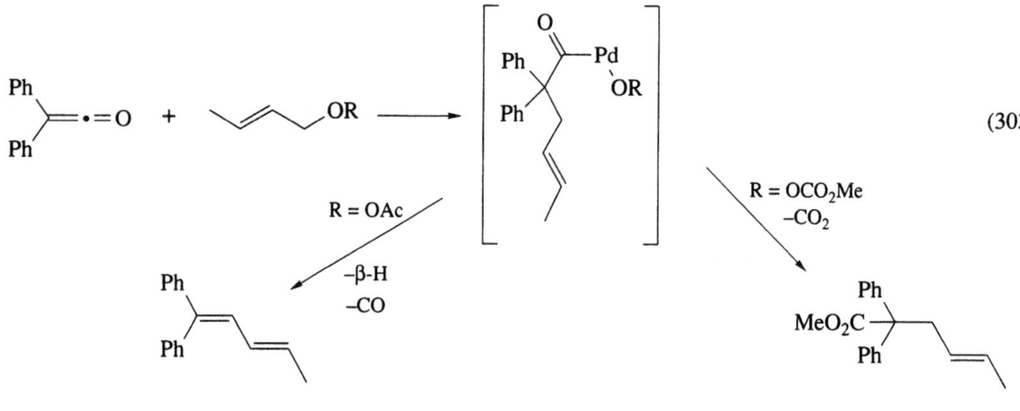

(303)

3.3.4.12.4 *Insertion of C=N into π-allylpalladium complexes*

The bis(π-allyl) complex derived from the linear dimerization of butadiene can be trapped with phenylhydrazones in a process involving C=N insertion into a π-allyl intermediate.[265] Predominant insertion into the more substituted terminus is observed (equation 304).

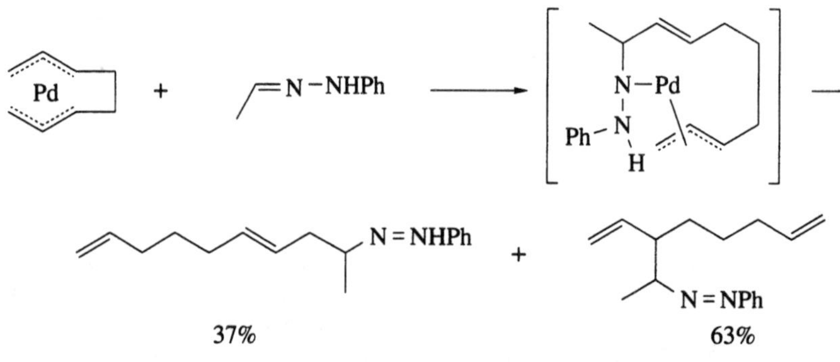

(304)

37% 63%

3.3.4.12.5 Oxidation of π-allylpalladium complexes

The oxidation of preformed π-allylpalladium complexes with MCPBA proceeds by attack at the metal to give oxidation at the less substituted allyl terminus (equation 305).[272] $MoO_2(acac)_2$–Bu^tO_2H oxidations also involve attack at palladium, but, in contrast to MCPBA, give oxidation preferentially at the more substituted allyl end (equations 306–308).[275]

MCPBA (305)

$MoO_2(acac)_2$ / Bu^tOOH (306)

$MoO_2(acac)_2$ / Bu^tOOH (307)

$MoO_2(acac)_2$ / Bu^tOOH (308)

Oxidation of preformed π-allylpalladium complexes with Cr^{VI} salts appears particularly capricious. $Na_2Cr_2O_7$ gave exclusive attack at the more substituted terminus (equation 309),[267] while a similar allyl substrate gave oxidation at the less substituted terminus with Collins reagent (Cr^{VI} oxide, pyridine; equation 310).[269] Photo-oxidation of π-allylpalladium complexes also shows inconsistent regioselectivity when directed by alkyl substituents on the terminus of an allyl ligand (equations 311 and 312).[276] A carbonyl or sulfonyl group appended on the allyl unit, however, gave exclusive oxidation at the allyl end remote to the functionality (equations 313 and 314).[277]

$Na_2Cr_2O_7$ (309)

Cr^{VI} (310)

$h\nu$ / O_2 (311)

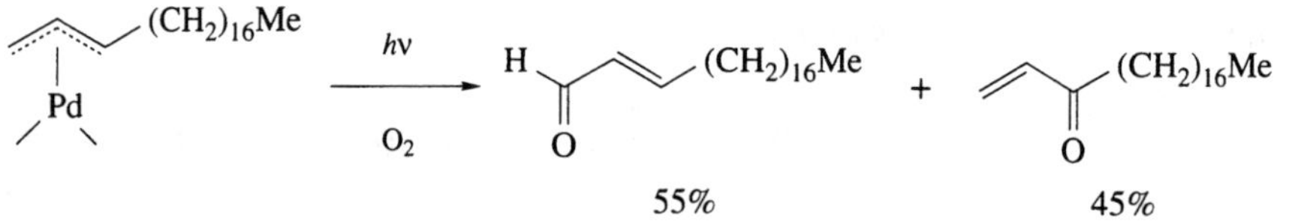

(312)

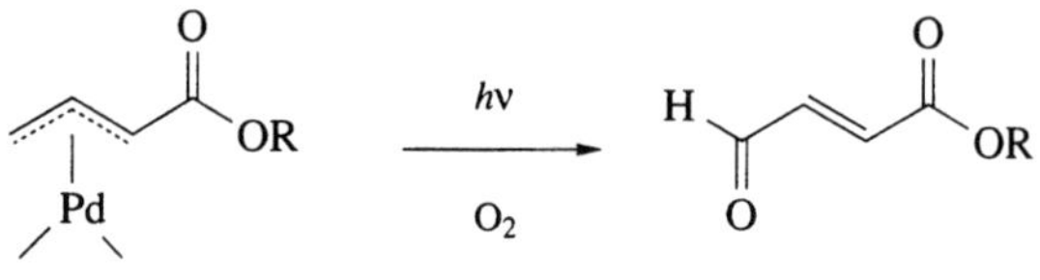

(313)

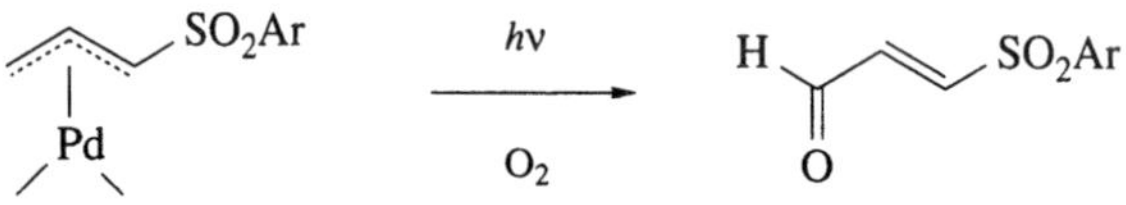

(314)

3.3.4.12.6 Reduction of π-allylpalladium complexes

A wide variety of methodologies have been developed for the reduction of a π-allylpalladium complex. Although these processes appear to uniformly involve attack at the metal center, the regiochemistry of the reduction is highly dependent on the reagent selected. For example, the regiodirection exerted by alkyl or aryl groups at the allyl terminus is clearly reagent dependent. KOH–MeOH can accomplish reduction of π-allyl complexes, and does so with hydride addition to the less substituted allyl terminus (equations 315 and 316).[281]

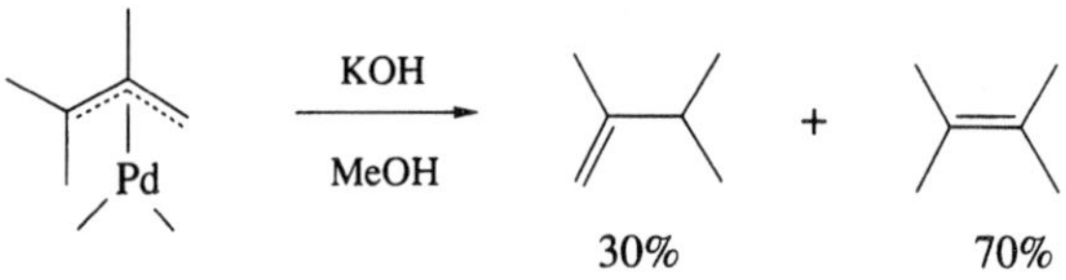

(315)

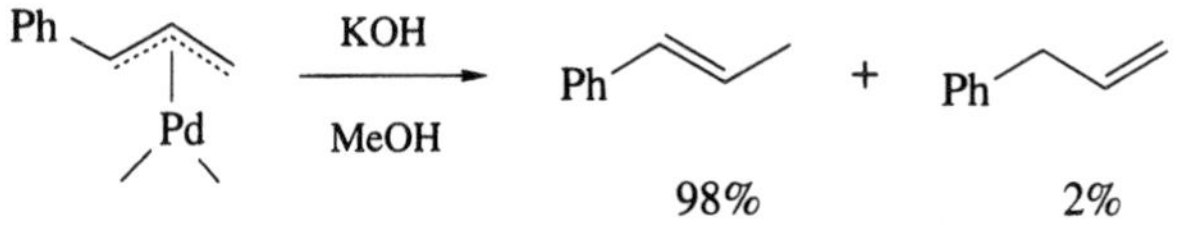

(316)

Reduction with $LiAlH(OBu^t)_3$[293] or LAH[292] also gives selective hydride addition to the less substituted allyl end (equations 317 and 318). In contrast, formate reductions selectively deliver hydride to the more substituted allyl terminus (equations 319 and 320).[302,303]

Si—H-mediated reduction, conveniently performed with polymethylhydrosiloxane (PMHS), demonstrates no clear pattern of regioselectivity (equation 321).[320]

$LiHBEt_3$ delivers hydride regioselectivity to the less substituted allyl terminus (equation 322).[289–291]

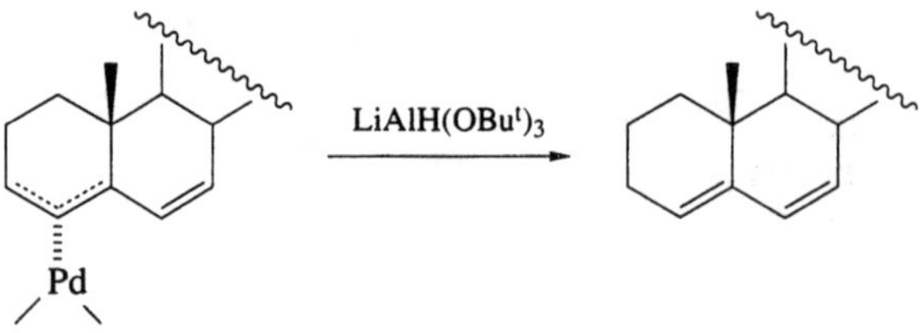

(317)

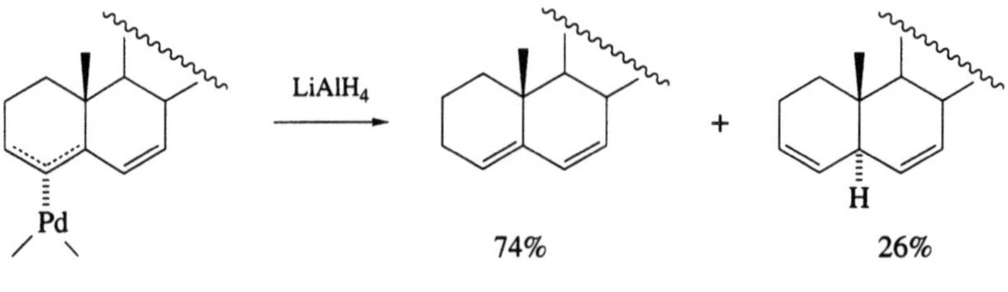

(318)

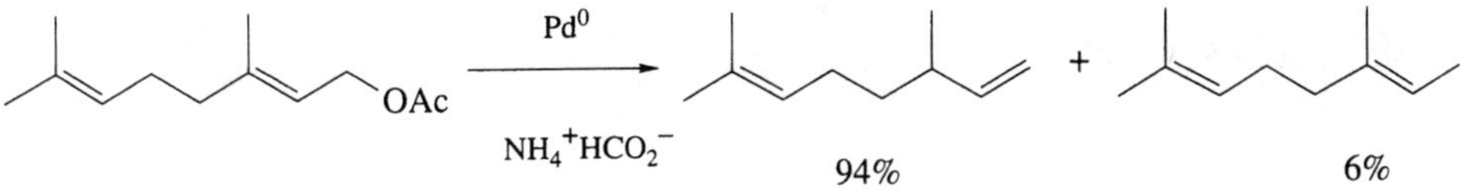

(319)

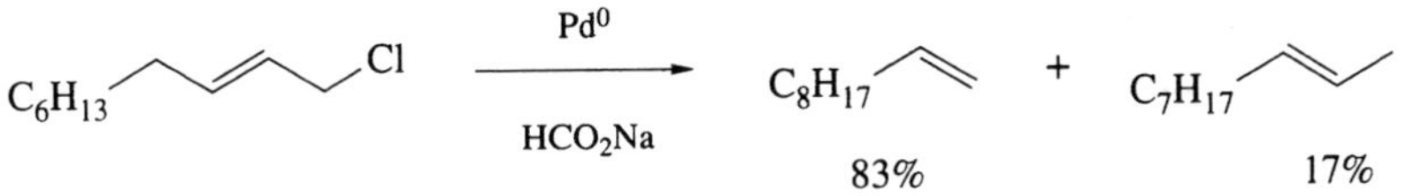

(320)

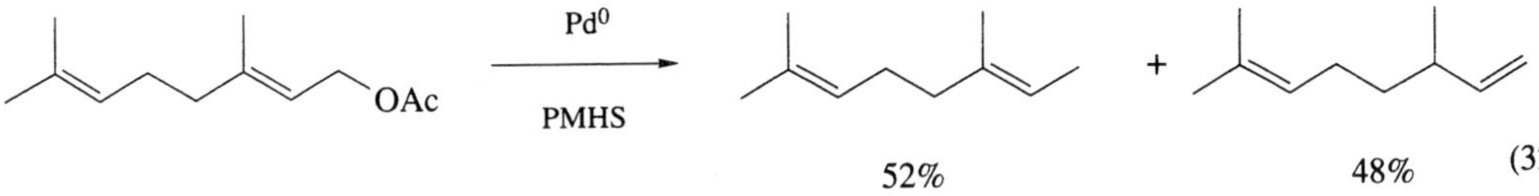

(321)

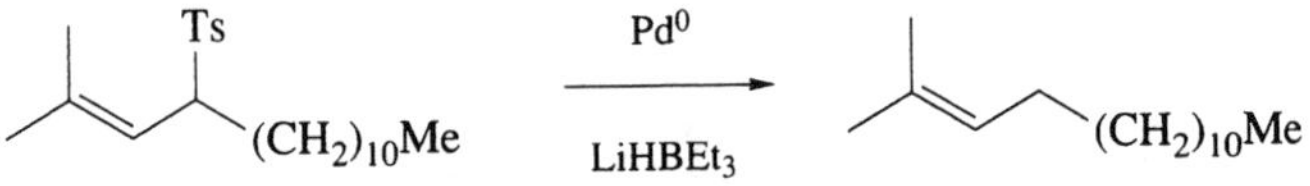

(322)

Electrochemical reduction of a π-allylpalladium complex results in preferred formation of the more substituted alkene (equation 323).[307] Conversely, reduction by SmI_2, followed by protonation, provides the less substituted alkene predominantly (equation 324).[305]

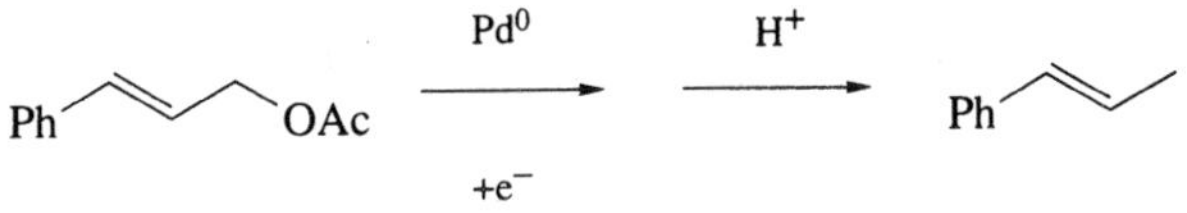

(323)

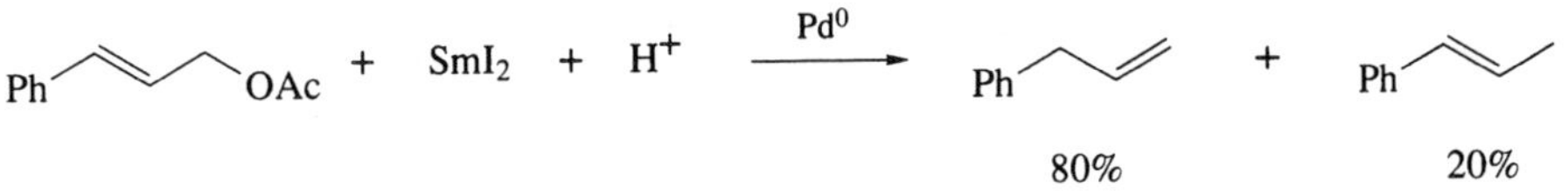

(324)

Delivery of hydride from a NAD(P)H model, *i.e.* *N*-Pr-1,4-dihydronictotinamide, occurs prefentially at the more substituted allyl end (equation 325).[306]

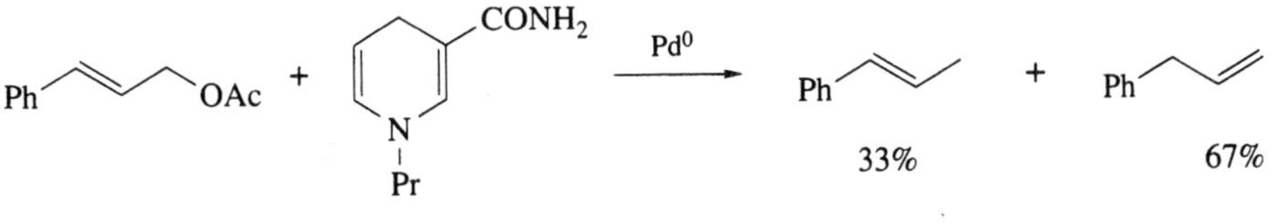

(325)

An informative comparison of a number of these reducing agents reacting with a single π-allyl substrate has been reported (Scheme 6).[291]

Some special directing effects have been observed in these reductions. For example, R_3SnH is directed by an OAc or Cl substituent on the allyl terminus to deliver hydride to the remote terminus,[297] but a CN directs delivery to the proximal terminus (equations 326–328).[287] Electrochemical reduction[307] also results in formation of an α,β-unsaturated organo product (equation 329). A remote OH functionality appears to direct $NaBH_4$ reduction to the proximal allyl terminus (equation 330).[288]

3.3.4.12.7 *Umpolung*

π-Allylpalladium complexes, which act as electrophiles, can be transformed into allyl-silanes and -stannanes, which typically act as nucleophiles. The allylsilanes, generated by electrochemical reduction and trapping with R_3SiCl,[307] and allylstannanes, formed by treatment with $Et_2AlSnBu_3$[309,310] or SmI_2 and

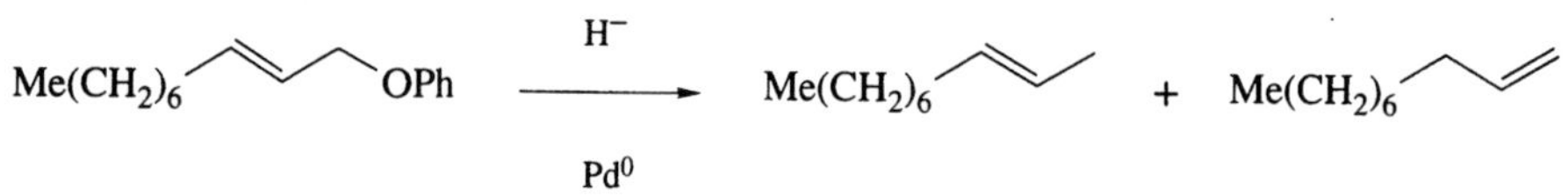

H$^-$, Pd^0

Reducing agent	*2-Alkene*	*1-Alkene*
$LiBHEt_3$	100	0
$LiAlH_4$	97	3
$Bu^t{}_2AlH$	100	0
$NaBH_4$	45	55
$NaBH_3CN$	71	29
Ph_2SiH_2	55	45
$NH_4{}^+HCO_2{}^-$	12	88

Scheme 6

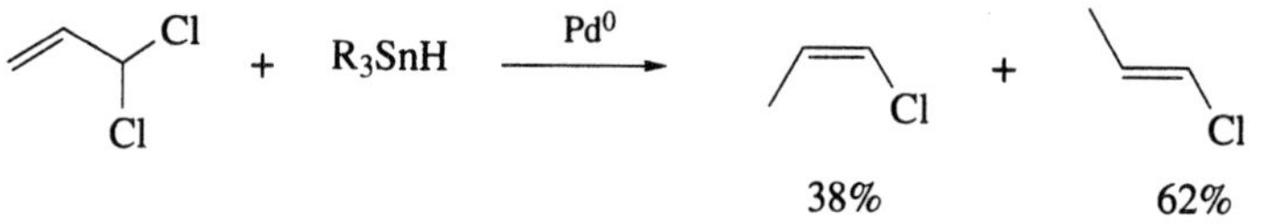

Cl, Cl + R_3SnH —Pd^0→ Cl (38%) + Cl (62%) (326)

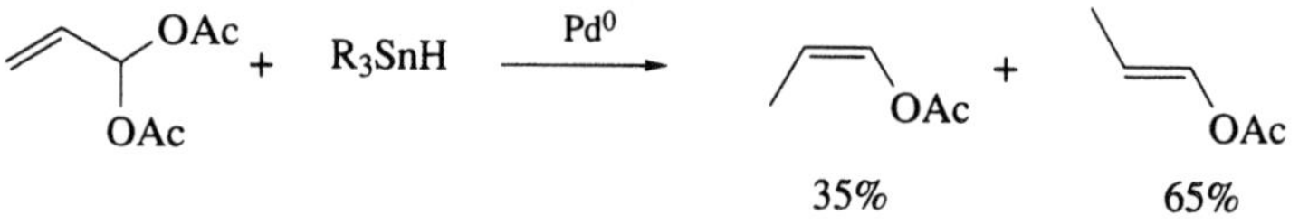

OAc, OAc + R_3SnH —Pd^0→ OAc (35%) + OAc (65%) (327)

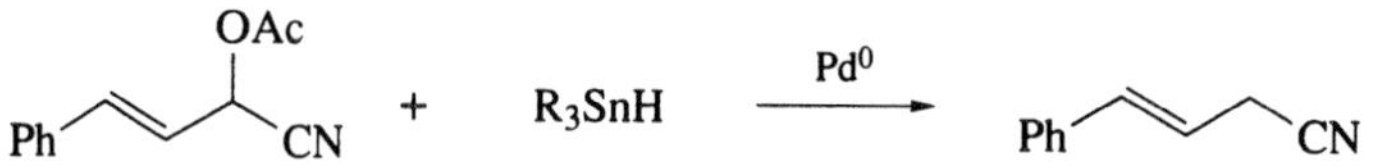

Ph, OAc, CN + R_3SnH —Pd^0→ Ph, CN (328)

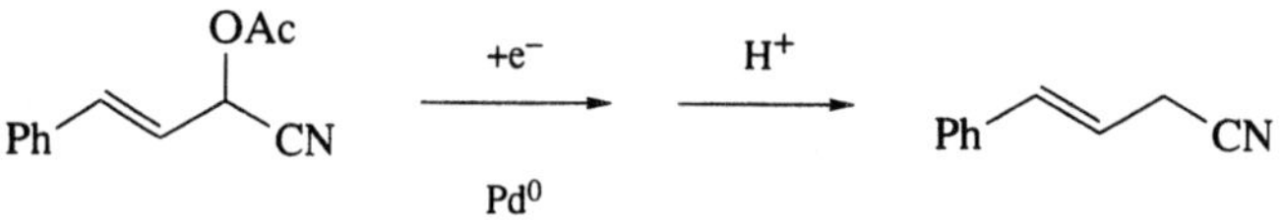

Ph, OAc, CN —+e$^-$, Pd^0→ —H$^+$→ Ph, CN (329)

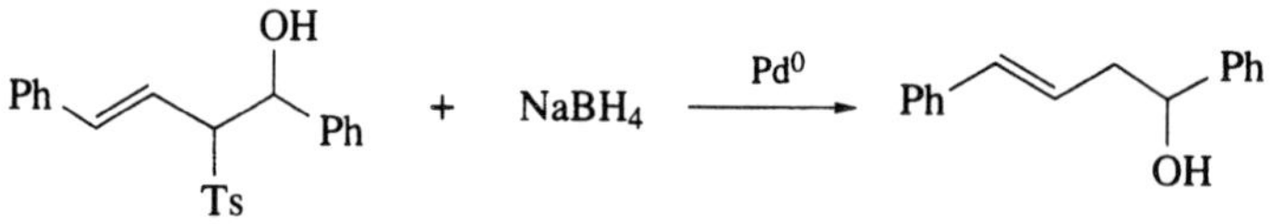

Ph, OH, Ph, Ts + $NaBH_4$ —Pd^0→ Ph, Ph, OH (330)

R_3SnCl,[311] have the heteroatom selectivity positioned at the less substituted allyl terminus (equations 331–334).

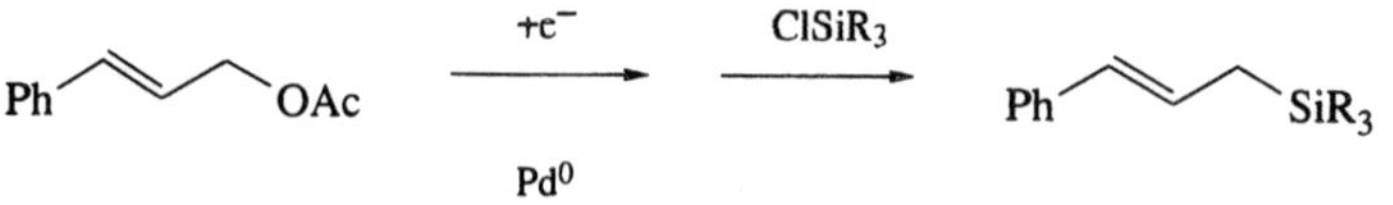

Ph, OAc —+e$^-$, Pd^0→ —$ClSiR_3$→ Ph, SiR_3 (331)

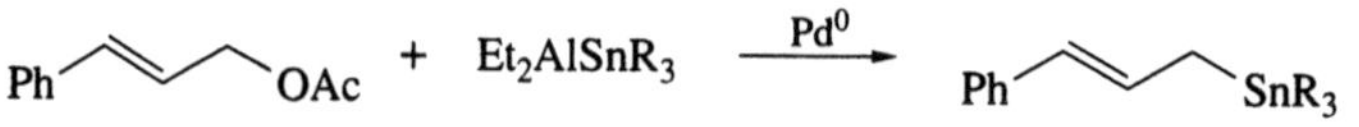

Ph, OAc + Et_2AlSnR_3 —Pd^0→ Ph, SnR_3 (332)

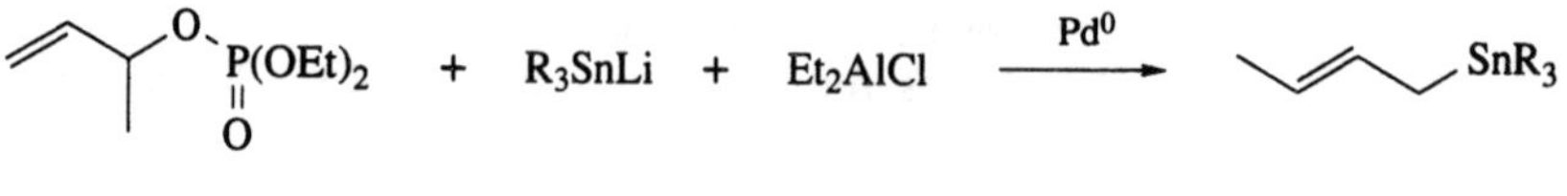

O, $P(OEt)_2$, O + R_3SnLi + Et_2AlCl —Pd^0→ SnR_3 (333)

OAc + SmI_2 + R_3SnCl —Pd^0→ SnR_3 (334)

3.3.4.12.8 Regioselectivity of addition reactions

Addition of disilanes to π-allylpalladium complexes proceeds selectively to yield the silyl group appended to the less substituted allyl terminus (equations 335 and 336).[338,339]

Ph, Pd + $PhCl_2SiSiMe_3$ → —EtOH→ Ph $SiPh(OEt)_2$ (335)

Pd —$PhCl_2SiSiMe_3$→ —MeMgBr→ $SiPhMe_2$ (336)

3.3.5 ENANTIOSELECTIVITY

3.3.5.1 Introduction

Three basic types of enantioselectivity have been investigated in conjunction with the reaction of π-allylpalladium complexes with nucleophiles. In the first type, the chirality exists in the allyl–X precursor. The stereospecificities associated with the oxidative addition to convert it to a π-allyl complex and the subsequent attack by the nucleophile are key in the chirality transfer (equation 337).

X, R, R' —Nuc, Pd^0→ Nuc, R, R' (337)

In the second type, the source of the chirality is the phosphine ligand on the palladium. Two scenarios have been evaluated. In the first, an achiral *meso* allyl ligand is employed. This ligand when complexed to an optically active PdL_2* creates a chiral π-allyl complex. The transition states leading to attack at one terminus or the other are then diastereomeric and can lead to optically active products. The second scenario employs a prochiral allyl ligand. The resulting π-allyl complexes with PdL_2* are now diastereomeric and the relative reactivity of these diastereomeric complexes with the nucleophile is the source of formation of optically active products. The third and last type involves the use of a chiral leaving group or chirality in the nucleophile in the allyl alkylation (equations 338 and 339).

R X* —Nuc, Pd^0→ R, *, H, Nuc (338)

Pd + Nuc* → Nuc* (339)

3.3.5.2 Chirality Transfer

Optically active methyl (*R*)-3-oxo-7-(methoxycarbonyloxy)-8-nonenoate, when converted to its sodium salt and treated with a Pd^0 catalyst, proceeds to form an optically active allylcyclohexanone with complete retention at the original chiral center.[424] For this chirality transfer from a C—O to a C—C bond to be successful, both the formation of the π-allyl complex and the attack of the nucleophile must be completely stereospecific. In addition, the attack by the nucleophile must be fast relative to a *syn–anti* type interconversion of the π-allyl intermediate (*via* an η^1-allyl, with the Pd attached to the less substituted terminus) which would result in racemization (Scheme 7).

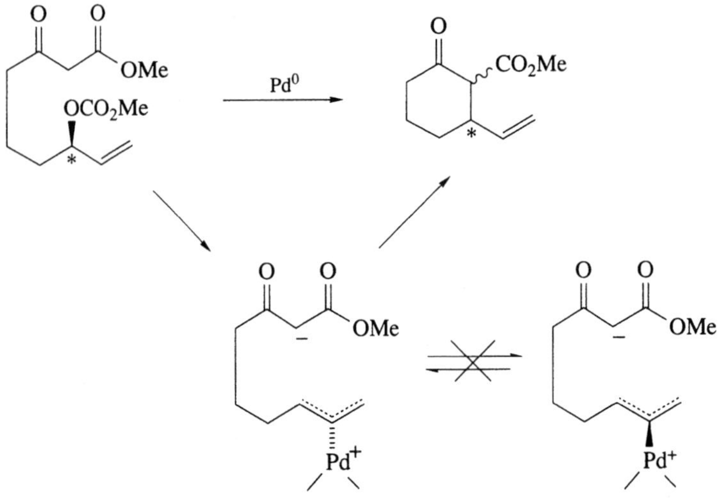

Scheme 7

An efficient cyclization is also observed with methyl (5*R*)-methoxycarbonyl-(3*E*)-decenyl malonate to yield an optically active product (equation 340).[425]

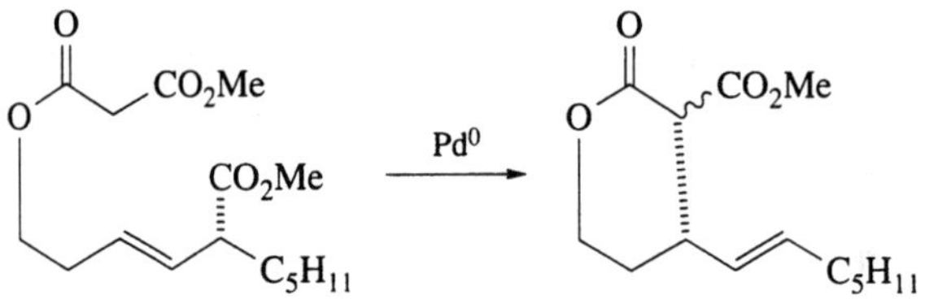

(340)

An intermolecular example of this process was also shown to proceed with a high degree of chirality transfer (equation 341).[426] The chirality transfer process can also function with oxygen nucleophiles as shown in equations (342) and (343).[427]

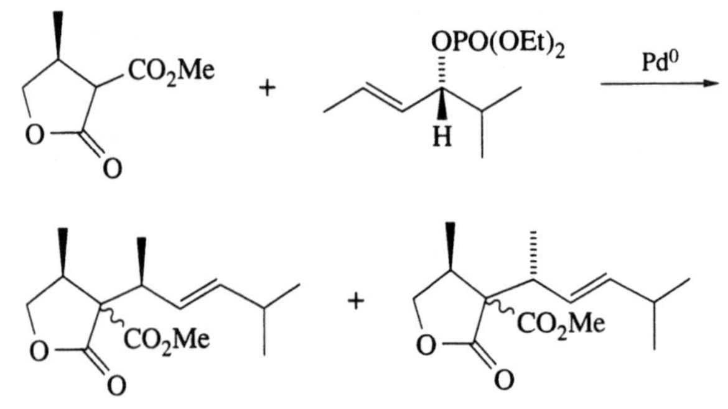

(341)

96.1% 3.9%

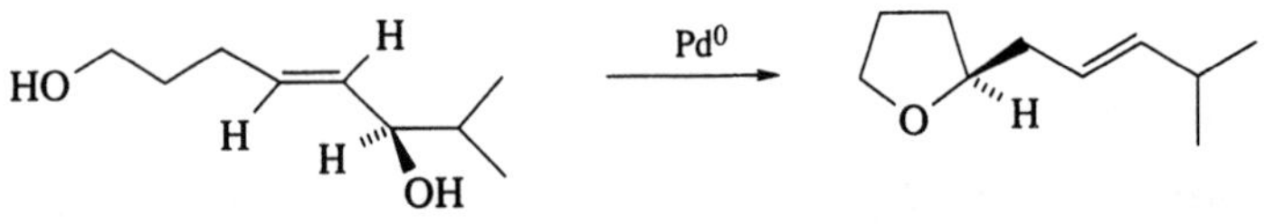

(342)

100% optical yield

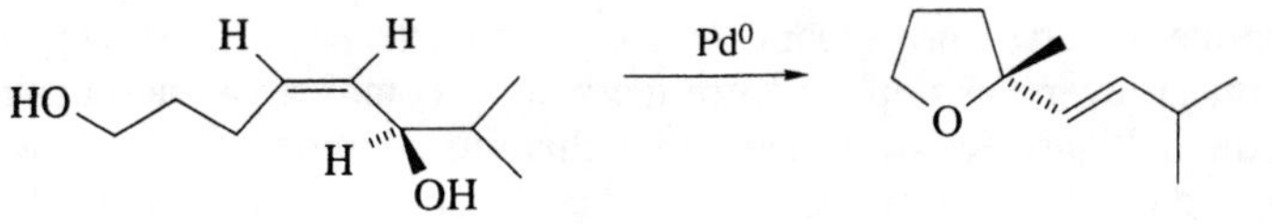

(343)

90% optical yield

A net C—O to C—C bond chirality transfer has been demonstrated in the preparation of a cyclopropane (equation 344).[428] Another C—O to C—C chirality transfer was accomplished starting with an optically active alkylidenetetrahydrofuran.[429] Not only was complete stereospecificity *via* a double inversion mechanism observed, but, in addition, regioselectivity had to be exercised to avoid the alternate cyclization to give a seven-membered ring product (equation 345).

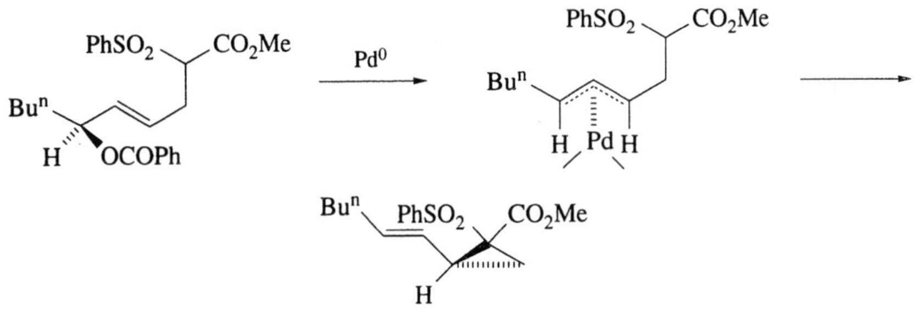

(344)

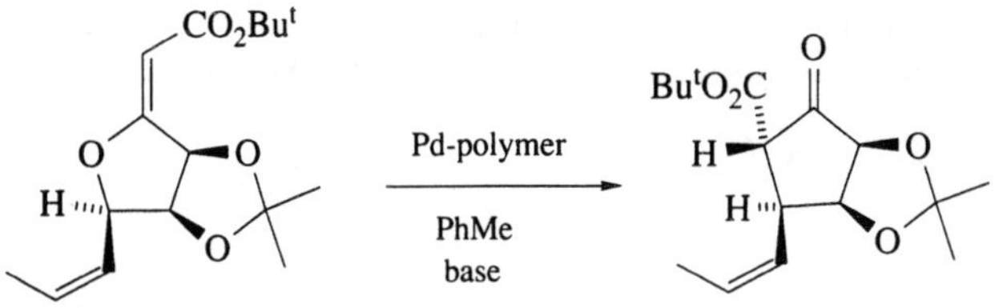

(345)

An elegant example of a stereo relay in a nonrigid system employed a chiral vinyl lactone.[430] For the relay to be completely successful (i) the ionization of the vinyl lactone must occur from one conformation, (ii) the intermediate π-allylpalladium complex must retain its stereochemistry, and (iii) the nucleophile must attack regioselectively at the carbon of the allyl system distal to the incipient carboxylate. This distal transfer of chirality was achieved as shown in equation (346). This same process has been utilized in a synthesis of an optically active vitamin E side chain from D-glucose.[431]

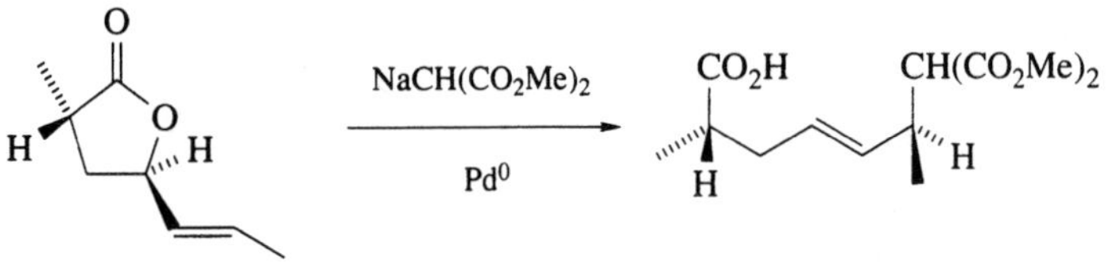

(346)

3.3.5.3 Enantioselectivity with Optically Active Phosphines

Significant progress has been made in enantioselective C—C bond formation in π-allylpalladium chemistry, where the source of the chirality is the phosphine ligand on the palladium. A number of potential difficulties must be addressed in order for this approach to result in synthetically useful optical yields. Racemization can occur by a variety of mechanisms. Intermediates not bearing substituents at both allyl termini can undergo *syn–anti* isomerization, which accomplishes racemization (equation 347).

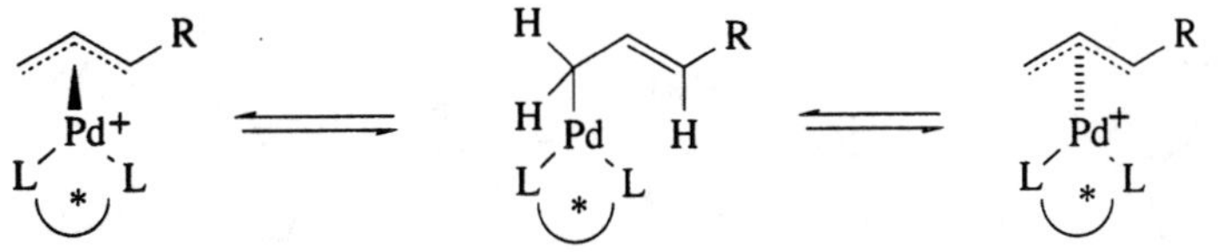

(347)

Allyl acetates are the preferred precursors for π-allyl complexes. The acetate liberated on oxidative addition (stereochemical inversion step) to form the π-allyl complex can add again to the allyl ligand, displacing the palladium. It has been demonstrated[167] that this addition can take place either *via* direct attack on the allyl ligand on the face opposite the palladium or at the palladium, followed by reductive elimination (equation 348). These processes have opposite stereochemical results, and with respect to enantioselectivity, attack at the metal will result in racemization of the starting allyl acetate (equation 349). It has been shown that this process can become competitive if the incoming nucleophile is particularly sterically bulky and slow to add to the allyl ligand.[167]

ligand addition — OAc + Pd^0; metal addition — Pd-OAc —reductive elimination→ OAc + Pd^0 (348)

Pd^0 ; ^-OAc (349)

A recently uncovered mode of racemization involves attack of the π-allylpalladium complex by an uncomplexed PdL_2 species present in the solution. This mode of racemization can be significant if high concentrations of PdL_4 catalyst are employed in the reaction (equation 350).

(350)

An inherent difficulty in achieving high optical yields in π-allylpalladium reactions is that with the soft carbon nucleophiles employed, attack occurs on the allyl ligand on the face opposite the palladium. This places the source of the chirality (the phosphine ligand) remote from where C—C bond formation is taking place. Despite these limitations, considerable success has been achieved in these reactions.

The initial study in this area employed the *meso*-1,3-dimethylallyl ligand.[432] The chloride dimer was generated from 2-pentene, and a variety of optically active phosphine ligands were added to form the chiral bisphosphine complexes. Reaction of these allyl complexes with sodiodiethyl malonate resulted in optical yields in the range of 2–29% (equation 351).

$NaCH(CO_2Me)_2$; CO_2Me ; CO_2Me (351)

Asymmetric induction in catalytic allylic alkylation was next evaluated.[433] Alkylation of a racemic ethylcyclopentenyl acetate with the sodium salt of methyl phenylsulfonylacetate with (+)-DIOP gave the best result (optical yield 46%; equation 352). The asymmetry of the allyl ligand in this case results in diastereomeric π-allyl complexes with the PdL_2* moieties. The optical yields observed demand a racemization of the starting allyl acetate or interconversion of the diastereomeric complexes. If the re-

OAc + Na^+ SO_2Ph CO_2Me —PdL_2^*→ SO_2Ph CO_2Me (352)

action proceeded without racemization or interconversion and strictly by a double inversion process from the racemic allyl acetate, then the product would of necessity be racemic (equation 352).

A variation on this asymmetric catalytic allylation scheme employed β-diketones or β-keto ester nucleophiles with allyl ethers and a palladium–DIOP catalyst.[434] The chiral center generated is now one carbon removed from the allyl ligand. As a result, somewhat lower optical yields were observed (≈10%; equation 353). A variety of chiral phosphines were evaluated in the cyclization of β-keto esters. Optical yields up to 48% were measured in these reactions.[435]

PdL_2^* (353)

Allylic alkylations using a benzophenone imine of glycine methyl ester as a prochiral nucleophile and chiral phosphine ligands on palladium produced optical yields up to 57% (equation 354).[436]

Ph, N, Ph, CO_2Me + OAc $\xrightarrow{PdL_2^*}$ N, Ph, MeO_2C, Ph (354)

Attempts to address the difficulty associated with the distance of the chirality in the Pd ligand to the site of C—C bond formation have been most fruitful. New chiral phosphine ligands containing an optically active functional group remote from the phosphines have been very effective, providing optical yields up to 52% in the case where the new chiral center is one carbon removed from the allyl ligand[437] (*cf.* 10% optical parity obtained previously;[434] equation 355).

X, PdL_2^* (355)

L^* = Ph_2P, Ph_2P, N, $(CH_2)_n$, N, H, Pr^i, CO_2Me $n = 2, 3$

'Chiral pockets' have been created, employing optically active ligands derived from 1,2′-binaphthol. Asymmetric alkylations with nearly 70% optical yield have been realized using this method.[438]

Optically active ferrocenylphosphines containing a functional group on the side chain, which may complex to the incoming nucleophile and induce asymmetric attack, have proven to be most effective in π-allyl allylations, with optical yields up to 92% being observed (equations 356 and 357).[439–442]

The most detailed and thorough studies on asymmetric catalytic allylation using palladium chiral phosphine complexes have come from the work of Bosnich *et al.*[443–445] Employing a combination of NMR, chemical and labelling studies on this process, this group was able to determine that with 1,1-diphenyl-3-

Ar, Ar, OAc + Na^+ CO_2Me, CO_2Me $\xrightarrow{PdL_2^*}$ Ar, Ar, MeO_2C, CO_2Me (356)

92% optical yield

$$\text{2-acetylcyclohexanone} + CH_2{=}CHCH_2OAc \xrightarrow{PdL_2^*} \text{2-acetyl-2-allylcyclohexanone (*)} \qquad (357)$$

73% optical yield

alkyl- or aryl-substituted allyls: (i) nucleophilic attack is the turnover limiting step and is also the enantioselective step; (ii) the major product enantiomer originates from the major four-coordinate π-allyl diastereomeric intermediate and that the diastereomeric complexes rapidly interconvert; (iii) the enantioselectivity is relatively insensitive to the nucleophile employed; (iv) a *syn* substituent at C-2 or C-3 in these allyl units is not significant in determining optical yields; and (v) a viable mechanism for racemization is attack by an uncomplexed PdL_2 on a π-allyl complex with inversion of configuration.

An X-ray structure of the major diastereomeric complex was also obtained to give further insight into the origin of asymmetry in the attack by nucleophiles.[446]

3.3.5.4 Chiral Nucleophiles and Leaving Groups

The final modes of enantioselective allyl alkylations catalyzed by palladium involve the use of chiral nucleophiles[447] and chiral leaving groups.[448,449] Chiral enamines were found to undergo allylation in 100% optical yield in an intramolecular case and in up to 50% optical yield in intermolecular reactions (equation 358).

$$\text{N-(cyclohexenyl)-2-}(CH_2OCO_2CH_2CH{=}CH_2)\text{pyrrolidine} \xrightarrow{Pd^0} \text{2-allylcyclohexanone (*, H)} \qquad (358)$$

A palladium-catalyzed allylic sulfonate–sulfone rearrangement with chiral sulfonates produced 92% optical yields (equation 359).[448,449] Presumably, the optically active sulfonate induces chirality in the oxidative addition by Pd^0 in these reactions.

$$p\text{-}MeC_6H_4\text{–}\overset{*}{S}(O)\text{–}O\text{–}CH_2CH{=}CHCH_3 \xrightarrow{Pd^0} p\text{-}MeC_6H_4\text{–}SO_2\text{–}\overset{*}{C}H(CH_3)CH{=}CH_2 \qquad (359)$$

92% optical yield

3.3.6 REFERENCES

1. J. Tsuji, *Acc. Chem. Res.*, 1969, **2**, 144.
2. B. M. Trost, *Tetrahedron*, 1977, **33**, 2615.
3. B. M. Trost, *Acc. Chem. Res.*, 1980, **13**, 385.
4. R. Baker, *Chem. Ind. (London)*, 1980, 816.
5. J. Tsuji, *Pure Appl. Chem.*, 1981, **53**, 2371.
6. B. M. Trost, *Pure Appl. Chem.*, 1981, **53**, 2357.
7. J. Tsuji, *Pure Appl. Chem.*, 1982, **54**, 197.
8. J. Tsuji, *Pure Appl. Chem.*, 1986, **58**, 869.
9. J. Tsuji, *Tetrahedron*, 1986, **42**, 4361.
10. B. M. Trost, *J. Organomet. Chem.*, 1986, **300**, 263.
11. J. Tsuji, *J. Organomet. Chem.*, 1986, **300**, 281.
12. J. Tsuji, 'Organic Synthesis with Palladium Compounds', Springer-Verlag, Heidelberg, 1980.
13. B. M. Trost and T. R. Verhoeven, in 'Comprehensive Organometallic Chemistry', ed. G. Wilkinson, Pergamon Press, Oxford, 1982, vol. 8, p. 799.

14. P. M. Maitlis, in 'Comprehensive Organometallic Chemistry', ed. G. Wilkinson, Pergamon Press, Oxford, 1982, vol. 6, p. 385.
15. J. Tsuji, *Tetrahedron*, 1986, **42**, 4361.
16. B. M. Trost, *Angew. Chem., Int. Ed. Engl.*, 1986, **25**, 1.
17. G. W. Parshall and G. Wilkinson, *Inorg. Chem.*, 1962, **1**, 896.
18. R. Hüttel and H. Christ, *Chem. Ber.*, 1963, **96**, 3101.
19. R. Hüttel and H. Christ, *Chem. Ber.*, 1964, **97**, 1439.
20. R. Hüttel, H. Dietl and H. Christ, *Chem. Ber.*, 1964, **97**, 2037.
21. R. Hüttel, H. Christ and K. Herzig, *Chem. Ber.*, 1964, **97**, 2710.
22. R. Hüttel and H. Dietl, *Chem. Ber.*, 1965, **98**, 1753.
23. R. Hüttel and H. Schmid, *Chem. Ber.*, 1968, **101**, 252.
24. H. C. Vogler, *Recl. Trav. Chim. Pays-Bas*, 1969, **88**, 225.
25. K. Dunne and F. J. McQuillin, *J. Chem. Soc. C*, 1970, 2196.
26. B. M. Trost and T. J. Fullerton, *J. Am. Chem. Soc.*, 1973, **95**, 292.
27. K. Kikukawa, K. Sakai, K. Asada and T. Matsuda, *J. Organomet. Chem.*, 1974, **77**, 131.
28. B. W. Howsam and F. J. McQuillin, *Tetrahedron Lett.*, 1968, 3667.
29. D. N. Jones and S. D. Knox, *J. Chem. Soc., Chem. Commun.*, 1975, 165.
30. S. Baba, T. Sobata, T. Ogura and S. Kawaguchi, *Bull. Chem. Soc. Jpn.*, 1974, **47**, 2792.
31. B. M. Trost and P. J. Metzher, *J. Am. Chem. Soc.*, 1980, **102**, 3572.
32. B. M. Trost and L. Weber, *J. Am. Chem. Soc.*, 1975, **97**, 1161.
33. E. V. Fischer and H. Werner, *Chem. Ber.*, 1962, **93**, 2075.
34. E. V. Fischer and H. Werner, *Chem. Ber.*, 1972, **95**, 695.
35. S. D. Robinson and B. L. Shaw, *J. Chem. Soc.*, 1963, 4806.
36. J. M. Rove and D. A. White, *J. Chem. Soc. A*, 1967, 1451.
37. R. Hüttel, H. Dietl and H. Christ, *Chem. Ber.*, 1964, **97**, 2037.
38. J. H. Lukas, P. W. N. M. Van Leeuwen, H. C. Vogler and A. P. Konwenhoven, *J. Organomet. Chem.*, 1973, **47**, 153.
39. J. H. Lukas, J. P. Visser and A. P. Konwenhoven, *J. Organomet. Chem.*, 1973, **50**, 349.
40. R. G. Schultz, *Tetrahedron*, 1964, **20**, 2809.
41. M. S. Lupin, J. Powell and B. L. Shaw, *J. Chem. Soc. A*, 1966, 1687.
42. A. D. Ketley and J. A. Braatz, *J. Organomet. Chem.*, 1967, **9**, P5.
43. T. Shono, T. Yoshimura, Y. Matsumura and R. Oda, *J. Org. Chem.*, 1968, **33**, 876.
44. R. P. Hughes, D. E. Hunton and K. Schumann, *J. Organomet. Chem.*, 1979, **169**, C37.
45. R. Noyori and H. Takaya, *Chem Commun.*, 1969, 525.
46. A. D. Ketley and T. A. Braatz, *Chem. Commun.*, 1968, 959.
47. A. D. Ketley, T. A. Braatz and J. Craig, *Chem. Commun.*, 1970, 1117.
48. M. A. Battiste, L. E. Friedrich and R. A. Fiato, *Tetrahedron Lett.*, 1975, 45.
49. R. A. Fiato, P. Mushak and M. A. Battiste, *J. Chem. Soc., Chem. Commun.*, 1975, 869.
50. R. C. Larock and M. A. Mitchell, *J. Am. Chem. Soc.*, 1976, **98**, 6718.
51. R. C. Larock and M. A. Mitchell, *J. Am. Chem. Soc.*, 1978, **100**, 180.
52. E. O. Fischer and G. Burger, *Z. Naturforsch., Teil B*, 1961, **16**, 702.
53. W. Dent, R. Long and J. Wilkinson, *J. Chem. Soc.*, 1964, 1585.
54. J. K. Nicholson, J. Powell and B. L. Shaw, *Chem. Commun.*, 1966, 174.
55. J. Tsuji and N. Iwamoto, *Chem. Commun.*, 1966, 828.
56. J. Sakakibara, Y. Takahashi, S. Sakai and Y. Ishii, *Chem. Commun.*, 1969, 396.
57. R. Jira and J. Sedlmeier, *Tetrahedron Lett.*, 1971, 1227.
58. J. H. Lukas and J. E. Blom, *J. Organomet. Chem.*, 1971, **26**, 465.
59. H. A. Quinn, W. R. Jackson and J. J. Rooney, *J. Chem. Soc., Dalton Trans.*, 1972, 180.
60. F. R. Hartley and S. R. Jones, *J. Organomet. Chem.*, 1974, **66**, 465.
61. K. H. Pannell, M. F. Lappert and K. Stanley, *J. Organomet. Chem.*, 1976, **112**, 37.
62. R. D. Rieke, A. V. Kavaliunas, L. D. Rhyne and D. J. J. Fraser, *J. Am. Chem. Soc.*, 1979, **101**, 246.
63. R. D. Rieke and A. V. Kavaliunas, *J. Org. Chem.*, 1979, **44**, 3069.
64. J. Smidt and W. Hafner, *Angew. Chem.*, 1959, **71**, 284.
65. S. Imamura and J. Tsuji, *Tetrahedron*, 1969, **25**, 4187.
66. R. Pietropaolo, F. Faraone and S. Sergi, *J. Organomet. Chem.*, 1977, **42**, 177.
67. T. Hayashi, M. Konishi and M. Kumada, *J. Chem. Soc., Chem. Commun.*, 1983, 736.
68. K. Fugami, K. Oshima, K. Utimoto and H. Nozaki, *Bull. Chem. Soc. Jpn.*, 1987, **60**, 2509.
69. B. M. Trost and T. J. Fullerton, *J. Am. Chem. Soc.*, 1973, **95**, 292.
70. B. M. Trost and P. E. Strege, *Tetrahedron Lett.*, 1974, 2603.
71. B. M. Trost and T. J. Dietsche, *J. Am. Chem. Soc.*, 1973, **95**, 8200.
72. B. M. Trost and P. E. Strege, *J. Am. Chem. Soc.*, 1975, **97**, 2534.
73. B. M. Trost and L. Weber, *J. Org. Chem.*, 1975, **40**, 3617.
74. B. M. Trost and T. R. Verhoeven, *J. Am. Chem. Soc.*, 1976, **98**, 630.
75. J. Tsuji, *Acc. Chem. Res.*, 1973, **6**, 8.
76. Y. Tanigawa, K. Nishimura, A. Kawasaki and S. Murahashi, *Tetrahedron Lett.*, 1982, **23**, 5549.
77. K. E. Atkins, W. E. Walker and R. M. Manyik, *Tetrahedron Lett.*, 1970, 3821.
78. T. Hirao, N. Yamada, Y. Ohshiro and T. Agawa, *J. Organomet. Chem.*, 1982, **236**, 409.
79. U. M. Dzhemilev, D. L. Minsker, L. M. Khalilov and A. G. Ibragimov, *Izv. Akad. Nauk SSSR, Ser. Khim.*, 1988, 378.
80. A. Hosomi, K. Hoashi, S. Kohra, Y. Tominaga, K. Otaka and H. Sakurai, *J. Chem. Soc., Chem. Commun.*, 1987, 570.
81. R. Tamura and L. S. Hegedus, *J. Am. Chem. Soc.*, 1982, **104**, 3727.
82. L. McGarry, unpublished results.

83. H. Okamura and H. Takei, *Tetrahedron Lett.*, 1979, 3425.
84. T. Yamamoto, M. Akimoto, O. Saito and A. Yamamoto, *Organometallics*, 1986, **5**, 1559.
85. X. Lu and L. Lu, *J. Organomet. Chem.*, 1986, **307**, 287.
86. B. M. Trost, N. R. Schmuff and M. J. Miller, *J. Am. Chem. Soc.*, 1980, **102**, 5981.
87. R. Tamura, K. Hayashi, Y. Kai and D. Oda, *Tetrahedron Lett.*, 1984, **25**, 4437.
88. R. Tamura, K. Hayashi, M. Kakihana, M. Tsuji and D. Oda, *Chem. Lett.*, 1985, 229.
89. X. Lu, X. Jiang and X. Tao, *J. Organomet. Chem.*, 1988, **344**, 109.
90. T. G. Schenck and B. Bosnich, *J. Am. Chem. Soc.*, 1985, **107**, 2058.
91. G. Balavione and F. Guibe, *Tetrahedron Lett.*, 1979, 3949.
92. M. Moreno-Manas and A. Trius, *Tetrahedron*, 1981, **37**, 3009.
93. X. Lu, L. Lu and J. Sun, *J. Mol. Catal.*, 1987, **41**, 245.
94. Y. Inoue, M. Toyofuku and H. Hashimoto, *Chem. Lett.*, 1984, 1227.
95. Y. Inoue, M. Toyofuku, M. Akimoto, O. Saito and A. Yamamoto, *Bull. Chem. Soc. Jpn.*, 1986, **59**, 885.
96. J. Tsuji, Y. Ohashi and I. Shimizu, *Tetrahedron Lett.*, 1985, **26**, 3825.
97. J. Tsuji, H. Takahashi and M. Morikawa, *Tetrahedron Lett.*, 1965, 4387.
98. J. Tsuji, H. Takahashi and M. Morikawa, *Kogyo Kagaku Zasshi*, 1966, **69**, 920.
99. H. Onoue, I. Moritani and S. Murahashi, *Tetrahedron Lett.*, 1977, 121.
100. J. Tsuji, *Bull. Chem. Soc. Jpn.*, 1973, **46**, 1897.
101. J. Tsuji, T. Yamakawa and T. Mandai, *Tetrahedron Lett.*, 1978, 565.
102. P. A. Wade, S. D. Morrow and S. A. Hardinger, *J. Org. Chem.*, 1982, **47**, 365.
103. P. Aleksandrowicz, H. Piotrowska and W. Sas, *Tetrahedron*, 1982, **38**, 1321.
104. P. Aleksandrowicz, H. Piotrowska and W. Sas, *Monatsh. Chem.*, 1982, **113**, 1221.
105. J. C. Fiaud and J. L. Malleron, *Tetrahedron Lett.*, 1980, **21**, 4437.
106. B. M. Trost and E. Keinan, *Tetrahedron Lett.*, 1980, **21**, 2591.
107. B. M. Trost and C. R. Self, *J. Org. Chem.*, 1984, **49**, 468.
108. N. Ono, I. Hamamoto and A. Kaji, *Bull. Chem. Soc. Jpn.*, 1985, **58**, 1863.
109. E. Negishi, H. Matsushita, S. Chatterjee and R. A. John, *J. Org. Chem.*, 1982, **47**, 3188.
110. J. C. Fiaud and J. L. Malleron, *J. Chem. Soc., Chem. Commun.*, 1981, 1159.
111. B. Akermark and A. Jutland, *J. Organomet. Chem.*, 1981, **217**, C41.
112. T. Tsuda, Y. Chujo, S. Nishi, K. Tawara and T. Saegusa, *J. Am. Chem. Soc.*, 1980, **102**, 6381.
113. I. Shimizu, T. Yamada and J. Tsuji, *Tetrahedron Lett.*, 1980, **21**, 3199.
114. J. C. Fiaud and L. Aribi-Zoviueche, *Tetrahedron Lett.*, 1982, **23**, 5279.
115. J. E. Backväll, R. E. Nordberg and J. O. Vagberg, *Tetrahedron Lett.*, 1983, **24**, 411.
116. I. Shimizu, Y. Ohashi and J. Tsuji, *Tetrahedron Lett.*, 1983, **24**, 3865.
117. J. Tsuji and I. Minami, *Acc. Chem. Res.*, 1987, **20**, 140.
118. J. Tsuji, J. Yamada, I. Minami, M. Yuhara, M. Nisar and I. Shimizu, *J. Org. Chem.*, 1987, **52**, 2988.
119. J. Tsuji, I. Minami and I. Shimizu, *Tetrahedron Lett.*, 1983, **24**, 1793.
120. J. Tsuji, Y. Ohashi and I. Minami, *Tetrahedron Lett.*, 1987, **28**, 2397.
121. J. Tsuji, I. Minami and I. Shimizu, *Chem. Lett.*, 1983, 1325.
122. J. Tsuji, I. Minami and I. Shimizu, *Tetrahedron Lett.*, 1983, **24**, 4713.
123. J. Tsuji, K. Takahashi, I. Minami and I. Shimizu, *Tetrahedron Lett.*, 1984, **25**, 4783.
124. J. Tsuji, I. Shimizu, I. Minami, Y. Ohashi, T. Sugiura and K. Takahashi, *J. Org. Chem.*, 1985, **50**, 1523.
125. I. Minami, Y. Ohashi, I. Shimizu and J. Tsuji, *Tetrahedron Lett.*, 1985, **26**, 2449.
126. T. Tsuda, M. Okada, S. Nishi and T. Saegusa, *J. Org. Chem.*, 1986, **51**, 421.
127. T. Tsuda, M. Tokai, T. Ishida and T. Saegusa, *J. Org. Chem.*, 1986, **51**, 5216.
128. B. M. Trost, *Angew. Chem., Int. Ed. Engl.*, 1986, **25**, 1.
129. B. M. Trost, S. M. Mignani and T. N. Nanninga, *J. Am. Chem. Soc.*, 1986, **108**, 6051.
130. B. M. Trost and S. Chen, *J. Am. Chem. Soc.*, 1986, **108**, 6053.
131. B. M. Trost and S. A. King, *Tetrahedron Lett.*, 1986, **27**, 5971.
132. B. M. Trost and S. M. Mignani, *Tetrahedron Lett.*, 1986, **27**, 4137.
133. B. M. Trost, S. A. King and T. N. Nanninga, *Chem. Lett.*, 1987, 15.
134. B. M. Trost and D. T. MacPherson, *J. Am. Chem. Soc.*, 1987, **109**, 3983; D. G. Cleary and L. A. Paquette, *Synth. Commun.*, 1987, **17**, 497; B. M. Trost, S. M. Mignani and T. N. Nanninga, *J. Am. Chem. Soc.*, 1988, **110**, 1602.
135. I. Shimizu, Y. Ohashi and J. Tsuji, *Tetrahedron Lett.*, 1984, **25**, 5183.
136. P. Brevilles and D. Uguen, *Tetrahedron Lett.*, 1987, **28**, 6053.
137. B. M. Trost and E. Keinan, *Tetrahedron Lett.*, 1980, **21**, 2595.
138. J. P. Godschalx and J. K. Stille, *Tetrahedron Lett.*, 1980, **21**, 2599.
139. A. Gollaszewski and J. Schwartz, *Organometallics*, 1985, **4**, 417.
140. B. M. Trost and K. M. Pietrosiewicz, *Tetrahedron Lett.*, 1985, **26**, 4039.
141. N. A. Bumagin, P. N. Kasatkin and I. P. Bevetskaya, *Dokl. Akad. Nauk SSSR*, 1982, **266**, 862.
142. F. K. Sheffy and J. K. Stille, *J. Am. Chem. Soc.*, 1983, **105**, 7173.
143. A. M. Echavarren, D. R. Tueting and J. K. Stille, *J. Am. Chem. Soc.*, 1988, **110**, 4039.
144. N. A. Bumagin, A. N. Kasatkin and I. P. Bevetskaya, *Izv. Akad. Nauk SSSR, Ser. Khim.*, 1983, 912.
145. F. K. Sheffy, J. P. Godschalx and J. K. Stille, *J. Am. Chem. Soc.*, 1984, **106**, 4833.
146. S. Katsumura, S. Fujiwara and S. Isoe, *Tetrahedron Lett.*, 1987, **28**, 1191.
147. J. H. Merrifield, J. P. Godschalx and J. K. Stille, *Organometallics*, 1984, **3**, 1108.
148. S. Numata, H. Kurosawa and R. Okawara, *J. Organomet. Chem.*, 1975, **102**, 259.
149. S. Numata and H. Kurosawa, *J. Organomet. Chem.*, 1977, **131**, 301.
150. T. Kitazume and N. Ishikawa, *Chem. Lett.*, 1982, 137.
151. S. Sasoka, T. Yamamoto, H. Kinoshita, K. Inomata and H. Kotake, *Chem. Lett.*, 1985, 315.
152. J. van der Louw, J. L. van der Baan, F. Bickelhaupt and G. W. Klumpp, *Tetrahedron Lett.*, 1987, **28**, 2889.
153. T. Hayashi, A. Yamamoto and T. Hagihara, *J. Org. Chem.*, 1986, **51**, 723.

154. H. Matsushita and E. Negishi, *J. Am. Chem. Soc.*, 1981, **103**, 2882.
155. J. S. Temple and J. Schwartz, *J. Am. Chem. Soc.*, 1980, **102**, 7382.
156. Y. Hayashi, M. Riediker, J. S. Temple and J. Schwartz, *Tetrahedron Lett.*, 1981, **22**, 2629.
157. J. E. McMurry, J. R. Matz and K. L. Kees, *Tetrahedron*, 1987, **43**, 5489.
158. Y. Castanet and T. Petit, *Tetrahedron Lett.*, 1979, 3221.
159. T. Hayashi, M. Konishi, K. Yokota and M. Kumada, *J. Chem. Soc., Chem. Commun.*, 1981, 313.
160. S. A. Godleski, K. B. Gundlach, H. Y. Ho, E. Keinan and F. Frolow, *Organometallics*, 1984, **3**, 21.
161. S. Holle, P. W. Jolly, R. Mynott and R. Salz, *Z. Naturforsch., Teil B*, 1982, **37**, 675.
162. Y. Takahashi, S. Sakai and Y. Ishii, *J. Organomet. Chem.*, 1969, **16**, 177.
163. Y. Takahashi, K. Tsukiyama, S. Sakai and Y. Ishii, *Tetrahedron Lett.*, 1970, 1913.
164. J. E. Backvall, R. E. Nordberg, E. E. Bjorkman and C. Moberg, *J. Chem. Soc., Chem. Commun.*, 1980, 943.
165. P. M. Henry, *J. Am. Chem. Soc.*, 1972, **94**, 5200.
166. L. E. Overman and F. M. Knoll, *Tetrahedron Lett.*, 1979, 321.
167. B. M. Trost, T. R. Verhoeven and J. M. Fortunak, *Tetrahedron Lett.*, 1979, 2301.
168. J. E. Backvall and R. E. Nordberg, *J. Am. Chem. Soc.*, 1981, **103**, 4959.
169. J. E. Backvall, R. E. Nordberg and D. Wilhelm, *J. Am. Chem. Soc.*, 1985, **107**, 6892.
170. J. E. Backvall and A. Gogoll, *Tetrahedron Lett.*, 1988, **29**, 2243.
171. J. E. Backvall, R. E. Nordberg, E. E. Bjorkman and C. Moberg, *J. Chem. Soc., Chem. Commun.*, 1980, 943.
172. J. Tsuji, K. Sakai, H. Nagashima and I. Shimizu, *Tetrahedron Lett.*, 1981, **22**, 131.
173. K. Ogura, N. Shibuya and H. Iida, *Tetrahedron Lett.*, 1981, **22**, 1519.
174. K. Ogura, N. Shibuya, K. Takahashi and H. Iida, *Bull. Chem. Soc. Jpn.*, 1984, **57**, 1092.
175. R. C. Larock, L. W. Harrison and M. H. Hsu, *J. Org. Chem.*, 1984, **49**, 3662.
176. D. R. Deardorff, D. C. Myles and K. O. MacFerrin, *Tetrahedron Lett.*, 1985, **26**, 5615.
177. R. C. Larock, D. J. Leuck and L. W. Harrison, *Tetrahedron Lett.*, 1987, **28**, 4977.
178. K. Takahashi, G. Hata and A. Miyake, *Bull. Chem. Soc. Jpn.*, 1973, **46**, 1012.
179. D. G. Brady, *Chem. Commun.*, 1970, 434.
180. S. A. Stanton, S. W. Felman, C. S. Parkhurst and S. A. Godleski, *J. Am. Chem. Soc.*, 1983, **105**, 1964.
181. K. E. Atkins, W. E. Walker and R. M. Manyik, *Tetrahedron Lett.*, 1970, 3821.
182. W. Kuran and A. Musco, *J. Organomet. Chem.*, 1972, **40**, C47.
183. J. P. Genet, M. Balabane, J. E. Backvall and J. E. Nystrom, *Tetrahedron Lett.*, 1983, **24**, 2745.
184. B. Akermark and K. Zetterberg, *Tetrahedron Lett.*, 1975, 3733.
185. B. Akermark, J. E. Backvall, A. Lowenberg and K. Zetterberg, *J. Organomet. Chem.*, 1979, **166**, C33.
186. F. G. Stakem and R. F. Heck, *J. Org. Chem.*, 1980, **45**, 3584.
187. J. E. Backvall, R. E. Nordberg and J. E. Nystrom, *J. Org. Chem.*, 1981, **46**, 3479.
188. B. Åkermark, G. Åkermark, L. S. Hegedus and K. Zetterberg, *J. Am. Chem. Soc.*, 1981, **103**, 3037.
189. Y. Tanigawa, K. Nishimura, A. Kawasaki and S. Murahashi, *Tetrahedron Lett.*, 1982, **23**, 5549.
190. J. E. Backvall, R. E. Nordberg, K. Zetterberg and B. Akermark, *Organometallics*, 1983, **2**, 1625.
191. R. Tamura, K. Hayashi, Y. Kai and D. Oda, *Tetrahedron Lett.*, 1984, **25**, 4437.
192. G. C. Ninokogu, *J. Org. Chem.*, 1985, **50**, 3900.
193. B. M. Trost and S. Chen, *J. Am. Chem. Soc.*, 1986, **108**, 6053.
194. K. Takahashi, A. Miyake and G. Hata, *Bull. Chem. Soc. Jpn.*, 1972, **45**, 230.
195. J. E. Backvall, J. E. Nystrom and R. E. Nordberg, *J. Am. Chem. Soc.*, 1985, **107**, 3666.
196. Y. Inoue, M. Taguchi and H. Hashimoto, *Bull. Chem. Soc. Jpn.*, 1985, **58**, 2721.
197. S. E. Bystrom, R. A. Slanian and J. E. Backvall, *Tetrahedron Lett.*, 1985, **26**, 1749.
198. S. Murahashi, Y. Tanigawa, Y. Imada and Y. Taniguchi, *Tetrahedron Lett.*, 1986, **27**, 227.
199. U. M. Dzhemilev, A. G. Ibragimov, D. L. Minsker and R. R. Muslukov, *Izv. Akad. Nauk SSSR, Ser. Khim.*, 1987, 406.
200. B. M. Trost and E. Keinan, *J. Org. Chem.*, 1979, **44**, 3452.
201. S. Murahashi, T. Shimamura and I. Mortiani, *J. Chem. Soc., Chem. Commun.*, 1974, 931.
202. B. M. Trost and E. Keinan, *J. Org. Chem.*, 1980, **45**, 2741.
203. B. M. Trost and J. P. Genet, *J. Am. Chem. Soc.*, 1976, **98**, 8516.
204. B. M. Trost, S. A. Godleski and J. P. Genet, *J. Am. Chem. Soc.*, 1978, **100**, 3930.
205. B. M. Trost, S. A. Godleski and J. L. Belletire, *J. Org. Chem.*, 1979, **44**, 2052.
206. R. Z. Andriamialisoa, N. Langlois and Y. Langlois, *Heterocycles*, 1980, **14**, 1457.
207. S. A. Godleski, D. J. Heacock, J. D. Meinhart and S. Van Wallendael, *J. Org. Chem.*, 1983, **48**, 2101.
208. H. H. Baer and Z. S. Hanna, *Can. J. Chem.*, 1981, **59**, 889.
209. N. Nikado, R. Aslanian, F. Scavo, P. Helquist, B. Akermark and J. E. Backvall, *J. Org. Chem.*, 1984, **49**, 4738.
210. S. A. Godleski, J. D. Meinhart, D. J. Miller and S. Van Wallendael, *Tetrahedron Lett.*, 1981, **22**, 2247.
211. M. Julia, M. Nel and L. Saussime, *J. Organomet. Chem.*, 1979, **181**, C17.
212. K. Inomata, T. Yamamoto and H. Kotake, *Chem. Lett.*, 1981, 1357.
213. R. V. Kunakova, R. L. Gaisin, M. M. Sirazova and U. M. Dzhemilev, *Izv. Akad. Nauk. SSSR, Ser. Khim.*, 1983, 157.
214. Y. Tamaru, Y. Yamada, M. Kagotani, H. Ochiai, E. Nakato, R. Suzuki and Z. Yoshida, *J. Org. Chem.*, 1983, **48**, 4669.
215. B. Akermark, J. E. Nystrom, T. Rein, J. E. Backvall, P. Helquist and R. Aslanian, *Tetrahedron Lett.*, 1984, **25**, 5719.
216. B. M. Trost and N. R. Schmuff, *J. Am. Chem. Soc.*, 1985, **107**, 396.
217. M. Julia, D. Lave, M. Mulhauser, M. Ramirez-Munoz and D. Uguen, *Tetrahedron Lett.*, 1983, **24**, 1783.
218. R. Tamaru, K. Hayashi, M. Tsuji and D. Oda, *Chem. Lett.*, 1985, 229.
219. N. Ono, I. Hamamoto, T. Kawai, A. Kaji, R. Tamuru and M. Kakihana, *Bull. Chem. Soc. Jpn.*, 1986, **59**, 405.
220. Y. Tamaru, Z. Yoshida, Y. Yamada, K. Mukai and H. Yoshioka, *J. Org. Chem.*, 1983, **48**, 1293.
221. P. R. Auburn, J. Whelan and B. Bosnich, *J. Chem. Soc., Chem. Commun.*, 1986, 146.

222. X. Lu and Z. Ni, *Synth. Commun.*, 1987, 66.
223. B. M. Trost and T. S. Scanlan, *Tetrahedron Lett.*, 1986, **27**, 4141.
224. J. C. Fiaud, *J. Chem. Soc., Chem. Commun.*, 1983, 1055.
225. B. M. Trost, J. Yoshida and M. Lautens, *J. Am. Chem. Soc.*, 1983, **105**, 4494.
226. L. S. Hegedus and R. Tamura, *Organometallics*, 1982, **1**, 1188.
227. J. Tsuji and T. Suzuki, *Tetrahedron Lett.*, 1965, 3027.
228. D. Medema, R. VanHelden and C. F. Kohll, *Inorg. Chim. Acta*, 1969, **3**, 255.
229. W. Dent, R. Long and G. H. Whitfield, *J. Chem. Soc.*, 1964, 1588.
230. J. Tsuji, J. Kiji and M. Murikawa, *Tetrahedron Lett.*, 1963, 1811.
231. J. Tsuji and S. Imamura, *Bull. Chem. Soc. Jpn.*, 1967, **40**, 197.
232. J. Tsuji, J. Kiji and S. Hosaka, *Tetrahedron Lett.*, 1964, 605.
233. J. Tsuji and S. Hosaka, *J. Am. Chem. Soc.*, 1965, **87**, 4075.
234. I. G. Dinulescu, S. Staticu, F. Chiraleu and M. Avram, *J. Organomet. Chem.*, 1977, **140**, 91.
235. D. Milstein, *Organometallics.*, 1982, **1**, 888.
236. J. Tsuji, K. Sato and H. Okomoto, *J. Org. Chem.*, 1984, **49**, 1341.
237. K. Yamamoto, R. Deguchi and J. Tsuji, *Bull. Chem. Soc. Jpn.*, 1985, **58**, 3397.
238. J. Tsuji, S. Hosaka, J. Kiji and T. Susuki, *Bull. Chem. Soc. Jpn.*, 1969, **39**, 141.
239. B. C. Soderberg, B. Akermark and S. S. Hall, *J. Org. Chem.*, 1988, **53**, 2925.
240. T. Kajimoto, H. Takahashi and J. Tsuji, *J. Organomet. Chem.*, 1979, **23**, 275.
241. Y. Ito, T. Hirao, N. Ohta and T. Saegusa, *Tetrahedron Lett.*, 1977, 1009.
242. T. Boschi and B. Crociani, *Inorg. Chim. Acta*, 1971, **5**, 477.
243. R. Santi and M. Marchi, *J. Organomet. Chem.*, 1979, **182**, 117.
244. T. Ito, Y. Kindaichi and Y. Takami, *Chem. Ind. (London)*, 1980, 83.
245. T. Hung, P. W. Jolly and G. Wilke, *J. Organomet. Chem.*, 1980, **190**, C5.
246. D. Medema, R. VanHelden and C. F. Kohll, *Inorg. Chim. Acta*, 1969, **3**, 255.
247. Y. Takahashi, S. Sakai and Y. Ishii, *J. Organomet. Chem.*, 1969, **16**, 177.
248. D. Medema and R. VanHelden, *Recl. Chim. Trav. Pays-Bas*, 1971, **90**, 304.
249. R. P. Hughes and J. Powell, *J. Am. Chem. Soc.*, 1972, **94**, 7723.
250. R. P. Hughes, T. Jack and J. Powell, *J. Organomet. Chem.*, 1973, **63**, 451.
251. J. Kiji, Y. Miura and J. Furukawa, *J. Organomet. Chem.*, 1977, **140**, 317.
252. B. M. Trost and J. I. Luengo, *J. Am. Chem. Soc.*, 1988, **110**, 8239.
253. R. C. Larock, J. P. Burkhart and K. Oertle, *Tetrahedron Lett.*, 1982, **23**, 1071.
254. R. C. Larock, K. Takagi, J. P. Burkhart and S. S. Hershberger, *Tetrahedron*, 1986, **42**, 3759.
255. R. P. Hughes and J. Powell, *J. Organomet. Chem.*, 1973, **60**, 427.
256. R. P. Hughes and J. Powell, *J. Organomet. Chem.*, 1973, **60**, 387.
257. M. Zocchi, G. Tieghi and A. Albinati, *J. Organomet. Chem.*, 1971, **33**, C47.
258. M. C. Gallazzi, T. L. Hanlon, G. Vittulli and L. Porri, *J. Organomet. Chem.*, 1971, **33**, C45.
259. R. P. Hughes and J. Powell, *J. Organomet. Chem.*, 1971, **30**, C45.
260. M. Zocchi and G. Tieghi, *J. Chem. Soc., Dalton Trans.*, 1979, 944.
261. M. Zocchi, G. Tieghi and A. Albinati, *J. Chem. Soc., Dalton Trans.*, 1973, 883.
262. W. Oppolzer and J. Gandin, *Helv. Chim. Acta*, 1987, **70**, 1477.
263. E. Negishi, S. Iyer and C. J. Rousset, *Tetrahedron Lett.*, 1989, **30**, 291.
264. T. Mitsudo, M. Kadokura and Y. Watanabe, *J. Org. Chem.*, 1984, **52**, 1695.
265. R. Baker, M. S. Nobbs and D. T. Robinson, *J. Chem. Soc., Chem. Commun.*, 1976, 723.
266. P. Baker, M. S. Nobbs and D. T. Robinson, *J. Chem. Soc., Perkin Trans. 1*, 1978, 543.
267. R. Hüttel and H. Christ, *Chem. Ber.*, 1964, **97**, 1439.
268. G. A. Gray, W. R. Jackson and J. J. Rooney, *J. Chem. Soc. C*, 1970, 1788.
269. E. Vedejs, M. F. Salomon and R. P. Weeks, *J. Organomet. Chem.*, 1972, **40**, 221.
270. J. Y. Satoh and C. A. Horiuchi, *Bull. Chem. Soc. Jpn.*, 1981, **54**, 625.
271. D. N. Jones and S. D. Knox, *J. Chem. Soc., Chem. Commun.*, 1975, 165.
272. C. Mahe, H. Patin and M. VanHulle, *J. Chem. Soc., Perkin Trans. 1*, 1981, 2504.
273. R. F. Heck, *J. Am. Chem. Soc.*, 1968, **90**, 5542.
274. S. Wolfe and P. G. C. Campbell, *J. Am. Chem. Soc.*, 1971, **93**, 1499.
275. K. Jitsukawa, K. Kaneda and S. Teranishi, *J. Org. Chem.*, 1983, **48**, 389.
276. J. Muzart, P. Pale and J. Pete, *J. Chem. Soc., Chem. Commun.*, 1981, 668.
277. J. Muzart, P. Pale and J. Pete, *Tetrahedron Lett.*, 1983, **24**, 4567.
278. K. Dunne and F. J. McQuillin, *J. Chem. Soc. C*, 1970, 2196.
279. J. W. Faller and K. J. Laffey, *Organomet. Chem. Synth.*, 1972, **1**, 471.
280. J. W. Faller, M. T. Tully and K. J. Laffey, *J. Organomet. Chem.*, 1972, **37**, 193.
281. H. Christ and R. Hüttel, *Angew. Chem., Int. Ed. Engl.*, 1963, **2**, 626.
282. R. Hüttel and P. Kochs, *Chem. Ber.*, 1968, **101**, 1043.
283. W. A. Donaldson and B. S. Taylor, *Tetrahedron Lett.*, 1985, **26**, 4163.
284. T. A. Schenach and F. F. Caserio, Jr., *J. Organomet. Chem.*, 1969, **18**, P17.
285. R. O. Hutchins, K. Learn, M. Markowitz and R. P. Fulton, in abstracts of 'Second Chemical Congress of the North American Continent, Las Vegas, 1980', Orgn-183.
286. S. A. Godleski, K. B. Gundlach, H. Y. Ho, E. Keinan and F. Frolow, *Organometallics*, 1984, **3**, 21.
287. E. Keinan and N. Greenspoon, *Tetrahedron Lett.*, 1982, **23**, 241.
288. H. Kotake, T. Yamamoto and H. Kinoshita, *Chem. Lett.*, 1982, 1331.
289. M. Mohri, H. Kinoshita, K. Inomata and H. Kotake, *Chem. Lett.*, 1985, 451.
290. R. O. Hutchins, K. Learn and R. P. Fulton, *Tetrahedron Lett.*, 1980, **21**, 27.
291. R. O. Hutchins and K. Learn, *J. Org. Chem.*, 1982, **47**, 4380.
292. D. N. Jones and S. D. Knox, *J. Chem. Soc., Chem. Commun.*, 1975, 165.
293. D. H. R. Barton and A. Patin, *J. Chem. Soc., Chem. Commun.*, 1977, 799.

294. F. Guibe, Y. T. Xian, A. M. Zigna and G. Balavione, *Tetrahedron Lett.*, 1985, **26**, 3559.
295. P. J. Ssebuwufu, F. Glocking and P. Harriott, *Inorg. Chim. Acta*, 1985, **98**, L35.
296. F. Guibe, H. X. Zang and G. Balavione, *Nouv. J. Chim.*, 1984, **8**, 611.
297. F. Guibe, A. M. Zigna and G. Balavione, *J. Organomet. Chem.*, 1986, **306**, 257.
298. F. Guibe, Y. T. Xian and G. Balavione, *J. Organomet. Chem.*, 1986, **306**, 267.
299. F. Guibe and Y. Saintmileux, *Tetrahedron Lett.*, 1981, **22**, 3591.
300. E. Keinan and N. Greenspoon, *J. Org. Chem.*, 1983, **48**, 3545.
301. H. Hey and H. J. Arpe, *Angew. Chem., Int. Ed. Engl.*, 1973, **12**, 928.
302. J. Tsuji and T. Yamakawa, *Tetrahedron Lett.*, 1979, 613.
303. T. Okano, Y. Moriyama, H. Konishi and J. Kiji, *Chem. Lett.*, 1986, 1463.
304. H. Matsushita and E. Negishi, *J. Org. Chem.*, 1982, **47**, 4161.
305. T. Tabuchi, J. Inanaga and M. Yamaguchi, *Tetrahedron Lett.*, 1986, **27**, 601.
306. K. Nakamura, A. Ohno and S. Oka, *Tetrahedron Lett.*, 1983, **24**, 3335.
307. S. Torii, H. Tanaka, T. Katoh and K. Morisaki, *Tetrahedron Lett.*, 1984, **25**, 3207.
308. Y. Tamaru, M. Kagotani and Z. Yoshida, *J. Chem. Soc., Chem. Commun.*, 1978, 367.
309. B. M. Trost and J. W. Herndon, *J. Am. Chem. Soc.*, 1984, **106**, 6835.
310. S. Matsubara, K. Wakamatsu, Y. Morizawa, N. Tsuboniwa, K. Oshima and H. Nozaki, *Bull. Chem. Soc. Jpn.*, 1985, **58**, 1196.
311. T. Tabuchi, J. Inanaga and M. Yamaguchi, *Tetrahedron Lett.*, 1987, **28**, 215.
312. Y. Masuyama, R. Hamashi, K. Otaka and Y. Kurusu, *J. Chem. Soc., Chem. Commun.*, 1988, 44.
313. T. Tabuchi, J. Inanaga and M. Yamaguchi, *Tetrahedron Lett.*, 1986, **27**, 1195.
314. Y. Masuyama, N. Kinugawa and Y. Kirusu, *J. Org. Chem.*, 1987, **52**, 3704.
315. S. Torii, H. Tanaka, T. Katoh and K. Morisaki, *Tetrahedron Lett.*, 1984, **25**, 3207.
316. H. Kurosawa and A. Urabe, *Chem. Lett.*, 1985, 1839.
317. H. Kurosawa, A. Urabe, K. Miki and N. Kasai, *Organometallics*, 1986, **5**, 2002.
318. R. Benn, G. Gabor, P. W. Jolly, R. Mynott and B. Raspel, *J. Organomet. Chem.*, 1985, **296**, 443.
319. R. Benn, P. W. Jolly, R. Mynott, B. Raspel, G. Schenker, K. P. Schick and G. Schroth, *Organometallics*, 1985, **4**, 1945.
320. M. Para-Hake, M. F. Rettig and R. M. Wing, *Organometallics*, 1983, **2**, 1013.
321. B. M. Trost and G. B. Tometzki, *J. Org. Chem.*, 1988, **53**, 915.
322. K. Dunne and F. J. McQuillin, *J. Chem. Soc. C*, 1970, 2200.
323. M. Donati and F. Conti, *Tetrahedron Lett.*, 1966, 4953.
324. B. M. Trost, T. R. Verhoeven and J. M. Fortunak, *Tetrahedron Lett.*, 1979, 2301.
325. I. T. Harrison, G. Kimura, E. Bohme and J. H. Frigd, *Tetrahedron Lett.*, 1969, 1589.
326. M. Suzuki, Y. Oda and R. Noyori, *J. Am. Chem. Soc.*, 1979, **101**, 1623.
327. D. J. Collins, W. R. Jackson and R. N. Timms, *Tetrahedron Lett.*, 1976, 495.
328. W. R. Jackson and J. V. Strauss, *Aust. J. Chem.*, 1978, **31**, 1073.
329. T. Mandai, J. Gotah, J. Utera and M. Kawada, *Chem. Lett.*, 1980, 313.
330. A. F. Noels, J. J. Herman and P. Teyssie, *J. Org. Chem.*, 1976, **41**, 2527.
331. J. Tsuji, T. Yamakawa, M. Kaito and T. Mandai, *Tetrahedron Lett.*, 1978, 2075.
332. H. Kumobayashi, S. Mitsuhashi, S. Akutagawa and S. Ohtsuka, *Chem. Lett.*, 1986, 157.
333. B. M. Trost and J. M. Fortunak, *J. Am. Chem. Soc.*, 1980, **102**, 2841.
334. J. Muzart and J. P. Pète, *J. Chem. Soc., Chem. Commun.*, 1980, 257.
335. J. Muzart, P. Pale and J. P. Pète, *J. Chem. Soc., Chem. Commun.*, 1981, 668.
336. B. DePorter, J. Muzart and J. P. Pète, *Organometallics*, 1983, **2**, 1494.
337. H. Matsumoto, T. Yako, S. Nagashima, T. Motegi and Y. Nagai, *J. Organomet. Chem.*, 1978, **148**, 97.
338. H. Orata, H. Sutuki, Y. Moro-Oka and T. Ikawa, *Bull. Chem. Soc. Jpn.*, 1984, **57**, 607.
339. T. Hayashi, A. Yamamoto, T. Iwata and Y. Ito, *J. Chem. Soc., Chem. Commun.*, 1987, 398.
340. N. A. Bumagin, A. N. Kasatkin and I. P. Beletskaya, *Izv. Akad. Nauk SSSR, Ser. Khim.*, 1984, 636.
341. A. N. Kashin, I. G. Bumagin, V. N. Bakunin and I. P. Beletskaya, *Zh. Org. Khim.*, 1981, **17**, 805.
342. S. Ali, S. Tanimoto and T. Okamoto, *J. Org. Chem.*, 1988, **53**, 3639.
343. S. Wolff and W. C. Agosta, *Synthesis*, 1976, 240.
344. E. Mincione, G. Ortaggi and A. Sirna, *Tetrahedron Lett.*, 1977, 773.
345. V. B. Bierling, K. Kirschke, H. Oberender and M. Schulz, *J. Prakt. Chem.*, 1972, **314**, 170.
346. T. Susuki and J. Tsuji, *Bull. Chem. Soc. Jpn.*, 1973, **46**, 655.
347. F. J. Theissen, *J. Org. Chem.*, 1971, **36**, 752.
348. K. Kirschke, H. Müller and D. Timm, *J. Prakt. Chem.*, 1975, **317**, 807.
349. J. M. Townsend, I. D. Reingold, M. C. R. Kendall and T. A. Spencer, *J. Org. Chem.*, 1975, **40**, 2976.
350. M. E. Kuehne and T. C. Hall, *J. Org. Chem.*, 1976, **41**, 2742.
351. R. D. Gillard and J. R. Lyons, *Transition Met. Chem.*, 1977, **2**, 19.
352. A. Kasahara, T. Izumi and M. Oshima, *Bull. Chem. Soc. Jpn.*, 1974, **47**, 2526.
353. T. Hosokawa, N. Shimo, K. Maeda, A. Sonoda and S. Murahashi, *Tetrahedron Lett.*, 1976, 383.
354. Y. Ito, T. Hirao, N. Ohta and T. Saegusa, *Tetrahedron Lett.*, 1977, 1009.
355. B. M. Trost, Y. Nishimura and K. Yamamoto, *J. Am. Chem. Soc.*, 1979, **101**, 1328.
356. P. A. Wender and J. C. Lechleiter, *J. Am. Chem. Soc.*, 1980, **102**, 6340.
357. G. J. Quallich and R. H. Schlessinger, *J. Am. Chem. Soc.*, 1979, **101**, 7627.
358. R. S. Lott, E. G. Breitholle and C. H. Stammer, *J. Org. Chem.*, 1980, **45**, 1151.
359. T. Tsuda, Y. Chujo, S. Nisai, K. Tamara and T. Saegusa, *J. Am. Chem. Soc.*, 1980, **102**, 6381.
360. I. Shimizu and J. Tsuji, *J. Am. Chem. Soc.*, 1982, **104**, 5844.
361. J. Tsuji, I. Minami, I. Shimizu and H. Kataoka, *Chem. Lett.*, 1984, 1133.
362. N. Kataoka, T. Yamada, K. Goto and J. Tsuji, *Tetrahedron*, 1987, **43**, 4107.
363. I. Minami, M. Nisar, M. Yuhara, I. Shimizu and J. Tsuji, *Synthesis*, 1987, 992.
364. I. Shimizu, I. Minami and J. Tsuji, *Tetrahedron Lett.*, 1983, **24**, 1797.

365. J. Tsuji, I. Minami and I. Shimizu, *Tetrahedron Lett.*, 1983, **24**, 5639.
366. J. Tsuji, I. Minami and I. Shimizu, *Tetrahedron Lett.*, 1983, **24**, 5635.
367. I. Minami, K. Takahashi, I. Shimizu, T. Kimura and J. Tsuji, *Tetrahedron*, 1986, **42**, 2971.
368. J. Tsuji, M. Nisar and I. Minami, *Tetrahedron Lett.*, 1986, **22**, 2483.
369. I. Minami, M. Yuhara, I. Shimizu and J. Tsuji, *J. Chem. Soc., Chem. Commun.*, 1986, 118.
370. Y. Ito, M. Nakatsuka, N. Kise and T. Saegusa, *Tetrahedron Lett.*, 1980, **21**, 2873.
371. N. Yoshimura, S. Murahashi and I. Moritani, *J. Organomet. Chem.*, 1973, **52**, C58.
372. Y. Ito, H. Aoyama and T. Saegusa, *J. Am. Chem. Soc.*, 1980, **102**, 4519.
373. Y. Ito, H. Aoyama, T. Hirao, A. Mochizuki and T. Saegusa, *J. Am. Chem. Soc.*, 1979, **101**, 494.
374. Y. Takahashi, K. Tsukiyama, S. Sakai and Y. Ishii, *Tetrahedron Lett.*, 1970, 1913.
375. A. S. Kende, B. Roth and P. J. Sanfilippo, *J. Am. Chem. Soc.*, 1982, **104**, 1784.
376. A. S. Kende, B. Roth, P. J. Sanfilippo and T. J. Blacklock, *J. Am. Chem. Soc.*, 1982, **104**, 5808.
377. A. S. Kende and D. J. Wustrow, *Tetrahedron Lett.*, 1985, **26**, 5411.
378. A. S. Kende, R. A. Battista and S. B. Sandoval, *Tetrahedron Lett.*, 1984, **25**, 1341.
379. A. S. Kende and P. J. Sanfilippo, *Synth. Commun.*, 1983, **13**, 715.
380. B. M. Trost, L. Weber, P. E. Strege, T. J. Fullerton and T. J. Dietsche, *J. Am. Chem. Soc.*, 1978, **100**, 3416.
381. P. K. Wong, K. S. Y. Lau and J. K. Stille, *J. Am. Chem. Soc.*, 1974, **96**, 5956.
382. T. Hayashi, T. Hagihara, M. Konishi and M. Kumara, *J. Am. Chem. Soc.*, 1983, **105**, 7767.
383. S. Sakari, M. Nishikawa and A. Ohyoshi, *J. Am. Chem. Soc.*, 1980, **102**, 4062.
384. H. Kurosawa, M. Emoto, H. Ohnishi, K. Miki, N. Kasai, K. Tatsumi and A. Nakamura, *J. Am. Chem. Soc.*, 1987, **109**, 6333.
385. E. Keinan and Z. Roth, *J. Org. Chem.*, 1983, **48**, 1772.
386. J. C. Fiaud and J. Y. Leuros, *J. Org. Chem.*, 1987, **52**, 1907.
387. T. Hayashi, M. Konishi and M. Kumada, *J. Chem. Soc., Chem. Commun.*, 1984, 107.
388. B. M. Trost and T. R. Verhoeven, *J. Am. Chem. Soc.*, **100**, 3435.
389. B. M. Trost and T. R. Verhoeven, *J. Org. Chem.*, 1976, **41**, 3215.
390. B. M. Trost and T. R. Verhoeven, *J. Am. Chem. Soc.*, 1980, **102**, 4730.
391. D. J. Collins, W. R. Jackson and R. N. Timms, *Aust. J. Chem.*, 1977, **30**, 2167.
392. D. A. Hunt, J. M. Quante, R. L. Tyson and L. W. Dasher, *J. Org. Chem.*, 1984, **49**, 5262.
393. M. Moreno-Manas, J. Ribas and A. Virgili, *J. Org. Chem.*, 1988, **53**, 5328.
394. H. Kirosawa, *J. Organomet. Chem.*, 1987, **334**, 243.
395. J. S. Temple, M. Riediker and J. Schwartz, *J. Am. Chem. Soc.*, 1982, **104**, 1310.
396. A. Goliaszewski and J. Schwartz, *Tetrahedron*, 1985, **41**, 5779.
397. A. Goliaszewski and J. Schwartz, *J. Am. Chem. Soc.*, 1984, **106**, 5028.
398. B. M. Trost and E. Keinan, *J. Am. Chem. Soc.*, 1978, **100**, 7779.
399. E. Keinan and M. Sahai, *J. Chem. Soc., Chem. Commun.*, 1984, 648.
400. B. M. Trost and T. R. Verhoeven, *J. Am. Chem. Soc.*, 1977, **99**, 3866.
401. B. M. Trost and T. R. Verhoeven, *J. Am. Chem. Soc.*, 1979, **101**, 1595.
402. B. M. Trost and T. R. Verhoeven, *J. Am. Chem. Soc.*, 1980, **102**, 4743.
403. W. R. Jackson and J. U. G. Strauss, *Aust. J. Chem.*, 1977, **30**, 553.
404. W. R. Jackson and J. U. G. Strauss, *Tetrahedron Lett.*, 1975, 2591.
405. J. Tsuji, H. Ueno, Y. Kobayashi and H. Okumoto, *Tetrahedron Lett.*, 1981, **22**, 2573.
406. S. A. Godleski and E. B. Villhauer, *J. Org. Chem.*, 1984, **49**, 2246.
407. S. A. Godleski and E. B. Villhauer, *J. Org. Chem.*, 1986, **51**, 486.
408. Z. Zhu and X. Lu, *Tetrahedron Lett.*, 1987, **28**, 1897.
409. D. P. Grant, N. W. Murrall and A. S. Welch, *J. Organomet. Chem.*, 1987, **333**, 403.
410. J. Tsuji, H. Kataoka and T. Kobayashi, *Tetrahedron Lett.*, 1981, **22**, 2575.
411. T. Takahashi, H. Kataoka and J. Tsuji, *J. Am. Chem. Soc.*, 1983, **105**, 147.
412. B. M. Trost and R. W. Warner, *J. Am. Chem. Soc.*, 1983, **105**, 5940.
413. B. M. Trost and R. W. Warner, *J. Am. Chem. Soc.*, 1982, **104**, 6112.
414. V. I. Ognyanov and M. Hesse, *Synthesis*, 1985, 645.
415. R. S. Valpey, D. J. Miller, J. M. Estes and S. A. Godleski, *J. Org. Chem.*, 1982, **47**, 4717.
416. D. S. Manchand, H. S. Wong and J. F. Blount, *J. Org. Chem.*, 1978, **43**, 4769.
417. J. P. Genet and D. Ferroud, *Tetrahedron Lett.*, 1984, **25**, 3579.
418. R. J. P. Corriu, N. Escudie and C. Guerin, *J. Organomet. Chem.*, 1984, **271**, C7.
419. T. Hirao, J. Enda, Y. Ohshiro and T. Agawa, *Tetrahedron Lett.*, 1981, **22**, 3079.
420. B. Akermark, K. Zetterberg, S. Hansson, B. Krakenberger and A. Vitagliano, *J. Organomet. Chem.*, 1987, **335**, 133.
421. B. Akermark, B. Krakenberger, S. Hansson and A. Vitagliano, *Organometallics*, 1987, **6**, 620.
422. B. Akermark and A. Vitagliano, *Organometallics*, 1985, **4**, 1275.
423. S. Murahashi, Y. Tanigawa, Y. Imada and Y. Taniguchi, *Tetrahedron Lett.*, 1986, **27**, 227.
424. K. Yamamoto, R. Deguchi, Y. Ogimura and J. Tsuji, *Chem. Lett.*, 1984, 1657.
425. T. Takahashi, Y. Jinbo, K. Kitamura and J. Tsuji, *Tetrahedron Lett.*, 1984, **25**, 5921.
426. F. E. Ziegler, A. Kreisley and R. T. Wester, *Tetrahedron Lett.*, 1986, **27**, 1221.
427. G. Stork and J. M. Poirier, *J. Am. Chem. Soc.*, 1983, **105**, 1073.
428. F. Colobert and J. P. Genet, *Tetrahedron Lett.*, 1985, **26**, 2779.
429. B. M. Trost and T. A. Runge, *J. Am. Chem. Soc.*, 1981, **103**, 2485.
430. B. M. Trost and T. P. Klun, *J. Am. Chem. Soc.*, 1979, **101**, 6756.
431. B. M. Trost and T. P. Klun, *J. Am. Chem. Soc.*, 1981, **103**, 1864.
432. B. M. Trost and T. J. Dietsche, *J. Am. Chem. Soc.*, 1973, **95**, 8200.
433. B. M. Trost and P. E. Strege, *J. Am. Chem. Soc.*, 1977, **99**, 1649.
434. J. C. Fiaud, A. H. DeGournay, M. Larchevêque and H. B. Kagan, *J. Organomet. Chem.*, 1978, **154**, 175.
435. K. Yamamoto and J. Tsuji, *Tetrahedron Lett.*, 1982, **23**, 3089.

436. J. P. Genet, D. Ferroud, S. Juge and J. R. Montes, *Tetrahedron Lett.*, 1986, **27**, 4573.
437. T. Hayashi, K. Kanehira, H. Tsuchiya and M. Kumada, *J. Chem. Soc., Chem. Commun.*, 1982, 1162.
438. B. M. Trost and D. J. Murphy, *Organometallics*, 1985, **4**, 1143.
439. T. Hayashi, A. Yamamoto and Y. Ito, *J. Chem. Soc., Chem. Commun.*, 1986, 1090.
440. T. Hayashi, A. Yamamoto, T. Hagihara and Y. Ito, *Tetrahedron Lett.*, 1986, **27**, 191.
441. T. Hayashi, Y. Yamamoto and Y. Ito, *Chem. Lett.*, 1987, 177.
442. T. Hayashi, K. Kanehira, T. Hagihara and M. Kumada, *J. Org. Chem.*, 1988, **53**, 113.
443. B. Bosnich and P. B. Mackenzie, *Pure Appl. Chem.*, 1982, **54**, 189.
444. P. R. Auburn, P. B. Mackenzie and B. Bosnich, *J. Am. Chem. Soc.*, 1985, **107**, 2033.
445. P. B. Mackenzie, J. Whelan and B. Bosnich, *J. Am. Chem. Soc.*, 1985, **107**, 2046.
446. D. H. Farrar and N. C. Payne, *J. Am. Chem. Soc.*, 1985, **107**, 2054.
447. K. Hiroi, K. Suya and S. Sato, *J. Chem. Soc., Chem. Commun.*, 1986, 469.
448. K. Hiroi, R. Kitayama and S. Sato, *J. Chem. Soc., Chem. Commun.*, 1984, 303.
449. K. Hiroi, R. Kitayama and S. Sato, *Chem. Lett.*, 1984, 929.

3.4
Nucleophiles with Cationic Pentadienyl–Metal Complexes

ANTHONY J. PEARSON
Case Western Reserve University, Cleveland, OH, USA

3.4.1 INTRODUCTION

This chapter includes those transition metal–pentadienyl cationic complexes that are quite stable, can be stored and handled easily, and are therefore useful as stoichiometric intermediates for organic synthesis. The dienyliron systems, which are readily available and inexpensive, have dominated this area of chemistry, and will occupy the larger part of the discussion. The chemistry of more expensive and less easily prepared dienylmetal complexes, such as those of manganese and cobalt, will be dealt with at the end of the chapter.

Although tricarbonylbutadieneiron (**1**) was prepared by Reihlen *et al.*[1] in 1930, some considerable time passed before the corresponding cyclohexadiene complex (**2**; equation 1) was reported.[2] Fischer and Fischer described the conversion of (**2**) to the cationic cyclohexadienyliron complex (**3**; equation 1) by reaction with triphenylmethyl tetrafluoroborate in dichloromethane.[3] This particular complex is extremely easy to prepare and isolate; as the hydride abstraction reaction proceeds the product (**3**) crystallizes out. Precipitation is completed by pouring the reaction mixture into 'wet' diethyl ether, the small amount of water present serving to destroy any excess triphenylmethyl tetrafluoroborate by conversion to triphenylmethanol. Filtration, followed by washing the residue with ether, gives pure dienyl complex.

The analogous cycloheptadienyl complex (**5**; equation 1) was similarly prepared by Dauben and Bertelli,[4] but the acyclic pentadienyl systems were a little more difficult to obtain. The triphenylmethyl cation does not remove hydride from tricarbonyl(*trans*-pentadiene)iron (**6**; equation 2). The corresponding *cis*-pentadiene complex (**7**; equation 3) cannot be prepared directly from the diene and an iron carbonyl,

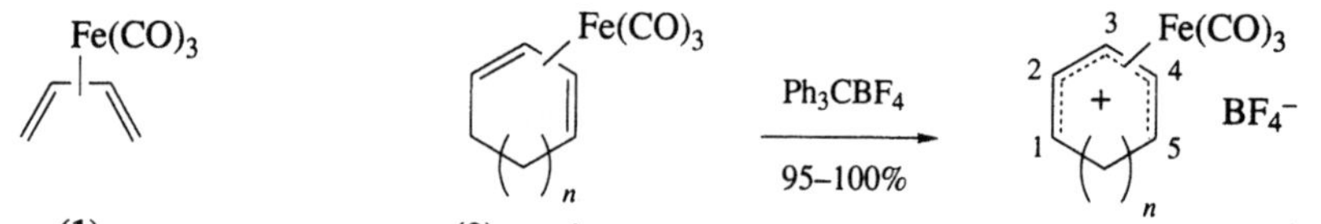

$Fe(CO)_3$ (1); (2) $n = 1$, (4) $n = 2$ —[Ph_3CBF_4, 95–100%]→ (3) $n = 1$, (5) $n = 2$ BF_4^- (1)

since this reaction always leads to (**6**), although it is now known that (**7**) can be converted to (**8**; equation 3) by treatment with triphenylmethyl tetrafluoroborate.

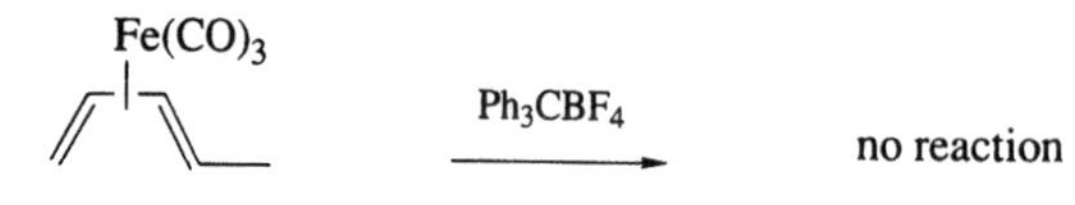

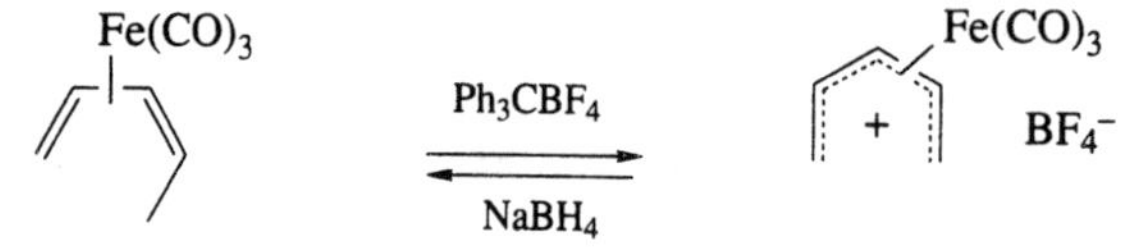

(**6**) —[Ph_3CBF_4]→ no reaction (2)

(**7**) ⇌[Ph_3CBF_4 / $NaBH_4$] (**8**) BF_4^- (3)

An elegant solution to this problem was devised by Mahler and Pettit,[5] who showed that treatment of the (pentadienol)$Fe(CO)_3$ complex (**9**; equation 4) with acid gave (**8**), which was isolated as the perchlorate in 50% yield. These two methods (hydride abstraction and dehydroxylation) form the basis for the majority of preparations of dienyliron complexes.

(**9**) —[$HClO_4$]→ (**8**) (4)

In these dienylmetal complexes, the metal is bound to all five carbon atoms, and these are almost coplanar. The methylene carbon of (**3**) is bent out of the dienyl plane, pointing away from the metal. The angle between the dienyl plane and the C(1)–C(6)–C(5) plane is approximately 39°, as measured by X-ray crystallography.[6] The complexes such as (**3**) can be drawn as shown in Figure 1, but are more conveniently depicted as in the above structures.

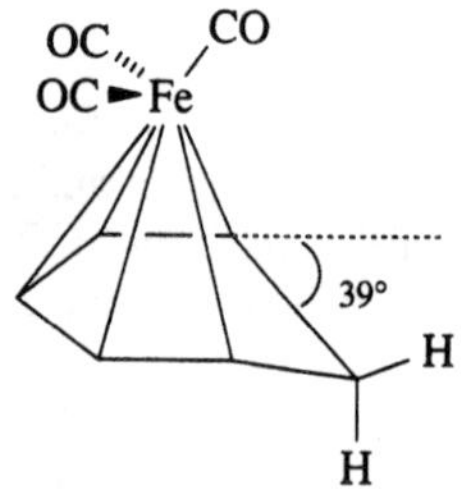

Figure 1 Structure of the tricarbonylcyclohexadienyliron cation

Generally, the dienyliron complexes discussed here are stable towards air and moisture (indeed, some of them can be recrystallized from hot water), and most can be stored for extended periods of time at room temperature. In spite of their stability, they are extremely reactive towards nucleophiles. The result of nucleophile addition to complex (**3**) is shown in equation (5), giving substituted cyclohexadieneiron complexes (**10**) resulting from *exo* attack. On the other hand, complexes (**5**) and (**8**) are more problematic; mixtures of regioisomers (**11**) and (**12**; equation 6) result from (**5**), while the initial products from (**8**) are prone to isomerization to the more stable systems (**14**; Scheme 1).

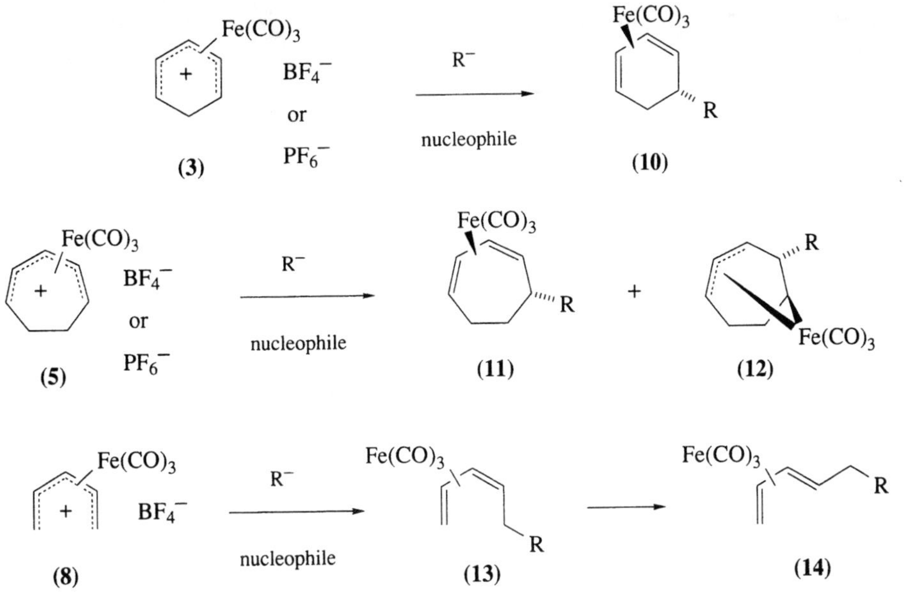

$Fe(CO)_3$ BF_4^- or PF_6^- (3) — R^- nucleophile → $Fe(CO)_3$ R (10) (5)

$Fe(CO)_3$ BF_4^- or PF_6^- (5) — R^- nucleophile → $Fe(CO)_3$ R (11) + R $Fe(CO)_3$ (12) (6)

$Fe(CO)_3$ BF_4^- (8) — R^- nucleophile → $Fe(CO)_3$ R (13) → $Fe(CO)_3$ R (14)

Scheme 1

Because of the highly efficient nucleophile additions to (**3**), substituted cyclohexadienyliron complexes have been exploited for natural products synthesis. The problems associated with (**5**) have been largely overcome and the seven-membered ring systems now show promise as synthetic building blocks. On the other hand, very little attention has been given to the acyclic systems, and this remains a virgin area. Finally, in order to be useful in organic synthesis, it is essential that the metal can be removed from the product dienes selectively and without detriment to any functional groups in the molecule.

3.4.2 DIENYL COMPLEXES OF IRON

3.4.2.1 Preparation of Substituted Tricarbonylcyclohexadienyliron Complexes

Since these complexes are all prepared from appropriate (cyclohexadiene)tricarbonyliron derivatives, we shall first briefly review the preparation of the diene complexes. As mentioned earlier, complex (**2**) was initially prepared by reaction of cyclohexa-1,3-diene with pentacarbonyliron (autoclave; 135–140 °C). Arnet and Pettit observed[7] that cyclohexa-1,4-diene also gave (**2**; equation 7) on reaction with pentacarbonyliron at elevated temperature, *via* a rearrangement of the diene which is promoted by the iron carbonyl. Birch and coworkers[8] applied this observation to the reaction of iron carbonyls with various dihydrobenzene derivatives that are readily prepared by metal–ammonia (Birch) reduction of the aromatic compound. Various iron carbonyls have been investigated as starting materials. Pentacarbonyliron in refluxing di-*n*-butyl ether, and dodecacarbonyltriiron [$Fe_3(CO)_{12}$] in hot benzene are reported[8] to convert cyclohexa-1,4-dienes to their (1,3-diene)$Fe(CO)_3$ complexes; since pentacarbonyliron is inexpensive and usually gives better yields, this is the reagent of choice. Nonacarbonyldiiron [$Fe_2(CO)_9$] may also be used, usually in an inert solvent at *ca.* 60 °C, but invariably is successful only with 1,3-dienes. For sensitive 1,3-dienes the heterodiene complex (benzylideneacetone)tricarbonyliron is a useful tricarbonyliron transfer reagent,[9] which can be employed under mild conditions.

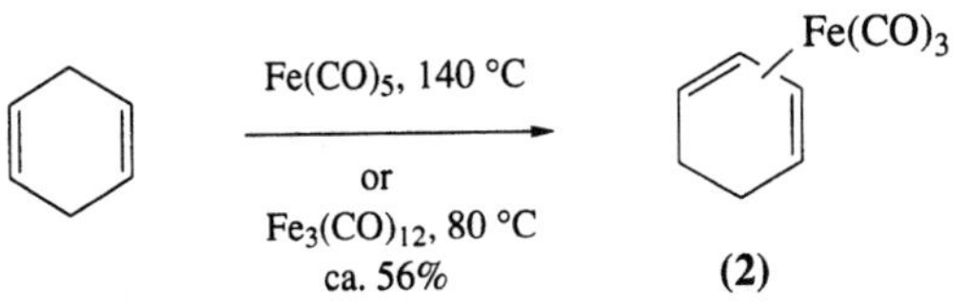

$Fe(CO)_5$, 140 °C or $Fe_3(CO)_{12}$, 80 °C ca. 56% → $Fe(CO)_3$ (2) (7)

The earlier work of Birch *et al.*[8,10,11] laid the foundations for much of the synthetic application of cyclohexadienyliron complexes. Of particular interest are the reactions of dihydroanisole derivatives[8,10] and dihydrobenzoic esters[11] with iron carbonyls, which are summarized in Schemes 2 and 3 and equations (8)–(10).

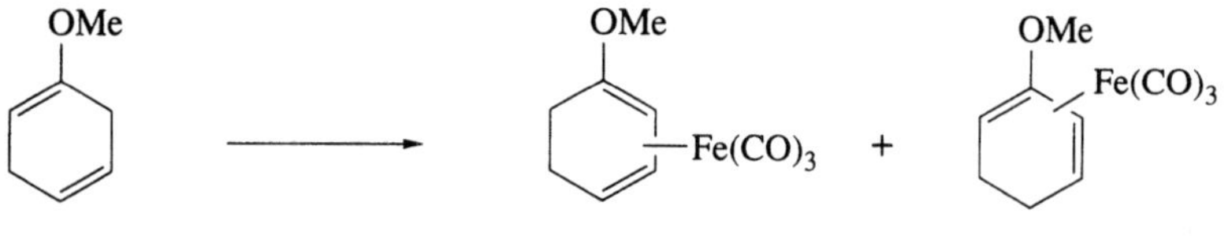

OMe → OMe $-Fe(CO)_3$ + OMe $Fe(CO)_3$

(15) $Fe_3(CO)_{12}$, Δ, 32% (16) 76% (17) 24%
$Fe(CO)_5$, Δ, 70–75% (16) *ca.* 50% (17) *ca.* 50%

Scheme 2

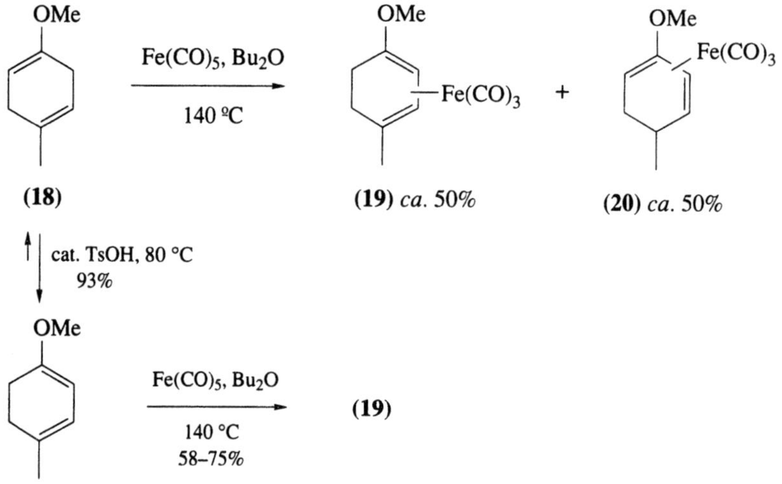

OMe → ($Fe(CO)_5$, Bu_2O; 140 °C) OMe $-Fe(CO)_3$ + OMe $Fe(CO)_3$

(18) (19) *ca.* 50% (20) *ca.* 50%

cat. TsOH, 80 °C 93%

OMe → ($Fe(CO)_5$, Bu_2O; 140 °C 58–75%) (19)

(21)

Scheme 3

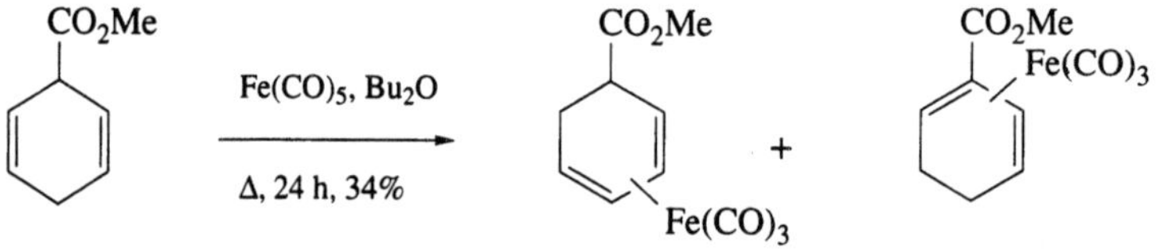

CO_2Me → ($Fe(CO)_5$, Bu_2O; Δ, 24 h, 34%) CO_2Me $Fe(CO)_3$ + CO_2Me $Fe(CO)_3$ (8)

(22) (23) 75% (24) 25%

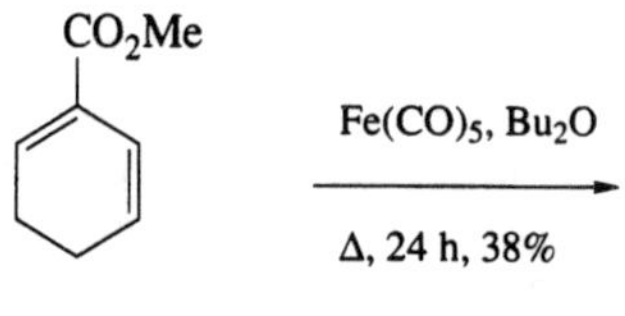

CO_2Me → ($Fe(CO)_5$, Bu_2O; Δ, 24 h, 38%) (24) (9)

(25)

(23) + (24) → (10% H_2SO_4/MeOH; Δ, 16 h) CO_2Me $-Fe(CO)_3$ (10)

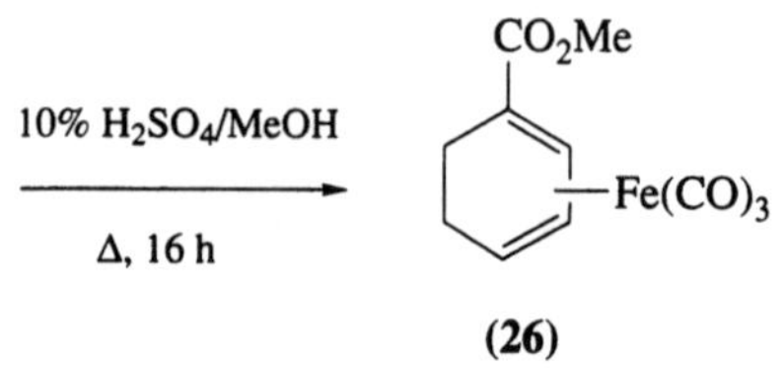

(26)

With substituted 1,4-dienes, such as (**15**), (**18**) and (**22**), the isomerization/complexation reaction invariably produces mixtures of complexes. Although these can often be separated chromatographically, it is better to preconjugate the diene for synthetic applications. A useful example is the conversion of diene (**18**) to an equilibrium mixture (*ca.* 1:4) of (**18**) and (**21**), followed by reaction with $Fe(CO)_5$ to give complex (**19**; Scheme 3) in 58% yield.[12]

The methoxy- and methoxycarbonyl-substituted complexes show pronounced electronic effects during the hydride abstraction process, leading to good regiocontrol in most cases. Examples are shown in equations (11)–(15).[10–13]

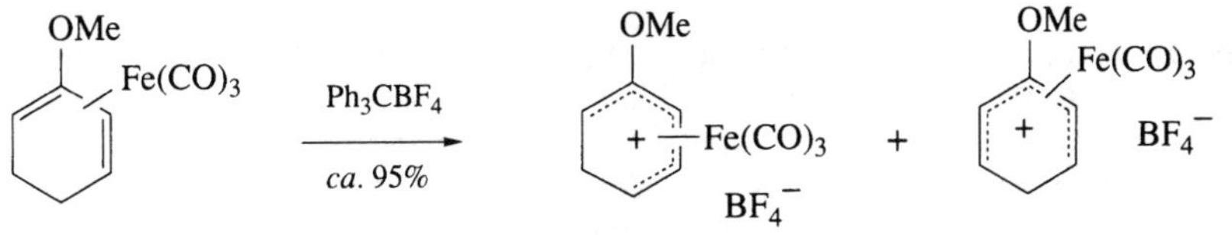

(**17**) $\xrightarrow[ca.\ 95\%]{Ph_3CBF_4}$ (**27**) 94% + (**28**) 6% (11)

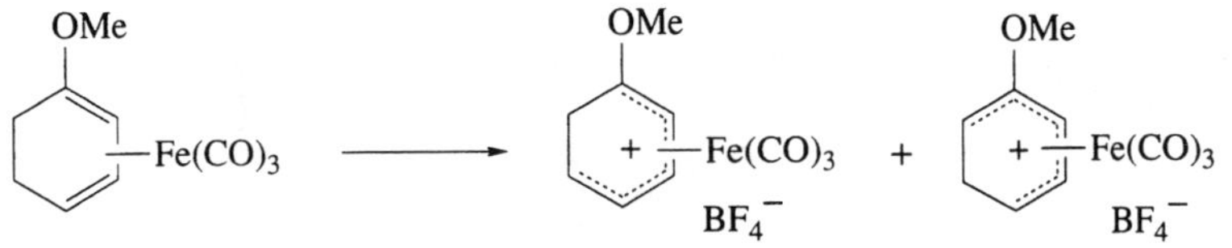

(**16**) ⟶ (**29**) 80% + (**27**) 20% (12)

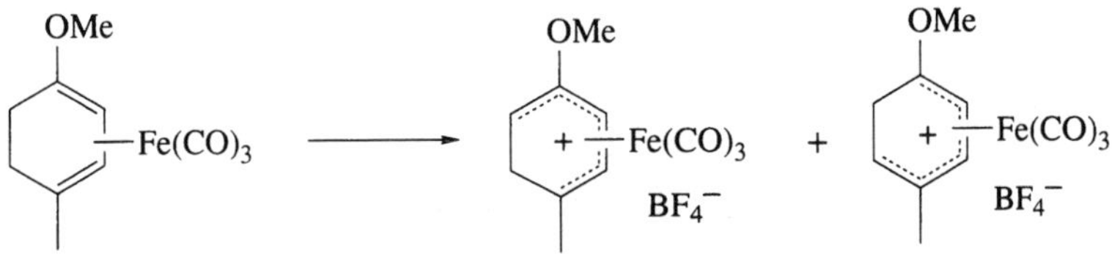

(**19**) ⟶ (**30**) 95% + (**31**) 5% (13)

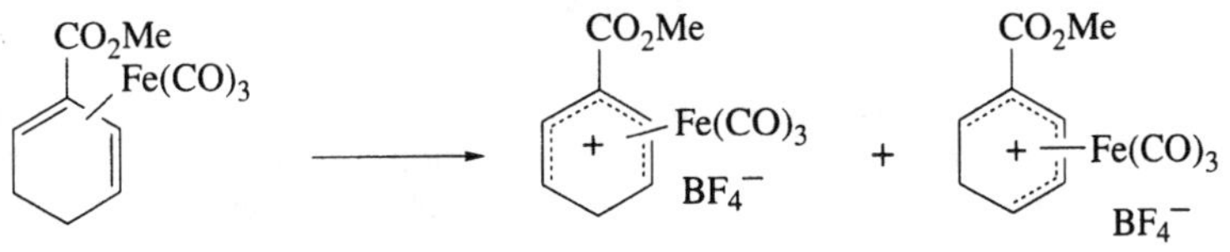

(**24**) ⟶ (**32**) 80% + (**33**) 20% (14)

(**26**) ⟶ (**34**) (15)

With complexes such as (**17**), (**24**) and (**35**) that do not have sterically demanding substituents adjacent to the point of hydride abstraction, the reaction is controlled by electronic effects from the substituent. The results of these particular reactions are somewhat unexpected by comparison with the anticipated stabilities of the corresponding noncomplexed dienyl cations. For example, (**27**; equation 11) corresponds to a less stable dienyl ligand than (**28**), and (**32**; equation 14) is expected to be less stable than (**33**). In the conversion of (**35**) to (**36**; equation 16) reported by Paquette *et al.*,[13] the trimethylsilyl group appears to act as an electron acceptor, and is therefore analogous to the methoxycarbonyl group. These results of hydride abstraction have been rationalized using molecular orbital calculations,[6b] the synergistic interaction between metal *d*-orbitals and dienyl HOMO and LUMO being the most important consideration. A detailed discussion will not be presented here. Suffice it to say that the dienyl with highest energy HOMO and lowest energy LUMO gives the strongest net bonding interaction with the $Fe(CO)_3$ fragment, and the hydride abstraction is thermodynamically driven.

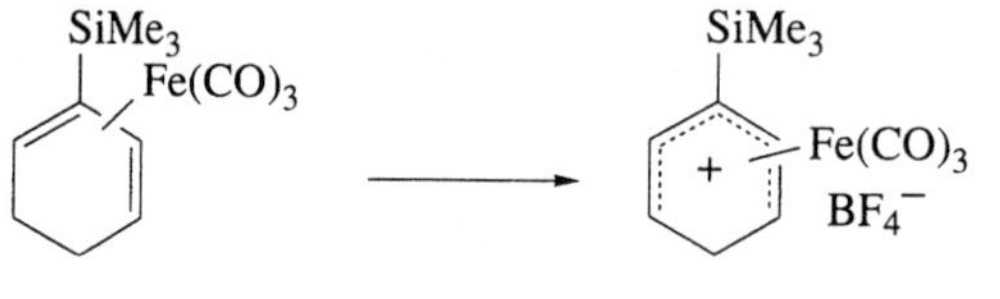

(35) → (36) (16)

In complexes (**16**) and (**26**), the situation is rather different, because there is now steric hindrance from the diene terminal substituent. As a result, the electronically less favorable products are predominant. However, when steric effects at both methylene groups are balanced, as in complex (**19**), the hydride abstraction proceeds with 'electronic' control to give (**30**; equation 13).

With methyl-substituted cyclohexadienes, very little selectivity is observed during the preparation of the complexes and the hydride abstraction reaction. Dihydrotoluene, on heating with pentacarbonyliron, gives a mixture of complexes (**37**) and (**38**; Scheme 4). These cannot be easily separated using standard chromatographic procedures, and little is known about hydride abstraction from the individual complexes. Treatment of the equimolar mixture with trityl tetrafluoroborate gives a mixture of all three possible products (**39–41**; Scheme 4).

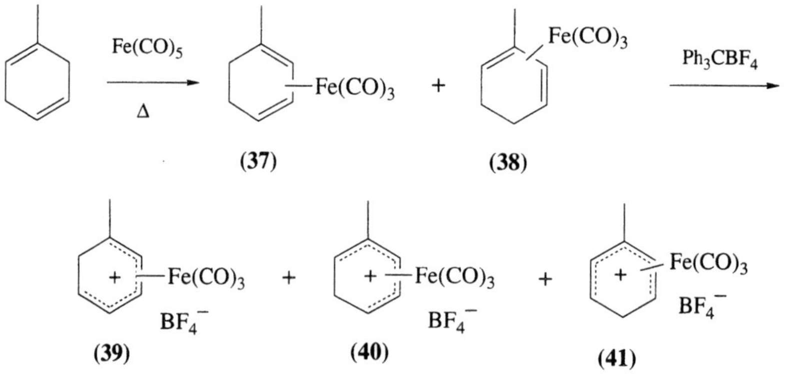

Scheme 4

A method for preparing specific alkyl-substituted dienyl complexes takes advantage of the propensity of (diene)$Fe(CO)_3$ complexes to rearrange under acidic conditions, coupled with the acid-promoted dehydration illustrated earlier for the conversion of (**9**) to (**8**). Birch and Haas[14] discovered that complexes derived from methylanisoles could be converted to methyl-substituted cyclohexadienyliron complexes, whose substitution pattern is defined by the relative positions of the methoxy and methyl substituents in the precursor. Several examples are given in Schemes 5 and 6 and equation (17).

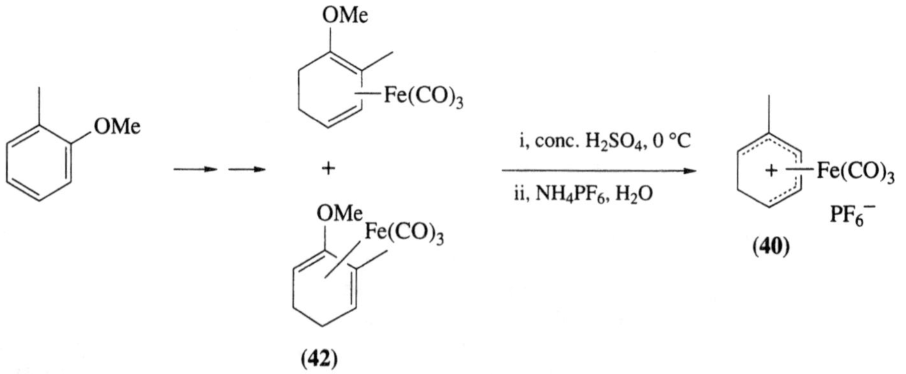

Scheme 5

$$(\mathbf{19}) + (\mathbf{20}) \xrightarrow[\text{ii, } NH_4PF_6,\ H_2O]{\text{i, conc. } H_2SO_4,\ 0\ ^\circ C} (\mathbf{40}) \qquad (17)$$

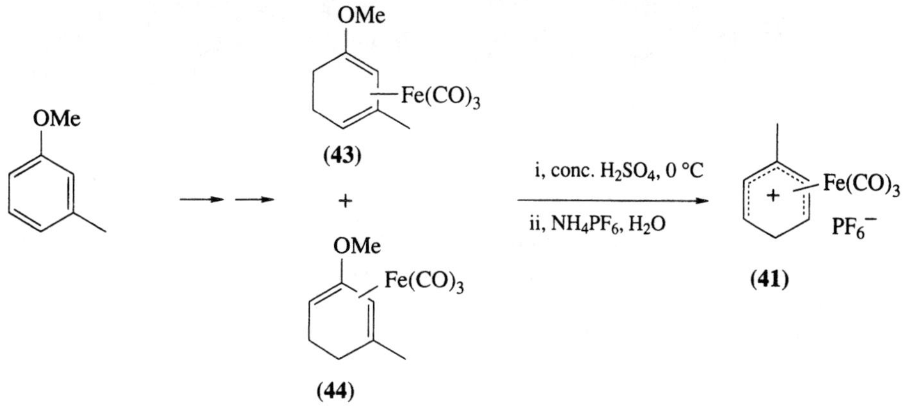

OMe (43) + OMe (44) → i, conc. H_2SO_4, 0 °C; ii, NH_4PF_6, H_2O → (41) $Fe(CO)_3$ PF_6^-

Scheme 6

Related to this reaction is the acid treatment of the (1-hydroxymethylcyclohexadiene)tricarbonyliron (**45**), prepared *via* DIBAL-H reduction of ester (**26**) or borane reduction of the corresponding carboxylic acid, which leads to the 1-methylcyclohexadienyliron complex (**46**; equation 18).[11] Using these methods, it is therefore possible to prepare a range of alkyl-substituted dienyl complexes having a defined substitution pattern.

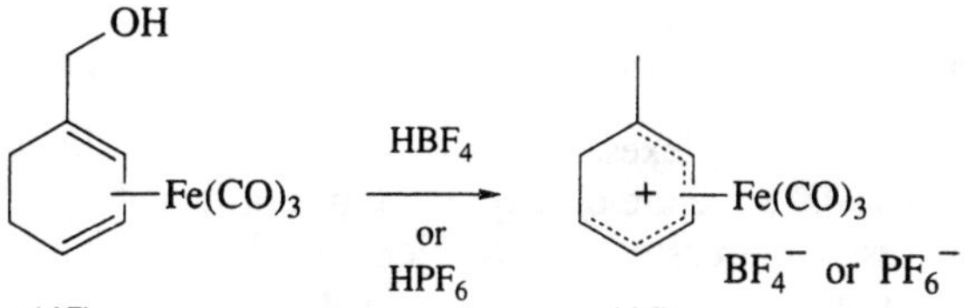

(45) $Fe(CO)_3$ —HBF_4 or HPF_6→ (46) $Fe(CO)_3$ BF_4^- or PF_6^- (18)

As mentioned earlier, steric effects can be important in determining the outcome of the hydride abstraction reaction. This is particularly vexing in cases where an alkyl substituent is present at the sp^3 carbon of the cyclohexadiene complex. For example, complexes such as (**47**; equation 19) are untouched by trityl cation, provided traces of acid are not present (these are formed by hydrolysis of the trityl tetrafluoroborate due to atmospheric moisture, and will cause rearrangement of the diene complex). This is due to the fact that only the hydride *trans* to the $Fe(CO)_3$ group can be removed, and the methyl substituent prevents close approach to this hydrogen.

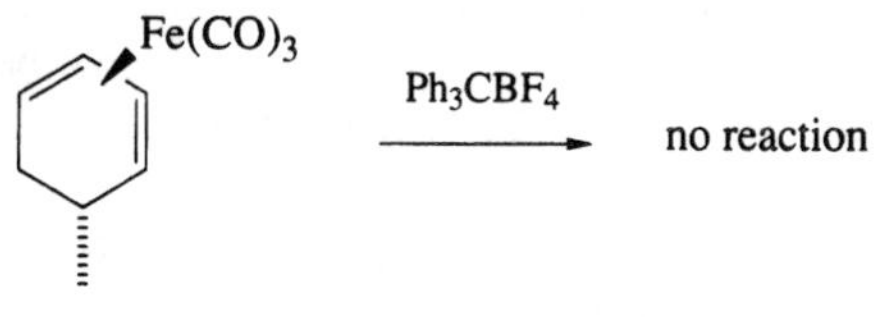

(47) $Fe(CO)_3$ —Ph_3CBF_4→ no reaction (19)

The trimethylsilyl substituent proves to be of considerable value in this respect. It appears to activate the diene towards hydride abstraction to such an extent that conversion of complexes such as (**48**) and (**50**) to the dienyl derivatives (**49**; equation 20) and (**51**; equation 21) proceeds quite readily.[13]

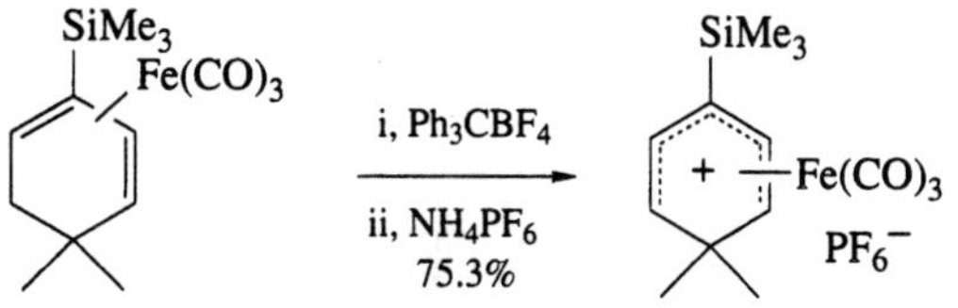

(48) $SiMe_3$ $Fe(CO)_3$ —i, Ph_3CBF_4; ii, NH_4PF_6, 75.3%→ (49) $SiMe_3$ $Fe(CO)_3$ PF_6^- (20)

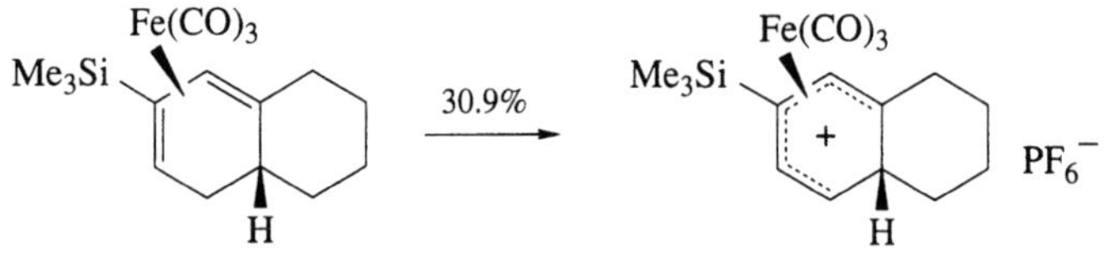

(21)

(50) (51)

In those cases where a hydroxyethyl substituent is attached to the cyclohexadiene ring, an oxidative cyclization procedure has been developed to effect conversion to cyclohexadienyliron complexes.[14] Thus, treatment of complex (**52**) with manganese dioxide gives the cyclic ether (**53**) which can be converted to the dienyl complex (**54**; Scheme 7) by treatment with tetrafluoroboric acid in acetic anhydride.

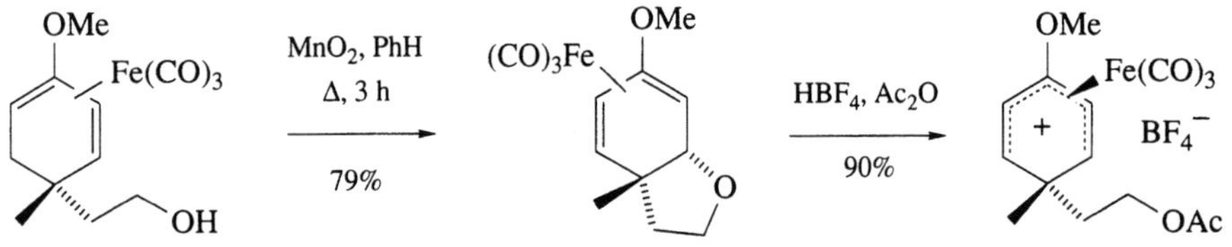

(52) (53) (54)

Scheme 7

3.4.2.2 Nucleophile Additions to Dienyliron Complexes; General Comments

Imagining tricarbonyldienyliron complexes to consist of a positively charged iron surrounded by dienyl and carbon monoxide ligands, there are clearly two types of ligand site at which nucleophile addition might take place. However, only in rare instances does attack occur at a carbonyl ligand, and so this leads to an extremely powerful method for preparing substituted dienes. A wide range of nucleophiles are available for this method, including alkoxides, hydroxide, amines, borohydride (as a source of hydride), enols of simple ketones and β-dicarbonyl compounds,[8,10] enamines,[10] aromatic amines and ethers,[15] silyl enol ethers,[16] allylsilanes,[17] organocuprates,[18] organo-zinc and -cadmium reagents[19] (but not dimethylzinc), phosphines,[20] phosphites and sulfites,[21] alkynylborates,[22] lithium[23] and tin[24] enolates, silylketene acetals,[25] as well as the enolates from arylsulfonyl esters,[26] malonates,[27] and β-keto esters.[27] Typical reactions of representative nucleophiles with the parent cyclohexadienyliron complex (**3**) are summarized in equations (22)–(37).

All of these reactions are chemospecific (only the dienyl ligand undergoes nucleophile addition), regiospecific (only the terminal dienyl carbon is attacked), and stereospecific (the nucleophile always adds *anti* to the metal). However, the addition of organolithium and Grignard reagents in ether solvents to (**3**) gives very low yields of alkylation products, the major product being the dimer (**71**; Scheme 8). This is presumed to arise by electron transfer from the RLi or RMgX to the dienyl complex, giving the free radical (**70**) which then dimerizes.

Use of less reactive alkylating agents such as dialkylcuprates, R_2Zn or R_2Cd, as described above, is one way to overcome this problem, but the yields are variable. The use of dichloromethane as solvent at −78 °C resulted in good yields of alkylation product (84–94%),[28] and a more general solution is achieved

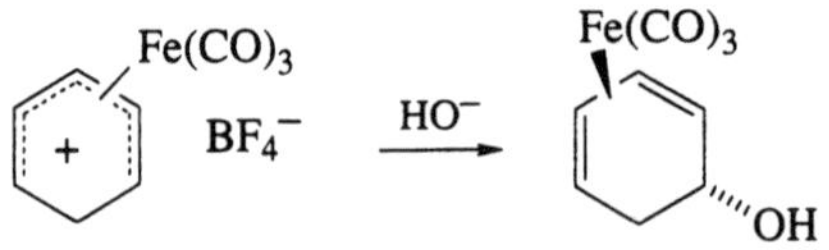

(22)

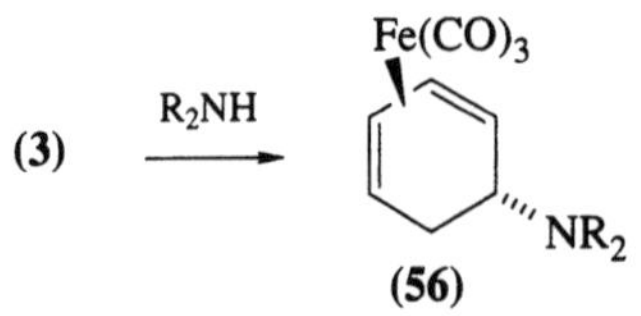

(3) (55)

(23)

(3) (56)

(3) —acetone, EtOH, Δ→ $Fe(CO)_3$ O **(57)** (24)

(3) —$MeCOCH_2CO_2Et$, Δ→ $Fe(CO)_3$ O CO_2Et **(58)** (25)

(3) —i, NR_2; ii, H_2O→ $Fe(CO)_3$ O **(59)** (26)

(3) —$NaBH_4$→ $Fe(CO)_3$ **(2)** (27)

(3) —$SiMe_3$, Δ→ $Fe(CO)_3$ **(60)** (28)

(3) —R_2Zn or R_2Cd→ $Fe(CO)_3$ R **(61)** (29)

R = Ph, $PhCH_2$, Me_2CH; CH=CHMe, $CH_2CH=CH_2$

(3) —Me_2CuLi→ $Fe(CO)_3$ **(62)** (30)

(3) —i, $OSiMe_3$ $OSiMe_3$; ii, H^+, MeOH→ $Fe(CO)_3$ **(63)** O (31)

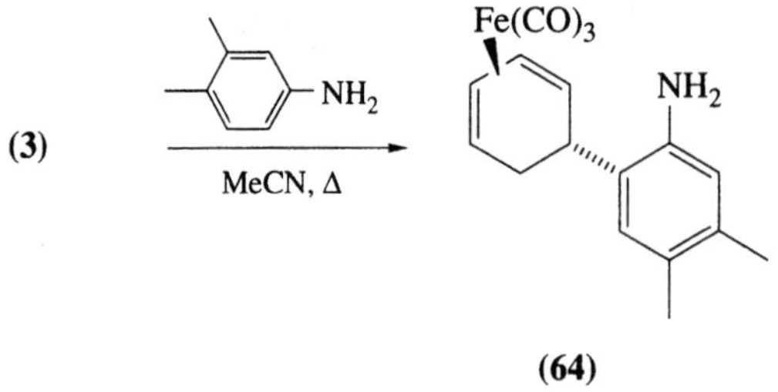

(3) $\xrightarrow[\text{MeCN, }\Delta]{}$ (64) (32)

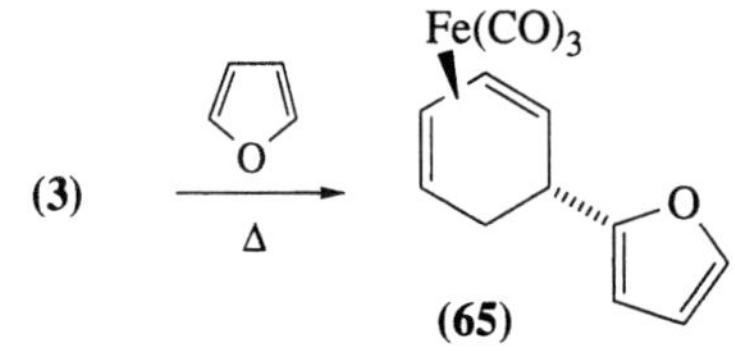

(3) $\xrightarrow[\Delta]{}$ (65) (33)

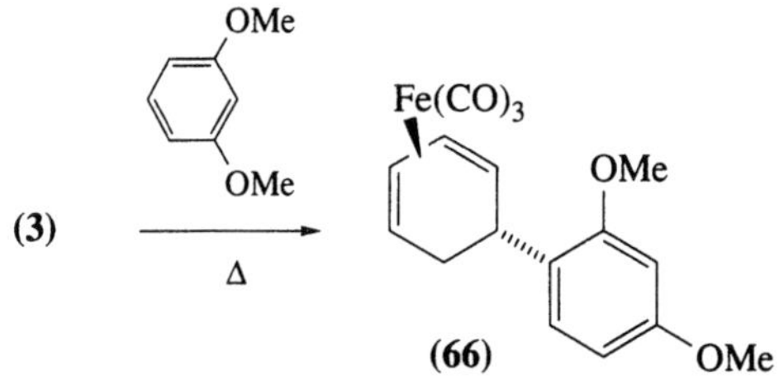

(3) $\xrightarrow[\Delta]{}$ (66) (34)

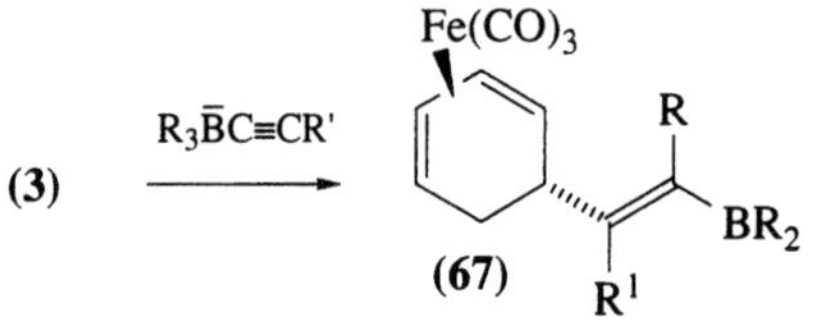

(3) $\xrightarrow{R_3\bar{B}C{\equiv}CR'}$ (67) (35)

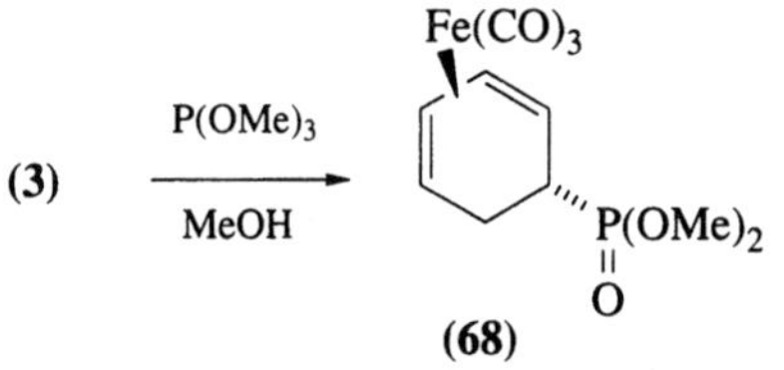

(3) $\xrightarrow[\text{MeOH}]{P(OMe)_3}$ (68) (36)

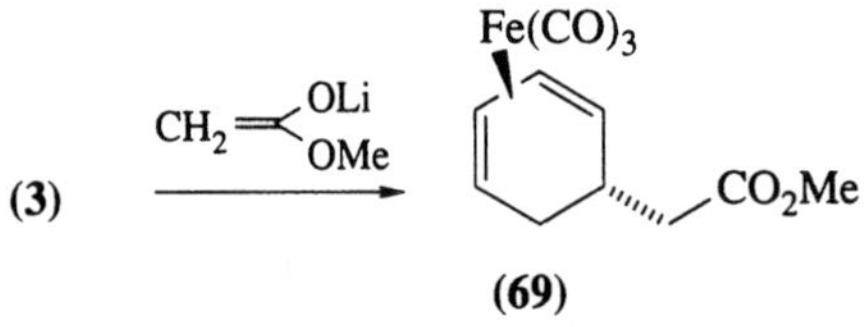

(3) $\longrightarrow$ (69) (37)

(3) $\xrightarrow[\text{THF or }Et_2O]{\text{RLi or RMgX}}$ (70) $\longrightarrow$ (71)

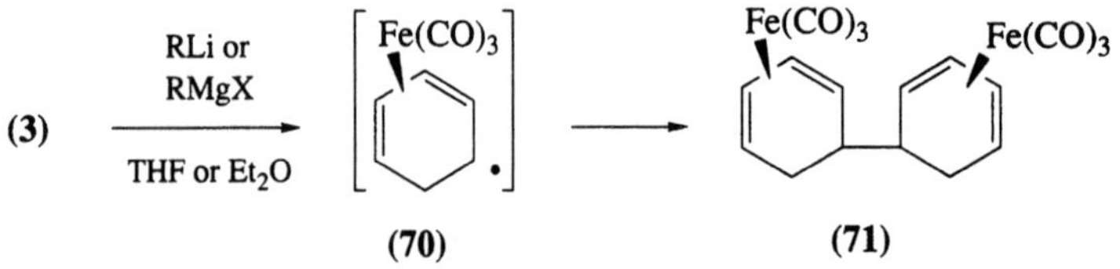

Scheme 8

by replacing a carbonyl ligand by, *e.g.* triphenylphosphine, a poorer π-acceptor. Very good yields of alkylation can be obtained using Grignard reagents under standard conditions.[29] This was especially noticeable with vinylmagnesium bromide, which converts complex **(3)** almost exclusively to the dimer

(**71**), even in dichloromethane at low temperature, but which gives exclusively the addition product (**73**) on reaction with complex (**72**; equation 38).

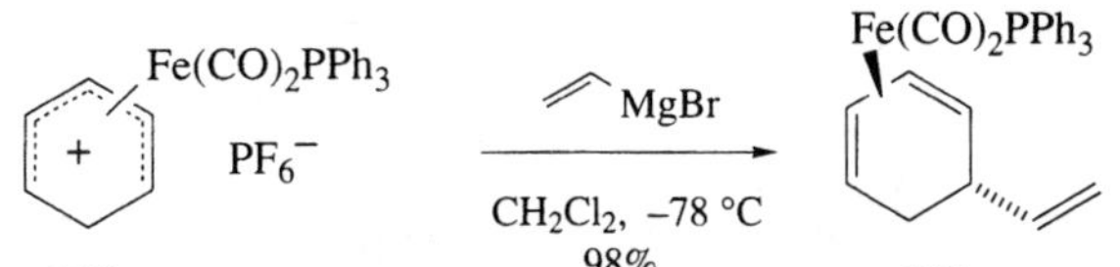

$Fe(CO)_2PPh_3$ PF_6^- (**72**) → MgBr, CH_2Cl_2, −78 °C, 98% → $Fe(CO)_2PPh_3$ (**73**) (38)

This observation has been used in a key step during a total synthesis of the alkaloid (±)-dihydrocannivonine (**77**) outlined in Scheme 9.[30] The final product is formed by intramolecular hetero-Diels–Alder reaction of the intermediate (**76**), which is not isolated.

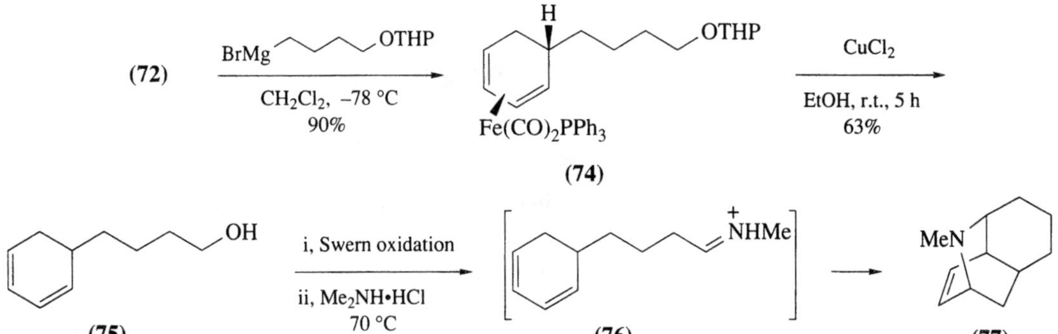

(**72**) → BrMg~OTHP, CH_2Cl_2, −78 °C, 90% → H, OTHP, $Fe(CO)_2PPh_3$ (**74**) → $CuCl_2$, EtOH, r.t., 5 h, 63%

OH (**75**) → i, Swern oxidation; ii, $Me_2NH{\cdot}HCl$, 70 °C → [$\overset{+}{N}HMe$] (**76**) → MeN (**77**)

Scheme 9 Grieco's synthesis of (±)-dihydrocannivonine (**77**)

Substitution of a CO ligand by a phosphine or phosphite derivative similarly leads to reactivity control in the cycloheptadienyliron system. The tricarbonyliron complex (**5**) invariably gives mixtures of products from nucleophile addition.[31] For example, treatment of (**5**) with sodium borohydride or sodium cyanide gives mixtures of products of general structure (**11**) and (**12**; equation 39), resulting from addition of nucleophile at C-1 and C-2, respectively, while addition of cyanide to the triphenylphosphine complex (**78**) gives exclusively (**80**; R = CN; equation 39).

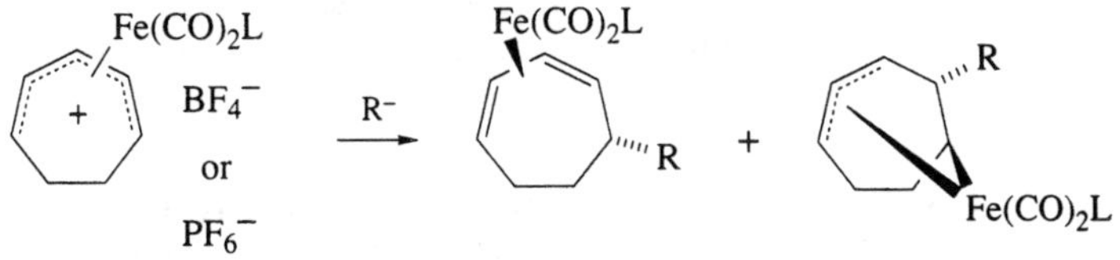

$Fe(CO)_2L$ BF_4^- or PF_6^- → R^- → $Fe(CO)_2L$, R + R, $Fe(CO)_2L$ (39)

(**5**) L = CO; (**78**) L = PPh_3 — (**11**) L = CO; (**79**) L = PPh_3 — (**12**) L = CO; (**80**) L = PPh_3

Similarly, reaction of lithium dimethylcuprate with (**5**) gave (**11**; R = Me) in low yield, together with significant amounts of the dimeric complex (**81**; equation 40) and substantial decomposition, but the phosphine analog (**78**) was more useful. Reaction of cuprate reagents with (**78**) gives (**79**; R = Me) in yields greater than 95%. The ligand exchange reaction on complex (**4**) does not proceed in good yield on a large scale using triphenylphosphine under thermal conditions (Bu_2O, reflux, 24 h). However, this same reaction using triphenyl phosphite proceeds well, giving complex (**82**) in *ca.* 95% yield.[32] Conversion to the dienyl complex (**83**) is essentially quantitative. Complex (**83**) behaves similarly to (**79**), and some of its reactions with nucleophiles are summarized in Scheme 10. Interestingly, dimethylcuprate reacts exclusively at C-1, while methyllithium adds exclusively to C-2. In general it appears that 'soft' nucleophiles add to the dienyl terminus, while harder nucleophiles attack C-2. (This contrasts with the reactions of cyclohexadienyliron complexes which always give C-1 addition products.) The high yields from these reactions, coupled with the ease of preparation of (**83**) on a large scale (>200 g batches), augurs well for this methodology as a means of stereoselectively functionalizing the seven-membered ring. We shall return to this in Section 3.4.2.4.

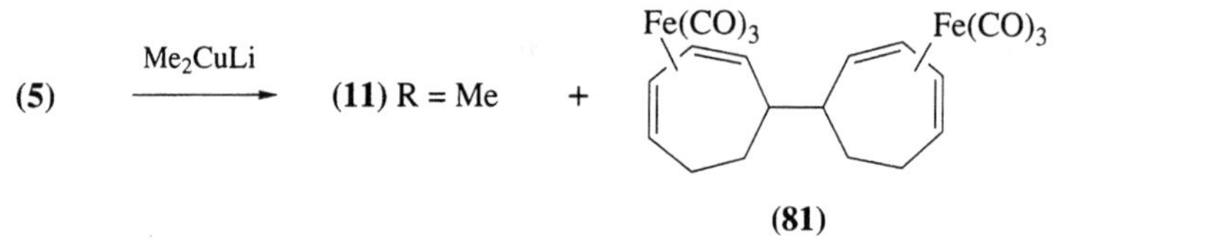

(40)

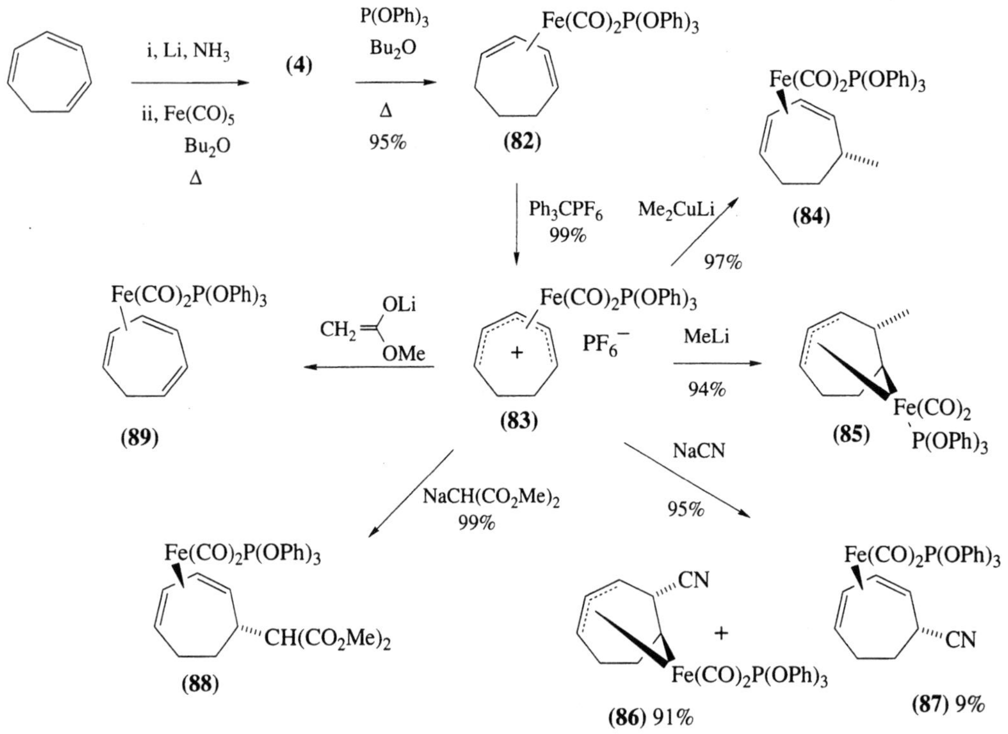

Scheme 10 Preparation of dicarbonyl(triphenyl phosphite)(cycloheptadienyl)iron hexafluorophosphate and its reactions with simple nucleophiles

The major limitation on the reactivity of cycloheptadienyliron complexes is the fact that treatment with hard enolates, such as methyl lithioacetate results in deprotonation to give, *e.g.* the η^4-triene complex (**89**). This appears to be less of a problem in the corresponding cyclohexadienyl systems.

3.4.2.3 Regiocontrol During Nucleophile Addition to Cyclohexadienyliron Complexes: Applications of Alkoxy- and Methoxycarbonyl-substituted Complexes in Organic Synthesis

A methoxy substituent at C-2 of the cyclohexadienyl ligand, as in complex (**27**), exerts a very powerful directing effect. The C-1 terminus appears to be electronically deactivated, so that nucleophiles add exclusively to C-5, giving products of general structure (**90**; Scheme 11).[8,10] The tricarbonyliron group can be removed quite easily using trimethylamine *N*-oxide[33] to give the dienes (**91**), which are readily hydrolyzed under mild acid conditions to give substituted cyclohexenones (**92**). Alternatively, direct conversion of (**90**) to (**92**) can be accomplished by treatment of the complex with ethanolic copper(II) chloride.[34]

Various other oxidizing agents have been used successfully for the conversion of iron complexes to the dienes or enones, including iron(III) chloride in ethanol,[8,35] Jones' reagent,[17] and cerium(IV) ammonium nitrate.[8] Overoxidation to give the aromatic compound is noted with methoxy-substituted complexes, and can be minimized by using trimethylamine *N*-oxide or copper(II) chloride. Sometimes, further oxidation of dienes such as (**91**) to aromatic compounds is a desirable transformation. This can be accomplished in good yield using either DDQ or palladium on charcoal, and this has been utilized in a formal synthesis of the natural product *O*-methyljoubertiamine (**93**; Scheme 12).[36]

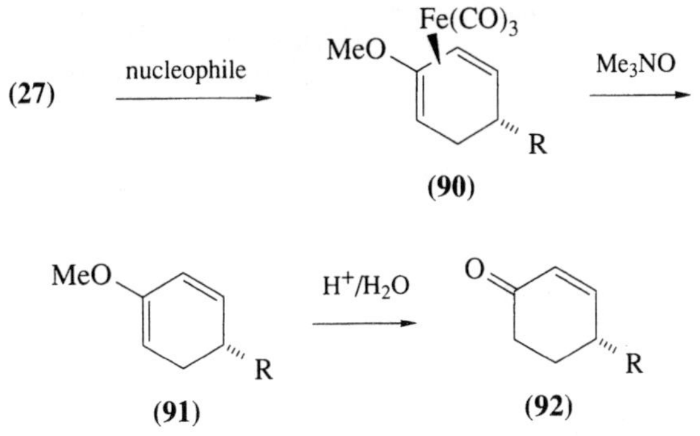

(27) —nucleophile→ (90) —Me_3NO→ (91) —H^+/H_2O→ (92)

Scheme 11

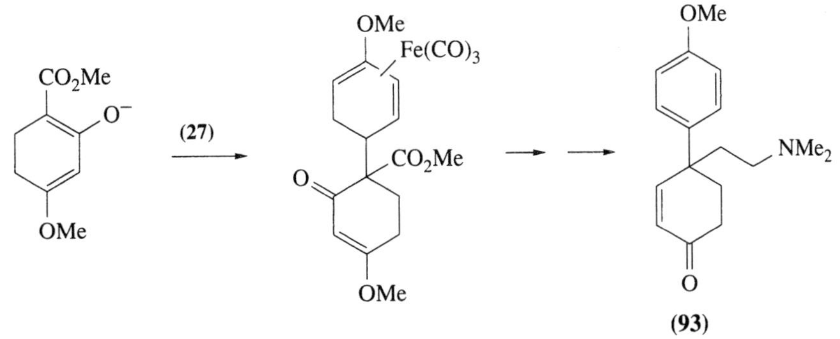

(27) → → → (93)

Scheme 12

The addition of carbon nucleophiles to complex (**27**), followed by demetallation, is equivalent to the γ-alkylation of cyclohexenone. This overall transformation can also be accomplished directly *via* addition of electrophiles to dienolsilanes, but it becomes nontrivial for cases where the cyclohexenone C-4 position is already substituted.[37] On the other hand, 1-substituted cyclohexadienyliron complexes, such as (**30**), react very cleanly with certain carbon nucleophiles, at the substituted dienyl terminus. This provides useful methodology for the construction of 4,4-disubstituted cyclohexenones, and has been employed in a variety of natural product syntheses.

Reaction of (**30**) with diethyl sodiomalonate gives exclusively complex (**94**), which can be demetallated to give the cyclohexenone derivative (**95**; Scheme 13) in good overall yield.[27] A range of nucleophiles can be used, including cyanide, and keto ester enolates, giving, *e.g.* (**96**) or (**97**; Scheme 13), respectively.

The steric bulk of the nucleophile does not seem to affect the regioselectivity during reaction with complex (**30**), but mixtures are obtained when the complex carries bulky substituents at the dienyl terminus. This is illustrated in equations (41)–(46), where some of the potential problems are highlighted.[27,38]

The regioselectivity during these reactions can be improved by using isopropoxy instead of methoxy as the directing group. In this case the electronic deactivation of the neighboring dienyl terminus is reinforced by the steric demand of the more bulky isopropyl group. For example, (**114**; Scheme 14) and (**117**; Scheme 15) give dienyl complexes (**115**) and (**118**) in high yield on treatment with triphenylmethyl hexafluorophosphate[39] and these add to dimethyl malonate anion with much better regiocontrol than the corresponding methoxy derivatives. Cleaner reaction occurs if potassium enolates are used in place of the sodium derivatives,[39] as summarized in Schemes 14 and 15.

Multiple substitution on the dienyl ligand has not been extensively investigated. The dimethyl-substituted complexes (**121** and **122**; equation 47) and (**123**; Scheme 16)[40,41] all react with malonate anion to give good yields of the desired products, although the isopropoxy derivative (**122**) gives better regioselectivity than the methoxy compound (**121**). A sterically more demanding substituent *para* to the alkoxy directing group has adverse effects though, and complexes (**124**) and (**125**; equation 48) react with nucleophiles exclusively or predominantly at the unsubstituted dienyl terminus.[40]

Steric hindrance to nucleophile addition becomes a major problem with 6-*exo*-substituted cyclohexadienyliron complexes, and the outcome of these reactions is highly dependent on the size of the substi-

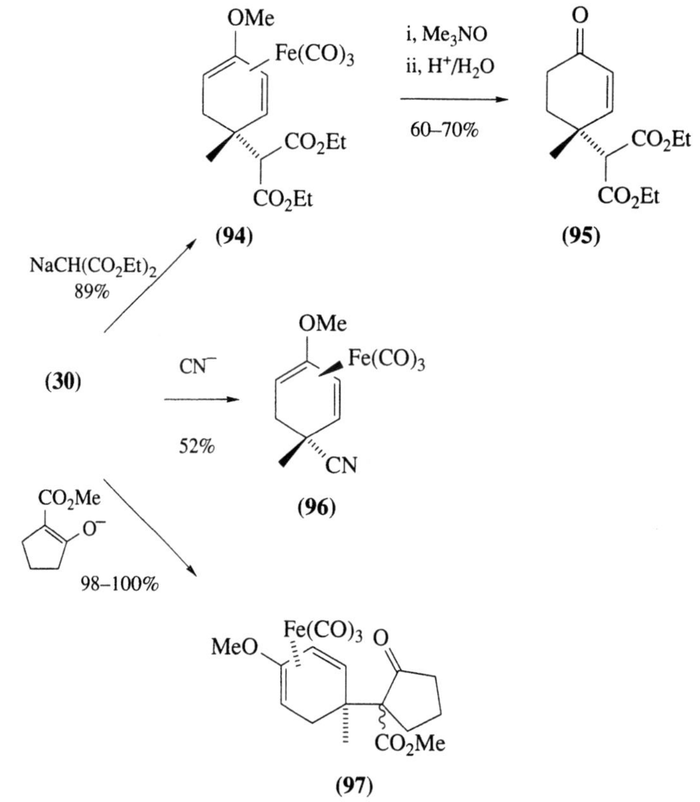

OMe
$Fe(CO)_3$
CO_2Et
CO_2Et
(94)

i, Me_3NO
ii, H^+/H_2O
60–70%

O
CO_2Et
CO_2Et
(95)

$NaCH(CO_2Et)_2$
89%

(30)

CN^-
52%

OMe
$Fe(CO)_3$
CN
(96)

CO_2Me
O^-
98–100%

$Fe(CO)_3$ O
MeO
CO_2Me
(97)

Scheme 13

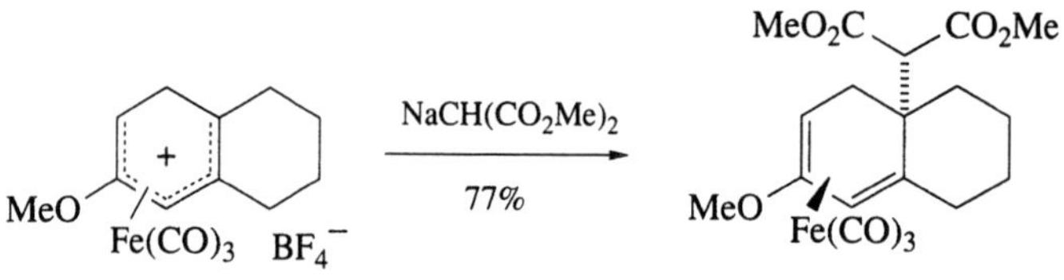

+
MeO
$Fe(CO)_3$ BF_4^-
(98)

$NaCH(CO_2Me)_2$
77%

MeO_2C CO_2Me
MeO
$Fe(CO)_3$
(99)

(41)

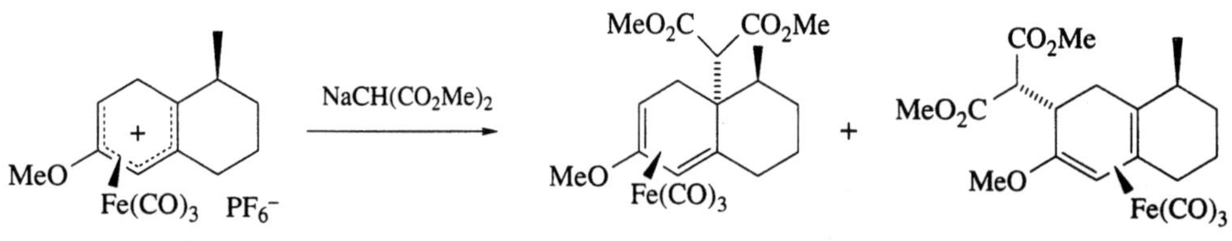

+
MeO
$Fe(CO)_3$ PF_6^-
(100)

$NaCH(CO_2Me)_2$

MeO_2C CO_2Me
MeO
$Fe(CO)_3$
(101) 50%

+

CO_2Me
MeO_2C
MeO
$Fe(CO)_3$
(102) 50%

(42)

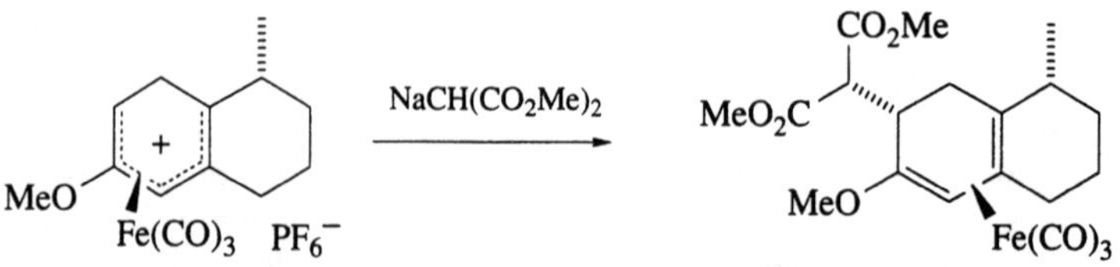

+
MeO
$Fe(CO)_3$ PF_6^-
(103)

$NaCH(CO_2Me)_2$

CO_2Me
MeO_2C
MeO
$Fe(CO)_3$
(104)

(43)

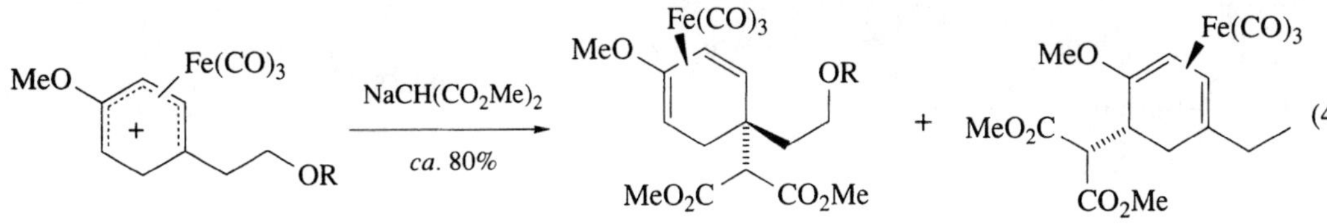

MeO, $Fe(CO)_3$, +, OR; $NaCH(CO_2Me)_2$, *ca.* 80%; MeO, $Fe(CO)_3$, OR, MeO_2C, CO_2Me + MeO, $Fe(CO)_3$, MeO_2C, CO_2Me (44)

(105) R = Me
(108) R = Ac

(106) R = Me, 72%
(109) R = Ac, 50%

(107) R = Me, 28%
(110) R = Ac, 50%

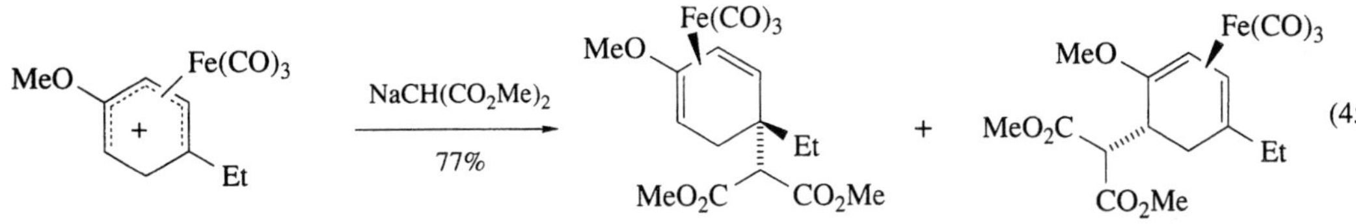

MeO, $Fe(CO)_3$, +, Et; $NaCH(CO_2Me)_2$, 77%; MeO, $Fe(CO)_3$, Et, MeO_2C, CO_2Me + MeO, $Fe(CO)_3$, MeO_2C, Et, CO_2Me (45)

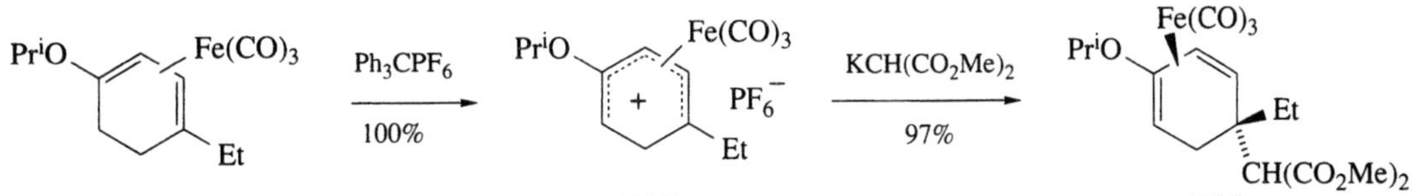

(111) **(112)** 82% **(113)** 18%

(111) —$KCH(CO_2Me)_2$, 96%→ **(112)** 85% + **(113)** 15% (46)

Pr^iO, $Fe(CO)_3$, Et **(114)** —Ph_3CPF_6, 100%→ Pr^iO, $Fe(CO)_3$, +, PF_6^-, Et **(115)** —$KCH(CO_2Me)_2$, 97%→ Pr^iO, $Fe(CO)_3$, Et, $CH(CO_2Me)_2$ **(116)**

Scheme 14

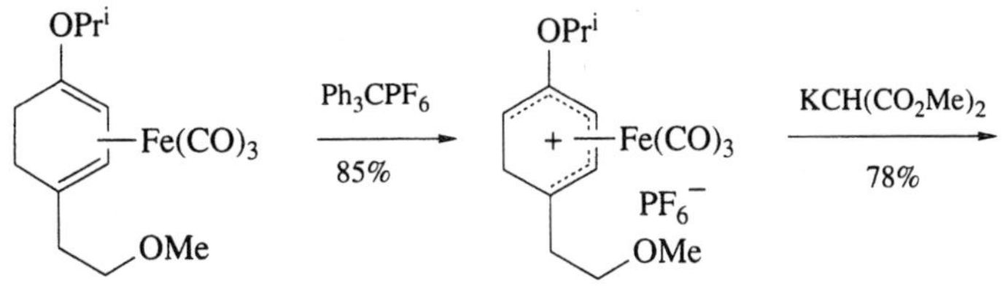

OPr^i, $Fe(CO)_3$, OMe **(117)** —Ph_3CPF_6, 85%→ OPr^i, +, $Fe(CO)_3$, PF_6^-, OMe **(118)** —$KCH(CO_2Me)_2$, 78%→

OPr^i, $Fe(CO)_3$, MeO, CO_2Me, CO_2Me + MeO_2C, MeO_2C, OPr^i, $Fe(CO)_3$, OMe

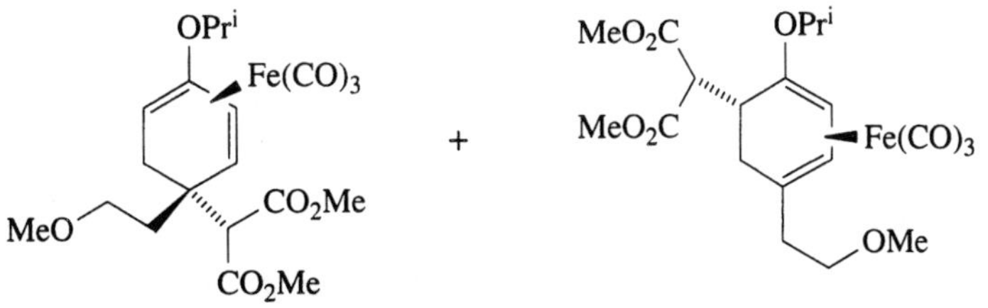

(119) 90% **(120)** 10%

Scheme 15

tuent. For example, complex (**54**; Scheme 17), which has a linear, sterically undemanding 6-*exo* substituent, reacts with malonate and cuprate nucleophiles with excellent stereocontrol to give complexes (**126**) or (**127**), and (**126**) is readily converted to the cyclohexenone (**128**).[14]

A range of enolate nucleophiles have been added to trimethylsilyl-substituted cyclohexadienyl complexes (**49**; Scheme 18) and (**51**; equation 49) and related compounds. Again, high yields are obtained and stereodirection by the $Fe(CO)_3$ group is very powerful.[13]

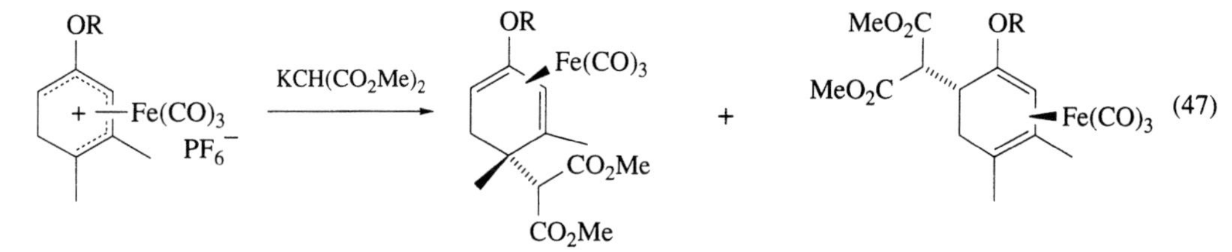

OR KCH(CO$_2$Me)$_2$ Fe(CO)$_3$ PF$_6^-$ (47)

(121) R = Me 76% 78% 22%
(122) R = Pr^i 86% 90% 10%

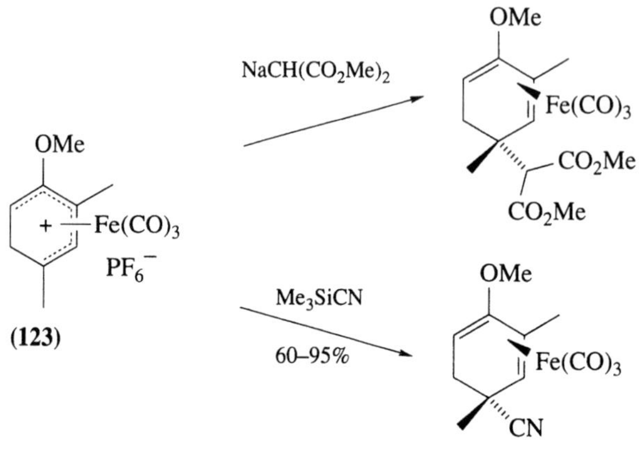

OMe NaCH(CO$_2$Me)$_2$ Fe(CO)$_3$ PF$_6^-$ CO$_2$Me Me$_3$SiCN 60–95% CN

(123)

Scheme 16

OR KCH(CO$_2$Me)$_2$ Fe(CO)$_3$ PF$_6^-$ CO$_2$Me MeO$_2$C (48)

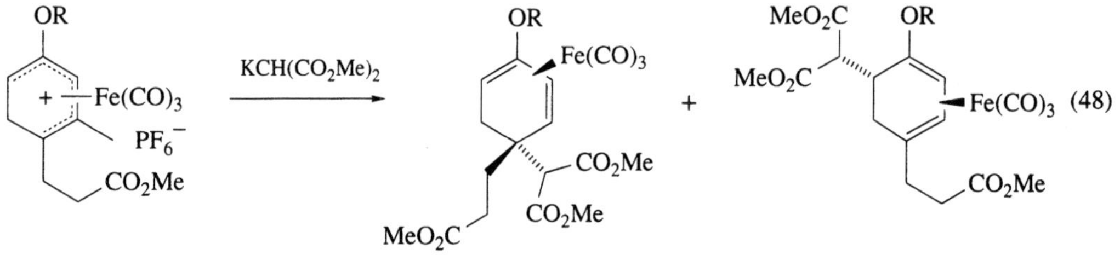

(124) R = Me 89% 0% 100%
(125) R = Pr^i 97% 22% 78%

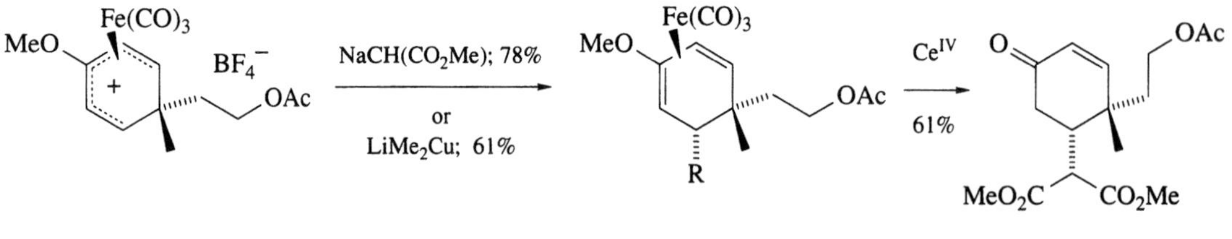

MeO Fe(CO)$_3$ BF$_4^-$ OAc NaCH(CO$_2$Me); 78% or LiMe$_2$Cu; 61% Ce^{IV} 61% O OAc MeO$_2$C CO$_2$Me

(54) (126) R = CH(CO$_2$Me)$_2$ (127) R = Me (128)

Scheme 17

Exo addition is always observed. When the 6-*exo* substituent is very bulky, as in complex (**132**; equation 50), nucleophile addition to the dienyl ligand is completely suppressed. Instead, decomplexation is observed giving the dienone (**133**), possibly *via* attack of nucleophile at the metal (*i.e.* ligand exchange).[42] *Endo* addition is not observed.

Regioselectivity problems, such as those encountered with complexes (**124**) or (**125**), may be overcome by using an intramolecular addition of the nucleophile. Spirocyclization of keto ester groups onto the dienyliron system proceeds in good yield to form a six-membered ring, *e.g.* (**134**) gives (**135**; equation 51), but *O*-alkylation occurs when five-membered ring formation is attempted, *e.g.* (**136**) gives (**137**;

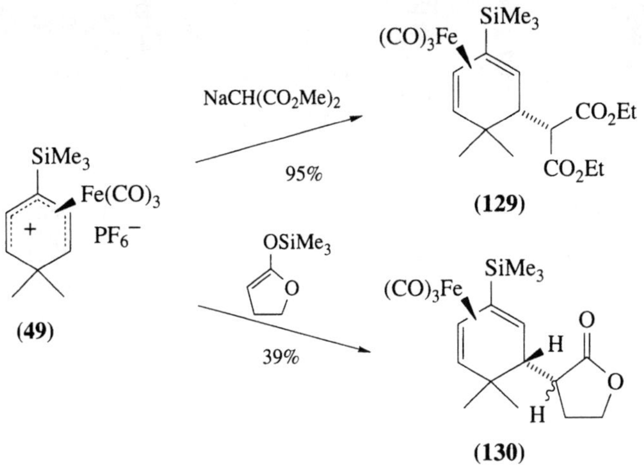

NaCH(CO$_2$Me)$_2$ 95%

OSiMe$_3$ 39%

(49) (129) (130)

Scheme 18

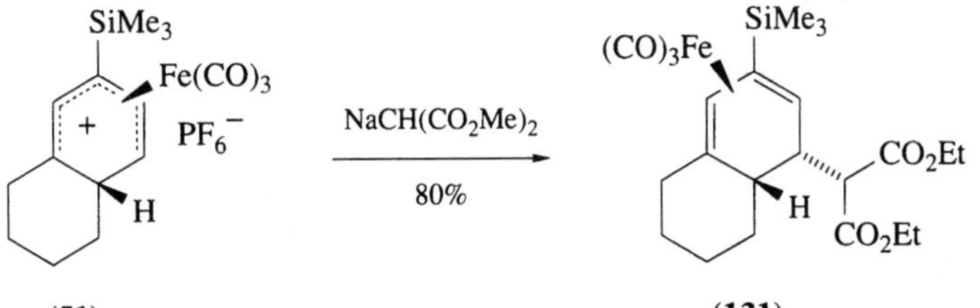

NaCH(CO$_2$Me)$_2$ 80%

(51) (131) (49)

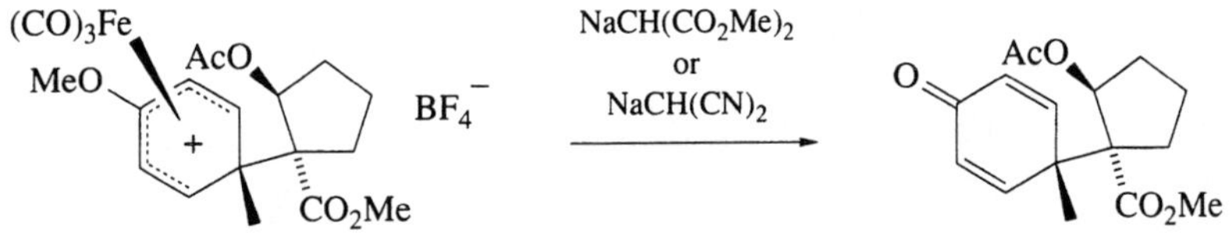

NaCH(CO$_2$Me)$_2$ or NaCH(CN)$_2$

(132) (133) (50)

equation 52).[43] A malonate group is insufficiently acidic for selective anion generation, and treatment of (**138**) with base results in deprotonation of the side chain to give (**139**; equation 53). On the other hand, the more acidic malononitrile or cyano ester groups can be deprotonated and cyclized to give spiro[4.5]decane derivatives such as (**141**; equation 54) and (**143**; equation 55).[40,44,45]

Regioselectivity during the addition of heteroatom nucleophiles to these complexes can be influenced in special cases by taking advantage of the reversibility of the reaction. For example, primary amine nucleophiles add reversibly to dienyliron complexes. When they are allowed to react with complexes such as (**144**) and (**149**), containing a leaving group in the side chain, the adducts (**146**) and (**151**) can undergo further reaction to generate azaspirocyclic systems (**147**) and (**152**), while the regioisomers (**145**) and (**150**) cannot (Scheme 19).[45] Since the formation of the amine adducts is reversible, the net result is that the azaspirocycles can be produced in high yield. These are readily converted to enones (**148**) and (**153**), and (**153**) has been used in a formal total synthesis of (±)-perhydrohistrionicotoxin (**154**; Scheme 20).[46]

The regiocontrol displayed by an alkoxy substituent at C-2 of the dienyl system has been exploited for the total synthesis of a number of relatively complex natural products. For example, the malonate ad-

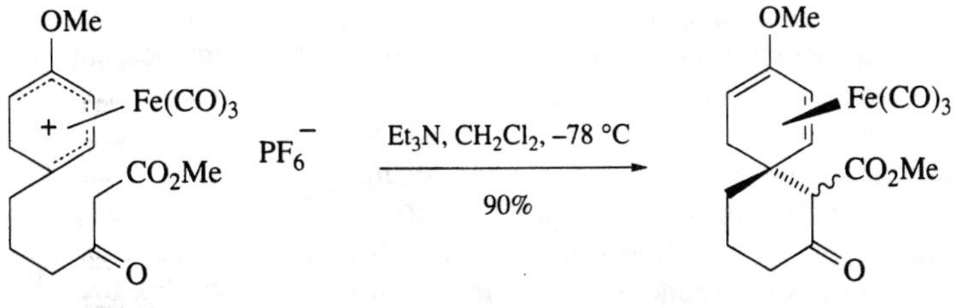

Et$_3$N, CH$_2$Cl$_2$, –78 °C 90%

(134) (135) (51)

OMe Fe(CO)$_3$ + PF_6^- O CO_2Me (136) — Et_3N, CH_2Cl_2, –78 °C, 100% → OMe Fe(CO)$_3$ O CO_2Me (137) (52)

OMe Fe(CO)$_3$ + PF_6^- CO_2Me CO_2Me (138) — base, 83% → OMe Fe(CO)$_3$ CO_2Me CO_2Me (139) (53)

OMe Fe(CO)$_3$ + PF_6^- CN CN (140) — Et_3N, CH_2Cl_2, 41% → OMe Fe(CO)$_3$ CN CN (141) (54)

OMe Fe(CO)$_3$ + PF_6^- CN CN (142) — Et_3N, CH_2Cl_2/MeCN, –78 °C, 96% → OMe Fe(CO)$_3$ CN CN (143) (55)

ducts (**112**) or (**116**) can be employed in a formal synthesis of the alkaloid aspidospermine (**158**; Scheme 21).[47,48] A diverse range of organic transformations can be accomplished in the presence of the (diene)Fe(CO)$_3$ unit; the tricarbonyliron group may be regarded as an enone protecting group in complexes such as (**155**) and (**156**).

Since the perhydroquinoline derivative (**157**) had previously been converted to (**158**) by Stork and Dolfini,[49] the conversion of (**112**) to (**157**; Scheme 21) constitutes a formal total synthesis of the natural product. However, this route to (**157**) is a longer sequence of reactions than that employed by Stork so that, while demonstrating the applications of dienyliron complexes in natural products synthesis, there is a need for considerable improvement in efficiency. Aspidosperma alkaloids bearing oxygen functionality at C-18 are less readily accessible *via* standard organic methodology,[50] and the application[48,51] of dienyl complex (**118**) in the total synthesis of limaspermine (**163**; Scheme 22) is of greater significance. This route parallels that described above for the construction of intermediate (**157**), and gives the analogous perhydroquinoline (**161**) in eight steps from (**118**). Conversion of (**161**) to (**163**) proceeds according to standard methods, resulting in a total synthesis of (±)-limaspermine in 16 steps from complex (**118**).

The trichothecenes, a group of sesquiterpenes having a reasonably complex tricyclic structure, represent useful targets for the synthetic application of cyclohexadienyliron complexes. Several members of

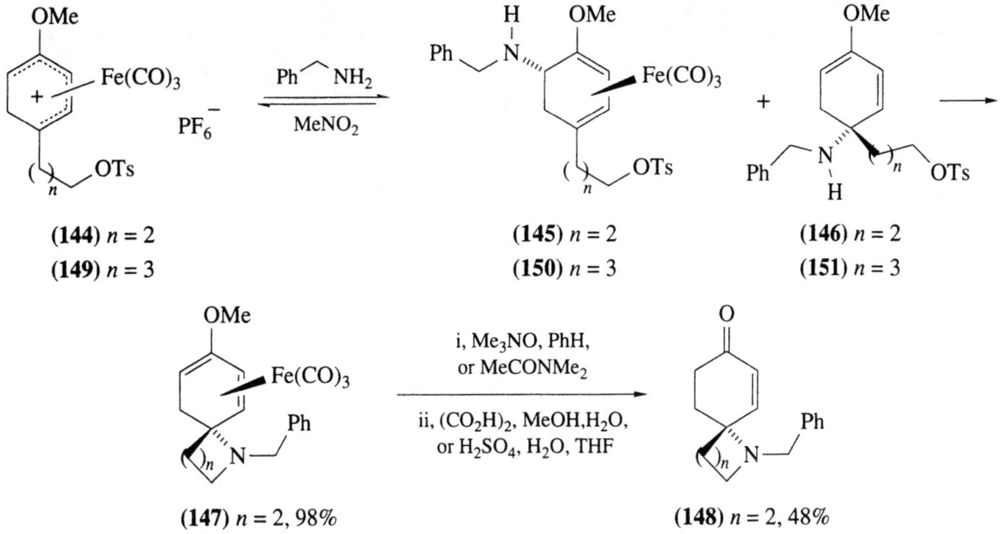

(144) $n = 2$ (145) $n = 2$ (146) $n = 2$
(149) $n = 3$ (150) $n = 3$ (151) $n = 3$

(147) $n = 2$, 98% (148) $n = 2$, 48%
(152) $n = 3$, 90% (153) $n = 3$, 75%

Scheme 19

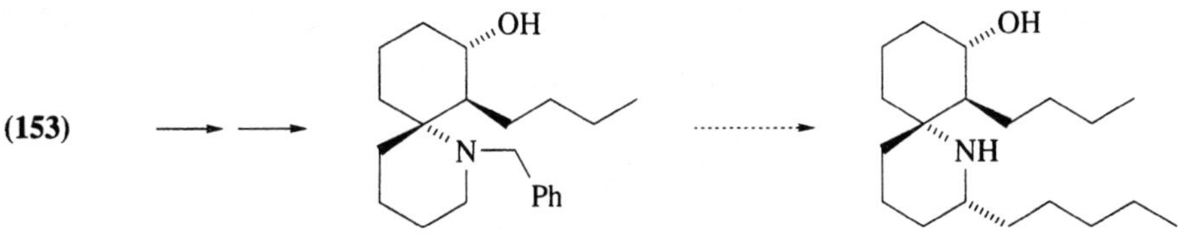

(154)

Scheme 20

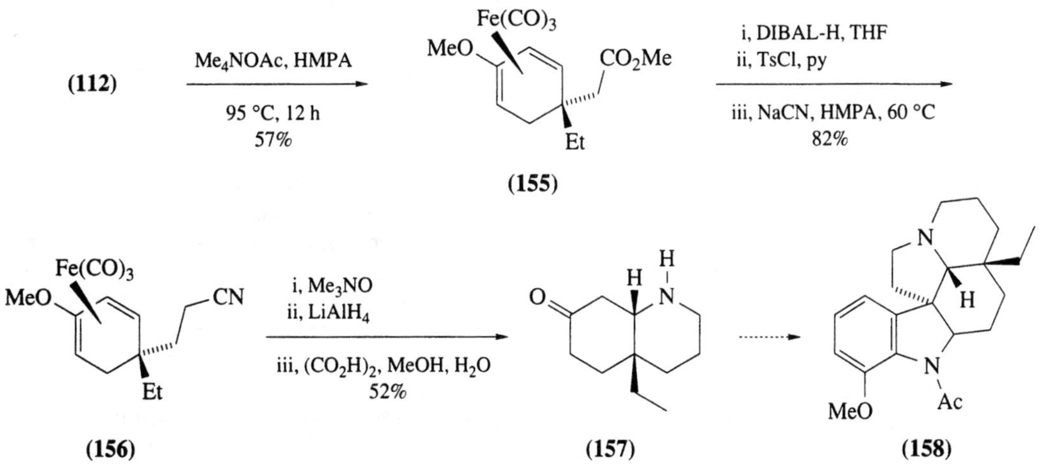

(155)

(156) (157) (158)

Scheme 21

this class, exemplified by trichodermol (**164**), verrucarol (**165**), and calonectrin (**166**), as well as their biogenetic precursor trichodiene (**167**), show interesting biological activity and have been the objects of a number of total syntheses.[52–54]

The reaction of complex (**30**) with the enolate from methyl 2-oxocyclopentanecarboxylate to give complex (**97**), provides a method of joining five- and six-membered rings at highly substituted positions, generating contiguous quaternary centers appropriate for trichothecene synthesis. An equimolar mixture of diastereomers is generated in almost quantitative yield, and these are readily separated by crystallization to give (**168**; Scheme 23), which has relative configurations about the two quaternary centers appropriate for trichothecene synthesis. This intermediate has been converted to the trichothecene analogs

i, $KCH(CO_2Me)_2$; ii, KCN, DMSO, H_2O, 100–120 °C, 22 h; iii, DIBAL-H; iv, TsCl, py; v, NaCN, HMPA, 60 °C; vi, Me_3NO; vii, $LiAlH_4$; viii, H^+, H_2O; ix, $ClCH_2COCl$, py, PhH; x, $KOBu^t$; xi, H^+, $(CH_2OH)_2$; xii, $LiAlH_4$; xiii, H^+, H_2O

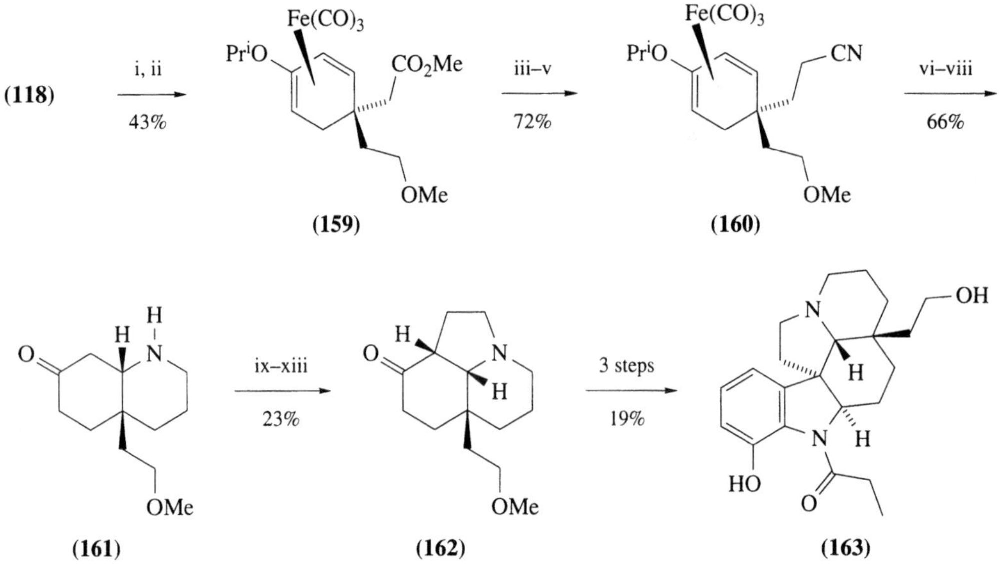

Scheme 22

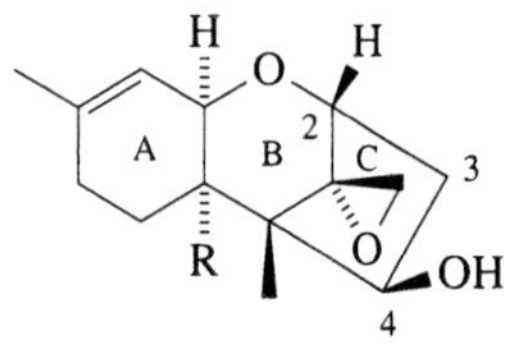

(164) R = Me
(165) R = CH_2OH

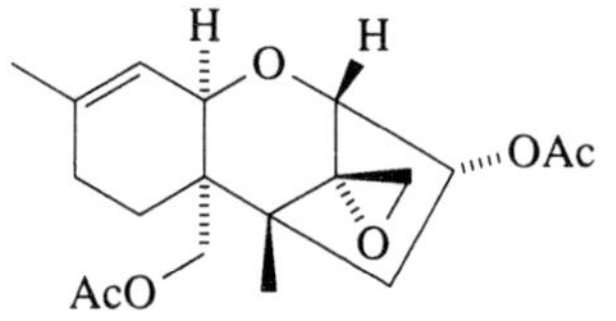

(166)

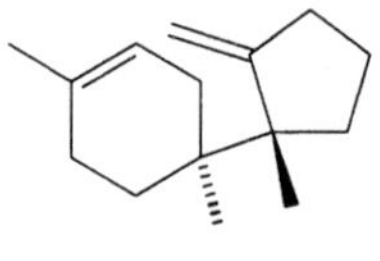

(167)

(**171**; Scheme 23) and (**172**; Scheme 24), and the two sequences of reactions illustrate the wide range of organic manipulations that can be performed without detriment to the (diene)$Fe(CO)_3$ unit.[12,55]

For an effective synthesis of trichodermol, there remain three points of concern. First, the formation of an equimolar mixture of diastereomers (**97**) leads to inefficiency, since realistically only half of the material can be used. Second, the ester group of complex (**168**) has to be reduced to methyl, a multistep operation which again is unattractive; it would be better to attach to complex (**30**) an enolate having methyl in place of the ester. Third, the introduction of the trichodermol 4-hydroxy group must be accomplished in some manner.

The direct coupling of the tin enolate (**175**; Scheme 25) with (**30**) gives complex (**176**) in good yield (87%). Fortuitously, the bond formation gives approximately a 5:1 mixture in favor of the diastereomer required for trichothecene synthesis. Further elaboration of the major isomer leads to (±)-trichodiene (**167**), representing an eight step diastereoselective total synthesis of the natural product from *p*-methylanisole, which compares very favorably with previous methods.[52,56]

In order to address the problem of 4-hydroxy group introduction, a tin enolate bearing a substituent that can be later converted to hydroxy must be employed. A suitable OH surrogate is the phenyldimethylsilyl group.[57] The readily prepared silyl-substituted tin enolate (**177**) reacts with complex (**30**) with a high degree of diastereoselectivity (*ca.* 6:1) to give complex (**178**) as the major product (72%).[58] Subsequent manipulation of (**178**) leads to a stereoselective total synthesis of (±)-trichodermol (**164**; Scheme 26) in 13 steps from complex (**30**), which is itself easily prepared from *p*-methylanisole on a large scale (>200 g batches).

The regioselectivity during hydride abstraction afforded by an ester substituent attached to cyclohexadiene has also been exploited for total synthesis. The synthesis of (+)- and (–)-gabaculine in homochiral

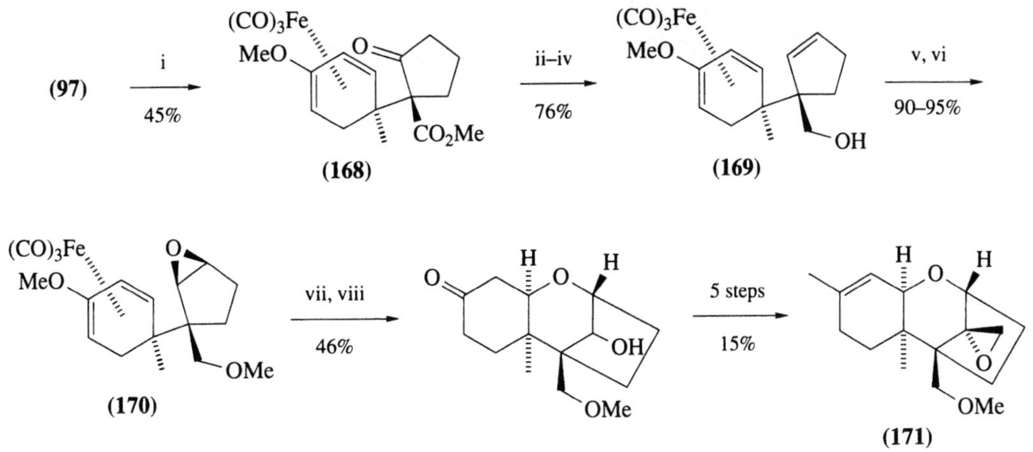

(97) → i, 45% → (168) → ii–iv, 76% → (169) → v, vi, 90–95% → (170) → vii, viii, 46% → 5 steps, 15% → (171)

i, recryst.; ii, $NaBH_4$; iii, $SOCl_2$, py; iv, DIBAL-H; v, Bu^tOOH, $VO(acac)_2$, PhH; vi, MeI, NaH, THF; vii, Me_3NO; viii, H^+, H_2O

Scheme 23

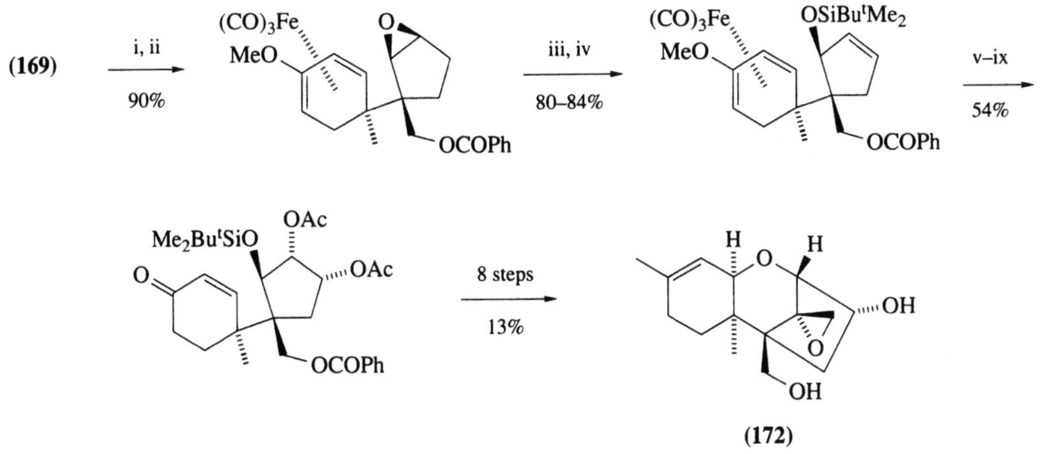

(169) → i, ii, 90% → iii, iv, 80–84% → v–ix, 54% → 8 steps, 13% → (172)

i, Bu^tOOH, $VO(acac)_2$, PhH; ii, PhCOCl, py; iii, Bu^tMe_2SiI, MeCN, 0 °C, 20 min; iv, DBU, THF, Δ, 48 h; v, OsO_4, py, 20 °C, 24 h; vi, $Na_2S_2O_5$, H_2O; vii, Ac_2O, py; viii, Me_3NO; ix, H^+, H_2O

Scheme 24

form takes advantage of the efficient resolution of the carboxylic acid derived from complex (**26**), which in turn allows the preparation of homochiral dienyl complex (**34**), as either enantiomer.[59] Gabaculine (**181**) is a naturally occurring unsaturated amino acid, which has been found to act as a potent inhibitor of 4-aminobutyrate:2-oxoglutarate aminotransferase, and which is therefore of interest in the treatment of certain neurochemical disorders, *e.g.* Parkinsonism, epilepsy and schizophrenia. The elegant synthesis described here illustrates two important features of dienyliron chemistry: (i) the stereospecific addition of nucleophiles, and (ii) the ability to remove the $Fe(CO)_3$ group from a delicate ligand. Reaction of optically pure (–)-(**34**) with *t*-butyl carbamate in the presence of Hunig's base leads to the formation of (–)-(**179**), which is decomplexed under mild conditions using trimethylamine *N*-oxide in dimethylacetamide. The resulting diene (**180**) is converted to (–)-gabaculine (**181**) in two steps (Scheme 27).

The ester-substituted complex (**34**) has been used in synthesis of (+)- and (–)-shikimic acid, an important intermediate in the biosynthesis of aromatic compounds, as well as stereospecifically deuterium labeled shikimic acid.[60] Addition of hydroxide anion to (+)-(**34**) gives the diene complex (+)-(**182**),

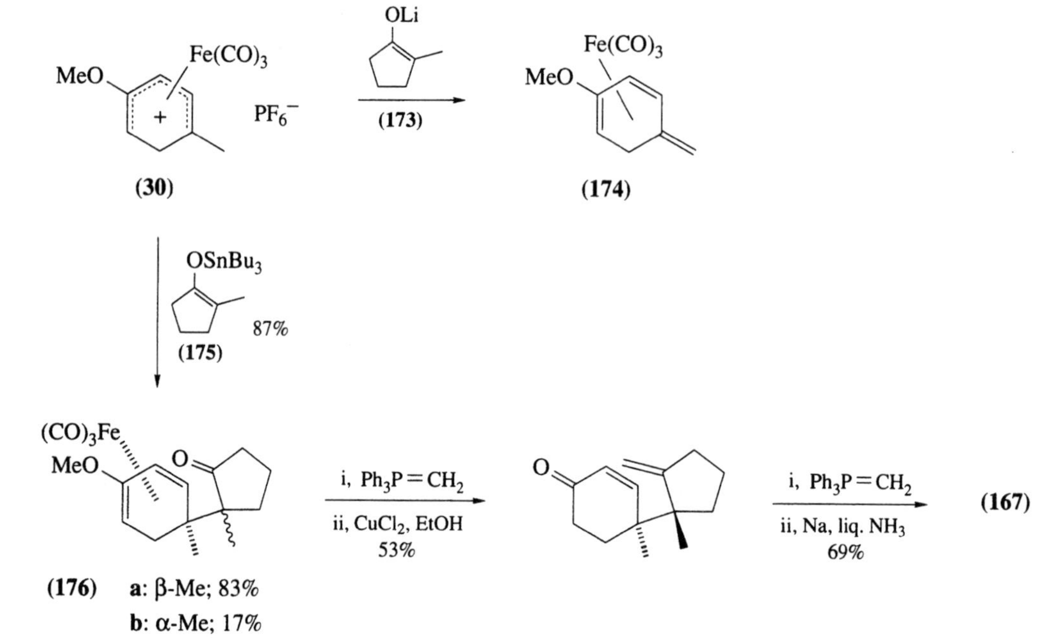

MeO Fe(CO)3 + PF6−
(30)

OLi
(173)

MeO Fe(CO)3
(174)

OSnBu3
(175)
87%

(CO)3Fe MeO O
(176) **a**: β-Me; 83%
b: α-Me; 17%

i, $Ph_3P{=}CH_2$
ii, $CuCl_2$, EtOH
53%

O

i, $Ph_3P{=}CH_2$
ii, Na, liq. NH_3
69%

(167)

Scheme 25

OSnBu3
SiMe2Ph
(177)

i
72%

(CO)3Fe MeO O SiMe2Ph
(178)

ii–iv
70%

OMs O O SiMe2Ph

v–vii
52%

O O OMs FMe2Si

viii–x
81%

H O H SiMe2OBu^t

xi, xii
56%

H O H OH

xiii
65%

(164)

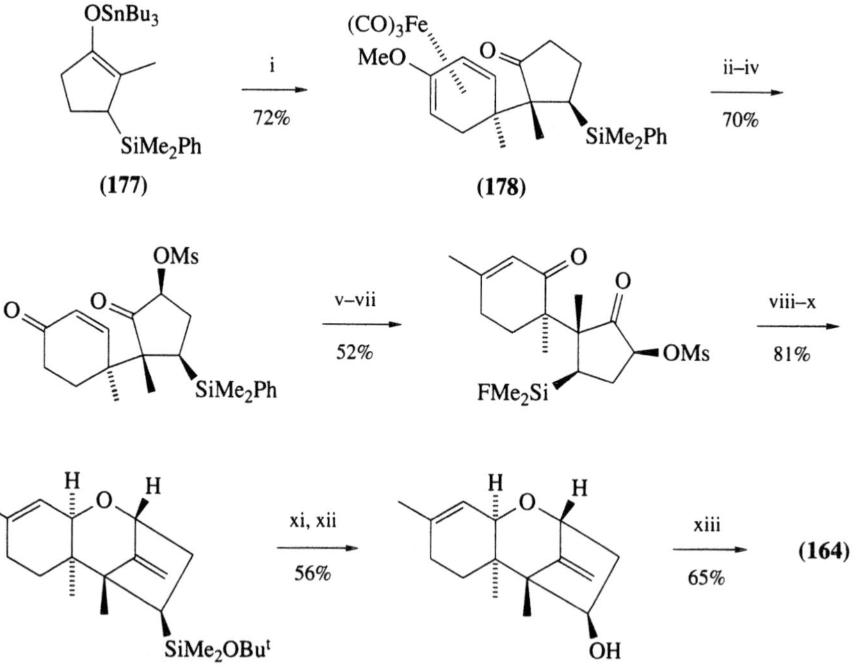

i, **(30)**; ii, LDA, THF, MoOPH; iii, MsCl, py, r.t.; iv, $CuCl_2$, EtOH; v, MeMgBr; vi, PCC; vii, $HBF_4{\bullet}Et_2O$, CH_2Cl_2; viii, DIBAL-H; ix, PCC; x, [Ph_3PMe]Br, KOBu^t, Et_2O, Bu^{t}OH; xi, excess Bu_4NF, THF, H_2O; xii, H_2O_2, KF, $NaHCO_3$, THF, MeOH, 60 °C; xiii, MCPBA

Scheme 26

which is protected and decomplexed to give the diene (**183**). This is converted to (–)-methyl shikimate (**184**) using standard procedures (Scheme 28).

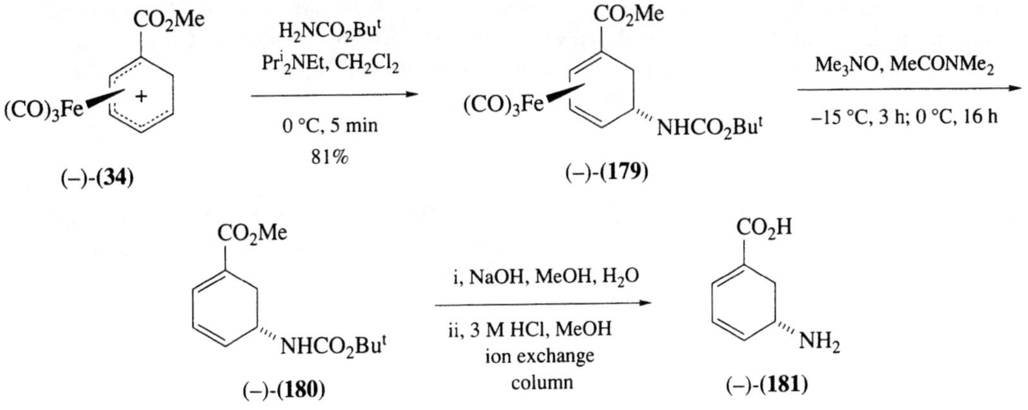

CO_2Me | $H_2NCO_2Bu^t$, Pr^i_2NEt, CH_2Cl_2, 0 °C, 5 min, 81% | $(CO)_3Fe$ | (–)-(34) → (–)-(179) $NHCO_2Bu^t$ | Me_3NO, $MeCONMe_2$, –15 °C, 3 h; 0 °C, 16 h → (–)-(180) | i, NaOH, MeOH, H_2O; ii, 3 M HCl, MeOH, ion exchange column → (–)-(181) CO_2H, NH_2

Scheme 27

(+)-(34) | $KHCO_3$, MeCN, H_2O, 95% → (+)-(182) HO, $Fe(CO)_3$ | i, Bu^tMe_2SiCl, DMF, Pr^i_2NEt; ii, Me_3NO, PhH, 82% → (183) Bu^tMe_2SiO | i, OsO_4, acetone; ii, Bu_4NF, THF, 57% → (–)-(184) HO, OH, OH

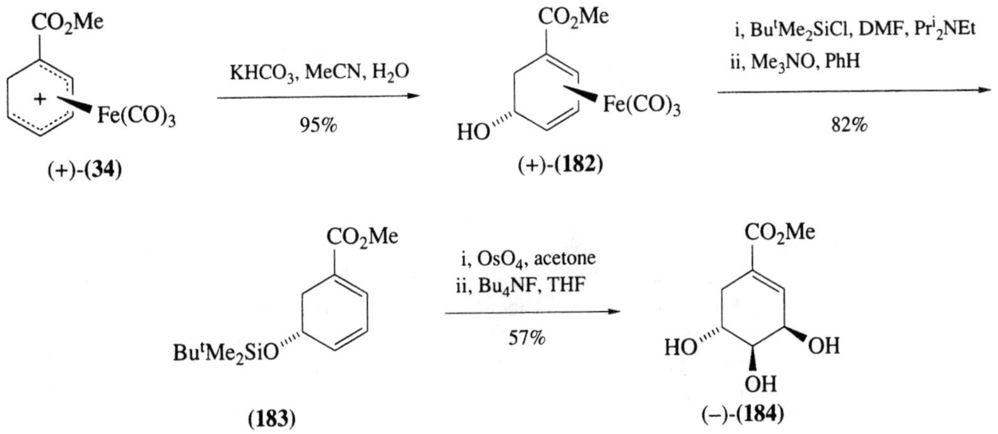

Scheme 28

3.4.2.4 Stereocontrol During Nucleophile Addition: Applications in Relative Stereocontrol During Multiple Functionalization of Six- and Seven-membered Rings

Given that nucleophile addition to (dienyl)Fe(CO)3 complexes proceeds stereospecifically *trans* to the metal, the question arises as to whether this can be used to control relative stereochemistry during multiple functionalization of cyclohexadienes and cycloheptadienes. A hypothetical example is shown in Scheme 29, where nucleophile addition is followed by a second hydride abstraction, or its equivalent, to generate a substituted dienyl complex. Addition of a second nucleophile, assuming stereocontrol from the metal, would generate a disubstituted derivative with defined relative stereochemistry.

$Fe(CO)_3$ → $(CO)_3Fe$, R → $(CO)_3Fe$, R → $(CO)_3Fe$, R, R'

$Fe(CO)_3L$ → $L(CO)_3Fe$, R → $L(CO)_3Fe$, R → $L(CO)_3Fe$, R, R'

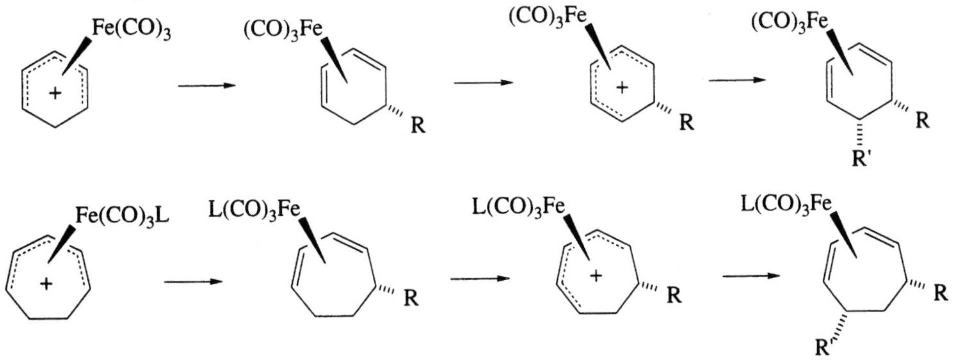

Scheme 29 Stereocontrolled multiple functionalization of dienyliron complexes

As mentioned earlier, direct hydride abstraction from 5-*exo*-substituted cyclohexadiene complexes is in general difficult, except for the 2-trimethylsilyl-substituted derivatives such as (**48**) and (**50**). Oxidative cyclization techniques have been developed to overcome this problem, exemplified by the conversion of (**52**) to (**53**) and thence to (**54**; Scheme 7). Stereocontrolled addition of a second nucleophile has already been illustrated by the conversion of (**54**) to (**126**) or (**127**), and the limitations imposed by a sterically demanding 6-*exo* substituent have been mentioned.

Hydride abstraction from cycloheptadienyl complex (**84**), having a sterically less encumbered allylic CH_2 group, occurs in excellent yield to generate complex (**185**; Scheme 30). Addition of nucleophiles to (**185**) provides complexes of general structure (**186**; Scheme 30) in high yield. The regioselectivity of this reaction is controlled only by steric hindrance from the methyl group in (**185**), but this is sufficient to give high selectivity.

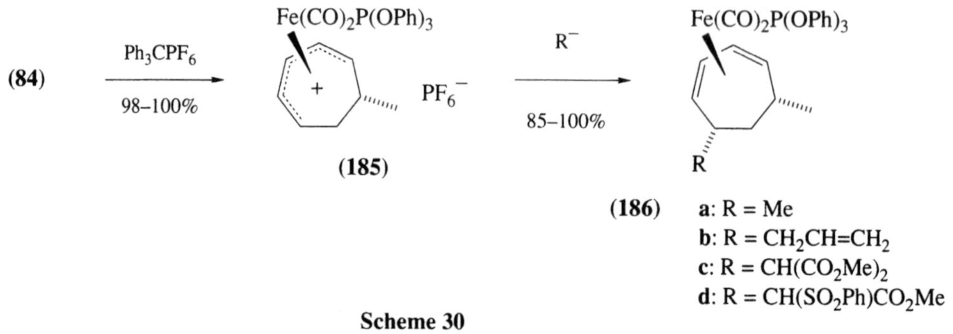

(**84**) $\xrightarrow[98–100\%]{Ph_3CPF_6}$ $Fe(CO)_2P(OPh)_3$ complex PF_6^- (**185**) $\xrightarrow[85–100\%]{R^-}$ $Fe(CO)_2P(OPh)_3$ complex (**186**)

(**186**) **a**: R = Me
b: R = $CH_2CH{=}CH_2$
c: R = $CH(CO_2Me)_2$
d: R = $CH(SO_2Ph)CO_2Me$

Scheme 30

The symmetrical dimethyl-substituted complex (**186a**) has been employed in an asymmetric synthesis of the Prelog–Djerassi lactone (Scheme 31).[61] Decomplexation of (**186a**) affords the diene (**187**), which undergoes stereospecific palladium-catalyzed 1,4-diacetoxylation to give (**188**). Enzymatic hydrolysis of this compound allows the preparation of optically pure hydroxy acetate (**189**) which is readily converted to the (+)-Prelog–Djerassi lactone (**190**), an important degradation product of several macrolide antibiotics and a building block for their chemical synthesis.

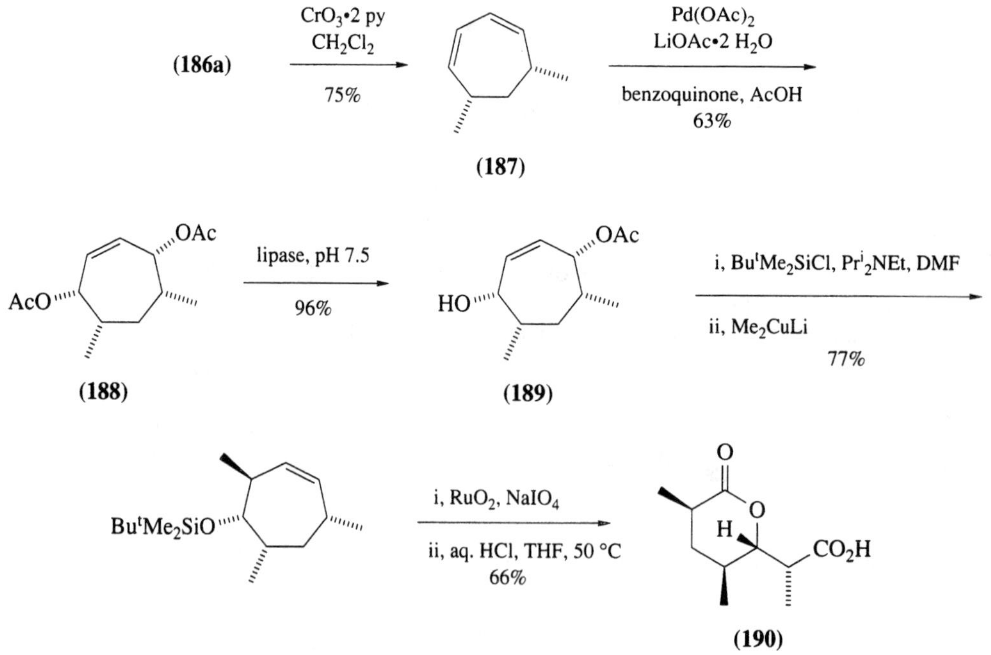

(**186a**) $\xrightarrow[75\%]{CrO_3{\cdot}2\,py,\ CH_2Cl_2}$ (**187**) $\xrightarrow[\text{benzoquinone, AcOH, } 63\%]{Pd(OAc)_2,\ LiOAc{\cdot}2\,H_2O}$

(**188**) [AcO, OAc] $\xrightarrow[96\%]{\text{lipase, pH 7.5}}$ (**189**) [HO, OAc] $\xrightarrow[\text{ii, } Me_2CuLi;\ 77\%]{\text{i, } Bu^tMe_2SiCl,\ Pr^i_2NEt,\ DMF}$

Bu^tMe_2SiO $\xrightarrow[\text{ii, aq. HCl, THF, 50 °C; } 66\%]{\text{i, } RuO_2,\ NaIO_4}$ (**190**) [O, O, H, CO_2H]

Scheme 31

The malonate adduct (**186c**) has been used[62] as an intermediate for the preparation of molecules corresponding to substructures of the macrolide antibiotics tylosin (**193**) and carbomycin B (**194**; Scheme 32). Decomplexation of (**186c**), followed by decarboxylation and ester hydrolysis, gives the carboxylic acid

(**191**), which undergoes conjugate bromolactonization to generate (**192**), and this can serve as an intermediate for approaches to the right-hand section of either (**193**) or (**194**), as shown in Scheme 32.

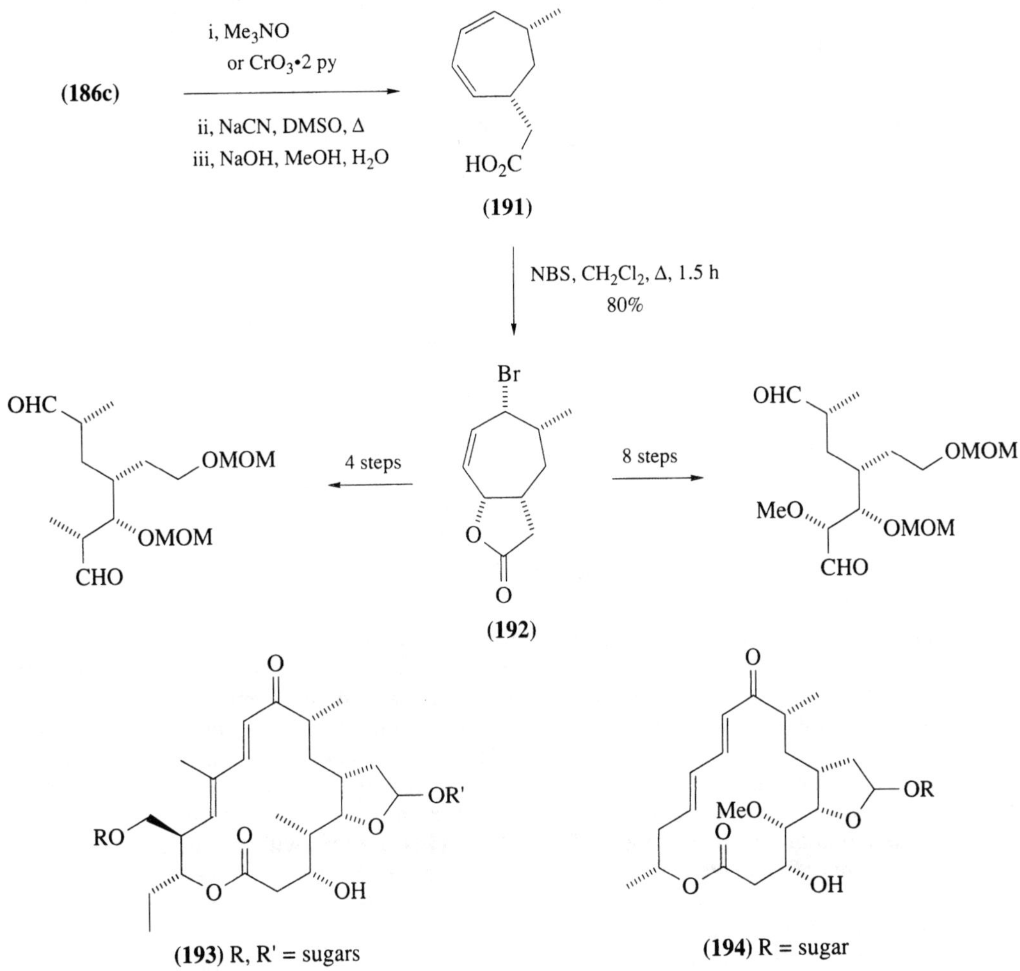

(**186c**) → i, Me_3NO or CrO_3•2 py; ii, NaCN, DMSO, Δ; iii, NaOH, MeOH, H_2O → HO_2C (**191**)

(**191**) → NBS, CH_2Cl_2, Δ, 1.5 h, 80% → Br (**192**)

(**192**) → 4 steps → OHC, OMOM, OMOM, CHO

(**192**) → 8 steps → OHC, OMOM, MeO, OMOM, CHO

(**193**) R, R' = sugars

(**194**) R = sugar

Scheme 32

Double functionalization appears to be quite general, since a variety of 5-substituted cycloheptadiene-iron complexes are readily converted to 6-substituted cycloheptadienyliron systems, which in turn undergo regio- and stereo-controlled nucleophile addition.[23]

3.4.2.5 Control of Absolute Stereochemistry During Carbon–Carbon Bond Formation Using Dienyliron Complexes

3.4.2.5.1 Preparation of enantiomerically enriched complexes

There are two approaches for obtaining enantiomerically pure, unsymmetrically substituted, dienyliron complexes: (i) Resolution of either the dienyl complex itself or its dieneiron complex precursor; and (ii) asymmetric synthesis of the diene or dienyl complex.

The first method, resolution, is unattractive unless both enantiomers are useful in synthesis. In some cases, such as the resolution of dienecarboxylic acid derivatives mentioned earlier (*via* the phenylethylammonium salt), the resolution is efficient and provides optically pure materials in good yield.[59,60,63] In certain cases, the dienyliron complex can be treated with a chiral nucleophile to give a mixture of diastereomers which are separated and then reconverted to enantiomerically pure dienyl complex.[64] An example of this method is the resolution of complex (**27**; Scheme 33), *via* the menthyl ethers (**195**) and

(**196**), which can be separated chromatographically. Treatment of the separated ethers with acid leads to regeneration of optically pure complexes (+)- and (–)-(**27**).

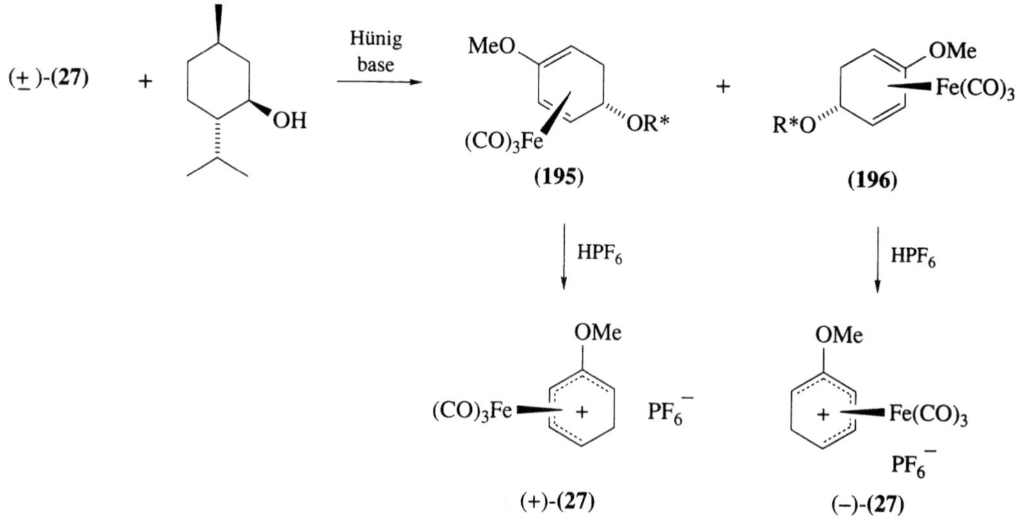

(±)-(**27**) + [menthol] —Hünig base→ (**195**) + (**196**); each —HPF_6→ (+)-(**27**) and (–)-(**27**), PF_6^-

Scheme 33

A modest degree of kinetic and/or thermodynamic discrimination has been observed during reactions of unsymmetrically substituted cyclohexadienyliron complexes with a deficiency of chiral nucleophile,[65] but this does not generate synthetically useful amounts of optically pure complexes and also involves loss of at least 50% of the complex (as does any classical resolution). A more promising approach,[66] which has not been fully refined to give high optical yields, is based on the knowledge that iron carbonyl complexes of α,β-unsaturated ketones, such as benzylideneacetone, readily transfer the $Fe(CO)_3$ group to 1,3-dienes.[9,35,67] The use of $Fe(CO)_3L_2$ complexes derived from chiral enones as asymmetric transfer reagents allows preparation of both enantiomers of complexes (**16**) and (**19**) with enantiomeric excesses of up to 43%. Assignments of absolute stereochemistry have been made, and it is noteworthy that the enantiomer which predominates is dependent on the chiral enone used to generate (*in situ*) the (enone)$Fe(CO)_3$ complex. For example, the use of (+)-pulegone (**197**; equation 56) gives (+)-(**19**), while the use of (–)-3β-(acetyloxy)pregna-5,16-dien-20-one (**198**; equation 57) gives (–)-(**19**).

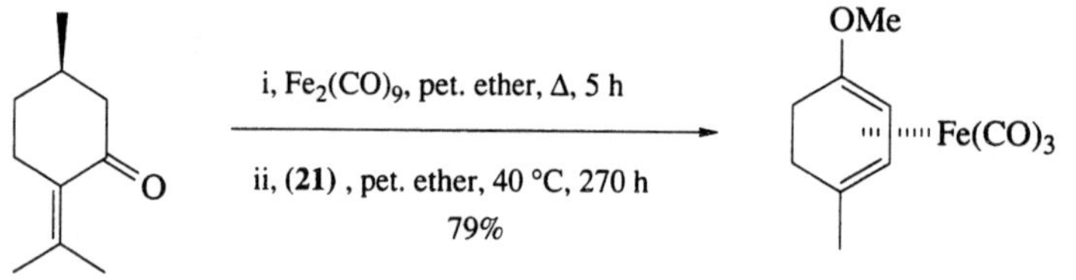

(**197**) —i, $Fe_2(CO)_9$, pet. ether, Δ, 5 h; ii, (**21**), pet. ether, 40 °C, 270 h; 79%→ (+)-(**19**) 10% *ee* (56)

(**198**) —i, $Fe_2(CO)_9$, pet. ether, 40 °C, 5 h; ii, (**21**), PhH, 65 °C, 88 h; 26%→ (–)-(**19**) 43% *ee* (57)

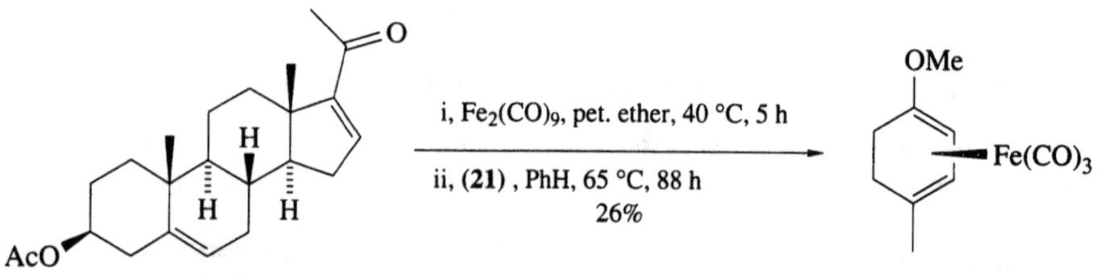

While the optical yields are rather low for useful synthetic application, this represents a promising lead for future work in the area of asymmetric complexation reactions.

3.4.2.5.2 Reactions of symmetrical dienyliron complexes with chiral nucleophiles

An alternative approach to asymmetric synthesis using dienyliron complexes involves addition of an enolate nucleophile bearing a chiral auxiliary to the complex.[68] Reaction of complex (**83**) with the enolate from optically pure sulfoximine-stabilized ester (**199**) gives the adduct (**200**), desulfonylation of which leads to the monoester derivative (**201**; Scheme 34), obtained in 50% enantiomeric excess. Since this complex undergoes hydride abstraction to give (**202**), which in turn reacts with a variety of carbon nucleophiles to generate complexes (**203**; Scheme 35), this represents a potentially useful approach to asymmetric multiple functionalization of the seven-membered ring. However, an enantiomeric excess of 50% is disappointing and there is much work to be done in seeking chiral auxiliaries that give higher asymmetric induction.

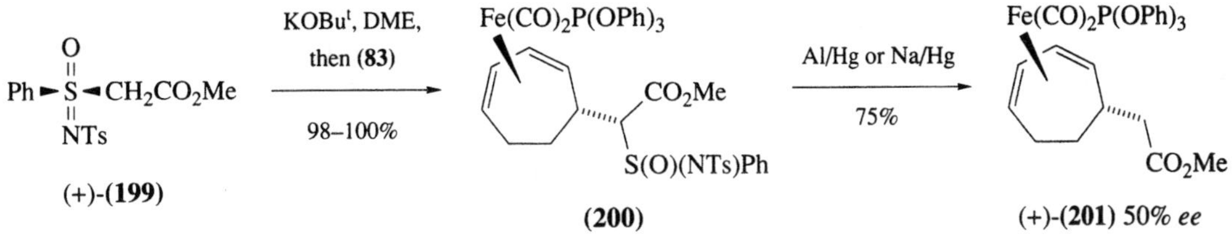

Ph–S(O)(=NTs)–CH_2CO_2Me (+)-(**199**) → [KOBu^t, DME, then (**83**); 98–100%] → $Fe(CO)_2P(OPh)_3$ complex, CO_2Me, S(O)(NTs)Ph (**200**) → [Al/Hg or Na/Hg; 75%] → $Fe(CO)_2P(OPh)_3$ complex, CO_2Me (+)-(**201**) 50% *ee*

Scheme 34

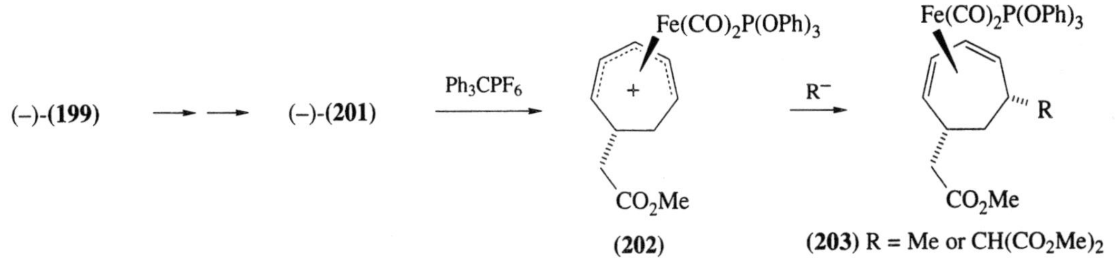

(–)-(**199**) → → (–)-(**201**) → [Ph_3CPF_6] → $Fe(CO)_2P(OPh)_3$, +, CO_2Me (**202**) → [R^-] → $Fe(CO)_2P(OPh)_3$, R, CO_2Me (**203**) R = Me or $CH(CO_2Me)_2$

Scheme 35

3.4.3 CATIONIC DIENYL COMPLEXES OF OTHER METALS

The chemistry of pentadienyl complexes of metals other than iron has not been explored in depth, and consequently few applications in synthesis have emerged. The most noteworthy systems are the complexes of manganese and cobalt, and these will be described briefly in this section. One of the major problems with the use of these metals stoichiometrically is the high cost of their carbonyl complexes compared to iron [at the time of writing $Fe(CO)_5$ was cataloged at *ca.* \$75.00 kg^{-1}, $Mn_2(CO)_{10}$ at \$79.60 for 10 g, and $CpCo(CO)_2$ \$142.75 for 25 g (Aldrich Catalog 1988/89)]. While this may reflect the rather low demand for these reagents, the high price is certainly a disincentive for extensive research that requires large amounts.

3.4.3.1 Dienylmanganese Complexes

Treatment of 1,3-cycloheptadiene with $Mn_2(CO)_{10}$ at high temperature (refluxing mesitylene) gives the (cycloheptadienyl)$Mn(CO)_3$ derivative (**204**) in high yield.[69] This complex bears no charge and is unreactive towards most nucleophiles. However, substitution of CO ligand by NO^+ gives complex (**205**), which reacts with nucleophiles to give substituted diene complexes (**206**; Scheme 36).[70] Stereocontrolled double functionalization of the cycloheptadiene ring is possible since (**204**) can be converted to the (triene)$Mn(CO)_3$ complex (**207**), which also undergoes nucleophile addition. Using these tactics, substituted (dienyl)$Mn(CO)_2NO$ cation complexes (**208**; Scheme 37) can be prepared and converted to diene complexes (**209**). In several cases, the addition of nucleophiles is improved by using the trialkylphosphine or trialkyl phosphite derivatives (**210**) or (**211**; Scheme 37).[70]

(Arene)tricarbonylmanganese complexes react with nucleophiles to give substituted (cyclohexadienyl)$Mn(CO)_3$ complexes[71] which can be activated towards nucleophile addition by CO/NO^+ exchange giving, *e.g.* (**214**).[72] Kinetic studies[73] indicate that the $[Mn(CO)_2NO]^+$ and $[Fe(CO)_3]^+$ groups are vir-

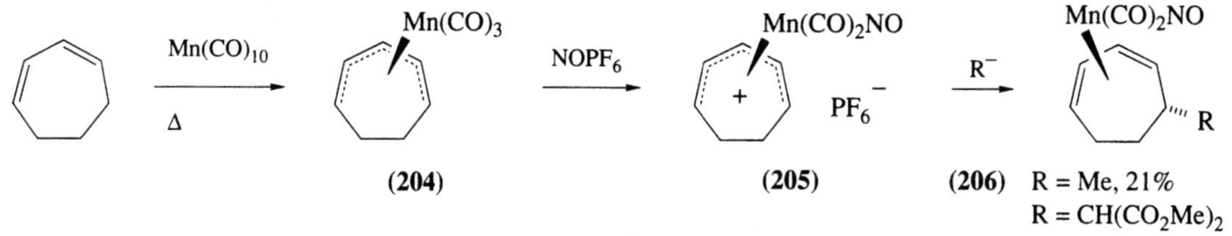

$Mn(CO)_{10}$, Δ; $Mn(CO)_3$ **(204)**; $NOPF_6$; $Mn(CO)_2NO$ PF_6^- **(205)**; R^-; $Mn(CO)_2NO$ **(206)** R = Me, 21%; R = $CH(CO_2Me)_2$

Scheme 36

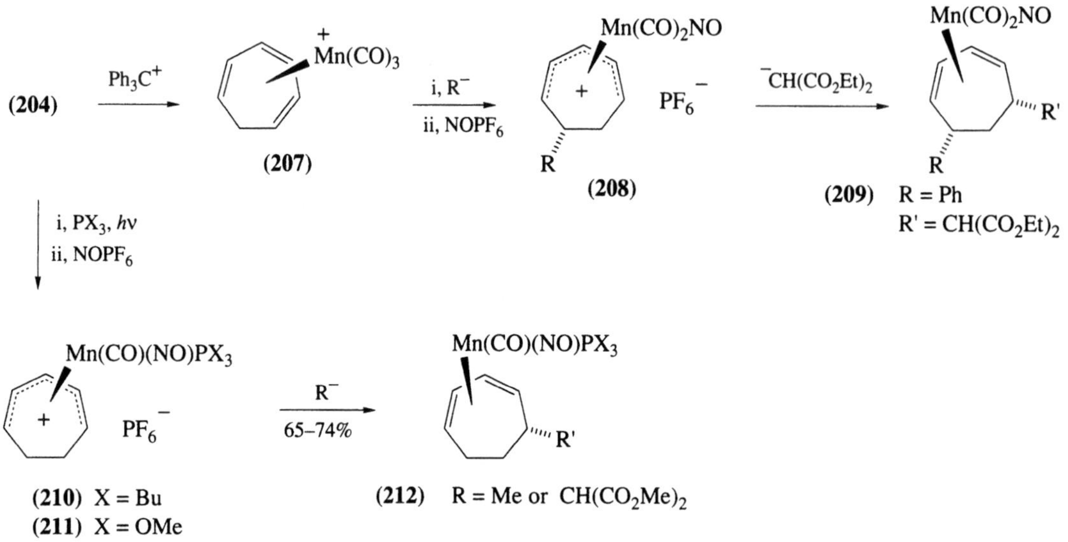

(204) Ph_3C^+; $Mn(CO)_3^+$ **(207)**; i, R^-; ii, $NOPF_6$; $Mn(CO)_2NO$ PF_6^- **(208)**; $^-CH(CO_2Et)_2$; $Mn(CO)_2NO$ **(209)** R = Ph; R' = $CH(CO_2Et)_2$

i, PX_3, $h\nu$; ii, $NOPF_6$

$Mn(CO)(NO)PX_3$ PF_6^- **(210)** X = Bu; **(211)** X = OMe; R^-, 65–74%; $Mn(CO)(NO)PX_3$ **(212)** R = Me or $CH(CO_2Me)_2$

Scheme 37

tually identical in their activating power, so that they can be expected to react with similar nucleophiles. From the reactions of (**214**; Scheme 38), it can be seen that all nucleophiles add *exo*, except borohydride (borodeuteride) which adds *endo*, possibly *via* initial attack on a CO or NO ligand. (It may be noted that hydride adds only *exo* to dienyliron complexes.)

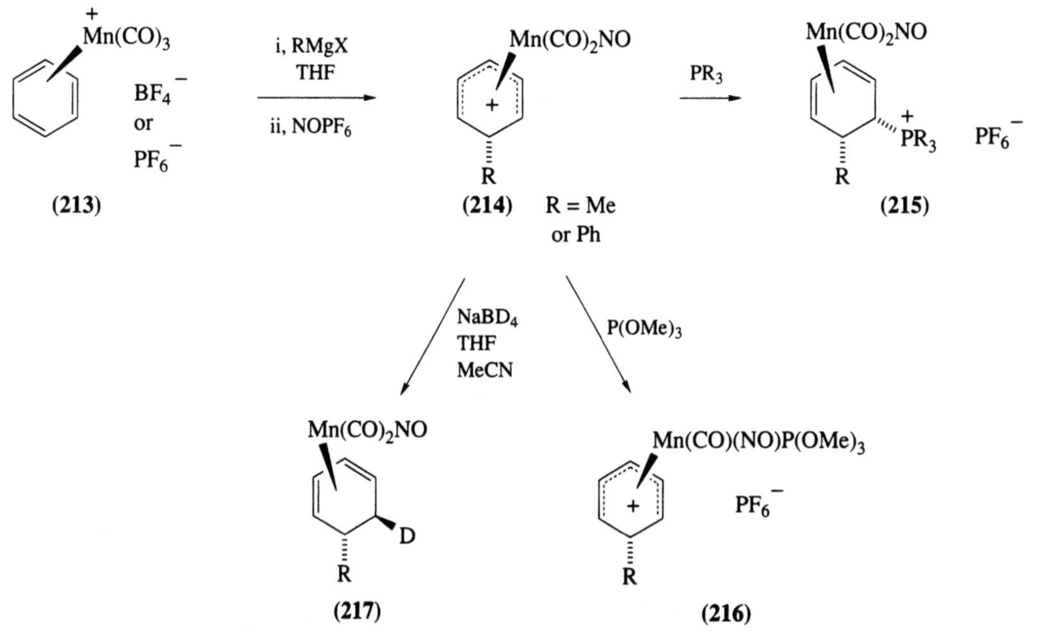

$Mn(CO)_3^+$ BF_4^- or PF_6^- **(213)**; i, RMgX, THF; ii, $NOPF_6$; $Mn(CO)_2NO$ **(214)** R = Me or Ph; PR_3; $Mn(CO)_2NO$, $^+PR_3$ PF_6^- **(215)**

$NaBD_4$, THF, MeCN; $Mn(CO)_2NO$ **(217)**; $P(OMe)_3$; $Mn(CO)(NO)P(OMe)_3$ PF_6^- **(216)**

Scheme 38

Of potentially greater interest are the preparation and reactions[74] of (dienyl)$Mn(CO)_2NO$ complexes such as (**219**; Scheme 39), which take advantage of the *meta*-directing effect of a methoxy substituent during nucleophile addition to the (arene)$Mn(CO)_3$ complex (**218**). The dienyl system of (**219**) is very similar to the organoiron derivatives such as (**30**), and so it is interesting to note that sodium borohydride reduction of this complex gives a 1.5:1 mixture of regioisomers (**220**) and (**221**; Scheme 39) (hydride again adds *endo*). While there is some regioselectivity, it is unlikely to be of sufficient interest to prompt synthetic applications of this complex. Perhaps the use of isopropoxy in place of the methoxy group will allow greater regiocontrol, as seen with the iron analogs, but no study has been reported. Similarly, the arenemanganese complex (**222**; equation 58) has been converted to (**223**; 9:1 regioselectivity is observed).[75] No reports of nucleophile addition to this complex have appeared.

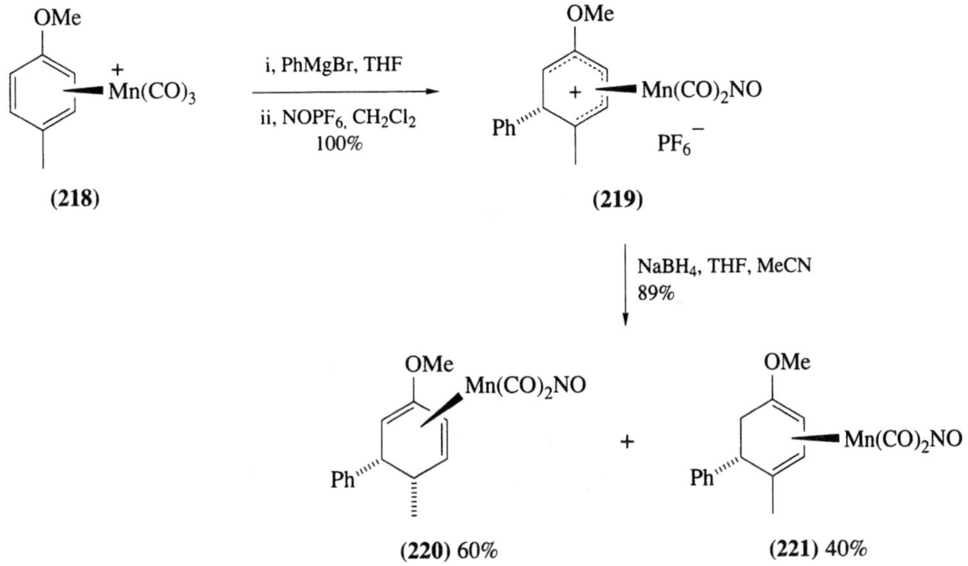

OMe, + $Mn(CO)_3$ (**218**) → i, PhMgBr, THF; ii, $NOPF_6$, CH_2Cl_2, 100% → OMe, Ph, + $Mn(CO)_2NO$ PF_6^- (**219**)

↓ $NaBH_4$, THF, MeCN, 89%

OMe, $Mn(CO)_2NO$, Ph (**220**) 60% + OMe, Ph, $Mn(CO)_2NO$ (**221**) 40%

Scheme 39

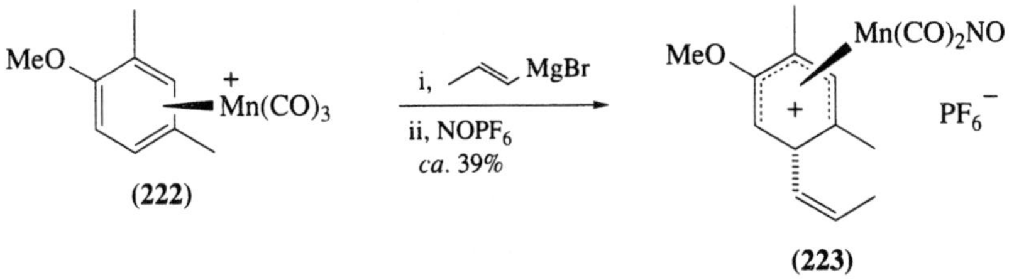

MeO, + $Mn(CO)_3$ (**222**) → i, MgBr; ii, $NOPF_6$, *ca.* 39% → MeO, $Mn(CO)_2NO$, +, PF_6^- (**223**) (58)

The range of nucleophiles that will react with complexes such as (**219**) and (**223**) has not been extensively investigated. The regiocontrolled addition of carbon nucleophiles, if successful, might provide new methodology for construction of highly functionalized cyclohexenones.

3.4.3.2 Dienylcobalt Complexes

The reactivity of the cobalt complexes has not been extensively investigated. The methods of preparation[76] of (dienyl)CoCp complexes are very similar to those for (dienyl)$Fe(CO)_3$ systems, including acid-promoted dehydroxylation, hydride abstraction, *etc.* and are illustrated here for complex (**224**; Scheme 40) and the analogous rhodium complex (**225**; Scheme 41). Somewhat surprisingly, the sterically congested (diene)CoCp complex (**226**) undergoes hydride abstraction, on treatment with trityl hexafluorophosphate under mild conditions, to give (**228**; Scheme 42).[77]

Nucleophilic addition to these complexes is a little less predictable than for the dienyliron systems. For example, the nucleophiles studied add to C-3 of the acyclic dienyl ligand in (**224**), but to C-1 of the cyclic system (**225**). Alkyllithiums add both to the cyclohexadienyl ligand, *e.g.* (**227**) gives (**229**), and to the cyclopentadienyl ligand, followed by interligand hydrogen transfer to give, *e.g.* (**228**). However, the fact that a hindered, very basic alkyllithium adds to a sterically hindered dienyl terminal carbon and does

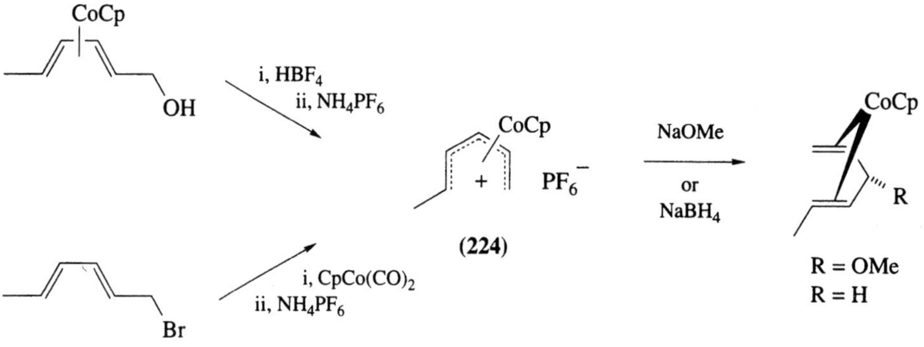

CoCp OH i, HBF_4 ii, NH_4PF_6 CoCp + PF_6^- (224) NaOMe or $NaBH_4$ CoCp R R = OMe R = H Br i, $CpCo(CO)_2$ ii, NH_4PF_6

Scheme 40

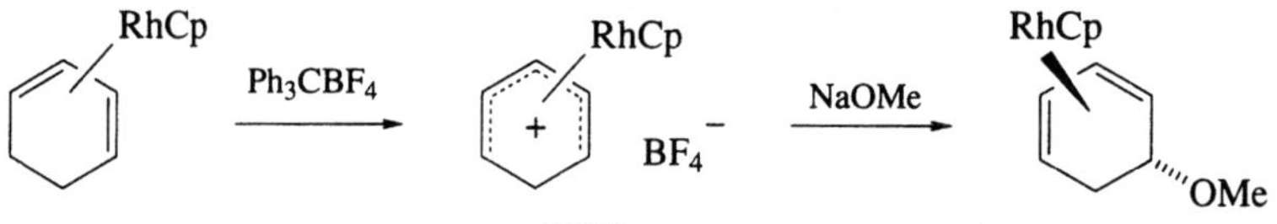

RhCp Ph_3CBF_4 RhCp + BF_4^- (225) NaOMe RhCp OMe

Scheme 41

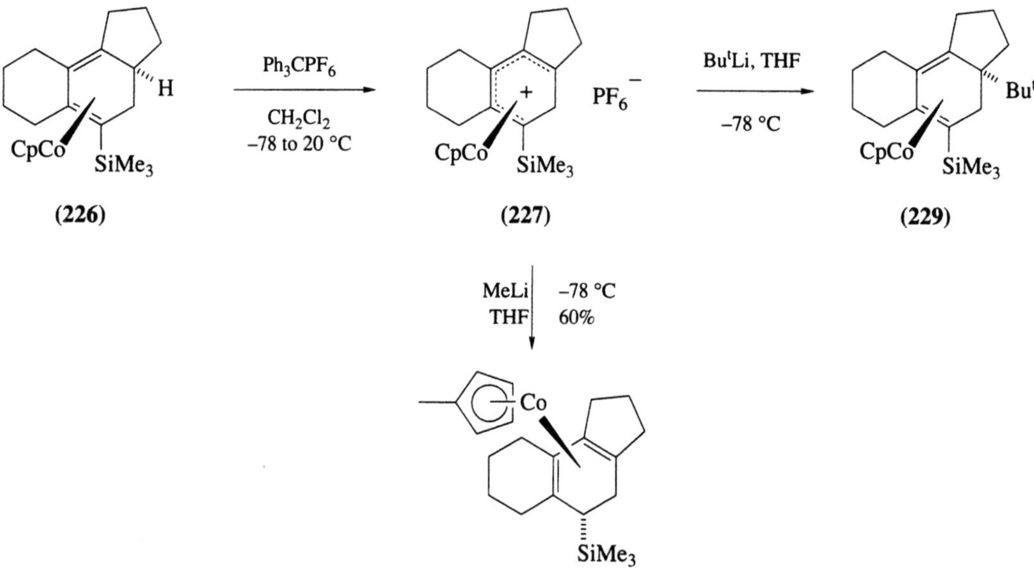

H CpCo $SiMe_3$ (226) Ph_3CPF_6 CH_2Cl_2 –78 to 20 °C + PF_6^- CpCo $SiMe_3$ (227) Bu^tLi, THF –78 °C Bu^t CpCo $SiMe_3$ (229) MeLi THF –78 °C 60% Co $SiMe_3$ (228)

Scheme 42

not deprotonate the complex is very encouraging and indicates that a set of reactions might be developed that are complementary to those already known for the organoiron derivatives.

3.4.4 REFERENCES

1. H. Reihlen, A. Gruhl, G. von Hessling and O. Pfrengle, *Justus Liebigs Ann. Chem.*, 1930, **482**, 161.
2. B. F. Hallam and P. L. Pauson, *J. Chem. Soc.*, 1958, 642; see also R. Burton, L. Pratt and G. Wilkinson, *J. Chem. Soc.*, 1961, 594.
3. E. O. Fischer and R. D. Fischer, *Angew. Chem.*, 1960, **72**, 919.
4. H. J. Dauben and D. J. Bertelli, *J. Am. Chem. Soc.*, 1961, **83**, 497.
5. J. E. Mahler and R. Pettit, *J. Am. Chem. Soc.*, 1963, **85**, 3955.
6. (a) R. Mason, *23rd IUPAC Congress, Boston*, 1971, **6**, 31; (b) O. Eisenstein, W. M. Butler and A. J. Pearson, *Organometallics*, 1984, **3**, 1150.

7. J. E. Arnet and R. Pettit, *J. Am. Chem. Soc.*, 1961, **83**, 2954.
8. A. J. Birch, P. E. Cross, J. Lewis, D. A. White and S. B. Wild, *J. Chem. Soc. A*, 1968, 332.
9. J. A. S. Howell, B. F. G. Johnson, P. L. Josty and J. Lewis, *J. Organomet. Chem.*, 1972, **39**, 329; G. Evans, B. F. G. Johnson and J. Lewis, *J. Organomet. Chem.*, 1975, **102**, 507; B. F. G. Johnson, J. Lewis, G. R. Stephenson and E. J. S. Vichi, *J. Chem. Soc., Dalton Trans.*, 1978, 829.
10. A. J. Birch, K. B. Chamberlain, M. A. Haas and D. J. Thompson, *J. Chem. Soc., Perkin Trans. 1*, 1973, 1882; see also R. E. Ireland, G. G. Brown, Jr., R. H. Stanford, Jr. and T. C. McKenzie, *J. Org. Chem.*, 1974, **39**, 51.
11. A. J. Birch and D. H. Williamson, *J. Chem. Soc., Perkin Trans. 1*, 1973, 1892.
12. A. J. Pearson and C. W. Ong, *J. Am. Chem. Soc.*, 1981, **103**, 6686.
13. L. A. Paquette, R. G. Daniels and R. Gleiter, *Organometallics*, 1984, **3**, 560.
14. A. J. Pearson and C. W. Ong, *J. Org. Chem.*, 1982, **47**, 3780; see also A. J. Birch, K. B. Chamberlain and D. J. Thompson, *J. Chem. Soc., Perkin Trans. 1*, 1973, 1900.
15. A. J. Birch, A. J. Liepa and G. R. Stephenson, *Tetrahedron Lett.*, 1979, 3565; G. R. John and L. A. P. Kane-Maguire, *J. Chem. Soc., Dalton Trans.*, 1979, 1196; L. A. P. Kane-Maguire and C. A. Mansfield, *J. Chem. Soc., Chem. Commun.*, 1973, 540; C. A. Mansfield, K. M. Al-Kathumi and L. A. P. Kane-Maguire, *J. Organomet. Chem.*, 1974, **71**, C11.
16. A. J. Birch, A. S. Narula, P. Dahler, G. R. Stephenson and L. F. Kelly, *Tetrahedron Lett.*, 1980, **21**, 979.
17. L. F. Kelly, A. S. Narula and A. J. Birch, *Tetrahedron Lett.*, 1980, **21**, 871, 2455.
18. A. J. Pearson, *Aust. J. Chem.*, 1976, **29**, 1101; 1977, **30**, 345.
19. A. J. Birch and A. J. Pearson, *J. Chem. Soc., Perkin Trans. 1*, 1976, 954; *Tetrahedron Lett.*, 1975, 2379.
20. M. Gower, G. R. John, L. A. P. Kane-Maguire, T. I. Odiaka and A. Salzer, *J. Chem. Soc., Dalton Trans.*, 1979, 2003, and references cited therein.
21. A. J. Birch, I. D. Jenkins and A. J. Liepa, *Tetrahedron Lett.*, 1975, 1723.
22. A. Pelter, K. J. Gould and L. A. P. Kane-Maguire, *J. Chem. Soc., Chem. Commun.*, 1974, 1029.
23. A. J. Pearson, S. L. Kole and J. Yoon, *Organometallics*, 1986, **5**, 2075.
24. A. J. Pearson and M. K. O'Brien, *J. Chem. Soc., Chem. Commun.*, 1987, 1445.
25. A. J. Pearson and M. K. O'Brien, *Tetrahedron Lett.*, 1988, **29**, 869.
26. L. F. Kelly, *J. Org. Chem.*, 1982, **47**, 3965.
27. A. J. Pearson, *J. Chem. Soc., Perkin Trans. 1*, 1977, 2069.
28. B. M. R. Bandara, A. J. Birch and T. C. Khor, *Tetrahedron Lett.*, 1980, **21**, 3625.
29. A. J. Pearson and J. Yoon, *Tetrahedron Lett.*, 1985, **26**, 2399.
30. P. A. Grieco and S. D. Larsen, *J. Org. Chem.*, 1986, **51**, 3553.
31. R. Aumann, *J. Organomet. Chem.*, 1973, **47**, C28; R. Edwards, J. A. S. Howell, B. F. G. Johnson and J. Lewis, *J. Chem. Soc., Dalton Trans.*, 1974, 2105.
32. A. J. Pearson, S. L. Kole and T. Ray, *J. Am. Chem. Soc.*, 1984, **106**, 6060.
33. Y. Shvo and E. Hazum, *J. Chem. Soc., Chem. Commun.*, 1974, 336.
34. D. J. Thompson, *J. Organomet. Chem.*, 1976, **108**, 381.
35. D. H. R. Barton, A. A. L. Gunatilaka, T. Nakanishi, H. Patin, D. A. Widdowson and B. R. Worth, *J. Chem. Soc., Perkin Trans. 1*, 1976, 821; D. H. R. Barton and H. Patin, *J. Chem. Soc., Perkin Trans. 1*, 1976, 829, and references cited therein.
36. A. J. Pearson, I. C. Richards and D. V. Gardner, *J. Org. Chem.*, 1984, **49**, 3887.
37. I. Fleming, J. Goldhill and I. Paterson, *Tetrahedron Lett.*, 1979, 3205, 3209.
38. (a) A. J. Pearson, *J. Chem. Soc., Perkin Trans. 1*, 1978, 495; (b) A. J. Pearson and M. Chandler, *J. Chem. Soc., Perkin Trans. 1*, 1980, 2238.
39. (a) A. J. Pearson, P. Ham, C. W. Ong, T. R. Perrior and D. C. Rees, *J. Chem. Soc., Perkin Trans. 1*, 1982, 1527; (b) A. J. Pearson, D. C. Rees and C. W. Thornber, *J. Chem. Soc., Perkin Trans. 1*, 1983, 619.
40. A. J. Pearson, T. R. Perrior and D. A. Griffin, *J. Chem. Soc., Perkin Trans. 1*, 1983, 625.
41. R. P. Alexander, C. Morley and G. R. Stephenson, *J. Chem. Soc., Perkin Trans. 1*, 1988, 2069; A. J. Pearson and A. D. White, unpublished results.
42. A. J. Pearson and C. W. Ong, *J. Chem. Soc., Perkin Trans. 1*, 1981, 1614.
43. A. J. Pearson, *J. Chem. Soc., Perkin Trans. 1*, 1980, 400.
44. A. J. Pearson and T. R. Perrior, *J. Organomet. Chem.*, 1985, **285**, 253.
45. A. J. Pearson, P. Ham and D. C. Rees, *J. Chem. Soc., Perkin Trans. 1*, 1982, 489.
46. A. J. Pearson and P. Ham, *J. Chem. Soc., Perkin Trans. 1*, 1983, 1421.
47. A. J. Pearson, *Tetrahedron Lett.*, 1981, **22**, 4033.
48. A. J. Pearson and D. C. Rees, *J. Chem. Soc., Perkin Trans. 1*, 1982, 2467.
49. G. Stork and J. E. Dolfini, *J. Am. Chem. Soc.*, 1963, **85**, 2872.
50. J. E. Saxton, A. J. Smith and G. Lawton, *Tetrahedron Lett.*, 1975, 4161; G. Lawton, J. E. Saxton and A. J. Smith, *Tetrahedron*, 1977, **33**, 1641; Y. Ban, I. Ijima, I. Inoue, M. Akagi and T. Oishi, *Tetrahedron Lett.*, 1969, 1067; I. Inoue and Y. Ban, *J. Chem. Soc. C*, 1970, 602.
51. A. J. Pearson and D. C. Rees, *J. Am. Chem. Soc.*, 1982, **104**, 1118; A. J. Pearson, D. C. Rees and C. W. Thornber, *J. Chem. Soc., Perkin Trans. 1*, 1983, 619.
52. For an excellent review of the total synthesis of trichothecenes, see P. G. McDougal and R. N. Schmuff, *Prog. Chem. Org. Nat. Prod.*, 1985, **47**, 153.
53. E. W. Colvin, R. A. Raphael and J. S. Roberts, *J. Chem. Soc., Chem. Commun.*, 1971, 858; E. W. Colvin, S. Malchenko, R. A. Raphael and J. S. Roberts, *J. Chem. Soc., Perkin Trans. 1*, 1973, 1989.
54. W. C. Still and M. Tsai, *J. Am. Chem. Soc.*, 1980, **102**, 3654.
55. A. J. Pearson and Y. S. Chen, *J. Org. Chem.*, 1986, **51**, 1939.
56. G. A. Kraus and P. J. Thomas, *J. Org. Chem.*, 1986, **51**, 503, and references cited therein; J. C. Gilberg and J. A. Kelly, *J. Org. Chem.*, 1986, **51**, 4485; F. L. Middlesworth, *J. Org. Chem.*, 1986, **51**, 5019.
57. T. Hayashi, Y. Ito and Y. Matsumoto, *J. Am. Chem. Soc.*, 1988, **110**, 5579; I. Fleming and P. E. J. Sanderson, *Tetrahedron Lett.*, 1987, **28**, 4229; K. Tamao, N. Ishida, T. Tanaka and M. Kunada, *Organometallics*, 1983, **2**, 1694; I. Fleming, D. J. Ager and S. K. Patel, *J. Chem. Soc., Perkin Trans. 1*, 1981, 2520.

58. M. K. O'Brien, A. J. Pearson, A. A. Pinkerton, W. Schmidt and K. Willman, *J. Am. Chem. Soc.*, 1989, **111**, 1499.
59. B. M. R. Bandara, A. J. Birch and L. F. Kelly, *J. Org. Chem.*, 1984, **49**, 2496.
60. A. J. Birch, L. F. Kelly and D. V. Weerasuria, *J. Org. Chem.*, 1988, **53**, 278.
61. A. J. Pearson and Y. S. Lai, *J. Chem. Soc., Chem. Commun.*, 1988, 442.
62. A. J. Pearson and T. Ray, *Tetrahedron Lett.*, 1986, **27**, 3111.
63. A. J. Birch and B. M. R. Bandara, *Tetrahedron Lett.*, 1980, **21**, 2981.
64. B. M. R. Bandara, A. J. Birch, L. F. Kelly and T. C. Khor, *Tetrahedron Lett.*, 1983, **24**, 2491; J. A. S. Howell and M. J. Thomas, *J. Chem. Soc., Dalton Trans.*, 1983, 1401; J. A. S. Howell and M. J. Thomas, *Organometallics*, 1985, **4**, 1054.
65. L. F. Kelly, A. S. Narula and A. J. Birch, *Tetrahedron Lett.*, 1979, 4107; L. F. Kelly, A. S. Narula and A. J. Birch, *Tetrahedron*, 1982, **38**, 1813; J. G. Atton, L. A. P. Kane-Maguire, P. A. Williams and G. R. Stephenson, *J. Organomet. Chem.*, 1982, **232**, C5; D. J. Evans, L. A. P. Kane-Maguire and S. B. Wild, *J. Organomet. Chem.*, 1982, **232**, C9; D. J. Evans and L. A. P. Kane-Maguire, *J. Organomet. Chem.*, 1982, **236**, C15.
66. A. J. Birch and G. R. Stephenson, *Tetrahedron Lett.*, 1981, **22**, 779; A. J. Birch, W. D. Raverty and G. R. Stephenson, *Tetrahedron Lett.*, 1980 **21**, 197; A. J. Birch, W. D. Raverty and G. R. Stephenson, *J. Org. Chem.*, 1981, **46**, 5166; A. J. Birch, W. D. Raverty and G. R. Stephenson, *Organometallics*, 1984, **3**, 1075.
67. M. Brookhart, G. O. Nelson, G. Scholes and R. A. Watson, *J. Chem. Soc., Chem. Commun.*, 1976, 195; C. C. Santini, J. Fischer, F. Mathey and A. Mitschler, *Inorg. Chem.*, 1981, **20**, 2848; M. Brookhart, C. R. Graham, G. O. Nelson and G. Scholes, *Ann. N.Y. Acad. Sci.*, 1977, **295**, 254.
68. A. J. Pearson and J. Yoon, *J. Chem. Soc., Chem. Commun.*, 1986, 1467; see also A. J. Pearson, S. L. Blystone and B. A. Roden, *Tetrahedron Lett.*, 1987, **28**, 2459; A. J. Pearson, S. L. Blystone, H. Nar, A. A. Pinkerton, B. A. Roden and J. Yoon, *J. Am. Chem. Soc.*, 1989, **111**, 134.
69. F. Haque, J. Miller, P. L. Pauson and J. B. P. Tripathi, *J. Chem. Soc. C*, 1971, 743.
70. E. D. Honig and D. A. Sweigart, *J. Chem. Soc., Chem. Commun.*, 1986, 691.
71. A. Mawby, P. J. C. Walker and R. J. Mawby, *J. Organomet. Chem.*, 1973, **55**, C39; G. Winkhaus, L. Pratt and G. Wilkinson, *J. Chem. Soc.*, 1961, 3807.
72. Y. K. Chung, H. S. Choi, D. A. Sweigart and N. G. Connelly, *J. Am. Chem. Soc.*, 1982, **104**, 4245; Y. K. Chung, D. A. Sweigart, N. G. Connelly and J. B. Sheridan, *J. Am. Chem. Soc.*, 1985, **107**, 2388.
73. Y. K. Chung and D. A. Sweigart, *J. Organomet. Chem.*, 1986, **308**, 223.
74. Y. K. Chung, E. D. Honig, W. T. Robinson, D. A. Sweigart, N. G. Connelly and S. D. Ittel, *Organometallics*, 1983, **2**, 1479.
75. R. P. Alexander, C. Morley and G. R. Stephenson, *J. Chem. Soc., Perkin Trans. 1*, 1988, 2069.
76. P. Powell and L. J. Russell, *J. Chem. Res. (S)*, 1978, 283; P. Powell, *J. Organomet. Chem.*, 1977, **165**, C43; B. F. G. Johnson, J. Lewis and D. J. Yarrow, *J. Chem. Soc., Dalton Trans.*, 1972, 2084; J. Evans, B. F. G. Johnson and J. Lewis, *J. Chem. Soc., Dalton Trans.*, 1972, 2868.
77. E. D. Sternberg and K. P. C. Vollhardt, *J. Am. Chem. Soc.*, 1980, **102**, 4841.

3.5

Carbon Electrophiles with Dienes and Polyenes Promoted by Transition Metals

MAURICE BROOKHART, ANTHONY F. VOLPE, JR. and JAEYON YOON

University of North Carolina, Chapel Hill, NC, USA

3.5.1 INTRODUCTION

Numerous synthetically useful carbon–carbon bond-forming reactions are based on the fact that unsaturated hydrocarbon ligands bound to electrophilic transition metal moieties are activated toward addition of nucleophiles. Normally the metal moiety in such complexes is a neutral or cationic metal carbonyl group. Prominent and well-studied examples include $[Cr(arene)(CO)_3]$ complexes (covered in Chapter 2.4, this volume),[1] $[Fe(dienyl)(CO)_3]^+$ complexes (covered in Chapter 3.4, this volume),[2] $[FeCp(CO)_2(alkene)]^+$ complexes[3] and $[M(CO)_n(diene)]$ complexes.[4]

The complementary approach, activation of unsaturated hydrocarbons toward electrophilic attack by complexation with electron-rich metal fragments, has seen limited investigation. Although there are certainly opportunities in this area which have not been exploited, the electrophilic reactions present a more complex problem relative to nucleophilic addition. For example, consider the nucleophilic *versus* electrophilic addition to a terminal carbon of a saturated 18-electron metal–diene complex. Nucleophilic addition generates a stable 18-electron saturated π-allyl complex. In contrast, electrophilic addition at carbon results in removal of two valence electrons from the metal and formation of an unstable π-allyl unsaturated 16-electron complex (Scheme 1).

The 16-electron species invariably must be stabilized by addition of an external ligand or by internal coordination. For example, as will be discussed later, acylation of a $[Fe(diene)(CO)_3]$ complex yields a

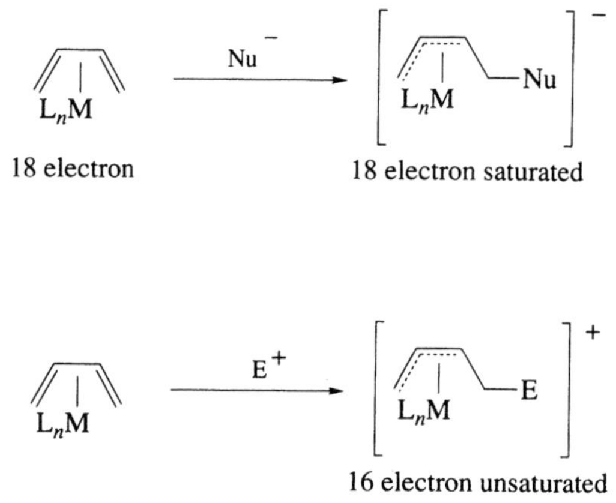

Nu⁻ → 18 electron → 18 electron saturated

E⁺ → 16 electron unsaturated

Scheme 1

carbonyl-coordinated complex, while alkylation of a $[Mn(diene)(CO)_3]^-$ complex gives an 18-electron agostic species (Scheme 2).

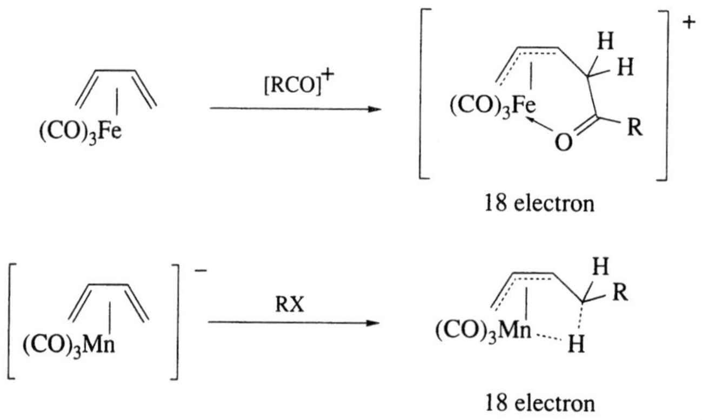

$(CO)_3Fe$ —$[RCO]^+$→ $(CO)_3Fe$ 18 electron

$(CO)_3Mn$ —RX→ $(CO)_3Mn$ 18 electron

Scheme 2

A common observation is that electrophiles add readily to complexes containing uncomplexed double bonds (normally at the uncomplexed double bond) to give 18-electron species as the initially generated complex. For example, acylation of $[Fe(cycloheptatriene)(CO)_3]$ yields an 18-electron dienyl complex, as shown in equation (1).

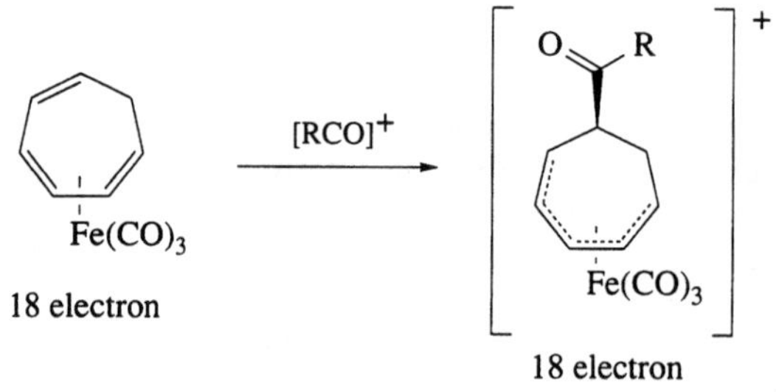

$Fe(CO)_3$ 18 electron —$[RCO]^+$→ $Fe(CO)_3$ 18 electron (1)

There are a few examples of additions of carbon electrophiles to metal complexes of allyl[5] and dienyl[6] moieties and arenes[7] which suggest many possible future extensions. However, the vast majority of systems examined involve additions of carbon electrophiles to electron-rich metal–diene complexes. This chapter will present a general discussion and several examples of such additions. Section 3.5.2 will examine simple η^4-diene complexes, while Section 3.5.3 will treat polyalkene complexes containing an η^4-diene unit and one or more uncomplexed double bonds.

3.5.2 REACTIONS OF η⁴-DIENE TRANSITION METAL COMPLEXES WITH ELECTROPHILES

3.5.2.1 η⁴-Diene Iron Tricarbonyl Complexes

Electrophilic substitution of [Fe(diene)(CO)₃] complexes was first described by Ecke who reported that acetylation of [Fe(butadiene)(CO)₃], (**1**), gives 1- and 2-acetyl derivatives.[8] Subsequent studies showed that acylation occurred only at the terminal carbons,[9–12] to give the *trans* and *cis* isomers (**2**) and (**3**), respectively (equation 2).

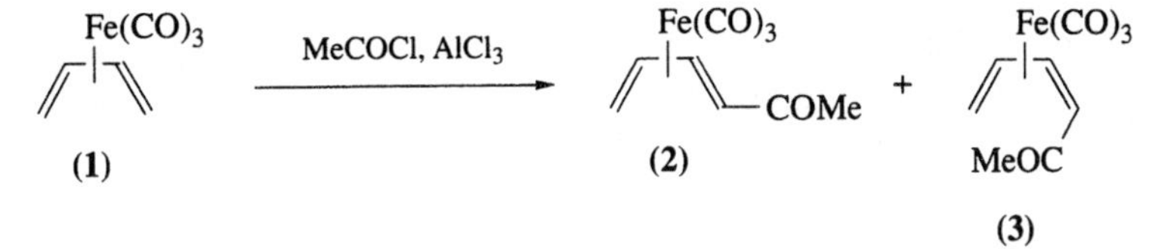

Fe(CO)₃ (**1**) —MeCOCl, AlCl₃→ Fe(CO)₃ COMe (**2**) + Fe(CO)₃ MeOC (**3**) (2)

The acylation is thought to occur as shown in Scheme 3. The π-allyl intermediate (**5**) was isolated in 86% yield and characterized by X-ray analysis.[13] The NMR spectrum and crystal structure of (**5**) indicate that only the *anti*-allyl structure is formed; no other positional isomers were detected. The iron atom in (**5**) obtains a closed-shell electron configuration by intramolecular σ-donation of a lone pair of electrons from the acyl oxygen atom, which is confirmed by a lowering the acyl CO bond order (ν_{CO} 1637 cm^{-1}).[12] There is no direct evidence for formation of (**4**); its intermediacy has been suggested primarily based on the *endo* stereochemistry of the acylation reaction (see below).

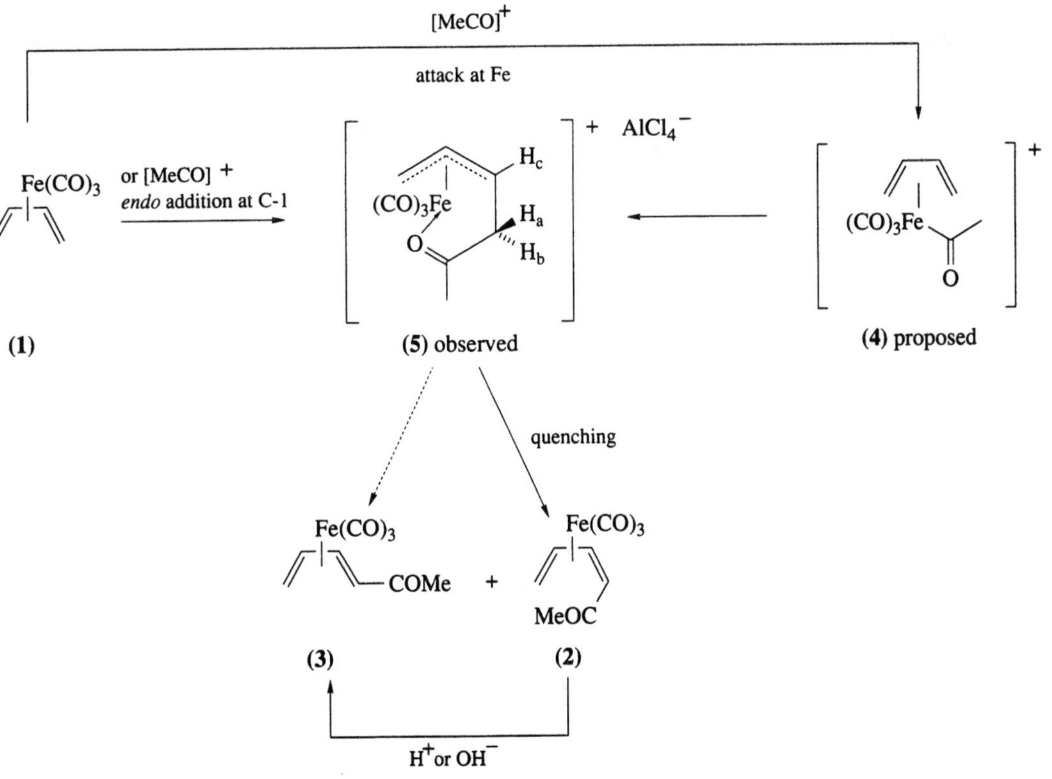

[MeCO]⁺
attack at Fe

Fe(CO)₃ (**1**) —or [MeCO]⁺ *endo* addition at C-1→ [(CO)₃Fe, O, Hc, Ha, Hb]⁺ AlCl₄⁻ (**5**) observed ← [(CO)₃Fe, O]⁺ (**4**) proposed

quenching

Fe(CO)₃ COMe (**3**) + Fe(CO)₃ MeOC (**2**)

H⁺ or OH⁻

Scheme 3

The kinetic product of deprotonation of (**5**) is the *cis* isomer (**2**); however, (**2**) is readily isomerized to the thermodynamically more stable *trans* isomer (**3**) under both acidic and basic conditions. Thus acylation reactions can often lead to a mixture of isomers and product ratios are dependent on the precise quenching conditions. Convincing evidence for *endo* addition of [RCO]⁺ to iron–diene complexes first came from the X-ray crystal structure determination of the cationic intermediate (**7**) formed from acylation of the 1,4-dimethylbutadiene complex (**6**; equation 3).[14]

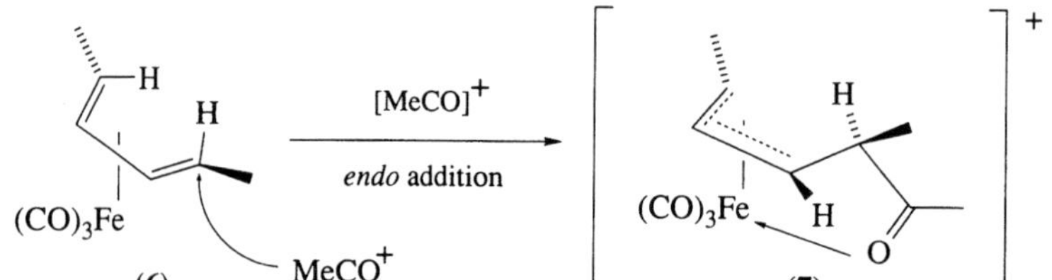

(3)

Using a competition method, Lillya *et al.*[15,16] obtained relative reactivities for a series of substituted diene–tricarbonyliron compounds toward the methyloxocarbonium tetrachloroaluminate ion pair in dichloromethane. The results are qualitatively summarized in Scheme 4.

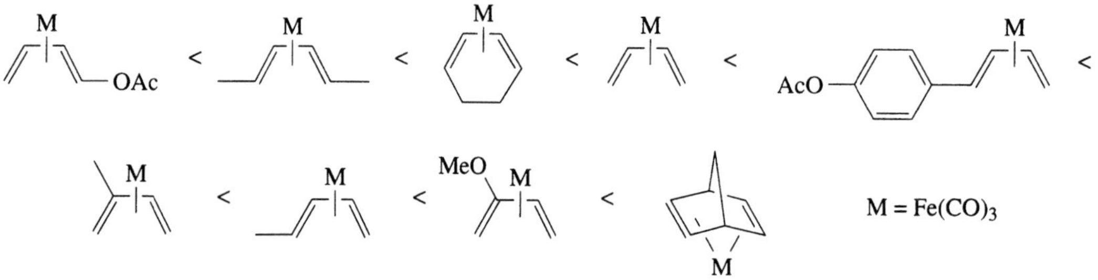

M = $Fe(CO)_3$

Scheme 4

Quantitative relative partial rate factors for acetylation of several of these iron–diene complexes were also obtained and are summarized in Scheme 5.[15,16]

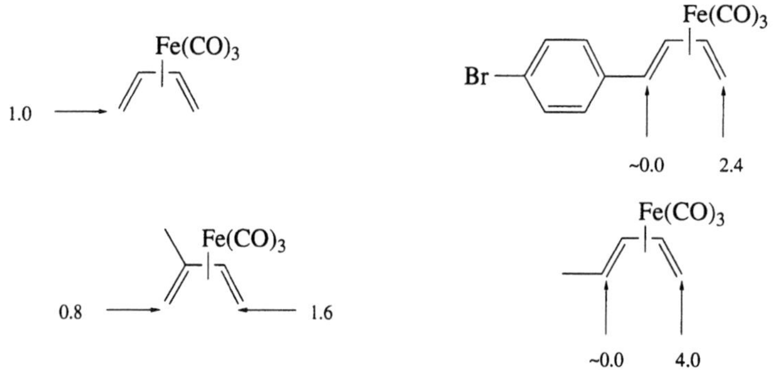

Scheme 5 Relative partial rate factors for acylation with acetyl chloride/$AlCl_3$ in dichloromethane at 25 °C

These results show that substitution of a terminal carbon decreases its reactivity dramatically presumably due to a steric effect. Substituent effects are moderate; for example, a 1-methyl group increases the rate of substitution at C-4 by a factor of 4. A 2-methyl group exerts a small directing effect to C-4;[16] however, substitution of more strongly electron-donating groups at C-2 (*e.g.* $SiEt_3$, $SiPr^i_3$) results in higher regioselectivity for substitution at C-4 (see below).[17]

As mentioned earlier, the *cis:trans* ratios of acylation products are highly dependent on the method of quenching.[18] Quenching with cold 28% aqueous ammonia gives high yields of *cis*-dienone complex. In contrast, quenching with potassium *t*-butoxide in *t*-butanol gives high yields of *trans*-dienone (equation 4 and Table 1).

Cis isomers can be isomerized to *trans* isomers by various methods. Illustrative examples are shown in equations (5) and (6), Tables 2 and 3.

Birch and Pearson[22] have studied electrophilic substitution of a triphenylphosphine-substituted system, $[Fe(cyclohexadiene)(CO)_2PPh_3]$, (**8**; equation 7). Several features of their results are instructive. First, substitution of CO by the better σ-donor, poorer π-acceptor PPh_3 ligand renders the complex more reac-

Table 1 Variation of *cis:trans* Isomer Ratios with Quenching Method

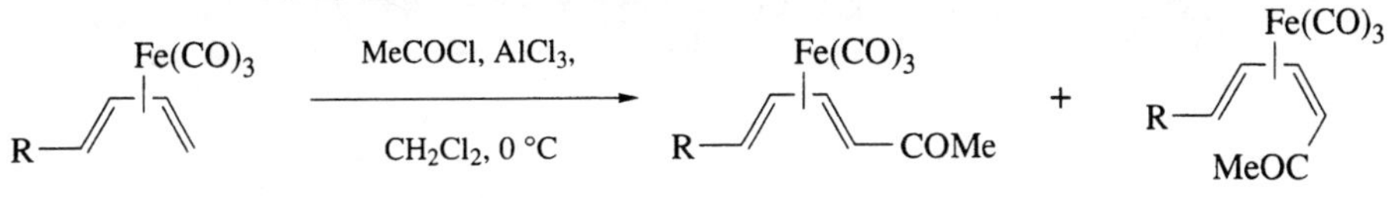

(4)

A **B**

R	*Quenching method*	*A:B*	*Yield (%)*
H	Aq. ammonia (28%)	1:0	86
Me	Aq. ammonia (28%)	1:0	84
p-BrC_6H_4	Aq. ammonia (28%)	1:0	90
p-$MeCOC_6H_4$	K_2CO_3 (5%)	0:1	90

Table 2 Isomerization of *cis* Isomers to *trans* Isomers Under Basic Conditions

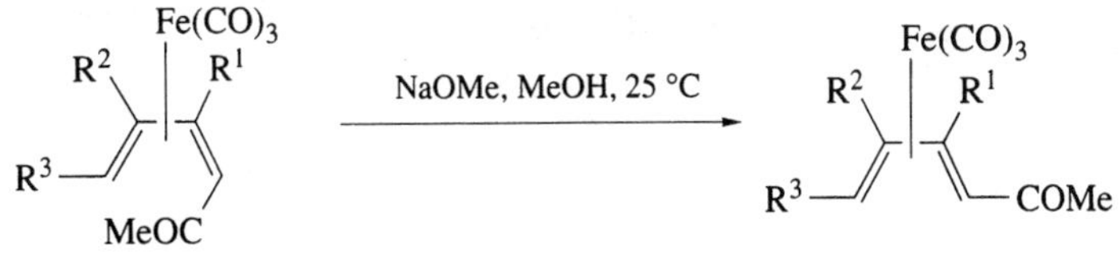

(5)

R^1	R^2	R^3	*Yield (%)*	*Ref.*
H	H	H	69	18
H	H	Me	71	18
H	H	*p*-BrC_6H_4	79	18
H	OMe	H	60	18
Me	H	Me	75	18
H	H	$SiMe_3$	85	19
Me	H	$SiMe_3$	95	19
MeCO	$SiEt_3$	H	92	20

Table 3 Isomerization of *cis* Isomers to *trans* Isomers in the Presence of Acetyl Chloride

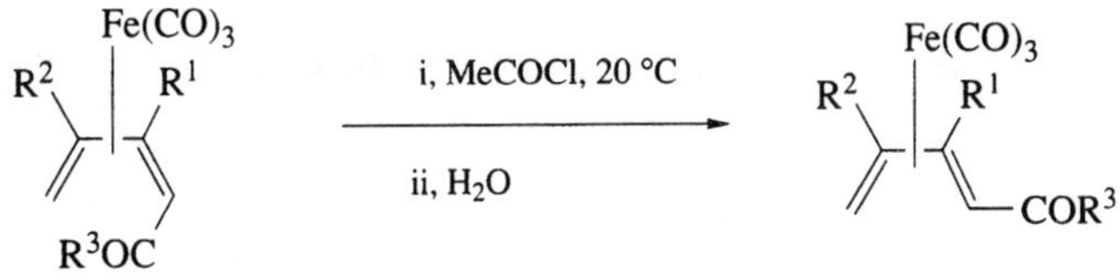

(6)

R^1	R^2	R^3	*Yield (%)*	*Ref.*
H	Me	Me	89	21
Me	Me	Me	99	21
Me	H	Me	89	21
H	$SiEt_3$	Me	95	20
H	$SiPr^i_3$	Me	95	20
$SiEt_3$	H	Ph	68	17
H	$SiEt_3$	Ph	68	17
$SiEt_3$	H	Bu^t	73	17
H	$SiEt_3$	Bu^t	73	17
$SiEt_3$	H	Pr^i	96	17
H	$SiEt_3$	Pr^i	96	17

tive toward electrophiles. Thus, substitutions can be carried out under milder conditions and in better yields.

Second, acylation reactions of (**8**) yield only isomer (**10**); (**11**) is not formed. This observation can be explained by considering intermediate (**9**; Scheme 6). The coordination between the acyl oxygen and the

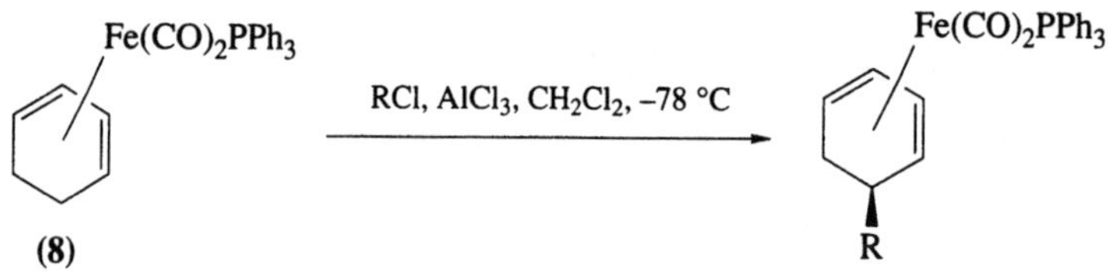

Fe(CO)$_2$PPh$_3$ (8) —RCl, AlCl$_3$, CH$_2$Cl$_2$, −78 °C→ Fe(CO)$_2$PPh$_3$ ... R (7)

R = COMe, 96%; R = COPh, 50%; R = CH$_2$OMe, 65%

iron atom in cationic intermediate (**10**) would result in the carbonyl π-orbital being orthogonal to a developing carbanion lone pair orbital at the α-position, thereby making the α-hydrogen, H_{1X}, less acidic than the alternative methylene hydrogen, H_{5N}.

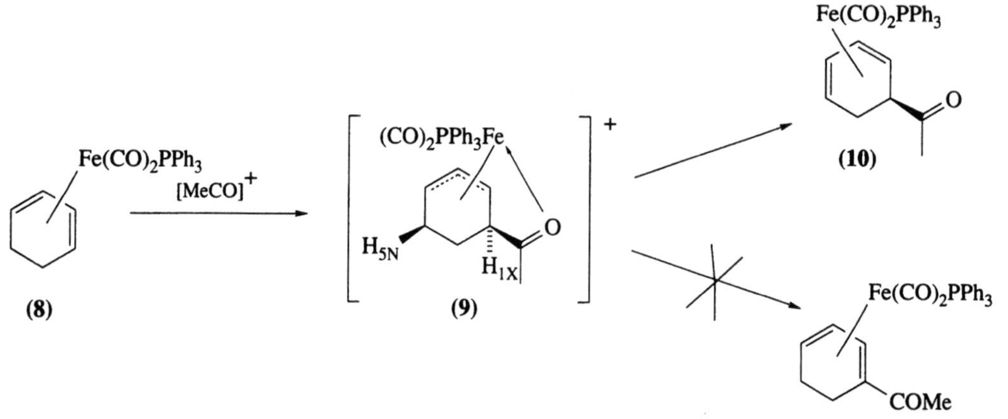

(8) —[MeCO]$^+$→ [(9)]$^+$ → (10); (9) ↛ (11)

Scheme 6

All of the examples cited above involve monoacylation of iron–diene complexes. Frank-Neumann has recently demonstrated that monoacylated derivatives are subject to a second acylation and that bis-1,4-diacylated complexes could be obtained in moderate to good yields. Examples are given in equations (8) and (9) and Table 4.[20,23]

Table 4 Diacylation Reactions of [Fe(2-Triethylsilylbutadiene)(CO)$_3$]

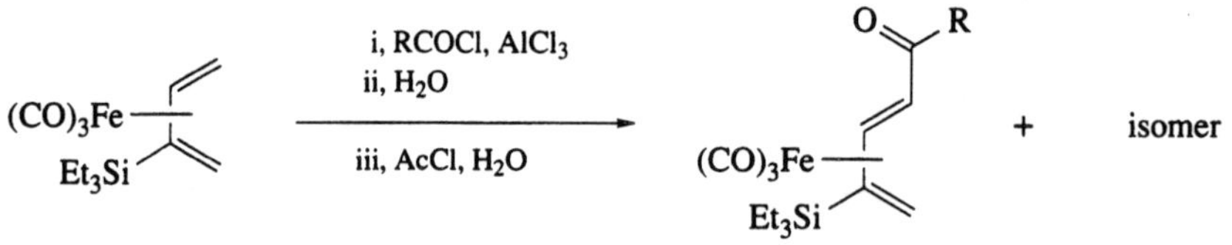

(CO)$_3$Fe, Et$_3$Si —i, RCOCl, AlCl$_3$; ii, H$_2$O; iii, AcCl, H$_2$O→ (CO)$_3$Fe, Et$_3$Si, O, R + isomer (8)

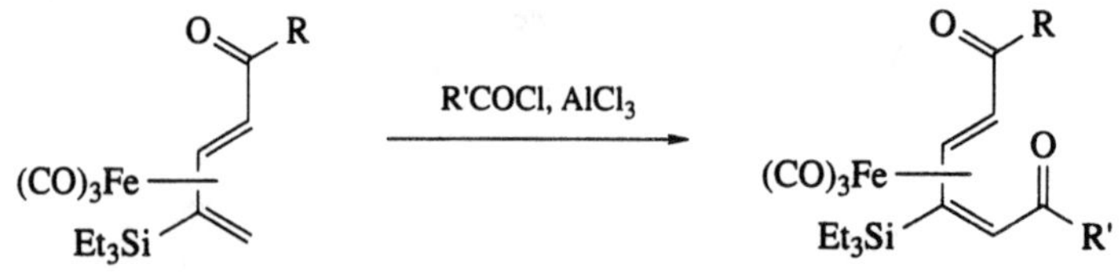

(CO)$_3$Fe, Et$_3$Si, O, R —R'COCl, AlCl$_3$→ (CO)$_3$Fe, Et$_3$Si, O, R, O, R' (9)

R	*R'*	*Yield (%)*
Me	Pr^i	50
Me	(CH$_2$)$_4$CO$_2$Me	76
Pr^i	Me	40
Bu^t	Me	32
Ph	Me	55
(CH$_2$)$_7$COMe	Et	90

Formylation of diene–tricarbonyliron complexes has also been reported by Lillya *et al.*[18,24] Examples are given in Scheme 7.

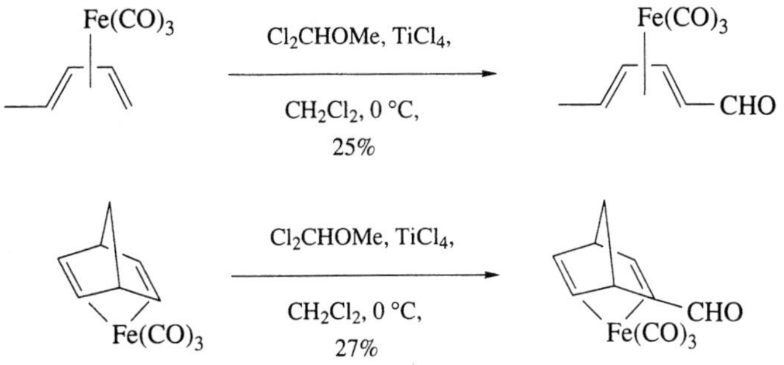

$Fe(CO)_3$ Cl_2CHOMe, $TiCl_4$, CH_2Cl_2, 0 °C, 25% $Fe(CO)_3$ CHO

Cl_2CHOMe, $TiCl_4$, CH_2Cl_2, 0 °C, 27% $Fe(CO)_3$ CHO $Fe(CO)_3$

Scheme 7

In a series of classic studies, Pettit *et al.* reported the synthesis of (cyclobutadiene)iron tricarbonyl together with a variety of electrophilic substitution reactions of this 'aromatic' system (Scheme 8).[25–27]

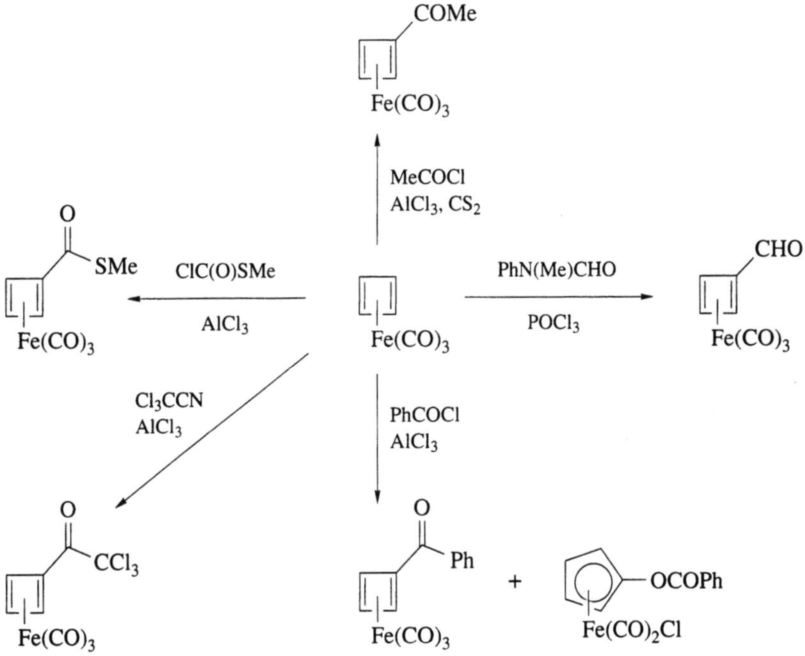

COMe $Fe(CO)_3$ MeCOCl $AlCl_3$, CS_2 O SMe ClC(O)SMe $AlCl_3$ $Fe(CO)_3$ $Fe(CO)_3$ PhN(Me)CHO $POCl_3$ CHO $Fe(CO)_3$ Cl_3CCN $AlCl_3$ PhCOCl $AlCl_3$ O CCl_3 $Fe(CO)_3$ O Ph $Fe(CO)_3$ + OCOPh $Fe(CO)_2Cl$

Scheme 8

There are numerous methods available for the cleavage of the iron tricarbonyl group from the diene moiety and recovery of the organic ligand in good yields. The most frequently used methods involve oxidative procedures including reaction of the diene complexes with Me_3NO, H_2O_2/OH^- and Fe^{III} and Ce^{IV} salts. Reactions of iron–diene complexes with $LiAlH_4$ normally leads to cleavage and complete reduction of the diene unit.[28,29]

The availability of such procedures coupled with several straightforward methods for synthesis of the diene complexes make these species attractive for use in organic synthesis. Several illustrative examples follow, which include acylation (or diacylation) of iron–diene complexes followed by cleavage and recovery of the free organic ligand.

Frank-Neumann has described a series of double acylations followed by oxidative cleavage reactions to yield 1,6-dione-2,4-dienes (Schemes 9 and 10).[21]

In an elegant synthetic application of iron–diene complexes, Knox[30] has reported acylation of a series of 1-alkyl-substituted diene complexes which after cleavage, reduction and esterification give a series of moth pheromones (Scheme 11).

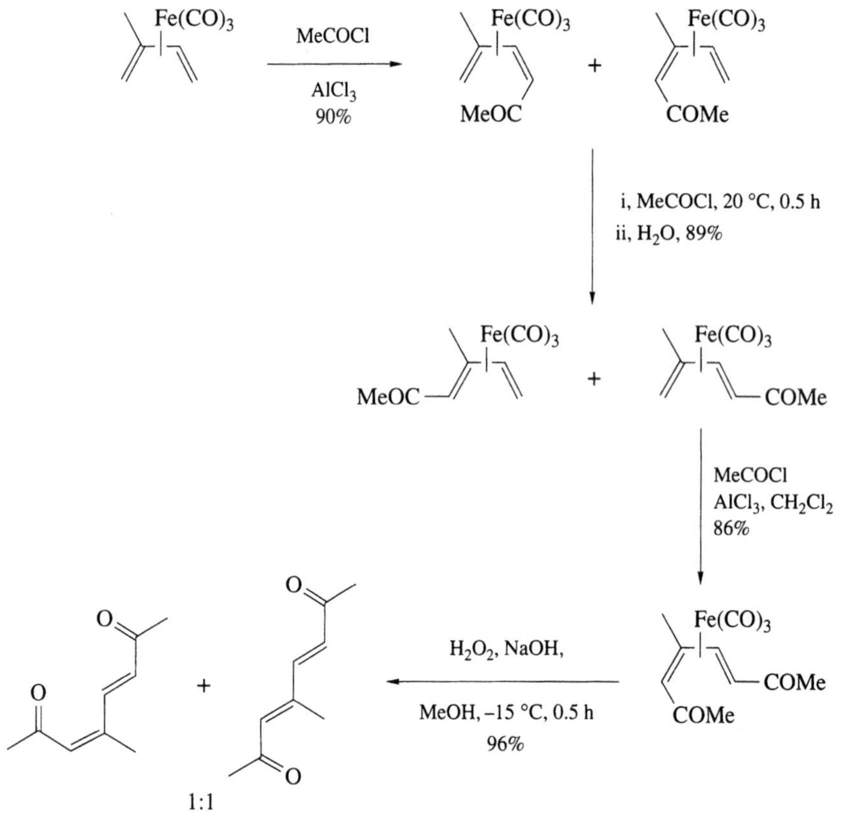

$Fe(CO)_3$ MeCOCl $AlCl_3$ 90% MeOC + COMe

i, MeCOCl, 20 °C, 0.5 h
ii, H_2O, 89%

MeOC + COMe

MeCOCl
$AlCl_3$, CH_2Cl_2
86%

H_2O_2, NaOH, MeOH, –15 °C, 0.5 h 96%

1:1

Scheme 9

i 90% MeOC ii, iii 99%

COMe iv 98% COMe COMe v, vi 82%

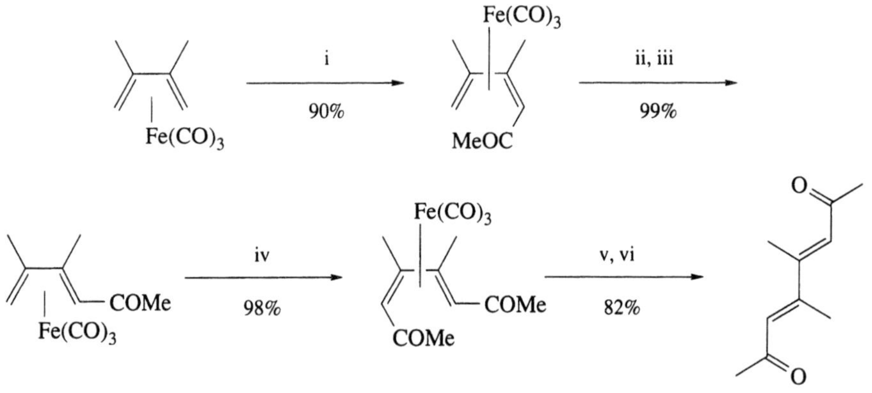

i, MeCOCl, $AlCl_3$; ii, MeCOCl, 20 °C, 5 h; iii, H_2O; iv, MeCOCl, $AlCl_3$;
v, NaOMe, MeOH, 20 °C, 15 h; vi, H_2O_2, NaOH, MeOH, -15 °C, 0.5 h

Scheme 10

Cleavage of acylated diene–iron complexes with $LiAlH_4$ is illustrated by the following reports of Nesmeyanov and Anisimov (Scheme 12).[28,29]

Pearson[31] has reported an example of intramolecular alkylation under CO to yield an allyliron tetracarbonyl cation. Hydride reduction yields the alkene (Scheme 13).

3.5.2.2 η^4-Diene Manganese Tricarbonyl Anions

$[Mn(diene)(CO)_3]^-$ complexes are isoelectronic with $[Fe(diene)(CO)_3]$ complexes but owing to their anionic character are much more reactive toward electrophiles than the neutral iron analogs. The first examples of these species to be reported were $[Mn(cyclohexadiene)(CO)_3]^-$ complexes prepared *via*

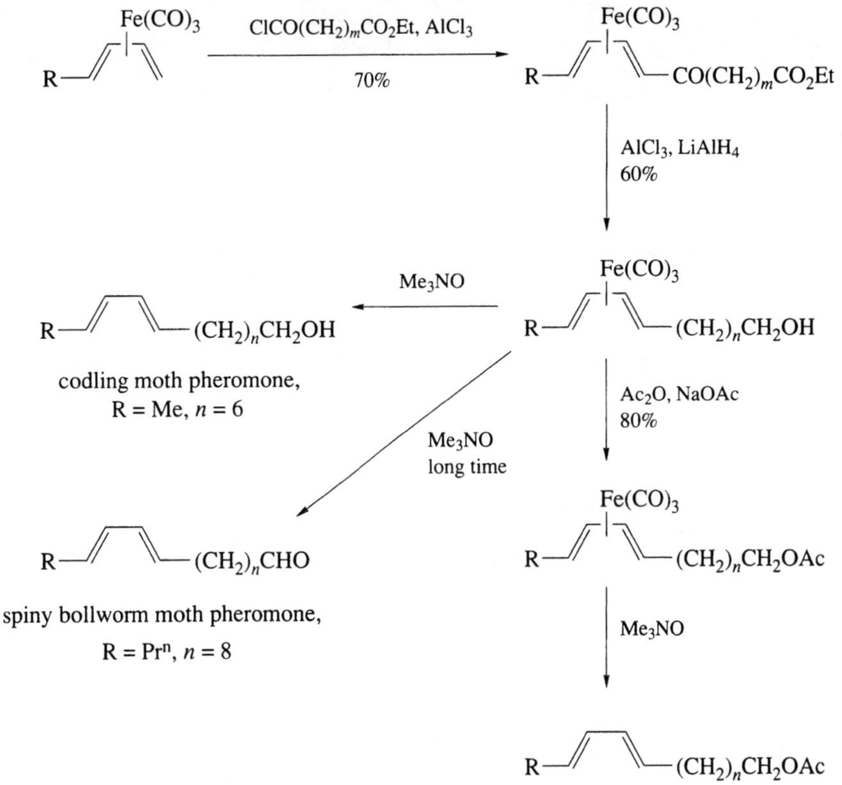

$Fe(CO)_3$ diene complexes: R–diene–$Fe(CO)_3$ → $ClCO(CH_2)_mCO_2Et$, $AlCl_3$ (70%) → R–diene($Fe(CO)_3$)–$CO(CH_2)_mCO_2Et$ → $AlCl_3$, $LiAlH_4$ (60%) → R–diene($Fe(CO)_3$)–$(CH_2)_nCH_2OH$

→ Me_3NO → R–diene–$(CH_2)_nCH_2OH$

codling moth pheromone, R = Me, *n* = 6

→ Me_3NO, long time → R–diene–$(CH_2)_nCHO$

spiny bollworm moth pheromone, R = Pr[n], *n* = 8

→ Ac_2O, NaOAc (80%) → R–diene($Fe(CO)_3$)–$(CH_2)_nCH_2OAc$ → Me_3NO → R–diene–$(CH_2)_nCH_2OAc$

R	*m*	*n*	*Moth species*
H	6	7	red bollworm moth
Me	5	6	pea moth; pitch pine tip moth
Et	6	7	light-brown apple moth

Scheme 11

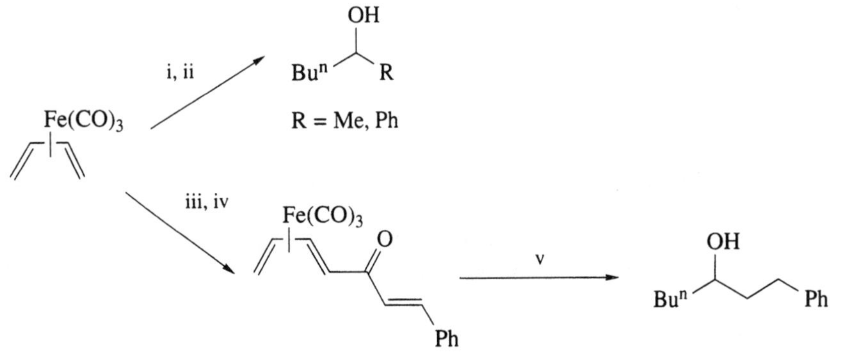

$Fe(CO)_3$(butadiene) → i, ii → $Bu^nCH(OH)R$, R = Me, Ph

$Fe(CO)_3$(butadiene) → iii, iv → $Fe(CO)_3$(dienyl–CO–CH=CH–Ph) → v → $Bu^nCH(OH)CH_2CH_2Ph$ (OH, Bu[n], Ph)

i, $RCOCl$, $AlCl_3$; ii, $LiAlH_4$, THF, reflux; iii, $MeCOCl$, $AlCl_3$; iv, PhCHO, NaOH; v, $LiAlH_4$, THF, reflux

Scheme 12

double hydride reduction of $[Mn(arene)(CO)_3]^+$ complexes (Scheme 14).[32] An acyclic example (**12**) was prepared by thermolysis of [1,1-dimethylallyl$Mn(CO)_4$], (**13**), to give agostic complex (**14**) followed by deprotonation with KH (Scheme 14).[33]

A more general route to these anionic manganese complexes has recently been described[34,35] and involves Red-Al reduction of $[Mn(2\text{-methylallyl})(CO)_4]$ to yield $[Mn(butene)(CO)_4]^-$, which reacts with a variety of both cyclic and acyclic dienes or polyenes to give $[Mn(\eta^4\text{-diene})(CO)_3]^-$ or $[Mn(\eta^4\text{-polyene})(CO)_3]^-$ complexes in moderate to good isolated yields (equation 10).

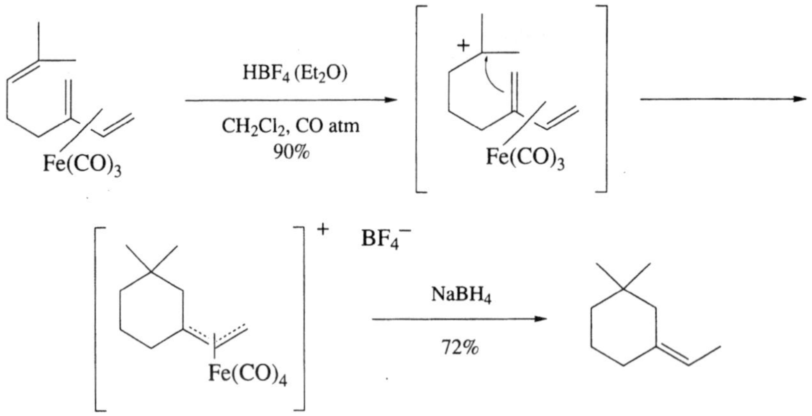

Scheme 13

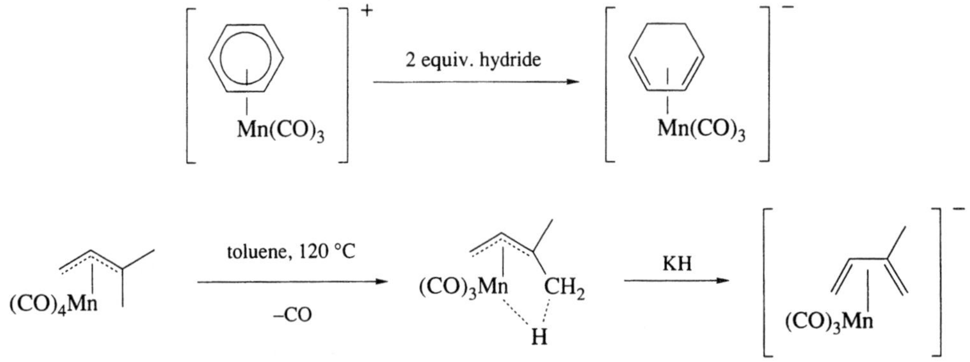

Scheme 14

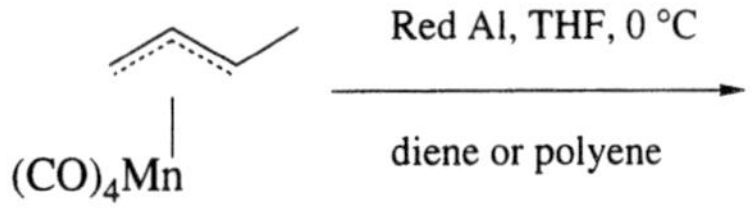

$$(CO)_4Mn(\text{allyl}) \xrightarrow[\text{diene or polyene}]{\text{Red Al, THF, } 0\ ^\circ C} [Mn(\eta^4\text{-diene})(CO)_3]^- \text{ or } [Mn(\eta^4\text{-polyene})(CO)_3]^- \qquad (10)$$

As indicated above, the anionic manganese species are quite nucleophilic and much more reactive toward electrophiles than are the neutral iron analogs. Although little has been done with regard to applying these reactions in synthesis, their potential is illustrated by the reactions of these complexes with methylating reagents. $[Mn(cyclohexadiene)(CO)_3]^-$ reacts with methyl iodide to give the equilibrating mixture of agostic isomers (**15a** and **15b**) possessing an *endo* methyl group (Scheme 15). Based on *endo* addition of the electrophile, the reaction likely proceeds *via* a methylmanganese intermediate (**16**). Following methyl migration, the 16-electron allyl intermediate aquires an 18-electron configuration *via* bridging to an *endo* C—H bond to give initially agostic isomer (**15a**) which can reversibly rearrange to isomer (**15b**). The isomers (**15a** and **15b**) can be deprotonated with KH to yield the anionic *endo*-methylcyclohexadiene complex (**17**). This anion can be methylated a second time or can be readily oxidatively cleaved with O_2 to give 5-methyl-1,3-cyclohexadiene.[32] The conversion of 1,3-cyclohexadiene to 5-methyl-1,3-cyclohexadiene thus represents a stepwise procedure for electrophilic substitution of cyclohexadiene *via* nucleophilic anionic diene complex (**18**; Scheme 15).

Reactions of acyclic derivatives with carbon electrophiles have also been examined.[33,34] An illustrative reaction involving methylation of the unsubstituted complex $[Mn(\eta^4\text{-butadiene})(CO)_3]^-$, (**19**), is shown in Scheme 16. Again, the reaction is presumed to occur *via* a methylmanganese species (**20**) and after methyl migration the unsaturated metal center is stabilized by formation of a Mn---H---C bridge (isomers **21a** and **21b**). Deprotonation of equilibrating (**21a** and **21b**) yields the $[Mn(1\text{-methylbutadiene})(CO)_3]^-$ complex (**22**), which has exclusively *trans* stereochemistry.[34] This sequence represents alkylation of the terminal carbon of butadiene and complements the iron carbonyl chemistry, where terminal acylation has been achieved as described above. Unpublished results indicate that a second methylation of (**22**) occurs

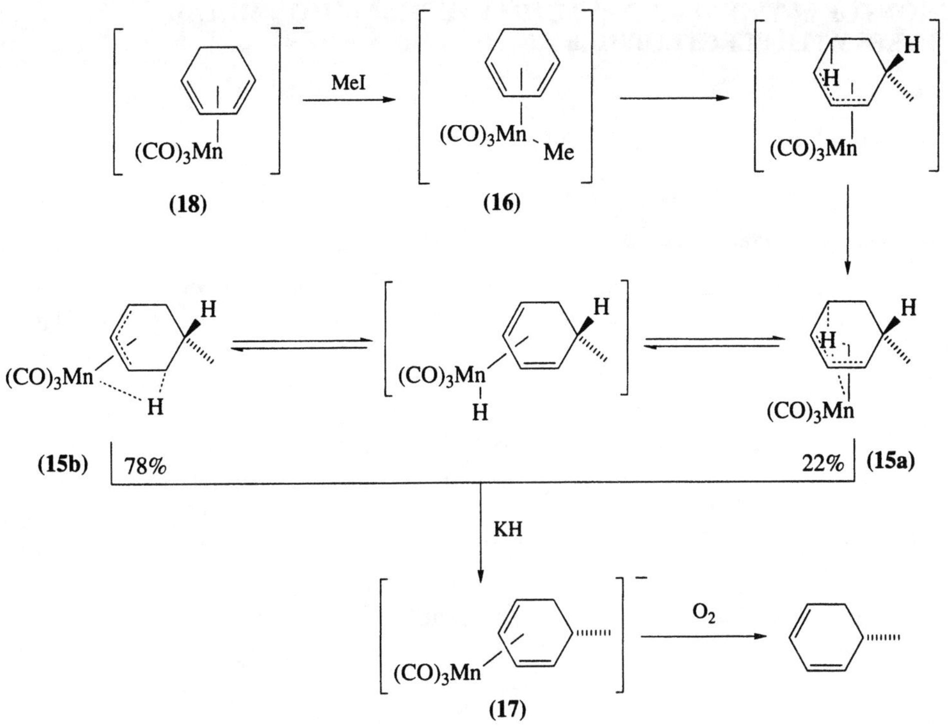

(CO)$_3$Mn (18) MeI (CO)$_3$Mn Me (16) (CO)$_3$Mn H H

(CO)$_3$Mn H H (15b) 78% (CO)$_3$Mn H H (CO)$_3$Mn H H 22% (15a)

KH

(CO)$_3$Mn (17) O_2

Scheme 15

exclusively at C-4.[34b] The basic reactivity patterns described here suggests that, using these manganese reagents, procedures may be available for multiple alkylation of butadiene with control of regio- and stereo-chemistry.

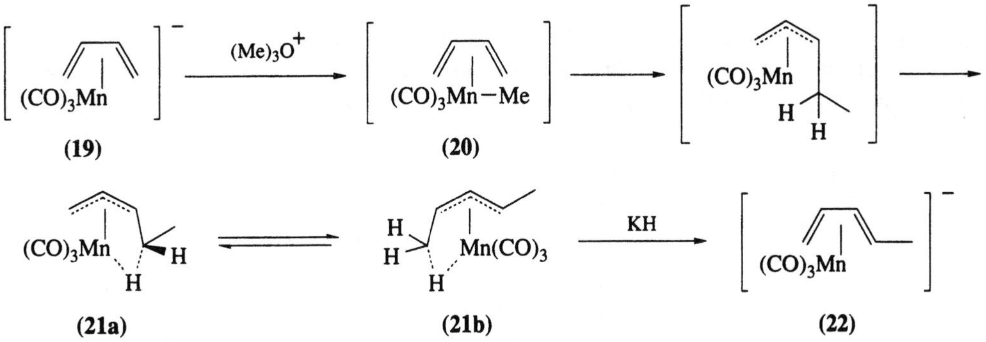

(CO)$_3$Mn (19) $(Me)_3O^+$ (CO)$_3$Mn—Me (20) (CO)$_3$Mn H H

(CO)$_3$Mn H H (21a) H H H Mn(CO)$_3$ (21b) KH (CO)$_3$Mn (22)

Scheme 16

3.5.3 REACTIONS OF η^4-TRIENE AND η^4-TETRAENE TRANSITION METAL COMPLEXES WITH ELECTROPHILES

3.5.3.1 η^4-Triene and η^4-Tetraene Iron Tricarbonyl Complexes

3.5.3.1.1 Acylation and alkylation reactions

[Fe(η^4-cyclooctatetraene)(CO)$_3$] can be formylated by treatment with $POCl_3$ in DMF.[36,37] The resulting aldehyde can then be cleaved from the metal with Ce^{IV} ion or further elaborated using standard procedures (Scheme 17).

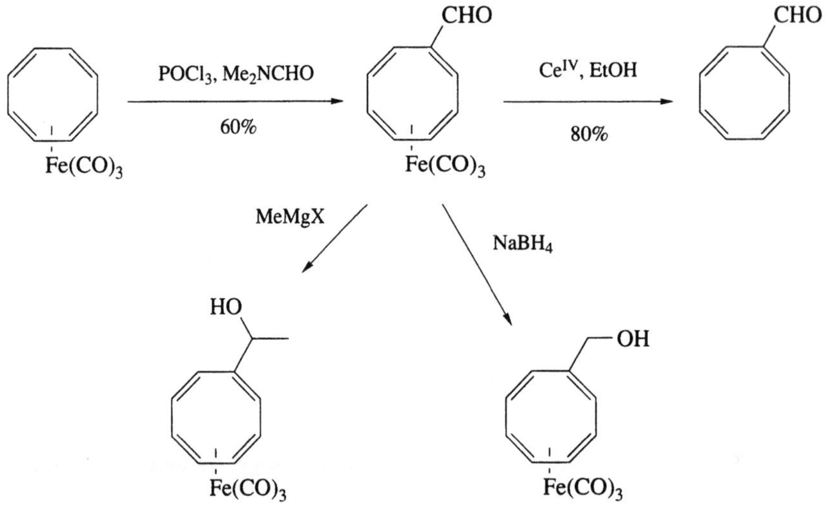

CHO CHO

POCl$_3$, Me$_2$NCHO Ce^{IV}, EtOH

60% 80%

Fe(CO)$_3$ Fe(CO)$_3$

MeMgX NaBH$_4$

HO OH

Fe(CO)$_3$ Fe(CO)$_3$

Scheme 17

A general mechanism for the formylation of [Fe(cyclooctatetraene)(CO)$_3$] is depicted in equation (11). The first step is electrophilic attack on the ring to form a cationic η^5-dienyl intermediate **(23)**. Subsequent proton loss yields the substituted product.

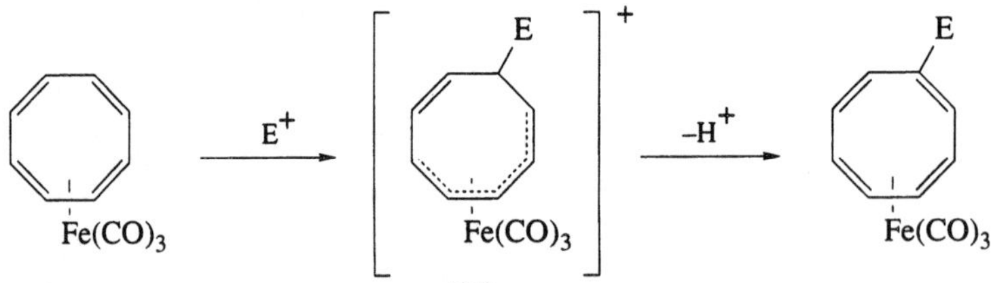

E + E

E$^+$ -H$^+$ (11)

Fe(CO)$_3$ Fe(CO)$_3$ Fe(CO)$_3$

(23)

[Fe(cyclooctatetraene)(CO)$_3$] reacts under Friedel–Crafts acetylation conditions (acetyl chloride/$AlCl_3$), but only low yields of the substituted product (**24**) are formed; the major product being the bicyclic cation (**25**) (equation 12).[36–38]

Bicyclic cation (**25**) is likely formed by the route shown in Scheme 18. The [5.1.0] cation (**26**) has been intercepted with methoxide to give (**27**), which can be oxidatively cleaved to give the free bicyclic diene (**28**).[37] The bicyclo[3.2.1] complex (**25**) reacts with a variety of anionic nucleophiles to give neutral [3.2.1] derivatives.[38]

[Fe(cycloheptatriene)(CO)$_3$] can be formylated in the same manner as [Fe(cyclooctatetraene)(CO)$_3$] and likely proceeds through the same mechanism. Reaction of [Fe(cycloheptatriene)(CO)$_3$] with $POCl_3$ in DMF gives the corresponding aldehyde (70% yield), which can be converted to the primary alcohol with $NaBH_4$ or secondary alcohols using Grignard reagents (Scheme 19).[39,40]

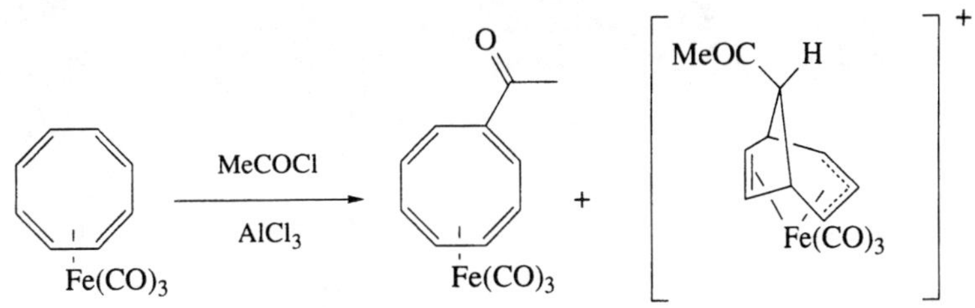

MeCOCl, $AlCl_3$ (12)

(24) 5% **(25)** 28%

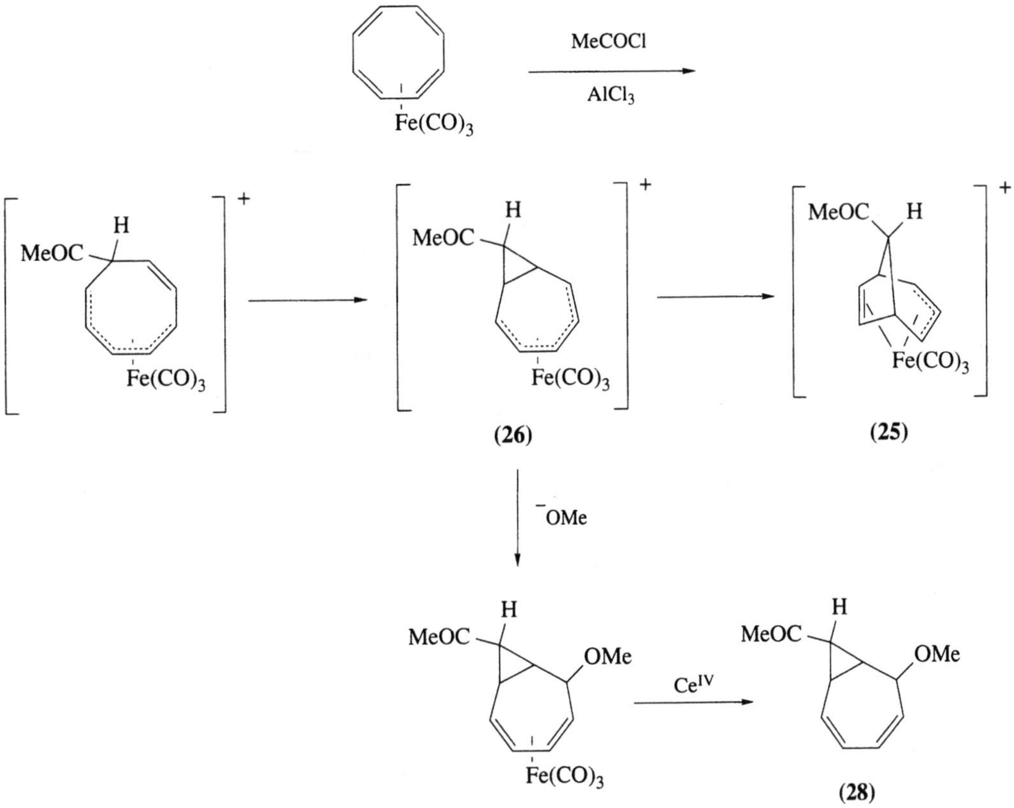

(26) **(25)** **(27)** **(28)**

Scheme 18

The reaction of [Fe(cycloheptatriene)$(CO)_3$] with acyl tetrafluoroborates yields only cationic complex (**29**), which can then be converted to the acylated product (**30**) in high yield *via* the methoxide adduct (**31**; Scheme 20).[40]

Complex (**29**) results from *exo* addition of $[RCO]^+$, and when treated with Et_3N does not yield the neutral acyl derivatives. In contrast the *endo* salt, (**32**) does deprotonate readily to give the acyl derivative (**30**; Scheme 21). These observations suggest *endo* (**32**) may be the intermediate responsible for formation of (**30**) under Friedel–Crafts conditions.

When various [Fe(heptafulvene)$(CO)_3$] complexes are treated with $POCl_3$ in DMF formylated products are produced.[41] The reactivity of the iron carbonyl complex is markedly different than the reactivity of uncoordinated heptafulvenes which react with electrophiles to form the tropylium ion.

Azepines do not undergo electrophilic substitution reactions and introduction of substituents at the 3- or 4-position is a difficult problem. Acylation at C-3 can be achieved *via* the iron tricarbonyl complex as shown in Scheme 22.[42] The *N*-ethoxycarbonyl group of the acylated product can be removed with methoxide methanol and the resulting 1*H* compound can then be methylated.

Franck-Neumann has reported the synthesis of naturally occurring tropolones β-thujaplicin and β-dolabrin *via* the readily available [Fe(tropone)$(CO)_3$] as shown in Scheme 23.[43] A key step in the sequence is regiospecific acetylation of the [Fe(tropone)$(CO)_3$] complex, a reaction which cannot be carried out on

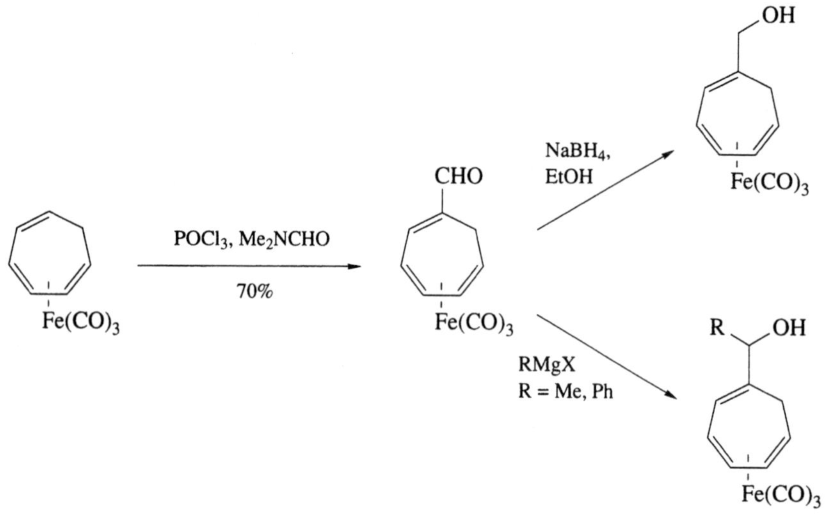

$POCl_3$, Me_2NCHO

70%

CHO

$Fe(CO)_3$

$NaBH_4$, EtOH

OH

RMgX
R = Me, Ph

R OH

Scheme 19

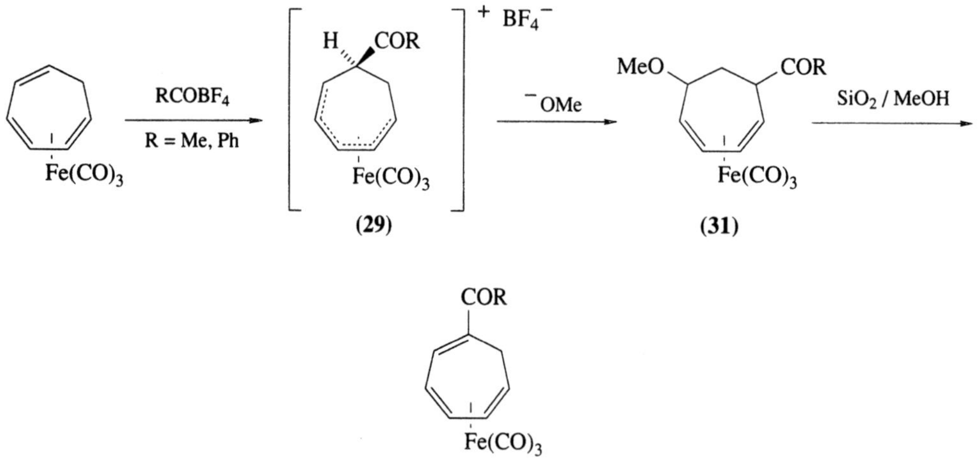

$RCOBF_4$
R = Me, Ph

H COR

BF_4^-

(29)

^-OMe

MeO COR

(31)

SiO_2 / MeOH

COR

(30)

Scheme 20

O R

(29)

(30)

Et_3N

HBF_4

(32)

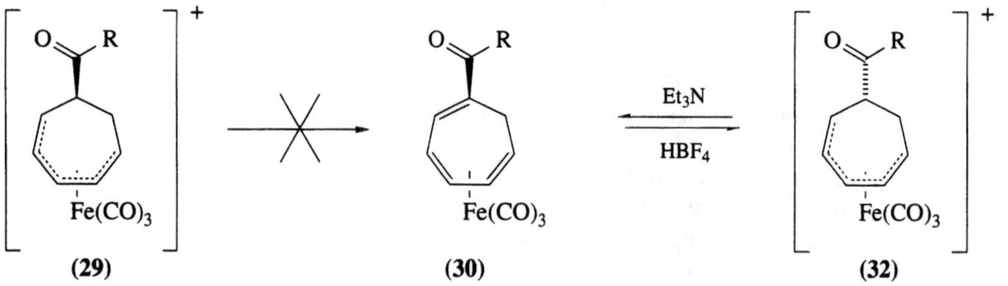

Scheme 21

tropone itself. Acetylation was necessary in order to induce the proper orientation of addition of the 1,3-dipolar dimethyldiazomethane.

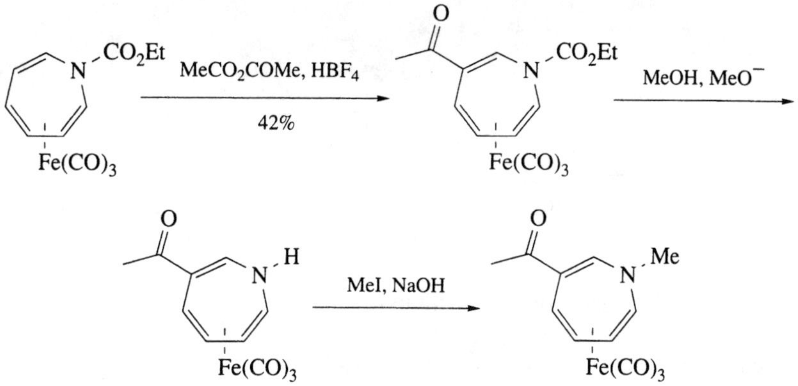

$MeCO_2COMe$, HBF_4 42%; MeOH, MeO^-; MeI, NaOH; $Fe(CO)_3$; CO_2Et

Scheme 22

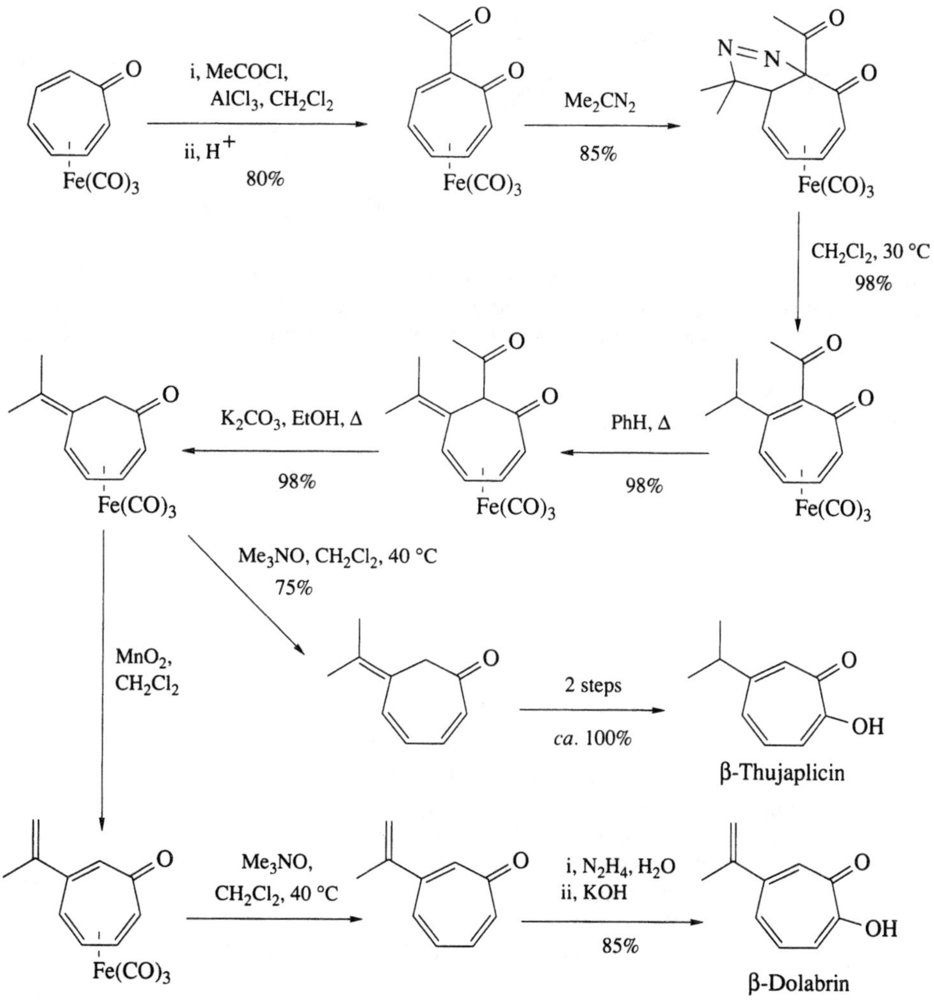

i, MeCOCl, $AlCl_3$, CH_2Cl_2; ii, H^+ 80%; Me_2CN_2 85%; CH_2Cl_2, 30 °C 98%; PhH, Δ 98%; K_2CO_3, EtOH, Δ 98%; Me_3NO, CH_2Cl_2, 40 °C 75%; 2 steps *ca.* 100%; β-Thujaplicin; MnO_2, CH_2Cl_2; Me_3NO, CH_2Cl_2, 40 °C; i, N_2H_4, H_2O ii, KOH 85%; β-Dolabrin

Scheme 23

3.5.3.1.2 Reactions with electrophilic alkenes, ketones and ketenes

[Fe(cyclooctatetraene)(CO)3] reacts with TCNE to form a cycloadduct having the bicyclic structure (**33**) shown in equation (13).[44–47] Addition is *exo* to the metal. Similar reactions with other electrophilic alkenes such as 1,1-dicyano-2,2-bis(trifluoromethyl)ethylene have also been observed.[46]

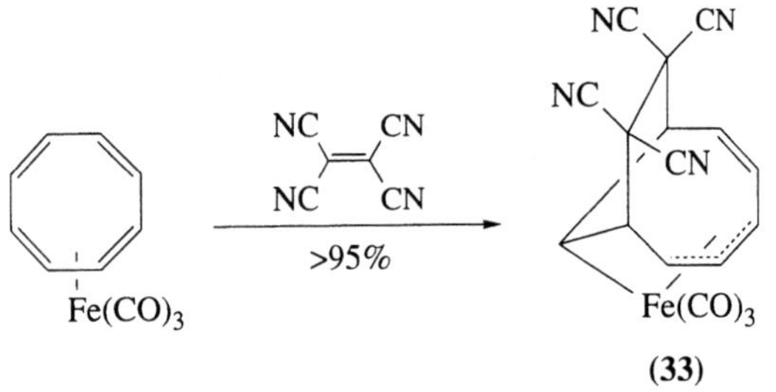

(13)

(33)

Initially, the mechanism for these types of reactions was thought to involve the formation of a Zwitterionic intermediate which would then collapse to form the product.[44,46–48] However, recent evidence suggests that such reactions proceed *via* concerted addition (equation 14).[49,50]

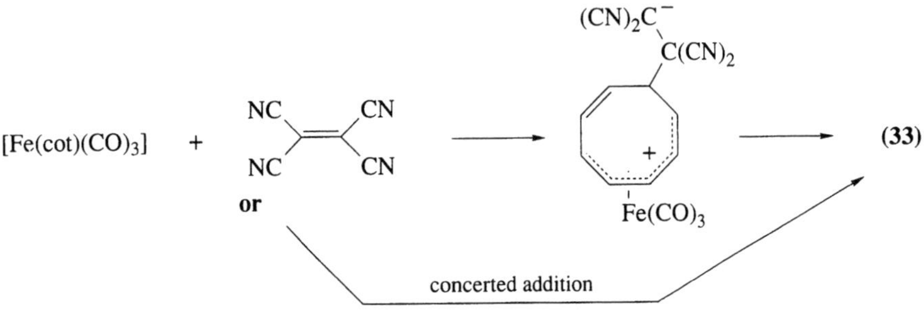

[Fe(cot)(CO)3] + ... **(33)** (14)

or ... concerted addition

Oxidative cleavage of the organic fragment from the metal is possible using Ce^{IV} and gives the tricyclic rearrangement product (**34**) in very high yields (equation 15).[45,46] The tetranitrile, (**34**), prepared in this manner has been used as a key intermediate in the synthesis of chiral 2-substituted triquinacene derivatives.[51]

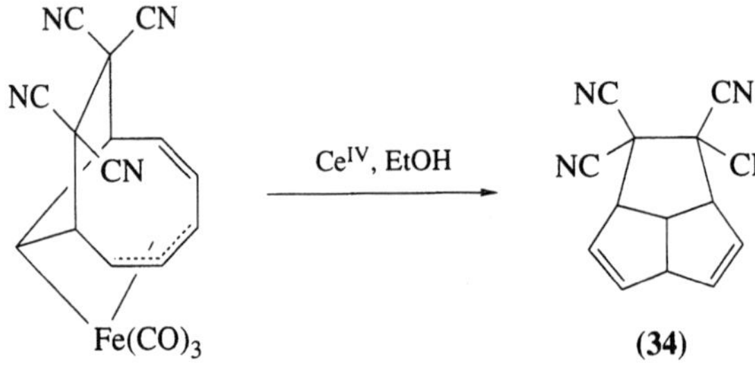

Ce^{IV}, EtOH (15)

(34)

TCNE addition to substituted derivatives of [Fe(cyclooctatetraene)$(CO)_3$] including benzocyclooctatetraene have been examined to gain information on regioselectivity.[47,52]

[Fe(cycloheptatriene)$(CO)_3$] and its ring-substituted derivatives react with a variety of strong dienophiles including TCNE,[46,53–56] 1,1-dicyano-2,2-bis(trifluoromethyl)ethylene,[47] 1,2-dicyano-1,2-bis(trifluoromethyl)ethylene,[47] carbomethoxymaleic anhydride,[57] hexafluoroacetone[47,49] and arylketenes.[58–60] For example, reaction with TCNE produces compound (**35**) as the major product (Scheme 24).[46,53,54] Treatment of (**35**) with Ce^{IV} or $FeCl_3$ cleaves the organic fragment from the metal to yield rearranged diene (**36**) in good yields.[46,54] Complex (**35**) can be converted to (**37**) by heating.[54,53] Cleavage of (**37**) with Ce^{IV} yields (**38**).

Reaction of $(CN)_2C{=}C(CF_3)_2$ with [Fe(cycloheptatriene)$(CO)_3$] yields adduct (**39**), which under CO pressure gives tricyclic ketone (**40**) presumably *via* iron–acyl complex (**41**; equation 16).[47]

Treatment of [Fe(cycloheptatriene)$(CO)_3$] with diphenylketene at room temperature yields the [2 + 2] product (**42**; Scheme 25).[58] Heating (**42**) leads to the rearrangement product, (**43**),[58,60] which can be cleaved from the metal with Ce^{IV} to yield the tricyclic ketone (**44**).[58]

The TCNE adducts of a variety of other $[\eta^4\text{-Fe(CO)}_3]^-$ complexed trienes and tetraenes have been investigated. In most cases the adducts have been oxidatively cleaved from the metal with Ce^{IV} and the free ligands recovered in good yields. Reactivity patterns generally follow those described above for cycloheptatriene and cyclooctatetraene complexes. Specific complexes investigated include [Fe(2,4,6-cyclooctatrieneone)$(CO)_3$],[61] [Fe(tropone)$(CO)_3$],[62,63] [Fe(heptafulvalene)$(CO)_3$] and various 8-substituted

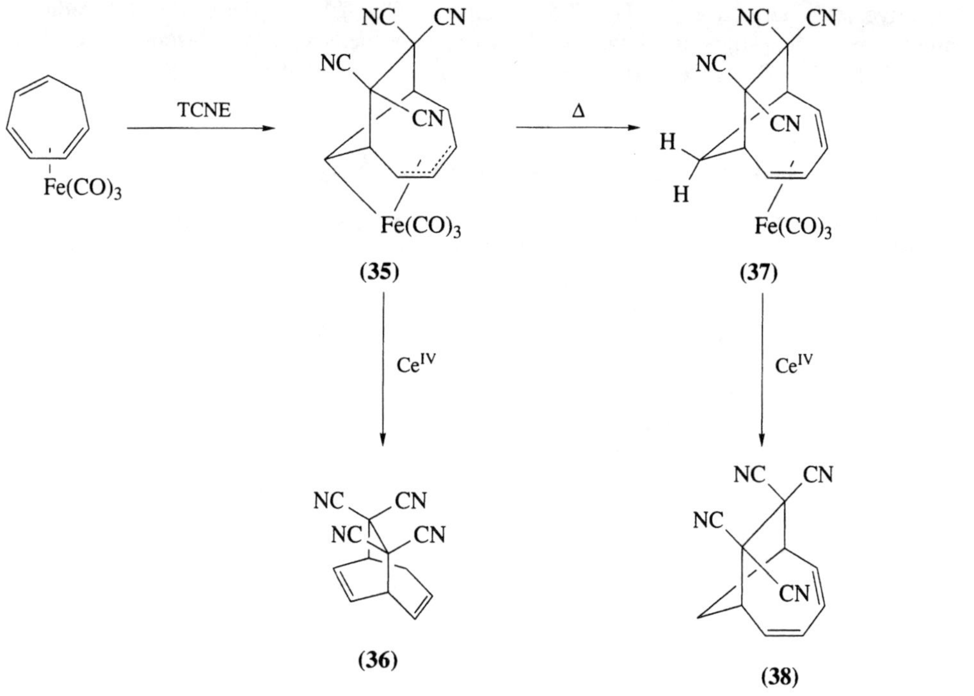

Scheme 24

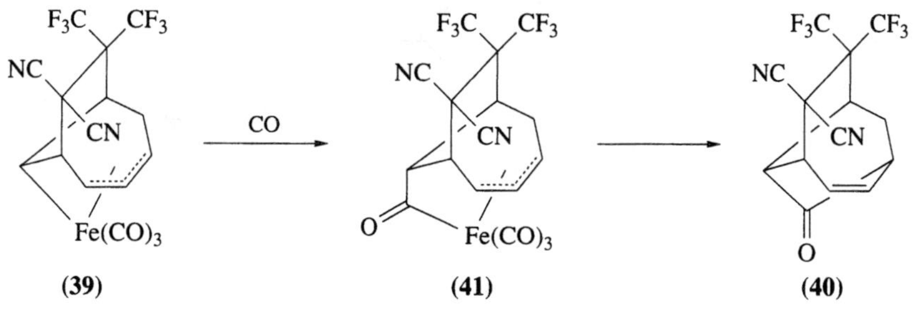

(16)

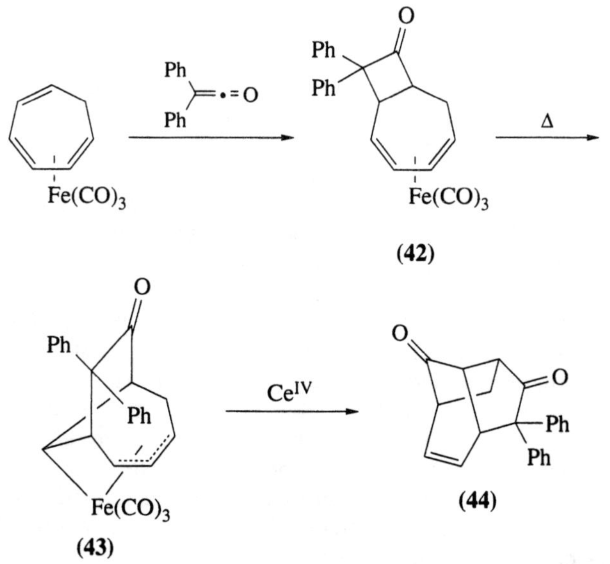

Scheme 25

derivatives,[64–67] derivatives of [Fe(1,3,5-heptatriene)$(CO)_3$][68] and [Fe(vinylcyclobutadiene)$(CO)_3$].[69] A comprehensive investigation of the reactions of [M(*N*-methoxycarbonylazepine)$(CO)_3$] (M = Fe, Ru) with TCNE, $(CN)_2C{=}C(CF_3)_2$ and $(CF_3)_2C{=}O$ has been reported.[49,70]

3.5.3.2 η^4-Triene and η^4-Tetraene Manganese Tricarbonyl Anions

As noted in Section 3.5.2.2, a general method has been recently reported for the synthesis of anionic $[Mn(\eta^4\text{-polyene})(CO)_3]^-$ complexes. While reactions with carbon electrophiles have not been extensively examined, it is clear these species are much more reactive than their neutral iron analogs. For example, whereas [Fe(cycloheptatriene)$(CO)_3$] and [Fe(cyclooctatetraene)$(CO)_3$] are inert toward alkyl halides, the corresponding anionic manganese derivatives (**45**) and (**46**) can be methylated under mild conditions[34,35] with methyl iodide or methyl triflate as shown in Scheme 26. X-Ray analysis of (**47**) establishes the *exo* stereochemistry of the methyl group and suggests attack of the methylating reagent at the uncoordinated double bond.[35] Similar to these reports is the reaction of $PhCH_2Br$ with $[Cr(\eta^4\text{-}C_6H_6)(CO)_3]^{2-}$ to give [Cr(*S*-*exo*-benzylcyclohexadienyl)$(CO)_3$]$^-$ reported by Cooper.[7(b)]

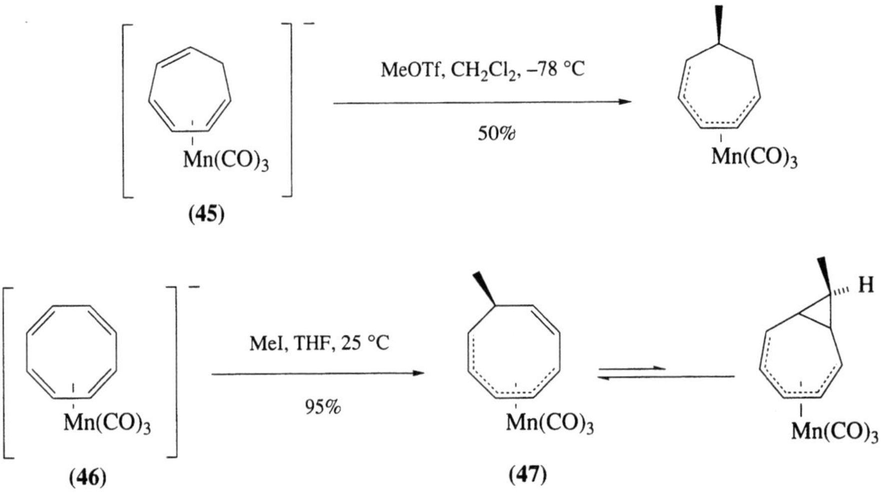

MeOTf, CH_2Cl_2, −78 °C; 50%

$Mn(CO)_3$ (**45**) → $Mn(CO)_3$

MeI, THF, 25 °C; 95%

$Mn(CO)_3$ (**46**) → $Mn(CO)_3$ (**47**) ⇌ $Mn(CO)_3$ (H)

Scheme 26

3.5.4 SUMMARY

This chapter illustrates that electron-rich transition metal–diene complexes can couple with carbon electrophiles and, thereby, provide unusual methods for carbon–carbon bond formation. These procedures are of interest from a synthetic viewpoint since normally uncomplexed dienes or polyenes are not reactive toward weak carbon electrophiles or, with strong electrophiles, undesirable reactions such as polymerization occur. Furthermore, the metal-mediated route often results in desirable regio- and/or stereo-selectivity. Important to the utility of these methods is the ability to free the organic ligand from the metal. In most instances efficient oxidative procedures have been developed for such cleavage reactions.

As noted in the introduction, in contrast to attack by nucleophiles, attack of electrophiles on saturated alkene–, polyene– or polyenyl–metal complexes creates special problems in that normally unstable 16-electron, unsaturated species are formed. To be isolated, these species must be stabilized by intramolecular coordination or *via* intermolecular addition of a ligand. Nevertheless, as illustrated in this chapter, reactions of significant synthetic utility can be developed with attention to these points. It is likely that this area will see considerable development in the future. In addition to refinement of electrophilic reactions of metal–diene complexes, synthetic applications may evolve from the coupling of carbon electrophiles with electron-rich transition metal complexes of alkenes, alkynes and polyenes, as well as allyl– and dienyl–metal complexes. Sequential addition of electrophiles followed by nucleophiles is also viable to rapidly assemble complex structures.

3.5.5 REFERENCES

1. (a) M. F. Semmelhack, *Pure Appl. Chem.*, 1981, **53**, 2379; (b) M. F. Semmelhack, in 'Comprehensive Organic Synthesis', ed. B. M. Trost, Pergamon Press, Oxford, 1991, vol. 4, chap. 2.4.
2. (a) A. J. Pearson, *Acc. Chem. Res.*, 1980, **13**, 463; (b) A. J. Birch, *Ann. N. Y. Acad. Sci.*, 1980, **333**, 101; (c) A. J. Pearson, in 'Comprehensive Organic Synthesis', ed. B. M. Trost, Pergamon Press, Oxford, 1991, vol. 4, chap. 3.4.
3. (a) M. Rosenblum, *Pure Appl. Chem.*, 1984, **56**, 129; (b) M. Rosenblum, *Acc. Chem. Res.*, 1974, **7**, 122; (c) L. S. Hegedus, in 'Comprehensive Organic Synthesis', ed. B. M. Trost, Pergamon Press, Oxford, 1991, vol. 4, chaps. 3.1 and 3.2.
4. (a) J. W. Faller, H. H. Murray, D. L. White and K. H. Chao, *Organometallics*, 1983, **2**, 400; (b) A. J. Pearson, M. A. Khan, J. C. Clardy and H. Cun-Heng, *J. Am. Chem. Soc.*, 1985, **107**, 2748; (c) M. F. Semmelhack, J. W. Herndon and J. P. Springer, *J. Am. Chem. Soc.*, 1983, **105**, 2497; (d) L. S. Barinelli, K. Tao and K. M. Nicholas, *Organometallics*, 1985, **5**, 588.
5. (a) M. Brookhart, J. Yoon and S. K. Noh, *J. Am. Chem. Soc.*, 1989, **111**, 4117; (b) G. M. Williams and D. E. Rudisill, *Inorg. Chem.*, 1989, **28**, 797.
6. E. P. Kundig, *Pure Appl. Chem.*, 1985, **57**, 1855.
7. (a) M. Brookhart, P. K. Rush and S. K. Noh, *Organometallics*, 1986, **5**, 1745; (b) V. S. Leong and N. J. Cooper, *J. Am. Chem. Soc.*, 1988, **110**, 2644.
8. G. G. Ecke, *US pat.* 3 149 135 (1964). (*Chem. Abstr.*, 1965, **62**, 4054e).
9. D. F. Hunt, C. P. Lillya and M. D. Rausch, *J. Am. Chem. Soc.*, 1968, **90**, 2561.
10. D. R. Falkowski, D. F. Hunt, C. P. Lillya and M. D. Rausch, *J. Am. Chem. Soc.*, 1967, **89**, 6387.
11. N. A. Clinton and C. P. Lillya, *J. Am. Chem. Soc.*, 1970, **92**, 3065.
12. E. O. Greaves, G. R. Knox and P. L. Pauson, *J. Chem. Soc., Chem. Commun.*, 1969, 1124.
13. A. D. U. Hardy and G. A. Sim, *J. Chem. Soc., Dalton Trans.*, 1972, 2305.
14. (a) E. O. Greaves, G. R. Knox, P. L. Pauson, S. Toms, G. A. Sim and D. I. Woodhouse, *J. Chem. Soc., Chem. Commun.*, 1974, 257; (b) G. A. Sim and D. I. Woodhouse, *Acta Crystallogr., Sect. B*, 1979, **35**, 1477.
15. R. E. Graf and C. P. Lillya, *J. Organomet Chem.*, 1979, **166**, 53.
16. R. E. Graf and C. P. Lillya, *J. Am. Chem. Soc.*, 1972, **94**, 8282.
17. M. Frank-Neumann, M. Sedrati and M. Mokhi, *J. Organomet. Chem.*, 1987, **326**, 389.
18. R. E. Graf and C. P. Lillya, *J. Organomet Chem.*, 1976, **122**, 377.
19. M. Frank-Neumann, M. Sedrati and A. Abdali, *J. Organomet. Chem.*, 1988, **339**, C9.
20. M. Frank-Neumann, M. Sedrati and M. Mokhi, *Tetrahedron Lett.*, 1986, **27**, 3861.
21. M. Frank-Neumann, M. Sedrati and M. Mokhi, *Angew. Chem., Int. Ed. Engl.*, 1986, **25**, 1131.
22. A. J. Birch, W. D. Raverty, S. Y. Hsu and A. J. Pearson, *J. Organomet. Chem.*, 1979, **166**, 53.
23. M. Frank-Neumann, M. Sedrati and A. Abdali, *J. Organomet. Chem.*, 1988, **339**, C9.
24. R. E. Graf and C. P. Lillya, *J. Chem. Soc., Chem. Commun.*, 1973, 271.
25. R. Pettit, *J. Organomet. Chem.*, 1975, **110**, 205.
26. J. D. Fitzpatrick, L. Watts, G. F. Emerson and R. Pettit, *J. Am. Chem. Soc.*, 1965, **87**, 3254.
27. A. Efraty, *Chem. Rev.*, 1977, **77**, 691.
28. A. N. Nesmeyanov, K. N. Anisimov and G. K. Magomedov, *Izv. Akad. Nauk SSSR, Ser. Khim.*, 1970, **11**, 715.
29. K. N. Anisimov, G. K. Magomedov, N. E. Kolobova and A. G. Trufanov, *Izv. Akad. Nauk SSSR, Ser. Khim.*, 1970, **11**, 2533.
30. G. R. Knox and I. G. Thom, *J. Chem. Soc., Chem. Commun.*, 1981, 373.
31. A. J. Pearson, *Aust. J. Chem.*, 1976, **29**, 1841.
32. (a) W. Lamanna and M. Brookhart, *J. Am. Chem. Soc.*, 1981, **103**, 989; (b) M. Brookhart, W. Lamanna and M. B. Humphrey, *J. Am. Chem. Soc.*, 1982, **104**, 2117; (c) M. Brookhart, W. Lamanna and A. R. Pinhas, *Organometallics*, 1983, **2**, 638; (d) P. Bladon, G. A. M. Munro, P. L. Pauson and C. A. L. Mahaffy, *J. Organomet. Chem.*, 1981, **221**, 79; (e) M. Brookhart and A. Lukacs, *J. Am. Chem. Soc.*, 1984, **106**, 4161.
33. F. J. Timmers and M. Brookhart, *Organometallics*, 1985, **4**, 1365.
34. (a) M. Brookhart, S. K. Noh and F. J. Timmers, *Organometallics*, 1987, **6**, 1829; (b) S. K. Noh and M. Brookhart, unpublished results.
35. M. Brookhart, S. K. Noh, F. J. Timmers and Y. H. Hong, *Organometallics*, 1988, **7**, 2458.
36. B. F. G. Johnson, J. Lewis, A. W. Parkins and G. L. P. Randall, *J. Chem. Soc., Chem. Commun.*, 1969, 595.
37. B. F. G. Johnson, J. Lewis and G. L. P. Randall, *J. Chem. Soc. A*, 1971, 422.
38. A. D. Charles, P. Diversi, B. F. G. Johnson, K. D. Karlin, J. Lewis, A. V. Rivera, and G. M. Sheldrick, *J. Organomet. Chem.*, 1977, **128**, C31.
39. B. F. G. Johnson, J. Lewis and G. L. P. Randall, *J. Chem. Soc., Chem. Commun.*, 1969, 1273.
40. B. F. G. Johnson, J. Lewis and G. L. P. Randall, *J. Chem. Soc., Dalton Trans.*, 1972, 456.
41. B. F. G. Johnson, J. Lewis, P. McArdle and G. L. P. Randall, *J. Chem. Soc., Dalton Trans.*, 1972, 2076.
42. G. B. Gill, N. Gourlay, A. W. Johnson and M. Mahendran, *J. Chem. Soc., Chem. Commun.*, 1969, 631.
43. M. Frank-Neumann, F. Brion and D. Martina, *Tetrahedron Lett.*, 1978, 5033.
44. M. Green and D. C. Wood, *J. Chem. Soc., A*, 1969, 1172.
45. L. A. Paquette, S. V. Ley, M. J. Broadhurst, D. Truesdell, J. Fayos and J. Clardy, *Tetrahedron Lett.*, 1973, 2943.
46. D. J. Ehntholt and R. C. Kerber, *J. Organomet. Chem.*, 1972, **38**, 139.
47. M. Green, S. Heathcock and D. C. Wood, *J. Chem. Soc., Dalton Trans.*, 1973, 1564.
48. G. Deganello, P. Uguagliati, L. Calligaro, P. L. Sandrini and F. Zingales, *Inorg. Chim. Acta*, 1975, **13**, 247.
49. M. Green, S. M. Heathcock, T. W. Turney and D. M. P. Mingos, *J. Chem. Soc., Dalton Trans.*, 1977, 204.
50. N. Hallinan, P. McArdle, J. Burgess and P. Guardado, *J. Organomet. Chem.*, 1987, **333**, 77.
51. L. A. Paquette, S. V. Ley and W. B. Farnham, *J. Am. Chem. Soc.*, 1974, **96**, 312.
52. L. A. Paquette, S. V. Ley, S. Maiorana, D. F. Schneider, M. J. Broadhurst and R. A. Boggs, *J. Am. Chem. Soc.*, 1975, **97**, 4658.

53. Z. Goldschmidt, H. E. Gottlieb, E. Genizi and D. Cohen, *J. Organomet. Chem.*, 1986, **301**, 337.
54. Z. Goldschmidt and E. Genizi, *Synthesis*, 1985, 949.
55. S. K. Chopra, M. J. Hynes and P. McArdle, *J. Chem. Soc. Dalton Trans.*, 1981, 586.
56. Z. Goldschmidt and H. E. Gottlieb, *J. Organomet. Chem.*, 1989, **361**, 207.
57. Z. Goldschmidt, S. Antebi, H. E. Gottlieb and D. Cohen, *J. Organomet. Chem.*, 1985, **282**, 369.
58. Z. Goldschmidt and S. Antebi, *Tetrahedron Lett.*, 1978, 271.
59. Z. Goldschmidt, S. Antebi, D. Cohen and I. Goldberg, *J. Organomet. Chem.*, 1984, **273**, 347.
60. Z. Goldschmidt and H. E. Gottlieb, *J. Organomet. Chem.*, 1987, **329**, 391.
61. Z. Goldschmidt and Y. Bakal, *Tetrahedron Lett.*, 1977, 955.
62. Z. Goldschmidt and Y. Bakal, *Tetrahedron Lett.*, 1976, 1229.
63. Z. Goldschmidt, H. E. Gottlieb and D. Cohen, *J. Organomet. Chem.*, 1985, **294**, 219.
64. P. McArdle, *J. Organomet. Chem.*, 1978, **144**, C31.
65. Z. Goldschmidt and Y. Bakal, *J. Organomet. Chem.*, 1979, **179**, 197.
66. Z. Goldschmidt and Y. Bakal, *J. Organomet. Chem.*, 1979, **160**, 215.
67. S. K. Chopra, M. J. Hynes, G. Moran, J. Simmie and P. McArdle, *Inorg. Chim. Acta*, 1982, **63**, 177.
68. Z. Goldschmidt and Y. Bakal, *J. Organomet. Chem.*, 1984, **191**, 269.
69. A. Efraty, *Chem. Rev.*, 1977, **72**, 691.
70. M. Green, S. Tolson, J. Weaver, D. C. Wood and P. Woodward, *J. Chem. Soc., Chem. Commun.*, 1971, 222.

4.1
Radical Addition Reactions

DENNIS P. CURRAN
University of Pittsburgh, PA, USA

4.1.1 INTRODUCTION

The formation of a new σ-bond at the expense of an existing π-bond is a transformation that rests at the heart of modern organic synthesis. It underlies such diverse reactions as nucleophilic additions to carbonyls and activated alkenes, transition metal catalyzed additions, and pericyclic reactions. The addition of a radical to a π-bond, to form a σ-bond and a new radical, is a fundamental reaction of organic radicals that organic polymer chemists have continuously applied to the synthesis of new materials. Applications in small molecule synthesis evolved more slowly for a time.[1,2] However, over the last decade, the application of radical addition reactions to problems in fine synthesis has grown from the level of a curiosity to that of a major reaction class. For recent reviews see refs. 3–5. This rapid growth can be attributed, in part, to the high level of understanding of organic radicals that has been provided by fundamental research on structure and reactivity and, in part, to the fact that the transformations that can be accomplished by radical additions are often those that are difficult to accomplish by more traditional means.

This chapter begins with an introduction to the basic principles that are required to apply radical reactions in synthesis, with references to more detailed treatments. After a discussion of the effect of substituents on the rates of radical addition reactions, a new method to notate radical reactions in retrosynthetic analysis will be introduced. A summary of synthetically useful radical addition reactions will then follow. Emphasis will be placed on how the selection of an available method, either chain or non-chain, may affect the outcome of an addition reaction. The addition reactions of carbon radicals to multiple bonds and aromatic rings will be the major focus of the presentation, with a shorter section on the addition reactions of heteroatom-centered radicals. Intramolecular addition reactions, that is radical cyclizations, will be covered in the following chapter with a similar organizational pattern. This second chapter will also cover the use of sequential radical reactions. Reactions of diradicals (and related reactive intermediates) will not be discussed in either chapter. Photochemical [2 + 2] cycloadditions are covered in Volume 5, Chapter 3.1 and diyl cycloadditions are covered in Volume 5, Chapter 3.1. Related functional group transformations of radicals (that do not involve π-bond additions) are treated in Volume 8, Chapter 4.2.

4.1.2 BASIC PRINCIPLES

4.1.2.1 Introduction

New organic compounds are synthesized by the chemical reactions of diverse molecules that range from stable reagents to highly reactive intermediates. These molecules usually have one common feature: their electrons are paired. In the accounting system that is used to follow electrons from reactants to products, 'double-headed' arrows symbolize two electrons (Scheme 1). Radicals are species that contain at least one unpaired electron.[6] This feature gives radicals a unique reactivity profile relative to closed shell species. And with an understanding of the basic principles of radicals, this reactivity profile can be exploited in synthesis. To account for electrons in radical reactions, 'single-headed' arrows symbolizing one electron are used by convention. Bear in mind that, as with double-headed arrows, the use of single-headed arrows helps one to quickly grasp the changes in electron distribution in a step or transformation but does not necessarily indicate the mechanism.

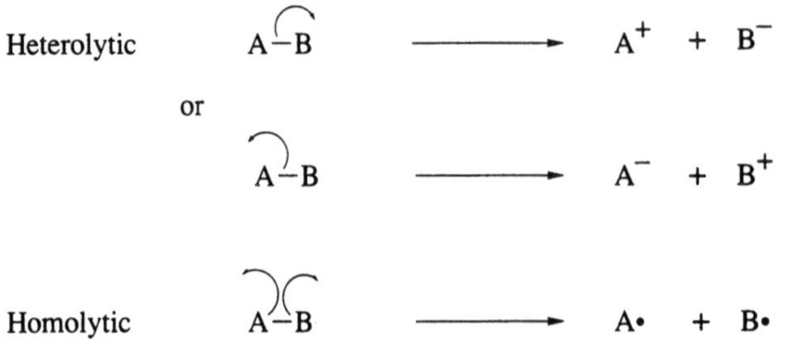

Heterolytic $\quad A{-}B \longrightarrow A^+ + B^-$

or

$A{-}B \longrightarrow A^- + B^+$

Homolytic $\quad A{-}B \longrightarrow A\bullet + B\bullet$

Scheme 1 Bond cleavage formalisms

This section will provide an introduction to the basic principles of organic radicals that are needed for synthetic application. Many of these principles have been understood and applied for more than 30 years[7] and a variety of good treatments are available.[8] The book of Prior still provides a thorough, readable introduction for newcomers to the field.[9] The first chapter of Giese's excellent book[3] and an article by Walling[10] are especially good introductions for the synthetic chemist. Although it was published in 1973, the two volume set edited by Kochi[11] still provides a wealth of information on many fundamental aspects of radical chemistry.

4.1.2.2 Transiency of Radicals

A major difference between radicals and other reactive species that are employed in synthesis is that virtually all radicals react rapidly with themselves. Compare *t*-butyllithium and *t*-butyl radical, for example. *t*-Butyllithium is one of the most reactive organometallic species that synthetic chemists employ. It reacts in some fashion with virtually all types of organic molecules, even most solvents, given long enough time and high enough temperature. However, *t*-butyllithium does not react readily with itself! As a result, one can purchase a bottle of *t*-butyllithium and admix it with a desired reactant. In many cases, a new reactive intermediate is formed (by deprotonation, for example) and another reagent is then added.

In contrast, one need not consult an Aldrich catalog to see if *t*-butyl radical is commercially available. Nor can one generate a significant quantity of *t*-butyl radical in solution prior to the addition of a reagent with which it is destined to react. This is because *t*-butyl radical, like nearly every other radical that is employed in synthesis, is transient. The lifetime of a transient radical rarely exceeds 1 μs and is limited by the rate of self reaction. This self reaction, which can be either a disproportionation or a recombination (Scheme 2), typically occurs at the diffusion-controlled limit; the enthalpy of activation of most radical–radical reactions $\approx$ 0 kcal mol^{-1}.

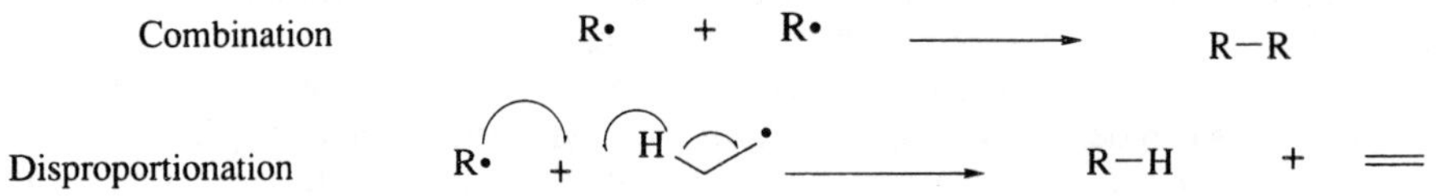

Combination $\quad R\bullet + R\bullet \longrightarrow R{-}R$

Disproportionation $\quad R\bullet + H{-}CH_2CH_2\bullet \longrightarrow R{-}H + CH_2{=}CH_2$

Scheme 2 Termination

Radicals can be stabilized by substituents and a simple and useful measure of radical stabilization is C—H bond dissociation energy [D(C—H)].[12] The approximate bond dissociation energies of some representative C—H bonds are given in Scheme 3.[13] However, these can be misleading: tertiary C—H bonds are weaker than primary C—H bonds, not so much because the alkyl groups stabilize the radical (although they do), but more because repulsion between these alkyl groups in the tetrahedral precursor is diminished in the radical. Conjugating groups and heteroatoms stabilize radicals and such electronically stabilized radicals sometimes react more slowly with neutral molecules than their non-stabilized counterparts. However, radical–radical reactions are so exothermic that the rates of the self reactions of stabilized radicals are not retarded. Thus, the benzyl radical is much less reactive than the methyl radical in common reactions such as C—H abstraction, yet it still reacts with itself or with another radical at the diffusion-controlled rate.

$PhCH_2\bullet$	<	$(CH_3)_3C\bullet$	<	$(CH_3)_2CH\bullet$	<	$CH_3CH_2\bullet$	<	$CH_3\bullet$	<	$=\dot{C}-H$	<	$\equiv C\bullet$	RO•
85		92		96		100		104		108		110	

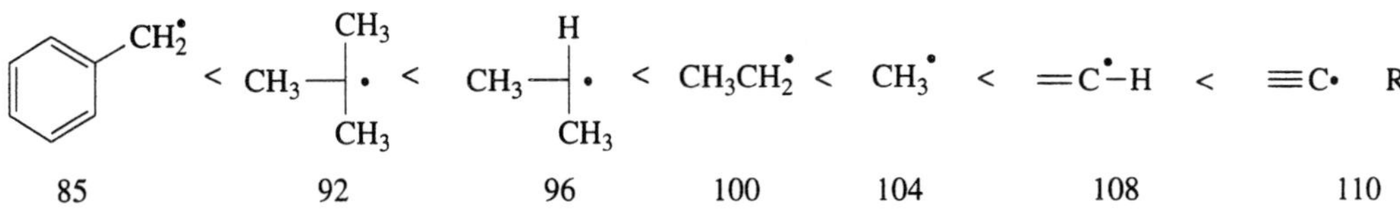

D[R—H] (kcal mol^{-1})

Scheme 3 Bond dissociation energies

In radical terminology, the opposite of transience is not stability but persistence.[14] Persistent radicals do not react with themselves at diffusion-controlled rates; however, they may still react readily with other radicals or with triplet oxygen. Thus, persistence is a kinetic property that is more often related to sterically hindered recombination than to electronic stabilization. Persistent radicals typically also lack β-C—H bonds and they cannot disproportionate. Several persistent radicals are illustrated in Scheme 4. Persistent radicals are rarely present in synthetic applications, but when they are, there will be important consequences.

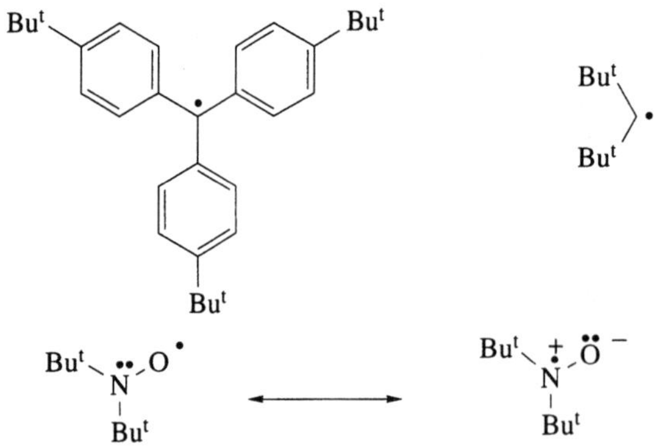

Scheme 4 Persistent radicals

The thermodynamic driving force notwithstanding, the coupling of two different radicals is not an especially practical preparative method to form a new bond. This is because the preparation of precursors that directly decompose to radicals is rarely convenient, because disproportionation can often compete effectively with recombination, and especially because chemoselectivity (that is, the selective-coupling of two different radicals to the exclusion of self-coupling) is difficult to achieve if all coupling reactions occur at the same rate (the diffusion-controlled limit).

The reactions of radicals with nonradicals (typically, neutral organic molecules[15]) have much greater use in synthesis. These reactions occur with a wide and predictable range of rate constants, and the concentration of a nonradical component is a readily controlled experimental variable that can greatly affect the chemoselectivity of a reaction. For radical–reagent reactions to be synthetically useful, they must be faster than radical–radical reactions and radical–solvent reactions. By estimating the rates of these destructive reactions, one can determine lower rate limits for synthetic procedures. The rate expression for the self reaction of two radicals (R·)[16] is shown in equation (1), while the rate expression for the addition of the same radical to an alkene (A═B) is shown in equation (2). For a useful addition reaction to occur, the rate of disappearance of R· by equation (2) must be significantly greater than that by equation (1). This comparison is stated in equation (3).[17] By substituting $k_t \approx 10^{10}\ M^{-1}\ s^{-1}$ (the approximate diffusion-controlled limit) and by assuming that 10^{-8} M is a typical radical concentration in a chain reaction, we find that the rate of formation of the adduct radical R—A—B· must be $>10^2\ s^{-1}$ (equation 3). Equation (4) shows the requirement for a first order reaction such as a cyclization.

$$R\bullet + R\bullet \xrightarrow{k_t} R{-}R \qquad (1)$$

$$\frac{d[R{-}R]}{dt} = k_t[R\bullet][R\bullet]$$

$$R\bullet + A{=}B \longrightarrow R{-}A{-}B\bullet \qquad (2)$$

$$\frac{d\,[R{-}A{-}B\bullet]}{dt} = k_a[R\bullet]\,[A{=}B]$$

$$k_a\,[A{=}B] > k_t\,[R\bullet] \qquad (3)$$

$$k_a\,[A{=}B] > 10^2\ s^{-1}$$

$$R\bullet \xrightarrow{k_c} R'\bullet \qquad (4)$$

$$k_c > 10^2\ s^{-1}$$

This simple example illustrates how the rates of two reactions competing for the same radical (R·) are compared. Provided they are not too rapid, the rates of termination steps are not of concern, but comparison of the rates of competing propagation steps is very informative. Fortunately, all propagation steps are first order in [R·] and this term (which is difficult to know accurately) cancels when two propagation steps are compared.

Concerns about nonproductive reactions of radicals apply not only to carbon–carbon bond forming steps, such as additions and cyclizations, but to every step in a sequence of radical reactions. For example, to obtain a good yield of product from the reaction in equation (2), not only must the addition of R· to A═B occur but the conversion of R—A—B· to a nonradical product must also be efficient. In a chain reaction, the slowest propagation step must still be rapid relative to loss of the radicals by radical–radical or radical–solvent reactions. In practice, reactions with rates of product formation of 10^2–10^3 s^{-1} are experimentally difficult to conduct. Reactions with rates of 10^4–10^5 s^{-1} are manageable and those with rates $>10^6$ s^{-1} are usually conducted with ease.

The reactions of radicals with solvents may also limit the types of radical–nonradical reactions that can be conducted. The two types of reactions encountered are addition and hydrogen atom abstraction. Alkyl radicals add to benzene, a popular solvent for radical reactions, with pseudo first order rate constants ≈ 10^2 s^{-1}.[18] This is roughly competitive with radical–radical reactions. Ethereal solvents provide a different limitation: alkyl radicals abstract an α-hydrogen atom from ether with a pseudo first order rate constant $k_H \approx 10^3$ s^{-1}. THF is a better H-donor than ether ($k_H \approx 10^4$ s^{-1}).[19]

4.1.2.3 Structure and Stereochemistry of Radicals

It is generally safe to assume that the complex radicals formed in synthesis will have predictable structural and stereochemical features that can be extrapolated from simple radicals.[20] The methyl radical is known to be planar (or perhaps very slightly pyramidalized with a low barrier to inversion). The replacement of hydrogen atoms by alkyl groups or heteroatoms causes pyramidalization but the inversion barrier remains very low. Conjugating substituents favor planar structures. In practice, this means that the stereochemistry of a reaction that occurs at an alkyl radical center cannot be dictated by the stereogenicity of the pro-radical center because stereoisomeric precursors generate the same radical. Instead, the stereochemistry of reaction at a radical center is controlled by the relative rates of the competing reactions. Scheme 5 provides examples where the competing reactions are axial and equatorial hydrogen atom transfer.[21]

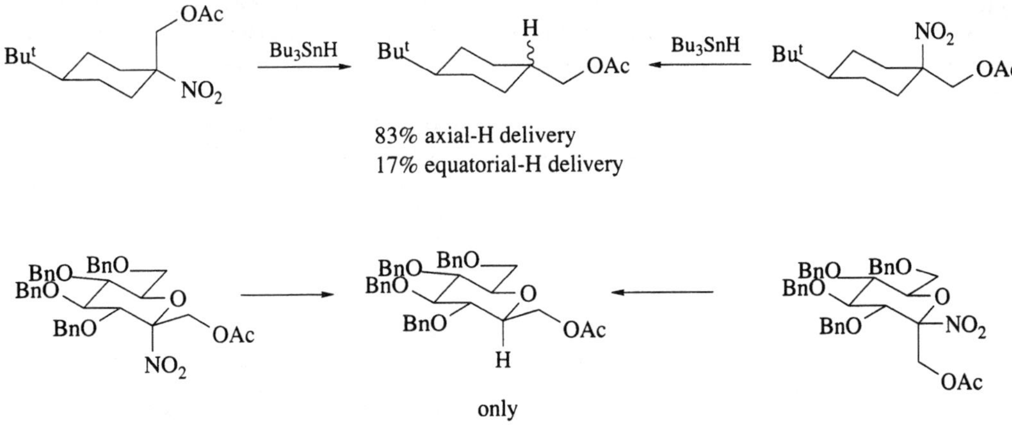

Scheme 5 Stereochemistry of alkyl radicals

Somewhat less is known about the structures of vinyl radicals. Two limiting structures, bent and linear, are possible (Scheme 6). The vinyl radical itself exists in a bent form (*sp*^{2}-hybridized) but with a very low barrier to inversion.[22] It is generally assumed that alkyl-substituted vinyl radicals behave similarly. Thus, as with alkyl radicals, the stereochemical outcome of a reaction of a vinyl radical does not generally depend on the stereochemistry of the precursor (Scheme 6).[23] However, there are several examples in which a subsequent reaction of a vinyl radical has been proposed to be more rapid than its inversion.[24] Heteroatom substituents (notably ethers[25]) substantially raise the barrier for inversion of vinyl radicals and it seems likely that stereoselective methods that capitalize on this barrier might be developed.[26] In contrast, π-conjugating substituents are usually thought to favor the linear (*sp*-hybridized) form to promote resonance overlap.[27] With a linear vinyl radical, there is again no correlation between precursor and product stereochemistry. However, to understand the stereochemical outcome of a reaction of a vinyl radical, it may be important to know whether it is bent and rapidly inverting or linear.

In return for the possible disadvantage that existing stereochemistry at the radical site is lost, radicals provide several advantages (see below). Not the least of these is that the synthesis of the radical precursor is often facilitated because stereochemistry at the pro-radical center is not a concern.

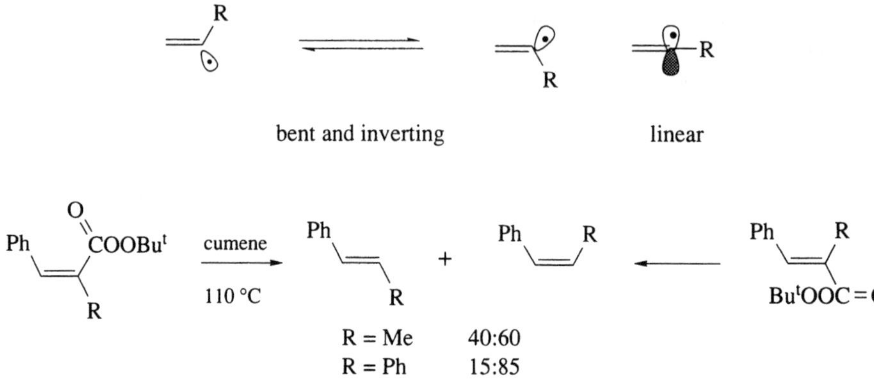

bent and inverting linear

cumene, 110 °C

R = Me 40:60
R = Ph 15:85

Scheme 6 Stereochemistry of vinyl radicals

4.1.2.4 General Considerations for the Use of Radical Reactions in Synthesis

This section presents some general chemoselectivity considerations for designing substrates and planning experimental conditions for radical reactions. More detailed concerns of chemo-, regio- and stereoselectivity are addressed for each specific technique in the individual sections below.

Synthetic chemists are well-accustomed to designing experimental conditions for reactions that are ionic.[28] The alkylation, hydroxyalkylation and acylation reactions of metal enolates provide typical examples. Such reactions generate intermediates that may be sensitive to oxygen, water or heat. The relative rates of competing reactions (and, hence, chemo-, regio- and stereo-selectivities) of such intermediates are sometimes strongly influenced by solvent (and cosolvent additives), metal counterions and additives, and temperature. Depending on the reaction, such characteristics can either be significant advantages or severe limitations. Many radical reactions have a very different profile.

Nearly all radicals add to triplet oxygen at rates approaching the diffusion-controlled limit. Unless this is a desired reaction, it is advisable to take some precautions to exclude atmospheric oxygen. However, rigorous measures, such as freeze–thaw degassing, are not usually required in either exploratory or preparative experiments. The evacuation of a reaction vessel by using a water aspirator or a house vacuum apparatus, followed by repressurization with nitrogen or argon, suffices for most purposes. Sometimes the solvent is purged by bubbling inert gas for several minutes. After such standard measures, the concentration of oxygen is too low to interfere with most radical chains.

Most researchers distil solvents for radical reactions in the same manner as they might for use with a reactive organometallic. This practice is recommended to ensure that solvents are sufficiently pure. That the solvents are simultaneously dried during purification is of little consequence. Water is a much poorer hydrogen atom donor than any common solvent (conversely, the hydroxyl radical is a powerful hydrogen atom abstractor). Thus, the presence of trace quantities of water will have no adverse effect on most radical reactions. As a corollary, water can be a useful solvent or cosolvent provided that the reagents or substrates are not susceptible to hydrolysis or protonolysis.

Few radicals live long enough to decompose the way ionic intermediates might. The lifetimes of radicals are strictly limited by radical–radical and radical–solvent reactions. Since such reactions are temperature dependent, changing the reaction temperature can be a productive technique. Temperature changes are commonly used to effect the partitioning between competing unimolecular and bimolecular processes. For entropic reasons, bimolecular processes usually exhibit a much greater temperature dependence.

Effects analogous to ion-pairing and aggregation are nonexistent for radicals; their structures are not especially solvent dependent. Solvent effects on the rates of radical–neutral reactions are often small and are frequently neglected (although the transition states of many radical reactions have some polar character that might be influenced by solvent). The solvent for a radical reaction is often chosen based on concerns about solubility of reagents and reactants, solvent boiling point, and rate of radical–solvent reactions. Restrictions in solvent selection apply mainly to radical reactions of slow or intermediate rates. Certain chains (such as tin hydride mediated cyclizations of some hexenyl radicals) comprise only rapid propagation steps and they could be conducted in a wide variety of common solvents.

Aromatic hydrocarbons that lack benzylic hydrogen atoms react especially slowly with radicals. Thus benzene is a popular solvent. One can conveniently conduct exploratory reactions in d_6-benzene in an NMR tube and monitor progress by periodically recording an NMR spectrum of the reaction mixture. Higher boiling alternatives to benzene include dichlorobenzene and *t*-butylbenzene. While benzene is an acceptable solvent for small scale reactions, its toxicity can be a concern for industrial applications. *t*-Butyl alcohol[29] is an inexpensive, general purpose solvent that is an excellent substitute for benzene in many radical reactions. Its boiling point (83 °C) is similar to that of benzene (80 °C) and is convenient for the decomposition of some common radical initiators. Because *t*-butyl alcohol contains strongly bonded hydrogen atoms (methyl and hydroxy), rates of H-atom transfer are lower than those for other alcohols (such as methanol and ethanol) and for ethers. Scrupulous drying of *t*-butyl alcohol is not critical.

Alcohols, water and even acetic acid are useful solvents for some radical reactions. However, they cannot be employed with organometallics that are basic. Conversely, halocarbon solvents are popular for many types of ionic reactions but they are not generally useful for radical reactions. Chloroform and carbon tetrachloride (and to a lesser extent, dichloromethane) can interfere by donating either hydrogen or halogen atoms to intermediate radicals and they are used only in atom transfer reactions where the solvent is also a reagent.

Radicals are highly reactive intermediates, but they are often very chemoselective. This is because the rates of reactions of radicals with nonradicals are a function of the structure of both components and vary over a wide range. The planned execution of chemoselective radical reactions will be a major theme of the remainder of this chapter. For the moment, recall that reactions proceeding at rates below the the solution limit cannot be conducted. This negative statement has a very positive side: functional groups that react at rates below the critical limit can be deemed stable to free radical reaction conditions. The homolytic strength of O—H bonds makes them inert to hydrogen transfer reactions. Hence, free acids and alcohols do not interfere with radical reactions. Amide N—H bonds are also relatively strong. Even most bimolecular abstractions of hydrogen atoms from C—H bonds are relatively slow, competing only with other slow reactions. Even though acidic O—H and N—H groups do not require protection in most radical reactions, they are often present in protected form to be compatible with other reactions in a multistep sequence. Virtually all common protecting groups are stable to typical radical reactions. Potential problems can usually be anticipated based on standard radical reactions: protecting groups should not have weak C—H bonds six atoms from any pro-radical site (problems with 1,5-H transfer may arise) nor should they have reactive π-bonds five or six atoms away from a pro-radical site (cyclization to the protecting group may occur).

Unless a substrate or reagent contains an acidic or basic site, the conditions for most radical reactions are neutral. Thus, ionic side reactions such as base-catalyzed epimerization are rarely a problem. While radical reactions are typically conducted at temperatures above ambient, this is often solely for experimental convenience: most commercially available initiators require heating to generate radicals. Many radical reactions should succeed at lower temperatures provided that the chain is maintained (in chain methods) or that the rate of generation of radicals is sufficiently rapid (in nonchain methods). Low temperature initiators are available.[30,31]

In contrast to ionic or organometallic intermediates, most radicals are inert to reactions that involve fragmentation of a β-bond. Elimination of β-oxygen and β-nitrogen groups is frequently a rapid reaction

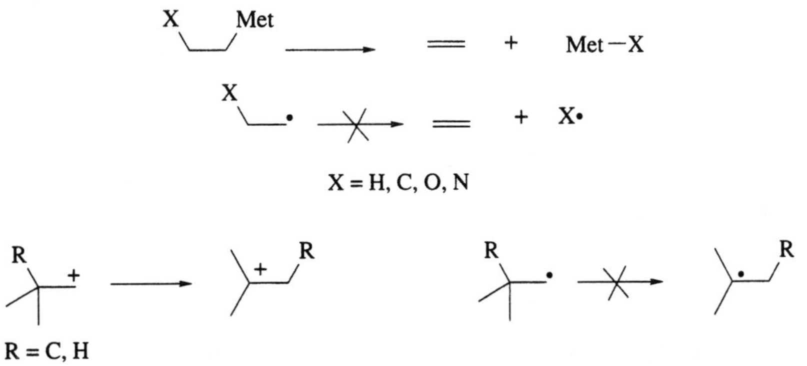

X Met → = + Met—X

X • ↛ = + X•

X = H, C, O, N

R = C, H

Scheme 7 β-Bond cleavage

that limits the use of many metals (Scheme 7). Transition metal alkyls are often prone to β-hydride eliminations. In contrast, the related elimination reactions of radicals are usually so slow in the normal reaction temperature range[32] that their occurrence can be discounted (for some useful exceptions, see Chapter 4.2, this volume). However, β-elimination reactions can be rapid and useful with halogens, sulfides and stannanes.

In carbocation chemistry, 1,2-alkyl and -hydride migrations are very common. In contrast, the direct 1,2-migration of hydrogen atoms to radicals is virtually unknown. While useful migrations of carbon substituents to radicals are known, they frequently proceed by a sequence of addition to a π-bond followed by elimination. The 1,2-migration of saturated alkyl groups to radicals is also virtually unknown.

4.1.2.5 Rate Constants and Synthetic Planning

Despite our increasing knowledge of the structures and reactions of organometallic intermediates, it is virtually impossible to plan complex ionic reactions by using known rate constants from simple model systems. Rate constants are potentially more useful when planning pericyclic reactions because they are easier to measure and because pericyclic reactions are much less susceptible to medium effects than ionic reactions. However, the need to evaluate the rates of competing pericyclic reactions is relatively rare (often there is only one reasonable possibility).

In contrast, the need to evaluate the relative rates of competing radical reactions pervades synthetic planning of radical additions and cyclizations. Further, absolute rate constants are now accurately known for many prototypical radical reactions over wide temperature ranges.[19,33–35] These absolute rate constants serve to calibrate a much larger body of known relative rates of radical reactions.[33] Because rates of radical reactions show small solvent dependence, rate constants that are measured in one solvent can often be applied to reactions in another, especially if the two solvents are similar in polarity. Finally, because the effects of substituents near a radical center are often predictable, and because the effects of substituents at remote centers are often negligible, rate constants measured on simple compounds can often provide useful models for the reactions of complex substrates with similar substitution patterns.

One of the goals of this and the following chapter is to show how a knowledge of a few of the most important rate constants of radical reactions and some substituent effect trends can be used to design successful experimental conditions for applications of known radical-based synthetic methods. The design of new synthetic methods is only a small step beyond this.

Table 1 contains representative absolute rate constants for prototypes of the most important radical reactions.[36] While some known rate constants (like those in Table 1) are very precise (<25% error), others are only rough estimates because they have not been measured directly but by competition against a reaction of known absolute rate. In such competition studies, any error in the original direct measurement is compounded by errors in the competition reaction. It is often difficult to estimate the error limits precisely (errors could exceed a factor of two). In some cases, rates measured by indirect methods become two (or even three) reactions removed from known absolute rate constants. These errors should not be ignored during synthetic planning. When estimating the rates of two competing radical reactions by using approximate rate constants, an order of magnitude is probably about the highest level of precision that is meaningful.

Rate constants are a function of temperature and should be compared at the same temperature whenever possible. Arrhenius equations for the reactions in Table 1 are provided and these equations can often be used as substitutes to calculate the temperature dependence for similar reactions. In practice, the error introduced by temperature uncertainties in the normal range of 25–80 °C is not particularly important in the comparison of two indirectly measured rate constants. Because competitive rate studies are so simple to conduct, synthetic chemists can (and sometimes do!) measure the rates of reactions that are required to plan a new synthetic method or total synthesis. Such 'quick and dirty' experiments are often conducted at only one temperature. A temperature in the middle of the normal range (~50 °C) is particularly useful because the rate constants that are obtained can be directly compared (with the above provisos) to rate constants in the normal range.

4.1.2.6 A Word about Concentrations

Include them! Synthetic chemists are accustomed to publishing communications or letters in which the conditions of relatively standard procedures are summarized in a scheme. Except for special reactions such as macrocyclizations, reaction concentrations are rarely included in these schemes. However, the

Table 1 Representative Rate Constants of Radical Reactions

Reaction	*Log A* ($M^{-1}\ s^{-1}$)	E_a (kcal mol^{-1})	*k at* 25 °C	50 °C	80 °C	*Ref.*
Fragmentation						
	13.15	7.05	0.91	2.3	$5.9 \times 10^{8}\ s^{-1}$	37
Cyclization						
	10.42	6.85	0.24	0.58	$1.5 \times 10^{6}\ s^{-1}$	38
Hydrogen transfer						
$Bu^{n}\bullet$ + $HSnBu_3$	9.06	3.65	2.3	3.8	$6.2 \times 10^{6}\ M^{-1}\ s^{-1}$	39
Iodine transfer						
n-$C_8H_{17}\bullet$ + ICH_2CO_2Et	10.4	4.4	1.4	2.6	$4.7 \times 10^{7}\ M^{-1}\ s^{-1}$	40
Thiohydroxamate transfer						
n-$C_8H_{17}\bullet$ + $C_8H_{17}C(O)O$–N (pyridine-2-thione)	9.04	3.95	1.3	2.3	$3.2 \times 10^{6}\ M^{-1}\ s^{-1}$	41
Alkene addition						
$Bu^{t}\bullet$ + $PhCH{=}CH_2$	7.60	3.42	1.2	1.9	$3.0 \times 10^{5}\ M^{-1}\ s^{-1}$	42
Pyridinium addition						
n-$C_7H_{15}\bullet$ + N^{+}–H	9.2	6.9	1.3	3.0	$8.1 \times 10^{4}\ M^{-1}\ s^{-1}$	18

relative rates of competing radical processes and the general efficiency of radical chain reactions are highly dependent on the concentration of the reagents. Omit the yield, the product ratio, the solvent, the temperature, the work-up procedure or the method of initiation, but never omit the concentration! For reactions (such as syringe pump additions) that are not conducted at a fixed concentration, a brief comment on the initial concentrations and the rate of addition is helpful.

4.1.3 METHODS TO CONDUCT RADICAL REACTIONS

Although the synthetic transformations that can be accomplished by radical reactions are virtually limitless, precious few general methods exist by which radical reactions can be conducted. These methods can be classified in two main groups: chain and nonchain.

As illustrated in Scheme 8, useful methods to conduct radical addition (and cyclization) reactions must: (i) generate an initial radical; (ii) provide this radical some lifetime in which to undergo the desired reaction(s); and (iii) productively remove the final radical before radical–radical, radical–solvent or undesired radical–reagent reactions intervene. High chemoselectivity is required at all stages. Radical generation must occur at a single site in the molecule to the exclusion of all other sites. The initially formed radical is usually a precursor for one or more subsequent radicals, hence, several intermediate radicals will be simultaneously present in a reaction. Ideally, each intermediate radical must suffer a single fate. This selectivity is imparted by the radicals themselves (structural variations in radicals can dramatically affect the rates of reactions with other molecules) and by the experimental conditions (especially concentration and temperature). Finally, the method employed must selectively remove the final radical in a sequence before it is destroyed, but it must not prematurely remove any prior radicals.

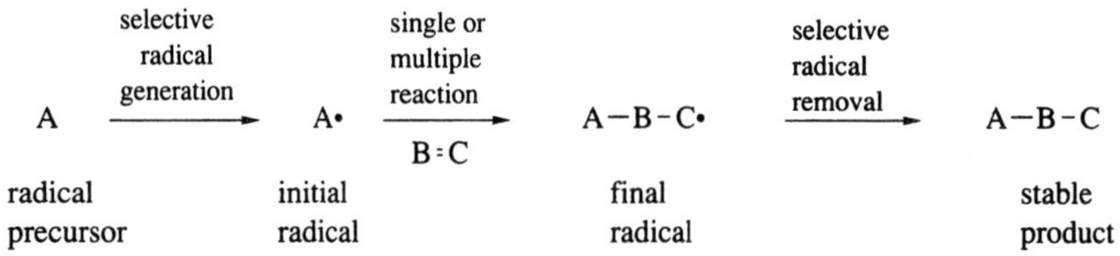

$$\underset{\text{radical precursor}}{A} \xrightarrow{\text{selective radical generation}} \underset{\text{initial radical}}{A\bullet} \xrightarrow[B=C]{\text{single or multiple reaction}} \underset{\text{final radical}}{A{-}B{-}C\bullet} \xrightarrow{\text{selective radical removal}} \underset{\text{stable product}}{A{-}B{-}C}$$

Potential problems: (1) radical–molecule or radical–solvent reactions of A• or A–B–C•; (2) removal of A• prior to reaction with B=C; (3) reaction of A–B–C with B=C prior to removal.

Scheme 8 Selectivity requirements

4.1.3.1 Chain Reactions

Chain reactions offer an ideal vehicle to conduct radical additions because the concentration of radicals is maintained at a low, steady level by nature (it is limited by the rates of initiation and termination) and because the first and last stages of the reaction sequence are directly linked. That is, the process of productively removing the final radical either directly or indirectly (*via* the intermediacy of a chain transfer agent such as $Bu_3Sn\cdot$) regenerates the initial radical.

Chain reactions comprise initiation, propagation and termination steps. Initiation steps generate radicals from nonradicals, while termination steps generate nonradicals by removing radicals. All of the desired transformations in a chain occur in propagation steps, which involve the inter- or intra-molecular reactions of radicals with nonradicals. Propagation steps effect no net gain or loss of radicals. In an ideal chain, initiation and termination steps are rare and propagation steps are common.

Initiation can be accomplished by photochemical or redox reactions, but it is most often accomplished by homolytic bond cleavage of a chemical initiator (promoted by heat or light) to give two radicals. A steady source of initiation is critical because, even under ideal conditions, the life of a single chain (from initiation to termination) rarely exceeds 1 s and can be much shorter if slow propagation steps are involved. The amount of initiator that is required depends on the efficiency of the chain (chain length) and on the temperature. Efficient chains (such as tin hydride reductions conducted at relatively high concentrations) require only trace amounts of initiator: 0.1–1% is more than sufficient. Low reagent concentrations or inherently slow reactions allow terminations to compete more effectively. To maintain the chain, additional initiator is required.

Most initiators are either peroxides (or peresters)[43] or azo compounds.[44] Several common initiators are listed in Scheme 9 along with their half-lives for unimolecular scission (first order decomposition to two radicals).[10] Although the peroxides shown are regarded as relatively stable, it is advisable to take appropriate safety precautions during their handling and use.[43] 2,2′-Azobisisobutyronitrile (AIBN) is probably the most generally useful and commonly employed initiator. It is a stable, free-flowing solid that decomposes in a useful temperature range for preparative experiments. This decomposition is well understood if not widely appreciated in the synthetic arena: AIBN decomposes to two free isobutyronitrile radicals[45] with only about 60% efficiency.[46] The remainder of the radicals undergo cage recombination.[47] The useful temperature range for AIBN is about 60–120 °C. At lower temperatures, too few radicals are produced. At higher temperatures, the decomposition of AIBN may be too rapid and the chain may break after the AIBN is consumed (if another initiation step is not available). To counter this rapid decomposition, AIBN can be added slowly to the reaction mixture.[48] Photolysis of initiators such as AIBN and benzoyl peroxide can generate radicals at room temperature or below and special low temperature initiators are not difficult to use.[30,31]

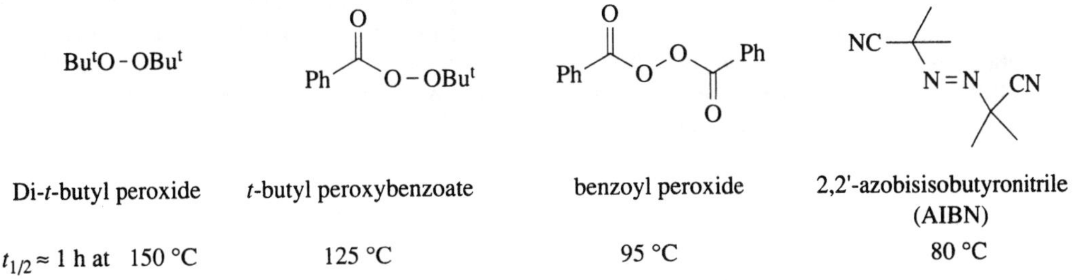

Bu^tO-OBu^t	PhC(O)O–OBu^t	PhC(O)O–OC(O)Ph	NC–C(Me)$_2$–N=N–C(Me)$_2$–CN
Di-*t*-butyl peroxide	*t*-butyl peroxybenzoate	benzoyl peroxide	2,2'-azobisisobutyronitrile (AIBN)
$t_{1/2} \approx$ 1 h at 150 °C	125 °C	95 °C	80 °C

Scheme 9 Useful initiators

Propagation steps are the heart of any chain and generally fall into two classes: atom or group transfer reactions and addition reactions to π-bonds (or the reverse: elimination). The rate of the chain transfer step is especially important in synthetic planning because, by fixing the maximum lifetime that radicals can exist, it determines what reactions will (or will not) be permitted. Termination steps are generally undesirable but are naturally minimized during chain reactions because initiation events are relatively uncommon.

The overall kinetics of a radical chain (that is, the rate at which a nonradical starting material is converted to a nonradical product) are well-modeled by the steady state approximation.[49] According to this treatment, the overall rate of a radical chain depends to the first order on the rate of the slowest propagation step and to the half order on the ratio of initiation to termination rates. Thus, the overall rate is not extremely sensitive to the initiator concentration. For synthetic purposes, a knowledge of the overall rate is not often especially useful; even slow propagation steps provide reasonable conversion rates. It is the knowledge of the kinetics of the individual, competing propagation steps that is useful.

4.1.3.2 Nonchain reactions

Nonchain processes differ significantly from chain processes in their methods of generating and removing radicals from a reaction. Nonetheless, useful reactions can be conducted between the generation and removal of a radical provided that the selectivity requirements are met (Scheme 8). In nonchain processes, radicals can be generated by stoichiometric bond homolysis and removed by selective radical–radical couplings. In effect, such reactions consist of only initiation and termination steps. Radicals can also be both generated and removed by redox methods (oxidative or reductive). Some fundamental redox reactions that can be used to generate both neutral and charged radicals are outlined in Scheme 10.

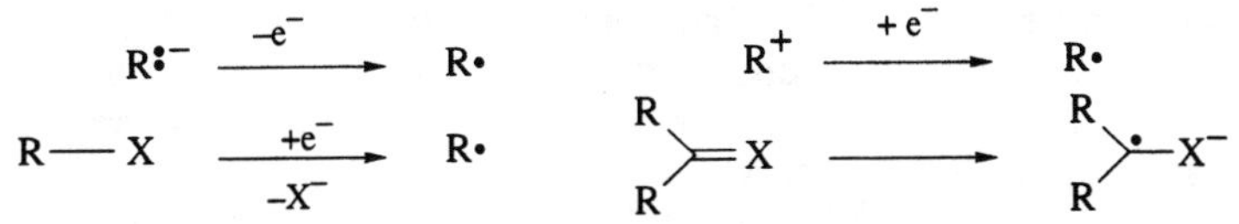

$$R{:}^- \xrightarrow{-e^-} R\bullet \qquad R^+ \xrightarrow{+e^-} R\bullet$$

$$R{-}X \xrightarrow[-X^-]{+e^-} R\bullet \qquad R_2C{=}X \longrightarrow R_2\dot{C}{-}X^-$$

Scheme 10 Redox generation of radicals

4.1.4 ELEMENTARY RADICAL REACTIONS

Almost all of the reactions of radicals can be grouped into three classes: redox reactions, atom (or group) transfer reactions and addition reactions. A detailed discussion of these reactions is beyond the scope of this chapter, but a summary of some important features (with references to more in-depth discussions) is essential. Although addition reactions will receive the most attention, redox and atom transfer reactions are important because nearly all radicals formed by addition reactions will be removed from the radical pool to give nonradical products by one of these methods.

4.1.4.1 Redox (Electron Transfer) Reactions

Radicals can be either reduced (to anions or organometallics) or oxidized to cations by formal single electron transfer (Scheme 11).[50] Such redox reactions can be conducted either chemically or electrochemically[51] and the rates of electron transfer are usually analyzed by the Marcus theory and related treatments.[50] These rates depend (in part) on the difference in reduction potential between the radical and the reductant (or oxidant). Thus a species such as an α-amino radical with high-lying singly occupied molecular orbital (SOMO) is more readily oxidized, while a species such as the malonyl radical with a low-lying SOMO is more readily reduced. The inherent difference in reduction potential of substituted radicals is an important control element in several kinds of reactions.

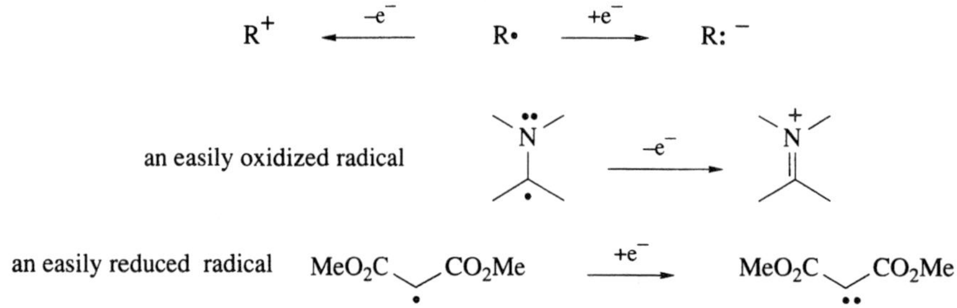

Scheme 11 Redox reactions of radicals

4.1.4.2 Atom and Group Transfer Reactions

In this very broad class of reactions, also called atom abstractions or homolytic substitutions, a univalent atom (hydrogen[52] or a halogen[53]) or a group (such as SPh, SePh) is transferred from a neutral molecule to a radical to form a new σ-bond and a new radical (Scheme 12). As with nucleophilic substitution reactions, the energy gained by the forming A—X bond helps to lower the transition state energy below that required for complete cleavage of the R—X bond. The site of the new radical is determined by the location of the abstracted atom or group X. The direction of the reaction is determined by the strength of the forming and breaking bonds and the rate of the reaction often roughly parallels its exothermicity. Endothermic atom or group transfers are rarely rapid enough to proceed at a useful rate so most transfers are irreversible under normal solution conditions. The reactions of alkyl radicals with alkyl iodides are an important exception to this generalization.

R–X + •A ⇌ R• + X–A

Scheme 12 Atom or group abstraction

Factors besides those associated with thermodynamics can also greatly effect the rates. For example, in the reaction of a carbon-centered radical with a univalent atom donor (alkyl halide or alkane, R—X), the following decreasing order of reactivity is observed: R—I > R—Br > R—H > R—Cl >> R—F.[53] The range of reactivity is impressive; alkyl iodides are outstanding iodine atom donors that can suffer even slightly endothermic atom transfers, while alkyl fluorides are virtually inert to even the most exothermic of reactions. In general, atoms connected by weaker bonds are transferred more rapidly, even though the forming bonds are also weaker. The unusual placement of hydrogen is thought to be due at least in part to repulsive interactions of the incoming radical with the lone pairs of chlorine and fluorine.

Polar effects can also be important in atom transfer reactions.[54] In an oft-cited example (Scheme 13), the methyl radical attacks the weaker of the C—H bonds of propionic acid, probably more for reasons of bond strength than polar effects. However, the highly electrophilic chlorine radical attacks the stronger of the C—H bonds to avoid unfavorable polar interactions. As expected, the hydroxy hydrogen remains intact in both reactions.

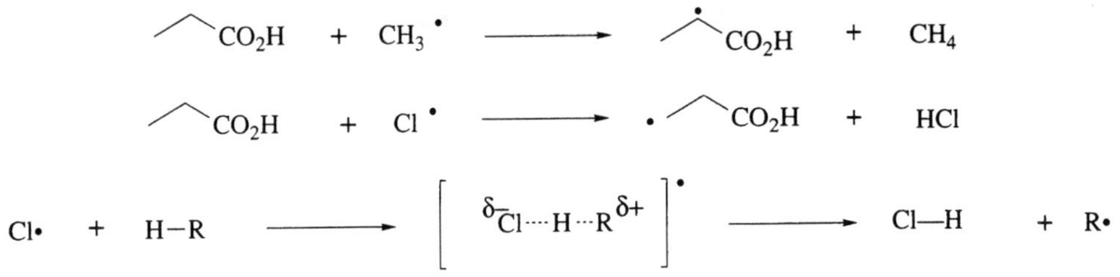

$$CH_3CH_2CO_2H + CH_3^{\bullet} \longrightarrow CH_3\dot{C}HCO_2H + CH_4$$

$$CH_3CH_2CO_2H + Cl^{\bullet} \longrightarrow {}^{\bullet}CH_2CH_2CO_2H + HCl$$

$$Cl^{\bullet} + H{-}R \longrightarrow \left[{}^{\delta-}Cl \cdots H \cdots R^{\delta+} \right]^{\bullet} \longrightarrow Cl{-}H + R^{\bullet}$$

Scheme 13 Polar effects

4.1.4.3 Addition Reactions

Redox and homolytic substitution reactions almost never directly form C—C, C—N and C—O bonds. Such bonds are generated in radical addition reactions (Scheme 14). Intermolecular addition reactions are presented in this chapter. Cyclization reactions have important similarities with, and differences from, bimolecular additions, and they are presented in Chapter 4.2 of this volume. Falling under the umbrella of addition reactions are radical eliminations (the reverse of addition) and radical migrations (which are usually, but not always, comprised of an addition and an elimination).

$$R^{\bullet} + A{=}B \rightleftharpoons R{-}A{-}B^{\bullet}$$

Scheme 14 Addition–elimination

Radical addition reactions are reversible in principle and sometimes in practice. The addition of a carbon-centered radical to a carbon–carbon double bond is not reversible at normal reaction temperatures because of its exothermicity; the conversion of a C—C π-bond to a C—C σ-bond is highly favorable. Additions can become reversible if the π-bond is strong or if the σ-bond is weak or if the starting radical is significantly more stable than the adduct radical. For example, additions of carbon-centered radicals to carbonyl groups are often reversible; fragmentation reactions of oxy radicals are more common than additions to carbonyls. Whether an addition or a fragmentation reaction occurs depends not only on the position of the equilibrium but also on the rates of subsequent reactions of the starting and product radicals with reactants. Thus, the choice of reaction method can greatly affect the product distribution when reversible reactions are involved. The additions of heteroatom-centered radicals to multiple bonds are frequently reversible (and the equilibrium may lie to the left) for elements (R) that are not in the first row. The fragmentation reactions of these radicals can be very useful, especially when R = SnR_3 or SPh.

The additions of carbon-centered radicals to C—C double and triple bonds are the most important reactions for preparative purposes and are also the most well understood. Because of the exothermicity of these additions, they are generally regarded as having early transition states. As such, thermodynamic considerations, such as stability of the adduct radical with respect to the starting radical, are not especially useful in predicting the rates of radical additions. However, Frontier Molecular Orbital (FMO) Theory[55] has provided an excellent framework to interpret and predict the rates of addition reactions of carbon-centered radicals.[3,56] The summary that follows is intended as an introduction of the most important concepts and conclusions. The authoritative review by Giese[56] is highly recommended as an in-depth treatment with references to the primary literature.

Radicals are often classified according to their rates of reactions with alkenes. Those radicals that react more rapidly with electron poor alkenes than with electron rich are termed nucleophilic radicals. Conversely, those that react more rapidly with electron rich alkenes than electron poor are termed electrophilic radicals. Recently, it has been found that this simple division does not suffice because certain radicals react more rapidly with both electron rich and electron poor alkenes than they do with alkenes of intermediate electron density. These radicals are termed ambiphilic. The appropriate pairing of a radical and an acceptor is important for the success of an addition reaction.

4.1.4.3.1 *Nucleophilic radicals*

Most synthetically useful radical addition reactions pair nucleophilic radicals with electron poor alkenes. In this pairing, the most important FMO interaction is that of the SOMO of the radical with the LUMO of the alkene.[56] Thus, many radicals are nucleophilic (despite being electron deficient) because they have relatively high-lying SOMOs. Several important classes of nucleophilic radicals are shown in Scheme 15. These include: heteroatom-substituted radicals, vinyl, aryl and acyl radicals, and most importantly, alkyl radicals.

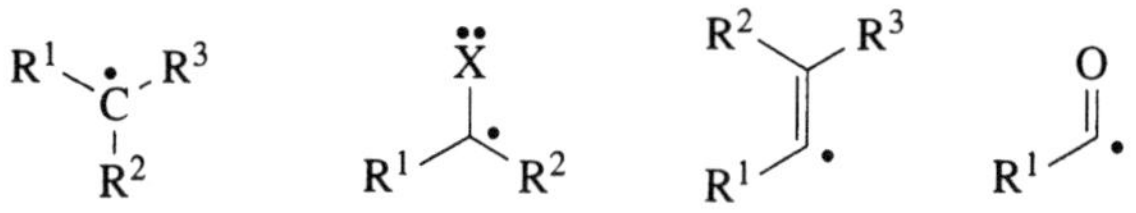

R^1, R^2, R^3, = H, alkyl, vinyl, aryl

Scheme 15 Nucleophilic radicals

A large number of accurate rate constants are known for addition of simple alkyl radicals to alkenes.[33–35] Table 2 summarizes some substituent effects in the addition of the cyclohexyl radical to a series of monosubstituted alkenes.[56] The resonance stabilization of the adduct radical is relatively unimportant (because of the early transition state) and the rate constants for additions roughly parallel the LUMO energy of the alkene. Styrene is selected as a convenient reference because it is experimentally difficult to conduct additions of nucleophilic radicals to alkenes that are much poorer acceptors than styrene. Thus, high yield additions of alkyl radicals to acceptors, such as vinyl chloride and vinyl acetate, are difficult to accomplish and it is not possible to add alkyl radicals to simple alkyl-substituted alkenes. Alkynes are slightly poorer acceptors than similarly activated alkenes but are still useful.[57]

Table 2 Substituent Effects — Alkene

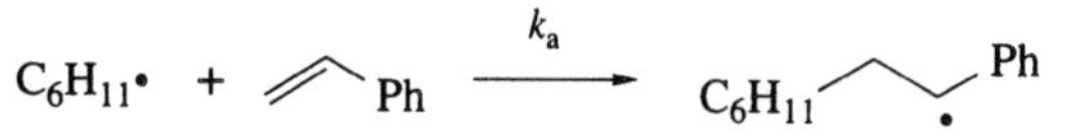

$C_6H_{11}\bullet$ + ═Ph $\xrightarrow{k_a}$ C_6H_{11}–CH₂–ĊH–Ph $k_{relative} = 1$ ($k_a \cong 4 \times 10^4\ M^{-1}s^{-1}$)

Alkene	k_{rel}	*Alkene*	k_{rel}	*Alkene*	k_{rel}
═CHO	34	═Cl	0.1	═CN	24
═OAc	0.02	═COMe	13	═Bu	<0.005

Other alkene substituent effects are summarized in Table 3 with reference to the addition of cyclohexyl radical to methyl acrylate, a good acceptor. β-Alkyl substituents have a powerful decelerating effect which is usually attributed to unfavorable steric interactions with the incoming radical. Indeed, this effect is very general; virtually any radical will add more rapidly to the less-substituted end of any multiple bond in any bimolecular reaction. Most preparatively useful addition reactions have been conducted with terminal alkene acceptors (methyl crotonate is less reactive than styrene). Also useful are alkenes with β-electron-withdrawing groups. These groups provide a very modest acceleration (compare methyl acrylate to dimethyl fumarate); apparently the electronic activation provided by the lowering of the LUMO slightly outweighs unfavorable steric interactions.

Table 3 Relative Rates of Addition of $C_6H_{11}\bullet$ to Acrylates

Compound	*Rate*	*Compound*	*Rate*
β═α CO₂Me	1	(crotonate) CO₂Me	0.01
═C(CO₂Me) CO₂Me	150	MeO₂C═CO₂Me	5

Substituent effects on the radical-bearing carbon are summarized by the examples in Table 4.[58] Although they must increase steric interactions between the radical and the alkene, alkyl substituents on the radical serve to raise the SOMO and a small acceleration of addition occurs with increasing substitution. This ability to tolerate a high degree of substitution at the radical-bearing carbon is an important asset for synthesis. The introduction of a powerful electron-withdrawing group on the radical generates an electronic mismatch and deceleration results. However, this 'mismatching' of electron-withdrawing substituents may not be as general as was once thought (see Ambiphilic Radicals, below). Conjugated substituents like aryl and vinyl thermodynamically stabilize the radical (by lowering the HOMO) without greatly affecting the position of the SOMO. Thus, most benzylic and allylic radicals will probably be nucleophilic but will exhibit reduced rates relative to simple alkyl radicals due to the reduced exothermicity of the addition.

Table 4 Substituent Effects — Radical

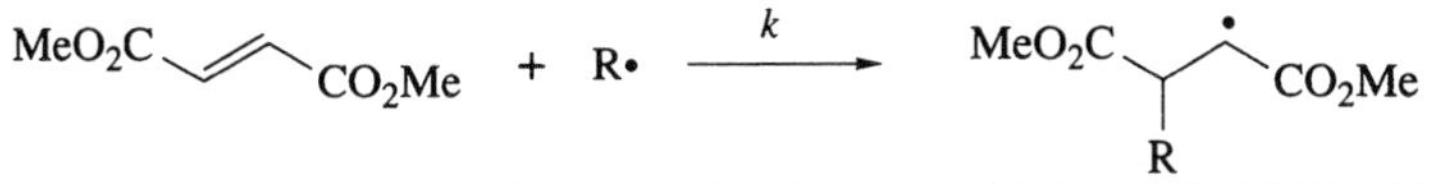

MeO_2C–CH=CH–CO_2Me + R• $\xrightarrow{k}$ MeO_2C–CH(R)–$\dot{C}$H–CO_2Me

Radical	k_{rel}	*Radical*	k_{rel}
n-C_7H_{15} •	1	Bu^t •	1.6
c-C_6H_{11} •	1.1	Alkyl-OCH_2 •	2.2

The transition states for additions of simple alkyl radicals to alkenes have been studied at high levels of theory[59] and the model that emerges (Figure 1) fits well with intuitive expectations based on an early transition state dominated by the SOMO(radical)–LUMO(alkene) interaction.[56] The forming C—C bond is very long yet still perfectly staggered, all the carbons involved are only slightly pyramidalized and the angle of attack of the radical is close to the ideal tetrahedral angle of 109°. There is little charge separation in the transition state, consistent with the small solvent effects that are usually observed. In summary, the reactions of nucleophilic radicals are favored by high radical SOMOs and low alkene LUMOs. Substituents are well tolerated at the radical-bearing carbon and the remote alkene carbon but strongly retard the addition when directly attached to the alkene carbon that is being attacked.

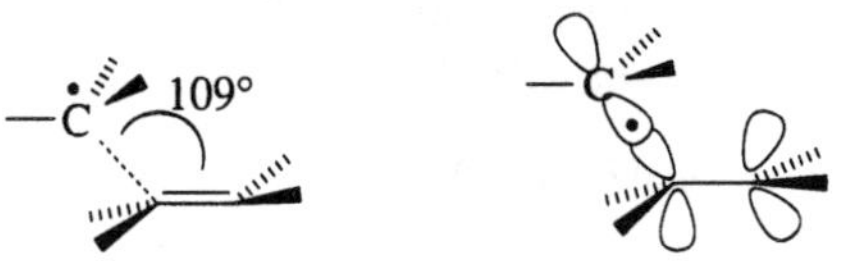

109°

Exothermic reaction, early transition state (TS)
Forming C—C bond staggered and >2.0 Å
All carbons only slightly pyramidalized

Figure 1 Transition state for addition of radicals to alkenes

4.1.4.3.2 Electrophilic radicals

Radicals that add more rapidly to electron rich alkenes and more slowly to electron poor alkenes are classified as electrophilic. Although the use of electrophilic radicals in synthesis is increasing, the body of rate data for electrophilic radical additions is presently much smaller than that for the nucleophilic counterparts.[60–62] Until very recently, virtually all radicals that possessed conjugative electron-withdrawing substituents were classed as electrophilic. The notion that ambiphilic radicals exist has only recently been introduced and the transition between nucleophilic, ambiphilic and electrophilic radicals is a continuum with rather ill-defined borders at present. The distinction between electrophilic radicals and ambiphilic radicals is especially gray. For the purposes of simplicity, radicals with two or more strong electron-withdrawing groups will be classed as electrophilic (perhalo-substituted radicals are also electrophilic),[63] while radicals with a single electron-withdrawing group will be classed as ambiphilic. However, certain radicals with one powerful electron-withdrawing group (like cyano) might best be thought of as electrophilic radicals. Future rate studies will no doubt define these borders more accurately.

Most of the information on substituent effects on the addition of electrophilic radicals to alkenes comes from several relative rate studies with substituted malonyl and malononitrile radicals. FMO theory predicts that such reactions will be accelerated by substituents that lower the SOMO of the radical or that raise the HOMO of the alkene. Experiments indicate that electron-donating alkene substituents modestly accelerate the additions of electrophilic radicals[60] and electron-withdrawing groups dramatically decelerate such additions. The relative rates of addition of the malononitrile radical to several monosubstituted alkenes are listed in Table 5.[61] The interplay between steric deactivation and electronic activation for substituents at the attacking carbon is less well defined. There are indications that electronic activating effects may outweigh steric effects for small, highly electrophilic radicals.[61,64]

Table 5 Substituent Effects — Electrophilic Radicals

$$\cdot CH(CN)_2 + CH_2{=}CHR \xrightarrow{k_a} (NC)_2CHCH_2\dot{C}HR$$

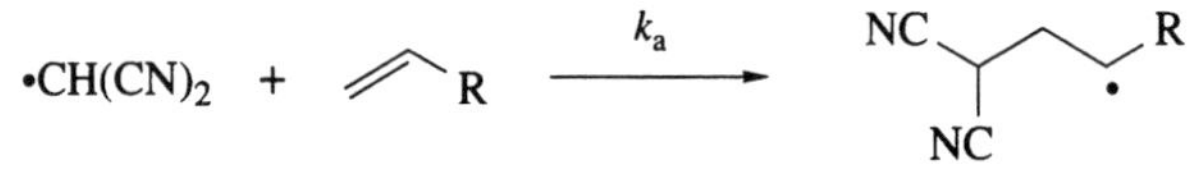

Alkene	k_{rel}	*Alkene*	k_{rel}
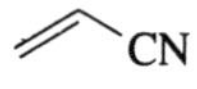 $CH_2{=}CHCN$	$<10^{-3}$	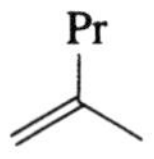 $CH_2{=}C(CH_3)Pr$	16
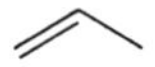 $CH_2{=}CHCH_3$	1		

4.1.4.3.3 *Ambiphilic radicals*

One of the triumphs of FMO theory is the ability to rationalize the reactivity profile of 1,3-dipoles.[65] Certain classes of dipoles prefer electron rich cycloaddition partners, while others prefer electron poor partners. There is a relatively large group of 1,3-dipoles that have a U-shaped reactivity curve; that is, their cycloaddition reactions are accelerated by both electron-donating and electron-withdrawing substituents on the alkene. Extending this cycloaddition analogy, there must be a certain number of radicals whose SOMOs lie in an intermediate range. The addition reactions of such ambiphilic radicals would be accelerated by either electron-withdrawing or electron-donating alkene substituents. Recent rate studies from the laboratories of Fischer[66] and of Giese[67] indicate that this is indeed the case. Fischer[66] has determined accurate rate constants for the addition of $\cdot CH_2CO_2Bu^t$ to 19 terminal alkenes. A classic U-shaped

Table 6 $\cdot CH_2CO_2Bu^t$ – An Ambiphilic Radical

$$RCH{=}CH_2 + \cdot CH_2CO_2Bu^t \longrightarrow R\dot{C}HCH_2CH_2CO_2Bu^t$$

R	k_{rel}	*R*	k_{rel}
CO_2Me	12	OEt	3
Et	1 ($5 \times 10^4\ M^{-1}s^{-1}$)	Ph	40

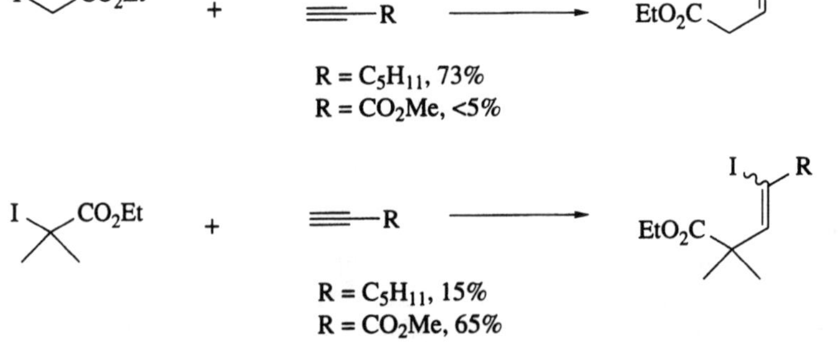

$$ICH_2CO_2Et + HC{\equiv}CR \longrightarrow EtO_2CCH_2CH{=}C(I)R$$

R = C_5H_{11}, 73%
R = CO_2Me, <5%

$$ICMe_2CO_2Et + HC{\equiv}CR \longrightarrow EtO_2CCMe_2CH{=}C(I)R$$

R = C_5H_{11}, 15%
R = CO_2Me, 65%

Scheme 16 Atom transfer addition to alkynes

reactivity curve is obtained (when log k is plotted against ionization potential) and this series of experiments is the most clear-cut evidence to date for the existence of ambiphilic radicals. Selected examples from this study are summarized in Table 6.

From the Fischer rate study, it appears that primary ester-substituted radicals are not electrophilic but ambiphilic and the borderline between ambiphilic and electrophilic radicals is not at all clear. Consider our results[68] (Scheme 16) on the atom transfer additions of ester-substituted radicals to alkynes (with the caution that it may be dangerous to compare yields in place of rate constants). The primary ester-substituted radical adds more efficiently to 1-heptyne but the tertiary ester-substituted radical prefers ethyl propiolate.

Because of the centrality of the carbonyl group in synthesis, carbonyl-substituted radicals are especially useful. The above results indicate that, if planned addition or cyclization reaction of a carbonyl-substituted radical fails due to lack of reactivity of the acceptor, one should consider activation of the alkene not only with electron donors but also with electron acceptors.

4.1.4.3.4 Heteroatom-centered radicals

The reactivity of heteroatom-centered radicals depends on both thermodynamic factors (stabilities of starting and product radicals and strengths of forming and breaking bonds) and kinetic factors. The electronegativity of the radical-bearing element is an important consideration. For example, radicals such as $Cl\cdot$, $RO\cdot$ and $R_3N^{+}\cdot$ are strongly electrophilic, while $R_3Sn\cdot$ is more nucleophilic.

4.1.5 RADICALS AND SYNTHETIC PLANNING

The notion of retrosynthetic analysis, first introduced by Corey,[69] is an excellent heuristic device that has become standard practice in synthetic planning.[70] Dissection of a molecule leads to imaginary synthons from which real reagents with the desired reactivity profiles are selected or designed. The notation of this thought process (and as a result, the process itself) has focused on the natural alternating polarity imposed on a carbon chain by a heteroatom. Bonds are dissected in a polar sense to give either normal or polarity-inverted (umpoled[71]) synthons.[72] The polarity of these synthons (nucleophilic or electrophilic) is often represented simply by plus (+) and minus (–) signs (Scheme 17). A more rigorous notation introduced by Seebach is also popular.[73] Here, nucleophilic sites are classed as donors (**d**) and electrophilic site are classed as acceptors (**a**). Such notations are useful not only in the planning of complex synthetic paths but also in the conceptualization of such paths (many steps can often be summarized with one picture) and in the conceptualization and design of multifunctional reagents.

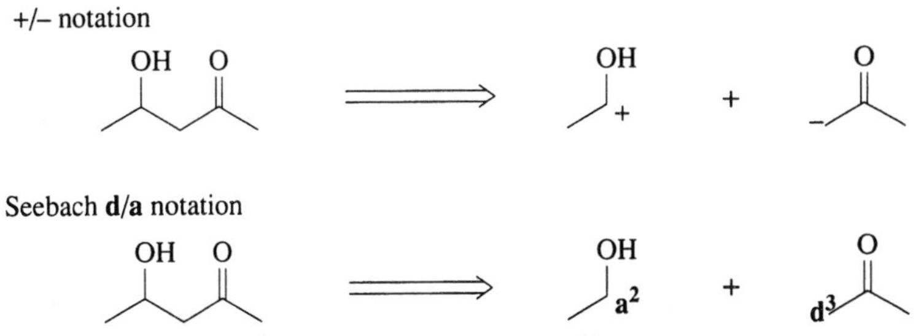

+/– notation

OH O ⟹ OH + O (synthons with + and – sites)

Seebach **d/a** notation

OH O ⟹ OH a^2 + O d^3

Scheme 17 Notations for retrosynthetic planning

As radical reactions are increasingly applied in synthesis, a convenient and informative method to notate these reactions in retrosynthetic analysis becomes desirable. Currently, two fragments united by a radical reaction are sometimes represented simply by dots '·'. Seebach has introduced the radical synthon, notated as '**r**' (Scheme 18).[74] Both of these notations have two related limitations: first, they imply that C—C bonds are formed by radical–radical coupling, and second, they do not indicate which site provided the radical and which site accepted it. I introduce here a notation for radicals in retrosynthetic analysis that strives to be in harmony with current ionic notations without artificially imposing[75] on radical reactions the features needed for the planning of ionic reactions.

In this notation, there are two kinds of sites that a synthon may possess: radical sites themselves and sites that react with radicals. In synthons, radical sites are designed simply by '·' to integrate with the

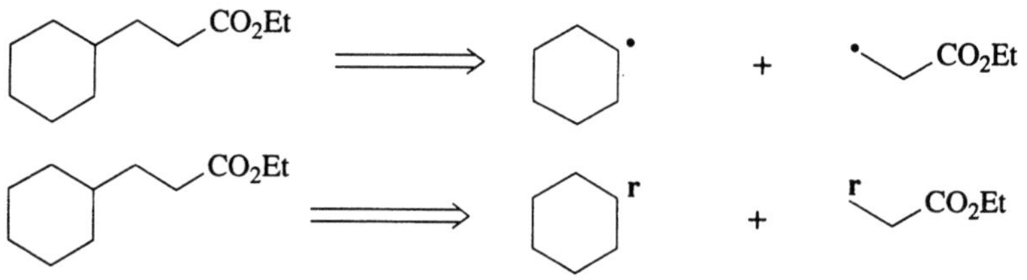

Scheme 18 Current notations for radical reactions

plus–minus convention and more precisely by '·r' to integrate with Seebach's donor–acceptor notation. Sites that react with radicals may be either atom (or group) donors or multiple bonds. Such sites are designated by '∘' in the plus–minus convention and by '∘' in the Seebach convention. This modification clearly shows the direction in which radical reactions are occurring and it is especially illustrative when sequences of reactions (either all radical or combined radical and ionic) are notated.

The simple addition reaction in Scheme 19 illustrates how the notation is used. Ester (**1**) can be dissected into synthons (**2**), (**3**) and (**4**). Synthons for radical precursors (pro-radicals) possess radical sites (·). A reagent that is an appropriate radical precursor for the cyclohexyl radical, such as cyclohexyl iodide, is the actual equivalent of synthon (**2**). By nature, alkene acceptors have one site that reacts with a radical (∘) and one adjacent radical site (·) that is created upon addition of a radical. Ethyl acrylate is a reagent that is equivalent to synthon (**3**). Atom or group donors are represented as sites that react with radicals (∘). Tributyltin hydride is a reagent equivalent of (**4**). In practice, such analysis will usually focus on carbon–carbon bond forming reactions and the atom transfer step may be omitted in the notation for simplicity.

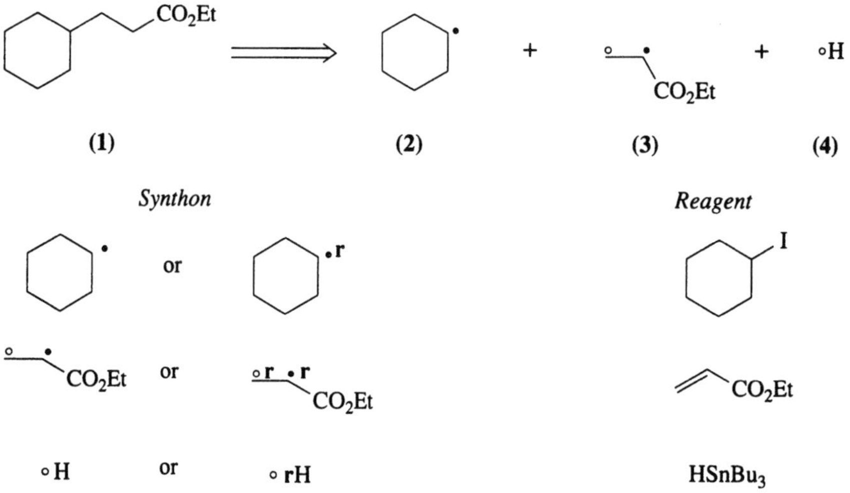

Scheme 19 Retrosynthetic notation of radical reactions

This notation blends smoothly with the plus–minus notation to indicate sequences of reactions or multifunctional reagents. In the simple, hypothetical transformation in Scheme 20, dibromomethane might serve as a reagent for the synthon (**5**). A slash (/) separates the two symbols (–/·) to indicate that (**5**) will behave separately as a nucleophilic site and a radical site, rather than as a radical anion. The more rigorous notation of Seebach avoids this confusion. Related transformations have actually been executed by Wilcox and Gaudino.[76]

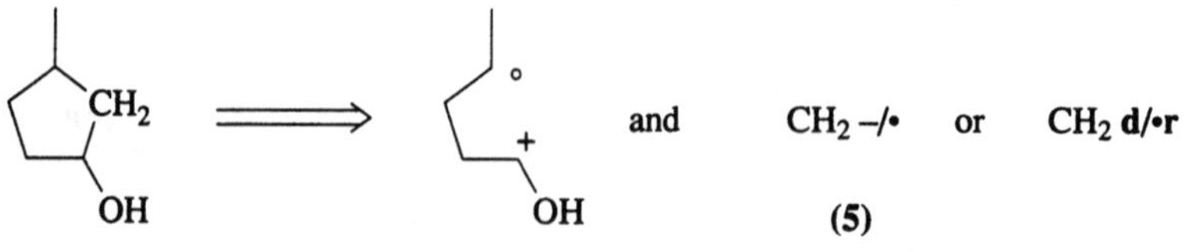

Scheme 20 A simple synthon

The notation becomes useful in the planning and conceptualization of a complex synthesis. A propellane triquinane such as modhephene presents almost limitless opportunities for the design of strategies based on radical cyclizations. Scheme 21 illustrates four of the 24 possible[77] strategies for the synthesis of modhephene that would conduct two radical cyclizations of a precursor that contains a pre-existing ring and two geminal side chains. In strategy A, both radical sites are generated in the side chains and both acceptor sites are on the ring. In strategy B, the roles are reversed; both radical sites are generated on the ring and cyclize to the chains. Strategies C and D have one site each on the ring and the chain. The notation systems used in Scheme 18 would have represented all four different strategies in Scheme 21 in the same way. In contrast, the new notation method clearly differentiates the four strategies.

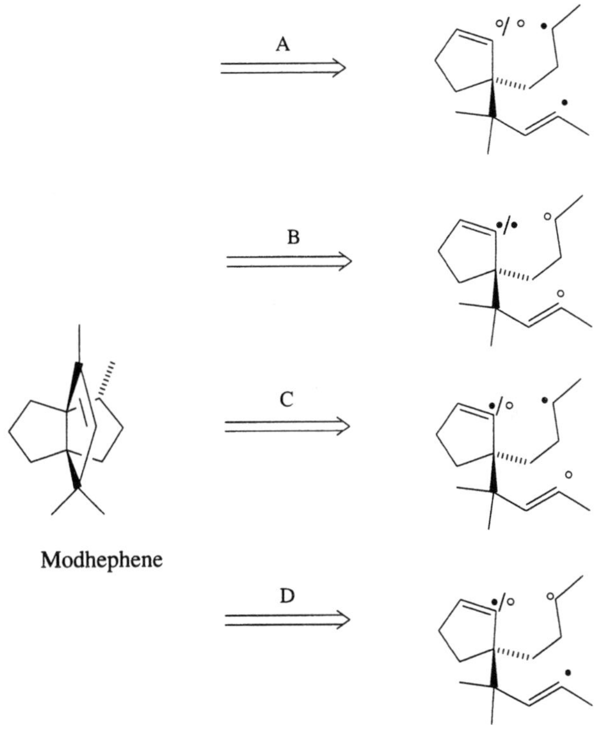

Modhephene

A

B

C

D

Scheme 21 Four strategies for modhephene

As with any retrosynthetic analysis, once potential strategies are identified, one must address the problems posed by each route. In the present synthesis, such questions arise as: Are the planned cyclizations likely to succeed? Will the two cyclizations be conducted together or separately? In what order will they occur? How will the stereochemistry of the methyl group be controlled? By what method will the cyclizations be conducted and what will the actual cyclization precursor(s) be? How will the precursors be prepared? The answers to these questions will help determine the selection of a practical synthetic pathway.

In practice, Jasperse has developed a synthesis of modhephene that is a variant of strategy A (Scheme 22).[78] Modhephene can be prepared from the enone (**6**). In turn, enone (**6**) is prepared by two separate radical cyclizations with complete control of regio- and stereo-chemistry. The vinylstannane serves as the synthetic equivalent of a double radical acceptor site (∘/∘). The whole synthetic route can be conceptualized in one simple diagram as shown in Scheme 23. Below the diagram, each synthon is associated with the actual starting material that was used to accomplish the required transformation in the synthesis.

Scheme 24 illustrates how this notation can be combined with that of Seebach.[73] Clive has formed a new ring by sequencing a Michael reaction, a carbonyl addition, and a radical cyclization.[79] Phenylselenoacrylonitrile is the actual reagent that accomplishes the transformation implied by synthon (**7**).

No heuristic device for retrosynthesis can replace a knowledge of the synthetic literature. This device aids the synthetic chemist in planning possible strategic routes to a target, but it remains to the chemist to evaluate the viability of these routes and to make an informed selection. A knowledge of the principles of radical addition and cyclization reactions and the methods to conduct these reactions is essential for the evaluation of synthetic strategies, however they are devised. To provide a foundation of this knowledge is the principle goal of this and the following chapter.

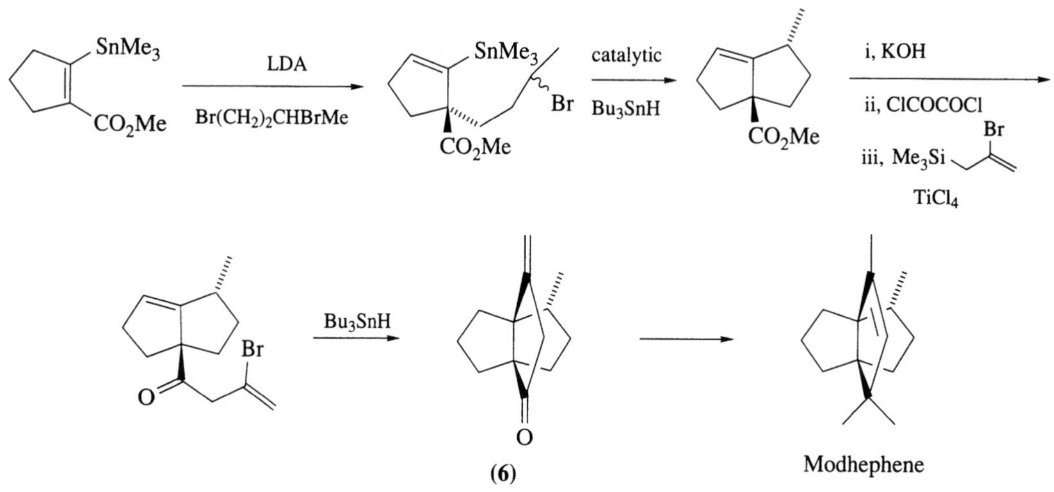
SnMe3
CO2Me
LDA
Br(CH2)2CHBrMe
SnMe3
Br
CO2Me
catalytic
Bu3SnH
CO2Me
i, KOH
ii, ClCOCOCl
iii, Me3Si
Br
TiCl4
Br
O
Bu3SnH
O
(6)
Modhephene

Scheme 22

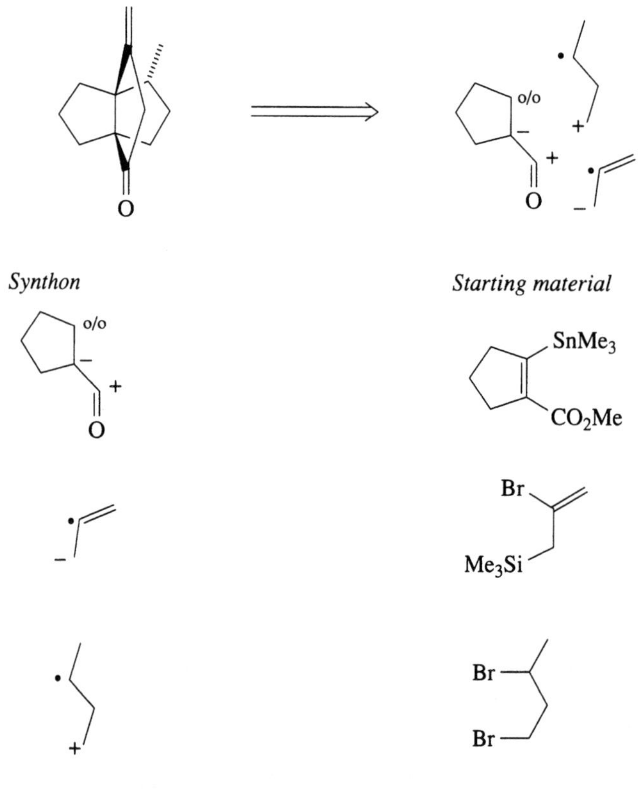
O
o/o
+
+
O
Synthon
Starting material
o/o
+
O
SnMe3
CO2Me
Br
Me3Si
+
Br
Br

Scheme 23 An analysis of the modhephene synthesis

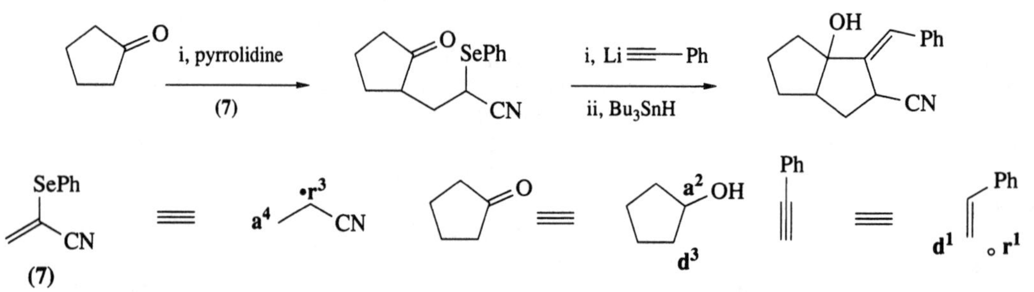
O
i, pyrrolidine
(7)
O
SePh
CN
i, Li≡—Ph
ii, Bu3SnH
OH
Ph
CN
SePh
CN
(7)
a4
•r3
CN
O
a2 OH
d3
Ph
Ph
d1
r1

Scheme 24 Seebach notation

4.1.6 ADDITIONS OF CARBON-CENTERED RADICALS TO ALKENES AND ALKYNES

One of the mildest general techniques to extend a carbon chain entails the addition of a carbon-centered radical to an alkene or alkyne. The method for conducting these addition reactions often determines the types of precursors and acceptors that can be used and the types of products that are formed. In the following section, synthetically useful radical additions are grouped into chain and non-chain reactions and then further subdivided by the method of reaction. Short, independent sections that follow treat the addition of carbon-centered radicals to other multiple bonds and aromatic rings and the additions of heteroatom-centered radicals.

4.1.6.1 Chain Methods

4.1.6.1.1 Metal hydrides (the Giese method)

Developed in large measure by Giese and coworkers, metal hydride mediated addition reactions are among the most useful and commonly employed synthetic reactions that involve radical intermediates. Two classes of metal hydrides have emerged as superior reagents to conduct addition reactions by radical chains: tin hydrides (and closely related germanium and silicon hydrides) and mercury(II) hydrides. This subject has been recently covered in detail in a number of specific[80–82] and general reviews[3–5,83,84] and the following treatment is intended only as an introduction.

(i) Tin hydride

The tin hydride method[85] generates radicals by abstraction of an atom or group from a pro-radical by $Bu_3Sn\cdot$ and removes radicals by hydrogen transfer from Bu_3SnH. Selectivity is often imparted by controlling reagent concentrations. Most of the applications of the tin hydride method pair nucleophilic radicals, generated from alkyl iodides, bromides or related precursors, with electron deficient alkenes (Scheme 25). This reaction is complementary to the more common tactic of conjugate addition of an organometallic reagent (such as a cuprate) to an electron deficient alkene. For many applications, the standard organometallic route will remain the method of choice because yields are consistently high and the conjugate addition reaction is more tolerant to the presence of β-substituents on the acceptor. However, there are situations in which the radical method may be advantageous. These include additions of secondary, tertiary and heteroatom-substituted radicals (where organometallic derivatives are often difficult to prepare or are unstable) and additions in which either the halide or the alkene contains functionality sensitive to organometallic methods. Scheme 26 provides some representative examples in which readily available, optically active precursors have been chain extended.[85c–87] Although very high yields have been obtained in favorable cases, isolated yields typically fall in the range of 40–75%. Reduction products or short telomers often account for the remainder of the mass balance (see below).

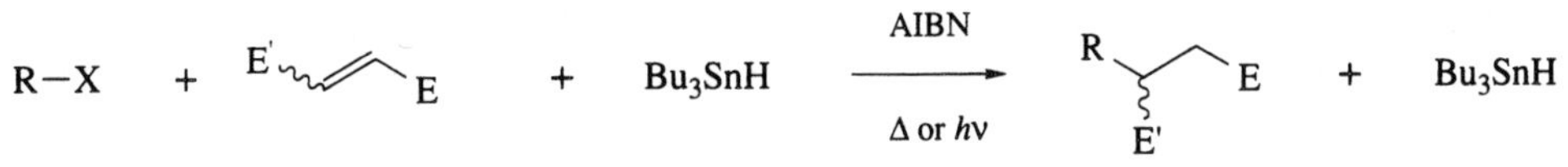

R—X + E'∼CH=CH–E + Bu_3SnH $\xrightarrow[\Delta \text{ or } h\nu]{\text{AIBN}}$ R–CH(E')–CH₂–E + Bu_3SnH

R = primary, secondary, tertiary or heteroatom-substituted alkyl, phenyl

X = I or Br, SePh, xanthate, NO_2

E = COR, CO_2R, CN, Ph, SO_2Ph

E' = H, activating group

Scheme 25

A detailed mechanistic analysis of the factors affecting the success of these tin hydride mediated addition reactions has been provided by Giese.[3] This analysis, which is especially illustrative of how experimental conditions for free radical reactions are planned, is summarized in Scheme 27. Three intermediate radicals, (**8**), (**9**) and (**10**), are involved. As is characteristic of all radical reactions, these radicals are simultaneously exposed to the same reagent pool and each can potentially undergo an addition reaction or an atom transfer reaction. The required reaction of the tributyltin radical (**8**) is atom ab-

optically active from Sharpless epoxidation

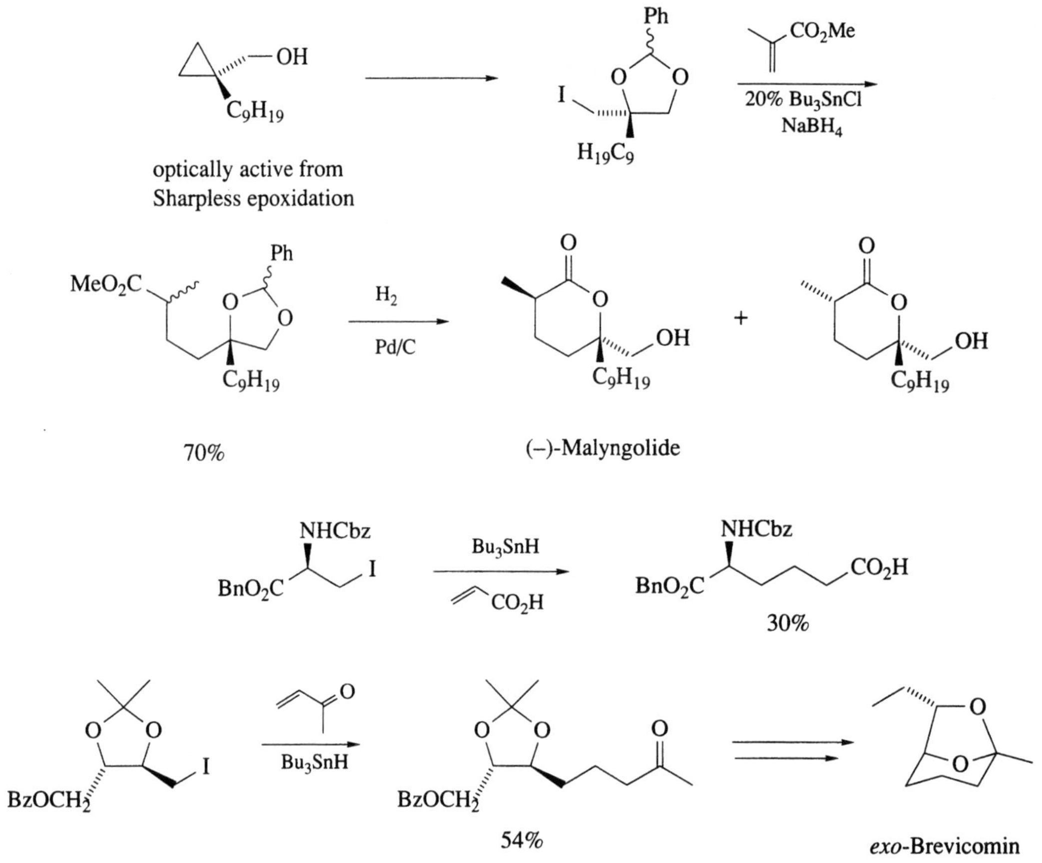

20% Bu_3SnCl $NaBH_4$

H_2 Pd/C

70%

(–)-Malyngolide

Bu_3SnH

30%

Bu_3SnH

54%

exo-Brevicomin

Scheme 26

straction to generate the initial carbon-centered radical (**9**); a possible competing reaction is the well-known hydrostannylation reaction of electron deficient alkenes. Because it is not possible to use low alkene concentrations to retard hydrostannylation (this also retards the desired addition), the selectivity problem in step 1 is solved by using a highly reactive atom donor. Iodides are often the precursors of choice because trialkyltin radicals abstract iodine from alkyl iodides at rates approaching the diffusion-controlled limit.[88] Iodide precursors of more stable radicals (tertiary alkyl, α-oxy, acyl) are often difficult to prepare but are not usually required. Substituents that stabilize radicals also facilitate atom transfer and less reactive bromides, phenyl selenides, nitro groups and xanthates often become acceptable precursors.

In steps 2 and 3, addition to the electron deficient alkene and hydrogen abstraction from tin hydride compete. Step 2 requires addition of R· to be faster than hydrogen atom transfer and step 3 requires precisely the reverse. That this reaction can be conducted at all is a result of the electronic properties of the intermediate radicals: alkyl radical (**9**) is a nucleophilic radical that pairs well in the addition to the electrophilic alkene, but the resulting adduct radical (**10**) does not add as rapidly in the corresponding reaction to give (**11**). Better yields are obtained as the alkene becomes more electron deficient because electron-withdrawing substituents increase k_{a1} and, at the same time, may decrease k_{a2}. The ratio of the two rate constants (k_{a1}:k_{a2}) may range anywhere from 10 in borderline cases up to 1000 in ideal cases, depending on radical and alkene substituents. The rate constants k_{H1} and k_{H2} are also important and it is often assumed that they are approximately equal.[89]

To successfully conduct the desired addition reaction, one must adopt conditions so that the partitioning in step 2 between reduction and addition of (**9**) favors addition, without going so far as to disrupt the natural rate constant bias in favor of reduction over addition in step 3. A specific example, the addition of cyclohexyl radical to methyl acrylate, best illustrates this point. Equation (5) states the requirement that addition of the cyclohexyl radical be more rapid than reduction with tin hydride. The rate constant for the addition of the cyclohexyl radical is about 3×10^5 M^{-1} s^{-1} while the rate constant for abstraction of hydrogen is about 2×10^6 M^{-1} s^{-1}. This slight bias in rate constant ratio in favor of hydrogen transfer must be offset by altering the concentrations of the reagents as indicated in Equation (6); the concentration of methyl acrylate must be greater than that of tin hydride. This alteration of reagent concentrations is ac-

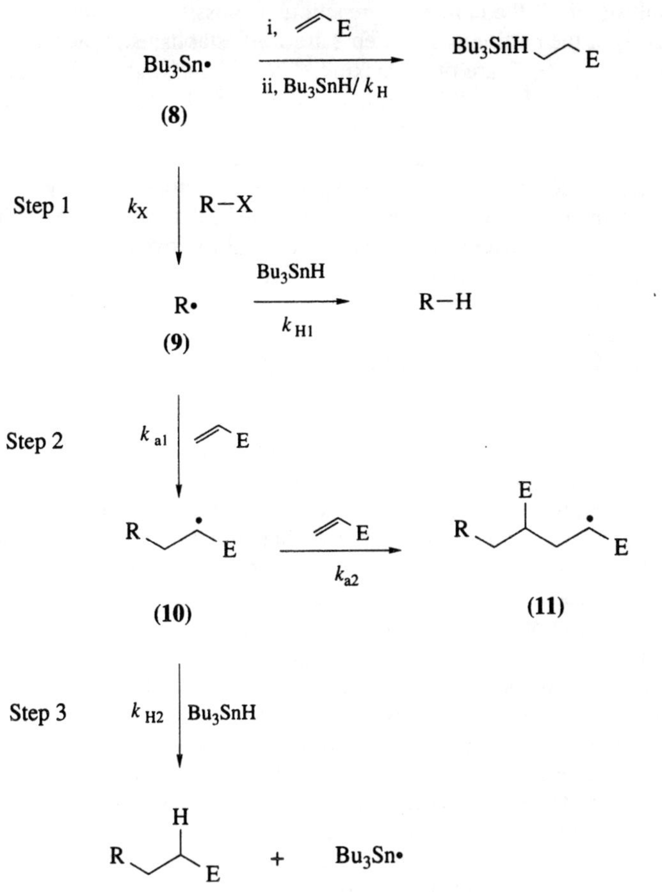

$$Bu_3Sn\bullet\ (\mathbf{8}) \xrightarrow[\text{ii, } Bu_3SnH/k_H]{\text{i, } =\!\!\!\diagup E} Bu_3SnH\!\!\diagdown\!\!\diagup E$$

Step 1 $\quad k_X \downarrow R{-}X$

$$R\bullet\ (\mathbf{9}) \xrightarrow[k_{H1}]{Bu_3SnH} R{-}H$$

Step 2 $\quad k_{a1} \downarrow =\!\!\!\diagup E$

$$R\diagdown\!\diagup\!\overset{\bullet}{C}\diagdown E\ (\mathbf{10}) \xrightarrow[k_{a2}]{=\!\!\!\diagup E} R\diagdown\!\diagup\!CH(E)\diagdown\!\diagup\!\overset{\bullet}{C}\diagdown E\ (\mathbf{11})$$

Step 3 $\quad k_{H2} \downarrow Bu_3SnH$

$$R\diagdown\!\diagup\!CH\diagdown E \quad + \quad Bu_3Sn\bullet$$

Scheme 27

ceptable provided that it is not sufficiently great to offset the inherent rate constant bias in favor of hydrogen transfer in step 3, as illustrated in equations (7) and (8).

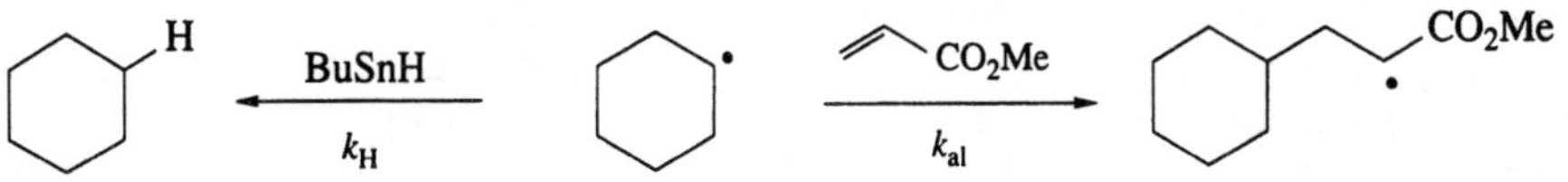

$$\text{cyclohexane (C–H)} \xleftarrow[k_H]{BuSnH} \text{cyclohexyl}\bullet \xrightarrow[k_{a1}]{=\!\!\!\diagup CO_2Me} \text{cyclohexyl}\diagdown\!\diagup\!\overset{\bullet}{C}H\diagdown CO_2Me$$

$$k_{a1}[=\!\!\!\diagup CO_2Me] \quad > \quad k_H[Bu_3SnH] \tag{5}$$

$$k_{a1} \approx 3\times10^5\ M^{-1}\,s^{-1}$$

$$k_H \approx 2\times10^6\ M^{-1}\,s^{-1}$$

$$\frac{[=\!\!\!\diagup CO_2Me]}{[Bu_3SnH]} \quad > \quad 7 \tag{6}$$

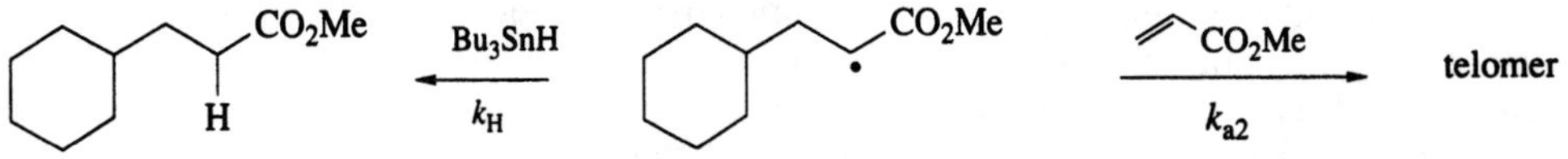

$$\text{cyclohexyl}\diagdown\!\diagup\!CH(H)\diagdown CO_2Me \xleftarrow[k_H]{Bu_3SnH} \text{cyclohexyl}\diagdown\!\diagup\!\overset{\bullet}{C}H\diagdown CO_2Me \xrightarrow[k_{a2}]{=\!\!\!\diagup CO_2Me} \text{telomer}$$

$$k_{a2}\left[=\!\!\!\diagup CO_2Me\right] \quad < \quad k_H[Bu_3SnH] \tag{7}$$

$$7k_{a2} \quad < \quad k_H \tag{8}$$

A quantitative analysis of all the competing reactions is possible in some simple systems. Rate constants to permit analysis of the partitioning at step 2 are well established; however, specific rate constants to model the partitioning at step 3 are often lacking.[89] Nonetheless, general guidelines are forthcoming from the mechanistic analysis. First and foremost, use the most reactive acceptor possible in order to favor addition in step 2 (by increasing k_{a1}) and reduction in step 3 (by decreasing k_{a2}). This simplifies the design of reaction conditions. Next, once an acceptor is selected, choose reaction conditions carefully to maximize the yield of addition. For reactive, inexpensive acceptors like acrylonitrile, the components can simply be mixed in an appropriate ratio and the reaction initiated.[90] However, because alkyl radicals react with most acceptors with rate constants equal to or slightly lower than with tin hydride, it is desirable to maintain a situation where the concentration of the alkene exceeds that of the tin hydride. This is best accomplished by using reaction conditions that minimize the tin hydride concentration and by employing a modest excess of an alkene acceptor. The two most popular ways to mininize the tin hydride concentration are by using syringe pump addition techniques or by using a limited quantity of a catalytic precursor of tin hydride.

In the syringe pump method, the halide and the alkene acceptor are typically refluxed in benzene and a solution of tributyltin hydride and AIBN in benzene is added slowly over a period of hours by syringe drive. The exact concentration of tin hydride is not known but it remains low, provided that the chain continues at a rate more rapid than syringe pump addition. Telomerization may intervene if the tin hydride concentration falls too low. To insure that the chain continues, it is important to use a reactive atom donor (to maximize the rate of step 1) and to insure that a constant source of initiator is present.[91]

Catalytic procedures (introduced by Kuivila and Menapace[92]) are easier to conduct and the tin hydride concentration is more easily controlled. A catalytic amount of tributyltin hydride or tributyltin chloride is mixed with the radical precursor, the alkene acceptor and a stoichiometric quantity of a coreductant such as sodium borohydride[93] or sodium cyanoborohydride.[29] Over the course of the reaction, the borohydride continuously converts the tin halide to tin hydride. The use of the catalytic procedure is probably restricted to halide precursors (tin products derived from other precursors may not be reduced to tin hydrides). This method has several advantages over the standard procedures: (i) it is simple to conduct; (ii) most functional groups are stable to the coreductants (especially sodium cyanoborohydride); (iii) the tin hydride concentration is known, is stationary (assuming that the tin halide is rapidly reduced to tin hydride), and can be varied by either changing the concentration of the reaction or the quantity of the tin reagent (10% is a typical value, but lower quantities can be used); and finally, (iv) the amount of tin hydride precursor that is added limits the amount of tin by-product that must be removed at the end of the reaction.

This last advantage should not be underestimated because the removal of large amounts of tin products on a preparative scale can be inconvenient. With relatively polar organic products, the non-polar tin products are often removed by chromatographic purification.[80] Partitioning between acetonitrile and hexane has also been used.[94] Nonpolar reaction products sometimes present more problems. Pretreatment of reaction mixtures with fluoride anion to convert tin halides to insoluble tin fluorides in a popular work-up technique.[95] Chang has recently introduced a work-up procedure with DBU that serves to remove much of the tin halide by-product from a crude product prior to chromatography.[96] These work-up procedures are designed to remove tin halides and it is recommended that any work-up procedure be prefaced by treatment of the crude reaction mixture with iodine to destroy any residual tin hydride.[97] Methods based on polymer-bound tin hydrides have not become popular, but offer the potential advantages of low concentration of tin hydride and easy removal of the tin products.[98]

Another strategy to improve the partitioning of the initial radical between direct reduction and desired addition is to design a reagent that is a poorer hydrogen donor than tributyltin hydride.[99] Bulky substituents on the tin atom do not significantly decrease the rate of hydrogen transfer and substituents like aryl groups and heteroatoms actually accelerate hydrogen transfer.[100] Germanium is above tin in the periodic table and forms stronger bonds to most elements, making it a poorer atom donor and a better atom abstractor. Indeed, tributylgermanium hydride is known to be about 10–20 times poorer as a donor of a hydrogen atom towards alkyl radicals than tin hydride[38,101] and it is also a powerful halogen atom abstractor.[38] Thus it would appear to offer significant advantages over tributyltin hydride in slow addition reactions by giving better partitioning in favor of addition over reduction.[99] However, there are two problems. First, hydrogermylation is more rapid than hydrostannylation and it can be a serious competing side reaction,[102] and second, there is evidence that reactions of germanium hydride with stabilized radicals (such as tertiary alkyl) may be too slow to propagate chains at low concentrations[101] (that is, what is gained by decreasing k_H might be lost by increasing the germanium hydride concentration in order to maintain the chain).[103]

A detailed study by Hershberger has evaluated the efficacy of germanium hydride compared to tin hydride under a standard set of conditions.[104] In general, tin hydride propagates chains more efficiently[105] and gives superior yields in additions to acrylonitrile. However, in the addition of a primary radical to acrylonitrile (Scheme 28), the germanium hydride was found to be superior provided that acetonitrile was used as a solvent rather than benzene[105] and that an iodide was used as the radical precursor.[102]

$C_{11}H_{23}I$ (0.1 M) + $Bu_3M{-}H$ (0.1 M) + CN (0.15 M) ⟶ $C_{13}H_{27}CN$ + $C_{11}H_{24}$

	$C_{13}H_{27}CN$	$C_{11}H_{24}$
Bu_3SnH/PhH	40%	47%
Bu_3GeH/MeCN	71%	11%

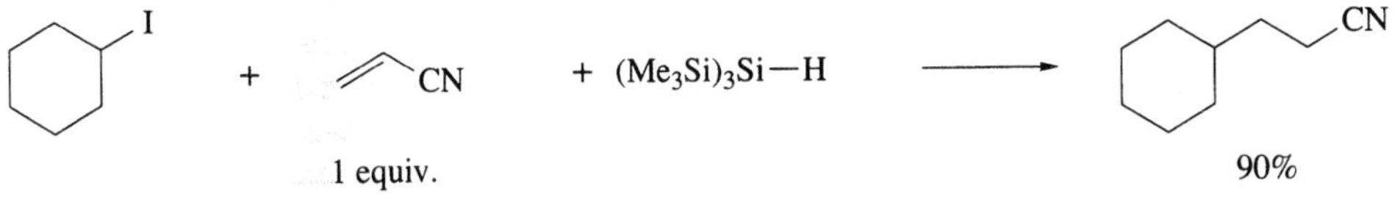

I + CN (1 equiv.) + $(Me_3Si)_3Si{-}H$ ⟶ CN 90%

Scheme 28

Silicon–hydrogen bonds in most trialkylsilanes are too strong to donate hydrogen atoms to alkyl radicals. However, Griller and Chatgilialoglu have recently shown that tris(trimethylsilyl)silane has a significantly weaker Si—H bond than normal silicon hydrides (but still stronger than tin hydrides) and that it can be used as a reagent to propagate chain reductions of halides.[106] Very recently, Giese and Chatgilialoglu have shown that this new reagent is a useful substitute for tin hydride in simple addition and cyclization reactions (Scheme 28).[107] Future work will be required to determine the general suitability of this promising reagent as a substitute for tin hydride.

The level of stereoselectivity in tin hydride mediated addition reactions often depends on whether cyclic or acyclic systems are involved. Two recent, representative examples are shown in Scheme 29.[108,109] Good selectivity is often observed with cyclic radicals in both the addition and hydrogen transfer steps. The reagent attacks the less-hindered face of the radical. Acyclic systems have been less frequently studied and typically show low asymmetric induction in either addition or hydrogen transfer reactions.[110] There is at least one example where reasonable stereoselectivity has been observed (albeit in only 13% yield) in the addition of an alkyl radical to an acceptor bearing a stereogenic center (Scheme 30).[111] There are currently no generally useful methods for absolute asymmetric induction although this is an ongoing area of research in several laboratories.

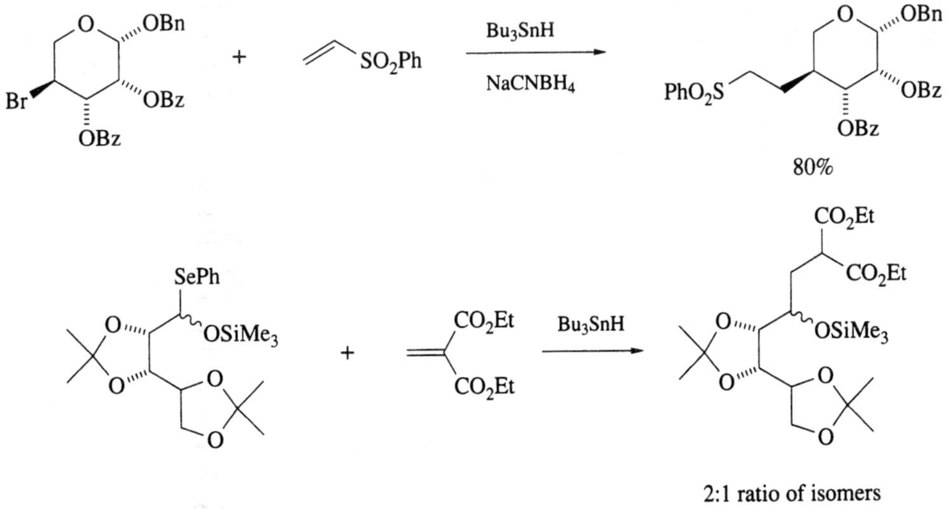

O OBn Br OBz OBz + SO_2Ph $\xrightarrow[NaCNBH_4]{Bu_3SnH}$ PhO_2S O OBn OBz OBz 80%

SePh O O $OSiMe_3$ O O + CO_2Et CO_2Et $\xrightarrow{Bu_3SnH}$ CO_2Et CO_2Et O O $OSiMe_3$ O O

2:1 ratio of isomers

Scheme 29

In addition to alkyl radicals, other nucleophilic radicals can participate in tin hydride mediated addition reactions (Scheme 31). Phenyl radicals are highly reactive[112] but give only modest yields of addition products[113] and one suspects that vinyl radicals should behave similarly.[114] α-Oxy radicals have been

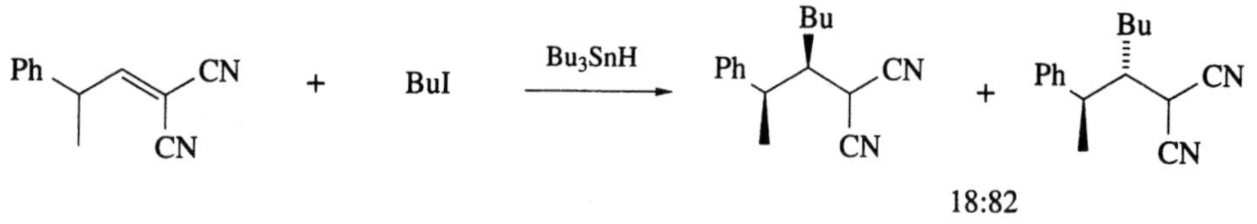

Ph CN CN + BuI $\xrightarrow{Bu_3SnH}$ Ph Bu CN CN + Ph Bu CN CN

18:82

Scheme 30

used frequently and with good success, especially at the anomeric center in carbohydrates. In such anomeric radicals, stereoelectronic effects can cause very high levels of asymmetric induction.[115,116] Very recently, acyl selenides have been demonstrated to be useful precursors for the addition of acyl radicals to alkenes.[117]

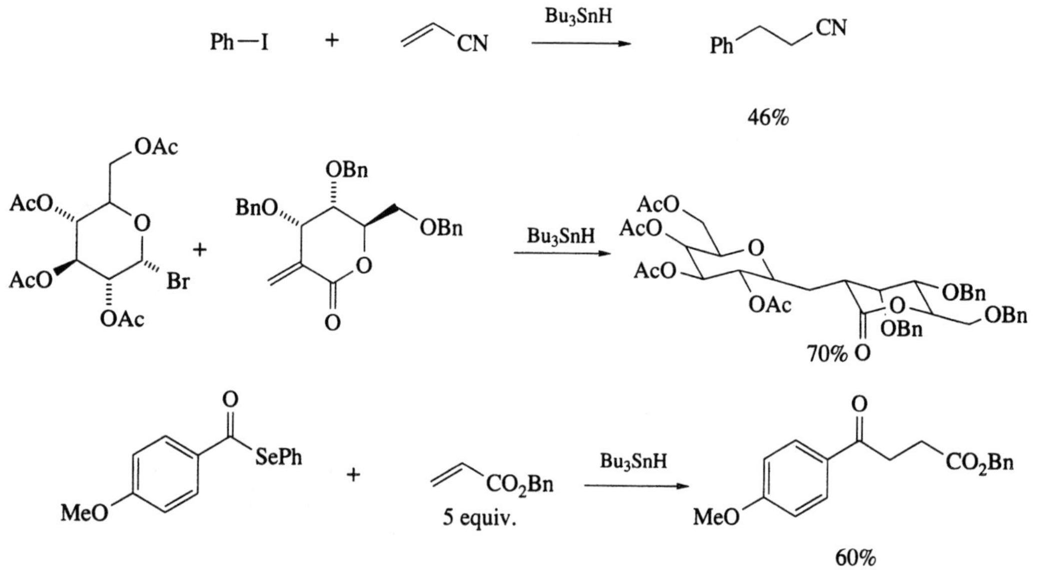

Ph—I + CN $\xrightarrow{Bu_3SnH}$ Ph CN

46%

AcO OAc O AcO OAc Br + BnO OBn OBn O O $\xrightarrow{Bu_3SnH}$ AcO AcO O AcO OAc O OBn OBn OBn O

70%

MeO SePh O + CO_2Bn 5 equiv. $\xrightarrow{Bu_3SnH}$ MeO O CO_2Bn

60%

Scheme 31

There are several examples of the addition reactions of carbonyl-substituted radicals to alkenes by the tin hydride method. The first reaction cited in Scheme 32 is a clear-cut example of reversed electronic requirement: an electrophilic radical pairing with a nucleophilic alkene.[60] Because enol ethers are not easily hydrostannylated, the use of a chloride precursor (which is activated by the esters) is possible. Indeed, the use of a bromomalonate results in a completely different product (Section 4.1.6.1.4). The second example is more intriguing (especially in light of the recent proposals on the existence of ambiphilic radicals) because it appears to go against conventional wisdom in the pairing of radicals and acceptors.[118,119]

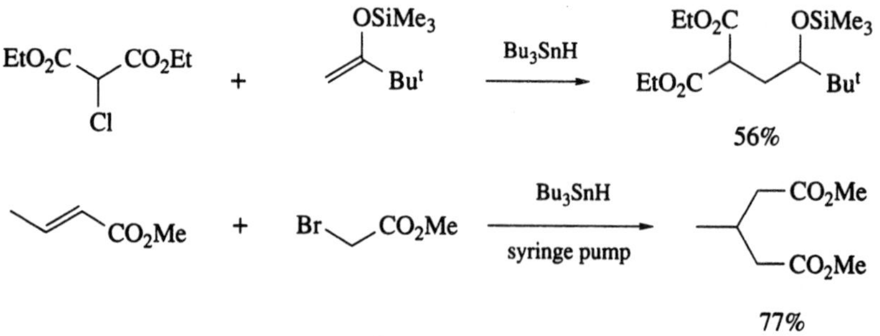

EtO_2C CO_2Et Cl + $OSiMe_3$ Bu^t $\xrightarrow{Bu_3SnH}$ EtO_2C EtO_2C $OSiMe_3$ Bu^t

56%

CO_2Me + Br CO_2Me $\xrightarrow[\text{syringe pump}]{Bu_3SnH}$ CO_2Me CO_2Me

77%

Scheme 32

(ii) Mercury(II) hydrides

Radical addition reactions conducted by the mercury(II) hydride method were also pioneered by Giese and are very similar in principle to the tin hydride method. The radical is generated from an organomercurial (rather than a halide) and removed by hydrogen transfer from a mercury(II) hydride (rather than a tin hydride). Mercury(II) hydride reductions have been covered in several recent reviews.[3,5,81,82]

The basic transformation and mechanism are outlined in Scheme 133. Addition of a reducing agent ($NaBH_4$, $NaCNBH_3$, Bu_3SnH) to a solution (methanol, THF, dichloromethane) of an organomercury(II) halide or acetate in the presence of an excess of a reactive acceptor provides addition products in modest to good yield. The directly reduced hydrocarbon is a common by-product. It is believed that the organomercurial is rapidly reduced to an alkylmercury(II) hydride and the accepted mechanism for addition is a mercury(II) hydride chain.[120] This chain is probably initiated by spontaneous decomposition of the alkylmercury(II) hydride. Alkyl radicals can compete for addition of the alkene (step 1) or hydrogen atom abstraction from the alkylmercury(II) hydride. Adduct radicals transfer the chain by abstracting hydrogen from the mercury(II) hydride (step 2).

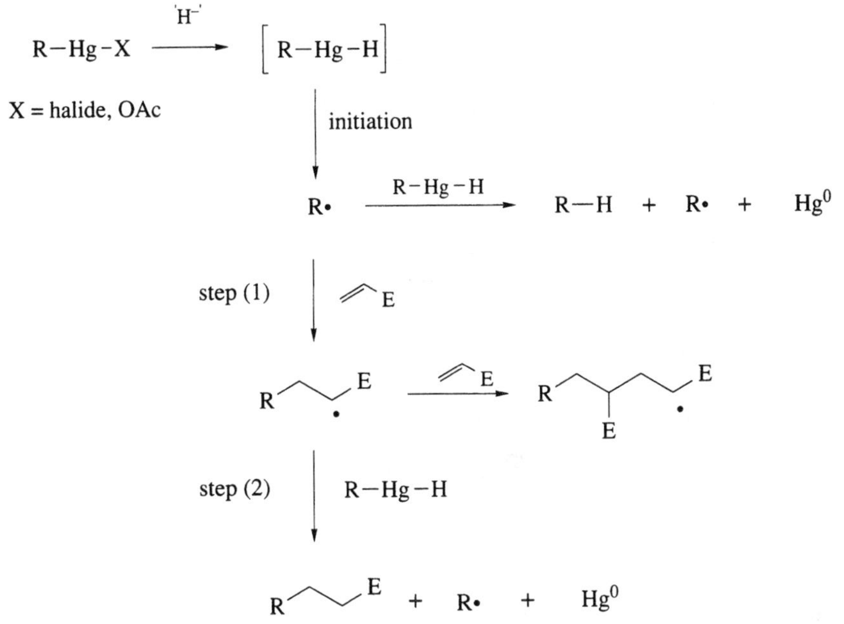

R–Hg–X —('H⁻')→ [R–Hg–H]

X = halide, OAc

initiation

R• —(R–Hg–H)→ R–H + R• + Hg^0

step (1) (alkene E)

R(CH₂CH•E) —(alkene E)→ R–CH₂CH(E)CH₂CH•E

step (2) R–Hg–H

RCH₂CH₂E + R• + Hg^0

Scheme 33

It is known that mercury(II) hydrides are better hydrogen donors towards alkyl radicals than tin hydride by about one order of magnitude.[121] This means that very reactive acceptors are required to attain useful yields in the mercury(II) hydride method. For many applications, tin hydride will be the reagent of choice. However, the mercury(II) hydride method does have advantages: (i) the mercury(II) hydride is both the radical trap and the radical precursor and thus there are only two radicals involved in the reaction (rather than three as in the tin hydride method);[122] (ii) the reaction is very easy to conduct and proceeds rapidly at room temperature; (iii) the organomercurials are often formed *in situ*; and (iv) the separation of products from reagents is very straightforward (metallic mercury is removed by filtration and the inorganic products are removed by water extraction).

A useful aspect of the mercury(II) hydride method is that it can be directly coupled with the many standard techniques for heteromercuration of alkenes and cyclopropanes. The resulting overall transformation adds a heteroatom and a carbon atom across the carbon–carbon double bond of an alkene or the carbon–carbon single bond of a cyclopropane. This is a difficult transformation to conduct by standard ionic techniques. An alkene thus becomes an equivalent of synthon (**12**) and a cyclopropane of synthon (**13**; Scheme 34). Many equivalent transformations (like haloetherification and phenylselenolactonization) are available to make precursors for tin hydride mediated additions.

The selection of examples collected in Scheme 35 illustrates just a few of the many transformations that are possible with mercurials.[123–126] (The comprehensive review of Barluenga is recommended as a complete source of references.)[82] The oxy- or amido-mercuration can be inter- or intra-molecular and the addition products are often converted to lactones or lactams. The last example illustrates a simple prep-

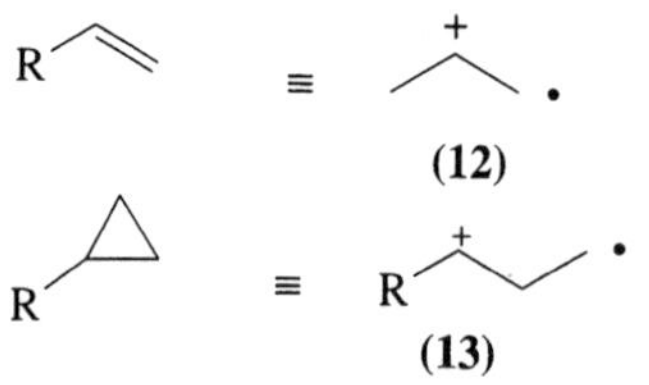

R
≡
(12)
R
≡
R
(13)

Scheme 34 Synthons in the mercury(II) hydride method

aration of a 1,6-dicarbonyl compound. The standard ionic approach to such molecules calls for umpolung; a reagent equivalent to homoenolate synthon (**14**) would need to be prepared. The preparation of a reagent for the radical synthon (**15**) and the execution of the coupling are very simple compared to many ionic methods for the equivalent transformation.

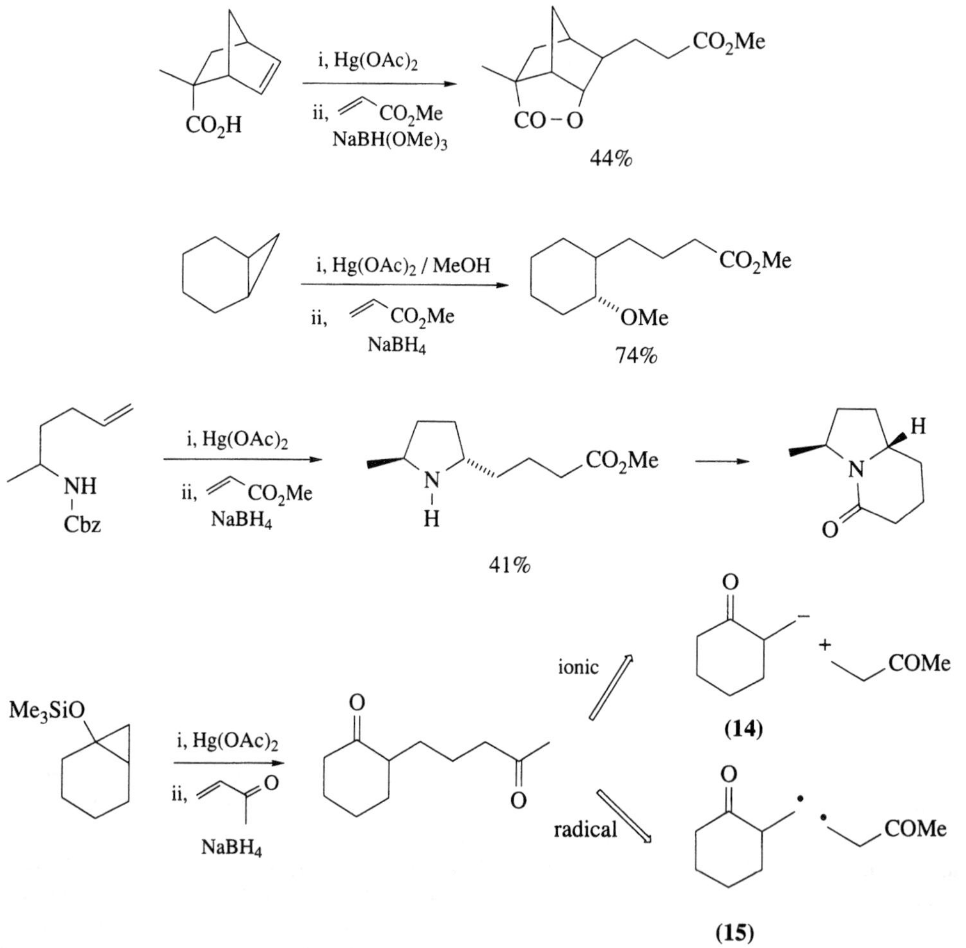

CO2H
i, Hg(OAc)2
ii, CO2Me
NaBH(OMe)3
CO2Me
CO–O
44%
i, Hg(OAc)2 / MeOH
ii, CO2Me
NaBH4
CO2Me
OMe
74%
NH
Cbz
i, Hg(OAc)2
ii, CO2Me
NaBH4
N
H
CO2Me
41%
N
H
O
Me3SiO
i, Hg(OAc)2
ii, O
NaBH4
O
O
ionic
radical
O
COMe
(14)
O
COMe
(15)

Scheme 35

4.1.6.1.2 *The fragmentation method*

While carbon and oxygen radicals add irreversibly to carbon–carbon double bonds, the fragmentation reaction is rapid (and often reversible) for elements like tin, sulfur, selenium and the halogens (Scheme 36). This elimination reaction can be very useful in synthesis if the eliminated radical Y· can either directly or indirectly react with a radical precursor to propagate a chain. Given this prerequisite, an addition chain can be devised with either an allylic or a vinylic precursor, as illustrated in Scheme 37. Carbon radicals are generated by the direct or indirect reaction with Y· and are removed by the β-elimination of Y·. Selectivity is determined by the concentration of the alkene acceptor and the rate of β-elimination (which is usually very fast).[127] The allylic variant (**16**) is more straightforward because the alkene substituent (CH_2Y) sterically (and perhaps also electronically) biases addition in the desired sense to form (**17**). For selectivity, the addition of R· to the allylic precursor (**16**) must of course be more rapid than radical–radical or radical–solvent reactions, but it must also be more rapid than addition of R· to the final product. In the vinylic variant, an activating group A must be introduced as in (**18**) to prevent the direct addition of R· to the carbon remote from Y.

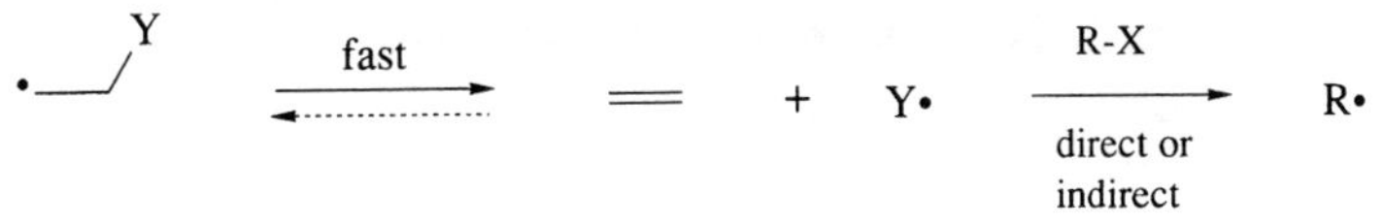

Y = halogen, SR, SeR, SnR_3

Scheme 36 Fragmentation reactions

A = activating group

A = H, cannot fragment

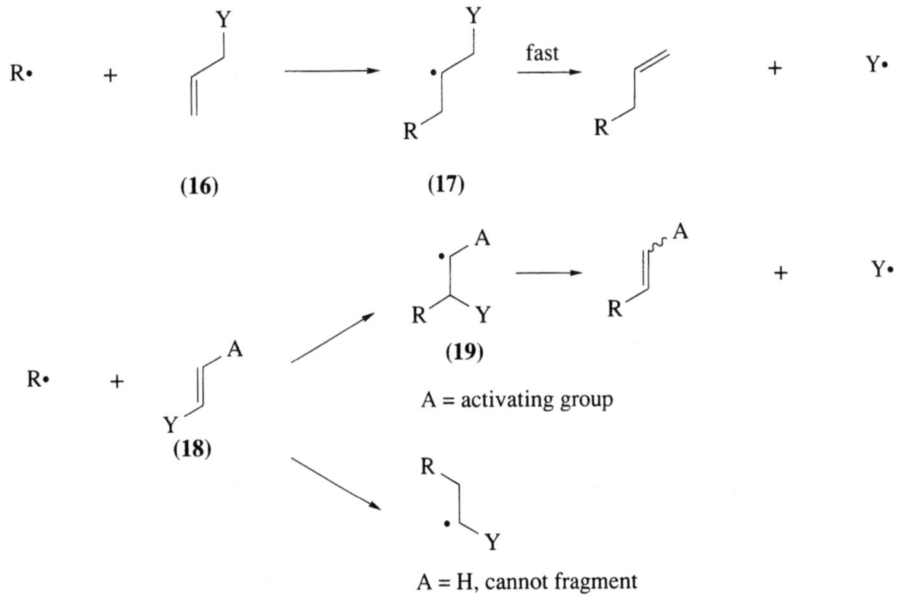

Scheme 37 Allyl and vinyl fragmentation chains

A strategy that propagates the chain by β-elimination has important advantages over metal hydride methods that propagate chains by hydrogen transfer. Because the overall transformation is a substitution and not a reduction, there is no metal hydride reagent present in a fragmentation chain. The problem of reduction of an initial radical by the reagent prior to addition does not exist. Relatively unreactive acceptors can be used and the reactions can (and generally should) be conducted at high concentrations. The differentiation of the initial radical from the final radical is also not difficult because most fragmentation reactions are very fast unimolecular processes with which few bimolecular processes can compete;[127] that is, radicals (**17**) and (**19**) fragment much more rapidly than they add to the starting alkenes (**16**) and (**18**), even if these alkenes are present in high concentration. The trade-off in using the fragmentation strategy is that it is presently more limited in the type and location of substituents that can be introduced compared to the metal hydride method (see below).

(i) Allyl- and vinyl-stannanes

The most popular fragmentation methods use allyl- or vinyl-stannane reagents. These reagents have all the advantages associated with using the trialkyltin radical (a reactive atom and group abstractor) as a chain transfer agent without the disadvantages of tin hydride (a reactive hydrogen donor). Several recent reviews provide sources of references in this field.[3–5]

Allylation reactions with allyltributylstannane[128] are now routine (Scheme 38) and have been used to advantage in complex synthetic endeavors. The experimental procedures developed by Keck are popular.[129] Thermal methods involve heating of allyltributylstannane (or allyltrimethylstannane), the radical precursor, and AIBN in degassed benzene for several hours. If lower temperatures are desired, AIBN can be omitted and the reaction can be irradiated through a Pyrex filter with a standard Hanovia lamp. Two equiv. of the stannane are typically employed and the initial concentration is about 0.5 M in radical precursor. The products are isolated simply by evaporation of the solvent and chromatography of the residue. Alternatively, methods to remove tin halides can be applied prior to work-up and chromatography. Keck has shown that this allylation is very powerful because it succeeds in complex synthetic settings, where more traditional synthetic methods would be inappropriate.[130]

R−X + $\diagup\!\!\diagdown$SnBu$_3$ ⟶ R$\diagdown\!\!\diagup$ + X−SnBu$_3$

X = I, Br, SeR, NO_2, xanthate, activated SPh, Cl

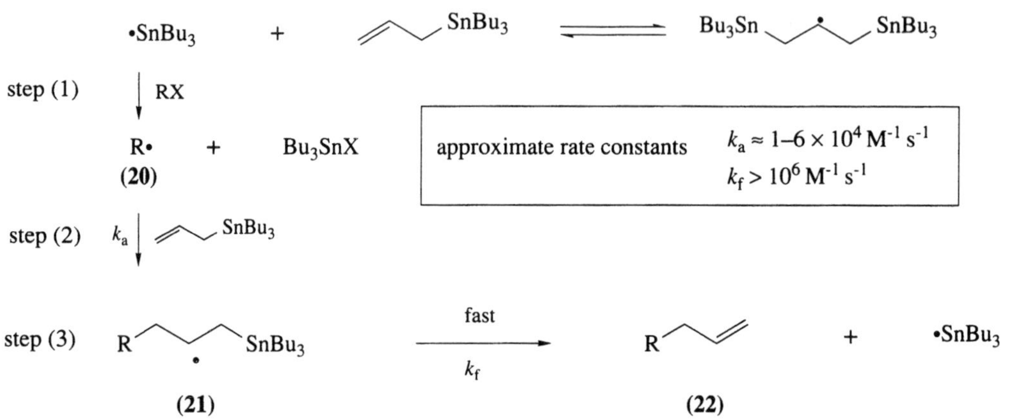

•SnBu$_3$ + $\diagup\!\!\diagdown$SnBu$_3$ ⇌ Bu$_3$Sn$\diagdown\!\!\overset{\bullet}{\diagup}\!\!\diagdown$SnBu$_3$

step (1) ↓ RX

R• + Bu$_3$SnX
(20)

approximate rate constants	$k_a \approx 1\text{–}6 \times 10^4\ M^{-1}\ s^{-1}$ $k_f > 10^6\ M^{-1}\ s^{-1}$

step (2) k_a ↓ $\diagup\!\!\diagdown$SnBu$_3$

step (3) R$\diagdown\!\!\diagup\!\!\diagdown$SnBu$_3$ (•) $\xrightarrow[k_f]{\text{fast}}$ R$\diagdown\!\!\diagup$ + •SnBu$_3$

(21) (22)

Scheme 38

The mechanism of this transformation is outlined in Scheme 38 and each step has important features. In step 1, the tributyltin radical abstracts the radical precursor X. A possible side reaction, the addition of the tributyltin radical to the allylstannane, is much slower than comparable additions to activated alkenes. Even if this addition occurs, the stannyl radical is simply eliminated to regenerate the starting materials. Thus, for symmetric allylstannanes, this reaction is of no consequence. As a result, the range of precursors X that can be used in allylation is more extensive than in the tin hydride method. Even relatively unreactive precursors like chlorides and phenyl sulfides can be used if they are activated by adjacent radical-stabilizing groups.

That step 2 succeeds implies that there must be some activating effect of the tin on the rate of radical addition to the allylstannane.[131,132] Otherwise the product (**22**) would be of equal reactivity to the allylstannane reagent. However, this activation may only be modest enough so that addition of radical (**20**) to allylstannane is more rapid than its addition to benzene or its reaction with another radical; there are no rapid alternative reactions for (**20**) to undergo. Although there are no accurate rate constants yet available for k_a in step 2, we have recently estimated the rate constant for addition of octyl radical to allyltributylstannane at 50 °C: $k_a \approx 1\text{–}6 \times 10^4\ M^{-1}\ s^{-1}$.[133] Likewise, there are no accurate rate constants known for the fragmentation reaction in step 3; however, this must be a very fast reaction.[127] In some cases (see below) fragmentation of a β-stannyl radical is even competitive with σ-bond rotation. It is very likely that $k_f > 10^6\ M^{-1}\ s^{-1}$. Addition of (**21**) to the starting allylstannane cannot possibly compete with fragmentation.

Illustrative examples of the kinds of transformations that can be conducted with allylstannane reagents are presented in Scheme 39. Example 1 emphasizes that activating groups on the alkene provide especially reactive partners for radicals.[134] Example 2 illustrates in the context of a prostaglandin synthesis that allylstannanes need not be paired with nucleophilic radicals.[135] Ambiphilic and electrophilic radicals should be equally useful, if not more. Example 3 illustrates that allenylations can be accomplished with the appropriate propargylstannane.[136] Finally, example 4 illustrates the kinds of new synthons and reagents that can be developed by combining this type of allylation with known transformations.[137] Silylstannane (**23**) becomes a useful reagent for the synthon (**24**).

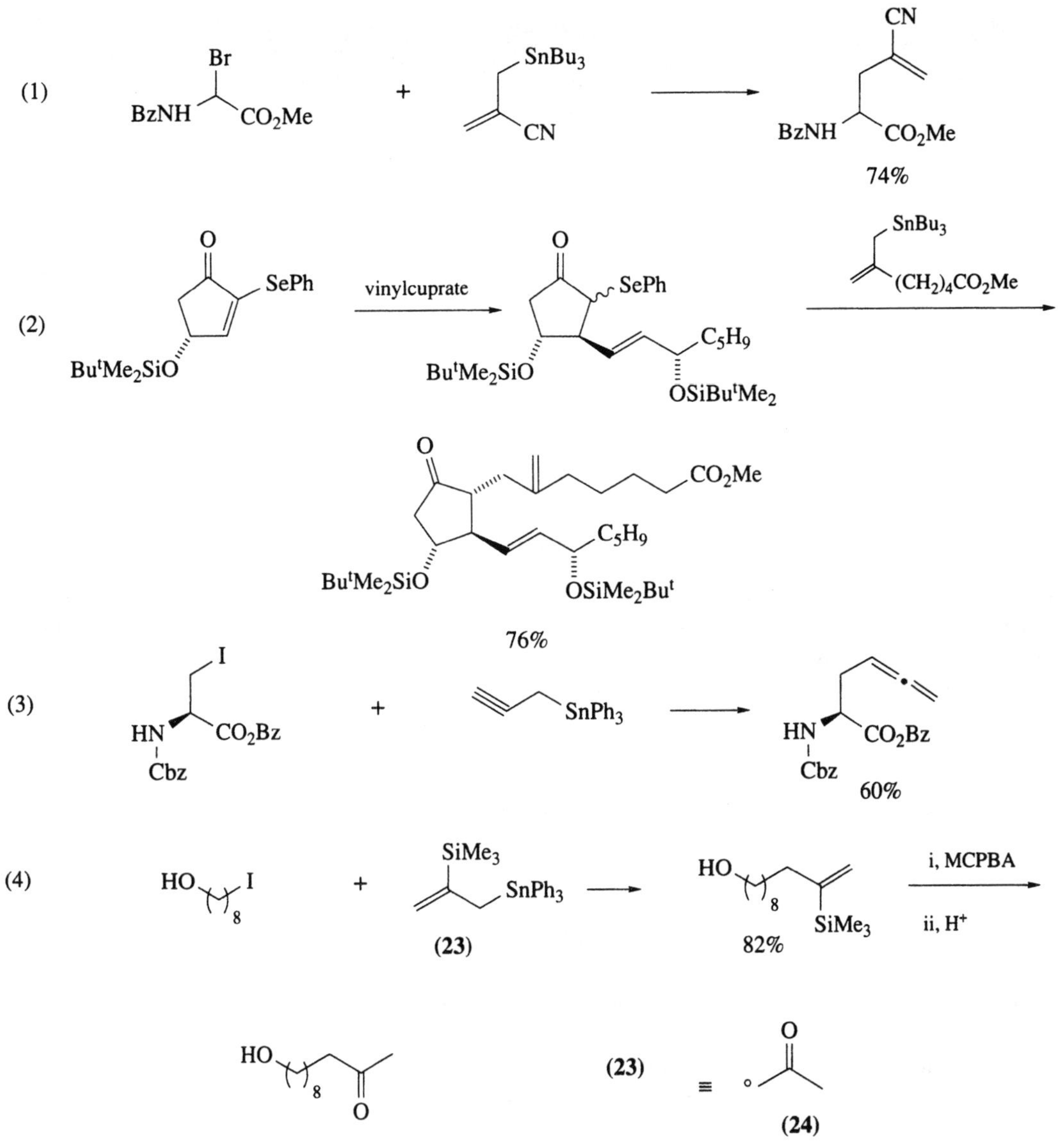

Scheme 39

A sequence of reactions that was recently reported by Hanessian and Alpegiani nicely illustrates how the allylstannane method is useful for functionalization of complex, sensitive substrates and, more generally, how stereochemistry can be controlled in radical addition reactions (Scheme 40).[138] Dibromo-β-lactam (**25**) can be monoallylated with a slight excess of allyltributylstannane and then reduced with tributyltin hydride to provide β-allylated β-lactam (**26**) (the acid salt of which shows some activity as a β-lactamase inhibitor). Stereochemistry is fixed in the reduction step: hydrogen is delivered to the less-hindered face of the radical. Alternatively, monodebromination, followed by allylation, now delivers the allyl group from the less-hindered face to provide stereoisomer (**27**). Finally, allylation of (**25**) with excess allylstannane produces the diallylated product (not shown).

At present, bimolecular allylation is limited to unsubstituted and 2-substituted allylstannanes (which are symmetrical so that addition and elimination of a tin radical are degenerate). 1-Substituted allylstannanes (for example, crotylstannane) are not useful reagents because the substituent decelerates the rate of the radical addition step below the useful limit.[139] The usefulness of 3-substituted allylstannanes has been limited by their relatively facile isomerization to the inert 1-substituted isomers under typical reaction conditions.[140]

A useful solution to the latter problem, introduced by Keck and Byers,[141] makes possible crotylation and isoprenylation. Irradiation of an allyl sulfide (**28**; 3 equiv.), a reactive radical precursor (**29**; 1 equiv.) and hexabutylditin (1.5 equiv.) produces substitution products (**30**) in good yields (Scheme 41). The

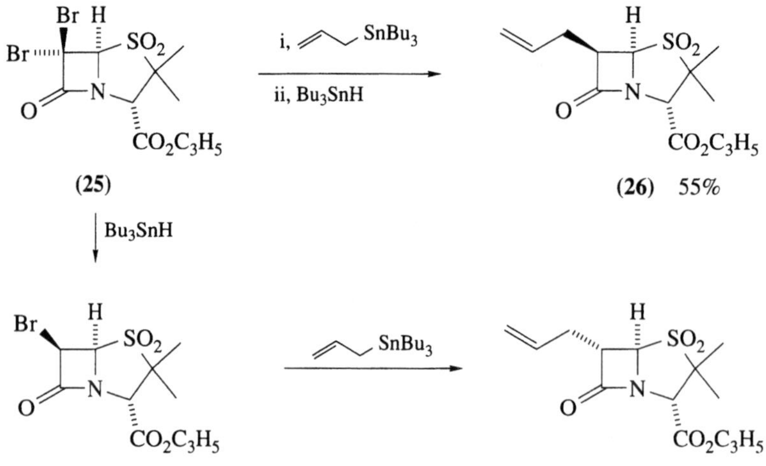

(25) (26) 55%

(27) 69%

Scheme 40

mechanism is similar to that with an allylstannane except that the thiophenyl radical is ejected in the fragmentation step. This thiophenyl radical cleaves hexabutylditin to give Bu_3SnSPh and the chain-carrying tributyltin radical. Reactive radical precursors must be used because addition of the tributyltin radical to (**28**) is a reasonably rapid competing reaction.

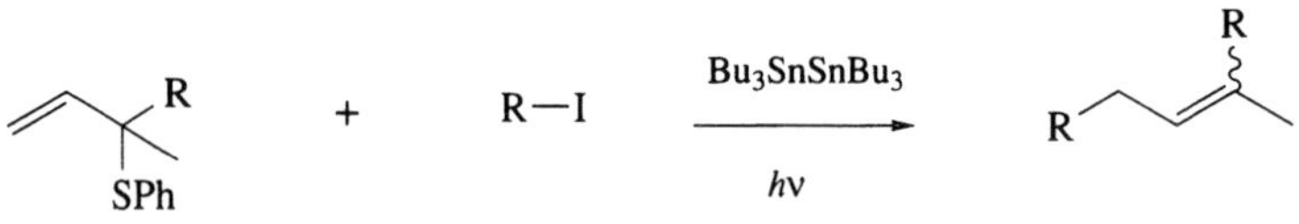

(28) R = H, Me (29) (30)

Scheme 41

Vinylstannanes can also be used in the fragmentation method provided that an activating group is introduced to accelerate and direct the addition of the radical to the carbon bearing the tin (Scheme 42). Useful activating groups include carbonyls,[142] phenyl,[143] phenylsulfonyl and diphenylphosphinyl.[144] Most preparative studies have reported the predominant or exclusive formation of the (*E*)-isomer of the product regardless of whether the starting stannane was (*E*) or (*Z*). However, careful studies by Russell[143] on the addition of alkyl radicals (generated from alkylmercury(II) chlorides) to the individual stereoisomers of β-stannylstyrene have shown that the stereochemistry of the vinylstannane precursor is retained to a significant extent in the product.[145] These studies demonstrate that fragmentations of β-stannyl radicals are very rapid. The success of these procedures again implies that there is at least a modest activating effect of the β-stannyl group on the rate of addition, otherwise the product would be equally reactive with the starting vinylstannane. Even so, β-stannyl groups must be decelerating relative to hydrogen and an excess of the acceptor is usually employed.

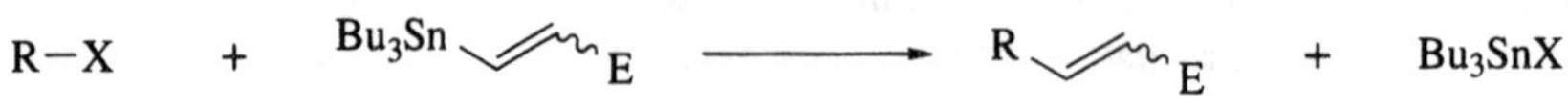

$$R{-}X + Bu_3Sn{\sim}E \longrightarrow R{\sim}E + Bu_3SnX$$

E = COR, Ph, SO_2Ph, $P(O)Ph_2$

Scheme 42

(ii) Other reagents

A variety of other fragmentation reactions are potentially useful in the formation of carbon–carbon bonds, although none has yet shown the generality of the allyl- and vinyl-stannanes.[146] In addition to the use of allyl sulfides discussed above (Scheme 41), allylcobalts have also been employed.[147] Extensive studies by Russell have shown that a variety of activated vinyl (and propargyl) acceptors will propagate chains when alkylmercury(II) halides are used as radical precursors (Scheme 43).[148]

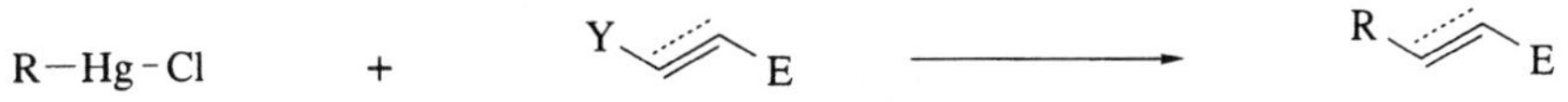

R—Hg-Cl + Y~E ——→ R~E

E = Ph, $PhSO_2$; Y = halogen, SPh, SO_2Ph

Scheme 43

4.1.6.1.3 Thiohydroxamate esters (the Barton method)

One of the most important chain methods to conduct radical addition reactions that does not revolve around the chemistry of the trialkyltin radical is Barton's thiohydroxamate method.[149] This is presently the most generally useful method to conduct addition reactions of nucleophilic radicals to electrophilic acceptors that are not terminated by hydrogen atom transfer. In the thiohydroxamate method, radicals are generated and removed in the same process: the addition of a radical to a thiohydroxamate ester and the fragmentation of the ensuing intermediate. Thus, the lifetimes of radicals are limited by their rate constants for addition to the radical precursor and by the concentration of this precursor. Several pertinent reviews of the chemistry of thiohydroxamate esters are available;[3–5] a thorough, succinct review by Crich is an excellent starting point for citations to the original literature.[150] In addition to serving as radical precursors for C—C bond-forming reactions, these esters are excellent precursors for the decarboxylative functionalization of carboxylic acids with a variety of heteroatomic groups. This aspect of thiohydroxamate chemistry is described in Volume 7, Chapter 5.4.

Scheme 44 summarizes an addition reaction by the Barton method. Thiohydroxamate esters (**32**) are readily prepared and isolated, but, more typically, they are generated *in situ*. Experimental procedures have been described in detail[148,151] and often entail the slow addition of an acid chloride to a refluxing chlorobenzene solution of the readily available sodium salt (**31**), dimethylaminopyridine (DMAP, to catalyze the esterification), and excess alkene. The products are usually isolated by standard aqueous work-up and chromatographic purification.

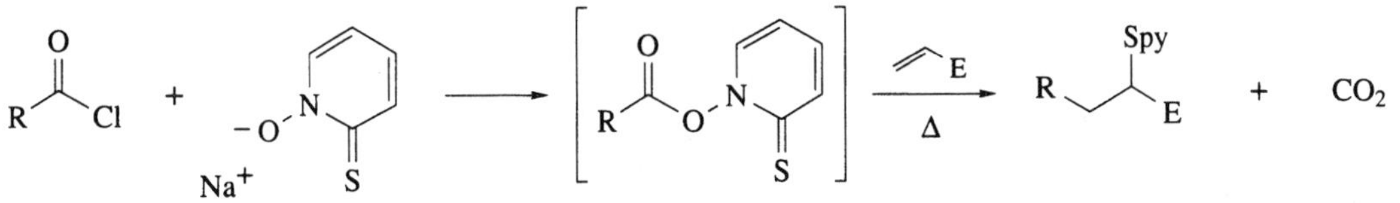

RCOCl + Na^+ (**31**) ——→ [(**32**)] —(alkene–E, Δ)→ R–CH₂–CH(Spy)–E + CO_2

Scheme 44 The Barton method

The basic transformation that underlies the Barton method is outlined in Scheme 45, steps 1 and 2.[152] Thermolysis in refluxing toluene or photolysis with a sunlamp rapidly converts a thiohydroxamate ester (**32**) to the decarboxylated pyridyl sulfide (**33**). This pyridyl sulfide is formed by addition of an alkyl radical R· to the thiohydroxamate (**32**) followed by fragmentation of (**34**) as indicated. In the planning of addition reactions by the Barton method, it is usually assumed that the addition step 1 is rate limiting. However, there is now evidence that step 1 may sometimes be reversible and step 2 may be rate limiting.[153]

In the presence of a reactive alkene acceptor, the intermediate radical R· can be intercepted by addition to a reactive alkene (step 3) prior to reaction with the starting thiohydroxamate (step 2). If the adduct radical (**35**) then adds to the thiohydroxamate (step 4) more rapidly than it adds to another molecule of alkene, a successful chain addition reaction results. In practice, a wide variety of addition reactions can be conducted by this method in yields that range from 30–90%.

Considerations for synthetic planning are remarkably similar to the tin hydride method. The initial radical can either add to the alkene or add to its own precursor (**32**). To maximize the addition, it is advisable to use a reactive alkene (typically in excess) and to keep a relatively low concentration of the thiohydroxamate (**32**); hence the slow addition of the acid chloride. At present, rate constants for the addition of a primary and a tertiary alkyl radical to a thiohydroxamate are known (Scheme 46)[38,40] and these are useful in selecting alkene acceptors and in planning reaction conditions.

Like the tin hydride method, this addition succeeds because of differences in reactivity of the starting and the adduct radicals towards the alkene acceptor.[154] Unlike the tin hydride method, there is a loss of

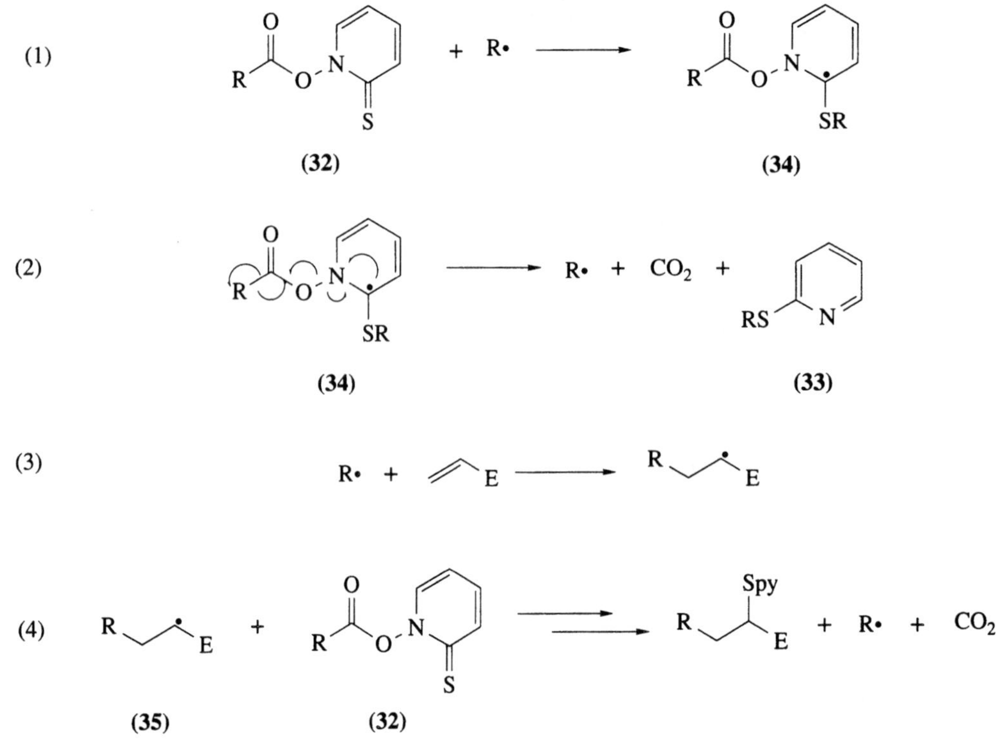

(1) (32) + R• → (34)

(2) (34) → R• + CO_2 + (33)

(3) R• + E-alkene → R-CH2-CH(•)-E

(4) (35) + (32) → R-CH2-CH(Spy)-E + R• + CO_2

Scheme 45

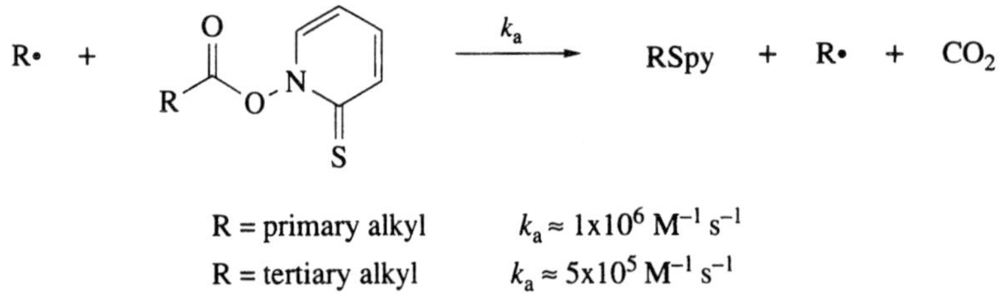

R• + thiohydroxamate $\xrightarrow{k_a}$ RSpy + R• + CO_2

R = primary alkyl $k_a \approx 1\text{x}10^6\ M^{-1}\ s^{-1}$

R = tertiary alkyl $k_a \approx 5\text{x}10^5\ M^{-1}\ s^{-1}$

Scheme 46 Rates of addition to thiohydroxamates

one carbon in the radical precursor (as carbon dioxide). Also unlike the tin hydride method, the addition reaction is not terminated by hydrogen atom transfer. Instead, the thiopyridyl group, a useful precursor for several other functionalities, is introduced. Several addition reactions, along with subsequent transformations of the adducts, are compiled in Scheme 47.[155–158]

Fragmentation reactions can also be executed within the framework of the thiohydroxamate method, as illustrated in Scheme 48. Addition of acid chloride (**36**) to a mixture of activated allyl sulfide (**37**) and sodium salt (**31**) provides allylated product (**38**) in 75% yield. The activating group insures that the initial radical adds to the acceptor more rapidly than to the thiohydroxamate. Rapid fragmentation of the resulting intermediate provides the product (**38**) and the thiobutyl radical. This last radical propagates the chain by addition to the starting thiohydroxamate.

Loss of a carbon atom from the precursor need not always result. Barton and Crich have introduced a related procedure based on the chemistry of mixed oxalates, an example of which is provided in Scheme 49.[159] Double decarboxylation is involved in the decomposition of oxalate precursors, such as (**39**). Unfortunately, there are indications that this method may to be limited to tertiary alcohols; one secondary alcohol derived mixed oxylate did not fragment completely to the alkyl radical.

In the thiohydroxamate method, activated acceptors are required for successful addition reactions because the thiohydroxamate is relatively reactive towards alkyl radicals. An ingenious method to circumvent this problem has recently been communicated by Zard.[160] Photolysis of benzyl xanthate (**40**) with a UV lamp in the presence of *N*-methylmaleimide (NMM) provided the adduct (**41**) in 40% yield, as illustrated in Scheme 50. In this reaction, the competition reaction between the desired addition to the acceptor and the addition to the precursor still exists. However, addition to the precursor is reversible and,

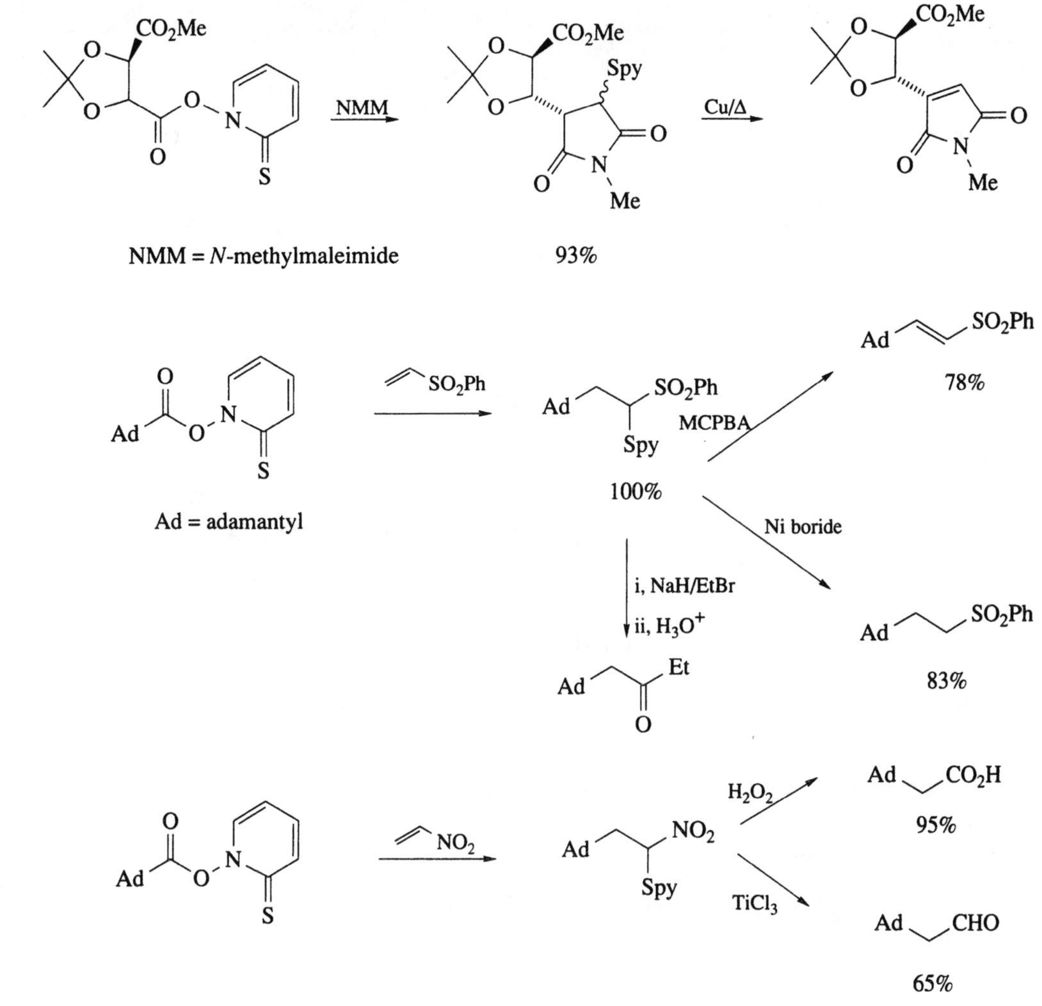

NMM = *N*-methylmaleimide

93%

Ad = adamantyl

Scheme 47

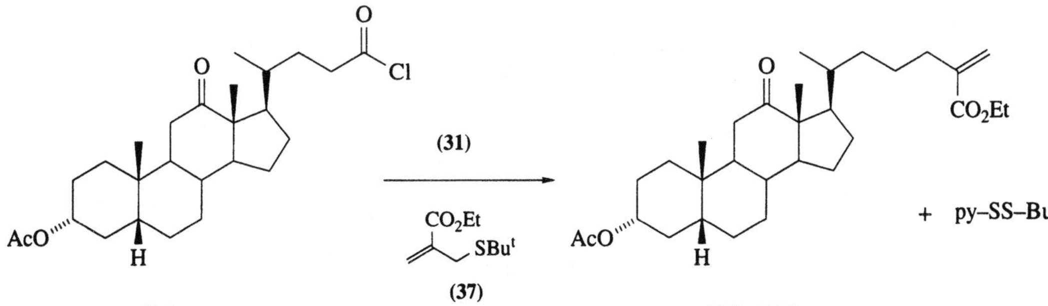

(36) (37) (38) 75%

Scheme 48

by the design of the method, it can only regenerate the starting radical (fragmentation to generate the unstable methyl radical is not rapid). One potential problem is that the product (**41**) contains the same reactive functional group as the starting material (**40**) and it is continuously recycled to the radical pool. This recycling may result in a reduced yield.

This problem is not so severe when acyl xanthates are used as precursors because these substrates absorb in the visible region, while the products do not (however, the products might still be recycled to the radical pool by radical addition–elimination). Visible light photolysis of benzoyl xanthane (**42**) and allyl acetate provides (**43**) in 60% yield. Standard (ionic) β-elimination of the xanthane is a facile reaction that gives (**44**). When the tertiary acyl xanthane (**45**) is irradiated in the presence of *N*-benzylmaleimide

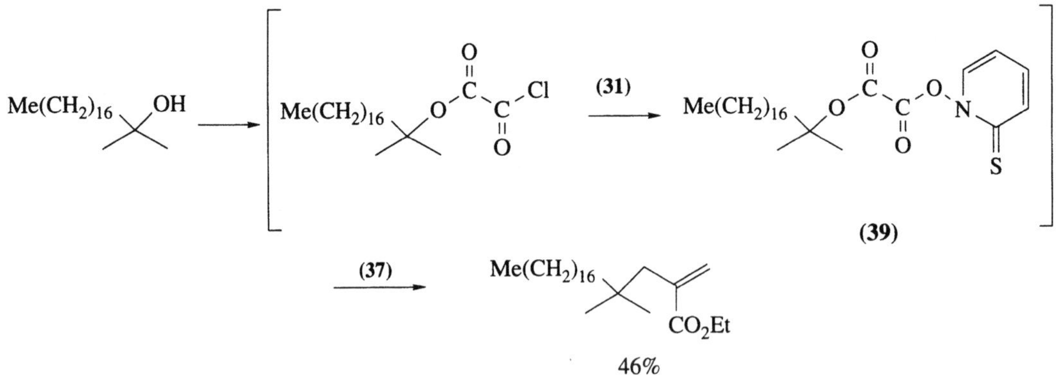

Scheme 49

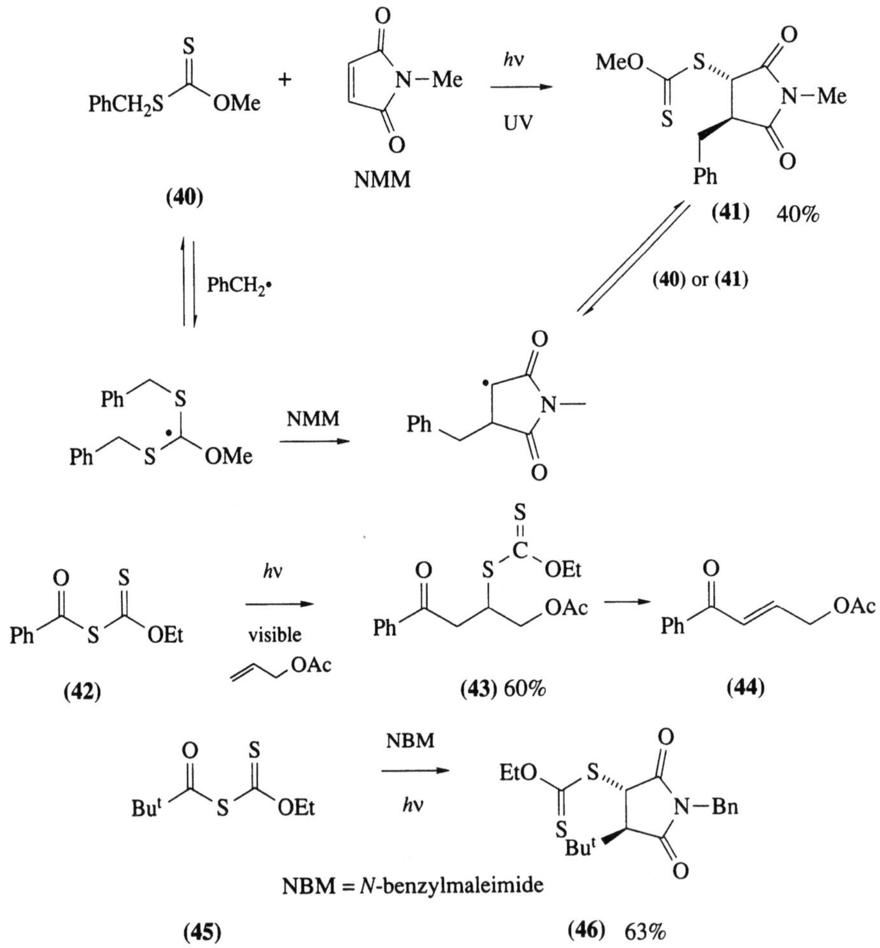

NBM = *N*-benzylmaleimide

Scheme 50

(NBM), decarbonylated adduct (**46**) is isolated in 63% yield. The initially formed acyl radical decarbonylates to give the *t*-butyl radical (in contrast, aroyl radicals are much more reluctant to decarbonylate). Now a true competition again exists because addition of the *t*-butyl radical to the precursor is not necessarily degenerate. Nonetheless, addition to the acceptor is more favorable and a useful yield is obtained. Although the scope and limitations remain to be determined, this new method shows the promise of being especially useful for conducting reactions of stabilized radicals and aroyl radicals.

4.1.6.1.4 Atom transfer reactions (the Kharasch method)

The addition of a single-bonded reagent across a multiple bond is one of the fundamental reactions of organic radicals. The basic principles of this reaction were first advanced by Kharasch in pioneering studies on the mechanism of the peroxide-initiated anti-Markovnikov addition of hydrogen bromide to alkenes.[1] In the atom transfer method, the generation and removal of radicals are coupled and occur in the key atom transfer step. Compared to other methods, the atom transfer method provides unique options for synthetic reactions. But there are also important limitations. Recently, there has been a renewed interest in the application of the characteristics of atom transfer reactions in synthesis and new developments have been reviewed.[5,161]

The basic mechanism for the addition of a reagent containing a C—X bond across an alkene acceptor is outlined in Scheme 51. In the formation of C—C bonds, the group X is usually limited to one of the univalent atoms (listed in increasing order of reactivity): Cl, H, Br, I. Many more groups X can be used in forming C–heteroatom bonds because of the weak nature of most interheteroatom bonds (Section 4.1.8). The first propagation step is the addition of the carbon radical (**47**) to an alkene. The adduct radical (**48**) then abstracts the univalent atom X from the starting reagent to transfer the chain. That the atom donor and the radical precursor are the same molecule (**49**) is a great advantage because the reaction of the initial radical (**47**) with the atom donor is degenerate (and often slow anyway because it is thermoneutral). Compare this to the tin hydride or thiohydroxamate methods, where reactions of the starting radical with the radical precursor can efficiently compete with the desired addition reaction. Thus, the atom transfer method should be ideal to conduct relatively slow addition reactions because there is no competing trap for the starting radical (of course it must still be trapped by the alkene faster than radical–radical or radical–solvent reactions). The limitations in the atom transfer method are imposed by step 2. This step must be sufficiently rapid, otherwise the adduct radical (**48**) can add to another molecule of acceptor as shown in step 3. For an atom transfer reaction to be rapid, it is usually important that it be exothermic: the more exothermic the better. Thus, the adduct radical (**48**) must be a more reactive (that is, less resonance stabilized) radical than the starting radical (**47**). This increase in radical reactivity is made possible because the gain in energy associated with conversion of a π-bond to a σ-bond in the addition step often offsets any loss in energy associated with decreased resonance stabilization of the product radical.

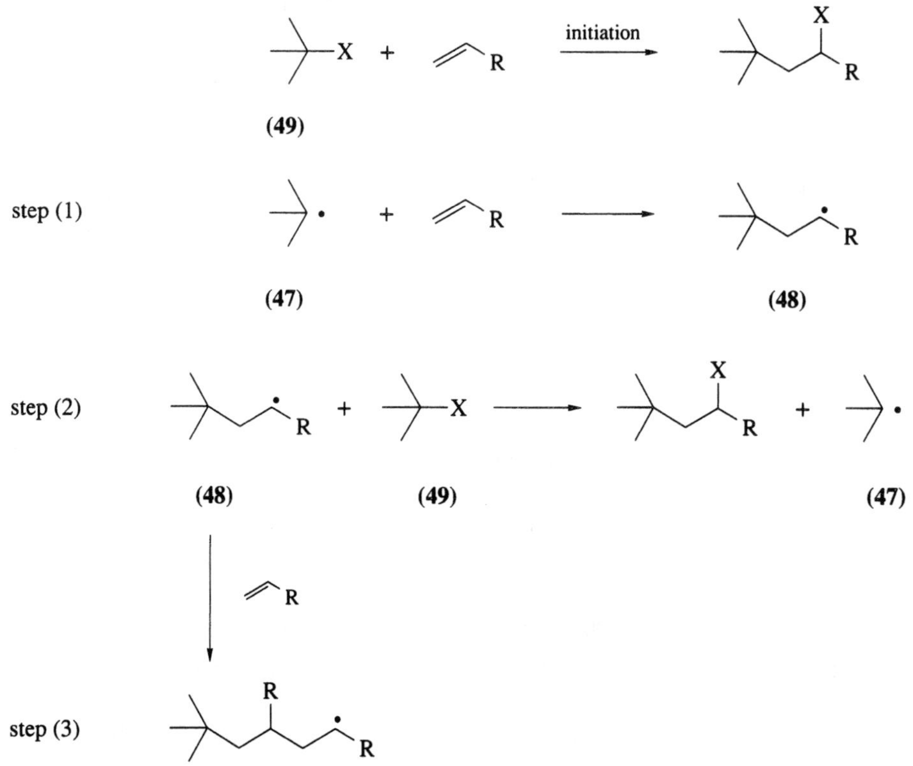

X R R initiation X R (**49**)

step (1) R R (**47**) (**48**)

step (2) R X X R (**48**) (**49**) (**47**)

R

step (3) R R

Scheme 51

To summarize, an atom transfer sequence has two basic requirements: (i) a rapid, exothermic addition, cyclization or fragmentation reaction must convert a starting radical into a more reactive (less resonance stabilized) counterpart; and (ii) atom transfer from the radical precursor to the adduct must occur more rapidly than any other competing reaction. Considering these requirements, one can conclude that the atom transfer method will not be particularly useful for conducting addition reactions of nucleophilic radicals to electron deficient alkenes because most electron-attracting groups that are used to activate alkenes also stabilize radicals. Thus, although addition might succeed, atom transfer will fail because the adduct radical is more stable than the starting radical. In contrast, the atom transfer method should be very useful for the additions of ambiphilic and electrophilic radicals (which are usually resonance-stabilized) to alkyl-substituted alkenes. Thus, the tin hydride and the thiohydroxamate methods are nicely complemented by the atom transfer method. In the former methods, it is the rate of the addition step that is important for success and not the relative stabilities of the starting and adduct radicals; in the atom transfer method, it is precisely the reverse.

(i) Hydrogen atom transfer additions

The addition of a C—H bond across a C—C multiple bond has often been applied to the synthesis of simple organic molecules, but has rarely been used in complex synthetic settings.[3,5,162] This is because the abstraction of hydrogen atoms from C—H bonds by carbon-centered radicals is a relatively slow process and reaction conditions must be designed around this serious limitation (Scheme 52). To serve as a hydrogen donor, a molecule should have one C—H bond that is significantly weaker than the rest. Almost any resonance-stabilizing group can serve as an activator for an adjacent C—H bond. In classical reaction conditions, a peroxide initiator is added slowly to an excess of the radical precursor and the acceptor (sometimes the acceptor is also added slowly with the initiator). There is usually no solvent other than the excess of the precursor and reaction temperatures over 100 °C often give the best results. Such conditions favor hydrogen transfer over telomerization of the initial adduct. Even so, the chains are very short in such processes because of the inherent inefficiency of hydrogen transfer between carbon radicals. Large amounts of initiator are required and the formation of short telomers as by-products is very common. The products are often isolated by fractional distillation or chromatography.

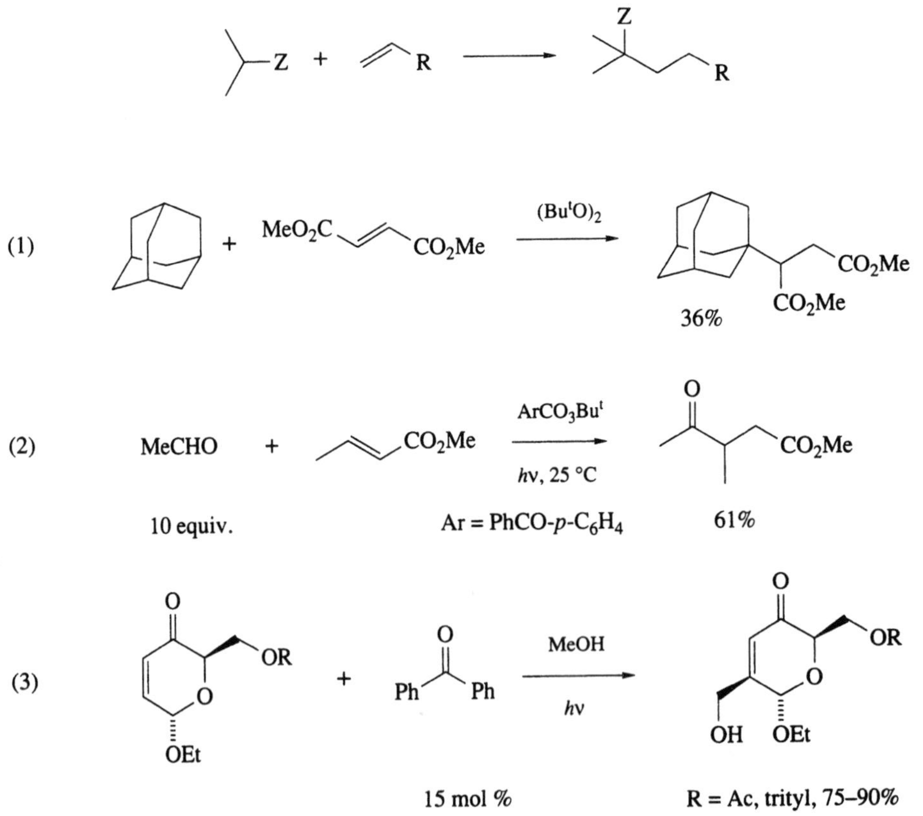

(1) + MeO_2C CO_2Me $(Bu^tO)_2$ CO_2Me CO_2Me 36%

(2) MeCHO + CO_2Me $ArCO_3Bu^t$, $h\nu$, 25 °C CO_2Me
10 equiv. Ar = PhCO-p-C_6H_4 61%

(3) OR + Ph Ph (15 mol %) MeOH, $h\nu$ OR, OH OEt
R = Ac, trityl, 75–90%

Scheme 52

Scheme 52 provides a sample of the types of reactions that have recently been conducted under a variety of different conditions. Example (1) illustrates the classical conditions,[163] while example (2) shows that modified initiators that permit low temperature reactions may prove beneficial.[29] Finally, example (3) illustrates that the photochemical addition of alcohols (and also acetals) to enones in the presence of benzophenone is a very useful preparative procedure.[164] Here, the carbinyl hydrogen is abstracted by an excited state benzophenone molecule rather than a peroxy radical.[165]

The problem of inefficient hydrogen atom transfer to carbon-centered radicals has been circumvented by Maillard as outlined in Scheme 53.[166] Heating of excess methyl propionate with allyl *t*-butyl peroxide provides a mixture of adducts in which (**49**) is the major component. In this reaction, the carbon-centered adduct radical (**50**) reacts in an intramolecular homolytic substitution (group transfer) reaction with a neighboring O—O bond. The highly reactive (but relatively unselective) alkoxy radical that is generated in this step then serves as the hydrogen abstractor. A net addition of a carbon and an oxygen across a C—C double bond is accomplished in modest yield. The synthetic price to pay is that the peroxide or perester must be built into one of the reagents.

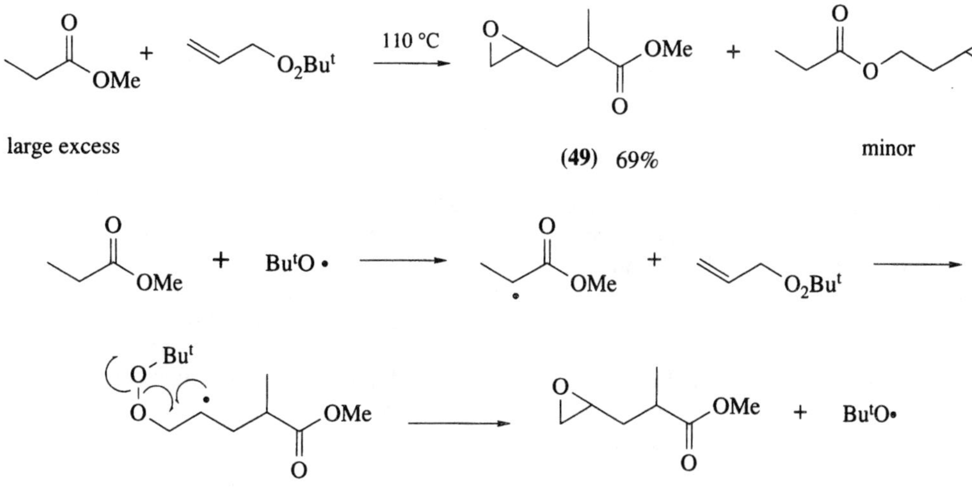

large excess; 110 °C; (**49**) 69%; minor; (**50**)

Scheme 53

Because it is inexpensive and very simple, the hydrogen atom transfer method is useful for the preparation of many kinds of simple molecules. Its use in complex synthetic tasks is limited because the yields are often modest, a large excess of the H-donor is required, and chemoselectivity is certain to be poor in polyfunctional molecules with several relatively weak C—H bonds. It can be best employed to advantage when the addition of a relatively small molecule to a complex acceptor is required.

(ii) Halogen atom transfer additions

Replacement of a C—H bond by a C—halogen bond may have relatively little effect on the overall thermodynamics of an atom transfer reaction, but it can dramatically influence the kinetics. While C—Cl bonds are often poorer atom donors than corresponding C—H bonds, C—Br and C—I bonds are dramatically better. Thus the introduction of a halogen can improve an atom transfer reaction by: (i) ensuring that the generation of the radical is site selective; (ii) allowing more practical reaction conditions by accelerating the atom transfer step; and (iii) providing a functional group handle (the C–halogen bond in the product) for subsequent synthetic transformations.

The addition of a polyhaloalkane across an alkene is a mild and practical way to form a carbon–carbon bond and some representative examples are shown in Scheme 54.[159,167–169] Polyhaloalkyl radicals are regarded as electrophilic and are usually paired with electron rich alkenes. The halogen substituents also accelerate atom transfer by stabilizing the starting radical relative to the adduct radical. The addition reactions of polyhaloalkanes are often dramatically facilitated and become very practical in the presence of certain metal additives[170] including complexes of palladium,[171] aluminum,[167] ruthenium,[172,173] samarium[168] and others. However, the function of the metal is not always clear. It may simply initiate an atom transfer chain or provide a reactive halogen source (through a metal–halogen intermediate) or provide a source of metal-complexed radicals or even initiate a catalytic cycle that does not involve radical intermediates.

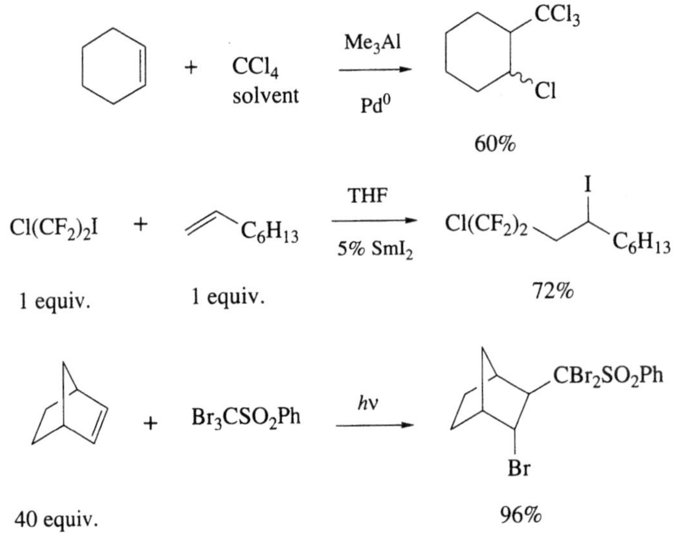

+ CCl_4 solvent $\xrightarrow[\text{Pd}^0]{\text{Me}_3\text{Al}}$ 60%

$Cl(CF_2)_2I$ + C_6H_{13} $\xrightarrow[\text{5\% SmI}_2]{\text{THF}}$ $Cl(CF_2)_2$ I C_6H_{13}

1 equiv. 1 equiv. 72%

+ Br_3CSO_2Ph $\xrightarrow{h\nu}$ CBr_2SO_2Ph Br

40 equiv. 96%

Scheme 54

The most common and useful additives are copper(I) salts (such as CuCl), which produce high yields of 1:1 adducts in many cases.[174] Several examples from the extensive work of the Ciba-Geigy group in Basel are compiled in Scheme 54, with an emphasis on subsequent conversions of the highly functionalized products into important heterocycles.[175] These procedures are very simple and have been conducted on a multigram scale. Typically, the halogen component and the acceptor are heated without solvent at 110 °C in the presence of 1–10% CuCl. After several hours, the copper salts are removed by filtration and the product is isolated by distillation. It is clear that the copper additive behaves as more than just an initiator; the additions of electrophilic radicals to electron deficient alkenes like those shown in Scheme 54 would not be likely to succeed otherwise.

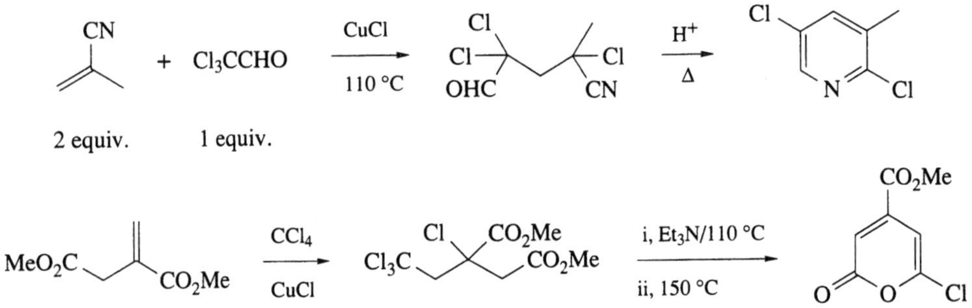

CN + Cl_3CCHO $\xrightarrow[\text{110 °C}]{\text{CuCl}}$ Cl Cl OHC Cl CN $\xrightarrow[\Delta]{\text{H}^+}$ Cl N Cl

2 equiv. 1 equiv.

MeO_2C CO_2Me $\xrightarrow[\text{CuCl}]{\text{CCl}_4}$ Cl_3C Cl CO_2Me CO_2Me $\xrightarrow[\text{ii, 150 °C}]{\text{i, Et}_3\text{N/110 °C}}$ CO_2Me O O Cl

Scheme 55

The move away from polyhalogenated precursors to simple monohalogenated compounds has recently accelerated. One need only retain a radical-stabilizing group to promote atom transfer and to substitute a reactive iodine atom as the donor halogen (with radicals possessing two stabilizing groups, bromine can often be used). From the examples shown in Scheme 56, it is beginning to appear as if the additions of iodocarbonyls across multiple bonds will be a very general reaction.[58,176–178] The γ-iodocarbonyl adducts are versatile precursors of several different functional groups and the subsequent transformations outlined in Scheme 56 are often conducted *in situ*.

These reactions are often initiated by sunlamp photolysis of the iodide and the acceptor in benzene containing 10% hexabutylditin.[175,179] Because the iodine donor is the radical precursor, high reaction concentrations are beneficial rather than detrimental: 0.3 M is a typical value. Initiation can occur either directly by photolytic cleavage of the C—I bond or indirectly by cleavage of the Sn—Sn bond (the resulting tin radicals will abstract iodine from the precursor). An important function of the ditin is as a trap for iodine (either atomic or molecular), which is invariably produced when iodides are photolyzed but which strongly suppresses radical chains. Isolated yields for these bimolecular reactions are rarely higher than 80% but often exceed 50%. By using a slight excess of either the iodide or the alkene (1.5 to 2 equiv.), a significant increase in yield is often observed.

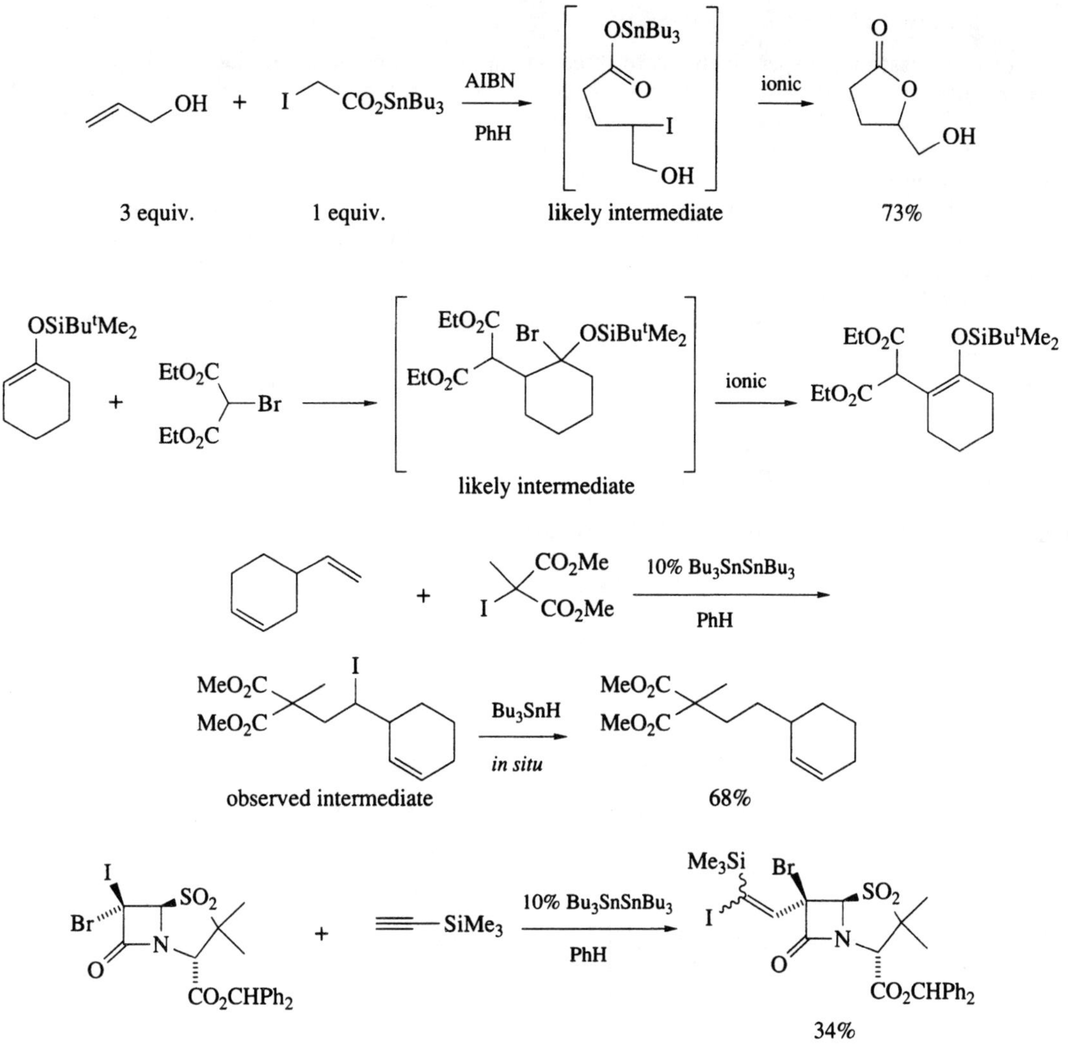

Scheme 56

Rate constants for the abstraction of halogens by primary alkyl radical from representative halocarbonyls have recently been measured and they are very high (considerably higher than the rate constant for the transfer of hydrogen from tin hydride, for example).[180] It is no wonder that telomerization is not as big a problem as in hydrogen transfer reactions. Of greater concern is the rate of the initial addition step; yields decrease as this step slows. A common by-product is the reduced iodide resulting from hydrogen transfer from the medium, and the yield of this reduction product increases as the efficiency of the addition step decreases.

It is more difficult to conduct the addition reactions of nucleophilic radicals to electron poor alkenes because the resulting atom transfer steps are often endothermic and are too slow to propagate chains, even with iodides. An exception is illustrated in Scheme 57: resonance-stabilized vinyl radicals (especially if they are secondary or tertiary) are reactive enough to abstract iodine from alkyl iodides.[178]

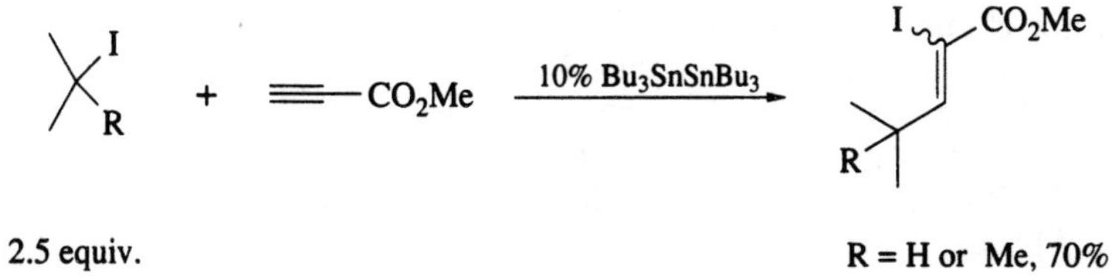

Scheme 57

(iii) Group transfer of organoboranes

There is at least one preparatively useful method for conducting addition reactions by using group transfer (rather than atom transfer). The transfer of alkyl substituents from organoboranes to enones is a chain reaction that was discovered by Brown in the late 1960s.[181] Since that time, the reaction has been neglected by synthetic practitioners; however, it has good synthetic potential and a recent development promises to significantly expand its current scope.

In the basic reaction, outlined in Scheme 58, an organoborane (**51**; usually formed by hydroboration of an alkene with diborane) is admixed with an enone or enal and the solution is exposed to atmospheric oxygen to initiate the reaction. After aqueous work-up, adducts (**52**) are isolated in good to excellent yield, even with some β-substituted acceptors. The proposed chain mechanism involves addition of an alkyl radical to the enone followed by (stepwise or concerted) group transfer of the organoborane to give a stable enol borane (**53**).[182] This product is hydrolyzed on work up. There are two main limitations to this method: (i) it fails with certain good radical traps like unsaturated esters (presumably because the adduct radicals bearing these substituents telomerize more rapidly than they add to the borane); and (ii) only one of the three initial borane substituents is transferred. The latter limitation can be circumvented for secondary- and tertiary-radicals by adjusting the substituents on boron as outlined in Scheme 59.[183] This transformation provides an alternative to the usual tactic of conversion of the borane to a halide or mercury(II) halide for use in a metal hydride mediated addition.

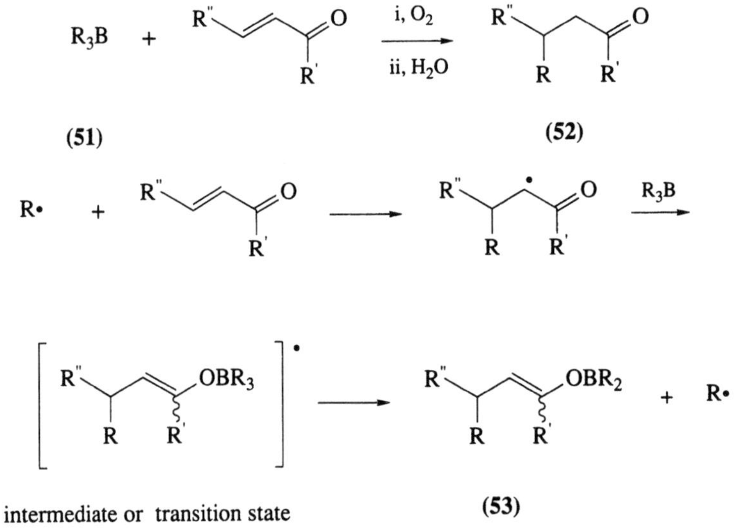

R_3B + R″ R′ O → (i, O_2; ii, H_2O) R″ R R′ O

(51) **(52)**

R• + R″ R′ O → R″ R R′ O → (R_3B)

[R″ R R′ OBR_3]• → R″ R R′ OBR_2 + R•

intermediate or transition state **(53)**

Scheme 58

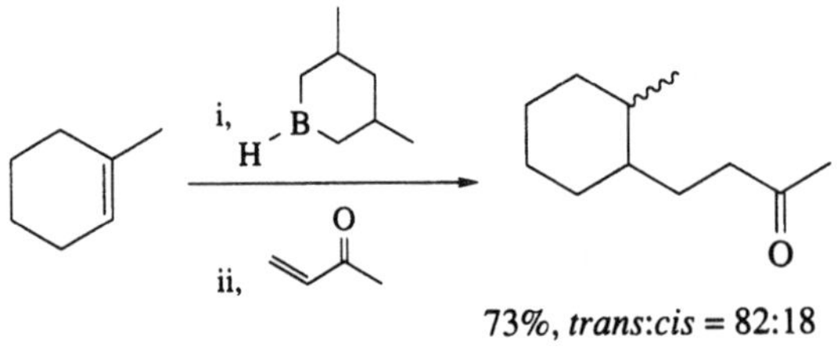

i, H B ; ii, O

73%, *trans*:*cis* = 82:18

Scheme 59

A beautiful extension of this reaction has recently been communicated by Nozaki, Oshima, and Utimoto.[184] These workers simply admixed *t*-butyl iodide (3 equiv.), benzaldehyde (1 equiv.), methyl vinyl ketone (1 equiv.) and triethylborane (1 equiv.) in benzene (Scheme 60). After 5 min at 25 °C, the reaction was subjected to standard extractive work-up and the crude product was purified by chromatography to give (**54**) in 63% yield. If methanol is substituted for benzaldehyde, the protonated product (**55**) is isolated in 79% yield. Although enones are equivalents of synthon (**56**), such a direct coupling of radical and ionic reactions had not been achieved previously.

The mechanism outlined in Scheme 61 is probably responsible for the observed transformation. Adduct radicals (**57**) react with the triethylborane to form the enol boronate (**58**) and release an ethyl radical. The enol borate is then trapped in a standard aldol reaction. The ethyl radical partitions between

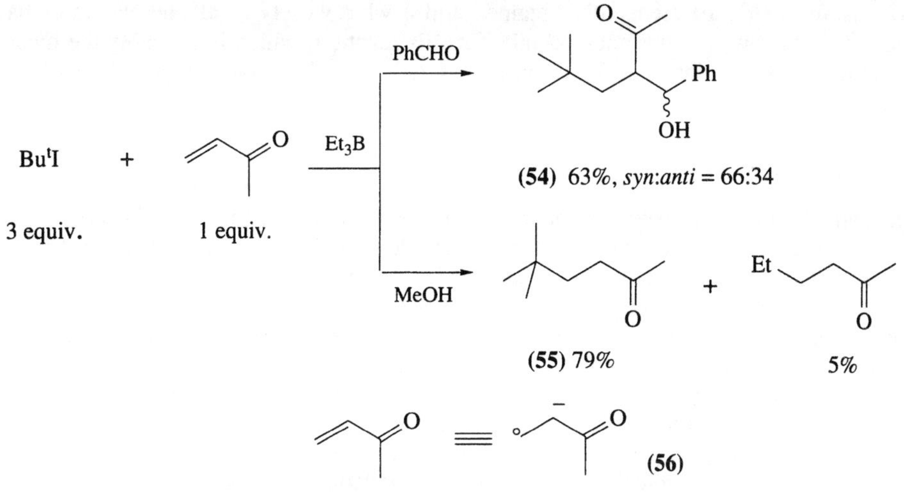

PhCHO

Ph

OH

Bu^tI + O Et_3B

(54) 63%, *syn*:*anti* = 66:34

3 equiv. 1 equiv.

Et

MeOH +

O O

(55) 79% 5%

(56)

Scheme 60

addition to the acceptor (a minor pathway) and abstraction of iodine from *t*-butyl iodide (the major pathway). This so-generated *t*-butyl radical (which is more reactive than the ethyl radical) adds to the enone to continue the chain. Accurate rate constants for the reaction of octyl radical (a good model for ethyl radical) with alkyl iodides are now available.[185] By comparing the estimated rate of iodine transfer (step 2) to that of addition to methyl vinyl ketone, one can plan successful reaction conditions to ensure a favorable partitioning in step 2.[186] This use of a reactive, easily generated alkyl radical (Et·) as a 'relay' in a reaction with an iodide to translocate a radical site has many potential applications, and it was first applied to synthesis by Minisci in addition reactions to protonated aromatics (Section 4.1.7.2).

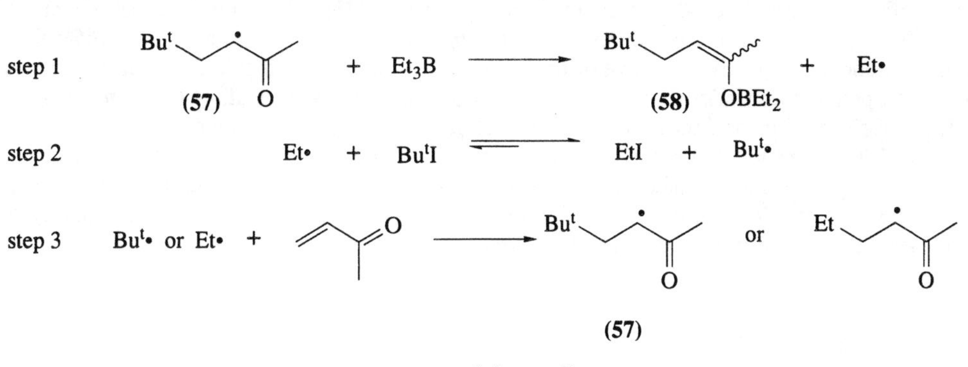

step 1 Bu^t **(57)** O + Et_3B ⟶ Bu^t **(58)** $OBEt_2$ + Et•

step 2 Et• + Bu^tI ⇌ EtI + Bu^t•

step 3 Bu^t• or Et• + (methyl vinyl ketone) ⟶ Bu^t (radical) O or Et (radical) O

(57)

Scheme 61

(iv) Meerwein arylation

The Meerwein arylation is at least formally related to the atom transfer method because a net introduction of an aromatic ring and a chlorine across a double bond is accomplished (Scheme 62). Facile elimination of HCl provides an efficient route to the kinds of substituted styrenes that are frequently prepared by Heck arylations. Standard protocol calls for the generation of an arene diazonium chloride *in situ*, followed by addition of an alkene (often electron deficient because aryl radicals are nucleophilic) and a catalytic quantity of copper(II) chloride. It is usually suggested that the copper salt operates in a catalytic redox cycle, reducing the diazonium salt to the aryl radical as Cu^I and trapping the adduct radical as Cu^{II}.

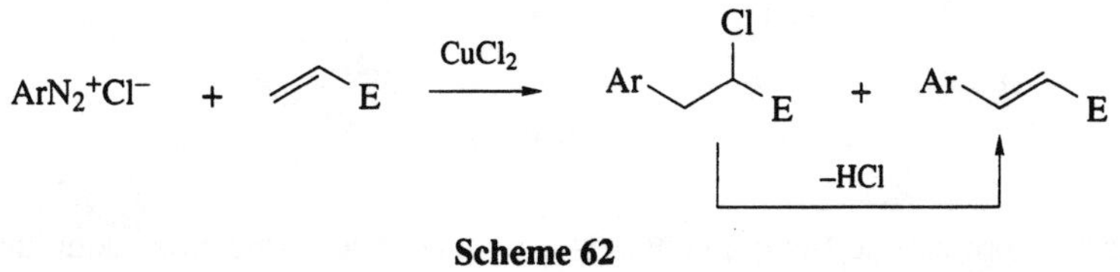

Cl

$ArN_2^+Cl^-$ + E $\xrightarrow{CuCl_2}$ Ar E + Ar E

–HCl

Scheme 62

Preparatively useful yields are frequently obtained and a wide variety of alkene activators have been used. Authoritative reviews provide more details for this reaction, which is probably the most general method available for the addition of aryl radicals across carbon–carbon double bonds.[187,188]

4.1.6.2 Nonchain Methods

Although chain methods have been the foundation of the recent resurgence of radical reactions in synthesis, many of the earliest preparatively useful radical addition reactions were based on nonchain processes. Of late, such nonchain processes have regained importance.

Nonchain reactions do not couple the formation and removal of carbon-centered radicals the way chain processes do. Instead of needing only a small quantity of initiator, nonchain additions need a method to generate and remove stoichiometric quantities of radicals. Generation of radicals poses special problems due to the transiency of radicals: if high concentrations of radicals are generated, the rates of radical–radical reactions will surpass those of radical–molecule reactions and addition reactions will not be possible. Removal of radicals poses the usual selectivity problem: how is an adduct radical removed without intercepting an initial radical prior to addition? Nonchain methods can be classified into two groups by the method in which the radicals are removed from the reaction: radical–radical coupling and redox.

4.1.6.2.1 Radical–radical coupling

In chain methods, it is important to avoid radical–radical reactions. However, radicals that are generated in a stoichiometric quantity by bond homolysis can be productively removed by radical–radical coupling. Despite the inherent problems in controlling reactions that occur at rates near the diffusion-controlled limit, radical–radical coupling reactions can be selective and preparatively useful.

The addition reactions to capto-dative alkenes developed by Viehe[189] illustrate one way to employ selective radical–radical coupling in product formation (Scheme 63).[190] Heating of cyclopentanone, capto-dative alkene (**59**) and a stoichiometric amount of di-*t*-butyl peroxide gives dimer (**60**) in 46% yield. Initial radicals are generated by slow cleavage of the peroxide O—O bond and subsequent hydrogen abstraction from cyclopentanone. Under the conditions of the reaction, these radicals (**61**) add to the reactive alkene (**59**) more rapidly than they undergo radical–radical reaction. However, the intermediate capto-dative radical (**62**) does not add rapidly to another molecule of (**59**). Instead, it suffers radical recombination to give (**60**). The reaction succeeds because capto-dative radicals are especially unreactive in radical–molecule reactions[191] (preventing telomerization) but they still react with each other (mainly by recombination rather than disproportionation) at diffusion-controlled rates.

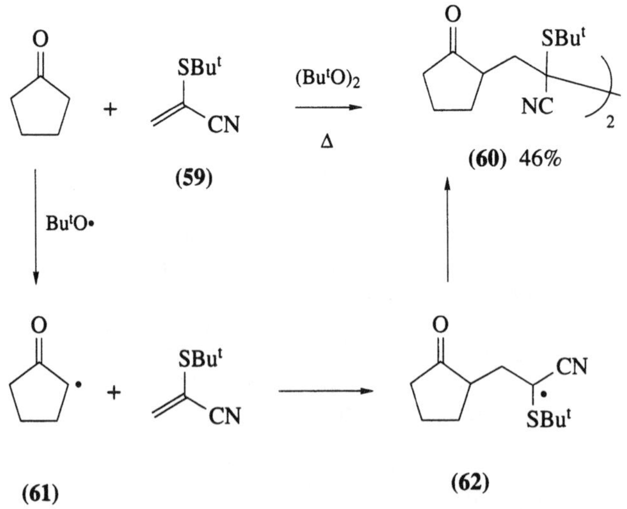

O SBu^t (Bu^{t}O)$_2$ CN Δ SBu^t NC 2 (**60**) 46% (**59**) Bu^{t}O• SBu^t CN CN SBu^t (**61**) (**62**)

Scheme 63

Such recombinations produce self-coupled dimers. In general, selective cross-coupling of two radicals (A· + B· → A—B) cannot be conducted because: (i) the two radicals are generated at the same rate; and

(ii) both possible self-coupling reactions (2A· → A—A; 2B· → B—B) and the cross reaction all proceed at the diffusion-controlled rate. Given these two conditions, cross-coupling and self-coupling must proceed at the same rate and no selectivity is possible. However, if either condition is violated, selective radical–radical couplings are possible.

Some selectivity can be enforced in cross-coupling by generating radicals at different rates. Schäfer has shown that Kolbe electrolysis of two acid salts, one major component (A) and one minor component (B), produces a near-statistical mixture of the self-coupled product of the major component (A—A) and the cross-coupled product of the major and minor components (A—B) (Scheme 64).[192] Provided that recombination is preferred over disproportionation and that the self-coupled product is easily removed, this is a useful method for coupling a precious component (B) with an inexpensive component (A); however, it is relatively energy inefficient since the self-coupled dimer is the major product. Scheme 65 illustrates how this principle has been applied to a simple addition reaction.[193] Presumably, methyl radicals generated in electrolysis add rapidly to the alkene and the resulting adduct radicals couple preferentially with methyl radicals (which are present in much higher concentration). Ethane is probably formed in large amounts by the self-coupling of methyl radicals, but it just evaporates.

$$\underset{\text{5–10 equiv.}}{A-CO_2^-} + B-CO_2^- \xrightarrow{-e^-} \underset{\text{major product}}{A-A} + \underset{\text{minor product but useful yield based on } B\text{-}CO_2^-}{B-B}$$

Scheme 64 Kolbe cross-coupling

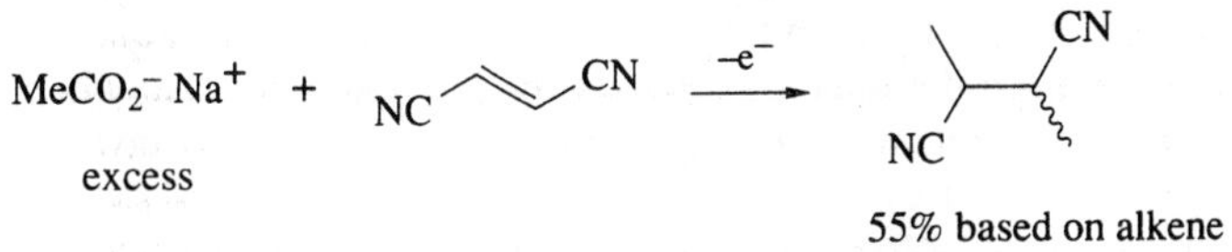

$MeCO_2^-\ Na^+$ (excess) + NC–CH=CH–CN $\xrightarrow{-e^-}$ product (NC, CN)

55% based on alkene

Scheme 65

A potentially more useful way to control cross-coupling is to mediate radical–radical reactions with a persistent radical (that is, one that does not react with itself at a diffusion-controlled rate). This strategy has not been widely recognized, but an excellent paper by Fischer has recently provided a quantitative theoretical framework for such reactions.[194] The basic strategy is qualitatively summarized in Scheme 66. Assume that a molecule A—P homolyzes to give two radicals A· and P·. Assume also that A· reacts with both itself and P· at diffusion-controlled rates but that P· is a persistent radical that couples with itself at a significantly lower rate. Throughout the reaction, A· and P· are produced at the same rate. However, transient radical A· is removed by self- and cross-couplings but P· is removed only by cross-coupling. Because P· is not removed by self-coupling, the concentration of P· soon exceeds that of A· and selective cross-coupling reactions become possible. For example, if A· is converted to B·, and if this new radical B· lives long enough to suffer radical–radical reactions, it may react exclusively with P·. The extent of the concentration gradient in favor of P· depends of its rate of self reaction (the more slowly it dimerizes, the higher the gradient).[192] More importantly, this concentration gradient can be estab-

$$A-P \longrightarrow A\bullet + P\bullet \quad \text{(a persistent radical)}$$

$$A\bullet + A\bullet \longrightarrow A-A$$

depletes concentration of A• at early reaction time

$$[P\bullet] >> [A\bullet] \text{ at steady state}$$

$$A\bullet \longrightarrow B\bullet \xrightarrow{+P\bullet} B-P$$

Scheme 66 Transient radical–persistent radical coupling

lished very early in the reaction and it is maintained at a steady state throughout the course of the reaction. From the preparative standpoint, this means that very high yields of radical–radical coupling product are theoretically possible because practically all radical–radical encounters involve P·.

Hart has recently developed an exciting new method to conduct radical addition reactions that probably operates by selective transient radical–persistent radical coupling.[195] When tin pinacolate[196] (TINPIN; **63**), methyl crotonate and cyclohexyl iodide are simply heated in benzene, adduct (**64**) is isolated in 60% yield after chromatography (Scheme 67).[197] This is a very respectable yield considering that: (i) the acceptor is not used in large excess (1–3 equiv.); (ii) it possesses a β-substituent; and (iii) the reaction is conducted at high concentration (0.3 M). Indeed, this reaction promises to be a first step towards the development of a second generation of addition reactions based on the trialkyltin radical that permit the addition of nucleophilic radicals to electron deficient alkenes at a high reaction concentration without special experimental techniques. One potential disadvantage of this method is that stoichiometric quantities of tin by-products and benzophenone must be separated from the desired product.

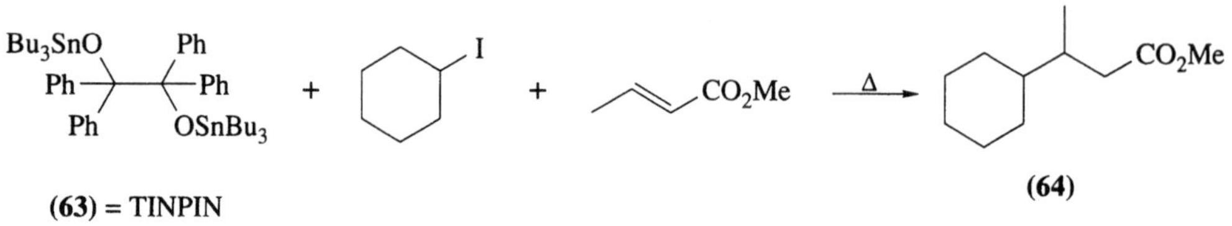

Bu₃SnO Ph Ph Ph Ph OSnBu₃ + I + CO₂Me Δ CO₂Me

(**63**) = TINPIN (**64**)

Scheme 67

Although uncertainties remain, the mechanism outlined in Scheme 68 is supported by observations of both Hart[197] and Neumann.[196] Neumann has shown that thermolysis of (**63**) causes homolytic cleavage to give the suspected persistent radical (**65**), which is in equilibrium with benzophenone and the trialkyltin radical. This radical (**65**) resembles a ketyl and it is stabilized both sterically and electronically. The trialkyltin radical abstracts iodine from the radical precursor to give a new radical, which in turn adds to the alkene to give (**66**). At very early reaction time, a concentration gradient in favor of (**65**) over all transient radicals develops because it is the only persistent radical. The most rapid reaction of (**66**) is radical–radical coupling, and this occurs selectively with (**65**) because it is present in the highest concentration. The coupled product (**67**) is then converted to (**64**) by thermal retroaldol reaction and hydrolysis. The use of a persistent radical does not eliminate the usual selectivity concerns. If the concentration of the persistent radical is too high, initial radicals may be trapped by coupling prior to addition, but if it is too low, the adduct radical may begin to telomerize. Thus, the concentration of the persistent radical is an important, yet ill-defined, variable in these reactions.

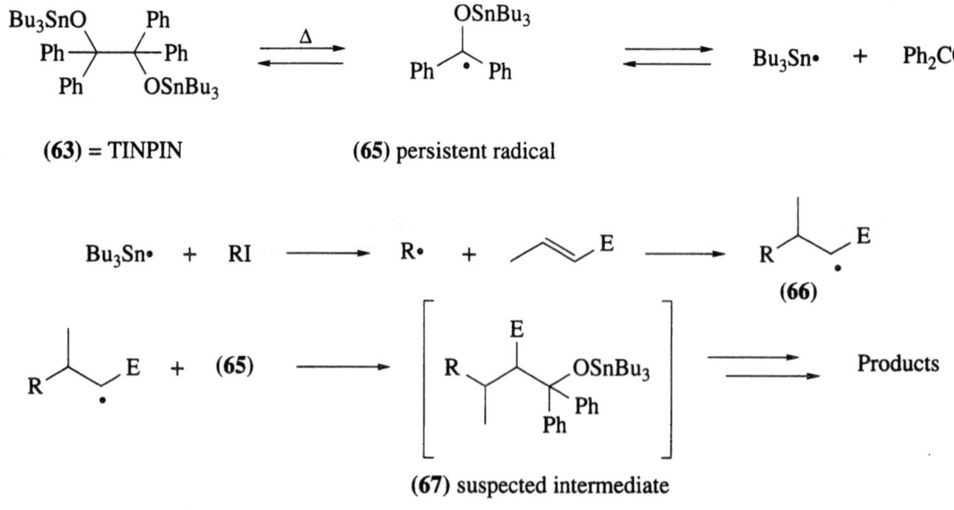

Bu₃SnO Ph Ph Ph Ph OSnBu₃ ⇌(Δ) OSnBu₃ Ph • Ph ⇌ Bu₃Sn• + Ph₂CO

(**63**) = TINPIN (**65**) persistent radical

Bu₃Sn• + RI → R• + E → R • E (**66**)

R • E + (**65**) → [R E OSnBu₃ Ph Ph] → Products

(**67**) suspected intermediate

Scheme 68

In contrast to the relatively small number of persistent carbon-centered radicals, there are many persistent metal-centered radicals, some of which are very stable. Indeed, it now appears that the emerging synthetic methods based on the chemistry of alkylcobalt compounds should be classed as reactions controlled by selective transient radical–persistent radical coupling.[198] In this case, the persistent radical is a cobalt(II) entity.

Alkyl- or acyl-cobalt(III) salen, salophen or dimethylglyoximato complexes are readily prepared by the displacement of a wide variety of halides and related leaving groups with highly nucleophilic cobalt(I) anions.[199,200] Because such alkylations may sometimes proceed *via* $S_{RN}1$ mechanisms, even cobalt complexes derived from unreactive (in an S_N2 sense) halides can be formed. The resulting carbon–cobalt bonds are very weak and Pattenden[201] and Branchaud[202] have shown that simple irradiation of these complexes in the presence of excess alkene produces adducts by the mechanism outlined in Scheme 69. This mechanism has been discussed in detail by Branchaud[202] and Giese,[203] and a key feature is that high concentrations of cobalt(II) are present in the reaction[204] because it does not readily dimerize or disproportionate (by electron transfer to give cobalt(III) and cobalt(I)). This permits the selective formation of products by radical–Co^{II} coupling. A high concentration of cobalt(II) can be both a blessing and a curse: a blessing because the adduct radicals are efficiently trapped and a curse because initial radicals can also be trapped. While this latter trapping only regenerates the starting complex, if it becomes too efficient, few of the radicals will live long enough to undergo addition.[204] A key feature of the reaction is that substituents that facilitate radical addition apparently also accelerate cobalt hydride elimination. Thus, the adducts are more prone to elimination of cobalt hydride than the starting complexes.

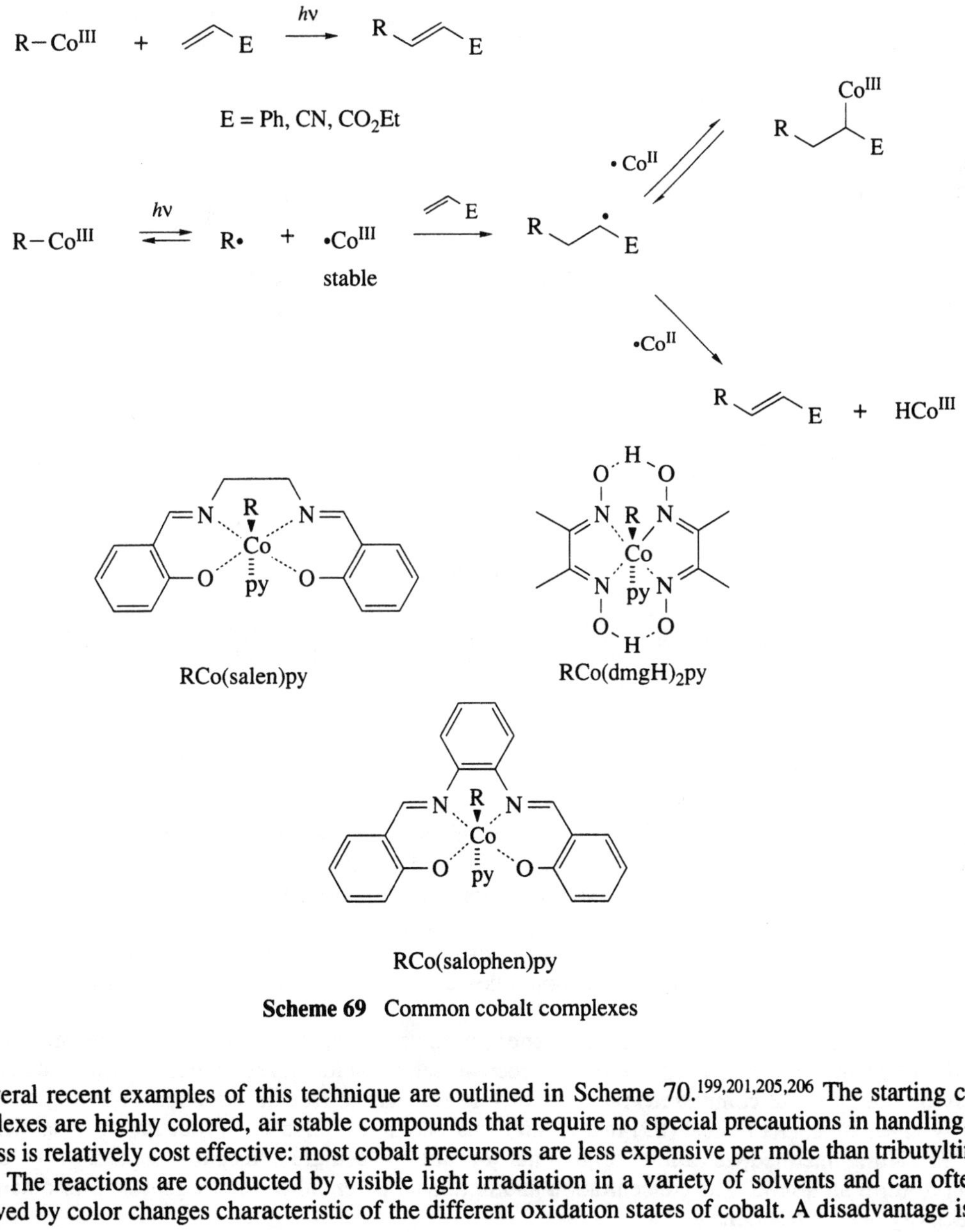

E = Ph, CN, CO_2Et

RCo(salen)py

RCo(dmgH)$_2$py

RCo(salophen)py

Scheme 69 Common cobalt complexes

Several recent examples of this technique are outlined in Scheme 70.[199,201,205,206] The starting cobalt complexes are highly colored, air stable compounds that require no special precautions in handling. The process is relatively cost effective: most cobalt precursors are less expensive per mole than tributyltin hydride. The reactions are conducted by visible light irradiation in a variety of solvents and can often be followed by color changes characteristic of the different oxidation states of cobalt. A disadvantage is that large excesses of alkene are often required for maximum yields. The products of the reaction are usually derived from apparent β-hydride elimination although sometimes protonated products are isolated.

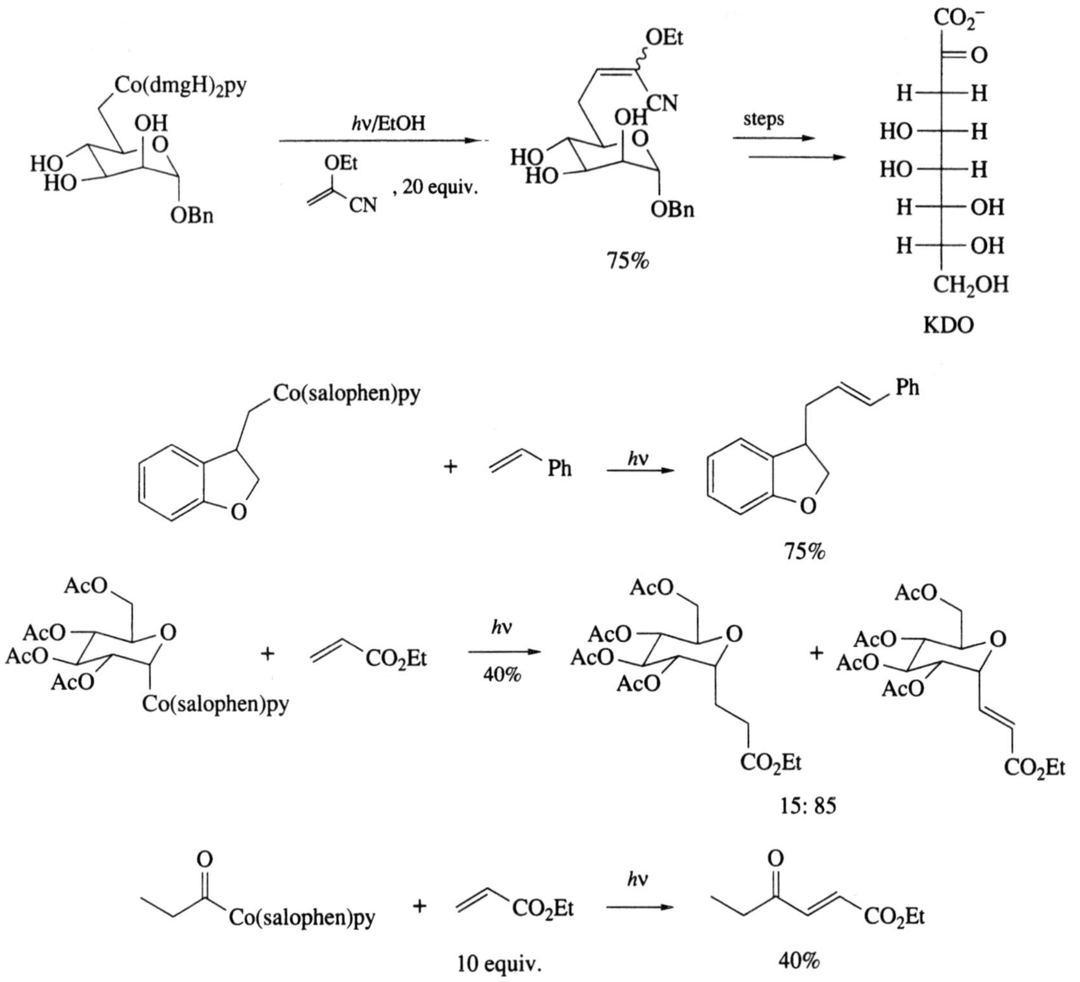

Co(dmgH)$_2$py ... hν/EtOH, OEt / CN, 20 equiv. ... 75% ... steps ... KDO

Co(salophen)py + Ph hν → Ph 75%

AcO, AcO, AcO, AcO, Co(salophen)py + CO$_2$Et hν 40% → CO$_2$Et + CO$_2$Et 15: 85

Co(salophen)py + CO$_2$Et (10 equiv.) hν → CO$_2$Et 40%

Scheme 70

The preliminary developments with TINPIN and organocobalt species are very encouraging and bode well for the future development of synthetic procedures that use radical–radical coupling methods controlled either by organic or organometallic persistent radicals.

4.1.6.2.2 Redox methods

In redox methods, radicals are generated and removed either by chemical or electrochemical oxidation or reduction. Initial and final radicals are often differentiated by their ability to be oxidized or reduced, as determined by substituents. In oxidative methods, radicals are removed by conversion to cations. Such oxidations are naturally suited for the additions of electrophilic radicals to alkenes (to give adduct radicals that are more susceptible to oxidation than initial radicals). Reductive methods are suited for the reverse: addition of alkyl radicals to electron poor alkenes to give adducts that are more easily reduced to anions (or organometallics).

The oxidative method is often conducted on enol (or enolate) derivatives and a simplified mechanism is shown in Scheme 71. Initial chemical or electrochemical oxidation gives an electrophilic radical (**68**; that may be free or metal-complexed) that is relatively resistant to further oxidation. Addition to an alkene now gives an adduct radical (**69**) that is more susceptible to oxidation. Products are often derived from the resulting intermediate cation (**70**) by inter- or intra-molecular nucleophilic capture or by loss of a proton to form an alkene. The concentration and oxidizing potential of the reagent help to determine the selectivity in such reactions.

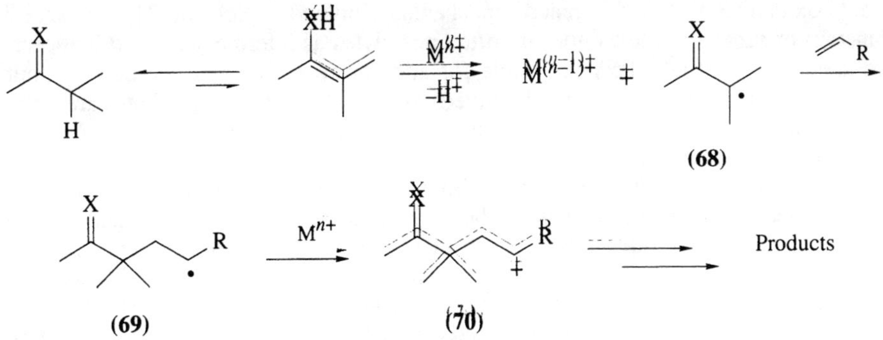

(68)

Products

(69) (70)

Scheme 71 Oxidative addition reactions

In addition to electrochemical methods,[192] copper(II),[207] iron(III)[208] and cerium(IV)[209] oxidants are useful for mediating addition reactions. During the last decade, manganese(III) acetate has emerged as the the most popular reagent for conducting such transformations.[210] The basic reaction, first studied by Heiba and Dessau,[211] is the oxidation of acetic acid in the presence of alkenes and 2 equiv. of Mn^{III} triacetate dihydrate (which is actually an oxo-centered manganese(III) trimer). Lactones are formed in good yields as shown in Scheme 72.

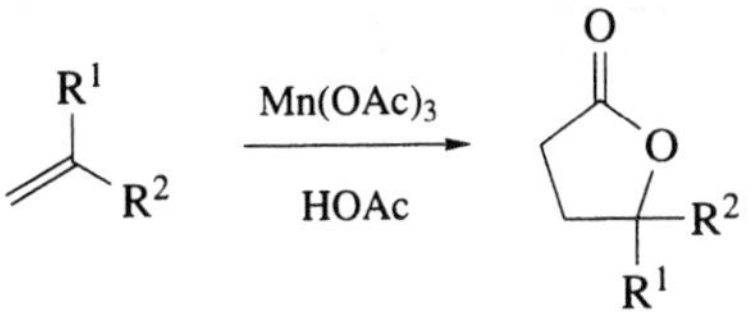

$Mn(OAc)_3$, HOAc

Scheme 72

Variants on this theme have provided many useful preparative procedures and a few representative examples are provided in Scheme 73.[212–214] Acetic acid is a relatively unreactive substrate and is often used as a solvent. However, more enolizable substrates, especially those possessing two activating groups, are readily oxidized on a stoichiometric basis at 25–50 °C in acetic acid. If the products contain enolizable hydrogens, over oxidation is a potential problem (see the first example). The reactions can often be monitored by the disappearance of the characteristic brown color of the oxidant. While it is clear

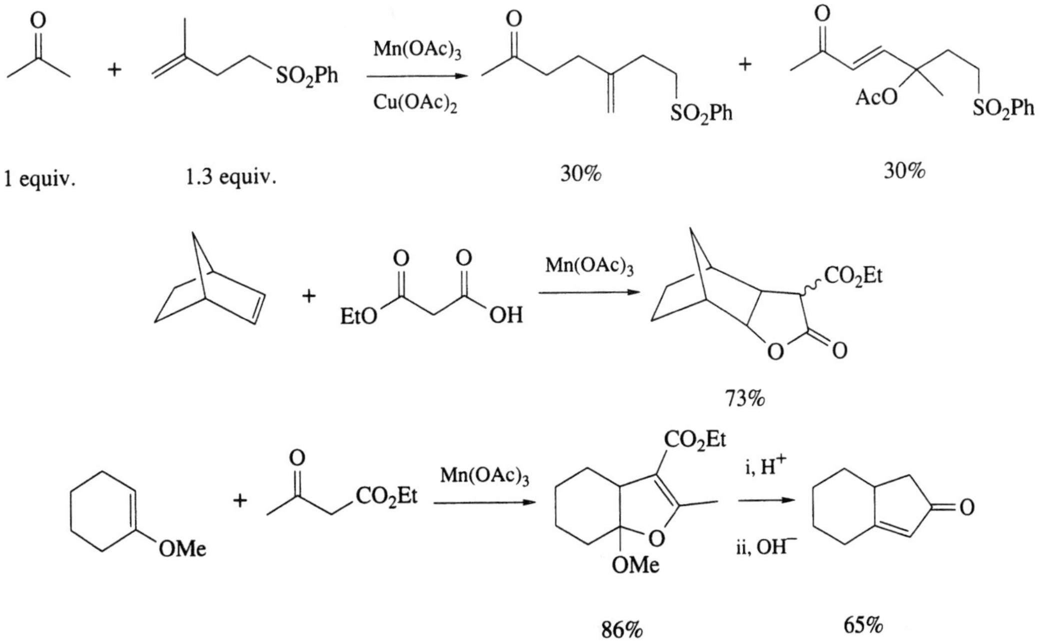

1 equiv. 1.3 equiv. $Mn(OAc)_3$, $Cu(OAc)_2$ 30% 30%

$Mn(OAc)_3$ 73%

$Mn(OAc)_3$ 86% i, H^+ ii, OH^- 65%

Scheme 73

that these Mn^{III} oxidations follow the general mechanism outlined in Scheme 71, the exact details can vary as a function of substrate. Mn^{III} enols are often postulated as intermediates and it appears that Mn-complexed radicals may actually be responsible for addition reactions in some cases,[215] but in others (particularly where more stable radicals are involved), true free radical intermediates are strongly implicated. A recent paper by Snider provides a detailed mechanistic discussion with complete references to previous synthetic and mechanistic studies.[216]

In contrast to oxidative methods, reductive methods are more suited to conduct additions of nucleophilic radicals to electron deficient alkenes as shown in Scheme 74. A useful reagent should reduce a radical precursor but give the resulting radical sufficient lifetime to add to an alkene prior to reduction. The adduct radical should then be readily reduced to remove it from the radical pool. This reaction is formally equivalent to the much more common conjugate addition of an organometallic reagent. In principle, the organometallic intermediate that is formed can be trapped by electrophiles, but in practice most reductions have been conducted in protic solvents and have formed only protonated products. Several examples of addition reactions in which the products are trapped by reduction are contained in Scheme 75. The first example illustrates a convenient reductive equivalent of the Meerwein arylation,[112,217] while the second example shows the reductive generation of alkoxy radicals for use in C—H abstraction.[218,219] The last example is particularly interesting because it appears on the surface to be a standard organometallic conjugate addition but experimental evidence gathered by Luche indicates that the reductive radical mechanism is operating.[220]

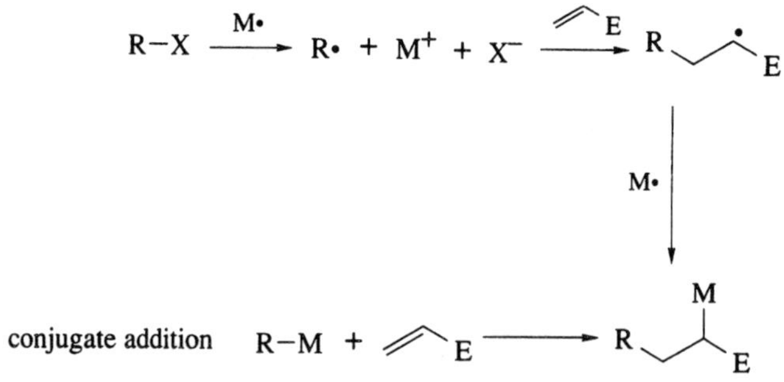

$$\text{R–X} \xrightarrow{\text{M}\bullet} \text{R}\bullet + \text{M}^+ + \text{X}^- \longrightarrow \text{R}\frown\dot{\text{C}}\text{H–E} \xrightarrow{\text{M}\bullet} \text{R}\frown\text{CH(M)–E}$$

conjugate addition $\text{R–M} + \text{CH}_2\text{=CH–E} \longrightarrow \text{R}\frown\text{CH(M)–E}$

Scheme 74 Reductive addition reaction

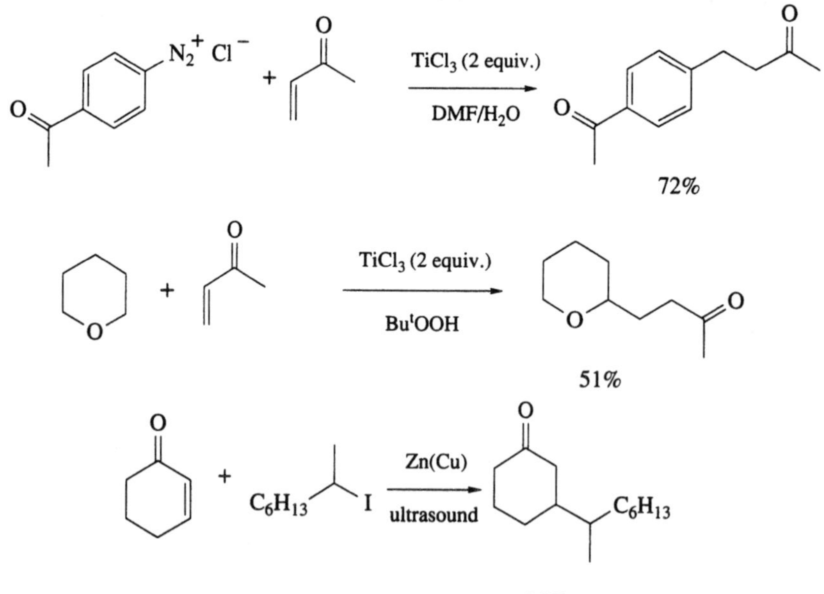

4-acetylbenzenediazonium chloride ($N_2^+ Cl^-$) + methyl vinyl ketone $\xrightarrow[\text{DMF/}H_2O]{TiCl_3 \text{ (2 equiv.)}}$ 4-(4-acetylphenyl)butan-2-one, 72%

tetrahydropyran + methyl vinyl ketone $\xrightarrow[\text{Bu}^t\text{OOH}]{TiCl_3 \text{ (2 equiv.)}}$ 4-(tetrahydropyran-2-yl)butan-2-one, 51%

cyclohexenone + $C_6H_{13}CH(CH_3)I$ $\xrightarrow[\text{ultrasound}]{\text{Zn(Cu)}}$ 3-(1-methylheptyl)cyclohexanone (C_6H_{13}), 90%

Scheme 75

Perhaps the most generally useful reaction of this type is the vitamin B_{12} catalyzed reductive addition shown in Scheme 76.[221] This reaction, pioneered by Scheffold,[222] is related to the organocobalt additions in Section 4.1.6.1, but only a catalytic quantity of vitamin B_{12} is required along with a chemical or electrochemical coreductant. This procedure is a very practical alternative to the metal hydride method for conducting addition reactions of nucleophilic radicals. The catalytic cycle outlined in Scheme 76 is a reasonable mechanism but variants are possible.[222] The coreductant recycles cobalt(II) products to the nucleophilic cobalt(I) complexes required for radical generation.

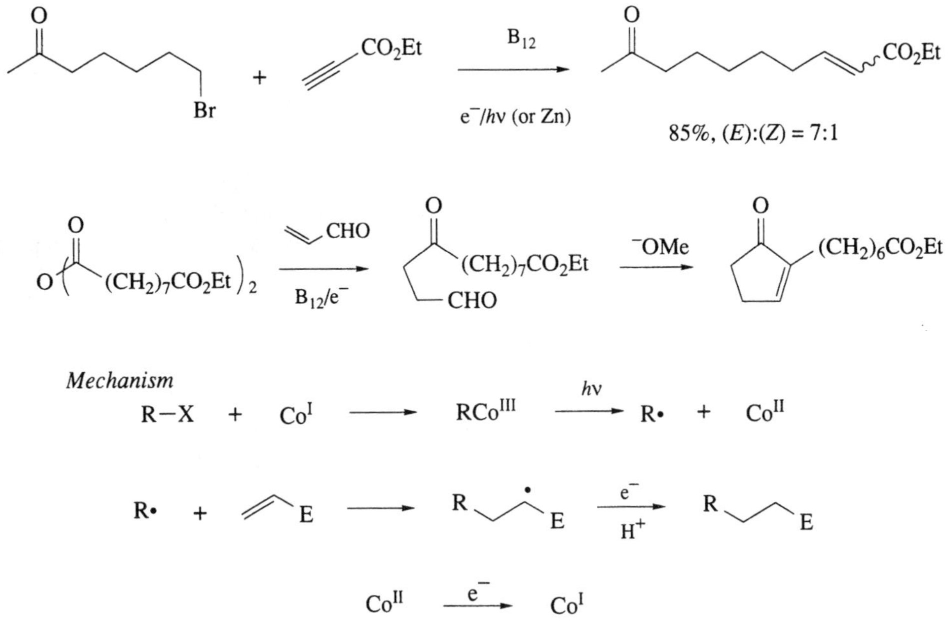

Scheme 76

4.1.7 ADDITIONS OF CARBON-CENTERED RADICALS TO OTHER MULTIPLE BONDS AND AROMATIC RINGS

4.1.7.1 Additions to Other Multiple Bonds

Besides alkenes and alkynes, other multiple bonds can be used as acceptors in addition reactions of carbon radicals provided the usual requirements of reactivity and selectivity are met. Other types of carbon–carbon multiple bonds that have been used as acceptors include dienes,[162] allenes,[61] enolates (and nitronates, see below) and quinones.[223] Even highly strained σ-bonds have served as acceptors on occasion.[224]

Addition reactions of carbon radicals to C—O and C—N multiple bonds are much less-favored than additions to C—C bonds because of the higher π-bond strengths of the carbon–heteroatom multiple bonds. This reduction in exothermicity (additions to carbonyls can even be endothermic) often reduces the rate below the useful level for bimolecular additions. Thus, acetonitrile and acetone are useful solvents because they are not subject to rapid radical additions. However, entropically favored cyclizations to C—N and C—O bonds are very useful, as are fragmentations (see Chapter 4.2, this volume).

Only a few additions to highly activated (electron deficient) carbonyls are known[222,225] and at present they provide little competition for standard ionic carbonyl addition reactions. But additions to C–N multiple bonds show more preparative potential. Both Stork and Sher,[29] and Barton[226] have developed useful additions to isonitriles; additions to oximes are also known.[227] The TINPIN method (Scheme 67) of adding nucleophilic radicals to oxime ethers recently reported by Hart and Seely appears to be one of the most useful reactions in this class.[197] This reaction oxaminomethylates radical precursors under very mild conditions in good yields and it might often be advantageous compared to the standard tactic of addition of an organometallic reagent to an oxime ether (Scheme 77).

Even though they are electron deficient, radicals do not react rapidly with simple anions and organometallic species because the resulting radical anions would place an electron in a very high energy orbital. However, conjugated anions with low lying LUMOs can react readily with radicals. Indeed, nitronates and certain enolates are good radical acceptors and a large class of chain reactions, termed $S_{RN}1$ substitutions, rely on this capability.[228] Several examples of $S_{RN}1$ reactions and a generic mechan-

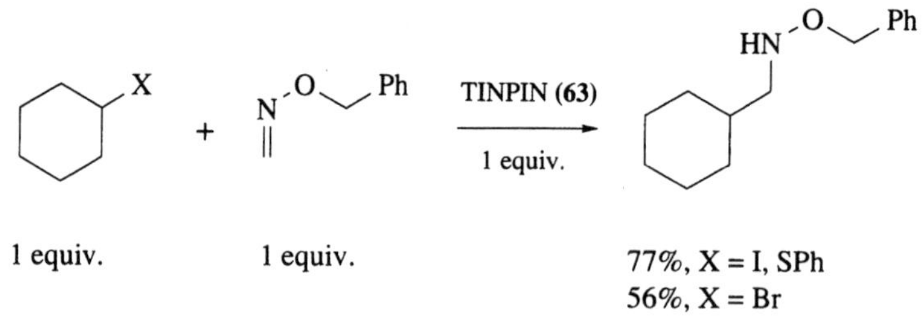

TINPIN (63)

1 equiv.

1 equiv. 1 equiv. 77%, X = I, SPh

56%, X = Br

Scheme 77

ism are collected in Scheme 78.[229,230] These chain reactions often require chemical or photochemical initiation. While the overall transformation is the same as that of an alkylation, $S_{RN}1$ reactions nicely complement standard enolate alkylations, working best with tertiary alkyl and aryl halides (and related precursors). The key step in Semmelhack's classic cephalotaxine synthesis is an $S_{RN}1$ reaction that illustrates this point very well (Scheme 78).[229]

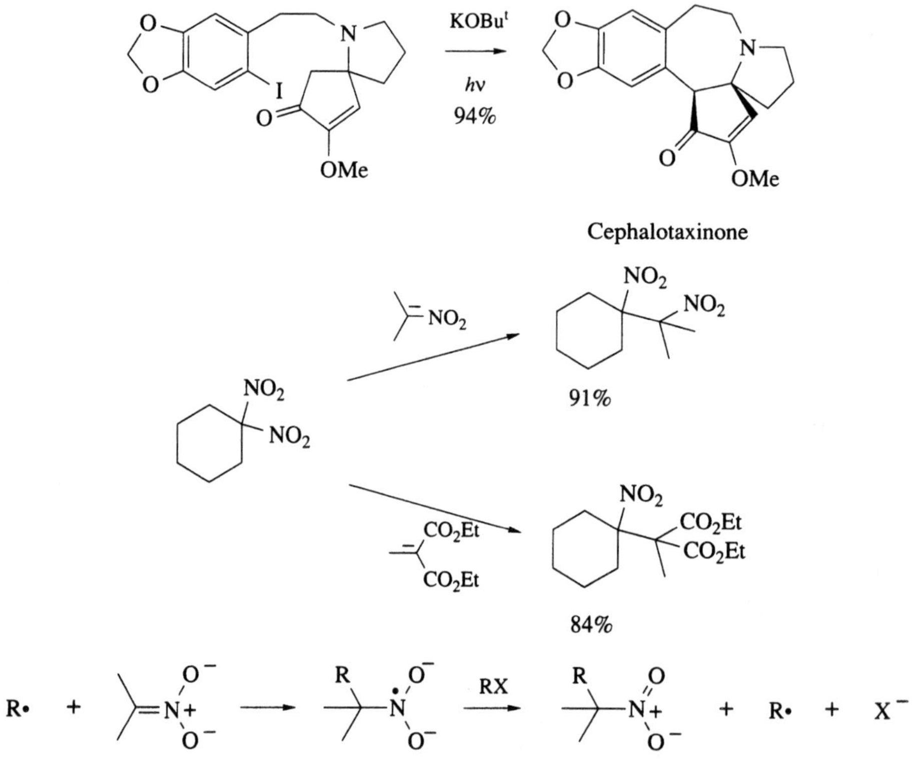

KOBu^t, $h\nu$, 94%

Cephalotaxinone

91%

84%

Scheme 78

4.1.7.2 Additions to Aromatic Rings

Most known addition reactions of radicals to aromatic rings involve net substitution for a hydrogen atom by the general mechanism outlined in Scheme 79. The cyclohexadienyl radical (**71**) that is produced on addition is a powerful reductant compared to most other radicals. It is important to have a mild oxidant present to trap this radical to prevent it from undergoing standard radical–radical reactions. Thus, the overall driving force for these substitutions is usually provided by the oxidant. Unlike additions to C—C multiple bonds, fragmentation can sometimes be important in these less exothermic additions to aromatic rings (because aromaticity is sacrificed on addition). The oxidant may play an important role in reversible additions by oxidizing the cyclohexadienyl radical more readily than the starting radical.

The addition reactions of alkyl and substituted alkyl radicals to simple aromatic rings are very slow (see Section 4.1.2.2). Hence, benzene is a good solvent. The additions of aryl radicals to aromatic rings are considerably faster and, while such reactions have been studied intensely,[231] their preparative utility

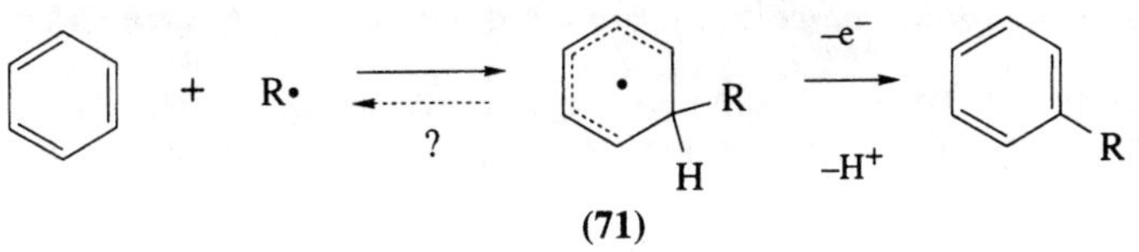

R• (71) H R −e⁻ −H⁺ ?

Scheme 79 Additions to aromatic rings

is not high because of low regio- and chemo-selectivity. However, the addition of an aryl radical to benzene can become a limiting side reaction when techniques like slow syringe pump addition of tin hydride are used and there is no good trap for the aryl radical. An example is provided in Scheme 80.

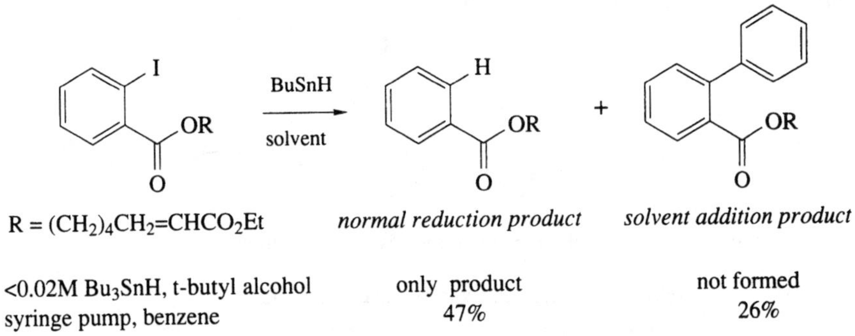

I OR O BuSnH solvent H OR O + OR O

R = $(CH_2)_4CH_2$=$CHCO_2Et$ *normal reduction product* *solvent addition product*

<0.02M Bu_3SnH, t-butyl alcohol only product not formed
syringe pump, benzene 47% 26%

Scheme 80

Additions to aromatic rings can become useful when radicals and acceptors are electronically paired. The additions of electrophilic radicals to electron rich aromatic rings are growing in importance and the additions of nucleophilic radicals to electron poor alkenes have long been of preparative value. This chapter can provide only a few representative examples of each class. Giese's book is recommended as a more thorough overview of additions to aromatic rings.[232]

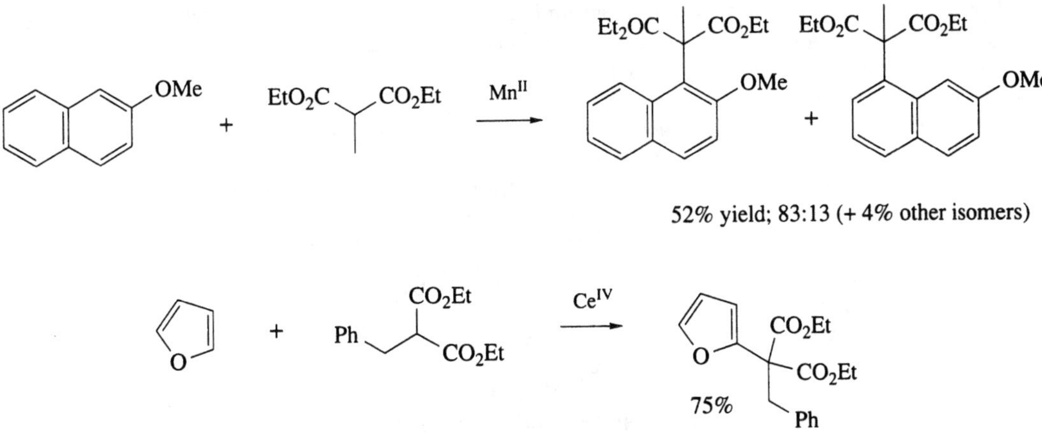

OMe + EtO_2C CO_2Et Mn^{II} Et_2OC CO_2Et OMe + EtO_2C CO_2Et OMe

52% yield; 83:13 (+ 4% other isomers)

O + Ph CO_2Et CO_2Et Ce^{IV} O CO_2Et CO_2Et Ph 75%

Scheme 81

4.1.7.2.1 Additions of electrophilic radicals to electron rich aromatic rings

Oxidation of enolizable nitro, carbonyl and dicarbonyl compounds with Fe^{III} Mn^{III} and Ce^{IV} reagents in the presence of electron rich aromatic (or heteroaromatic) rings often provides modest to good yields of substituted products. Typical examples are shown in Scheme 81.[233,234] The oxidant functions both to generate the initial radical (Scheme 71) and to trap the adduct radical. Products of *ortho* substitution usually predominate but significant amounts of *para* and *meta* products are often formed, and in some cases, reversibility in the addition step may influence the product distribution. A recent paper by Citterio and Santi provides a nice introduction to these types of reactions.[219]

4.1.7.2.2 *Addition of nucleophilic radicals to protonated heteroaromatics (the Minisci reaction)*

The functionalization of electron rich aromatics rings is often accomplished by electrophilic aromatic substitution. However, electrophilic substitutions require stringent conditions or fail entirely with electron deficient aromatic rings. Nucleophilic aromatic substitutions are commonly used but must usually be conducted under aprotic conditions. In contrast, nucleophilic radicals can add to electron deficient aromatic rings under very mild conditions.

The addition of radicals to protonated nitrogen heterocycles is by far the most important reaction in this class. Pioneered by the group of Minisci, this reaction was recognized as being preparatively useful well before the recent resurgence of radical reactions in synthesis. In addition, associated mechanistic studies contributed significantly to our present understanding of polar effects in free radical reactions. The general reaction is outlined in Scheme 82 and, because there are many important variants, several reviews are recommended for more comprehensive coverage.[235]

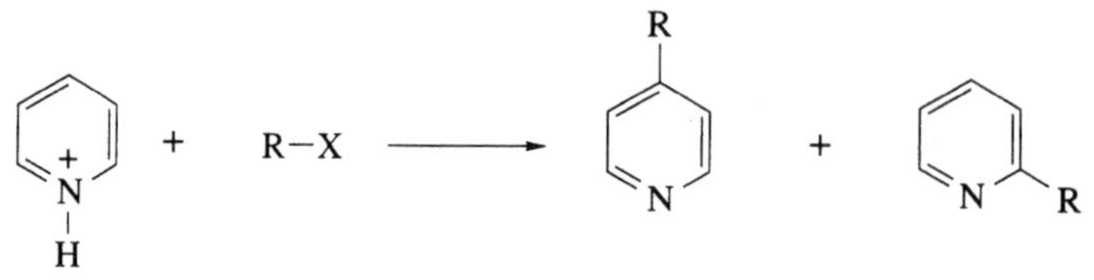

X = H, I, Co, Hg, CO_2H; R = alkyl, hydroxyalkyl, aminoalkyl, acyl

Scheme 82 Additions to protonated pyridines

Alkyl radicals add to pyridine relatively slowly; however, the rate of addition is dramatically accelerated by simple protonation, and alkyl radicals add to pyridinium salts at about the same rate as they add to styrene ($k \approx 5 \times 10^4$ M^{-1} s^{-1}).[236] While pyridine itself gives mixture of 2- and 4-substituted products, substituted pyridines and polycyclic *N*-heterocycles often exhibit good regioselectivity. Electronegative substituents on the ring accelerate additions of nucleophilic radicals and rates comparable to those with reactive alkenes like acrylonitrile are not uncommon.[236] Radicals can be generated by a variety of methods from C—H and C—I bonds (by abstraction), from C—Co[237] and C—Hg[238] bonds, and from carboxylic acids (by oxidative decarboxylation or from thiohydroxamates[239]). Representative examples are provided in Scheme 83.[240–242] The Minisci reaction has two potential advantages over nucleophilic aromatic substitution: (i) it readily introduces acyl and related functional groups that are not usually directly available by nucleophilic substitutions; and (ii) it proceeds under mild, protic conditions and does not require organometallic intermediates. Activation of *N*-heterocycles for organometallic additions is often beneficial but simple protonation does not suffice because addition reactions are usually precluded by simple acid–base reactions with the nucleophile.

A representative example from the recent literature illustrates that a redox chain substitution is one of the most practical means to conduct these reactions.[243] Simple heating of lepidine (**72**) and *t*-butyl hydroperoxide in aqueous acidic methanol with 5% iron sulfate provides hydroxymethyllepidine (**73**) in 93% yield. The accepted mechanism is a redox chain outlined in Scheme 84. Iron(II) reduces the *t*-butyl hydroperoxide to generate the *t*-butoxyl radical (**74**) in the classical Fenton reaction. This radical (**74**) abstracts a hydrogen from the solvent to give the hydroxymethyl radical (**75**), which in turn adds to the lepidinium salt to form (**76**). Such addition reactions may sometimes be reversible but subsequent loss of a proton (from carbon) is not. The resulting radical (**77**) is a very weak base compared to lepidine. But it is a potent reductant because of its high-lying HOMO and it reduces iron(III) to iron(II).

These redox chain reactions, which cycle iron(II) and iron(III), have advantages over methods that use stoichiometric quantities of oxidants because the hydroxymethyl radical is also a good reductant and, at high oxidant concentrations, it may be oxidized more rapidly than it adds to (**72**). The disadvantage of this type of reaction is that the initial radical is generated by a relatively non-selective hydrogen atom abstraction reaction. To be efficient, the H-donor must be used in large excess; it is often a cosolvent. Nonetheless, this is a very practical method to prepare hydroxyalkylated and acylated heteroaromatic and related derivatives.

The addition of functionalized alkyl radicals to protonated heteroaromatics was more difficult (because the radicals could not be generated by H-atom abstraction), but a recent development holds promise to resolve this problem. Generation of a methyl radical in the presence of an alkyl iodide sets up a relatively rapid equilibrium as indicated in Scheme 85. This equilibrium will favor any more highly substituted alkyl radical over methyl, and further, this latter radical will be significantly more nucleophilic. Thus when methyl radicals are generated in the presence of cyclohexyl iodide and a protonated quinaldine, the

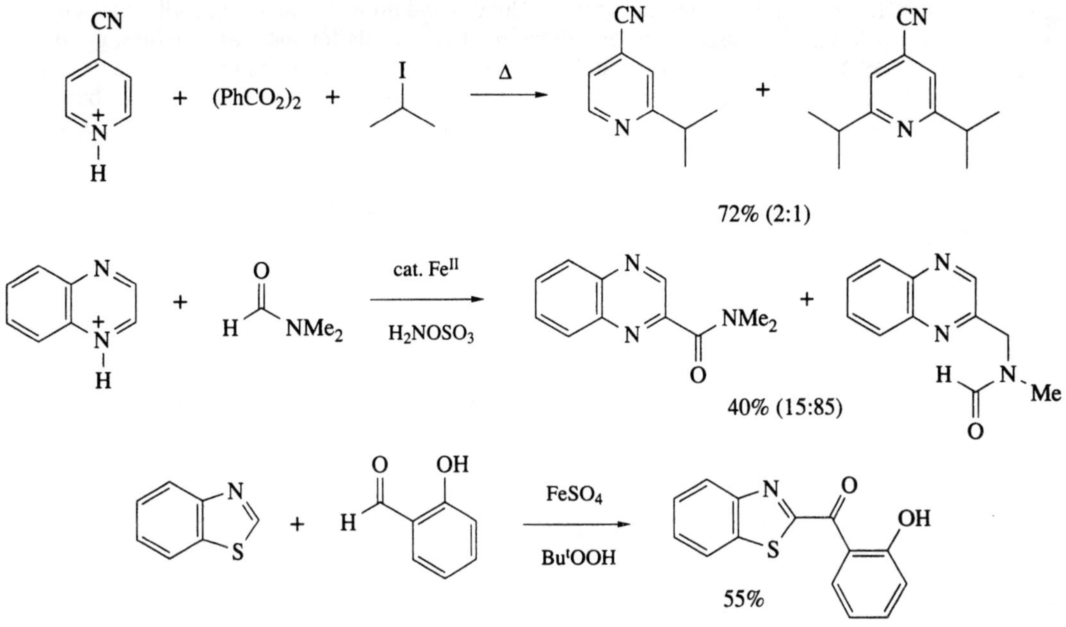
CN
+ (PhCO2)2 +
I
Δ
72% (2:1)
cat. FeII
H2NOSO3
NMe2
H
Me
40% (15:85)
FeSO4
ButOOH
OH
55%

Scheme 83

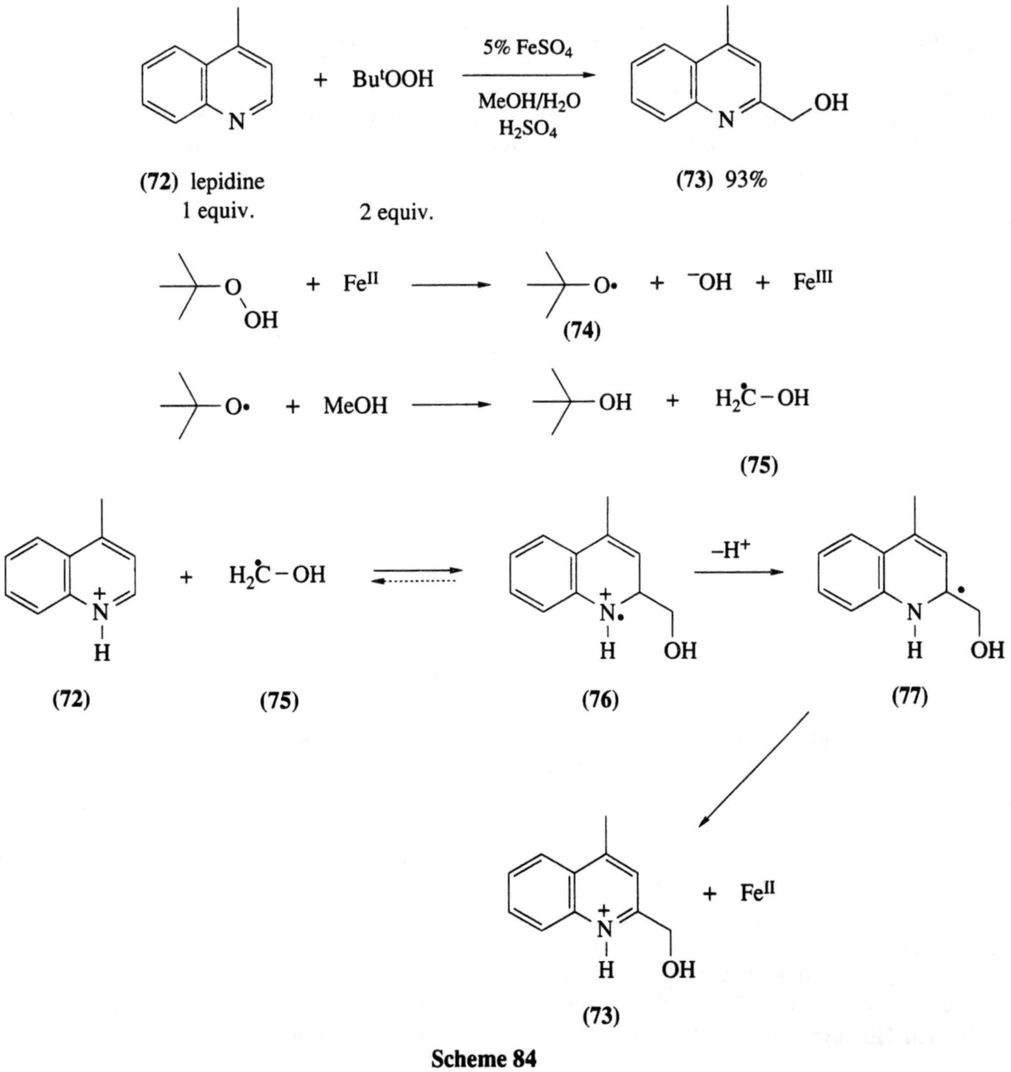
5% FeSO4
ButOOH
MeOH/H2O
H2SO4
(72) lepidine
1 equiv.
2 equiv.
(73) 93%
+ FeII
(74)
+ −OH + FeIII
+ MeOH
H2C−OH
(75)
−H+
(72)
(75)
(76)
(77)
(73)

Scheme 84

product results not from methyl addition but from cyclohexyl addition.[244] Scheme 85 illustrates one of several methods that Minisci has used to generated the methyl radicals for such a procedure: the oxidative cleavage of DMSO. As above, the iron catalyst mediates a redox chain and the ultimate oxidant is hydrogen peroxide. In principle then, any alkyl radical that is more stable than methyl might be generated by this elegant method and added not only to protonated heteroaromatics but to other acceptors as well.

$$CH_3\bullet + R{-}I \rightleftharpoons MeI + R\bullet$$

R = substituted alkyl

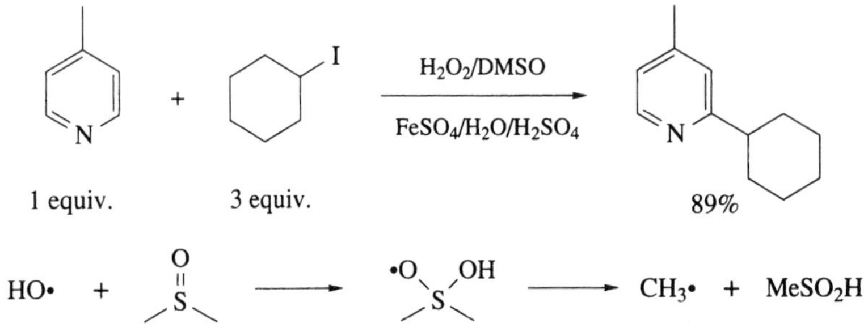

1 equiv. + 3 equiv. — H_2O_2/DMSO, $FeSO_4/H_2O/H_2SO_4$ → 89%

$$HO\bullet + \text{DMSO} \longrightarrow \bullet O\text{–}S(OH)Me_2 \longrightarrow CH_3\bullet + MeSO_2H$$

Scheme 85

4.1.8 ADDITIONS OF HETEROATOM-CENTERED RADICALS

Addition and substitution reactions of heteroatom-centered radicals with multiple bonds have been extensively studied and are sometimes preparatively useful.[11] This section will briefly consider the addition reactions of H—Y and X—Y reagents (Kharasch reactions) and substitution reactions (Scheme 86).[245]

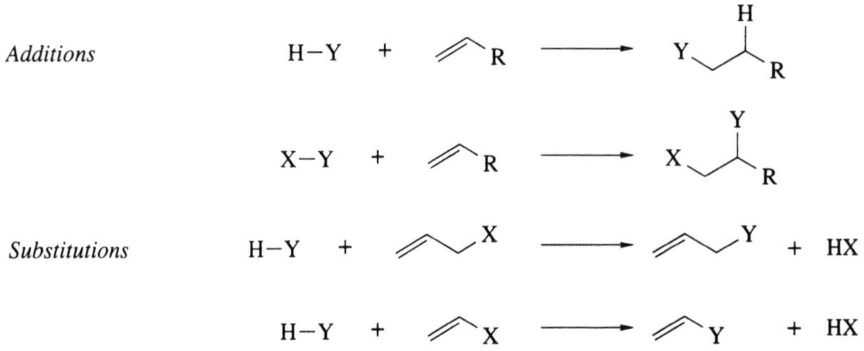

Additions

H–Y + alkene–R ⟶ Y–CH₂–CH(H)–R

X–Y + alkene–R ⟶ X–CH₂–CH(Y)–R

Substitutions

H–Y + allyl–X ⟶ allyl–Y + HX

H–Y + vinyl–X ⟶ vinyl–Y + HX

Scheme 86 Reactions of heteroatom radicals with alkenes

The anti-Markovnikov addition of HBr to alkenes is an historically important reaction in the development of organic chemistry, but its synthetic value for alkene functionalization has waned with the advent of reactions such as hydroboration. Radical additions of thiols and selenols to alkenes and alkynes are important preparative reactions for the synthesis of sulfides and selenides.[246] Acid- and base-catalyzed additions of thiols and selenols are also popular. Hydrostannylation and hydrogermylation (but not hydrosilylation) can also be accomplished by radical chain addition.[80,83,84] Metal-catalyzed additions and ionic conjugate additions of stannyl metals are also popular. Like most radical additions, radical hydrostannylation is not stereospecific (that is, alkene geometry is not conserved in the product) but it can be highly stereoselective in cyclic systems. Hydrostannylations are usually conducted at high concentrations because the addition of a tin radical is reversible and a high tin hydride concentration favors trapping of the adduct. Scheme 87 provides several recent examples.[247–249]

Many variants on the addition of two heteroatoms (X—Y) across an alkene or alkyne have been developed.[245] These reactions always proceed by the atom transfer method and require a reactive atom or group donor (X = halogen, SPh, SePh). Many atoms and groups Y can be introduced including oxygen and nitrogen. However, such additions are only occasionally advantageous when compared to ionic

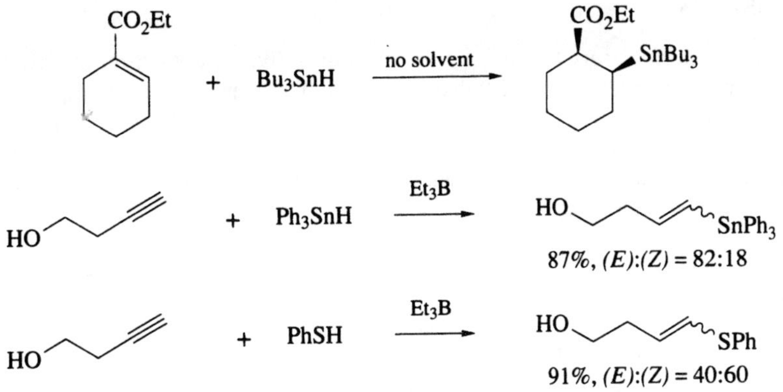

Scheme 87

equivalents like oxy- and amido-mercuration, sulfenylation, iodination and the like. These ionic additions are often stereospecific, high yielding and very general.

Among the many variants that add two heteroatoms, those that add phenylsulfonyl groups are particularly useful because they proceed in high yield under mild conditions and provide functionality for subsequent synthetic transformations. The examples contained in Scheme 88 illustrate some recent applications.[250–252] In contrast to most radical additions to alkynes, the additions of $ArSO_2X$ are often highly stereoselective.

The interchange of heteroatom groups by free radical substitution of allyl or vinyl groups (see Section 4.1.6.2) is also a common and preparatively useful reaction that occurs by an addition–elimination mechanism. Those groups that participate in fragmentation reactions (halogen, tin, sulfur, selenium) are susceptible to interchange. Depending on reaction conditions, such substitutions can sometimes be conducted in either direction. Representative examples are provide in Scheme 89.[253,254]

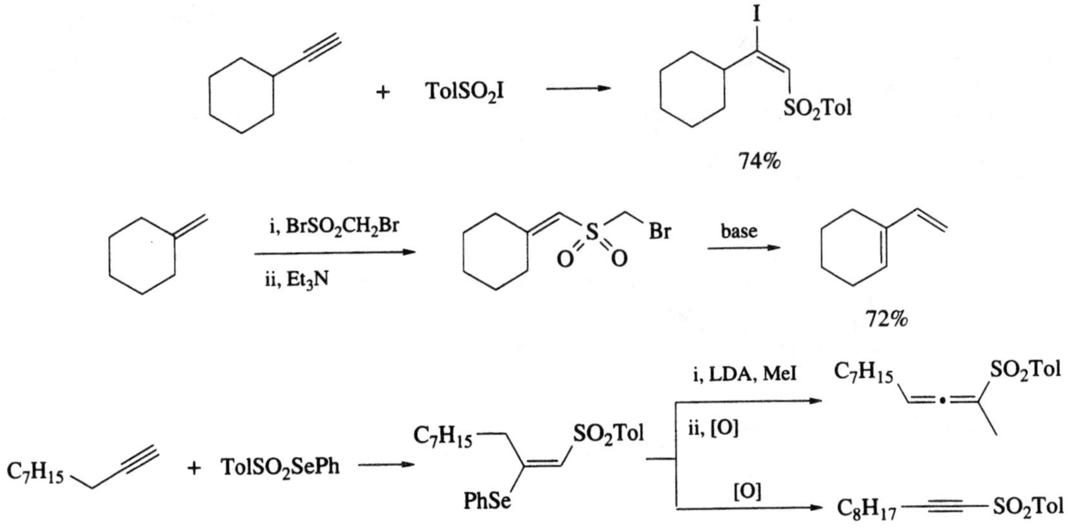

Scheme 88

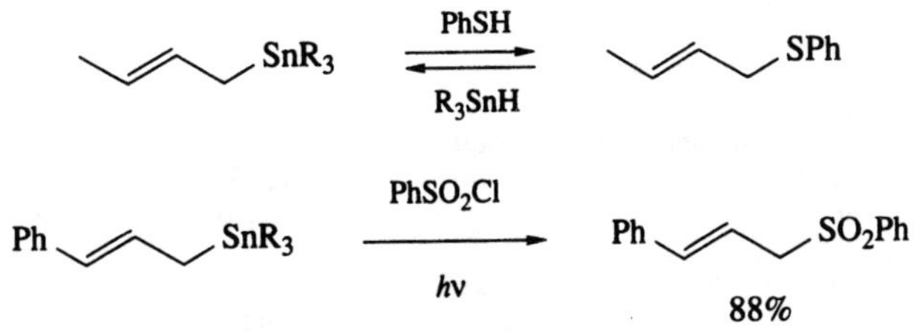

Scheme 89

4.1.9 CONCLUSIONS

Radical addition reactions are useful in the formation of a diverse collection of structural motifs and they often offer advantages over standard transformations. This chapter has attempted to show that the barrier to becoming familiar with the principles and techniques of radical reactions is small compared to the rewards, both practical and intellectual, that derive from this familiarity. The following chapter (Chapter 4.2) assumes that the reader has now acquired this familiarity and it presents radical cyclization reactions and sequences of radical reactions.

ACKNOWLEDGMENTS

I am very grateful to Jean Rock, Churl-Min Seong, Wang Shen, and M. Tottleben for their help in preparing the manuscript and I also thank Drs. C. Jasperse, E. Schwartz and R. Wolin, and H. Liu and P. S. Ramamoorthy. Our contributions in the free radical area have been supported by the National Institutes of Health.

4.1.10 REFERENCES AND NOTES

1. F. R. Mayo, *J. Chem. Educ.*, 1986, **63**, 97; C. Walling, *J. Chem. Educ.*, 1986, **63**, 99; *Chem. Br.*, 1987, 767.
2. D. J. Hart, *Science (Washington, D.C.)*, 1984, **223**, 883.
3. B. Giese, 'Radicals in Organic Synthesis: Formation of Carbon–Carbon Bonds', Pergamon Press, Oxford, 1986.
4. M. Ramaiah, *Tetrahedron*, 1987, **43**, 3541.
5. D. P. Curran, *Synthesis*, 1988, part 1, 417; part 2, 489.
6. In the original usage, the term 'radical' meant 'disconnected substituent'. What we now represent as 'Me—' was called a 'methyl radical'. When it was discovered that real methyl radicals existed, these were termed 'free radicals'. In English, the original usage of 'radical' has completely disappeared. Currently, the terms 'radical' and 'free radical' are often used interchangeably although a 'free radical' is sometimes distinguished from a 'caged radical' or a 'metal-complexed radical'. Most synthetic reactions involve 'free' radicals that are not complexed and are free to diffuse in solution. In this chapter, these are simply called 'radicals'.
7. W. A. Waters, 'The Chemistry of Free Radicals', Clarendon Press, Oxford, 1946; C. Walling, 'Free Radicals in Solution', Wiley, New York, 1957.
8. D. C. Nonhebel, J. M. Tedder and J. C. Walton, 'Radicals', Cambridge University Press, 1979; D. I. Davies and M. J. Parrott, 'Free Radicals in Organic Synthesis', Springer-Verlag, Berlin, 1978; J. M. Hay, 'Reactive Free Radicals', Academic Press, New York, 1974; E. S. Huyser, 'Free Radical Chain Reactions', Wiley, New York, 1970; W. A. Pryor, 'Free Radicals', McGraw-Hill, New York, 1966; F. A. Carey and R. J. Sundberg, in 'Advanced Organic Chemistry' part A, 2nd edn., Plenum Press, New York, 1984, chap. 12; T. H. Lowry and K. S. Richardson, in 'Mechanism and Theory in Organic Chemistry', 3rd edn., Harper and Row, New York, 1987, chap. 9.
9. W. A. Pryor, 'Introduction to Free Radical Chemistry', Prentice-Hall, Englewood Cliffs, NJ, 1971.
10. C. Walling, *Tetrahedron*, 1985, **41**, 3887. This article introduces a Tetrahedron Symposium in print edited by B. Giese and entitled: 'Selectivity and Synthetic Application of Radical Reactions'.
11. J. K. Kochi, in 'Free Radicals', Wiley, New York, 1973, vols. 1 and 2.
12. C. Rüchardt and H.-D. Beckhaus, *Top. Curr. Chem.*, 1986, **130**, 1.
13. A. L. Castelhano and D. Griller, *J. Am. Chem. Soc.*, 1984, **106**, 3655.
14. D. Griller and K. U. Ingold, *Acc. Chem. Res.*, 1976, **9**, 13.
15. Radicals can also react with charged molecules; reactions with stabilized anions are especially important.
16. This analysis is simplified by assuming that all termination is by recombination.
17. Equation (3) is derived by setting equation (2) > equation (1) and canceling a '[R·]' from both sides.
18. A. Citterio, F. Minisci, O. Porta and G. Sesana, *J. Am. Chem. Soc.*, 1977, **99**, 7960; K. Münger and H. Fischer, *Int. J. Chem. Kinet.*, 1985, **17**, 809.
19. M. Newcomb and D. P. Curran, *Acc. Chem. Res.*, 1988, **21**, 206.
20. J. K. Kochi, *Adv. Free Radical Chem.*, 1975, **5**, 189.
21. F. Baumberger and A. Vasella, *Helv. Chim. Acta*, 1983, **66**, 2210.
22. R. W. Fessenden and R. H. Schuler, *J. Chem. Phys.*, 1963, **39**, 2147; E. L. Cochran, F. J. Adrian and V. A. Bowers, *J. Chem. Phys.*, 1964, **40**, 213.
23. R. M. Kopchik and J. A. Kampmeier, *J. Am. Chem. Soc.*, 1968, **90**, 6733; L. A. Singer and N. P. Kong, *J. Am. Chem. Soc.*, 1966, **88**, 5213.
24. O. Simamura, *Top. Stereochem.*, 1969, **4**, 1.
25. M. S. Liu, S. Soloway, D. K. Wedegaertner and J. A. Kampmeier, *J. Am. Chem. Soc.*, 1971, **93**, 3809.
26. T. Ohnuki, M. Yoshida and O. Simamura, *Chem. Lett.*, 1972, 797, 999.
27. D. Griller, J. W. Cooper and K. U. Ingold, *J. Am. Chem. Soc.*, 1975, **97**, 4269.
28. The term ionic is intended only in the most general sense to indicate overall 'two-electron' processes. It is not meant to imply a mechanism.
29. G. Stork and P. M. Sher, *J. Am. Chem. Soc.*, 1986, **108**, 303.
30. L. Thijs, S. N. Gupta and D. C. Neckers, *J. Org. Chem.*, 1979, **44**, 4123.

31. K. Miura, Y. Ichinose, K. Nozaki, K. Fugami, K. Oshima and K. Utimoto, *Bull. Chem. Soc. Jpn.*, 1989, **62**, 143.
32. β-Cleavage reactions of C—C bonds occur at higher temperatures and are important reactions in processes like hydrocarbon cracking. See: K. Klenke, J. O. Metzger and S. Lübben, *Angew. Chem., Int. Ed. Engl.*, 1988, **27**, 1168.
33. H. Fischer (ed.), 'Radical Reaction Rates in Liquids', Springer-Verlag, West Berlin, 1983–85; Landolt-Börnstein, new series, vols. II/13a–e. This is the single most comprehensive source of rate information currently available.
34. D. Griller and K. U. Ingold, *Acc. Chem. Res.*, 1980, **13**, 193, 317.
35. H. Fischer and H. Paul, *Acc. Chem. Res.*, 1987, **20**, 200.
36. Temperature dependence is usually calculated with the Arrhenius equation: $\log k = \log A - E_a/\theta$ where $\theta = 2.3RT$ (= 1.36, 1.47, 1.61 kcal mol^{-1} at 25 °C, 50 °C, 80 °C)
37. M. Newcomb and A. G. Glenn, *J. Am. Chem. Soc.*, 1989, **111**, 275.
38. J. Lusztyk, B. Maillard, S. Deycard, D. A. Lindsay and K. U. Ingold, *J. Org. Chem.*, 1987, **52**, 3509.
39. C. Chatgilialoglu, K. U. Ingold and J. C. Scaiano, *J. Am. Chem. Soc.*, 1981, **103**, 7739.
40. D. P. Curran, E. Bosch, J. Kaplan and M. Newcomb, *J. Org. Chem.* 1989, **54**, 1826.
41. M. Newcomb and S. U. Park, *J. Am. Chem. Soc.*, 1986, **108**, 4132; M. Newcomb and J. Kaplan, *Tetrahedron Lett.*, 1987, **28**, 1615.
42. K. Münger and H. Fischer, *Int. J. Chem. Kinet.*, 1985, **17**, 809.
43. D. Swern, in 'Comprehensive Organic Chemistry', ed. D. H. R. Barton and W. D. Ollis, Pergamon Press, Oxford, 1979, vol. 1, p. 909.
44. P. S. Engel, *Chem. Rev.*, 1980, **80**, 99.
45. These isobutyronitrile radicals are relatively unreactive compared to the alkoxyl radicals that are generated from peroxides. Thus, the types of chains that are initiated are more limited in scope.
46. H. P. Waits and G. S. Hammond, *J. Am. Chem. Soc.*, 1964, **86**, 1911.
47. Most caged isobutyronitrile radicals undergo C—C coupling to give the stable tetramethylfumaronitrile. A small percentage undergo C—N coupling to give a product that is still subject to slow homolytic cleavage on heating.
48. In reactions such as tin hydride reductions, excess tin hydride must be added if large amounts of initiator are employed because the initiation steps consume tin hydride.
49. J. W. Moore and R. G. Pearson, 'Kinetics and Mechanism', Wiley, New York, 1981.
50. L. E. Eberson, 'Electron Transfer Reactions in Organic Chemistry', Springer-Verlag, Berlin, 1987.
51. M. M. Baizer and H. Lund, 'Organic Electrochemistry: an Introduction and a Guide', Dekker, New York, 1983.
52. G. A. Russell, in ref. 11, vol. 1, p. 275.
53. W. C. Danen, in 'Methods in Free Radical Chemistry', ed. E. L. S. Huyser, Dekker, New York, 1974, vol. 5, p. 1; M. Poutsma, in ref. 11, vol. 2, p. 113.
54. J. M. Tedder, *Angew. Chem., Int. Ed. Engl.*, 1982, **21**, 401.
55. I. Fleming, 'Frontier Orbitals and Organic Chemical Reactions', Wiley, Chichester, 1976.
56. B. Giese, *Angew. Chem., Int. Ed. Engl.*, 1983, **22**, 753.
57. B. Giese and S. Lachhein, *Angew. Chem., Int. Ed. Engl.*, 1982, **21**, 768.
58. B. Giese, J. Dupuis, T. Hasskerl and J. Meixner, *Tetrahedron Lett.*, 1983, **24**, 703.
59. K. N. Houk, M. N. Paddon-Row, D. C. Spellmeyer, N. G. Rondan and S. Nagase, *J. Org. Chem.*, 1986, **51**, 2874.
60. B. Giese, H. Horler and M. Leising, *Chem. Ber.*, 1986, **119**, 444.
61. K. Riemenschneider, H. M. Bartels, R. Dornow, E. Drechsel-Grau, W. Eichel, H. Luthe, Y. M. Matter, W. Michaelis and P. Boldt, *J. Org. Chem.*, 1987, **52**, 205.
62. V. Ghodoussi, G. J. Gleicher and M. Kravetz, *J. Org. Chem.*, 1986, **51**, 5007.
63. J. M. Tedder and J. C. Walton, *Tetrahedron*, 1980, **36**, 701.
64. For example, the dicyanomalonyl radical adds more rapidly to Δ^9-octalin than to 1-hexene: G. J. Gleicher, B. Mahiou and A. J. Aretakis, *J. Org. Chem.*, 1989, **54**, 308.
65. A. Padwa (ed.), '1,3-Dipolar Cycloaddition Chemistry', Wiley, New York, 1984, vols. 1 and 2.
66. I. Beranek and H. Fischer, in 'Free Radicals in Synthesis and Biology', ed. F. Minisci, Kluwer, Dordrecht, 1989, p. 303.
67. B. Giese, J. He and W. Mehl, *Chem. Ber.*, 1988, **121**, 2063.
68. D. Kim, Ph.D. Thesis, University of Pittsburgh, 1988.
69. E. J. Corey, *Pure Appl. Chem.*, 1967, **14**, 19; *Chem. Soc. Rev.*, 1988, **17**, 111.
70. S. G. Warren, 'Organic Synthesis: The Disconnection Approach', Wiley, New York, 1982.
71. The word synthon is now commonly used to mean 'reagent' or 'starting material'. However, I subscribe here to the original use of 'synthon' as a conceptual entity created by homolysis or heterolysis of a bond.
72. T. A. Hase, 'Umpoled Synthons', Wiley, New York, 1987.
73. D. Seebach, *Angew. Chem., Int. Ed. Engl.*, 1979, **18**, 239.
74. See ref. 1c in ref. 73.
75. Even though radicals can be nucleophilic or electrophilic, the chemistry of radicals is not naturally intertwined with the carbonyl group and related functionalites in the same way that the chemistry of ionic reactions is. This notation does not now account for the frequent need for radicals and alkenes to be electronically paired for a successful reaction. This is in part because I could not envision a simple, unambiguous notation device and more importantly because such a device could be artificial and misleading in the realm of radical cyclizations where arrangement of the double bond and the radical site in the molecule is often much more important than any electronic pairing.
76. C. S. Wilcox and J. J. Gaudino, *J. Am. Chem. Soc.*, 1986, **108**, 3102.
77. Each of three rings can be dissected in two places and each of these six dissections has four possible cyclization arrangements (Scheme 21 shows one set of four possibilities). If one considers the ordering of the

cyclizations as another variable (two orders for each route), there are actually 48 variants on this one basic strategy.
78. D. P. Curran and C. Jasperse, *J. Am. Chem. Soc.*, 1990, **112**, 5601.
79. D. L. J. Clive, T. L. B. Boivin and A. G. Angoh, *J. Org. Chem.*, 1987, **52**, 4943.
80. Review of tin hydride chemistry: W. P. Neumann, *Synthesis*, 1987, 665.
81. Review of tin and mercury(II) hydride additions: B. Giese, *Angew. Chem., Int. Ed. Engl.*, 1985, **24**, 553.
82. Review of radical reactions of organomercurials: J. Barluenga and M. Yus, *Chem. Rev.*, 1988, **88**, 487.
83. M. Pereyre, J.-P. Quintard and A. Rahm, 'Tin in Organic Synthesis', Butterworths, London, 1987.
84. A. G. Davies, in 'Comprehensive Organometallic Chemistry', ed. G. Wilkinson, F. G. A. Stone and E. W. Abel, Pergamon Press, Oxford, 1982, vol. 2, p. 519.
85. Tin hydride additions were introduced by three groups in the early 1980s (a) S. D. Burke, W. F. Fobare and D. M. Armistead, *J. Org. Chem.*, 1982, **47**, 3348; (b) B. Giese and J. Dupuis, *Angew. Chem., Int. Ed. Engl.*, 1983, **22**, 622; (c) R. M. Adlington, J. E. Baldwin, A. Basak and R. P. Kozyrod, *J. Chem. Soc., Chem. Commun.*, 1983, 944.
86. B. Giese and R. Rupaner, *Liebigs Ann. Chem.*, 1987, 231.
87. B. Geise and R. Rupaner, *Synthesis*, 1988, 219.
88. For leading references to rate constants of atom and group transfer to tin, germanium and silicon radicals, see: K. U. Ingold, J. Lusztyk and J. C. Scaiano, *J. Am. Chem. Soc.*, 1984, **106**, 343; A. L. J. Beckwith and P. E. Pigou, *Aust. J. Chem.*, 1986, **39**, 77; A. L. J. Beckwith and P. E. Pigou, *Aust. J. Chem.*, 1986, **39**, 1151.
89. It is well known that k_H is similar for all alkyl-substituted radicals but rate constants for reaction of tin hydride with carbonyl-substituted radicals are not known. Substituents can effect the rate constant for hydrogen transfer. For example, the benzyl radical is about 50 times less reactive than a primary alkyl radical. J. A. Franz, N. K. Suleman and M. S. Alnajjar, *J. Org. Chem.*, 1986, **51**, 19.
90. B. Giese, J. Dupuis and M. Nix, *Org. Synth.*, 1987, **65**, 236.
91. Even though alkyl iodides and many alkyl bromides do not require an added reagent for reduction reactions conducted at high concentration, it is advisable to provide an initiator to insure that chains continue at low concentration.
92. H. G. Kuivila and L. W. Menapace, *J. Org. Chem.*, 1963, **28**, 2165.
93. E. J. Corey and J. W. Suggs, *J. Org. Chem.*, 1975, **40**, 2554; D. B. Gerth and B. Giese, *J. Org. Chem.*, 1986, **51**, 3726.
94. D. P. G. Hamon and K. R. Richards, *Aust. J. Chem.*, 1983, **36**, 2243.
95. D. Milstein and J. K. Stille, *J. Am. Chem. Soc.*, 1978, **100**, 3636; J. E. Leibner and J. Jacobus, *J. Org. Chem.*, 1979, **44**, 449.
96. D. P. Curran and C.-T. Chang, *J. Org. Chem.*, 1989, **54**, 3140.
97. Trialkyltin hydrides are converted rapidly to trialkyltin iodides with molecular iodine. An ether solution of iodine is added dropwise until the iodine color just persists.
98. Y. Ueno, O. Moriya, K. Chino, M. Watanabe and M. Okawara, *J. Chem. Soc., Perkin Trans. 1*, 1986, 1351; Y. Ueno, K. Chino, M. Watanabe, O. Moriya and M. Okawara, *J. Am. Chem. Soc.*, 1982, **104**, 5564; N. M. Weinshenker, G. A. Crosby and J. Y. Wong, *J. Org. Chem.*, 1975, **40**, 1966; H. Schumann and B. Pachaly, *Angew. Chem., Int. Ed. Engl.*, 1981, **20**, 1043; D. E. Bergbreiter and J. R. Blanton, *J. Org. Chem.*, 1987, **52**, 472.
99. For addition reactions, the absolute magnitude of k_{H1} is not the real concern. Instead it is the k_{a1}:k_{H1} ratio compared to the k_{a2}:k_{H2} ratio that is important. Reducing both k_{H1} and k_{H2} may not provide a significant advantage. An ideal reagent would donate hydrogen slowly to nucleophilic initial radicals but rapidly to electrophilic product radicals.
100. P. W. Pike, V. Gilliatt, M. Ridenour and J. W. Hershberger, *Organometallics*, 1988, **7**, 2220.
101. J. Lusztyk, B. Maillard, D. A. Lindsay and K. U. Ingold, *J. Am. Chem. Soc.*, 1983, **105**, 3578.
102. In fact, additions of tributylgermyl radical and tributyltin radical to activated alkenes occur at about the same rate (see refs. 38 and 101). This addition reaction is probably more readily reversible in the case of tin (because a weaker bond is formed) and therefore hydrostannylation is a less serious problem than hydrogermylation. Thus, very reactive precursors (preferably iodides) are required as precursors if germanium hydride is used with an electron deficient alkene but this is not because the germanium radical is less reactive towards halides than the tin radical.
103. This leads to the conclusion that germanium hydride might be a valuable reagent when the trapping of reactive radicals such as vinyl or alkoxyl in the presence of less reactive alkyl radicals is desired.
104. P. Pike, S. S. Hershberger and J. W. Hershberger, *Tetrahedron*, 1988, **44**, 6295.
105. Many workers have observed difficulties in propagating chains with germanium hydrides, especially at low concentrations. The observation that germanium hydride chains were propagated more efficiently in acetonitrile than in benzene leads to the speculation that the germyl radical may react more rapidly with benzene than the stannyl radical. This suggests that solvents like *t*-butyl alcohol might be useful for low concentration germanium hydride reductions.
106. J. M. Kanabus-Kaminska, J. A. Hawari, D. Griller and C. Chatgilialoglu, *J. Am. Chem. Soc.*, 1987, **109**, 5267; C. Chatgilialoglu, D. Griller and M. Lesage, *J. Org. Chem.*, 1988, **53**, 3641.
107. B. Giese, B. Kopping and C. Chatgilialoglu, *Tetrahedron Lett.*, 1989, **30**, 681.
108. M. Vaman Rao and M. Nagarajan, *J. Org. Chem.*, 1988, **53**, 1432.
109. B. Giese, T. Linker and R. Muhn, *Tetrahedron*, 1989, **45**, 935.
110. D. J. Hart and H.-C. Huang, *Tetrahedron Lett.*, 1985, **26**, 3749.
111. Y. Yamamoto, S. Nishii and T. Ibuka, *J. Am. Chem. Soc.*, 1988, **110**, 617.
112. A. Citterio, F. Minisci and E. Vismara, *J. Org. Chem.*, 1982, **47**, 81.
113. See p. 242 in ref.3.
114. Because phenyl and vinyl radicals are highly reactive, side reactions like inter- and intra-molecular hydrogen transfer and addition to the solvent are of greater concern.
115. B. Giese, M. Hoch, C. Lamberth and R. R. Schmidt, *Tetrahedron Lett.*, 1988, **29**, 1375.

116. B. Giese, J. Dupuis, M. Leising, M. Nix and H. J. Linder, *Carbohydr. Res.*, 1987, **171**, 329; B. Giese, *Pure Appl. Chem.*, 1988, **60**, 1655.
117. D. L. Boger and R. J. Mathvink, *J. Org. Chem.*, 1989, **54**, 1777.
118. G. Sacripante, C. Tan and G. Just, *Tetrahedron Lett.*, 1985, **26**, 5643.
119. G. Sacripante and G. Just, *J. Org. Chem.*, 1987, **52**, 3659.
120. G. A. Russell and D. Guo, *Tetrahedron Lett.*, 1984, **25**, 5239.
121. B. Giese and G. Kretzschmar, *Chem. Ber.*, 1984, **117**, 3160.
122. This means that there is no potential problem equivalent to the hydrostannation in the tin hydride method.
123. B. Giese and K. Heuck, *Chem. Ber.*, 1981, **114**, 1572.
124. B. Giese and W. Zwick, *Chem. Ber.*, 1982, **115**, 2526.
125. S. J. Danishefsky, E. Taniyama and R. R. Webb, II, *Tetrahedron Lett.*, 1983, **24**, 11.
126. B. Giese, H. Horler and W. Zwick, *Tetrahedron Lett.*, 1982, **23**, 931.
127. P. J. Wagner and M. J. Lindstom, *J. Am. Chem. Soc.*, 1987, **109**, 3057.
128. H. Kosugi, K. Kurino, K. Takayama and T. Migita, *J. Organomet. Chem.*, 1973, **56**, C11; J. Grignon and M. Pereyre, *J. Organomet. Chem.*, 1973, **61**, C33; J. Grignon, C. Servens and M. Pereyre, *J. Organomet. Chem.*, 1975, **96**, 225.
129. G. E. Keck, E. J. Enholm, J. B. Yates and M. R. Wiley, *Tetrahedron*, 1985, **41**, 4079.
130. G. E. Keck and E. H. Enholm, *Tetrahedron Lett.*, 1985, **26**, 3311; G. E. Keck and J. B. Yates, *J. Org. Chem.*, 1982, **47**, 3590.
131. The origin of this activating effect is not clear. An allylic C—Sn bond should not lower the LUMO of an alkene. Crystal structures of several allylic stannanes have been determined and several appear to show unusually short C=C bonds. See: W. Kitching, K. G. Penman, G. Valle, G. Tagliavini and P. Ganis, *Organometallics*, 1989, **8**, 785.
132. For relative rate studies involving allyltributylstannane, see: G. A. Russell, P. Ngoviwatchai and H. I. Tashtoush, *Organometallics*, 1988, **7**, 696.
133. D. P. Curran, P. A. van Elbury, B. Giese and S. Gilges, *Tetrahedron Lett.*, 1990, **31**, 2861.
134. J. E. Baldwin, R. M. Adlington, C. Lowe, I. A. O'Neil, G. L. Sanders, C. J. Schofield and J. B. Sweeney, *J. Chem. Soc., Chem. Commun.*, 1988, 1030; C. J. Easton, I. M. Scharfbillig and E. W. Tan, *Tetrahedron Lett.*, 1988, **29**, 1565.
135. T. Toru, Y. Yamada, T. Ueno, E. Maekawa and Y. Ueno, *J. Am. Chem. Soc.*, 1988, **110**, 4815.
136. J. E. Baldwin, R. M. Adlington and A. Basak, *J. Chem. Soc., Chem. Commun.*, 1984, 1284; G. A. Russell and L. L. Herold, *J. Org. Chem.*, 1985, **50**, 1037.
137. E. Lee, S.-G. Yu, C.-U. Hur and S.-M. Yang, *Tetrahedron Lett.*, 1988, **29**, 6969.
138. S. Hanessian and M. Alpegiani, *Tetrahedron*, 1989, **45**, 941.
139. G. E. Keck and J. B. Yates, *J. Organomet. Chem.*, 1983, **248**, C21.
140. J. E. Baldwin, R. M. Adlington, D. J. Birch, J. A. Crawford and J. B. Sweeney, *J. Chem. Soc., Chem. Commun.*, 1986, 1339.
141. G. E. Keck and J. H. Byers, *J. Org. Chem.*, 1985, **50**, 5442; A. Yanagisawa, Y. Noritake and H. Yamamoto, *Chem. Lett.*, 1988, 1899.
142. J. E. Baldwin and D. R. Kelly, *J. Chem. Soc., Chem. Commun.*, 1985, 682; J. E. Baldwin, D. R. Kelly and C. B. Ziegler, *J. Chem. Soc., Chem. Commun.*, 1984, 133.
143. G. A. Russell, H. I. Tashtoush and P. Ngoviwatchai, *J. Am. Chem. Soc.*, 1984, **106**, 4622.
144. G. E. Keck, J. H. Byers and A. M. Tafesh, *J. Org. Chem.*, 1988, **53**, 1127.
145. When activated (*Z*)-vinylstannanes are used, it is possible that any kinetically formed (*Z*)-adducts are isomerized to (*E*)-adducts by addition–elimination of the tributyltin radical to the product (which is also an activated alkene).
146. Allyl-germanes and -silanes are inferior to stannanes: J. P. Light II, M. Ridenour, L. Beard and J. W. Hershberger, *J. Organomet. Chem.*, 1987, **326**, 17.
147. M. D. Johnson, *Acc. Chem. Res.*, 1983, **16**, 343; A. Gaudemer, K. Nguyen-van-Duong, N. Shahkarami, S. S. Achi, M. Frostin-Rio and D. Pujol, *Tetrahedron*, 1985, **41**, 4095.
148. G. A. Russell, *Acc. Chem. Res.*, 1989, **22**, 1; G. A. Russell, P. Ngoviwatchai, H. I. Tashtoush, A. Pla-Dalmau and R. K. Khanna, *J. Am. Chem. Soc.*, 1988, **110**, 3530.
149. D. H. R. Barton, D. Crich and W. B. Motherwell, *Tetrahedron*, 1985, **41**, 3901.
150. D. Crich, *Aldrichimica Acta*, 1986, **20**, 35.
151. D. H. R. Barton, D. Crich and G. Kretzschmar, *J. Chem. Soc., Perkin Trans. 1*, 1986, 39.
152. D. H. R. Barton, D. Crich and P. Potier, *Tetrahedron Lett.*, 1985, **26**, 5943.
153. D. H. R. Barton, D. Bridon, I. Fernandez-Picot and S. Z. Zard, *Tetrahedron*, 1987, **43**, 2733.
154. There may also be differences in the rate of addition of the initial radical (R·) and the adduct radical (**35**) to (**32**) but rates for the addition of (**35**) to (**32**) are not yet known.
155. D. H. R. Barton, *Pure Appl. Chem.*, 1988, **60**, 1549.
156. D. H. R. Barton, A. Gateau-Olesker, S. D. Gero, B. Lacher, C. Tachdjian and S. Z. Zard, *J. Chem. Soc., Chem. Commun.*, 1987, 1790.
157. D. H. R. Barton, H. Togo and S. Z. Zard, *Tetrahedron*, 1985, **41**, 5507.
158. D. H. R. Barton, J. Guilhem, Y. Hervé, P. Potier and J. Thierry, *Tetrahedron Lett.*, 1987, **28**, 1413.
159. D. H. R. Barton and D. Crich, *J. Chem. Soc., Perkin Trans. 1*, 1986, 1603.
160. P. Delduc, C. Tailhan and S. Z. Zard, *J. Chem. Soc., Chem. Commun.*, 1988, 308; C. Tailhan and S. Z. Zard, in 'Free Radicals in Synthesis and Biology', ed. F. Minisci, Kluwer, Dordrecht, 1989, p. 263.
161. D. P. Curran, in 'Free Radicals in Synthesis and Biology', ed. F. Minisci, Kluwer, Dordrecht, 1989, p. 37.
162. C. Walling and E. S. Huyser, *Org. React. (N.Y.)*, 1963, **13**, 91.
163. K. Fukunishi and I. Tabushi, *Synthesis*, 1988, 826.
164. B. Fraser-Reid, *Acc. Chem. Res.*, 1975, **8**, 192; 1985, **18**, 347; K. Ogura, A. Yanagisawa, T. Fujino and K. Takahashi, *Tetrahedron Lett.*, 1988, **29**, 5387; K. Inomata, H. Suhara, H. Kinoshita and H. Kotake, *Chem. Lett.*, 1988, 813.

165. Z. Benko, B. Fraser-Reid, P. S. Mariano and A. L. J. Beckwith, *J. Org. Chem.*, 1988, **53**, 2066.
166. E. Montaudon, M. Agorrody, F. Rakotomanana and B. Maillard, *Bull. Chim. Soc. Belg.*, 1987, **96**, 769; B. Maillard, A. Kharrat and C. Gardrat, *Nouv. J. Chim.*, 1986, 259.
167. K. Maruoka, H. Sano, Y. Fukutani and H. Yamamoto, *Chem. Lett.*, 1985, 1689.
168. X. Lu, S. Ma and J. Zhu, *Tetrahedron Lett.*, 1988, **29**, 5129.
169. D. L. Fields, Jr. and H. Shechter, *J. Org. Chem.*, 1986, **51**, 3369.
170. F. Minisci, in 'Fundamental Research in Homogeneous Catalysis', ed. M. Graziani and M. Giongo, Plenum Press, New York, 1984, vol. 4, p. 173.
171. J. Tsuji, K. Sato and H. Nagashima, *Tetrahedron*, 1985, **41**, 5003, 5645.
172. M. Kameyama and N. Kamigata, *Bull. Chem. Soc. Jpn.*, 1987, **60**, 3687.
173. R. Grigg, J. P. Devlin, A. Ramasubbu, R. M. Scott and P. Stevenson, *J. Chem. Soc., Perkin Trans. 1*, 1987, 1515.
174. D. Bellus, *Pure Appl. Chem.*, 1985, **57**, 1827.
175. P. Martin, E. Steiner, J. Streith, T. Winkler and D. Bellus, *Tetrahedron*, 1985, **41**, 4057.
176. G. A. Kraus and K. Landgrebe, *Tetrahedron*, 1985, **41**, 4039; M. Degueil-Castaing, B. de Jéso, G. A. Kraus, K. Landgrebe and B. Maillard, *Tetrahedron Lett.*, 1986, **27**, 5927.
177. D. P. Curran and C.-T. Chang, *J. Org. Chem.*, 1989, **54**, 3140.
178. D. P. Curran, D. Kim and C. Ziegler, Jr., manuscript in preparation.
179. D. P. Curran, M.-H. Chen and D. Kim, *J. Am. Chem. Soc.*, 1989, **111**, 6265.
180. For example, octyl radical abstracts iodine for dimethyl methyliodomalonate with $k_I \approx 2 \times 10^9$ s^{-1} at 50 °C; see ref. 40.
181. H. C. Brown and M. M. Midland, *Angew. Chem., Int. Ed. Engl.*, 1972, **11**, 692.
182. The enol borane can be trapped by aldehydes (Scheme 60): T. Mukaiyama, K. Inomata and M. Muraki, *J. Am. Chem. Soc.*, 1973, **95**, 967.
183. H. C. Brown and E. Negishi, *J. Am. Chem. Soc.*, 1971, **93**, 3777.
184. K. Nozaki, K. Oshima and K. Utimoto, *Tetrahedron Lett.*, 1988, **29**, 1041.
185. M. Newcomb, R. M. Sanchez and J. Kaplan, *J. Am. Chem. Soc.*, 1987, **109**, 1195.
186. About 5% of the ethyl radical adduct was isolated in the reaction with *t*-butyl iodide and about 20% of the ethyl radical adduct was formed in the reaction with *i*-propyl iodide. These ratios compare quite well with those calculated by using the rate constants for halogen transfer in the above reference and the rate constants for addition of a primary radical to methyl vinyl ketone.
187. C. Galli, *Chem. Rev.*, 1988, **88**, 765.
188. C. S. Rondestvedt, Jr. *Org. React. (N.Y.)*, 1976, **24**, 225; 1960, **11**, 189.
189. H. G. Viehe, R. Merényi and Z. Janousek, *Pure Appl. Chem.*, 1988, **60**, 1635; Z. Janousek, R. Merényi and L. Stella, *Acc. Chem. Res.*, 1985, **18**, 148.
190. S. M. Mignani, R. Merényi, Z. Janousek and H. G. Viehe, *Bull. Soc. Chim. Belg.*, 1984, **93**, 991; S. M. Mignani, M. Beaujean, Z. Janousek, R. Merényi and H. G. Viehe, *Tetrahedron*, 1981, **37** (suppl. 1, Woodward issue), 111.
191. H. G. Viehe, Z. Janousek and R. Merényi, in 'Free Radicals in Synthesis and Biology', ed. F. Minisci, Kluwer, Dordrecht, 1989, p. 1; H. Birkhofer, J. Hädrich, J. Pakusch, H.-D. Beckhaus, C. Rückhardt, K. Peters and H.-G. V. Schnering, in 'Free Radicals in Synthesis and Biology', ed. F. Minisci, Kluwer, Dordrecht, 1989, p. 27. Capto-dative radicals are probably unreactive with molecules because the captor and donor substituents stabilize the radical (making reactions less exothermic) without dramatically increasing its electrophilicity (as with two captors) or nucleophilicity (as with two donors).
192. H. J. Schäfer, *Angew. Chem., Int. Ed. Engl.*, 1981, **20**, 911; J. Yoshida, K. Sakaguchi and S. Isoe, *J. Org. Chem.*, 1988, **53**, 2525.
193. R. N. Renaud and P. J. Champagne, *Can. J. Chem.*, 1979, **57**, 990.
194. H. Fischer, *J. Am. Chem. Soc.*, 1986, **108**, 3925.
195. D. J. Hart and F. L. Seely, *J. Am. Chem. Soc.*, 1988, **110**, 1631.
196. H. Hillgärtner, W. P. Neumann and B. Schroeder, *Justus Liebigs Ann. Chem.*, 1975, 586; W. P. Neumann, H. Hillgärtner, K. M. Baines, R. Dicke, K. Vorspohl, U. Kobs and U. Nussbutel, *Tetrahedron*, 1989, **45**, 951.
197. D. J. Hart and F. L. Seely, unpublished results.
198. However, Samsel and Kochi concluded from a detailed mechanistic study that a chain mechanism analogous to group transfer (rather than radical–cobalt coupling) was operative in the cyclization of hexenylcobalt compounds to cyclopentylmethyl isomers. E. G. Samsel and J. K. Kochi, *J. Am. Chem. Soc.*, 1986, **108**, 4790.
199. G. N. Schrauzer, *Angew. Chem., Int. Ed. Engl.*, 1976, **15**, 417.
200. Precursors can also be prepared by hydrocobaltation. H. Bhandal and G. Pattenden, *J. Chem. Soc., Chem. Commun.*, 1988, 1111.
201. G. Pattenden, *Chem. Soc. Rev.*, 1988, **17**, 361; V. F. Patel and G. Pattenden, *J. Chem. Soc., Chem. Commun.*, 1987, 871.
202. B. P. Branchaud, M. S. Meier and Y. Choi, *Tetrahedron Lett.*, 1988, **29**, 167.
203. A. Ghosez, T. Göbel and B. Giese, *Chem. Ber.*, 1988, **121**, 1807.
204. Indeed the cobalt hydride disproportionates to Co^{II} and hydrogen and at high conversions the addition is suppressed because cobalt(II) couples with starting radicals (to regenerate the starting complex). This is one reason why excess alkene is required. Hydrocobaltation of the acceptor by the cobalt hydride intermediate can also cause problems, see: W. M. Bandaranayake and G. Pattenden, *J. Chem. Soc., Chem. Commun.*, 1988, 1179.
205. B. P. Branchaud and M. S. Meier, *J. Org. Chem.*, 1989, **54**, 1320.
206. D. J. Coveney, V. F. Patel and G. Pattenden, *Tetrahedron Lett.*, 1987, **28**, 5949; V. F. Patel and G. Pattenden, *Tetrahedron Lett.*, 1988, **29**, 707.
207. F. Minisci, *Acc. Chem. Res.*, 1975, **8**, 165; J. K. Kochi, *Acc. Chem. Res.*, 1974, **7**, 351.
208. A. Citterio, A. Cerati, R. Sebastiano, C. Finzi and R. Santi, *Tetrahedron Lett.*, 1989, **30**, 1289.

209. E. Baciocchi and R. Ruzziconi, in 'Free Radicals in Synthesis and Biology', ed. F. Minisci, Kluwer, Dordrecht, 1989, p. 155.
210. W. J. DeKlein, in 'Organic Synthesis by Oxidation with Metal Compounds', ed. W. J. Mijs and C. R. H. de Jonge, Plenum Press, New York, 1986, p. 261.
211. E. I. Heiba and R. M. Dessau, *J. Org. Chem.*, 1974, **39**, 3456; E. I. Heiba, R. M. Dessau and P. G. Rodewald, *J. Am. Chem. Soc.*, 1974, **96**, 7977.
212. P. Brevilles and D. Uguen, *Bull. Soc. Chim. Fr.*, 1988, 705.
213. E. J. Corey and A. W. Gross, *Tetrahedron Lett.*, 1985, **26**, 4291; W. E. Fristad and J. R. Peterson, *J. Org. Chem.*, 1985, **50**, 10.
214. E. J. Corey and A. K. Ghosh, *Tetrahedron Lett.*, 1987, **28**, 175; *Chem. Lett.*, 1987, 223.
215. W. E. Fristad, J. R. Peterson, A. B. Ernst and G. B. Urbi, *Tetrahedron*, 1986, **42**, 3429.
216. B. B. Snider, J. J. Patricia and S. A. Kates, *J. Org. Chem.*, 1988, **53**, 2137.
217. A. Citterio, *Org. Synth.*, 1984, **62**, 67.
218. A. Citterio, A. Arnoldi and A. Griffini, *Tetrahedron*, 1982, **38**, 393.
219. A. Citterio and R. Santi, 'Free Radicals in Synthesis and Biology', ed. F. Minisci, Kluwer, Dordrecht, 1989, p. 187.
220. J.-L. Luche and C. Allavena, *Tetrahedron Lett.*, 1988, **29**, 5369; J. L. Luche, C. Allavena, C. Petrier and C. Dupuy, *Tetrahedron Lett.*, 1988, **29**, 5373.
221. R. Scheffold and R. Olinski, *J. Am. Chem. Soc.*, 1983, **105**, 7200.
222. R. Scheffold, S. Abrecht, R. Orlinski, H.-R. Ruf, P. Stamouli, O. Tinembart, L. Walder and C. Weymuth, *Pure Appl. Chem.*, 1987, **59**, 363.
223. D. H. R. Barton, D. Bridon and S. Z. Zard, *Tetrahedron*, 1987, **43**, 5307.
224. P. Kaszynski and J. Michl, *J. Org. Chem.*, 1988, **53**, 4593.
225. See ref. 3, p. 109.
226. D. H. R. Barton, N. Ozbalik and B. Vacher, *Tetrahedron*, 1988, **44**, 3501.
227. A. Citterio and L. Filipini, *Synthesis*, 1986, 473.
228. N. Kornblum, *Angew. Chem., Int. Ed. Engl.*, 1975, **14**, 734; R. A. Rossi and R. H. de Rossi, 'Aromatic Substitution by the $S_{RN}1$ Mechanism', American Chemical Society (monograph 178), Washington, 1983; J. F. Bunnett, *Acc. Chem. Res.*, 1978, **11**, 413.
229. M. F. Semmelhack, B. P. Chong, R. D. Stauffer, T. D. Rogerson, A. Chong and L. D. Jones, *J. Am. Chem. Soc.*, 1975, **97**, 2507.
230. N. Kornblum, W. J. Kelly and M. M. Kestner, *J. Org. Chem.*, 1985, **50**, 4720.
231. M. J. Perkins, in ref. 11, vol. 2, p. 231.
232. B. Giese, in ref. 3, chap. 5.
233. A. Citterio, R. Santi, D. Fancelli, A. Pagani and S. Bonsignore, *Gazz. Chim. Ital.*, 1988, **118**, 408.
234. L. M. Weinstock, E. Corley, N. L. Abramson, A. O. King and S. Karady, *Heterocycles*, 1988, 2627.
235. F. Minisci, E. Vismara and F. Fontana, *Heterocycles*, 1989, **28**, 489; F. Minisci, A. Citterio and C. Giordano, *Acc. Chem. Res.*, 1983, **16**, 27; F. Minisci, in 'Substitutent Effects in Radical Chemistry', ed. H. G. Viehe, Reidel, Dordrecht, 1986, p. 391; F. Minisci, *Top. Curr. Chem.*, 1976, **62**, 1; H. Vorbrüggen and M. Maas, *Heterocycles*, 1988, **27**, 2659.
236. A. Citterio, F. Minisci and V. Franchi, *J. Org. Chem.*, 1980, **45**, 4752.
237. B. P. Branchaud and Y. L. Choi, *J. Org. Chem.*, 1988, **53**, 4638.
238. G. A. Russell, D. Guo and R. K. Khanna, *J. Org. Chem.*, 1985, **50**, 3423.
239. D. H. R. Barton, B. Garcia, H. Togo and S. Z. Zard, *Tetrahedron Lett.*, 1986, **27**, 1327.
240. F. Minisci, E. Vismara, F. Fontana, G. Morini, M. Serravalle and C. Giordano, *J. Org. Chem.*, 1986, **51**, 4411.
241. A. Citterio, A. Gentile, F. Minisci, M. Serravalle and S. Venturas, *J. Org. Chem.*, 1984, **49**, 3364.
242. F. Minisci, *Synthesis*, 1973, 1.
243. F. Minisci, E. Visamara, F. Fontana and D. Redaelli, *Gazz. Chim. Ital.*, 1987, **117**, 363.
244. F. Fontana, F. Minisci and E. Vismara, *Tetrahedron Lett.*, 1988, **29**, 1975.
245. F. W. Stacey and J. F. Harris, Jr., *Org. React. (N.Y.)*, 1963, **13**, 150.
246. K. Griesbaum, *Angew. Chem., Int. Ed. Engl.*, 1970, **9**, 273.
247. A. D. Ayala, N. Giagante, J. C. Podestá and W. Neumann, *J. Organomet. Chem.*, 1988, **340**, 317.
248. Y. Ichinose, K. Wakamatsu, K. Nozaki, J.-L. Birbaum, K. Oshima and K. Utimoto, *Chem. Lett.*, 1987, 1647.
249. K. Nozaki, K. Oshima and K. Utimoto, *Tetrahedron*, 1989, **45**, 923.
250. W. E. Truce and G. C. Wolf, *J. Org. Chem.*, 1971, **36**, 1727.
251. E. Block, M. Aslam, V. Eswarakrishnan, K. Gebreyes, J. Hutchinson, R. S. Iyer, J.-A. Laffitte and A. Wall, *J. Am. Chem. Soc.*, 1986, **108**, 4568.
252. T. G. Back, M. V. Krishna and K. R. Muralidharan, *Tetrahedron Lett.*, 1987, **28**, 1737.
253. Y. Ichinose, K. Oshima and K. Utimoto, *Chem. Lett.*, 1988, 669.
254. G. A. Russell, P. Ngoviwatchai, H. I. Tashtoush and J. W. Hershberger, *Organometallics*, 1987, **6**, 1414.

4.2
Radical Cyclizations and Sequential Radical Reactions

DENNIS P. CURRAN
University of Pittsburgh, PA, USA

4.2.1 INTRODUCTION

Reactions that form rings are the central steps in the synthesis of cyclic organic compounds, and the development of cyclization reactions that are mild and general has been a recurring theme since the emergence of organic synthesis as a discipline. Although they have been recognized only recently, intramolecular addition reactions of radicals (hereafter called cyclizations) are among the most powerful tools at the disposal of the synthetic chemist. These radical cyclization reactions have all the advantages of their bimolecular counterparts, such as predictability and functional group tolerance; furthermore, because of the entropic advantages, cyclization reactions are of much broader scope.

The purpose of this chapter is to provide an introduction to the scope and limitations of radical cyclization reactions. Emphasis will be placed on the reactivity profile of radicals with respect to chemo-, regio- and stereo-selectivity. Because most sequential radical reactions include at least one cyclization, they are also presented in this chapter. The organization of this chapter is similar to the previous chapter on radical additions. However, the basic principles of radical reactions, selectivity requirements, methods to conduct radical reactions (including experimental techniques), and mechanisms are extensively discussed in the previous chapter, and these aspects will be reiterated rather sparingly. A reader who is not familiar with the principles of radical reactions as applied to synthesis should read the addition chapter (Chapter 4.1, this volume) first.

The general requirements to conduct a selective radical cyclization are summarized in Scheme 1. Like an addition reaction, one must have selective radical generation, cyclization and selective radical removal. Each step must be more rapid than the loss of radicals by (nonselective) radical/radical or radical/solvent reactions, and the method which is chosen must convert the cyclic radical, but not the initial radical, to a stable product. Cyclization reactions are often easier to conduct than additions. Indeed, some cyclizations are so rapid that it is difficult to trap initial radicals with standard reagents (like tin hydride) prior to closure. If bimolecular trapping is a problem, the standard experimental approach is to reduce the concentration of the trapping reagent, thus slowing bimolecular reactions relative to unimolecular competitors. In intermolecular additions, reaction of the product radical with the starting alkene is an important competing reaction, and the electronic differences between the initial and the product radicals are of great concern. However, in cyclizations this bimolecular reaction is rarely a problem because (for certain ring sizes) entropy greatly favors cyclization relative to addition. Indeed, many potential competing bimolecular addition reactions are so slow that they do not occur at all. Thus, the selectivity concerns arise mainly at the stage of the initial radical. The cyclic radical need only be converted to a stable product more rapidly than it is lost by radical/radical or radical/solvent reactions. In practice, this means that cyclizations can often be conducted down to the lower solution rate limit (10^2–10^3 s^{-1}), provided that the chains are maintained.

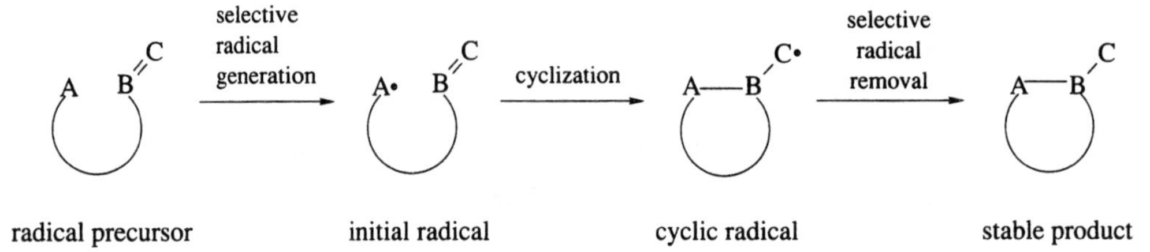

radical precursor —selective radical generation→ initial radical —cyclization→ cyclic radical —selective radical removal→ stable product

Potential problems: i, reactions of radicals with other radicals or solvent;
ii, removal of initial radical before cyclization

Scheme 1 Selectivity requirements for cyclization

The following introduction will briefly recount some of the key features of radical cyclizations with an emphasis on basic concepts that control regio- and stereo-selectivity. More details will be provided in the following sections, which describe specific types of reactions. The factors affecting the cyclization reactions to carbon–heteroatom multiple bonds are treated separately in Section 4.2.5, and the cyclizations of heteroatom-centered radicals are contained in Section 4.2.4.

Several excellent reviews[1–4] on the subject of radical cyclizations have stimulated synthetic advances in the past and will continue to do so in the future; they are highly recommended reading. As with radical additions, the book by Giese[5] provides a very good overview of the whole field (and it also includes cyclizations that do not occur by additions to π-bonds).

4.2.1.1 Hexenyl Radical Cyclizations

Hexenyl radical cyclizations are typically more rapid than higher homologs, and, unlike lower homologs, they are irreversible. Therefore, they are the most generally useful class of radical cyclizations. Thanks to intensive mechanistic study, they are also the most well understood.

4.2.1.1.1 The 5-hexenyl radical

The behavior of the 5-hexenyl radical merits special attention not only because it is one of the best understood reactive intermediates in organic chemistry, but also because it is a representative parent of a larger class of cyclizations. Because it is a cyclization of intermediate rate, it provides a convenient (if arbitrary) dividing point for synthetic planning: cyclizations that are more rapid than that of the hexenyl radical are easily conducted, but those that are significantly slower may present experimental difficulties.

As shown in Scheme 2, the hexenyl radical partitions between 5-exo cyclization (to give the cyclopentylmethyl radical) and 6-endo cyclization (to give the cyclohexyl radical) with the indicated rate constants (the terms 'exo' and 'endo' derive from the orientation of the reacting alkene with respect to the forming rings). At standard operating temperatures (25–80 °C), this partitioning translates to a ratio of about 98:2 in favor of 5-exo over 6-endo closure. This cyclization is highly exothermic, and strictly kinetically controlled (in this example, the less stable product is preferred). The contrast between this cyclization and an electronically equivalent bimolecular addition is dramatic. The addition of an alkyl radical to an unactivated alkene is not sufficiently rapid to be a preparatively useful reaction; its bimolecular rate constant is about three orders of magnitude below that of cyclization. However, when such an addition occurs, the alkyl radical adds exclusively to the terminus of the alkene. The bimolecular addition of an alkyl radical to the internal carbon of a simple alkene is virtually an unknown reaction. The hexenyl radical cyclization, a rapid addition to the more substituted end of an alkene, is indeed a special transformation.

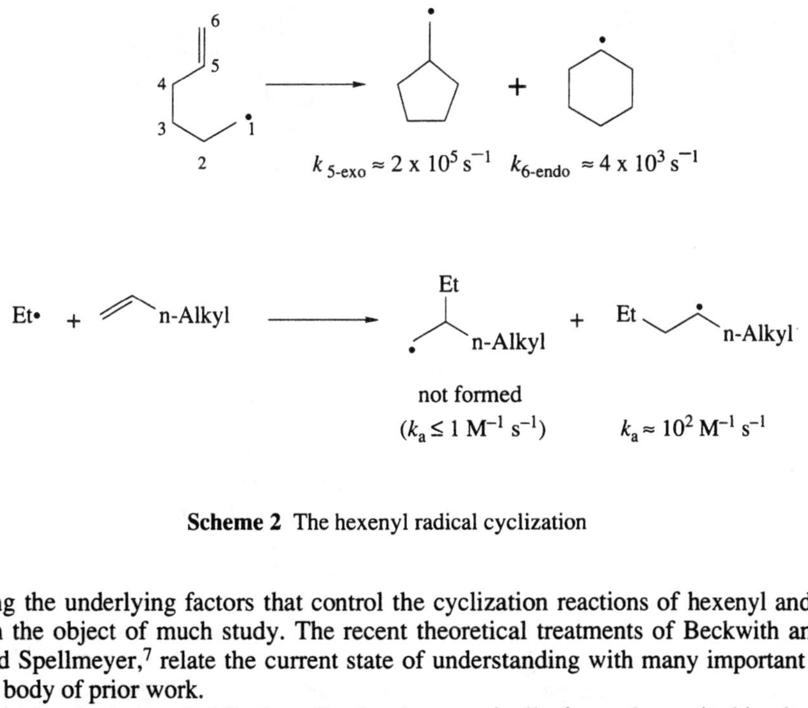

6 5 4 3 2 1

$k_{5\text{-exo}} \approx 2 \times 10^5\ s^{-1}$ $k_{6\text{-endo}} \approx 4 \times 10^3\ s^{-1}$

Et• + n-Alkyl → Et n-Alkyl + Et n-Alkyl

not formed
($k_a \leq 1\ M^{-1}\ s^{-1}$) $k_a \approx 10^2\ M^{-1}\ s^{-1}$

Scheme 2 The hexenyl radical cyclization

Determining the underlying factors that control the cyclization reactions of hexenyl and related radicals has been the object of much study. The recent theoretical treatments of Beckwith and Schiesser,[6] and Houk and Spellmeyer,[7] relate the current state of understanding with many important references to the extensive body of prior work.

Not surprisingly, the hexenyl radical cyclization is entropically favored over its bimolecular counterpart (enthalpies of activation are similar). However, entropy is not the whole story in the preferential formation of the less stable 5-exo product by attack at the more substituted carbon. To understand this key feature, we must turn to the transition state model outlined in Figure 1. At least three low energy transition states are important to consider: the 5-exo chair (**1**), the 5-exo boat (**2**), and the 6-endo chair (**3**).[8] This model, which will be used to interpret both regio- and stereo-chemical features, was proposed and developed by Beckwith in pioneering work that spans more than a decade.[6] The recognition by Houk and Spellmeyer[7] that the 5-exo boat transition state is relatively low in energy and cannot be neglected is a very significant advance that aids in the understanding of stereoselectivity.

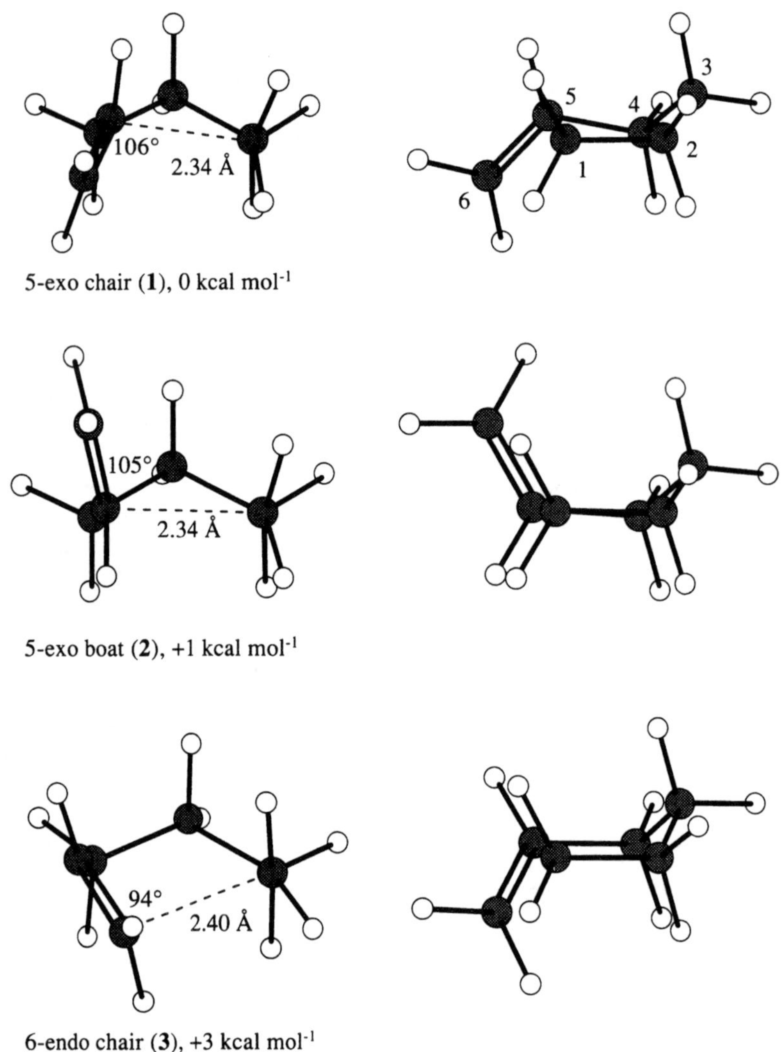

Figure 1 Transition states for hexenyl radical cyclization

The energetic differences between these three transition states are mainly enthalpic in origin, and the relative energy values calculated by Houk/Spellmeyer (STO-3G level) are included in Figure 1.[9] Provided that the 5-exo boat transition state (**2**) is not neglected, the Houk computational model closely reproduces the experimentally observed regioselectivity. The 5-exo chair transition state (TS) bears a striking resemblance to a cyclohexane chair, as originally proposed by Beckwith. Reflective of the early TS, the forming C—C bond is very long (its length is similar to the distance between C-1 and C-3 in cyclohexane), and the radical and alkene carbons have deviated little from their sp^2 geometries. The angle of attack (106°) is near to the preferred angle for unconstrained bimolecular additions (109°) and, because the forming bond is long, the angle and torsional strain of the connecting chain resemble more a cyclohexane ring than the product cyclopentane ring. The forming bond is nearly eclipsed, but it is so long that the energetic penalty is minimal. The length of the forming bond in the 6-endo TS (**3**) is similar to that of the 5-exo TS (**1**), but the angle of attack is constrained to be much smaller (94°) because of the endo orientation of the alkene. Thus, the overlap between the SOMO of the radical and the LUMO of the alkene is poorer, and energy is sacrificed. This stereoelectronic rationale, first proposed by Beckwith, is generally regarded as the single most important factor that favors 5-exo over 6-endo cyclization. Baldwin's guidelines for ring closure integrated these stereoelectronic preferences for radicals with those for other types of cyclizations.[10] Because of the importance of SOMO–LUMO overlap, cyclizations of radicals are more closely related to those of anions than of cations.

In addition to this important stereoelectronic effect, it is less widely recognized that the 6-endo transition state (**3**) is believed to have a higher degree of torsional and/or bending strain that its 5-exo counterpart (**1**).[11] This added strain may be due to deformations in the chain, which are necessary to accommodate the best possible angle of attack, or to the elongated forming bond, which stretches the 'cyclohexane ring' away from its ideal geometry.

The comparison of exo and endo transition states has emphasized the similarity of the 5-exo transition state (**1**) with a cyclohexane chair. The comparison of the two 5-exo transition states, (**1**) and (**2**), illustrates the all-important differences. The two transition states are very similar, but only five of the six atoms have analogies in chair or boat cyclohexane. The terminal alkene carbon (C-6) is well out of place compared to its partner in cyclohexane. Thus, the transition states really resemble stretched cyclopentane rings more than cyclohexanes. A boat cyclohexane is disfavored relative to a chair by two eclipsed carbon–carbon bonds and by the 'flagpole' interaction. In the 5-exo boat transition state, all of these interactions are greatly reduced: the flagpole interaction because the terminal alkene carbon is not appropriately placed, and the eclipsing interactions because of the presence of the sp^2 atoms that have yet to rehybridize. The small energy difference disfavoring the boat (**2**) relative to the chair (**1**) probably results from the remnants of the flagpole interaction (if any), and from the presence of a *gauche* butene arrangement (allylic C—H eclipsing vinylic C—H) rather than the more favored skew butene (allylic C—H eclipsing C=C). That (**1**) and (**2**) are comparable in energy is important for understanding stereoselectivity (see below).

The analogy between 5-exo transition states and the chair/boat cyclohexane is simple to apply and has useful predictive and interpretive value, provided that one recalls the differences as well as the similarities.

Substituent effects on cyclizations of simple nucleophilic hexenyl radicals have been well studied, and much quantitative rate data is available.[12] The trends that emerge from this data can often be translated to qualitative predictions in more complex settings. Once the large preference for 5-exo cyclization is understood, other substituent effects can often be interpreted in the same terms as for addition reactions. For example, electronegative substituents activate the alkene towards attack, and alkyl substituents retard attack at the carbon that bears them. The simple hexenyl radical provides a useful dividing point: $k_c \approx 2 \times 10^5\ s^{-1}$. More rapid cyclizations are easily conducted by many methods, but slower cyclizations may cause difficulties. Like the hexenyl radical, most substituted analogs undergo irreversible 5-exo closure as the predominate path. However, important examples of kinetic 6-endo closure and reversible cyclization will be presented.

4.2.1.1.2 Accelerating substituents

Fortunately for the synthetic chemist, most substituents accelerate 5-exo cyclizations, and the complete absence of 6-endo products is not uncommon. Several important types of accelerating groups are summarized in Scheme 3. Like addition reactions, activating groups on the alkene (C-6) provide a large rate enhancement due to favorable FMO interactions. Unlike additions, such activating substituents are not required for successful cyclizations; however, they can prove beneficial to offset deactivating substituents in hexenyl radical cyclizations, and they are often essential when forming larger rings. Most substituents on the chain also accelerate the hexenyl radical cyclization, probably by raising the energy of the ground state relative to the transition state (as in the *gem* dimethyl effect). The presence of additional rings in the cyclization substrate is usually beneficial, provided that a relatively unstrained ring is forming. In general, fused rings are preferred over bridged; however, bridged rings have been formed on many occasions. Substitutions of oxygen or nitrogen for C-3 are powerfully accelerating because they provide better overlap in the 5-exo transition state. Most alkyl substituents on the radical carbon (C-1) have relatively little effect on the rate: simple primary, secondary and tertiary alkyl radicals all cyclize at about the same rate. Vinyl and phenyl radicals are much more reactive in cyclizations than alkyl radicals. Acyl radicals are also excellent substrates.

4.2.1.1.3 Decelerating substituents

The most common reason that 6-endo products are observed in radical cyclizations is because 5-exo cyclizations are decelerated by substituent effects (see Scheme 4). Just as in addition reactions, the substitution of any group for hydrogen at the β-alkene carbon (C-5) slows 5-exo cyclization significantly, and 6-endo products are formed, sometimes predominantly. This effect also appears to be sensitive to radical substitution: 5-alkyl-substituted tertiary radicals give increased amounts of 6-endo products compared to primary radicals. The introduction of a radical-stabilizing heteroatom endocyclic to the forming ring (substitution of X for C-2) also retards 5-exo closure due either to its stabilizing effect, or (more

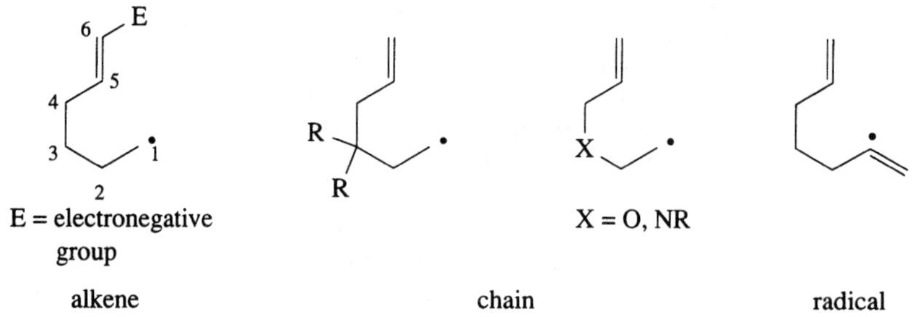

E = electronegative group

alkene

X = O, NR

chain

radical

Scheme 3 Accelerating substituents

likely) to geometric constraints imposed on the radical by resonance.[13] Relatively little is known about the effect of heteroatom substituents attached to the radical (C-1) exocyclic to the forming ring. One suspects that they will be at least modest rate depressors because they stabilize the radical.

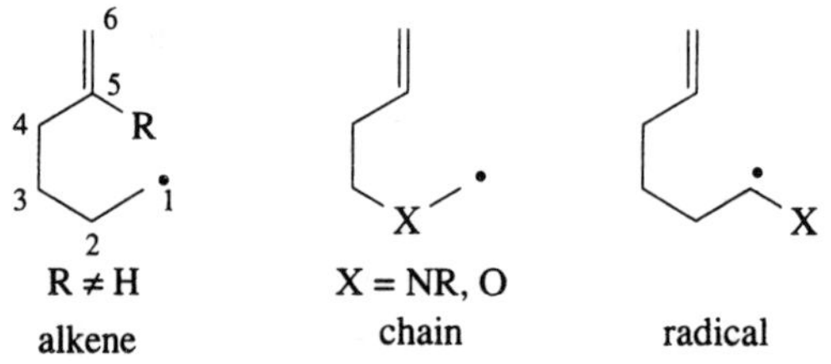

$R \neq H$

alkene

X = NR, O

chain

radical

Scheme 4 Decelerating substituents

All of these decelerating effects are easily overridden by appropriate alkene substitution. Indeed, any terminal alkene substituent will decelerate 6-endo cyclization. To accelerate 5-exo cyclization, one requires only an electronegative alkene substituent at C-6.

The appropriate placement of activating groups can also be used to accelerate 6-endo cyclizations, as illustrated by the examples in Scheme 5.[14–16] However this prescription is not a guaranteed cure to overcome 5-exo cyclizations when 6-endo products are desired. As illustrated by the last example in Figure 6,[17] the preference for 5-exo cyclization can often be so high that it cannot be overridden by an activating group.

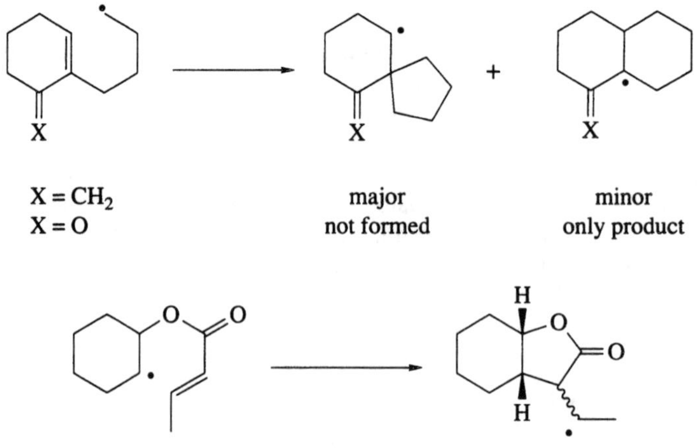

$X = CH_2$ major minor

X = O not formed only product

Scheme 5 Activation of 6-*endo* cyclizations

4.2.1.1.4 *Carbonyl-substituted radicals*

Because of the centrality of the carbonyl group in synthesis, the cyclizations of carbonyl-substituted radicals are of special importance. At present, there are no accurately known rate constants for the cyclizations of any carbonyl-substituted radicals, and quantitative statements are risky. Because carbonyl-substituted radicals add more rapidly to simple alkenes than alkyl radicals add, it might be suspected that they cyclize more rapidly as well. However, qualitative evidence does not support this analogy.

Cyclizations of carbonyl-substituted radicals can sometimes provide very high levels of 6-endo products by either kinetic or thermodynamic methods. Reversible hexenyl radical cyclizations are presently restricted to radicals with two carbonyl or cyano groups.[18] The discussion of reversible cyclizations is reserved for the hydrogen atom transfer section (Section 4.2.2.1.4) because this is the only method by which they have been conducted. Some kinetic substituent effects are summarized in Scheme 6.[19] If the carbonyl group is outside the forming ring, there is a slight increase in the amount of 6-endo product with monosubstituted alkenes (R = H); however, when R is an alkyl group, the 6-endo cyclization is often the only pathway. The placement of a ketone (but not an ester or an amide) inside the forming ring also greatly favors 6-endo closure.[19–21] It has been proposed that 5-exo cyclizations of the ketone-substituted radicals are retarded because overlap between the radical and the carbonyl group must be sacrificed to attain the transition state geometry.[19] Even without knowing rate constants, it is clear from experiment that the cyclizations of these latter types of radicals are very slow (although still useful).

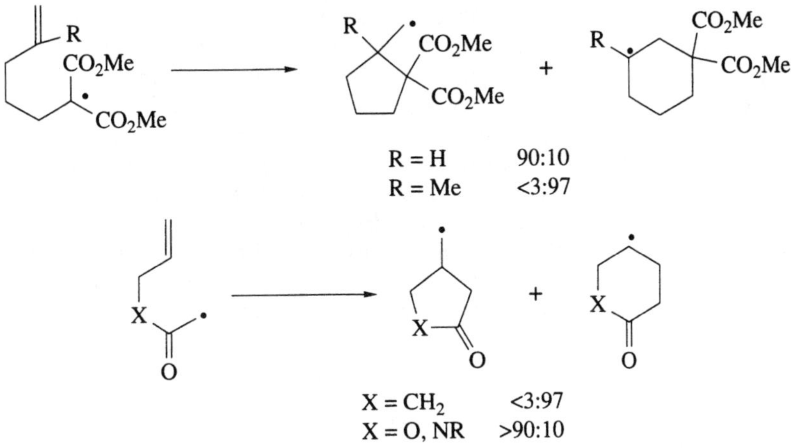

R = H 90:10
R = Me <3:97

$X = CH_2$ <3:97
X = O, NR >90:10

Scheme 6 Carbonyl-substituted radicals

4.2.1.2 Butenyl, Pentenyl, Heptenyl and Other Radical Cyclizations

Although cyclizations of hexenyl radicals are the best understood and most common, other cyclizations have unique features that can be exploited in synthesis. The basics for cyclization of butenyl and pentenyl radicals are summarized in Scheme 7. In accord with the stereoelectronic rationale, these radicals are very reluctant to cyclize in an endo mode due to the very small angle of attack that is required. The butenyl radical undergoes 3-exo cyclization with a rate that is near the lower limit of synthetic utility.[22] Like the hexenyl radical, this rate may be dramatically increased by substituents (for example, *gem* dimethyl substituents on the chain dramatically accelerate the closure).[23] Unlike the hexenyl radical, the cyclization is reversible and, due to ring strain, the equilibrium usually lies far to the side of the open radical. Accordingly, 3-exo cyclizations that synthesize cyclopropanes are only useful if there is a means to rapidly trap the cyclic radical (but not the initial radical) or if substituents dramatically shift the unfavorable equilibrium. Selective trapping of equilibrating radicals can be planned within the framework of the Curtin–Hammett principle, as described for cyclizations to carbonyls (see Section 4.2.5). 3-Exo fragmentations are very useful reactions, either alone, or in combination with 3-exo cyclizations (this combination results in rearrangement if the forming and fragmenting bond are not the same; see Section 4.2.6.2.1).

Like many ionic cyclizations, the formation of cyclobutane rings by radical cyclizations is very slow, and powerful accelerating effects will be required for any synthetic applications (see Scheme 7).[2] These cyclizations are again reversible, and, in the absence of any other effects, the equilibrium lies to the side of the open radical due to ring strain. The fragmentations of simple cyclobutyl carbinyl radicals are slow,

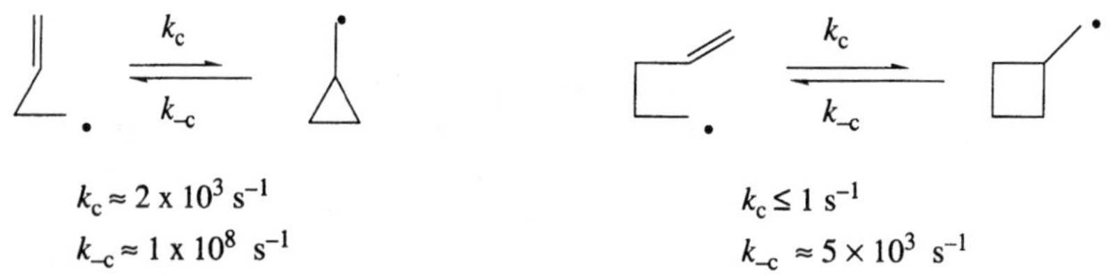

$k_c \approx 2 \times 10^3\ s^{-1}$
$k_{-c} \approx 1 \times 10^8\ s^{-1}$

$k_c \leq 1\ s^{-1}$
$k_{-c} \approx 5 \times 10^3\ s^{-1}$

Scheme 7 Butenyl and pentenyl radical cyclizations

but still potentially useful. Any stabilization of the open radical relative to the cyclic radical should accelerate such fragmentations.

Cyclizations of higher homologs of the hexenyl radical are useful, and they are generally irreversible. As the size of the forming ring increases, the rate of ring closure begins to decrease for entropic reasons. In addition, larger rings can better accommodate the favored trajectory angle for an endo transition state, and the large preference for exo cyclization that is seen in smaller rings begins to erode.

Heptenyl radical cyclizations are very useful in synthesis, and the behavior of the parent radical is summarized in Scheme 8.[24] Rate constants of 6-exo and 7-endo closure are near the lower limit of synthetic utility, and the ratio of cyclohexylcarbinyl radical to cycloheptyl radical is about 85:15 at normal operating temperatures. Besides cyclization, there is another competing unimolecular reaction that is unique to heptenyl radicals: 1,5-allylic hydrogen transfer. This intramolecular H-transfer is always thermodynamically favorable, but is strongly disfavored kinetically in both smaller rings (for stereoelectronic reasons) and larger rings (for entropic reasons). In the parent heptenyl radical, the rate of 1,5-hydrogen transfer is comparable to that of 7-endo cyclization, but there are examples where it can even exceed the rate of 6-exo cyclization. Thus, there are at least three problems to consider when planning a heptenyl radical cyclization; (i) the rates of cyclization are relatively slow, (ii) an overriding bias for exo cyclization does not exist, and (iii) competitive 1,5-hydrogen transfer may occur if there are accessible allylic hydrogens. Fortunately, these problems can often be solved by placement of appropriate substituents. For example, the introduction of an activating group on the alkene terminus often accelerates 6-exo cyclization sufficiently so that all three problems are simultaneously solved. It is also possible to effect selective 7-endo cyclizations in some cases.

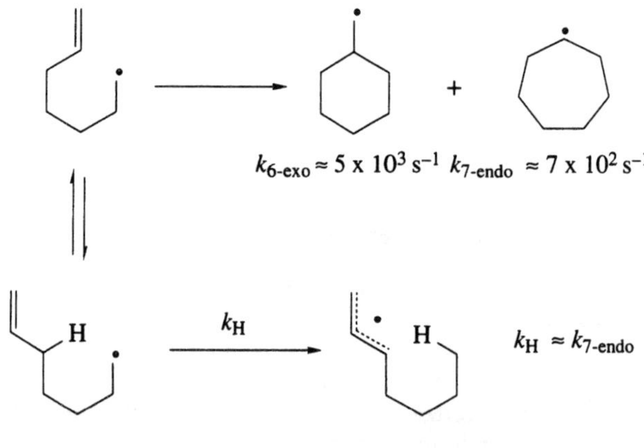

$k_{6\text{-exo}} \approx 5 \times 10^3\ s^{-1}$ $k_{7\text{-endo}} \approx 7 \times 10^2\ s^{-1}$

k_H $\qquad$ $k_H \approx k_{7\text{-endo}}$

Scheme 8 The heptenyl radical cyclization

The formation of medium and large rings by radical cyclizations is no different from most other standard methods. The formation of medium-sized rings by radical cyclizations is difficult. For example, the octenyl radical undergoes selective 8-endo closure rather than 7-exo closure[6] but the rate constant for cyclization is very low ($k \approx 1 \times 10^2\ s^{-1}$). However, one suspects that the kinds of substituent effects that favor medium ring formation in standard ionic reactions (like the introduction of (*Z*)-double bonds in the connecting chain)[25] should also accelerate radical cyclizations. As the ring size becomes larger and transannular interactions decrease, unimolecular rate constants for cyclization recover and begin to approach those of bimolecular addition. Pioneering work of Porter[26] has shown that, with appropriate electronic pairing and reaction concentration, large rings can be formed with a facility comparable to typical ionic processes (see Section 4.2.2.1.1).

4.2.1.3 Stereoselectivity

Although radical cyclizations have a reputation of exhibiting poor stereoselectivity, there are now many examples where highly stereoselective transformations have occurred, and both empirical and theoretical guidelines are beginning to emerge. As with regiochemistry, the papers of Beckwith[6] and Houk[7] are highly recommended starting points for stereochemical information. When predicting stereochemistry, it is important to keep in mind that transition states for radical cyclizations are early, and steric interactions in the final product may not be important in the transition state. More important are local conformation effects about the alkene, and allylic substituents and *cis*-substituted alkenes have large effects on stereoselectivity.

4.2.1.3.1 Hexenyl radical cyclizations

Stereoselectivity in the cyclization of substituted hexenyl radicals often follows the guidelines originally proposed by Beckwith.[27] The Beckwith model, shown in Figure 2, is probably a good representation of the lowest energy transition state, and it predicts the major product by placing a chain substituent in an equatorial-like orientation. Thus, 3-substituted hexenyl radicals give *cis* products, and 2- and 4-substituted radicals give *trans* products. The stereoselectivity of C-1-substituted radicals requires special comment because this center is sp^2 hybridized, and thus the substituent (R^1) is not in an equatorial-like orientation but actually eclipses the alkene CH_2 about the forming bond. (However, this eclipsing interaction is relatively small because of the long forming bond and the obtuse angle of attack.) The Beckwith model correctly predicts that C-1 alkyl substituents will give mainly *cis* products, although the underlying cause for the *cis* selectivity has been the source of considerable speculation. Houk and Spellmeyer[7] have suggested that van der Waals attraction is a likely cause. Although simple alkene substituents give 1,2-*cis* products, larger R^1 groups are *trans* selective, as are most ethers (R^1 = *O*-alkyl or *O*-silyl). Esters and ketones show virtually no selectivity in simple examples.[19]

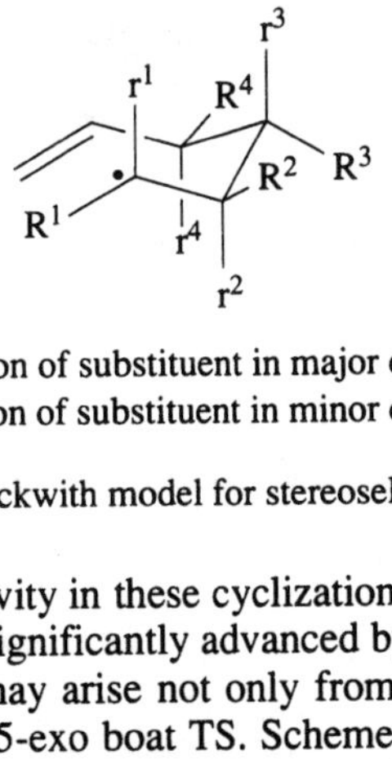

R = location of substituent in major diastereomer
r = location of substituent in minor diastereomer

Figure 2 Beckwith model for stereoselectivity

The understanding of the stereoselectivity in these cyclizations (and hence the prediction of which cyclizations will be highly selective) was significantly advanced by Houk and Spellmeyer.[7] They proposed that the minor diastereomeric product may arise not only from the 5-exo chair transition state with an axial-like substituent, but also from the 5-exo boat TS. Scheme 9 gives a specific example: the cyclization of the 3-methylhexenyl radical. The relative energies (rounded to the nearest 0.5 kcal mol^{-1} (1 kcal = 4.18 kJ)) are derived from force field calculations. The Beckwith model correctly predicts the formation of the major *cis* product, which arises *via* transition state A (chair, methyl equatorial). However, the calculations indicate that the minor product arises not only from the chair transition state C with the methyl group axial-like, but also from the boat transition state D with the methyl group equatorial-like. According to the force field model, this is a general phenomenon for simple systems: a single low energy transition state accounts for the major product but two transition states of intermediate level combine to produce the minor product.

The important studies of RajanBabu provide solid experimental support for the postulated energetic importance of boat transition states (Scheme 10).[28,29] Cyclization of (**4**α,β) gives a stereochemical outcome that is at first glance surprising: the stereochemistry of the methyl group is apparently controlled by the remote *t*-butyl group, and not by the neighboring residues on C-1 and C-2. The *t*-butyl group not only locks the conformation of the cyclohexane ring, but it also locks the conformation of the forming ring because the forming ring must be *cis* fused (the orbital overlap to form a *trans* ring is extremely poor). This is akin to locking a *cis* decalin with a *t*-butyl group (see the triple Neumann projections in Scheme 10), and instead of four transition states, only two (one chair and one boat) are now accessible. In each case, the major diastereoisomer (**5**α) or (**6**β) results from the chair-like transition state of the forming ring. Be-

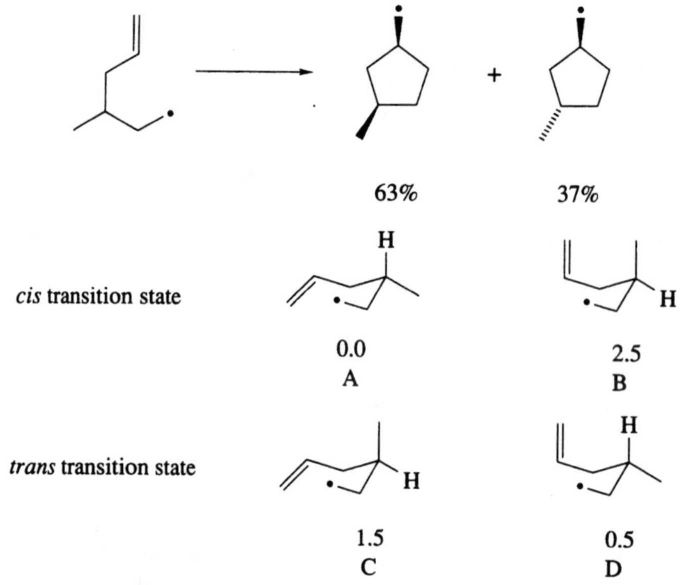

63% 37%

cis transition state

0.0 A | 2.5 B

trans transition state

1.5 C | 0.5 D

Scheme 9 Houk/Spellmeyer model for cyclization of a 3-substituted hexenyl radical

cause the other chair is prohibitively high in energy, it is very likely that the minor products (**5**β) and (**6**α) result from boat-like transition states.

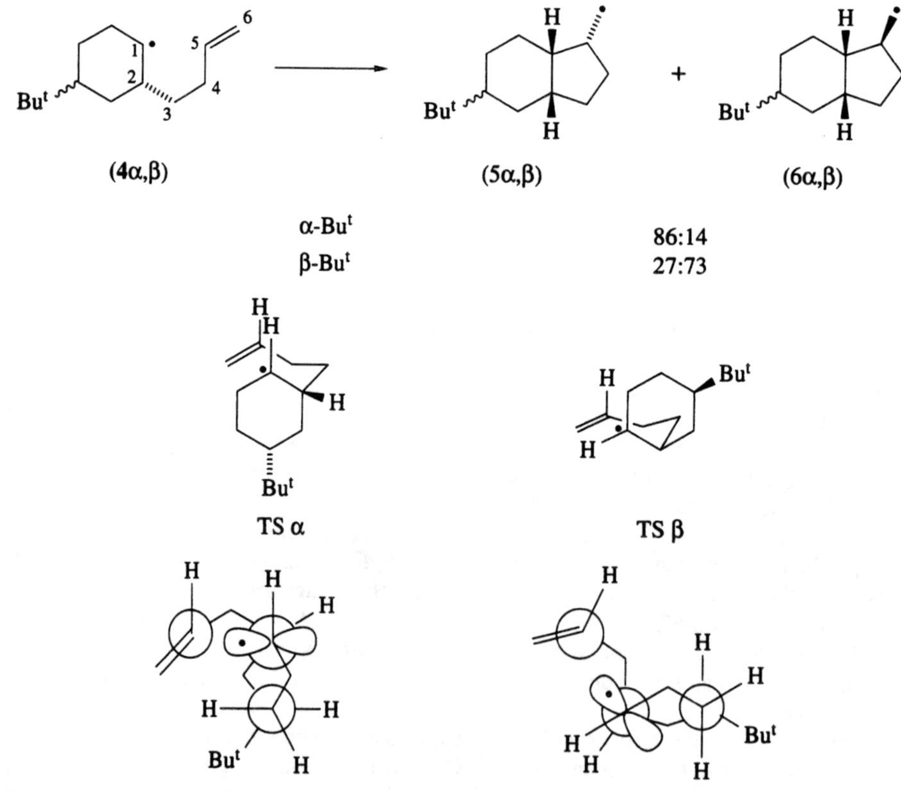

Scheme 10 Evidence for boat transition states

Indeed, RajanBabu has shown that the introduction of allylic substituents at C-4 can either reinforce or completely override the preference for a chair transition state, depending on the configuration. Two examples are provided in Scheme 11. In the framework of the stereochemical model, high selectivity results in these examples because three of the four normally accessible transition states have been eliminated by unfavorable interactions (the locked dioxolane ring eliminates one chair and one boat, and A-strain caused by the allylic substituent selects between the remaining chair and boat). The recognition that boat transition states are important, and can even dominate, should greatly facilitate the design of stereoselective hexenyl radical cyclizations.

derived from boat TS

derived from chair TS

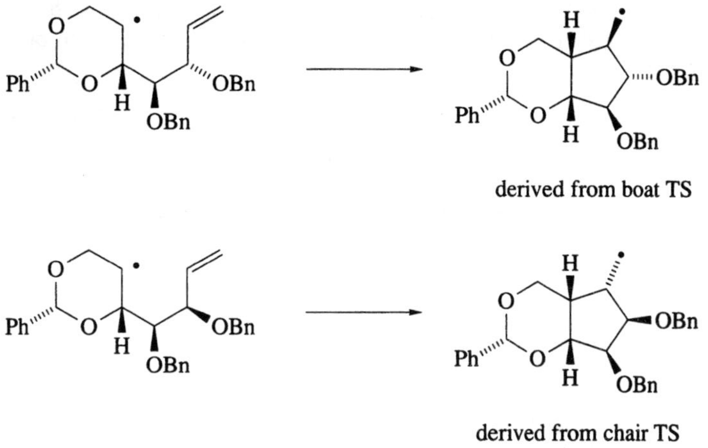

Scheme 11 Highly stereoselective cyclizations

4.2.1.3.2 Heptenyl radical cyclizations

Stereoselectivity in the cyclization of heptenyl radicals has recently been studied and the analogy to chair cyclohexane has again been drawn.[30] Figure 3 illustrates the two chair-like transition states that have been considered (in most cases, an activating group is present to facilitate the cyclization). The assumptions that model A is of lower energy than B, and that equatorial-like substituents are preferred, suffice to rationalize some of the literature examples; however, the situation is less straightforward than with hexenyl radical cyclizations. For example, this model predicts that 2- and 4-substituted precursors will give *cis* products, and that 3- and 5-substituted precursors will give *trans*. However, Hanessian has shown that 2-, 3- and 4-methyl-substituted heptenyl radicals (*E* = *trans* CO_2Me) all show a very slight preference for formation of the *trans* product.[30] This serves as a reminder that the forming bond in the transition state for the heptenyl radical cyclization is so long that distortions are inevitable, and the connecting chain cannot closely resemble a cyclohexane chair. Indeed, if the connecting chain in a hexenyl radical reaction resembles a cyclohexane ring, then cycloheptane-like conformers might be of considerable importance in the cyclizations of heptenyl radicals.

Figure 3 Stereoselectivity in heptenyl radical cyclizations

4.2.2 CYCLIZATIONS OF CARBON-CENTERED RADICALS TO CARBON–CARBON MULTIPLE BONDS

Most of the recent synthetic developments in the field of radical cyclization have involved the reactions of carbon-centered radicals with alkenes and alkynes. Other useful acceptors include allenes,[31] dienes[30] and vinyl epoxides.[32] The same methods are used for cyclizations to these acceptors as for radical additions, and the preceding chapter should be consulted for specific details on an individual method (the organization of this section parallels that of Section 4.1.6). Selection of a particular method to conduct a proposed cyclization is based on a variety of criteria, including the availability of the requisite precursor, the electronic properties of the radical, the expected rate of the cyclization, the type of functionality that is desired in the product, and the ease of purification of the products. These aspects will be briefly featured with each cyclization method.

The literature on radical cyclizations has grown so rapidly over the last decade that an introductory review cannot even sample all of the important types of reactions. In broad areas such as tin hydride cyclizations, this chapter will try to focus on the most commonly used types of cyclizations. With lesser used techniques, the selected examples will attempt to illustrate the synthetic possibilities and limitations. Several recent reviews that cover major parts of the radical cyclization field can be consulted for more detailed coverage.[3–5,33,34]

4.2.2.1 Chain Methods

Most radical cyclizations are conducted by one of the common chain methods. Kinetic analysis and synthetic planning are usually more straightforward than in addition reactions because the cyclizations are intramolecular.

4.2.2.1.1 Metal hydrides

(i) Tin hydride, general considerations

The use of tin hydride is by far the most popular and generally applicable method to conduct radical cyclizations. Because the tin radical is such a powerful atom and group abstractor, a wide variety of radical precursors can be used. A partial list of precursors (in rough order of decreasing reactivity) includes iodides, bromides, phenyl selenides, xanthates (and related derivatives),[35] nitro groups,[36] thio acetals,[37] phenyl sulfides and chlorides. The less reactive precursors are more useful for the production of more stable radicals. Although the tin hydride method is most commonly used for the cyclizations of nucleophilic radicals, it is appropriate for electrophilic (and ambiphilic) radicals as well. The kinds of radicals that can be generated for cyclization range from the most reactive aryl and vinyl radicals, through alkyl radicals, to stable radicals such as allyl[38] and benzyl.[39]

The tin hydride method is reductive, and the cyclic radical is almost always trapped by a hydrogen atom. In simple cyclizations, both the radical precursor and the alkene are lost during tin hydride reduction, and this sometimes results in underfunctionalized products, necessitating the introduction of extra functional groups for subsequent transformations. However, in the synthesis of simple molecules, this is often an advantage as steps to remove residual alkenes, carbonyl groups and the like, left by ionic methods of C—C bond formation, are not required. Work-up requires separation of the desired products from the tin by-products (see Section 4.1.6.2.1).

The most important consideration in conducting a tin hydride cyclization is not the electronic nature of the radical but the rate of cyclization. Equation 1 expresses the requirement for the successful cyclization of the initial radical prior to trapping with tin hydride. Cyclizations more rapid than that of the hexenyl radical are easily conducted by mixing the precursors with a slight excess of tin hydride (1.1 equiv. is common) at 0.01–1.0 M, and then initiating the chain. Slow cyclizations require high dilution conditions: syringe pump addition or catalytic tin hydride techniques are the most popular. With reactive radical precursors and proper conditions, cyclizations approaching the lower rate limit in solution can be conducted with the tin hydride method.[40] To maintain the chains, excess initiator and a greater excess of reducing agent (either tin hydride in the syringe pump method or the coreductant in the catalytic method) are often required.

$$k_c > k_H\ [Bu_3SnH] \tag{1}$$

As the rate of cyclization becomes slower, the reactivity of the precursor becomes more important. To ensure that the radical generation step does not break the chain, it is important to use the most reactive precursor available. For very slow cyclizations, the advice is simple: use iodides whenever possible. The purity of the precursor is also critical for slow cyclizations because tin hydride can sometimes react with impurities to generate hydrogen atom sources that are much more reactive than itself. Any impurities that might generate thiols or selenols may cause undue amounts of reduction (thus, the purity of phenyl sulfides and selenides is especially important). Metal impurities, which may form transition metal hydrides, can be devastating, even for fast cyclizations.[41] Empirically, it seems that breaking of the chain is less of a concern in the hydrogen transfer step, probably because radicals often react with each other or with the medium by hydrogen transfer. Thus, even if hydrogen transfer from tin hydride fails at low concentration, many of the product radicals will nevertheless end up with a hydrogen. Finally, high temperatures usually favor cyclization over atom transfer, and the use of a higher boiling solvent can be a productive means to favor the formation of cyclic products. Rapid cyclizations can be conducted at temperatures as low as –78 °C by using triethylborane initiation as prescribed by Oshima and Utimoto.[42]

The following sections provide selected examples of tin hydride cyclizations that are organized by the nature of the radical. More detailed coverage is provided in several general reviews,[5,33,34] and in the excellent specialized review of Neumann.[43]

(ii) Substituted alkyl radicals (carbocycles)

Many synthetic methods are currently available to form simple, isolated rings, and radical cyclizations, although certainly applicable, are not often used. However, as the complexity of the molecules is increased, radical cyclizations begin to offer advantages even in the preparation of isolated rings; preparations of highly oxygenated carbocycles from sugars are prime examples.[44–46] For the synthesis of polycyclic molecules, radical cyclizations offer a powerful alternative to traditional means for ring closure, and tin hydride cyclizations are useful for the preparation of fused, bridged and spiro rings.

Recent representative examples in which fused rings have been formed are shown in Scheme 12. As illustrated by the conversion of (**7**) to (**8**),[47] the forming of bonds to preexisting small rings results in *cis* ring fusion (5,5-, 5,6- and 6,6-bicycles). However, in extensive studies that provide insight into both stereo- and regio-selectivity in complex cyclizations,[41] Hart has shown that smaller rings take precedence over larger in formation of *cis*-fused rings. For example, cyclization of (**9**) provides (**10α**) and (**10β**), both of which contain a *trans* 6,5-ring fusion and a *cis* 5,5-fusion.[48] Likewise, when the activating group is relocated, a *trans* 6,6/*cis* 6,5 fusion is preferred [see (**11**) → (**12**)].[41] The stereoselectivity in the first two examples is noteworthy; in each case the C-1 alkyl group is *cis* to the alkene acceptor in the transition state leading to the major product. As larger rings are involved, overlap to form *trans*-fused rings can improve.[49] Winkler and Sridar have shown that high selectivity in the formation of *trans*-fused 5,8-rings can be obtained by using ring allylic substituents to dictate the local conformation of the alkene acceptor.[50] For example, cyclization of (**13**) provides exclusively (**14**), but the hydroxy epimer of (**13**) gives about a 1:1 mixture of products derived from *cis* and *trans* cyclization.

Examples of bridged and spiro ring formation are provided in Scheme 13. In the formation of bridged rings[51] the conformation of the radical is important, as illustrated by two examples from Giese.[52] Cyclization of (**15**) gives (**16**) in excellent yield but the related compound (**17**) is reduced to (**18**) under the same conditions (however, this unsuccessful cyclization is 6-exo, not 5-exo). These results are taken as evidence that the intermediate radicals adopt conformations like (**19**), in which a group at C-3 is appropriately oriented for cyclization, but one at C-5 is not. The formation of spiro rings has been extensively investigated by Clive,[53] and the conversion of (**20**) to (**21**) provides an example that also illustrates that alkynes are very useful acceptors for cyclization. The resulting alkene then provides a handle for subsequent transformations.

The combination of standard synthetic transformations with radical cyclizations can often result in efficient, stereocontrolled methods to build rings. For example, protocols that form cyclization precursors by Michael additions[54] and aldol reactions[55] have been developed. Scheme 14 illustrates a stereocontrolled method of ring formation developed by Clive that relies on the Ireland–Claisen rearrangement.[56] One of the most elegant examples of this approach to ring construction is the combination of Diels–Alder reactions with radical cyclizations, as recently described by Ghosh and Hart.[57] In the example provided in Scheme 14, an aryl radical cyclization is conducted (see below for aryl radical cyclizations).

Cyclizations to form larger rings are also possible as the two examples contained in Scheme 15 illustrate. 6-Exo cyclizations usually require activating groups unless other substituents accelerate the cyclization.[30,58] A few 7-endo cyclizations are known, but intermediate size rings (8–10) will be difficult to form unless special conformational advantages are built into the precursors. The formation of larger rings has been accomplished with great success by Porter.[21,40] Substituent effects in these cyclizations parallel those in bimolecular additions, and electronic pairing of the radical and the acceptor is essential. Because the effective molarities of macrocyclization are favorable, low concentrations (0.01–0.001 M) promote cyclization over addition. Scheme 15 features a recent application of this macrocyclization by Pattenden with several interesting features.[59] An allylic radical undergoes cyclization, but, due to rotation, some alkene stereochemistry is lost. The product is converted to the cembranolide mukulol by reduction of the carbonyl.

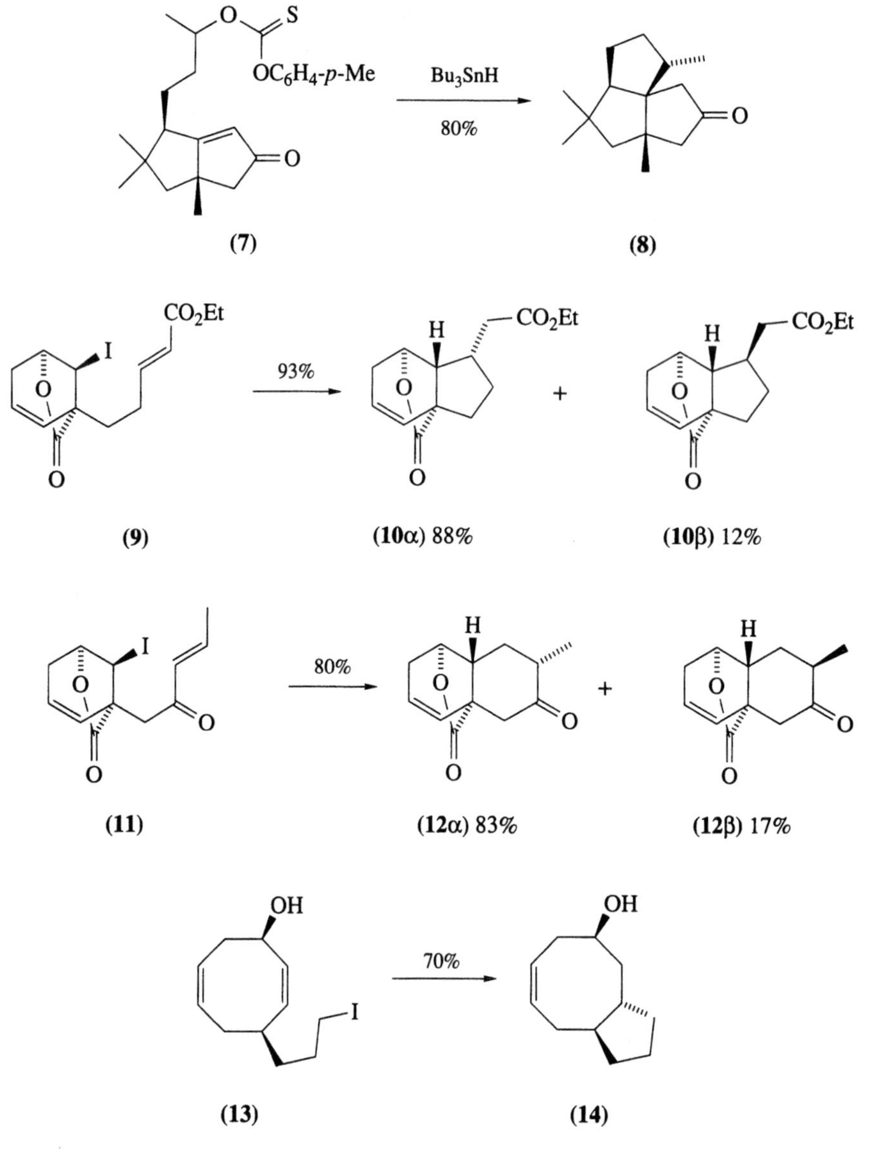

OC_6H_4-*p*-Me

Bu_3SnH

80%

(7) **(8)**

CO_2Et

93%

(9) **(10α)** 88% **(10β)** 12%

80%

(11) **(12α)** 83% **(12β)** 17%

OH

70%

(13) **(14)**

Scheme 12

(iii) Substituted alkyl radicals (heterocycles)

As indicated above, nuclear substitution of C-3 by oxygen or nitrogen accelerates 5-exo cyclizations, and some of the best radical cyclizations form heterocycles. Several examples are provided in Scheme 16. As indicated by the conversion of (**22**) to (**23**), the directing effect of the heteroatom is often powerful enough to overcome an otherwise detrimental C-5 methyl substituent.[60] Several research groups[61,62] have developed mild methods to form fused carbohydrates, and the conversion of (**24**) to (**25**) is a representative example.[63] Although the stereochemistry of most such cyclizations follows the Beckwith guidelines, Watanabe and Endo have recently shown that dichloroalkyl radicals give the opposite stereoselectivity when compared to simple alkyl radicals.[64] This observation appears to have some generality,[65] and it is synthetically significant because the perchlorinated products can be reductively dechlorinated.

The cyclizations of halo acetals, introduced by Stork[66] and Ueno,[67] are among the most popular of all radical methods because the precursors are easy to prepare, the cyclizations are very rapid, and the products can be converted to a variety of different heterocycles. The conversion of (**26**) to (**28**; Scheme 17) illustrates the basic procedure.[68] Standard bromoacetalization of an alcohol provides the precursor (**27**). After tin hydride cyclization and acidic oxidation, lactone (**28**) is formed in excellent yield with 97%

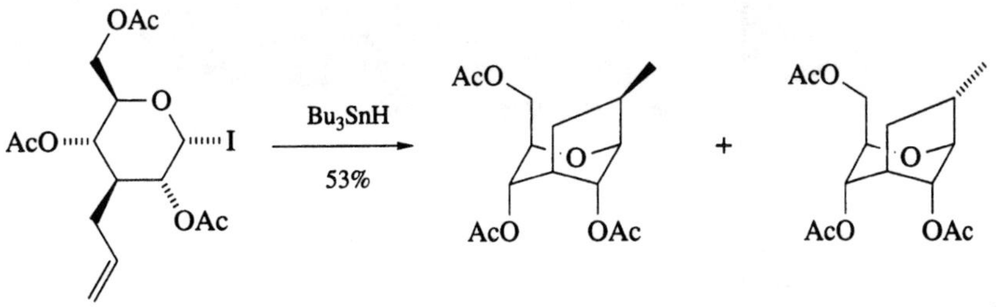

(15) **(16β)** 90% **(16α)** 10%

(17) **(18)** **(19)**

(20) **(21)**

Scheme 13

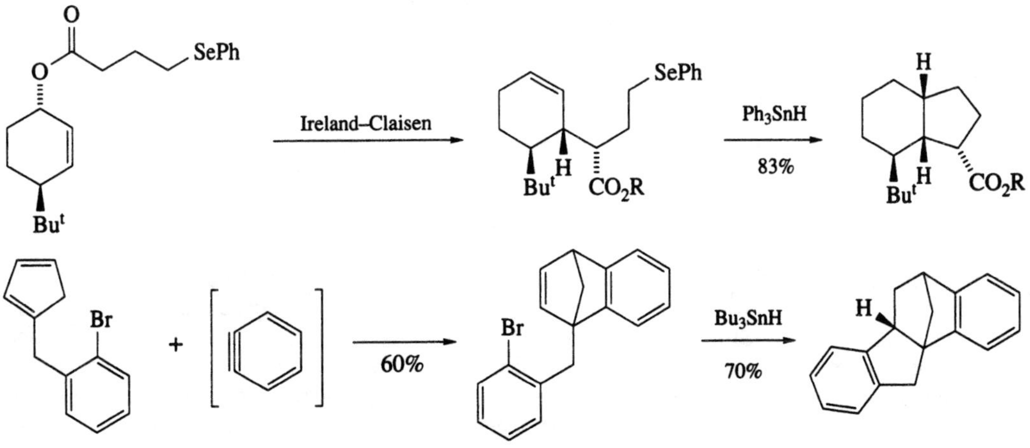

Scheme 14

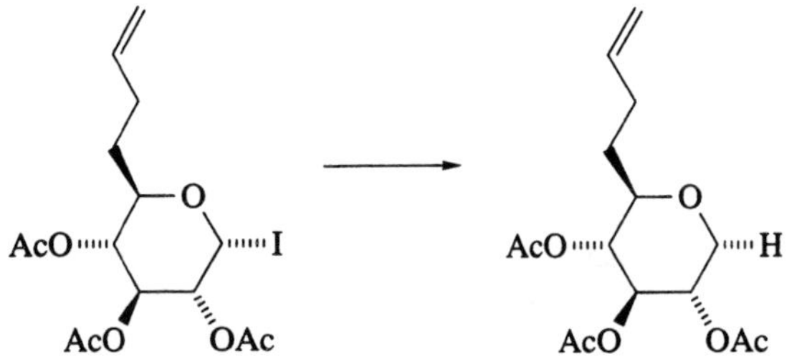

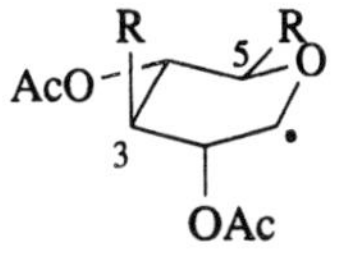

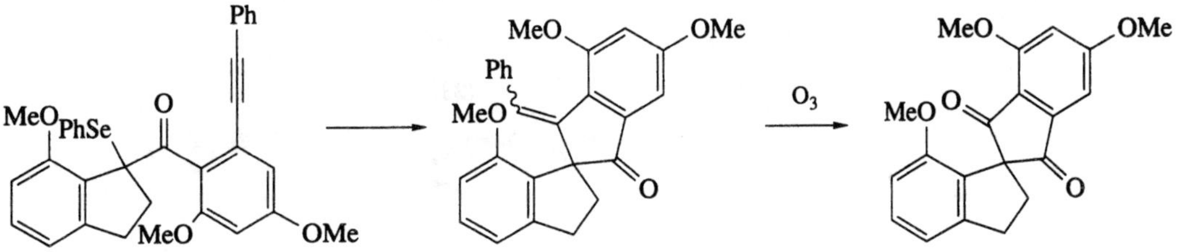

trans selectivity. The cyclizations of halo esters provide a potentially more direct route from alcohols to lactones; however, these cyclizations are much slower than the hexenyl radical benchmark, while the acetal cyclizations are much more rapid. Recent variants on this theme include the formation of 6-exo products,[69] furans[70] and unsaturated lactones,[71,72] as illustrated in Scheme 17.

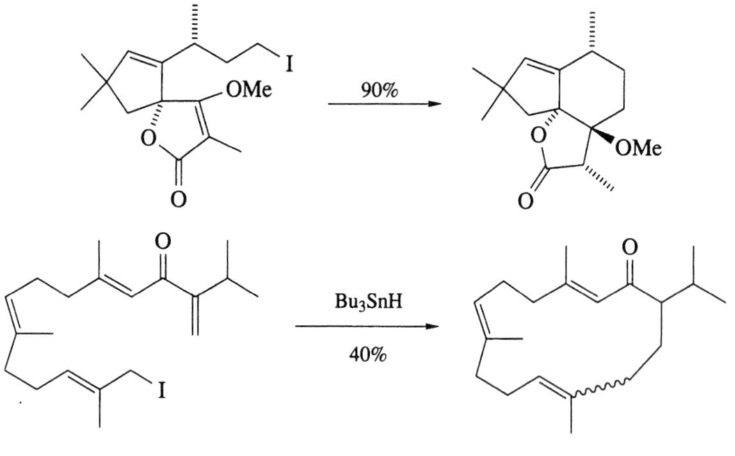

90%

Bu_3SnH

40%

(*E*)-only

4:1 (*E*):(*Z*)

Scheme 15

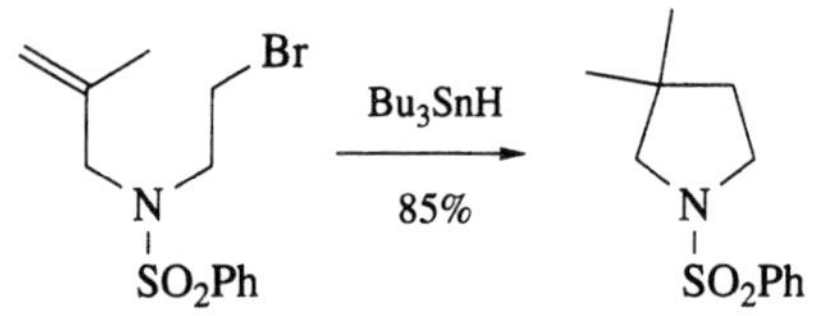

Bu_3SnH

85%

(22) **(23)**

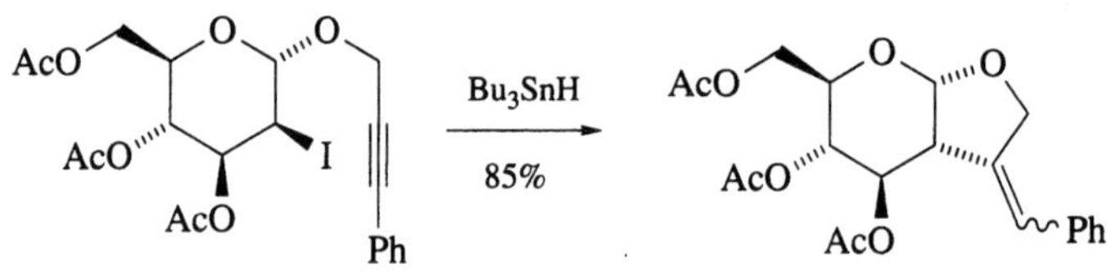

Bu_3SnH

85%

(24) **(25)** 1:3 (*E*):(*Z*)

Bu_3SnH

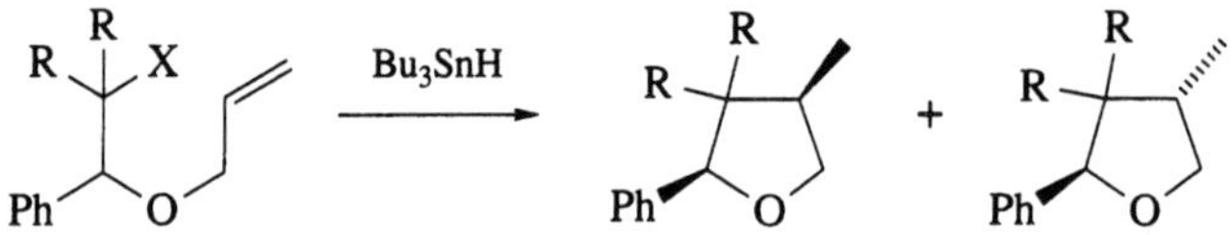

X = Br, Cl

R = H,	83%	34:66
R = Cl,	92%	90:10

Scheme 16

Silylmethyl radical cyclizations are a related class of reactions in which the heterocycle serves a temporary purpose to direct the introduction of an alkyl or hydroxyalkyl group adjacent to an alcohol. These reactions were introduced by Stork[73] and Nishiyama,[74] and a recent example from Crimmins[75] nicely shows the usefulness of the method (Scheme 18). Talaromycin A (**31**) is a challenging target because it is the less stable of the two possible spiroacetal diastereomers. The problem is to introduce the axial hydroxymethyl group after the formation of the spiroacetal (if it is introduced before, the diastereomeric spiroacetal will form). Cyclization of (**29**) produces (**30**) which, upon direct oxidation, gives talaromycin A (**31**) in 78% yield. Silylmethyl radical cyclizations can provide stereocontrol in acyclic systems,[74,76] and 6-endo examples are also known.[77] Finally, in addition to oxidative cleavage, protiodesilylation is also possible.[73]

The placement of heteroatoms adjacent to the radical center often slows the rate of cyclization and, although low concentration techniques are required, good yields of cyclic products can still be obtained. Hart's systematic studies on the cyclizations of acylamino radicals were among the first to demonstrate the preparative utility of radical cyclizations.[78] Scheme 19 provides some recent examples. In general,

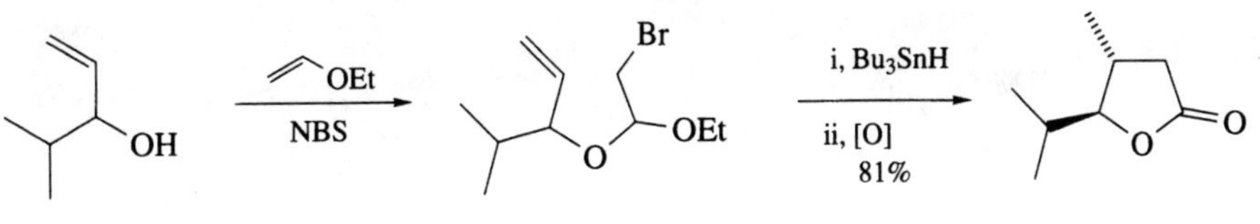

(26) (27) (28)

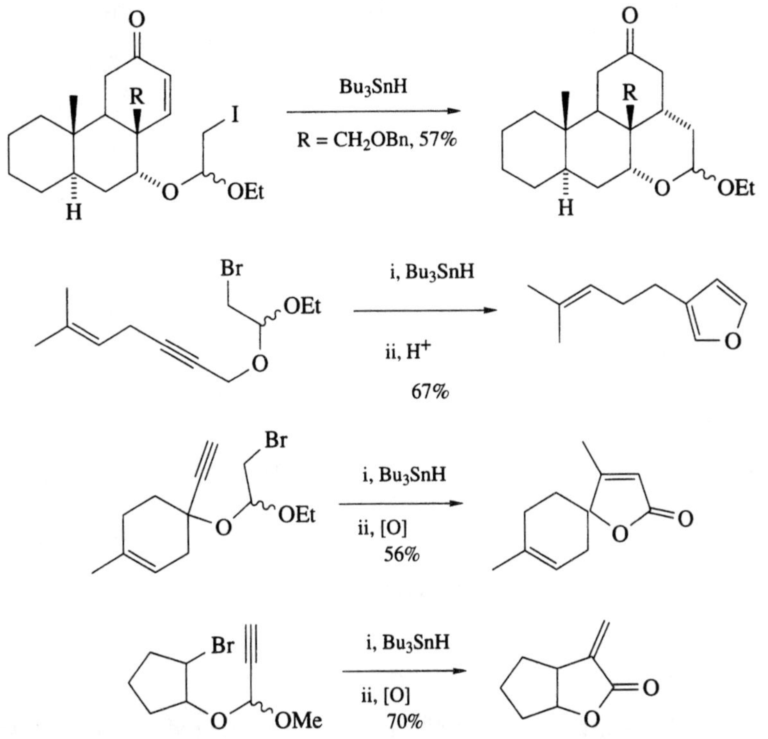

Scheme 17

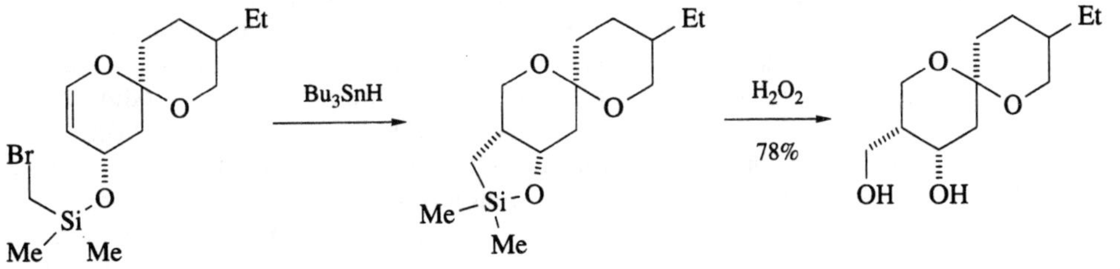

(29) (30) (31) Talaromycin A

Scheme 18

phenylthio precursors are preferred because acylamino halides are prone to solvolysis.[79] The first example in Scheme 19 shows that these cyclizations are dramatically improved by an accelerating substituent.[80] The cyclization of (**32**) to (**33**) succeeds even though a relatively high energy conformer is required. The cyclization of the unactivated alkene acceptor fails. The second example illustrates that even 6-exo cyclizations are possible with activated acceptors.[81] In the absence of directing alkene substituents, acylamino radicals often give significant amounts of endo products. This is especially true in the β-lactam series,[82,83] as shown by the third example in Scheme 19.[84] There are even examples of 7-endo cyclizations to fused β-lactams.[85] Although Scheme 19 shows only acylamino radicals, amino radicals[86] and ether-substituted radicals have also been studied.[37,87–89]

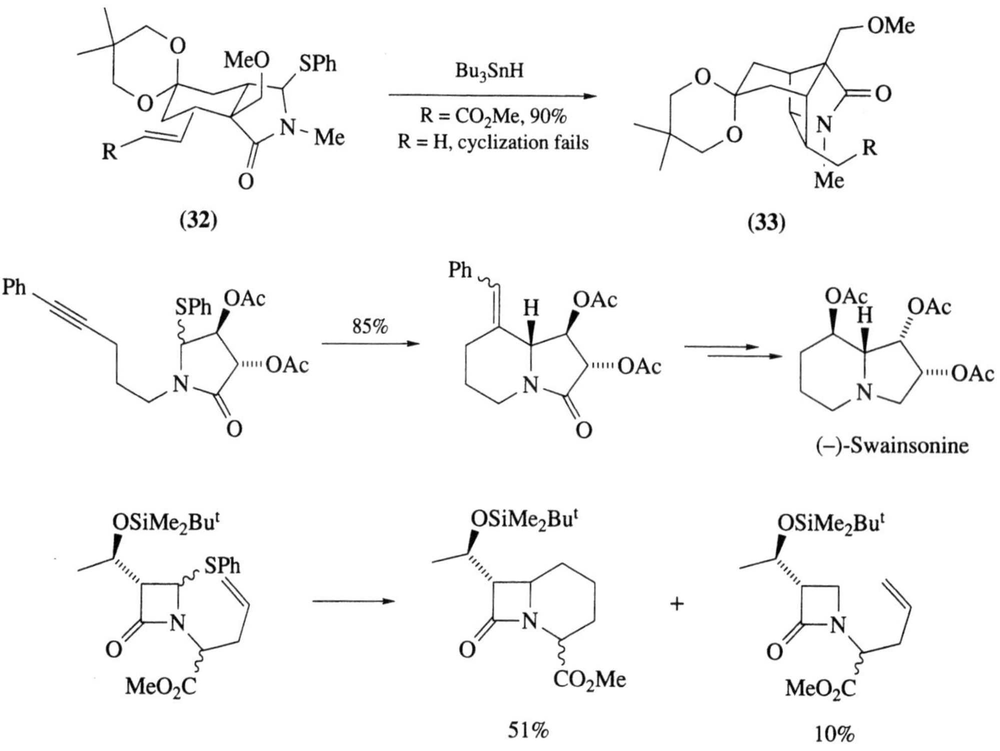

Bu_3SnH
R = CO_2Me, 90%
R = H, cyclization fails

(32) (33)

85%

(–)-Swainsonine

51% 10%

Scheme 19

(iv) Vinyl and aryl radicals

The use of vinyl radicals (pioneered by Stork)[90] and aryl radicals (introduced by Beckwith)[91] in the tin hydride method is attractive because these radicals are extremely reactive,[92] and often provide excellent yields of cyclic products that contain useful functionality for subsequent transformations.

Each of the syntheses of seychellene summarized in Scheme 20 illustrates one of the two important methods for generating vinyl radicals. In the more common method, the cyclization of vinyl bromide (**34**) provides tricycle (**35**).[93] Because of the strength of sp^2 bonds to carbon, the only generally useful precursors of vinyl radicals in this standard tin hydride approach are bromides and iodides. Most vinyl radicals invert rapidly, and therefore the stereochemistry of the radical precursor is not important. The second method, illustrated by the conversion of (**36**) to (**37**),[94] generates vinyl radicals by the addition of the tin radical to an alkyne.[95–98] The overall transformation is a hydrostannylation, but a radical cyclization occurs between the addition of the stannyl radical and the hydrogen transfer. Concentration may be important in these reactions because direct hydrostannylation of the alkyne can compete with cyclization. Stork has demonstrated that the reversibility of the stannyl radical addition step confers great power on this method.[95] For example, in the conversion of (**38**) to (**39**), the stannyl radical probably adds reversibly to all of the multiple bond sites. However, the radicals that are produced by additions to the alkene, or to the internal carbon of the alkyne, have no favorable cyclization pathways. Thus, all the product (**39**) derives from addition to the terminal alkyne carbon. Even when cyclic products might be derived from addition to the alkene, followed by cyclization to the alkyne, they often are not found because β-stannyl alkyl radicals revert to alkenes so rapidly that they do not close.

The cyclizations of vinyl, aryl, and acyl radicals differ from those of alkyl radicals because there is a mechanism by which cyclic radicals can interconvert that does not entail reversal of the original cyclization. This mechanism is shown in Scheme 21. Reduction of (**40**) with tin hydride at low concentration gives a mixture of 5-exo and 6-endo products (**41**) and (**42**).[99] However, only the 5-exo product (**41**) is obtained at high concentration, indicating that the 6-endo product is not kinetic. (Direct reduction of the vinyl radical cannot compete with cyclization even at 1.7 M!) Beckwith[100] and Stork[99] have convincingly demonstrated that the cyclic radicals (**43**) and (**45**) equilibrate, not through reversible 5-exo cyclization, but through (**44**), which arises from reversible 3-exo cyclization. At high tin hydride concentrations, (**43**) is trapped, but at lower concentrations, equilibration can compete. This isomerization, which is a

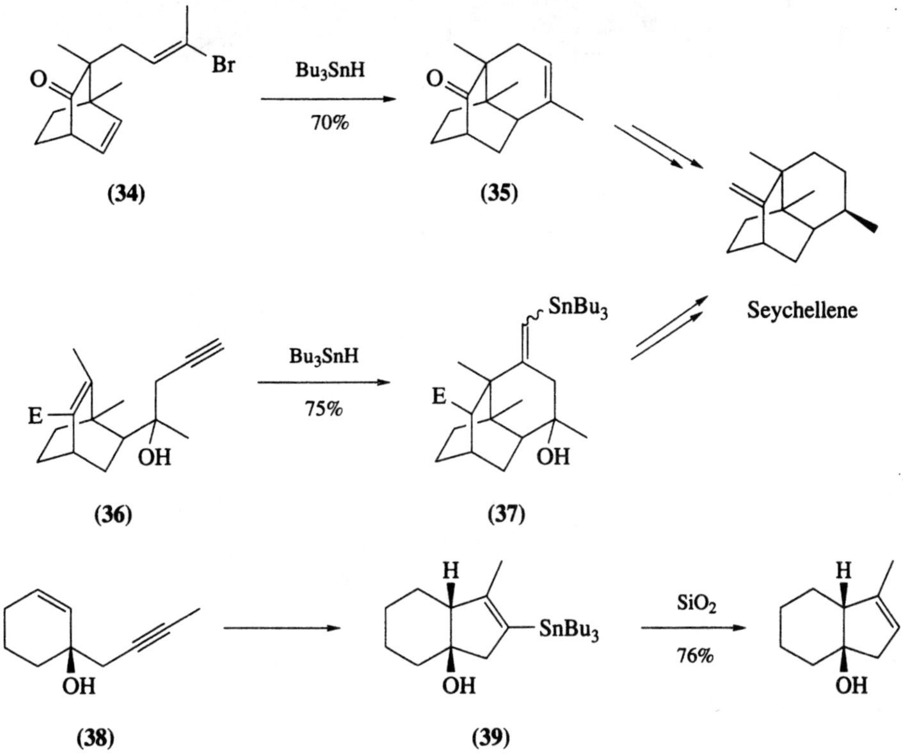

(34) Bu_3SnH 70% (35)

Seychellene

(36) Bu_3SnH 75% (37) $SnBu_3$ E OH

(38) (39) $SnBu_3$ SiO_2 76% H OH

Scheme 20

general phenomenon for radicals oriented β to unsaturated groups, was not widely recognized by practitioners of vinyl radical cyclizations prior to the work of Stork and Beckwith. To date, conditions have usually been designed to trap kinetic 5-exo products, but equilibration to 6-endo products has potential utility. Such isomerizations are well developed in cases where the acceptor for the cyclization is a carbonyl group (see Section 4.2.6).

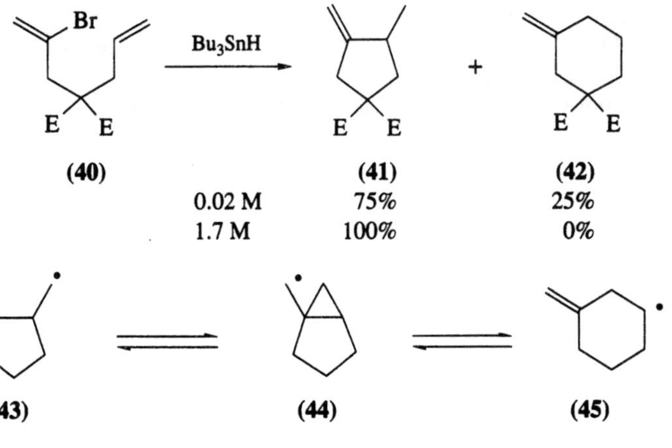

(40) Br Bu_3SnH (41) + (42)

	(41)	(42)
0.02 M	75%	25%
1.7 M	100%	0%

(43) (44) (45)

Scheme 21

Scheme 22 presents some selected aryl radical cyclizations. The first example shows that isomerization of the intermediate radicals is again a concern. This isomerization, which is usually called the neophyl rearrangement,[101] is promoted by the activating aldehyde group in the case at hand.[102,103] The other examples illustrate that aryl radical cyclizations provide practical routes to benzo-fused heterocycles.[104,105]

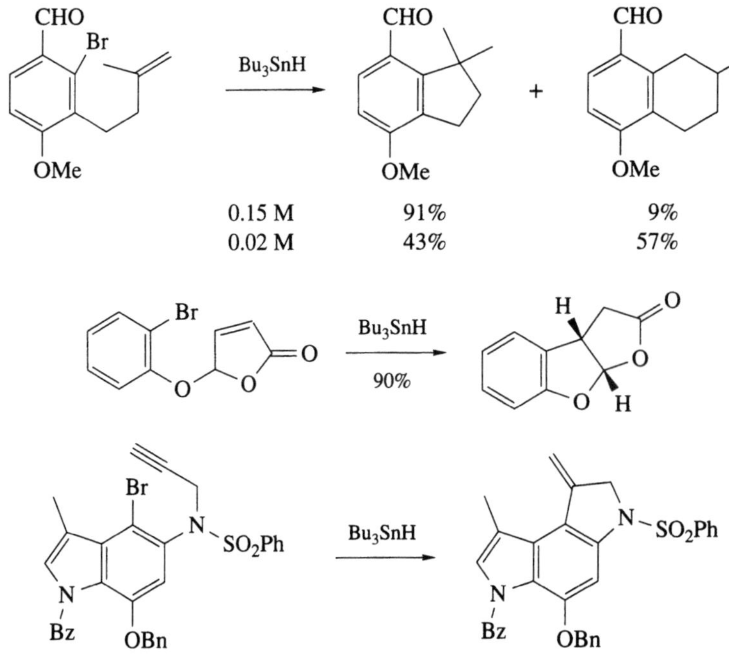

	Bu_3SnH		
	0.15 M	91%	9%
	0.02 M	43%	57%

Scheme 22

(v) Acyl radicals

Acyl, carbonyloxy and related radical cyclizations are rapidly gaining prominence and appear to have excellent scope. Examples of 5-exo (bridged and fused), 6-endo (possibly through equilibration as in Scheme 21) and 6-exo cyclization[106] have appeared. It is not presently clear whether the favorable ratios of cyclic to reduced products obtained with acyl radicals are due to inherently high reactivity of acyl radicals towards cyclization or to low reactivity towards tin hydride (due to the relative weakness of the forming C—H bond).

Two recent examples are contained in Scheme 23. The acyl radical cyclization, selected from the work of Boger and Mathvink,[107] illustrates that bridged rings can be formed in good yields (the corresponding alkyl radical cyclizes very slowly). Cyclizations of carbonyloxy[108] and related[109] radicals have been pioneered by Bachi, and the example in Scheme 23, from the work of Corey,[110] is a key step in the synthesis of atractyligenin. The precursor of choice for such radicals is often the phenyl selenide; halides are susceptible to ionic reactions. If a stable radical will be generated, decarboxylation or decarbonylation can sometimes compete with cyclization.

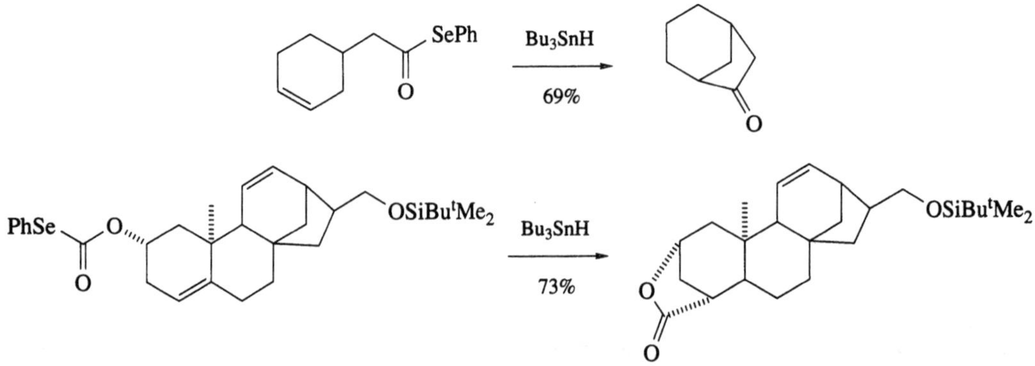

Scheme 23

(vi) Mercuric hydride

Although most reductive cyclizations have been conducted by the tin hydride method, cyclizations of mercuric hydrides are also straightforward to conduct. Precursors are usually prepared by heteromercuration of an alkene, and substituted alkyl radicals (nucleophilic) often result. Cyclizations are conducted like addition reactions: oxymercuration is followed by the direct addition of a hydride source to convert the intermediate mercuric halide or acetate to a mercuric hydride. It may be difficult to conduct slow cyclizations because mercuric hydrides are very good hydrogen atom donors. In principle, the concentration of the mercuric hydride can be controlled, but mercuric hydrides are unstable and their exact concentrations are not usually known. Purification of the products is straightforward (there are no byproducts that require chromatographic separation), but the toxicity of metallic mercury may cause problems on a larger scale. The recent review of Barluenga and Yus provides comprehensive coverage of the radical cyclization reactions of mercurials,[111] and two representative transformations are contained in Scheme 24.[112,113]

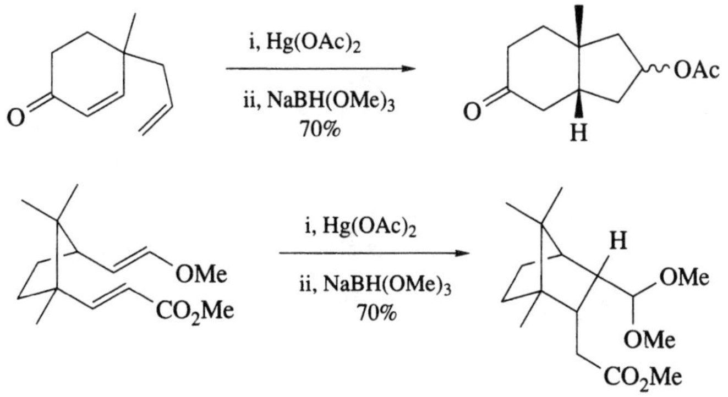

i, $Hg(OAc)_2$; ii, $NaBH(OMe)_3$; 70%

i, $Hg(OAc)_2$; ii, $NaBH(OMe)_3$; 70%

Scheme 24

4.2.2.1.2 The fragmentation method

Conducting radical cyclizations by fragmentation reactions offers a powerful alternative to the tin hydride method. Instead of obtaining reduced products, one obtains products of substitution; an alkene is regenerated in the fragmentation step.

Even though there are relatively few examples, the use of allyl- and vinyl-stannanes in radical cyclizations is probably very general.[114] This should be a very good method to conduct relatively slow cyclizations. Since high dilution is not necessary, a wide range of radical precursors (including the less reactive types) can be used. Although most of the examples in the literature use nucleophilic radicals, all classes of radicals (nucleophilic, electrophilic, ambiphilic) should be amenable to cyclization by this method. Scheme 25 provides examples of addition reactions to both allyl- and vinyl-stannanes. The first example is a convenient route to enol acetates.[115] The second example is particularly interesting because it illustrates that control over the stereochemistry of exocyclic alkenes is possible because fragmentation of the intermediate β-stannyl radical is more rapid than σ-bond rotation[116] (however, there are also examples where rotation is faster than elimination, see Section 4.2.6). The complete control of relative stereochemistry in the third example is noteworthy.[117]

Allyl sulfones offer the possibility to conduct cyclization reactions that are isomerizations, and investigations by Smith and Whitham indicate that this is a very promising technique for the preparation of functionalized products.[118] The mechanism consists of reversible addition of a sulfonyl radical to a terminal alkene or alkyne, cyclization, and fragmentation of the resulting β-sulfonyl radical. Two examples of these isomerizations are provided in Scheme 26.

4.2.2.1.3 The Barton thiohydroxamate method

Cyclizations conducted by the thiohydroxamate method are even less common than those conducted by the fragmentation method, but there is every reason to believe that thiohydroxamate cyclizations should be generally applicable.[119] Precursors are formed from carboxylic acids, and require the inclusion of an extra carbon since decarboxylation occurs. This feature may be very useful when normal precursors containing carbon–heteroatom bonds are unstable or difficult to prepare. The thiohydroxamate

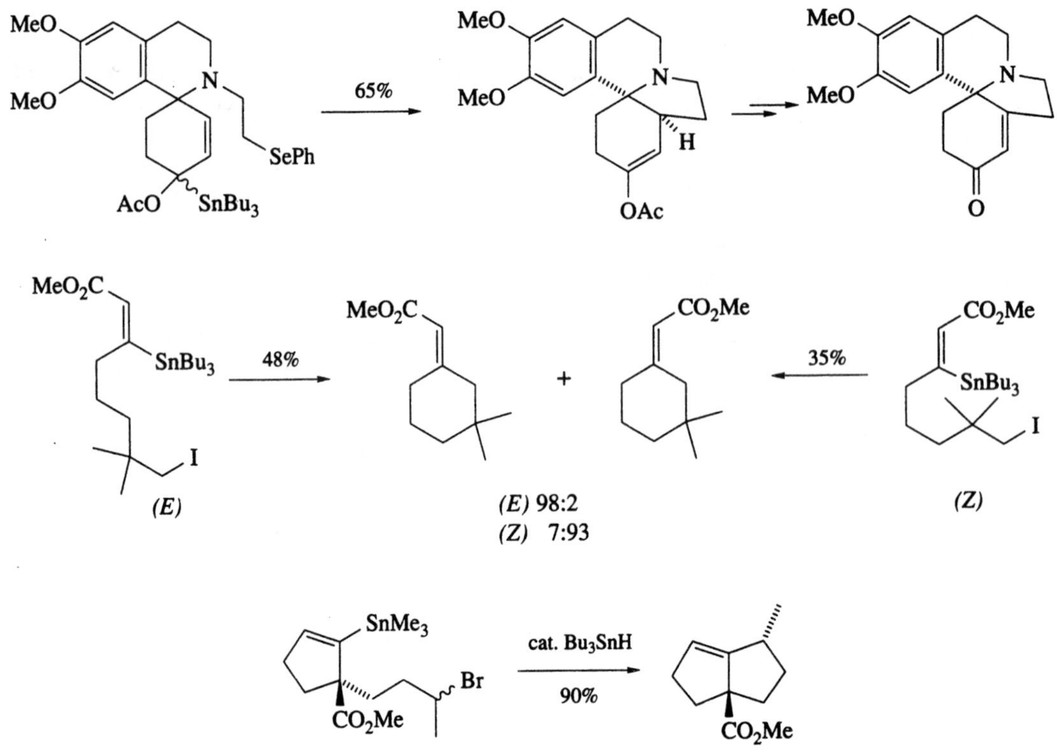

MeO MeO N SePh AcO SnBu3 65% MeO MeO N H OAc MeO MeO N O

MeO2C SnBu3 I (E) 48% MeO2C + CO2Me (E) 98:2 (Z) 7:93 35% CO2Me SnBu3 I (Z)

SnMe3 Br CO2Me cat. Bu3SnH 90% CO2Me

Scheme 25

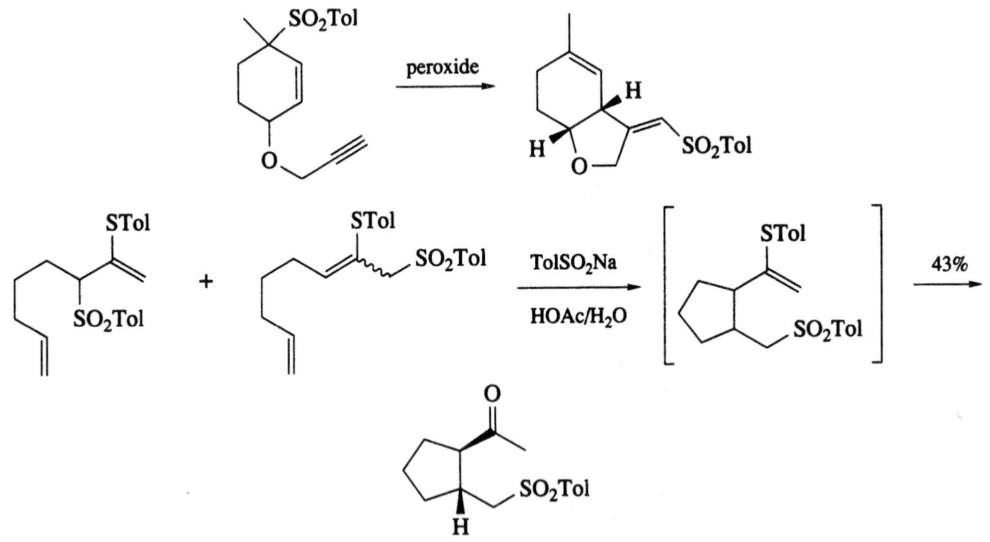

SO2Tol O peroxide H H O SO2Tol

STol SO2Tol + STol SO2Tol TolSO2Na HOAc/H2O STol SO2Tol 43%

O SO2Tol H

Scheme 26

method is usually used to generate nucleophilic radicals, and, if the standard techniques are used to minimize the thiohydroxamate concentration, it should be possible to conduct relatively slow cyclizations. The most important feature of the thiohydroxamate method is that it is not reductive (although reduction products are readily obtained if desired). In the standard procedure, cyclic radicals are trapped by addition to the thiohydroxamate, and a thiopyridyl group results. Equation 2 presents an example from the work of Barton in which optically active indolizidines are prepared from glutamic acid.[120,121]

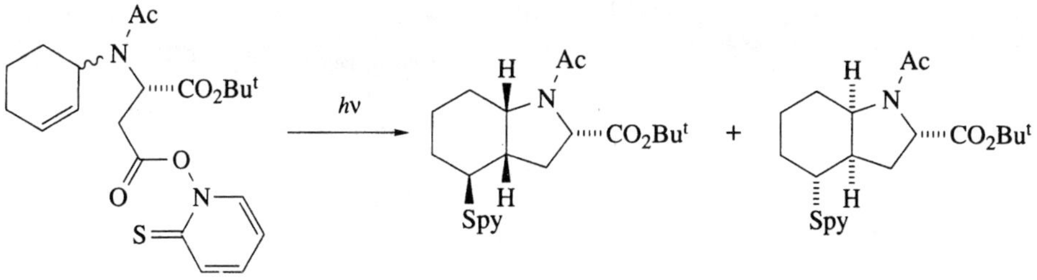

(2)

4.2.2.1.4 *The atom transfer method*

Cyclizations conducted by the atom transfer method offer a powerful alternative to tin hydride chemistry. Both hydrogen and halogen atom transfer cyclizations are possible; however, the scope of halogen atom transfer cyclizations is far greater. Although the mechanisms of hydrogen and halogen atom transfer cyclizations are similar, the same types of products are not always obtained. Halogen atom transfer cyclizations virtually always provide kinetically controlled products, but thermodynamic control can operate in hydrogen atom transfer cyclizations.

(i) Hydrogen atom transfer cyclizations

Preparative and mechanistic aspects of hydrogen atom transfer cyclizations were studied extensively by the group of Julia,[122] and the review of Surzur[4] provides an excellent overview of these pioneering contributions. Although most hexenyl radical cyclizations are not reversible, Julia demonstrated that the rate of reverse cyclization can be rapid enough to be preparatively useful if multiple radical-stabilizing groups are present on the initial radical center. These radical-stabilizing groups also facilitate hydrogen transfer, and the most important preparative use of the hydrogen atom transfer method is to obtain thermodynamically controlled products in reversible cyclizations. If kinetically controlled products are desired, it is usually better to prepare a halide (or related precursor) and to conduct the cyclization by either the tin hydride method or by halogen atom transfer.

The hydrogen atom transfer method is most useful for electrophilic radicals (for example, malonate, acetoacetate, *etc.*). Because radicals are generated from C—H bonds, the preparation of cyclization precursors by alkylation is routine. The hydrogen atom transfer method is very good for conducting slow cyclizations. In addition reactions, the hydrogen donor is typically used in large excess relative to the acceptor to facilitate H-transfer; however, cyclizations must use different conditions because the H-donor and the alkene acceptor are in the same molecule.

An example from the work of Julia,[123] which illustrates how reversible cyclizations can be conducted, is presented in Scheme 27. A solution of 2 equiv. of benzoyl peroxide in cyclohexane is added in portions over 2 d to a refluxing solution of cyanomalonate (**46**) in cyclohexane. After an additional 3 d at reflux, the solvent is removed and the 6-exo product (**47**) is isolated in 75% yield. Because a benzoyloxy radical produced by cleavage of the peroxide is highly reactive (and very nonselective), it probably abstracts a hydrogen from the solvent cyclohexane. The cyclohexyl radical is less reactive (and more selective), and it may abstract a hydrogen from (**46**) to form the initial radical (**48**). Thus, the selectivity of the cyclohexyl radical helps to offset the usual requirement for a large excess of hydrogen atom donor; however, the inefficiency of such C–H transfers necessitates the use of a very large amount of peroxide. It has recently been shown that (**48**) closes almost exclusively by 5-exo cyclization to (**49**),[19] but, because the ensuing hydrogen transfer is very slow, equilibration occurs to produce (**50**).[124] The source of the hydrogen atom in the final product is not certain. It may derive from the cyclohexane, or from (**46**), or from disproportionation. If chains are involved, they must be very short.

Substrates for hydrogen atom transfer cyclization are limited in the functional groups that they can bear because of the hydrogen atom transfer step. Even the allylic hydrogens in the acceptor are at risk of being abstracted. Further, such cyclizations are only useful when the kinetic cyclization product is different for the thermodynamic product (sometimes they are the same), and when the reverse cyclization is sufficiently rapid so that equilibrium is possible. Many reverse cyclizations are just too slow to be useful. Cyano groups appear to be good at increasing the rate of reverse cyclization, but even so, the conversion of (**49**) to (**48**) is very slow (see k_{-c} in Scheme 27). All the limitations not withstanding, hydrogen transfer cyclization can sometimes be the best method to obtain product ratios that are not kinetically controlled.[125]

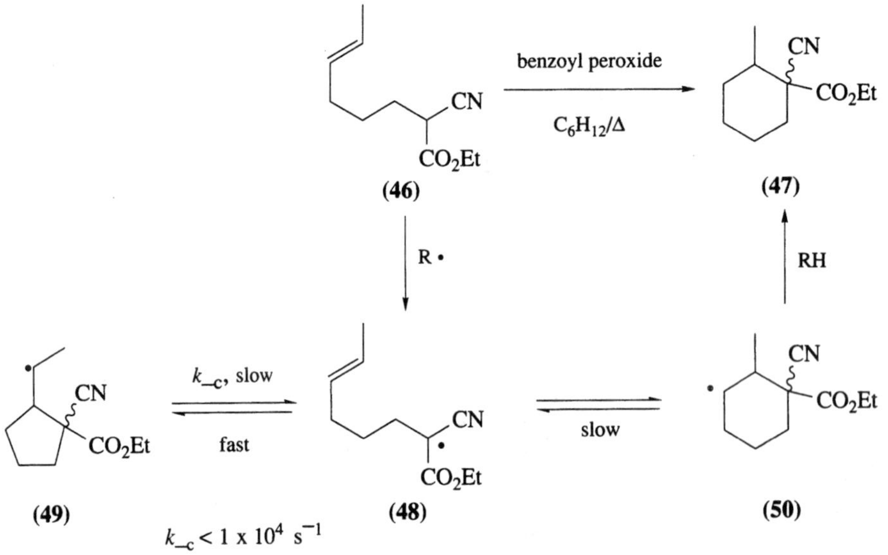

Scheme 27

(ii) Halogen atom transfer cyclizations

Halogen atom transfer cyclizations are conceptually similar to their hydrogen atom transfer counterparts, but, because halogen transfers are much more rapid, these cyclizations have a much broader scope and usually trap kinetic products efficiently.[34] Early examples of halogen transfer cyclization used perhalo-substituted radicals. In such cases, any halogen can serve as the radical precursor, and, while certain functional groups (like trichloroacetates) are very readily prepared, the incorporation of a polyhalogenated radical precursor in a complex molecule can sometimes be difficult. Polyhalogenated radicals are electrophilic and their cyclizations are usually promoted by metal additives. Any exothermic iodine atom transfer reaction will be very fast, and the recent development of methods based on iodine atom transfer has considerably expanded the types of atom transfer cyclizations that can be conducted.[126] The iodine atom transfer method is excellent for iodocarbonyl and related derivatives, but it can also be used for some nucleophilic radical cyclizations. The halogen atom transfer method is excellent for conducting slow cyclizations at high concentrations because there is nothing present to intercept the initial radical. However, for the reaction to succeed, it is critical that the halogen atom transfer step be sufficiently rapid to propagate the chain. Final products either retain the original halogen atom (in a new location) or are derived from subsequent ionic reactions of the initial products.

Several recent examples of metal-promoted cyclizations of perchlorocarbonyl compounds are presented in Scheme 28, and a full paper by Weinreb is recommended as an excellent source of references to prior work in this area (including mechanistic studies on the role of the metal).[127] The first two examples illustrate that the choice of substrates can dictate the types of products that are formed: the initially formed γ-chloro esters are stable to subsequent ionic reactions, but the *cis*-γ-chloro acids form lactones. Interestingly, Weinreb has shown that the metal can equilibrate the *cis*- and *trans*-γ-chloro esters by reversible chlorine atom transfer. The third example[128] illustrates a general feature of the atom transfer method: yields at high concentration are comparable to (and sometimes better than) those provided by using tin hydride at low concentrations. Indeed, in the third example, the three chlorines on the ester provided three opportunities for cyclization during the tin hydride reduction, but 40% of the product still failed to cyclize. (Unfortunately, the tin hydride concentration was not specified.)

Iodocarbonyls are excellent substrates for atom transfer cyclization, as shown by examples from our recent work in Scheme 29.[19,129] When two carbonyl (or cyano) groups are present, bromides can also serve as radical precursors. Photolysis with 10% ditin usually provides excellent yields of kinetic products at high concentration, and alkene substituents often dictate the regioselectivity. The γ-iodo ester products are particularly versatile for subsequent transformations, which can often be conducted *in situ*. Although tertiary iodine products sometimes go on to give lactones or alkenes, primary and secondary iodides can often be isolated if desired. The last example is particularly noteworthy: the kinetic product from the cyclization presented in Scheme 27 is trapped, because bromine atom transfer is much more rapid that reverse cyclization.

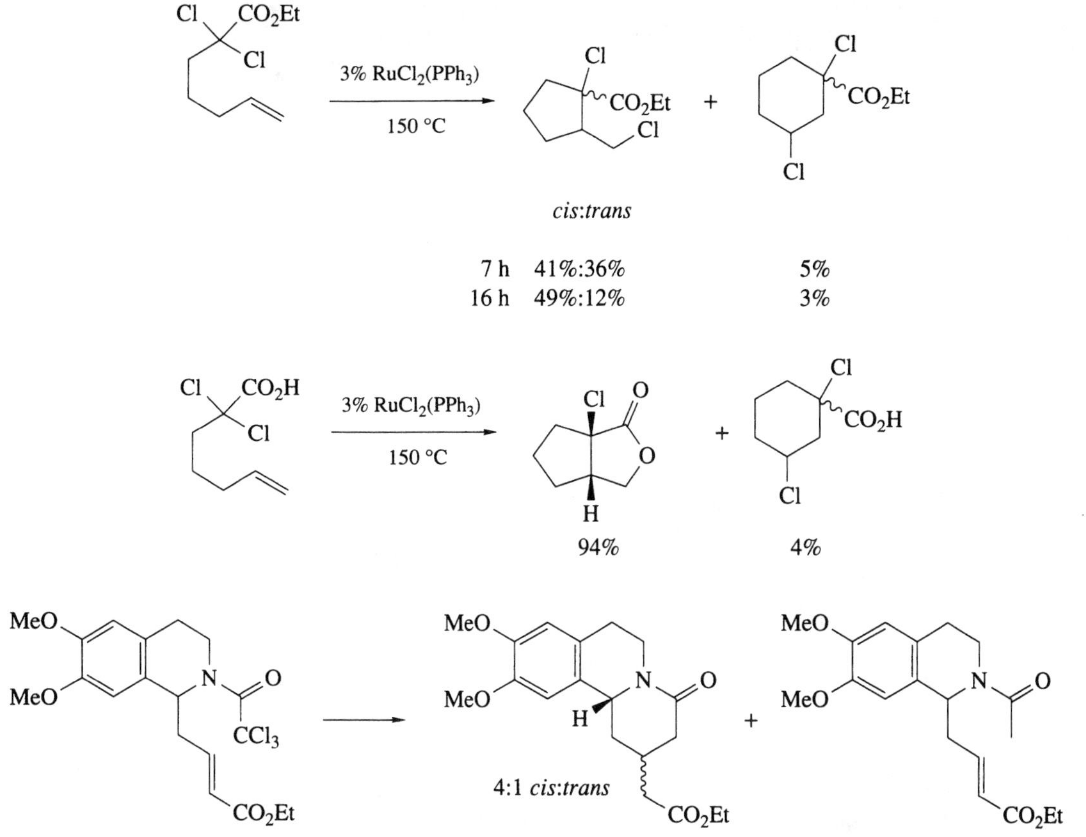

3% $RuCl_2(PPh_3)$, 150 °C

cis:*trans*

7 h	41%:36%	5%
16 h	49%:12%	3%

3% $RuCl_2(PPh_3)$, 150 °C

94% 4%

4:1 *cis*:*trans*

Bu_3SnH (unspecified concentration)	97%,	60:40
or		
i, CuCl, 140 °C; ii, Bu_3SnH	98%,	~100:0

Scheme 28

Iodine atom transfer reactions between alkyl radicals and iodocarbonyls are very rapid (10^7 M^{-1} s^{-1} to 10^9 M^{-1} s^{-1}).[130] This means that, even when these iodides are cyclized by the tin hydride method, iodine atom transfer may supersede hydrogen transfer, and the reductively cyclized product will ultimately be derived from the reduction of a cyclic iodide. Tin hydride cyclizations of halocarbonyls also often require very low concentration to avoid reduction of the initial radical prior to cyclization. For these reasons, reductively cyclized products are best formed by atom transfer cyclization at high concentration, followed by reduction of the product *in situ*. In a recent full paper, we have described in detail the preparative and mechanistic features of these cyclizations,[19] and Jolly and Livinghouse have reported a modification of our reaction conditions that appears to be especially useful for substrates that cyclize very slowly.[131] Cyclizations of α-iodocarbonyls can also be promoted by palladium.[132]

Cyclizations of nucleophilic radicals by the atom transfer method are more restricted in scope. Iodine transfer usually is too slow to be useful when the cyclic radicals are more stable than the initial radicals (as is often the case with alkyl radical cyclizations), and iodide precursors of stabilized nucleophilic radicals (α-iodo ethers, for example) are often too unstable to use. However, Scheme 30 illustrates that there are at least two types of atom transfer cyclizations that are preparatively useful. In the first type, we have found that the isomerization of hex-5-ynyl iodides to iodomethylenecyclopentanes is a very general reaction, which is made possible by the rapid, exothermic abstraction of iodine atoms from alkyl iodides by vinyl radicals.[133,134] Indeed, this transfer is so rapid that normal tin hydride reductive cyclizations of hexenyl iodides can proceed through iodomethylenecyclopentane intermediates. In the second type, the exchange of iodine atoms among alkyl radicals[135] is sufficiently rapid that isomerizations based on slightly exothermic (or even thermoneutral) iodine atom transfers are possible. The second example in Scheme 30 is representative of this type of reaction.[136] Even though the iodine transfer is reversible, the cyclization is not, and so the acyclic iodide is eventually isomerized to the cyclic iodide. Modest yields

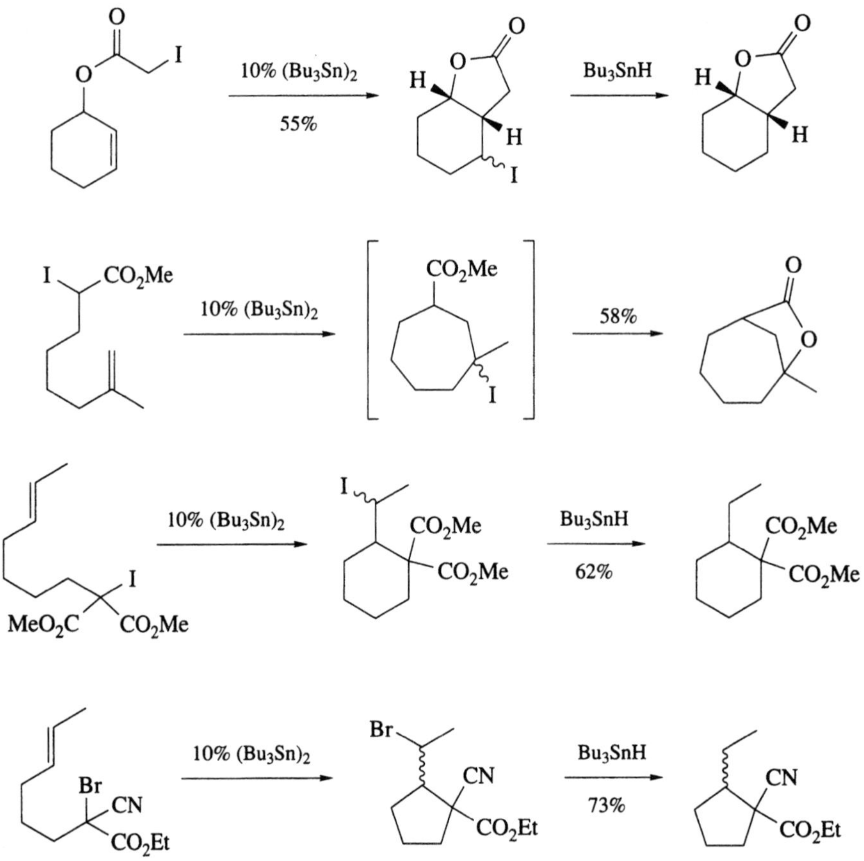

Scheme 29

are the rule because the product is continuously recycled to the radical pool, subjecting it to radical/radical and radical/solvent reactions. In such cases, higher yields of reduced products can probably be obtained by the tin hydride method (by using low concentration, if necessary).[137] But the atom transfer method may still be useful because the iodide in the product is suitable for further functionalization. Cyclizations like those outlined in Scheme 30 must be conducted with iodides because alkyl bromides are not sufficiently good bromine atom donors.

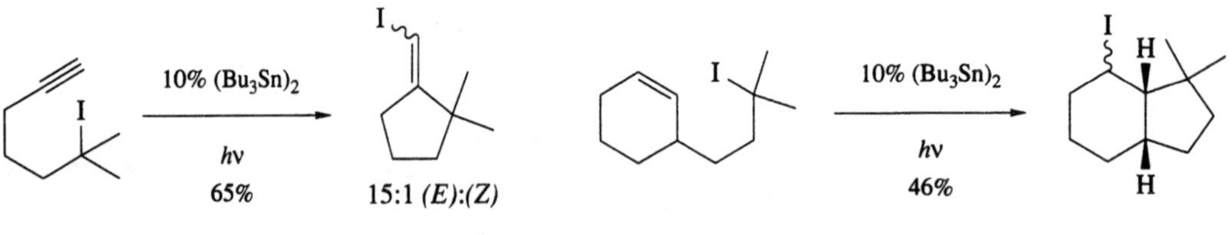

Scheme 30

(iii) Aromatic diazonium salts

Cyclizations of aromatic diazonium salts[138] (intramolecular Meerwein arylations) are preparatively related to atom transfer reactions because a radical cyclization is terminated by the transfer of an atom or group other than hydrogen. However, the two methods are not mechanistically related. In the atom transfer method, the atom that is transferred to the cyclic product always derives from the radical precursor, but in the cyclizations of aryldiazonium salts, the atom or group transferred derives from an added reagent. This means that many different products can be prepared from a single diazonium precursor, but it also means that the standard competition is operative, and initial radicals are at risk of being intercepted by reagents prior to cyclization. Illustrative examples of cyclizations of aromatic diazonium salts taken from the recent work of Beckwith are presented in equation (3).[139–141]

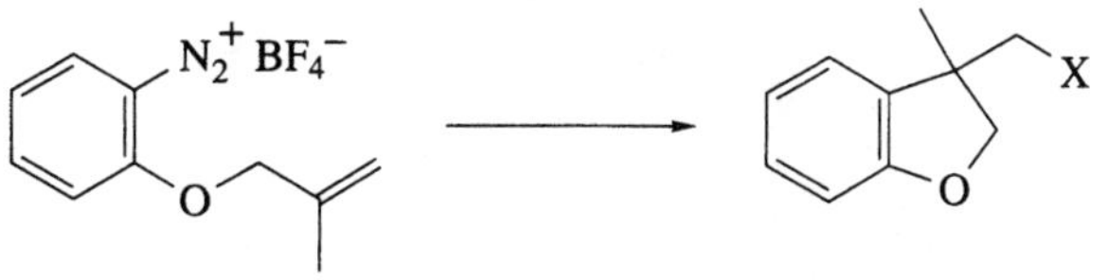

(3)

	NaI	CuCl	CuBr	CuCN	NaSPh
X	I	Cl	Br	CN	SPh
%	89	63	60	40	60

4.2.2.2 Nonchain Methods

Cyclization reactions can be conducted by methods that remove cyclic radicals by selective radical/radical coupling, oxidation, or reduction. The usual selectivity concerns are operative: initial radicals must cyclize, and cyclic radicals must be productively removed.

4.2.2.2.1 Selective radical/radical coupling

Becking and Schäfer have shown that mixed Kolbe coupling reactions can provide useful yields (40–60%) of cyclic products.[142] In the example provided in equation (4), 1 equiv. of acid (**51**) and 4 equiv. of acid (**52**) are electrochemically cooxidized, and the cyclic cross adduct (**53**) is formed in 53% yield. Because the rates of oxidation of (**51**) and (**52**) are similar, the concentration of radicals derived from (**52**) is higher. Thus, radicals derived from (**51**) are more likely to cross couple than to self couple. The strength of the mixed Kolbe method is that two carbon–carbon bonds are formed rather than one because the cyclic radical is removed by radical/radical coupling.

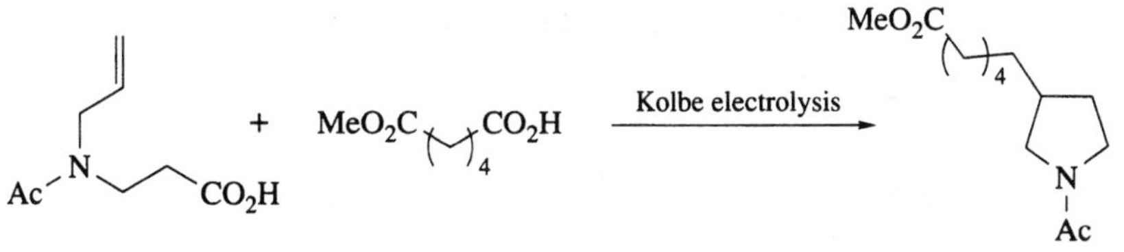

$$\text{(51)} + MeO_2C(CH_2)_4CO_2H \xrightarrow{\text{Kolbe electrolysis}} \text{(53)} \quad (4)$$

(**51**) 1 equiv. (**52**) 4 equiv. (**53**) 53% based on (**51**)

The cyclization reactions of organocobalt complexes are very useful, and they offer an excellent alternative to the tin hydride method when reduced products are not desired. Most cobalt cyclizations have been conducted with nucleophilic radicals. Precursors are prepared by alkylation of cobalt(I) anions, and are usually (but not always) isolated. One suspects that alkylcobalt precursors should be useful for slow cyclizations because there are no rapid competing reactions that would consume the initial radical (coupling of the initial radical with cobalt(II) regenerates the starting complex).

The most important feature of organocobalt cyclizations is that a variety of functionalized products can be obtained, depending on the nature of the substrate and the reaction conditions. The most common transformation has been formation of an alkene by cobalt hydride elimination. Alkenes are often formed *in situ* during the photolysis, and with activated alkene acceptors the formation of these products by cobalt hydride elimination is very facile. Scheme 31 provides a representative example from the work of Baldwin and Li.[143] The alkene that is formed by cobalt hydride elimination maintains the correct oxidation state in the product (**54**) for formation of the pyrimidone ring of acromelic acid. Under acidic conditions, protonation of the cyclic organocobalt compound may compete;[144] however, if protonated products are desired, the cyclization can probably be conducted by the reductive method with only catalytic quantities of cobalt (see Section 4.2.2.2.2).

When simple terminal alkenes are used as acceptors, the cyclic primary alkyl cobalt species are stable, and can often be isolated and purified by standard techniques.[145] Scheme 32 shows some of the transformations that Pattenden has accomplished with the cyclic alkylcobalt complex (**55**).[146] In addition to standard elimination to an alkene, the complexes can be converted to alcohols, halides, oximes, and phe-

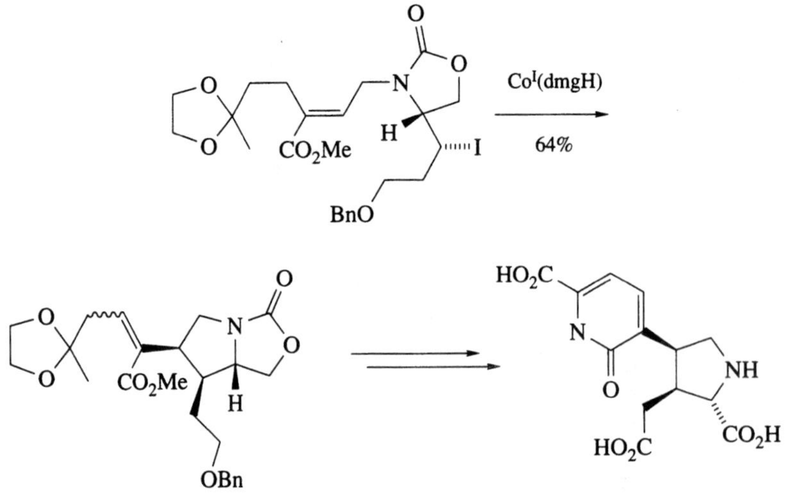
Co^I(dmgH)
64%
CO$_2$Me
BnO
OBn
HO$_2$C
NH
CO$_2$H

(54) Acromelic acid

Scheme 31

nyl sulfides and selenides. Of course, complexes like (**55**) are also precursors for subsequent radical reactions.

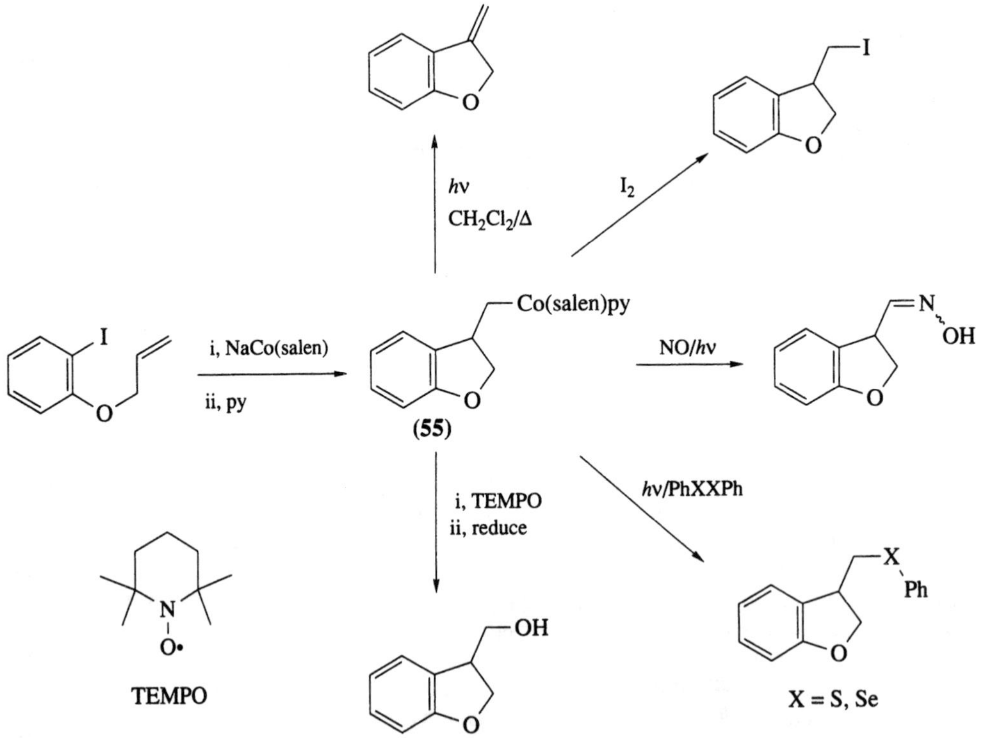
i, NaCo(salen)
ii, py
Co(salen)py
(55)
hν
CH$_2$Cl$_2$/Δ
I$_2$
NO/hν
hν/PhXXPh
i, TEMPO
ii, reduce
TEMPO
X = S, Se

Scheme 32

4.2.2.2.2 Redox methods

Among the oxidants that have been used to generate radicals, manganese(III) acetate has emerged as a powerful reagent to mediate radical cyclizations.[147] The manganese(III) acetate-mediated oxidation of enolizable carbonyl compounds is one of the best methods available for the cyclization of electrophilic radicals. The substrates are very easily prepared by standard alkylation and acylation reactions. Radicals are formed with high selectivity by oxidation of acidic C—H bonds, and, because the reaction is an oxidation, the product actually has a higher level of functionality than the starting material. While cyclic radicals are readily oxidized by manganese(III) if they are tertiary or heteroatom-substituted, the initial electrophilic radicals are inert to oxidation. Therefore, the manganese(III) method is excellent for slow cyclizations. In addition to the usual 5-exo cyclizations, 6-endo, 6-exo, 7-endo, and even 8-endo[148] cyclizations can be conducted depending on substituents. Substrates should be somewhat resistant to acid because the usual solvent is acetic acid. Standard extractive work-up procedures are the norm.

When one of the activating groups is an ester or acid, the cyclic cation is often trapped as a lactone.[149,150] As illustrated by the first two examples in Scheme 33,[151,152] this results in the simultaneous formation of two rings. The oxidation of secondary radicals to cations by Mn^{III} is much slower than that of tertiary radicals. However, addition of Cu^{II} salts smoothly (and often regioselectively) promotes the formation of alkenes, probably through copper hydride elimination from an intermediate alkylcopper species. The last two examples in Scheme 33, from the extensive studies of Snider,[148,153] show that this is an excellent technique for conducting other cyclizations besides the most common 5-exo variety. It appears that the regiochemistry of these cyclizations is controlled mainly by the alkene substitution pattern. As would be expected for free acetoacetate radicals, terminal alkenes provide significant amounts of endo-cyclized products, and Snider has provided evidence that tertiary radicals (but not secondary) formed by Mn^{III} oxidation are indeed free, rather than metal complexed.[153] The ratios of cyclic products are in line with expectations for irreversible cyclizations.

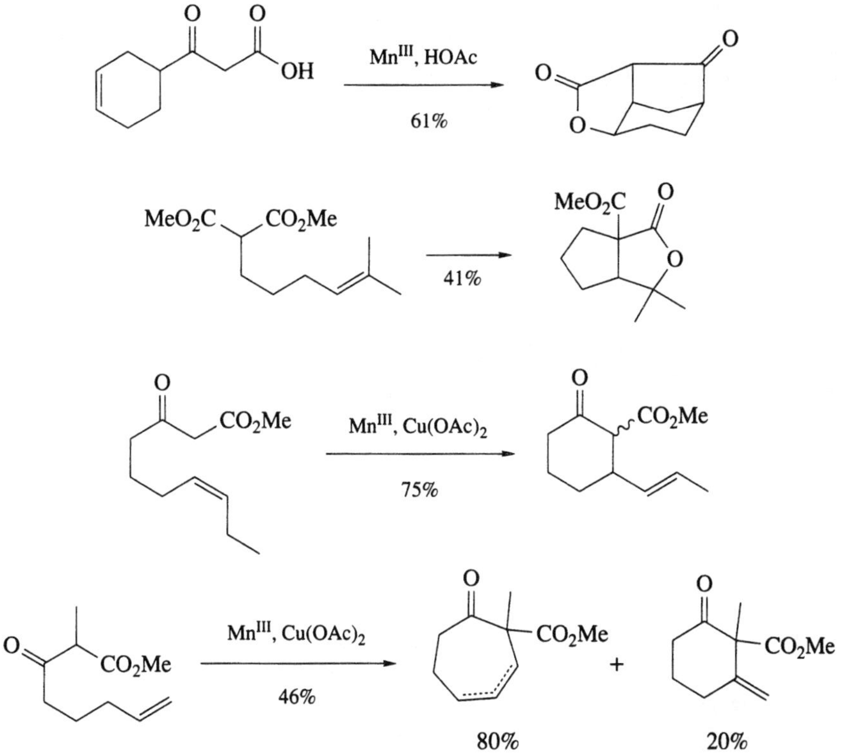

Mn^{III}, HOAc; 61%

MeO_2C, CO_2Me; 41%

Mn^{III}, $Cu(OAc)_2$; 75%

Mn^{III}, $Cu(OAc)_2$; 46%; 80% + 20%

Scheme 33

If the products contain an enolizable hydrogen, overoxidation is a concern. When cyclopentanones are involved, the cyclic products are often more rapidly oxidized than the starting materials. However, further oxidation of the cyclic products is not necessarily undesirable. For example, Snider has developed a simple synthesis of phenols in which a manganese(III)-mediated 6-endo radical cyclization is followed by oxidative aromatization.[154] An example of this process is provided in Scheme 34. Snider has also shown that the overoxidation can be blocked with a chlorine atom, as shown by the second example in Scheme 34.[153]

Even though radical intermediates are involved in many reductive metallations, cyclizations of alkyl radicals by reductive trapping methods are relatively uncommon. When protonated products are formed, the overall transformations are similar to a tin hydride reductions; however, reductive methods may facilitate purification of the cyclic product because tin by-products are absent. Two examples of simple reductive cyclizations are shown in Scheme 35. The first example illustrates that Sheffold's procedure, which uses catalytic quantities of vitamin B_{12} with a chemical or electrochemical coreductant, is attrac-

i, Mn^{III}, LiCl

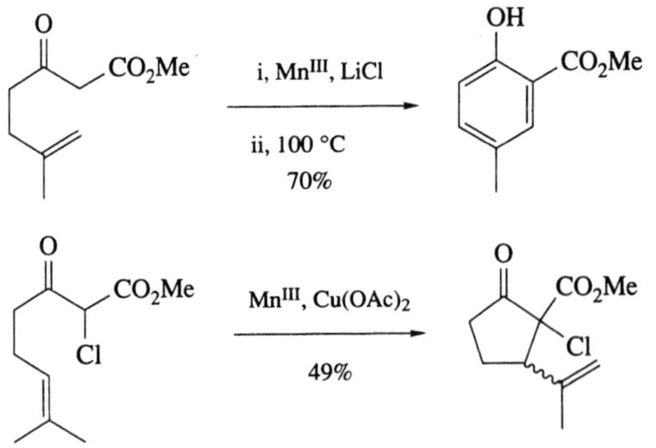

ii, 100 °C

70%

Mn^{III}, $Cu(OAc)_2$

49%

Scheme 34

tive for preparative cyclizations.[155] The second example illustrates that Cr^{II} salts are excellent reductants.[156] In this case, the $Cr(ClO_4)_2$ must be added slowly to the substrate so that the initial radical is not reduced to an alkylchromium(III) complex prior to cyclization (this is a general problem when powerful reductants are employed). Although alkyl radicals derived from such cyclizations can be reduced to alkylmetal complexes,[157] vinyl (and phenyl) radicals (such as those produced in the second example) often abstract a hydrogen atom from the solvent more rapidly than they are reduced.

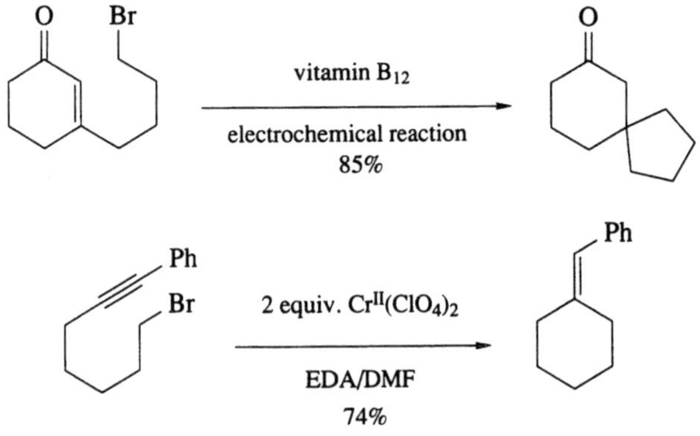

vitamin B_{12}

electrochemical reaction

85%

2 equiv. $Cr^{II}(ClO_4)_2$

EDA/DMF

74%

EDA = ethylenediamine

Scheme 35

If reduction of the cyclic radical is more rapid than hydrogen abstraction from the solvent, then organometallic intermediates are produced. That these can be trapped by electrophiles other than protons significantly extends the power of the reductive method. Two examples of sequences of (i) radical cyclization, (ii) reduction of the cyclic radical to an organometallic, and (iii) electrophilic trapping are provided in Scheme 36.[158,159]

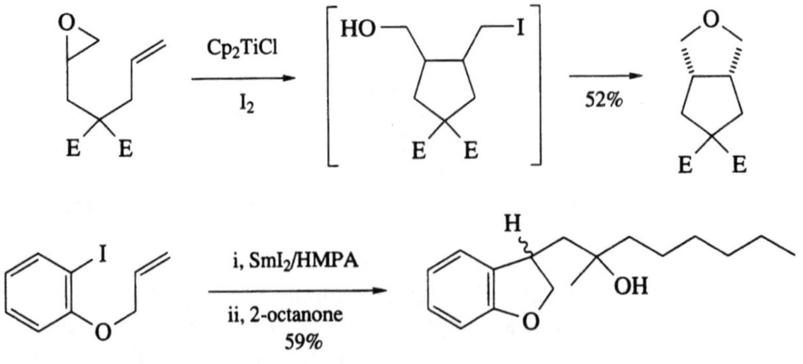

Cp_2TiCl

I_2

52%

i, SmI_2/HMPA

ii, 2-octanone

59%

Scheme 36

The most common type of reductive cyclization involves the conversion of an aldehyde or a ketone to a ketyl. Ketyls are radical anions, and, while dimerization to form pinacols is a common reaction, it does not occur at the diffusion-controlled rate because of charge repulsion. Nonetheless, pinacol coupling can sometimes compete with cyclization, as can reduction to give an alcohol. As illustrated by the examples in Scheme 37, ketyl cyclizations can be conducted in several ways. Chemical reductions are the most common techniques, and these are conducted like standard metal reductions.[160] Samarium(II) iodide is a particularly useful reagent due to its high reducing power and its solubility in common solvents. Molander has shown that the degree of asymmetric induction in samarium-mediated ketyl cyclizations is often very high.[161] Electrochemical reductions are also useful, and can be further facilitated by the presence of soluble catalysts that aid in electron transfer.[162] Cossy has shown that photolysis of carbonyls in the presence of HMPA or triethylamine is an especially simple preparative method to conduct ketyl cyclizations.[163] Levels of stereocontrol in ketyl cyclizations are usually much higher than in the corresponding cyclizations of α-oxy radicals; the alkene-derived substituent is oriented *trans* to the ketyl oxygen in the final product.

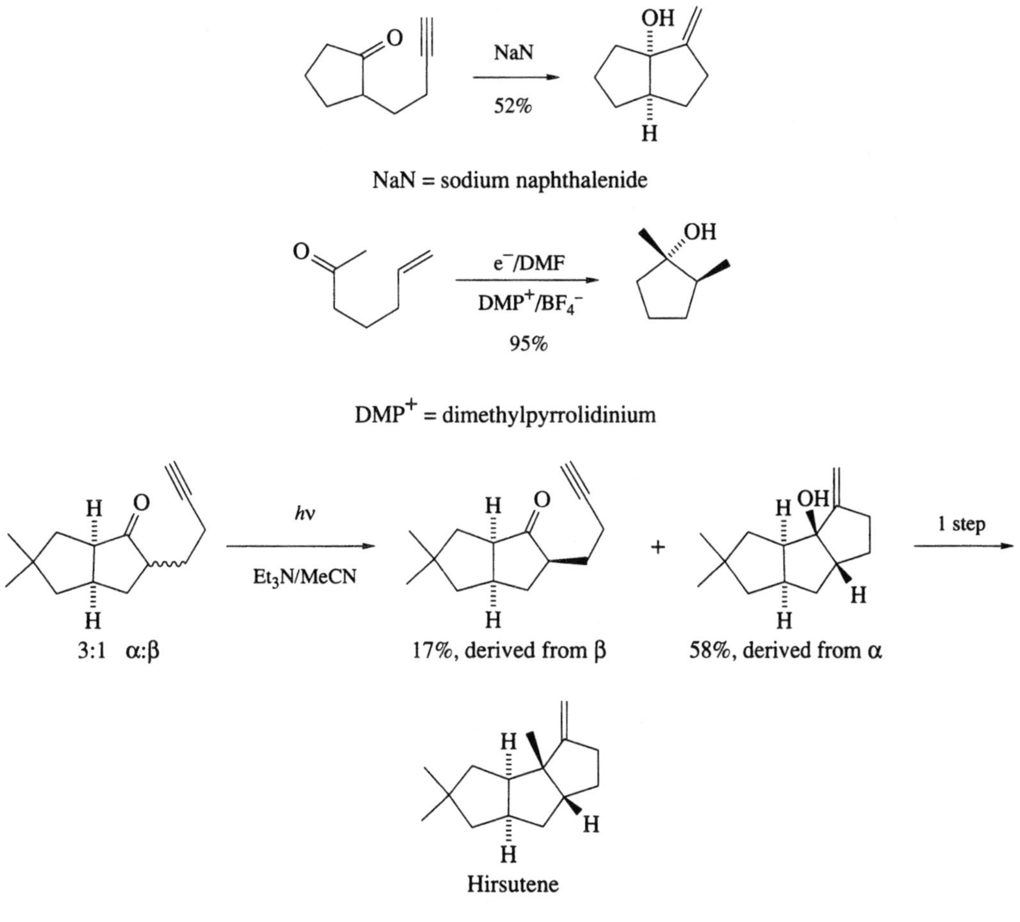

NaN
52%

NaN = sodium naphthalenide

e^-/DMF
DMP^+/BF_4^-
95%

DMP^+ = dimethylpyrrolidinium

3:1 α:β
hν
Et_3N/MeCN
17%, derived from β
58%, derived from α
1 step

Hirsutene

Scheme 37

Cyclizations of carbonyl-containing α,β-unsaturated esters like those illustrated in Scheme 38[164,165] are conceptually very similar to ketyl cyclizations, but they may be very different mechanistically. At least for electrochemical cyclizations, Little has proposed, based on reduction potentials, that the unsaturated ester is reduced first.[164] Even though questions remain about the timing of electron transfer and proton transfer steps, such cyclizations often provide good yields of products.

4.2.3 CYCLIZATIONS OF CARBON-CENTERED RADICALS TO AROMATIC RINGS

The recent resurgence in the application of radical reactions to organic synthesis has provided only a few examples of additions to aromatic rings; however, radical cyclizations should be applicable for con-

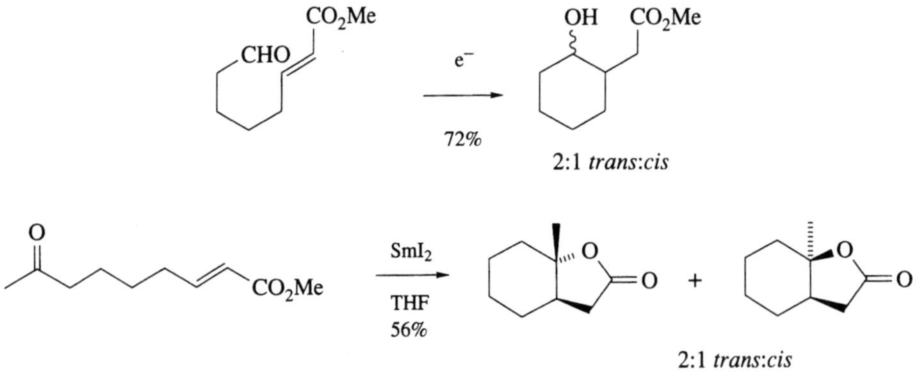

CO_2Me CHO e⁻ 72% OH CO_2Me 2:1 *trans*:*cis*

O CO_2Me SmI_2 THF 56% O O + O O 2:1 *trans*:*cis*

Scheme 38

structing many types of fused aromatic rings. Most such cyclizations occur under oxidative conditions that effect the rearomatization of intermediate cyclohexadienyl radicals. Several recent examples of additions to aromatic rings are provided in Scheme 39.[166,167] Unlike cyclizations with alkenes, cyclizations to aromatic nuclei that form six-membered rings are very good. Indeed, Citterio has shown that malonyl radicals add more rapidly to aromatic rings if six-membered rings are formed than if five-membered rings are formed.[168] As with addition reactions, one must be concerned that cyclizations to aromatic rings could be reversible, especially when stabilized initial radicals are involved.

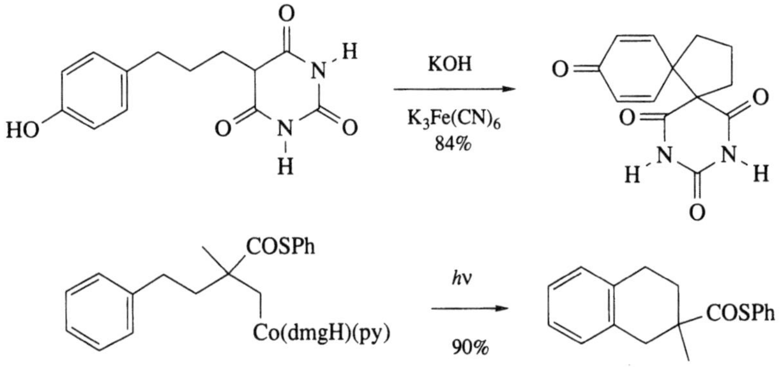

HO O N H O N H O KOH $K_3Fe(CN)_6$ 84% O O O O N H N H O

COSPh Co(dmgH)(py) *h*ν 90% COSPh

Scheme 39

Reductive cyclizations to aromatic rings are less common. The first example in Scheme 40 illustrates that cyclizations of ketyls to aromatic rings can be performed in good yields with high levels of stereocontrol.[169] The intermediate cyclohexadienyl radical is reduced and protonated to form a 1,4-cyclohexadiene. Tin hydride can also be used to conduct cyclizations to aromatic rings. And even though tin hydride is a reducing agent, rearomatized products are usually isolated, as indicated by the other example in Scheme 40.[170,171] The mechanism that converts the presumed intermediate cyclohexadienyl radical to the aromatized product is not clear.

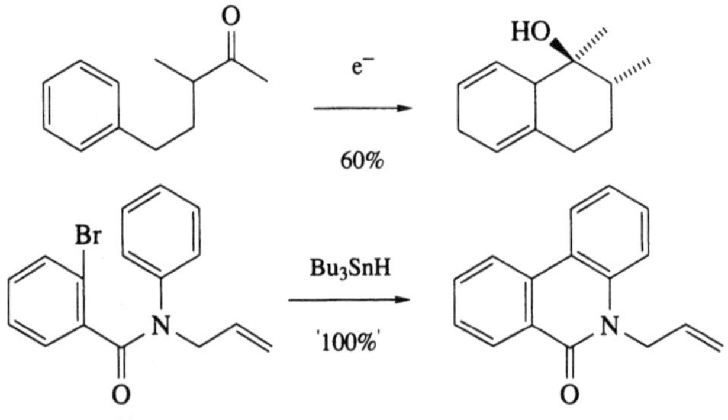

O e⁻ 60% HO

Br N O Bu_3SnH '100%' N O

Scheme 40

Although the bimolecular additions of aryl radicals to aromatic rings are not useful due to a lack of selectivity, the corresponding cyclizations are useful for constructing biaryl bonds. Equation (5) provides a generalized example of the Pschorr reaction.[172] This cyclization of a diazonium salt is one of the classical methods for biaryl bond formation, and there are a variety of related reactions.[138]

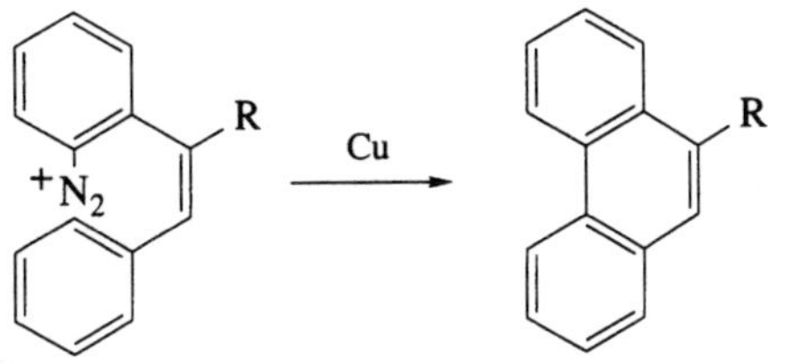

$$\xrightarrow{\text{Cu}} \qquad (5)$$

4.2.4 CYCLIZATIONS OF OXYGEN- AND NITROGEN-CENTERED RADICALS

Cyclizations of heteroatom-centered radicals offer alternatives to standard ionic methods for forming heterocyclic rings. Among the many heteroatom-centered radicals, those centered on nitrogen and oxygen are the most important for synthetic applications. There are significant differences between carbon-centered radicals and their nitrogen- and oxygen-centered counterparts, and these differences must be considered in synthetic planning. In general, oxy radicals are much more reactive than carbon radicals. In contrast, aminyl radicals are less reactive than carbon radicals, but their reactivity can be increased by protonation or by metal complexation. As always, consideration of the factors that will permit selective trapping of the cyclic radical without premature trapping of the initial radical is essential.

An important factor that provides insight into the reactivity of oxy and aminyl radicals is reactivity towards tin hydride. Figure 4 lists approximate rate constants for abstraction of a hydrogen atom from tributylin hydride by alkyl, aminyl,[173] and alkoxyl radicals.[174] The differences in reactivity are striking, and they do not parallel the order of elements in the periodic table: aminyl radicals are much less reactive than alkyl radicals, but alkoxyl radicals are much more reactive. This reactivity pattern follows the X—H bond strengths. Due to the high electronegativity of oxygen, alkoxyl radicals are very reactive in hydrogen abstraction, and the resulting OH bonds are very strong. Even though nitrogen is more electronegative than carbon, it forms weaker bonds to hydrogen. This is because electron–electron repulsions between bonded electron pairs and lone pairs are greater than repulsions between bonded electron pairs and other bonded electron pairs. This effect, which raises the ground state energy of an amine (as opposed to lowering the energy of an aminyl radical), is apparently more important than electronegativity in determining reactivity of aminyl radicals. Such large differences in reactivity can be very important for preparative applications. It is not uncommon (see below) to have alkyl radicals present in equilibrium with aminyl or alkoxyl radicals. If tin hydride is present, it provides a factor of 50–100 bias in rate that favors the trapping of alkoxyl radicals over carbon radicals, or the trapping of carbon radicals over aminyl radicals.

$R_3C\cdot$	$R_2\ddot{N}\cdot$	$R\ddot{O}\cdot$	
$k_H \approx 3 \times 10^6\ M^{-1}\ s^{-1}$	$8 \times 10^4\ M^{-1}s^{-1}$	$2 \times 10^8\ M^{-1}\ s^{-1}$	at 50 °C

Figure 4 Rates of hydrogen abstraction from Bu_3SnH

4.2.4.1 Nitrogen-centered Radicals

In contrast to the large body of kinetic data that is available on hexenyl radical cyclizations, relatively little is known about the cyclizations of 'azahexenyl' analogs. Only very recently have the detailed studies of Newcomb permitted a complete analysis of a useful parent cyclization.[175] As illustrated in equation (6), aminyl radical (**55**) cyclizes to (**56**) with a rate constant of about $3 \times 10^3\ s^{-1}$, but (**56**) reverts to (**55**) with a slightly greater rate constant of $7 \times 10^3\ s^{-1}$. That the cyclization of (**55**) is very slow, and that (**55**) is actually slightly preferred at equilibrium over (**56**), can be understood in terms of the above explanations for N—H bond strengths.

The slow rate of cyclization and the unfavorable equilibrium may cause problems for aminyl radical cyclizations. However, most chain substituents should accelerate such cyclizations, and in so doing, shift

Bu N• (55) $\underset{k_{-c}}{\overset{k_c}{\rightleftharpoons}}$ Bu N (56) (6)

$$k_c \approx 3 \times 10^3\ s^{-1};\ k_{-c} \approx 7 \times 10^3\ s^{-1}$$

the equilibrium to the cyclic product. Further, cyclic carbon-centered radicals can be trapped either kinetically (if the trapping reaction is faster than reverse cyclization), or at equilibrium. In either case, a reaction that traps alkyl radicals more rapidly than aminyl radicals is required. Aminyl radicals can be generated and cyclized by the halogen atom transfer method (N–halogen bonds are very weak), by the Barton method,[176] or by oxidation of metal amides.[177]

To facilitate the cyclizations of nitrogen-centered radicals, one need only increase the electronegativity at nitrogen. This makes the radical more electrophilic (which accelerates the cyclization), and strengthens the forming bonds (which shifts the equilibrium to the cyclic product). Amminium radical cations are more well-behaved in cyclizations than aminyl radicals, and these radical cations can either be formed directly from ammonium salts, or by protonation of aminyl radicals (which are relatively weak bases and require strong acids for protonation). Metal-complexed aminyl radicals are highly reactive, and amidyl radicals are also useful.[178]

The atom transfer method is most commonly employed for the cyclizations of *N*-haloamines, and pioneering work by the groups of Minisci[179] (in arylations), and Surzur and Stella[180] (in cyclizations) has shown that this is a powerful method for construction of N-heterocycles. These reactions are usually promoted by metals: popular procedures use $TiCl_3$, $FeSO_4$, or CuCl and $CuCl_2$. It is proposed that the aminyl radicals formed in some of these reactions are complexed rather than free.[180] The atom transfer mechanism operates, but it is not always clear whether the donor is the starting *N*-halide or a metal halide additive (for example, $CuCl_2$ is an excellent chlorine donor). Because *N*-haloamines are reactive sources of positive halogen, ionic products are also sometimes formed. The first example in Scheme 41 illustrates a typical radical reaction and the basic mechanism. The products of such reactions are nitrogen mustards that can be used to form additional bonds through the intermediacy of aziridinium ions (sometimes *in situ*). Recent work of Broka also shown in Scheme 41 illustrates how such combinations can lead to rapid construction of complex heterocyclic skeletons like the morphans.[181]

Recent work by Newcomb has resulted in the emergence of a second method for the cyclizations of amminium radical cations that shows good synthetic potential.[175,176] As illustrated in Scheme 42, *N*-hydroxypyridine-2-thione carbamates can be generated and cyclized by analogy to the Barton method (which uses thione esters). In Scheme 42, the preparation of the precursors and two variants on the cyclization, reductive trapping and thiopyridyl trapping are illustrated.

4.2.4.2 Oxygen-centered Radicals

In contrast to aminyl radicals, alkoxyl (and acyloxy) radicals are highly reactive. As illustrated in equation (7), their cyclization reactions are extremely rapid and irreversible. However, the rapidity of such cyclizations does not guarantee success because alkoxyl radicals are also reactive in inter- and intramolecular hydrogen abstractions, and β-fragmentations (see Section 4.2.5.2). This lack of selectivity may limit the use of alkoxyl radicals in cyclizations, but 5-exo cyclizations are so rapid that they should succeed in many cases, and other types of cyclizations may also be possible.

Alkoxyl radicals can be generated by a variety of methods including peroxide reduction, nitrite ester photolysis, hypohalite thermolysis, and fragmentation of epoxyalkyl radicals (for additional examples of alkoxyl radical generation, see Section 4.2.5.2). Hypohalites are excellent halogen atom donors to carbon-centered radicals, and a recent example of this type of cyclization from the work of Kraus is illustrated in Scheme 43.[182] Oxidation of the hemiketal (**57**) presumably forms an intermediate hypoiodite, which spontaneously cyclizes to (**58**) by an atom transfer mechanism. Unfortunately, the direct application of the Barton method for the generation of alkoxyl radicals fails because the intermediate pyridinethione carbonates are sensitive to hydrolytic reactions. However, in a very important recent development, Beckwith and Hay have shown that alkoxyl radicals are formed from *N*-alkoxypyridinethiones.[183] Al-

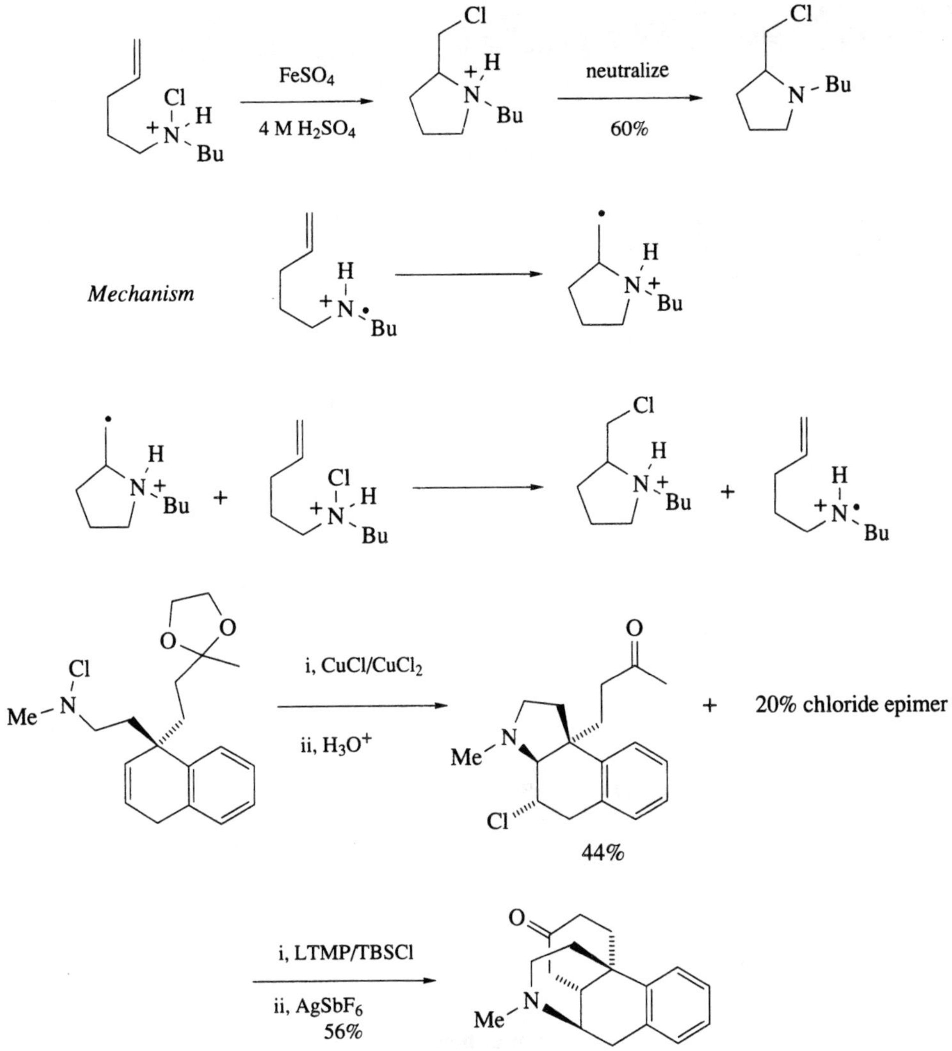

FeSO$_4$ / 4 M H_2SO_4 ; neutralize 60%

Mechanism

i, CuCl/CuCl$_2$; ii, H_3O^+ ; 44% + 20% chloride epimer

i, LTMP/TBSCl ; ii, AgSbF$_6$; 56%

Scheme 41

hν TFA/Bu^{t}SH 50% ; hν TFA 70%

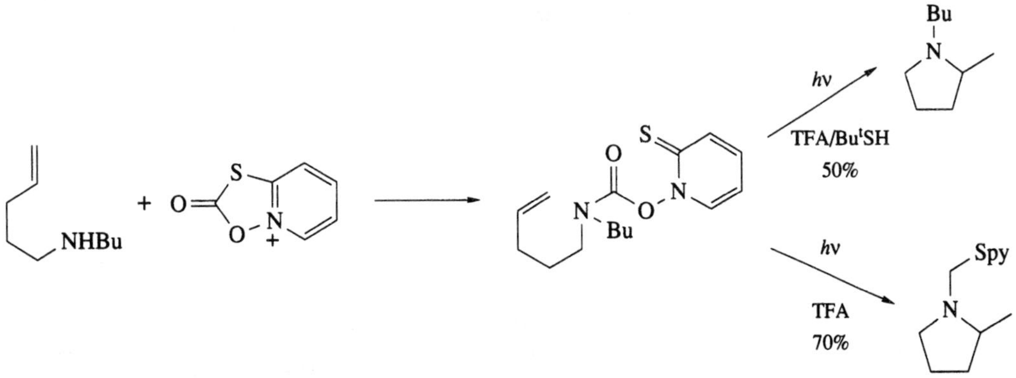

Scheme 42

though the only current example of a cyclization is the closure of the simplest precursor (**59**) to give methyl tetrahydrofuran (**60**), this method should allow access to Barton-like products in alkoxyl radical cyclizations. Precursors are formed by *O*-alkylation of *N*-hydoxypyridinethione salts, and a potential problem with such amibident nucleophiles is *S*-alkylation. In a related method, Fraser-Reid has recently shown that reductions of nitrate esters with tin hydride give products derived from alkoxyl radicals.[184,185]

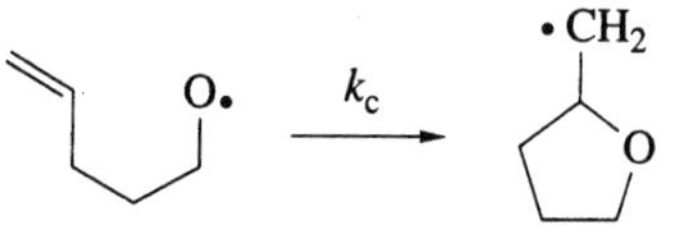

$k_c > 6 \times 10^8\ s^{-1}$ (7)

This is rather surprising because primary nitroalkanes (which would give more stable alkyl radicals) are not readily reduced by tin hydride. The third example in Scheme 43 illustrates a cyclization of an alkoxyl radical derived from a nitrate ester.

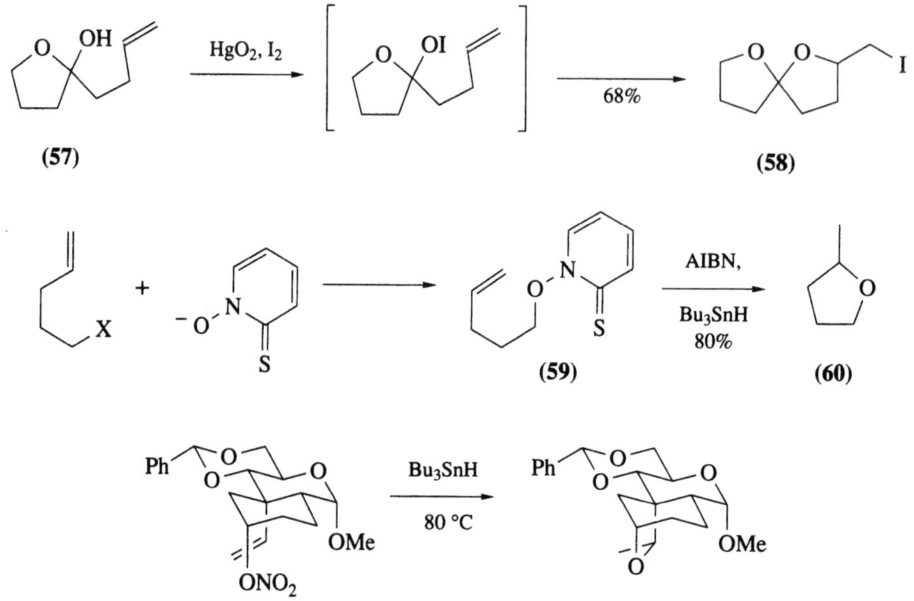

Scheme 43

There is another standard method for formation of heterocyclic rings from nitrogen- and oxygen-centered radicals that does not involve cyclization of a radical to a π-bond. Instead, δ-halo ethers or amines are formed by a chain reaction comprised of hydrogen and halogen atom transfer steps, and the ring closure occurs by ionic (S_N1 or S_N2) substitution. Two recent examples of this broad class of reactions are provided in Scheme 44.[186,187] When nitrogen-centered radicals are involved, this transformation is usually called the Hofmann–Löffler–Freytag reaction.[188]

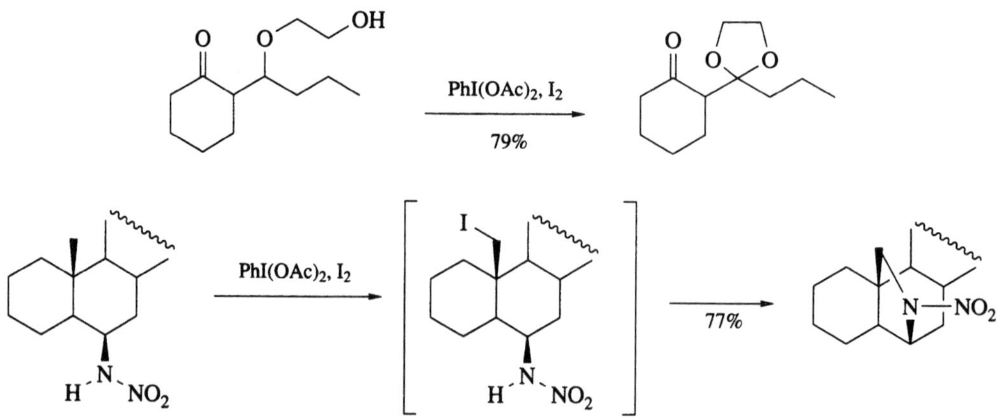

Scheme 44

4.2.5 ADDITIONS TO CARBON–NITROGEN AND CARBON–OXYGEN MULTIPLE BONDS, AND β-FRAGMENTATIONS

When constructing five-membered rings by radical cyclizations, it is required that the multiple bond acceptor be exocyclic to the forming ring. Thus, with C—C multiple bond acceptors, an exocyclic ring residue of at least one carbon atom is always produced. When this residue is not required, an alkyne acceptor is often used, and the resulting alkylidene chain is excised by ozonolysis (see Scheme 13). A more direct route to rings that do not possess carbon substituents at the site of cyclization uses a carbon–nitrogen or carbon–oxygen multiple bond as the acceptor.

4.2.5.1 Carbon–Nitrogen Multiple Bonds

Radical cyclizations to carbon–nitrogen multiple bonds resemble additions to carbon–carbon multiple bonds in that they usually give products of irreversible exo cyclization. To date, the most useful acceptors have been oximes[189] and nitriles,[190] and one example of each type of cyclization is given in Scheme 45.[191] Nitriles are useful because the intermediate imines are readily hydrolyzed by mild acid to ketones. Although this route to ketones is shorter than the two-step sequence of alkyne cyclization/ozonolysis, nitriles are slightly poorer acceptors than terminal alkynes, and much poorer acceptors than activated alkynes. Thus, when slow cyclizations are involved, the two-step protocol is preferable.

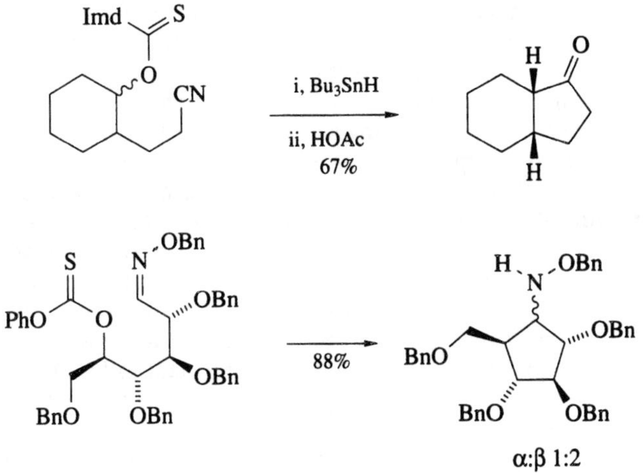

Imd S O CN i, Bu_3SnH ii, HOAc 67% H O H

S N OBn PhO O OBn OBn BnO OBn 88% H N OBn BnO OBn BnO OBn α:β 1:2

Scheme 45

4.2.5.2 Carbon–Oxygen Multiple Bonds

Radical cyclizations to carbon–oxygen multiple bonds are considerably different from cyclizations to C—C and C—N double bonds. Indeed, before the recent discoveries of Fraser-Reid and Tsang,[185,192–194] cyclizations to C–O multiple bonds were not considered generally useful, and only the reverse reaction (β-fragmentation of an alkoxyl radical) was common. However, a much better understanding of these reactions is beginning to emerge. As illustrated in Scheme 46, the cyclizations of radicals to carbonyl groups are best interpreted in the framework of the Curtin–Hammett principle.[195] A semiquantitative analysis of these reactions (introduced in a recent review)[196] has been greatly facilitated by the recent rate studies of Beckwith and Hay.[197]

Analysis of the behavior of the parent substrates for 5-exo cyclization (**61a**) and 6-exo cyclization (**61b**) is an instructive point of departure. As indicated in Scheme 46, these cyclizations are approximately equal in rate,[198] and both are more rapid than the cyclization of the hexenyl radical (see Scheme 2). The high rate of cyclization and the complete absence of *endo* products (which would be much more stable radicals) indicate the importance of FMO effects on rates: the carbonyl group has a low-lying LUMO with the large orbital coefficient on carbon. Unlike the hexenyl radical, the products of cyclization (**62a**) and (**62b**) can revert to the starting radicals. Indeed, both reverse cyclizations are faster than the forward cyclizations; that is, the starting radicals (**61a**) and (**61b**) are favored at equilibrium. That (**61a,b**) are more stable than (**62a,b**) is not surprising: the cyclization sacrifices a C═O bond (which is much stronger than a C═C bond that is lost in a hexenyl radical cyclization) and generates an alkoxyl radical (which is much less stable than the starting carbon radical). Because of the increased strain in the cyclopentane ring relative to cyclohexane, alkoxyl radical (**62a**) fragments much more rapidly than (**62b**).

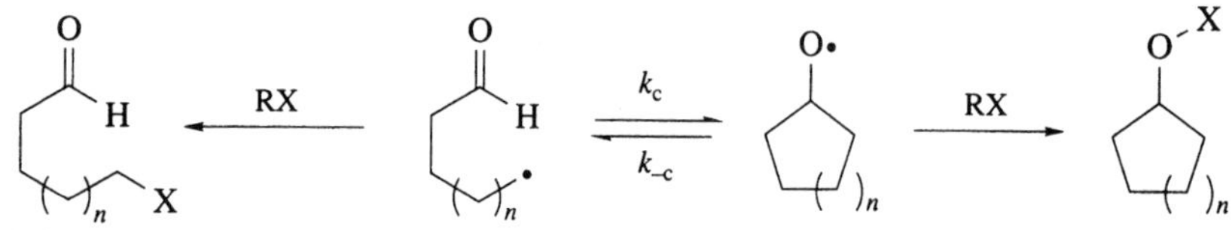

(**63a,b**) ← RX — (**61a**) $n = 1$, (**61b**) $n = 2$ $\underset{k_{-c}}{\overset{k_c}{\rightleftharpoons}}$ (**62a**), (**62b**) — RX → (**64a,b**)

at 80 °C, $n = 1$, $k_c = 9 \times 10^5\ s^{-1}$ $n = 2$, $k_c = 1 \times 10^6\ s^{-1}$

$k_{-c} = 5 \times 10^8\ s^{-1}$ $k_{-c} = 1 \times 10^7\ s^{-1}$

$K_{eq} \approx 0.002$ $K_{eq} \approx 0.1$

Scheme 46 Cyclization to carbonyls

Whether products derived from cyclization or fragmentation are isolated depends not only on the rates of forward and reverse cyclization, but also on the rates of trapping of the intermediate radicals to give (**63a,b**) or (**64a,b**). Several different scenarios are possible (the reader is referred to more quantitative discussions).[196,197] If cyclization is the desired pathway, a reagent (RX) that reacts much more rapidly with alkoxyl radicals than with carbon radicals is required. As mentioned in Figure 4, tin hydride is just such a reagent (RX = Bu_3SnH).[199] When forming six-membered rings, it is often possible to trap cyclic radicals (**62b**) with tin hydride more rapidly than they revert to (**61b**). Even if partial equilibrium occurs, an excellent yield of cyclic product is still possible because tin hydride reacts about 100 times more rapidly with (**62b**) than with (**61b**). Therefore, (**62b**) will be selectively trapped at equilibrium provided that the equilibrium constant is not too low (that is k_c is not too much smaller than k_{-c}). The formation of five-membered ring products is more difficult because the reverse cyclization of (**62a**) is so rapid. When tin hydride is used, its concentration must be sufficiently low to allow most of (**61a**) to close before it is directly reduced to (**63a**). At such concentrations (<0.1M), most of (**62a**) will revert to (**61a**) before it is trapped to give (**64a**). The increased reactivity of tin hydride toward (**62a**) over (**61a**) will not be sufficient to offset the equilibrium constant, which strongly favors (**61a**) because of the magnitude of k_{-c}. Thus, both (**63a**) and (**64a**) will be produced.

Even though there are no reaction conditions that can promote the high yield formation of (**64a**) in this simple example, the prospects for forming five-membered rings are far from hopeless. Most substituents along the forming ring should increase the equilibrium constant either by increasing k_c or decreasing k_{-c}, thus facilitating the trapping of cyclic products. The important experimental lesson from Scheme 46 is that cyclizations to aldehydes are fast, and low tin hydride concentrations are not necessarily required to obtain good yields of cyclic products.

Because equilibrium constants often favor open products, fragmentation reactions of alkoxyl radicals are easy to conduct and are generally useful. The atom transfer method is excellent for fragmentations. Alkoxyl radicals are formed from *in situ* generated hypohalites (RX = O—halogen). Hypohalite bonds are very weak, and thus oxygen-centered radicals (**62**) are not efficiently trapped, but carbon-centered radicals (**61**) abstract halogen very rapidly to give (**63**) (X = halogen). Peroxides are also good precursors for alkoxyl radical fragmentations. When radical-stabilizing groups are present, or when strained rings are involved, the fragmentation reactions are often so rapid that no reagent can trap the alkoxyl radical prior to its scission. Indeed, fragmentation reactions are especially valuable when conducted in sequence with other radical reactions because group transfers or ring expansions can result (see Section 4.2.6.2.1).

In cases where the fragmentation reactions are not prohibitively rapid, it may be possible to conduct either cyclization or fragmentation by selection of appropriate reagents and conditions. Scheme 47 provides an example where the same two intermediate radicals are involved.[200,201] The cyclized product is formed in the presence of tin hydride, but the fragmented product is produced under the atom transfer conditions.

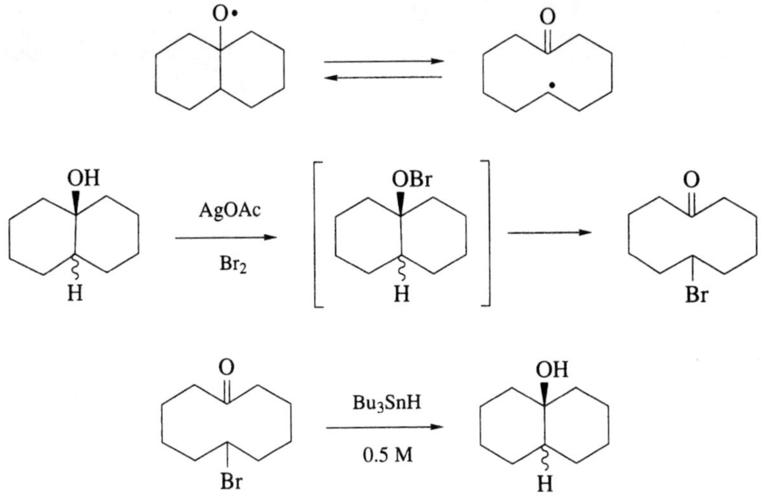

O•
O
OH
AgOAc
Br_2
H
OBr
H
O
Br
O
Br
Bu_3SnH
0.5 M
OH
H

Scheme 47

4.2.5.2.1 Cyclizations to aldehydes and ketones

Additions to carbonyls are probably restricted to aldehydes and ketones, and ketones will be less reactive than aldehydes because the extra alkyl group on a ketone is at the carbon that the radical is attacking. Carbonyl groups (like esters and amides) with resonance-stabilizing substituents are usually too unreactive to serve as acceptors. In contrast, fragmentations to give resonance-stabilized carbonyls will be very rapid. Two examples of cyclization to aldehydes from the work of Frasier-Reid and Tsang are presented in Scheme 48.[193,194] There is good evidence to indicate that the first cyclization is not reversible; tin hydride traps the cyclic radical before it can revert.[185]

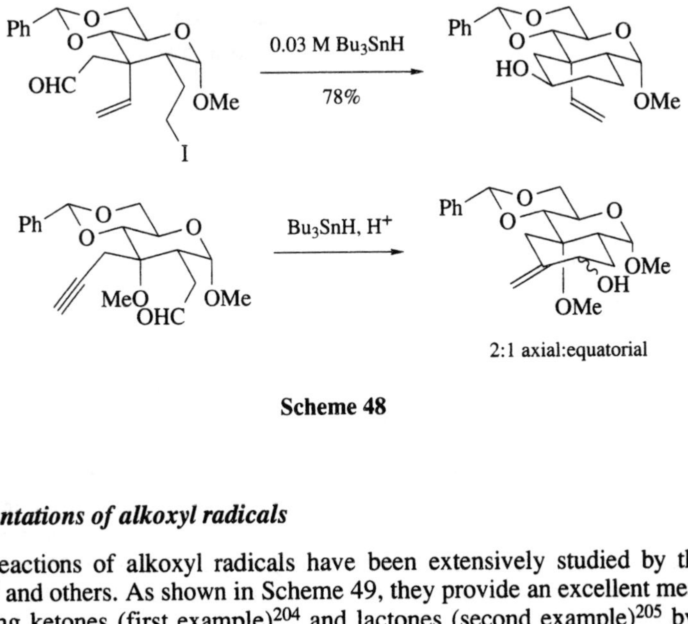

Ph O O O OHC OMe I
0.03 M Bu_3SnH
78%
Ph O O O HO OMe

Ph O O O MeO OHC OMe
Bu_3SnH, H^+
Ph O O O OMe OH OMe

2:1 axial:equatorial

Scheme 48

4.2.5.2.2 Fragmentations of alkoxyl radicals

Fragmentation reactions of alkoxyl radicals have been extensively studied by the groups of Suginome,[202] Suarez[203] and others. As shown in Scheme 49, they provide an excellent means for the preparation of medium ring ketones (first example)[204] and lactones (second example)[205] by scission of a ring fusion bond. Schreiber has shown that alkoxyl radicals generated from peroxides can fragment, and the resulting radicals can be trapped by Cu^{II}. The ensuing copper hydride elimination often proceeds with very high regio- and stereo-selectivity (third example).[206] The last example in Scheme 49 illustrates an important variant: when readily eliminated groups are present β to the cleaving bond, an additional fragmentation occurs to introduce an alkene whose stereochemistry is dictated by the configuration of the precursor.[207,208]

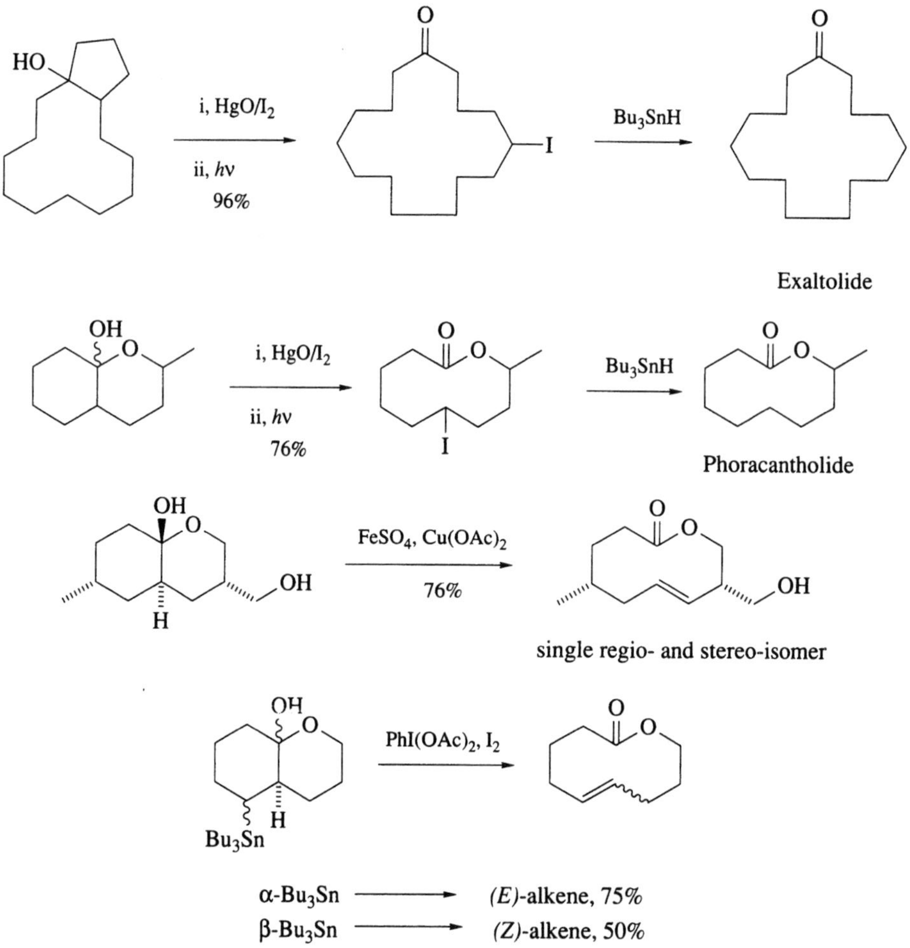

i, HgO/I_2
ii, $h\nu$
96%

Bu_3SnH

Exaltolide

i, HgO/I_2
ii, $h\nu$
76%

Bu_3SnH

Phoracantholide

$FeSO_4$, $Cu(OAc)_2$
76%

single regio- and stereo-isomer

$PhI(OAc)_2$, I_2

α-Bu_3Sn ⟶ (*E*)-alkene, 75%
β-Bu_3Sn ⟶ (*Z*)-alkene, 50%

Scheme 49

4.2.6 SEQUENTIAL RADICAL REACTIONS

Radical reactions are naturally suited to sequencing, because a new radical results from every radical addition, cyclization, or fragmentation. In a real sense, virtually every radical chain reaction (and many nonchain reactions as well) is a sequence of reactions in which several intermediate radicals are involved. However, for synthetic purposes, we define a radical sequence more narrowly as any reaction in which *more than one radical transformation occurs in the substrate, excluding steps that involve radical generation and radical removal.* By this definition, simple addition or cyclization reactions are not sequences because only one reaction (the addition or cyclization) occurs between radical generation in the substrate and radical removal. Within this definition, we classify two groups of radical sequences: simple sequences and tandem (and higher) sequences. In simple sequences, there are two or more radical precursors present in the radical substrate(s), and the two radical reactions occur separately. In tandem (and higher) sequences, the radical reactions are intimately linked because the precursor for the second radical reaction is generated by the first reaction. Aspects of sequential radical chain reactions have been discussed in a recent review.[34]

4.2.6.1 Simple Sequences

In simple sequences, reactions are conducted by repeating at least once the three stages of a radical reaction: (i) radical generation, (ii) addition, cyclization, or fragmentation, and (iii) radical removal. This means that two (or more) radical precursors will be required, and the only difference between a simple sequence and a standard cyclization or addition is that one must control the timing and location of the first set of radical reactions relative to the second. Because the transformations occur individually, there is no requirement that the two reactions of a simple sequence be conducted in the same step rather than in separate steps; however, the potential to conduct two important synthetic transformations together is attractive.

There are only a few examples of simple sequences in the literature, and the two provided in Scheme 50 illustrate different ways in which the timing of the individual steps in the sequence can be controlled. The first example from Bellus sequences an addition reaction prior to a cyclization by using the atom transfer method.[209] Because the initial radicals for both the addition and the cyclization are generated on the same carbon atom, there is no problem with selective radical generation. Selection between the two alkenes for the initial addition step is interesting: the dichloroacetate radical adds to the terminal alkene (even though this alkene is electron deficient) rather than the disubstituted alkene. The second example from our laboratory[210] illustrates the sequencing of an addition reaction, conducted by the Keck/Byers fragmentation method,[211] with a cyclization reaction, conducted by the tin hydride method. To ensure that the radicals are generated in the correct order, a highly reactive alkyl iodide is used as the radical precursor for the first step while a less reactive vinyl bromide serves as the precursor for the second step. The two reactions of this sequence can be conducted either together in one pot, or individually.

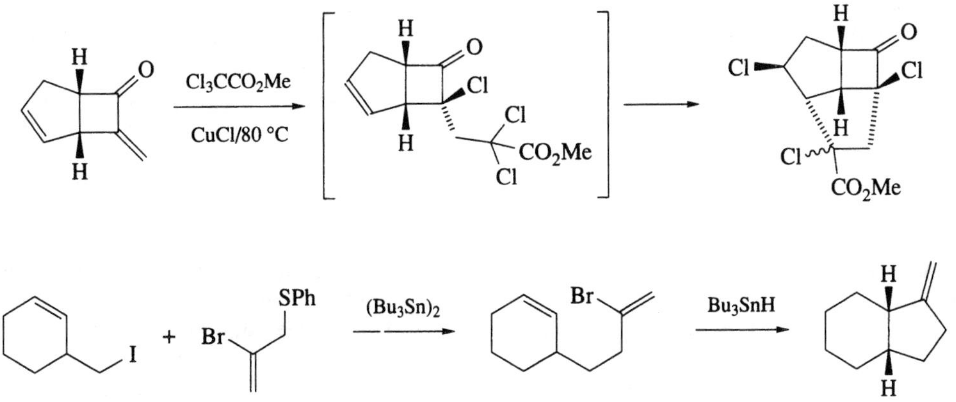

Cl_3CCO_2Me, CuCl/80 °C; $(Bu_3Sn)_2$; Bu_3SnH

Scheme 50

4.2.6.2 Tandem (and Higher) Sequences

Tandem sequences are far more common than simple sequences, and they are far more powerful. In a tandem sequence, the radical precursor for the second transformation is generated not from a standard precursor (like a halide), but from the preceding transformation in the sequence. There is only one step that generates an initial radical and one step that removes a final radical. In effect, all of the transformations that are conducted must occur during the solution lifetime of a single radical. General considerations for tandem sequences are shown schematically in Scheme 51. The radical generation step provides the initial radical A·. This undergoes an inter- or intra-molecular reaction to form B·, the precursor for the second step in the sequence. The final radical in the sequence, C·, must be converted to a stable product. In many respects, designing tandem sequences is no different from simple additions and cyclizations. All the reactions must be faster than radical/radical reactions or radical/solvent reactions, and the final radical must be selectively removed from the radical pool (but now in the presence of two other substrate radicals rather than one). In principle, many reactions can be conducted between radical generation and radical removal. The radical polymerization of alkenes is a prime example: thousands of addition steps take place between radical generation and radical removal. The problem in synthesis is how to have several different intermediate radicals (A·, B·, C·), which are simultaneously present, avoid suffering the same fate.

The types of reactions that are sequenced and the order in which they occur often dictate the choice of an appropriate method. Thus, sequential radical reactions are classified by transformation, and they are presented below in roughly increasing order of difficulty. With careful substrate design and selection of an appropriate method and reaction conditions, a diverse collection of tandem radical reactions has been executed. Yet the known reactions only begin to show the potential of such radical sequences for the rapid assemblage of complex molecules.

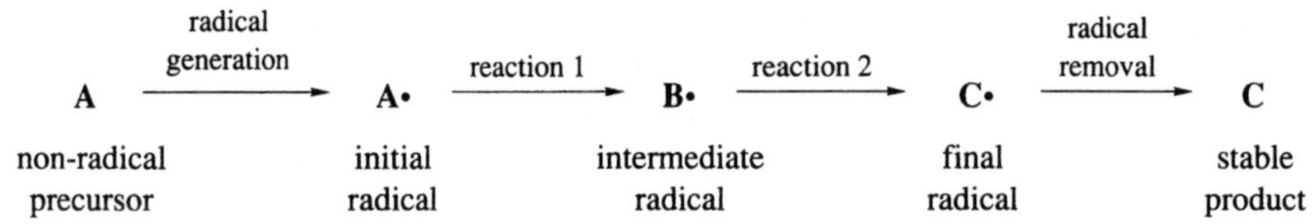

A	$\xrightarrow{\text{radical generation}}$	A•	$\xrightarrow{\text{reaction 1}}$	B•	$\xrightarrow{\text{reaction 2}}$	C•	$\xrightarrow{\text{radical removal}}$	C
non-radical precursor		initial radical		intermediate radical		final radical		stable product

Potential problems: all reactions must be faster than radical–radical or radical–solvent reactions
radicals **A•**, **B•** and **C•** must be differentiated
radical **C•** must be selectively removed in the presence of **A•** and **B•**

Scheme 51 Sequential radical reactions

4.2.6.2.1 Sequences involving only intramolecular transformations

Sequences that involve all intramolecular transformations are by far the easiest to conduct, yet they are among the most powerful. The requirements for success are similar to those for standard radical cyclizations. In general, the slowest intramolecular reaction must still be more rapid than the reaction that converts that radical to a nonradical product (this ensures that the initial and intermediate radicals are not intercepted prior to intramolecular reaction). The differentiation of the intermediate radicals is provided by the structure of the substrate itself: radicals A· and B· should have only one reasonably rapid intramolecular option, and radical C· should have none.

(i) Cyclization/cyclization

The sequencing of two cyclizations is a simple but very powerful strategy to construct polycyclic rings. Any method that is useful for a single cyclization can be used for a tandem cyclization as well, and standard techniques are used to allow intermediate radicals to live long enough to cyclize (for example, in the tin hydride method, syringe pump additions or catalytic methods are common). Scheme 52 provides just a sampling of the types of tandem cyclizations that have recently been conducted by the tin hydride method,[192,212–216] hydrogen[50] and halogen[133] atom transfer methods, oxidation by Mn^{III},[217] and reduction by SmI_2.[218] Provided that the cyclizations are rapid, it should be possible to sequence more than two reactions, and there are already several examples of triple radical cyclizations.[219,220]

(ii) Intramolecular 1,5-hydrogen atom transfer/cyclization (radical translocation)

Intramolecular 1,5-hydrogen atom transfer reactions of radicals are well known and have often been used to introduce functional groups at unactivated centers. For example, the Barton photolysis of nitrite esters is a preparatively useful method to make oximes,[221] and the Hoffman–Löffler–Freytag and related reactions are also important methods to construct heterocycles (see Scheme 44). These reactions use highly reactive alkoxyl or amminium radicals. 1,5-Hydrogen transfer reactions of alkoxyl radicals have also been used to generate radicals for subsequent cyclizations[222] and additions.[168]

We have recently introduced a variant of this class of reaction, termed radical translocation, that will be especially useful because it operates within the confines of the commonly used tin hydride method.[223] In this variant, the initial radical is generated at a remote location by standard halogen atom abstraction, and then translocated by a 1,5-hydrogen atom shift. The intermediate radical that results then undergoes standard cyclization. The two approaches illustrated in Scheme 53 have a common theme: the radical is translocated from a site where it is easily generated to another site where a standard radical precursor might be difficult either to prepare or to carry through a multistep synthetic sequence. In the first example, the initial radical center is in a 'protecting group', and in the second, it is on the alkene that is destined to become the acceptor. In an indirect way, this allows a simple C—H bond to be a radical precursor in the tin hydride method. It is important that the 1,5-hydrogen transfer be as exothermic as possible so that it is sufficiently rapid to compete with direct tin hydride reduction.

(iii) Cyclization/fragmentation

When a cyclization reaction precedes a fragmentation, carbon–carbon bonds are exchanged rather than created. Nonetheless, such sequences are very useful for reorganizing the carbon skeleton of a molecule, especially when easily fragmentable alkoxyl radicals are formed in the cyclization step. Depending on

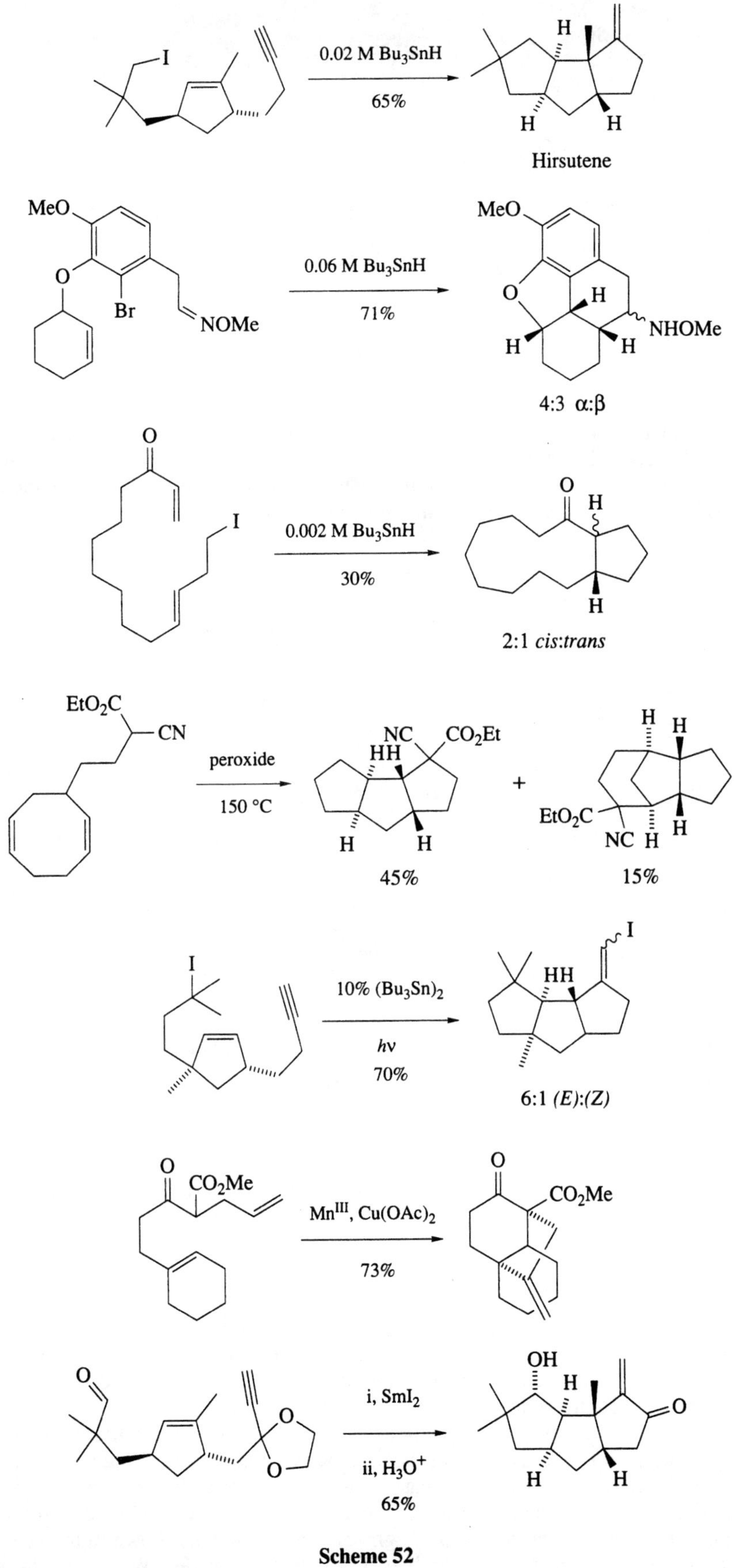

0.02 M Bu_3SnH
65%
H
H H
Hirsutene

MeO
O
Br
NOMe
0.06 M Bu_3SnH
71%
MeO
O
H
H
H
NHOMe
4:3 α:β

O
I
0.002 M Bu_3SnH
30%
O
H
H
2:1 *cis*:*trans*

EtO_2C
CN
peroxide
150 °C
NC
HH
CO_2Et
H H
45%
+
H H
EtO_2C
NC H
H
15%

I
10% $(Bu_3Sn)_2$
$h\nu$
70%
HH
I
6:1 *(E)*:*(Z)*

O
CO_2Me
Mn^{III}, $Cu(OAc)_2$
73%
O
CO_2Me

O
O
O
i, SmI_2
ii, H_3O^+
65%
OH
H
O
H H

Scheme 52

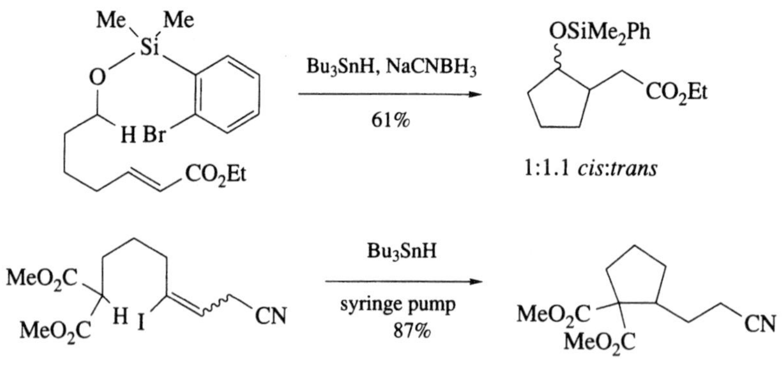

Me Me Si O H Br CO_2Et — Bu_3SnH, $NaCNBH_3$ 61% → $OSiMe_2Ph$ CO_2Et 1:1.1 *cis:trans*

MeO_2C MeO_2C H I CN — Bu_3SnH syringe pump 87% → MeO_2C MeO_2C CN

Scheme 53

the structure of the substrate, group migration or ring expansion can occur. Scheme 54 provides examples from the work of Beckwith,[224] Dowd and Choi[225] and Baldwin.[226] In these sequences, a radical-stabilizing group often controls the direction of fragmentation, and thus the final radical is usually more stable than the starting radical. Methods like atom transfer will not be useful for such sequences, but the tin hydride method and the tin/sulfur fragmentation method are appropriate.

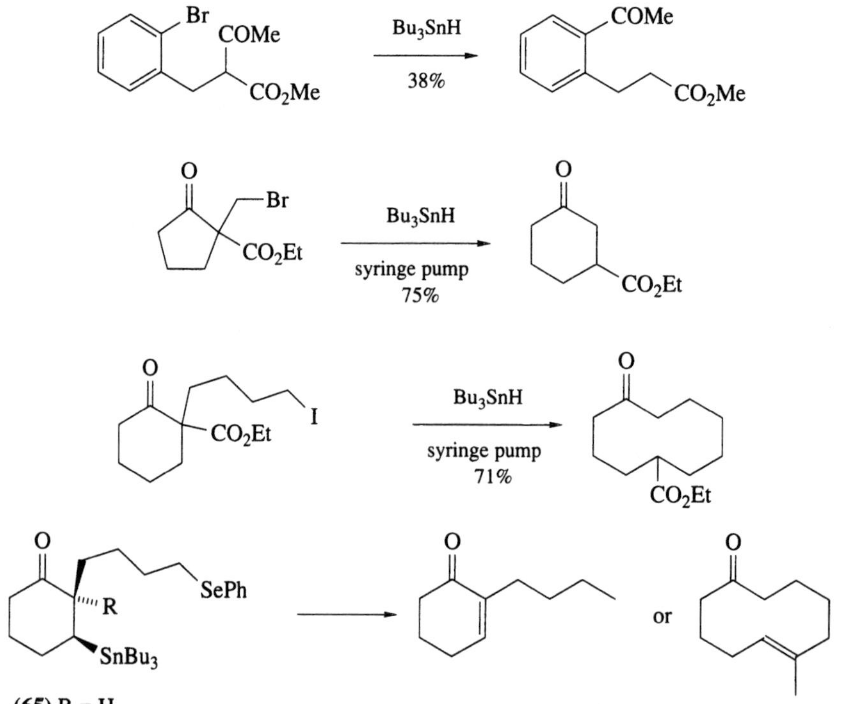

Br COMe CO_2Me — Bu_3SnH 38% → COMe CO_2Me

O Br CO_2Et — Bu_3SnH syringe pump 75% → O CO_2Et

O CO_2Et I — Bu_3SnH syringe pump 71% → O CO_2Et

O R $SnBu_3$ SePh → O or O

(65) R = H
(66) R = Me

(67) R = H, 86% or **(68)** R = Me, 89%

Scheme 54

The first example in Scheme 54 illustrates the migration of an acetyl group while the second example illustrates the expansion of a ring by one carbon atom (see also Scheme 21). As shown in the third example, rings can also be expanded by three and four atoms (but not two atoms because the 4-exo closure is too slow). Thus, although medium size rings are not directly available by radical cyclization of an acyclic precursor, they are readily prepared by ring expansion of a standard small ring. In the second and third examples, the ester group not only controls the direction of fragmentation, but it also accelerates the initial cyclization to the carbonyl. Even so, syringe pump methods must be used to minimize direct reduction of the starting halide. As usual, this difficulty can be overcome by building the tin group into the precursor and conducting the cyclization by the fragmentation method, as shown in the last example of Scheme 54. One of the pitfalls of this approach is 1,5-hydrogen transfer. When **(65)** is heated with

AIBN, (**67**) forms in high yield because the initial radical abstracts the hydrogen α to the carbonyl (with concomitant elimination of $Bu_3Sn\cdot$) more rapidly than it cyclizes. However, blocking this pathway results in smooth ring expansion, as illustrated by the conversion of (**66**) to (**68**). The alkene stereochemistry is dictated by the configuration of the precursor (see Scheme 49).

(iv) Fragmentation/cyclization

Very useful transformations can also be designed when the fragmentation precedes the cyclization, as shown in Scheme 55. Motherwell and Harling have demonstrated that the stereochemistry of a spirocyclic ring can be controlled by a fragmentation of a cyclopropyl carbinyl radical, followed by cyclization, as illustrated by the first two examples.[227] The hydroxy group ultimately dictates the stereochemistry of the spirocycle by directing the Simmons–Smith cyclopropanation. Such fragmentations of cyclopropyl carbinyl radicals are useful even when they are not conducted in a sequence,[228] and the kinetic cleavage of a lateral bond of a fused cyclopropane is usually observed unless radical-stabilizing groups accelerate cleavage of the ring fusion bond. The third example, taken from the studies of Citerrio, illustrates how a fragmentation/cyclization sequence conducted by a reductive method can result in ring contraction.

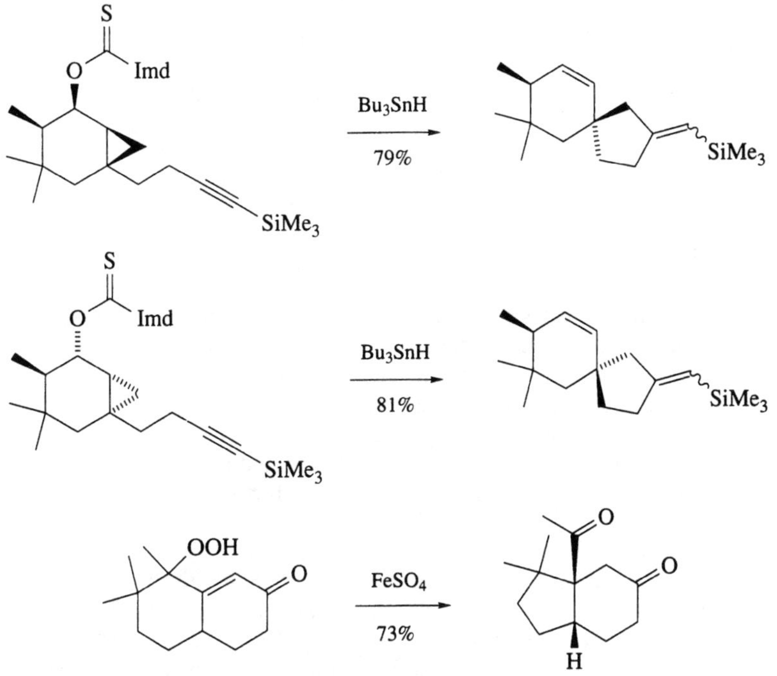

S
Imd
O
Bu_3SnH
79%
$SiMe_3$
$SiMe_3$

S
Imd
O
Bu_3SnH
81%
$SiMe_3$
$SiMe_3$

OOH
O
$FeSO_4$
73%
O
O
H

Scheme 55

4.2.6.2.2 Sequences combining inter- and intra-molecular reactions

Sequences that combine inter- and intra-molecular reactions are very powerful, but they require more careful planning to ensure selectivity. The order in which such reactions are conducted has a significant impact on the design of substrates and the selection of a method.

(i) Cyclization/addition

When combining cyclizations and additions, design is simpler if the cyclization precedes the addition. The usual considerations apply for the success of the addition step (see the preceding chapter). Any cyclization that is more rapid than a given addition can be sequenced before that addition. Thus, rapid cyclizations are easily conducted prior to additions, but slow cyclizations may be problematic since the initial radical might competitively add to the acceptor. However, even slow cyclizations can probably be conducted by electronically mismatching the initial radical and the acceptor, or by using relatively unreactive acceptors like allylstannanes. Examples of cyclization/addition sequences conducted by the tin hydride method,[229,230] the fragmentation method,[231,232] and the reductive method[168] are presented in Scheme 56.

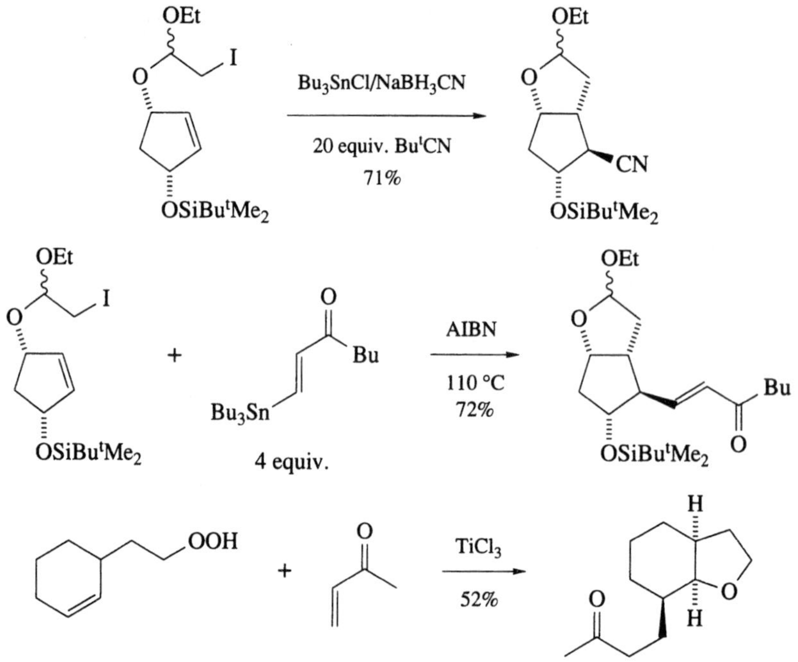

Scheme 56

(ii) Addition/cyclization

Sequences in which addition precedes cyclization are not as straightforward to conduct as the reverse; however, they are very important because a net annulation results (that is, a new ring is formed by the union of two acyclic precursors in one experimental step). The intermediate radical is differentiated from the other radicals provided that the cyclization reaction is rapid, but it can be difficult to differentiate the initial radical from the final radical. As illustrated in Scheme 57, this is particularly true in the tin hydride method because many different types of radicals react with tin hydride at similar rates. Reaction of **(69)** under standard radical addition conditions produces **(70)**, which results from a sequence of addition/cyclization/addition.[233] That the last C—C bond is formed actually results from a lack of selectivity: the initial and final radicals are not differentiated and they must undergo the same reaction. Of course, this 'lack of selectivity' is of no consequence if the product contains the desired skeleton and the needed functionality for subsequent transformations. Such sequences are very useful for forming three carbon–carbon bonds, and they can also be conducted by Barton's thiohydroxamate method.[234] Structural modifications are required to differentiate the initial and final radicals, and, as illustrated by the conversion of **(71)** to **(72)**, phenyl groups can provide the needed differentiation (probably by retarding the rate of addition more than they retard the rate of hydrogen abstraction). Clive has demonstrated that phenyl-substituted vinyl radicals also provide the needed selectivity, as illustrated by the second example in Scheme 57.[235]

Work from our group has shown that the atom transfer method[236,237] and the fragmentation method[238] are especially useful for conducting these radical annulations. As illustrated by the first two examples in Scheme 58, atom transfer annulations can be designed by pairing nucleophilic radicals with electrophilic acceptors, or by the reverse. In either case, the substrates are designed such that the least resonance stabilized radical appears last in the sequence. This radical rapidly abstracts iodide from either the starting iodide or from an intermediate iodide, thereby terminating the sequence before the final radical can undergo a second addition reaction. The last example in Scheme 58 illustrates that fragmentation reactions should be especially powerful in controlling such annulations: whenever a β-stannyl radical arises, its rapid fragmentation will terminate the sequence and transfer the chain. In contrast to other cyclization and addition reactions of vinylstannanes (see Scheme 25), the example in Scheme 58 provides a high level of inversion of alkene stereochemistry, and an explanation for this phenomenon has been proposed.[238]

Feldman[239,240] and Oshima and Utimoto[241] have introduced a very sophisticated sequence of reactions that effects a radical annulation starting from a vinylcyclopropane. Scheme 59 provides two examples and a likely mechanism for this sequence, which involves fragmentation, addition and cyclization. The

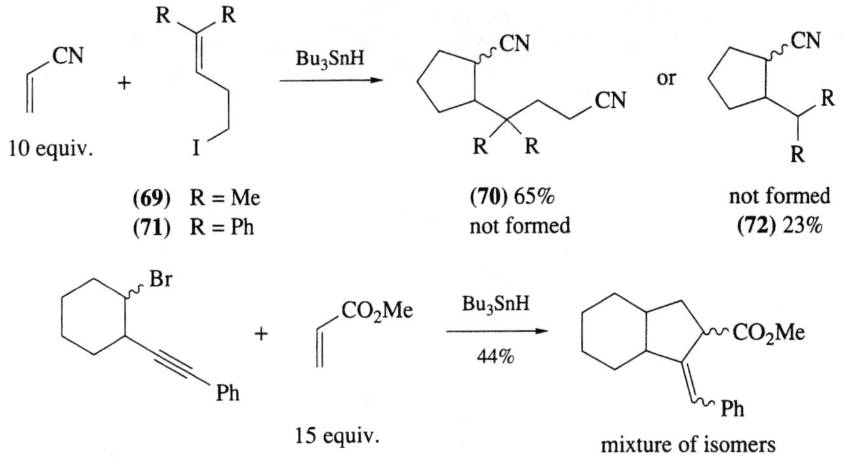

Bu_3SnH

10 equiv.

(69) R = Me
(71) R = Ph

(70) 65%
not formed

or

not formed
(72) 23%

Bu_3SnH
44%

15 equiv.

mixture of isomers

Scheme 57

10% $(Bu_3Sn)_2$
$h\nu$
68%

2 equiv.   1 equiv.

10% $(Bu_3Sn)_2$
$h\nu$
71%

1 equiv.   2 equiv.

85:15 *cis*:*trans*

10% $(Bu_3Sn)_2$
$h\nu$
56%

98:2 *(E)*:*(Z)*

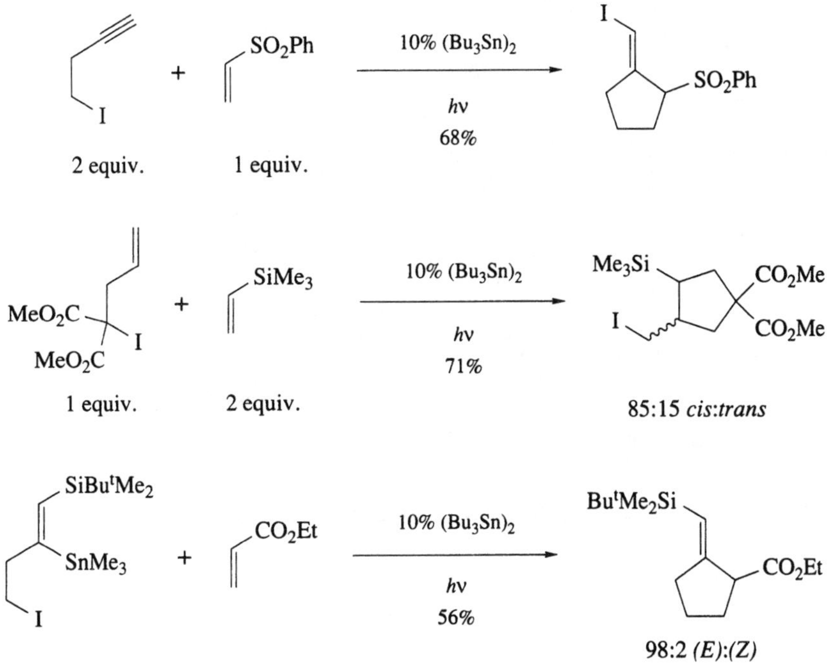

Scheme 58

irreversibility of the addition and 5-exo cyclization steps drives the sequence forward, and a rapid fragmentation reaction (of a β-phenylthio radical) plays a key role in terminating the sequence before the final radical can undergo another addition.

With the goal of sequencing several addition reactions, Feldman and Lee have recently introduced a beautiful protocol, termed template-controlled oligomerization, that allows a controlled number of addition reactions to take place prior to a radical macrocyclization.[242] As illustrated in equation (8), the number of additions is dictated by the distance of the gap between the 'initiator' and the 'terminator' in this strictly controlled oligomerization. The requirement here is that the macrocyclization be more rapid than the bimolecular addition. A rapid fragmentation of a phenylthio radical is again used to terminate the sequence, and this prevents the final radical from continuing to polymerize. When the oligomer is detached from the template, it appears as if four sequential additions (three to the external acceptor and one to the acceptor on the template) have occurred. This type of reaction is virtually impossible to conduct in a controlled fashion by sequencing addition reactions (see below).

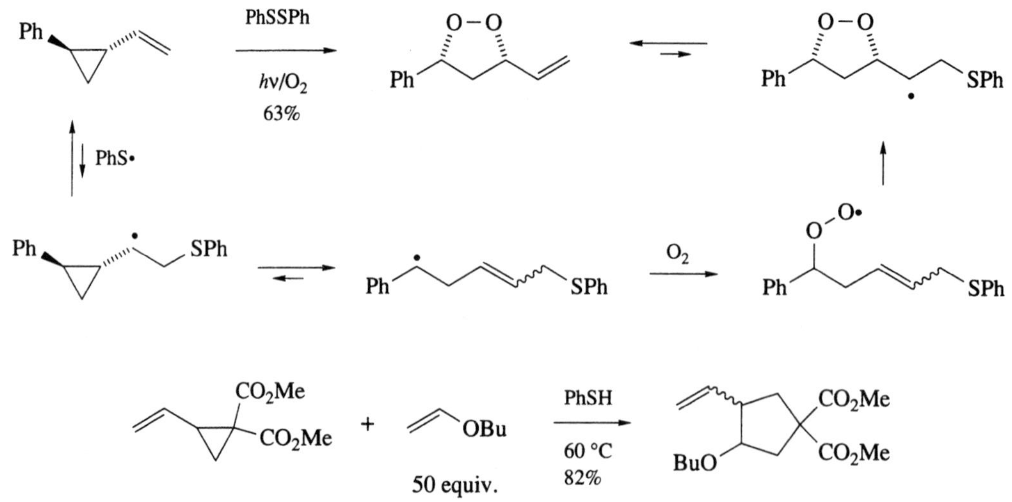

Ph
PhSSPh
hν/O2
63%
O–O
Ph
O–O
Ph
SPh
PhS•
Ph
SPh
Ph
SPh
O2
O•
O
Ph
SPh
CO2Me
CO2Me
+
OBu
50 equiv.
PhSH
60 °C
82%
CO2Me
BuO
CO2Me

Scheme 59

O
O
CCl3
O
O
SPh
CO2Me
Mo(CO)6, 80 °C
41%
O
O
Cl
Cl
E
E
E
O
O

(8)

4.2.6.2.3 Sequences involving only intermolecular reactions

The sequencing of addition reactions is more difficult to design because polymerization is a natural outcome. Two addition reactions can be sequenced by alternating the reactivity of the radicals: one nucleophilic, the other electrophilic, or the reverse. This approach is reminiscent of radical copolymerization, except that reaction conditions are designed to intercept the radical after only one pair of additions. The first example in Scheme 60 provides a very slick example from the pioneering work of Minisci[243] in which two electronically reversed additions occur in sequence. Although the prior radicals are not susceptible to oxidation, the final allylic radical is removed by oxidation (and the allylic cation lactonizes), thus preventing polymerization. The second example again shows the power of the fragmentation method for conducting difficult sequences; when the β-stannyl radical arises, its rapid elimination shuts down the sequence.[244]

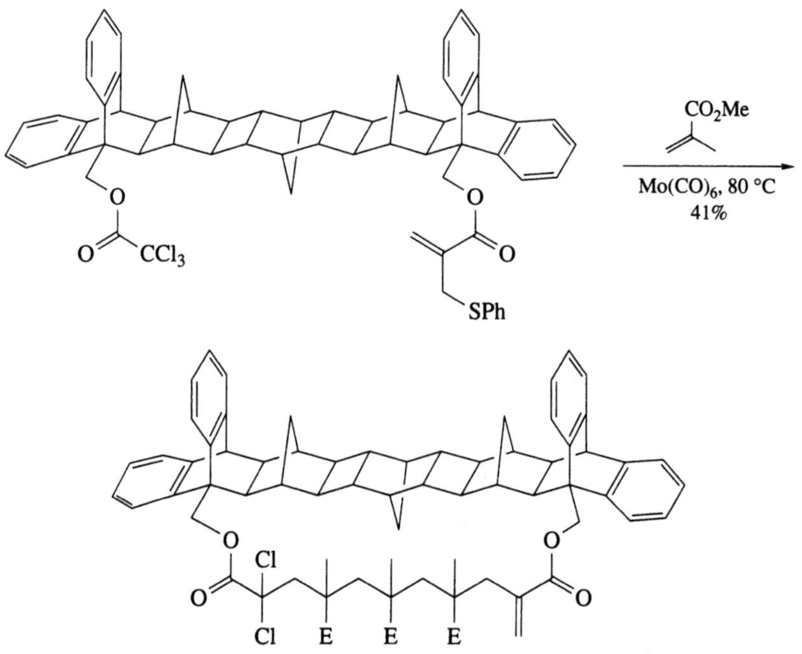

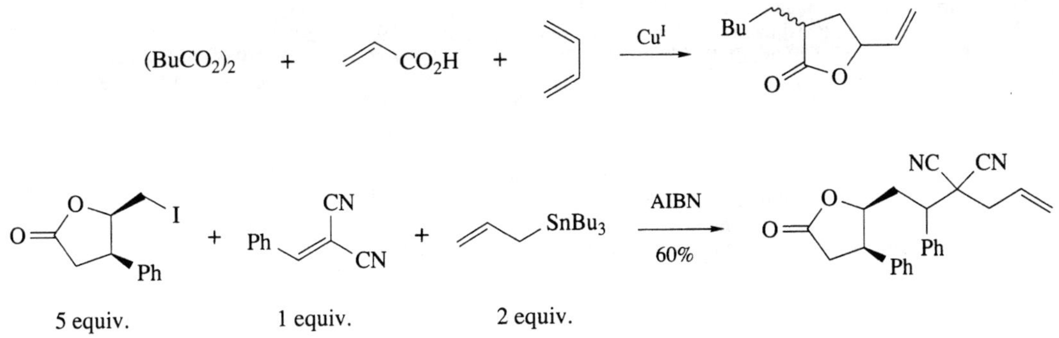

$(BuCO_2)_2$ + CO_2H + → Cu^I Bu O O

I O O + Ph CN CN + $SnBu_3$ AIBN 60% → NC CN O O Ph Ph

5 equiv. 1 equiv. 2 equiv.

Scheme 60

4.2.7 CONCLUSIONS

Radical cyclization reactions offer a valuable synthetic tool with which to construct both carbocyclic and heterocyclic rings. The success and the diversity of recent advances have dispelled the old notion that radical cyclizations are only useful for making five-membered rings, and with poor stereoselectivity at that. With the exception of three- and four-membered rings, almost any ring size can be built either by direct radical cyclization or by ring expansion, and the design of highly stereoselective radical cyclizations is now possible. The general principles that have emerged from this work facilitate the planning of new radical reactions, whether individual or sequenced. Radical additions and cyclizations that are chemo- and regio-selective, and exhibit control of relative stereochemistry are now appearing regularly in synthetic journals; can the development of new radical reactions that control absolute stereochemistry be far behind?

ACKNOWLEDGEMENTS

I am very grateful to Jean Rock, Churl Min Seong, Wang Shen and Scott Gothe for their help in preparing the manuscript, and I also thank Drs. C. Jasperse, E. Schwartz, R. Wolin and J. Stack, and H. T. Liu, M. Totleben and P. S. Ramamoorthy. Our contributions in the free radical area have been supported by the National Institutes of Health.

4.2.8 REFERENCES

1. D. J. Hart, *Science (Washington, D.C.)*, 1984, **223**, 883.
2. A. L. J. Beckwith and K. U. Ingold, in 'Rearrangements in Ground and Excited States', ed. P. de Mayo, Academic Press, New York, 1980, vol. 1, p. 162.
3. A. L. J. Beckwith, *Tetrahedron*, 1981, **37**, 3073.
4. J.-M. Surzur, in 'Reactive Intermediates', ed. R. A. Abramovitch, Plenum Press, New York, 1982, vol. 2, chap. 3.
5. B. Giese, 'Radicals in Organic Synthesis: Formation of Carbon–Carbon Bonds', Pergamon Press, Oxford, 1986.
6. A. L. J. Beckwith and C. H. Schiesser, *Tetrahedron Lett.*, 1985, **26**, 373; *Tetrahedron*, 1985, **41**, 3925.
7. D. C. Spellmeyer and K. N. Houk, *J. Org. Chem.*, 1987, **52**, 959.
8. In some cases, other conformations of the 6-endo transition state may be important, see ref. 7.
9. The diagrams in Figure 1 were produced from the coordinates of the transition states in the Houk/Spellmeyer force field model. I am grateful to Professor K. N. Houk for providing these coordinates.
10. J. E. Baldwin, *J. Chem. Soc., Chem. Commun.*, 1976, 734.
11. The two computational models differ on this point. The Beckwith model (ref. 6) indicates increased bending and torsional strain in the 6-endo TS. The Houk model (ref. 7) indicates increased bending strain only, and states this strain is due in part to the bending of the allylic carbon out of the alkene plane to accommodate a better angle of attack.
12. H. Fischer (ed.), 'Radical Reaction Rates in Liquids', Springer-Verlag, West Berlin, 1983–85, Landolt-Börnstein, new series, vols. II/13a–e.
13. A. L. J. Beckwith and S. A. Glover, *Aust. J. Chem.*, 1987, **40**, 157.
14. A. L. J. Beckwith, G. E. Gream and D. L. Struble, *Aust. J. Chem.*, 1972, **25**, 1081.
15. D. Wehle, N. Scholmann and L. Fitjer, *Chem. Ber.*, 1988, **121**, 2171.
16. T. Satoh, M. Itoh and K. Yamakawa, *Chem. Lett.*, 1987, 1949.
17. D. L. J. Clive and P. L. Beaulieu, *J. Chem. Soc., Chem. Commun.*, 1983, 307.
18. M. Julia, *Acc. Chem. Res.*, 1971, **4**, 386.

19. D. P. Curran and C.-T. Chang, *J. Org. Chem.*, 1989, **54**, 3140.
20. D. L. J. Clive and D. R. Cheshire, *J. Chem. Soc., Chem. Commun.*, 1987, 1520.
21. N. A. Porter, V. H.-T. Chang, D. R. Magnin and B. T. Wright, *J. Am. Chem. Soc.*, 1988, **110**, 3554.
22. A. Effio, D. Griller, K. U. Ingold, A. L. J. Beckwith and A. K. Serelis, *J. Am. Chem. Soc.*, 1980, **102**, 1734.
23. M. Newcomb and W. G. Williams, *Tetrahedron Lett.*, 1985, **26**, 1179.
24. A. L. J. Beckwith and G. Moad, *J. Chem. Soc., Chem. Commun.*, 1974, 472.
25. G. Illuminati and L. Mandolini, *Acc. Chem. Res.*, 1981, **14**, 95.
26. N. A. Porter and V. H.-T. Chang, *J. Am. Chem. Soc.*, 1987, **109**, 4976.
27. A. L. J. Beckwith, C. J. Easton, T. Lawrence and A. K. Serelis, *Aust. J. Chem.*, 1983, **36**, 545.
28. T. V. RajanBabu, T. Fukanaga and G. S. Reddy, *J. Am. Chem. Soc.*, 1989, **111**, 1759.
29. T. V. RajanBabu and T. Fukanaga, *J. Am. Chem. Soc.*, 1989, **111**, 296.
30. S. Hanessian, D. S. Dhanoa and P. L. Beaulieu, *Can. J. Chem.*, 1987, **65**, 1859.
31. M. Apparu and J. K. Crandall, *J. Org. Chem.*, 1984, **49**, 2125.
32. Y. Ichinose, K. Oshima and K. Utimoto, *Chem. Lett.*, 1988, 1437.
33. M. Ramaiah, *Tetrahedron*, 1987, **43**, 3541.
34. D. P. Curran, *Synthesis*, 1988, 417, 489.
35. D. H. R. Barton and W. B. Motherwell, *Pure Appl. Chem.*, 1981, **53**, 15; in 'Organic Synthesis, Today and Tomorrow', ed. B. M. Trost and C. R. Hutchinson, Pergamon Press, New York, 1981, p. 1; *Heterocycles*, 1984, **21**, 1.
36. N. Ono, H. Miyake, A. Kamimura, I. Hamamoto, R. Tamura and A. Kaji, *Tetrahedron*, 1985, **41**, 4013; N. Ono, M. Fujii and A. Kaji, *Synthesis*, 1987, 532.
37. V. K. Yadav and A. G. Fallis, *Tetrahedron Lett.*, 1988, **29**, 897.
38. G. Stork and M. E. Reynolds, *J. Am. Chem. Soc.*, 1988, **110**, 6911.
39. J.-M. Fang, H.-T. Chang and C.-C. Lin, *J. Chem. Soc., Chem. Commun.*, 1988, 1385.
40. N. A. Porter, D. R. Magnin and B. T. Wright, *J. Am. Chem. Soc.*, 1986, **108**, 2787.
41. C.-P. Chuang, J. C. Gallucci, D. J. Hart and C. Hoffman, *J. Org. Chem.*, 1988, **53**, 3218.
42. K. Miura, Y. Ichinose, K. Nozaki, K. Fugami, K. Oshima and K. Utimoto, *Bull. Chem. Soc. Jpn.*, 1989, **62**, 143.
43. W. P. Neumann, *Synthesis*, 1987, 665.
44. C. S. Wilcox and J. J. Guadino, *J. Am. Chem. Soc.*, 1986, **108**, 3102.
45. T. V. RajanBabu, *J. Org. Chem.*, 1988, **53**, 4522.
46. M. F. Jones and S. M. Roberts, *J. Chem. Soc., Perkin Trans. 1*, 1988, 2927.
47. Y. K. Rao and M. Nagarajan, *Tetrahedron Lett.*, 1988, **29**, 107.
48. C.-P. Chuang, J. C. Gallucci and D. J. Hart, *J. Org. Chem.*, 1988, **53**, 3210.
49. D. L. J. Clive, D. R. Cheshire and L. Set, *J. Chem. Soc., Chem. Commun.*, 1987, 353.
50. J. D. Winkler and V. Sridar, *Tetrahedron Lett.*, 1988, **29**, 6219.
51. F. MacCorquodale and J. C. Walton, *J. Chem. Soc., Perkin Trans. 1*, 1989, 347.
52. K. S. Gröniger, K. F. Jäger and B. Giese, *Liebigs Ann. Chem.*, 1987, 731.
53. D. L. J. Clive, *Pure Appl. Chem.*, 1988, **60**, 1645.
54. D. L. J. Clive, T. L. B. Boivin and A. G. Angoh, *J. Org. Chem.*, 1987, **52**, 4943.
55. W. R. Leonard and T. Livinghouse, *Tetrahedron Lett.*, 1985, **26**, 6431.
56. A. Y. Mohammed and D. L. J. Clive, *J. Chem. Soc., Chem. Commun.*, 1986, 588.
57. T. Ghosh and H. Hart, *J. Org. Chem.*, 1988, **53**, 2396.
58. M. Ladlow and G. Pattenden, *Tetrahedron Lett.*, 1985, **26**, 4413.
59. N. J. G. Cox, G. Pattenden and S. D. Mills, *Tetrahedron Lett.*, 1989, **30**, 621.
60. A. Padwa, H. Nimmesgern and G. S. K. Wong, *J. Org. Chem.*, 1985, **50**, 5620; *Tetrahedron Lett.*, 1985, **26**, 957.
61. Y. Chapleur and N. Moufid, *J. Chem. Soc., Chem. Commun.*, 1989, 39.
62. C. Audin, J.-M. Lancelin and J.-M. Beau, *Tetrahedron Lett.*, 1988, **29**, 3691.
63. A. De Mesmaeker, P. Hofmann and B. Ernst, *Tetrahedron Lett.*, 1989, **30**, 57.
64. Y. Watanabe and T. Endo, *Tetrahedron Lett.*, 1988, **29**, 321.
65. Y. Watanabe, Y. Ueno, C. Tanaka, M. Okawara and T. Endo, *Tetrahedron Lett.*, 1987, **28**, 3953.
66. G. Stork, in 'Current Trends in Organic Synthesis', ed. H. Nozaki, Pergamon Press, Oxford, 1983, p. 359; G. Stork, in 'Selectivity—A Goal For Synthetic Efficiency', ed. W. Bartmann and B. M. Trost, Springer Verlag, Florida, 1984, p. 281.
67. Y. Ueno, O. Moriya, K. Chino, M. Watanabe and M. Okawara, *J. Chem. Soc., Perkin Trans. 1*, 1986, 1351; Y. Ueno, K. Chino, M. Watanabe, O. Moriya and M. Okawara, *J. Am. Chem. Soc.*, 1982, **104**, 5564.
68. G. Stork, R. Mook, Jr., S. A. Biller and S. D. Rychnovsky, *J. Am. Chem. Soc.*, 1983, **105**, 3741.
69. M. Kim, R. S. Gross, H. Sevestre, N. K. Dunlap and D. S. Watt, *J. Org. Chem.*, 1988, **53**, 93.
70. A. Srikrishna and G. SunderBabu, *Chem. Lett.*, 1988, 371.
71. A. Srikrishna and G. V. R. Sharma, *Tetrahedron Lett.*, 1988, **29**, 6487.
72. J. P. Dulcere, M. N. Mihoubi and J. Rodriguez, *J. Chem. Soc., Chem. Commun.*, 1988, 237.
73. G. Stork and M. Kahn, *J. Am. Chem. Soc.*, 1985, **107**, 500; G. Stork and M. J. Sofia, *J. Am. Chem. Soc.*, 1986, **108**, 6826.
74. H. Nishiyama, T. Kitajima, M. Matsumoto and K. Itoh, *J. Org. Chem.*, 1984, **49**, 2298.
75. M. T. Crimmins and R. O'Mahony, *J. Org. Chem.*, 1989, **54**, 1157.
76. E. Magnol and M. Malacria, *Tetrahedron Lett.*, 1986, **27**, 2255.
77. M. Koreeda and I. A. George, *J. Am. Chem. Soc.*, 1986, **108**, 8098.
78. D. A. Burnett, J.-K. Choi, D. J. Hart and Y.-M. Tsai, *J. Am. Chem. Soc.*, 1984, **106**, 8201; D. J. Hart and Y.-M. Tsai, *J. Am. Chem. Soc.*, 1984, **106**, 8209; J.-K. Choi and D. J. Hart, *Tetrahedron*, 1985, **41**, 3959.
79. Reduction of imines is a useful route to related α-amino radicals; S. F. Martin, C.-P. Yang, W. L. Laswell and H. Rüeger, *Tetrahedron Lett.*, 1988, **29**, 6685.
80. J.-K. Choi, D.-C. Ha, D. J. Hart, C.-S. Lee, S. Ramesh and S. Wu, *J. Org. Chem.*, 1989, **54**, 279.

81. J. M. Dener, D. J. Hart and S. Ramesh, *J. Org. Chem.*, 1988, **53**, 6022.
82. M. D. Bachi, A. De Mesmaeker and N. Stevenart-De Mesmaeker, *Tetrahedron Lett.*, 1987, **28**, 2637, 2887.
83. A. L. J. Beckwith and D. R. Boate, *Tetrahedron Lett.*, 1985, **26**, 1761.
84. T. Kametani, S.-D. Chu, A. Itoh, S. Maeda and T. Honda, *J. Org. Chem.*, 1988, **53**, 2683.
85. M. D. Bachi, F. Frolow and C. Hoornaert, *J. Org. Chem.*, 1983, **48**, 1841.
86. A. Padwa, W. Dent, H. Nimmesgern, M. K. Venkatramanan and G. S. K. Wong, *Chem. Ber.*, 1986, **119**, 813.
87. K. C. Nicolaou, D. G. McGarry, P. K. Somers, C. A. Veale and G. T. Furst, *J. Am. Chem. Soc.*, 1987, **109**, 2504.
88. A. L. J. Beckwith and P. E. Pigou, *J. Chem. Soc., Chem. Commun.*, 1986, 85.
89. J. P. Marino, E. Laborde and R. S. Paley, *J. Am. Chem. Soc.*, 1988, **110**, 966.
90. G. Stork and N. H. Baine, *J. Am. Chem. Soc.*, 1982, **104**, 2321.
91. A. L. J. Beckwith and W. B. Gara, *J. Chem. Soc., Perkin Trans. 2*, 1975, 593, 795; A. L. J. Beckwith and G. F. Meijs, *J. Chem. Soc., Chem. Commun.*, 1981, 136.
92. For an analysis of the relative reactivity of vinyl and aryl radicals towards cyclization and hydrogen abstraction from tin hydride, see ref. 34, p. 429.
93. G. Stork and N. H. Baine, *Tetrahedron Lett.*, 1985, **26**, 5927.
94. K. Vijaya Bhaskar and G. S. R. Subba Rao, *Tetrahedron Lett.*, 1989, **30**, 225.
95. G. Stork and R. Mook, Jr., *J. Am. Chem. Soc.*, 1987, **109**, 2829.
96. K. Nozaki, K. Oshima and K. Utimoto, *J. Am. Chem. Soc.*, 1987, **109**, 2547.
97. J. Ardisson, J. P. Férézou, M. Julia and A. Pancrazi, *Tetrahedron Lett.*, 1987, **28**, 2001.
98. R. Mook, Jr. and P. M. Sher, *Org. Synth.*, 1987, **66**, 75.
99. G. Stork and R. Mook, Jr., *Tetrahedron Lett.*, 1986, **27**, 4529.
100. A. L. J. Beckwith and D. M. O'Shea, *Tetrahedron Lett.*, 1986, **27**, 4525.
101. J. A. Franz, R. D. Barrows and D. M. Camaioni, *J. Am. Chem. Soc.*, 1984, **106**, 3964.
102. A. N. Abeywickrema, A. L. J. Beckwith and S. Gerba, *J. Org. Chem.*, 1987, **52**, 4072.
103. K. A. Parker, D. M. Spero and K. C. Inman, *Tetrahedron Lett.*, 1986, **27**, 2833.
104. C. P. Sloan, J. C. Cuevas, C. Quesnelle and V. Snieckus, *Tetrahedron Lett.*, 1988, **29**, 4685; S. Wolff and H. M. R. Hoffmann, *Synthesis*, 1988, 760.
105. D. L. Boger and R. S. Coleman, *J. Am. Chem. Soc.*, 1988, **110**, 4796; J. P. Dittami and H. Ramanathan, *Tetrahedron Lett.*, 1988, **29**, 45.
106. D. Crich and S. M. Fortt, *Tetrahedron Lett.*, 1988, **29**, 2585.
107. D. L. Boger and R. J. Mathvink, *J. Org. Chem.*, 1988, **53**, 3377.
108. M. D. Bachi and E. Bosch, *Tetrahedron Lett.*, 1988, **29**, 2581; *Tetrahedron Lett.*, 1986, **27**, 641.
109. M. D. Bachi and E. Bosch, *J. Chem. Soc., Perkin Trans. 1*, 1988, 1517; M. D. Bachi and D. Denemark, *J. Am. Chem. Soc.*, 1989, **111**, 1886.
110. A. K. Singh, R. K. Bakshi and E. J. Corey, *J. Am. Chem. Soc.*, 1987, **109**, 6187; it is not entirely clear that this example is kinetically controlled, see footnote 16 in this paper for a comment.
111. J. Barluenga and M. Yus, *Chem. Rev.*, 1988, **88**, 487.
112. S. J. Danishefsky, S. Chackalamannil and B.-J. Uang, *J. Org. Chem.*, 1982, **47**, 2231.
113. K. Weinges and W. Sipos, *Chem. Ber.*, 1988, **121**, 363.
114. G. E. Keck and E. J. Enholm, *Tetrahedron Lett.*, 1985, **26**, 3311.
115. S. J. Danishefsky and J. S. Panek, *J. Am. Chem. Soc.*, 1987, **109**, 917.
116. F. L. Harris and L. Weiler, *Tetrahedron Lett.*, 1987, **28**, 2941.
117. D. P. Curran and C. Jasperse, *J. Am. Chem. Soc.*, 1990, **112**, 5601.
118. T. A. K. Smith and G. H. Whitham, *J. Chem. Soc., Perkin Trans. 1*, 1989, 313, 319.
119. D. Crich, *Aldrichimia Acta*, 1986, **20**, 35.
120. D. H. R. Barton, J. Guilhem, Y. Hervé, P. Potier and J. Thierry, *Tetrahedron Lett.*, 1987, **28**, 1413.
121. A. J. Bloodworth, D. Crich and T. Melvin, *J. Chem. Soc., Chem. Commun.*, 1987, 786.
122. M. Julia, *Acc. Chem. Res.*, 1971, 4, 386; *Pure Appl. Chem.*, 1974, **40**, 553.
123. M. Julia and M. Maumy, *Org. Synth., Coll Vol.*, 1988, **6**, 586.
124. If all the cyclizations and cleavages are rapid relative to H-atom transfer, then the product distribution depends on the equilibrium constant between (**49**) and (**50**) and the rate of hydrogen atom transfer to (**49**) and (**50**) (Curtin–Hammett kinetics).
125. J. D. Winkler and V. Sridar, *J. Am. Chem. Soc.*, 1986, **108**, 1708.
126. D. P. Curran, in 'Free Radicals in Synthesis and Biology', ed. F. Minisci, Kluwer, Dordrecht, 1989, p. 37.
127. T. K. Hayes, R. Villani and S. M. Weinreb, *J. Am. Chem. Soc.*, 1988, **110**, 5533.
128. Y. Hirai, A. Hagiwara, T. Terada and T. Yamazaki, *Chem. Lett.*, 1987, 2417.
129. D. P. Curran and C.-T. Chang, *Tetrahedron Lett.*, 1987, **28**, 2477.
130. D. P. Curran, E. Bosch, J. Kaplan and M. Newcomb, *J. Org. Chem.*, 1989, **54**, 1826.
131. R. S. Jolly and T. Livinghouse, *J. Am. Chem. Soc.*, 1988, **110**, 7536.
132. M. Mori, N. Kanda, Y. Ban and K. Aoe, *J. Chem. Soc., Chem. Commun.*, 1988, 12.
133. D. P. Curran, M.-H. Chen and D. Kim, *J. Am. Chem. Soc.*, 1986, **108**, 2489; *J. Am. Chem. Soc.*, 1989, **111**, 6265.
134. G. Haaima and R. T. Weavers, *Tetrahedron Lett.*, 1988, **29**, 1085.
135. M. Newcomb and S.-U. Park, *J. Am. Chem. Soc.*, 1986, **108**, 4132; M. Newcomb and J. Kaplan, *Tetrahedron Lett.*, 1987, **28**, 1615.
136. D. P. Curran and D. Kim, *Tetrahedron Lett.*, 1986, **27**, 5821.
137. M. T. Crimmins and S. W. Mascarella, *Tetrahedron Lett.*, 1987, **28**, 5063.
138. C. Galli, *Chem. Rev.*, 1988, **88**, 765.
139. G. F. Meijs and A. L. J. Beckwith, *J. Am. Chem. Soc.*, 1986, **108**, 5890.
140. A. N. Abeywickrema and A. L. J. Beckwith, *J. Am. Chem. Soc.*, 1986, **108**, 8227.
141. A. L. J. Beckwith and G. F. Meijs, *J. Org. Chem.*, 1987, **52**, 1922.
142. L. Becking and H. J. Schäfer, *Tetrahedron Lett.*, 1988, **29**, 2797, 3001.

143. J. E. Baldwin and C.-S. Li, *J. Chem. Soc., Chem. Commun.*, 1988, 261.
144. A. Ghosez, T. Göbel and B. Giese, *Chem. Ber.*, 1988, **121**, 1807.
145. E. G. Samsel and J. K. Kochi, *J. Am. Chem. Soc.*, 1986, **108**, 4790.
146. G. Pattenden, *Chem. Soc. Rev.*, 1988, **17**, 361.
147. W. Klein, in 'Organic Synthesis by Oxidation with Metal Compounds', ed. W. J. Mijs and C. R. H. de Jonge, Plenum Press, New York, 1986, p. 261.
148. J. E. Merritt, M. Sasson, S. A. Kates and B. B. Snider, *Tetrahedron Lett.*, 1988, **29**, 5209.
149. J.-M. Surzur and M. P. Bertrand, *Pure Appl. Chem.*, 1988, **60**, 1659.
150. Other types of products can also be formed. For example, see H. Oumar-Mahamat, C. Moustrou, J.-M. Surzur and M. P. Bertrand, *Tetrahedron Lett.*, 1989, **30**, 331.
151. E. J. Corey and M. Kang, *J. Am. Chem. Soc.*, 1984, **106**, 5384.
152. A. B. Ernst and W. E. Fristad, *Tetrahedron Lett.*, 1985, **26**, 3761.
153. B. B. Snider, J. J. Patricia and S. A. Kates, *J. Org. Chem.*, 1988, **53**, 2137.
154. B. B. Snider and J. J. Patricia, *J. Org. Chem.*, 1989, **54**, 38.
155. R. Scheffold, S. Abrecht, R. Orlinski, H.-R. Ruf, P. Stamouli, O. Tinembart, L. Walder and C. Weymuth, *Pure Appl. Chem.*, 1987, **59**, 363.
156. J. K. Crandall and W. J. Michaely, *J. Org. Chem.*, 1984, **49**, 4244.
157. J. K. Kochi and J. W. Powers, *J. Am. Chem. Soc.*, 1970, **92**, 137.
158. W. A. Nugent and T. V. RajanBabu, *J. Am. Chem. Soc.*, 1988, **110**, 8561.
159. D. P. Curran, T. L. Fevig and M. Totleben, *Synlett*, 1990, 773.
160. J. K. Crandall and M. Mualla, *Tetrahedron Lett.*, 1986, **27**, 2243; E. J. Corey and S. G. Pyne, *Tetrahedron Lett.*, 1983, **24**, 2821; G. Pattenden and G. M. Robertson, *Tetrahedron*, 1985, **41**, 4001.
161. G. A. Molander and C. Kenny, *Tetrahedron Lett.*, 1987, **28**, 4367.
162. E. Kariv-Miller and T. J. Mahachi, *J. Org. Chem.*, 1986, **51**, 1041; J. E. Swartz, T. J. Mahachi and E. Kariv-Miller, *J. Am. Chem. Soc.*, 1988, **110**, 3622; T. Shono, I. Nishiguchi, H. Ohmizu and M. Mitani, *J. Am. Chem. Soc.*, 1978, **100**, 545.
163. J. Cossy, D. Belloto and J. P. Pète, *Tetrahedron Lett.*, 1987, **28**, 4547.
164. R. D. Little, D. P. Fox, L. V. Hijfte, R. Dannecker, G. Sowell, R. L. Wolin, L. Moëns and M. M. Baizer, *J. Org. Chem.*, 1988, **53**, 2287.
165. S. Fukuzawa, A. Nakanishi, T. Fujinami and S. Sakai, *J. Chem. Soc., Perkin Trans. 1*, 1988, 1669.
166. A. S. Kende, K. Koch and C. A. Smith, *J. Am. Chem. Soc.*, 1988, **110**, 2210.
167. M. Tada, T. Nakamura and M. Matsumoto, *Chem. Lett.*, 1987, 409.
168. A. Citterio and R. Santi, in 'Free Radicals in Synthesis and Biology', ed. F. Minisci, Kluwer, Dordrecht, 1989, p. 187.
169. T. Shono, N. Kise, T. Suzumoto and T. Morimoto, *J. Am. Chem. Soc.*, 1986, **108**, 4676.
170. H. Togo and O. Kikuchi, *Heterocycles*, 1989, **28**, 373.
171. N. S. Narasimhan and I. S. Aidhen, *Tetrahedron Lett.*, 1988, **29**, 2987.
172. A. J. Floyd, S. F. Dyke and S. E. Ward, *Chem. Rev.*, 1976, **76**, 509.
173. M. Newcomb, S.-U. Park, J. Kaplan and D. J. Marquardt, *Tetrahedron Lett.*, 1985, **26**, 5651.
174. J. C. Scaiano, *J. Am. Chem. Soc.*, 1980, **102**, 5399.
175. M. Newcomb, M. T. Burchill and T. M. Deeb, *J. Am. Chem. Soc.*, 1988, **110**, 6528.
176. M. Newcomb and T. M. Deeb, *J. Am. Chem. Soc.*, 1987, **109**, 3163.
177. M. Tokuda, Y. Yamada, T. Takagi and H. Suginome, *Tetrahedron Lett.*, 1985, **26**, 6085; *Tetrahedron*, 1987, **43**, 281.
178. R. Sutcliffe and K. U. Ingold, *J. Am. Chem. Soc.*, 1982, **104**, 6071.
179. F. Minisci, *Synthesis*, 1973, 1.
180. L. Stella, *Angew. Chem., Int. Ed. Engl.*, 1983, **22**, 337.
181. C. A. Broka and J. F. Gerlits, *J. Org. Chem.*, 1988, **53**, 2144.
182. G. A. Kraus and J. Thurston, *Tetrahedron Lett.*, 1987, **28**, 4011.
183. A. L. J. Beckwith and B. P. Hay, *J. Am. Chem. Soc.*, 1988, **110**, 4415.
184. G. D. Vite and B. Fraser-Reid, *Synth. Commun.*, 1988, **18**, 1339.
185. B. Fraser-Reid, G. D. Vite, B.-W. A. Yeung and R. Tsang, *Tetrahedron Lett.*, 1988, **29**, 1645.
186. K. Furuta, T. Nagata and H. Yamamoto, *Tetrahedron Lett.*, 1988, **29**, 2215.
187. P. de Armas, C. G. Francisco, R. Hernández, J. A. Salazar and E. Suárez, *J. Chem. Soc., Perkin Trans. 1*, 1988, 3255.
188. M. E. Wolff, *Chem. Rev.*, 1963, **63**, 55.
189. P. A. Bartlett, K. L. McLaren and P. C. Ting, *J. Am. Chem. Soc.*, 1988, **110**, 1633.
190. D. L. J. Clive, P. L. Beaulieu and L. Set, *J. Org. Chem.*, 1984, **49**, 1313.
191. Azo compounds have also been used as acceptors; A. L. J. Beckwith, S. Wang and J. Warkentin, *J. Am. Chem. Soc.*, 1987, **109**, 5289.
192. R. Tsang and B. Fraser-Reid, *J. Am. Chem. Soc.*, 1986, **108**, 2116.
193. R. Tsang and B. Fraser-Reid, *J. Am. Chem. Soc.*, 1986, **108**, 8102.
194. R. Tsang, J. K. Dickson, Jr., H. Pak, R. Walton and B. Fraser-Reid, *J. Am. Chem. Soc.*, 1987, **109**, 3484.
195. J. I. Seeman, *Chem. Rev.*, 1983, **83**, 83.
196. See pages 427, 428 and 508 in ref. 34.
197. A. L. J. Beckwith and B. P. Hay, *J. Am. Chem. Soc.*, 1989, **111**, 230, 2674.
198. Because these cyclizations are less exothermic than the hexenyl radical cyclization, their TSs must be later. Thus, at the transition state, radical (**61a**) 'feels' some of the cyclopentane strain energy. This is probably why the 6-exo cyclization of (**61a**) is not more rapid than the 6-exo cyclization of (**61b**).
199. Germanium and silicon hydrides should be even more selective toward alkoxyl radicals relative to carbon radicals. Indeed, while silicon hydrides react rapidly with alkoxyl radicals, reactions with carbon radicals are too slow to propagate chains. The rapid addition of silyl and germyl radicals to C=O bonds is a possible complicating reaction.

200. A. L. J. Beckwith, R. Kazlauskas and M. R. Syner-Lyons, *J. Org. Chem.*, 1983, **48**, 4718.
201. D. E. O'Dell, J. T. Loper and T. L. Macdonald, *J. Org. Chem.*, 1988, **53**, 5225.
202. H. Suginome, C. F. Liu, S. Seko, K. Kobayashi and A. Furusaki, *J. Org. Chem.*, 1988, **53**, 5952.
203. R. Freire, J. J. Marrero, M. S. Rodriguez and E. Suárez, *Tetrahedron Lett.*, 1986, **27**, 383; C. G. Francisco, R. Freire, M. S. Rodriguez and E. Suarez, *Tetrahedron Lett.*, 1987, **28**, 3397.
204. H. Suginome and S. Yamada, *Tetrahedron Lett.*, 1987, **28**, 3963.
205. H. Suginome and S. Yamada, *Tetrahedron*, 1987, **43**, 3371.
206. S. L. Schreiber, B. Hulin and W.-F. Liew, *Tetrahedron*, 1986, **42**, 2945.
207. M. Ochiai, S. Iwaki, T. Ukita and Y. Nagao, *Chem. Lett.*, 1987, 133.
208. M. G. O'Shea and W. Kitching, *Tetrahedron*, 1989, **45**, 1177.
209. D. Bellus, *Pure Appl. Chem.*, 1985, **57**, 1827.
210. C. Jasperse and B. Yoo, unpublished results, University of Pittsburgh.
211. G. E. Keck and J. H. Byers, *J. Org. Chem.*, 1985, **50**, 5442.
212. G. Stork and R. Mook, Jr., *J. Am. Chem. Soc.*, 1983, **105**, 3720.
213. D. P. Curran and D. M. Rakiewicz, *J. Am. Chem. Soc.*, 1985, **107**, 1448; *Tetrahedron*, 1985, **41**, 3943; D. P. Curran and M.-H. Chen, *Tetrahedron Lett.*, 1985, **26**, 4991; D. P. Curran and S.-C. Kuo, *J. Am. Chem. Soc.*, 1986, **108**, 1106; *Tetrahedron*, 1987, **43**, 5653.
214. K. A. Parker, D. M. Spero and J. Van Epp, *J. Org. Chem.*, 1988, **53**, 4628.
215. S. Kanno and T. Ohya, *Chem. Pharm. Bull.*, 1988, **36**, 4095.
216. P. J. Parsons, P. A. Willis and S. C. Eyley, *J. Chem. Soc., Chem. Commun.*, 1988, 283.
217. B. B. Snider and M. A. Dombroski, *J. Org. Chem.*, 1987, **52**, 5487.
218. T. L. Fevig, R. L. Elliott and D. P. Curran, *J. Am. Chem. Soc.*, 1988, **110**, 5064.
219. A. L. J. Beckwith, D. H. Roberts, C. H. Schiesser and A. Wallner, *Tetrahedron Lett.*, 1985, **26**, 3349.
220. L. Stella, D. R. Boate and E. Guittet, in 'Free Radicals in Synthesis and Biology', ed. F. Minisci, Kluwer, Dordrecht, 1989, p. 145.
221. D. H. R. Barton, *Pure Appl. Chem.*, 1968, **16**, 1.
222. A. Johns and J. A. Murphy, *Tetrahedron Lett.*, 1988, **29**, 837; Z. Cekovic and D. Ilijev, *Tetrahedron Lett.*, 1988, **29**, 1441.
223. D. P. Curran, D. Kim, H. T. Liu and W. Shen, *J. Am. Chem. Soc.*, 1988, **110**, 5900; see also D. Lathbury, P. J. Parsens and I. Pinto, *J. Chem. Soc., Chem. Commun.*, 1988, 81.
224. A. L. J. Beckwith, D. M. O'Shea and S. W. Westwood, *J. Am. Chem. Soc.*, 1988, **110**, 2565.
225. P. Dowd and S.-C. Choi, *Tetrahedron*, 1989, **45**, 77.
226. J. Baldwin, R. M. Adlington and J. Robertson, *Tetrahedron*, 1989, **45**, 909.
227. W. B. Motherwell and J. D. Harling, *J. Chem. Soc., Chem. Commun.*, 1988, 1380.
228. M. C. M. de C. Alpoim, A. D. Morris, W. B. Motherwell and D. M. O'Shea, *Tetrahedron Lett.*, 1988, **29**, 4173.
229. G. Stork and P. M. Sher, *J. Am. Chem. Soc.*, 1983, **105**, 6765.
230. H. Togo and O. Kikuchi, *Tetrahedron Lett.*, 1988, **29**, 4133.
231. O. Moriya, M. Kakihana, Y. Urata, T. Sugizaki, T. Kageyama, Y. Ueno and T. Endo, *J. Chem. Soc., Chem. Commun.*, 1985, 1401.
232. G. E. Keck and D. A. Burnett, *J. Org. Chem.*, 1987, **52**, 2958.
233. Z. Cekovic and R. Saicic, *Tetrahedron Lett.*, 1986, **27**, 5893.
234. D. H. R. Barton, S. Z. Zard and E. daSilva, *J. Chem. Soc., Chem. Commun.*, 1988, 285.
235. D. L. J. Clive and A. G. Angoh, *J. Chem. Soc., Chem. Commun.*, 1985, 980.
236. D. P. Curran and M.-H. Chen, *J. Am. Chem. Soc.*, 1987, **109**, 6558.
237. D. P. Curran, C.-T. Chang, E. Spletzer, C. M. Seong and M.-H. Chen, *J. Am. Chem. Soc.*, 1989, **111**, 8872. .
238. D. P. Curran and P. van Elburg, *Tetrahedron Lett.*, 1989, **30**, 2501.
239. K. S. Feldman, R. E. Simpson and M. Parvez, *J. Am. Chem. Soc.*, 1986, **108**, 1328.
240. K. S. Feldman, A. L. Romanelli, R. E. Ruckle, Jr. and R. F. Miller, *J. Am. Chem. Soc.*, 1988, **110**, 3300.
241. K. Miura, K. Fugami, K. Oshima and K. Utimoto, *Tetrahedron Lett.*, 1988, **29**, 5135.
242. K. S. Feldman and Y. B. Lee, *J. Am. Chem. Soc.*, 1987, **109**, 5850.
243. F. Minisci, *Synthesis*, 1973, 1.
244. K. Mizuno, M. Ikeda, S. Toda and Y. Otsuji, *J. Am. Chem. Soc.*, 1988, **110**, 1288.

4.3
Vinyl Substitutions with Organopalladium Intermediates

RICHARD F. HECK
University of Delaware, Newark, DE, USA

4.3.1 INTRODUCTION

4.3.1.1 Basic Reaction

Organopalladium halides, acetates, triflates and similar derivatives lacking β sp^3-bonded hydrogens generally react easily with unhindered alkenes to form new alkenes in which an original, vinyl hydrogen is replaced by the organic group of the palladium reactant. The reactions occur in solution. Other Group VIII metals are ineffective in bringing about this transformation in acceptable yields. In the palladium reaction the organopalladium(II) reactant is reduced to the zero valent state while the halide, acetate, triflate or other, initial anionic ligand is converted into the corresponding acid (equation 1). If alkyl groups are attached directly to the double-bond carbons, double-bond rearrangement may occur and allylically substituted products can be produced. Normally, substitution occurs before any double-bond rearrangement occurs in the reactant alkene and the substituent is placed on one of the initial double-bond carbons (equation 2). Rearrangement of the double bond along a hydrocarbon chain to positions more distant from the substituent than the allylic position is occasionally observed, most often when X is a halogen and the halogen acid produced by the reaction is not neutralized as it is formed by inclusion of a base.

RPdX + H(alkene) → R(alkene) + Pd + HX (1)

RPdX + H, H, H (alkene) → R, H, H (alkene) and/or R, H, H (alkene) + Pd + HX (2)

If two hydrogen atoms on a β-carbon atom are available for elimination, (*E*)-alkenes are strongly preferred. If only one hydrogen atom is present the product stereochemistry will be predictable on the basis of a *syn* addition of the organopalladium group to the double bond followed by a *syn* elimination of a palladium hydride group, provided the reaction is conducted under the proper conditions as described in Section 4.3.5.1.2.i. Yields of substituted alkenes from these reactions generally decline as the number and size of the substituents on the vinyl carbons increase.

Other atoms or groups than hydrogen may be lost in the final elimination step. Chlorine (or bromine or iodine) especially is lost easily, in fact in preference to hydrogen when it occurs in positions β to the palladium group.

Some isolated organopalladium compounds, particularly cyclopalladated complexes, have been employed in the vinyl substitution, but most often simple organopalladium compounds are unstable and must be formed *in situ*. Even then, alkylpalladium derivatives with β-hydrogens (on saturated carbon) generally decompose by elimination of Pd—H too rapidly to be useful except in some favorable intramolecular examples. The situation is somewhat complicated by the fact that palladium(II) compounds are usually four coordinate and the organopalladium derivatives will normally have two, two-electron donor ligands associated with them in addition to the organic group and the anionic ligand. These ligands may be various species such as solvent, halide ions, organophosphines or amines. The basic chemistry of the organopalladium derivatives is not usually greatly altered by the ligands but they can and do affect the stability of intermediates, reaction rates and selectivity. Under the proper conditions vinyl substitution can be carried out with aryl, heterocyclic, vinyl and occasionally alkyl derivatives lacking β-hydrogens. A very few examples of vinyl substitution with acyl groups are known, also.

Three methods are commonly employed for the *in situ* preparation of organopalladium derivatives: (i) direct metallation of an arene or heterocyclic compound with a palladium(II) salt; (ii) exchange of the organic group from a main group organometallic to a palladium(II) compound; and (iii) oxidative addition of an organic halide, triflate or aryldiazonium salt to palladium(0) or a palladium(0) complex.

The vinyl substitution reaction often may be achieved with catalytic amounts of palladium. Catalytic reactions are carried out in different ways depending on how the organopalladium compound is generated. Usually copper(II) chloride or *p*-benzoquinone is employed to reoxidize palladium(0) to palladium(II) in catalytic reactions when methods (i) or (ii) are used for making the organopalladium derivative. The procedures developed for making these reactions catalytic are not completely satisfactory, however. The best catalytic reactions are achieved when the organopalladium intermediates are obtained by the oxidative addition procedures (method iii), where the halide is both the reoxidant and a reactant. Reviews of some aspects of these reactions have been published.[1a–1e]

Methods (i) and (ii) require palladium(II) salts as reactants. Either palladium acetate, palladium chloride or lithium tetrachloropalladate(II) usually are used. These salts may also be used as catalysts in method (iii) but need to be reduced *in situ* to become active. The reduction usually occurs spontaneously in reactions carried out at 100 °C but may be slow or inefficient at lower temperatures. In these cases, zero valent complexes such as bis(dibenzylideneacetone)palladium(0) or tetrakis(triphenylphosphine)palladium(0) may be used, or a reducing agent such as sodium borohydride, formic acid or hydrazine may be added to reaction mixtures containing palladium(II) salts to initiate the reactions. Triarylphosphines are usually added to the palladium catalysts in method (iii), but not in methods (i) or (ii). Normally, 2 equiv. of triphenylphosphine, or better, tri-*o*-tolylphosphine, are added per mol of the palladium compound. Larger amounts may be necessary in reactions where palladium metal tends to precipitate prematurely from the reaction mixtures. Large concentrations of phosphines are to be avoided, however, since they usually inhibit the reactions.

4.3.1.2 Chapter Coverage

This chapter will be concerned only with substitutions of vinyl hydrogens with groups forming carbon–carbon bonds to one of the vinyl carbons. Replacements of halogens or other heteroatom-bound groups at vinylic positions will not be included. Likewise, replacements of vinyl hydrogens by groups forming heteroatom bonds to vinyl carbons will not be covered. Substitutions at alkynyl carbons by any substituents and additions to alkynes also will not be discussed.

4.3.1.3 Advantages and Disadvantages of the Various Procedures

Normally, the most practical vinyl substitutions are achieved by use of the oxidative additions of organic bromides, iodides, diazonium salts or triflates to palladium(0)–phosphine complexes *in situ*. The organic halide, diazonium salt or triflate, an alkene, a base to neutralize the acid formed and a catalytic amount of a palladium(II) salt, usually in conjunction with a triarylphosphine, are the usual reactants at about 25–100 °C. This method is useful for reactions of aryl, heterocyclic and vinyl derviatives. Acid chlorides also react, usually yielding decarbonylated products, although there are a few exceptions. Likewise, arylsulfonyl chlorides lose sulfur dioxide and form arylated alkenes. Aryl chlorides have been reacted successfully in a few instances but only with the most reactive alkenes and usually under more vigorous conditions. Benzyl iodide, bromide and chloride will benzylate alkenes but other alkyl halides generally do not alkylate alkenes by this procedure.

The main group organometallic exchange with palladium salts is useful for vinylic substitution with all types of groups: aryl, heterocyclic, vinyl and alkyl groups, but the only effective way to make the reaction catalytic is to use a stoichiometric amount of copper(II) chloride with a catalytic quantity of palladium. This procedure is acceptable on a small scale but is a problem on a large scale because of the large quantities and thick slurry of copper(II) salts involved. Other reoxidants, such as *p*-benzoquinone, silver salts, peracids or oxygen, are often not effective or give only low turnovers of the palladium.

The direct palladation procedure is limited to substitution with aryl and heterocyclic groups. The metallation is an electrophilic process and therefore does not work well with deactivated aromatics. When possible, mixtures of isomers may be obtained. Since the palladium salts employed for the metallation are moderately strong oxidizing agents, the reaction cannot be used with easily oxidizable alkenes or aromatics. The only effective method for making this procedure catalytic is to reoxidize the palladium *in situ* with oxygen under pressure; an inconvenient and potentially dangerous procedure.

Stable, cyclopalladated complexes sometimes may be used in vinyl substitution but these reactions cannot be carried out catalytically. A major limitation is that few cyclopalladated structures are available and only a few types of vinyl substitution products can be made by this procedure.

4.3.2 VINYL SUBSTITUTION WITH ARENES AND HETEROCYCLES

A variety of arenes and heterocyclic compounds will react with alkenes in the presence of palladium(II) salts to produce vinyl substitution products (equation 3).

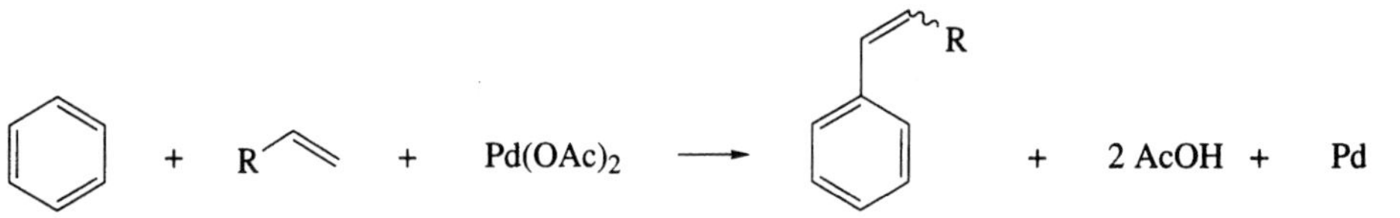

+ R + $Pd(OAc)_2$ ⟶ + 2 AcOH + Pd (3)

4.3.2.1 Mechanism

Two mechanisms of reaction appear possible depending upon the reactants. Initial palladation of either the alkene or the arene may occur: alkene palladation is proposed to occur initially in reactions of chlorinated alkenes.[2,3] Alkenes without strongly electron-withdrawing chloride substituents and moderately reactive arenes generally react *via* an initial arene palladation, and the reaction takes the route shown in Scheme 1.[4]

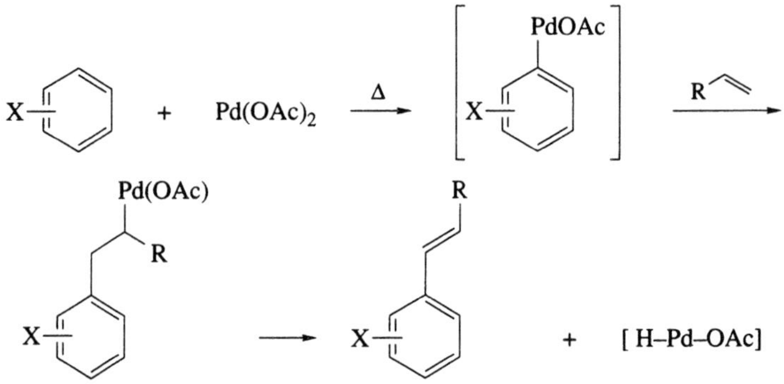

X + $Pd(OAc)_2$ $\xrightarrow{\Delta}$ [PdOAc, X] $\xrightarrow{R}$

Pd(OAc), R, X ⟶ R, X + [H–Pd–OAc]

Scheme 1

4.3.2.2 Limitations

The mechanism shown in Scheme 1 seems to be the one operating in most of the known examples of the reaction. The initial aromatic palladation is an electrophilic substitution, but unfortunately it is not a very selective one.[4] Therefore, isomeric mixtures of products may be expected in cases where palladation can occur at different aromatic positions.

4.3.2.3 Reaction Scope and Conditions

4.3.2.3.1 Stoichiometric reactions

The majority of examples reported have been carried out in refluxing acetic acid with stoichiometric amounts of palladium acetate.[5] Occasionally lower temperatures give higher yields, however.

Several examples of the cyclization of indole derivatives with alkenic side chains in the 3-position have been reported.[6] In these examples, palladium chloride in combination with silver tetrafluoroborate is the cyclizing agent. The palladium tetrafluoroborate, presumably formed, should be a very reactive palladating species and probably is the reason why these reactions proceed at room temperature, although the mechanism is not yet completely clear. These reactions were worked up reductively (by addition of sodium borohydride) in order to reduce the expected alkenic product or any relatively stable organopalladium complexes that may have been formed (equation 4).[6]

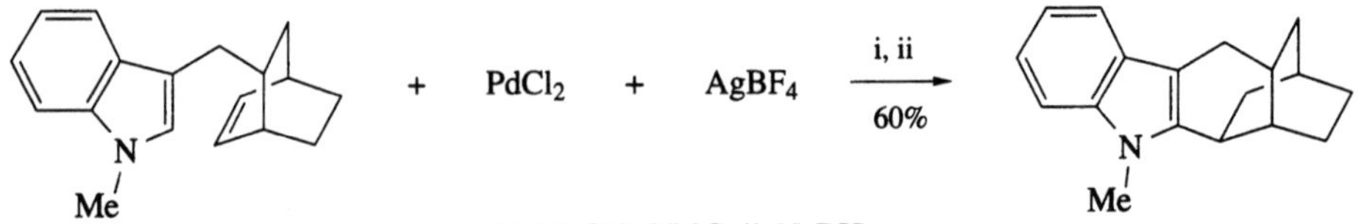

N Me + $PdCl_2$ + $AgBF_4$ $\xrightarrow[60\%]{\text{i, ii}}$ N Me (4)

i, Et_3N, MeCN, 25 °C; ii, $NaBH_4$

Several other heterocyclic compounds have been used in the vinyl substitution. Quinones, for example, are readily substituted with a variety of heterocycles as well as with benzene.[7] Heterocycles which react include 2-acetylfuran (5-position), furfural (5-position) (equation 5) methyl furoate (5-position), 2-acetylthiophene (5-position), *N*-benzenesulfonylpyrrole (2-position), *N*-benzenesulfonylindole (3-position), 4-pyrone (3-position) and *N*-methyl-2-pyridone (3- and 5-positions).

$$\text{1,4-naphthoquinone} + \text{furfural} + Pd(OAc)_2 \xrightarrow[\Delta,\ 79\%]{AcOH} \text{2-(5-formyl-2-furyl)-1,4-naphthoquinone} + 2AcOH + Pd \quad (5)$$

Acrylate esters have been substituted with *N*-methyl-2-pyridone (5-position), furfural (5-position) and 2-thiophenecarbaldehyde (5-position).[8] Benzo[*b*]furan reacts similarly with alkenes (2-position).[9] Alkene substitutions with furan,[10] thiophene,[10] benzothiophene[10] and *N*-acetylindoles,[10,11] also, have been reported. Furan and thiophene yield mainly 2,5-dialkenylated products.[9] α-Substituted chalcones with strongly electron-withdrawing substituents substitute with benzene.[12] Ferrocene[13] and η^4-cyclobutadienetricarbonyliron[14] will substitute alkenes as well.

4.3.2.3.2 Catalytic methods

Silver acetate has a small catalytic effect on the alkene substitution reaction but 5 equiv. of the salt only give 140% of stilbene in the styrene phenylation, based upon palladium.[15] The same reaction carried out at 80 °C under 300 lbf in^{-2} (1 lbf in^{-2} = 6.89 kPa) of oxygen gives stilbene in 248% yield, based upon palladium.[16] The best reoxidation reagent is *t*-butyl perbenzoate, which yields 10–14 turnovers of the palladium in the vinyl substitution of cinnamaldehyde and similar alkenes with benzene.[17]

Methyl acrylate and benzene under the above conditions give a mixture of methyl cinnamate and methyl 3,3-diphenylacrylate.[17]

4.3.3 VINYL SUBSTITUTION WITH CYCLOPALLADATED COMPLEXES

A variety of isolable cyclopalladated complexes are readily available by direct reaction of palladium(II) salts, usually the chloride or acetate, with aromatic derivatives. The aromatic compound needs only to have a side chain with a good coordinating atom, usually nitrogen (although sulfur, oxygen, arsenic and phosphorus atoms also will work), in a position such that a five- or rarely a four- or six-membered chelate ring can form when orthopalladation occurs.[18] Reactions are slowed or even totally inhibited by strongly electron-withdrawing substituents in the ring being palladated.

The cyclopalladated complexes have limited practical value in organic synthesis because of the high cost of the stoichiometric amount of palladium required. Some compensation for their expense may be obtained, however, if several steps can be saved compared with an alternate method and, of course, the palladium may be recovered relatively easily.[1a]

Vinyl substitutions with *N*,*N*-dialkylbenzylamine-cyclopalladated complexes have been studied most thoroughly. In the presence of triethylamine the substitution occurs quite selectively with styrene derivatives[19] and α,β-unsaturated carbonyl compounds.[20] For example, chloride-bridged cyclopalladated *N*,*N*-dimethylbenzylamine dimer and methyl vinyl ketone give 92% of *o*-dimethylaminomethylbenzalacetone in 1 h at 110 °C (equation 6).

$$\text{[cyclopalladated } N,N\text{-dimethylbenzylamine–PdCl]}_2 + 2\ \text{methyl vinyl ketone} + 2\,Et_3N \xrightarrow{92\%} 2\ o\text{-}(Me_2NCH_2)C_6H_4CH{=}CHCOMe + 2\,Et_3NH^+Cl^- + 2\,Pd \quad (6)$$

4.3.4 VINYL SUBSTITUTION WITH MAIN GROUP ORGANOMETALLICS

A variety of main group organometallics and some nonmetal organic derivatives have served as the source of organic groups, transferred *via* palladium to the vinyl carbons of alkenes.

4.3.4.1 Organomercury Compounds

Organomercury compounds are often obtainable from the reaction of Grignard reagents with mercury(II) chloride or in the aromatic series by direct mercuration of the aromatic compound. They are air and water stable but have the disadvantages of being highly toxic, having high molecular weights and being relatively expensive. Some arylmercury(II) chlorides also have quite low solubility in organic solvents, which may limit their reactivity. Exchange of the organic group from an organomercury species to a palladium salt (usually chloride or acetate) frequently occurs at or below room temperature, allowing vinyl substitutions to be carried out at or below 25 °C.

4.3.4.1.1 Alkylation

Vinyl substitution with aliphatic groups has been achieved with several mercurials in which the organic substituents are lacking sp^3-bound hydrogens on a carbon one-removed from the mercury. Other aliphatic organomercury reagents generally yield intermediate palladium derivatives which eliminate palladium hydride faster than vinyl substitution takes place, except where favorable intramolecular reactions can occur. Methyl,[21] alkoxycarbonyl,[21] neopentyl[22] and neophyl[22] groups have been substituted for vinyl hydrogens by this reaction (equation 7).

$$\mathrm{Bu^t{\diagup\!\!\diagdown}HgOAc} + \mathrm{Pd(OAc)_2} + \mathrm{{=\!\!\diagdown}CO_2Me} \xrightarrow[25\ ^\circ C]{\mathrm{MeCN}}$$

$$\underset{94\%}{\mathrm{Bu^t{\diagup\!\!\diagdown}{=}CO_2Me}} + \mathrm{Pd} + \mathrm{HOAc} + \mathrm{Hg(OAc)_2} \qquad (7)$$

Usually, lithium tetrachloropalladate(II) or palladium acetate have served as the palladium reagents because of their solubilities in organic solvents such as acetonitrile and methanol. This reaction has only been carried out with stoichiometric quantities of the palladium salt. The neophyl group has been observed to rearrange or partially rearrange in its reactions with methyl acrylate or styrene.[22] The regiochemistry of the addition of aliphatic groups to unsymmetrical alkenes is known in only a few cases where reactions are quite selective. The results are consistent with the general rule that vinyl substitution takes place (predominantly or exclusively) at the least-substituted vinyl carbon and/or at the vinyl carbon having the smallest substituents. Substitution, also, usually occurs β in α,β-unsaturated esters and other α,β-unsaturated carbonyl compounds. It is to be expected that vinyl substitution with electron-donating aliphatic groups on alkenes with electron-donating substituents will give mixtures of regioisomeric products in the absence of significant steric effects.

4.3.4.1.2 Arylation

Vinyl substitution with arylmercury reagents is a very general reaction.[21] Arylmercury(II) chlorides and acetates as well as diarylmercury compounds with various substituents — diethylamino, nitro, carboxy, chloro, amide, hydroxy, and aldehyde groups — have been employed in the reaction. Naphthylmercury reagents and very hindered compounds such as mesitylmercury(II) chloride react normally. The regiochemistry of the arylation with arylmercury(II) acetates follows the usual rules as stated in Section 4.3.4.1.1. (*E*)-Alkene products are strongly favored if two β-hydrogens on one carbon are available for elimination, while if only one is present the product sterochemistry is predictable on the basis of a *syn* addition of the organopalladium species followed by a *syn* elimination of the palladium hydride. Arylmercury(II) chlorides, however, do not give very selective reactions. Both *cis–trans* isomerization and double-bond migration are seen in the chloride reactions.[23] Quantitative data on regiochemistry for the *p*-methoxyphenylation of propene indicates solvent also can play a significant role.[23]

All of the organomercury reactions may be carried out catalytically in palladium by use of copper(II) chloride as a reoxidant for the palladium. This modification, however, often reduces the yield of product and lowers the stereospecificity of the arylation.[21,23]

4.3.4.1.3 Vinyl substitution of dienes

Vinyl substitution occurs with conjugated dienes as well as with alkenes, employing aryl-, vinyl-, methyl-, alkoxycarbonyl- or benzyl-mercury reagents and lithium tetrachloropalladate(II), but the products are usually π-allylpalladium complexes if the reactions are carried out under mild conditions (equation 8).[24,25] The π-allylic complexes may be decomposed thermally to substituted dienes[26] or reacted with nucleophiles to form allylic derivatives of the nucleophile. Secondary amines, for example, react to give tertiary allylic amines in modest yields, along with dienes and reduced dienes (equation 9).[25]

PhHgCl + [isoprene] + $Li_2[PdCl_4]$ ⟶ [π-allyl Pd Cl dimer] + 2 LiCl + $HgCl_2$ (8)

[π-allyl Pd Cl dimer] + piperidine —(r.t., 3 h)→ (9)

30% 43% 13%

4.3.4.1.4 Formation of heterocyclic derivatives

It is to be expected that organomercury reagents of all of the heterocycles that are employed in the palladation–vinyl substitution reaction (Section 4.3.2.3.1) could be used in the exchange palladation–substitution, also. Few of these organomercury reagents have been tried in the reaction, however. 2-Chloromercuriothiophene[21] and both chloromercurio- and 1,1-bischloromercurio-ferrocene[27] react as expected with styrene and with α,β-unsaturated carbonyl compounds. 4-Acetoxymercurio-*N*-methylisocarbostyril also yields the expected products with styrene and methyl acrylate.[28] Pyrimidinylmercury(II) acetates have found application in the synthesis of *C*-nucleosides in their reactions with chiral furanoid glycals[29] and other unsaturated cyclic ethers.[30]

4.3.4.1.5 Vinylation

In the absence of base, vinylmercury reagents and lithium tetrachloropalladate(II) react with alkenes to form π-allylpalladium complexes arising from addition of the 'vinylpalladium chloride' to the alkene followed by palladium hydride elimination, a reverse readdition and π-allyl formation (equation 10).[31]

Bu^tCH=CHHgCl + $Li_2[PdCl_4]$ + CH_2=CHCO_2Et —(THF, 0 °C, 90%)→ [π-allyl Pd Cl dimer] (10)

In the presence of triethylamine, decomposition of the π-allylic complexes to conjugated dienes may occur, particularly when electron-withdrawing substituents are present on a methyl group in the 1- or 3-position of the π-allyl system.[31] Cyclic alkenes and vinylpalladium chlorides also yield π-allylic complexes in the absence of an amine. If a tertiary amine is present, however, 1,4-dienes are obtained (equation 11).[32]

HgCl + + $Li_2[PdCl_4]$ + Et_3N →

PdCl —84%→ (11)

4.3.4.2 Silanes and Silyl Enol Ethers

4.3.4.2.1 Vinyl substitution with silanes

One organic group is readily transferred from tetraorganosilanes (and some other silanes) to palladium(II). Tetramethylsilane, lithium tetrachloropalladate(II) and styrene at 120 °C in acetonitrile solution form 1-phenyl-1-propene in 65% yield along with *ca.* 1.5% 2-phenyl-1-propene.[33] Trimethylphenylsilane transfers phenyl and with styrene under the above conditions gives *trans*-stilbene in 94% yield.[33] Similar vinyl substitution reactions have been achieved with potassium (*E*)-alkenyl pentafluorosilicates.[34]

4.3.4.2.2 Vinyl substitution with silyl enol ethers

Palladium(II) enolates are produced by the exchange reaction of trimethylsilyl enol ethers with palladium chloride or acetate. If β-hydrogens are present the enolates usually decompose by eliminating palladium hydride, forming α,β-unsaturated ketones (equation 12).[35] *p*-Benzoquinone (50%) was added in this example as a reoxidant for the palladium. Palladium enolates have been isolated in cases where β-hydrogens are absent.[36] The isolated complexes react with ethylene to yield the α-vinylated ketones.[36] It may be presumed that palladium enolates possessing β-hydrogens do not successfully substitute alkenes intermolecularly, since no examples have been reported. Intramolecular examples are well known, however, and these reactions provide a useful method for preparing five- and six-membered ring ketones (equation 13).[37]

Me_3SiO + $Pd(OAc)_2$ —*p*-benzoquinone; MeCN, 30 h, r.t.→ O (94%) + O (5%) (12)

$OSiMe_3$ + $Pd(OAc)_2$ —MeCN; 10 h, r.t.; 87%→ O + $AcOSiMe_3$ + Pd + AcOH (13)

Bicyclic ketones, also, have been prepared intramolecularly from silyl enol ethers. Six-membered rings are formed more easily in these reactions than in the reactions forming monocyclic products, described earlier.[38]

The mechanism of the trimethylsilyl enol ether cyclization may involve formation of a palladium enolate which adds to the double bond. However, another mechanism is also possible involving attack of a palladium(II)–alkene complex upon the silyl enol ether double bond.

4.3.4.3 Organothallium Reagents

Thallation of aromatic compounds with thallium tris(trifluoroacetate) proceeds more readily than mercuration and in many instances more selectively. *Ortho* derivatives are obtained frequently when functional groups with unshared electron pairs are present to direct the metallation.[39]

Several examples of the arylation of vinyl ketones with arylthallium bis(trifluoroacetate) and lithium tetrachloropalladate(II) have been reported (equation 14).[40]

$$PhTl(O_2CCF_3)_2 + \text{(methyl vinyl ketone)} + Li_2[PdCl_4] \xrightarrow[\text{25 °C, 98\%}]{\text{THF}} PhCH{=}CHC(O)CH_3 \qquad (14)$$

Benzoic acid and its derivatives orthothallate. Reaction of *o*-carboxyphenylthallium bis(trifluoroacetate) with palladium chloride and alkenes followed by base treatment yields isocoumarin derivatives (equation 15).[41] Cyclization does not occur until base is added in the second step.

$$o\text{-}(CO_2H)C_6H_4Tl(O_2CCF_3)_2 + Li_2[PdCl_4] + PhCH{=}CH_2 \xrightarrow[\text{79\%}]{\text{i, ii}} \text{3-phenylisocoumarin} \qquad (15)$$

i, $PdCl_2$, MeCN, 25 °C; ii, Et_3N, Na_2CO_3

Several similar reactions have been carried out with other orthothallated derivatives to produce various oxygen and nitrogen heterocycles.[42]

4.3.4.4 Vinyl Substitution with Other Organometal and Nonmetal Derivatives

A very few examples of the use of aryl-tin,[21] -lead,[21] -lithium[43] and -magnesium[21,43,44] derivatives as the source of the aryl group in vinyl substitutions have been reported. Tin derivatives have been used with palladium dichloride bis(benzonitrile) and a copper(II) chloride reoxidant in a regioselective synthesis of oxygen heterocycles from unsaturated alcohols.[45]

Organo-iron and -cobalt phthalocyanines are reported to transfer the organic groups to palladium salts, and they have been used to substitute alkenes.[46]

There is one report of vinyl substitution with an organosodium derivative.[47] Organoboronic acids, also, have been used.[48]

Triarylphosphines, which are often employed with palladium in catalysts, also can transfer aryl groups to the palladium and cause vinyl substitution reactions with alkenes.[49,50] Fortunately, this reaction is usually slower than other methods for generating arylpalladium derivatives so that it usually is not a problem, but there are exceptions (equation 16).

$$PhCH{=}CH_2 + [Pd(PPh_3)_2(OAc)_2] \xrightarrow[\text{60 °C, 123 h}]{\text{HOAc}} PhCH{=}CHCH_2OAc + (Z)\text{-}PhCH{=}CHPh\ (28\%) + (E)\text{-}PhCH{=}CHPh\ (82\%) \qquad (16)$$

4.3.5 VINYL SUBSTITUTION WITH ORGANIC HALIDES, DIAZONIUM SALTS, ACID CHLORIDES, SULFINATE SALTS AND TRIFLATES

These reactions are grouped together because they all involve replacement of a leaving group by palladium(0) to form the organopalladium reactant. The ease of replacement varies significantly with the leaving group. The diazonium group is most easily replaced and reactions of these salts may be carried out at room temperatures. Triflate esters are less reactive and generally require temperatures of 50 °C or more to form the palladium derivatives. Under some conditions some vinyl iodides will react at room temperature, but generally higher temperatures (*ca.* 100 °C) are employed. Vinyl bromides and aryl iodides and bromides also are usually reacted at 100 °C. Carboxylic acid chlorides, sulfonyl chlorides and sulfinate salts also generally need a higher temperature.

These reactions do not require reoxidants since the organic compounds reoxidize the palladium(0) in the initial replacement step. Most results have been obtained with vinyl, aryl and heterocyclic iodides and bromides because of the easy and convenient availability of the required halides, and because this reaction was discovered for these reactants first.

It is necessary to have a base present in all of these reactions since strong acids are formed in every case. Secondary or tertiary amines, bicarbonates, carbonates and acetates have been the usual bases employed.

4.3.5.1 Vinyl Substitution with Alkyl, Aryl and Vinyl Halides

4.3.5.1.1 Alkylation and benzylation of alkenes

Alkyl halides with β-hydrogens generally undergo only elimination reactions under the conditions of the vinyl substitution (100 °C in the presence of an amine or other base). Exceptions are known only in cases where intramolecular reactions are favorable. Even alkyl halides without β-hydrogens appear not to participate in the intermolecular alkene substitution since no examples have been reported, with the exception of reactions with benzyl chloride and perfluoroalkyl iodides.

Benzyl chloride reacts easily with methyl acrylate in the presence of tri-*n*-butylamine and palladium acetate (1 mol %) as catalyst.[51] The product is a mixture of (*E*)-methyl 4-phenyl-3-butenoate (67%) and (*E*)-methyl 4-phenyl-2-butenoate (9%), arising from elimination–addition reactions of the palladium hydride group which largely isomerize the initial elimination product.

Several perfluoroalkyl iodides add to alkenes in the presence of 1 mol % of tetrakis(triphenylphosphine)palladium(0) at room temperature in fair to good yields forming fluoroalkyl iodides.[52] Palladium hydride elimination is less favorable than substitution of the palladium by iodide in these examples. This is due to the reaction proceeding by an unusual radical chain mechanism (equation 17).

$$CF_3CF_2I + \text{CH}_2{=}\text{CHBu}^n \xrightarrow[\text{r.t., 20 min}]{[Pd(PPh_3)_4]} CF_3CF_2CH_2CH(I)Bu^n \quad (17)$$

(18) [Scheme: N-allyl-N-benzyl iodoacetamide + 1,8-bis(dimethylamino)naphthalene, [Pd(PPh₃)₄], HMPA, 65 °C, 6 h → N-benzyl-4-methylene-2-pyrrolidinone (41%) + N-benzyl-4-(iodomethyl)-2-pyrrolidinone (26%) + $Me_2N\cdots H^+\cdots NMe_2$ naphthalene I^-]

[Pd(PPh3)4]

HMPA, 65 °C, 6 h

O, N, Ph, Me2N, NMe2, I

41% 26%

Several (*N*-alkenyl)iodoacetamides undergo cyclization when they are reacted with tetrakis(triphenylphosphine)palladium(0). The highest yields (which are generally only moderate) are obtained in DMF solution with 1,8-dimethylaminonaphthalene (Proton Sponge) to take up the hydrogen iodide formed. This reaction has been used to prepare piperidones, oxindoles, indolizidines, quinolizidines, pyrrolidines, indoles, quinolines[53] and pyrrolizidines (equation 18).[54]

Similar reactions in which β-hydrogens are available in the presumed intermediate palladium complex also give cyclized products.[55] Quite possibly these are radical chain reactions also, and that is why the hydride elimination is not observed (equation 19).

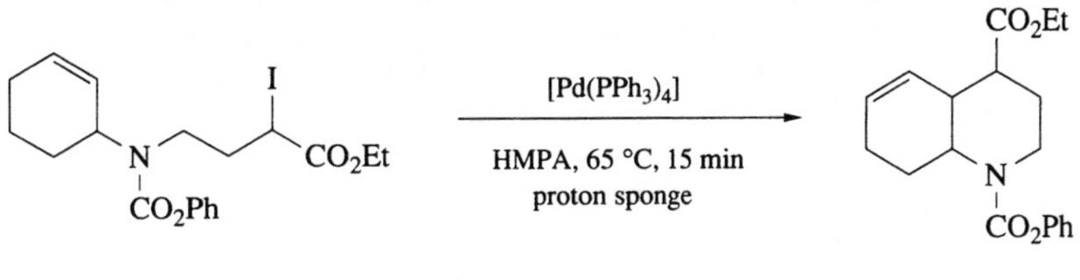

I, N, CO_2Et, CO_2Ph —[$[Pd(PPh_3)_4]$; HMPA, 65 °C, 15 min; proton sponge]→ CO_2Et, N, CO_2Ph (19)

4.3.5.1.2 Vinyl substitution with aryl halides

(i) Arylation of alkenes

Alkene arylation has been the most-studied of the several vinyl substitution reactions. Not only is the reaction convenient to carry out but it is easily made catalytic in palladium, it is often quite regio- and stereo-specific and yields generally are good.

(a) Mechanism and catalyst formation.
The many known examples of this reaction support a mechanism involving an oxidative addition of the aryl halide to the palladium(0) catalyst as the initial step. The catalyst actually added to the reaction mixtures is often a palladium(II) salt or complex. This must be reduced before the catalytic cycle can begin. Reduction usually occurs simply by heating the palladium salts with the alkene in the reaction mixture at about 80–100 °C. The alkene is oxidized to a vinyl or allyl derivative of the anion of the salt and in the process the palladium is reduced. Alternatively, the catalyst may be reduced with reagents such as sodium borohydride, formic acid or hydrazine. The use of an added reducing reagent sometimes allows the reaction to be carried out at lower temperatures than when reduction is achieved by reaction with an alkene. Some aryl iodides add easily to unhindered alkenes even at room temperature when silver salts are added to remove the apparently inhibiting halide ion from the reaction mixture or when quaternary ammonium chlorides are introduced, especially if potassium carbonate is the base. In the second step of the reaction, the arylpalladium halide adds *syn* to the alkenic double bond. A final *syn* elimination of a hydridopalladium halide yields the arylated alkene, probably by way of an alkene π-complex. A base must be added to neutralize the hydrogen halide produced by the reaction, if it is to be catalytic. The free hydrogen halide appears to inactivate the palladium by keeping it in the hydridopalladium halide form. Palladium(0) or (II) are normally tetracoordinated, but tri- and even di-coordinated complexes are known. Scheme 2 summarizes the proposed mechanism, presuming a dicoordinated palladium(0) complex to be involved in the catalytic cycle.

$$PdL_2 + ArX \longrightarrow RPdL_2X$$

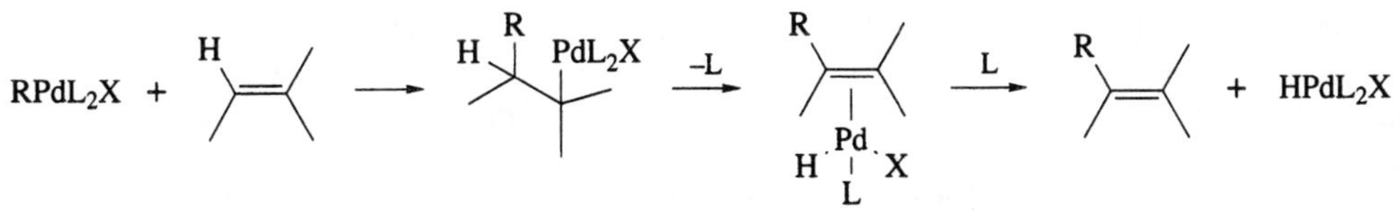

$RPdL_2X$ + (H-alkene) → (H, R, PdL_2X adduct) —(−L)→ (R-alkene π-complex with Pd, H, X, L) —(L)→ (R-alkene) + $HPdL_2X$

$$HPdL_2X \rightleftharpoons HX + PdL_2 \qquad L = \text{a ligand}$$

$$HX + \text{base} \longrightarrow \text{salt}$$

Scheme 2

Ligands: The ligands associated with the catalyst are a halogen anion, often added triphenyl-, or tri-*o*-tolyl-phosphine, possibly an amine, solvent, reactants or products. Arylphosphine ligands (two or more equivalents) are generally required when aryl bromides are to be reacted, in order to keep the palladium metal from precipitating, while with iodides no special ligand needs to be added. In the latter case, reactants, solvent or product(s), no doubt, function as ligands. The situation with aryl bromides is not clearly defined, however. In some examples at least, aryl bromide reactions proceed homogeneously without phosphines and with only sodium carbonate or bicarbonate as base. In these reactions, tetra-*n*-butylammonium chloride is frequently added as a solid–liquid phase transfer agent. There is evidence that tetraalkylammonium chlorides may have other beneficial effects upon palladium-catalyzed reactions than simply acting as phase transfer agents.[56] The effect may be due to coordination of chloride ions with palladium(0) keeping it in solution in catalytic reactions. The secondary amine, piperidine, is also capable of keeping palladium metal in solution in the absence of phosphines.[57]

Aryl chlorides: Aryl chlorides will substitute alkenes only under very special conditions, and then catalyst turnover numbers are generally not very high. Palladium on charcoal in the presence of triethylphosphine catalyzes the reaction of chlorobenzene with styrene,[58] but the catalyst becomes inactive after one use.[59] Examples employing an activated aryl chloride and highly reactive alkenes, such as acrylonitrile, with a palladium acetate–triphenylphosphine catalyst in DMF solution at 150 °C with sodium acetate as base react to the extent of only 51% or less.[60] Similar results have been reported for the combination of chlorobenzene with styrene in DMF–water at 130 °C, using sodium acetate as the base and palladium acetate–diphos as a catalyst.[61] Most recently, a method for reacting chlorobenzene with activated alkenes has been claimed where, in addition to the usual palladium dibenzilideneacetone–tri-*o*-tolylphosphine catalyst, nickel bromide and sodium iodide are added. It is proposed that an equilibrium concentration of iodobenzene is formed from the chlorobenzene–sodium iodide–nickel bromide catalyst and the iodobenzene then reacts in the palladium-catalyzed alkene substitution. Moderate to good yields were reported from reactions carried out in DMF solution at 140 °C.[62]

The problem with aryl chlorides seems to be that they do not compete effectively with reactant alkenes or even many solvents as ligands or reactants at palladium(0). Even if the arylpalladium chloride is formed its equilibrium concentration under the usual reaction conditions is probably very low.

The catalyst: The amount of catalyst required in an aryl bromide or iodide alkene substitution varies widely with the reactants and the reaction conditions. Most examples reported have used 1–2 mol % of palladium salt relative to the aryl halide, but much lower amounts are sufficient in some instances. In an extreme case, where very reactive *p*-nitrobromobenzene was added to the very active alkene, ethyl acrylate and sodium acetate was the base in DMF solution at 130 °C with a palladium acetate–tri-*o*-tolylphosphine catalyst; in 6 h the palladium turned over 134 000 times and ethyl *p*-nitrocinnamate was obtained in 67% yield.[63]

The phosphorus ligand most often used with the palladium catalyst has been triphenyl- or tri-*o*-tolylphosphine. The second phosphine is more stable under the reaction conditions because it does not undergo the palladium-catalyzed arylation of the phosphorus as triphenylphosphine does.[64] Reactions employing tri-*o*-tolyphosphine are often faster than the triphenylphosphine reactions, also. Other triarylphosphines, phosphites, trialkylphosphines and chelating diphosphines do not usually give as good catalysts. The molar ratio of the phosphine to the palladium is usually 2:1; however, larger amounts of phosphines are sometimes advantageous because this often prevents premature precipitation of the metal from the reaction mixtures. High concentrations of triarylphosphines are usually to be avoided, however, because they inhibit the reaction.

The base: One equivalent, at least, of a base relative to the aryl halide must be present to achieve the alkene substitution catalytically. Most often a tertiary amine is employed. Secondary amines also appear to be suitable but primary amines usually are not. The base strength of the amine is important since only quite basic amines such as triethylamine work well. Acetate salts, carbonates and bicarbonates also are suitable bases but solubility may cause difficulties in some instances. The addition of a phase transfer agent such as a quaternary ammonium salt has often solved this problem. The inorganic bases, of course, may cause other problems such as ester hydrolysis, aldol condensations and other undesired side reactions.

Solvent and temperature: The alkene substitution may be carried out in a variety of solvents or without an added solvent if the reactants form a homogeneous solution. DMF and acetonitrile are the most popular solvents but DMSO, HMPA, THF and methanol have also been used.

Reaction temperatures range from room temperature to about 140 °C. Higher temperatures generally cause premature precipitation of the palladium and at lower temperatures than about 20 °C reactions are usually very slow.

(b) Regio- and stereo-chemistry.

The regiochemistry of the alkene substitution reaction is primarily controlled by steric factors in noncyclic cases. The aryl group of the organopalladium halide adds preferentially to the least-substituted alkenic carbon. There is a secondary electronic influence which directs the aryl group to the most electron deficient vinyl carbon. Thus, α,β-unsaturated carbonyl compounds usually react quite selectively at the β-carbon, while terminal alkenes yield about 80:20 mixtures of 1- and 2-arylated alkenes in the absence of major steric influences. The regiochemistry is only slightly influenced by the substituents on the aryl halide. Aryl bromides and iodides give similar reaction mixtures and the substituents in any phosphines used with the catalyst have little influence on the regiochemistry of the reaction. Iodides are much more reactive than bromides in the substitution in the absence of phosphines, but the factor is usually less than about 10 when phosphines are present.

Isomeric alkene mixtures are often obtained in the alkene substitution reaction when different hydrogen atoms are available for the final elimination step. The factors controlling the direction of elimination are obscure. It is clear that a *syn* palladium hydride elimination is much preferred over an *anti* elimination. In the arylation of cyclic alkenes, therefore, allylically arylated compounds are major products. Rearrangements of the allylic isomers to 4-aryl-1-cycloalkenes also occur often, presumably because the neighboring hydrogen atoms are relatively favorably placed for multiple palladium hydride elimination–readdition sequences. This isomerization may be prevented by performing the reactions at the lowest possible temperatures and by employing either potassium acetate[65] or silver carbonate[66] as the base. The acetate anion or silver cation presumably decompose the intermediate product/hydridopalladium iodide complex before it has time to readd to the double bond and cause isomerization. Reactions of noncyclic alkenes generally do not yield rearranged arylated alkenes. In general, the conjugated alkenes are favored in these cases with the (*E*)-isomer as the main or sole product, if there is a choice. The stereochemistry of the products is determined by the *syn* addition–*syn* elimination mechanism. However, more or less equilibration may occur by multiple hydride addition–elimination steps if the reaction conditions cannot be kept mild enough (equation 20).[67]

PhI + Ph(CH=CHCH$_3$) + Et_3N —[$Pd(OAc)_2$, PPh_3; 60 °C, 142 h]→

80% + 15% + 5% (20)

(c) Intermolecular reactions.

The alkene arylation reaction is tolerant of a wide variety of substituents.[1b] A partial listing of substituents which may be on the aromatic halide includes: carboxy, alkoxycarbonyl, formyl, nitrile, methylthio, alkyl, nitro, hydroxy, alkoxy, acetoxy, amino, dimethylamino, acetamido, aminocarbonyl, chloro, trifluoromethyl, acyl and methylenedioxy. These substituents may be in the *o*-, *m*- or *p*-position and more than one or a combination may be present. A strong chelating effect of an *o*-carboxy group inhibits reactions of *o*-bromobenzoic acid but the methyl ester reacts normally.[68] It should be noted that free carboxy groups are not actually tolerated by the reaction since they are acidic, but an extra equivalent of base should be present to neutralize the acid and achieve catalytic reaction. Of course, if the reacting halogen group is hindered the reaction rate will decrease, but even 2,5-diisopropylbromobenzene reacts normally at 125 °C.[68] As noted above, aryl halides with strongly electron-supplying substituents often do not undergo the substitution reaction in high yield. The use of tri-*o*-tolylphosphine with these halides eliminates the phosphine quaternization problem[64] but product palladation and oxidation, as well as dehalogenation of the aromatic reactant, are other problems that may be encountered.

Very few examples of vinyl substitution with polynuclear aromatic halides have been reported but indications are that they generally react like the halobenzenes.

Essentially the same substituents as listed above may be present in the alkene being substituted, with the possible exception of chloro, alkoxy and acetoxy groups on vinyl or allyl carbons. These groups, especially chloro, may be lost or partially lost with palladium when the final elimination step occurs. For example, vinyl acetate, iodobenzene and triethylamine with a palladium acetate–triphenylphosphine catalyst at 100 °C form mainly (*E*)-stilbene, presumably *via* phenylation of styrene formed in the first arylation step (equation 21).[69]

PhI + CH_2=CHOAc + Et_3N —[$Pd(OAc)_2$, PPh_3; 8 h, 100 °C]→ PhCH=CHPh (52%) + PhCH=CHOAc (10%) (21)

Trimethylsilyl groups also may be lost from vinylic or allylic positions in the arylation reaction.[70] Halide ion facilitates the desilylation as well as the loss of alkoxy and acetoxy groups. Inclusion of soluble silver salts in reactions where trimethylsilyl groups are lost prevents this reaction.[70] The reaction in the absence of silver ion is useful for preparing styrene derivatives. However, they also may be prepared directly from aryl halides and ethylene in fair to good yields.[71]

The large difference in reactivity between aryl iodides and bromides in the absence of phosphines allows the iodo group to be reacted in the presence of a bromo group in the same molecule. Subsequently the bromide may be reacted with another alkene if a phosphine is added.[72] Aryl chloride groups are usually inert in the presence of bromo or iodo substituents.

Numerous methods for forming heterocycles and carbocycles by means of the alkene substitution reaction have been published. There are a few examples of direct intermolecular ring-forming reactions. *o*-Iodoaniline and dimethyl maleate, for example, form 4-methoxycarbonyl-2-quinolone in 71% yield (equation 22).[73] As a result of the *syn* addition of the organopalladium complex and the *syn* hydride elimination, the *cis* ester (maleate) yields the correct stereoisomer for cyclization. However, the *trans* ester, dimethyl fumarate, also gives the quinolone but only in 47% yield. Isomerization must be occurring at some stage of the reaction.

o-Iodoaniline (NH_2, I) + MeO_2CCH=CHCO_2Me + Et_3N —[$Pd(OAc)_2$; 100 °C]→ [intermediate: CO_2Me, CO_2Me, NH_2] —71%→ 4-methoxycarbonyl-2-quinolone (CO_2Me, N–H, O) 71% (22)

Further examples of the vinyl substitution reaction with aryl halides are given in Table 1.

(d) Intramolecular reactions.

Intramolecular vinyl substitutions are commonly used to form cyclic products. In a typical reaction, β-indoleacetic acid has been prepared in up to 43% yield from an *o*-bromoaniline derivative, as shown in equation (23).[74]

o-Bromoaniline derivative (Br, N–CO_2Me, CO_2Me) + Me_2NCH$_2$CH$_2$NMe$_2$ —[$Pd(OAc)_2$, PPh_3; 125 °C, 5 h]→ indole (CO_2Me, N–CO_2Me) —[H_3O^+; 43%]→ indole (CO_2H, N–H) (23)

Cyclizations occur with secondary amine derivatives, also. *N*-Allyl-*o*-iodoaniline produces 3-methylindole, for example, in high yield at 25 °C using sodium carbonate as the base with tetrabutylammonium chloride in DMF solution (equation 24).[75] Similar procedures have been applied to the synthesis of indolines, oxindoles, quinolines, isoquinolines and isoquinolones.[75] Some *N*-vinyl-*o*-bromoaniline derivatives cyclize to substituted indoles, also.[76]

Similar reactions with 3-(*o*-bromoarylamino)cyclohexenones yield carbazole derivatives.[77] Spiropyrrolidinones are obtained from *N*-(cyclohexylcarbonyl)-*o*-iodoaniline derivatives.[66] The use of silver nitrate in reactions significantly reduces the amount of double-bond isomerization observed. This system is useful for preparing six- and seven-membered ring analogs, also. Even quaternary centers can be

Table 1 Palladium-catalysed Arylations with Aryl Halides

Aryl halide	*Alkene*	*Product*	*Yield (%)*	*Ref.*
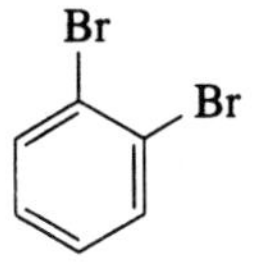 Br, Br	$H_2C{=}CH_2$	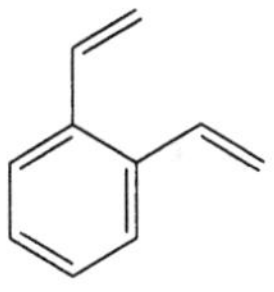	78	
		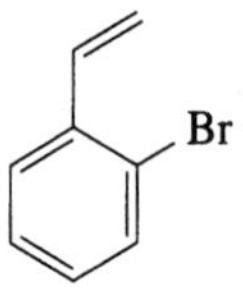 Br	12	71
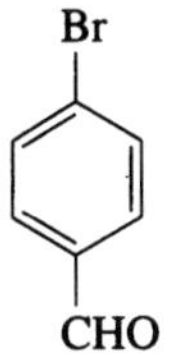 Br, CHO	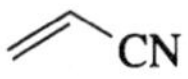 CN	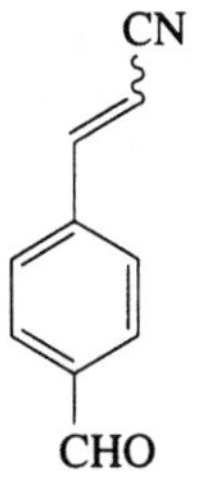 CN, CHO	79	63
PhI		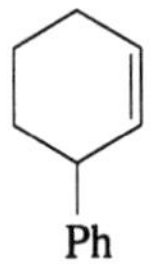 Ph	72	73
 I, I	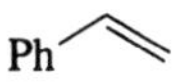 Ph	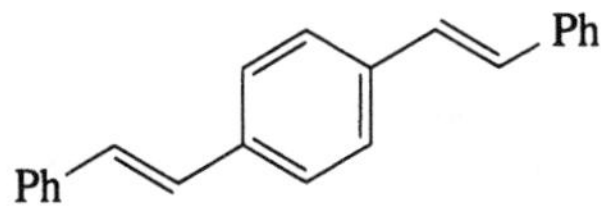 Ph, Ph	67	51
PhBr	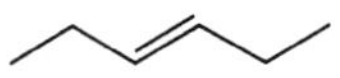	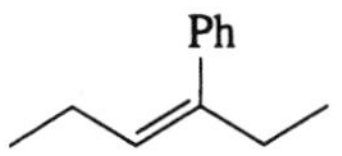 Ph	23	
		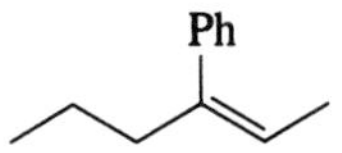 Ph	23	73
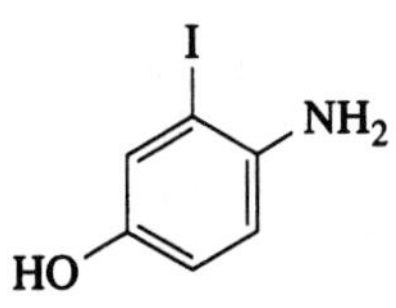 I, NH_2, HO	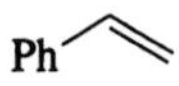 Ph	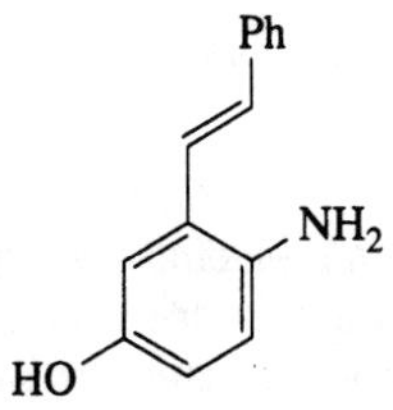 Ph, NH_2, HO	50	68

Table 1 *(continued)*

Aryl halide	*Alkene*	*Product*	*Yield (%)*	*Ref.*
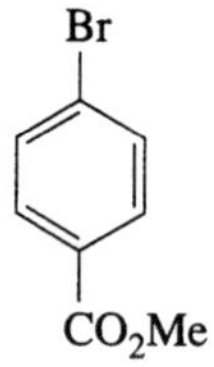 4-Br-C6H4-CO2Me	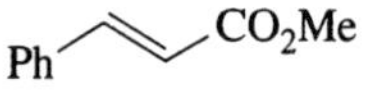 Ph-CH=CH-CO2Me	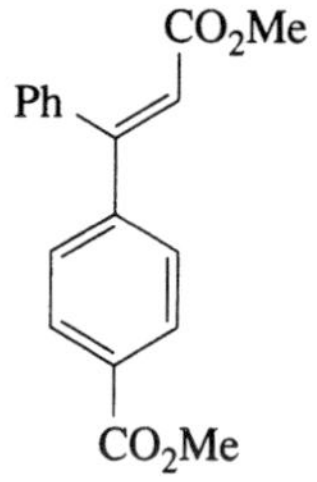 Ph(4-MeO2C-C6H4)C=CH-CO2Me	52	73

$$\text{2-iodo-}N\text{-allylaniline} + Na_2CO_3 \xrightarrow[\text{DMF, 25 °C, 24 h; 97\%}]{Pd(OAc)_2,\ Bu^n{}_4NCl} \text{3-methylindole} \quad (24)$$

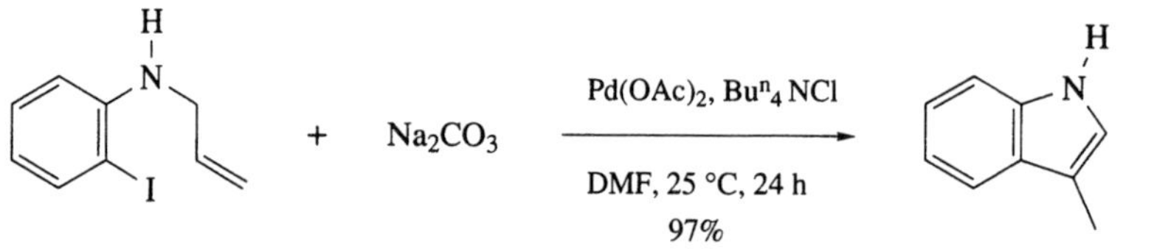

formed from tetrasubstituted double bonds. Carbocycles are formed by an analogous procedure (equation 25).[66,78] In this example, silver carbonate functions as both a base and a precipitant for the halogen to minimize double-bond isomerization. Another variation of this reaction permits the synthesis of bridged ring systems (equation 26).[66] Other polynuclear systems have been prepared with the potassium carbonate–tetra-*n*-butylammonium chloride phase transfer system.[79]

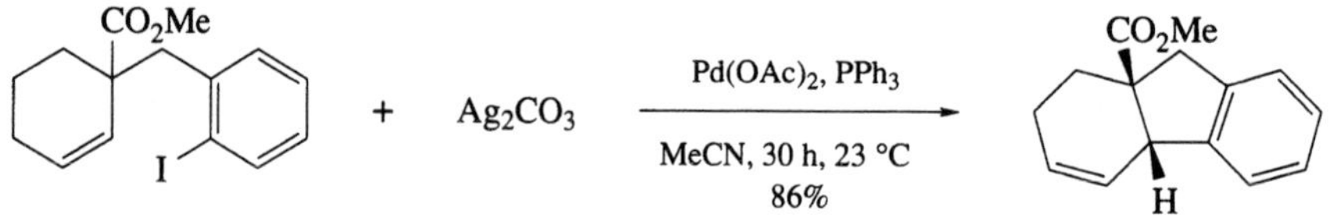

$$+ Ag_2CO_3 \xrightarrow[\text{MeCN, 30 h, 23 °C; 86\%}]{Pd(OAc)_2,\ PPh_3} \quad (25)$$

$$+ Et_3N \xrightarrow[\text{MeCN, 36 h, 82 °C; 78\%}]{Pd(OAc)_2,\ PPh_3} \quad (26)$$

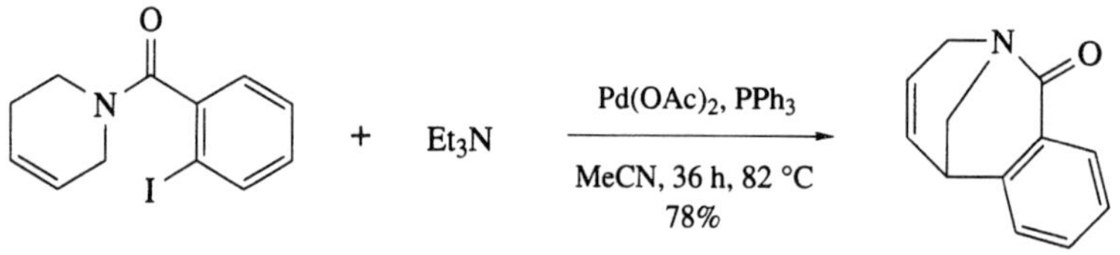

Some success has been achieved in the tandem formation of two rings when two double bonds are appropriately situated on an *ortho* side chain of iodobenzene (equation 27).[80]

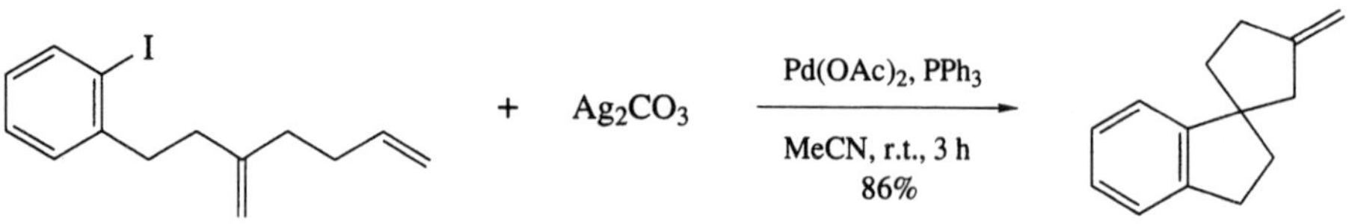

$$+ Ag_2CO_3 \xrightarrow[\text{MeCN, r.t., 3 h; 86\%}]{Pd(OAc)_2,\ PPh_3} \quad (27)$$

(ii) Arylation of allylic and other unsaturated alcohols

Vinyl substitution of primary or secondary allylic alcohols with aryl halides usually produces 3-aryl aldehydes or ketones, respectively. The reaction is believed to involve an addition of the intermediate arylpalladium halide to the double bond, placing the aryl group mainly on the more distant carbon from the hydroxy group, followed by palladium hydride elimination, a reverse readdition and another elimination with a hydrogen atom on the carbon bearing the hydroxy group. The product is probably a π-complex of the enol which ultimately either dissociates or collapses to a σ-complex with palladium on the carbon bearing the hydroxy group. In the latter case, palladium hydride elimination with the hydroxy hydrogen will yield the carbonyl product (Scheme 3). The procedure for this reaction is the same as that for the vinyl substitution of simple alkenes. Triethylamine, sodium bicarbonate or potassium carbonate are the usual bases employed. When the products are aldehydes with two α-hydrogens potassium carbonate may cause aldol condensations, so it probably should be avoided in these cases.

Mixtures of regioisomers are frequently obtained in these reactions.[80] The problem is most serious with primary allylic alcohols without α- or β-substituents. Even the 2-arylated products generally rearrange to saturated aldehydes. Allyl alcohol itself, when reacted with iodobenzene and triethylamine, with palladium acetate as catalyst, for example, produces a 71% yield of an 84:16 mixture of 3-phenyl- and 2-phenyl-propanal (equation 28).

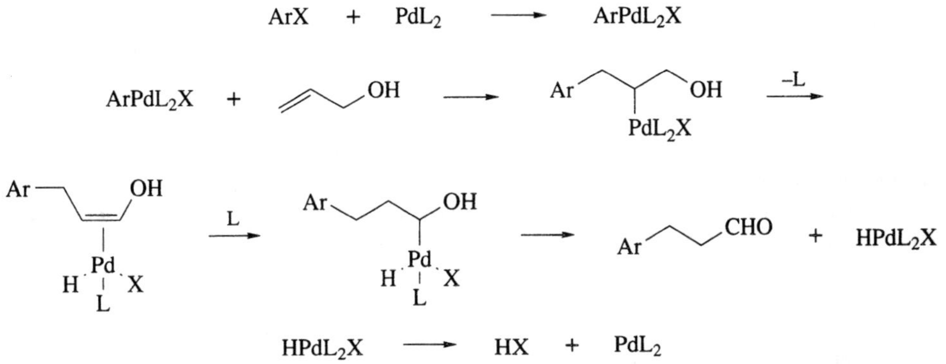

$$ArX + PdL_2 \longrightarrow ArPdL_2X$$

$$ArPdL_2X + CH_2{=}CHCH_2OH \longrightarrow ArCH_2CH(PdL_2X)CH_2OH \xrightarrow{-L}$$

$$ArCH_2CH{=}CHOH\,(\pi\text{-}HPdLX) \xrightarrow{L} ArCH_2CH_2CH(OH)PdHLX \longrightarrow ArCH_2CH_2CHO + HPdL_2X$$

$$HPdL_2X \longrightarrow HX + PdL_2$$

Scheme 3

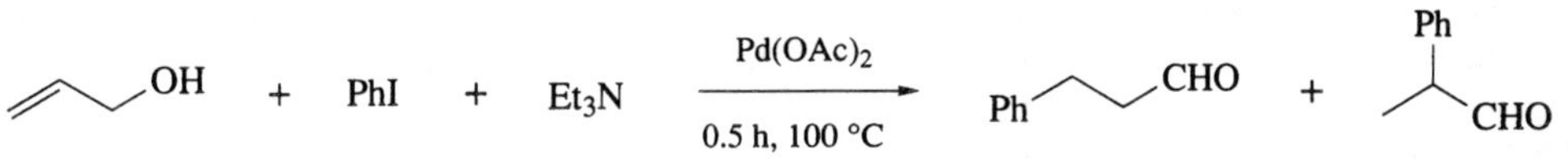

$$CH_2{=}CHCH_2OH + PhI + Et_3N \xrightarrow[0.5\ h,\ 100\ ^\circ C]{Pd(OAc)_2} PhCH_2CH_2CHO + CH_3CH(Ph)CHO \quad (28)$$

Allylic alcohols also may be arylated in *N*-methylpyrrolidinone or DMF solution with sodium bicarbonate as the base.[81] The use of this base improves the yields of aldehydes obtained compared with tertiary amine bases in the case of aryl bromides with electron-withdrawing substituents, where reduction to an arene is sometimes a problem.

(iii) Arylation of dienes and trienes

(a) Arylation with elimination.

As noted in Section 4.3.4.1.3, organopalladium halides add to conjugated (and some nonconjugated) dienes to give isolable π-allylpalladium halide dimers at room temperature. At higher temperatures (100 °C) some of these complexes are stable and some eliminate palladium hydride to form dienes. The elimination does not appear to be base catalyzed since the presence of tertiary amines has little effect upon the reaction. The elimination is most facile in cases where three or more conjugated double bonds are formed in the reaction. Thus, arylated dienes may be prepared catalytically by the reaction of aryl halides with conjugated dienes, a tertiary amine and a palladium catalyst at about 100 °C (equation 29).[82] The reaction proceeds in better yield if an additional conjugating group such as a carboxy group is in the diene. Vinyl substitution generally takes place at the least-substituted end of the diene system. 1,3-Butadiene may be diarylated easily. In fact, 1-phenyl-1,3-butadiene is more reactive than butadiene itself in the reaction. With a 1:2 ratio of butadiene to iodobenzene only 2% 1-phenyl-1,3-butadiene is formed along with 43% 1,4-diphenyl-1,3-butadiene, while if the ratio is 2.5:1 only 1-phenyl- and 1,4-diphenyl-1,3-butadiene are obtained in 24% and 34% yields, respectively.[83] In this reaction it is necessary to use 10 mol % palladium acetate and 20 mol % of tri-*o*-tolylphosphine as catalyst, since the reaction is very slow with the usual 1–2 mol %. Aryl halides with strongly electron-withdrawing substituents react with dienes much more readily and require only 1–2 mol % catalyst. They yield only 1,4-diarylated products even with a diene to aryl halide ratio of only 2:1. Butadiene yields products with *trans* double bonds while mixtures of isomers may be obtained from substituted butadienes. The phenylation of (*E*)-3-methyl-1-phenyl-1,3-butadiene, for example, yields 37% (*E,E*)- and 9% (*E,Z*)-1,4-diphenylisoprene.

$$\text{PhI} + \text{isoprene} + \text{Et}_3\text{N} \xrightarrow[\text{18 h, 100 °C; 52\%}]{\text{Pd(OAc)}_2\text{, PPh}_3} \text{product} \quad (29)$$

1,3,5-Hexatriene may be arylated similarly to 1,3-butadiene. Iodobenzene and aryl halides with electron-donating substituents require more catalyst (as in the butadiene reactions) than aryl halides with electron-withdrawing substituents to give 1,6-diarylated hexatrienes in a reasonable time.

(b) Arylation with substitution.

π-Allylpalladium complexes are well known to undergo nucleophilic attack at the terminal π-allylic carbons. This reaction sometimes can be coupled with the catalytic vinyl substitution reaction of conjugated dienes. In these cases the nucleophile attacks the intermediate π-allylic complex forming the allylated nucleophile, and the catalyst is recycled. This variation is successful even with simple dienes not possessing conjugating substituents which sometimes fail to react in good yield in the usual diene substitution reaction.

The reaction proceeds well with unhindered secondary amines as both nucleophiles and bases. The yield of allylic amine formed depends upon how easily palladium hydride elimination occurs from the intermediate. In cases such as the phenylation of 2,4-pentadienoic acid, elimination is very facile and no allylic amines are formed with secondary amine nucleophiles, while phenylation of isoprene in the presence of piperidine gives 29% phenylated diene and 69% phenylated allylic amine (equation 30).[84] Arylation occurs at the least-substituted and least-hindered terminal diene carbon and the amine attacks the least-hindered terminal π-allyl carbon. If one of the terminal π-allyl carbons is substituted with two methyl groups, however, then amine substitution takes place at this carbon. The reasons for this unexpected result are not clear but perhaps the intermediate reacts in a σ- rather than a π-form and the tertiary center is more accessible to the nucleophile. Primary amines have been used in this reaction also, but yields are only low to moderate.[85] A cyclic version occurs with *o*-iodoaniline and isoprene.[85]

$$\text{PhBr} + \text{isoprene} + \text{piperidine} \xrightarrow[\text{100 °C, 48 h}]{\text{Pd(OAc)}_2\text{, P}(o\text{-tolyl})_3} \text{Ph-diene (29\%)} + \text{Ph-allyl-piperidine (69\%)} + \text{piperidinium } \text{Br}^- \quad (30)$$

Other nucleophiles than amines which have been employed in the reaction are malononitrile and cyanoacetic ester anions. Both of these anions undergo a preliminary reaction with the aryl halide to form the *C*-aryl derivatives before they attack the π-allylpalladium intermediate, so that diarylated products are formed (equation 31).[86] Phenylmalononitrile anion reacts with iodobenzene and butadiene to give the same product in 70% yield.

$$\text{PhI} + \text{butadiene} + \text{NaCH(CN)}_2 \xrightarrow[\text{THF, 70 °C, 76 h; 62\%}]{[\text{PdCl}_2(\text{PPh}_3)_2]} \text{PhCH=CHCH}_2\text{C(CN)}_2\text{Ph} \quad (31)$$

4.3.5.1.3 Vinyl substitution with heterocyclic halides

A variety of *N*-, *O*- and *S*-heterocyclic halides have been found to undergo the palladium-catalyzed alkene substitution reaction. Halogen derivatives of furan, thiophene, pyridine, pyrazine, uracil, indole, quinoline and isoquinoline, for example, undergo the reaction. Iodoferrocene also reacts normally.

4.3.5.1.4 Vinyl substitution with vinyl halides

(i) Alkene reactions

(a) Variations and mechanism.
Alkenes generally are believed to react with vinylpalladium halides to form π-allylpalladium halide dimers.[87] This reaction also is proposed to occur in the vinylation of alkenes with vinyl halides and a palladium catalyst. The stability of the π-allylic complexes varies greatly with structure. In the vinylation reaction, if the thermal decomposition of the π-allylic intermediate to diene takes place easily, a useful catalytic diene synthesis occurs. In other cases the reaction may occur only very slowly or not at all. As described in Section 4.3.5.1.2.iii, the decomposition (elimination of palladium hydride) takes place easily when the diene formed in the reaction will have one or more additional groups conjugated with it, such as an aromatic ring or a carbonyl group. In other cases, the π-allylic complex may be decomposed by nucleophilic substitution analogously to the diene arylation reactions (Section 4.3.5.1.2.iii) and catalytic reactions are produced. Unhindered secondary amines have usually been employed as the nucleophile and the base in these reactions, in which case allylic amines are the products. Even in reactions where decomposition of the π-allylic complex occurs thermally, inclusion of a secondary amine in the reaction mixture often leads to the formation of allylic amines as major products. The π-allylpalladium complexes formed under conditions giving catalytic reactions usually undergo equilibration of *syn* and *anti* isomers through σ-allylpalladium derviatives and the stereochemistry of the starting vinyl halide or alkene is not retained in the products. Generally, (*E*,*E*)-dienes or (*E*)-allylic amines are the products of this reaction. However, it is sometimes possible to intercept the initial palladium hydride–diene complex before it forms a π-allylic complex. Conducting the reactions at as low a temperature as possible increases the stereoselectivity, as does addition of triarylphosphines even at higher temperatures. Under favorable conditions the stereochemistry of the diene should be the same as that of the vinyl halide at one double bond and reflect the *syn* addition–*syn* elimination mechanism at the other. The reactions believed to be involved in the vinyl substitution with vinyl halides are summarized in Scheme 4. The scheme is shown with several substituents present but it should be remembered that, like other vinyl substitution reactions, steric effects are quite important and too many or too large substituents can inhibit the reaction or even prevent it entirely.

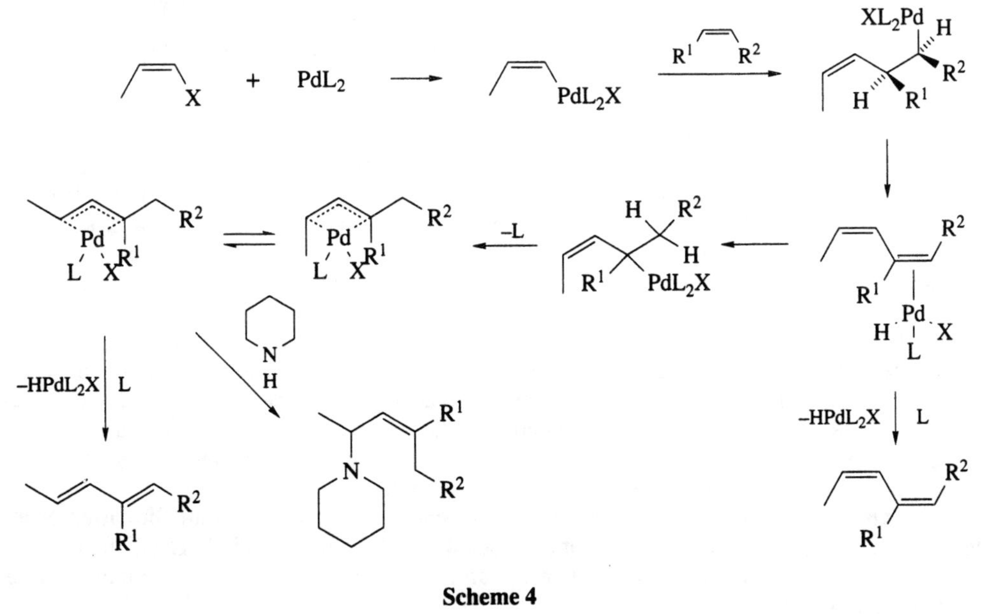

Scheme 4

(b) Intermolecular reactions.

Vinyl iodides are considerably more reactive than bromides in the vinylations. It may be presumed that chlorides are not generally useful, with one exception noted below, since they have not been employed in the reaction. The bromides are usually reacted with a palladium acetate–triphenyl- or tri-*o*-tolyl-phosphine catalyst at about 100 °C. The reaction will occur without the phosphine if a secondary amine is present. Vinyl iodides will react in the absence of a phosphine even with only a tertiary amine present.[48,57] The iodides are so reactive, in fact, that reactions occur even at room temperature if potassium carbonate is the base and tetra-*n*-butylammonium chloride is used as phase transfer agent in DMF solution when palladium acetate is the catalyst.[88]

The regiochemistry of the vinylation is quite structure dependent. Additions to alkenes with strongly electron-withdrawing substituents on one of the double-bond carbons causes addition of the vinyl group, regardless of structure, to the other double-bond carbon, in the absence of large steric effects. Additions to 1-alkenes and other alkenes with electron-donating substituents at the double bond may give mixtures of regioisomers. Vinyl bromide itself and 2-bromo-1-propene (and very probably other 2-bromo-1-alkenes) add quite selectively to the terminal carbon of 1-hexene. 2-Alkyl-1-bromo-1-alkenes usually yield mixtures. Stereospecific vinylations of alkenes only have been reported for α,β-unsaturated carbonyl compounds. If vinyl bromides are used, elevated temperatures are needed and the bromide or the diene product may isomerize under the reaction conditions. Stereospecific reactions have been obtained with the 1-iodo-1-hexenes at low temperatures (18–27 °C) using the phase transfer DMF procedure mentioned above. Under these conditions even acrolein and methyl vinyl ketone, as well as methyl acrylate, give better than 95% stereospecific reactions with the 1-iodo-1-hexenes (equation 32).[88]

$$\text{I} \quad + \quad \text{CHO} \quad + \quad K_2CO_3 \xrightarrow[\text{DMF, 18 °C, 5 h}]{Pd(OAc)_2,\ Bu^n{}_4NCl} \text{CHO} \quad + \quad \text{CHO} \qquad (32)$$

89% 1%

2-Chlorotropone, apparently, is the only example of a vinyl-type chloride which is known to undergo the substitution reaction. This chloride and styrene with triethylamine gives a 25% yield of 2-styryltropone (equation 33).[89]

$$\text{O, Cl} \quad + \quad \text{Ph} \quad + \quad Et_3N \xrightarrow[\text{THF, 100 °C, 20 h}]{[Pd(PPh_3)_4]} \text{O, Ph} \qquad (33)$$

Vinyl substitutions on alkenes not having their double bonds conjugated with carbonyl groups often proceed more rapidly and give better product yields when the reactions are conducted in the presence of an unhindered secondary amine. Conjugated and nonconjugated dienes are usually only minor products in these cases. The major products normally are allylic amines obtained by nucleophilic attack of the secondary amine upon the π-allylpalladium intermediates. Since allylic amines may be quaternized and subjected to the Hoffmann elimination, this is a two-step alternative to the direct vinyl substitution reaction.[90]

As expected from the results obtained in the arylation of dienes with secondary amines (Section 4.4.5.1.2.iii), the amine attacks the least-substituted (hindered) end of the π-allylic group. An exception to this behavior occurs in these reactions as it did in the diene arylation case where there are two methyl substituents on one terminal allylic carbon, in which case the amine attacks this tertiary carbon (equation 34).[57] If groups larger than methyls are present on one terminal allylic carbon, steric hindrance to attack at that carbon causes reaction at the other end of the allyl system.

Reactions of vinyl halides with acrolein acetals and secondary amines lead to the formation of minor amounts of dienal acetals and major amounts of aminoenal acetals (equation 35).[53] The diene product retains the sterochemistry in the vinyl halide while the aminoenal acetal loses it through equilibration of the π-allylpalladium intermediate.

Br + + morpholine (H–N) → Pd(OAc)$_2$, P(*o*-tolyl)$_3$; 100 °C, 48 h

13% + 15% + 31% (34)

Br + OMe/OMe + piperidine (H–N) → Pd(OAc)$_2$, P(*o*-tolyl)$_3$; 100 °C, 8 h

OMe, MeO 22% + N, OMe, OMe 73% (35)

(c) Intramolecular reactions.

Several bromodienes have been cyclized by use of the palladium–tri-*o*-tolylphosphine catalyst with a secondary amine as nucleophile and base. The regiochemistry of the substitution may be different from the intermolecular cases. For example, (*Z*)-1-bromo-1,5-hexadiene and piperidine form exclusively five-membered rather than six-membered ring products as might have been expected since they would arise from vinylation at the least-substituted double-bond carbon (equation 36).[91] It is surprising that even 9% of the very hindered *N*-tertiary alkylamine is formed in this example. The ratio of products obtained varies significantly with solvent. In benzene solution, for example, the yield of the tertiary alkylamine is 29% and the allylic isomer is 50%. Ring closure is not observed with 2-bromo-1,5-hexadiene and only polymers are formed. Six-membered ring products are obtained from 2-bromo-1,7-octadiene, along with a minor amount of a five-membered ring product (equation 37). In the absence of acetonitrile, the reaction in equation (37) gives a much lower yield (28%) of the six-membered ring amine. The origin of the five-membered ring product is not clear but it is proposed to be formed by a double-bond shift in the intermediate palladium complex followed by ring closure. An attempt to form a seven- or eight-membered ring from 2-bromo-1,8-nonadiene and piperdine gave only 8% of the seven-membered ring amine

Br + piperidine (H–N) → Pd(OAc)$_2$, P(*o*-tolyl)$_3$; MeCN, 100 °C, 2 h

[Pd(PPh$_3$)$_2$Br] → [Pd(PPh$_3$)$_2$Br] →

N 68% + N 9% (36)

and 4% of the rearranged six-membered ring amine. The reaction, therefore, in the carbocyclic series appears to be useful for forming only five- and possibly six-membered rings.

Br + piperidine —Pd(OAc)$_2$, P(*o*-tolyl)$_3$; MeCN, 100 °C, 8 h→ 8% + 10% + 72% + 6% (37)

Further examples of the formation of carbocyclic and carbopolycyclic compounds from (*Z*)-vinyl iodides with triethylamine as base have been reported and as noted in the related bromide reactions, double-bond isomerization is commonly observed.[92]

The same reactions, carried out with potassium carbonate as base in place of a secondary amine, yield exocyclic dienes in good yield, although double-bond isomerization sometimes occurs (equation 38).[93] Inclusion of tetra-*n*-butylammonium chloride in the reaction mixture stops the double-bond isomerization. Thus, the reaction in equation (38) with the chloride yields only the bis(exomethylene) product in 45% yield in a slow reaction. Some *N*- and *O*-heterocyclic products, also, have been prepared by the intramolecular vinyl substitution reaction.[94] A 16-membered ring lactone was made by the ring closure of a vinylic iodide group with a vinyl ketone group. The yield, based upon the reactant, was 55% but a stoichiometric amount of bis(acetonitrile)palladium dichloride was employed. The 'catalyst' was prereduced with formic acid so that the reaction proceeded at 25 °C (equation 39).[95]

Ph, Br, EtO_2C CO_2Et + K_2CO_3 —Pd(OAc)$_2$, PPh$_3$, MeCN; 30 °C, 3.5 h→ Ph, EtO_2C CO_2Et 24% + Ph, EtO_2C CO_2Et (38)

O, I, O, O + Et_3N —[PdCl$_2$(MeCN)$_2$], HCO$_2$H; MeCN, 25 °C, 11 h; 55%→ O, O, O (39)

(ii) Allylic alcohols and allylic amines

Vinylation of allylic alcohols with only a tertiary amine as base is often a very slow reaction in which product (if any) decomposition may be a serious side reaction. Exceptions to this behavior occur when β-halo-α,β-unsaturated esters (and other β-halo-α,β-unsaturated carbonyl compounds and nitriles if they are stable under the reaction conditions) are reacted with allylic alcohols. The presence of the conjugating substituent in the adduct is usually necessary to facilitate the hydridopalladium halide elimination. In most other cases, at least part of the reaction proceeds by way of a π-allylpalladium complex and these complexes with the β-hydroxy substituents are often unusually stable, probably as a result of coordination of the hydroxy group with the palladium. In these cases, the reactions proceed easily if a nucleophilic secondary amine is added to decompose the intermediate π-allylic complexes to amino enols. Usually, hydridopalladium halide elimination from the initial vinylpalladium adduct with the allylic alcohol takes place in both possible directions and two products are formed. Elimination of the hydrogen atom on the carbon bearing the hydroxy group leads to formation of an enone, while elimination in the other direction gives a conjugated diene π-complex which usually goes on to a π-allylpalladium complex and finally reacts with the amine to yield an amino enol. The mechanism proposed is illustrated in Scheme 5 with the reaction between 2-bromopropene and 3-buten-2-ol, where only one regioisomer is formed.[96]

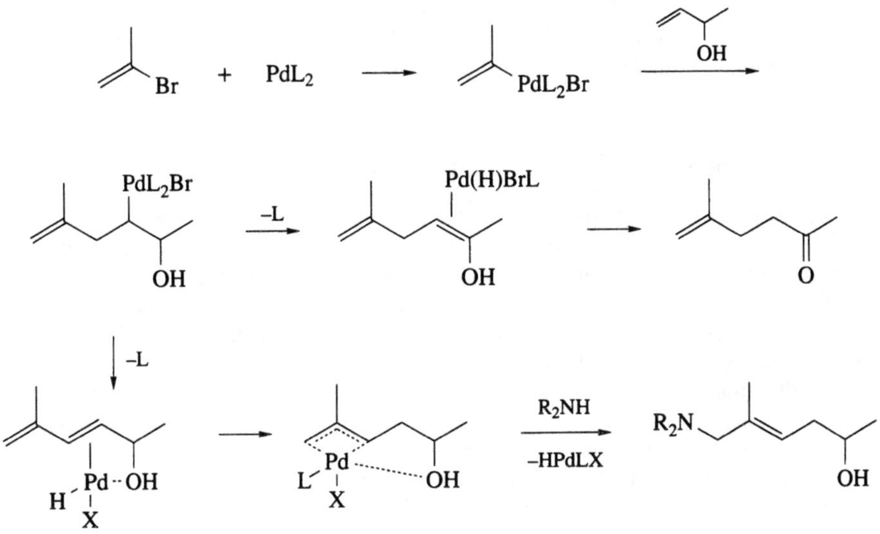

Br + PdL_2 → PdL_2Br —(OH)→

PdL_2Br / OH —(–L)→ Pd(H)BrL / OH → O

–L

Pd··OH / H / X → Pd / L / X ····OH —(R_2NH / –HPdLX)→ R_2N / OH

Scheme 5

(iii) Dienes and trienes

Vinylation of dienes in the presence of piperidine or morpholine yields aminodienes as major products. Sometimes trienes are minor products. The reaction is believed to proceed by way of a π-allylpalladium complex formed by addition of the vinylpalladium halide to the least-substituted diene double bond. Nucleophilic attack of the amine upon the π-allylic complex gives the aminodienes, while hydridopalladium halide elimination yields trienes (Scheme 6).[97]

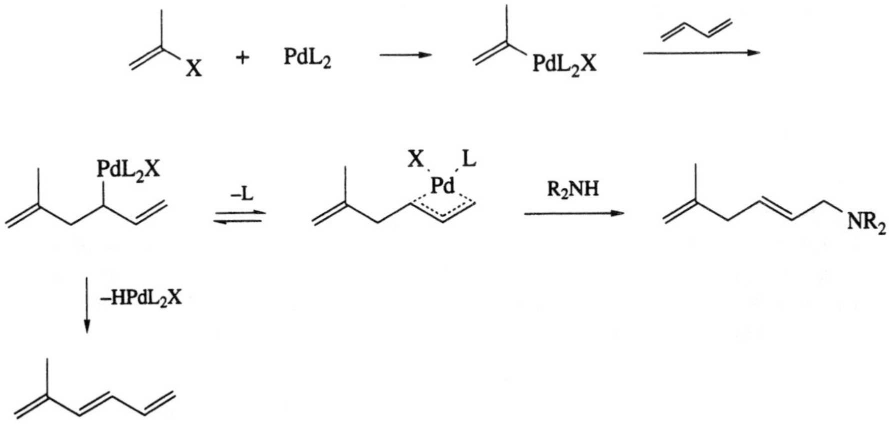

X + PdL_2 → PdL_2X →

PdL_2X ⇌(–L) X L Pd —(R_2NH)→ NR_2

–$HPdL_2X$

Scheme 6

Stereochemistry is largely lost in the diene vinylation, at least at 100 °C, as evidenced by the reaction of (*Z,E*)-3-methyl-2,4-pentadienoic acid shown in equation (40).

MeO_2C Br + CO_2H + 2 Et_3N —$Pd(OAc)_2$, P(*o*-tolyl)$_3$ / 100 °C, 1.5 h→

MeO_2C CO_2H 16% + MeO_2C CO_2H 39% (40)

Many conjugated dienes react twice with methyl 3-bromo-2-propenoate and triethylamine, making this procedure convenient for preparing a variety of dimethyl 2,4,6,8-decatetraenedioates.[98] 1,3,5-Hexatriene and its substituted derivatives undergo vinylation also, to form tetraenes and pentaenes in low yields.[83]

4.3.5.2 Vinyl Substitutions with Aryldiazonium Salts and Arylamines

Aryldiazonium salts react with bis(dibenzylideneacetone)palladium to form arylpalladium salts and nitrogen. Therefore, diazonium salts may be employed to catalytically arylate alkenes under mild conditions. Since many aryl halides are made from diazonium salts this variation could even be more convenient than using aryl halides. The reaction proceeds in good to excellent yields in nonaqueous solvents, using sodium acetate as the base at room temperature with terminal alkenes and cyclopentene.[99] Internal alkenes usually give poor yields, however.

Palladium(II) salts apparently oxidize arylamines to arylpalladium salts since alkenes are arylated by reaction with only an aromatic amine and a palladium salt. However, yields are generally low.[100] Much better yields are obtained if *t*-butyl nitrite is added and, of course, this forms the diazonium salt *in situ*. This not only saves a step but some diazonium salts which are too unstable to be isolated may be used as well. The reactions are carried out in the presence of acetic or chloroacetic acid with 5–10% bis(dibenzylideneacetone)palladium as catalyst (equation 41).[101]

NH_2 / NO_2 (arene) + styrene + Bu^tONO —[Pd(dba)$_2$], AcOH / $ClCH_2CO_2H$, 50 °C, 0.5 h→ O_2N (stilbene) 79% (41)

4.3.5.3 Vinyl Substitutions with Carboxylic Acid Halides, Aryl Sulfinates and Arylsulfonyl Chlorides

4.3.5.3.1 Vinyl acylation

Carboxylic acid chlorides and chloroformate esters add to tetrakis(triphenylphosphine)palladium(0) to form acylpalladium derivatives (equation 42).[102] On heating, the acylpalladium complexes can lose carbon monoxide (reversibly). Attempts to employ acid halides in vinylic acylations, therefore, often result in obtaining decarbonylated products (see below). However, there are some exceptions. Acylation may occur when the alkenes are highly reactive and/or in cases where the acylpalladium complexes are resistant to decarbonylation and in situations where intramolecular reactions can form five-membered rings.

$$MeCOCl + [Pd(PPh_3)_4] \xrightarrow[PhH]{25\ ^\circ C} [MeCOPd(PPh_3)_2Cl] + 2PPh_3 \quad (42)$$

Vinyl ethers yield acylated vinyl ethers when they are reacted with aroyl chlorides or 3-thienylcarbonyl chloride, triethylamine and a palladium acetate catalyst at 60–70 °C (equation 43).[103] The products of this reaction are useful 1,3-dicarbonyl equivalents. At higher temperatures the reaction yields arylated vinyl ethers. It is interesting that the acylations of vinyl ethers are regioselective while the direct arylations are usually not.[104]

COCl, OMe + OBu^n + Et_3N —[$Pd(OAc)_2$; 60–70 °C, 24 h; 62%]→ O, OBu^n, OMe (43)

Homoallyl chloroformates cyclize catalytically to α-methylene-γ-butyrolactones in moderate yields at 130 °C with tetrakis(triphenylphosphine)palladium(0) (equation 44).[105] This reaction only gives 1–2% of product when it is carried out intermolecularly. The presumed intermediate in the homoallyl chloroformate cyclization has been isolated and kinetic measurements show that that cyclization is inhibited by an excess of triphenylphosphine. A chelated π-alkene intermediate is proposed.[106]

Bu^n, O, Cl, O + $NaHCO_3$ —[$[Pd(PPh_3)_4]$, PPh_3; xylene, 130 °C, 16 h; 58%]→ Bu^n, O, O (44)

Since vinylic iodides (and bromides) can be catalytically carbonylated with a palladium catalyst, intermolecular acylations sometimes can be carried out with appropriate halodienes and carbon monoxide as, for example, is shown in equation (45).[107]

I + CO + Et_3N —[$[Pd(PPh_3)_4]$, THF; 60 °C, 18 h, 1.1 atm, 51 %]→ O (45)

4.3.5.3.2 Arylation with carboxylic acid chlorides

Aromatic acid chlorides are decarbonylated to aryl chlorides when they are heated to 300–360 °C with palladium on carbon. The reaction proceeds by way of an aroylpalladium chloride, then to an arylpalladium chloride and finally through a reductive elimination to the aryl chloride. If the reaction is conducted in the presence of a reactive alkene under mild conditions the aroylpalladium chloride intermediate will sometimes acylate the alkene, as noted in Section 4.3.5.3.1. More usually, however, decarboxylation is more rapid than acylation, especially at higher temperatures (>100 °C), and decarbonylation occurs. The

COCl, MeO + $H_2C{=}CH_2$ + NMe_2 —[toluene, 10 atm; 100 °C, 4 h, 48%]→ MeO (46)

COCl, Br + CN + NMe_2 —[$Pd(OAc)_2$; 130 °C, 25 h, 70%]→

Br, CN —[$[Pd(OAc)_2\{P(o\text{-tolyl})_3\}_2]$, NaOAc, CO_2Et; DMF, 130 °C, 3.5 h; 75%]→ EtO_2C, CN (47)

arylpalladium chloride formed is relatively stable and usually will arylate the alkene. Many examples of this reaction are known (equation 46). The reaction only appears to go well with ethylene, styrene, vinyl ethers and α,β-unsaturated carbonyl compounds and nitriles. Since aromatic carboxylic acid chlorides are often more readily available than the corresponding aryl bromides or iodides this variation will be useful for some syntheses. Bromobenzoyl chlorides give aryl bromides as products which can be used in a subsequent vinyl substitution (equation 47).[108]

4.3.5.3.3 *Arylation with sulfinate salts and sulfonyl chlorides*

Sodium arylsulfinates and palladium salts react to form arylpalladium salts and sulfur dioxide. When carried out in the presence of some dienes, stable organopalladium complexes have been obtained. For example, 1,5-cyclooctadiene and sodium *p*-toluenesulfinate with palladium chloride yields the σ,π-complex shown in equation (48).[109]

SO_2Na + + $PdCl_2$ —MeOH, reflux, 2 h, 86%→ SO_2 + Pd C_6H_4Me-*p* Cl 2 (48)

Arylsulfonyl chlorides react similarly with palladium(0) catalysts, producing sulfur dioxide and arylpalladium chlorides. In the presence of activated alkenes, the combination forms arylated alkenes in modest yields (equation 49).[110]

SO_2Cl Cl + CO_2Et + Et N O —$Pd(OAc)_2$, xylene, 130 °C, 8 h, 48%→

Cl CO_2Et + SO_2 (49)

4.3.5.4 Vinyl Substitution with Triflate Esters and Related Esters

4.3.5.4.1 *Arylation with aryl triflates*

Only a few examples of vinyl substitution with aryl fluorosulfonate esters have been reported but the reaction will surely turn out to be general.[111] A tetrafluoroethoxytetrafluoroethanesulfonate ester was used in the example shown in equation (50)

$OSO_2CF_2CF_2OCF_2CF_2H$ OMe + CO_2Et + Et_3N —$[PdCl_2(PPh_3)_2]$, DMF, 90 °C, 18 h, 87%→

CO_2Et OMe (50)

4.3.5.4.2 Vinylation with enol triflates

Many examples of the vinylation of alkenes with enol triflates have been reported. In general, the reactions proceed in good yields under similar conditions to the vinyl iodide vinylations. Most examples have been carried out in DMF solution with triethylamine as base at 60–75 °C. At these temperatures reduction of the palladium(II)–triphenylphosphine catalyst is sometimes slow and the use of tetrakis(triphenylphosphine)palladium(0) is preferred. The reaction appears to be relatively tolerant of steric congestion around the triflate group since 2,5,5-trimethyl-1-cyclopentenyl triflate reacts easily in good yields. The stereochemistry of the vinyl triflate is largely retained in the reaction while the second double bond of the product should have sterochemistry resulting from a *syn* addition of the vinylpalladium derivative followed by *syn* elimination of the palladium hydride group, although no reactions with 1,2-disubstituted alkenes have been reported to confirm this prediction. This reaction appears to be a very useful variation of the vinyl substitution since yields are generally good and the vinyl triflates are easily available from ketones and trifluoromethanesulfonic anhydride. Examples of the reaction are shown in equation (51) and Table 2.

OTf

AcO

+ CO_2Me + Et_3N $\xrightarrow[\text{DMF, 60 °C, 3 h; 82\%}]{Pd(OAc)_2,\ PPh_3}$

CO_2Me

AcO

(51)

Table 2 Vinyl Substitution with Vinyl Triflates

Triflate	*Alkene*	*Catalyst*	*Product*	*Yield (%)*	*Ref.*
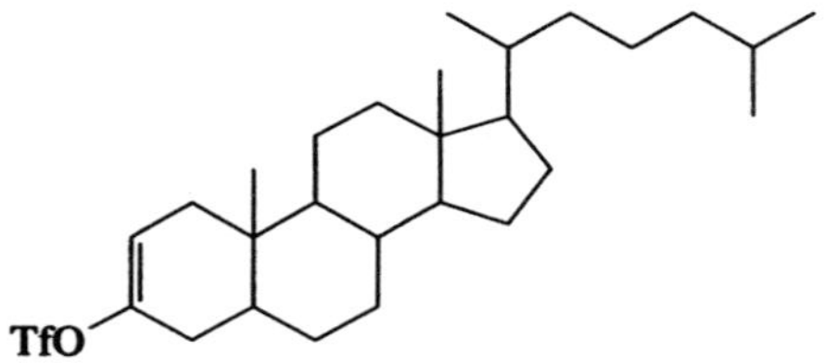 TfO	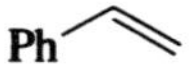 Ph	$[Pd(OAc)_2(PPh_3)_2]$	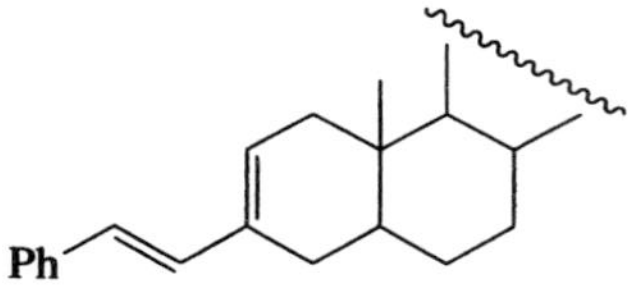 Ph	77	112
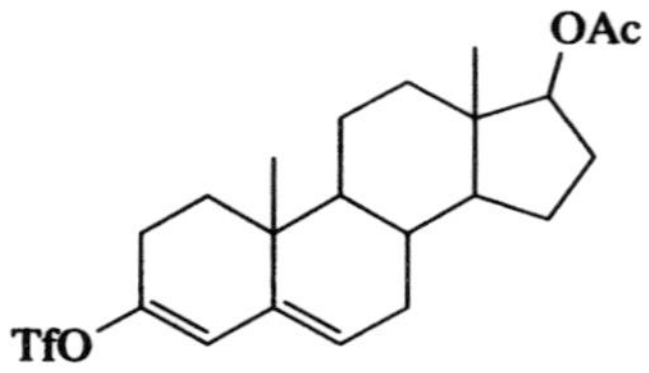 TfO, OAc	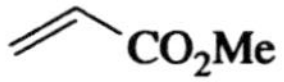 CO_2Me	$[Pd(OAc)_2(PPh_3)_2]$	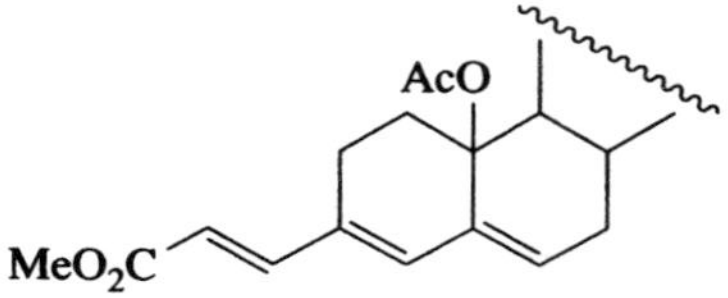 MeO_2C, AcO	82	112
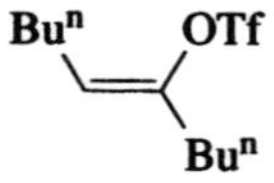 Bu^n, OTf, Bu^n	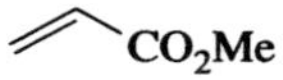 CO_2Me	$[Pd(OAc)_2(PPh_3)_2]$	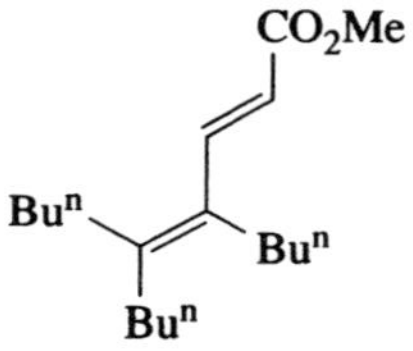 CO_2Me, Bu^n, Bu^n, Bu^n	20	
			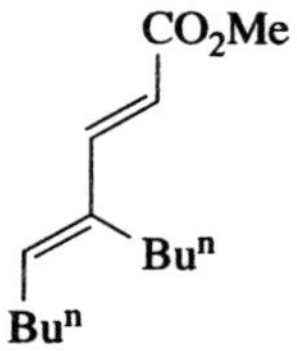 CO_2Me, Bu^n, Bu^n	5	112

Table 2 *(continued)*

Triflate	*Alkene*	*Catalyst*	*Product*	*Yield (%)*	*Ref.*
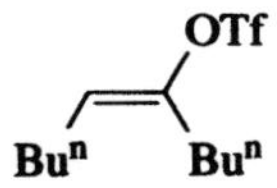 OTf, Bu^n, Bu^n	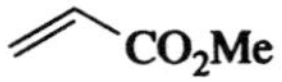 CO_2Me	$[Pd(OAc)_2(PPh_3)_2]$	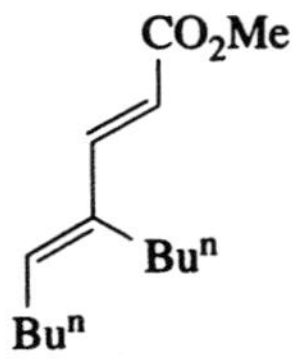 CO_2Me, Bu^n, Bu^n	24	112
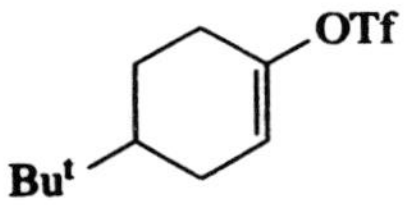 OTf, Bu^t	 CHO	$[PdCl_2(PPh_3)_2]$	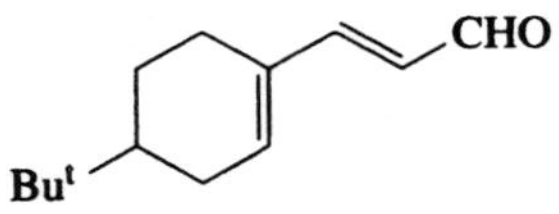 CHO, Bu^t	86	113
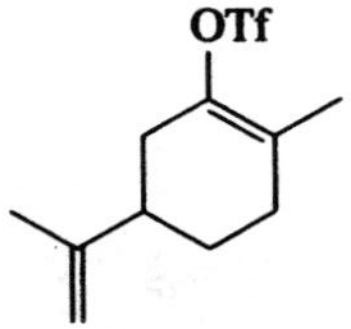 OTf	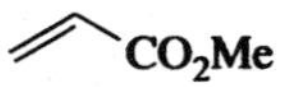 CO_2Me	$[PdCl_2(PPh_3)_2]$	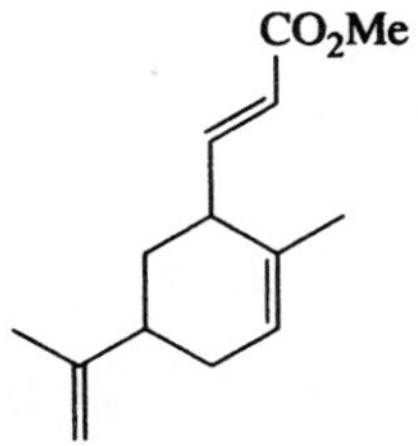 CO_2Me	88	113
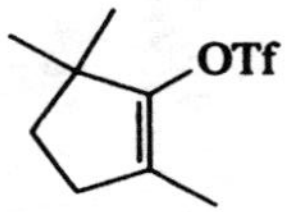 OTf	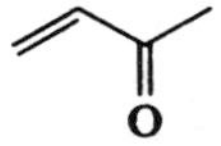 O	$[PdCl_2(PPh_3)_2]$	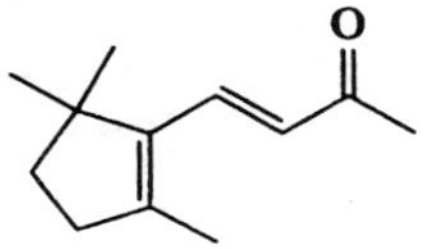 O	83	113
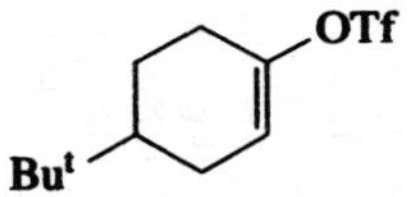 OTf, Bu^t	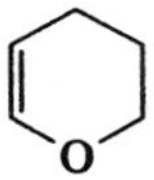 O	$[Pd(PPh_3)_4]$	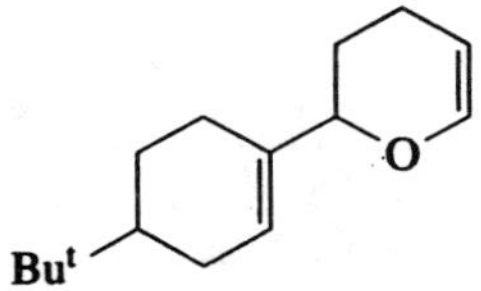 O, Bu^t	53	113

4.3.6 REFERENCES

1. (a) R. F. Heck, 'Palladium Reagents in Organic Syntheses', Academic Press, London, 1985, chap. 6; (b) R. F. Heck, *Org. React.*, 1982, **27**, 345; (c) L. S. Hegedus, *Tetrahedron*, 1984, **40**, 2415; (d) P. M. Henry, 'Palladium Catalyzed Oxidation of Hydrocarbons', Reidel, Dordrecht, 1980; (e) P. M. Maitlis, 'The Organic Chemistry of Palladium', Academic Press, London, 1971.
2. I. Moritani, S. Danno, Y. Fujiwara and S. Teranishi, *Bull. Chem. Soc. Jpn.*, 1971, **44**, 578.
3. I. Moritani, Y. Fujiwara and S. Danno, *J. Organomet. Chem.*, 1971, **27**, 279.
4. Y. Fujiwara, R. Asano, I. Moritani and S. Teranishi, *J. Org. Chem.*, 1976, **40**, 1681
5. Y. Fujiwara, I. Moritani, R. Asano and S. Teranishi, *Tetrahedron Lett.*, 1968, 6015.
6. B. M. Trost and J. M. Fortunak, *Organometallics*, 1982, **1**, 7.
7. T. Itahara, *J. Org. Chem.*, 1985, **50**, 5546.
8. T. Itahara and F. Ouseto, *Synthesis*, 1984, 488.
9. A. Kasahara, T. Izumi, M. Yodono, R. Saito, T. Takeda and T. Sugawara, *Bull. Chem. Soc. Jpn.*, 1973, **46**, 1220.
10. Y. Fujiwara, O. Maruyama, M. Yoshidomi and H. Taniguchi, *J. Org. Chem.*, 1981, **46**, 851.
11. T. Itahara, M. Ikeda and T. Sakakibara, *J. Chem. Soc., Perkin Trans. 1*, 1983, 1361.
12. K. Yamamura, *J. Org. Chem.*, 1978, **43**, 724.
13. R. Asano, I. Moritani, A. Sonoda, Y. Fujiwara and S. Teranishi, *J. Chem. Soc. C*, 1971, 3691.
14. Y. Fujiwara, R. Asano, I. Moritani and S. Teranishi, *Chem. Lett.*, 1975, 1061.
15. Y. Fujiwara, I. Moritani, M. Matsud and S. Teranishi, *Tetrahedron Lett.*, 1968, 3863.
16. R. S. Shue, *J. Chem. Soc., Chem. Commun.*, 1971, 1510.
17. J. Tsuji and H. Nagashima, *Tetrahedron*, 1984, **40**, 2699.
18. E. C. Constable, *Polyhedron*, 1984, **3**, 1037.
19. B. J. Brisdon, P. Nair and S. F. Dyke, *Tetrahedron*, 1981, **37**, 173.
20. R. A. Holton, *Tetrahedron Lett.*, 1977, 355.
21. R. F. Heck, *J. Am. Chem. Soc.*, 1968, **90**, 5518.
22. R. F. Heck, *J. Organomet. Chem.*, 1972, **37**, 389.
23. R. F. Heck, *J. Am. Chem. Soc.*, 1969, **91**, 6707.
24. R. F. Heck, *J. Am. Chem. Soc.*, 1968, **90**, 5542.
25. F. G. Stakem and R. F. Heck, *J. Org. Chem.*, 1980, **45**, 3584.
26. M. Donati and F. Conti, *Tetrahedron Lett.*, 1966, 4953.
27. A. Kasahara, T. Izumi, G. Saito, M. Yodono, R. Saito and Y. Goto, *Bull. Soc. Chem. Jpn.*, 1972, **45**, 895.
28. S. F. Dyke and M. McCartney, *Tetrahedron*, 1981, **37**, 431.
29. U. Hacksell and G. Daves, Jr., *J. Org. Chem.*, 1983, **48**, 2870.
30. I. Arai and G. Daves, Jr., *J. Org. Chem.*, 1978, **43**, 4110.
31. R. Larock and M. Mitchall, *J. Am. Chem. Soc.*, 1978, **100**, 180.
32. R. Larock, K. Takagi, S. S. Hershberger and M. Mitchell, *Tetrahedron Lett.*, 1981, **22**, 5231.
33. I. Akhrem, N. Chistovalova, E. Mysov and M. E. Vol'pin, *J. Organomet. Chem.*, 1974, **72**, 163.
34. J. Yoshida, K. Tamao, H. Yamamoto, T. Kakui, T. Uchida and M. Kumada, *Organometallics*, 1982, **1**, 542.
35. Y. Ito, T. Hirao and T. Saegusa, *J. Org. Chem.*, 1978, **43**, 1011.
36. Y. Ito, M. Nakatsuka, N. Kise and T. Saegusa, *Tetrahedron Lett.*, 1980, **21**, 2873.
37. Y. Ito, H. Aoyama, T. Hirao, A. Mochizuki and T. Saegusa, *J. Am. Chem. Soc.*, 1979, **101**, 494.
38. A. S. Kende, B. Roth and P. J. Sanfilippo, *J. Am. Chem. Soc.*, 1982, **104**, 1784.
39. R. Larock and C. Fellows, *J. Am. Chem. Soc.*, 1982, **104**, 1900.
40. R. A. Kjonaas, *J. Org. Chem.*, 1986, **51**, 3708.
41. R. Larock, S. Varaprath, H. Lau and C. Fellows, *J. Am. Chem. Soc.*, 1984, **106**, 5274.
42. R. Larock, C. Liu, H. Lau and S. Veraprath, *Tetrahedron Lett.* 1984, **25**, 4459.
43. L. S. Hegedus and M. A. McGuire, *Organometallics*, 1982, **1**, 1175.
44. N. Luong-Thi and H. Riviere, *J. Chem. Soc., Chem. Commun.*, 1978, 918.
45. Y. Tamaru, M. Hojo, H. Higashimura and Z. Yoshida, *Angew. Chem., Int. Ed. Engl.*, 1986, **25**, 735.
46. M. E. Vol'pin, R. Taube, H. Drevs, L. Volkova, G. Levitin and T. Ushakova, *J. Organomet. Chem.*, 1972, **39**, C79.
47. T. Hayashi and L. S. Hegedus, *J. Am. Chem. Soc.*, 1977, **99**, 7093.
48. H. Dieck and R. F. Heck, *J. Org. Chem.*, 1975, **40**, 1083.
49. T. Yamane, K. Kikukawa, M. Takagi and T. Matsuda, *Tetrahedron*, 1973, **29**, 955.
50. R. Asano, I. Moritani, Y. Fujiwara and S. Teranishi, *Bull. Chem. Soc. Jpn.*, 1973, **37**, 2320.
51. R. F. Heck and J. Nolley, Jr., *J. Org. Chem.*, 1972, **37**, 2320.
52. Q. Chen, Z. Yang, C. Zhao and Z. Qiu, *J. Chem. Soc., Perkin Trans. 1*, 1988, 563.
53. M. Mori, N. Kubo, N. Kanda, I. Oda and Y. Ban, *Heterocycles*, 1985, **23**, 220.
54. M. Mori, N. Kanda, I. Oda and Y. Ban, *Heterocycles*, 1985, **41**, 5465.
55. M. Mori, Y. Kubo and Y. Ban, *Tetrahedron Lett.*, 1985, **26**, 1519.
56. H. Dieck, R. Laine and R. F. Heck, *J. Org. Chem.*, 1975, **40**, 2819.
57. B. Patel and R. F. Heck, *J. Org. Chem.*, 1978, **43**, 3898.
58. M. Julia and M. Duteil, *Bull. Soc. Chim. Fr.*, 1973, 2790.
59. Author's observation.
60. A. Spencer, *J. Organomet. Chem.*, 1984, **270**, 115.
61. J. Davison, N. Simon and S. Sojka, *J. Mol. Catal.*, 1984, **22**, 349.
62. J. J. Bozell and C. Vogt, *J. Am. Chem. Soc.*, 1988, **110**, 2655.
63. A. Spencer, *J. Organomet. Chem.*, 1983, **258**, 101.
64. C. Ziegler, Jr. and R. F. Heck, *J. Org. Chem.*, 1978, **43**, 2941.
65. R. Larock and B. Baker, *Tetrahedron Lett.*, 1988, **29**, 905.
66. M. Abelman, T. Oh and L. E. Overman, *J. Org. Chem.*, 1987, **52**, 4130.

67. H. Dieck and R. F. Heck, *J. Am. Chem. Soc.*, 1974, **96**, 1133.
68. B. Patel, C. Ziegler, Jr., N. Cortese, J. Plevyak, T. Zebovitz, M. Terpko and R. F. Heck, *J. Org. Chem.*, 1977, **42**, 3903.
69. A. Kasahara, T. Izumi and N. Fukuda, *Bull. Chem. Soc. Jpn.*, 1977, **50**, 551.
70. K. Karabelas and A. Hallberg, *J. Org. Chem.*, 1986, **51**, 5286.
71. J. Plevyak and R. F. Heck, *J. Org. Chem.*, 1978, **43**, 2454.
72. J. Plevyak, J. Dickerson and R. F. Heck, *J. Org. Chem.*, 1979, **44**, 4078.
73. N. Cortese, C. Ziegler, Jr., B. Hrnjez and R. F. Heck, *J. Org. Chem.*, 1978, **43**, 2952.
74. M. Mori, K. Chiba and Y. Ban, *Tetrahedron Lett.*, 1977, 1037.
75. R. Larock and S. Babu, *Tetrahedron Lett.*, 1987, **28**, 5291.
76. A. Kasahara, T. Izumi, S. Murakami, H. Yanai and M. Takatori, *Bull. Chem. Soc. Jpn.*, 1986, **59**, 927.
77. H. Iida, Y. Yuasa and C. Kibayashi, *J. Org. Chem.*, 1980, **45**, 2938.
78. R. Larock, H. Song, B. Baker and W. Gong, *Tetrahedron Lett.*, 1988, **29**, 2919; R. Grigg, V. Sridharan, P. Stevenson and T. Worakun, *J. Chem. Soc., Chem. Commun.*, 1986, 1697.
79. M. Abelman and L. E. Overman, *J. Am. Chem. Soc.*, 1988, **110**, 2328.
80. J. Melpolder and R. F. Heck, *J. Org. Chem.*, 1976, **41**, 265.
81. A. Chalk and S. Magennis, *J. Org. Chem.*, 1976, **41**, 1206.
82. B. Patel, J. Dickerson and R. F. Heck, *J. Org. Chem.*, 1978, **43**, 5018.
83. T. Mitsudo, W. Fischetti and R. F. Heck, *J. Org. Chem.*, 1984, **49**, 1640.
84. F. G. Stakem and R. F. Heck, *J. Org. Chem.*, 1980, **45**, 3584.
85. J. O'Connor, B. Stallman, W. Clark, A. Shu, R. Spada, T. Stevenson and H. Dieck, *J. Org. Chem.*, 1983, **48**, 807.
86. M. Uno, T. Takahashi and S. Takahaski, *J. Chem. Soc., Chem. Commun.*, 1987, 785.
87. R. Larock and M. Mitchell, *J. Am. Chem. Soc.*, 1976, **98**, 6718.
88. T. Jeffery, *Tetrahedron Lett.*, 1985, **26**, 2667.
89. H. Horino, N. Inone and T. Asao, *Tetrahedron Lett.*, 1981, **22**, 741.
90. B. Patel, J. Kim, D. Bender, L. Kao and R. F. Heck, *J. Org. Chem.*, 1981, **46**, 1061.
91. C. Narula, K. Mak and R. F. Heck, *J. Org. Chem.*, 1983, **48**, 2792.
92. E. Negishi, Y. Zhang and B. O'Conner, *Tetrahedron Lett.*, 1988, **29**, 2915.
93. R. Grigg, P. Stevenson and T. Worakun, *Tetrahedron*, 1988, **44**, 2033.
94. L. Shi, C. Narula, K. Mak, L. Kao, Y. Xu and R. F. Heck, *J. Org. Chem.*, 1983, **48**, 3894.
95. F. Ziegler, V. Chakraborty and R. B. Weisenfeld, *Tetrahedron*, 1981, **37**, 1267.
96. L. Kao, F. G. Stakem, B. Patel and R. F. Heck, *J. Org. Chem.*, 1982, **47**, 1267.
97. B. Patel, L. Kao, N. Cortese, J. Minkiewicz and R. F. Heck, *J. Org. Chem.*, 1979, **44**, 918.
98. W. Fischetti, K. Mak, F. G. Stakem, J. Kim, A. L. Rheingold and R. F. Heck, *J. Org. Chem.*, 1983, **48**, 948.
99. K. Kikukawa, K. Nagira, F. Wada and T. Matsuda, *Tetrahedron*, 1981, **37**, 31.
100. F. Akiyama, H. Miyazaki, K. Kaneda, S. Teranishi, Y. Fujiwara, M. Abe and H. Taniguchi, *J. Org. Chem.*, 1980, **45**, 2359.
101. K. Kikukawa, K. Maemura, Y. Kliseki, F. Wada and T. Matsuda, *J. Org. Chem.*, 1981, **46**, 4885.
102. P. Fitton, M. Johnson and J. E. McKeon, *J. Chem. Soc., Chem. Commun.*, 1968, 6.
103. C. Anderson and A. Hallberg, *Tetrahedron Lett.*, 1983, **24**, 4215.
104. K. Hori, M. Ando, N. Takaishi and Y. Inamoto, *Tetrahedron Lett.*, 1987, **28**, 5883.
105. F. Henin and J. Pete, *Tetrahedron Lett.*, 1983, **24**, 4687.
106. E. G. Samsel and J. Norton, *J. Am. Chem. Soc.*, 1984, **106**, 5506.
107. E. Negishi and J. Miller, *J. Am. Chem. Soc.*, 1983, **105**, 6761.
108. A. Spencer, *J. Organomet. Chem.*, 1984, **265**, 323.
109. Y. Tamaru and Z. Yoshida, *Tetrahedron Lett.*, 1978, 4527.
110. A. Kasahara, T. Izumi, N. Kudon, H. Azami and S. Yamamoto, *Chem. Ind. (London)*, 1988, 51.
111. Q. Chen and Z. Yang, *Tetrahedron Lett.*, 1986, **27**, 1171.
112. S. Cacchi, E. Morera and G. Ortar, *Tetrahedron Lett.*, 1984, **25**, 2271.
113. W. Scott, M. Pena, K. Sward, A. Stoessel and J. Stille, *J. Org. Chem.*, 1985, **50**, 2302.

4.4
Carbometallation of Alkenes and Alkynes

PAUL KNOCHEL
University of Michigan, Ann Arbor, MI, USA

4.4.1 INTRODUCTION

Reactions which result in the addition of the carbon–metal bond of an organometallic (**1**) across a carbon–carbon multiple bond (**2**) leading to a new organometallic (**3**) are called carbometallation reactions (equation 1).[1–3] Only those reactions in which the newly formed carbon–metal bond of (**3**) can be used for further synthetic transformations will be considered here. Since the first carbometallation discovered by Ziegler and Bähr[4] in 1927 (equation 2), an ever-increasing number of additions of organometallics to carbon–carbon multiple bonds have been reported.[3] To be synthetically useful, a carbometallation reaction must show good chemo-, regio- and stereo-selectivity.

$$\underset{(\mathbf{1})}{\text{R–M}} \quad + \quad \underset{(\mathbf{2})}{\text{C}\equiv\text{C}} \quad \longrightarrow \quad \underset{(\mathbf{3})}{\text{R–C=C–M}} \qquad (1)$$

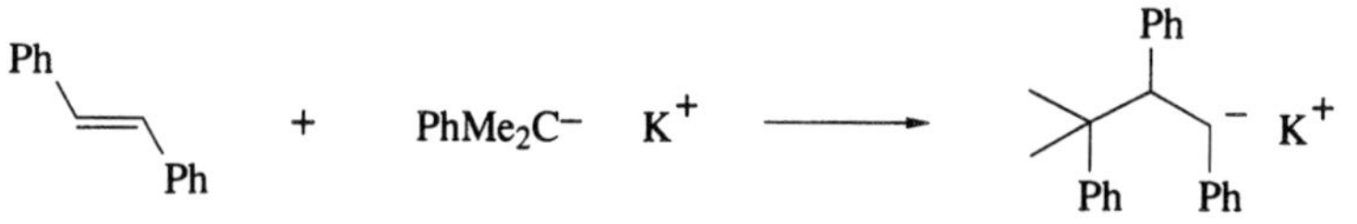

$$\text{PhCH=CHPh} \quad + \quad \text{PhMe}_2\text{C}^- \ \text{K}^+ \quad \longrightarrow \quad \text{PhMe}_2\text{C–CH(Ph)–CH(Ph)}^- \ \text{K}^+ \qquad (2)$$

Most organometallics which undergo additions to carbon–carbon bonds are polar reagents and they may react with several functional groups before they add to a relatively nonpolar carbon–carbon multiple bond. In practice, only a limited number of functional groups of low reactivity (such as ethers, sulfides, acetals, hydroxy and amino groups) can be present in the unsaturated organic substrate submitted to a carbometallation. Organometallics can also act as bases, especially organolithium and organomagnesium derivatives, abstracting acidic allylic, propargylic or alkynic protons. The deprotonation of terminal alkynes, instead of their carbometallation, is a common side reaction.[3] Furthermore, the addition of an organometallic (**1**) to a multiple bond (**2**) affords a new organometallic compound (**3**; Scheme 1), which can participate in a subsequent carbometallation if its reactivity is too similar to that of (**1**). In this case, a polymerization of the unsaturated substrate will result. In order to realize a controlled monoaddition, the carbometallation ability of (**1**) must be higher than that of (**3**). In the case of an intramolecular carbometallation, this does not necessarily have to be true because the entropy effect will favor the monoaddition even if the starting organometallic and the product have similar reactivities.

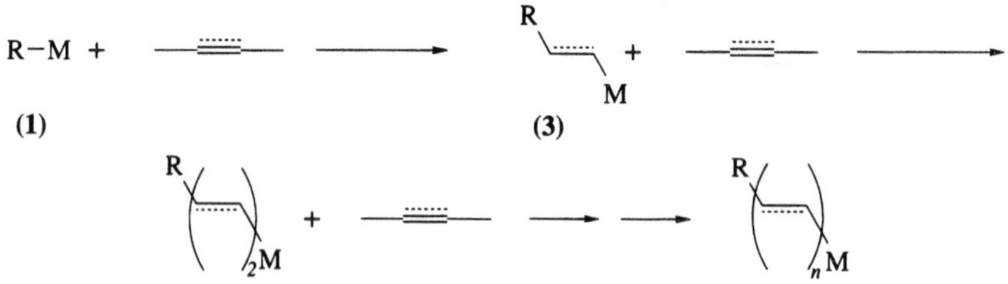

$$\underset{(\mathbf{1})}{\text{R–M}} + \text{C}\equiv\text{C} \longrightarrow \underset{(\mathbf{3})}{\text{R–C=C–M}} + \text{C}\equiv\text{C} \longrightarrow \text{R–(C=C)}_2\text{–M} + \text{C}\equiv\text{C} \longrightarrow \longrightarrow \text{R–(C=C)}_n\text{–M}$$

Scheme 1

The carbometallation of an unsymmetrical alkyne or alkene can lead to two regioisomers (**4**) and (**5**) (equation 3). The ratio of (**4**) to (**5**) depends on the nature of the organometallic, the organic substrate and the reaction conditions used (presence of a catalyst, solvent, cosolvent, *etc.*).

$$\text{R–M} \quad + \quad \text{C}\equiv\text{C–R}^1 \quad \longrightarrow \quad \underset{(\mathbf{4})}{\text{R–C=C–R}^1} \quad + \quad \underset{(\mathbf{5})}{\text{RR}^1\text{C=C}} \qquad (3)$$

The addition to an alkyne can either be *syn* or *anti*, leading respectively to the two stereoisomers (**6**) and (**7**) (Scheme 2). Most carbometallations of alkynes are *syn* additions and certain organometallics such as organocopper reagents show an almost perfect *syn* selectivity.[3] With others, such as organomagnesium derivatives, an *anti* addition is preferred in the case of some functionalized alkenes and alkynes. Since most C(sp^3)—M bonds of polar organometallics are not configurationally stable, the *syn* or the *anti* character of the addition to alkenes often cannot be established. However, in the case of cyclopropenes, which leads, after carbometallation, to configurationally stable cyclopropyl organometallics, a *syn* addition is always observed.

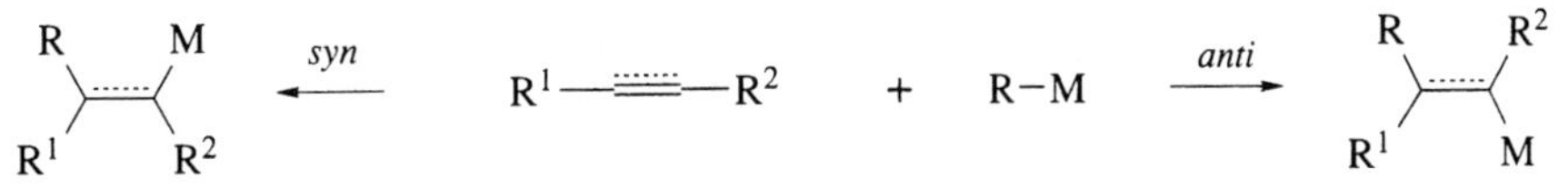

R^1—≡—R^2 + R—M ; *syn* → (**6**) ; *anti* → (**7**)

Scheme 2

More severe reaction conditions are required for the carbometallation of alkenes and dienes than for alkynes. Most carbometallations of alkenes strongly depend on the structure of the organic substrate, and generally show a narrow synthetic scope. More successful are additions to functionalized alkenes in which the heteroatom, by precoordination of the organometallic, directs the carbometallation reaction. Also, intramolecular carbometallation reactions proceed more readily and may show very high regio- and diastereo-selectivities. Alkynes, which are more reactive, are carbometallated with several organometallics under mild conditions. The carbocupration discovered by Normant[3] has the largest synthetic possibilities. It shows an excellent chemo-, regio- and stereo-selectivity with nonfunctionalized alkynes. A very high regioselectivity and almost perfect stereoselectivity can be reached with functionalized alkynes by the fine tuning of the reaction conditions (solvent, ligand, cosolvent, *etc.*). Of special interest is also the zirconium-catalyzed carboalumination discovered by Negishi,[1,2,5] which complements the carbocupration reaction and allows the performance of methylaluminations with excellent regio- and stereoselectivity. These carbometallations give an easy access to a wide range of stereo-defined alkenyl-copper and -aluminum reagents, of great synthetic interest. Both alkenyl organometallics are able to form new carbon–carbon bonds with a variety of electrophiles and have found numerous synthetic applications.[1–3,5]

4.4.2 THE CARBOLITHIATION REACTION

4.4.2.1 General Considerations[6,7]

Due to its high ionic character, the carbon–lithium bond is very reactive and adds under mild conditions to ethylene or dienes and under more severe conditions to other alkenes. Some functionalized alkenes can be used, and high regio- and stereo-selectivity is usually observed in these carbolithiation reactions, especially if a precoordination of the lithium organometallic with the alkene is possible. Intramolecular carbolithiations of alkenes proceed under mild conditions and allow the preparation of several stereochemically well defined mono- and bi-cyclic compounds. Alkynes are too reactive, and can lead, with organolithium derivatives, to several side reactions, and seldom afford the desired carbolithiated product in good yield.

4.4.2.2 The Carbolithiation of Alkenes and Dienes

4.4.2.2.1 Intermolecular reactions

The controlled addition of organolithium reagents to isolated double bonds is of limited preparative interest. The carbolithiation of ethylene under pressure by primary lithium organometallics (*e.g.* BuLi, EtLi) leads to new lithium derivatives, RCH_2CH_2Li, which have a similar reactivity toward ethylene as their precursor, RLi, and thus low molecular weight polymers are formed.[8] In the presence of ligands like DABCO or TMEDA, which enhance the reactivity of organolithium derivatives by reducing their aggregation,[9] extensive polymerization is observed, and alkyllithium reagents are in fact excellent initiators for anionic polymerization, especially of styrene and conjugated dienes.[10] Secondary and tertiary alkyllithium reagents are far more reactive toward addition to ethylene (Bu^sLi is at least 10^6 times more reactive than Bu^nLi), and afford primary alkyllithium compounds in good yields under mild conditions. These organometallics are inert toward further addition of ethylene (equation 4).[11] Other isolated alkenes[12] like 1-octene are far less reactive and the addition will occur in moderate yield only at a higher temperature (80 °C). At higher temperatures, allylic proton abstraction and decomposition of the lithium reagent (β-hydride elimination) become important side reactions. Noteworthy is the high regioselectivity of the carbolithiation. In the case of 1-octene, only the primary alkyllithium (**8**) is formed, in strong contrast to the corresponding aluminum and magnesium organometallics, which afford the opposite regioisomer (equation 5).[12]

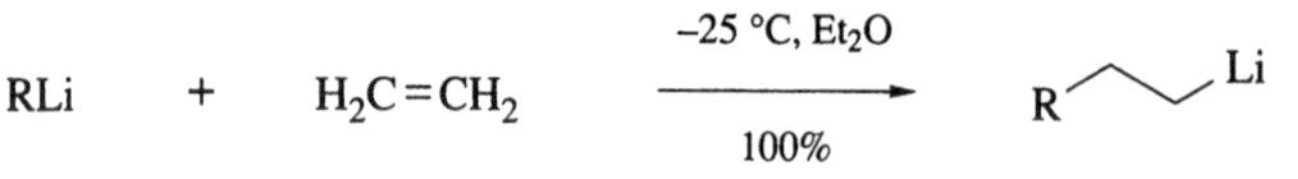

RLi + $H_2C{=}CH_2$ —(–25 °C, Et_2O; 100%)→ R–CH₂CH₂–Li (4)

R = Bu^t, Bu^s, Pr^i, cyclohexyl

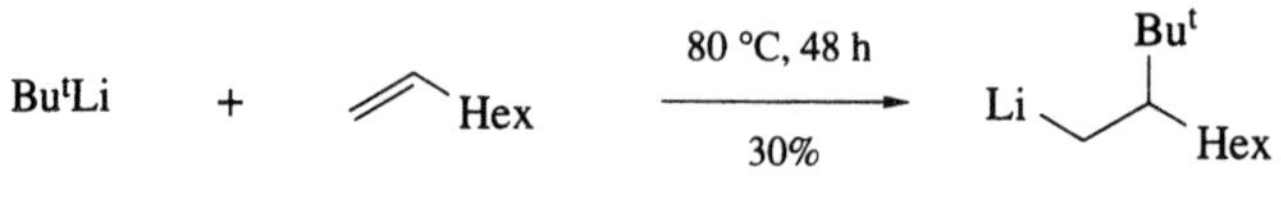

Bu^tLi + CH₂=CH–Hex —(80 °C, 48 h; 30%)→ (**8**) (5)

Li, Bu^t, Hex

(8)

The addition to aryl-substituted alkenes and to dienes occurs readily in Et_2O or THF, but controlled reaction conditions have to be used in order to avoid polymerization.[13] Alkyllithium reagents are more reactive than allyllithiums,[14] and the fast addition of Bu^tLi to butadiene[16] provides a quantitative formation of 5,5-dimethyl-2-hexenyllithium (**9**). The subsequent addition of (**9**) to butadiene is more sluggish and occurs with low regioselectivity (Scheme 3). The addition of lithium organometallics to enynes[15] is a useful reaction for the preparation of allenes. Strained alkenes such as norbornene,[17] methylcyclopropene[18] or *trans*-cyclooctene[19] react under moderate reaction conditions (Scheme 4). In the case of

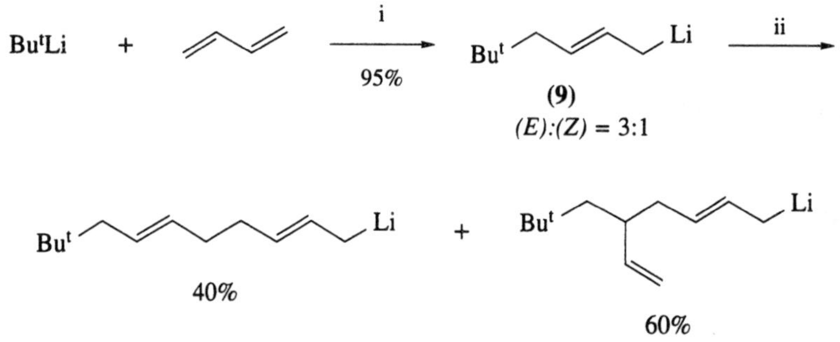

Bu^tLi + butadiene —(i; 95%)→ Bu^t–CH₂CH=CHCH₂–Li (**9**) —(ii)→

(9)
(E):(Z) = 3:1

Bu^t ... Li 40% + Bu^t ... Li 60%

i, pentane, –78 °C to 20 °C, 0.5 h; ii, butadiene, 25 °C, 8 h

Scheme 3

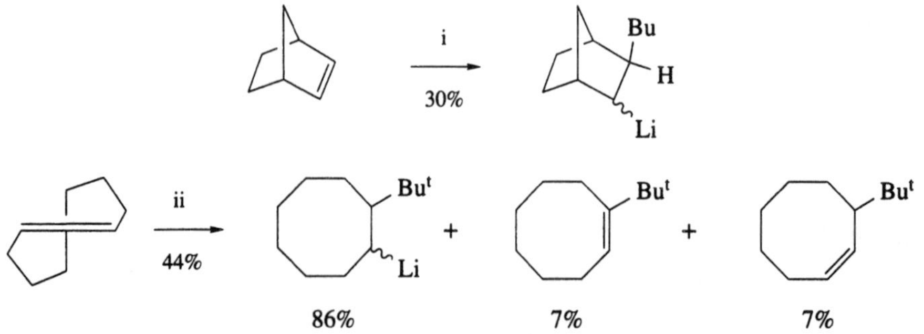

norbornene —(i; 30%)→ Bu, H, Li

trans-cyclooctene —(ii; 44%)→ Bu^t, Li 86% + Bu^t 7% + Bu^t 7%

i, BuLi, TMEDA, hexane, 25 °C, 36 h; ii, Bu^tLi, Et_2O, 25 °C, 18 h

Scheme 4

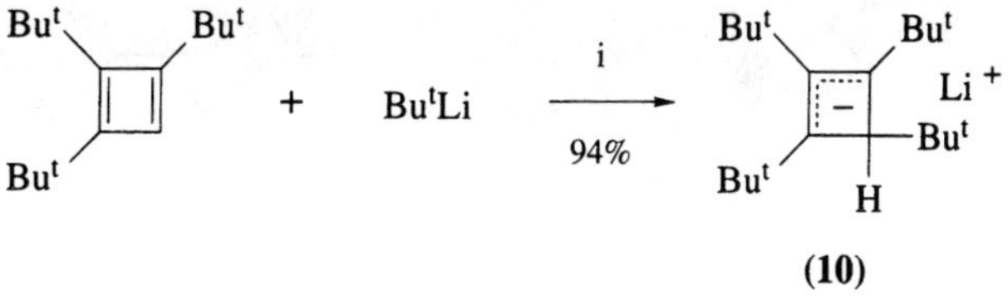

(6)

(10)

i, TMEDA, Et_2O, –78 °C

48% 32% 6% 62%

92%

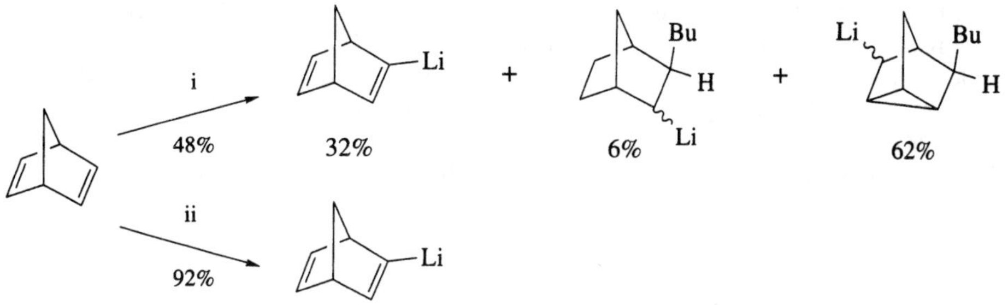

i, BuLi, Et_2O; ii, BuLi, Bu^tOK, THF, –80 °C to –60 °C, then LiBr

Scheme 5

norbornene, only the product of *exo* attack is observed.[17] The addition of Bu^tLi to a cyclobutadiene derivative[20] occurs readily and affords the addition product (**10**) in a very high yield (equation 6). In some cases, such as that of norbornadiene,[21] a proton abstraction can compete with the carbolithiation reaction and, depending on the reaction conditions, can become the predominant reaction pathway (Scheme 5).[22]

The presence of a donor group at the proximity of the double bond of the alkene promotes the carbolithiation reaction, as has been shown by Wittig,[21] Bickelhaupt,[23] Klumpp[23] and their respective coworkers (equation 7). The precoordination of the lithium reagent with the heteroatom explains both the high rate of the carbometallation and the high regio- and stereo-selectivity (*syn* addition) observed in these reactions. The addition proceeds well with several acyclic alkenyl ethers and amines,[24,25] regiospecifically affording the stabilized lithium derivatives of type (**11**) and (**12**) (Scheme 6). Similarly, the addition of organolithium compounds to β-substituted homoallylic sulfides provides an approach to γ-lithiated sulfides (**13**; equation 8).[26]

The addition of organolithium compounds[27] to allylic alcohols in the presence of TMEDA is a preparatively useful method which allows a high regio- and stereo-selective synthesis of alcohols. Several side reactions can be observed depending on the structure of the allylic alcohol (**14**; Scheme 7). If the

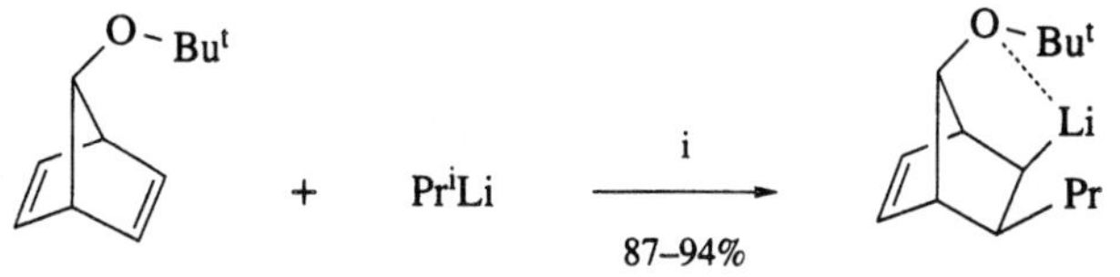

(7)

87–94%

i, petroleum ether, –20 °C

95–100% 97%

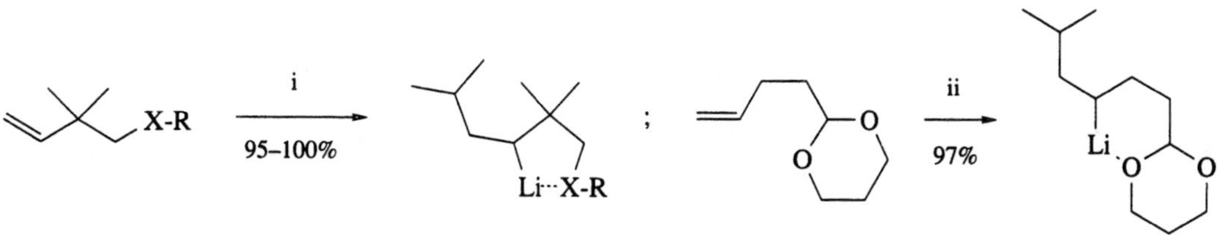

X = OMe or NMe_2 (**11**) (**12**)

i, Pr^iLi (1.5 equiv.), pentane/Et_2O, 25 °C, 0.3 h; ii, Pr^iLi (3 equiv.), no solvent, 0 °C, 2 h

Scheme 6

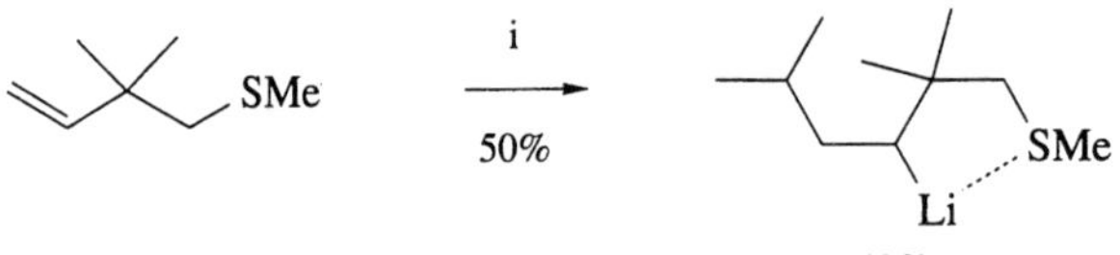

(8)

(13)

i, Pr^iLi, Et_2O/pentane, 25 °C, 2 h

substituent R^2 of (**14**) is a methyl group, then the carbolithiation is disfavored and an allylic deprotonation occurs, leading to the allylic lithium derivative (**15**).[28] If R^1 or R^3 is a phenyl group, another allylic deprotonation can occur, furnishing the dilithiated compound (**16**). Finally, if the carbolithiation leads to a β-lithioalcoholate of type (**17**), a fast elimination of lithium oxide will occur, affording the alkene (**18**; Scheme 7). This reaction pathway is the most important if R^1 and R^3 are part of a five- or six-membered ring, or if a bulky organolithium reagent is used.[27] Representative synthetic applications, as developed by Crandall and coworkers, are shown in Scheme 8. Felkin and coworkers[27] demonstrated that these car-

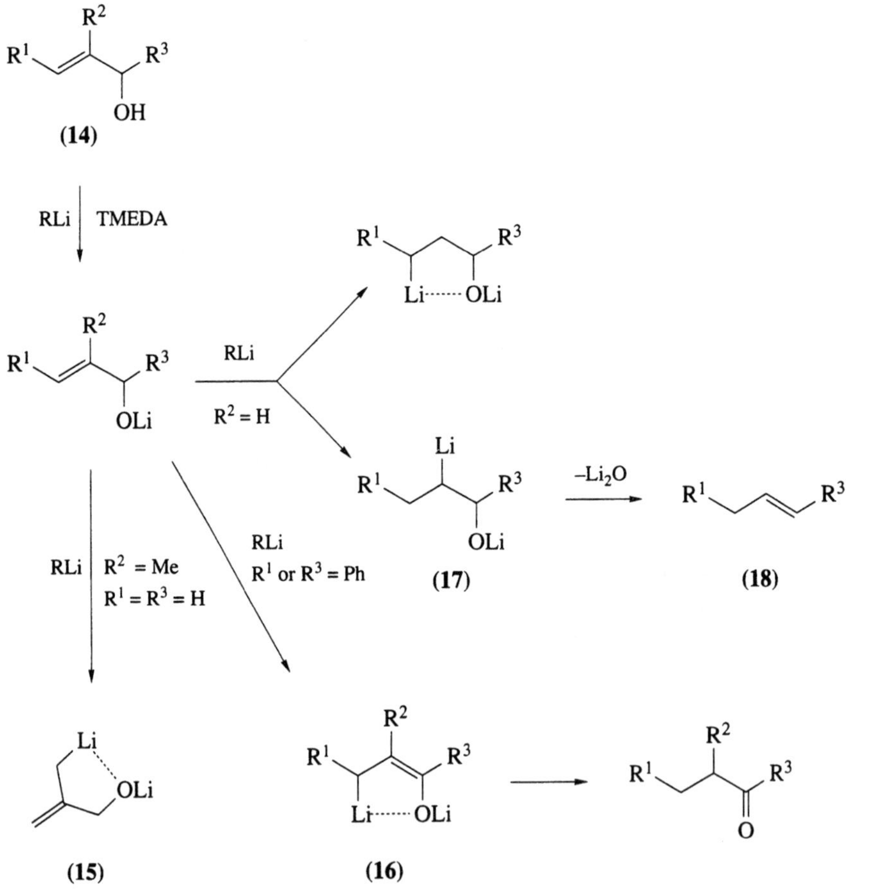

Scheme 7

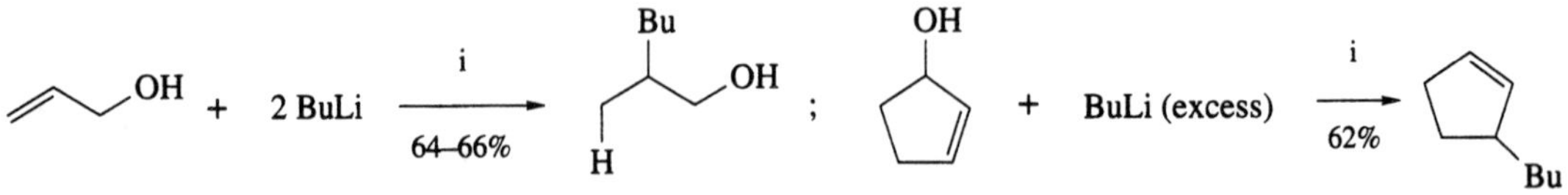

i, TMEDA, pentane, 25 °C, then H_3O^+

Scheme 8

bolithiations can be highly diastereoselective, and a chelate-type transition state (**19**) has been proposed to account for the observed stereochemistry (Scheme 9).[27] Other related compounds like allylic amines[29] and α-amino- or α-hydroxy-enynes[30] are also able to add various lithium organometallics across the double bond, but in moderate yields (30–60%). The direct addition of organolithium reagents to aromatic compounds such as naphthalene or phenanthrene occurs only under severe reaction conditions (decalin, 165 °C).[31]

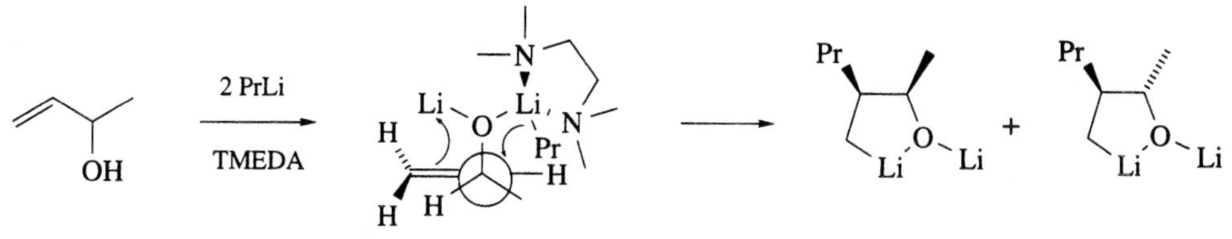

2 PrLi, TMEDA

(**19**) >98% <2%

Scheme 9

4.4.2.2.2 Intramolecular reactions

Intramolecular carbolithiations proceed more readily than intermolecular carbolithiations.[32] Thus, 5-hexen-1-yllithium (**20**) is indefinitely stable at –78 °C, but undergoes a rapid cyclization upon warming to higher temperatures (at 23 °C, $t_{1/2}$ = 5.5 min; see Scheme 10). However, it has been estimated that the rate of cyclization of (**20**) is 10^8 times slower than that of the corresponding radical.[33] Bailey and coworkers showed that several intramolecular organolithium cyclizations occur in good yields and with very high diastereoselectivities, as indicated in Scheme 11.[34]

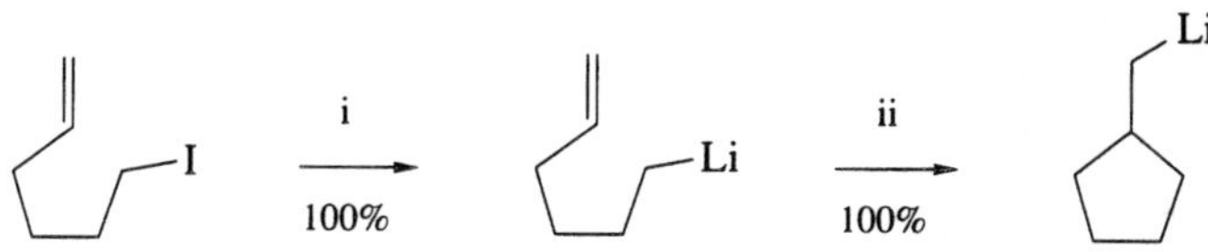

i 100% (**20**) ii 100%

i, Bu^tLi (2 equiv.), pentane/Et_2O, –78 °C, 5 min; ii, –78 °C to 25 °C

Scheme 10

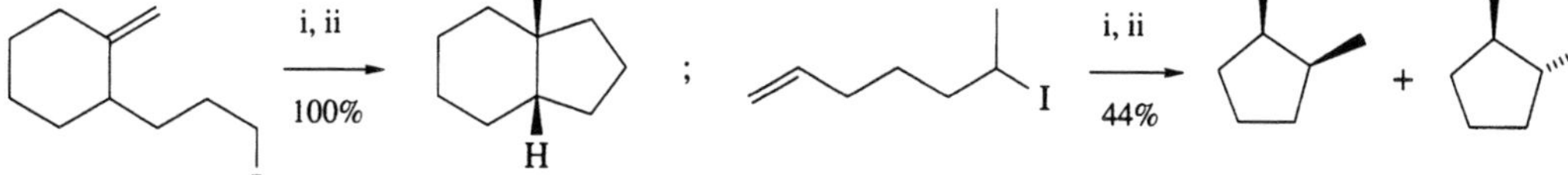

i, ii 100% ; i, ii 44% >99% <1%

i, Bu^tLi (2.2 equiv.), pentane/Et_2O, 5 min; ii, 5 h at –20 °C, then H_3O^+

Scheme 11

The ring closures can be reversible, as in the case of the diphenylcyclopropylcarbinyllithium species (**21**), which is stable in THF but completely undergoes opening to 4,4-diphenylbuten-3-yllithium (**22**) in ether (equation 9).[35] The equilibrium between unsaturated primary and secondary organolithium organometallics such as (**23**) and (**24**) proceeds *via* a reversible intramolecular carbolithiation reaction[36] which favors the more stable primary organolithium (**23**) in over 99% (equation 10). Even the relatively unreactive aryllithium reagent (**25**) cyclizes to the corresponding indanyllithium (**26**) in good yield (Scheme

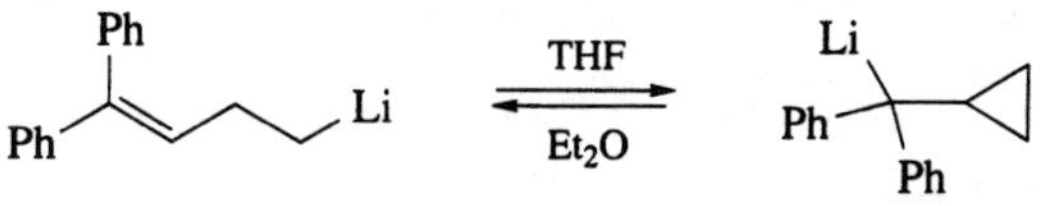

(**22**) THF ⇌ Et_2O (**21**) (9)

12).[37] Chamberlin and coworkers[38] showed that vinyllithium reagents derived from ketone (2,4,6-triisopropylphenyl)sulfonylhydrazones undergo intramolecular carbolithiation reactions with high diastereoselectivity (Scheme 13).

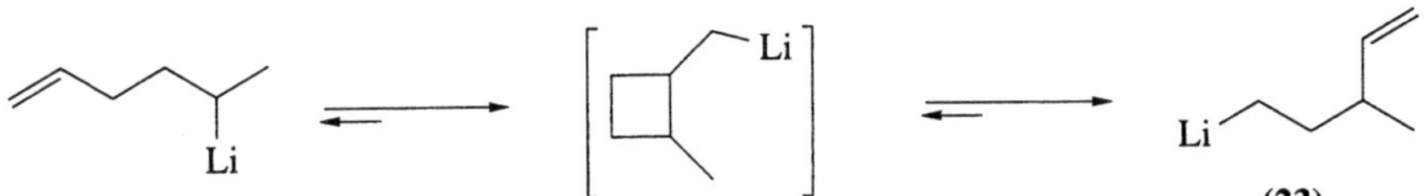

Li (24) ⇌ [cyclobutane Li] ⇌ Li (23) (10)

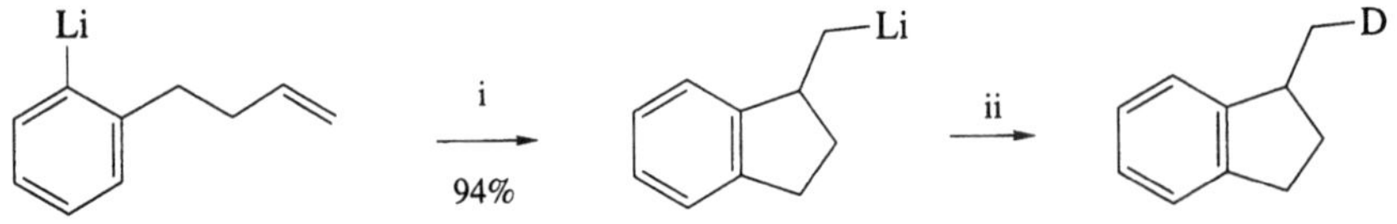

Li (25) —i, 94%→ Li (26) —ii→ D 85% d_1

i, Et_2O/TMEDA, 23 °C, 0.5 h, ii, D_2O

Scheme 12

Major diastereoisomers — Minor diastereomers

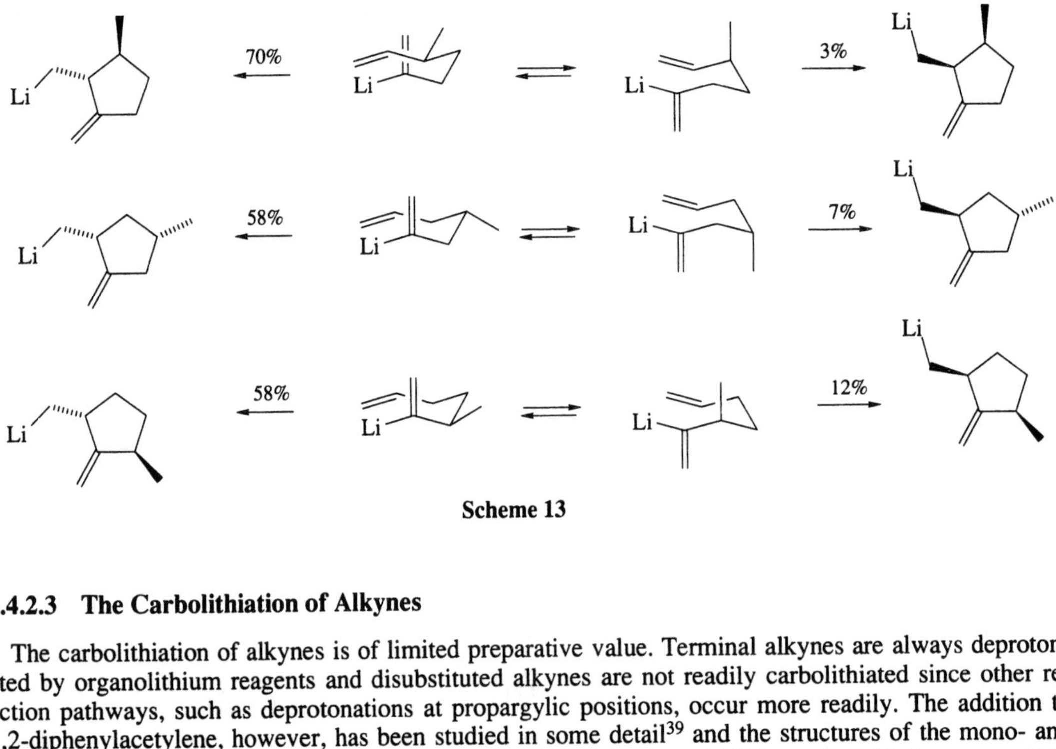

70% — Li — Li — 3% — Li

58% — Li — Li — 7% — Li

58% — Li — Li — 12% — Li

Scheme 13

4.4.2.3 The Carbolithiation of Alkynes

The carbolithiation of alkynes is of limited preparative value. Terminal alkynes are always deprotonated by organolithium reagents and disubstituted alkynes are not readily carbolithiated since other reaction pathways, such as deprotonations at propargylic positions, occur more readily. The addition to 1,2-diphenylacetylene, however, has been studied in some detail[39] and the structures of the mono- and di-lithiated products (**27**) and (**28**) (Scheme 14) have been elucidated.[40] The carbolithiation proceeds in various solvents and is accelerated by the presence of TMEDA. An initial *syn* addition product (**29**) has been postulated, since it has been found that Bu^tLi reacts with 1,2-diphenylacetylene to afford mainly the *syn* addition product. A rapid isomerization of (**29**) leads to the only detectable *trans* isomer (**27**), which is again lithiated to afford (**28**). *Ab initio* calculations[41] on the addition of lithium hydride to acetylene also predict a *syn* addition and support the postulated mechanism of Mulvaney (Scheme 14).[39b] Functionalized alkynes, such as 3-phenylpropargyl alcohol, readily add BuLi to provide, after hydrolysis, the stereoisomerically pure alcohol (**30**; Scheme 15).[42] The addition of organolithium reagents to benzyne and benzyne derivatives is of preparative interest.[43] Meyers and coworkers[44] showed that the reaction of the functionalized aryl chloride (**31**) with BuLi allows the *in situ* generation of the benzyneoxazoline (**32**), which undergoes a regioselective carbolithiation with lithium organometallics to afford mainly the aryllithium derivatives (**33**). The carbocupration of the benzyne (**32**) furnishes the opposite regioisomer

(**34**; Scheme 16). This regioisomer is also obtained in appreciable amounts if bulky lithium organometallics are used. Intramolecular carbolithiations have been reported, such as the cyclization of 5-alkynyllithium compounds to furnish the cyclized products in low yields.[3]

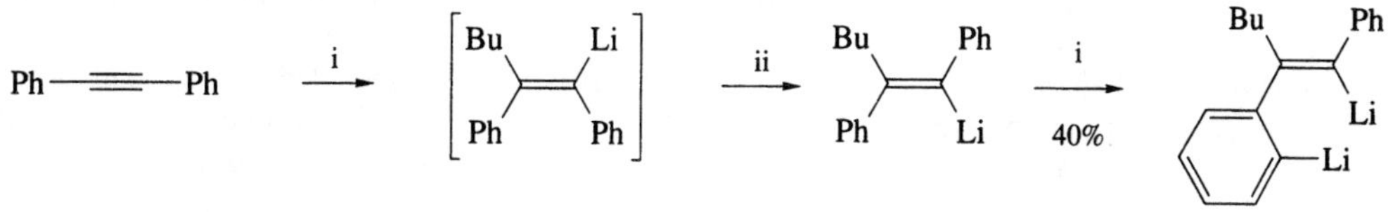

i, BuLi, hexane/TMEDA or Et_2O; ii, fast isomerization

Scheme 14

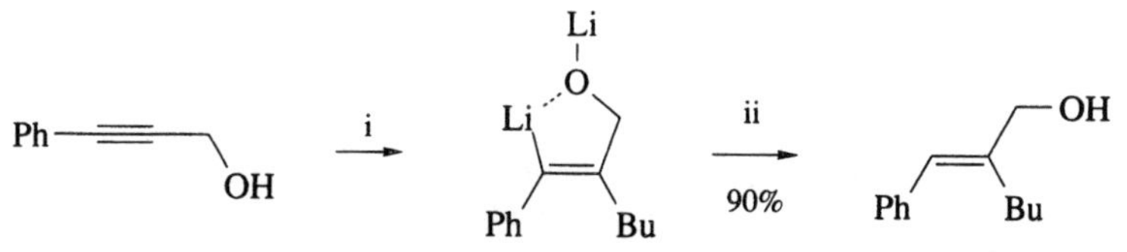

i, BuLi (2.5 equiv.), Et_2O, TMEDA (0.2 equiv.), –30 °C to 20 °C, 3 h, ii, H_2O

Scheme 15

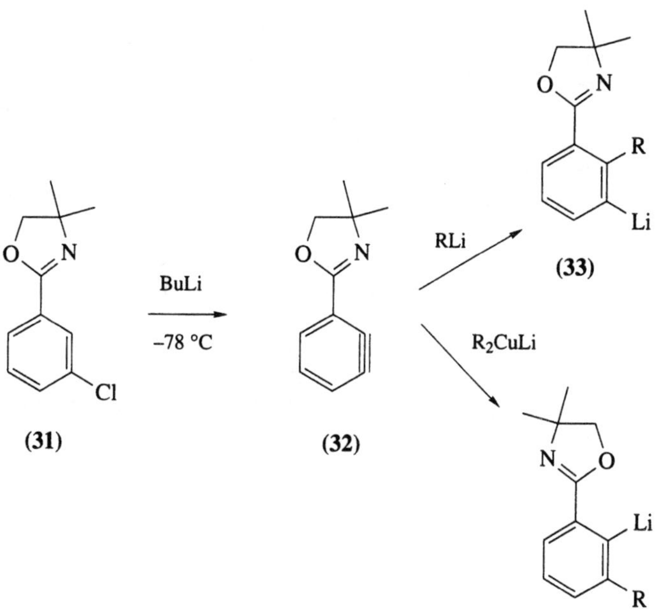

Scheme 16

4.4.3 THE CARBOMAGNESIATION REACTION

4.4.3.1 General Considerations[45]

The carbomagnesiation of alkenes has been extensively investigated in the last 40 years and several reactions with high synthetic potential have been discovered. The addition of Grignard reagents to nonfunctionalized alkenes usually requires severe reaction conditions and only the zirconium-catalyzed addition proceeds under mild conditions, but its scope seems limited. The introduction of a heteroatom in the proximity of the double bond considerably facilitates the addition by a precoordination of the Grignard reagent to the heteroatom. In these cases, high regio- and stereo-selectivity of the addition usually results. The intramolecular carbomagnesiation requires more severe conditions than the corresponding carbolithiation, but if the Grignard reagent is allylic the reaction proceeds well and is of high synthetic utility (see Volume 5, Chapter 1.2). In contrast to many carbometallations, the addition of Grignard reagents to functionalized alkynes shows an *anti* selectivity. It is highly regioselective and leads to useful functionalized alkenyl organomagnesium derivatives. The addition of magnesium derivatives to alkynes can, in several cases, be catalyzed by salts of transition metals such as copper or nickel.

4.4.3.2 The Carbomagnesiation of Alkenes

4.4.3.2.1 Addition to nonfunctionalized alkenes

(i) Intermolecular reactions

Intermolecular and uncatalyzed additions of Grignard reagents to nonfunctionalized alkenes proceed only under very severe conditions and are of limited synthetic interest. In order to avoid polymerization reactions, the reactivity of the starting magnesium organometallic has to be higher than that of the carbomagnesiated product. Lehmkuhl[46] and coworkers have shown that secondary and tertiary alkyl,[47] benzylic,[48] and allylic[49] magnesium derivatives add to alkenes, whereas the less reactive primary alkylmagnesium compounds do not. In most cases, forcing reaction conditions have to be used. Ethylene reacts under pressure with allylic magnesium halides[49,50] in ether (40–70 atm, 20–80 °C, 2–100 h) to afford unsaturated Grignard derivatives of type (**35**; equation 11). Ether is the best solvent for these reactions; stronger Lewis bases like THF or dioxane diminish the reaction rate. Lehmkuhl and coworkers have shown that the reaction is first order both for the alkene and the allylic Grignard reagent. A cyclic six-membered transition state of type (**36**) has been proposed since relatively high activation entropies have been observed (between –17 and –24 cal °C^{-1} mol^{-1}; 1 cal = 4.2 J). The activation parameters ($\Delta H^{\ddagger}$ = 20 kcal mol^{-1} and $\Delta G^{\ddagger}$ = 26–30 kcal mol^{-1}) are very similar to those of Diels–Alder or ene reactions.[46] The reaction usually proceeds with almost complete allylic rearrangement (Scheme 17).[51] The reactivity of alkenes increases in the order: 1-alkene < styrene < butadiene < ethylene. Alkenes with an internal double bond do not react. Strained alkenes such as cyclopropenes[50,52] or norbornene[50–52] are exceptions and add readily to a variety of Grignard reagents under mild conditions and in good yields. The stereospecific *syn* addition of 2-methylpropenylmagnesium chloride to 3,3-dimethylcyclopropene has been used by Lehmkuhl[50,52,53] for a short and efficient approach to (*Z*)-chrysanthemic acid (Scheme 18).

The regioselectivity of the carbomagnesiation of 1-alkenes depends strongly on the structure of the Grignard reagent.[54] Primary organomagnesium halides attack the alkene preferentially at C-2, leading to a new primary organometallic, whereas secondary and tertiary magnesium derivatives give more C-1 attack (formation of a secondary Grignard reagent; see Table 1). These reactions require temperatures between 70 and 120 °C and afford only moderate yields of the addition products.[50] Allylic magnesium halides[53,55] react with more than 90% regioselectivity with butadiene (Mg at C-2) to afford the 2,6-heptadienylmagnesium derivatives (**37**) as the major products (Scheme 19). In the presence of an excess of bu-

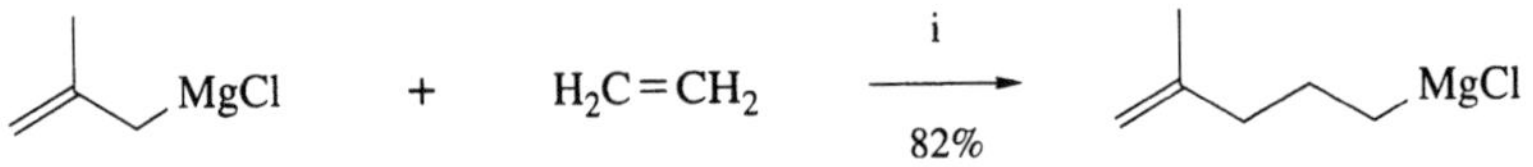

MgCl + $H_2C{=}CH_2$ —(i, 82%)→ MgCl (**35**) (11)

i, Et_2O, 40–70 atm, 20–62 °C

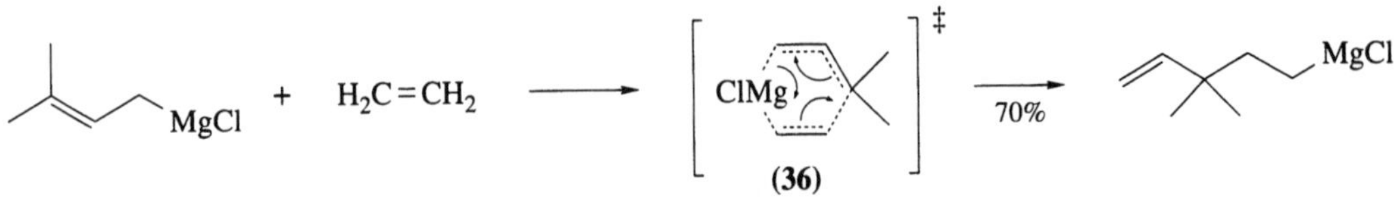

MgCl + $H_2C{=}CH_2$ → [ClMg … (**36**)]‡ —(70%)→ MgCl

Scheme 17

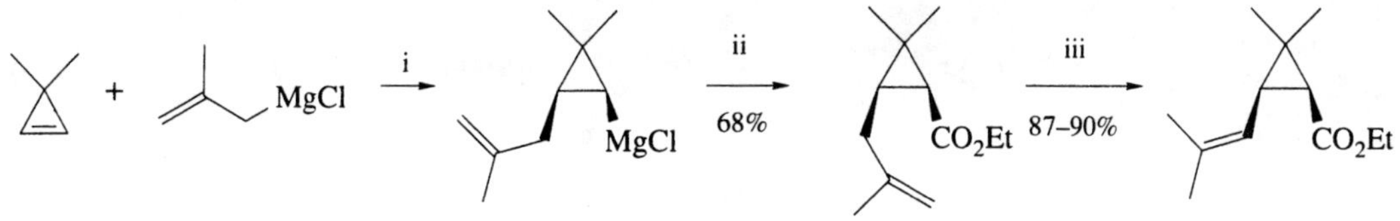

i, Et_2O, 0–20 °C; ii, $ClCO_2Et$, Et_2O; iii, H^+ (cat.), dioxane

Scheme 18

tadiene, a second carbomagnesiation takes place leading to the 1:2 adduct (**38**), which undergoes a fast intramolecular ring closure to give the cyclic Grignard reagent (**39**). Isoprene shows a similar reactivity pattern, and the initial carbomagnesiation occurs predominantly at C-4 (Mg at C-3). Allene also reacts with 2-methylallylmagnesium chloride[47] (10 atm, Et_2O) to provide a cyclic 1:2 adduct in 44% yield (equation 12). Allylic magnesium compounds add across the double bond of enynes to afford mixtures of alkynes and the corresponding isomeric allenes.[30b]

Table 1 Proportion of Mg at C-2 (Formation of a Secondary Carbomagnesiation Adduct) in the Reaction of Organomagnesium Chloride with 1-Octene

Organo group	*Mg at C-2 (%)*
$—CH_2CH{=}CH_2$	0
$—CH_2CMe{=}CH_2$	0.6
$—CH_2Ph$	0
$—CHMeCH{=}CH_2$	31
Pr^i	81–88
Bu^t	98–99

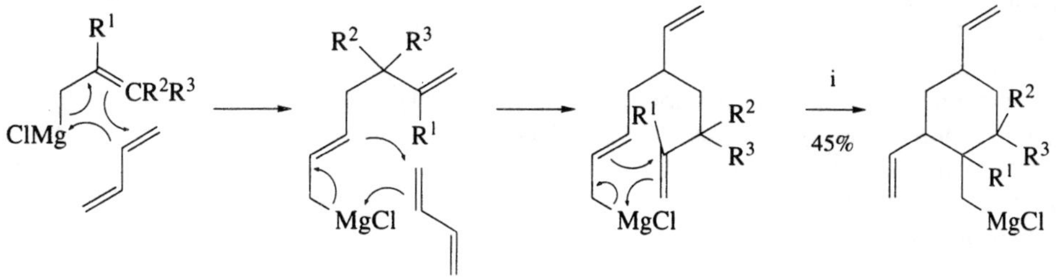

(**37**) (**38**) (**39**)

i, $R^1 = R^2 = R^3 = H$, Et_2O, 85 °C, 24 h

Scheme 19

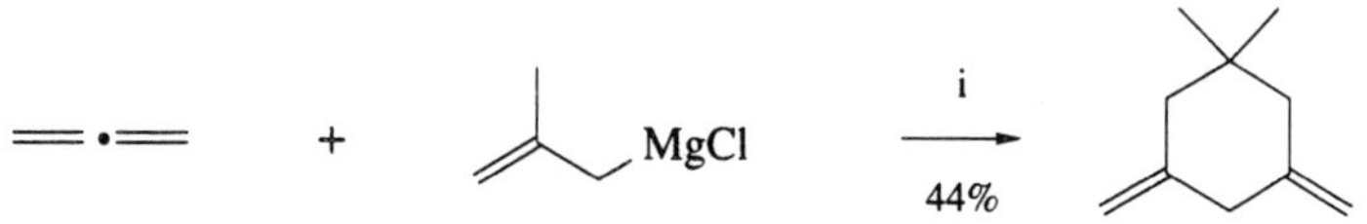

(12)

i, 10 atm, Et_2O, 80–88 °C

In strong contrast, the transition metal catalyzed carbomagnesiation occurs under much milder conditions. Dzhemilev and coworkers[56] have shown that Et_2Mg adds regiospecifically to terminal alkenes in the presence of a catalytic amount of Cp_2ZrCl_2 (equation 13). Several functional groups, such as a double bond, a dialkylamino, a ketal, a trimethylsilyl, an alkoxy or a hydroxy group, are tolerated under these mild reaction conditions (equation 14). The use of higher homologs of Et_2Mg, such as Pr_2Mg or Bu_2Mg, is complicated by the formation of appreciable amounts of elimination and dimerization products. Several other ethylmagnesium derivatives (EtMgX; X = alkyl, halide, NR_2, $SiMe_3$ or Cp) can also be used, but slower addition rates are observed.

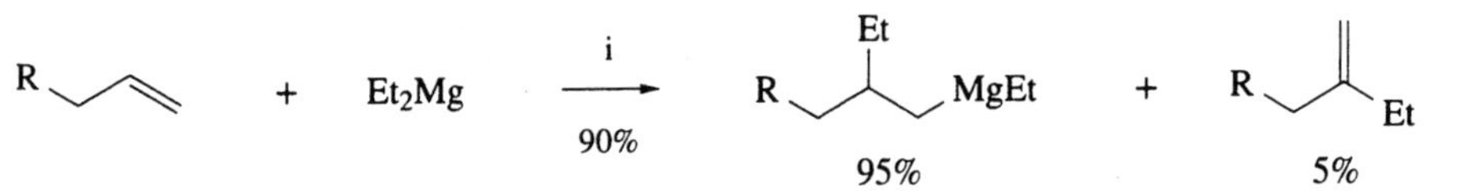

R-CH2CH=CH2 + Et_2Mg —i (90%)→ R-CH2CH(Et)CH2MgEt (95%) + R-CH2C(=CH2)Et (5%) (13)

i, Cp_2ZrCl_2 (1 mol %), Et_2O, 2–4 h

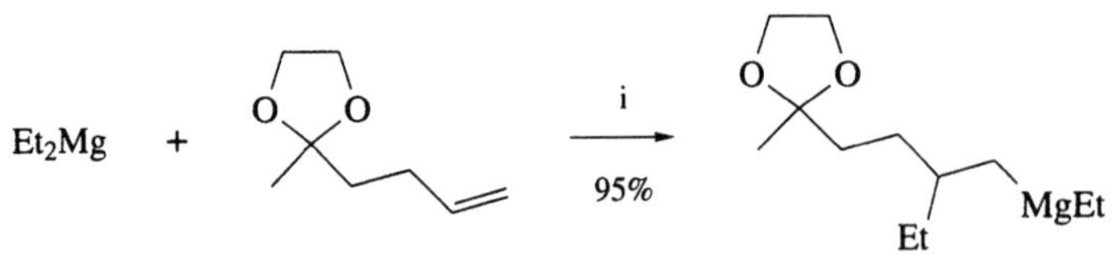

Et_2Mg + (dioxolane-protected alkene) —i (95%)→ (dioxolane product with Et and MgEt) (14)

i, Cp_2ZrCl_2 (1 mol %), Et_2O, 4 h

(ii) Intramolecular reactions

The intramolecular carbomagnesiation[57] of alkenes has been extensively studied in the past 40 years and reversible intramolecular carbomagnesiation reactions have been postulated in isomerizations (Scheme 20).[58] All cyclizations proceed by an exo-trig (or dig) mode[59] leading to the smaller ring (Scheme 21).[60,61] A similar preference has been observed in radical cyclizations.[62] The ring closure rates can be modified by several factors.[57] For example, the use of the more basic solvent THF in place of ether leads to slower cyclization rates by factors ranging from 2 to over 100. Hydrocarbon cosolvents, as well as aprotic solvents like HMPA, generally increase the reaction rate. Higher Grignard concentrations usually lead to a rate increase, possibly by formation of more reactive ate complexes.[35b] In several rearrangement studies, an increased rate has been observed with dialkylmagnesium derivatives (prepared by precipitation in dioxane).[57] The relative rates for the formation of various ring sizes follows the order $C_3 > C_5 > C_4 > C_6$. The effects of substituents have been well studied[57] and substituents at the double bond lead to a rate decrease. Allylic and secondary Grignard compounds are more reactive than primary alkyl Grignard compounds, as shown in Scheme 22.[57,60,63] Interestingly, the cyclized product (**40**) has a *cis*-configuration, whereas the cyclizations of secondary alkylmagnesium compounds mainly afford the *trans* product (*trans*:*cis* > 10:1).[57] The high stereoselectivity in the ring closure of allylic Grignard reagents has been used by Oppolzer and coworkers[64] for the formation of five-, six-, and seven-membered rings (Volume 5, Chapter 1.2).

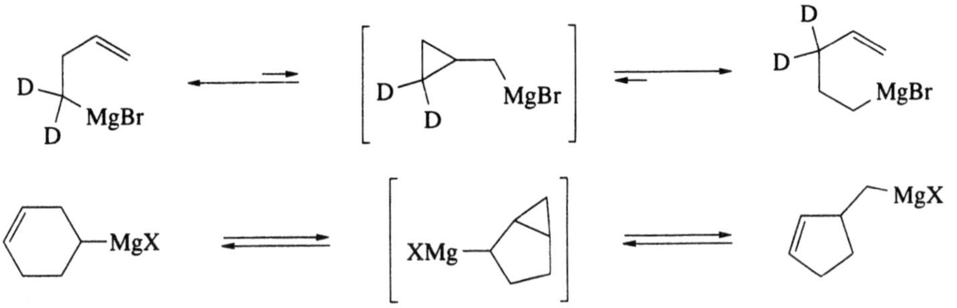

D D MgBr ⇌ [D D cyclopropyl MgBr] ⇌ D D MgBr

MgX (cyclohexenyl) ⇌ [XMg bicyclic] ⇌ cyclopentenyl MgX

Scheme 20

R MgX (cyclohexane) ✕← R MgX (alkene) → R cyclopentane MgX

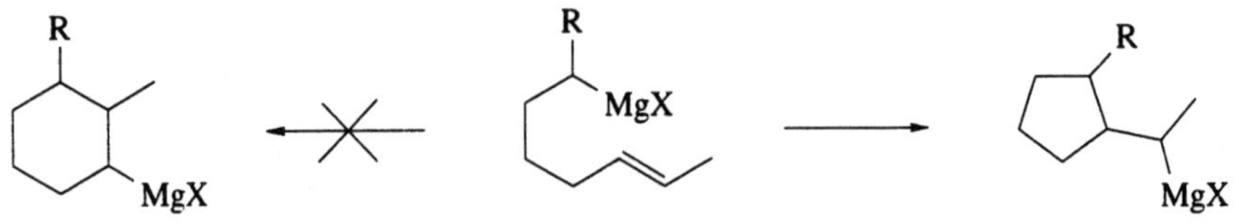

Scheme 21

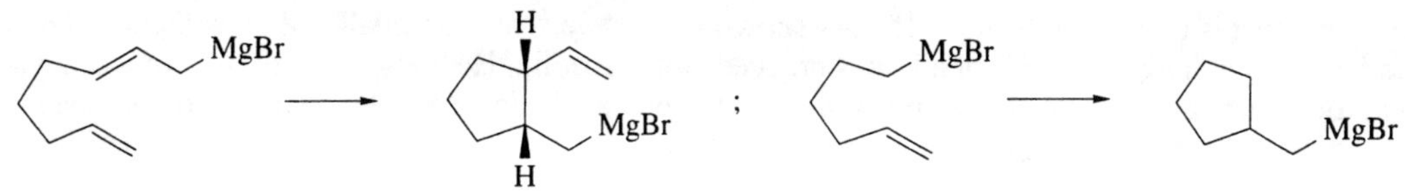

MgBr → H MgBr H ; MgBr → MgBr

$k \approx 10^{-4}$ s^{-1} (refluxing ether) **(40)** $k \approx 10^{-5}$ s^{-1} (THF, 100 °C)

Scheme 22

4.4.3.2.2 Addition to functionalized alkenes and dienes

The presence of a neighboring hydroxy, alkoxy or amino group facilitates the addition of reactive Grignard reagents (allylic and benzylic) to alkenes. Eisch,[65] Felkin,[27d,66] Richey[67] and coworkers found that allylic magnesium organometallics add in a *syn* fashion to alkenols and some unsaturated amines under mild conditions. The reaction proceeds by a precoordination of the organomagnesium derivative[68] to the heteroatom followed by a *syn* addition (Scheme 23).[65,67] The addition of diallylmagnesium to 3-cyclopentenol provides only *cis*-3-allylcyclopentanol.[65] Felkin and coworkers[27d,66] showed that a high diastereoselectivity can also be observed for open chain systems. The addition of allylmagnesium bromide to 3-buten-2-ol affords mainly (R^*,R^*)-3-methyl-5-hexen-2-ol. In this case, a different mechanism has been proposed. The allylic magnesium reagent adds in an *anti* fashion to the double bond (Scheme 24). Functionalized enynes also react under mild conditions with allylic and vinylic magnesium bromides in ether to afford mixtures of isomeric allenes and alkynes in good yields.[30] Oxygen and nitrogen heterocycles have been prepared by Klumpp and coworkers *via* an intramolecular carbomagnesiation reaction (Scheme 25).[69]

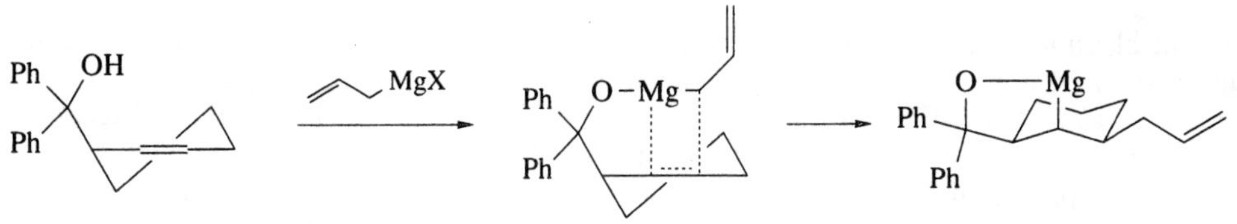

Ph OH Ph MgX → Ph O–Mg Ph → O—Mg Ph Ph

Scheme 23

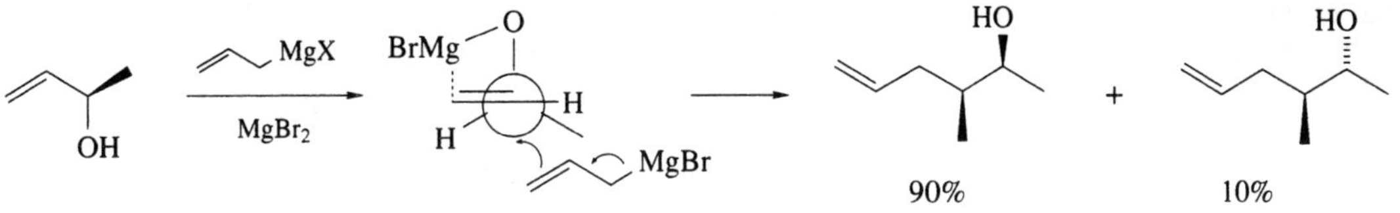

OH MgX MgBr$_2$ → BrMg O H H MgBr → HO 90% + HO 10%

Scheme 24

Me N MgBr → THF → Me N MgBr → Me_3SnCl 76% → Me N $SnMe_3$

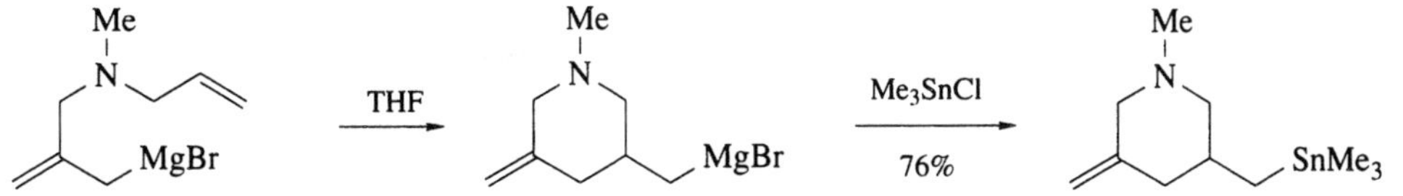

Scheme 25

4.4.3.3 The Carbomagnesiation of Alkynes

4.4.3.3.1 Addition to nonfunctionalized alkynes

Terminal alkynes are readily deprotonated by Grignard reagents, and no further addition occurs to alkynylmagnesium halides. In the presence of transition metal complexes of titanium,[70] iron,[70] rhodium,[71] nickel,[70,72] palladium[70] or copper,[73] the carbomagnesiation takes place in moderate yields. The regio- and stereo-selectivity of the additions are variable. In the presence of a copper(I) salt, however, only the *syn* addition product is obtained (equation 15). As shown by Richey,[74] and Crandall and coworkers,[75] the intramolecular carbomagnesiation[57] of alkynes proceeds more readily. Here also, the addition of a catalytic amount of a copper salt facilitates the reaction. *In situ* generated benzynes add to various organomagnesium derivatives,[76] representing a useful method for the generation of a wide range of arylmagnesium organometallics (Scheme 26).

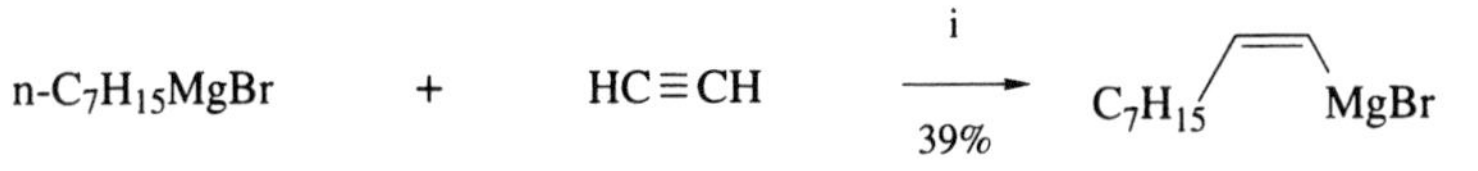

$$n\text{-}C_7H_{15}MgBr + HC{\equiv}CH \xrightarrow[39\%]{i} C_7H_{15}CH{=}CHMgBr \quad (15)$$

i, CuBr (5 mol %), Et_2O, –10 °C, 1 h

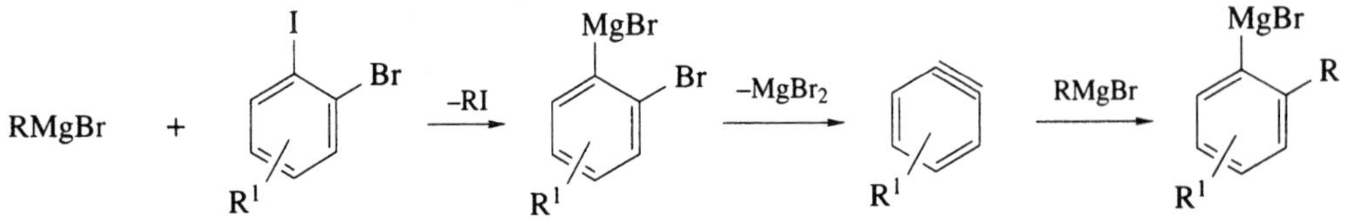

RMgBr + (2-bromo-1-iodoarene, R^1) —(–RI)→ (2-bromophenylmagnesium bromide, R^1) —($-MgBr_2$)→ (benzyne, R^1) —(RMgBr)→ (2-R-phenylmagnesium bromide, R^1)

R = alkenyl, aryl, alkynyl

Scheme 26

4.4.3.3.2 Addition to functionalized alkynes

Propargylic alcohols react regiospecifically with allylic magnesium halides by an *anti* addition to furnish relatively unreactive cyclic magnesium compounds of type (**41**; R^1 = allyl, R^2 = alkyl) in good yields (equation 16).[77] Other Grignard reagents add efficiently to propargyl alcohol and higher alkynols in the presence of a catalytic amount of copper(I) iodide. Only the *anti* addition product is obtained in ether.[78] Metallated propargylic alcohol[59,79] is also able to add allylmagnesium bromide in the presence of a copper(I) salt, and leads to the 1,1-dimetallic alkenyl derivative (**42**) which undergoes a further carbomagnesiation reaction leading to (**43**) in 50% yield (Scheme 27). The carbomagnesiation of propargylic amines proceeds less effectively.[80] Functionalized propargylic derivatives[81] add a wide range of Grignard reagents in an *anti* fashion (equation 17). The addition of the dimagnesium alcoholate of 2-butyne-1,4-diol and related derivatives has found several synthetic applications (equation 18).[82,83]

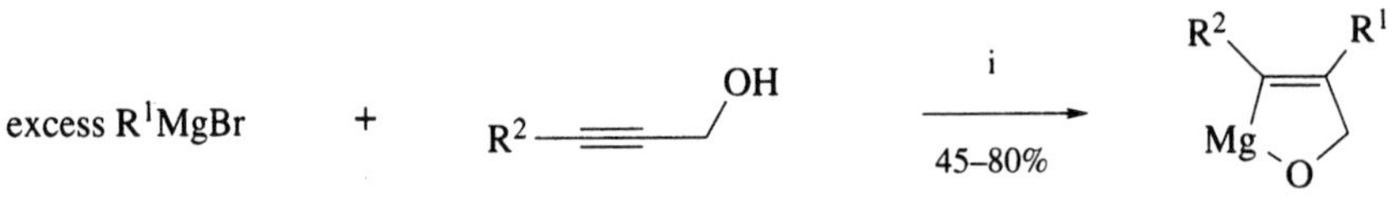

$$\text{excess } R^1MgBr + R^2C{\equiv}CCH_2OH \xrightarrow[45\text{–}80\%]{i} (\mathbf{41}) \quad (16)$$

(**41**)

R^1 = alkyl, aryl, allyl, benzyl

R^2 = H, alkyl, aryl, alkenyl, trimethylsilyl

i, 10% CuI, Et_2O, 0 °C, 1 h

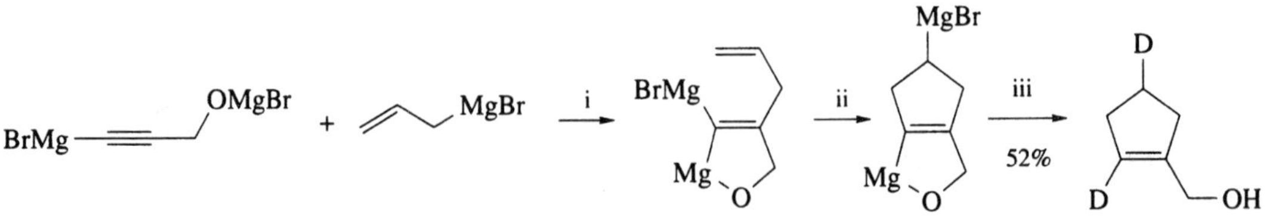

BrMg–C≡C–CH₂OMgBr + CH₂=CHCH₂MgBr —i→ (**42**) —ii→ (**43**) —iii (52%)→ D-substituted cyclopentenylmethanol (OH)

(**42**) (**43**)

i, 10 mol % CuI, Et_2O, 25 °C, 3h, ii, 5 h reflux; iii, D_2O

Scheme 27

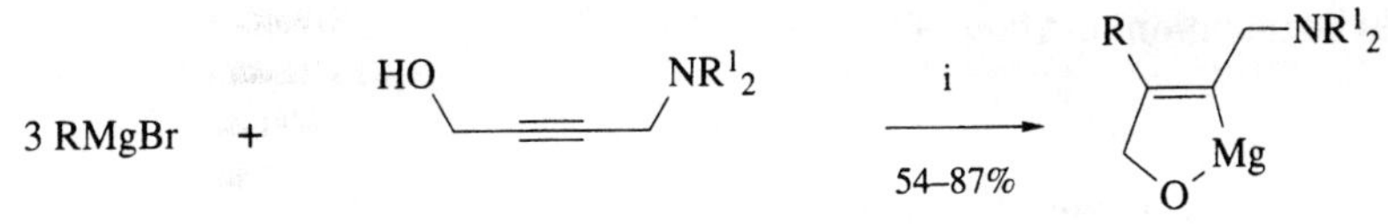

3 RMgBr + HO–CH$_2$–C≡C–CH$_2$–NR$^1_2$ →(i, 54–87%) (17)

i, Et_2O, reflux 4–30 h

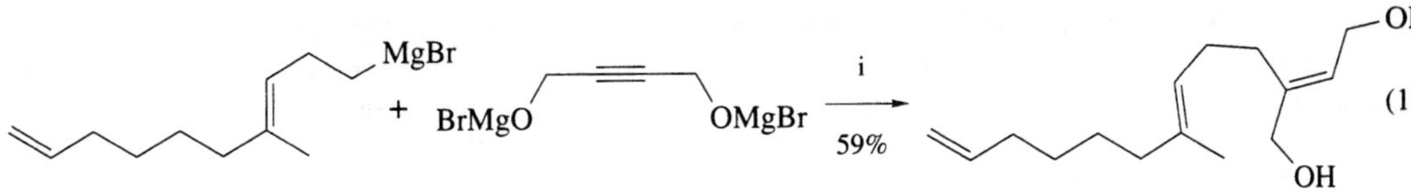

(18)

i, THF, Et_2O, reflux, 5 h, then aq. NH_4Cl

In contrast, the addition of allylmagnesium bromides to homopropargylic alcohols proceeds with low regioselectivity.[65,77] Snider and coworkers found that methylmagnesium bromide adds to silylalkynes in the presence of a 1:1 mixture of nickel acetylacetonate and trimethylaluminum (10 mol %) to afford mainly the *syn* addition product. The use of ethylmagnesium bromide leads only to hydromagnesiation and coupling products.[84] Utimoto and coworkers[85] have described an intramolecular version of this reaction and have formed cyclic tetrasubstituted vinylsilanes in high yields (Scheme 28). The intramolecular carbomagnesiation of an alkenylsilane is also possible.[85] The reaction proceeds specifically in a suprafacial and 5-exo-trig manner and produces, after allylation, the silane (**44**) with complete control of the stereochemistry of the three adjacent chiral centers (equation 19).

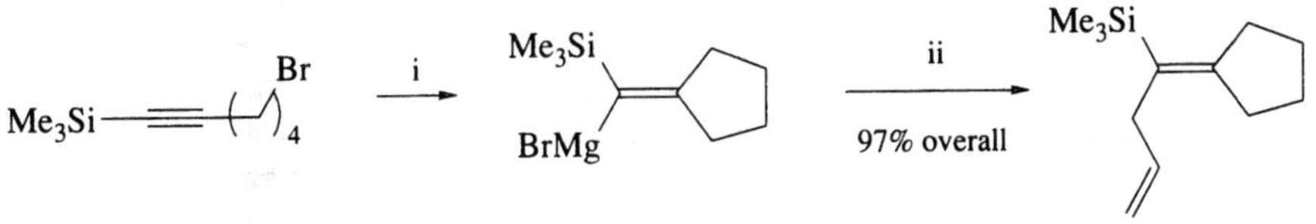

i, Mg, Et_2O, reflux, 5 h; ii, allyl bromide

Scheme 28

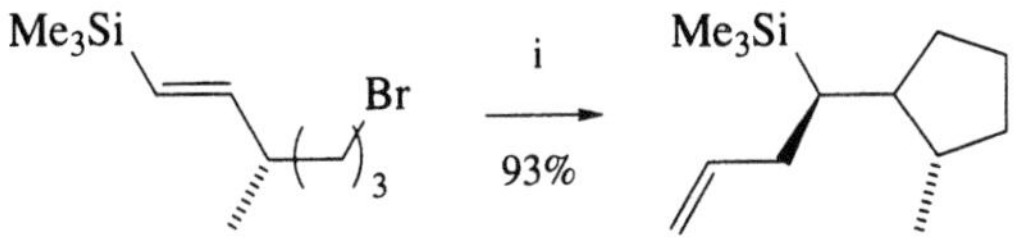

(19)

(**44**)

i, Mg, THF, 67 °C, 6 h, then allyl bromide

4.4.4 THE CARBOZINCATION REACTION

4.4.4.1 General Considerations[86]

Di-*t*-butyl- and diallyl-zinc compounds are able to add to alkenes, whilst other types of organozinc derivatives are usually inert toward nonfunctionalized alkene addition. In strong contrast, allylic zinc halides add readily to alkenyl organometallics under mild conditions to give mixed 1,1-diorganometallics. Alkynes are more reactive toward carbozincation than alkenes, and dialkylzinc and allylic zinc halides add to alkynes in the presence of transition metal salts. The addition of allylic zinc halides to alkynyl organometallics allows a unique approach to 1,1,1-triorganometallics. Allylic zinc halides show a higher reactivity toward the addition to alkynes than any other main group allylic organometallic.

4.4.4.2 The Carbozincation of Alkenes

4.4.4.2.1 Addition to nonfunctionalized alkenes

As shown by Lehmkuhl,[87] the addition of di-*t*-butylzinc to alkenes and 1,3-dienes proceeds between −20 and 75 °C and affords the addition products in good yields (Scheme 29), the reaction showing a good regioselectivity. The addition to octene or propene furnishes a secondary dialkylzinc with over 90% selectivity.[87] Several diallylic zinc derivatives[88] add to ethylene, giving the corresponding dialk-4-enylzinc compounds in high yields (>90%) and under mild conditions (20 °C, toluene, 36–96 h). The addition to octene is again highly regioselective (opposite regioselectivity than that with di-*t*-butylzinc) and furnishes primary dialk-4-enylzinc derivatives of type (**45**; equation 20). Diallylic zinc compounds add in a *syn* fashion to cyclopropenes in almost quantitative yields and under very mild conditions.[88] In all these cases, the new carbon–carbon bond is formed from the most-substituted end of the allylic organometallic. The addition to various substituted styrenes proceeds with moderate regioselectivity. A Hammett correlation between the observed regioselectivity and substituent constant σ^* has been found.[89] Diallylzinc and allylzinc bromide have only a moderate stability[90] and undergo a carbozincation reaction leading to 1,3-diorganometallic compounds of type (**46**). The reaction of (**46**) with acetaldehyde furnishes, after β-hydride elimination, the corresponding addition product (Scheme 30).

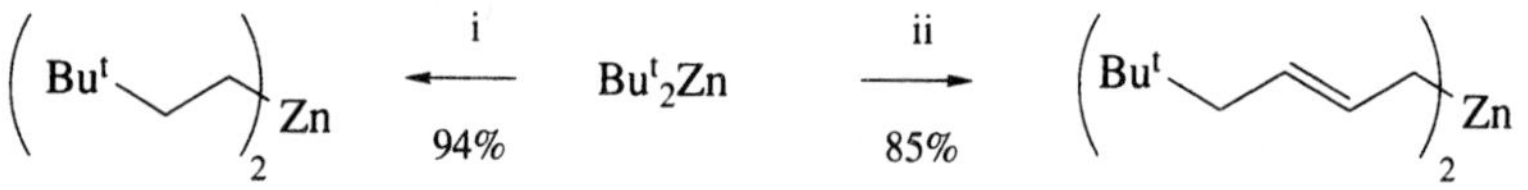

Bu^t_2Zn —(i, 94%)→ $(Bu^t\text{CH}_2\text{CH}_2\text{CH}_2)_2Zn$; —(ii, 85%)→ $(Bu^t\text{-alkenyl})_2Zn$

i, ethylene, 60 atm, 50–70 °C, 48 h; ii, butadiene, pentane, −20 to 10 °C, 21 h

Scheme 29

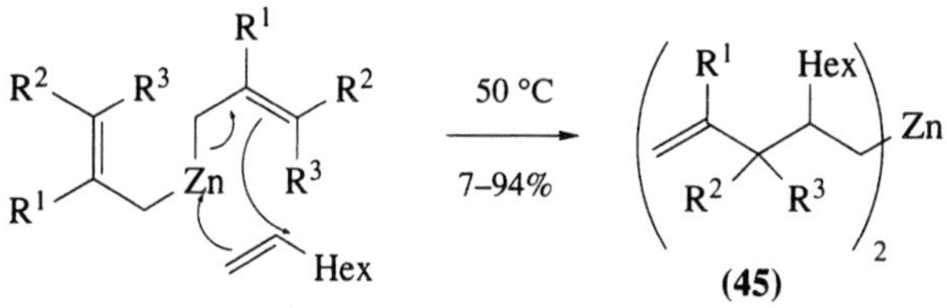

$$\text{diallylic zinc (R}^1\text{, R}^2\text{, R}^3\text{)} + \text{Hex-CH=CH}_2 \xrightarrow[\text{7–94\%}]{\text{50 °C}} (\mathbf{45}) \qquad (20)$$

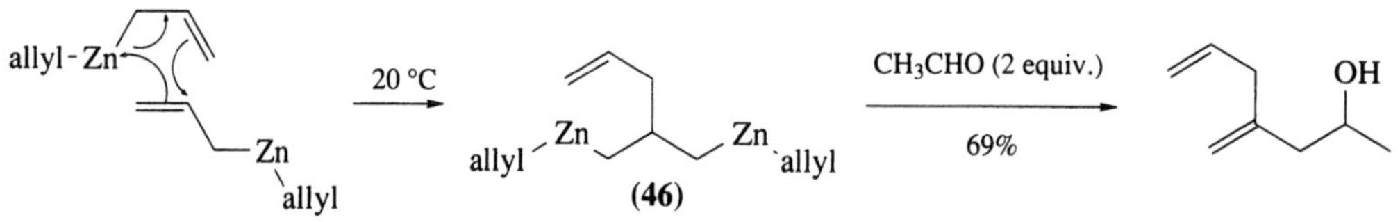

allyl-Zn (allyl) —(20 °C)→ allyl-Zn~~Zn-allyl (**46**) —(CH_3CHO (2 equiv.), 69%)→ OH product

Scheme 30

4.4.4.2.2 Addition to functionalized alkenes

In 1971, Gaudemar[91] discovered that allylzinc bromide was able to add to alkenylmagnesium bromides to furnish the mixed 1,1-diorganometallics of zinc and magnesium of type (**47**). Knochel and coworkers[92] showed that this reaction was quite general and that various alkenyl organometallics of magnesium, lithium and aluminum showed similar reactivity. The reaction is believed to proceed through the formation of a mixed allylic, vinylic zinc compound of type (**48**) which then undergoes a 3,3-sigmatropic shift (metallo-Claisen reaction; compare with Volume 5, Chapter 1.2) to give the corresponding 1,1-diorganometallic (**47**) in fair to good (60–95%) yields (Scheme 31). The substitution pattern of both allylic and alkenylorganometallics (**49** and **50**, respectively) can be quite varied and many new and, in some cases, highly functionalized mixed 1,1-diorganometallics of zinc, lithium, magnesium and aluminum can be obtained. The compounds (**51**) to (**56**), which were all prepared in good yields, illustrate the synthetic potential of the method. The cyclic transition state leading to compounds of type (**47**; Scheme 31) confers a complete regioselectivity and a high stereoselectivity to this carbozincation reaction.[92,93] Thus, the alkoxy-substituted allylic zinc compound (**57**) reacts with alkenylmagnesium bromides stereospecifically through chair transition state intermediates of types (**58a**) and (**58b**), in which the configuration of both starting organometallics is retained. The diastereomeric diorganometallics (**59a**) and (**59b**) formed in this reaction can then be converted stereospecifically by a mild oxidation[93,94] reaction into the aldol products (**60a**) and (**60b**), the ratio of (**60a**) to (**60b**) being the same as the ratio between the starting (*E*)- and (*Z*)-octenylmagnesium bromides (Scheme 32).

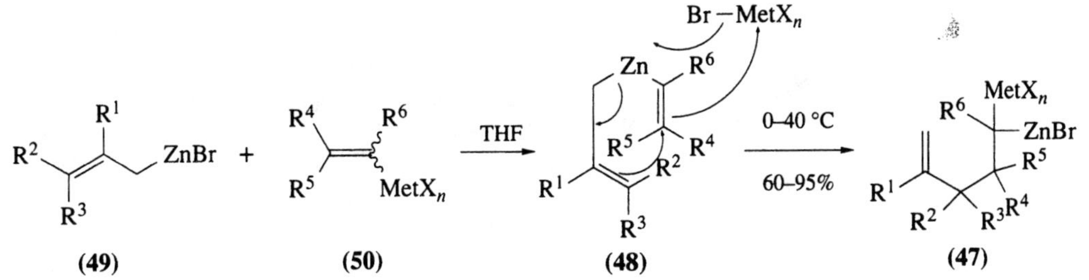

(49) + (50) —THF→ (48) —0–40 °C, 60–95%→ (47)

$MetX_n$ = Li, MgBr, AlR_2

Scheme 31

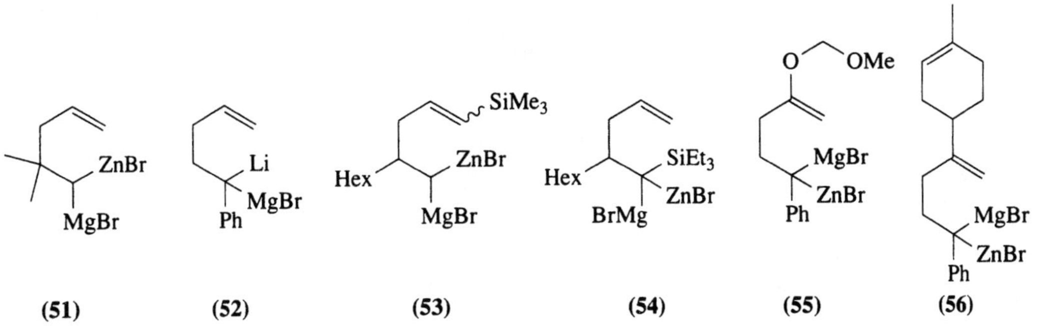

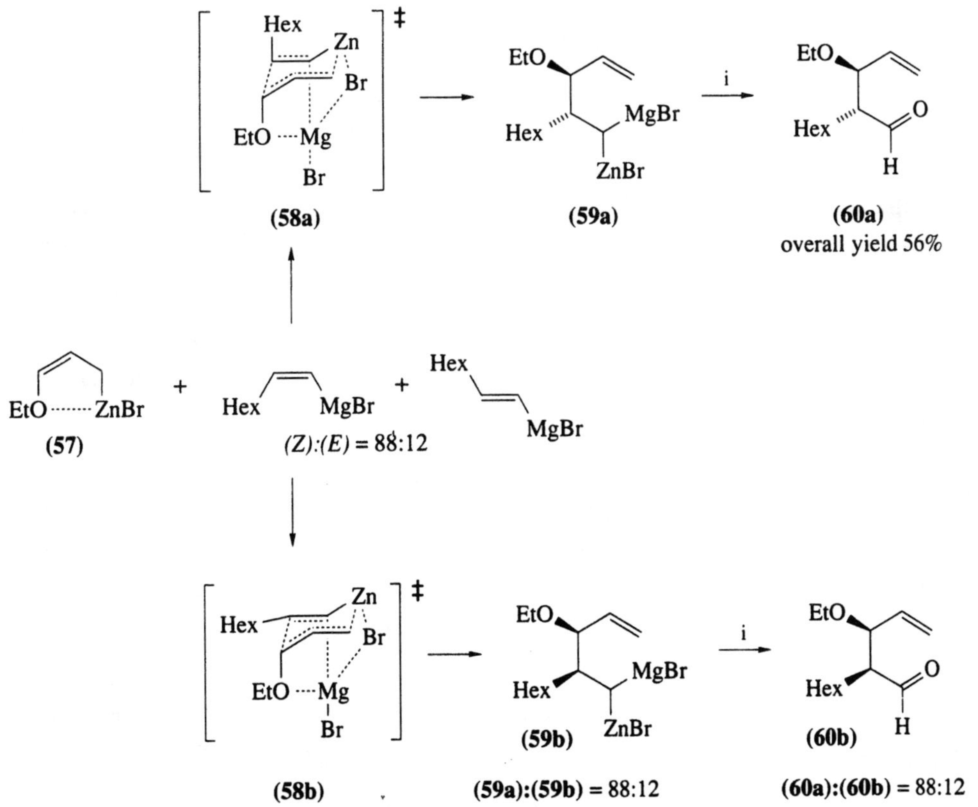

(51) (52) (53) (54) (55) (56)

(57) + Hex–CH=CH–MgBr (*Z*):(*E*) = 88:12 → (58a) → (59a) —i→ (60a) overall yield 56%

→ (58b) → (59b) —i→ (60b)

(**59a**):(**59b**) = 88:12 (**60a**):(**60b**) = 88:12

i, Me_3SnCl (1 equiv.), –40 to 0 °C, air (–15 °C, 6 h)

Scheme 32

The reactivity of the 1,1-diorganometallics (**61**) has been extensively investigated by Knochel and co-workers.[92,93] Since the reactivity of a carbon–lithium or a carbon–magnesium bond is considerably different from that of a carbon–zinc bond, a selective reaction of (**61**) with two different electrophiles is often possible. The general reaction pathways are summarized in Scheme 33. The addition of allylic zinc

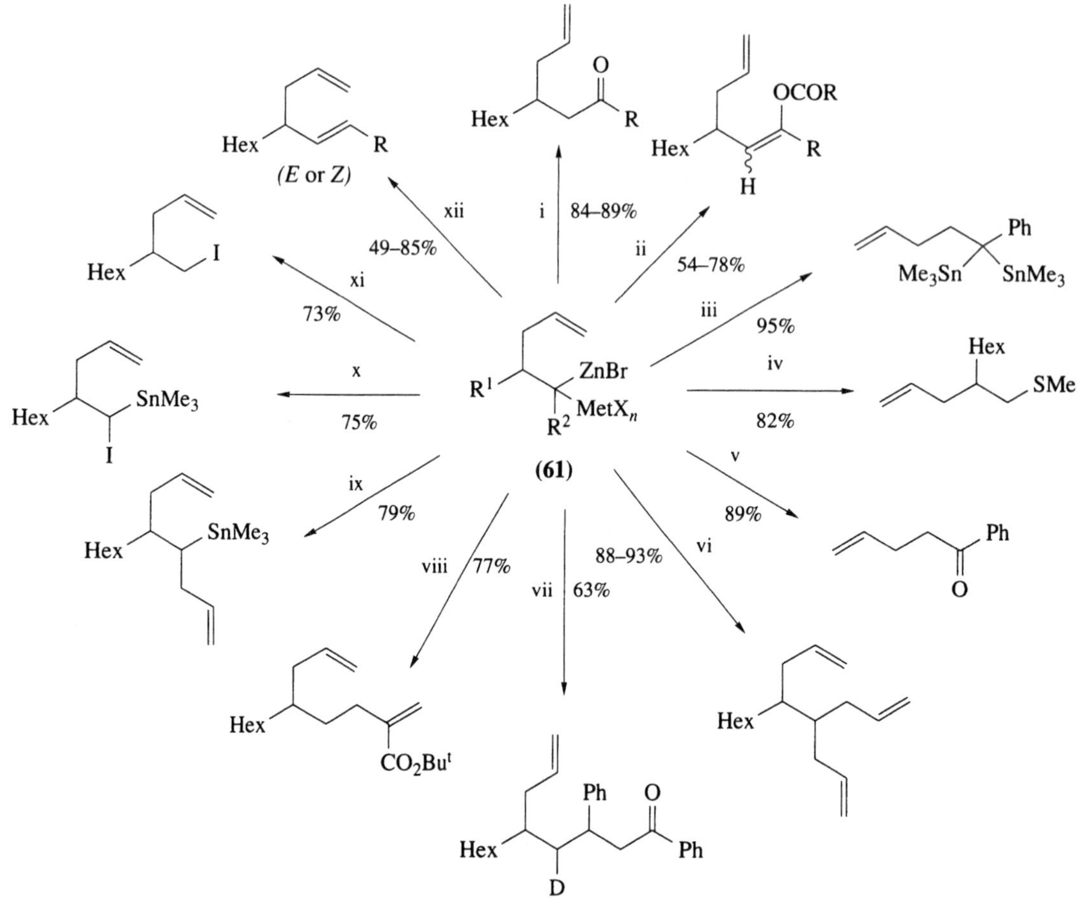

i, R^1 = Hex, R^2 = H, $MetX_n$ = Li, Pr^iOH (1.01 equiv.), –70 to –30 °C, RCOCl (2.5 equiv.), $[Pd(PPh_3)_4]$ (0.05 equiv.), 0 °C, 0.5–2 h; ii, R^1 = Hex, R^2 = H, $MetX_n$ = MgBr; CuCN (1 equiv.), –35 °C, 0.25 h, RCOCl (3 equiv.), –10 °C, 2 h; iii, R^1 = H, R^2 = Ph, $MetX_n$ = MgBr; CuCN (1 equiv.), –20 °C, 0.5 h, then Me_3SnCl (2 equiv.), –40 to 0 °C; iv, R^1 = Hex, R^2 = H, $MetX_n$ = MgBr; MeSSMe, –40 to 25 °C, then AcOH; v, R^1 = H, R^2 = Ph, $MetX_n$ = MgBr; Me_3SnCl (1 equiv.), –40 to 0 °C, 0.3 h, then dry air, –10 to 0 °C, 0.5 h; vi, R^1 = Hex, R^2 = H, $MetX_n$ = MgBr or Li; CuCN (1 equiv.), –35 °C, 0.25 h, then allyl bromide (excess), –40 to 0 °C, 0.5 h; vii, R^1 = Hex, R^2 = H, $MetX_n$ = Li; benzylideneacetophenone (1.0 equiv.), –78 to 0 °C, 1 h, D_3O^+; viii, R^1 = Hex, R^2 = H, $MetX_n$ = Li; MeOH (1 equiv.), –78 to 30 °C, then CuCN (1 equiv.), –40 °C, 0.5 h, *t*-butyl bromomethylacrylate, –78 to 0 °C, 0.5 h; ix, R^1 = Hex, R^2 = H, $MetX_n$ = Li; Me_3SnCl (1 equiv.), –40 to 0 °C, 0.5 h, CuCN (1 equiv.), –30 °C, 0.5 h, allyl bromide (3.0 equiv.), –30 to –10 °C, 0.5 h; x, R^1 = Hex, R^2 = H, $MetX_n$ = MgBr; Me_3SnCl (1 equiv.), –25 to 0 °C, 0.5 h, I_2 (1 equiv.), –78 °C, 0.5 h; xi, R^1 = Hex, R^2 = H, $MetX_n$ = Li, MeOH (1.05 equiv.), –70 to –30 °C, then I_2 (1 equiv.), –78 to –40 °C, 0.1 h; xii, R^1 = Hex, R^2 = H, $MetX_n$ = MgBr, $BF_3{\bullet}OEt_2$ (1 equiv.), –90 °C, RCHO, –90 to –50 °C, 0.25 h (formation of the *E*-isomer in over 90% selectivity), or $RCH{=}C(CO_2Et)_2$, –78 °C, 0.5 h (formation of the *Z*-isomer in over 84% selectivity; R = alkyl)

Scheme 33

bromides to α,β-unsaturated esters proceeds, in the presence of $[NiBr_2(PPh_3)_2]$, with moderate regioselectivity.[95]

4.4.4.3 The Carbozincation of Alkynes

4.4.4.3.1 Addition to nonfunctionalized alkynes

Gaudemar[86,90,91,96] found that allylic zinc bromides readily add to terminal alkynes. The first step in these reactions is always the deprotonation of the alkyne, leading to an alkynylzinc bromide (**62**), which subsequently adds the allylic zinc bromide, leading to vinylic 1,1-diorganometallics of type (**63**). Depending on the reaction conditions and on the substitution pattern of the allylic zinc reagent, a further carbozincation can occur, leading to 1,1,1-triorganometallics of type (**64**). The deuterolysis of (**64**) affords trideuterated dienes of type (**65**; Scheme 34). Miginiac and coworkers[86,97] showed that these additions are reversible, so that the amount of the thermodynamically more stable adduct (**66**) can be increased by longer reaction times (equation 21). Whereas disubstituted nonfunctionalized alkynes do not react, an intramolecular allylzincation reaction[98] is possible and affords, after hydrolysis, the *syn* addition product (**67**) stereospecifically (Scheme 35).

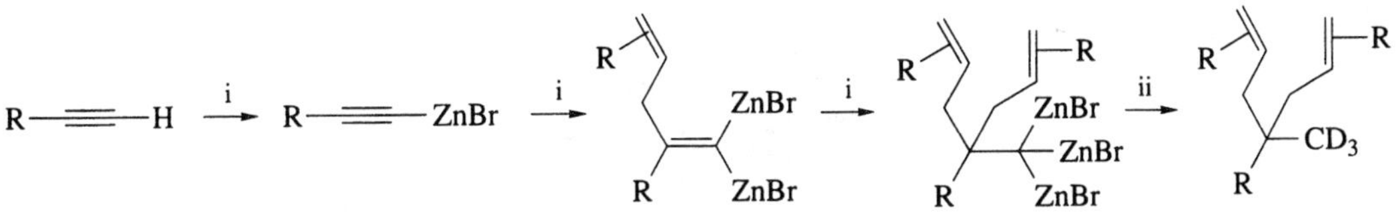

(**62**) (**63**) (**64**) (**65**)

i, allyl zinc bromide, THF; ii, D_2O

Scheme 34

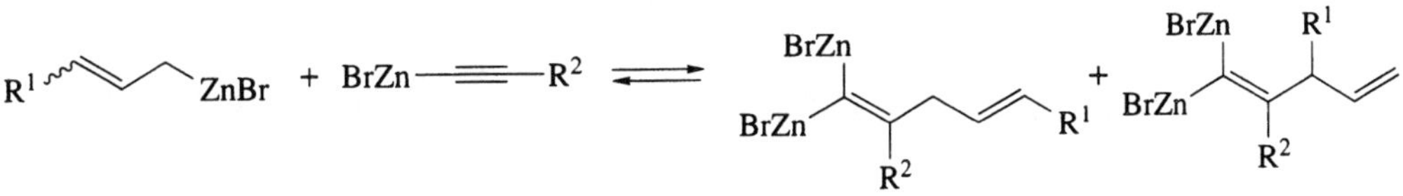

(**66**) (21)

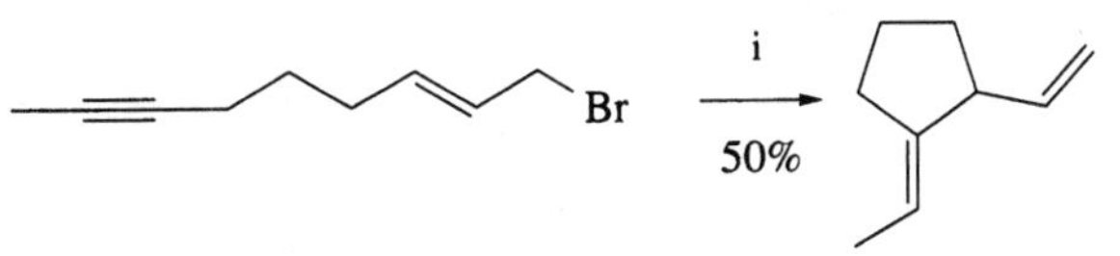

(**67**)

i, Zn, THF, 24 h, 65 °C, then H_3O^+

Scheme 35

Although several less reactive organozinc derivatives add to alkynes, the scope of these reactions is more limited. Di-*t*-butylzinc, for example, adds only to phenylacetylene, affording the *anti* addition product, (*Z*)-3,3-dimethyl-1-phenylbutene.[99] More interestingly, zinc malonates of type (**68**) add to various alkynes[100] in fair yields to give derivatives of type (**69**; Scheme 36). Propargylic zinc bromides[101] add to terminal alkynes and afford 1,4-enynes in good yields (equation 22). Negishi and coworkers[5,102] showed that dialkylzinc reagents, in the presence of I_2ZrCp_2, add to terminal and internal alkynes in good yields to furnish the *syn* adducts with good regioselectivity (70–90%) and very high stereoselectivity (>98%). The reaction seems to involve a direct addition of the carbon–zinc bond across the alkyne (equation 23).

The zirconium-promoted addition of diethylzinc to an alkynylzinc derivative is also possible,[102,103] and allows for several interesting synthetic applications (Scheme 37).[104]

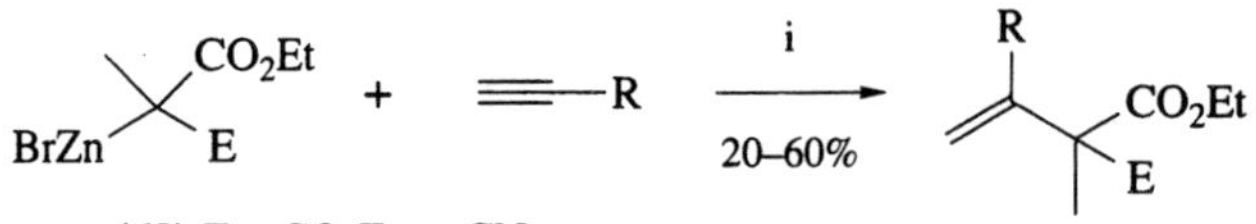

(68) E = CO_2Et or CN

(69)

i, THF, 4–24 h, 42 °C, then H_2O

20–60%

Scheme 36

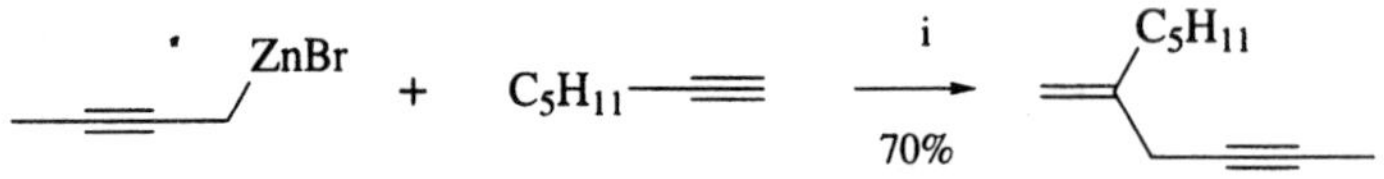

(22)

i, THF, 24 h, 30 °C, then H_3O^+

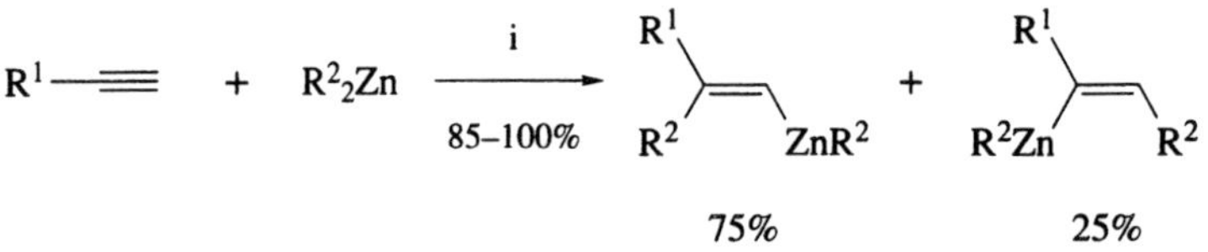

70%

(23)

85–100%; 75% : 25%

i, I_2ZrCp_2 (0.1 to 1 equiv.), $ClCH_2CH_2Cl$, 20 °C, 3–48 h

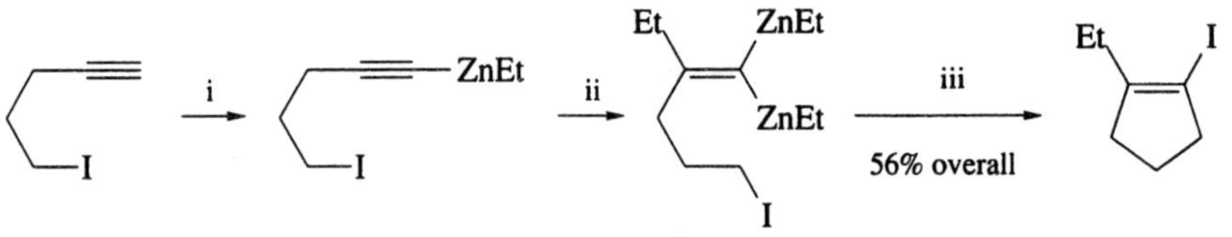

56% overall

i, BuLi, then EtZnCl, CH_2Cl_2; ii, Et_2Zn, I_2ZrCp_2; iii, evaporation, then THF and I_2

Scheme 37

4.4.4.3.2 Addition to functionalized alkynes

A number of functionalized alkynes containing halides or hydroxy, alkoxy and amino groups react with allylic[80,90,96,97,105] and propargylic zinc halides,[101,106] zinc malonates,[100] and di-*t*-butylzinc.[99] The presence of a functional group in the alkyne allows new synthetic possibilities, such as cyclization reactions leading to the highly functionalized cyclopropanes (**70**)[101,106] and (**71**)[101,106] or to the unsaturated lactone (**72**; Scheme 38).[100] The addition of zinc organometallics to enynes has also been extensively studied by Miginiac and coworkers.[15] The *syn* addition of allylic zinc bromides to 1-trimethylsilylalkynes has been reported by Molander,[107] Negishi,[107] Klumpp[108] and coworkers and has led to several new cyclization reactions (Scheme 39).[103,104,107b,108]

4.4.5 THE CARBOBORATION REACTION

4.4.5.1 General Considerations

Trialkylboranes generally show a low reactivity toward addition to alkenes and alkynes. Only triallylboranes react with alkynes under mild conditions. This reaction is often complicated by further intramolecular carboboration reactions and has found only limited synthetic application.

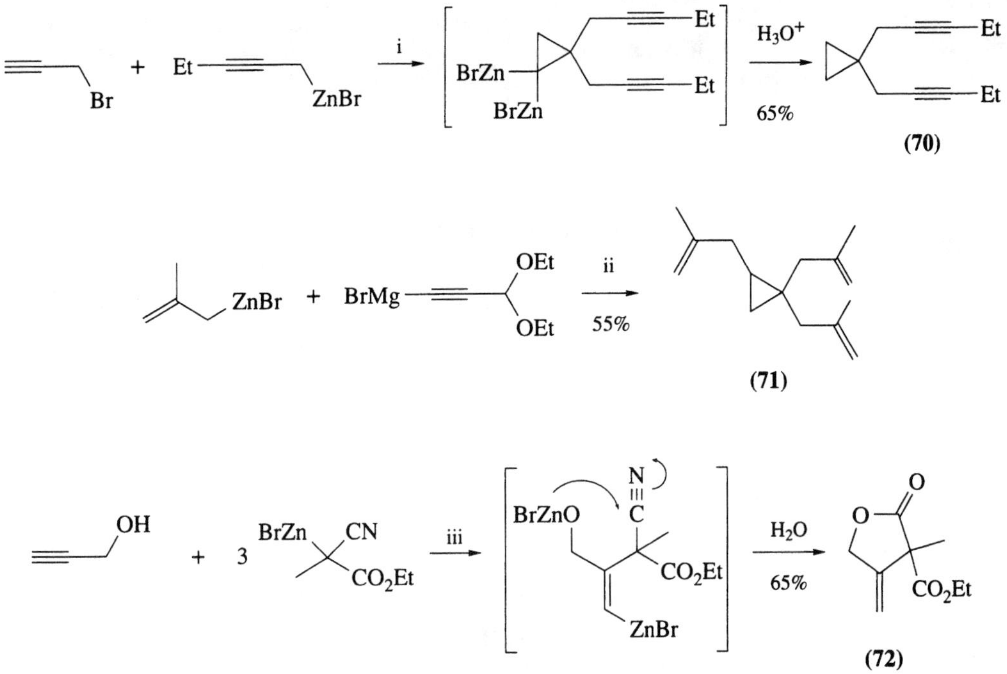

i, 40 °C, 24 h, THF; ii, 65 °C, 1.5 h, THF; iii, 42 °C, 23 h, THF

Scheme 38

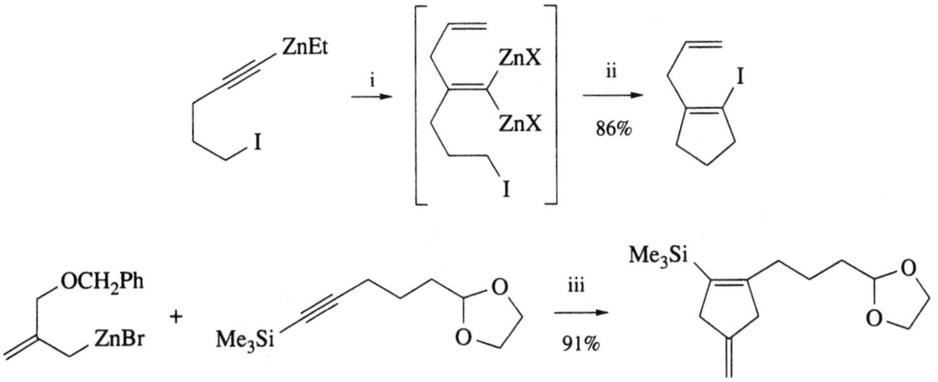

i, allylzinc bromide, THF; ii, cyclization, then I_2; iii, 100 °C, 30 h, then $[Pd(PPh_3)_4]$ (10 mol %), 24 h, 65 °C

Scheme 39

4.4.5.2 The Carboboration of Alkenes

As has been shown by Köster,[109] trialkylboranes react with alkenes and dienes at high temperatures (>140 °C) to give hydroboration adducts by a dehydroboration/hydroboration mechanism. It is only under special reaction conditions that carboboration reactions can be observed. Triethylborane adds slowly to 1-decene at 160–170 °C to give diethyl(2-ethyldecyl)borane.[109] In the presence of trialkylaluminum, trialkylboranes react with ethylene at 150–170 °C.[110] Under irradiation, trialkylboranes give a formal *syn* addition to cycloalkenes.[111] The highly strained *trans*-cyclohexenes (**73**) have been assumed to be the reactive intermediates in this reaction (Scheme 40). The more reactive triallylboranes add at 120–140 °C to alkyl enol ethers.[113] The reaction with 2-methyl-4,5-dihydrotetrahydrofuran (**74**) proceeds readily at 80 °C and affords, after hydrolysis, (*E*)-4-methylhepta-1,4-dien-1-ol (**75**) stereospecifically (Scheme 41). Strained alkenes like 1-methylcyclopropene react readily with trialkylboranes in two ways:[114] *syn* addition of the organoborane leading to the cyclopropylborane (**76**; the predominant mode of reaction), and cleavage of the C(2)—C(3) bond of the cyclopropene ring leading to the allylic borane (**77**; equation 24). Triallylboranes also react with allenes at high temperatures.[115]

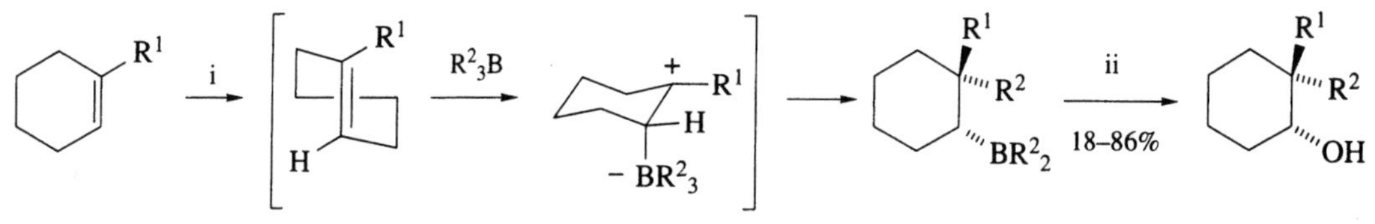

(73)

R^1 = H, Et; R^2 = alkyl

i, $h\nu$, xylene, benzene; ii, H_2O_2

Scheme 40

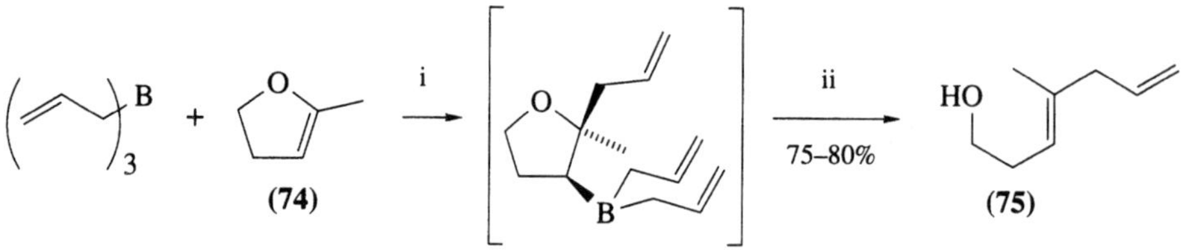

(74) (75)

i, 80–100 °C, 5–10 min; ii, OH^-

Scheme 41

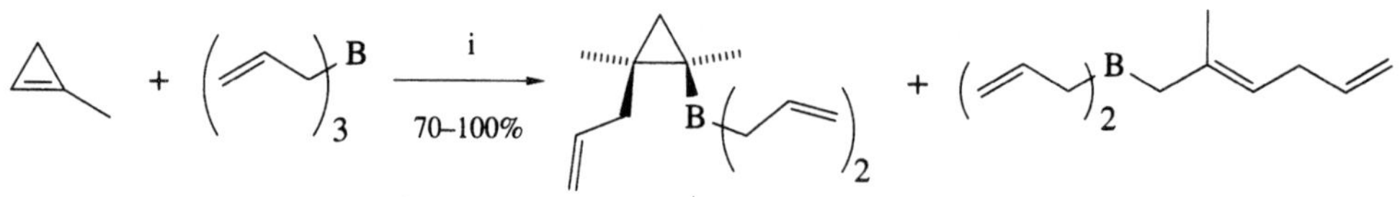

(24)

(76) 50–60% (77) 20–40%

i, –70 to 0 °C, heat

4.4.5.3 The Carboboration of Alkynes

Like alkenes, alkynes do not react readily with trialkylboranes. Under severe reaction conditions, only hydroboration products are obtained.[116] Mikhailov[115] showed that triallylboranes react with various alkynes (20 °C) to afford *syn* addition products of type (**78**) which rapidly cyclize (40–60 °C) to give the cyclic boranes of type (**79**; Scheme 42). In the case of trimethylsilylacetylene and ethoxyacetylene, the reaction affords compounds of type (**78**; R = Me_3Si or OEt) which do not cyclize further.[115] Recently, a transition metal silylboration reaction has been described by Oshima and coworkers.[117] A formal *syn* carboboration reaction leading to a variety of alkenylboranes has been reported by Suzuki and coworkers[118] (Scheme 43). Hexamethyldistannylacetylene (**80**) reacts readily with various trialkylboranes[119] to afford *syn* addition products of type (**81**; equation 25).

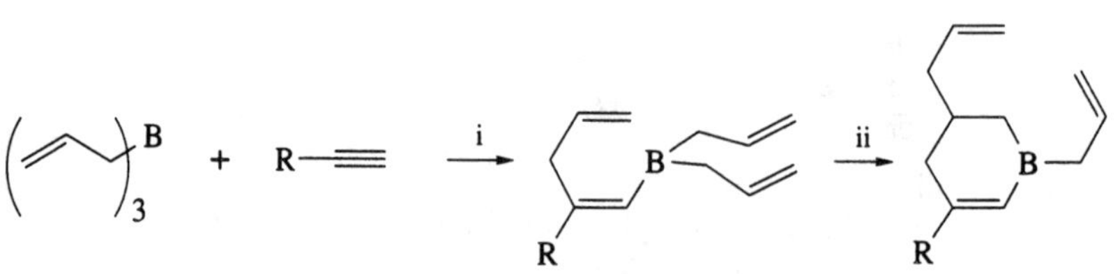

(78) (79)

i, 20 °C; ii, 40–60 °C

Scheme 42

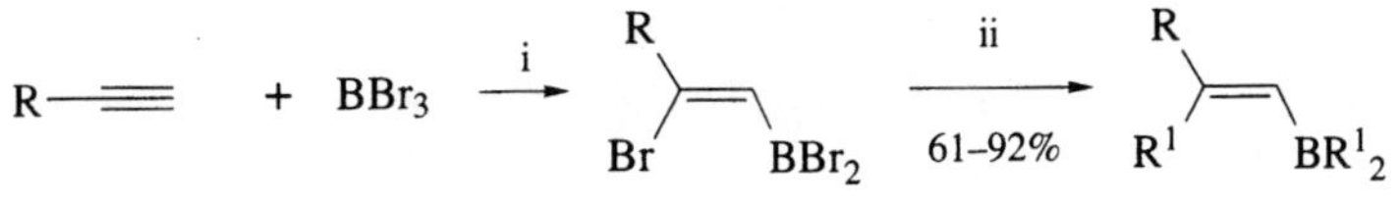

$$\text{R—≡} + \text{BBr}_3 \xrightarrow{\text{i}} \text{(}Z\text{)-R(Br)C=CH(BBr}_2\text{)} \xrightarrow[\text{61–92\%}]{\text{ii}} \text{R(R}^1\text{)C=CH(BR}^1{}_2\text{)}$$

i, CH_2Cl_2, –78 to 25 °C, 30 min; ii, $[PdCl_2(PPh_3)_2]$ (5 mol %), 3 equiv. R^1ZnCl

Scheme 43

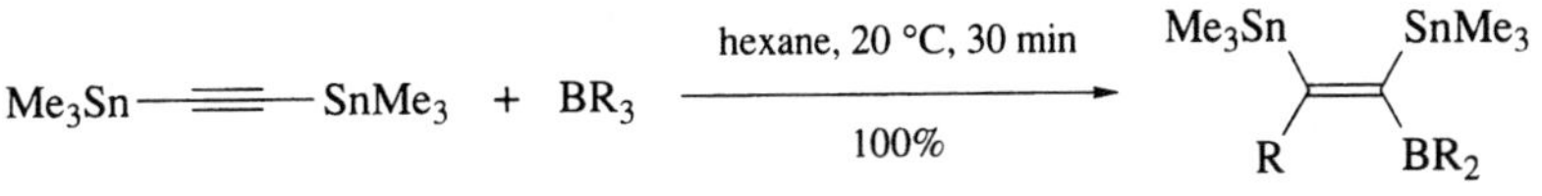

$$\underset{\textbf{(80)}}{\text{Me}_3\text{Sn—≡—SnMe}_3} + \text{BR}_3 \xrightarrow[\text{100\%}]{\text{hexane, 20 °C, 30 min}} \underset{\textbf{(81)}}{\text{(Me}_3\text{Sn)(R)C=C(SnMe}_3\text{)(BR}_2\text{)}} \quad (25)$$

4.4.6 THE CARBOALUMINATION REACTION

4.4.6.1 General Considerations

The uncatalyzed carboalumination[120] of nonfunctionalized alkenes and alkynes occurs generally under quite severe conditions. The reaction is of limited synthetic interest because of the low regioselectivity observed and the numerous side reactions, such as polymerization, oligomerization and dehydroalumination, which can occur. The intramolecular carboalumination of alkenes is easier and good stereoselectivities may be observed. Transition metal catalyzed carboalumination of alkynes is far more useful. These reactions occur under mild conditions, are highly regioselective and stereoselective (*syn* addition), and represent a powerful synthetic tool, especially the zirconium-catalyzed methylalumination reaction.

4.4.6.2 The Carboalumination of Alkenes

The reaction of organoaluminum compounds with alkenes was discovered and studied in great detail by Ziegler and coworkers.[121,122] Triethylaluminum adds to ethylene to give a Poisson distribution[122] of oligomeric trialkylaluminum organometallics (equation 26). A stepwise carboalumination reaction is not possible, and a clean synthesis of defined oligomers cannot be achieved by this method. However, with substituted alkenes, only monoaddition occurs. Carboaluminations of 1-alkenes are regioselective, and the new carbon–carbon bond is formed in 90–95% at the 2-position.[123] Secondary trialkylaluminum derivatives are very reactive toward addition to double bonds, but show a low regioselectivity.[12] Tertiary aluminum organometallics are even more reactive, and, in this case, a selective monoaddition is observed (equation 27).[124] The addition of triphenylaluminum to conjugated dienes, styrenes and strained alkenes occurs more readily.[125] More synthetically useful is the intramolecular carboalumination of 1,5-dienes, which allows a high yield preparation of bicyclic systems[126,127] of type (**82**) or (**83**) under mild conditions. The dehydroalumination of (**83**) in the presence of ethylene affords the diene (**84**) in 90% overall yield (Scheme 44).[126] Since the elimination of Et_2AlH occurs at similar temperatures as the carboalumination step, a catalytic amount of this hydride can be used to promote cyclization reactions (equation 28).[128] Donor solvents like ether can prevent intramolecular carboaluminations,[129] although exceptions[130] are known. The presence of ether also may prevent the formation of dehydroalumination products.[130b] If substituted 1,5-dienes are used, a very high diastereoselectivity is observed in the ring closure (Scheme 45).[131]

$$\text{Et}_3\text{Al} + 3n\,\text{H}_2\text{C=CH}_2 \longrightarrow [\text{Et(CH}_2\text{CH}_2)_n]_3\text{Al} \quad (26)$$

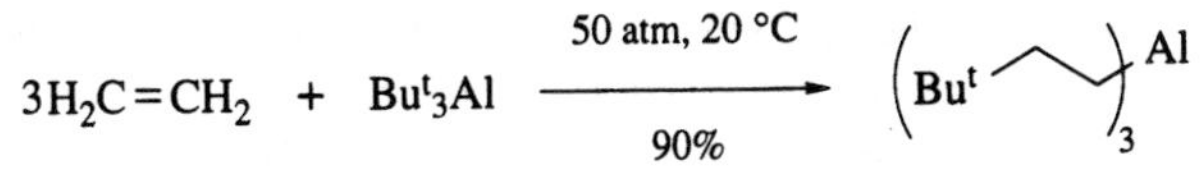

$$3\text{H}_2\text{C=CH}_2 + \text{Bu}^t{}_3\text{Al} \xrightarrow[\text{90\%}]{\text{50 atm, 20 °C}} (\text{Bu}^t\text{CH}_2\text{CH}_2)_3\text{Al} \quad (27)$$

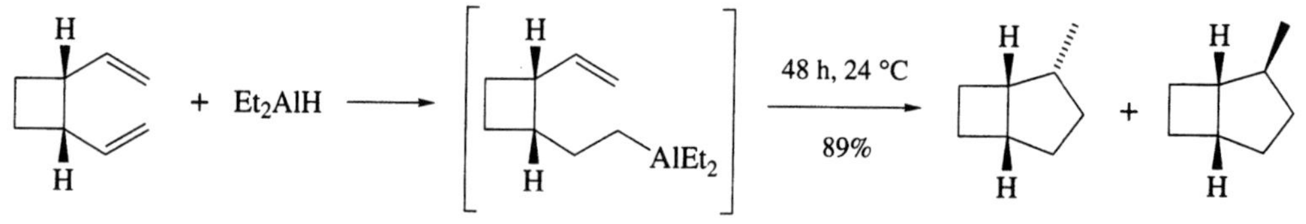

(82a) 48% (82b) 52%

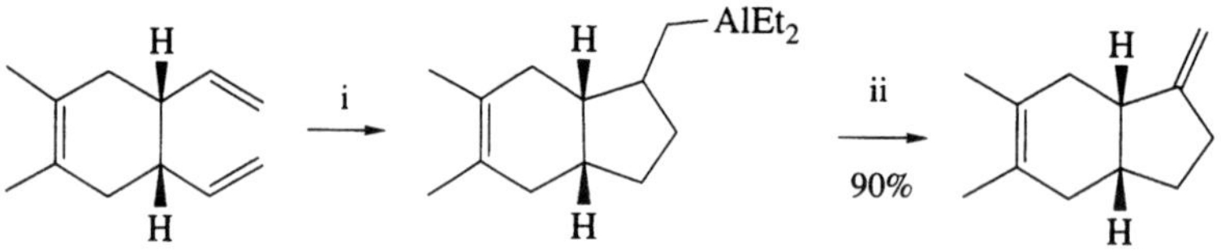

48 h, 24 °C
89%

(83) (84)

i, Et_2AlH, benzene, 24 h, heat; ii, ethylene, 70 atm, 18 h, 50 °C

Scheme 44

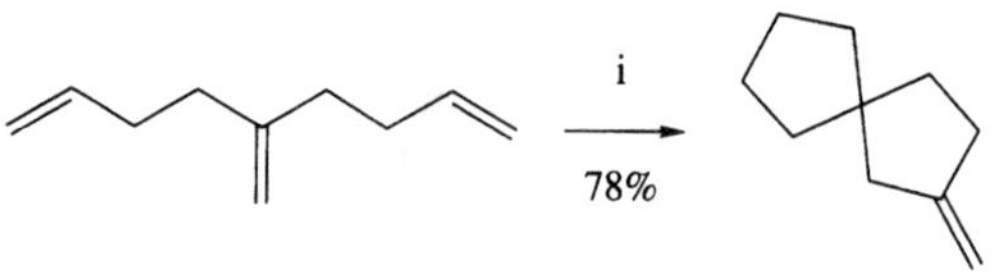

i
78%

(28)

i, $Bu^i{}_2AlH$, mineral oil, 110 °C, 48 h

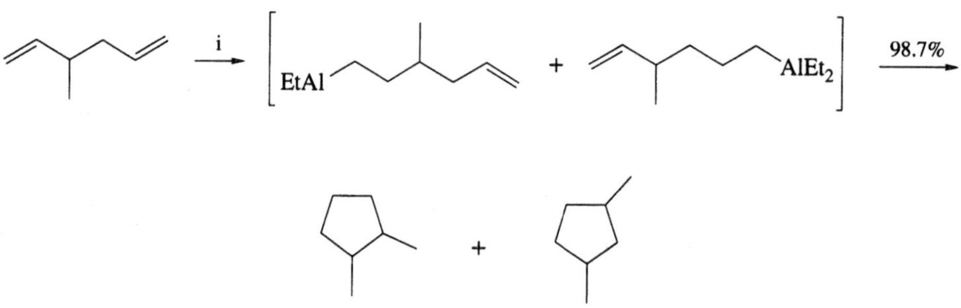

98.7%

53.1%
cis:trans = 2.3:97.7

45.6%
cis:trans = 91.1:8.9

i, Et_2AlH, toluene, 24 h, 25 °C

Scheme 45

4.4.6.3 The Carboalumination of Alkynes

4.4.6.3.1 Addition to nonfunctionalized alkynes

Acetylene[132] itself reacts readily with trialkylaluminum compounds and affords the *syn* addition products in good yields and under mild conditions (20 to 60 °C compared to 150 °C if ethylene is the substrate; see equation 29). The reaction with higher alkynes[133] is more complex. The metallation of the alkyne leading to an alkynylaluminum organometallic of type (**85**) is the major pathway. The reaction shows a low regioselectivity, and a mixture of the two regioisomers (**86**) and (**87**) is generally obtained (equation 30). Thus, the addition of triethylaluminum to phenylacetylene affords a 1:1 mixture of the two possible regioisomers.[133] If an excess of alkyne is used, then the corresponding alkynylaluminum derivative (**85**) can be obtained in excellent yields.[123] Interestingly, the carboalumination reaction proceeds with retention of the configuration of the alkyl group (equation 31).[134] Disubstituted alkynes react only at temperatures higher than 80 °C, where the adducts of type (**86**) and (**87**) are no longer stable and give further reactions.[132,134] Only diphenylacetylene gives a clean *syn* addition (equation 32). The reaction conditions required for the addition to disubstituted alkynes are sufficiently severe for the alkenylaluminum reagent (**88**) to add again to the starting alkyne, leading to the dienylalane (**89**). The organometallics of type (**89**) can be further elaborated into stereodefined 1,3-dienes (Scheme 46).[137]

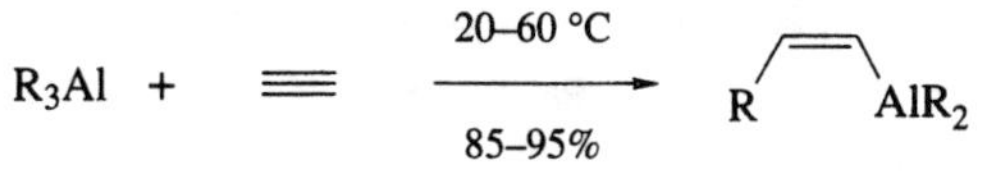

R_3Al + ≡ —(20–60 °C; 85–95%)→ R–CH=CH–AlR_2 (29)

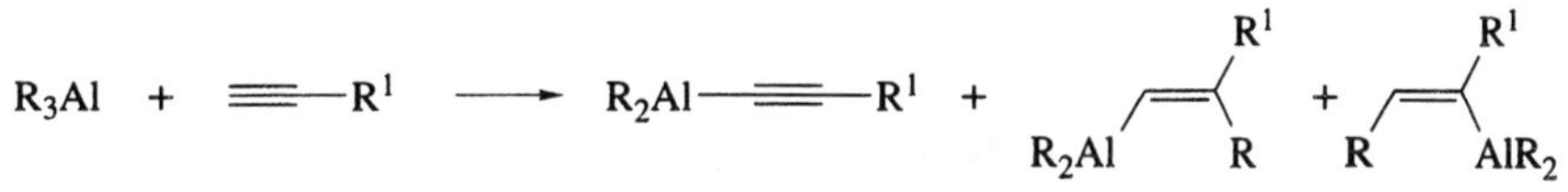

R = alkyl

R_3Al + ≡–R^1 → R_2Al–≡–R^1 (**85**) + (**86**) + (**87**) (30)

(31)

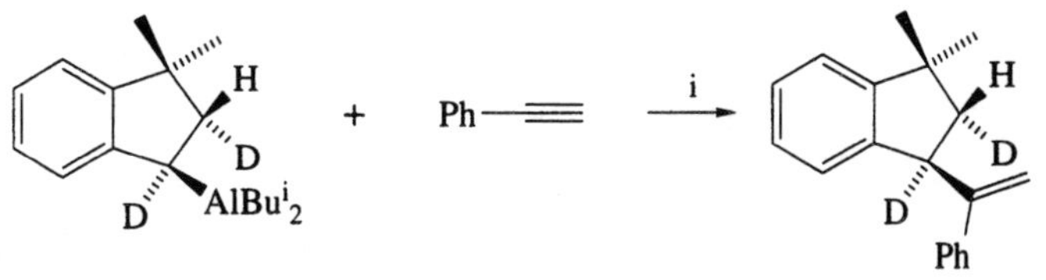

i, Et_2O, 110 °C, several days, then H_3O^+

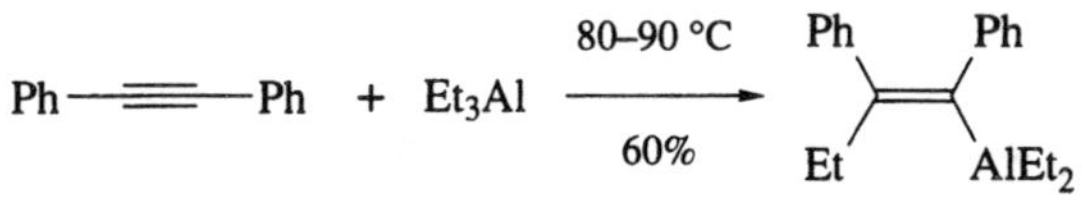

Ph–≡–Ph + Et_3Al —(80–90 °C; 60%)→ (32)

Et–≡–Et + Et_3Al —(80 °C)→ [(**88**)] —(Et–≡–Et)→ (**89**)

Et–≡–Et —(i)→ —(ii; 63%)→

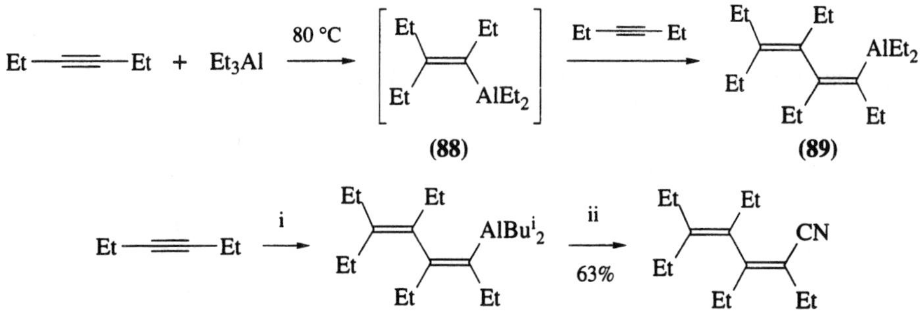

i, $Bu^i{}_2AlH$, 3-hexyne, 70 °C; ii, MeLi, –78 °C, then $(CN)_2$, then H_3O^+

Scheme 46

The low chemo- and regio-selectivity of the carboalumination reaction can be dramatically improved by the use of transition metal catalysts. Negishi and coworkers[135] showed that Me_3Al—Cl_2TiCp_2 reacts with diphenylacetylene to give a single organometallic species (**90**) which, after hydrolysis or iodolysis, affords, respectively, (*Z*)-1,2-diphenylpropene and (*E*)-1-iodo-1,2-diphenylpropene (Scheme 47). The scope of this reaction is limited, and the use of other alkynes leads to the formation of side products[135,136] due to the high reactivity of the Al–Ti reagent. The corresponding Al–Zr reagent (Me_3Al—Cl_2ZrCp_2) has been shown by Negishi to have much better synthetic utility. The reaction of phenylacetylene with Me_3Al—Cl_2ZrCp_2 gives 2-phenylpropene (**91**) and (*E*)-1-phenylpropene (**92**) in a 96:4 ratio and a 98% overall yield (Scheme 48). The reaction can be performed in the presence of only a catalytic amount of Cl_2ZrCp_2 (10 mol %), but the reaction then proceeds at a slower rate and a slight yield decrease is observed.[136] A wide range of alkynes containing alkyl, aryl, conjugated alkenyl or isolated alkenyl groups give the methylalumination products in excellent yields. The regioselectivity observed with terminal alkynes is always better than 96% and the *syn* stereoselectivity is always higher than 98%.[136] Internal alkynes such as 5-decyne give a relatively smooth carboalumination with $Me_3Al—Cl_2ZrCp_2$ and furnish *syn* addition products in good yields (equation 33).[136]

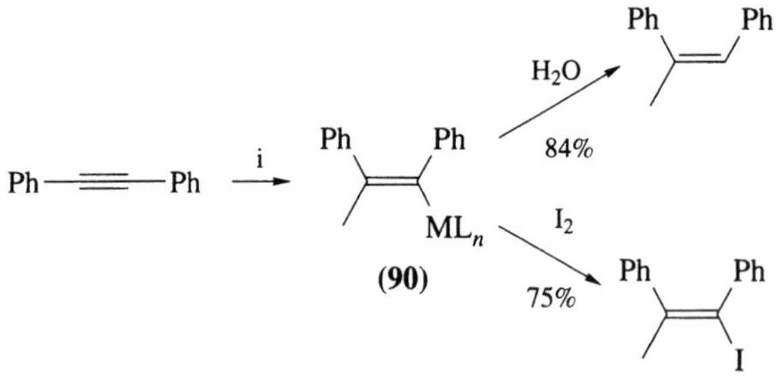

Ph—≡—Ph →(i) Ph(Ph)C=C(ML$_n$) **(90)**; H_2O, 84%; I_2, 75%

i, $Me_3Al–Cl_2TiCp_2$, $ClCH_2CH_2Cl$, 20 to 25 °C
ML_n = $–Ti(Cp)_2Cl$ or $–AlMe_2$

Scheme 47

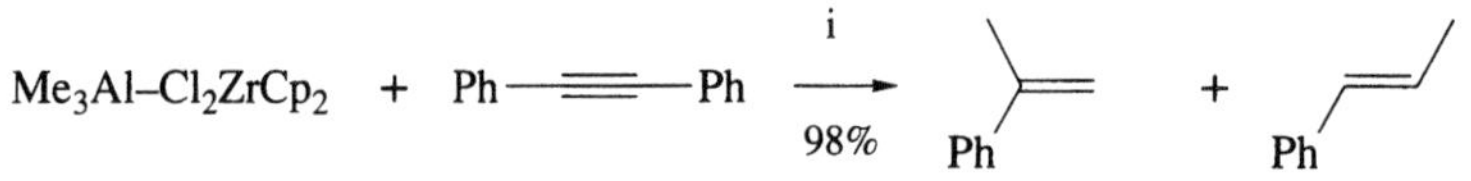

$Me_3Al–Cl_2ZrCp_2$ + Ph—≡—Ph →(i, 98%) **(91)** + **(92)**

i, Me_3Al (2 equiv.), Cl_2ZrCp_2 (1 equiv.), 20–25 °C, 24 h, then H_3O^+

Scheme 48

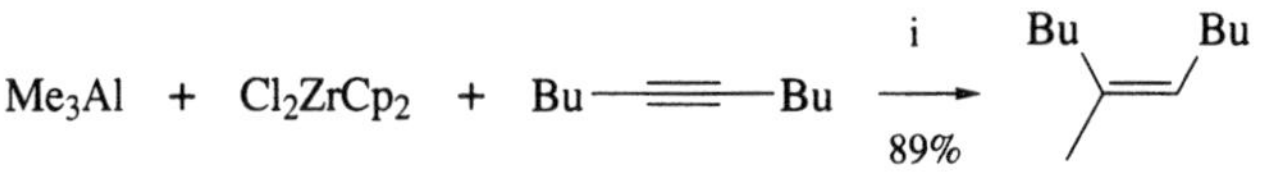

Me_3Al + Cl_2ZrCp_2 + Bu—≡—Bu →(i, 89%) (33)

i, $ClCH_2CH_2Cl$, 60 °C, 6 h, then H_3O^+

Higher homolog organoaluminum reagents can also carboaluminate alkynes, but the reaction is less regioselective, and the reagent R_2AlCl has to be used instead of R_3Al in order to avoid competitive hydrometallations (equation 34). The mechanism of the reaction has been carefully studied by Negishi and coworkers.[136] All data are consistent with a zirconium-assisted carboalumination reaction in which an intermediate of type (**93**) is involved (Scheme 49). The Lewis acid–base interactions between the aluminum and the zirconium reagent enhance the carbometallation ability of (**93**) and favor the addition to the alkyne. The following facts support this mechanism: the reaction of a 1:1 mixture of Et_3Al and $Cl(Me)ZrCp_2$ with 1-heptyne affords, after hydrolysis, only ethylalumination products. No methylalumination products were detected. Also, the reaction of 1-heptyne with $CD_3(Cl)ZrCp_2$ and Me_3Al affords, after iodolysis, only nondeuterated products (less than 4% of deuterium incorporation; see equation 35). An aluminum-assisted carbozirconation has also been observed by Negishi and coworkers.[138] The reaction of $Cl(Me)ZrCp_2$ with 1-pentynyldimethylalane (**94**) gives the mixed vinylic 1,1-diorganometallic of zirconium and aluminum (**95**) which, after iodolysis, affords a 1,1-diiodoalkene in good yield (Scheme 50).

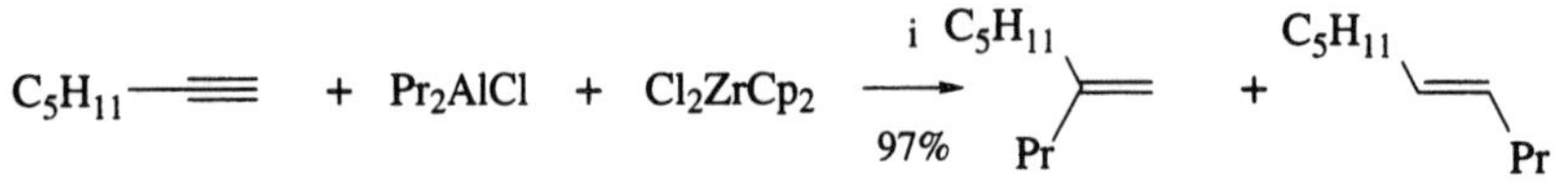

C_5H_{11}—≡ + Pr_2AlCl + Cl_2ZrCp_2 →(i, 97%) 78% + 22% (34)

i, $ClCH_2CH_2Cl$, r.t., 20 h, then H_3O^+

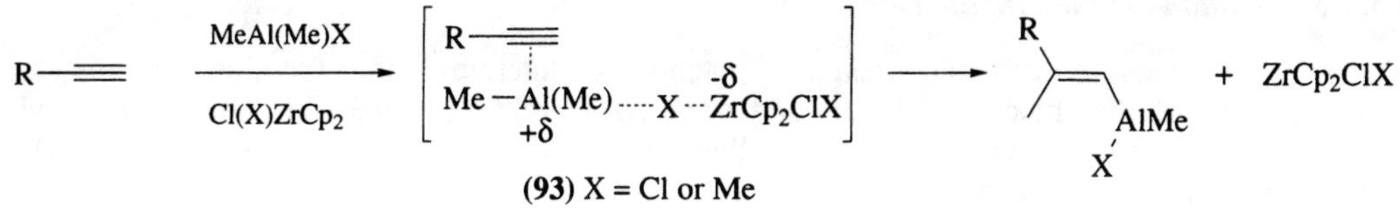

R—≡ —[MeAl(Me)X / Cl(X)ZrCp$_2$]→ [R—≡ ⋮ Me—Al(Me)····X···ZrCp$_2$ClX, +δ, −δ] → (R)(Me)C=CH–AlMe(X) + ZrCp$_2$ClX

(93) X = Cl or Me

Scheme 49

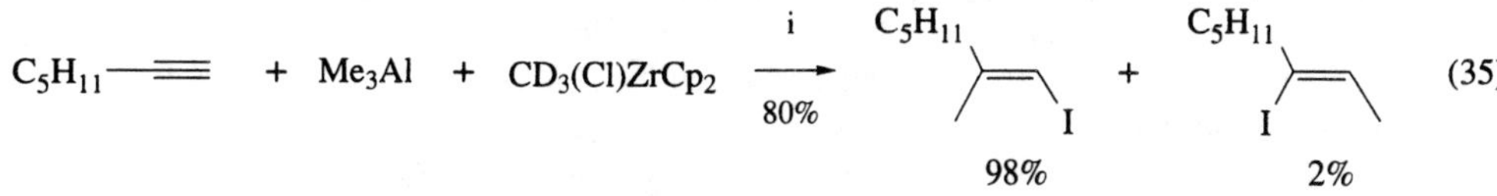

$$C_5H_{11}\text{—≡} + Me_3Al + CD_3(Cl)ZrCp_2 \xrightarrow[80\%]{i} \text{98\%} + \text{2\%} \quad (35)$$

i, $ClCH_2CH_2Cl$, 25 °C, 6 h, then −10 °C, I_2

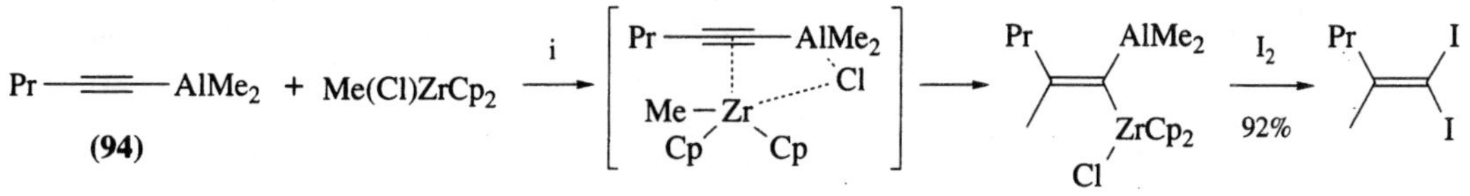

Pr—≡—AlMe$_2$ **(94)** + Me(Cl)ZrCp$_2$ —i→ [Pr—≡—AlMe$_2$ ··· Cl ··· Me—Zr(Cp)(Cp)] → Pr(Me)C=C(AlMe$_2$)(ZrCp$_2$Cl) —I$_2$, 92%→ Pr(Me)C=CI$_2$

i, CH_2Cl_2, 25 °C, 3 h

Scheme 50

Allylalanes and benzylalanes[139] also react with terminal and internal alkynes in the presence of Cl_2ZrCp_2 to produce *syn* allylation products (>98% *syn* addition) with a moderate regioselectivity (equation 36). Interestingly, in the case of substituted allylic alanes, the addition product (**96**), formed without allylic rearrangement *via* a four-membered transition state, is obtained as the major isomer. This is in contrast with most allylmetallations, which afford the allylic rearrangement regioisomer of type (**97**)[107b] *via* a six-membered transition state (Scheme 51).

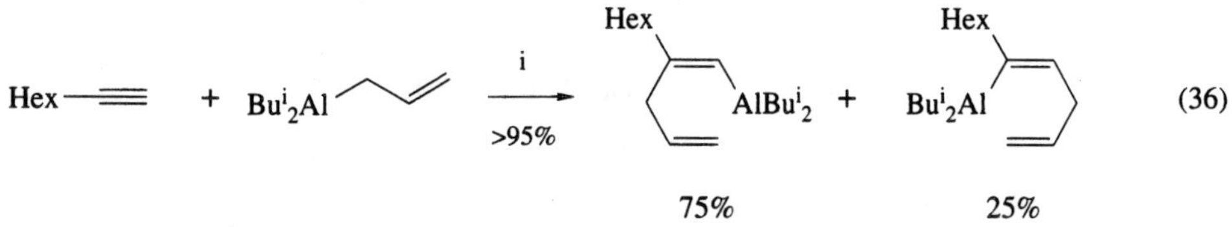

$$Hex\text{—≡} + Bu^i_2Al\text{(allyl)} \xrightarrow[>95\%]{i} \text{(AlBu}^i_2\text{ adduct) 75\%} + \text{(Bu}^i_2\text{Al adduct) 25\%} \quad (36)$$

i, Cl_2ZrCp_2 (1 equiv.), $ClCH_2CH_2Cl$, 25 °C, 1–2 h

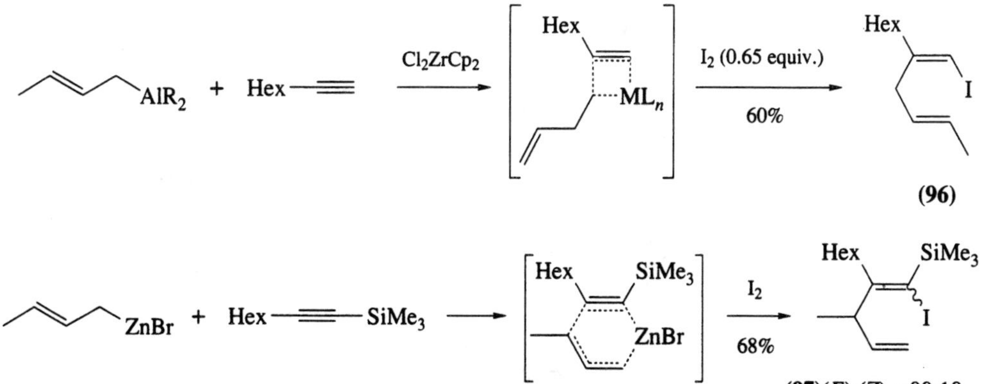

(crotyl)AlR$_2$ + Hex—≡ —Cl$_2$ZrCp$_2$→ [Hex, ML$_n$ four-membered TS] —I$_2$ (0.65 equiv.), 60%→ **(96)**

(crotyl)ZnBr + Hex—≡—SiMe$_3$ → [Hex, SiMe$_3$, ZnBr six-membered TS] —I$_2$, 68%→ **(97)**(*E*):(*Z*) = 90:10

Scheme 51

4.4.6.3.2 Addition to functionalized alkynes

The zirconium-catalyzed carboalumination[140] reaction can tolerate various functional groups such as hydroxy, OTBDMS, SPh, halogens, alkenes and arenes. The addition of Me_3Al to phenyl propynyl sulfide[141] affords the *syn* addition product with excellent regio- and stereo-selectivity (equation 37). The addition to alkynols in the presence of Cl_2TiCp_2 proceeds in good yields with moderate regioselectivity. However, the dimethylaluminum alkoxide of 3-hexyn-1-ol undergoes a regio- and stereo-specific carbotitanation reaction and leads to the *syn* adduct (**98** Scheme 52). The addition of various organoaluminum compounds to alkynylsilanes in the presence of titanium or zirconium complexes gives mainly the *syn* carboalumination products (equation 38).[136,144] A high yield regiospecific dialkylmagnesium-promoted carboalumination of alkynylsilanes has been reported by Oshima and coworkers.[143] It was found that alkynylsilanes furnish mainly the *syn* addition product. Negishi and coworkers demonstrated that the vinylic 1,1-diorganometallics obtained by carbometallation can participate in various cyclization reactions.[104] The zirconium-catalyzed methylalumination of 4-haloalkynylsilanes gives cyclobutene derivatives in excellent yields (Scheme 53). Although three- and six-membered rings are formed by the same method, five-membered rings cannot be prepared directly.[104,145] An intramolecular carboalumination reaction of alkynylsilanes can also be used to produce five- and six-membered rings.[104,146] Although the

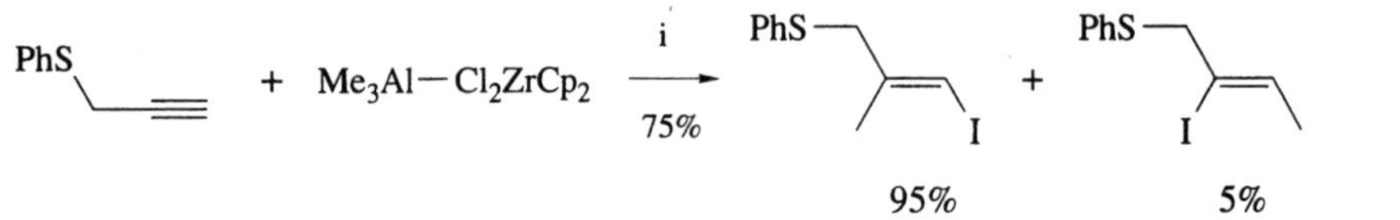

PhS + $Me_3Al-Cl_2ZrCp_2$ —i, 75%→ PhS … I (95%) + PhS … I (5%) (37)

i, $ClCH_2CH_2Cl$, 25 °C, 3 h

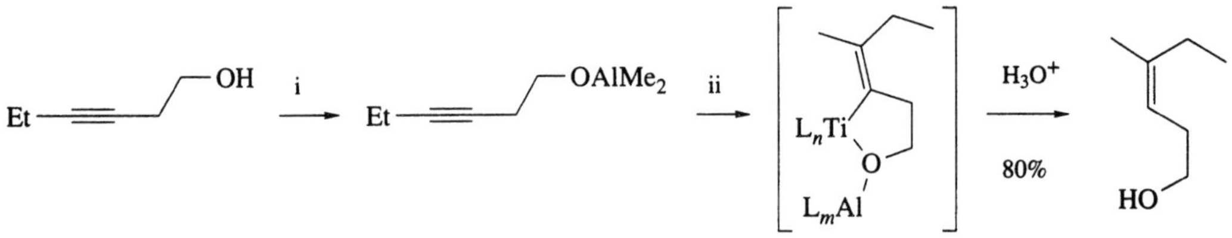

Et—≡—OH —i→ Et—≡—$OAlMe_2$ —ii→ [L_nTi, O, L_mAl] —H_3O^+, 80%→ HO (**98**)

i, Me_3Al, CH_2Cl_2, 0 to 25 °C; ii, $TiCl_4$, –78 °C, 6 h

Scheme 52

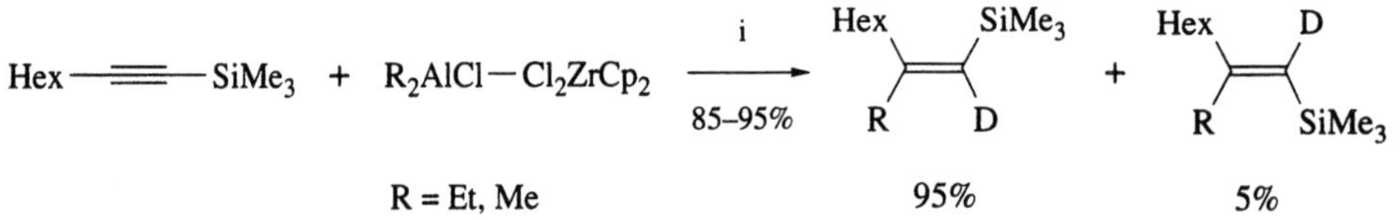

Hex—≡—$SiMe_3$ + $R_2AlCl-Cl_2ZrCp_2$ —i, 85–95%→ Hex, $SiMe_3$, R, D (95%) + Hex, D, R, $SiMe_3$ (5%) (38)

R = Et, Me

i, CH_2Cl_2, 25 °C, 2 h, then $D_2O/NaOD$

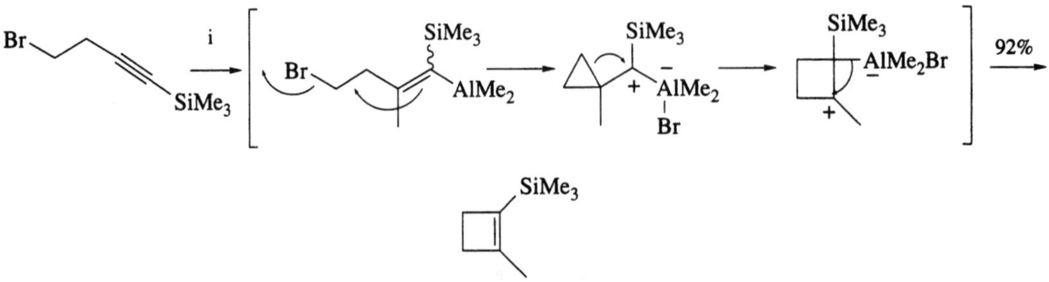

Br, $SiMe_3$ —i→ [Br, $SiMe_3$, $AlMe_2$ → $SiMe_3$, $^+$ $\bar{A}lMe_2$, Br → $SiMe_3$, $\bar{A}lMe_2Br$, +] —92%→ $SiMe_3$

i, Me_3Al, Cl_2ZrCp_2, $ClCH_2CH_2Cl$, 25 °C, 6 h

Scheme 53

carboalumination of nonfunctionalized alkynylsilanes does not produce stereoisomerically pure α-silylalkenylalanes of type (**99**; Scheme 54), since these are configurationally unstable,[147] it is possible to achieve a highly stereoselective chelation-controlled cyclization by using the α-hydroxyalkynylsilane (**100**) as a substrate.

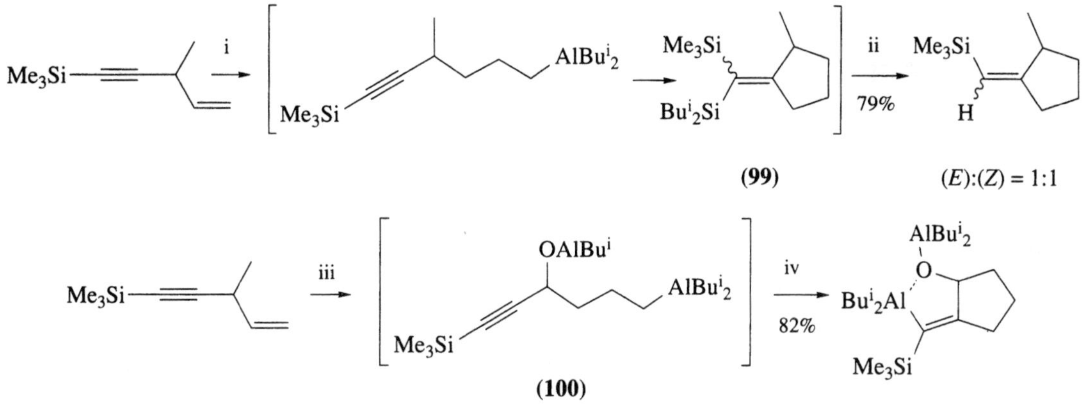

i, Bu^i_3Al, Cl_2ZrCp_2 (10 mol %), CH_2Cl_2, 0 °C, 2 h, then 25 °C, 18 h; ii, H_3O^+; iii, Bu^i_3Al, $ClCH_2CH_2Cl$, 0 to 25 °C; iv, Cl_2ZrCp_2 (5 mol %), 0 °C, 2 h, then 25 °C, 4 h

Scheme 54

4.4.6.4 Synthetic Applications of Alkenylaluminum Organometallics

As mentioned in the previous section, alkenylaluminum reagents of type (**101**) obtained by methylalumination are synthetically very useful reagents which react with a variety of electrophiles. The carbon–aluminum bond can be readily converted in good yields into carbon–hydrogen, –deuterium, –iodine, –mercury, –boron, –zirconium and –carbon bonds[107b,136,148] (Scheme 55). The initial reactivity pattern of organometallics (**101**) can be considerably increased by the transmetallation to more reactive derivatives, such as organozinc halides, which undergo a variety of cross-coupling reactions[136] in the presence of palladium(0) complexes. With other types of less reactive electrophiles the use of the corresponding alanates (**102**), prepared by the addition of one equivalent of an organolithium reagent to (**101**), is required and further expands their synthetic application (Scheme 56). Applications to the synthesis of a number of natural products such as geraniol,[149] farnesol,[150] monocyclofarnesol,[151] ocimene,[152] α-farnesene,[152] dendrolasin (**103**),[150] mokupalide (**104**),[150] vitamin A (**105**),[153] brassinolide,[154] milbemycin,[155] verrucarin[156] and zoapatanol[157] have been reported.

4.4.7 THE CARBOCUPRATION REACTION

4.4.7.1 General Considerations

The carbocupration[3,158,159] of isolated double bonds is generally not possible, but some strained alkenes, as well as conjugated 1,2- and 1,3-dienes, react with several organocopper reagents. The carbocupration of alkynes is certainly the most versatile carbometallation reaction. It proceeds with very high stereoselectivity and affords the *syn* addition product (>99.5% *syn*) regardless of the substrate used. The regioselectivity of the addition is usually very good, but may be complicated by the presence of heteroatoms at α- or β-positions to the triple bond. The scope of the reaction is very broad. Besides alkynyl- and, surprisingly, allyl-copper reagents, most organic groups can be added across alkynes by a carbocupration reaction. The best substrates are certainly primary alkylcopper derivatives; however, secondary, tertiary, alkenyl and even the less reactive methylcopper can undergo carbocuprations. The alkenylcopper reagents formed after carbocupration have found a wide range of applications. They have proven to be very useful intermediates for the synthesis of a variety of natural products.[3,158,159]

i, D_3O^+; ii, X = I, I_2; X = Br, NBS or Br_2; iii, $ClCO_2Et$; iv, allyl chloride or bromide, $[Pd(PPh_3)_4]$ (5 mol %), 0 to 25 °C, 3 h; v, $HgCl_2$ (1.5 equiv.), THF, 0 °C; vi, $[Pd(PPh_3)_4]$ (5 mol %), $ZnCl_2$, (I)BrCH=CHR[1], 20–25 °C; vii, $[Pd(PPh_3)_4]$ (5 mol %), $ZnCl_2$, R^1—≡—I, 20–25 °C; viii, $ArCH_2Cl$, $[Pd(PPh_3)_4]$ (3 mol %), THF, 25 °C, 3 h; ix, *B*-methoxy-9-borabicyclo[3.3.1]nonane, hexane, 25 °C, 1 h; x, R^1Li, THF, –78 °C, then I_2, –78 to 25 °C; xi, BuLi, –78 °C, then $ZnCl_2$, evaporation, THF, Pd^0L_n (1 mol %), RCOCl (1.2 equiv.); xii, ArI, $[Pd(PPh_3)_4]$ (5 mol %), THF, 25 °C, 5 h

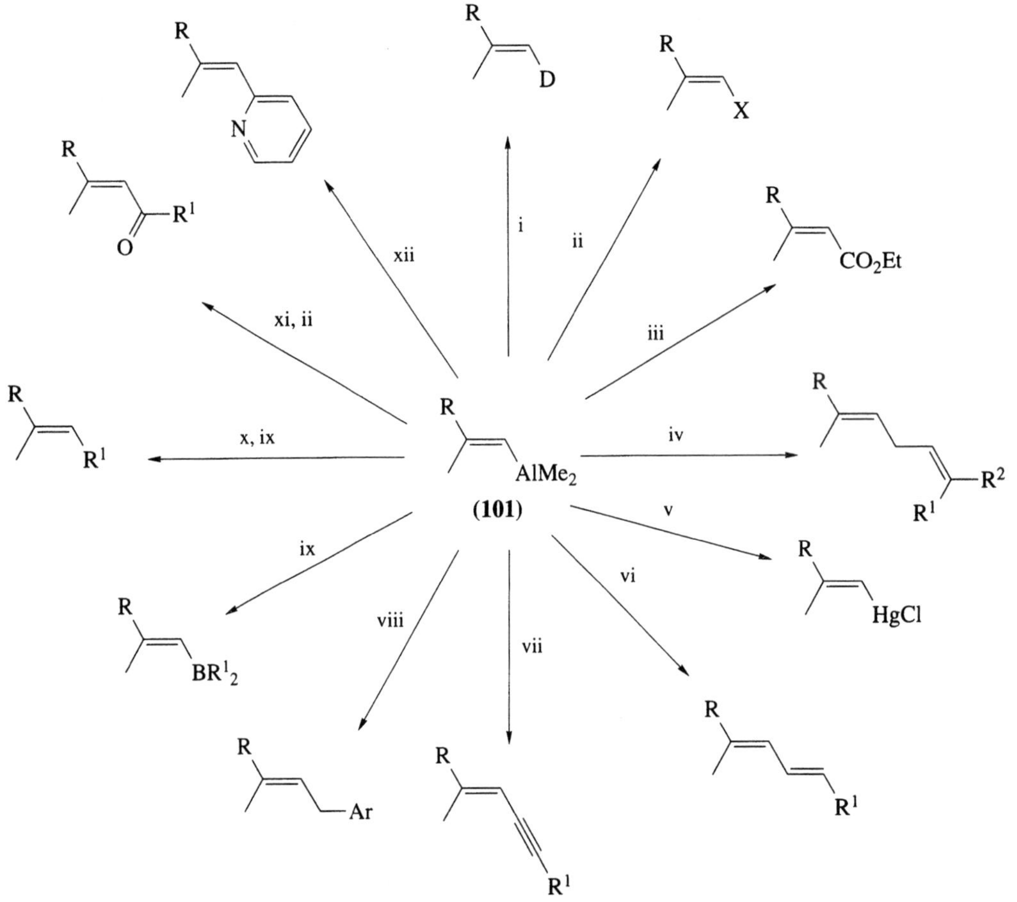

Scheme 55

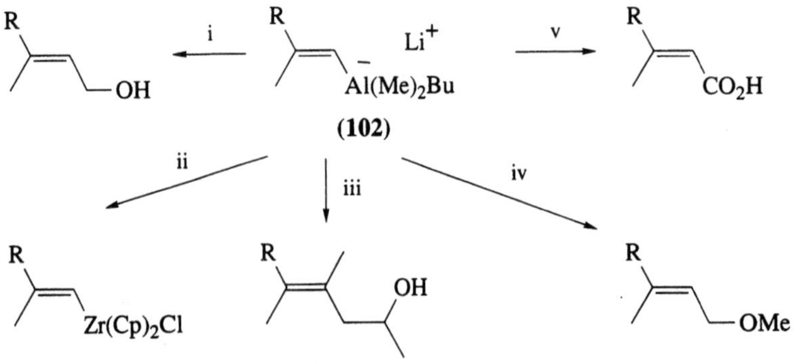

i, $(CH_2O)_n$, THF, 25 °C, 3 h; ii, Cl_2ZrCp_2, –78 to 25 °C; iii, 1,2-epoxypropane, –30 °C; iv, $MeOCH_2Cl$, 0 °C, 2 h; v, CO_2, –30 to –10 °C

Scheme 56

(103)

(105)

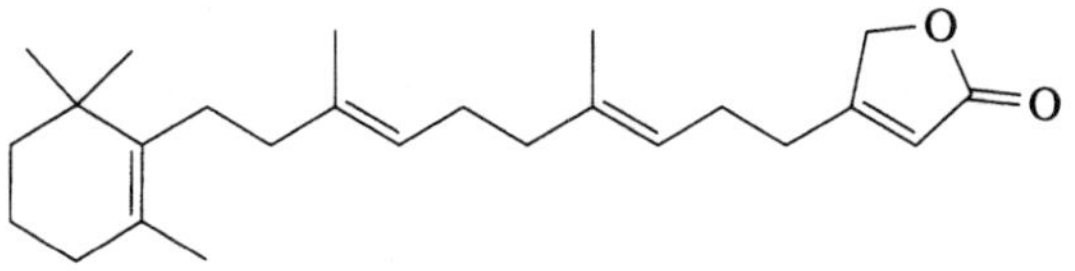

(104)

4.4.7.2 The Carbocupration of Alkenes, 1,2- and 1,3-Dienes

The addition of organocopper reagents to isolated double bonds does not occur readily and only some classes of transition metal–alkene complexes add copper organometallics in good yields (see Volume 4, Chapter 3.2).[160] Strained alkenes such as cyclopropenes are more prone to undergo addition reactions. The *in situ* generated cyclopropene derivative (**106**) adds $PrCu{\cdot}MgBr_2$ at –25 °C and affords, after hydrolysis, the *syn* adduct (**107**) stereospecifically.[161] Functionalized cyclopropenes can also be used and the cyclopropene ketal (**108**) adds a wide range of organocopper reagents[162] in a *syn* fashion to afford stereodefined cyclopropylcopper derivatives of type (**109**), which can be trapped by various electrophiles with retention of configuration (Scheme 57).

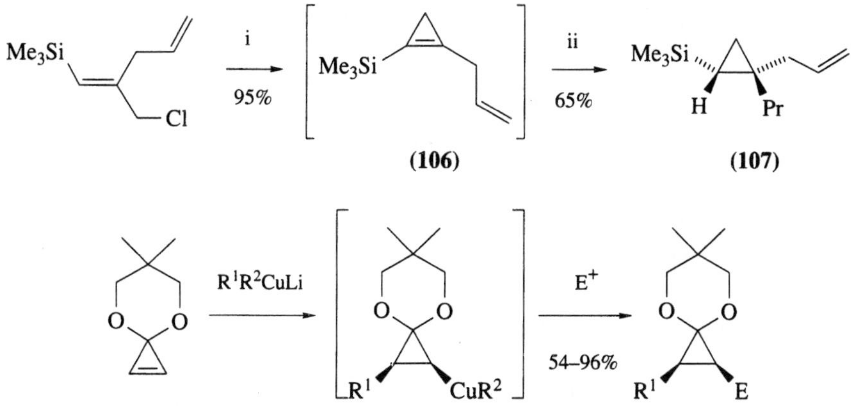

Me_3Si Cl —i, 95%→ [Me_3Si (106)] —ii, 65%→ Me_3Si H Pr (107)

(108) —R^1R^2CuLi→ [O O R^1 CuR^2 (109)] —E^+, 54–96%→ O O R^1 E

i, BuLi, Et_2O, –78 °C; ii, $PrCuMgBr_2$, Et_2O, –25 °C, 1–2 h, then H_2O

Scheme 57

It has been shown that magnesium organocopper and cuprate reagents[163] add to 1-methoxy-1,2-propadiene with low stereoselectivity. In strong contrast, Alexakis and Normant[164] showed that lithium organocopper and cuprate reagents allow a very high control of the newly formed double-bond stereochemistry. Thus, lithium diorganocuprates in ether furnish mostly the (*E*)-substituted enol ethers (**110**), whereas lithium organocopper reagents in THF afford the corresponding (*Z*)-substituted enol ethers (**111**; Scheme 58). The observed stereoselectivity has been explained by a delicate balance between chelate and steric effects. In a solvent of low basicity like ether, the organometallic reagent will complex with the methoxy group of 1-methoxy-1,2-propadiene and will lead to a transition state like (**112**). In the more basic solvent THF, only steric effects are important and the attack of the organocopper will come from the less-hindered side (*anti* to the methoxy group), leading to a transition state like (**113**).

Pulido and coworkers[165] showed that lithium bis(phenyldimethylsilyl)cuprate adds readily to 1,2-propadiene to afford allylic copper reagents, which react with various electrophiles, leading to functionalized alkenylsilanes (Scheme 59). Normant and coworkers[166] have shown that secondary and tertiary organocopper reagents ($R_2CuMgBr$) readily add to butadiene in HMPA/THF to afford regiospecifically the C-1 addition product (**114**). This allylic organometallic can be trapped in good yields with several

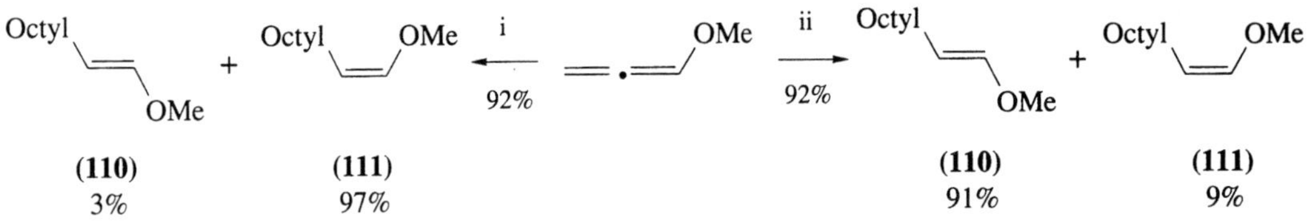

Octyl OMe (110) 3% + Octyl OMe (111) 97% ← i 92% ← =•= OMe → ii 92% → Octyl OMe (110) 91% + Octyl OMe (111) 9%

i, heptylCu, LiX, THF, –30 °C, then H_2O; ii, heptyl$_2$CuLi, Et_2O, –40 °C, 0.5 h, then H_2O

Scheme 58

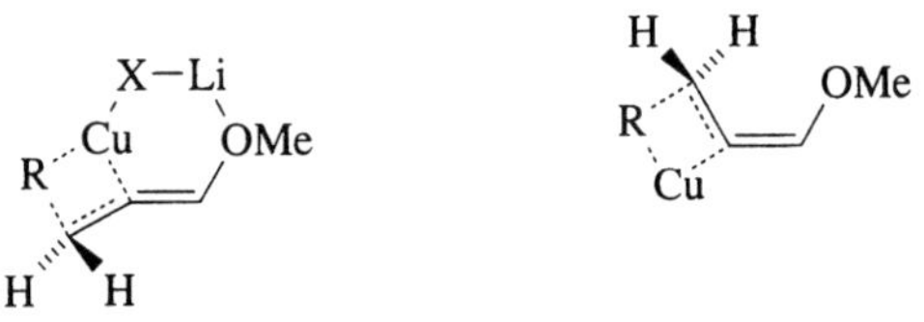

X–Li Cu OMe R H H **(112)**

H H OMe R Cu **(113)**

electrophiles (Scheme 60). The scope of the reaction seems to be limited, as primary copper reagents do not give an addition and 1-substituted dienes such as 1,3-pentadiene are inert.

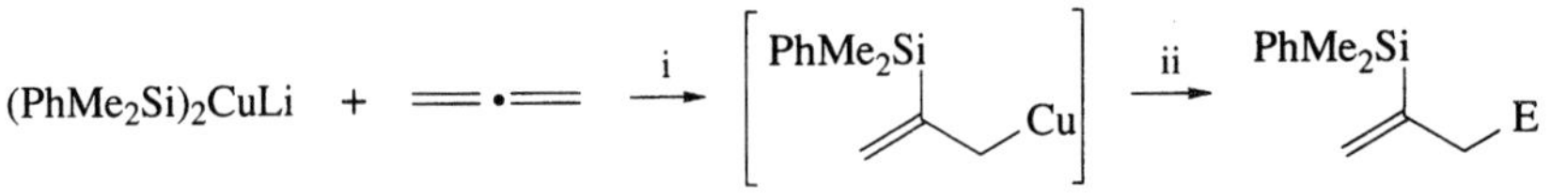

$(PhMe_2Si)_2CuLi$ + =•= →(i) [$PhMe_2Si$ Cu] →(ii) $PhMe_2Si$ E

i, THF, –78 °C, 1 h; ii, E^+, –78 to 0 °C, 1 h

Scheme 59

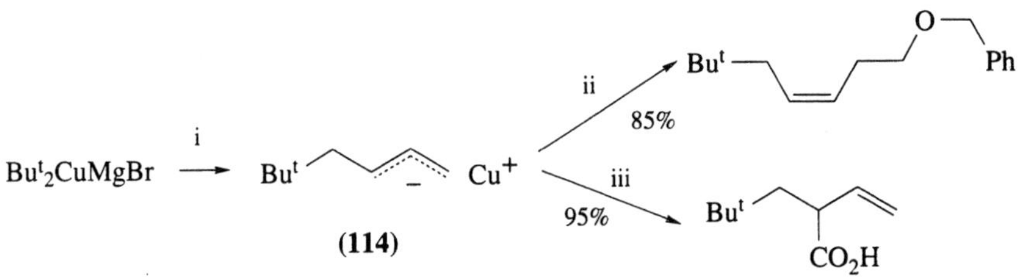

$Bu^t{}_2CuMgBr$ →(i) Bu^t (allyl anion) Cu^+ **(114)** →(ii, 85%) Bu^t (Z-alkene) O Ph; →(iii, 95%) Bu^t (CO_2H)

i, butadiene (excess), HMPA, THF, –15 °C; ii, $PhCH_2OCH_2Cl$; iii, CO_2

Scheme 60

4.4.7.3 The Carbocupration of Alkynes

4.4.7.3.1 Addition to nonfunctionalized alkynes

In 1971, Normant and Bourgain[167] discovered that organocopper reagents were able to add in a *syn* fashion, with almost complete stereoselectivity (>99.5%) and very high regioselectivity, to terminal alkynes and acetylene itself, leading to vinylcopper reagents of type (**115**; equation 39). The reaction was found to be quite general,[3] and of very high synthetic utility. Several organocopper or organocuprate reagents can be used, such as $RCu{\cdot}MgX_2$, R_2CuMgX, R_2CuLi or $RCu{\cdot}(X)Li$ (X = OBu^t, SPh, CN), but for each reagent the choice of the reaction conditions (solvent, cosolvents, ligands) is critical for the success of the reaction. Acetylene is the most reactive nonfunctionalized alkyne[168] and adds all types of copper reagents. Lithium cuprates react particularly well and insert two acetylene units to afford vinylcuprates of type (**116**; equation 40).

A controlled oligomerization[3,169,170] of acetylene is possible by carefully controlling the reaction temperature and adding the two first equivalents of acetylene at –50 °C and the remaining acetylene at 0 °C.

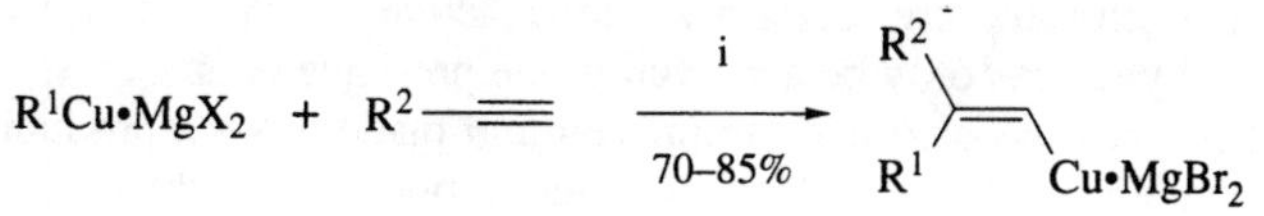

$$R^1Cu{\bullet}MgX_2 + R^2{-}{\equiv} \xrightarrow[70\text{–}85\%]{\text{i}} R^1R^2C{=}CH{-}Cu{\bullet}MgBr_2 \quad \mathbf{(115)} \tag{39}$$

i, addition at –35 °C, then warm to –15 °C

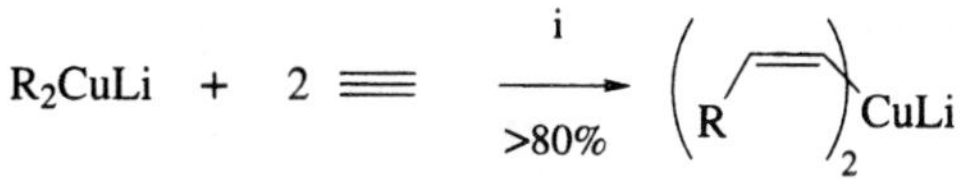

$$R_2CuLi + 2 \equiv \xrightarrow[>80\%]{\text{i}} (R{-}CH{=}CH)_2CuLi \quad \mathbf{(116)} \tag{40}$$

i, Et_2O, –50 °C to –20 °C, 15 min

Under these conditions, the (*Z*,*Z*)-dienylcuprates (**117**) are formed in fair yields and can be trapped by various electrophiles (Scheme 61). α-Substituted alkenylcuprates react even more readily with acetylene and dienylcuprates are obtained in high yields (Scheme 62).[170]

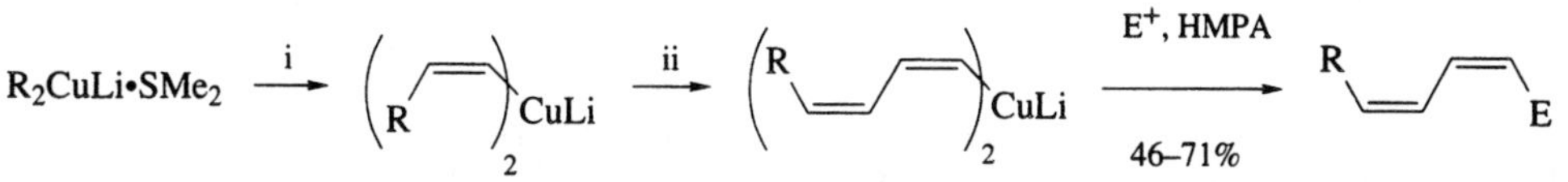

$$R_2CuLi{\bullet}SMe_2 \xrightarrow{\text{i}} (R{-}CH{=}CH)_2CuLi \xrightarrow{\text{ii}} (R{-}CH{=}CH{-}CH{=}CH)_2CuLi \ \mathbf{(117)} \xrightarrow[46\text{–}71\%]{E^+,\ HMPA} R{-}CH{=}CH{-}CH{=}CH{-}E$$

i, acetylene (2 equiv.), –50 °C, then 30 min at –25 °C; ii, acetylene (4 equiv.), 0 °C

Scheme 61

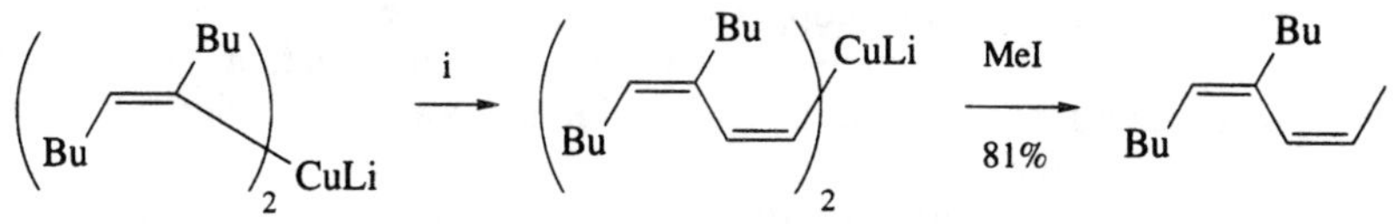

$$(BuCH{=}C(Bu))_2CuLi \xrightarrow{\text{i}} (BuCH{=}C(Bu){-}CH{=}CH)_2CuLi \xrightarrow[81\%]{MeI} BuCH{=}C(Bu){-}CH{=}CHMe$$

i,acetylene (2 equiv.), Et_2O, 20 °C, 15 min

Scheme 62

Organocuprates in which the R group is phenyl, allyl, alkynyl or vinyl do not add to acetylene or to other alkynes. Phenylacetylene and propyne, although less reactive than acetylene, show a higher reactivity toward carbocupration than most higher alkynes and Vermeer[171] showed that secondary and tertiary magnesium alkylcopper reagents and dialkylcuprates add in THF to phenylacetylene under very mild conditions, albeit with a low regioselectivity (equation 41).

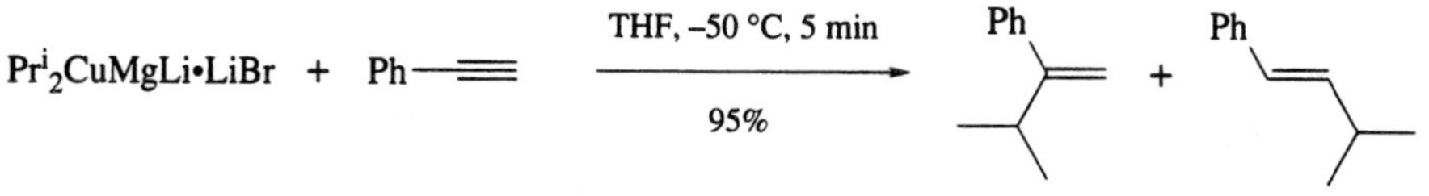

$$Pr^i_2CuMgLi{\bullet}LiBr + Ph{-}{\equiv} \xrightarrow[95\%]{THF,\ -50\ °C,\ 5\ min} Ph(Pr^i)C{=}CH_2\ (68\%) + PhCH{=}CHPr^i\ (32\%) \tag{41}$$

As indicated previously, higher terminal alkynes react with copper reagents derived from magnesium organometallics in ether and afford the *syn* addition products with high regio- and stereo-selectivity (equation 39). The first step in these carbocuprations seems to be the formation of a π-complex between the alkyne and the copper reagent, since it has been found by Normant and coworkers[172] that the stability of an alkylcopper in ether can be increased substantially by the presence of an alkyne. The presence of magnesium salts is also essential and no carbocupration takes place in their absence, indicating that mixed magnesium–copper clusters are certainly the reactive species. Similarly, the addition of RMgX/CuCl to terminal alkynes can only be achieved in the presence of an excess of $MgBr_2$.[173] The carbocupration of terminal alkynes has also been improved and made more reproducible by the use of dimethyl sulfide–copper halide complexes instead of crude copper salts, which did not always have the required purity. Helquist and coworkers have used dimethyl sulfide as a cosolvent to achieve higher yields and a broader range of applicability (Scheme 63).[174] Of special interest is that this procedure allows one to perform methylcupration reactions, which are usually difficult due to the low reactivity of methylcopper derivatives.[3,171] This gives access to trisubstituted alkenes which are of widespread occurrence among natural products (Scheme 64).[174]

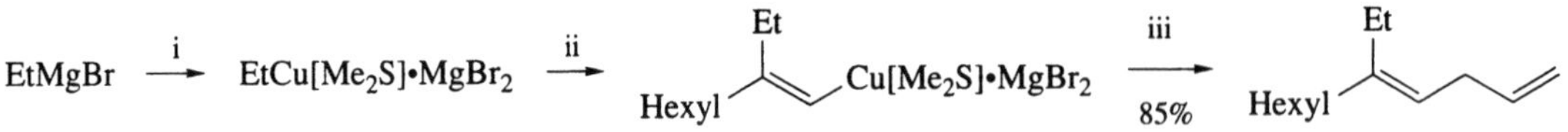

EtMgBr →(i) $EtCu[Me_2S]{\bullet}MgBr_2$ →(ii) Hexyl(Et)C=CH–$Cu[Me_2S]{\bullet}MgBr_2$ →(iii, 85%) Hexyl(Et)C=CH–CH₂CH=CH₂

i, $CuBr{\bullet}Me_2S$ (1 equiv.), Et_2O:dimethyl sulfide (1:1), –45 °C, 2 h; ii, 1-octyne (0.9 equiv.), –45 °C, 2 h; iii, HMPA, allyl bromide (1 equiv.), –78 °C to –30 °C, 12 h, then aq. NH_4Cl

Scheme 63

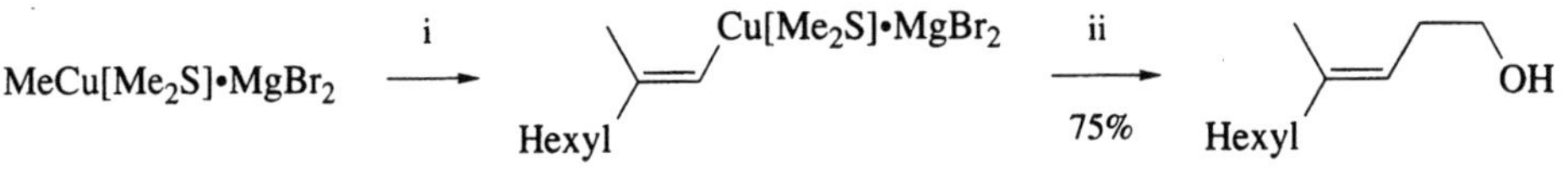

$MeCu[Me_2S]{\bullet}MgBr_2$ →(i) Hexyl(Me)C=CH–$Cu[Me_2S]{\bullet}MgBr_2$ →(ii, 75%) Hexyl(Me)C=CH–CH₂CH₂OH

i, octyne (0.9 equiv.), Et_2O, dimethyl sulfide,–25 °C, 120 h; ii, 1-lithiopentyne (1 equiv.), HMPA (2 equiv.), Et_2O, ethylene oxide (1 equiv.), –78 °C, 2 h, then –25 °C, 24 h, then aq. NH_4Cl

Scheme 64

The addition of secondary alkylcopper derivatives often proceeds with low stereoselectivity. Only the magnesium diisopropylcuprate adds regioselectively to 1-octyne.[171] A procedure developed recently by Periasamy, using RMgX/CuCl in the presence of an excess of $MgBr_2$ in THF, seems to allow a regiospecific addition of secondary and tertiary alkylcopper reagents.[173] While nonfunctionalized disubstituted alkynes usually do not undergo carbocuprations,[175] the intermolecular reaction, as shown by Crandall and coworkers,[75,176] proceeds under mild conditions and with high yields (equation 42). Some functionalized organocopper reagents bearing various protected hydroxy and thiol functions have been added to acetylene and propyne. The optimum conditions for these reactions have been determined (equations 43 and 44).[177] Normant and coworkers[178] also showed that trimethylsilylmethylcopper adds, under forcing

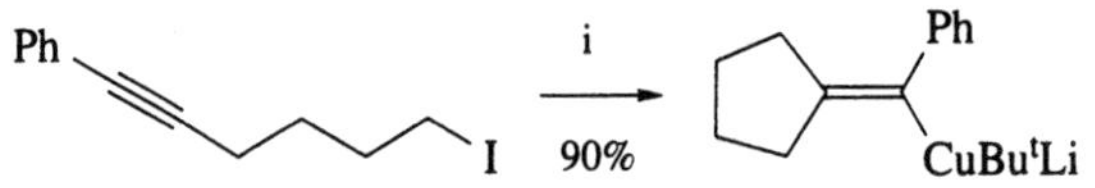

PhC≡C(CH₂)₄I →(i, 90%) cyclopentylidene=C(Ph)(CuBu^{t}Li) (42)

i, Bu^t_2CuLi, PBu_3, pentane:ether (7:1)

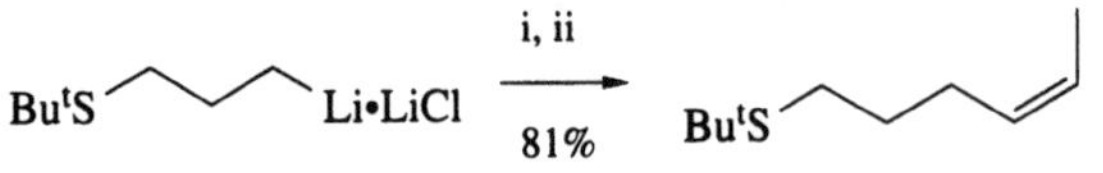

$Bu^tS(CH_2)_4Li{\bullet}LiCl$ →(i, ii, 81%) Bu^tS–(CH₂)₄–CH=CH–Me (43)

i, CuI, then acetylene, –50 °C, warm to –30 °C, 0.5 h; ii, MeI, HMPA, –40 °C to 25 °C, then H_3O^+

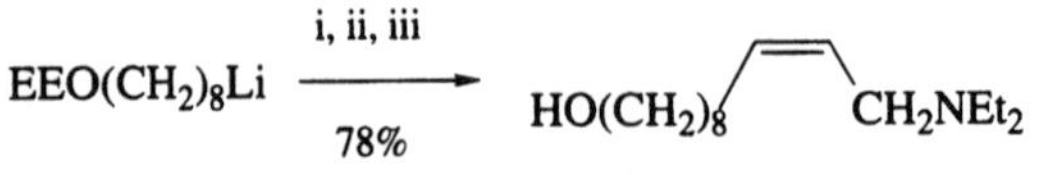

$EEO(CH_2)_8Li$ →(i, ii, iii, 78%) $HO(CH_2)_8$–CH=CH–CH_2NEt_2 (44)

i, $CuBr{\bullet}Me_2S$, Et_2O, –35 °C, 0.5 to 1 h; ii, acetylene, –50 °C, then warm to –30 °C, 0.5 h; iii, $PhSCH_2NEt_2$, THF, 20 °C, 2 h, then H_3O^+

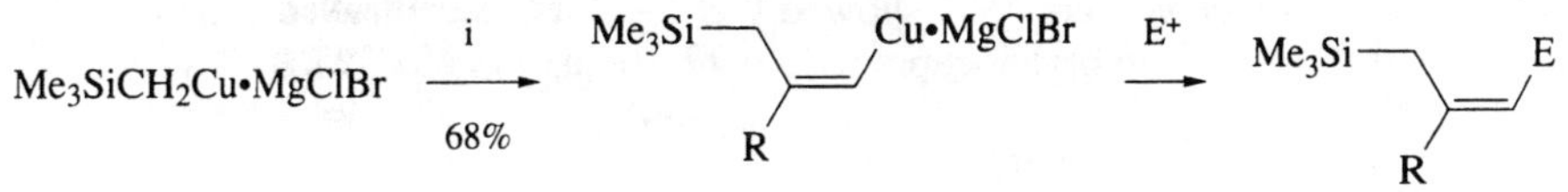

$Me_3SiCH_2Cu{\bullet}MgClBr$ —i, 68%→ (118) —E^+→ (119)

i, R—≡, $P(OEt)_3$, Et_2O, −35 °C, 62 h

Scheme 65

conditions, to alkynes to afford functionalized copper derivatives of type (**118**) which react with a wide range of electrophiles to afford stereoisomerically pure allylsilanes (**119**; Scheme 65).

In summary, lithium diorganocuprates[3] are the reagents of choice for the carbocupration of acetylene itself. For higher alkynes, the use of copper organometallics made from Grignard reagents[3] ($RCu{\cdot}MgX_2$ in ether and R_2CuMgX in THF[171]) is recommended. The use of dimethyl sulfide[174] as a cosolvent significantly improves the yields.

4.4.7.3.2 Addition to functionalized alkynes

Whereas nonfunctionalized alkynes undergo carbocupration with high, predictable regioselectivity, the introduction of a heteroatom at the propargylic position, the homopropargylic position or directly at the triple bond has a strong influence on the regioselectivity of the carbocupration. It has been shown by Alexakis and Normant[3] that the observed regioselectivity is determined by complexation and steric effects. The use of a solvent of low basicity usually favors the complexation between the organocopper derivative and the heteroatoms present in the alkyne, whereas in a more basic solvent like THF steric interactions are more important. These points are illustrated in equations (45) and (46). The addition of $BuCu{\cdot}MgBr_2$ to the homopropargylic acetal (**120**) in ether is nonregioselective due to a partial complexation of the copper reagent to the acetal, and affords equal amounts of (**121**) and (**122**) (equation 45). In THF, the copper reagent, already strongly complexed by the solvent, does not coordinate to the acetal function of (**120**) and steric effects lead to the formation of (**121**; equation 45). With the cyclic acetal (**123**), a better chelation of the copper reagent is possible, leading mostly to the regioisomer (**125**; equation 46).

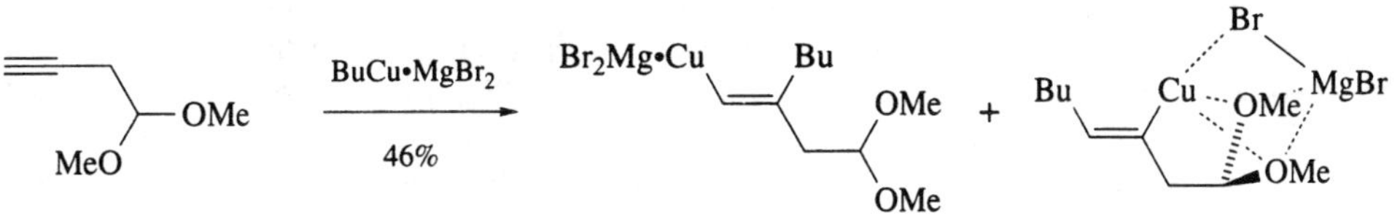

(120) —$BuCu{\bullet}MgBr_2$, 46%→ (121) + (122) (45)

	(121)	(122)
in Et_2O	49%	51%
in THF	98%	2%

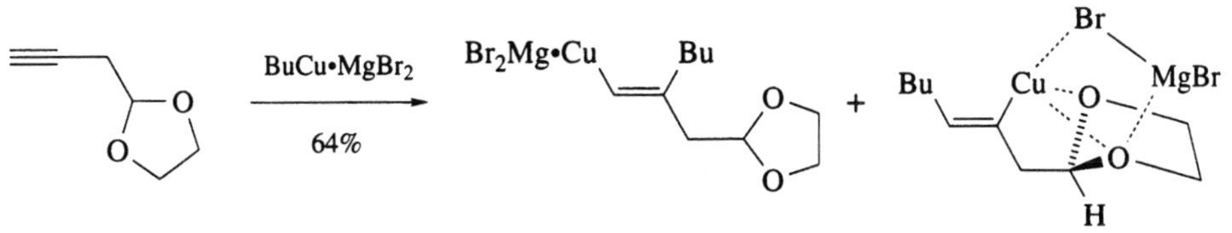

(123) —$BuCu{\bullet}MgBr_2$, 64%→ (124) + (125) (46)

	(124)	(125)
in Et_2O	13%	87%

For propargylic substrates, a similar rationale is possible to explain the regiochemistry of the carbocupration. A propargylic trimethylsilyloxy group does not seem to undergo any complexation with copper reagents, and steric interactions are dominant, leading to the formation of only one regioisomer (equation 47). Propargylic ethers show a higher reactivity compared to nonfunctionalized alkynes and the addition of methylcopper[3] proceeds smoothly leading to the interesting reagent (**126**). In contrast, propargylic acetals give low regioselectivities with lithium- or magnesium-derived organocopper reagents,

$RCu{\cdot}MX_n$, but Alexakis and coworkers[165,179] showed that the chelate-controlled product (**127**) was obtained regiospecifically with lithium organocuprates, R_2CuLi (equation 48).

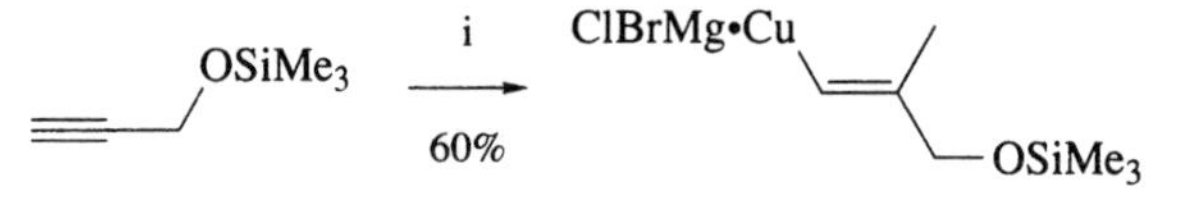

(47)

(**126**)

i, MeCu•MgClBr, THF, –25 °C, 24 h

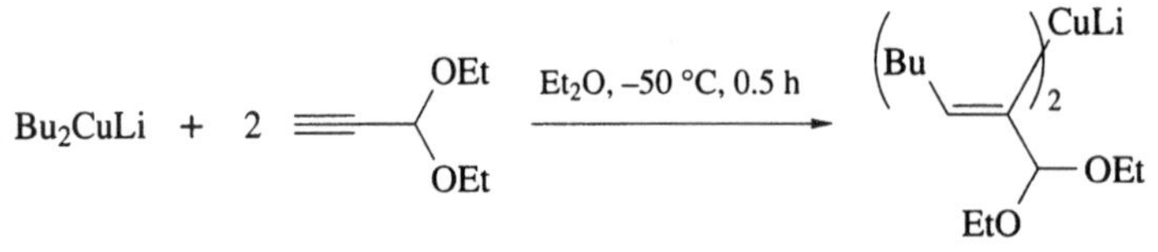

(48)

(**127**)

Secondary and tertiary dialkylcuprates, lithium dialkenyl-, and even diphenyl-cuprates, add in very good yields to the reactive propionaldehyde diethyl acetal. The *syn* addition products may be trapped with a variety of electrophiles such as alkyl, alkenyl, alkynyl and aryl halides. The method has been used for the synthesis of several natural products. Substituted alkynic acetals also react with lithium dialkylcuprates in ether to furnish stable dialkenylcuprates of type (**128**) which do not eliminate to the corresponding alkoxy allenes (**129**) if the temperature is maintained below –20 °C.[164,179]

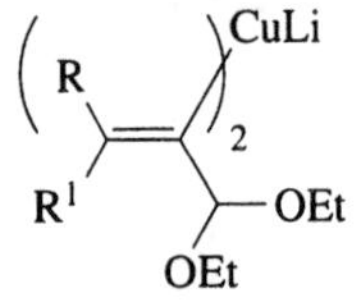

(**128**) 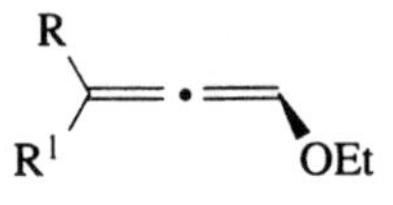 (**129**)

Heteroatom-substituted alkynes of type (**130**) react regio- and stereo-specifically with various organocopper reagents[3,180,181] to afford the regioisomer (**131**) if X = OR, NR_2 or $SiMe_3$, and the regioisomer (**132**) if X = SR or PR_2 (Scheme 66). Some adducts, (**131**) and (**132**), have limited thermal stability. Thus, the organocopper derivatives (**131**), where X = OR and NR_2, undergo an elimination reaction at temperatures higher than –20 °C and +20 °C, respectively. However, the vinylic carbenoid (**132**), where X = SR, is stable up to 60 °C. Vermeer[181] and coworkers showed that the alkenylcopper reagents derived from the addition to ethoxyacetylene are useful intermediates for the preparation of 1,4-diketones (Scheme 67). The addition of $RCu{\cdot}MgBr_2$ to alkynylsilanes[182] allows an easy access to various regio- and stereo-isomerically pure alkenylsilanes (Scheme 68).

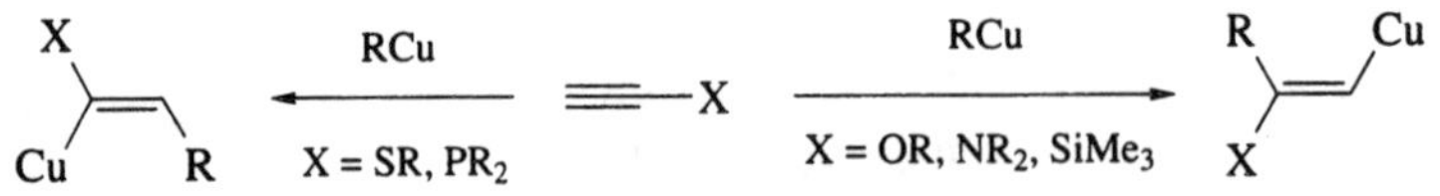

(**132**) (**130**) (**131**)

Scheme 66

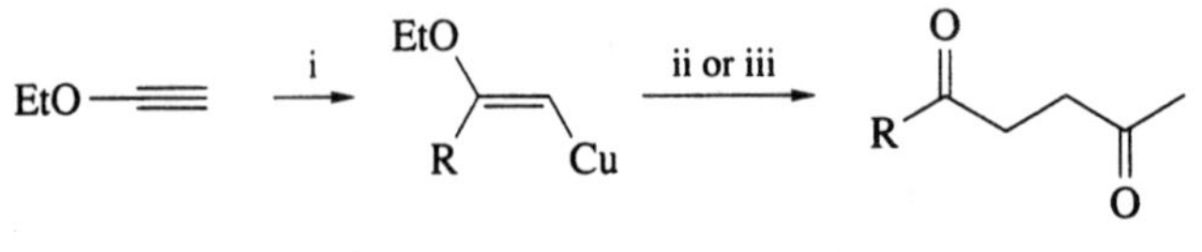

i, $RCu{\cdot}MgX_2$ (1.0 equiv.), THF, –50 °C, 1 h, several hours at –25 °C; ii, allyl bromide (1.0 equiv.), HMPA, –20 °C, 3 h, then 3 M HCl followed by Wacker oxidation ($PdCl_2$, CuCl, O_2 in DMF, H_2O);
iii, 2,3-dibromopropene (1.0 equiv.), HMPA, –20 °C, 4 h, then $Hg(OAc)_2$, HCO_2H

Scheme 67

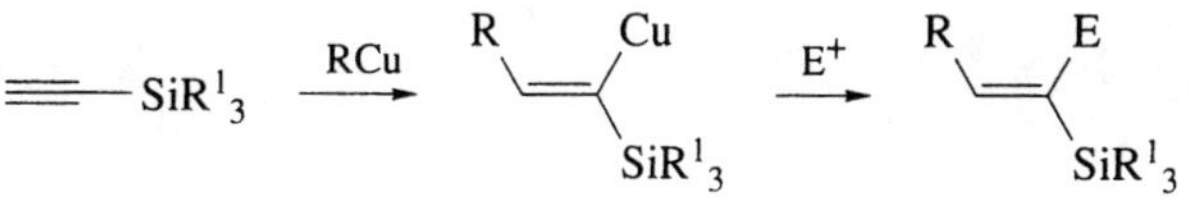

$$\equiv\!-\mathrm{SiR^1_3} \xrightarrow{\mathrm{RCu}} \mathrm{R(Cu)C{=}CH(SiR^1_3)} \xrightarrow{\mathrm{E^+}} \mathrm{R(E)C{=}CH(SiR^1_3)}$$

E = Cl, Br, I, CN, SMe, Me, alkyl, allyl

Scheme 68

Finally, silyl,[183,184] germyl,[185] and stannyl[184,186] cuprations of alkynes have also been described. Fleming[183,184] and coworkers showed that silylcuprate reagents of type (**133**; $R^1 = R^2$ = Me or R^1 = Me, R^2 = Ph) add readily to various alkynes with high regioselectivity, affording the *syn* addition products. After reaction with electrophiles, such as the proton, iodine, acyl and alkyl halides, enones and epoxides, highly functionalized vinylsilanes of defined stereochemistry are formed in good yields. It is noteworthy that the regiochemistry of the silylcupration is opposite to that observed with other organocopper reagents (see Section 4.4.7.3.1 and Scheme 69). Oshima and coworkers[185] showed that germylcopper reagents add to alkynes, but that the regiochemistry of the addition depends on the nature of the copper reagent and of the alkynic compound (equation 49). Due to their reversibility,[185,186] some germylcuprations and stannylcuprations need to be performed in the presence of a proton source, such as an alcohol, in order to drive the addition to completion, as has been shown by Piers and coworkers.[186a] With some reactive alkynes, the *syn* addition proceeds even in the absence of a proton source, but shows only a moderate regioselectivity (equation 50). Related reactions have been developed by Fleming, Oshima and Oehlschlager and coworkers.[186]

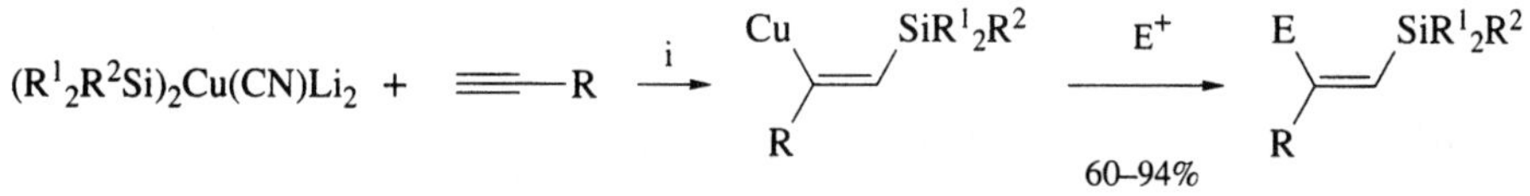

$$\mathrm{(R^1_2R^2Si)_2Cu(CN)Li_2} + \equiv\!-\mathrm{R} \xrightarrow{\mathrm{i}} \mathrm{Cu(R)C{=}CH(SiR^1_2R^2)} \xrightarrow[60\text{–}94\%]{\mathrm{E^+}} \mathrm{E(R)C{=}CH(SiR^1_2R^2)}$$

R^1 = Me; R^2 = Me or Ph

i, THF, –50 °C to 0 °C, 15 min

Scheme 69

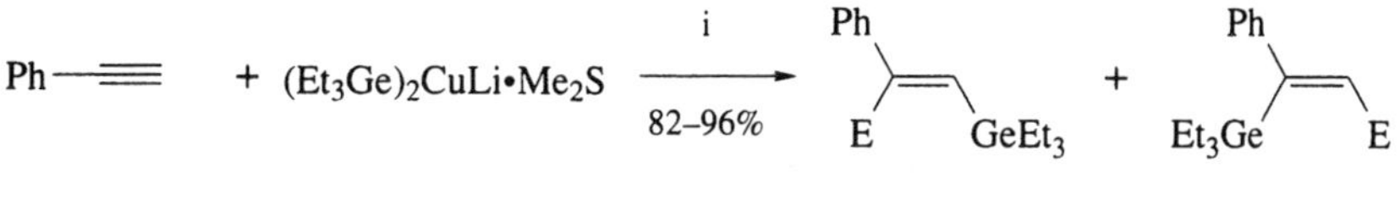

$$\mathrm{Ph}\!-\!\equiv + \mathrm{(Et_3Ge)_2CuLi{\bullet}Me_2S} \xrightarrow[82\text{–}96\%]{\mathrm{i}} \underset{7\text{–}17\%}{\mathrm{Ph(E)C{=}CH(GeEt_3)}} + \underset{83\text{–}93\%}{\mathrm{Ph(Et_3Ge)C{=}CH(E)}} \quad (49)$$

i, $E^+ = D_2O$, MeI, or allyl bromide

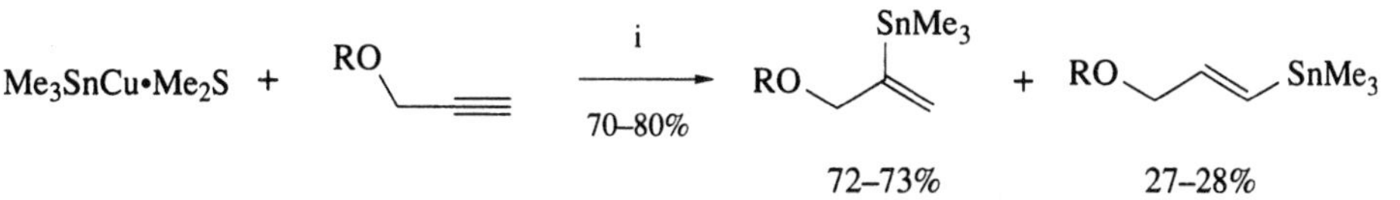

$$\mathrm{Me_3SnCu{\bullet}Me_2S} + \mathrm{RO\!-\!CH_2\!-\!C{\equiv}CH} \xrightarrow[70\text{–}80\%]{\mathrm{i}} \underset{72\text{–}73\%}{\mathrm{RO\!-\!CH_2\!-\!C(SnMe_3){=}CH_2}} + \underset{27\text{–}28\%}{\mathrm{RO\!-\!CH_2\!-\!CH{=}CH\!-\!SnMe_3}} \quad (50)$$

R = THP, TBDMS

i, THF, –78 °C, 6 h, then MeOH, aq.NH_4Cl

4.4.7.4 Synthetic Applications of Alkenylcopper Organometallics

The synthetic applications of alkenylcopper reagents have been reviewed in detail by Normant and Alexakis.[3] The high synthetic potential of these reagents has been summarized in Scheme 70. It is important to notice that the reactivity of alkenylcoppers and dialkenylcuprates is not the same, the latter being more reactive. Thus, some electrophiles[3] do not react readily with alkenylcoppers, but only with

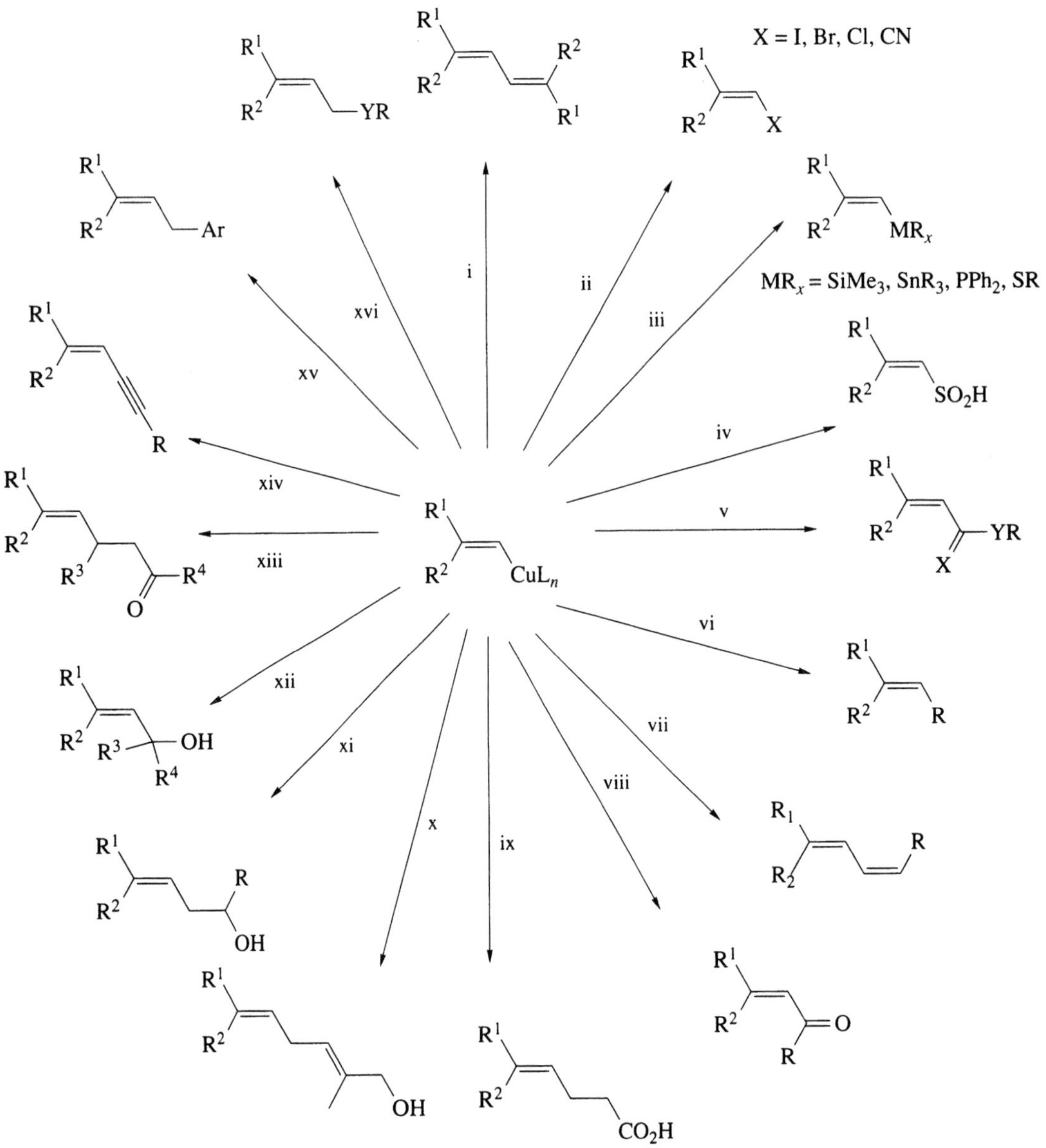

X = I, Br, Cl, CN

MR_x = $SiMe_3$, SnR_3, PPh_2, SR

i, thermal decomposition (10–20 °C) or O_2, –20 °C (60–80%); ii, X = I: I_2, –78 °C, 0.5 h (90–100%); X = Br: HgX_2, HMPA, Br_2, pyridine or NBS (50–90%); X = Cl: NCS, –45 °C (50–79%); X = CN: NCCl, $NCSO_2Ph$ or $NCSO_2Tol$, –50 °C to –20 °C (90–97%); iii, Me_3SiCl (80%); $ClSnR_3$ or $ClPPh_2$, –40 °C, 1.5 h (80–90%); RSSR, HMPA, –50 °C to 25 °C, 3–4 h (58–84%); iv, SO_2, –40 °C to 10 °C (70%); v, X=C=Y: X = O, Y = NPh (50%); X = Y = S, then MeI (R = Me) (98%); X = Y = O (R = H) (80–100%); vi, allylic and benzylic halides, HMPA (80–98%); primary iodides and bromides, HMPA, $P(OEt)_3$, –20 °C (77–85%); vii, for magnesium copper reagents: $[Pd(PPh_3)_4]$ (5 mol %) (50–78%); for cuprates (R_2CuM): $ZnCl_2$ (1 equiv.), THF, $[Pd(PPh_3)_4]$ (5 mol %), –10 °C to 20 °C, 30 min (80–94%);[188] viii, $ZnCl_2$, RCOCl, $[Pd(PPh_3)_4]$ (5 mol %), 25 °C, 1 h (70–100%);[188,189] ix, propiolactone,[190] –70 °C to –10 °C, then 5 M HCl (40–95%);

Scheme 70

x, 1,2-epoxyisoprene,[191] –20 °C, 1 h (70–90%), *(E)*:*(Z)* = 95:5; xi, epoxide,[191] HMPA, –78 °C, 3 h, –33 °C, 24 h (60–90%);[3,174,192] xii, lithium cuprate, $BF_3{\bullet}OEt_2$ (1 equiv.), aldehyde or ketone (0.7 equiv.) (60–70%);[3,193] xiii, enone or α,β-unsaturated aldehyde (65–95%);[3,174,194] xiv, R–≡–X (X = Br or I), TMEDA (1.2 to 2 equiv.), –15 °C, 1–2 h, then H_3O^+;[3,195] xv, aryl iodide, $ZnBr_2$, THF, $[Pd(PPh_3)_4]$ (5 mol %), –15 °C to 20 °C, 0.5 h (65–85%);[188] xvi, $RYCH_2X$: (a) $MeOCH_2Cl$[3] or $ClCH_2SR$[196] (1 equiv.), HMPA or $P(OEt)_3$, 25 °C, 4 h, H_3O^+ (60–99%); (b) $ClCH_2N(Me)CHO$ or *N*-chloromethylphthalimide,[197] THF, 25 °C, 4 h, H_3O^+ (50–89%)

Scheme 70 *(continued)*

dialkenylcuprates. The addition of polar cosolvents, such as HMPA, or of ligands, such as 1-pentynyllithium or magnesium salts, can often enhance the reactivity of an alkenylcopper. Of special interest is the transmetallation, described by Alexakis and coworkers,[188,189] from dialkenylcuprates to the corresponding zinc derivatives, since it is known that these alkenylzinc reagents react efficiently with alkenyl and aryl halides, as well as acid chlorides,[198] under palladium(0) catalysis (equation 51). This method allows an easy access to a variety of dienes with defined stereochemistry, and has been used extensively by Alexakis and coworkers for the synthesis of insect sex pheromones.[199]

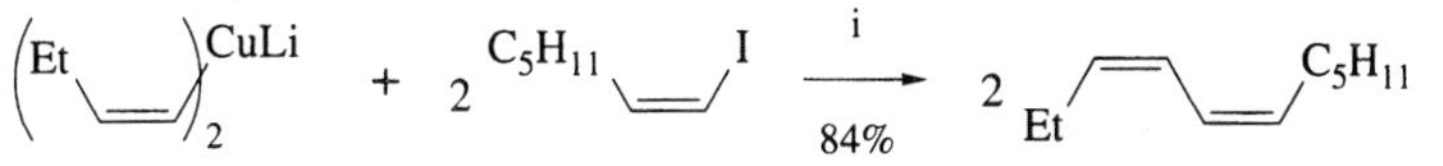

$$\left(\text{Et}\diagdown\!\!=\!\!\diagup\text{CuLi}\right)_2 + 2\ C_5H_{11}\diagdown\!\!=\!\!\diagup\text{I} \xrightarrow[84\%]{\text{i}} 2\ \text{Et}\diagup\!\!=\!\!\diagdown\!\!=\!\!\diagup C_5H_{11} \qquad (51)$$

i, $MgCl_2$ (1 equiv.), $ZnBr_2$ (1 equiv.), $[Pd(PPh_3)_4]$ (5 mol %), –20 °C to 15 °C, 1 h

4.4.8 OTHER CARBOMETALLATION REACTIONS

4.4.8.1 General Considerations

A number of transition metal complexes react with alkenes, alkynes and dienes to afford insertion products (see Volume 4, Part 3). A general problem is that the newly formed carbon–metal bond is usually quite reactive and can undergo a variety of transformations, such as β-hydride elimination or another insertion reaction, before being trapped by an electrophile.[200] Usually, a better stability and lower reactivity is observed if the first carbometallation step leads to a metallacycle. It is worthy to note that the carbometallation of perfluorinated alkenes and alkynes constitutes a large fraction of the substrates investigated with transition metal complexes.[200b]

4.4.8.2 Addition to Alkenes and Dienes

The organomercurial-mediated[201] addition of organopalladium derivatives across alkenes is a common reaction (Heck reaction[201b]). The new carbon–carbon bond is formed mainly from the less-hindered end of the double bond (Scheme 71).[201] The initial carbopalladation product (**134**) is usually not stable and undergoes a β-hydride elimination, leading to the alkene (**135**). However, the *syn* adducts (**134**) can be isolated in certain systems where there are no β-hydrogens available for elimination, or where the palladium complex is stabilized by coordination with a neighboring π-system or a heteroatom, as shown in Scheme 72.[201,202] Besides arylpalladium compounds, the addition of π-allyl, heterocyclic, alkenyl and

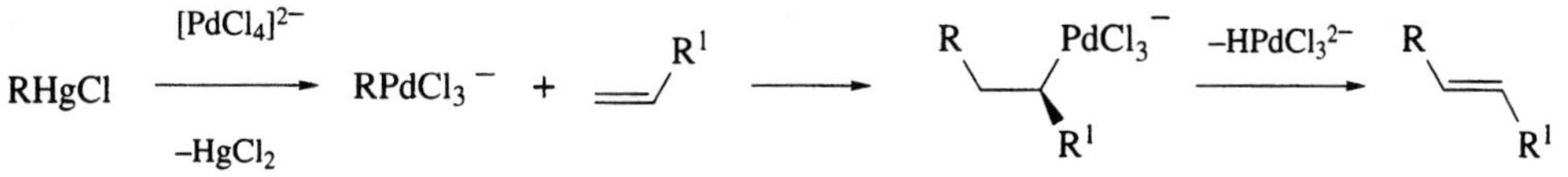

$$RHgCl \xrightarrow[-HgCl_2]{[PdCl_4]^{2-}} RPdCl_3^- + \diagup\!\!=\!\!\diagup R^1 \longrightarrow \underset{(\mathbf{134})}{R\diagdown\!\!\diagup\!CH(PdCl_3^-)R^1} \xrightarrow{-HPdCl_3^{2-}} \underset{(\mathbf{135})}{R\diagdown\!\!=\!\!\diagdown R^1}$$

Scheme 71

benzylic organopalladium compounds to bicyclic alkenes is well known.[202,203] These *syn* palladium adducts can undergo a variety of synthetically useful transformations which all occur with retention of the *exo* stereochemistry and complement the Diels–Alder approach to bicyclic systems (Scheme 73).

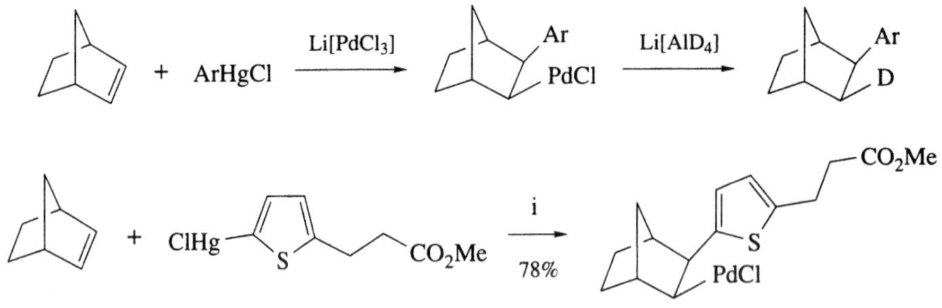

+ ArHgCl —Li[PdCl₃]→ Ar / PdCl —Li[AlD₄]→ Ar / D

+ ClHg–(thienyl)–CO₂Me —i, 78%→ PdCl / S / CO₂Me

i, Li[PdCl₄], MeCN, 0 to 25 °C

Scheme 72

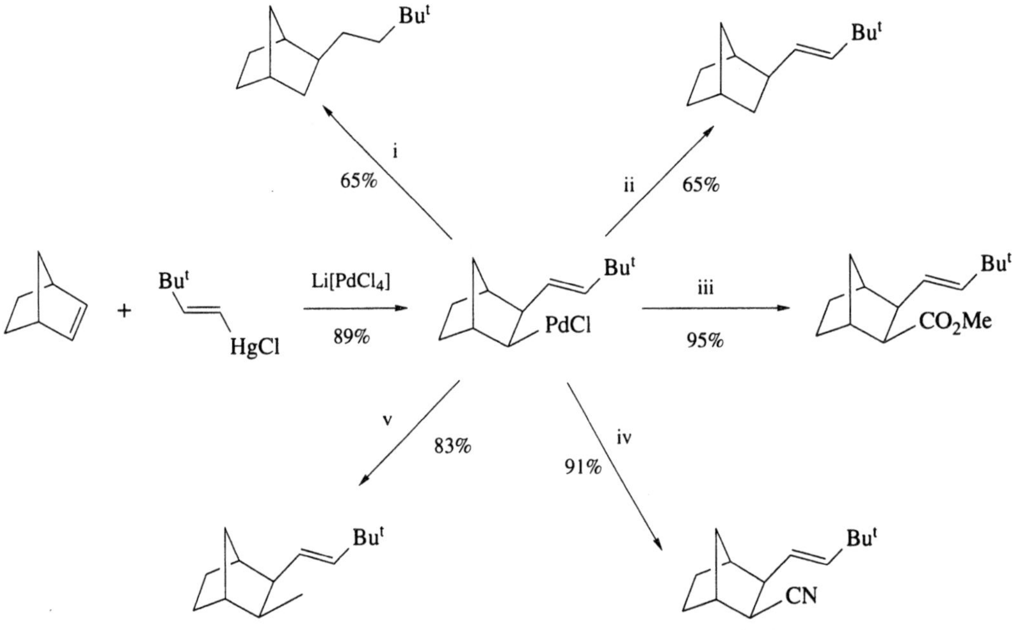

+ Bu^t / HgCl —Li[PdCl₄], 89%→ Bu^t / PdCl

i, 65% → Bu^t
ii, 65% → Bu^t
iii, 95% → Bu^t / CO₂Me
iv, 91% → Bu^t / CN
v, 83% → Bu^t

i, H_2; ii, NaOMe/MeOH; iii, CO/MeOH/Et_3N; iv, CuCN; v, MeLi, $2PPh_3$

Scheme 73

A functionalized mercury(II) compound like ethyl (acetoxymercurio)acetate (**136**) allows an easy approach to prostaglandin endoperoxide analogs (equation 52).[204] Several organomercury(II) compounds, RHgCl (R = Me, aryl, benzylic), are able to add to 1,3-dienes in the presence of a stoichiometric amount of a palladium(II) salt and afford π-allylpalladium compounds of type (**137**) in variable yields (equation 53).[205] A related intramolecular carbomercuration has been reported by Snider.[206] It allows a stereospecific approach to the chloromercury compound (**138**; equation 54). Similar palladium-mediated reactions

$AcOHgCH_2CO_2Et$ + —i, 40%→ O / OEt / O (52)

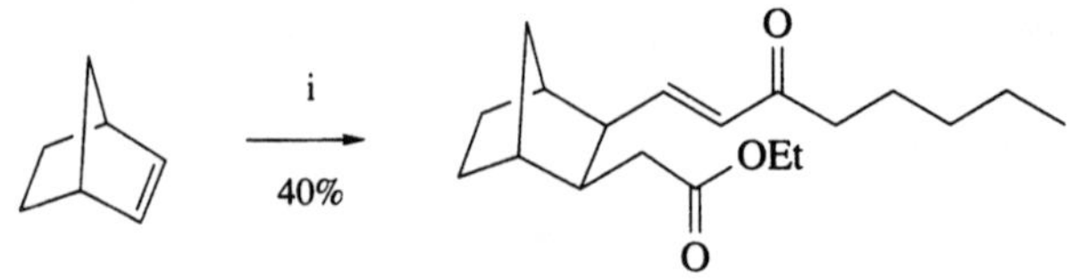

(**136**)

i, $Li_2[PdCl_4]$ (1 equiv.), $H_2C{=}CHCOC_5H_{11}$ (10 equiv.), THF, 0 °C, 1 h, then 25 °C, 4 d

have been studied by Saegusa,[207] Kende[208] and coworkers. Of interest is also the carbometallation of enynes with stabilized organosilver[209] compounds RAg·2LiBr. These organometallics show a high tendency to add to the double bond of enynes (**139**) to afford, after hydrolysis, the allene (**140**), whereas the corresponding copper reagent mainly adds to the triple bond to give the 1,3-diene (**141**; Scheme 74).

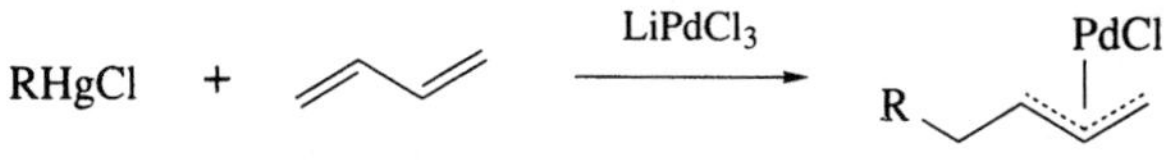

RHgCl + [diene] —$LiPdCl_3$→ (**137**) (53)

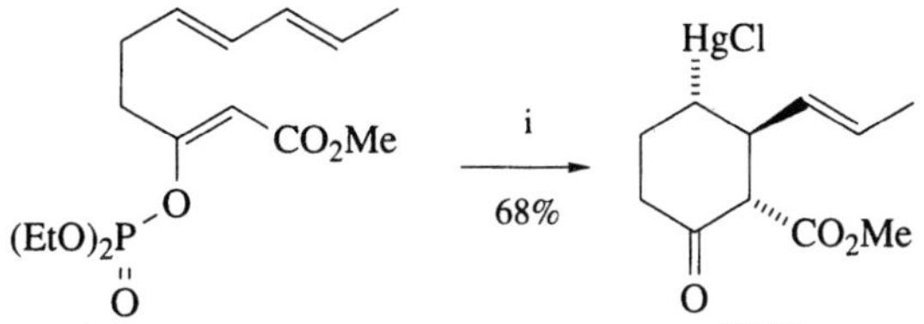

(**138**) (54)

i, $Hg(OCOCF_3)_2$, $MeNO_2$, –20 °C, NaCl

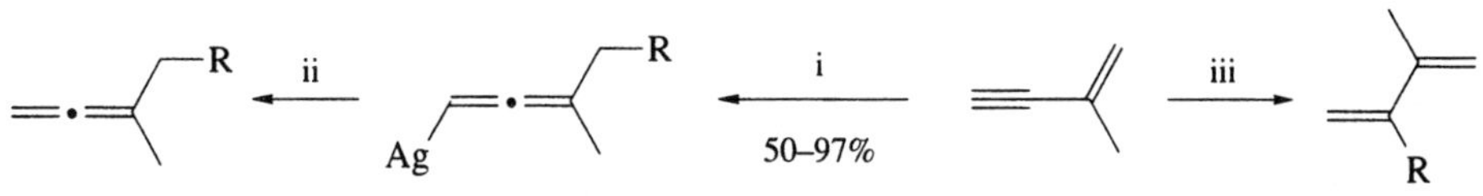

(**140**) (**139**) (**141**)

i, RAg•2LiBr, THF/HMPA, –40 to –10 °C, 3–6 h; ii, H_3O^+; iii, $R_2CuMgCl$, THF, then H_3O^+

Scheme 74

4.4.8.3 Addition to Alkynes

Drouin and Conia[210] have shown that alkynic silyl enol ethers undergo a highly stereoselective (>95% *syn* addition) carbomercuration, leading to five- or six-membered rings in high yields (72–90%). The intermediate alkenylmercury derivative can be trapped by various electrophiles (Scheme 75). Several transition metal complexes[200b,211] can effect the methylmetallation of alkynes in moderate yields, low selectivity and no special advantage compared with main group organometallics. Acylmetallation reactions, however, which are very difficult with main group organometallics, proceed far more readily with transition metal complexes and are of some synthetic utility. Hoberg and coworkers[212] showed that several symmetrical alkynes react with CO_2 in the presence of nickel(0) complexes to give five-membered cyclic nickel complexes of type (**142**) in fair to good yields, which can be further converted into a variety of organic products (Scheme 76). DeShong and coworkers[213] showed that the sequential insertion of carbon monoxide and alkynes into various alkylmanganese pentacarbonyl complexes of type (**143**)

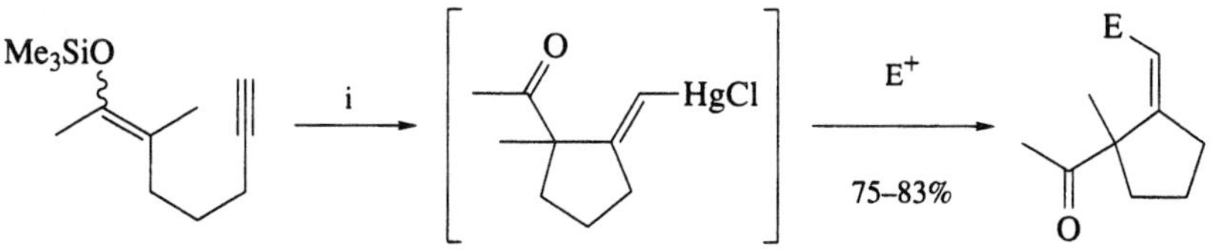

E = COMe, CO_2Me, Br

i, $HgCl_2$, CH_2Cl_2, hexamethyldisilazane, 30 °C, 0.5 h

Scheme 75

proceeds readily at high pressures (2–10 kbar), and provides the manganacycles (**144**) in good yields. These manganese complexes are valuable intermediates for the preparation of organic compounds. By demetallation under acidic conditions, (*E*)-enones are obtained. Alternatively, hydride reduction of complexes (**144**) affords the butenolides (**145**) by an intramolecular Reppe reaction (Scheme 77). The presence of an electron-withdrawing group in the alkyne makes the addition easier, but nonfunctionalized alkynes can also add with excellent regioselectivity.

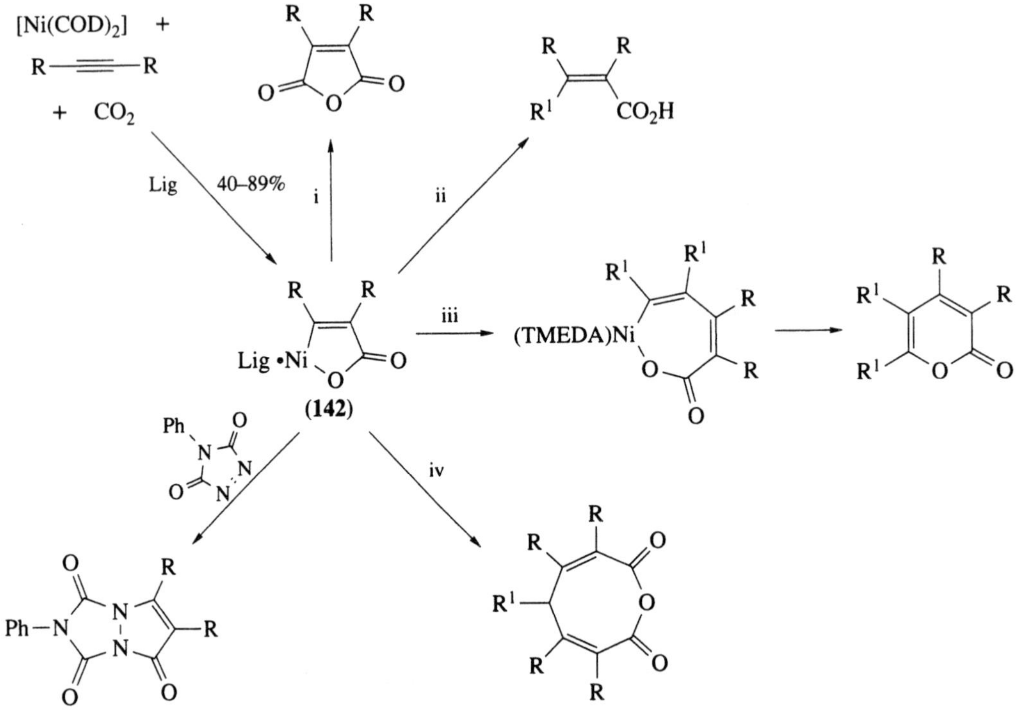

[Ni(COD)$_2$] + R—≡—R + CO_2 →(Lig, 40–89%) Lig•Ni (**142**)

R = Me (H, Ph); Lig = TMEDA, 1,2-bis(dicyclohexylphosphino)ethane or 2,2'-bipyridine

i, CO; ii, R^1X, then H_3O^+; iii, TMEDA, R^1CCR^1; iv, R^1CHX_2

Scheme 76

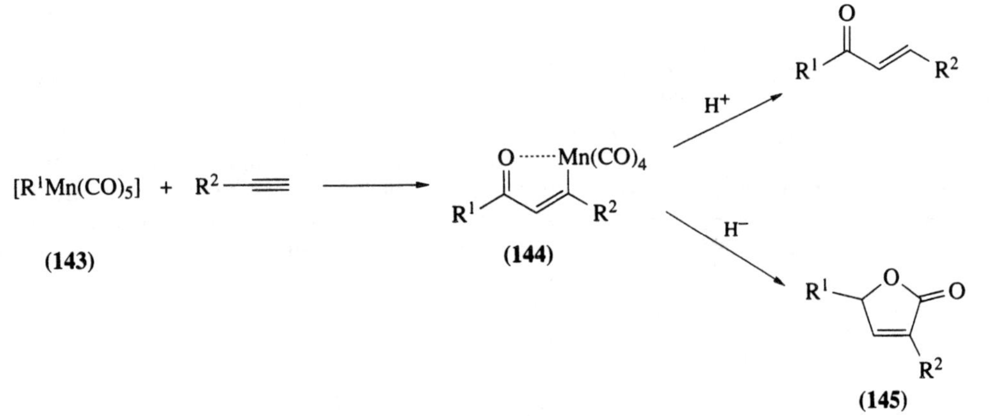

[$R^1Mn(CO)_5$] (**143**) + R^2—≡ → (**144**) →(H^+) enone; →(H^-) (**145**)

Scheme 77

4.4.9 REFERENCES

1. D. E. Van Horn and E. Negishi, *J. Am. Chem. Soc.*, 1978, **100**, 2252.
2. E. Negishi, *Pure Appl. Chem.*, 1981, **53**, 2333.
3. J. F. Normant and A. Alexakis, *Synthesis*, 1981, 841.
4. K. Ziegler and K. Bähr, *Chem. Ber.*, 1928, **61**, 253.
5. E. Negishi and T. Takahashi, *Synthesis*, 1988, 1.
6. B. J. Wakefield, 'The Chemistry of Organolithium Compounds', Pergamon Press, New York, 1974.
7. (a) J. L. Wardell and E. S. Paterson, in 'The Chemistry of the Metal–Carbon Bond', ed. F. R. Hartley and S. Patai, Wiley, New York, 1985, vol. 2, p. 219; (b) J. L. Wardell, in 'Inorganic Reactions and Methods', ed. J. J. Zuckerman, VCH, Weinheim, 1988, vol. 11, p. 129.
8. K. Ziegler and H. G. Gellert, *Justus Liebigs Ann. Chem.*, 1950, **567**, 195.
9. (a) D. Seebach, *Angew. Chem.*, 1988, **100**, 1685; (b) D. Seebach, *Angew. Chem., Int. Ed. Engl.*, 1988, **27**, 1624.
10. M. Morton, 'Anionic Polymerization: Principles and Practice', Academic Press, New York, 1983.
11. P. D. Bartlett, S. J. Tauber and W. P. Weber, *J. Am. Chem. Soc.*, 1969, **91**, 6362.
12. (a) H. Lehmkuhl, O. Olbrysch, D. Reinehr, G. Schomburg and D. Henneberg, *Justus Liebigs Ann. Chem.*, 1975, 145; (b) A. Maercker and J. Troesch, *J. Organomet. Chem.*, 1975, **102**, C1.
13. K. Ziegler, F. Dersch and H. Wollthan, *Justus Liebigs Ann. Chem.*, 1934, **511**, 13.
14. R. Waack and M. A. Doran, *J. Org. Chem.*, 1967, **32**, 3395.
15. L. Miginiac, *J. Organomet. Chem.*, 1982, **238**, 235.
16. (a) W. H. Glaze, J. E. Hanicak, M. L. Moore and J. Chandhuri, *J. Organomet. Chem.*, 1972, **44**, 39; (b) W. H. Glaze, J. E. Hanicak, D. J. Berry and D. P. Duncan, *J. Organomet. Chem.*, 1972, **44**, 49.
17. (a) H. G. Richey, Jr., C. W. Wilkins, Jr. and R. M. Bension, *J. Org. Chem.*, 1980, **45**, 5042; (b) J. A. Marshall and H. Faubl, *J. Am. Chem. Soc.*, 1970, **92**, 948; (c) J. E. Mulvaney and Z. G. Gardlund, *J. Org. Chem.*, 1965, **30**, 917; (d) R. Caple, G. M. S. Chen and J. D. Nelson, *J. Org. Chem.*, 1971, **36**, 2874.
18. S. Wawzonek, B. J. Studnicka and A. R. Zigman, *J. Org. Chem.*, 1969, **34**, 1316.
19. R. D. Bach, K. W. Bair and C. L. Willis, *J. Organomet. Chem.*, 1974, **77**, 31.
20. (a) G. Maier and F. Köhler, *Angew. Chem.*, 1979, **91**, 327; (b) G. Maier and F. Köhler, *Angew. Chem., Int. Ed. Engl.*, 1979, **18**, 308.
21. G. Wittig and J. Otten, *Tetrahedron Lett.*, 1963, 601.
22. (a) H. D. Verkruijsse and L. Brandsma, *Recl. Trav. Chim. Pays-Bas*, 1986, **105**, 66; (b) M. Stähle, R. Lehmann, J. Kramar and M. Schlosser, *Chimia*, 1985, **39**, 229; (c) E. Moret, P. Schneider, C. Margot, M. Stähle and M. Schlosser, *Chimia*, 1985, **39**, 231.
23. G. W. Klumpp, A. H. Veefkind, W. L. de Graaf and F. Bickelhaupt, *Justus Liebigs Ann. Chem.*, 1967, **706**, 47.
24. A. H. Veefkind, F. Bickelhaupt and G. W. Klumpp, *Recl. Trav. Chim. Pays-Bas*, 1969, **88**, 1058.
25. G. W. Klumpp, *Recl. Trav. Chim. Pays-Bas*, 1986, **105**, 1.
26. A. H. Veefkind, J. v. d. Schaaf, F. Bickelhaupt and G. W. Klumpp, *J. Chem. Soc., Chem. Commun.*, 1971, 722.
27. (a) J. K. Crandall and A. C. Clark, *Tetrahedron Lett.*, 1969, 325; (b) J. K. Crandall and A. C. Clark, *J. Org. Chem.*, 1972, **37**, 4236; (c) J. K. Crandall and A. C. Rojas, *Org. Synth., Coll. Vol.*, 1988, **6**, 786; (d) H. Felkin, G. Swierczewski and A. Tambute, *Tetrahedron Lett.*, 1969, 707; (e) D. R. Dimmel and S. Huang, *J. Org. Chem.*, 1973, **38**, 2756.
28. B. M. Trost, D. M. T. Chan and T. N. Nanninga, *Org. Synth.*, 1984, **62**, 58.
29. H. G. Richey, Jr., A. S. Heyn and W. F. Erickson, *J. Org. Chem.*, 1983, **48**, 3821.
30. (a) G. Courtois, B. Mauze and L. Miginiac, *J. Organomet. Chem.*, 1974, **72**, 309; (b) B. Mauze, *J. Organomet. Chem.*, 1977, **131**, 321; (c) B. Mauze, *J. Organomet. Chem.*, 1977, **134**, 1; (d) D. Mesnard and L. Miginiac, *J. Organomet. Chem.*, 1976, **117**, 99; (e) G. Courtois and L. Miginiac, *J. Organomet. Chem.*, 1976, **117**, 201.
31. R. L. Eppley and J. A. Dixon, *J. Am. Chem. Soc.*, 1968, **90**, 1606 and refs. therein.
32. W. F. Bailey and J. J. Patricia, *J. Organomet. Chem.*, 1988, **352**, 1.
33. W. F. Bailey, J. J. Patricia, V. C. Del Gobbo, R. M. Jarret and P. J. Okarma, *J. Org. Chem.*, 1985, **50**, 1999.
34. W. F. Bailey, T. T. Nurmi, J. J. Patricia and W. Wang, *J. Am. Chem. Soc.*, 1987, **109**, 2442 and refs. therein.
35. (a) A. Maercker and J. D. Roberts, *J. Am. Chem. Soc.*, 1966, **88**, 1742; (b) A. Maercker and K. Weber, *Justus Liebigs Ann. Chem.*, 1972, **756**, 43.
36. (a) E. A. Hill, H. G. Richey, Jr. and T. C. Rees, *J. Org. Chem.*, 1963, **28**, 2161; (b) E. A. Hill and J. A. Davidson, *J. Am. Chem. Soc.*, 1964, **86**, 4663.
37. G. A. Ross, M. D. Koppang, D. E. Bartak and N. F. Woolsey, *J. Am. Chem. Soc.*, 1985, **107**, 6742.
38. A. R. Chamberlin, S. H. Bloom, L. A. Cervini and C. H. Fotsch, *J. Am. Chem. Soc.*, 1988, **110**, 4788 and refs. therein.
39. (a) J. E. Mulvaney, Z. G. Gardlund and S. L. Gardlund, *J. Am. Chem. Soc.*, 1963, **85**, 3897; (b) J. E. Mulvaney and D. J. Newton, *J. Org. Chem.*, 1969, **34**, 1936.
40. W. Bauer, M. Feigel, G. Müller and P. von R. Schleyer, *J. Am. Chem. Soc.*, 1988, **110**, 6033.
41. K. N. Houk, N. G. Rondan, P. von R. Schleyer, E. Kaufmann and T. Clark, *J. Am. Chem. Soc.*, 1985, **107**, 2821.
42. (a) L.-I. Olsson and A. Claesson, *Tetrahedron Lett.*, 1974, 2161; (b) L.-I. Olsson and A. Claesson, *Acta Chem. Scand., Ser. B*, 1976, **30**, 521.
43. (a) G. Wittig, H. J. Schmidt and H. Renner, *Chem. Ber.*, 1962, **95**, 2377; (b) R. Huisgen and H. Rist, *Justus Liebigs Ann. Chem.*, 1955, **594**, 137.
44. (a) A. I. Meyers and P. D. Pansegrau, *J. Chem. Soc., Chem. Commun.*, 1985, 690; (b) A. I. Meyers and P. D. Pansegrau, *Tetrahedron Lett.*, 1983, **24**, 4935.
45. J. V. N. Vara Presad and C. N. Pillai, *J. Organomet. Chem.*, 1983, **259**, 1.
46. H. Lehmkuhl, *Bull. Soc. Chim. Fr., Part 2*, 1981, 87.

47. H. Lehmkuhl and D. Reinehr, *J. Organomet. Chem.*, 1972, **34**, 1.
48. H. Lehmkuhl, D. Reinehr, J. Brandt and G. Schroth, *J. Organomet. Chem.*, 1973, **57**, 39.
49. H. Lehmkuhl and D. Reinehr, *J. Organomet. Chem.*, 1970, **25**, C47.
50. (a) H. Lehmkuhl, D. Reinehr, G. Schomburg, D. Henneberg, H. Damen and G. Schroth, *Justus Liebigs Ann. Chem.*, 1975, 103; (b) H. Lehmkuhl and K. Mehler, *Justus Liebigs Ann. Chem.*, 1978, 1841, 1846.
51. H. Lehmkuhl, D. Reinehr, D. Henneberg, G. Schomburg and G. Schroth, *Justus Liebigs Ann. Chem.*, 1975, 119.
52. (a) H. Lehmkuhl and E. Janssen, *Justus Liebigs Ann. Chem.*, 1978, 1854; (b) H. Lehmkuhl and K. Mehler, *Liebigs Ann. Chem.*, 1982, 2244; (c) H. G. Richey, Jr. and E. K. Watkins, *J. Chem. Soc., Chem. Commun.*, 1984, 772; (d) A. M. Moiseenkov, B. A. Czeskis and A. V. Semenovsky, *J. Chem. Soc., Chem. Commun.*, 1982, 109.
53. H. Lehmkuhl, in 'Organometallics in Organic Synthesis', ed. A. de Meijere and H. tom Dieck, Springer-Verlag, Berlin, 1987, p. 185.
54. H. Lehmkuhl, W. Bergstein, D. Henneberg, E. Janssen, O. Olbrysch, D. Reinehr and G. Schomburg, *Justus Liebigs Ann. Chem.*, 1975, 1176.
55. H. Lehmkuhl, D. Reinehr, K. Mehler, G. Schomburg, H. Kötter, D. Henneberg and G. Schroth, *Justus Liebigs Ann. Chem.*, 1978, 1449.
56. (a) U. M. Dzhemilev and O. S. Vostrikova, *J. Organomet. Chem.*, 1985, **285**, 43; (b) U. M. Dzhemilev, O. S. Vostrikova and G. A. Tolstikov, *J. Organomet. Chem.*, 1986, **304**, 17 and refs. therein.
57. (a) E. A. Hill, *Adv. Organomet. Chem.*, 1977, **16**, 131; (b) E. A. Hill, *J. Organomet. Chem.*, 1975, **91**, 123.
58. (a) M. S. Silver, P. R. Shafer, J. E. Nordlander, C. Rüchardt and J. D. Roberts, *J. Am. Chem. Soc.*, 1960, **82**, 2646; (b) D. J. Patel, C. L. Hamilton and J. D. Roberts, *J. Am. Chem. Soc.*, 1965, **87**, 5144; (c) A. Maercker and R. Geuss, *Chem. Ber.*, 1973, **106**, 773.
59. For an exception, see: J. G. Duboudin and B. Jousseaume, *Synth. Commun.*, 1979, **9**, 53.
60. W. C. Kossa, Jr., T. C. Rees and H. G. Richey, Jr., *Tetrahedron Lett.*, 1971, 3455.
61. E. A. Hill, R. J. Theissen, A. Doughty and R. Miller, *J. Org. Chem.*, 1969, **34**, 3681.
62. B. Giese, 'Radicals in Organic Synthesis: Formation of Carbon–Carbon Bonds', Pergamon Press, Oxford, 1986.
63. H. Felkin, J. D. Umpleby, E. Hagaman and E. Wenkert, *Tetrahedron Lett.*, 1972, 2285.
64. W. Oppolzer, 'Selectivity—A Goal for Synthetic Efficiency', ed. W. Bartmann and B. M. Trost, Verlag Chemie, Weinheim, 1984, vol. 14, p. 137.
65. (a) J. J. Eisch and G. R. Husk, *J. Am. Chem. Soc.*, 1965, **87**, 4194; (b) J. J. Eisch and J. H. Merkley, *J. Am. Chem. Soc.*, 1979, **101**, 1148; (c) J. J. Eisch, J. H. Merkley and J. E. Galle, *J. Org. Chem.*, 1979, **44**, 587.
66. (a) M. Cherest, H. Felkin, C. Frajerman, C. Lion, G. Roussi and G. Swierczewski, *Tetrahedron Lett.*, 1966, 875; (b) H. Felkin and C. Kaeseberg, *Tetrahedron Lett.*, 1970, 4587.
67. (a) H. G. Richey, Jr. and C. W. Wilkins, Jr., *J. Org. Chem.*, 1980, **45**, 5027; (b) H. G. Richey, Jr. and R. M. Bension, *J. Org. Chem.*, 1980, **45**, 5036; (c) H. G. Richey, Jr., L. M. Moses, M. S. Domalski, W. F. Erickson and A. S. Heyn, *J. Org. Chem.*, 1981, **46**, 3773; (d) H. G. Richey, Jr. and M. S. Domalski, *J. Org. Chem.*, 1981, **46**, 3780.
68. R. Lazzaroni, D. Pini, S. Bertozzi and G. Fatti, *J. Org. Chem.*, 1986, **51**, 505.
69. (a) J. van der Louw, J. L. van der Baan, H. Stieltjes, F. Bickelhaupt and G. W. Klumpp, *Tetrahedron Lett.*, 1987, **28**, 5929; (b) J. v. d. Louw, J. L. van der Baan, H. Stichter, G. J. J. Out, F. Bickelhaupt and G. W. Klumpp, *Tetrahedron Lett.*, 1988, **29**, 3579.
70. J. R. C. Light and H. H. Zeiss, *J. Organomet. Chem.*, 1970, **21**, 517.
71. M. Michman and M. Balog, *J. Organomet. Chem.*, 1971, **31**, 395.
72. J. G. Duboudin and B. Jousseaume, *J. Organomet. Chem.*, 1972, **44**, C1.
73. A. Alexakis, G. Cahiez and J. F. Normant, *J. Organomet. Chem.*, 1979, **177**, 293.
74. H. G. Richey, Jr. and A. M. Rothman, *Tetrahedron Lett.*, 1968, 1457.
75. J. K. Crandall, P. Battioni, J. T. Wehlacz and R. Bindra, *J. Am. Chem. Soc.*, 1975, **97**, 7171.
76. (a) R. W. Hoffmann, 'Dehydrobenzene and Cycloalkynes', Academic Press, New York, 1967, chap. 2; (b) H. Hart, K. Harada and C.-J. F. Du, *J. Org. Chem.*, 1985, **50**, 3104; (c) K. Harada, H. Hart and C. J. F. Du, *J. Org. Chem.*, 1985, **50**, 5524; (d) C.-J. F. Du, H. Hart and K.-K. D. Ng, *J. Org. Chem.*, 1986, **51**, 3162; (e) C. J. F. Du and H. Hart, *J. Org. Chem.*, 1987, **52**, 4311; (f) H. Hart and T. Ghosh, *Tetrahedron Lett.*, 1988, **29**, 881; (g) T. K. Vinod and H. Hart, *Tetrahedron Lett.*, 1988, **29**, 885; (h) J. G. Duboudin, B. Jousseaume and M. Pinet-Vallier, *J. Organomet. Chem.*, 1979, **172**, 1.
77. (a) J. J. Eisch and J. H. Merkley, *J. Organomet. Chem.*, 1969, **20**, P27; (b) J. J. Eisch, J. H. Merkley and J. E. Galle, *J. Org. Chem.*, 1979, **44**, 587; (c) H. G. Richey, Jr. and F. W. von Rein, *J. Organomet. Chem.*, 1969, **20**, P32; (d) H. G. Richey, Jr. and F. W. von Rein, *Tetrahedron Lett.*, 1971, 3777; (e) R. B. Miller and T. Reichenbach, *Synth. Commun.*, 1976, **6**, 319.
78. (a) J. G. Duboudin and B. Jousseaume, *J. Organomet. Chem.*, 1975, **91**, C1; 1979, **168**, 1.
79. (a) J. G. Duboudin, B. Jousseaume and A. Bonakdar, *J. Organomet. Chem.*, 1979, **168**, 227; (b) J. G. Duboudin and B. Jousseaume, *J. Organomet. Chem.*, 1979, **168**, 233.
80. (a) R. Mornet and L. Gouin, *Bull. Soc. Chim. Fr.*, 1977, 737; (b) H. G. Richey, Jr., W. F. Erickson and A. S. Heyn, *Tetrahedron Lett.*, 1971, 2183; (c) C. Nivert, B. Mauze and L. Miginiac, *J. Organomet. Chem.*, 1972, **44**, 69.
81. (a) R. Mornet and L. Gouin, *J. Organomet. Chem.*, 1975, **86**, 57, 297; (b) G. Bouet, R. Mornet and L. Gouin, *J. Organomet. Chem.*, 1977, **135**, 151.
82. (a) R. Mornet and L. Gouin, *Tetrahedron Lett.*, 1977, 167;
83. (a) Y. Ishino, K. Wakamoto and T. Hirashima, *Chem. Lett.*, 1984, 765; (b) R. Whitby, C. Yeates, P. Kocienski and G. Costello, *J. Chem. Soc., Chem. Commun.*, 1987, 429.
84. (a) B. B. Snider, M. Karras and R. S. E. Conn, *J. Am. Chem. Soc.*, 1978, **100**, 4624; (b) B. B. Snider, R. S. E. Conn and M. Karras, *Tetrahedron Lett.*, 1979, 1679; (c) R. S. E. Conn, M. Karras and B. B. Snider, *Isr. J. Chem.*, 1984, **24**, 108.

85. (a) S. Fujikura, M. Inoue, K. Utimoto and H. Nozaki, *Tetrahedron Lett.*, 1984, **25**, 1999; (b) K. Utimoto, K. Imi, H. Shiragami, S. Fujikura and H. Nozaki, *Tetrahedron Lett.*, 1985, **26**, 2101.
86. L. Miginiac, in 'The Chemistry of the Metal–Carbon Bond', ed. F. R. Hartley and S. Patai, Wiley, New York, 1985, vol. 3, p. 99.
87. H. Lehmkuhl and O. Olbrysch, *Justus Liebigs Ann. Chem.*, 1975, 1162.
88. H. Lehmkuhl, I. Döring and H. Nehl, *J. Organomet. Chem.*, 1981, **221**, 123.
89. (a) H. Lehmkuhl and R. McLane, *Liebigs Ann. Chem.*, 1980, 736; (b) H. Lehmkuhl and H. Nehl, *J. Organomet. Chem.*, 1981, **221**, 131.
90. (a) G. Courtois and L. Miginiac, *J. Organomet. Chem.*, 1973, **52**, 241; (b) H. Lehmkuhl, I. Döring, R. McLane and H. Nehl, *J. Organomet. Chem.*, 1981, **221**, 1.
91. (a) M. Gaudemar, *C. R. Hebd. Seances Acad. Sci., Ser. C*, 1971, **273**, 1669; (b) Y. Frangin and M. Gaudemar, *C. R. Hebd. Seances Acad. Sci., Ser. C*, 1974, **278**, 885; (c) M. Bellasoued, Y. Frangin and M. Gaudemar, *Synthesis*, 1977, 205.
92. (a) P. Knochel and J. F. Normant, *Tetrahedron Lett.*, 1986, **27**, 1035, 1043, 4427, 4431, 5727; (b) P. Knochel, C. Xiao and M. C. P. Yeh, *Organometallics*, 1989, **8**, 2831.
93. (a) P. Knochel, C. Xiao and M. C. P. Yeh, *Tetrahedron Lett.*, 1988, **29**, 6697; (b) P. Knochel and M. C. P. Yeh, unpublished results.
94. H. G. Chen and P. Knochel, *Tetrahedron Lett.*, 1988, **29**, 6701.
95. A. Yanagisawa, S. Habaue and H. Yamamoto, *J. Am. Chem. Soc.*, 1989, **111**, 366.
96. (a) G. Courtois and L. Miginiac, *J. Organomet. Chem.*, 1974, **69**, 1 and refs. therein; (b) F. Bernardou, B. Mauze and L. Miginiac, *C. R. Hebd. Seances Acad. Sci., Ser. C*, 1973, **276**, 1645; (c) F. Bernardou and L. Miginiac, *C. R. Hebd. Seances Acad. Sci., Ser. C*, 1975, **280**, 1473.
97. F. Bernardou and L. Miginiac, *Tetrahedron Lett.*, 1976, 3083 and refs. therein.
98. G. Courtois, A. Masson and L. Miginiac, *C. R. Hebd. Seances Acad. Sci., Ser. C*, 1978, **286**, 265.
99. (a) J. Auger, G. Courtois and L. Miginiac, *J. Organomet. Chem.*, 1977, **133**, 285; (b) G. Courtois and L. Miginiac, *C. R. Hebd. Seances Acad. Sci., Ser. C*, 1977, **285**, 207.
100. (a) K. E. Schülte, G. Rücker and J. Feldkamp, *Chem. Ber.*, 1972, **105**, 24; (b) H. T. Bertrand, G. Courtois and L. Miginiac, *Tetrahedron Lett.*, 1974, 1945 and 1975, 3147.
101. M. Bellasoued, Y. Frangin and M. Gaudemar, *J. Organomet. Chem.*, 1979, **166**, 1.
102. E. Negishi, D. E. Van Horn, T. Yoshida and C. L. Rand, *Organometallics*, 1983, **2**, 563.
103. E. Negishi, H. Sawada, J. M. Tour and Y. Wei, *J. Org. Chem.*, 1988, **53**, 913.
104. E. Negishi, *Acc. Chem. Res.*, 1987, **20**, 65.
105. (a) Y. Frangin and M. Gaudemar, *J. Organomet. Chem.*, 1977, **142**, 9; (b) F. Bernardou and L. Miginiac, *J. Organomet. Chem.*, 1977, **125**, 23.
106. M. Bellasoued and Y. Frangin, *Synthesis*, 1978, 838.
107. (a) G. A. Molander, *J. Org. Chem.*, 1983, **48**, 5409; (b) E. Negishi and J. A. Miller, *J. Am. Chem. Soc.*, 1983, **105**, 6761.
108. J. van der Louw, J. L. van der Baan, F. Bickelhaupt and G. W. Klumpp, *Tetrahedron Lett.*, 1987, **28**, 2889.
109. R. Köster, *Justus Liebigs Ann. Chem.*, 1958, **618**, 31.
110. R. Köster, *Methoden Org. Chem. (Houben–Weyl), 4th Ed.*, 1982, **13**, 44.
111. (a) N. Miyamoto, S. Isiyama, K. Utimoto and H. Nozaki, *Tetrahedron Lett.*, 1971, 4597; (b) N. Miyamoto, S. Isiyama, K. Utimoto and H. Nozaki, *Tetrahedron*, 1973, **29**, 2365.
112. (a) B. M. Mikhailov, *Organomet. Chem. Rev. Sect. A*, 1972, **8**, 1; (b) B. M. Mikhailov, *Pure Appl. Chem.*, 1974, **39**, 505; (c) J. J. Eisch, *Adv. Organomet. Chem.*, 1977, **16**, 67.
113. (a) B. M. Mikhailov and Y. N. Bubnov, *Tetrahedron Lett.*, 1971, 2127; (b) B. M. Mikhailov and Y. N. Bubnov, *Zh. Obshch. Khim.*, 1971, **41**, 2039.
114. Y. N. Bubnov, O. A. Nesmeyanova, T. Y. Budashevskaya, B. M. Mikhailov and B. A. Kazansky, *Tetrahedron Lett.*, 1971, 2153.
115. (a) B. M. Mikhailov and V. N. Smirnow, *Bull. Acad. Sci. USSR, Div. Chem. Sci.*, 1974, **23**, 1079; (b) B. M. Mikhailov, Y. N. Bubnov, M. S. Grigoryan and V. S. Bobdanov, *J. Gen. Chem. USSR (Engl. Transl.)*, 1974, **44**, 2669; (c) Y. N. Bubnov and M. Y. Etinger, *Tetrahedron Lett.*, 1985, **26**, 2797; (d) Y. N. Bubnov and L. I. Lavrinovich, *Tetrahedron Lett.*, 1985, **26**, 4551.
116. A. J. Hubert, *J. Chem. Soc.*, 1965, 6669.
117. K. Nozaki, K. Wakamatsu, T. Nonaka, W. Tückmantel, K. Oshima and K. Utimoto, *Tetrahedron Lett.*, 1986, **27**, 2007.
118. Y. Satoh, H. Serizawa, N. Miyaura, S. Hara and A. Suzuki, *Tetrahedron Lett.*, 1988, **29**, 1811.
119. B. Wrackmeyer and H. Nöth, *J. Organomet. Chem.*, 1976, **108**, C21.
120. T. Mole and E. A. Jeffery, 'Organoaluminum Compounds', Elsevier, Amsterdam, 1972.
121. K. Ziegler, *Adv. Organomet. Chem.*, 1968, **6**, 1.
122. H. Lehmkuhl, K. Ziegler and H. G. Gellert, *Methoden Org. Chem. (Houben–Weyl), 4th Ed.*, 1970, **13**, 184.
123. J. R. Zietz, Jr., G. C. Robinson and K. L. Lindsay, in 'Comprehensive Organometallic Chemistry', ed. G. Wilkinson, Pergamon Press, Oxford, 1982, vol. 7, p. 365.
124. H. Lehmkuhl, *Justus Liebigs Ann. Chem.*, 1969, **719**, 40.
125. (a) J. J. Eisch, N. E. Burlinson and M. Boleslawski, *J. Organomet. Chem.*, 1976, **111**, 137; (b) J. J. Eisch and N. E. Burlinson, *J. Am. Chem. Soc.*, 1976, **98**, 753.
126. R. Schimpf and P. Heimbach, *Chem. Ber.*, 1970, **103**, 2122.
127. J. J. Eisch and G. R. Husk, *J. Org. Chem.*, 1966, **31**, 3419.
128. P. W. Chum and S. E. Wilson, *Tetrahedron Lett.*, 1976, 1257.
129. R. Rienäcker and D. Schwengers, *Justus Liebigs Ann. Chem.*, 1977, 1633 and refs. therein.
130. (a) G. Zweifel, G. M. Clark and R. Lynd, *J. Chem. Soc., Chem. Commun.*, 1971, 1593; (b) M. J. Smith and S. E. Wilson, *Tetrahedron Lett.*, 1982, **23**, 5013.
131. A. Stefani, *Helv. Chim. Acta*, 1974, **57**, 1346.
132. G. Wilke and H. Müller, *Justus Liebigs Ann. Chem.*, 1960, **629**, 222.

133. H. Lehmkuhl, K. Ziegler and H. G. Gellert, *Methoden Org. Chem. (Houben–Weyl), 4th Ed.*, 1970, **13**, 141.
134. J. J. Eisch and K. C. Fichter, *J. Am. Chem. Soc.*, 1974, **96**, 6815.
135. D. E. Van Horn, L. F. Valente, M. J. Idacavage and E. Negishi, *J. Organomet. Chem.*, 1978, **156**, C20.
136. (a) E. Negishi, D. E. Van Horn and T. Yoshida, *J. Am. Chem. Soc.*, 1985, **107**, 6639; (b) E. Negishi, *Pure Appl. Chem.*, 1981, **53**, 2333; (c) E. Negishi and T. Takahashi, *Aldrichimica Acta*, 1985, **18**, 31.
137. (a) G. Wilke and H. Müller, *Chem. Ber.*, 1956, **89**, 444; (b) G. Wilke and H. Müller, *Justus Liebigs Ann. Chem.*, 1958, **618**, 267; (c) G. Zweifel, J. T. Snow and C. C. Whitney, *J. Am. Chem. Soc.*, 1968, **90**, 7139; (d) J. J. Eisch and W. C. Kaska, *J. Am. Chem. Soc.*, 1966, **88**, 2213, 2976.
138. T. Yoshida and E. Negishi, *J. Am. Chem. Soc.*, 1981, **103**, 1276.
139. J. A. Miller and E. Negishi, *Tetrahedron Lett.*, 1984, **25**, 5863.
140. C. L. Rand, D. E. Van Horn, M. W. Moore and E. Negishi, *J. Org. Chem.*, 1981, **46**, 4093.
141. E. Negishi, D. E. Van Horn, A. O. King and N. Okukado, *Synthesis*, 1979, 501.
142. (a) L. C. Smedley, H. E. Tweedy, R. A. Coleman and D. W. Thompson, *J. Org. Chem.*, 1977, **42**, 4147; (b) D. C. Brown, S. A. Nichols, A. B. Gilpin and D. W. Thompson, *J. Org. Chem.*, 1979, **44**, 3457; (c) M. D. Schiavelli, J. J. Plunkett and D. W. Thompson, *J. Org. Chem.*, 1981, **46**, 807.
143. H. Hayami, K. Oshima and H. Nozaki, *Tetrahedron Lett.*, 1984, **25**, 4433.
144. (a) J. J. Eisch, R. J. Manfre and D. A. Komar, *J. Organomet. Chem.*, 1978, **159**, C13; (b) B. B. Snider and M. Karras, *J. Organomet. Chem.*, 1979, **179**, C37; (c) J. J. Eisch, A. M. Piotrowski, S. K. Brownstein, E. J. Gabe and F. L. Lee, *J. Am. Chem. Soc.*, 1985, **107**, 7219.
145. (a) E. Negishi, L. D. Boardman, J. M. Tour, H. Sawada and C. L. Rand, *J. Am. Chem. Soc.*, 1983, **105**, 6344; (b) L. D. Boardman, V. Bagheri, H. Sawada and E. Negishi, *J. Am. Chem. Soc.*, 1984, **106**, 6105; (c) E. Negishi, L. D. Boardman, H. Sawada, V. Bagheri, A. T. Stoll, J. M. Tour and C. L. Rand, *J. Am. Chem. Soc.*, 1983, **110**, 5383.
146. J. A. Miller and E. Negishi, *Isr. J. Chem.*, 1984, **24**, 76.
147. E. Negishi and T. Takahashi, *J. Am. Chem. Soc.*, 1986, **108**, 3402.
148. G. Zweifel and J. A. Miller, *Org. React.*, 1984, **32**, 375.
149. N. Okukado and E. Negishi, *Tetrahedron Lett.*, 1978, **19**, 2357.
150. E. Negishi, L. F. Valente and M. Kobayashi, *J. Am. Chem. Soc.*, 1980, **105**, 3298.
151. H. Kobayashi, L. F. Valente, E. Negishi, W. Patterson and A. Silveira, Jr., *Synthesis*, 1980, 1034.
152. H. Matsushita and E. Negishi, *J. Am. Chem. Soc.*, 1981, **103**, 2882.
153. A. O. King, Ph.D. Thesis, Syracuse University, 1975.
154. (a) S. Fung and J. B. Siddall, *J. Am. Chem. Soc.*, 1980, **102**, 6580; (b) K. Mori, M. Sakakibara and K. Okada, *Tetrahedron*, 1984, **40**, 1767.
155. D. R. Williams, B. A. Barner, K. Nishitani and J. G. Phillips, *J. Am. Chem. Soc.*, 1982, **104**, 4708.
156. (a) W. R. Roush and A. P. Spada, *Tetrahedron Lett.*, 1983, **24**, 3693; (b) W. R. Roush and T. A. Blizzard, *J. Org. Chem.*, 1983, **48**, 758; 1984, **49**, 1772, 4332.
157. R. C. Cookson and N. J. Liverton, *J. Chem. Soc., Perkin Trans. 1*, 1985, 1589.
158. J. F. Normant, *Pure Appl. Chem.*, 1978, **50**, 709.
159. J. F. Normant and A. Alexakis, *Mod. Synth. Methods*, 1983, **3**, p. 139.
160. J. P. Collman, L. S. Hegedus, J. R. Norton and R. G. Finke, 'Principles and Applications of Organotransition Metal Chemistry', University Science Books, Mill Valley, CA, 1987.
161. A. T. Stoll and E. Negishi, *Tetrahedron Lett.*, 1985, **26**, 5671.
162. E. Nakamura, M. Isaka and S. Matsuzawa, *J. Am. Chem. Soc.*, 1988, **110**, 1297.
163. H. Kleijn, H. Eijsurga, H. Westmijze, J. Meijer and P. Vermeer, *Tetrahedron Lett.*, 1976, 947.
164. (a) A. Alexakis and J. F. Normant, *Isr. J. Chem.*, 1984, **24**, 113; (b) A. Alexakis, P. Mangeney, A. Ghribi, I. Marek, R. Sedrani, C. Guir and J. F. Normant, *Pure Appl. Chem.*, 1988, **60**, 49.
165. P. Cuadrado, A. M. Gonzalez, F. J. Pulido and I. Fleming, *Tetrahedron Lett.*, 1988, **29**, 1825.
166. J. F. Normant, G. Cahiez and J. Villieras, *J. Organomet. Chem.*, 1975, **92**, C28.
167. J. F. Normant and M. Bourgain, *Tetrahedron Lett.*, 1971, 2583.
168. A. Alexakis, G. Cahiez and J. F. Normant, *Org. Synth.*, 1984, **62**, 1.
169. M. Furber, R. J. K. Taylor and S. C. Burford, *Tetrahedron Lett.*, 1985, **26**, 3285.
170. A. Alexakis and J. F. Normant, *Tetrahedron Lett.*, 1982, **23**, 5151.
171. (a) H. Westmijze, J. Meijer, T. J. T. Bos and P. Vermeer, *Recl. Trav. Chim. Pays-Bas*, 1976, **95**, 299, 304; (b) H. Westmijze, H. Kleijn and P. Vermeer, *Tetrahedron Lett.*, 1977, 2023.
172. (a) J. F. Normant, G. Cahiez, M. Bourgain, C. Chuit and J. Villieras, *Bull. Soc. Chim. Fr.*, 1974, 1656; (b) E. C. Ashby, R. S. Smith and A. B. Goel, *J. Org. Chem.*, 1981, **46**, 5133.
173. S. Achyutha Rao and M. Periasamy, *Tetrahedron Lett.*, 1988, **29**, 4313.
174. (a) R. S. Iyer and P. Helquist, *Org. Synth.*, 1986, **64**, 1; (b) A. Marfat, P. R. McGuirk and P. Helquist, *J. Org. Chem.*, 1979, **44**, 1345, 3888.
175. J. K. Crandall and F. Collonges, *J. Org. Chem.*, 1976, **41**, 4089.
176. E. D. Sternberg and K. P. C. Vollhardt, *J. Org. Chem.*, 1984, **49**, 1574.
177. (a) A. Alexakis, G. Cahiez and J. F. Normant, *Synthesis*, 1979, 826; (b) M. Gardette, A. Alexakis and J. F. Normant, *Tetrahedron Lett.*, 1982, **23**, 5155; (c) M. Gardette, A. Alexakis and J. F. Normant, *Tetrahedron*, 1985, **41**, 5887; (d) T. Fujisawa, T. Sato, T. Kawara and H. Tago, *Bull. Chem. Soc. Jpn.*, 1983, **56**, 345.
178. (a) J. P. Foulon, M. Bourgain-Commerçon and J. F. Normant, *Tetrahedron*, 1986, **42**, 1398, 1399; (b) H. Kleijn and P. Vermeer, *J. Org. Chem.*, 1985, **50**, 5143.
179. (a) A. Alexakis, A. Commerçon, C. Coulentianos and J. F. Normant, *Tetrahedron*, 1984, **40**, 715; (b) I. Marek, P. Mangeney, A. Alexakis and J. F. Normant, *Tetrahedron Lett.*, 1986, **27**, 5499.
180. (a) P. Vermeer, J. Meijer and C. de Graaf, *Recl. Trav. Chim. Pays-Bas*, 1974, **93**, 24; (b) J. Meijer, H. Westmijze and P. Vermeer, *Recl. Trav. Chim. Pays-Bas*, 1976, **95**, 102; (c) A. Alexakis, G. Cahiez, J. F. Normant and J. Villieras, *Bull. Soc. Chim. Fr.*, 1977, 693.
181. P. Wijkens and P. Vermeer, *J. Organomet. Chem*, 1986, **301**, 247.

182. (a) M. Obayashi, K. Utimoto and H. Nozaki, *J. Organomet. Chem.*, 1979, **177**, 145; (b) M. Obayashi, K. Utimoto and H. Nozaki, *Tetrahedron Lett.*, 1977, 1805; (c) H. Westmijze, J. Meijer and P. Vermeer, *Tetrahedron Lett.*, 1977, 1823; (d) H. Westmijze, H. Kleijn and P. Vermeer, *J. Organomet. Chem.*, 1984, **276**, 317.
183. (a) I. Fleming, T. W. Newton and F. Roessler, *J. Chem. Soc., Perkin Trans. 1*, 1981, 2527; (b) I. Fleming and T. W. Newton, *J. Chem. Soc., Perkin Trans. 1*, 1984, 1805.
184. I. Fleming and M. Taddei, *Synthesis*, 1985, 899.
185. H. Oda, Y. Morizawa, K. Oshima and H. Nozaki, *Tetrahedron Lett.*, 1984, **25**, 3217, 3221.
186. (a) E. Piers and J. M. Chong, *Can. J. Chem.*, 1988, **66**, 1425 and refs. therein; (b) I. Fleming and M. Taddei, *Synthesis*, 1985, 889; (c) J. Hibino, S. Matsubara, Y. Morizawa, K. Oshima and H. Nozaki, *Tetrahedron Lett.*, 1984, **25**, 2151; (d) S. Matsubara, J. Hibino, Y. Morizawa, K. Oshima and H. Nozaki, *J. Organomet. Chem.*, 1985, **285**, 163; (e) S. Sharma and A. C. Oehlschlager, *Tetrahedron Lett.*, 1988, **29**, 261; (f) G. Zweifel and W. Leong, *J. Am. Chem. Soc.*, 1987, **109**, 6409.
187. A. Alexakis and J. F. Normant, *Synthesis*, 1985, 72.
188. (a) N. Jabri, A. Alexakis and J. F. Normant, *Tetrahedron Lett.*, 1982, **23**, 1589; (b) N. Jabri, A. Alexakis and J. F. Normant, *Tetrahedron Lett.*, 1981, **22**, 959, 3851; (c) N. Jabri, A. Alexakis and J. F. Normant, *Bull. Soc. Chim. Fr., Part 2*, 1983, 321, 332.
189. (a) N. Jabri, A. Alexakis and J. F. Normant, *Tetrahedron*, 1986, **42**, 1369.
190. (a) J. F. Normant, A. Alexakis and G. Cahiez, *Tetrahedron Lett.*, 1980, 935; (b) T. Fujisawa, T. Sato, T. Kawara and K. Naruse, *Chem. Lett.*, 1980, 1123.
191. (a) G. Cahiez, A. Alexakis and J. F. Normant, *Synthesis*, 1978, 528; (b) A. Alexakis, G. Cahiez and J. F. Normant, *Tetrahedron Lett.*, 1978, 2027.
192. (a) B. H. Lipschutz, D. A. Parker, J. A. Kozlowski and S. L. Nguyen, *Tetrahedron Lett.*, 1984, **25**, 5959; (b) A. Alexakis, D. Jachiet and J. F. Normant, *Tetrahedron*, 1986, **42**, 5607.
193. P. Knochel and J. F. Normant, unpublished results.
194. A. Alexakis, G. Cahiez and J. F. Normant, *Tetrahedron*, 1980, **36**, 1961.
195. (a) A. Commerçon, J. F. Normant and J. Villieras, *Tetrahedron*, 1980, **36**, 1215; (b) P. J. Stang and T. Kitamura, *J. Am. Chem. Soc.*, 1987, **107**, 7561.
196. C. Germon, A. Alexakis and J. F. Normant, *Synthesis*, 1984, 43.
197. C. Germon, A. Alexakis and J. F. Normant, *Synthesis*, 1984, 40.
198. (a) E. Negishi, *Acc. Chem. Res.*, 1982, **15**, 340; (b) E. Negishi, V. Bagheri, S. Chatterjee, F.-T. Luo, J. A. Miller and A. T. Stoll, *Tetrahedron Lett.*, 1983, **24**, 5181.
199. (a) M. Gardette, N. Jabri, A. Alexakis and J. F. Normant, *Tetrahedron*, 1984, **40**, 2741 and refs. therein; (b) B. O'Conner and G. Just, *J. Org. Chem.*, 1987, **52**, 1801.
200. (a) L. S. Hegedus, *Tetrahedron*, 1984, **40**, 2415; (b) J. J. Alexander, in 'The Chemistry of the Metal–Carbon Bond', ed. F. R. Hartley and S. Patai, Wiley, New York, 1985, vol. 2, p. 339.
201. (a) R. C. Larock, 'Organomercury Compounds in Organic Synthesis', Springer-Verlag, Berlin, 1985, p. 263; (b) R. F. Heck, *Org. React.*, 1982, **27**, 345.
202. (a) H. Horino, M. Arai and N. Inoue, *Tetrahedron Lett.*, 1974, 647; (b) R. C. Larock, D. R. Leach and S. M. Bjorge, *Tetrahedron Lett.*, 1982, **23**, 715; (c) A. Kasahara, T. Izumi, K. Endo, T. Takeda and M. Ookita, *Bull. Chem. Soc. Jpn.*, 1974, **47**, 1967.
203. (a) R. C. Larock, S. S. Hershberger, K. Takagi and M. A. Mitchell, *J. Org. Chem.*, 1986, **51**, 2450 and refs. therein; (b) R. C. Larock, K. Takagi, S. S. Hershberger and M. A. Mitchell, *Tetrahedron Lett.*, 1981, **22**, 5231.
204. R. C. Larock, K. Narayanan, R. K. Carlson and J. A. Ward, *J. Org. Chem.*, 1987, **52**, 1364.
205. (a) R. F. Heck, *J. Am. Chem. Soc.*, 1968, **90**, 5542; (b) F. G. Stakem and R. F. Heck, *J. Org. Chem.*, 1980, **45**, 3584; (c) R. C. Larock and K. Takagi, *Tetrahedron Lett.*, 1983, **24**, 3457.
206. R. Cordova and B. B. Snider, *Tetrahedron Lett.*, 1984, **25**, 2945.
207. (a) Y. Ito, H. Aoyama, T. Hirao, A. Mochizuki and T. Saegusa, *J. Am. Chem. Soc.*, 1979, **101**, 494; (b) Y. Ito, H. Aoyama and T. Saegusa, *J. Am. Chem. Soc.*, 1980, **102**, 4519.
208. (a) A. S. Kende, B. Roth, P. J. Sanfilippo and T. J. Blacklock, *J. Am. Chem. Soc.*, 1982, **104**, 1784, 5808.
209. (a) H. Westmijze, H. Kleijn and P. Vermeer, *J. Organomet. Chem.*, 1979, **172**, 377; (b) H. Kleijn, M. Tigchelaar, J. Meijer and P. Vermeer, *Recl. Trav. Chim. Pays-Bas*, 1981, **100**, 337; (c) H. Kleijn, M. Tigchelaar, R. J. Bullee, C. J. Elsevier, J. Meijer and P. Vermeer, *J. Organomet. Chem.*, 1982, **240**, 329.
210. (a) M. A. Boaventura, J. Brouin and J. M. Conia, *Synthesis*, 1983, 801; (b) J. Drouin, M. A. Boaventura and J. M. Conia, *J. Am. Chem. Soc.*, 1985, **107**, 1726; (c) M. A. Boaventura, F. Theobald and J. Drouin, *Bull. Soc. Chim. Fr.*, 1987, 1006, 1015.
211. (a) W. H. Boon and M. D. Rausch, *J. Chem. Soc., Chem. Commun.*, 1977, 397; (b) M. Michman and M. Balog, *J. Organomet. Chem.*, 1971, **31**, 395.
212. (a) H. Hoberg, D. Schaefer, G. Burkhart, C. Krüger and M. J. Romao, *J. Organomet. Chem.*, 1984, **266**, 203 and refs. therein; (b) H. Hoberg and B. Apotecher, *J. Organomet. Chem.*, 1984, **270**, C15.
213. (a) P. DeShong, D. R. Sidler, P. J. Rybczynski, G. A. Slough and A. L. Rheingold, *J. Am. Chem. Soc.*, 1988, **110**, 2575 and refs. therein; (b) P. DeShong and D. R. Sidler, *J. Org. Chem.*, 1988, **53**, 4892.

4.5 Hydroformylation and Related Additions of Carbon Monoxide to Alkenes and Alkynes

JOHN K. STILLE
Colorado State University, Fort Collins, CO, USA

4.5.1 INTRODUCTION

The hydroformylation reaction — the addition of dihydrogen and carbon monoxide to an alkene or alkyne to produce an aldehyde — is one of the most important industrial chemical reactions. It is remarkable in that it converts simple, inexpensive starting materials into useful products. A new carbon–carbon bond is formed along with a very useful and reactive functionality.

The following discussion deals not only with this reaction, but related reactions in which a transition metal complex achieves the addition of carbon monoxide to an alkene or alkyne to yield carboxylic acids and their derivatives. These reactions take place either by the insertion of an alkene (or alkyne) into a metal–hydride bond (equation 1) or into a metal–carboxylate bond (equation 2) as the initial key step. Subsequent steps include carbonyl insertion reactions, metal–acyl hydrogenolysis or solvolysis and metal–carbon bond protonolysis.

Reactions of alkenes and alkynes that generate a carbon–metal bond by nucleophilic addition to a metal π-complex and subsequently undergo carbon monoxide insertion to yield a carbonyl product are

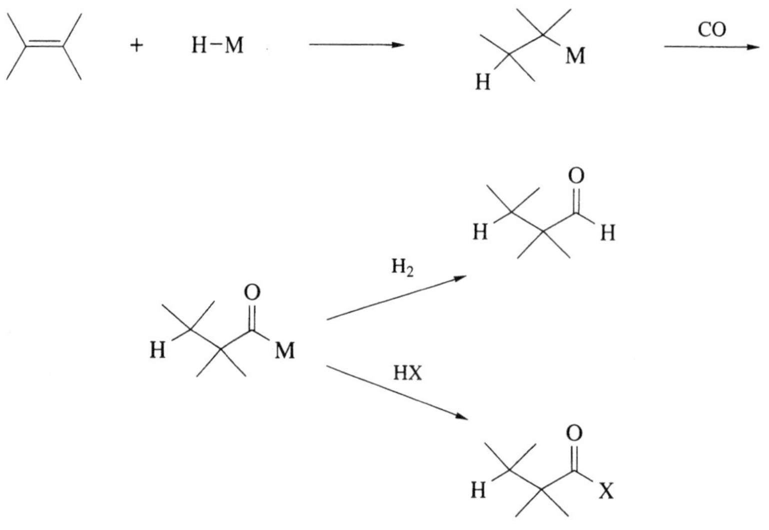

(1)

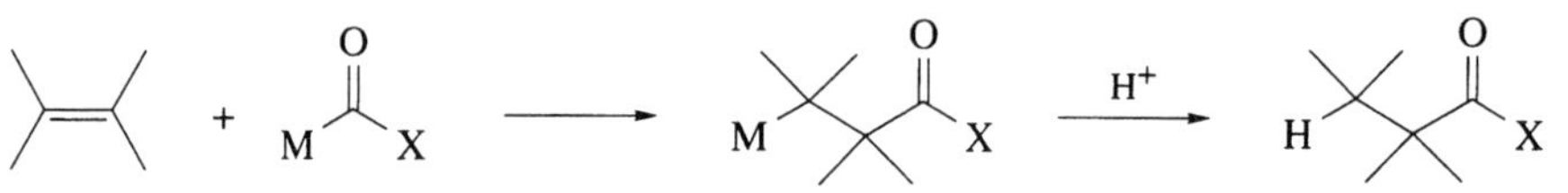

(2)

covered in Volume 4, Chapter 3.1. Reactions of carbon monoxide with alkenes and alkynes generating quinones, cyclopentanones and their derivatives that proceed *via* a metal-assisted cycloassembly (Pauson–Khand reaction) are discussed in Volume 5, Chapter 9.1.

The hydroformylation of alkenes generally has been considered to be an industrial reaction unavailable to a laboratory scale process. Usually bench chemists are neither willing nor able to carry out such a reaction, particularly at the high pressures (200 bar) necessary for the hydrocarbonylation reactions utilizing a cobalt catalyst. (Most of the previous literature reports pressures in atmospheres or pounds per square inch. All pressures in this chapter are reported in bars (SI); the relationship is 14.696 p.s.i. = 1 atm = 101 325 Pa = 1.013 25 bar.) However, hydroformylation reactions with rhodium require much lower pressures and related carbonylation reactions can be carried out at 1–10 bar. Furthermore, pressure equipment is available from a variety of suppliers and costs less than a routine IR instrument. Provided a suitable pressure room is available, even the high pressure reactions can be carried out safely and easily. The hydroformylation of cyclohexene to cyclohexanecarbaldehyde using a rhodium catalyst is an *Organic Syntheses* preparation (see Section 4.5.2.5).

Several excellent comprehensive reviews that include the hydroformylation reaction and related reactions of carbon monoxide have appeared in the past decade.[1,2] Because of the industrial importance of these reactions, much of the literature is process oriented and has focused on the carbonylation of simple alkenes derived from petroleum feedstock.

4.5.2 HYDROFORMYLATION[1–5]

The reaction of alkenes (and alkynes) with synthesis gas ($CO + H_2$) to produce aldehydes, catalyzed by a number of transition metal complexes, is most often referred to as a hydroformylation reaction or the oxo process. The discovery was made using a cobalt catalyst, and although rhodium-based catalysts have received increased attention because of their increased selectivity under mild reaction conditions, cobalt is still the most used catalyst on an industrial basis. The most industrially important hydrocarbonylation reaction is the synthesis of *n*-butanal from propene (equation 3). Some of the butanal is hydrogenated to butanol, but most is converted to 2-ethylhexanol *via* aldol and hydrogenation sequences.

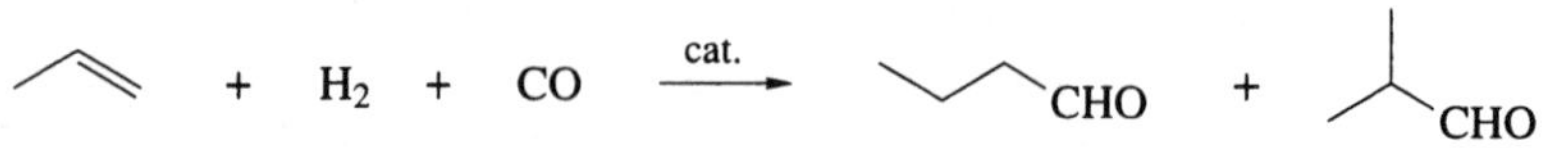

(3)

4.5.2.1 The Catalysts

All Group VIII, IX and X transition metals show some catalytic activity for hydroformylation, although cobalt and rhodium are the most active, rhodium catalysts being 10^4 times more reactive. More recently, platinum catalysts containing the trichlorostannate ligand have been shown to be selective catalysts that effect hydroformylation under mild conditions.[6]

The treatment of a cobalt(II) salt with synthesis gas generates sequentially $Co_2(CO)_8$ then $HCo(CO)_4$. This catalyst is generated only at 120–140 °C; for the carbonylation to proceed smoothly 200–300 bar is required to stabilize the catalyst. If the hydridocobalt catalyst is prepared separately and then introduced into the reaction, temperatures as low as 90 °C can be used for the hydrocarbonylation. An important consideration in industrial reactions is the normal to branched *n*/*b* ratio to give the desired straight chain aldehyde, the hydridocobalt catalyst providing an *n*/*b* ratio of ~4 in the hydroformylation of propene under the lower temperature conditions. This catalyst will stoichiometrically hydroformylate 1-alkenes under ambient conditions.

Rhodium carbonyl catalysts effect hydrocarbonylation under very mild reaction conditions at higher rates and selectivity. Rhodium introduced as $Rh_4(CO)_{12}$, for example, probably is converted to $RhH(CO)_n$, but there is no direct evidence for such species. These rhodium carbonyls are not particularly attractive since they produce mostly branched aldehydes.

Both cobalt and rhodium catalysts containing phosphine ligands have the advantages that they operate at lower pressures and are more selective. The cobalt catalyst is stabilized by phosphines, *e.g.* $HCo(CO)_3(PBu_3)$, and therefore lower CO pressures (100 bar) can be utilized. The catalyst is less active for hydroformylation than $HCo(CO)_4$ but gives an *n*/*b* ratio of 7 instead of 4 for $HCo(CO)_4$. This phosphine-stabilized catalyst is active for the hydrogenation of butanal, so instead of aldehydes, alcohols are generated under the reaction conditions. Propene can be hydroformylated under 10–20 bar at 100 °C with a rhodium catalyst, $HRh(CO)(PPh_3)$, to give predominately *n*-butanal (*n*/*b* = 9). In this case, little or no hydrogenation to butanol takes place.

Until recently, the use of platinum complexes for hydroformylation reactions was less encouraging because of the lower reaction rates and the tendency for the substrate to undergo competitive hydrogenation. However, the complex $HPt(CO)(PPh_3)_2SnCl_3$ hydroformylates alkenes under mild reaction conditions selectively without hydrogenating the alkene directly.[6,7] The function of the $SnCl_3^-$ is apparently to stabilize the ligand *trans* to it and to provide an easily dissociating ligand for coordination of an alkene.[8,9] Indeed, the use of certain phosphine ligands in place of triphenylphosphine yields catalysts that give even higher reaction rates than $HRh(CO)(PPh_3)$ and provide *n*/*b* ratios of 99.[10] Hydroformylations using $HPt(CO)(PPh_3)_2SnCl_3$ generally give high yields of aldehyde (90%) at 60–80 °C and 65–100 bar.[7]

4.5.2.2 Mechanisms

Although the overall reaction mechanisms (catalytic cycles) written for hydroformylation reactions with an unmodified cobalt catalyst (Scheme 1) and the rhodium catalyst (Scheme 2) serve as working models for the reaction, the details of many of the steps are missing and there are many aspects of the reaction that are not well understood.

The key features of both catalytic cycles are similar. Alkene coordination to the metal followed by insertion to yield an alkyl–metal complex and CO 'insertion' to yield an acyl–metal complex are common to both catalytic cycles. The oxidative addition of hydrogen followed by reductive elimination of the aldehyde regenerates the catalyst (Scheme 2 and middle section of Scheme 1). The most distinct departure in the catalytic cycle for cobalt is the alternate possibility of a dinuclear elimination occurring by the intermolecular reaction of the acylcobalt intermediate with hydridotetracarbonylcobalt to generate the aldehyde and the cobalt(0) dimer.[11,12] In the cobalt catalytic cycle, therefore, the valence charges can be from +1 to 0 or +1 to +3, while the valence charges in the rhodium cycles are from +1 to +3.

In both catalytic cycles none of the complexes exceed 18 electrons. In order to account for the effect of phosphine on the *n*/*b* ratio in rhodium catalysis (*vide infra*), an associative mechanism has been proposed in which alkene coordinates directly to the 18-electron $HRh(CO)_2L_2$ complex. Since this gives a 20-electron complex, this mechanism is not particularly attractive. A number of the intermediates in the rhodium catalytic cycle have been verified by various spectroscopic techniques.[13,14]

Much less is known concerning the platinum-catalyzed hydroformylations. However, a reasonable catalytic cycle can be constructed (Scheme 3) from the available information on the generation and reactions of many of the intermediate complexes shown.[6,8,9,15] The ability of platinum to catalyze hydroformylation reactions while palladium is not a good catalyst could be due to the ability of platinum to achieve the +4 oxidation state more readily.

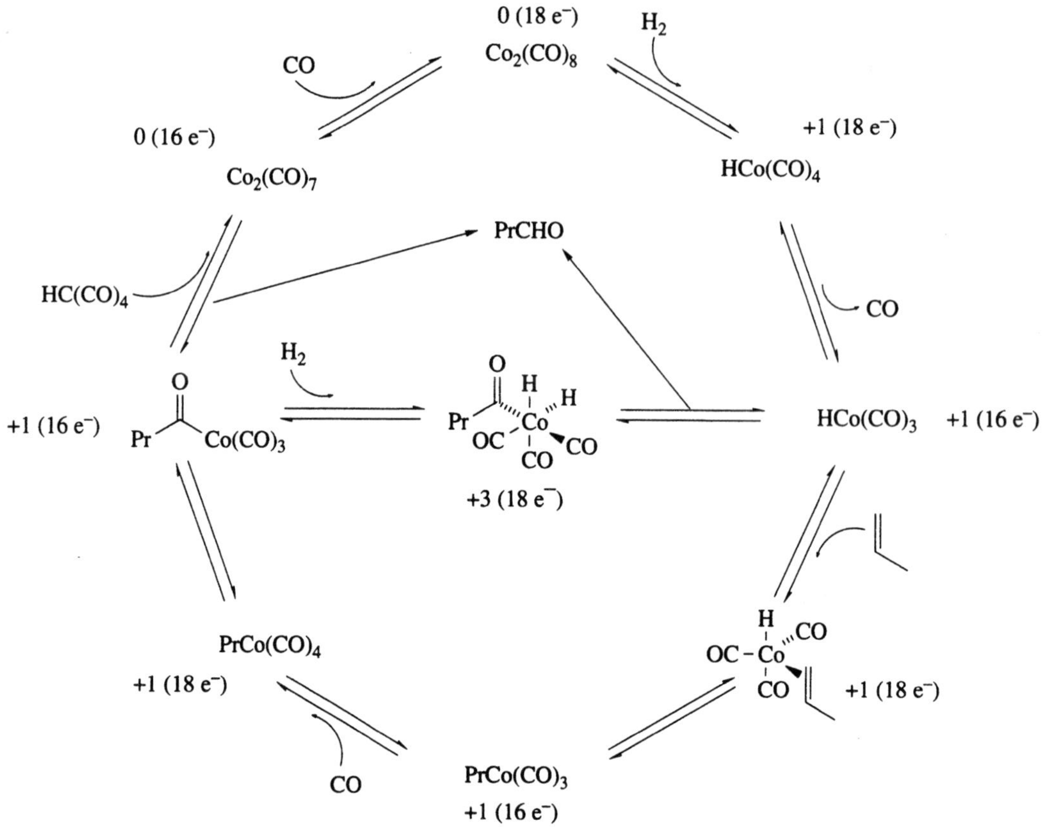

Scheme 1

In the unmodified catalyst system (Scheme 1), the rate shows a first-order dependence on hydrogen pressure and an inverse first-order dependence on carbon monoxide pressure, so that the rate is nearly independent of total pressure. The reaction is first order in alkene and first order in cobalt at higher CO pressures. With phosphine-modified cobalt catalysts, the rate-determining step depends on the ligand and the alkene.

Catalysis by phosphine-modified rhodium is even more complex and the rate-limiting steps under different conditions are uncertain. Addition of excess phosphine to the system or a high CO pressure slows the rate, a result consistent either with the rate-determining step being dihydrogen oxidative addition (**6**) → (**7**) or alkene complexation (**2**) → (**3**), both steps requiring coordinatively unsaturated species. At high alkene concentration the reaction is insensitive to alkene concentration, the reaction being saturated with acyl complex (**6**), but at low alkene concentration or with highly substituted (unreactive) alkene, the rate is first order in alkene concentration, the rate-limiting step changing to the complexation of alkene (**2**) → (**3**).

4.5.2.3 Regio- and Stereo-chemistry

The direction of insertion of the alkene into the metal–hydride bond can take place to yield the normal or branched metal alkyl, which ultimately determines the *n*/*b* ratio of aldehydes (equation 4).

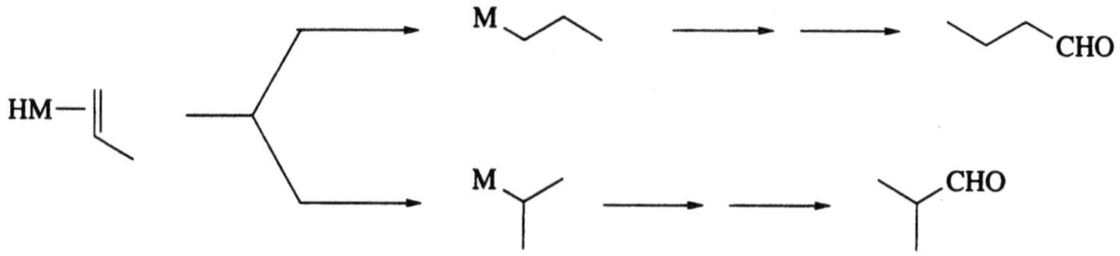

(4)

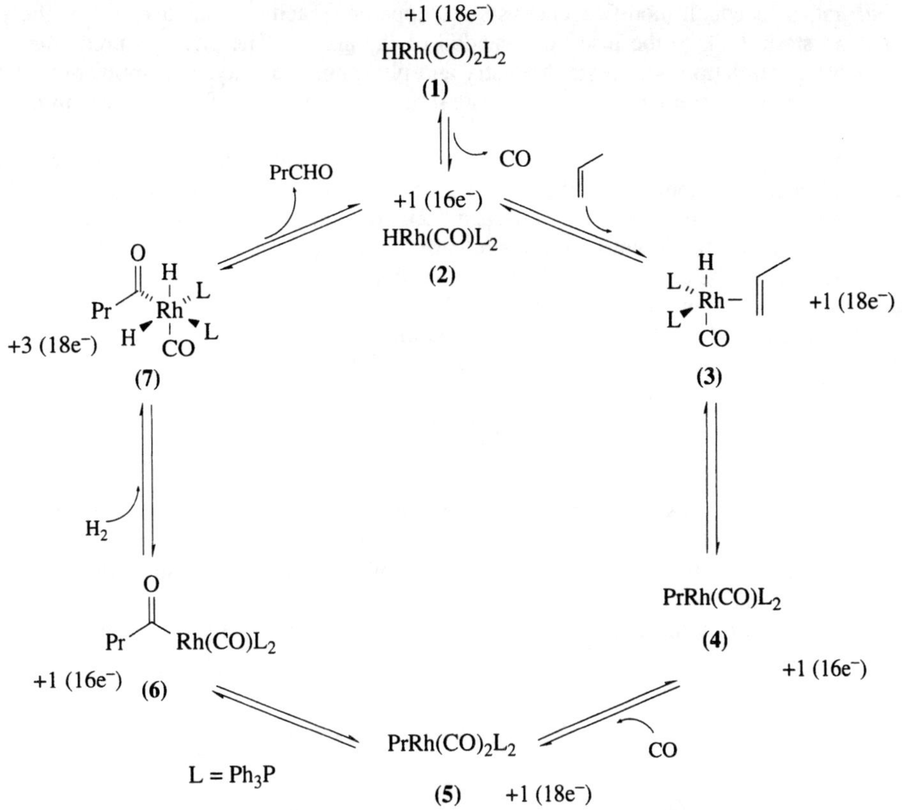

+1 (18e⁻)
$HRh(CO)_2L_2$
(1)

CO

PrCHO

+1 (16e⁻)
$HRh(CO)L_2$
(2)

Pr(C=O)Rh(H)(H)(CO)L$_2$ +3 (18e⁻) **(7)**

H–Rh(L)(L)(CO)(propene) +1 (18e⁻) **(3)**

H_2

Pr(C=O)–$Rh(CO)L_2$ +1 (16e⁻) **(6)**

$PrRh(CO)L_2$ **(4)** +1 (16e⁻)

$PrRh(CO)_2L_2$ **(5)** +1 (18e⁻)

CO

L = Ph_3P

Scheme 2

$L_2PtCl(SnCl_3)$ —(H_2, –HCl)→ H–Pt(L)(L)–$SnCl_3$

CO

+2 (18e⁻)
$[HPtL_2(CO)]^+SnCl_3^-$

PrCHO

+4 (18e⁻) Pr(C=O)(H)(H)$PtL_2(CO)$

$SnCl_3^-$

$L_2(CO)$–Pt^+(H)–(propene) +2 (18e⁻)

H_2

+2 (16e⁻) Pr(C=O)$PtL_2(CO)^+$

Pr–Pt^+(L)(L)–CO +2 (16e⁻)

Pr–Pt^+(CO)$(CO)L_2$
+2 (18e⁻)

CO

L = Ph_3P

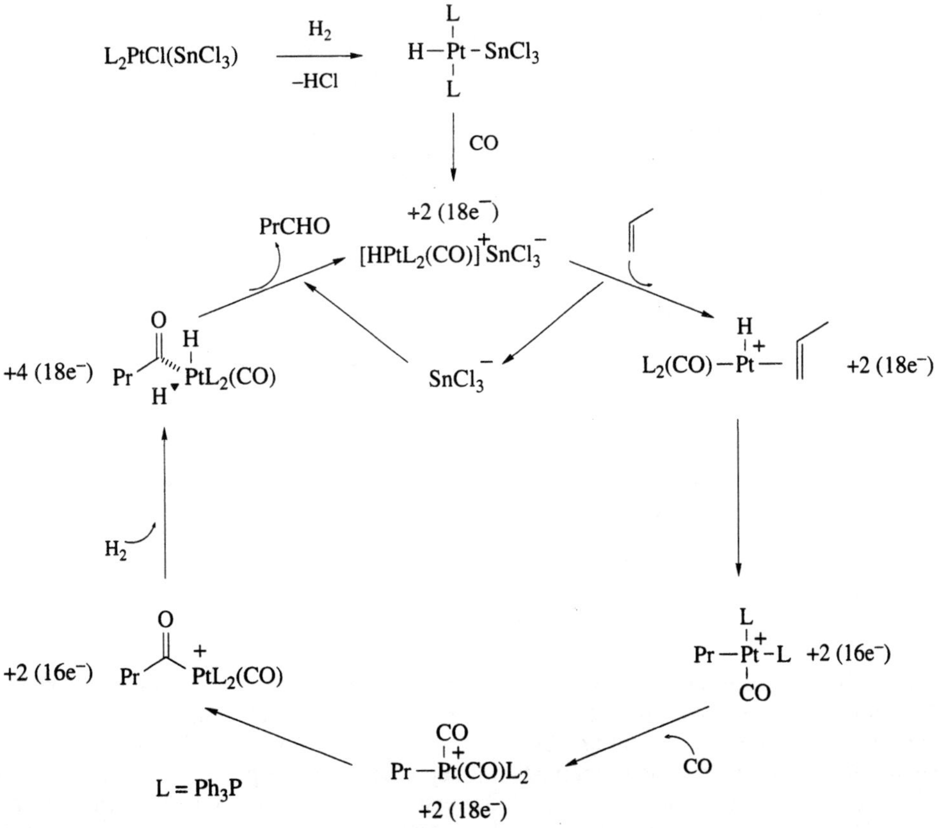

Scheme 3

Higher *n*/*b* ratios in cobalt-modified catalysts are apparently achieved as a result of the phosphine presenting more steric bulk to the insertion reaction of the alkene. The effect is predominately steric, since there is little correlation with regiochemistry and phosphine bacisity. A combination of electronic (χ_i) and steric (cone angle θ) effects provide a more rational explanation of the increase in *n*/*b* ratio with added phosphines.[16,17]

The addition of excess triphenylphosphine to rhodium catalysts can improve the *n*/*b* ratio to 9. The concentration of added phosphine and the partial pressure of CO may determine whether the species undergoing a reaction with the alkene is (**2**) (Scheme 2), containing two phosphine ligands and one carbonyl, or $HRh(CO)_2L$. Certainly with a more sterically hindered complex containing two phosphine ligands (**3**), insertion could be expected to take place to provide the linear metal alkyl. At lower phosphine concentrations the alkene complex derived from $HRh(CO)_2L$, containing only one phosphine ligand, should lead to more branched aldehyde. There are major uncertainties as to the numbers of phosphine and carbonyl ligands on rhodium and certainly mixtures of these complexes are present in the catalytic reaction leading to *n*/*b* mixtures of products.

High *n*/*b* ratios (5.7–13) are obtained in the platinum-catalyzed hydroformylation reaction; *n*/*b* ratios as high as 99 have been obtained utilizing a chelating phosphine, *trans*-1,2-diphenylphosphinocyclobutane.[10]

The *n*/*b* ratios are also affected by the alkene structure and are sensitive to steric hindrance. For example, dicobaltoctacarbonyl hydroformylates longer chain 1-alkenes more slowly than the shorter alkenes. Methyl substitution on the alkene chain not only slows the reaction as the methyl group is moved closer to the double bond, but the *n*/*b* ratio increases (Table 1). A quaternary carbon does not undergo CO insertion and, as a result, the aldehyde is almost never attached at a fully substituted sp^2 carbon of an alkene.

Table 1 Hydroformylation of Alkenes: Relative Rate and Isomer Ratios

Alkene	*Relative rate constant*	*Substitution (%)* Carbon 1	*Substitution (%)* Carbon 2
	1	82	18
	0.97	86	14
	—	96	4
	0.11	100	0
	0.088	—	—

Once metal hydride addition (alkene insertion) has taken place, for example (**3**) → (**4**), β-elimination (**4**) → (**3**) and readdition can occur (Scheme 2). Accordingly, alkene isomerization can take place in the hydroformylation process (equation 5).

$$* \ + \ M{-}H \ \rightleftharpoons \ {-}M \ \rightleftharpoons \ * \ + \ M{-}H \qquad (5)$$

This becomes especially apparent in hydroformylation reactions of internal alkenes, since not only does (*E*)/(*Z*)-isomerization take place, but *n*-aldehydes are obtained. Thus, in the hydroformylation of (*E*)-4-octene by $Co_2(CO)_8$, *n*-nonanal (78%), 2-methyloctanal (10%), 2-ethylheptanal (6%) and 2-propylhexanal (6%) are obtained. This isomerization is supressed with the phosphine-modified catalysts, in the presence of excess phosphine and at high CO pressures. Both carbon monoxide and phosphine can react with a 16-electron complex to provide an 18-electron complex (*e.g.* **4** → **5**; Scheme 2), the reverse β-hydride elimination is prevented, a requirement for this elimination being the presence of a vacant coordination site (16-electron complex). Hydroformylation of 2,3-dimethyl-2-butene yields only 3,4-dimethylpentanal (equation 6).

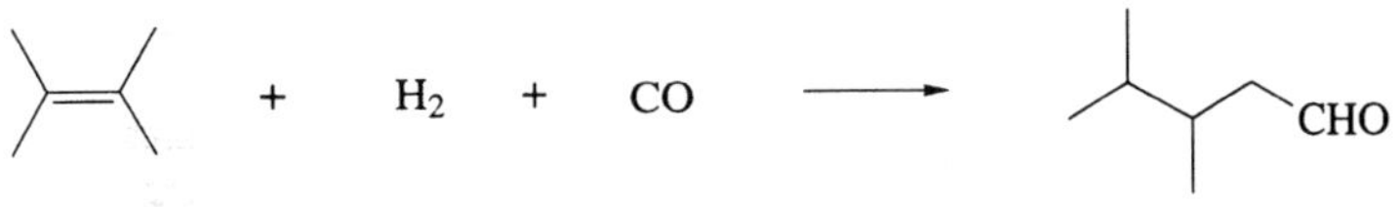

$$\text{Me}_2\text{C=CMe}_2 + \text{H}_2 + \text{CO} \longrightarrow \text{3,4-dimethylpentanal (CHO)} \qquad (6)$$

The overall stereochemistry of the hydroformylation reaction exhibits *syn* addition of the hydride and the formyl groups, both with cobalt and rhodium catalysts. Thus in the hydroformylation of 1-methylcyclohexene, where (*E*)/(*Z*)-isomerization cannot occur, the predominate product is *trans*-2-methylcyclohexanecarbaldehyde (equation 7) resulting from the delivery of the aldehyde and the formyl groups from the same face of the alkene.[18] Similarly the hydroformylation of (*Z*)-3-methyl-2-pentene gives mainly the *erythro*-aldehyde (equation 8).[19]

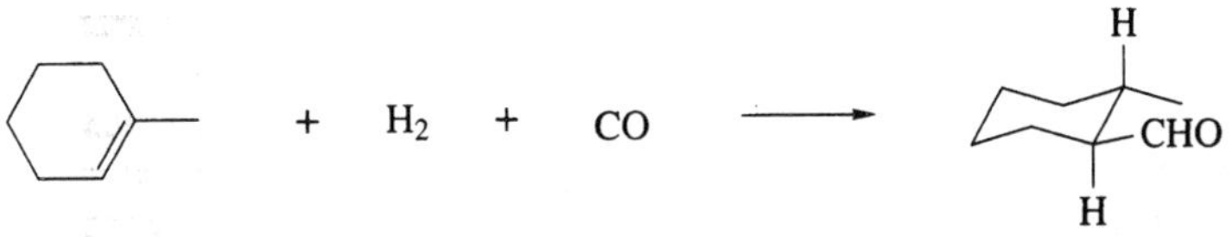

$$\text{1-methylcyclohexene} + \text{H}_2 + \text{CO} \longrightarrow \textit{trans}\text{-2-methylcyclohexanecarbaldehyde (H, CHO, H)} \qquad (7)$$

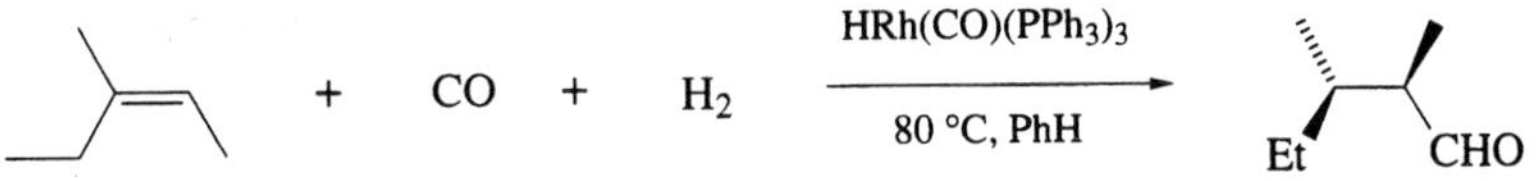

$$(Z)\text{-3-methyl-2-pentene} + \text{CO} + \text{H}_2 \xrightarrow[\text{80 °C, PhH}]{\text{HRh(CO)(PPh}_3)_3} \textit{erythro}\text{-aldehyde (Et, CHO)} \qquad (8)$$

This stereochemistry is a result of *syn* metal hydride addition across the alkene followed by CO 'insertion' with retention of configuration at the carbon bound to the metal (*e.g.* steps **3** → **4** and **5** → **6**; Scheme 2).

4.5.2.4 Hydroformylation of Unfunctionalized Alkenes

The hydroformylation of 1-alkenes and some of the simpler alkenes has been discussed (*vide supra*). In many cases these aldehydes are commercially available or are not of particular interest as fine chemicals for organic syntheses. The hydroformylation of alkenes is especially useful when isomerization is difficult or only one regioisomer can be obtained.

The hydroformylation of cyclohexene, as mentioned earlier, provides good yields of cyclohexanecarbaldehyde (entry 1, Table 2). The regiochemistry of the hydroformylation of vinyl aromatics is strongly influenced by electronic effects, the formyl group being incorporated at the most electron-deficient carbon. Consequently the hydroformylation of indene places the formyl group predominately at the carbon adjacent to the aromatic ring (entry 2). The indanealdehyde and aldehyde derived from acenaphthalene (entry 3) can be converted to β-phenylethylamines (**8**) and (**9**), respectively, that have interesting hypotensive properties.

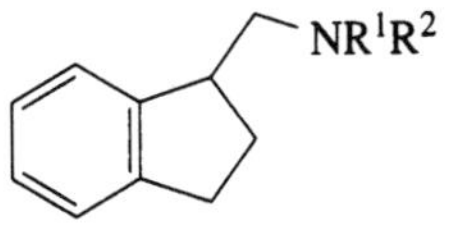

NR^1R^2 (**8**) $\qquad$ NR^1R^2 (**9**)

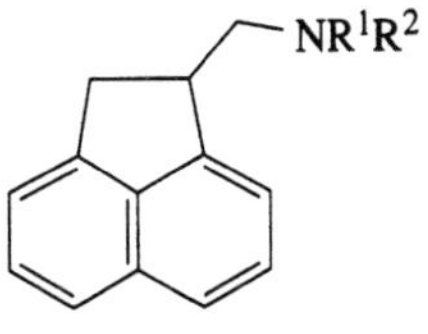

Similarly, hydroformylation of substituted styrenes yields the corresponding aldehydes, the β-substituted styrene providing the α-phenyl aldehyde (entry 4). This aldehyde can be converted to the pharmacologically active amine.

(–)-α-Pinene (entry 5) from the cluster pine (*Pinus pinaster* Sol.) is diastereoselectively hydroformylated to (–)-3-pinanecarbaldehyde. Hydroformylation of (+)-α-pinene from Aleppo pine (*Pinus pinaster* Mill.) yields the diastereomer. This selective hydroformylation takes place from the least-hindered alkene face. Depending on the reaction conditions, selectivities as high as 85% can be achieved.

Table 2 Hydroformylation of Alkenes

Entry	*Alkene*	*Catalyst*	*Temperature* (°C)	*Pressure* (bar)	*Aldehyde*	*Yield* (%)	*Ref.*
1		Rh_2O_3	100	150	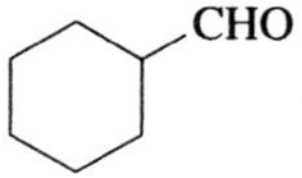 CHO	82–84	20
2	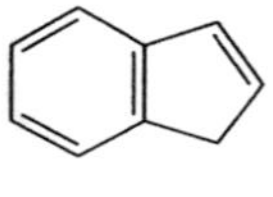	$[RhCl(COD)]_2$	100	700	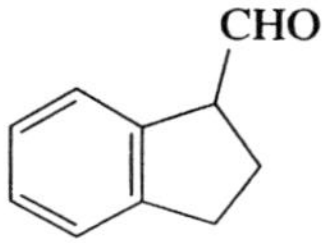 CHO	a	22
3	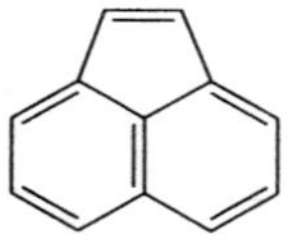	$[RhCl(COD)]_2$	100	700	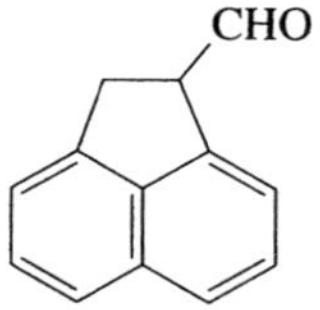 CHO	54	22
4	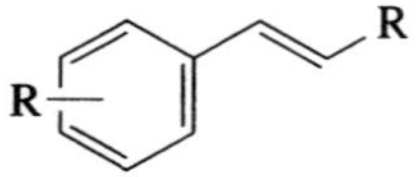 R, R	$[RhCl(COD)]_2$	100	160	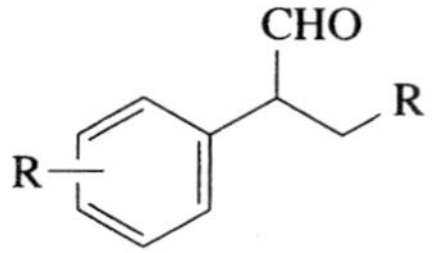 CHO, R, R	83	22
5		$[RhCl(COD)]_2$	70	600	 OHC	a	22–24
6	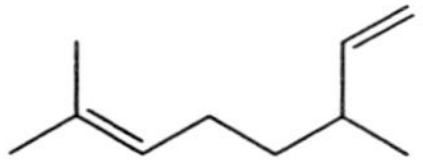	$[RhCl(COD)]_2$	70	600	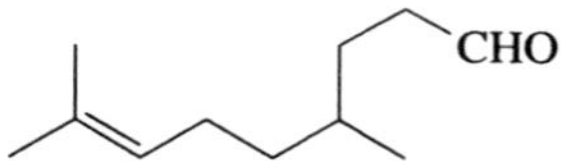 CHO	a	22

Table 2 *(continued)*

Entry	*Alkene*	*Catalyst*	*Temperature* (°C)	*Pressure* (bar)	*Aldehyde*	*Yield* (%)	*Ref.*
7		$[RhCl(COD)]_2$	70	600	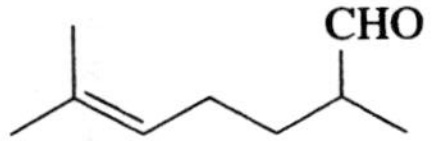 CHO	37	22
8	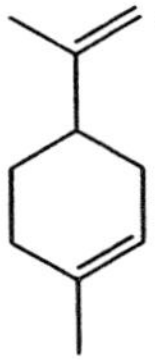	[Rh(OAc)COD] 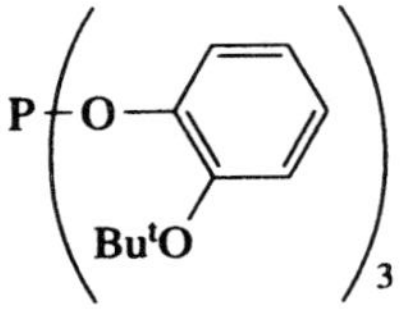 P(O–C₆H₄–OBu^t)$_3$	90	14	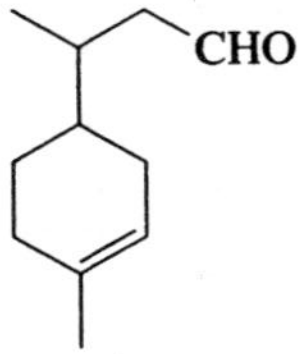 CHO	b	21
9	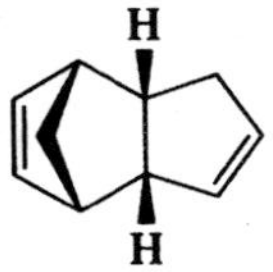 H, H	$RhH(CO)(PPh_3)_3$ Et_3N	70	100	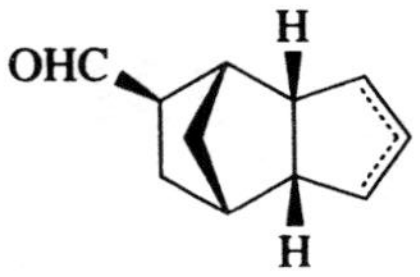 OHC, H, H	82	26
			130	82	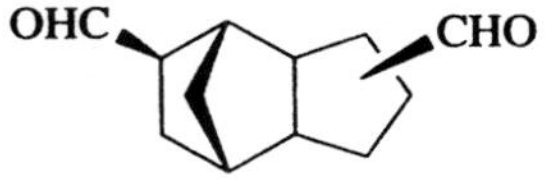 OHC, CHO	75	27
			180	150	 $HOCH_2$, CH_2OH	67	28

[a] Yields not provided; conversions varied, depending on reaction time. [b] Conversion run to 50% to determine rates.

The enantiomeric purity of the 3-pinanecarbaldehyde corresponds to the α-pinane utilized (70–85%). Enantiomerically pure aldehyde can be obtained by the acid-catalyzed trimerization of the aldehyde, with only one enantiomer being preferentially cyclotrimerized to a crystalline compound.[23b] Cleavage of the trimer results in enantiomerically pure aldehyde. If cobalt catalysts are employed in the cyclization, rearrangement to the bornane structure takes place (equation 9).[25]

$$+ \; CO \; + \; H_2 \xrightarrow[\text{205 bar, 65\%}]{Co_2(CO)_8,\ 110\text{–}120\ °C} \text{OHC-bornane} \qquad (9)$$

Selective hydroformylation of nonconjugated dienes can be carried out such that the least-hindered double bond reacts. Thus citronellene (entry 6) and 2,6-dimethyl-1,5-heptadiene both undergo preferential reaction at the least-substituted double bond, the hydroformylation of citronellene patterned after that of 3-methyl-1-pentene (Table 1) and the hydroformylation of 2,6-dimethyl-1,5-heptadiene to citronellal following that of 2-methyl-1-pentene, avoiding placing the formyl group at the quaternary carbon. The reaction of limonene (entry 8) shows similar behavior.

Because the norbornene double bond of dicyclopentadiene is more strained, its selective hydroformylation can be achieved under mild reaction conditions (entry 9). Under more vigorous reaction conditions both double bonds are hydroformylated. The *exo* faces of both rings are the least hindered, accounting for the selectivity.

The hydroformylation of conjugated dienes with unmodified cobalt catalysts is slow, since the insertion reaction of the diene generates an η^3-cobalt complex by hydride addition at a terminal carbon (equation 10).[5] The stable η^3-cobalt complex does not undergo facile CO insertion. Low yields of a mixture of *n*- and iso-valeraldehyde are obtained. The use of phosphine-modified rhodium catalysts gives a complex mixture of C_5 monoaldehydes (58%) and C_6 dialdehydes (42%). A mixture of mono- and di-aldehydes are also obtained from 1,3- and 1,4-cyclohexadienes with a modified rhodium catalyst (equation 11).[29] The 3-cyclohexenecarbaldehyde, an intermediate in the hydrocarbonylation of both 1,3- and 1,4-cyclohexadiene, is converted in 73% yield, to the same mixture of dialdehydes (*cis*:*trans* = 35:65) as is produced from either diene.

$$HCo(CO)_4 \; + \; \text{butadiene} \xrightarrow{-CO} [(\eta^3\text{-allyl})-Co(CO)_3] \qquad (10)$$

$$\text{1,3- or 1,4-cyclohexadiene} + H_2 + CO \xrightarrow[\text{100 bar, 100 °C, 6 h}]{Rh(CO_2Me)(CO)(PPh_3)_2} \text{(CHO)}_2\text{-cyclohexane} + \text{CHO-cyclohexene} + \text{CHO-cyclohexane} \qquad (11)$$

	dialdehyde	cyclohexenecarbaldehyde	cyclohexanecarbaldehyde
1,3-	29%	5%	25%
1,4-	73%	7%	20%

Phosphine-modified rhodium catalysts hydroformylate alkynes to saturated aldehydes.[1] The reaction most likely proceeds by a rapid hydrogenation to yield the alkene, followed by hydroformylation.

4.5.2.5 Hydroformylation of Functionalized Alkenes

The hydroformylation of alkenes containing a variety of functional groups can be carried out without poisoning the catalyst and without reaction of the functionality taking place. Exceptions are those in which the functional group reacts with the newly formed aldehyde. Alcohols, ethers, acetals and ketals are tolerated in hydroformylations. If the double bond is remote enough from the functional group, then the hydroformylation regiochemistry and rate are similar to the analogous alkene. If the double bond is closer to the functionality, then alkene isomerization may be favored and the regiochemistry may be altered or even reversed.

Hydroformylation of 2,6-dimethyl-6-hepten-2-ol produces hydroxycitronellal (equation 12).[22] Subjecting allyl alcohol to hydroformylation reaction conditions with $HCo(CO)_4$ yields only propanal, isomerization taking place more rapidly than hydroformylation.[2] Phosphine-modified rhodium catalysts will convert allyl alcohol to butane-1,4-diol under mild conditions in the presence of excess phosphine, however (equation 13).[5,30,31] When isomerization is blocked, hydroformylation proceeds normally (equation 14). An elegant synthesis of the Prelog–Djerassi lactone has been accomplished starting with the hydroformylation of an allylic alcohol (equation 15).[32]

OH + H_2 + CO —(100 °C, 705 bar; $[RhCl(COD)]_2$; 90%)→ HO ... CHO (12)

OH + H_2 + CO —($HRh(CO)(PPh_3)_3$, excess PPh_3; 30 °C, 1.4 bar)→ HO...OH (77%) + HO...OH (19.5%) (13)

OH + H_2 + CO —($HCo(CO)_4$)→ CHO OH ⇌ O (14)

BnO ... OH + CO + H_2 —($Rh_2(OAc)_4$, Ph_3P; 100 °C, 24 bar; 6 h)→ [BnO ... O ... OH] —PCC→

BnO ... O ... O → → (15)

86% overall

The electron-withdrawing ability of an ether oxygen alters the regiochemistry of hydroformylation. This effect is present even in allyl ethers such as ethyl allyl ether, which gives predominately 2-methyl-3-ethoxypropanol.[2] The reaction with vinyl ethers is regioselective (equation 16)[5] or in some examples regiospecific (equation 17).[22] The products of hydroformylation of phenyl vinyl ethers can be converted into pharmacologically active aryloxypropylamines by reductive amination.

O + H_2 + CO —($Co_2(CO)_8$; 180 °C; 300 bar)→ O ... OH (78%) + O ... OH (8%) (16)

The regiochemistry of hydroformylation of acrolein cyclic acetals can be controlled to some extent by the presence or absence of phosphine. Excess phosphine provides predominately the straight chain alde-

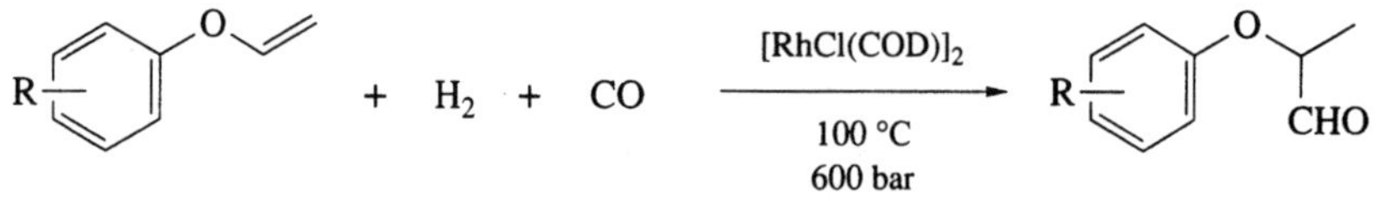

(17)

hyde with a number of catalysts (equation 18).[5] In the absence of phosphine, the major product is the branched aldehyde (equation 19).[22]

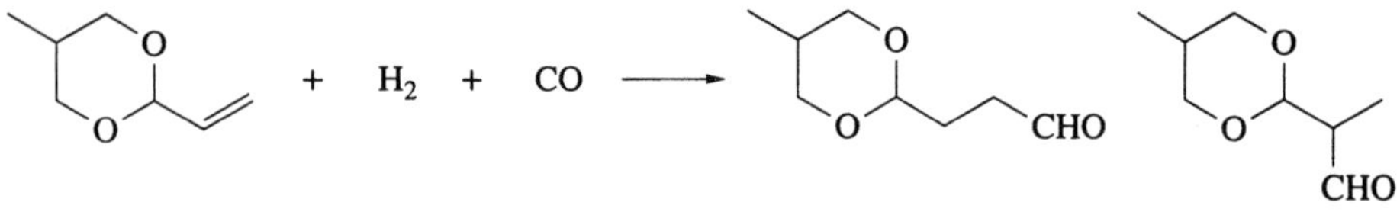

(18)

$Co_2(CO)_8$, $(C_8H_{17})_3P$ 150 °C, 68–82 bar, 86%	81%	19%
$Rh_6(CO)_{11}$, $(MeO)_3P$ 110 °C, 7.2 bar, 97%	80%	12%
$HRh(CO)(PPh_3)_3$ 140 °C, 21–41 bar, 89%	81%	19%

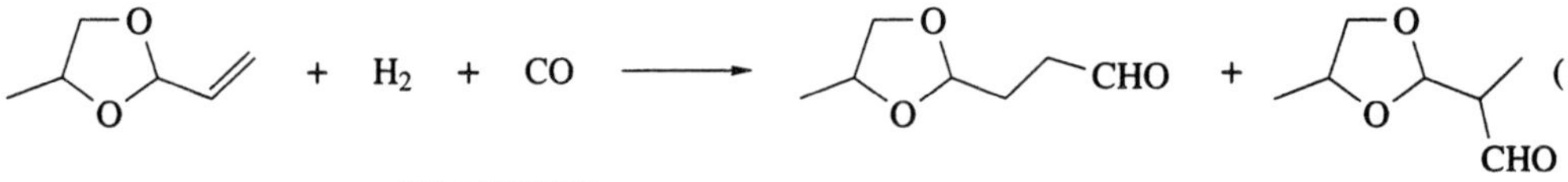

(19)

$[RhCl(COD)]_2$ 100 °C, 600 bar	33%	57%
$[RhCl(COD)]_2$, 50 equiv. PPh_3 140 °C, 20 bar	70%	30%

The hydroformylation of allyl and vinyl acetals yields some useful intermediates. Allyl acetate undergoes partial double bond migration prior to hydroformylation with a cobalt catalyst in the absence of phosphine (equation 20).[2,5] Rhodium catalysts containing chelating phosphines are more selective to the linear aldehyde.[31]

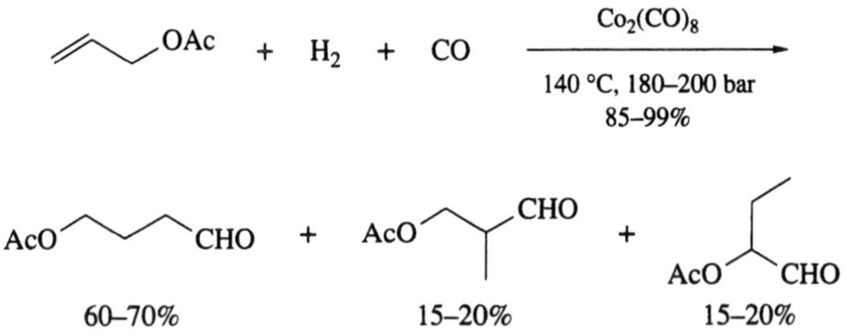

(20)

Industrial (BASF) syntheses of vitamin A and vitamin A aldehyde have been accomplished utilizing the aldehydes obtained from allyl acetate hydroformylation.[22] Either aldehyde (**10**) or (**11**) reacts with the same phosphorus ylide to give vitamin A or retinal (Scheme 4). Hydroformylation of 3-methyl-2-butenyl acetate gives a high yield of 2-formyl-3-methylbutyl acetate. Elimination of acetic acid followed by isomerization provides trimethylacrylaldehyde, which is an intermediate in the synthesis of irones (Scheme 5).

The hydroformylation of vinyl acetate has been accomplished by a number of different catalysts.[30,31,33,34] The branched isomer is the major product, the regioselectivity being attributed both to the electron-withdrawing effect of oxygen and the chelating effect of the ester carbonyl.[4] High regioselectivity can be achieved in a polar solvent with a rhodium catalyst (equation 21).[33,34] Vinyl pivalate shows nearly a five times greater selectivity for the branched aldehyde.[34]

Unsaturated aldehydes, ketones and esters undergo hydroformylation of the double bond without reaction at the carbonyl group. For example, dienes yield aldehydes in a stepwise fashion (*vide supra*); however, α,β-unsaturated aldehydes and ketones undergo rapid hydrogenation to the saturated aldehyde

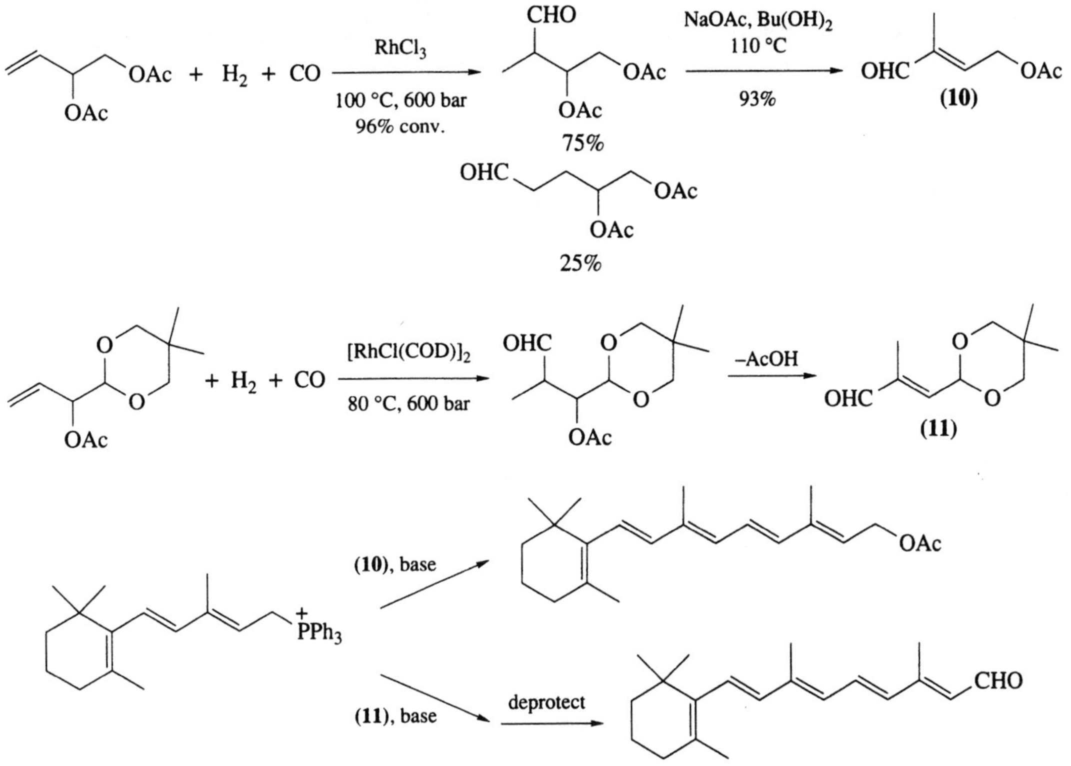

RhCl$_3$ 100 °C, 600 bar 96% conv. 75% 25% NaOAc, Bu(OH)$_2$ 110 °C 93% (10)

[RhCl(COD)]$_2$ 80 °C, 600 bar –AcOH (11)

(10), base (11), base deprotect

Scheme 4

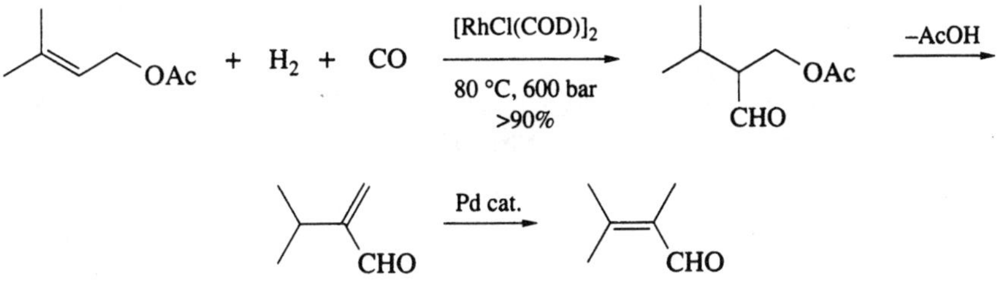

[RhCl(COD)]$_2$ 80 °C, 600 bar >90% –AcOH Pd cat.

Scheme 5

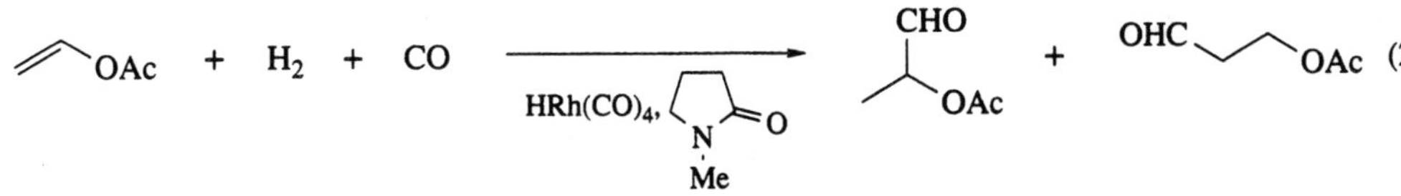

HRh(CO)$_4$, (21)

Conditions	CHO (branched)	OHC (linear)
60 °C, 200 bar, 83%	95%	5%
Rh(CO)$_2$(acac), 15 equiv. Ph$_3$P DMSO or DMF, 8 bar, 86%	86%	14%

or ketone instead of undergoing hydroformylation.[1,2,5] Hydroformylation of an unsaturated ketone (in which the carbonyl is not conjugated) takes place selectively at the more reactive double bond to give an aldehyde that can be converted to guaiazulene (equation 22).[22]

α,β-Unsaturated esters are not hydrogenated as readily with hydroformylation catalysts, so aldehydes can be obtained (equation 23).[5,30,31,35,36] The regioselectivity is sensitive to substitution and reaction conditions. Formylation at the β-position is driven by conjugation of the double bond with the carbonyl, substitution at the α-position, and excess phosphine (equations 24 and 25). Similarly, the selectivity in the hydroformylation of dimethyl itaconate can be reversed by phosphine-modified catalysts (equation 26).

$$+ \; H_2 \; + \; CO \xrightarrow[100\ ^\circ C,\ 600\ bar]{[RhCl(COD)]_2} \qquad (22)$$

$$R^1CH{=}C(R^2)CO_2R \; + \; H_2 \; + \; CO \longrightarrow R^1CH(CHO)CH(R^2)CO_2R \; + \; R^1CH_2C(R^2)(CO_2R)CHO \qquad (23)$$

$$CH_2{=}CHCO_2R \; + \; H_2 \; + \; CO \longrightarrow OHC\!\!\diagup\!\!\diagdown CO_2R \; + \; CH_3CH(CHO)CO_2R \qquad (24)$$

	OHC–CH$_2$CH$_2$–CO$_2$R	CH$_3$CH(CHO)CO$_2$R
Rh(acac), Ph$_3$P 1 bar, 40 °C, 60%	100%	–
RhH(CO)PPh$_3$, Ph$_3$P 2 equiv. Ph$_2$P(CH$_2$)$_4$PPh$_2$ 4 bar, 100% conv.	13%	34%

(53% CH$_3$CH$_2$CO$_2$R)

$$CH_2{=}C(CH_3)CO_2Me \; + \; H_2 \; + \; CO \longrightarrow OHC\,CH_2CH(CH_3)CO_2Me \; + \; OHC\,C(CH_3)_2CO_2Me \qquad (25)$$

	OHC–CH$_2$CH(CH$_3$)CO$_2$Me	OHC–C(CH$_3$)$_2$CO$_2$Me
Rh$_4$(CO)$_{12}$, 100 °C, 15 bar, 98%	6%	94%
[Rh(nbd)Cl]$_2$, 150 °C, 80 bar, 100%	53%	47%
[Rh(nbd)Cl]$_2$, Bu$_3$P, 100 °C, 15 bar, 58%	82%	18%

$$MeO_2C\,C(=CH_2)CH_2CO_2Me \; + \; H_2 \; + \; CO \xrightarrow{RhH(CO)PPh_3} MeO_2C\,CH(CH_2CHO)CH_2CO_2Me \; + \; OHC\,C(CH_3)(CO_2Me)CH_2CO_2Me \qquad (26)$$

Acrylonitrile yields predominately the linear cyano aldehyde, although considerable double bond reduction takes place (equation 27).[37] The reaction has received considerable attention since it is a precursor to glutamic acid *via* a Strecker reaction.[1,2]

$$CH_2{=}CHCN \; + \; H_2 \; + \; CO \longrightarrow OHC\,CH_2CH_2CN \longrightarrow NC\,CH(NH_2)CH_2CH_2CN \longrightarrow HO_2C\,CH(NH_2)CH_2CH_2CO_2H \qquad (27)$$

Nitro groups are tolerated in hydroformylation as are amines, provided they are protected. The reaction of *o*-nitrostyrene gives an intermediate for the synthesis of 3-methylindole (equation 28).[38] With a phosphine-modified rhodium catalyst, the reaction is regioselective, placing the formyl group in the α-position.

Vinylamides are not regioselectively hydroformylated but rhodium catalysts tend to cause formylation at the carbon adjacent to the nitrogen (Table 3).[2] *N*-Allylamides exhibit a regiochemistry that indicates some electronic influence of nitrogen even though it is one methylene removed from the double bond. The aldehydes obtained from the regioselective hydroformylation of *N*-vinylamides are of interest because they represent intermediates for amino acid synthesis.[40]

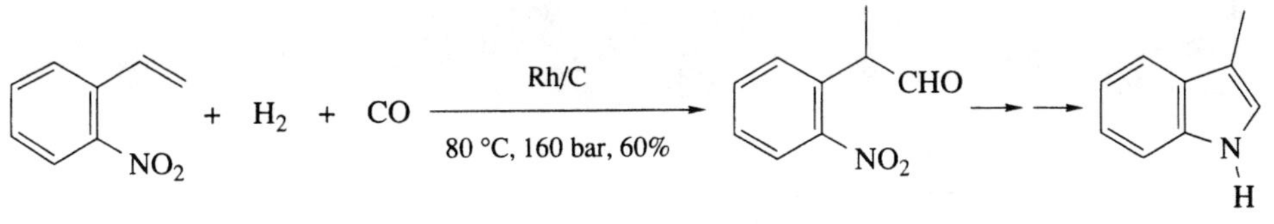

$$\text{2-nitrostyrene} + H_2 + CO \xrightarrow[\text{80 °C, 160 bar, 60\%}]{\text{Rh/C}} \text{2-(2-nitrophenyl)propanal (CHO, NO}_2\text{)} \rightarrow\rightarrow \text{3-methylindole (N-H)} \quad (28)$$

Although *N*-methyltetrahydropyridines are not selectively hydroformylated, an equal mixture of the two aldehydes being obtained from *N*-methyl-1,2,3,6-tetrahydropyridine, *N*-methyltropidenes are selectively formylated at the 3-position (equation 29). The difference in the selectivity can be attributed to the substitution in tropidine (bridge head carbon).

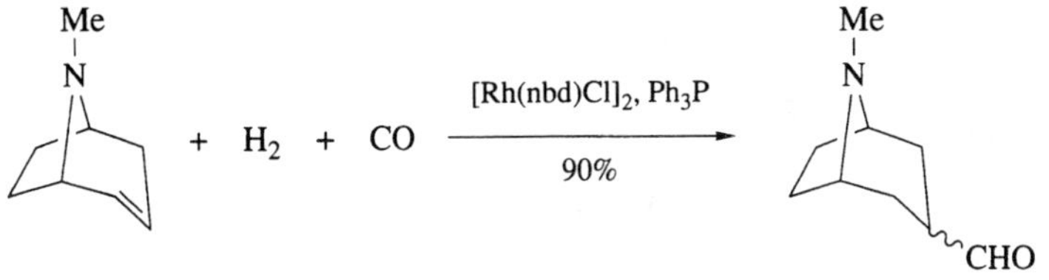

$$\text{tropidine (N-Me)} + H_2 + CO \xrightarrow[\text{90\%}]{[Rh(nbd)Cl]_2,\ Ph_3P} \text{3-formyltropane (N-Me, CHO)} \quad (29)$$

Fluoroalkenes such as trifluoropropene and pentafluorostyrene show unique regioselectivities in hydroformylation reactions.[41] While cobaltoctacarbonyl catalysts give *n*/*b* ratios of about 93/7 and 90/10 for trifluoropropene and pentafluorostyrene, respectively, ruthenium and rhodium catalysts show selectivities of about 3/97, regardless of the ligands (the exception being that a ruthenium carbonyl catalyst gave a *n*/*b* ratio of 26/74 for pentafluorostyrene). These selectivities are in marked contrast to those obtained for the unfluorinated alkenes: propene, $Co_2(CO)_8$, 80/20; $Ru_3(CO)_{12}$, 74/26; styrene, $Co_2(CO)_8$, 41/59; $Rh_2Cl_2(CO)_4/PPh_3$, 57/43. The hydroformylation of these fluorinated alkenes has been utilized in the synthesis of fluorinated amino acids (Scheme 6).

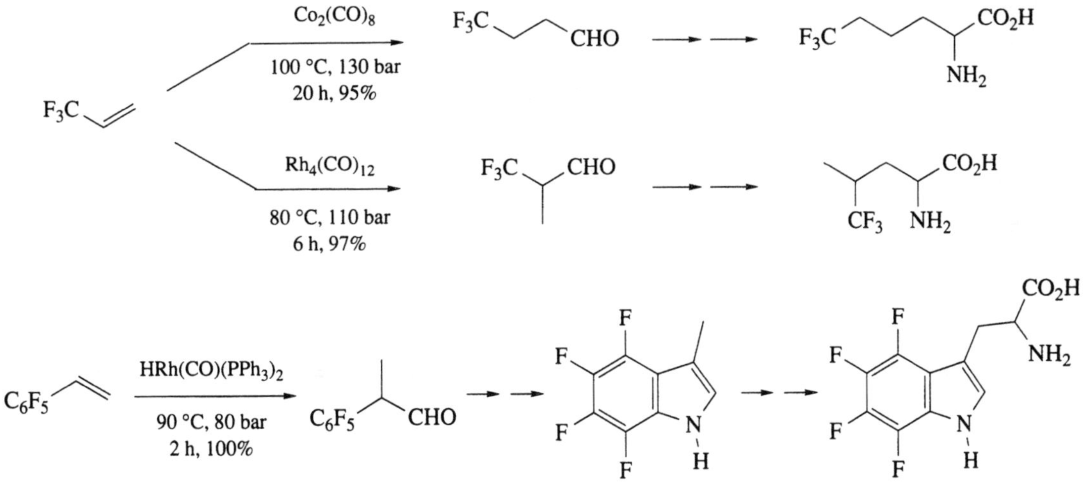

$$F_3C\text{CH=CH}_2 \xrightarrow[\text{100 °C, 130 bar, 20 h, 95\%}]{Co_2(CO)_8} F_3C\text{CH}_2\text{CH}_2\text{CHO} \rightarrow\rightarrow F_3C\text{CH}_2\text{CH}_2\text{CH}_2\text{CH(NH}_2\text{)CO}_2\text{H}$$

$$F_3C\text{CH=CH}_2 \xrightarrow[\text{80 °C, 110 bar, 6 h, 97\%}]{Rh_4(CO)_{12}} F_3C\text{CH(CH}_3\text{)CHO} \rightarrow\rightarrow \text{CH}_3\text{CH(CF}_3\text{)CH}_2\text{CH(NH}_2\text{)CO}_2\text{H}$$

$$C_6F_5\text{CH=CH}_2 \xrightarrow[\text{90 °C, 80 bar, 2 h, 100\%}]{HRh(CO)(PPh_3)_2} C_6F_5\text{CH(CH}_3\text{)CHO} \rightarrow\rightarrow \text{4,5,6,7-tetrafluoro-3-methylindole (N-H)} \rightarrow\rightarrow \text{4,5,6,7-tetrafluorotryptophan (CO}_2\text{H, NH}_2\text{)}$$

Scheme 6

4.5.2.6 Asymmetric Hydroformylation[42–44]

The generation of a chiral center as a result of alkene hydroformylation can take place either by formylation or hydride addition at the prochiral carbon (equations 30 and 31). Kinetic resolution in the hydroformylation reaction of a racemic alkene containing a chiral center could also occur, but in this example the chiral center is not generated as a result of the hydroformylation reaction.

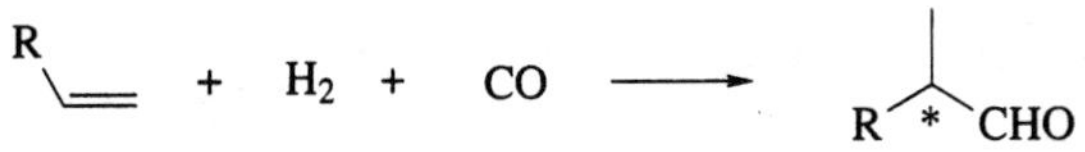

$$R\text{CH=CH}_2 + H_2 + CO \longrightarrow R\text{C}^*\text{H(CH}_3\text{)CHO} \quad (30)$$

Table 3 Hydroformylation of Unsaturated Amides

Vinylimide	*Catalyst*	*Temperature* (°C)	*Pressure* (bar)	*Products*			*Yield* (%)	*Ref.*
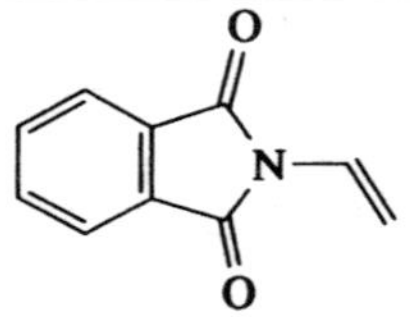 O, N, O	$Co_2(CO)_8$	120	170	O, N, CHO, O 15	:	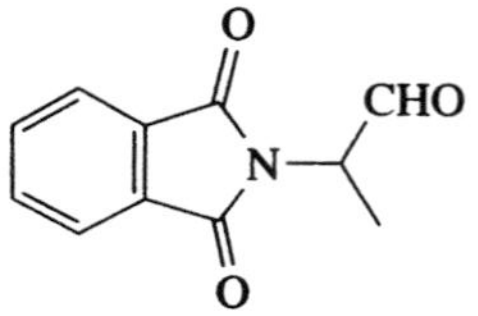 O, N, CHO, O 1	78	38
	$HRh(CO)(PPh_3)_3$	65	34	None		Only	100	39
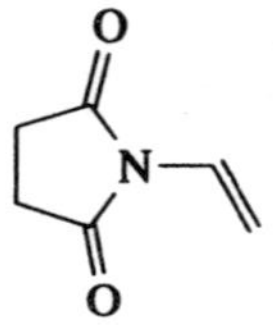 O, N, O	$HRh(CO)(PPh_3)_3$	65	34			O, N, CHO, O	100	39
	$[RhCl(COD)]_2$	90	700			Only	90	2
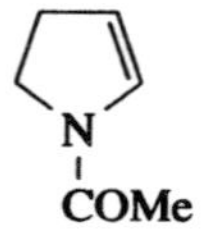 N, COMe	$HRh(CO)(PPh_3)_3$, DIPHOL[a]	60	34			N, CHO, COMe	100	39
 AcNH	$HRh(CO)(PPh_3)_3$	40–45	34	OHC, NHAc 46	:	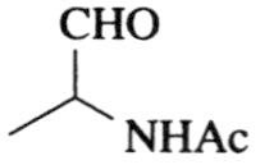 CHO, NHAc 54	70	39
 NAc_2	$Co_2(CO)_8$	120	170	OHC, NAc_2 6.7	:	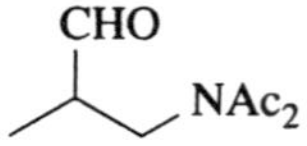 CHO, NAc_2 1	90	38

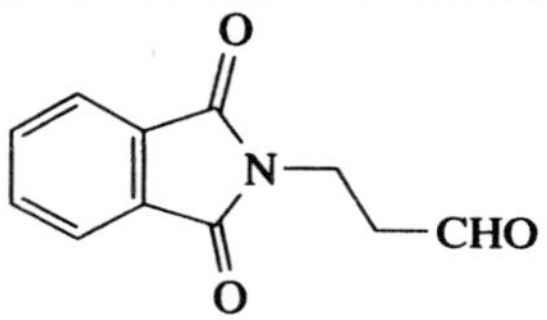

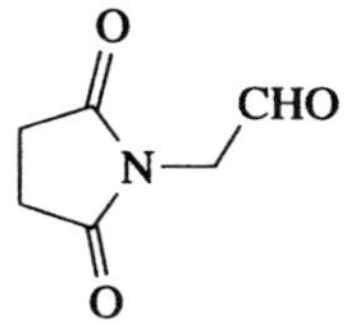

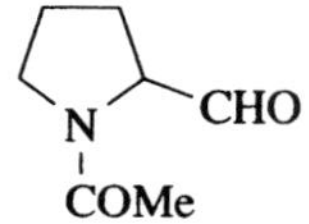

Table 3 *(continued)*

Vinylimide	*Catalyst*	*Temperature* (°C)	*Pressure* (bar)	*Products*	*Yield* (%)	*Ref.*
NHAc	$HRh(CO)(PPh_3)_3$	40	34	[OHC NAc$_2$] → N-Ac : CHO NHAc 46 : 54	100	39

[a] See structure **(13)**.

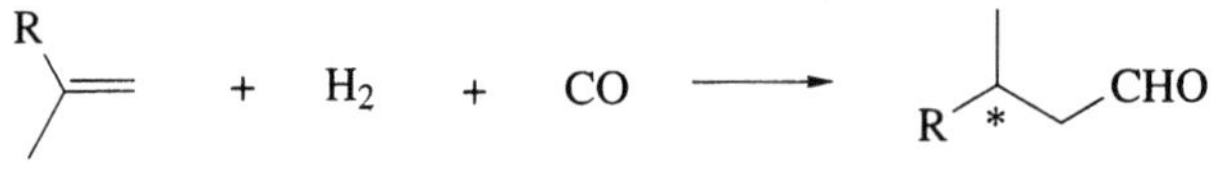

$$\text{R(Me)C=CH}_2 + \text{H}_2 + \text{CO} \longrightarrow \text{R}\overset{*}{\text{C}}\text{H(Me)CH}_2\text{CHO} \qquad (31)$$

The enantioselective step in rhodium-catalyzed hydroformylation reactions (refer to Scheme 2) can take place either in the formation of the alkene complex (**3**), at the insertion of the alkene to yield (**4**), or in the CO insertion step (**5**), since these steps are all reversible. For example, if the diastereomers (**3**), obtained from the complexation of the alkene *re* and *si* faces, are converted in a fast step to (**4**), then the ratios of the diastereomeric complexes (**3**) will determine the enantiomeric excess. If, however, one diastereomer (for example, the highest energy complex, present in a minor amount) undergoes the insertion reaction several orders of magnitude more rapidly than the other diastereomer, then it is possible, depending on the subsequent step, that the diastereomeric ratios of (**4**) are representative of the enantiomeric excess. This question has been addressed by carrying out the asymmetric hydroformylation of 1-butene, (*Z*)-2-butene and (*E*)-2-butene, which yield the same product, 2-methylbutanol. This means that all three substrates must pass through a common alkylmetal intermediate (Scheme 7).[45] Since with both rhodium and platinum catalysts, the prevailing chirality of the product and the enantiomeric excesses depend on substrate structure, this suggests that the enantioselection occurs before or during the formation of the alkyl metal intermediate.

(a) 1-butene–M → Et/M alkyl (wedge)
(b) (Z)-2-butene–M → + → Et–CH(Me)–CHO
(c) (E)-2-butene–M → Et/M alkyl (hash)

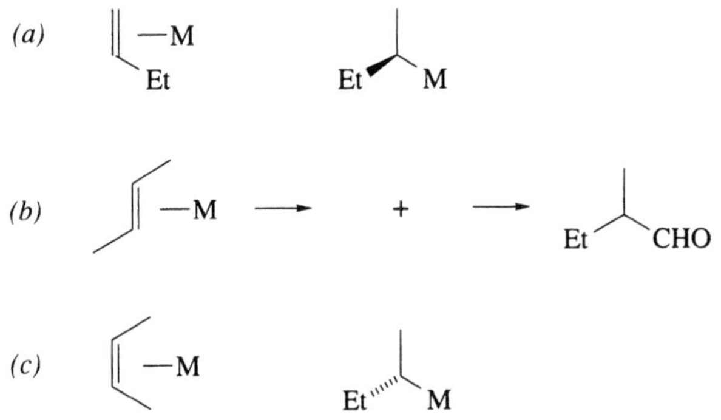

S.M.	*M*	*Absolute configuration*	*% ee*
a	Rh	*(R)*	18.8
	Pt	*(R)*	46.7
b	Rh	*(S)*	32.0
	Pt	*(S)*	24.2
c	Rh	*(S)*	27.0
	Pt	*(S)*	14.5

Scheme 7

By using a stereochemical model for the transition state of the insertion reaction with nonbonded interactions between a complex of given chirality and the alkene the enantiomeric excesses as well as the regioisomeric excesses of aldehydes from monosubstituted, and 1,1- and 1,2-disubstituted ethylenes can be predicted. The prediction is correct in more than 80% of the cases studied.[46,47]

Until recently, only moderate enantiomeric excesses of branched aldehydes had been realized in hydroformylation reactions with rhodium and platinum catalysts. Utilizing the chelating ligand (4*R*,5*R*)-2,2-dimethyl-4,5-bis(diphenylphosphinomethyl)-1,3-dioxolane (**12**; (–)-DIOP) in the presence of the appropriate rhodium complex generates HRh(CO)[(–)-DIOP], which catalyzes the hydroformylation of styrene to hydrotropaldehyde in 25% *ee*.[42,43] Replacing the diphenylphosphino groups with dibenzophosphole gives the (–)-DBP-DIOP ligand (**13**), which yields a rhodium catalyst that hydroformylates styrene in 33% *ee*.[45] Other chelating ligands such as phosphinites, phosphines and phospholes of 1,2-dimethylene-cyclohexane and -cyclobutane gave lower enantiomeric excesses (2–30%).[41] Low enantiomeric excesses (~25%) have also been obtained with the use of CHIRAPHOS (**14**)[48] and EPHOS (**15**).[49] A variety of other alkene substrates have been hydroformylated but enantiomeric excesses have never been higher than about 50% with rhodium as a catalyst,[42,50,51] the highest enantiomeric excess being achieved in the hydroformylation of vinyl acetate with Rh/(–)-DBP-DIOP (equation 32).[50] In a number of reactions, the optical yields were observed to be dependent on the temperature, as expected, and on the partial pressures of carbon monoxide and hydrogen.[42,52] This reaction is particularly attractive since 2-acetoxypropanol is a precursor for the Strecker synthesis of threonine.

Until recently, the use of platinum complexes in asymmetric hydroformylation had been less encouraging because of the lower reaction rates and the tendency for the substrate to undergo competitive hydrogenation. In addition, relatively high normal to branched (*n*/*b*) ratios have been observed in the hydroformylation of monosubstituted alkenes. Hydroformylation of 1- and 2-butenes in the presence of [(–)-DIOP]$PtCl_2/SnCl_2$ was reported by two groups to give nonreproducible (and sometimes contradictory) results.[53] In contrast, the use of the preformed [(–)-DIOP]$Pt(SnCl_3)Cl$ complex as a catalyst precursor did give reproducible results.[54]

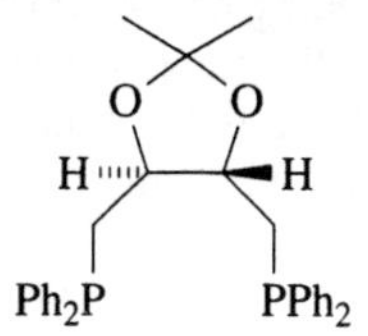

(12) 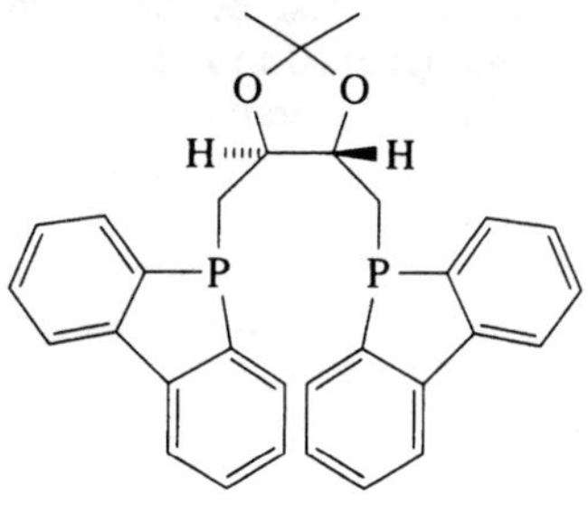 **(13)** 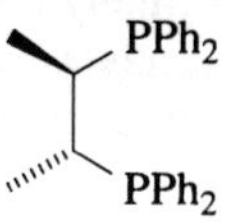 **(14)** 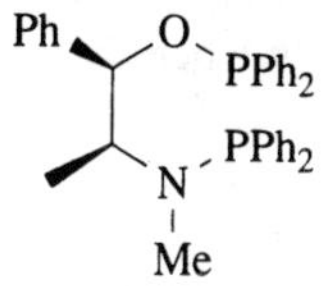 **(15)**

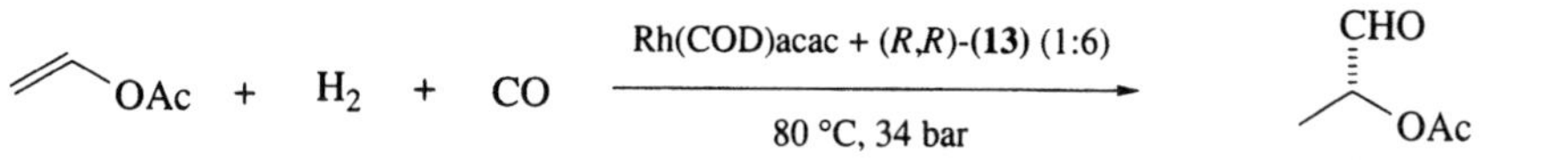

$$\text{CH}_2\text{=CHOAc} + H_2 + CO \xrightarrow[\text{80 °C, 34 bar}]{\text{Rh(COD)acac + }(R,R)\text{-}\mathbf{(13)}\text{ (1:6)}} \text{CH}_3\text{CH(CHO)OAc} \quad (32)$$

51% *ee*
75–95% selectivity

Although the enantiomeric excesses, selectivity and *n*/*b* ratios are dependent on such variables as the H_2/CO ratio, the total pressure, the substrate concentration and the ratio of phosphorus to platinum, no consistent pattern emerges to reveal much information as to the reaction mechanism.[55] The hydroformylation of a variety of substrates has been carried out, with enantiomeric excesses ranging from 20 to 44%,[56] usually with DIOP (**12**) or CHIRAPHOS (**14**) as ligands.[48,56] The highest enantiomeric excesses (73% and 85%, respectively) obtained from the hydroformylation of styrene were with [(–)-DBP-DIOP]$PtCl_2$/$SnCl_2$ at 40–60 °C, 220 bar, H_2/CO = 2.4[57] and (*R*,*R*)-2,3-bicyclo[2.2.2]octanediylbis(methylene)bis(diphenylphosphine)$PtCl_2$/$SnCl_2$ (equation 33).[58]

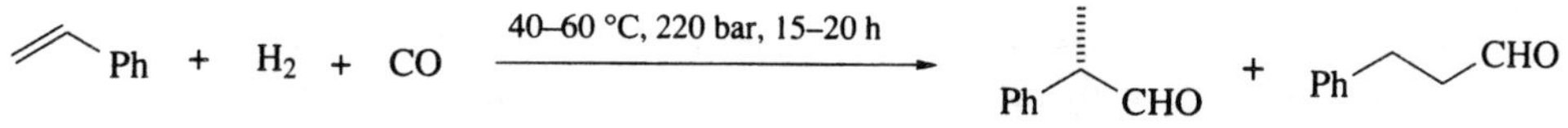

$$\text{CH}_2\text{=CHPh} + H_2 + CO \xrightarrow{\text{40–60 °C, 220 bar, 15–20 h}} \text{PhCH(CH}_3\text{)CHO} + \text{PhCH}_2\text{CH}_2\text{CHO} \quad (33)$$

5:1

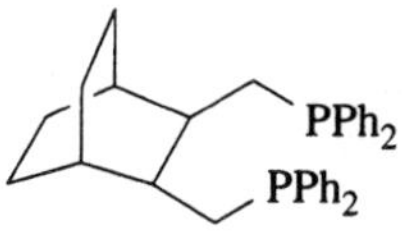

(16)	$PtCl_2$•$SnCl_2$ (**13**)	73% *ee*
	(**16**)•$PtCl_2$•$SnCl_2$	85% *ee*

Thus the enantiomeric excesses obtained in the platinum-catalyzed hydroformylation reactions of certain alkenes were achieved at a level (75–85% *ee*) necessary for facile enrichment to optically pure compounds, presenting an opportunity to establish this reaction as a viable asymmetric synthesis of aldehydes.

Some of the highest enantiomeric excesses in hydroformylation reactions of styrene (~70–80% *ee*) have been obtained with a platinum catalyst containing the chiral ligand (2*S*,4*S*)-4-(diphenylphosphino)-2-[(diphenylphosphino)methyl]pyrrolidine (**17**; (–)-BPPM).[59]

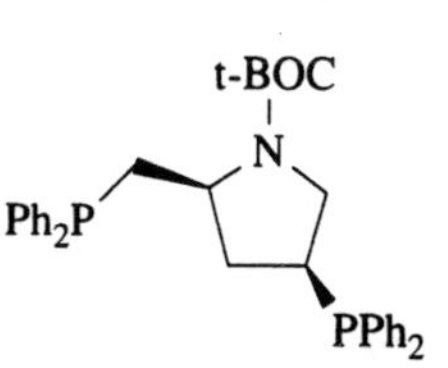

(17)

Styrene hydroformylation with [(–)-BPPM]$PtCl_2$ and added $SnCl_2$ to yield a mixture of 2- and 3-phenylpropanal has been carried out in benzene at different reaction pressures (100–180 bar), temperatures (50–95 °C) and reaction times (2–15 h; equation 34).[59] The normal to branched ratio was relatively high

($n/b = 2$), but the selectivity to aldehyde was high, less than 2% of ethylbenzene being obtained in each case. Branched aldehyde, 2-phenylpropanal, was obtained in relatively high enantiomeric excess, particularly when the reaction times were short and the temperature was low, 78–80% *ee* being obtained at 56–57 °C after 2–4 h, and low conversion. The lower enantiomeric excesses obtained at longer reaction times are a result of product racemization under the reaction conditions.

The hydroformylation reaction of vinyl aromatics (Table 4)[60] lends itself to the synthesis of a number of 2-arylpropionic acids in high enantiomeric excess that are nonsteroidal antiinflammatory agents.[61] Previous asymmetric syntheses of these acids required the use of stoichiometric amounts of chiral auxiliaries, which in most cases are not easily recovered. The branched aldehyde was oxidized to (*S*)-(+)-naproxen,[62] in 84% yield.

$$\text{CH}_2{=}\text{CHPh} + H_2 + CO \xrightarrow[\text{170 bar, 55–60 °C}]{PtCl_2{\bullet}SnCl_2\ (\mathbf{17})} \text{PhCH(CH}_3\text{)CHO} + \text{PhCH}_2\text{CH}_2\text{CHO} \quad (1:2) \tag{34}$$

An analgesic with greater antiinflammatory and antipyretic activity than Naproxen is α-methyl-4-(2-thienylcarbonyl)benzeneacetic acid, Suprofen.[63] This compound has been tested only as a racemic mixture, and any difference in the activity of the enantiomers is not known. Hydroformylation of 4-(2-thienylcarbonyl)styrene (entry 5, Table 4) to the 2-arylpropanal was achieved in 78% *ee* for the branched aldehyde [(+)-enantiomer in excess], which can be converted to Suprofen by oxidation.

Hydroformylations of several other vinyl substrates were carried out (Table 4). Vinyl acetate is a particularly interesting substrate, since, as mentioned earlier, it gives 2-acetoxypropanal, a precursor in the Strecker synthesis of threonine. The product can be converted to 2-hydroxypropanal, a useful intermediate in the synthesis of steroids, pheromones, antibiotics and peptides.

Hydroformylation of norbornene proceeds slowly in spite of the expected reactivity of the strained double bond. Racemization of the product aldehyde does not occur under the reaction conditions. The hydroformylation of methyl methacrylate gives only one isomer, which is a useful chiral synthon. Dimethyl itaconate also can be hydroformylated using (**12**) as the chiral ligand to give aldehyde product in 82% *ee* accompanied by hydrogenated ester.[64]

In the hydroformylation reactions of vinyl aromatics, racemization of the products takes place under the reaction conditions. Presumably racemization of the aldehydes also can take place in hydroformylations carried out with substrates other than vinyl aromatics, although this was not verified in all examples.

The hydroformylation of styrene in triethyl orthoformate is slower than that observed in benzene, but a 98% *ee* is obtained, since racemization of the product acetal does not occur. Hydrolysis of the acetal to the aldehyde can be accomplished without racemization. A number of other substrates are hydroformylated in the presence of triethyl orthoformate. The reactions are slower, but with all substrates tried except norbornene, enantiomerically pure products can be obtained.

4.5.3 CARBOXYLATION[1,2,65]

The hydrocarboxylation reaction of alkenes and alkynes is one which utilizes carbon monoxide to produce carboxylic acid derivatives. The source of hydrogen is a protic solvent (equation 35); dihydrogen is not usually added to the reaction. There are a number of variations to this reaction, since the solvent can be water, alcohols, amines, acids, *etc*. The catalysts can be Group VIII–X transition metals, but cobalt, rhodium, nickel, palladium and platinum have found the most use.

$$\text{R}_2\text{C}{=}\text{CR}_2 + CO + HX \longrightarrow \text{H–CR}_2\text{–CR}_2\text{–C(=O)X} \tag{35}$$

X = OH, OR, SR, NR_2, OCOR, Cl

Industrially, low molecular weight acids can be made from water, carbon monoxide and an alkene, but higher molecular weight acids are produced as esters. In water a phase-transfer agent is required for the

Table 4[60] Hydroformylation of Alkenes[a]

Entry	*Alkene*	*Reaction time* (h)	*Solvent*	*Product*	*Conversion* (%)	*n/b*	*ee* (%)
1	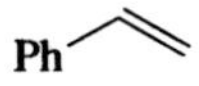 Ph	4	PhH	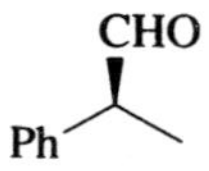 CHO, Ph	50–90	2	80
2		150	$HC(OEt)_3$	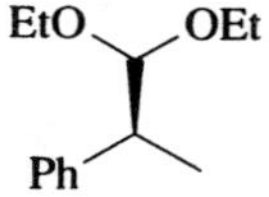 EtO, OEt, Ph	100	2	>96
3	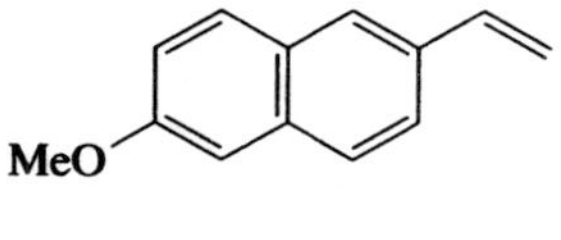 MeO	9	PhH	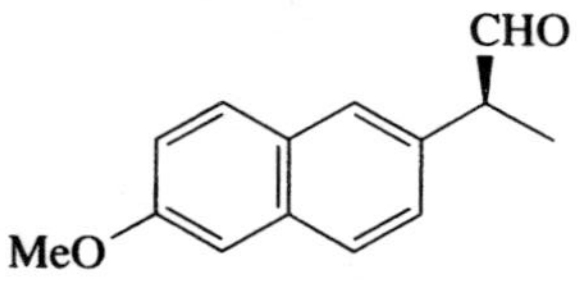 CHO, MeO	90	1.4	81
4		200	$HC(OEt)_3$	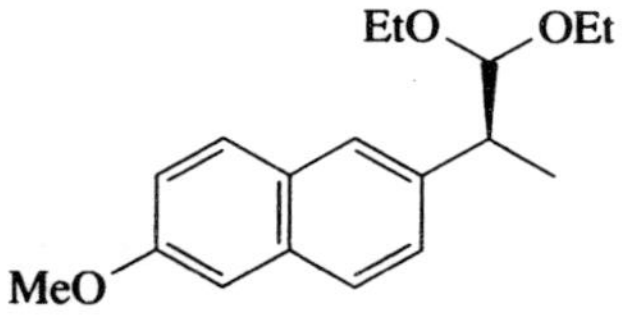 EtO, OEt, MeO	55	1.4	>96
5	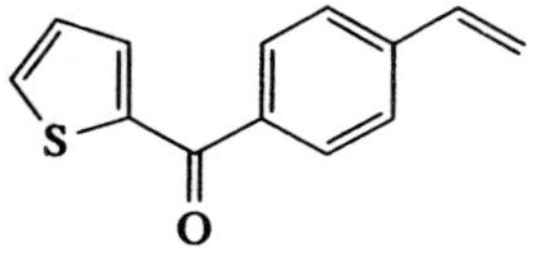 S, O	9	PhH	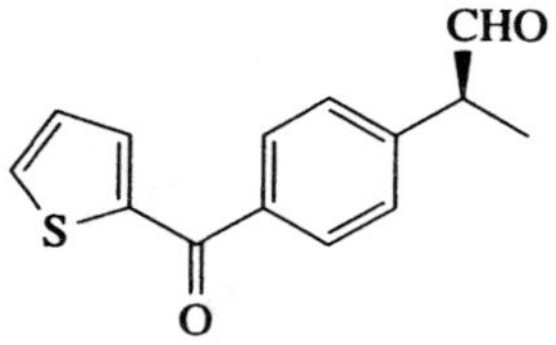 CHO, S, O	73	2	78
6	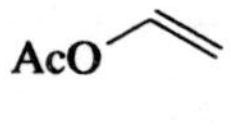 AcO	40	PhH	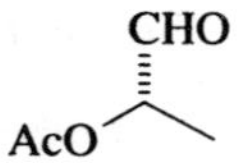 CHO, AcO	50	2.3[b]	82

Table 4 *(continued)*

Entry	*Alkene*	*Reaction time* (h)	*Solvent*	*Product*	*Conversion* (%)	*n/b*	*ee* (%)
7	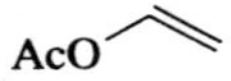 AcO	240	$HC(OEt)_3$	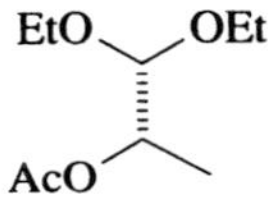 EtO, OEt, AcO	25	0.67[b]	>98
8	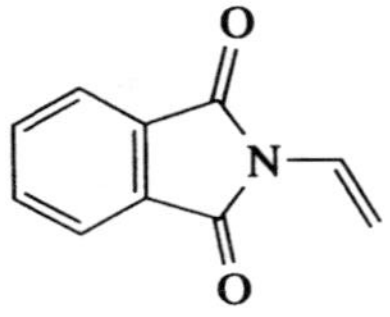 O, N, O	46	PhH	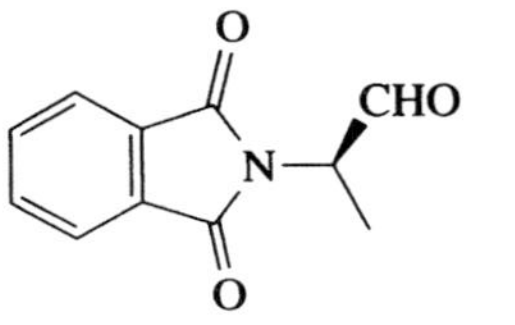 O, N, CHO, O	52	2	73
9		240	$HC(OEt)_3$	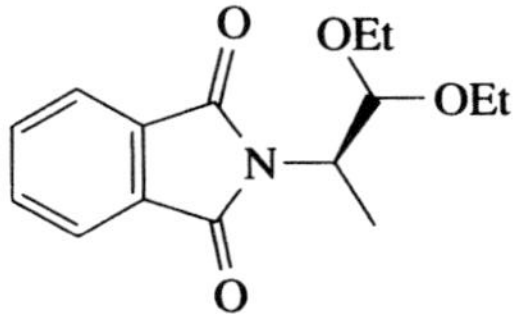 O, OEt, OEt, N, O	46	2	>96
10		20	PhH	 CHO, H	84	c	60
11		140	$HC(OEt)_3$	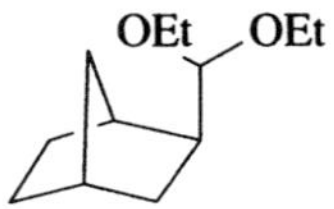 OEt, OEt	100	c	60
12	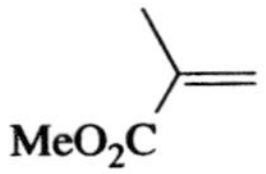 MeO_2C	50	PhH	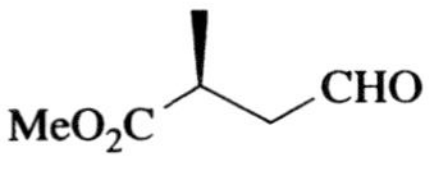 MeO_2C, CHO	36	c	60

Table 4 *(continued)*

Entry	*Alkene*	*Reaction time* (h)	*Solvent*	*Product*	*Conversion* (%)	*n/b*	*ee* (%)
13	MeO_2C–C(=CH_2)–CH_2CO_2Me	45	PhMe[d,e]	MeO_2C–CH(CH_2CHO)–CH_2CO_2Me	28	f	82

[a] Entries 1–12: **(17)**•$PtCl_2$•$SnCl_2$; $H_2/CO = 1$; temperature = 60 °C (except entry 1, 55 °C and entries 10 and 11, 30 °C); pressure = 165–185 bar; selectivities > 98% except entry 8, 85%. [b] Linear aldehyde and acetal partially decomposed. [c] Only one aldehyde (or acetal) obtained. [d] $H_2/CO = 3$. [e] From ref. 64: **(12)**•$PtCl_2$•$SnCl_2$; $H_2/CO = 1$; temperature = 50 °C; pressure = 40 bar. [f] 45% of product was $MeCH(CO_2Me)CH_2CO_2Me$, 35% *ee*.

higher molecular weight alkenes. A problem with aqueous hydrocarboxylation is that most of the catalysts also effect the water gas shift reaction.

4.5.3.1 Mechanisms[66]

There are two different reaction mechanisms that are possible for the overall transformation shown in equation (35). It is not always apparent which of the two mechanisms may be operative, the mechanistic pathway apparently depending on the reaction conditions (particularly added acids or bases), the transition metal and the substrate.

The hydrocarboxylation can take place by insertion of the alkene into a metal–hydride bond followed by CO insertion and finally reaction of the acyl complex with solvent as illustrated in equation (36). Alternatively, a transition metal–carboxylate complex can be generated initially. Insertion of the alkene into the metal–carbon bond of this carboxylate complex followed by cleavage of the metal–carbon bond by solvent completes the addition, as shown in equation (37). Both sequences provide the same product.

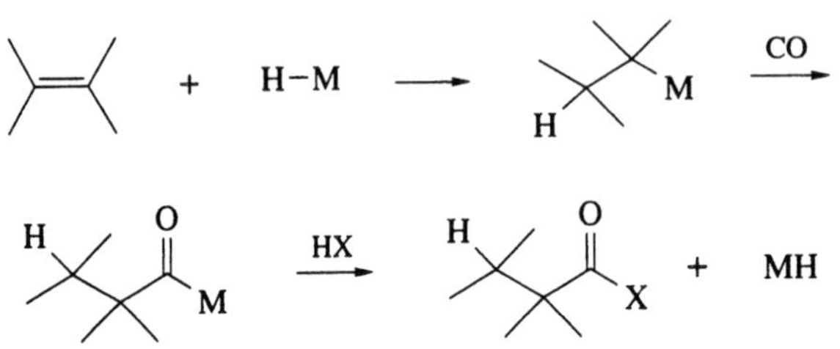

+ H–M → (H, M) —CO→

(H, O, M) —HX→ (H, O, X) + MH (36)

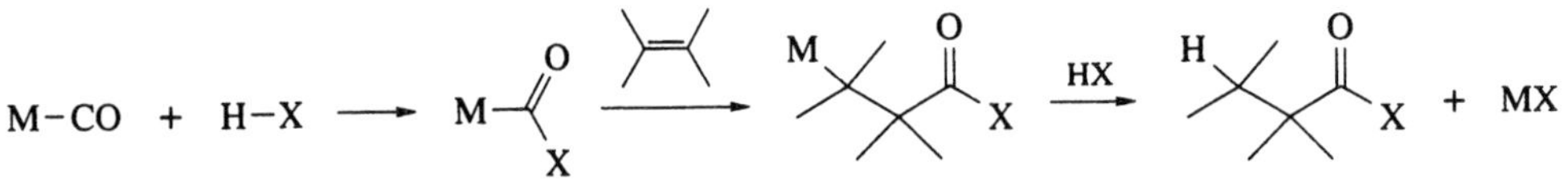

M–CO + H–X → M–C(=O)X → (M, O, X) —HX→ (H, O, X) + MX (37)

The metal hydride mechanism has been written particularly for hydrocarboxylation reactions with a palladium catalyst.[67,68] In the reactions of propene in the presence of $(Ph_3P)_2PdCl_2$, the acyl complex (**18**) was isolated from the reaction mixture, and also shown to be a catalyst for the reaction.

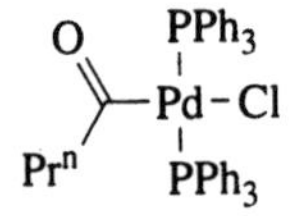

Pr^n–C(=O)–Pd(PPh_3)$_2$–Cl

(18)

Both mechanisms are predicted to show *syn* addition of hydride and carboxylate to the alkene. In the metal hydride mechanism (equation 36) alkene insertion is *syn* and CO insertion proceeds with retention of configuration at carbon. In the metal carboxylate mechanisms (equation 37) alkene insertion is *syn* and cleavage of the metal–carbon bond can take place with retention at carbon. The palladium-catalyzed hydroesterification reaction produces the *erythro* ester from (*Z*)-3-methyl-2-pentene (equation 38) and the *threo* ester from (*E*)-3-methyl-2-pentene (equation 39).[69]

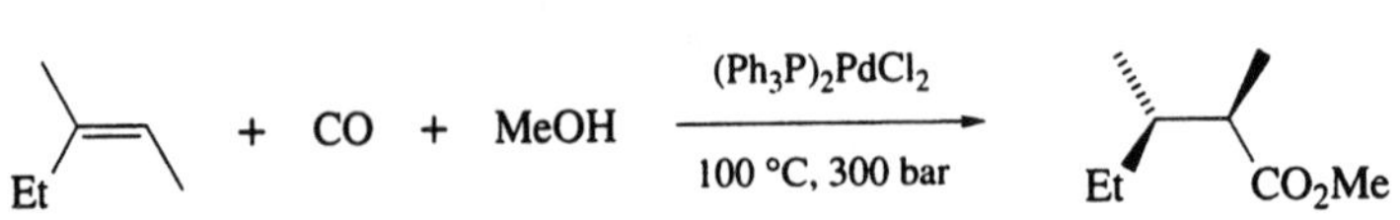

Et + CO + MeOH —$(Ph_3P)_2PdCl_2$, 100 °C, 300 bar→ Et, CO_2Me (38)

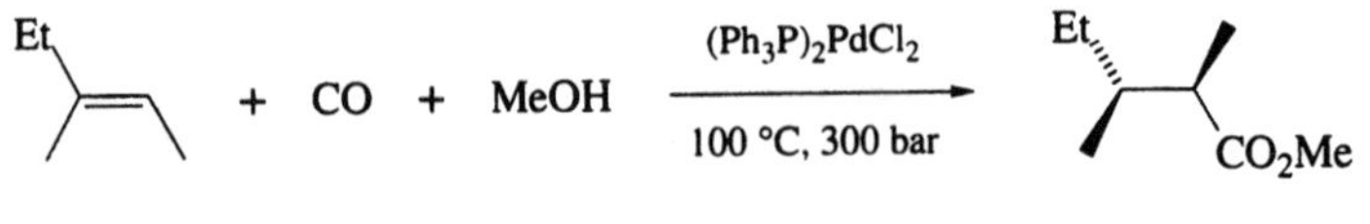

Et + CO + MeOH —$(Ph_3P)_2PdCl_2$, 100 °C, 300 bar→ Et, CO_2Me (39)

In certain other systems, there is compelling evidence for the insertion into a metal–carboxylate complex (equation 37). For example, in the synthesis of α-methylene-γ-lactones from alkynic alcohols,[70,71] no double bond rearrangement to a butenolide occurs, a reaction shown to take place in the presence of transition metal hydrides. The source of the vinyl proton (deuterium) on the α-methylene group is indeed the alcohol function. Finally, palladium carboxylate complexes containing alkynic (equation 40) or vinyl tails (equation 41) can be isolated and the corresponding insertion reaction can be observed.

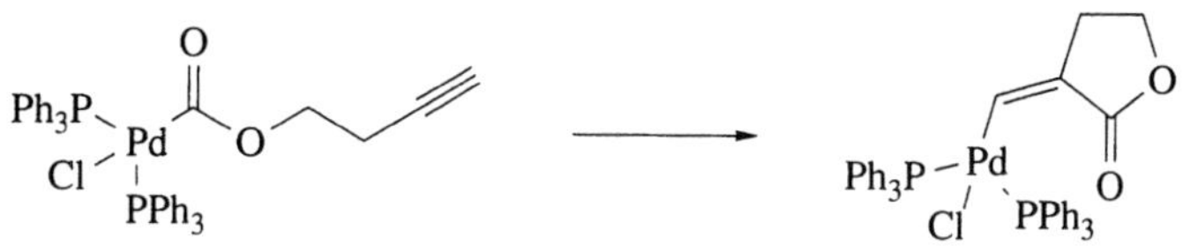

(40)

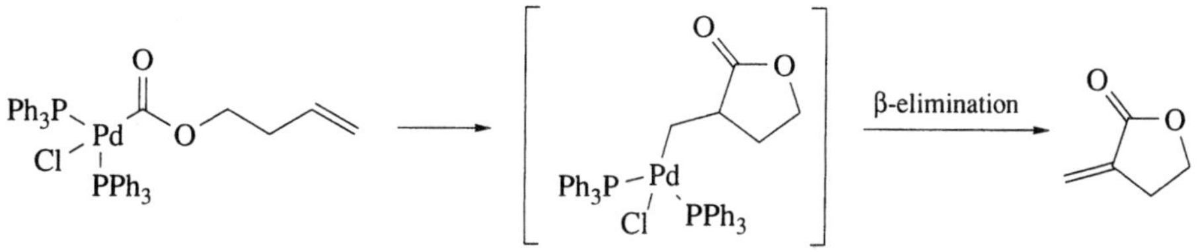

β-elimination

(41)

The metal carboxylate insertion mechanism has also been demonstrated in the dicobaltoctacarbonyl-catalyzed carbomethoxylation of butadiene to methyl 3-pentenoate.[66,72] The reaction of independently synthesized cobalt–carboxylate complex (**19**) with butadiene (Scheme 8) produced η^3-cobalt complex (**20**) *via* the insertion reaction. Reaction of (**20**) with cobalt hydride gives the product. The pyridine–CO catalyst promotes the reaction of methanol with dicobalt octacarbonyl to give (**19**) and $HCo(CO)_4$.

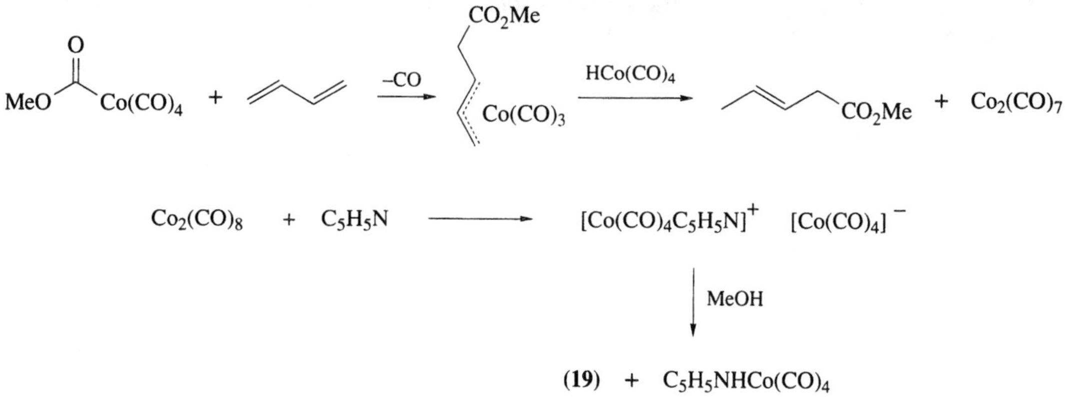

$MeOC(O)Co(CO)_4$ + butadiene $\xrightarrow{-CO}$ (η^3-allyl, CO_2Me)$Co(CO)_3$ $\xrightarrow{HCo(CO)_4}$ $CH_3CH{=}CHCH_2CO_2Me$ + $Co_2(CO)_7$

$Co_2(CO)_8$ + C_5H_5N ⟶ $[Co(CO)_4C_5H_5N]^+$ $[Co(CO)_4]^-$ $\xrightarrow{MeOH}$ (**19**) + $C_5H_5NHCo(CO)_4$

Scheme 8

Thus the catalytic cycles that have been written reflect these mechanistic findings. In the hydride mechanism, illustrated with cobalt (Scheme 9), the hydride is generated either by the alcohol *vide supra* or by adventitious hydrogen (*e.g.* from a water gas shift reaction). In the case of a palladium catalyst introduced as $(Ph_3P)_2PdCl_2$, the palladium is first reduced to palladium(0), which then undergoes the rapid oxidative addition of HCl to yield $(Ph_3P)_2Pd(Cl)H$. Alkene insertion and carbon monoxide insertion steps are the characteristic sequence of events in such catalysis. The hydride $HCo(CO)_4$ is acidic and readily decomposes as cobalt metal in the absence of carbon monoxide; high CO pressures are necessary to stabilize $HCo(CO)_4$ and the intermediates in the catalytic cycle.

The mechanism of the reaction of the alcohol (or water) with the acyl complex to produce ester (or acid) and regenerate the cobalt hydride complex is not known. Because the reaction of the analogous manganese complex with alcohols is known to proceed through a hemiacetal-like complex, this mechanism has been written for the carboxylation reaction (equation 42).

The transition metal carboxylate mechanism (Scheme 10), illustrated with palladium, requires the generation of this type of complex in the initial steps. This can occur by the alcoholysis of dicobalt octacarbonyl (Scheme 6) or by, for example, the reaction of methanol with a palladium(II) carbonyl (equa-

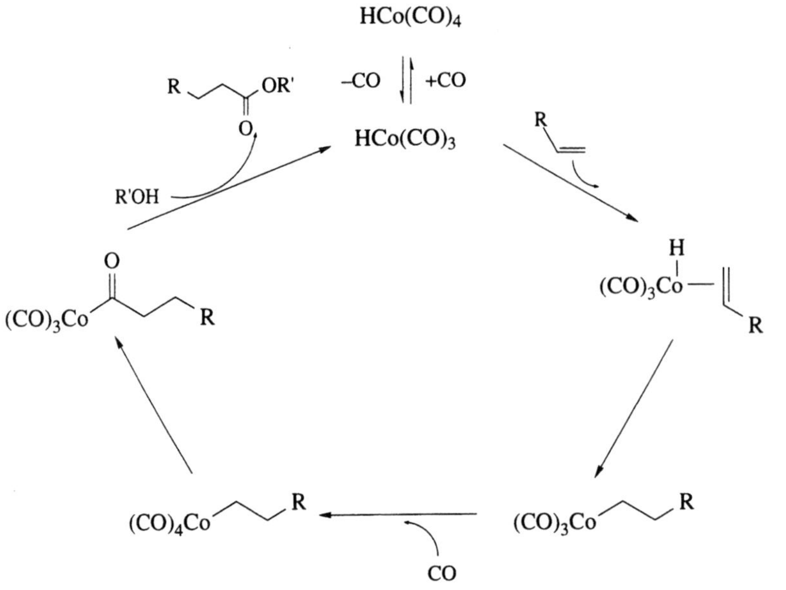

$HCo(CO)_4$

$-CO$ ⇌ $+CO$

$HCo(CO)_3$

R'OH

$(CO)_3Co$ ... R

$(CO)_4Co$ ... R

CO

Scheme 9

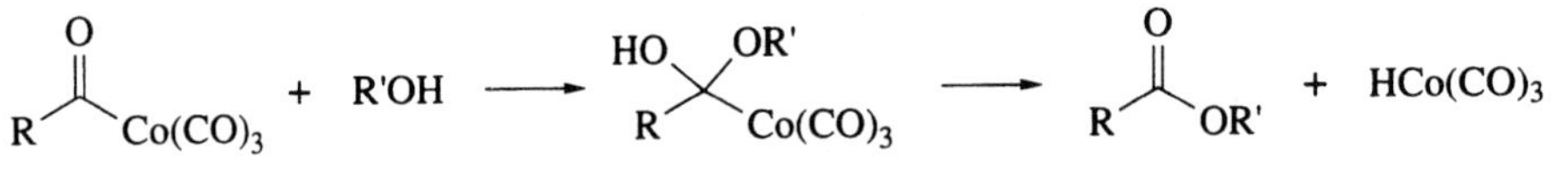

$$RC(O)Co(CO)_3 + R'OH \longrightarrow R(HO)(OR')C\text{-}Co(CO)_3 \longrightarrow RC(O)OR' + HCo(CO)_3 \quad (42)$$

tion 43). The palladium catalyst containing a trichlorostannate ligand, an excellent catalyst for the hydrocarboxylation, has been shown to proceed by the palladium carboxylate insertion mechanism.[70,71]

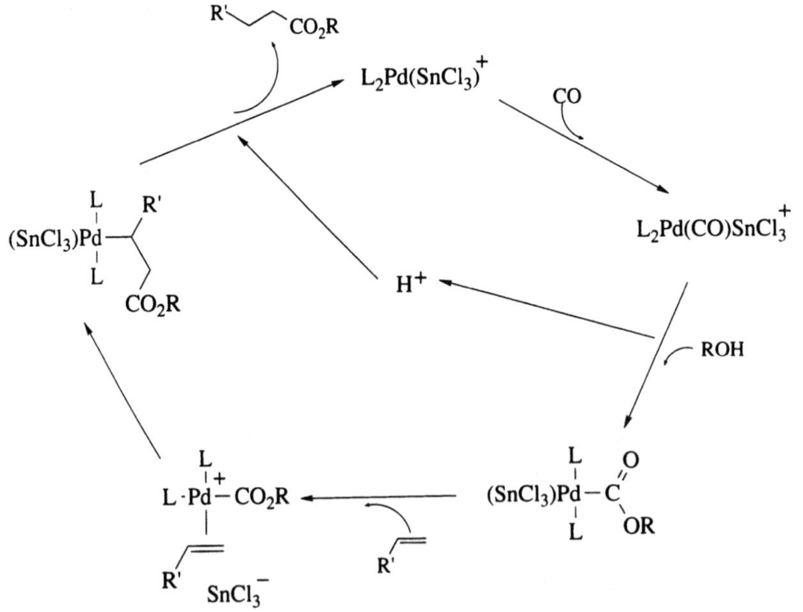

$R'CH_2CH_2CO_2R$

$L_2Pd(SnCl_3)^+$

CO

$L_2Pd(CO)SnCl_3^+$

H^+

ROH

$(SnCl_3)PdL_2\text{-}C(O)OR$

$L_3Pd^+\text{-}CO_2R$ (alkene R'), $SnCl_3^-$

$(SnCl_3)PdL_2\text{-}CH(R')CH_2CO_2R$

Scheme 10

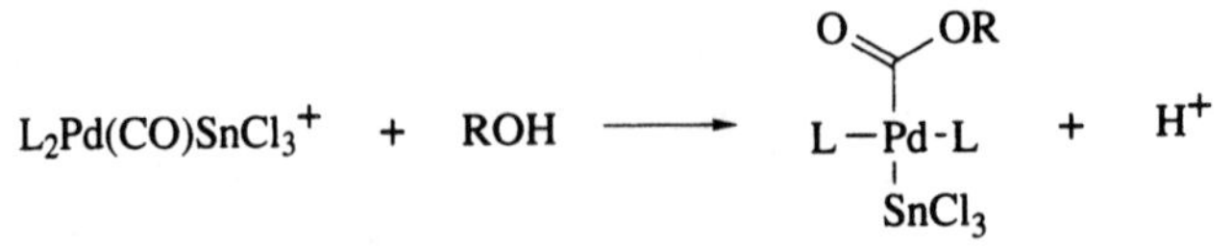

$$L_2Pd(CO)SnCl_3^+ + ROH \longrightarrow L\text{-}Pd(CO_2R)(SnCl_3)\text{-}L + H^+ \quad (43)$$

The final step in the catalytic cycle is the cleavage of the metal–alkyl bond with acid, which must take place faster in the hydrocarboxylation of alkenes than β-elimination.

4.5.3.2 Catalysts and Products

Nickel catalysts are utilized for the industrial synthesis of acrylic acid or esters either in a semicatalytic process with $Ni(CO)_4$ or a catalytic process with $NiBr_2$ (equation 44).[73] The reaction is carried out in THF containing water or alcohol (to avoid acetylene detonation at 60 bar).

$$\mathrm{H{-}C{\equiv}C{-}H + CO + H_2O \xrightarrow[\text{200 °C, 60 bar, THF}]{NiBr_2} CH_2{=}CH{-}CO_2H} \quad (44)$$

Most industrial carboxylation reactions utilize cobalt catalysts, which produce predominately linear acids or esters from 1-alkenes. The hydroesterification of 1-pentene produces 70% methyl hexanoate and 30% branched esters (entry 1, Table 5). Hydroesterification reactions utilizing the palladium catalyst $(Ph_3P)_3PdCl_2$[74] take place under milder conditions than required for a cobalt catalyst. However, the branched isomer is produced in larger amounts (entry 2, Table 5). As in the hydroformylation reaction, the addition of excess phosphine improves the *n*/*b* ratio.[75] The hydrocarboxylation reaction of styrene under these conditions yields the branched ester almost exclusively. Chelating phosphines such as 1,4-bis(diphenylphosphino)butane shift the *n*/*b* ratio for styrene to 1/1.[76] The addition of tin(II) chloride to the palladium catalyst shifts the isomer ratio dramatically.[77] Thus 87% selectivity to the straight chain product is achieved. This combined with the relatively mild conditions makes the palladium-catalyzed hydroesterification particularly attractive for the production of straight chain acids (entry 3, Table 5). It is likely that under these latter conditions a metal carboxylate mechanism is taking place, while without added tin(II) chloride, the metal hydride mechanism is operative. Hydroesterification reactions of $C_6F_5CH{=}CH_2$ and $CF_3CH{=}CH_2$ show approximately the same regioselectivities with these palladium catalysts.[41]

Platinum also catalyzes hydrocarboxylation, under comparable reaction conditions (240 bar, 80 °C). The best ligand in this case is triphenylarsine and a 10-fold excess of tin(II) chloride provides even higher selectivity to the linear ester (entry 4, Table 5).[78]

Branched acids and esters are obtained from the palladium-catalyzed reaction in the absence of phosphines, and in the presence of copper chloride and HCl.[79] The mild reaction conditions and the regiospecificity make this a very attractive carboxylation procedure (entry 5, Table 5). Internal straight chain alkenes can be hydrocarboxylated, but the rates are slower and the reaction is not regiospecific.

4.5.3.3 Hydrocarboxylation of Alkenes

Although the hydrocarboxylation of 1-alkenes is not of interest for the synthesis of more complex organic molecules, the information obtained from the hydrocarboxylation reactions with various catalysts can be applied to the synthesis and reactions of other alkene substrates.

The reaction with norbornene gives the *exo* acid (equation 45).[2] Under mild conditions, vinylcyclohexene is carbonylated at the least-substituted double bond; higher temperatures (120 °C) are necessary to attain the diester (equation 46).[74,80]

$$\text{norbornene} + \mathrm{CO + H_2O \xrightarrow[\text{HCl, 80\%}]{(Ph_3P)_2PdCl_2}} \textit{exo}\text{-norbornane-2-carboxylic acid } (\mathrm{CO_2H}) \quad (45)$$

$$\text{4-vinylcyclohexene} + \mathrm{CO + MeOH \xrightarrow[\text{60 °C, 300–700 bar; 85–98\%}]{(Ph_3P)_2PdCl_2}} \text{cyclohexenyl ester } (\mathrm{CO_2Me}) \quad (46)$$

Table 5 Hydroesterification Reaction of 1-Alkenes with Methanol

Entry	*Alkene*	*CO/pressure* (bar)	*Temperature* (°C)	*Catalyst*	*Yield* (%)	*Products*	*Relative yield* (%)
1		100–200	140–170	$Co_2(CO)_8$	80–90	CO_2Me	70
						CO_2Me	30
2		35	80–110	$(Ph_3P)_2PdCl_2$	80 (conv.)	CO_2Me	60
						CO_2Me	40
3		138	80	$(Ph_3P)_2PdCl_2$, $10SnCl_2$	96 (conv.)	CO_2Me	87
						CO_2Me	14
4		240	80	$(Ph_3As)_2PtCl_2$, $10SnCl_2$	95 (conv.)	CO_2Me	98
5		1	25	$PdCl_2$, $10CuCl_2$ O_2, HCl	100	CO_2Me	100

The reaction of nonconjugated dienes such as 1,4-pentadiene and 1,5-hexadiene yields cyclopentanone derivatives, while 1,6-heptadiene gives only traces of ketone (equation 47).[1,2,80] The yields in these reactions are low, 1,4-pentadiene giving only 10% of cyclic ketone product.

$$+ \ CO + MeOH \xrightarrow[50\%]{(Ph_3P)_2PdCl_2,\ HCl} \ (CO_2Me) \quad (47)$$

The hydrocarboxylation reaction of simple alkenes and alkynes in the presence of primary or secondary amines or ammonia yields amides (equations 48 and 49). The fact that formamides can be used in place of amines suggests that a key intermediate in the reaction is the hydride metal carboxamide (**20**).

$$R\text{–CH=CH}_2 + CO + R^1R^2NH \longrightarrow R\text{–CH}_2\text{CH}_2\text{–C(O)NR}^1R^2 + R\text{–CH(CH}_3)\text{–C(O)NR}^1R^2 \quad (48)$$

$$R\text{–C}\equiv\text{CH} + CO + R^1R^2NH \longrightarrow R\text{–CH=CH–C(O)NR}^1R^2 + R\text{–C(=CH}_2)\text{–C(O)NR}^1R^2 \quad (49)$$

$$HM\text{–C(O)–}NR_2$$

(20)

4.5.3.4 Alkenes and Alkynes Containing Functional Groups

The hydrocarboxylation of *N*-vinylphthalimide catalyzed by palladium takes place regioselectively, under mild conditions, to give the α-amino acid derivative (equation 50).[39]

$$\text{N-vinylphthalimide} + CO + MeOH \xrightarrow[70\ ^\circ C,\ 94\ bar,\ 27\ h,\ 70\%]{(Ph_3P)_2PdCl_2} \text{PhthN–CH(CH}_3)\text{–CO}_2Me \quad (50)$$

When either an alcohol or an amine function is present in the alkene, the possibility for lactone or lactam formation exists. Cobalt or rhodium catalysts convert 2,2-dimethyl-3-buten-1-ol to 2,3,3-trimethyl-γ-butyrolactone, with minor amounts of the δ-lactone being formed (equation 51).[2] In this case, isomerization of the double bond is not possible. The reaction of allyl alcohols catalyzed by cobalt or rhodium is carried out under reaction conditions that are severe, so isomerization to propanal occurs rapidly. Running the reaction in acetonitrile provides a 60% yield of lactone, while a rhodium carbonyl catalyst in the presence of an amine gives butane-1,4-diol in 60–70% (equation 52).[81] A mild method of converting allyl and homoallyl alcohols to lactones utilizes the palladium chloride/copper chloride catalyst system (Table 6).[79,82,83]

The synthesis of α-methylenelactones from alkynic alcohols can be effected in 84% yield utilizing $HPt(SnCl_3)(PPh_3)_2$ (equation 53).[84] This is essentially the same α-methylenelactone synthesis elegantly developed earlier utilizing a palladium catalyst (Table 7).[85] The starting *trans* alkynic alcohols are ob-

$$\text{OH} + CO \xrightarrow[125\text{–}250\ ^\circ C,\ 70\text{–}300\ bar]{Co_2(CO)_8} \text{lactone (51\%)} \quad (51)$$

Table 6 Hydrocarboxylation of Alkenols[a]

Entry	*Substrate*	*Product*	*Yield (%)*	*cis:trans*
1	 OH	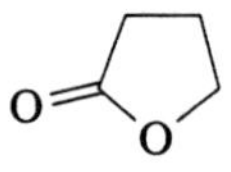 O O	35	
2	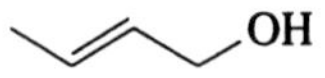 OH	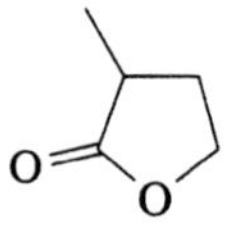 O O	50	
3	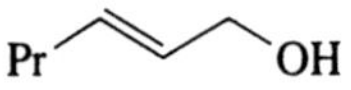 Pr OH	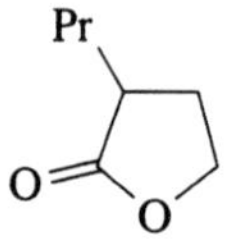 Pr O O	35	
4	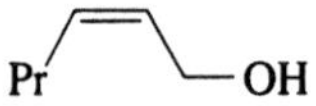 Pr OH	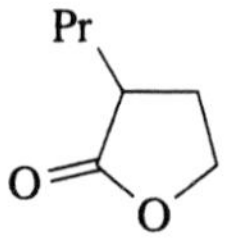 Pr O O	60	
5	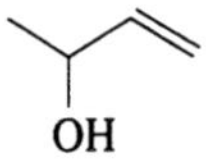 OH	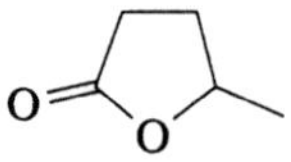 O O	70	
6	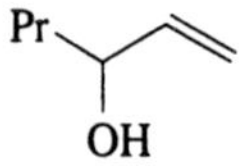 Pr OH	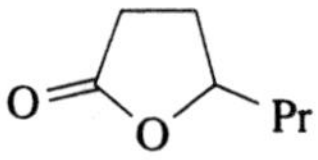 O O Pr	40	
7	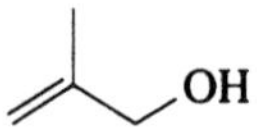 OH	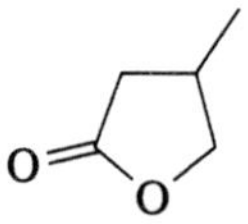 O O	15	
8	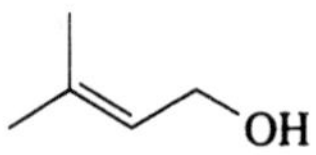 OH	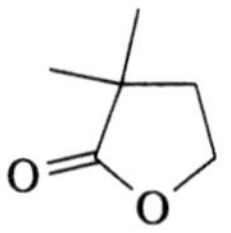 O O	45	
9	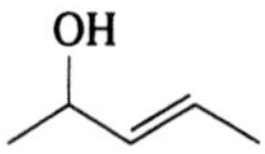 OH	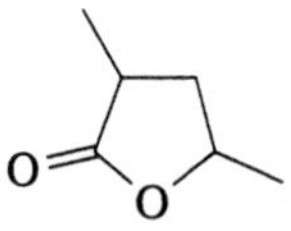 O O	65	1:1
10	 OH	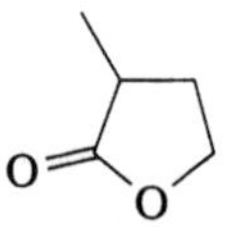 O O	60	
11	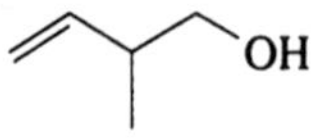 OH	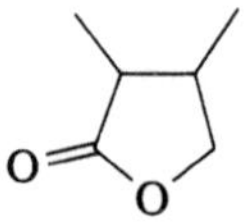 O O	42	1:2
12	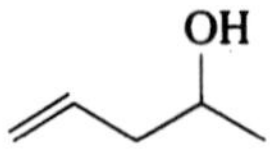 OH	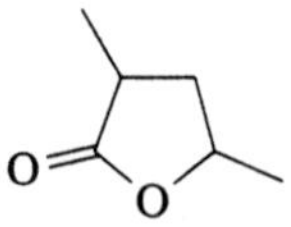 O O	80	1:1

Table 6 *(continued)*

Entry	*Substrate*	*Product*	*Yield (%)*	*cis:trans*
13	OH	O O	75	
14	OH	O O	50	2:3

[a] $PdCl_2$:$CuCl_2$ (<1:5 or 1:10), HCl, O_2, 25 °C, 1 bar.

OH + CO + H_2O —[$Rh_6(CO)_{11}$; 10 bar, 60 °C]→ HO OH; EtO OH; Me_2N NMe_2; 60–70% (52)

tained by the reaction of the corresponding epoxides with aluminum acetylides, the *cis* alcohols are obtained from the *trans* by an oxidation–reduction sequence (equation 54). The reaction does not work well for 3-butenol, only a 26% yield being obtained.

HO + CO —[$HPt(SnCl_3)(PPh_3)_2$; 100 °C, 80 bar; MeCN, 84%]→ O O (53)

$(CH_2)_n$ H O H —[$Et_2O•AlMe_2$]→ $(CH_2)_n$ OH —[i, Cr_2O_7, H^+; ii, L-selectride]→ $(CH_2)_n$ OH (54)

Allylamines cyclize readily with a dicobalt octacarbonyl catalyst (equation 55).[1,2] Rhodium catalysis generally allows the carbonylative cyclization to be carried out under milder conditions.[86] Application of this reaction to unsaturated amides yields the corresponding imides, the best yields arising when R^1 = H and R^2 = allyl (equation 56).[1,2]

NHR + CO —[$Co_2(CO)_8$, 125–250 °C, 60–300 bar; or $Rh(CO)(PPh_3)_2Cl$, 120 °C, 136 bar]→ N(R) O (55)

R = H, alkyl, aryl; R = Me, 78%; R = H, 68%

R^1 NHR^2 O + CO —[$Co_2(CO)_8$; 200 °C, 300 bar]→ R^1; O N O; R^2 (56)

Table 7 Synthesis of α-Methylenelactones[a]

Entry	*Substrate*	*Phosphine*	*Product*	*Yield (%)*
1	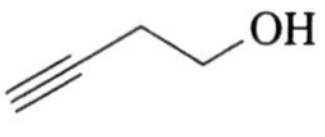 OH	PPh_3	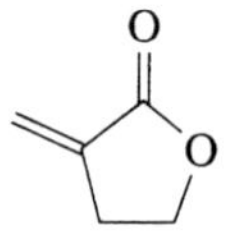 O, O	100
2	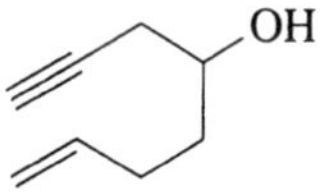 OH	PPh_3	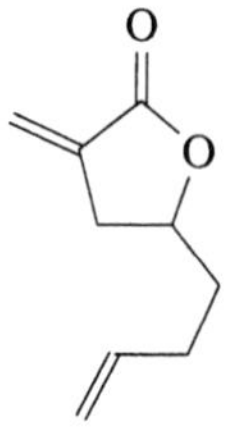 O, O	43
3	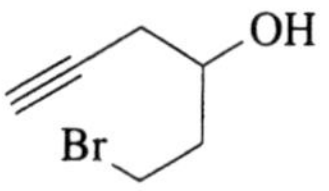 OH, Br	PBu_3	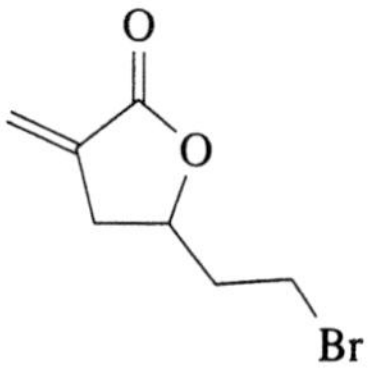 O, O, Br	77
4	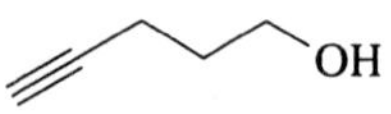 OH	PPh_3[b]	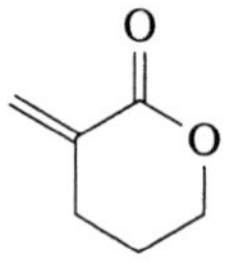 O, O	52
5	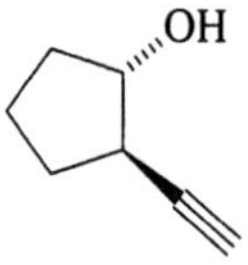 OH	PPh_3[c]	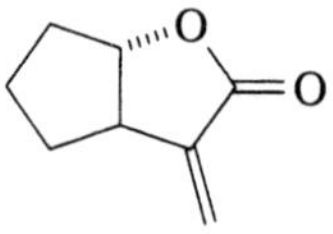 O, O	27
6	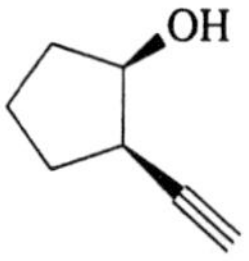 OH	PPh_3[c]	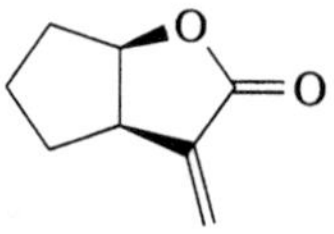 O, O	93
7	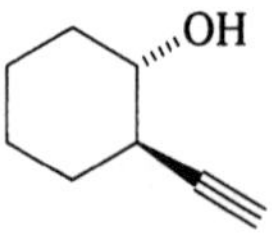 OH	PBu_3	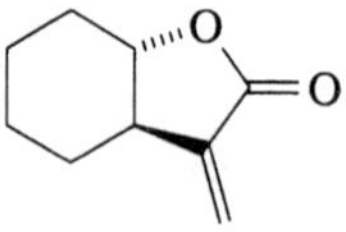 O, O	85
8	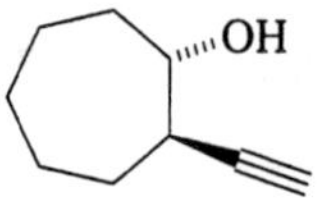 OH	PBu_3	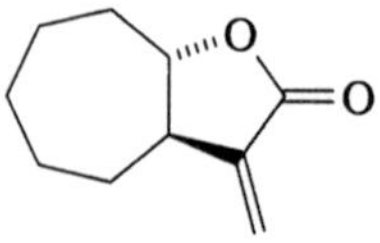 O, O	71
9	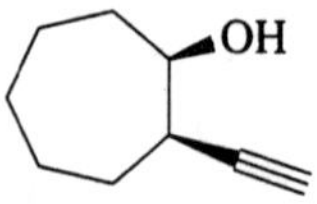 OH	PBu_3	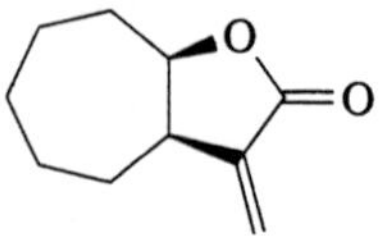 O, O	83

[a] $PdCl_2$:PR_3:$SnCl_2$ 1:1:2 in MeCN, 7.8 bar CO, 75 °C. [b] 5.1 bar CO. [c] 5.7 bar CO, 65 °C.

4.5.3.5 Conjugated Diene Hydrocarboxylation

The hydrocarboxylation of conjugated dienes such as butadiene can yield monocarboxylate, dicarboxylate or diene dimerized carboxylated products. The carboxylation reaction is important because it is a potential route to adipic acid.

Palladium catalysts that are free of halide ions effect the dimerization and carboxylation of butadiene to yield 3,8-nonadienoate esters. Palladium acetate, solubilized by a tertiary amine or an aromatic amine, gives the best yields and selectivities (equation 57).[87] Palladium chloride catalyzes the hydrocarboxylation to yield primarily 3-pentenoates.[88] The hydrocarboxylation of isoprene and chloroprene is regioselective, placing the carboxy function at the least-hindered carbon (82% and 71% selectively); minor amounts of other products are obtained (equation 58). Cyclic dienes such as 1,3-cyclohexadiene and 1,3-cyclooctadiene are similarly hydrocarboxylated.

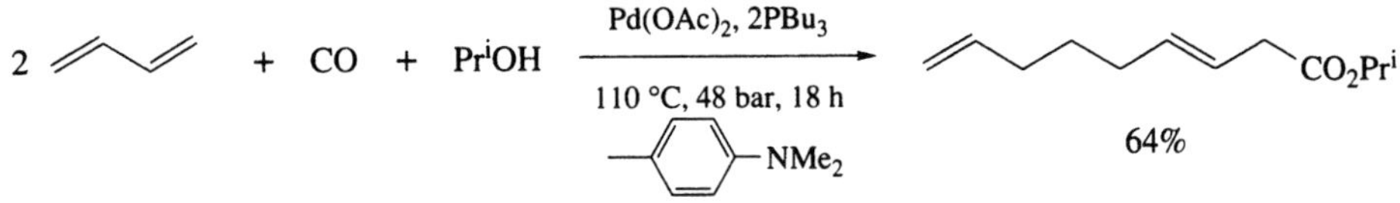

2 [butadiene] + CO + Pr^{i}OH —[$Pd(OAc)_2$, $2PBu_3$; 110 °C, 48 bar, 18 h; p-tolyl-NMe_2]→ [3,8-nonadienoate] CO_2Pr^i (57)

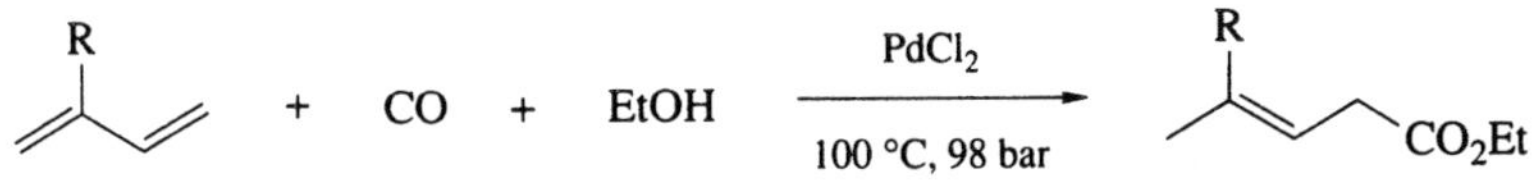

64%

R + CO + EtOH —[$PdCl_2$; 100 °C, 98 bar]→ R CO_2Et (58)

R = H, 96%
R = Me, 82%
R = Cl, 71%

Cobalt is the catalyst of choice for the hydrocarboxylation of butadiene to adipic esters.[89] The reaction is carried out in two steps, the first of which yields methyl-3-butenoate. This product can either be isolated or carried on to dimethyl adipate at high temperatures (Scheme 11). The first hydrocarboxylation occurs by the metal carboxylate insertion mechanism (*vide supra*).

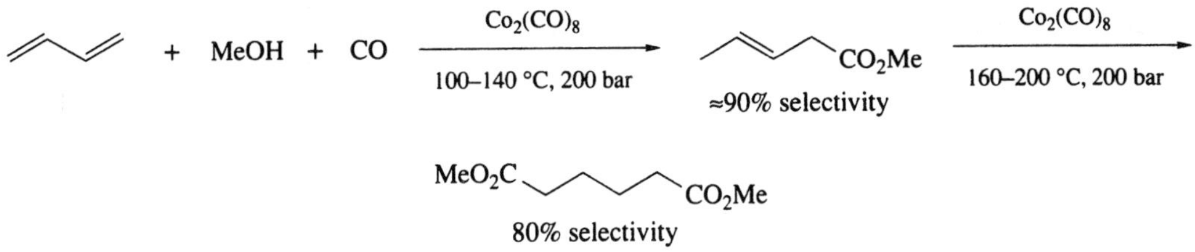

[butadiene] + MeOH + CO —[$Co_2(CO)_8$; 100–140 °C, 200 bar]→ [pentenoate] CO_2Me (≈90% selectivity) —[$Co_2(CO)_8$; 160–200 °C, 200 bar]→

MeO_2C[adipate chain]CO_2Me

80% selectivity

Scheme 11

4.5.3.6 Asymmetric Hydrocarboxylation[42–44]

High enantiomeric excesses in the hydroesterification of alkenes have not been achieved. The reaction carried out with unbranched 1-alkenes or internal alkenes gives products in less than 20% enantiomeric excess (equation 59). The low enantiomeric excesses are due, in part, to the relatively high temperatures necessary to effect the reaction.

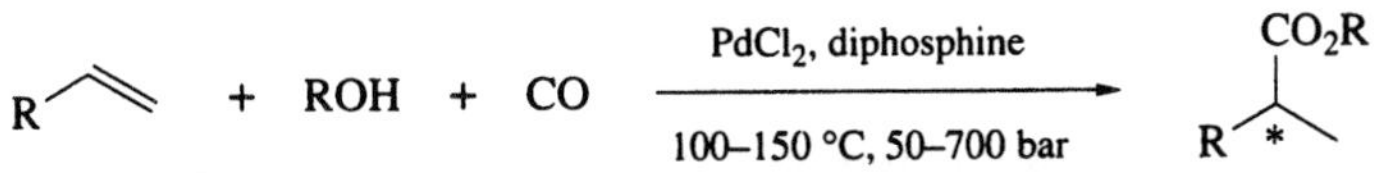

R[alkene] + ROH + CO —[$PdCl_2$, diphosphine; 100–150 °C, 50–700 bar]→ R–*CH(CH$_3$)–CO_2R (59)

Much higher enantiomeric excesses have been achieved in the hydroesterification of α-methylstyrene.[90,91] In this case, the asymmetric center is generated by hydrogen addition at the tertiary site. The reaction is sensitive to the alcohol, the highest enantiomeric excesses being obtained with isopropanol or *t*-butanol (equation 60).

$$\text{PhC(Me)=CH}_2 + \text{ROH} + \text{CO} \xrightarrow{\text{PdCl}_2,\ \text{diphosphine}} \text{PhCH(Me)CH}_2\text{CO}_2\text{R} \quad (60)$$

R	*Solvent*	*Diphosphine*	*Temperature* (°C)	*Pressure* (bar)	*ee* (%)
Bu^t	PhH	DIOP (**12**)	100	45–400	58
Pr^i	dioxane	(**21**)	100	220	52

(**21**)

4.5.4 DICARBOXYLATION[92]

Dicarboxylation reactions of alkenes can be carried out such that predominately 1,2-addition of the two ester functions occurs (equation 61). The reaction takes place under mild conditions (1–3 bar, 25 °C) in alcohol. It is stoichiometric in palladium, since the palladium(II) catalyst is reduced to palladium(0) in the process, but by use of an oxidant (stoichiometric copper chloride or catalytic copper chloride plus oxygen; equation 62 and 63) the reaction becomes catalytic in palladium. In the reoxidation process, water is generated and the build-up of water increases the water gas shift reaction at the expense of the carboxylation. Thus a water scavenger such as triethyl orthoformate is necessary for a smooth reaction.

$$PdCl_2 + R_2C{=}CR_2 + 2CO + 2ROH \longrightarrow RO_2C{-}CR_2{-}CR_2{-}CO_2R + Pd^0 + 2HCl \quad (61)$$

$$2CuCl_2 + Pd^0 \longrightarrow CuCl_2 + PdCl_2 \quad (62)$$

$$CuCl_2 + 2HCl + 1/2O_2 \longrightarrow 2CuCl_2 + H_2O \quad (63)$$

4.5.4.1 Mechanisms

The key palladium intermediate is a carboalkoxypalladium complex formed through the nucleophilic attack by alcohol on carbonyl coordinated to palladium. The addition of a base with the appropriate pK_a [sodium butyrate; $pK_a = 4.82$ (H_2O)] promotes the formation of the palladium carboxylate (**22**).[93] The reaction is a general method for formation of inorganic alkoxycarbonyl derivatives.

$$-Pd^{II}-CO_2Me$$

(**22**)

The sequence of events is: (i) the insertion of the alkene into the palladium–carboxylate bond; followed by (ii) CO insertion into the newly generated palladium–carbon bond; followed by (iii) the reaction with solvent to give a palladium hydride that undergoes reductive elimination to palladium(0) (Scheme 12).

The formation of dimethyl succinates is stereospecific, *syn* addition of the two carbomethoxy functions taking place, as illustrated by the dicarboxylation of (*Z*)-2-butene to the *meso* diester and (*E*)-2-butene to the (±)-diester (equations 64 and 65). This occurs since the alkene insertion to yield intermediate (**23**) is *syn* and the CO insertion (**23**) → (**24**) takes place with retention of configuration at carbon.

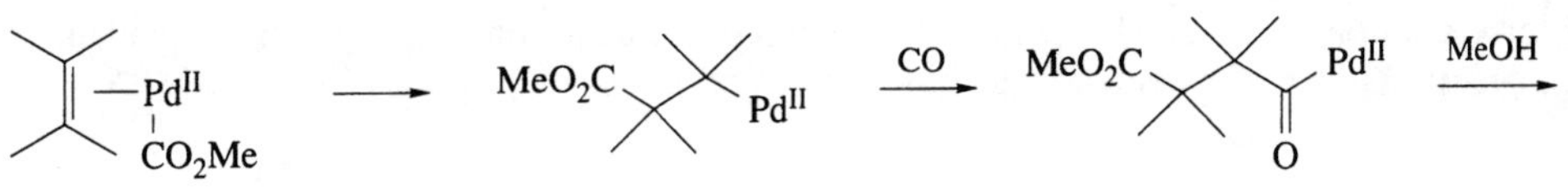

(23) (24)

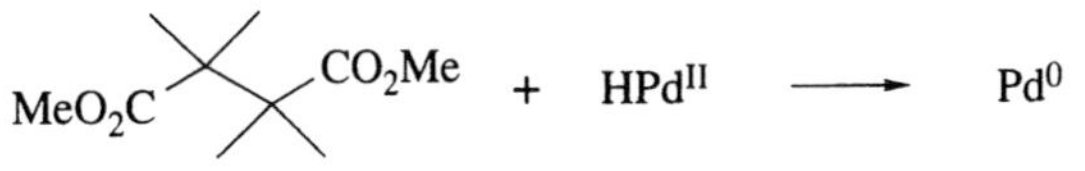

Scheme 12

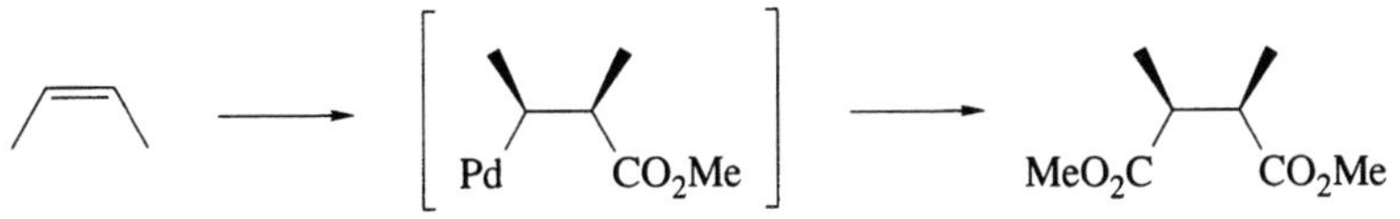

(64)

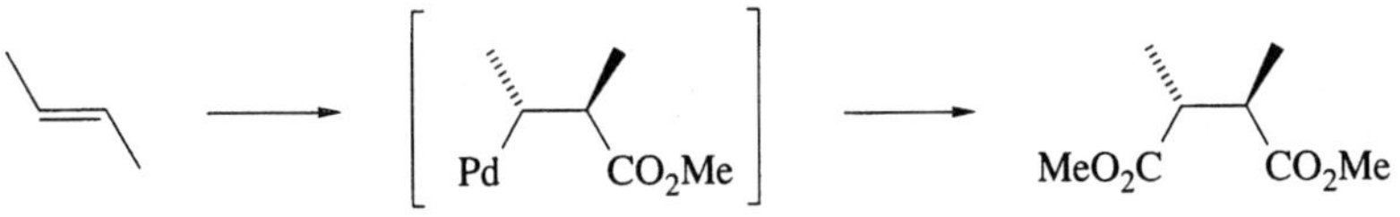

(65)

4.5.4.2 Dicarboxylation of Alkenes

The catalytic dicarbonylation of ethylene to dimethyl succinate can be carried out in 90% conversion.[94] High reaction temperatures and low carbon monoxide pressures can lead to unsaturated esters as a result of a faster β-hydride elimination from the intermediate (**23**) than carbon monoxide insertion. This later reaction path has been termed oxidative carboxylation.

The reaction of norbornene yields the *cis exo* diester (equation 66).[93] This *exo* isomer is not obtained directly by Diels–Alder chemistry. Other cyclic alkenes such as cyclopentene yield *cis* diesters, but isomers are obtained as a result of β-hydride elimination–readdition from intermediates such as (**23**) prior to CO insertion (equation 67). Thus the palladium walks around the ring to some extent, but always stays on the same face. The extent of rearrangement can be minimized by higher CO pressures since CO insertion becomes more competitive with β-elimination. This rearrangement becomes a critical problem in the dicarboxylation of 1-alkenes, since a variety of diesters are formed and the reaction is not particularly useful. These reactions were carried out with catalytic amounts of palladium and stoichiometric amounts of copper chloride.

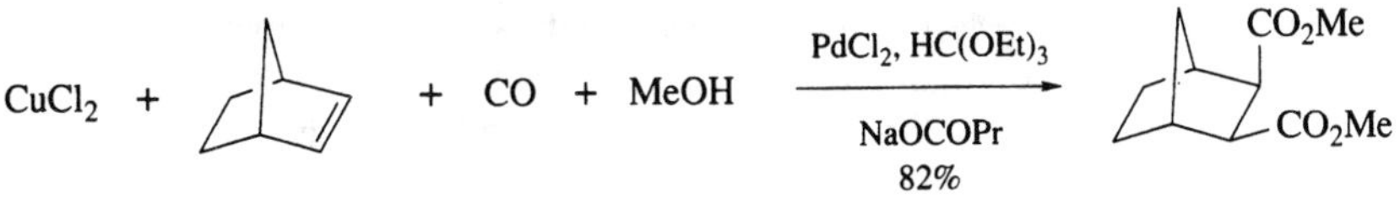

(66)

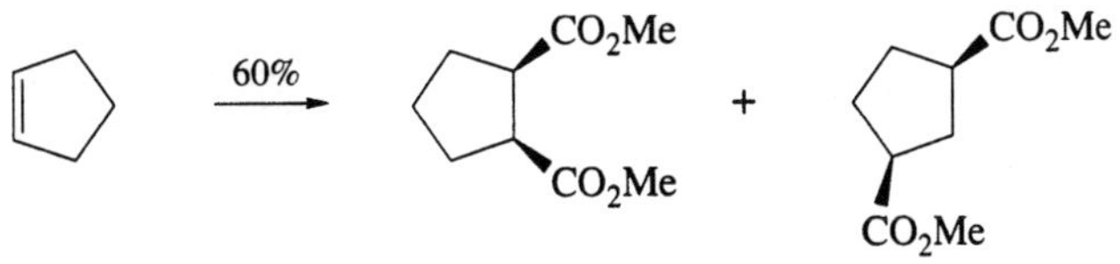

(67)

Tertiary carbons are not carboxylated, however, so the rearrangement reaction can be of advantage for the synthesis of certain symmetrical glutaric esters (equation 68).

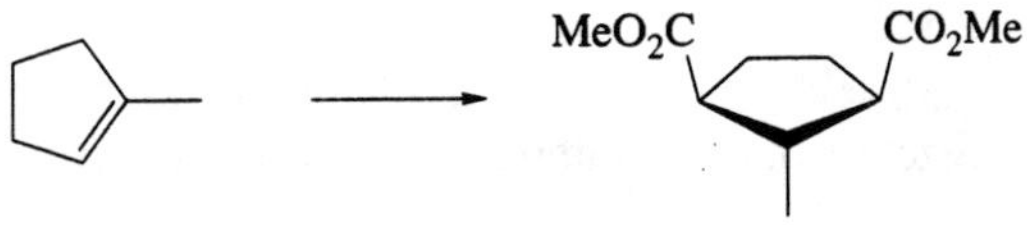

(68)

The dicarboxylation of 2-methyl-4-penten-2-ol provides an interesting example in that palladium carboxylate is generated from the pentenol alcohol function producing a γ-lactone (Scheme 13).[95]

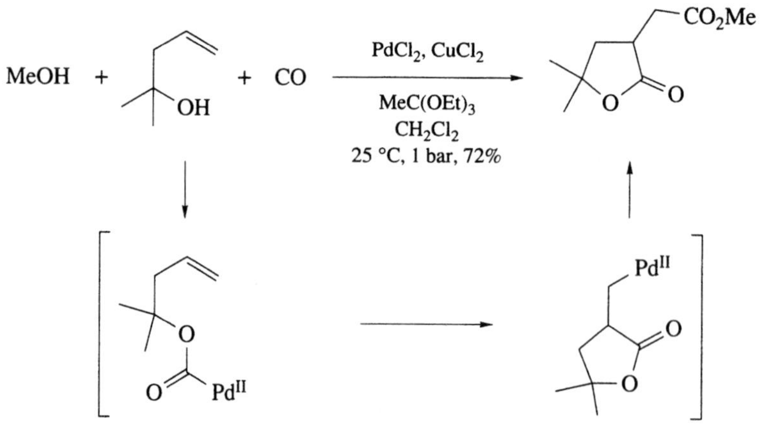

MeOH + (2-methyl-4-penten-2-ol) + CO —[$PdCl_2$, $CuCl_2$ / $MeC(OEt)_3$, CH_2Cl_2, 25 °C, 1 bar, 72%]→ γ-lactone bearing CH_2CO_2Me

Pd^{II} carboxylate intermediate → lactone–CH_2Pd^{II} intermediate

Scheme 13

Under these reaction conditions, cyclopentenone, allyl acetate, and butenyl acetate are dicarboxylated in high yields (equations 69–71).[96]

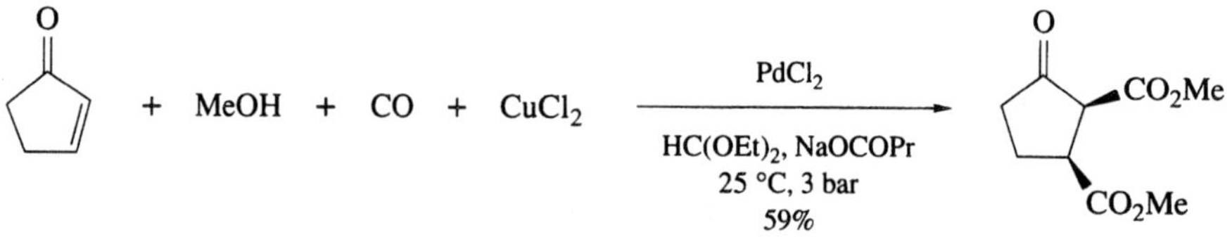

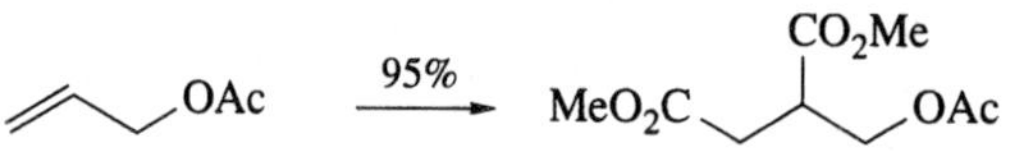

$$\text{cyclopentenone} + MeOH + CO + CuCl_2 \xrightarrow[\text{HC(OEt)}_2\text{, NaOCOPr; 25 °C, 3 bar; 59\%}]{PdCl_2} \text{2,3-bis(CO}_2\text{Me)cyclopentanone} \quad (69)$$

$$\text{allyl acetate} \xrightarrow{95\%} MeO_2C\text{–CH}_2\text{–CH(CO}_2\text{Me)–CH}_2\text{OAc} \quad (70)$$

$$\text{butenyl acetate} \xrightarrow{81\%} MeO_2C\text{–CH}_2\text{–CH(CO}_2\text{Me)–CH}_2\text{CH}_2\text{OAc} \quad (71)$$

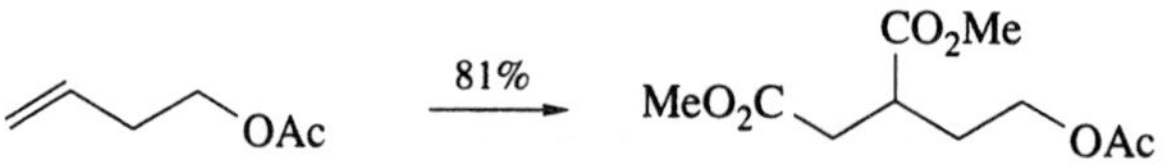

When the substrate contains two double bonds, either di- or tetra-carboxylation can be achieved.[96] The reaction of 1,5-hexadiene at 5 bar CO gives the diester, while at 1–3 bar the tetraester is obtained (equation 72). These somewhat surprising results are consistent with a pressure-dependent competition of the alkene and carbon monoxide for coordination sites on palladium. Since 1,5-hexadiene is not easily displaced by CO, even at higher pressures, the carboxylation of one double bond is ensured. Competition of the resulting monoalkenic diester with carbon monoxide is dependent on the CO pressure, lower pressures allowing monoalkene coordination and thus dicarboxylation.

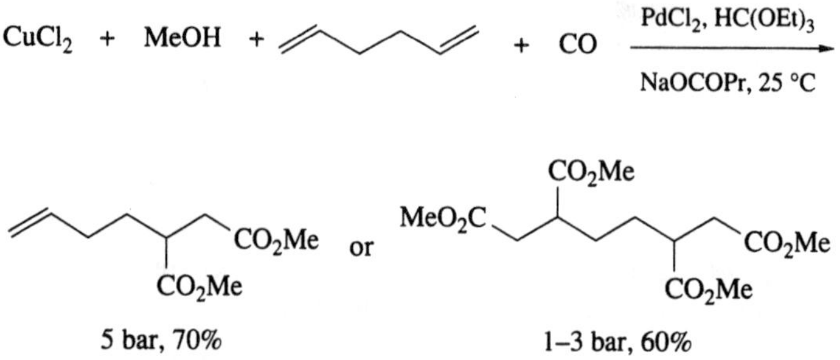

$$CuCl_2 + MeOH + \text{1,5-hexadiene} + CO \xrightarrow[\text{NaOCOPr, 25 °C}]{PdCl_2\text{, HC(OEt)}_3}$$

diester (5 bar, 70%) or tetraester (1–3 bar, 60%) (72)

Vinylcyclohexene can be dicarboxylated under these conditions at the more reactive vinyl double bond, which is 100 times more reactive than the internal one (equation 73).

$$\text{(cyclohexenyl)-CH=CH}_2 \xrightarrow{85\%} \text{(cyclohexenyl)-CH(CO}_2\text{Me)CH}_2\text{CO}_2\text{Me} \quad (73)$$

Butadiene carboxylation requires somewhat higher reaction temperatures, producing (*Z*)- and (*E*)-dimethylhex-3-ene-1,6-dioate (equation 74).

$$\text{CH}_2\text{=CH-CH=CH}_2 + \text{MeOH} + \text{CO} \xrightarrow{75\%} \text{MeO}_2\text{C}\text{CH}_2\text{CH=CHCH}_2\text{CO}_2\text{Me} \quad (74)$$

$Z{:}E = 32{:}68$

4.5.5 REFERENCES

1. I. Tkatchenko, in 'Comprehensive Organometallic Chemistry', ed. G. Wilkinson, F. G. A. Stone and E. W. Abel, Pergamon Press, Oxford, 1982, vol. 8, p. 101.
2. J. Falbe (ed.), 'New Syntheses With Carbon Monoxide', Springer-Verlag, Berlin, 1980.
3. P. Piño, *J. Organomet. Chem.*, 1980, **200**, 223.
4. P. J. Davidson, R. R. Hignett and D. T. Thompson, in 'Catalysis', ed. C. Kemball, The Chemical Society, London, 1977, vol. 1, p. 369.
5. R. L. Pruett, *Adv. Organomet. Chem.*, 1979, **17**, 1.
6. (a) C.-Y. Hsu and M. Orchin, *J. Am. Chem. Soc.*, 1975, **97**, 3553; (b) M. Orchin, *Prepr., Div. Pet. Chem., Am. Chem. Soc.*, 1976, **21(3)**, 482.
7. I. Schwager and J. F. Knifton, *J. Catal.*, 1976, **45**, 256.
8. H. C. Clark and J. A. Davies, *J. Organomet. Chem.*, 1981, **213**, 503.
9. H. J. Ruegg, P. S. Pregosin, A. Scrivanti, L. Toniolo and C. Botteghi, *J. Organomet. Chem.*, 1986, **316**, 233.
10. Y. Kawabata, T. Hayashi and I. Ogata, *J. Chem. Soc., Chem. Commun.*, 1979, 462.
11. N. H. Alemdaroglu, J. L. Penninger and E. Oltay, *Monatsh. Chem.*, 1976, **107**, 1153.
12. J. R. Norton, *Acc. Chem. Res.*, 1979, **12**, 139.
13. C. D. Tolman and J. W. Falls, in 'Homogeneous Catalysis with Metal Phosphine Complexes', ed. L. H. Pingolet, Plenum Press, New York, 1983.
14. W. R. Moser, C. J. Papile and S. J. Weininger, *J. Mol. Catal.*, 1987, **41**, 293.
15. (a) R. Bardi, A. M. Piazzesi, A. Del Pra, G. Cavinato and L. Toniolo, *J. Organomet. Chem.*, 1982, **234**, 107; (b) R. Bardi, A. M. Piazzesi, G. Cavinato, P. Cavoli and L. Toniolo, *J. Organomet. Chem.*, 1982, **224**, 407; (c) A. Scrivanti, C. Botteghi, L. Toniolo and A. Berton, *J. Organomet. Chem.*, 1988, **344**, 261; (d) G. K. Anderson, H. C. Clark and J. A. Davies, *Organometallics*, 1982, **1**, 64; (e) A. B. Goel and S. Goel, *Inorg. Chim. Acta*, 1983, **77**, L53.
16. C. A. Tolman, *Chem. Rev.*, 1977, **77**, 313.
17. B. Fell and H. Bahrmann, *J. Mol. Catal.*, 1977, **2**, 211.
18. A. Stefani, G. Consiglio, C. Botteghi and P. Piño, *J. Am. Chem. Soc.*, 1977, **99**, 1058.
19. A. Stefani. G. Consiglio, C. Botteghi and P. Piño, *J. Am. Chem. Soc.*, 1973, **95**, 6504.
20. P. Piño and C. Botteghi, *Org. Synth.*, 1977, **57**, 11.
21. P. W. N. M. Van Leeuwen and C. F. Roobeeck, *J. Organomet. Chem.*, 1983, **258**, 343.
22. (a) H. Siegel and W. Himmele, *Angew. Chem., Int. Ed. Engl.*, 1980, **19**, 178; (b) H. Pummer and A. Neürrenbach, *Pure Appl. Chem.*, 1975, **43**, 527.
23. (a) W. Himmele and H. Siegel, *Tetrahedron Lett.*, 1976, 907; (b) W. Himmele and H. Siegel, *Tetrahedron Lett.*, 1976, 911.
24. W. H. Clement and M. Orchin, *Ind. Eng. Chem. Prod. Res. Dev.*, 1965, **4**, 283.
25. J. C. LoCierco and R. T. Johnson, *J. Am. Chem. Soc.*, 1952, **74**, 2094.
26. Y. Fujikura, Y. Inamoto, N. Takaishi and H. Ikeda, *Synth. Commun.*, 1976, **6**, 199.
27. R. L. Pruett, *Ann. N. Y. Acad. Sci.*, 1977, **295**, 239.
28. J. Falbe and N. Huppes, *Brennst. Chem.*, 1967, **48**, 182.
29. A. Spencer, *J. Organomet. Chem.*, 1977, **124**, 85.
30. A. M. Trzeciak and J. J. Ziólkowski, *J. Mol. Catal.*, 1987, **43**, 15.
31. M. Matsumoto and M. Tamura, *J. Mol. Catal.*, 1982, **16**, 195.
32. P. G. M. Wuts, M. L. Obrzut and P. A. Thompson, *Tetrahedron Lett.*, 1984, **25**, 4051.
33. B. Fell and M. Barl, *J. Mol. Catal.*, 1977, **2**, 301.
34. A. G. Abatjoglou, D. R. Bryant and L. C. D'Esposito, *J. Mol. Catal.*, 1983, **18**, 381.
35. C. U. Pittman, Jr., W. D. Honnick and J. J. Yang, *J. Org. Chem.*, 1980, **45**, 684.
36. K. Prokai-Tatrai, S. Toros and B. Heil, *J. Organomet. Chem.*, 1987, **332**, 331.
37. L. Kollar, G. Consiglio and P. Piño, *Chimica*, 1986, **40**, 428.
38. E. Ucciani and A. Bonfand, *J. Chem. Soc., Chem. Commun.*, 1981, 82.
39. S. Sato, *J. Chem. Soc. Jpn.*, 1969, **90**, 404.
40. Y. Becker, A. Eisenstadt and J. K. Stille, *J. Org. Chem.*, 1980, **45**, 2145.
41. K. Prokai-Tatrai, S. Toros and B. Heil, *J. Organomet. Chem.*, 1986, **315**, 231.
42. G. Consiglio and P. Piño, *Top. Curr. Chem.*, 1982, **105**, 77.

43. I. Ojima and K. Hirai, in 'Asymmetric Synthesis', ed. J. D. Morrison, Academic Press, New York, 1985, vol. 5, p. 103.
44. B. Bosnich (ed.), 'Asymmetric Catalysis', Nijhoff, Boston, 1986.
45. P. Haelg, G. Consiglio and P. Piño, *J. Organomet. Chem.*, 1985, **296**, 281.
46. M. Tanaka, Y. Ikeda and I. Ogata, *Chem. Lett.*, 1975, 1115.
47. (a) T. Hayashi, M. Tanaka, Y. Ikeda and I. Ogata, *Bull. Chem. Soc. Jpn.*, 1979, **52**, 2605; (b) T. Hayashi, M. Tanaka and I. Ogata, *Tetrahedron Lett.*, 1978, 3925.
48. G. Consiglio, F. Morandini, M. Scalone and P. Piño, *J. Organomet. Chem.*, 1985, **279**, 193.
49. M. Petit, A. Mortreux, F. Petit, G. Buono and G. Peiffer, *Nouv. J. Chim.*, 1983, **7**, 593.
50. (a) C. Botteghi, M. Branca and A. Saba, *J. Organomet. Chem.*, 1980, **184**, C17; (b) C. Botteghi, M. Branca, G. Micera, F. Piacenti and G. Menchi, *Chim. Ind. (Milan)*, 1978, **60**, 16; (c) Y. Becker, A. Eisenstadt and J. K. Stille, *J. Org. Chem.*, 1980, **45**, 2145.
51. (a) C. F. Hobbs and W. S. Knowles, *J. Org. Chem.*, 1981, **46**, 4422; (b) H. B. Tinker and A. J. Solodar, *US Pat.* 4 268 688 (1981).
52. M. Tanaka, Y. Watanabe, T. Mitsudo and Y. Takegami, *Bull. Chem. Soc. Jpn.*, 1974, **47**, 1698.
53. G. Consiglio and P. Piño, *Helv. Chim. Acta*, 1976, **59**, 642.
54. P. S. Pregosin and S. N. Sze, *Helv. Chim. Acta*, 1978, **61**, 1848.
55. T. Hayashi, Y. Kawabata, T. Isoyama and I. Ogata, *Bull. Chem. Soc. Jpn.*, 1981, **54**, 3438.
56. (a) G. Consiglio and P. Piño, *Isr. J. Chem.*, 1976/77, **15**, 221; (b) G. Consiglio, W. Arber and P. Piño, *Chim. Ind. (Milan)*, 1978, **60**, 396.
57. (a) C. U. Pittman, Jr., Y. Kawabata and L. I. Flowers, *J. Chem. Soc., Chem. Commun.*, 1982, 473; (b) G. Consiglio, P. Piño, L. I. Flowers and C. U. Pittman, Jr., *J. Chem. Soc., Chem. Commun.*, 1983, 612.
58. G. Consiglio and A. Borer, Abstracts of Papers Presented at the Third International Conference on The Chemistry of the Platinum Group Metals, Sheffield, UK, 12–17 July 1987.
59. J. K. Stille and G. Parrinello, *J. Mol. Catal.*, 1983, **21**, 203.
60. G. Parrinello and J. K. Stille, *J. Am. Chem. Soc.*, 1987, **109**, 7122.
61. T. Y. Shen, *Angew. Chem., Int. Ed. Engl.*, 1972, **11**, 460.
62. I. T. Harrison, B. Lewis, P. Nelson, W. Rooks, A. Roszkowski, A. Tomolonis and J. H. Fried, *J. Med. Chem.*, 1970, **13**, 203.
63. (a) P. A. Janssen, *Arzneim.-Forsch.*, 1975, **25**, 1495; (b) R. J. Capetola, D. A. Shriver and M. E. Rosenthale, *J. Pharmacol. Exp. Ther.*, 1980, **214**, 16; (c) R. S. Pujalte, E. Valdez and R. De la Paz, *Curr. Ther. Res.*, 1984, **36**, 245.
64. L. Kollár, G. Consiglio and P. Piño, *J. Organomet. Chem.*, 1987, **330**, 305.
65. G. P. Chiusoli, *Transition Met. Chem.*, 1987, **12**, 89.
66. D. Milstein, *Acc. Chem. Res.*, 1988, **21**, 428.
67. J. F. Knifton, *J. Org. Chem.*, 1976, **41**, 793.
68. R. Bardi, A. del Pra, A. M. Piazzesi and L. Toniolo, *Inorg. Chim. Acta*, 1979, **35**, L345.
69. G. Consiglio and P. Piño, *Gazz. Chim. Ital.*, 1975, **105**, 1133.
70. T. F. Murray and J. R. Norton, *J. Am. Chem. Soc.*, 1979, **101**, 4107.
71. E. G. Samsel and J. R. Norton, *J. Am. Chem. Soc.*, 1984, **106**, 5505.
72. D. Milstein and J. L. Huckaby, *J. Am. Chem. Soc.*, 1982, **104**, 6150.
73. P. W. Jolly, in 'Comprehensive Organometallic Chemistry', ed. G. Wilkinson, F. G. A. Stone and E. W. Abel, Pergamon Press, Oxford 1982, vol. 8, p. 773.
74. K. Bittler, N. von Kutepow, D. Neubauer and H. Reis, *Angew. Chem., Int. Ed. Engl.*, 1968, **7**, 329.
75. D. M. Fenton, *J. Org. Chem.*, 1973, **38**, 3192.
76. (a) Y. Sugi, K. Bando and S. Shin, *Chem. Ind. (London)*, 1975, **9**, 397; (b) Y. Sugi and K. Bando, *Chem. Lett.*, 1976, **7**, 727; (c) G. Consiglio and M. Marchetti, *Chimia*, 1976, **30**, 26.
77. J. F. Knifton, *J. Org. Chem.*, 1976, **41**, 2885.
78. J. F. Knifton, *J. Org. Chem.*, 1976, **41**, 793.
79. (a) B. Despeyroux and H. Alper, *Ann. N.Y. Acad. Sci.*, 1983, **415**, 148; (b) H. Alper, J. B. Woell, B. Despeyroux and D. J. H. Smith, *J. Chem. Soc., Chem. Commun.*, 1983, 1270.
80. R. Hüttel, *Synthesis*, 1970, **5**, 225.
81. K. Kaneda, T. Imanaka and S. Teranishi, *Chem. Lett.*, 1983, 1465.
82. H. Alper and D. Leonard, *Tetrahedron Lett.*, 1985, **26**, 5639.
83. H. Alper and D. Leonard, *J. Chem. Soc., Chem. Commun.*, 1985, 511.
84. Y. Tsuji, T. Kondo and Y. Watanabe, *J. Mol. Catal.*, 1987, **40**, 295.
85. T. F. Murray, E. G. Samsel, V. Varma and J. R. Norton, *J. Am. Chem. Soc.*, 1981, **103**, 7520.
86. J. F. Knifton, *J. Organomet. Chem.*, 1980, **188**, 223.
87. (a) J. F. Knifton, *J. Catal.*, 1979, **60**, 27; (b) J. F. Knifton, *Ann. N.Y. Acad. Sci.*, 1980, **333**, 264.
88. S. Hosaka and J. Tsuji, *Tetrahedron*, 1971, **27**, 3821.
89. (a) A. Matsuda, *Bull. Chem. Soc. Jpn.*, 1973, **46**, 524; (b) F. J. Waller, *J. Mol. Catal.*, 1985, **31**, 123; (c) E. I. du Pont de Nemours and Co., *Jpn. Pat.* 61 172 850 (1985).
90. (a) G. Consiglio and P. Piño, *Chimia*, 1976, **30**, 193; (b) G. Consiglio, *J. Organomet. Chem.*, 1977, **132**, C26.
91. (a) T. Hayashi, M. Tanaka and I. Ogata, *Tetrahedron Lett.*, 1978, 3925; (b) T. Hayashi, M. Tanaka and I. Ogata, *J. Mol. Catal.*, 1984, **26**, 17.
92. J. K. Stille and D. E. James, in 'The Chemistry of Functional Groups, Supplement A. Double Bonded Functional Groups', ed. S. Patai, Wiley, London, 1976, p. 1099.
93. D. E. James and J. K. Stille, *J. Am. Chem. Soc.*, 1976, **98**, 1810.
94. D. M. Fenton and P. J. Steinwand, *J. Org. Chem.*, 1972, **37**, 2034.
95. Y. Tamara, M. Hojo and Z. Yoshida, *Tetrahedron Lett.*, 1987, **28**, 325.
96. J. K. Stille and R. Divakaruni, *J. Org. Chem.*, 1979, **44**, 3474.

4.6
Methylene and Nonfunctionalized Alkylidene Transfer to Form Cyclopropanes

PAUL HELQUIST

University of Notre Dame, IN, USA

4.6.1 INTRODUCTION

4.6.1.1 Scope of Chapter

Cyclopropanes occur among several classes of natural products, they have a number of commercial applications, and they serve as useful synthetic intermediates leading to other classes of cyclic and acyclic compounds. Methods for the synthesis of cyclopropanes have been the subject of several earlier reviews.[1] The principal methods may be divided into two broad categories: (i) addition of a one-carbon unit (*e.g.* a carbene) to a two-carbon unit, most commonly an alkene; and (ii) 1,3-coupling of difunctional compounds. Much rarer are examples of direct joining of three one-carbon units.

The coverage of this chapter emphasizes category (i) reactions in which a simple methylene group is transferred, or added, to an alkene substrate. Also covered is transfer of nonfunctionalized alkylidene groups when these reactions may be regarded as simple extensions of the methylene transfer reactions. Whenever appropriate, aspects of the stereoselectivity, enantioselectivity, regioselectivity and chemoselectivity of these reactions will be emphasized by means of specific examples.

Methods for cyclopropane formation which fall outside of the defined scope of this chapter and which are not specifically covered herein are the following: (i) transfer of 1,1-dihalo-carbenes or -carbenoids (see Volume 4, Chapter 4.7); (ii) transfer of other heterosubstituted alkylidene groups;[2] (iii) cyclopropanations using diazocarbonyl compounds (see Volume 4, Chapter 4.8);[3] (iv) transfer of other functionalized alkylidene groups when these reactions are dependent upon this functionalization (although reactions are included when the presence of a functional group in the alkylidene unit instead emphasizes aspects of chemoselectivity); (v) direct 1,3-coupling reactions;[4] (vi) intramolecular C—H insertions of carbenes (see Volume 3, Chapter 4.2);[5] (vii) ring contraction and other rearrangements; (viii) generation of vinyl- or divinyl-cyclopropanes as intermediates directed specifically toward the synthesis of cyclopentenes or cycloheptadienes (see Volume 5, Chapters 8.1 and 8.3);[6] (ix) synthesis of alkylidenecyclopropanes; (x) synthesis of cyclopropenes; and (xi) intramolecular alkylidene transfer, since most examples involve the use of diazocarbonyl compounds (see Volume 4, Chapter 4.8).[7]

4.6.1.2 Stereochemical Definitions

There are several aspects of stereochemistry that must be considered when cyclopropanes are produced by the usual methods of alkylidene group addition to alkenic substrates. In turn, several terms have been devised to express the extent to which different types of stereoisomers are formed selectively, but unfortunately, the meanings of these terms have become confused because of their inconsistent usage in the literature. In the context of this chapter, the term stereoselectivity refers to the degree of selectivity for formation of cyclopropane products having *endo versus exo* or, alternatively, *syn versus anti* stereochemistry of the substituents originating in the alkylidene group relative to substituents originating in the alkene substrate. The term stereospecificity refers to the stereochemistry of vicinal cyclopropane substituents originating as double-bond substituents in the starting alkene, *i.e.* a cyclopropane-forming reaction is stereospecific if the *cis/trans* relationship of the double-bond substituents is retained in the cyclopropane product. Diastereofacial selectivity refers to the face of the alkene to which addition occurs relative to other substituents in the alkene substrate. Finally, enantioselectivity refers to the formation of a specific enantiomer of the cyclopropane product. These stereochemical principles are illustrated by general examples in equations (1) to (4).

Stereoselectivity: R[1] R[2] *versus* R[2] R[1] (1)

Stereospecificity: R[1] H R[2] H → R[1] H R[2] H *versus* H R[1] R[2] H (2)

Diastereofacial selectivity: R → R H H *versus* R H H (3)

Enantioselectivity: 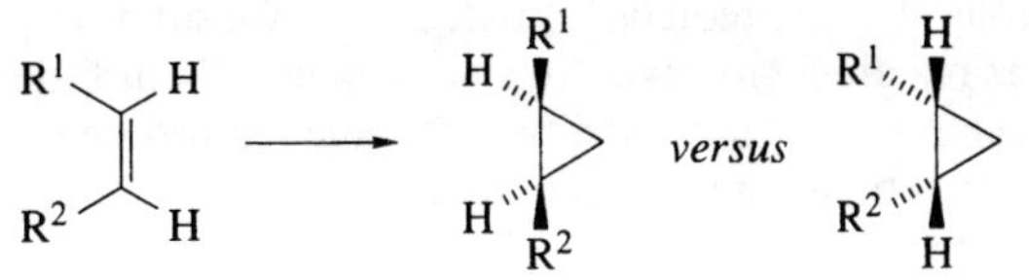 (4)

4.6.2 USE OF SIMPLE DIAZOALKANES

The most generally employed approach for the formation of cyclopropanes is the addition of a carbene or carbenoid to an alkene. In many cases, a free carbene is not involved as an actual intermediate, but instead the net, overall transformation of an alkene to a cyclopropane corresponds, in at least a formal sense, to carbene addition. In turn, the most traditional method for effecting these reactions is to employ diazo compounds, $R^1R^2C{=}N_2$, as precursors. Thermal, photochemical and metal-catalyzed reactions of these diazo compounds have been studied thoroughly and are treated separately in the discussion below. These reactions have been subjects of several comprehensive reviews,[8] to which the reader is referred for further details and literature citations. Emphasis in the present chapter is placed on recent examples.

In considering the possible use of diazoalkanes in synthetic applications, the prospective user must bear in mind that the handling of these compounds is very hazardous. They are highly toxic, and they are susceptible to undergoing violent explosions. Therefore, they must be handled with proper precautions and by properly informed laboratory personnel.

4.6.2.1 Thermal Reactions

The thermal reactions of diazo compounds with alkenes generally do not proceed *via* direct formation of cyclopropanes. Rather, the diazo compounds serve as 1,3-dipoles that first undergo [3 + 2] cycloaddition to form 1-pyrazolines (**1**) that in a separate step undergo extrusion of nitrogen to give the cyclopropane products (Scheme 1).[1h,8a–8g,8i] The extrusion may occur under thermal or under either direct or triplet-sensitized photochemical conditions. The initially formed 1-pyrazolines may first undergo double-bond positional isomerization to give 2-pyrazolines (**2**), especially when a hydrogen substituent next to nitrogen is reasonably acidic due to the presence of electron-withdrawing substituents. In many cases, the pyrazolines are unstable under the conditions of their formation and are consequently not observed, but instead, the cyclopropanes are isolated directly from the two-step reaction sequence.

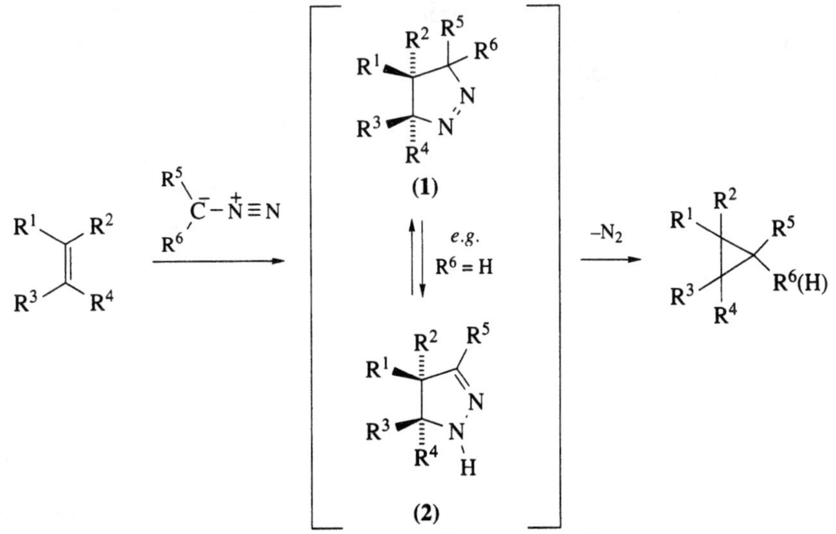

(**1**)

e.g. $R^6 = H$

$-N_2$

(**2**)

Scheme 1

Addition to simple, unstrained alkenes is usually quite slow, but these reactions become more useful when electron-withdrawing and electron-donating substituents are present on the double bond. On the other hand, the reactivity of diazo compounds is reduced when they bear conjugating substituents (*e.g.* aryl or carbonyl groups). Acceleration of these reactions by high pressure has been reported.[9] The initial cycloaddition is generally stereospecific and is considered to be concerted. The regioselectivity of these dipolar additions is not as consistently predictable, based upon substitution patterns, as other cycloadditions such as the Diels–Alder reaction. However, this point is usually of little consequence in the overall transformation leading to cyclopropanes since the two possible regioisomers are converted into the same final product upon loss of nitrogen. Several types of side reactions of diazo compounds frequently compete with the cycloadditions. Insertions into various C—H and heteroatom X—H bonds are well known, but especially commonly encountered are insertions into O—H bonds of alcohols and carboxylic acids, leading to ether and ester derivatives, respectively.

The subsequent extrusion of nitrogen from pyrazolines to produce cyclopropanes is also subject to complications. The cyclopropanes may be formed along with alkenes, resulting from rearrangements and eliminations, especially under thermal and direct photolysis conditions, whereas triplet-sensitized photochemical conditions often lead to much smaller amounts of alkenes. Nitrogen extrusion and carbon–carbon bond formation proceed largely with retention of configuration of the pyrazoline under direct photochemical conditions, but scrambling of stereochemistry occurs to a large extent in many cases of triplet-sensitized photochemical reactions and thermal reactions. Many examples of these reactions have shown evidence for the formation of singlet or triplet diradical intermediates upon extrusion of nitrogen, although zwitterionic intermediates have also been proposed. The formation of triplet diradicals, having longer lifetimes than their singlet counterparts, would explain the greater extent of stereochemical scrambling in the triplet-sensitized photochemical reactions.

A variation on the thermal reactions of diazo compounds with alkenes is the decomposition of salts of sulfonylhydrazones. This procedure, known as the Bamford–Stevens reaction, is believed to occur *via* the formation of diazo compounds. Subsequent 1,2-hydrogen shifts generally lead to the formation of alkenes as the final products. Cyclopropanes may also be formed as the result of intramolecular 1,3-C—H insertion reactions or when the original hydrazone substrate contains a remote alkenic group as a site for intramolecular cyclopropanation.[10]

Examples of the overall conversion of alkenes to cyclopropanes under the above sets of conditions are summarized in Table 1.

4.6.2.2 Photochemical Reactions

The reaction of diazoalkanes with alkenes under either direct or sensitized photochemical conditions has been investigated extensively.[1h,8a,8b,8d–8f,8i,8l,8m,8p] These reactions proceed *via* formation of carbenes as reactive intermediates. Side reactions, such as C—H insertions, compete with the cyclopropanations. The relative reactivities of singlet and triplet carbenes have been studied. Direct photolysis generally produces singlet carbenes, but these may in turn undergo intersystem crossing to produce lower energy triplet carbenes. Alternatively, triplet carbenes may be produced through use of sensitizers. Singlet carbenes add stereospecifically (equation 2) to alkenes, whereas triplet carbenes generally undergo nonstereospecific addition. Interesting stereoselectivity (equation 1) is seen in several cases in which there is at least a small predominance of the thermodynamically less stable *syn* (or *cis* or *endo*) product. Even though singlet carbenes are generally of higher reactivity than the triplets, there is evidence that, unlike the usual intermolecular reactions, certain intramolecular additions of arylcarbenes to alkenes occur more readily *via* triplets than singlets.[29] As is also true of thermal reactions (see previous section), hydrazones may be used as substrates in photochemical reactions. A brief compilation of examples, with emphasis on more recent work, is found in Table 2.

4.6.2.3 Metal-catalyzed Reactions

The use of metal catalysts for the addition of diazoalkanes to alkenes has become increasingly common,[1h,8a–8e,8i,8k,8m–8o] while the use of thermal and photochemical reactions has waned. Many generalizations have been made to describe several different aspects of the metal-catalyzed reactions, but one must bear in mind that many of the detailed studies of these reactions have involved primarily diazocarbonyl compounds and not necessarily the simpler diazoalkanes that pertain to the general topic of the present chapter.

Both heterogeneous and homogeneous catalysts have been used in these reactions. These catalysts have consisted of compounds or complexes of many different metals, but the most common have been copper derivatives or metallic copper itself. In recent years, palladium and rhodium compounds have been shown to be particularly effective catalysts and have increased in popularity relative to many other

Table 1 Thermal Reactions of Diazoalkanes with Alkenes

Alkene	*Diazoalkane*	*Method of pyrazoline decomposition*	*Products*[a] *and yields*		*Ref.*
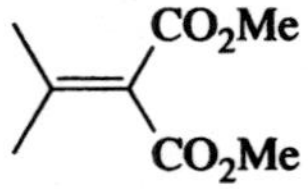 CO_2Me, CO_2Me	CH_2N_2	Xylene at reflux	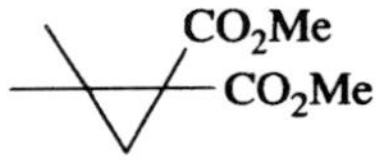 CO_2Me, CO_2Me 67%		11
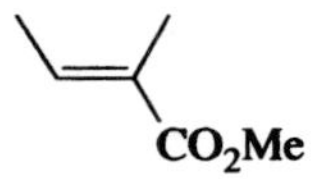 CO_2Me	$MeCHN_2$	120 °C	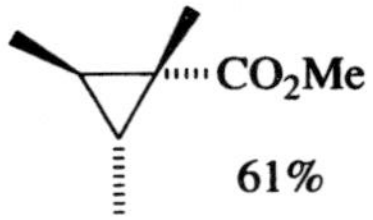 CO_2Me 61%	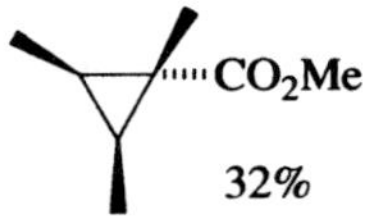 CO_2Me 32%	12
		Photolysis	88%	10%	
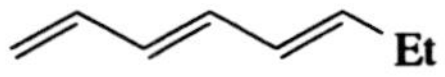 Et	CH_2N_2	140 °C or photolysis	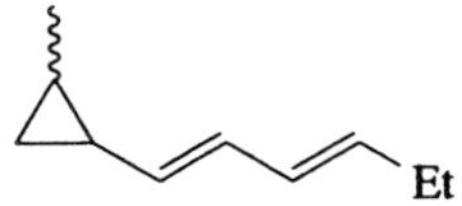 Et	*ca.* 60%	13
	$MeCHN_2$	140 °C	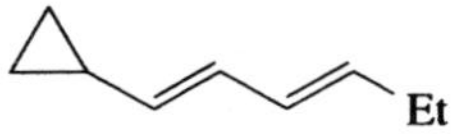 Et (*cis*:*trans* = 0.5:1.0)	*ca.* 82%	13
	$MeCHN_2$	Photolysis	(*cis*:*trans* = 0.5:1.0)		13
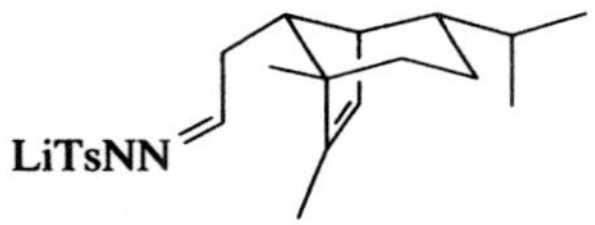 LiTsNN		Photolysis	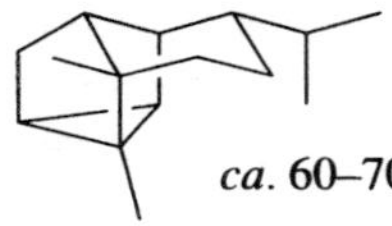 *ca.* 60–70%		14
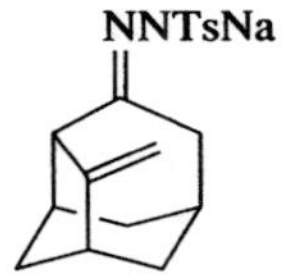 NNTsNa		200 °C	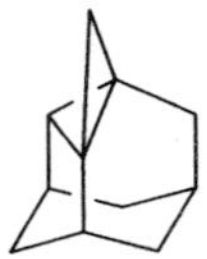 75%		15

Table 1 *(continued)*

Alkene	*Diazoalkane*	*Method of pyrazoline decomposition*	*Products*[a] *and yields*	*Ref.*
CO_2Me	Me_2CN_2	110 °C	CO_2Me *ca.* 44%	16
			(*cis*:*trans* = 2:1)	
	Me_2CN_2	Photolysis (direct)	(*cis*:*trans* = 4:1) *ca.* 44%	16
	Me_2CN_2	Photolysis (sensitized)	(*cis*:*trans* = 0.25:1) *ca.* 44%	16
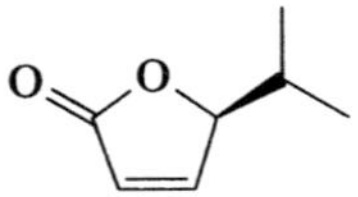 O, O	Me_2CN_2	Photolysis (sensitized)	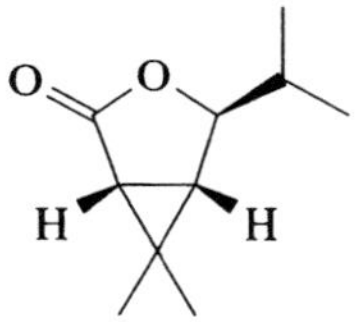 O, O, H, H 81%	17
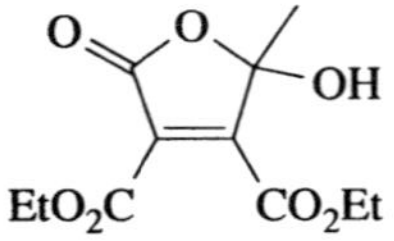 O, O, OH, EtO_2C, CO_2Et	CH_2N_2	?	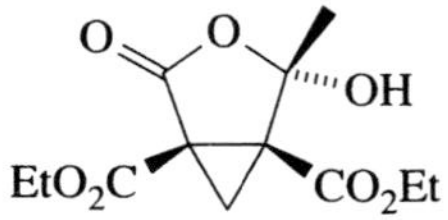 O, O, OH, EtO_2C, CO_2Et 70%	18
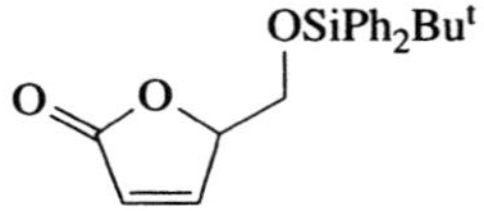 $OSiPh_2Bu^t$, O, O	Me_2CN_2	Photolysis (sensitized)	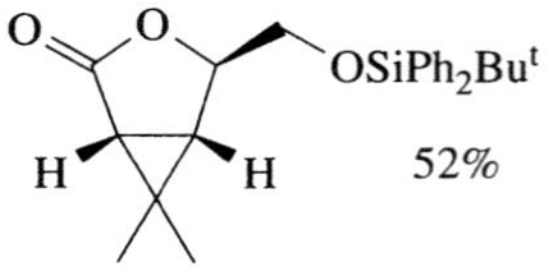 O, O, $OSiPh_2Bu^t$, H, H 52%	19
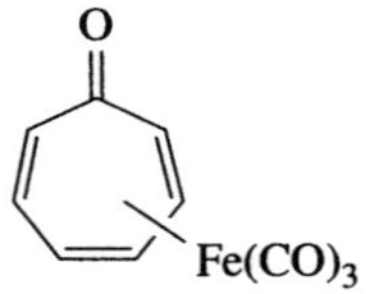 O, $Fe(CO)_3$	Me_2CN_2	80 °C	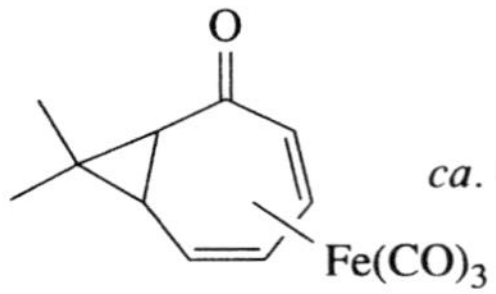 O, $Fe(CO)_3$ *ca.* 45%	20

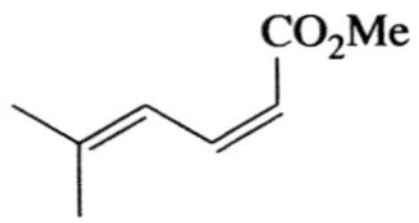

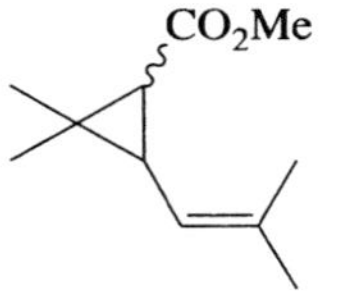

Table 1 *(continued)*

Alkene	*Diazoalkane*	*Method of pyrazoline decomposition*	*Products*[a] *and yields*	*Ref.*
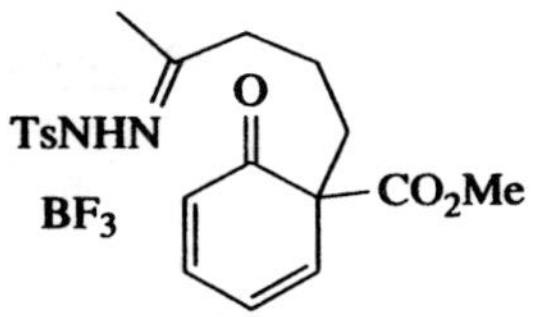 TsNHN, O, CO_2Me, BF_3		110 °C	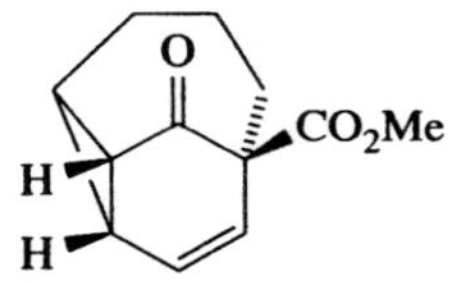 O, H, H, CO_2Me 39%	21
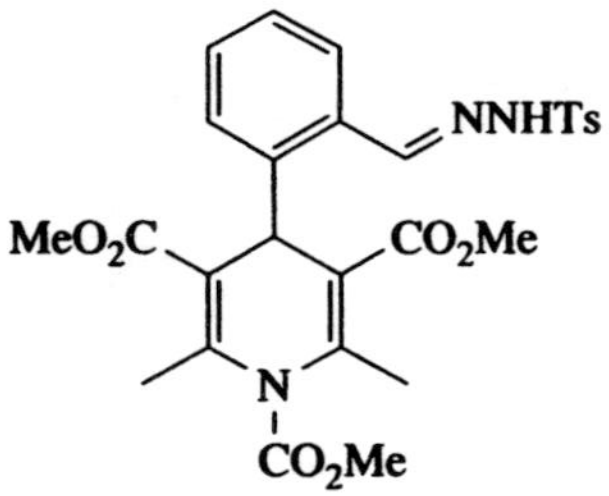 NNHTs, MeO_2C, CO_2Me, N, CO_2Me		Benzene at reflux	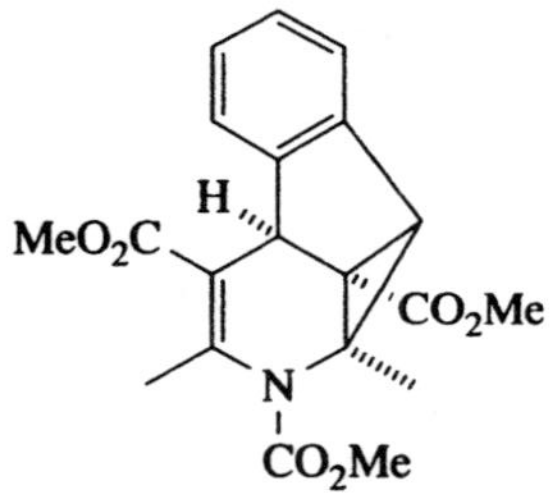 H, MeO_2C, CO_2Me, N, CO_2Me 32%	22
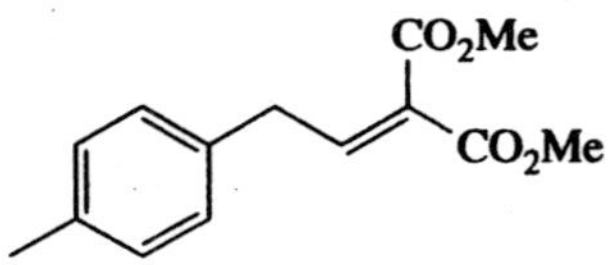 CO_2Me, CO_2Me	CH_2CN_2	$Ce(NH_4)_2(NO_3)_6$	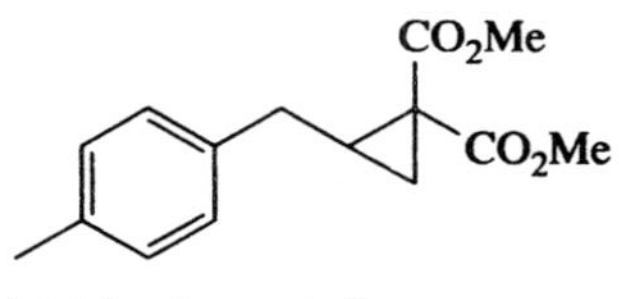 CO_2Me, CO_2Me (yield not reported)	23
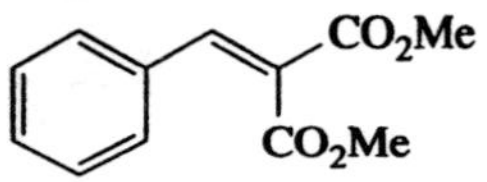 CO_2Me, CO_2Me	$(MeO)_2CHCHN_2$	Photolysis	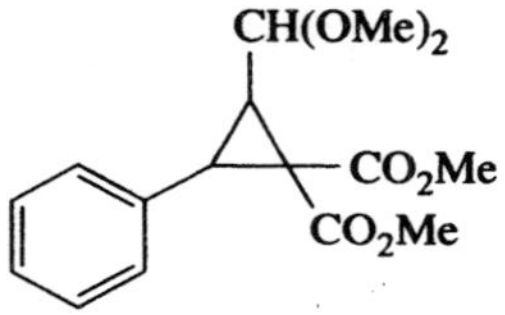 $CH(OMe)_2$, CO_2Me, CO_2Me *ca.* 80%	24

Table 1 *(continued)*

Alkene	*Diazoalkane*	*Method of pyrazoline decomposition*	*Products*[a] *and yields*	*Ref.*
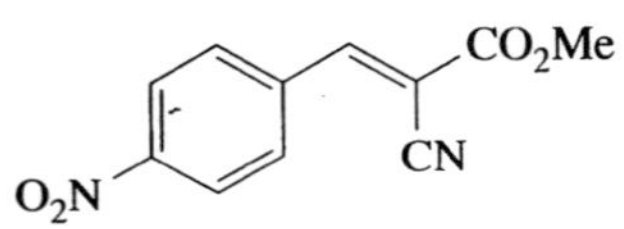 O_2N, CO_2Me, CN	$PhCMeN_2$	25 °C	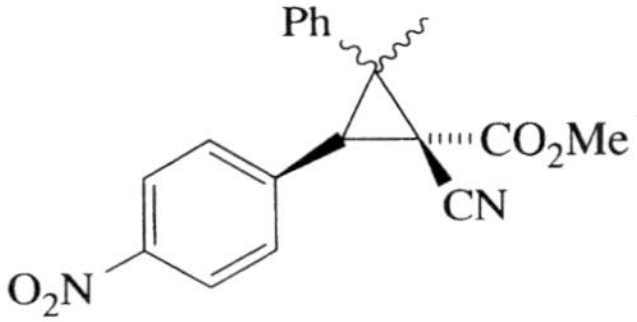 Ph, CO_2Me, CN, O_2N 38% (*cis*:*trans*-diaryl = 9:29)	25
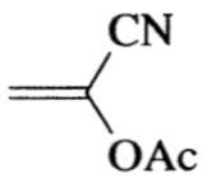 CN, OAc	$PhCMeN_2$	25 °C	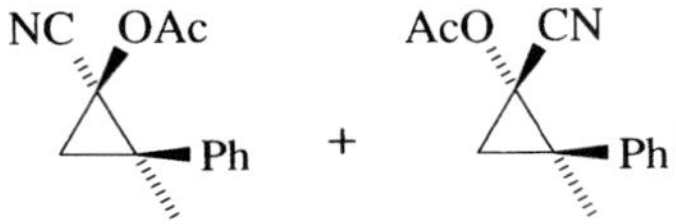 NC, OAc, Ph + AcO, CN, Ph 81% (1.6:1)	26
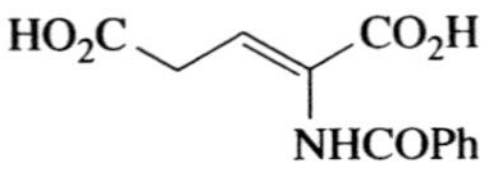 HO_2C, CO_2H, NHCOPh	CH_2N_2	Photolysis	 MeO_2C, CO_2Me, NHCOPh 78%	27
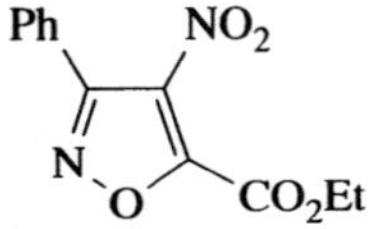 Ph, NO_2, N, O, CO_2Et	Me_2CN_2	25 °C	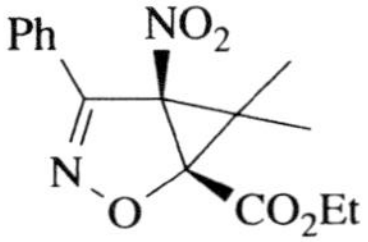 Ph, NO_2, N, O, CO_2Et 73%	28

[a] Only the cyclopropane products are shown, although alkenes and other side products may also have been produced.

Table 2 Photochemical Reactions of Diazoalkanes with Alkenes

Alkene	*Diazoalkane*	*Method of photolysis*	*Products*[a]	*Yield (%)*	*Ref.*
	CH_2N_2	Direct	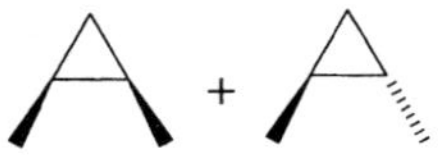 + 46:1	Not reported	30
		Sensitized (Ph_2CO)	1.9:1		31
	CH_2N_2	Direct	0:1		30
		Sensitized (Ph_2CO)	Trace Major isomer		31
	p-$MeOC_6H_4CHN_2$	Direct	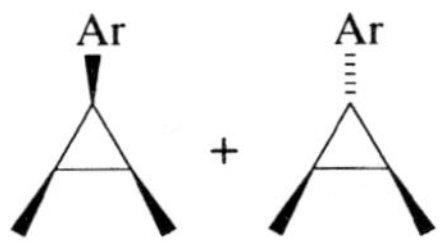 Ar + Ar 2.8:1 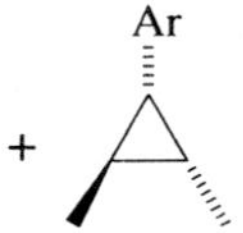 + Ar (trace)	21	32
	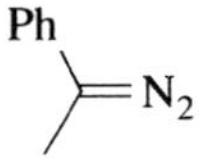 Ph $=N_2$	Direct	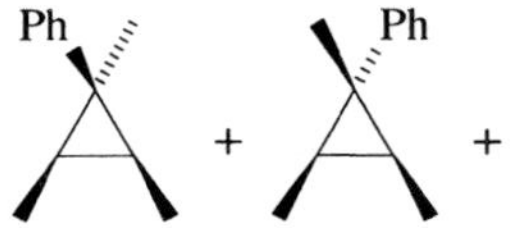 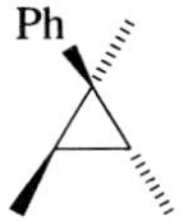 Ph + Ph + Ph 44:52:4		33
		Sensitized (Ph_2CO)	47:38:15		
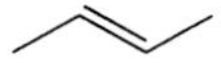		Direct	6:3:91		

Table 2 *(continued)*

Alkene	*Diazoalkane*	*Method of photolysis*	*Products*[a]	*Yield (%)*	*Ref.*
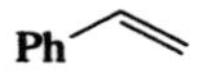 Ph	$PhCHN_2$	Direct	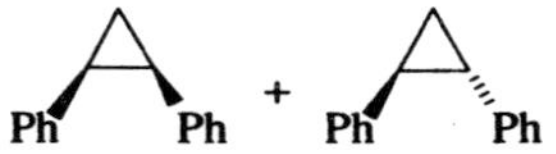 Ph Ph + Ph Ph 1.0:1.4	73	34
		Direct (but in crystalline state)	1.0:5.0	54	
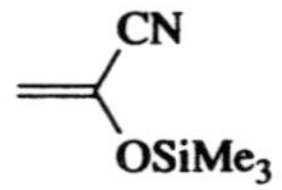 CN $OSiMe_3$	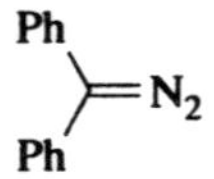 Ph $=N_2$ Ph	Direct	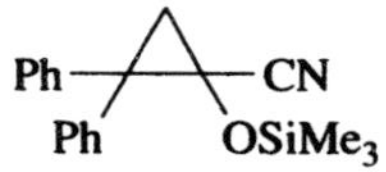 Ph CN Ph $OSiMe_3$	63	26
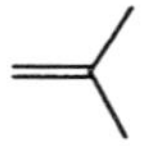	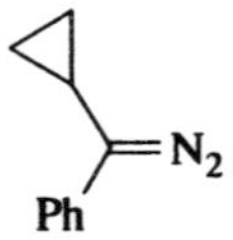 $=N_2$ Ph	Direct (128 °C)	 Ph	42	35
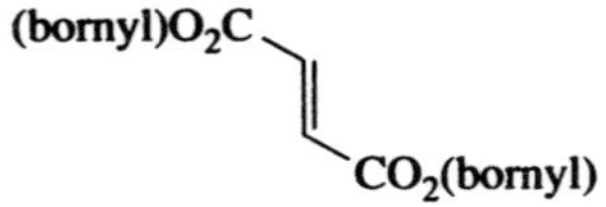 $(bornyl)O_2C$ $CO_2(bornyl)$	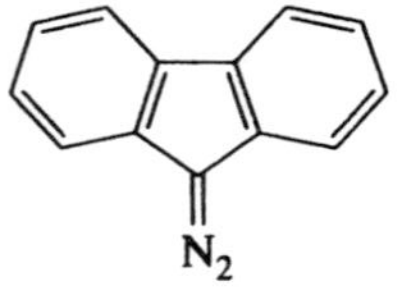 N_2	Direct	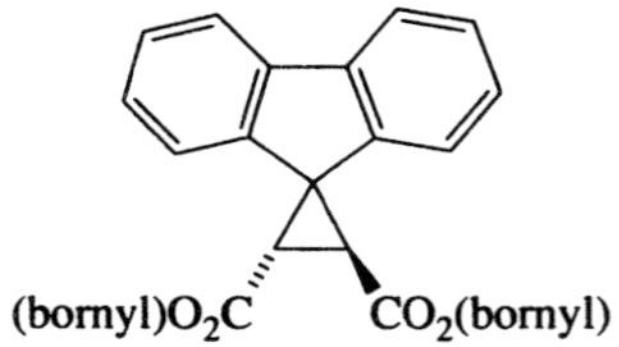 $(bornyl)O_2C$ $CO_2(bornyl)$ + diastereomer (ratio = 1.3)		36
 N_2		Direct		16	37

[a] Only the cyclopropane products are shown, although alkenes and other side products may also have been produced.

metals. Evidence has been reported for the intermediacy of metal carbene complexes in these reactions. (The discussion of well-defined examples of metal carbene complexes is reserved for Section 4.6.5.)

The range of alkenes that may be used as substrates in these reactions is vast. Suitable catalysts may be chosen to permit use of ordinary alkenes, electron deficient alkenes such as α,β-unsaturated carbonyl compounds, and very electron rich alkenes such as enol ethers. These reactions are generally stereospecific, and they often exhibit *syn* stereoselectivity, as was also mentioned for the photochemical reactions earlier. Several optically active catalysts and several types of chiral auxiliaries contained in either the alkene substrates or the diazo compounds have been studied in asymmetric cyclopropanation reactions, but diazocarbonyl compounds, rather than simple diazoalkanes, have been used in most of these studies. When more than one possible site of cyclopropanation exists, reactions of less highly substituted alkenes are often seen, whereas the photochemical reactions often occur predominantly at more highly substituted double bonds. However, the regioselectivity of the metal-catalyzed reactions can be very dependent upon the particular catalyst chosen for the reaction.

The occurrence of competing side reactions is generally reduced when metal catalysts are employed instead of photochemical conditions. For example, insertions into C—H bonds are of diminished importance, although insertions into O—H and other X—H bonds are still commonplace. The extent to which any of these other reactions compete with cyclopropanation is, however, again dependent upon the choice of catalyst.

Examples of metal-catalyzed reactions are compiled in Table 3.

4.6.3 USE OF CARBENES FROM OTHER PRECURSORS

Carbenes may be generated by many other methods for use in cyclopropanation reactions. However, relatively few of these methods are used very commonly in actual applications in synthetic organic chemistry and, therefore, the discussion that follows is limited to just these few methods.

4.6.3.1 α-Elimination Reactions

Treatment of an appropriate substrate with a base can lead to a carbanionic intermediate that bears a leaving group on the carbanionic center. Upon loss of this group, a carbene may be formed, the net overall process being called an α-elimination since both the hydrogen substituent and the leaving group are lost from the same carbon center (equation 5). The resulting carbene may then participate in cyclopropanation reactions.

$$>C(H)X \xrightarrow{\text{base}} \left[>\bar{C}-X \right] \xrightarrow{-X^-} \left[>C: \right] \qquad (5)$$

Most often, the starting materials bear additional substituents, especially electronegative groups, on the prospective carbene centers. Consequently, there are only a small number of examples in which this approach has been used to add a simple alkylidene group to an alkene (Table 4).

A related approach is the treatment of geminal dihalides with metal reagents to give an intermediate α-haloalkylmetal species, from which halide may in principle be lost in a net overall α-elimination of the two halide substituents (see Section 4.6.4.2).

4.6.3.2 Carbenes from Other Sources

Other methods for the formation of carbenes that have seen at least limited application in cyclopropanation reactions include the use of diazirines, epoxides, ketenes and *N*-nitrosoureas.[1h,8e,8f,8l,8m,8p]

4.6.4 USE OF GEMINAL DIHALOALKANES AND RELATED COMPOUNDS

Geminal dihaloalkanes are very commonly used reagents for the addition of alkylidene groups to alkenes. The simpler dihalides, especially the dihalomethanes, are readily available and are certainly much less hazardous to use than the corresponding diazoalkanes. Nevertheless, any reactive alkylating agent should be regarded as a toxic substance.

Table 3 Metal-catalyzed Reactions of Diazoalkanes with Alkenes

Alkene	*Diazoalkane*	*Catalyst*	*Products*[a]	*Yield (%)*	*Ref.*
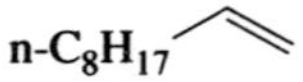 $n-C_8H_{17}$	CH_2N_2	$Pb(OAc)_2$	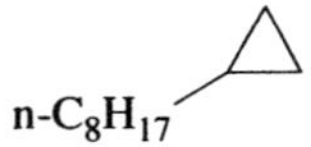 $n-C_8H_{17}$	89	37
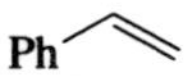 Ph	CH_2N_2	ZnI_2[b]	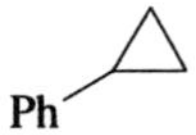 Ph	85	38
	CH_2N_2	Et_2AlCl[b]	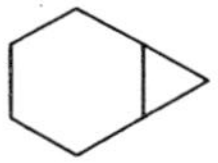	83	39
	$PhCHN_2$	$ZnCl_2$[b]	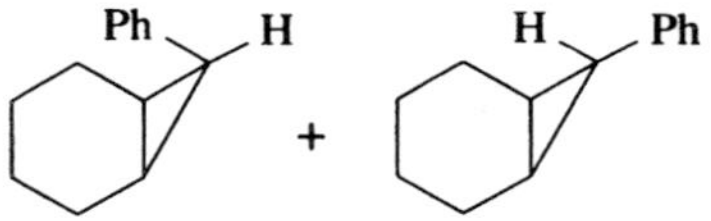 Ph H + H Ph 3:1	90	40
	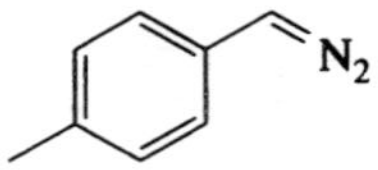 N_2	ZnI_2[b]	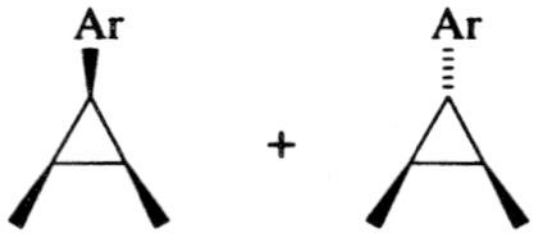 Ar + Ar 21:1	60	40
	CH_2N_2	$Cu(acac)_2$	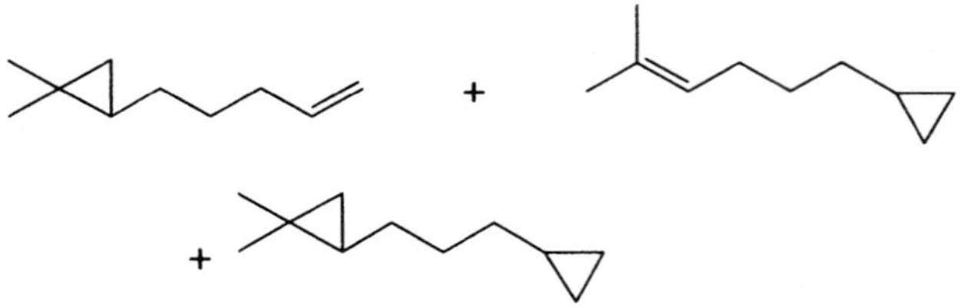 + + 1.0:0.1:0.6	90	41
		$Cu(OTf)_2$	1.0:2.5:0.4	50	41

Table 3 *(continued)*

Alkene	*Diazoalkane*	*Catalyst*	*Products*[a]	*Yield (%)*	*Ref.*
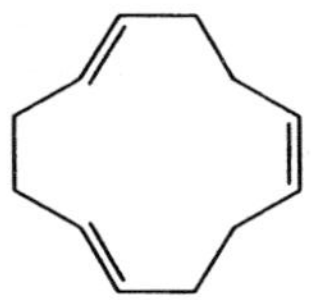	CH_2N_2	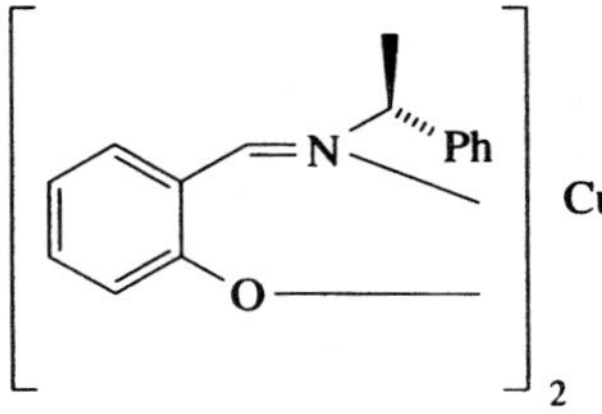 [=N Ph O]$_2$ Cu (optically active)	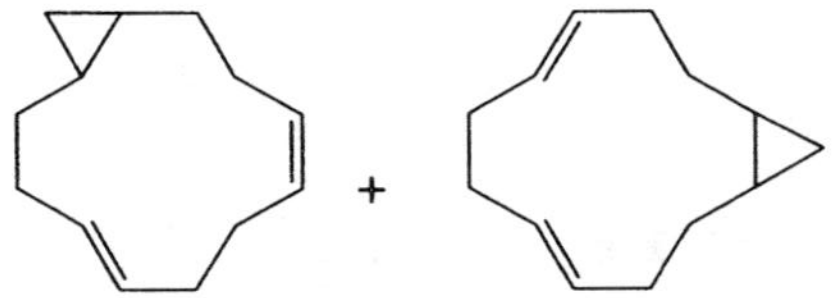 + 9:1 (low *ee*)	40	42
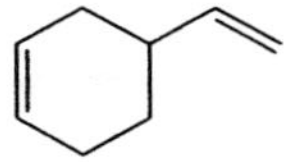	CH_2N_2	$Pd(OAc)_2$	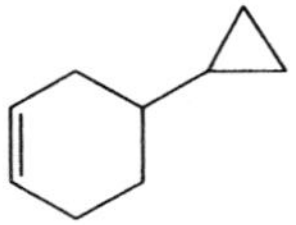	77	37
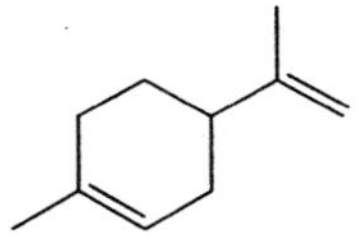	CH_2N_2	$Pd(OAc)_2$	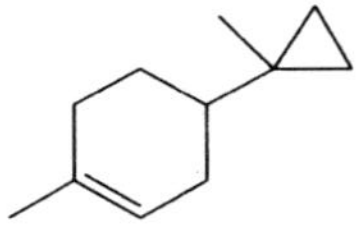	82	37
	CH_2N_2	$[(PhCN)_2PdCl_2]$	 + >30:1	90	43
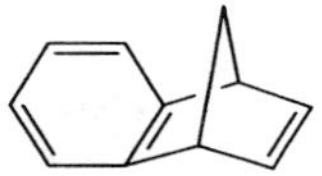	CH_2N_2	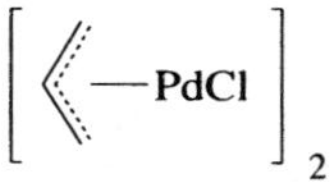 [—PdCl]$_2$		100	44

Table 3 *(continued)*

Alkene	*Diazoalkane*	*Catalyst*	*Products*[a]	*Yield (%)*	*Ref.*
	Ph_2CN_2	$Rh_2(OAc)_4$	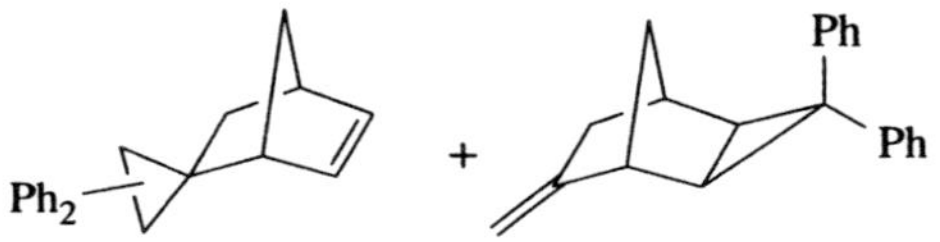 Ph_2 + Ph Ph >100:1		45
	Ph_2CN_2	$[(PhCN)_2PdCl_2]$	<1:50		45
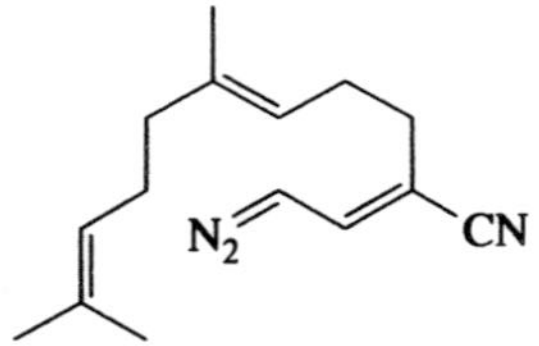 N_2 CN		CuI	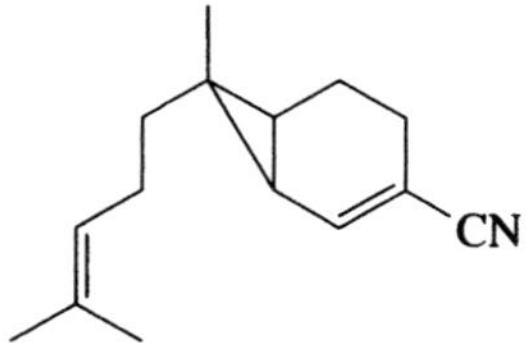 CN	>45	46
 OH	CH_2N_2	$[(PhCN)_2PdCl_2]$	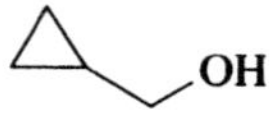 OH	*ca.* 70	47
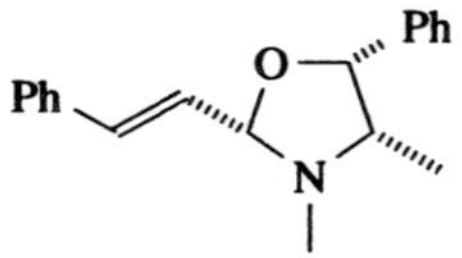 Ph O Ph N	CH_2N_2	$Pd(OAc)_2$	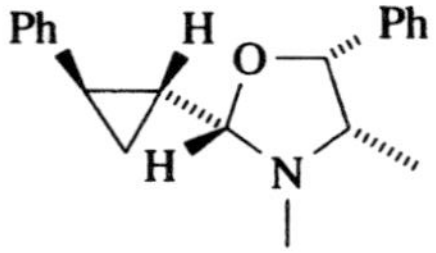 Ph H O Ph H N (>90% *de*)	*ca.* 100	48
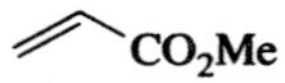 CO_2Me	CH_2N_2	$Ni(Bu^tNC)_2$	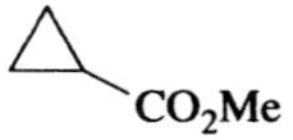 CO_2Me	66	49
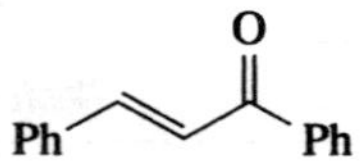 O Ph Ph	CH_2N_2	$Pd(OAc)_2$	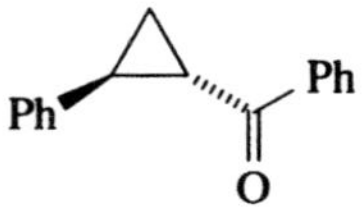 Ph Ph O	98	50

Table 3 *(continued)*

Alkene	*Diazoalkane*	*Catalyst*	*Products*[a]	*Yield (%)*	*Ref.*
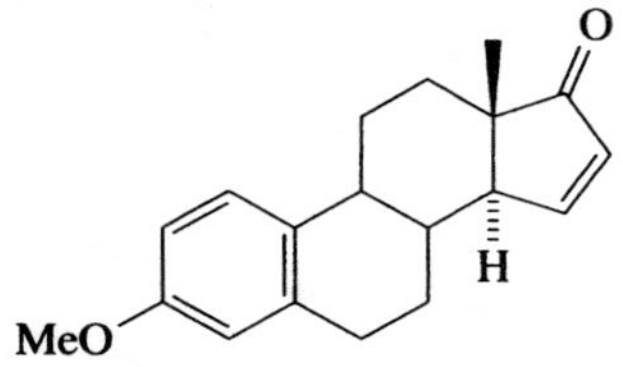 MeO	CH_2N_2	$Pd(OAc)_2$	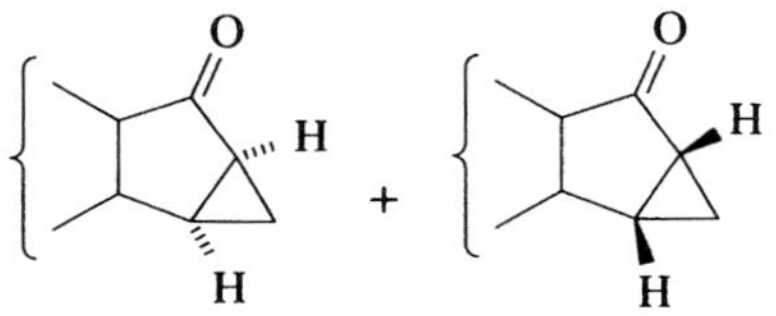 7:3	100	50
 OBu^n	$PhCHN_2$	$Rh_2(OAc)_4$	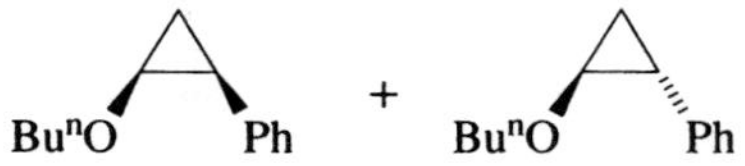 Bu^{n}O, Ph + Bu^{n}O, Ph 2.5:1	92	51

[a] Only the cyclopropane products are shown, although alkenes and other side products may also have been produced. [b] Stoichiometric or near stoichiometric rather than catalytic quantities of metal species employed.

Table 4 Cyclopropanations *via* α-Eliminations

Alkene	*Substrate*	*Base*	*Products*[a]	*Yield (%)*	*Ref.*
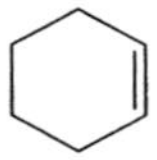	$PhCH_2Cl$	Bu^nLi	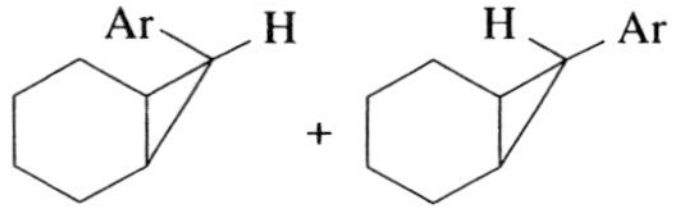 Ar H + H Ar	14[b]	52
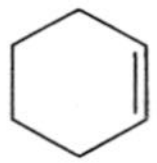	$PhCH_2Cl$	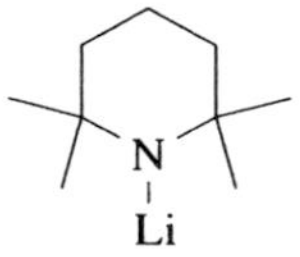 N Li	2.2:1	53	53
	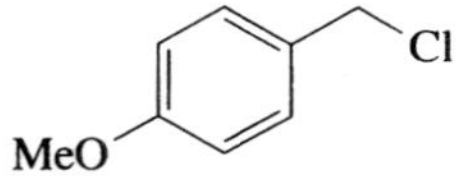 MeO Cl	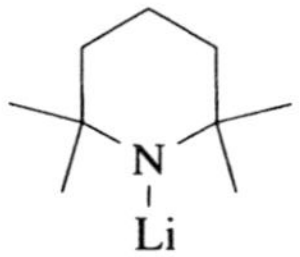 N Li	10:1	82	53
	$PhCH_2OPh$	Bu^nLi		2[c]	54
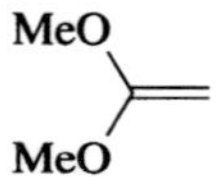 MeO MeO	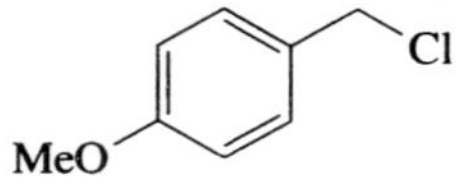 MeO Cl	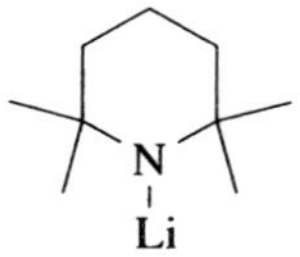 N Li	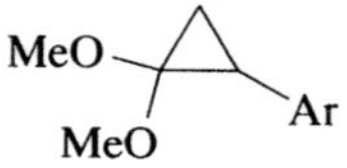 MeO MeO Ar	87	53
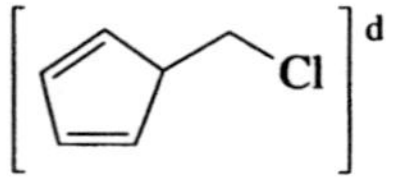 [Cl][d]		MeLi		29	55

Table 4 *(continued)*

Alkene	*Substrate*	*Base*	*Products*[a]	*Yield (%)*	*Ref.*
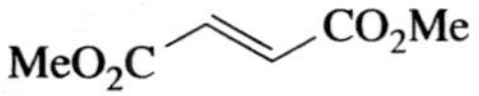 MeO_2C–CH=CH–CO_2Me	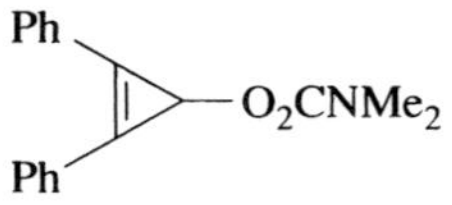 Ph, Ph-cyclopropene–O_2CNMe_2	$KOBu^t$	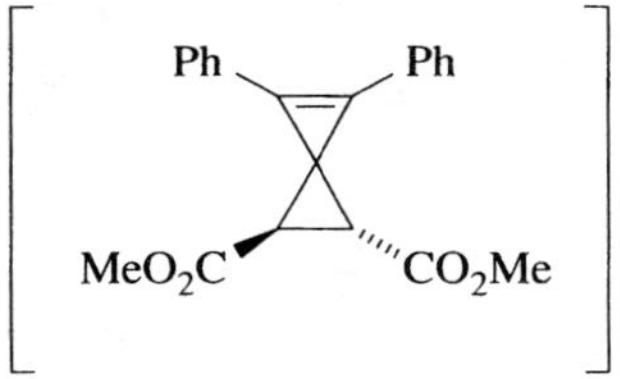 [Ph, Ph spiro-cyclopropene-cyclopropane, MeO_2C, CO_2Me][e]	>70	56

[a] Only the cyclopropane products are shown, although alkenes and other side products may also have been produced. [b] Ratio of isomers not reported. [c] A 3:1 mixture was obtained, but the isomers were not assigned. [d] Generated as an intermediate from cyclopentadiene + CH_2Cl_2 + MeLi. [e] Unstable — rearranges to an alkylidene cyclopropene.

4.6.4.1 Simmons–Smith Reaction

The Simmons–Smith reaction has become a very popular method for the addition of a methylene or certain other simple alkylidene groups to alkenes in the course of normal synthetic organic research applications.[1,8e,57] The originally developed procedure employed the reaction of diiodomethane with a zinc–copper couple in the presence of an alkene to give a cyclopropane as a methylene transfer product (equation 6). The mechanism of this reaction is not well understood, but an α-iodomethylmetal species, such as ICH_2ZnI, is generally regarded as being a key intermediate. Similar species may be involved in reactions of diazoalkanes promoted by zinc halides and related salts (see Section 4.6.2.3).

$$\text{alkene} + I\frown I \xrightarrow{Zn/Cu} \text{cyclopropane} \quad (6)$$

Many modifications have been made in the Simmons–Smith reaction over the years. A disadvantage of the original procedure was that the preparation of the required zinc–copper couple was a rather messy operation. Therefore, some of the modifications have been directed at developing other metal reagents. Among the several that have been reported, diethylzinc[58] and trialkylaluminums[59] are quite useful, but these reagents are, of course, quite air and moisture sensitive. A disadvantage of using diiodomethane is its expense, but the much less expensive dibromomethane has been found to be a useful substitute under various conditions, including sonication with zinc–copper couple.[60] A photochemical variant of the reaction using diiodomethane has also been reported which avoids the use of metal reagents altogether and which is less sensitive to steric hindrance.[61] The extension of the Simmons–Smith reaction to transfer of alkylidene groups other than just the simple methylene group has been limited. Ethylidene and benzylidene transfers have been the most thoroughly studied extensions of the method.

The Simmons–Smith reaction is stereospecific. Also, when other than just the simple methylene group is transferred, a low to modest level of *syn* stereoselectivity is typically seen. One of the most useful stereochemical features of the reaction is the directing effect of various heteroatomic functional groups. Experimental studies of this effect have been followed by theoretical studies.[62] The effect has been studied most commonly for allylic hydroxy groups. In these cases, the general outcome is diastereoselective delivery of the incoming methylene or alkylidene group to the face of the double bond having the closer proximity to the hydroxy group. The diastereoselectivity can be reversed by introducing an appropriate blocking or protecting group on the free hydroxy function. Also, the effect of the hydroxy group can be exploited in cases of substrates containing more than one alkene unit; regioselective delivery of the incoming group to a double bond having an associated allylic hydroxy function occurs in preference to double bonds not in close proximity to the directing group. Asymmetric versions of the Simmons–Smith reaction have also been studied, with the most common involving the use of alkene substrates containing chiral auxiliaries.

The intermediates in the Simmons–Smith reaction show much more controlled reactivity than, for example, diazoalkanes. Competing side reactions are greatly reduced, and insertion reactions are generally not seen. For example, the free hydroxy group is tolerated as indicated by the above discussion of directing effects. The reaction is also compatible with many other functional groups.

A wide range of alkenes may be used as substrates. The reaction is most commonly performed with alkenes of normal electronic nature, but electron deficient alkenes, such as α,β-unsaturated carbonyl compounds, and very electron rich alkenes, such as enol ethers and enamines, have also been used successfully. Not surprisingly, the cyclopropylcarbinyl ethers and amines that are formed in the latter reactions (see Table 5) are subject to facile rearrangements.

Examples of the Simmons–Smith reaction are presented in Table 5, with emphasis on more recent work. For additional examples, the reader is referred to extensive tabulations in earlier, comprehensive reviews.[1,8e,57]

4.6.4.2 Related Procedures

Organomercury compounds such as ICH_2HgI and $Hg(CH_2Br)_2$ may be prepared by the reaction of metallic mercury with diiodomethane or by reaction of mercury(II) salts with diazoalkanes. These mercury derivatives are related to the intermediate that has been proposed in the Simmons–Smith reaction, and likewise, they have been found to convert alkenes into cyclopropanes.[92]

A procedure which combines characteristics of the Simmons–Smith reaction with the previously discussed α-eliminations (see Section 4.6.3.1) is the reaction of a geminal dihalide with an alkyllithium re-

Table 5 Simmons–Smith Reactions and Modifications

Alkene	*Dihalide*	*Reagent*	*Products*	*Yield (%)*	*Ref.*
	CH_2I_2	Zn(Cu) (or Zn/CuCl)	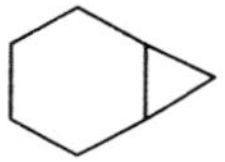	86–92	63
	CH_2Br_2	Zn powder		94	64
	CHBrI	Cu powder		69	65
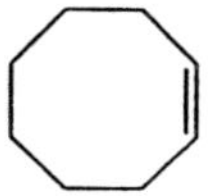	CH_2Br_2	Zn/CuCl, sonication	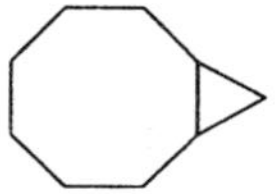	72	60
	CH_2I_2	Me_3Al		96	59
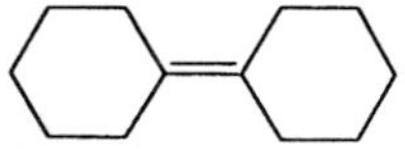	CH_2I_2	$h\nu$		86	61
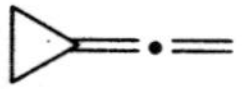	CH_2I_2	Zn/CuCl, sonication	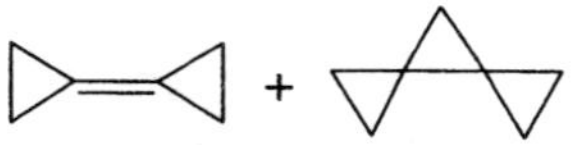 + 30–40% 10–15%		66
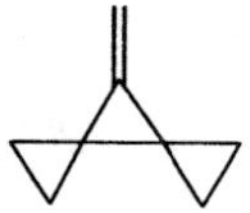	CH_2I_2	Zn(Ag)	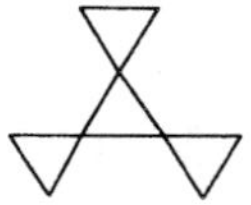	100	67
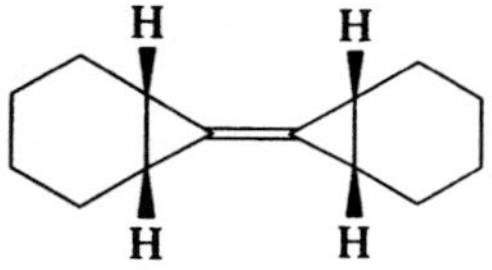 H H H H	CH_2I_2	Et_2Zn	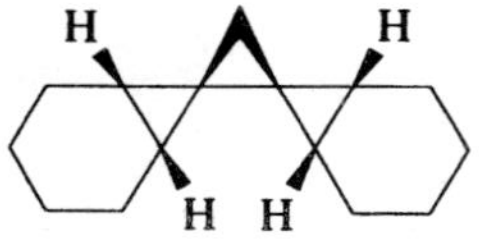 H H H H	92	68

Table 5 *(continued)*

Alkene	*Dihalide*	*Reagent*	*Products*	*Yield (%)*	*Ref.*
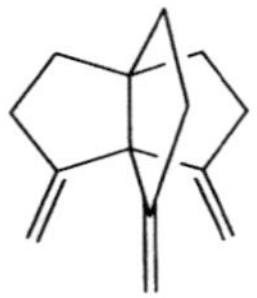	CH_2I_2	Et_2Zn		Not reported	69
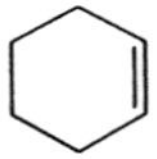	$MeCHI_2$	Et_2Zn	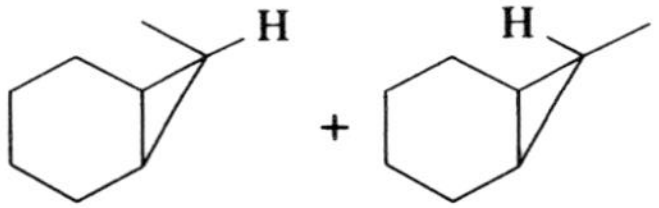 H + H 1.5:1	66	58b
	$PhCHI_2$	Et_2Zn	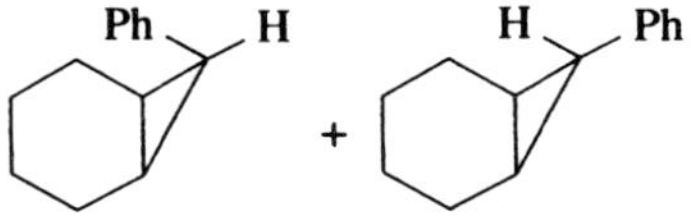 Ph H + H Ph 17:1	69	70
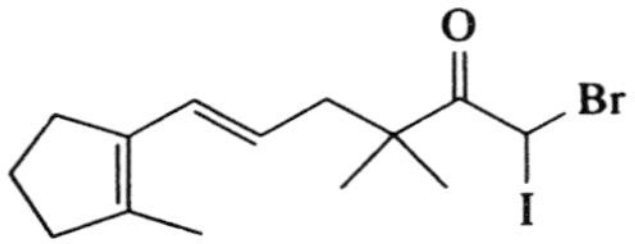 O Br I		Bu_3SnH, AIBN	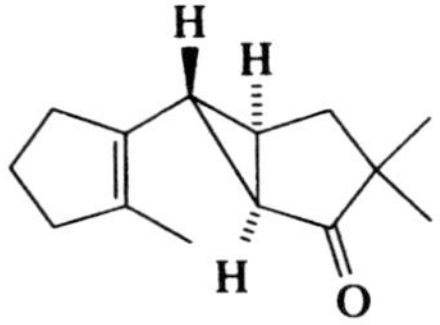 H H H O	45	71
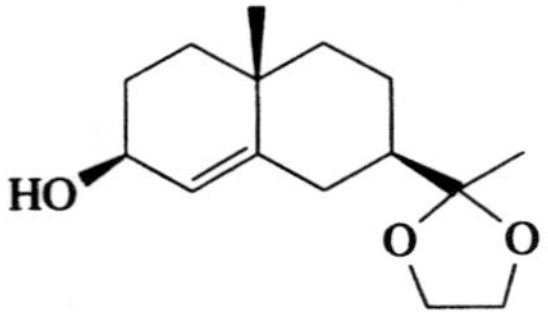 HO O O	CH_2I_2	Zn(Cu)	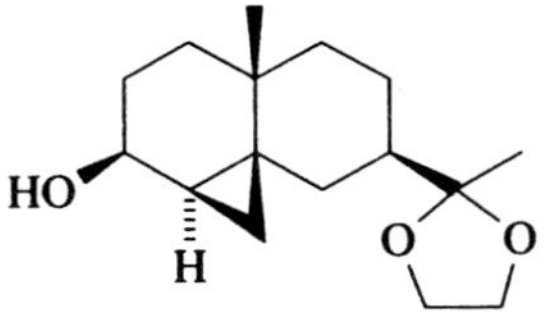 HO H O O	78	72

Table 5 *(continued)*

Alkene	*Dihalide*	*Reagent*	*Products*	*Yield (%)*	*Ref.*
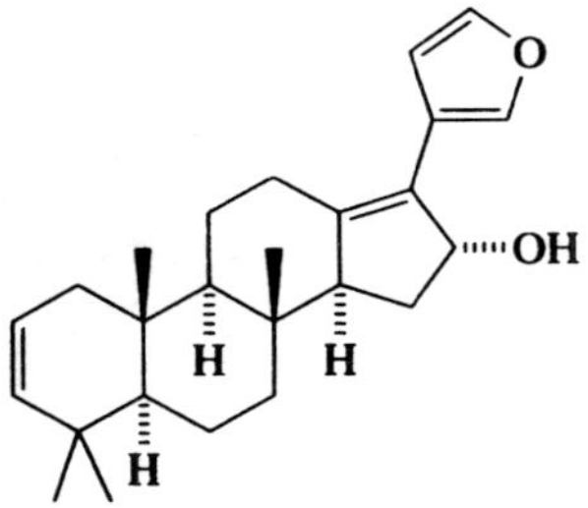 O, OH, H, H, H	CH_2I_2	Zn(Ag)	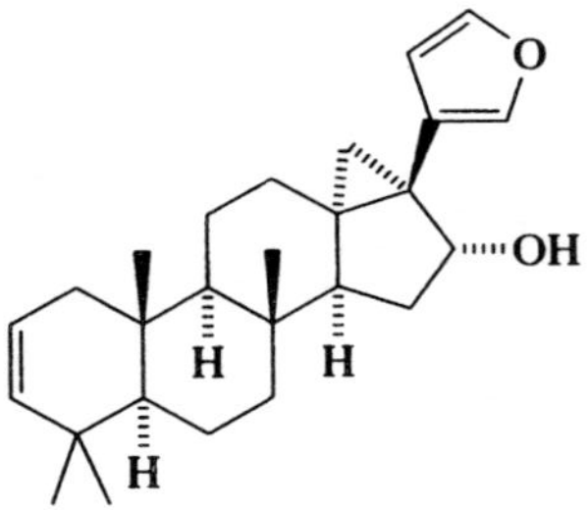 O, OH, H, H, H	89	73
 Bu_3Sn, OH	CH_2I_2	Zn(Cu)	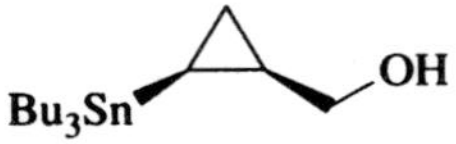 Bu_3Sn, OH	70	74
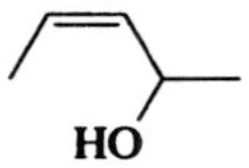 HO	CH_2I_2	Zn(Cu)	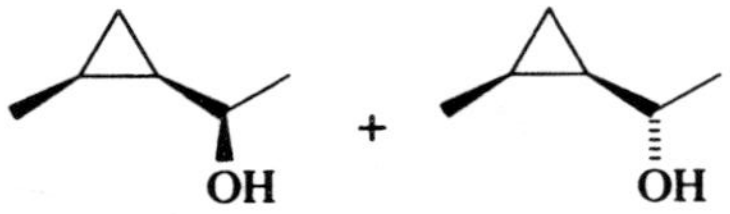 OH + OH >99:1		75
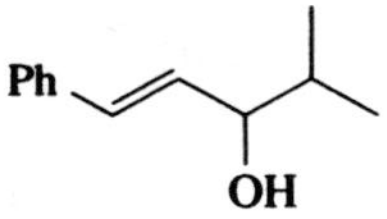 Ph, OH	CH_2I_2	Sm(Hg)	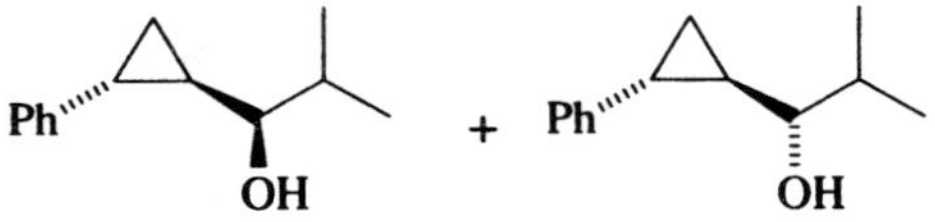 Ph, OH + Ph, OH >200:1	88	76
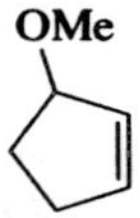 OMe	CH_2I_2	Cu	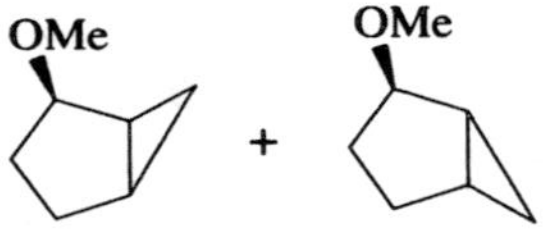 OMe + OMe 88:12	75	77

Table 5 *(continued)*

Alkene	*Dihalide*	*Reagent*	*Products*	*Yield (%)*	*Ref.*
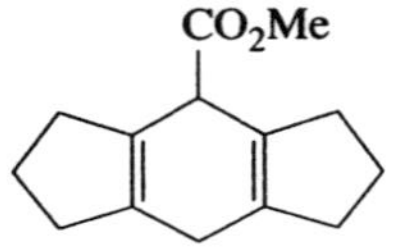	CH_2I_2	Zn(Cu)	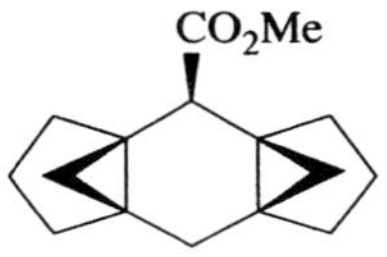	>45	78
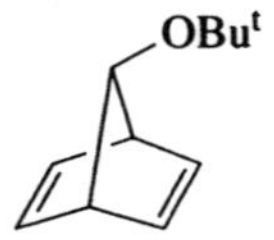		$Zn(CH_2I)_2$	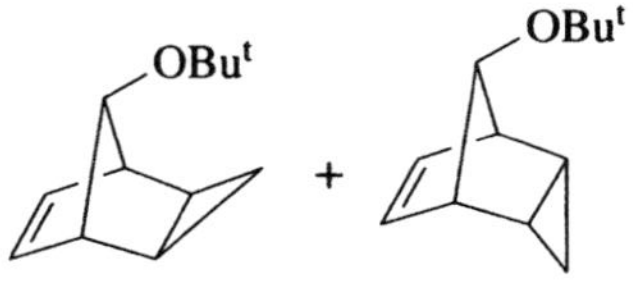 + 1.8:1	Not reported	79
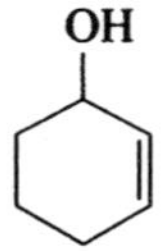	$MeCHI_2$	Et_2Zn	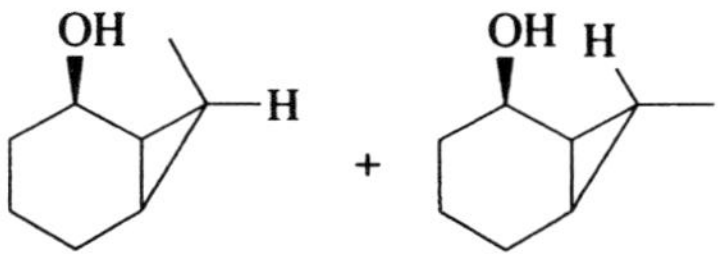 + 1:1.6	79	80
		Sm(Hg)	1:5	100	76
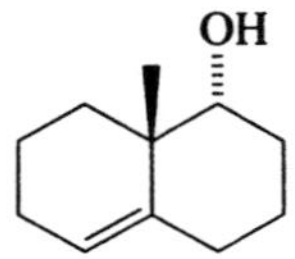	CH_2I_2	Zn(Cu)	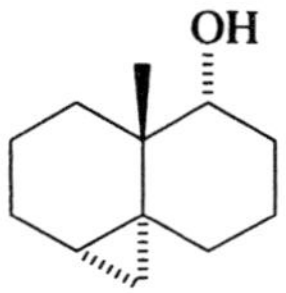	63	81
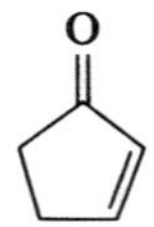	CH_2I_2	Zn(Cu)	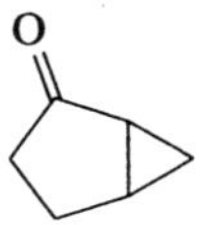	80	82

Table 5 *(continued)*

Alkene	*Dihalide*	*Reagent*	*Products*	*Yield (%)*	*Ref.*
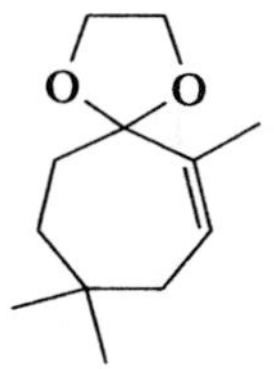 O O	CH_2I_2	EtZnI	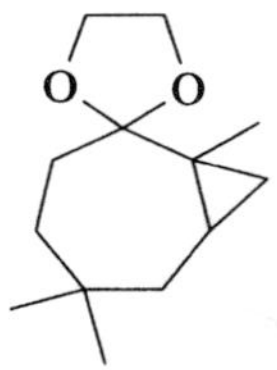 O O	92	83
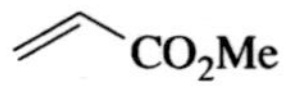 CO_2Me	CH_2Br_2	$[Ni(PPh_3)_4]/ZnBr_2$	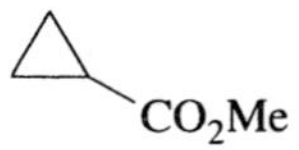 CO_2Me	75	84
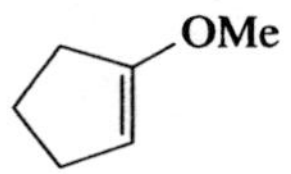 OMe	CH_2I_2	Zn(Cu)	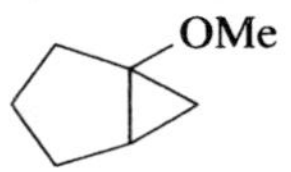 OMe	72	85
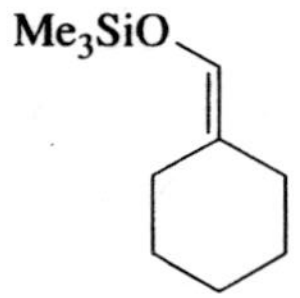 Me_3SiO	CH_2I_2	Zn(Cu)	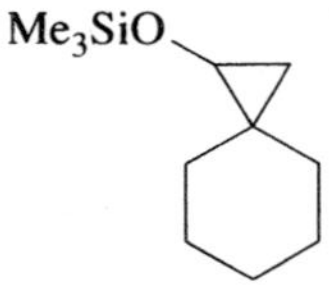 Me_3SiO	86	85
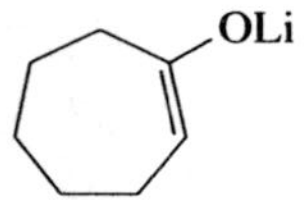 OLi	CH_2I_2	SmI_2 (then HCl)	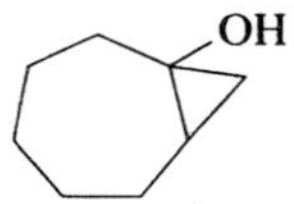 OH	58	86
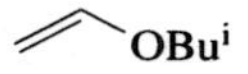 OBu^i	$MeCHI_2$	Et_2Zn	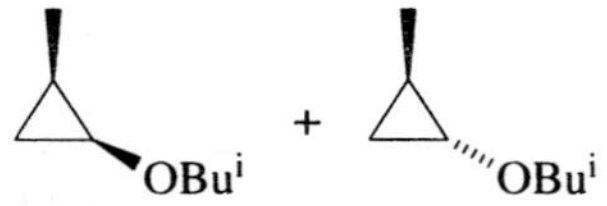 OBu^i + OBu^i 2.3:1	96	58b

Table 5 *(continued)*

Alkene	*Dihalide*	*Reagent*	*Products*	*Yield (%)*	*Ref.*
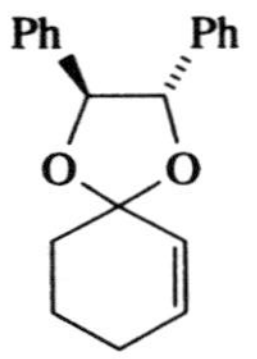 Ph Ph O O (optically active)	CH_2I_2	Zn(Cu)	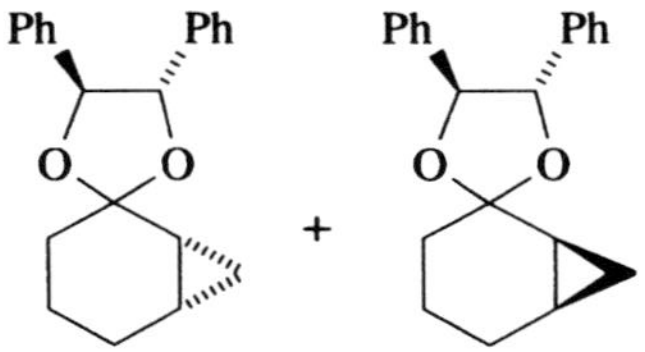 Ph Ph O O + Ph Ph O O 19:1	90	87
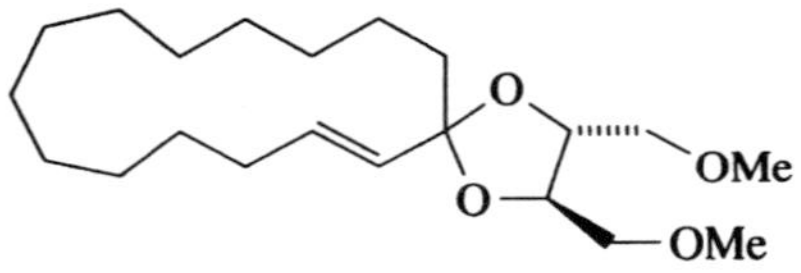 O OMe O OMe (optically active)	CH_2I_2	Zn(Cu)	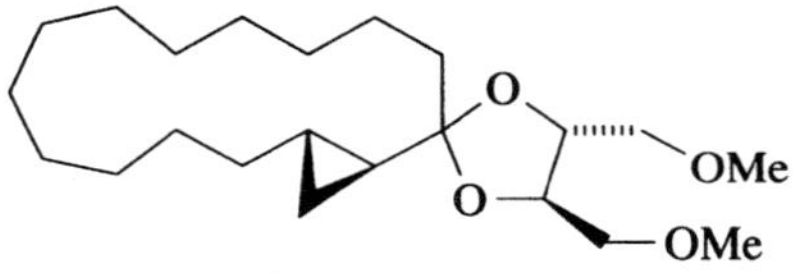 O OMe O OMe + diastereomer (>20:1)	95	88
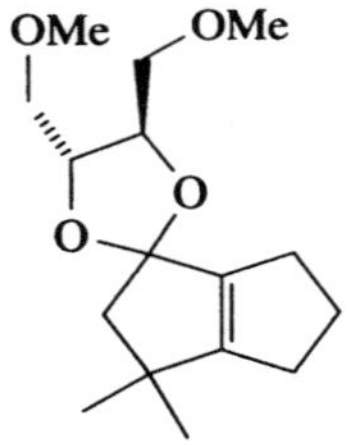 OMe OMe O O (optically active)	CH_2I_2	Zn(Cu)	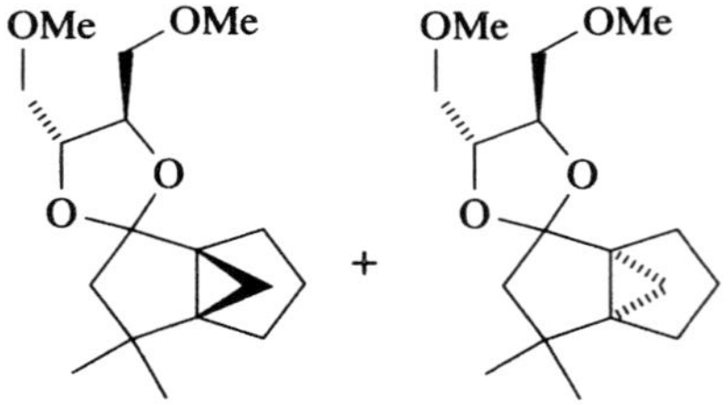 OMe OMe O O + OMe OMe O O 8:1	84	89
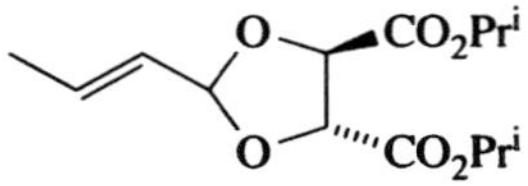 O CO_2Pr^i O CO_2Pr^i	CH_2I_2	Et_2Zn	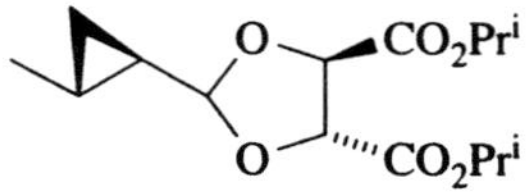 O CO_2Pr^i O CO_2Pr^i + diastereomer (97:3)	90	90

Table 5 *(continued)*

Alkene	*Dihalide*	*Reagent*	*Products*	*Yield (%)*	*Ref.*
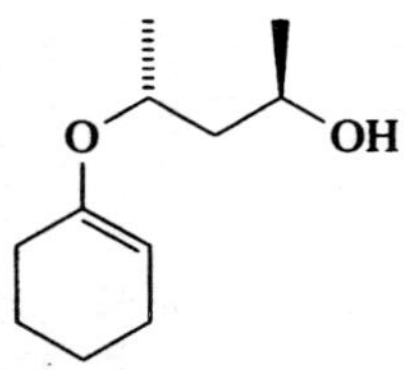 OH	CH_2I_2	Et_2Zn	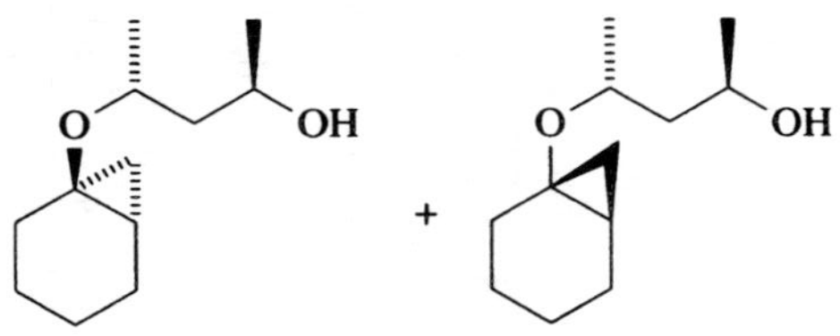 OH + OH 97:3	65	91

agent in the presence of an alkene. Halogen–lithium exchange occurs to give again an α-haloalkylmetal species as an apparent intermediate which then leads to alkylidene transfer to the alkene (equation 7).

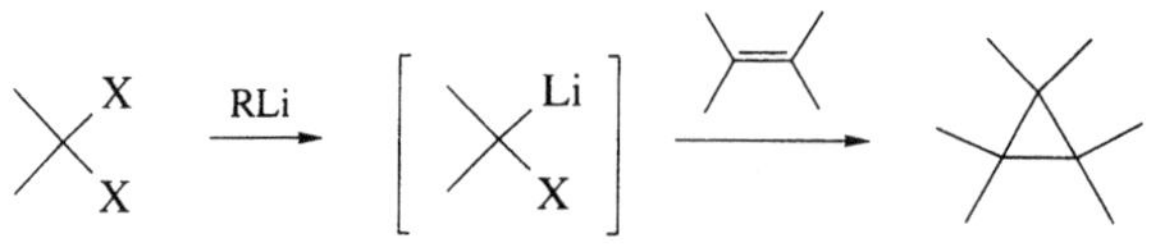

(7)

Finally, aldehydes have also been reported to serve as alkylidene transfer reagents.[93] Examples of these further cyclopropanation procedures are summarized in Table 6.

4.6.4.3 Dialkylation of 1,2-Dicarbanionic Species by Geminal Dihaloalkanes

In principle, a vicinal or 1,2-dicarbanion, as at least a formal intermediate, should be capable of undergoing a double alkylation by an appropriate 1,1-difunctionalized electrophile to give a cyclopropane. Indeed, examples of this general strategy have been reported. In one case, a 1,2-bis(oxazolinyl)ethane derivative undergoes dilithiation followed by reaction with bromochloromethane to give the corresponding disubstituted cyclopropane (equation 8), although the stereochemistry was not explicitly reported.[96] A related asymmetric cyclopropanation occurs with optically active (–)-dimenthyl succinate (equation 9).[97] Finally, a β-tosyloxy-α,β-unsaturated ketone undergoes a related cyclopropanation *via* an apparent double Michael addition pathway (equation 10).[98] This last reaction was said to produce only one stereoisomer, but the *endo*/*exo* stereochemistry was not stated explicitly.

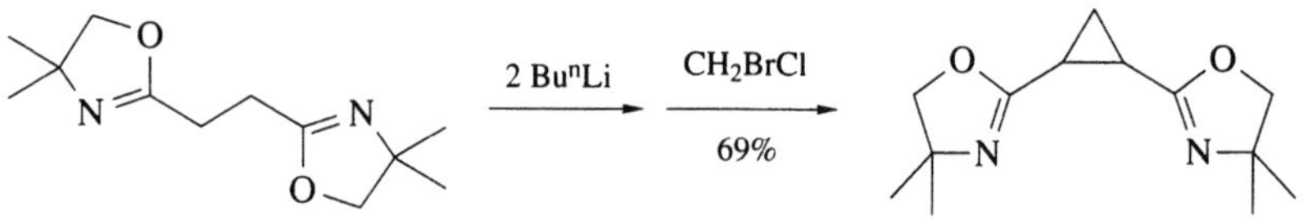

2 Bu^nLi; CH_2BrCl, 69% (8)

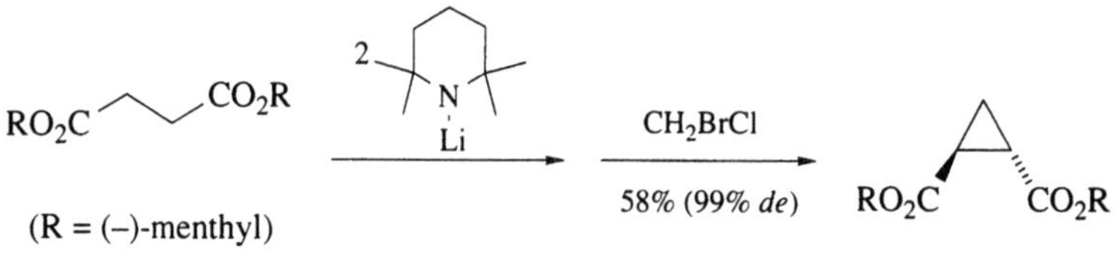

$RO_2C\sim CO_2R$ (R = (–)-menthyl); 2 LiN(2,2,6,6-tetramethylpiperidide); CH_2BrCl, 58% (99% *de*) → RO_2C, CO_2R cyclopropane (9)

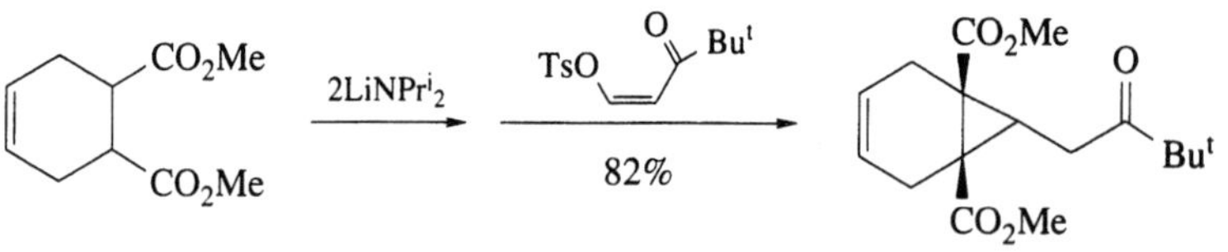

CO_2Me, CO_2Me; $2LiNPr^i_2$; TsO, O, Bu^t, 82% → CO_2Me, O, Bu^t, CO_2Me (10)

4.6.5 USE OF TRANSITION METAL CARBENE COMPLEXES

Complexes containing carbenes formally coordinated to transition metals were first reported by E. O. Fischer and his coworkers in the early 1960s.[99] The particular compounds that they studied were of the general structure $[(OC)_5M{=}C(OR^1)R]$ (M = Cr, Mo, W), having alkoxycarbenes coordinated to the Group VI metals. Note that the structures of these complexes are written with a double bond between the carbene center and the metal. This notation signifies the combination of donation of electron density from carbon to the metal to form a normal coordinative, or dative, σ-bond and donation from a *d*-like orbital of the metal to the empty *p*-orbital of the carbene, resulting in back-bonding to form a retrodative π-bond.

Since the pioneering studies of Fischer, carbene complexes have become the subjects of very extensive investigations in several laboratories, and they have become known for a large number of the transition metals.[100] A fairly general statement about their chemistry is that these compounds usually are neither prepared directly from free carbenes nor are they to be considered as sources of free carbenes, *i.e.* free carbenes are generally not in equilibrium with the carbene complexes. Instead, they are prepared by indi-

Table 6 Further Cyclopropanation Procedures Related to the Simmons–Smith Reaction

Alkene	*Reagents*	*Products*	*Yield (%)*	*Ref.*
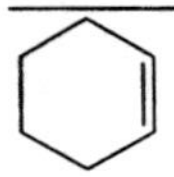	ICH_2HgI/Ph_2Hg	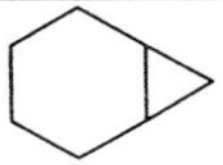	64	92
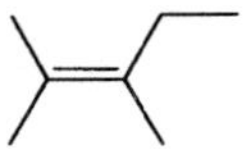	$Hg(CH_2Br)_2$		72	92
	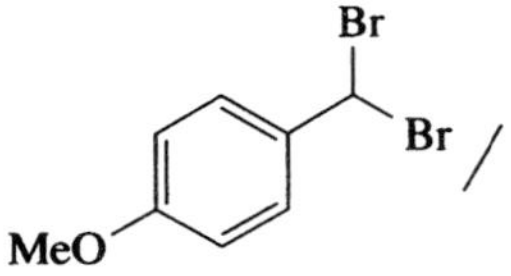 MeO, Br, Br / MeLi	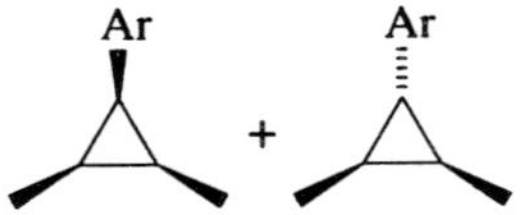 Ar + Ar 8.1:1	55	32
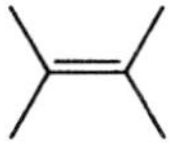	Me_2CBr_2/Bu^nLi	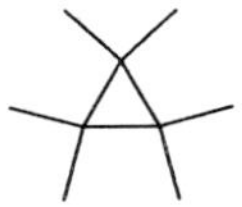	46[a]	94
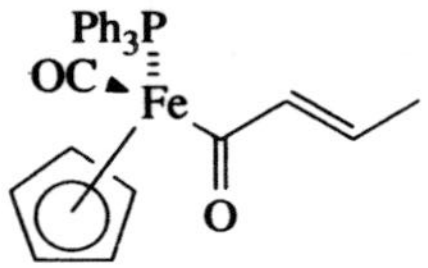 Ph_3P OC Fe O	$CH_2ClI/MeLi$	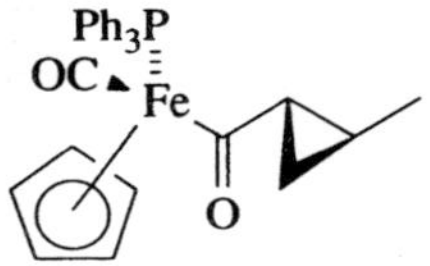 Ph_3P OC Fe O + diastereomer (>50:1)	85	95
	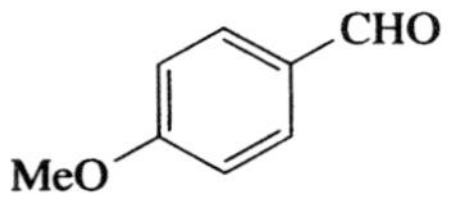 MeO, CHO $Zn(Hg)/BF_3•OEt_2$	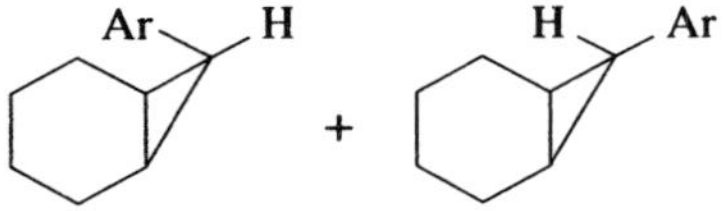 Ar H + H Ar 95:5	60	93
 OAc	$PhCHO/Zn(Hg)/BF_3•OEt_2$	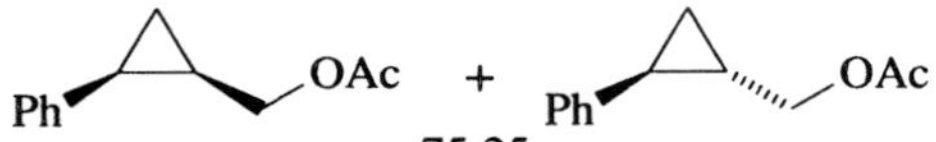 Ph OAc + Ph OAc 75:25	50	93

Table 6 *(continued)*

Alkene	*Reagents*	*Products*	*Yield (%)*	*Ref.*
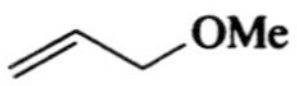 OMe	PhCHO/Zn(Hg)/BF_3•OEt_2	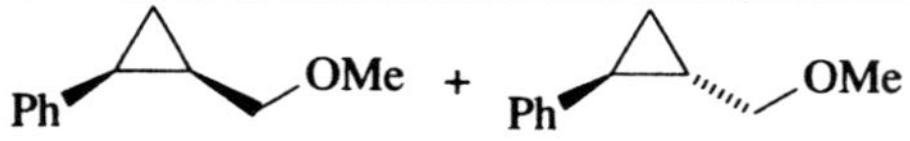 Ph, OMe + Ph, OMe 33:67	56	93

[a] Product yield corrected for unreacted starting material.

rect routes not proceeding *via* free carbenes. Furthermore, the vast majority of these complexes do not exhibit typical carbene-like reactivity. For example, relatively few of them to date have shown reactivity toward alkenes to give cyclopropanes. Rather, for the most part, these complexes behave as a distinctly new class of chemical species showing their own unique types of reactivity.

Another broad class of compounds are the bridged carbene complexes. These compounds contain two identical or two different metal centers with the carbene centers bonded to both of the metal atoms in a bridging relationship. However, these binuclear complexes generally do not show classical carbene reactivity and will therefore not be discussed further, except to mention briefly the special case of the titanium–aluminum complex (**3**) developed by Tebbe and Grubbs and their coworkers.[101] This, and related complexes, has proven to be particularly useful in organic synthesis, although its principal importance is in reactions other than cyclopropanations.

Ti AlMe$_2$ Cl

(**3**)

A special note about nomenclature is in order. Both the terms 'carbene' and 'alkylidene' complex have come into use, and attempts have been made to use these terms to differentiate various types of reactivity of these species. However, the term 'carbene' complex is used throughout this chapter, primarily because of the emphasis on cyclopropanation reactivity.

Several methods have been reported for the formation of carbene complexes, but three of these methods have been used especially commonly. The first method, as used by Fischer in his original studies, involves the addition of an organolithium reagent to a transition metal carbonyl complex to give an anionic intermediate which upon *O*-alkylation produces an alkoxycarbene complex (equation 11). In turn, the alkoxy group may undergo substitution by other heteroatomic groups. The second method involves the use of transition metal alkyls which bear a prospective leaving group on the carbon atom bonded to the metal. Upon either direct loss of this group or upon its prior activation, a carbocationic species is formed for which a resonance structure can be written based upon retrodative π-bonding from the metal to the carbon (equation 12). The resulting species may be regarded as a metal-stabilized carbocation or, in the context of this chapter, as a cationic metal–carbene complex. The third method to be discussed here involves addition of an electrophilic reagent, most commonly just a proton source ($E^+ = H^+$), to an alkenylmetal complex. The resulting adduct may again be regarded as a cationic carbene complex (equation 13).

$$(L)_nM(CO) \xrightarrow{R\text{—}Li} \left[(L)_nM{=}C(OLi)R \right] \xrightarrow{R^1X} (L)_nM{=}C(OR^1)R \qquad (11)$$

$$(L)_nM\text{—}C(X)(R)R^1 \xrightarrow{-X^-} \left[(L)_nM\text{—}\overset{+}{C}(R^1)R \longleftrightarrow (L)_n\overset{+}{M}{=}C(R^1)R \right] \qquad (12)$$

$$(L)_nM\text{—}CH{=}CH\text{—}R \xrightarrow{E^+} \left[(L)_nM\text{—}\overset{+}{C}H\text{—}CH(E)\text{—}R \longleftrightarrow (L)_n\overset{+}{M}{=}CH\text{—}CH(E)\text{—}R \right] \qquad (13)$$

Several carbene complexes prepared by the first route are sufficiently stable to be isolated and characterized. However, the cationic complexes prepared by the second and third methods are quite reactive species that usually cannot be isolated and for which only partial characterization has been possible through special spectroscopic techniques. Exceptions are those complexes bearing substituents R or R^1 that are good cation-stabilizing groups. Consequently, these cationic carbene complexes are most commonly generated *in situ* from their precursors in the presence of the substrates desired for further reactions.

Of the several carbene complexes that are presently known, only a relatively small fraction of them have shown synthetically useful reactivity in cyclopropanations of alkenes. Comprehensive reviews of this aspect of their chemistry appeared in 1987 and 1989,[100j,100k] and the reader is referred to these articles for more detailed treatment of the topic than is given in the present chapter.

The original Fischer-type carbene complexes react with electron deficient alkenes such as α,β-unsaturated esters[102] and especially electron rich alkenes such as enol ethers[103] and enamines,[104] but reactions with simple alkenes have been limited in their usefulness. Furthermore, because these particular carbene complexes most commonly contain heterosubstituted carbene units, they formally fall outside the scope of this chapter. A few nonheterosubstituted examples that have proven to be useful are benzylidene[105] and diphenylcarbene[106] complexes of tungsten.

The most generally useful carbene complexes for transfer of methylene and other simple alkylidene groups are the cationic complexes of the general types shown in equations (12) and (13). In particular, the complexes containing the cyclopentadienyldicarbonyliron group, commonly written as $Cp(CO)_2Fe$ or simply Fp, have been the most useful to date. Also, derivatives of this system have been studied in which at least one of the carbonyl ligands has been replaced by another ligand such as a phosphine or a phosphite.

The first examples of these iron complexes arose from early work of Pettit[107] and Green[108] and their coworkers in the middle 1960s. A key finding was that the reaction of the methoxymethyliron complex (**4**) with acid in the presence of cyclohexene produced a modest yield of the cyclopropane norcarane (equation 14). The cationic methylene complex (**5**) was proposed as an intermediate. This species is very reactive and has not been studied by direct means in solution, although it has been detected in gas phase studies.[109]

$$\left[Cp(CO)_2FeCH_2OMe \right] \xrightarrow[\text{HBF}_4,\ -\text{MeOH}]{\text{cyclohexene}} \left[\begin{matrix} Cp(CO)_2\overset{+}{Fe}{=}CH_2 \\ BF_4^- \end{matrix} \right] \xrightarrow{46\%} \text{norcarane} \qquad (14)$$

(**4**) (**5**)

Since the time of this early work, many other investigators have studied the parent methylene complex (**5**), several substituted derivatives, and various precursors of these species.[100j,100k] The sulfonium salt (**6**) has proven to be particularly useful (equation 15) because of its unusually high stability. It is a highly crystalline solid that is stable in air and even in hot water.[110] Substituted derivatives have also been developed. In addition, several α-alkoxyalkyliron derivatives related to the parent compound (**4**) have been studied extensively.

$$\left[Cp(CO)_2FeCH_2SMe_2 \right]^+ BF_4^- \xrightarrow[\text{polar solvent, heat}]{\text{>C=C<}} \text{>C–C< (cyclopropane)} \qquad (15)$$

(**6**)

The reactions of these iron carbene reagents with alkenes to give cyclopropanes are stereospecific. They also exhibit high *syn* stereoselectivity in many cases. Optically active derivatives have been reported that have chiral ligands on iron or chiral alkoxy groups on the prospective carbene center and which have been resolved with the iron itself as a chiral center. Resulting from this work have been some highly enantioselective cyclopropanations.

The cyclopropanation reactions of the cationic iron carbene complexes occur most efficiently with alkenes of normal electronic characteristics. Very electron deficient alkenes such as α,β-unsaturated carbonyl compounds are very poor substrates. Very electron rich alkenes such as enol ethers react rapidly, but the expected cyclopropanes generally cannot be isolated; if they are indeed formed, they apparently undergo further reactions, perhaps promoted by the metallic species present in the reaction mixtures.

Specific examples of cyclopropanations using carbene complexes in general are summarized in Table 7, but the reader is referred to other recent reviews for much more extensive compilations.[100j,100k]

A recent modification of these reactions that appears to have significant potential in organic synthesis is a tandem sequence of alkyne insertion and cyclopropanation (Scheme 2).[115] One particularly impressive, fully intramolecular case is shown in Table 7 (ref. 115).

Table 7 Cyclopropanations Using Transition Metal–Carbene Complexes

Alkene	*Carbene complex or precursor*	*Products*	*Yield (%)*	*Ref.*
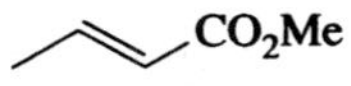 CO_2Me	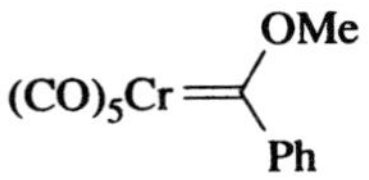 $(CO)_5Cr{=}C(OMe)Ph$	 MeO, Ph, CO_2Me + Ph, OMe, CO_2Me (2.5:1)	60	102a
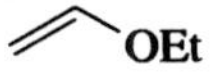 OEt		 MeO, Ph, OEt + Ph, OMe, OEt (3.2:1)	61	103
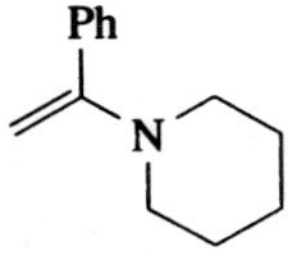 Ph, N		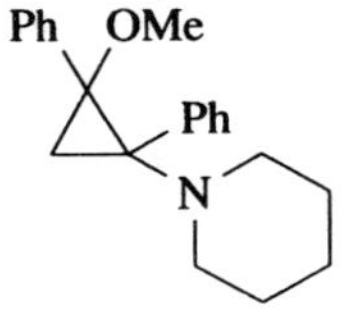 Ph, OMe, Ph, N	16	104
 $OSiMe_2Bu^t$, OMe	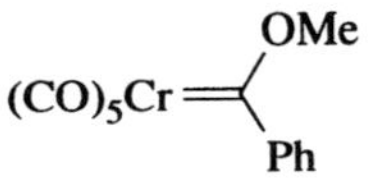 $(CO)_5Cr{=}C(OMe)Ph$	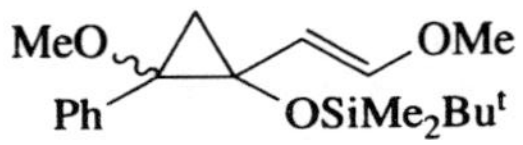 MeO, Ph, OMe, $OSiMe_2Bu^t$ 2:1	67	111
	$(CO)_5W{=}CHPh$	 Ph + Ph (41:1)	60	105
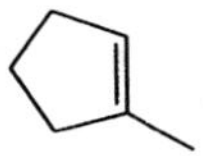		 Ph, H + H, Ph (8:1)	50	105
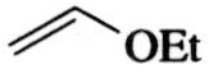 OEt	$(CO)_5W{=}CPh_2$	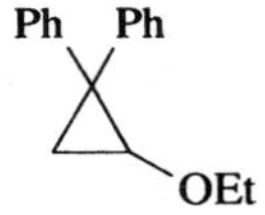 Ph, Ph, OEt	65	106

Table 7 *(continued)*

Alkene	*Carbene complex or precursor*	*Products*	*Yield (%)*	*Ref.*
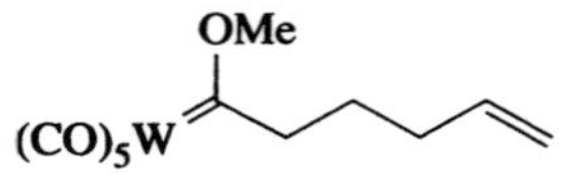 $(CO)_5W$ (OMe)		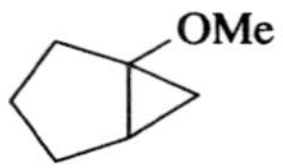 OMe	*ca.* 100	112
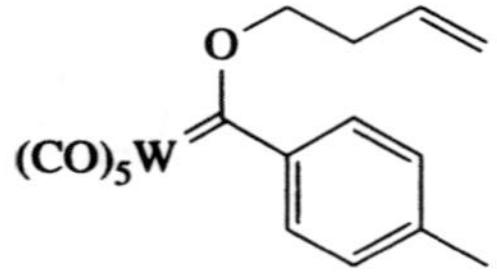 $(CO)_5W$ (O)		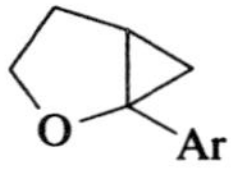 O, Ar	*ca.* 95	113
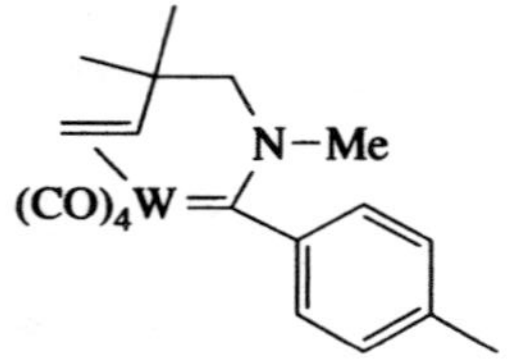 $(CO)_4W$ (N–Me)		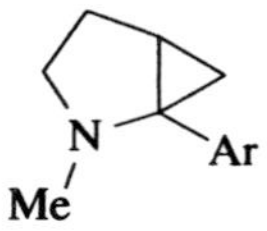 N, Me, Ar	*ca.* 90	114
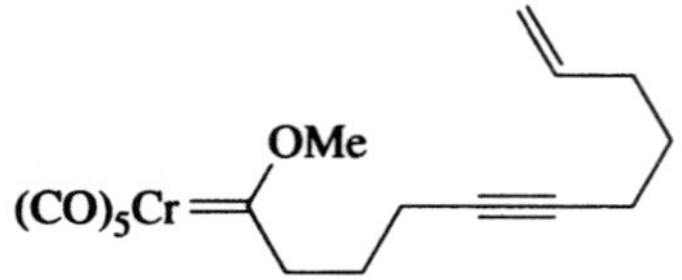 $(CO)_5Cr$ (OMe)		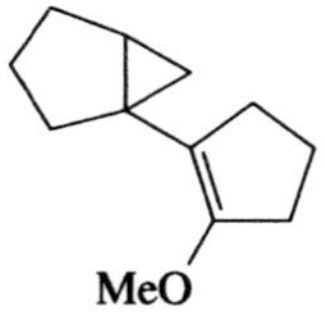 MeO	97	115
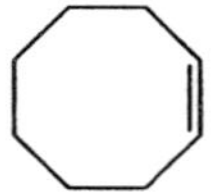	$[Cp(CO)_2FeCH_2SMe_2]^+ BF_4^-$	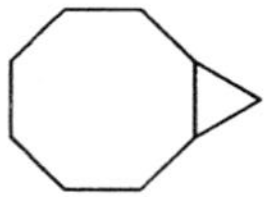	92	110
 Bu^n Bu^n		 Bu^n Bu^n	87[a]	110

Table 7 *(continued)*

Alkene	*Carbene complex or precursor*	*Products*	*Yield (%)*	*Ref.*
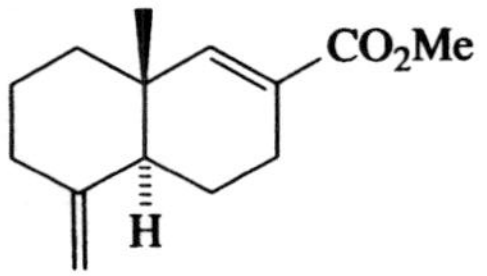 CO_2Me; H	$[Cp(CO)_2FeCH_2SMe_2]^+ BF_4^-$	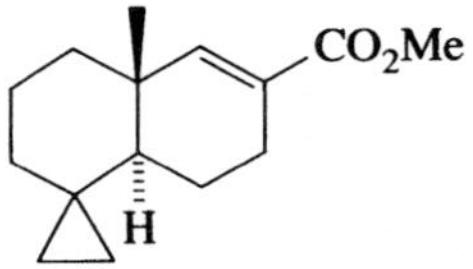 CO_2Me; H	86	116
 Et, Et	$[Cp(CO)_2FeCH(OMe)Me]$, $CF_3SO_3SiMe_3$	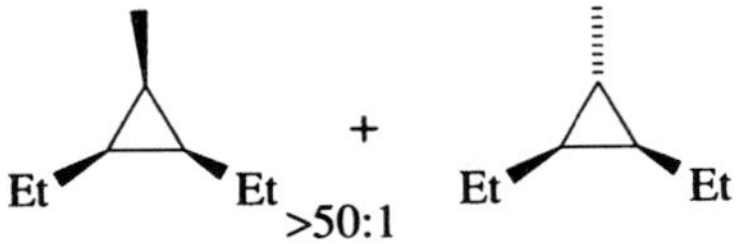 Et, Et + Et, Et (>50:1)	58	117
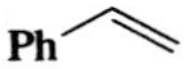 Ph		 Ph + Ph (4.7:1)	75	117
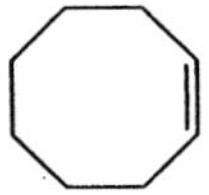	$[Cp(CO)_2FeCH(SPh)Me]$, $Me_3O^+ BF_4^-$	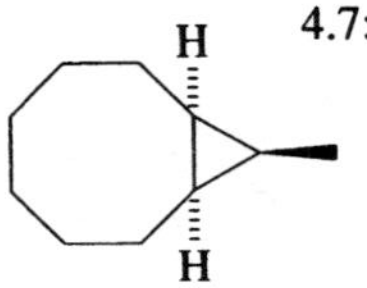 H, H	73	118
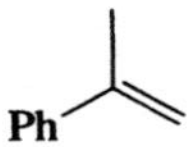 Ph		 Ph + Ph (2.4:1)	58	118
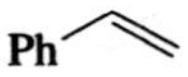 Ph	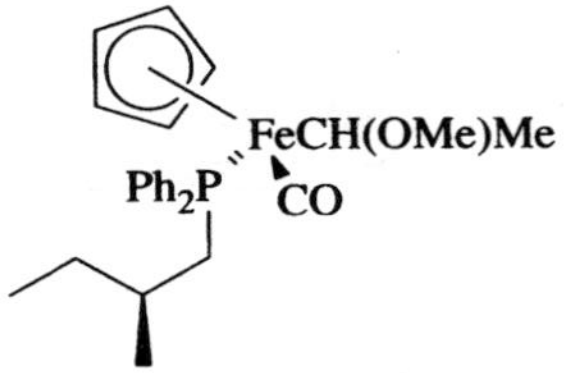 FeCH(OMe)Me, Ph_2P, CO (optically active)	 Ph + Ph (3.5:1)	75 (*ca.* 90% *ee*)	119

Table 7 *(continued)*

Alkene	*Carbene complex or precursor*	*Products*	*Yield (%)*	*Ref.*
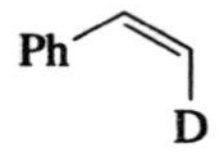 Ph, D	[Cp(CO)$_2$FeCH(OMe)Me], $CF_3SO_3SiMe_3$	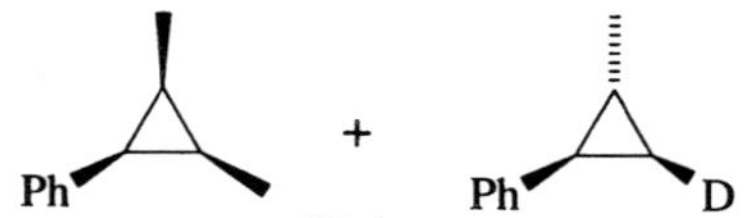 Ph + Ph, D 6.7:1	Not reported	120
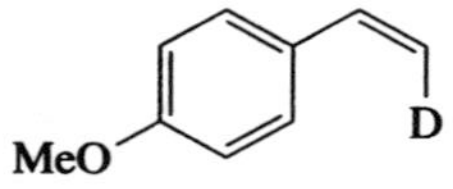 MeO, D		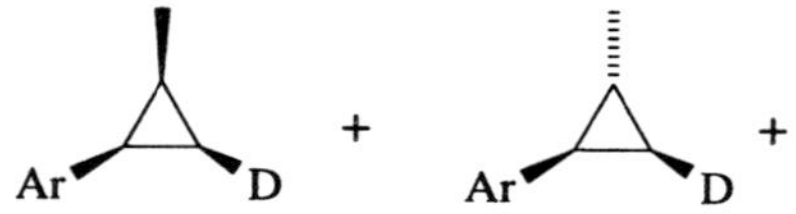 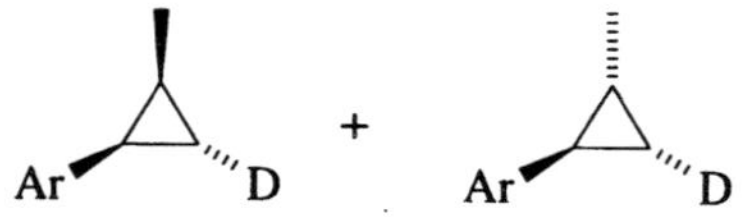 Ar, D + Ar, D + Ar, D + Ar, D 6.8:2.2:6.7:1	Not reported	120
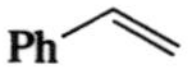 Ph	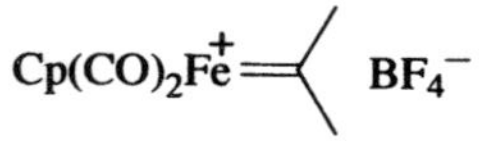 $Cp(CO)_2\overset{+}{Fe}$= BF_4^-	 Ph	45	121
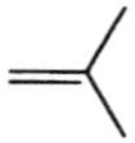	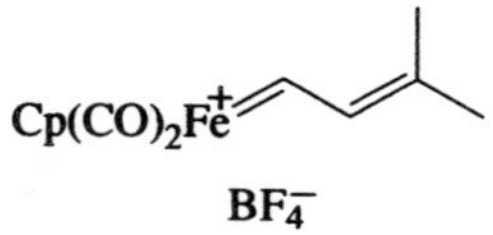 $Cp(CO)_2\overset{+}{Fe}$= BF_4^-	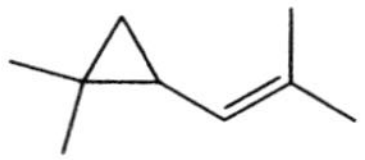	56	121, 122
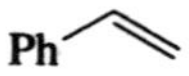 Ph	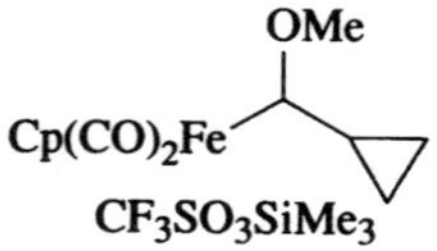 OMe, $Cp(CO)_2Fe$, $CF_3SO_3SiMe_3$	 Ph + Ph 1:3.4	66	123
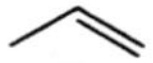	 OMe, $Cp(CO)_2Fe$, F, $CF_3SO_3SiMe_3$	 Ar + Ar 9.4:1	80	124

Table 7 *(continued)*

Alkene	*Carbene complex or precursor*	*Products*	*Yield (%)*	*Ref.*
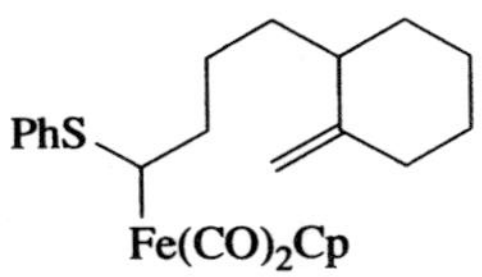 PhS, $Fe(CO)_2Cp$; $Me_3O^+BF_4^-$		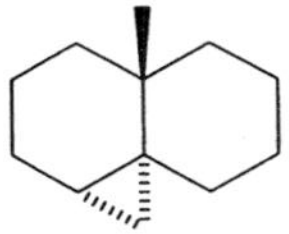	50	125
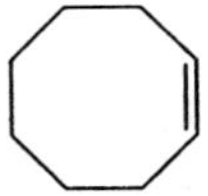	$[Cp(CO)_2FeCH_2SPh_2]^+ BF_4^-$	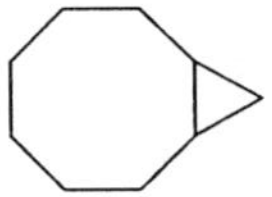	85	126
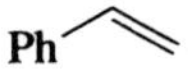 Ph	$[(C_5Me_5)(CO)_2MCH_2OMe]$ (M = Fe, Ru) $CF_3SO_3SiMe_3$	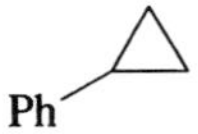 Ph	100	127
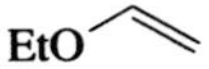 EtO	$(CO)_5WCH(Ph)SEt_2$	  Ph, OEt + Ph, OEt	Not reported	128
	$(H_2C{=}CH_2)_2NiCH_2S(O)Me_2$		16	129
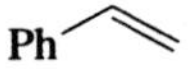 Ph	$[Cp(CO)_2(PPh_3)WCH_2OC(O)Bu^t]$ $CF_3SO_3SiMe_3$	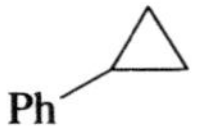 Ph	50	130
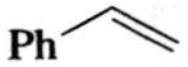 Ph	$[Cp(CO)_2(PPh_3)WMe]$, $Ph_3C^+ BF_4^-$	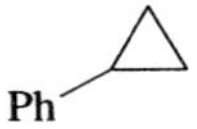 Ph	52	130
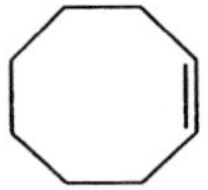	$[Cp(PPh_3)NiCH_2SMe_2]^+ BF_4^-$	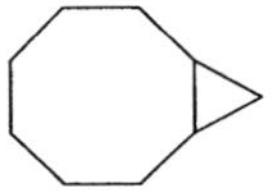	49	131

[a] Product yield corrected for unreacted starting material.

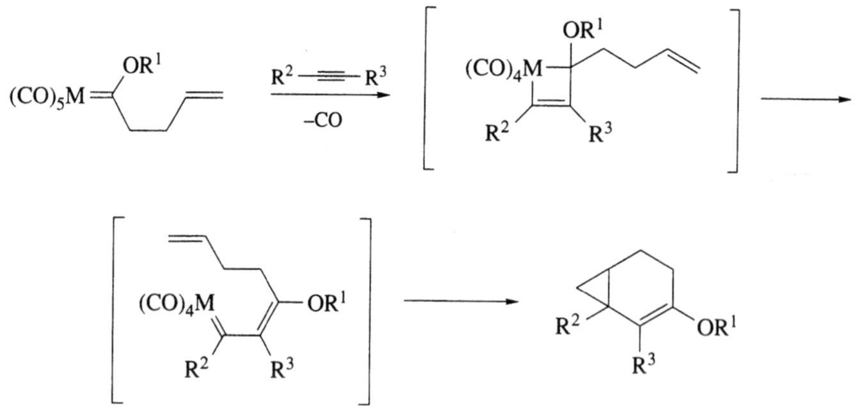

Scheme 2

4.6.6 TANDEM CARBANIONIC ADDITION/INTRAMOLECULAR ALKYLATIONS

A number of methods for forming cyclopropanes are based upon the general reaction sequence shown in Scheme 3. In the first step, a carbanionic intermediate (**7**) is formed which bears a substituent G. This substituent generally serves as both a carbanion-stabilizing group and, eventually, as a leaving group. The intermediate (**7**) then undergoes addition to an alkene double bond bearing an electron-withdrawing group (EWG). The resulting adduct (**8**) is another carbanionic species which undergoes intramolecular alkylation, with displacement of the group G, to produce a cyclopropane.

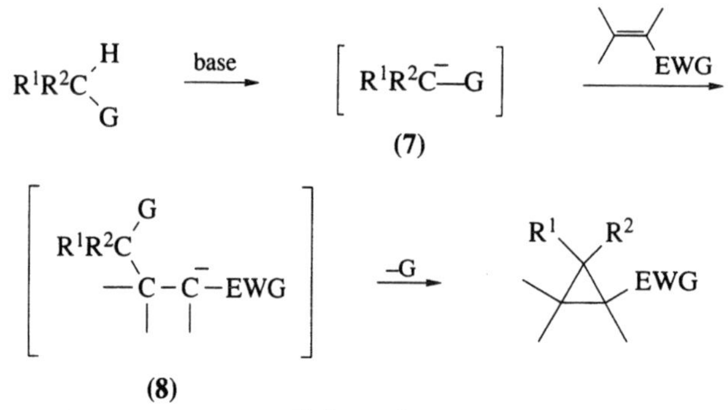

(**7**)

(**8**)

Scheme 3

This set of methods is inherently related to some of the carbene-based approaches to cyclopropanes. The relationship is especially close to the α-elimination pathways (Section 4.6.3.1) with a principal difference being the ordering of steps with respect to loss of the leaving group.

Because of the stepwise formation of the two new carbon–carbon bonds of the cyclopropane product, and because of the intermediacy of the carbanionic adduct (**8**), the double-bond configuration of the alkene substrate may be lost, depending upon the lifetime of the intermediate relative to carbon–carbon single-bond rotations. Consequently, the stereospecificity of these cyclopropanations varies from one case to another.[132]

4.6.6.1 Use of Ylides

With reference back to Scheme 3, the substituent G in these reactions is most commonly an onium group bearing a positive charge. Examples are sulfonium and phosphonium groups. The intermediates (**7**) are thus various types of ylides, and the adducts (**8**) are defined as being betaines.

4.6.6.1.1 Sulfur ylides

Ylides based upon sulfur are the most generally useful in these cyclopropane-forming reactions.[133] Early work in this area was done with the simple dimethyloxysulfonium methylide (**9**) derived from dimethyl sulfoxide. The even simpler dimethylsulfonium methylide (**10**) was studied at the same time as a reagent primarily for the conversion of carbonyl compounds into epoxides.[134] Somewhat later, other types of sulfur ylides were developed, among which the nitrogen-substituted derivatives such as (**11**) are particularly important.[135]

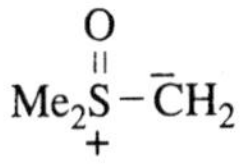

$Me_2\overset{+}{S}(=O)-\overset{-}{C}H_2$ **(9)** $Me_2\overset{+}{S}-\overset{-}{C}H_2$ **(10)** $MeS(=O)(NMe_2)\overset{\pm}{}\overset{-}{C}H_2$ **(11)**

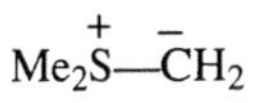

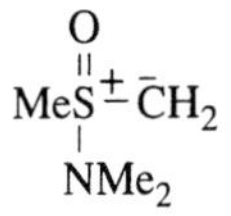

Despite the stepwise pathway for these reactions, some cyclopropanations using ylides may be stereospecific.[132] Successful efforts have also been reported for the development of enantioselective cyclopropanations through use of, for example, optically active ylides related to (**11**)[136] and of alkene substrates bearing chiral auxiliaries.

The use of sulfur ylides has been covered by thorough reviews[133] to which the reader is referred for extensive listings of examples. A few representative cases are given in Table 8.

The reactions are normally limited to the use of electron deficient alkenes as substrates. However, there have been some reports of copper-catalyzed reactions of sulfur ylides with simple alkenes, as exemplified in equation (16).[147]

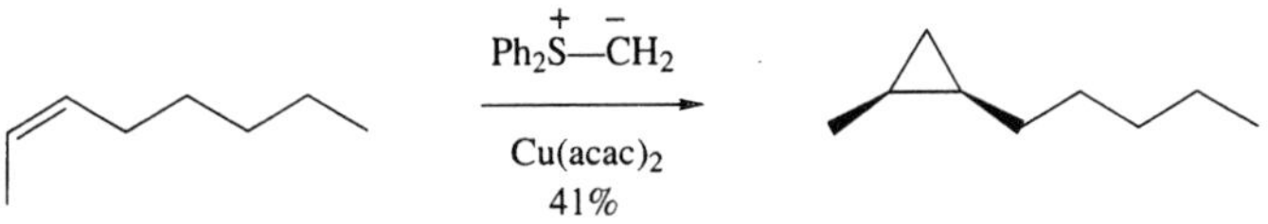

$$\xrightarrow[\text{Cu(acac)}_2,\ 41\%]{Ph_2\overset{+}{S}-\overset{-}{C}H_2} \quad (16)$$

4.6.6.1.2 Other ylides

Ylides of other elements have been used much less commonly than sulfur ylides in cyclopropanations. Rather, other ylides are better known for their uses in other types of reactions, the best example being the use of phosphonium ylides in the Wittig reaction with carbonyl compounds to give alkenes. Nonetheless, some cases of cyclopropanations have been reported with phosphonium ylides and the related arsenic derivatives. Examples are given in Table 9.

4.6.6.2 Addition of Other Carbanionic Species Bearing Leaving Groups

Cyclopropanations are known for several other carbanionic intermediates of the general type (**7**), in which the substituent G is ultimately lost as an anionic leaving group in the last step of the ring-forming pathway (see Scheme 3 above). The substituent G is most often a functional group based upon sulfur, selenium or nitrogen. Halide-substituted derivatives probably react *via* the α-elimination pathway in most cases (see Section 4.6.3.1), but in some reactions with electron deficient alkenes as substrates, the normal order of steps may be altered (*e.g.* Table 10, ref. 162).

Representative examples of these further carbanionic cyclopropanations are compiled in Table 10.

4.6.7 CONCLUSION

A large number of useful synthetic methods are available for the addition of methylene and other simple alkylidene groups to alkenes to give cyclopropanes. The methods that are most commonly used with alkenes of normal electronic characteristics are based upon the intermediacy of carbenes and related species. Very electron rich alkenes bearing electron-donating substituents are also good substrates for many of these carbene-based methods. Some of these same methods may be used with electron deficient alkenes, but the carbanion-based methods have been developed more specifically for these latter substrates.

Table 8 Cyclopropanations Using Sulfur Ylides

Alkene	*Ylide*	*Products*	*Yield (%)*	*Ref.*
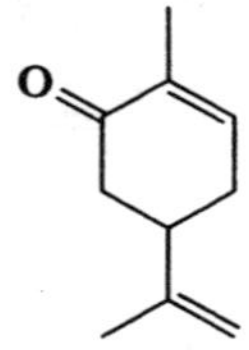 O	$Me_2\overset{+}{S}(=O)-\overset{-}{C}H_2$	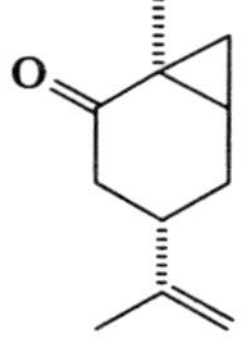 O	81	134
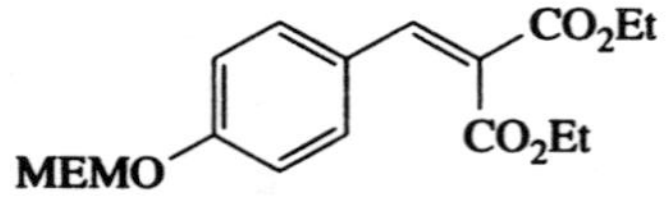 MEMO, CO_2Et, CO_2Et		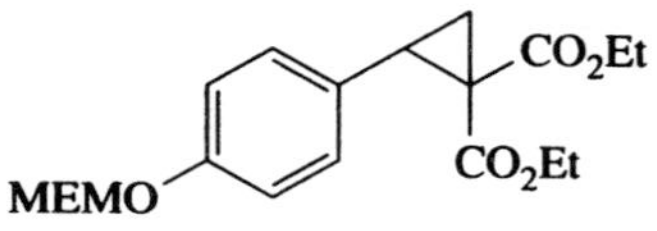 MEMO, CO_2Et, CO_2Et	76	137
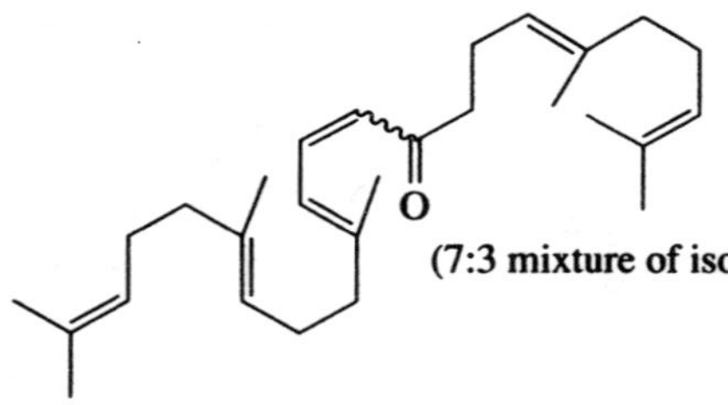 O (7:3 mixture of isomers)		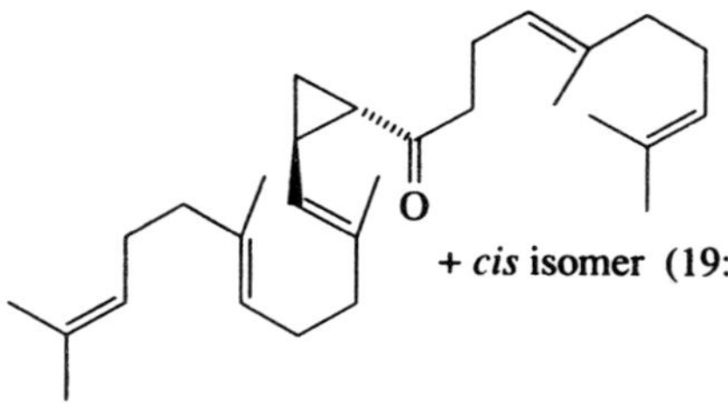 O + *cis* isomer (19:1)	Not reported	138
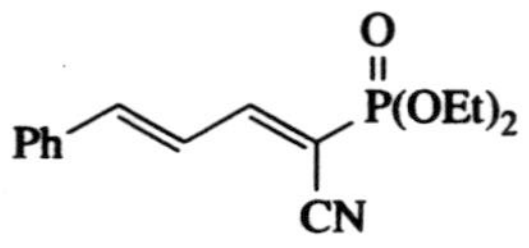 Ph, $P(OEt)_2$ (=O), CN		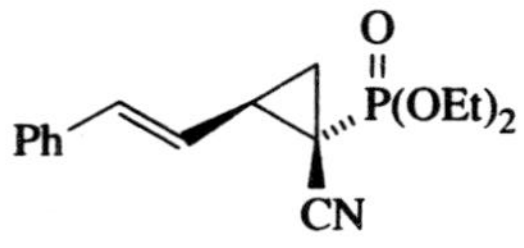 Ph, $P(OEt)_2$ (=O), CN + Ph, CN, $P(OEt)_2$ (=O); 9:1 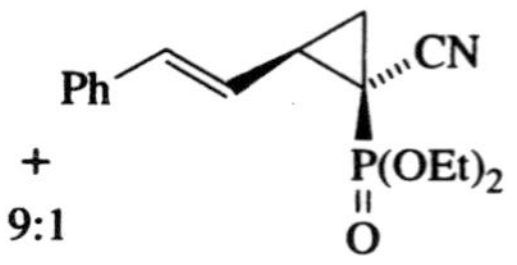	87	139
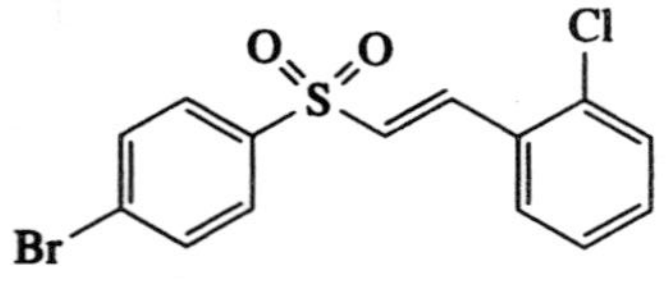 Br, O=S=O, Cl	$Me_2\overset{+}{S}-\overset{-}{C}H_2$	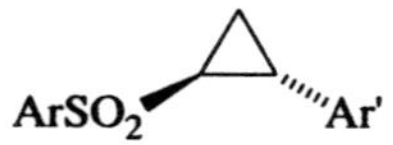 $ArSO_2$, Ar'	80	140

Table 8 *(continued)*

Alkene	*Ylide*	*Products*	*Yield (%)*	*Ref.*
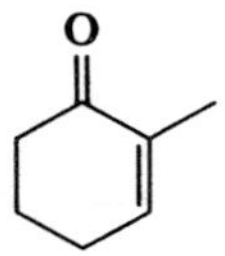 O	$Ph_2\overset{+}{S}-\overset{-}{C}Me_2$	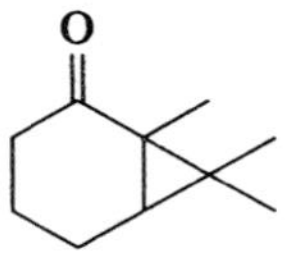 O	86	141
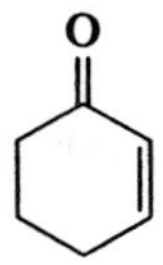 O	$Me_2\overset{+}{S}-\overset{-}{C}HSiMe_3$	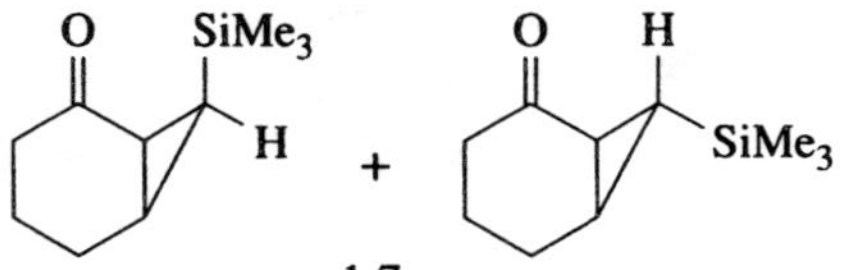 O, SiMe$_3$, H + O, H, SiMe$_3$ 1:7	55	142
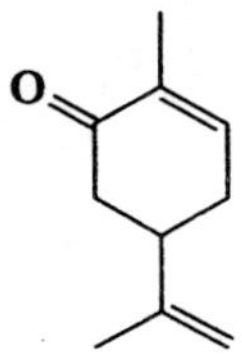 O	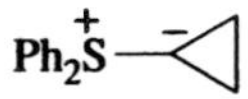 $Ph_2\overset{+}{S}-$ (cyclopropylide)	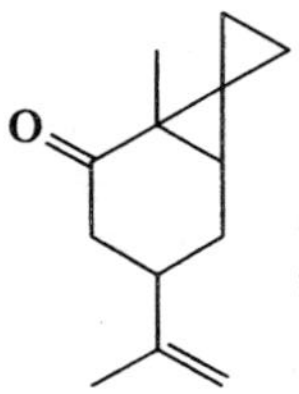 O (stereochemistry not reported)	75	143
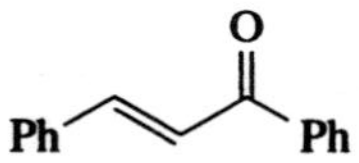 Ph, O, Ph	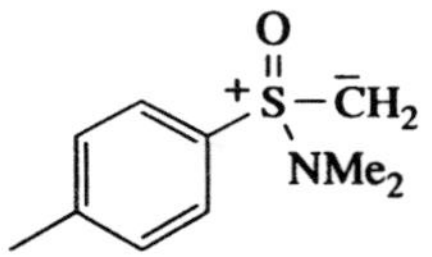 O, +S, $\overset{-}{C}H_2$, NMe$_2$ (optically active)	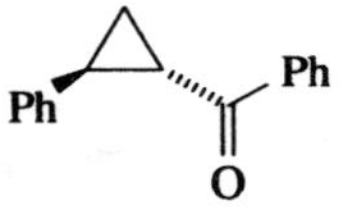 Ph, Ph, O	94 (35% optical yield)	136
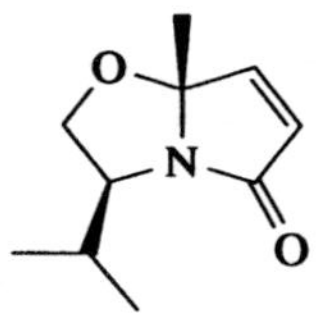 O, N, O (optically active)	$Me_2\overset{O}{\overset{\parallel}{\underset{+}{S}}}-\overset{-}{C}H_2$	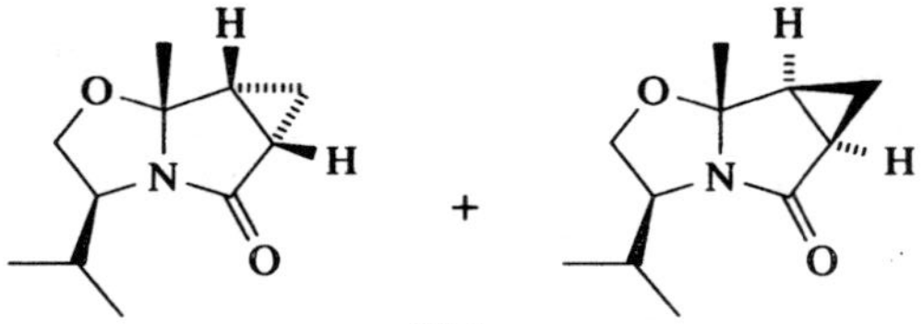 O, N, H, H, O + O, N, H, H, O 39:1	64	144

Table 8 *(continued)*

Alkene	*Ylide*	*Products*	*Yield (%)*	*Ref.*
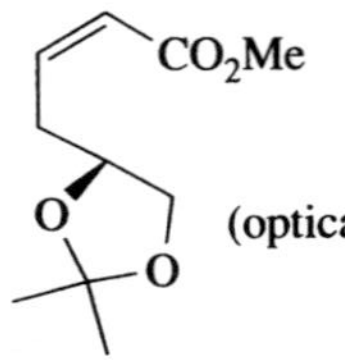 CO_2Me, O, O (optically active)	$Ph_2\overset{+}{S}-\overset{-}{C}Me_2$	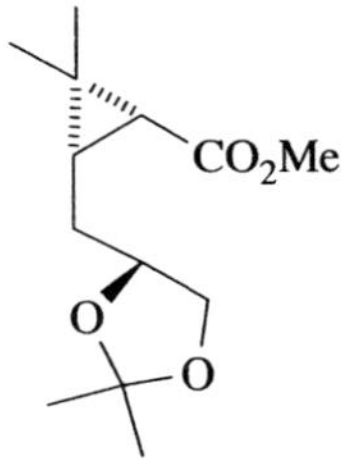 CO_2Me, O, O	84 (>96% *de*)	145
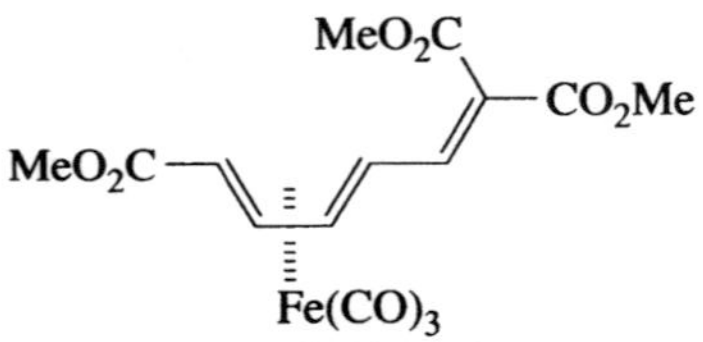 MeO_2C, MeO_2C, CO_2Me, $Fe(CO)_3$ (optically active)	$Me_2\overset{+}{S}-\overset{-}{C}HCO_2Me$	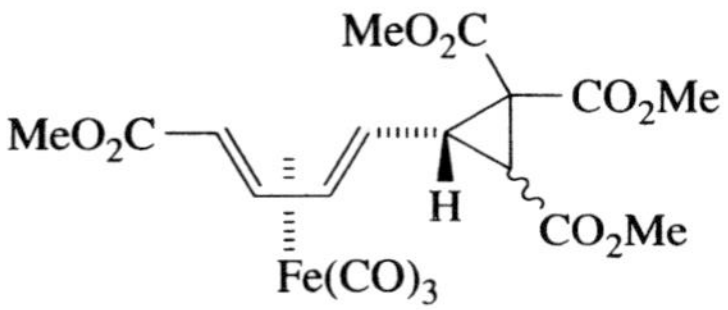 MeO_2C, MeO_2C, CO_2Me, H, CO_2Me, $Fe(CO)_3$	Not reported (≥90% *de*)	146

Table 9 Cyclopropanations Using Other Ylides

Alkene	*Ylide*	*Products*	*Yield (%)*	*Ref.*
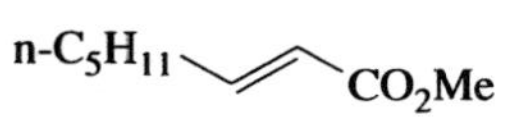 n-C_5H_{11} CO_2Me	$Ph_3\overset{+}{P}-\overset{-}{C}Me_2$	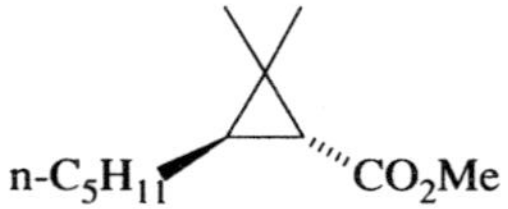 n-C_5H_{11} CO_2Me	75	148
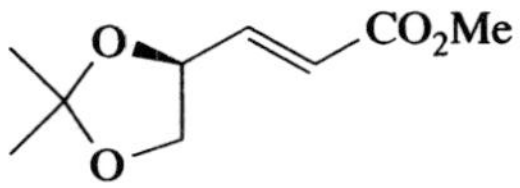 O O CO_2Me (optically active)		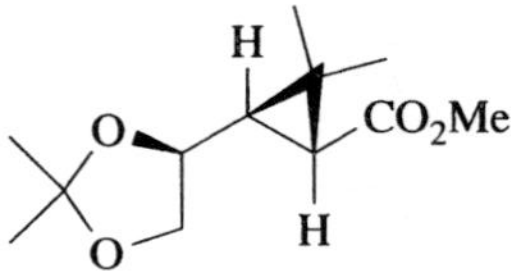 O O H H CO_2Me	60 (>96% *de*)	149
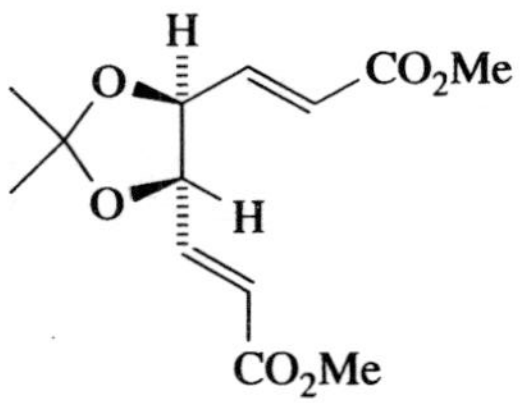 O O H H CO_2Me CO_2Me		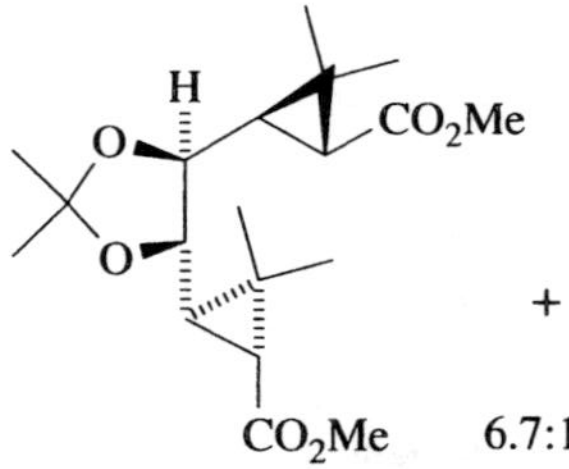 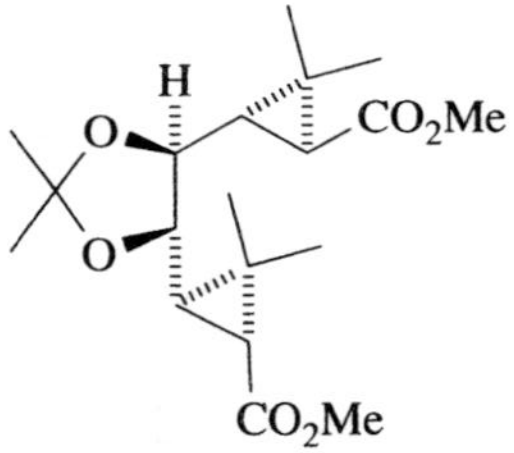 O O H CO_2Me CO_2Me + O O H CO_2Me CO_2Me 6.7:1	75	150
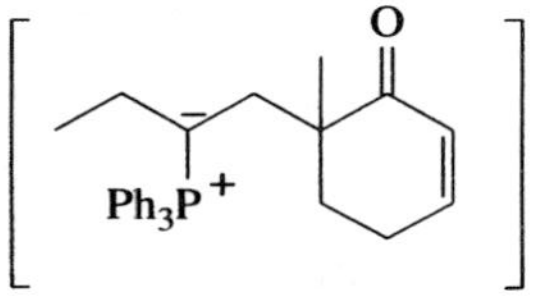 [Ph_3P^+ O]		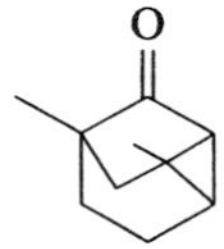 O	44	151
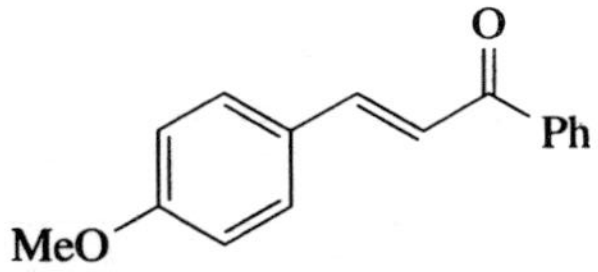 MeO O Ph	$Ph_3\overset{+}{As}-\overset{-}{C}HSiMe_3$	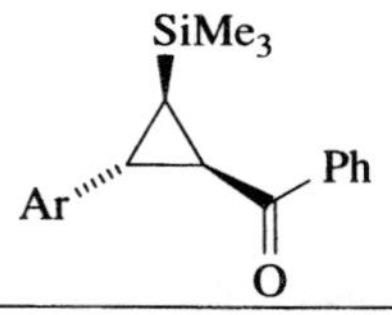 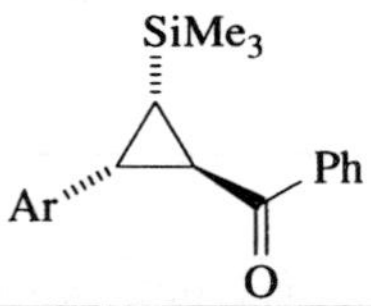 Ar $SiMe_3$ Ph O + Ar $SiMe_3$ Ph O 1:22	47	152

Table 10 Other Carbanionic Cyclopropanations

Alkene	*Reagent*	*Products*	*Yield (%)*	*Ref.*
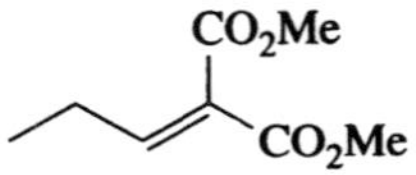 CO_2Me, CO_2Me	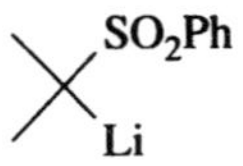 SO_2Ph, Li	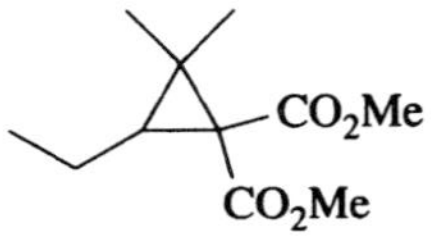 CO_2Me, CO_2Me	63	153
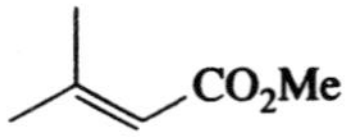 CO_2Me	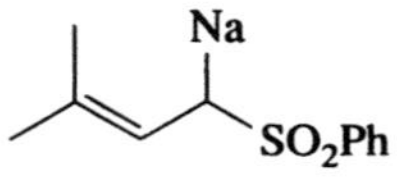 Na, SO_2Ph	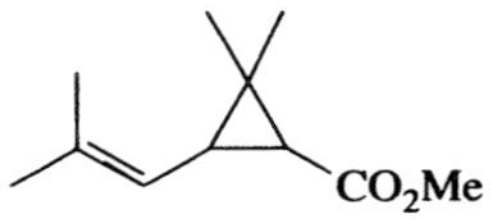 CO_2Me (stereochemistry not reported)	Not reported	154
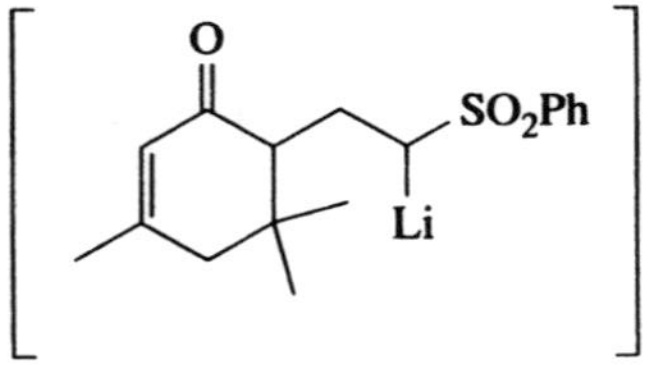 [O, SO_2Ph, Li]		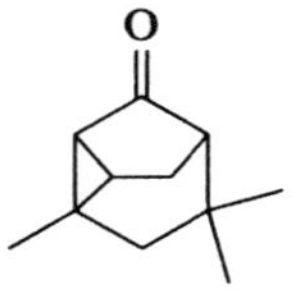 O	38	155
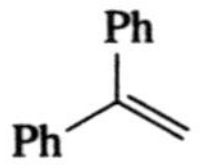 Ph, Ph	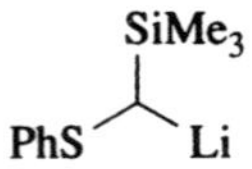 $SiMe_3$, PhS, Li	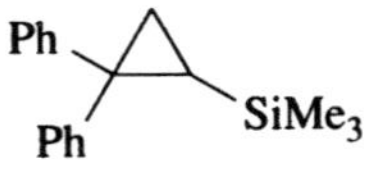 Ph, Ph, $SiMe_3$	75	156
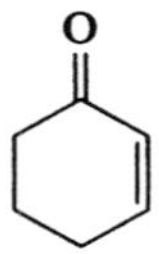 O	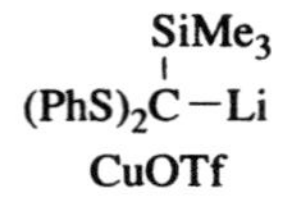 $SiMe_3$, $(PhS)_2C-Li$, CuOTf	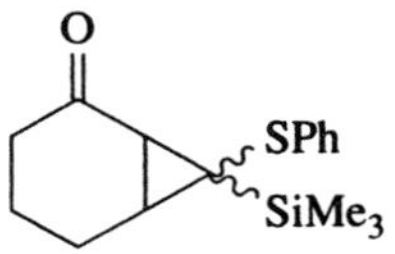 O, SPh, $SiMe_3$	76	157
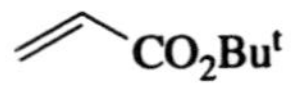 CO_2Bu^t	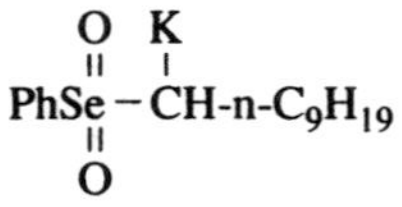 O K, PhSe–CH-n-C_9H_{19}, O	 n-C_9H_{19}, CO_2Bu^t	76	158

Table 10 *(continued)*

Alkene	*Reagent*	*Products*	*Yield (%)*	*Ref.*
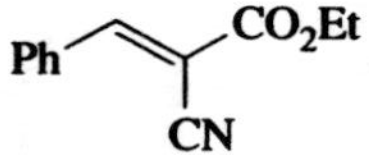 Ph, CO_2Et, CN	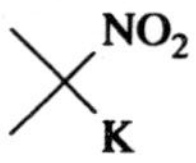 NO_2, K	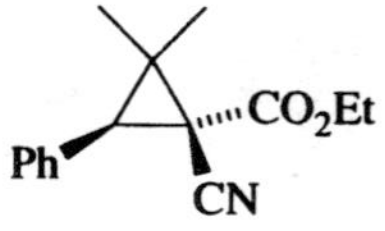 Ph, CO_2Et, CN	81	159a
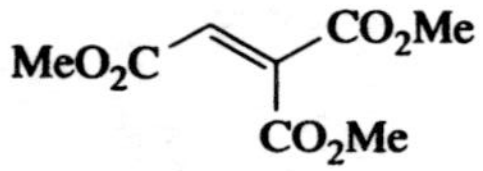 MeO_2C, CO_2Me, CO_2Me	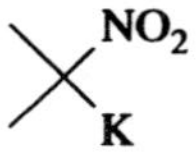 NO_2, K	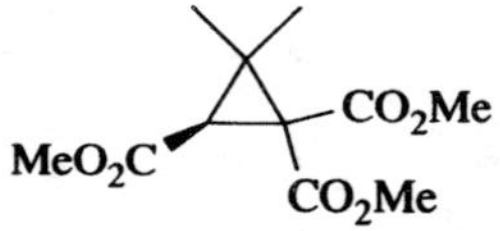 MeO_2C, CO_2Me, CO_2Me	72	160
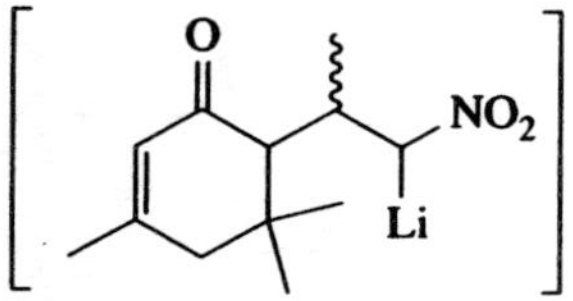 O, NO_2, Li		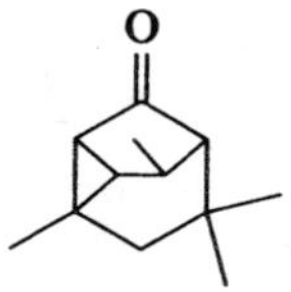 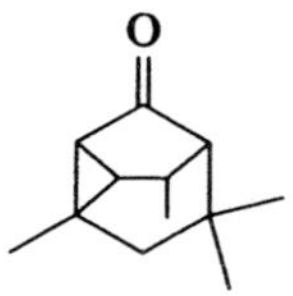 O + O (12:1)	63	161
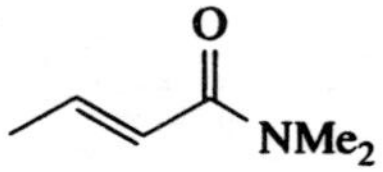 O, NMe_2	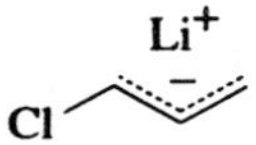 Li^+, Cl	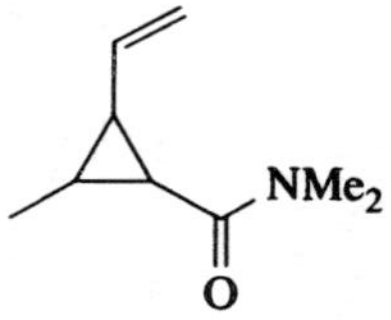 NMe_2, O (mixture of stereoisomers)	80	162

With the wide choice of methods now available, typical cyclopropanations may be performed with excellent stereospecificity, stereoselectivity, diastereofacial selectivity, regioselectivity and chemoselectivity. Good progress has also been made with respect to enantioselective cyclopropanations, but much further work in this area during the coming years would be appropriate. Ideal methods would not only exhibit high enantioselectivity with a wide range of alkenes, but they would also use very readily available and inexpensive starting materials as the sources of the methylene or alkylidene groups that undergo the alkene additions, and if catalysts are needed, they should be reasonably inexpensive, exhibit high catalytic turnovers and be readily recyclable. In terms of availability, aldehydes and ketones would appear to be particularly ideal starting materials, but methods for their use in direct, deoxygenative alkylidene transfer reactions as suggested in equation (17) are still relatively rare.[93] Further studies of reactions of this type are surely justified, especially with recyclable, oxygenophilic metal species that are also chiral.

$$R^1R^2C{=}CR^3R^4 + R^5R^6C{=}O \xrightarrow{(L)_nM^*} \text{cyclo-}C(R^1)(R^2)C(R^3)(R^4)C(R^5)(R^6) + (L)_nM^*{=}O \qquad (17)$$

(optically active)

4.6.8 REFERENCES

1. (a) D. Wendisch, '*Methoden Org. Chem. (Houben–Weyl)*', 1971, **4** (3), 15; (b) L. N. Ferguson, 'Highlights of Alicyclic Chemistry', Franklin, Palisade, NJ, 1973, p. 21; (c) P. H. Boyle, in 'Rodd's Chemistry of Carbon Compounds', 2nd edn., ed. M. F. Ansell, Elsevier, Amsterdam, 1974, vol. 2A suppl., p. 9, and the earlier volumes of this series; (d) L. A. Yanovskaya and V. A. Dombrovskii, *Russ. Chem. Rev. (Engl. Transl.)*, 1975, **44**, 154; (e) L. N. Ferguson and D. R. Paulson, 'Highlights of Alicyclic Chemistry', Franklin, Palisade, NJ, 1977, part 2, p. 89; (f) F. J. McQuillin and M. S. Baird, 'Alicyclic Chemistry', 2nd edn., Cambridge University Press, Cambridge, 1983, p. 92; (g) L. A. Paquette, *Chem. Rev.*, 1986, **86**, 733; (h) T. Tsuji and S. Nishida, in 'The Chemistry of the Cyclopropyl Group', ed. Z. Rappoport, Wiley, New York, 1987, part 1, chap. 7, p. 307.
2. R. A. Moss, *Acc. Chem. Res.*, 1980, **13**, 58.
3. T. Aratani, *Pure Appl. Chem.*, 1985, **57**, 1839.
4. R. Kh. Freidlina, A. A. Kamyshova and E. Ts. Chukovskaya, *Russ. Chem. Rev. (Engl. Transl.)*, 1982, **51**, 368.
5. (a) R. H. Shapiro, J. H. Duncan and J. C. Clopton, *J. Am. Chem. Soc.*, 1967, **89**, 1442; (b) S. C. Welch, C.-Y. Chou, J. M. Gruber and J.-M. Assercq, *J. Org. Chem.*, 1985, **50**, 2668.
6. T. Hudlicky, T. M. Kutchan and S. M. Naqvi, *Org. React. (N. Y.)*, 1985, **33**, 247.
7. S. D. Burke and P. A. Grieco, *Org. React. (N. Y.)*, 1979, **26**, 361.
8. (a) R. Huisgen, R. Grashey and J. Saver, in 'The Chemistry of Alkenes', ed. S. Patai, Interscience, London, 1964, p. 821; (b) B. Eistert, M. Regitz, G. Heck and H. Schwall, '*Methoden Org. Chem. (Houben–Weyl)*, 1968, **10** (4), p. 471; (c) M. F. Lappert and J. S. Poland, *Adv. Organomet. Chem.*, 1970, **9**, 397; (d) G. W. Cowell and A. Ledwith, *Q. Rev. Chem. Soc.*, 1970, **24**, 119; (e) W. Kirmse, 'Carbene Chemistry', Academic Press, New York, 1971, p. 267; (f) 'Carbenes', ed. M. Jones, Jr. and R. A. Moss, Wiley, New York, 1973, vol. 1; 1975, vol. 2; (g) K. Mackenzie, in 'The Chemistry of the Hydrazo, Azo, and Azoxy Groups', ed. S. Patai, Wiley, London, 1975, part 1, p. 329; (h) R. J. Drewer, in 'The Chemistry of the Hydrazo, Azo, and Azoxy Groups', ed. S. Patai, Wiley, London, 1975, part 2, p. 935; (i) D. S. Wulfman, G. Linstrumelle and C. F. Cooper, in 'The Chemistry of Diazonium and Diazo Groups', ed. S. Patai, Wiley, Chichester, 1978, part 2, p. 821; (j) J. T. Sharp, in 'Comprehensive Organic Chemistry', ed. D. H. R. Barton and W. D. Ollis, Pergamon Press, Oxford, 1979, vol. 1, p. 455; (k) D. S. Wulfman and B. Poling, in 'Reactive Intermediates', ed. R. A. Abramovitch, Plenum Press, New York, 1980, vol. 1, p. 321; (l) C. Wentrup, 'Reactive Molecules', Wiley, New York, 1984, p. 162; (m) R. A. Moss and M. Jones, Jr., in 'Reactive Intermediates', ed. M. Jones, Jr., and R. A. Moss, Wiley, New York, 1985, vol. 3, p. 45, and the earlier volumes of this series; (n) M. P. Doyle, *Chem. Rev.*, 1986, **86**, 919; (o) G. Maas, *Top. Curr. Chem.*, 1987, **137**, 75; (p) R. A. Aitken, in 'Organic Reaction Mechanisms 1986', ed. A. C. Knipe and W. E. Watts, Wiley, Chichester, 1988, p. 227, and the earlier volumes of this series.
9. H. de Suray, G. Leroy and J. Weiler, *Tetrahedron Lett.*, 1974, 2209.
10. (a) J. Casanova and B. Waegell, *Bull. Soc. Chim. Fr.*, 1975, 922; (b) R. H. Shapiro, *Org. React. (N. Y.)*, 1976, **23**, 405.
11. R. Danion-Bougot and R. Carrie, *C. R. Hebd. Seances Acad. Sci., Ser. C*, 1968, **266**, 645.
12. D. E. McGreer, N. W. K. Chiu, M. G. Vinje and K. C. K. Wong, *Can. J. Chem.*, 1965, **43**, 1407.
13. M. Schneider and A. Rau, *Angew. Chem., Int. Ed. Engl.*, 1979, **18**, 231.
14. E. Piers, M. B. Geraghty and M. Soucy, *Synth. Commun.*, 1973, **3**, 401.
15. Z. Majerski and M. Zuanic, *J. Am. Chem. Soc.*, 1987, **109**, 3496.
16. M. Franck-Neumann and M. Miesch, *Bull. Soc. Chim. Fr.*, 1984, 362.
17. M. Franck-Neumann, M. Sedrati, J.-P. Vigneron and V. Bloy, *Angew. Chem., Int. Ed. Engl.*, 1985, **24**, 996.
18. G. Adembri, D. Donati, L. R. Lampariello, M. Scotton and A. Sega, *Tetrahedron Lett.*, 1984, **25**, 6055.
19. J. Mann and A. Thomas, *J. Chem. Soc., Chem. Commun.*, 1985, 737.

20. M. Saha, B. Bagby and K. M. Nicholas, *Tetrahedron Lett.*, 1986, **27**, 915.
21. A. G. Schultz, K. K. Eng and R. K. Kullnig, *Tetrahedron Lett.*, 1986, **27**, 2331.
22. D. C. Remy, S. W. King, D. Cochran, J. P. Springer and J. Hirshfield, *J. Org. Chem.*, 1985, **50**, 4120.
23. J. Martelli and R. Gree, *J. Chem. Soc., Chem. Commun.*, 1980, 355.
24. H. Abdallah and R. Gree, *Tetrahedron Lett.*, 1980, **21**, 2239.
25. W. Nagai and Y. Hirata, *J. Org. Chem.*, 1989, **54**, 635.
26. A. Oku, T. Yokoyama and T. Harada, *J. Org. Chem.*, 1983, **48**, 5333.
27. T. Wakamiya, Y. Oda, H. Fujita and T. Shiba, *Tetrahedron Lett.*, 1986, **27**, 2143.
28. R. Nesi, D. Giomi, S. Papaleo and S. Bracci, *J. Org. Chem.*, 1989, **54**, 706.
29. G. Hömberger, A. E. Dorigo, W. Kirmse and K. N. Houk, *J. Am. Chem. Soc.*, 1989, **111**, 475.
30. (a) R. C. Woodworth and P. S. Skell, *J. Am. Chem. Soc.*, 1959, **81**, 3383; (b) W. von E. Doering and P. LaFlamme, *J. Am. Chem. Soc.*, 1956, **78**, 5447.
31. K. R. Kopecky, G. S. Hammond and P. A. Leermakers, *J. Am. Chem. Soc.*, 1962, **84**, 1015.
32. G. L. Closs and R. A. Moss, *J. Am. Chem. Soc.*, 1964, **86**, 4042.
33. I. Moritani, Y. Yamamoto and S.-I. Murahashi, *Tetrahedron Lett.*, 1968, 5697.
34. H. Tomioka, Y. Ozaki, Y. Koyabu and Y. Izawa, *Tetrahedron Lett.*, 1982, **23**, 1917.
35. R. A. Moss and W. P. Wetter, *Tetrahedron Lett.*, 1981, **22**, 997.
36. L. M. Tolbert and M. B. Ali, *J. Am. Chem. Soc.*, 1985, **107**, 4589.
37. M. Suda, *Synthesis*, 1981, 714.
38. G. Wittig and K. Schwarzenbach, *Justus Liebigs Ann. Chem.*, 1962, **650**, 1.
39. H. Hoberg, *Justus Liebigs Ann. Chem.*, 1962, **656**, 1.
40. S. H. Goh, L. E. Closs and G. L. Closs, *J. Org. Chem.*, 1969, **34**, 25.
41. R. G. Salomon and J. K. Kochi, *J. Am. Chem. Soc.*, 1973, **95**, 3300.
42. H. Nozaki, H. Takaya, S. Moriuti and R. Noyori, *Tetrahedron*, 1968, **24**, 3655.
43. Yu. V. Tomilov, V. G. Bordakov, I. E. Dolgii and O. M. Nefedov, *Bull. Acad. Sci. USSR, Div. Chem. Sci.*, 1984, **33**, 533.
44. I. G. Dinulescu, L. N. Enescu, A. Ghenciulescu and M. Avram, *J. Chem. Res. (S)*, 1978, 456.
45. M. P. Doyle, L. C. Wang and K.-L. Loh, *Tetrahedron Lett.*, 1984, **25**, 4087.
46. T. Mandai, K. Hara, M. Kawada and J. Nokami, *Tetrahedron Lett.*, 1983, **24**, 1517; E. J. Corey and K. Achiwa, *Tetrahedron Lett.*, 1970, 2245.
47. Yu. V. Tomilov, A. B. Kostitsyn and O. M. Nefedov, *Bull. Acad. Sci. USSR, Div. Chem. Sci.*, 1987, **36**, 2678.
48. H. Abdallah, R. Gree and R. Carrie, *Tetrahedron Lett.*, 1982, **23**, 503.
49. A. Nakamura, T. Yoshida, M. Cowie, S. Otsuka and J. A. Ibers, *J. Am. Chem. Soc.*, 1977, **99**, 2108.
50. U. Mende, B. Radüchel, W. Skuballa and H. Vorbrüggen, *Tetrahedron Lett.*, 1975, 629.
51. M. P. Doyle, J. H. Griffin, V. Bagheri and R. L. Dorow, *Organometallics*, 1984, **3**, 53.
52. G. L. Closs and L. E. Closs, *Tetrahedron Lett.*, 1960 (24), 26.
53. R. A. Olofson and C. M. Dougherty, *J. Am. Chem. Soc.*, 1973, **95**, 581.
54. U. Schöllkopf and M. Eisert, *Justus Liebigs Ann. Chem.*, 1963, **664**, 76.
55. T. J. Katz, E. J. Wang and N. Acton, *J. Am. Chem. Soc.*, 1971, **93**, 3782.
56. W. M. Jones, M. E. Stowe, E. E. Wells, Jr. and E. W. Lester, *J. Am. Chem. Soc.*, 1968, **90**, 1849.
57. (a) H. E. Simmons, Jr., T. L. Cairns, S. A. Vladuchick and C. M. Hoiness, *Org. React. (N.Y.)*, 1973, **20**, 1; (b) J. Furukawa and N. Kawabata, *Adv. Organomet. Chem.*, 1974, **12**, 83; (c) B. J. Wakefield, in 'Comprehensive Organic Chemistry', ed. D. H. R. Barton and W. D. Ollis, Pergamon Press, Oxford, 1979, vol. 3, p. 989.
58. (a) J. Furukawa, N. Kawabata and J. Nishimura, *Tetrahedron Lett.*, 1966, 3353; (b) J. Nishimura, N. Kawabata and J. Furukawa, *Tetrahedron*, 1969, **25**, 2647.
59. K. Maruoka, Y. Fukutani and H. Yamamoto, *J. Org. Chem.*, 1985, **50**, 4412.
60. E. C. Friedrich, J. M. Domek and R. Y. Pong, *J. Org. Chem.*, 1985, **50**, 4640.
61. P. J. Kropp, N. J. Pienta, J. A. Sawyer and R. P. Polniaszek, *Tetrahedron*, 1981, **37**, 3229.
62. S. D. Kahn, C. F. Pau, A. R. Chamberlin and W. J. Hehre, *J. Am. Chem. Soc.*, 1987, **109**, 650.
63. (a) H. E. Simmons and R. D. Smith, *J. Am. Chem. Soc.*, 1958, **80**, 5323; (b) E. P. Blanchard and H. E. Simmons, *J. Am. Chem. Soc.*, 1964, **86**, 1337; (c) R. J. Rawson and I. T. Harrison, *J. Org. Chem.*, 1970, **35**, 2057.
64. R. D. Rieke, P. T.-J. Li, T. P. Burns and S. T. Uhm, *J. Org. Chem.*, 1981, **46**, 4324.
65. N. Kawabata, I. Kamemura and M. Naka, *J. Am. Chem. Soc.*, 1979, **101**, 2139.
66. N. S. Zefirov, K. A. Lukin, S. F. Politanskii and M. A. Margulis, *J. Org. Chem. USSR*, 1987, **23**, 1603.
67. (a) L. Fitjer, *Chem. Ber.*, 1982, **115**, 1047; (b) I. Erden, *Synth. Commun.*, 1986, **16**, 117.
68. Y. Fukuda, Y. Yamamoto, K. Kimura and Y. Odaira, *Tetrahedron Lett.*, 1979, 877.
69. J. E. Maggio, H. E. Simmons, III and J. K. Kouba, *J. Am. Chem. Soc.*, 1981, **103**, 1579.
70. J. Nishimura, J. Furukawa, N. Kawabata and H. Koyama, *Bull. Chem. Soc. Jpn.*, 1971, **44**, 1127.
71. T. Hudlicky, B. C. Ranu, S. M. Naqvi and A. Srnak, *J. Org. Chem.*, 1985, **50**, 123.
72. M. Ando, S. Sayama and K. Takase, *J. Org. Chem.*, 1985, **50**, 251.
73. E. J. Corey, J. G. Reid, A. G. Myers and R. W. Hahl, *J. Am. Chem. Soc.*, 1987, **109**, 918.
74. E. J. Corey and T. M. Eckrich, *Tetrahedron Lett.*, 1984, **25**, 2415.
75. M. Ratier, M. Castaing, J.-Y. Godet and M. Pereyre, *J. Chem. Res. (S)*, 1978, 179.
76. G. A. Molander and J. B. Etter, *J. Org. Chem.*, 1987, **52**, 3942.
77. N. Kawabata and N. Ikeda, *Bull. Chem. Soc. Jpn.*, 1980, **53**, 563.
78. T. Hanafusa, L. Birladeanu and S. Winstein, *J. Am. Chem. Soc.*, 1965, **87**, 3510.
79. G. W. Klumpp, A. H. Veefkind, W. L. de Graaf and F. Bickelhaupt, *Justus Liebigs Ann. Chem.*, 1967, **706**, 47.
80. N. Kawabata, T. Nakagawa, T. Nakao and S. Yamashita, *J. Org. Chem.*, 1977, **42**, 3031.
81. M. Ando, S. Sayama and K. Takase, *Chem. Lett.*, 1981, 377.
82. J.-C. Limasset, P. Amice and J.-M. Conia, *Bull. Soc. Chim. Fr.*, 1969, 3981.
83. L. A. Paquette, W. H. Ham and D. S. Dime, *Tetrahedron Lett.*, 1985, **26**, 4983.

84. H. Kanai and N. Hiraki, *Chem. Lett.*, 1979, 761.
85. I. Ryu, T. Aya, S. Otani, S. Murai and N. Sonoda, *J. Organomet. Chem.*, 1987, **321**, 279.
86. T. Imamoto and N. Takiyama, *Tetrahedron Lett.*, 1987, **28**, 1307.
87. E. A. Mash and D. S. Torok, *J. Org. Chem.*, 1989, **54**, 250.
88. K. A. Nelson and E. A. Mash, *J. Org. Chem.*, 1986, **51**, 2721.
89. E. A. Mash, S. K. Math and C. J. Flann, *Tetrahedron Lett.*, 1988, **29**, 2147.
90. I. Arai, A. Mori and H. Yamamoto, *J. Am. Chem. Soc.*, 1985, **107**, 8254.
91. T. Sugimura, T. Futagawa and A. Tai, *Tetrahedron Lett.*, 1988, **29**, 5775.
92. D. Seyferth, R. M. Turkel, M. A. Eisert and L. J. Todd, *J. Am. Chem. Soc.*, 1969, **91**, 5027.
93. I. Elphimoff-Felkin and P. Sarda, *Tetrahedron*, 1975, **31**, 2785.
94. P. Fischer and G. Schaefer, *Angew. Chem., Int. Ed. Engl.*, 1981, **20**, 863.
95. P. W. Ambler and S. G. Davies, *Tetrahedron Lett.*, 1988, **52**, 6983.
96. K. Furuta, N. Ikeda and H. Yamamoto, *Tetrahedron Lett.*, 1984, **25**, 675.
97. A. Misumi, K. Iwanaga, K. Furuta and H. Yamamoto, *J. Am. Chem. Soc.*, 1985, **107**, 3343.
98. E. J. Corey, W. Su and I. N. Houpis, *Tetrahedron Lett.*, 1986, **27**, 5951.
99. (a) E. O. Fischer and A. Maasböl, *Angew. Chem., Int. Ed. Engl.*, 1964, **3**, 580; (b) E. O. Fischer, *Pure Appl. Chem.*, 1970, **24**, 407; (c) E. O. Fischer, *Pure Appl. Chem.*, 1972, **30**, 353.
100. (a) F. Brown, *Prog. Inorg. Chem.*, 1980, **27**, 1; (b) K. H. Dötz, H. Fischer, P. Hofmann, F. R. Kreissl, U. Schubert and K. Weiss, 'Transition Metal Carbene Complexes', Verlag Chemie, Weinheim, 1983; (c) K. H. Dötz, *Angew. Chem., Int. Ed. Engl.*, 1984, **23**, 587; (d) U. Schubert, *Coord. Chem. Rev.*, 1984, **55**, 261; (e) J. E. Hahn, *Prog. Inorg. Chem.*, 1984, **31**, 205; (f) D. B. Pourreau and G. L. Geoffroy, *Adv. Organomet. Chem.*, 1985, **24**, 249; (g) C. P. Casey, in 'Reactive Intermediates', ed. M. Jones and R. A. Moss, Wiley, New York, 1985, vol. 3, and the earlier volumes of this series; (h) M. J. Winter, *Organomet. Chem.*, 1986, **14**, 225; (i) A. K. Smith, *Organomet. Chem.*, 1986, **14**, 278, and the earlier volumes of this series; (j) M. Brookhart and W. B. Studabaker, *Chem. Rev.*, 1987, **87**, 411; (k) P. Helquist, in 'Advances in Metal–Organic Chemistry', ed. L. S. Liebeskind, JAI Press, Greenwich, CT, 1991, vol. 2, in press; see also refs. 3 and 8n.
101. K. A. Brown-Wensley, S. L. Buchwald, L. Cannizzo, L. Clawson, S. Ho, D. Meinhardt, J. K. Stille, D. Straus and R. H. Grubbs, *Pure Appl. Chem.*, 1983, **55**, 1733.
102. (a) E. O. Fischer and K. H. Dötz, *Chem. Ber.*, 1970, **103**, 1273; (b) K. H. Dötz and E. O. Fischer, *Chem. Ber.*, 1972, **105**, 1356; (c) M. D. Cooke and E. O. Fischer, *J. Organomet. Chem.*, 1973, **56**, 279; (d) A. Wienand and H.-U. Reissig, *Tetrahedron Lett.*, 1988, **29**, 2315.
103. E. O. Fischer and K. H. Dötz, *Chem. Ber.*, 1972, **105**, 3966.
104. K. H. Dötz and I. Pruskil, *Chem. Ber.*, 1981, **114**, 1980.
105. C. P. Casey, S. W. Polichnowski, A. J. Shusterman and C. R. Jones, *J. Am. Chem. Soc.*, 1979, **101**, 7282.
106. (a) C. P. Casey and T. J. Burkhardt, *J. Am. Chem. Soc.*, 1974, **96**, 7808; (b) K. Weiss and K. Hoffmann, *J. Organomet. Chem.*, 1983, **255**, C24.
107. (a) P. W. Jolly and R. Pettit, *J. Am. Chem. Soc.*, 1966, **88**, 5044; (b) P. E. Riley, C. E. Capshew, R. Pettit and R. E. Davis, *Inorg. Chem.*, 1978, **17**, 408.
108. M. L. H. Green, M. Ishaq and R. N. Whiteley, *J. Chem. Soc. A*, 1967, 1508.
109. A. E. Stevens and J. L. Beauchamp, *J. Am. Chem. Soc.*, 1978, **100**, 2584.
110. (a) S. Brandt and P. Helquist, *J. Am. Chem. Soc.*, 1979, **101**, 6473; (b) E. J. O'Connor, S. Brandt and P. Helquist, *J. Am. Chem. Soc.*, 1987, **109**, 3739; (c) M. N. Mattson, J. P. Bays, J. Zakutansky, V. Stolarski and P. Helquist, *J. Org. Chem.*, 1989, **54**, 2467.
111. W. D. Wulff, D. C. Yang and C. K. Murray, *J. Am. Chem. Soc.*, 1988, **110**, 2653; M. Buchert and H.-U. Reissig, *Tetrahedron Lett.*, 1988, **29**, 2319.
112. C. A. Toledano, H. Rudler, J.-C. Daran and Y. Jeannin, *J. Chem. Soc., Chem. Commun.*, 1984, 574.
113. C. P. Casey and A. J. Shusterman, *Organometallics*, 1985, **4**, 736; C. P. Casey, N. L. Hornung and W. P. Kosar, *J. Am. Chem. Soc.*, 1987, **109**, 4908.
114. C. P. Casey, N. W. Vollendorf and K. J. Haller, *J. Am. Chem. Soc.*, 1984, **106**, 3754.
115. P. F. Korkowski, T. R. Hoye and D. B. Rydberg, *J. Am. Chem. Soc.*, 1988, **110**, 2676; A. Parlier, H. Rudler, N. Platzer, M. Fontanille and A. Soum, *J. Chem. Soc., Dalton Trans.*, 1987, 1041.
116. P. A. Wender and S. L. Eck, *Tetrahedron Lett.*, 1982, **23**, 1871.
117. M. Brookhart, J. R. Tucker and G. R. Husk, *J. Am. Chem. Soc.*, 1983, **105**, 258.
118. K. A. M. Kremer and P. Helquist, *J. Organomet. Chem.*, 1985, **285**, 231.
119. M. Brookhart, D. Timmers, J. R. Tucker, G. D. Williams, G. R. Husk, H. Brunner and B. Hammer, *J. Am. Chem. Soc.*, 1983, **105**, 6721.
120. M. Brookhart, S. E. Kegley and G. R. Husk, *Organometallics*, 1984, **3**, 650.
121. C. P. Casey, W. H. Miles and H. Tukada, *J. Am. Chem. Soc.*, 1985, **107**, 2924; K. A. M. Kremer, G.-H. Kuo, E. J. O'Connor, P. Helquist and R. C. Kerber, *J. Am. Chem. Soc.*, 1982, **104**, 6119.
122. G.-H. Kuo, P. Helquist and R. C. Kerber, *Organometallics*, 1987, **6**, 1141.
123. M. Brookhart, W. B. Studabaker and G. R. Husk, *Organometallics*, 1987, **6**, 1141.
124. M. Brookhart, W. B. Studabaker, M. B. Humphrey and G. R. Husk, *Organometallics*, 1989, **8**, 132.
125. R. S. Iyer, G.-H. Kuo and P. Helquist, *J. Org. Chem.*, 1985, **50**, 5898.
126. E. K. Barefield, P. McCarten and M. C. Hillhouse, *Organometallics*, 1985, **4**, 1682.
127. (a) V. Guerchais and D. Astruc, *J. Chem. Soc., Chem. Commun.*, 1985, 835; (b) V. Guerchais, C. Lapinte, J.-Y. Thepot and L. Toupet, *Organometallics*, 1988, **7**, 604.
128. H. Fischer, J. Schmid and S. Zeuner, *Chem. Ber.*, 1987, **120**, 583.
129. K.-R. Pörschke, *Chem. Ber.*, 1987, **120**, 425.
130. S. E. Kegley, M. Brookhart and G. R. Husk, *Organometallics*, 1982, **1**, 760.
131. (a) J. G. Davidson, E. K. Barefield and D. G. Van Derveer, *Organometallics*, 1985, **4**, 1178; (b) E. K. Barefield, P. McCarten and M. C. Hillhouse, *Organometallics*, 1985, **4**, 1682.
132. Y. Apeloig, M. Karni and Z. Rappoport, *J. Am. Chem. Soc.*, 1983, **105**, 2784.

133. (a) B. M. Trost and L. S. Melvin, Jr., 'Sulfur Ylides', Academic Press, New York, 1975; (b) E. Block, 'Reactions of Organosulfur Compounds', Academic Press, New York, 1978, p. 91; (c) C. R. Johnson, in 'Comprehensive Organic Chemistry', ed. D. H. R. Barton and W. D. Ollis, Pergamon Press, Oxford, 1979, vol. 3, p. 247; (d) Yu. G. Gololobov, A. N. Nesmeyanov, V. P. Lysenko and I. E. Boldeskul, *Tetrahedron*, 1987, **43**, 2609.
134. E. J. Corey and M. Chaykovsky, *J. Am. Chem. Soc.*, 1965, **87**, 1353.
135. C. R. Johnson, *Acc. Chem. Res.*, 1973, **6**, 341.
136. (a) C. R. Johnson and C. W. Schroeck, *J. Am. Chem. Soc.*, 1973, **95**, 7418; (b) C. R. Johnson, C. W. Schroeck and J. R. Shanklin, *J. Am. Chem. Soc.*, 1973, **95**, 7424.
137. C. Mapelli, G. Turocy, F. L. Switzer and C. H. Stammer, *J. Org. Chem.*, 1989, **54**, 145.
138. E. J. Corey and P. R. Ortiz de Montellano, *Tetrahedron Lett.*, 1968, 5113.
139. T. Minami, T. Yamanouchi, S. Tokumasu and I. Hirao, *Bull. Chem. Soc. Jpn.*, 1984, **57**, 2127.
140. D. B. Reddy, B. Sankaraiah and T. Balaji, *Indian J. Chem., Sect. B*, 1980, **19**, 563.
141. E. J. Corey and M. Jautelat, *J. Am. Chem. Soc.*, 1967, **89**, 3912.
142. F. Cooke, P. Magnus and G. L. Bundy, *J. Chem. Soc., Chem. Commun.*, 1978, 714.
143. B. M. Trost and M. J. Bogdanowicz, *J. Am. Chem. Soc.*, 1973, **95**, 5307.
144. A. I. Meyers, J. L. Romine and S. A. Fleming, *J. Am. Chem. Soc.*, 1988, **110**, 7245.
145. A. Krief and W. Dumont, *Tetrahedron Lett.*, 1988, **29**, 1083.
146. A. Monpert, J. Martelli, R. Gree and R. Carrie, *Tetrahedron Lett.*, 1981, **22**, 1961.
147. (a) B. M. Trost, *J. Am. Chem. Soc.*, 1967, **89**, 138; (b) T. Cohen, G. Herman, T, M. Chapman and D. Kuhn, *J. Am. Chem. Soc.*, 1974, **96**, 5627; M. Clagett, A. Gooch, P. Graham, N. Holy, B. Mains and J. Strunk, *J. Org. Chem.*, 1976, **41**, 4033.
148. P. A. Grieco and R. S. Finkelhor, *Tetrahedron Lett.*, 1972, 3781.
149. J. Mulzer and M. Kappert, *Angew. Chem., Int. Ed. Engl.*, 1983, **22**, 63.
150. A. Krief, W. Dumont and P. Pasau, *Tetrahedron Lett.*, 1988, **29**, 1079.
151. R. M. Cory, D. M. T. Chan, Y. M. A. Naguib, M. H. Rastall and R. M. Renneboog, *J. Org. Chem.*, 1980, **45**, 1852.
152. Y. Shen, Z. Gu, W. Ding and Y. Huang, *Tetrahedron Lett.*, 1984, **25**, 4425.
153. A. Krief and M. J. De Vos, *Tetrahedron Lett.*, 1985, **26**, 6115.
154. L. N. Andreeva, T. G. Perlova, V. V. Negrebetskii, S. F. Dymova, T. I. Koroleva, K. D. Shvetsova-Shilovskaya and V. K. Promonenkov, *J. Org. Chem. USSR*, 1981, **17**, 1800.
155. R. M. Cory and R. M. Renneboog, *J. Chem. Soc., Chem. Commun.*, 1980, 1081.
156. E. Schaumann, C. Friese and C. Spanka, *Synthesis*, 1986, 1035.
157. T. Cohen and M. Myers, *J. Org. Chem.*, 1988, **53**, 457.
158. A. Krief, W. Dumont and A. F. De Mahieu, *Tetrahedron Lett.*, 1988, **29**, 3269.
159. (a) J. H. Babler and K. P. Spina, *Tetrahedron Lett.*, 1985, **26**, 1923; (b) K. Annen, H. Hofmeister, H. Laurent, A. Seeger and R. Wiechert, *Chem. Ber.*, 1978, **111**, 3094; (c) N. Ono, T. Yanai, I. Hamamoto, A. Kamimura and A. Kaji, *J. Org. Chem.*, 1985, **50**, 2806.
160. A. Krief, M. J. Devos and M. Sevrin, *Tetrahedron Lett.*, 1986, **27**, 2283.
161. R. M. Cory, P. C. Anderson, F. R. McLaren and B. R. Yamamoto, *J. Chem. Soc., Chem. Commun.*, 1981, 73.
162. P. Ongoka, B. Mauze and L. Miginiac, *J. Organomet. Chem.*, 1987, **322**, 131.

4.7
Formation and Further Transformations of 1,1-Dihalocyclopropanes

VIJAY NAIR
American Cyanamid, Pearl River, NY, USA

4.7.1 INTRODUCTION

In this chapter we will be concerned with only 1,1-dihalocyclopropanes and these will be referred to as dihalocyclopropanes throughout the text.

The first description of a dihalocyclopropane[1] was made nearly a century ago and the idea that dichlorocarbene is an intermediate in the basic hydrolysis of chloroform dates back to 1862.[2] The pioneering work of Hine confirmed the existence of dichlorocarbene[3,4] and Doering then demonstrated its generation and its addition to alkenes leading to the formation of dichlorocyclopropanes (Doering reaction).[5] Undoubtedly, this work laid the foundation for dihalocyclopropane chemistry and has opened a new vista in organic chemistry. The formation and transformations of dihalocyclopropanes including synthetic applications have received considerable attention during the last three decades. Excellent reviews on aspects of dihalocyclopropanes also have been published.[6–10] Effort has been made to expose all the original work with special attention being paid to publications that have appeared during the last ten years.

Dihalocyclopropanes containing all possible combinations of halogens have been synthesized. From the vantage point of the synthetic chemist, dibromo- and dichloro-cyclopropanes elicit the most useful and fascinating chemistry, and therefore this discussion will be centered around the formation and transformations of these two groups of compounds. For the sake of completeness, dihalocyclopropenes have been discussed where appropriate. To emphasize the synthetic potential, separate subsections are devoted to certain topics, *e.g.* formation of heterocycles.

4.7.2 SYNTHESIS OF DIHALOCYCLOPROPANES

Dihalocyclopropanes are generally prepared by the addition of dihalocarbenes to alkenic substrates. As indicated in the introduction, the first synthesis of a dihalocyclopropane was accomplished by Doering and Hoffmann by the addition of dichlorocarbene, generated from chloroform and potassium *t*-butoxide (Bu^tOK), to cyclohexene giving dichloronorcarane (**1**), as shown in equation (1).[5]

$$\text{cyclohexene} \xrightarrow[\text{CHCl}_3/\text{Bu}^t\text{OK}]{:\text{CCl}_2} \text{7,7-dichloronorcarane} \quad \textbf{(1)} \qquad (1)$$

Subsequently, other dihalocarbenes (**2**) have been generated and their addition to alkenes investigated in varying detail.

$$:\text{CXY} \quad \textbf{(2)}$$

X = Y = F, Cl, Br, I
X = F, Y = Cl
X = Cl, Y = Br
X = F, Y = Br
X = F, Y = I

4.7.2.1 Methods for the Generation of Dihalocarbenes

The original procedure for the generation of dichlorocarbene from chloroform by Bu^tOK (equation 2) still finds limited use.[5] In one variation[11] dichlorocarbene is generated by the reaction of an alkyllithium with bromotrichloromethane (equation 3). However, these procedures have largely been replaced by milder and synthetically more useful methods. Thermal decomposition of sodium trichloroacetate in dimethoxyethane has found extensive use in the generation of dichlorocarbene (equation 4).[12] Similarly, difluorochloroacetate has been used as a convenient source of difluorocarbene (equation 5).[13] Trichloroacetic acid itself undergoes thermal decomposition to give dichlorocarbene (equation 6).[14]

$$\text{CHCl}_3 \xrightarrow{\text{Bu}^t\text{OK}} :\text{CCl}_2 + \text{KCl} + \text{Bu}^t\text{OH} \qquad (2)$$

$$\text{BrCCl}_3 \xrightarrow{\text{Bu}^n\text{Li}} :\text{CCl}_2 + \text{LiCl} + \text{Bu}^n\text{Br} \qquad (3)$$

$$\text{Cl}_3\text{CCO}_2\text{Na} \xrightarrow{\Delta} :\text{CCl}_2 + \text{NaCl} + \text{CO}_2 \qquad (4)$$

$$\text{ClF}_2\text{CCO}_2\text{Na} \xrightarrow{\Delta} :\text{CF}_2 + \text{NaCl} + \text{CO}_2 \qquad (5)$$

$$\text{Cl}_3\text{CCO}_2\text{H} \xrightarrow{\Delta} :\text{CCl}_2 + \text{HCl} + \text{CO}_2 \qquad (6)$$

The observation that phenyl(trichloromethyl)mercury undergoes dissociation at 150 °C to give phenylmercury(II) chloride and dichlorocarbene has led to a wide range of organomercury compounds ($PhHgCX_3$ — Seyferth reagents)[15–18] which serve as efficient dihalocarbene transfer reagents (equation

7). Seyferth reagents are ideal for the dihalocyclopropanation of deactivated or base sensitive substrates such as allyl halides, allyl isocyanates, polychloroalkenes, α,β-unsaturated ketones, esters, nitriles and sulfones.[19]

$$PhHgCX_2Y \xrightarrow{\Delta} :CX_2 + PhHgY \quad (7)$$

X = Cl, Br or F, Y = Cl or Br

Undoubtedly the most important and widely used procedure for the generation of dichlorocarbene involves the reaction of chloroform with aqueous sodium hydroxide under the conditions of phase transfer catalysis (PTC), introduced by Makosza.[20–22] Under these conditions chloroform reacts with sodium hydroxide to form sodium trichloromethylide which on exchange with a quaternary ammonium salt, usually benzyltriethylammonium chloride, is converted to the unstable quaternary ammonium methylide which dissociates in the organic phase to give dichlorocarbene. The dichlorocarbene irreversibly adds to the alkene (Scheme 1).

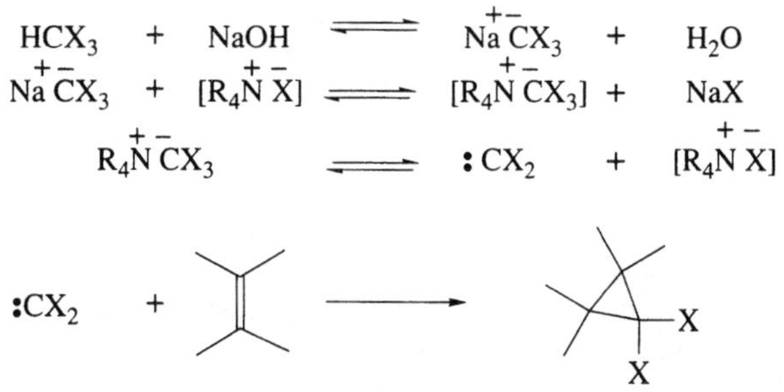

$$HCX_3 + NaOH \rightleftharpoons \overset{+}{Na}\overset{-}{C}X_3 + H_2O$$
$$\overset{+}{Na}\overset{-}{C}X_3 + [R_4\overset{+}{N}\overset{-}{X}] \rightleftharpoons [R_4\overset{+}{N}\overset{-}{C}X_3] + NaX$$
$$R_4\overset{+}{N}\overset{-}{C}X_3 \rightleftharpoons :CX_2 + [R_4\overset{+}{N}\overset{-}{X}]$$

:CX$_2$ + (alkene) ⟶ (1,1-dihalocyclopropane, X, X)

X = Cl or Br

Scheme 1

Although illustrated for the generation of dibromo- and dichloro-carbenes, the PTC method has found application in the generation of other dihalocarbenes. Indeed the PTC method has assumed great importance in recent years in the synthesis of a variety of organic compounds and the procedure is well documented in excellent reviews[21–25] and books.[26–28] The procedure has been used extensively[21–28] for the addition of dihalocarbenes to a variety of substrates such as alkenes, alkynes, conjugated polyalkenes, allenes, α,β- and γ,δ-unsaturated esters, imines, enamines, enol acetates, enol ethers, electron rich aromatic compounds, sterically hindered arylated alkenes, ferrocene-substituted alkenes, methylfurans, thiophenes, indenes, strained alkenes such as bicyclo[2.2.1]heptenes and a number of other π-systems. Dehmlow has investigated[24,25,28] the PTC method for the generation of dichlorocarbene in great detail and his prescription calls for the use of a 4 mol excess of $CHCl_3$, 50% aqueous sodium hydroxide and 1 mol % of the catalyst in proportion to the alkenic substrate. Quaternary ammonium salts such as benzyl triethylammonium chloride and tetra-*n*-butylammonium chloride are the most efficient catalysts.

Improvements in the generation of dibromocarbene have also been reported.[29,30] Although diiodocarbene has been generated and added to alkenes under PTC conditions,[31] the CHI_3/Bu^tOK procedure seems to be superior.[32,33] Dihalocarbenes carrying two different halogens, *viz.* :CBrCl,[34,36] :CBrF,[37,38] :CClF,[37] :CClI[36] and :CFI[39] have been generated under PTC conditions. A number of modifications, such as solid/liquid PTC involving $Cl_3CCO_2Na/CHCl_3$/catalyst,[40] powdered $NaOH/CHCl_3$/catalyst[41] with ultrasonic irradiation,[42] have enhanced the usefulness of the PTC procedure. Tertiary amines have also been used as catalysts for the generation of dichlorocarbene from chloroform in two-phase systems consisting of aqueous organic solvents.[43,44]

Dihalocarbenes generated by PTC exhibit enhanced reactivity which may be attributed to the fact that the carbene is formed in the proximity of the substrate and a large excess of it is generated, albeit in small quantities at a time. In addition, the dihalocarbene generated in this fashion is completely ‘free’ and not associated with any metallic species, thereby attenuating the reactivity. The use of crown ethers and CH_2Cl_2 has been shown to suppress halogen exchange during the generation of mixed dihalocarbenes.[34–36]

In comparison to other dihalocarbenes, difluorocarbene reactions are less well investigated. Invariably, difluorocarbene is generated thermally and a number of reagents have been investigated.[19,45] The method of choice appears to be the reaction of $PhHgCF_3$ with NaI, which gives excellent results with a variety of alkenes[45]

The low valent titanium species obtained from $TiCl_4$ and LAH has been used[46] to generate dichlorocarbene from CCl_4. Finally, it has been shown that certain dihalocyclopropanes dissociate thermally[47] or photochemically[48] to generate the carbene. Such procedures may be useful under special circumstances.

4.7.2.2 Addition of Dihalocarbenes to π-Bonds

The electronic description and hybridization of dihalocarbenes (**3**) are similar to those of carbocations. Not surprisingly, therefore, dihalocarbenes behave as electrophiles in their reactivity towards alkenic substrates and this is discussed in the following sections.

X X

(3)

4.7.2.2.1 Alkenes

By virtue of their electrophilic nature, dihalocarbenes add to alkenes smoothly. Predictably, the following order of reactivity is observed: tetraalkyl- > trialkyl- > unsymmetrical dialkyl- > symmetrical dialkyl- > monoalkyl-substituted alkenes. It has long been known that dihalocarbene addition to alkenes is stereospecific (equations 8 and 9).[49,50] The stereospecificity has its origin in the spin state of the carbene, a ground state singlet,[51] which allows concerted reaction. The ground state singlet is stabilized by the interaction of one of the nonbonding orbitals of the carbene with orbitals on the halogen.

$:CBr_2$ Br Br (8)

$:CBr_2$ Br Br (9)

The general methods outlined (Section 4.7.2.1) are suitable for the dihalocyclopropanation of simple alkenes, although the PTC procedure is preferable in the vast majority of cases, with the exception of alkenes carrying electron-withdrawing groups. Seyferth reagents are ideal in such cases, as illustrated by the conversion of acrylonitrile to 1-cyano-2,2-dihalocyclopropane (equation 10).[52] Dihalocyclopropanes are generally stable compounds and are readily isolable. However, those resulting from the addition of dihalocarbenes to strained alkenes are unstable and often undergo rearrangements (equation 11).[10,53,54] Selected examples of the addition of dihalocarbenes to alkenes are given in equations (12) to (16). The addition of dichlorocarbene to methylenecyclopropane, giving a spiro compound (equation 12),[55] is similar to the earlier work on the addition of dibromo- and dichloro-carbenes to cyclopropylidenecycloalkenes.[56] A monoadduct can be conveniently prepared from isotetralin by reaction with dihalocarbenes (equation 13).[57] However, with a large excess of $:CCl_2$, the reaction gives a mixture of di- and tri-adducts.[58] Normally 1,3-dienes add dichlorocarbene to give the 1,2-adduct (equation 14),[59] but there are instances where both the 1,2- and the 1,4-adducts are obtained.[60] Certain metal complexes of alkenes have been shown to add dichlorocarbene (equation 15).[61] Dichlorocarbene generated from sodium trichloroacetate adds to 2-bromovinyltrimethylsilane (equation 16).[62]

CN $:CCl_2/PhHgCCl_2Br$ 78% Cl Cl NC (10)

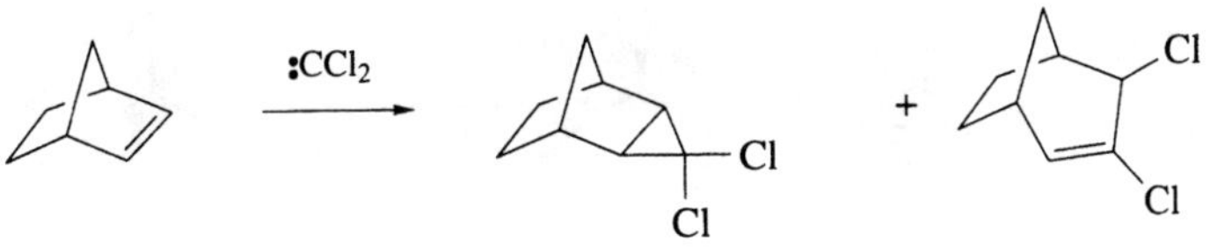

(11)

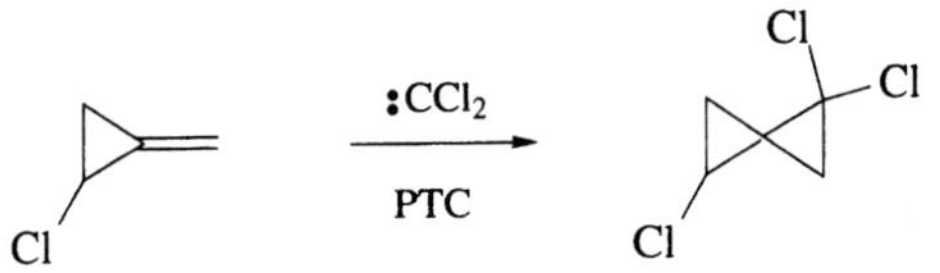

(12)

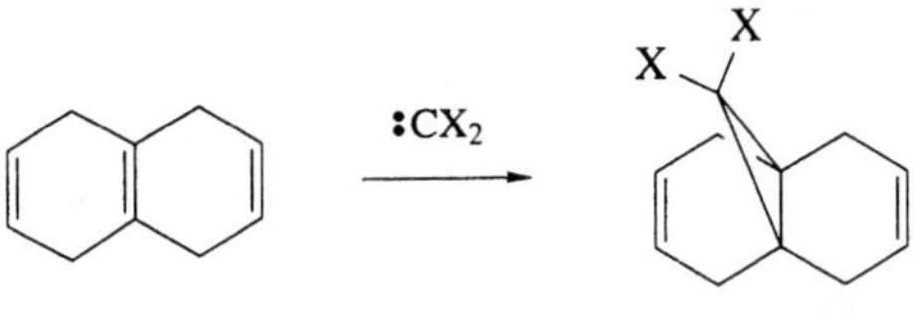

(13)

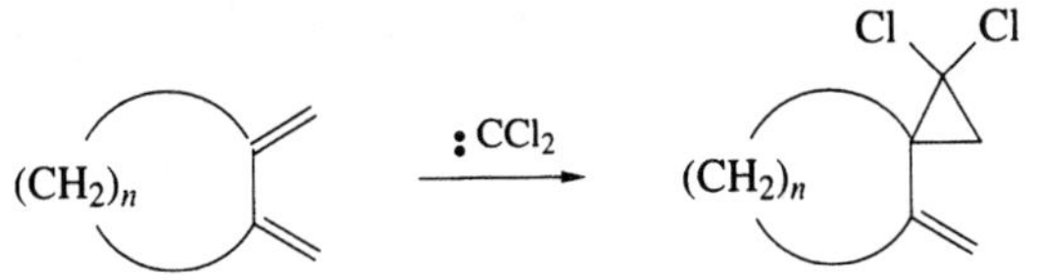

(14)

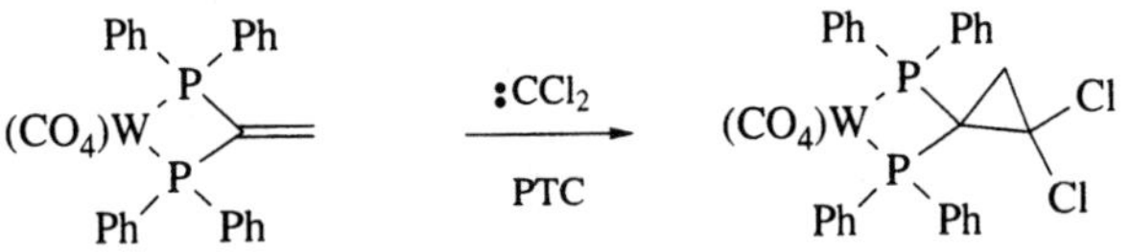

(15)

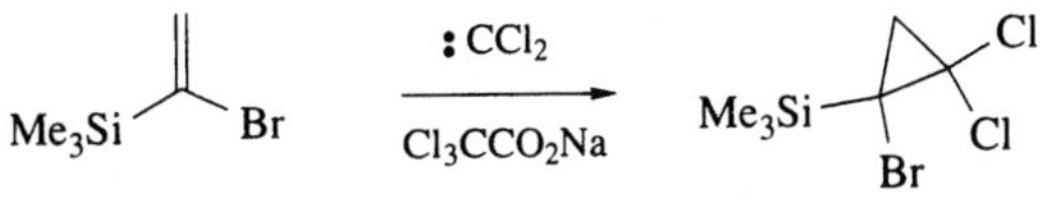

(16)

Predictably, with enynes dihalocarbene addition occurs preferentially with the double bond (equation 17).[63,64] The reaction of dichlorocarbene with allenes usually gives rise to spiro-linked dichlorocyclopropanes (equation 18),[65] but rearranged products are obtained in some cases (equation 19).[65,66] Somewhat surprisingly, a measure of stereoselectivity has been observed in the formation of dichlorocyclopropylcarbinols from secondary allylic alcohols and dichlorocarbene (equation 20).[67]

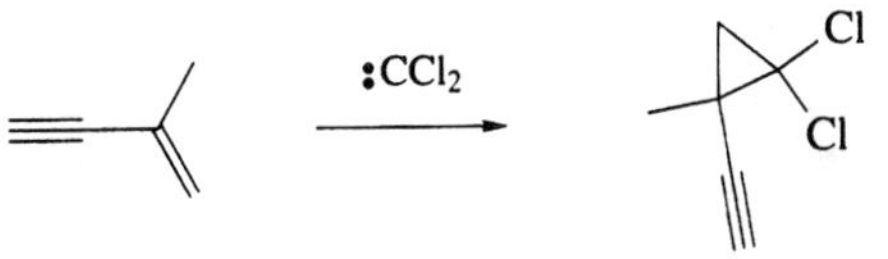

(17)

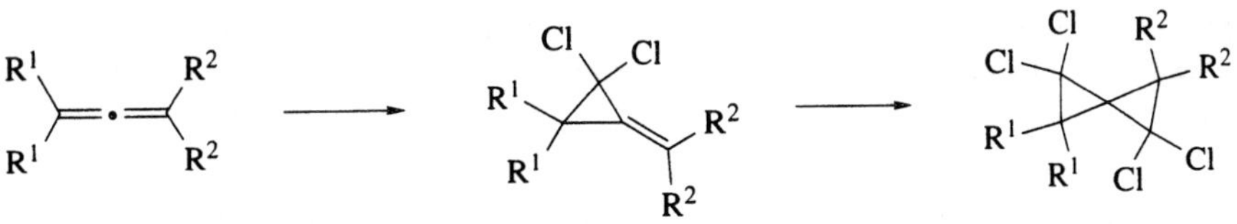

(18)

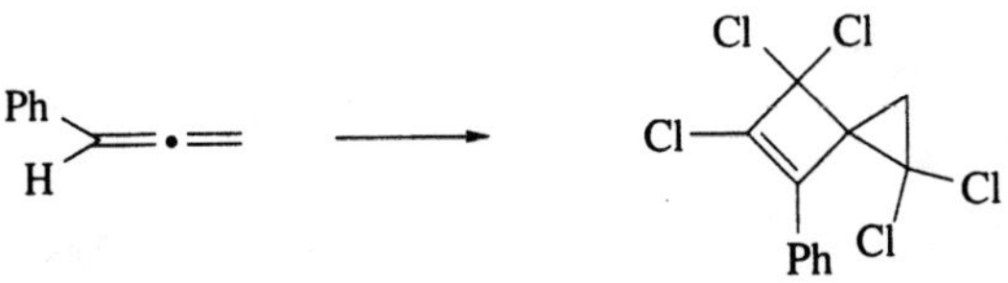

(19)

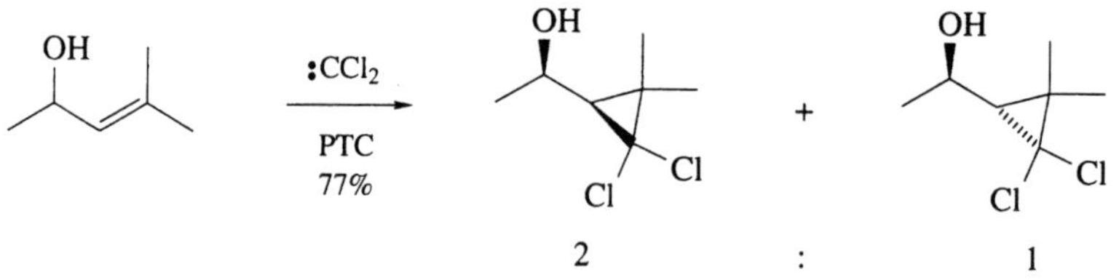

:CCl$_2$ PTC 77% (20)

2 : 1

4.7.2.2.2 Electron rich π-systems

The electrophilic nature of dihalocarbenes makes them highly reactive towards enamines, enamides, enol ethers, vinyl sulfides, vinyl selenides and electron rich aromatic and heteroaromatic systems. Unlike in the case of simple alkenes, the products are often unstable and readily undergo rearrangement. The reaction of dichlorocarbene with the enamine of cyclohexanone (**4a**) affords an adduct (**5a**) which on hydrolysis gives 2-chloromethylenecyclohexanone (Scheme 2).[68] Subsequently, the reactions of dihalocarbenes with a variety of enamines have been studied.[69–74] The dichlorocarbene adduct of cyclohexanone enamine is stable, whereas the one resulting from cyclopentanone enamine (**5b**) undergoes ring expansion (Scheme 2).[69] Such ring expansions are discussed in a later section (4.7.3.7.1). As with other substrates, the dihalocarbene is generated by PTC, by the thermal decomposition of sodium trihaloacetate or from a Seyferth reagent.

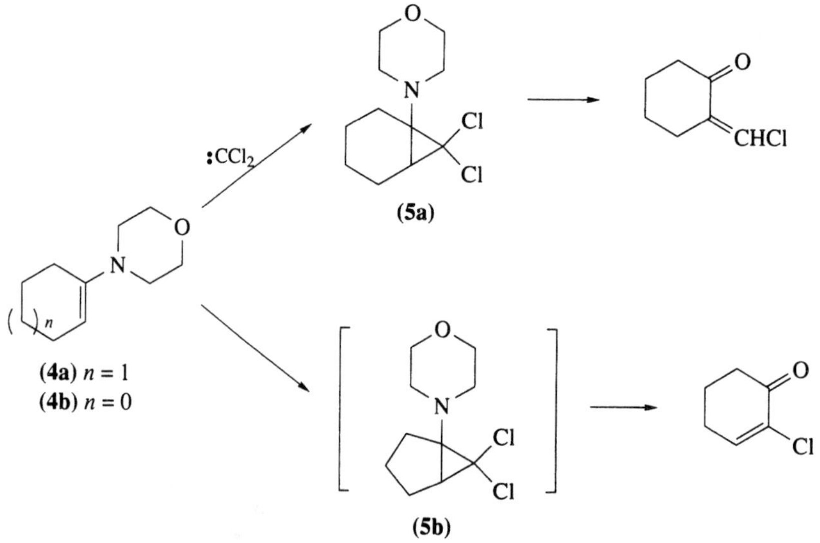

:CCl$_2$

(**4a**) n = 1
(**4b**) n = 0

(**5a**)

(**5b**)

Scheme 2

In addition to acyclic and monocyclic enamines, bicyclic and tricyclic enamines also undergo cycloaddition with dihalocarbenes. 'Endocyclic enamines', such as pyrrole and indole, add dichlorocarbene and the adducts rapidly undergo ring cleavage to afford 3-chloropyridine and 3-chloroquinoline, respectively, in moderate yields (*cf.* Section 4.7.3.9).[75–77]

Although less reactive than enamines, enamides also undergo cycloaddition with dihalocarbenes and the adducts are usually stable.[73,74] A variety of other nitrogen heterocycles, in which the nitrogen lone pair is in conjugation with a π-system, have been shown to undergo dihalocyclopropanation leading to synthetically useful intermediates.[74] An example from isoquinoline alkaloid synthesis[78] is shown in equation (21).

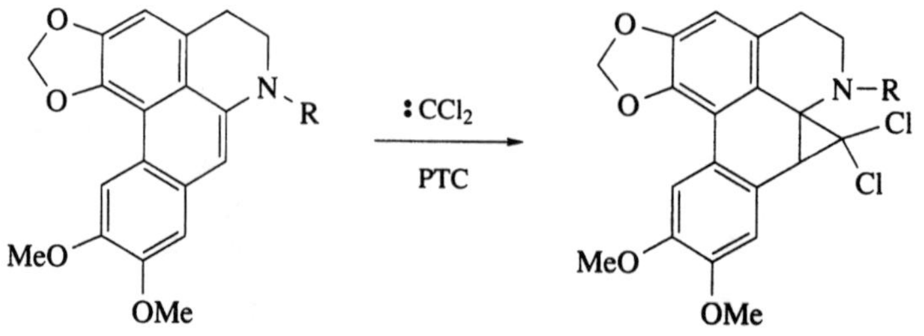

:CCl$_2$ PTC (21)

Like enamines, dihalocarbenes add smoothly to enol ethers and in many cases it is possible to isolate the dihalocyclopropyl intermediates which are valuable synthons for chloroenones (*cf.* Section 4.7.3.7.1). The earliest example of the addition of a dihalocarbene to an enol ether was provided by Parham,[6,79] who studied the addition of dichlorocarbene to dihydropyran (equation 22). An example which illustrates the synthetic potential of the process is the conversion of the cyclohexanone enol ether (**6**) to the dichlorocyclopropane (**7**; equation 23).[80] The latter served as a useful intermediate in a stereospecific synthesis of Prelog–Djerassi lactonic acid.

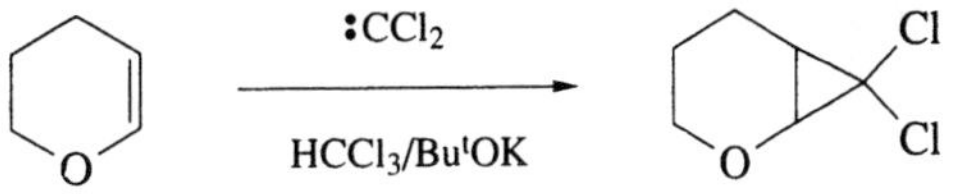

:CCl_2 ; $HCCl_3$/Bu^{t}OK — Cl, Cl, O (22)

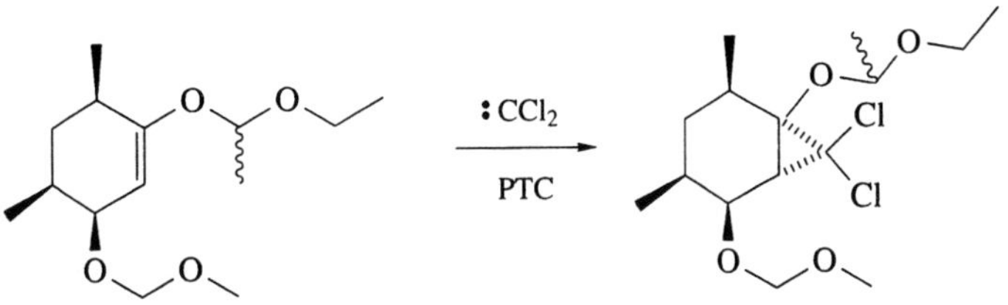

:CCl_2 ; PTC — O, O, Cl, Cl (23)

(6) **(7)**

Enol silyl ethers,[81–83] enol acetates,[84,85] and enol lactones[86] also undergo facile dihalocyclopropanation with dihalocarbenes generated by the thermal decomposition of Seyferth reagents or sodium trihaloacetates. Ketene acetals give unstable dihalocarbene adducts which are synthons for α-haloacrolein and other valuable intermediates.[87,88] Vinyl sulfides[89] and selenides[90] have been reported to undergo addition with dichlorocarbene.

4.7.2.2.3 Alkynes

Interest in the reaction of dihalocarbenes with alkynes has been limited. In comparison with alkenes, dihalocarbenes react sluggishly with alkynes (*cf.* equation 17) and the resulting dihalocyclopropenes are unstable (Scheme 3). These are, however, valuable precursors for cyclopropenones (equations 24 and 25).[91–95]

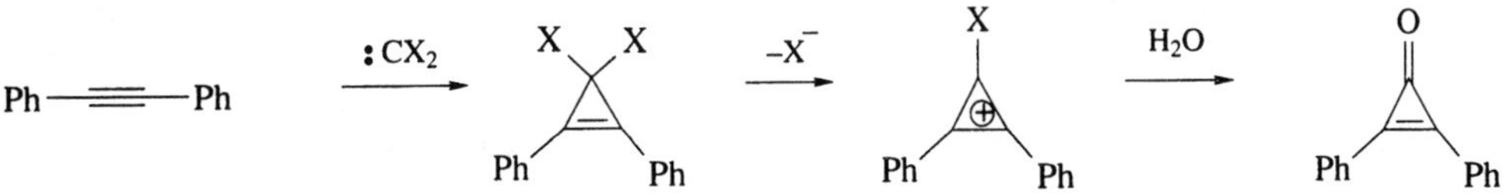

Ph—≡—Ph —(:CX_2)→ X X, Ph Ph —($-X^-$)→ X, ⊕, Ph Ph —(H_2O)→ O, Ph Ph

Scheme 3

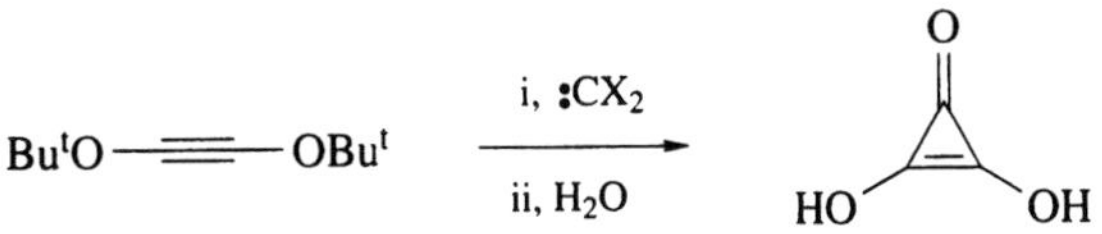

Bu^{t}O—≡—OBu^t —(i, :CX_2; ii, H_2O)→ O, HO OH (24)

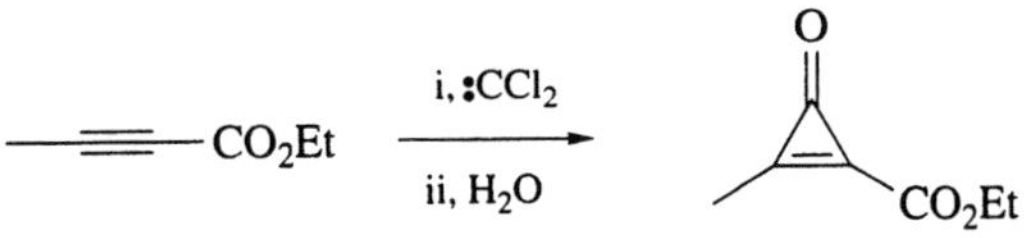

—≡—CO_2Et —(i, :CCl_2; ii, H_2O)→ O, CO_2Et (25)

4.7.2.3 Other Methods for the Formation of Dihalocyclopropanes

Indirect methods, which are primarily essentially of academic interest, nevertheless may find use under special circumstances. One of these, involving the brominative decarboxylation (Borodin–Hunsdiecker reaction) of silver α-fluorocyclopropanecarboxylate, provides a convenient and stereospecific route to the synthesis of fluorobromocyclopropyl compounds (equation 26).[96] Another approach makes use of the Diels–Alder reaction between tetrachlorocyclopropene and dienes (equation 27).[97,98] The adducts can undergo facile dehydrohalogenation or ring opening.

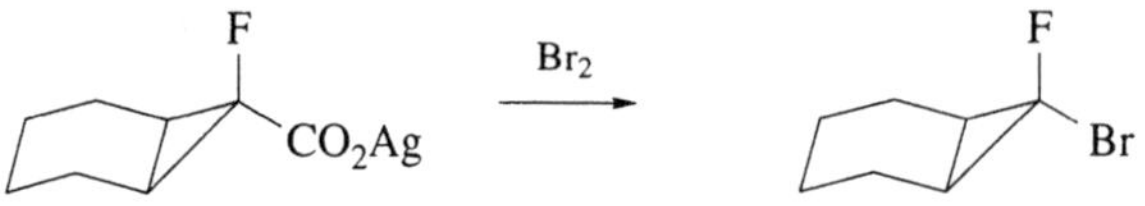

(26)

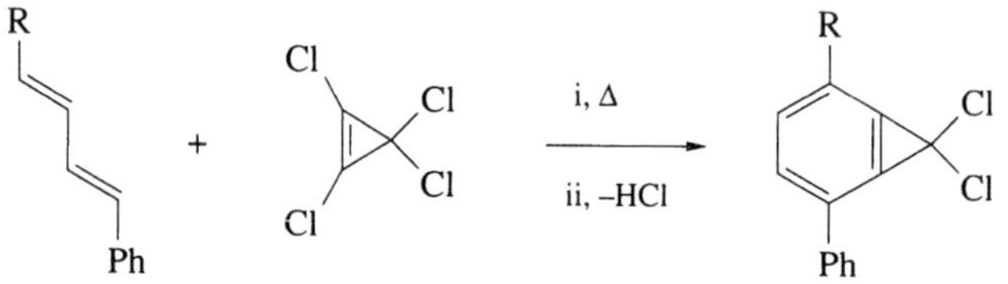

(27)

4.7.3 TRANSFORMATIONS OF 1,1-DIHALOCYCLOPROPANES

The electronic and geometric features of dihalocyclopropanes make them susceptible to a variety of chemical transformations. Some of these are manifested in simple peripheral changes involving only the halogen atoms while others involve complex transformations resulting from highly reactive species — carbenes and carbenoids — unleashed by the removal of the halogens. Yet another set of reactions emerge from the electrocyclic opening of the cyclopropane ring attended by the departure of a halide ion. These transformations are described in the following sections with emphasis on synthetic utility.

4.7.3.1 Reductive Dehalogenation

Dihalocyclopropanes readily undergo reductive dehalogenation under a variety of conditions. Suitable choice of reagents and reaction conditions will allow the synthesis of monohalocyclopropanes or the parent cyclopropanes.[19,99] The ease of reduction follows the expected order: I > Br > Cl > F. In general, complete reduction of dibromo and dichloro compounds is accomplished by alkali metal in alcohol,[99–102] liquid ammonia[103] or tetrahydrofuran (equations 28 and 29).[104] The dihalocyclopropanes can be reduced conveniently with LAH (equation 30).[105] LAH reduction is particularly suited for difluoro compounds which are resistant to dissolving metal reductions.[19,106] It is noteworthy that the sequence of dihalocarbene addition to an alkene followed by the reduction of the dihalocyclopropyl compounds (equation 31) provides a convenient and powerful alternative to Simmons–Smith cyclopropanation, which is not always reliable.

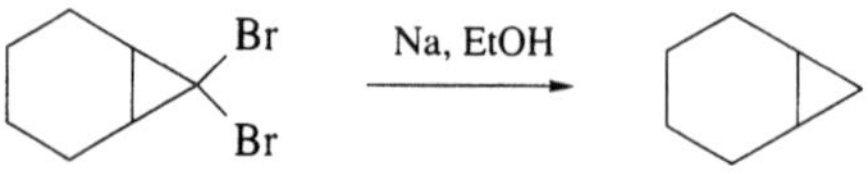

(28)

(29)

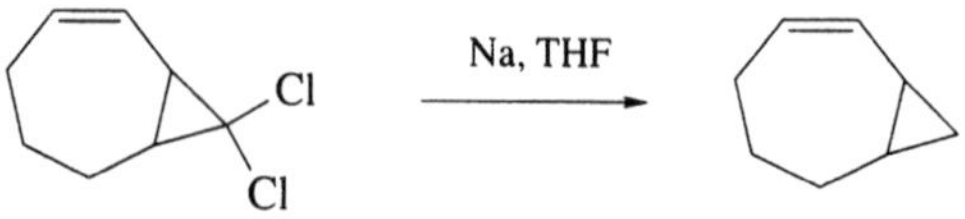

(30)

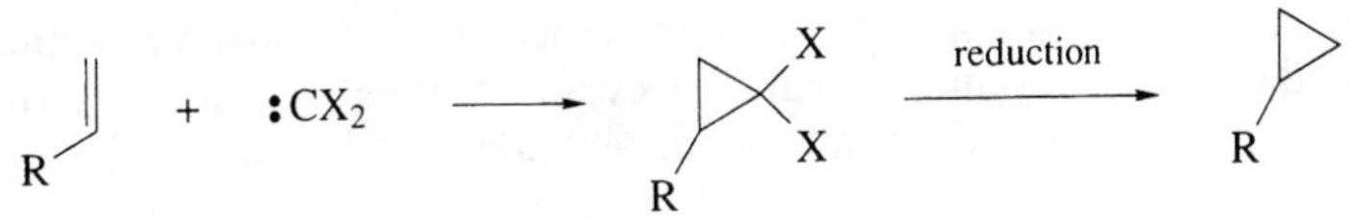

(31)

The selective or partial reduction of dihalocyclopropanes to monohalocyclopropanes[107] can be achieved with LAH[108] or Bu_3SnH (equations 32 to 34).[109,110] A number of other reducing agents also lead to partial reduction.[19] Reduction of dihalocyclopropanes proceeds *via* radical intermediates and the cyclopropyl radical thus generated can be trapped by an adjacent homoallylic double bond leading to a bicyclo[3.1.0]hexane system (equation 35).[111,112]

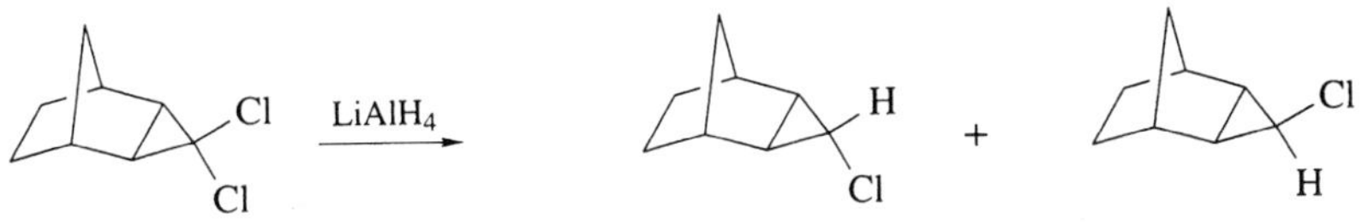

(32)

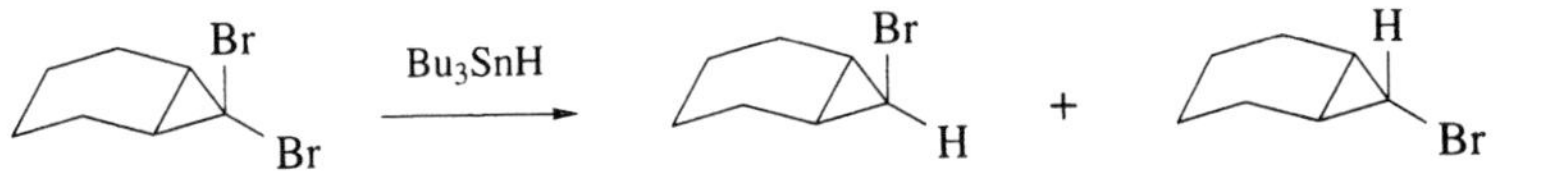

(33)

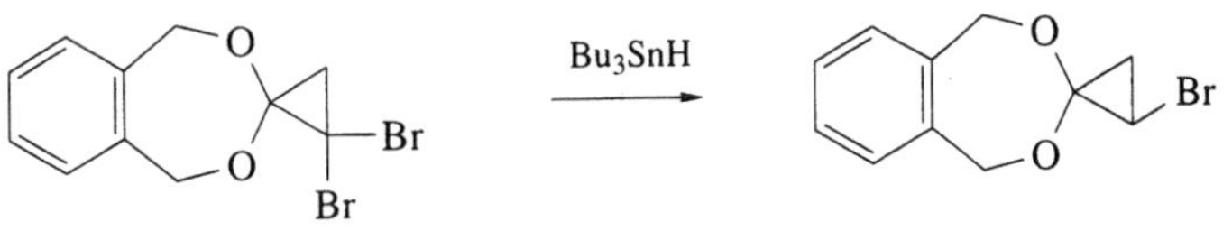

(34)

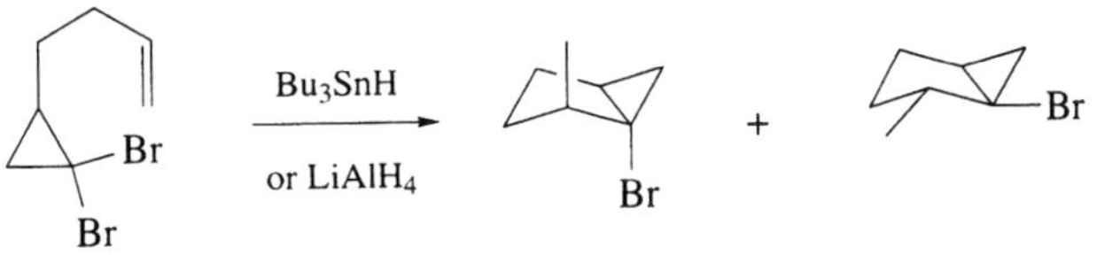

(35)

Halogen–lithium exchange (*cf.* Section 4.7.3.2) at low temperatures with an alkyllithium followed by quenching with methanol or water results in stereospecific reduction of dihalocyclopropyl compounds.[113] Stereospecific reduction has also been achieved by reaction with potassium diphenylphosphide (equation 36).[114]

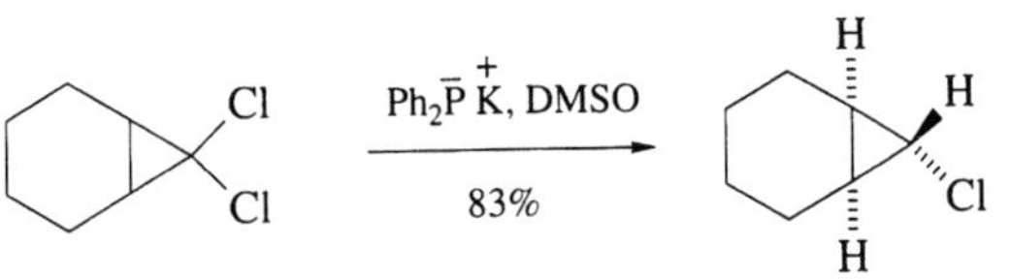

(36)

4.7.3.2 Lithium–Halogen Exchange and the Chemistry of 1-Lithio-1-halocyclopropanes

Dihalocyclopropanes undergo rapid metal–halogen exchange with alkyllithiums at low temperature.[19,115] The exchange is most convenient with dibromocyclopropanes and the resulting lithium carbenoids[116,117] are versatile intermediates which can be trapped by electrophiles (equation 37). Alkyl halides,[118–120] aldehydes,[121,122] ketones,[121,122] disulfides,[123] iminium salts,[124] trimethylsilyl chloride,[125,126] trimethylstannyl chloride,[127] carbon dioxide[115,128] and sulfur dioxide[129] react smoothly giving the corresponding 1-bromocyclopropyl derivatives (Scheme 4). In many cases the products themselves serve as valuable intermediates in synthesis by virtue of their ability to undergo a variety of substitution reactions.[19,130] For example, methanol reacts with 1-bromo-1-methylthiocyclopropane (**8**) to give (**9**) and subsequently (**10**) with methanethiol (equation 38).[131] Other nucleophiles react in an analogous manner (equation 39) to provide useful synthetic intermediates. The lithio compound resulting from (**8**) on reaction with an aldehyde gives a carbinol which is a cyclobutanone synthon.[123] 1-Bromo-1-trimethylsilylcyclopropanes have been used in the synthesis of alkylidenecyclopropanes by Peterson alkenation (equation 40).[126,132] A stereoselective cyclopropanol synthesis is based on the sequential reaction of lithiobromocyclopropane with catechol borane followed by alkaline hydrogen peroxide oxidation (equation 41).[133] With cyclopentenone, 1,2-addition occurs predominantly (equation 42) and the product is a precursor for fulvene derivatives.[134]

It was noted earlier that the lithium carbenoids on reaction with alkyl halides afford l-alkyl-1-halocyclopropanes. It is possible to effect *endo* selective monoalkylation, as illustrated with a benzobarrelene

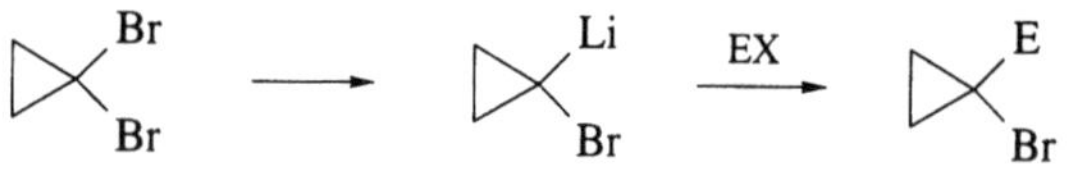

Br, Br → Li, Br —EX→ E, Br (37)

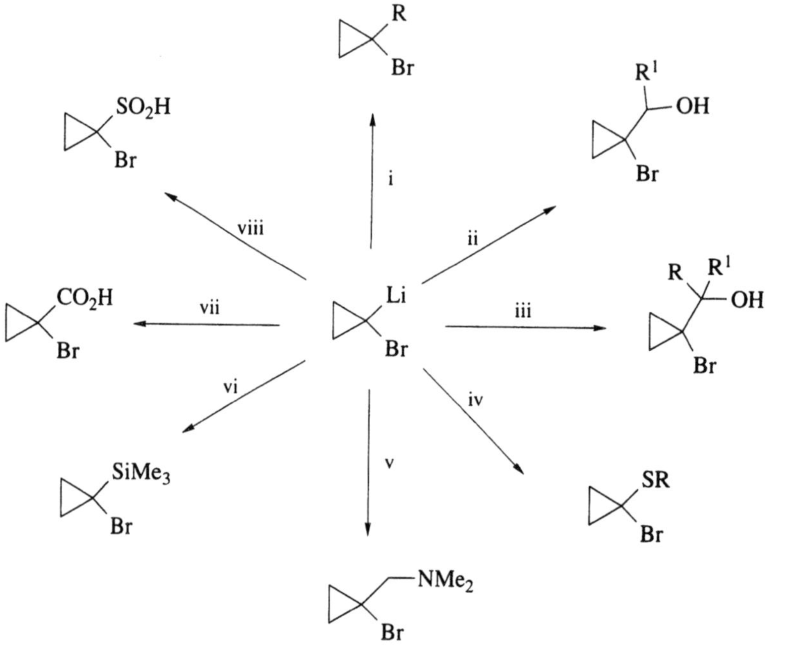

R, Br; SO₂H, Br; R¹, OH, Br; i; viii; ii; R, R¹, OH; CO₂H, Br; vii; Li, Br; iii; Br; vi; iv; SiMe₃, Br; v; SR, Br; NMe₂, Br

i, RX; ii, R¹CHO; iii, RR¹CO; iv, RSSR; v, $H_2C{=}\overset{+}{N}Me_2\,X^-$, vi, Me_3SiCl; vii, CO_2; viii, SO_2

Scheme 4

SMe, Br → SMe, OMe → SMe, SMe (38)

(**8**) (**9**) (**10**)

SMe, Br —MX, DMF, 0 °C→ SMe, X (39)

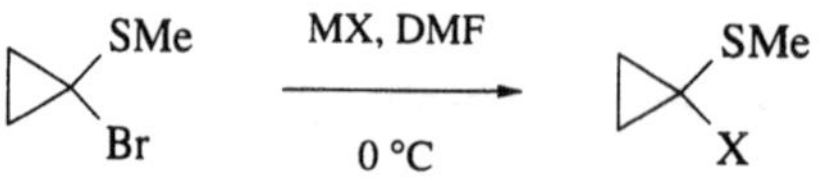

$X = O_2CH$, I, N_3, SH

SiMe₃, Br —i, BuLi; ii, R¹COR²→ SiMe₃, R², R¹, OH —ButOK or KF→ R¹, R² (40)

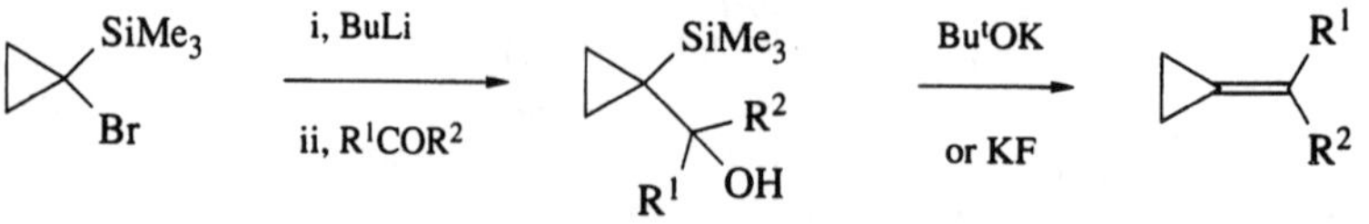

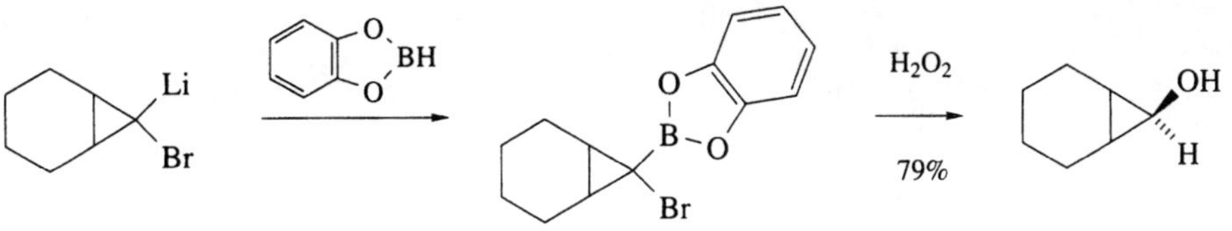

(41)

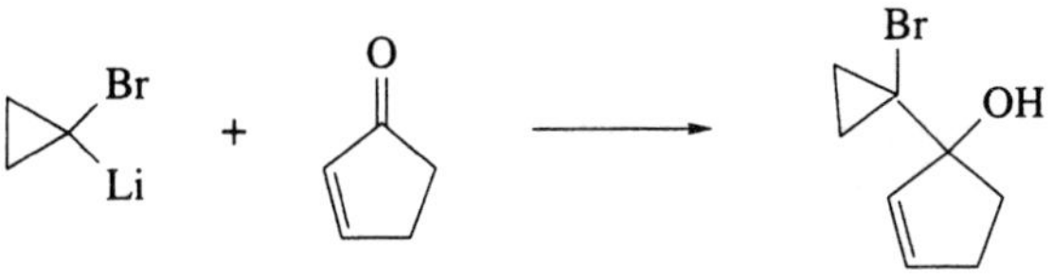

(42)

derivative in equation (43).[135] Stereoselective dialkylation is also possible.[136,137] Dialkylation with lithium dialkylcuprate, involving copper carbenoids, is a synthetically useful reaction (Scheme 5),[138,139] and has been employed in the synthesis of sesquicarene and (±)-sirenin.[118,136] An efficient conversion of dihalocyclopropanes to *gem* dialkyl compounds utilizes higher order organocuprates (equation 44).[140–142] This procedure has been used for the introduction of *gem* dimethyl groups in the phorbol skeleton.[142]

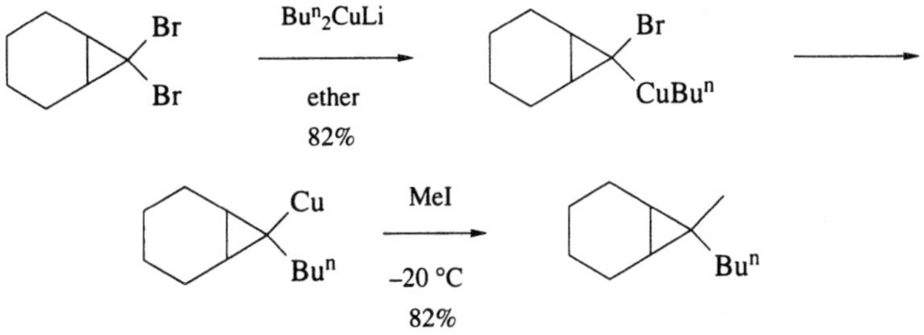

Scheme 5

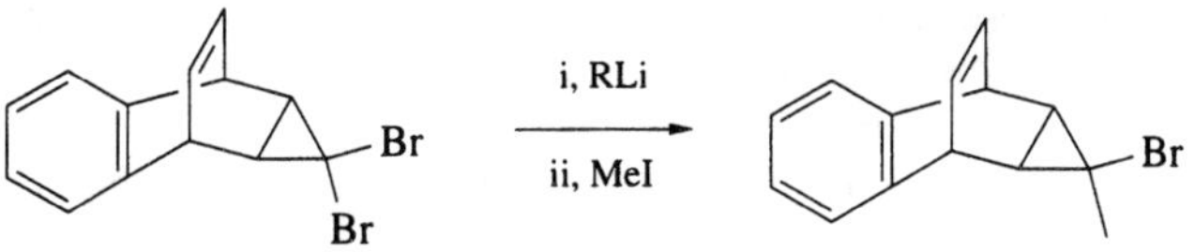

(43)

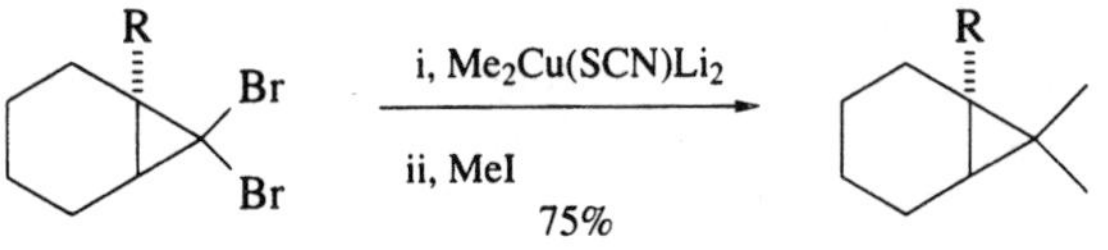

(44)

Intramolecular alkylation of a lithiated species has been exploited in the synthesis of [1.1.0]- and [1.1.1]-propellanes (equation 45).[143–145] In what appears to be a related reaction, the bis(dibromocarbene) adduct (**11**) on treatment with methyllithium undergoes a ring closure by 1,6-elimination, presumably *via* a four-center transition state, to give (**12**; equation 46), which was subsequently converted to a bis(homobenzene).[146]

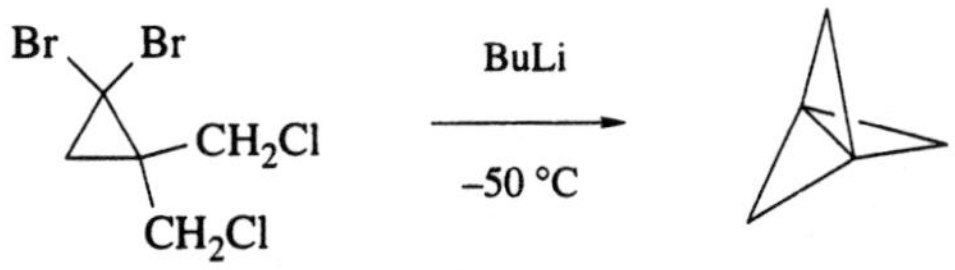

(45)

4.7.3.3 Doering–Moore–Skattebøl Allene Synthesis

Dehalogenation of dihalocyclopropanes with sodium or magnesium gives allenes (equation 47).[147] Since the original observation, a number of reagents have been introduced for the allene synthesis *via* de-

(46)

(11) **(12)**

halogenation of dihalocyclopropanes, but the greatest improvement has resulted from the use of alkyllithiums.[148,149] In general, methyllithium is the reagent of choice and the reaction is usually carried out in ether at low temperature. Dibromocyclopropanes are preferred over the dichloro analogs. The reaction can be viewed as a multistep process initiated by metal–halogen exchange to produce an α-haloalkyllithium intermediate. The latter then eliminates lithium halide to generate a carbene which rearranges to allene by cyclopropane ring scission (Scheme 6). The overall process represents the most versatile and practical synthesis of acyclic mono-, di-, and tri-substituted allenes, cyclic allenes, allenes containing functional groups, and acyclic and cyclic cumulenes. An important feature of the reaction is the absence of alkynic side products, which beset other procedures for the synthesis of terminal allenes.

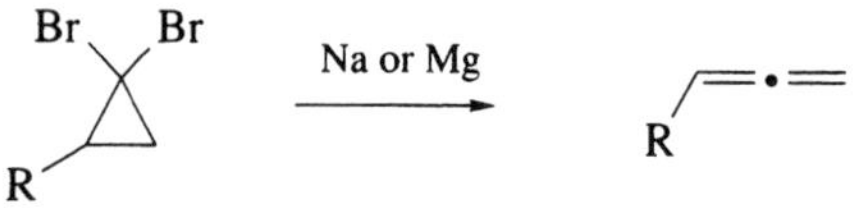

Na or Mg (47)

RLi –LiX

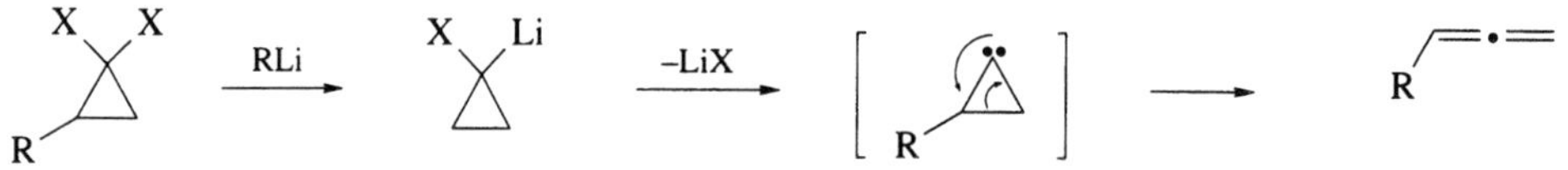

Scheme 6

Since the synthesis of allenes from dihalocyclopropanes has been reviewed,[150–152] only representative examples highlighting the process, along with the more recent results, are presented in this section. The formation of diallenes (equation 48) illustrates the usefulness of this process.[153,154] Two other interesting examples are given in equations (49) and (50).[155–157]

Cyclic allenes have been obtained in high yields, as illustrated by the synthesis of 1,2-cyclononadiene from the dibromocarbene adduct of the readily available cyclooctene (equation 51).[158] The smallest stable cyclic allene known to date is (**14**): it was prepared from the dibromocyclopropane (**13**) in high yield.[159] A small amount of the tricyclic compound (**15**) was also obtained (equation 52). The cyclic allene (**14**) did not undergo dimerization even on prolonged standing at ambient temperatures. In contrast, the unsubstituted analog was detected only at –60 °C by ^{1}H NMR. It should also be noted that cyclohexa-1,2-diene was generated by the reaction of methyllithium on dibromobicyclo[3.1.0]hexane and trapped as the Diels–Alder adduct.[160]

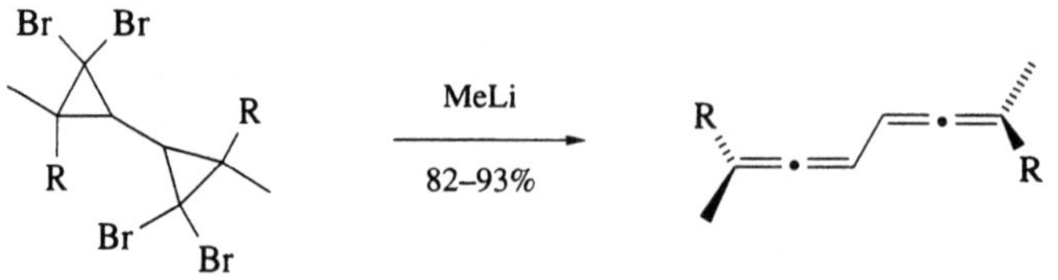

MeLi, 82–93% (48)

R = Me, Et, Ph, Bu^t

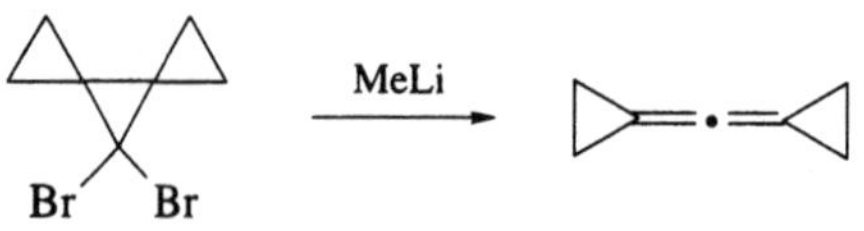

MeLi (49)

The conversion of dihalocyclopropane to allene can be extended to the synthesis of cumulenes by repeating the sequence, as illustrated by the synthesis of 1,2,3-cyclononatriene (Scheme 7).[161]

(50)

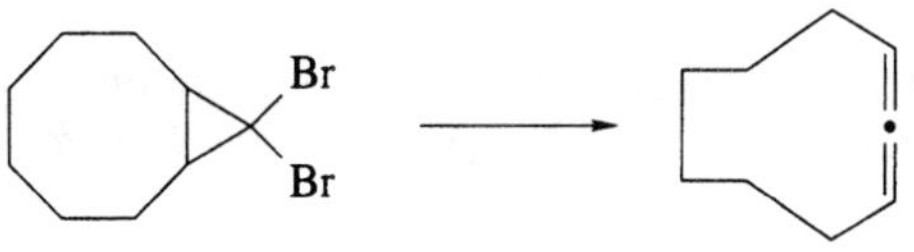

(51)

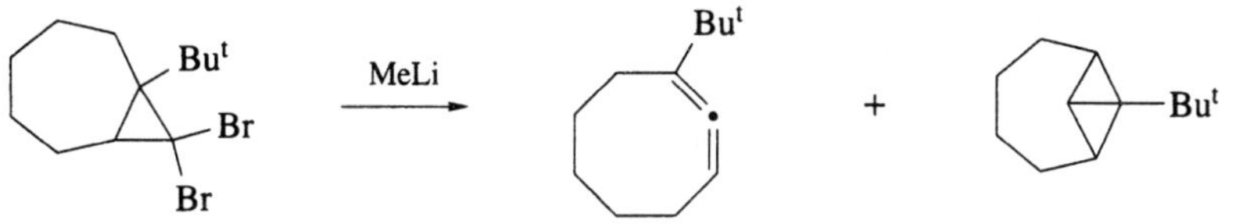

MeLi

(52)

(13) **(14)** **(15)**

MeLi CBr$_4$/MeLi MeLi

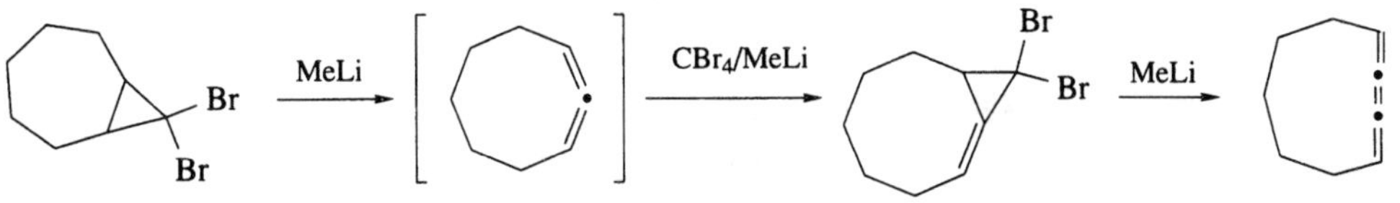

Scheme 7

The rigidity of the allene and its facile formation were imaginatively exploited in a synthesis of the vitamin E side chain alcohol, in which an overall transfer of 1,4 to 1,5 'acyclic stereoselection process' was achieved (Scheme 8).[162]

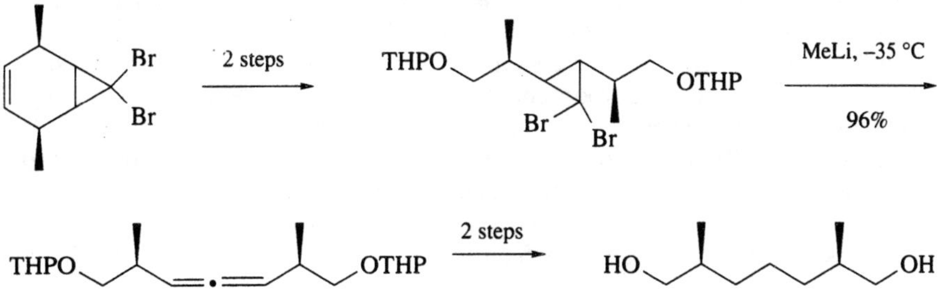

2 steps MeLi, –35 °C 96%

2 steps

Scheme 8

Interestingly, the alkene to allene conversion can be carried out directly without isolation of the intermediate dihalocyclopropane. This process involves the treatment of the alkene with 1 equiv. of carbon tetrabromide and 2 equiv. of methyllithium in ether at –65 °C.[163] Ultrasonic irradiation facilitates the formation of cyclopropylidenes, and therefore the allenes, from dihalocyclopropanes under the influence of Li, Na or Mg.[164] The reactions are usually complete in 5–15 min. A report[165] on the use of *n*-butyllithium complexed with the chiral tertiary amine (–)-sparteine, leading to optically active allenes, seems to be of questionable value.

It should be pointed out that while the dihalocyclopropane–allene conversion discussed here represents the best general synthesis of allenes, occasional problems due to the formation of side products have been reported. An example is the formation of the bicyclobutane (**16**) along with the expected allene (equation 53).[166] The method is especially unsuitable for the synthesis of tetrasubstituted allenes. Sterically demanding substituents on the dihalocyclopropanes are believed to be responsible for the change of course of the reaction.

Endocyclic allenes which contain other double bonds lead to rearranged products. An example is the reaction of dibromobicyclononatriene (**17**) with methyllithium to form initially a carbenoid (**18**) which

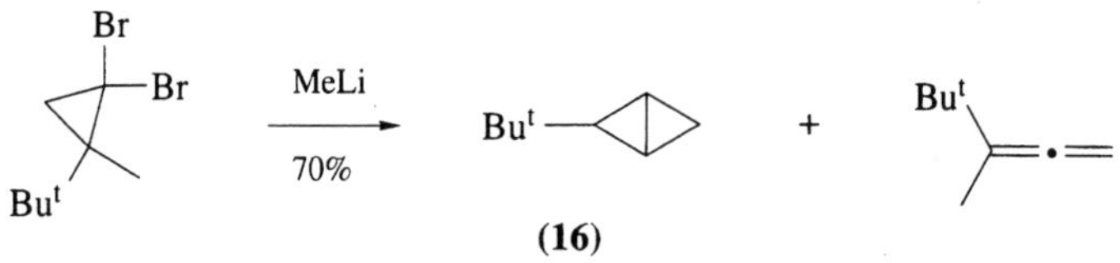

Br Br MeLi 70% Bu^t (16) + Bu^t (53)

rearranges to the nine-membered allenic triene (**19**). The latter rearranges to the bicyclic compound (**20**) and finally a [1,5] sigmatropic rearrangement of (**20**) yields indene (Scheme 9).[167]

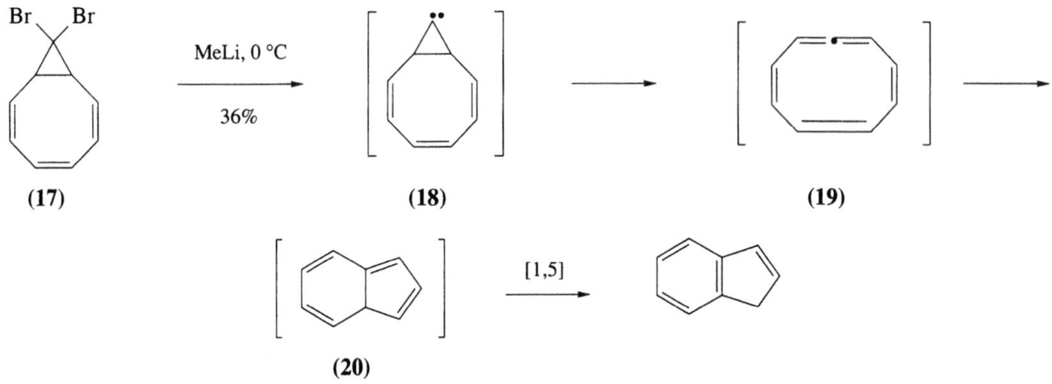

Br Br MeLi, 0 °C 36% (17) (18) (19)

[1,5] (20)

Scheme 9

4.7.3.4 Skattebøl Rearrangement

Vinyldihalocyclopropanes on treatment with methyllithium give rise to cyclopentadienes, formally resulting from a vinylcyclopropylidene–cyclopentylidene rearrangement followed by a rapid 1,2-hydrogen shift (Scheme 10). This unique carbene–carbene rearrangement is known as the Skattebøl rearrangement.[168] It occurs in competition with the cyclopropylidene–allene rearrangement. The Skattebøl rearrangement has been investigated thoroughly[169–172] and the earlier results as well as the mechanistic interpretations have been the subject of critical discussions.[173,174] It is now recognized that carbenoid species rather than free carbenes are intermediates in the rearrangement.[117,175] The conversion of the dibromobicyclo[4.1.0]heptene (**21**) to the norbornene derivative (**22**) is a good illustration of the Skattebøl rearrangement (equation 54).[169] Mechanistically the reaction is formulated as proceeding *via* the ionic rearrangement of the carbenoid (**23**) to the ion pair (**24**), and the conversion of the latter by sequential reaction with methyllithium and methyl bromide (Scheme 11).[175] The stereospecificity in the product formation is consistent with this interpretation and argues strongly against the involvement of free carbene. Two other representative examples of the Skattebøl rearrangement are given in equations (55)[176] and (56).[177]

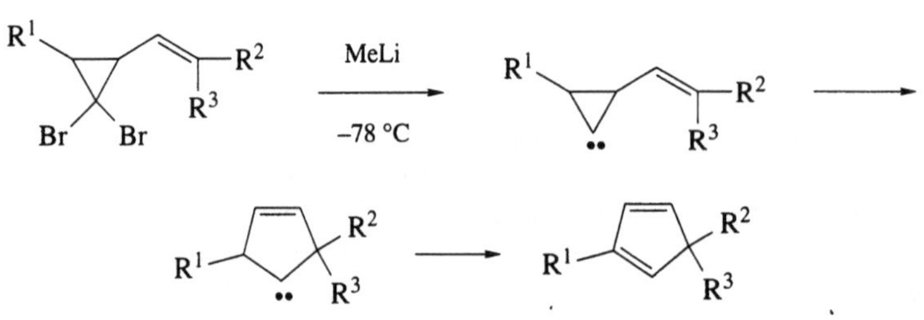

R^1 R^2 R^3 Br Br MeLi −78 °C R^1 R^2 R^3

R^1 R^2 R^3 R^1 R^2 R^3

Scheme 10

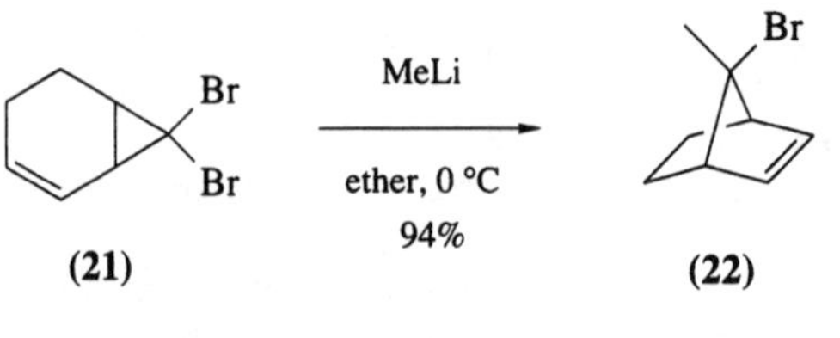

Br Br MeLi ether, 0 °C 94% Br (21) (22) (54)

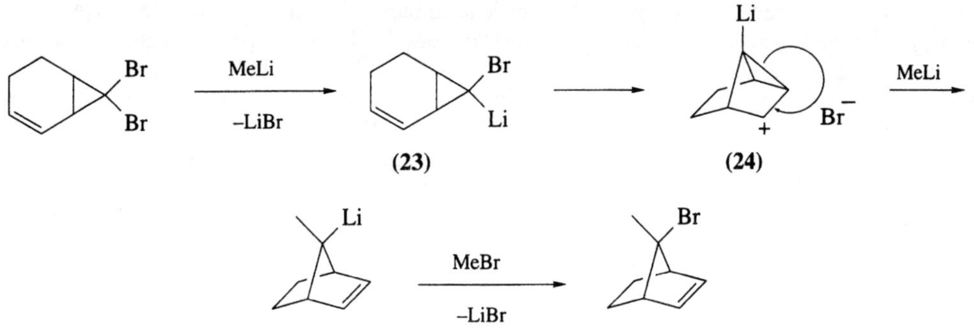

MeLi, –LiBr

(23)

(24)

MeLi

MeBr, –LiBr

Scheme 11

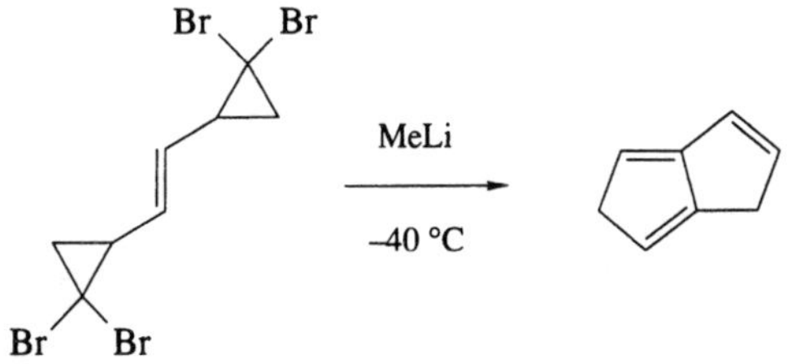

MeLi, –40 °C (55)

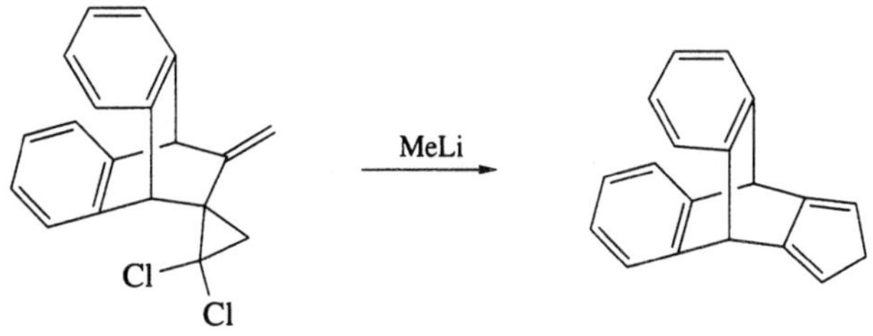

MeLi (56)

4.7.3.5 Other Carbenoid Reactions

When the carbene or carbenoid resulting from a dihalocyclopropane is unable to rearrange to the allene due to steric or other factors, insertion or addition reactions characteristic of carbenes take place. Thus dibromonorcarane on reaction with methyllithium gives a bicyclobutane derivative by insertion of the carbene into a β-C—H bond (equation 57).[178] Allene formation is sterically unfavorable in this case. Similarly, dibromotetramethylcyclopropane gives 1,2,2-trimethylbicyclo[1.1.0]butane instead of tetramethylallene (equation 58).[179–181] An example involving a tricyclic dibromocyclopropane is given in equation (59).[182]

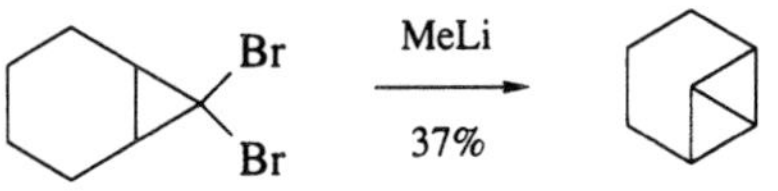

MeLi, 37% (57)

(58)

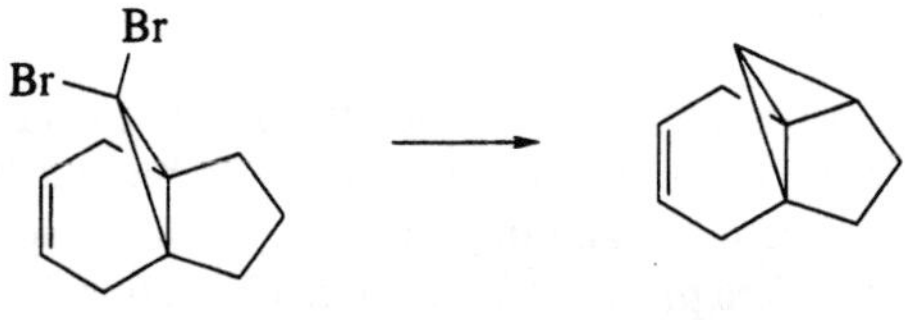

(59)

In general, tetrasubstituted dibromocyclopropanes lead to bicyclobutanes predominantly, whereas mixtures of bicyclobutanes and allenes result from trisubstituted dihalocyclopropanes and those bearing bulky substituents (see also Section 4.7.3.3).

In addition to insertion into β-C—H bonds, cyclopropylidenes can undergo other reactions such as alkylation (*cf.* Section 4.7.3.2), dimerization, insertion into C—H bonds of the ether solvent (equation 60)[183] or reaction with alkenes to afford spirocyclopropanes (equation 61).[184] Addition of stoichiometric amounts of Bu^{t}OK has been shown to promote the reactions of lithium carbenoids, even at –85 °C, with THF to give the insertion product (equation 62).[185] Addition to alkenes is also promoted under these conditions. Intramolecular addition of the carbenoid to double bonds has been exploited in the synthesis of spirotricyclic compounds (equation 63).[186]

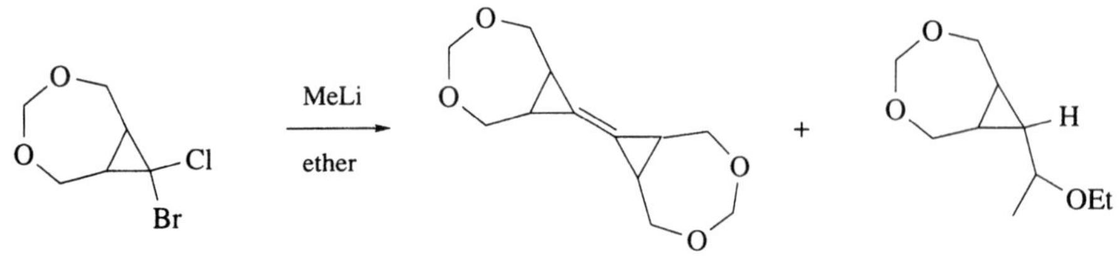

MeLi, ether (60)

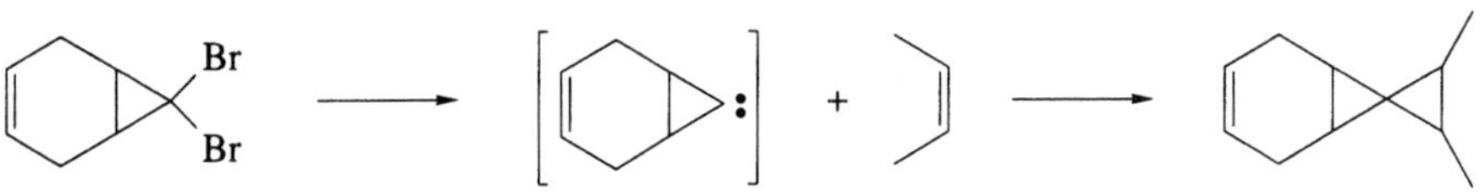

(61)

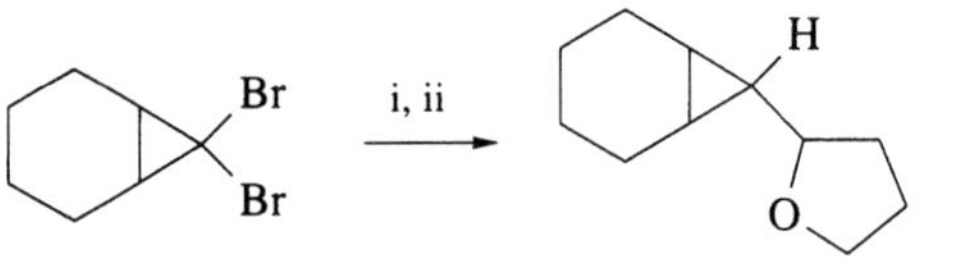

i, ii (62)

i, BuLi, THF, –85 °C; ii, Bu^{t}OK

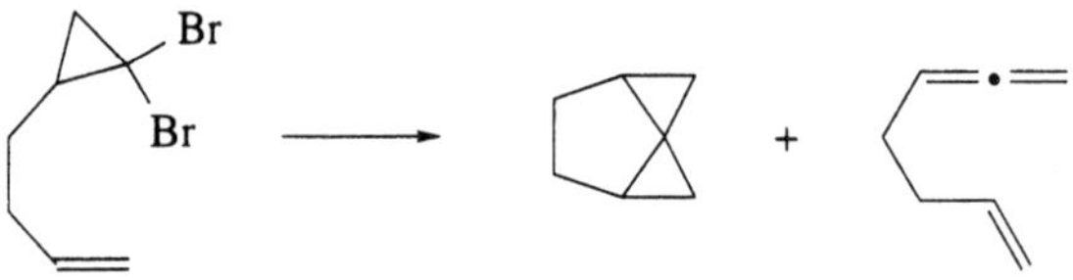

(63)

Insertions into O—H bonds, N—H bonds and C—H bonds adjacent to oxygen and nitrogen have found use in the synthesis of a number of heterocycles (see Section 4.7.3.9).[187–189] The reaction of alcohol (**25**) with methyllithium in ether, leading to the ketone (**26**), has been interpreted[190] as involving the insertion of an intermediate carbenoid into a β-C—H bond followed by ring opening (equation 64).

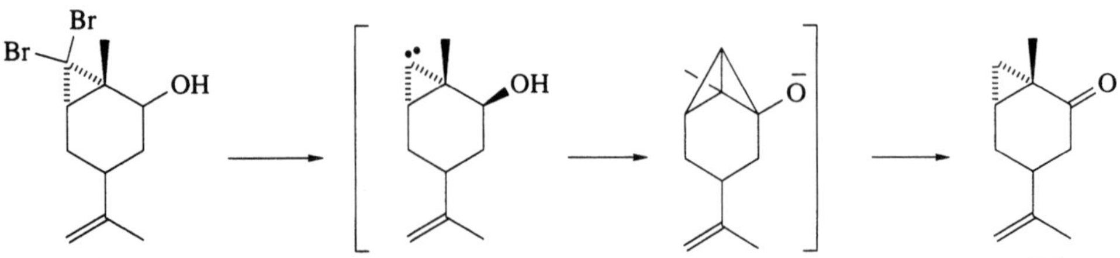

(**25**) (**26**) (64)

4.7.3.6 Elimination and Elimination–Addition

Elimination of hydrogen halide from dihalocyclopropanes occurs under the influence of strong bases, leading to halocyclopropenes. Simple halocyclopropenes are unstable but can be trapped by nucleophiles (Scheme 12).[191,192] In situations where hydrogen migration is possible, isomerization of the chlorocyclopropene to alkylidenecyclopropane takes place readily. Thus (**27**; equation 65) is an excellent precursor for vinylmethylenecyclopropane.[193–195] The latter is a valuable synthon for methylenecyclopentene. The double-bond migration attending dehydrohalogenation by Bu^{t}OK has been utilized in the synthesis of benzocyclopropene,[196] a remarkably stable compound, and its analogs (equation 66). The dehydrohalogenation of (**28**) gives a mixture of (**29**) and (**30**) in a ratio of 3:2, presumably *via* a carbene intermediate (equation 67).[197] A related reaction occurs with (**31**), as shown in equation (68) (*cf.* Section 4.7.3.9).[198]

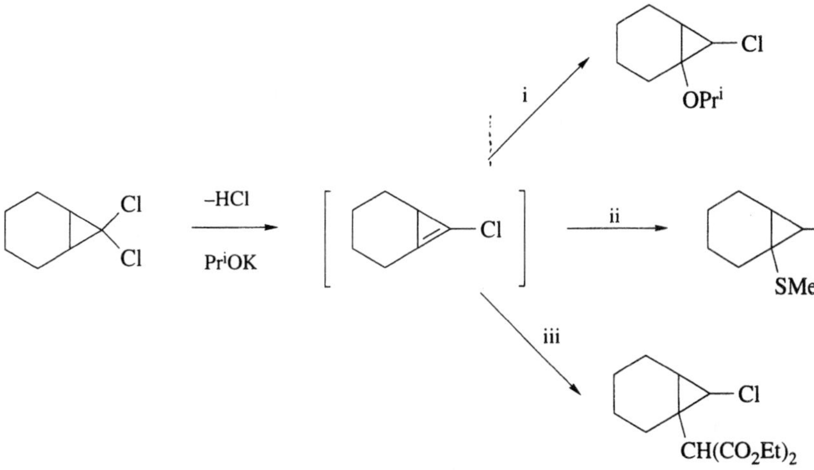

–HCl, Pr^{i}OK; i; ii; iii

Cl Cl Cl Cl OPr^i SMe $CH(CO_2Et)_2$

i, Pr^{i}OH; ii, MeSH; iii, $^-CH(CO_2Et)_2$

Scheme 12

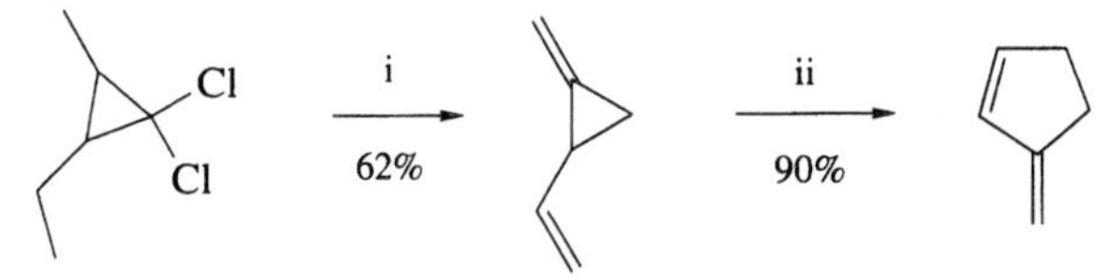

Cl Cl i 62% ii 90% (65)

(**27**)

i, Bu^{t}OK, DMSO; ii, 80 °C, 13 h

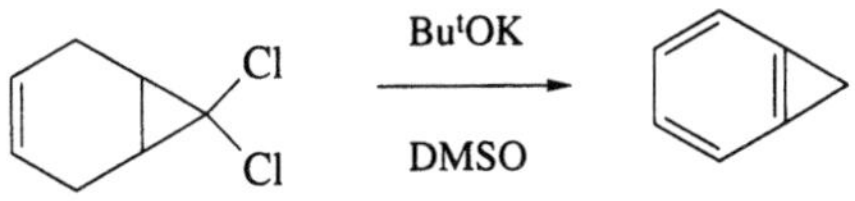

Cl Cl Bu^{t}OK DMSO (66)

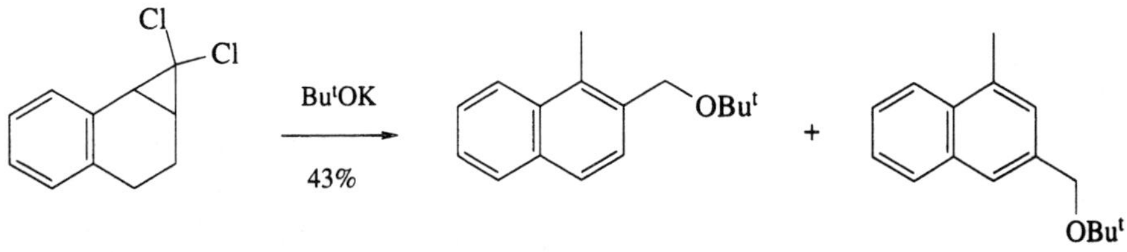

Cl Cl Bu^{t}OK 43% OBu^t + OBu^t (67)

(**28**) (**29**) (**30**)

In cases such as the one shown in equation (69), what appears to be displacement is more likely a multistep elimination–addition process.[37] Carbonyl or sulfonyl groups in the 2-position greatly facilitate each of these two steps. Thus (**32**) on treatment with alkoxide or thiolate gave the cyclopropane acetal or thioacetal, respectively (equation 70).[199] In related work, tetrachlorocyclopropene, which has interesting synthetic potential, is formed from pentachlorocyclopropane by treatment with potassium hydroxide (equation 71).[200] Certain 1,1,2-trihalocyclopropanes on reaction with methyllithium in ether at low temperature afford the corresponding halocyclopropenes by 1,2-dehalogenation (equation 72).[201] Similarly,

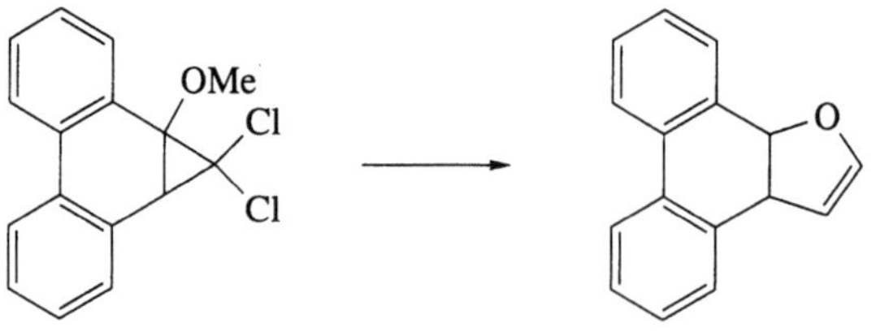

(68)

(31)

2-bromo-2-trimethylsilyldibromocyclopropane on reaction with tetra-*n*-butylammonium fluoride undergoes elimination to give 1,2-dibromocyclopropene (equation 73).[202]

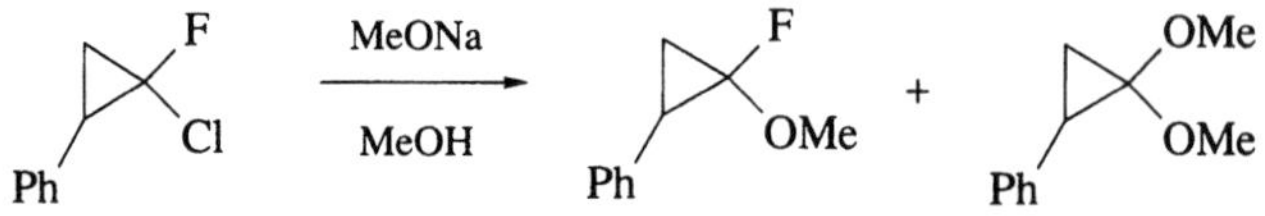

(69)

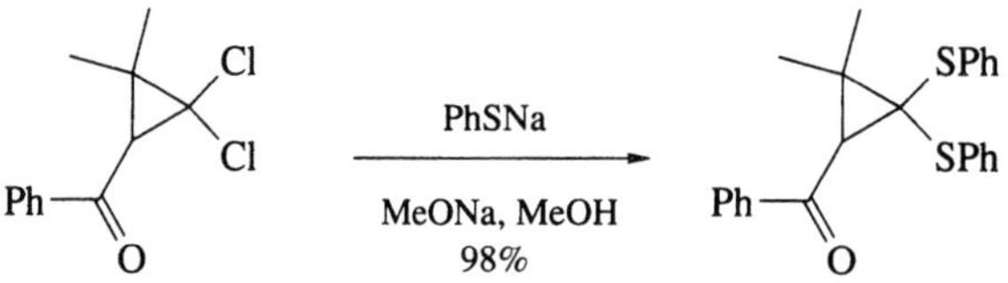

(70)

(32)

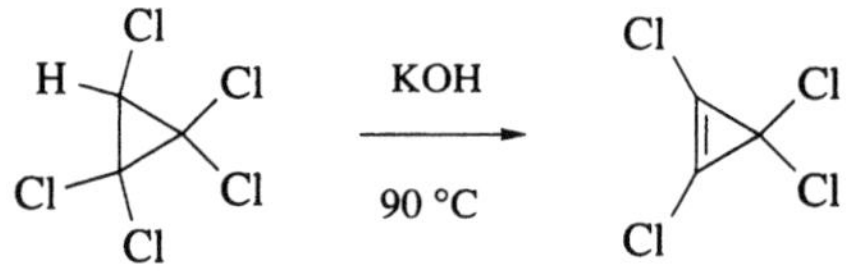

(71)

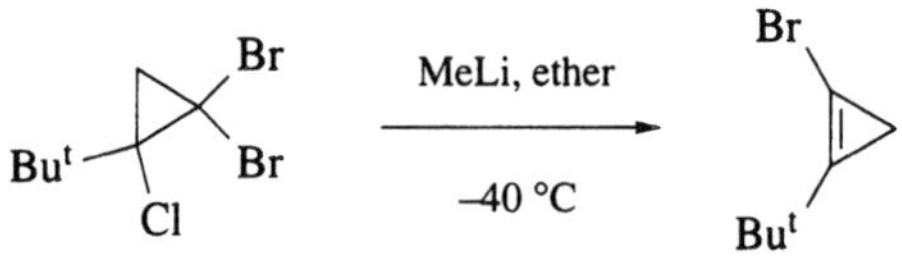

(72)

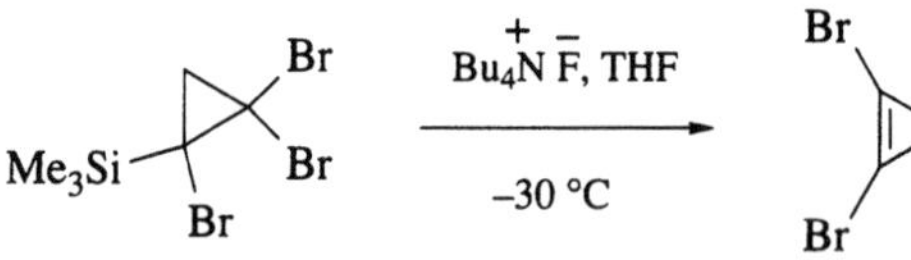

(73)

4.7.3.7 Electrocyclic Ring Opening

Shortly after the advent of the Doering reaction, it was recognized that dihalocyclopropanes derived from certain cyclic alkenes can undergo facile thermal rearrangement to ring-opened products.[6,9,10] The first example of this rearrangement was the rapid conversion of the dichlorocarbene adduct of indene to 2-chloronaphthalene, observed by Parham (equation 74).[6,203] A number of dihalocyclopropanes, however, are thermally stable and the facility of the rearrangement depends on factors such as ring size, stereochemical disposition and the nature of the halogens. In addition to the thermal rearrangement, the cyclopropyl–allyl conversion can be facilitated by electrophiles such as silver ion, protic and Lewis acids, and by nucleophiles. On the basis of theoretical considerations, the rearrangement has been characterized as a concerted process proceeding *via* stereospecific disrotatory opening.[204–209] The ring opening is concomitant with the departure of the leaving group, leading to a transition state where the positive charge is delocalized (equation 75). Important sterochemical consequences of the concerted nature of the rearrangement include the formation of the *cis* and *trans* allyl cations from the *endo* and *exo* derivatives (**33**; equation 76) and (**34**; equation 77), respectively. The electrocyclic ring opening of dihalocyclopropanes has been exploited in the synthesis of one-carbon ring-expanded carbocycles and heterocycles and chain-extended compounds.

(74)

(75)

(33) (76)

(34) (77)

4.7.3.7.1 Ring expansion/chain extension

It has been already noted (Section 4.7.3.7) that the facility of the dihalocyclopropane–allyl cation rearrangement depends on a number of factors, including the stability of the allyl cation, which in turn is influenced by electronic factors. The addition of dichlorocarbene to cyclopentadiene to afford chlorobenzene *via* the unstable adduct (**35**) and the relatively stable cation (**36**) is illustrative (equation 78).[210] In the absence of external nucleophiles, the departing halide ion is usually captured by the allyl cation to form a haloene. Alternatively, the cation loses a proton to form a halodiene. Even the substitution product itself is thermally unstable and can lose hydrogen halide at higher temperatures to give the halodiene. For instance, when *endo*-6-chloro-*exo*-6-fluorobicyclo[3.1.0]hexane is heated in quinoline at 110 °C 3-chloro-2-fluorocyclohexene is obtained which at higher temperatures loses HCl to give 2-fluoro-1,3-cyclohexadiene (equation 79).[211] This example also illustrates the effect of stereochemistry and the nature of the halogens on the rearrangement. Not surprisingly, the cyclopropyl–allyl rearrangement of dihalocyclopropanes has found a number of synthetic applications. Selected examples are given in equations (80) to (92). The adducts resulting from the addition of a dihalocarbene to the internal double bond of bicyclic systems lead to rearrangement products *via* bridgehead alkenes. This process has been exploited in the synthesis of [6]- and even [5]-metacyclophanes (equation 80).[212,213] The dibromocarbene adduct (**37**) on heating in acetonitrile gives (**38**), which has found use in the synthesis of 3-phenyl-5-bromooctavalene (equation 81).[214] The diadduct (**39**) on thermal rearrangement affords a mixture of dichloroazulenes (equation 82).[215] Such rearrangements have been employed in a number of similar systems.[216]

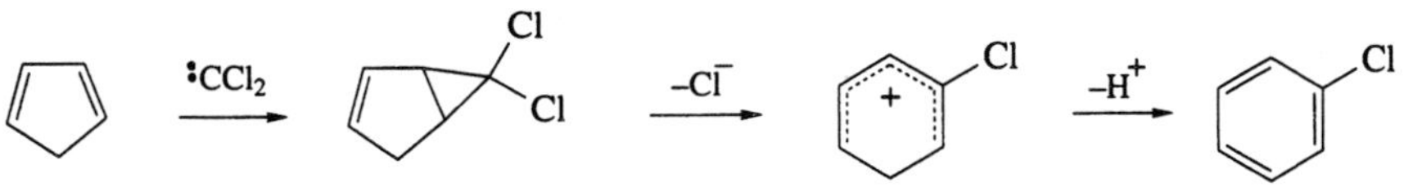

(35) (36) (78)

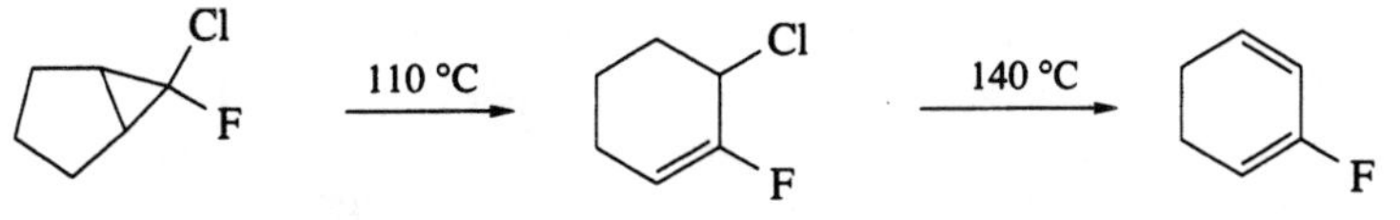

(79)

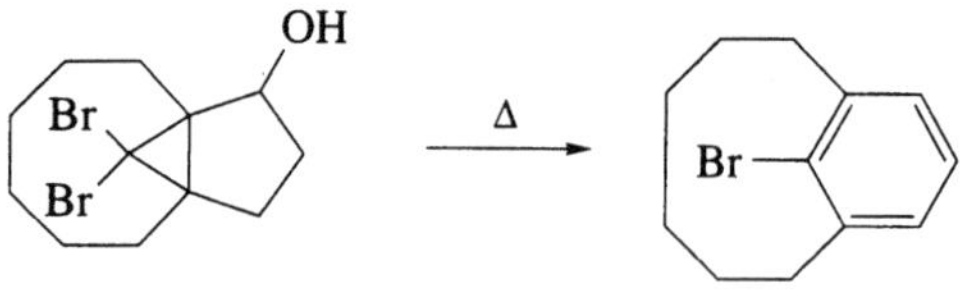

(80)

(81)

(37) **(38)**

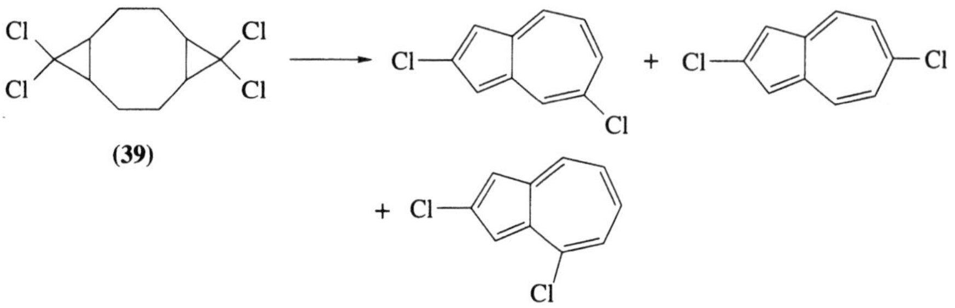

(82)

(39)

Acid-catalyzed hydrolysis followed by ring opening of the dichlorocyclopropane adduct (**7**) proceeded smoothly on treatment with aqueous acetic acid buffered with sodium acetate to afford the chlorocycloheptenone (**40**), an intermediate in the stereoselective synthesis of Prelog–Djerassi lactonic acid (equation 83).[80] Similar ring expansion has been used in the synthesis of muscone and tropolones such as thujaplicins.[81–83] A recent synthesis of α-tropolone[217,218] involves the ring expansion attended by the loss of hydrogen halide when a dihalocarbene adduct, such as (**41**), is subjected to Albright–Swern oxidation conditions (equation 84). The method is also quite suitable for the synthesis of open chain α-chloroenones. An efficient and somewhat related cyclopentenone synthesis[219] involves the acid-catalyzed rearrangement of the dichlorocarbene adduct (**42**), as shown in equation (85).

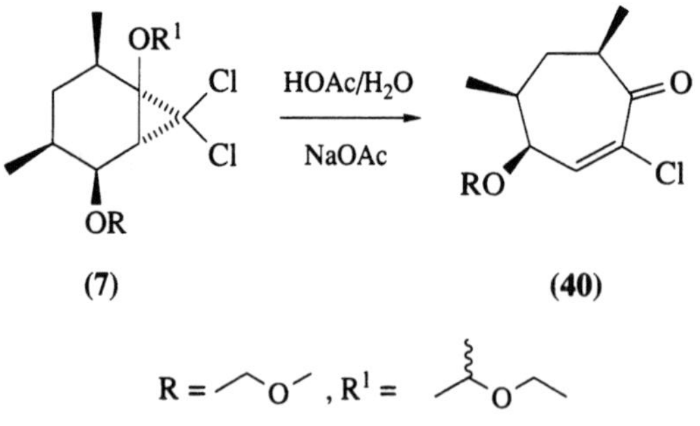

(83)

(7) **(40)**

(84)

(41)

The silver ion assisted carbon–halogen bond cleavage and the unraveling of the cyclopropane ring by the cyclopropyl–allyl rearrangement was first noted in the formation of 2-bromocyclohexen-1-ol from dibromobicyclo[3.1.0]hexane under solvolytic conditions (equation 86).[220] The silver ion assisted solvolysis of the dihalocyclopropane adduct (**43**), derived from a Birch reduction product, smoothly rearranges to the tropone (equation 87).[221] A number of other synthetic applications[222–226] have been reported

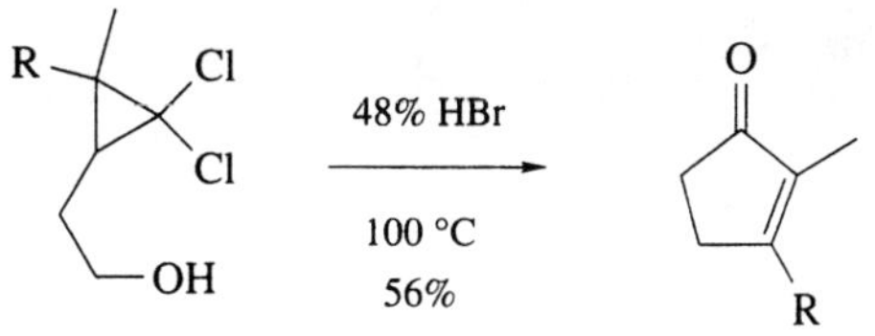

(85)

(42)

recently. Analogous to the formation of a [5]metacyclophane,[213,223] the dibromocarbene adduct (**44**) undergoes silver ion assisted solvolysis to give (**45**) in high yield (equation 88).[224] Similarly, (**46**) has been reported[225] to give (**47**), albeit in low yield (equation 89). It should be noted that (**46**) is thermally stable.

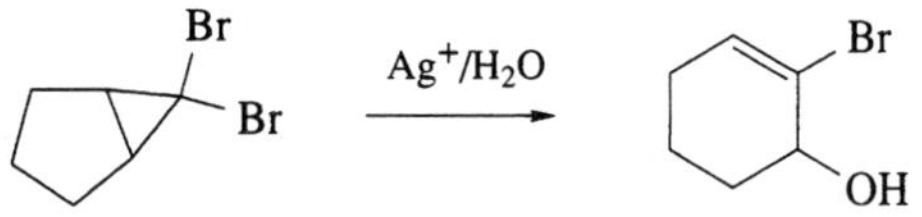

(86)

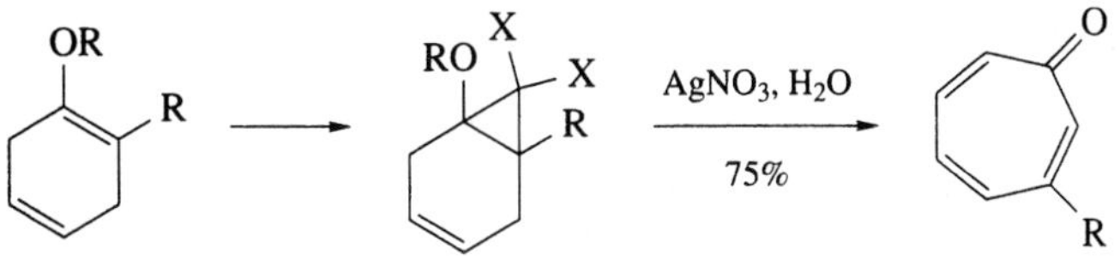

(87)

(43)

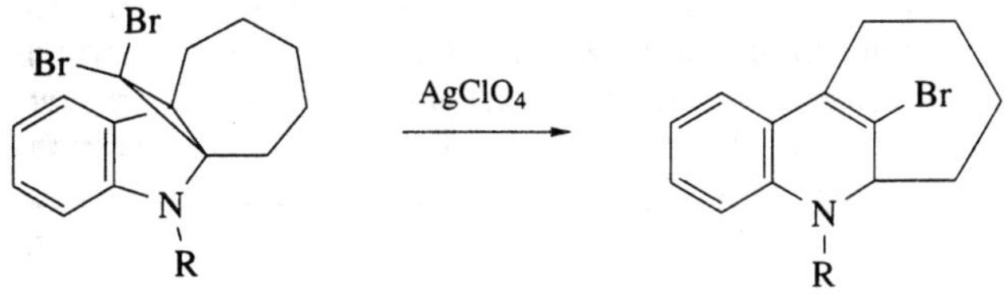

(88)

(44) R = t-BOC **(45)**

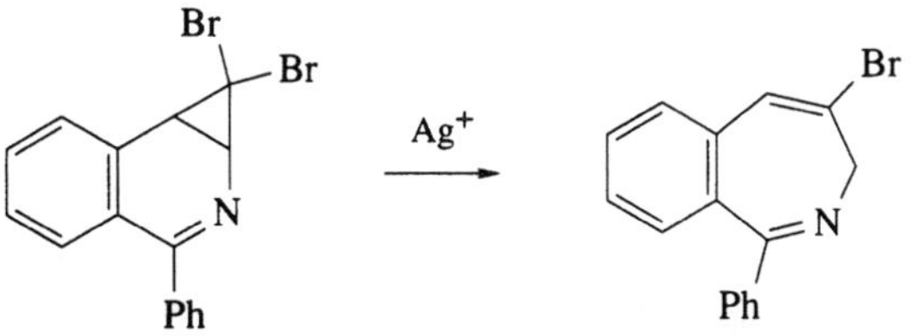

(89)

(46) **(47)**

In certain cases, the silver ion assisted solvolysis fails to initiate electrocyclic reaction to give ring-expanded compounds, leading instead to a cleavage product (equation 90).[227] The dichlorocarbene adduct (**48**) on $FeCl_3$-promoted cleavage in ether affords exclusively α-trichloromethylcyclohexanone, whereas in DMF a small amount of chlorocycloheptenone also was obtained (equation 91).[228]

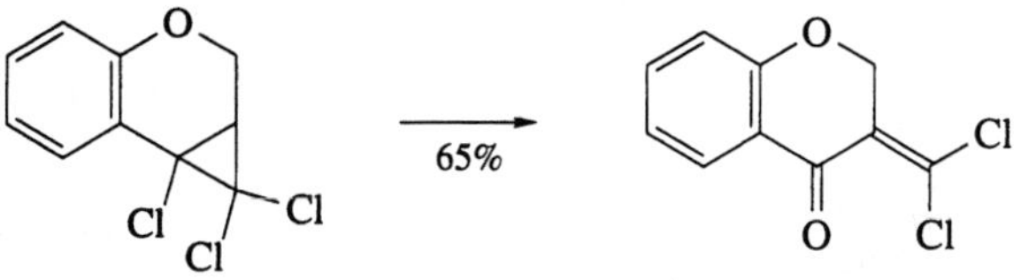

(90)

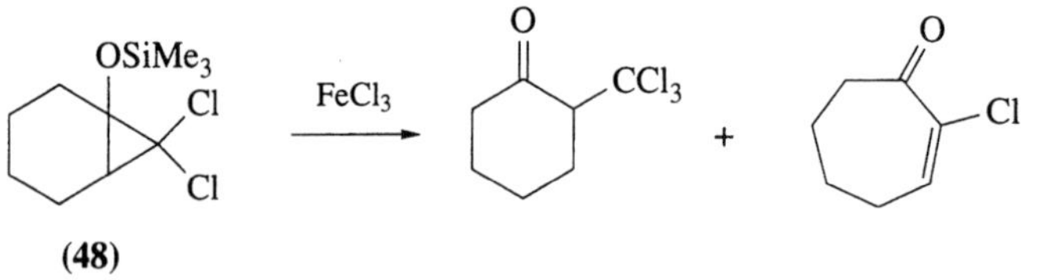

(48) (91)

Not much is known about the opening of dihalocyclopropanes by Lewis acids beyond reports on the reaction of dibromo- and dichloro-cyclopropanes with aromatic hydrocarbons under the influence of $AlCl_3$ or $FeCl_3$ leading to indenes.[229–231]

The cyclopropyl–allyl rearrangement has been shown to proceed with nucleophilic assistance,[87,232,233] and the intermediate allyl cation can be trapped by nucleophiles leading to synthetically useful derivatives. An example is the formation of an unsaturated acetal and the propiolic acid ortho ester (equations 92 and 93).[232]

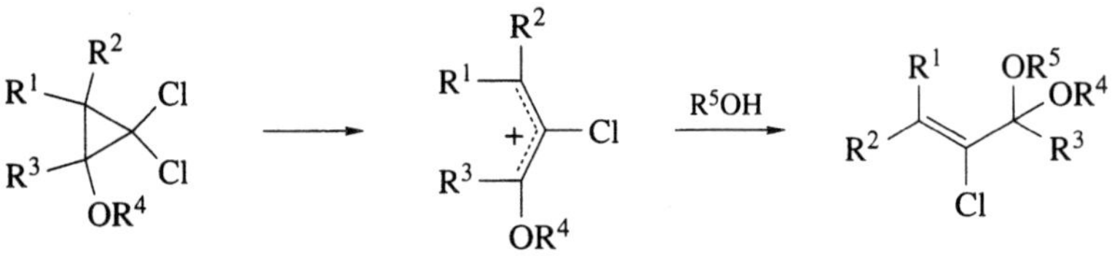

(92)

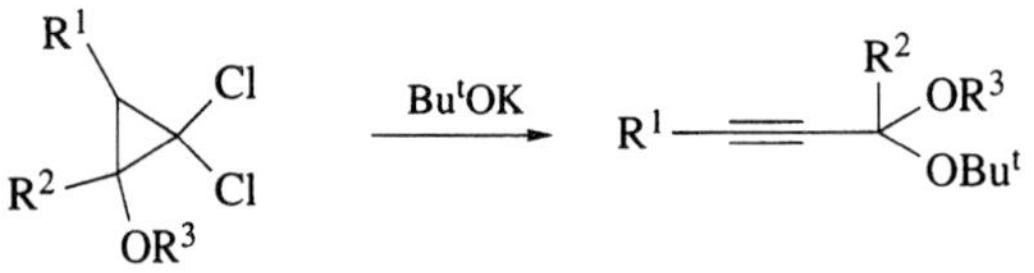

(93)

A stereospecific synthesis of conjugated dienes involving the ring opening of difluorocyclopropanes has been reported.[234] In an extension of this work, 1-acetoxy-3-alkyldifluorocyclopropanes undergo ring opening with LAH to give β-fluoroallylic alcohols with high stereospecificity and in good yields (equation 94).[235] Similarly, chlorofluorocyclopropyl ethers undergo rearrangement to give α-fluoroacrolein[236] *via* its acetal and chlorofluoro-2-(trimethylsilyl)methylcyclopropane gives 2-fluoro-1,3-butadiene in high yield.[237]

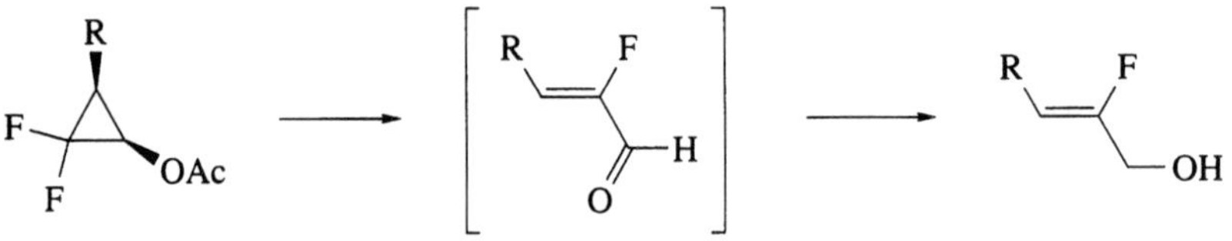

(94)

The C-ring expansion of aporphines *via* the dichlorocyclopropane ring opening[238] induced by LAH has been utilized in a synthesis of homoaporphine alkaloids (equation 95). Reductive cleavage of 2-silyloxy-dibromocyclopropanes gives α,β-unsaturated ketones directly (equation 96).[239]

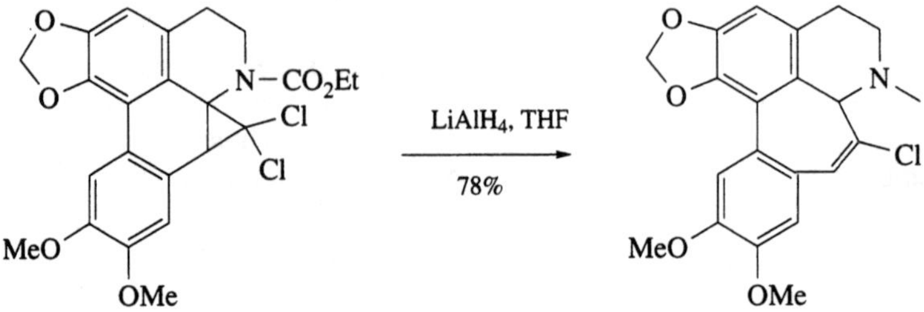

(95)

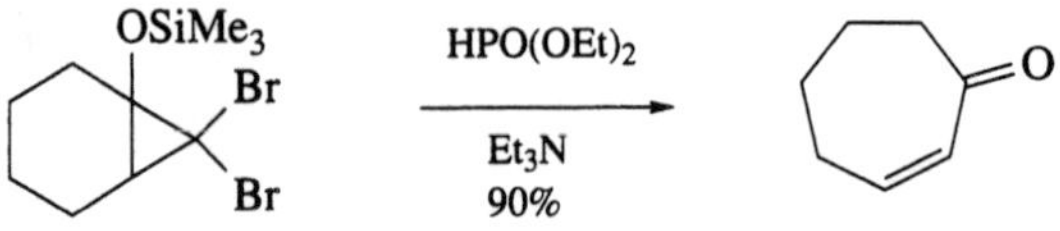

(96)

4.7.3.8 Solvolysis without Ring Opening

It is clear from the discussion in the previous section that generally dihalocyclopropanes undergo solvolysis with concomitant electrocyclic ring opening. However, dihalocyclopropanes with a barrier to ring opening have been shown to undergo silver ion assisted solvolysis to give both substituted cyclopropanes and ring-opened products (equation 97).[211,212,240] The [4.4.1]propellane undergoes solvolysis without ring opening (equation 98). Similar results were obtained with [3.3.1]propellanes.[241] Dihalocyclopropanes that carry a β-TMS group undergo a unique solvolytic displacement of the halogens without rupture of the cyclopropane ring when treated with silver trifluoroacetate and alcohol (equation 99).[242] Finally, it has been shown that halogens can be displaced by nucleophiles under photolytic conditions.[243] A mixture of mono- and di-substitution products are obtained (equation 100).

Br; Br; CF_3CO_2Ag; Br; $OCOCF_3$; +; $OCOCF_3$; Br (97)

Br; Br; Br; $OCOCF_3$ (98)

$SiMe_3$; Br; Br; CF_3CO_2Ag; MeOH; $SiMe_3$; OMe; OMe (99)

Br; Br; $PhZ^-/h\nu$; H; ZPh; +; ZPh; ZPh (100)

Z = S or Se

4.7.3.9 Formation of Heterocycles

Although the use of dihalocyclopropanes in the synthesis of heterocycles has not been investigated in great detail, there are a number of examples which illustrate the usefulness in heterocyclic synthesis. The earliest example is the conversion of pyrrole to 3-chloropyridine with chloroform and strong base (*cf.* Section 4.7.2.2.2), a reaction now recognized as involving the formation and electrocyclic rearrangement of a dihalocyclopropane (equation 101).[75,76,244,245] The yield of 3-chloropyridine is influenced by the source of dichlorocarbene and the formation of significant amounts of side products limits the usefulness of the reaction. The formation of 3-chloroquinolines from indoles by the reaction of chloroform and sodium ethoxide can be considered to proceed in an analogous manner.[75,77,246–248] This reaction is also complicated by side products resulting from the Reimer–Tiemann reaction (equation 102).[246] Pyrrole reacts with dichlorocarbene in the gas phase to give a mixture of 2-chloro- and 3-chloro-pyridines.[249] Imidazole under similar conditions affords a mixture of chloropyrazine and 5-chloropyrimidine (equation 103).[250]

N–H; $:CCl_2$; [Cl; Cl; N]; Cl; N (101)

In what can be viewed as a heterocyclic version of the Skattebøl rearrangement (*cf.* Section 4.7.3.4), the dibromocarbene adduct (**49**) on treatment with methyllithium affords the pyrrole (**50**), albeit in low yield (equation 104).[251] The propensity of the carbenoid species resulting from dihalocyclopropanes to insert into C—H bonds adjacent to oxygen and nitrogen was mentioned previously (Section 4.7.3.5).

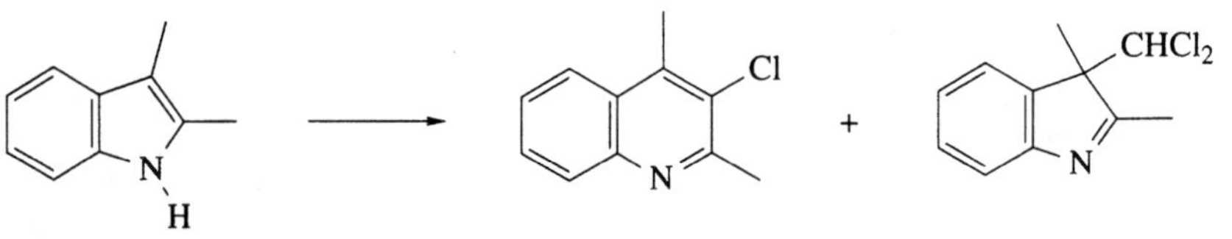

(102)

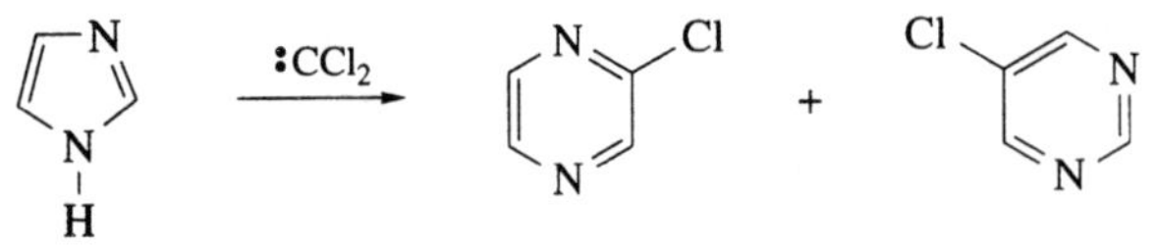

$:CCl_2$ (103)

This process has found occasional use in the synthesis of heterocycles. The major product resulting from the reaction of MeLi on the dibromocarbene adduct (**51**) is the tetrahydrofuran (**52**), as shown in equation (105).[189] In a novel approach to the synthesis of spiroacetal pheromones, carbene insertion into an acetal C—H bond was studied (equation 106).[252] Unfortunately, the reaction proceeds in low yield and the approach is further hampered by the lack of stereocontrol in the ring opening of the cyclopropane. Carbene insertions into N—H and C—H bonds adjacent to nitrogen have been shown to give azabicyclo systems, as shown in equations (107) and (108).[188]

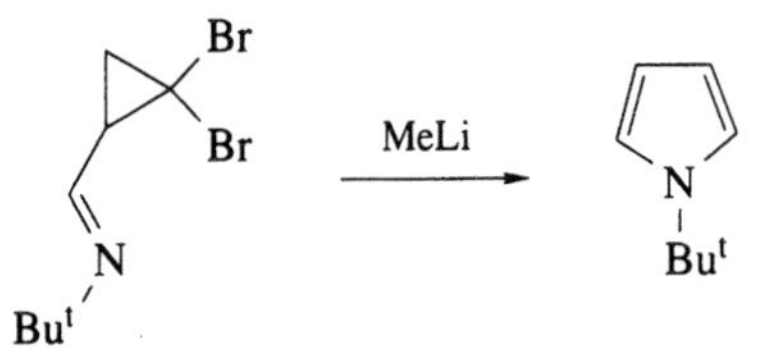

MeLi (104)

(**49**) (**50**)

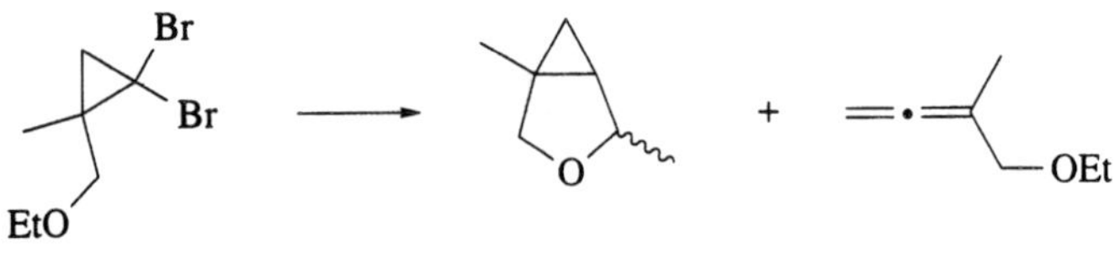

(105)

(**51**) (**52**)

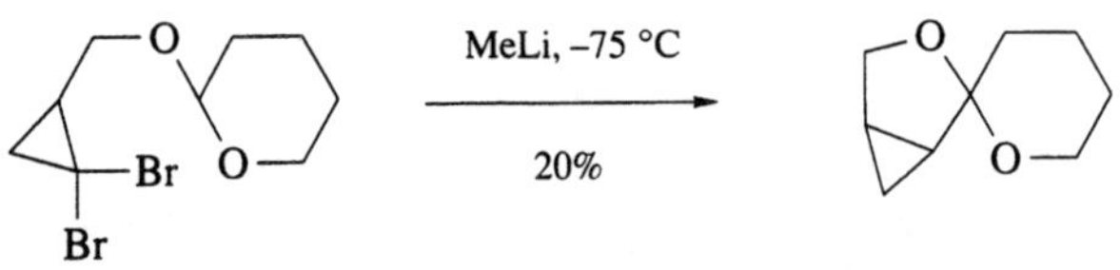

MeLi, –75 °C; 20% (106)

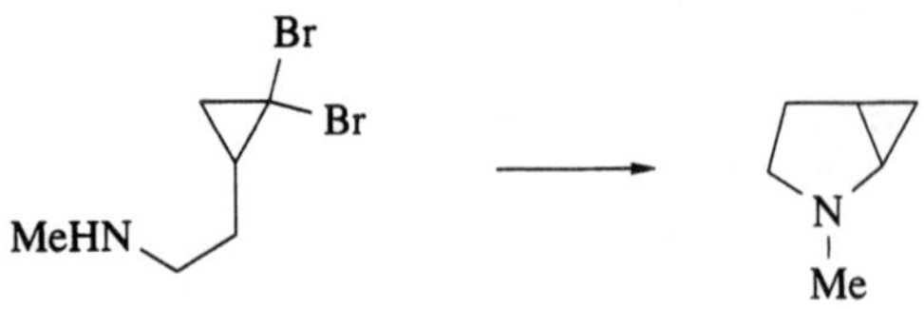

(107)

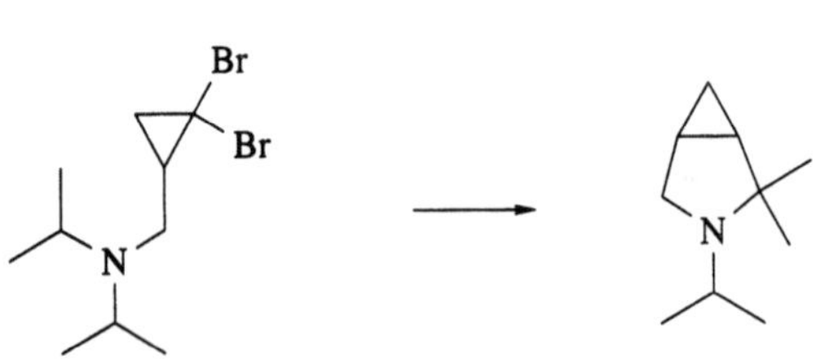

(108)

A crucial step in the synthesis of a methyleneaziridine involves the nucleophile-induced ring opening of a dibromocyclopropane (equation 109).[253] Nucleophile-assisted opening of a dichlorocyclopropyl ketone leads to 2,2-dimethoxy-2,3-dihydrofurans which are useful γ-keto aldehyde synthons (equation

110).[254] The reaction of acyloxydihalocyclopropanes with hydrazine has been reported to yield pyrazoles.[255]

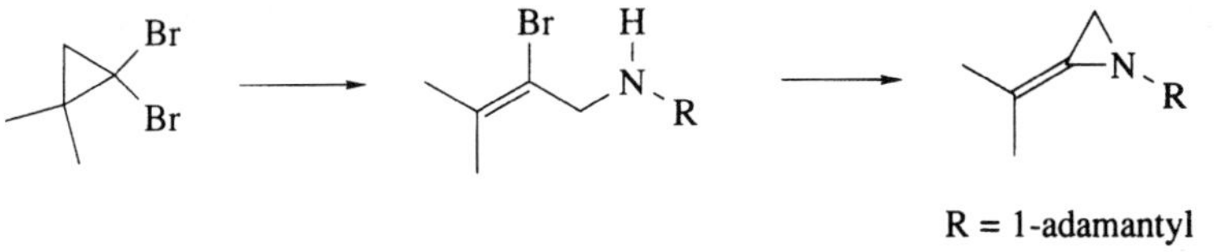

(109)

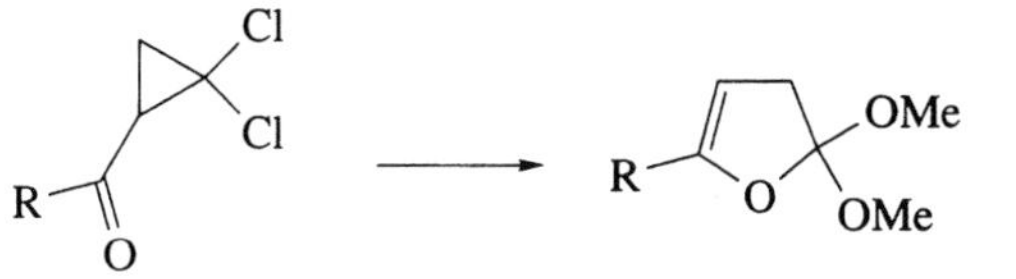

R = 1-adamantyl

(110)

The allyl cation generated by the electrocyclic cleavage of dibromocyclopropanes (*cf.* Section 4.7.3.7.1) has been trapped by a carboxyl group in a highly efficient synthesis of furanones (equation 111) and pyranones.[256] Acid-catalyzed openings of dihalocyclopropanes also give similar results.[257]

(111)

An elegant synthesis of the neurotoxic alkaloid anatoxin has exploited the electrocyclic opening of the dibromobicyclo[5.1.0]octane followed by transannular cyclization (Scheme 13).[258] Similarly, the thermal electrocyclic opening of the dichlorocyclopropane followed by intramolecular trapping of the developing allylic cation by a suitably positioned amine has been used in a homoaporphine synthesis.[238]

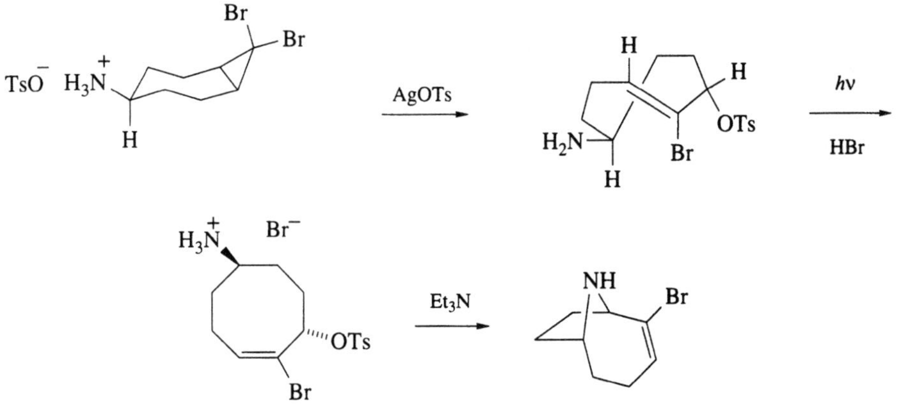

Scheme 13

4.7.3.10 Miscellaneous Reactions

Dihalocyclopropanes elicit a number of reactions which cannot be conveniently accommodated in the above categories and these are outlined in this section. Photolysis of the bis(dichlorocarbene) adduct (**53**) gave only the ring cleavage products (equation 112).[259] The vinyldichlorocyclopropane (**54**) has been shown to undergo thermal rearrangement to give the dichlorocyclopentene (**55**; equation 113).[260] Electrolysis of dihalocyclopropanes has been reported to give ring-opened products (equation 114).[261]

Dibromocyclopropanes with vicinal chloromethoxy or mesyloxymethyl substituents undergo [$Ni(CO)_4$]-induced ring opening–carbonylation in the presence of alcohol or amine, leading to γ,δ-unsaturated carboxylic acid derivatives selectively *via* intermediate nickel enolates (equation 115).[262] Di-

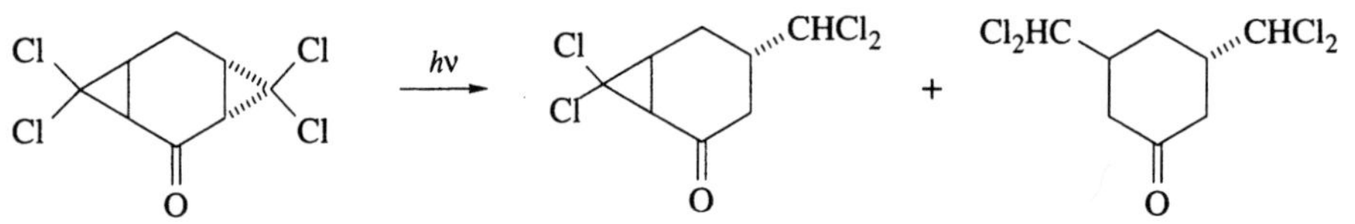

(53)

(112)

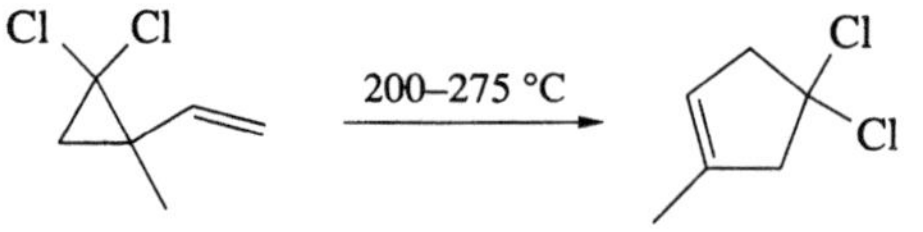

(54) (55)

(113)

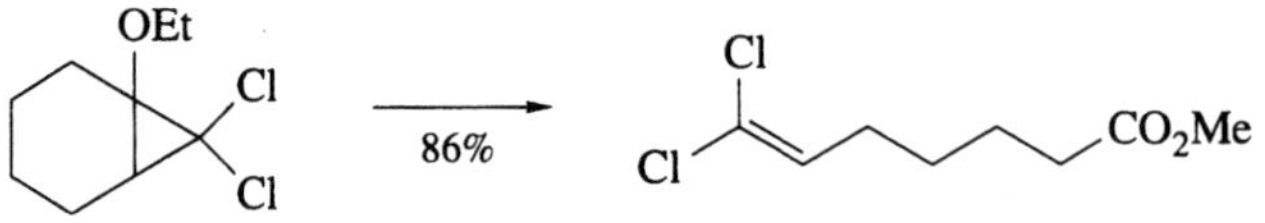

(114)

bromo-2-chlorocyclopropanes also undergo a similar ring-opening carbonylation with [$Ni(CO)_4$] to give a mixture of β,γ-unsaturated esters, along with a small amount of the reduced cyclopropane derivative (equation 116).[263] The same authors have shown that [$Ni(CO)_4$]-induced reductive carbonylation of dibromocyclopropanes is a versatile method for the synthesis of cyclopropanecarboxylic acid derivatives (equation 117).[264]

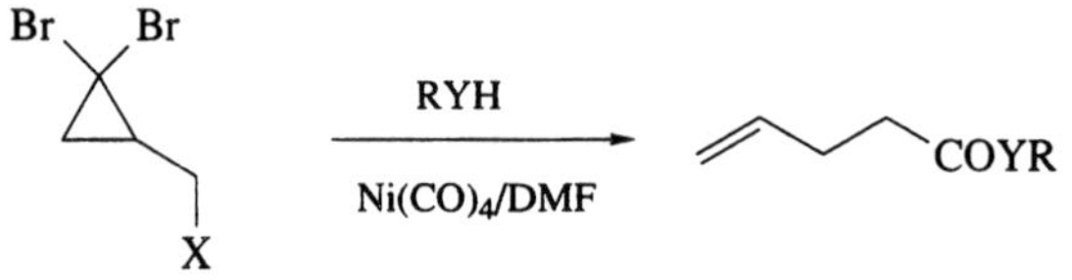

X = Cl, OMs Y = O, NH, NR

(115)

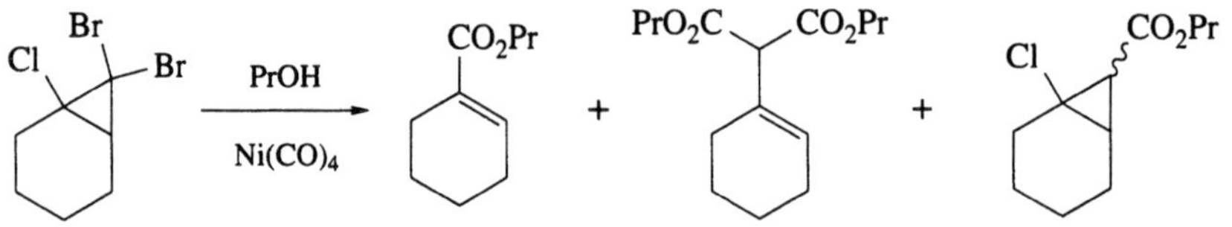

(116)

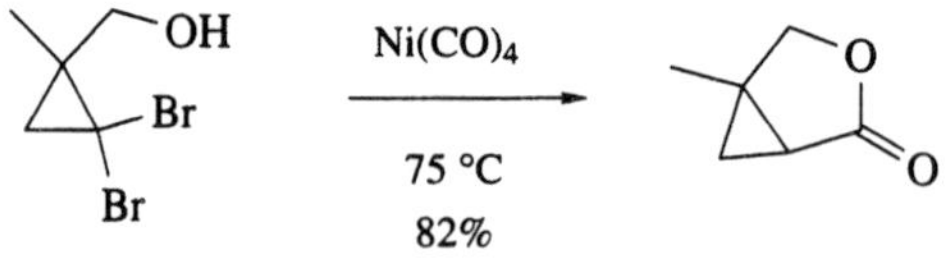

(117)

In a somewhat intriguing sequence of reactions, the organometallic compound (**56**) on treatment with methyllithium afforded the rearranged product (**57**), as shown in equation (118).[265] While the mechanism of this transformation is not known, it has been speculated that the initially formed carbene rearranges to a spirocyclopropene intermediate *via* a novel ring contraction.

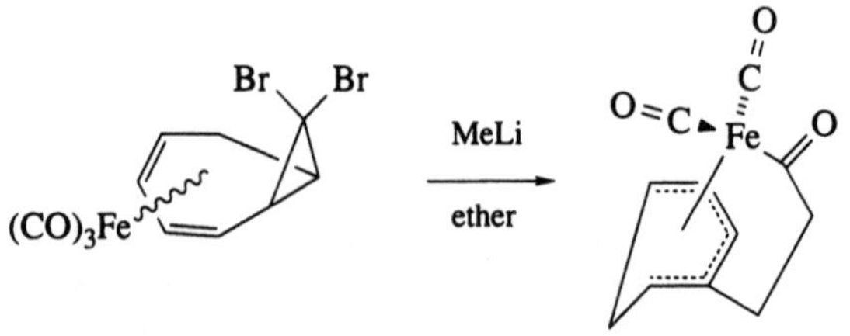

(56) (57)

(118)

4.7.4 CONCLUSION

It is abundantly clear from the preceding discussion that dihalocyclopropanes are versatile intermediates in organic synthesis. Although a wealth of chemistry has already been uncovered, prospects remain bright for interesting developments in the future. Areas such as the application of dihalocyclopropanes in heterocyclic synthesis *via* carbene insertion into C—H bonds adjacent to heteroatoms, reactions of dihalocyclopropanes with organometallics and the synthetic applications of metallated derivatives deserve further exploration. The chemistry of difluoro-, diiodo- and mixed dihalo-cyclopropanes can be expected to attract some attention. Finally, other heteroatom-substituted cyclopropanes derived from dihalocyclopropanes will also invoke further investigation.

4.7.5 REFERENCES

1. G. Gustavson, *J. Prakt. Chem.*, 1890, **42**, 496.
2. A. Geuther, *Justus Liebigs Ann. Chem.*, 1862, **123**, 121.
3. J. Hine, *J. Am. Chem. Soc.*, 1950, **72**, 2438.
4. J. Hine, 'Divalent Carbon', Ronald Press, New York, 1964, chap. 3.
5. W. E. Doering and A. K. Hoffmann, *J. Am. Chem. Soc.*, 1954, **76**, 6162.
6. W. E. Parham and E. E. Schweizer, *Org. React. (N. Y.)*, 1963, **13**, 55.
7. R. Barlet and Y. Vo-Quang, *Bull. Soc. Chim. Fr.*, 1969, 3729.
8. W. Kirmse, 'Carbene Chemistry', 2nd edn., Academic Press, New York, 1971, chap. 8.
9. D. Wendisch, *Methoden Org. Chem. (Houben–Weyl), 4th Ed.*, 1971, **4/3**, 150.
10. P. Weyerstahl, in 'The Chemistry of Functional Groups', suppl. D., ed. S. Patai and Z. Rappoport, Wiley, New York, 1983, p. 1451.
11. W. T. Miller, Jr. and C. S. Y. Kim, *J. Am. Chem. Soc.*, 1959, **81**, 5008.
12. W. M. Wagner, J. Kloosterziel and S. Van der Ven, *Recl. Trav. Chim. Pays-Bas*, 1961, **80**, 740.
13. J. M. Birchall, G. W. Gross and R. N. Haszeldine, *Proc. Chem. Soc., London*, 1960, 81.
14. K. Nanjo, K. Suzuki and M. Sekiya, *Chem. Lett.*, 1977, 553.
15. D. Seyferth, J. M. Burlitch, R. J. Minasz, J. Y.-P. Mui, H. D. Simmons, Jr., A. J. H. Treiber and S. R. Dowd, *J. Am. Chem. Soc.*, 1965, **87**, 4259.
16. D. Seyferth and C. K. Haas, *J. Org. Chem.*, 1975, **40**, 1620.
17. D. Seyferth, *Acc. Chem. Res.*, 1972, **5**, 65.
18. D. Seyferth, C. K. Haas and D. Dagani, *J. Organomet. Chem.*, 1976, **104**, 9.
19. T. Tsuji and S. Nishida, in 'The Chemistry of the Cyclopropyl Group' ed. Z. Rappoport, Wiley, New York, 1987, part 1, chap. 7.
20. M. Makosza and W. Wawrzyniewicz, *Tetrahedron Lett.*, 1969, 4659.
21. M. Makosza, *Pure Appl. Chem.*, 1975, **43**, 439.
22. M. Makosza, *Russ. Chem. Rev. (Engl. Transl.)*, 1977, **46**, 1151.
23. J. Dockx, *Synthesis*, 1973, 441.
24. E. V. Dehmlow, *Angew. Chem., Int. Ed. Engl.*, 1974, **13**, 170.
25. E. V. Dehmlow, *Angew. Chem., Int. Ed. Engl.*, 1977, **16**, 493.
26. W. P. Weber and G. W. Gokel, 'Phase Transfer Catalysis in Organic Synthesis', Springer-Verlag, Berlin, 1977.
27. C. M. Starks and C. Liotta, 'Phase Transfer Catalysis: Principles and Techniques', Academic Press, New York, 1978.
28. E. V. Dehmlow and S. S. Dehmlow, 'Phase Transfer Catalysis', 2nd edn., Verlag Chemie, Weinheim, 1983.
29. M. Makosza and M. Fedorynsky, *Synth. Commun.*, 1973, **3**, 305.
30. L. Skattebøl, G. A. Abskharoun and T. Greibrokk, *Tetrahedron Lett.*, 1973, 1367.
31. R. Mathias and P. Weyerstahl, *Angew. Chem., Int. Ed. Engl.*, 1974, **13**, 132.
32. M. S. Baird, *J. Chem. Soc., Perkin Trans. 1*, 1976, 54.
33. R. J. Kricks and A. A. Volpe, *Synthesis*, 1976, 313.
34. M. Fedorynsky, *Synthesis*, 1977, 783.
35. E. V. Dehmlow and M. Slopianka, *Liebigs Ann. Chem.*, 1979, 1465.
36. E. V. Dehmlow and J. Stütten, *Liebigs Ann. Chem.*, 1989, 187.
37. P. Weyerstahl, G. Blume and C. Müller, *Tetrahedron Lett.*, 1971, 3869.
38. L. V. Chau and M. Schlosser, *Synthesis*, 1973, 112.
39. P. Weyerstahl, R. Mathias and G. Blume, *Tetrahedron Lett.*, 1973, 611.
40. E. V. Dehmlow, *Tetrahedron Lett.*, 1976, 91.
41. S. Julia and A. Ginebreda, *Synthesis*, 1977, 682.
42. S. L. Regen and A. Singh, *J. Org. Chem.*, 1982, **47**, 1587.
43. K. Isagawa, Y. Kimura and S. Kwon, *J. Org. Chem.*, 1974, **39**, 3171.
44. M. Makosza, A. Kacprowicz and M. Fedorynsky, *Tetrahedron Lett.*, 1975, 2119.
45. D. Seyferth, in 'Carbenes', ed. R. A. Moss and M. Jones, Jr., Wiley, New York, 1975, vol. 2, p. 101.
46. T. Mukaiyama, M. Shiono, K. Watanabe and M. Onaka, *Chem. Lett.*, 1975, 711.
47. V. Rautenstrauch, H.-J. Scholl and E. Vogel, *Angew. Chem., Int. Ed. Engl.*, 1968, **7**, 288.
48. J. F. Hartwig, M. Jones, Jr., R. A. Moss and W. Tawrynowics, *Tetrahedron Lett.*, 1986, **27**, 5907.
49. P. S. Skell and A. Y. Garner, *J. Am. Chem. Soc.*, 1956, **78**, 3409.
50. E. V. Dehmlow and R. A. Kramer, *Angew. Chem., Int. Ed. Engl.*, 1984, **23**, 706.

51. J. T. Sharp, in 'Comprehensive Organic Chemistry', ed. D. H. R. Barton and W. D. Ollis, Pergamon Press, Oxford, 1979, vol. 1, p. 467.
52. D. Seyferth, J. Y.-P. Mui, M. E. Gordon and J. M. Burlitch, *J. Am. Chem. Soc.*, 1965, **87**, 681.
53. C. W. Jefford, *Proc. Chem. Soc., London*, 1963, 64.
54. C. W. Jefford, U. Burger and F. Delay, *Helv. Chim. Acta*, 1973, **56**, 1083.
55. W. E. Billups and L.-J. Lin, *Tetrahedron*, 1986, **42**, 1575.
56. M. Bertrand, A. Tubul and C. Ghiglione, *J. Chem. Res. (S)*, 1983, 251.
57. E. Vogel and H. D. Roth, *Angew. Chem., Int. Ed. Engl.*, 1964, **3**, 228.
58. M. A. Hashem, H. Marschall-Weyerstahl and P. Weyerstahl, *Chem. Ber.*, 1986, **119**, 464.
59. L. W. Jenneskens, L. A. M. Turkenburg, W. H. De Wolf and F. Bickelhaupt, *Recl. Trav. Chim. Pays-Bas*, 1985, **104**, 184.
60. H. Mayr and U. W. Heigl, *Angew. Chem., Int. Ed. Engl.*, 1985, **24**, 579.
61. G. R. Cooper, F. Hassan, B. L. Shaw and M. Thornton-Pett, *J. Chem. Soc., Chem. Commun.*, 1985, 614.
62. W. E. Billups, L.-J. Lin, B. E. Arney, Jr., W. A. Rodin and E. W. Casserly, *Tetrahedron Lett.*, 1984, **25**, 3935.
63. I. A. Dyankonov, *J. Gen. Chem. USSR (Engl. Transl.)*, 1960, **30**, 3475.
64. L. Vo-Quang and P. Cadot, *C. R. Hebd. Seances Acad. Sci., Ser. C*, 1961, **252**, 3827.
65. E. V. Dehmlow, *Tetrahedron Lett.*, 1975, 203.
66. T. Greibrokk, *Acta Chem. Scand., Ser. B*, 1973, **27**, 3207.
67. F. Mohamadi and W. C. Still, *Tetrahedron Lett.*, 1986, **27**, 893.
68. G. Stork, in 'Enamine Symposium', 140th National Meeting of the American Chemical Society, Chicago, Sept. 1961, abstr. 45Q.
69. M. Ohno, *Tetrahedron Lett.*, 1963, 1753.
70. M. Makosza and A. Kacprowicz, *Bull. Acad. Pol. Sci., Ser. Sci. Chim.*, 1974, **22**, 467.
71. U. K. Pandit, S. A. G. De Graaf, C. T. Brams and J. S. T. Raaphorst, *Recl. Trav. Chim. Pays-Bas*, 1972, **91**, 799.
72. S. A. G. De Graaf and U. K. Pandit, *Tetrahedron*, 1974, **30**, 1115.
73. A. G. Cook, in 'Enamines', 2nd edn., ed. A. G. Cook, Dekker, New York, 1988, p. 388.
74. E. Vilsmaier, in 'The Chemistry of the Cyclopropyl Group', ed. Z. Rappoport, Wiley, New York, 1987, part 2, p. 1356.
75. H. C. van der Plas, in 'Ring Transformations of Heterocycles', Academic Press, New York, 1973, p. 215.
76. A. H. Jackson, in 'Comprehensive Organic Chemistry', ed. D. H. R. Barton and W. D. Ollis, Pergamon Press, Oxford, 1979, vol. 4, p. 291.
77. R. T. Brown, J. A. Joule and P. G. Sammes, in 'Comprehensive Organic Chemistry', ed. D. H. R. Barton and W. D. Ollis, Pergamon Press, Oxford, 1979, vol. 4, p. 445.
78. J. L. Castro, L. Castedo and R. Riguera, *Tetrahedron Lett.*, 1985, **25**, 1561.
79. W. E. Parham and E. E. Schweizer, *J. Org. Chem.*, 1959, **24**, 1733.
80. G. Stork and V. Nair, *J. Am. Chem. Soc.*, 1979, **101**, 1315.
81. G. Stork and T. L. Macdonald, *J. Am. Chem. Soc.*, 1975, **97**, 1264.
82. P. Amice, L. Blanco and J. M. Conia, *Synthesis*, 1976, 196.
83. T. L. Macdonald, *J. Org. Chem.*, 1978, **43**, 3621.
84. G. Stork, M. Nussim and B. August, *Tetrahedron, Suppl.*, 1966, part 1, 105.
85. R. C. DeSelms, *Tetrahedron Lett.*, 1966, 1965.
86. K. J. Shea, W. M. Fruscella and W. P. England, *Tetrahedron Lett.*, 1987, **28**, 5623.
87. S. M. McElvain and P. L. Weyna, *J. Am. Chem. Soc.*, 1959, **81**, 2579.
88. T. Taguchi, T. Takigawa, Y. Tawara, T. Morikawa and Y. Kobayashi, *Tetrahedron Lett.*, 1984, **25**, 5689.
89. W. E. Parham, S. Kajigaeshi and S. H. Groen, *Bull. Chem. Soc. Jpn.*, 1972, **45**, 509.
90. B. R. Dent and B. Halton, *Tetrahedron Lett.*, 1984, **25**, 4279.
91. M. E. Vol'pin, Y. D. Koreshkov and D. N. Kursanov, *Izv. Akad. Nauk SSSR, Ser. Khim.*, 1959, 560 (*Chem. Abstr.*, 1959, **53**, 21 799f).
92. R. Breslow and R. Peterson, *J. Am. Chem. Soc.*, 1960, **82**, 4426.
93. E. V. Dehmlow, *Chem. Ber.*, 1968, **101**, 427.
94. M. A. Pericas and F. Serratosa, *Tetrahedron Lett.*, 1977, **17**, 4437.
95. E. V. Dehmlow, S. S. Dehmlow and F. Marschner, *Chem. Ber.*, 1977, **110**, 154.
96. T. Ishihara, K. Hayashi, T. Ando and H. Yamanaka, *J. Org. Chem.*, 1975, **40**, 3264.
97. R. Neidlein, V. Poignee, W. Kramer and C. Glück, *Angew. Chem., Int. Ed. Engl.*, 1986, **25**, 731.
98. G. Seitz and R. Van Gemmern, *Synthesis*, 1987, 953.
99. A. R. Pinder, *Synthesis*, 1980, 444.
100. J. T. Groves and K. W. Ma, *J. Am. Chem. Soc.*, 1974, **96**, 6528.
101. W. E. Doering and P. M. LaFlamme, *J. Am. Chem. Soc.*, 1956, **78**, 5447.
102. E. Vogel, *Angew. Chem.*, 1960, **72**, 7.
103. E. E. Schweizer and W. E. Parham, *J. Am. Chem. Soc.*, 1960, **82**, 4085.
104. H. Meier, C. Antony-Mayer, C. Schulz-Popitz and G. Zerban, *Liebigs Ann. Chem.*, 1987, 1087.
105. R. E. Ireland, D. Häbich and D. W. Norbeck, *J. Am. Chem. Soc.*, 1984, **107**, 3273.
106. E. Funakubo, I. Moritani, S. Murahashi and T. Tsuji, *Tetrahedron Lett.*, 1962, 539.
107. G. Boche and H. M. Walborsky, in 'The Chemistry of the Cyclopropyl Group', ed. Z. Rappoport, Wiley, New York, 1987, part 1, p. 715, 720.
108. C. W. Jefford, D. Kirkpatrick and F. Delay, *J. Am. Chem. Soc.*, 1972, **94**, 8905.
109. D. Seyferth, H. Yamasaki and D. L. Alleston, *J. Org. Chem.*, 1963, **28**, 703.
110. P. Dowd, C. Kaufman and R. H. Abeles, *J. Am. Chem. Soc.*, 1984, **106**, 2703.
111. C. Descoines, M. Julia and H. V. Sang, *Bull. Soc. Chim. Fr.*, 1971, 4087.
112. M. A. McKinney, S. W. Anderson, M. Keyes and R. Schmidt, *Tetrahedron Lett.*, 1982, **23**, 3443.
113. K. G. Taylor and J. Chaney, *J. Am. Chem. Soc.*, 1976, **98**, 4158.
114. G. G. Meijs, *J. Org. Chem.*, 1987, **52**, 3923.

115. G. Köbrich, *Angew. Chem., Int. Ed. Engl.*, 1972, **11**, 473.
116. D. Seebach, H. Siegel, K. Müllen and K. Hiltbrünner, *Angew. Chem., Int. Ed. Engl.*, 1979, **18**, 784.
117. K. G. Taylor, *Tetrahedron*, 1982, **38**, 2751.
118. K. Kitatani, T. Hiyama and H. Nozaki, *J. Am. Chem. Soc.*, 1975, **97**, 949.
119. K. Kitatani, H. Yamamoto, T. Hiyama and H. Nozaki, *Bull. Chem. Soc. Jpn.*, 1977, **50**, 2158.
120. A. Weber, U. Stämpfli and M. Neuenschwander, *Helv. Chim. Acta*, 1989, **72**, 29.
121. M. Braun, R. Dammann and D. Seebach, *Chem. Ber.*, 1975, **108**, 2368.
122. T. Hiyama, S. Takehara, K. Kitatani and H. Nozaki, *Tetrahedron Lett.*, 1974, 3295.
123. D. Seebach, M. Braun and N. DuPreez, *Tetrahedron Lett.*, 1973, 3509.
124. T. Hiyama, H. Saimoto, K. Nishio, M. Shinoda, H. Yamamoto and H. Nozaki, *Tetrahedron Lett.*, 1979, 2043.
125. S. Halazy, W. Dumont and A. Krief, *Tetrahedron Lett.*, 1981, **22**, 4737.
126. T. Hiyama, A. Kanakura, Y. Morizawa and H. Nozaki, *Tetrahedron Lett.*, 1982, **23**, 1279.
127. D. Seyferth and R. L. Lambert, Jr., *J. Organomet. Chem.*, 1975, **91**, 31.
128. A. Schmidt and G. Köbrich, *Tetrahedron Lett.*, 1974, 2561.
129. E. Vogel, H. Wieland, L. Schmalstieg and J. Lex, *Angew. Chem., Int. Ed. Engl.*, 1984, **23**, 717.
130. H.-U. Reissig, in 'The Chemistry of the Cyclopropyl Group', ed. Z. Rappoport, Wiley, New York, 1987, part 1, chap. 8.
131. M. Braun and D. Seebach, *Chem. Ber.*, 1976, **109**, 669.
132. R. Hässig, H. Siegel and D. Seebach, *Chem. Ber.*, 1982, **115**, 1990.
133. R. L. Danheiser and A. C. Savoca, *J. Org. Chem.*, 1985, **50**, 2403.
134. A. Weber, R. Galli, G. Sabbioni, U. Stämpfli, S. Walther and M. Neuenschwander, *Helv. Chim. Acta*, 1989, **72**, 41.
135. J. Krebs, D. Guggisberg, U. Stämpfli and M. Neuenschwander, *Helv. Chim. Acta*, 1986, **69**, 835.
136. K. Kitatani, T. Hiyama and H. Nozaki, *J. Am. Chem. Soc.*, 1976, **98**, 2362.
137. K. Kitatani, T. Hiyama and H. Nozaki, *Bull. Chem. Soc. Jpn.*, 1977, **50**, 1600, 3288.
138. E. J. Corey and G. H. Posner, *J. Am. Chem. Soc.*, 1967, **89**, 3911.
139. F. Scott, B. G. Mafunda, J. F. Normant and A. Alexakis, *Tetrahedron Lett.*, 1983, **24**, 5767.
140. T. Harayama, H. Fukushi, K. Ogawa and F. Yoneda, *Chem. Pharm. Bull.*, 1985, **33**, 3564.
141. J. H. Rigby and A.-R. Bellemin, *Synthesis*, 1989, 188.
142. P. A. Wender, R. M. Keenan and H. Y. Lee, *J. Am. Chem. Soc.*, 1987, **109**, 4391.
143. N. O. Nilsen, L. Skattebøl, M. S. Baird, S. R. Buxton and P. D. Slowey, *Tetrahedron Lett.*, 1984, **25**, 2887.
144. K. B. Wiberg and J. V. McClusky, *Tetrahedron Lett.*, 1987, **28**, 5411.
145. K. Semmler, G. Szeimies and J. Belzner, *J. Am. Chem. Soc.*, 1985, **107**, 6410.
146. U. H. Brinker, J. Wüster and G. Maas, *J. Chem. Soc., Chem. Commun.*, 1985, 1812.
147. W. E. Doering and P. M. LaFlamme, *Tetrahedron*, 1958, **2**, 75.
148. W. R. Moore and H. R. Ward, *J. Org. Chem.*, 1960, **25**, 2073.
149. L. Skattebøl, *Tetrahedron Lett.*, 1961, 167.
150. H. F. Schuster and G. M. Coppola, 'Allenes in Organic Synthesis', Wiley, New York, 1984, p. 20.
151. P. D. Landor, in 'The Chemistry of the Allenes' ed. S. R. Landor, Academic Press, New York, 1982, vol. 1, chap. 2.
152. M. Murray, *Methoden Org. Chem. (Houben–Weyl) 4th Ed.*, 1977, **5/2a**, 985.
153. L. Skattebøl *J. Org. Chem.*, 1966, **31**, 2789.
154. K. Kleveland and L. Skattebøl, *Acta Chem. Scand., Ser. B*, 1975, **29**, 191.
155. Ya. M. Slobodin, V. E. Maiorova and P. A. Khitrov, *Zh. Obshch. Khim.*, 1969, **5**, 851.
156. P. L. Perchec and J. M. Conia *Tetrahedron Lett.*, 1970, 1587.
157. L. A. Paquette, K. E. Green, R. Gleiter, W. Schafer and J. C. Gallucci, *J. Am. Chem. Soc.*, 1984, **106**, 8233.
158. L. Skattebøl and S. Solomon, *Org. Synth. Coll. Vol.*, 1973, **5**, 306.
159. J. D. Price and R. P. Johnson, *Tetrahedron Lett.*, 1986, **27**, 4679.
160. M. Christl and M. Schreck, *Angew. Chem., Int. Ed. Engl.*, 1987, **26**, 449.
161. R. O. Angus, Jr. and R. P. Johnson, *J. Org. Chem.*, 1984, **49**, 2880.
162. G. Bérubé and P. Deslongchamps, *Can. J. Chem.*, 1984, **62**, 1558.
163. K. G. Untch, D. J. Martin and N. T. Castellucci, *J. Org. Chem.*, 1965, **30**, 3572.
164. L. Xu, F. Tao and T. Yu, *Tetrahedron Lett.*, 1985, **26**, 4231.
165. H. Nozaki, T. Aratani, T. Toraya and R. Noyori, *Tetrahedron*, 1971, **27**, 905.
166. D. W. Brown, M. E. Hendrick and M. Jones, Jr., *Tetrahedron Lett.*, 1973, 3951.
167. E. E. Waali and N. T. Allison, *J. Org. Chem.*, 1979, **44**, 3266.
168. L. Skattebøl, *Chem. Ind. (London)*, 1962, 2146.
169. L. Skattebøl, *Tetrahedron*, 1967, **23**, 1107.
170. K. H. Holm and L. Skattebøl, *Tetrahedron Lett.*, 1977, 2347.
171. M. S. Baird and C. B. Reese, *J. Chem. Soc., Chem. Commun.*, 1972, 523.
172. M. S. Baird and C. B. Reese, *Tetrahedron Lett.*, 1976, 2895.
173. W. M. Jones and U. H. Brinker, *Org. Chem. (N. Y.)*, 1977, **35/1**, 110.
174. R. A. Moss and M. Jones, Jr., *React. Intermed.*, 1981, **2**, 113; 1985, **3**, 78.
175. P. Warner and S.-C. Chang, *Tetrahedron Lett.*, 1978, 3981.
176. U. H. Brinker and I. Fleischauer, *Tetrahedron*, 1981, **37**, 4495.
177. D. N. Butler and I. Gupta, *Can. J. Chem.*, 1978, **56**, 80.
178. W. R. Moore, H. R. Ward and R. F. Merritt, *J. Am. Chem. Soc.*, 1961, **83**, 2019.
179. S. Hoz, in 'The Chemistry of the Cyclopropyl Group', ed. Z. Rappoport, Wiley, New York, 1987, part 2, p. 1137.
180. L. Skattebøl, *Tetrahedron Lett.*, 1970, 2361.
181. W. R. Moore, K. G. Taylor, P. Muller, S. S. Hall and Z. L. F. Gaibel, *Tetrahedron Lett.*, 1970, 2365.
182. L. A. Paquette, E. Chamot and A. R. Browne, *J. Am. Chem. Soc.*, 1980, **102**, 637.
183. K. G. Taylor, J. Chaney and J. C. Deck, *J. Am. Chem. Soc.*, 1976, **98**, 4163.

184. M. Jones, Jr. and E. W. Petrillo, Jr., *Tetrahedron Lett.*, 1969, 3953.
185. A. Oku, T. Harada, Y. Homoto and M. Iwamoto, *J. Chem. Soc., Chem. Commun.*, 1988, 1490.
186. M. S. Baird, *J. Chem. Soc., Chem. Commun.*, 1974, 197.
187. A. R. Allan and M. S. Baird, *J. Chem. Soc., Chem. Commun.*, 1975, 172.
188. M. S. Baird and A. C. Kaura, *J. Chem. Soc., Chem. Commun.*, 1976, 356.
189. M. S. Baird, *Chem. Commun.*, 1971, 1145.
190. M. S. Baird, S. R. Buxton and P. Sadler, *J. Chem. Soc., Perkin Trans. 1*, 1984, 1379.
191. T. C. Shields and P. D. Gardner, *J. Am. Chem. Soc.*, 1967, **89**, 5425.
192. J. Arct, B. Migaj and J. Leonczynski, *Tetrahedron*, 1981, **37**, 3689.
193. W. E. Billups, T. C. Shields, W. Y. Chow and N. C. Deno, *J. Org. Chem.*, 1972, **37**, 3676.
194. T. C. Shields, W. E. Billups and A. R. Lepley, *J. Am. Chem. Soc.*, 1968, **90**, 4749.
195. A. S. Kende and E. F. Riecke, *J. Am. Chem. Soc.*, 1972, **94**, 1397.
196. W. E. Billups, A. J. Blakeney and W. Y. Chow, *Chem. Commun.*, 1971, 1461.
197. W. E. Billups, L. E. Reed, E. W. Casserly and L. P. Lin, *J. Org. Chem.*, 1981, **46**, 1326.
198. P. Müller, H. C. N. Thi and J. Pfyffer, *Helv. Chim. Acta*, 1986, **69**, 855.
199. W. E. Parham, W. D. McKown, V. Nelson, S. Kajigaeshi and N. Ishikawa, *J. Org. Chem.*, 1973, **38**, 1361.
200. S. W. Tobey and R. West, *J. Am. Chem. Soc.*, 1966, **88**, 2461.
201. M. S. Baird, H. H. Hussain and W. Nethercott, *J. Chem. Soc., Perkin Trans. 1*, 1986, 1845.
202. I. J. Anthony and D. Wege, *Tetrahedron Lett.*, 1987, **28**, 4217.
203. W. E. Parham and H. E. Reiff, *J. Am. Chem. Soc.*, 1955, **77**, 1177.
204. R. B. Woodward and R. Hoffmann, *J. Am. Chem. Soc.*, 1965, **87**, 395.
205. H. C. Longuet-Higgins and E. W. Abrahamson, *J. Am. Chem. Soc.*, 1965, **87**, 2045.
206. C. H. DePuy, L. G. Schmack, J. W. Hauser and W. Wiedeman, *J. Am. Chem. Soc.*, 1965, **87**, 4006.
207. S. J. Cristol, R. M. Sequeira and C. H. DePuy, *J. Am. Chem. Soc.*, 1965, **87**, 4007.
208. P. von R. Schleyer, T. M. Su, M. Saunders and J. C. Rosenfeld, *J. Am. Chem. Soc.*, 1969, **91**, 5174.
209. P. von R. Schleyer, W. F. Sliwinski, G. W. Van Dine, U. Schöllkopf, J. Paust and K. Fellenberger, *J. Am. Chem. Soc.*, 1972, **94**, 125.
210. A. P. ter Borg and A. F. Bickel, *Recl. Trav. Chim. Pays-Bas*, 1961, **80**, 1217.
211. T. Ando, H. Hosaka, H. Yamanaka and W. Funasaka, *Bull. Chem. Soc. Jpn.*, 1969, **42**, 2018.
212. D. Ginsburg, in 'The Chemistry of the Cyclopropyl Group', ed. Z. Rappoport, Wiley, New York, 1987, part 2, chap. 20.
213. S. Hirano, H. Hara, T. Hiyama, S. Fujita and H. Nozaki, *Tetrahedron*, 1975, **31**, 2219.
214. M. Christl, R. Lang and C. Herzog, *Tetrahedron*, 1986, **42**, 1585.
215. E. V. Dehmlow and D. Balschukat, *Chem. Ber.*, 1985, **118**, 3805.
216. E. V. Dehmlow, C. Gröning, H. Bögge and A. Müller, *Chem. Ber.*, 1988, **121**, 621.
217. C. M. Amon, M. G. Banwell and G. L. Gravatt, *J. Org. Chem.*, 1987, **52**, 4851.
218. M. G. Banwell, K. A. Herbert, J. R. Buckleton, G. R. Clark, C. E. F. Rickard, C. M. Lin and E. Hamel, *J. Org. Chem.*, 1988, **53**, 4945.
219. T. Hiyama, M. Shinoda, M. Tsukanaka and H. Nozaki, *Bull. Chem. Soc. Jpn.*, 1980, **53**, 1010.
220. P. S. Skell and S. R. Sandler, *J. Am. Chem. Soc.*, 1958, **80**, 2024.
221. A. J. Birch, J. M. H. Graves and F. Stansfield, *Proc. Chem. Soc., London*, 1962, 282.
222. E. C. Friedrich, in 'The Chemistry of the Cyclopropyl Group', ed. Z. Rappoport, Wiley, New York, 1987, part 1, p. 640.
223. P. Grice and C. B. Reese, *J. Chem. Soc., Chem. Commun.*, 1980, 424.
224. D. Dhanak, R. Kuroda and C. B. Reese, *Tetrahedron Lett.*, 1987, **28**, 1827.
225. I. Lantos, D. Bhattacharjee and D. S. Eggleston, *J. Org. Chem.*, 1986, **51**, 4147.
226. L. W. Jenneskens, F. J. J. De Kanter, L. A. M. Turkenburg, H. J. R. De Boer, W. H. De Wolf and F. Bickelhaupt, *Tetrahedron*, 1984, **40**, 4401.
227. P. E. Brown and Q. Islam, *Tetrahedron Lett.*, 1987, **28**, 3047.
228. L. Blanco and A. Mansouri, *Tetrahedron Lett.*, 1988, **29**, 3239.
229. J. Buddrus and F. Nerdel, *Tetrahedron Lett.*, 1965, 3197.
230. J. Buddrus, *Chem. Ber.*, 1968, **101**, 4152.
231. L. Skattebøl and B. Boulette, *J. Org. Chem.*, 1966, **31**, 81.
232. L. Skattebøl, *J. Org. Chem.*, 1966, **31**, 1554.
233. S. R. Sandler, *J. Org. Chem.*, 1968, **33**, 4537.
234. Y. Kobayashi, T. Morikawa, A. Yoshizawa and T. Taguchi, *Tetrahedron Lett.*, 1981, **22**, 5297.
235. T. Taguchi, T. Takigawa, Y. Tawara, T. Morikawa and Y. Kobayashi, *Tetrahedron Lett.*, 1984, **25**, 5689.
236. H. Molines, T. Nguyen and C. Wakselman, *Synthesis*, 1985, 754.
237. M. Schlosser, R. Daham and S. Cottens, *Helv. Chim. Acta*, 1984, **67**, 284.
238. J. L. Castro, L. Castedo and C. Riguera, *J. Org. Chem.*, 1987, **52**, 3579.
239. T. Hirao, T. Masunaga, K. Hayashi, Y. Ohshiro and T. Agawa, *Tetrahedron Lett.*, 1983, **24**, 399.
240. J. T. Groves and K. W. Ma, *Tetrahedron Lett.*, 1974, 909.
241. P. Warner and S.-L. Lu, *J. Am. Chem. Soc.*, 1976, **98**, 6752.
242. T. Ishihara, T. Kudaka and T. Ando, *Tetrahedron Lett.*, 1984, **25**, 4765.
243. R. A. Rossi and A. N. Santiago, *J. Chem. Res. (S)*, 1988, 172.
244. E. R. Alexander, A. B. Herrick and T. M. Rodes, *J. Am. Chem. Soc.*, 1950, **72**, 2760.
245. R. L. Jones and C. W. Rees, *J. Chem. Soc. C*, 1969, 2249, 2255.
246. F. DeAngelis, A. Gambacosta and R. Nicoletti, *Synthesis*, 1976, 798.
247. C. W. Rees and C. E. Smithen, *Adv. Heterocycl. Chem.*, 1964, **3**, 57.
248. H. E. Dobbs, *Tetrahedron*, 1968, **24**, 491.
249. F. S. Baker, R. E. Busby, M. Iqbal, J. Parrick and C. J. G. Shaw, *Chem. Ind. (London)*, 1969, 1344.
250. R. E. Busby, M. Iqbal, J. Parrick and C. J. G. Shaw, *Chem. Commun.*, 1969, 1344.
251. J. Arct and L. Skattebøl, *Tetrahedron Lett.*, 1982, **23**, 113.

252. U. H. Brinker, A. Haghani and K. Gomann, *Angew. Chem., Int. Ed. Engl.*, 1985, **24**, 230.
253. H. Quast, R. Jakob, K. Peters, E.-M. Peters and H. G. von Schnering, *Chem. Ber.*, 1984, **117**, 840.
254. O. G. Kulinkovich, I. G. Tischenko and N. V. Masalov, *Synthesis*, 1984, 886.
255. W. E. Parham and J. F. Dooley, *J. Org. Chem.*, 1968, **33**, 1476.
256. R. L. Danheiser, J. M. Morin, Jr., M. Yu and A. Basak, *Tetrahedron Lett.*, 1981, **22**, 4205.
257. K. Sugahara, K. Suga, T. Fujita, S. Watanabe and K. Sugimoto, *Synthesis*, 1985, 342.
258. R. L. Danheiser, J. M. Morin, Jr. and E. J. Salaski, *J. Am. Chem. Soc.*, 1985, **107**, 8066.
259. P. Weyerstahl and M. A. Hashem, *Chem. Ber.*, 1987, **120**, 449.
260. A. D. Ketley, A. J. Berlin, E. Gorman and L. P. Fisher, *J. Org. Chem.*, 1966, **31**, 305.
261. M. Klehr and H. J. Schäfer, *Angew. Chem., Int. Ed. Engl.*, 1975, **14**, 247.
262. T. Hirao, S. Nagata and T. Agawa, *Tetrahedron Lett.*, 1985, **26**, 5795.
263. T. Hirao, S. Nagata and T. Agawa, *Chem. Lett.*, 1985, 1625.
264. T. Hirao, Y. Harano, Y. Yamana, Y. Hamada, S. Nagata and T. Agawa, *Bull. Chem. Soc. Jpn.*, 1986, **59**, 1341.
265. P. Skarstad, P. J. Vuuren, J. Meinwald and R. E. Hughes, *J. Chem. Soc., Perkin Trans. 2*, 1975, 88.

4.8
Addition of Ketocarbenes to Alkenes, Alkynes and Aromatic Systems

HUW M. L. DAVIES
Wake Forest University, Winston-Salem, NC, USA

4.8.1 INTRODUCTION

The reactions of keto-carbenes (**1**) or -carbenoids with C=C π-bonds[1–7] offer a direct entry to highly functionalized cyclopropanes (**2**),[8] which, due to their lability, are readily transformed to a range of products (Scheme 1). The cyclopropanation is often highly stereoselective and further reaction can lead to products of defined stereochemistry, controlled by the compact arrangement of functionality within the cyclopropane ring. Although free carbenes undergo additions to alkenes, these reactions are unselective and competing carbene rearrangements tend to occur. Consequently, metal-stabilized ketocarbenoids, generally formed by catalyzed decomposition of diazo compounds, are of much greater synthetic utility and will be the emphasis of this chapter.

Extensive synthetic developments have occurred recently in this field, primarily because of improvements in the catalysts used and the procedures for preparation of the carbenoid precursors. Consequently,

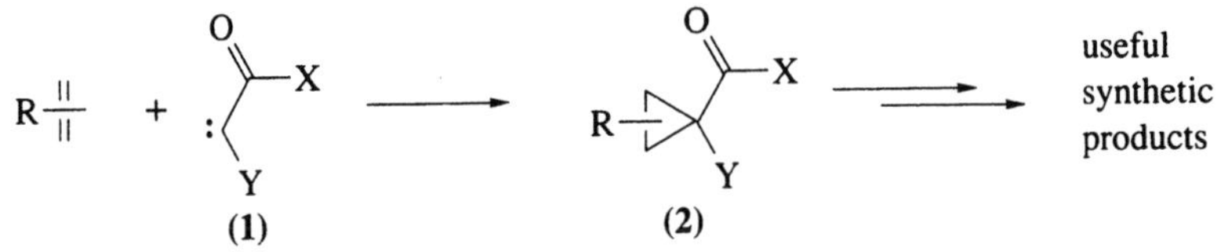

(1) (2)

useful synthetic products

Scheme 1

previously inefficient yet intriguing carbenoid reactions can often now be carried out in synthetically viable yields.

In order to outline the scope of this chemistry, Sections 4.8.2 and 4.8.3 will discuss the catalysts and carbenoid precursors used. This will be followed by reactions of carbenoids with π-systems, organized according to the π-system involved, alkenes (Section 4.8.4), alkynes (Section 4.8.5), benzenes and electron-rich heterocycles (Section 4.8.6). Particular emphasis will be placed on the stereochemical outcome of these reactions with reference to applications in organic synthesis.

4.8.2 GENERATION OF KETOCARBENOIDS

Ketocarbenes (**1**) are usually generated from the corresponding diazo compounds (**3**).[5] Other sources which are occasionally used are α,α-dibromo compounds (**4**),[9] sulfur ylides (**5**)[10] and iodonium ylides (**6**; Scheme 2).[11] The thermal or photochemical decomposition of diazo compounds in the presence of π-systems is often complicated by indiscriminate side reactions, such as Wolff rearrangements,[12] C—H insertions and hydride migrations. To avoid such problems, the use of metal-catalyzed decomposition of diazo compounds is generally preferred.[1,2]

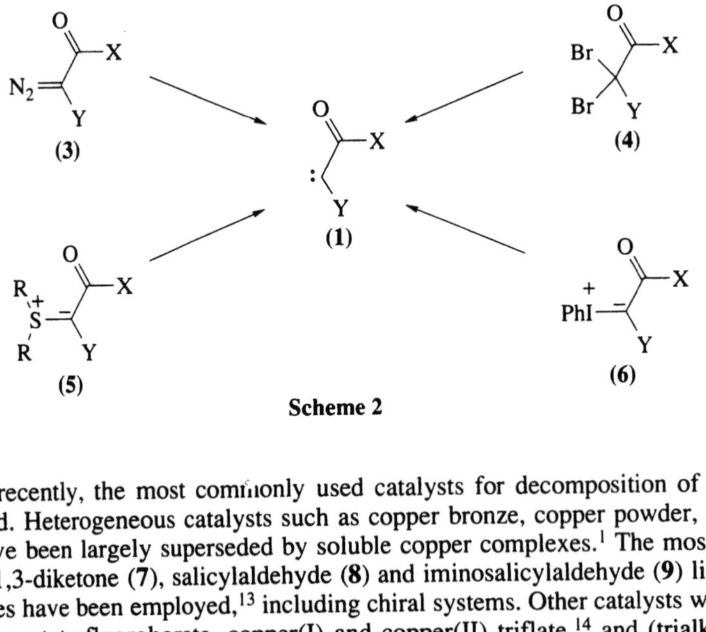

(3) (4) (1) (5) (6)

Scheme 2

Up until quite recently, the most commonly used catalysts for decomposition of diazo compounds were copper based. Heterogeneous catalysts such as copper bronze, copper powder, copper halide and copper sulfate have been largely superseded by soluble copper complexes.[1] The most extensively used are based on the 1,3-diketone (**7**), salicylaldehyde (**8**) and iminosalicylaldehyde (**9**) ligands, and a wide range of complexes have been employed,[13] including chiral systems. Other catalysts which are occasionally used are copper tetrafluoroborate, copper(I) and copper(II) triflate,[14] and (trialkyl phosphite)copper(I) iodide.[15] Even though copper sources ranging in oxidation state from zero to two have been used, it has often been found[14] that the active catalytic species is copper(I), which is generated under the reaction conditions.

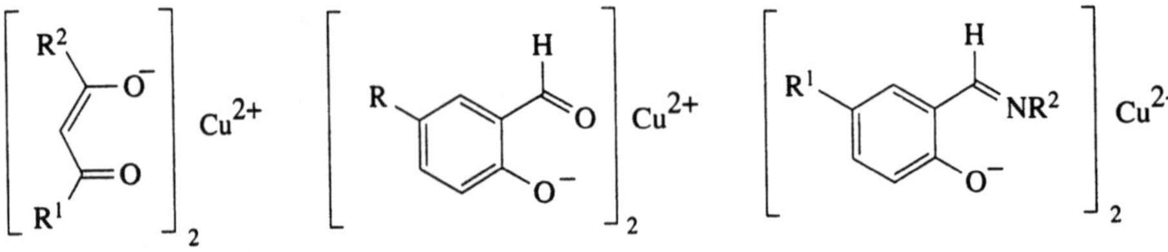

(7) (8) (9)

The development of the dimeric rhodium(II) carboxylates (**10**)[16] as extremely mild catalysts for decomposition of diazo compounds has greatly enhanced the whole field of carbenoid chemistry, and these have quickly become the preferred catalysts in most instances. Commercially available rhodium(II) acetate (**10a**) has been most extensively used, but rhodium(II) pivalate (**10b**) and rhodium(II) hexanoate (**10c**) often offer advantages due to their greater solubility. The ligands can also be varied to fine tune the catalyst. Complexes with electron-withdrawing ligands such as trifluoroacetate (**10d**) result in less back-bonding to the carbenoid, producing a more reactive electrophilic carbenoid.[16d] Alternatively, the use of rhodium(II) acetamide, $Rh_2(NHCOMe)_4$, generates a carbenoid with greater back-bonding to the metal than in the case of rhodium(II) acetate, resulting in a system with greater selectivity.[17] Sterically hindered ligands do not have a great effect on reactivity except in extreme cases, such as (**10e**)[18] and the porphyrin complex (**11**).[19] Rhodium–carbonyl complexes such as $Rh_6(CO)_{16}$ and $[Rh(CO)_2Cl]_2$ are also effective catalysts.[20]

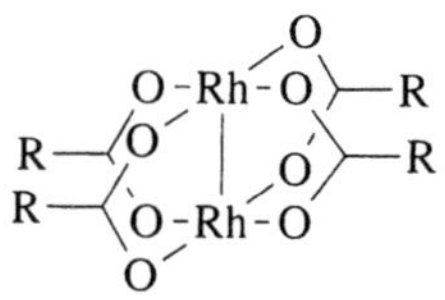

(**10**) **a**: R = Me
b: R = Bu^t
c: R = $(CH_2)_4Me$
d: R = CF_3
e: R = 2,4,6-triarylphenyl

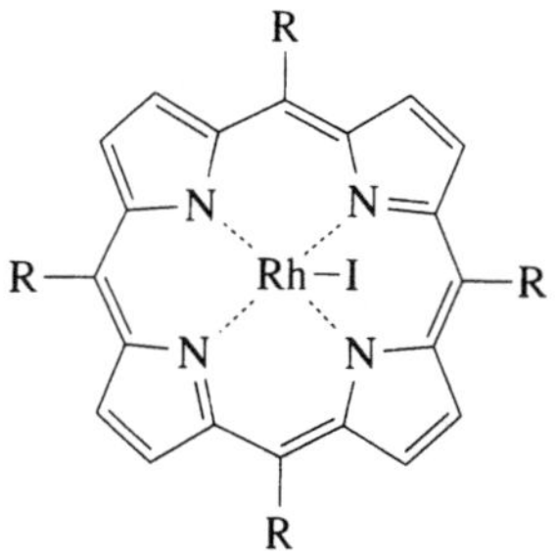

(**11**) R = 2,4,6-trimethylphenyl

The rhodium(II) catalysts and the chelated copper catalysts are considered to coordinate only to the carbenoid, while copper triflate and tetrafluoroborate coordinate to both the carbenoid and alkene and thus enhance cyclopropanation reactions through a template effect.[14] Palladium-based catalysts, such as palladium(II) acetate and bis(benzonitrile)palladium(II) chloride,[16e] are also believed to be able to coordinate with the alkene. Some chiral complexes based on cobalt have also been developed,[21] but these have not been extensively used.

4.8.3 CLASSES OF KETOCARBENOIDS

One of the most attractive features about the application of carbenoid chemistry in organic synthesis is that their diazo precursors are readily prepared. Diazoacetates (**12a**) have been most extensively used, but a variety of other systems including diazoacetamides (**12b**),[17,22,23] diazo ketones (**12c**),[24] diazomalonates (**12d**),[25] diazoacetoacetates (**12e**) and others with two electron-withdrawing groups, diazopyruvates (**12f**)[26] and vinyldiazomethanes (**12g**),[27] have also been exploited in useful synthetic processes. A particular advantage of the diazocarbonyl compounds is that the presence of an electron-withdrawing group causes the system to be more stable than simple diazoalkanes.

The diazo transfer reaction with sulfonyl azides has been extensively used for the preparation of diazo compounds with two electron-withdrawing groups (equation 1).[28] Toluenesulfonyl azide (**13a**)[29] is the standard reagent used, but due to problems of safety and ease of product separation, several alternative reagents have been developed recently. *n*-Dodecylbenzenesulfonyl azide (**13b**)[30] is very effective for the preparation of crystalline diazo compounds, while *p*-acetamidobenzenesulfonyl azide (**13c**)[31] or naphthalenesulfonyl azide (**13d**)[30] are particularly useful with fairly nonpolar compounds. Other useful reagents are methanesulfonyl azide (**13e**)[32] and *p*-carboxybenzenesulfonyl azide (**13f**).[33]

The diazo transfer reaction can be used to prepare, indirectly, compounds with only one electron-withdrawing group because base-induced deacylation is very favorable (equation 2;[28] Scheme 3[22a]). Diazo ketones are also conveniently prepared by reaction of acid chlorides with diazoalkanes (equation 3),[34] while diazoacetates are readily formed through reaction of the toluenesulfonylhydrazone (**14**)[35] or the corresponding triisopropylphenylhydrazone[36] with alcohols (equation 4). Other classical methods for preparation of diazocarbonyl compounds include the nitrosation of amines and dehydrogenation of hydrazones.[28]

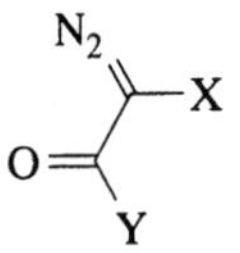

RSO_2N_3

(12) **a**: Y = OEt, X = H
b: Y = NR_2, X = H
c: Y = alkyl, X = H, Me
d: Y = OEt, X = CO_2Et
e: Y = OEt, X = COMe
f: Y = CO_2Et, X = H
g: Y = OEt, X = CH=CHR

(13) **a**: R = 4-MeC_6H_4–
b: R = 4-($C_{12}H_{25}$)C_6H_4–
c: R = 4-(MeCONH)C_6H_4–
d: R = 1-naphthyl–
e: R = Me–
f: R = 4-(HO_2C)C_6H_4–

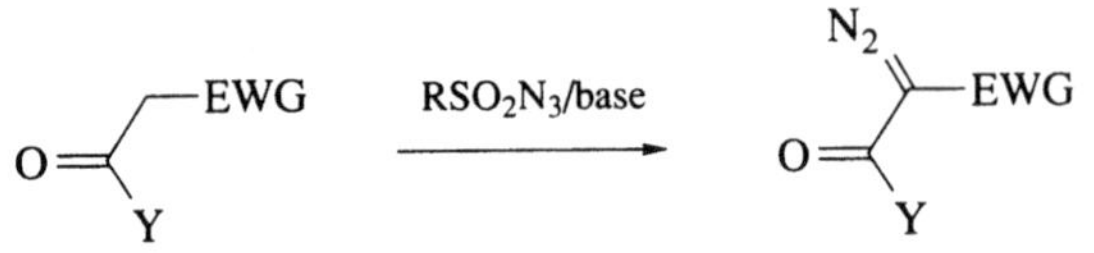

RSO_2N_3/base (1)

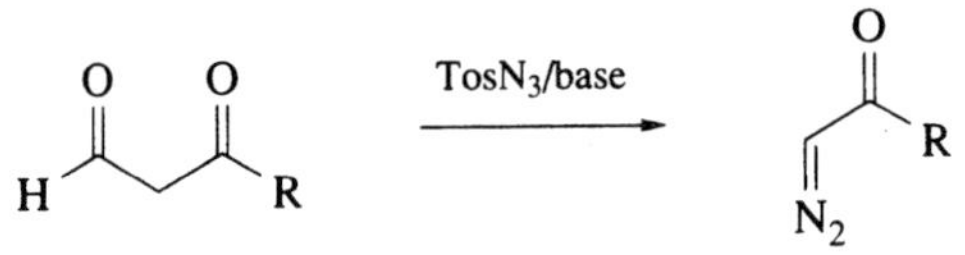

$TosN_3$/base (2)

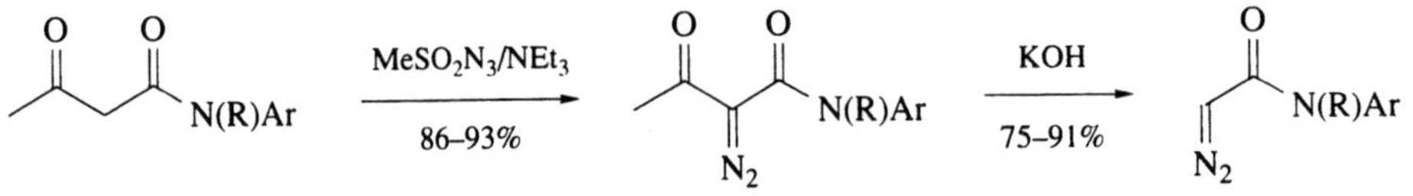

$MeSO_2N_3/NEt_3$ 86–93% ; KOH 75–91%

Scheme 3

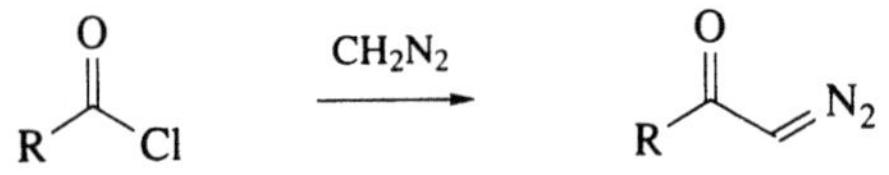

CH_2N_2 (3)

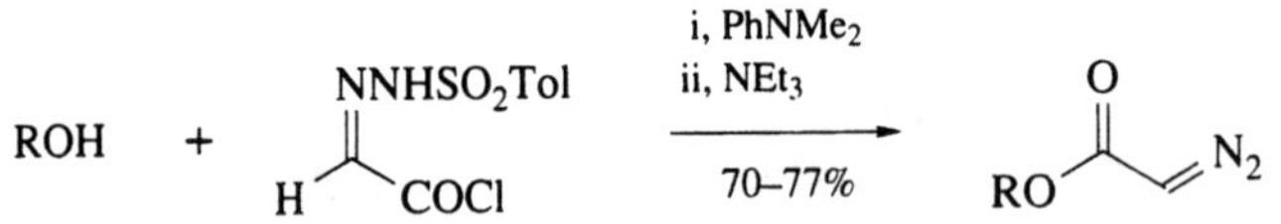

ROH + (14) → i, $PhNMe_2$; ii, NEt_3 70–77% (4)

4.8.4 ADDITION OF KETOCARBENOIDS TO ALKENES

4.8.4.1 Intermolecular Reactions

The addition of carbenoids derived from alkyl diazoacetates to alkenes has been extensively studied. As two thorough reviews on the subject,[1,2] dealing with a detailed comparison of the various catalysts, have recently appeared, only a general summary concerning regioselectivity, competing reactions, diastereoselectivity and enantioselectivity will be presented here.

4.8.4.1.1 Regioselectivity

Rhodium(II) acetate appears to be the most generally effective catalyst, and most of this discussion will center around the use of this catalyst with occasional reference to other catalysts when significant synthetic advantages can be gained. Cyclopropanation of a wide range of alkenes is possible with alkyl diazoacetate, as is indicated with the examples shown in Table 1.[16e,37] The main limitations are that the alkene must be electron rich and not too sterically crowded. Poor results were obtained with *trans*-alkenes. Comparison studies have been carried out with copper and palladium catalysts and commonly the yields were lower than with rhodium catalysts. Cyclopropanation of styrenes and strained alkenes, however, proceeded extremely well with palladium(II) acetate, while copper catalysts are still often used for cyclopropanation of vinyl ethers.[38–40]

Table 1 Yield of Cyclopropanation with Ethyl Diazoacetate

R-alkene + N_2=CHCO_2Et —$Rh_2(OAc)_4$→ R-cyclopropane-CO_2Et

Alkene	*Yield (%)*
$PhCH{=}CH_2$	93
$EtOCH{=}CH_2$	88
$AcOCH{=}CH_2$	77
$Bu^tCH{=}CH_2$	87
$ClCH_2CH{=}CH_2$	90
$Me_2C{=}CMe_2$	70
cis-$MeCH{=}CH(CH_2)_4Me$	65
trans-$MeCH{=}CH(CH_2)_4Me$	24
trans-$Me(CH_2)_2CH{=}CH(CH_2)_2Me$	7
$Me_2C{=}CHCH{=}CMe_2$	81
Cyclohexene	90
Cyclohexadiene	90
2,3-Dihydropyran	91

Another series of experiments which nicely demonstrates the regioselectivity of cyclopropanation reactions has been carried out with substituted dienes.[41] In general, cyclopropanation occurs at the least-substituted double bond with terminal dienes (equation 5), but electronic factors control the position of cyclopropanation with 2-substituted dienes (equation 6). With the fluorodiene (**15**) the predominant product was (**16a**; equation 7), which arose through cyclopropanation at the fluorine-bound double bond.[42] This result and related observations have been rationalized by assuming a nonsynchronous cyclopropanation mechanism in which the adverse electron-withdrawing effect of the fluoride in the somewhat zwitterionic transition state is compensated by a mesomeric electron-releasing effect. Even though extensive studies of the rhodium-catalyzed cyclopropanation with diazoacetates have been reported,[1,2] similar studies with other carbenoids are less common[26a,43] and most of the available information on regioselectivity is based on earlier work using less efficient copper catalysts.[5]

The structure of the carbenoid has considerable effect on the outcome of the reaction with vinyl ethers. Unlike the case with diazoacetate, reaction with diazopyruvate resulted in the formation of a dihydrofuran (**17**) rather than a cyclopropane (equation 8).[26d] The reaction is a formal [2 + 3] cycloaddition but it

1-methoxybutadiene + N_2=CHCO_2Et —$Rh_2(OAc)_4$, 90%→ 2-(2-methoxyvinyl)cyclopropane-CO_2Et (89%) + 2-methoxy-3-vinylcyclopropane-CO_2Et (11%) (5)

2-methoxybutadiene (OMe) + N_2=CHCO_2Et —$Rh_2(OAc)_4$, 88%→ 2-methoxy-2-vinylcyclopropane-CO_2Et (MeO) (6)

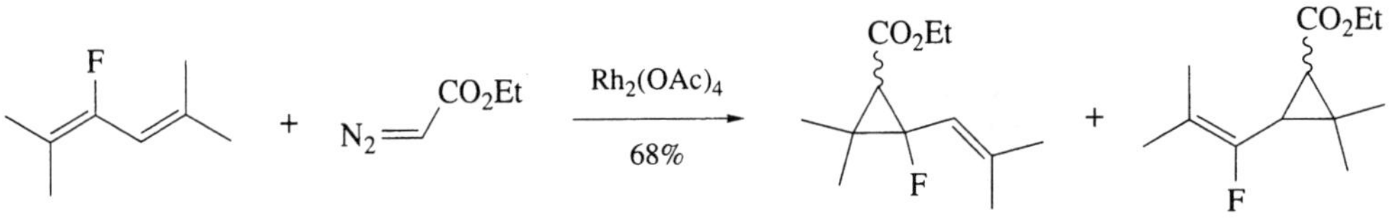

(15) + N_2=CHCO$_2$Et $\xrightarrow[68\%]{Rh_2(OAc)_4}$ (16a) 56% + (16b) 44% (7)

has been proposed[26c] that the mechanism involves a dipolar intermediate. Similar reactions were observed with diazoacetoacetates[44] and also with diazo ketones reacting with ketene acetals.[45]

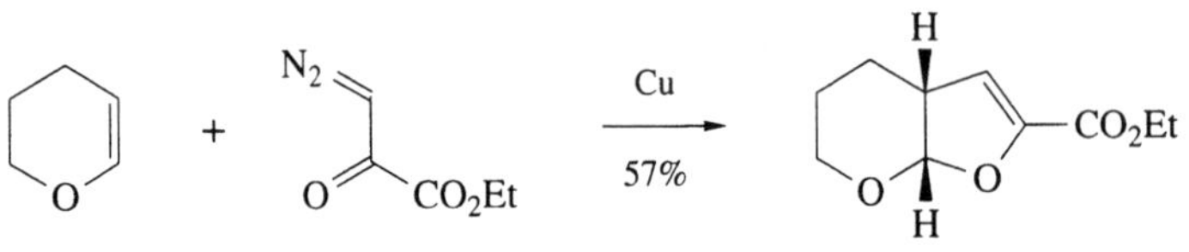

$\xrightarrow[57\%]{Cu}$ (17) (8)

4.8.4.1.2 Competing reactions

One of the main side reactions that can occur in cyclopropanation reactions is a competitive capture of the carbenoid with unreacted diazo compound rather than with the alkene, which results in the formation of dimeric products such as (**18**) and (**19**) (equation 9).[1,46] This reaction is usually avoided by using a five- to ten-fold excess of alkene and ensuring that a low concentration of the diazo compound is maintained. Indeed, Doyle has shown that by very slow addition of the diazo compound it is possible to obtain reasonable yields of cyclopropane even when stoichiometric quantities of the carbenoid precursor and alkene were used.[47]

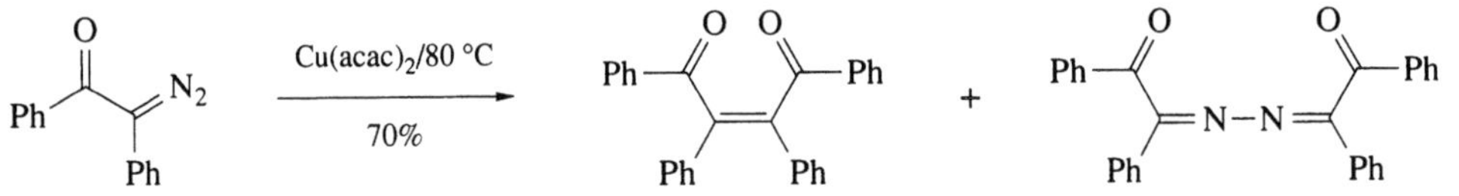

$\xrightarrow[70\%]{Cu(acac)_2/80\ °C}$ (**18**) 86% + (**19**) 14% (9)

Insertion into an O—H bond is generally favored over cyclopropanation, and consequently protection of hydroxy functionality is normally required. The ease of O—H insertion is nicely illustrated in a recent synthesis of chorismic acid derivatives, where the alkene functionality in (**20**) was totally unaffected by the carbenoid (Scheme 4).[48]

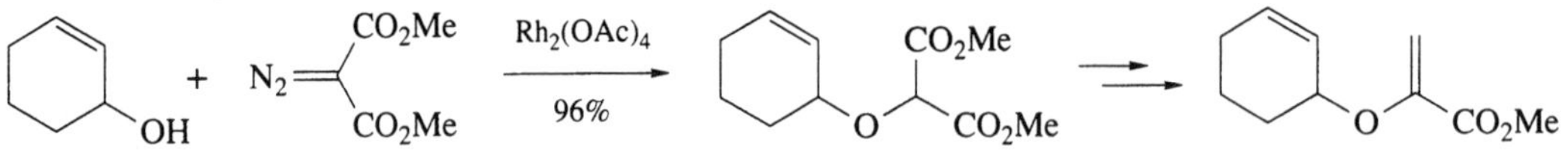

(**20**) + N_2=C(CO$_2$Me)$_2$ $\xrightarrow[96\%]{Rh_2(OAc)_4}$

Scheme 4

The carbenoid can also be competitively captured by halogens, as demonstrated in a series of experiments on allyl halides (Scheme 5).[49] Capture of the carbenoid with the halide generated an ylide capable of a 2,3-sigmatropic rearrangement, forming (**21**). As would be expected, the order of reactivity was iodide > bromide > chloride, but the carbenoid structure was also crucial because halogen insertion was much more favorable with diazomalonate than diazoacetate. Other nucleophilic sites such as oxygen,[50] nitrogen and sulfur can also effectively compete for the carbenoid.

Although C—H insertion reactions rarely occur in intermolecular reactions with diazoacetates, these are common side reactions with diazomalonates[51,52] (equation 10) and diazo ketones (with α-allyl vinyl ethers).[53] Several mechanistic pathways are available to generate the products of an apparent direct C—H insertion reaction and these include dipolar intermediates, π-allyl complexes and ring opening of cyclopropanes.[1] Oxidative problems due to the presence of oxygen are common with copper catalysts, but these are rarely encountered with rhodium catalysts except in systems where the carbenoid is ineffectively captured.[54]

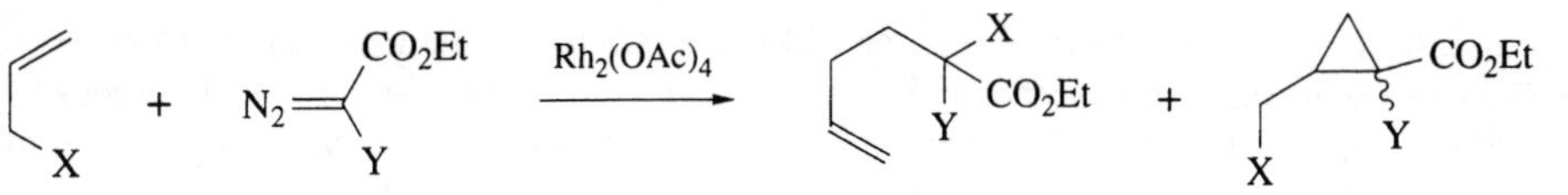

X	Y	Yield (%)	Ratio (21):(22)
I	H	98	100:0
Br	H	76	28:72
Cl	H	95	5:95
Br	CO_2Et	92	93:7

Scheme 5

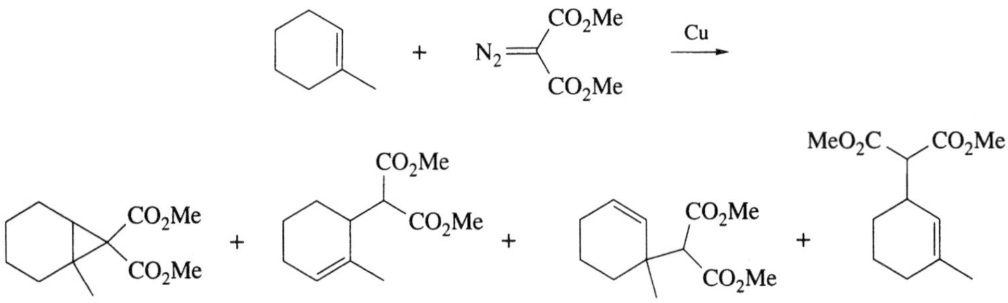

(10)

4.8.4.1.3 *Diastereoselectivity*

Considering that the ring opening of cyclopropanes can proceed in a stereodefined manner, very valuable synthetic sequences would be possible if the original cyclopropanation was also stereoselective. Cyclopropanation of alkenes generally proceeds stereospecifically from the point of view of alkene geometry, although this is not the case with cobalt catalysis.[21] The stereoselectivity with respect to the arrangement of the alkene and carbenoid, however, is not usually particularly impressive with ethyl diazoacetate, as can be seen in some representative examples shown in Table 2.[37] A slight preference for the thermodynamically more stable *trans* isomer (**23a**) was observed. Further enhancement occurred when *cis*-disubstituted alkenes were used as substrates. This tendency was greater for copper catalysts than the rhodium(II) salts, while palladium catalysts were less selective. Ligands on the catalyst do not exert a great effect but with extremely hindered systems such as (**11**)[18] it was possible to change the distribution to slightly favor the *cis* isomer.

Table 2 Stereoselectivity of Cyclopropanation Reactions with Ethyl Diazoacetates

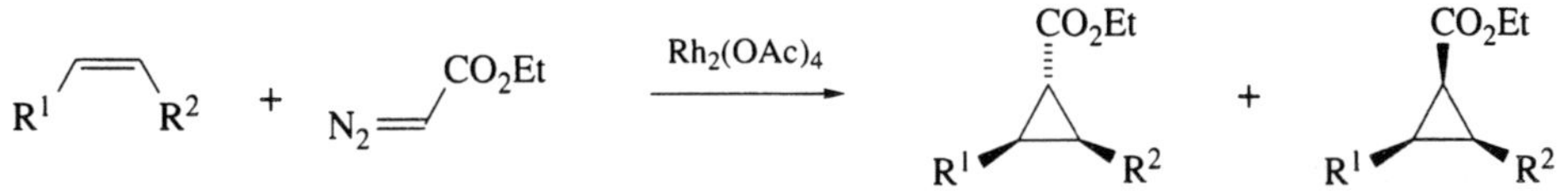

R^1	R^2	Ratio (**23a**):(**23b**)
$Me(CH_2)_3O$	H	1.7
Ph	H	1.4
$Me(CH_2)_3$	H	1.5
Me_2CH	H	2.0
Bu^t	H	4.2
—c-$(CH_2)_4$—		3.8
—c-$(CH_2)_3$—		2.1
—c-$(CH_2)_3O$—		6.5

Some recent studies suggest that considerable improvement in the stereoselectivity of the cyclopropanation reaction may be possible by altering the carbenoid structure. For example, Doyle[17] has shown that the use of sterically bulky diazoacetates, such as (**24a**) and (**24b**) resulted in a much more selective cyclopropanation, favoring the *trans* isomer (**25a**), particularly when rhodium(II) acetamide was used as catalyst (Scheme 6). Unfortunately, (**24b**) may be of limited utility because cyclopropanation proceeded in low yield due to competing intramolecular C—H insertion. Entirely different carbenoid structures may be much more selective, as can be seen with the vinyldiazomethane (**26**), whose reaction with styrene under standard conditions produced an isomer ratio of 13:1 (equation 11).[55]

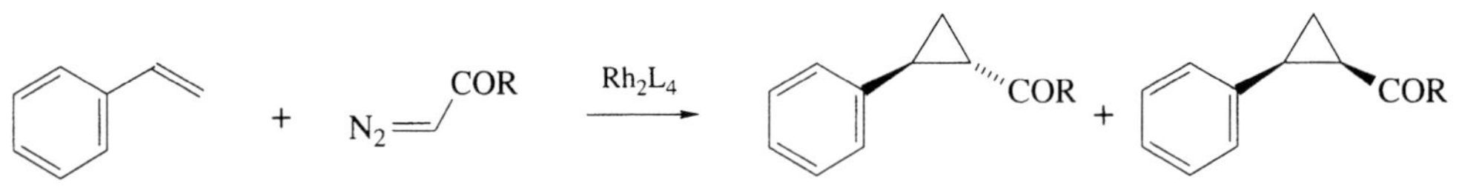

(**24**) (**25a**) (**25b**)

Substrate	*R*	*L*	*Yield (%)*	*Ratio* (**25a**):(**25b**)
(**24a**)	$OCMePr^i_2$	OAc	95	2.4:1
(**24a**)	$OCMePr^i_2$	NHCOMe	87	4.4:1
(**24b**)	NPr^i_2	OAc	53	64:1
(**24b**)	NPr^i_2	NHCOMe	47	114:1

Scheme 6

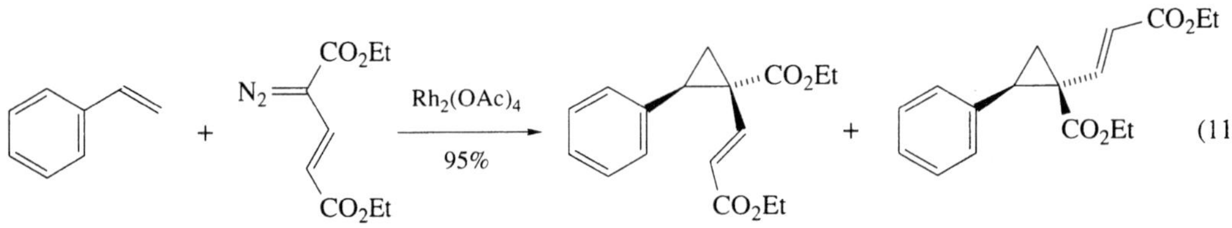

$Rh_2(OAc)_4$, 95% (11)

(**26**) 92% 8%

The effect of the alkene is rather subtle and is not simply caused by steric factors. This is clearly demonstrated in the cyclopropanation of alkenes substituted with halide in the side chain, *e.g.* with (**27**) the predominant product was the thermodynamically unfavored *cis* isomer (**28a**; equation 12).[56]

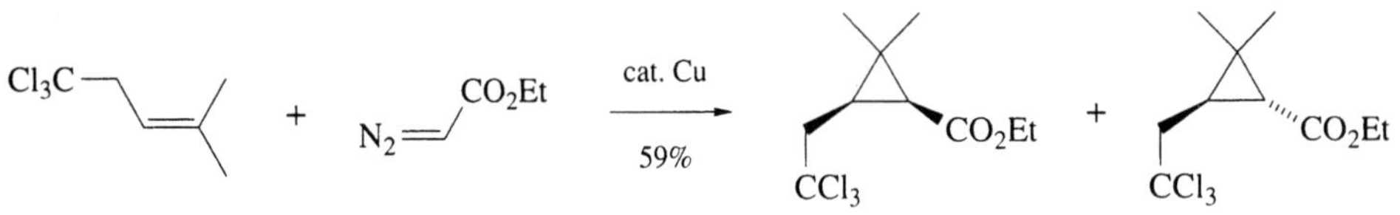

cat. Cu, 59% (12)

(**27**) (**28a**) 85% (**28b**) 15%

4.8.4.1.4 Enantioselectivity

The development of asymmetric cyclopropanation protocols has been actively studied and in recent years remarkable progress has been made. The extent of chiral induction that can now be obtained in these reactions approaches the level of other classic catalytic asymmetric reactions on alkenes, such as catalytic hydrogenation and the Sharpless' epoxidation.[57]

The use of ketocarbenoids with chiral auxiliaries has not been terribly effective at chiral induction. Menthol and borneol esters of diazoacetates resulted in very low enantioselectivity.[58] Some improvements were obtained by using the chiral amide (**29**; equation 13), but low overall yields were obtained due to competing intramolecular side reactions.[59] Related studies with other types of carbenoids, however, have resulted in high enantioselectivity.[60]

In contrast, much more effective asymmetric reactions have been obtained by using chiral copper catalysts. Since the pioneering work of Nozaki and coworkers[61] with a chiral salicylamide catalyst (**30**), a wide variety of other chiral complexes has been developed, the most significant of which are (**31**)–(**34**). Another useful catalyst is the cobalt complex (**35**).

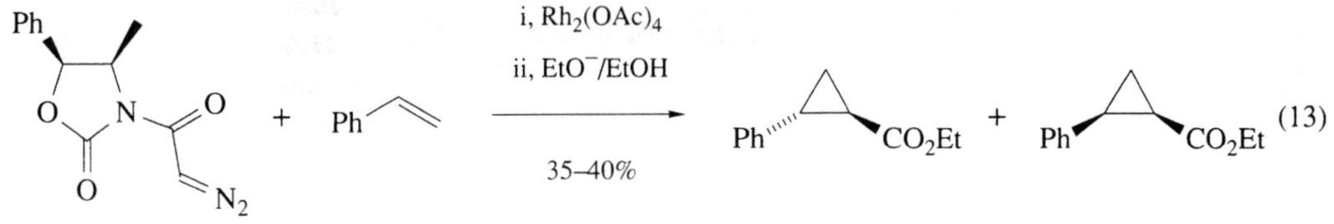

(29) + Ph–CH=CH$_2$ → (i, $Rh_2(OAc)_4$; ii, EtO^-/EtOH; 35–40%) → Ph–cyclopropane–CO_2Et (64%, 14% *ee* (1*R*,2*R*)) + Ph–cyclopropane–CO_2Et (36%, 13% *ee* (1*R*,2*S*)) (13)

(29) (30) 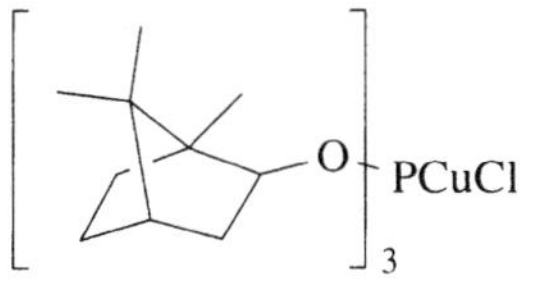 (31) 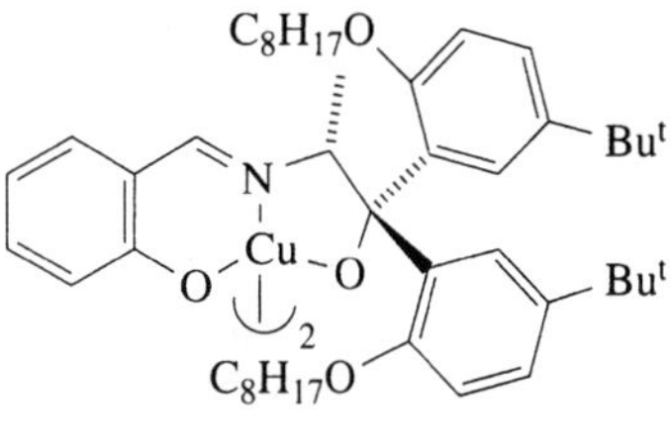 (32) 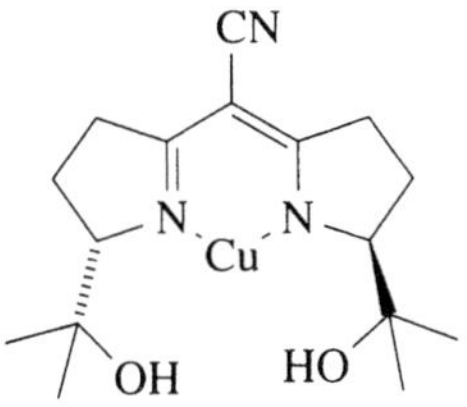 (33) 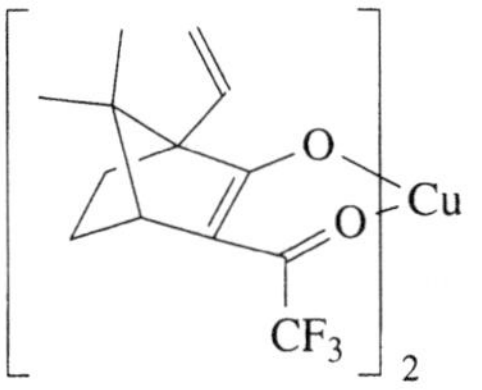 (34) 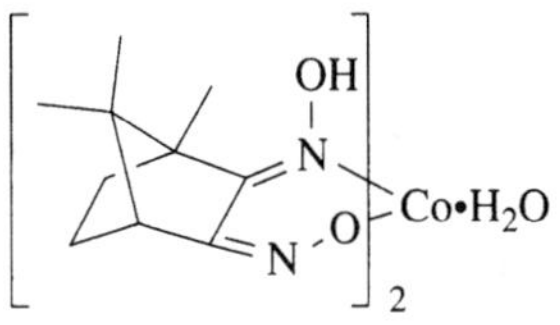 (35)

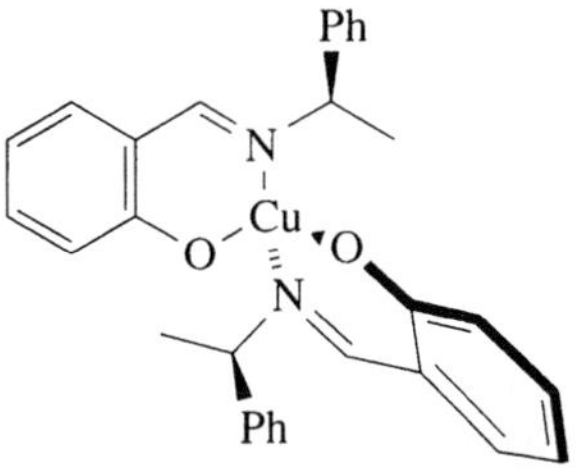

The initial studies in the 1960s with (**30**) and Moser's catalyst (**31**)[15] resulted in rather low levels of enantioselectivity. Great enhancements were obtained by Aratani and coworkers[56,58] using complex (**32**), which is closely related to Nozaki's catalyst except that the chiral ligand is tridentate. Much of Aratani's work was directed towards the enantioselective synthesis of pyrethroid derivatives and a representative example is shown in equation (14). The extent of enantioselectivity was dependent on the nature of the ester functionality and the highest value (90% *ee*) was obtained using the (1*S*,3*S*,4*R*)-menthyl group. Even though (1*R*,3*R*,4*S*)-menthyl diazoacetate produced minimal chiral induction with achiral catalysts, the interaction between (**36**) and (**32**) significantly enhanced the selectivity. Further applications of (**32**) to the synthesis of pyrethroid derivatives[62] and deuterated cyclopropanes[63] have been reported.

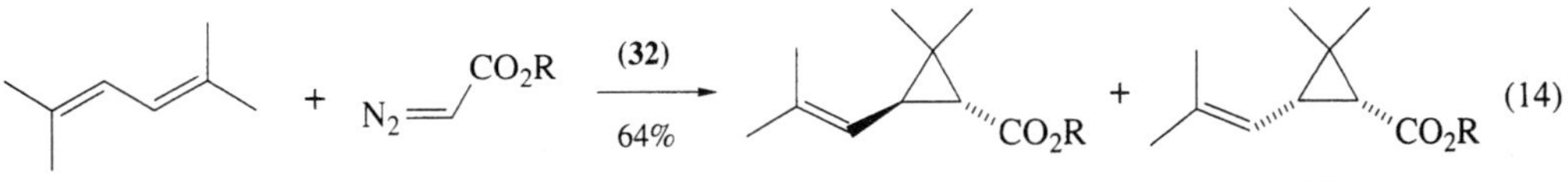

diene + N_2=CH–CO_2R (**36**) → ((**32**), 64%) → cyclopropane–CO_2R (81%, 90% *ee* (1*R*,3*R*)) + cyclopropane–CO_2R (19%, 59% *ee* (1*R*,3*S*)) (14)

R = (1*S*,3*S*,4*R*)-menthyl

A remarkable complex (**33**) with a C_2-symmetric semicorrin ligand has been recently developed by Pfaltz and coworkers.[64] A copper(II) complex was used as a procatalyst, but (**33**) was shown to be the active cyclopropanation catalyst. As shown in Table 3, this complex resulted in spectacular enantioselectivities in the range of 92–97% *ee*. Once again, the (1*S*,3*S*,4*R*)-menthyl group attenuated the selectivity. Unfortunately, even though respectable yields were obtained with dienes and styrenes, the reaction with 1-heptene was rather inefficient.

Another efficient chiral catalyst (**34**) has been developed by Matlin.[65] The only reaction of (**34**) reported so far has been between 2-diazodimedone (**38**) and styrene, which resulted in the exclusive production of only one enantiomer of (**39**) as determined by NMR (equation 15). An immobilized derivative of (**34**) was also found to be an effective catalyst, generating (**39**) in 98% enantiomeric excess (43% yield). These results are remarkable because they represent efficient chiral induction at the prochiral al-

Table 3 Asymmetric Induction with (**33**)

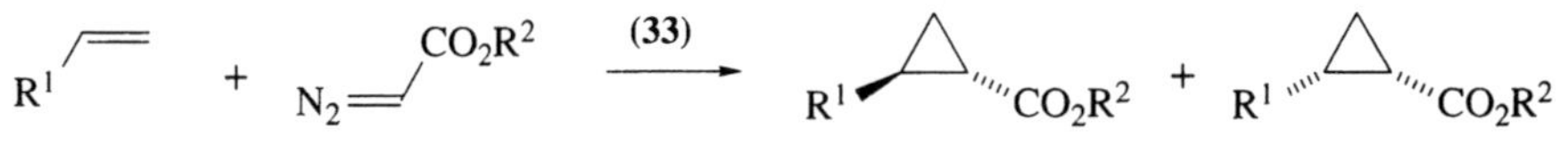

R^1	R^2	Yield (%)	(**37a**):(**37b**)	ee (%)	(**37a**)	ee (%)	(**37b**)
Ph	(1*R*,3*R*,4*S*)-Menthyl	65–75	85:15	91	(1*S*,2*S*)	90	(1*S*,2*R*)
Ph	(1*S*,3*S*,4*R*)-Menthyl	60–70	82:18	97	(1*S*,2*S*)	95	(1*S*,2*R*)
CH_2=CH	(1*S*,3*S*,4*R*)-Menthyl	60	63:37	97	(1*S*,2*R*)	97	(1*S*,2*S*)
Me_2C=CH	(1*S*,3*S*,4*R*)-Menthyl	77	63:37	97	(1*S*,2*R*)	97	(1*R*,2*S*)
C_5H_{11}	(1*S*,3*S*,4*R*)-Menthyl	30	82:18	92	(1*S*,2*S*)	92	(1*S*,2*R*)

kene, rather than induction at the prochiral carbenoid, which occurs with all the other asymmetric cyclopropanation systems. Further examination of the scope of this type of induction would be very valuable.

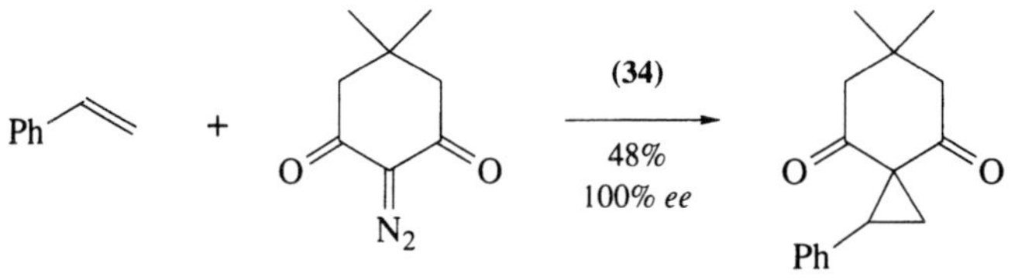

(15)

Although cobalt catalysts have been rarely used in cyclopropanation reactions, Nakamura and coworkers[21] have developed the camphor-based complex (**35**) as a useful asymmetric catalyst, as shown in a typical example in equation (16). High yields were obtained with dienes and styrenes but cyclopropanation did not occur with simple alkenes. Studies with *cis*-d_2-styrene showed that, unlike other catalytic systems, the reaction was not stereospecific with respect to alkene geometry.

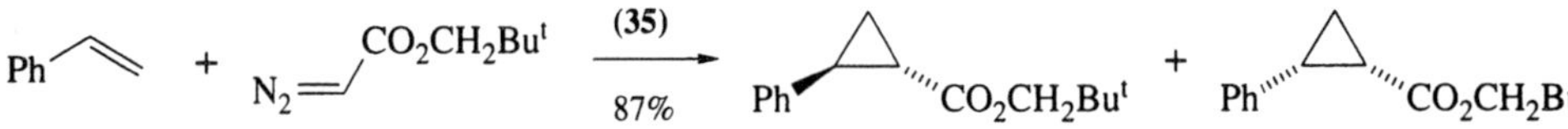

(16)

4.8.4.2 Intramolecular Reactions

The intramolecular cyclopropanation of alkenes has been extensively employed in the synthesis of natural products and molecules of theoretical interest. Reactions which generate [3.1.0]-[1,7,66–94] and [4.1.0]-bicyclic[1,7,95–98] systems are the most common, while longer tethers tend to result in inefficient capture of the carbenoid (equation 17).[99] The main side reactions are due to C—H insertions,[100] but even these can be minimized by appropriate substitution, as shown in equation (18).[101]

With β,γ-unsaturated α′-diazo ketones, the resulting [2.1.0]-bicyclic systems (**40**) were quite unstable and underwent a [2 + 2] cycloreversion to generate ketenes (**41**), which were then trapped by nucleophiles (Scheme 7). The overall scheme has been named a vinylogous Wolff rearrangement and offers a novel entry to products usually derived from a Claisen rearrangement.[102] A recent report describes its application for functionalized angular alkylation in fused ring systems.[103] In contrast, the intramolecular re-

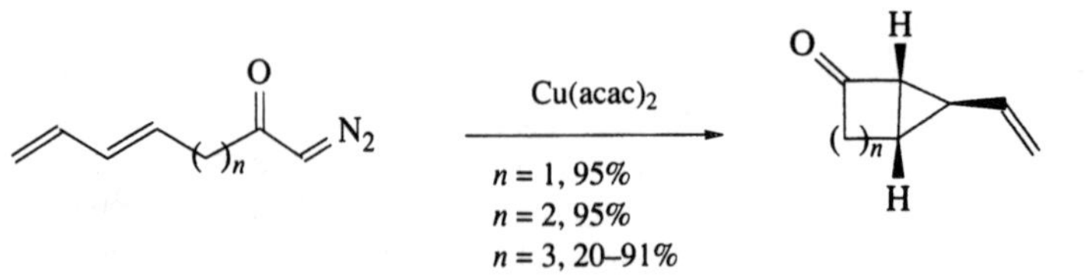

(17)

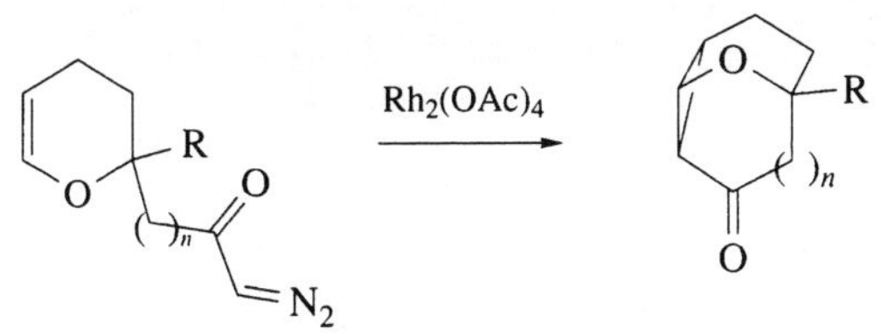

$Rh_2(OAc)_4$

$n = 0$, R = H; 58%
$n = 1$, R = H; 61%
$n = 2$, R = H; 0%
$n = 2$, R = Me; 95%
$n = 3$, R = Me; 95%
$n = 4$, R = H; 0%

(18)

action to cyclopropenes such as (**42**) leads to the highly strained but isolable tricyclic ketone (**43**; equation 19).[104] Other novel structures that have been derived through intramolecular cyclopropanations include barbaralone and 4,4,4-propellane.[7] A recent example includes the rhodium(II) acetate catalyzed decomposition of (**44**), which generated the novel structure (**45**; equation 20).[105] The reaction was considered to proceed through an unusual 1,6-addition, although cyclopropanation followed by a 1,5-sigmatropic rearrangement would be a reasonable alternative mechanism.

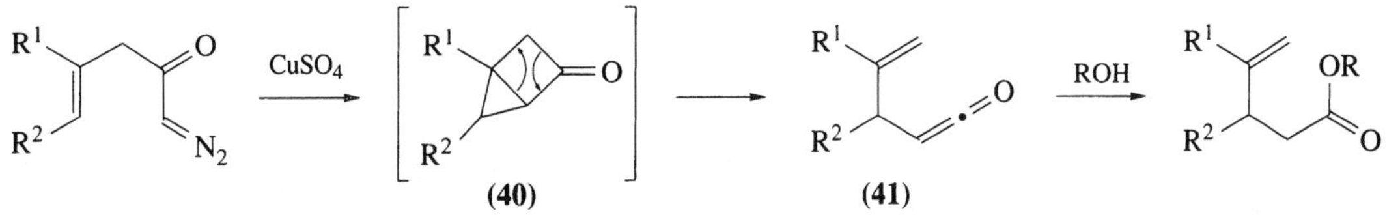

$CuSO_4$ (**40**) (**41**) ROH

Scheme 7

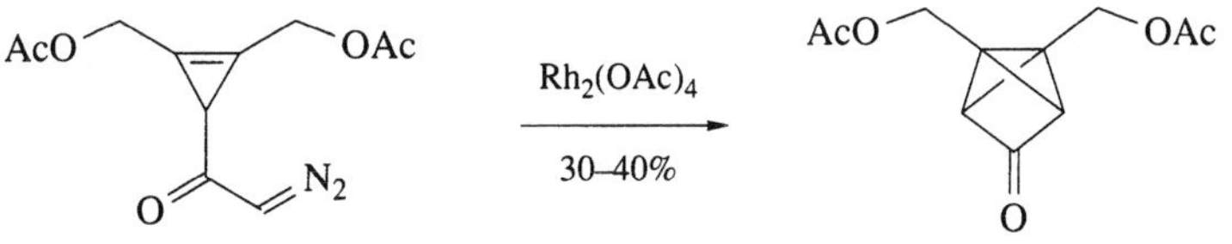

(**42**) $Rh_2(OAc)_4$, 30–40% (**43**) (19)

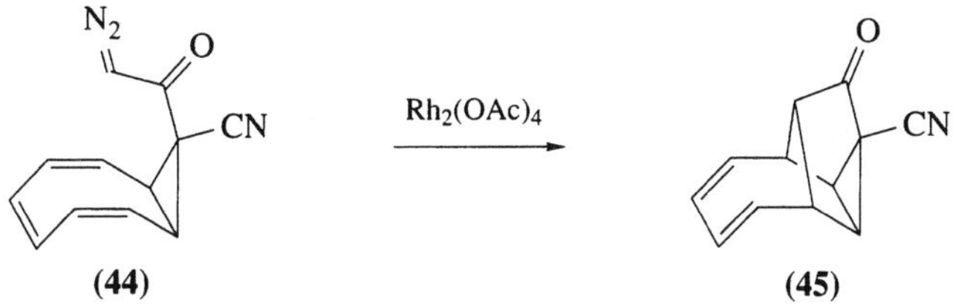

(**44**) $Rh_2(OAc)_4$ (**45**) (20)

Unlike their intermolecular counterparts, intramolecular cyclopropanations are stereospecific with respect to both alkene and carbenoid structure, as can be seen in equations (21) and (22).[87] The influence of other stereogenic centers can also be significant. Numerous examples are known where excellent stereocontrol is obtained when the stereogenic center is part of a tether constrained within a ring, as illustrated in equation (23).[96] Stereocontrol is less certain in acyclic systems. With the extremely bulky grouping in (**46**) excellent stereocontrol was observed to generate only (**47**; equation 24).[95a] In the case of (**48**), which has a longer and more flexible tether, only moderate stereoselectivity was observed (equation 25),[95b] while with (**49**) the methyl group has no influence on the product distribution (equation 26).[66] Only moderate asymmetric inductions by chiral catalysts[93] or chiral auxiliaries[94] have been reported so far for intramolecular cyclopropanations. More effective chiral catalysts such as (**33**) and (**34**) are now available and it is reasonable to predict that high asymmetric inductions should be feasible in the near future.

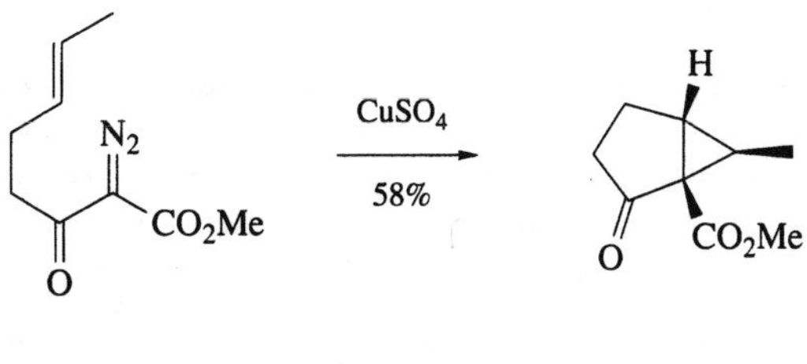

$CuSO_4$, 58% (21)

$CuSO_4$, 60% (22)

$[\ldots NBu^t]_2 Cu^{2+}$, 84% (23)

Cu bronze/110 °C, 48% (24)

(46) **(47)**

(48)

$Cu(acac)_2$, 35%

84% 16% (25)

(49) $Cu(acac)_2$, 92% 50% 50% (26)

A spectacular example showing the effect of catalyst on product distribution is shown in Scheme 8.[92] Palladium(II) acetate catalyzed decomposition of (**50**) resulted in cyclopropanation to form (**51**). When rhodium(II) acetate was used as catalyst, however, a different reaction occurred, whereby the carbenoid was predominantly captured by the carbonyl to generate ultimately (**52**).

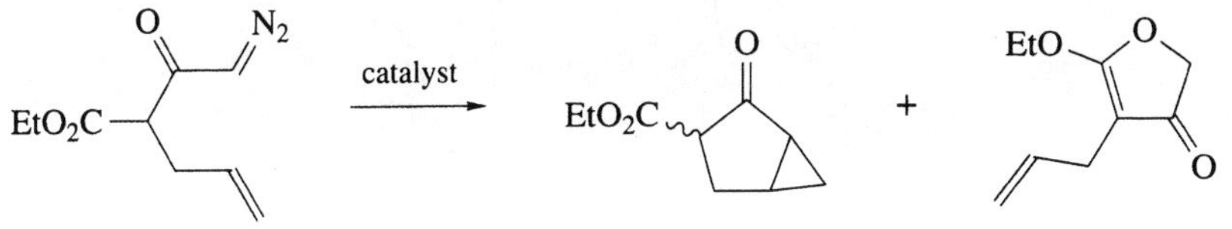

catalyst

(50) **(51)** **(52)**

	(51)	(52)
$Pd(OAc)_2$	53%	–
$Rh_2(OAc)_4$	1%	58%

Scheme 8

Cyclopropanes are present in a variety of natural products and the intramolecular cyclopropanation sequence allows ready access to such compounds. The sesquiterpenoid antibiotic (±)-cycloeudesmol (**53**) was readily prepared from the monocyclic system (**54**; Scheme 9).[91] Copper sulfate catalyzed decomposition of (**54**) generated the tricyclic system (**55**) with full control of stereochemistry. Further conversion of (**55**) to (±)-cycloeudesmol was achieved in four steps in 81% overall yield. A second example shown in equation (27)[73] allowed access to (**56**), an important substructure of the antibiotic CC-1066.

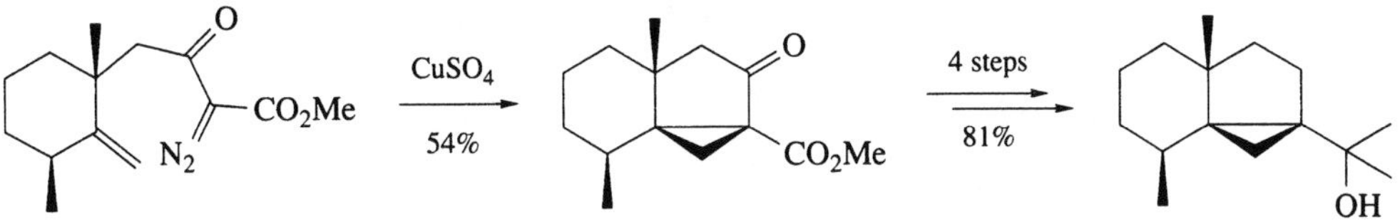

$CuSO_4$, 54%; 4 steps, 81%

(54) **(55)** **(53)**

Scheme 9

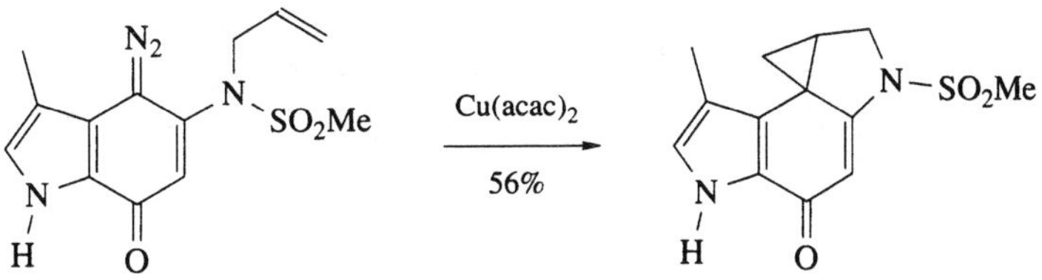

$Cu(acac)_2$, 56% (27)

(56)

4.8.4.3 Further Transformations of Cyclopropanes

Due to the strain associated with the cyclopropane system, a variety of ring-opening transformations can occur. As this material has been extensively reviewed in recent years,[6–8,106–108] this section will concentrate only on the most useful synthetic processes by means of some illustrative examples.

Cyclopropanes in rigid systems have been selectively cleaved by catalytic hydrogenation[80] or lithium/liquid ammonia reduction.[81,97,98] Only the bond with greater overlap to the carbonyl was ruptured. A particularly attractive application of this transformation is the synthesis of spiro compounds such as (**57**) by cleavage of the tricyclic system (**58**; Scheme 10).[97] Alternatively, under acidic conditions, ring opening to the bicyclic system (**59**) was observed. The flexibility of this approach to spiro compounds has been nicely demonstrated through the synthesis of (±)-α-chamigrene (**60**) and (±)-acorenone (**61**).[98]

The cyclopropane ring can also be cleaved by a retro-Michael reaction. By appropriate positioning of the carbonyl groups by means of an intramolecular cyclopropanation, controlled ring opening to spiro systems (**62**),[84] bicyclo[3.2.1]octanes (**63**)[82] or bicyclo[2.2.2]octanes (**64**),[82] has been achieved (equations 28–30).

α-Hydroxyalkylcyclopropanes readily ring open under elimination conditions. An example of this sequence is the synthesis of (±)-hinesol (**65**), as shown in Scheme 11.[83,84] Corey and coworkers have reported several elegant applications of this ring-opening sequence, as shown in the synthesis of (±)-cafestol (**66**; Scheme 12).[71] An intramolecular cyclopropanation generated (**67**) stereoselectively. On conversion of (**67**) to the triflate (**68**) a remarkable transformation occurred to generate the polycyclic

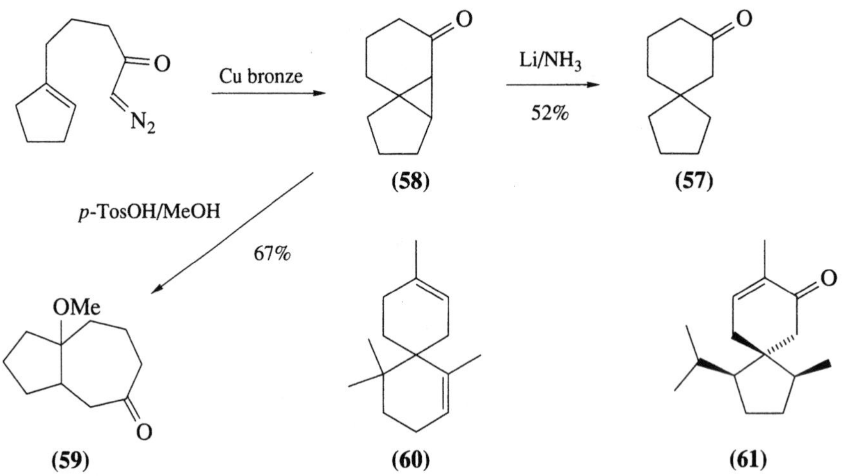

Cu bronze

Li/NH_3

52%

(58) **(57)**

p-TosOH/MeOH

67%

OMe

(59) **(60)** **(61)**

Scheme 10

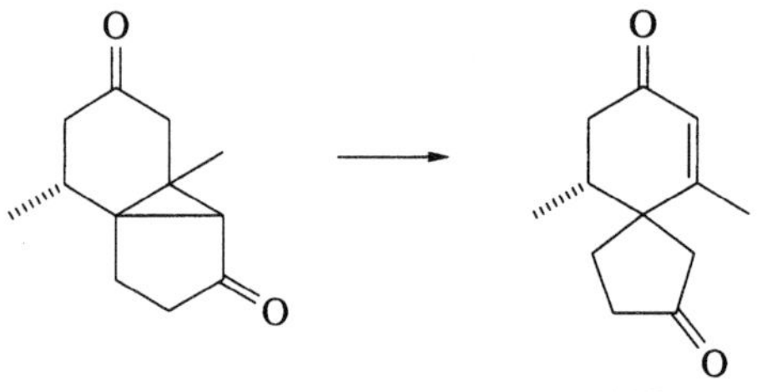

(28)

(62)

(29)

(63)

(30)

(64)

system (**69**) which was then modified to cafestol (**66**). Variations of this strategy have been used for the synthesis of (±)-atractyligenin[72] and (±)-kahweol.[70]

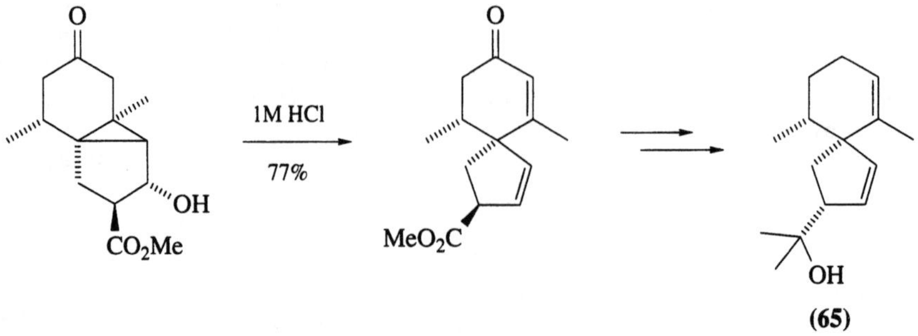

1M HCl

77%

CO_2Me MeO_2C OH

(65)

Scheme 11

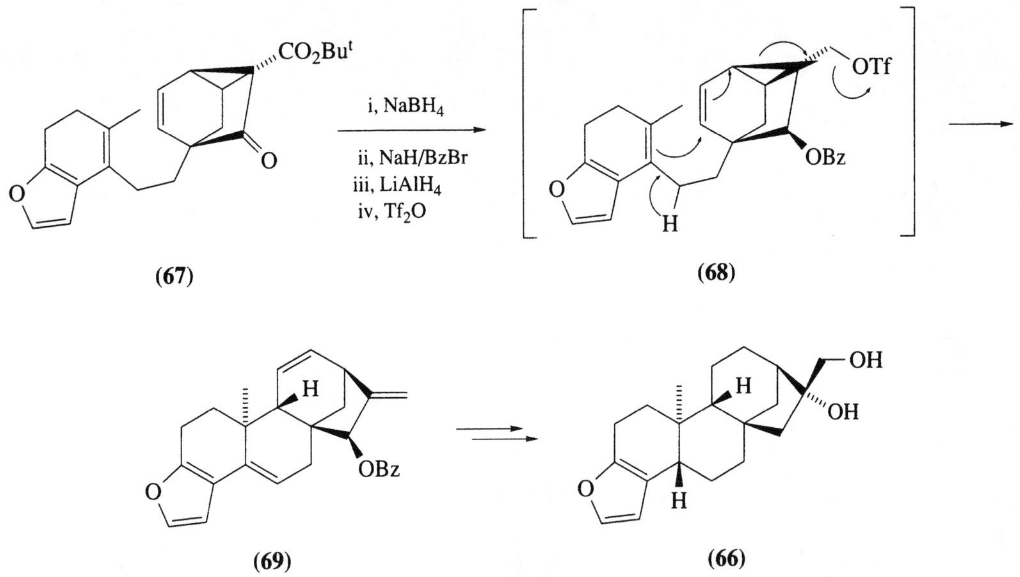

i, $NaBH_4$
ii, NaH/BzBr
iii, $LiAlH_4$
iv, Tf_2O

(67) (68)

(69) (66)

Scheme 12

Nucleophile-induced ring opening of cyclopropanes generally proceeds with inversion of configuration.[8a] Only very powerful soft nucleophiles will react with monoactivated cyclopropanes, but a much wider range of nucleophiles is possible if two electron-withdrawing groups are present. An early example of the use of this chemistry was reported by Corey and Fuchs (Scheme 13).[85] Copper powder induced decomposition of the diazo compound (**70**) generated the tricyclic system (**71**). Nucleophilic ring opening of the doubly activated cyclopropane with divinylcuprate followed by decarboxylation generated the advanced prostanoid precursor (**72**). Another study in the prostaglandin field was carried out by Kondo and coworkers (Scheme 14).[86] Copper(acetylacetonate)-catalyzed decomposition of (**73**) generated an inseparable mixture (2:1) of (**74a**) and (**74b**). Fortunately, (**74a**) and (**74b**) were distinguishable in their reaction with benzenethiol, because under controlled conditions only the desired isomer (**74a**) underwent ring opening with backside displacement to form (**75**; 85% yield based on **74a** in the mixture). In a series of further transformations (**75**) was converted to (±)-prostaglandin $F_{2\alpha}$ (**76**). A key step in a recent synthesis of (±)-quadrone also involved a nucleophilic ring opening of a cyclopropane.[89,90]

Stereoselective synthesis of medium-sized rings was possible from compounds such as (**77**; Scheme 15),[101a] which were readily derived from intramolecular cyclopropanations (see equation 18). Treatment of (**77**) with 1 equiv. of ethanethiol in the presence of a catalytic amount of zinc chloride generated the hemithioacetal (**78**), which in the presence of boron trifluoride etherate formed the cycloheptane system (**79**), in which the two carbonyls are differentiated.

Ring opening of cyclopropanes by nitrogen nucleophiles has been extensively used by Danishefsky and coworkers for the synthesis of alkaloids.[109] The synthesis of (±)-hastanicine (**80**; Scheme 16)[95a] is described to illustrate the strategy. The stereodefined system (**47**) was obtained through an intramolecular cyclopropanation (equation 24). Deprotection of the amine in (**47**) was achieved by treatment with hydrazine, which then opened the cyclopropane to form (**81**), in which the correct stereochemistry was obtained for further conversion to (**80**). A more ambitious target, based on the same overall strategy, was the synthesis of a potential mytomycine precursor.[95b]

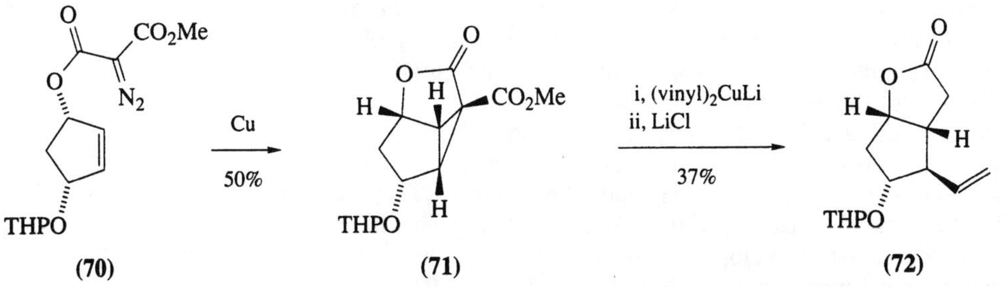

Cu
50%

i, $(vinyl)_2CuLi$
ii, LiCl
37%

(70) (71) (72)

Scheme 13

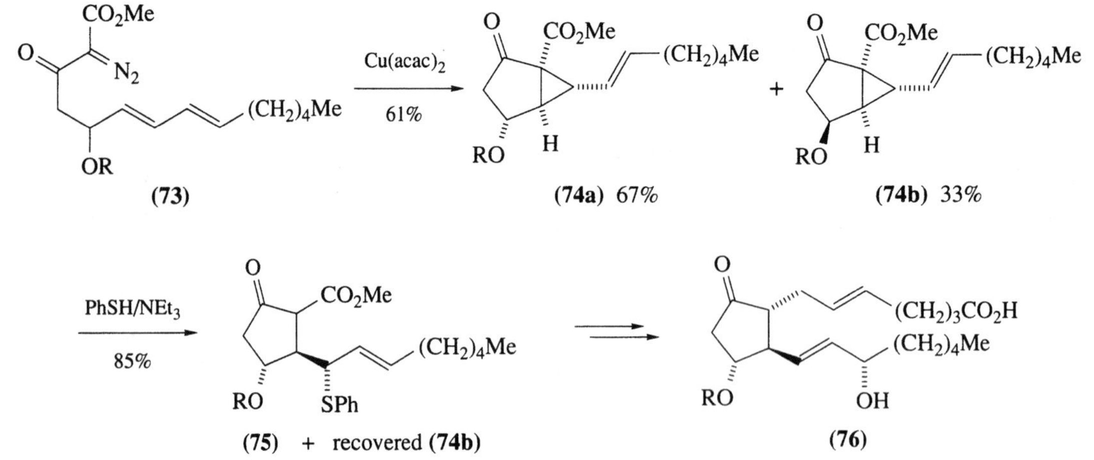

(73) (74a) 67% (74b) 33%

(75) + recovered (74b) (76)

Scheme 14

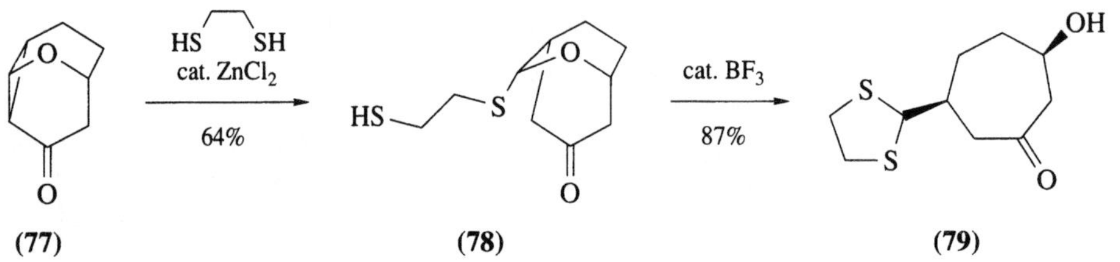

(77) (78) (79)

Scheme 15

(47) (81) (80)

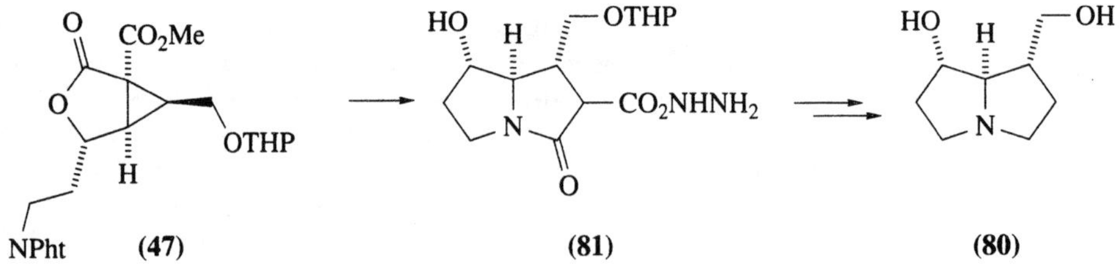

Scheme 16

Vinyl ethers are very efficiently cyclopropanated and the resulting donor–acceptor-substituted cyclopropanes (**82**) have been widely used in organic synthesis.[8a,106] Ring opening can be achieved under very mild conditions resulting in a versatile approach to 1,4-difunctionalized compounds. On treatment with mild acid ring opening of (**82**) occurred to form 1,4-dicarbonyl compounds (**83**),[110] which are useful precursors for cyclopentenones (**84**)[110] or furans (**85**).[26a,b,110,111] Alternatively, treatment of (**82**) with phenylselenyl chloride,[112] or bromine,[113] generated the substituted systems (**86**) and (**87**), respectively. Another mild procedure involved treatment of (**82**) with a catalytic amount of trimethylsilyl chloride and bis(trimethylsilyl)amine, which resulted in the formation of the silylated product (**88**).[114] If (**82**) was first reduced to the corresponding alcohol (**89**), the resulting mild acid hydrolysis generated β,γ-unsaturated ketones (**90**; Scheme 17).[115]

Some recent studies have shown that the enolate of the cyclopropane (**91**) was readily formed, and the resulting aldol product (**92**) underwent a remarkable diversity of ring-opening reactions, as shown in Scheme 18.[39] Treatment with acid generated the β,γ-unsaturated ketone (**93**), which under more vigorous conditions cyclized to the furanone (**94**). On treatment of (**92**) with acidic methanol the tetrahydrofuran (**95**) was formed, which on thermolysis was converted to the dihydrofuran (**96**).

Wenkert and coworkers have reported several applications of this chemistry to the synthesis of alkaloids and terpenes.[110] A recent example, leading to the methyl ether of tetrahydropyrethrolone (**97**), is illustrated in Scheme 19.[116] Reaction of 1,2-dimethoxypropene with 1-diazo-2-heptanone (**98**) resulted in the formation of the labile cyclopropane (**99**), which was directly converted to the 1,4-diketone (**100**). Base-induced aldol condensation of (**100**) readily formed (**97**). Stereoselective syntheses of prostaglandins[117] and dicranenone A[118] have also been developed using acid-induced ring-opening reactions.

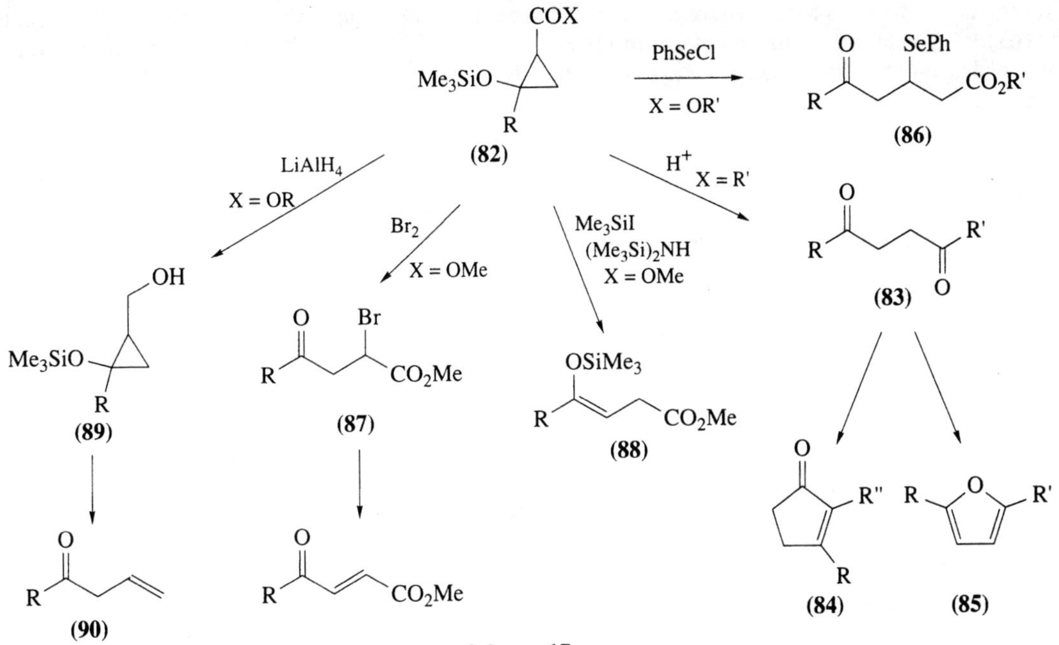

Scheme 17

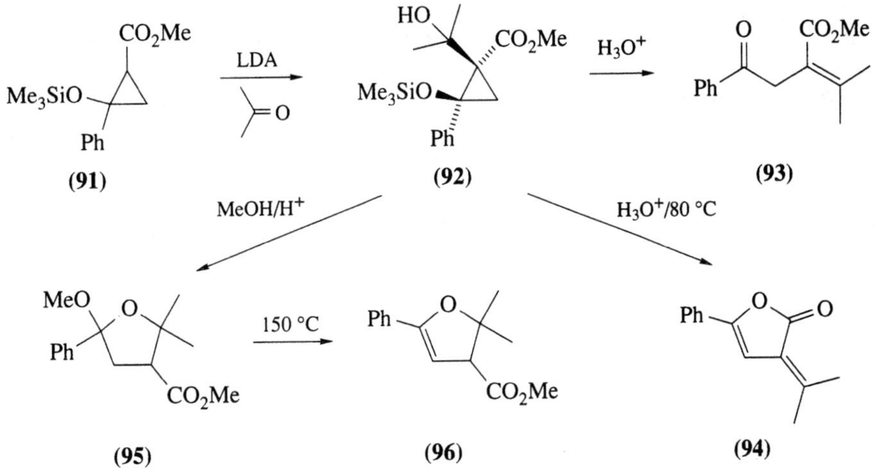

Scheme 18

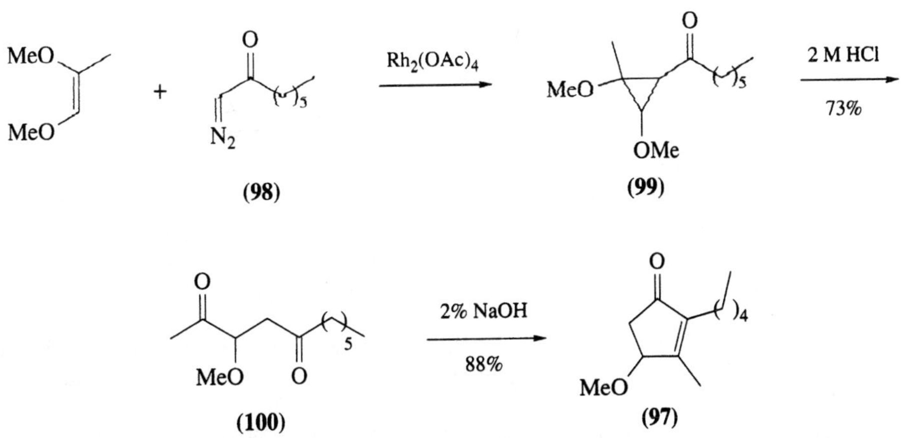

Scheme 19

Reissig and coworkers recently reported a rather direct route to the intramolecular Diels–Alder precursor (**101**).[119] Alkylation of the enolate from (**102**) followed by fluoride-induced ring opening generated (**101**), which on standing underwent a smooth transformation to the cycloadducts (**103a**) and (**103b**) (Scheme 20).

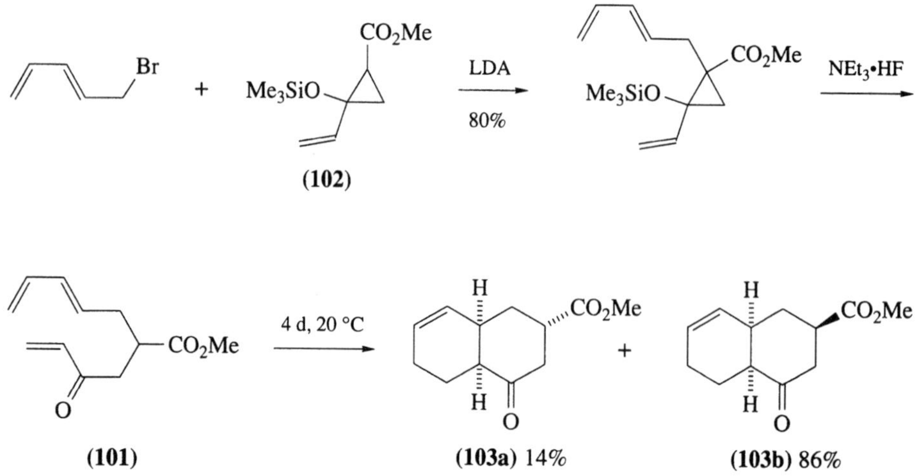

Br + Me_3SiO CO_2Me (**102**) LDA 80% Me_3SiO CO_2Me $NEt_3{\cdot}HF$

CO_2Me O (**101**) 4 d, 20 °C H H O CO_2Me (**103a**) 14% + H H O CO_2Me (**103b**) 86%

Scheme 20

An elegant strategy for the synthesis of fused cyclopentanoids has been reported by Marino and coworkers (Scheme 21).[38] Reaction of (**104**) with the phosphonium salt (**105**) generated the bicyclic system (**106**). Further conversion of (**106**) to (**107**) enabled the annulation sequence to be repeated to form triquinane derivatives such as (**108**).

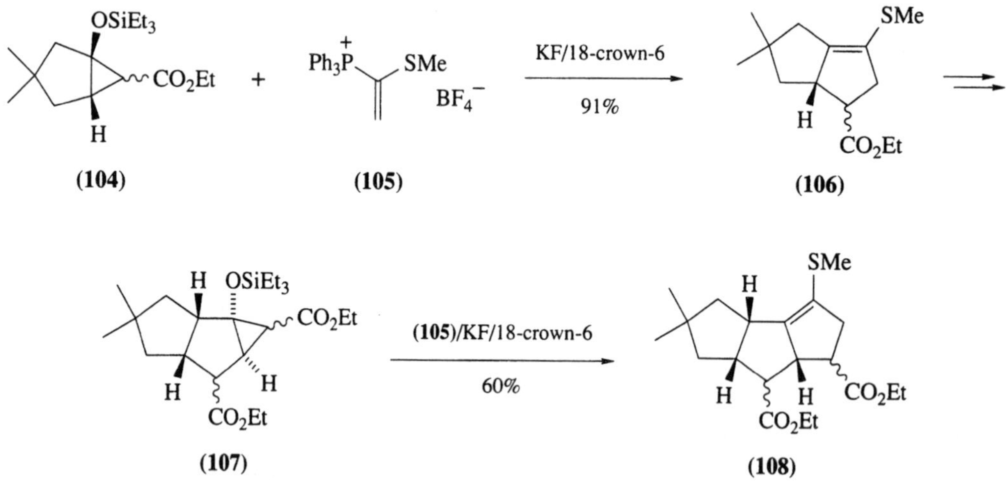

$OSiEt_3$ CO_2Et H (**104**) + $Ph_3\overset{+}{P}$ SMe BF_4^- (**105**) KF/18-crown-6 91% SMe H CO_2Et (**106**)

H $OSiEt_3$ CO_2Et H H CO_2Et (**107**) (**105**)/KF/18-crown-6 60% SMe H H H CO_2Et CO_2Et (**108**)

Scheme 21

Thermolysis of vinylcyclopropanes offers a direct approach to the synthesis of cyclopentenes and several imaginative applications have been developed.[107,108] Hudlicky and coworkers[88] have used this chemistry to synthesize a range of compounds, as demonstrated in Scheme 22.[88a] Decomposition of (**109**) by a mixture of copper salts generated (**110**) in 75% yield. This particular example of cyclopropanation is quite significant because it showed that dienes with electron-withdrawing groups were tolerated and also that two contiguous quaternary carbons were readily formed. Flash vacuum pyrolysis of (**110**) generated the tricyclic system (**111**), which was further transformed to the triquinane derivative (**112**). This strategy has been applied to the synthesis of (±)-pentalene,[88c] (±)-sakromycin,[88g] a guaiane ring system[88b] and the plant hormone antheridiogen-An.[96]

The Cope rearrangement of *cis*-divinylcyclopropanes is thermally allowed and offers an attractive stereoselective approach to cycloheptadienes. Cyclopropanation reactions can be used to prepare divinylcyclopropanes, as shown in Scheme 23.[120] Reaction of ethyl diazopyruvate with butadiene generated

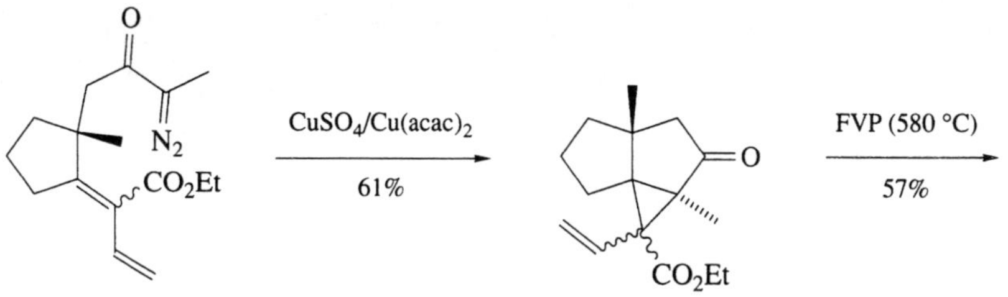

Scheme 22

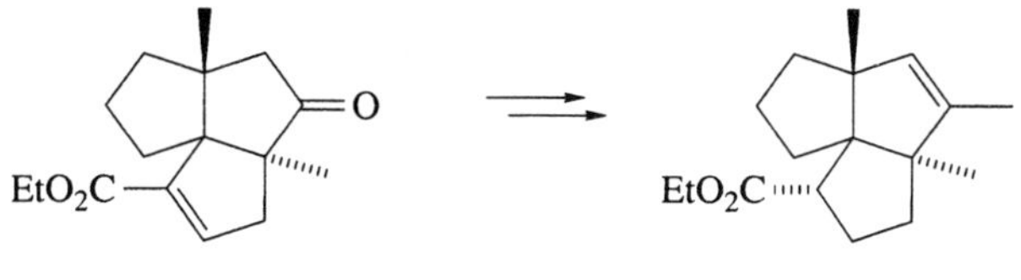

the *trans*-cyclopropane (**113**), the dihydrooxepine (**114**; presumably derived from the corresponding *cis*-cyclopropane) and the dihydrofuran (**115**). The *trans*-cyclopropane (**113**) was readily converted to the *trans*-divinylcyclopropane (**116**) by a Wittig reaction. Pyrolysis of (**116**) led to the cycloheptadiene (**117**), presumably *via* initial isomerization of (**116**) to the *cis*-divinylcyclopropane. Further transformations on (**117**) generated the tropone natural product, nezukone (**118**).

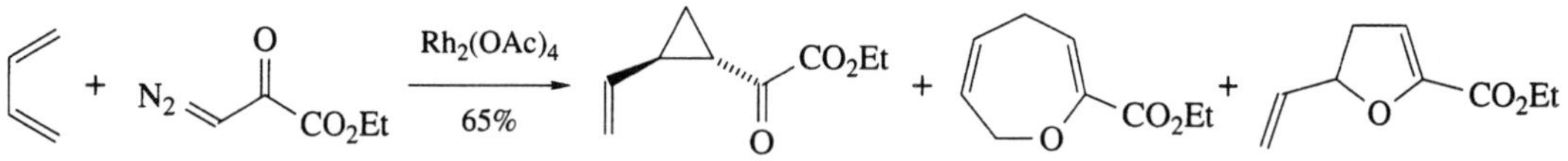

Scheme 23

An elegant application of this chemistry to the formal synthesis of (±)-quadrone has been reported by Piers and coworkers.[78a,121a] Decomposition of ethyl diazoacetate in the presence of the bicyclic structure (**119**) resulted in selective cyclopropanation to form (**120**). Further modification of (**120**) generated the *trans*-divinylcyclopropane (**121**), which on thermolysis followed by desilylation produced the tricyclic system (**122**). Conversion of (**122**) to the ketone (**123**) completed the formal synthesis because (**123**; Scheme 24) had been previously converted to (±)-quadrone.[121b]

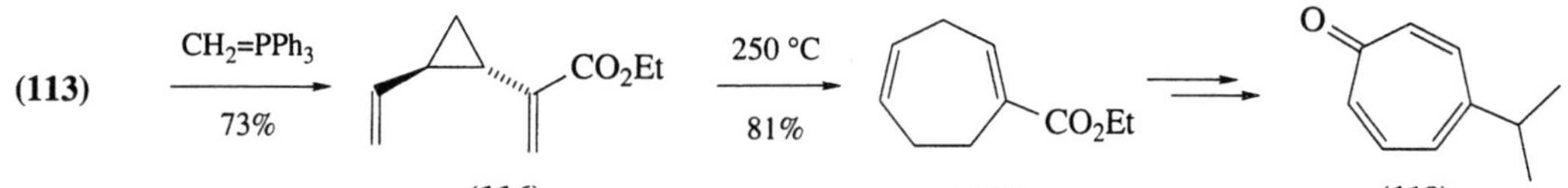

Recently, Davies and coworkers have shown that direct formation of *cis*-divinylcyclopropanes is possible by reaction of dienes with vinylcarbenoids, as illustrated in Scheme 25.[27] The reaction of (**26**) with cyclopentadiene gave exclusively the *cis* isomer (**124**) which rearranged to the bicyclo[3.2.1]octadiene system (**125**) under the reaction conditions. The involvement of a divinylcyclopropane intermediate in this reaction was confirmed by using more sterically congested vinylcarbenoids, as this enabled the presumed intermediate to be isolated. The remarkable *cis* selectivity for the cyclopropanation of vinylcarbenoids is in stark contrast to the poor selectivity that is generally observed in cyclopropanations with ethyl diazoacetate.[1,2] The reaction has been applied to intramolecular examples and the results were dependent on diene geometry.[79] A system with a *trans* double bond nearest the tether (**126**) gave rise to a fused cycloheptadiene (**127**) *via* rearrangement of the *cis*-divinylcyclopropane intermediate (**128**), while the system with a *cis* geometry (**129**) gave exclusively the *trans*-divinylcyclopropane (**131**; Scheme 26).

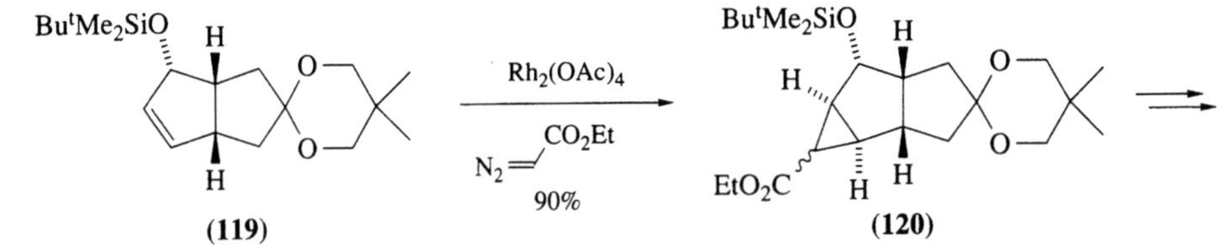

(119) (120)

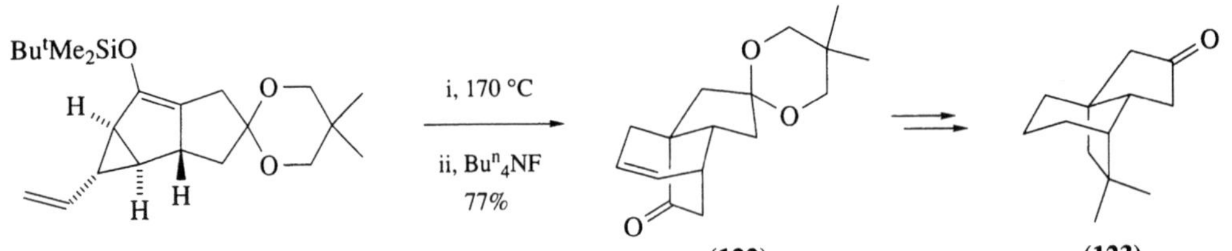

(121) (122) (123)

Scheme 24

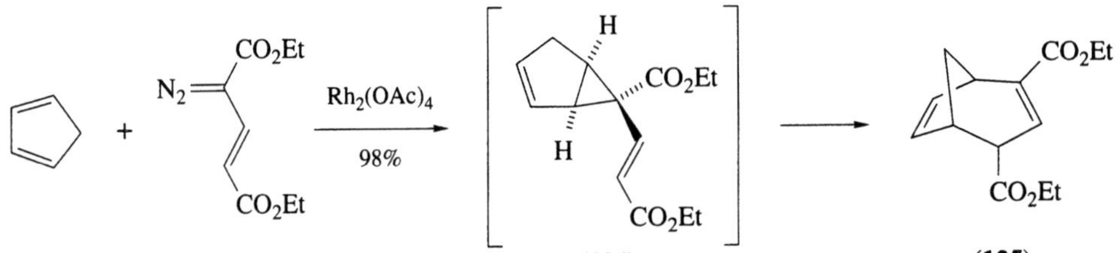

(26) (124) (125)

Scheme 25

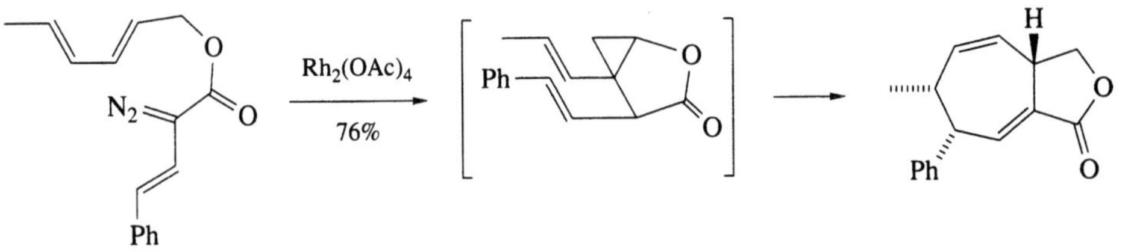

(126) (127) (128)

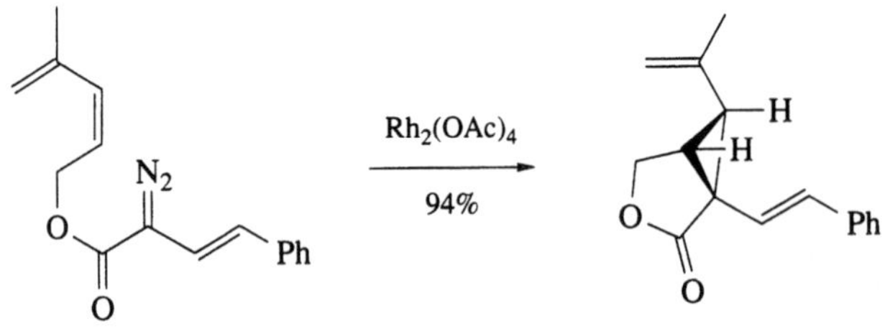

(129) (130)

Scheme 26

4.8.5 ADDITION OF KETOCARBENOIDS TO ALKYNES

The reaction of ketocarbenoids with alkynes is a direct method for the synthesis of functionalized cyclopropenes.[1,122,123] Until quite recently copper catalysis was generally used and the reactions proceeded in fairly moderate yields, except with terminal alkynes, which failed to generate cyclopropenes due to competing C—H insertions.[1] This limitation could be circumvented, however, by using trimethylsilyl derivatives. This approach is illustrated in the synthesis of (**131**), the unsaturated analog of 1-aminocyclopropanecarboxylic acid, the biosynthetic precursor to ethylene in plants (Scheme 27).[124] The initial cyclopropene (**132**) could be readily desilylated by stirring in aqueous potassium carbonate solution at room temperature.

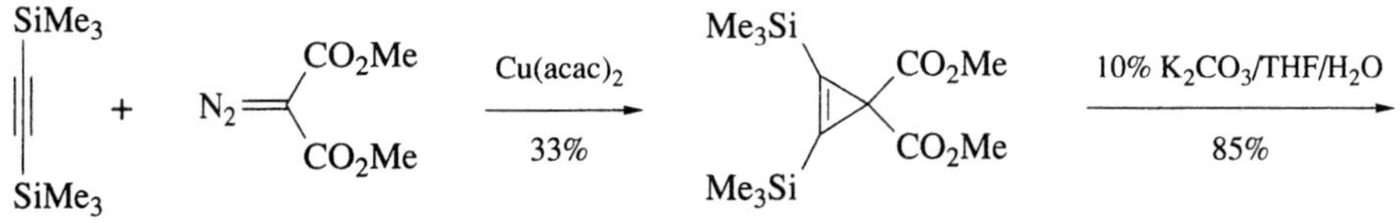

$SiMe_3$–C≡C–$SiMe_3$ + N_2=C(CO_2Me)$_2$ —[$Cu(acac)_2$, 33%]→ (**132a**) —[10% K_2CO_3/THF/H_2O, 85%]→ cyclopropene(CO_2Me)$_2$ → (**131**) (NH_3Cl, CO_2H)

Scheme 27

Over the last few years it has become clear that rhodium(II) acetate is more effective than the copper catalysts in generating cyclopropenes.[125,126] As shown in Scheme 28,[125] a range of functionality, including terminal alkynes, can be tolerated in the reaction with methyl diazoacetate. Reactions with phenylacetylene and ethoxyacetylene were unsuccessful, however, because the alkyne polymerized under the reaction conditions.

R^1–C≡C–R^2 + N_2=CHCO_2Me —[$Rh_2(OAc)_4$, 38–86%]→ cyclopropene(R^1, R^2)–CO_2Me

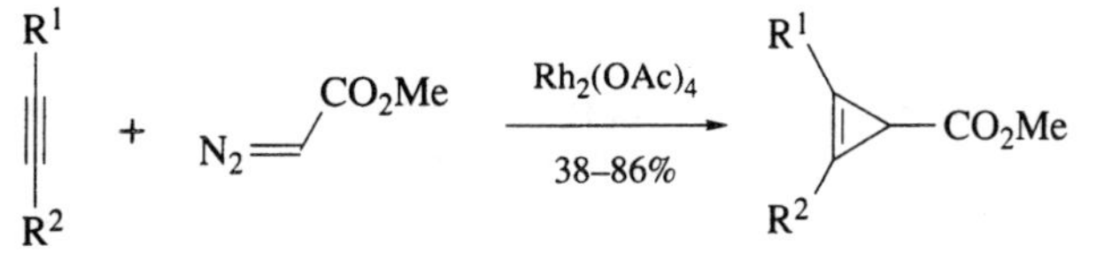

R^1 = Bu^n, Bu^t, c-C_6H_{11}, $MeCO_2CH_2$, $MeOCH_2$, Et; R^2 = H, Et, $MeOCH_2$

Scheme 28

Depending on the carbenoid and alkyne used, furans rather than cyclopropenes may become the predominant products. Indeed, copper salts are known to catalyze the thermal rearrangement of cyclopropenyl esters to furans (equation 31).[127] In systems where a dipolar intermediate is stabilized, furan formation is prevalent. Copper-catalyzed decomposition of ethyl diazopyruvate in the presence of 3-hexyne generated the furan (**133**) in moderate yield (equation 32).[26a] With an unsymmetrical alkyne, a mixture of furans was formed. In contrast, carbenoids with two electron-withdrawing groups are more regioselective in their reaction with unsymmetrical alkynes. Reaction of ethyl diazoacetoacetate with phenylacetylene produced the furan (**134**), in which the keto carbonyl has been incorporated into the ring (equation 33).[128] Alternatively, fused furans (**135**) could be prepared from the copper-catalyzed decomposition of the dibromo compound (**136**) in the presence of alkynes (equation 34).[9] The formation of furans could conceivably be derived directly or *via* an unstable cyclopropene precursor and unless the cyclopropene has been observed or its stability under the reaction conditions determined, definitive conclusions about the actual reaction pathway cannot be made.

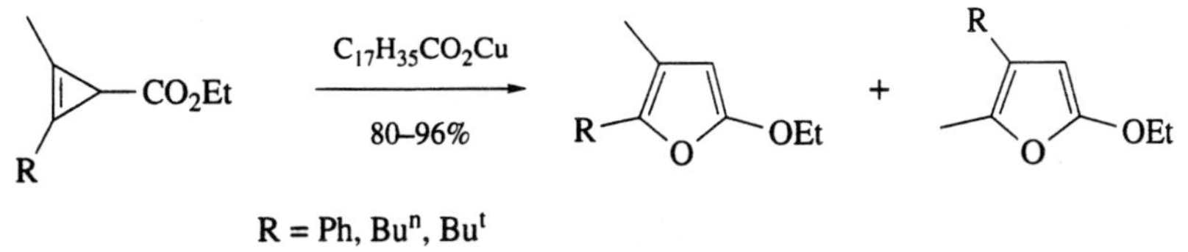

cyclopropene(R)–CO_2Et —[$C_{17}H_{35}CO_2Cu$, 80–96%]→ furan(R, OEt) + furan(R, OEt) (31)

R = Ph, Bu^n, Bu^t

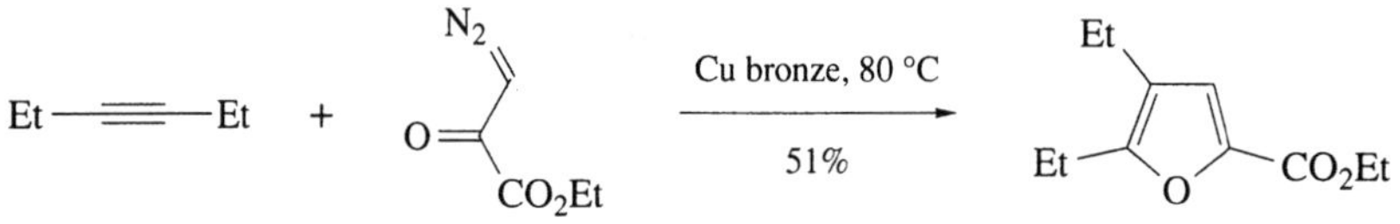

(133) (32)

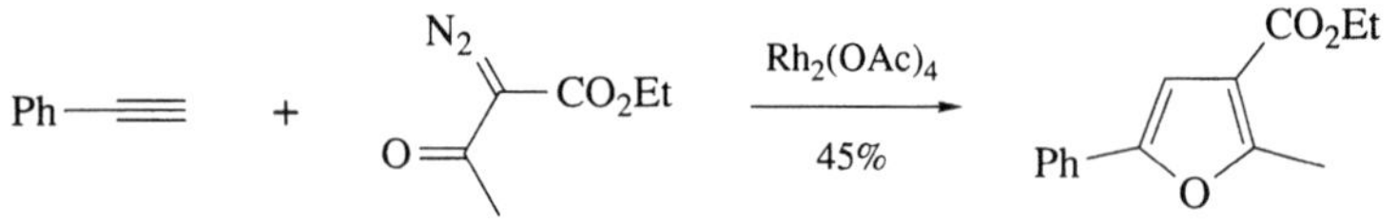

(134) (33)

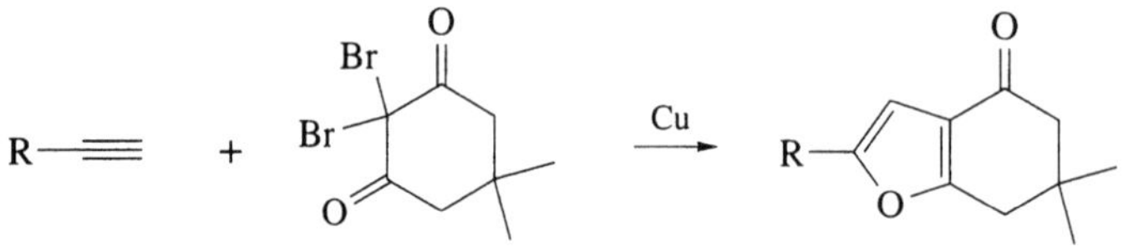

(136) (135) (34)

4.8.6 ADDITION OF KETOCARBENOIDS TO AROMATIC SYSTEMS

4.8.6.1 Benzenes

The reaction of carbenoids with aromatic systems was first reported by Buchner and coworkers in the 1890s.[6] The reaction offers a direct entry to cycloheptatrienes and has been used to synthesize tropones, tropolones and azulenes.[6] Neither the thermal nor copper-catalyzed reactions, however, proceed in good yield. The problems associated with these transformations were clearly demonstrated in a recent reexamination of the thermal decomposition of ethyl diazoacetate in excess anisole (**137**).[129] A careful analysis of the reaction mixture revealed the presence of seven components (**138–144**) in 34% overall yield (Scheme 29). The cycloheptatrienes (**138**)–(**142**) were considered to be formed by cyclopropanation followed by electrocyclic ring opening of the resulting norcaradienes. A mixture of products arose because the cyclopropanation was not regioselective and, also, the initially formed cycloheptatrienes were labile under the reaction conditions.

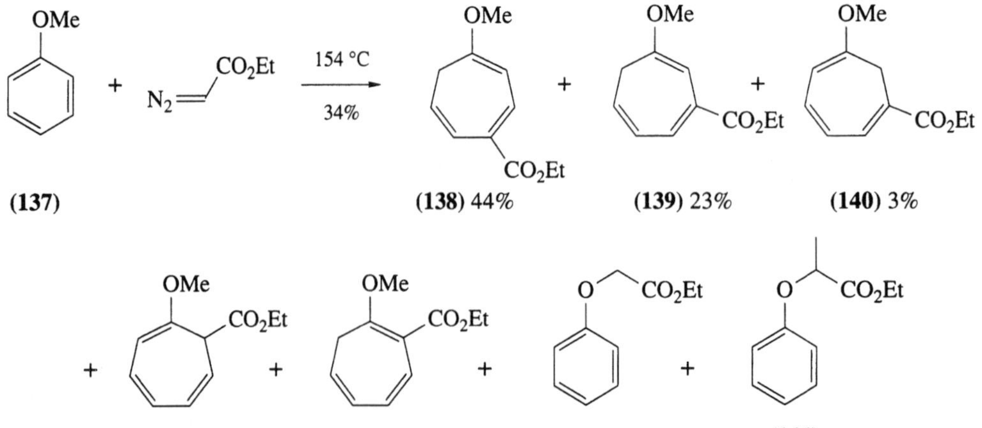

(**137**) + N_2CHCO_2Et —154 °C, 34%→ (**138**) 44% + (**139**) 23% + (**140**) 3% + (**141**) 6% + (**142**) trace + (**143**) 24% + (**144**) trace

Scheme 29

A considerable improvement in the efficiency of the reaction of alkyl diazoacetates with benzenoid systems occurred with the development of rhodium(II) carboxylates as catalysts.[16d] As can be seen in the reaction with benzene, rhodium(II) salts with electron-withdrawing ligands were far superior (Scheme

30), which is presumably because a highly electrophilic carbenoid is required for good interaction with the aromatic system. A particular advantage of rhodium catalysis is that the reaction conditions are quite mild and sigmatropic rearrangement of the rather labile cycloheptatrienes (**145**) usually can be avoided. The carbenoid in these reactions was not particularly selective. With substituted benzenes mixtures were invariably formed (Scheme 31) and selectivity between electronically different aromatic systems was minimal. Alkyl, methoxy and halo substituents could be tolerated on the aromatic ring. Another interesting feature of this system is that steric effects on the carbenoid are significant because as the ester increases in size, from methyl to ethyl to *t*-butyl, a steady decrease in yield was observed.

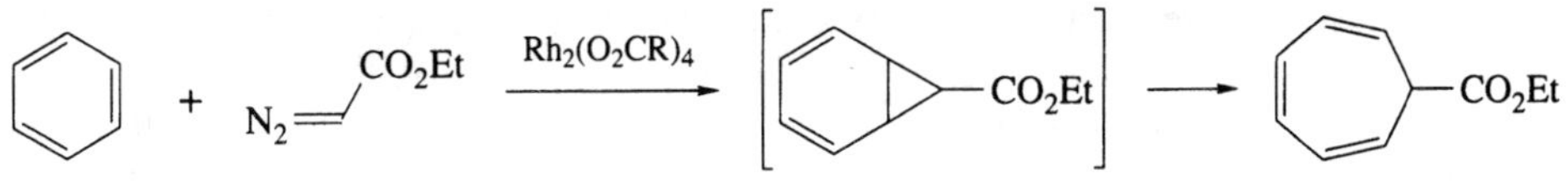

CO_2Et N_2 $Rh_2(O_2CR)_4$ CO_2Et CO_2Et

(**145**)

R	*Yield (%)*
CF_3	100
Ph	87
$MeOCH_2$	30
Me	7
Bu^t	5

Scheme 30

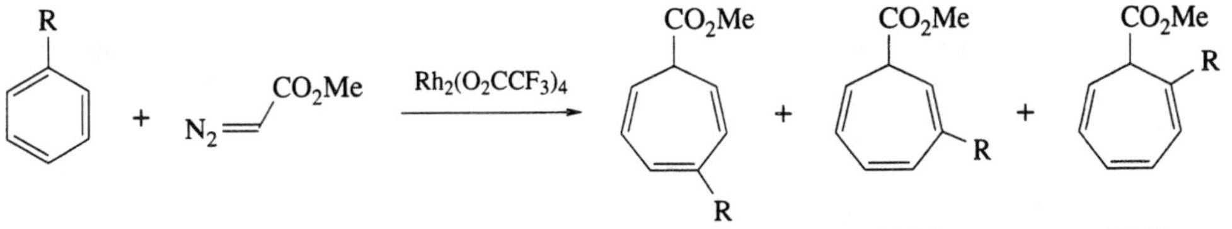

R CO_2Me N_2 $Rh_2(O_2CCF_3)_4$ CO_2Me R CO_2Me R CO_2Me R

(**146a**) (**146b**) (**146c**)

R	*Yield (%)*	(**146a**):(**146b**):(**146c**)
Me	95	58:24:18
OMe	73	60:9:31
Cl	72	80:15:5
F	46	80:12:8

Scheme 31

A similar transformation was observed with the rhodium trifluoroacetate catalyzed decomposition of diazo ketones in the presence of benzene (Scheme 32).[130] The cycloheptatrienes (**147**) formed in this case were acid labile and could be readily rearranged to benzyl ketones (**148**) on treatment with TFA. The reaction was effective even when the side chain contained reactive halogen and cyclopropyl functionality, but competing intramolecular reactions occurred with benzyl diazomethyl ketone. A more exotic example of this reaction is the rhodium(II) trifluoroacetate catalyzed decomposition of the diazopenicillinate (**149**) in the presence of anisole, which resulted in the formation of two cycloheptatriene derivatives (**150**) and (**151**) (equation 35).[131]

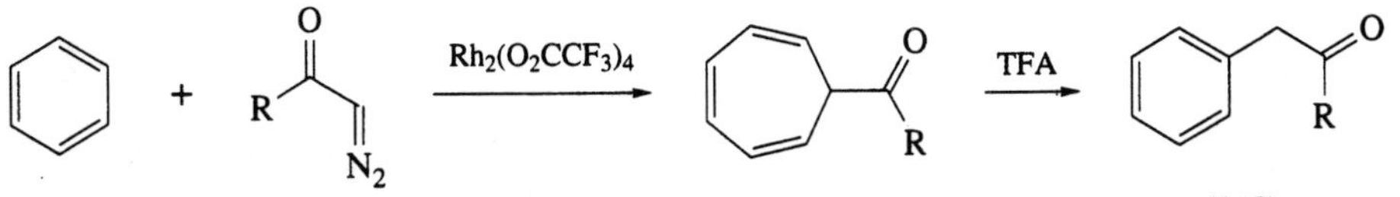

O R N_2 $Rh_2(O_2CCF_3)_4$ O R TFA O R

(**147**) (**148**)

Scheme 32

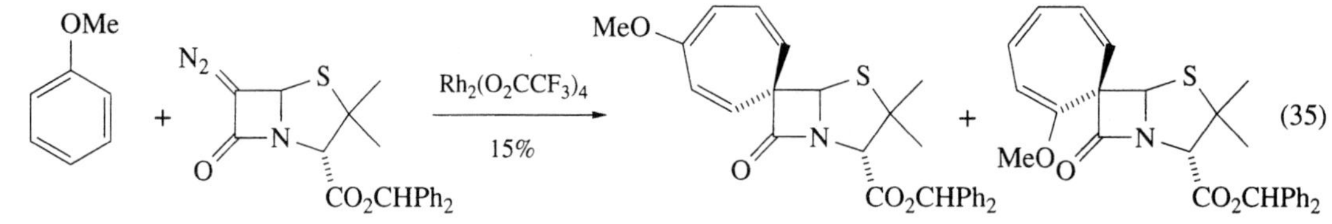

(35)

(149) **(150)** 87% **(151)** 13%

The products formed in these reactions are very sensitive to the functionality on the carbenoid. A study of Schechter and coworkers[132] using 2-diazo-1,3-indandione (**152**) nicely illustrates this point. The resulting carbenoid would be expected to be more electrophilic than the one generated from alkyl diazoacetate and consequently rhodium(II) acetate could be used as catalyst. The alkylation products (**153**) were formed in high yields without any evidence of cycloheptatrienes (Scheme 33). As can be seen in the case for anisole, the reaction was much more selective than the rhodium(II)-catalyzed decomposition of ethyl diazoacetate (Scheme 31), resulting in the exclusive formation of the *para* product. Application of this alkylation process to the synthesis of a novel *p*-quinodimethane has been reported.[133] Similar alkylation products were formed when dimethyl diazomalonate was decomposed in the presence of aromatic systems, but as these earlier studies[134] were carried out either photochemically or by copper catalysis, side reactions also occurred, as can be seen in the reaction with toluene (equation 36).

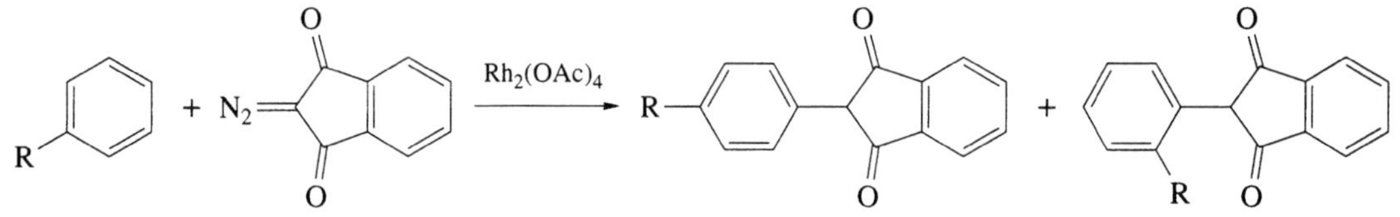

(152) **(153a)** **(153b)**

R	*Yield* (%)	**(153a)**:**(153b)**
H	95	–
OMe	73	100:0
Me	86	75:25
Cl	62	75:25
Br	21	–
NO_2	0	–

Scheme 33

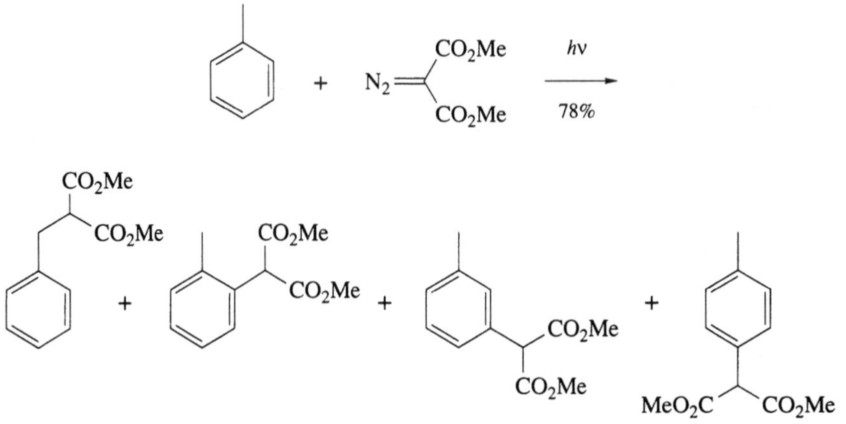

(36)

33% 40% 4% 23%

The intramolecular reaction between diazo ketones and benzenes is an effective way to generate a range of bicyclic systems.[7] The earlier copper-based catalysts have largely been superseded by rhodium(II) salts. Unlike the case in the intermolecular reactions, rhodium(II) acetate is the catalyst that has been most commonly used. Studies by McKervey,[135,136] however, indicated that rhodium(II) mandelate, which would be expected to generate a slightly more electrophilic carbenoid than rhodium(II) acetate, often gave improved yields.

Decomposition of 1-diazo-4-arylbutan-2-ones offers a direct entry to bicyclo[5.3.0]decatrienones and the approach has been extensively used by Scott and coworkers to synthesize substituted azulenes.[137] Respectable yields were obtained with copper catalysis,[137] but a more recent study[24] showed that rhodium(II) acetate was much more effective, generating bicyclo[5.3.0]decatrienones (**154**) under mild conditions in excess of 90% yield (Scheme 34). The cycloheptatrienes (**154**) were acid labile and on treatment with TFA rearranged cleanly to 2-tetralones (**155**), presumably *via* norcaradiene intermediates (**156**). Substituents on the aromatic ring exerted considerable effect on the course of the reaction. With *m*-methoxy-substituted systems the 2-tetralone was directly formed. Thus, it appeared that rearrangement of (**156**) to (**154**) was kinetically favored, but under acidic conditions or with appropriate functionality, equilibration to the 2-tetralone (**155**) occurred.

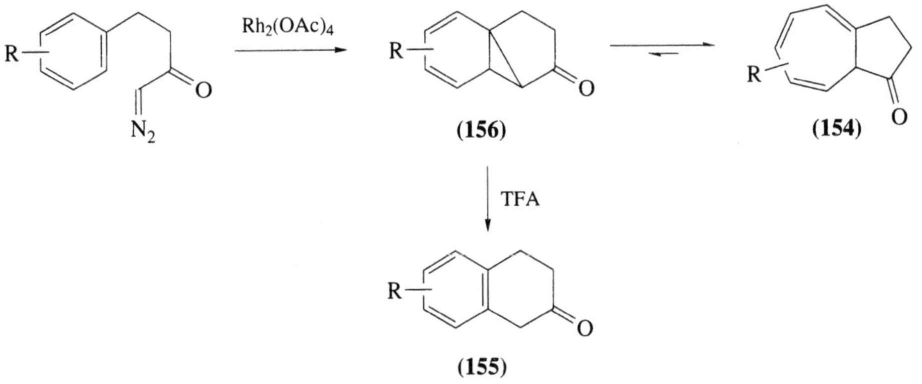

$Rh_2(OAc)_4$

(**156**) (**154**)

TFA

(**155**)

R = H, 4-Me, 4-OMe, 4-OAc, 2-Me, 2-OMe, 3-OAc, 2,3-$(OMe)_2$, 3,4,5,-$(OMe)_3$

Scheme 34

This methodology has been applied to the synthesis of the bicyclic system (**157**),[135] an advanced intermediate previously used in the synthesis of confertin.[138] Reaction of the diazo compound (**158**) with rhodium(II) mandelate gave an equilibrating mixture of the norcaradiene (**159**) and the fused cycloheptatriene (**160**) in essentially quantitative yield (Scheme 35). The mixture of (**159**) and (**160**) was converted to (**157**) in six steps in 20% overall yield.

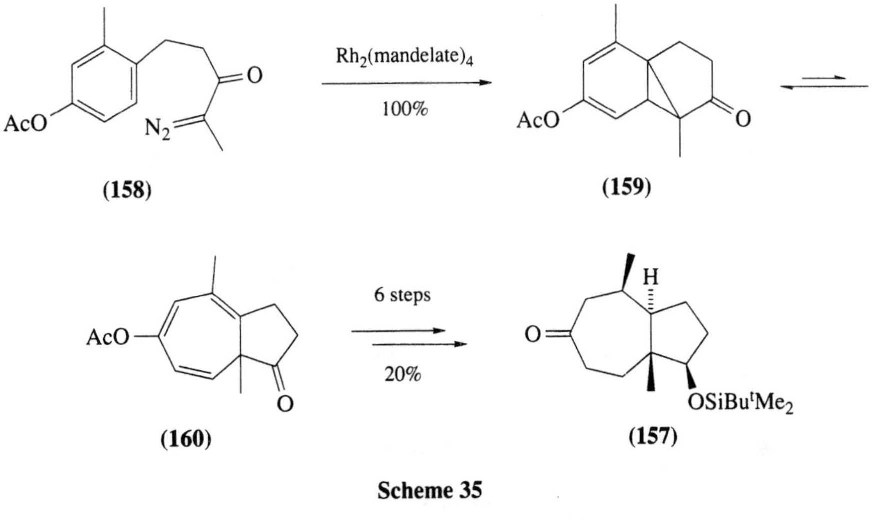

$Rh_2(mandelate)_4$, 100%

(**158**) (**159**)

6 steps, 20%

(**160**) (**157**)

Scheme 35

Further extensions of this chemistry have been reported, leading to more complex aromatic systems. Decomposition of the biphenyl (**161**) led to the trienone (**162**; Scheme 36).[136] Treatment of (**162**) with acid generated the fused cycloheptanone (**163**), but an alternative aromatization to (**164**) occurred on exposure of (**162**) to triethylamine. A similar reaction with the lower homolog efficiently generated a phenanthrol, but the higher homolog failed to produce a colchicine framework. Naphthalene derivatives have also been generated through intramolecular reactions.[139]

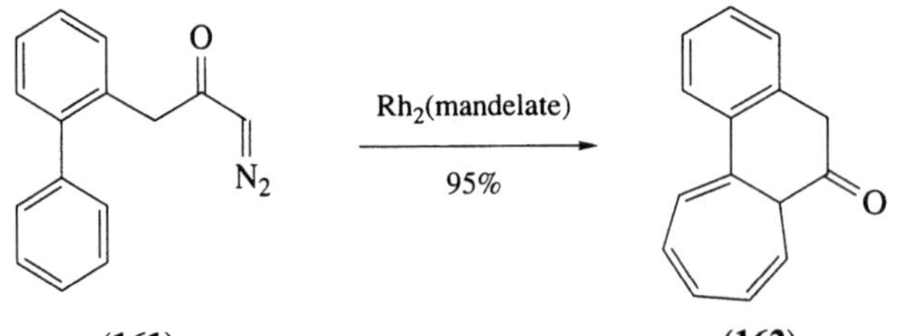

O N$_2$ Rh$_2$(mandelate) 95% O

(**161**) (**162**)

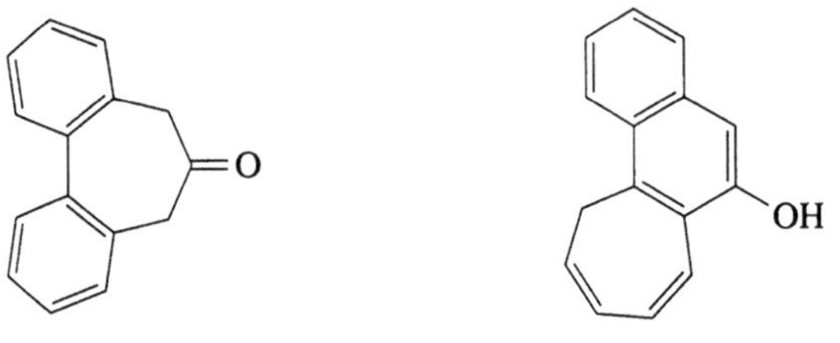

O OH

(**163**) (**164**)

Scheme 36

Intramolecular reactions using lower homologs invariably result in substitution products because competitive formation of bicyclo[5.2.0]nonatrienones is unfavorable due to the strained ring. Consequently, reaction of (**165**) with rhodium(II) acetate cleanly generated (**166**) in 98% yield (equation 37).[140] Aliphatic C—H insertion can be a competing reaction with aromatic substitution, and which reaction occurs appears to be very sensitive to steric and electronic effects.[100] Alkylation also occurred with phenolic diazo ketones (**167**), leading to the formation of spirodienones (**168**; Scheme 37).[141] Further reactions on (**170**) allowed ready access to (±)-solavetivone (**169**).[141b]

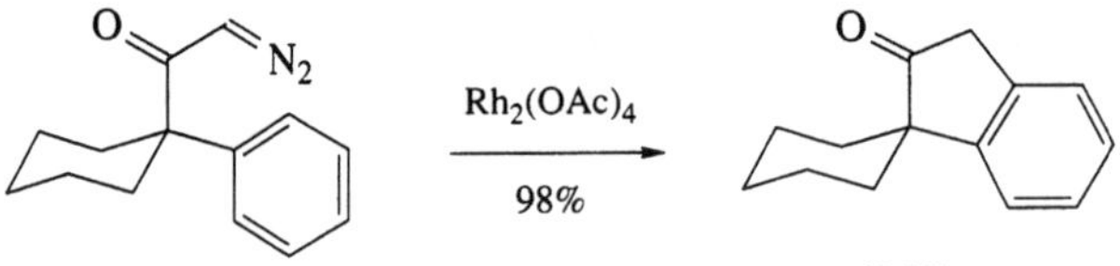

O N$_2$ Rh$_2$(OAc)$_4$ 98% O (37)

(**165**) (**166**)

OH O N$_2$ Rh$_2$(OAc)$_4$ O O O

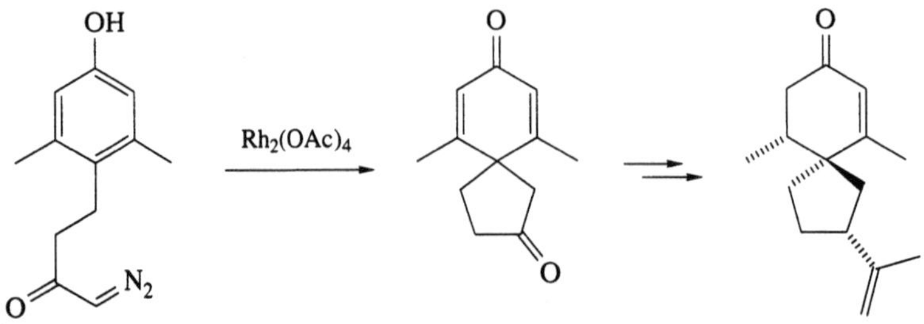

(**167**) (**168**) (**169**)

Scheme 37

The role of the catalyst is often crucial in determining the outcome of this chemistry and this is nicely illustrated in Scheme 38. Decomposition of (**170**) with rhodium(II) acetate resulted in insertion into the benzene ring, forming (**171**), but with palladium(II) acetate a different insertion occurred to give (**172**).[142]

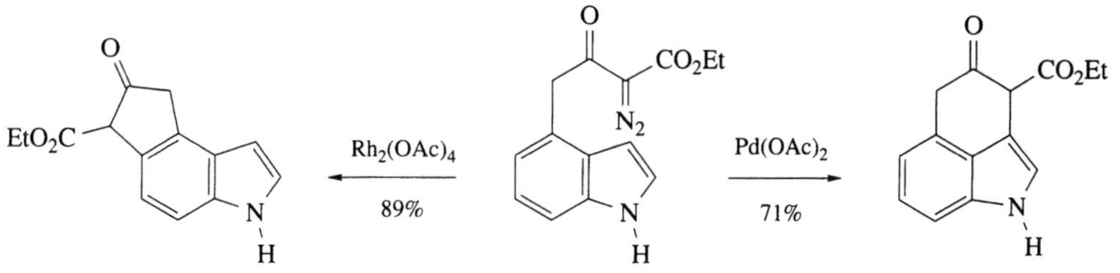

EtO_2C (**171**) ← $Rh_2(OAc)_4$, 89% — (**170**) — $Pd(OAc)_2$, 71% → (**172**)

Scheme 38

Cyclizations can occur with heteroatoms present in the tether as long as the groups are not strongly nucleophilic. Decomposition of α-diazo-β-arylmethanesulfonyl esters (**173**) resulted in the formation of 1,3-dihydrobenzo[*c*]thiophene 2,2-dioxides (**174**; equation 38),[143] which are valuable precursors to *o*-quinodimethanes. Reaction with *N*-aryldiazoamides (**175**) has been shown to be a useful method for preparing 2(3*H*)-indolinones (**176**; equation 39),[22a] while reaction of α-diazo-β-keto esters (**177**) has been developed as a process to synthesize 3-acetylbenzofuran-2(3*H*)-ones (**178**; equation 40).[144]

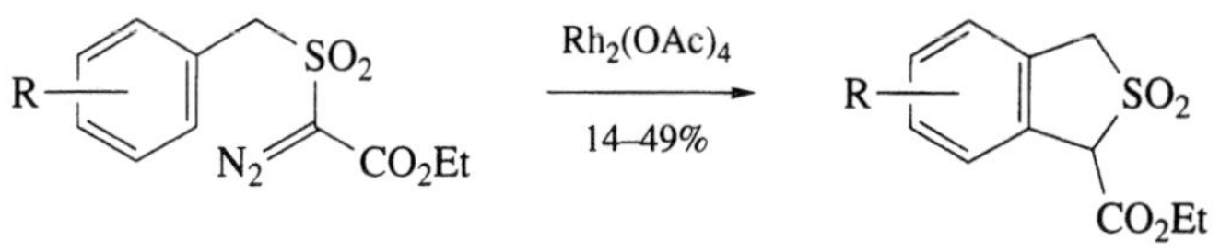

(**173**) $\xrightarrow[14\text{–}49\%]{Rh_2(OAc)_4}$ (**174**) (38)

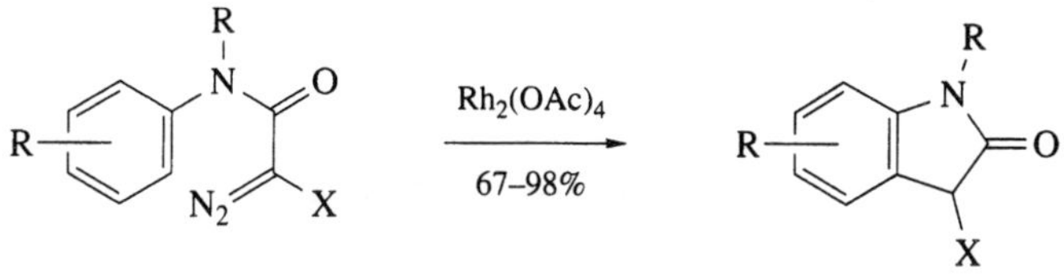

(**175**) X = H, MeCO $\xrightarrow[67\text{–}98\%]{Rh_2(OAc)_4}$ (**176**) (39)

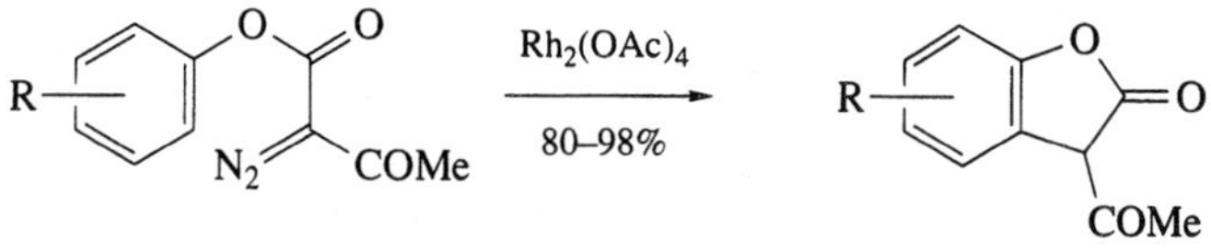

(**177**) $\xrightarrow[80\text{–}98\%]{Rh_2(OAc)_4}$ (**178**) (40)

When the heteroatom is contained within a longer tether, fused cycloheptatrienes once again become viable products. Rhodium(II) acetate catalyzed decomposition of *N*-benzyldiazoamides (**179**; X = H)[22b] formed cycloheptatrienes (**180**) in excellent yield as long as the R group was bulky such as *t*-butyl (Scheme 39). The carbenoid structure played an important role as reaction with *N*-benzyldiazoacetoacetamides (**179**; X = COMe)[22c] resulted in an entirely different reaction pathway, generating β-lactams (**181**) by aliphatic C—H insertion. Subtle conformational effects were also shown to be significant in this reaction because with (**182**) the predominant product (**183a**) arose from capture of the carbenoid by the inherently less reactive benzene ring (equation 41).[22b] Similar reactions have been carried out on systems with an oxygen tether[145,146] and these generated a mixture of fused cycloheptatrienes and chromanones, formal C—H insertion products.

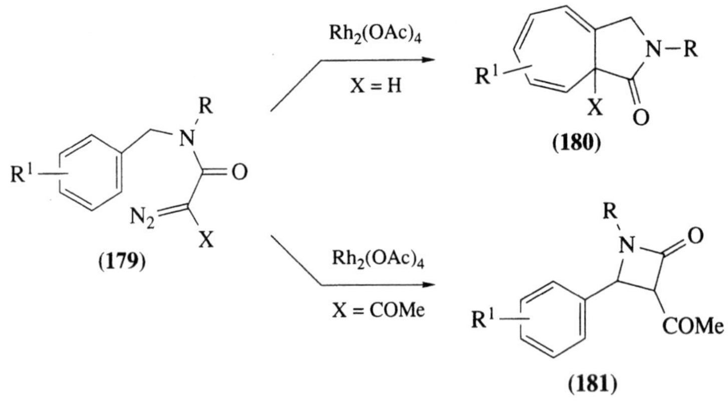

Scheme 39

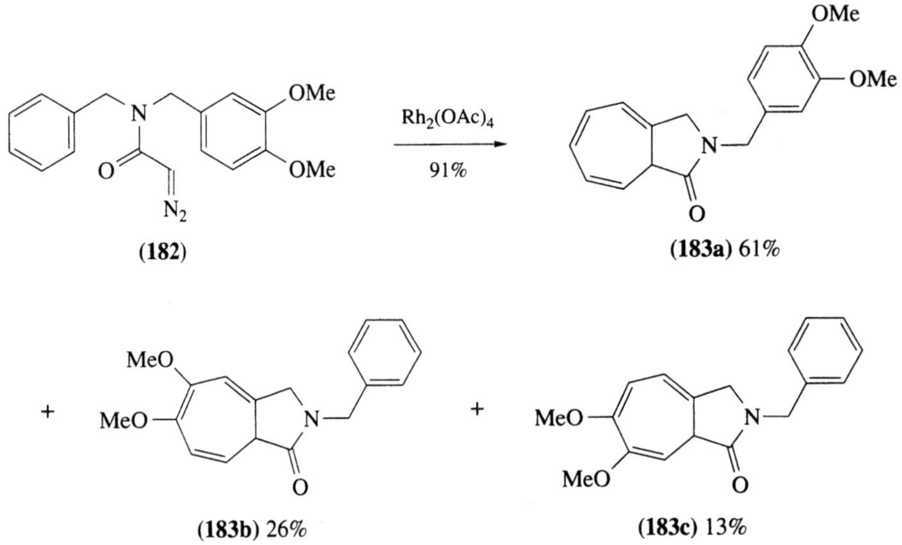

(41)

The overall mechanistic picture of these reactions is poorly understood, and it is conceivable that more than one pathway may be involved. It is generally considered that cycloheptatrienes are generated from an initially formed norcaradiene, as shown in Scheme 30. Equilibration between the cycloheptatriene and norcaradiene is quite facile and under acidic conditions the cycloheptatriene may readily rearrange to give a substitution product, presumably *via* a norcaradiene intermediate (Schemes 32 and 34). When alkylated products are directly formed from the intermolecular reaction of carbenoids with benzenes (Scheme 33 and equation 36) a norcaradiene considered as an intermediate; alternatively, a mechanism may be related to an electrophilic substitution may be involved leading to a zwitterionic intermediate. A similar intermediate has been proposed[143] in the intramolecular reactions of carbenoids with benzenes, which result in substitution products (equations 37–40). It has been reported,[144] however, that a considerable kinetic deuterium isotope effect was observed in some of these systems. Unless the electrophilic attack is reversible, this would indicate that a C—H insertion mechanism is involved in the rate-determining step.

4.8.6.2 Furans

The reaction of carbenoids with furans usually leads to the unravelling of the heterocycle, resulting in the formation of differentially functionalized dienes in good yield.[147–151] A recent reexamination of the reaction between ethyl diazoacetate with furan revealed the formation of four products, a furanocyclopropane (**184**), two isomeric dienes (**185** and **186**) and a trace of an alkylated product (**187**), in 65% overall yield (Scheme 40).[152] The furanocyclopropane (**184**) was unstable, however, and on prolonged standing rearranged cleanly to the (*Z,E*)-diene (**185**). Also, both the (*Z,E*)- and (*Z,Z*)-dienes (**185** and **186**) cleanly rearranged to the (*E,E*)-isomer (**188**) on treatment with iodine. Thus, the reaction could be used to prepare predominantly the (*Z,E*)- or (*E,E*)-products (**185** or **188**) by appropriate choice of work-up conditions. Furans with electron-withdrawing groups such as esters and unsaturated esters could be tolerated, but with vinylfurans competing cyclopropanation of the vinyl group occurred.[153]

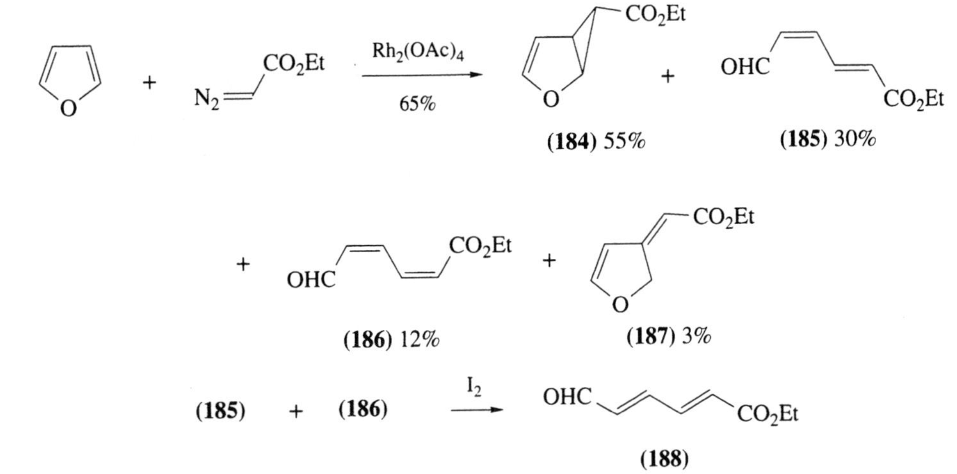

CO_2Et N_2 $Rh_2(OAc)_4$ 65% CO_2Et OHC CO_2Et

(184) 55% **(185)** 30%

OHC CO_2Et CO_2Et

(186) 12% **(187)** 3%

(185) + **(186)** I_2 → OHC CO_2Et

(188)

Scheme 40

Numerous applications of this chemistry to the synthesis of leukotrienes have been reported, as illustrated for the preparation of 12-hydroxyeicosatetraenoic acid (**189**; Scheme 41).[154] Reaction of the carbenoid precursor (**190**) with furan in the presence of rhodium(II) acetate generated a furanocyclopropane, which on standing reverted to predominantly the (*Z,E*)-isomer (**191**). Reduction of the dicarbonyl compound (**191**) gave the diol (**192**), which was then selectively converted to the bromide (**193**). Subsequently, (**193**) was coupled with the alkynide (**194**), followed by Lindlar reduction and deprotection to produce (**189**). The overall procedure is quite general and has been applied to a range of related compounds (**195–199**)[155–158] and useful synthetic fragments (**200** and **201**).[152]

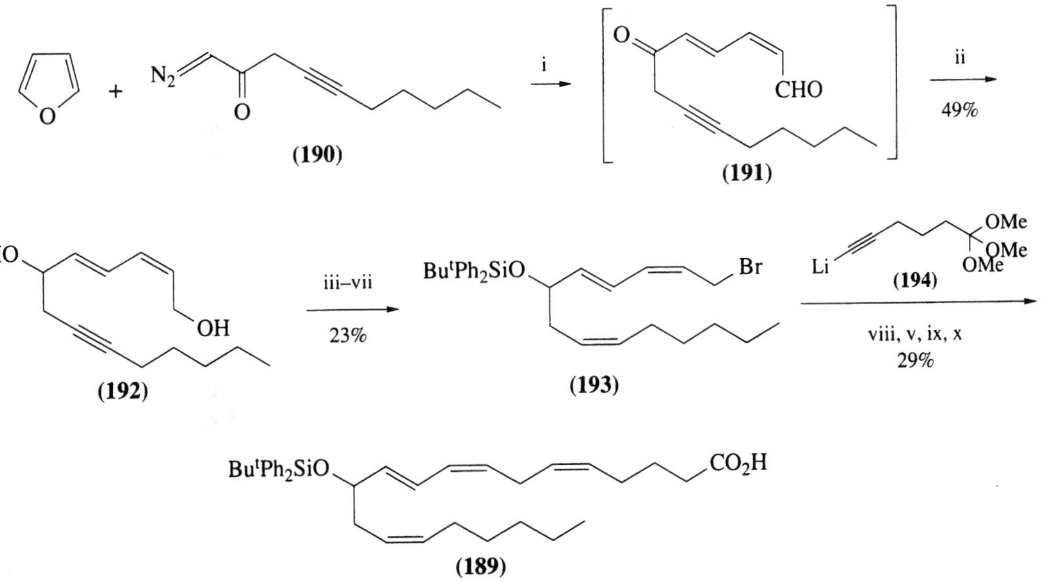

N_2 O **(190)** i → O CHO **(191)** ii 49%

HO OH **(192)** iii–vii 23% Bu^tPh_2SiO Br **(193)** Li OMe OMe OMe **(194)** viii, v, ix, x 29%

Bu^tPh_2SiO CO_2H **(189)**

i, $Rh_2(OAc)_4$; ii, $NaBH_4$, $CeCl_3$; iii, (*p*-(MeO)C_6H_4)Ph_2Cl, pyridine; iv, Bu^tPh_2SiCl, imidazole, DMF; v, Lindlar H_2; vi, 80% AcOH; vii, DIPHOS, CBr_4; viii, (**194**), CuI, THF, HMPA; ix, $Bu^n{}_4NF$, THF; x, HO^-

Scheme 41

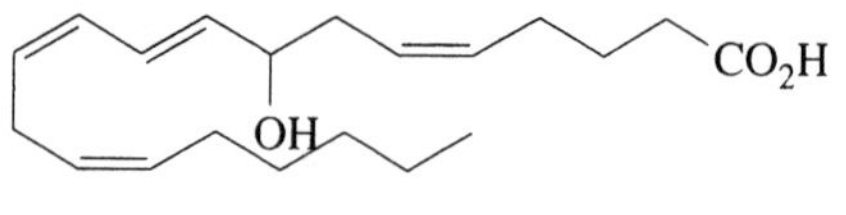

(195)

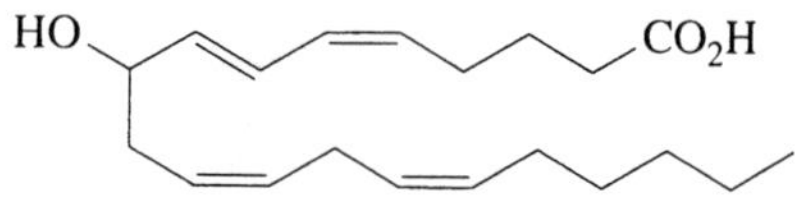

(196)

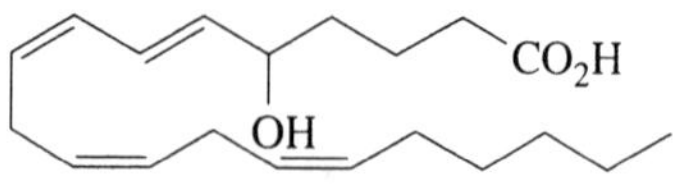

(197)

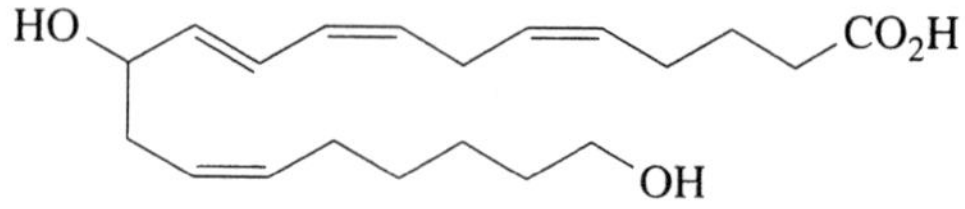

(198)

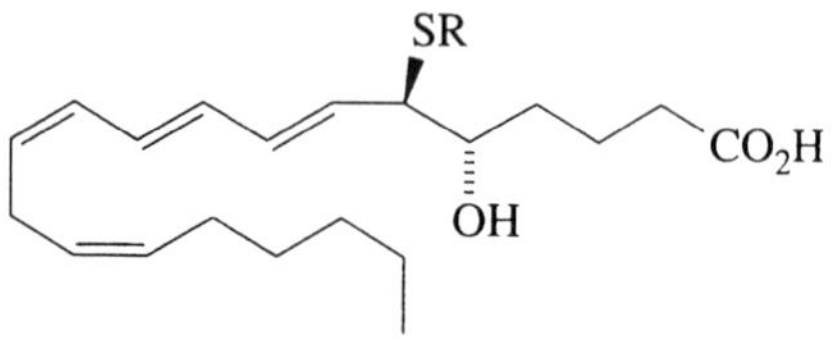

(199)

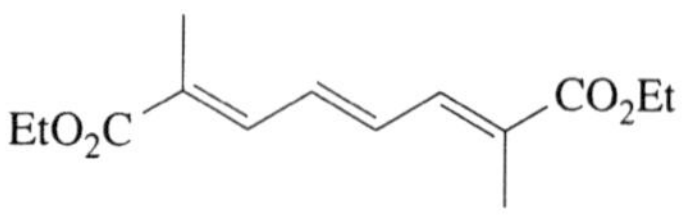

(200)

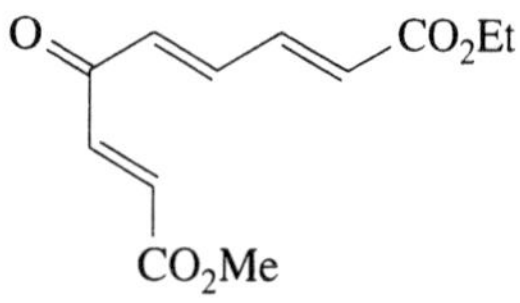

(201)

Reaction of vinylcarbenoids with furans offers another level of complexity because the furanocyclopropanes in this case would be divinylcyclopropanes capable of a Cope rearrangement as well as electrocyclic ring opening to trienes.[27b,c] As illustrated in Scheme 42, the product distribution was dependent on the furan structure. With 2,5-disubstituted furans [3.2.1]-bicyclic systems (**202**) were exclusively formed, but with furan or 2-substituted furans, trienes (**203**) were also produced.

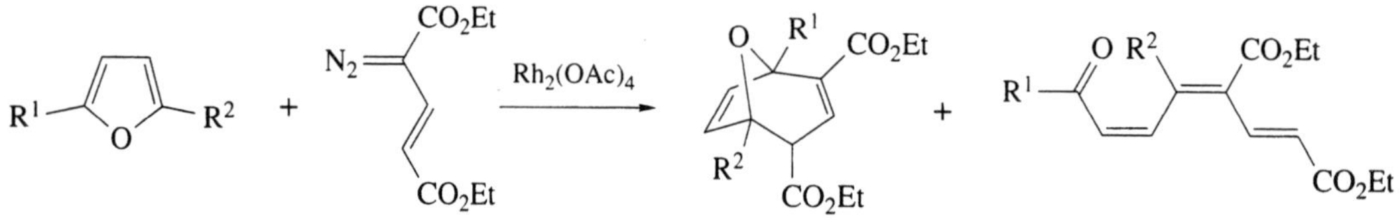

$Rh_2(OAc)_4$

(202)

(203)

R[1]	*R*[2]	(**202**) (%)	(**203**) (%)
Me	Me	70	0
H	H	62	26
Me	H	8	74
MeO	H	0	92

Scheme 42

A similar unravelling of the heterocyclic ring occurred in intramolecular reactions with furans,[79b,159–161] leading to cyclic 1,4-diacyl-1,3-butadienes (**204**; equation 42)[159] and (**205**; equation 43).[79b] The ideal size for the tether was five, but less efficient capture of the carbenoid was also possible with six- and seven-membered rings. With the 3-furanyl derivative (**206**), the acid labile structure (**207**) was formed, which readily rearranged to *p*-hydroxyphenylacetaldehyde (**208**; Scheme 43).[161]

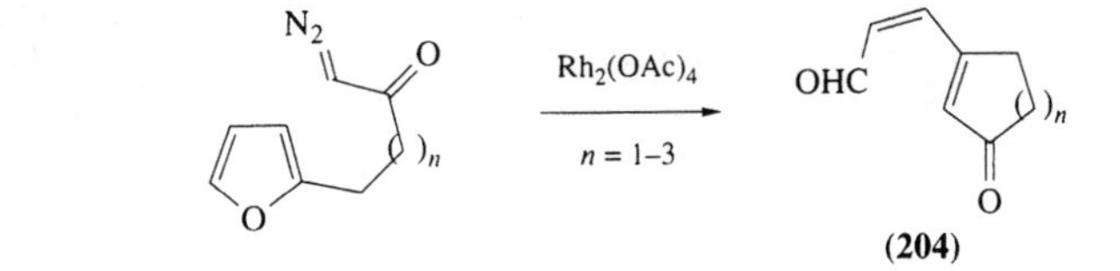

$Rh_2(OAc)_4$, n = 1–3 (42)

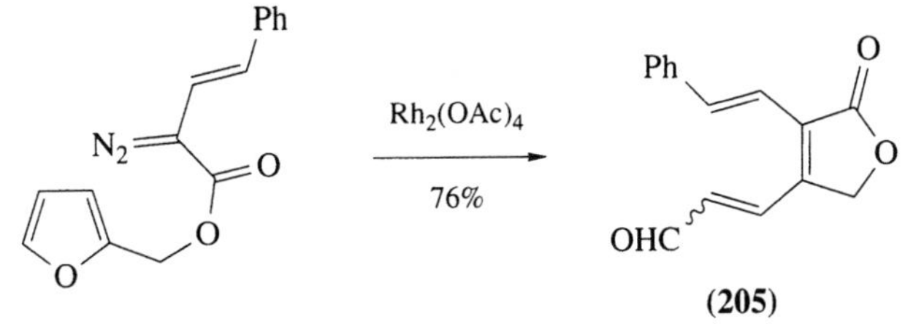

(204)

$Rh_2(OAc)_4$, 76% (43)

(205)

$Rh_2(OAc)_4$, 88%; H^+

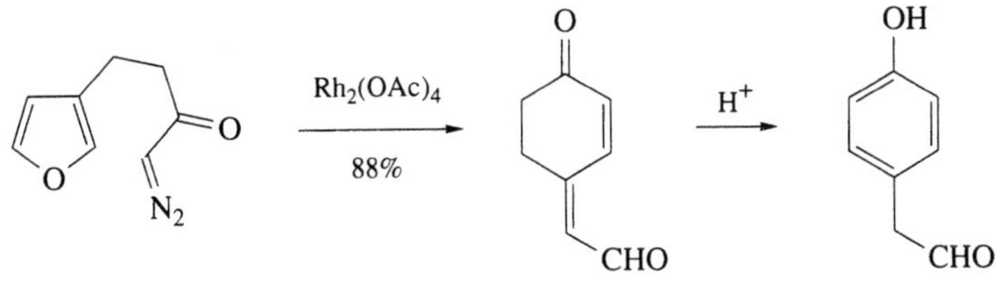

(206) (207) (208)

Scheme 43

4.8.6.3 Pyrroles

The reaction of ketocarbenoids with pyrroles leads to either substitution or cyclopropanation products, depending on the functionality on nitrogen. With *N*-acylated pyrrole (**209**) reaction of ethyl diazoacetate in the presence of copper(I) bromide generated the 2-azabicyclo[3.1.0]hex-3-ene system (**210**) and some of the diadduct (**211**; Scheme 44).[162,163] On attempted distillation of (**210**) in the presence of copper(I) bromide rearrangement to the 2-pyrrolylacetate (**212**) occurred, which was considered to proceed through the dipolar intermediate (**213**). In contrast, on flash vacuum pyrolysis (**210**) was transformed to the dihydropyridine (**214**). A plausible mechanism for the formation of (**214**) involved rearrangement of (**210**) to the acyclic imine (**215**), which then underwent a 6π-electrocyclization.

The reaction of *N*-alkylated pyrroles with carbenoids leads exclusively to substitution products. Due to the pharmaceutical importance of certain pyrrolylacetates, the reaction with alkyl diazoacetates (Scheme 45) has been systematically studied using about 50 different catalysts.[13] Both the 2- and 3-alkylated products (**216**) and (**217**) could be formed and the ratio was dependent on the size of the *N*-alkyl group and ester and also on the type of catalyst used. This has been interpreted as evidence that transient cyclopropane intermediates were not involved because if this were the case, the catalyst should not have influenced the isomer distribution. Instead, the reaction was believed to proceed by dipolar intermediates, whereby product control is determined by the position of electrophilic attack by the carbenoid. Similar alkylations with dimethyl diazomalonate gave greater selectivity and yields.[164]

The intramolecular reaction of carbenoids with pyrroles is extremely effective and leads to the formation of heterobicyclic systems in high yield, as can be seen in the simple systems shown in equation (44).[165] The reaction is so favorable that even an alkylcarbenoid capable of hydride migration could be utilized to generate a bicyclic system (**218**), but formation of (**219**) was a competing reaction in this case (equation 45).[166]

Intramolecular carbenoid reactions of pyrrole derivatives have recently been applied in a total synthesis of ipalbidine (**220**; Scheme 46).[167] The carbenoid generated from (**221**) was preferentially captured by the pyrrole ring, producing the indolizidinone (**222**) in 82% isolated yield, with only a trace of the indanone side product (**223**) formed. In four steps (**222**) was readily converted to (**220**) in 13% overall yield.

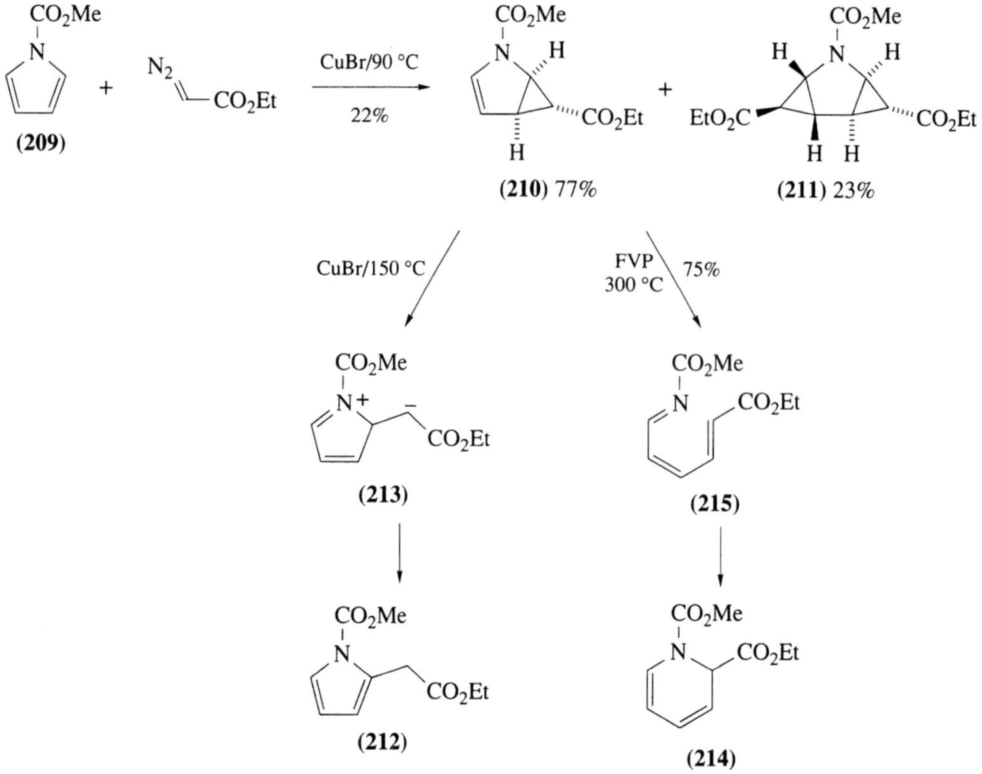

CuBr/90 °C, 22%

(209) **(210)** 77% **(211)** 23%

CuBr/150 °C; FVP 300 °C 75%

(213) **(215)** **(212)** **(214)**

Scheme 44

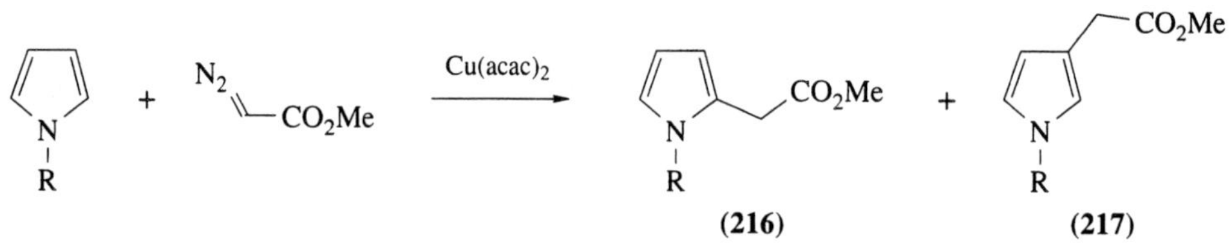

Cu(acac)$_2$

(216) **(217)**

R	*Yield* (%)	**(216)**:**(217)**
H	53	94:6
Me	42	92:8
Bu^t	34	0:100

Scheme 45

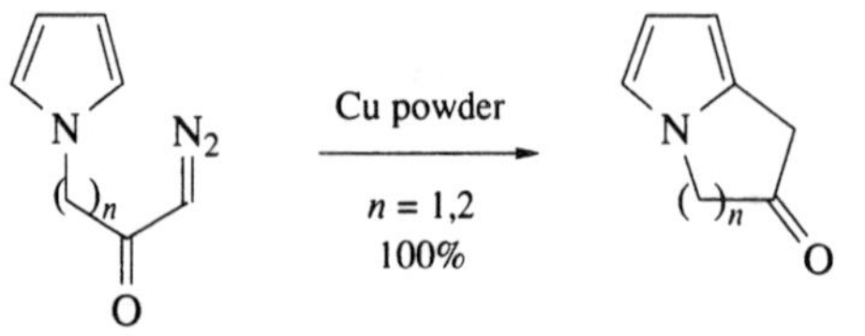

Cu powder, n = 1,2, 100% (44)

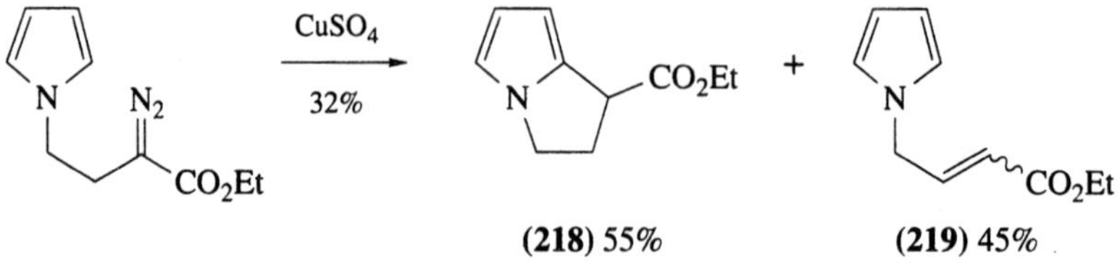

$CuSO_4$, 32% (45)

(218) 55% **(219)** 45%

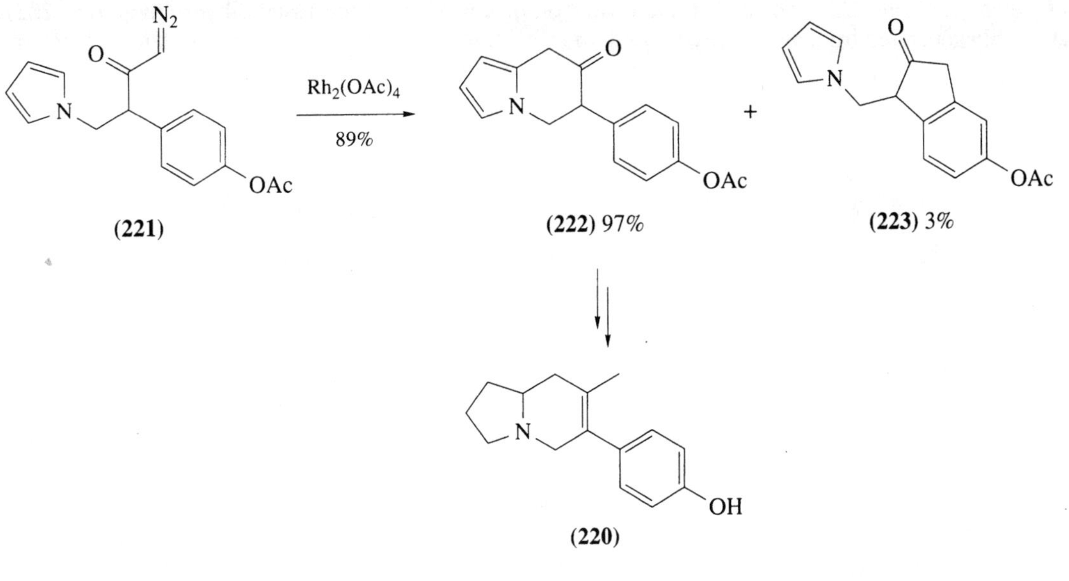

$Rh_2(OAc)_4$ 89%

(221) (222) 97% (223) 3%

(220)

Scheme 46

4.8.6.4 Thiophenes

Usually, the initial reaction between ketocarbenoids and thiophenes is formation of a sulfur ylide (**224**).[1,6,7,168–170] Further rearrangement of (**224**) may then lead to products which formally could have been considered to have been derived from direct interaction of the carbenoid with the π-system. Depending on the functionality, rearrangement of (**224**) to the thiophenocyclopropane (**225**),[170] the dihydrothiopyran (**226**)[168,171] or the substitution product (**227**; Scheme 47)[10a] has been reported. Cross-over experiments have shown[168] that the formation of (**227**) was by an intramolecular process and theoretical calculations have indicated[169] that all three products were probably formed from the common intermediate (**228**). 2-Alkenylthiophenes undergo competing cyclopropanation at the vinyl group.[172] The ylide from 2,5-dichlorothiophene has been used as a carbenoid precursor,[10] but depending on the carbenoid, a 2,3-sigmatropic rearrangement to a 1,4-oxathiocine has been shown to be an alternative reaction pathway.[173] A variety of carbenoid systems have been examined including a penicillin derivative.[131]

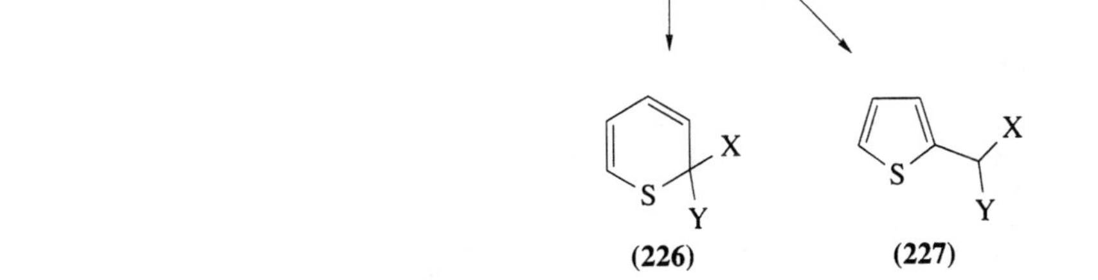

(224) (228) (225) (226) (227)

Scheme 47

In intramolecular reactions, direct interaction with the thiophene ring is possible by appropriate positioning of the tether as shown for the 3-substituted thiophene (**229**), which resulted in alkylation to form (**230**) and (**231**) (equation 46).[161a] In contrast, the rhodium(II) acetate catalyzed decomposition of the 2-

substituted thiophene (**232**) resulted in the formation of a mixture of the fused dihydrothiopyran (**233**), presumably generated from an initially formed ylide, and the alkylated product (**234**; equation 47).[174]

$Rh_2(OAc)_4$, 73%

(**229**) → (**230**) 87% + (**231**) 13% (46)

$Rh_2(OAc)_4$, 93%

(**232**) → (**233**) 73% + (**234**) 27% (47)

4.8.7 SUMMARY

The novel and varied chemistry of carbenoids has intrigued the practicing synthetic chemist for several decades. The development of rhodium(II) salts as extremely mild catalysts for the generation of metal-stabilized carbenoids has made a remarkable impact on this area of research because much more efficient synthetic pathways based on carbenoids have now become available. With the diverse range of rearrangements possible for cyclopropanes whose generation can now be achieved with high enantioselectivity, the future application of carbenoids to organic synthesis is assured to be an active field.

4.8.8 REFERENCES

1. G. Maas, *Top. Curr. Chem.*, 1987, **137**, 75.
2. M. P. Doyle, *Chem. Rev.*, 1986, **86**, 919.
3. M. P. Doyle, *Acc. Chem. Res.*, 1986, **19**, 348.
4. A. Demonceau, A. F. Noels and A. J. Hubert, in 'Aspects of Homogeneous Catalysis', ed. R. Ugo, Reidel, Dordrecht, 1988, vol. 6, p. 199.
5. D. S. Wulfman, G. Linstrumelle and C. F. Cooper, in 'The Chemistry of the Diazonium and Diazo Groups', ed. S. Patai, Wiley, New York, 1978, part 2, p. 821.
6. V. Dave and E. W. Warnhoff, *Org. React. (N.Y.)*, 1970, **18**, 217 and references cited therein.
7. S. D. Burke and P. A. Grieco, *Org. React. (N.Y.)*, 1979, **26**, 361.
8. (a) H.-U. Reissig, in 'The Chemistry of the Cyclopropyl Group', ed. Z. Rappoport, Wiley, New York, 1987, part 1, p. 375; (b) R. Vehre and N. De Kimpe, in 'The Chemistry of the Cyclopropyl Group', ed. Z. Rappoport, Wiley, New York, 1987, part 1, p. 445; (c) T. T. Tidwell, in 'The Chemistry of the Cyclopropyl Group', ed. Z. Rappoport, Wiley, New York, 1987, part 1, p. 565.
9. J. Yoshida, S. Yano, T. Ozawa and N. Kawabata, *J. Org. Chem.*, 1985, **50**, 3467.
10. (a) R. J. Gillespie, A. E. A. Porter and W. E. Willmott, *J. Chem. Soc., Chem. Commun.*, 1978, 85; (b) J. Cuffe, R. J. Gillespie and A. E. A. Porter, *J. Chem. Soc., Chem. Commun.*, 1978, 641; (c) R. J. Gillespie and A. E. A. Porter, *J. Chem. Soc., Chem. Commun.*, 1979, 50.
11. L. Hatjiarapoglou, A. Varvoglis, N. W. Alcock and G. A. Pike, *J. Chem. Soc., Perkin Trans. 1*, 1988, 2839.
12. H. Meier and K.-P. Zeller, *Angew. Chem., Int. Ed. Engl.*, 1975, **14**, 32.
13. B. E. Maryanoff, *J. Org. Chem.*, 1979, **44**, 4410.
14. R. G. Salomon and J. K. Kochi, *J. Am. Chem. Soc.*, 1973, **95**, 3300.
15. (a) W. R. Moser, *J. Am. Chem. Soc.*, 1969, **91**, 1135; (b) W. R. Moser, *J. Am. Chem. Soc.*, 1969, **91**, 1141.
16. (a) R. Paulissen, H. Reimlinger, E. Hayez, A. J. Hubert and P. Teyssie, *Tetrahedron Lett.*, 1973, 2233; (b) A. J. Hubert, A. F. Noels, A. J. Anciaux and P. Teyssie, *Synthesis*, 1976, 600; (c) A. J. Anciaux, A. Demonceau, A. J. Hubert, A. F. Noels, N. Petiniot and P. Teyssie, *J. Chem. Soc., Chem. Commun.*, 1980, 765; (d) A. J. Anciaux, A. Demonceau, A. F. Noels, A. J. Hubert, R. Warin and P. Teyssie, *J. Org. Chem.*, 1981, **46**, 873; (e) A. J. Anciaux, A. J. Hubert, A. F. Noels, N. Petiniot and P. Teyssie, *J. Org. Chem.*, 1980, **45**, 695; (f) A. F. Noels, A. Demonceau, N. Petiniot, A. J. Hubert and P. Teyssie, *Tetrahedron*, 1982, **38**, 2733.
17. M. P. Doyle, R. L. Dorow, W. H. Tamblyn and W. E. Buhro, *Tetrahedron Lett.*, 1982, **23**, 2261.
18. H. J. Callot and F. Metz, *Tetrahedron*, 1985, **41**, 4495.
19. H. J. Callot, F. Metz and C. Piechocki, *Tetrahedron*, 1982, **38**, 2365.
20. W. H. Tamblyn, S. R. Hoffmann and M. P. Doyle, *J. Organomet. Chem.*, 1981, **216**, C64.

21. (a) A. Nakamura, A. Konishi, Y. Tatsuno and S. Otsuka, *J. Am. Chem. Soc.*, 1978, **100**, 3443; (b) A. Nakamura, A. Konishi, R. Tsujitani, M. Kudo and S. Otsuka, *J. Am. Chem. Soc.*, 1978, **100**, 3449; (c) B. Scholl and H. J. Hansen, *Helv. Chim. Acta*, 1986, **69**, 1936.
22. (a) M. P. Doyle, M. S. Shanklin, H. Q. Pho and S. N. Mahapatro, *J. Org. Chem.*, 1988, **53**, 1017; (b) M. P. Doyle, M. S. Shanklin and H. Q. Pho, *Tetrahedron Lett.*, 1988, **29**, 2639; (c) M. P. Doyle, M. S. Shanklin, S. M. Oon, H. Q. Pho, F. R. Heide and W. R. Veal, *J. Org. Chem.*, 1988, **53**, 3384.
23. A. Jeganathan, S. K. Richardson, R. S. Mani, B. E. Haley and D. S. Watt, *J. Org. Chem.*, 1986, **51**, 5362.
24. M. A. McKervey, S. M. Tuladhar and M. F. Twohig, *J. Chem. Soc., Chem. Commun.*, 1984, 129.
25. B. W. Peace and D. S. Wulfman, *Synthesis*, 1973, 137.
26. (a) E. Wenkert, M. E. Alonso, B. L. Buckwalter and E. L. Sanchez, *J. Am. Chem. Soc.*, 1983, **105**, 2021; (b) M. E. Alonso, P. Jano, M. Hernandez, R. S. Greenberg and E. Wenkert, *J. Org. Chem.*, 1983, **48**, 3047; (c) M. E. Alonso, A. Morales and A. W. Chitty, *J. Org. Chem.*, 1982, **47**, 3747.
27. (a) H. M. L. Davies, H. D. Smith and O. Korkor, *Tetrahedron Lett.*, 1987, **28**, 1853; (b) H. M. L. Davies, D. M. Clarke and T. K. Smith, *Tetrahedron Lett.*, 1985, **26**, 5659; (c) H. M. L. Davies, D. M. Clarke, D. B. Alligood and G. R. Eiband, *Tetrahedron*, 1987, **43**, 4265.
28. (a) M. Regitz, in 'The Chemistry of the Diazonium and Diazo Groups', ed. S. Patai, Wiley, New York, 1978, part 2, p. 751; (b) M. Regitz and G. Maas, 'Diazo Compounds. Properties and Synthesis', Academic Press, New York, 1986.
29. M. Regitz, J. Hocker and A. Liedhegener, *Org. Synth., Coll. Vol.*, 1973, **5**, 197.
30. G. G. Hazen, L. M. Weinstock, R. Connell and F. W. Bollinger, *Synth. Commun.*, 1981, **11**, 947.
31. J. S. Baum, D. A. Shook, H. M. L. Davies and H. D. Smith, *Synth. Commun.*, 1987, **17**, 1709.
32. D. F. Taber, R. E. Ruckle, Jr. and M. J. Hennessy, *J. Org. Chem.*, 1986, **51**, 4077.
33. J. B. Hendrickson and W. A. Wolf, *J. Org. Chem.*, 1968, **33**, 3610.
34. T. H. Black, *Aldrichimica Acta*, 1983, **16**, 3.
35. E. J. Corey and A. G. Myers, *Tetrahedron Lett.*, 1984, **25**, 3559.
36. H. Ok, C. Caldwell, D. R. Schroeder and K. Nakanishi, *Tetrahedron Lett.*, 1988, **29**, 2275.
37. M. P. Doyle, R. L. Dorow, W. E. Buhro, J. H. Griffin, W. H. Tamblyn and M. L. Trudell, *Organometallics*, 1984, **3**, 44.
38. (a) J. P. Marino and E. Laborde, *J. Org. Chem.*, 1987, **52**, 1; (b) J. P. Marino and E. Laborde, *J. Am. Chem. Soc.*, 1985, **107**, 734.
39. (a) C. Bruckner and H.-U. Reissig, *J. Org. Chem.*, 1988, **53**, 2440; (b) C. Bruckner, H. Holzinger and H.-U. Reissig, *J. Org. Chem.*, 1988, **53**, 2450.
40. P. Kolsaker, A. Kvarsnes and H. J. Storesund, *Org. Mass Spectrom.*, 1986, **21**, 535.
41. M. P. Doyle, R. L. Dorow, W. H. Tamblyn and W. E. Buhro, *Tetrahedron Lett.*, 1982, **23**, 2261.
42. S. Cottens and M. Schlosser, *Tetrahedron*, 1988, **44**, 7127.
43. K. Burgess, *J. Org. Chem.*, 1987, **52**, 2046.
44. E. Wenkert, M. E. Alonso, B. L. Buckwalter and K. J. Chou, *J. Am. Chem. Soc.*, 1977, **99**, 4778.
45. M. L. Graziano and R. Scarpati, *J. Chem. Soc., Perkin Trans. 1*, 1985, 289.
46. S. B. Singh, *Indian J. Chem., Sect. B*, 1981, **20**, 810.
47. M. P. Doyle, D. van Leusen and W. H. Tamblyn, *Synthesis*, 1981, 787.
48. J. J. Gajewski, J. Jurayj, D. R. Kimbrough, M. E. Gande, B. Ganem and B. K. Carpenter, *J. Am. Chem. Soc.*, 1987, **109**, 1170.
49. (a) M. P. Doyle, W. H. Tamblyn and V. Bagheri, *J. Org. Chem.*, 1981, **46**, 5094; (b) M. P. Doyle, J. H. Griffin, M. S. Chinn and D. van Leusen, *J. Org. Chem.*, 1984, **49**, 1917.
50. M. P. Doyle, V. Bagheri and N. K. Harn, *Tetrahedron Lett.*, 1988, **29**, 5119.
51. D. S. Wulfman, R. S. McDaniel, Jr. and B. W. Peace, *Tetrahedron*, 1976, **32**, 1241.
52. M. E. Alonso and M. C. Garcia, *J. Org. Chem.*, 1985, **50**, 988.
53. M. P. Doyle, J. H. Griffin, V. Bagheri and R. L. Dorow, *Organometallics*, 1984, **3**, 53.
54. J. Elzinga, H. Hogeveen and E. P. Schudde, *J. Org. Chem.*, 1980, **45**, 4337.
55. H. M. L. Davies, T. J. Clark and L. A. Church, *Tetrahedron Lett.*, 1989, **30**, 5057.
56. T. Aratani, Y. Yoneyoshi and T. Nagase, *Tetrahedron Lett.*, 1982, **23**, 685.
57. J. D. Morrison (ed.), 'Asymmetric Synthesis, Chiral Catalysts', Academic Press, Orlando, FL, 1985, vol. 5.
58. T. Aratani, Y. Yoneyoshi and T. Nagase, *Tetrahedron Lett.*, 1977, 2599.
59. M. P. Doyle, R. L. Dorow, J. W. Terpstra and R. A. Rodenhouse, *J. Org. Chem.*, 1985, **50**, 1663.
60. U. Schollkopf, B. Hupfeld, S. Kuper, E. Egert and M. Dyrbusch, *Angew. Chem., Int. Ed. Engl.*, 1988, **27**, 433.
61. H. Nozaki, H. Takaya, S. Moriuti and R. Noyori, *Tetrahedron*, 1968, **24**, 3655.
62. A. Becalski, W. R. Cullen, M. D. Fryzuk, G. Herb, B. R. James, J. P. Kutney, K. Piotrowska and D. Tapiolas, *Can. J. Chem.*, 1988, **66**, 3108.
63. (a) J. E. Baldwin and C. G. Carter, *J. Am. Chem. Soc.*, 1982, **104**, 1362; (b) J. E. Baldwin, T. W. Patapoff and T. C. Barden, *J. Am. Chem. Soc.*, 1984, **106**, 1421; (c) J. E. Baldwin and T. C. Barden, *J. Am. Chem. Soc.*, 1984, **106**, 6364.
64. (a) H. Fritschi, U. Leutenegger, K. Siegmann, A. Pfaltz, W. Keller and C. Kratky, *Helv. Chim. Acta*, 1988, **71**, 1541; (b) H. Fritschi, U. Leutenegger and A. Pfaltz, *Helv. Chim. Acta*, 1988, **71**, 1553; (c) H. Fritschi, U. Leutenegger and A. Pfaltz, *Angew. Chem., Int. Ed. Engl.*, 1986, **25**, 1005.
65. S. A. Matlin, W. J. Lough, L. Chan, D. M. H. Abram and Z. Zhou, *J. Chem. Soc., Chem. Commun.*, 1984, 1038.
66. M. Laabassi and R. Gree, *Tetrahedron Lett.*, 1988, **29**, 611.
67. M. Fujita, T. Hiyama and K. Kondo, *Tetrahedron Lett.*, 1986, **27**, 2139.
68. S. H. Kang, W. J. Kim and Y. B. Chae, *Tetrahedron Lett.*, 1988, **29**, 5169.
69. R. A. Roberts, V. Schüll and L. A. Paquette, *J. Org. Chem.*, 1983, **48**, 2076.
70. E. J. Corey and Y. B. Xiang, *Tetrahedron Lett.*, 1987, **28**, 5403.
71. E. J. Corey, G. Wess, Y. B. Xiang and A. K. Singh, *J. Am. Chem. Soc.*, 1987, **109**, 4717.
72. A. K. Singh, R. K. Bakshi and E. J. Corey, *J. Am. Chem. Soc.*, 1987, **109**, 6187.

73. R. J. Sundberg, E. W. Baxter, W. J. Pitts, R. A. Schofield and T. Nishiguchi, *J. Org. Chem.*, 1988, **53**, 5097.
75. B. Tinant, J. P. DeClercq, M. van Meerssche, P. Kok, G. Rozing, F. Scott, M. DeMuynck, N. D. Shinna, P. DeClercq and E. M. Vandewalle, *Bull. Soc. Chim. Belg.*, 1987, **96**, 81.
76. M. Ramaiah and T. L. Nagabhushan, *Synth. Commun.*, 1986, **16**, 1049.
77. P. Ceccherelli, M. Curini, M. C. Marcotullio and E. Wenkert, *J. Org. Chem.*, 1986, **51**, 738.
78. (a) E. Piers, G. L. Jung and N. Moss, *Tetrahedron Lett.*, 1984, **25**, 3959; (b) E. Piers and E. H. Ruediger, *J. Org. Chem.*, 1980, **45**, 1725.
79. (a) H. M. L. Davies, C. E. M. Oldenburg, M. J. McAfee, J. G. Nordahl, J. P. Henretta and K. R. Romines, *Tetrahedron Lett.*, 1988, **29**, 975; (b) H. M. L. Davies, M. J. McAfee and C. E. M. Oldenburg, *J. Org. Chem.*, 1989, **54**, 930.
80. U. R. Ghatak and S. C. Roy, *J. Chem. Res. (S)*, 1981, 5.
81. L. N. Mander, R. H. Prager and J. V. Turner, *Aust. J. Chem.*, 1974, **27**, 2645.
82. D. J. Beames, J. E. Halleday and L. N. Mander, *Aust. J. Chem.*, 1972, **25**, 137.
83. M. Mongrain, J. Lafontaine, A. Belanger and P. Deslongchamps, *Can. J. Chem.*, 1970, **48**, 3273.
84. J. Lafontaine, M. Mongrain, M. Sergent-Guay, L. Ruest and P. Deslongchamps, *Aust. J. Chem.*, 1980, **58**, 2460.
85. E. J. Corey and P. L. Fuchs, *J. Am. Chem. Soc.*, 1972, **94**, 4014.
86. K. Kondo, T. Umemoto, K. Yako and D. Tunemoto, *Tetrahedron Lett.*, 1978, 3927.
87. D. Tunemoto, N. Araki and K. Kondo, *Tetrahedron Lett.*, 1977, 109.
88. (a) R. P. Short, J.-M. Revol, B. C. Ranu and T. Hudlicky, *J. Org. Chem.*, 1983, **48**, 4453; (b) T. Hudlicky, S. V. Govindan and J. O. Frazier, *J. Org. Chem.*, 1985, **50**, 4166; (c) T. Hudlicky, M. G. Natchus and G. Sinai-Zingde, *J. Org. Chem.*, 1987, **52**, 4641; (d) T. Hudlicky, D. B. Reddy, S. V. Govindan, T. Kulp, B. Still and J. P. Sheth, *J. Org. Chem.*, 1983, **38**, 3422; (e) T. Hudlicky, L. D. Kwart, M. H. Tiedje, B. C. Ranu, R. P. Short, J. O. Frazier and H. L. Rigby, *Synthesis*, 1986, 716; (f) T. Hudlicky and R. P. Short, *J. Org. Chem.*, 1982, **47**, 1522; (g) S. V. Govindan, T. Hudlicky and F. J. Koszyk, *J. Org. Chem.*, 1983, **48**, 3581.
89. T. Imanishi, M. Matsui, M. Yamashita and C. Iwata, *Tetrahedron Lett.*, 1986, **27**, 3161.
90. T. Imanishi, M. Matsui, M. Yamashita and C. Iwata, *J. Chem. Soc., Chem. Commun.*, 1987, 1802.
91. E. Y. Chen, *J. Org. Chem.*, 1984, **49**, 3245.
92. (a) S. Bien, A. Gillon and S. Kohen, *J. Chem. Soc., Perkin Trans. 1*, 1976, 489; (b) A. Gillon, D. Ovadia, M. Kapon and S. Bien, *Tetrahedron*, 1982, **38**, 1477.
93. H. Hirai and M. Matsui, *Agric. Biol. Chem.*, 1974, **40**, 169.
94. D. F. Taber, J. C. Amedio and K. Raman, *J. Org. Chem.*, 1988, **53**, 2984.
95. (a) S. J. Danishefsky, R. McKee and R. K. Singh, *J. Am. Chem. Soc.*, 1977, **99**, 7711; (b) S. J. Danishefsky, J. Regan and R. Doehner, *J. Org. Chem.*, 1981, **46**, 5255.
96. E. J. Corey and A. G. Myers, *J. Am. Chem. Soc.*, 1985, **107**, 5574.
97. J. F. Ruppert and J. D. White, *J. Am. Chem. Soc.*, 1981, **103**, 1808.
98. J. D. White, J. F. Ruppert, M. A. Avery, S. Torii and J. Nokami, *J. Am. Chem. Soc.*, 1981, **103**, 1813.
99. T. Hudlicky, J. P. Sheth, V. Gee and D. Barnvos, *Tetrahedron Lett.*, 1979, 4889.
100. D. F. Taber and R. E. Ruckle, Jr., *J. Am. Chem. Soc.*, 1986, **108**, 7686.
101. (a) J. Adams, R. Frenette, M. J. Belley, F. Chibante and J. P. Springer, *J. Am. Chem. Soc.*, 1987, **109**, 5432; (b) J. Adams and M. Belley, *Tetrahedron Lett.*, 1986, **27**, 2075; (c) J. Adams and M. J. Belley, *J. Org. Chem.*, 1986, **51**, 3878.
102. (a) A. B. Smith, III, B. H. Toder and S. J. Branca, *J. Am. Chem. Soc.*, 1984, **106**, 3995; (b) A. B. Smith, III, B. H. Toder, R. E. Richmond and S. J. Branca, *J. Am. Chem. Soc.*, 1984, **106**, 4001.
103. B. Saha, G. Bhattacharjee and U. R. Ghatak, *J. Chem. Soc., Perkin Trans. 1*, 1988, 939.
104. P. Dowd, P. Garner, R. Schappert, H. Irngartinger and A. Goldman, *J. Org. Chem.*, 1982, **47**, 4240.
105. M. Brikhahn, E. V. Dehmlow and H. Bogge, *Angew. Chem., Int. Ed. Engl.*, 1987, **26**, 72.
106. H.-U. Reissig, *Top. Curr. Chem.*, 1988, **144**, 73.
107. T. Hudlicky, T. M. Kutchan and S. M. Naqvi, *Org. React. (N.Y)*, 1985, **33**, 247.
108. Z. Goldschmidt and B. Crammer, *Chem. Soc. Rev.*, 1988, **17**, 229.
109. S. Danishefsky, *Acc. Chem. Res.*, 1979, **12**, 66.
110. E. Wenkert, *Acc. Chem. Res.*, 1980, **13**, 27.
111. K. Saigo, H. Kurihara, H. Miura, A. Hongo, N. Kubota and N. Nohira, *Synth. Commun.*, 1984, **14**, 787.
112. H.-U. Reissig and I. Reichelt, *Tetrahedron Lett.*, 1984, **25**, 5879.
113. I. Reichelt and H.-U. Reissig, *Liebigs Ann. Chem.*, 1984, 820.
114. H.-U. Reissig, *Tetrahedron Lett.*, 1985, **26**, 3943.
115. W. Bretsch and H.-U. Reissig, *Liebigs Ann. Chem.*, 1987, 175.
116. E. Wenkert, R. S. Greenberg and M. S. Raju, *J. Org. Chem.*, 1985, **50**, 4681.
117. J. P. Marino, R. F. de la Pradilla and E. Laborde, *J. Org. Chem.*, 1984, **49**, 5279.
118. J. Ollivier and J. Salaun, *J. Chem. Soc., Chem. Commun.*, 1985, 1269.
119. R. Zschiesche, E. L. Grimm and H.-U. Reissig, *Angew. Chem., Int. Ed. Engl.*, 1986, **25**, 1086.
120. E. Wenkert, R. S. Greenberg and H.-S. Kim, *Helv. Chim. Acta*, 1987, **70**, 2159.
121. (a) E. Piers and N. Moss, *Tetrahedron Lett.*, 1985, **26**, 2735; (b) S. D. Burke, C. W. Murtiashaw, J. O. Saunders, J. A. Oplinger and M. S. Dike, *J. Am. Chem. Soc.*, 1984, **106**, 4558.
122. B. Halton and M. G. Banwell, in 'The Chemistry of the Cyclopropyl Group', ed. Z. Rappoport, Wiley, New York, 1987, part 2, p. 1223.
123. M. S. Baird, *Top. Curr. Chem.*, 1988, **144**, 137.
124. T. N. Wheeler and J. Ray, *J. Org. Chem.*, 1987, **52**, 4875.
125. N. Petiniot, A. J. Anciaux, A. F. Noels, A. J. Hubert and P. Teyssie, *Tetrahedron Lett.*, 1978, 1239.
126. I. N. Domnin, E. F. Zhuravleva and N. V. Pronina, *Zh. Org. Khim.*, 1978, **14**, 2323.
127. M. I. Komendantov, I. N. Domnin and E. V. Bulecheva, *Tetrahedron*, 1975, **31**, 2495.
128. H. M. L. Davies and K. R. Romines, *Tetrahedron*, 1988, **44**, 3343.
129. M. E. Garst and V. A. Roberts, *J. Org. Chem.*, 1982, **47**, 2188.

130. M. A. McKervey, D. N. Russell and M. F. Twohig, *J. Chem. Soc., Chem. Commun.*, 1985, 491.
131. L. Chan and S. A. Matlin, *Tetrahedron Lett.*, 1981, **22**, 4025.
132. M. J. Rosenfeld, B. K. R. Shankar and H. Shechter, *J. Org. Chem.*, 1988, **53**, 2699.
133. M. Toda, M. Hattori, K. Okada and M. Oda, *Chem. Lett.*, 1987, 1263.
134. H. Ledon, G. Linstrumelle and S. Julia, *Bull. Soc. Chim. Fr., Part 2*, 1973, 2065.
135. M. Kennedy and M. A. McKervey, *J. Chem. Soc., Chem. Commun.*, 1988, 1028.
136. H. Duddeck, M. Kennedy, M. A. McKervey and M. F. Twohig, *J. Chem Soc., Chem. Commun.*, 1988, 1586.
137. L. T. Scott, P. Grutter and R. E. Chamberlain, *J. Am. Chem. Soc.*, 1984, **106**, 159.
138. G. Quinkert, H.-G. Schmalz, E. Walzer, T. Kowalczyk-Przewloka, G. Dürner and J. W. Bats, *Angew. Chem., Int. Ed. Engl.*, 1987, **26**, 61.
139. E. C. Taylor and H. M. L. Davies, *Tetrahedron Lett.*, 1983, **24**, 5453.
140. K. Nakatani, *Tetrahedron Lett.*, 1987, **28**, 165.
141. (a) C. Iwata, K. Miyashita, T. Imao, K. Masua, N. Kondo and S. Uchida, *Chem. Pharm. Bull.*, 1985, **33**, 853; (b) C. Iwata, T. Fusaka, T. Fujiwara, K. Tomita and M. Yamada, *J. Chem. Soc., Chem. Commun.*, 1981, 463.
142. M. Matsumoto, N. Watanabe and H. Kubayashi, *Heterocycles*, 1987, **26**, 1479.
143. M. Hrytsak, N. Etkin and T. Durst, *Tetrahedron Lett.*, 1986, **27**, 5679.
144. M. Hrytsak and T. Durst, *J. Chem. Soc., Chem. Commun.*, 1987, 1150.
145. A. Pusino, A. Saba and V. Rosnati, *Tetrahedron*, 1986, **42**, 4319.
146. A. Saba, *Synthesis*, 1984, 268.
147. J. Nvak and F. Sorm, *Collect. Czech. Chem. Commun.*, 1958, **23**, 1126.
148. G. O. Schenck and R. Steinmetz, *Justus Liebigs Ann. Chem.*, 1963, **668**, 19.
149. E. Wenkert, M. L. F. Bakuzis, B. L. Buckwalter and P. D. Woodgate, *Synth. Commun.*, 1981, **11**, 533.
150. O. M. Nefedov, V. M. Shostakovsky, M. Y. Samoilova and M. I. Kravchenko, *Izv. Akad. Nauk SSSR, Ser. Khim.*, 1972, 2342 (*Chem. Abstr.*, 1973, **78**, 43 165k).
151. O. M. Nevedov, L. E. Saltykova, L. E. Vasilvitskii and A. E. Shostakovsky, *Izv. Akad. Nauk SSSR, Ser. Khim.*, 1986, 2625 (*Chem. Abstr.*, 1987, **107**, 115 235a).
152. E. Wenkert, 'New Trends in Natural Product Chemistry', ed. Atta-ur-Rahman and P. W. Le Quesne, 'Studies in Organic Chemistry', 1986, Elsevier, Amsterdam, vol. 26, p. 557.
153. O. M. Nefedov, V. M. Shostakovsky and A. E. Vasilvitskii, *Angew. Chem., Int. Ed. Engl.*, 1977, **16**, 646.
154. Y. Leblanc, B. J. Fitzsimmons, J. Adams, F. Perez and J. Rokach, *J. Org. Chem.*, 1986, **51**, 789.
155. J. Rokach, Y. Girard, Y. Guindon, J. G. Atkinson, M. Larue, R. N. Young, P. Masson and G. Holme, *Tetrahedron Lett.*, 1980, **21**, 1485.
156. J. Rokach, J. Adams and R. Perry, *Tetrahedron Lett.*, 1983, **24**, 5185.
157. J. Adams and J. Rokach, *Tetrahedron Lett.*, 1984, **25**, 35.
158. J. Rokach and J. Adams, *Acc. Chem. Res.*, 1985, **18**, 87.
159. M. N. Nwaji and O. S. Onyiriuka, *Tetrahedron Lett.*, 1974, 2255.
160. E. Wenkert, M. Gua, F. Pizzo and K. Ramachandran, *Helv. Chim. Acta*, 1987, **70**, 1429.
161. (a) A. Padwa, T. J. Wisnieff and E. J. Walsh, *J. Org. Chem.*, 1989, **54**, 299; (b) A. Padwa, T. J. Wisnieff and E. J. Walsh, *J. Org. Chem.*, 1986, **51**, 5036.
162. S. R. Tanny, J. Grossman and F. W. Fowler, *J. Am. Chem. Soc.*, 1972, **94**, 6495.
163. J. F. Biellmann and M. P. Goeldner, *Tetrahedron*, 1971, **27**, 2957.
164. B. E. Maryanoff, *J. Org. Chem.*, 1982, **47**, 3000.
165. C. W. Jefford and W. Johncock, *Helv. Chim. Acta*, 1983, **66**, 2666.
166. E. Galeazzi, A. Guzman, A. Pinedo, A. Saldana, D. Torre and J. M. Muchowski, *Can. J. Chem.*, 1983, **61**, 454.
167. C. W. Jefford, T. Kubota and A. Zaslona, *Helv. Chim. Acta*, 1986, **69**, 2048.
168. T. Bowles, R. J. Gillespie, A. E. A. Porter, J. A. Rechka and H. S. Rzepa, *J. Chem. Soc., Perkin Trans. 1*, 1988, 803.
169. A. E. A. Porter and H. S. Rzepa, *J. Chem. Soc., Perkin Trans. 1*, 1988, 809.
170. O. Meth-Cohn and G. Vuuren, *J. Chem. Soc., Perkin Trans. 1*, 1986, 233.
171. R. J. Gillespie and A. E. A. Porter, *J. Chem. Soc., Perkin Trans. 1*, 1979, 2624.
172. V. M. Shostakovsky, V. L. Zlatkina, V. L. Vasilvitskii and O. M. Nefedov, *Izv. Akad. Nauk SSSR, Ser. Khim.*, 1982, 2126.
173. H. Storflor, J. Skramstad and S. Nordenson, *J. Chem. Soc., Chem. Commun.*, 1984, 208.
174. O. Meth-Cohn and E. Vuorinen, *J. Chem. Soc., Chem. Commun.*, 1988, 138

4.9
Intermolecular 1,3-Dipolar Cycloadditions

ALBERT PADWA
Emory University, Atlanta, GA, USA

4.9.1 INTRODUCTION

Cycloaddition reactions have figured prominently in both synthetic and mechanistic organic chemistry.[1,2] Current understanding of the underlying principles in this area has grown from a fruitful interplay between theory and experiment.[3] The monumental work of Huisgen and coworkers in the early 1960s led to the general concept of 1,3-dipolar cycloaddition.[4–7] Few reactions rival this process in the number of bonds that undergo transformation during the reaction, producing products considerably more complex than the reactants. Over the years this reaction has developed into a generally useful method of five-membered heterocyclic ring synthesis, since many 1,3-dipolar species are readily available and react with a wide variety of dipolarophiles. 1,3-Dipolar cycloadditions are bimolecular in nature and involve the addition of a 1,3-dipole (**1**) to a multiple π-bond system (**2**) leading to five-membered heterocycles (**3**; equation 1).

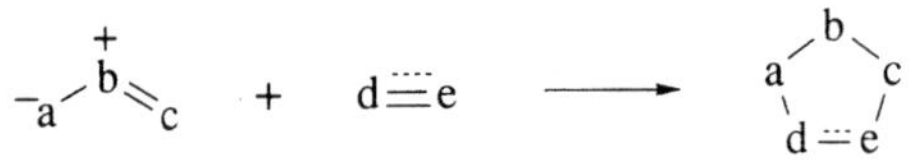

$$\overset{-}{a}\diagup\overset{+}{b}\diagdown\!\!\!=c \quad + \quad d\overset{\cdots}{=}e \longrightarrow \text{five-membered ring } a\!-\!b\!-\!c\!-\!e\overset{\cdots}{=}d\!-\!a \qquad (1)$$

(1) **(2)** **(3)**

4.9.2 NATURE OF THE DIPOLE

A 1,3-dipole is basically a system of three atoms over which are distributed four π-electrons, as in the allyl anion system. The three atoms can be a wide variety of combinations of C, O and N. Table 1 provides a list of the more common 1,3-dipoles possessing carbon, nitrogen and oxygen centers which are known to undergo the dipolar cycloaddition reaction. The dipolarophiles can be virtually any double or triple bond. The term 1,3-dipole arose because in valence bond theory such compounds can only be described in terms of dipolar resonance contributors, as shown for diazomethane (Scheme 1). The extreme 1,3-dipolar forms with their complementary nucleophilic and electrophilic centers readily explain the tendency to undergo addition to π-bonds. Indeed, it must be possible to write 1,3-dipolar forms for all such species that undergo this type of addition. Significant computational efforts have been directed toward characterization of 1,3-dipoles and the literature in this area has recently been reviewed by Houk and Yamaguchi.[8]

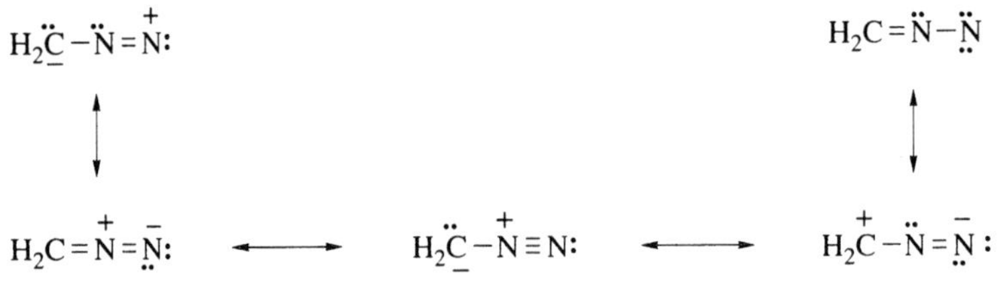

$$H_2\ddot{\underline{C}}-\ddot{N}=\overset{+}{N}: \qquad\qquad H_2C=\ddot{N}-\ddot{\underset{\cdot\cdot}{N}}$$

$$\updownarrow \qquad\qquad\qquad\qquad \updownarrow$$

$$H_2C=\overset{+}{N}=\overset{-}{\underset{\cdot\cdot}{N}}: \longleftrightarrow H_2\ddot{\underline{C}}-\overset{+}{N}\equiv N: \longleftrightarrow H_2\overset{+}{C}-\ddot{N}=\overset{-}{\underset{\cdot\cdot}{N}}:$$

Scheme 1

4.9.3 MECHANISTIC CONSIDERATIONS

Huisgen and coworkers have systematically studied the mechanism of 1,3-dipolar cycloadditions.[4–7] In the majority of 1,3-cycloadditions, the reaction rate is not markedly influenced by the dielectric constant of the solvent medium in which the reaction is conducted. The independence of solvent polarity, the very negative entropies of activation and the stereospecificity and regiospecificity point to a highly ordered transition state.[7] In most 1,3-dipolar cycloaddition reactions, when two isomers are possible as a result of the use of unsymmetrical reagents, one isomer usually predominates, often to the exclusion of the other. The principal question that arises when considering the regiospecificity of the 1,3-dipolar additions is whether the two new σ-bonds formed on addition of the 1,3-dipole to the dipolarophile are formed simultaneously, or one after the other. The mechanism that has emerged from Huisgen's group is that of a single-step, four-center, 'no mechanism' cycloaddition, in which the two new bonds are both partially formed in the transition state, although not necessarily to the same extent.[9] A symmetry–energy correlation diagram reveals that such a thermal cycloaddition reaction is an allowed process.[10] A proposed alternative mechanism is a two-step process involving a spin-paired diradical intermediate.[11]

The controversy between Huisgen and Firestone concerning the mechanism for 1,3-dipolar cycloaddition is longstanding.[9,11] For nitrile oxide cycloadditions, experimental data have been interpreted either as supportive of a concerted mechanism[9] or in favor of a stepwise mechanism with diradical intermediates.[11] Theory has compounded, rather than resolved, this problem. *Ab initio* calculations on the reaction of fulmonitrile oxide with acetylene predict a concerted mechanism at the molecular orbital level,[12,13] but a stepwise mechanism after inclusion of extensive electron correlation.[14] MNDO predicts a stepwise mechanism with a diradical intermediate.[15] The existence of an extended diradical intermediate such as (**4**; Scheme 2) has been postulated by Firestone in order to account for the occasional formation of 1,4-addition products such as the oxime (**5**).[11] Of course, the intermediates (**4**) and (**5**) for the Firestone mechanism do not correspond to the initial transition states in Firestone's theory. These are attained prior to the formation of, and at higher energy than, the intermediates.

Table 1 Classification of 1,3-Dipoles Consisting of Carbon, Nitrogen and Oxygen Centers

		Propargyl–Allenyl Type		
Nitrilium betaines	$-C\equiv\overset{+}{N}-\overset{-}{\ddot{C}}<$	⟷	$-\overset{-}{\ddot{C}}=\overset{+}{N}=C<$	Nitrile ylides
	$-C\equiv\overset{+}{N}-\overset{-}{\ddot{N}}\diagdown$	⟷	$-\overset{-}{\ddot{C}}=\overset{+}{N}=N\diagdown$	Nitrile imines
	$-C\equiv\overset{+}{N}-\overset{-}{\ddot{O}}$	⟷	$-\overset{-}{\ddot{C}}=\overset{+}{N}=O$	Nitrile oxides
Diazonium betaines	$N\equiv\overset{+}{N}-\overset{-}{\ddot{C}}<$	⟷	$\overset{-}{\ddot{N}}=\overset{+}{N}=C<$	Diazoalkanes
	$N\equiv\overset{+}{N}-\overset{-}{\ddot{N}}\diagdown$	⟷	$\overset{-}{\ddot{N}}=\overset{+}{N}=N\diagdown$	Azides
	$N\equiv\overset{+}{N}-\overset{-}{\ddot{O}}$	⟷	$\overset{-}{\ddot{N}}=\overset{+}{N}=O$	Nitrous oxide
		Allyl Type		
Nitrogen function as middle center	$>C=\overset{+}{N}(-)-\overset{-}{\ddot{C}}<$	⟷	$>\overset{-}{\ddot{C}}-\overset{+}{N}(-)=C<$	Azomethine ylides
	$>C=\overset{+}{N}(-)-\overset{-}{\ddot{N}}\diagdown$	⟷	$>\overset{-}{\ddot{C}}-\overset{+}{N}(-)=N\diagdown$	Azomethine imines
	$>C=\overset{+}{N}(-)-\overset{-}{\ddot{O}}$	⟷	$>\overset{-}{\ddot{C}}-\overset{+}{N}(-)=O$	Nitrones
	$\diagdown N=\overset{+}{N}(-)-\overset{-}{\ddot{N}}\diagdown$	⟷	$\diagdown\overset{-}{\ddot{N}}-\overset{+}{N}(-)=N\diagdown$	Azimines
	$\diagdown N=\overset{+}{N}(-)-\overset{-}{\ddot{O}}$	⟷	$\diagdown\overset{-}{\ddot{N}}-\overset{+}{N}(-)=O$	Azoxy compounds
	$O=\overset{+}{N}(-)-\overset{-}{\ddot{O}}$	⟷	$\overset{-}{\ddot{O}}-\overset{+}{N}(-)=O$	Nitro compounds
Oxygen atom as middle center	$>C=\overset{+}{O}-\overset{-}{\ddot{C}}<$	⟷	$>\overset{-}{\ddot{C}}-\overset{+}{O}=C<$	Carbonyl ylides
	$>C=\overset{+}{O}-\overset{-}{\ddot{N}}\diagdown$	⟷	$>\overset{-}{\ddot{C}}-\overset{+}{O}=N\diagdown$	Carbonyl imines
	$>C=\overset{+}{O}-\overset{-}{\ddot{O}}$	⟷	$>\overset{-}{\ddot{C}}-\overset{+}{O}=O$	Carbonyl oxides
	$\diagdown N=\overset{+}{O}-\overset{-}{\ddot{N}}\diagdown$	⟷	$\diagdown\overset{-}{\ddot{N}}-\overset{+}{O}=N\diagdown$	Nitrosimines
	$\diagdown N=\overset{+}{O}-\overset{-}{\ddot{O}}$	⟷	$\diagdown\overset{-}{\ddot{N}}-\overset{+}{O}=O$	Nitrosoxides
	$O=\overset{+}{O}-\overset{-}{\ddot{O}}$	⟷	$\overset{-}{\ddot{O}}-\overset{+}{O}=O$	Ozone

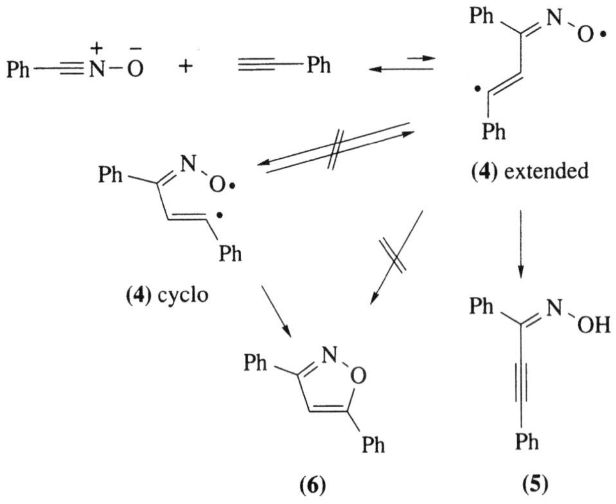

Ph—≡N⁺—O⁻ + ≡—Ph

(4) extended

(4) cyclo

(6)

(5)

Scheme 2

4.9.3.1 Stepwise Mechanism

The stereospecificity observed in many 1,3-dipolar cycloadditions is often considered to be compelling, if not conclusive, evidence for concert in these reactions.[9] However, if the rate constant for rotation (k_r) about bond 'a' in diradical intermediate, (**7a**) or (**7b**), was much smaller than the rate constant for cyclization (k_c), high stereospecificity would still be observed (Scheme 3).[11] The reported examples of stereospecific 1,3-dipolar cycloadditions involve di-, tri- or tetra-substituted alkenes. Barriers to rotation of simple primary, secondary and tertiary alkyl radicals are only 0–1.2 kcal mol^{-1} (1 kcal = 4.18 kJ),[16] but more highly substituted radicals have barriers to rotation estimated to be as high as 4 kcal mol^{-1}.[17] The rate ratio of rotation to ring closure is highly dependent on the degree of substitution at the termini. A deuterium-labeled methylene rotor offers the highest chance to bring about an intermediate in the cycloaddition reaction. This was the reason why Houk and Firestone studied the 1,3-dipolar cycloadditions of 4-nitrobenzonitrile oxide to *cis*- and *trans*-dideuterioethylene (Scheme 3; R = D).[18] These experiments establish that the reaction is ≥98% stereospecific. If a diradical intermediate were formed, the barrier to rotation about bond a would have to be at least 2.3 kcal mol^{-1} higher than the barrier to cyclization. Since the rotational barrier of bond a is that expected for a normal primary radical (*i.e.* ≤0.4 kcal mol^{-1}), there can be no barrier to cyclization for the predominant cycloaddition pathway. Thus, the most reasonable mechanism for 1,3-dipolar cycloadditions appears to be a concerted one.

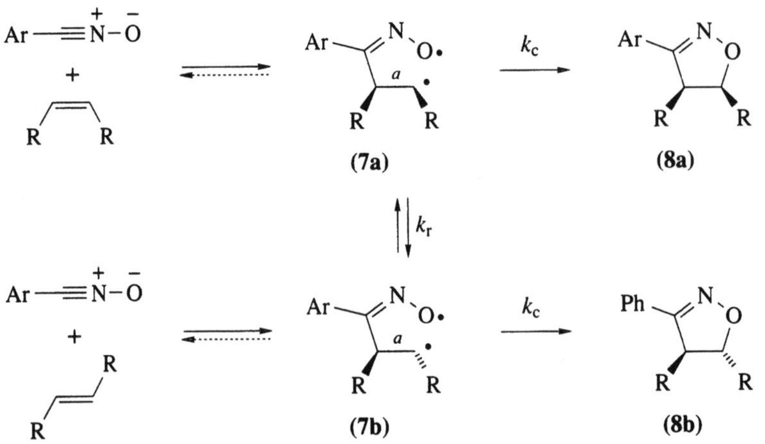

(7a) (8a)

k_c, k_r

(7b) (8b)

Scheme 3

4.9.4 FRONTIER MOLECULAR ORBITAL THEORY

It was only through the application of frontier molecular orbital (FMO) theory that a coherent picture of a mechanism based on a concerted nonsynchronous process emerged.[19,20] This theoretical solution to the mechanistic dilemma went far beyond the usual rationalization of steric and electronic effects in the transition state of a concerted process. The reactivity and regiochemical control in the [3 + 2] dipolar cycloaddition of 1,3-dipole to a substituted alkene can now be understood by consideration of the interactions of the corresponding molecular orbitals. Sustmann has categorized cycloaddition processes into three types.[21] The first (Type 1, Figure 1) involves a dominant interaction between the highest occupied molecular orbital of the dipole [HOMO (dipole)] and the lowest unoccupied molecular orbital of the dipolarophile [LUMO (dipolarophile)]. The majority of Diels–Alder reactions are accommodated by this classification. A second possibility (Type 3) is dominated by the interaction between the LUMO (dipole) and the HOMO (dipolarophile). The ozonization of alkenes may be a reaction that fits this description.[22] In the case of a Type 2 cycloaddition reaction, the similarity of LUMO and HOMO energies in dipole and dipolarophile implies that both HOMO (dipole)–LUMO (dipolarophile) and LUMO (dipole)–HOMO (dipolarophile) interactions may be important in determining reactivity and regiochemistry.

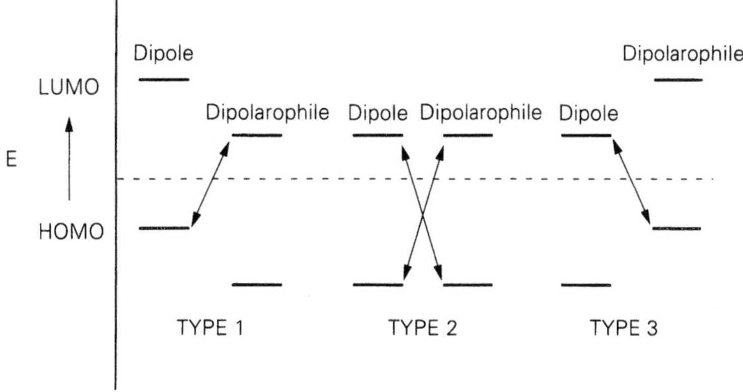

Figure 1 Frontier molecular orbital classification of cycloaddition reactions

The frontier orbital energies may be obtained from empirical ionization potential (IP) data to give the HOMO energies. The LUMO energies are derived from electron affinity (EA) data. The electron affinities may be estimated by EA = IP – ΔE ($\pi\pi^*$) – 4.5 eV.[23] As the two addends in a cycloaddition reaction approach each other in proper orientation for reaction, an interaction occurs between the reactant frontier molecular orbitals. According to second-order perturbation theory, this interaction leads to a stabilization of the occupied molecular orbital that is inversely proportional to the energy separation between the orbitals involved.[24] The more proximate the orbitals are in energy, the more extensive the interaction. Additionally, the magnitude of interaction depends on the degree of overlap involved, the larger the overlap, the greater the interaction. When reactant HOMO–LUMO interactions occur, the result of the process will be to stabilize the bonding orbital (HOMO) and to destabilize the antibonding orbital (LUMO). When the interaction involves one filled HOMO and one vacant LUMO orbital, the result of the interaction will be to stabilize the system as a whole, since both electrons will enter the orbital of lower energy. Sustmann first applied FMO theory to the 1,3-dipolar cycloadditions,[25] while Houk[19] and Bastide and Henri-Rousseau[26] used it to explain reactivity and regioselectivity of the reaction. In order to predict regioselectivity, it becomes necessary to determine the relative magnitudes of the coefficients in the HOMO and LUMO of the 1,3-dipole and dipolarophile. The favored cycloadduct will be that formed by union of the atoms with the largest coefficients.

4.9.5 NONCONCERTED CYCLOADDITIONS

Recently, Huisgen and coworkers have reported on the first unequivocal example of a nonconcerted 1,3-dipolar cycloaddition.[27] Sustmann's FMO model of concerted cycloadditions envisions two cases in which the stepwise mechanism might compete with the concerted one.[21] Two similar HOMO–LUMO interaction energies correspond to a minimum of rate and a diradical mechanism is possible, especially if stabilizing substituents are present. A second case is when the HOMO (1,3-dipole)–LUMO (dipolarophile) is strongly dominant in the transition state. The higher the difference in π-MO energies of reactant, the lower would be the energy contribution by the second HOMO–LUMO interaction. In the extreme case, the interaction of the LUMO of the dipole with the HOMO of the dipolarophile would no longer contribute to the attractive forces in the transition state. The unidirectional electron flow from HOMO (1,3-dipole) to LUMO (dipolarophile) would establish one bond between the reactants and a zwitterionic intermediate would result.

Thiocarbonyl ylides are the 1,3-dipoles with the highest π-MO energies.[28,29] Huisgen described the cycloadditions of 2,2,4,4-tetramethyl-1-oxocyclobutane-3-thione *S*-methylide (**10**) and of adamantanethione *S*-methylide (**11**) to dimethyl 2,3-dicyanofumarate which proceed in a nonstereospecific manner (Scheme 4).[27] The presence of four electron-attracting substituents in the dipolarophile significantly lowers the MO energy of the ethylenic dipolarophile. Thus, in this particular case, the pair of reactants ful-

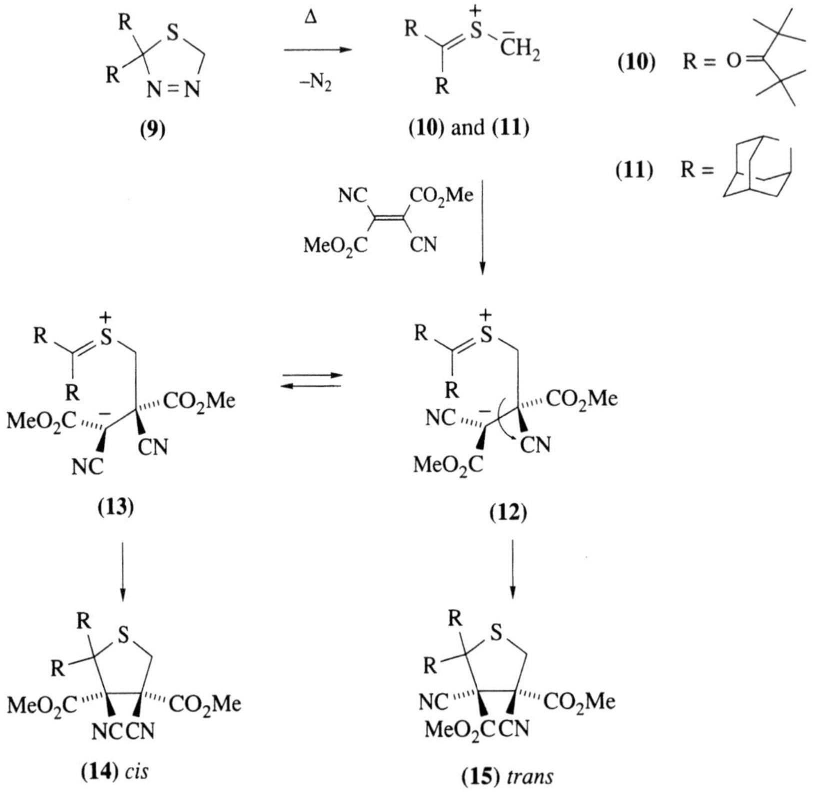

(**9**) (**10**) and (**11**) (**10**) R = (**11**) R = (**13**) (**12**) (**14**) *cis* (**15**) *trans*

Scheme 4

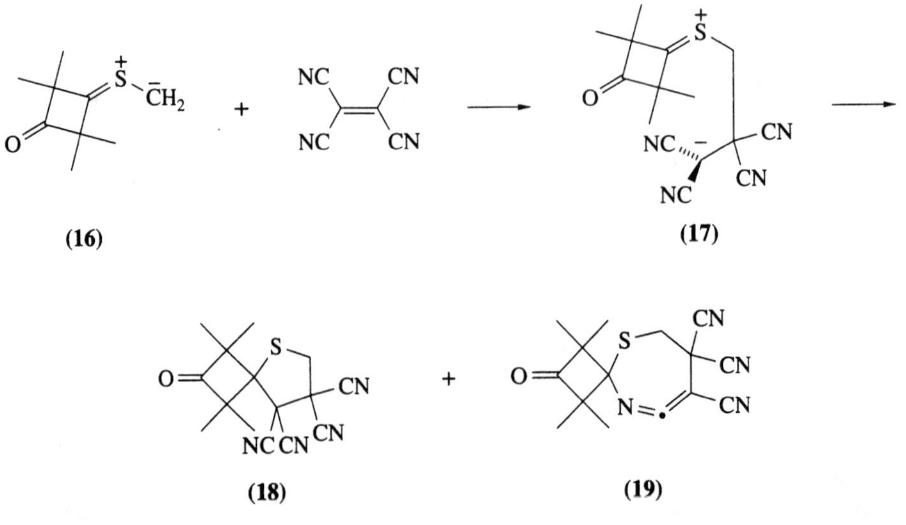

(**16**) (**17**) (**18**) (**19**)

Scheme 5

fills the prerequisite of zwitterion formation with its extremely different π-MO energies. Huisgen has also reported that the zwitterionic intermediate (**17**) derived from thiocarbonyl ylide (**10**) and tetracyanoethylene can also undergo competing cyclizations to give the normal cycloadduct (**18**) as well as a seven-membered ketenimine (**19**; Scheme 5). The existence of a discrete intermediate with these thiocarbonyl ylide cycloadditions also testifies to the fact that normal stereospecific 1,3-dipolar cycloadditions follow a fundamentally different mechanism that does not involve any intermediates.

4.9.5.1 [4 + 3] Cycloaddition

The principles of orbital symmetry conservation provide a permissive, although not obligatory, theoretical basis for the [3 + 2] concerted mechanism,[10] as do the observations of [$\pi^4s + \pi^6s$] to the exclusion of [$\pi^4s + \pi^4s$] dipolar cycloadditions in the reactions of 1,3-dipoles with trienes.[30] In accordance with the orbital symmetry rules,[10] 1,3-dipoles have been found to react with 1,3-dienes to give vinyl-substituted five-membered rings exclusively.[31] Corresponding [4 + 3] cycloadditions have only been observed in the ozonization of anthracenes,[32] and in one case as an intramolecular variant.[33] Recent work by Mayr has uncovered the first example of a [4 + 3] cycloaddition of a 1,3-dipole with a diene (Scheme 6).[34] The reaction of hexamethylcyclopentadiene (**20**) with *C*,*N*-diphenylnitrone (**21**) afforded a mixture of diastereomeric [3 + 2] cycloadducts (**22**; 32%) as well as a [4 + 3] cycloadduct (**23**; 18%). If the applicability of orbital symmetry is taken for granted, then a concerted [$\pi^4s + \pi^4s$] process cannot account for the formation of (**23**). Mayr has suggested that the isolation of (**23**) requires the intervention of a diradical intermediate (*i.e.* **24**). He also claims that the activation energy for the stepwise mechanism is only 2–3 kcal mol^{-1} greater than that of the concerted process. It remains to be seen whether other examples of nonconcerted dipolar cycloadditions will be discovered in the future.

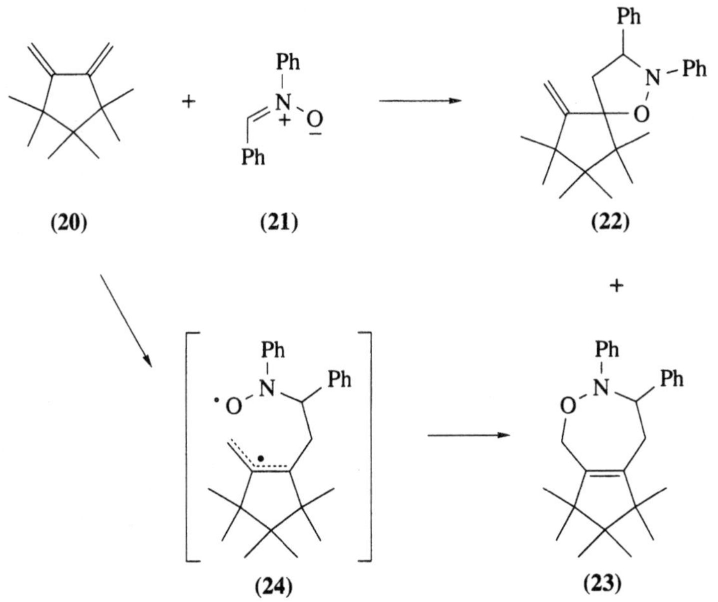

Scheme 6

The aim of the present chapter is to review the characteristics of the dipolar cycloaddition reaction of the most common 1,3-dipoles. The scope, generality and limitations of the cycloaddition reaction will be summarized and the general outline and potential synthetic use for the cycloadditions noted. It is hoped that general and specific points in need of study will be revealed, stimulating further work in this field.

4.9.6 1,3-DIPOLAR CYCLOADDITIONS

4.9.6.1 Nitrones

Nitrones represent a long-known and thoroughly investigated class of 1,3-dipoles.[35] Through the use of nitrone cycloaddition chemistry, numerous isoxazolidines have been prepared with excellent stereochemical control.[36,37] The 1,3-dipolar cycloaddition reaction of nitrones has attracted considerable attention as a convenient tool for the rapid construction of widely varied classes of natural products (Scheme 7).[38–43] The presence of a nitrogen atom within the isoxazolidine ring has made this heterocycle moiety especially attractive for the synthesis of the β-lactam ring (Scheme 8).[42,43] The key feature of this approach involves a reductive cleavage of the isoxazolidine ring to give a γ-amino alcohol, which undergoes subsequent cyclization with a neighboring methoxycarbonyl group.[44–46]

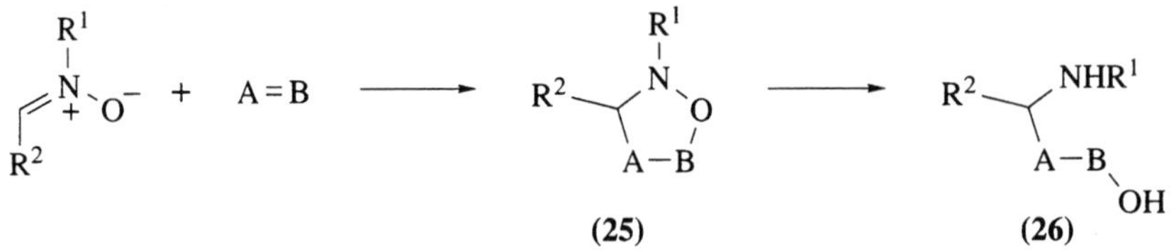

Scheme 7

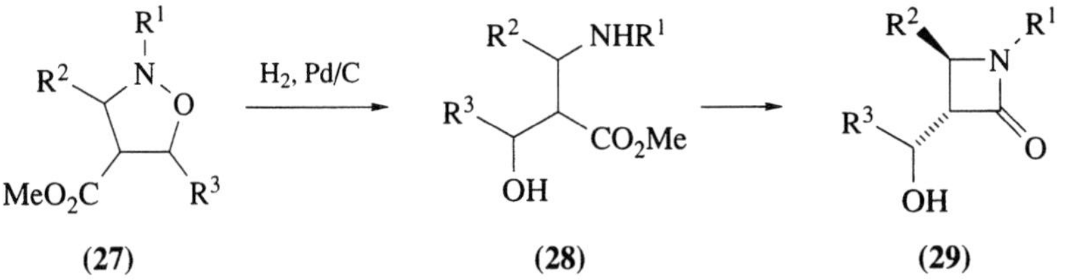

Scheme 8

1,3-Dipolar cycloadditions of nitrones have been reported with a variety of alkenes, alkynes, isocyanates, isothiocyanates, thiocarbonyl compounds, phosphoranes, sulfenes and sulfinyl groups.[35] All these reactions are usually carried out by heating the two reagents together in an inert solvent, and the products are often easily isolated in high yield. It should be pointed out that the cycloadducts are not stable in all cases and sometimes undergo interesting transformations. The regiochemical control exhibited by nitrone–alkene cycloadditions has been accommodated by the development of a FMO treatment.[19–21] Nitrone cycloadditions are believed to be Type 2 processes (Scheme 9). As such, both possible FMO interactions may be significant. The interaction that dominates in a particular case will depend on the nature of both the dipole and the dipolarophile. Most dipolarophiles undergo cycloaddition to give 5-substituted isoxazolidines with high regioselectivity. As the ionization potential of the nitrone decreases or the electron affinity of the dipolarophile increases, an increasing tendency toward production of the 4-substituted isoxazolidines is found.[47] At some point, there must be a switch over from LUMO to HOMO control as one increases the electron-withdrawing power of the substituents on the alkene. That point is apparently approached with methyl acrylate, since regioisomeric mixtures of adducts are encountered with this dipolarophile.[48,49]

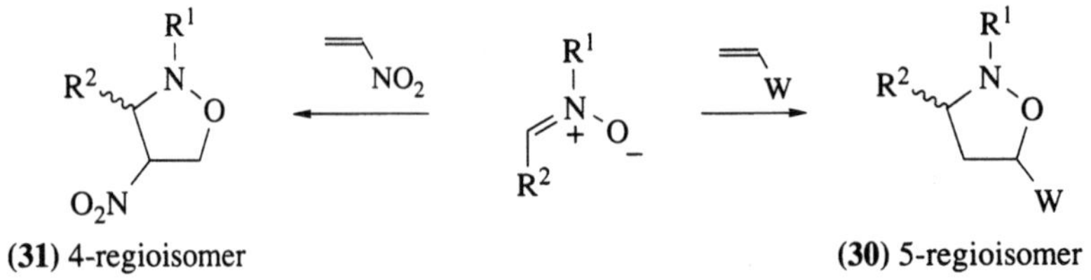

Scheme 9

Monosubstituted alkenes react with nitrones to give rise to a diastereomeric ratio of *cis*- and *trans*-disubstituted isoxazolidines (**32**) and (**33**). The *exo* approach gives the *cis* product, provided that the NR^1 and CR^2 substituents are in a *trans* relationship; the *endo* approach gives the *trans* product (Scheme 10). Inspection of the available data tends to show that the *exo* approach is favored in most cases when secondary orbital interactions are negligible. When X = CO_2Me, CN or Ph such secondary interactions could stabilize the transition state and give predominantly the *trans* adduct. These secondary interactions are analogous to those directing the *endo*/*exo* approach in the Diels–Alder reaction. However, there are data available that are inconsistent with this interpretation.[48] Predictions are therefore unreliable, and it is advisable to determine structures thoroughly for every reaction. The unpredictable behavior of nitrones is demonstrated by the reactions of *N*,*C*-diphenylnitrone and *N*-methylnitrone with diethyl methylenemalonate; which in the former case gives only the 4,4-disubstituted regioisomer (**34**), and in the latter case the 5,5-disubstituted regioisomer (**35**).

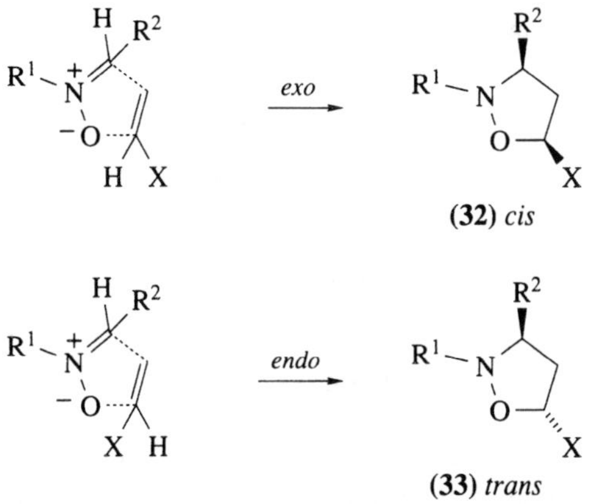

exo → (**32**) *cis*

endo → (**33**) *trans*

Scheme 10

(**34**) *versus* (**35**)

Entropic factors allow the intramolecular version of the nitrone–alkene cycloaddition reaction to proceed under milder conditions than the corresponding bimolecular equivalent.[51,52] The regiochemistry of the internal reaction is complicated by a complex interplay of such factors as alkene polarity, ring strain and other nonbonded interactions. In general, the intramolecular situation can be assessed as a competition between the bridged and fused modes of cycloaddition. Since LeBel first reported the facile intramolecular cycloaddition of unsaturated nitrones to form bicyclic isoxazolidines,[53] there has been considerable interest in this reaction (Scheme 11). When the dipolarophile is substituted at the 2-position with an alkyl group and the nitrone is substituted at the α-position, there is a tendency for the predominant formation of a fused bicyclic system (**38**) as opposed to the bridged system.[51,52] The intramolecular cycloaddition of *C*-alkenyl-substituted nitrones represents an efficient and general route to five-, six- and higher-membered carbocyclic and heterocyclic rings. This methodology has been used to synthesize prostanoids, terpenoids, various bridged and condensed cycloalkanes, alkaloids, carbohydrates and miscellaneous natural products.[51] Some typical examples are outlined in Scheme 12.

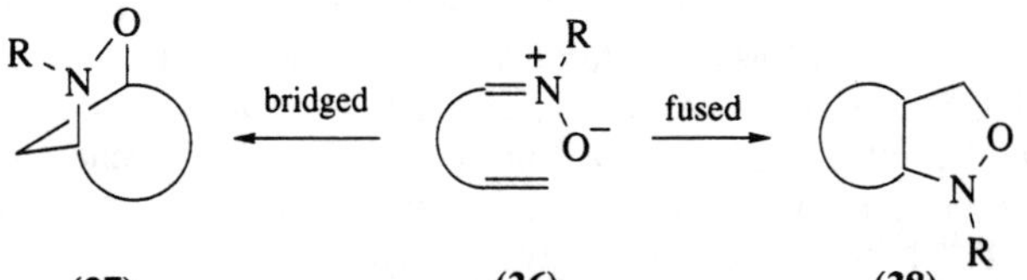

(**37**) ← bridged — (**36**) — fused → (**38**)

Scheme 11

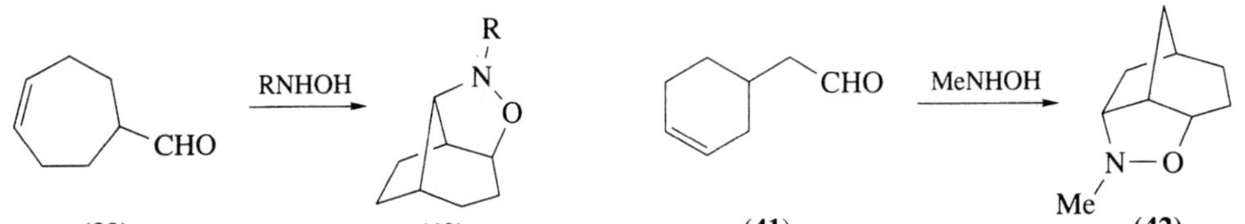

(39) $\xrightarrow{\text{RNHOH}}$ (40) (41) $\xrightarrow{\text{MeNHOH}}$ (42)

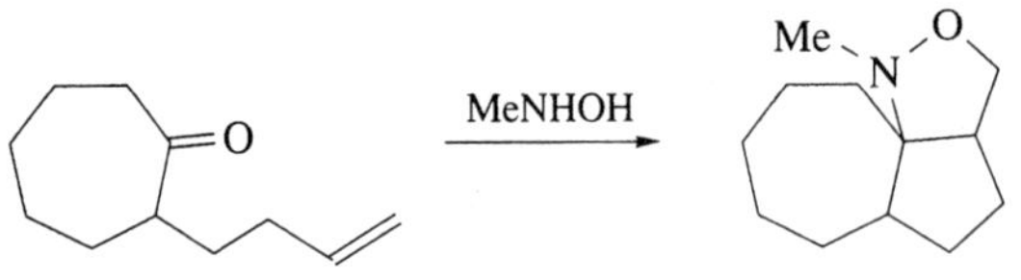

(43) $\xrightarrow{\text{MeNHOH}}$ (44)

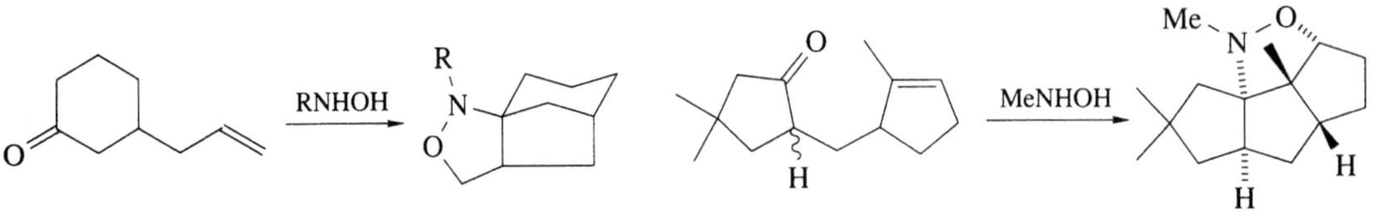

(45) $\xrightarrow{\text{RNHOH}}$ (46) (47) $\xrightarrow{\text{MeNHOH}}$ (48)

Scheme 12

The utility of nitrone–alkene [3 + 2] cycloadditions in the total synthesis of natural products arises not from the presence of the isoxazolidine moiety in naturally occurring substances but from the various transformation products of this saturated heterocycle. The most important of these is the conversion of the isoxazolidine ring system to an *N*-substituted 1,3-amino alcohol by a reduction step (equation 2). The formation of these 1,3-amino alcohols from readily available aldehydes and alkenes represents a very powerful synthetic method. Since this transformation not only yields new carbon–carbon and carbon–oxygen bonds but also incorporates a nitrogen atom in the molecule, the obvious synthetic targets have been nitrogenous substances, particularly alkaloids. Applications of this method to nonnitrogenous targets have also been reported.[48–52]

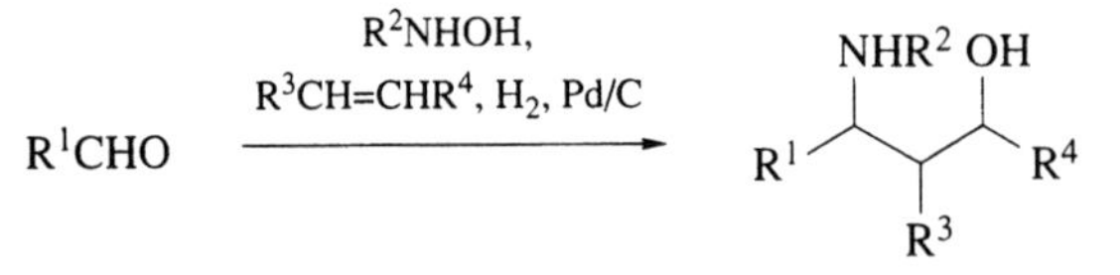

$$R^1CHO \xrightarrow[R^3CH{=}CHR^4,\ H_2,\ Pd/C]{R^2NHOH,} R^1CH(NHR^2)CH(R^3)CH(OH)R^4 \quad \textbf{(49)} \qquad (2)$$

4.9.6.2 Nitrile Oxides

1,3-Dipoles can be classified into two major types: (i) those with internal octet stabilization, where a mesomeric formula can be drawn such that the central atom of the dipole has a positive charge and all centers have completely filled valences; and (ii) those without internal octet stabilization, where each mesomeric form has an electron sextet.[4] By far the more common group of dipoles is the former, mainly because the dipoles in the second group are all unstable and must be prepared *in situ*. In recent years a good deal of attention has focused on the chemistry of the octet-stabilized class of dipoles known as the nitrile oxides.[54,55] Nitrile oxides readily cycloadd to a wide variety of alkenes to generate isoxazolines which represent an extremely useful class of heterocycles (Scheme 13). The presence of a nitrogen atom within the isoxazoline ring has made this heterocycle moiety especially attractive for the synthesis of a wide assortment of natural products. The key feature of this approach generally involves a reductive cleavage of the isoxazoline ring to give either a γ-amino alcohol (**52**) or a β-hydroxy ketone (**53**) which can be further manipulated.[56] Nitrile oxides are conveniently generated by the base treatment of hydroxamic acid chlorides,[54] by the oxidation of aldoximes[55] or by the dehydrogenation of primary nitroalkanes.[57] Most nitrile oxides (**50**) are highly reactive and, in the absence of added trapping agents, undergo rapid dipolar cycloaddition with themselves to give furoxans (**54**; equation 3) Steric bulk increases the stability of the nitrile oxide. For example, *t*-butyl nitrile oxide is readily generated and handled in solution, whereas mesityl nitrile oxide is a normal crystalline organic compound that is indefinitely stable in the pure form. The cycloadditions of nitrile oxides with electron-rich and conjugated alkenes are LUMO controlled.[58] Union of the larger coefficients leads to 5-substituted isoxazolines. Electron-deficient dipolarophiles also react rapidly due to the influence of both dipole HOMO and LUMO interactions. With this class of dipolarophiles, regioselectivity is decreased, since the dipole LUMO tends to produce 5-substituted isoxazolines, whereas the dipole HOMO produces 4-substituted isoxazolines.[59]

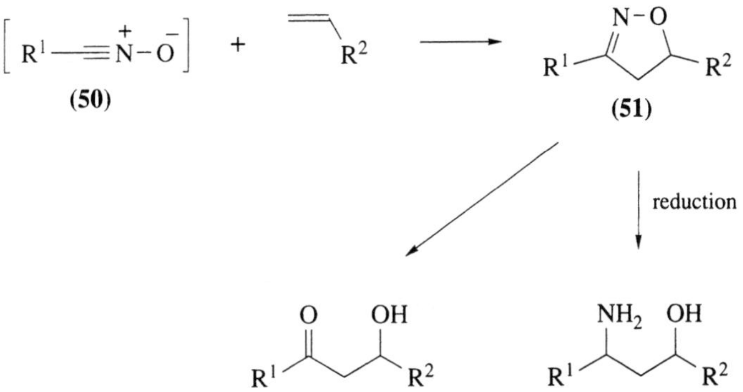

(50) (51) reduction (53) (52)

Scheme 13

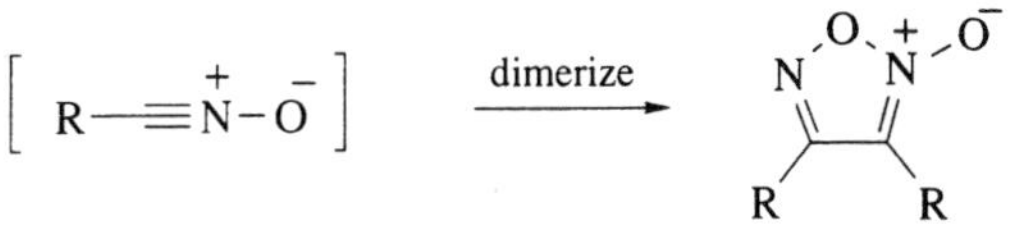

dimerize (3)

(50) (54)

Asymmetric induction has become one of the cornerstones in modern synthetic methodology, leading to a host of enantiomerically enriched compounds.[60] The use of chiral auxiliaries to induce biased stereochemical changes has allowed investigators to form a variety of C—X bonds with high enantioselectivity. Recently, the asymmetric Diels–Alder reaction has received a great deal of attention.[61] Some of these [4 + 2] cycloadditions proceed with very high, and sometimes near perfect, diastereoselection.[62] In contrast to the well-documented asymmetric Diels–Alder reactions, asymmetric 1,3-dipolar cycloadditions have not been extensively explored. There are, however, several reports of stereoselective cycloadditions involving nitrile oxides and nitrones. Nitrones which carry a chiral substituent on nitrogen often display high diastereoselectivity in dipolar cycloadditions with a variety of achiral dipolarophiles.[63–66] In certain cases, the reaction of chiral dipolarophiles with nitrile oxides leads to reasonably high diastereoselectivity in the formation of the cycloadducts. Since the N—O bonds of the cycloadducts are readily cleaved to produce acyclic molecules, the diastereoselectivity displayed in the cycloadditions serves as a means of controlling acyclic stereochemistry (equation 4).[56]

(4)

C* = stereogenic center

Curran and coworkers[67] have reported high diastereoselectivity (9:1) in the cycloaddition reactions of Oppolzer's chiral sultam[68] (**55**) with nitrile oxides (equation 5). The major diastereomer (**56**) results from 'top-side' attack of the incoming nitrile oxide and it has been suggested that the α-oxygen atom of the sultam provides a steric or electronic encumbrance to 'bottom-side' attack. These results point the way to the design of new chiral auxiliaries in which both the facial selectivity and the acrylate conformer are controlled in the transition state for cycloaddition.

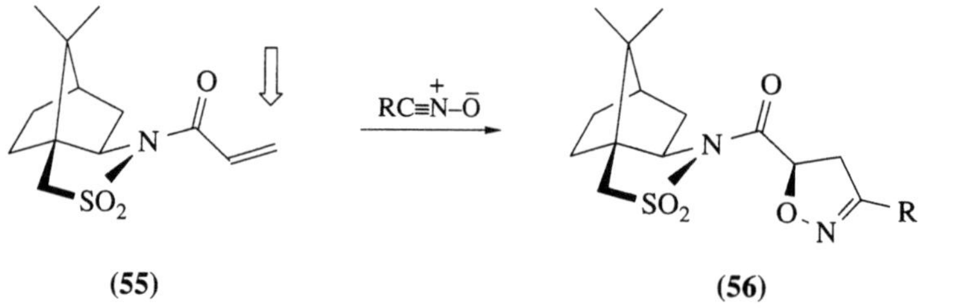

$RC\equiv\overset{+}{N}-\overline{O}$

(55) **(56)** (5)

The application of intramolecular dipolar cycloaddition reactions to the synthesis of complex natural products has recently come to be recognized as a very powerful synthetic tool, one equally akin to the intramolecular Diels–Alder reaction in its potential scope of application.[69] This is particularly the case with nitrile oxides and the INOC reaction has been extensively utilized in total synthesis.[70] The intramolecular nitrile oxide cycloaddition (INOC) generally displays exceptional regio- and stereo-chemical control which undoubtedly accounts for the popularity of this reaction. Internal cycloadditions of nitrile oxides have been found to offer a powerful solution to many problems in complex natural product synthesis.[48] For example, Confalone and coworkers have utilized the INOC reaction for the stereospecific synthesis of the key amino alcohol (**60**), which was converted in five subsequent steps to (±)-biotin (**61**; Scheme 14).[71]

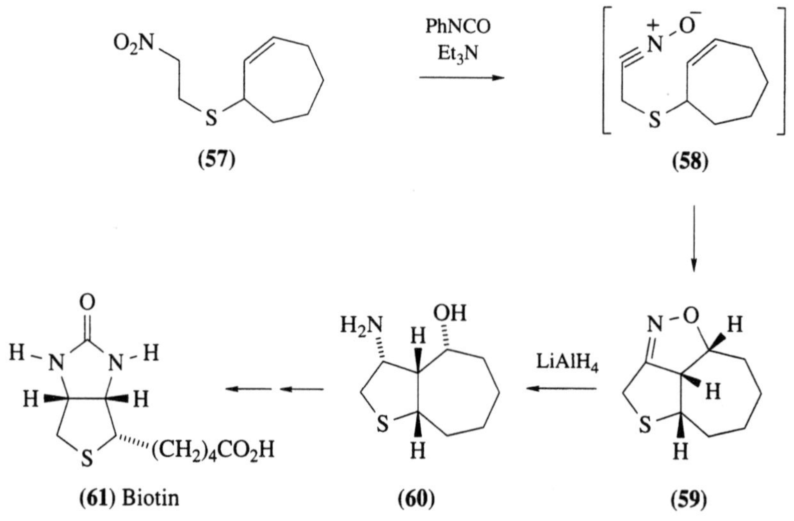

PhNCO
Et_3N

(57) **(58)**

$LiAlH_4$

(61) Biotin **(60)** **(59)**

Scheme 14

Kozikowski's group has been particularly active in the application of the INOC reaction toward the construction of a variety of natural products. One of the many examples from his laboratory involves the synthesis of tetracyclic compounds possessing suitably functionalized C rings for elaboration to a diverse number of ergot alkaloids *via* the INOC reaction. A total synthesis of chanoclavine I (**65**) was accomplished by this chemistry (Scheme 15). The key step in the synthesis involved the conversion of the nitro group of indole (**62**) into the corresponding nitrile oxide using the phenyl isocyanate procedure developed by Mukaiyama.[57] The major product corresponded to isoxazoline (**64**). The isoxazoline nucleus was converted into chanoclavine I (**65**) in a series of subsequent steps. The application of nitrile oxide cycloaddition chemistry to the construction of other natural products can be expected to be an active area in future years.

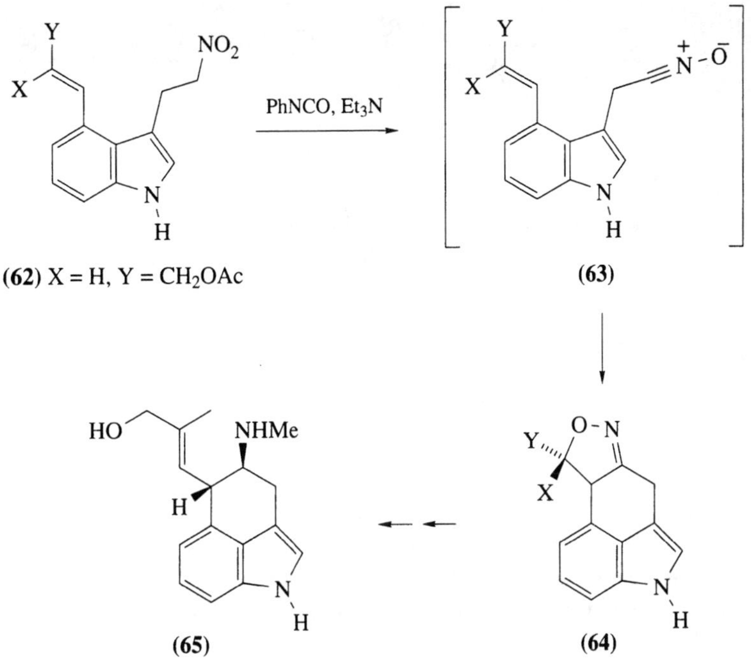

PhNCO, Et_3N

(62) X = H, Y = CH_2OAc

(63)

(65)

(64)

Scheme 15

4.9.6.3 Nitrile Ylides

One of the more interesting members of the nitrilium betaine family is the nitrile ylides.[72–74] This class of dipoles has traditionally been prepared by: (i) treatment of imidoyl halides with base,[75] (ii) thermal or photochemical elimination of phosphoric acid esters from 4,5-dihydro-1,3,5-oxazaphospholes,[76] and (iii) photolysis of aryl-substituted azirines.[77,78] Nitrile ylides have also been observed to be formed upon photolysis of carbene precursors in nitrile solvents.[79,80] FMO theory correctly rationalizes the regioselectivity of nitrile ylide cycloadditions (equation 6).[3] When nitrile ylides are used as 1,3-dipoles, the dipole highest occupied and dipolarophile lowest unoccupied orbital interaction stabilizes the transition state. MNDO calculations of simple nitrile ylides indicate that the coefficient size in both the HOMO and LUMO is slightly greater at the unsubstituted carbon atom.[81] The photoconversion of arylazirines to alkoxyimines in alcoholic solvents (Scheme 16) indicates that in the highest occupied orbital of the nitrile ylide, the electron density at the disubstituted carbon is greater than at the trisubstituted carbon atom.[82] The preferred regioisomeric transition state will be that in which the larger terminal coefficients of the interacting orbitals are united. With this conclusion, all of the regiochemical data found in the literature can be rationalized (Scheme 17).[72–74]

Inductive effects exerted by substituents on the nitrile ylide also have an important effect on the regioselectivity of the cycloaddition. Benzonitriliohexafluoro-2-propanide (**73**) and methyl acrylate yield products with inverse regioselectivity, as compared with the reactions of the related benzonitrilio-2-propanide (**76**; Scheme 18).[83] The difference in regioselectivity has been attributed to the larger coefficient

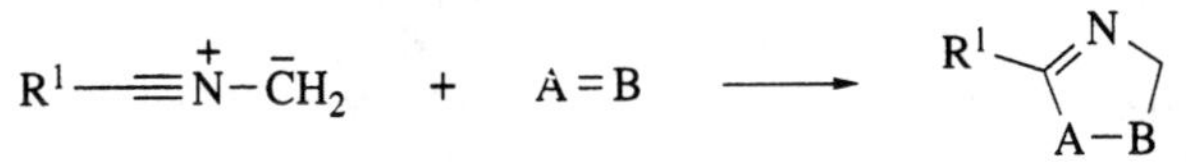

$$R^1-\!\!\equiv\overset{+}{N}-\overset{-}{C}H_2 \quad + \quad A{=}B \quad \longrightarrow \quad \text{(cycloadduct)} \qquad (6)$$

hν, MeOD

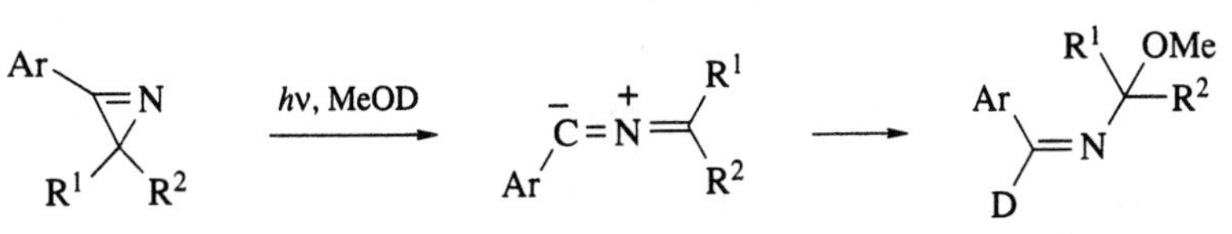

(66) (67) (68)

Scheme 16

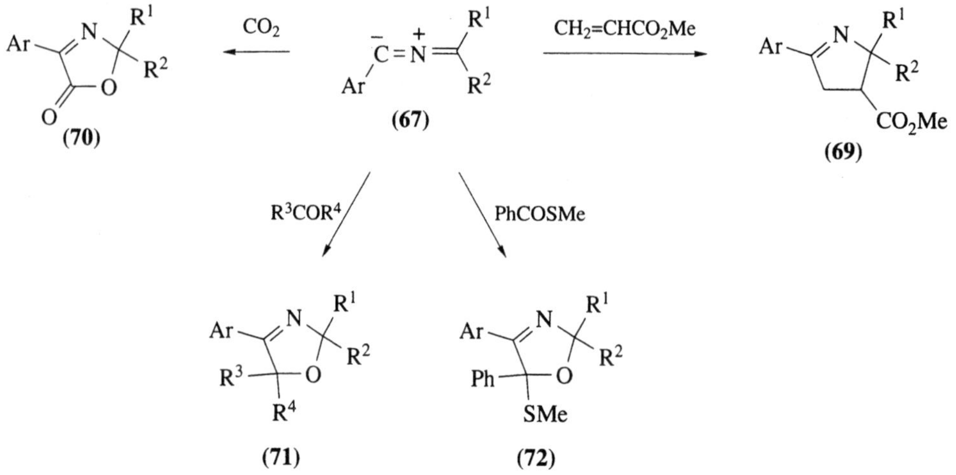

Scheme 17

at the trisubstituted carbon atom of the gem-trifluoromethyl-substituted nitrile ylide (**73**). This result also parallels the different mode of addition of alcohols to nitrile ylides (**73**) and (**76**).

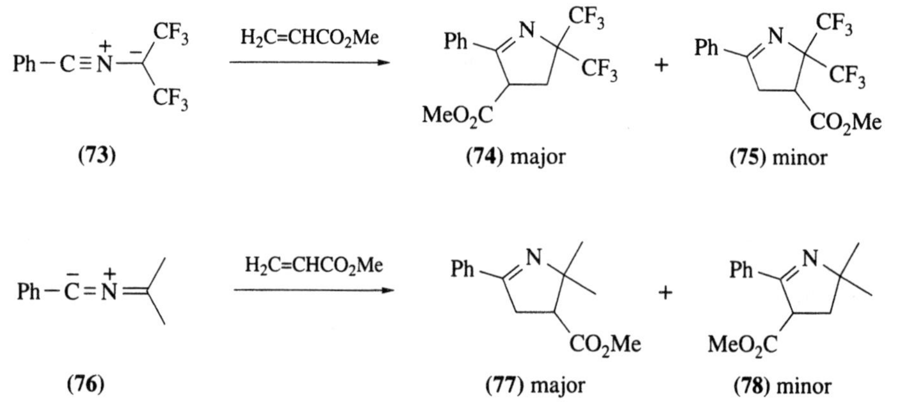

Scheme 18

Among the possible geometric forms of a nitrile ylide, a carbene structure (**79**) can be envisaged,[4] which makes conceivable a 1,1-cycloaddition of this 1,3-dipole (Scheme 19). Molecular orbital calculations have shown that nitrile ylides exist preferentially in the bent form with an HCN angle of 114–116°.[84–86] The nitrile ylide HOMO is heavily concentrated at C-1, but still resembles the normal three-orbital, four-electron system present in other 1,3-dipoles, so that concerted cycloaddition can still occur. In a series of papers, Padwa and coworkers have shown that there are two pathways by which nitrile ylides react with multiple π-bonds.[87] The most frequently encountered path involves a 'parallel-plane' approach of addends and can be considered to be an orbital symmetry allowed [4 + 2] concerted process.[88] With this path, the relative reactivity of the nitrile ylide toward a series of dipolarophiles will be determined primarily by the extent of stabilization afforded the transition state by interaction of the dipole HOMO and dipolarophile LUMO orbitals. Substituents which lower the dipolarophile LUMO energy will accelerate the 1,3-dipolar cycloaddition reaction. Bimolecular reactions of nitrile ylides with electron-rich alkenes have never been observed, thereby indicating that the dipole LUMO–dipolarophile HOMO interaction is never large. Because of their high nucleophilicities, nitrile ylides generally undergo reactions with their precursors, dimerize or isomerize faster than they undergo reactions with electron-rich alkenes.[89–91]

The other path by which nitrile ylides react with π-bonds occurs only in certain intramolecular cases and has been designated as a 1,1-cycloaddition reaction.[87] It occurs when the *p*-orbitals of the alkenic group have been deliberately constrained to attack perpendicular to the nitrile ylide plane. Houk and Caramella have suggested that the 1,1-cycloaddition reaction is initiated by interaction of the terminal carb-

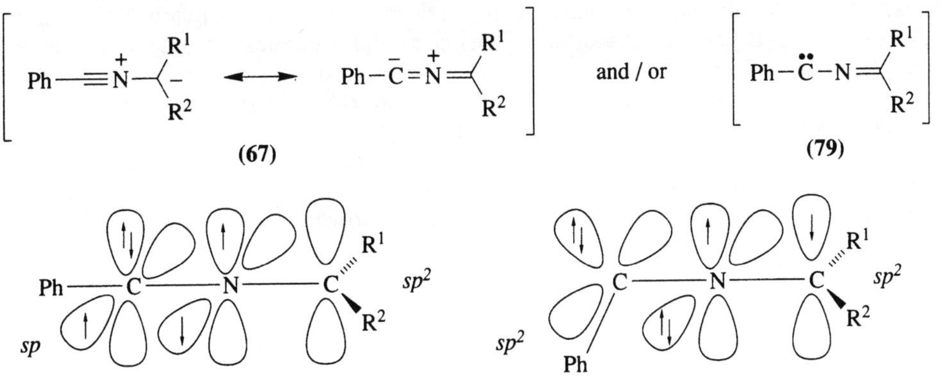

Scheme 19

on of the alkene with the second LUMO of the nitrile ylide.[81] The second LUMO of the dipole is perpendicular to the ylide plane and presents a large vacancy at C-1 of the dipole for attack by the terminus of the neighboring double bond, without the possibility of simultaneous bonding at the C-3 carbon. In fact, the HOMO and second LUMO of the bent nitrile ylide bear a strong resemblance to the HOMO and LUMO of a singlet carbene. Placement of electron-releasing substituents on the π-bond will raise both the HOMO and LUMO orbital energies of the alkene, and consequently facilitate the rate of the 1,1-cycloaddition reaction.[87]

As was clearly pointed out by Huisgen,[4,7] 1,3-dipolar additions generally proceed *via* a two-plane orientation complex, in which the dipole and dipolarophile approach each other in parallel planes. Inspection of molecular models of the nitrile ylides (**81**) and (**85**) derived from azirines (**80**) and (**84**), indicates that the normal two-plane orientation approach of the nitrile ylide and the allyl π-system is impossible as a result of the geometric restrictions imposed on the system. Consequently, the normal mode of 1,3-dipolar cycloaddition cannot occur here. Instead, attack of the carbene carbon on the neighboring double bond generates a six-membered ring trimethylene intermediate (Scheme 20). Collapse of this species results in the formation of the observed azabicyclohexene system (**83**) and (**87**). This sequence proceeds in a nonconcerted manner and bears a strong resemblance to the stepwise diradical mechanism suggested by Firestone to account for bimolecular 1,3-dipolar cycloadditions.[11] It is evident from the available data that, unless the dipole and dipolarophile approach each other in parallel planes, an alternative nonconcerted mechanism for dipolar cycloaddition can occur.

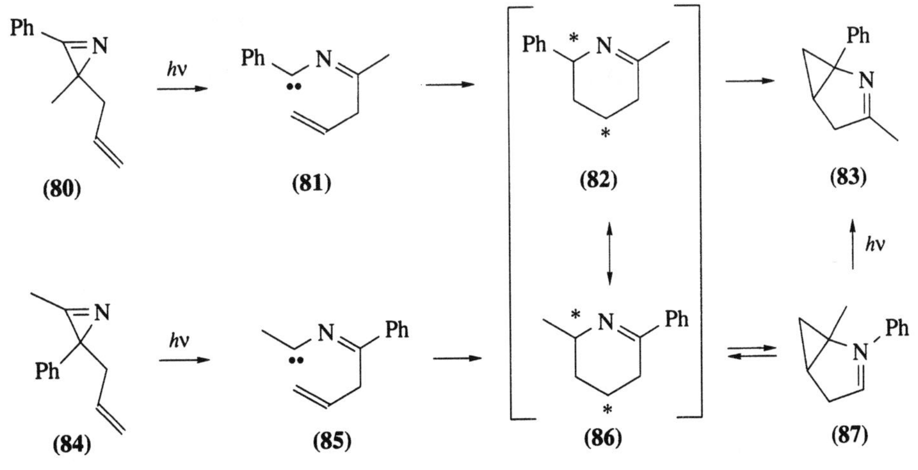

Scheme 20

4.9.6.4 Nitrilimines

Nitrilimines represent a well investigated class of 1,3-dipoles.[55] Access to this group of dipoles can be realized by: (i) treatment of hydrazonyl halides with base,[92] (ii) thermal or photochemical decomposition of tetrazoles,[93,94] (iii) photolysis of sydnones,[95] and (iv) thermal elimination of carbon dioxide from oxadiazolin-5-ones.[96,97] 1,3-Dipolar cycloaddition of this class of 1,3-dipoles has been widely investigated, and in many cases has led to the synthesis of a variety of interesting heterocyclic compounds (Scheme 21). The cycloadditions of simple nitrilimines with electron-rich dipolarophiles are LUMO (dipole)–HOMO (dipolarophile) controlled.[55] For conjugate dipolarophiles, both HOMO and LUMO interactions are important, but the greater difference in LUMO coefficients leads to a preference for 5-substituted Δ^2-pyrazolines. With electron-deficient dipolarophiles, the regioselectivity is reversed since the cycloaddition becomes HOMO (1,3-dipole)–LUMO (dipolarophile) controlled.

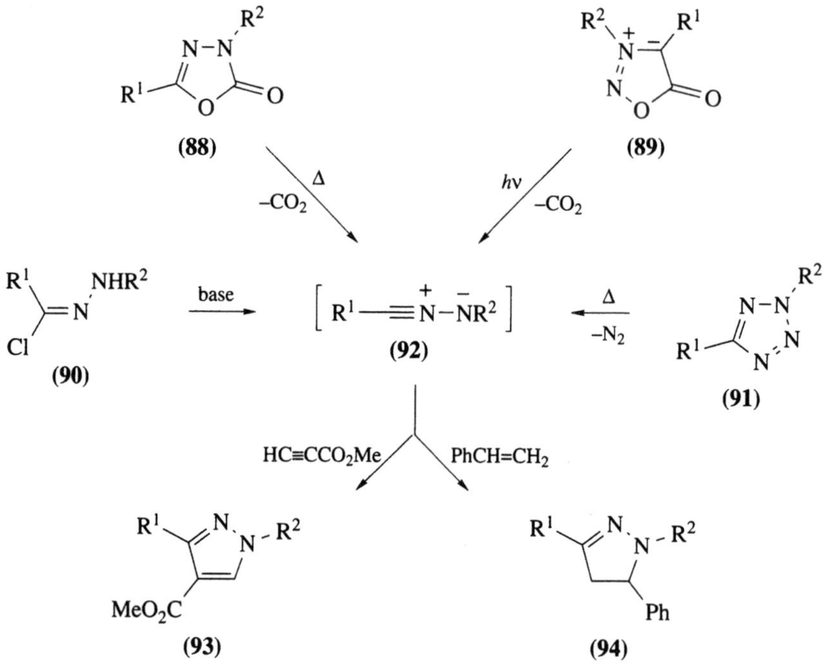

(88) **(89)** **(90)** **(92)** **(91)** **(93)** **(94)**

Δ, $-CO_2$; $h\nu$, $-CO_2$; base; Δ, $-N_2$; $HC{\equiv}CCO_2Me$; $PhCH{=}CH_2$

Scheme 21

(95) base **(96)**

(97) base **(98)**

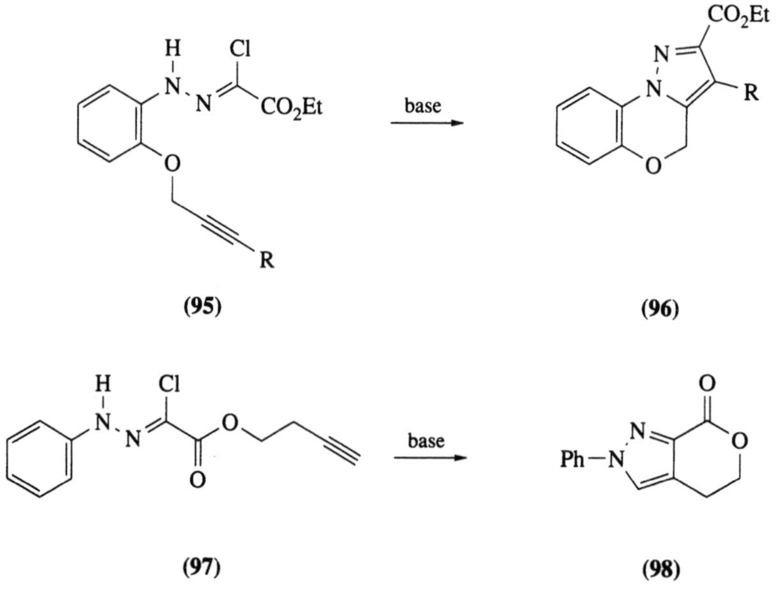

Scheme 22

During the past few years a large number of studies dealing with the intramolecular 1,3-dipolar cycloaddition reactions of nitrilimines have been reported.[51] The reaction products are generally complex polycyclic, annelated and fused-ring molecules that are very difficult to prepare by other methods. A typical example involves the treatment of hydrazidryl chloride (**95**) with base to give a nitrilimine that can be readily intercepted by the adjacent alkynic functionality present in the molecule to give pyrazole (**96**).[98] Similarly, treatment of (**97**) with base affords cycloadduct (**98**) in high yield.[99] These ring closures are particularly interesting in that they involve cycloadditions with unconjugated alkynes, substrates that are generally unreactive toward nitrilimines.[100] Furthermore, although the usual orientation of monosubstituted alkynes in the bimolecular cycloadducts with nitrilimines is such as to afford 5-substituted pyrazoles,[100–102] obtainment of the 4-substituted pyrazole (**98**) reveals that in a properly chosen substrate the regiospecificity of monosubstituted alkynes can be inverted (Scheme 22).

MO calculations by Houk and Caramella suggest that nitrilimine is a flexible 1,3-dipole able to adapt its geometry according to the nature of the reaction.[85] Electrophilic reagents tend to favor a planar nitrilimine structure possessing a relatively high-lying HOMO, whereas nucleophilic reagents promote the bent azo carbene-like form, which possesses a low-lying LUMO. Independent results by Garanti[103] and Padwa[104] have shown that the 1,1-cycloaddition of nitrilimines to π-bonds can occur when the *p*-orbitals of the dipolarophile have been deliberately constrained to attack perpendicular to the nitrilimine plane. Thus, treatment of *o*-vinylphenyl-substituted chloroglyoxylate phenylhydrazones with base leads to nitrilimines that undergo intramolecular 1,1-cycloaddition to give cyclopropa[*c*]cinnolines (Scheme 23). The results with a set of *cis* and *trans* methyl-substituted chlorohydrazones indicate that complete retention of stereochemistry about the π-bond has occurred in the cycloaddition reaction.

H Cl N N CO_2Et base N N CO_2Et R^1 R^1 R^2 R^2

(**99**) (**100**)

N + N CO_2Et N N CO_2Et R^1 R^2 R^1 R^2

(**101**) (**102**)

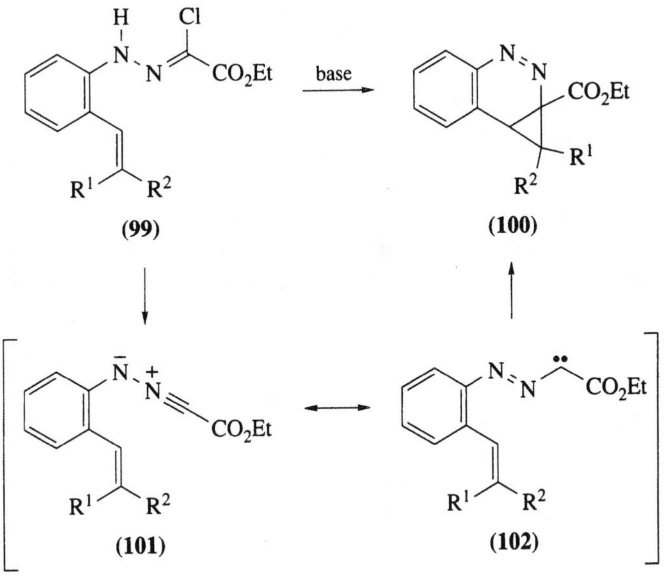

Scheme 23

4.9.6.5 Azomethine Ylides

Azomethine ylides have represented attractive synthetic building blocks for pyrrolidine synthesis since the discovery that they can be generated by pyrolysis of aziridines.[105] The reaction of 1,2,3-triphenylaziridine with electron-deficient alkenes or alkynes to yield five-membered nitrogen rings was initially reported by Heine[106] and similar findings were described by Padwa[107] and by Huisgen[108] shortly thereafter. Due to the systematic investigations by Huisgen,[109] it is well known that the thermolysis of 1-phenyl-2,3-bis(methoxycarbonyl)aziridine involves conrotatory ring opening to the carbonyl-stabilized ylides (**104**) or (**105**; Scheme 24). Trapping products of the S-dipole (**104**) are obtained from the *cis*-aziridine (**103**), while adducts of the isomeric W-dipole (**105**) result from the *trans*-aziridine (**106**). The S-dipole (**104**) is trapped by several dipolarophiles without loss of dipole geometry. In contrast, the W-dipole (**105**) reacts cleanly only with the most reactive of trapping agents such as dimethyl acetylenedicarboxylate (DMAD). With less reactive dipolarophiles, products derived from the S-dipole (**104**) are also observed due to dipole interconversion. These topics have been extensively reviewed, and the concepts have also been extended to nonstabilized azomethine ylides.[110–113] Many other examples of aziridine thermolysis have been reported in the literature.[105] Although N-arylaziridines have been investigated most intensively, cycloadditions using N-alkyl, N-H and N-acyl derivatives have also been encountered. Other routes to stabilized azomethine ylides include the thermolysis of benzaldimines,[114] the related method of thermal N-alkylamino ester/aldehyde condensation[115,116] and carbene insertion into an imine nitrogen lone pair.[117]

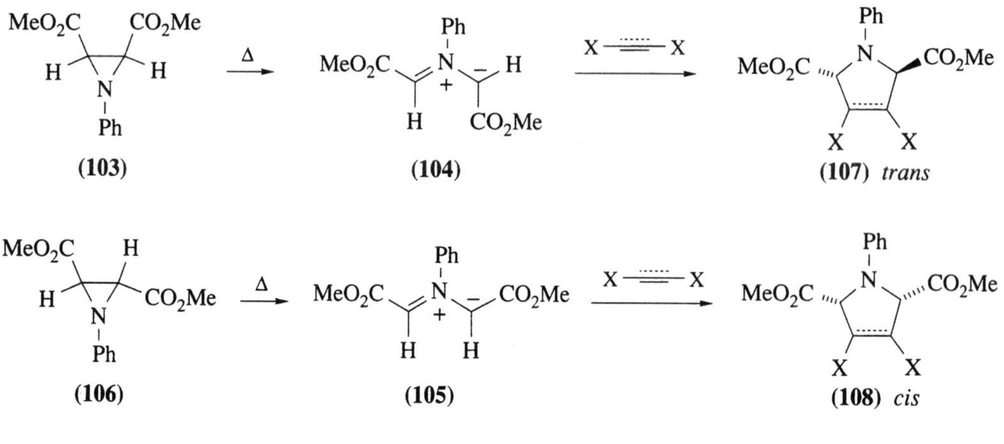

(103) **(104)** **(107)** *trans*

(106) **(105)** **(108)** *cis*

Scheme 24

In frontier molecular orbital terms, the reaction of an azomethine ylide with an electron-deficient dipolarophile is suggested to be a dipole HOMO controlled reaction.[118] Thus, the dominant FMO involves the HOMO of the dipole and the LUMO of the dipolarophile, and factors that decrease the HOMO–LUMO gap increase the efficiency of the reaction. Generally, donor and/or conjugative groups (alkyl, alkoxy, aryl) on the dipole raise the HOMO energy, while acceptor groups on the dipolarophile lower the LUMO energy. This combination of effects facilitates the reaction. In the case of stabilized azomethine ylides, the dipole contains electron-deficient groups which tend to lower the HOMO and moderate the dipole reactivity. These stabilized dipoles are still reactive enough to participate in cycloadditions with highly electron-deficient dipolarophiles such as acrylate, DMAD and maleimide.

Although ring openings of aziridines to azomethine ylides work well when the substituent groups are capable of stabilizing the dipole centers,[105] the ring cleavage generally fails when simple alkyl substituents are used. Recent studies by Vedejs and coworkers have shown that the desilylation of α-trimethylsilylonium salts represents a convenient method for azomethine ylide generation.[119] Cycloaddition of these transient species with various dipolarophiles gives rise to dipolar cycloadducts in high yield (Scheme 25). The advantage of using cesium fluoride for these desilylations is related to its reasonable

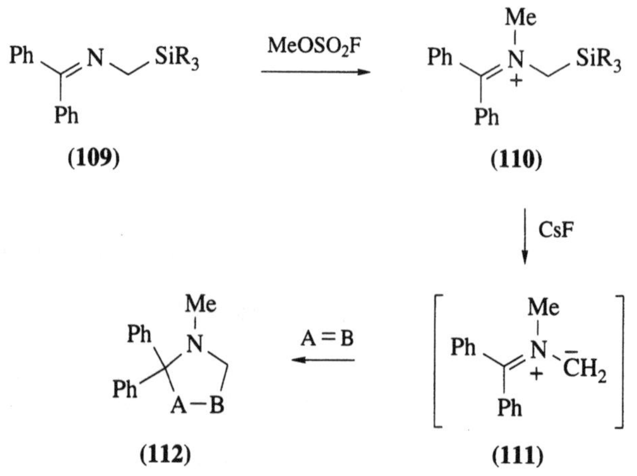

(109) **(110)**

(112) **(111)**

Scheme 25

thermal stability, which allows thorough drying under vacuum to remove water of hydration. Similar results can be achieved more tediously with scrupulously dried KF/18-crown-6, but organic fluoride donors are less effective because they are contaminated with water. The presence of water results in the ready protonation of the nonstabilized ylides and the net result is protodesilylation.

Generation of intermediates having azomethine ylide reactivity has also been achieved by the cesium fluoride-induced desilylation reaction of several immonium salts derived from amides, thioamides and vinylogous amides. A variety of dipolarophiles have been used to trap the azomethine ylide (Scheme 26). Methyl acrylate or α-chloroacrylonitrile are especially efficient trapping agents, and in both cases the more hindered product is formed.[110] The same regiochemical preference is seen in virtually all of the trapping experiments as long as unsymmetrical dipolarophiles derived from acrylic acid are used.

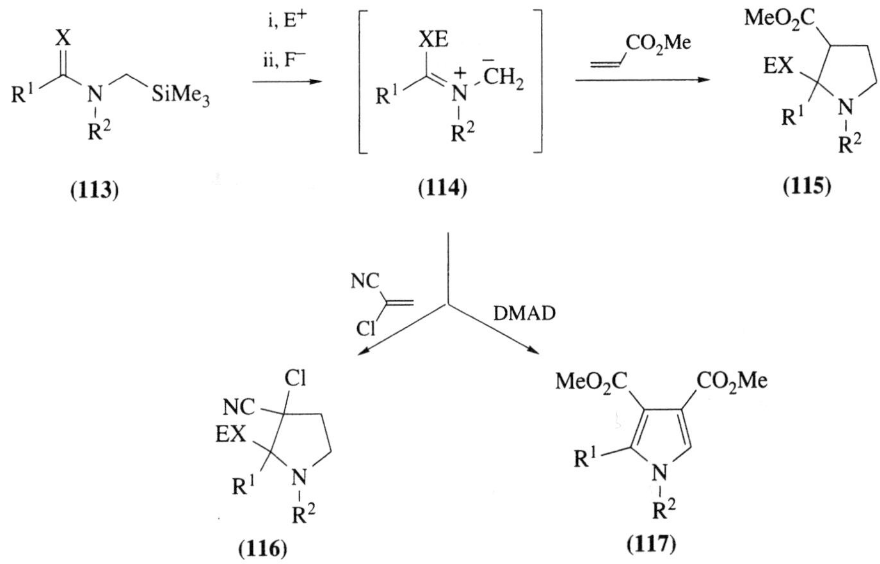

(113) **(114)** **(115)** **(116)** **(117)**

Scheme 26

Imidate-derived dipoles have played a prominent role in the synthesis of the pyrrolizidine alkaloid retronecine (**121**).[119] The imidate salt derived from lactam (**118**) was found to undergo a smooth desilylation reaction to produce azomethine ylide (**119**). Trapping of this dipole with methyl acrylate affords

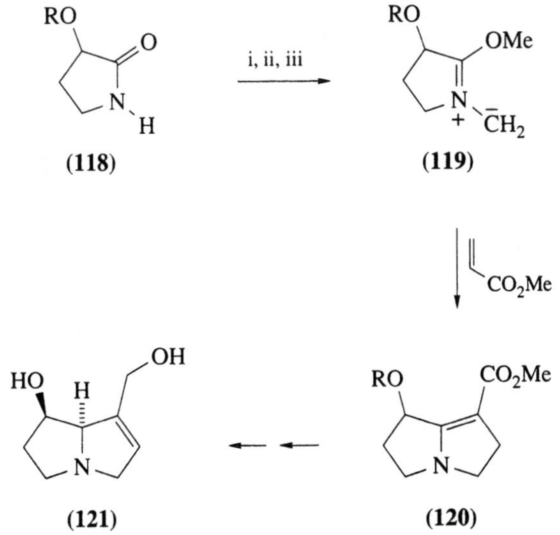

(118) **(119)** **(121)** **(120)**

i, Me_3SiCH_2I; ii, MeOTf; iii, F^-

Scheme 27

the conjugate enamine ester (**120**). This material was converted to retronecine in several additional steps (Scheme 27).

Amidine-derived azomethine ylides can also be used in dipolar cycloaddition reactions. Livinghouse has applied the cycloaddition of this class of dipoles to the synthesis of eserethiole (**125**).[120] The starting amidine (**123**) is made by coupling the imidate salt (**122**) with trimethylsilylmethylamine. *N*-Alkylation followed by desilylation affords the azomethine ylide which readily undergoes intramolecular dipolar cycloaddition to give the natural product (Scheme 28). Other attempts to trap nonstabilized azomethine ylides by intramolecular cycloaddition across unactivated π-bonds using the desilylation technique have so far been unsuccessful.

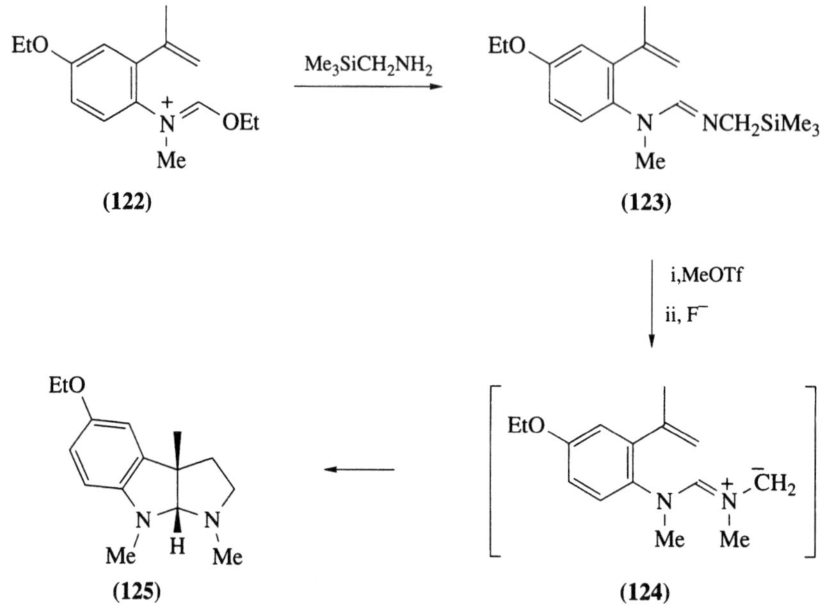

EtO, N+, Me, OEt **(122)** — $Me_3SiCH_2NH_2$ → EtO, N, Me, NCH_2SiMe_3 **(123)** — i, MeOTf; ii, F^- → EtO, N, Me, N+, $\bar{C}H_2$, Me **(124)** → EtO, N, H, N, Me, Me **(125)**

Scheme 28

There have been a number of additional approaches used for the synthesis of nonstabilized azomethine ylides. Padwa and Chen have reported that α-cyanosilylamines (**126**) are useful and convenient synthons for azomethine ylides.[121] The development of this strategy was based on literature reports that cyanomethylamines can function as convenient iminium ion precursors.[122] The propensity of silicon to transfer to a silylophile[123] when bound to an electronegative carbon strongly inferred that the treatment of α-cyanosilylamines with silver fluoride will generate azomethine ylides. Further reports by Padwa[118] and Sakurai[124] have described a mechanistically related method using the aminal (**127**) as starting material for TMSI-induced dipole generation (Scheme 29).

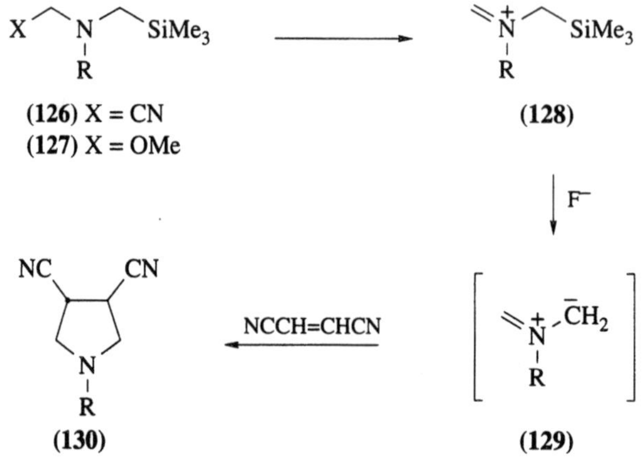

X, N, R, $SiMe_3$ — **(126)** X = CN; **(127)** X = OMe → N+, R, $SiMe_3$ **(128)** — F^- → N+, R, $\bar{C}H_2$ **(129)** — NCCH=CHCN → NC, CN, N, R **(130)**

Scheme 29

The examples of desilylation discussed so far all assume that the reactive intermediates are silicon-free azomethine ylides. It is conceivable, however, that pentacoordinated silicon-containing intermediates may be able to mimic the reactivity of nucleophilic azomethine ylides. Two groups have probed this question and have reported evidence in favor of the dipole pathway. If silicon is still present as bond formation occurs, then it should matter which of the isomeric starting materials (**131**) or (**132**) is used in the desilylation experiment (Scheme 30). Work by Padwa and Dent has shown that treatment of either isomeric aminonitrile with AgF affords the same mixture of product regioisomers using methyl propiolate as the trapping agent.[118] Achiwa has observed that the deuterated ylide precursor (**133**) affords cycloadducts with complete scrambling of the label as expected in a silicon-free azomethine ylide.[125]

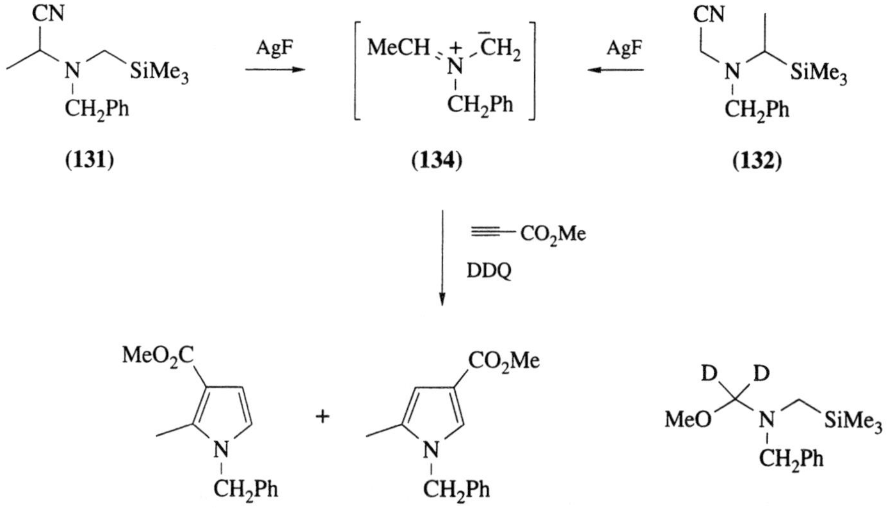

CN N $SiMe_3$ CH_2Ph AgF MeCH + $\bar{C}H_2$ N CH_2Ph AgF CN N $SiMe_3$ CH_2Ph

(**131**) (**134**) (**132**)

$\equiv$—CO_2Me DDQ

MeO_2C N CH_2Ph + CO_2Me N CH_2Ph D D MeO N $SiMe_3$ CH_2Ph

(**135**) 89% (**136**) 11% (**133**)

Scheme 30

Finally, the LDA deprotonation of amine *N*-oxides has been reported to generate azomethine ylides that can be trapped in [2 + 3] cycloadditions with simple alkenes.[126] For example, *N*-methylpyrrolidine *N*-oxide (**137**) reacts with LDA in the presence of cyclopentene to give adduct (**139**; Scheme 31). A variety of other *N*-oxides behave similarly. Interestingly, there are no examples published to date where nonstabilized azomethine ylides generated by the desilylation procedure can be trapped by simple, unactivated alkenes. It is not clear whether these discrepancies are due to some fundamental difference in the reactive intermediate being generated, or whether the differences in environment are responsible for differing behavior. Further work is needed to establish this point.

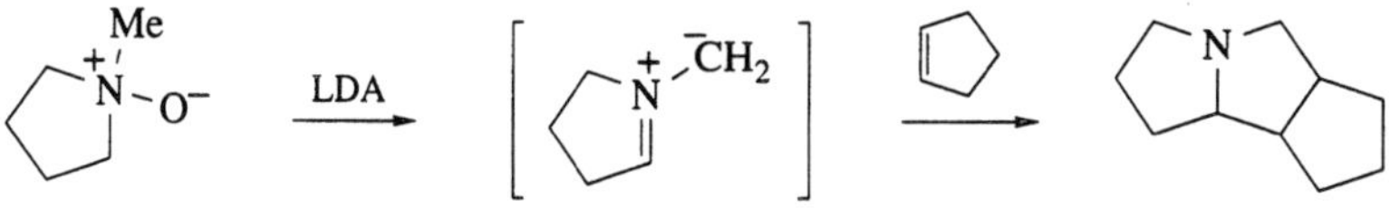

Me + N O^- LDA + N $\bar{C}H_2$ N

(**137**) (**138**) (**139**)

Scheme 31

4.9.6.6 Carbonyl Ylides

In recent years there has been a growing interest in the use of carbonyl ylides as 1,3-dipoles for total synthesis.[127–130] Their dipolar cycloaddition to alkenic, alkynic and hetero multiple bonded dipolarophiles has been well documented.[6] Methods for the generation of carbonyl ylides include the thermal and photochemical opening of oxiranes,[131] the thermal fragmentation of certain heterocyclic structures such as Δ^3-1,3,4-oxadiazolines (141) or 1,3-dioxolan-4-ones[132–134] (**142**) and the reaction of carbenes or carbenoids with carbonyl derivatives.[135–138] Formation of a carbonyl ylide by attack of a rhodium carbenoid intermediate onto the lone pair of electrons of the carbonyl group represents a particularly useful approach to this dipole (Scheme 32).[137] The regiochemistry of the cycloaddition reaction of carbonyl ylides can be successfully predicted by FMO theory. Of the three categories described by Sustmann,[21] type 2 is particularly common for carbonyl ylides since they have one of the smallest HOMO–LUMO energy gaps of the common 1,3-dipoles. For carbonyl ylides, the HOMO of the dipole is dominant for reactions with electron-deficient dipolarophiles such as acrolein, while the LUMO becomes important for cycloaddition to electron-rich species such as ethyl vinyl ether.

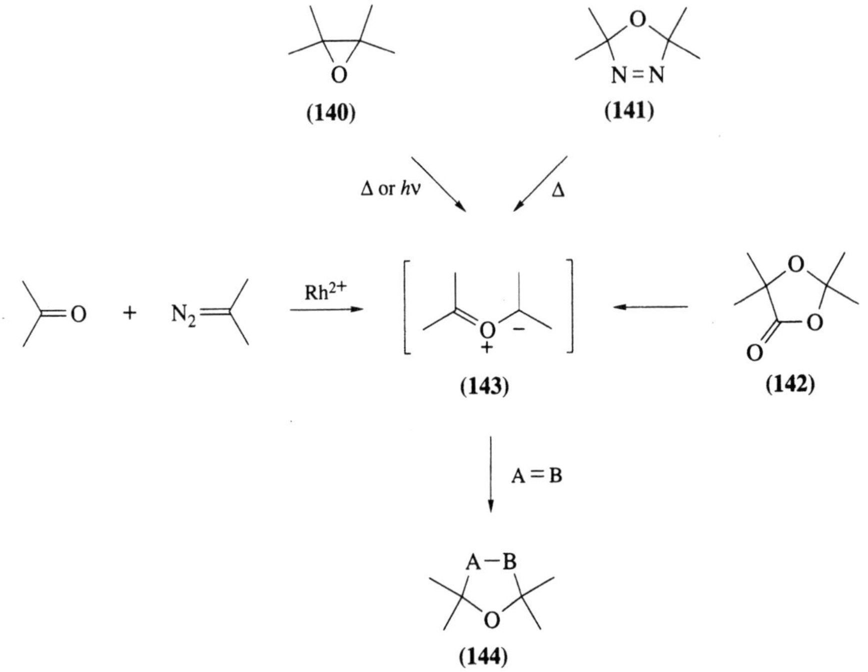

(140) (141)

Δ or $h\nu$ Δ

Rh^{2+}

(143) (142)

A=B

(144)

Scheme 32

Linn and Benson discovered in 1965 the ability of tetracyanoethylene oxide (**145**) to react with alkenes and alkynes at elevated temperatures in [3 + 2] cycloadditions.[139] The kinetics of the reaction of (**145**) with styrene revealed that the formation of (**147**) is preceded by a first-order step consisting of the electrocyclic ring opening to the carbonyl ylide (**146**; Scheme 33).

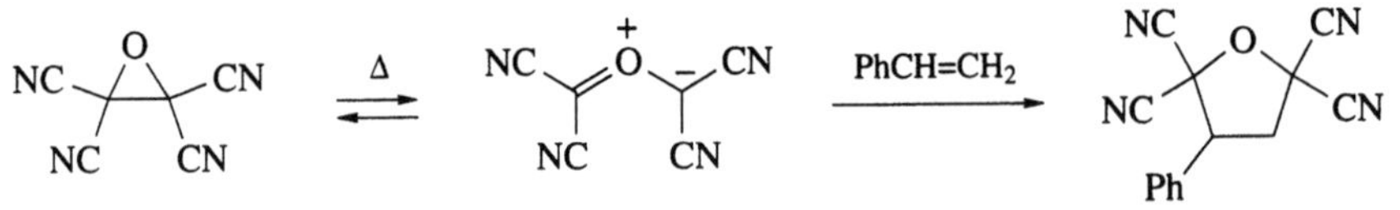

Δ PhCH=CH$_2$

(145) (146) (147)

Scheme 33

1,2-Diaryloxiranes undergo a photofragmentation giving carbonyl compounds and a carbene.[140] Do-Minh, Trozzolo and Griffin noticed an orange color on irradiation of (**148**) at –196 °C and ascribed it to carbonyl ylide (**149**; Scheme 34)[141] Photogenerated carbonyl ylides can be captured *in situ* by α,β-unsaturated carboxylic esters as dipolarophiles.[142] Disrotation predicted by orbital symmetry control appears to be the predominant mode involved in the photoconversion of oxiranes to carbonyl ylides. An investigation of the thermal behavior of α-cyano-*trans*- and α-cyano-*cis*-stilbene oxide revealed that the symmetry-allowed conrotation is the favored pathway of electrocyclic ring opening (Scheme 34); however, steric hindrance can induce a partial switch to disrotation.[143]

The reaction of dimethyl diazomalonate with an excess of benzaldehyde with or without metal catalyst furnished the diastereomeric 1,3-dioxolanes (**156**) and the oxirane (**155**; Scheme 35). The reaction was proposed to proceed *via* the intermediacy of a carbonyl ylide dipole.[135] Benzaldehyde plays a double role

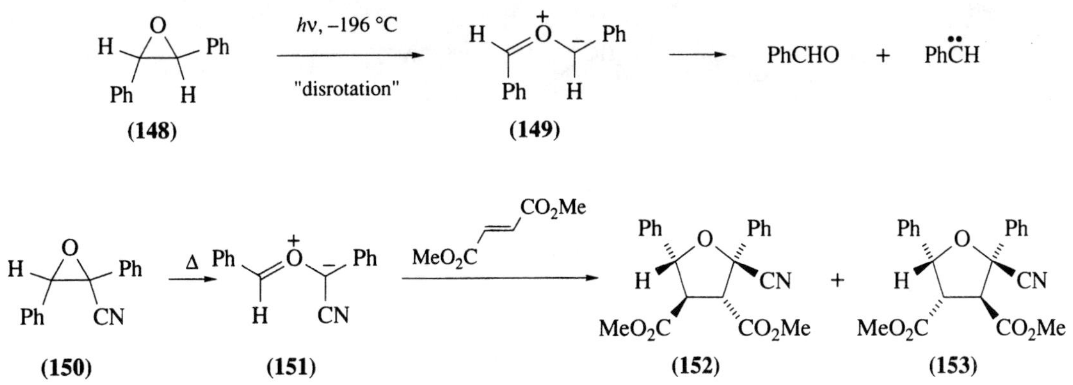

Scheme 34

here, first as a constituent of the carbonyl ylide and, subsequently, as a dipolarophile in the trapping of (**154**).

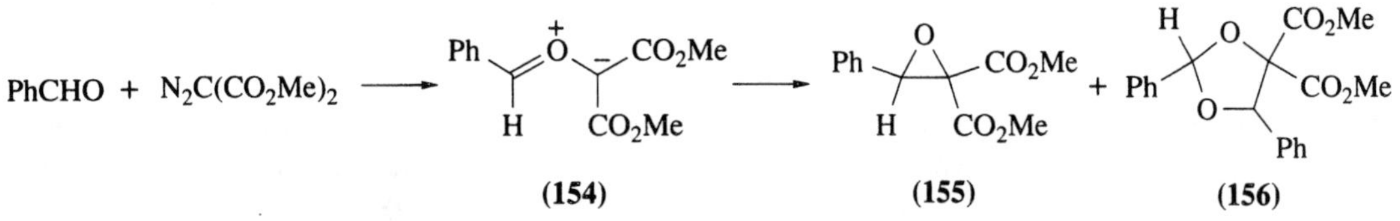

Scheme 35

The intramolecular reaction of carbenes with a neighboring carbonyl group results in the formation of cyclic carbonyl ylides. This approach, pioneered largely by Ibata and coworkers,[138] allows for the convenient generation of various five- or six-membered carbonyl ylides which can be trapped by π-bonds. Generation of the carbene center involves treating a diazo compound with an appropriate transition metal catalyst. An interesting example of this involves the tandem cyclization–cycloaddition reactions of α-diazoacetophenone derivatives which occur upon treatment of *O*-alkyl-2-enoxycarbonyl-α-diazoacetophenones (**157**) with rhodium(II) acetate. Initial cyclization gives a six-ring carbonyl ylide which undergoes a subsequent intramolecular dipolar cycloaddition across the neighboring double bond. The process is illustrated by the conversion of diazoketone (**157**) with rhodium(II) acetate to cycloadduct (**159**) in 87% yield (Scheme 36).[137] The initial work involved systems in which the ketorhodium carbenoid and the remote carbonyl were attached in a 1,2-fashion on a benzene ring. This arrangement provides interatomic distances and bond angles that are ideal for dipole formation. The conformationally unencumbered diazo ketone (**160**) was also found to undergo this same reaction when treated with a catalytic amount of rhodium(II) acetate. The cycloaddition proceeded quite smoothly in the presence of DMAD and produced cycloadduct (**161**; Scheme 37). The intermediate carbonyl ylide could also be trapped with benzaldehyde and methyl propiolate. The orientation observed with these systems can readily be rationalized in terms of maximum overlap of the dipole HOMO–dipolarophile LUMO. MNDO calculations of the carbonyl ylide derived from (**160**) clearly indicate that the largest coefficient in the HOMO resides on the enolate carbon.[137] This site becomes linked with the less-substituted carbon atom of the alkyne.

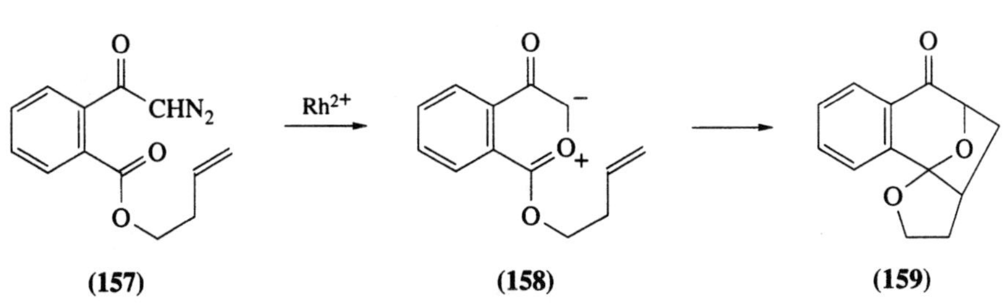

Scheme 36

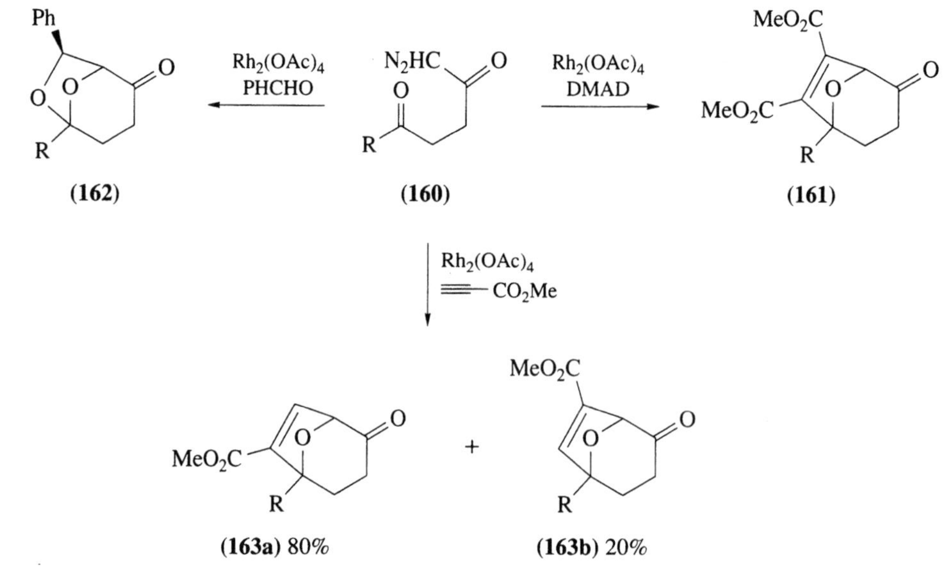

Scheme 37

The cycloaddition chemistry of the 3-oxidopyrylium system (**165**; Scheme 38) has been studied in some detail by Sammes and coworkers.[128] Pyranulose acetate (**164**) is readily available from simple furan derivatives and thus a route is available for the preparation of a variety of substituted 3-oxidopyrylium compounds.[127] These reactive species can be trapped with a wide range of unsaturated materials including alkenes, which cycloadd across the 2,6-positions. 3-Oxidopyrylium (**165**) reacts sluggishly with electron-deficient dipolarophiles giving reasonable yields of adducts (*i.e.* **166**) with only the most reactive traps (*i.e.* acrolein). Both experimental studies[144] and theoretical interpretations[21] have shown that 1,3-dipoles can behave as ambident species, reacting faster with either electron-deficient or electron-rich alkenes than neutral, unsubstituted ones. Huisgen, for example, obtained U-shaped curves when plotting

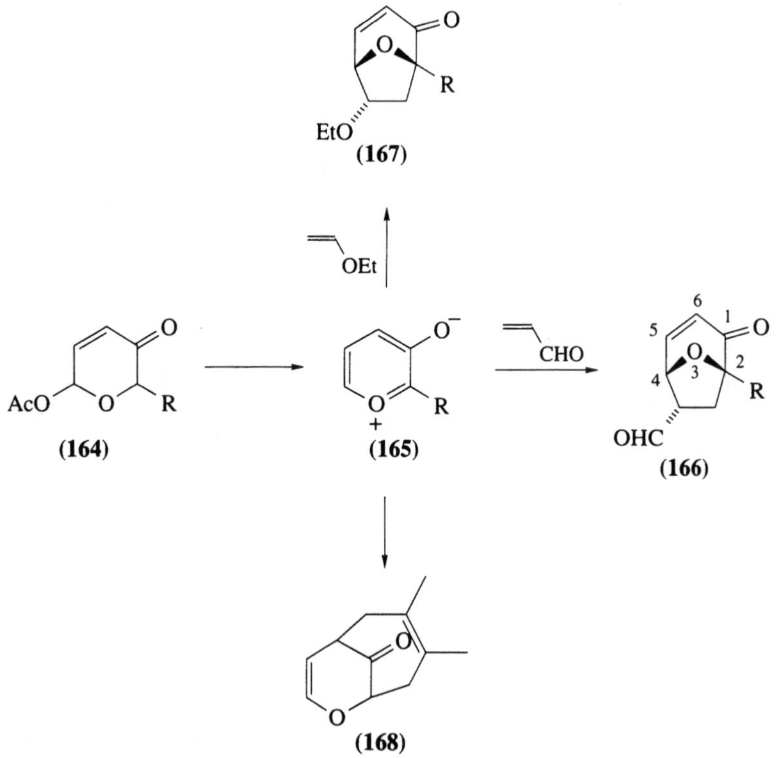

Scheme 38

reaction rates of a given dipole with different dipolarophiles against the ionization potential of the latter.[144] Thus, the formation of adduct (**167**) when ethyl vinyl ether was used as a trapping agent is perfectly understandable. Ylides of type (**165**) are also expected to undergo pericyclic additions across other centers depending on electronic demands. 4π-Electron systems such as dienes would be expected to add across the 2,4-positions, as observed with certain 3-oxidopyridinium derivatives.[145] In fact, 2,3-dimethylbutadiene reacted with (**165**) to give adduct (**168**) as the principal product. Sammes also studied the chemistry of oxidopyrylium ylides which contain an unsaturated side chain. These molecules (*i.e.* **169**) undergo an intramolecular dipolar cycloaddition reaction upon thermolysis to give bicyclic adducts which provide a simple entry into the perhydroazulene system (Scheme 39). This method was used to synthesize a number of natural products.[128]

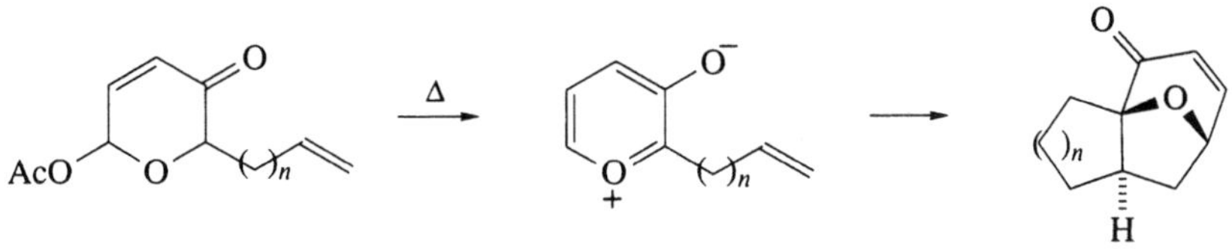

(**169**) n = 3 or 4 (**170**) (**171**)

Scheme 39

4.9.6.7 Thiocarbonyl Ylides

The first recognition of the existence of the thiocarbonyl ylide appears to have been made by Knott in 1955 in conjunction with considerations of the contributions of this segment to the electronic make-up of certain dyestuffs.[146] Knott very likely generated the unstable thiocarbonyl ylide (**173**) *via* the route shown in Scheme 40.

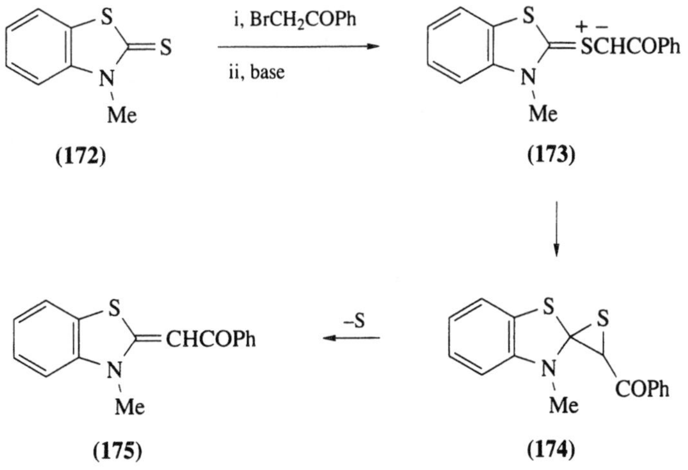

i, $BrCH_2COPh$; ii, base

(**172**) (**173**) (**174**) (**175**)

Scheme 40

The most flexible and general route to thiocarbonyl ylides involves the Δ^3-1,3,4-thiadiazoline (**177**) ring system as the dipole precursor.[28] Mild thermolysis of (**177**) releases nitrogen and produces the thiocarbonyl ylide (**178**). The principle involved is that of a retro-1,3-dipolar cycloaddition.[147] Two generalized routes to the precursor (**177**) are shown in Scheme 41.[148–150] The thiocarbonyl ylide (**178**) generated as a reactive and short-lived intermediate can be trapped by various dipolarophiles. Kinetic analysis of the reaction establishes unambiguously the transitory existence of intermediate (**178**). The predicted conrotatory ring closure occurs to give thiirane (**180**) in the absence of trapping agents. Cycloaddition of (**178**) proceeds with retention of configuration as predicted.[10]

Thiocarbonyl ylides can also be generated *via* the electrocyclic ring closure of divinyl sulfides. Schultz has studied this transformation in some detail.[151] The formation of (**182**) from the irradiation of (**181**) has been rationalized in terms of an electrocyclic ring closure to generate thiocarbonyl ylide (**183**) which then undergoes a 1,4-hydrogen shift to give (**182**; Scheme 42). The successful trapping of the dipole intermediate with *N*-phenylmaleimide, to produce (**184**) provides strong support for the suggested mechan-

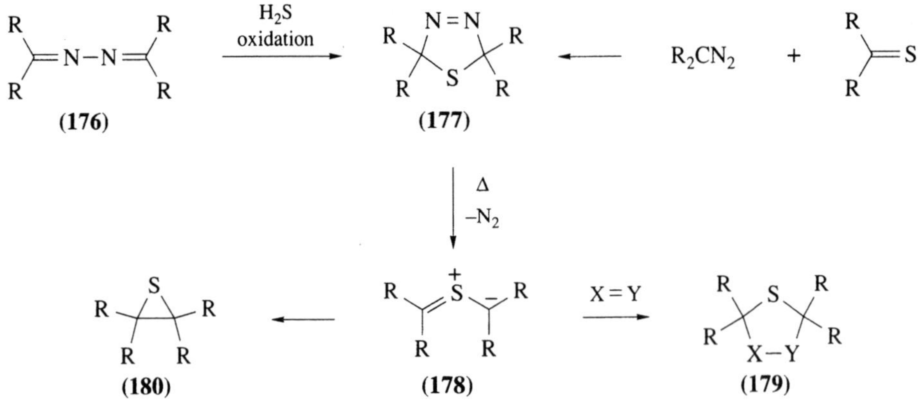

Scheme 41

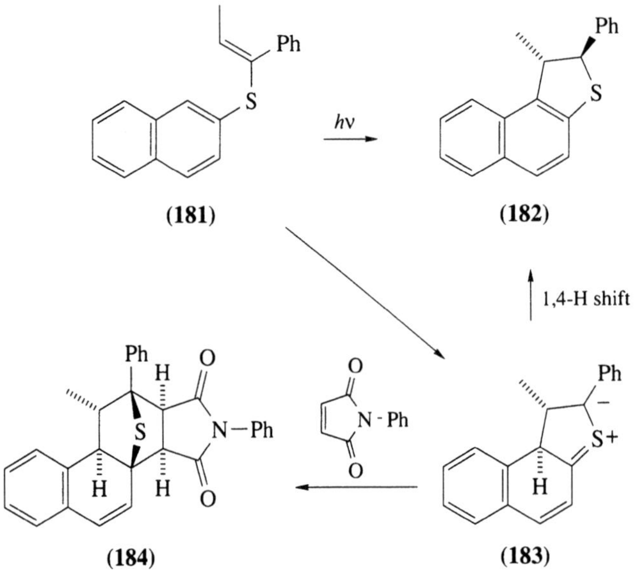

Scheme 42

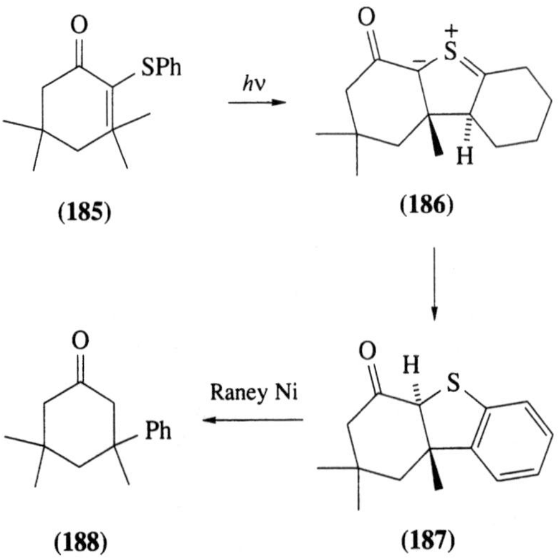

Scheme 43

ism. A clever variation of the above reaction was used by Schultz for the preparation of 3-phenylcyclohexanones.[151] Essentially the same reaction sequence was used here except that the final product was desulfurized. The overall effect is that an aryl group has been added in Michael fashion to an enone (Scheme 43).

Another route which has recently been used to generate thiocarbonyl ylides involves the bromodesilylation reaction of α-bromosilyl silyl sulfide (**189**). This process is related to the methodology first developed by Padwa for the preparation of azomethine ylides. Thus, heating a sample of (**189**) generates thiocarbonyl ylide (**190**) which can be trapped with various dipolarophiles to give cycloadducts of type (**191**; Scheme 44).

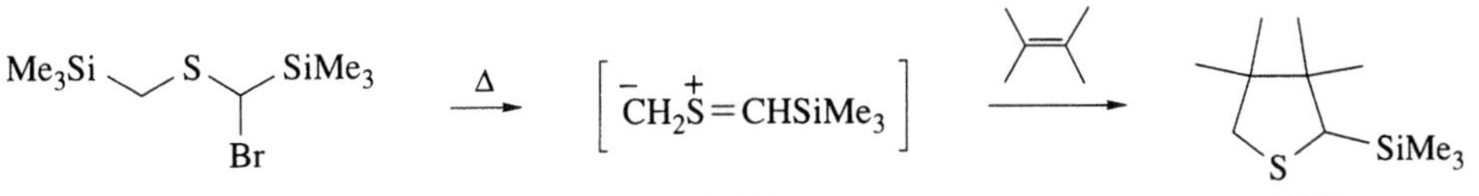

(**189**) (**190**) (**191**)

Scheme 44

Further work by the Achiwa group has demonstrated that this desilylation protocol can be used to generate the simple trimethylsilylthioaldehyde *S*-methylide (**193**).[152] The strategy employed the release of disiloxane from bis(trimethylsilylmethyl) sulfoxide (**192**) through a pathway related to a sila-Pummerer rearrangement. Heating a sample of sulfoxide (**192**) with a dipolarophile in HMPA at 100 °C gave the corresponding cycloadduct (**195**; Scheme 45) in 55–80% yield. Both alkenes and alkynes were used as dipolarophiles.

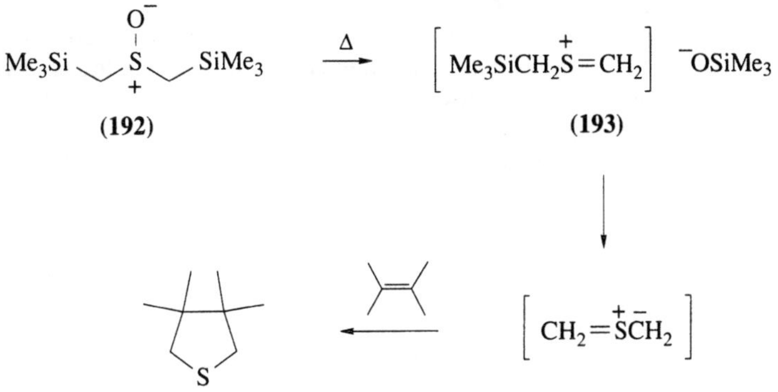

(**192**) (**193**) (**195**) (**194**)

Scheme 45

4.9.6.8 Azomethine Imines

Azomethine imines possess the two-nitrogen 4π-dipolar system (**196**; equation 7) and therefore fall into the category of 1,3-dipoles without an orthogonal double bond.[153] The resonance form (**196a**) is expected to be more important as a result of the higher electronegativity of nitrogen relative to carbon. The condensation of aldehydes with *N,N'*-disubstituted hydrazines is a frequently employed method for generating azomethine imines.[154] Placing an acyl group on the nitrogen end of the dipole, as in (**197**), lowers both the HOMO and the LUMO of the unsubstituted azomethine imine. Reaction of (**197**) with conjugated alkenes like styrene turns out to be a dipole LUMO-controlled process, thereby accounting for the preferential formation of cycloadduct (**198**; Scheme 46).[155] When unsaturated aldehydes are used, the

(7)

(**196a**) (**196b**)

reactions first give the 1,3-dipole, and are then followed by an intramolecular dipolar cycloaddition. An interesting application of this method involves the conversion of (**199**) to the diaza[3.2.1]octane system (**200**; equation 8) in 69% overall yield.[156]

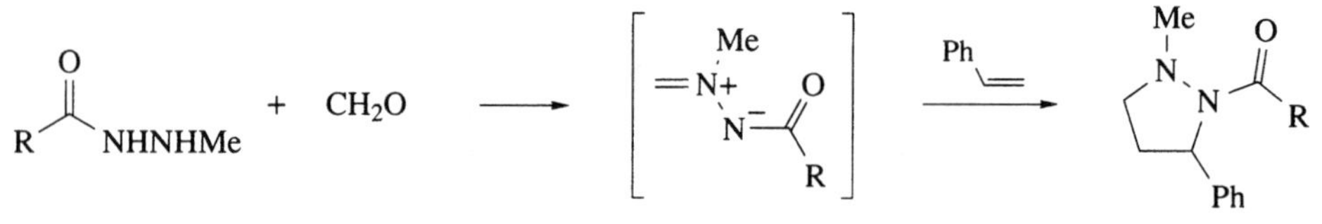

(**197**) (**198**)

Scheme 46

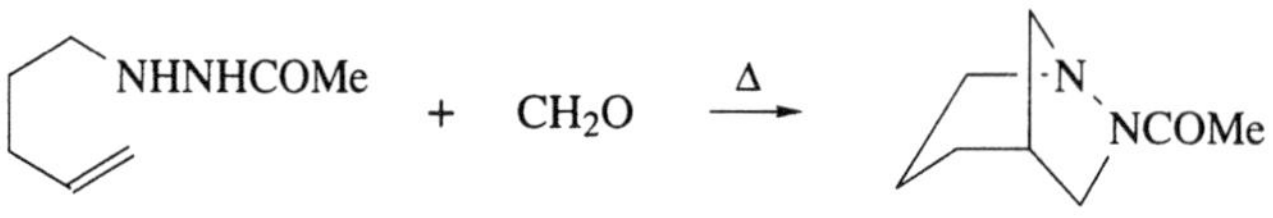

(**199**) (**200**) (8)

Hydrazones have also been used as azomethine imine precursors to achieve cycloadditions.[157] Protonated hydrazones act under suitable conditions as quasi-azomethine imines in polar [3^+ + 2] cycloadditions. Thus, acetaldehyde phenylhydrazone (**201**) was found to react with styrene in the presence of sulfuric acid in a regiospecific manner to give pyrazolidine (**203**; Scheme 47) as a diastereomeric mixture.[157] The most commonly used azomethine imine has a phenyl group attached to one end of the dipole and hence has a raised HOMO relative to the unsubstituted system. Because the coefficients at the terminal atoms of the dipole are smaller in the LUMO than they are in the HOMO, the phenyl group does not lower the energy of the LUMO as much as it raises the energy of the HOMO. With electron-deficient dipolarophiles like methyl acrylate, the reaction is dipole HOMO-controlled, and mixtures can be expected. In fact, a 1:1 mixture of regioisomers was obtained in the reaction of (**201**) with acrylonitrile (equation 9).[157]

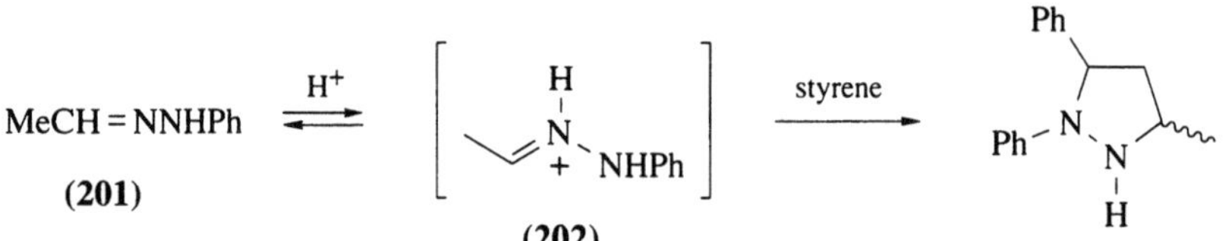

MeCH=NNHPh (**201**) (**202**) (**203**)

Scheme 47

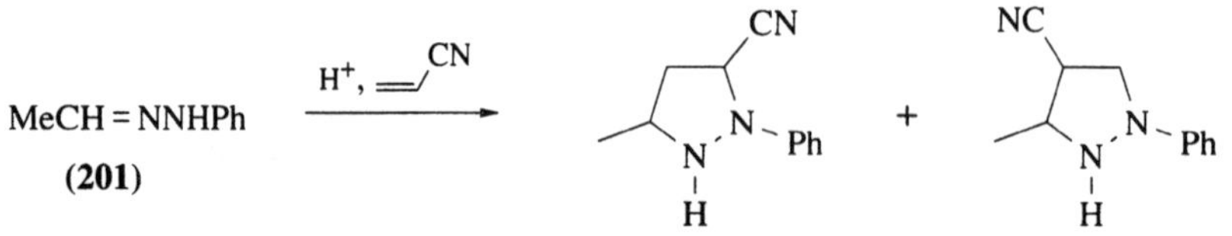

MeCH=NNHPh (**201**) (**204**) (**205**) (9)

4.9.6.9 Mesoionic Ring Systems

Mesoionic compounds have been known for many years and have been extensively utilized as substrates in 1,3-dipolar cycloadditions.[158–160] Of the known mesoionic heterocycles, munchnones and sydnones have generated the most interest in recent years. These heterocyclic dipoles contain a mesoionic aromatic system (*i.e.* **206**) which can only be depicted with polar resonance structures.[158] Although sydnones were extensively investigated after their initial discovery in 1935,[160] their 1,3-dipolar character was not recognized until the azomethine imine system was spotted in the middle structure of (**206**). *C*-Methyl-*N*-phenylsydnone (**206**) combines with ethyl phenylpropiolate to give the tetrasubstituted pyrrole (**208**; Scheme 48). Kinetic studies revealed that the rate-determining cycloaddition to give the bicyclic intermediate (**207**) is succeeded by a fast elimination of CO_2.[161] In the analogous reaction of *N*-phenylsydnone with styrene, the intermediate (**209**) suffers a 1,3-dipolar cycloreversion giving a new azomethine imine (**210**), which is no longer stabilized by aromaticity as was the case with the sydnone. This species either undergoes a 1,4-hydrogen shift to give 2-pyrazoline (**211**; Scheme 49), or adds a second molecule of the dipolarophile.[162]

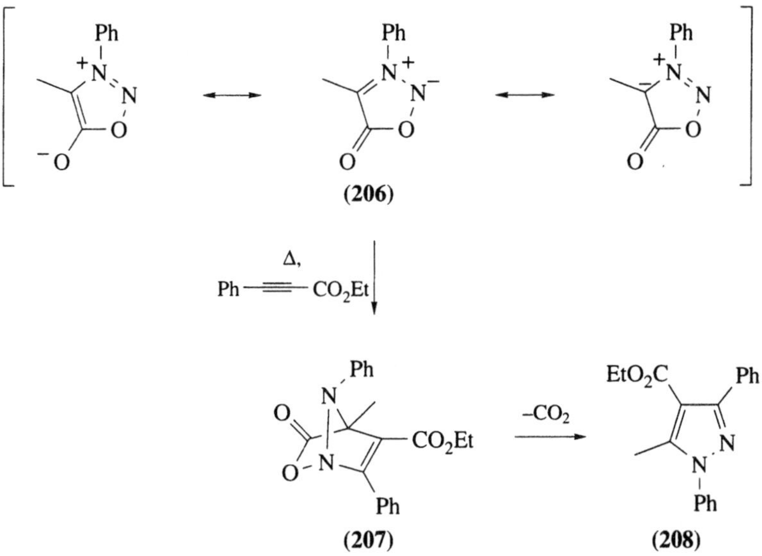

(206)

(207) (208)

Scheme 48

(209) (210) (211)

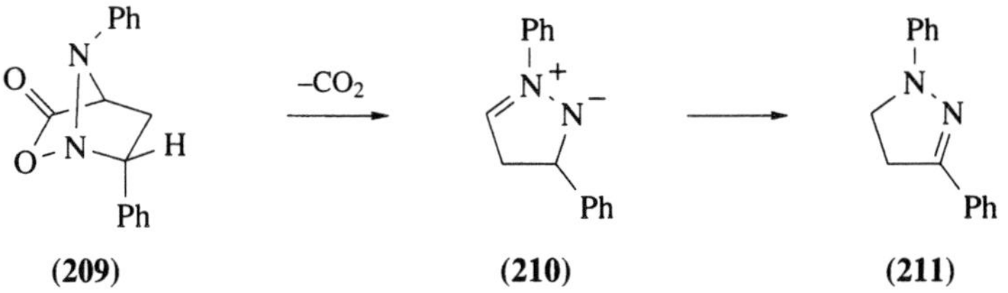

Scheme 49

When sydnones are used as 1,3-dipoles, the dipole LUMO and dipolarophile HOMO interaction has been suggested to be the controlling term.[163] Calculations by Houk indicate that the terminal coefficients of the azomethine imine system are almost identical in the LUMO.[164] Although LUMO control of reactivity will dominate, a decrease in regioselectivity of sydnone cycloaddition with respect to that observed with simpler azomethine imines is expected. In fact, sydnones do undergo regioselective reactions, but the degree of regioselectivity appears to be smaller than is observed with simpler azomethine imines.[163] For example, *N*-phenylsydnone is known to react with all three classes of dipolarophiles to give predominantly the products resulting from intermediate adduct (**212**).[165] Methyl propiolate gives a 4:1 mixture of adducts arising from (**212**) and the other regioisomer (**213**), respectively (Scheme 50).

Huisgen and coworkers have also described the cycloaddition behavior of the 'munchnones', unstable mesoionic Δ^2-oxazolium 5-oxides with azomethine ylide character.[166] Their reactions closely parallel those of the related sydnones. These mesoionic dipoles are readily prepared by cyclodehydration of *N*-acyl amino acids (**216**) with reagents such as acetic anhydride. The reaction of munchnones with alkynic dipolarophiles constitutes a pyrrole synthesis of broad scope.[158–160] 1,3-Dipolar cycloaddition of alkynes to the Δ^2-oxazolium 5-oxide (**217**), followed by cycloreversion of carbon dioxide from the initially formed adduct (**218**), gives pyrrole derivative (**219**; Scheme 51) in good yield. Cycloaddition studies of munchnones with other dipolarophiles have resulted in practical, unique syntheses of numerous functionalized monocyclic and ring-annulated heterocycles.[167–169]

Houk has suggested that unsymmetrically substituted azomethine ylides such as munchnones will react readily with both electron-deficient and electron-rich dipolarophiles due to the narrow frontier orbital

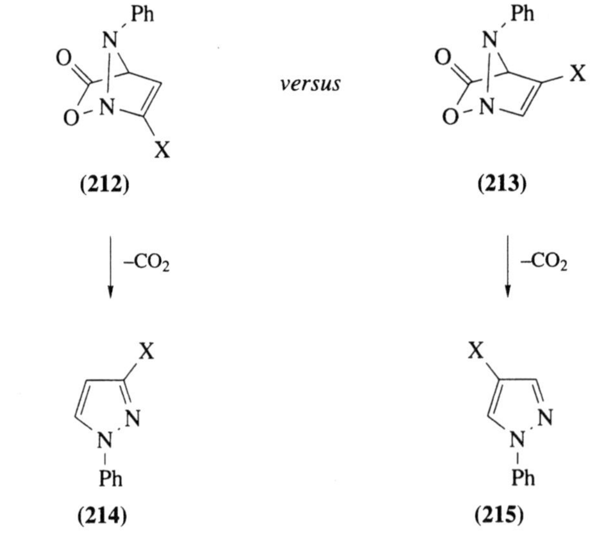

(212) versus (213)

$-CO_2$ $-CO_2$

(214) (215)

Scheme 50

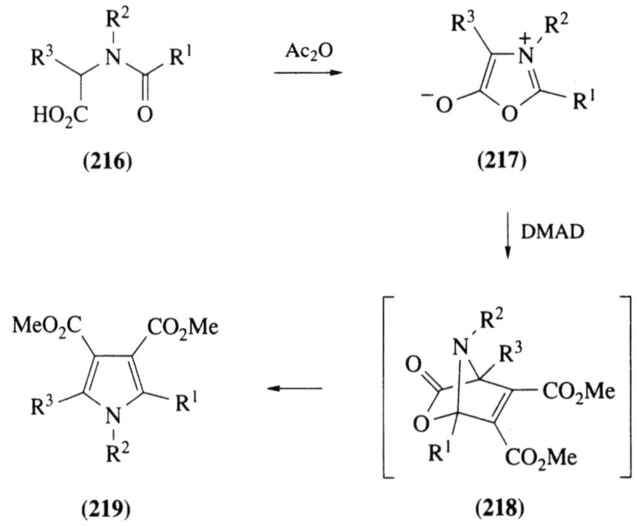

Ac_2O

(216) (217)

DMAD

(219) (218)

Scheme 51

separation[170] (*i.e.* Sustmann Type 2 classification).[21] The regiochemistry of the cycloaddition should be controlled by asymmetry in the dipole frontier orbitals caused by the substituent groups. Examples of dipolar cycloadditions of other mesoionic compounds can be found in several reviews.[158–160]

4.9.6.10 Ozone

Combination of an oxygen atom with the oxygen molecule yields the triatomic modification ozone. Despite the high oxidation potential of ozone, its reactions with organic compounds display considerable specificity. Reaction of ozone with alkenes has been formulated as a 1,3-dipolar cycloaddition.[7] The basic mechanism that describes the ozonolysis of an alkene to produce a secondary ozonide evolved during the 1950s and is now well recognized as the Criegee mechanism.[171–175] It consists of a 1,3-dipolar cycloaddition of ozone, a 1,3-dipolar cycloreversion to carbonyl oxide (**221**) and a carbonyl compound, followed by a renewed 1,3-cycloaddition producing the ozonide (Scheme 52). Criegee assumed a two-step cleavage of the 1,2,3-trioxolane (**220**) and formulated the carbonyl oxide as a zwitterion. Carbonyl oxides (Criegee zwitterions) have never been isolated, but indirect evidence for their existence is abundant. Although these species are usually generated by ozonolysis of a suitable alkene, evidence has been presented for their formation by reaction of a carbene with oxygen.[176] Carbonyl oxide reactions are LUMO-controlled and cycloadditions should be facile only with electron-rich species.

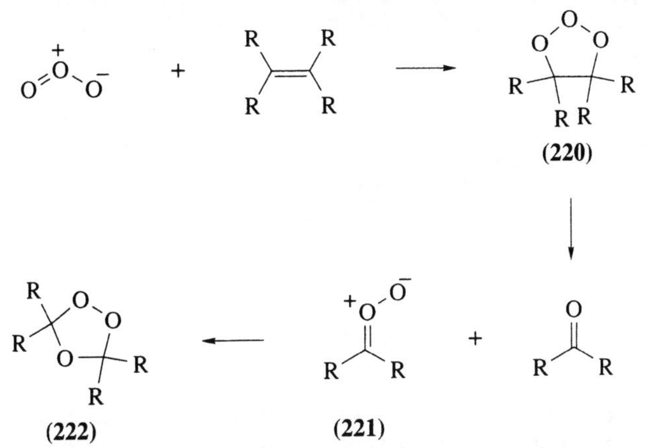

(220)

(222) (221)

Scheme 52

Additional work by Criegee's group has shown that the ozonolysis of 1,2-disubstituted 1-cyclopentenes (*e.g.* **223**) containing different substituents in the 4-position of the ring produces carbonyl oxides such as (**224**) as reaction intermediates.[177] This dipole undergoes intramolecular 1,3-dipolar cycloaddition with each of the two functional groups present. Thus two different ozonides are formed, in the present case (**225**) and (**226**; Scheme 53). This reaction strategy was further pursued to evaluate the relative reactivity of nonequivalent carbonyls with the carbonyl oxide intermediate. In this way the 1,3-dipolarophilic character of carbonyl groups was estimated to be formyl > acetyl > benzoyl.[177]

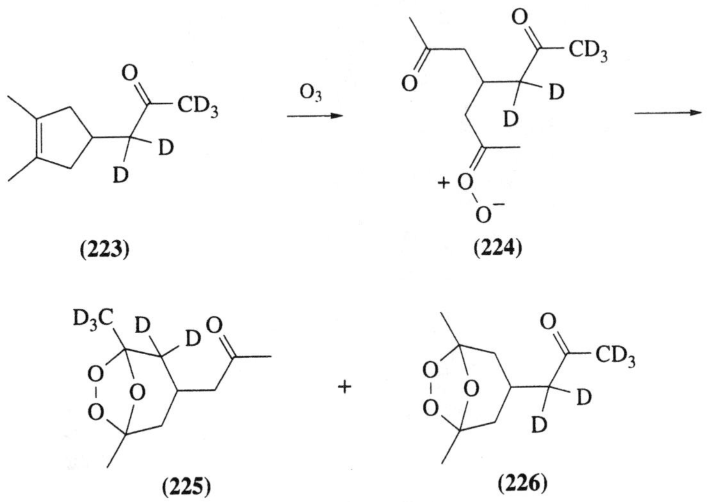

(223) (224)

(225) (226)

Scheme 53

4.9.6.11 Azides

The cycloaddition of azides to multiple π-bonds is an old and widely used reaction. Organic azides are well known to behave as 1,3-dipoles in thermal cycloaddition reactions.[178] The first example of this reaction was observed by Michael in 1893.[179] Since then the addition of azide to carbon–carbon double and triple bonds has become the most important synthetic route to 1,2,3-triazoles, -triazolines and their derivatives.[180–184] The cycloadditions of simple organic azides with electron-rich dipolarophiles are LUMO controlled.[3] Since the larger terminal coefficients are on the unsubstituted nitrogen in the azide and unsubstituted terminus in the dipolarophiles, the 5-substituted Δ^2-triazolines are favored, in agreement with experiment.[185–187] Reactions with electron-deficient dipolarophiles are HOMO controlled, and union of the substituted azide N with the unsubstituted dipolarophile C leads to 4-substituted Δ^2-triazolines (Scheme 54).[178] Sustmann correlated the reactivity of 21 alkenes with phenyl azide by means of the alkene ionization potentials.[188] The plot of the rate constants forms a curve, resembling a parabola, with a minimum at 9.8 eV. This is exactly what FMO theory predicts for a system in which the mechanism changes from dipole LUMO to dipole HOMO control.

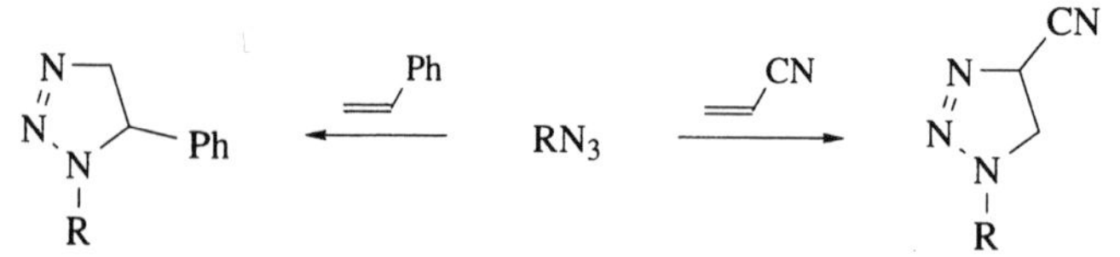

(227) 5-regioisomer **(228)** 4-regioisomer

Scheme 54

Simple carbon–carbon double bonds react slowly with azides and frequently take more than a week to react at 25 °C. Increasing the reaction temperature is restricted by the thermal lability of most triazolines. In fact, some classes of triazolines decompose spontaneously at room temperature with the extrusion of nitrogen.[189] The principal decomposition modes involve the formation of aziridines and imines and are outlined in Scheme 55. Usually, just a few of these modes operate in any particular system.[190]

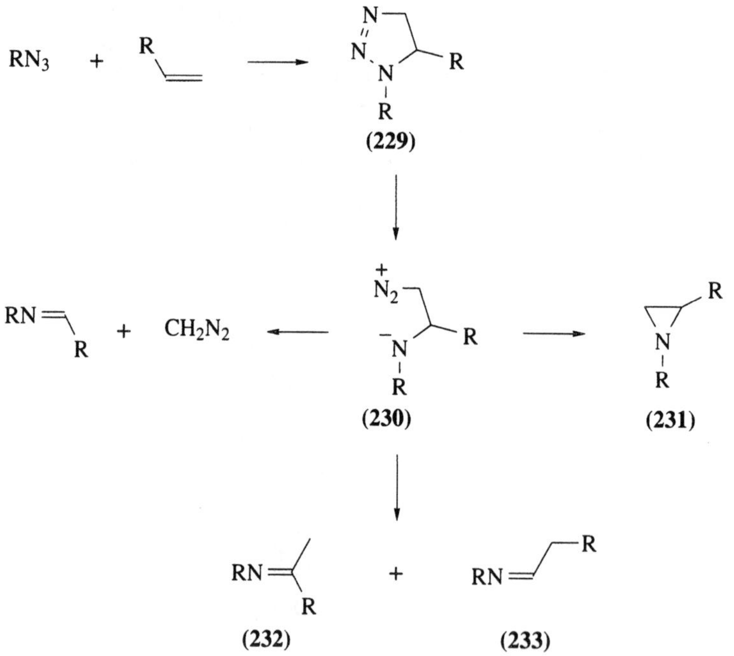

(229) **(230)** **(231)** **(232)** **(233)**

Scheme 55

Alkynes add to azides to form 1,2,3-triazoles (equation 10). In contrast to the azide additions to alkenes, the regioselectivity with alkynes is low and usually mixtures of triazoles are formed.[178] Alkynes have HOMO energy levels lower than those of the corresponding alkenes, although the LUMOs are ap-

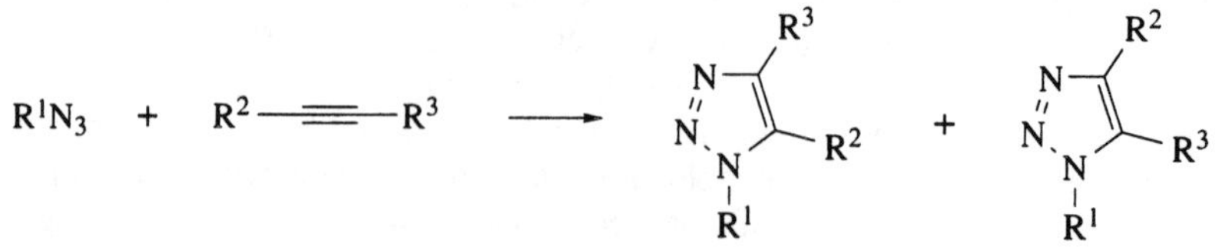

$$R^1N_3 + R^2-\equiv-R^3 \longrightarrow \textbf{(234)} + \textbf{(235)} \quad (10)$$

proximately at the same level. Consequently, most alkynes undergo simultaneously both azide LUMO- and azide HOMO-controlled cycloadditions.

Although less reactive than C—C triple bonds, C—N triple bonds are also known to undergo 1,3-dipolar cycloadditions with some organic azides. An interesting example of this reaction which occurs intramolecularly involves the conversion of 2′-azido-2-phenylcarbonitrile (**236**) into tetrazolophenanthridine (**237**; equation 11).[191]

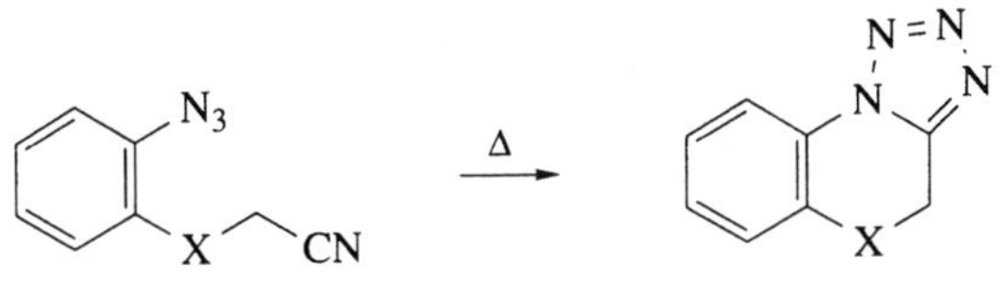

(**236**) X = O, CH_2 → (**237**) (11)

The intramolecular 1,3-dipolar cycloaddition reaction of azides has become an increasingly useful process for the construction of natural products and molecules of theoretical interest.[192,193] For example, 2-substituted azido enone (**238**) was prepared from the corresponding bromide by treatment with sodium azide. Thermolysis of this material afforded aziridinyl ketone (**240**) presumably *via* a transient dipolar cycloadduct (**239**).[193] Ketone (**240**) was subsequently converted to an intermediate previously used to prepare histrionicotoxin (**241**; Scheme 56).

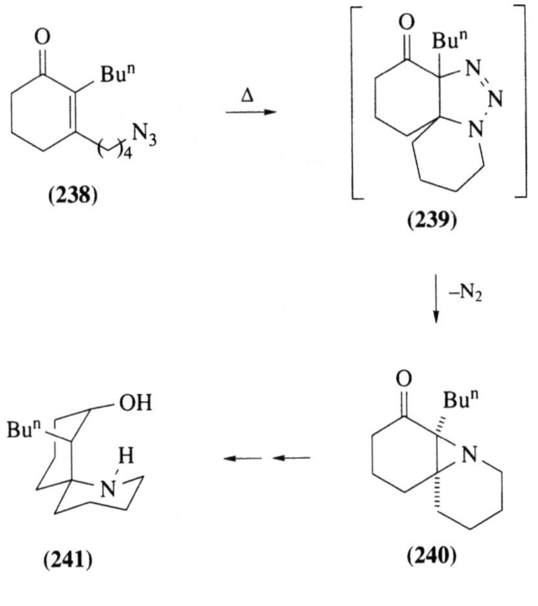

(**238**) → (**239**) → $-N_2$ → (**240**) → (**241**)

Scheme 56

A very interesting sequence of intramolecular dipolar cycloaddition–nitrogen extrusion retro-Diels–Alder reactions that lead to pyrrole (**247**) was uncovered by Shultz and Sha.[194] These workers found that the thermolysis of azido dienone (**242**) in refluxing benzene gave triazoline (**243**) in high yield. Brief irradiation of (**243**) in methanol produced pyrrole carboxylic ester (**247**) in quantitative yield. The mechanism suggested for the formation of (**247**) involves homolytic extrusion of nitrogen from (**243**) to give diradical (**244**), from which recombination yields the bridged intermediate (**245**). Tricyclic (**245**) is formally an intramolecular pyrrole ketene Diels–Alder adduct and retro-Diels–Alder reaction of (**245**) would lead to the pyrrole ketene. Reaction of (**246**) with methanol nicely accounts for the formation of pyrrole carboxylic ester (**247**; Scheme 57).

4.9.6.12 Diazoalkanes

Apart from the significance of diazoalkanes for the generation of carbenes,[195] these compounds also play a dominant role in dipolar cycloaddition chemistry.[196,197] Recent advances in the synthesis of di-

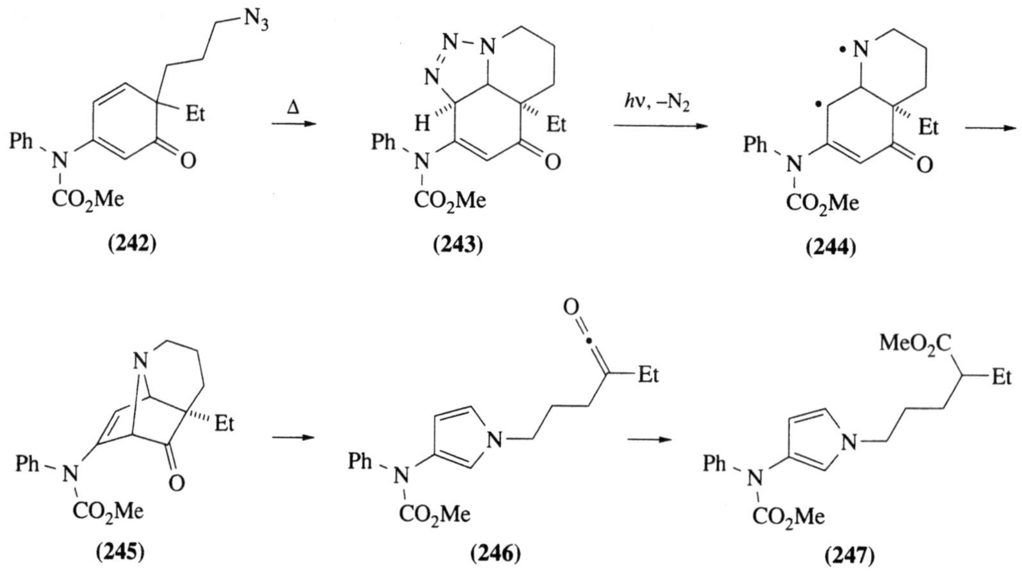

(242) (243) (244)

(245) (246) (247)

Scheme 57

azoalkanes have frequently led to new applications in cycloaddition chemistry.[198] The additions of diazoalkanes to alkenes are among the most thoroughly studied 1,3-dipolar cycloadditions. Tosylhydrazones are commonly used as precursors to generate diazoalkanes. The cycloadditions of simple diazoalkanes are HOMO (dipole)–LUMO(dipolarophile) controlled (equation 12).[196] Both conjugating and electron-attracting groups accelerate reactions of dipolarophiles with diazoalkanes as compared to ethylene. With these dipolarophiles, 3-substituted Δ′-pyrazolines are favored, a result of the union of the larger diazoalkane HOMO coefficient on carbon with that of the larger dipolarophile LUMO coefficient on the unsubstituted carbon. Huisgen has shown that introduction of a methoxycarbonyl group into diazomethane shifts the 1,3-dipole to a Type 2 (Sustmann's classification)[21] in methyl diazoacetate and further toward a Type 3 for dimethyl diazomalonate and methyl diazo(phenylsulfonyl)acetate.[199] Electron-releasing substituents in the diazoalkane, on the other hand, raise the HOMO energy and enhance the cycloaddition rate.[200] 3-Substituted pyrazolines are formed as the major products in the 1,3-dipolar cycloaddition of diazomethane with 1-alkenes.[201,202] With these systems, the difference between the two frontier orbital interactions is quite small, but the nearly equal magnitude of the terminal coefficients in the diazomethane LUMO suggests that the diazomethane HOMO determines product regiochemistry. As is frequently observed for cycloadditions of diazo compounds to alkenes, the initial cycloadducts are unstable and readily undergo further reactions. Thus, the pyrazoline (**248**), generated from diazomethane and 3-cyanopropene, escapes isolation by undergoing a very facile hydrogen shift to give Δ²-pyrazoline (**249**; Scheme 58).[196]

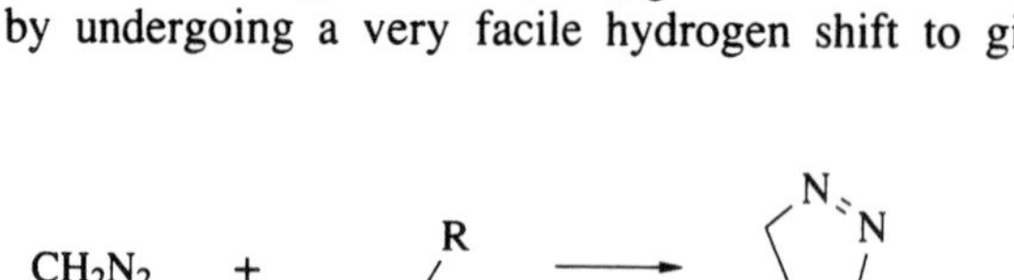

CH_2N_2 + (R-alkene) → (pyrazoline) (12)

CH_2N_2 + (CN-alkene) → [(248)] → (249)

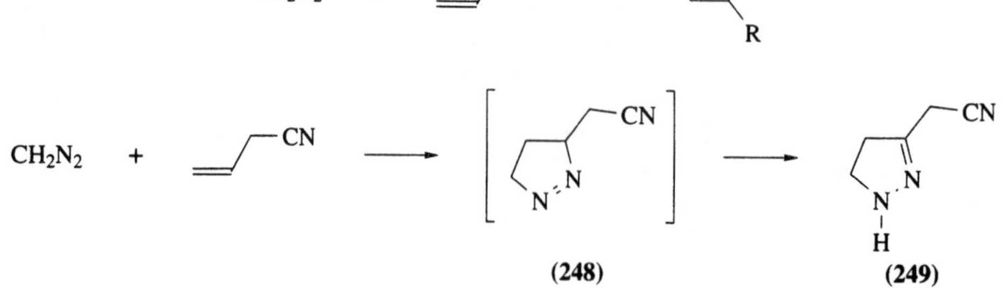

Scheme 58

Elimination of nitrogen from the Δ¹-pyrazoline to give cyclopropanes frequently occurs, even during the cycloaddition step.[196] This reaction profits considerably from an increase in temperature. Thus, although Δ^1-pyrazoline (**251**), which arises from 9-diazoxanthene (**250**) and methyl acrylate at –20 °C, can be isolated, it is transformed into a mixture of Δ^2-pyrazoline (**252**) and the spirocyclopropane (**253**) on warming to room temperature (Scheme 59).[203]

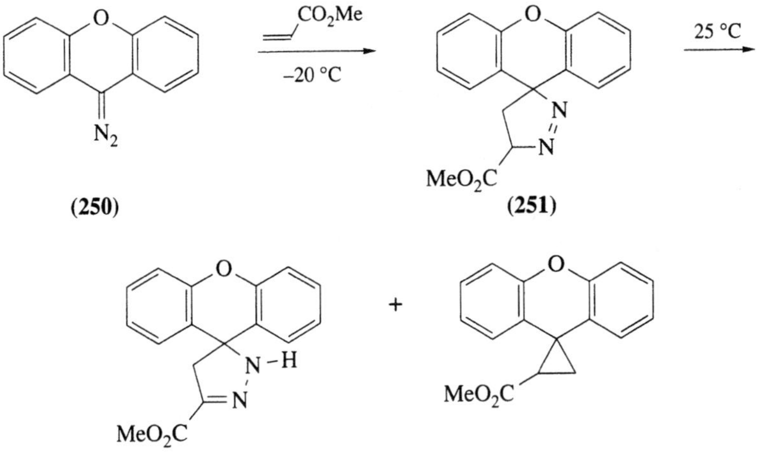

CO_2Me –20 °C 25 °C

N_2 N N MeO_2C

(**250**) (**251**)

N–H N MeO_2C + MeO_2C

(**252**) (**253**)

Scheme 59

Although it has been established that the HOMO (diazoalkane)–LUMO (alkene) controlled concerted cycloaddition occurs without intervention of any intermediate for the reactions of simple diazoalkanes with alkenes, Huisgen once proposed a mechanistic alternative;[4] namely an initial hypothetical nitrene-type 1,1-cycloaddition reaction of phenyldiazomethane to styrene followed by a vinylcyclopropane–cyclopentene-type 1,3-sigmatropic rearrangement. Control experiments, however, excluded this hypothesis for the bimolecular 1,3-dipolar cycloaddition reaction of diazomethane (Scheme 60).[204]

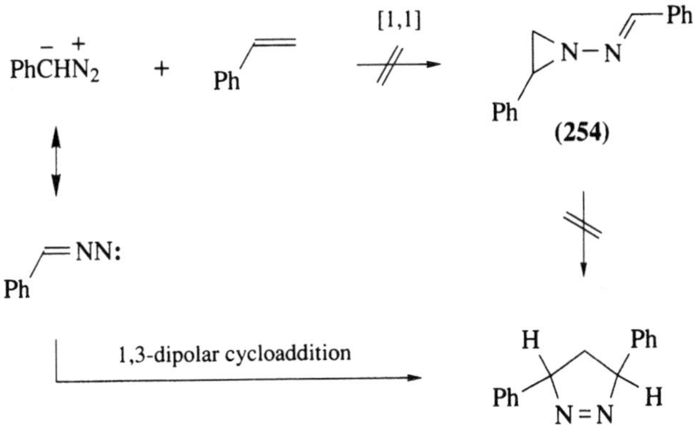

$Ph\overset{-}{C}H\overset{+}{N}_2$ + Ph [1,1] N–N Ph Ph

(**254**)

=NN: Ph

1,3-dipolar cycloaddition

H Ph Ph H N=N

(**255**)

Scheme 60

The hypothetical nitrene reactivity of the terminal nitrogen of diazomethane was simultaneously uncovered by Miyashi[205] and Padwa.[206] Various allyl-substituted diazomethanes generated by pyrolyses of the sodium salts of the corresponding tosylhydrazones were found to undergo a reversible intramolecular nitrene-type 1,1-cycloaddition to give 1,2-diazabicyclo[3.1.0]hex-2-enes (Scheme 61). Further work established that these allyl-substituted diazoalkenes undergo the intramolecular 1,1-cycloaddition with complete retention of configuration. It was concluded by both groups that the 1,1-cycloaddition reaction occurs concertedly without intervention of a zwitterionic intermediate, being controlled by the electrophilic LUMO energy of diazomethane. The results parallel the stereospecific addition of singlet nitrenes[207] and the intramolecular 1,1-cycloaddition of nitrile ylides[87] and nitrile imines.[104] Formation of the 1,2-diazabicyclohexene ring seems to be limited to α-phenyl-substituted derivatives.

α-Diazo ketones represent an interesting subclass of diazoalkanes since several discrete modes of bimolecular cycloaddition are possible.[208] Among these are those involving the diazoalkane moiety or a re-

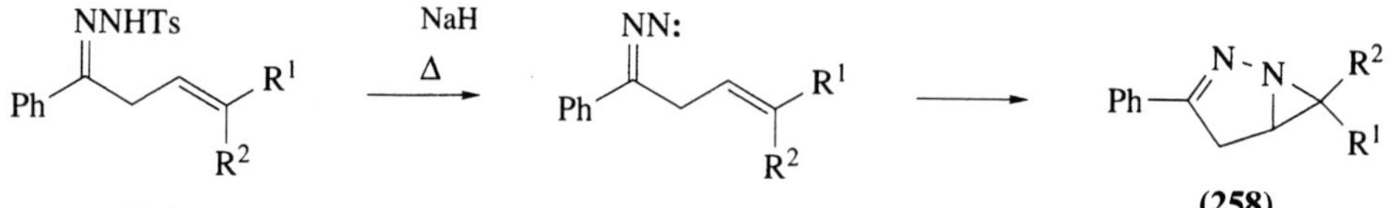

Scheme 61

active intermediate possessing the stoichiometry of a keto carbene species derived from an initial loss of nitrogen (Scheme 62). Much less common modes of addition involving the extended 6π-electron 1,5-dipolar system are also observed with certain quinonoid α-diazo ketones.[209] The use of extended diazoalkanes with six or more electrons has not received much attention despite the obvious synthetic and theoretical interest in such processes. When the diazoketone moiety is incorporated into a heterocyclic ring, dipolar cycloaddition processes can provide ready access to more elaborate and rare heterocyclic ring systems (Scheme 63).[210–213]

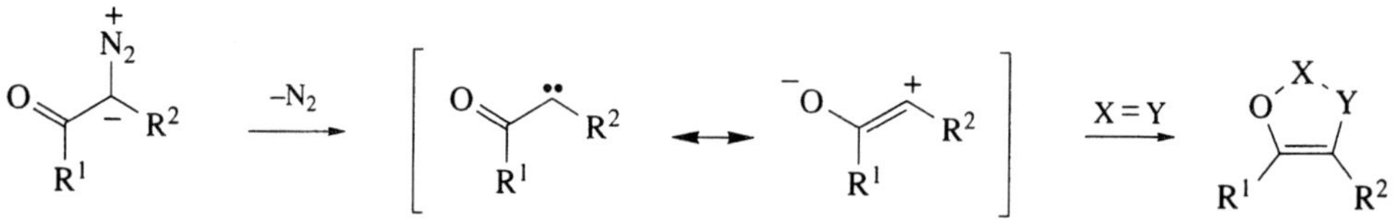

Scheme 62

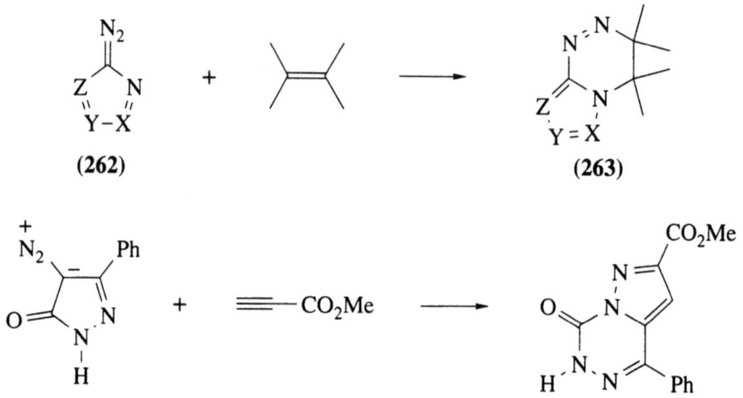

Scheme 63

4.9.7 CONCLUSION

The prominent role that 1,3-dipolar cycloaddition reactions play in the elaboration of a variety of organic molecules has become increasingly apparent in recent years. The ease of the cycloaddition, the rapid accumulation of polyfunctionality in a relatively small molecular framework, the high stereochemical control of the cycloaddition, and the fair predictability of its regiochemistry have contributed to the popularity of the reaction. In the realm of synthesis, in which a premium is put on the rapidity of construction of polyfunctional, highly bridged carbon and heteroatom networks, the dipolar cycloaddition reaction has now emerged as a prominent synthetic method. In the past decade, interest in the intramolecular 1,3-dipolar cycloaddition reaction, where the dipole and dipolarophile are constrained in the same molecule, has increased exponentially. When the reacting components are themselves cyclic or have ring substituents, complex multicyclic arrays, such as those contained in drugs and natural products, can be constructed in a single step. Often the syntheses of molecules of this complexity are more difficult and lengthy by other routes. Future developments will no doubt continue to enhance the synthetic utility of the dipolar cycloaddition reaction. After a century of research in this field, there is no end in sight.

ACKNOWLEDGEMENT

We thank the National Institute of Health for generous support of our research program in the area of dipolar cycloaddition chemistry. Dedicated with respect and admiration to Professor Rolf Huisgen, the leading authority in the area of dipolar cycloaddition chemistry, on the occasion of his 70th birthday.

4.9.8 REFERENCES

1. A. Wasserman, 'Diels–Alder Reactions', Elsevier, New York, 1965.
2. J. Sauer, *Angew. Chem., Int. Ed. Engl.*, 1967, **6**, 16.
3. I. Fleming, 'Frontier Orbitals and Organic Chemical Reactions', Wiley, New York, 1976.
4. R. Huisgen, *Angew. Chem., Int. Ed. Engl.*, 1963, **2**, 565, 633.
5. R. Huisgen, R. Grashey and J. Sauer, 'The Chemistry of Alkenes', ed. S. Patai, Interscience, New York, 1964, p. 739.
6. A. Padwa (ed.), '1,3-Dipolar Cycloaddition Chemistry', Wiley-Interscience, New York, 1984.
7. R. Huisgen, in '1,3-Dipolar Cycloaddition Chemistry', ed. A. Padwa, Wiley-Interscience, New York, 1984, p 1.
8. K. N. Houk and K. Yamaguchi, in '1,3-Dipolar Cycloaddition Chemistry', ed. A. Padwa, Wiley-Interscience, New York, 1984.
9. R. Huisgen, *J. Org. Chem.*, 1968, **33**, 2291; 1976, **41**, 1976.
10. R. B. Woodward and R. Hoffmann, in 'The Conservation of Orbital Symmetry', Academic Press, New York, 1970.
11. R. A. Firestone, *J. Org. Chem.*, 1968, **33**, 2285; 1972, **37**, 2181; *J. Chem. Soc. A*, 1970, 1570; *Tetrahedron*, 1977, **33**, 3009.
12. D. J. Poppinger, *J. Am. Chem. Soc.*, 1975, **97**, 7468; *Aust. J. Chem.*, 1976, **29**, 465.
13. A. Komornicki, J. Goddard and H. F. Schaefer, *J. Am. Chem. Soc.*, 1980, **102**, 1763.
14. P. C. Hiberty, G. Ohanessian and H. B. Schlegel, *J. Am. Chem. Soc.*, 1983, **105**, 719.
15. M. J. S. Dewar, S. Olivella and H. S. Rzepa, *J. Am. Chem. Soc.*, 1978, **100**, 5650.
16. J. Pacansky and W. Schubert, *J. Chem. Phys.*, 1982, **76**, 1459.
17. S. W. Benson, K. W. Egger and D. M. Golden, *J. Am. Chem. Soc.*, 1965, **87**, 468.
18. K. N. Houk, R. A. Firestone, L. L. Munchausen, P. H. Mueller, B. H. Arison and L. A. Garcia, *J. Am. Chem. Soc.*, 1985, **107**, 7227.
19. K. N. Houk, *Acc. Chem. Res.*, 1975, **8**, 361.
20. K. Fukui, *Acc. Chem. Res.*, 1971, **4**, 57; *Fortschr. Chem. Forsch.*, 1970, **15**, 1.
21. R. Sustmann and R. Schubert, *Tetrahedron Lett.*, 1972, 2739; R. Sustmann and H. Trill, *Tetrahedron Lett.*, 1972, 4271; R. Sustmann, *Pure Appl. Chem.*, 1974, **40**, 569.
22. R. L. Kuczkowski, in '1,3-Dipolar Cycloaddition Chemistry', ed. A. Padwa, Wiley-Interscience, New York, 1984, p. 197.
23. K. N. Houk, in 'Pericyclic Reactions', ed. A. P. Marchand and R. E. Lehr, Academic Press, New York, vol. 2, 1977.
24. W. C. Herndon, *Chem. Rev.*, 1972, **72**, 157.
25. R. Sustmann, *Tetrahedron Lett.*, 1971, 2717.
26. J. Bastide and O. Henri-Rousseau, *Bull. Soc. Chim. Fr.*, 1973, 2290.
27. R. Huisgen, R. Mloston and E. Langhals, *J. Org. Chem.*, 1986, **51**, 4085, 6401.
28. R. M. Kellogg, *Tetrahedron*, 1976, **32**, 2165.
29. R. Huisgen, C. Fulka, I. Kalwinsch, X. Li, G. Mloston, J. R. Moran and A. Probstl, *Bull. Soc. Chim. Belg.*, 1984, **93**, 511.
30. K. N. Houk and C. R. Watts, *Tetrahedron Lett.*, 1970, 4025; K. N. Houk and L. J. Luskus, *Tetrahedron Lett.*, 1970, 4029.
31. J. N. Crabb and R. C. Storr, in '1,3-Dipolar Cycloaddition Chemistry', ed. A. Padwa, Wiley-Interscience, New York, 1984, p. 543.
32. P. S. Bailey, in 'Ozonization in Organic Chemistry'; Academic Press, New York, 1982, vol. 2.
33. M. Boshar, H. Heydt, G. Maas, H. Gumbel and M. Regitz, *Angew. Chem., Int. Ed. Engl.*, 1985, **24**, 597.
34. J. Baran and H. Mayr, *J. Am. Chem. Soc.*, 1987, **109**, 6519.
35. J. J. Tufariello, in '1,3-Dipolar Cycloaddition Chemistry', ed. A. Padwa, Wiley-Interscience, New York, 1984, p. 83.
36. J. J. Tufariello, *Acc. Chem. Res.*, 1979, **12**, 396.
37. D. St. C. Black, R. F. Crozier and V. C. Davis, *Synthesis*, 1975, **7**, 205.
38. T. Kametani, S. D. Huang, A. Nakayama and T. Hondu, *J. Org. Chem.*, 1982, **47**, 2328.
39. W. Oppolzer, J. I. Grayson, H. Wegmann and M. Urea, *Tetrahedron*, 1983, **39**, 3695.
40. P. M. Wovkulich and M. R. Uskokovic, *J. Am. Chem. Soc.*, 1981, **103**, 3956.
41. P. DeShong and J. M. Leginus, *J. Am. Chem. Soc.*, 1983, **105**, 1686.
42. J. E. Baldwin, M. F. Chan, G. Gallacher, P. Monk and K. Prout, *J. Chem. Soc., Chem. Commun.*, 1983, 250.
43. S. W. Baldwin and J. Aube, *Tetrahedron Lett.*, 1987, **28**, 179.
44. A. P. Kozikowski and Y. Y. Chen, *J. Org. Chem.*, 1981, **46**, 5248.
45. D. P. Curran, *J. Am. Chem. Soc.*, 1982, **104**, 4024.
46. V. Jager and W. Schwab, *Tetrahedron Lett.*, 1978, 3129.
47. K. N. Houk, A. Bimanand, D. Mukherjee, J. Sims, Y. M. Chang, D. C. Kaufman and L. N. Domelsmith, *Heterocycles*, 1977, **7**, 293.
48. K. B. G. Torssell, in 'Nitrile Oxides, Nitrones and Nitronates in Organic Synthesis', VCH, Weinheim, 1988.

49. P. N. Confalone and E. M. Huie, *Org. React.*, 1988, **36**, 1.
50. S. A. Ali, P. A. Senaratne, C. R. Illig, H. Meckler and J. J. Tufariello, *Tetrahedron Lett.*, 1979, 4167.
51. A. Padwa, *Angew. Chem., Int. Ed. Engl.*, 1976, **15**, 123; A. Padwa (ed.), in '1,3-Dipolar Cycloaddition Chemistry', Wiley-Interscience, New York, 1984, p. 277.
52. W. Oppolzer, *Angew. Chem., Int. Ed. Engl.*, 1977, **16**, 10.
53. N. A. LeBel and J. J. Whang, *J. Am. Chem. Soc.*, 1959, **81**, 6334.
54. C. Grundmann and P. Grunanger, in 'The Nitrile Oxides', Springer-Verlag, New York, 1971.
55. P. Caramella and P. Grunanger, in '1,3-Dipolar Cycloaddition Chemistry', ed. A. Padwa, Wiley-Interscience, New York, 1984, p. 291.
56. D. P. Curran, in 'Advances in Cycloaddition', JAI Press, Greenwich, CT, 1988, vol. 1.
57. T. Mukaiyama and T. Hoshino, *J. Am. Chem. Soc.*, 1960, **82**, 5339.
58. K. Bast, M. Christl, R. Huisgen, W. Mack and R. Sustmann, *Chem. Ber.*, 1973, **106**, 3258.
59. M. Christl and R. Huisgen, *Tetrahedron Lett.*, 1968, 5209.
60. J. D. Morrison, in 'Asymmetric Synthesis, a Multivolume Treatise', Academic Press, New York, 1984.
61. S. Masamune, L. A. Reed, J. T. Davis and W. Choy, *J. Org. Chem.*, 1983, **48**, 4441.
62. W. Choy, L. A. Reed and S. Masamune, *J. Org. Chem.*, 1983, **48**, 1137.
63. A. Vasella, *Helv. Chim. Acta*, 1977, **60**, 1273; 1982, **65**, 1134.
64. V. Jager, R. Schone and E. F. Paulus, *Tetrahedron Lett.*, 1983, **24**, 5501.
65. K. N. Houk, S. R. Moses, Y.-D. Wu, N. G. Rondan, V. Jager, R. Schohe and F. R. Fronczek, *J. Am. Chem. Soc.*, 1984, **106**, 3880.
66. A. P. Kozikowski and A. K. Ghosh, *J. Org. Chem.*, 1984, **49**, 2762; *J. Am. Chem. Soc.*, 1982, **104**, 5788.
67. D. P. Curran, B. H. Kim, J. Daugherty and T. A. Heffner, *Tetrahedron Lett.*, 1988, **29**, 3555.
68. W. Oppolzer, *Angew. Chem., Int. Ed. Engl.*, 1984, **23**, 876.
69. E. Ciganek, *Org. React.*, 1984, 32, 1.
70. A. P. Kozikowski, *Acc. Chem. Res.*, 1984, **17**, 410.
71. P. N. Confalone, E. D. Lolla, G. Pizzolato and M. R. Uskokovic, *J. Am. Chem. Soc.*, 1978, **100**, 6291; 1980, **102**, 1954.
72. A. Padwa, *Acc. Chem. Res.*, 1976, **9**, 371; A. Padwa, in 'Reactive Intermediates', ed. R. A. Abramovitch, Plenum Press, New York, vol. 2, 1982, p. 55.
73. P. Gilgen, H. Heimgartner, H. Schmid and H. J. Hansen, *Heterocycles*, 1977, **6**, 143.
74. H. J. Hansen and H. Heimgartner, in '1,3-Dipolar Cycloaddition Chemistry', ed. A. Padwa, Wiley-Interscience, New York, 1984, vol. 1, chap. 2, p. 177.
75. R. Huisgen, H. Stangle, H. J. Sturm and H. Wagenhofer, *Angew. Chem.*, 1962, **74**, 31.
76. K. Burger and J. Fehn, *Chem. Ber.*, 1972, **105**, 3814.
77. A. Padwa, M. Dharan, J. Smolanoff and S. I. Wetmore, *J. Am. Chem. Soc.*, 1973, **95**, 1945.
78. W. Sieber, P. Gilgen, S. Chaloupka, H. J. Hansen and H. Schmid, *Helv. Chim. Acta*, 1973, **56**, 1679.
79. A. Padwa, J. Gasdaska, N. J. Turro, Y. C. Cha and I. R. Gould, *J. Am. Chem. Soc.*, 1986, **108**, 6739; *J. Org. Chem.*, 1985, **50**, 4417.
80. D. Griller, C. R. Montgomery, J. C. Scaiano, M. S. Platz and L. Hadel, *J. Am. Chem. Soc.*, 1982, **104**, 6813.
81. P. Caramella and K. N. Houk, *J. Am. Chem. Soc.*, 1976, **98**, 6397.
82. A. Padwa and J. Smolanoff, *J. Chem. Soc., Chem. Commun.*, 1973, 342.
83. K. Burger, J. Albanbauer and F. Manz, *Chem. Ber.*, 1974, **107**, 1823; K. Burger, J. Fehn and E. Muller, *Chem. Ber.*, 1973, **106**, 1; K. Burger and S. Einhellig, *Chem. Ber.*, 1973, **106**, 3421.
84. L. Salem, *J. Am. Chem. Soc.*, 1974, **96**, 3486.
85. P. Caramella, R. W. Gandour, J. A. Hall, C. G. Deville and K. N. Houk, *J. Am. Chem. Soc.*, 1977, **99**, 385.
86. B. Bigot, A. Sevin and A. Devaquet, *J. Am. Chem. Soc.*, 1978, **100**, 6924.
87. A. Padwa and P. H. J. Carlsen, *J. Am. Chem. Soc.*, 1975, **97**, 3862; 1976, **98**, 2006; 1977, **99**, 1514; A. Padwa, A. Ku, A. Mazzu and S. I. Wetmore, *J. Am. Chem. Soc.*, 1976, **98**, 1048; A. Padwa and P. H. J. Carlsen, *J. Org. Chem.*, 1978, **43**, 3757; A. Padwa, P. H. J. Carlsen and A. Ku, *J. Am. Chem. Soc.*, 1978, **100**, 3494.
88. A. Padwa, M. Dharan, J. Smolanoff and S. I. Wetmore, *J. Am. Chem. Soc.*, 1973, **95**, 1954.
89. A. Padwa, M. Dharan, J. Smolanoff, S. I. Wetmore and S. Clough, *J. Am. Chem. Soc.*, 1972, **94**, 1395.
90. A. Padwa, J. Smolanoff and S. I. Wetmore, *J. Chem. Soc., Chem. Commun.*, 1972, 409.
91. R. Huisgen, R. Sustmann and K. Bunge, *Chem. Ber.*, 1972, **105**, 1324.
92. R. Huisgen, M. Seidel, J. Sauer, J. W. McFarland and G. Wallbillich, *J. Org. Chem.*, 1959, **24**, 892.
93. P. Scheiner and J. F. Dinda, *Tetrahedron*, 1970, **26**, 2619; P. Scheiner and W. M. Litchmann, *J. Chem. Soc., Chem. Commun.*, 1972, 781.
94. J. S. Clover, A. Eckell, R. Huisgen and R. Sustmann, *Chem. Ber.*, 1967, **100**, 60.
95. M. Marky, H. Meier, A. Wunderli, H. Heimgartner, H. Schmid and H. J. Hansen, *Helv. Chim. Acta*, 1978, **61**, 1477.
96. C. Wentrup, A. Damerius and W. Richen, *J. Org. Chem.*, 1978, **43**, 2037.
97. A. Padwa, T. Caruso and S. Nahm, *J. Org. Chem.*, 1980, **45**, 4065.
98. R. Fusco, L. Garanti and G. Zecchi, *Tetrahedron Lett.*, 1974, 269.
99. L. Garanti and G. Zecchi, *Synthesis*, 1974, 814.
100. R. Huisgen, H. Knupfer, R. Sustmann, G. Wallbillich and V. Weberndorfer, *Chem. Ber.*, 1967, **100**, 1580.
101. R. Huisgen, M. Seidel, G. Wallbillich and H. Knupfer, *Tetrahedron*, 1962, **17**, 3.
102. S. Morrocchi, A. Ricca and A. Zanarotti, *Tetrahedron Lett.*, 1970, 3215.
103. L. Garanti, A. Vigevani and G. Zecchi, *Tetrahedron Lett.*, 1976, 1527; L. Garanti and G. Zecchi, *J. Chem. Soc., Perkin Trans. 1*, 1977, 2092; *J. Heterocycl. Chem.*, 1979, **16**, 377.
104. A. Padwa and S. Nahm, *J. Org. Chem.*, 1979, **44**, 4746; 1981, **46**, 1402.
105. J. W. Lown, in '1,3-Dipolar Cycloaddition Chemistry', ed. A. Padwa, Wiley-Interscience, New York, 1984, p. 653.
106. H. W. Heine and R. E. Peavy, *Tetrahedron Lett.*, 1965, 3123.
107. A. Padwa and L. Hamilton, *Tetrahedron Lett.*, 1965, 4363.

108. R. Huisgen, W. Scheer, G. Szeimies and H. Huber, *Tetrahedron Lett.*, 1966, 397.
109. R. Huisgen and H. Mader, *J. Am. Chem. Soc.*, 1967, **89**, 1753; R. Huisgen, R. Scheer and H. Huber, *J. Am. Chem. Soc.*, 1967, **89**, 1753.
110. E. Vedejs and F. G. West, *Chem. Rev.*, 1986, 941; E. Vedejs, in 'Advances in Cycloaddition Chemistry', ed. D. P. Curran, JAI Press, Greenwich, CT, 1988, p. 33.
111. E. Vedejs and G. R. Martinez, *J. Am. Chem. Soc.*, 1979, **101**, 6542; E. Vedejs and J. W. Grissom, *J. Am. Chem. Soc.*, 1988, **110**, 3238.
112. J. W. Lown, *Rec. Chem. Prog.*, 1971, **32**, 51.
113. R. Huisgen, *Spec. Publ. Chem. Soc.*, 1970, **21**, 1.
114. M. Joucla and J. Hamelin, *Tetrahedron Lett.*, 1978, 2885; R. Grigg, *Bull. Soc. Chim. Belg.*, 1984, **93**, 593; O. Tsuge, K. Ueno, S. Kanemasa and K. Yorozu, *Bull. Chem. Soc. Jpn.*, 1986, **59**, 1809; M. Joucla, B. Fouchet and J. Hamelin, *Tetrahedron*, 1985, **41**, 2707.
115. O. Tsuge and K. Ueno, *Heterocycles*, 1983, **20**, 2133; O. Tsuge, S. Kanemasa, M. Ohe, K. Yorozu, S. Takenaka and K. Ueno, *Chem. Lett.*, 1986, 127, 973.
116. P. Armstrong, R. Grigg, M. Jordan and J. F. Malone, *Tetrahedron*, 1985, **41**, 3547; P. N. Confalone and R. A. Earl, *Tetrahedron Lett.*, 1986, **27**, 2695.
117. R. Bartnik and G. Mloston, *Tetrahedron*, 1984, **40**, 2569.
118. A. Padwa and W. Dent, *J. Org. Chem.*, 1987, **52**, 235; A. Padwa, Y. Y. Chen, W. Dent and H. Nimmesgern, *J. Org. Chem.*, 1985, **50**, 4006.
119. E. Vedejs and F. G. West, *J. Org. Chem.*, 1983, **48**, 4773; E. Vedejs and G. R. Martinez, *J. Am. Chem. Soc.*, 1980, **102**, 7993; E. Vedejs, S. Larsen and F. G. West, *J. Org. Chem.*, 1985, **50**, 2170.
120. T. Livinghouse and R. Smith, *J. Chem. Soc., Chem. Commun.*, 1983, 210; *J. Org. Chem.*, 1983, **48**, 1554; *Tetrahedron*, 1985, **41**, 3559.
121. A. Padwa and Y. Y. Chen, *Tetrahedron Lett.*, 1983, **24**, 3447.
122. L. E. Overman and E. J. Jacobsen, *Tetrahedron Lett.*, 1982, **23**, 2741.
123. E. W. Colvin, *Chem. Soc. Rev.*, 1978, **7**, 15.
124. A. Hosomoi, Y. Sakata and H. Sakurai, *Chem. Lett.*, 1984, 1117.
125. Y. Terao, H. Kotaki, N. Imai and K. Achiwa, *Chem. Pharm. Bull.*, 1985, **33**, 896.
126. R. Beugelmans, G. Negron and G. Roussi, *J. Chem. Soc., Chem. Commun.*, 1983, 31; R. Beugelmans, L. B. Iguertsira, J. Chastanet, G. Negron and G. Roussi, *Can. J. Chem.*, 1985, **63**, 724; J. Chastanet and G. Roussi, *Heterocycles*, 1985, **23**, 653; *J. Org. Chem.*, 1985, **50**, 2910.
127. J. B. Hendrickson and J. S. Farina, *J. Org. Chem.*, 1980, **45**, 3359.
128. P. G. Sammes and R. Whitby, *J. Chem. Soc., Chem. Commun.*, 1984, 702; P. G. Sammes and L. J. Street, *J. Chem. Res.*, 1984, 196; *J. Chem. Soc., Perkin Trans. 1*, 1983, 1261; *J. Chem. Soc., Chem. Commun.*, 1983, 666; *J. Chem. Soc., Perkin Trans. 1*, 1987, 195; S. M. Bromidge, P. G. Sammes and L. J. Street, *J. Chem. Soc., Perkin Trans. 1*, 1985, 1715.
129. M. E. Garst, B. J. McBride and J. G. Douglass, III, *Tetrahedron Lett.*, 1983, **24**, 1675.
130. K. S. Feldman, *Tetrahedron Lett.*, 1983, **24**, 5585.
131. G. W. Griffin and A. Padwa, in 'Photochemistry of Heterocyclic Compounds', ed. O. Buchart, Wiley, New York, 1976, chap. 2, p. 41; P. K. Das and G. W. Griffin, *J. Photochem.*, 1985, **27**, 317.
132. M. Bekhazi, P. J. Smith and J. Warkentin, *Can. J. Chem.*, 1984, **62**, 1646; J. Warkentin, *J. Org. Chem.*, 1984, **49**, 343; M. Bekhazi and J. Warkentin, *J. Am. Chem. Soc.*, 1983, **105**, 289.
133. N. Shimizu and P. D. Bartlett, *J. Am. Chem. Soc.*, 1978, **100**, 4260.
134. R. W. Hoffmann and H. Luthardt, *Chem. Ber.*, 1968, **101**, 3861.
135. P. de March and R. Huisgen, *J. Am. Chem. Soc.*, 1982, **104**, 4952.
136. R. P. L'esperance, T. M. Ford and M. Jones, *J. Am. Chem. Soc.*, 1988, **110**, 209.
137. A. Padwa, S. P. Carter and H. Nimmesgern, *J. Org. Chem.*, 1985, **50**, 4417; *J. Am. Chem. Soc.*, 1988, **110**, 2894; A. Padwa, G. E. Fryxell and L. Zhi, *J. Org. Chem.*, 1988, **53**, 2875.
138. T. Ibata, T. Motoyama and M. Hamaguchi, *Bull. Chem. Soc. Jpn.*, 1976, **49**, 2298; T. Ibata and J. Toyoda, *Chem. Lett.*, 1983, 1453; T. Ibata, J. Toyoda, M. Sawada and T. Tanaka, *J. Chem. Soc., Chem. Commun.*, 1986, 1266; T. Ibata, M. Liu and J. Toyoda, *Tetrahedron Lett.*, 1986, **27**, 4383; K. Ueda, T. Ibata and M. Takebayashi, *Bull. Chem. Soc. Jpn.*, 1972, **45**, 2779; T. Ibata, K. Jitsuhiro and Y. Tsubokura, *Bull. Chem. Soc. Jpn.*, 1981, **54**, 240.
139. W. J. Linn and R. E. Benson, *J. Am. Chem. Soc.*, 1965, **87**, 1965; W. J. Linn, *J. Am. Chem. Soc.*, 1965, **87**, 3665.
140. G. W. Griffin, *Angew. Chem., Int. Ed. Engl.*, 1971, **10**, 537.
141. T. DoMinh, A. M. Trozzolo and G. W. Griffin, *J. Am. Chem. Soc.*, 1970, **92**, 1402.
142. I. J. Lev, K. Ishikawa, N. S. Bhacca and G. W. Griffin, *J. Org. Chem.*, 1976, **41**, 2654; G. A. Lee, *J. Org. Chem.*, 1976, **41**, 2656; V. Markowski and R. Huisgen, *Tetrahedron Lett.*, 1976, 4643.
143. A. Dahmen, H. Hamberger, R. Huisgen and V. Markowski, *J. Chem. Soc., Chem. Commun.*, 1971, 1192.
144. W. Bihlmaier, R. Huisgen, H.-U. Reissig and S. Voss, *Tetrahedron Lett.*, 1979, 261.
145. N. Dennis, B. E. Ibrahim, A. R. Katritzky, I. G. Taulov and Y. Takeuchi, *J. Chem. Soc., Perkin Trans. 1*, 1974, 1883; 1979, 399.
146. E. B. Knott, *J. Chem. Soc.*, 1955, 916.
147. G. Bianchi and R. Gandolfi, in '1,3-Dipolar Cycloaddition Chemistry', ed. A. Padwa, Wiley-Interscience, New York, 1984, p. 451.
148. A. Schonberg, B. Konig and E. Singer, *Chem. Ber.*, 1967, **100**, 767.
149. D. H. R. Barton and B. J. Willis, *J. Chem. Soc., Chem. Commun.*, 1970, 1225; *J. Chem. Soc., Perkin Trans. 1*, 1972, 305.
150. R. M. Kellogg and S. Wassenaer, *Tetrahedron Lett.*, 1970, 1987, 4689; J. Buter, S. Wassenaer and R. M. Kellogg, *J. Org. Chem.*, 1972, **40**, 2573; R. M. Kellogg, M. Noteboom and J. K. Kaiser, *J. Org. Chem.*, 1975, **40**, 2573; R. M. Kellogg and J. K. Kaiser, *J. Org. Chem.*, 1975, **40**, 2575.
151. A. G. Schultz and M. B. DeTar, *J. Am. Chem. Soc.*, 1974, **96**, 296; *J. Org. Chem.*, 1974, **39**, 3185.

152. Y. Terao, M. Tanaka, N. Imai and K. Achiwa, *Tetrahedron Lett.*, 1985, **26**, 3011; M. Aono, Y. Terao and K. Achiwa, *Heterocycles*, 1986, **24**, 313; M. Aono, C. Hyodo, Y. Terao and K. Achiwa, *Tetrahedron Lett.*, 1986, **27**, 4039.
153. R. Grashey, in '1,3-Dipolar Cycloaddition Chemistry', ed. A. Padwa, Wiley-Interscience, New York, 1984, p. 733.
154. H. Neunhoeffer and P. F. Wiley, *Chem. Heterocycl. Comp.*, 1978, **33**, 1.
155. W. Oppolzer, *Tetrahedron Lett.*, 1970, 2199.
156. W. Oppolzer, *Tetrahedron Lett.*, 1972, 1707.
157. G. Le Fevre, S. Sinbandhit and J. Hamelin, *Tetrahedron*, 1979, **35**, 1821.
158. K. T. Potts, in '1,3-Dipolar Cycloaddition Chemistry', ed. A. Padwa, Wiley-Interscience, New York, 1984.
159. H. L. Gingrich and J. S. Baum, in 'Oxazoles', ed. N. J. Turchi, Wiley, New York, 1987, p. 35.
160. W. D. Ollis and C. Ramsden, *Adv. Heterocycl. Chem.*, 1976, **19**, 1.
161. R. Huisgen and H. Gotthardt, *Chem. Ber.*, 1968, **101**, 1059.
162. R. Huisgen, H. Gotthardt and R. Grashey, *Angew. Chem., Int. Ed. Engl.*, 1962, **1**, 49; H. Gotthardt and R. Huisgen, *Chem. Ber.*, 1968, **101**, 552.
163. H. Gotthardt and F. Reiter, *Chem. Ber.*, 1979, **112**, 1193.
164. K. N. Houk, J. Sims, C. R. Watts and L. J. Luskus, *J. Am. Chem. Soc.*, 1973, **95**, 7301.
165. R. Huisgen, H. Gotthardt and R. Grashey, *Chem. Ber.*, 1968, **101**, 536, 839, 1059.
166. R. Huisgen, H. Gotthardt, H. O. Bayer and F. C. Schaefer, *Chem. Ber.*, 1970, **103**, 2611; *Angew. Chem., Int. Ed. Engl.*, 1964, **3**, 136; *Tetrahedron Lett.*, 1964, 487; *Chem. Ber.*, 1970, **103**, 2625.
167. K. T. Potts and U. P. Singh, *J. Chem. Soc. D*, 1969, 66; K. T. Potts and J. Baum, *J. Chem. Soc., Chem. Commun.*, 1973, 833.
168. F. M. Hershenson, *J. Org. Chem.*, 1975, **40**, 1260.
169. A. Padwa, E. Burgess, H. L. Gingrich and D. Roush, *J. Org. Chem.*, 1982, **47**, 786.
170. K. N. Houk, J. Sims, R. E. Duke, R. W. Strozier and J. K. George, *J. Am. Chem. Soc.*, 1973, **95**, 7287.
171. R. Criegee, *Rec. Chem. Prog.*, 1957, **18**, 111; *Justus Liebigs Ann. Chem.*, 1953, **583**, 1; *Angew. Chem., Int. Ed. Engl.*, 1975, **14**, 745; *Chimia*, 1968, **22**, 392.
172. P. S. Bailey, *Chem. Rev.*, 1958, **58**, 926.
173. R. L. Kuczkowski, in '1,3-Dipolar Cycloaddition Chemistry', ed. A. Padwa, Wiley-Interscience, New York, 1984, p. 197.
174. P. S. Bailey, in 'Ozonization in Organic Chemistry', Academic Press, New York, vol. 1, 1978.
175. R. W. Murray, *Acc. Chem. Res.*, 1968, **1**, 313.
176. P. D. Bartlett and T. G. Traylor, *J. Am. Chem. Soc.*, 1962, **84**, 3408.
177. R. Criegee, A. Bancin and H. Keul, *Chem. Ber.*, 1975, **108**, 1642.
178. W. Lwowski, in '1,3-Dipolar Cycloaddition Chemistry', ed. A. Padwa, Wiley-Interscience, New York, 1984, p. 559.
179. A. Michael, *J. Prakt. Chem.*, 1893, **48**, 94.
180. F. R. Benson and W. L. Savel, *Chem. Rev.*, 1950, **46**, 1.
181. J. H. Boyer, in 'Heterocyclic Compounds', ed. R. C. Elderfield, Wiley, New York, vol. 7, 1961.
182. G. L'Abbe, *Chem. Rev.*, 1969, **69**, 345.
183. T. L. Gilchrist and G. E. Gymer, in 'Advances in Heterocyclic Chemistry', ed. A. R. Katritzky and A. J. Boulton, Academic Press, New York, 1974.
184. K. T. Finley, *Chem. Heterocycl. Comp.*, 1980, **39**, 1.
185. P. Scheiner, J. H. Schomaker, S. Deming, W. J. Libbey and G. P. Nowack, *J. Am. Chem. Soc.*, 1965, **87**, 306.
186. R. Huisgen, L. Mobius and G. Szeimies, *Chem. Ber.*, 1965, **98**, 1138; R. Huisgen and G. Szeimes, *Chem. Ber.*, 1965, **98**, 1153.
187. R. Fusco, G. Bianchetti and D. Pocar, *Gazz. Chim. Ital.*, 1961, **91**, 849.
188. R. Sustmann and H. Trill, *Angew. Chem., Int. Ed. Engl.*, 1972, **11**, 838.
189. R. Huisgen, G. Szeimies and L. Mobius, *Chem. Ber.*, 1967, **100**, 2494.
190. G. Szeimies and R. Huisgen, *Chem. Ber.*, 1966, **99**, 491.
191. P. A. Smith, J. M. Clegg and J. H. Hall, *J. Org. Chem.*, 1958, **23**, 524.
192. A. G. Schultz and K. K. Eng, *Tetrahedron Lett.*, 1984, **25**, 1255; A. G. Schultz and S. Puig, *J. Org. Chem.*, 1985, **50**, 915; A. G. Schultz, J. P. Dittami, S. O. Myong and C.-K. Sha, *J. Am. Chem. Soc.*, 1983, **105**, 3273; A. G. Schultz and S. O. Myong, *J. Org. Chem.*, 1983, **48**, 2432; A. G. Schultz and G. W. McMahon, *J. Org. Chem.*, 1984, **49**, 1676; A. G. Schultz, in 'Advances in Cycloaddition Chemistry', ed. D. P. Curran, JAI Press, Greenwich, CT, 1988.
193. C.-K. Sha, K.-S. Chuang and J.-J. Young, *J. Chem. Soc., Chem. Commun.*, 1984, 1552; C. Y. Tsai and C.-K. Sha, *Tetrahedron Lett.*, 1987, **28**, 1419; C.-K. Sha, S.-L. Ouyang, D.-Y. Hsieh, R.-C. Chang and C.-C. Chang, *J. Org. Chem.*, 1986, **51**, 1490.
194. A. G. Schultz and C.-K. Sha, *J. Org. Chem.*, 1980, **45**, 2040.
195. W. Kirmse, 'Carbene Chemistry', Academic Press, New York, 1971; R. A. Moss and M. Jones, 'Carbenes'; Wiley-Interscience, New York, 1975.
196. M. Regitz and H. Heydt, in '1,3-Dipolar Cycloaddition Chemistry', ed. A. Padwa, Wiley-Interscience, New York, 1984, p. 393; M. Regitz, 'Synthesis of Diazoalkanes in the Chemistry of Diazonium and Diazo Groups', Wiley, New York, 1978, vol. 2; M. Regitz, *Synthesis*, 1972, 351.
197. G. W. Cowell and Q. Ledwith, *Q. Rev., Chem. Soc.*, 1970, **24**, 119.
198. D. S. Wulfman, G. Linstrumelle and C. F. Cooper, 'Chemistry of Diazonium and Diazo Groups', Wiley-Interscience, New York, 1978, vol. 2.
199. W. Bihlmaier, R. Huisgen, H.-U. Reissig and S. Voss, *Tetrahedron Lett.*, 1979, 2621.
200. R. Huisgen and J. Geittner, *Heterocycles*, 1978, **11**, 105.
201. R. Huisgen, J. Koszinowski, A. Ohta and R. Schiffer, *Angew. Chem., Int. Ed. Engl.*, 1980, **19**, 202.
202. R. A. Firestone, *Tetrahedron Lett.*, 1980, **21**, 2209.
203. G. W. Jones, K. T. Chang and H. Schechter, *J. Am. Chem. Soc.*, 1979, **101**, 3906.

204. R. Huisgen, R. Sustmann and K. Bunge, *Tetrahedron Lett.*, 1966, 3603; *Chem. Ber.*, 1972, **105**, 1324.
205. Y. Nishizawa, T. Miyashi and T. Mukai, *J. Am. Chem. Soc.*, 1980, **102**, 1176; T. Miyashi, Y. Fujii, Y. Nishizawa and T. Mukai, *J. Am. Chem. Soc.*, 1981, **103**, 725; T. Miyashi, K. Yamakawa, M. Kamata and T. Mukai, *J. Am. Chem. Soc.*, 1983, **105**, 6342; T. Miyashi, Y. Nishizawa, Y. Fujii, K. Yamakawa, M. Kamata, S. Akao and T. Mukai, *J. Am. Chem. Soc.*, 1986, **108**, 1617.
206. A. Padwa and H. Ku, *Tetrahedron Lett.*, 1980, **21**, 1009; A. Padwa and R. Rodriguez, *Tetrahedron Lett.*, 1981, **22**, 187; A. Padwa, A. Rodriguez, M. Tohidi and T. Fukanaga, *J. Am. Chem. Soc.*, 1983, **105**, 933.
207. W. Lwowski, 'Nitrenes', Wiley-Interscience, New York, 1970.
208. S. D. Burke and P. A. Grieco, *Org. React.*, 1979, **26**, 361.
209. W. Reid and R. Dietrich, *Justus Liebigs Ann. Chem.*, 1963, **666**, 113.
210. J. N. Crabb and R. C. Storr, in '1,3-Dipolar Cycloaddition Chemistry', ed. A. Padwa, Wiley-Interscience, New York, 1984, p. 543.
211. H. Durr and H. Schmitz, *Chem. Ber.*, 1978, **111**, 2258.
212. W. L. Magee and H. Shechter, *J. Am. Chem. Soc.*, 1977, **99**, 633; W. L. Magee, C. B. Rao, J. Glinka, H. Jui, T. J. Amick, D. Fiscus, S. Kakodkar, M. Nair and H. Shechter, *J. Org. Chem.*, 1987, **52**, 5538.
213. A. Padwa, A. D. Woolhouse and J. J. Blount, *J. Org. Chem.*, 1983, **48**, 1069; A. D. Woolhouse, T. Caruso and A. Padwa, *Tetrahedron Lett.*, 1982, **23**, 2167.

4.10
Intramolecular 1,3-Dipolar Cycloadditions

PETER A. WADE

Drexel University, Philadelphia, PA, USA

4.10.1 INTRODUCTION

The most general approach to synthesis of five-membered heterocyclic compounds involves cycloaddition of a 1,3-dipole to an appropriate unsaturated substrate, the dipolarophile. Intermolecular cycloadditions result in the formation of one new ring only. When the 1,3-dipole and the substrate are part of the same molecule, cycloaddition is intramolecular and leads to a new bicyclic system. Thus, intramolecular cycloadditions are amenable to the construction of inherently more complex products than intermolecular cycloadditions. Markedly different regioselectivity, controlled by the geometric constraints of bringing the 1,3-dipole into correct internal alignment for reaction with the dipolarophile, is often observed in an intramolecular cycloaddition. Sometimes these geometric constraints overwhelm the normal regiochemical preference dictated by electronic factors. The greater steric constraint inherent to intramolecular cycloaddition often affords higher diastereofacial discrimination: accordingly, these reactions can exhibit very high stereoselectivity. Periselectivity can also be different for an intramolecular cycloaddition, leading to formation of unusual bicyclic systems, containing other than five-membered heterocycles.

With all of these advantages, intramolecular cycloaddition is certainly a powerful synthetic tool. However, prior assembly of the 1,3-dipole and the dipolarophile within the same molecule is required. Careful consideration of transition state geometries is also essential: an accurate molecular model should always be constructed, preferably prior to carrying out the intramolecular cycloaddition. It is even better to do a calculational treatment prior to attempting the reaction. The recent development of routine molecular mechanics programs portends a much more accurate approach to predicting the outcome of an intramolecular cycloaddition.

There is an excellent recent review on intramolecular dipolar cycloadditions.[1] Reviews on mechanistic aspects of intermolecular 1,3-dipolar cycloadditions abound[2] and a treatise on intramolecular Diels–Alder cycloadditions is available.[3]

4.10.2 NITRONE CYCLIZATIONS

4.10.2.1 General

The earliest and to date most extensively studied class of intramolecular cycloadditions involves unsaturated nitrones.[4] These are most readily available from condensation of an unsaturated aldehyde with a hydroxylamine or an unsaturated hydroxylamine with an aldehyde. Another approach is simply to oxidize an unsaturated hydroxylamine. Nitronic esters are nitrones containing an alkoxy substituent attached to the N-atom: they can be prepared from nitro compounds. Frequently an unsaturated nitrone can be isolated and purified, although much work has been done with the nitrone generated *in situ*; eventual cyclization can provide three new contiguous chiral centers, often with only one diastereomer actually formed.

The outcome of cyclization depends on the intervening chain (bridge) separating the nitrone from the dipolarophile and whether this bridge is attached at C or N of the nitrone. Also important are the nonbonding interactions and entropic differences for competing cycloaddition pathways. Incorporation of the nitrone and/or dipolarophile in an existing ring system offers possibilities for the construction of a wide variety of polycyclics.

4.10.2.2 Open-chain Alkenylnitrones

4.10.2.2.1 C-(5-Alkenyl)nitrones

Cyclization of *C*-(5-alkenyl)nitrones typically produces an eight-membered 5,5-fused product and no bridged product as originally demonstrated by LeBel *et al.*[4] As a recent example, cyclization of the nitrone derived from aldehyde (**1**) led to isomeric fused isoxazolidines (**2**) and (**3**) in a 96:4 *anti:syn* ratio: no bridged isomers were detected (Scheme 1).[5,6] Similarly, aldehyde (**4**) afforded the fused isoxazolidines (**5**) and (**6**) in a 92:8 *anti:syn* ratio. These results illustrate three other important features of intramolecular nitrone cycloaddition. First, the stereochemistry of the alkenyl group is retained. Thus, (*Z*)-(**1**) led to *cis* products, whereas (*E*)-(**4**) led to *trans* products. Second, only *cis*-fused cyclization products were observed, presumably because a much higher energy transition state would be required to obtain *trans*-fused five-membered rings. Third, there was a high degree of asymmetric induction: the allylic chiral center, here outside of the incipient bicyclic system, controlled facial selectivity.

Intramolecular cycloaddition of nitrone (**7**) led to the diastereomeric fused isoxazolidines (**8a**) and (**8b**) in an 82:18 ratio (Scheme 2).[7a] The authors suggested cyclization through competing conformational isomers A and B as a rationale of the cyclization outcome. Notably, absence of the *trans*-alkenyl methyl

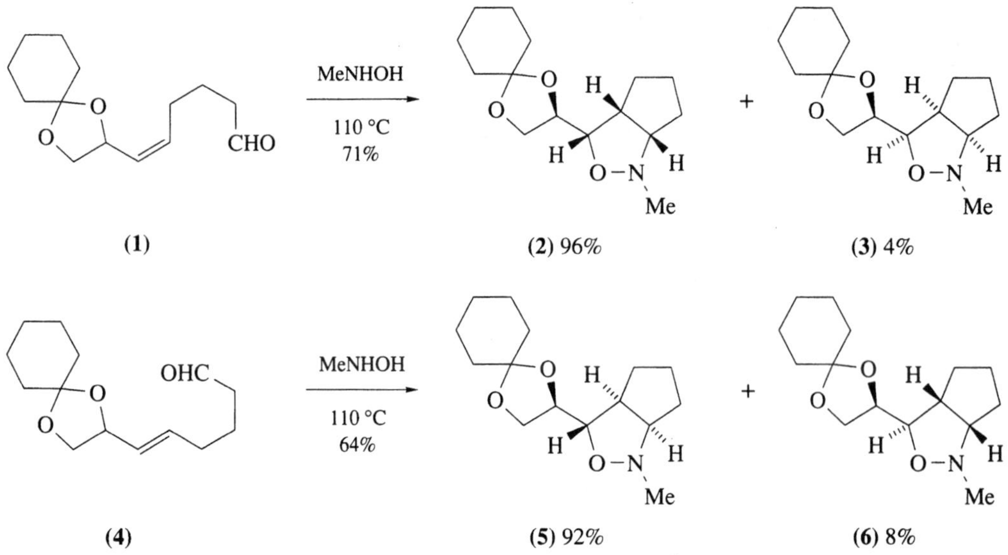

MeNHOH 110 °C 71%

(1) (2) 96% (3) 4%

MeNHOH 110 °C 64%

(4) (5) 92% (6) 8%

Scheme 1

group led to lower diastereoselection. The major product, isoxazolidine (**8a**) was transformed in three steps to acosamine. Reductive cleavage of the N—O bond is a very common transformation of isoxazolidines: γ-amino alcohols are typically obtained. Here hydrogenolysis over Pd—C was employed and concomitant debenzylation was accomplished. Nitrones are easily derived from sugars. For example, lactol (**9**) was prepared from D-ribose; reaction of (**9**) with *N*-methylhydroxylamine afforded the nitrone which cyclized *in situ* to the 5,5-fused bicyclic isoxazolidine (**10**).[7b]

4.10.2.2.2 C-(6-Alkenyl)nitrones

Cyclization of *C*-(6-alkenyl)nitrones leads to competitive formation of the nine-membered 6,5-fused ring system and bridged bicyclo[4.2.1]nonyl system. For example, NMR analysis of nitrone-derived products obtained from ketone (**11**) indicated a 62:38 mixture of fused isoxazolidine (**12a**) and bridged isoxazolidine (**12b**), respectively (Scheme 3).[8] The preference for the 6,5-fused ring system is typical: it is entropic in nature. Frontier MO theory suggests a more developed C—C bond than C—O bond in the typical nitrone cycloaddition transition state. Formation of the fused product occurs through a six-membered carbocyclic transition state, whereas the bridged product is formed *via* a seven-membered carbocyclic transition state. This factor is usually, but not always, more important than opposing factors. Thus, nitrones (**13a**) and (**13b**) gave the 6,5-fused isoxazolidines (**14a**) and (**14b**) as the only detectable cyclization products.[9] Conversely, nitrones (**15a**) and (**15b**) gave the bridged 7-aza-8-oxabicyclo[4.2.1]nonanes (**16a**) and (**16b**) as the only detectable products. The major reason for the abrupt change in regiochemistry is presumably steric: the phenyl substituent in (**15a–b**) would interact unfavorably with the larger nitrone terminus, the substituted C- rather than the unsubstituted O-atom. Another contributory factor (with **15a** only) should be the electronic preference: nitrones prefer to cycloadd to alkenes affording five-rather than four-substituted isoxazolidines.

Whether *cis* or *trans* fusion is observed in nitrone cycloadditions can depend on reaction conditions as first determined by LeBel *et al.*[8] At lower temperature where cycloaddition is irreversible, kinetic control prevails and this usually favors *cis* fusion. However, at higher temperature equilibration can occur through retrocycloaddition and the more stable product will predominate (*i.e.* thermodynamic control). The nitrone may also undergo (*E*)/(*Z*) isomerization, particularly at elevated temperature, and this complicates the analysis: a different kinetically favored ratio might prevail. A recent example of temperature dependence involves formation of isoxazolidines (**18**) and (**19**) from aldehyde (**17a**; Scheme 4). At 90 °C *cis*-fused (**18**) and *trans*-fused (**19**) were formed in 74% and 9% yield, respectively. At 140 °C, however, (**18**) and (**19**) were formed in 31% and 34% yield.

The isoxazolidines derived from aldehydes (**17a–b**) were converted to eudesmane sesquiterpenes. Thus, quaternization of isoxazolidine (**18**) was followed by ring expansion *via* a Stevens rearrangement.

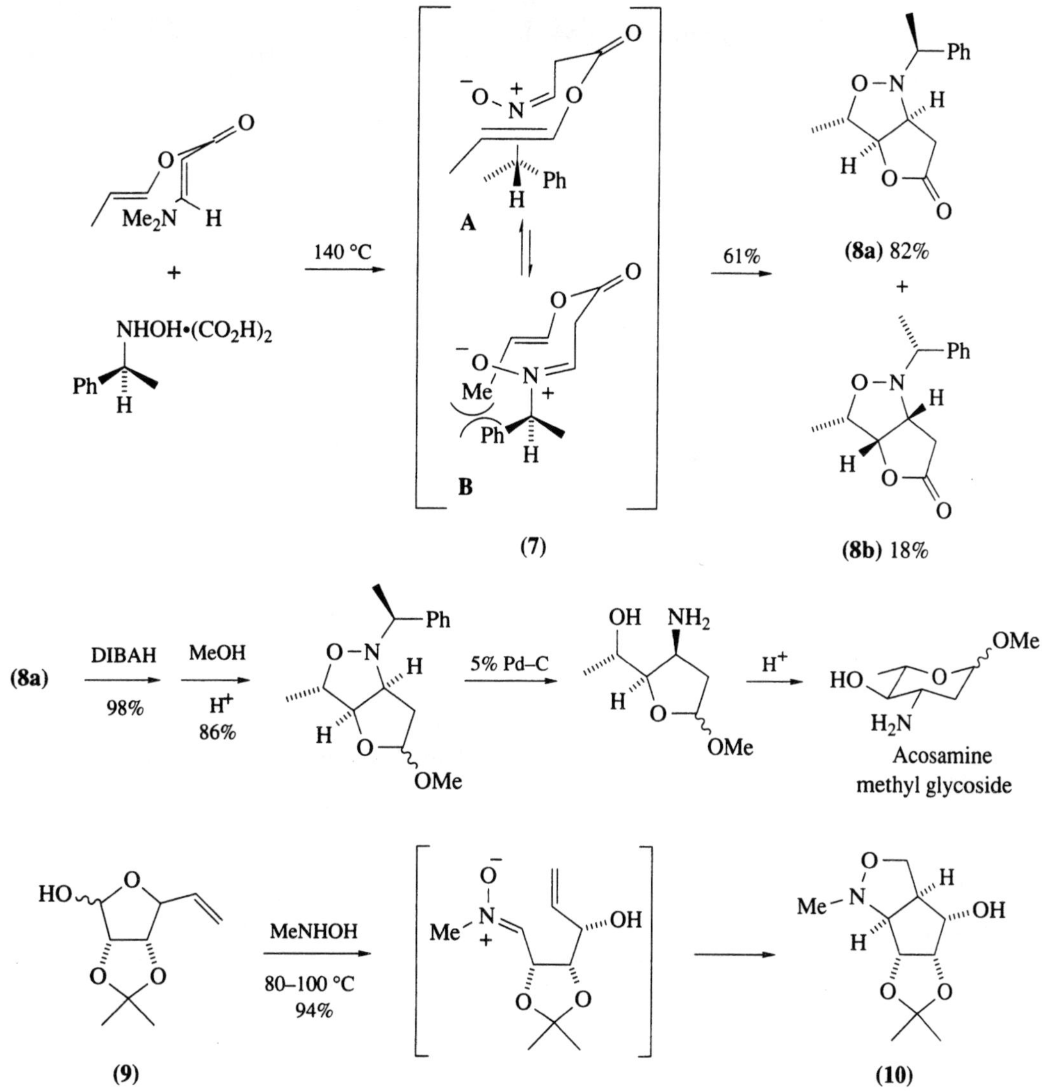

Scheme 2

Quaternization of the resulting tetrahydro-1,3-oxazine (**20**) and subsequent hydrogenolysis with lithium in liquid ammonia provided the nitrogen-free (±)-5-epi-α-eudesmol.

4.10.2.2.3 N-(Alkenyl)nitrones

N-(Alkenyl)nitrones must always cyclize to give a new bridged bicyclic system. For example, nitrones (**21a–b**) gave the 1-aza-2-oxabicyclo[2.2.1]heptanyl derivatives (**22a–b**) in 76% and 85% yield, respectively (Scheme 5).[11] Regioselectivity depends heavily on the length of the intervening bridge. Thus, nitrones (**23a–b**) gave regioisomeric 1-aza-8-oxabicyclo[3.2.1]octanyl derivatives (**24a–b**) in 95% and 69% yield, respectively. In all of these cases, only one isomer was detected. Formation of (**22a–b**) occurred *via* a six-membered cyclic transition state avoiding angle strain inherent to formation of the other regioisomer. Formation of (**24a–b**) occurred *via* six-membered transition state A, entropically favored over seven-membered transition state B leading to the regioisomer. These rationalizations are predicated on the assumption that C—C bond formation is further developed in the transition state than C—O bond formation, in agreement with frontier molecular orbital theory. Accordingly, nitrone (**25**) with a longer bridge afforded a 75:25 mixture of isomers through competing seven- and eight-membered transition states, respectively.

Relatively low stereoselectivity in the cyclization of *N*-(alkenyl)nitrones (**26**) and (**27**) has been reported (Scheme 6).[12a] Cyclization of (**26**) led to an 80:11:9 ratio of the *exo,exo-*, *exo,endo-* and *endo,exo-*

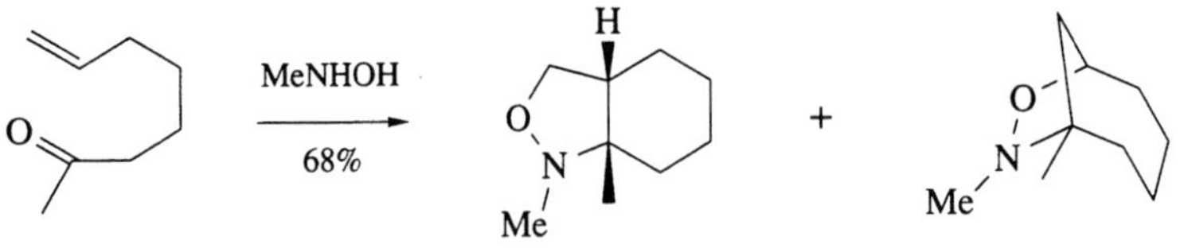

MeNHOH, 68%

(11) → **(12a)** 62% + **(12b)** 38%

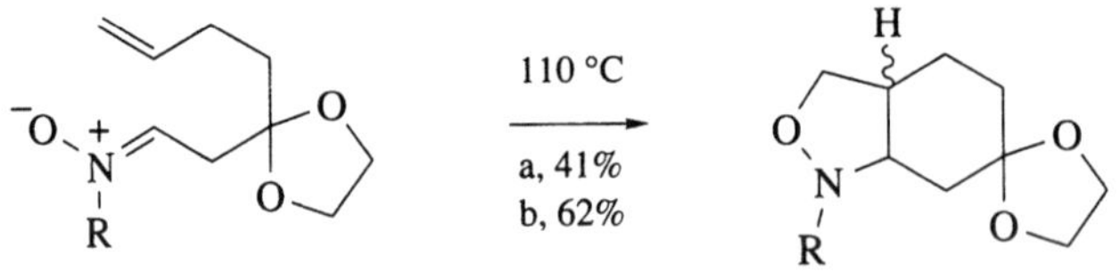

110 °C; a, 41%; b, 62%

(13a) R = Me
(13b) R = Bn

(14a) R = Me
(14b) R = Bn

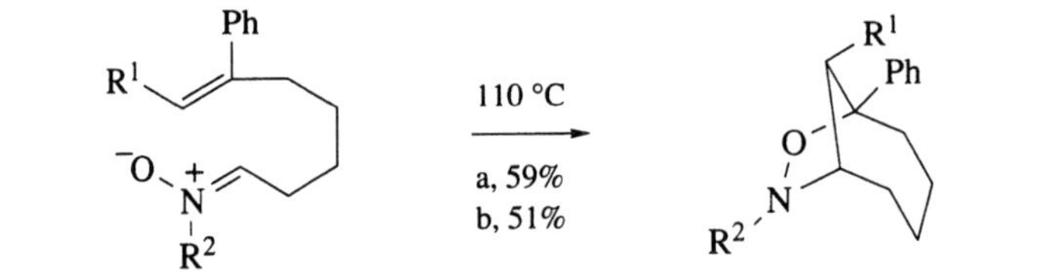

110 °C; a, 59%; b, 51%

(15a) R^1 = H; R^2 = Bn
(15b) R^1 = CO_2Et; R^2 = Me

(16a) R^1 = H; R^2 = Bn
(16b) R^1 = CO_2Et; R^2 = Me

Scheme 3

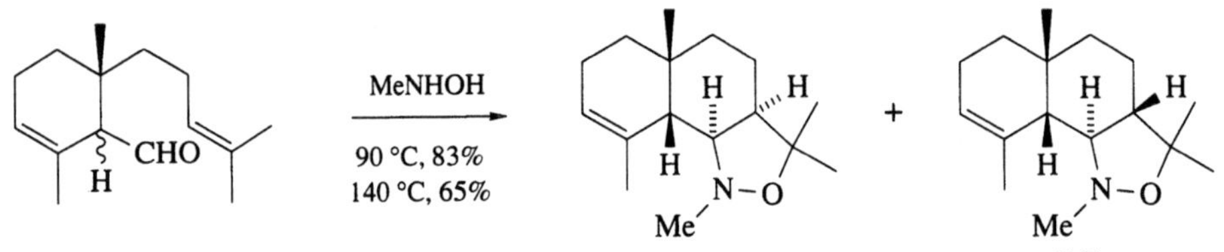

MeNHOH; 90 °C, 83%; 140 °C, 65%

(17a) ◄H
(17b) ''''''H

	(18)		**(19)**
	89%	90 °C	11%
	48%	140 °C	52%

(18) 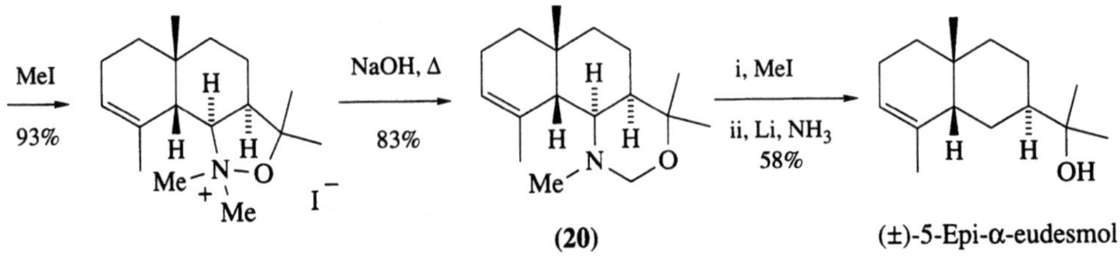 → MeI, 93% → NaOH, Δ, 83% → **(20)** → i, MeI; ii, Li, NH_3, 58% → (±)-5-Epi-α-eudesmol

Scheme 4

isoxazolidines. A very similar ratio was obtained for cyclization of nitrone (**27**). These results do not arise from product interconversion: the *exo,exo* product gave none of the isomers on heating. Perhaps interconversion of (**26**) and (**27**) occurred while heating *via* an aza-Cope rearrangement. The modest stereoselectivity could also have arisen simply from nitrone (*Z*)/(*E*) isomerization and competing cycloaddition, although in this case the nearly identical results are coincidental.

A synthesis of (±)-lasubine II was carried out by an analogous nitrone cyclization.[12a] The *exo,exo*, *exo,endo* and *endo,exo* ratio was 84:10:6 and the *exo,exo* product (**28**) was isolated pure in 60% yield.

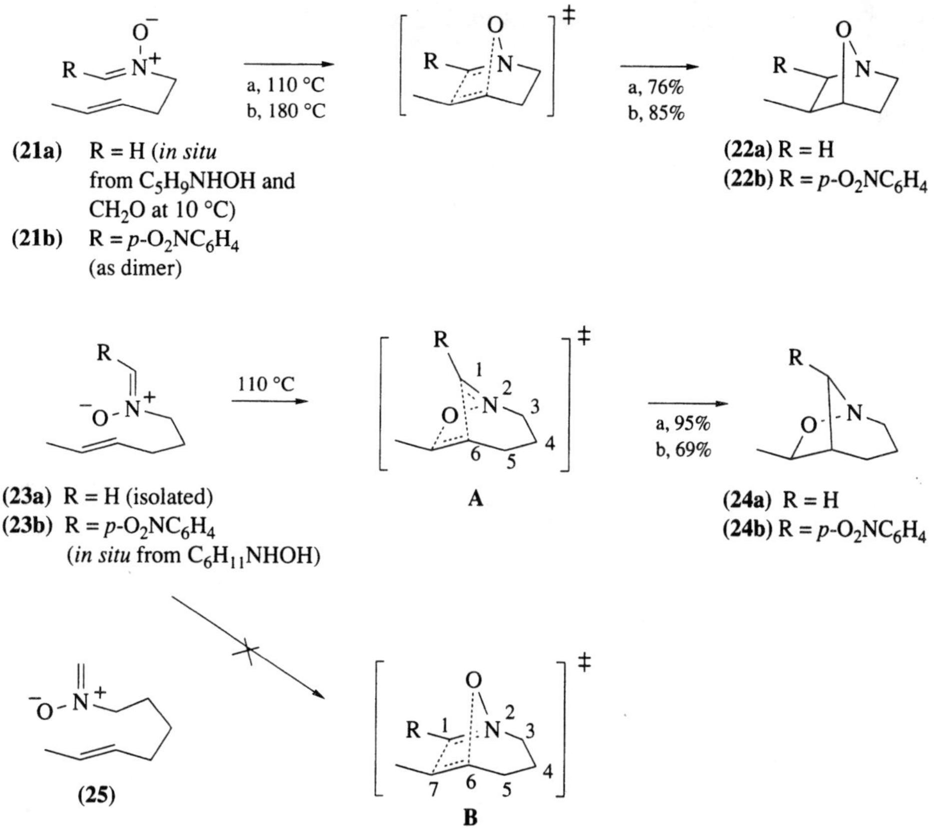

a, 110 °C
b, 180 °C

a, 76%
b, 85%

(21a) R = H (*in situ* from C_5H_9NHOH and CH_2O at 10 °C)
(21b) R = *p*-$O_2NC_6H_4$ (as dimer)

(22a) R = H
(22b) R = *p*-$O_2NC_6H_4$

110 °C

a, 95%
b, 69%

(23a) R = H (isolated)
(23b) R = *p*-$O_2NC_6H_4$ (*in situ* from $C_6H_{11}NHOH$)

A

(24a) R = H
(24b) R = *p*-$O_2NC_6H_4$

(25)

B

Scheme 5

Reduction of (**28**) with zinc–acetic acid provided the all-*cis*-piperidinol which was converted in several steps to (±)-lasubine II.

A very recent synthesis of (±)-pumiliotoxin C employed the nitrone cyclization product (**29**), obtained in 74% overall yield from the nitrone precursors.[12b] Reduction of (**29**) with zinc–acetic acid provided a trisubstituted piperidine which was converted in several steps to (±)-pumiliotoxin C.

4.10.2.3 Cyclic Alkenylnitrones

Cyclic substrates incorporating alkenyl and nitrone functional groups also undergo intramolecular cycloaddition. Examples of three simple types are known: the alkene may be part of a ring, the bridge only may include a ring or the nitrone may be part of a ring. Cyclizations involving complex combinations where more than one of the three elements are part of a ring are also numerous, including a case where all three elements are in rings. Typically, nonbonded interactions tend to be more severe in these cyclic substrates, often leading to greater selection of competing isomers and occasionally to unexpected preferences.

4.10.2.3.1 C-(Cycloalkenyl)nitrones

Cyclization of *C*-(cycloalkenyl)nitrones has been extensively investigated.[13–15] For example, the cyclohexenylnitrone (**30**) (an equilibrated 2:1 (*E*)/(*Z*) isomer mixture) afforded (**31**) in 56% yield (Scheme 7).[13] Presumably the (*Z*)-nitrone cyclized exclusively based on the strained transition state required for cyclization of the (*E*)-nitrone. None of the isomeric bridged [4.2.1] product (**32**) was formed: this is attributed to unfavorable nonbonded interaction between the carbonyl group and a ring methylene group in the required transition state.

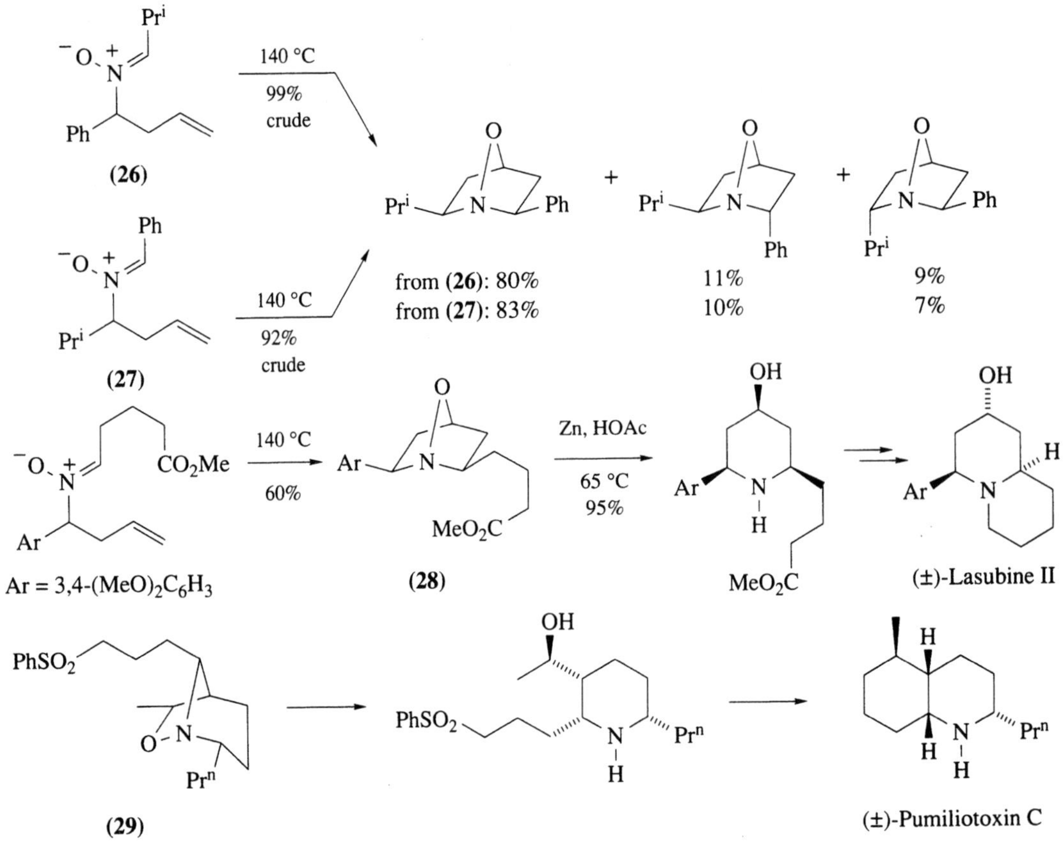

(26) 140 °C, 99% crude

(27) 140 °C, 92% crude

from (26): 80% / 11% / 9%

from (27): 83% / 10% / 7%

$Ar = 3,4\text{-}(MeO)_2C_6H_3$

140 °C, 60% → (28) → Zn, HOAc, 65 °C, 95% → (±)-Lasubine II

(29) → (±)-Pumiliotoxin C

Scheme 6

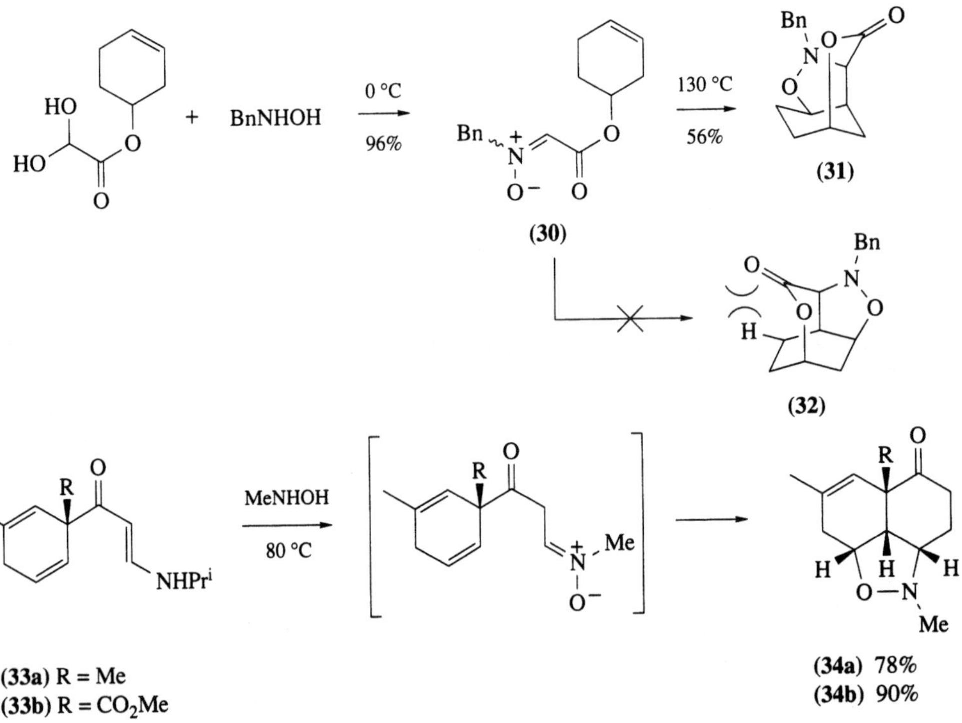

+ BnNHOH → 0 °C, 96% → (30) → 130 °C, 56% → (31)

(30) ✕→ (32)

(33a) R = Me

(33b) $R = CO_2Me$

MeNHOH, 80 °C →

(34a) 78%

(34b) 90%

Scheme 7

The nitrones derived from dienes (**33a–b**) cyclize exclusively at the disubstituted double bond to give the isoxazolidines (**34a–b**) in 78% and 90% yield, respectively.[14] Chemoselection here arises from the marked rate difference in cycloaddition to di- *vs.* tri-substituted alkenyl double bonds.

Nitrones have been reported to undergo intramolecular cycloaddition to quinones[15a] and to thiophenes.[15b]

4.10.2.3.2 N-(Cycloalkenyl)nitrones

Intramolecular cycloaddition of *N*-(cycloalkenyl)nitrones has also been extensively investigated. Thus, stepwise oxidation of the amine (**35**) and cyclization of the ensuing nitrone provided isoxazolidine (**36**) in 64% overall yield as the only observed regioisomer (Scheme 8).[16] Isoxazolidine (**36**) was then desulfonated with buffered sodium amalgam and further reduced with zinc–acetic acid to provide the corresponding γ-amino alcohol; subsequent dehydration provided (–)-hobartine.

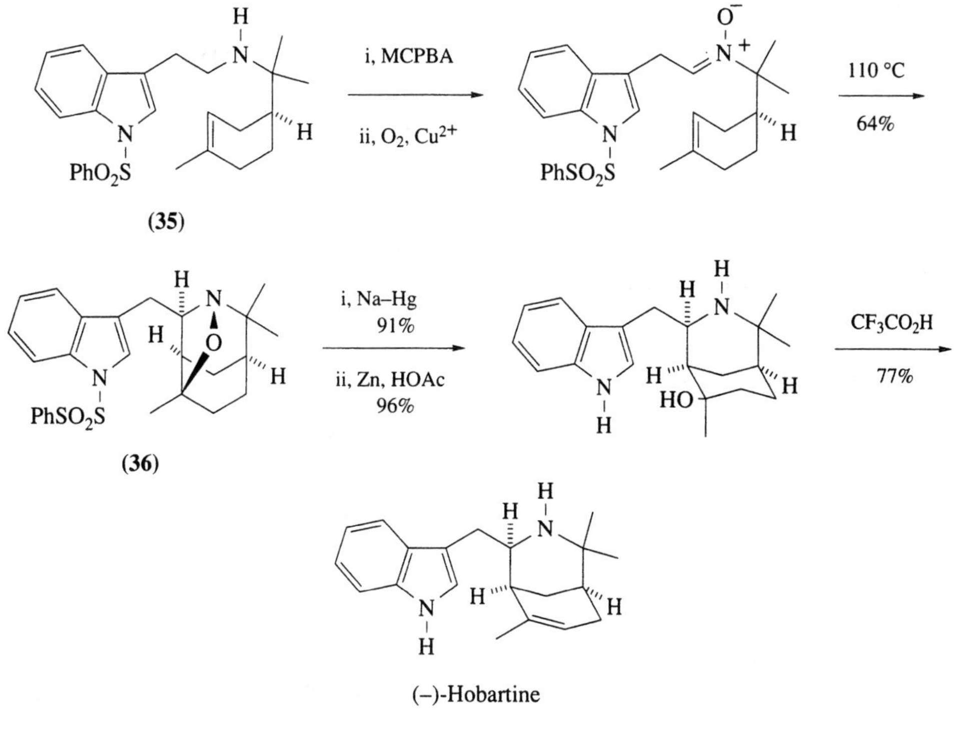

(**35**) i, MCPBA; ii, O_2, Cu^{2+} → 110 °C, 64% → (**36**) i, Na–Hg, 91%; ii, Zn, HOAc, 96% → CF_3CO_2H, 77% → (–)-Hobartine

Scheme 8

The *N*-(bicycloalkenyl)nitrone (**37**) afforded the tetracyclic isoxazolidine (**38**) in 67% yield: none of the regioisomeric isoxazolidine was observed (Scheme 9).[17] The authors attributed this to non-bonded C—H interactions as shown, but the observed isoxazolidine was also entropically favored, forming *via* a six-membered, as opposed to a seven-membered, carbocyclic transition state. The *N*-(bicycloalkenyl)nitrone formed from (**39**) and furfural cyclized in 45% yield predominantly to the tetracyclic product (**40b**), but some (**41b**) was also produced (95:5 ratio).[18] Reaction of (**39**) and formaldehyde gave a mixture of (**40a**) and (**41a**) (62:38 ratio). The authors attribute the somewhat higher regioselectivity for (**40b**) in part to non-bonded interaction of the 2-furyl substituted methylene with the C-8 endocyclic C—H bond.

4.10.2.3.3 Aryl-bridged alkenylnitrones

Many intramolecular nitrone cycloadditions have been carried out on substrates containing a cyclic intervening bridge. Perhaps the simplest possibility here is an *o*-disubstituted aromatic substrate, as, for example, the nitrone derived from aldehyde (**42**), which cyclizes to the 5,6-fused isoxazolidine (**43**; Scheme 10).[19] There are many additional possibilities: the nitrones derived from biphenyl aldehyde

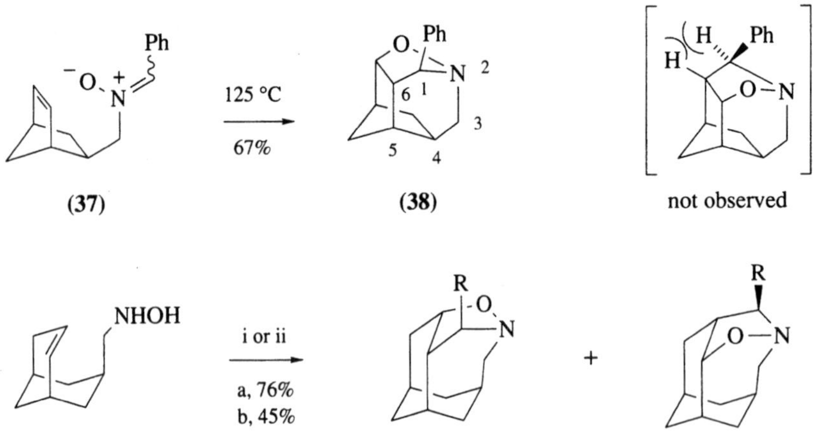

125 °C, 67%

(37) **(38)** not observed

i or ii; a, 76%; b, 45%

(39) **(40a)** 62% R = H **(41a)** 38%

(40b) 95% R = 2-furyl **(41b)** 5%

i, $(CH_2O)_n$, 100 °C; ii, furfural, 140 °C

Scheme 9

(**44**)[20] and indolyl nitrile (**45**)[21] cyclize uneventfully to fused isoxazolidines. Isoxazolidine (**46**) has been converted to (±)-chanoclavene I in a multi-step synthesis.

4.10.2.3.4 Alicylic-bridged alkenylnitrones

The isoxazolidines (**47**) and (**48**) were obtained by cyclization of the corresponding ketone-derived exocyclic nitrones (Scheme 11).[22] Such reactions have been used for natural product sysnthesis, as in the conversion of isoxazolidine (**49**) to (±)-hirsutene. Methylation and catalytic hydrogenolysis provided a γ-dimethylamino alcohol, which underwent Cope elimination to provide an alkenyl alcohol in a key step. A related synthesis of (±)-7,12-sechoishwaran-12-ol is also reported.

4.10.2.3.5 Cyclic nitrones

Cyclic nitrones also undergo intramolecular cycloaddition to an alkenyl group with very high regioselectivity. For example, (±)-cocaine was synthesized *via* cyclization of the cyclic nitrone (**50**), obtained from reduction of a nitroacetal or from thermolysis of an isoxazolidine, to isoxazolidine (**51**; Scheme 12).[23] Sequential methylation, reductive N—O cleavage and benzoylation of (**51**) provided (±)-cocaine.

Cyclic nitrone (**52**) has a cyclic bridge and cyclic enol ether dipolarophile: thus all three elements are present in rings (Scheme 13).[24a] Models show that direct cyclization of (**52**) is not possible because of the relative positions of the dipole and dipolarophile. However, a pentacyclic cage isoxazolidine was formed by heating (**52**); presumably epimerization at the phenyl-bearing bridgehead carbon preceded cycloaddition.

4.10.2.4 Oxime Cyclization Reactions

Unsaturated oximes can be cyclized simply by heating.[24b] It has been suggested that nitrone intermediates are involved in this process. Thus, oxime (**53**) on heating at 180 °C cyclized to a pyrrolizidine in 60% yield: presumably tautomerization to nitrone (**54**) first occurred followed by [3 + 2] cycloaddition (Scheme 14). Likewise, quinolizidine (**55**) was obtained in 69% yield from the corresponding oxime.

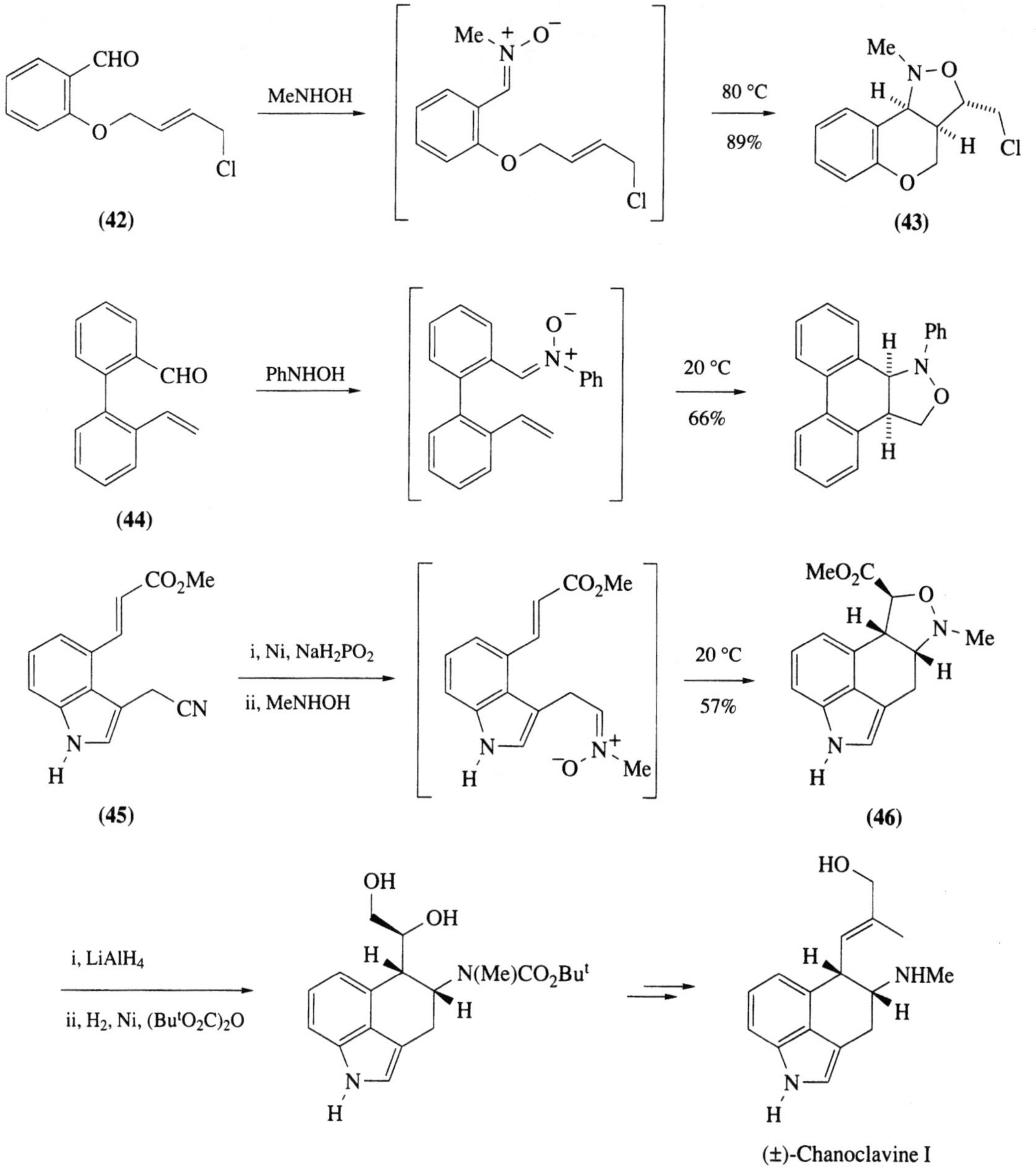

Scheme 10

4.10.2.5 Tandem Michael–Nitrone Cyclization Reactions

Nitrones can be generated by Michael reaction of oximes with appropriate conjugated substrates.[25] If a generated nitrone has a built-in dipolarophile, cyclization can ensue (Scheme 15). There are three synthetic variations on this theme.[25a,b] First, the oxime may contain the dipolarophile as in (**56**). Reaction of (**56**) with phenyl vinyl sulfone provided a quantitative yield of the tandem product as one stereoisomer. Alternatively, the dipolarophile can reside in the Michael substrate as in (**57**). Reaction of (**57**) with cyclohexanone oxime produced two isoxazolidines from competitive cyclization of the intermediate nitrone through six- and seven-membered carbocyclic transition states. It is also possible to carry out an intramolecular Michael addition followed by an intramolecular nitrone cyclization as in thermolysis of (**58**) to produce a tricyclic isoxazolidine. Very recently several examples of a tandem Diels–Alder, Michael addition, nitrone cyclization sequence have been reported.[25c]

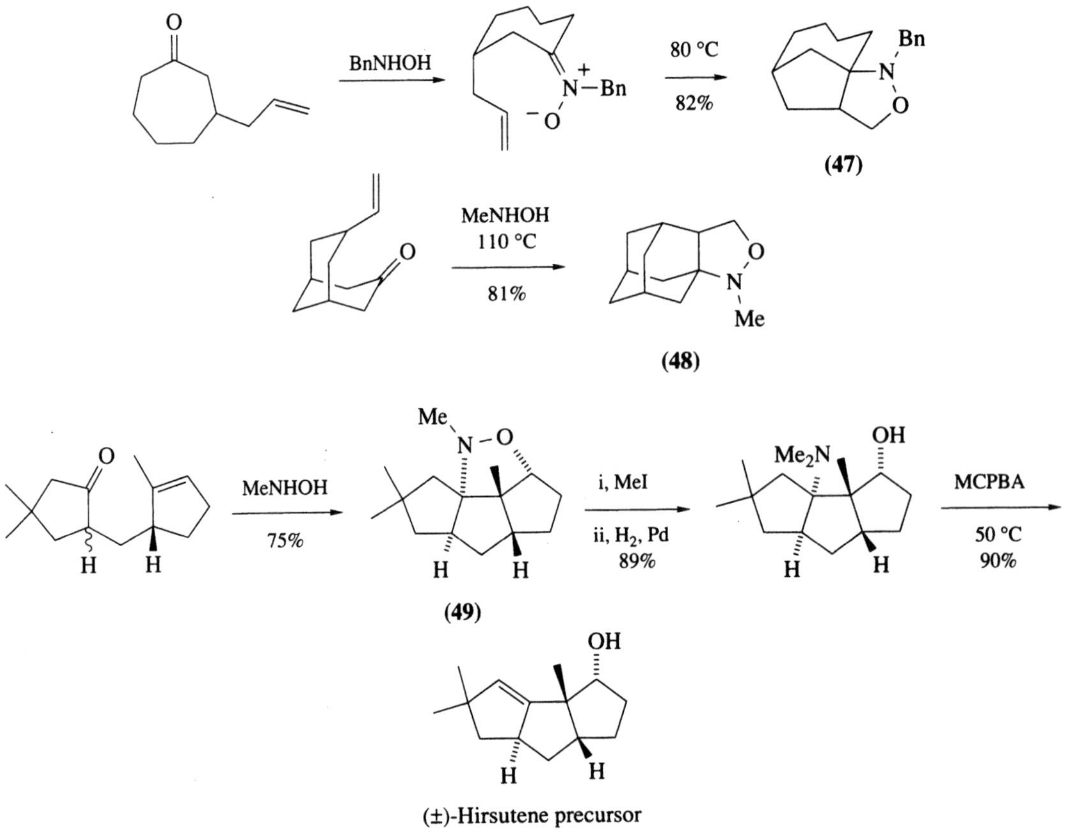

BnNHOH

80 °C 82%

(47)

MeNHOH 110 °C 81%

(48)

MeNHOH 75%

(49)

i, MeI ii, H_2, Pd 89%

MCPBA 50 °C 90%

(±)-Hirsutene precursor

Scheme 11

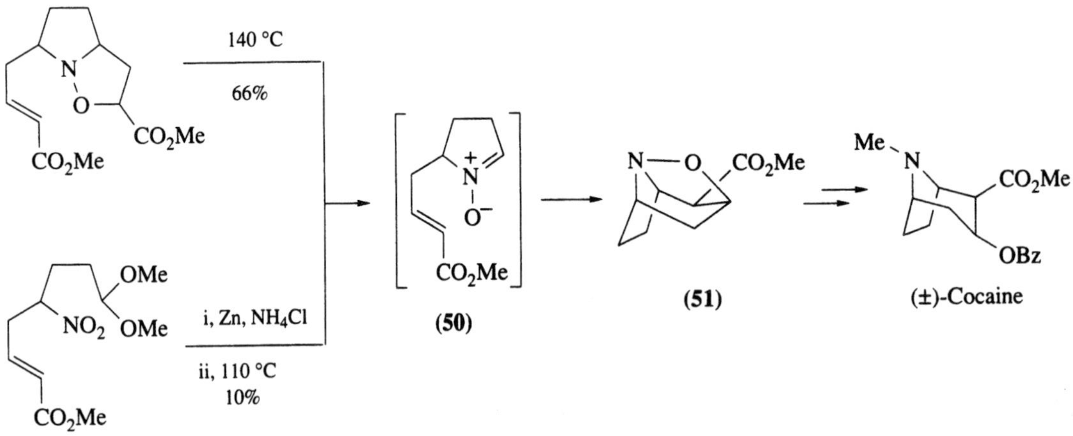

140 °C 66%

i, Zn, NH_4Cl ii, 110 °C 10%

(50)

(51)

(±)-Cocaine

Scheme 12

4.10.2.6 Tandem Diels–Alder–Nitronic Ester Cyclization Reactions

Allylic nitro compounds containing a suitable dipolarophile undergo Diels–Alder cycloaddition to alkenes in the presence of tin(IV) chloride affording cyclic nitronic esters (Scheme 16).[26a] Nitronic ester (**59**) could not be isolated but spontaneously cyclized to the 5,5-fused cyclic product (**60**), isolated in 68% yield. The nitronic esters (**61a**) and (**61b**) were isolated from the Diels–Alder reaction and could be separated. Heating (**61a**) in refluxing benzene afforded the 5,6-fused dipolar cyclization product (**62a**) in 93% (68% overall) yield; (**61b**) likewise afforded (**62b**) in 62% (11% overall) yield. Either (**62a**) or (**62b**) could be converted to the tricyclic lactam (**63**) by catalytic hydrogenolysis followed by lactamiza-

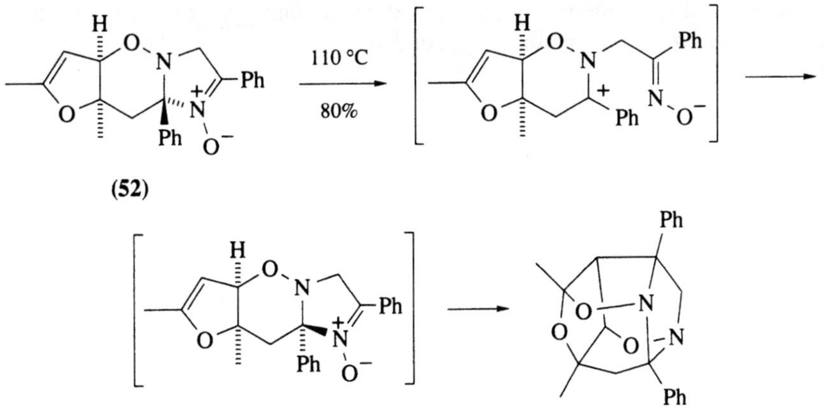
110 °C
80%
Ph
(52)

Scheme 13

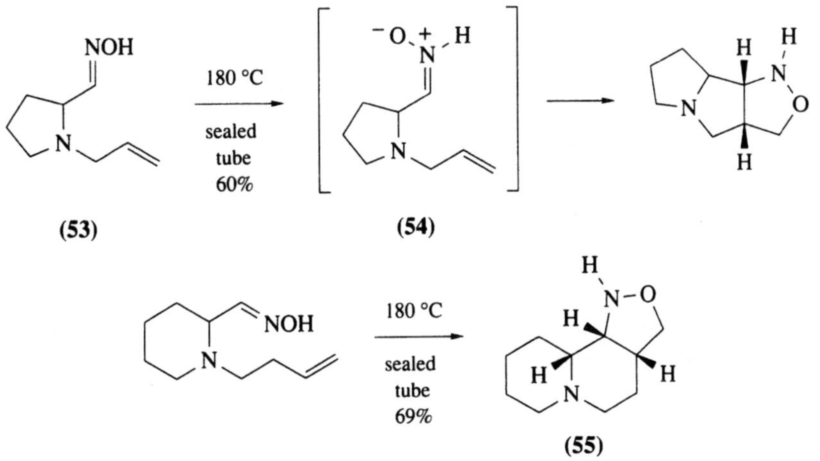
NOH
180 °C
sealed
tube
60%
(53)
(54)
180 °C
sealed
tube
69%
(55)

Scheme 14

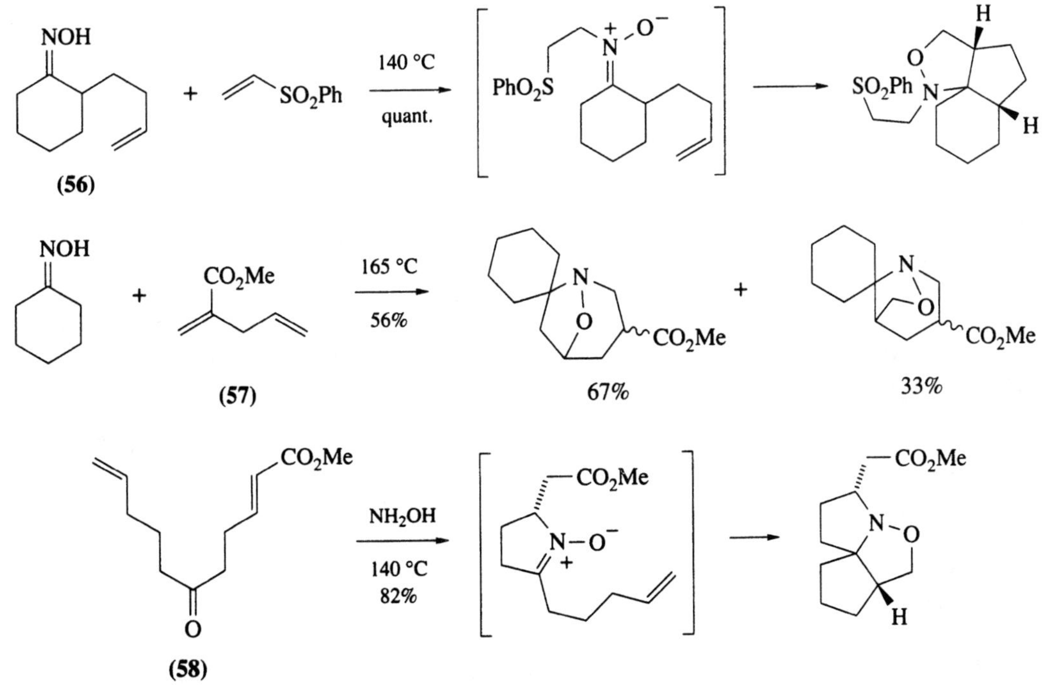
NOH
SO2Ph
140 °C
quant.
PhO2S
(56)
165 °C
56%
CO2Me
(57)
67%
33%
NH2OH
140 °C
82%
(58)

Scheme 15

tion. Nitronic ester (**64**) did not undergo intramolecular 1,3-dipolar cycloaddition, presumably because formation of the 5,7-fused product would be less favorable.

Scheme 16

4.10.2.7 Alkynylnitrones and Allenylnitrones

In constrast with intermolecular nitrone cycloadditions to alkynes and allenes, very little work has been done on the corresponding intramolecular cycloadditions. The bicyclic isoxazolidines (**65a–b**) were reported as products from reaction of an alkynone with methylhydroxylamine in ethanol.[26b] Presumably the initial strained bridgehead C—C double bond of the Δ^4-isoxazoline added ethanol under the reaction conditions. Cyclization of an allenyl ketone with methylhydroxylamine in ethanol solution also led to isoxazolidines (**65a–b**) as the major products and isoxazolidine (**66**) as a minor product.[26b] Thus, preferential cyclization to the internal C—C double bond of the allene occurred followed by addition of ethanol to the exocyclic C—C double bond of the methyleneisoxazolidine intermediate.

4.10.3 NITRILE OXIDE CYCLIZATIONS

4.10.3.1 General

Intermolecular nitrile oxide cycloadditions have been known for a very long time.[27] However, it was not until the mid-1970s that intramolecular nitrile oxide cycloaddition (INOC) reactions were studied.[28]

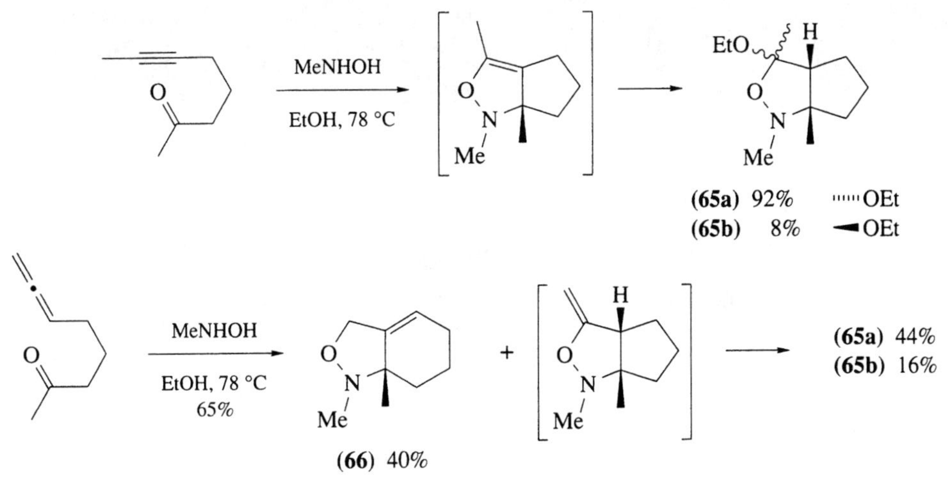

MeNHOH, EtOH, 78 °C

(**65a**) 92% ·····OEt
(**65b**) 8% ◂OEt

MeNHOH, EtOH, 78 °C, 65%

(**66**) 40%

(**65a**) 44%
(**65b**) 16%

Scheme 17

Successful INOC reactions have typically been run on nitrile oxides generated *in situ*. It is necessary to avoid strong halogenating reagents to generate the nitrile oxide because of the presence of the dipolarophile. Consequently, an INOC procedure typically utilizes an isocyanate to generate the nitrile oxide from a nitro compound[29] or sodium hypochlorite to generate the nitrile oxide from an oxime.[30]

The entropically enhanced cycloaddition rate of an intramolecular cycloaddition permits alkenyl nitrile oxides to cyclize in the presence of free amino groups, in contrast to intermolecular reactions. Since nitrile oxides have a wide spectrum of possible intermolecular reactions, they also react well in macrocyclic cyclizations.

INOC reactions are occasionally synthetic alternatives to nitrone cyclizations. The isoxazolines produced can substitute for isoxazolidines in a number of further transformations. Thus, isoxazolines, like isoxazolidines, can be converted to γ-amino alcohols.[31] Isoxazolines can be converted to isoxazolidines and *vice versa*.[32]

Nitrile oxides are linear as opposed to nitrones which are planar with an approximately 120° C—N—O bond angle. Consequently, the cycloaddition transition states are rather different for these two dipoles. The nitrone usually, although not always, offers more promise of stereoselective cyclization; the nitrile oxide can offer more promise of regiospecific cyclization, especially in forming 5,6-, 5,7- and larger fused bicyclic systems.

4.10.3.2 Open-chain Alkenyl Nitrile Oxides

4.10.3.2.1 5-Alkenyl nitrile oxides

Cyclization of nitrile oxides with a three-atom intervening chain to the alkene always leads to 5,5-fused bicylic isoxazolines possessing a bridgehead C—N double bond. Thus, the alkenyl oximes (**67**) and (**68**) cyclized exclusively to fused bicyclic isoxazoline products (Scheme 18).[33] These reactions illustrate other important features of nitrile oxide cyclizations. Like nitrone cyclizations, the C—C double bond stereochemistry is retained. Thus, (*Z*)-(**67**) gave only C(4)—C(5) *cis* products, while (*E*)- gave only C(4)—C(5) *trans* products. Yields in these reactions are moderate, and this is attributed to some difficulty in the dipole reaching the dipolarophile. The stereoselectivity of these reactions was moderate, determined by the chiral center which remains exocyclic to the developing ring system. Similar results were obtained using nitro compounds as the nitrile oxide precursors.

Higher stereoselectivity has been observed where the chiral center is included in the developing ring system. Thus, cyclization of (*Z*)-nitroalkene (**69**) led solely to *anti* isomer (**70**; Scheme 19).[34] This was explained based on the differing degree of $A^{1,3}$-strain in the transition state A leading to (**70**) and transition state B leading to the unobserved *syn* isomer. The much more severe Me—Me interaction in B precluded formation of the *syn* isomer.

Cyclizations of the nitrile oxide derived from nitroalkene (**71**) also led to bicyclic isoxazoline (**72**) as one stereoisomer, presumably through a transition state that minimized $A^{1,3}$-strain (Scheme 20).[35] Re-

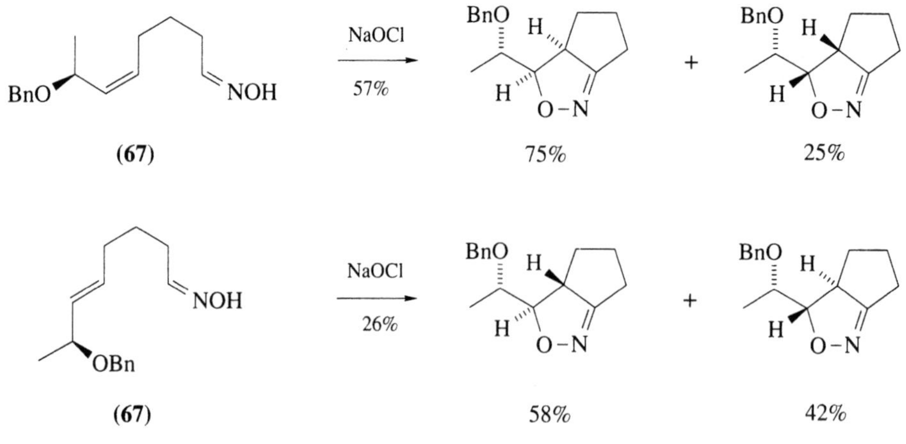

(67) NaOCl 57% → 75% + 25%

(67) NaOCl 26% → 58% + 42%

Scheme 18

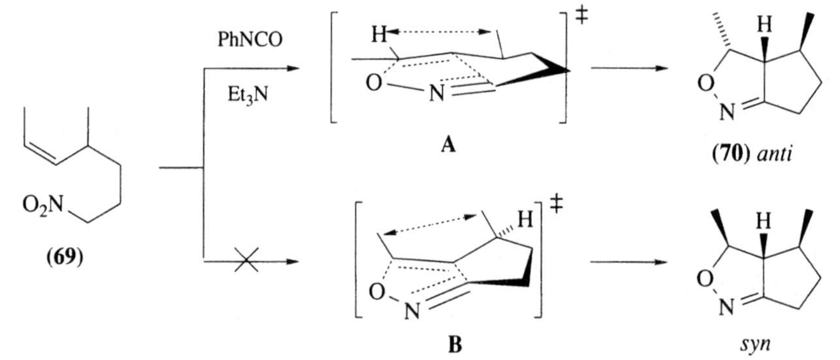

PhNCO, Et_3N → A → (70) *anti*

(69) → B → *syn*

Scheme 19

ductive cleavage of (**72**) under hydrolytic conditions afforded a β-hydroxy ketone which underwent elimination to give (±)-sarkomycin on treatment with mesyl chloride.

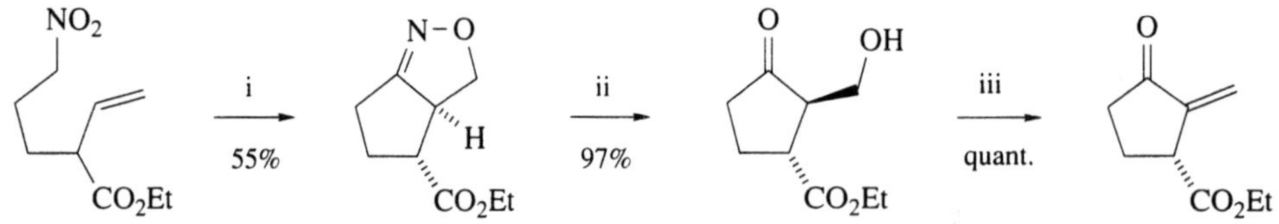

(71) i 55% → (72) ii 97% → iii quant. → (±)-Sarkomycin

i, p-$ClC_6H_4N{=}C{=}O$, Et_3N, 20 °C; ii, H_2, W-2 Raney Ni, $MeOH–H_2O$ (5:1), $B(OH)_3$; iii, MsCl, Et_3N

Scheme 20

4.10.3.2.2 6-Alkenyl nitrile oxides

Cyclization of nitrile oxides with a four-atom intervening chain to the alkene always leads to 5,6-fused bicylic isoxazolines possessing a bridgehead C—N double bond. This is in contrast to nitrone cyclizations where competition to form bridged bicyclic isoxazolidines is observed. The alkenyl oximes (**73**) and (**74**) cyclize in typical fashion *via* nitrile oxide intermediates (Scheme 21).[33a,36] The stereochemistry of cyclization here was studied both experimentally and by calculation. The higher stereoselectivity observed with the (*Z*)-alkene is typical. (*Z*)-Alkenes cycloadd much slower than (*E*)-alkenes in intermolecular reactions: this is attributed to greater crowding in the transition state. Thus, intramolecular cycloaddition of (*Z*)-alkenes depends on a transition state that is heavily controlled by steric factors.

MM2 calculations suggest that transition state A where the small group (H) is on the 'inside' is favored for cyclization of (**73**); attack is then *anti* to the benzyloxy group. (*E*)-Alkenyl ethers in general undergo cycloaddition with the alkoxy group preferentially on the 'inside' arising from an electronic effect; it is suggested that transition state B is favored for cyclization of (**74**), although MM2 calculations were unable to substantiate this without modification of a torsional parameter.[33a] The additional transition state C with the benzyloxy group 'outside' was predicted to be second most stable (C also minimizes overlap of $\sigma_{C—O}*$ overlap with $\pi_{C—C}$, a major problem if the benzyloxy group were *anti*[37]) resulting in the lower stereoselectivity observed with (**74**) compared to (**73**).

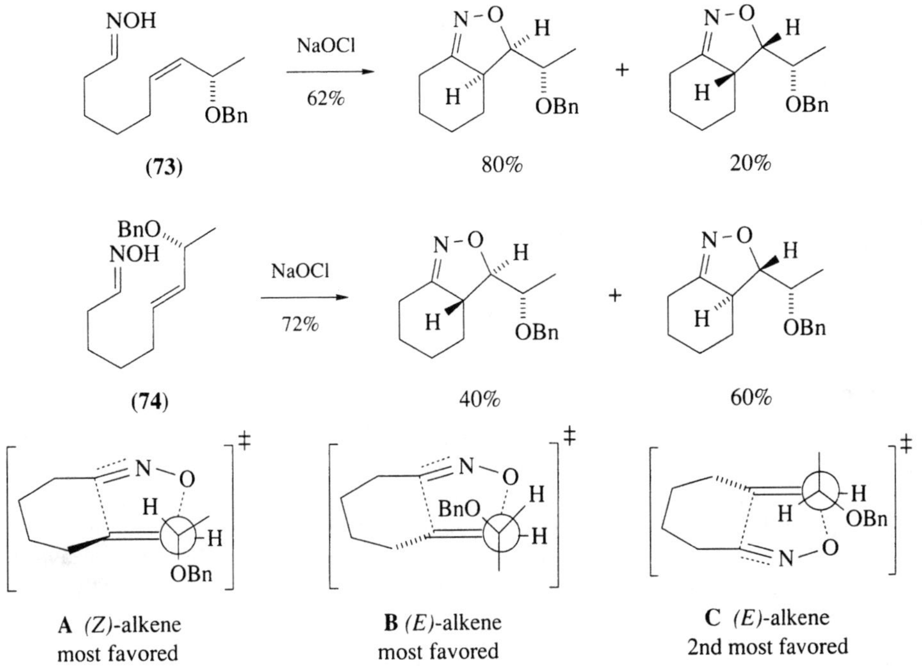

NOH OBn NaOCl 62% N-O H H OBn + N-O H H OBn

(**73**) 80% 20%

BnO NOH NaOCl 72% N-O H H OBn + N-O H H OBn

(**74**) 40% 60%

N-O H H OBn ‡ N-O BnO H H ‡ H H OBn N-O ‡

A (*Z*)-alkene most favored

B (*E*)-alkene most favored

C (*E*)-alkene 2nd most favored

Scheme 21

4.10.3.2.3 Long-chain alkenyl nitrile oxides

Macro carbocyclic rings can be constructed by cyclization of nitrile oxides derived from ω-nitro-1-alkenes (Scheme 22). If the intervening bridge is not longer than seven atoms, only fused bicyclic products are obtained. Thus, the nitrile oxide derived from nitro compound (**75a**) is cyclized in 44% yield to the 5,9-fused bicyclic isoxazoline (**76a**).[38] 10-Nitro-1-decene (**75b**) also cyclized to (**76b**) in unspecified yield.[39] It should be noted that these results go counter to the usual regiochemistry of an intermolecular nitrile oxide cycloaddition where the five-substituted isoxazoline is usually,[27] although not always,[40] heavily preferred from reaction of a terminal alkene. Thus, geometric constraints have won out over the normal electronic control.

In the case of a longer intervening chain, a bridged bicyclic isoxazoline will form. Cyclization of the nitrile oxide derived from nitro compounds (**77a–b**) afforded only the bridged products (**78a–b**) in 67% and 50% yield, respectively.[38] None of the isomeric fused products were detected. Here the intervening chain has 13 atoms. Cases where the chain has 12 and 10 intervening atoms have also been studied: a mixture of bridged and fused bicyclic isoxazolines was obtained with the bridged product predominating in both cases. Isoxazoline (**78b**) was converted to an enone lactone and eventually to racemic antitumor agent A26771B.

4.10.3.3 Cyclic Alkenyl Nitrile Oxides

Intramolecular cycloaddition can be carried out on a variety of cyclic nitrile oxide substrates. Two simple possibilities exist: either the alkene may be part of a ring or the intervening bridge may include a ring. Both possibilities may also coexist.

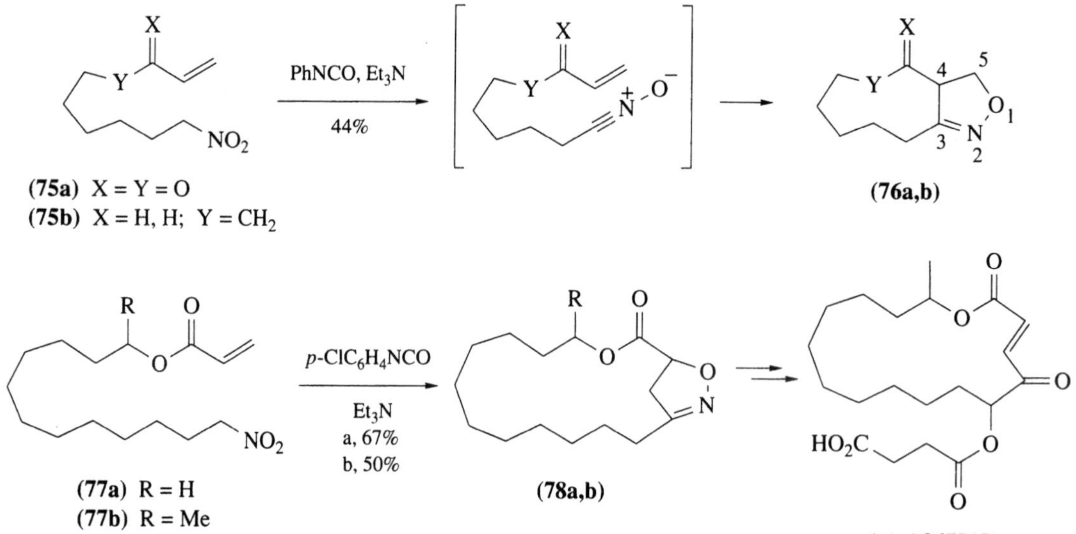

(75a) X = Y = O
(75b) X = H, H; Y = CH_2

PhNCO, Et_3N 44%

(76a,b)

(77a) R = H
(77b) R = Me

p-ClC_6H_4NCO, Et_3N, a, 67%; b, 50%

(78a,b)

(±)-A26771B

Scheme 22

4.10.3.3.1 Cycloalkenyl nitrile oxides

Cyclizations of nitrile oxides derived from the nitro compounds (**79**) and (**80**) are typical examples where the dipolarophile is part of a ring (Scheme 23).[41] In both cases only a fused polycyclic system was formed, typical of most nitrile oxide cyclizations. Nitro compound (**79**) led to tricyclic isoxazoline (**81**), which possesses a new 5,5-fused bicyclic system, doubly fused to the existing seven-membered ring. Isoxazoline (**81**) was the sole diastereomer obtained: all ring fusions were *cis*. Reductive cleavage of (**81**) afforded a γ-amino alcohol, which was converted to (±)-biotin in several additional synthetic steps.[41a] Nitro compound (**80**) led to tricyclic isoxazoline (**82**), which possesses a new 5,7-fused bicyclic system, doubly fused to the existing five-membered ring. Again, only one diastereomer, the all-*cis*-fused product, was obtained: models suggest that the transition state leading to *trans* fusion at the carbocyclic 5,7-ring junction would involve considerable nonbonded interaction absent in the transition state leading to *cis* fusion. Reductive cleavage of (**82**) with Raney nickel in the presence of water afforded a β-hydroxy ketone, which was dehydrated to provide a conjugated ketone in 80% yield. This procedure provides a good alternative to aldol strategies leading to polycyclic conjugated ketones.[41b]

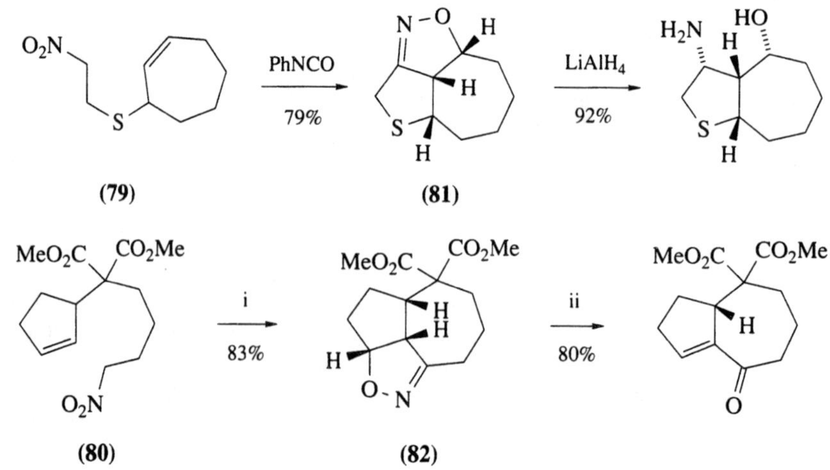

(79) —PhNCO, 79%→ (81) —$LiAlH_4$, 92%→

(80) —i, 83%→ (82) —ii, 80%→

i, PhNCO added dropwise at 80 °C, cat. Et_3N; ii, O_3, –78 °C then TsOH, MeOH

Scheme 23

The hexahydronaphthalene ring system of compactin has been constructed by a nitrile oxide cyclization strategy (Scheme 24).[42] Thus, cyclization of either (**83**) or (**84**) *via* nitrile oxide intermediates led to the corresponding tricyclic isoxazoline in which a new 5,6-fused bicyclic system was appended to the existing six-membered ring. Again, the all-*cis*-fused diastereomer was obtained exclusively. These new isoxazolines were reductively cleaved with Raney nickel in the presence of water and the ensuing β-hydroxy ketones were dehydrated to provide conjugated ketones (**85a–b**). 1,2-Reduction then provided allylic alcohols that underwent dehydration to provide hexahydronaphthalenes (**86**) and (**87**). Using alumina for the dehydration provided dienes (**86a–b**) as major products, whereas other dehydrating agents gave more of the isomeric dienes (**87a–b**). The diene (**86b**) obtained from oxime (**84**) was converted in several steps to compactin.

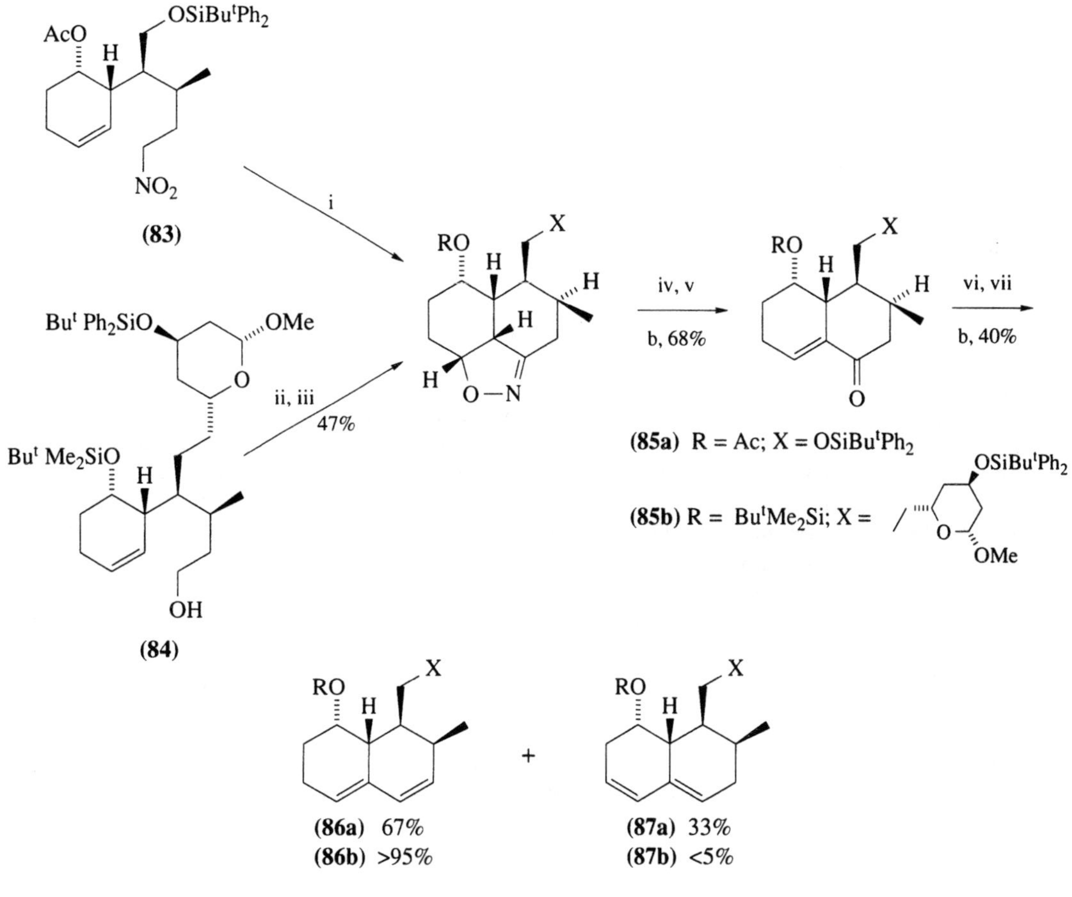

(**83**)

(**84**)

(**85a**) R = Ac; X = $OSiBu^tPh_2$

(**85b**) R = Bu^tMe_2Si; X =

(**86a**) 67%
(**86b**) >95%

(**87a**) 33%
(**87b**) <5%

i, PhNCO, Et_3N; ii, DMSO, $(COCl)_2$, Et_3N then NH_2OH; iii, NaOCl, Et_3N; iv, H_2, Raney Ni, aq. MeOH; v, Al_2O_3, 80 °C; vi, $NaBH_4$, $CeCl_3$; vii, Al_2O_3, 120 °C

Scheme 24

The spirocyclic alkaloid sibirine was prepared from the nitrile oxide cyclization product (**88**; Scheme 25).[43a] Here the intervening bridge of the nitrile oxide precursor is attached directly to the cycloalkenyl C—C double bond: cyclization to the new 5,6-fused bicyclic system afforded the azacarbocyclic ring attached spiro to the former cycloalkenyl ring. The furanylnitrile oxide (**89**), generated from the corresponding nitro compound, cyclized to (**90**) in 79% overall yield.[43b] Furan is typically sluggish in nitrile oxide cycloaddition reactions. Cycloaddition to the more substituted double bond of furan derivatives is highly atypical: therefore, the INOC reaction of (**89**) must be entropically driven.

4.10.3.3.2 Alicyclic-bridged nitrile oxides

Many substrates containing a cyclic bridge have been investigated in intramolecular nitrile oxide cycloadditions. Typical examples include cyclization of the nitrile oxides derived from nitro compounds (**91**) and (**92**; Scheme 26).[44] MM2 calculations are consistent with these results: a chair-like transition state led to formation of the 5,6-fused product (**93**). Cyclization of (**92**) occurred through a more flexible

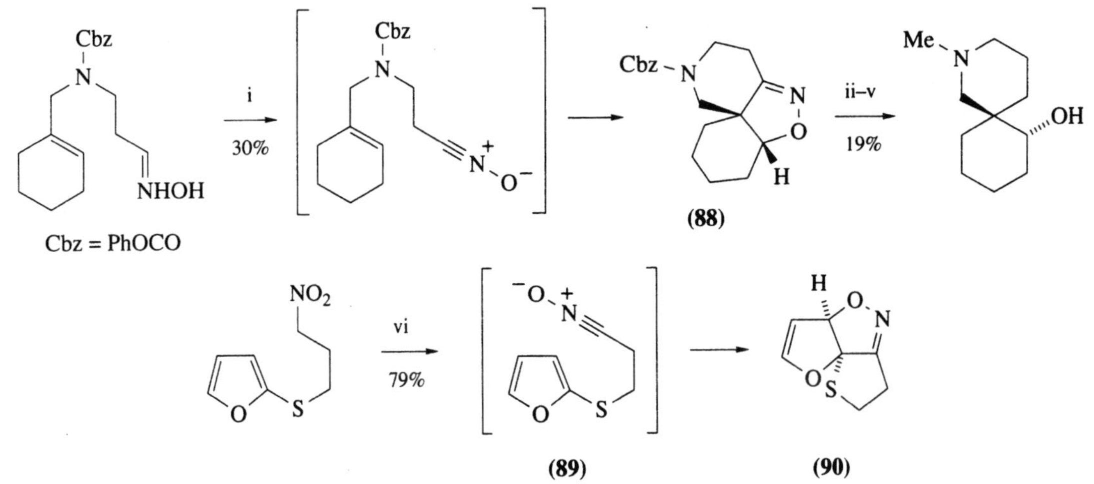

Cbz = PhOCO

(88)

(89) (90)

i, 30% aq. NaOCl; ii, H_2, Raney Ni, aq. MeOH, AcOH; iii, $HSCH_2CH_2SH$, BF_3; iv, Bu_3SnH, AIBN; v, H_2, Raney Ni, MeOH; vi, p-ClC_6H_4NCO, Et_3N, 80 °C

Scheme 25

transition state, affording both *syn* and *anti* products. Cyclization of the nitrile oxide (**94**) (generated from nitro and oximino precursors) also led to a mixture of *syn* and *anti* products with the additional observation that only products with phenyl *cis* to methyl were obtained, although a 1:1 mixture of diastereomers was present in the precursor in both cases. It was suggested that $A^{1,3}$-strain present in the transition state leading to *syn* product (**95**) was responsible for the preferred formation of *anti* product. Presumably, the *cis* phenyl–methyl relationship arose from a severe nonbonded interaction between phenyl and one of the azetidine C(4)—H bonds. Interestingly, calculations based on product geometries were essentially equivalent to calculations based on transition state geometries: this is in accordance with a late transition state.

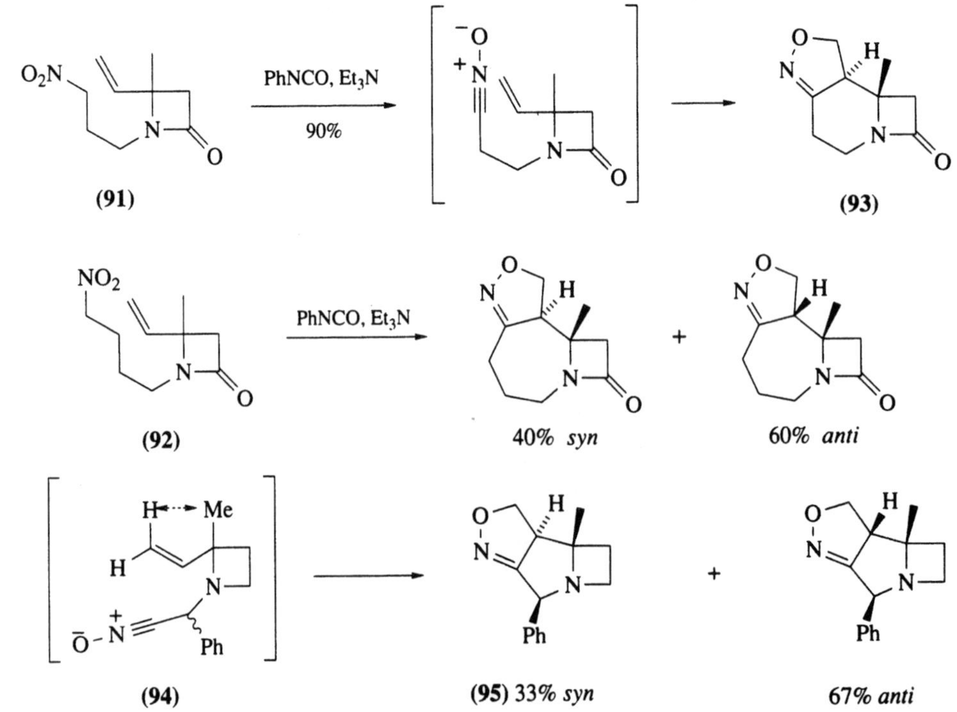

(91) (93)

(92) 40% *syn* 60% *anti*

(94) (95) 33% *syn* 67% *anti*

Scheme 26

The *cis*-dioxane (**97**) underwent cyclization to afford a single tricyclic isoxazoline in which the new 5,5-fused bicyclic system formed rapidly at 0 °C (Scheme 27).[45] The isomeric *trans*-dioxane also gave a single tricyclic isoxazoline (**96**) as product but only at 20 °C. The results were rationalized as follows: cyclization of the nitrile oxide derived from (**96**) occurred with the vinyl group equatorial, while the *cis* isomer required an axial vinyl group in the transition state. The isoxazoline (**96**) was converted in several steps to a known $PGF_{2\alpha}$ precursor.

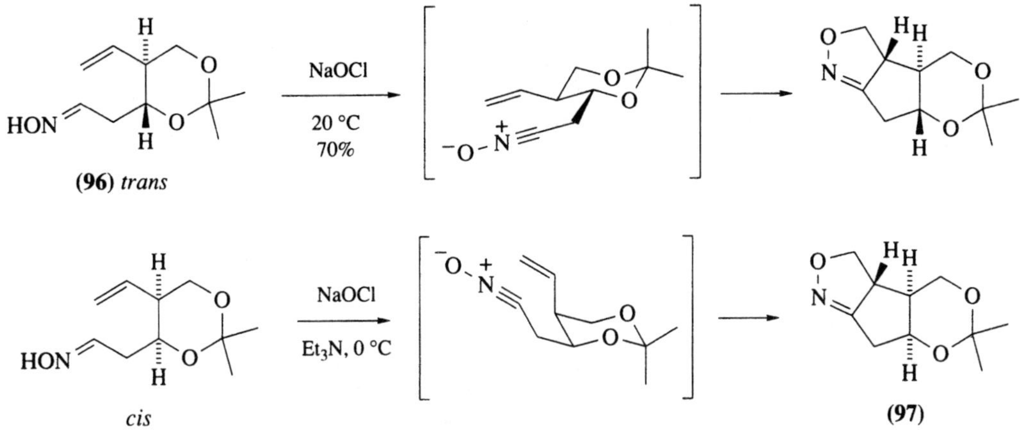

NaOCl, 20 °C, 70%

(**96**) *trans*

NaOCl, Et_3N, 0 °C

cis

(**97**)

Scheme 27

4.10.3.3.3 Aryl-bridged nitrile oxides

The indolyl nitro compound (**98**) was converted to the corresponding nitrile oxide, which cyclized to afford an inseparable mixture of isoxazolidines (**99**) and (**100**; Scheme 28).[46] These were acetylated and the THP group replaced by a mesyl group at which point the desired β-isomer could be separated. Elimination *via* a selenide intermediate provided the alkenylisoxazoline (**101**), which was converted to an isoxazolinium salt and reduced with LAH to give an *N*-methylisoxazolidine. Reductive cleavage with aluminum amalgam provided (+)-paliclavine.

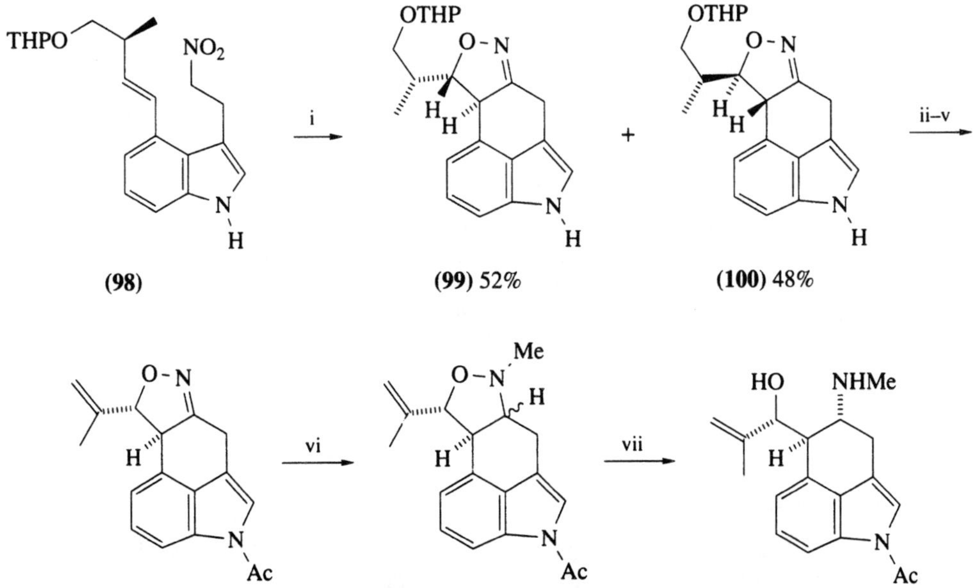

(**98**) (**99**) 52% (**100**) 48%

(**101**) (+)-Paliclavine

i, PhNCO, Et_3N; ii, Ac_2O, 4-DMAP, Et_3N then H^+; iii, MsCl, Et_3N then chromatography to separate mesylates; iv, $Na^{+}\ {}^{-}SePh$ then Ac_2O, 4-DMAP; v, $NaIO_4$; vi, $MeO_3{}^{+}BF_4{}^{-}$ then $LiAlH_4$; vii, Al–Hg

Scheme 28

In an early example of the INOC reaction, the oxime (**102**) with an *o*-aryl five-atom intervening bridge was cyclized to a tricyclic isoxazoline (Scheme 29).[47] The macrocylic bisisoxazoline (**103**) was formed as a byproduct, apparently *via* a tandem intermolecular–intramolecular cycloaddition sequence. The homologous oxime with a four-atom intervening bridge gave only intramolecular cycloaddition whereas from reaction of the homologs with six- and seven-atom intervening bridges, only the macrocyclic products corresponding to (**103**) were isolated. The effect of high-dilution conditions on cyclization was apparently not studied.

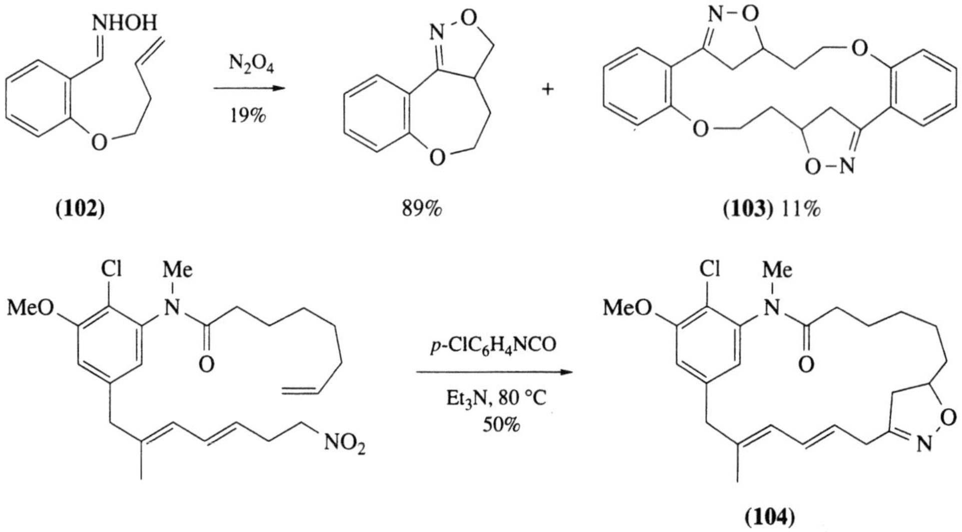

Scheme 29

Confalone and Ko have reported formation of the macrocyclic INOC product (**104**) in 50% yield from cyclization of an aryl-bridged nitrotriene containing a 17-atom intervening bridge.[48] Cyclization to the conjugated diene system did not occur, presumably because of geometric constraints and the fact that the more accessible double bond is trisubstituted. Isoxazoline (**104**) is a potential maytansine precursor.

4.10.3.4 Tandem Reaction Sequences Involving Nitrile Oxide Cyclizations

4.10.3.4.1 Tandem Diels–Alder–INOC reactions

Nitrodienes undergo intermolecular Diels–Alder reactions with appropriate dienophiles. The resulting nitro compounds can then be cyclized *via* a nitrile oxide intermediate.[49] Thus, the 2-chloroacrylonitrile Diels–Alder adduct of 8-nitro-1,3-octadiene was prepared and cyclized to give (**105**) as a 3:1 mixture of diastereomers (Scheme 30). The Diels–Alder adduct of dimethyl acetylenedicarboxylate and 8-nitro-1,3-octadiene cyclized exclusively at the conjugated double bond, activated by the ester groups. Similarly, the quinone Diels–Alder adduct (**106**) cyclized at the conjugated double bond: reduction of the conjugated double bond permitted cyclization on the cycloalkenyl double bond.

4.10.3.4.2 Tandem Michael–INOC reactions

Allylic stannanes condense with nitro dienes in the presence of titanium tetrachloride to give alkenyl α-chlorooximes; treatment of the α-chlorooximes with base then affords bicyclic isoxazolines.[50] Thus, cyclization of the intermediates (**107a–c**), obtained from appropriate 1-nitro-1,5-hexadienes and (3-propenyl)trimethyltin, gave bicyclic isoxazolines (Scheme 31). Cyclization occurred exclusively on the double bond with the longer (three-atom *vs.* two-atom) intervening bridge, even when that double bond was trisubstituted.

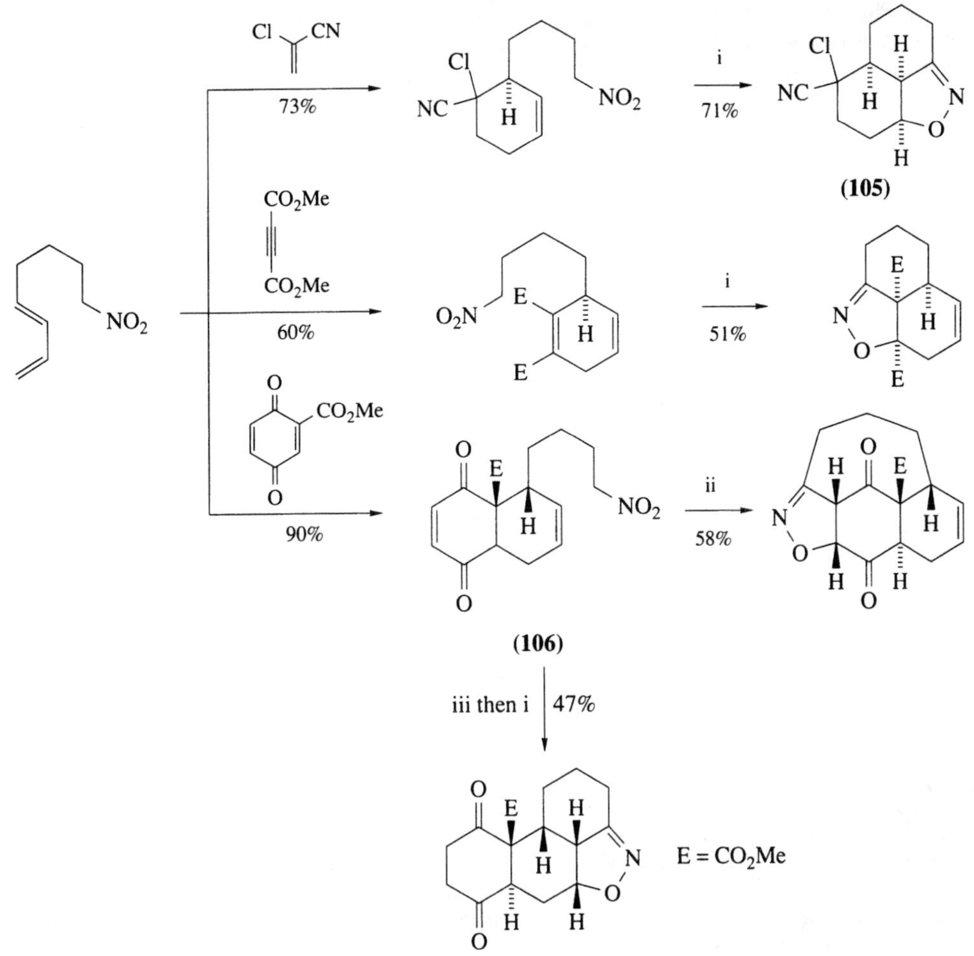

i, PhNCO, Et_3N, 80 °C; ii, p-ClC_6H_4NCO, Et_3N, 80 °C; iii, L-selectride, THF, –78 °C

Scheme 30

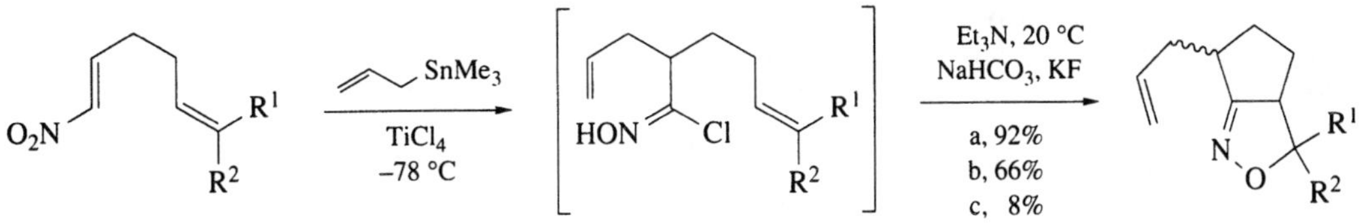

(107a) $R^1 = R^2 = H$
(107b) $R^1 = H$; $R^2 = Ph$
(107c) $R^1 = R^2 = Me$

Scheme 31

4.10.3.5 Alkynyl Nitrile Oxides

Only a few INOC reactions have been run on alkynyl substrates.[51,52] The reactions do proceed well, as for example, cyclization of the alkynyl nitrile oxide (**108**), derived *in situ* from the corresponding oxime.[51] The oxime itself was formed *in situ* by reaction of an α-bromosilyloxime with propargyl alcohol, presumably *via* a vinylnitroso intermediate (Scheme 32). The alkynyl acylnitrile oxide (**109**) also cyclized to afford a bicyclic isoxazole.[52]

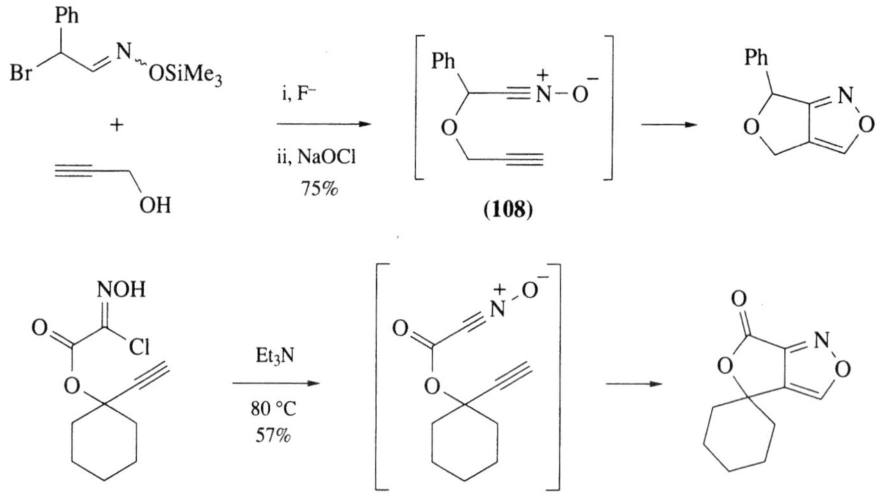

Scheme 32

4.10.4 AZOMETHINE YLIDE CYCLIZATIONS

4.10.4.1 General

Azomethine ylides are not typically isolable but must be used *in situ*. They undergo cycloaddition reactions that produce highly functionalized pyrrolidines, dihydropyrroles and pyrroles. The success of these reactions often depends on a judicious choice of dipole and dipolarophile. Azomethine ylides are reluctant to cycloadd to nonactivated alkenes, in large part owing to electronic considerations. The LUMO of most azomethine ylides is high in energy and there is a large gap with the HOMO of a nonactivated alkene.

Intramolecular cycloaddition of an azomethine ylide was first documented in 1977.[53] One frequently employed method for *in situ* generation of the azomethine ylide involves thermolysis of an aziridine, a reaction demonstrated to occur exclusively in a conrotatory fashion.[54] Other methods involve deprotonation of an iminium salt, either directly with base or preferably indirectly *via* desilylation.[55a] Oxazolium salts can be converted to 4-oxazolines, which undergo ring opening to afford azomethine ylides.[55b] Tautomerization of the imines of α-amino acid esters also gives azomethine ylides, presumably at low concentration.[56] Very recently it has been found that nonstabilized azomethine ylides can be generated by deprotonation of amine *N*-oxides.[57a] The ylides are formed in the absence of imines and have a wide reactivity range in intermolecular cycloadditions. This generation procedure has not yet been applied to intramolecular cycloadditions. Tandem Michael addition–intermolecular azomethine ylide cycloaddition to a tetrahydropyrazinone has also been carried out.[57b]

4.10.4.2 Alkenyl Azomethine Ylides

4.10.4.2.1 Open-chain C-alkenyl azomethine ylides

Independent thermolysis (80 °C) of the aziridines (**110a**) and (**110b**) led to the same 5,5-fused bicylic pyrrolidine (Scheme 33).[58] This was rationalized as follows: stereospecific conrotatory aziridine ring-opening afforded the doubly-stabilized azomethine ylides (**111a–b**). These azomethine ylides were in equilibrium but only (**111a**) was able to cyclize. The aziridines (**110c**) and (**110d**) underwent *cis–trans* isomerization, quite possibly *via* (**111c–d**), but gave no dipolar cyclization products. This was attributed to the lower reactivity of the unactivated dipolarophile. The doubly stabilized azomethine ylide (**112**), generated by thermolysis (140 °C) of the corresponding imine, did cyclize, affording a mixture of *cis*-

and *trans*-5,5-fused bicyclic pyrrolidines.[56] An attempt to form a homologous 5,6-fused bicyclic pyrrolidine was unsuccessful.

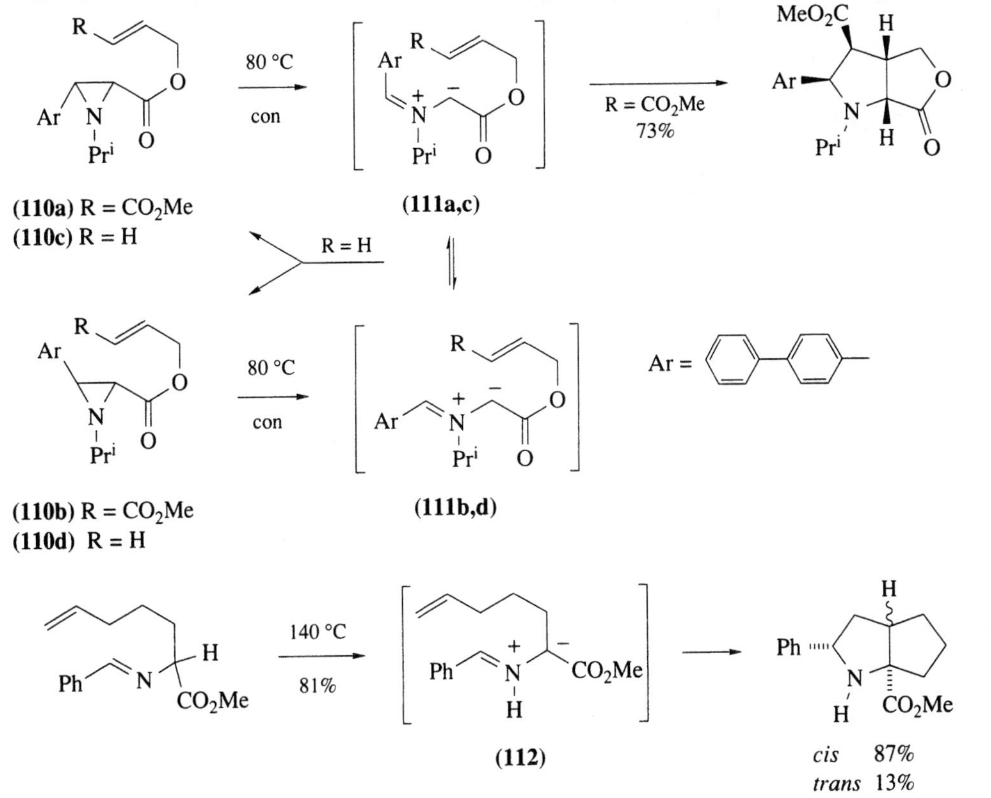
80 °C
con
R = CO_2Me
73%
(110a) R = CO_2Me
(110c) R = H
(111a,c)
R = H
80 °C
con
(110b) R = CO_2Me
(110d) R = H
(111b,d)
Ar =
140 °C
81%
(112)
cis 87%
trans 13%

Scheme 33

Unactivated dipolarophiles readily participate in intramolecular azomethine ylide cycloadditions with a more reactive azomethine ylide. Thus, flash vacuum pyrolysis of aziridine (**113**) afforded a 67% yield of the 5,5-fused bicyclic pyrrolidine (Scheme 34).[59] A singly stabilized azomethine ylide was the apparent intermediate. Similarly, cyclization of the azomethine ylides derived from (**114a–c**) gave the corresponding *cis*-fused 6,6-bicyclic pyrrolidines in 69%, 26% and 16% yield, respectively; the original double bond stereochemistry was retained in the latter two cases.

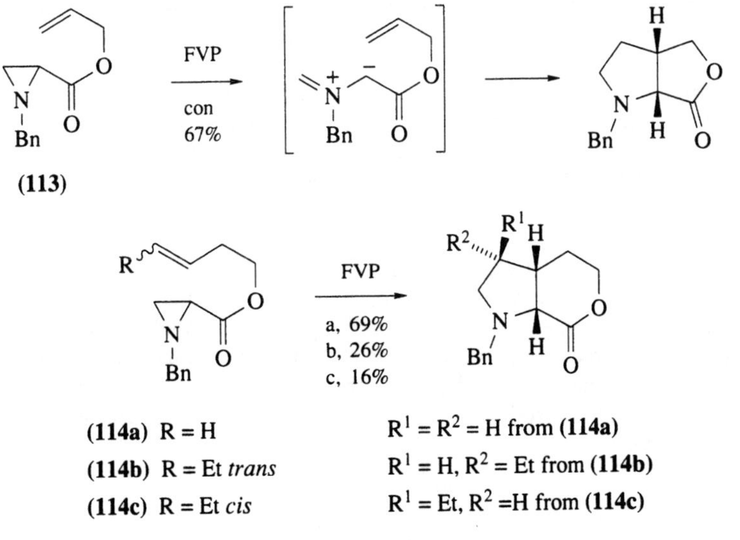
FVP
con
67%
(113)
FVP
a, 69%
b, 26%
c, 16%
(114a) R = H
(114b) R = Et *trans*
(114c) R = Et *cis*
$R^1 = R^2$ = H from (114a)
R^1 = H, R^2 = Et from (114b)
R^1 = Et, R^2 = H from (114c)

Scheme 34

Confalone and Earl have reported a series of intramolecular azomethine ylide cyclizations employing α-dithiolanyl aldehydes (Scheme 35).[60] Thus, reaction of the aldehydes (**115a–b**) with ethyl sarcosinate at 140 °C provided the 5,6-fused bicyclic pyrrolidines (**116a–b**), presumably *via* cyclization of the singly stabilized azomethine ylides. The dithiolanyl group could be removed by standard reactions: this route then provides bicyclic pyrrolidines which cannot be prepared directly from enolizable aldehydes.

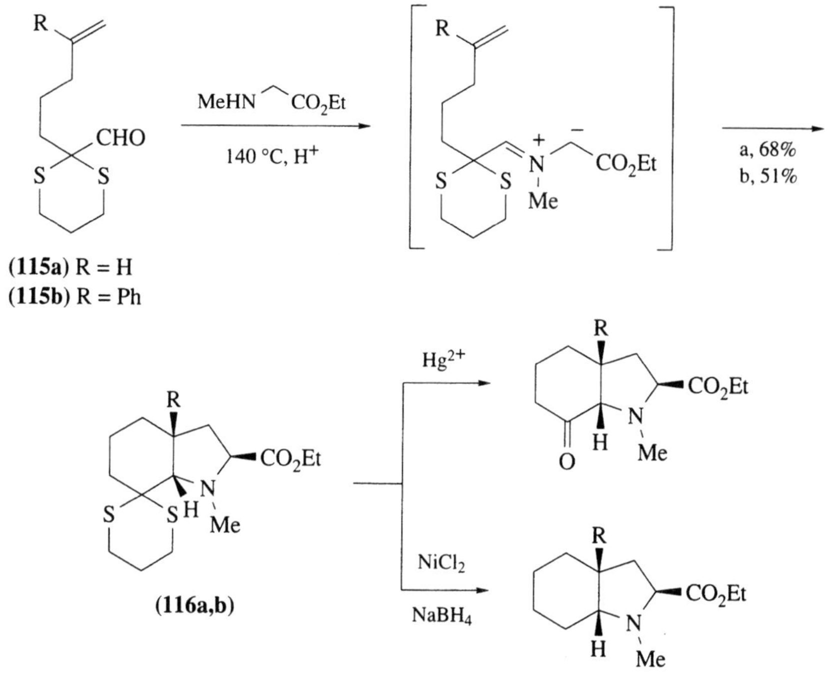

(**115a**) R = H
(**115b**) R = Ph

MeHN⌒CO$_2$Et, 140 °C, H$^+$

a, 68%
b, 51%

(**116a,b**)

Hg^{2+}

NiCl$_2$, NaBH$_4$

Scheme 35

4.10.4.2.2 Cyclic C-alkenyl azomethine ylides

A variety of cyclic azomethine ylides undergo cyclization. The intervening bridge may be part of an aryl ring system in the most common case. In principal, the alkene can be part of a ring, but there are no known examples. The azomethine ylide itself may also be cyclic.

A number of aryl-bridged azomethine ylides have been shown to undergo intramolecular cyclization. Thus, the imine (**117a**) was converted to a triflate salt and then desilylated to provide an azomethine ylide (Scheme 36).[61a] The ylide cyclized under high dilution to provide (±)-eserethole. Reaction of (**117b**) with benzoyl fluoride provided a similar cyclization product. These desilylation approaches to azomethine ylide generation show much promise for adaptation to intramolecular cycloadditions. The imine (**118**), containing an α-ester group, affords a tricyclic pyrrolidine on strong heating.[56] Presumably (**118**) tautomerizes to the corresponding azomethine ylide which then cyclizes. In similar fashion, the azomethine ylide (**119**) is presumably an intermediate in the formation of (**120a**), formed as a mixture of two isomers in 50% overall yield from the aldehyde and aminomethylphosphonate.[61b] Cycloadduct (**120a**) was converted to the corresponding lactam (**120b**) by sequential reaction with butyllithium and oxygen.

Cyclization of the azomethine ylide derived from aziridine (**121**) gave two diastereomeric pyrrolidinometacyclophanes (Scheme 37).[62a] Here an 11-atom chain separated the dipole from the dipolarophile. The homologous aziridine with an eight-atom intervening chain afforded (**122**) in 5% yield. Larger homologs were also studied: when a 17-atom chain was present, a small amount of bridged product accompanied the fused product. The stabilized cyclic azomethine ylide (**123**), formed *via* the desilylation route, afforded a pyrrolidine cyclization product in 70% yield as a single stereoisomer.[67b] This product possesses an erythrinane skeleton.

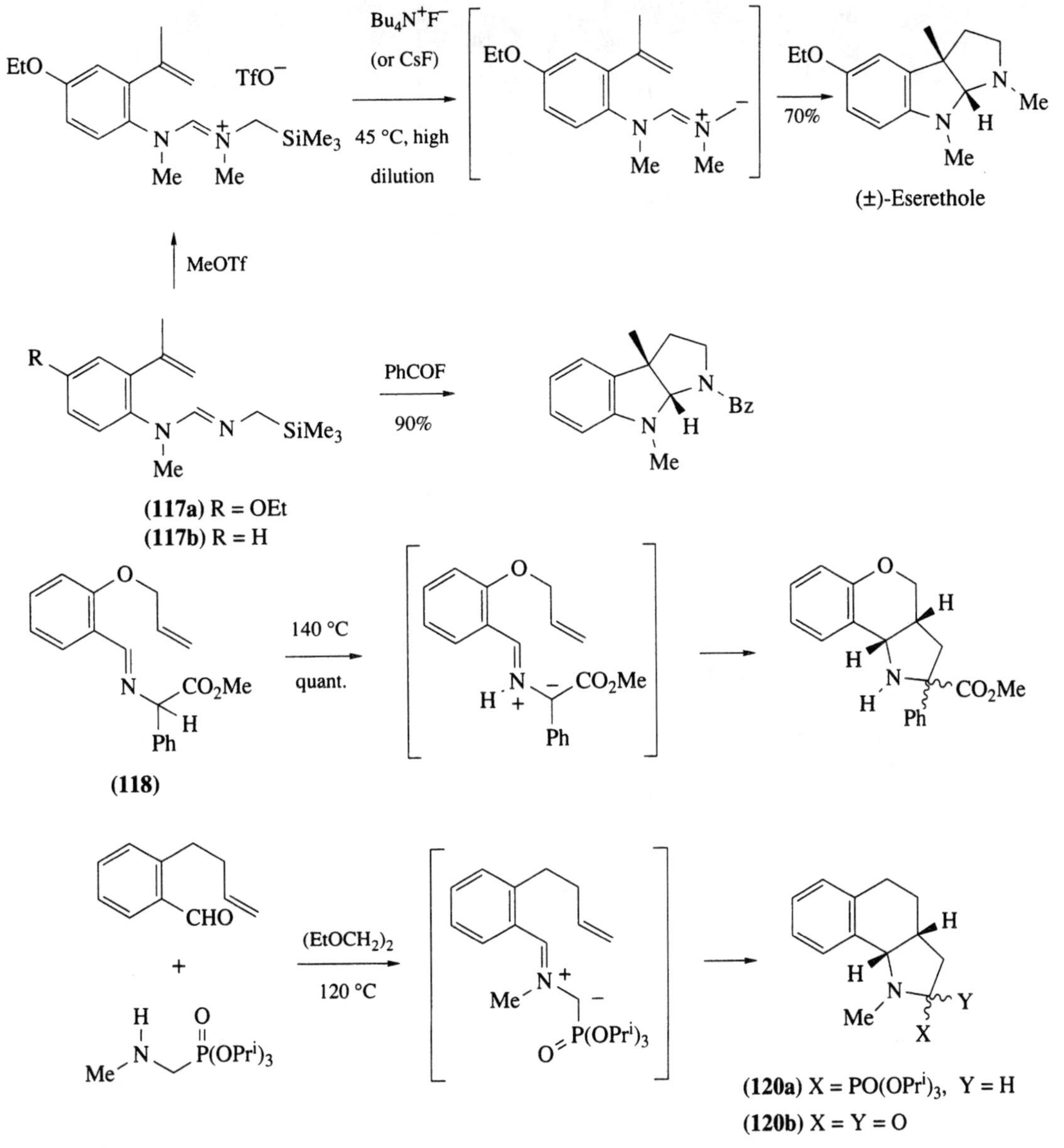

(±)-Eserethole

(117a) R = OEt
(117b) R = H

(118)

(120a) X = $PO(OPr^i)_3$, Y = H
(120b) X = Y = O

Scheme 36

4.10.4.3 Tandem Michael–Azomethine Ylide Cyclization Reactions

Very recently examples of tandem Michael–azomethine ylide cyclization reactions have been presented.[62b] Thus, divinyl sulfone reacted with imine (**124**) in the presence of lithium bromide and triethylamine to give (**126**) in 40% yield (Scheme 38). Presumably formation of Michael adduct (**125**), tautomerization to an azomethine ylide and ensuing intramolecular [3 + 2] cycloaddition afforded (**126**). Indeed, (**125**) could be independently synthesized and converted to (**126**) under the reaction conditions. The preference for initial Michael addition, rather than cycloaddition, was variable. When (**124**) and divinyl sulfone were treated with silver acetate and triethylamine in DMSO, intermolecular azomethine cycloaddition occurred giving (**127**) in 27% yield.

4.10.4.4 Munchnones

The mesoionic heterocycles called munchnones function as cyclic azomethine ylides in cycloaddition reactions.[63] The alkenyl moiety can be attached at carbon or at nitrogen.

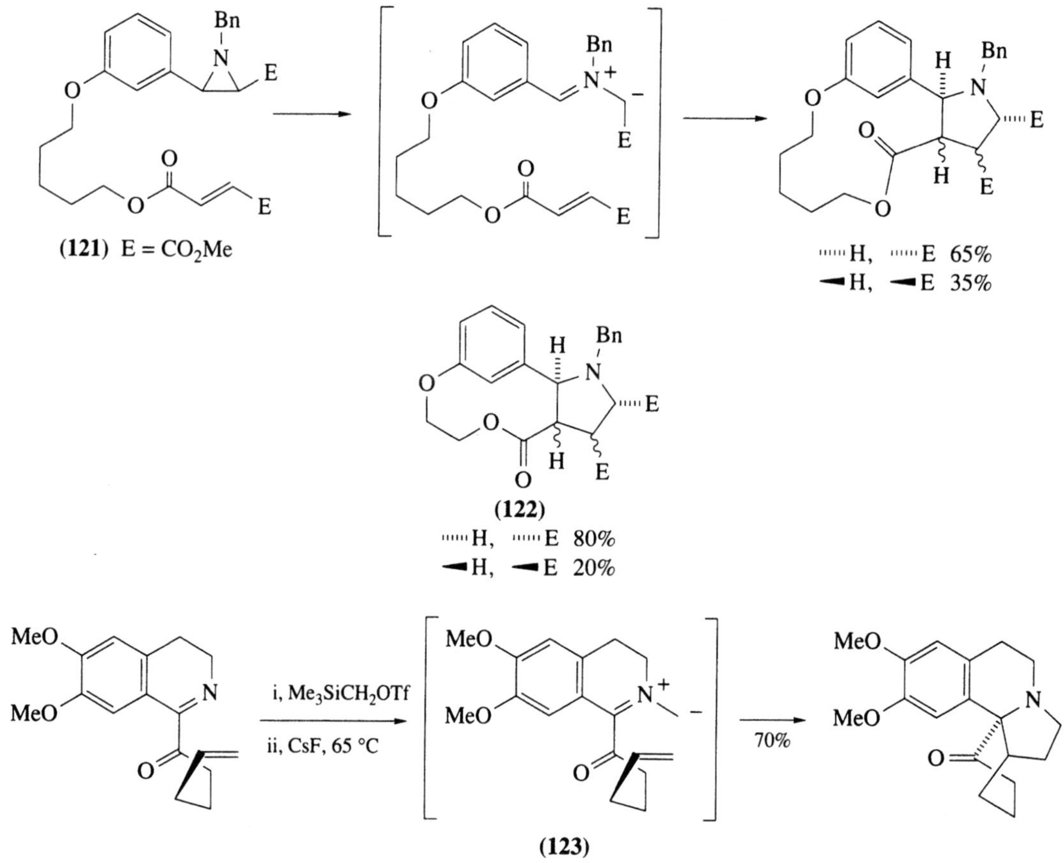

(121) E = CO_2Me

"""""H, """""E 65%
◄H, ◄E 35%

(122)
"""""H, """""E 80%
◄H, ◄E 20%

i, Me_3SiCH_2OTf
ii, CsF, 65 °C

(123)

70%

Scheme 37

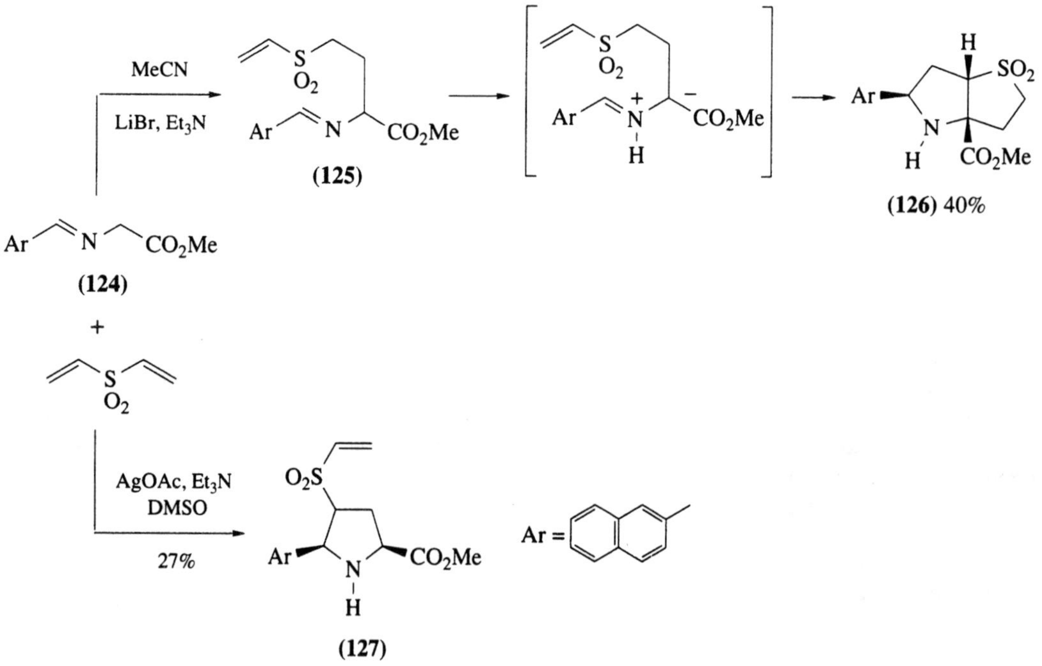

MeCN
LiBr, Et_3N

(125)

(126) 40%

(124)

AgOAc, Et_3N
DMSO
27%

(127)

Ar =

Scheme 38

4.10.4.4.1 C-Alkenyl munchnone azomethine ylides

Acetylation of the amide (**128**) produced a fused tricyclic pyrrole in 17% yield (Scheme 39).[64] A munchnone intermediate was postulated and could be trapped *via* intermolecular cycloaddition with diethyl acetylenedicarboxylate. Presumably the direct cyclization product lost carbon dioxide, a known reaction of munchnone cycloadducts.[65] An homologous munchnone with an *o*-(propenyl)phenyl group, rather than *o*-(butenyl)phenyl group, failed to cyclize, although it could again be trapped with diethyl acetylenedicarboxylate.

HO_2C NPh Ph O OMe (**128**) Ac_2O 55 °C

$-CO_2$ [Ox] 17%

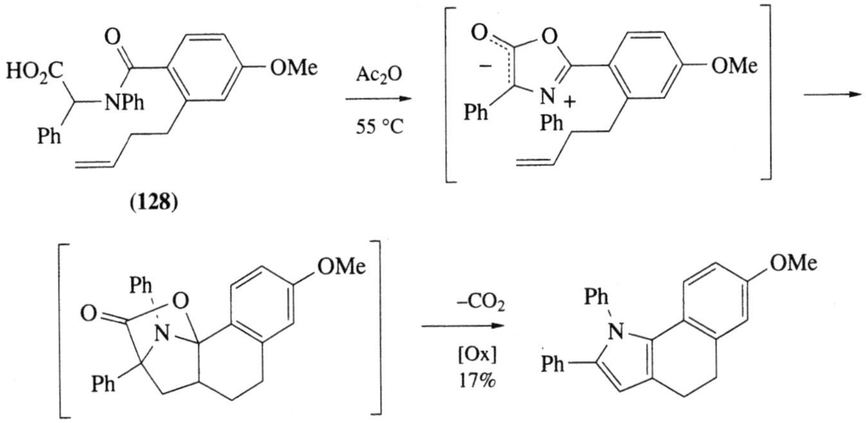

Scheme 39

4.10.4.4.2 N-Alkenyl munchnone azomethine ylides

Munchnone (**129a**), prepared from the corresponding formamide and cyclized *in situ*, afforded a 75% yield of one product (Scheme 40).[66] Trifluoroacetylmunchnone (**129b**), formed directly from the corresponding amine *via* trifluoroacetylation, similarly gave a 70% yield of one cyclization product. The munchnone (**130a**) gave a mixture of two regioisomeric cyclization products in 92% yield, indicating the importance of substituent effects in determining regioselectivity. Munchnone (**130b**), presumably formed *via in situ* trifluoroacetylation of (**130a**), gave a 33% yield of one cyclization product, the opposite regioisomer to the one preferred from (**130a**). The homologous munchnones with a vinylic, rather than allylic, dipolarophile failed to undergo cyclization. However, munchnone (**131**), also with a vinylic dipolarophile but a one-atom longer bridge, did cyclize in 85% yield to afford two regioisomers.

4.10.4.5 Alkynyl and Allenyl Azomethine Ylides

Intramolecular cycloaddition can be carried out on a variety of alkynyl azomethine ylides analogous to reactions of the alkenyl azomethine ylides. There is also one report of cyclization with an allene.

4.10.4.5.1 Open-chain C-alkynyl azomethine ylides

Flash vacuum pyrolysis of aziridine (**132**) led to a 35% yield of the bicyclic 2,5-dihydropyrrole, accompanied by varying amounts of the pyrrole and traces of a fused tricyclic pyrrole (Scheme 41).[59] The dihydropyrrole was easily converted to the pyrrole simply on exposure to oxygen. The homologous aziridine with a one-atom longer bridge directly afforded the aromatized product (**133a**) in 63% yield. Pyrrole (**133b**) was formed analogously but only in 34% yield. Very recently the generation of azomethine ylide (**135**) from oxazole (**134**) and its ensuing intramolecular cycloaddition have been reported.[67a] Thus, (**134**) was converted to the isoxazolium salt with methyl triflate. Reaction of the crude salt with trimethylsilyl cyanide/cesium fluoride afforded a mixture of (**136a**) and (**136b**); treatment with tetrabutylammonium fluoride prior to isolation afforded (**136b**) in 68% overall yield from (**134**).

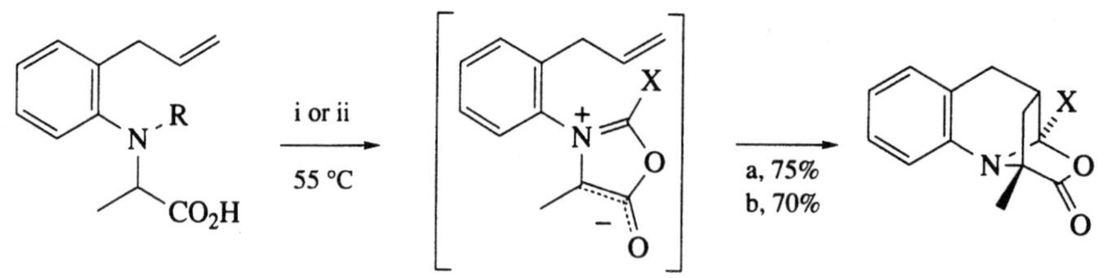

i or ii, 55 °C; a, 75%; b, 70%

R = COH or H

(129a) X = H
(129b) X = CF_3

i, Ac_2O with R = COH; ii, $(CF_3CO)_2O$ with R = H

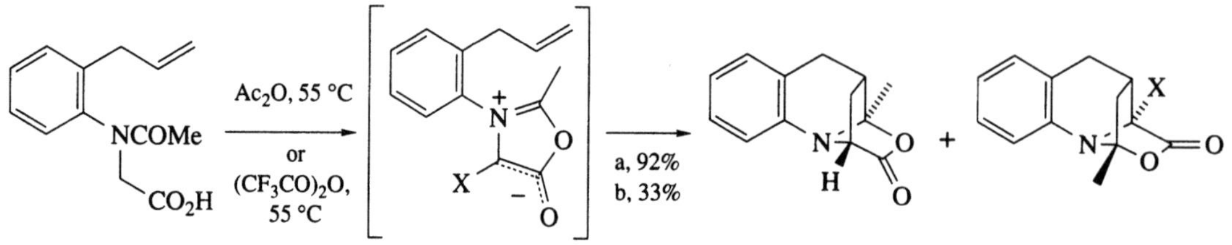

Ac_2O, 55 °C or $(CF_3CO)_2O$, 55 °C; a, 92%; b, 33%

(130a) X = H
(130b) X = $COCF_3$

from **(130a)**: 85% — 15%
from **(130b)**: 0% — 100%

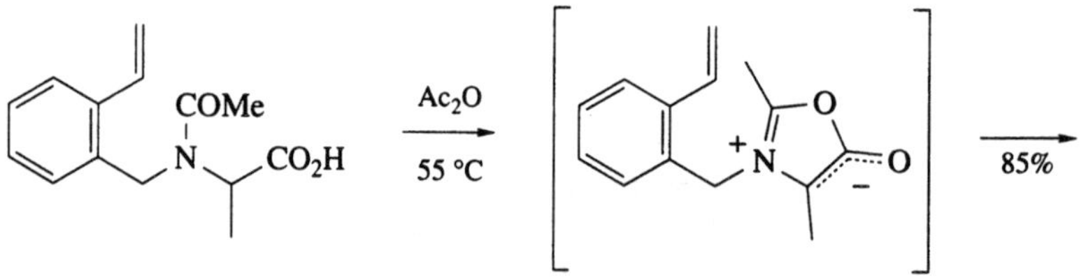

Ac_2O, 55 °C; 85%

(131)

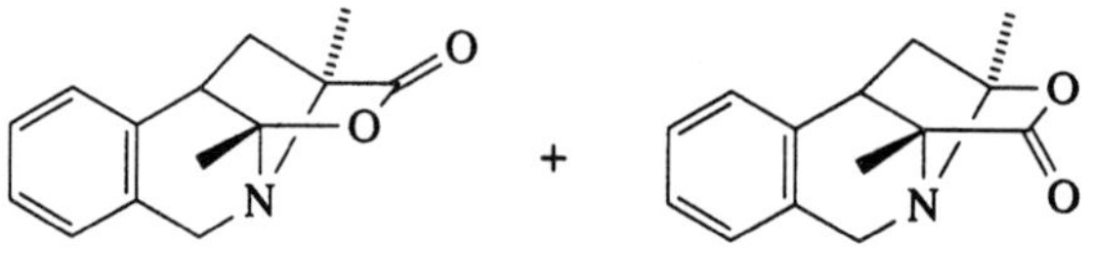

40% + 60%

Scheme 40

4.10.4.5.2 Cyclic C-alkynyl azomethine ylides

Alkynyl azomethine ylides containing an aryl bridge have been shown to undergo cyclization. Thus, tautomerization of imine (**137**) afforded an azomethine ylide which cyclized to give two diastereomeric dihydropyrroles and traces of aromatized product (Scheme 42).[56] The activated cyclic ylide (**138**), generated *via* the desilylation route, underwent cyclization to afford dihydropyrrole (**139**) in 42% yield.[67b] The structure of (**139**) is closely related to erythramine. The structurally related non-stabilized azomethine ylide (**140**) did not cyclize, tautomerizing to the corresponding enamine instead.

4.10.4.5.3 C-Allenyl azomethine ylides

Azomethine ylide (**141**), formed from condensation of the corresponding aldehyde with ethyl sarcosinate, gave a mixture of the regioisomeric cyclization products (Scheme 43).[60] The major product has two of the three aliphatic rings of lycorenine.

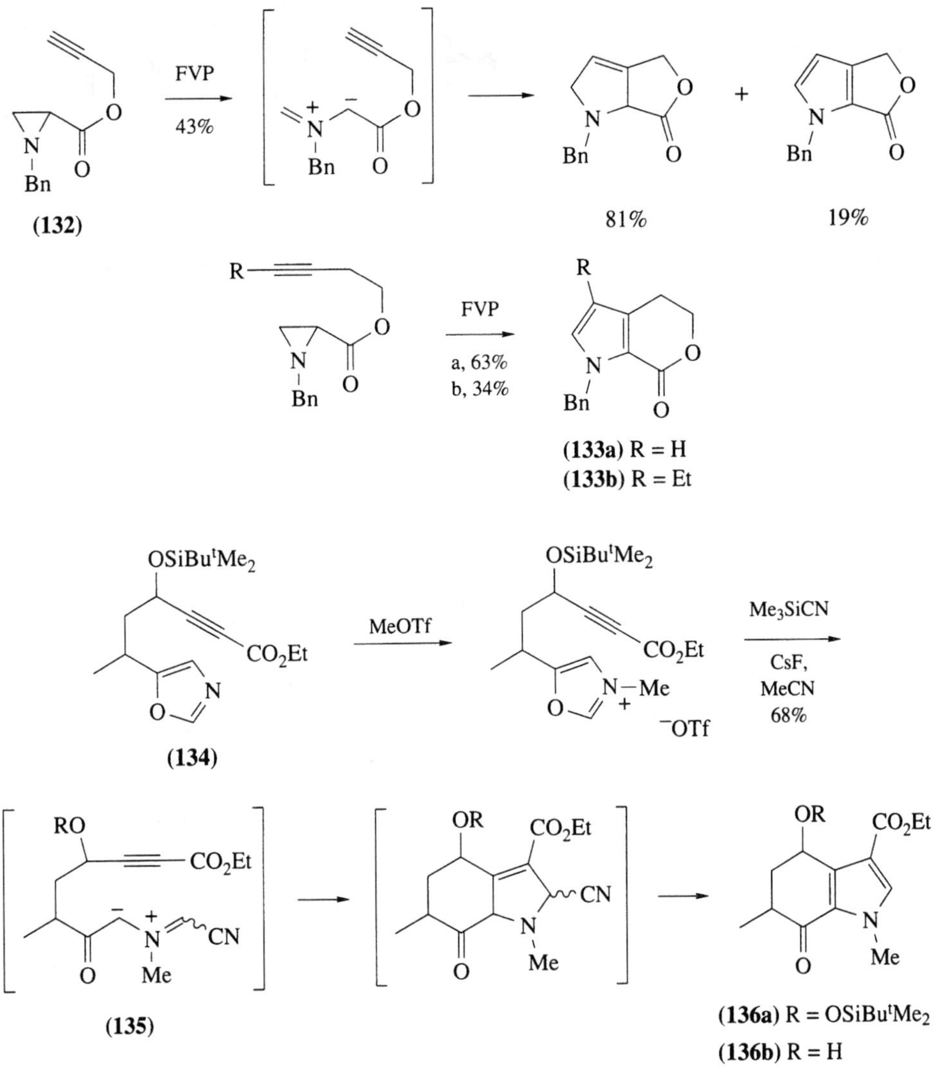

Scheme 41

4.10.5 NITRILE YLIDE CYCLIZATIONS

4.10.5.1 General

Nitrile ylides are not readily isolable but must be used *in situ*. They are typically bent, not linear like nitrile oxides.[68] As bent species containing six π-electrons, nitrile ylides have some carbene character. Consequently, they undergo both intramolecular 1,3- and 1,1-cycloaddition reactions. These lead to a variety of bicyclic dihydropyrroles, pyrroles and cyclopropanes. The LUMO of most azomethine ylides is high in energy. Consequently, the dipole LUMO–dipolarophile HOMO gap is large, leading to poor reaction with nonconjugated alkenes. Electron-attracting groups on the dipolarophile permit successful intermolecular cycloaddition.

Intramolecular azomethine ylide cycloaddition to the C—O double bond of an aldehyde was reported in 1973[69] and cycloaddition to the C—C double bond was first reported in 1975.[70] Competition between 1,1- and 1,3-cycloaddition is observed in intramolecular reactions, although intermolecular reactions give only 1,3-cycloaddition. Photolysis of 2*H*-azirines is one generation method of nitrile ylides applicable to intramolecular cycloaddition.[70] Another method involves the base-catalyzed 1,3-elimination of hydrogen halide from alkenyl imidoyl halides. Still other procedures involve thermolytic and photolytic cycloreversions of oxazolinones and dihydrooxazaphospholes.

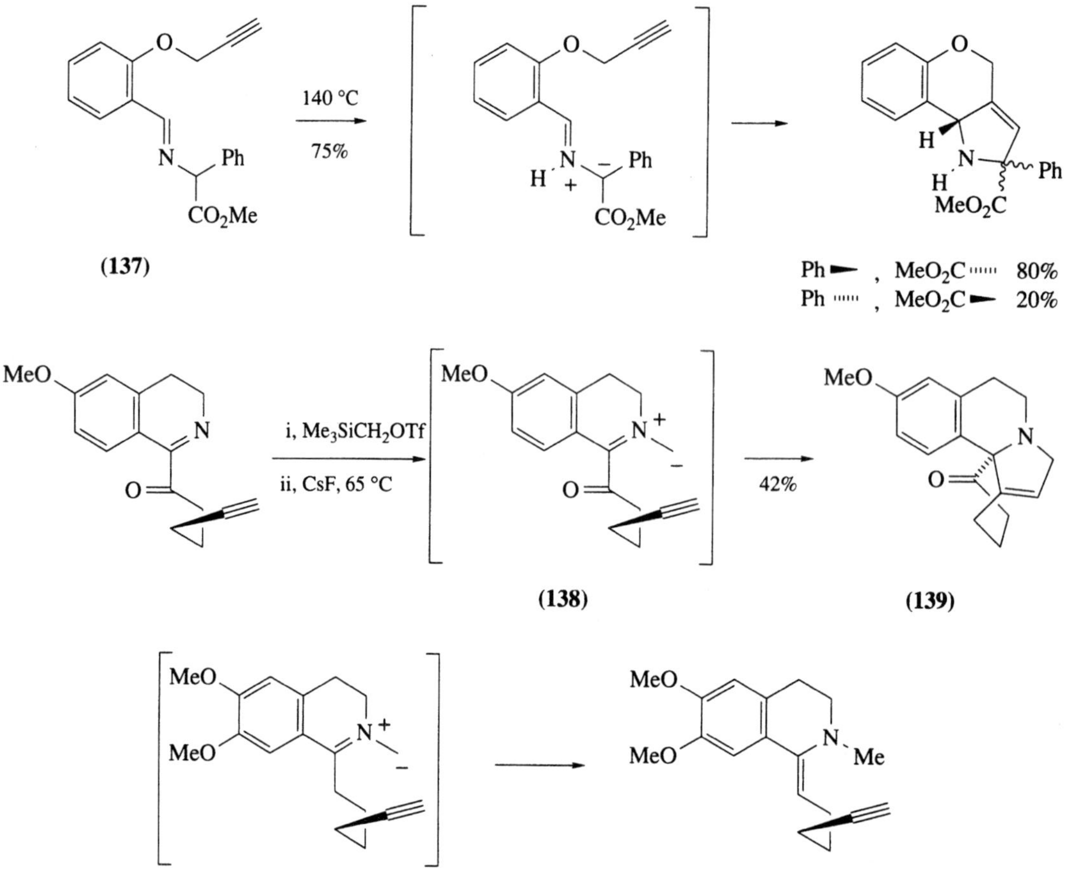

140 °C
75%

(137)

Ph, MeO_2C 80%
Ph, MeO_2C 20%

i, Me_3SiCH_2OTf
ii, CsF, 65 °C

42%

(138)

(139)

(140)

Scheme 42

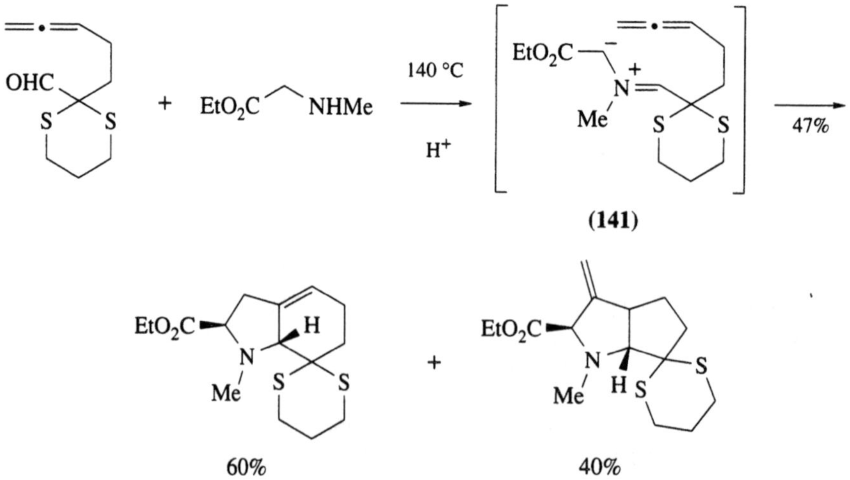

140 °C
H^+

47%

(141)

60%

40%

Scheme 43

4.10.5.2 Alkenyl Nitrile Ylides

Nitrile ylides can have the alkenyl group attached at either end, at the disubstituted or monosubstituted carbon.

4.10.5.2.1 Open-chain nitrile ylides

Open-chain intramolecular systems have been examined with the chain attached at the disubstituted carbon. If the intervening bridge is sufficiently long (three atoms), only 1,3-cycloaddition is observed.[71] Thus, photolysis of azirine (**142**) gave a 5,5-fused bicyclic dihydropyrrole in 81% yield (Scheme 44). This result is rationalized based on the normal concerted cycloaddition of an intermediate azomethine ylide. Proof for the intermediate dipole was obtained by trapping as a pyrrole with dimethyl acetylenedicarboxylate. Approach of the dipole to the dipolarophile on parallel planes is geometrically possible. However, the regiochemistry normally dictated by electronic factors (which should favor a bridged product) was reversed, presumably by geometric constraints.

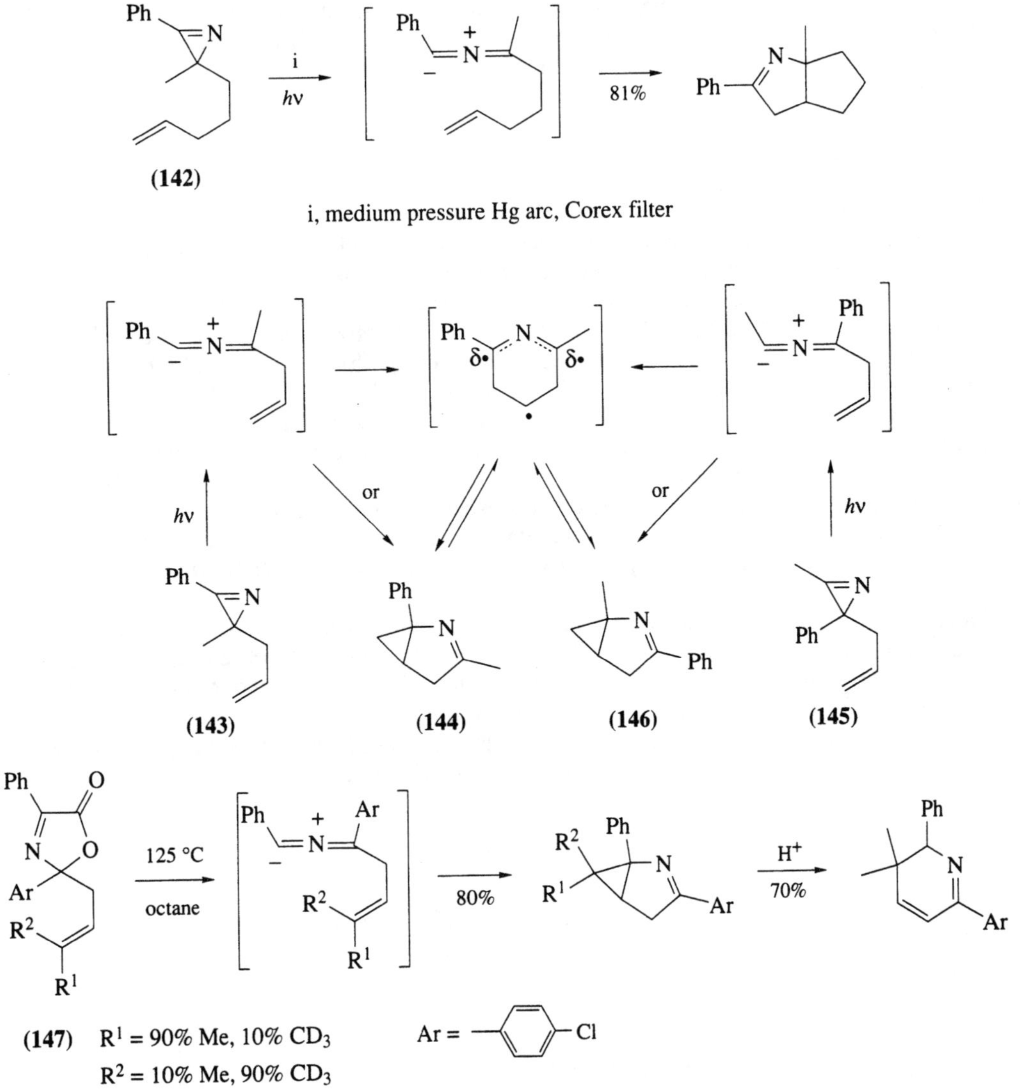

(**142**)

i, medium pressure Hg arc, Corex filter

(**143**) (**144**) (**146**) (**145**)

(**147**) R^1 = 90% Me, 10% CD_3
R^2 = 10% Me, 90% CD_3

Ar = 4-ClC$_6$H$_4$

Scheme 44

The homologous azirine (**143**) with a one-atom bridge gave quite different results.[70] Photolysis led to the 3,5-fused bicyclic dihydropyrrole (**144**). The isomeric azirine (**145**) also led to (**144**), although the initial products included dihydropyrrole (**146**) which apparently converted to (**144**) as photolysis continued. Azirines (**143**) and (**145**) were shown to not interconvert and the postulated two discrete azomethine ylides were trapped with methyl trifluoroacetate. Formation of dihydropyrrole (**144**) was explained based on a two-step cycloaddition process involving a common diradical intermediate. The observation of (**146**) from photolysis of (**145**) but not (**143**) can be explained based on extinction coefficient differences. Azirine (**145**) has a high extinction coefficient as does (**146**). The initial product (**146**) can then be optically pumped to (**144**) with a low extinction coefficient. Azirine (**143**) also has a low extinction coefficient and any (**146**) that formed from it would be optically pumped to (**144**) before observation. It has also been suggested that the cycloaddition is concerted: according to this viewpoint the diradical is a secondary photoproduct, only formed from photolysis of (**146**) and (**144**).[72]

The complete change in periselectivity can be explained based on geometric constraints. The azomethine ylide generated from (**142**) can undergo approach of the dipole to the internal dipolarophile in parallel planes, the required geometry for a concerted [$4\pi + 2\pi$] process.[1a] However, the azomethine ylides generated from (**143**) and (**145**) *cannot* undergo approach of the dipole to the dipolarophile in parallel planes. Instead, it has been suggested that the dipolarophile HOMO must approach perpendicular to the nitrile ylide plane, interacting with the second LUMO, and resulting in a step-wise process.[73]

The 1,1-cycloaddition process also occurs in nonphotolytic reactions involving azomethine ylides. Thermolysis of oxazolinone (**147**) led to a 3,5-fused bicyclic dihydropyrrole in 80% yield.[72] The alkene stereochemistry was maintained in the product, although subsequent photolysis scrambled the methyl and trideuteromethyl groups. Nondeuterated oxazolinone gave the cyclization product which was converted to a dihydropyridine on warming with acid.[74]

4.10.5.2.2 Aryl-bridged nitrile ylides

The only cyclic nitrile ylides reported to undergo cyclization have been aryl-bridged species. The intervening bridge has typically been attached at the monosubstituted nitrile ylide carbon. Attachment at this site permits a large variation in the substituents attached at the disubstituted carbon. Changes in these groups lead to changes in periselectivity: the 1,1- *vs.* 1,3-cycloaddition ratio is altered.

The azomethine ylide (**148a**), formed by photolysis of the corresponding azirine, cyclized exclusively to the 1,3-cycloaddition product, obtained in 75% yield (Scheme 45).[75] The ylide (**148b**) cyclized exclusively to the 1,1-cycloaddition product, isolated in 55% yield as the amine. The azomethine ylide (**148c**) gave a 1:1 mixture of 1,1- and 1,3-cycloaddition products. These results apparently arise from electronic effects: electron-releasing groups at the disubstituted ylide carbon increase the amount of 1,1-cycloaddition while electron-attracting groups favor 1,3-cycloaddition. Thus, azomethine ylides (**148d–e**), generated from an imidoyl chloride and a dihydrooxazaphosphole, gave only the corresponding 1,3-cycloaddition products in 64% yield and quantitative yield, respectively. When the cyclization result for (**148b**) is compared to (**148d**), it is apparent that periselectivity was primarily controlled by electronic rather then steric factors.

The electronic effect has been rationalized based on the degree of linearity of the azomethine ylide.[75] Electron-donating groups should make the dipole more bent while electron-attracting groups should make it more linear. It has been reasoned that the more linear the dipole, the better the concerted 1,3-cycloaddition process. An electron-attracting group on the dipolarophile terminal carbon also favors 1,3-cycloaddition, presumably by speeding up the concerted process. Thus, azomethine ylide (**149a**) gave 1,1-cycloaddition while azomethine ylide (**149b**) gave 1,3-cycloaddition.[76]

4.10.5.3 Alkynyl Nitrile Ylides

Cyclization of the nitrile ylide generated from amide (**150a**) *via* an imidoyl chloride led to an 18% yield of a fused tricyclic pyrrole (Scheme 46).[77] Similarly, the nitrile ylide derived from (**150b**) cyclized in 6% yield.

4.10.6 AZOMETHINE IMINE CYCLIZATIONS

4.10.6.1 General

Azomethine imine cycloadditions provide access to pyrazolidines, pyrazolines and pyrazoles. Intramolecular cyclizations were first reported in 1970.[78] The main method for generation of azomethine imines involves reaction of a 1,2-disubstituted hydrazine with an aldehyde or an aldehyde precursor.

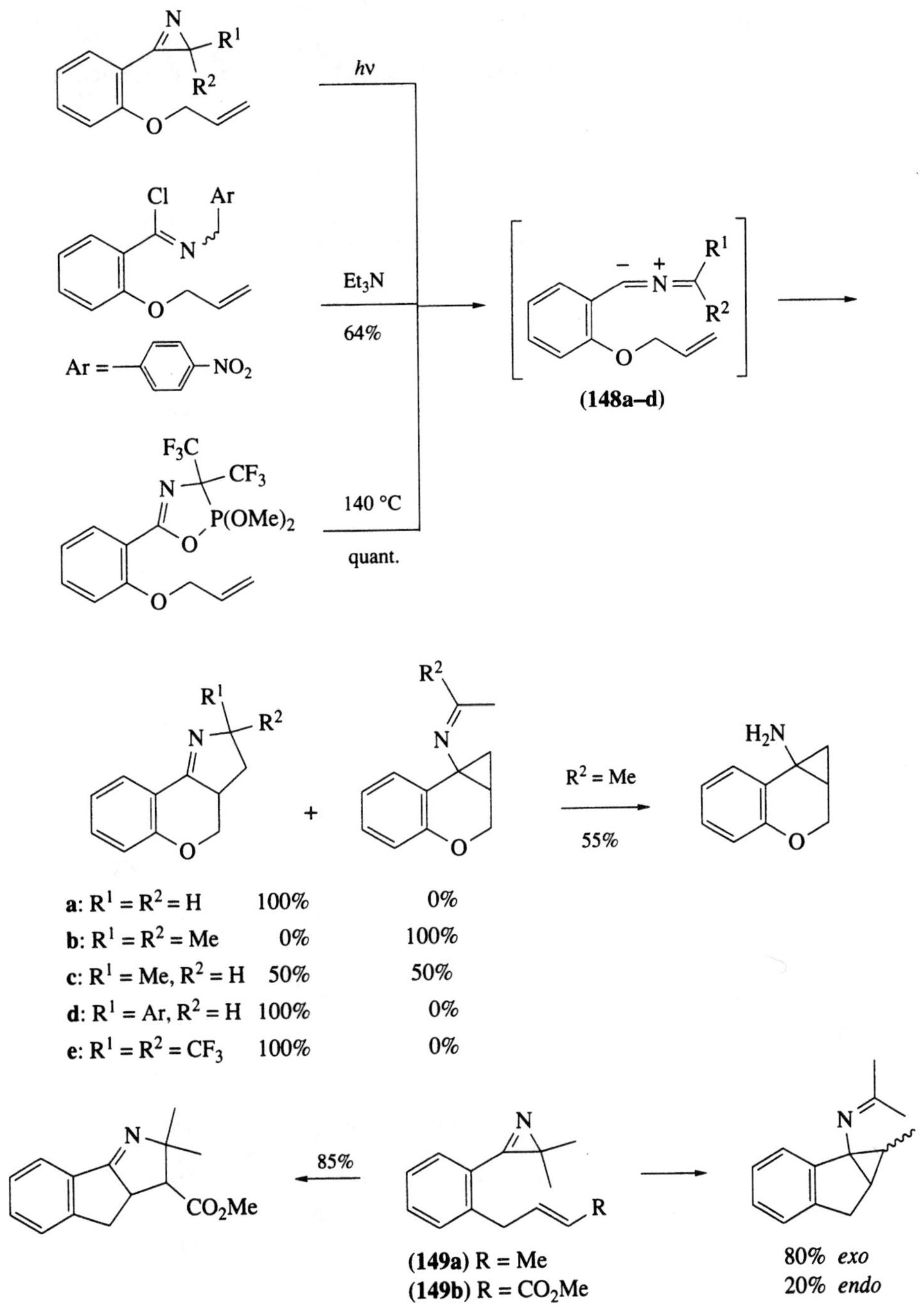

$h\nu$

Cl Ar

N

Et_3N

64%

Ar = —⟨⟩—NO_2

F_3C

CF_3

N

$P(OMe)_2$

O

140 °C

quant.

R[1]

R[2]

(148a–d)

a: $R^1 = R^2 = H$	100%	0%
b: $R^1 = R^2 = Me$	0%	100%
c: $R^1 = Me, R^2 = H$	50%	50%
d: $R^1 = Ar, R^2 = H$	100%	0%
e: $R^1 = R^2 = CF_3$	100%	0%

$R^2 = Me$

55%

H_2N

85%

CO_2Me

(149a) R = Me

(149b) R = CO_2Me

80% *exo*

20% *endo*

Scheme 45

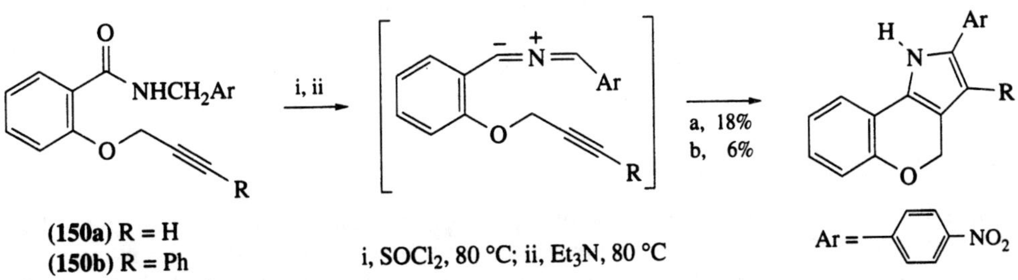

$NHCH_2Ar$

(150a) R = H

(150b) R = Ph

i, ii

i, $SOCl_2$, 80 °C; ii, Et_3N, 80 °C

a, 18%

b, 6%

Ar = —⟨⟩—NO_2

Scheme 46

4.10.6.2 Alkenyl Azomethine Imines

Azomethine imines can have the alkenyl group attached at carbon, the terminal nitrogen or the internal nitrogen.

4.10.6.2.1 Open-chain azomethine imines

The azomethine imine (**151**), having the alkene attached to the terminal dipole nitrogen, was generated *in situ* from the corresponding hydrazine by reaction with benzaldehyde (Scheme 47).[79] As is typical in these reactions, condensation at the more basic benzyl-substituted nitrogen occurred, rather than at the acyl-substituted nitrogen. Cyclization of (**151**) afforded the 5,5-fused pyrazolidine and no bridged product.

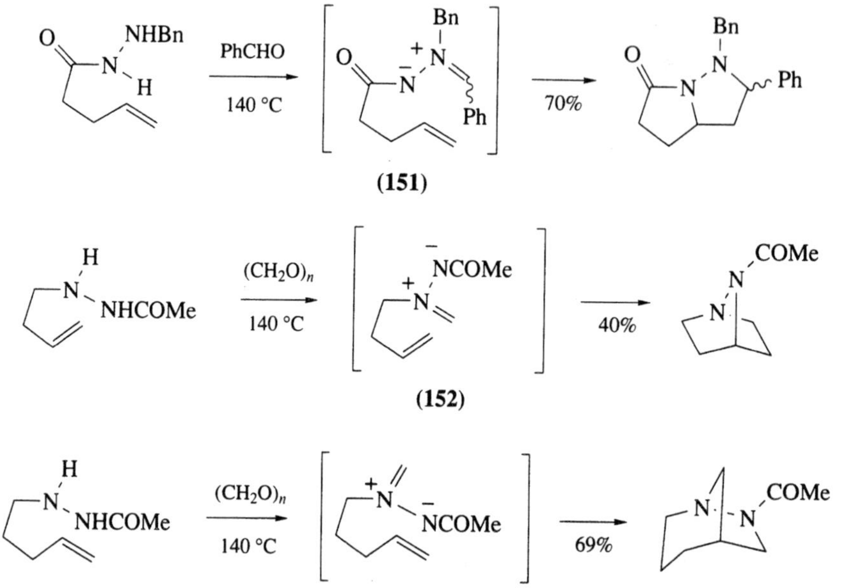

PhCHO, 140 °C → (**151**) → 70%

$(CH_2O)_n$, 140 °C → (**152**) → 40%

$(CH_2O)_n$, 140 °C → (**153**) → 69%

Scheme 47

Cyclization of azomethine imines substituted at the internal nitrogen with an alkenyl group must give bridged products. Thus, imines (**152**) and (**153**) cyclized to bridged pyrazolidines with a different regioisomer obtained as the sole reported product in each case: (**152**) cyclized with the dipole bonding through nitrogen to the terminal dipolarophile carbon whereas (**153**) cyclized in reverse fashion.[79] The imine corresponding to (**153**) but derived from benzaldehyde gave a mixture of regioisomers.

4.10.6.2.2 Aryl-bridged azomethine imines

Azomethine imines which contain an intervening aromatic ring between the dipole and dipolarophile are accessible from aryl aldehydes and readily undergo cyclization. Thus, imine (**154**), where the dipolarophile is attached at the carbon of the dipole, afforded *cis*- and *trans*-fused tricyclic pyrazolidines in which the alkene stereochemistry was retained (Scheme 48).[78a]

4.10.6.2.3 Cycloalkenyl azomethine imines

The azomethine imine (**155**), containing a cyclic dipolarophile, cyclized to a tetracyclic pyrazolidine in 48% yield.[78a] Aldehyde (**156**), containing a furan ring, reacted with *N*-methyl-*N'*-(phenylacetyl)hydrazine to afford an azomethine imine; the imine cyclized to one product, probably with the two new rings *cis* fused.[80]

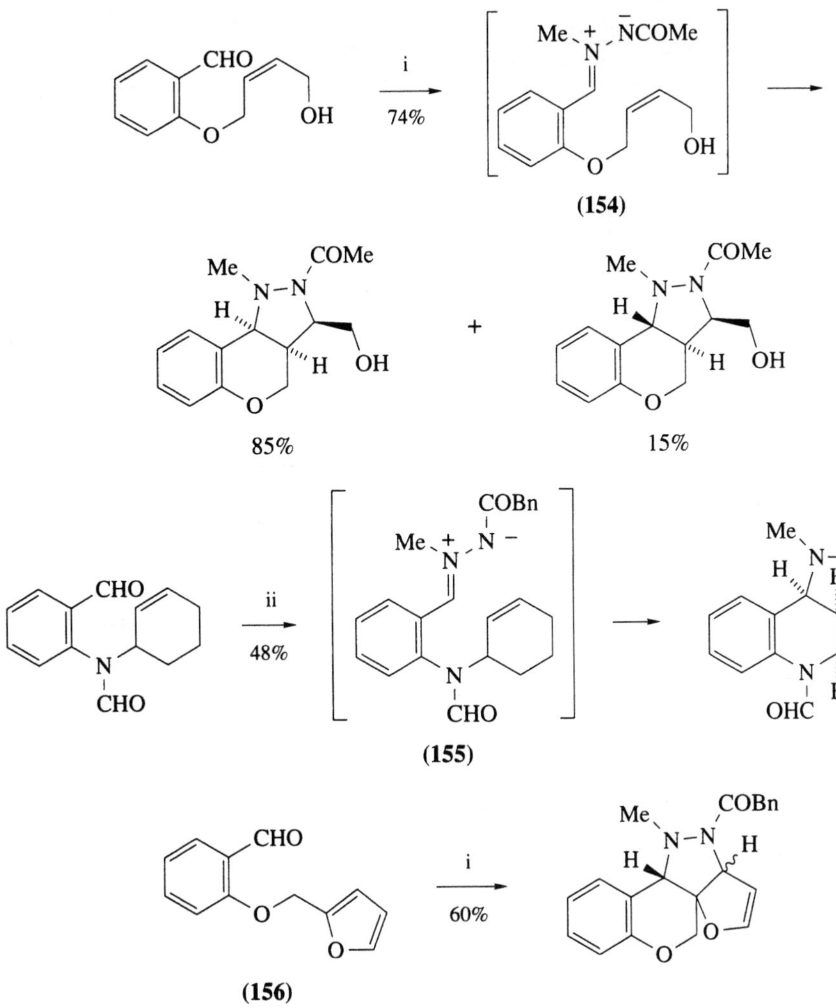

(154)

(155)

(156)

i, MeNHNHCOMe, 110 °C; ii, MeNHNHCOBn, 110 °C

Scheme 48

Jacobi *et al.* have reported the preparation of azomethine imines (**157a–b**) by the acid-catalyzed reaction of the corresponding hydrazines with acetals (Scheme 49).[81] The imine (**157a**) cyclized to a single product as did (**157b**). The stereochemistry of cyclization is apparently controlled by nonbonded interactions between the benzyl group and R group. The pyrazolidine obtained from (**157a**) was converted in several steps to (**160**) and hence to (±)-saxitoxin. First, it was necessary to epimerize the methoxycarbonyl group (thermodynamic control) which was then reduced *in situ* affording (**158**). Direct cleavage of the pyrazolidine ring of (**158**) could be accomplished with sodium in liquid ammonia. However, to complete the synthesis, it was found necessary first to prepare (**159**). Reductive cleavage of the pyrazolidine ring and *in situ* thioacylation afforded (**160**) with the requisite six-membered ring of saxitoxin.

4.10.6.3 Alkynyl Azomethine Imines

Azomethine imines can cycloadd to an internal C—C triple bond producing new bicyclic pyrazoline systems. Thus, imine (**161**), prepared *in situ* from the corresponding aldehyde, cyclized to a tricyclic pyrazoline (Scheme 50).[78a]

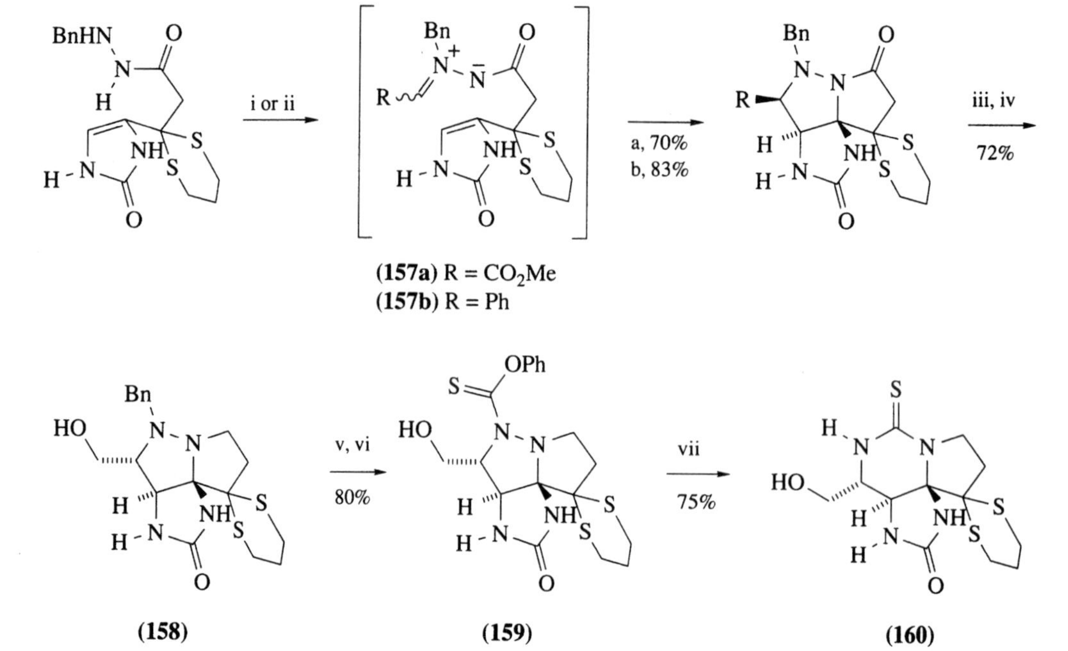

(157a) R = CO_2Me
(157b) R = Ph

(158) (159) (160)

i, $PhCH(OEt)_2$, TsOH, 80 °C; ii, $MeOCH(OH)CO_2Me$, BF_3•OEt_2, 82 °C; iii, NaOMe; iv, $NaBH_4$ then BH_3•SMe_2; v, Pd, HCO_2H; vi, PhOCSCl, pyridine; vii, Na, NH_3, –78 °C

Scheme 49

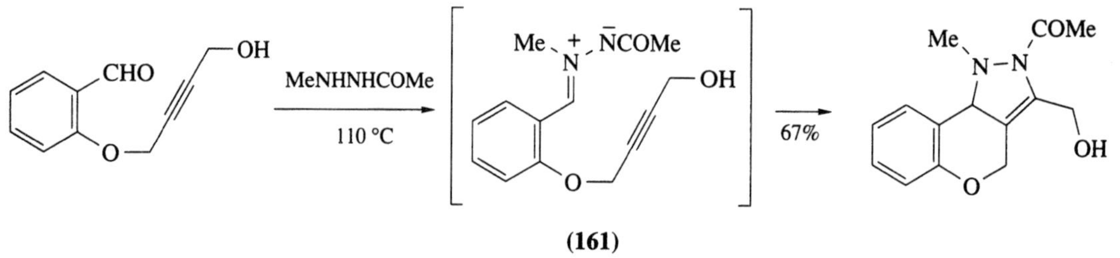

(161)

Scheme 50

4.10.6.4 Hydrazone Cyclization Reactions

Alkenyl hydrazones undergo cyclization to pyrazolines on heating, presumably *via* an azomethine imine tautomer; cyclization to a pyrazolidine is presumably followed by *in situ* oxidation.[82] Thus, **(162b)** afforded a 36% yield of the tricyclic pyrazoline (**163b**; Scheme 51). Several similar cyclizations have been reported.[82,83] However, a synthetically more useful procedure is to treat the hydrazone with strong acid.[84,82] Under these conditions (**162a**) was converted to a mixture of pyrazoline (**163a**) and pyrazolidine (**164a**), which could be completely converted at 140 °C to the corresponding pyrazole. Hydrazone (**165**) cyclized to a tricyclic pyrazolidine under similar conditions. In these acid-catalyzed reactions, a protonated azomethine imine, rather then the free azomethine imine, is the likely intermediate and the process is formally a $[3^+ + 2]$ cycloaddition.

Aldehyde (**166**) reacted with excess hydrazine hydrochloride to afford a pyrazoline in 63% yield: a mixture of pyrazoline and pyrazolidine was obtained under an inert atmosphere.[85] This reaction presumably involves the hydrazone and protonated azomethine imine as intermediates. Using (**166**) in excess afforded a bisintramolecular adduct in 87% yield.

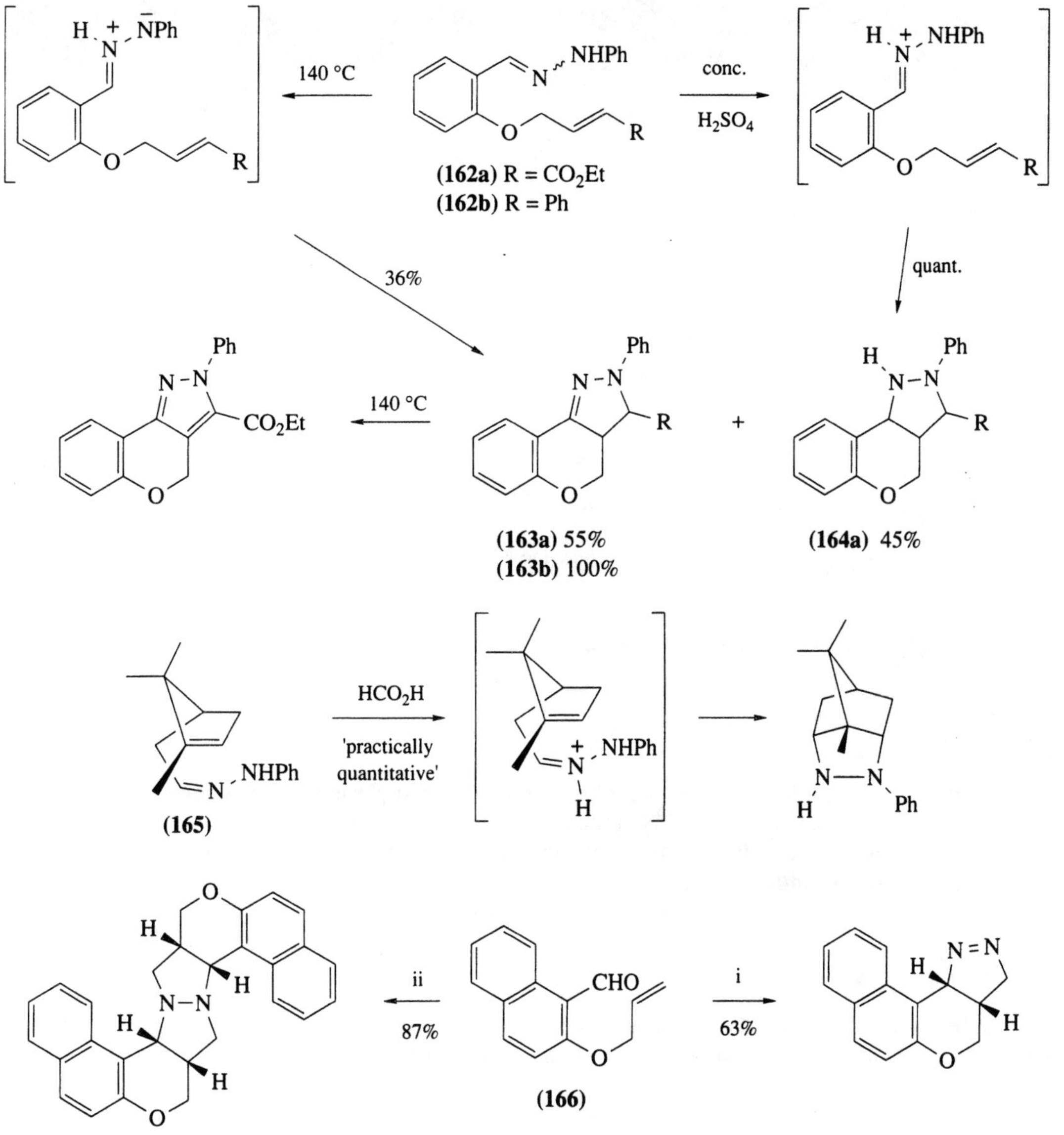

140 °C; conc. H_2SO_4; 36%; quant.; 140 °C

(162a) R = CO_2Et
(162b) R = Ph

(163a) 55%
(163b) 100%

(164a) 45%

HCO_2H 'practically quantitative'

(165)

ii 87%; i 63%

(166)

i, 20 equiv. $NH_2NH_2{\bullet}2HCl$, 78 °C; ii, 0.5 equiv. $NH_2NH_2{\bullet}2HCl$, 78 °C then Et_3N

Scheme 51

4.10.6.5 Azomethine Imine Cyclizations Employing Sydnones

The mesoionic compounds known as sydnones serve as cyclic azomethine imines. Thus, sydnone (**167**), isolable after preparation from the corresponding nitrosamine, underwent cyclization as an azomethine imine at 20–35 °C (Scheme 52).[86] Photolysis of sydnones also results in cyclization but through nitrile imine intermediates (*vide infra*).

4.10.6.6 Tandem Sydnone Intermolecular–Intramolecular Cycloadditions

The products derived from intermolecular cycloaddition of sydnones readily lose carbon dioxide. If the original cycloaddition is carried out on a diene, the resulting azomethine imine intermediate can be trapped intramolecularly. Thus, 1,5-cyclooctadiene cycloadded to a sydnone to afford the cycloadduct (**168**).[78b] It is theorized that the intermolecular monocycloadduct (**169**) was formed, lost carbon dioxide, and cycloadded as a cyclic azomethine imine to the remaining C—C double bond. Conjugated dienes have been shown to undergo a similar sequence.[87]

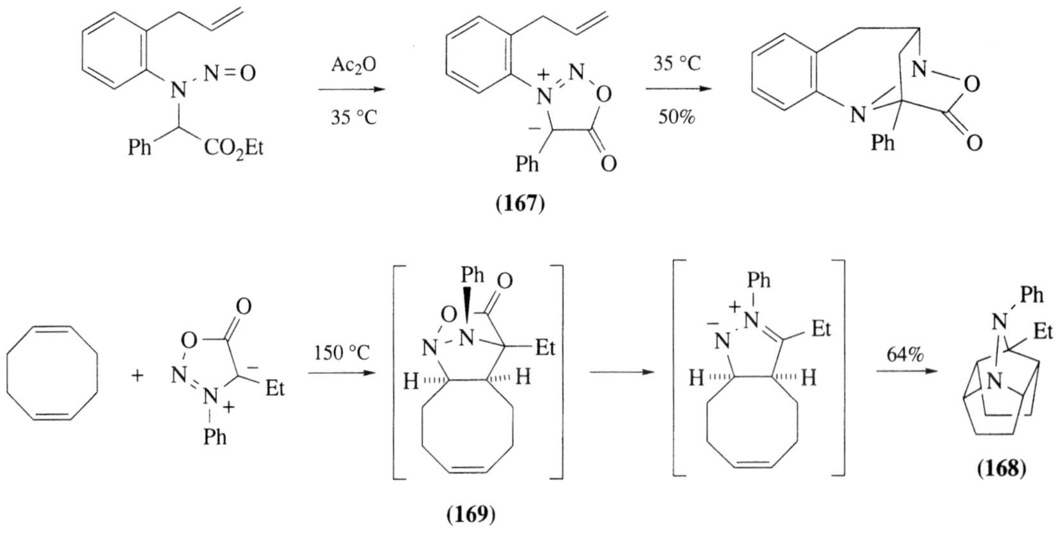

(167)

(169)

(168)

Scheme 52

4.10.7 NITRILE IMINE CYCLIZATIONS

4.10.7.1 General

Nitrile imine cycloadditions provide access to pyrazolines and pyrazoles. Intramolecular cyclizations to alkynes were first reported in 1974.[88] Perhaps the most useful method for generation of nitrile imines involves 1,3-elimination of hydrogen chloride from an α-chlorohydrazone. Tetrazoles and sydnones are also precursors to nitrile imines.

4.10.7.2 Alkenyl Nitrile Imines

Nitrile imines can have the alkenyl group attached at either end of the 1,3-dipole, through carbon or nitrogen. Predominantly 1,3-cyclizations have been observed although a few 1,1-cyclizations have been reported.

4.10.7.2.1 Open-chain nitrile imines

The nitrile imines (**170a–b**), having the alkene attached to carbon, were generated *in situ* from the corresponding chlorohydrazone by reaction with triethylamine (Scheme 53).[89] Cyclization afforded the 5,5-fused pyrazolines and no bridged product. Alkene stereochemistry present in (**170b**) was retained in the cyclization product. It is noteworthy that nitrile imines typically fail to cycloadd intermolecularly to unactivated alkenes. The nitrile imine (**171**), presumably formed by reaction of the corresponding chlorohydrazone with base, failed to cyclize, probably owing to a reduced entropic advantage. The nitrile imine (**172**) also failed to cyclize, presumably because the dipole and dipolarophile could not approach on parallel planes.[90] Nor did (**172**) give a 1,1-cycloadduct, although it did undergo intermolecular cycloaddition.

4.10.7.2.2 Aryl-bridged nitrile imines

A large number of nitrile imines which contain an intervening aromatic ring between the dipole and dipolarophile have been cyclized. Thus, the nitrile imines (**173a–b**), where the dipolarophile is attached at the nitrogen of the dipole, afforded tricyclic pyrazolines (Scheme 54).[89,86] Imine (**173a**) was generated from the chlorohydrazone, while (**173b**) was generated by photorearrangement of a sydnone. Nitrile imines (**174a–c**), where the dipolarophile is attached at the carbon of the dipole, also afforded tricyclic py-

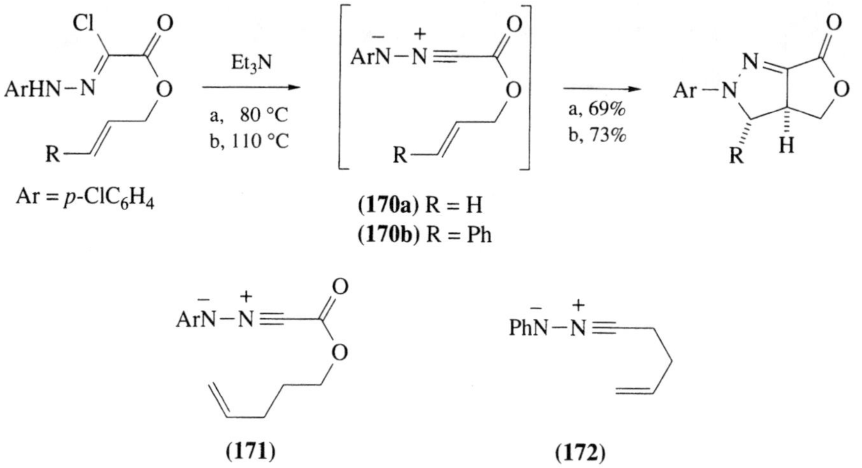

Cl O
ArHN–N O
R
Ar = p-ClC_6H_4

Et_3N
a, 80 °C
b, 110 °C

ArN–N≡ O O R

(170a) R = H
(170b) R = Ph

a, 69%
b, 73%

Ar–N N O O R H

ArN–N≡ O O

PhN–N≡

(171) **(172)**

Scheme 53

razolines.[91] Nitrile imine (**174a**) was formed from a tetrazole[91b] and (**174c**) by lead tetraacetate oxidation of a sulfonyl hydrazone.[91c] Other hydrazones also afford nitrile imines on lead tetraacetate oxidation but oxidation of the pyrazoline products to pyrazoles is a side reaction.[92]

Nitrile imines can cyclize to 1,1-cycloadducts if the dipole–dipolarophile parallel plane approach is unfavorable. Thus, cyclopropanes (**175a–b**) were independently obtained on short-term reaction of the chlorohydrazone precursors with silver carbonate (Scheme 55).[93] Longer reaction times afforded a 3:1 b:a ratio from either precursor: the products were shown to equilibrate. Heating of cyclopropanes (**175a–b**) afforded a benzodiazepine.

4.10.7.2.3 Cycloalkenyl nitrile imines

The nitrile imine (**176**), containing a cyclic dipolarophile, cyclized to a tricyclic pyrazoline in 73% yield.[89]

4.10.7.3 Alkynyl Nitrile Imines

Numerous alkynyl nitrile imines have been shown to undergo cyclization. Typical examples are the preparation of pyrazoles (**177**) and (**178**; Scheme 56).[90,94]

4.10.8 CYCLIZATION OF DIAZO COMPOUNDS

4.10.8.1 General

Diazomethane and its simple analogs undergo cycloaddition to unsaturated compounds both directly and after conversion to carbenes. The direct cycloadditions are 1,3-dipolar for the most part and provide access to pyrazolines and pyrazoles. Intramolecular cyclizations were recognized as early as 1965.[95] The two main methods used in generation of diazo compounds for subsequent intramolecular cycloaddition include thermolysis of tosylhydrazone salts and thermolysis of iminoaziridines. Decomposition of nitrosamines has also been employed.

4.10.8.2 Cyclization of Diazoalkenes

Diazoalkenes always have the intervening chain attached on carbon. The dipole HOMO–dipolarophile LUMO interaction is usually dominant in intermolecular cycloadditions. Consequently, nonactivated alkenes which have a relatively high-energy LUMO react poorly. Nevertheless, intramolecular cycloaddi-

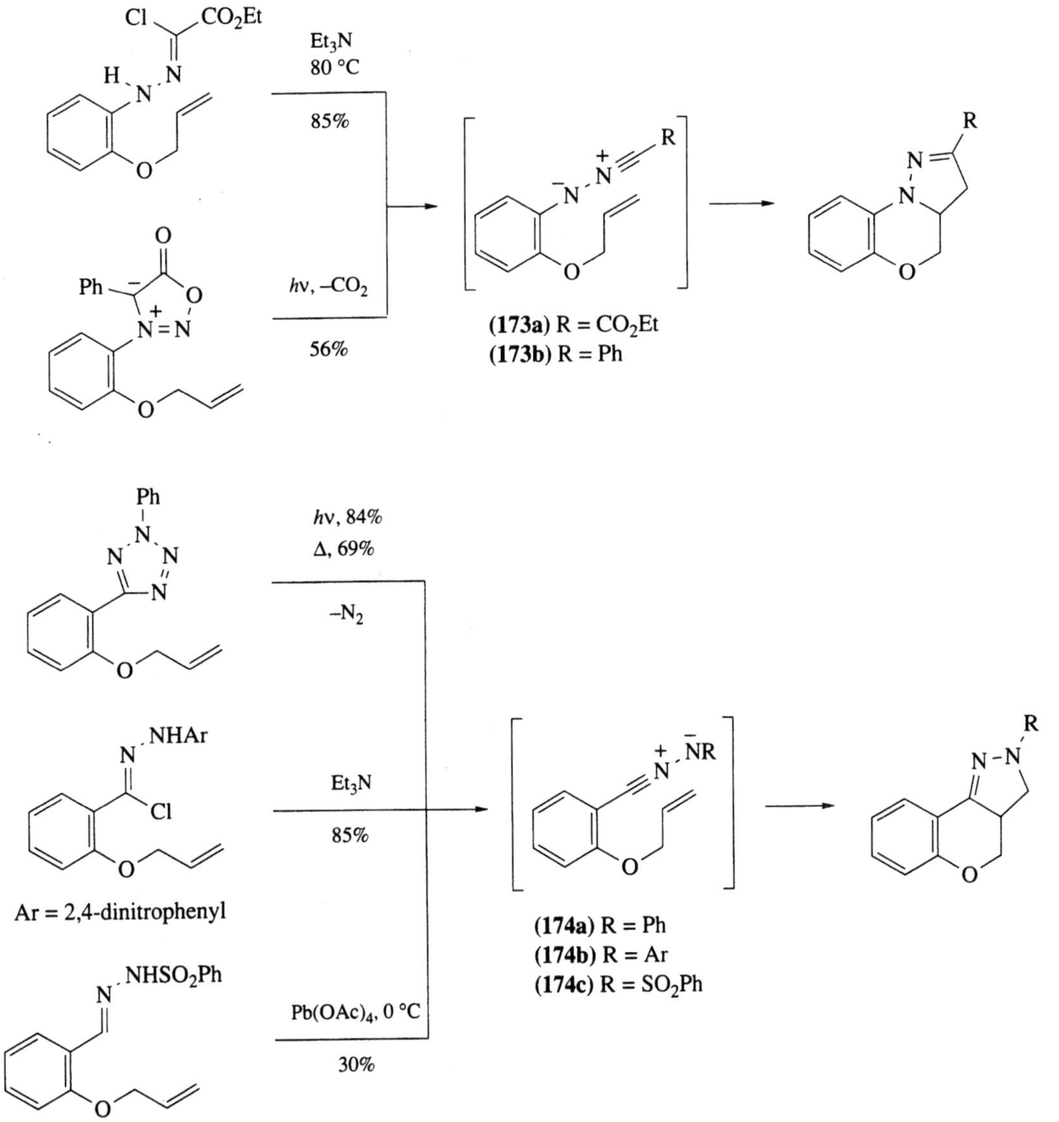

Et$_3$N
80 °C
85%

$h\nu$, $-CO_2$
56%

(**173a**) R = CO_2Et
(**173b**) R = Ph

$h\nu$, 84%
Δ, 69%
$-N_2$

Ar = 2,4-dinitrophenyl

Et$_3$N
85%

Pb(OAc)$_4$, 0 °C
30%

(**174a**) R = Ph
(**174b**) R = Ar
(**174c**) R = SO_2Ph

Scheme 54

tion of nonactivated alkenes does occur; indeed, activated alkenes have been largely ignored in intramolecular diazo cycloadditions.

Diazo compounds can often be isolated, although many intramolecular cycloadditions occur at room temperature. Thus, the diazo compounds have typically been treated as reactive intermediates. Both 1,3-cycloaddition and 1,1-cycloaddition reactions have been observed, depending on the substrate geometry.

4.10.8.2.1 Open-chain diazoalkenes

Open-chain diazoalkenes can conceivably cyclize to give either fused or bridged bicyclic products. Only fused products have been reported. Thus, the diazoalkene (**179a**) cyclized to a 5,5-fused bicyclic pyrazoline in 76% overall yield when (**179a**) was formed from an iminoaziridine precursor (Scheme 57).[96] A tosylhydrazone salt was also cyclized in unstated yield *via* (**179a**). The analogous diazoalkene (**179b**), formed from a tosylhydrazone precursor, cyclized in 69% overall yield.[97] The homolog (**180**) with a two- rather than three-carbon intervening chain did not cyclize at room temperature: it was stable unless exposed to air. The homolog with a four-carbon intervening chain cyclized in only 16% yield. Longer intervening chains have not been investigated, but the reduced entropic activation would make these compounds unattractive candidates for high-yield cyclization.

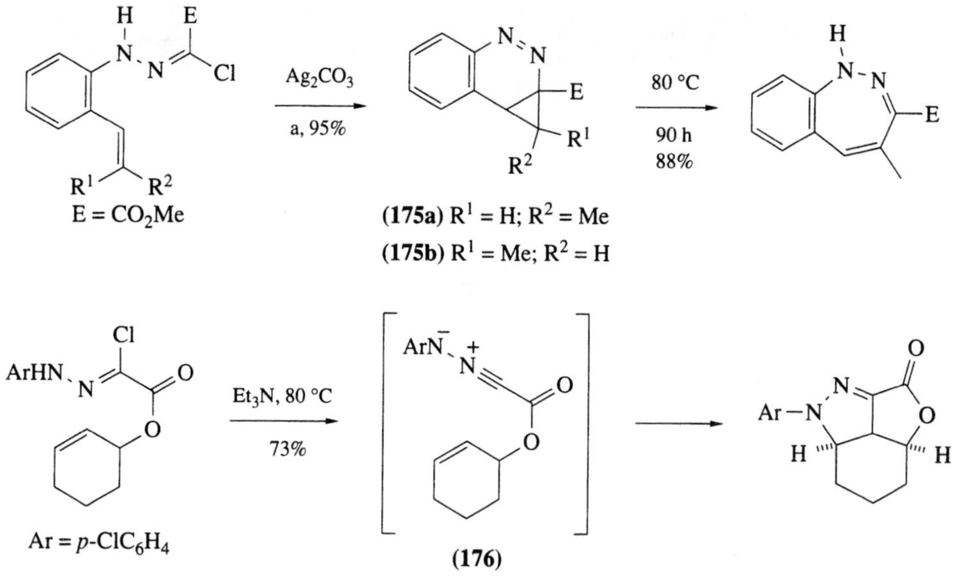

$E = CO_2Me$

(175a) $R^1 = H$; $R^2 = Me$

(175b) $R^1 = Me$; $R^2 = H$

$Ar = p\text{-}ClC_6H_4$

(176)

Scheme 55

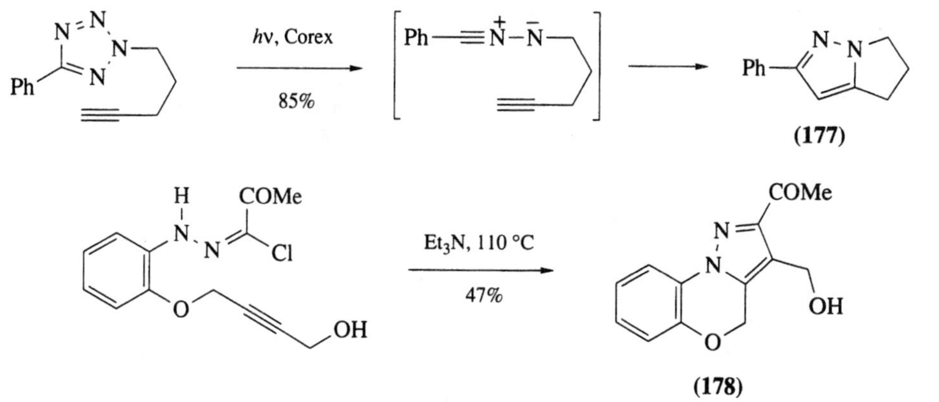

(177)

(178)

Scheme 56

The diazo compound (**181a**), prepared from the nitrosamine, cyclized to a pyrrolopyrazoline in 80% yield.[98] The diazo compound (**181b**), prepared from diethyl diazomalonate and allylamine, cyclized similarly but at a much more rapid rate. This is consistent with the lowered LUMO of the dipole of (**181b**), substituted with an ester group: here the dipole LUMO–dipolarophile HOMO is the likely dominant interaction. The N—N double bond of the pyrrolopyrazoline products was readily isomerized to afford Δ^2-isomers.

In contrast to the above results, the diazo compounds (**182a–b**) with a one-carbon intervening chain underwent exclusive 1,1-cycloaddition.[97] These cyclizations were stereospecific: the alkene stereochemistry was retained.[97,99]

4.10.8.2.2 Aryl-bridged diazoalkenes

A number of aryl-bridged diazoalkenes have been shown to undergo cyclization. Thus, (**183a**) cyclized to pyrazoline (**184a**) and (**183b**) cyclized to (**184b**; Scheme 58).[96] The alkene stereochemistry was retained in these cyclizations as in 1,1-cyclizations. The pyrazolines (**184a–b**) undergo nitrogen extrusion to afford cyclopropanes: this reaction, however, is nonstereospecific. Both (**184a**) and (**184b**) gave mixtures of the *endo*- and *exo*-methylcyclopropanes, although the product ratios differed (1:3 *endo:exo* from **184b** and 6:1 *endo:exo* from **184a**). These different ratios presumably reflect the rapid closure of the diradical intermediates and require a preference for inversion.

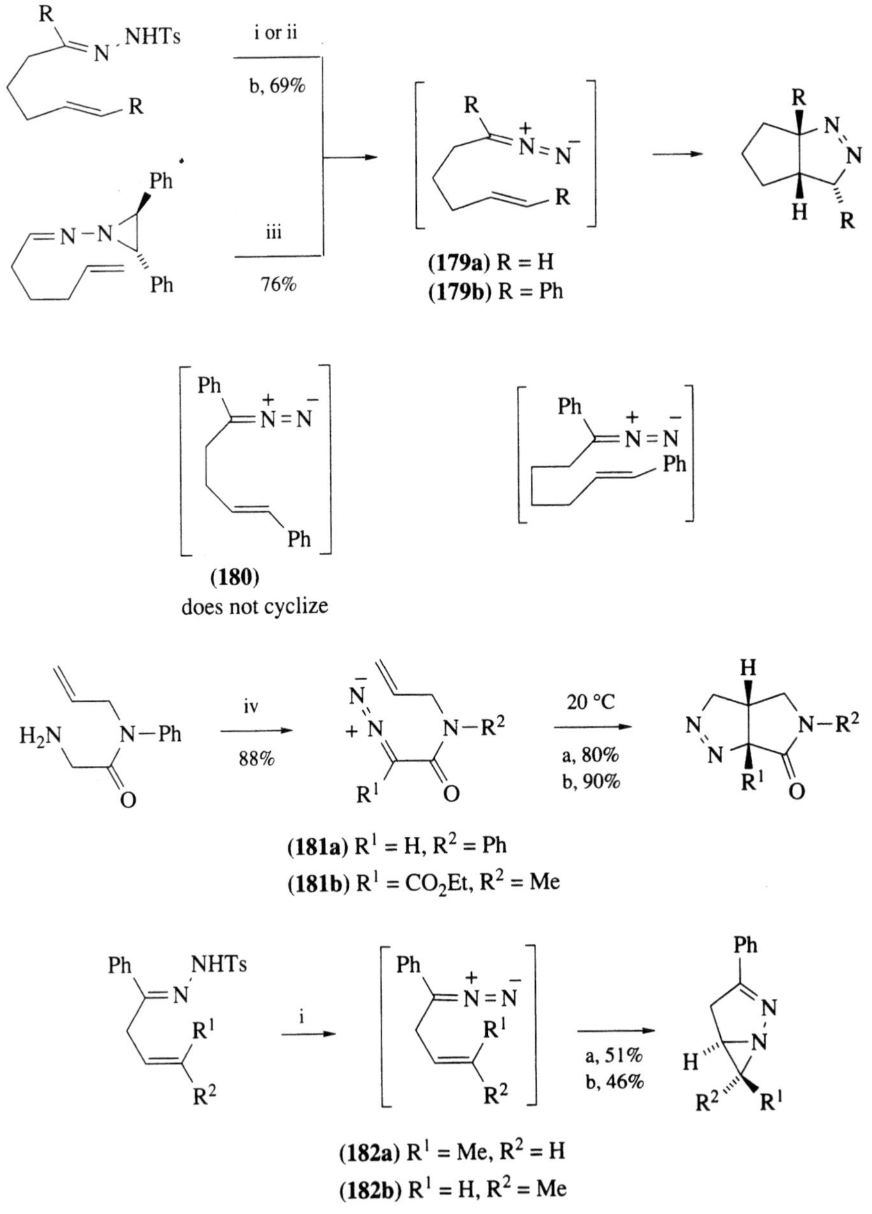

i or ii

b, 69%

iii

76%

(179a) R = H
(179b) R = Ph

(180)
does not cyclize

iv

88%

20 °C

a, 80%
b, 90%

(181a) R^1 = H, R^2 = Ph
(181b) R^1 = CO_2Et, R^2 = Me

i

a, 51%
b, 46%

(182a) R^1 = Me, R^2 = H
(182b) R^1 = H, R^2 = Me

i, NaH, 80 °C, CCl_4 then 20 °C; ii, NaH, THF then 80 °C, C_6H_6; iii, 80 °C, C_6H_6; iv, $NaNO_2$, HOAc, 20 °C

Scheme 57

4.10.8.2.3 Diazocycloalkenes

A variety of cyclic diazo substrates have been shown to undergo cyclization. Thus, the diazodiene derived from thermolysis of iminoaziridine (**185**) underwent intramolecular 1,3-cycloaddition in 43% yield (Scheme 59).[100] A nitrile side product was formed by elimination of diphenylaziridine from (**185**). The cycloaddition product was converted in a number of steps to (–)-longifolene. The bicyclic tosylhydrazone salt (**186**) afforded a tetracyclic pyrazoline in 81% yield from which nitrogen could be extruded to afford (–)-cyclocopacamphene.[101] The diazocyclopentene (**187**), with a one-carbon intervening chain, underwent intramolecular 1,1-cycloaddition affording a tricyclic pyrazoline in 73% yield.[97] The cyclohexenyl and cycloheptenyl homologs of (**187**) reacted similarly to give 1,1-cycloadducts. Presumably the 1,1-cycloadduct (**188**) was formed from cycloheptatriene (**189**), but the actual product, isolated in 66% yield, was isomeric.[102,97] It was hypothesized that (**188**) underwent a 3,3-sigmatropic rearrangement and a subsequent 1,3-hydride shift.

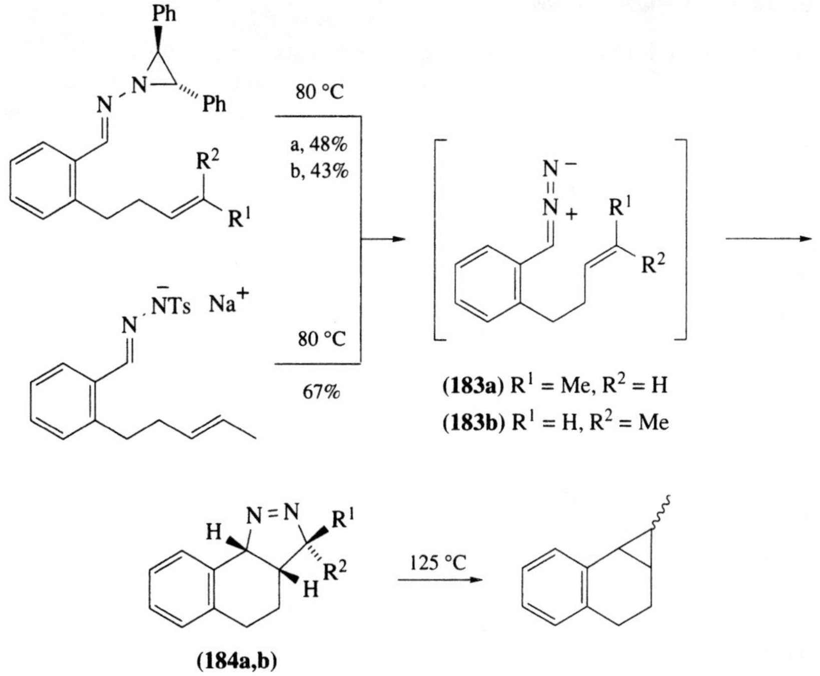

80 °C

a, 48%
b, 43%

80 °C

67%

(183a) R^1 = Me, R^2 = H
(183b) R^1 = H, R^2 = Me

125 °C

(184a,b)

Scheme 58

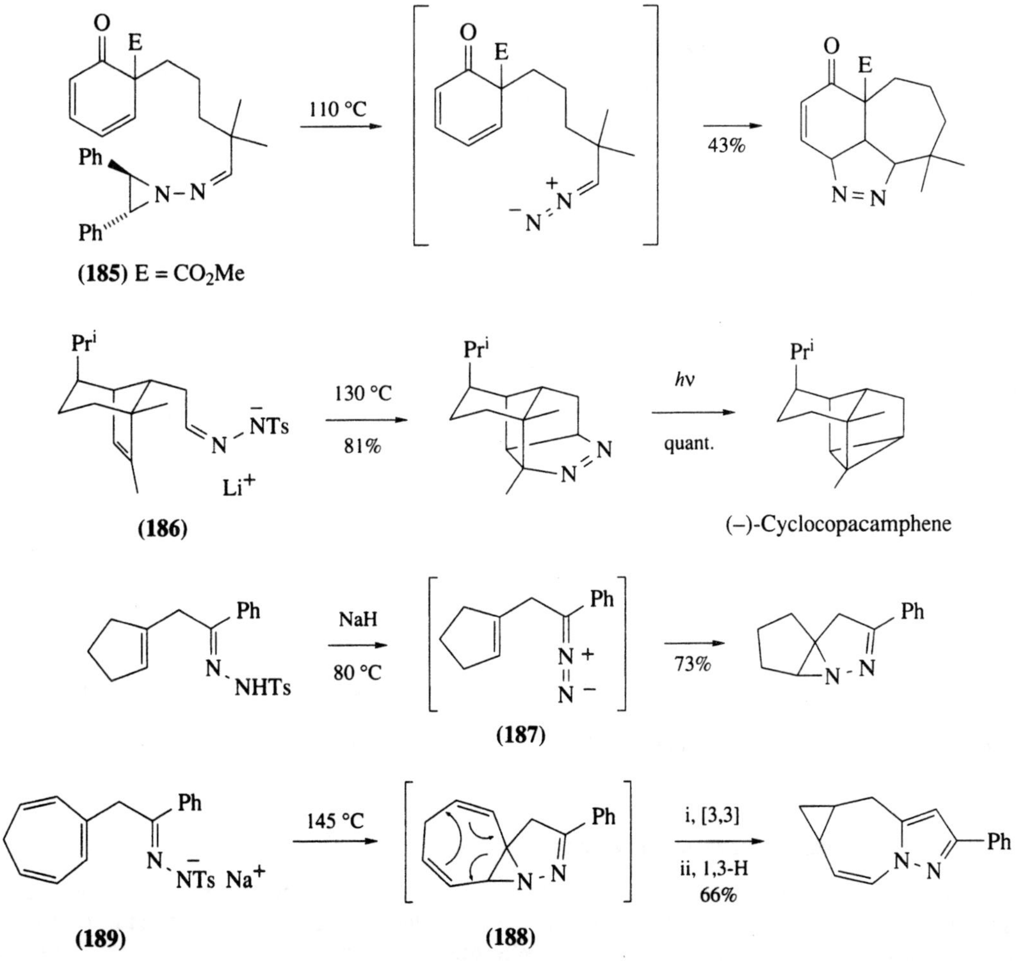

110 °C

43%

(185) E = CO_2Me

130 °C
81%

$h\nu$
quant.

(186)

(–)-Cyclocopacamphene

NaH
80 °C

73%

(187)

145 °C

i, [3,3]
ii, 1,3-H
66%

(189)

(188)

Scheme 59

4.10.8.3 Cyclization of 3-Diazoalkenes

3-Diazoalkenes have been known to cyclize to pyrazoles since 1935. This reaction has been rationalized as another variation of intramolecular 1,3-dipolar cycloaddition.[103] The 3-diazoalkenes, derived from nitrosamines (**190a–b**), gave typical results: cyclization afforded pyrazoles in 89% and 87% yield, respectively (Scheme 60).

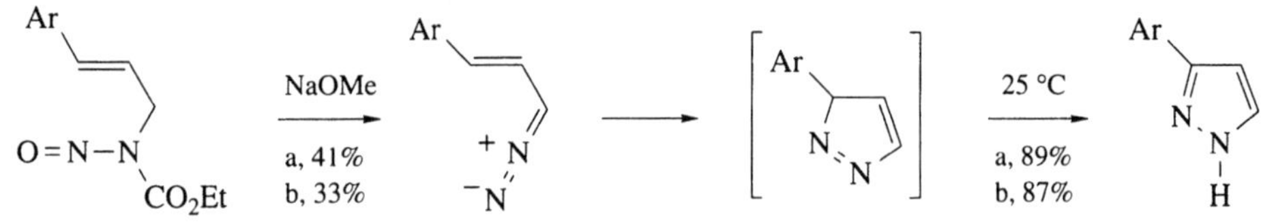

NaOMe; a, 41%; b, 33%

25 °C; a, 89%; b, 87%

(**190a**) Ar = *m*-$C_6H_4NO_2$
(**190b**) Ar = *p*-C_6H_4Cl

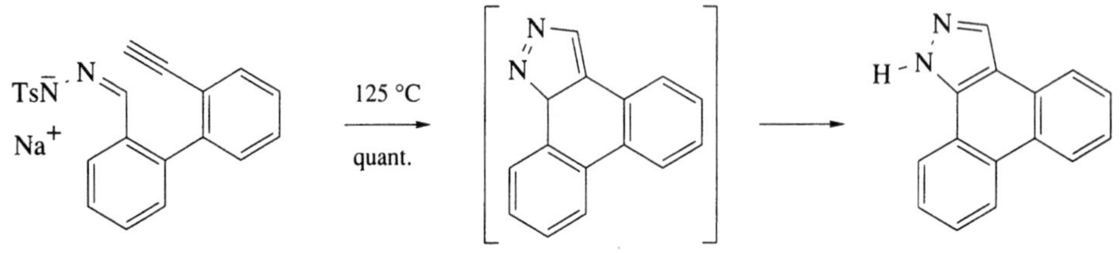

125 °C; quant.

(**191**)

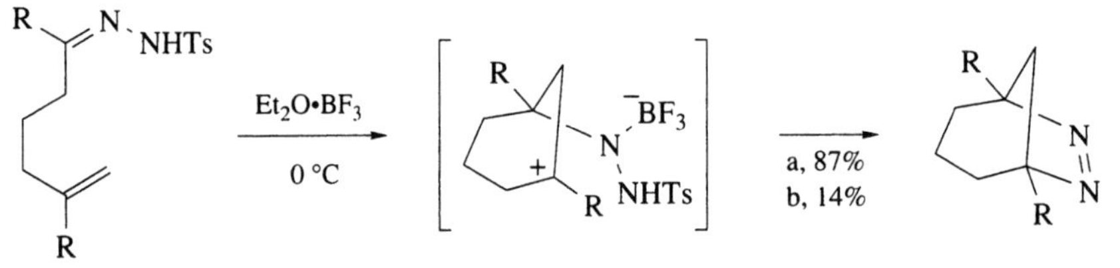

$Et_2O•BF_3$; 0 °C

a, 87%; b, 14%

(**192a**) R = Me
(**192b**) R = H

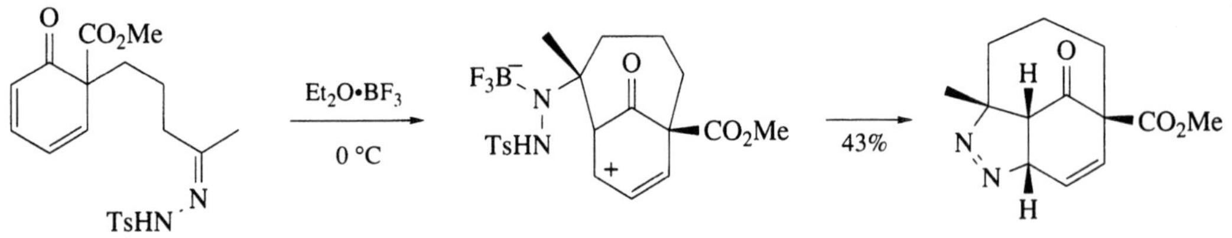

$Et_2O•BF_3$; 0 °C

43%

(**193**)

Scheme 60

4.10.8.4 Cyclization of Diazoalkynes

Intramolecular cycloaddition of diazoalkynes has been little studied. Cyclization of tosylhydrazone salt (**191**) illustrates the potential of these reactions (Scheme 60).[104] The actual product was isomeric with the expected 1,3-cycloadduct: presumably a 1,3-hydride shift followed the initial cyclization.

4.10.8.5 Acid-catalyzed Cyclization of Tosylhydrazones

A number of tosylhydrazones containing an alkene have been shown to undergo intramolecular cycloaddition in the presence of acid.[105,96] Boron trifluoride etherate appears to be the acid of choice.[105] Bridged, rather than fused, bicyclic pyrazolines are formed under these conditions. The mechanism almost certainly involves cationic intermediates. Thus, tosylhydrazone (**192a**) cyclized in 87% yield to the bridged bicyclic pyrazoline. Tosylhydrazone (**192b**), which lacks a methyl group on the C—C double bond, cyclized in only 14% yield while tosylhydrazone (**193**) cyclized in 43% yield.

4.10.9 AZIDE CYCLIZATIONS

4.10.9.1 General

Azides undergo both 1,3-dipolar cycloadditions and nitrene cycloadditions after loss of nitrogen. Triazolines and triazoles are the products of 1,3-dipolar cycloaddition. Triazoles are readily isolable, but triazolines differ markedly in stability as a function of substituents: their isolation sometimes provides an insurmountable challenge. Triazolines frequently lose nitrogen, giving aziridines: thus, the formal product of a nitrene cycloaddition can be obtained indirectly. Intramolecular cyclization of an azide to a nitrile was reported as early as 1958,[106] although it was not until 1965 that intramolecular azide–alkene cycloadditions (IAAC) were recognized.[107]

Azides are normally stable compounds and can be isolated prior to cycloaddition. Occasionally, intramolecular cycloaddition is so favored, however, that the azide cannot be isolated.

Azide cycloaddition to electron-deficient dipolarophiles is normally HOMO–dipole LUMO–dipolarophile controlled, whereas the reverse is true for electron-rich dipolarophiles. Products with an electron-deficient group at the 5-position or an electron-rich group at the 4-position are favored electronically; in intramolecular cycloadditions, steric constraints can be expected to outweigh these considerations.

4.10.9.2 Azidoalkenes

The alkenyl group of an azidoalkene is always attached to the end of the 1,3-dipole. Only 1,3-cyclizations have been observed with azides, typically giving fused bicyclic systems.

4.10.9.2.1 Open-chain azidoalkenes

Relatively few examples of aliphatic open-chain azide cyclizations have been reported in which the triazoline was isolated. In one example, the azidoalkene (**194a**) cyclized at 50 °C to give the 5,5-fused bicyclic triazoline (**195a**; Scheme 61).[107] At 80 °C, triazoline (**195a**) extruded nitrogen, affording a cyclic imine and a lesser amount of a bicyclic aziridine. The azidoalkene (**194b**) cyclized similarly, giving a diasteromeric mixture of (**195b–c**), which, however, was of very limited stability.[108] Silica gel chromatography converted (**195b–c**) to diastereomeric bicyclic aziridines. Thermolysis of (**195b–c**) at 80 °C afforded a Δ^3-oxazoline. A zwitterionic intermediate (**196**) was presumably responsible for formation of the Δ^3-oxazoline; the authors suggested a silica gel coordinated zwitterion as the intermediate in bicyclic aziridine formation.

4.10.9.2.2 Aryl-bridged azidoalkenes

Aryl azides to which an alkenyl group is attached can be readily cyclized. Thus, the azide (**197a**) afforded the tricyclic triazoline (**198a**) at 60 °C, which could be isolated in 12% yield by chromatography (Scheme 62).[109] Silica gel chromatography converted (**198a**) to diazo compound (**199**), used to synthesize a 1,2-dihydroisoquinoline. Azide (**197b**) was more reactive than (**197a**), cyclizing at 0 °C upon formation from bromide (**197b**).[110] At 50 °C, triazoline (**198b**) extruded nitrogen to afford a conjugated alkene.

The bromomethylindole (**200a**) cyclized on treatment with sodium azide to give a tetracyclic triazoline.[111] Treatment with a catalytic amount of *p*-toluenesulfonic acid afforded 2,4-dihydropyrrolo[3,4-*b*]indole (**201**), presumably *via* loss of diethyl diazomalonate by cycloreversion and subsequent tautomerization.

Kozikowski *et al.* have described a synthesis of clavicipitic acids where a key step involved preparation of the imine (**202**).[112] Azide cyclization at 190 °C presumably afforded a triazoline which, however, was nonisolable affording (**202**) in 62% yield.

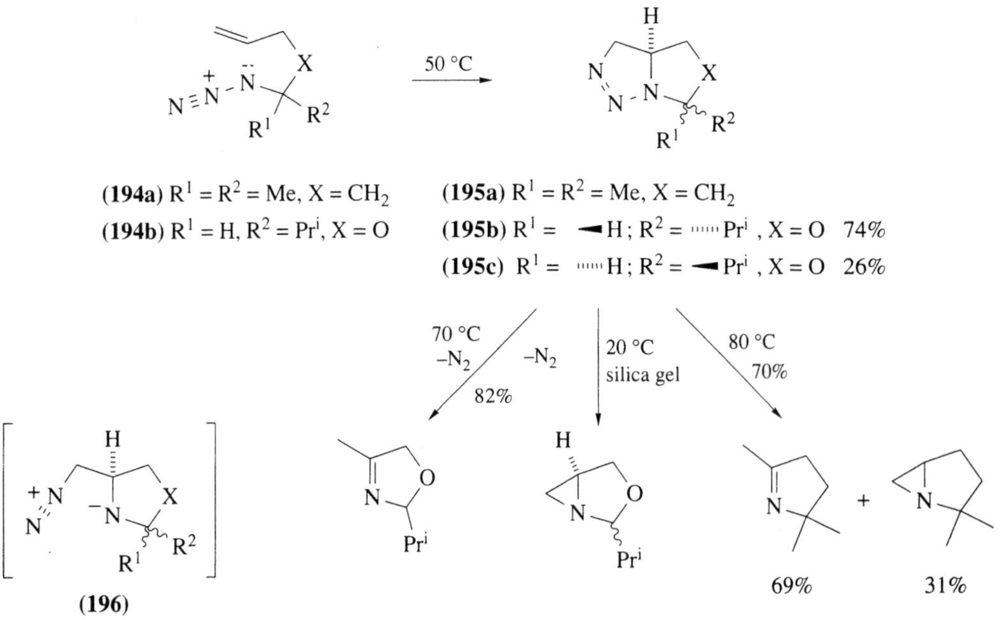

(**194a**) $R^1 = R^2 = Me$, $X = CH_2$
(**194b**) $R^1 = H$, $R^2 = Pr^i$, $X = O$

50 °C

(**195a**) $R^1 = R^2 = Me$, $X = CH_2$
(**195b**) R^1 = H; R^2 = Pr^i, X = O 74%
(**195c**) R^1 = H; R^2 = Pr^i, X = O 26%

70 °C $-N_2$ $-N_2$ 82%; 20 °C silica gel; 80 °C 70%

(**196**)

69% 31%

Scheme 61

4.10.9.2.3 Alicyclic-bridged azidoalkenes

There have been a number of reports where alicyclic-bridged precursors underwent an IAAC reaction. Thus, the dioxolane (**203b**), formed from triflate (**203a**), cyclized *in situ* to a tricyclic triazoline (Scheme 63).[113] Treatment of this triazoline with sodium ethoxide converted it to a diazopyrrolidine in 86% yield, which underwent smooth catalytic hydrogenation in 89% yield. The (*Z*)-azidoalkene (**204**), bridged by a β-lactam, cyclized at 20 °C to triazoline (**205**).[114] The triazoline (**205**) extruded nitrogen at 80 °C providing a tricyclic aziridine. The (*E*)-isomer of (**204**) did not cyclize to a triazoline but instead produced an azirine, presumably *via* a nitrene intermediate.

4.10.9.2.4 Cycloalkenyl azides

Azides undergo intramolecular cycloaddition to C—C double bonds present in rings to provide a variety of novel products. Schultz *et al*., for example, reported thermolysis of azide (**206**) at 110 °C to afford an isolable triazoline as a single diastereomer (Scheme 64).[115] A related reaction, however, gave an isomer mixture. Photochemical nitrogen extrusion from the triazoline provided a dihydropyrrole in 97% yield.

Treatment of the bromide (**207a**) with sodium azide led to a mixture of the bicyclic products (**209**) and (**210**).[116] It was postulated that tricyclic triazoline (**208**) was formed *via* an IAAC reaction: cleavage of the strained system to a common zwitterion followed by competing rearrangement paths might then be responsible for product formation. The bromide (**207b**) presumably gave similar triazoline and zwitterion intermediates but extruded nitrogen to afford the tricyclic aziridine (**211**) without rearrangement. An approach to (±)-desamylperhydrohistrionicotoxin based on this chemistry was also presented.

The (azidofuranosyl)uracil (**212**) cyclized at 110 °C to afford a mixture of mostly a tricyclic triazole and lesser amounts of a tricyclic pyrimidine.[117] A triazoline intermediate was hypothesized; indeed, the triazoline could be intercepted by DDQ-effected aromatization to afford the corresponding tetracyclic triazole. Evidence was presented that the major product was in equilibrium with the triazoline at the reaction temperature.

4.10.9.3 Azidoalkynes

Intramolecular azide cycloaddition to an alkyne produces a triazole. Triazoles are typically much more stable than triazolines and are most often directly isolable. Thus, azide (**213**) cyclized in 89% yield to a triazole (Scheme 65).[118] The azido enyne (**214**) presents an interesting case: intramolecular cycloaddition

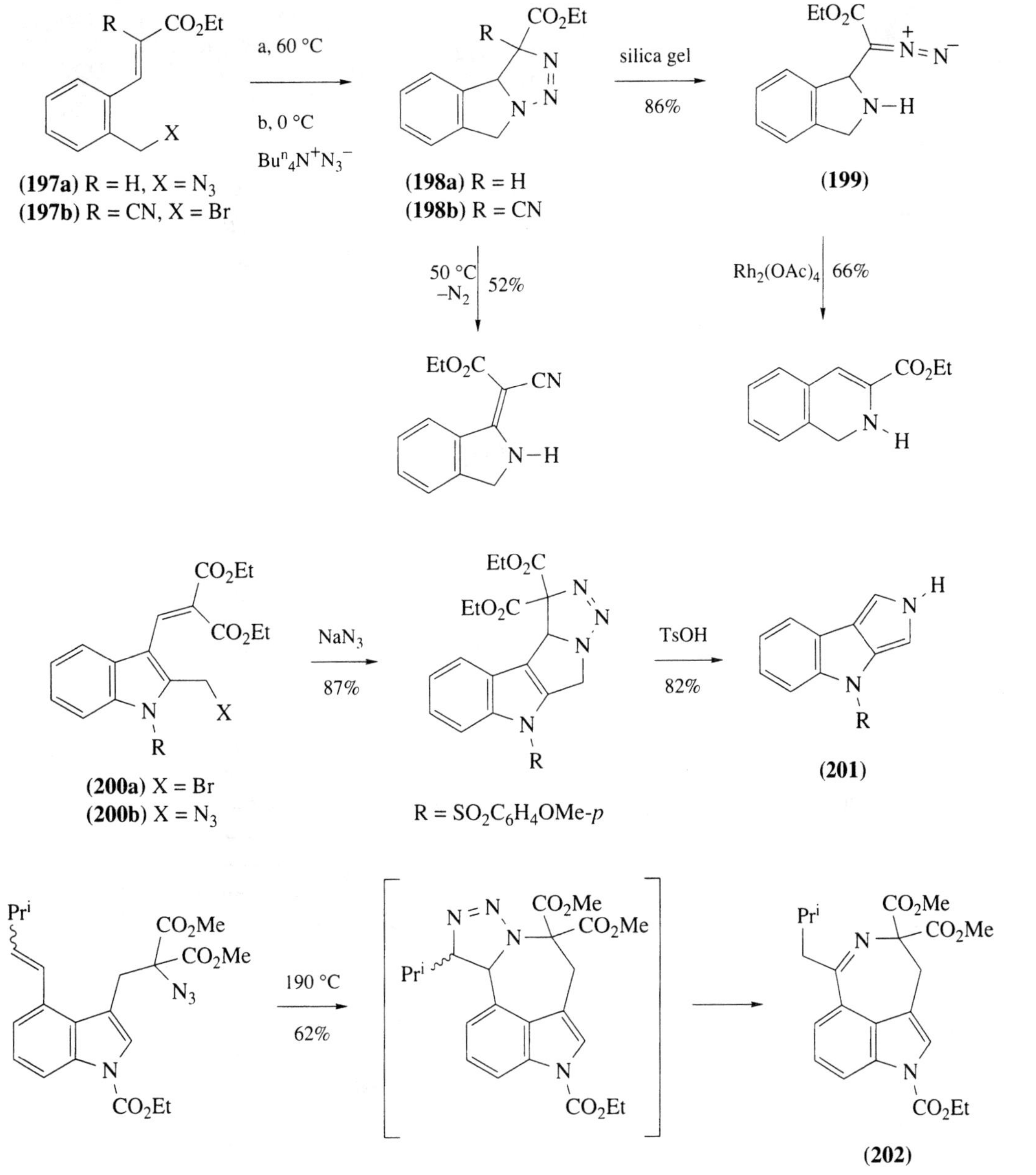

(197a) R = H, X = N_3
(197b) R = CN, X = Br

a, 60 °C; b, 0 °C, $Bu^n{}_4N^+N_3^-$

(198a) R = H
(198b) R = CN

silica gel, 86%

(199)

50 °C, $-N_2$, 52%

$Rh_2(OAc)_4$, 66%

(200a) X = Br
(200b) X = N_3

NaN_3, 87%

R = $SO_2C_6H_4OMe$-*p*

TsOH, 82%

(201)

190 °C, 62%

(202)

Scheme 62

afforded the triazole where the C—C triple bond served exclusively as the dipolarophile, presumably because of geometric constraints.[119] The alkynyl β-lactam (**215**) cyclized to afford triazolocephem (**216**) in 49% yield and an azirine in 12% yield.[120] The authors suggested that (**216**) was derived from (*Z*)-isomer (**215a**), while the azirine was derived from (*E*)-(**215b**) *via* a nitrene. A triazolopenam was synthesized in similar fashion.

4.10.10 CARBONYL YLIDE CYCLIZATIONS

4.10.10.1 General

Carbonyl ylides were first reported to undergo intramolecular 1,3-dipolar cycloaddition to C—C double bonds in 1980.[121] New polycyclic ring systems containing a tetrahydrofuran are formed. Dipola-

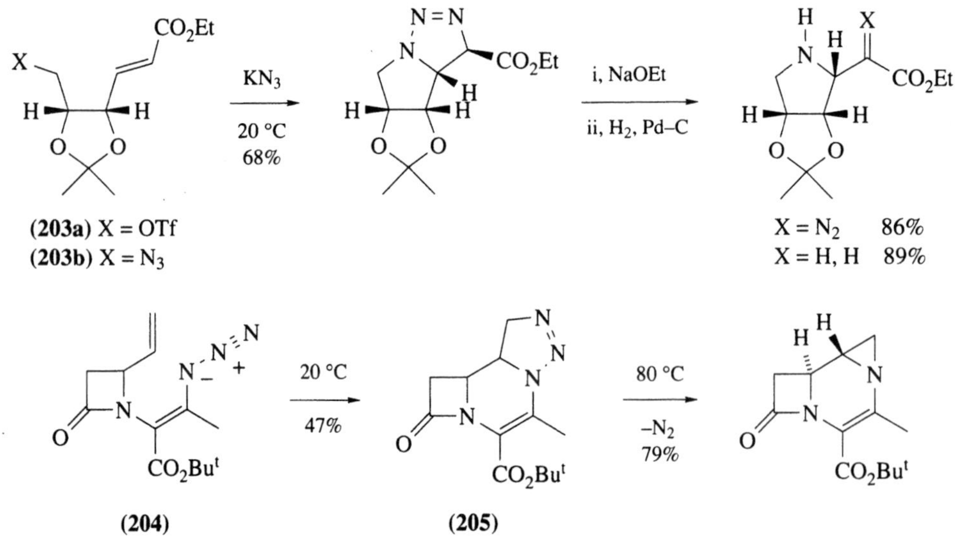

(203a) X = OTf
(203b) X = N_3

KN_3, 20 °C, 68%

i, NaOEt; ii, H_2, Pd–C

X = N_2 86%
X = H, H 89%

(204) 20 °C, 47% (205) 80 °C, $-N_2$, 79%

Scheme 63

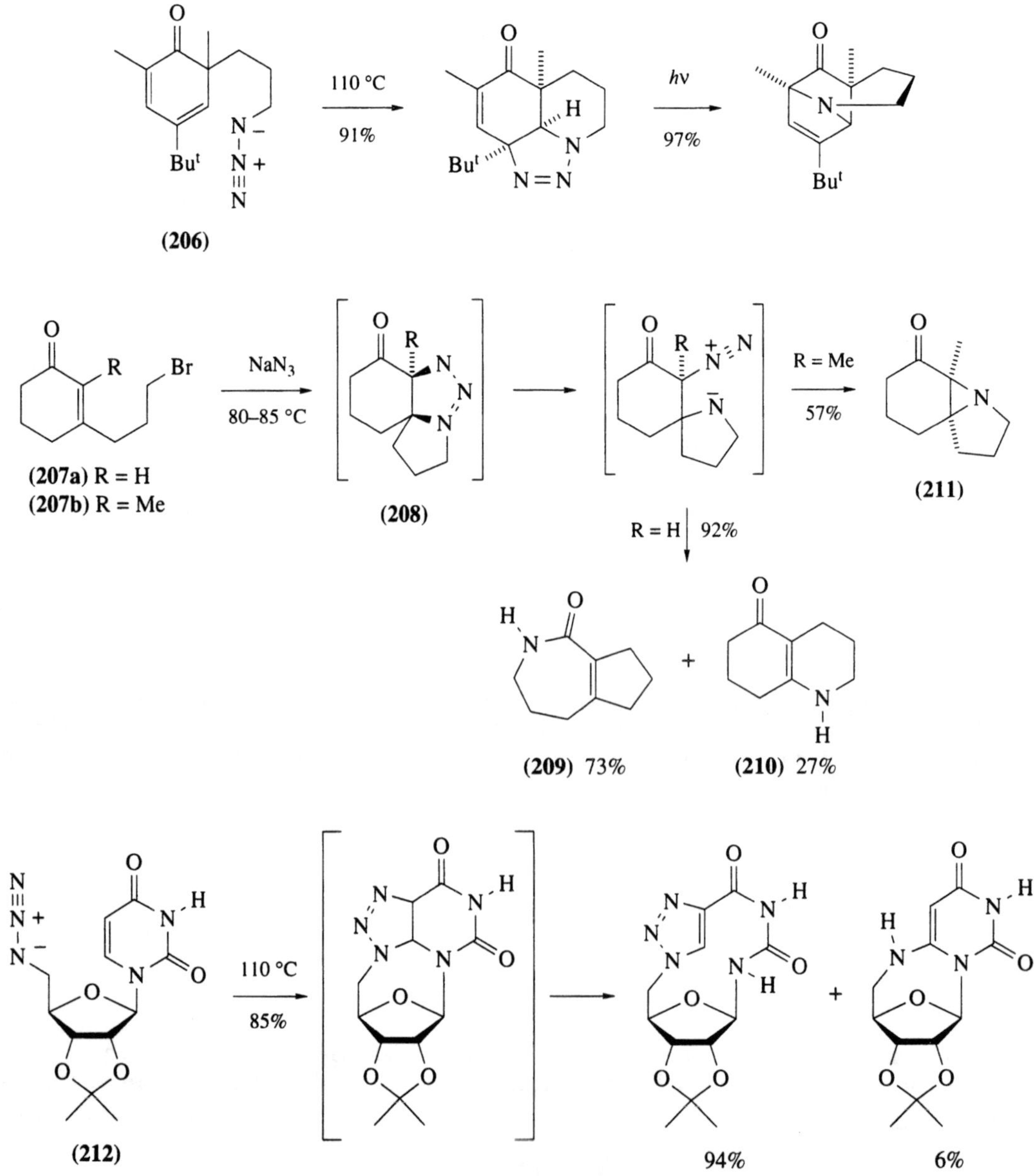

(206) 110 °C, 91%; hν, 97%

(207a) R = H
(207b) R = Me

NaN_3, 80–85 °C → (208) → R = Me, 57% → (211)

R = H 92%

(209) 73% + (210) 27%

(212) 110 °C, 85% → 94% + 6%

Scheme 64

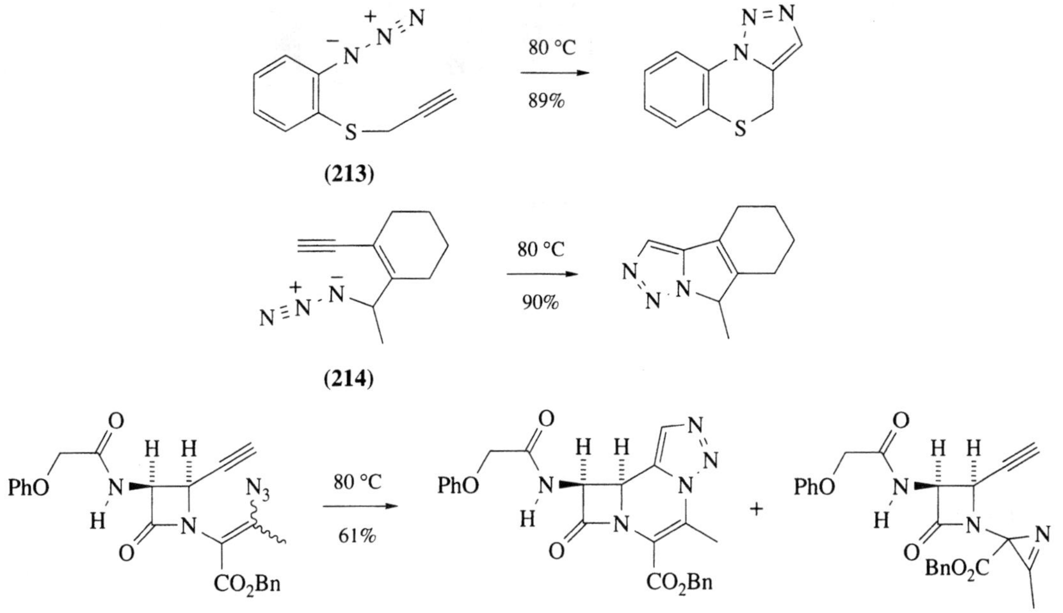

(213)

(214)

(215a) (*Z*)-isomer
(215b) (*E*)-isomer

(216) 80%

20%

Scheme 65

rophiles activated by an electron-attracting group are somewhat more succesful than nonactivated dipolarophiles.

Carbonyl ylides are highly reactive and must be used only *in situ*. There are a number of procedures applicable to their generation. The most extensively studied approach involves thermal or photochemical cleavage of ene oxiranes, predominantly those with a cyano group on the ring. Rhodium acetate catalyzed decomposition of α-diazo ketones containing an additional carbonyl group can afford cyclic carbonyl ylides and has been applied to intramolecular cycloadditions. Oxazolin-4-ones are mesoionic carbonyl ylides: they have also been generated from α-diazo ketones containing an additional carbonyl group. Oxidopyrylium ylides can also be classified as carbonyl ylides and are available from pyranones.

4.10.10.2 Open-chain Carbonyl Ylide–Alkene Cyclizations

An alkenyl group can be attached at either carbon atom of the carbonyl ylide. Cyclization normally affords fused bicyclic tetrahydrofurans, although with large rings bridged products have been observed.

4.10.10.2.1 Open-chain dipolarophiles

Cyclization of aryl carbonyl ylides substituted in the *ortho* position with an open-chain alkenyl group has been systematically studied. Thus, the *exo,exo*-carbonyl ylide (**217**), generated by thermal conrotatory opening of an oxirane precursor, cyclized at 175 °C to afford the tricyclic tetrahydrofuran (**218**) as a single stereoisomer (Scheme 66).[121,122] Despite the presumed entropic advantage, (**217**) cycloadded only well above the temperature of its formation: when *N*-phenylmaleimide was added, intermolecular cycloaddition occurred at 120 °C in preference to formation of (**218**). The homologous carbonyl ylide with a one-carbon shorter chain gave a mixture of *cis*- and *trans*-fused (**219a–b**). The homolog with a one-carbon longer chain underwent cycloaddition only at higher temperature (240 °C) where the stereointegrity of the carbonyl ylide was lost: only 21% yield of a mixture of (**220a–b**) at 60% conversion was observed. A trace (1%) of the *cis*-fused isomer of (**220a**) was also noted.

A wider range of cyclizations occurred to conjugated ester dipolarophiles. Thus, (**221**) underwent thermolysis to the corresponding carbonyl ylide, which cyclized to give mainly bridged cycloadduct.[122] The homologous oxirane with a five-carbon chain was also studied and gave only the *trans*-fused product (**222**) in 73% yield at 55% conversion.

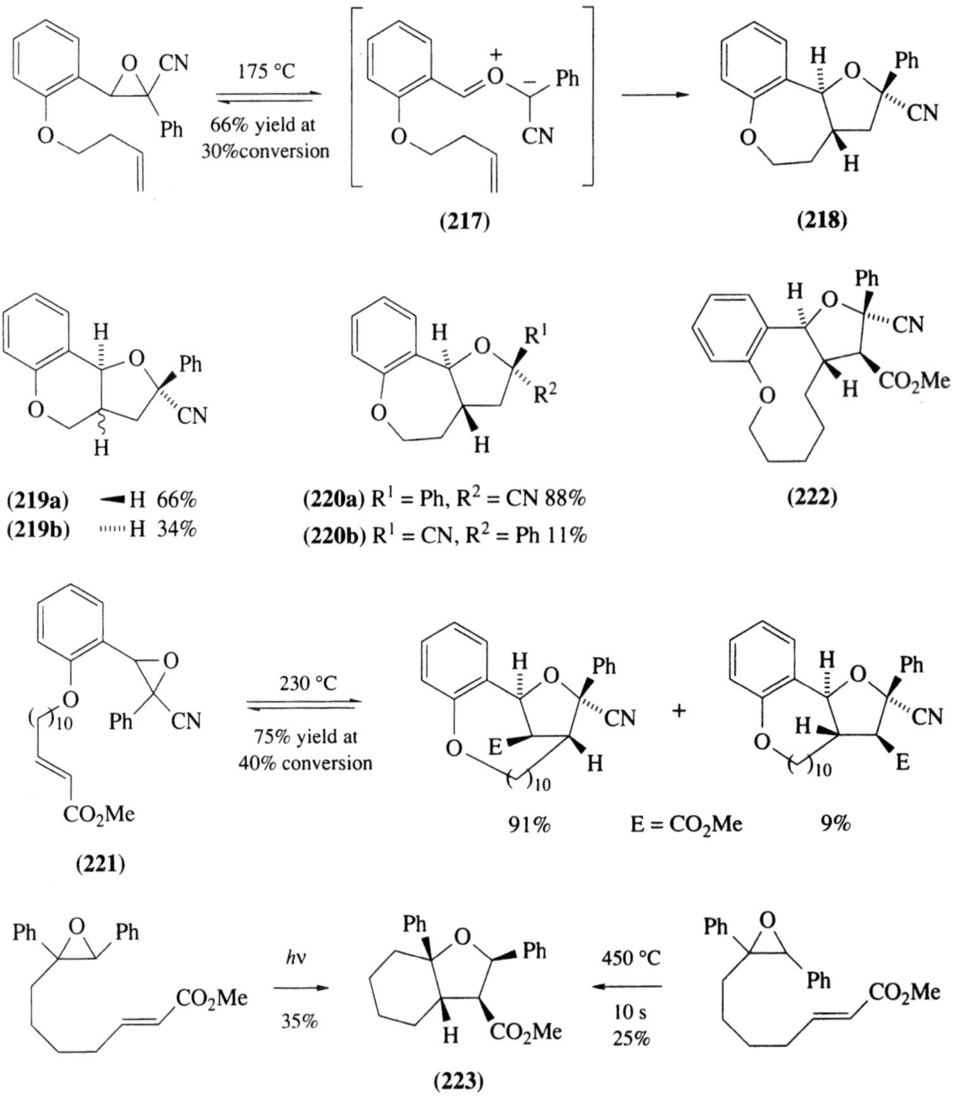

175 °C

66% yield at 30%conversion

(217)

(218)

(219a) ◀H 66%
(219b) ''''H 34%

(220a) R^1 = Ph, R^2 = CN 88%
(220b) R^1 = CN, R^2 = Ph 11%

(222)

230 °C

75% yield at 40% conversion

(221)

91% E = CO_2Me 9%

hν 35%

450 °C 10 s 25%

(223)

Scheme 66

Photolysis of ene oxiranes not possessing an α-cyano group has also led to intramolecular carbonyl ylide cycloaddition, but in low yield.[123] Thus, bicyclic tetrahydrofuran (**223**) was formed in 35% yield by photolytic disrotatory opening of the corresponding *trans*-diphenyloxirane. Thermolytic conrotatory opening of the isomeric *cis*-oxirane also afforded tetrahydrofuran (**223**) in 25% yield accompanied by an aldehyde byproduct.

4.10.10.2.2 Cyclic dipolarophiles

The diasteromeric tetracyclic products (**224a–b**) were formed from a cyclohexenyloxirane precursor (Scheme 67).[124] Similar cyclizations were carried out on cyclopentenyl, cycloheptenyl and cyclooctenyl substrates giving rise to diastereomeric mixtures in all cases. Interestingly, the cycloheptenyl substrate gave mostly (**225b**), corresponding to the minor product (**224b**).

4.10.10.3 Cyclic Carbonyl Ylide–Alkene Cyclizations

A variety of cyclic carbonyl ylides and related mesoionic species have been found to undergo intramolecular cycloaddition.

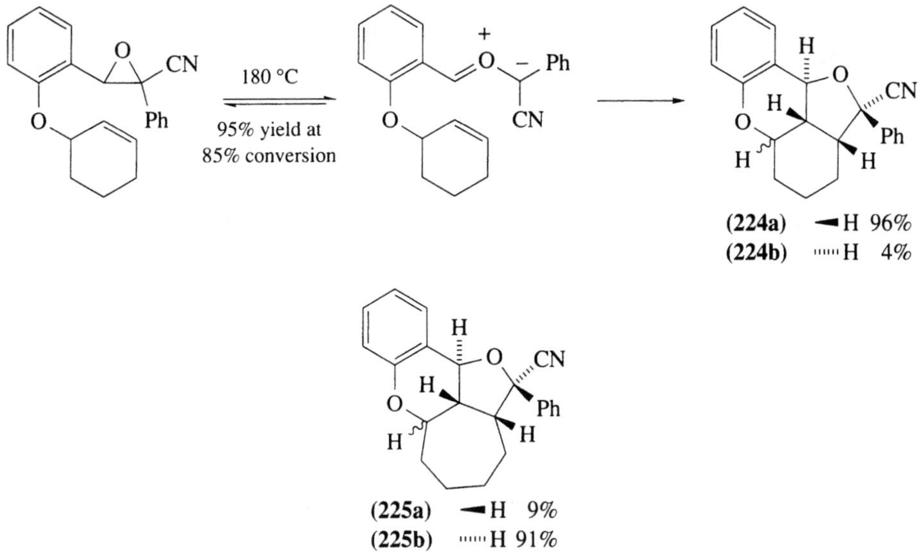

180 °C

95% yield at 85% conversion

(224a) ◄H 96%
(224b) ''''''H 4%

(225a) ◄H 9%
(225b) ''''''H 91%

Scheme 67

4.10.10.3.1 Oxidopyrylium ylide cycloadditions

Rhodium acetate catalyzed cyclization of the diazo compounds (**226a–b**) led to the tetracyclic tetrahydrofurans (**227a–b**) in high yield (Scheme 68).[125] It was suggested that the reaction to produce (**227a–b**) involved a tandem carbene addition–carbonyl ylide intramolecular cycloaddition sequence. The carbonyl ylide here is actually an oxidobenzopyrylium ylide, formally a carbonyl ylide. The oxidobenzopyrylium ylide intermediate formed from (**226b**) was trapped by intermolecular cycloaddition to dimethyl acetylenedicarboxylate. A related tandem carbene addition–carbonyl ylide intramolecular cycloaddition sequence has been reported where the dipole is present in a five- rather than six-membered ring.[126]

Thermolysis of the pyranyl acetate (**228**) gave rise to the isomeric perhydroazulenes (**229a–b**), presumably *via* intramolecular cycloaddition of an intermediate oxidopyrylium ylide.[127] A similar cyclization occurred for the pyran-4-one (**230**), leading to (**231a**), which was isolated as the acetate (**231b**).[128] No evidence was presented for an oxidopyrylium intermediate and the timing of carbonyl–enol tautomerization is unknown. Thus, another mechanism may apply.

4.10.10.3.2 Oxazolin-4-one (isomunchnone) cycloadditions

A number of appropriate diazo precursors have been subjected to tandem carbene cyclization–isomunchnone intramolecular cycloaddition.[129] Thus, (**232a**) was cyclized with rhodium acetate to provide a tricyclic tetrahydrofuran; (**232b**) cyclized similarly and produced just one stereoisomer.

4.10.10.4 Carbonyl Ylide–Alkyne Cyclizations

Dihydrofurans are valuable synthetic compounds. Nevertheless, little is known about intramolecular cycloaddition of carbonyl ylides to alkynes. In one example, alkynyl pyran-4-one (**233**) was cyclized to dihydrofuran (**234**; Scheme 69).[128] Possibly this reaction proceeds *via* an oxidopyrylium ylide intermediate as shown.

4.10.11 THIOCARBONYL YLIDE CYCLIZATIONS

There are two mesoionic sulfur heterocycles which have been shown to undergo intramolecular cycloaddition as thiocarbonyl ylides: 1,3-dithiolones[130] and 1,3-thiazolones.[131] Thus, the alkenyl 1,3-dithiolone (**235**) gave a 90% yield of cyclization product (Scheme 70). The analogous alkynyl

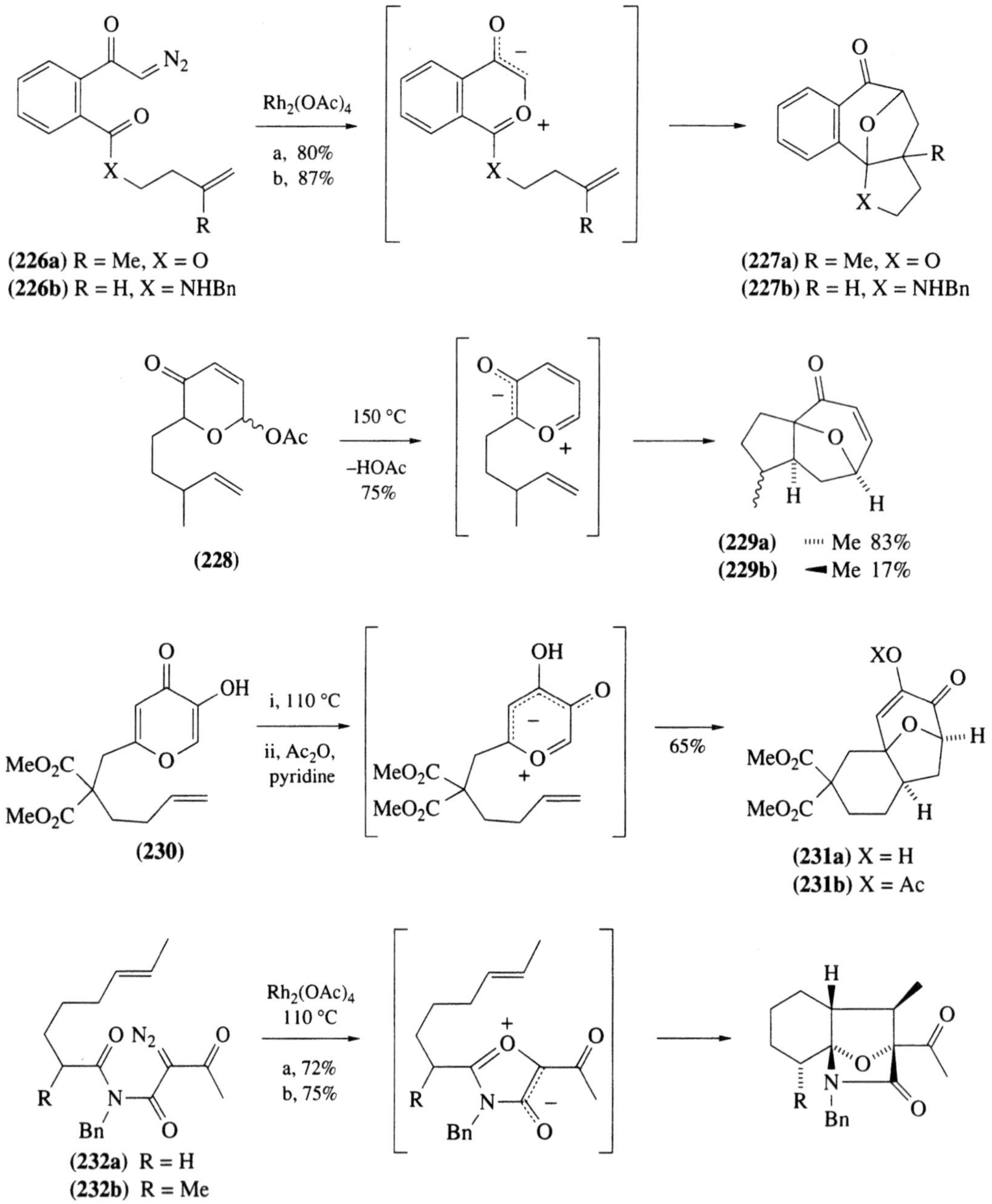

(**226a**) R = Me, X = O
(**226b**) R = H, X = NHBn

$Rh_2(OAc)_4$
a, 80%
b, 87%

(**227a**) R = Me, X = O
(**227b**) R = H, X = NHBn

(**228**)

150 °C
–HOAc
75%

(**229a**) Me 83%
(**229b**) Me 17%

(**230**)

i, 110 °C
ii, Ac_2O, pyridine

65%

(**231a**) X = H
(**231b**) X = Ac

(**232a**) R = H
(**232b**) R = Me

$Rh_2(OAc)_4$
110 °C
a, 72%
b, 75%

Scheme 68

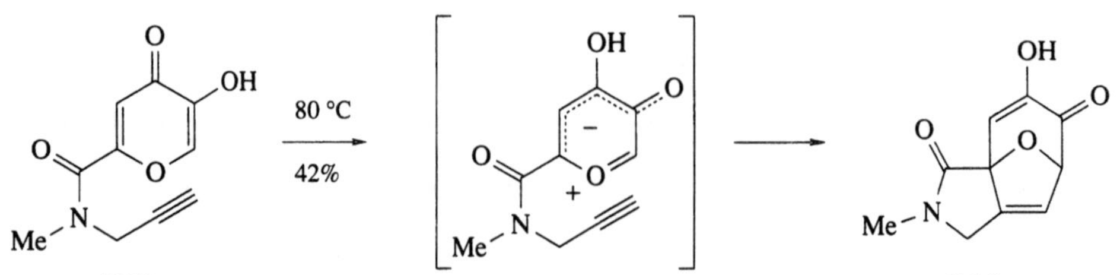

(**233**)

80 °C
42%

(**234**)

Scheme 69

1,3-dithiolone gave thiophene (**236**) in 92% yield, presumably formed from the corresponding cycloadduct after extrusion of carbonyl sulfide. The 1,3-thiazolone (**237**) reacted with 3-bromocyclohexene to afford pentacyclic (**238**) in 23% overall yield as a single isomer.[131] The intermediate alkylation product, tethered at the thiazolone 2-position, was partially cyclized to (**238**) during preparation. Similarly, thiazolone (**239**), tethered at the 5-position and generated from an α-hydroxy acid, also partially cyclized during preparation affording a 33% overall yield of (**240**). The thiazolone (**241**), tethered at the N-atom (3-position), was obtained pure: cyclization to (**242**) occurred only after eight days in refluxing xylene. Transition state analysis was consistent with the observed difference in reactivity depending on the position of the tether: 2-substituted > 5-substituted >> 3-substituted.

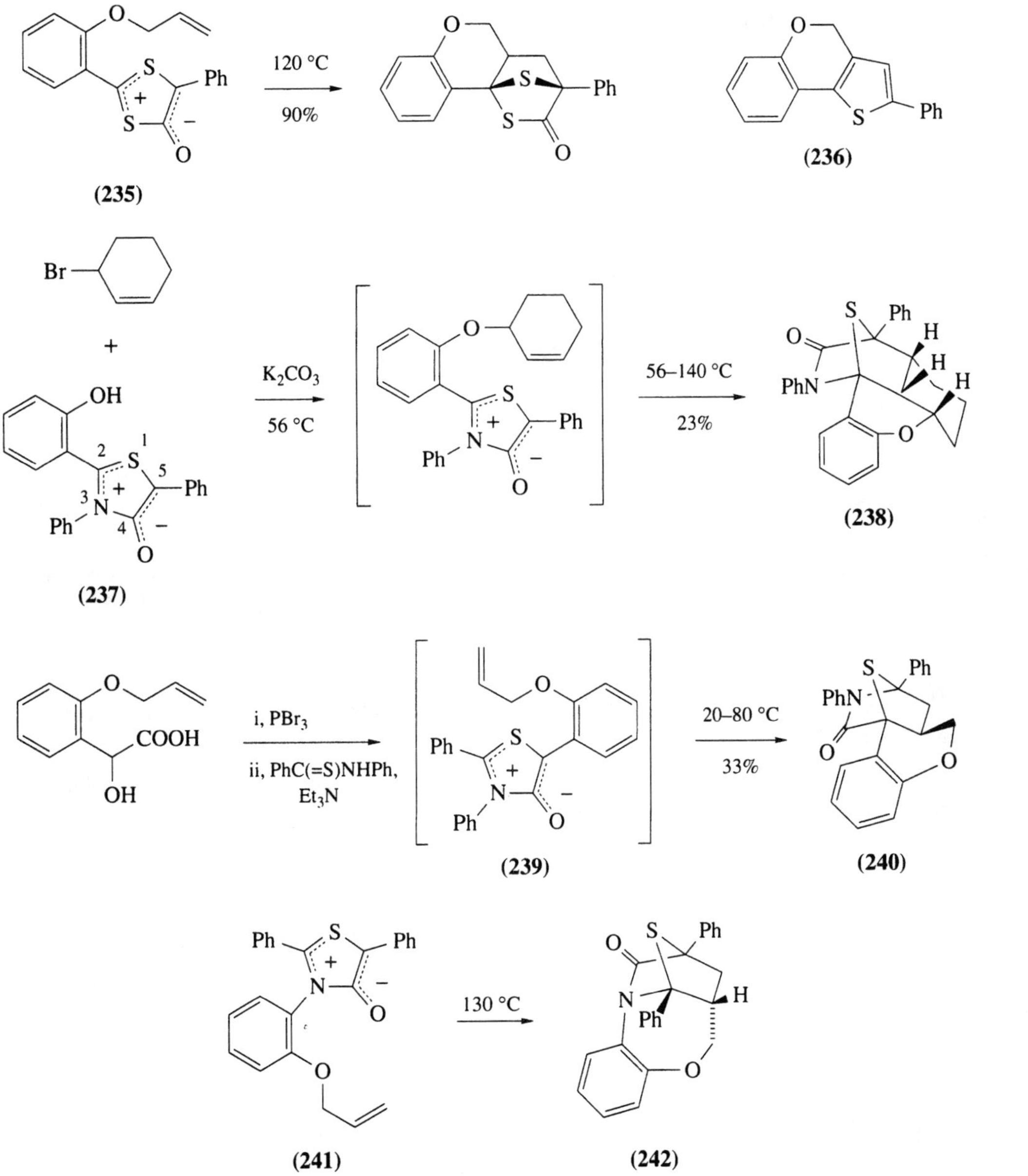

120 °C, 90%

(**235**) (**236**)

Br, +, OH, K_2CO_3, 56 °C, 56–140 °C, 23%

(**237**) (**238**)

i, PBr_3; ii, PhC(=S)NHPh, Et_3N; 20–80 °C, 33%

(**239**) (**240**)

130 °C

(**241**) (**242**)

Scheme 70

4.10.12 NITRILE SULFIDE CYCLIZATIONS

Intermolecular nitrile sulfide cycloadditions have been known for some time but the first intramolecular cycloadditions have just been reported.[132] The oxathiazolone (**243**) gave (**244a**) in 70% yield when heated in refluxing xylene (Scheme 71). Extrusion of carbon dioxide presumably afforded the nitrile sulfide as an intermediate. The oxathiazolone (**245**) gave (**244b**) when heated in refluxing xylene; here, aromatization of the initial cycloadduct presumably occurred.

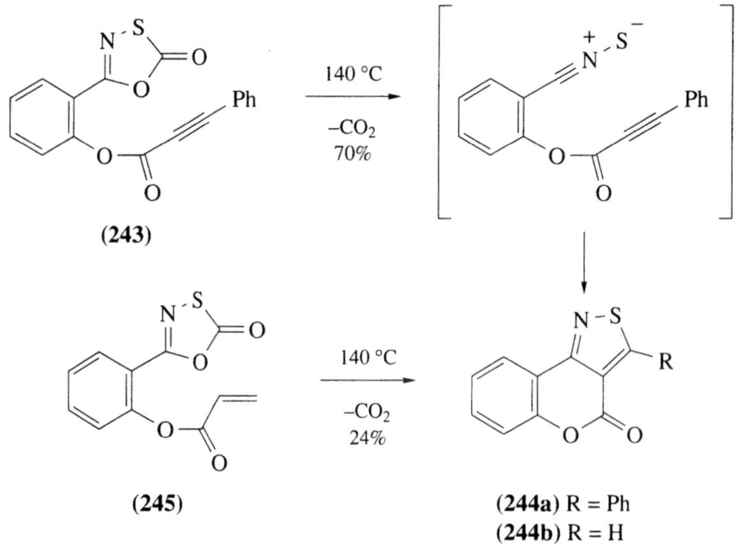

(243)

140 °C, $-CO_2$, 70%

(245)

140 °C, $-CO_2$, 24%

(244a) R = Ph
(244b) R = H

Scheme 71

4.10.13 REFERENCES

1. A. Padwa, in '1,3-Dipolar Cycloaddition Chemistry', ed. A. Padwa, Wiley-Interscience, New York, 1984, vol. 2, p. 277; see also: J. J. Tufariello, in '1,3-Dipolar Cycloaddition Chemistry', ed. A. Padwa, Wiley-Interscience, New York, 1984, vol. 2, p. 87.
2. (a) G. Bianchi, C. De Micheli and R. Gandolfi, in 'The Chemistry of Double-bonded Functional Groups', ed. S. Patai and H. Rapoport, Wiley-Interscience, New York, 1983, suppl. C, part 1, p. 752; (b) G. Bianchi, C. De Micheli and R. Gandolfi, in 'The Chemistry of Double-bonded Functional Groups', ed. S. Patai, Wiley-Interscience, New York, 1977, suppl. A, part 1, p. 369.
3. D. F. Taber, 'Intramolecular Diels–Alder and Ene Reactions', Springer-Verlag, New York, 1984; see also: W. R. Roush, in 'Comprehensive Organic Synthesis', ed. L. A. Paquette, Pergamon Press, New York, 1991, vol. 5, chap. 4.
4. (a) N. A. LeBel and J. J. Whang, *J. Am. Chem. Soc.*, 1959, **81**, 6334; (b) N. A. Le Bel, M. E. Post and J. J. Whang, *J. Am. Chem. Soc.*, 1964, **86**, 3759.
5. R. Annunziata, M. Cinquini and F. Cozzi, *Tetrahedron*, 1987, **43**, 4051.
6. For diastereoselective nitrone intramolecular cycloaddition to conjugated esters, see: R. Annunziata, M. Cinquini, F. Cozzi and L. Raimondi, *Tetrahedron Lett.*, 1988, **29**, 2881.
7. (a) P. M. Wovkulich and M. R. Uskovic, *Tetrahedron*, 1985, **41**, 3455; (b) T. K. M. Shing, D. A. Elsley and J. G. Gillhouley, *J. Chem. Soc., Chem. Commun.*, 1989, 1280.
8. N. A. LeBel and E. G. Banucci, *J. Org. Chem.*, 1971, **36**, 2440.
9. S. W. Baldwin, J. D. Wilson and J. Aubé, *J. Org. Chem.*, 1985, **50**, 4432.
10. M. A. Schwartz and A. M. Willbrand, *J. Org. Chem.*, 1985, **50**, 1359.
11. W. Oppolzer, S. Siles, R. L. Snowden, B. H. Bakker and M. Petrzilka, *Tetrahedron*, 1985, **41**, 3497.
12. (a) R. W. Hoffmann and A. Endesfelder, *Anal. Chem.*, 1986, 1823; (b) N. A. LeBel and N. Balasubramanian, *J. Am. Chem. Soc.*, 1989, **111**, 3363.
13. D. M. Tschaen, R. R. Whittle and S. M. Weinreb, *J. Org. Chem.*, 1986, **51**, 2604.
14. D. Stanssens, D. De Keukeliere and M. Vandewalle, *Bull. Soc. Chim. Belg.*, 1987, **96**, 813.
15. (a) A. G. Schultz and W. G. McMahon, *J. Org. Chem.*, 1987, **52**, 3905; (b) D. Prajapati and J. S. Sandhu, *Synthesis*, 1988, 342.
16. G. W. Gribble and T. C. Barden, *J. Org. Chem.*, 1985, **50**, 5900.
17. S. Eguchi, Y. Furukawa, T. Suzuki, K. Kondo and T. Sasaki, *J. Org. Chem.*, 1985, **50**, 1895.
18. S. Eguchi, Y. Furukawa, T. Suzuki and T. Sasaki, *J. Chem. Soc., Perkin Trans. 1*, 1988, 719.
19. W. Oppolzer and H. P. Weber, *Tetrahedron Lett.*, 1970, 1121.
20. A. Padwa, H. Ku and A. Mazzu, *J. Org. Chem.*, 1978, **43**, 381.
21. W. Oppolzer and J. I. Grayson, *Helv. Chim. Acta*, 1980, **63**, 1706.
22. R. L. Funk, G. L. Bolton, J. U. Daggett, M. M. Hansen and L. M. M. Horcher, *Tetrahedron*, 1985, **41**, 3479.
23. (a) J. J. Tufariello and G. B. Mullen, *J. Am. Chem. Soc.*, 1978, **100**, 3638; (b) J. J. Tufariello, J. T. Tegeler, S. C. Wong and S. A. Ali, *Tetrahedron Lett.*, 1978, 1733.
24. (a) D. Mackay and K. N. Watson, *J. Chem. Soc., Chem. Commun.*, 1982, 777; (b) A. Hassner and R. Maurya, *Tetrahedron Lett.*, 1989, **30**, 2289 and refs. therein.
25. (a) P. Armstrong, R. Grigg and W. J. Warnock, *J. Chem. Soc., Chem. Commun.*, 1987, 1325; (b) P. Armstrong, R. Grigg, S. Surendrakumar and W. J. Warnock, *J. Chem. Soc., Chem. Commun.*, 1987, 1327; (c) G. Donegan, R. Grigg, F. Heaney, S. Surendrakumar and W. J. Warnock, *Tetrahedron Lett.*, 1989, **30**, 609.
26. (a) S. E. Denmark, Y.-C. Moon, and C. B. W. Senanayake, *J. Am. Chem. Soc.*, 1990, **112**, 311; (b) N. A. LeBel and E. G. Banucci, *J. Am. Chem. Soc.*, 1970, **92**, 5278.
27. C. Grundmann and P. Grünanger, 'The Nitrile Oxides', Springer-Verlag, New York, 1971.

28. R. Fusco, L. Garanti and G. Zecchi, *Chim. Ind. (Milan)*, 1975, **57**, 16.
29. T. Mukaiyama and T. Hoshino, *J. Am. Chem. Soc.*, 1960, **82**, 5339.
30. G. A. Lee, *Synthesis*, 1982, 508.
31. (a) V. Jäger *et al.*, *Lect. Heterocycl. Chem.*, 1985, **8**, 79 and refs. therein; (b) I. Müller and V. Jäger, *Tetrahedron Lett.*, 1982, **23**, 4777.
32. (a) Isoxazolines → isoxazolidines: A. Vasella, *Helv. Chim. Acta*, 1977, **60**, 1273; (b) isoxazolidines → isoxazolines: A. Barco, S. Benetti, G. P. Pollini and P. G. Baraldi, *Synthesis*, 1977, 837.
33. (a) R. Annunziata, M. Cinquini, F. Cozzi, C. Gennari and L. Raimondi, *J. Org. Chem.*, 1987, **52**, 4674; (b) R. Annunziata, M. Cinquini, F. Cozzi, G. Dondio and L. Raimondi *Tetrahedron*, 1987, **43**, 2369.
34. A. P. Kozikowski and Y. Y. Chen, *Tetrahedron Lett.*, 1982, **23**, 2081.
35. A. P. Kozikowski and P. D. Stein, *J. Am. Chem. Soc.*, 1982, **104**, 4023.
36. R. Annunziata, M. Cinquini, F. Cozzi and L. Raimondi, *J. Chem. Soc., Chem. Commun.*, 1987, 529.
37. (a) K. N. Houk, S. R. Moses, Y.-D. Wu, N. G. Rondan, V. Jäger, R. Schohe and F. R. Fronczek, *J. Am. Chem. Soc.*, 1984, **106**, 3880; (b) S. D. Kahn and W. J. Hehre, *Tetrahedron Lett.*, 1985, **26**, 3647.
38. (a) M. Asaoka, M. Abe, T. Mukuta and H. Takei, *Chem. Lett.*, 1982, 215; (b) M. Asaoka, M. Abe and H. Takei, *Bull. Chem. Soc. Jpn.*, 1985, **58**, 2145.
39. A. P. Kozikowski and B. B. Mugrabe, *J. Chem. Soc., Chem. Commun.*, 1988, 198.
40. P. A. Wade and H. R. Hinney, *Tetrahedron Lett.*, 1985, **26**, 3647.
41. (a) P. N. Confalone, G. Pizzolato, D. L. Confalone and M. R. Uskovic, *J. Am. Chem. Soc.*, 1980, **102**, 1954; (b) A. P. Kozikowski and B. B. Mugrabe, *Tetrahedron Lett.*, 1983, **24**, 3705.
42. A. P. Kozikowski and C.-S. Li, *J. Org. Chem.*, 1987, **52**, 3541.
43. (a) A. P. Kozikowski and P.-W. Yuen, *J. Chem. Soc., Chem. Commun.*, 1985, 847; (b) R. Annunziata, M. Cinquini, F. Cozzi and L. Raimondi, *Tetrahedron Lett.*, 1989, **30**, 5013.
44. A. Hassner, K. S. Keshava Murthy, A. Padwa, W. H. Bullock and P. D. Stull, *J. Org. Chem.*, 1988, **53**, 5063.
45. A. P. Kozikowski and P. D. Stein, *J. Org. Chem.*, 1984, **49**, 2301.
46. A. P. Kozikowski, Y. Y. Chen, B. C. Wang and Z.-B. Xu, *Tetrahedron*, 1984, **40**, 2345.
47. L. Garanti, A. Sala and G. Zecchi, *J. Org. Chem.*, 1975, **40**, 2403.
48. S. S. Ko and P. N. Confalone, *Tetrahedron*, 1985, **41**, 3511.
49. A. P. Kozikowski, K. Hiraga, J. P. Springer, B. C. Wang and Z.-B. Xu, *J. Am. Chem. Soc.*, 1984, **106**, 1845.
50. H. Uno, N. Watanabe, S. Fujiki and H. Suzuki, *Synthesis*, 1987, 471.
51. A. Padwa, U. Chiacchio, D. C. Dean, A. M. Schoffstall, A. Hassner and K. S. K. Murthy, *Tetrahedron Lett.*, 1988, **29**, 4169.
52. L. Garanti, A. Sala and G. Zecchi, *Synthesis*, 1975, 666.
53. C. I. Deyrup, J. A. Deyrup and M. Hamilton, *Tetrahedron Lett.*, 1977, 3437.
54. R. Huisgen, W. Scheer and H. Huber, *J. Am. Chem. Soc.*, 1967, **89**, 1753.
55. (a) E. Vedejs and F. G. West, *Chem. Rev.*, 1986, **86**, 941; (b) E. Vedejs and J. W. Grissom, *J. Am. Chem. Soc.*, 1988, **110**, 3238.
56. P. Armstrong, R. Grigg, M. W. Jordan and J. F. Malone, *Tetrahedron*, 1985, **41**, 3547.
57. (a) J. Chastanet and G. Roussi, *J. Org. Chem.*, 1988, **53**, 3808 and refs. therein; (b) P. Garner, F. Arya and W.-B. Ho, *J. Org. Chem.*, 1990, **55**, 412.
58. A. Padwa and H. Ku, *J. Org. Chem.*, 1979, **44**, 255.
59. P. DeShong, D. A. Kell and D. R. Sidler, *J. Org. Chem.*, 1985, **50**, 2309.
60. P. N. Confalone and R. A. Earl, *Tetrahedron Lett.*, 1986, **27**, 2695.
61. (a) R. Smith and T. Livinghouse, *Tetrahedron*, 1985, **41**, 3559; (b) S. F. Martin and T. M. Cheavens, *Tetrahedron Lett.*, 1989, **30**, 7017.
62. (a) W. Eberbach, H. Fritz, I. Heinze, P. von Laer and P. Link, *Tetrahedron Lett.*, 1986, **27**, 4003; (b) D. A. Barr, G. Donegan and R. Grigg, *J. Chem. Soc., Perkin Trans. 1*, 1989, 1550.
63. R. Huisgen, *Angew. Chem., Int. Ed. Engl.*, 1963, **2**, 565.
64. A. Padwa, H. L. Gingrich and R. Lim, *J. Org. Chem.*, 1982, **47**, 2447.
65. H. Gotthardt, R. Huisgen and H. O. Bayer, *J. Am. Chem. Soc.*, 1970, **92**, 4340.
66. A. Padwa, R. Lim and J. G. MacDonald, *J. Org. Chem.*, 1985, **50**, 3816.
67. (a) E. Vedejs and S. L. Dax, *Tetrahedron Lett.*, 1989, **30**, 2627; (b) M. Westling, R. Smith and T. Livinghouse, *J. Org. Chem.*, 1986, **51**, 1159.
68. (a) L. Salem, *J. Am. Chem. Soc.*, 1974, **96**, 3486; (b) P. Caramella, R. W. Gandour, J. A. Hall, C. G. Deville and K. N. Houk, *J. Am. Chem. Soc.*, 1977, **99**, 385.
69. H. Giezendanner, H. Heimgartner, B. Jackson, T. Winkler, H. J. Hansen and H. Schmid, *Helv. Chim. Acta*, 1973, **56**, 2611.
70. (a) A. Padwa and P. H. J. Carlsen, *J. Am. Chem. Soc.*, 1975, **97**, 3862; (b) A. Padwa and P. H. J. Carlsen, *J. Am. Chem. Soc.*, 1976, **98**, 2006.
71. A. Padwa and N. Kamigata, *J. Am. Chem. Soc.*, 1977, **99**, 1871.
72. J. Fischer and W. Steglich, *Angew. Chem., Int. Ed. Engl.*, 1979, **18**, 167.
73. P. Caramella and K. N. Houk, *J. Am. Chem. Soc.*, 1976, **98**, 6397.
74. N. Engel, J. Fischer and W. Steglich, *J. Chem. Res. (S)*, 1977, 162.
75. A. Padwa, P. H. J. Carlsen and A. Ku, *J. Am. Chem. Soc.*, 1978, **100**, 3494.
76. A. Padwa and A. Ku, *J. Am. Chem. Soc.*, 1978, **100**, 2181 .
77. L. Garanti, G. Padova and G. Zecchi, *J. Heterocycl. Chem.*, 1977, **14**, 947.
78. (a) W. Oppolzer, *Tetrahedron Lett.*, 1970, 3091; (b) P. M. Weintraub, *Chem. Commun.*, 1970, 760.
79. W. Oppolzer, *Tetrahedron Lett.*, 1972, 1707.
80. O. Tsuge, K. Ueno and S. Kanemasa, *Chem. Lett.*, 1984, 285.
81. (a) P. A. Jacobi, M. J. Martinelli and (in part) S. Polanc, *J. Am. Chem. Soc.*, 1984, **106**, 5594; (b) P. A. Jacobi, A. Brownstein, M. J. Martinelli and K. Grozinger, *J. Am. Chem. Soc.*, 1981, **103**, 239.
82. T. Shimuzu, Y. Hayashi, Y. Kitora and K. Teramura, *Bull. Chem. Soc. Jpn.*, 1982, **55**, 2450.
83. R. Grigg, M. Jordan and J. F. Malone, *Tetrahedron Lett.*, 1979, 3877.

84. B. Fouchet, M. Joucla and J. Hamelin, *Tetrahedron Lett.*, 1981, **22**, 1333.
85. T. Shimuzu, Y. Hayashi, M. Miki and K. Teramura, *J. Org. Chem.*, 1987, **52**, 2277.
86. H. Meier and H. Heimgartner, *Helv. Chim. Acta*, 1986, 927 and refs. therein.
87. A. Haneda, T. Imagawa and M. Kawanisi, *Bull. Chem. Soc. Jpn.*, 1976, **49**, 748.
88. R. Fusco, L. Garanti and G. Zecchi, *Tetrahedron Lett.*, 1974, 269.
89. L. Garanti, A. Sala and G. Zecchi, *J. Org. Chem.*, 1977, **42**, 1389.
90. A. Padwa, S. Nahm and E. Sato, *J. Org. Chem.*, 1978, **43**, 1664.
91. (a) G. Schmitt and B. Laude, *Tetrahedron Lett.*, 1978, 3727; (b) H. Meier and H. Heimgartner, *Helv. Chim. Acta*, 1977, **60**, 3035; (c) T. Shimuzu, Y. Hayashi, Y. Nagano and K. Teramura, *Bull. Chem. Soc. Jpn.*, 1980, **53**, 429.
92. T. Shimuzu, Y. Hayashi, S. Ishikawa and K. Teramura, *Bull. Chem. Soc. Jpn.*, 1982, **55**, 2456.
93. A. Padwa and S. Nahm, *J. Org. Chem.*, 1981, **46**, 1402.
94. D. Janietz, K. Khoudary and W.-D. Rudorf, *J. Prakt. Chem.*, 1987, **329**, 343.
95. M. Schwarz, A. Besold and E. R. Nelson, *J. Org. Chem.*, 1965, **30**, 2425.
96. A. Padwa and H. Ku, *J. Org. Chem.*, 1980, **45**, 3756.
97. T. Miyashi, Y. Nishizawa, Y. Fujii, K. Yamakawa, M. Kamata, S. Akao and T. Mukai, *J. Am. Chem. Soc.*, 1986, **108**, 1617.
98. H. J. Sturm, K.-H. Ongania, J. J. Daly and W. Klötzer, *Chem. Ber.*, 1981, **114**, 190.
99. A. Padwa and A. Rodriguez, *Tetrahedron Lett.*, 1981, **22**, 187.
100. A. G. Schultz and S. Puig, *J. Org. Chem.*, 1985, **50**, 915.
101. (a) E. Piers, R. W. Britton, R. J. Keziere and R. D. Smillie, *Can. J. Chem.*, 1971, **49**, 2623; (b) in a possibly related sequence, photolysis of a diazo compound led to a nitrogen extruded product: B. M. Trost, R. M. Cory and (in part) P. H. Scudder and H. B. Neubold, *J. Am. Chem. Soc.*, 1973, **95**, 7812.
102. T. Miyashi, Y. Nishizawa, T. Sugiyama and T. Mukai, *J. Am. Chem. Soc.*, 1977, **99**, 6109.
103. J. L. Brewbaker and H. Hart, *J. Am. Chem. Soc.*, 1969, **91**, 711 and refs. therein.
104. J. P. Mykytka and W. M. Jones, *J. Am. Chem. Soc.*, 1975, **97**, 5933.
105. (a) R. M. Wilson, J. W. Rekers, A. B. Packard and R. C. Elder, *J. Am. Chem. Soc.*, 1980, **102**, 1633; (b) A. G. Schultz, K. K. Eng and R. K. Kullnig, *Tetrahedron Lett.*, 1986, **27**, 2331.
106. P. A. Smith, J. M. Clegg and J. H. Hall, *J. Org. Chem.*, 1958, **23**, 524.
107. A. Logothetis, *J. Am. Chem. Soc.*, 1965, **87**, 749.
108. A. Hassner, A. S. Amarasekara and D. Andisik, *J. Org. Chem.*, 1988, **53**, 27.
109. J.-M. Liu, J.-J. Young, Y.-J. Li and C.-K. Sha, *J. Org. Chem.*, 1986, **51**, 1120.
110. P. Kolsaker, P. O. Ellingsen and G. Wøien, *Acta Chem. Scand., Ser. B*, 1978, **32**, 683.
111. C.-K. Sha, K.-S. Chuang and S.-J. Wey, *J. Chem. Soc., Perkin Trans. 1*, 1987, 977.
112. A. P. Kozikowski and M. N. Greco, *J. Org. Chem.*, 1984, **49**, 2310.
113. J. G. Buchanan, A. R. Edgar and B. D. Hewitt, *J. Chem. Soc., Perkin Trans. 1*, 1987, 2371.
114. M. J. Pearson and J. W. Tyler, *J. Chem. Soc., Perkin Trans. 1*, 1985, 1927.
115. (a) A. G. Schultz, J. P. Dittami, S. O. Myong and C.-K. Sha, *J. Am. Chem. Soc.*, 1983, **105**, 3273; (b) see also: A. G. Schultz, R. R. Staib and K. K. Eng, *J. Org. Chem.*, 1987, **52**, 2968.
116. C.-K. Sha, S.-L. Ouyang, D.-Y. Hsieh, R.-C. Chang and S.-C. Chang, *J. Org. Chem.*, 1986, **51**, 1490.
117. (a) T. Sasaki, K. Minamoto, T. Suzuki and T. Sugiura, *J. Org. Chem.*, 1979, **44**, 1424; (b) see also: T. Sasaki, K. Minamoto, T. Suzuki and S. Yamashita, *Tetrahedron*, 1980, **36**, 865.
118. L. Garanti, A. Locatelli and G. Zecchi, *J. Heterocycl. Chem.*, 1976, **13**, 657.
119. M. Bertrand, J. P. Dulcere and M. Santelli, *Tetrahedron Lett.*, 1977, 1783.
120. C. L. Branch, S. C. Finch and M. J. Pearson, *Tetrahedron Lett.*, 1982, **23**, 4381.
121. W. Eberbach, J. Brokatsky and H. Fritz, *Angew. Chem., Int. Ed. Engl.*, 1980, **19**, 47.
122. J. Brokatsky-Geiger and W. Eberbach, *Chem. Ber.*, 1984, **117**, 2157.
123. J. Brokatsky-Geiger and W. Eberbach, *Tetrahedron Lett.*, 1984, **25**, 1137.
124. J. Brokatsky-Geiger and W. Eberbach, *Chem. Ber.*, 1983, **116**, 2383.
125. A. Padwa, S. P. Carter, H. Nimmesgern and P. D. Stull, *J. Am. Chem. Soc.*, 1988, **110**, 2894.
126. A. Gillon, D. Ovadia, M. Kapon and S. Bien, *Tetrahedron*, 1982, **38**, 1477.
127. S. M. Bromidge, P. G. Sammes and L. J. Street, *J. Chem. Soc., Perkin Trans. 1*, 1985, 1725.
128. M. E. Garst, B. J. McBride and J. G. Douglass, III, *Tetrahedron Lett.*, 1983, **24**, 1675.
129. M. E. Maier and K. Evertz, *Tetrahedron Lett.*, 1988, **29**, 1677.
130. H. Gotthardt and O. M. Huss, *Justus Liebigs Ann. Chem.*, 1981, 347.
131. K. T. Potts, M. O. Dery and W. A. Juzukonis *J. Org. Chem.*, 1989, **54**, 1077.
132. P. A. Brownsort, R. M. Paton and A. G. Sutherland, *J. Chem. Soc., Perkin Trans. 1*, 1989, 1679.

Author Index

This Author Index comprises an alphabetical listing of the names of over 7000 authors cited in the references listed in the bibliographies which appear at the end of each chapter in this volume.

Each entry consists of the author's name, followed by a list of numbers, each of which is associated with a superscript number. For example

Abbott, D. E., 6[12,12c], 10[40], 573[53,54]

The numbers indicate the text pages on which references by the author in question are cited; the superscript numbers refer to the reference number in the chapter bibliography. Citations occurring in the text, tables and chemical schemes and equations have all been included.

Although much effort has gone into eliminating inaccuracies resulting from the use of different combinations of initials by the same author, the use by some journals of only one initial, and different spellings of the same name as a result of transliteration processes, the accuracy of some entries may have been affected by these factors.

Subject Index